Y0-AAZ-061

Carpentry contains procedures commonly practiced in industry and the trade. Specific procedures vary with each task and must be performed by a qualified person. For maximum safety, always refer to specific manufacturer recommendations, insurance regulations, specific job site and plant procedures, applicable federal, state, and local regulations, and any authority having jurisdiction. The material contained herein is intended to be an educational resource for the user. American Technical Publishers, Inc. assumes no responsibility or liability in connection with this material or its use by any individual or organization.

American Technical Publishers, Inc., Editorial Staff

Editor in Chief:
Jonathan F. Gosse
Vice President—Production:
Peter A. Zurlis
Art Manager:
Jennifer M. Hines
Multimedia Manager:
Carl R. Hansen
Technical Editor:
Charles A. Vescoso Jr.
Copy Editor:
Catherine A. Mini
Jeana M. Platz
Cover Design:
Jennifer M. Hines

Illustration/Layout:
Melanie G. Doornbos
Thomas E. Zabinski
Joshua P. Hugo
Robert M. McCarthy
DVD Development:
Robert E. Stickley
Daniel Kundrat
Nicole S. Polak
Kathleen A. Moster
Amanda N. Sidorowicz
Cory S. Butler
Kathryn C. Deisinger

6 7 8 9 – 13 – 9 8 7 6 5 4 3 2 1

Printed in the United States of America

ISBN 978-0-8269-0809-4

Carpentry

sixth edition

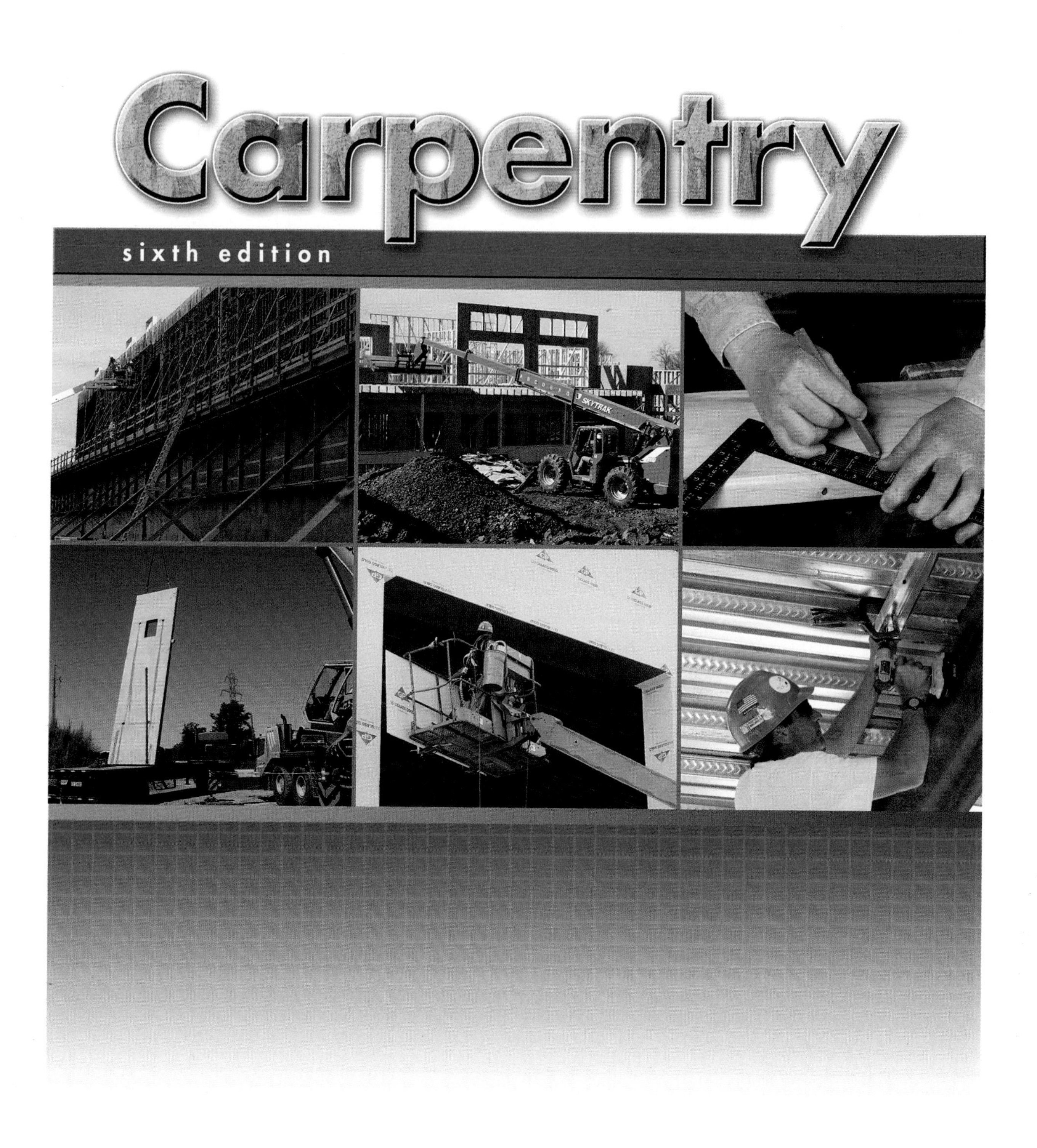

atp AMERICAN TECHNICAL PUBLISHERS
Orland Park, Illinois 60467-5756

Leonard Koel

ACKNOWLEDGMENTS

The author and publisher are grateful for the support provided by the following organizations:

United Brotherhood of Carpenters and Joiners of America (UBC)
- Carpenter Millwright Trades College, Sackville, Nova Scotia
- Carpenters District Council of Ontario, Mississauga, Ontario
- Carpenters' Local 27 Training Centre, Woodbridge, Ontario
- Carpenters Training Committee for Northern California, Pleasanton, CA
- Carpenters Training Institute, Paradise, Newfoundland
- Chicago Regional Council of Carpenters Apprentice & Training Program, Elk Grove Village, IL
- Detroit Carpentry Joint Apprenticeship Training Committee, Ferndale, MI
- International Training Center, Las Vegas, NV
- Southeast Wisconsin Carpentry Joint Apprenticeship and Training Committee, Milwaukee, WI

Construction Careers Center Charter High School, St. Louis, MO
North Iowa Area Community College—Building Trades Program, Mason City, IA

The author and publisher are grateful to the following companies and organizations for providing information, photographs, and technical assistance.

Adjustable Clamp Company
AHI Roofing
Alcoa Building Products, Inc.
Aluminum Association, Inc.
American Hardwood Export Council
American Saw & Mfg. Company
Anderson Corporation
APA—The Engineered Wood Association
Baldwin
Ballymore Company, Inc.
California Redwood Association
Case Foundation
Cedar Shake & Shingle Bureau
CertainTeed Corporation
Chestnut Tool Co.
Classic Products, Inc.
Classic Wood & Beam
Con-Tech Systems Ltd.
Crystal Window & Door Systems, Ltd.
David White Instruments
Dayton Superior
DC Roofing/Cedar Shake & Shingle Bureau
Deere & Company
DeWALT Industrial Tool Co.
Dunigan Custom Woodworking
ECO-Block, LLC
ELE International, Inc.
Empire Level
ET&F Fastening Systems, Inc.
Fastenal Company
Festool USA
Formica Corporation
The Garlinghouse Company
Gates & Sons, Inc.
Genie Industries
Hamilton Form Company, Ltd.
Hilti, Inc.
Icynene, Inc.
Ingersoll-Rand Material Handling
IR Security & Safety
ITW Paslode
ITW Ramset/Redhead
James Hardie Building Products
JCB Inc.
Jet
JLG Industries, Inc.
KIP America
Klein Tools, Inc.
Knape & Vogt
Kohler Co.
Kolbe & Kolbe Millwork Co., Inc.
Leica Geosystems Inc.
Lift-All Company, Inc.
Lignomat U.S.A., Ltd.
The Lincoln Electric Company
LiteGuard
L&S Technical Associates
Malco Products, Inc.
Manitowoc Crane Group
Marvin Windows and Doors
Merillat Industries
MEVA Formwork Systems, Inc.
MFG Construction Products
Miller Electric Manufacturing Company
Miller Fall Protection
Milwaukee Electric Tool Corporation
MiTek Industries, Inc.
Muro North America
National Gypsum Company
National Wood Flooring Association
The Original Saw Company
Owens-Corning Fiberglass Corp.
Ox Engineered Products
Patent Construction Systems
Pella® Windows and Doors
Pergo, Inc.
PERI Formwork Systems, Inc.
PLS-Pacific Laser Systems
Pocopson Industries, Inc.
Porter-Cable Corp.
Portland Cement Association
Powermatic
Quad-Lock Building Systems Ltd.
Quik Drive U.S.A., Inc.
Retrotec Inc.
Roseland Stair Works, Inc.
Safway Steel Products, Inc.
SawStop LLC
Senco Products, Inc.
Shakertown
Simpson Strong-Tie Company Inc.
Skil Corporation
The Sinco Group, Inc.
South Bend Lathe Co.
Southern Forest Products Association
Stabila, Inc.
StairWorld, Inc.
The Stanley Works
STIHL, Inc.
Structural Insulated Panel Association
Symons Corporation
Therma-Tru
Tierra de Zia Construction
Tower Manufacturing Corporation
Trus Joist, A Weyerhaeuser Business
TrusSteel Div. of Alpine Engineered Products, Inc.
TUFF-N-DRI®
United States Steel Corporation
U.S. Green Building Council
U.S. Tape
Vaughan & Bushnell Mfg. Co.
Vico Software, Inc.
WATCHDOG WATERPROOFING
Werner Ladder Co.
Western Forms, Inc.
Western Wood Products Association
Western Wood Structures, Inc.
Wick Homes
Wind-Lock Corporation
Wood Truss Council of America

CONTENTS

SECTION

4 POWER TOOLS

SECTION

5 CONSTRUCTION EQUIPMENT, JOB SITE SAFETY, AND WORKING CONDITIONS

SECTION

6 BUILDING DESIGN AND PRINTREADING *(continued on next page)*

CONTENTS

CONTENTS

SECTION

16 HEAVY CONCRETE CONSTRUCTION

INTERACTIVE DVD CONTENTS

INTRODUCTION

Carpentry, 6th edition, provides a comprehensive approach to step-by-step carpentry skill development with an overview of tools and equipment, materials, and proven trade practices. This textbook contains 70 units designed to help develop and upgrade the skills and competencies required in the trade and introduces related concepts.

This edition of *Carpentry* has been thoroughly updated while retaining the easy-to-use approach and design of the previous edition. In addition, *Carpentry* includes expanded coverage of rigging and lifting equipment and procedures, adhesives, hand tools, portable and stationary power tools, building materials, hazardous material safety, green building and certifications, prefabricated panel forming systems, and column forms. New content areas include the following:

- energy auditing
- high-definition surveying
- building information modeling (BIM)
- micropiles
- wind turbine foundations
- composite and noncomposite concrete slabs
- structural insulated sheathing
- advanced framing
- lead-based paint renovation, repair, and painting program

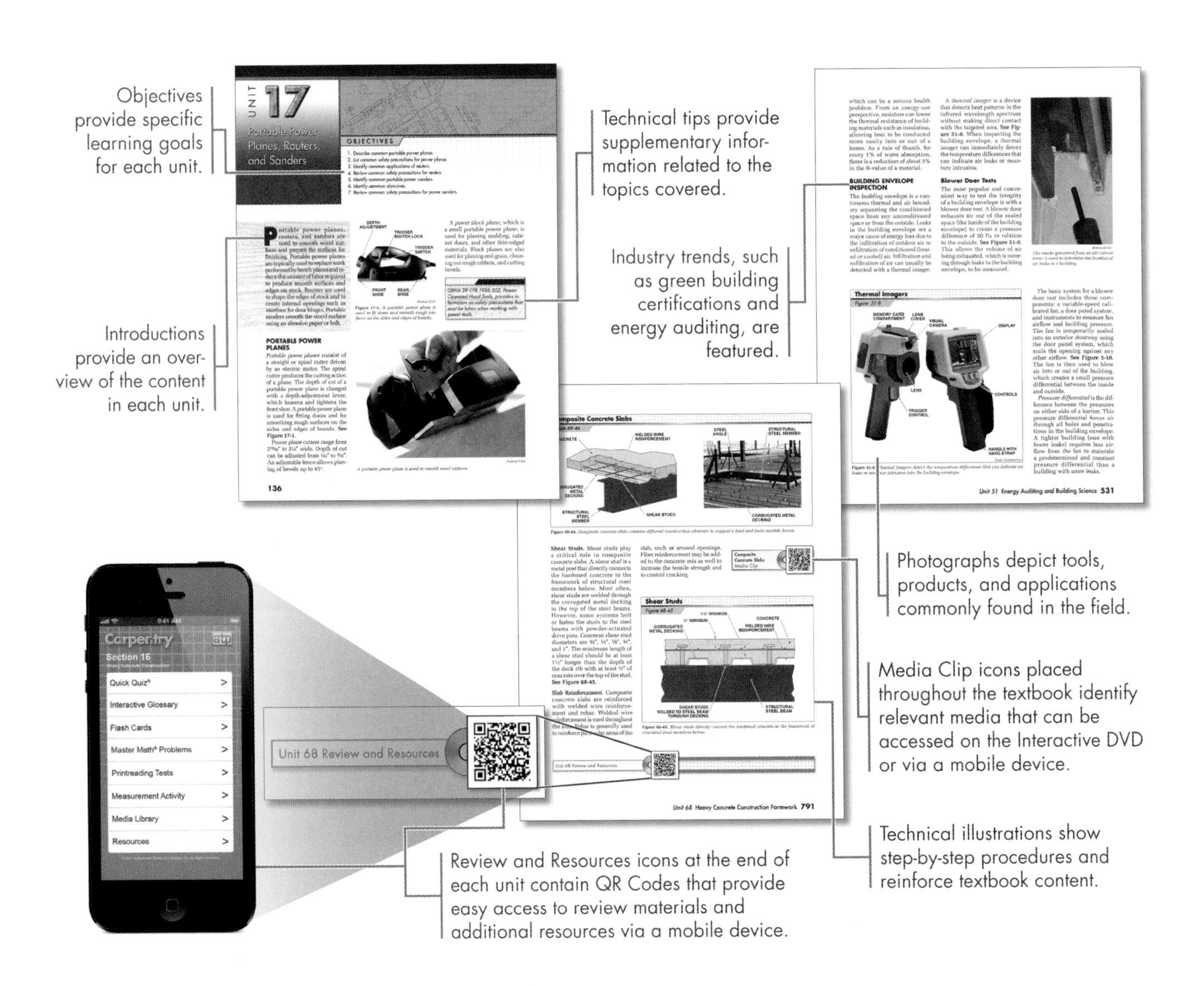

INTERACTIVE DVD FEATURES

The *Carpentry* Interactive DVD is a self-study aid designed to supplement content and learning activities in the textbook. The Interactive DVD includes expanded Quick Quizzes®, an Illustrated Glossary, Flash Cards, Master Math® problems, Printreading Tests, a Measurement Activity, a Media Library, and a link to ATPeResoureces.com.

The *Carpentry* Interactive DVD enhances content in the textbook with the following features:

- Quick Quizzes® that provide 20 interactive questions for each section and include links to highlighted content from the textbook and to the Illustrated Glossary
- An Illustrated Glossary that defines commonly used terms and links to selected illustrations and media clips
- Flash Cards that reinforce an understanding of the common industry terms and definitions found in the textbook
- Master Math Problems that provide practice applications of trade-related math
- Printreading Tests that reinforce comprehension of construction drawings using interactive printreading activities
- A Measurement Activity that teaches the basic use of a tape measure through a tutorial and interactive activity
- A Media Library containing media clips that reinforce and expand upon book content using videos and animated graphics
- ATPeResources.com, which provides a comprehensive array of instructional resources

SECTION

1

Carpentry and Construction

CONTENTS

UNIT 1 Types of Construction

OBJECTIVES

1. Identify common methods of residential and light construction.
2. List common methods of heavy construction.
3. Identify common types of prefabricated building units.
4. Discuss the importance of green building.

Carpenters work on two major types of building projects—residential and other light construction, and heavy construction. In each of these categories, many different construction methods are used to erect the buildings.

RESIDENTIAL AND LIGHT CONSTRUCTION

Residential construction is a type of light construction that includes houses, condominiums, and small multifamily dwellings (apartment buildings). **See Figure 1-1.** Residential and light construction employs the greatest number of carpenters. Other light construction projects include small- to medium-size commercial buildings such as stores, restaurants, and warehouses.

APA—The Engineered Wood Association

Figure 1-1. *Residential construction is a category of light construction.*

The framing method used for most light construction is *platform framing.* Platform framing consists of studs, plates, joists, bracing, and other structural members. Studs for a platform-framed structure are one story high. Double top plates over the studs support the floor joists for the next level. A subfloor is fastened to the joists, providing a platform for the next level. **See Figure 1-2.**

Balloon framing may also be used for light construction. In balloon framing, the studs extend from the sill plate to the roof. Second-floor joists are fastened to the studs, and a ribbon notched into and fastened to the studs provides their main support. Balloon framing gained in popularity in recent years due to the use of engineered lumber and metal framing components, which provide longer lengths at reasonable costs.

Post-and-beam construction is another wood-framing method. As its name implies, post-and-beam construction relies on posts and beams for its basic structure. **See Figure 1-3.** In most cases the beams are exposed, providing an attractive and open appearance to the building interior.

In *masonry construction,* the exterior walls of the building are constructed of bricks, hollow concrete masonry units (blocks), stone, or a combination of masonry products. The interior walls and floors of the building are usually constructed of wood and metal products. Carpenters work closely with bricklayers and stonemasons when building masonry structures, which are popular in the eastern and midwestern states.

Southern Forest Products Association

Figure 1-2. *In platform framing, each floor unit provides a working platform for the walls above.*

California Redwood Association

Figure 1-4. *Stone-veneer foundation and chimney combined with resawn redwood siding create an attractive effect for the house.*

In *brick-veneer* or *stone-veneer construction,* brick or stone is used as an outside covering (veneer) over a conventionally framed stud wall. **See Figure 1-4.** Brick-veneer and stone-veneer buildings give the appearance of having masonry exterior walls, but the primary support is provided by wood and metal structural members.

California Redwood Association

Figure 1-3. *Post-and-beam construction is characterized by exposed members, in this case redwood members, that also serve as the basic structure of the building.*

Another type of light construction is *alteration work,* or *remodeling,* in which a change or addition is made to a previously built structure. Examples of alteration work include adding a new room to an existing house, modernizing a kitchen or bathroom, changing locations of walls in an office, and updating an existing storefront. In large cities with limited space for new construction, alteration work is a significant portion of the carpentry trade.

HEAVY CONSTRUCTION

Most heavy construction uses reinforced concrete and structural steel. *Reinforced concrete* is concrete that contains steel reinforcement (rebar) or fiberglass reinforcing rod to strengthen it. Large office buildings, hospitals, bridges, freeways, and dams are types of heavy construction. **See Figure 1-5.**

Figure 1-5. *Reinforced concrete is typically used for most heavy construction.*

Steel is a primary building material in heavy construction.

Monolithic Concrete Structures

Monolithic concrete refers to the traditional method of concrete construction in which each major element of a building, such as a wall, is cast as a single continuous piece. Monolithic concrete construction is primarily used for small- to medium-size buildings. **See Figure 1-6.** Prefabricated forms are erected and/or wood forms are built to the shapes of the walls, beams, columns, and floors of the building. Rebar is placed and secured inside the forms and concrete is *placed* (poured) in the forms. When the concrete has *set* (hardened) sufficiently, the forms are stripped.

Figure 1-6. *In monolithic concrete construction, each major element of a structure is cast as a single piece.*

Concrete Construction Using Precast Units

Many large concrete buildings are constructed with *precast concrete units.* Precast concrete units, made with reinforced concrete, are manufactured at a precast plant and then transported to the job site where they are erected into the proper position. Many small- to medium-size concrete buildings are also constructed with precast units.

Steel-Framed Concrete Structures. Most high-rise buildings (skyscrapers) and some smaller buildings are erected with a steel framework. **See Figure 1-7.** In the past, carpenters built wood forms around the steel framework and concrete was placed in the forms. Today, however, precast concrete units are lifted into place by crane and fastened to the steel frame. The use of precast concrete units is more efficient and less costly than the older construction method. In many cases, carpenters are involved in attaching the precast concrete units to the steel framework.

Figure 1-7. *Most multistory commercial buildings have a steel framework.*

Concrete Structures without Steel Framework. Many small- to medium-size concrete buildings are constructed of precast units that are structurally tied together without steel framework. The *tilt-up* and *lift-slab* construction methods do not require a steel framework.

In tilt-up construction, concrete wall units are cast on the floor or transported to a job site after being cast at a precast plant. The wall units are raised into position with a crane and properly braced. **See Figure 1-8.** In lift-slab construction, floor slabs are stack-cast around columns at the first floor and raised into place by hydraulic jacks. Carpenters perform key operations in both tilt-up and lift-slab construction.

Figure 1-8. *Precast concrete wall panels of a building are raised into position with a crane.*

Heavy Timber Construction

Heavy timber construction is a type of heavy construction used to erect buildings, bridges, railroad trestles, and waterfront piers and docks. Heavy timber construction is one of the oldest construction methods used in North America. Today, heavy timber construction is often seen in large buildings such as domes.

PREFABRICATED BUILDING UNITS

A *prefabricated building unit* is a building unit constructed or built in a manufacturing facility and delivered to the job site. Precast concrete units, roof trusses, floor trusses, glued laminated (glulam) beams, box beams, and stressed-skin panels are *prefabricated structural units.* Packaged window assemblies, prehung doors, and exterior soffit systems are prefabricated *nonstructural* units. All site- constructed buildings today include a variety of prefabricated units.

Completely prefabricated buildings, typically used in residential construction, consist of *panel systems* or *modular units.* Over 15% of new single-family residential structures are manufactured housing. *Manufactured housing* refers to homes that are built in a manufacturing facility and transported to a building lot where they are installed or assembled. Manufactured housing includes panel, modular, and mobile homes.

Panel Systems

In addition to residential structures, light commercial structures such as office buildings and schools may be constructed with panel systems. The basic units of panel systems are the wall sections, which are constructed on an automated assembly line framing station. **See Figure 1-9.** When delivered by truck to the job site, the wall sections are installed by carpenters. **See Figure 1-10.** In addition to wall sections, a panel system may also include a roof system that includes roof trusses. Floor sections may also be provided, although the joists and subfloor are typically installed on the job site.

Figure 1-10. *Carpenters install prefabricated panel wall sections.*

Open Panel Systems. In an *open panel system,* the outside surface of the exterior wall panel is covered with sheathing or insulation board, but the inside surface of the wall is left exposed (open). In many cases, finish siding and completed window and door units are also installed at the manufacturing facility. **See Figure 1-11.** The plumbing and electrical systems are installed on the job.

Closed Panel Systems. In a *closed panel system,* the electrical and plumbing systems, insulation, and interior wallboard are installed at a manufacturing facility. As a result, the only work required on the job is usually some finish work where the panels butt together and connection of plumbing and electrical fixtures and appliances.

Some closed panel systems are available with a modular unit called a *mechanical core.* The mechanical core contains the kitchen and bathroom fixtures and appliances.

Building materials in manufactured homes are the same as those used in site-built homes.

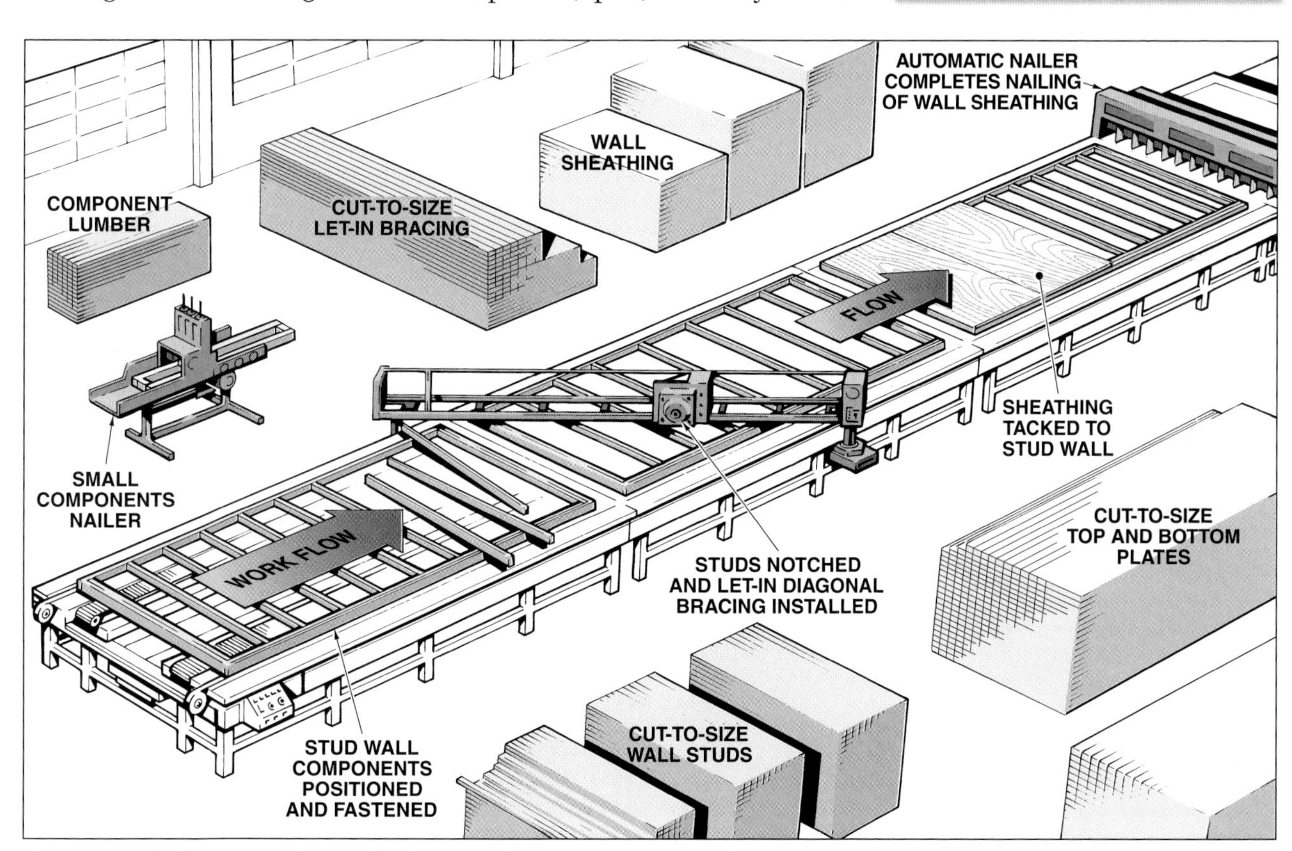

Figure 1-9. *A fully automated assembly line is used to construct panel wall sections.*

APA—The Engineered Wood Association

Figure 1-11. *An exterior wall panel is lifted after completion and prepared for shipment to the job site.*

Fold-Out Panel Systems. A *fold-out panel system* features panelized floors, walls, and roof sections hinged together. The units fold flat during shipping and are quickly unfolded and erected at the job site.

Modular Units

Modular construction is the most sophisticated and highly developed type of manufactured housing. A modular, or *sectional,* house is 95% complete when it is delivered to the job site. The house is prefabricated in three-dimensional units, each consisting of a floor, walls, and a ceiling or roof. The walls are finished on both sides and have electrical and plumbing systems installed.

A modular unit can be compared to a rectangular block. When all the blocks are fastened together, the house is completed. All modular housing packages include a mechanical core containing the kitchen and bathroom fixtures and appliances and most of the mechanical equipment needed in a house.

The installation procedure for a modular house is called *setting* the house. Modular units are placed on top of a foundation that has already been constructed on the job site. The modular units are set in place using a crane or slid into place on greased rails. As the modular units are placed, carpenters bolt the units together and fasten them to the foundation walls. Setting a house usually takes no more than a few hours.

A small amount of finish work is required after the modular units are set. Carpenters add exterior siding to the gable ends of the building to cover areas where the units are joined. Trim boards may be necessary where the house rests on the foundation. Inside the house, wall joints must be finished and floor covering installed where the modular units have been joined. Electrical and plumbing systems for the house must also be hooked up to the municipal systems.

Mobile Homes

Mobile homes are the primary type of manufactured housing. Mobile homes are built in accordance with federal standards contained in a building code established by the Department of Housing and Urban Development (HUD). The HUD code regulates manufactured home design and construction, strength and durability, energy efficiency, and fire resistance. The code also mandates wind resistance in areas that are subject to high-velocity winds.

The base of a mobile home rests on a steel frame. The floor, walls, and roof are wood-framed. Insulation and electrical and plumbing systems are set in place and floor coverings are installed. Heaters, furnaces, and heating ducts are

Southern Forest Products Association

Prefabricated building units, such as roof trusses, are raised into position using a crane.

installed. Mobile homes are available as single- or double-wide units. Double-wide mobile homes are transported to a job site in two sections where they are placed and bolted to a wall foundation or piers.

GREEN BUILDING

Green building, also known as *sustainable design*, refers to building design and construction methods that efficiently use materials, energy, water, and other natural resources. In comparison to traditional design, green building places a greater emphasis on occupant health and productivity; efficient energy, water, and other resource utilization; and reducing the overall impact on the environment. Green building principles are achieved through optimum building location on a building site and better design, construction, operation, maintenance, and removal—in other words, the complete building life cycle. **See Figure 1-12.** The green building initiative was developed due to concerns about diminishing natural resources, increasing pollution levels, and other factors that may cause harm to the environment.

SPECIALIZATION IN THE TRADE

Improved production and scheduling methods in construction have led to increased specialization of work. On housing tracts, for example, a large number of homes are constructed at the same time by crews of carpenters specializing in various divisions of carpentry work. **See Figure 1-13.** One crew of carpenters installs all the foundations of the houses while other crews separately place the floor units, walls, roofs, and ceilings.

In some areas, carpenters or subcontractors specialize in installing gypsum board (drywall) on walls and ceilings. Others work exclusively on installing acoustical tile ceilings. Still other carpenters install only certain types of patented partitions (interior walls) in office buildings. The number of such single-specialty operations is growing.

As a result of increased specialization, carpenters with versatile skills are becoming the exception rather than the rule. Still, it is possible for a person entering the carpentry trade to become an exceptional carpenter if he or she gains experience in a wide variety of construction work. This means learning such skills as building forms for concrete structures, erecting the framework for wood-framed buildings, and performing the trim work required in all types of construction. A variety of skills will give a person greater job security and earn the respect of employers and coworkers alike.

Figure 1-12. *Sustainable design and construction yields environmental, economic, and health benefits for the building occupants and the surrounding community.*

Figure 1-13. *Mass production methods are used on new subdivision housing tracts. Separate crews install the foundations, wall and roof framing, exterior trim, and interior trim.*

Unit 1 Review and Resources

UNIT 2

The Building Trades

OBJECTIVES

1. List and describe functions of carpenters and general contractors.
2. List other trades involved in the construction industry.
3. List and describe functions of industry and standards organizations.
4. Discuss functions of apprenticeship programs.

Carpenters work with a variety of building tradesworkers when working on a construction project. Apprentices attend classroom training to learn the principles of the trade, and then work closely with journeymen to continue to learn the trade. Cooperation and interaction between apprentices and journeymen, and between all tradesworkers, is vital to the success of a project.

CARPENTERS AND GENERAL CONTRACTORS

Carpentry is the largest craft in the construction industry. Most carpenters, unless they are self-employed, work for a *general contractor*. A general contractor, or building contractor, is a licensed individual or firm that can enter into legal contracts to perform construction work. In some cases, a general contractor is a single carpenter who either works alone to complete a project or hires extra carpenters to help. A general contractor may also be a large corporation involved in massive construction projects throughout the world. Whether the construction project is large or small, a general contractor is responsible for the overall organization, supervision, and execution of a construction project.

Carpenters work with tools, equipment, and a wide variety of materials to build structures such as one-family and multifamily dwellings, offices, bridges, and highways. **See Figure 2-1.** Carpenters may specialize in certain aspects of construction such as concrete formwork, residential carpentry, or interior systems commonly installed in commercial and public buildings.

Figure 2-1. *Carpenters work with tools, equipment, and a wide variety of materials. The decks being fabricated on the ground are lifted into position with the material lift shown in the background.*

OTHER BUILDING TRADESWORKERS

In addition to carpenters, workers from many other crafts work on construction projects. Most of the tradesworkers are employed by *subcontractors* who are licensed to perform work in their specialty area. For example, electrical work in a building under construction is performed by journeyman and master electricians who are employed by an electrical subcontractor.

Some of the major trades involved in the construction industry and examples of the work they perform follow.

Construction craft laborers (CCL), or laborers, are often directly employed by a general contractor. Activities carried out by construction craft laborers are preparing and carrying materials to carpenters; mixing, handling, and placing concrete in forms; excavating, placing, and compacting earth materials; and cleaning and maintaining work areas.

Bricklayers work with masonry materials such as brick, concrete block, and structural tile to construct or repair walls, partitions, arches, and fireplaces and chimneys. Bricklayers also work with glass, gypsum, and terra cotta block.

Stonemasons work on buildings that have solid stone or stone-veneer walls. Stonemasons set stone to build structures such as piers, walls, and abutments, or lay walks, curbstones, and special types of masonry.

Cabinetmakers and *millworkers* lay out, build, and install cabinets and built-in furniture, including kitchen cabinets, bathroom vanities, and store fixtures, using a variety of woodworking equipment and hand tools. Cabinetmakers also fabricate and install countertops, moldings, and panels.

Cement masons, or cement finishers, produce the finish on freshly placed concrete floor slabs, walls, sidewalks or curbs using hand tools or power equipment including floats, trowels, and screeds. **See Figure 2-2.** Cement masons may also set forms that are no more than one board high.

Figure 2-2. *Cement masons place and finish concrete to produce the desired texture.*

Structural-steel workers, or structural-iron workers, erect the steel framework for steel-framed buildings. Structural-steel workers raise, place, and join girders, columns, and other structural-steel members. **See Figure 2-3.**

Reinforcing-metal workers, or reinforcing-iron workers, place and secure rebar and welded wire fabric reinforcement used in reinforced concrete. Based on information provided in the construction prints, reinforcing-metal workers cut steel reinforcement to the proper dimensions and join the reinforcement using wire ties or welding equipment.

Electricians lay out and install electrical conductors (wiring), electrical fixtures, and control equipment for the electrical systems of a building. For many construction projects, electricians install temporary electrical systems at the beginning of construction projects so that tradesworkers can operate their power tools and equipment.

Figure 2-3. *Structural-steel workers erect the steel framework for steel-framed buildings.*

Plumbers assemble and install pipes, fittings, fixtures, and appliances connected to the water supply and drainage, waste, and vent (DWV) systems of a building according to the construction specifications and plumbing codes in effect in their jurisdiction. In addition, plumbers install pipe supplying gas or fuel to gas-operated appliances such as furnaces or kitchen appliances.

Pile-driver operators, or pile drivers, work primarily on heavy construction projects such as highways, high-rise buildings, dams, and bridges. Pile-driving rigs, mounted on skids, barges, crawler treads, or cranes, drive piling as foundations for structures and shoring.

Pipefitters typically work on industrial and commercial buildings where they lay out, assemble, install, and maintain pipe systems for hot water, steam, production, and processing equipment. In addition, pipefitters maintain pipe supports and related hydraulic and pneumatic equipment used in these systems.

Sheet-metal workers plan, lay out, fabricate, assemble, and install sheet-metal products such as ducts used to distribute air for heating, ventilating, and air conditioning (HVAC) systems. Sheet-metal workers also fabricate and install gutters and flashing for buildings.

Elevator constructors assemble and install electric and hydraulic freight and passenger elevators, escalators, and dumbwaiters. In some areas, escalators are installed by millwrights.

Operating engineers, or heavy-equipment operators, operate and maintain heavy construction equipment such as bulldozers, power shovels, and motor graders to excavate, move, and grade earth. In addition, operating engineers operate cranes to unload and distribute materials on the job and place concrete at higher elevations.

Millwrights install, set up, and maintain machinery and equipment such as pumps, turbines, generators, and conveyors. In some cases, millwrights construct foundations for machinery or equipment using wood, steel, and concrete.

Lathers install the basic framework for plaster, consisting of wire or metal mesh or perforated gypsum board. Plasterers apply plaster or stucco over the framework. In the past, lathers installed wood lath to which plaster was applied.

Drywallers, or drywall installers, plan gypsum board (drywall) installations, erect metal framing and furring channels for fastening gypsum board, and install gypsum board to cover walls, ceilings, soffits, shafts, and movable partitions in residential, commercial, and industrial buildings. **See Figure 2-4.**

Figure 2-4. *A drywaller attaches and finishes gypsum board.*

Plasterers apply plaster and stucco finishes to interior or exterior walls of buildings. Exterior insulation and finish systems (EIFS), used for residential and commercial buildings, are typically installed by plasterers.

Painters and *paperhangers* apply finishes to walls and ceilings. Painters apply coats of paint, varnish, or stain to finish and protect interior and exterior surfaces, trim, and fixtures. Paperhangers apply wallcoverings, such as wallpaper and fabric, to interior walls and ceilings.

Glaziers cut, fit, and install glass in windows, doors, skylights, and storefronts. In addition, glaziers may apply adhesive film to glass or spray glass with a liquid solution to prevent glare from the glass surface.

Floor layers, or floor-covering installers, install resilient flooring materials such as carpet, linoleum, and vinyl products over underlayment. Flooring materials provide shock-absorbing, sound-deadening, or decorative characteristics to the floor surface. In some areas, carpenters install resilient flooring.

Tilesetters apply ceramic and other types of tile to floors, walls, and ceilings following design specifications. Tilesetters may specialize in placing marble or terrazzo tile.

Roofers apply roofing materials, other than sheet metal, to buildings to waterproof roofs. Roofers work with composition shingles or sheets, wood shingles, rubber membranes, and asphalt and gravel roof coverings.

WORKING TOGETHER: CARPENTERS AND OTHER TRADESWORKERS

A smoothly functioning construction project depends on cooperation among the different tradesworkers assigned to the job. This is particularly true of carpenters in relation to other tradesworkers. Carpenters are the backbone of most construction projects, as all other trades depend on the work done by carpenters. Before an electrician can install wiring in a house or the plumber can install and hang pipe, carpenters must first frame the floors, walls, and ceilings.

During all stages of construction, carpenters should keep in mind the work of the trades that will follow. A good understanding of the work performed by other trades will help carpenters perform their work in a manner that facilitates the work of the other trades.

INDUSTRY AND STANDARDS ORGANIZATIONS

The construction industry has evolved over the years through the efforts of many industry and standards organizations. These organizations have sought to establish quality standards, provide quality and consistency between manufacturers, and provide a vehicle for improvement of the construction industry, including the carpentry trade. Contractors and industry professionals use the resources of these organizations to ensure product safety, quality, and efficiency.

United Brotherhood of Carpenters and Joiners of America

The *United Brotherhood of Carpenters and Joiners of America (UBC)* is the carpenters' union. Founded in 1881, the UBC now has approximately 520,000 members. Local unions number more than 1000. Every state of the United States and every province of Canada has at least one local union. More than 50,000 UBC members are enrolled in apprentice courses and over 100,000 journeymen are involved in upgrading their skills annually.

Most UBC members are construction carpenters. However, the UBC also includes cabinetmakers, pile drivers, floor layers, millwrights, lathers, and drywallers. *Millworkers* are industrial workers employed in lumber, panel product, and sawmill production, as well as in factories that produce prefabricated housing.

The UBC includes members from every racial, ethnic, and religious background. More than 27,000 women UBC members are employed in industrial plants or work as carpenters on construction sites. The number of women working in carpentry as well as the other construction trades increases each year.

The purpose of the UBC is to advance the interests of its membership. The union negotiates agreements with contractors' associations concerning wages, fringe benefits, working conditions, and provisions for apprenticeship training, hourly wage scale, and the number of hours in the work week. In addition, the agreement usually includes provisions for vacation pay and for a health and welfare plan that covers medical, dental, hospitalization, and prescription costs. Most agreements between the UBC and the employers also provide for pension plans.

The UBC has over 3500 full- and part-time instructors at over 200 training centers across North America.

Contractors' Associations

Many general contractors belong to industry associations that represent their interests as employers. Two influential contractor associations are the *Associated General Contractors (AGC)* and the *National Association of Home Builders (NAHB)*. The AGC and NAHB are national organizations with state and local chapters. The AGC primarily represents contractors that are involved with heavy construction work. The AGC represents more than 36,000 firms, including 8000 of the leading general contractors and 14,000 specialty-contracting firms. The NAHB typically represents contractors in residential and light construction. The NAHB represents more than 800 state and local builder associations throughout the United States. The UBC negotiates its working agreement with the AGC and NAHB.

American National Standards Institute

The *American National Standards Institute (ANSI)* is a national organization that helps identify industrial and public needs for national standards. Standards are commonly produced and co-published with ANSI and member technical societies, trade associations, and United States and Canadian governments. **See Figure 2-5.** A *technical society* is an organization composed of engineers and other technical personnel that are united by a professional interest. The Roof Consultants Institute is an example of a technical society. A *trade association* is an organization that represents the producers of specific products. An example of a trade association is APA—The Engineered Wood Association. A *government department* is a federal government department often responsible for developing specifications such as the United States Military Standards (MIL STD).

APA—The Engineered Wood Association

Engineered wood products are used to create a dramatic stairway.

APA—The Engineered Wood Association

APA—The Engineered Wood Association is a nonprofit trade association of the North American engineered wood products industry. The association represents manufacturers of structural plywood, oriented strand board (OSB), structural composite panels, glued laminated (glulam) timber, wood I-joists, and laminated veneer lumber (LVL).

APA—The Engineered Wood Association was founded in 1933 as the Douglas Fir Plywood Association to represent the Pacific Northwest plywood industry. After adhesive and technological improvements led to the manufacture of structural plywood from Southern pine and other species, the association changed its name to the American Plywood Association (APA) in 1964. The association expanded again in the early 1980s with the introduction of oriented strand board (OSB). A decade later, APA formed a related nonprofit organization called Engineered Wood Systems (EWS) to represent manufacturers of nonpanel engineered wood products such as glulam timber, wood I-joists, and laminated veneer lumber. The association once again changed its name in 1994 to APA—The Engineered Wood Association.

INDUSTRY AND STANDARDS ORGANIZATIONS	
United Brotherhood of Carpenters and Joiners of America 101 Constitution Avenue, NW Washington, DC 20005 www.carpenters.org	**Cedar Shake and Shingle Bureau** P.O. Box 1178 Sumas, WA 98295 www.cedarbureau.org
Associated General Contractors of America 2300 Wilson Blvd, Suite 400 Arlington, VA 22201 www.agc.org	**Composite Panel Association** 19465 Deerfield Avenue, Suite 306 Leesburg, VA 20176 www.pbmdf.com
National Association of Home Builders 1201 15th Street, NW Washington, DC 20005 www.nahb.com	**National Hardwood Lumber Association** P.O. Box 34518 Memphis, TN 38184-0518 www.nhla.com
Associated Builders and Contractors, Inc. 4250 N. Fairfax Drive, 9th Floor Arlington, VA 22203 www.abc.org	**Occupational Safety and Health Administration** 200 Constitution Avenue Washington, DC 20210 www.osha.gov
American National Standards Institute 1899 L Street, NW, 11th Floor Washington, DC 20036 www.ansi.org	**Portland Cement Association** 5420 Old Orchard Road Skokie, IL 60077 www.cement.org
APA—The Engineered Wood Association 7011 S. 19th Street Tacoma, WA 98466 www.apawood.org	**Southern Forest Products Association** 2900 Indiana Avenue Kenner, LA 70064-1700 www.sfpa.org www.southernpine.com
American Institute of Timber Construction 7012 S. Revere Parkway, Suite 140 Centennial, CO 80112 www.aitc-glulam.org	**Wood Truss Council of America** One WTCA Center 6300 Enterprise Lane Madison, WI 53719 www.woodtruss.com
American Wood Council 222 Catoctin Circle SE, Suite 201 Leesburg, VA 20175 www.awc.org	**Western Wood Products Association** 522 SW Fifth Avenue, Suite 500 Portland, OR 97204-2122 www.wwpa.org
California Redwood Association 818 Grayson Rd., Suite 201 Pleasant Hill, CA 94532 www.calredwood.org	**Canadian Centre for Occupational Health and Safety** 135 Hunter Street East Hamilton, Ontario, Canada L8N 1M5 www.ccohs.ca

Figure 2-5. *Industry and standards organizations establish quality standards, provide quality and consistency between manufacturers, and provide a vehicle for improvement of the construction industry.*

American Institute of Timber Construction

The *American Institute of Timber Construction (AITC)* is the national technical trade association of the structural glulam timber industry. AITC represents a majority of the glulam timber manufacturers in the United States, as well as several installers, suppliers, sales representatives, engineers, architects, designers, and researchers.

American Wood Council

The *American Wood Council (AWC),* formerly the National Forest Products Association, is the wood products division of the American Forest & Paper Association. AWC promotes the use of wood by ensuring that wood products are widely accepted by model codes and regulations, develops guidelines for wood construction, and influences the development of public policies affecting the use of wood products.

Many industry and standards organizations offer a wealth of technical information and training materials on their web sites. For example, the American Wood Council web site (www.awc.org) provides a variety of publications and on-line courses such as information related to building materials and model codes.

California Redwood Association

The *California Redwood Association (CRA)* is the trade association for redwood lumber producers. CRA members produce quality redwood products and are dedicated to responsible use of their private redwood forests. CRA members work to ensure there will be a continuous supply of redwood products from their mills.

Southern Forest Products Association

Industry and standards organizations perform tests on wood samples to ensure consistent performance and reliability of their products.

Cedar Shake and Shingle Bureau

The *Cedar Shake and Shingle Bureau* represents the manufacturers, distributors, installers, and maintenance technicians of Certi-label® cedar shake and shingle products for roofing, walls, and interior design. Responsibilities of the Cedar Shake and Shingle Bureau include building code and product standards development, marketing, testing, and education for technicians and consumers.

Composite Panel Association

The *Composite Panel Association (CPA)* is the North American trade association for particleboard (PB), medium-density fiberboard (MDF), and other compatible product manufacturers. CPA members include 34 of the leading U.S. and Canadian manufacturers, which represent more than 80% of the total North American production capacity. CPA was formed in 1997 by consolidating the National Particleboard Association and Canadian Particleboard Association memberships. The CPA represents the North American industry on technical, regulatory, and quality assurance concerns.

National Hardwood Lumber Association

The *National Hardwood Lumber Association (NHLA)* maintains order, structure, and ethics in the hardwood marketplace; ensures timber availability to meet society's needs; and builds positive business relationships within the hardwood production community. NHLA publishes technical information on lumber grading, forestry management, and harvesting.

Occupational Safety and Health Administration

The *Occupational Safety and Health Administration (OSHA)* is a federal agency that requires all employers to provide a safe environment for their employees. OSHA was established under the Occupational Safety and Health Act of 1970, which requires all employers to provide work areas free from recognized hazards likely to cause serious harm.

OSHA administers and enforces compliance with the act through inspection by trained OSHA inspectors. Under OSHA guidance, states may develop and administer state occupational safety and health plans. Currently, there are 23 states with their own occupational safety and health plans. State plans may include the private and/or public sector and must be revised as necessary in order to comply with minimum OSHA federal standards.

The Office of the Federal Register publishes all adopted OSHA standards and required amendments, corrections, insertions, and deletions. Current OSHA standards are reproduced in the Code of Federal Regulations (CFR). OSHA standards are included in Title 29 of CFR Parts 1900–1999. OSHA 29 CFR Part 1926, *Safety and Health Regulations for Construction,* provides regulations specific to the construction industry. OSHA documents are available at many libraries, from the Government Printing Office (GPO) in major cities, and on-line.

Portland Cement Association

Since its founding in 1916, the *Portland Cement Association (PCA)* has had the goal of improving and expanding the uses of portland cement and concrete. To promote the uses of cement and concrete, the PCA provides a wide range of research, testing, and consulting services. Support programs are offered, supplying informative data on cement use and market potential, training, and educational programs for the cement, concrete, and construction industries.

Southern Forest Products Association

The *Southern Forest Products Association (SFPA),* formerly the Southern Pine Association, was founded in 1915. SFPA is a nonprofit trade association of southern pine lumber manufacturers from 12 states: Virginia, Tennessee, North Carolina, South Carolina, Georgia, Florida, Alabama, Mississippi, Louisiana, Arkansas, Oklahoma, and Texas. SFPA members produce about 55% of the southern pine lumber in the United States.

Wood Truss Council of America

The *Wood Truss Council of America (WTCA)* is a trade association that consists of truss manufacturers, material suppliers, and industry professionals. WTCA promotes the use of wood trusses to contractors involved in the construction industry. WTCA allows members to stay current with trends in the field by distributing information through educational seminars and industry publications.

Trade associations, such as the Wood Truss Council of America, offer seminars to allow carpenters and supervisors to upgrade their skills.

Western Wood Products Association

The *Western Wood Products Association (WWPA)* is a trade association representing softwood lumber manufacturers in 12 western states, including Washington, Oregon, Montana, California, Idaho, and Alaska. WWPA mills produce lumber from western softwood species, including Douglas fir, western larch, western hemlock, ponderosa pine, and western red cedar. Products manufactured from these species include structural lumber, appearance lumber, and factory lumber.

ENTERING THE CARPENTRY TRADE

The carpentry trade offers many opportunities for a young person. Carpentry is a common offering at postsecondary career and technical institutes and community colleges. Career and technical and community college programs are strongly recommended for persons interested in entering the carpentry trade. However, these programs are only preparatory courses; a person must actually work in the trade to learn it fully.

Apprenticeship Programs

Apprenticeship is on-the-job training combined with other, related instruction. An apprentice carpenter works with experienced journeymen while learning the trade. An apprentice enters into an agreement with an employer and Joint Apprenticeship and Training Committee (JATC) for a required period of time to receive instruction and to learn a trade.

Origins of Apprenticeship. Apprenticeship began in ancient times when the master of a skilled craft taught the craft to his sons or other young men. The young workers were completely under the direction of the master for the period of the apprenticeship. This method of learning a craft evolved into the *indenture* system practiced in Europe during the Middle Ages. An indenture was a written contract between the master craftsman and an apprentice. Apprenticeships often lasted as long as eight years. The indenture system was brought over to the New World and was common practice in this country until the development of craft labor organizations in the construction trades.

From 1881 to 1939, the carpentry apprenticeship in the United States was mainly a union-sponsored effort to preserve the craft. During this period, the UBC developed a comprehensive apprenticeship program consisting of on-the-job training and related instruction.

Labor-Management Apprenticeship Programs. The passage of the National Labor Relations Act in 1935 brought about improved collective bargaining between trade unions and employers. As a result, apprenticeships became the joint responsibility of both labor and management. Programs were set up wherever negotiated agreements existed between the UBC and employers.

Locally, apprenticeship programs are directed by JATCs consisting of union and employer representatives. An apprenticeship program in a given area works closely with the local school district as well as with the Department of Education for that state. On a national level, all apprenticeship programs must conform to the standards established by the Bureau of Apprenticeship and Training, which is a division of the United States Department of Labor. Information on apprenticeship programs can be obtained from the following address:

Bureau of Apprenticeship and Training
Office of Apprenticeship Training,
Employer and Labor Services
U.S. Department of Labor
200 Constitution Avenue, NW
Washington, DC 20001

Over 1600 instructors teach at regional JATC training centers across the United States and Canada. These instructors are required to periodically attend "train-the-trainer" programs at the UBC International Training Center in Las Vegas, Nevada, to improve their instructional skills and trade knowledge.

Carpentry apprentices are required to attend the training centers full time for two to four weeks per year. Apprentices participate in classes that help them develop and improve their skills in many trade-related areas such as print-reading and trade math. In addition, a major part of the program focuses on hands-on construction.

Merit Shop Apprenticeship Programs. Apprenticeship programs for 20 trades (including carpentry) have also been established by the *Associated Builders and Contractors, Inc. (ABC)*. ABC apprenticeship programs are approved by the U.S. Department of Labor, Bureau of Apprenticeship and Training (BAT). ABC apprenticeship programs meet all federal and state requirements for formal apprenticeship and include employer-sponsored classroom training and on-the-job training.

Labor-Management Journeymen Programs. In addition to supervising apprenticeship programs, many labor-management JATCs throughout North America also conduct journeymen skill advancement courses. Skill advancement courses provide journeyman carpenters an opportunity to improve their skills in all areas of the trade, including printreading, estimating, level and transit applications, and welding.

Unit 2 Review and Resources

SECTION 2

Construction Materials

CONTENTS

UNIT 3

The Nature of Wood

1. Explain the basic structure of wood.
2. Describe the effects of the moisture content of wood.

Wood used in carpentry and other trades is obtained primarily from the trunk of a tree. An understanding of the different parts of a tree trunk and of how a tree grows is useful to anyone who works with wood.

STRUCTURE OF WOOD

Wood is composed of tiny cells (fibers) that are held together with a natural cement called *lignin*. The cells are tubular in shape and are about as thick as a human hair. **See Figure 3-1.** Cells in softwood trees are about 1/8″ long. Cells in hardwood trees are about 1/24″ long. The walls of each cell are composed of *cellulose* matter. A tree grows by forming new wood cells. When new cells stop forming, the tree reaches full maturity and stops growing.

Annual rings begin at the center of the trunk and continue outward to the bark. **See Figure 3-2.** Each ring represents a year of cellular growth. Therefore, the age of a tree is very close to the number of annual rings. In drier seasons there is less growth, so some rings are narrower than others. A close look at each annual ring shows that it is composed of an inner, light-colored section and an outer, darker section. The light-colored section is the early growth of a year and is known as *springwood*. The darker section develops later in the growing season and is known as *summerwood*. Springwood is usually weaker and less dense than summerwood.

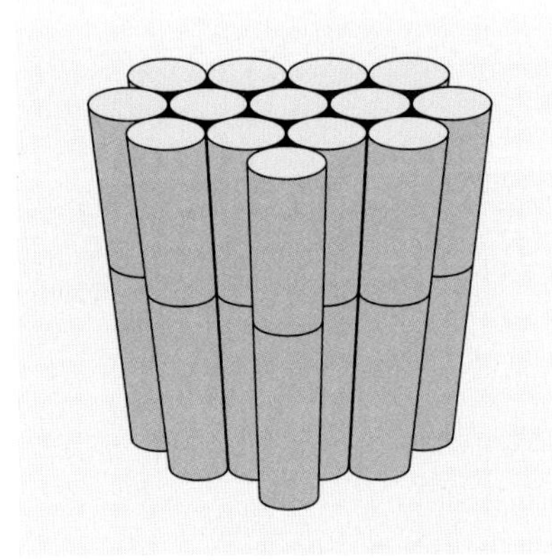

Figure 3-1. *Wood is composed of tiny cells that are held together with a natural cement called lignin. (The cells are drawn here many times larger than their actual size.) Trees grow by forming new wood cells, and reach full maturity and stop growing when new cells stop forming.*

The outside covering of a tree consists of *bark*. A tree has two layers of bark. The outer layer is dry, dead tissue. The purpose of the outer bark is to protect a tree from exterior damage. The inner bark is moist and soft, and helps transport food from the leaves to all the growing areas of the tree.

Directly underneath the bark is a very thin layer called the *cambium*. The light-colored section under the cambium is *sapwood*. The darker layer that extends from the sapwood to the *pith* (center) of the trunk is *heartwood*. The *medullary rays*, or wood rays, extend radially from the pith to the outer bark.

Color variation within wood is influenced by many conditions including soil types, minerals, water levels, available sunlight, and temperature.

Cambium

New cells are formed in the cambium. The inner part of the cambium develops wood cells that become sapwood. The outer part of the cambium produces new cells that form the bark.

Sapwood

Sapwood is the growing portion of a tree. Food is stored and absorbed in the sapwood. *Sap*, the watery fluid that circulates through a tree, travels from the

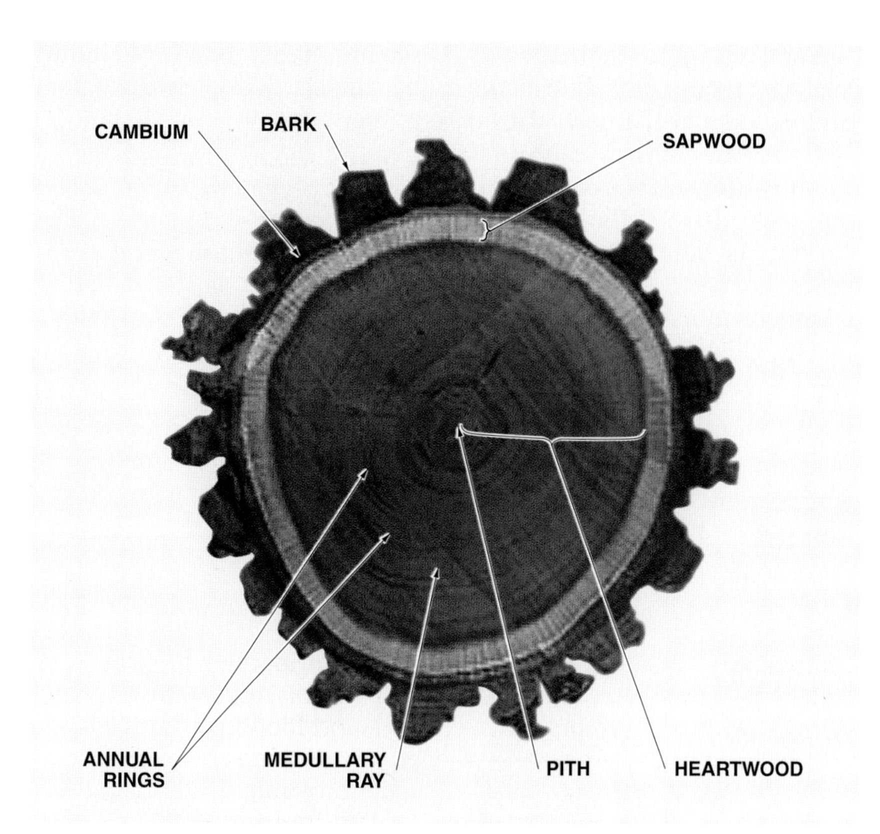

Figure 3-2. *Annual rings begin at the center of the trunk and continue outward to the bark. Each ring represents a year of cellular growth.*

roots, up through the sapwood, and to the leaves. A young tree consists entirely of sapwood.

Heartwood

As a tree grows, the number of annual rings increases. In addition, the layers of wood nearest the center of the trunk undergo certain changes. The wood cells become inactive and no longer conduct sap and food. When this occurs, the sapwood in the central part of the trunk changes into heartwood and usually begins to darken in color.

Lumber cut from sapwood and lumber cut from heartwood are about equal in strength. However, heartwood is more decay-resistant. As a result, heartwood is more durable than sapwood when exposed to weather and is a better exterior finish material.

Pith and Medullary Rays

The *pith* is the small central core of a tree. Pith is a soft, spongy material and does not produce a good structural grade of lumber. Medullary rays start at the pith area and extend outward toward the outside of the trunk. Medullary rays store and transport food for the tree.

Sapwood commonly ranges from 1½" thick to 2" thick. Maple, hickory, and ash may have 3" to 6" thick sapwood.

MOISTURE CONTENT OF WOOD

Water accounts for a large percentage of the weight of a living tree. Water is present in the cell cavities of the wood and cell walls. *Green lumber* (recently cut lumber) consequently has a high moisture content.

Lumber begins to dry out as water in the wood cells evaporates. The water evaporates first from the cell cavities, then from the cell walls. When water has been depleted from the cell cavities but the cell walls still contain water, the fiber saturation point has been reached. Wood does not begin to shrink until after it reaches the *fiber saturation point*. Only when water begins to leave the cell walls do the cells begin to decrease in size, which causes the wood to shrink.

Lumber gives off moisture until the amount of moisture in the wood is the same as the amount of moisture in the surrounding air. When this occurs, the lumber has reached a state of *equilibrium moisture content*. When the equilibrium moisture content is reached, lumber stops shrinking.

Moisture content can be tested in a laboratory by the oven-drying method shown in **Figure 3-3.** Moisture content can also be tested in the field with a moisture meter. **See Figure 3-4.** A moisture meter indicates an instant moisture content reading by measuring the resistance to current flow between two points driven into the wood. A moisture content reading obtained with a moisture meter is not as accurate as the oven-drying method, but it is accurate enough for most construction purposes.

Lumber should have a moisture content compatible with the air that will surround it after it is installed. In drier climates, lumber moisture content should be no more than 15%. In damp climates, a lumber moisture content up to 19% is acceptable. Interior finish materials in most areas should have a moisture content between 6% and 12%.

Effects of Moisture Content

If lumber has too high a moisture content when used on a job, problems may develop from additional shrinkage. Framing members inside the walls may shrink as the wood dries out, causing plaster to crack or nails to pop out if gypsum board (drywall) is installed. Most wood shrinkage occurs across the grain. A piece of lumber measuring 2″ thick, 4″ wide, and 8′ long will shrink very little along its 8′ length. The piece will, however, noticeably shrink across its 2″ thickness and 4″ width. **See Figure 3-5.**

Wood strength increases as moisture content decreases because the cell fibers stiffen and become more compact as they dry out. Wood will not decay (rot) if the moisture content is below 20%. Wood installed under conditions where the moisture content will remain higher than 20% should be treated with chemicals that prevent decay.

Many wood products today (such as plywood) consist of layers of wood glued together. The glue bond of these wood products improves as the moisture content of the wood decreases.

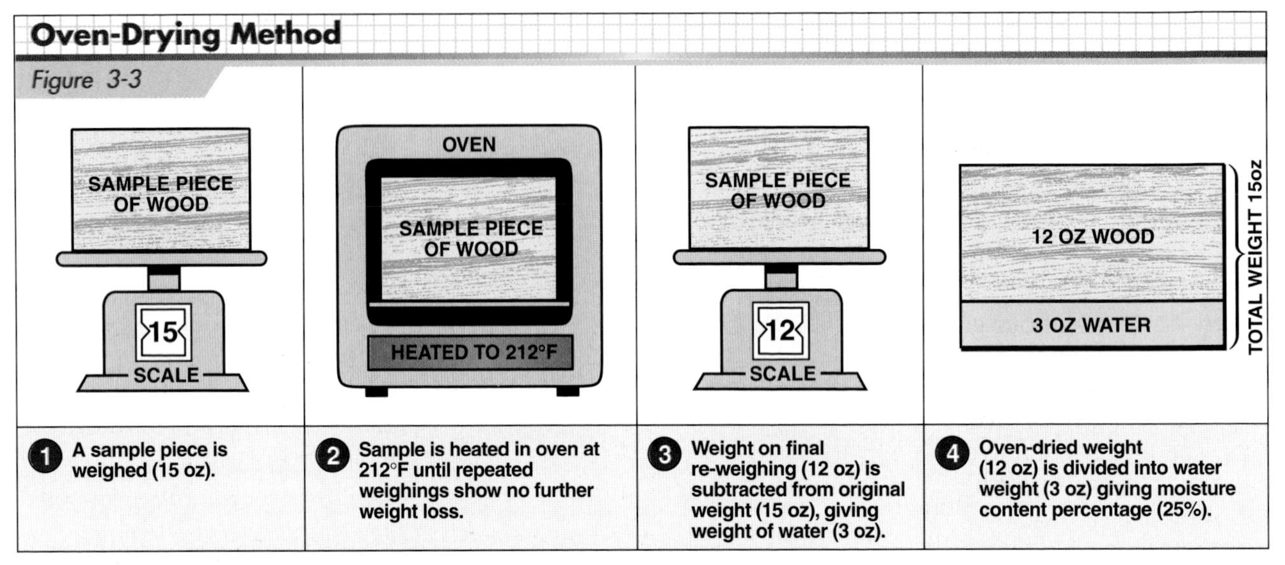

Figure 3-3. *The moisture content of wood is the percentage of its weight that is water. The oven-drying method provides an accurate measure of moisture content.*

Lignomat U.S.A., LTD.

Figure 3-4. *A moisture meter is a convenient tool to use to check the moisture content of wood.*

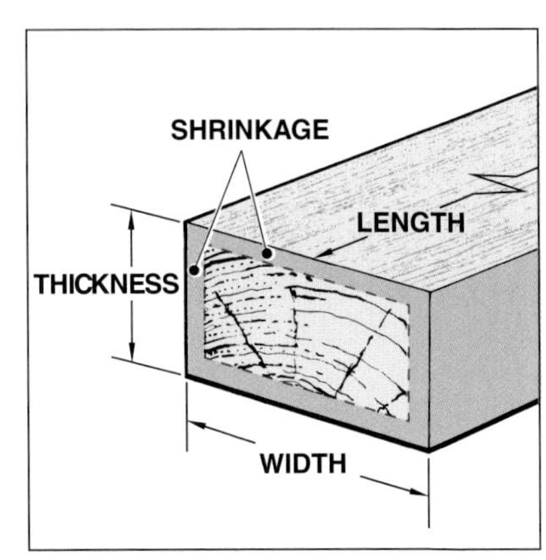

Figure 3-5. *Most lumber shrinkage occurs across the grain. For example, a 2″ × 4″ × 8′ piece of lumber will shrink more across its 2″ thickness and 4″ width than along its 8′ length.*

Unit 3 Review and Resources

UNIT 4 Lumber Manufacture

OBJECTIVES

1. Describe the procedure for lumber manufacture.
2. Discuss common methods of sawing.
3. Describe common methods of seasoning.
4. Explain the procedure for planing and grading.
5. Identify common types of lumber defects.
6. List common types of wood preservatives.
7. Identify the conditions that determine proper use of pressure-treated lumber.

Trees are one of our greatest natural resources. As the population of our country continues to grow, however, more trees are needed to supply the lumber required for housing and other types of construction. Improved forest management and waste reduction in sawmill operations provide a partial solution to the increasing wood needs of the industry.

SAWMILL OPERATIONS

Approximately one-half of the bulk of a log processed in a modern sawmill exits as usable construction lumber. The other half of the log is *residue,* consisting of chips, bark, trimmings, shavings, and sawdust. In the past, residue was considered of no value and was burned as waste. Today, however, most residue is converted into useful products. Wood chips are a major source of wood fiber used by paper mills. Planer shavings and chips are ingredients in the manufacturing of panels used for sheathing and insulation.

The manufacture of lumber begins in the forest where trees are cut. **See Figure 4-1.** Limbs and branches are removed, and the tree is cut into sections (logs) small enough to be transported by truck to a sawmill. **See Figure 4-2.** The logs are processed into the different sizes of lumber used in construction work. **See Figure 4-3.**

Photo Courtesy of Deere & Company

Figure 4-1. *Trees are cut using a feller buncher that grasps a tree as it is sawn off at the base. Trees sustain less damage when felled in this manner.*

Photo Courtesy of Deere & Company

Figure 4-2. *Log loaders stack logs on a truck for transport to a sawmill.*

APA—The Engineered Wood Association

Engineered wood products, such as wood I-joists and glulam beams, make efficient use of wood and wood by-products.

Western Wood Products Association

Figure 4-3. *A log can be cut into many shapes and sizes of lumber.*

At a sawmill, logs are stored in ponds or are stored on the ground and continually sprayed with water to prevent shrinkage caused by drying. The logs are then transported to the sawmill from the merchandising deck as shown in **Figure 4-4.** Initial sawmill operations include *debarking,* which is a process of stripping the bark from a log. **See Figure 4-5.** The debarked logs are cut into smaller sections, which are then cut into boards by a bandsaw. The boards travel by conveyor belts into a *trimmer,* which cuts the boards to standard lengths and also cuts off pieces containing defects. **See Figure 4-6.**

Southern Forest Products Association

Figure 4-4. *Logs are conveyed into the sawmill from a log merchandising deck.*

Southern Forest Products Association

Figure 4-5. *The logs pass through a debarker to remove bark and dirt, which may dull saw blades used in subsequent operations.*

SAWING METHODS

Two primary methods are used to cut logs into lumber—*plainsawing* and *quartersawing.* **See Figure 4-7.** Another sawing method, known as *rift sawing,* is used only for hardwood lumber. Each sawing method gives lumber a distinct grain pattern and also affects the performance characteristics of the lumber.

Plainsawing Lumber

When plainsawing lumber, the annual rings of the log are at an angle of 45° or less to the wide surface of the boards being cut. Plainsawing is also referred to as *tangential cutting.* Most lumber is produced using the plain-sawing method since it provides the widest boards and minimizes waste. When the plainsawing method is used for hardwood, the resulting lumber is referred to as *plainsawn.* When the plainsawing method is used for softwood, the lumber is referred to as *flat-grained.*

Color variation within a wood species is influenced by many factors including minerals, soil type, available sunlight, water levels, and temperature. Actual color variations are caused by natural chemical extractives found in the cell walls of wood.

Southern Forest Products Association

Figure 4-6. *Boards are fed into the trimmer by conveyor belts. A trimmer consists of a series of saw blades that remove defects and cut the boards to standard lengths.*

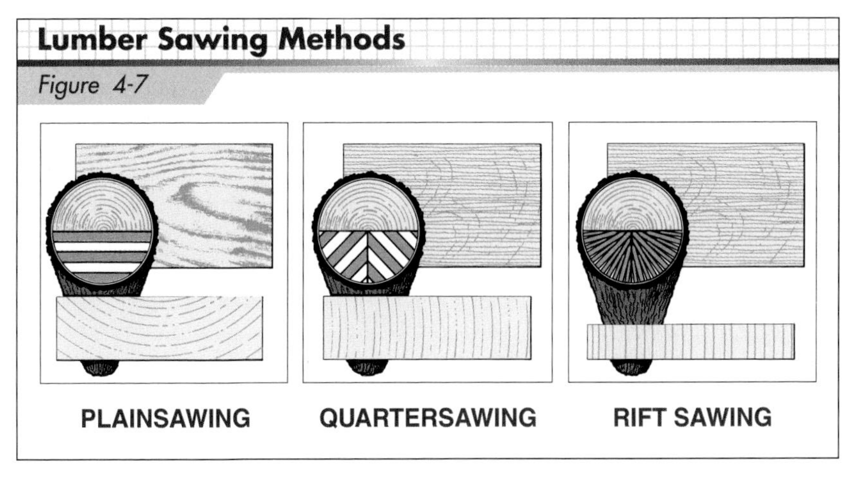

Figure 4-7. *Lumber sawing methods include plainsawing, quartersawing, and rift sawing.*

Quartersawing Lumber

When quartersawing lumber, the log is first quartered lengthwise. Lumber is cut from each quartered section, with the cuts made at a 90° angle to the annual rings. The quartersawing method is more expensive than plainsawing, as it produces narrower boards and creates more waste. However, quartersawing produces a more attractive grain pattern in some hardwoods. In addition, there is less warpage in quartersawn wood. When the quartersawing method is used for hardwood, the resulting lumber is referred to as *quartersawn.* When the quartersawing method is used for softwood, the lumber is referred to as *edge-grained.*

Rift Sawing Lumber

When rift sawing lumber, the log is first quartered lengthwise, similar to quartersawing. Lumber is cut from each quartered section, with the cuts made at a 30° or greater angle to the annual rings. Rift sawing results in relatively narrow boards. Rift-sawn lumber is available in limited quantities and is only available in certain wood species.

SEASONING METHODS

Lumber must be *seasoned* (dried) properly before it is placed on the market. The two seasoning methods are *air drying* and *kiln drying.*

Air Drying

When air drying lumber, the newly produced lumber is stacked outside with wood strips placed between the layers of wood so that air can circulate around each piece. A shed or flat roof is often placed over the stack so rain cannot fall directly on the lumber. It takes several months for the lumber to season adequately. Most softwood used for rough construction is seasoned by air drying.

Kiln Drying

When kiln drying lumber, the newly produced lumber is placed in a temperature-controlled building called a *kiln,* which acts like a large oven. **See Figure 4-8.** First, steam is introduced into the kiln to increase the humidity (amount of moisture in the air) in the kiln. As the temperature in the kiln is gradually increased, the humidity level is decreased. Kiln-dried lumber is stamped with a "KD." Kiln drying is usually used for higher grades of hardwood lumber used for finish work. Kiln-dried lumber is more expensive than air-dried lumber.

Southern Forest Products Association

Figure 4-8. *Lumber is stacked on tram cars and transported into a kiln for seasoning. Note the stickers placed between rows of lumber to allow for complete circulation of air.*

PLANING AND GRADING

Lumber has a rough surface when it is cut from a log. After the lumber is properly seasoned, it must be finished in a planing mill.

At the mill, the rough lumber passes through *planers,* which are machines with rotating knives that *surface* (smooth off) the sides and edges of the lumber. As the planed lumber moves along a conveyor belt, highly skilled workers called *graders* examine the boards for defects and mark each piece according to grade. The graded pieces are later sorted according to thickness, width, and length. **See Figure 4-9.**

LUMBER DEFECTS

Lumber defects may affect the strength, stiffness, or appearance of the lumber. Most lumber has some defects. The number and type of defects determine the lumber grade. Lumber with serious defects cannot be used for structural purposes.

Natural Defects

Many lumber defects occur from natural causes during the growth of the tree. *Knots* are the most common natural defect. As a tree grows, its upper limbs and limbs of surrounding trees cast shadows upon its lower limbs. The lack of light causes some lower limbs to die, decay, and fall away. However, a small piece of the dead limb may remain attached to the tree. As the tree continues to grow and expand, new sapwood is added to the trunk. The pieces of dead limb are covered over and become knots.

Knots are found in most lumber. **See Figure 4-10.** If knots are *sound* (remain firmly in place), a few of them will not significantly affect the lumber strength. Knots are classified by their diameter as follows:

- *pin knot*—½″ or less
- *small knot*—more than ½″ but less than ¾″
- *medium knot*—more than ¾″ but less than 1½″
- *large knot*—more than 1½″

Other natural defects occurring in lumber are as follows:

- *wane*—absence of wood or the presence of bark on the edge or corner of a piece of lumber. **See Figure 4-11.**
- *shake*—lengthwise separation of wood fibers between or through the annual rings
- *check*—separation of wood fibers across the annual rings
- *split*—separation of wood fibers across the annual rings (similar to a check) but extending entirely through a piece of lumber
- *pitch pocket*—opening in the wood that contains solid or liquid pitch
- *pitch streak*—section of wood fibers saturated with enough pitch to be visible

Southern Forest Products Association

Figure 4-9. *Lumber is graded and sorted according to thickness, width, and length as it moves along a conveyor belt.*

Warping

Warping is a distortion that occurs during the evaporation of water from the wood cells. Uneven shrinkage in the wood produces warping and results in twisted and uneven shapes of lumber. Common warpage shapes are the crown, bow, twist, and cup. **See Figure 4-12.** A *crown* (crook) is a deviation from a flat plane of the edge (narrow face) of a piece of lumber from end to end. A *bow* is a deviation from a flat plane of the wide face of a piece of lumber from end to end. A *twist* is a deviation from the flat planes of all four faces by a spiraling or torsional action, which is usually the result of improper seasoning. A *cup* is a deviation from a flat plane, edge to edge.

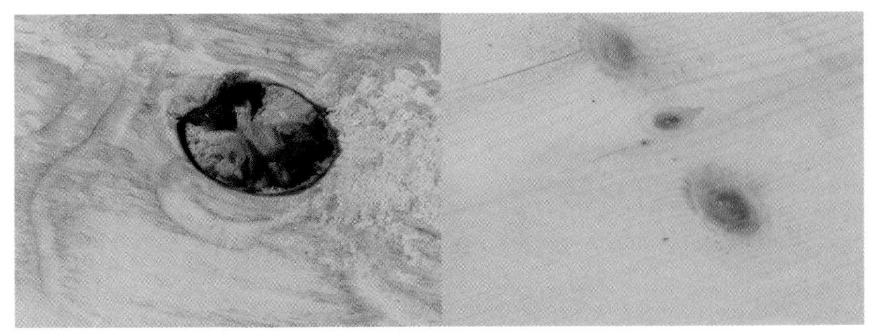

Figure 4-10. *Knots are found in most grades of lumber. A small number of sound knots will not significantly affect the lumber strength.*

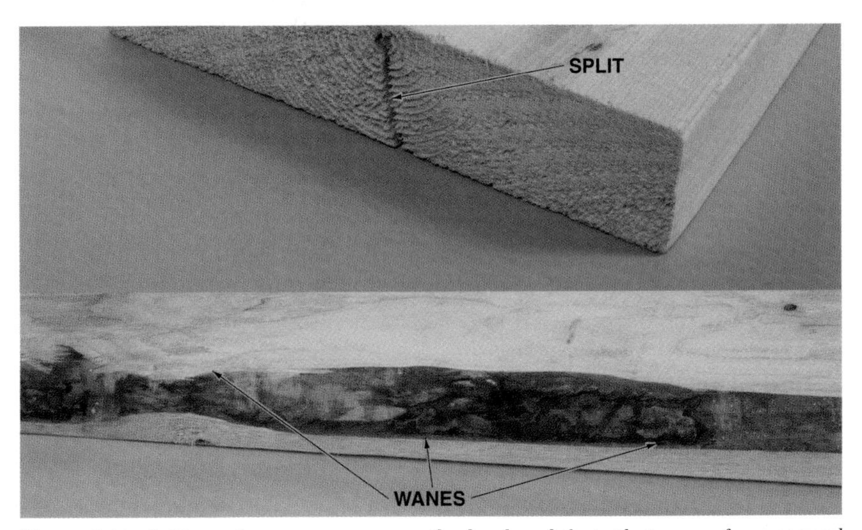

Figure 4-11. *Splits and wanes are among the lumber defects that occur from natural causes.*

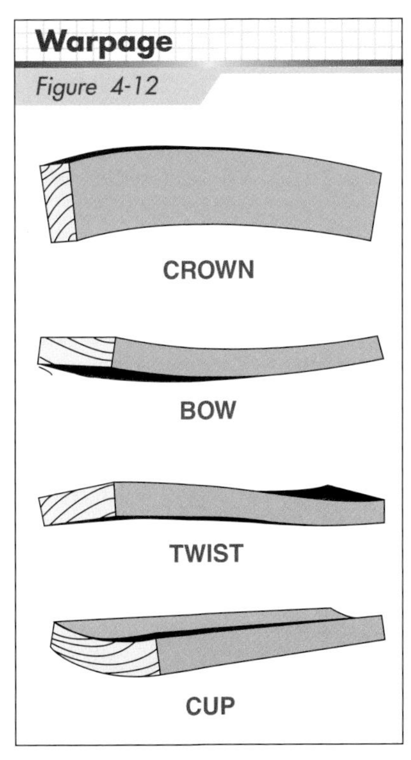

Figure 4-12. *Warpage can occur during the evaporation of water from wood cells.*

Damage from Manufacturing Processes

Defects that mar the appearance of lumber may be caused during sawmill operations. A list of these defects follows:

- chipped, torn, raised, or loosened grain
- skip marks that occur during surfacing
- burns caused by machines
- bite or knife marks

WOOD PROTECTIVE TREATMENT

Wood can be treated to make it resistant to fire, decay, and attacks from insects. Most wood treatments are designed to protect against fungi and termites since they are the main causes of serious damage to lumber. The damage usually begins in the wood members near ground level.

Damage Caused by Fungi

Dry rot causes wood tissue to deteriorate, reducing the strength of a wood member. **See Figure 4-13.** Dry rot is the most common type of damage caused by a fungus. The fungus that causes dry rot lives in the wood and can be seen only with a microscope. Since dry rot fungus must have water to live, it can survive only in wood with a moisture content of at least 20%. The term *dry rot* is misleading, since the deterioration begins under damp conditions. However, it is often not detected until after the wood has dried out.

Other types of fungi cause specks, molds, and stains. These fungi are fairly harmless, since they damage only the surface of the wood, affecting its appearance but not its structural quality. *White speck* and *honeycomb* are examples of surface damage caused by fungi. Wood with white speck has small white spots or pits in its surface. Honeycomb is similar, but the spots are larger or the pits are deeper. Another type of fungus causes *blue stain,* which is a blue-gray discoloration.

A key to preventing dry rot is controlling the exposure of wood to moisture. Most wood decay fungi grow only on wood with a high moisture content, usually 20% or above.

Wood Preservatives

When most woods are exposed to common weather conditions, excess moisture or humidity levels, or ground contact, the woods will deteriorate. Four conditions are required for insect attack and deterioration to occur—moisture, oxygen, 50°F to 90°F temperature, and a food source such as wood fiber. If any of these conditions is missing, deterioration will not occur.

Figure 4-13. *Dry rot is the breakdown of wood caused by a wood-destroying fungus.*

In some areas of the United States, moisture and oxygen levels and temperatures are more favorable to wood deterioration than in other areas. **See Figure 4-14.** Therefore, wood preservatives are used to eliminate wood fiber as a food source. Wood preservatives contain chemicals that protect the wood against fungus decay and insect attack. Preservatives are divided into three major types—waterborne, oil-borne, and creosote.

Waterborne Preservatives. Waterborne preservatives are used to treat lumber and panel products for residential, commercial, marine, agricultural, recreational, and industrial construction applications. Waterborne preservatives are the most common preservative specified for residential, commercial, and marine construction applications. Waterborne preservatives are clean, odorless, and easy to paint. Waterborne preservatives do not contain any arsenical or chromium compounds, but they provide excellent decay and termite resistance. Common waterborne preservatives are listed in the table in **Figure 4-15.**

American Hardwood Export Council

Proper forest management techniques must be employed to ensure that an adequate supply of lumber can be provided to the construction industry.

Chromated copper arsenate (CCA) is a waterborne preservative that is no longer permitted to be used in residential and other consumer-related applications. However, CCA is an approved preservative for industrial, marine, commercial, and agricultural applications such as piles and posts, permanent wood foundations, and wood shakes and shingles.

Inhalation of sawdust from waterborne preservative-treated wood should be avoided. When sawing or machining treated wood, an approved particulates mask (dust mask) and goggles should be worn. If possible, sawing and machining operations should be performed outdoors to avoid accumulation of sawdust within a building. Wood treated with waterborne preservatives may be used inside residential structures as long as all sawdust and scrap are cleaned up and properly disposed of after construction. Waterborne preservative-treated wood should not be burned; rather, it should be disposed of by ordinary trash collection or burial. After working with treated wood, exposed body areas should be washed thoroughly.

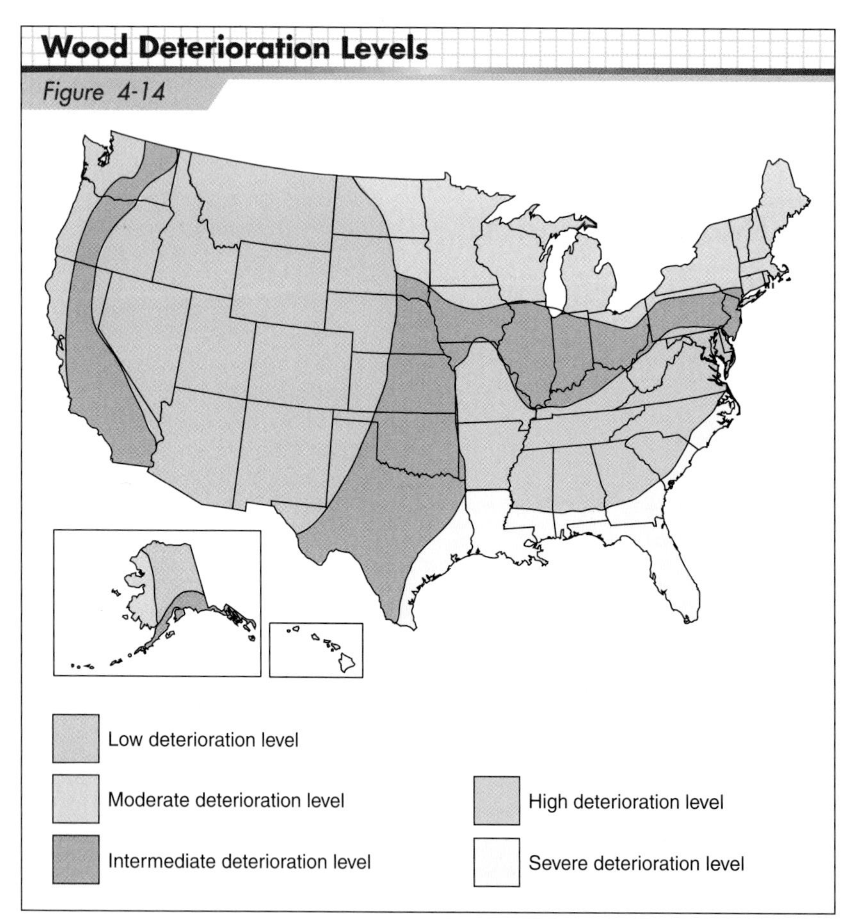

Figure 4-14. *Levels of wood deterioration vary depending on humidity levels and temperature.*

Oil-borne Preservatives. Best known in the form of *pentachlorophenol,* oil-borne preservatives are highly toxic to fungi and insects. However, pentachlorophenol may affect the surface color of the treated material.

WATERBORNE PRESERVATIVES		
Exposure Conditions	**Non-Copper-Based**	**Copper-Based**
Interior, dry or damp	Inorganic Boron (SBX) Propiconazole-Tebuconazole-Imidacioprid (PTI)	Copper Azole (CA-B) Copper HDO (CX-A) Copper Naphthenate (CuN-W) Copper Quat (ACQ, Micronized Copper)
Exterior, above ground	Propiconazole-Tebuconazole-Imidacioprid (PTI)	Copper Azole (CA-B) Copper HDO (CX-A) Copper Naphthenate (CuN-W) Copper Quat (ACQ, Micronized Copper)
Ground contact or fresh water		Copper Azole (CA-B) Copper Naphthenate (CuN-W) Copper Quat (ACQ, Micronized Copper)

Figure 4-15. *Waterborne preservatives are used for residential, commercial, marine, agricultural, recreational, and industrial construction applications.*

Inhalation of sawdust from pentachlorophenol-treated wood should be avoided. When sawing or machining pentachlorophenol-treated wood, an approved particulates mask (dust mask) and goggles should be worn. If possible, sawing and machining operations should be performed outdoors to avoid accumulation of sawdust within a building. Direct contact with pentachlorophenol-treated wood should be avoided; wear long-sleeved shirts, long pants, and gloves impervious to chemicals. Wood treated with oil-borne preservatives may be used inside residential structures as long as all sawdust and scrap are cleaned up and properly disposed of after construction. Pentachlorophenol-treated wood should not be burned; rather, it should be disposed of by ordinary trash collection or burial. After working with treated wood, exposed body areas should be washed thoroughly.

Creosote. Creosote is one of the oldest preservatives, but has largely been replaced with other wood preservatives. It comes in a variety of mixtures and leaves a slight odor after it has been applied. Surfaces coated with creosote cannot be painted.

Methods of Application. Wood preservatives may be applied by pressure or non-pressure processes. The pressure process is considered to be the most effective since preservatives penetrate deeply into the wood. **See Figure 4-16.** Untreated lumber is loaded onto tram cars and rolled into long steel cylinders, which are then sealed. The cylinders are filled with the preservative. Intense pressure builds up inside the tank and forces the preservative deeply into the wood.

The non-pressure process provides less protection than the pressure-treating process, but it is simpler and less expensive to apply. Untreated lumber is submerged in an open tank filled with preservative and allowed to soak for at least 3 min or the preservative is spray-applied. Lumber treated with this type of surface barrier should only be used for enclosed interior framing applications. Surface-coated wood cannot be substituted where pressure-treated wood is required by the building code.

Southern Forest Products Association

Figure 4-16. *Untreated lumber is placed in large cylinders and impregnated with preservative.*

Pressure-treated sill plates are typically required by building codes. Note the use of a sill sealer below the sill plate.

Exposure Conditions

Proper use of pressure-treated lumber is determined by its exposure to the weather and exterior conditions. Most pressure-treated lumber is either rated for *ground contact* or *above-ground* exposure conditions. Ground contact exposure conditions include situations when pressure-treated lumber is in contact with the ground or freshwater. Above-ground exposure conditions include applications such as sill plates, beams, joists, and decks. A grade mark used for pressure-treated lumber is shown in **Figure 4-17.**

Pressure-treated lumber is required for many structural applications, including the following:

- wood-framing members that rest on exterior foundation walls and are less than 8″ from exposed earth
- sill plates and sleepers on a concrete or masonry slab that is in direct contact with the soil
- joists or the bottom of a wood structural floor without joists closer than 18″ to exposed soil
- girders that are within 12″ of exposed soil in a crawl space

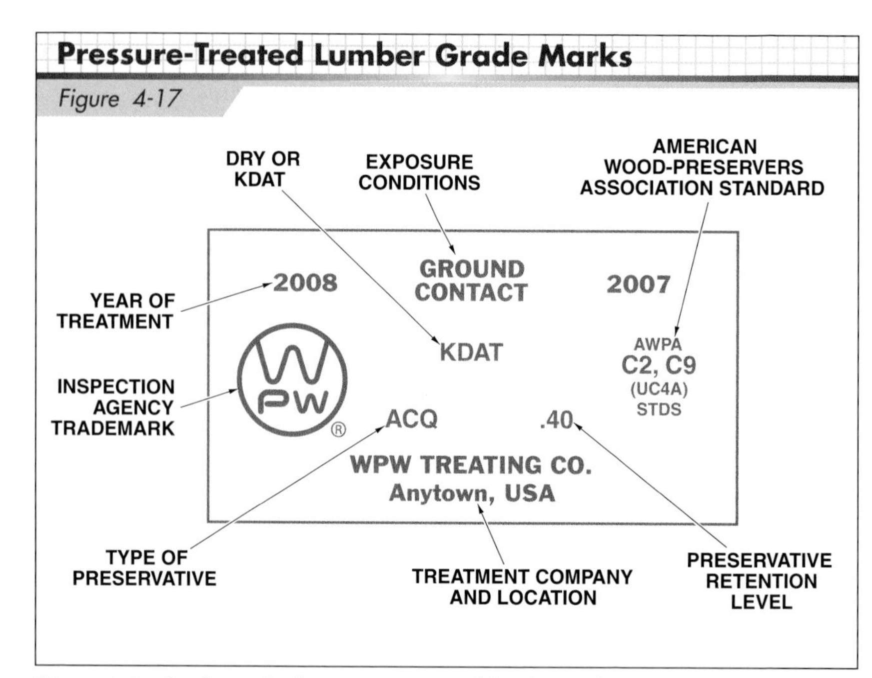

Figure 4-17. *Grade marks for pressure-treated lumber include information about the exposure conditions for proper use of the lumber, and also indicate the type of preservative used on the lumber.*

Fire-Retardant Treatment

Wood is a highly combustible material and methods are continually being developed to provide greater fire protection for construction lumber. Lumber that has been treated to make it fire-retardant is available for roof systems, beams, posts, studs, doors, hardwood paneling, and other interior trim products.

Fire retardants are applied the same way preservatives are applied. In a pressure process, wood is impregnated with the fire-retardant chemicals. In a non-pressure process, the wood receives a fire-retardant coating.

Fire-retardant chemicals react to heat slightly below the temperature required to ignite the wood. When heated, the chemicals release a vapor that surrounds the wood fibers and sets off a reaction in the wood by forming a protective insulation *char* on the surface of the wood. The char prevents the wood from igniting and reduces the amount of smoke and toxic fumes caused by the fire. Fire-retardant wood products are also resistant to termites and decay.

Per the International Building Code, only hot-dipped zinc-coated galvanized steel, stainless steel, silicon bronze, or copper fasteners are permitted for use with pressure-treated and fire-retardant-treated wood due to the chemicals used in the treatment process.

Southern Forest Products Association

Wood framing members and plywood used for permanent wood foundations must be pressure-treated.

Inhalation of sawdust from fire-retardant-treated wood should be avoided. When sawing or machining fire-retardant-treated wood, an approved particulates mask (dust mask) should be worn. If possible, sawing or machining operations should be performed outdoors. Direct contact with fire-retardant-treated wood should be avoided; wear long-sleeved shirts, long pants, and gloves impervious to chemicals. Fire-retardant-treated wood should not be burned; rather, it should be disposed of by ordinary trash collection or burial.

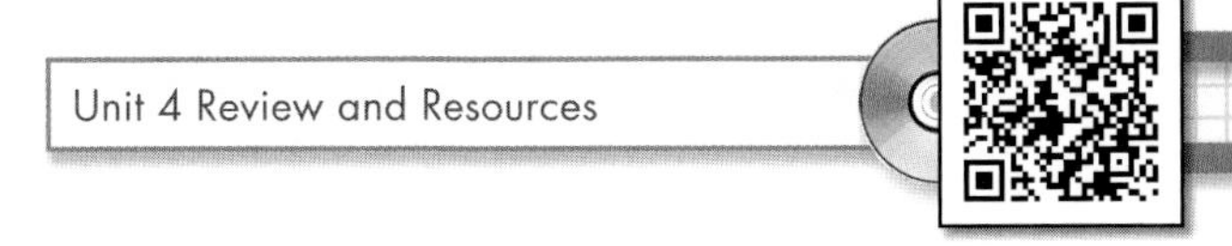

UNIT 5

Softwood and Hardwood

OBJECTIVES

1. Identify common applications of softwood and hardwood trees.
2. Describe grading systems of softwood.
3. Identify common applications of framing lumber.
4. List the classifications of Dimension framing lumber.
5. List the classifications of Special Dimension framing lumber.
6. List and describe the classifications of Timber framing lumber.
7. Describe applications and categories of Appearance grade lumber.
8. Identify common applications of lumber in the Select grade group.
9. Describe types of Finish grades.
10. Describe types of Special Western Red Cedar Pattern grades.
11. Identify common applications of General-purpose appearance category lumber.
12. Identify common applications of the Alternate Boards grade group.
13. Identify common applications of industrial lumber grades.
14. Explain the purpose and types of grade marks.
15. List and describe the primary grades of hardwood lumber.

Lumber is referred to by the same name as the tree it is produced from. For example, Douglas fir lumber is produced from a Douglas fir tree and walnut lumber is produced from a walnut tree. Trees and lumber are divided into two main classes—*softwood* and *hardwood*. The terms *softwood* and *hardwood* can be confusing since certain softwood lumber is physically harder than some hardwood lumber. In general, however, hardwoods are more dense and harder than softwoods.

SOFTWOOD AND HARDWOOD TREES

Softwood trees are called *conifers*. Conifers have thin needles and bear cones from which seeds germinate and grow. **See Figure 5-1.** Conifers are commonly referred to as *evergreens* since they bear green needles throughout the year. Over 75% of the wood used for construction applications is softwood.

Most hardwood trees in North America are broad-leaved, *deciduous* trees, which lose their leaves in the autumn. **See Figure 5-2.** Hardwood lumber is typically used for furniture and cabinetry where attractive grain patterns are desired.

Figure 5-1. *Softwood lumber comes from evergreen trees, which bear cones and have needle-shaped leaves.*

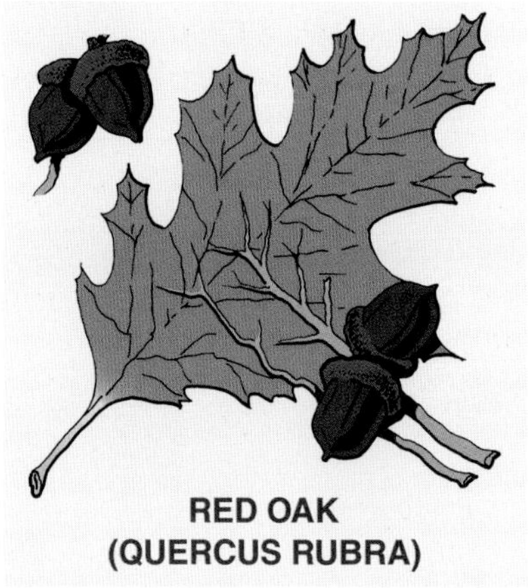

Figure 5-2. *Hardwood lumber comes from broad-leaved, deciduous trees, which lose their leaves in the fall.*

SOFTWOOD LUMBER

Lumber used for rough construction, such as framing or concrete forms, is softwood. Softwood may also be used for finish products such as moldings, doors, and cabinets. In general, softwood products used for finish applications are painted rather than stained. Some of the more frequently used softwoods are as follows:

- Douglas fir
- white fir
- white pine
- ponderosa pine
- sugar pine
- southern pine
- hemlock
- spruce
- cypress
- redwood
- western red cedar

Many softwood species have similar strength and stress characteristics and are used for general applications. However, redwood and western red cedar are more decay-resistant than other species and are recommended for exterior trim, siding, decks, and fences. **See Figure 5-3.**

Availability of softwood lumber species, especially framing lumber, varies from one part of the country to another. Framing lumber species are typically selected based on the species native to the area in which the lumber is being used, since lumber is expensive to transport long distances. In the western United States, Douglas fir lumber is commonly used for rough construction since Douglas fir trees are abundant along the Pacific coast. In the southeastern states, various species of southern pine lumber (longleaf, slash, shortleaf, loblolly) are widely used for rough construction since these species are grown in forests ranging from Virginia to Texas. However, a certain species of framing lumber may be specified for a particular construction project due to its physical characteristics.

California Redwood Association

Figure 5-3. *Redwood is commonly used for exterior applications, such as fences and siding, since it is decay-resistant.*

Softwood Lumber Standards

Voluntary Product Standard DOC PS 20-99, *American Softwood Lumber Standard,* establishes the principal trade classifications and lumber sizes for yard, structural, and factory and shop use. The *Standard* is primarily based on the input of softwood lumber-producing industry associations including the Western Wood Products Association, Southern Pine Inspection Bureau, West Coast Lumber Inspection Bureau, and Northern Softwood Lumber Bureau. The *Standard* encourages uniform grading rules, definitions, and markings. In addition, the *Standard* provides information to all areas of the U.S. and Canada about lumber sizes, design values, and other specifications.

Canadian softwood lumber standards are governed by the National Lumber Grading Association's publication, Standard Grading Rules for Canadian Lumber.

Softwood Grading Systems

The grade of lumber used on a construction project is usually stated in the specifications of the construction prints. Lumber grade is based on its strength, stiffness, and appearance. A high grade of lumber has few knots or other defects. A lower grade of lumber may have knotholes and loose knots. The lowest grades may have splits, checks, honeycombs, and warpage. Based on the *Western Lumber Product Use Manual,* published by the Western Wood Products Association, the basic softwood end-use categories are *Framing Lumber, Appearance Lumber,* and *Industrial Lumber.* The end-use categories are further divided into various general classifications and lumber grades.

Framing Lumber. Framing lumber is primarily used for structural load-bearing applications and is graded for its strength. Appearance of framing lumber is secondary unless the lumber will be visible, as in exposed posts, ceiling beams, and roof beams. Framing lumber includes *Dimension lumber, Special Dimension lumber,* and *Timbers* classifications.

The Dimension lumber classification includes products that are 2″ to 4″ thick by 2″ and wider and includes studs, joists, planks, roof rafters, trusses, and other components that form the framework of a building. **See Figure 5-4.** Dimension lumber is further classified as follows:

- *Structural Light Framing (SLF).* Structural Light Framing lumber applications include concrete forms, trusses, and engineered applications. Lumber that is 2 × 2 through 4 × 4 and used where higher strength design values are required is included in this classification. Structural Light Framing grades are Select Structural, No. 1, No. 2, and No. 3.
- *Light Framing (LF).* Lumber in the Light Framing classification includes 2 × 2s through 4 × 4s that are not intended for load-bearing applications. Light Framing lumber is used for non-load-bearing wall framing, plates, sills, and blocking. Light Framing grades are Construction, Standard, and Utility.
- *Stud.* The Stud classification includes 2 × 2s through 4 × 18s that are commonly used for vertical load-bearing walls. Lumber in the Stud classification is graded as Stud.
- *Structural Joists and Planks (SJ&P).* Structural Joists and Planks are used for horizontal load-bearing members such as floor joists, rafters, headers, and trusses, and include 2 × 5s through 4 × 18s and lumber 5″ and wider. Structural Joists and Planks grades are Select Structural, No. 1, No. 2, and No. 3.

The Special Dimension lumber classification includes Structural Decking and Machine Stress-Rated (MSR) lumber classifications. Special Dimension lumber is further classified as follows:

- *Structural Decking.* The Structural Decking, or Roof Decking, classification includes 2 × 4s through 4 × 12s that are used as roof covering. However, because of its load-bearing capacities, Structural Decking can also be used as floor decking and solid sidewall construction.

- *Machine Stress-Rated (MSR) lumber.* MSR lumber is Dimension lumber that has been evaluated by mechanical stress-rating equipment, which measures its stiffness and sorts it into various *modulus of elasticity* classes. Modulus of elasticity is a ratio of the amount a material bends in proportion to an applied load. The primary application of MSR lumber is trusses, but it is also used for floor and ceiling joists, rafters, and other applications where strength capabilities are a primary consideration.

The Timbers classification is a general category as well as a classification for the larger sizes of structural framing lumber. The Timbers classification includes two grade groups—*Beams and Stringers* and *Posts and Timbers*. Descriptions of the grade groups follow:

- *Beams and Stringers.* The Beams and Stringers grade group includes lumber that is 5″ and thicker, with the width more than 2″ greater than the thickness. Nominal 6 × 10s and 8 × 12s are included in this grade group. Beam and Stringers grades include Dense Select Structural, Dense No. 1, Dense No. 2, Select Structural, No. 1, and No. 2.
- *Posts and Timbers.* The Posts and Timbers grade group includes lumber that is 5″ × 5″ and larger, with the width not more than 2″ greater than the thickness. Nominal 6 × 6s and 6 × 8s are included in this grade group. **See Figure 5-5.** Post and Timbers grades include Dense Select Structural, Dense No. 1, Dense No. 2, Select Structural, No. 1, and No. 2.

Examples of Dimension lumber grading charts published by the Western Wood Products Association and Southern Pine Inspection Bureau are shown in the Appendix. Even though the charts show some variations in their sequence, they both follow the basic outlines of the American Softwood Lumber Standard.

Southern Forest Products Association

Figure 5-4. *Dimension lumber is used for studs, plates, trusses, and other components that form the framework of a building.*

Southern Forest Products Association

Figure 5-5. *Timbers are heavy structural members and are commonly used for posts, beams, and stringers.*

Appearance Grade Lumber. Good appearance is the main consideration for most grades of Appearance lumber. Appearance grade, or Board lumber, is visually inspected. The highest grades of Appearance lumber are typically not stamped with a grade mark as the stamping would deface the lumber. Differences in grade designations exist for different species of lumber. For example, cedar and redwood have different grade designations due to the difference in color between heartwood and sapwood in the lumber.

In most cases, appearance grades apply to lumber 1″ and thicker by 2″ and wider. The highest grades are *clear* (without knots or other defects) or nearly clear and are used where the beauty of exposed lumber is the primary consideration. Appearance grade lumber is typically used for applications such as moldings, window and door frames, interior paneling, exterior siding, and floor decking. **See Figures 5-6** and **5-7.** Appearance grades range from clear lumber to rustic blue-stained boards that contain knotholes.

Appearance grade lumber is classified into two main categories—*high-quality appearance grades* and *general-purpose appearance grades.* High-quality appearance grades include *Select, Finish,* and *Special Western Red Cedar Pattern* grade groups. General-purpose appearance grades include *Common Boards, Alternate Boards,* and *Special Western Red Cedar Pattern* grade groups. Each of the grade groups is divided into specific grades.

Lumber in the Select grade group is used for applications where only the best appearance is desired. Select grades are determined from the better side or face of the lumber. The Select grade group includes the following grades:

- *B & BTR.* Lumber graded as B & BTR contains very few defects, and is very limited in availability. B & BTR lumber is used for fine furniture, exposed cabinetry, trim, and flooring. **See Figure 5-8.**
- *C Select.* C Select lumber contains small, tight knots, and may be nearly perfect on one side. C Select lumber is used for most furniture, shelving, and some trim and flooring.
- *D Select.* D Select lumber contains many pin knots and other small defects. D Select lumber is used for furniture interiors, shelving, and some trim and flooring.

Lumber in the Finish grade group is graded from the better side or face and from both edges on pieces 5″ and narrower, and from the better side or face and one edge on pieces 6″ and wider. Finish grades are typically only available in Douglas fir and hemlock fir. The Finish grade group includes the following grades:

- *Superior.* Lumber with the Superior grade is clear of most defects and imperfections.

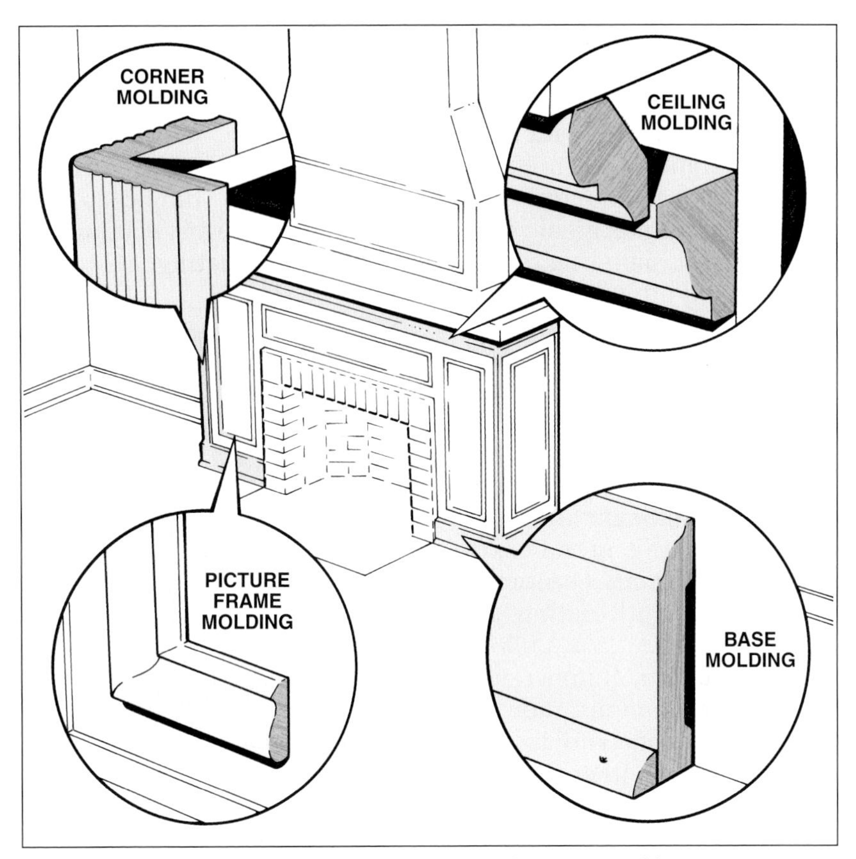

Figure 5-6. *Appearance grade lumber is used to make interior molding.*

- *Prime.* Lumber with the Prime grade includes a few minor defects or surface blemishes.
- *E.* Lumber with an E grade is typically used for ripping and crosscutting to obtain smaller pieces of Prime or better quality lumber.

California Redwood Association

Figure 5-7. *Appearance grade lumber, including redwood, is used for siding.*

The Special Western Red Cedar Pattern grade group is associated only with the highest cedar grades, which are used for siding or paneling products. Lumber with the Clear Heart grade is used only where the highest quality is desired. The exposed surface of Clear Heart lumber is all heartwood and free from imperfections. The A grade only allows minor imperfections.

The general-purpose appearance category includes the *Common Boards, Alternate Boards,* and *Special Western Red Cedar Pattern* grade groups. Common Board grades are graded from the better face of the lumber and include various types of knots. Common Board grades are used primarily for grading pines, spruces, and cedars. The Common Board grades are as follows:

- *1 Common* and *2 Common.* The 1 Common and 2 Common grades are typically sold as a 2 & BTR Common grade, and are primarily used for paneling, shelving, and other applications where a knotty material with a fine appearance is desired.

- *3 Common.* Lumber with the 3 Common grade is also used for paneling, shelving, and other applications where a knotty surface is desired, as well as for fences, crating, sheathing, and industrial applications.
- *4 Common.* Lumber with the 4 Common grade is the most commonly used grade for bracing members of concrete forms, economical fencing, and crating. **See Figure 5-9.**
- *5 Common.* Lumber with the 5 Common grade has many knots and is used only where other Common grades cannot be afforded.

California Redwood Association

Figure 5-8. *Finish appearance grades of redwood lumber may be used for interior paneling.*

The Alternate Boards grade group includes Select Merchantable, Construction, Standard, Utility, and Economy grades, and is primarily used for Douglas fir and hemlock fir. Lumber in the Alternate Boards grade group is graded from the better face. *Select Merchantable* lumber is used in housing and light construction where it is exposed as paneling, in shelving, and where knotty lumber is desired. *Construction grade* lumber is used for spaced sheathing, let-in bracing, fences, boxes, and industrial applications. *Standard grade* lumber is used for concrete forms, economical fencing, and crating.

Figure 5-9. *Concrete forms are commonly braced with 4 Common grade lumber.*

The Special Western Red Cedar Pattern General-Purpose grade group includes *Select Knotty* or *Quality Knotty* grades. Select Knotty and Quality Knotty grades are similar to 2 Common and 3 Common grades, and are used for siding and landscape applications.

Appearance lumber grade charts published by the Western Wood Products Association and the Southern Pine Inspection Bureau are found in the Appendix.

Industrial Lumber. Industrial lumber is seldom available in retail stores since the lumber is typically bought and sold at the wholesale level. Industrial products are used and graded for *structural, nonstructural,* and *remanufacturing* purposes. Structural examples include timbers used to shore mine and tunnel ceilings and walls, scaffold planks, and foundation lumber. Nonstructural grades are used for fence pickets, lath, batten, and gutters.

Finger-jointed lumber is made of short pieces of lumber which have been machined on the ends with a finger profile and glued together, and is accepted by most building codes.

Remanufacturing lumber products, commonly referred to as *factory and shop grades,* are used in the manufacture of molding, shutters, windows, doors, furniture, cabinets, and case goods. Factory and shop grades are primarily used for remanufacturing, and are shipped directly to mills and factories.

Grade Marks

Grade marks are stamped on lumber to provide grading information for end users. A typical grade mark includes the official certification mark of the grading association, such as the Western Wood Products Association or Southern Pine Inspection Bureau, lumber grade, mill identification number, wood species, and surfacing and moisture designations. **See Figure 5-10.**

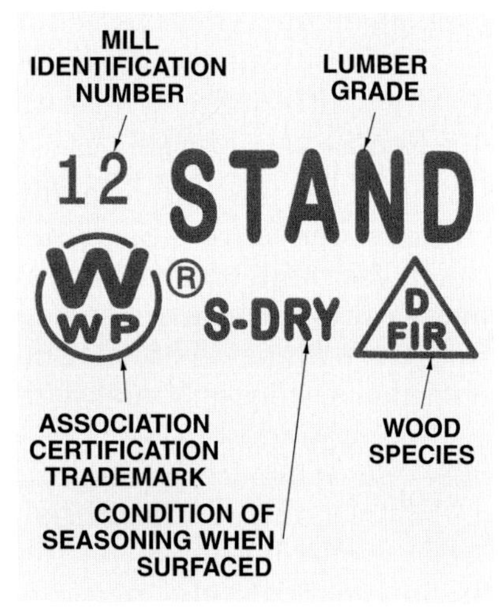

Western Wood Products Association

Figure 5-10. *Grade marks stamped on lumber provide grading and other relevant information.*

HARDWOOD LUMBER

Hardwood accounts for approximately 25% of total lumber production. Most hardwood species suitable for lumber are grown in the eastern United States. Hardwood lumber is typically more expensive than softwood. Hardwood lumber is used for moldings, stair treads, outside veneers of doors and wall paneling, and flooring. High-quality furniture and kitchen cabinets are often constructed from hardwood. **See Figure 5-11.**

Merillat Industries

Figure 5-11. *Hardwood lumber has an attractive grain pattern and is commonly used for high-quality furniture and cabinets.*

Since many hardwoods have an attractive grain pattern, hardwood trim is typically used when a natural or stained finish is desired. Oak and poplar are the most commonly used hardwood species. They account for over 60% of total hardwood production. Other frequently used hardwoods are:

- walnut
- birch
- white ash
- beech
- elm
- maple
- mahogany
- basswood
- butternut
- chestnut
- yellow poplar
- gum

Hardwood Grading System

The *National Hardwood Lumber Association (NHLA)* is the main industry-wide organization that developed and maintains the grades and applications for hardwood lumber in the United States and Canada. Appearance is the primary factor in determining most grades of hardwood lumber. Appearance is affected by lumber defects such as knots, knotholes, bird peck, bark, decay, splits, and warp. Other factors that define the grade of a piece of hardwood lumber are the percentage of high-grade yield from a piece and the number of cuttings (pieces) that can be produced. The primary grades of hardwood lumber are as follows:

- *First and Seconds (FAS)*—kiln dried, highest grade of hardwood lumber
- *First and Seconds 1-Face (F1F)*—select lumber that is 6″ and wider
- *Select*—the better side is graded FAS and the poorer side is graded No. 1 Common
- *No. 1 Common*—standard furniture grade
- *No. 2A Common*—standard grade for cabinets and millwork and is also used for medium to short cuttings
- *No. 2B Common*—the same as No. 2A Common, but stains and other defects are allowed in the clear cuttings; good lumber for a paint grade
- *No. 3A Common*—widely used for flooring and pallets; often combined with No. 3B common and sold as No. 3 common
- *No. 3B Common*—not an appearance grade; widely used for pallets and crating.

Unit 5 Review and Resources

UNIT 6

Size, Shapes, and Dimensions of Lumber

OBJECTIVES

1. Describe the principles of lumber size.
2. Explain how lumber is ordered.
3. Identify common types of lumber surfacing.
4. Compare actual and nominal lumber sizes.
5. Discuss the fundamentals of metric measurement.
6. Describe common methods of converting metric lumber and panel sizes.

Working with lumber, including ordering lumber, requires identifying the size, shape, and dimensions of the materials. In the United States, the English, or Imperial, system of measurement is predominantly used. Units of linear measurement in the English system include yards, feet, inches, and fractions of an inch. The SI metric system may also be used for building materials and components. Units of linear measurement in the SI metric system include meters, centimeters, and millimeters.

LUMBER SIZE

Lumber is usually referred to by its *nominal size*, which differs from its *actual*, or dressed, size. **See Figure 6-1.** A 2 × 4, for example, is 2″ thick and 4″ wide when it is cut from a log at the sawmill. However, lumber shrinks after being seasoned (air-dried or kiln-dried), and surfacing at the planing mill further reduces its measurements. When a 2 × 4 is placed on the market, its actual size is 1½″ × 3½″. Nevertheless, the lumber is referred to as a 2 × 4 (its nominal size) because its original dimensions were 2″ × 4″. See the Appendix for a listing of nominal and actual lumber dimensions.

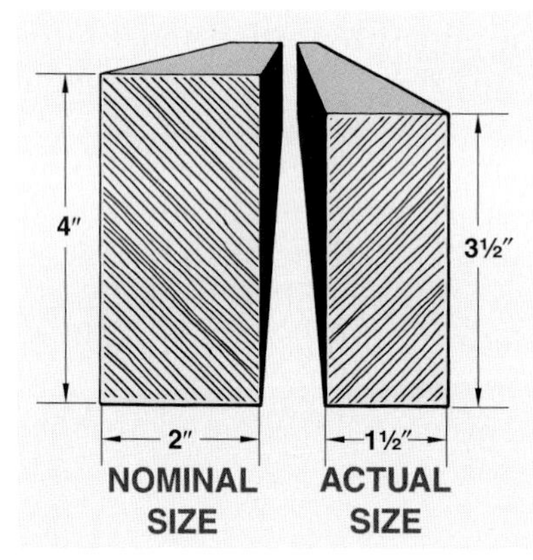

Figure 6-1. *Nominal thickness and width compared to actual thickness and width.*

Nominal Dimensions

Lumber measurements are stated in the following order: thickness, width, and length. For example, a piece of lumber 2″ (nominal size) thick, 4″ (nominal size) wide, and 16′ long is referred to as a 2 × 4 × 16. **See Figure 6-2.**

Softwood lumber is typically sold in even lengths ranging from 6′ to 24′. An extra (premium) charge is made for lumber over 20′ in length. Hardwood lumber is usually sold in random widths and lengths (RWL).

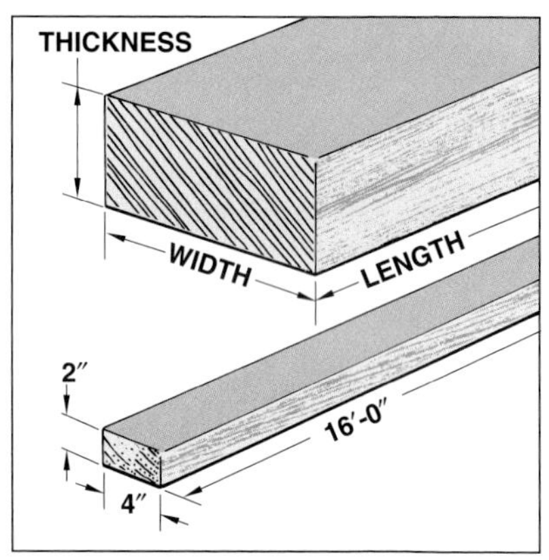

Figure 6-2. *The abbreviated way of referring to a piece of lumber 2″ thick by 4″ wide by 16′ long is 2 × 4 × 16.*

ORDERING LUMBER

Specifications included in a set of construction prints usually provide the species and grade of lumber to be used on a project. Specifications may also include the stress rating and moisture content of the lumber. Unless otherwise indicated, it is assumed that the lumber for a project is surfaced on all four sides (S4S). When ordering materials from a lumberyard or material supplier, the lumber species, grade, dimensions, and quantity are required. The quantity (except for molding) is stated in *board feet.* An example of a lumber order is:

Douglas fir
Standard grade
S4S, 2 × 6 × 16
4500 board feet

Board Foot

Lumber is usually priced by the *board foot (BF).* A board foot is 1″ thick (nominal size), 12″ wide (nominal size), and 12″ long (1″ × 12″ × 12″), or the equivalent. For example, a piece of lumber 1″ thick, 6″ wide, and 24″ long equals 1 BF. **See Figure 6-3.**

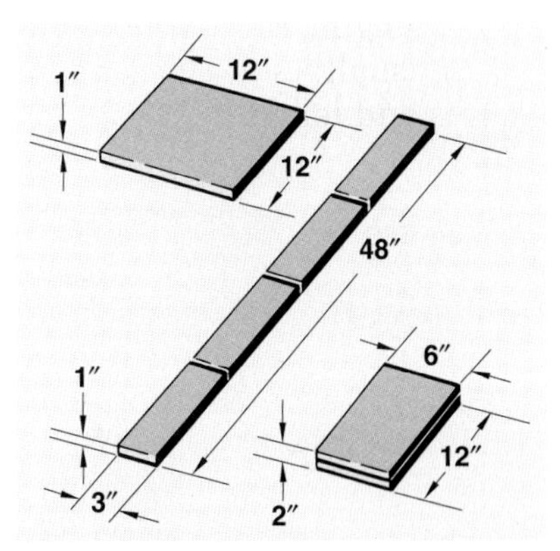

Figure 6-3. *A board foot is equal to a piece of lumber 1″ × 12″ × 12″, or any other measurement that contains 144 cubic inches.*

The number of board feet in a piece of lumber is not the same as the number of lineal feet. For example, a 2 × 4 × 16 is 16 lineal feet long and is approximately equal to 10.7 BF. To determine the number of board feet in a piece of lumber, the following formula is used:

$$BF = \frac{(T \times W \times L)}{12}$$

where

BF = board feet (in BF)
T = thickness (in in.)
W = width (in in.)
L = length (in ft)
12 = constant

For example, determine the number of board feet in a piece of lumber measuring 2″ thick by 4″ wide by 16′ long:

$$BF = \frac{(T \times W \times L)}{12}$$

$$BF = \frac{(2 \times 4 \times 16)}{12}$$

$$BF = \frac{128}{12}$$

BF = **10.67** or **10⅔ BF**

On most construction jobs, many pieces of the same-size lumber are ordered. The following formula is used to determine the total number of board feet when multiple pieces of the same-size lumber are ordered:

$$BF = \frac{(No. \times T \times W \times L)}{12}$$

where

BF = board feet (in BF)
No. = number of pieces
T = thickness (in in.)
W = width (in in.)
L = length (in ft)
12 = constant

For example, determine the number of board feet in 35 pieces of lumber measuring 2″ thick by 4″ wide by 16′ long:

$$BF = \frac{(No. \times T \times W \times L)}{12}$$

$$BF = \frac{(35 \times 2 \times 4 \times 16)}{12}$$

$$BF = \frac{4480}{12}$$

BF = **373.33** or **373⅓ BF**

To determine the cost of lumber, multiply the number of board feet by the cost per board foot as in the following formula:

Total cost = BF × Cost per BF

where

Total cost = total cost (in dollars)
BF = board feet (in BF)
Cost per BF = cost per board foot (in dollars)

The total cost of the lumber in the preceding example is determined by rounding the number of board feet to the next highest whole number (374 BF) and multiplying by the cost per board foot as follows:

Total cost = *BF* × *Cost per BF*
Total cost = 374 × $1.79
Total cost = **$669.46**

TYPES OF SURFACING

Most softwood lumber used in construction is surfaced on both sides and both edges. The *edge* of a piece of lumber is its narrowest dimension. The *side* is its widest dimension. A 2 × 4 has 2″ edges and 4″ sides. Lumber can also be ordered with all rough surfaces, or with a combination of smooth and rough surfaces. Mill-cabinet shops often order materials that have rough edges and smooth sides. A board with unsurfaced edges is wider, so more pieces can be cut from it.

A special type of rough surfacing is applied to *resawn* lumber. The pieces are run through a special bandsaw that produces a coarse, textured pattern on the surface of the wood. **See Figure 6-4.** Resawn lumber is most often used for exterior trim, siding, or paneling.

APA—The Engineered Wood Association

Figure 6-4. *Resawn lumber has an attractive textured surface.*

Abbreviations are used to indicate the type of surfacing required. For example, lumber surfaced on all four sides is specified as *S4S* lumber. Common abbreviations for lumber surfacing are as follows:

- S1S—surface on one side
- S2S—surface on two sides
- S4S—surface on four sides

- S1E—surface on one edge
- S2E—surface on two edges
- S1S1E—surface on one side and one edge
- S1S2E—surface on one side and two edges
- S2S1E—surface on two sides and one edge
- S/S—saw sized (resawn)

STANDARD SIZES

Boards, Dimension lumber, and Timbers are available in standard sizes. The actual size is always smaller than the nominal size. For example, the actual size of a 2 × 4 is 1½″ thick and 3½″ wide. **See Figure 6-5.** Additional standard sizes are found in the Appendix.

METRIC MEASUREMENT

In 1975, the United States Congress passed the Metric Conversion Act. In 1998, the Metric Conversion Act was amended by the Omnibus Trade and Competitiveness Act, which established the metric system as the preferred measurement system in the United States. The purpose of this act was to encourage U.S. businesses to replace the English measurement system with the *SI metric system.* Metric measurements are specified on federal construction projects. However, English measurements are typically specified for residential and light commercial construction projects. In recent years, national code books, such as the National Electrical Code®, have begun providing both English and SI metric measurements. Many manufacturers of lumber and nonstructural building materials also include metric dimensions in their specifications.

Metric Linear Measurement

The *meter* is the basic unit of linear measurement in the SI metric system. A meter is divided into *decimeters (dm), centimeters (cm),* and *millimeters (mm).* **See Figure 6-6.** Millimeters are the smallest division of a meter used in carpentry and are the standard unit of measurement for materials and metric construction plans. Following are comparisons of English and SI metric units:

39.37″ or 3′-3⅜″ = 1 m
1′ = .3048 m
1′ = 3.048 dm
1′ = 30.48 cm
1′ = 304.8 mm
1″ = .0254 m
1″ = .254 dm
1″ = 2.54 cm
1″ = 25.4 mm

Metric measurement is based on a system of 10. To convert to the next smaller unit, multiply by 10. For example, to determine the number of millimeters in 2.5 cm, multiply by 10 (2.5 cm × 10 = 25 mm). To convert to the next larger unit, divide by 10. For example, to determine the number of meters in 150 dm, divide by 10 (150 dm ÷ 10 = 15 m).

STANDARD LUMBER SIZES

Type	Nominal Size*		Actual Size*	
	Thickness	Width	Thickness	Width
Common Boards	1	2	¾	1½
	1	4	¾	3½
	1	6	¾	5½
	1	8	¾	7¼
	1	10	¾	9¼
	1	12	¾	11¼
Dimension	2	2	1½	1½
	2	4	1½	3½
	2	6	1½	5½
	2	8	1½	7¼
	2	10	1½	9¼
	2	12	1½	11¼
Timbers	5	5	4½	4½
	6	6	5½	5½
	6	8	5½	7½
	6	10	5½	9½
	8	8	7½	7½
	8	10	7½	9½

Common Boards: 1″ × 4″ Nominal; ¾″ × 3½″ Actual
Dimension: 2″ × 4″ Nominal; 1½″ × 3½″ Actual
Timbers: 6″ × 6″ Nominal; 5½″ × 5½″ Actual

* in in.

Figure 6-5. *The actual size of lumber is always smaller than the nominal size.*

Methods of converting English and metric measurement are shown in **Figure 6-7.** Conversion tables for area, volume, liquid, weight, pressure, and temperature are found in the Appendix.

Metric Lumber and Panel Sizes

Metric lumber and panel sizes are based on the actual standard lumber sizes and are expressed in millimeters (mm). **See Figure 6-8.** For example, the metric lumber size of a 2 × 8, which is 38 × 184, is based on the actual size of 1½″ × 7¼″. Actual lumber sizes are converted to their metric equivalents using *soft conversion* in which little rounding of the equivalents occur. Panel sizes are soft converted or *hard converted*, depending on the application.

Soft conversion is an exact or nearly exact conversion of English (U.S. Customary) measurements to their metric equivalents by multiplying by a metric conversion factor and rounding to a practical level of precision.

Hard conversion is an approximate conversion of an English measurement. Hard conversion measurements are convenient to work with and remember. Hard conversion is often preferred to eliminate odd metric values of soft conversion.

Metric Linear Measurement

Figure 6-6

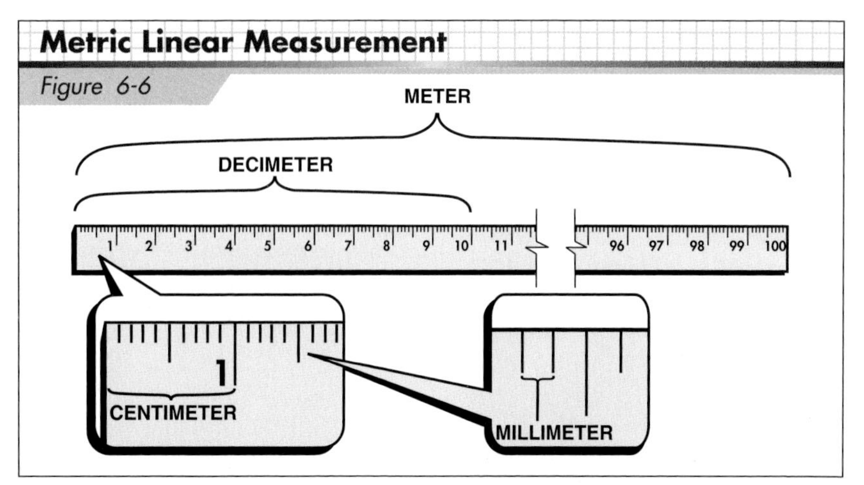

Figure 6-6. *A meter is divided into decimeters, centimeters, and millimeters.*

ENGLISH/METRIC CONVERSIONS

To Convert	To	Multiply By	Example
inches (in.)	millimeters (mm)	25.4	8″ × 25.4 = 203.2 mm
inches (in.)	centimeters (cm)	2.54	5″ × 2.54 = 12.7 cm
feet (ft)	centimeters (cm)	30.48	12′ × 30.48 = 365.76 cm
feet (ft)	meters (m)	.3048	7′ × .3048 = 2.1336 m
yards (yd)	centimeters (cm)	91.44	2 yd × 91.44 = 182.88 cm
yards (yd)	meters (m)	.9144	3 yd × .9144 = 2.7432 m
millimeters (mm)	inches (in.)	.03937	274 mm × .03937 = 10.7874″
centimeters (cm)	inches (in.)	.3937	18 cm × .3937 = 7.0866″
meters (m)	feet (ft)	3.281	4.5 m × 3.281 = 14.7645′
meters (m)	yards (yd)	1.0937	9 m × 1.0937 = 9.8433 yd

Figure 6-7. *English and metric measurements may be required to be converted. Additional conversions are included in the Appendix.*

METRIC LUMBER AND PANEL SIZES

Lumber Sizes		
Nominal*	Actual*	Metric†
2 × 4	1½ × 3½	38 × 89
2 × 6	1½ × 5½	38 × 140
2 × 8	1½ × 7¼	38 × 184
2 × 10	1½ × 9¼	38 × 235
2 × 12	1½ × 11¼	38 × 286

Panel Sizes			
Nominal‡	Actual*	Metric†	
		Soft	Hard
4 × 8	48 × 96	1220 × 2440	1200 × 2400

* in in.
† in mm
‡ in ft

Figure 6-8. *Metric lumber and panel sizes are based on the actual sizes and are expressed in millimeters.*

Unit 6 Review and Resources

UNIT 7

Engineered Wood Products

OBJECTIVES

1. Describe uses and guidelines for engineered wood panels.
2. Explain the procedure for manufacturing plywood.
3. List classifications of overlaid plywood.
4. Identify overlay materials of fiberglass-reinforced-plastic plywood.
5. Describe the procedure for manufacturing oriented strand board.
6. Describe the functions and appearance of composite panels.
7. List categories of APA performance rated panels.
8. Identify common types of APA Performance Rated Panel trademarks.
9. Describe common types of nonstructural panels.
10. Review common types of engineered lumber.

Engineered wood products are a class of building materials manufactured from solid and reconstituted wood products, which are combined with waterproof adhesives, resins, and other binders and subjected to extreme heat and pressure to form panels, timbers, and other structural or non-structural products. Engineered wood products may be as strong and dimensionally stable as traditional solid lumber products.

Engineered wood products may be manufactured from mill residue and short lumber pieces that were previously considered waste, resulting in environmental benefits. Long-span lumber, traditionally harvested from old-growth forests, can now be manufactured as engineered wood products.

Engineered wood products are a major material in the construction industry, and can be used for most structural applications. Engineered wood panels are commonly used for exterior walls and roof sheathing, and for subfloors in wood-framed floors. Engineered lumber products are frequently used for heavy beams, headers, studs, joists, rafters, and other framing members. **See Figure 7-1.**

APA—The Engineered Wood Association

Trus Joist, A Weyerhaeuser Business

Figure 7-1. *Engineered wood products, such as oriented strand board and wood I-joists, are commonly used for structural applications.*

ENGINEERED WOOD PANELS

Although plywood is still widely used in construction, especially for finish applications, engineered wood panels, such as oriented strand board and composite panels, are less expensive and provide similar performance results.

A *performance rated panel* is a structural wood panel that conforms to performance-based standards such as dimensional stability, bond durability, and structural integrity. Structural panels used for a specific application must meet the same minimum performance requirements, regardless of the composition of the panels or manufacturing process used to make them. In the past, plywood manufacturing techniques and materials varied little from mill to mill. However, as new techniques and materials became available, mills used a variety of techniques and a wider range of wood species. Therefore, it became necessary to develop performance-based standards for wood panels.

Guidelines for performance rated panels are based on product standards recommended by the U.S. Department of Commerce using input from major industry organizations such as APA—The Engineered Wood Association. Voluntary Product Standard PS 1-95, *Construction and Industrial Plywood,* and PS 2-92, *Performance Standard for Wood-Based Structural-Use Panels,* detail the minimum performance standards for performance rated panels.

Performance rated panels are manufactured as conventional veneered plywood, panels composed of wood strands and fibers such as oriented strand board and hardboard, and structural composite panels composed of veneer faces and reconstituted wood cores. **See Figure 7-2.**

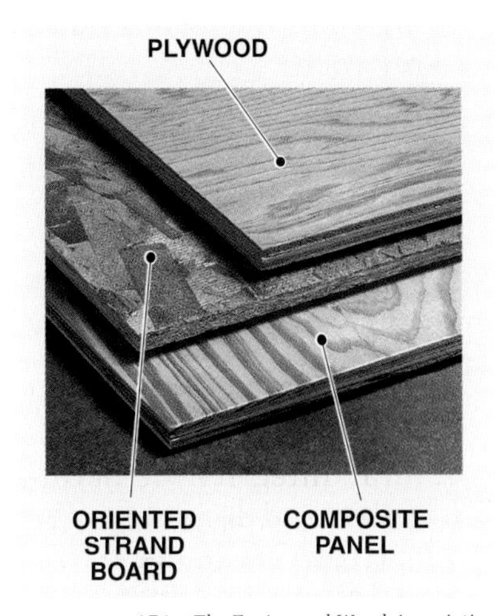

APA—The Engineered Wood Association

Figure 7-2. *Performance rated panels include plywood, oriented strand board, and composite panels.*

Plywood

Plywood is the original engineered wood product. It was first manufactured in the early 1900s, primarily for furniture production. Around 1940, plywood began to replace board lumber as wall and roof sheathing and subfloors.

Structural plywood panels are manufactured from softwood lumber. The particular softwood species used for plywood affects the strength and stiffness of plywood panels. Most structural plywood is manufactured from Douglas fir and various species of pine. More than 70 other wood species are also used in plywood manufacture. Wood species are classified into five groups, with species in Group 1 being the strongest and Group 5 being the weakest. **See Figure 7-3.**

Plywood panels consist of an odd number of *layers* such as three, five, or seven. Each layer consists of one or more plies. A *ply* is a single veneer sheet. The layers of a plywood panel are *cross-laminated,* a process in which each layer is placed with its grain at a 90° angle to the adjacent layer. Cross-lamination provides greater strength and stiffness in both directions while minimizing shrinkage and swelling in each direction. The outer layers of the panel are the *face veneer* and *back veneer.* The grain direction of the outer layer is always along the longest dimension of a plywood panel. Beneath the outer layers are the *crossbands* and the *core* (center layer). **See Figure 7-4.**

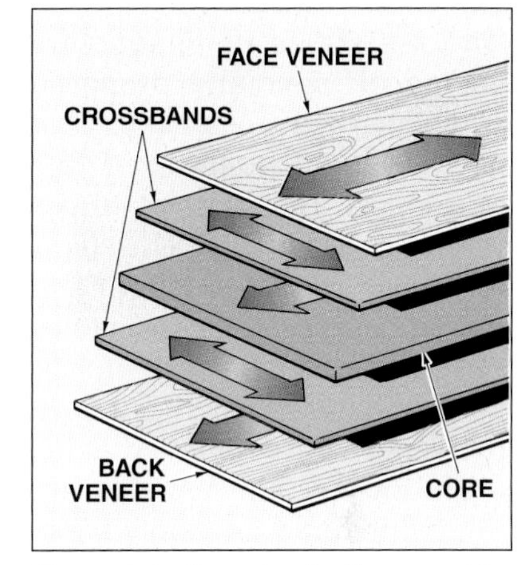

Figure 7-4. *Cross-lamination provides dimensional stability to plywood panels.*

Plywood Panel Sizes. Most plywood panels are manufactured in 4′ × 8′ sheets, but 4′ × 10′ sheets are also available. Some mills also produce plywood panels measuring up to 5′ wide by 12′ long. Common thicknesses of plywood panels used for wall

CLASSIFICATION OF WOOD SPECIES

Group 1	Group 2		Group 3	Group 4	Group 5
Apitong	Cedar, Port Orford	Maple, Black	Alder, Red	Aspen	Basswood
Beech, American	Cypress	Mengkulang	Birch, Paper	Bigtooth	Poplar, Balsam
Birch	Fir	Meranti, Red	Cedar, Alaska	Quaking	
Sweet	Balsam	Mersawa	Fir, Subalpine	Cativo	
Yellow	California Red	Pine	Hemlock, Eastern	Cedar	
Douglas fir	Grand	Pond	Maple, Bigleaf	Incense	
Kapur	Noble	Red	Pine	Western Red	
Keruing	Pacific Silver	Virginia	Jack	Cottonwood	
Larch, Western	White	Western White	Lodgepole	Eastern	
Maple, Sugar	Hemlock, Western	Spruce	Ponderosa	Black (Western Poplar)	
Pine	Lauan	Black	Spruce	Pine	
Carribean	Almon	Red	Redwood	Eastern White	
Ocote	Bagtikan	Sitka	Spruce	Sugar	
Pine, Southern	Mayapis	Sweetgum	Engelmann		
Loblolly	Red Lauan	Tamarack	White		
Longleaf	Tangile	Yellow Poplar			
Shortleaf	White Lauan				
Slash					
Tanoak					

APA—The Engineered Wood Association

Figure 7-3. *Strength and stiffness properties are affected by wood species.*

and roof sheathing and subfloors are 5/16″, 3/8″, 1/2″, 5/8″, and 3/4″. A 1 1/8″ thick plywood panel is often required for post-and-beam floors. Plywood panels are available with tongue-and-groove or square edges.

Plywood Manufacture. Continuous strips of veneer are peeled off specially prepared and debarked *peeler* logs that have been cut from longer logs to a length that will fit into a veneer-cutting lathe. The strips of veneer are cut to length and placed on a revolving table to be sorted by grade. Defects, such as knots and other imperfections in the veneers, are removed using a die cutter. The veneers are then glued and assembled into plywood sheets and subjected to intense heat and pressure in a hot press. **See Figure 7-5.** The sheets are then trimmed to size and stored in a warehouse for shipment to distributors.

Plywood veneers are graded according to their appearance, natural growth characteristics (such as knots and splits), and the size and number of repairs made during manufacture. High-grade veneers are used as face and back veneers on panels that will be exposed. If both sides of the panel will be exposed, high-grade veneers are used on both sides. If only one side will be exposed, a lower grade veneer is used on the unexposed side. Veneer grades and their descriptions are listed in **Figure 7-6.**

> *Over 731 million acres of the United States are covered by forests. Approximately 480 million acres of this forestland are suitable for planting and harvesting lumber. Landowners plant more than two billion trees each year. Approximately 27% more timber is grown than harvested.*

Overlaid Plywood

Overlaid plywood is a plywood panel with factory-applied, resin-treated fibers on one or both sides of the panel. Four common types of overlaid plywood are as follows:

- *medium density overlay (MDO) plywood*
- *high density overlay (HDO) plywood*
- *Plyform®*
- *fiberglass-reinforced-plastic plywood*

MDO and HDO Plywood. MDO and HDO plywood panels have opaque, resin-treated fiber overlays bonded to plywood with waterproof glue under heat and pressure. Overlays for MDO plywood are less dense than HDO overlays, and are applied on one or both sides of panels. Overlays for HDO plywood are applied to both sides of panels and have high chemical and abrasion

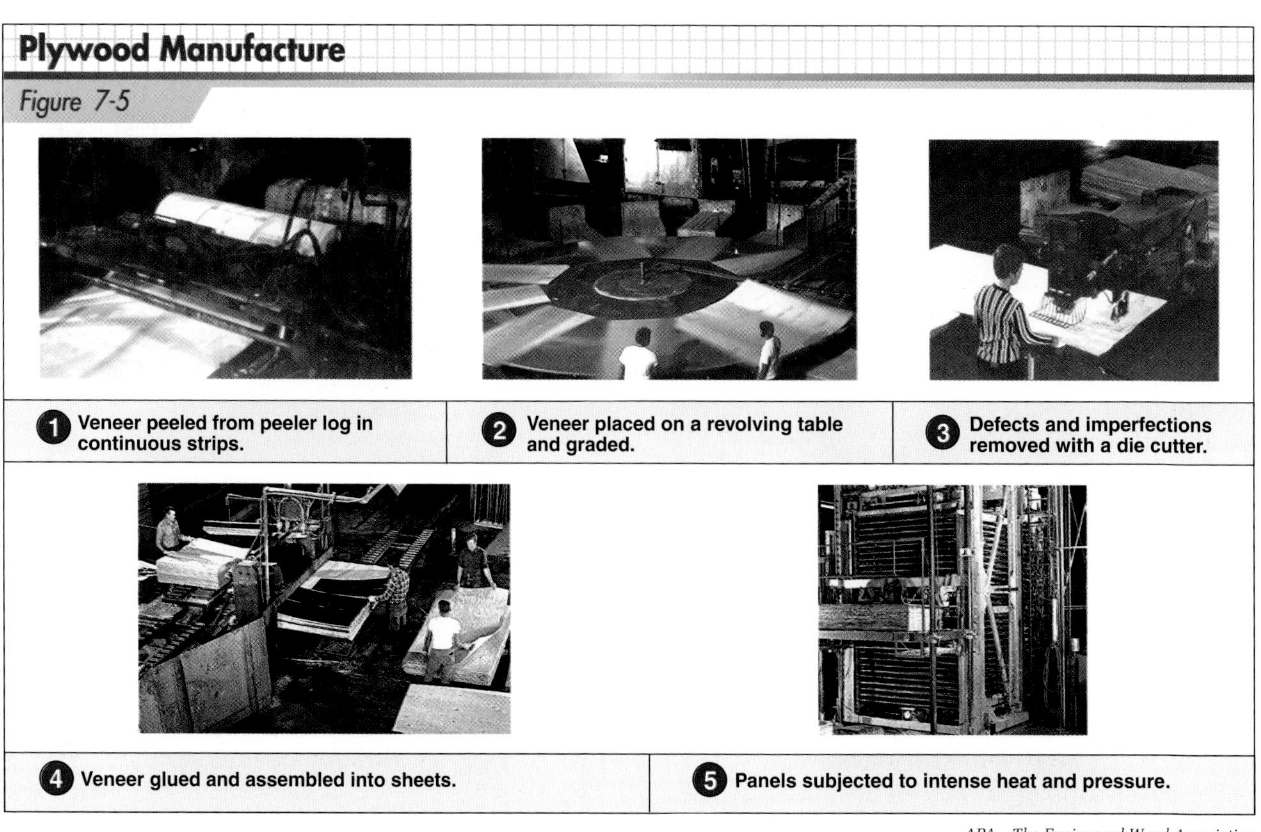

Figure 7-5. *The manufacture of plywood begins with peeling veneer strips from peeler logs. The veneer strips are then cross-laminated to produce plywood panels.*

resistance. HDO plywood is typically available in natural colors, but brown, black, and olive drab colors are also available.

MDO plywood provides a smooth, toothed surface for a paint base, and is recommended for siding and other exterior applications. **See Figure 7-7.** MDO plywood is also used for built-in furniture and cabinets, and signs. HDO plywood is used for concrete forms, highway signs, countertops, and other demanding applications. HDO plywood concrete forms can be reused 20 to 50 times provided they are properly maintained. HDO plywood is recommended where a smooth concrete finish is desired.

Plyform. Plyform is a performance rated plywood panel used for concrete forms. Even though other types of exterior plywood panels can be used as concrete forms, Plyform panels have an overlay which provides additional surface protection. The surface protection, composed of thermosetting resins, increases water, chemical, and abrasion resistance of the forms and increases panel stability, resulting in a smoother, more durable forming surface. **See Figure 7-8.**

The two basic grades of Plyform are *Plyform Class I* and *Plyform Class II.* Each grade can be ordered with an HDO surface on one or both sides. The primary difference between the Plyform grades is the wood species used in the manufacture of the panels. The strength of the wood species in the different groups varies, with Group 1 species being the strongest and Group 5 being the weakest.

Plyform Class I panels have Group 1 species on the face and back for high strength and stiffness properties. Plyform Class I panels are also available as *Structural I Plyform* when additional strength is required.

VENEER GRADES

Grade	Description
N	Smoothly cut 100% heartwood or 100% sapwood. Free from knots, knotholes, pitch pockets, stain, and other defects. Not more than six neatly made repairs. Intended for natural finish.
A	Smooth, paintable. Not more than eighteen neatly made repairs, boat, sled, or router type, and parallel to grain, permitted. May be used for natural finish in less demanding applications.
B	Solid surface. Shims, circular repair plugs and tight knots to 1″ across grain permitted. Some minor splits permitted.
C Plugged	Improved C veneer with splits limited to ⅛″ width and knotholes and borer holes limited to ¼″ x ½″. Admits some broken grain. Synthetic repairs permitted.
C	Tight knots to 1½″. Knotholes to 1″ across grain and some to 1½″ if total width of knots and knotholes is within specified limits. Synthetic or wood repairs. Discoloration and sanding defects that do not impair strength permitted. Limited splits allowed. Stitching permitted.
D	Knots and knotholes to 2½″ width across grain and ½″ larger within specified limits. Limited splits are permitted. Stitching permitted. Limited to interior panels.

Figure 7-6. *Veneers of higher grade are used on the faces and backs of panels that will be exposed.*

APA—The Engineered Wood Association

Figure 7-7. *Medium density overlay plywood is used for siding and other exterior applications, and provides a smooth, toothed surface for a paint base.*

Structural I Plyform panels use Group 1 species throughout the panel, and are recommended where face grain is parallel to its supports. Plyform Class II panels may have Group 2 faces and backs and are adequate for most concrete-forming applications.

Concrete formwork represents approximately one-half the cost of a concrete structure.

Nonoverlaid Plyform, referred to as *B-B Plyform,* is made of B-grade veneer, and is available as Class I, Class II, and Structural I grades. B-B Plyform is sanded on both sides and is treated with a release agent at the mill. Unless B-B Plyform panels have been recently manufactured, additional release agent may be required at the job site, and an edge sealer should be applied prior to first use. B-B Plyform panels may be reused 5 to 10 times if properly maintained.

APA—The Engineered Wood Association

Figure 7-8. *Plyform is a performance rated plywood panel that provides a smooth and durable concrete-forming surface.*

Fiberglass-Reinforced-Plastic Plywood. Fiberglass-reinforced-plastic (FRP) plywood is an engineered panel product that consists of a tough glass fiber-reinforced overlay bonded to plywood. **See Figure 7-9.** Overlay materials are as follows:

- fiberglass-woven fabric saturated with resin and cured under heat and pressure
- glass fiber mats saturated with resin, which are initially cured and bonded to the plywood surface and finally cured in a hot press under heat and pressure
- chopped glass strands and resin sprayed on the plywood and cured under heat

Overlay thicknesses range from 25 mils to 60 mils and are bonded to ¼″ to 1⅛″ thick plywood panels. Surfaces vary from extremely smooth to textured skid-resistant and aggregate-filled surfaces.

For construction applications, such as concrete forms, FRP plywood is typically available in 4′ × 8′ panels. However, larger panels are available for other applications such as shipping containers and walk-in coolers.

Glass content for FRP plywood ranges from ⅔ oz to 3 oz/sq ft.

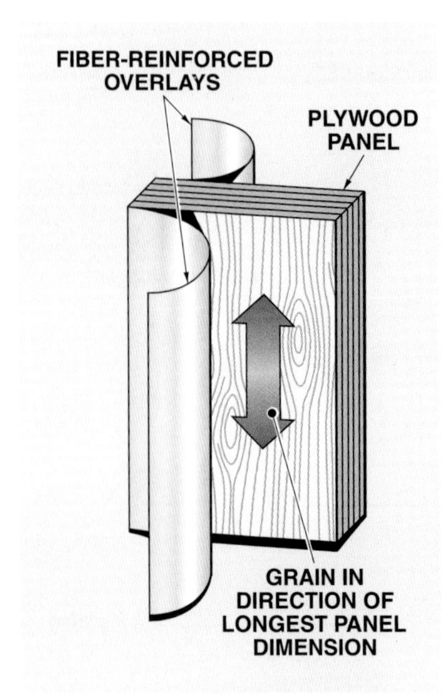

Figure 7-9. *Fiberglass-reinforced-plastic plywood panels consist of fiber-reinforced overlays bonded to a plywood panel.*

Oriented Strand Board

Oriented strand board (OSB) is structural wood panels manufactured from reconstituted, mechanically oriented wood strands that are bonded with a waterproof adhesive under heat and pressure. OSB is commonly used for roof and wall sheathing, for subfloors, and as the webs of wood I-joists. **See Figure 7-10.** The most commonly used OSB panel is 4′ wide by 8′ long. Panels measuring 4′ × 9′, 4′ × 10′, 4′ × 12′, and 4′ × 16′ can also be ordered. Thicknesses range from ¼″ to 1⅛″ and are available in both square-edged and tongue-and-groove (T&G) edges. Some T&G panels have notches cut into the tongue that allow the water to drain through. This helps prevent water from collecting under the panels before the house is enclosed. OSB can be obtained in the following grades:

- OSB/1—general-purpose applications in dry conditions
- OSB/2—load-bearing applications in dry conditions
- OSB/3—load-bearing applications in humid conditions
- OSB/4 — heavy duty load-bearing applications in dry or humid conditions

OSB is manufactured from first- and second-growth trees. After logs are shredded into uniform strands up to 6″ long, they are mixed with a waterproof adhesive and a small amount of wax. The strands are then formed into layers with the strands oriented in the same direction. Three to five layers are compressed together, with the strands in each layer at a right angle to the strands of the layer above or below. **See Figure 7-11.** The cross-lamination of the strands distributes the wood's strength in both directions of the panel. Most OSB sheathing panels have a nonskid surface on one side for construction site safety.

Figure 7-10. *Oriented strand board is a structural wood panel commonly used for wall sheathing.*

OSB has strength comparable to structural plywood and offers the following advantages:

- OSB has expanded the use of relatively weaker species, including aspen and western red cedar, making it possible to utilize a much broader timber resource.
- OSB panels are less expensive than plywood panels.

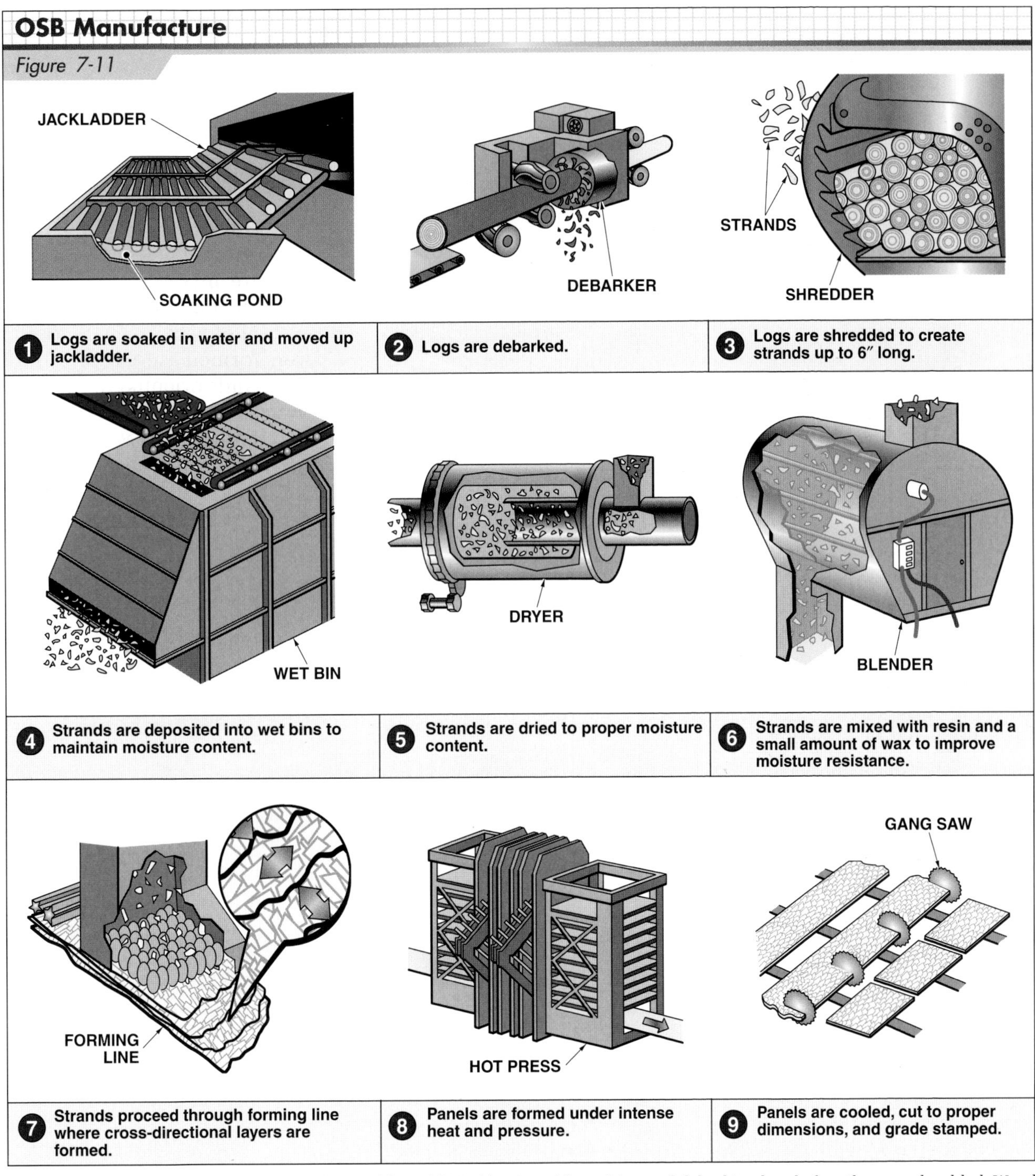

Figure 7-11. *The manufacture of oriented strand board (OSB) begins with soaking and debarking logs before they are shredded. Wood strands up to 6″ long are formed into layers and cross-laminated to provide OSB with its strength and dimensional stability.*

Composite Panels

Composite panels consist of veneer faces that are bonded to a wood fiber core, such as OSB, allowing an efficient use of wood resources while retaining the wood grain appearance on the panel face and back. Composite panels offer the advantages of structural panels while providing an appearance-grade finish. Composite panel applications include sheathing and siding.

Composite panels are available in three and five layers. Three-layer panels have a wood fiber core and veneer faces and backs. Five-layer panels have a wood veneer core with wood crossbands and veneers on each side. **See Figure 7-12.**

For every ton of wood grown, a young forest absorbs approximately 1.5 tons of carbon dioxide.

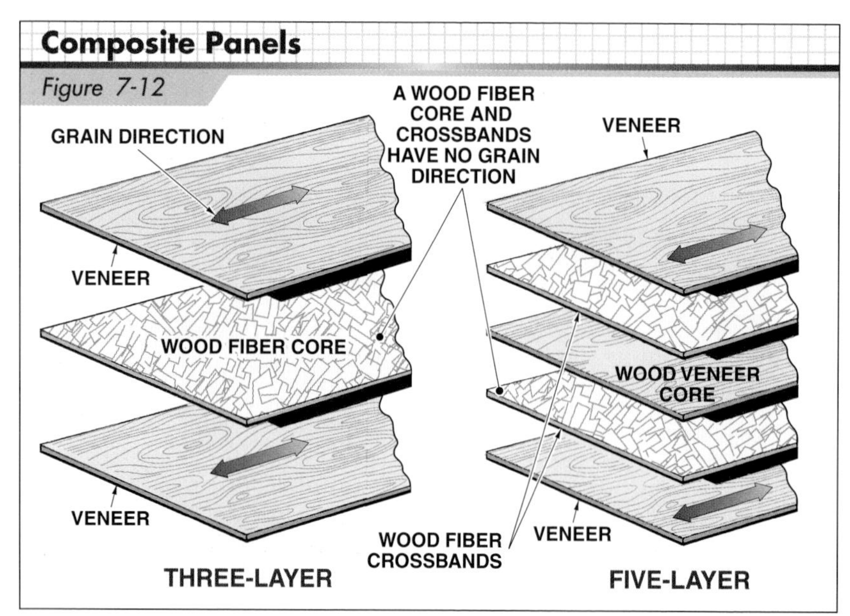

Figure 7-12. *Composite panels offer the advantages of structured panels while providing an appearance-grade finish.*

APA Performance Rated Panel Categories

Performance rated panels are principally categorized based on their end uses—roof, floor, and wall sheathing, single-layer flooring, or exterior siding. Other considerations when determining performance rated panel categories are wood species, span ratings, resins and other adhesives, bending strength, and stiffness. APA performance rated panel categories are as follows:

- *APA Rated Sheathing.* Used for applications where the highest strength and stiffness are required, such as subfloor and wall or roof sheathing. Rated Sheathing panels used for exterior applications should have an exterior exposure durability classification. Rated Sheathing panels are also available as *APA Rated Sheathing/Ceiling Deck* where one surface has an overlay, texture, or grooves applied to it. Common thicknesses for APA Rated Sheathing panels are 5⁄16″, 3⁄8″, 7⁄16″, 15⁄32″, 1⁄2″, 19⁄32″, 5⁄8″, 23⁄32″, and 3⁄4″.
- *Structural I.* Structural I panels are APA Rated Sheathing panels with increased racking and cross-panel strength properties. Structural I panels are commonly used for applications such as structural diaphragms and panelized roofs.
- *APA Rated Wall Bracing.* Used for wall sheathing where only diagonal bracing would typically be required. Some APA Rated Wall Bracing panels also have special fire-resistance properties.
- *APA Rated Sturd-I-Floor®* Combination subfloor/underlayment panel commonly used for flooring under a carpet and pad. The panels also provide extra resistance to punctures. When resilient flooring, such as vinyl, is applied over Sturd-I-Floor panels, an additional layer of thin, sanded under-layment is recommended. APA Rated Sturd-I-Floor panels are available with square or tongue-and-groove (T&G) edges. T&G panels are typically 47½″ wide. APA Rated Sturd-I-Floor is available in Exterior and Exposure 1 classifications in 19⁄32″, 5⁄8″, 23⁄32″, 3⁄4″, 7⁄8″, 1″, and 1⅛″ thicknesses.
- *APA Rated Siding.* APA Rated Siding panels are used as panel and lap siding, and can be placed over sheathing or directly over studs using special construction techniques. A variety of surface textures are available including rough sawn, reverse board and batten, channel groove, and brushed. Lap siding is available in widths up to 12″ and in lengths up to 16′. Panel siding is available in 11⁄32″, 3⁄8″, 7⁄16″, 15⁄32″, 1⁄2″, 19⁄32″, and 5⁄8″ thicknesses and in 4′ × 8′, 4′ × 9′, and 4′ × 10′ widths and lengths.

Square-edge panels measure 48″ × 96″ with a +0″, −1⁄8″ tolerance.

Each APA performance rated panel category is divided into exposure durability classifications as follows:

- *Exterior panels* have a waterproof bond and are designed for applications where they may be permanently exposed to weather or moisture.
- *Exposure 1 panels* have a waterproof bond and are designed for applications where long construction delays, high humidity, or water leakage may be experienced prior to providing protection. Approximately 95% of performance rated panels are Exposure 1 durability classification.
- *Exposure 2 panels* are suitable for interior applications where exposure to high humidity or water leakage may be experienced.
- *Interior panels* are manufactured with interior glue and are intended only for interior use.

APA Performance Rated Panel Trademarks

A *trademark* is stamped on one face and one edge of a panel at the time of its manufacture. When different grades of veneer are on the face and back, the trademark is stamped on the side of lesser quality. Examples of typical APA trademarks are shown in **Figure 7-13.** Descriptions of trademark elements are as follows:

- *Panel grade.* For plywood or other panels with veneer faces, panel grades are identified by their veneer grades. Panel grades for other panel materials are identified by the suggested end-use application, such as APA Rated Sheathing.
- *Span rating.* Span ratings indicate on-center spacing, in inches, of supports, such as studs, joists, and roof rafters, over which a panel can be placed. Some panels, such as Sturd-I-Floor panels, are used for only one application and contain one span rating. Other panels, such as APA Rated Sheathing, may be used for sheathing and subfloors and carry two span ratings. For example, span ratings on APA Rated Sheathing are shown as two numbers separated by a slash, such as 48/24. The number preceding the slash (48) indicates the maximum recommended spacing between supports when the panel is used for roof sheathing with the long dimension of the panel across three or more supports. The number following the slash (24) indicates the maximum recommended spacing between supports when the

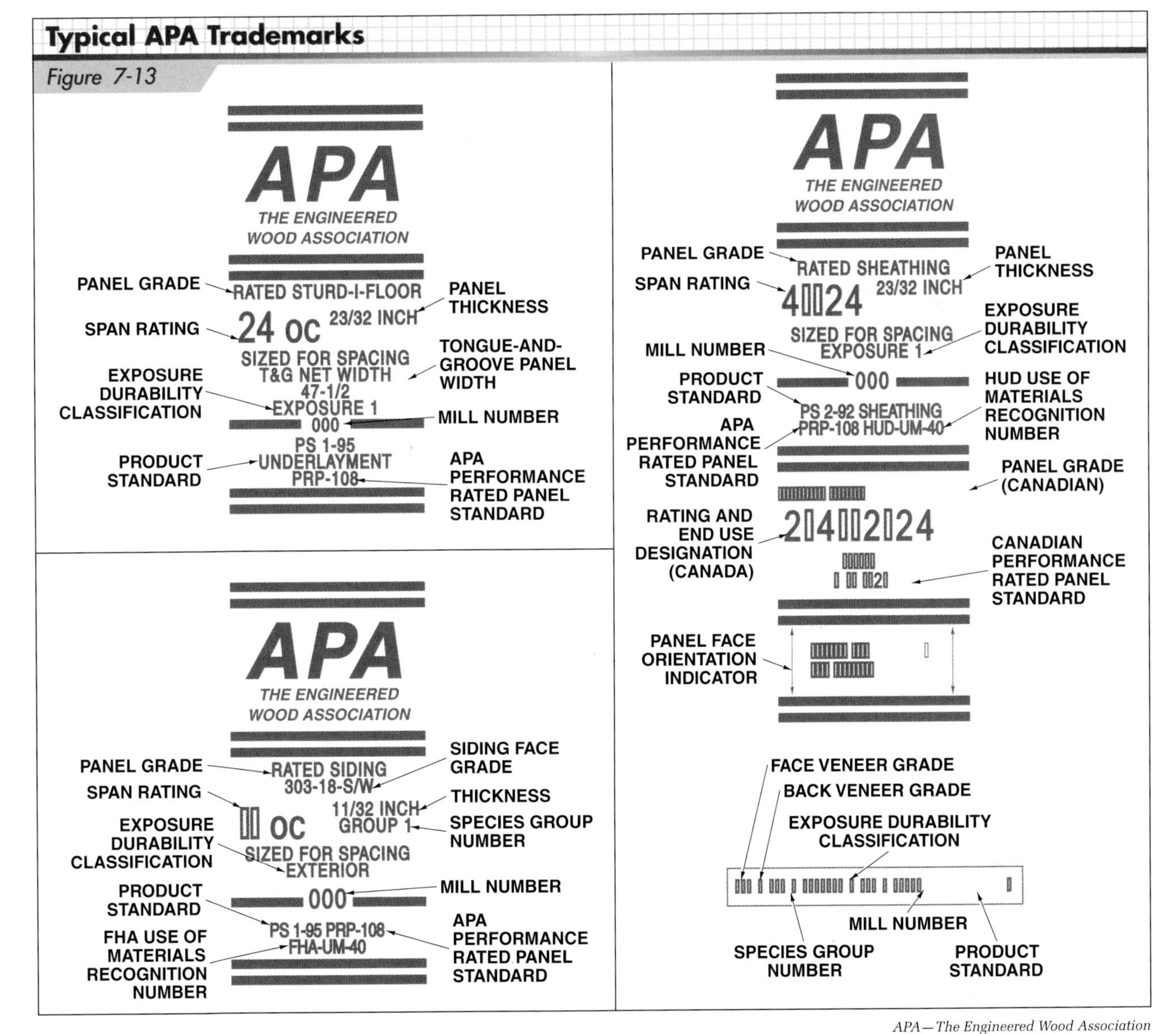

APA—The Engineered Wood Association

Figure 7-13. *Performance rated panels are identified by trademarks. Additional trademarks and descriptions are shown in the Appendix.*

panel is used for subfloors with the long dimension of the panel across three or more supports.

- *T&G net width.* The T&G net width (in inches) is indicated for a T&G panel. The T&G net width is omitted if the panel has square edges.
- *Exposure durability classification.* The exposure durability, or bond, classification identifies the applications for which the panel is recommended.
- *Product standard.* Identifies the voluntary standard developed cooperatively by the U.S. Department of Commerce and the panel industry association. The examples referred to in the figure are PS 1-95, *Construction and Industrial Plywood* and PS 2-92, *Performance Standard for Wood-Based Structural-Use Panels.*
- *Thickness.* Actual thickness dimension of the panel, in inches.
- *Mill number.* Identifies the mill where the panel was manufactured. Each mill where APA performance rated panels are manufactured is assigned its own mill number.
- *APA performance rated panel standard.* Indicates that the panel satisfies the APA quality auditing program requirements.
- *Siding face grade.* For APA Rated Siding, a siding face grade number is applied to the panel. In this example, "303" is a code for the type of siding and "18" indicates the maximum allowable repairs or patches. An "S" indicates the patch is synthetic or a "W" indicates the patch is wood.
- *Species group number.* Identifies the species group from which the panel veneer was manufactured.

APA—The Engineered Wood Association

Plywood sheathing is fastened to the tops of trusses made of 2 × 4s and tubular steel webs.

- *HUD-UM/FHA-UM recognition number.* Indicates the Use of Materials Bulletin number issued by the U.S. Department of Housing and Urban Development or Federal Housing Administration.
- *Canadian standard panel grade.* Panels manufactured in Canada include a Canadian standard panel grade number or description. In this example, the Canadian panel grade designation is Construction Sheathing.
- *Rating and end-use designation.* Panels manufactured in Canada include a Canadian standard rating and end-use designation, which indicates the allowable span between supports. In the APA Rated Sheathing example, roof panels will require edge supports such as panel clips or blocking along the unsupported edges. The number "48" is the maximum distance, in inches, between rafters (R). The number "24" is the maximum distance, in inches, between floor joists (F).
- *Canadian performance rated panel standard.* Indicates that the panel satisfies the Canadian Standards Association (CSA) quality auditing program requirements.
- *Panel face orientation indicator.* Identifies the strongest direction (axis) of the panel.

Nonstructural Panels

Nonstructural panels are manufactured from hardwood veneers or reconstituted wood products, and are used for nonstructural applications such as paneling, siding, or molding. Nonstructural panels are typically cut or trimmed, sanded, and finished as desired.

Hardwood plywood panels are produced in a manner similar to performance rated panels, by peeling veneers from peeling logs. However, the internal composition of hardwood plywood panels is different from that of performance rated panels.

Nonstructural reconstituted panels, such as *fiberboard, hardboard,* and *particleboard,* are widely used in the construction industry. Logs not suitable for dimension lumber and waste wood products are ground up and combined with a variety of resins, additives, and binders to produce nonstructural panels. The manufacturing process for nonstructural panels is similar to the process used for OSB.

Hardwood Plywood. Hardwood plywood is composed of hardwood face and back veneers with lumber, particleboard, MDF, or hardboard cores joined with an adhesive. The grain of alternate layers is typically at right angles. Hardwood plywood is commonly used for interior finish, such as wall paneling and cabinet and door exteriors. **See Figure 7-14.** Hardwood plywood is considerably more expensive than construction-grade plywood.

American Hardwood Export Council

Figure 7-14. *Hardwood plywood is commonly used for interior finish applications.*

Hardwood plywood is typically selected for its attractive grain pattern and receives a natural or stained finish. Most hardwood plywood is prefinished before placement. Common hardwood plywood face and veneer species are as follows:

- lauan
- walnut
- red and white oak
- maple
- cherry
- ash
- mahogany
- birch
- teak

Hardwood plywood panels are available in various sizes including 4′ × 4′, 4′ × 8′, and 4′ × 10′. Panel thicknesses are ¼″, ⅜″, ½″, and ¾″.

Hardwood plywood panels are manufactured in accordance with the Hardwood Plywood and Veneer Association (HPVA) standard HP-1-2000, *Hardwood and Decorative Plywood.* Hardwood plywood panel grades are based on the number of color streaks or spots, color variations, mineral streaks, burls, pin knots, and small repairs in the panel face and back. **See Figure 7-15.**

Medium Density Fiberboard. Medium density fiberboard (MDF) is a nonstructural reconstituted panel product used for a variety of exterior and interior products including siding, molding, furniture, shelving, and cabinets. **See Figure 7-16.** MDF is flat, smooth, uniform, dense, and free of knots and grain patterns, allowing MDF to be machined and finished easily. MDF may be sealed with a clear finish, painted, or wrapped with a plastic or paper laminate. Thicknesses range from ⅛″ to 1½″, although other thicknesses are available upon request. Standard panel sizes are 3′ × 8′ and 4′ × 8′.

HARDWOOD PLYWOOD PANEL GRADES

Grade	Description
A-1	Smooth face veneer running full panel length. Book-matched veneer edges if more than one veneer is used on face. May contain small burls and pin knots or small repairs. Sound veneer back, but may not be color matched.
B-2	Sound face veneer, but may have small burls, pin knots, and repairs. If face veneer consists of more than one piece, color and grain matching required. Solid back veneer with repaired voids.
C-2	Sound face veneer with more small burls, pin knots, and other growth characteristics. Sound repaired back.
C-3	Sound, repaired face. Back veneer contains more and larger repairs.
D-3	Economy-grade panel. Natural growth characteristics allowed on face and back, but no open areas allowed.

Figure 7-15. *Hardwood plywood panel grades are based on natural growth characteristics, including color variations, burls, and pin knots, and the number of repairs in the panel face and back.*

APA—The Engineered Wood Association

Figure 7-16. *Medium density fiberboard is used for a variety of exterior and interior applications including siding. Note the spacer block being used to uniformly space the courses of siding.*

MDF utilizes waste wooden chips and pieces generated by sawmill operations. The chips are reduced to fibers, which are then ground into smaller particles. The smaller particles adhere naturally to each other and are bonded together with water containing synthetic resins such as glue. The wet mixture is placed on a mesh-bottomed form to form panels and the excess water is squeezed out. The panels are dried by rolling, and then are cut into sheets, pressed, and heat treated.

Hardboard. *Hardboard* is a nonstructural reconstituted panel product commonly used for paneling, cabinet backs, floor underlayment, and exterior siding. **See Figure 7-17.** Hardboard does not have a grain pattern and has a uniform density, thickness, and appearance. Hardboard panels are manufactured with one side smooth (S1S) or both sides smooth (S2S). Thicknesses range from 1⁄12″ to 1⅛″, with ¼″, ⅜″ and ⅝″ thicknesses used most often. Hardboard panels are available in 4′ widths, and standard lengths of 8′, 10′, 12′, and 16′.

Figure 7-17. *Hardboard is a nonstructural panel product that has no grain pattern and has a uniform density, thickness, and appearance.*

Hardboard is manufactured from sawmill residue or logs that are ground up in a large chopper. The wood chips are broken down into fibers that are combined with natural and synthetic binders and other additives. Hardboard panels are formed using a wet or dry process. The wet forming process is similar to the forming process used for MDF. In the dry forming process, relatively dry fibers are formed into thick mats using air. Regardless of the forming process, the fibers in the mats are permanently bonded under heat and pressure into panels.

Hardboard panels are available in four grades—*standard, tempered, service,* and *service-tempered.* Standard hardboard panels have high strength and water resistance. Tempered hardboard panels are impregnated with resin, and provide greater stiffness, strength, hardness, and water-resistant properties than standard hardboard panels. Service hardboard panels have good strength, but are less strong than standard hardboard panels. Service-tempered hardboard panels are service hardboard panels impregnated with resin to provide better stiffness, strength, hardness, and water resistance than service hardboard panels.

Particleboard. *Particleboard* is a nonstructural reconstituted panel product consisting of particles of various sizes that are bonded together using a synthetic resin or binder under heat and pressure. Particleboard is commonly used for floor underlayment, shelving, and furniture, and as a core material for composite panels. **See Figure 7-18.** Particleboard panels are available in thicknesses ranging from ½″ to 1¾″, 4′ widths, and standard lengths of 8′, 10′, 12′, and 16′.

The production of particleboard begins with milling or grinding of wood shavings, chips, sawdust, or even whole logs into uniform small particles. The particles are then dried, combined with resin and other binders, and pressed into large panels using heat and pressure.

Figure 7-18. *Particleboard is a nonstructural reconstituted panel consisting of particles that are bonded together using a synthetic resin or binder. A plastic laminate is applied to the particleboard for certain applications.*

ENGINEERED LUMBER

Engineered lumber is manufactured using solid wood members or veneers, wood strands and fibers, or a combination of solid wood members and wood strand members. Engineered lumber, such as glued laminated (glulam) timbers and finger-jointed studs, is manufactured using trees from second- and third-growth forests. Laminated veneer lumber (LVL) is produced by combining and bonding wood veneers together.

Parallel strand lumber (PSL) and oriented strand lumber (OSL) consist of long wood strands oriented along the length of the lumber and bonded together using an adhesive. Wood I-joists are composed of OSB or plywood webs and dimension lumber or LVL flanges.

Glued Laminated Timber

Glued laminated (glulam) timber is an engineered lumber product composed of wood laminations (lams) that are bonded together with adhesives. The grain of the lams runs parallel to the length of the glulam member. Lams are 1⅜″ thick for southern pine species and 1½″ for western wood species. Lams up to 2″ thick may also be used to form glulam timbers. Glulam products are available in standard widths including 3⅛″, 3½″, 5⅛″, 5½″, and 6¾″, although almost any width can be custom produced.

Glulam dates back to the early 1900s when the first patents for glulam were obtained in Switzerland and Germany. One of the first glulam structures in the United States was built for the USDA Forest Products Laboratory in 1934. Since that time, glulam manufacture has evolved, making glulam one of the most widely used engineered lumber products in the construction industry. In the past, glulam was mainly associated with heavy timber construction such as industrial roofing systems, bridges, and marine piers. Glulam timbers are used in residential construction as well as in commercial and industrial construction. **See Figure 7-19.**

Glulam Manufacture. Douglas fir, larch, and southern yellow pine are most commonly used to produce glulam timber. In addition, Alaska cedar, redwood, ponderosa pine, spruce, and western red cedar may be used for framing and industrial-grade glulam timber. Hardwood species such as oak, maple, and poplar are used for glulam when appearance is important.

A special grade of lumber is used for glulam manufacture. The lumber is dried to a moisture content not exceeding 15% and is planed to close tolerances similar to dimension lumber. The lumber is graded or machine tested to determine its strength. The lumber is end-jointed and arranged into laminations with the ends of the lumber staggered so joints from adjacent lams are not aligned. The strongest lams are placed on the top and bottom of the beams where most of the tension and compression stresses occur. The lams are clamped and bonded together with a waterproof adhesive.

APA—The Engineered Wood Association

Figure 7-19. *Glulam timbers are commonly used for beams in commercial and residential construction.*

Glulam timbers are typically manufactured in lengths up to 20′. Longer glulam timbers are produced by finger-jointing the ends of the lams and gluing them together. Continuous straight beams spanning as much as 140′ have been installed on heavy timber projects. Glulams can be curved or arched. **See Figure 7-20.**

Glulam Appearance Classifications. All glulam appearance classifications have similar structural characteristics, but differ in their appearance. Appearance classifications allow natural growth characteristics, including limited open voids that can be repaired with inserts or wood fillers. Descriptions of glulam appearance classifications are as follows:

- *Framing.* Used only when glulam members are concealed. The widths of beams with a framing appearance classification are designed to fit flush with 2 × 4 and 2 × 6 wall framing members.
- *Industrial.* Used for concealed applications, such as warehouses and garages, where appearance is not important.
- *Architectural.* Available in stock sizes where the glulam member is exposed to view and must have a smooth and attractive finish.
- *Premium.* Available only as a custom order, but has the same appearance qualities as Architectural.

Laminated Veneer Lumber

Laminated veneer lumber (LVL) is an engineered lumber product composed of layered veneers and waterproof adhesive. LVL is

APA—The Engineered Wood Association

Figure 7-20. *Glulam timbers can be formed to produce curves or arches.*

commonly used for beams, headers, hip and valley rafters, and flanges for wood I-joists. **See Figure 7-21.** LVL is available in thicknesses ranging from ¾″ to 2½″, with 1¾″ thickness commonly used to produce built-up beams. LVL is available in lengths up to 80′, but is most commonly cut into lengths of 48′ and shorter.

APA—The Engineered Wood Association

Figure 7-21. *Laminated veneer lumber (LVL) is composed of layered veneers and waterproof adhesive and is commonly used for beams. Note the wood I-joists attached to the LVL beam.*

Materials commonly used for LVL are southern yellow pine, Douglas fir, larch, and poplar, which are cut into veneers ranging in thickness from 1⁄10″ to 3⁄16″. The veneers are dried, graded, and layered to form *billets* (blocks of material), with the grain of the veneers running lengthwise along the billet. The veneers are bonded together with waterproof adhesives and cured in a heated press. The cured billets are then cut to standard dimensions comparable to dimension lumber.

Parallel Strand Lumber

Parallel strand lumber (PSL) is an engineered lumber product manufactured from strands or elongated flakes of wood that are blended with a waterproof adhesive and cured under pressure. PSL is compatible with common wood-framing materials and is available in standard dimensions. PSL is used for beams and headers, and is frequently used for load-bearing columns. **See Figure 7-22.** PSL beams are available in 1¾″, 2 11⁄16″, 3½″, 5¼″, and 7″ thicknesses.

PSL and LVL design values conform to ASTM D5456-01ae1, *Standard Specification for Evaluation of Structural Composite Lumber Products.* PSL structural properties are comparable to, or exceed, similar properties of high-quality dimension lumber.

Trus Joist, A Weyerhaeuser Business

Figure 7-22. *Parallel strand lumber (PSL) is used for beams, headers, and load-bearing columns. PSL is used for the main beam and supporting column in this application.*

Oriented Strand Lumber

Similar to parallel strand lumber, *oriented strand lumber (OSL)* is made from flaked wood strands that have a high length-to-thickness ratio. The wood strands are combined with a waterproof adhesive and oriented and formed into a large mat or billet. The mat or billet is pressed and cured in a heated press. OSL is used in a variety of applications ranging from studs to millwork components.

Wood I-Joists

Wood I-joists, or I-beams, are load-bearing structural members consisting of a web placed between top and bottom flanges. *Webs* are OSB or plywood panels cut to size. *Flanges* are dimension lumber or LVL. Wood I-joists are used for floor and ceiling joists and rafters in both residential and commercial construction. **See Figure 7-23.**

One of the biggest advantages of wood I-joists over dimension lumber products is its high strength-to-weight ratio. Wood I-joists are lighter and easier to handle than solid lumber joists required for the same spans and subjected to similar loads. The flanges of wood I-joists provide a wide surface for attachment of sheathing and gypsum board.

Wood I-joists may have knockout holes in their webs to allow mechanical services, such as electrical wiring, to be easily installed. The knockout holes also provide ventilation when wood I-joists are used in cathedral ceilings. When knockout holes are not in the webs, holes can be drilled through the webs, but they must be located according to manufacturer recommendations.

Common wood I-joist depths range from 9½″ to 20″, and depths up to 30″ can be custom ordered. Wood I-joists are available in lengths up to 66′. For long wood I-joists, the flanges are finger-jointed. Flanges are typically 1½″ thick with widths varying from 1¾″ to 3½″. Web thicknesses range from ⅜″ to ½″. The upper and lower edges of the webs are butt jointed, toothed, tongue and grooved, or scarfed for fastening to the flanges.

APA—The Engineered Wood Association

Figure 7-23. *Wood I-joists are composed of oriented strand board or plywood webs with dimension lumber or laminated veneer lumber flanges.*

OSB Rim Boards

OSB rim boards are an engineered wood product used primarily with wood I-joists in floor construction where the ends of the I-joists are fastened to the face of the rim boards. Rim boards fill the space between a sill plate and bottom plate of a wall, or in two-story construction fill the space between the top plate and bottom plate of two wall sections. **See Figure 7-24.** In addition, rim boards tie together the floor joists. Rim boards serve the same purpose as header joists in traditional solid lumber construction.

Common rim board thicknesses are 1″ and 1⅛″, with depths of 9½″, 11⅞″, 14″, and 16″ to match the depths of wood I-joists and other engineered wood framing products. Rim boards are available in lengths up to 24′, depending on the material used.

APA—The Engineered Wood Association

Figure 7-24. *Rim boards tie together wood I-joists and fill the space between the top plate and bottom plate of two wall sections.*

Finger-Jointed Lumber

Finger-jointed lumber is an engineered lumber product composed of short pieces of dimension lumber joined end-to-end. The ends of the short pieces are machined with a fingerlike profile and glued together. **See Figure 7-25.** Longer lengths of finger-jointed lumber can be produced that have strength equal to conventional dimension lumber. Structural applications of finger-jointed lumber include studs, plates, joists, and rafters.

Southern Forest Products Association

Figure 7-25. *The ends of short pieces of lumber are milled to produce fingers. The short pieces are then attached to each other to form finger-jointed lumber, which is used for structural applications such as studs, plates, joists, and rafters.*

Douglas fir, larch, hemlock, spruce, and pine are commonly used for finger-jointed lumber. Finger-jointed lumber used for structural purposes is manufactured from lumber seasoned to less than 19% moisture content and is stamped "S-DRY" or "KD." The following grades of finger-jointed lumber have been established by *The National Grading Rule for Dimension Lumber (NGR)*:

- Stud
- Select Structural
- No. 1
- No. 2 & Better
- Standard and Better

Stud and No. 2 & Better are most often used for construction applications.

Finger-jointed lumber grades are further divided into two end-use applications. Finger-jointed lumber containing "CERT. EXT. JNTS." in the trademark is acceptable for use in all structural applications, and is manufactured with a waterproof, exterior-type adhesive. Finger-jointed lumber containing "VERTICAL USE ONLY" or "STUD USE ONLY" in the trademark is only used for vertical structural applications such as studs. **See Figure 7-26.**

Finger-jointed lumber dimensions range from 2 × 2 to 2 × 12 in standard dimension lumber sizes. Finger-jointed lumber with 2 × 4, 2 × 8, and 2 × 10 nominal dimensions is available in common standard lengths. Lengths of 32′ or more are also available.

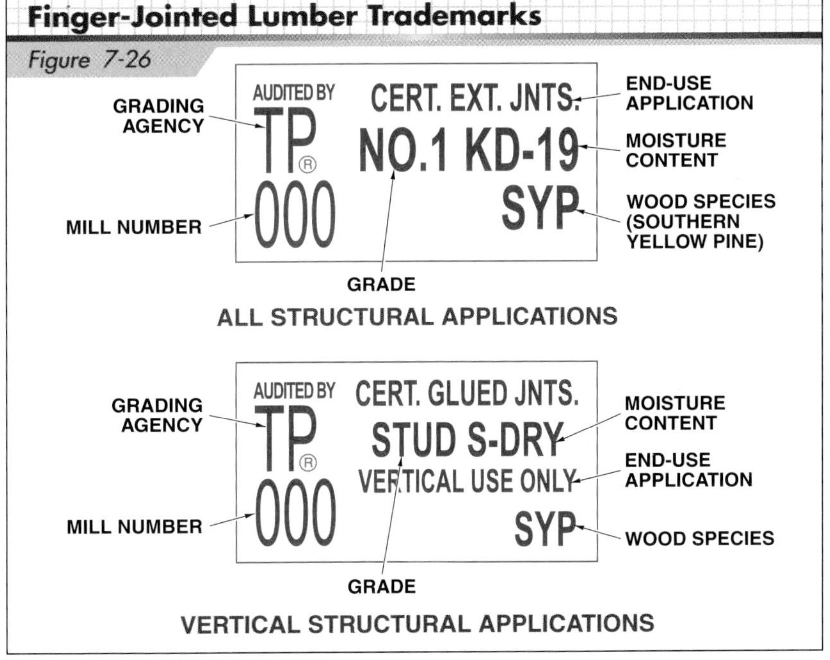

Southern Forest Products Association

Figure 7-26. *Finger-jointed lumber is divided into two end-use applications. Finger-jointed lumber acceptable for all structural applications is stamped with "CERT. EXT. JNTS." while finger-jointed lumber acceptable only for vertical applications is stamped with "VERTICAL USE ONLY" or "STUD USE ONLY."*

Unit 7 Review and Resources

UNIT 8

Fastening Systems

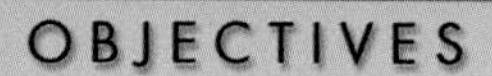

OBJECTIVES

1. List common types of nails.
2. Describe common types of staples.
3. Identify common types of screws.
4. Describe common types of bolts.
5. Discuss common types of hollow-wall fasteners.
6. Describe common types of solid concrete and solid masonry wall anchors.
7. List common types of drive pins, studs, and metal connectors.
8. Review common applications of adhesives.
9. Identify common types of caulk.

Many types of metal fastening devices are used for construction purposes, including nails, staples, screws, and bolts. *Nails* are the primary type of wood fasteners. *Staples* are frequently used for lighter applications. *Screws* have gained greater acceptance in construction and are specified for certain applications. *Bolts* may be used for applications where the ability to remove the fasteners is required. A variety of anchoring devices are also used to fasten materials to concrete, masonry, and steel. *Adhesives* (glues and mastics) are used in combination with nails, staples, and screws for applications where additional holding power is required.

The proper tools must be used to prevent injuries and to avoid damaging fasteners during installation or removal. In addition, appropriate personal protective equipment (PPE), such as safety glasses or goggles, must be worn to prevent or minimize injuries.

NAILS

Nails are available in many shapes and sizes, with a variety of heads, shanks, and points. **See Figure 8-1.** Some nails, typically those with special finishes or deformed shanks (barbs, spirals, rings), have greater holding power than other nails. Aluminum, stainless steel, and galvanized steel nails are used to fasten finish materials to a building exterior since they are corrosion-resistant and will not cause rust streaks on the surface of wood materials.

Special fasteners may be required when installing engineered wood products. Ring- or screw-shank nails are recommended when installing Sturd-I-Floor. Common cement-coated or ring-shank nails are used when installing subfloors. Common smooth-, ring-, or screw-shank nails or galvanized box nails are used when installing wall sheathing. Common smooth-, ring-, or screw-shank nails are recommended when installing roof sheathing, while hot-dipped galvanized box, siding, or casing nails are used to install Rated Siding directly to studs or over nonstructural sheathing.

This unit covers the types of nails used most often for rough and finish carpentry work. Specialty nails are discussed as appropriate in later units.

Simpson Strong-Tie Company, Inc.

Powder-actuated fasteners are commonly used to fasten bottom plates to concrete slabs.

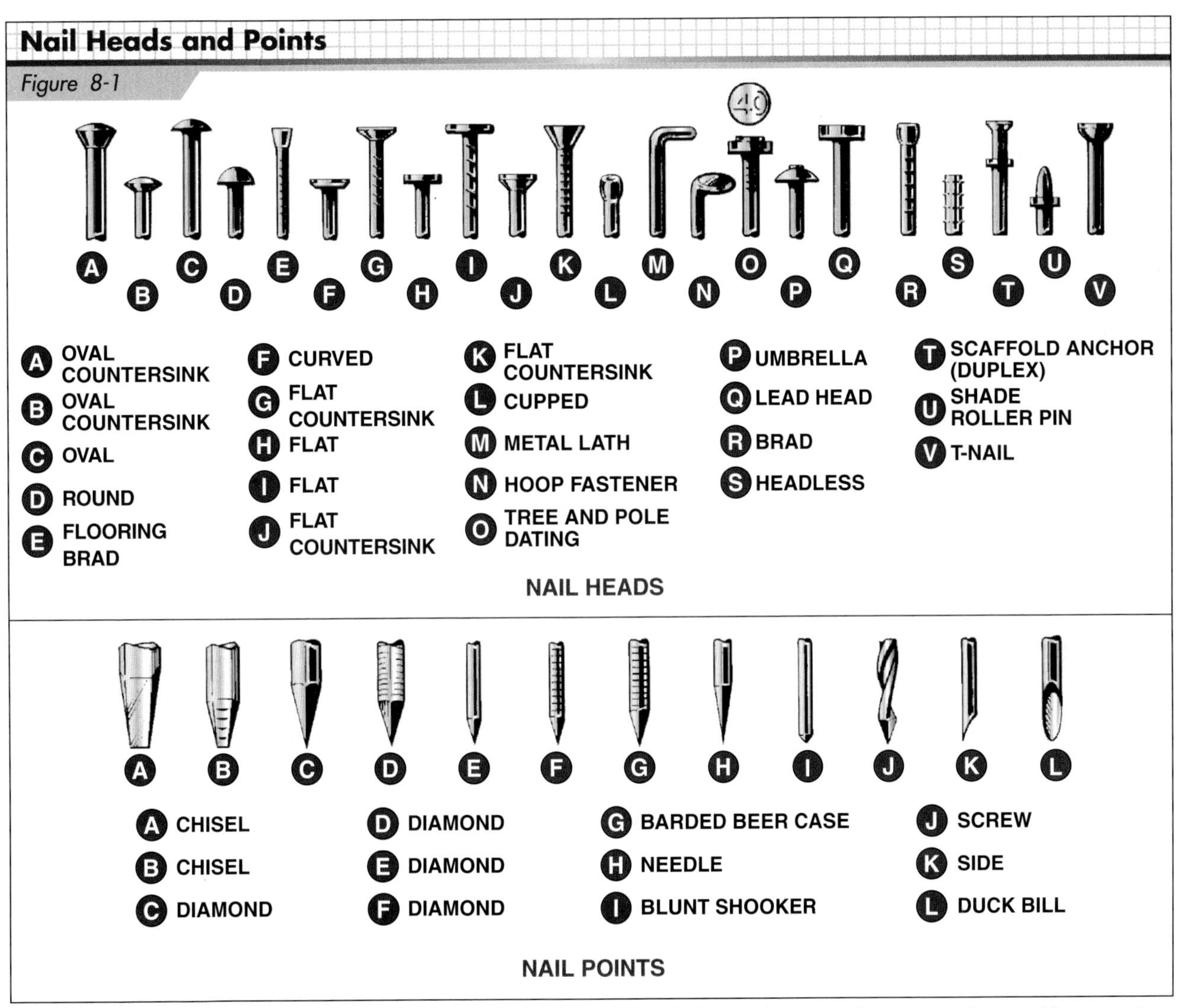

Figure 8-1. *Nails are available with a variety of heads and points. Flat-head, diamond-point nails are most often used by carpenters.*

Penny System of Nail Size

A number and the letter *d*, which represents "penny," are commonly used to designate nail sizes. Typical sizes are 6d, 8d, and 16d. A 6d nail is 2″ long, an 8d nail is 2½″ long, and a 16d nail is 3½″ long. **See Figure 8-2.** The letter *d* stands for *denarius*, an ancient Roman word for coin (or penny). The penny system originated hundreds of years ago in England where nails were priced by the cost (in pennies) per hundred nails. Smaller nail sizes cost less per hundred than larger sizes. The *gauge*, or diameter, of a nail depends on the type and length of the nail. The gauge of a nail increases as the length increases.

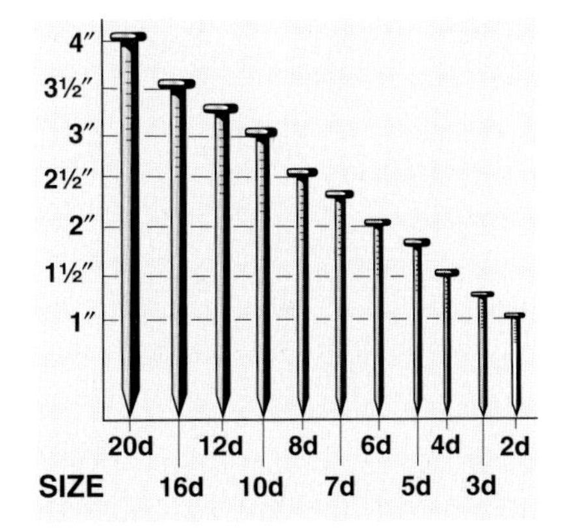

Figure 8-2. *Nail sizes are designated by a number and the letter* d.

Nail sizes may also be designated as the gauge of the nail and the length. For example, a 0.131 × 1½″ nail may be specified to fasten metal connectors to framing members.

Nails for Rough Work

Common nails are used most often in wood-frame construction. **See Figure 8-3.** A common nail is cut from wire and given a head and a point. Common nails are available in sizes from 2d (1″ long) to 60d (6″ long).

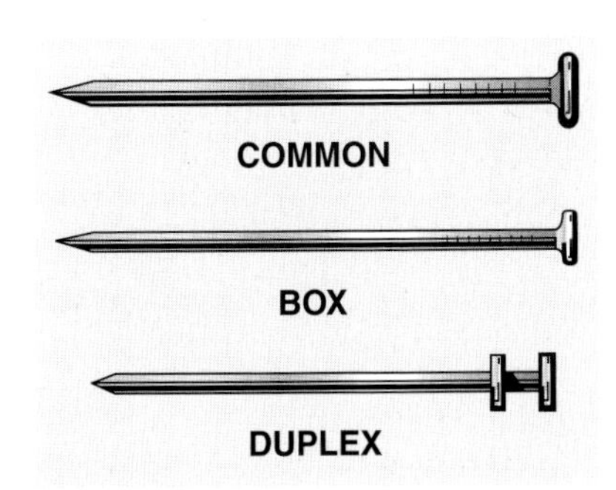

Figure 8-3. *Common nails are used most often for rough work.*

Box nails are similar in appearance to common nails, but their heads and shanks are thinner, making them less likely to split framing members. Box nails are often used to fasten exterior insulation board and siding and are available in sizes from 2d (1″ long) to 40d (5″ long). A disadvantage of box nails is that they bend more easily than common nails when driven.

Duplex nails, also known as double-headed or staging nails, are used for temporary construction such as formwork or scaffolding. The double head on a duplex nail makes it easy to pull out when dismantling forms or scaffolding.

Masonry nails are made with a special hardened steel. **See Figure 8-4.** A masonry nail is used to fasten wood to masonry (solid concrete, concrete block, brick, or stone). Masonry nails must be driven in perfectly perpendicular or they may chip the masonry.

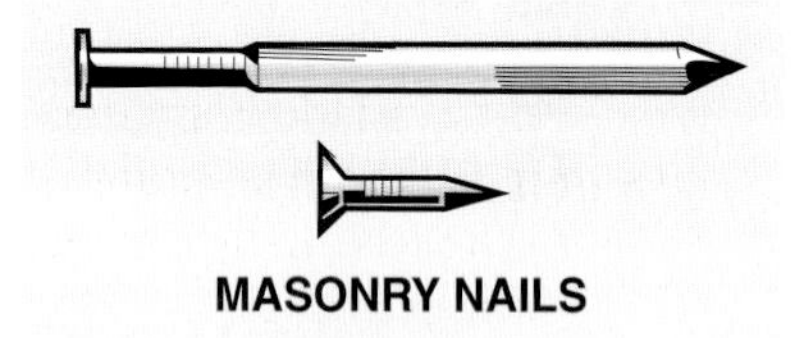

Figure 8-4. *Masonry nails can be driven into concrete or masonry.*

Nails for Finish Work

Nails used for finish work are thinner than common nails, making them easier to drive and less likely to split wood. **See Figure 8-5.** Finish work nails are used for applications, such as trim and moldings, where a nice final appearance is important. *Finish nails* have small, tulip-shaped heads, which are easily driven below the wood surface with a nail set. Finish nails are available in sizes from 2d (1″ long) to 20d (4″ long).

A *casing nail* is a thick version of a finish nail. The head of a casing nail is slightly larger than the head of a finish nail and is tapered toward the bottom. Casing nails are used to fasten heavier pieces of trim material.

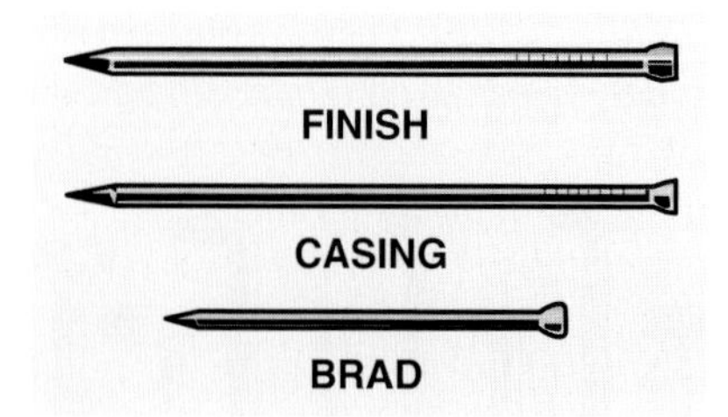

Figure 8-5. *Nails used for finish work are thinner than nails used for rough work.*

Wire brads are identified by their length in inches and gauge (diameter) rather than by the penny system. Wire brad sizes range from 3⁄16″ to 3″ long and from #14 to #20 gauge. Wire brads are thinner than finish or casing nails and are used with light trim materials.

Holding Power of Nails

When a nail is driven into wood, the shank compresses and pushes aside the wood fibers. When the nail is in place, the wood fibers spring back toward their original position. The pressure of the wood fibers against the surface of the nail gives the nail its holding power. **See Figure 8-6.**

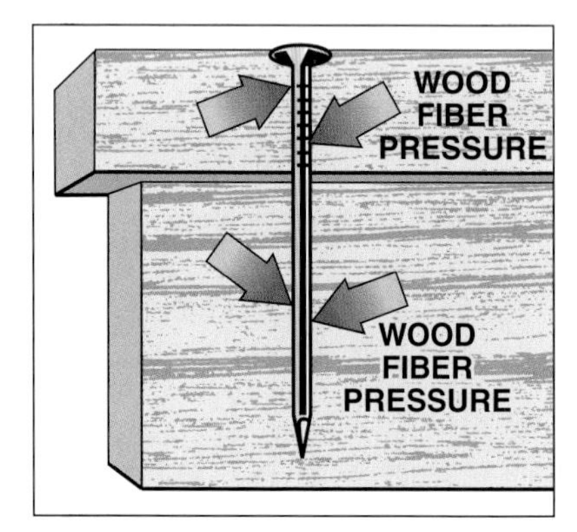

Figure 8-6. *Wood fibers apply pressure against the shank of a nail to give a nail its holding power.*

Smooth-shank nails have sufficient holding power for most construction applications. Deformed-shank nails are recommended when greater holding power is required. **See Figure 8-7.**

Spiral-shank and ring-shank nails are two commonly used deformed-shank nails. *Spiral-shank nails* have a spiral thread that causes them to rotate as they are driven into the wood (similar to driving a screw). Spiral-shank nails are specifically designed for use with hardwoods and are commonly used for siding, flooring, and roof trusses. *Ring-shank nails* have a series of small rings along the shank, which help to resist removal from the wood. Ring-shank nails are commonly used for plywood, underlayment, and roofing applications. The holding power of nails is also increased with cement-, resin-, or zinc-coated nails.

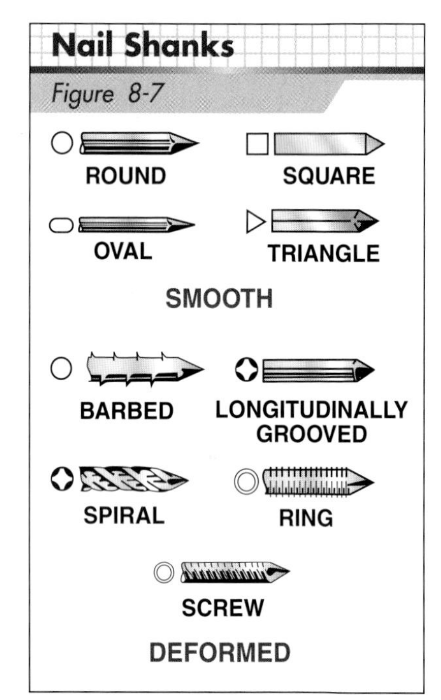

Figure 8-7. *Nail shanks are smooth or deformed. Deformed-shank nails provide greater holding power than smooth-shank nails.*

STAPLES

Staples are available in a variety of shapes and sizes. **See Figure 8-8.** Staples may be used to fasten subflooring, sheathing, and paneling. Heavy-duty staples are driven in by electric or pneumatic tools. Smaller staples may be driven in by hand-operated tools.

SCREWS

For most applications, screws provide greater holding power than nails. In general, carpenters use wood screws, sheet-metal screws, and machine screws for fastening hardware to wood or metal members, attaching cabinets to walls, and fastening trim to metal surfaces. Specialty screws are discussed in later units.

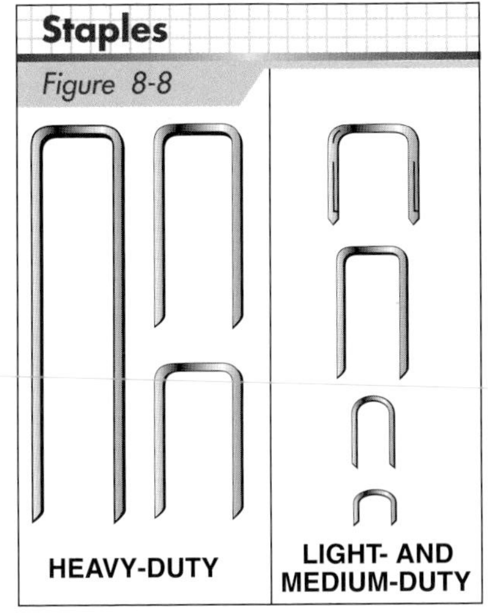

Figure 8-8. *Heavy-duty staples may be used to fasten plywood sheathing and subflooring. Light-duty and medium-duty staples are used for attaching molding and other interior trim.*

Most screws are made of soft steel. Brass, bronze, and copper screws are also available. For decorative purposes and for matching different hardware finishes, steel screws are available in many finishes, including nickel, chromium, silver plate, and gold plate.

Wood Screws

Wood screws have flat, round, or oval heads. Screw heads have a single slot, a recessed cross slot (Phillips), a square (Robertson) recess, or six-point star-shaped (Torx®) recess. **See Figure 8-9.** Phillips and square screw heads provide better grip than single-slot screw heads when driven by an electric screwdriver. They are also more attractive when in place.

Wood screws range in size from $\frac{1}{4}$" to 5" long. The diameter of the screw shank is identified by a gauge number. A higher gauge number indicates a thicker screw. **Figure 8-10** shows screws with gauges ranging from #1 to #14.

Screw Heads and Slots

Figure 8-9

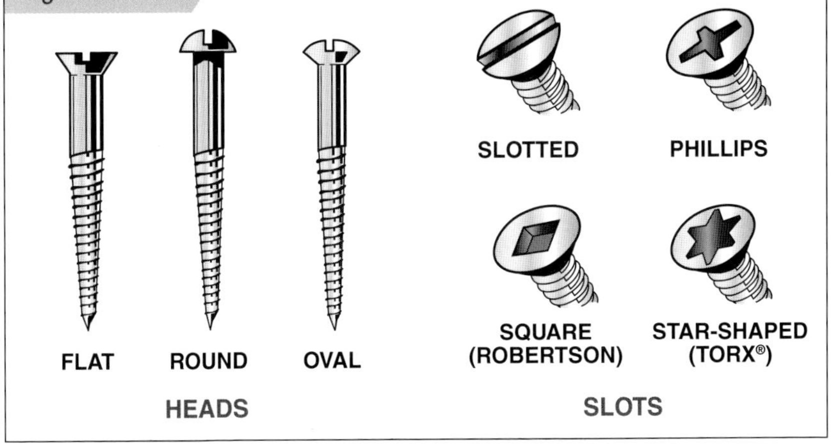

FLAT ROUND OVAL — HEADS

SLOTTED PHILLIPS SQUARE (ROBERTSON) STAR-SHAPED (TORX®) — SLOTS

Figure 8-9. *The three basic screw heads are flat, round, and oval. Screw heads have a single slot or a recessed cross slot.*

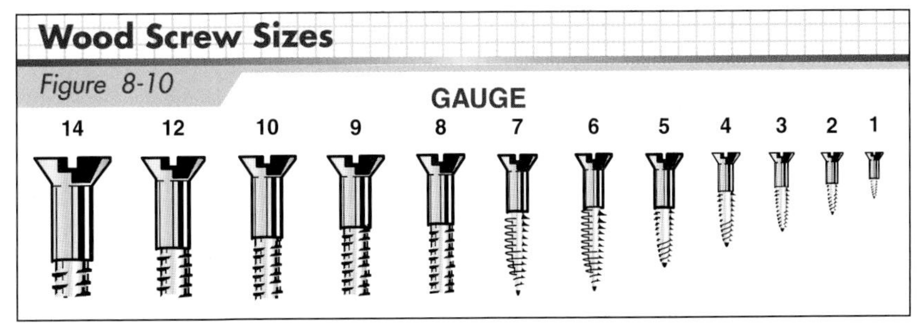

Figure 8-10. *A higher gauge number for screws indicates a thicker screw shank.*

When using wood screws to fasten wood pieces together, a shank hole should be drilled in the piece being fastened. An undersized pilot hole for the threads going into the receiving piece will make driving the screw easier and prevent splitting. **See Figure 8-11.** Flat-head screws may be countersunk or counter-bored. *Countersinking* places the top of the screw head flush with the wood surface. *Counterboring* is used when the screw head is to be concealed with a wood plug.

Self-Tapping Screws

Self-tapping screws are used to fasten metal framing members to each other and to fasten other materials to metal framing members. Self-tapping screws tap their own threads as they are being driven into the metal members. Self-tapping screws are available as self-drilling and self-piercing screws. **See Figure 8-12.**

Self-drilling screws are the most frequently used metal-to-metal fasteners. The points of self-drilling screws drill through the metal layers before the screw threads engage. The drill point of the screw must be sharp and long enough to penetrate the steel being fastened together. Coarse screw threads are commonly used for light-gauge steel-framing operations. Finer threads may be used when tapping into thicker steel material.

Self-piercing screws have a sharp point capable of penetrating and tapping thin metal. Screw diameters are identified by gauge numbers, which range from #6 to #14. The most frequently used self-piercing screw diameters are #6, #8, and #10, which are 0.138", 0.164", and 0.190", respectively.

Shank and Pilot Hole Sizes

Figure 8-11

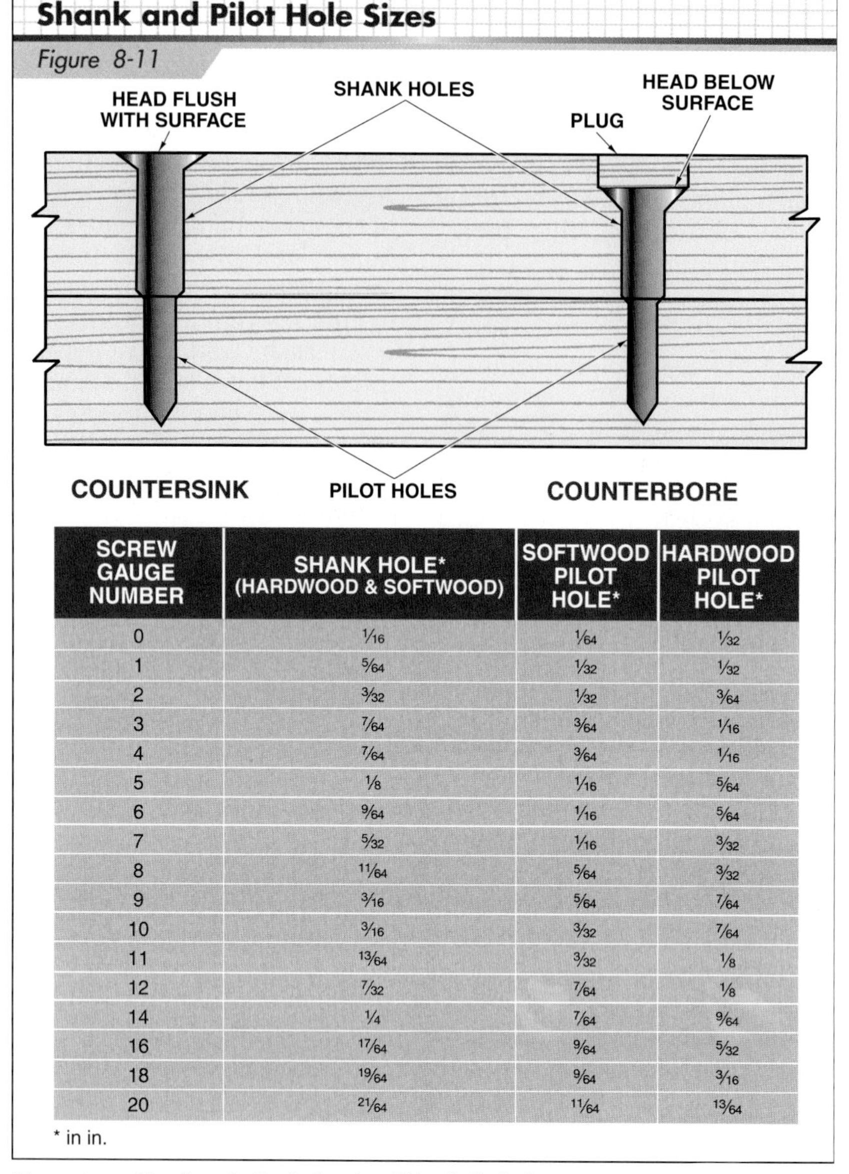

SCREW GAUGE NUMBER	SHANK HOLE* (HARDWOOD & SOFTWOOD)	SOFTWOOD PILOT HOLE*	HARDWOOD PILOT HOLE*
0	1/16	1/64	1/32
1	5/64	1/32	1/32
2	3/32	1/32	3/64
3	7/64	3/64	1/16
4	7/64	3/64	1/16
5	1/8	1/16	5/64
6	9/64	1/16	5/64
7	5/32	1/16	3/32
8	11/64	5/64	3/32
9	3/16	5/64	7/64
10	3/16	3/32	7/64
11	13/64	3/32	1/8
12	7/32	7/64	1/8
14	1/4	7/64	9/64
16	17/64	9/64	5/32
18	19/64	9/64	3/16
20	21/64	11/64	13/64

* in in.

Figure 8-11. *Shank and pilot holes should be drilled when using wood screws to fasten wood pieces together.*

Self-Tapping Screws

Figure 8-12

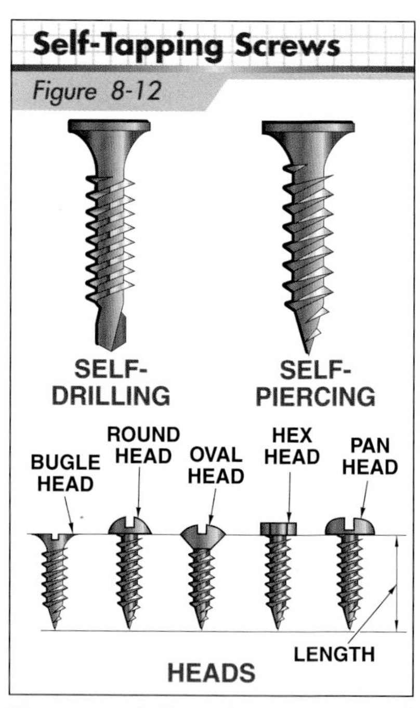

Figure 8-12. *Self-tapping screws are used to fasten metal framing members to each other by tapping their own threads in the metal members.*

Concrete Screws

Concrete screws are used to fasten items, such as plates, furring strips, and electrical boxes, to concrete, brick, or block. Concrete screws are cold-formed fasteners with twin-lead threads and a nail-point tip. Concrete screws are available in 3/16″, 1/4″, and 5/16″ diameters, and lengths ranging from 1 1/4″ to 4″. Concrete screws are available with Phillips, square (Robertson), and hex heads. Phillips-head concrete anchors are used when appearance is important or when the head must be flush with the surface. Hex-head screws are typically used since the head will engage better with the driver, making them easier to drive. Concrete screws provide excellent corrosion resistance in dry environments and can be removed without damaging the base material.

When installing concrete screws, first determine the proper length of screw to be installed. **See Figure 8-13.** Add the thickness of the material to be fastened to the minimum depth of embedment and select a screw of that length or longer. Concrete screws must be embedded a minimum of 1″. Maximum strength is achieved when the screw is embedded approximately 1 3/4″. Drill a hole for the screw; 3/16″ screws require a 5/32″ diameter bit, 1/4″ screws require a 3/16″ diameter bit, and 5/16″ screws require a 1/4″diameter bit. The hole should be drilled 1/4″ deeper than the length of the screw embedment. Hammer drills are commonly used to drill into masonry. Remove excess grit from the hole.

Drive the screw into the predrilled hole until the screw is fully seated. Use a slow to moderate drill speed and apply firm and even pressure when driving the screw.

Machine Screws

Machine screws are available with a variety of head shapes. **See Figure 8-14.** Machine screws screw into threaded holes in metal and have greater holding power than other types of screws that fasten to metal. Machine screws are commonly used for fastening door hinges, push plates, locks, and door closers to metal jambs and doors.

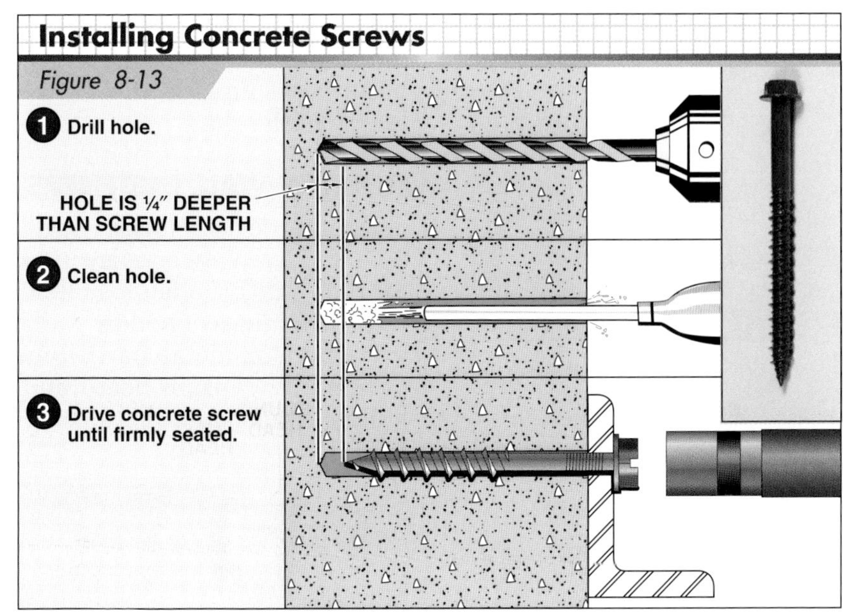

Figure 8-13. *Concrete screws are used to fasten items to concrete, brick, or block without the use of additional anchors.*

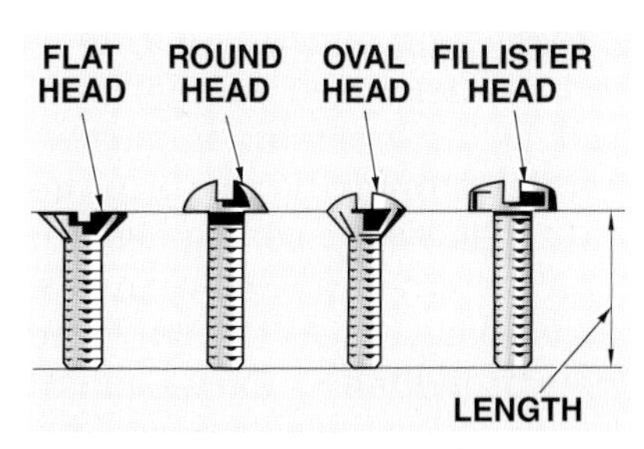

Figure 8-14. *Machine screws have greater holding power than other types of screws that fasten to metal.*

The size of a machine screw or machine bolt is expressed as the length, diameter (in inches), and number of threads per inch if the screw or bolt is ¼″ diameter or larger. For example, a 2″ × ¼–20 designates a screw or bolt that is 2″ long, ¼″ in diameter, and has 20 threads per inch. For smaller machine screws or bolts, a gauge number is used to specify the diameter.

Construction-Grade Screws

Construction-grade screws, driven by power screwdrivers or automatic screwguns, are used for many framing operations. Construction-grade screws are commonly used to fasten subfloor, roof sheathing, and wall sheathing to framing members. Construction-grade screws are self-driving and do not require a pilot hole. Care should be taken not to overtorque screws when driving them into panels with power tools. When the head is flush with the panel surface, the driving bit should be released from the head. Special construction-grade screws, which are self-drilling and self-tapping, are used in metal-frame construction and for attaching gypsum board to wood or metal framing members.

Construction-grade screws differ from conventional screws in several ways. Construction-grade panel screws have coarser and more steeply pitched threads, allowing them to be driven through harder materials such as particleboard and other reconstituted wood panels. **See Figure 8-15.** The high glue content of reconstituted panels makes them harder to penetrate with screws or nails.

Construction-grade panel screws are typically manufactured of corrosion-resistant materials, such as stainless steel, or have a corrosion-resistant finish, such as gray phosphate or zinc. Panel screws are typically #8 gauge and are available with Phillips or square drive recesses. Panel screws are commonly driven with a coil or strip-type screwgun. **See Figure 8-16.**

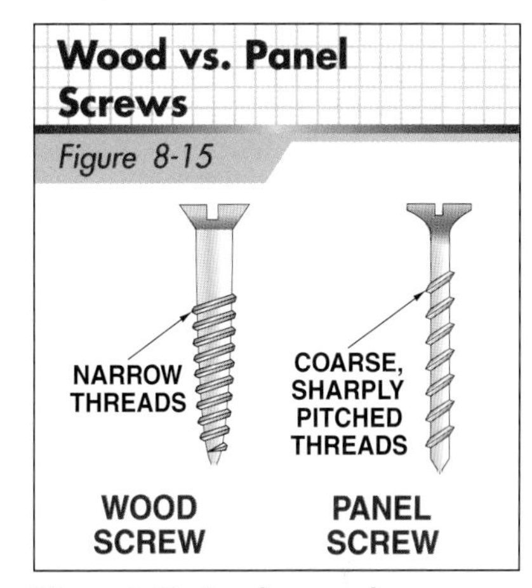

Figure 8-15. *Panel screws have coarse, steeply pitched threads in comparison to a standard wood screw.*

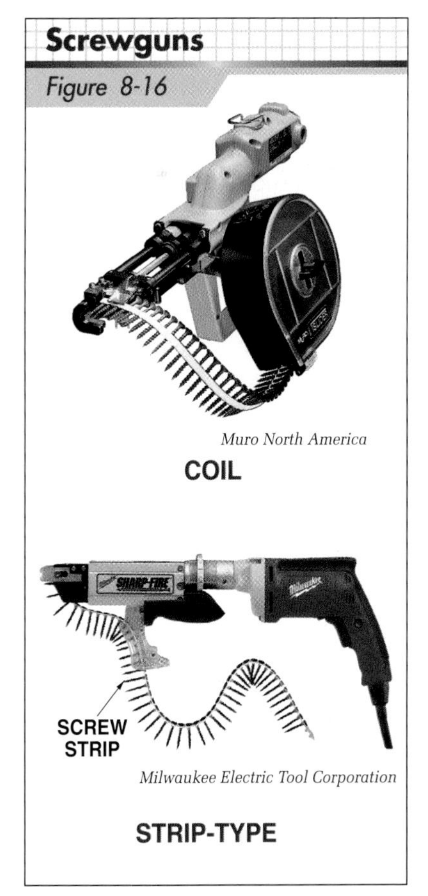

Figure 8-16. *Construction-grade panel screws are driven with coil or strip-type screwguns. An extension handle may be added to screwguns to allow carpenters to install screws from an upright position.*

BOLTS

Bolts are used to fasten together heavy wood and metal materials. **See Figure 8-17.** Most types of bolts require nuts. Whenever a nut bears against wood, a washer should be used to distribute pressure over a wider area and prevent the nut from digging into the wood. With the exception of carriage bolts and flat-head stove bolts, a washer should also be installed under the bolt head.

Carriage bolts are used in wood or metal. The square shank below the rounded head of a carriage bolt is embedded in the material as the nut is drawn up to prevent the bolt from turning as the nut is tightened. **See Figure 8-18.**

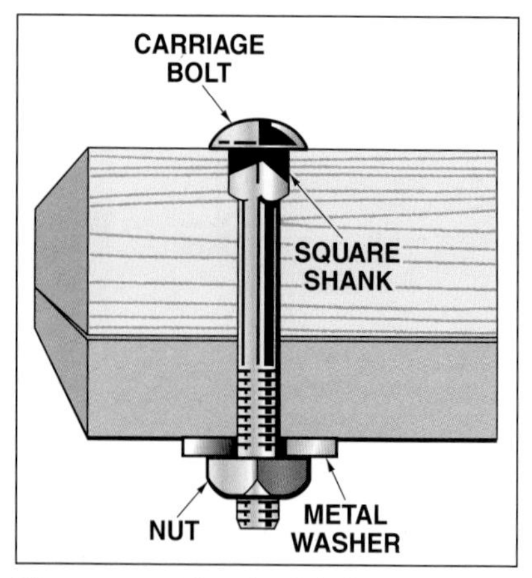

Figure 8-18. *The shank below a carriage bolt head is embedded in the wood.*

Machine bolts have square or hexagonal heads. They are used to fasten together wood or metal pieces.

Lag bolts are not true bolts; they are actually heavy screws with square or hexagonal heads. **See Figure 8-19.** Lag bolts are used to fasten heavy material to wood when a regular bolt-and-nut arrangement is impractical or inconvenient. Shank and pilot holes are drilled for the lag bolt, which is then screwed in with a wrench.

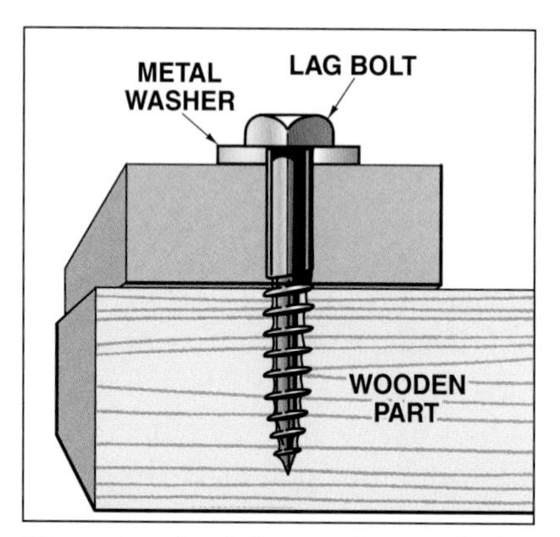

Figure 8-19. *Lag bolts are often used when it is inconvenient or impossible to use a nut-and-bolt arrangement. A washer should be used under the head.*

Stove bolts are used for lighter work. Stove bolts have a smaller size range than other types of bolts and are available in lengths from ⅜″ to 6″ and in diameters from ⅛″ to ⅜″. Unlike other bolts, stove bolts have slotted flat or round heads. Stove bolts have threads the entire length of the shaft.

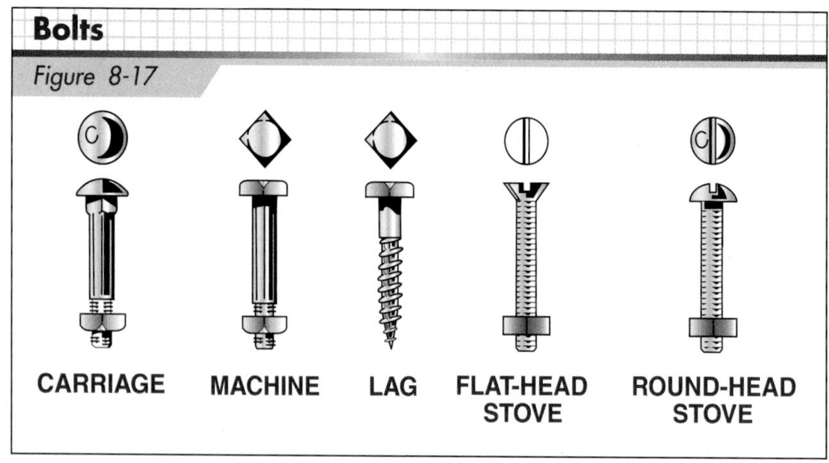

Figure 8-17. *Bolts are used to fasten together heavy wood and metal materials.*

HOLLOW-WALL FASTENERS

Anchoring devices, such as toggle bolts and screw anchors, are used to fasten light materials to hollow walls. Examples of hollow walls are wood or metal partitions covered by gypsum board and hollow-block masonry walls.

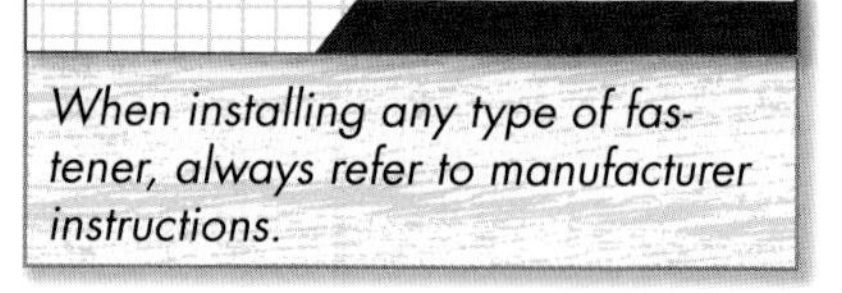

Toggle Bolts

A *toggle bolt* consists of a stove bolt with a winged nut that folds back as the assembly is pushed through a predrilled hole in the wall. The wings spring back to their original position inside the wall cavity. As the machine screw is tightened, the wings are drawn against the inner surface of the finish wall material. **See Figure 8-20.** Toggle bolts are available with a variety of machine screw combinations and range in size from ⅛″ to ⅜″ in diameter and 2″ to 6″ in length.

Screw Anchors

A *screw anchor* is used to fasten small cabinets, towel bars, drapery hangers, mirrors, electrical fixtures, and other lightweight items to hollow walls. A screw anchor is inserted into a predrilled hole in the wall so the prongs on the outside of the shield contact the wall surface. The prongs grip the wall surface to prevent the shield from turning. As the screw is tightened, the shield spreads and flattens against the inner surface of the wall. **See Figure 8-21.** Screw anchors are sized for the thickness of hollow walls from ⅛″ to 1¾″ thick.

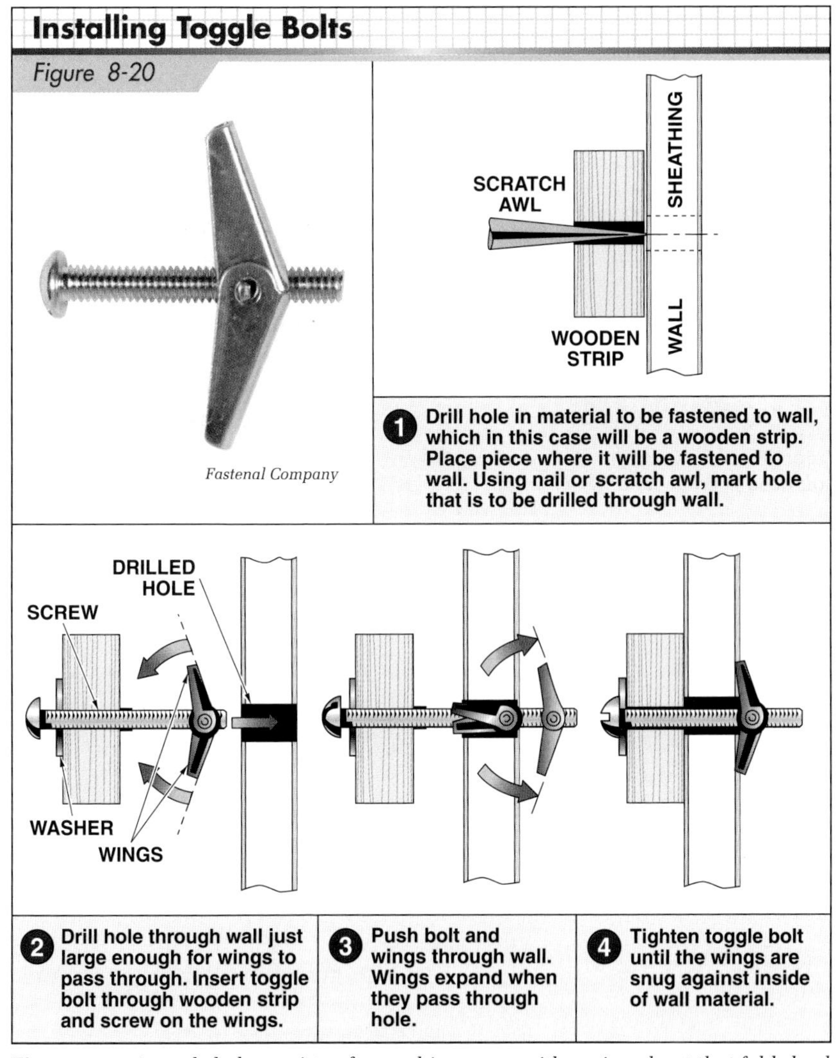

Figure 8-20. *A toggle bolt consists of a machine screw with a winged nut that folds back as the assembly is pushed through a predrilled hole in the wall.*

Wallboard Anchor

A *wallboard anchor* is designed for use in gypsum board and does not require predrilling prior to installation. Deep-cutting threads on the anchor allow it to be driven with a Phillips screwdriver or cordless screwdriver with a Phillips bit and resist stripping from the gypsum board. Wallboard anchors are used to support light loads such as pictures, curtain rods, and bathroom fixtures. Wallboard anchors are available in plastic, nylon, or zinc.

A wallboard anchor is installed by placing the point of the anchor against the wall and driving the anchor until it is firmly seated. **See Figure 8-22.** The fixture is installed over the anchor and a screw is driven into the anchor to tighten the fixture into place. A 1″ minimum engagement is required between the anchor and screw.

Neoprene sleeves and vibration-proof polypropylene screw anchors can be used to isolate vibrations from equipment (such as stereo speakers) from their surrounding surfaces.

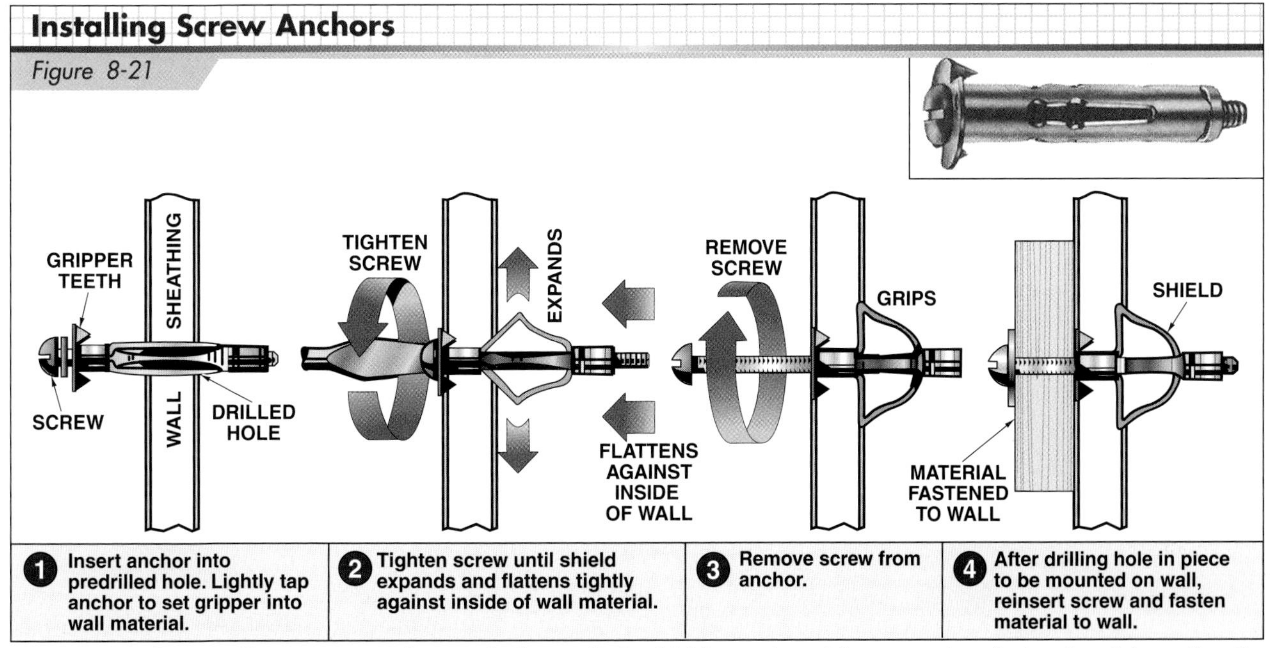

Figure 8-21. *When installing a screw anchor in a hollow wall, the shield spreads and flattens against the interior of the wall as the screw is tightened.*

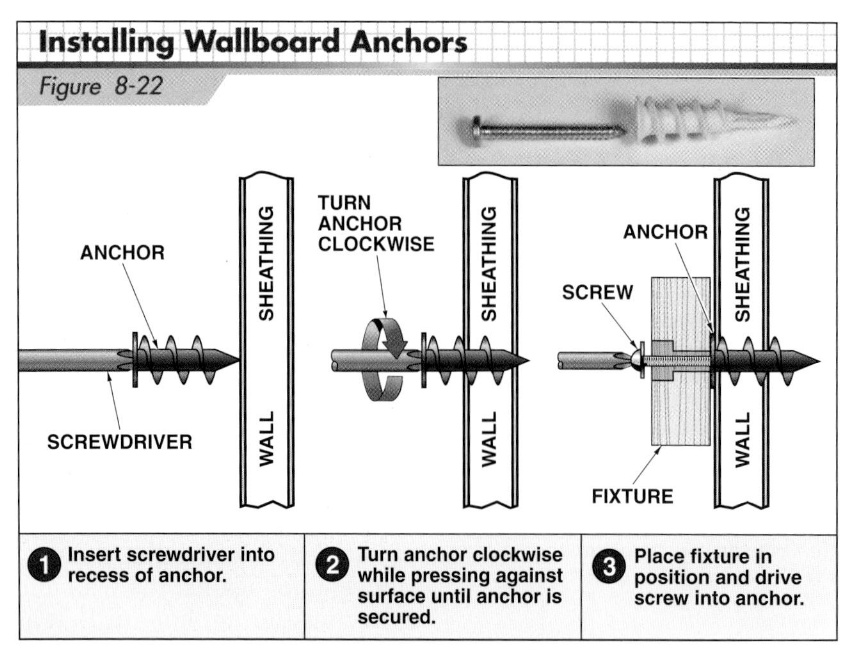

Figure 8-22. *A wallboard anchor does not require predrilling prior to installation.*

SOLID CONCRETE OR SOLID MASONRY WALL ANCHORS

Expansion anchors allow heavy materials and equipment to be securely and quickly fastened to solid concrete (or other solid masonry) walls, floors, and ceilings. Expansion anchors range from light-duty lead, plastic, and fiber plugs to heavy-duty expansion shields.

Most expansion anchors require that a hole equal to the outside diameter of the plug or shield be drilled in the concrete. The hole should be slightly deeper than the length of the plug or shield. Holes in concrete can be drilled with a standard electric drill or *rotary hammer.* **See Figure 8-23.**

Milwaukee Electric Tool Corporation

Figure 8-23. *A rotary hammer may be used to drill holes in concrete.*

Electric drills use carbide-tipped masonry bits or core bits. Masonry bits are available in 1⁄8″ to 1½″ diameters and 2½″ to 18″ lengths. Core bits are available in 5⁄8″ to 6″ diameters and 6″ to 18″ lengths. **See Figure 8-24.** Certain core bits cut through reinforcing bar and concrete. Smaller core bits have ports (holes) on the side to allow masonry debris to escape from the interior and be carried to the surface. Standard-, SDS-, or spline-drive masonry bits are used in rotary hammers.

After a hole has been drilled for an expansion anchor, the hole is cleaned out and the anchor is placed inside the hole. As the screw or bolt is driven into the anchor, it expands and presses against the concrete. The pressure holds the anchor in place.

Light-Duty Anchors

Light-duty anchors are typically used to secure light items such as electric fixtures, towel brackets, hooks, and other household accessories to solid masonry walls. Light-duty anchors are often made of plastic or lead. The plastic anchors in **Figure 8-25** are used with wood or sheet-metal screws. A *molded anchor* has a tapered cavity, and can be used in masonry and gypsum board. The *tubular anchor* has longitudinal slots that aid expansion and prevent turning or twisting of the anchor in the predrilled hole in the wall.

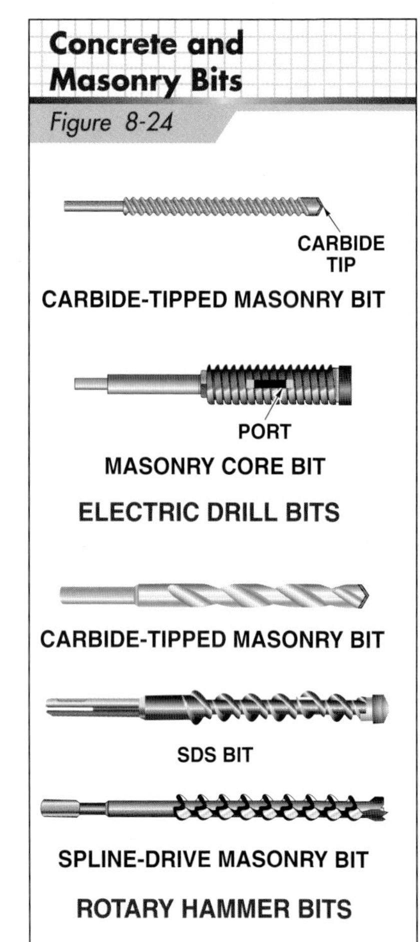

Figure 8-24. *A variety of masonry and core bits are used with electric drills or rotary hammers.*

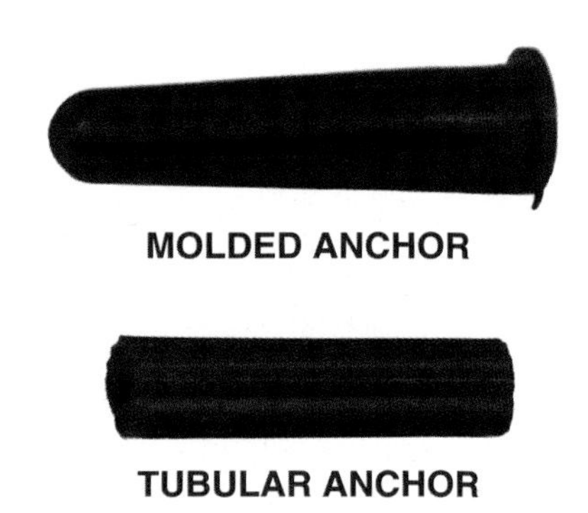

Figure 8-25. *Light-duty plastic anchors are used with wood or sheet-metal screws.*

The *lead-alloy anchor* in **Figure 8-26** can be used with lag, wood, or sheet-metal screws. The longitudinal ribs at the tip of the anchor grip the sides of the predrilled hole as a screw is driven into the anchor.

Figure 8-26. *Light-duty lead-alloy anchors are used with lag, wood, or sheet-metal screws.*

Hammer-Driven Anchors. A hammer-driven anchor does not receive a screw or bolt; a special nail called an expander pin expands the anchor as it is hammered in. **Figure 8-27** shows a steel hammer-driven anchor. The anchor expands as the expander pin is driven in.

To install a hammer-driven anchor, first drill a hole equal to the outside diameter of the anchor using the object to be fastened as a template for placement of the hole. Place the anchor, lead band first, through the object being fastened and into the hole in the masonry. The anchor flange should rest flush against the surface of the object being fastened. Place the expander pin into the anchor and drive it in.

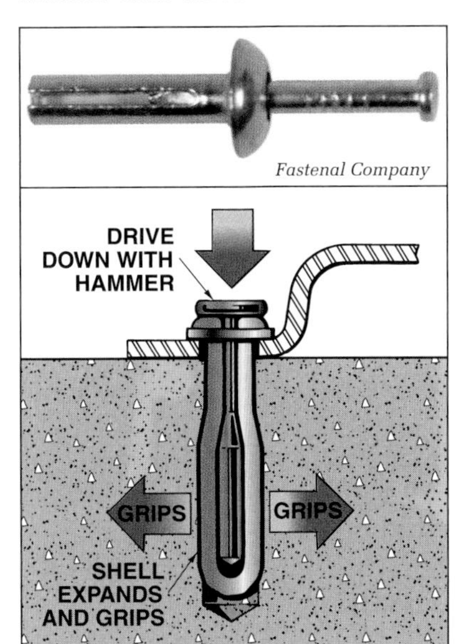

Figure 8-27. *The anchor of a hammer-driven anchor expands as the expander pin is driven in.*

Figure 8-28 shows a plastic hammer-driven anchor. Installation of a plastic hammer-driven anchor is similar to a steel hammer-driven anchor. The steel expander pin is threaded and has a screwdriver slot in the head to allow the pin to be removed with a screwdriver.

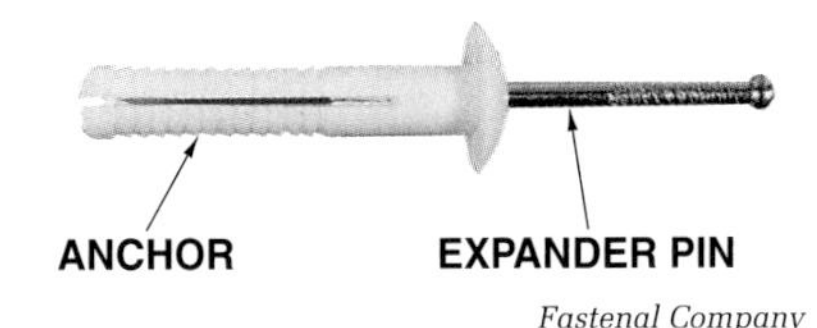

Fastenal Company

Figure 8-28. *The slotted head of a hammer-driven nylon anchor allows the expander pin to be removed with a screwdriver.*

Expansion Shields

A variety of expansion shields are available for medium to heavy loads. Expansion shields are used with lag or machine bolts. The shield expands as the bolt is tightened and grips the inside surface of a predrilled hole. **See Figure 8-29.** This medium-duty anchor is made of zinc alloy and is available in sizes accommodating lag screws of ¼″ to ¾″ diameter.

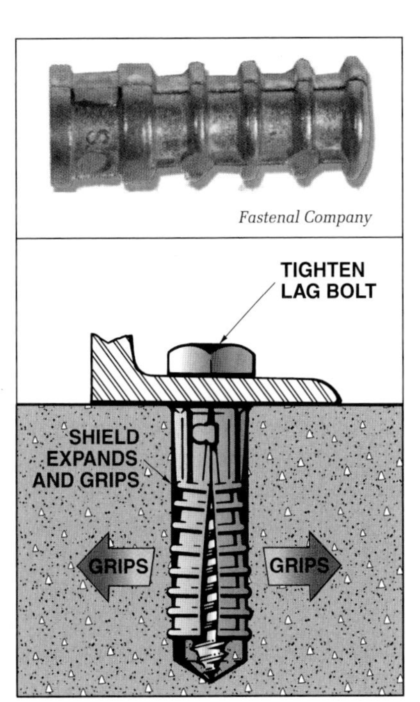

Figure 8-29. *When installing a lag bolt expansion shield, the shield expands as the bolt is tightened.*

To install an expansion shield, first drill a hole equal to the outside diameter of the shield. The hole should be as deep as the expansion shield is long, plus an additional ½″ or more. Place the expansion shield, ribbed end first, into the hole. A portion of the shield will protrude above the surface; therefore, hammer the shield into the hole until it is flush with the surface. Position the object being fastened and begin to screw in the lag bolt. If the lag bolt binds before the head of the bolt is against the object being fastened, drive the bolt by hammering against the head until it is flush with the object and finish tightening the lag bolt.

Figure 8-30 shows two machine bolt expansion shields. These shields require holes equal to their outside diameters and deep enough so that the shields will be flush or slightly below the concrete surface. The shields expand as the bolts are tightened. This type of fastener can be removed if desired, and the hole grouted to refinish the concrete surface.

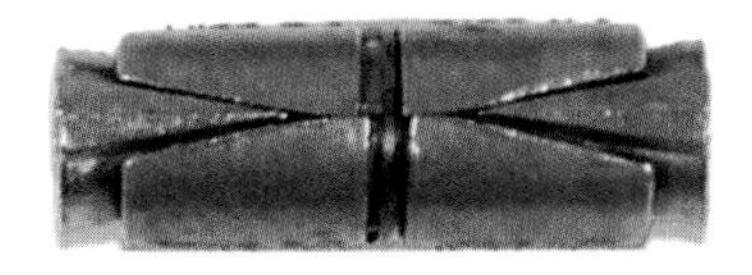

Fastenal Company

Figure 8-30. *A heavy-duty machine bolt expansion shield expands at both ends to distribute the anchor load.*

Medium-duty machine bolt anchors are available in short lengths of 1½″ to 2½″ and in long lengths of 1¼″ to 3½″. Short lengths are used for anchors in good grades of concrete when the thickness of the concrete layer limits the depth of the hole. Short-length anchors are available in sizes accommodating bolts of 5⁄16″ to 5⁄8″ diameter. Long lengths are used for anchoring in poorer grades of concrete (where extra anchoring strength is required) when the layer of concrete allows a deep enough hole for the longer bolt. Long-length anchors are available in sizes accommodating bolts of 1⁄4″ to 3⁄4″ diameter.

Figure 8-31 shows a steel anchor with an expansion plug and a setting tool. The steel anchor and expansion plug are placed in a predrilled hole. The setting tool is used to drive the anchor onto the plug and expand the anchor. Steel anchors with expansion plugs are available in sizes accommodating bolts of 1⁄4″ to 3⁄4″ diameter.

To install an anchor with an expansion plug, first drill a hole equal to the outside diameter of the anchor. Preassemble and place the plug and anchor into the hole. With the setting tool and hammer, drive the anchor over the expander plug. Position the object to be fastened over the anchor and fasten it into place.

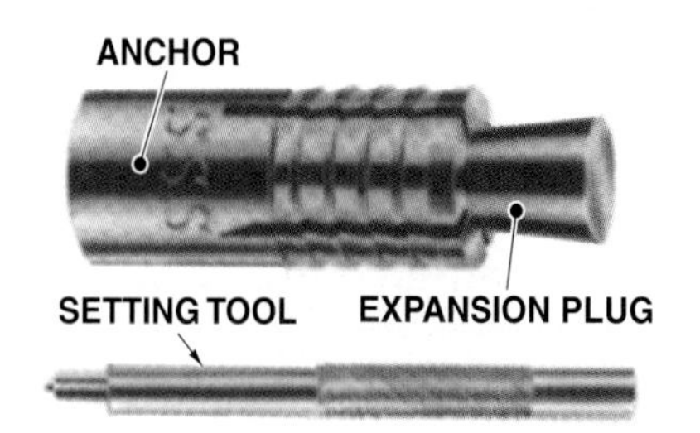

Figure 8-31. *The expansion plug expands the steel anchor as the anchor is driven onto the plug with a setting tool.*

Another type of expansion shield is a *self-drilling anchor.* The anchor is placed in a special chuck head adapted for a rotary hammer. The sharp teeth at the end of the anchor drill a hole in the concrete. The anchor is secured with an expansion plug.

Figure 8-32 shows the procedure for installing a self-drilling anchor. The chuck end of the anchor is snapped off with a quick lateral movement of the rotary hammer or by using a snap-off tool. Self-drilling anchors are available in sizes accommodating bolts of 1⁄4″ to 3⁄4″ diameter.

Stud-Bolt Anchors

Stud-bolt anchors are typically used to fasten large equipment and machinery in place. The equipment is set in its final position, and the holes are drilled in the concrete through the mounting lugs of the equipment. Stud-bolt anchors are then inserted and secured.

Three types of stud-bolt anchors are available—*expansion plug, sleeve,* and *wedge.* **See Figure 8-33.** Stud-bolt anchors are available in 1⁄4″ to 3⁄4″ diameters.

An expansion plug stud-bolt anchor has a tapered expansion plug located at the end opposite the external threads. A lead shield surrounding the expansion plug expands as the anchor is driven down on to the plug within the predrilled hole.

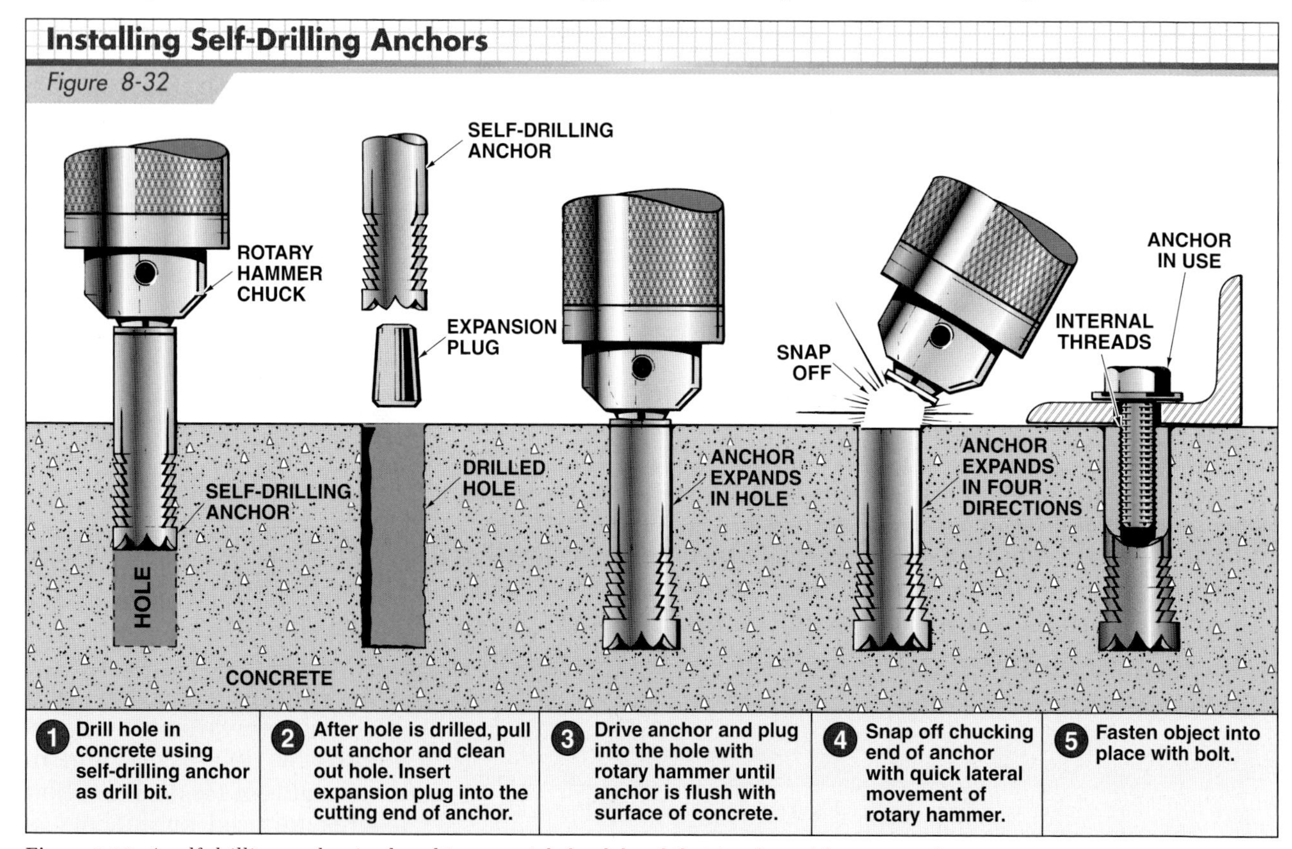

Figure 8-32. *A self-drilling anchor is placed in a special chuck head that is adapted for a rotary hammer.*

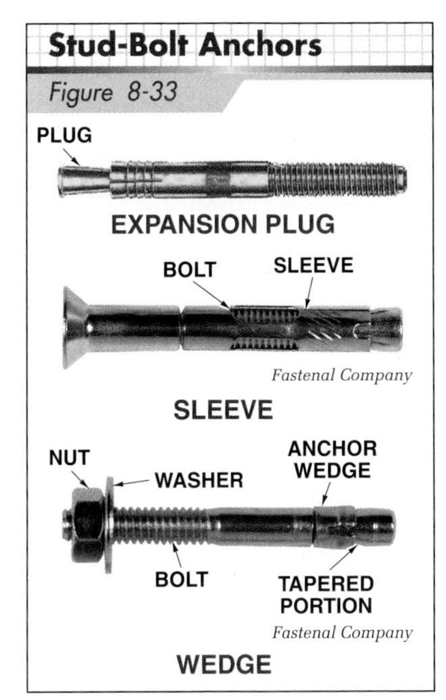

Figure 8-33. *Stud-bolt anchors are used to fasten large equipment and machinery in place.*

Sleeve stud-bolt anchors are available in a variety of preas-sembled units. When the nut or screw head of the bolt is tightened, the sleeve wedges outward and provides a grip for the anchor.

The anchor wedge of a wedge-type stud-bolt anchor is banded around the lower portion of the bolt. The wedge allows the anchor to be driven into the hole yet offer resistance to it being pulled out. A wedge stud-bolt anchor is driven into a hole drilled in the wall. The wedge expands outwardly as the tapered portion of the stud-bolt is drawn into it when the nut is tightened.

To install a stud-bolt anchor, first drill holes through the mounting lugs of the equipment to be fastened or use the mounting lugs provided. The holes should be the same diameter as the anchor to be used. Place the stud-bolt anchor, anchor end first, into the hole. Using a hammer or appropriate anchor-setting tool, drive the stud-bolt into position and secure the equipment in place with a washer and nut. When driving anchors into position, protect the end of the threads by installing a nut at the end or by placing a piece of wood over the end.

Adhesive Anchors

Adhesive anchors include adhesives mixed and dispensed from tubes and adhesives stored in glass capsules. The adhesive is placed in a drilled hole and a stud-bolt is placed in the proper position. *Tube-dispensed adhesives* are typically a two-part epoxy mixed to the proper proportions as it is dispensed. When using tube-dispensed adhesives, the hole is drilled and properly prepared by removing particles with a brush and low-pressure compressed air. **See Figure 8-34.** When starting a new cartridge or tube, dispense and discard enough adhesive until a uniform dark gray color is achieved. Insert the nozzle tip into the bottom of the hole and fill to approximately one-half the hole depth. The stud-bolt is inserted into the hole and slowly rotated to provide a good bond between the concrete and the bolt. After the recommended cure time, the equipment can be positioned and secured.

A *glass capsule adhesive anchor* consists of a sealed glass capsule containing an adhesive mixture, a stud-bolt, and a nut and washer. The capsules are available in a variety of sizes to accommodate various bolt diameters. Each capsule may contain synthetic resin and quartz aggregate filler. After a hole is drilled and properly prepared, the glass capsule is inserted into the hole. **See Figure 8-35.** The stud-bolt, attached to a rotary hammer, is inserted into the hole, breaking the capsule and agitating the mixture. As the capsule is crushed, a chemical reaction occurs and the mixture fills the gap between the stud-bolt and sides of the hole and penetrates the pores of the concrete. When the adhesive hardens full holding power is achieved.

DRIVE PINS AND STUDS

Drive pins and *studs* are also widely used devices for fastening materials to concrete and masonry walls. A drive pin is a hardened steel nail that is driven directly into the concrete to secure an object into place. A stud consists of a nail end that is embedded in concrete and a threaded end to receive a nut. **See Figure 8-36.** Drive pins and studs are driven into the concrete with a powder-actuated tool.

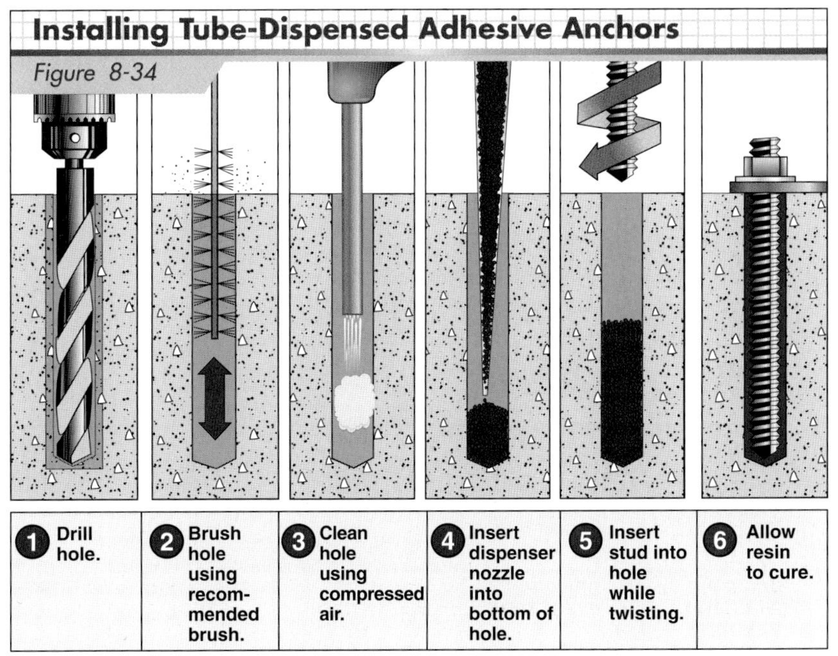

Figure 8-34. *A tube-dispensed adhesive is mixed as it is forced down the nozzle.*

A powder-actuated tool quickly embeds drive pins or studs into concrete or other masonry. **See Figure 8-37.** The tool can also be used to drive pins and studs into steel beams and columns. A *powder-actuated tool* is a gun with a powder-filled round (shell) that drives the pin or stud into the wall. Additional information on powder-actuated tools is found in Unit 18.

METAL CONNECTORS

Metal connectors are commonly used in wood- and metal-framed construction and are extremely important in areas subject to earthquakes, tornadoes, and hurricanes. Studies have shown that, when structural damage occurs, it is typically the result of framing members being pulled loose at their joints rather than being broken along their length. Metal connectors are available in many shapes and sizes for a variety of structural wood and metal connections. Some of the most commonly used metal connectors are joist and beam hangers, post bases, framing anchors, holddowns, and tiedowns. **See Figure 8-38.**

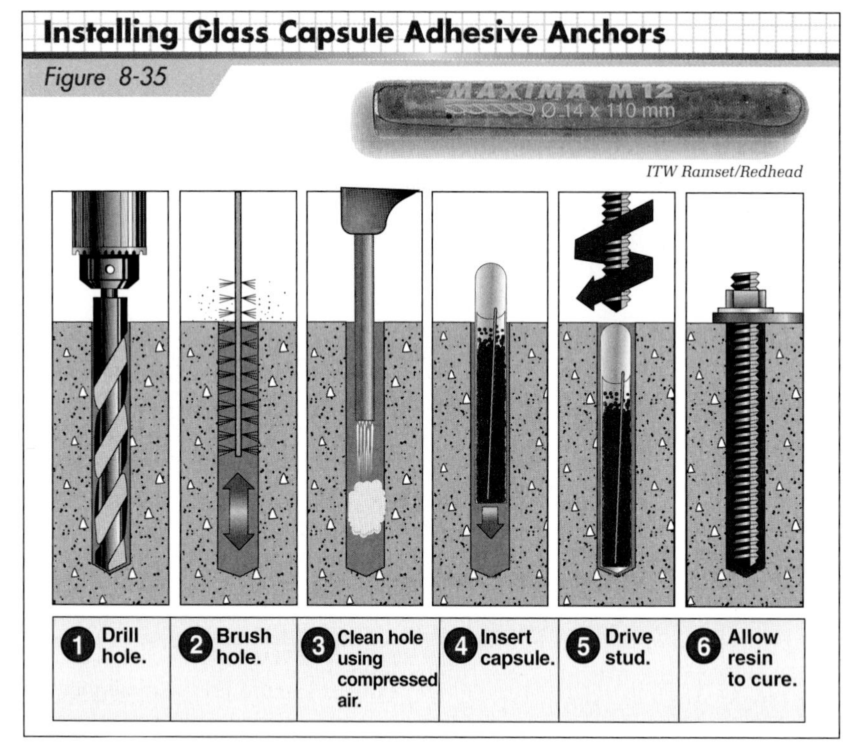

Figure 8-35. *A glass capsule adhesive anchor consists of a sealed glass capsule containing an adhesive mixture, a stud-bolt, and a nut and washer.*

Figure 8-37. *A powder-actuated tool is used to drive pins and studs into concrete or other masonry.*

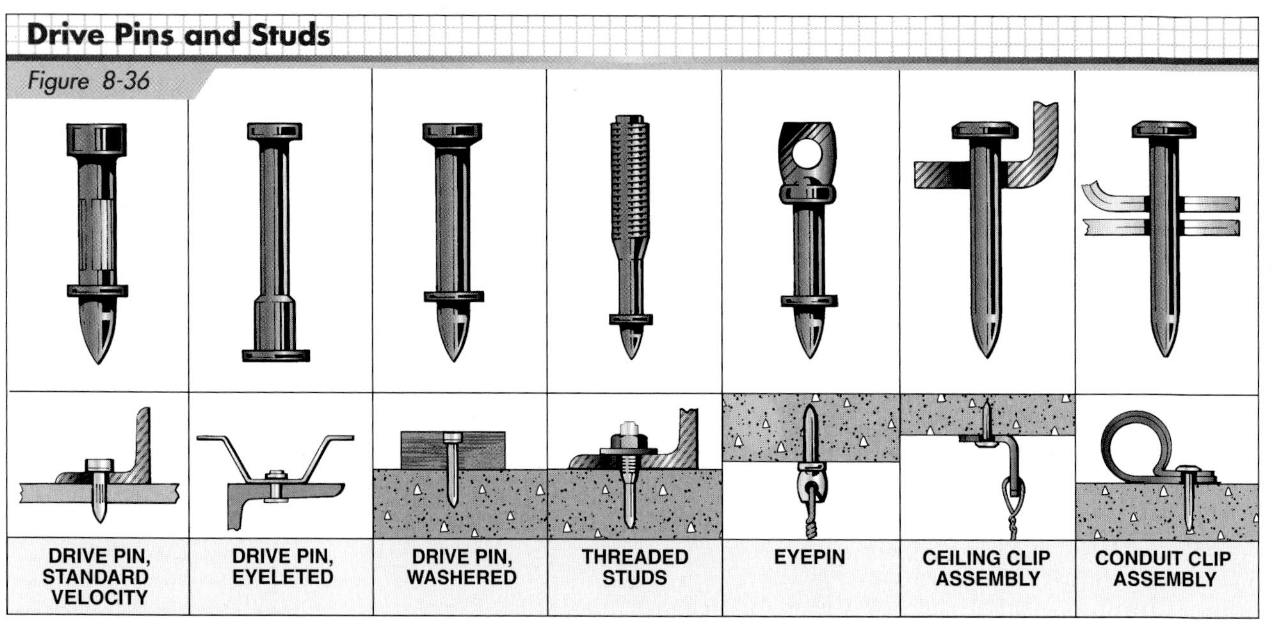

Figure 8-36. *Many types of drive pins and studs can be driven into concrete or other masonry.*

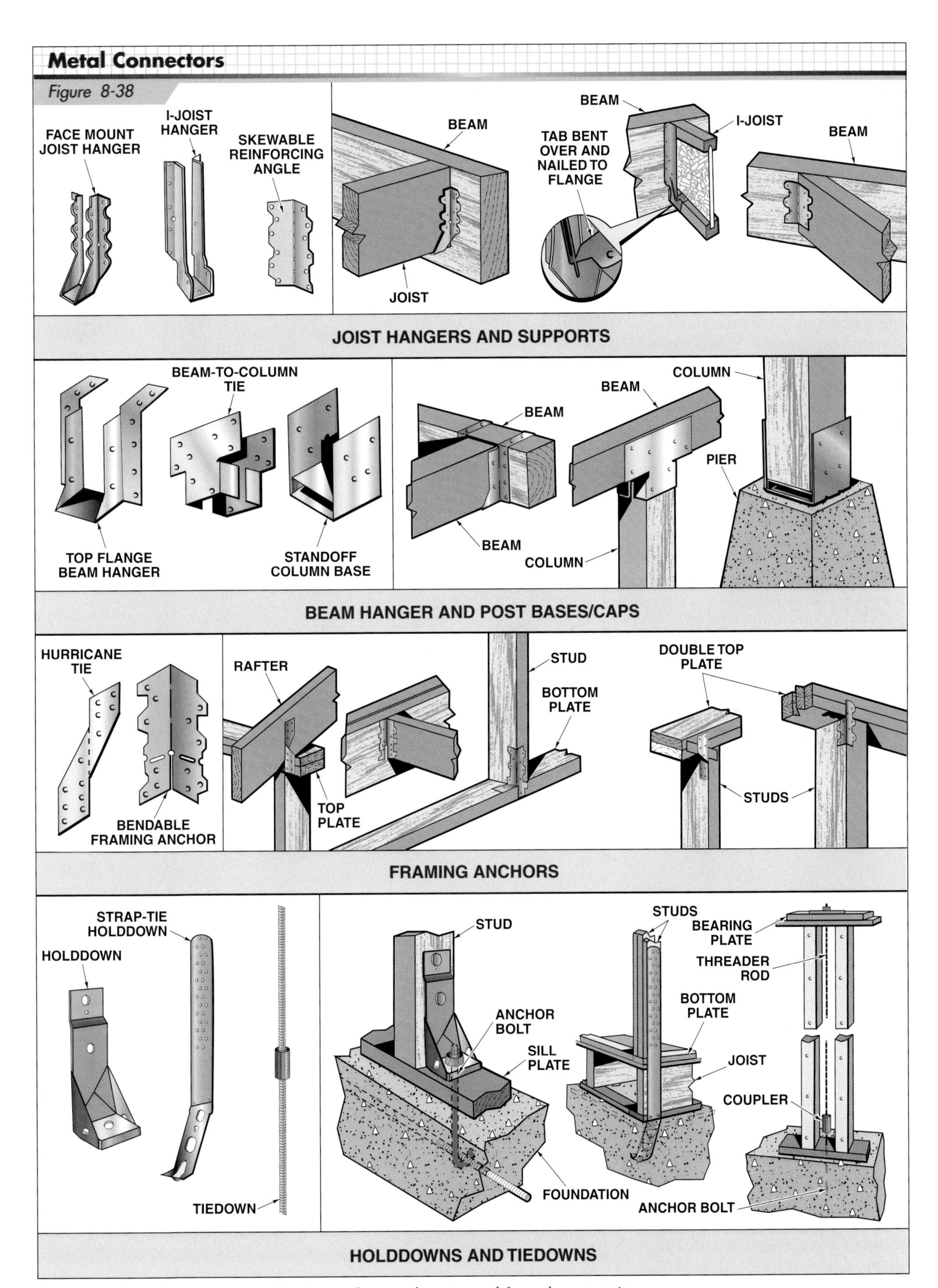

Figure 8-38. *Metal fasteners are used to tie together members in wood-framed construction.*

Metal connectors are manufactured from light-gauge stamped or punched sheet metal and heavy-gauge forged steel. The connectors may have a finish coating to resist corrosion. Metal connectors and fasteners will corrode and may lose load-carrying capability when installed in corrosive environments or when exposed to corrosive materials. In the past few years, the formulas for wood preservatives have changed, and many of the new wood preservatives contain higher copper contents, which are more corrosive to metal connectors and fasteners than traditional wood preservatives. Therefore, the metal connectors and fasteners with the proper finish are required to ensure proper load-carrying capacity. Always refer to manufacturer recommendations to determine the proper connector and fastener finish for a particular application.

When installing metal connectors, only approved fasteners, such as nails, screws, and bolts, should be used. Approved fasteners include triple-galvanized fasteners, stainless steel fasteners, and fasteners with specialized coatings. Always refer to the project specifications and/or manufacturer recommendations to determine the proper fasteners to use.

To prevent corrosion, metal connectors, washers, and bolts should not be in direct contact with pressure-treated wood. When installing metal connectors with pressure-treated wood, place a small section of self-adhesive polyethylene film, also known as *Peel 'N Stick™*, between the connector and the wood. For washers, cut a small piece of Peel 'N Stick, cut a small X in the center of it, remove the backing, and place it over the bolt before installing the washer. Bolts can be isolated from the surrounding pressure-treated wood by drilling a hole approximately ⅛″ larger than the bolt, placing the bolt through the hole, and applying epoxy between the bolt and wood. Refer to the local building code to ensure that oversizing the hole is approved.

The most effective installation of metal connectors is a vertical line of connectors from the foundation to the roof rafters. **See Figure 8-39.** Information regarding metal connector installation is detailed in later units covering different stages of wood- and metal-framed construction.

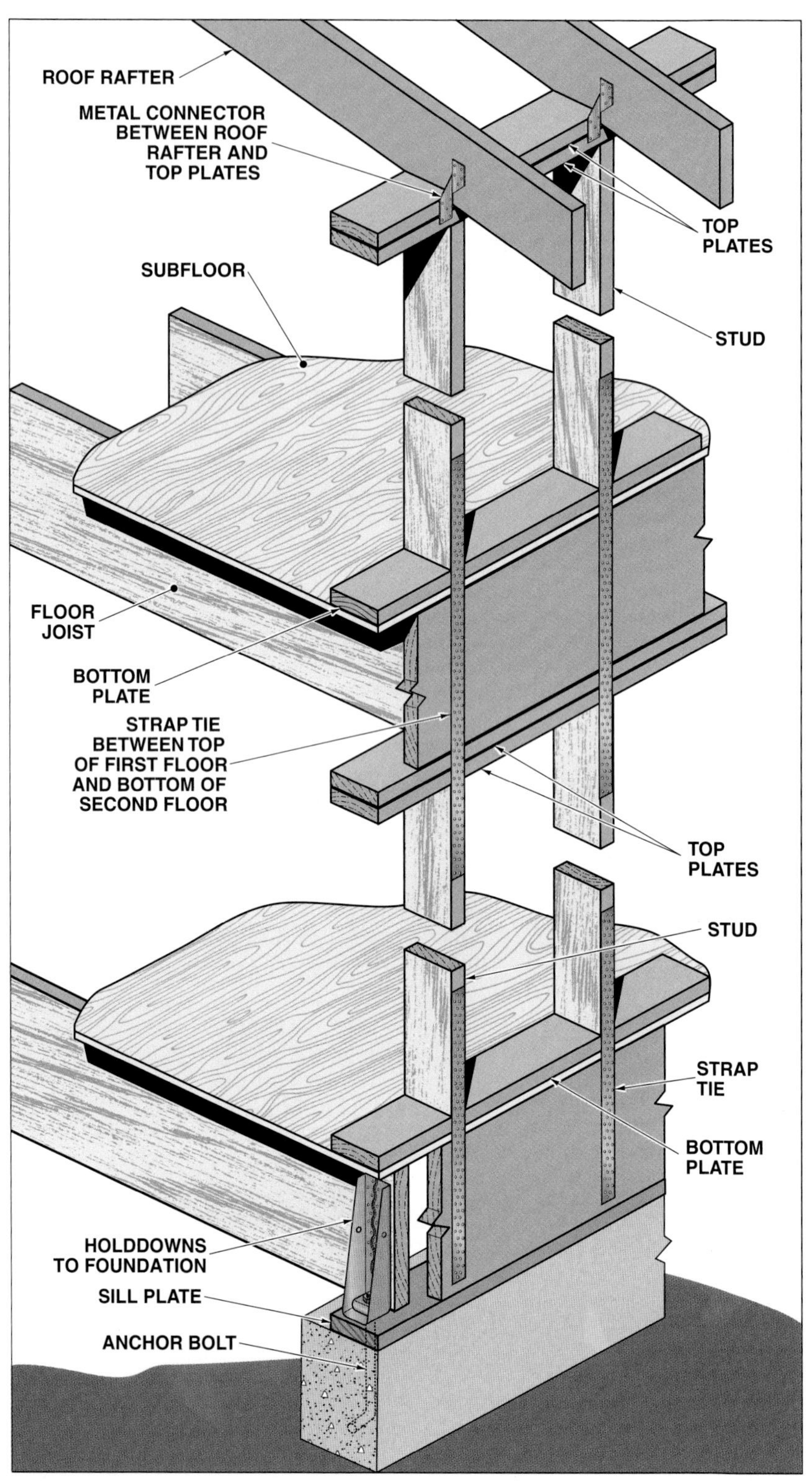

Figure 8-39. *Metal connectors aligned vertically provide the most effective protection against wind damage.*

Metal connectors are fastened to wood and metal structural members using nails, screws, or bolts, depending on the fastening requirements. Stainless steel fasteners should be used when installing stainless steel connectors.

Nails

Holes are provided in metal connectors through which nails are driven to fasten the connectors to structural members. Common nail sizes used with metal connectors range from 4d (1½″) to 16d (3½″). The nail size and diameter used for metal connectors depends on the thickness of the structural member and manufacturer recommendations. Nails with a diameter thinner than recommended by the manufacturer do not provide enough shear strength and should not be used. Some manufacturers stamp the recommended nail size on the metal connector. **See Figure 8-40.** Nail holes with domes or tabs guide nails into the structural members at a 45° angle. In most cases, all fastener holes in metal connectors must be filled with fasteners to ensure the rated load-carrying capacity. Nail heads should be seated flush with the surface of the metal connector to ensure maximum shear strength for the nail.

Figure 8-40. *Manufacturers may stamp the recommended nail size on metal connectors.*

Nails for metal connectors can be driven using a hammer, palm nailer, or pneumatic nailer. When using a pneumatic nailer for fastening metal connectors, the nailer must be fitted with a hole-locating mechanism. Specialized pneumatic nailers, referred to as *positive-placement nailers,* are designed specifically for driving nails for metal connectors. **See Figure 8-41.**

ITW Paslode

Figure 8-41. *A pneumatic nailer must be fitted with a hole-locating mechanism or be designed to expose the nail tip when driving nails for metal connectors.*

Screws

Screws may also be used to fasten metal connectors to structural members. Metal connectors that require screws are typically more expensive and require more time for installation. However, screws provide a strong connection between metal structural members when properly installed. Only screws specified for the metal connectors should be used. Do not substitute drywall screws for the specified type of screw as drywall screws do not provide the required shear strength.

Bolts

Metal connectors for larger framing members may require machine bolts as fasteners. **See Figure 8-42.** Holes for the bolts should not be more than 1⁄16″ larger than the bolt diameter, and the holes should be drilled from only one side of the member. When the nut for the machine bolt is to be placed against a wood member, a washer must be installed between the nut and member.

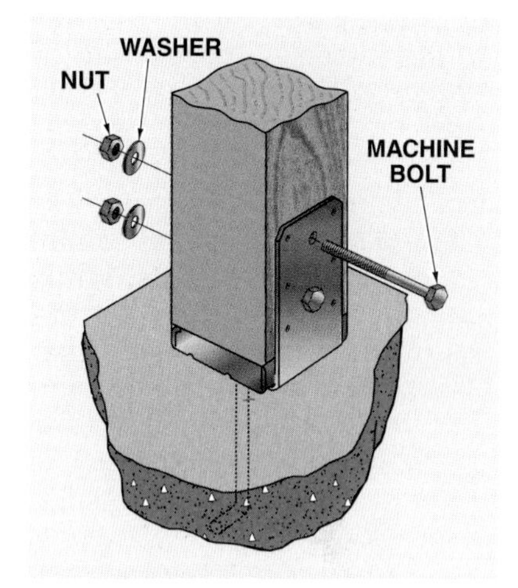

Figure 8-42. *Machine bolts may be used to fasten metal connectors to larger framing members.*

ADHESIVES

Several types of adhesives are available for construction purposes. Some are *glues,* which have a plastic base, and others are *mastics,* which have an asphalt, rubber, or resin base. The term "glue," however, is often used for mastic systems. Epoxies are adhesives that are comprised of a resin and a hardener.

Method of application, drying time, and bonding characteristics vary among adhesives. Some adhesives are more resistant to moisture and to hot and cold temperature extremes than others. Also, certain adhesives are toxic and flammable, so the work area must be well-ventilated. Other adhesives are highly irritating to the skin, so skin contact must be avoided. Manufacturer recommendations should always be followed when using adhesives.

Appropriate personal protective equipment must be worn when applying adhesives.

Glues

Glues, in conjunction with nails, staples, or screws, are commonly used to hold together joints in mill and cabinet work. **See Figure 8-43.** Glues are sold in a liquid form or as a powder to which water must be added.

Polyvinyl acetate (PVA) and *aliphatic resin glues* are available in different sizes of ready-to-use plastic squeeze bottles. Polyvinyl acetate and aliphatic resin glues have a good rating for bonding wood together and set up (dry) quickly after being applied. However, some formulations are not waterproof and should not be used on work that will be subject to constant moisture.

Urea resin is a plastic resin glue that is available in a powder form. The powder is mixed with water at the time it is needed. Urea resin makes an excellent bond for wood and has fair water resistance.

Phenolic resin has excellent water resistance and resistance to extreme temperature fluctuations. Phenolic resin is commonly used to bond the veneer layers of exterior grade plywood.

Resorcinol resin has excellent water resistance and resistance to extreme temperature fluctuations. Resorcinol resin creates a very strong bond and is frequently used for bonding the wood layers of glulam timbers.

Cyanoacrylate (CA) glue, also known as super glue, is quick-bonding glue that uses the moisture in the air or in the materials to be bonded for curing. CA glue can be used to bond wood, metal, glass, or rubber. CA glue comes in several viscosities from a very thin liquid to a thick gel. It has a high bond strength but can be brittle. Caution should be used when working with cyanoacrylate glue since the glue can easily bond skin together and emits hazardous vapors.

Polyurethane glue is a waterproof glue that cures by exposure to moisture. However, unlike cyanoacrylate glue, polyurethane glue has a long set-up time and may take 6 hr to 24 hr to reach full strength. Wetting the areas to be glued and clamping the work pieces is recommended to achieve the strongest bond.

Contact cement is primarily used to bond plastic laminates to wood surfaces. Contact cement has a neoprene rubber base. Since contact cement sets so quickly, it is very useful for joining parts that are difficult to clamp together.

Apply contact cement to both surfaces being joined. Allow contact cement to dry completely before pressing the pieces together.

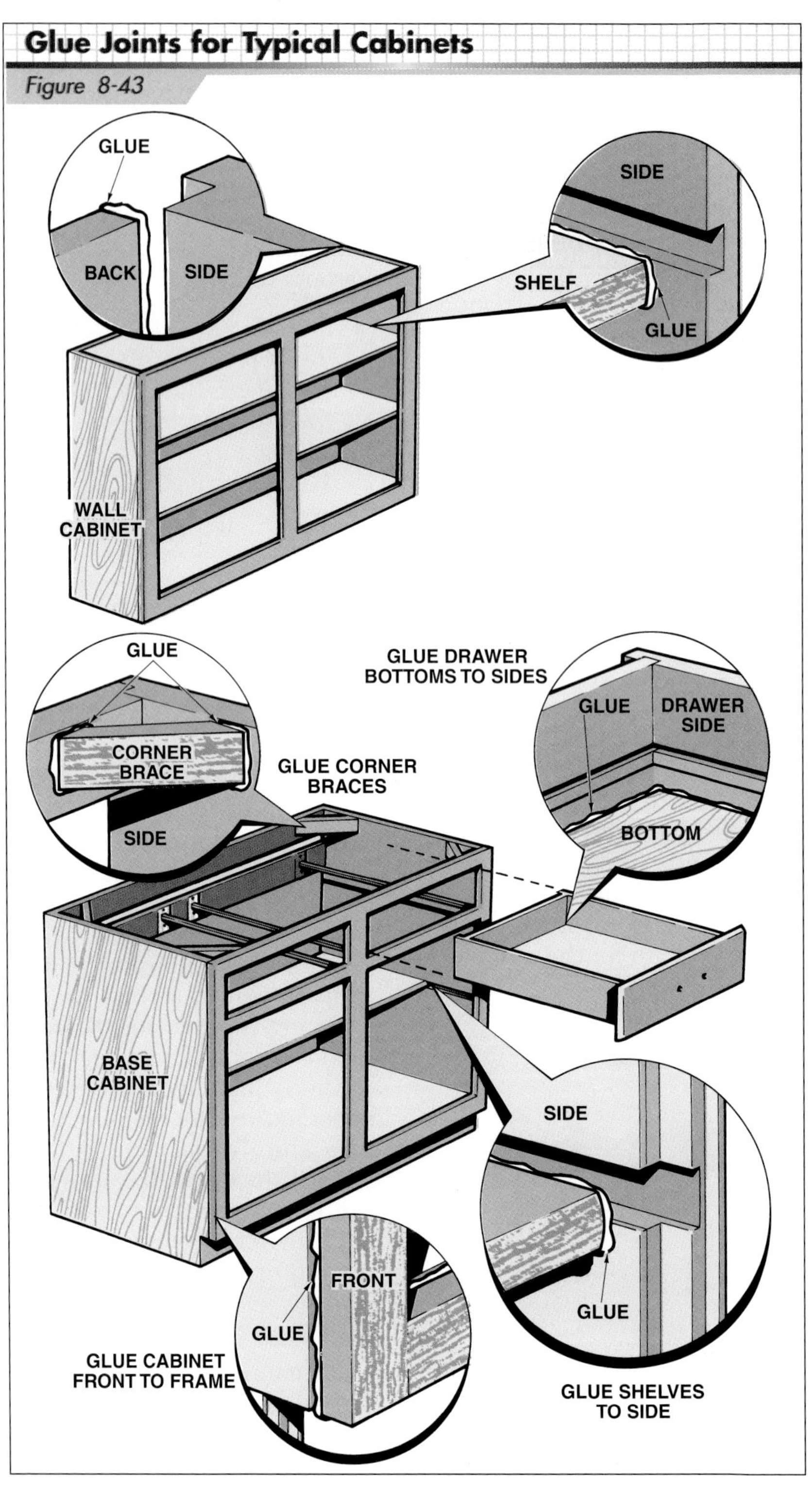

Figure 8-43. *Glue is used to help fasten the joints of cabinets.*

Mastics

Mastics have a thicker consistency than glues. Mastics are typically sold in cans, tubes, or canisters that fit into hand-operated or pneumatic caulking guns. **See Figure 8-44.**

Southern Forest Products Association

Figure 8-44. *Tubes that fit into caulking guns are the most convenient way of applying mastics to the upper surfaces of floor joists.*

Masonry and Concrete. Mastics bond materials directly to masonry or concrete walls. If furring strips are required on an uneven wall, they can be fastened with mastic rather than with concrete nails. In addition, insulation board can be permanently fastened to masonry and concrete walls with a mastic.

Gypsum Board. Mastics also bond gypsum board directly to wall studs, furring strips, and concrete or masonry walls. No nail indentations remain after installation, simplifying gypsum board finishing operations.

Wall Paneling. Prefinished wall panels used in commercial and residential construction have a neater appearance if nails are not driven through them. Mastics make it possible to apply paneling with very few or no nails at all. The panels can be bonded to studs, furring strips, or concrete or masonry walls.

Floor Systems. Mastics provide an important structural function for *glued floor systems* by increasing the stiffness and strength of the floor unit. Using an adhesive in addition to nailing or screwing produces a bond strong enough to cause the floor and joists to behave like integral T-beam units. Glued tongue-and-groove edges provide improved sound control by reducing the number of gaps between the panels. Glued floor systems also help to eliminate squeaks, bounce, and nail popping. Glued floors can be laid down quickly and efficiently, even during cold weather conditions, using standard construction materials and procedures.

In glued floor systems, a bead of mastic is applied with a caulking gun to the upper surface of the joists before each panel is placed. When tongue-and-groove panels are installed, another bead of mastic is applied into the grooved edges of a row of floor panels before the tongues of the next row of panels are inserted in the grooves. The floor panels are then nailed or screwed down before the adhesive sets. **See Figure 8-45.** The setting time of mastics varies, and manufacturer recommendations should be followed. As a general rule, setting time accelerates during warm weather. Additional information regarding glued floor systems is provided in Unit 42.

Mastic adhesives are also used to fasten wood sleepers to concrete floors to provide a nailing surface for wood flooring materials. This procedure is described and illustrated in Unit 62.

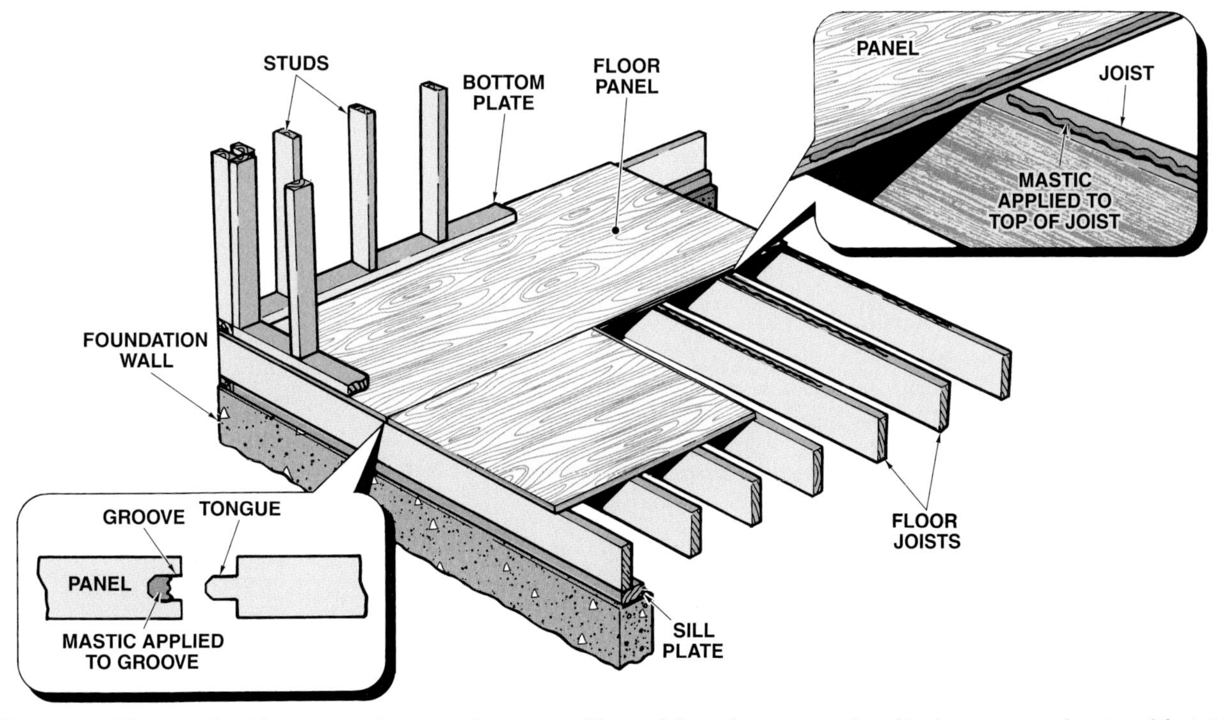

Figure 8-45. *When panels with tongue-and-groove edges are used for a subfloor, the mastic is placed in the grooves and on top of the joists.*

Epoxies

Epoxy is a two-part adhesive that consists of a resin and a hardener. When epoxy resin is combined with the hardener, the two parts chemically react to form a solid material with adhesive qualities. The working time for epoxies varies, but epoxy hardens quickly at normal temperatures. However, it may take several hours to weeks for the epoxy to fully cure and reach maximum strength depending on the formulation. Epoxies are used for bonding wood and other materials on the job site; for anchoring fasteners, threaded rods, reinforcing bars, or dowels in hardened concrete; for concrete repair; and as a floor coating. Once hardened, epoxy adheres well to most materials and provides outstanding strength and durability. Epoxies are resistant to heat and cold, and most epoxies are resistant to most acids, alkalis, and solvents.

The surfaces that will contact the epoxy must be free from contaminants such as dust, dirt, oil, grease, or other foreign materials. The surfaces must also be free of water, frost, and ice. In order to provide the best bonding, the surfaces should be cleaned right before use. Failure to properly prepare the surface could lead to bonding failure.

The epoxy must be properly mixed and dispensed in accordance with the manufacturer's instructions. Some epoxy manufacturers, especially manufacturers of epoxies used for anchoring fasteners to concrete, have proprietary mixing and dispensing equipment. Other manufacturers supply bulk resin and hardener, which places the responsibility of proper measuring and mixing of the two parts on the tradesperson. Proper PPE is required when using epoxy. The epoxy manufacturer should be consulted for instructions and safety precautions before use.

CAULKING

Caulk, also known as joint sealant, is a resilient construction material that incorporates synthetic polymers and is used to seal joints and cracks to prevent moisture and air leakage. Caulk adheres to surfaces and remains flexible to allow for joint movement. Caulking is used for residential and commercial construction where wall cladding meets with windows and doors, where foundation walls meet the sill plate, where a basement slab meets the foundation wall, and where the rim joist meets the sill plate and subfloor. Caulking is also used for utility penetrations such as water and gas pipes, vents, and electrical conduit. Caulking is also used to secure bathtubs, sinks, and solid surface counter tops and to fill in any gaps where various trim moldings meet a wall, ceiling, or floor.

The majority of caulking types used for residential and commercial applications are made with one of four different types of synthetic polymer: silicone, polyurethane, acrylic latex, and elastomeric polymer. Each type of caulk has advantages and limitations. **See Figure 8-46.** Some caulks contain two types of polymer that allow the caulk to overcome some of the limitations of a single polymer formula. For example, some painter's caulk is comprised of silicone and acrylic latex that gives the caulk the increased elasticity of silicone caulk and the paintability of acrylic latex caulk.

CAULK FORMULATIONS						
Caulk Type	Expected Life Span*	Elasticity	Adhesion	Application Temperature	Paintable	Other Characteristics
Acrylic Latex	10	Good	Good	Above Freezing	Yes	Good for use on doors, windows, siding, and trim
Silicone-Acrylic Latex	20	Good	Excellent	Above Freezing	Yes	More flexible than acrylic latex, waterproof, mold and mildew resistant once cured
Silicone	20 to 50	Excellent	Excellent except on vinyl or masonry	Above Freezing	No	Waterproof once cured, low shrinkage, available in multiple colors
Urethane	20 to 50	Excellent	Excellent	Above Freezing	Yes	Waterproof once cured, may have strong odor during cure, may need primer on certain substrates
Elastomeric Polymers	50	Excellent	Excellent	Can be applied below freezing	Yes	Will stick to damp surfaces, UV resistant, easy to work with, available in multiple colors
Butyl Rubber	3 to 10, although new formulations may last considerably longer	Excellent	Excellent	Some formulations may be installed below freezing	Yes	Used for exterior applications such as gutters, downspouts, flashing, and roof vents

Figure 8-46. *Caulk comes in several formulations and each formulation has advantages and limitations.*

To ensure proper adhesion of the caulk to the substrate, the substrate must be dry and free from dust, oils, debris, and other foreign substances. The areas to be caulked should be properly cleaned. The substrate may need to be cleaned with a clean cloth moistened with mineral spirits, denatured alcohol, or water depending on the substrate material and the location. Precautions must be taken with mineral spirits and denatured alcohol because they can pose a health and fire risk if used improperly, especially indoors.

A piece of painter's tape or masking tape can be run alongside the joint before caulking to allow a neat and straight caulk edge. Before the caulk cures, it should be tooled to improve the adhesion to the substrate and result in an aesthetically pleasing joint. Tooling ensures uniform sealant contact with the substrate and works air bubbles from the sealant. The tape should be removed after tooling but before the caulk has a chance to skin over.

Joints that are too deep or wide based on the caulk manufacturer's recommendations may require foam backer rod to be inserted into the joints before caulking. **See Figure 8-47.** *Backer rod* is a flexible polyethylene foam material that is used to control caulk joint depth and prevent three-sided adhesion.

Backer rod comes in a variety of diameters from ¼″ to 4″ and in a variety of lengths. The most common shape of backer rod is round, but quarter-round rod and triangular-shaped rod are also used. The size of backer rod required for a particular joint width should be available from the backer rod manufacturer. However, a good rule of thumb is that the size of the backer rod used should be about 25% larger than the joint width. Backer rod prevents the caulk from adhering to the back of a butt joint or fillet joint. This is known as three-sided adhesion. Caulk adhering to the back of a butt or fillet joint can limit the caulk's flexibility and cause the sealant to fail if there is differential movement between the substrates.

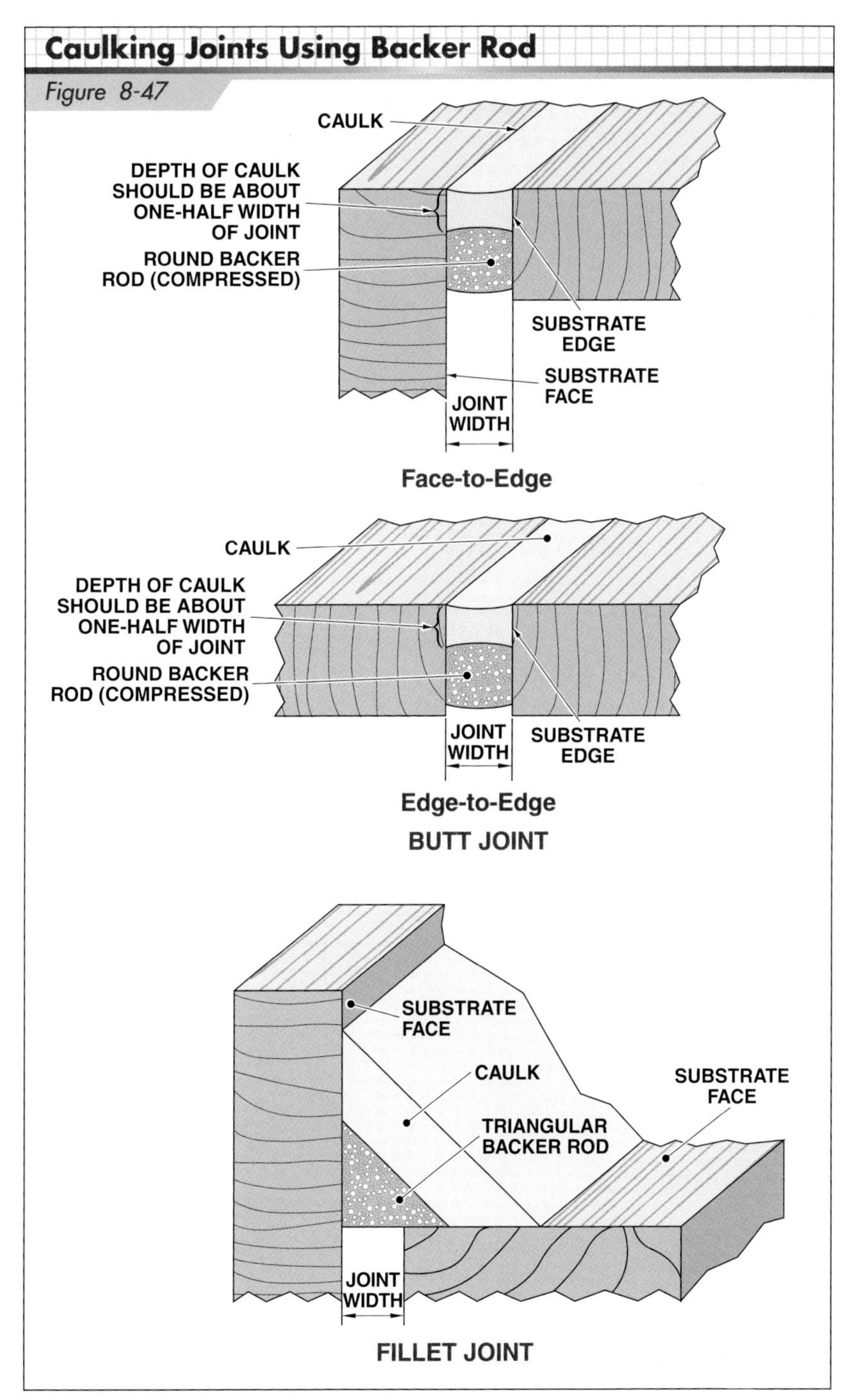

Figure 8-47. *Joints that are too deep or wide may need to have backer rod inserted before the joints can be caulked.*

Unit 8 Review and Resources

SECTION

3

Hand Tools

CONTENTS

Measuring and Layout Tools

OBJECTIVES

1. Describe functions of tapes and rules.
2. Identify common applications of leveling and plumbing tools.
3. List and describe common squaring tools.
4. List and describe common marking and scribing tools.
5. Describe common elements of a framing square.
6. Review layout procedures for a framing square.

Accurate measurement and layout are vital in construction work to minimize material waste. Adhering to the adage "measure twice, cut once" ensures that materials are properly measured before they are cut, drilled, or otherwise modified. Various tapes, squares, levels, and marking devices are available for accurate measurement and layout.

TAPES AND RULES

Steel tapes and rules are used to measure the lengths and widths of materials to be cut. In addition, they are used to establish the locations of and distances between walls, floors, ceilings, and other structural parts of a building. A laser distance meter may be used in place of tapes and rules in many construction applications.

Tape Measures

A *tape measure,* or pocket tape, is used more often than any other measuring tool on the job. **See Figure 9-1.** Tape measures come in various lengths, with 12′, 16′, 20′, 25′, 30′, and 55′ lengths being the most popular sizes. The blade of a tape measure is usually ½″ or ¾″ wide. Some 25′ tapes feature a 1″ blade, allowing the blade to remain rigid for an unsupported distance up to 7′. One edge of the blade of most tape measures is marked off in inches and the other edge in feet and inches. A special identifying mark is placed every 16″ on the blade for the 16″ OC (on-center) layout of studs and floor or ceiling joists. Many tape measures also have an identifying mark every 19.2″ on the blade for the 19.2″ OC layout of trusses and engineered lumber joists. Some tape measures have English (customary) measurement (feet, inches, and fractions) on one edge of the blade and metric measurement (meters and millimeters) on the other edge. **See Figure 9-2.**

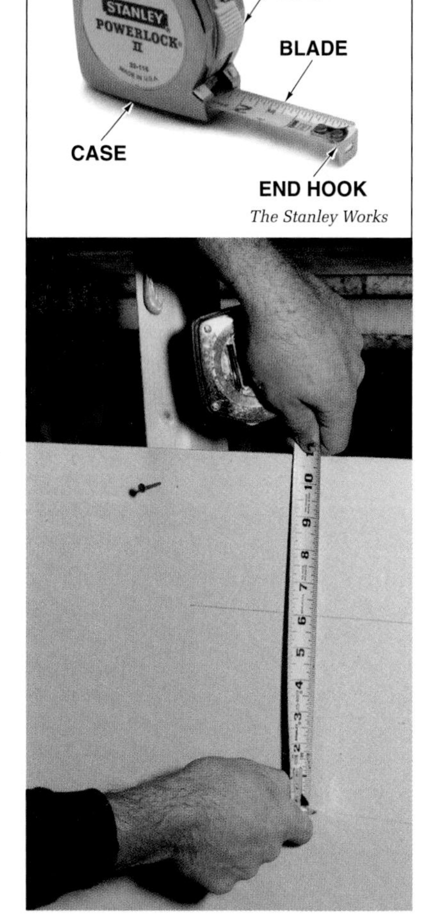

The Stanley Works

Figure 9-1. *A tape measure is used for many measuring and layout jobs.*

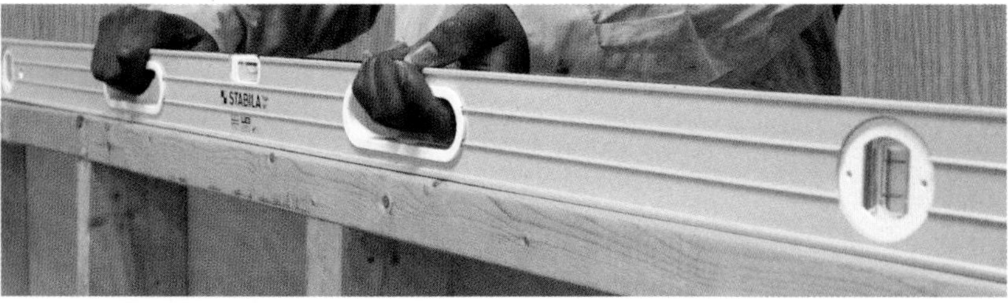

Stabila, Inc.

A carpenter's level is used to establish a level plane.

How to Read and Use a Tape Measure Media Clip

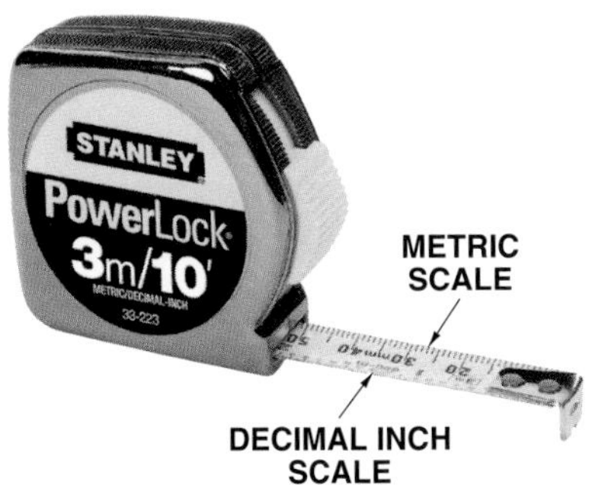

The Stanley Works

Figure 9-2. *Tape measures are available that show English measurement on one edge and metric measurement on the other edge.*

Steel Tapes

Steel tapes are used to measure long distances. **See Figure 9-3.** Carpenters typically use 50′ or 100′ steel tapes with blades incremented in feet, inches, and eighths of an inch. Most steel tapes have a steel ring with a fold-out hook. The hook can be placed over the edge of an object, or the ring can be placed over a nail when extending the blade.

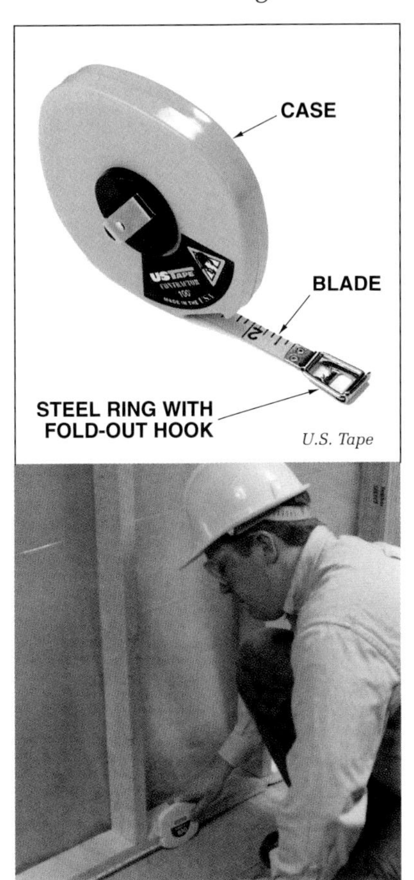

U.S. Tape

Figure 9-3. *A steel tape is used to measure long distances.*

Laser Distance Meters

A *laser distance meter* is a handheld measuring tool that uses a laser to aim an ultrasonic signal that measures the distance from the meter to an object or surface. **See Figure 9-4.** Some meters have the ability to store readings and calculate area and volume. Typical laser distance meters have a maximum range from 100′ to 350′ and are accurate to within ±1⁄16″ up to their maximum range. More expensive models have a maximum range up to 650′ and are accurate to within ±1⁄25″.

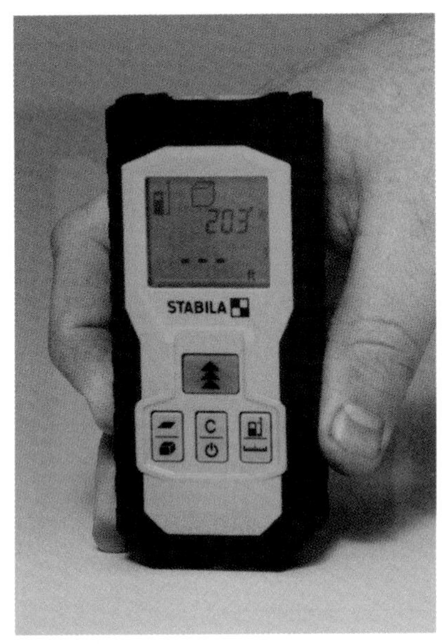

Figure 9-4. *A handheld laser distance meter uses a laser to accurately measure the distance from the meter to an object or surface.*

LEVELING AND PLUMBING TOOLS

The term *level* refers to horizontal planes. A perfectly level surface is a flat horizontal plane without high and low points. The term *plumb* refers to a vertical position. A plumb line is always at a right angle (90°) to a level surface. **See Figure 9-5.**

Carpenter's Levels

A *carpenter's level,* or hand level, is a hand tool used to establish level and plumb lines. **See Figure 9-6.** Most carpenters prefer lightweight aluminum or magnesium levels to the older wood levels as they are more durable than wood levels.

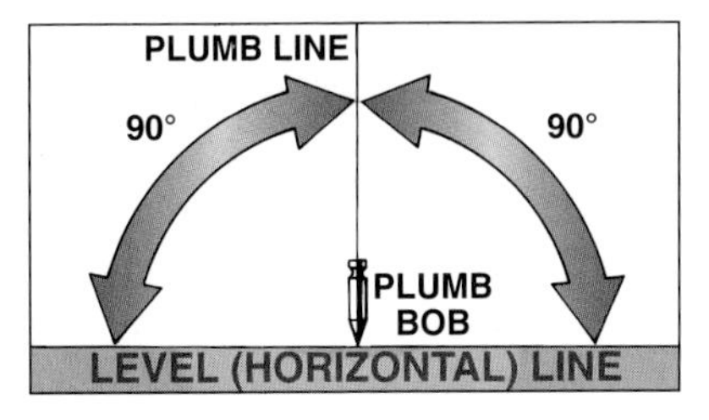

Figure 9-5. *A plumb (vertical) line is always at a 90° angle to a level (horizontal) line.*

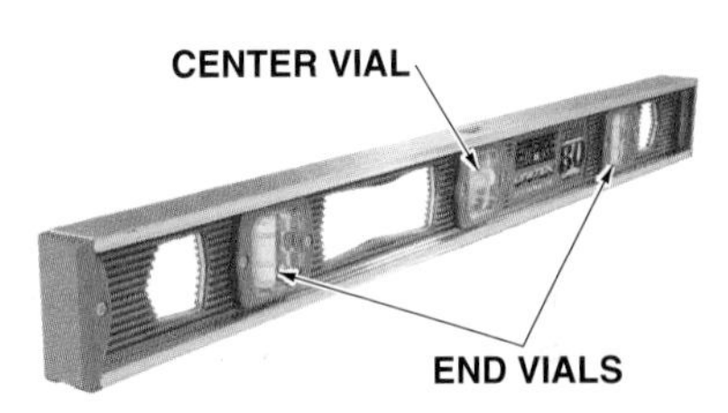

Empire Level

Figure 9-6. *Lightweight aluminum or magnesium carpenter's levels are generally preferred over older wood types.*

Three slightly curved vials are set in the frame of a carpenter's level and are protected by glass or plastic covers. The vials are partially filled with spirit alcohol that contains an air bubble. The vials at each end of a level are used to plumb objects. When a level is held in the vertical position and the air bubbles are centered between the two lines marked on the end vials, the level is exactly plumb. The center vial of the level is used to level objects. The air bubble will be centered between the two lines when the level is exactly level. **See Figure 9-7.**

Carpenter's levels are available in different lengths with the 24″ or 28″ level most often preferred by carpenters since they fit in a standard-size toolbox. Levels in 36″, 48″, 72″, and 78″ sizes are also available. Some levels have a magnetic base to secure the tool to a steel member or surface, freeing both hands of the carpenter to make adjustments to the member.

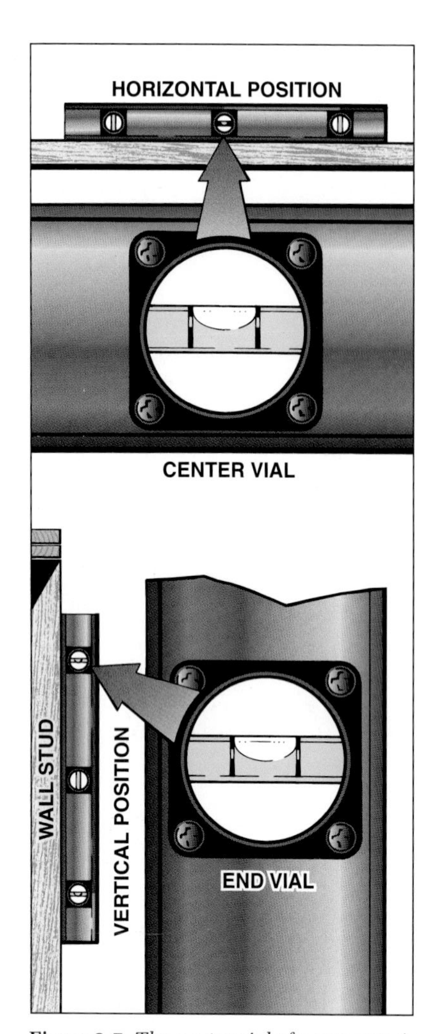

Figure 9-7. *The center vial of a carpenter's level is used for leveling. The end vials are used for plumbing. To ensure an accurate level reading, look straight into the vial and not at an angle.*

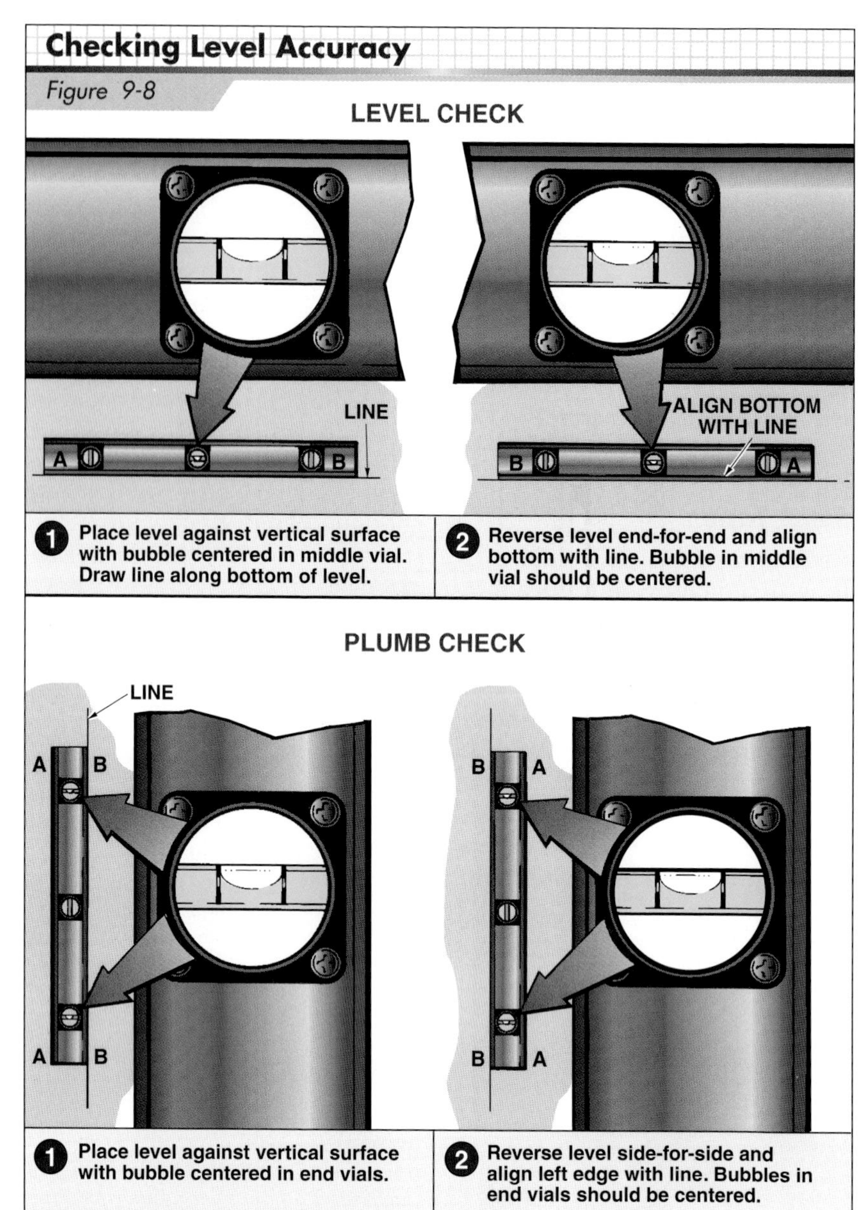

Figure 9-8. *A carpenter's level should be periodically checked for accuracy.*

A carpenter's level should be handled with care, since its vials and their protective covers are easily broken. The vials on most levels can be replaced if they are broken. A new level should always be checked for accuracy. Levels should be rechecked periodically since they can become inaccurate over a period of time. **Figure 9-8** shows how to test a level for accuracy.

A *plate level* is similar to a carpenter's level, but has an extendable arm to provide a longer reach for the level. **See Figure 9-9.** Plate levels are commonly used to plumb walls or level across longer horizontal distances without setting up additional leveling instruments.

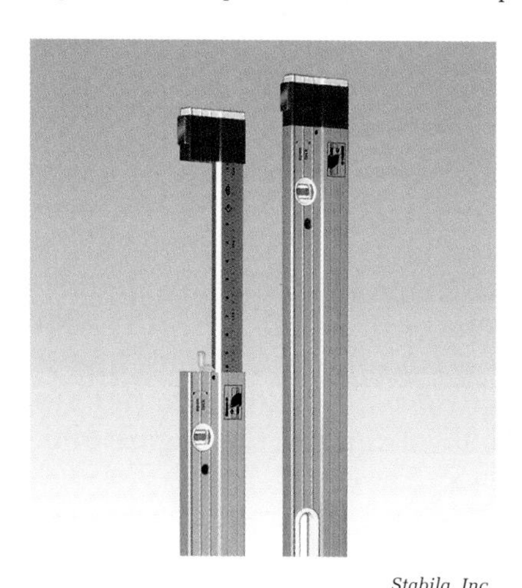

Stabila, Inc.

Figure 9-9. *The arm of a plate level can be extended to plumb a wall from the bottom plate to the top plate.*

Using a Straightedge with a Level. A carpenter's level is accurate only up to its length. Use a straightedge and carpenter's level when leveling or plumbing over long distances. A *straightedge* is usually a piece of lumber that is perfectly straight and is the same width from one end to the other. A piece of ¾″ plywood with a block at each end is often used for this purpose. **See Figure 9-10.**

Laser Hand Levels

A *laser hand level* can be used as a standard carpenter's level or it can be switched on to emit a laser beam. **See Figure 9-11.** Laser hand levels are mounted on a tripod or placed on a flat surface. A tripod mount allows the level to be easily rotated. A laser hand level accurately projects a laser beam 50′ to 100′ with minimum beam deflection.

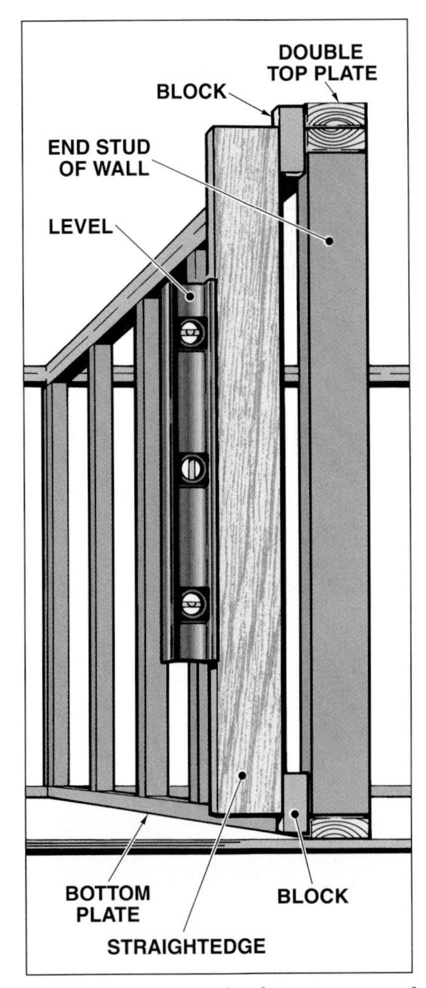

Figure 9-10. *A straightedge consisting of a piece of ¾″ plywood with blocks at the ends and a carpenter's level is often used when leveling or plumbing over a long distance.*

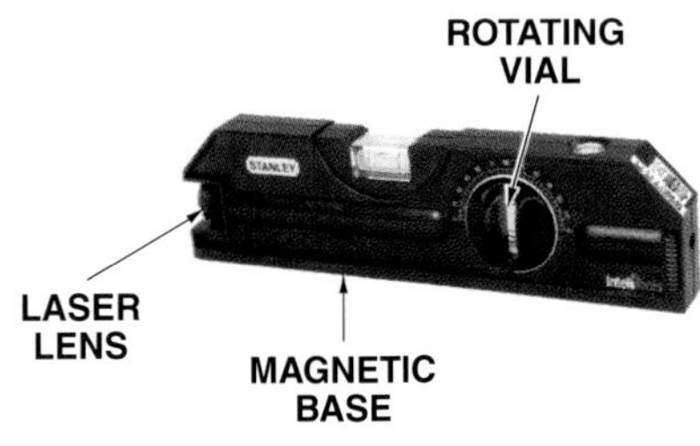

The Stanley Works

Figure 9-11. *A laser hand level is used to transfer locations from one surface to another surface which is nearby.*

Depending on the laser hand level manufacturer, the laser beam exits the level frame approximately 1″ above the base. When transferring marks from one surface to another, the distance between the laser beam and base must be taken into account to ensure accurate measurements.

Line Levels

Another way to level long distances is to use a *line level.* **See Figure 9-12.** A line level is hooked over a tightly stretched string. However, this is not always an accurate method since there may be some sag in the line from which the level is hung.

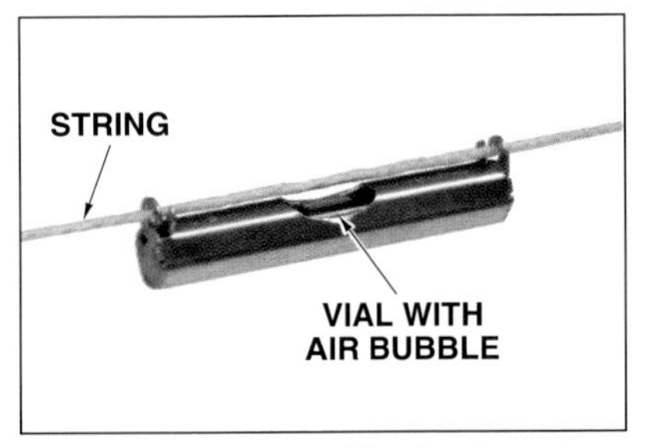

Figure 9-12. *A line level has hooks so that it can be attached to a tightly stretched string.*

Water Levels

A *water level* is an accurate leveling tool that is based on the principle that water will find its own level in a system open to atmospheric pressure. **See Figure 9-13.** While water levels have largely been replaced by laser levels and other leveling instruments, water levels are a convenient tool to use to determine a level location in another room or on the opposite side of a wall. Some water levels consist of a long section of clear plastic tubing with a clear reservoir on each end, while other water levels consist of a large water reservoir at one end that is connected to clear plastic tubing. Some models of water levels are equipped with a digital readout display.

Dye can be added to the water in a water level to make it easier to see. If the water level is going to be used in below-freezing temperatures, antifreeze should be added or another liquid, such as windshield washer fluid, can be used.

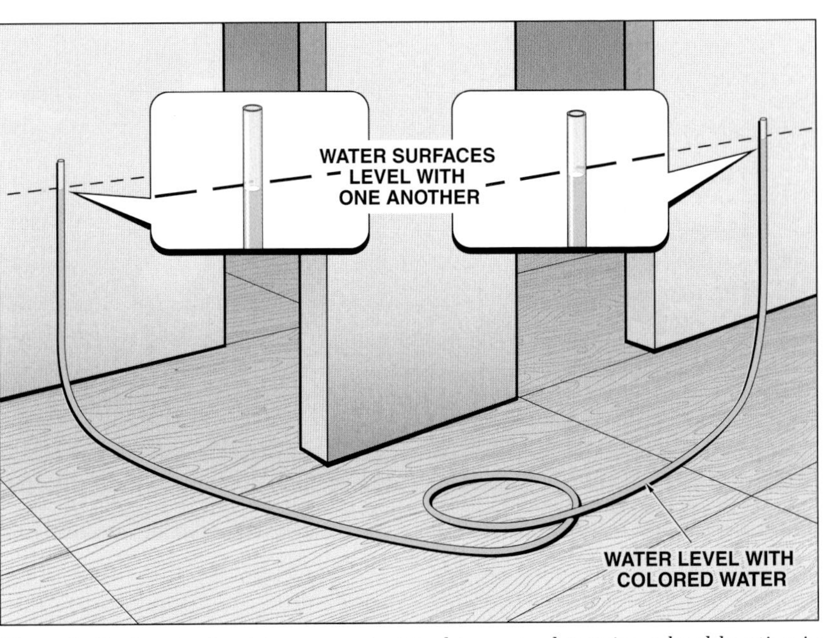

Figure 9-13. *A water level is a convenient tool to use to determine a level location in another room or otherwise obstructed location. Water levels are based on the principle that water will find its own level in a system open to atmospheric pressure.*

A water level is only accurate if no air bubbles are present in the system and if both ends of the plastic tubing (or plastic tubing and reservoir) are open to the atmosphere. Colored water is used in a water level so that level marks can be easily noted. One end of the plastic tubing is placed at a predetermined mark. The worker then moves around the area to be leveled with the other end of the tubing and places marks on the wall when the water level stabilizes.

Plumb Bob and Line

A *plumb bob* and line are used for plumbing greater heights than can be handled by a level and straightedge. **Figure 9-14** shows how a plumb bob is used. The wall is plumb when the distance between the line and the top of the wall is the same as the distance between the point of the plumb bob and the bottom of the wall. Plumb bobs range in weight from 4 oz to 48 oz.

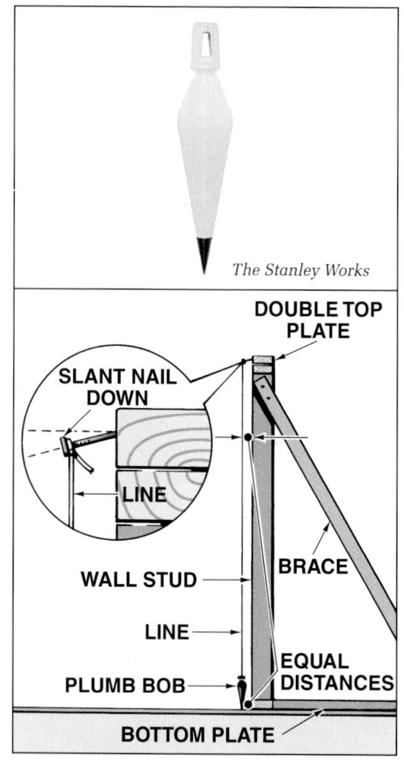

Figure 9-14. *When using a plumb bob to plumb a wall, the distance between the line and the top of the wall is equal to the distance between the point of the plumb bob and the bottom of the wall.*

SQUARING TOOLS

A 90° (right) angle is used most often for cutting and fastening together construction materials. Wall framing members and joists have square cuts at their ends. **See Figure 9-15.** A 45° angle is commonly used when fitting finish materials, such as the trim around a window or door opening. **See Figure 9-16.**

Figure 9-15. *Square cuts are required when framing walls and ceilings.*

Figure 9-16. *Trim pieces, such as this casing, are typically mitered at 45°.*

Combination Square

A *combination square* is used to lay out 90° and 45° angles. **See Figure 9-17.** Combination square blades are 4″ to 24″ long, with 12″ being the most common, and are marked off in inches and fractions. Metric blades are also available. Combination squares typically come with a square head, but protractor and center heads are also available.

A combination square can also be used as a marking gauge, as shown in **Figure 9-18.** The head of the combination square is locked in the desired position and placed flush against the edge of the surface to be marked. The tip of a pencil is held firmly against the blade while the combination square is pulled along the edge.

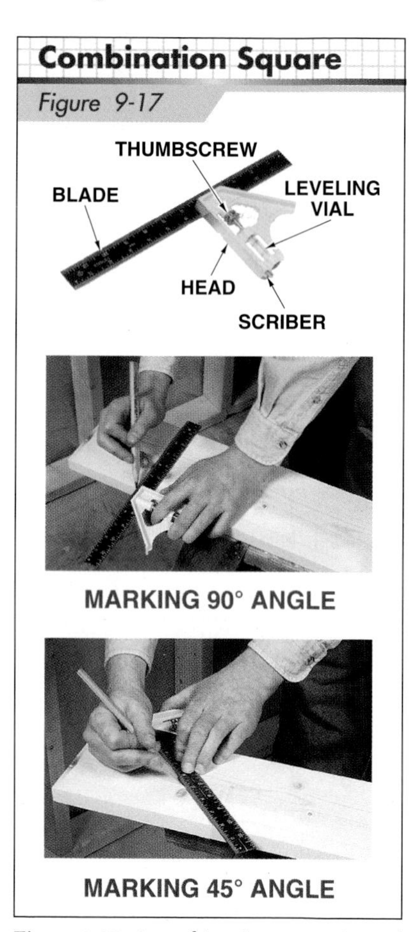

Figure 9-17. *A combination square is used to lay out 90° and 45° angles.*

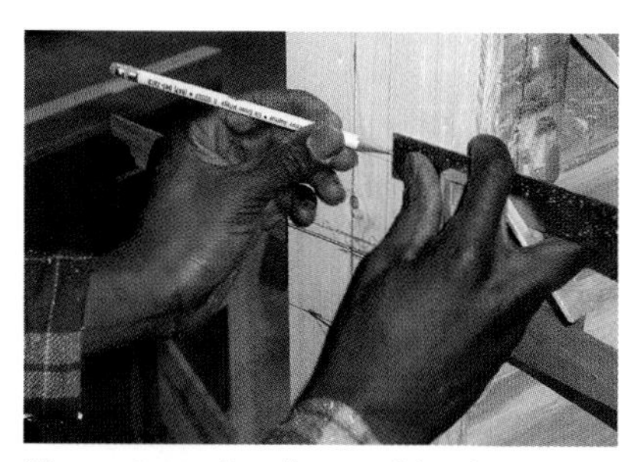

Figure 9-18. *Set the combination square to the desired position and guide the head along the edge of a board while holding a pencil along the end of the blade.*

Framing Square

A *framing square,* also known as a steel square or rafter square, is a valuable measuring and layout tool. The *blade,* or body, of a framing square is 2″ wide and 24″ long. The *tongue* is 1½″ wide and 16″ long. The outside corner of the square is the *heel.* The two sides of a framing square are the *face* and *back.*

Face of Square. The inches on the outside edges of the face of a framing square are divided into 1⁄16″ graduations. The inches along the inside edges of the face are divided into ⅛″ graduations. The face of a framing square typically includes the manufacturer name at the corner, *rafter tables* on the blade, and an *octagon table* on the tongue. **See Figure 9-19.**

The rafter tables enable a carpenter to determine lengths of roof rafters using the length per foot of run for unit rises ranging from 2″ to 18″. The procedure for using the rafter tables is explained in Section 10. The octagon table, used to lay out octagonal shapes, is located at the center of the face of the tongue.

Back of Square. The inches at the outside edges of the back of the square are divided into 1⁄12″ graduations, which is useful when making scale layouts of 1″=1′-0″. The inside edge of the blade of the back of the square is divided into 1⁄16″ graduations. However, the inside edge of the tongue is divided into 1⁄10″ graduations, which is also convenient for some types of layout. **See Figure 9-20.**

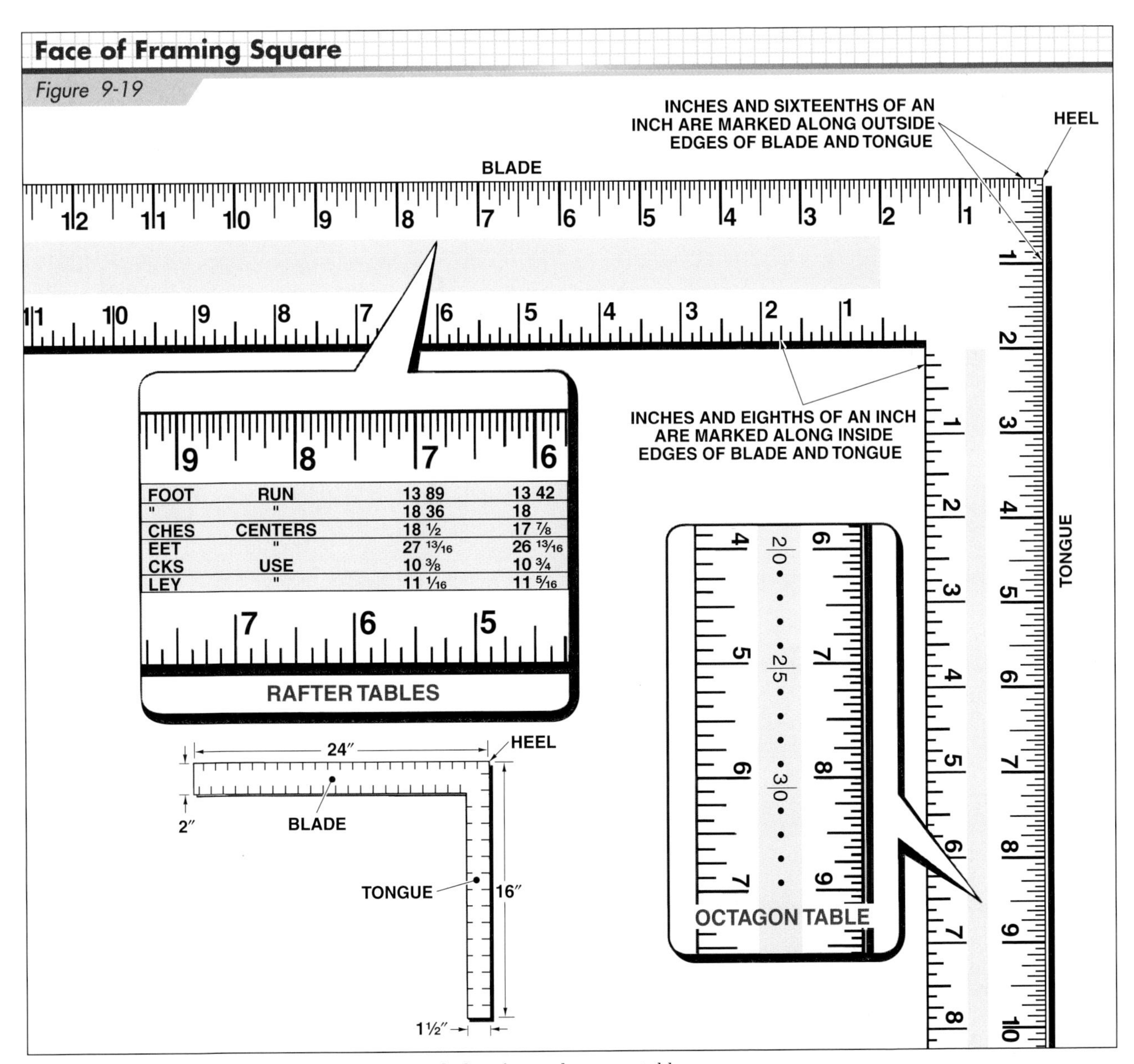

Figure 9-19. *The face of a framing square may include rafter and octagon tables.*

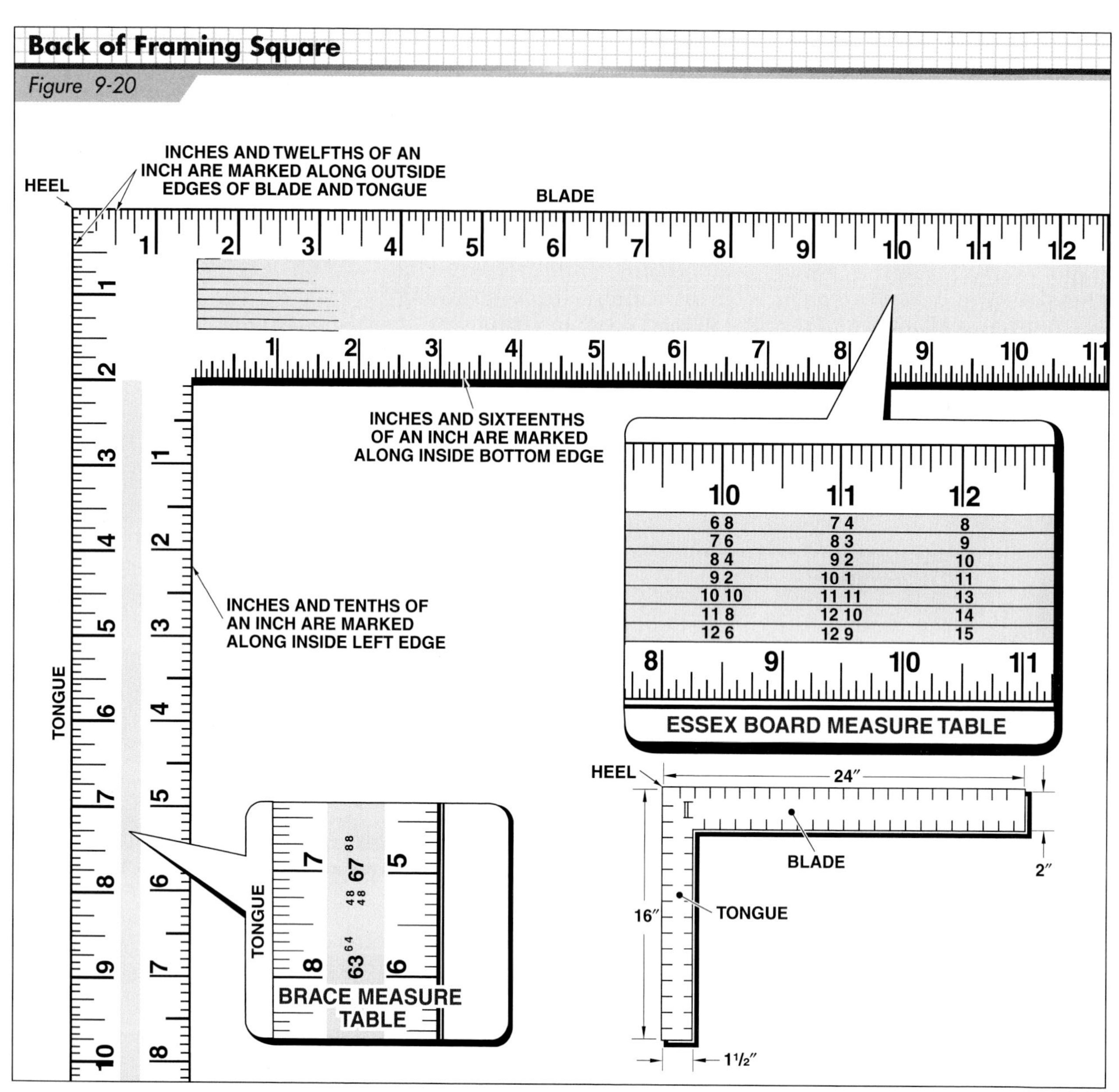

Figure 9-20. *The back of a framing square may include Essex board measure and brace measure tables.*

An *Essex board measure table* and a *brace measure table* are also found on the back of a framing square. The Essex board measure table, used to calculate and convert lumber size measurement to board foot measurement, is on the back of the blade. The brace measure table, found on the back of the tongue, identifies the diagonal lengths of braces based on common vertical and horizontal measurements.

Layout Procedures. Framing squares are used to check or square lines across wider boards or when greater accuracy is required than is possible with a combination square. A framing square is also used to mark 45° angles across wide boards and check inside corners for squareness. **See Figure 9-21.** To mark a 45° angle across a wide board, line up a number on the blade with the edge of the board and then line up the same number on the tongue with the edge of the board.

Framing squares can also be used to lay out and mark the spacing of short runs of studs and joists spaced 16″ and 24″ OC. To lay out marks 16″ OC, place the end of the tongue at the starting point of the layout and square a line at the heel. Move the square in the direction of the layout, place the end of the tongue on the first 16″ OC mark, and square the next line at the heel. Continue this process until the layout is completed. To lay out marks 24″ OC, follow the same procedure using the blade of the square.

A framing square is also used to lay out and mark the angle

cuts for roof rafters by lining up figures corresponding to the unit rise and unit run of the roof on the tongue and blade. Layout of the angle cuts is covered in detail in Section 10. Another important function of the framing square is for laying out and marking the risers and treads of stair stringers. This is done by marking the unit rise and run of the stairway by aligning the corresponding figures on the tongue and blade of the square. Stair layout is covered in greater detail in Section 14.

The octagon table, located on the face of a framing square, is used to shape squared lumber or timbers into an octagonal (eight-sided) shape. The octagon table consists of a series of dots with every fifth dot numbered. A carpenter may be required to lay out an octagonal newel post on a square piece of stock for a stairway, thus requiring the use of the octagonal table.

To lay out an octagonal shape on a piece of squared stock, first lay out the center of each side of the stock. **See Figure 9-22.** In this example, the stock is 8″ along each side. Using the octagonal table and dividers, set the dividers to 8 units since the stock measures 8″ along each side. Transfer the 8-unit distance to the stock, laying out the distance on each side of the centerline. Using a pencil, connect these points to form an octagon.

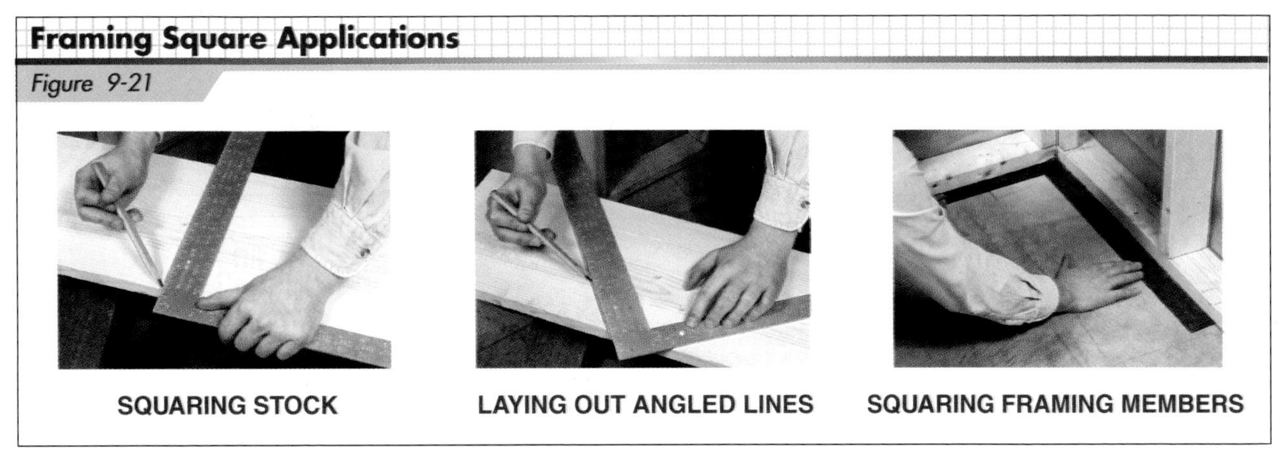

Figure 9-21. *Framing square applications include squaring a line across a wide board, marking a 45° angle across a wide board, and checking an inside corner for squareness.*

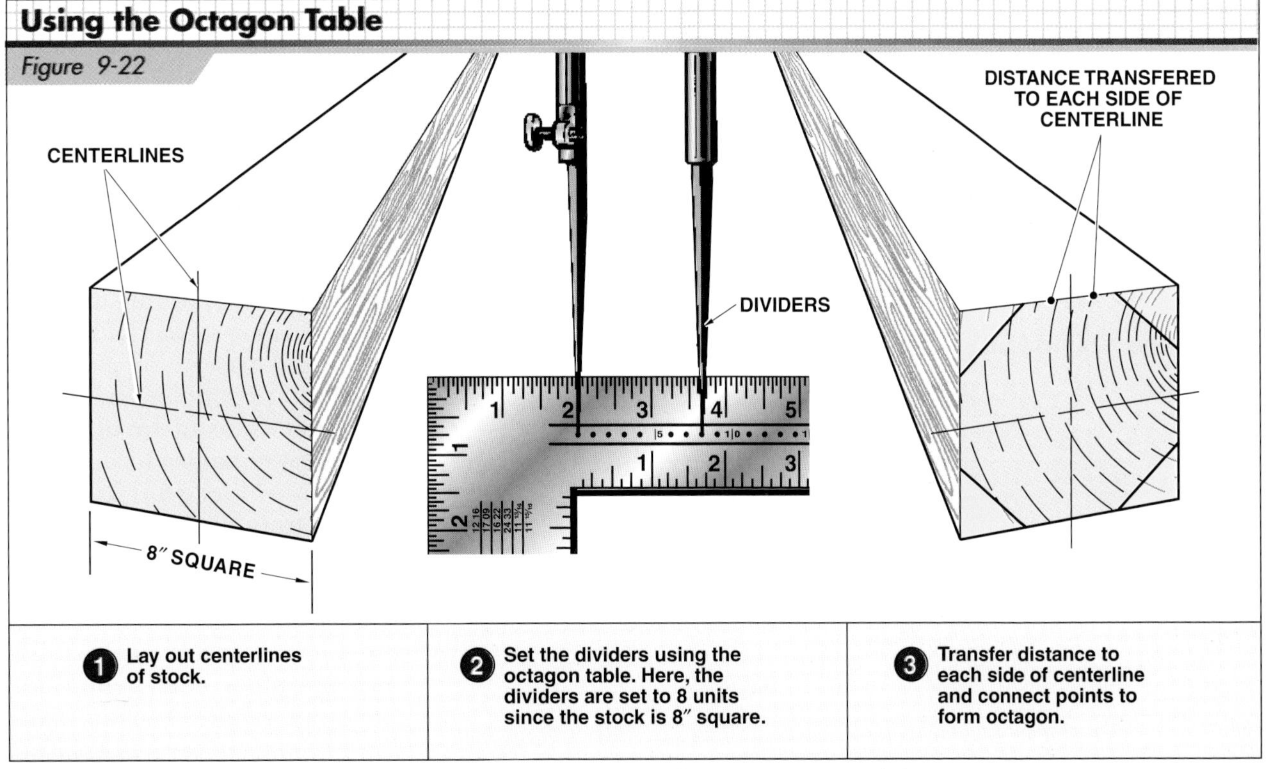

Figure 9-22. *The octagon table is used to shape squared lumber or timbers into an octagonal shape.*

Speed® Square

A *Speed® Square,* also referred to as a pocket square or rafter angle square, is used to lay out angles from 0° to 90°. **See Figure 9-23.** A Speed Square can also be used in place of a framing square to lay out plumb and seat cuts for common, hip, and valley rafters. Another use for Speed Squares is as saw guides when crosscutting lumber.

Speed® Squares are manufactured in 7", 12", and 25 cm versions.

Sliding T-Bevel

The *sliding T-bevel,* or bevel square, is used to transfer and test angles. The blade is adjusted to the desired angle and is locked into place with a wing nut. **See Figure 9-24.** A sliding T-bevel is considered essential by experienced carpenters.

Other Squaring Tools

Another type of squaring tool used by carpenters is the *try square.* **See Figure 9-25.** A try square has a short, fixed blade and is used to lay out 90° angles.

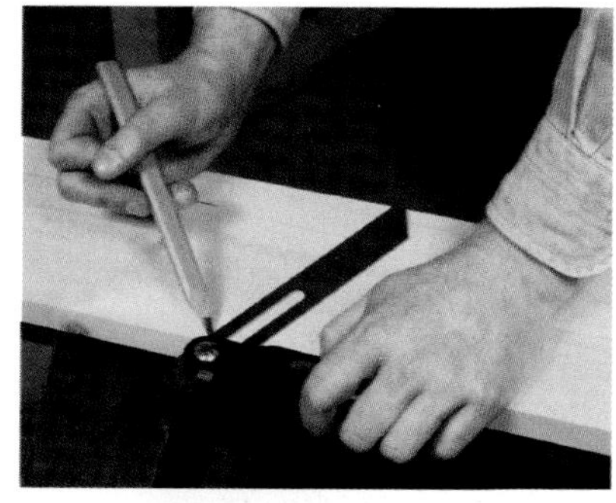

Figure 9-24. *A sliding T-bevel is used to transfer angles. The wing nut tightens the blade in place.*

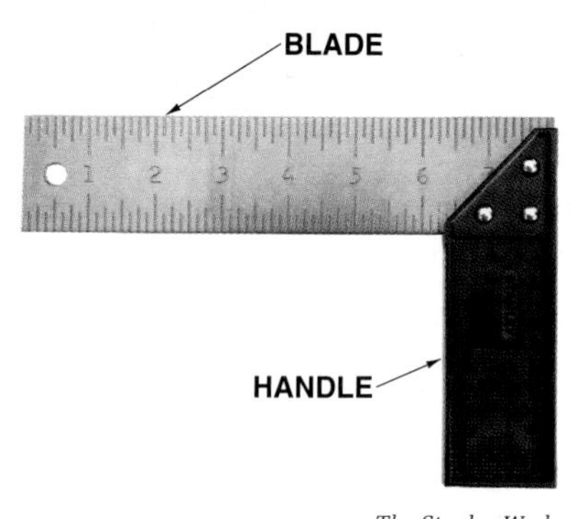

The Stanley Works

Figure 9-25. *A try square is used to lay out 90° angles.*

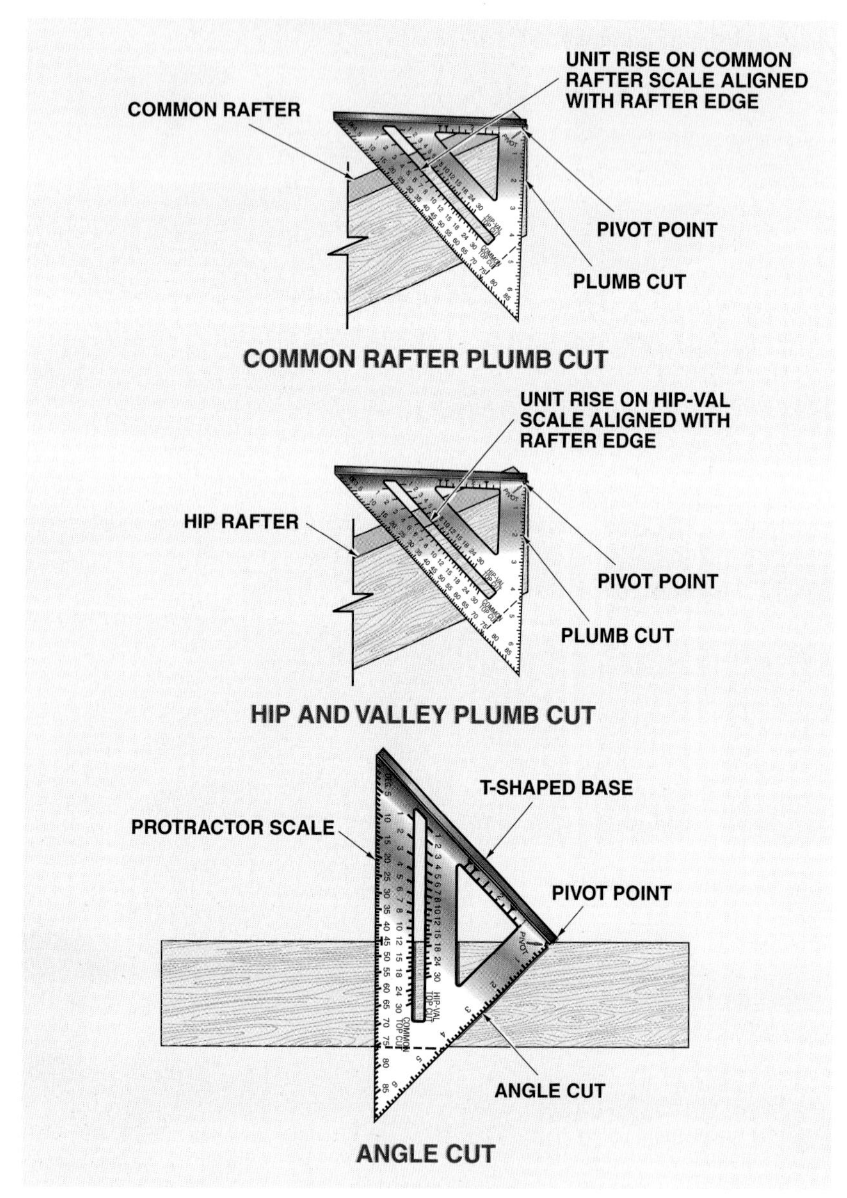

Figure 9-23. *The protractor scale of a Speed® Square is used to lay out angles ranging from 0° to 90°. The common, hip, and valley scales are used to lay out the plumb and seat cuts for common, hip, and valley roof rafters.*

An *angle divider* is a layout tool used to lay out joints that meet at angles other than 90°. **See Figure 9-26.** When laying out miter cuts on wall molding, an angle divider is first placed in the corner in which the molding will fit and locked into position using the thumbscrew. The angle divider is then positioned on the molding and the desired angle is marked along one of the legs.

A *miter protractor* is a layout tool that is used to measure and transfer angles for mitered trim work. Miter protractors are made of aluminum or plastic. The protractor has two scales. One scale corresponds to the angle required when making miter cuts for a mitered joint and the other corresponds to the angle required when making a single cut for a butt joint. **See Figure 9-27.**

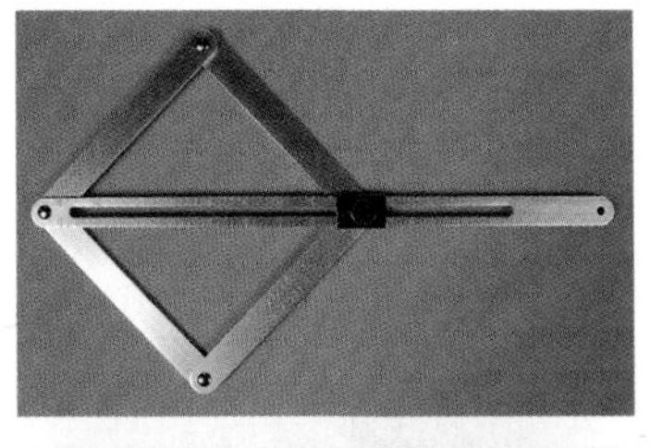

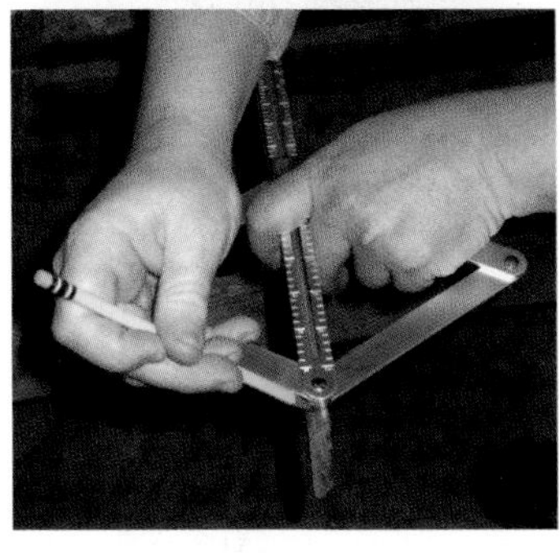

Figure 9-26. *An angle divider is used to lay out miter cuts on molding for walls that meet at angles other than 90°. First, the angle divider is adjusted to fit the inside corner. Then it is used to mark the miter cut on the molding.*

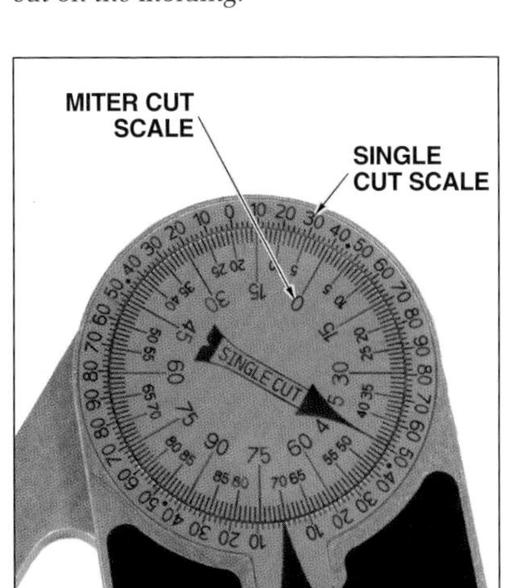

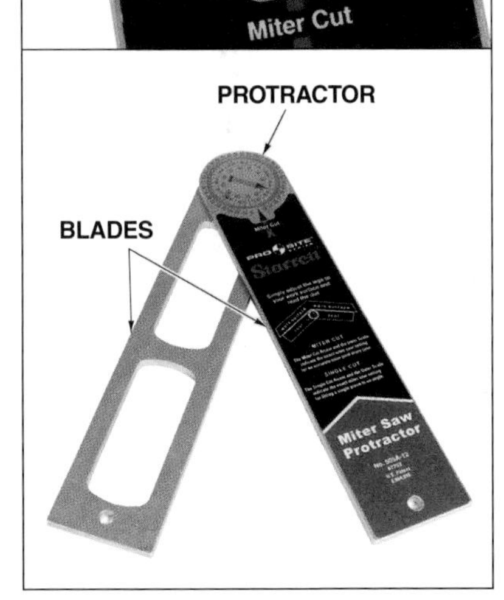

Figure 9-27. *A miter protractor is used to measure and transfer angles for mitered trim work.*

MARKING AND SCRIBING TOOLS

Many tools are available for marking materials for layout purposes including pencils, scribers, wing dividers, scratch awls, center punches, and trammel points. Flat pencils are commonly used by carpenters since they cannot roll when placed on a flat surface.

A *scriber* has two legs—one with a steel point and one holding a pencil. **See Figure 9-28.** A scriber is used when a close fit is required between two pieces of material when one of the surfaces is irregular. The steel point is placed against the irregular surface and is pulled along the surface, transferring the irregularities to the mating surface.

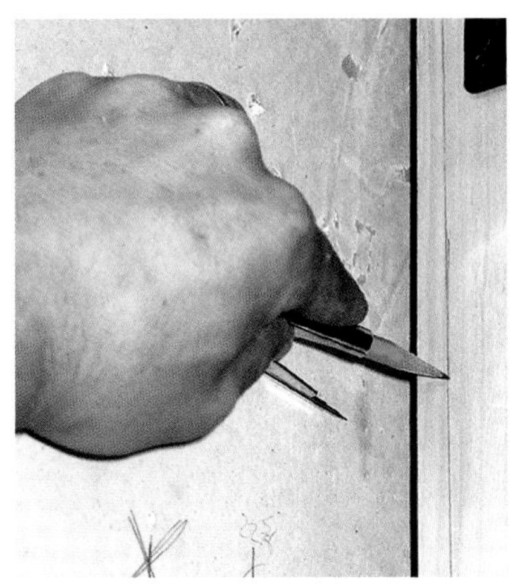

Figure 9-28. *A scriber is used to mark a line when a close fit is required between the edge of a piece of material and an irregular surface.*

Dividers are used to draw arcs and circles or lay out equal spaces. One metal leg is easily replaced with a pencil if desired. **See Figure 9-29.** Dividers can also be used when a close fit is required between two mating surfaces. When using dividers for this purpose, bend the point of the removable leg slightly outward.

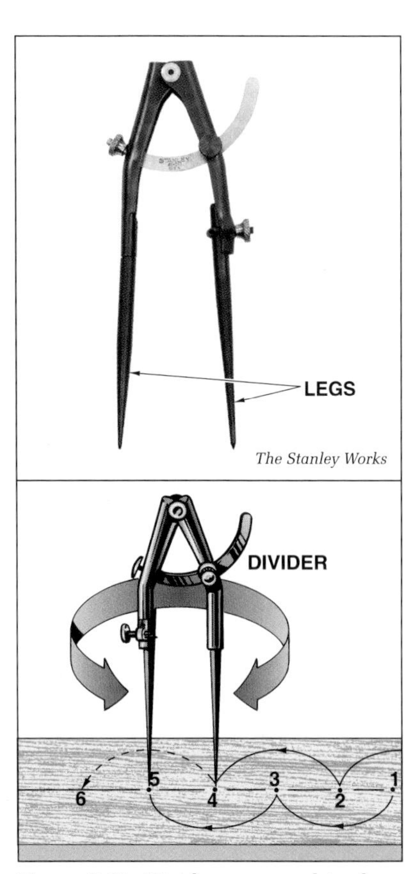

Figure 9-29. *Dividers are used to draw arcs and circles and can also be used as a scriber.*

A *scratch awl* is used to mark lines on materials that will not show a pencil line. Scratch awls are also used to start holes for wood screws, especially in hardwood lumber. **See Figure 9-30.**

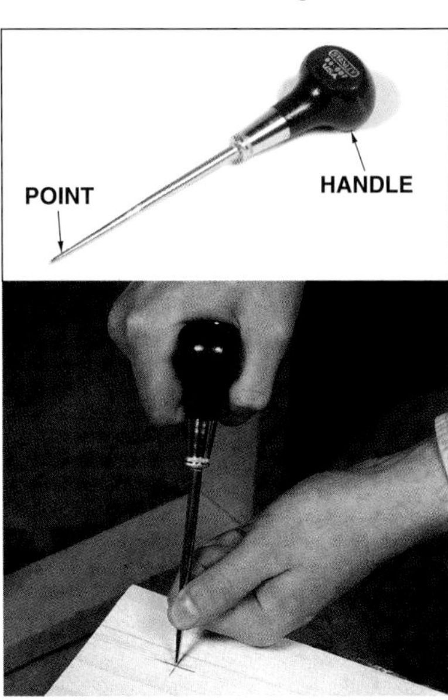

Figure 9-30. *A scratch awl is used to start holes for wood screws or mark lines on materials that will not show a pencil line.*

A *center punch* is struck with a hammer to mark holes to be drilled in hard wood or metal. The indentation in the metal will prevent the drill bit from "skating" across the surface. **See Figure 9-31.**

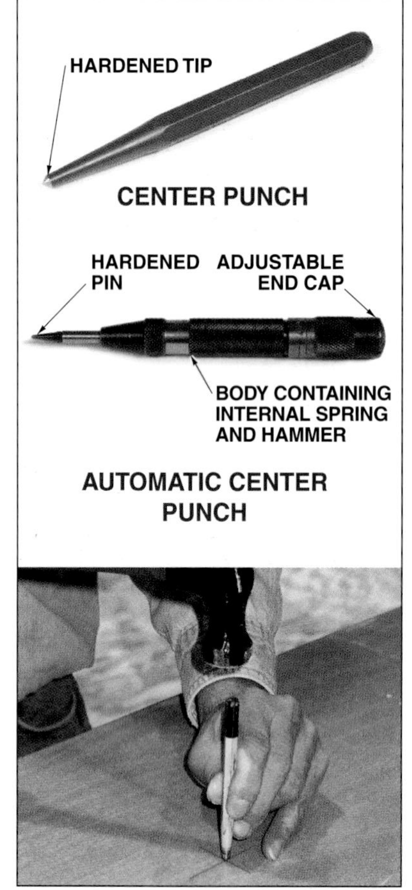

Figure 9-31. *A center punch is used to mark holes to be drilled in hard wood or metal.*

An *automatic center punch* does not require striking the punch with a hammer to mark holes. When the top of the punch is pressed towards the workpiece, an internal spring drives an internal hammer that strikes a hardened pin. The force is transferred from the spring to the surface of the workpiece. The amount of force can be adjusted by rotating the end cap.

A set of *trammel points* is used to lay out circles of any size. A pencil is clamped on to one of the trammel points. Both points are clamped to a piece of straight and narrow stock. The distance between the trammel points is equal to the radius of the circle.

Square gauges, or stair gauges, are used with a framing square to lay out roof rafters and stair stringers. The square gauges are clamped to the blade and tongue of a framing square to ensure the same angle is laid out for the rafters and stringers. **See Figure 9-32.**

Figure 9-32. *Square gauges are attached to a framing square to lay out angle cuts for roof rafters as well as tread and riser cuts for stair stringers.*

A *chalk line reel,* also known as a chalk box, is used to snap straight lines on flat surfaces. **See Figure 9-33.** The reel is filled with colored powdered chalk, which coats the line (string) wound up in it. A tapered ring at the end of the line can be hooked on a nail while the line is unwound. The line is stretched taut and then pulled straight up and released to snap a chalk line on the material beneath it. After snapping a few lines, the line will need to be rewound to recoat it.

Large circles can be formed by connecting a series of dashes laid out from a common center point using a tape measure.

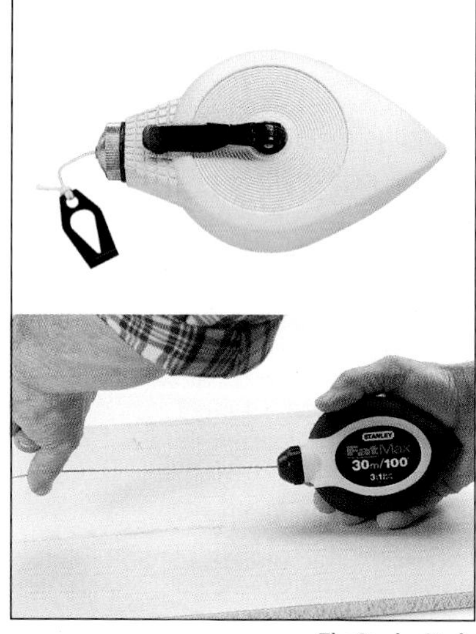

The Stanley Works

Figure 9-33. *A chalk line reel is used to snap straight lines on flat surfaces.*

Unit 9 Review and Resources

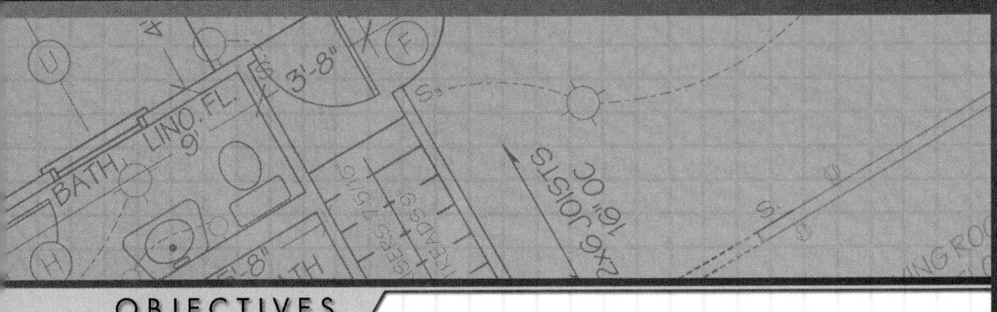

UNIT 10 Fastening and Prying Tools

OBJECTIVES

1. List and describe the various types of hammers and their uses.
2. Explain the proper hammering technique.
3. Discuss proper nailing methods.
4. List and describe common hatchets used in carpentry.
5. List and describe common staplers.
6. Identify different types of screwdrivers and the function of each.
7. Describe the proper procedure for driving wood screws.
8. Identify common safety guidelines for wrenches and pliers.
9. List basic prying tools and the function of each.

Fastening and prying tools are used to install and remove fasteners such as nails, bolts, and screws. The proper tools must be used to prevent injuries and to avoid damage to the fasteners during installation or removal. In addition to using the proper tool for the job, appropriate personal protective equipment (PPE), such as safety glasses or goggles, should be worn to prevent or minimize injuries. OSHA 29 CFR 1926.301, *Hand Tools,* provides general recommendations for proper hand tool usage for construction. OSHA 29 CFR 1926.102, *Eye and Face Protection,* details the proper eye protection for use with fastening and prying tools.

FASTENING TOOLS

Fastening tools are used to drive or install fasteners that hold together building materials. Nails, screws, staples, and bolts require different tools to drive or install the fasteners. Appropriate PPE must be worn by the worker using the fastening tool and other workers in the area where the work is being performed.

Hammers

Hammers are striking tools used to drive and remove nails. The main parts of a hammer are the *head, face, claw,* and *handle.* **See Figure 10-1.** A hammer head is forged of high-quality steel or titanium and its face may be bell-shaped or flat. A bell-shaped face is slightly curved so nails can be driven flush to the surface, leaving only light marks on the lumber.

The surface of the face may be smooth or milled (checkered). Milled hammer faces slip off a nail head less easily. However, a milled face will scar the wood and should only be used when doing rough work.

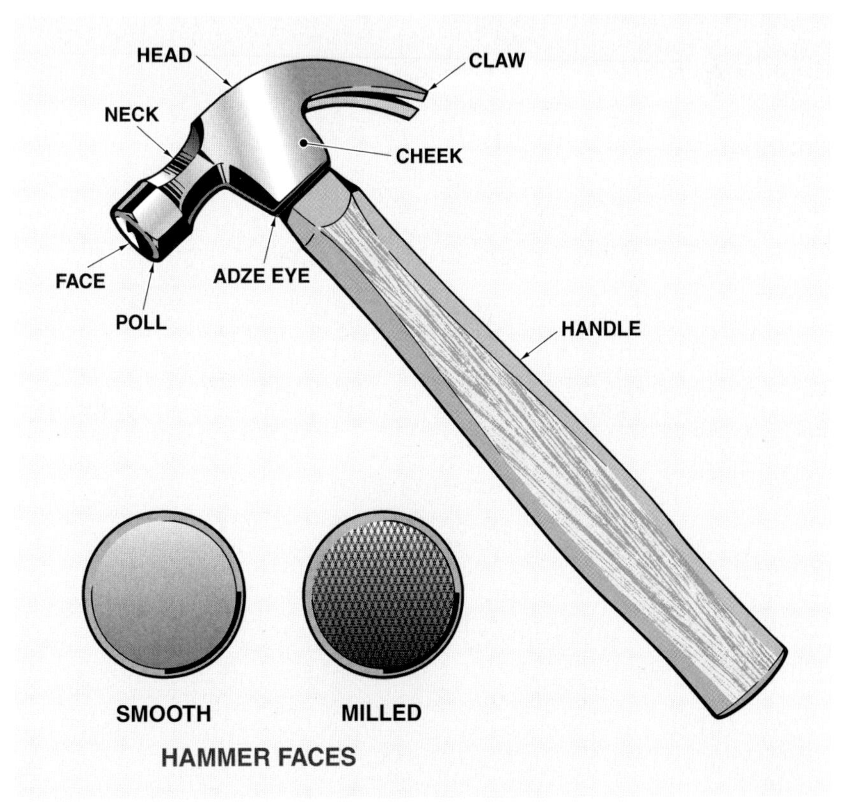

Figure 10-1. *The main parts of a hammer are the head, face, claw, and handle. The hammer face may be smooth or milled.*

Different materials are used for hammer handles, including fiberglass, steel, titanium and wood. Hammers with fiberglass or steel handles are cushioned with a vinyl or neoprene grip to absorb the blows when striking a nail. Hardwood hammer handles can be easily replaced if broken. Select a hammer with a handle diameter ranging from 1¼″ to 2″ to reduce the risk of injury.

Common types of hammers used by carpenters include straight-claw hammers, curved-claw hammers, ball-peen hammers, tack hammers, dead blow hammers, mallets, and sledgehammers. **See Figure 10-2.** The size of a hammer is determined by the weight of its head. Always use the proper weight hammer for the job.

Straight-Claw Hammers. A *straight-claw,* or *ripping,* hammer is used for rough work such as framing or concrete form construction. In addition, a straight-claw hammer can be used to pry boards apart and to split pieces of lumber. **See Figure 10-3.**

Straight-claw hammer weights include 16 oz, 20 oz, 22 oz, 24 oz, 28 oz, and 30 oz. A 22-oz hammer is common for rough work. Some carpenters who specialize in framing work use 28-oz hammers since a heavier hammer provides more driving power than lighter hammers. While 28-oz hammers are commonly used for framing work, care must be taken to avoid wrist and elbow injuries. When selecting a hammer, the proper head-to-handle weight distribution is very important. Good balance results in less stress on muscles and tendons.

Figure 10-3. *In addition to driving nails, straight-claw hammers are used to pry boards apart and to split pieces of lumber.*

Curved-Claw Hammers. A *curved-claw hammer* is commonly used for finish work. Some carpenters feel that the curved-claw design provides better balance and control when driving finish nails. Curved-claw hammers also provide better leverage for pulling nails. **See Figure 10-4.** Curved-claw hammers range in weight from 7 oz to 20 oz, with a 16-oz hammer being the most popular choice.

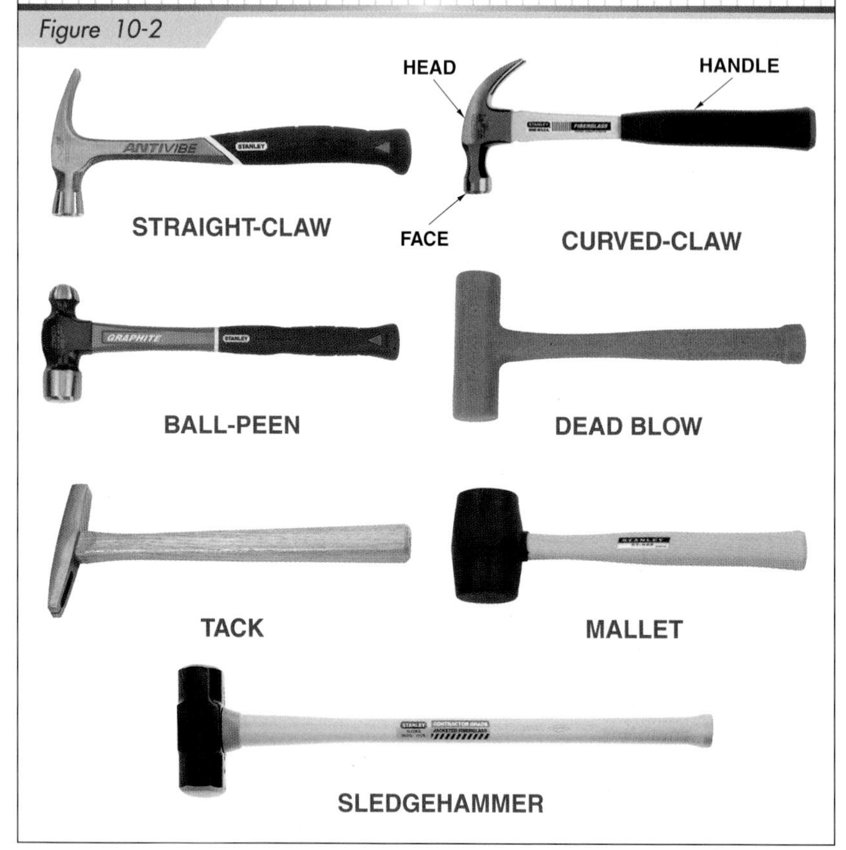

Figure 10-2. *Carpenters may use several types of hammers on the job site.*

Figure 10-4. *Curved-claw hammers provide better leverage when pulling nails than straight-claw hammers. A block of wood placed under the hammer head increases leverage and helps in removing the fastener.*

Ball-Peen Hammers. A *ball-peen hammer,* or machinist's hammer, is a hammer with a round, slightly curved face and a round head. A ball-peen hammer is a general-purpose hammer and is well suited for striking chisels and punches. Ball-peen hammers can also be used for riveting, shaping, and straightening unhardened metal.

Tack Hammers. A *tack hammer* is a lightweight hammer used to drive small nails into finish work. Tack hammers have magnetized heads that can hold small nails so that each nail can be started easily. After a nail is started, it is driven in with the opposite head of the hammer.

Dead Blow Hammers. A *dead blow hammer* is a striking tool with a head made of steel pellets encased in a plastic coating. This design prevents rebounding of the head when striking an object. A dead blow hammer is good for assembling various woodworking joints and other millwork.

Mallets. A *mallet* is a soft, double-faced striking tool. Mallets are used to prevent damage when the use of steel hammers would deface the work or damage the tool being struck. Mallets are frequently abused because they are often used to perform tasks they were not designed for. A mallet may be made of quality rubber, hardwood, or hard fiber and will last a long time if used in the correct manner. Mallets are useful for striking chisels and adjusting plane irons.

Sledgehammers. A *sledgehammer* is a heavy, double-faced striking tool used for heavy-duty striking work. Medium-sized sledgehammers (5 lb to 8 lb) are used for driving stakes and for other heavy-duty applications. Sledgehammers require a large working area for safety. Extreme caution must be taken while swinging a sledgehammer when others are near.

Hammering Technique

The proper hammering technique results in faster work with less effort. **See Figure 10-5.** Keep your wrist loose at all times while driving nails. The blow is delivered through the wrist, elbow, and/or shoulder, depending on the strength of blow to be struck. Rest the hammer face on the nail and draw the hammer back. Start the nail with a light tap and drive the nail. Strike the nail squarely to avoid bending it and to prevent the hammer head from bouncing off the nail and scarring the wood surface.

Nailing Method

Nails provide a strong tie between building materials. Improper nailing methods result in weak ties between materials. The following nailing methods should be used:

- When possible, always drive a nail from the thinner piece of material into the thicker piece. The nail should be long enough so the upper third of the nail is in the thinner piece and the rest of the nail is driven into the thicker piece. **See Figure 10-6.**
- Stagger the nails when driving nails near the end of a board. **See Figure 10-7.** Aligning the nails may split the board.
- To avoid splits in harder wood, either blunt the end of the nail with a hammer, cut off the point of the nail, or drill a pilot hole for the nail. **See Figure 10-8.** The blunt end of a nail forces the wood fibers ahead of it and reduces the possibility of splitting the wood.

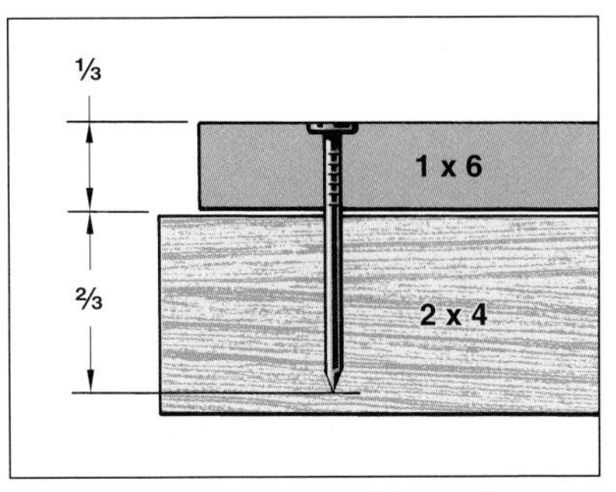

Figure 10-6. *When possible, nail from a thinner piece into the thicker piece.*

Titanium hammers are 45% lighter than steel hammers but drive nails much more efficiently with less recoil.

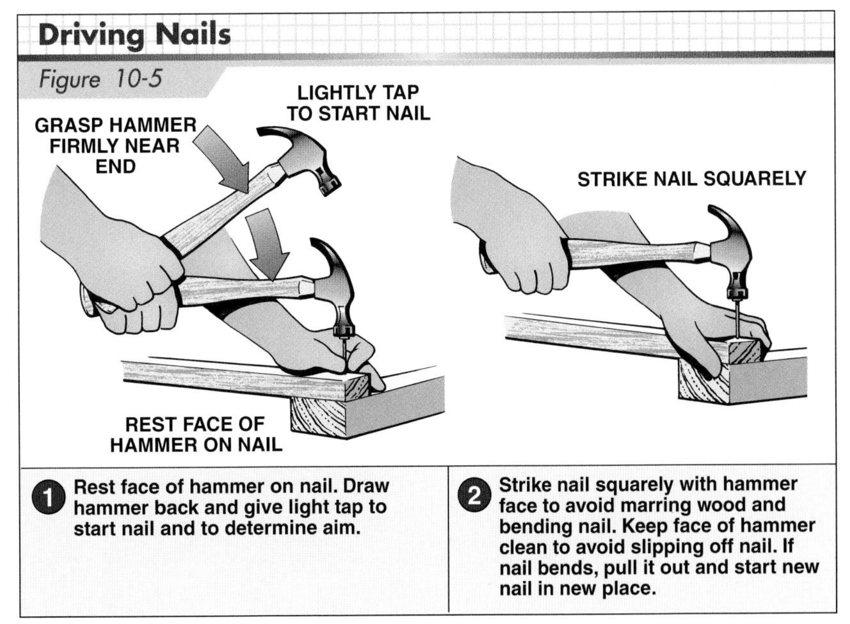

Figure 10-5. *Proper use of a hammer results in more efficient driving of nails.*

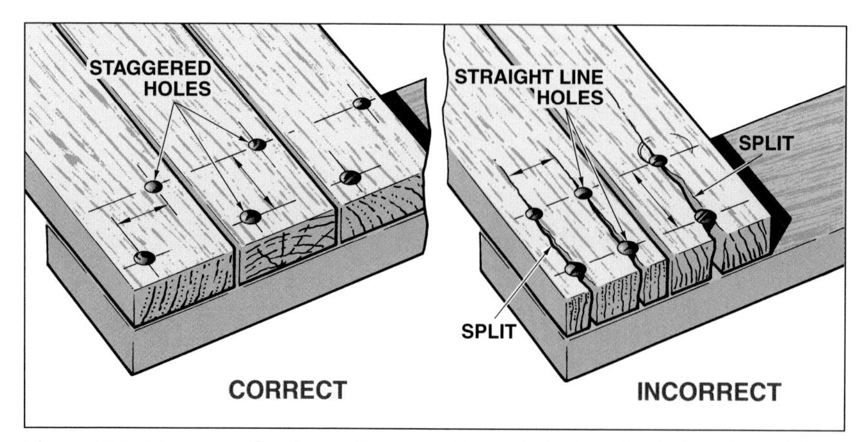

Figure 10-7. *Stagger nails when nailing near the end of a board. Placing the nails in a straight line may split the board.*

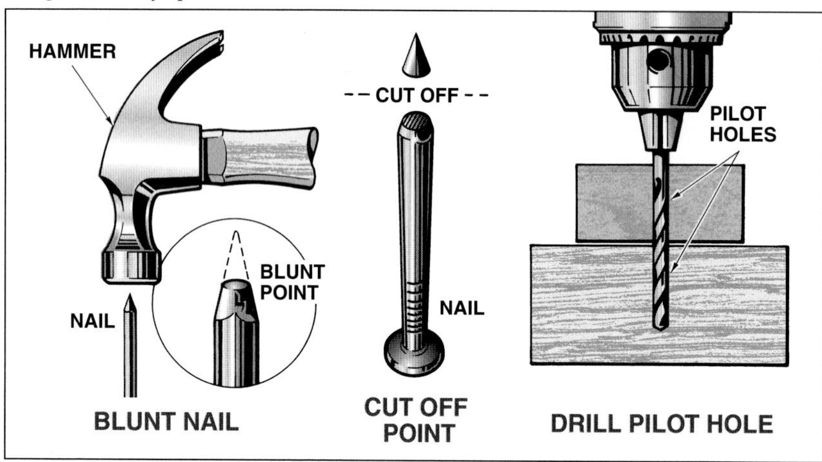

Figure 10-8. *To avoid splits in harder wood, blunt the end of the nail with a hammer, cut off the point of the nail, or drill a pilot hole for the nail.*

- Whenever possible, drive nails across the grain rather than into the end grain. Nails driven into end grain have less holding power. If it is necessary to nail into the end grain, drive nails in at an angle to increase their holding power. **See Figure 10-9.**
- When it is not possible to end-nail two pieces together, *toenail* them by driving the nail at an angle so approximately half the nail is in each piece of wood. **See Figure 10-10.**
- For temporary nailing, tack the nails so they stick out from the material and can be easily withdrawn. **See Figure 10-11.**
- To increase holding power between two pieces fastened face-to-face, drive nails at an angle. **See Figure 10-12.**

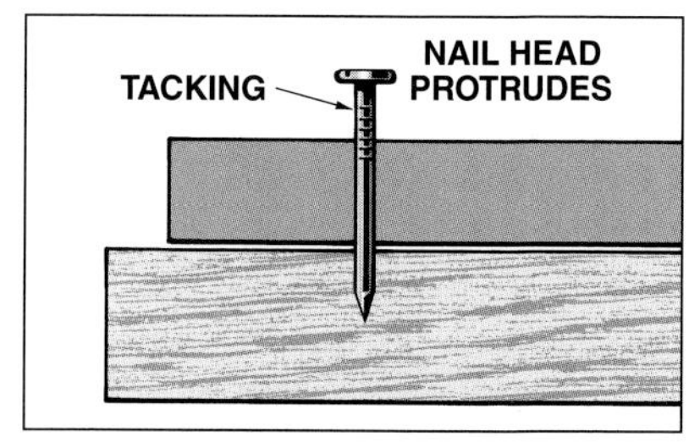

Figure 10-11. *Tacking is a procedure for temporary nailing. The head of the nail should stick out so that it can be easily withdrawn.*

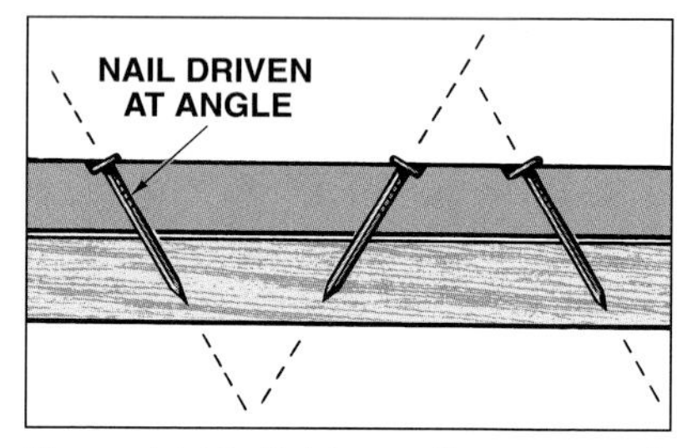

Figure 10-12. *Driving nails at an angle increases the holding power between two pieces fastened face-to-face.*

Nail Sets. A *nail set* is used to drive a nail head below the wood surface. **See Figure 10-13.** Nails are usually set when installing finish materials such as molding, paneling, or siding. Holes left by the set nails are filled by a painter. Nail sets are typically 4″ long and their tips range in diameter from 1/32″ to 5/32″.

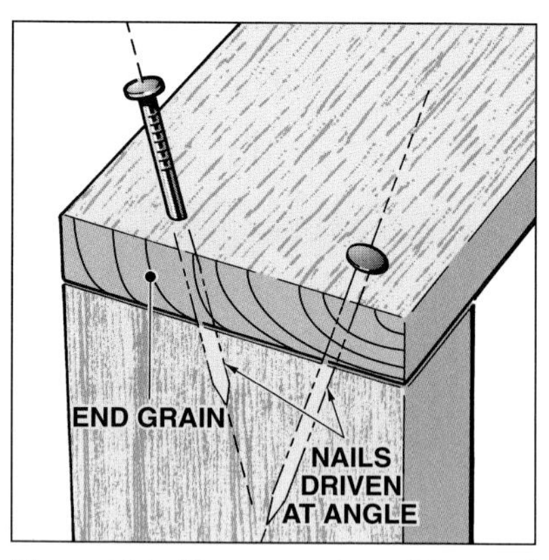

Figure 10-9. *If necessary to nail into end grain, drive the nails in at an angle to increase their holding power.*

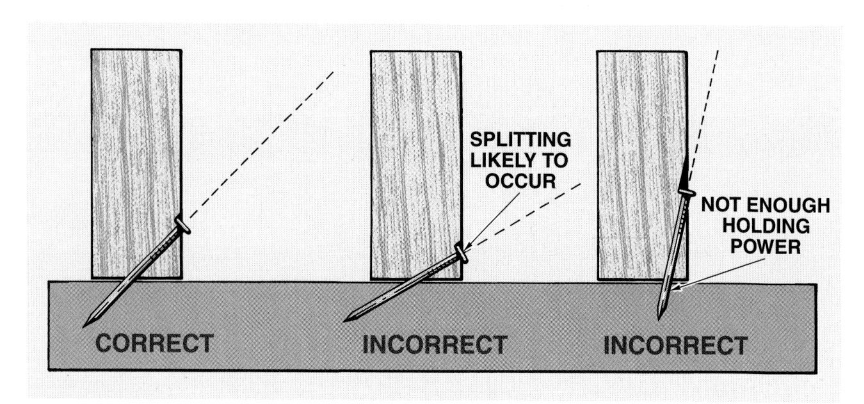

Figure 10-10. *Toenails should be driven in at an angle so approximately half of the nail is in each piece of wood.*

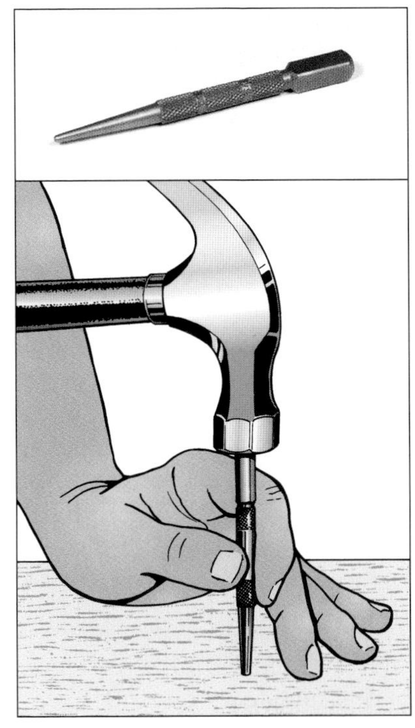

Figure 10-13. *A nail set is used to drive a nail head below the surface of the wood.*

Hatchets

Hatchets are striking tools that have a nailing face as well as a cutting edge. Three types of carpenter hatchets are the *half-hatchet, wallboard hatchet,* and *shingling hatchet.* **See Figure 10-14.** All carpenter hatchets have slots at the side of the blade for pulling nails.

A half-hatchet is used to sharpen and drive stakes when constructing wood concrete forms. A half-hatchet has a beveled single or double blade that is usually 3½″ wide. A wallboard hatchet, or drywall hammer, has a curved, milled nailing face, which permits the face to dimple the gypsum board without breaking through the paper covering. A shingling hatchet is used to install wood, composition, and fiberglass roof shingles. The blade of a shingling hatchet is usually 2½″ wide.

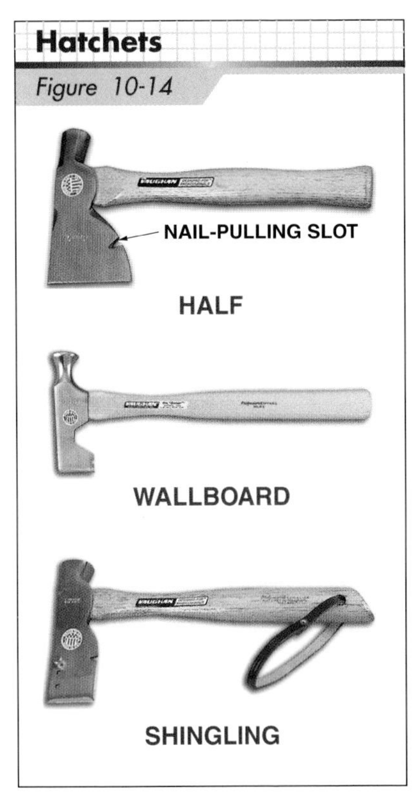

Vaughan & Bushnell Mfg. Co.

Figure 10-14. *Hatchets used in the carpentry trade have a nail-pulling slot at the side of the blade.*

Use and Care of Striking Tools

Finger injuries are the most frequent accidents caused by striking tools. Proper care and use of striking tools is essential to injury-free operation. Consider the following items to avoid injuries when using striking tools:

- When using a hammer or hatchet with a wood handle, ensure the head of the tool fits tightly on the handle.
- Do not use a hammer if the handle is loose or damaged. Replace a cracked wood handle.
- Use a flat-faced hammer for driving nails.
- Remove hammers from service that have signs of excessive wear, chips, cracks, or mushrooming.
- Hammer heads should be of proper hardness. Soft heads will mushroom and chips may break off.
- When using a hammer or hatchet, ensure the backswing path is clear.
- Do not use one hammer to strike another.

Staplers

As a rule, hand-operated mechanical staplers, or *tackers,* are not part of the carpenter's basic tool collection; they are usually provided by a contractor. Heavy-duty staplers can perform many operations previously performed with a hammer and nails.

A *strike tacker* is operated by striking the plunger with a rubber mallet. A strike tacker can be used to fasten floor underlayment to the subfloor. Strike tackers drive 18-ga staples between ⅞″ and 1⅛″ long. **See Figure 10-15.**

Figure 10-15. *A strike tacker may be used to fasten floor underlayment.*

A *hammer tacker* allows one-hand operation, since it releases a staple when it is struck against a surface. Hammer tackers are commonly used to install carpet padding, insulation, roofing paper, and vapor barriers. **See Figure 10-16.** Hammer tackers use 20-ga staples between ¼″ and 9⁄16″ long.

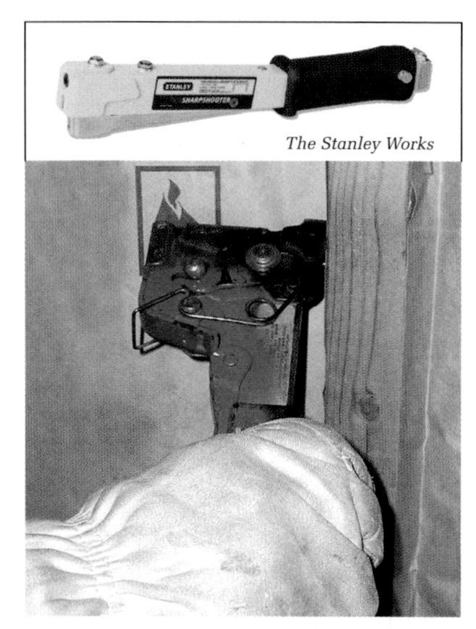

Figure 10-16. *A hammer tacker is used to fasten insulation batts.*

A *staple gun tacker* is a heavy-duty stapler that also allows one-hand operation. It is used to fasten a variety of materials, including vinyl flooring, insulation, roofing paper, screening, carpet padding, carpet, and ceiling tile. **See Figure 10-17.** Staple gun tackers typically drive 20-ga staples between ¼″ and 9/16″ long.

LEVER HANDLE

The Stanley Works

Figure 10-17. *Heavy-duty staple guns are used to fasten vinyl flooring, insulation, roofing paper, carpet padding, screening, carpet, and ceiling tile.*

Screwdrivers

Screwdrivers are used to drive or withdraw screws and should not be used to pry pieces apart. The parts of a screwdriver are the *head, handle, blade* (shank), and *tip.* **See Figure 10-18.** The size of a screwdriver is identified by the length of its blade. More frequently used lengths are 3″, 4″, 6″, 8″, and 10″. Screwdrivers with longer blades have larger handles allowing greater torque to be applied, which is required when driving larger screws. Common types of screwdrivers include standard (flat head), Phillips (cross-slot), square-drive (Robertson), and six-point star-shaped (Torx®) screwdrivers. Multifunction and offset screwdrivers are specialized screwdrivers that also may be used on the job site.

Screwdrivers are used for power tasks, such as driving screws into hardwood, and for precision tasks, such as adjusting the calibration of an instrument. For power tasks, select a screwdriver with a handle diameter ranging from 1¼″ to 2″. If necessary, a snug sleeve can be placed over the screwdriver handle to increase its diameter. In addition, for power tasks, select a screwdriver with a handle longer than the widest part of your hand—usually 4″ to 6″. If the handle is too short, the end of the handle will press against the palm of your hand and may cause an injury. For precision tasks, select a screwdriver with a handle diameter ranging from ¼″ to ½″.

Standard Screwdrivers. The tip of a standard (flat head) screwdriver fits into a single slot in the screw head. Tips range in width from 1/8″ to 3/8″. Longer screwdrivers typically have wider tips, but narrower tips are also available. For best results, use a screwdriver with a tip equal in width and thickness of the size of the screw slot.

Phillips Screwdrivers. The tip of a Phillips (cross-slot) screwdriver is shaped like a cross and is used to drive the cross slot head of a Phillips screw. The size of a Phillips screwdriver is determined by the length of the blade as well as by the tip size. Common tip sizes range from 0 to 4.

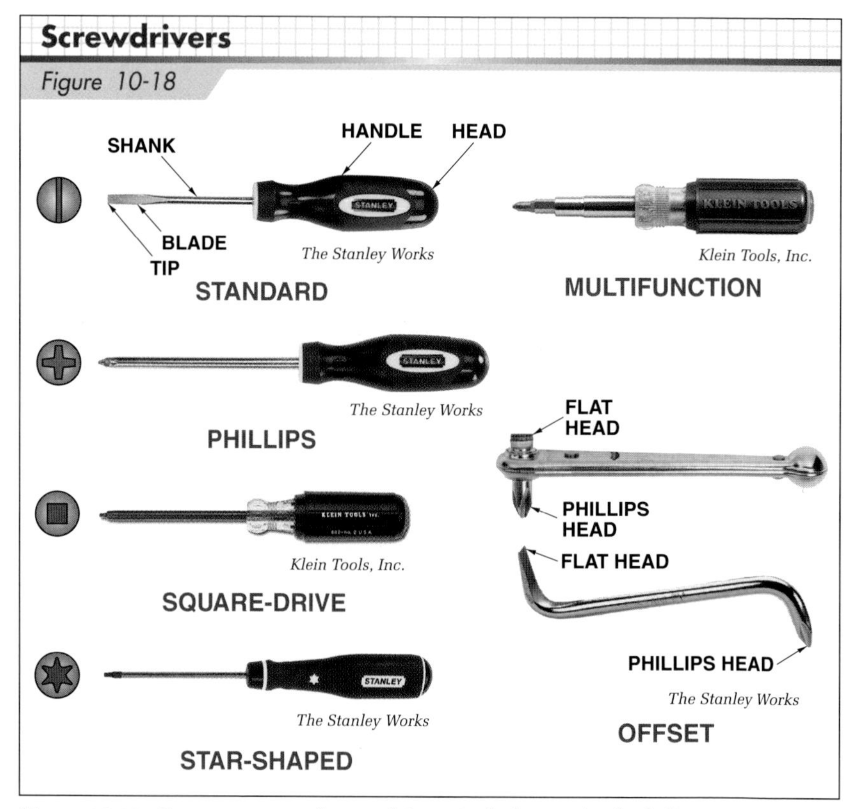

Figure 10-18. *Common types of screwdrivers include standard, Phillips, square-drive, star-shaped, multifunction, and offset screwdrivers.*

Square-Drive Screwdrivers. The tip of a square-drive screwdriver is square and fits into the square recess of a square-drive (Robertson) screw. The tip of a square-drive screwdriver fits deeper into the screw head and tends to lock into the screw recess, providing better torque. Square-drive screwdriver tip sizes include #1, #2, and #3.

Star-Shaped Screwdrivers. A star-shaped (Torx®) screwdriver has six points rather than four, as with a Phillips screwdriver. The design of the Torx head prevents the tip from slipping out as the fastener is tightened. Torx screwdriver sizes are indicated by a capital "T" followed by a number. Sizes range from T1 to T100.

Multifunction Screwdrivers. A *multifunction screwdriver* is a hand tool used to tighten or loosen a variety of sizes and types of fasteners. A multifunction screwdriver can replace several other single-function screwdrivers and nut drivers, saving space in a tool pouch or toolbox. Multifunction screwdrivers come in 5-in-1, 6-in-1, 10-in-1, and 11-in-1 models. For example, a 6-in-1 screwdriver includes 3⁄16″ and 1⁄4″ straight bits, 1⁄4″ and 5⁄16″ nut drivers, and #1 and #2 Phillips bits. A 10-in-1 screwdriver includes 3⁄16″ and 1⁄4″ straight bits, 1⁄4″ and 5⁄16″ nut drivers, #1 and #2 Phillips bits, T10 and T15 Torx® bits, and #1 and #2 square bits.

Offset Screwdrivers. Offset screwdrivers provide a means for tightening or loosening difficult to reach screws. Offset screwdrivers can be flat head or Phillips and are also available as a combination of the two. Some offset screwdrivers also allow the use of replaceable bits.

Driving Wood Screws. When driving screws into harder woods, shank and thread pilot holes must be drilled to prevent the wood from splitting. Combination countersink and drill bits are commonly used to drill the pilot hole, shank hole, and countersink at the same time. **Figure 10-19** describes the proper procedure for driving screws in hard wood. If a combination bit is not available, drill a shank hole equal in diameter to the diameter of the screw shank and a pilot hole slightly smaller in diameter than the thread diameter. If a flat-head screw is being installed, countersink the shoulder to accommodate the screw head.

When driving screws into softer woods, pilot holes should also be drilled. In addition, flat-head screws driven into softer woods tend to countersink themselves.

Rubbing wax or soap on screw threads also makes it easier to drive a screw. A piece of wax can be kept in the toolbox for lubricating screws and other applications.

Use and Care of Screwdrivers. Screwdriver accidents can cause puncture wounds in the hand. To avoid hand injuries, hold the screwdriver properly, as shown in **Figure 10-20.**

The pilot hole for a wood screw should be about 75% of the size of the total diameter of the screw.

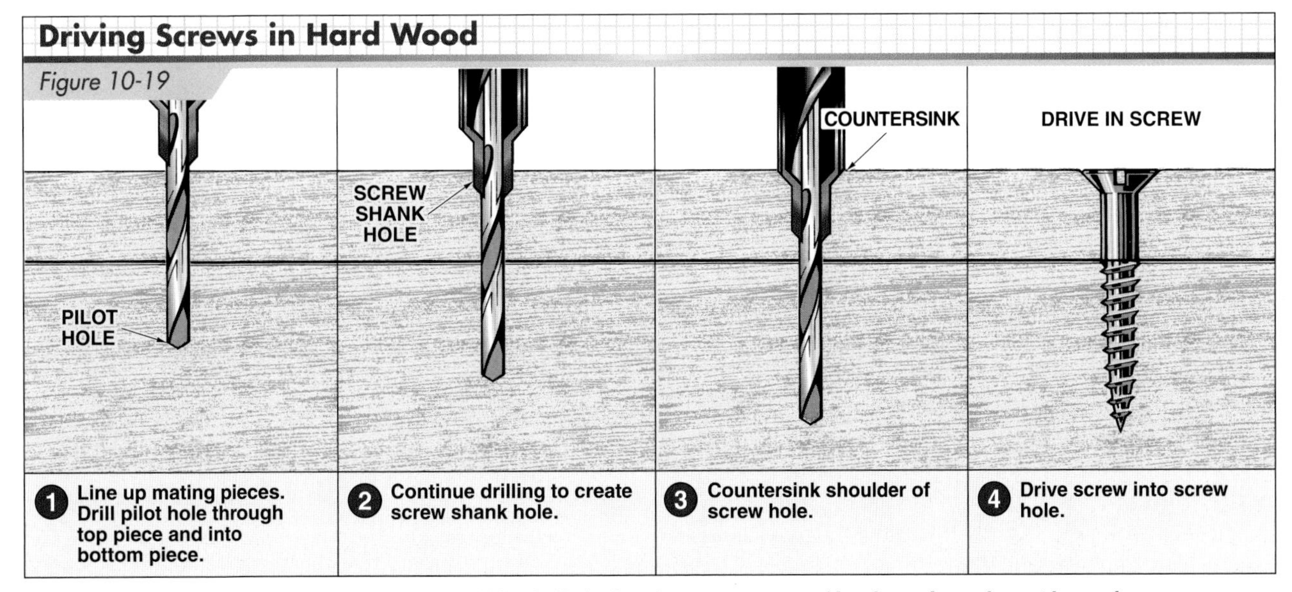

Figure 10-19. *Screw shank and pilot holes should be drilled when fastening pieces of hard wood together with wood screws.*

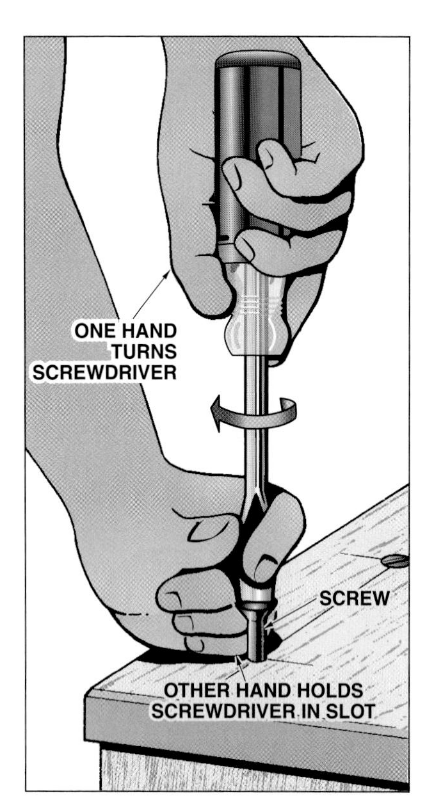

Figure 10-20. *When driving screws, one hand turns the screwdriver while the other hand holds it in position.*

Another important work and safety factor is the condition of the screwdriver tip and blade. A straight screwdriver tip should be straight and square-cornered, not rounded or excessively tapered. **See Figure 10-21.** The tip of a Phillips screwdriver should not be chipped or flattened. Other safety rules are as follows:

- Screwdrivers are designed to apply a certain amount of torque to a screw. Do not attach a wrench to a screwdriver to increase leverage or torque.
- Most screwdriver handles are plastic or wood. Do not drive a screwdriver with a hammer.
- Do not use a screwdriver as a punch, chisel, pry bar, or nail-puller.
- Always use a screwdriver tip that properly fits the slot.
- Do not carry a screwdriver in pants pocket.

Wrenches

Various types of bolts are used to fasten structural members together. Gripping tools, such as wrenches and pliers, are used to install bolts or tighten nuts. A *wrench* is a hand tool designed to turn bolts, nuts, or pipes to either fasten or loosen them. Wrenches are available in standard sizes, or they can be adjustable. Common types of wrenches include combination, socket, adjustable, and hex key (Allen) wrenches. **See Figure 10-22.**

Use and Care of Wrenches. Wrenches should be used only for their intended purpose. Using the wrong type of wrench or improperly using a wrench can cause scraped knuckles, injured muscles, or a fall if the wrench slips. The following safety rules should be followed when using wrenches:

- Check for worn, cracked, or sprung jaws on the wrench. Remove the wrench from service if such conditions exist.

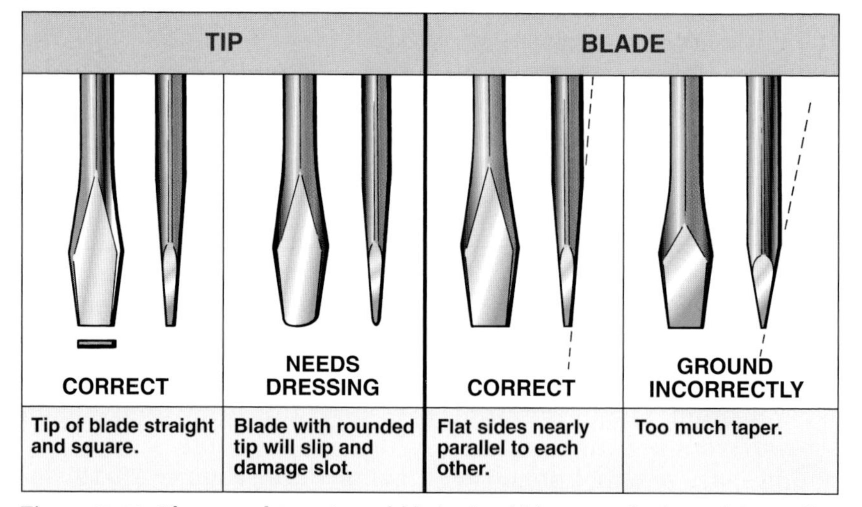

Figure 10-21. *The screwdriver tip and blade should be properly dressed (ground) to prevent the screwdriver from slipping from the slot.*

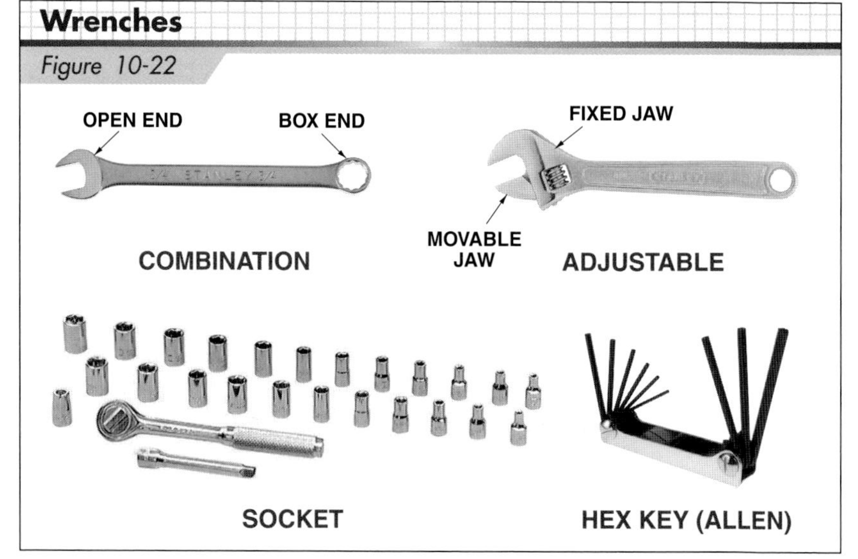

Figure 10-22. *Common types of wrenches include combination, socket, adjustable, and hex key (Allen) wrenches.*

- Use the proper size wrench for the job. Too short of a wrench will require additional pressure to turn the nut or bolt and may result in slippage and injury. Too long of a wrench may provide too much torque, resulting in a damaged nut or bolt.
- Always try to pull, rather than push, on a wrench. **See Figure 10-23.** A greater danger of a wrench slipping and causing a hand injury exists when pushing a wrench.
- Never use a wrench as a hammer.
- Inspect wrenches for damage, including cracks, wear, or distortion.
- Avoid using an extension on a wrench to improve the leverage.

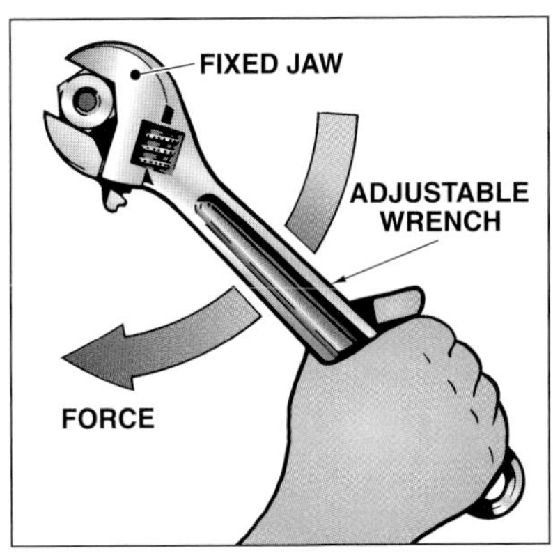

Figure 10-23. *When using an adjustable wrench, be sure it is tightly adjusted to the nut. Pull the wrench so that the force is on the side of the fixed jaw.*

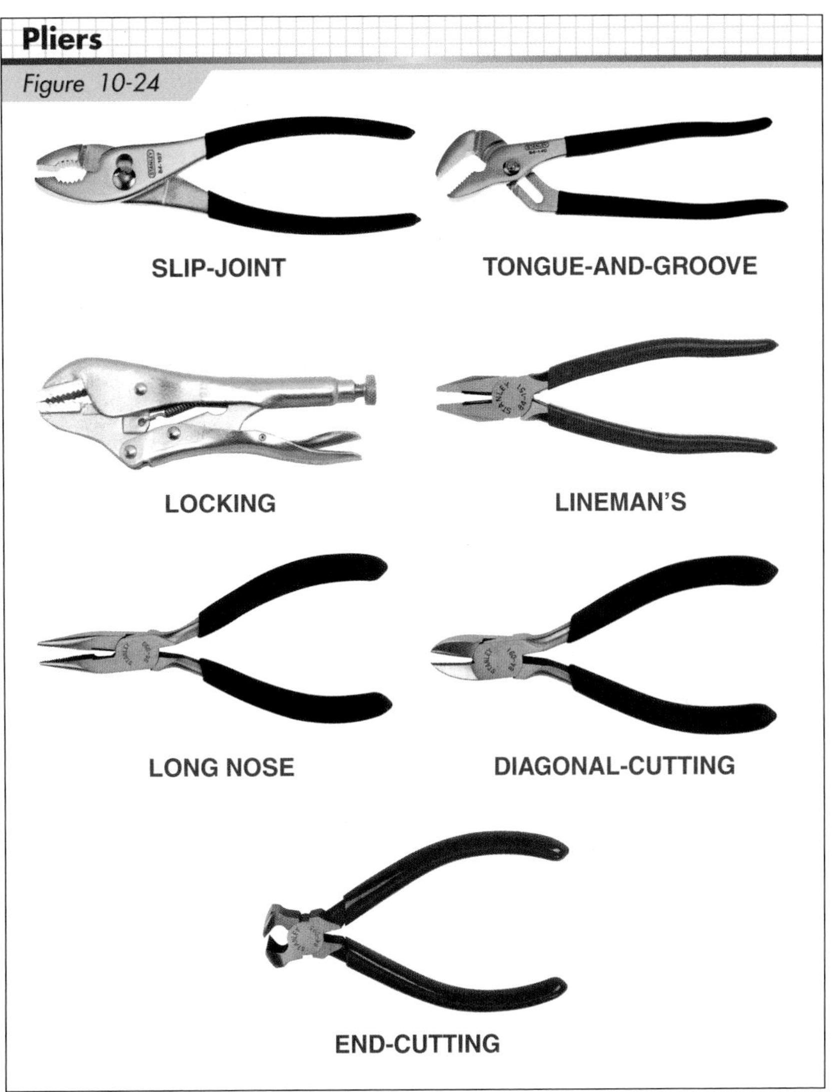

The Stanley Works

Figure 10-24. *Common types of pliers include slip-joint, tongue-and-groove (C-joint), locking, lineman's, long nose (needle nose), diagonal-cutting (side-cutting), and end-cutting pliers.*

Pliers

Pliers are hand tools that have opposing jaws for gripping or cutting. Carpenters use pliers for various gripping, turning, cutting, positioning, and bending tasks. Common types of pliers include slip-joint, tongue-and-groove (C-joint), locking, lineman's, long nose (needle nose), diagonal-cutting (side-cutting), and end-cutting (end nipper) pliers. **See Figure 10-24.**

Use and Care of Pliers. Pliers are a versatile tool used for gripping, turning, and bending. The following safety rules should be followed when using pliers:

- Cut hardened wire only with pliers designed for that purpose.
- Do not use pliers for tightening nuts and bolts; instead, use the proper size wrench.
- To gain more leverage when using pliers, use a larger size pliers.
- Never use pliers as a hammer.

Pliers must be carefully selected to minimize the risk of injury. For power tasks, select a pliers with an open grip span no greater than 3½″ and a closed grip span of at least 2″. The *grip span* is the distance between the thumb and fingers when the tool jaws are open or closed. **See Figure 10-25.** If continuous force is required, consider using a clamp or locking pliers. If available, select a pliers with handles that are spring-loaded to return the handles to the open position. In addition, select a pliers without sharp edges or finger grooves on the handle.

Pliers Grip Spans

Figure 10-25

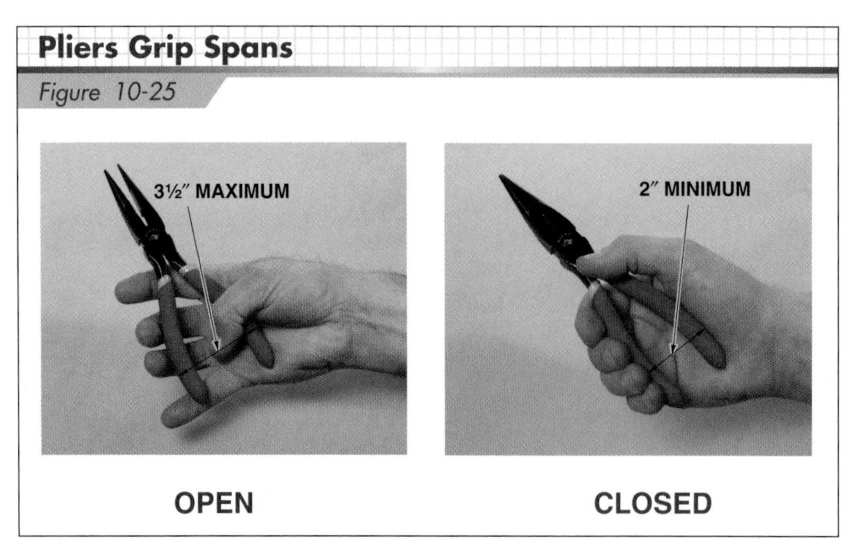

Figure 10-25. *For power tasks, a pair of pliers with an open grip span no greater than 3½″ and a closed grip span of at least 2″ should be used.*

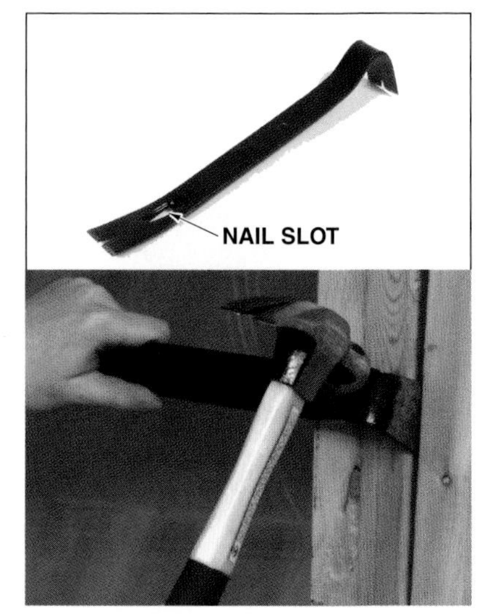

Figure 10-27. *A ripping chisel may be used to pry materials apart.*

PRYING TOOLS

Prying tools are designed to pull apart materials that have been fastened. Carpentry work sometimes requires tearing apart structural members, especially in remodeling work. Small or large sections of a building may have to be torn down and removed. In concrete form construction, wood forms must be stripped away after the concrete has hardened.

Most accidents related to the use of prying tools occur when the tool slips and the worker falls. Workers should maintain a balanced footing and a firm grip on the tool. Proper use of prying tools not only reduces the chance of accidents but also reduces damage to materials that must be reused.

One type of prying tool is the *ripping bar* or *pry bar.* **See Figure 10-26.** Ripping bars are available in lengths ranging from 12″ to 60″, with a 30″ bar being the most popular length since it conveniently fits in most toolboxes.

A *ripping chisel,* or *flat bar,* has a nail slot at the end to pull nails from tightly enclosed areas. A ripping chisel may also be used as a small pry bar. **See Figure 10-27.**

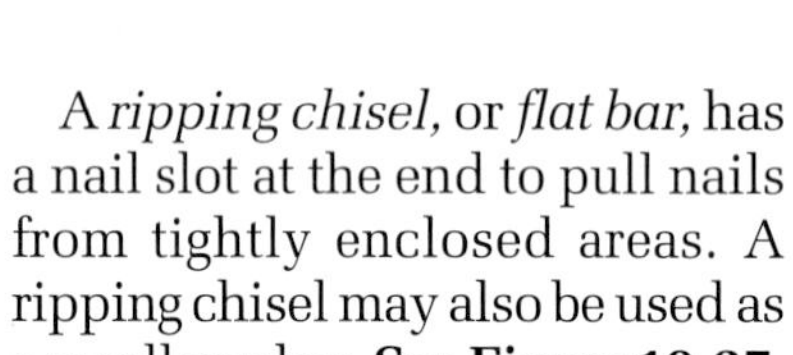

Figure 10-26. *A ripping bar is used to pry boards apart and remove large nails or spikes.*

A *nail claw* is used solely for nail removal. **See Figure 10-28.** The sharpened nail slot is driven under a nail head to pull the nail above the lumber surface. The nail is then pulled completely out with a hammer or ripping bar.

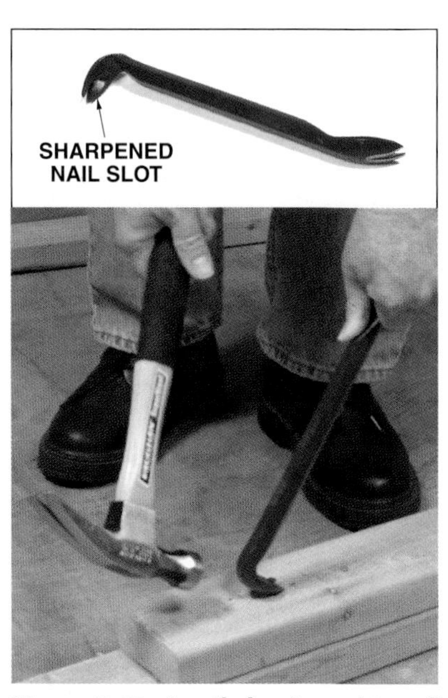

Figure 10-28. *A nail claw is used to pull nails driven flush with the lumber surface so they can be pulled out with a hammer or ripping bar.*

Unit 10 Review and Resources

UNIT 11

Sawing and Cutting Tools

OBJECTIVES

1. List and describe common uses of handsaws.
2. Describe the proper sawing technique.
3. Explain the proper use and care of handsaws.
4. List and describe common cutting tools.
5. Discuss the proper use and care of wood chisels.

Sawing and cutting tools are used to cut and trim construction materials to their proper dimensions. Handsaws, knives, chisels, and snips are commonly used on construction jobs. Sawing and cutting tools have sharp cutting edges that must be protected when the tool is not in use. In addition, sawing and cutting tools should be handled carefully to avoid injury. Eye protection must also be worn.

HANDSAWS

The main parts of a handsaw are the *blade* (including the *toe* and *heel* of the blade), *teeth, back,* and *handle.* **See Figure 11-1.** Although the basic construction of all handsaws is similar, there are many differences in the length and shape of the blade and the number and shape of the teeth. Most handsaws have a straight back, but *skewback saws* (curved-back) are also available.

The *kerf* (cut made by a handsaw) is wider than the thickness of a saw blade to allow the blade to move freely through the material being cut. If this were not the case, the wood fibers pressing against the blade would cause the saw to bind, making the cutting action difficult. Teeth of a saw are *set* (alternately bent from side to side), which provides a wider kerf than the blade thickness. High-quality saw blades are *taper ground.* **See Figure 11-2.** The blade back is thinner than the blade at the cutting edge, requiring less set in the teeth.

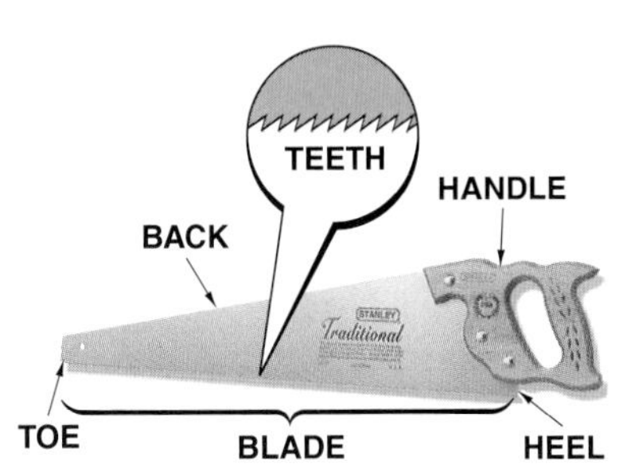

The Stanley Works

Figure 11-1. *Although the basic construction of handsaws is similar, many differences are found in the length and shape of the blade and the number and shape of the teeth.*

A saw usually has a number printed on its blade indicating the *number of teeth* or *points* per inch. **See Figure 11-3.** The lower the number, the larger the teeth, and the rougher the cut that is made. For example, an 8-point saw has larger teeth than an 11-point saw and produces a rougher cut than an 11-point saw.

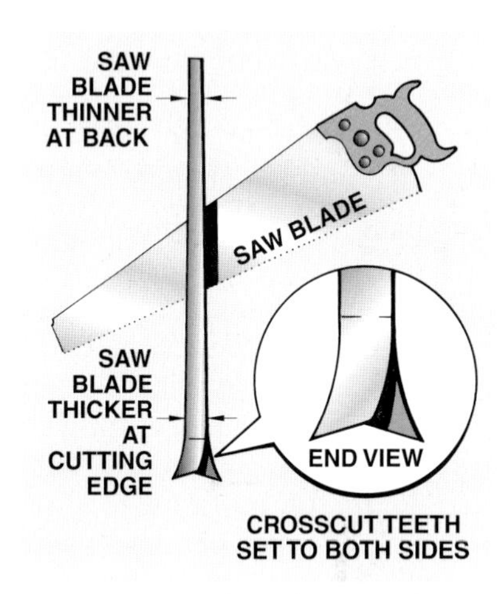

Figure 11-2. *High-quality saws have a taper-ground blade. The blade is thinner along its back than at its cutting edge.*

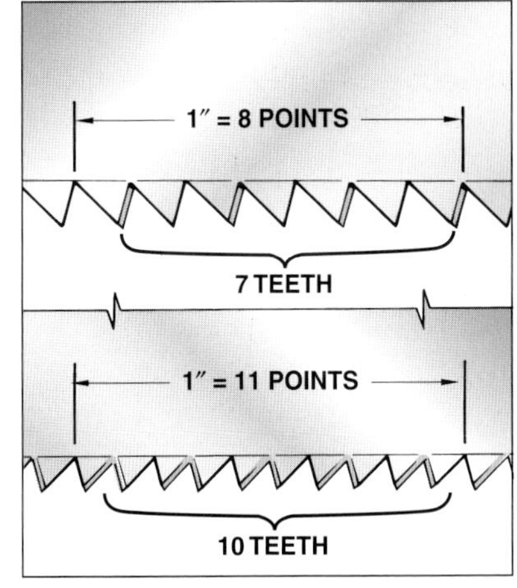

Figure 11-3. *An 8-point saw has larger teeth and fewer teeth per inch than an 11-point saw.*

When using a handsaw, position the piece being sawn face up to avoid splintering on the face. The back of the workpiece splinters due to the cutting action of the blade.

Crosscut Saws

Most saws used in carpentry are crosscut saws. The teeth of *crosscut saws* are shaped like knives to cut across the wood fibers and grain of the wood. **See Figure 11-4.** Crosscut saws should be held at a 45° angle to the work surface to cut efficiently.

The most popular type of crosscut saw for rough work has a 26″ blade with 8 points per inch. Crosscut saws for finish work usually have a shorter blade, such as 20″ or 22″, with 10 or 12 points per inch.

Compass Saws. A *compass saw* is used to saw curves, holes, and other internal openings. **See Figure 11-5.** A compass saw has a 12″ or 14″ blade with 8 or 10 points per inch. When cutting an internal opening, a hole is first drilled in the material to start the compass saw. A compass saw can also be used to start saw cuts in tight spaces where a regular saw will not fit.

Figure 11-5. *Compass saws are used to cut curves or internal openings.*

Several advancements in handsaws have been made that provide greater user productivity. Handsaws are now available with aggressive triple-bevel teeth, which cut on both the pull and push stroke. Induction-hardened compact-blade handsaws are also available for fast crosscutting in confined spaces.

Crosscut Saw Cutting Action

Figure 11-4

KERF

BOTTOM VIEW OF TEETH

FRONT VIEW CUTTING ACTION

SIDE VIEW CUTTING ACTION

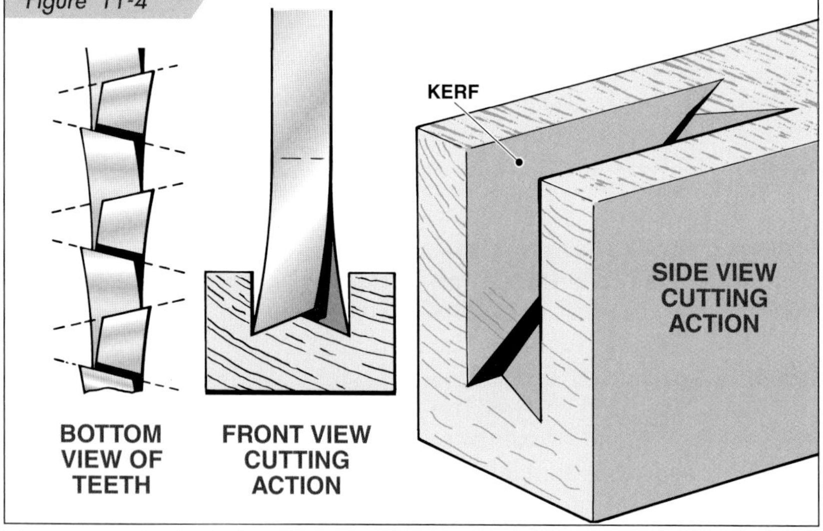

Figure 11-4. *The knife-shaped teeth of a crosscut saw are effective for cutting across the grain.*

Keyhole Saws. A *keyhole saw* is similar in appearance to a compass saw, but it has a narrower and shorter blade and its teeth are finer. Keyhole saws are used to make curved cuts in areas too small for a compass saw to be used in.

Backsaws. A *backsaw* is used with a miter box to make fine cuts in finish work. **See Figure 11-6.** The blade of a backsaw is 10″ to 26″ long and 3¼″ to 6″ wide, and has 10 to 14 points per inch. The back of a backsaw is reinforced to stiffen the blade for accurate straight cuts. A large version of a backsaw is a *miter saw.*

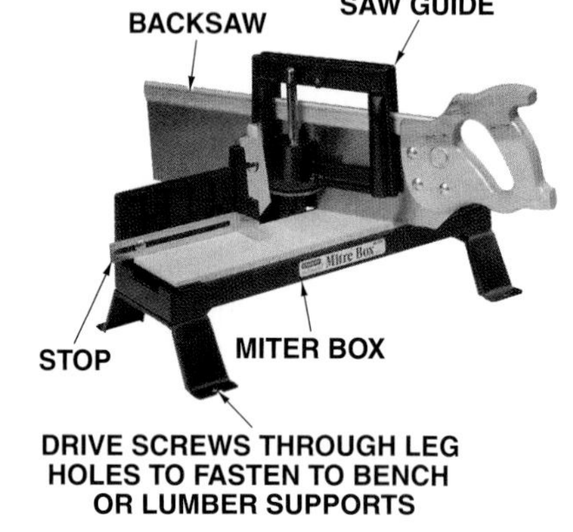

Figure 11-6. *Backsaws may be used with a miter box to cut accurate angled or straight cuts in finish work.*

A *miter box* may be used with a backsaw or miter saw to guide the saw for accurate angled cuts. A miter box can be set to the desired angle and locked in position. A miter box has built-in stops to cut accurate 90°, 67½°, 60°, 45°, and 22½° angles.

Dovetail Saws. A *dovetail saw* has a reinforced back, round handle, and narrow blade. **See Figure 11-7.** Dovetail saws are used to make fine cuts in molding and other small trim materials. The type of dovetail saw used most often has a 2″ × 10″ blade, with 15 points per inch.

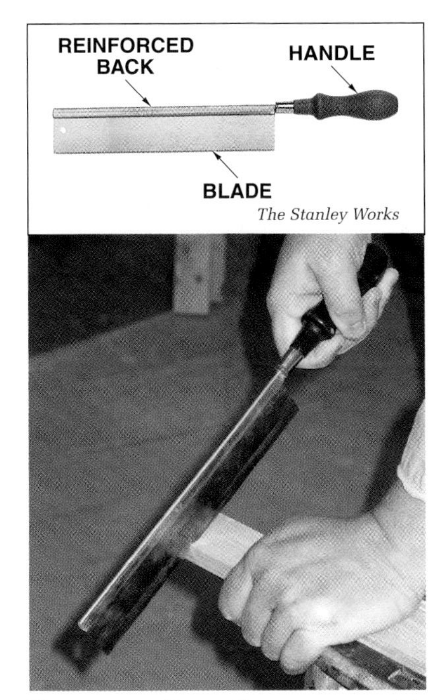

Figure 11-7. *Dovetail saws are used to make fine cuts in molding or other small trim materials.*

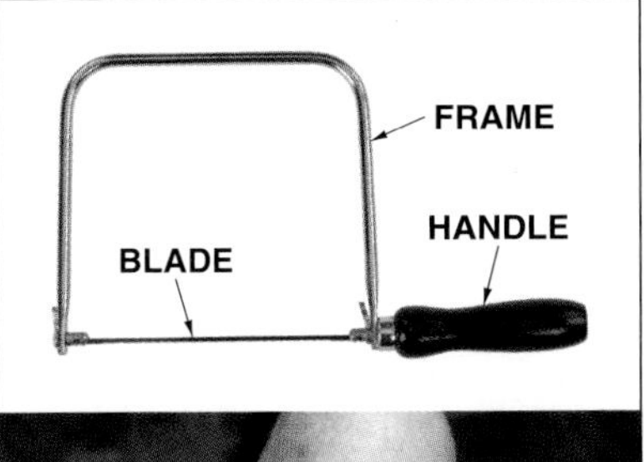

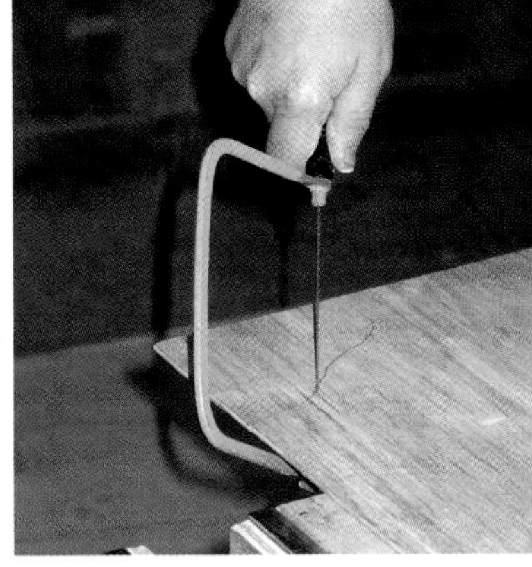

Figure 11-8. *Coping saws are used to make fine, irregular cuts in thin materials.*

Coping Saws. A *coping saw* is used to cut curves and irregular lines in thin material. **See Figure 11-8.** Coping saws are frequently used to cut coped joints when fitting the inside corners of molding. The coping saw preferred by most carpenters has a 1/8″ × 6 3/8″ blade. The blade can be adjusted to make angled cuts easier.

Ripsaws

Ripsaws are designed to cut with the wood grain. Ripsaw teeth are shaped like chisels, to be effective for cutting with the wood fibers. **See Figure 11-9.** Most ripsaws have a 26″ blade with 5½ points per inch. A ripsaw should be held at a 60° angle to the work surface for efficient cutting action. Sawing with the grain takes more time and effort than sawing across the grain.

If the kerf closes behind the cut when ripping a long board, use a wood shim or nail to keep the kerf open.

Sawing Technique

The proper sawing technique ensures quick, clean, and accurate cuts in wood or other material with minimal effort. **See Figure 11-10.** In addition, the proper sawing technique prevents hand lacerations from the saw teeth. Use the following technique when cutting with a crosscut saw or ripsaw:

1. Place the saw teeth along the waste side of the line. Press your thumb or side of hand lightly against the blade when starting the cut.
2. Pull back on the saw, holding up on the handle so the blade moves lightly back and forth until an initial saw kerf has been made.
3. Move the nonsawing hand a safe distance away from the blade and continue cutting. Adjust the cutting angle for crosscut saws and ripsaws to 45° and 60°, respectively.

Hand Sawing Techniques
Media Clip

Use and Care of Handsaws

Cuts are the most frequent accidents caused by sawing tools. Proper use and care of handsaws is essential to injury-free operation. Consider the following items when using and storing handsaws:

- Handsaws must be kept sharp. Dull saws cut poorly and can cause injuries.

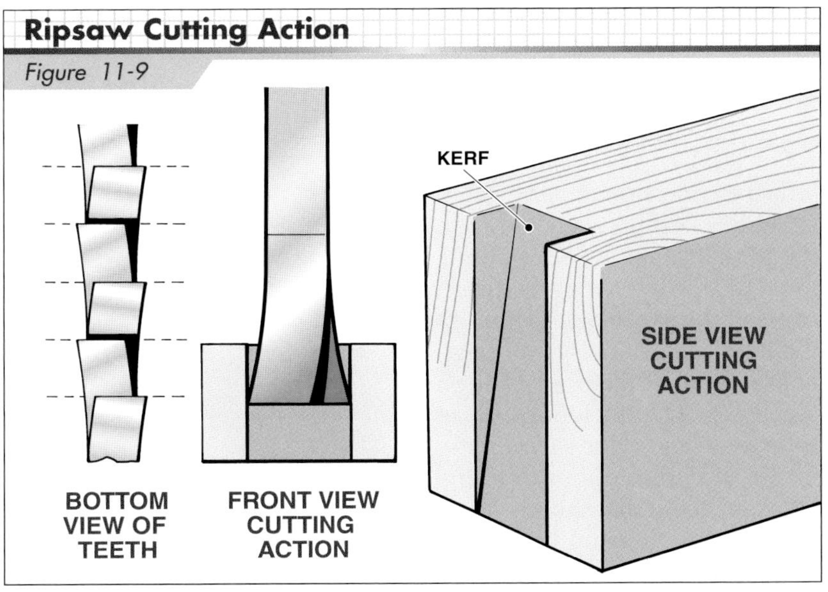

Figure 11-9. *The chisel-shaped teeth of ripsaws are effective for cutting with the grain.*

Hand Sawing

Figure 11-10

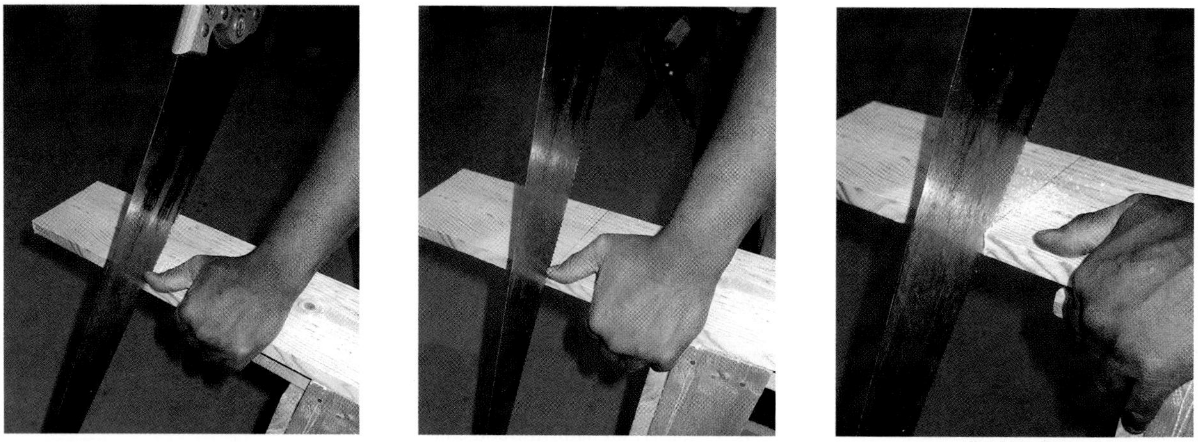

1. Align saw teeth along line and press thumb against blade.
2. Pull back on saw, holding up on handle.
3. Move nonsawing hand away from kerf and continue cut.

Figure 11-10. *The proper sawing technique results in accurate cuts while minimizing accidents.*

- Wipe a thin film of oil on a saw blade at the end of a work day to prevent the blade from rusting and to prolong tool life.
- Do not *ride* (dig in with) the blade when sawing. A sharp blade will cut quickly and accurately with very little pressure.
- Store handsaws in a toolbox so the teeth are protected from contact with other metal objects. Many carpenters fasten a slotted hardwood saw-block at the bottom and to one side of the toolbox. Handsaws can then be placed in the slots of the sawblock.

The teeth of a handsaw are *set* (offset to one side or another) to provide a kerf that is wider than the blade thickness. Over time, the set of the teeth may decrease, causing the saw blade to bind in the kerf. Occasionally, the original set must be restored. Using a *saw set,* each cutter tooth is offset from the plane of the saw slightly in the direction opposite to the adjacent teeth. The tooth offset must be the same on both sides of the saw or the saw will go off course during use.

Metal-Cutting Saws

Two metal-cutting saws are the *hacksaw* and *nail saw.* A hacksaw is used to cut metal materials such as framing members and door thresholds. **See Figure 11-11.** A nail saw is used to cut nails holding together framing members. **See Figure 11-12.** The nail saw blade fits into a metal or plastic handle similar to that of a compass saw.

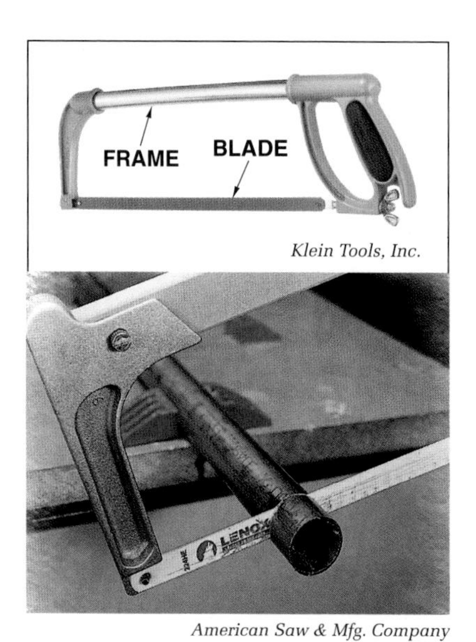

Klein Tools, Inc.

American Saw & Mfg. Company

Figure 11-11. *Hacksaws are used to cut metals.*

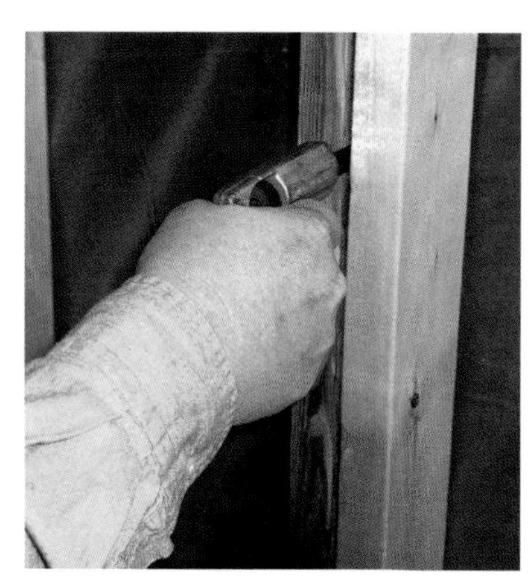

Figure 11-12. *Nail saws are used to cut nails so that framing members can be more easily separated.*

OTHER CUTTING TOOLS

In addition to handsaws, other cutting tools used by carpenters include chisels, knives, and specialty tools for cutting sheet metal and wire. Each tool should be used as designed for safe and efficient operation.

Wood Chisels

Wood chisels are hand tools used for rapid removal of waste stock. **See Figure 11-13.** Some chisels, such as *flooring chisels,* are designed for rough work. Other chisels are designed for finish work.

A *butt chisel* is a wood chisel used for mortising door hinges, flush bolts, and other kinds of finish hardware. Butt chisels have a plastic or wood handle that holds the blade. The end of a chisel handle is typically protected by a steel cap, which receives the direct hammer blow.

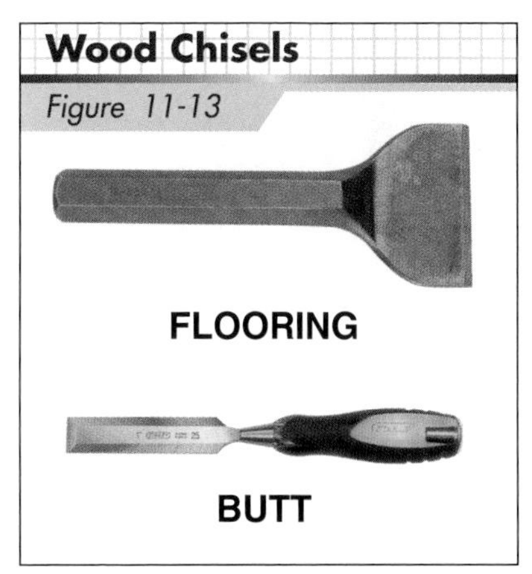

Figure 11-13. *Wood chisels include flooring chisels and butt chisels. A flooring chisel is used for rough work. A butt chisel is used to mortise lumber for door hinges and other types of finish hardware.*

Chisel blades are available in 1⁄8″ to 2″ widths and 3″ to 6″ lengths. Finish carpenters typically carry a wide selection of chisel sizes for a variety of applications. Chisels are often stored in a plastic roll to protect the cutting edges.

Use and Care of Wood Chisels. Wood chisels must be kept sharp. A dull chisel requires greater effort to use, results in poor work, and provides greater potential for injury. Consider the following safety precautions when using wood chisels:

- Carry wood chisels by the handle with the blade pointing downward. Do not carry chisels in pockets.
- Remove wood chisels from service if handles are loose or cracked.
- A wood chisel should only be used for its designed purpose. Do not use a chisel as a wedge or pry bar.
- When using a wood chisel, always keep both hands behind the cutting edge of the chisel. Always cut away from your body. **See Figure 11-14.**

Figure 11-14. *When using a wood chisel, keep both hands behind the cutting action of the chisel, and cut away from the body.*

Cold Chisels

Cold chisels are forged from special hardened and tempered alloy steel, and are used to cut through nails and other metals, stucco, and plaster, and to chip concrete. Cold chisels are available in 1⁄4″ to 1″ widths and 6″ to 12″ lengths. **See Figure 11-15.**

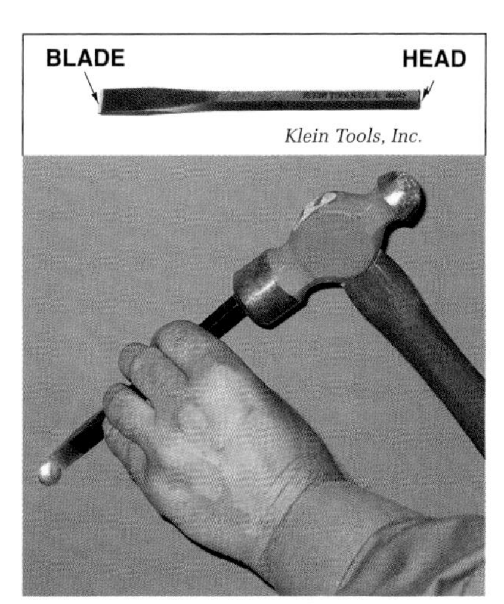

Figure 11-15. *Cold chisels are used to cut metals and chip concrete. Cold chisel heads must be properly ground to prevent chips from becoming detached from a mushroomed head.*

Wear protective goggles when working with cold chisels. Cold chisels may develop mushroomed heads during use. Steel chips from a mushroomed head create a safety hazard. Rough edges should be ground down as soon as mushrooming develops.

Knives

Carpenters use knives for cutting drywall, insulation, and many other construction materials. There are many different types of knives and each has its purpose. **See Figure 11-16.** One of the most popular knives is the *utility knife*. Utility knives can have either retractable or non-retractable blades. A retractable utility knife is a frequently used tool and is usually kept in a carpenter's tool pouch. For safety, the blade is retracted into the knife body when it is not being used. Additional blades are stored in the handle.

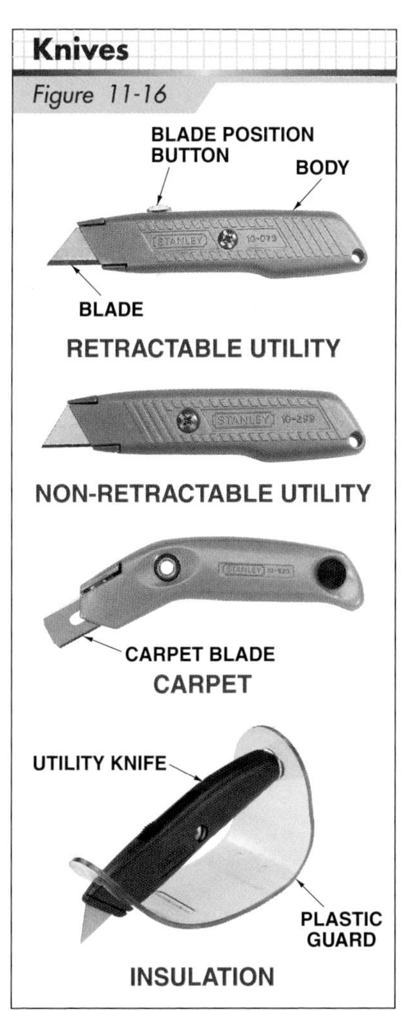

Figure 11-16. *Utility knives are used to cut materials such as gypsum board, fiberboard, and insulation.*

A non-retractable utility knife is just like a retractable utility knife except the blade cannot be retracted into the knife body. Non-retractable utility knives are often used for cutting drywall because they allow for a more accurate cut when scoring the paper. However, care must be taken when using a non-retractable utility knife since the blade is always exposed.

A carpet knife is a knife primarily used for cutting carpet and carpet pad. The carpet knife has a retractable blade and extra blades can be stored in the body. It also has an angled handle allowing it to be easily used in hard-to-reach areas. Carpet knives use slotted, double-edged blades that are very thin and extremely sharp.

An insulation knife is used to cut fiberglass or rock-wool batt insulation. A standard utility knife is used with a transparent shield. The transparent shield is used to compress the insulation and protect the user's hand from the abrasive material.

Some utility knives may use blades other than straight blades. Hook blades, which are used to cut roofing material such as shingles and underlayment, are used in retractable utility knives. Scoring blades, which are used to score materials such as laminate and Plexiglass®, and linoleum blades, which are used to cut sheet flooring, are used in non-retractable utility knives. **See Figure 11-17.**

Before cutting with a utility knife, adjust the blade so it barely clears through the opposite side of the material being cut. When using a utility knife, keep the free hand out of the way of the knife blade.

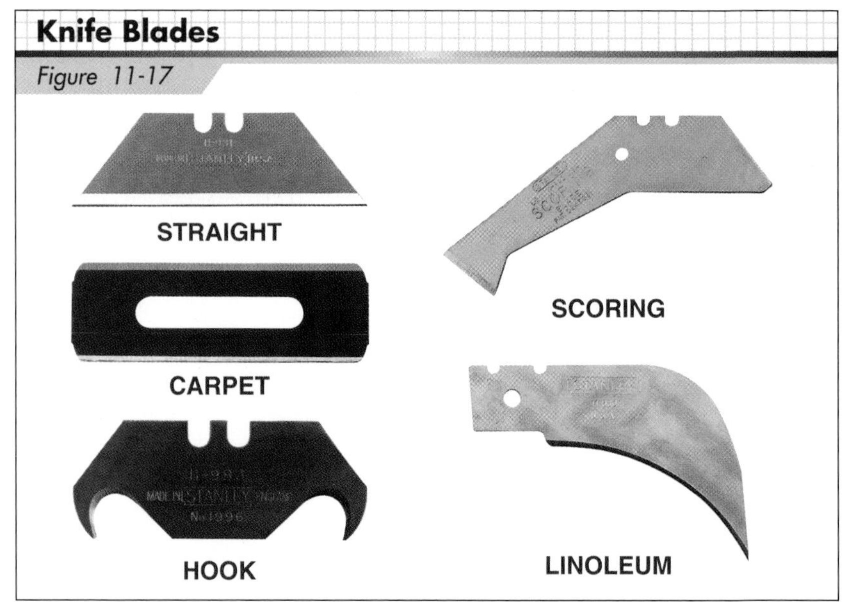

Figure 11-17. *Some utility knives may use blades other than straight blades.*

Sheet-Metal Snips

Sheet-metal snips, or *tin snips,* are available in a variety of sizes and designs. Sheet-metal snips are used to make straight cuts in metal framing members, screening, flashing, or other lightweight metal.

Aviation snips are used to make straight or angled cuts in light-gauge sheet metal, and are available in a 10½″ length. The compound action of aviation snips results in maximum cutting power with minimal effort. **See Figure 11-18.**

Aviation snips have color-coded handles to identify the direction of cut the snips are designed for. Yellow handles indicate a straight-cutting snip, red handles indicate a left-curve cutting snip, and green handles indicate a right-curve cutting snip.

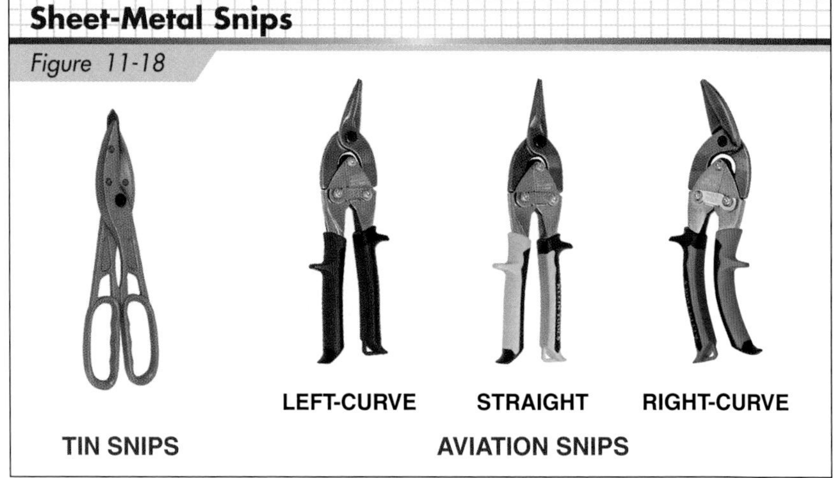

Figure 11-18. *Sheet metal snips are available in variety of sizes and designs.*

Unit 11 Review and Resources

UNIT 12

Clamping and Boring Tools

OBJECTIVES

1. Identify common clamping tools.
2. Identify common vises.
3. Describe the functions of boring tools.

Clamping tools, such as clamps and vises, securely hold and support materials being worked on or fastened together. Clamps and vises are used in rough and finish carpentry work. Clamps and vises are available in a variety of designs and sizes to perform a variety of tasks. A well-designed clamp or vise is easy to tighten and adjust, and will hold items securely without marring the surface.

Boring tools include ratchet braces and hand drills, which can be fitted with a variety of bits for performing a wide range of tasks.

Clamps

Figure 12-1

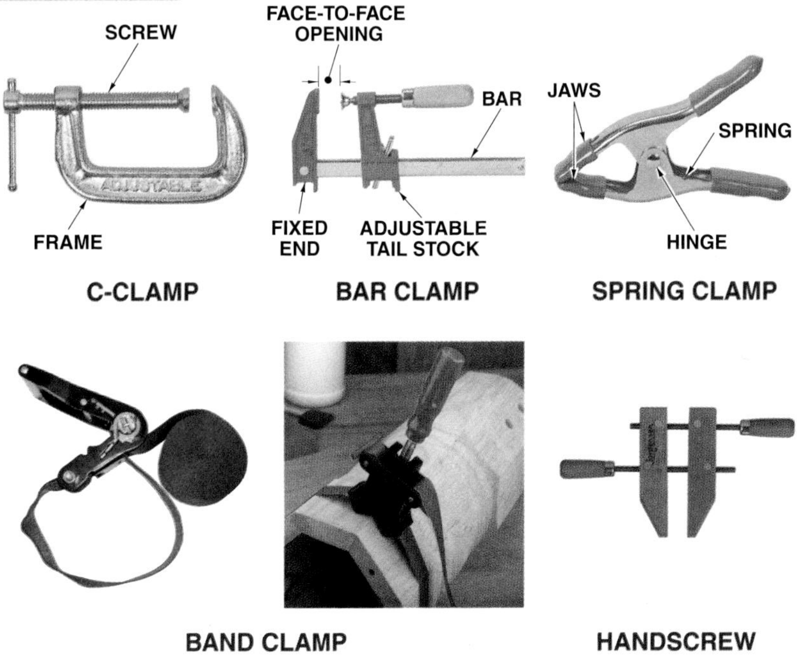

SCREW
FRAME
C-CLAMP
FACE-TO-FACE OPENING
BAR
FIXED END
ADJUSTABLE TAIL STOCK
BAR CLAMP
JAWS
SPRING
HINGE
SPRING CLAMP
BAND CLAMP
HANDSCREW

Adjustable Clamp Company

Figure 12-1. *Clamps are used to hold or secure pieces together to prevent movement or separation.*

CLAMPS

A *clamp* is a fastening tool used to hold or secure pieces together to prevent movement or separation. Clamps hold and apply pressure to materials while they are being worked on while glue or other adhesives set. Some of the more commonly used clamps are shown in **Figure 12-1.**

C-clamps are general-purpose clamps used for a variety of carpentry and welding tasks. When using C-clamps to hold together finish materials, protective pieces such as wood scraps should be used to prevent indentation. *Bar clamps* consist of a solid bar (typically aluminum or steel) with a fixed head at one end and an adjustable tail stop. **See Figure 12-2.**

Figure 12-2. *Bar clamps should be selected based on the size of the workpieces being joined.*

The "fixed" heads may also have an adjusting screw to make minor adjustments to the clamp. Bar clamps are available in many sizes with the face-to-face opening ranging from 6″ to 6′-0″. *Pipe clamps* are similar to bar clamps, but the solid bar is replaced with a pipe. Rubber pad protectors are commonly installed on the jaws of bar and pipe clamps to protect the surface being clamped from being marred.

Spring clamps consist of two hinged jaws and handles that are held in tension with a spring. Spring clamps apply moderate pressure to localized areas. *Band clamps* consist of a web belt that is adjusted in length using a crank- or ratchet-type handle. Band clamps are used for a variety of applications including cabinetmaking and SIP construction. Handscrews are constructed of two maple jaws and handles with connecting screws. *Handscrews* are capable of applying even pressure along tapered pieces of stock.

Specialty clamps, such as corner clamps and three-way edging clamps, are also available. **See Figure 12-3.** *Corner clamps* are lightweight clamps used for the assembly and glue-up of lightweight parts being joined at a 90° angle. *Three-way edging clamps* consist of three screws and are commonly used for securing countertop edging or edge veneer. *Clam clamps* are specialty clamps used to hold mitered pieces, such as door casing, in position until glue or other adhesive sets. Four pins along each side of the clamp engage the outer edges of the casing and exert pressure on the joint as the handle is turned. A perfect and tight 90° joint is formed. The pins leave small indentations in the casing edges, which can be filled. For stain-grade casing, two of the pins on each side can be removed, thus minimizing the indentations.

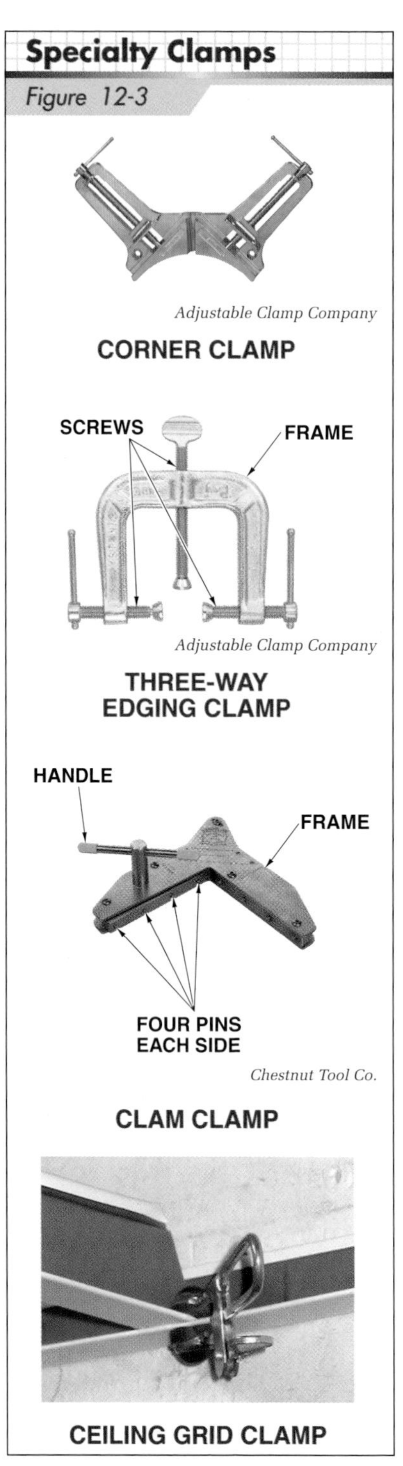

Figure 12-3. *Specialty clamps may be employed to perform certain operations more easily.*

Use the proper size clamp for the clamping load. Large clamps on small workpieces exert bending strain on the screw and frame.

Ceiling grid clamps are compact specialty clamps used to temporarily secure suspended ceiling members together. Ceiling grid clamps have rubber-padded jaws to ensure that the ceiling members are not marred or otherwise damaged. An integral cam is used to apply pressure to the ceiling members to secure them in position.

Trim clamps are a type of spring clamp with sharp points instead of jaws. Trim clamps are used to secure miter joints in position while the joint adhesive sets.

Use and Care of Clamping Tools

As with any other tool, proper use and care of clamping tools is a key to proper operation. Basic tips for the proper and safe use of clamping tools are as follows:

- Do not use a wrench, pipe, hammer, or pliers to tighten a clamp.
- Select the proper size clamp for the work being performed.
- Discard any clamp if its frame, screw, or spindle is bent or otherwise defective. Tremendous pressure is exerted on the clamp when securing items and may result in damage to the clamp.
- Do not use a C-clamp for hoisting operations.
- Store clamps by hanging them in racks.

VISES

Vises are used to hold or secure pieces together to prevent movement or separation, but they are typically attached to a solid base such as a workbench. Light-duty vises can be secured to the ends of sawhorses. Since a vise securely holds the workpiece, a worker can use both hands to perform the intended operation with hand or power tools. Vises typically have one fixed jaw and a movable parallel jaw that is adjusted using a screw mechanism. Common vises are shown in **Figure 12-4.**

Vises

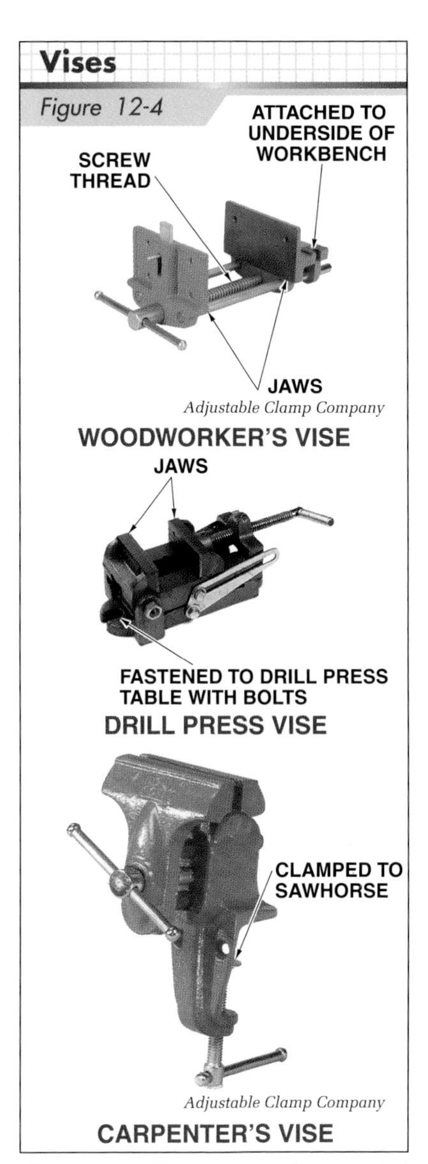

Figure 12-4. *Vises are usually mounted on or attached to a solid base such as a workbench or sawhorse.*

Woodworker's vises are attached to the underside of a workbench. A screw thread extending the length of the vise is used to adjust the distance between the faces. Some woodworker's vises have quick-release levers that bypass the screw thread to make fast adjustments. The faces of woodworker's vises are typically covered with wood to prevent marring of the workpiece.

Drill press vises allow accurate drill press work by securely clamping the stock to the vise. Slots along the edges of the vise allow it to be fastened to the drill press table. Similar to corner clamps, *miter vises* secure pieces together at a 90° angle to one another; however, a miter vise is fastened to a workbench. Some models of miter vises pivot 360° to allow irregularly shaped pieces to be properly secured. A *carpenter's vise* can be temporarily clamped to a sawhorse or piece of stock up to 2″ thick. Carpenter's vises are convenient to use for medium-duty operations on a job site.

Use and Care of Vises

As with clamping tools, proper use and care of vises is a key to proper operation. Basic tips for the proper and safe use of vises are as follows:

- Ensure a vise is properly fastened to a workbench or sawhorse before use.
- Select a vise of proper size and capacity to hold the object being worked on.
- Do not weld the base of a vise to a metal object such as a metal bench.
- Do not use an extension pipe to tighten the jaws or exert additional pressure on the jaws.
- When using a swivel-base vise, hand tighten the locknut handles at the sides of the vise. Do not use a wrench to tighten the locknut handles.
- Do not move the movable jaw beyond the maximum specified opening of the vise.
- Always wear proper eye and face protection when using striking tools or power tools with a vise.

BORING TOOLS

Hand boring tools, including ratchet braces and hand drills, may be used in interior finish work. **See Figure 12-5.** Although power tools and equipment have largely replaced manual boring tools on the job site, certain tasks may be better suited to hand tools.

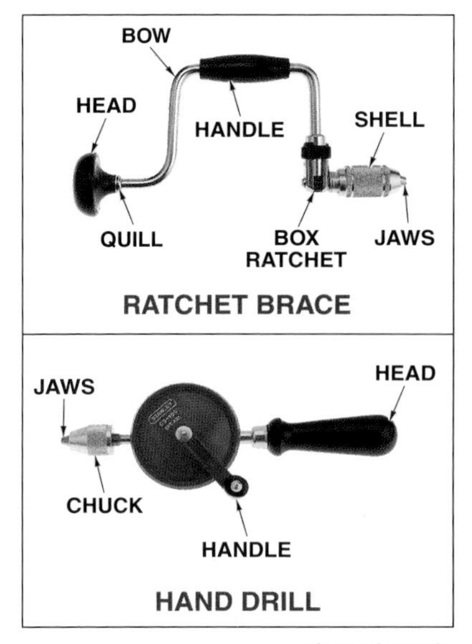

The Stanley Works

Figure 12-5. *A ratchet brace or hand drill can be used to manually drill holes.*

Ratchet Braces

A *ratchet brace* is used with auger bits and other specialized bits to bore holes in wood. The sweep of a ratchet brace is the diameter of the circle made when the handle is turned. A brace with a 10″ or 12″ sweep is used for general work.

The box ratchet is adjusted to operate the brace in a clockwise or counterclockwise direction, permitting holes to be bored close to walls or corners where it is not possible to make a complete turn of the handle.

Hand Drills

Hand drills use the same type of high-speed steel bits used with portable electric drills. Hand drills are typically used to drill holes that are 1/4″ diameter or less, although some have a 3/8″ chuck capacity.

Unit 12 Review and Resources

UNIT 13

Smoothing Tools

OBJECTIVES

1. List and describe common bench planes.
2. Describe the proper use and care of bench planes.
3. Describe common scrapers.
4. Describe functions of rasps and serrated-blade forming tools.

Most hand tools used to cut, saw, or bore lumber leave a rough and uneven surface when the operation is complete. Smoothing tools are used to smooth and flatten the uneven surfaces. Smoothing tools include bench planes, scrapers, rasps, and serrated-blade forming tools.

BENCH PLANES

Bench planes are used for smoothing and jointing lumber. *Jointing* is the process of truing the edges of boards before they are fitted (joined) together.

Many types of bench planes are available. The bench plane selected for a particular operation is often based on personal preference, since planes of more than one type may be equally effective. The primary parts of a bench plane are the *knob, lever cap, cam, plane iron,* and *handle.* **See Figure 13-1.** Since the basic construction of most bench planes is similar, the major difference is size.

Jointer and Fore Planes

Jointer and fore planes are longer and heavier than other planes and are used for jointing long boards and fitting doors. **See Figure 13-2.** *Jointer planes* range from 20″ to 24″ long with 2⅜″ to 2⅝″ wide plane irons. *Fore planes* are shorter versions of jointer planes and are 18″ long with 2⅜″ wide plane irons. The bottoms of jointer and fore planes may be smooth or grooved. Fore planes are preferred by carpenters since they are easier to handle and provide similar results to jointer planes.

Jack Planes

A *jack plane* is a general-purpose tool used for smoothing and fitting mating surfaces. Jack planes are commonly used for fitting doors. The most common size of jack plane is 14″ long with a 2″ wide plane iron.

The Stanley Works

Figure 13-2. *Jointer planes are 20″ to 24″ long and are used to joint long boards and fit doors.*

Smooth Planes

A *smooth plane,* or smoothing plane, is a general-purpose bench plane shorter than the jack plane. **See Figure 13-3.** Smooth planes are available in 9¼″, 9½″, and 9¾″ lengths with 1¾″ or 2″ wide plane irons. Smooth planes are useful for smoothing surfaces but are not as effective as larger planes for straightening surfaces.

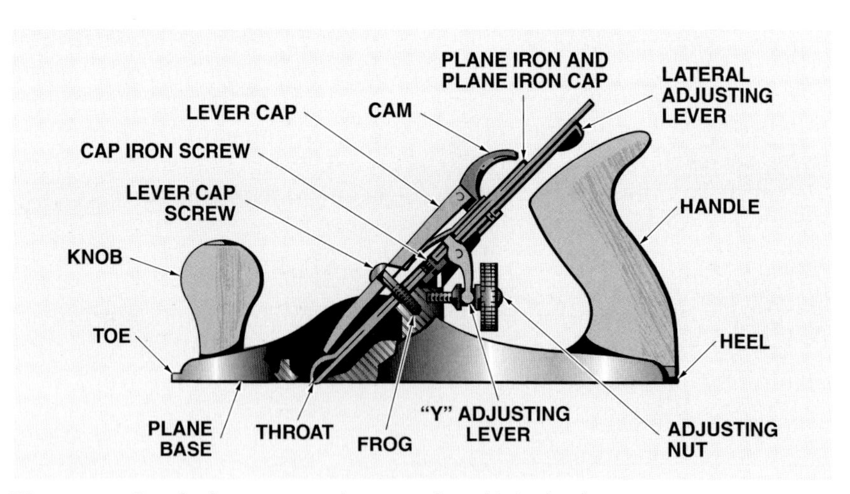

Figure 13-1. *Bench planes are used to smooth and joint lumber.*

The Stanley Works

Figure 13-3. *Smooth planes are general-purpose planes that are 9¼″, 9½″, or 9¾″ long.*

Block Planes

Block planes are used on small, narrow surfaces. A block plane is more simply constructed than other types of planes. Many block planes are adjusted for the depth of cut by properly positioning the plane iron in the plane mouth. **See Figure 13-4.** Block planes are 6″ or 7″ long with 1⅜″ or 1⅝″ wide plane irons. Plane irons of block planes rest at a 12° to 21° angle. Lower angles produce cleaner cuts across the end grain of lumber and plywood edges.

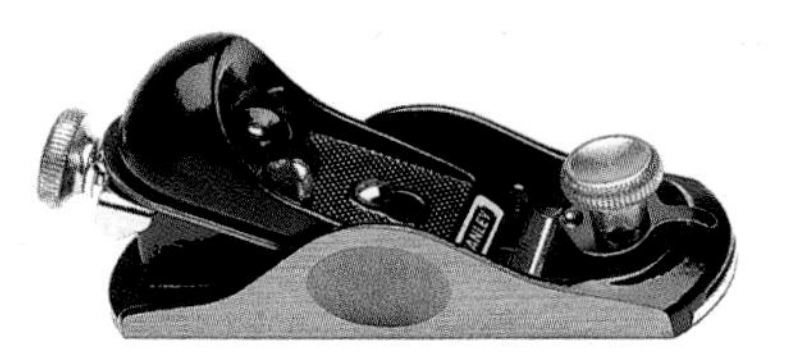

The Stanley Works

Figure 13-4. *Block planes are 6″ or 7″ long and are used on small, narrow surfaces.*

Rabbet Planes

Rabbet planes are used to make rabbet joints on the ends of boards. Two types of rabbet plane are the *bullnose* and *adjustable bullnose rabbet planes.* **See Figure 13-5.**

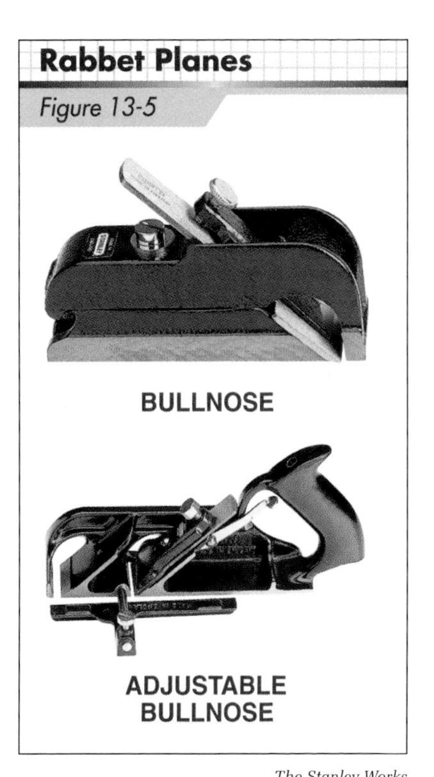

The Stanley Works

Figure 13-5. *A bullnose rabbet plane is used to reach into corners. An adjustable bullnose rabbet plane can be adjusted for regular or bullnose work.*

An adjustable throat on a plane allows the throat to be quickly opened or closed for fine or coarse work. Fine work requires the throat to be closed.

A bullnose rabbet plane is 4″ long, with a 1⅛″ wide plane iron. The plane iron is located toward the front of the plane for close work into corners.

An adjustable bullnose rabbet plane is 10″ long with a 1½″ wide plane iron. The plane iron is placed in the rear seat for regular work or in the front seat for bullnose work, such as cutting rabbets in corners. The fence on an adjustable bullnose rabbet plane is used to produce a rabbet parallel to the edge of the stock.

Use and Care of Bench Planes

The proper procedure for using a bench plane is shown in **Figure 13-6.** The effectiveness of a plane is determined by the condition and the sharpness of its plane iron. **See Figure 13-7.** If a plane iron is in good condition, it may require only touching up on an oilstone. If the cutting edge is nicked or the bevel has worn down, a bench grinder can restore the plane iron to its proper condition. **See Figure 13-8.** A plane iron should be pressed lightly to the wheel and dipped frequently in water to cool the steel and prevent it from burning. Since burning softens steel, a burned plane iron will not retain a sharp edge when it is used. A grinding attachment is used to properly position the plane iron on the adjustable tool rest.

Block planes are used on small, narrow surfaces. A fine shaving should be produced.

Exposed (nonpainted) parts of bench planes should be lightly oiled if the planes are to be stored for an extended period. Remove all dust and wood chips prior to oiling.

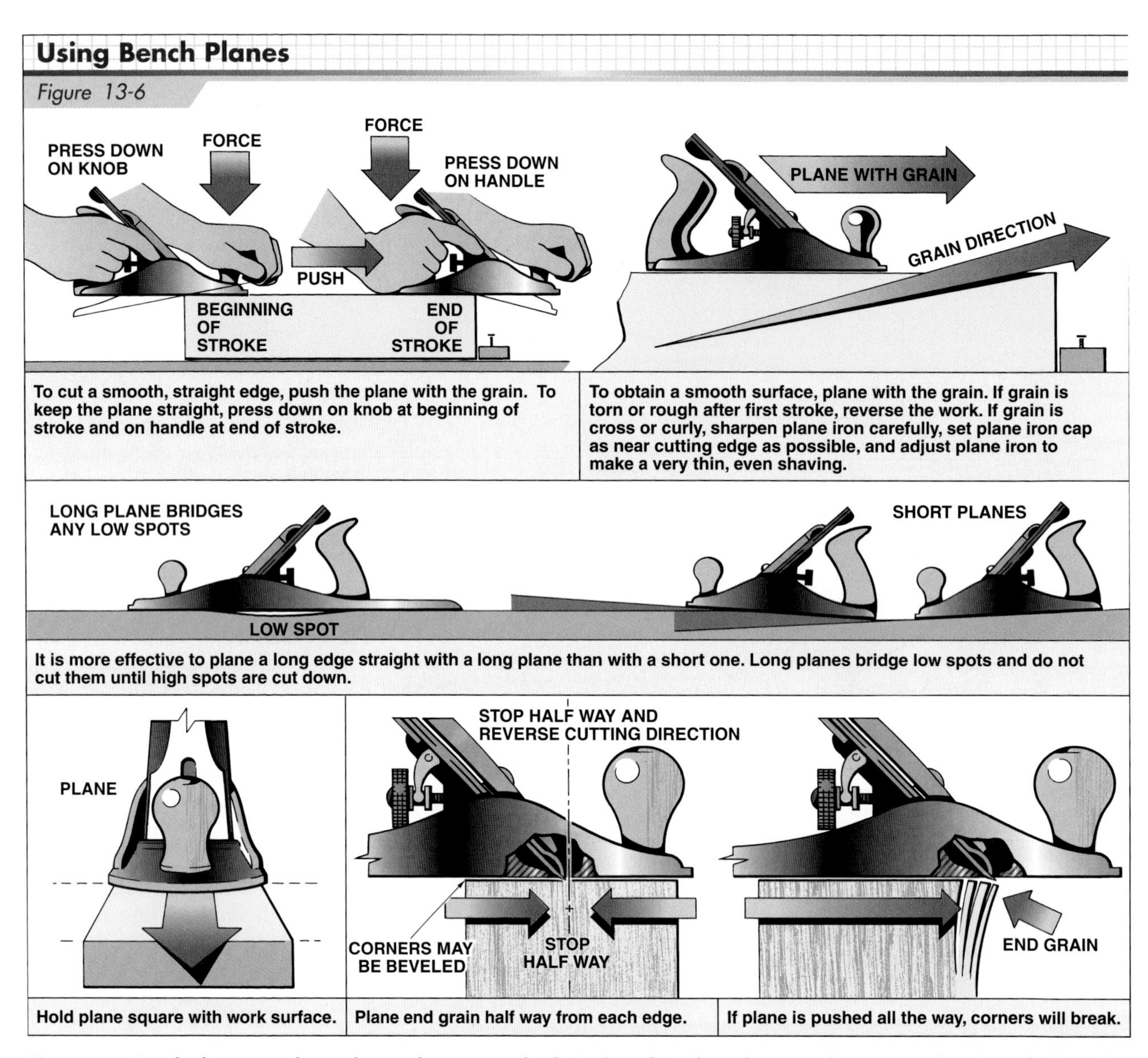

Figure 13-6. *Bench planes must be used properly to ensure the desired results. When planing end grain, a 45° bevel may be formed at the corner to eliminate the corners breaking if the plane is pushed all the way across the end grain.*

When joining pieces of stock, a bench plane may be initially used to remove excess wood prior to sanding.

After a plane iron is ground on a bench grinder, its fine edge is produced on an oilstone. **See Figure 13-9.** The plane iron is guided using a honing guide or by hand to ensure the edge is square to the plane iron. When guiding the plane iron, the hands should move parallel to the oilstone so the angle between the plane iron and oilstone remains constant throughout the stroke. Use enough oil on the oilstone to keep the surface moist. Oil prevents the steel particles from filling the pores of the oilstone.

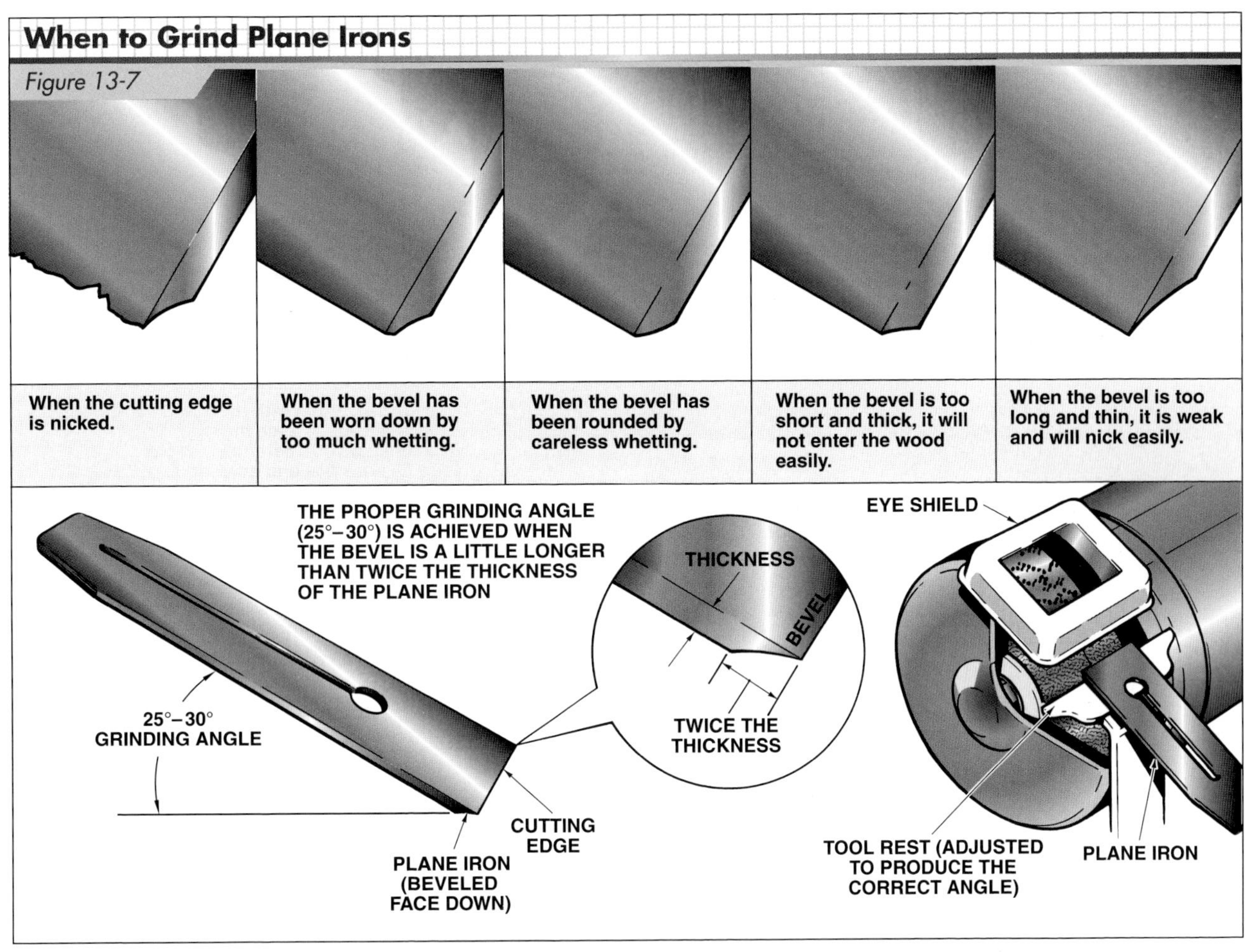

Figure 13-7. *A plane iron must be properly ground when it is dull or the cutting edge is deformed.*

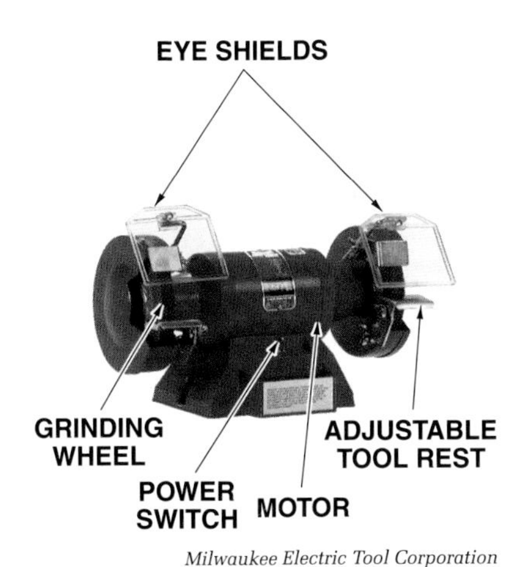

Milwaukee Electric Tool Corporation

Figure 13-8. *A bench grinder is used to restore a plane iron to its proper condition.*

When using a bench plane a fine shaving can only be produced with a sharp plane iron cutting edge, which is ground at the proper angle to allow the edge to slide smoothly across the wood.

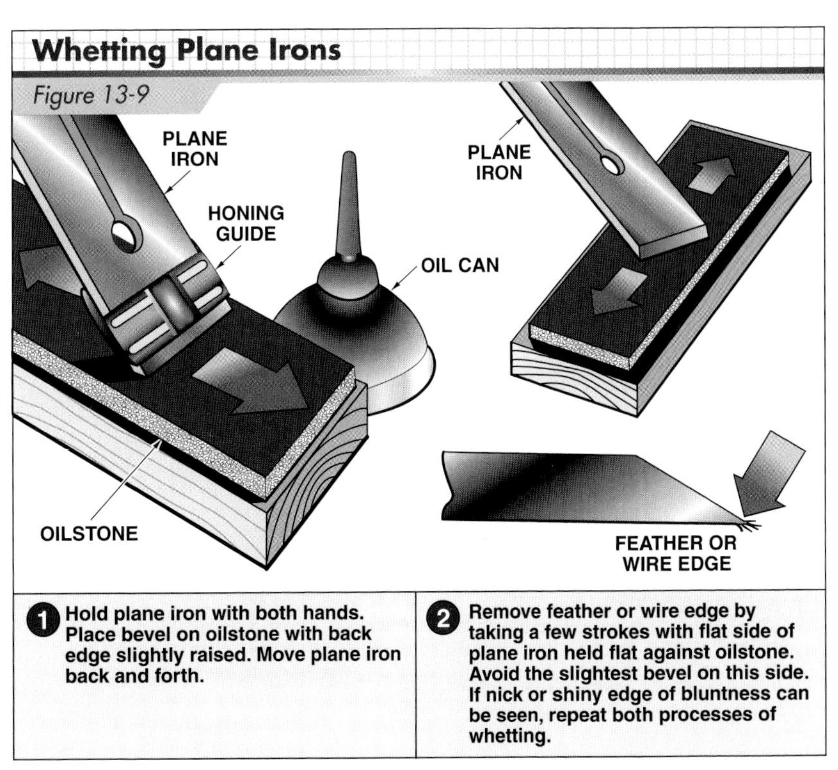

Figure 13-9. *Oilstones are used to produce a final, sharp cutting edge on plane irons.*

SCRAPERS

Scrapers are used to remove mill marks and blemishes, or to smooth grain that may have been torn during planing. A very fine finish can be produced with scrapers. Scrapers are also used to remove light scratches in surface veneers of doors and paneling.

A *hand scraper* is a thin piece of hardened steel with a slightly beveled edge. **See Figure 13-10.** The cutting action of a hand scraper is caused by a slight burr on the edge of the blade. The burr is produced using a burnishing tool, which turns the edge of the blade. A hand scraper should produce a fine shaving. Dust, instead of a shaving, indicates a dull scraper.

A *cabinet scraper* removes ridges that remain after planing and is used for final smoothing of a surface prior to sanding. A cabinet scraper is also used to smooth surfaces that are difficult to plane because of curly or irregular grain. **See Figure 13-11.** The cutting edge of a cabinet scraper blade has a greater angle than a hand scraper.

Light scratches in cabinet doors and drawer fronts may be removed prior to finishing using a cabinet scraper.

RASPS

Rasps are used for rapid removal of waste material. **See Figure 13-12.** The cutting action of the triangular teeth of a rasp produces a rough surface, which must be finished by another smoothing tool. Rasps may be rectangular or half-round in cross section. *Combination rasps* have coarse and fine teeth.

SERRATED-BLADE FORMING TOOLS

A *serrated-blade forming tool,* or Surform®, has hundreds of preset, sharpened steel teeth that act as tiny chisels. The holes between the teeth permit the shavings to pass through the blade to prevent clogging. **See Figure 13-13.** Serrated blades cannot be sharpened, so they must be replaced when dull.

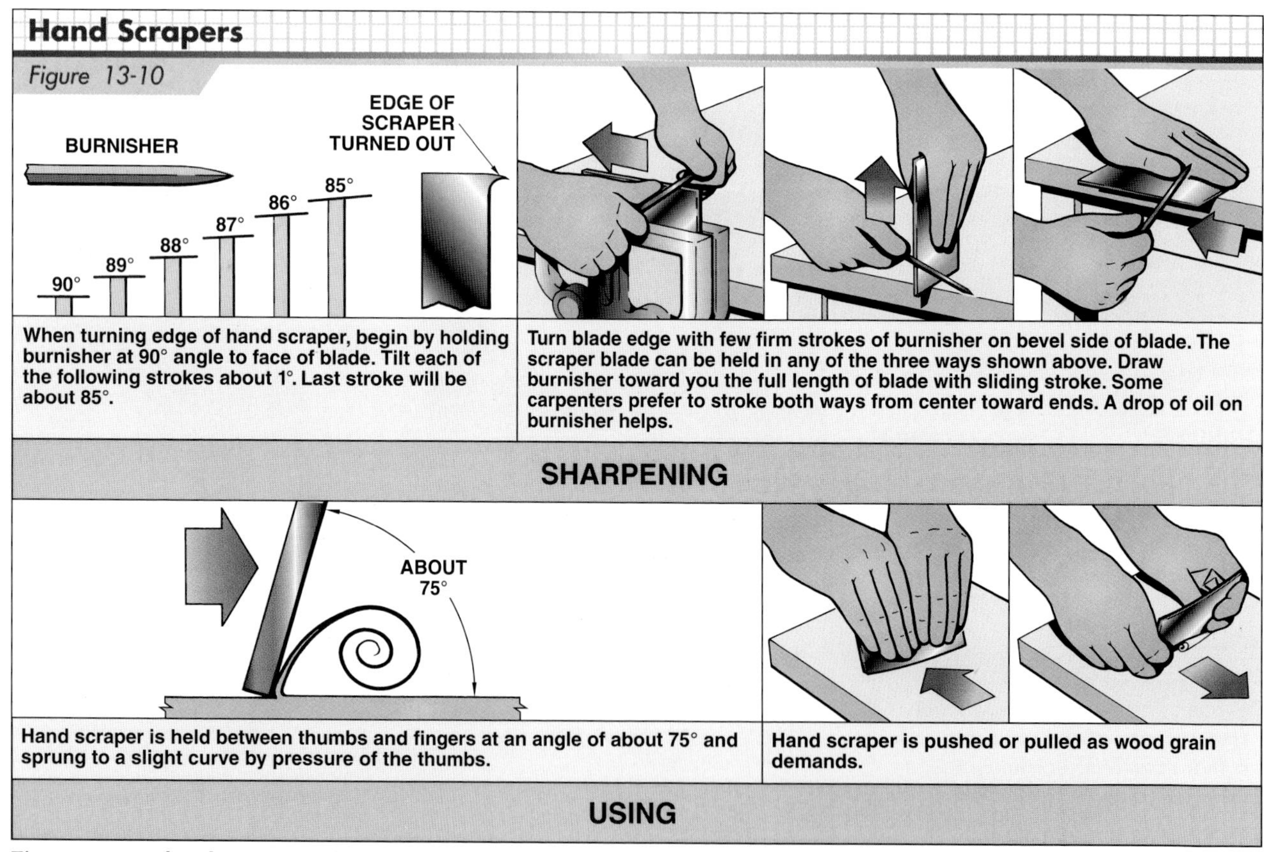

Figure 13-10. *A hand scraper produces a fine finish.*

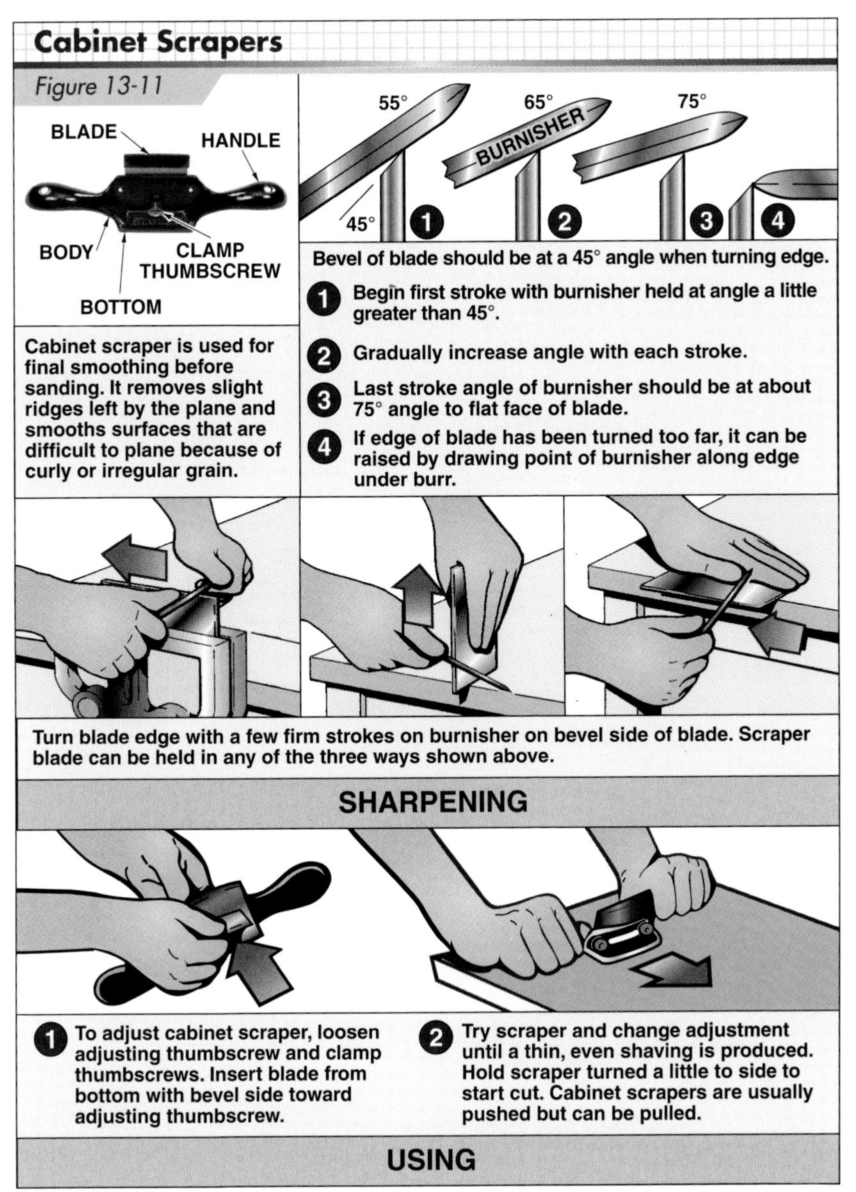

Figure 13-11. *The cutting action of a cabinet scraper is produced by a slight burr on the edge of the blade.*

Serrated-blade forming tools consist of a *body, handle,* and *serrated blade.* These tools can be adapted for rough cutting or for final smoothing by changing the angle at which they are held. **See Figure 13-14.** Serrated-blade forming tools are used on wood, plastic, vinyl, rubber, fiberglass, composition board, and soft metals such as aluminum.

Figure 13-12. *Rasps are used to rapidly remove waste material.*

Figure 13-13. *Serrated-blade forming tools have hundreds of sharp teeth with holes between them to permit the shavings to pass through the blade. This tool is specially designed for curved surfaces.*

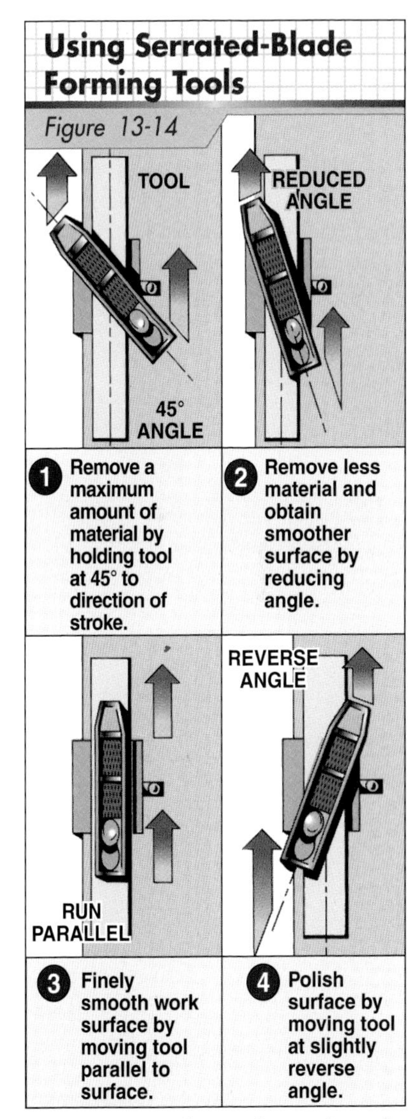

Figure 13-14. *The cutting action of a serrated-blade forming tool is determined by the angle of the tool.*

Unit 13 Review and Resources

SECTION

4

Power Tools

CONTENTS

UNIT 14

Portable Power Saws

1. Identify common safety precautions for power tools.
2. Explain grounding principles.
3. Identify common electrical safety precautions.
4. List and describe common circular saw blades.
5. Describe proper circular saw operation.
6. Identify common safety precautions for circular saws.
7. List and describe other types of portable power saws.
8. Identify common safety precautions for chain saws.

Safe work practices are required at a job site to minimize the potential for injury. The Occupational Safety and Health Administration (OSHA) requires all employers to provide a safe work environment for employees. A safe work environment is free of hazards and has safety practices in place to protect employees from injury or death. Employers are responsible for safety training and for ensuring that employees follow OSHA regulations. Safety meetings and/or toolbox talks should be frequently conducted to discuss current safety topics and address employee safety concerns.

POWER TOOL SAFETY

Power tools and equipment are powered by electric motors, pneumatic systems, or powder actuation. Tools and equipment should be used only for the purpose for which they are designed, and must be operated safely to reduce accidents. OSHA 29 CFR 1926.302, *Power-Operated Hand Tools,* and OSHA 29 CFR 1910.243, *Hand and Portable Power Tools and Other Hand-Held Equipment,* provide information on the safe operation of power tools used for construction. Power tool safety rules include the following:

- Inspect power tools for defects daily before initial use.
- Store power tools so as to protect them from damage, dampness, and dirt.
- Wear proper personal protective equipment (PPE), including eye, head, and/or hearing protection where required.
- Do not carry corded or cordless power tools with a finger on the power switch.
- Ensure the power switch is in the OFF position before connecting equipment to a power source.
- Ensure that all rotating parts of equipment are properly guarded.
- Do not allow hair to become entangled in moving parts.
- Avoid wearing loose clothing, ties, rings, or other items that could be caught by moving parts.
- Ensure all safety guards are in position before starting.

Electrical Safety

The greatest hazard related to electric tools is *electric shock,* which can cause serious injury or even death. Shock occurs when a defect in the electrical system, such as a loose or exposed wire, causes electrical current to pass through the tool housing.

Trus Joist, A Weyerhaeuser Business

Engineered wood products are easily cut using traditional portable power saws. A 90° saw guide designed for cutting wood I-joists is being used.

Electrical current seeks the easiest path to the ground, and the human body is a good conductor of electricity. Therefore, the current flows through the body to the ground.

Most electric power tools are available as *cordless* or *corded* models. Cordless tools permit a worker to perform operations in most locations without having the concerns associated with alternating-current (AC) equipment. The battery life of a cordless power tool must be considered when selecting a tool for an operation. Thick stock, pressure-treated lumber, and hardwood lumber require a considerable amount of power to cut and might best be cut or otherwise fabricated with corded tools.

Per OSHA 29 CFR 1926.404, *Wiring Design and Protection,* all electric corded tools, equipment, or devices must use either *ground fault circuit interrupter (GFCI)* outlets or acceptable means of *grounding* to ensure worker safety. A GFCI protects against electrical shock by detecting an imbalance of current in the normal conductor pathway and then quickly opening the circuit (in as little as 1⁄40 of a second). The most common type of GFCI outlet used on a job site is a portable GFCI. **See Figure 14-1.** Portable GFCIs should be inspected and tested before each use. GFCIs have a built-in test circuit to ensure that the ground fault protection is operational.

Extension cords are commonly used on a job site for corded tools and equipment. Worn or frayed extension cords must not be used for connection of electrical tools and equipment. In addition, extension cords must not be fastened with staples, hung from nails, or suspended by wire.

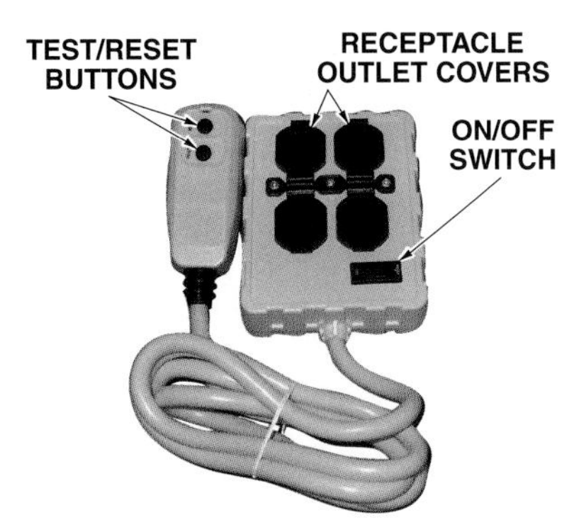

Tower Manufacturing Corporation

Figure 14-1. *A portable GFCI compares the amount of current in the hot or ungrounded conductor with the amount of current in the common or grounded conductor and immediately breaks the circuit if a difference greater than 5 mA exists.*

If GFCIs are not available, an acceptable means of grounding must be established to prevent shock. In order to be grounded, a power tool must have a three-conductor cord with a three-prong plug that fits into a grounded outlet. **See Figure 14-2.** A grounded outlet has a ground wire that is connected to a water pipe or a ground rod driven into the earth. If a fault occurs in the tool, the electrical current travels through the ground wire in the cord and then to the ground wire connected to the outlet. **Figure 14-3** shows a typical grounding system.

Double-insulated tools do not require a grounding system for safe operation. The electric motor components are covered by extra insulation that prevents current from reaching the surface of the tool and creating a safety hazard.

A properly grounded or double-insulated corded tool can still be dangerous under wet conditions since water is an electrical conductor. Special precautions should be taken when operating power tools near water or dampness. Electrical safety rules include the following:

- Periodically inspect electric cords for cuts, kinks, worn insulation, exposed wires, and any other defects.
- Place tool and extension cords so they do not present a tripping hazard. Do not expose cords to damage from mobile equipment or welding or burning operations.
- Carry a tool by its handle or housing, not by its cord. Do not remove a plug from a receptacle by pulling on its cord.
- When operating electric tools in damp locations cannot be avoided, use insulating platforms, rubber mats, and rubber gloves.
- Always disconnect electric corded tools when they are not in use.
- Keep the cord of a power tool safely away from the blade or other rotating part of the equipment.

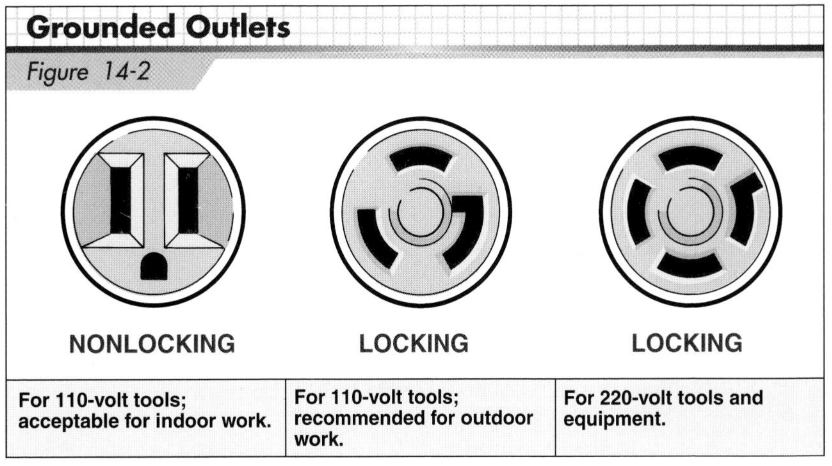

Figure 14-2. *Various configurations of approved grounded outlets are available for construction power tools and equipment.*

Typical Grounding System

Figure 14-3

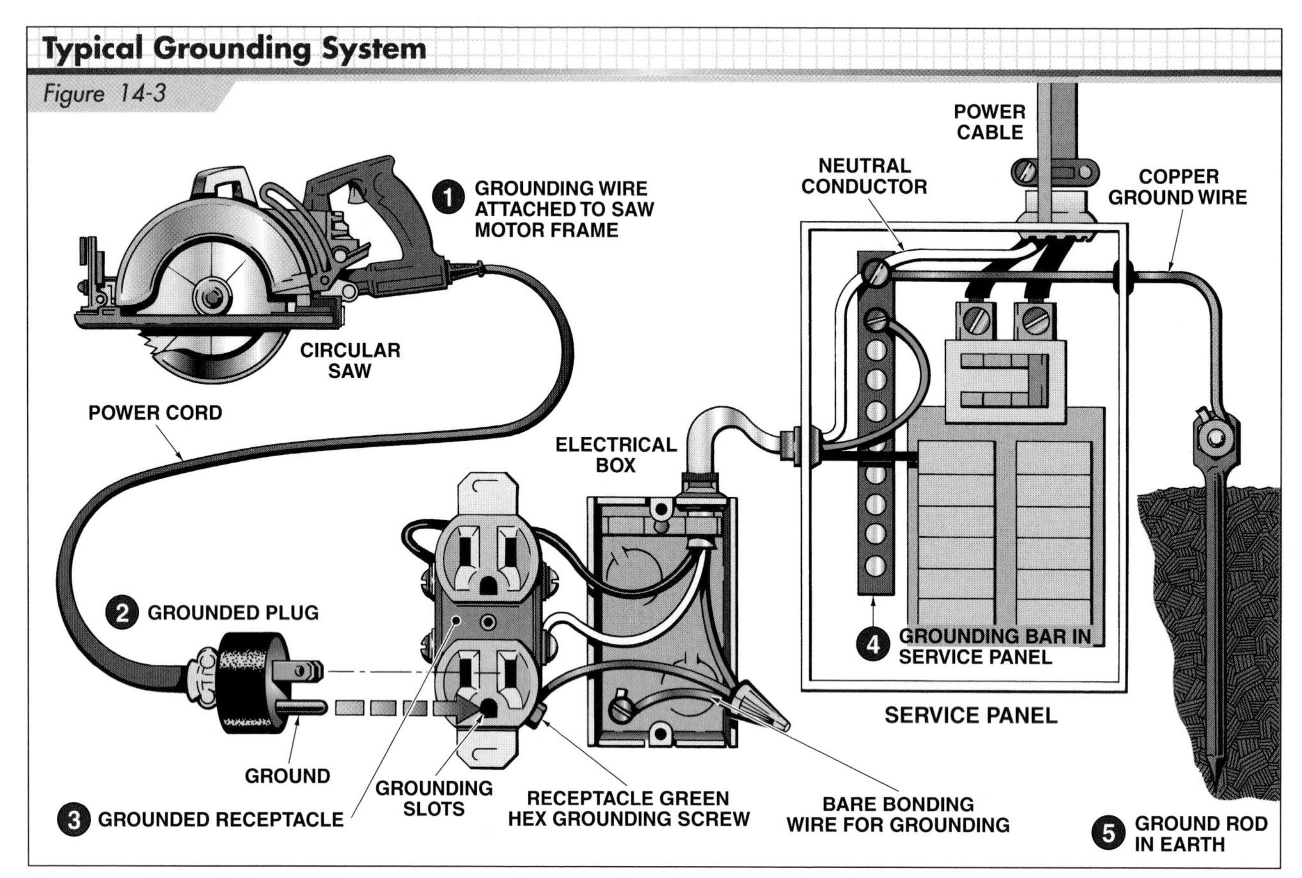

Figure 14-3. *In a typical grounding system, a ground wire runs from the power tool (in this case, a circular saw) to the grounded plug. Another ground wire runs from the grounded receptacle to a grounding bar in the servcie panel. A copper ground wire extends from the service panel to a ground rod or other grounding means in the earth.*

CIRCULAR SAWS

The *circular saw* is the portable power tool used most often on a construction site. Circular saws are used by carpenters to cut lumber to length or panels to length and width. A variety of saw blades are available for cutting wood and nonwood materials.

A circular saw blade cuts the material from the underside through the top. Circular saws are equally efficient for crosscutting and ripping lumber and panel products, and can be adjusted to cut angles ranging from 90° to 40°.

Two types of circular saws are used in construction—*side-drive* and *worm-drive* saws. **See Figure 14-4.** The size of a circular saw is based on the largest diameter blade that can be properly installed. Circular saw blade diameters range from 4½″ to 12″, with 7¼″, 7½″, or 8¼″ diameter blades commonly used by carpenters. **Figure 14-5** lists the different sizes of blades that are available and the depths to which they can cut at 45° or 90° angles.

Milwaukee Electric Tool Corporation

Use two hands to operate a circular saw—one hand on the handle and trigger switch and the other hand on the front knob or forward handle.

Circular Saw Blades

Many types of blades are used with circular saws. The proper circular saw blade to use for an application is based on the material being cut and the grain direction in which the cut is being made. **See Figure 14-6.** *Rip blades* have teeth shaped for cutting in the direction of the grain. Rip blades should not be used for plywood. *Crosscut blades* are used to cut across the grain and can be used to cut plywood. *Combination blades* are used for both crosscut and ripping operations. Combination blades are used more often than rip or crosscut blades since most jobs require sawing with and across the grain. Combination blades produce a rough cut.

Hollow-ground planer blades provide a smoother (though slower) crosscut or ripping cut than standard crosscut or rip blades. Hollow-ground planer blades are commonly used for mitering trim members. *Flat-ground plywood blades* cut plywood and engineered wood panels without tearing or splintering them. *Crosscut flooring blades* make smooth cross-grain cuts and can be used as a rip or cutoff blade on extremely hard woods. *Chisel-tooth combination blades* are commonly used to cut tempered laminates, exterior plywood, and other materials that dull blades rapidly. With any type of blade, a larger number of teeth per inch produces a cleaner cut.

Many types of circular saw blades have carbide-tips. *Carbide-tipped blades* have tungsten-carbide tips brazed to the teeth, and are effective for cutting hard materials. Carbide-tipped blades remain sharp longer than conventional blades. A major disadvantage of carbide-tipped blades, however, is the expense of repair if the tips are broken off by nails or other metals embedded in wood. Before cutting with a carbide-tipped blade, always check the material being cut for any material that could damage the tips.

Abrasive blades cut a variety of masonry materials and metals. *Silicon carbide abrasive blades* cut concrete, marble, granite, glazed and ceramic tile, slate, and terrazzo. *Aluminum oxide abrasive blades* cut metals such as stainless steel, aluminum, bronze, and brass.

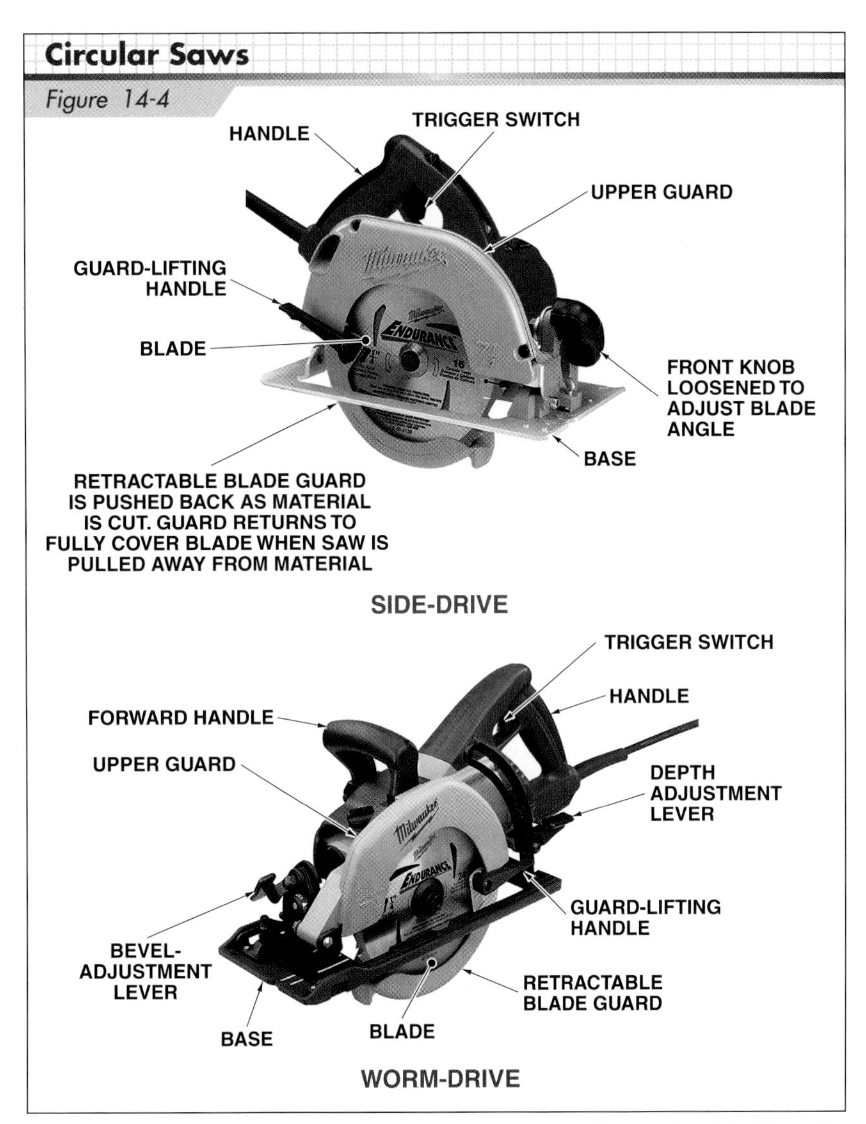

Milwaukee Electric Tool Corporation

Figure 14-4. *Side-drive and worm-drive circular saws are frequently used in construction work.*

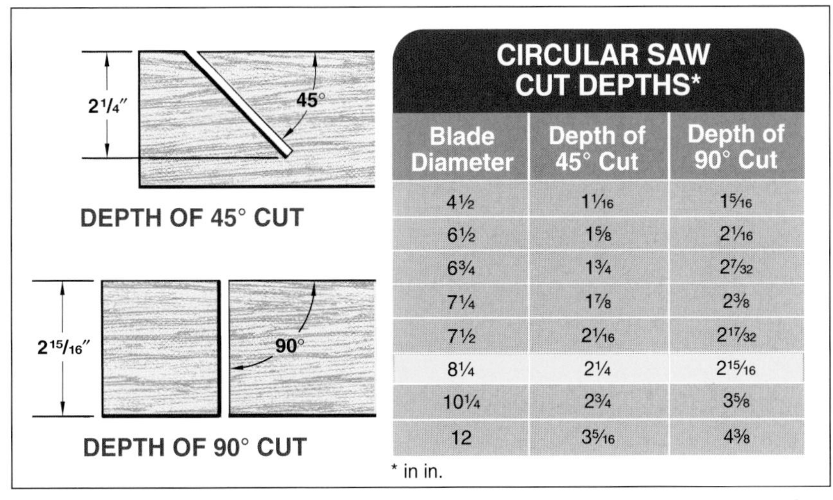

CIRCULAR SAW CUT DEPTHS*

Blade Diameter	Depth of 45° Cut	Depth of 90° Cut
4½	$1\frac{1}{16}$	$1\frac{5}{16}$
6½	$1\frac{5}{8}$	$2\frac{1}{16}$
6¾	$1\frac{3}{4}$	$2\frac{7}{32}$
7¼	$1\frac{7}{8}$	$2\frac{3}{8}$
7½	$2\frac{1}{16}$	$2\frac{17}{32}$
8¼	$2\frac{1}{4}$	$2\frac{15}{16}$
10¼	$2\frac{3}{4}$	$3\frac{5}{8}$
12	$3\frac{5}{16}$	$4\frac{3}{8}$

* in in.

Figure 14-5. *Blade diameter affects the maximum depth of cut. When the blade angle is set to 45°, the depth of cut is decreased.*

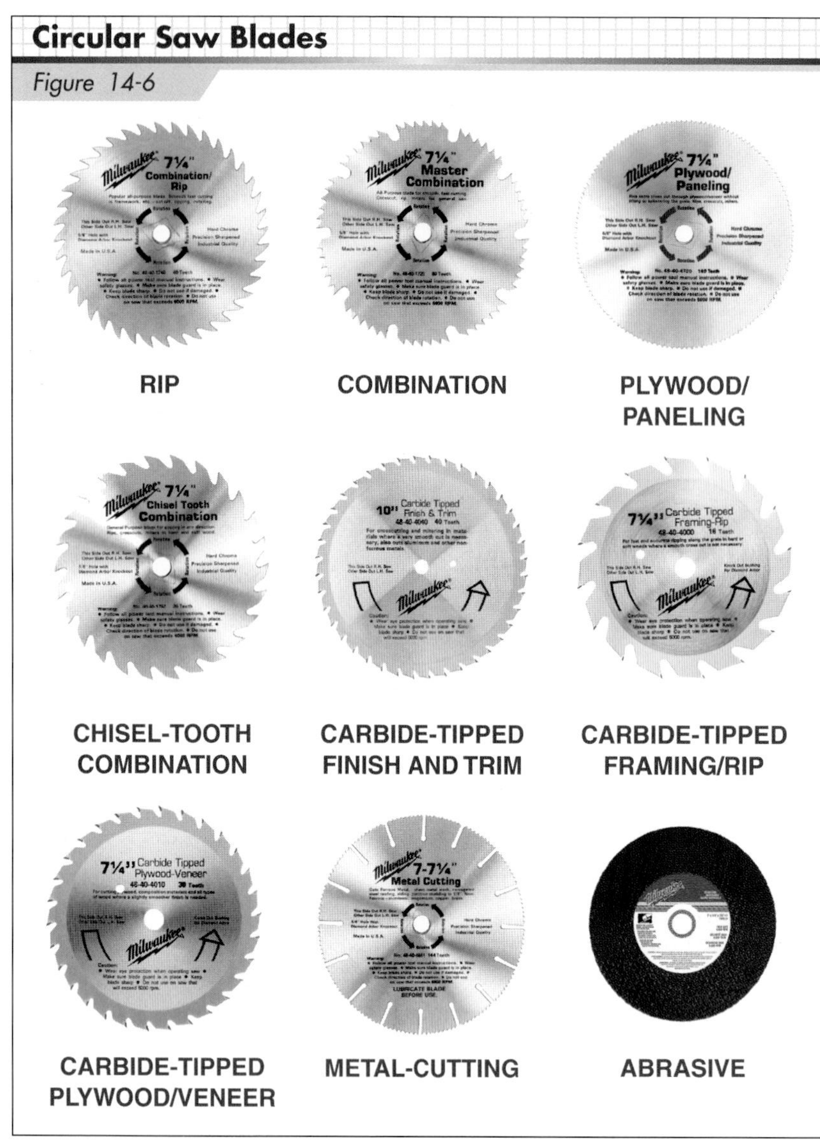

Milwaukee Electric Tool Corporation

Figure 14-6. *Different blades are used with a circular saw for different operations.*

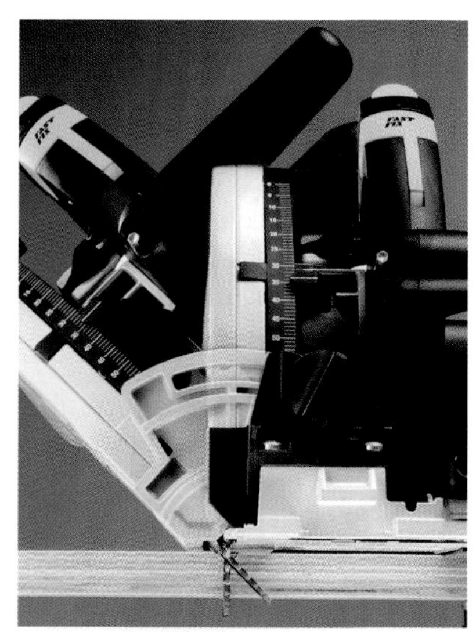

Festool USA

Figure 14-7. *When cutting a compound angle with a circular saw, set the depth of the blade slightly more than the thickness of the material.*

Other blades are available for making rabbet and dado cuts, for cutting fabrics (such as carpet) and rubber, and for cutting glass, asbestos, and cement. Blades with abrasive grit make it possible to saw and sand in one operation.

Circular Saw Operation

When cutting a piece of material with a circular saw, set the blade so the blade gullets clear the wood. The depth of the cut is adjusted by raising or lowering the blade in relation to the saw base. The blade is locked into place with a lever or knob. When adjusting the blade to cut at an angle, loosen the *bevel-adjustment knob,* or tilt knob or lever, to tilt the base to the desired angle of the cut. **See Figure 14-7.** The knob or lever is then retightened.

When beginning any cut, hold the blade slightly back from the material. Start the saw and let the blade attain full speed before pushing the saw ahead. The retractable (telescoping) guard pushes back as the blade advances into the material. If the guard becomes stuck when cutting material at an angle, pull the guard up with the *guard-lifting handle.*

When cutting freehand along a straight line, follow the guide slot on the tool or watch the blade. Many carpenters prefer to watch the blade instead of following the guide slot since guide slots may not be accurate.

When crosscutting long pieces, support the material with sawhorses. **See Figure 14-8.** Position the material with the waste piece overhanging the sawhorse. Do not make cuts between sawhorses; this causes the saw to bind and kick back.

Milwaukee Electric Tool Corporation

Figure 14-8. *When crosscutting long boards supported by sawhorses, do not cut between the sawhorses. Instead, make a cut past the end of a sawhorse.*

Plywood panels are supported by sawhorses or by other panels. When using sawhorses to support plywood panels, some carpenters place 2″ boards beneath the panels to prevent the cut pieces from dropping down and binding the saw. When using other panels for support, position the panel to be cut with the waste piece overhanging the supporting panels. **See Figure 14-9.**

Milwaukee Electric Tool Corporation

Figure 14-9. *Position the panel to be cut with the waste piece overhanging the supporting panels.*

Most circular saws can accommodate a fence attachment to make an accurate narrow rip cut. In **Figure 14-10,** a fence attachment is used to guide a circular saw along the bottom of a door when trimming it to size. To make an accurate cut for a wider piece, tack down a straightedge as shown in **Figure 14-11.**

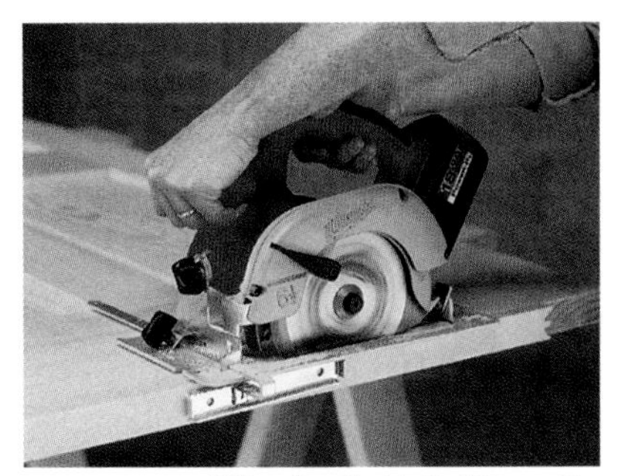

Milwaukee Electric Tool Corporation

Figure 14-10. *An adjustable fence attachment aids in making a narrow rip cut with a circular saw.*

Circular Saw Safety

More serious job site work-related injuries occur from the use of a circular saw than from any other portable power tool. Many accidents are due to carelessness or tool malfunction. One of the biggest causes of circular saw injuries is the retractable guard failing to snap back into position after the cut is completed. The guard may also jam due to a wood chip lodging between the saw guard and blade. Intentionally wedging the guard is a highly dangerous practice and should not be done under any circumstances.

Many safety features are integrated into circular saws to ensure carpenter safety when the tool is used properly. Circular saws are equipped with blade guards above and below the base. The upper guard is stationary and must cover the saw to the depth of the teeth. The lower guard covers the saw to the depth of the teeth and must automatically and instantly return to the covering position when the saw is retracted from the work. Never wedge the guard of a circular saw for any reason! Regulations regarding the proper guarding and operation of circular saws is located in OSHA 29 CFR 1926.304, *Woodworking Tools.*

Circular saws are equipped with a constant pressure switch or control that shuts off the power when the pressure is released. A circular saw must be properly grounded or double-insulated to ensure carpenter safety. Safety rules to observe when using a circular saw include the following:

- Wear appropriate eye protection such as safety glasses or goggles.
- Verify the retractable guard is operational before using a circular saw.
- Wait until the blade stops rotating before removing the saw from the material after completing a cut.
- Stand to one side of a circular saw in case of kickbacks. **See Figure 14-12.**

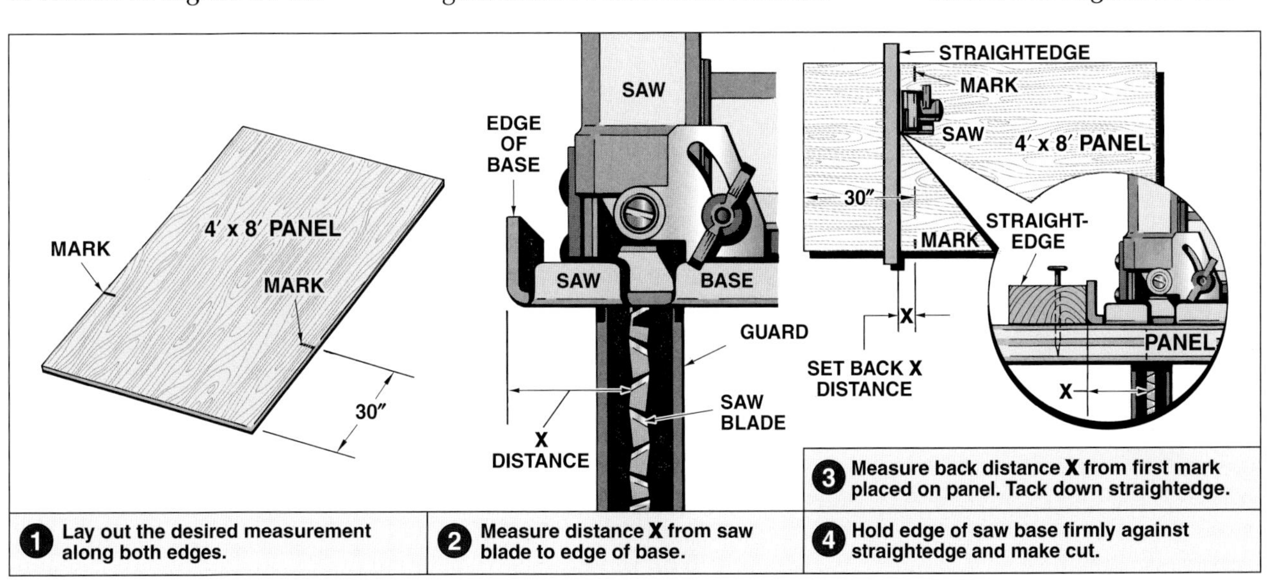

Figure 14-11. *A straightedge aids in making a straight cut across a panel.*

Festool USA

Figure 14-12. *Stand to one side of a circular saw during use.*

- Always disconnect the electrical plug before changing blades or making adjustments.
- Remove all damaged or cracked saw blades from service.

RECIPROCATING SAWS

A *reciprocating saw* is a multipurpose cutting tool in which the blade reciprocates (quickly moves back and forth) to create the cutting action. A reciprocating saw is particularly useful for remodeling work where sections of framing members, sheathing, or inside wall covering must be cut out. **See Figure 14-13.**

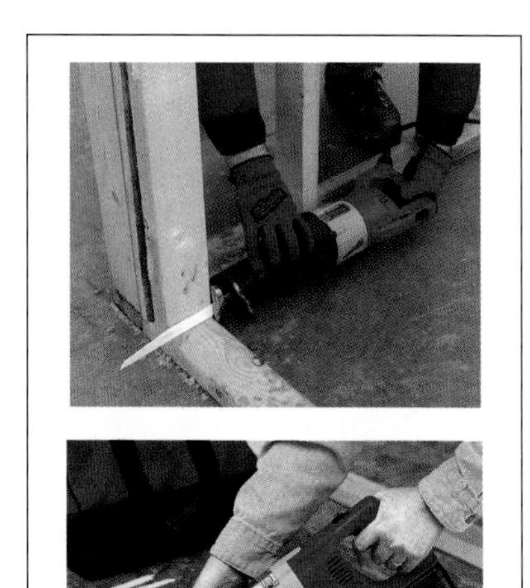

Milwaukee Electric Tool Corporation

Figure 14-13. *A reciprocating saw is used for a variety of applications such as cutting a door jamb or notching a stud.*

Many different blades are available for cutting different materials and for various types of cutting operations. Shapes of some of the more common blades are shown in **Figure 14-14.** In general, the more teeth per inch, the smoother the cut. Fewer teeth per inch result in a rougher, but quicker cut.

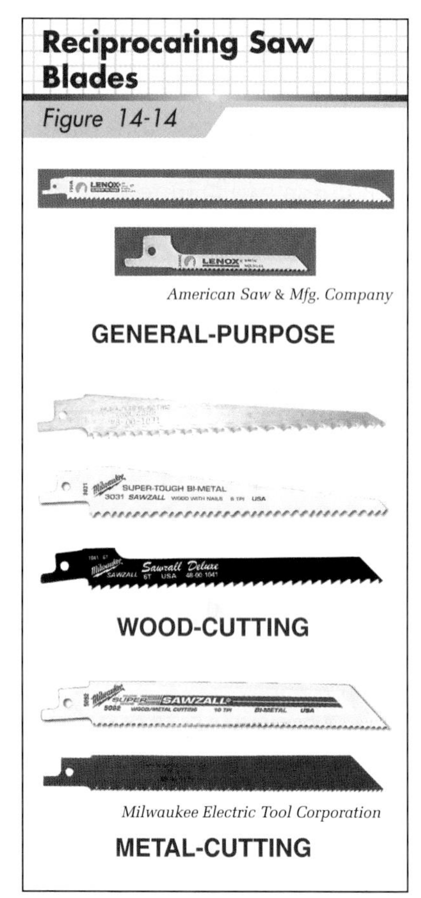

Figure 14-14. *Different blades are used with a reciprocating saw to cut different materials.*

JIGSAWS

A *jigsaw,* or saber saw, is a portable power saw used to cut thin wood and nonwood products. **See Figure 14-15.** The narrow blade of a jigsaw extends from the base and cuts with an orbital (circular) movement. The blade cuts on the upstroke and moves slightly away from the material on the downstroke. **See Figure 14-16.**

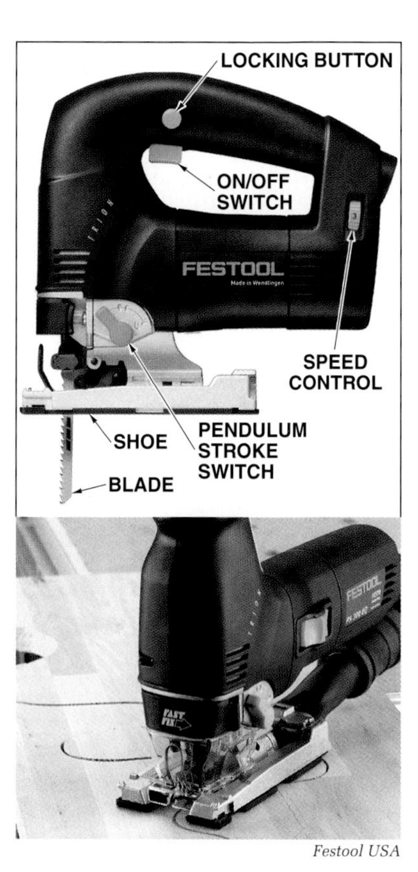

Festool USA

Figure 14-15. *A jigsaw is used to saw along curved lines and to cut circular and rectangular openings.*

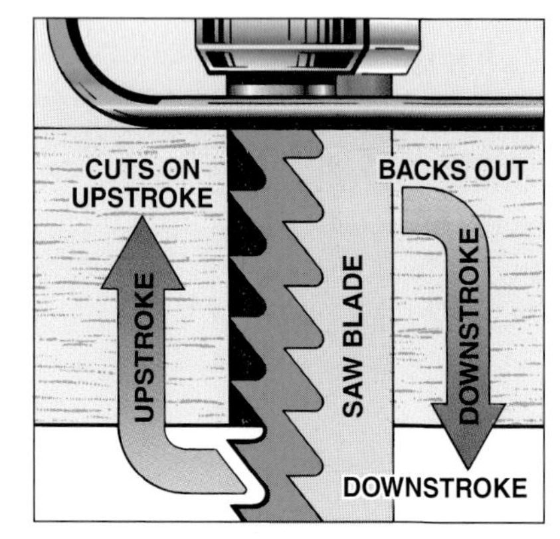

Figure 14-16. *The jigsaw cuts with an orbital movement. The blade cuts on the upstroke and moves slightly away from the material on the downstroke.*

The saber saw is used to cut a variety of thinner materials, and is well-suited for sawing along curved lines and for cutting circular and rectangular openings. Blades are available for cutting wood and metals and range in purpose from cutting nails in wood to plastic laminates. Some blades cut on the upstroke and downstroke.

Reciprocating and Jigsaw Safety

Regulations regarding the proper guarding and operation of reciprocating and jigsaws is located in OSHA 29 CFR 1926.304, *Woodworking Tools.* Reciprocating saws and jigsaws are equipped with a positive ON-OFF control, a constant pressure switch similar to a circular saw, or a variable-speed switch. Safety rules to observe when using reciprocating saws or jigsaws include the following:

- Do not reach under the material being cut.
- After completing a cut, wait until the motor stops before removing the saw from the material.
- Always disconnect the electrical plug before changing blades or making adjustments.
- Remove all damaged saw blades from service.

CUTOUT SAWS

A *cutout saw,* or spiral-cut saw, uses a ⅛″ or ¼″ bit to cut through wood and nonwood materials up to 1″ thick. **See Figure 14-17.** While especially useful for cutting openings in gypsum board, a cutout saw can also be used to cut laminate countertop sink cutouts, cement and backer board, and other materials. The tip of the bit is slowly plunged into the material until the base rests against the surface. The bit is then guided either freehand or against a surface such as the interior edges of a receptacle box.

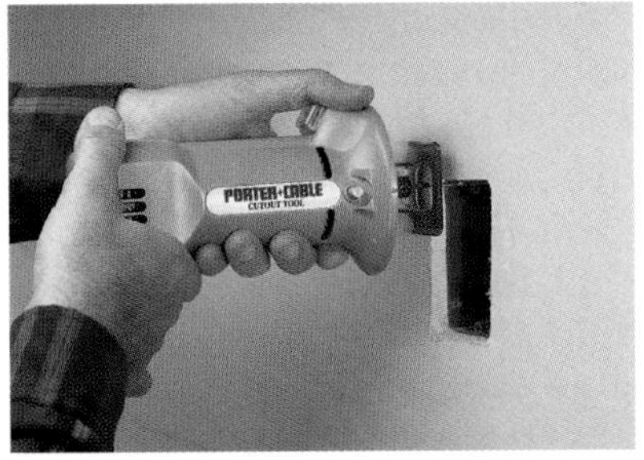

Porter-Cable Corp.

Figure 14-17. *A cutout saw uses a spiral bit to cut openings in wood and nonwood materials such as an opening in gypsum board for a receptacle box.*

Cutout Saw Safety

Basic safety rules to observe when using a cutout saw include the following:

- Ensure the work surface is free from nails or other foreign objects.
- Always hold a cutout saw with two hands during startup.
- Hold a cutout saw by the insulated gripping surface when performing an operation where the cutting tool may contact obscured electrical wiring.
- Ensure the tool is turned OFF before plugging it into an electrical receptacle.
- Ensure the material being cut is either clamped or securely fastened in order to allow you to operate the cutout saw with two hands.
- Allow the motor to come to a complete stop before laying the tool down.
- Do not start a cutout saw when the bit is in contact with the material to be cut.
- Never use dull or damaged bits.

CHAIN SAWS

A *chain saw* is used to cut heavy timbers and pilings and is useful equipment on demolition projects. **See Figure 14-18.** Chain saws are electrically or gasoline powered. Gasoline-powered chain saws are typically more powerful and easily used in remote locations.

Chain Saw Safety

Gasoline, a highly flammable substance, requires special care on a job site to ensure the safety of carpenters and other workers. Use only gasoline from properly labeled storage containers and use care not to overfill the chain saw tank. ANSI B175.1, *Safety Requirements for Gasoline Powered Chain Saws* and UL 1662, *Electric Chain Saws,* provide additional information regarding the safe use of gasoline-powered and electric chain saws, respectively. Basic safety rules to observe when using a chain saw include the following:

- Grip the handle bar when transporting a chain saw.
- Carry a chain saw with the guide bar and chain pointing behind you.
- Do not transport a chain saw with the engine running.
- Do not add gasoline to a gasoline-powered saw when the engine is hot. After pouring gasoline into the tank, wipe off any spilled fuel before starting the engine. Do not start a chain saw in the area where it was refueled.
- Do not smoke while refueling a gasoline-operated chain saw.
- Wear gloves when checking the saw chain for tension or when oiling the chain before or after use. Do not touch the cutters with your fingers or bare hands.
- Ensure the chain brake functions properly.
- Wear appropriate equipment when operating a chain saw. Legs may be protected using Kevlar® chaps or pants.

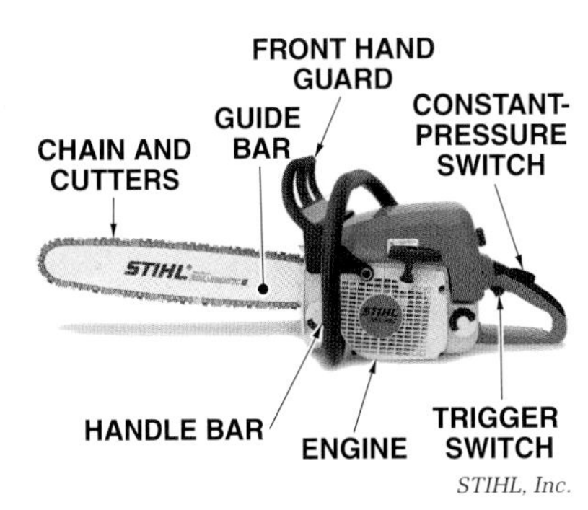

STIHL, Inc.

Figure 14-18. *A chain saw is used to cut heavy timbers and pilings.*

PORTABLE BAND SAWS

A *portable band saw* is a handheld power saw that has a flexible metal saw blade forming a continuous loop around two parallel pulleys. **See Figure 14-19.** Portable band saws use a continuous cutting action to slice through metals. Portable band saws are used in the field to cut all-thread rod, strut channel, metal studs, and other metal components and come in both corded and cordless versions.

Most portable band saws have two speeds. The lower speed is used when cutting hard materials, and the higher speed is used when cutting soft materials. When using a portable band saw, it is important to let the blade do the cutting and to not put too much pressure on the material with the blade. Excessive pressure can cause the blade to bind or break, leading to tool damage or worker injury.

CONCRETE SAWS

A *concrete saw* is a power saw with an abrasive or diamond rotating blade that is used to score and cut concrete. Concrete saws may be handheld or self-propelled. **See Figure 14-20.** Concrete saws are used to make control joints in concrete slabs or to cut into concrete slabs, floors, and walls for demolition or renovation. Cutting concrete can create excessive dust that may contain harmful materials such as silica. Many saws use water to reduce the amount of dust and debris created and to cool the blade. The proper respiratory, eye, and hearing protection must be worn when using a concrete saw.

ANGLE GRINDERS

An *angle grinder* is a handheld tool that removes material by abrasive action. **See Figure 14-21.** Angle grinders are used to sand, grind, and cut various materials such as concrete and metal and come in both corded and cordless versions. Angle grinders are often used to clean and smooth welds.

Different sanding, grinding, and cutting wheels can be used with an angle grinder. The wheel must be rated at or above the maximum rpm listed on the tool and must be inspected for damage. A cracked or damaged wheel can cause serious injury when the tool is powered. Angle grinders range in size from 4½″ to 9″ and may have more than one speed. The correct size wheel must be used with the appropriate grinder. WARNING: A full face shield and gloves must be worn when operating a grinder.

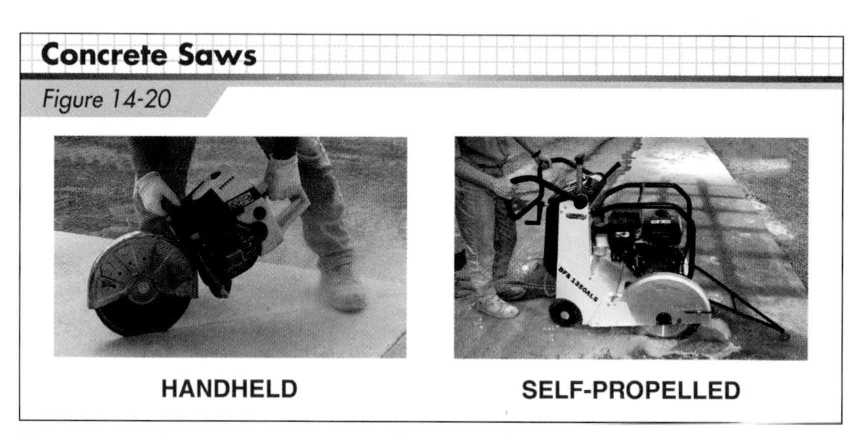

Figure 14-20. *A concrete saw uses an abrasive or diamond rotating blade to score and cut concrete.*

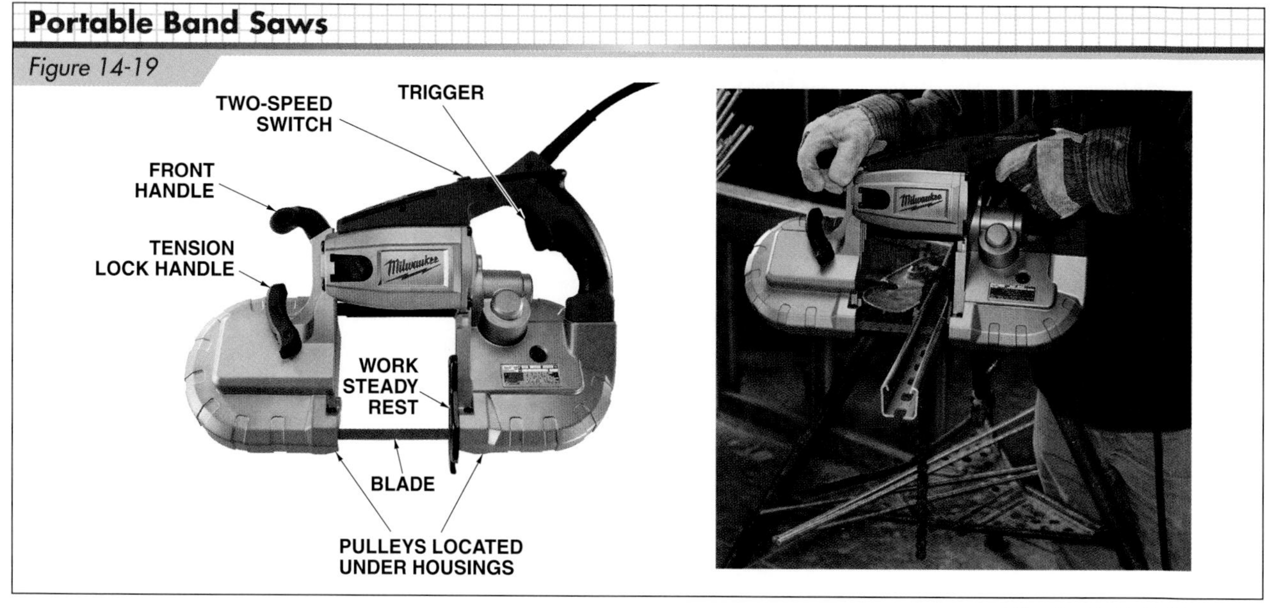

Milwaukee Electric Tool Corporation

Figure 14-19. *Portable band saws use a continuous cutting action to slice through metals.*

Angle Grinders

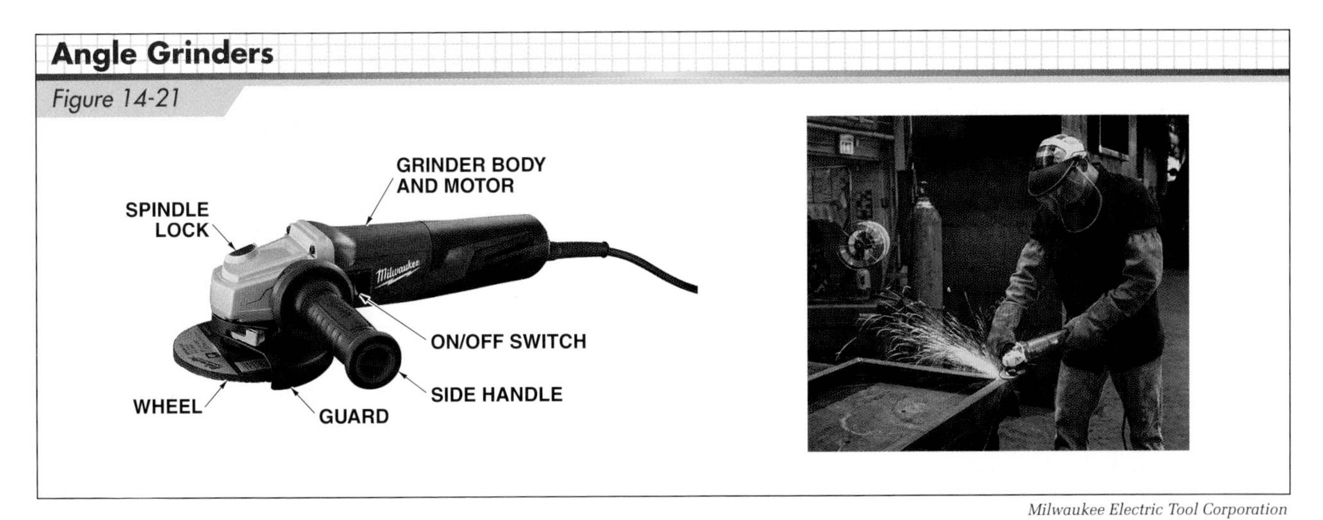

Milwaukee Electric Tool Corporation

Figure 14-21. *Angle grinders are used to sand, grind, and cut various types of materials such as concrete and metal.*

UNIT 15 Stationary Power Tools

OBJECTIVES

1. List common applications of radial arm saws.
2. Identify common safety precautions for radial arm saws.
3. Describe common components of table saws.
4. Identify common safety precautions of table saws.
5. List common applications of compound miter saws and frame-and-trim saws.
6. Describe the common components of band saws.
7. List common applications of drill presses and lathes.

Stationary power tools are usually furnished by the contractor. Radial arm saws are most often used on projects where crosscutting of a large amount of material is required. Table saws are convenient for ripping stock to width. Compound miter saws are commonly used for cutting trim members in finish work. Band saws are used to make curved cuts. Drill presses are convenient for drilling holes. Lathes are specialized tools used for turning wood or metal. Guarding and other safety requirements are detailed in OSHA 29 CFR 1926.304, Woodworking Tools, and ANSI O1.1, *Safety Requirements for Woodworking Machinery*.

The radial arm saw was invented in the early 1900s by Raymond E. DeWalt, the production manager for a woodworking mill. DeWalt designed a yoke and attached it to a motor and saw, which was then mounted on an overhead arm. The saw could be raised, lowered, slid back and forth, and moved or tilted to any angle.

RADIAL ARM SAWS

A *radial arm saw,* or cutoff saw, is typically used to crosscut lumber to length, but can also be set up for ripping and angled cuts. **See Figure 15-1.** On major construction projects where continuous use of a radial arm saw may be required, one carpenter (saw-person) is often assigned to operate the saw. **See Figure 15-2.** A radial arm saw is placed in a convenient location on the job, and an extension table is constructed on both sides to support long pieces of material.

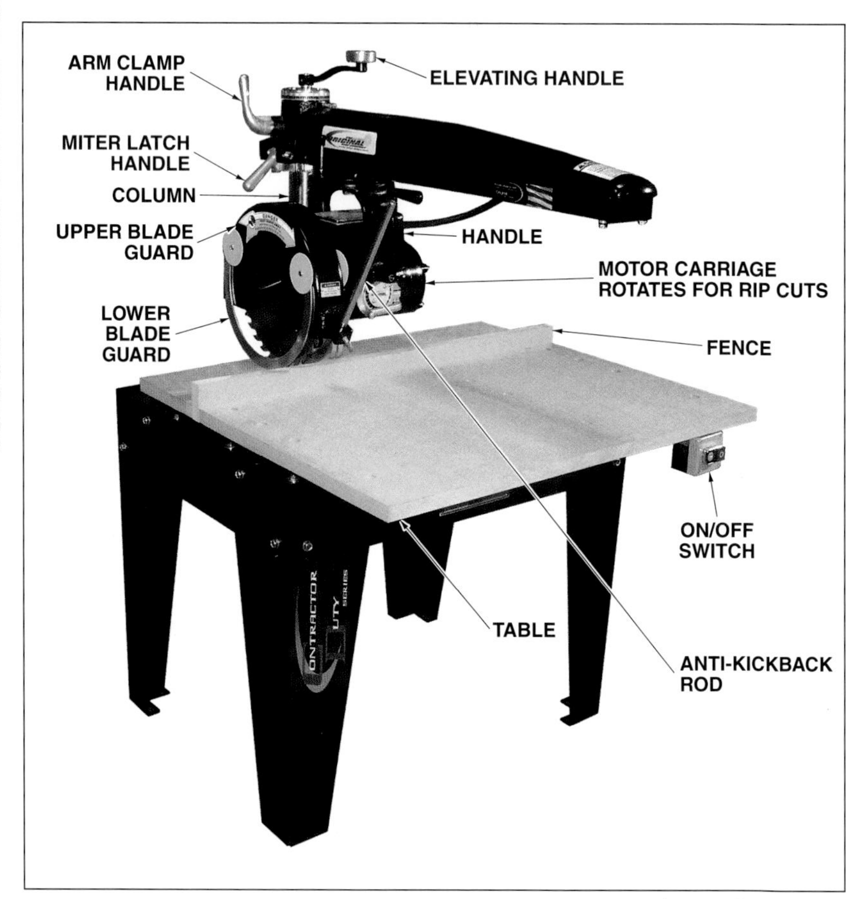

The Original Saw Company

Figure 15-1. *A radial arm saw is used to crosscut, rip, and make angled cuts on lumber.*

Figure 15-2. *A radial arm saw is typically used to crosscut lumber to length.*

The size of a radial arm saw is determined by the largest blade it will accommodate. Blades range from 8″ to 20″ in diameter. The 14″ and 16″ diameter blades are recommended for heavy-duty construction work. Many types of blades are available for different cutting operations. **See Figure 15-3.**

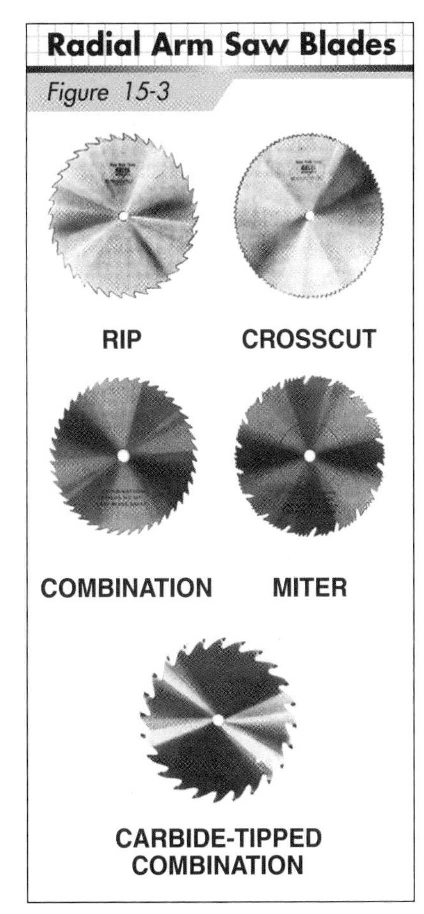

Figure 15-3. *Different blades are used with a radial arm saw for different operations.*

Radial Arm Saw Safety

Radial arm saws are equipped with upper and lower blade guards which protect the operator from the rotating saw blade and from flying debris generated by the cutting operation. The guards should be checked regularly for breakage or other defects. The upper blade guard should completely enclose the upper portion of the blade (down to the saw arbor). The lower blade guard automatically adjusts itself to the thickness of the stock and remains in contact with stock being cut to give maximum protection possible for the operation being performed.

Due to the rotation of the blade, radial arm saws have a tendency to "crawl" forward (away from the fence) when the power is on but material is not being cut. Crawl can usually be prevented if the saw table is slightly tilted back when the unit is set up. Be certain that a spring-loaded return device is operational when using a radial arm saw to ensure that the blade returns to the original position.

Radial arm saw safety rules to observe include the following:

- Wear appropriate eye protection.
- Inspect a radial arm saw daily before each use.
- Firmly grasp the yoke handle when pulling the saw blade through material. When making crosscuts, keep the hand holding the material at least 8″ away from the blade.
- Prior to making a cut and before turning on the saw, clear away scraps or other debris in the work area using a brush or scrap stick.
- Antikickback fingers must be installed on both sides of a blade for ripping operations.
- When ripping, push the material through the blade as it rotates toward the operator. Use a push stick when ripping.
- Do not force the saw through the material faster than the proper saw cutting speed.

The arm track on a radial arm saw pivots on the overarm to position the blade parallel to the fence for ripping operations. When the arm track is in the proper position, the track locking lever must be locked to secure the arm track position. The only part of a radial arm saw that can be allowed to move during a cutting operation is the arm track.

TABLE SAWS

A *table saw* is used for straight sawing, and is of great value in interior finish work such as cutting sheet goods and trim materials. **See Figure 15-4.** A *miter gauge* is used to guide a piece of material for crosscutting operations. A *rip fence* is mounted on a table saw for ripping operations. Freehand cutting is highly dangerous and is not permitted on table saws.

The portion of the blade below the table is protected by the saw housing. The portion of the blade above the table is protected with a self-adjusting guard, which adjusts to the thickness of material being cut.

A dado head is used to cut a dado or rabbet. **See Figure 15-5.** Multipiece dado heads consist of saw blades on the outside with chippers between them. Chippers are inserted or removed to adjust the cut width. Paper inserts are used to make fine adjustments. Single-piece dado heads are also available.

Table Saw Safety

Hand injuries are typically associated with table saw accidents. A hand may slip as material is being fed into the saw or if an operator is holding the stock too close to the blade while cutting. In addition, kickbacks may occur if material becomes jammed between the blade and fence or guard. When a kickback occurs, the material is ejected toward the operator at a high rate of speed. Table saw safety rules to observe include the following:

- Wear appropriate eye protection.
- When performing a ripping operation, always have a push stick available to use.
- Use a rip fence for ripping operations or a miter gauge for crosscutting operations. **See Figure 15-6.**
- Clear the work area around the saw of wood scraps.
- Set the saw blade height so the blade protrudes no more than ⅛″ to ¼″ above the stock being cut.
- Allow the blade to reach full speed before starting a cut.
- Do not reach over the saw blade when it is running.
- Use an extension bench when ripping long stock. If an extension bench is not available, have another worker assist in supporting the end of the material.
- Ensure that all belts are properly guarded.

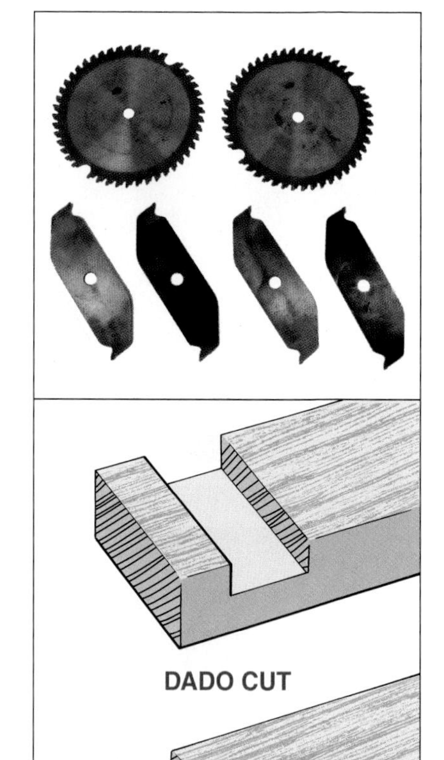

Figure 15-5. *Dado and rabbet cuts are made using a dado head. A multipiece set consists of two hollow-ground blades and chippers.*

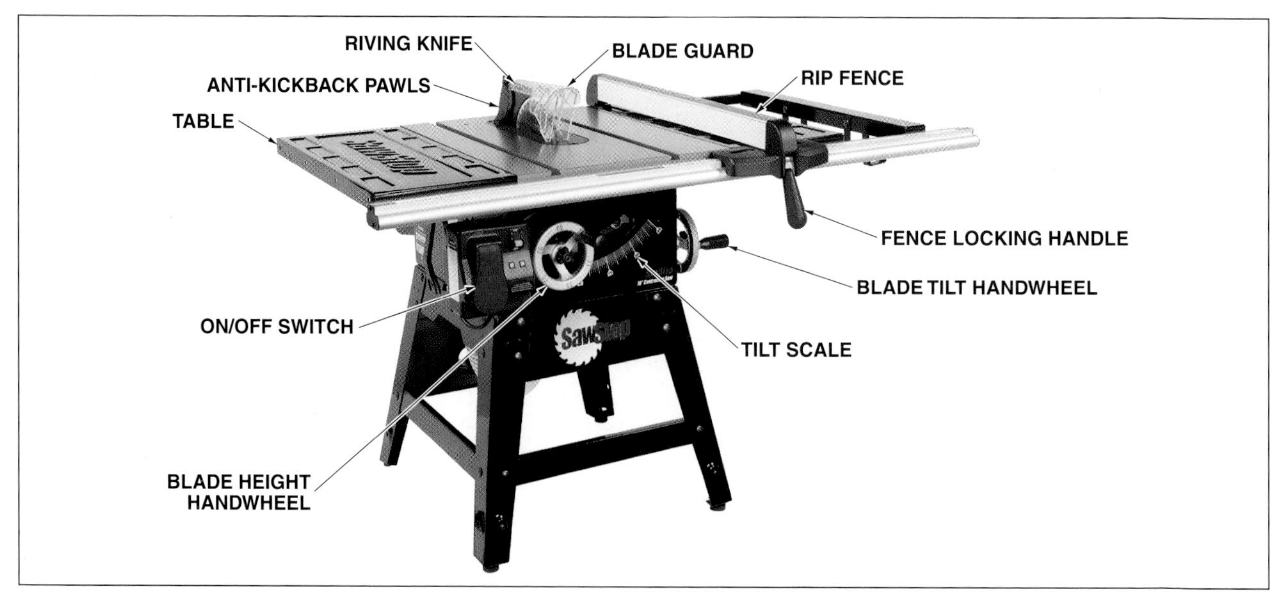

SawStop LLC

Figure 15-4. *A table saw is used to make straight and angle cuts in sheet goods and other finish materials.*

Using a Table Saw

Figure 15-6

Figure 15-6. *A guide such as a rip fence or miter gauge must be used when cutting stock on a table saw.*

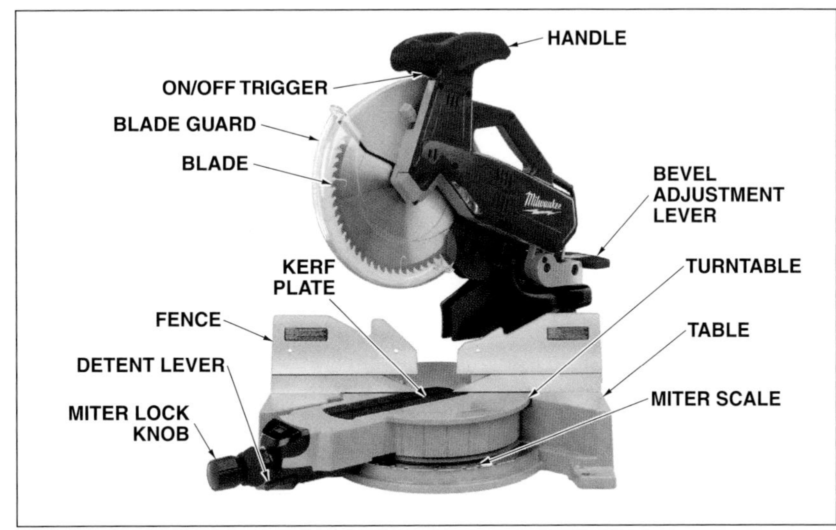

Milwaukee Electric Tool Corporation

Figure 15-7. *A compound miter saw is commonly used to cut trim materials.*

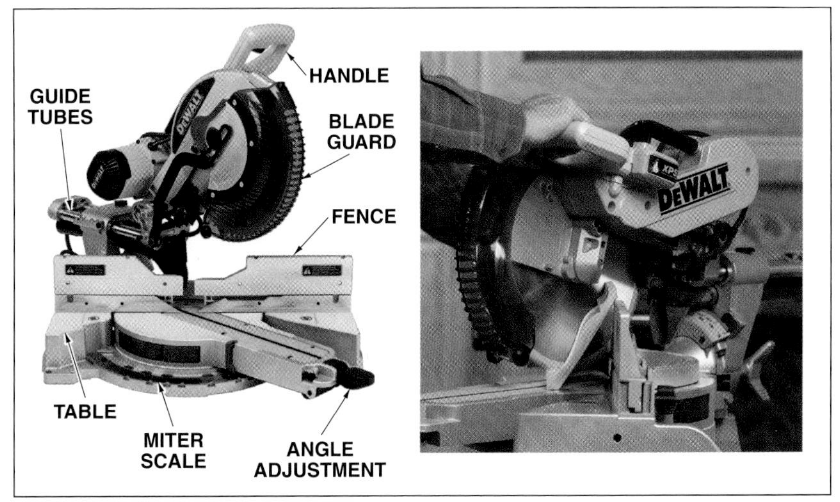

DeWalt Industrial Tool Co.

Figure 15-8. *Wider trim can be cut with a sliding compound miter saw.*

COMPOUND MITER SAWS

A *compound miter saw* is a stationary power tool primarily used to cut molding at angles for finish work. **See Figure 15-7.** Depending on the type of blade installed, a miter saw can be used for cutting wood, composition materials, plastic, and lightweight aluminum. A compound miter saw can rotate up to approximately 50° angles with positive stops at 0°, 15°, 22.5°, 30°, and 45° and also allows the blade to be angled up to 45°. Some compound miter saws, called sliding compound miter saws, are equipped with guide tubes or rails, allowing cuts to be made to wider stock. **See Figure 15-8.**

A compound miter saw is typically mounted on a stand or a couple of sawhorses at a convenient height. **See Figure 15-9.** Most compound miter saws are lightweight (less than 50 lb) and can be easily moved around the job site.

Miter Saw Safety. Miter saw safety rules to observe include the following:

- Wear appropriate eye and hearing protection.
- Attach a compound miter saw firmly to a workbench or other rigid frame at waist height. For job sites, securely mount the miter saw on a thick piece of plywood.
- Keep one hand on the trigger switch and handle and use the other hand to hold the stock against the fence.
- Keep hands out of the blade path.
- Remove the adjusting keys and wrenches before operation.
- Ensure that the blade and collars are clean and secure. Recessed sides of collars should be placed against the blade.
- Allow the motor to reach full speed before starting a cut.
- Clear the work area around the saw of wood scraps.
- Ensure the switch is in the OFF position before plugging in the miter saw.

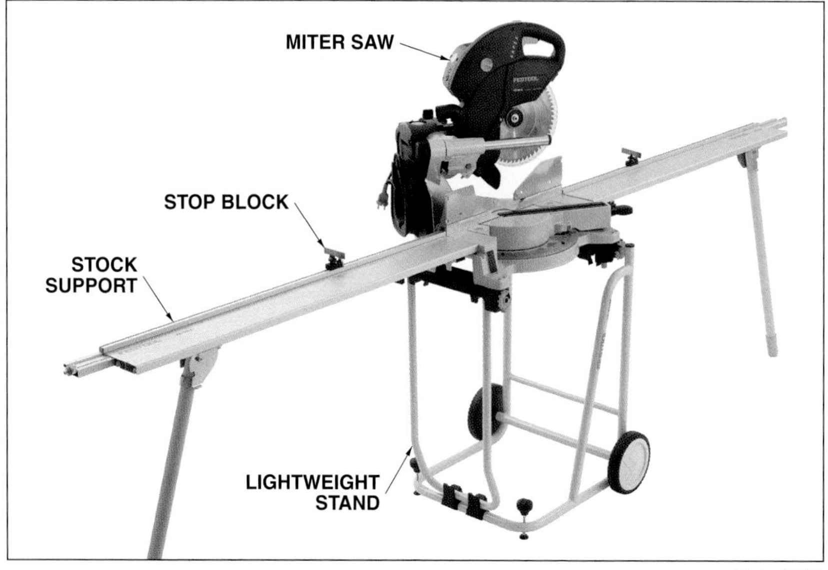

Festool USA

Figure 15-9. *A lightweight and sturdy stand is commonly used as a base for a power miter saw on a job site.*

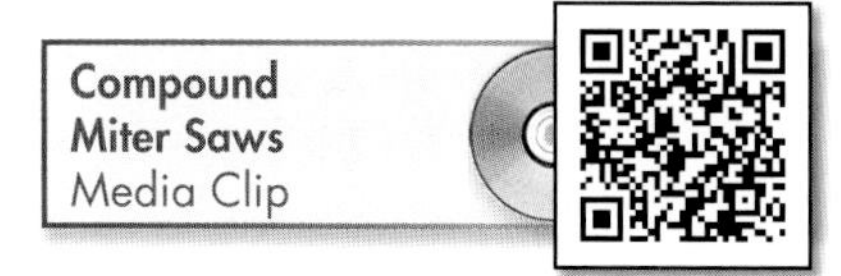

FRAME-AND-TRIM SAWS

A *frame-and-trim saw* combines many of the features of a radial arm saw and a miter saw. A frame-and-trim saw can perform crosscuts, miters, and beveled cuts on stock up to 2″ thick and 12″ wide, but it can not be used for ripping.

A frame-and-trim saw is a valuable tool on remodeling jobs and limited framing operations since it is easily moved and set up. The saw folds to a compact traveling and storage size of approximately 22″ deep, 32″ wide, and 50″ long.

BAND SAWS

A *band saw* is a stationary power saw that has a flexible saw blade forming a continuous loop around two parallel wheels. The saw blade is called a band due to the long, looping configuration of the blade. Some band saws have variable-speed motors that can be changed to accommodate the cutting of various types of materials.

Band saws are available in both vertical and horizontal models and in a wide range of sizes. **See Figure 15-10.** The size of the band saw is measured by its throat width (the clearance between the column and the saw blade), which in turn determines the size of the workpiece the machine can handle.

Band Saw Safety. The principal hazard to the band saw operator is bodily injury. The injury may be caused by contact with the rotating saw band, machine parts or tools, flying chips entering the eyes, or material falling on fingers or toes. The following are recommended safety tips for the use of band saws:

- Close all machine access doors prior to operating the machine.
- Adjust the saw blade guard opening slightly above the thickness of the material to be cut when the machine power is off.
- Wear safety goggles at all times when sawing.
- Fasten small workpieces in a fixture and keep fingers at least 6″ to 8″ away from the cutting band.
- Use the proper saw band, blade tension, and cutting speed for the material being cut.
- Use the proper saw band width for the radius being cut.
- Provide sufficient working space around the machine (4′ clearance in front and back and 3′ clearance on each side).
- Replace dull saw bands. Saw bands used with heavy feed pressures can break if not sharp.
- Wear gloves when handling saw bands.

DRILL PRESSES

A *drill press* is a stationary power tool for drilling or modifying holes of various depths. The drill press is one of the most frequently used machine tools. Its principal purpose is the cutting of round holes into or through materials. This machine employs a variety of cutting tools, with the twist drill being the most common. Holes that do not need to be accurately sized may be drilled without finishing. When close dimensional tolerances are required, secondary operations that can also be performed on the drill press are necessary.

Drill presses are available as bench or floor models. Successful operation of a drill press requires selection of the correct speed, using the correct coolant or lubricant when drilling metals, and a working knowledge of the machine. While drill presses may vary with respect to power options and controls, they typically have a head, spindle, column, table, and base. **See Figure 15-11.**

Band Saws

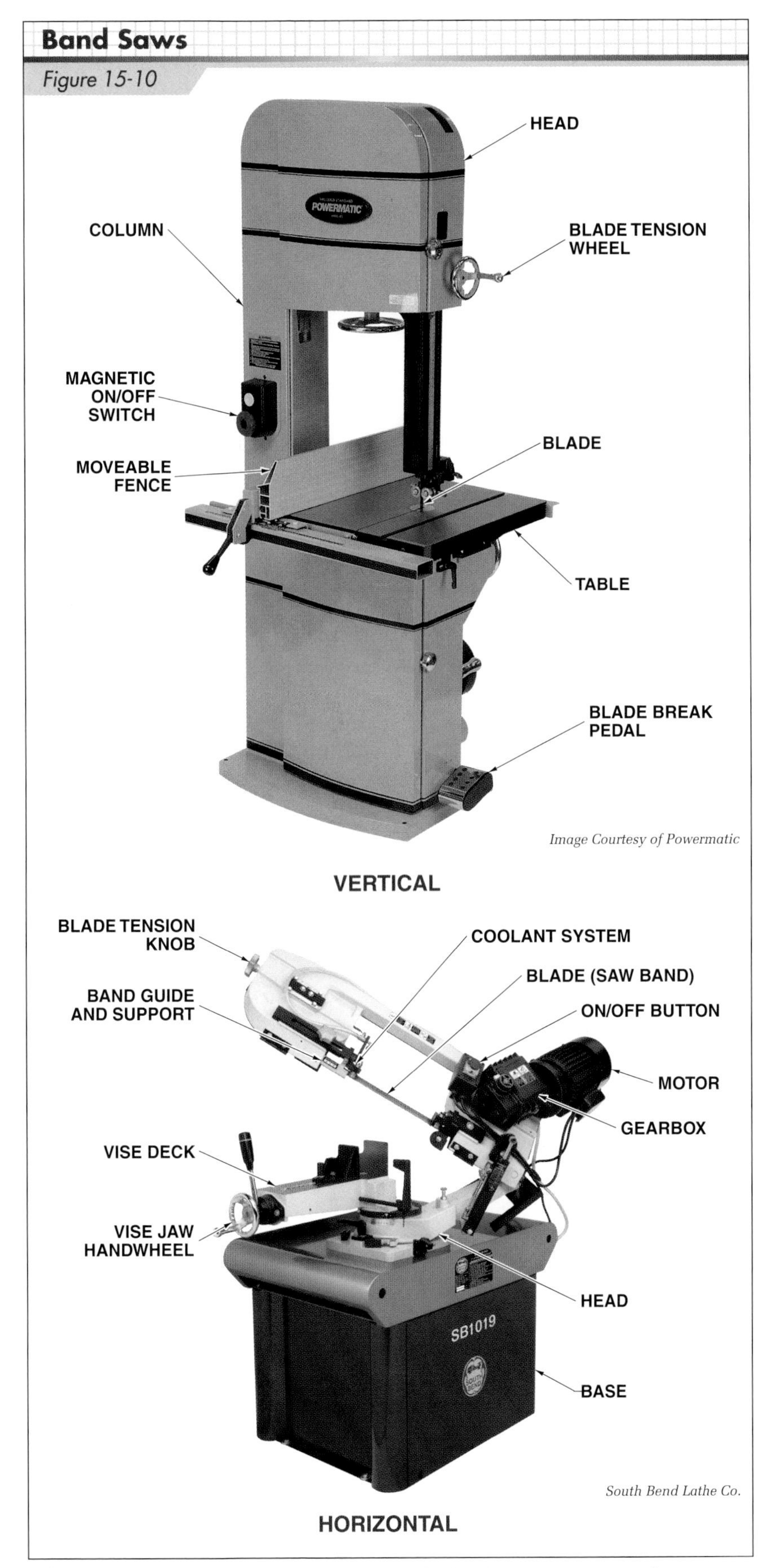

Figure 15-10. *Band saws use a flexible saw blade that forms a continuous loop to cut a variety of materials.*

Drill Presses

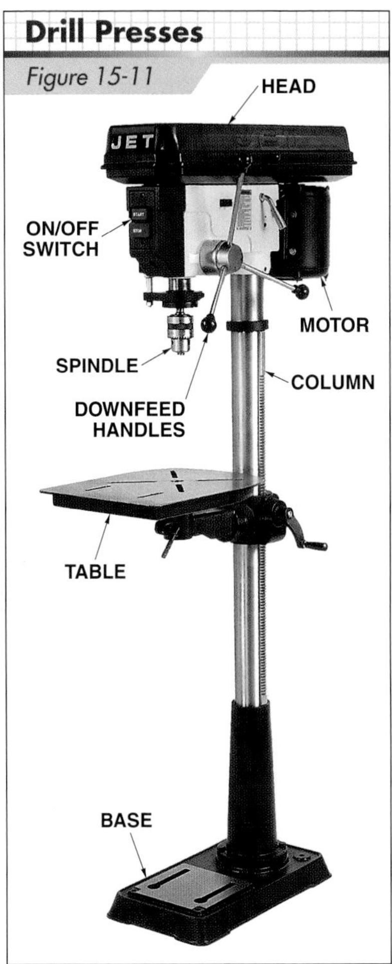

Image Courtesy of Jet

Figure 15-11. *A drill press is used to drill holes of various depths and diameters.*

Drill Press Safety

The principal hazard to a drill press operator is bodily injury. Injury may be caused by contact with moving machine parts or tools, flying chips entering the eyes, or material falling on fingers or toes. The following are recommended safety tips for the use of a drill press:

- Always wear appropriate eye and hearing protection.
- Do not wear gloves, rings, aprons, neckties, or loose or torn clothing. Long hair should be secured to prevent entanglement in rotating machinery.
- Use a drill press vise, clamps, or other means for properly holding the work in place while it is being machined.

- Maintain clean floors around drill presses and, if necessary, cover with an antislip material to safeguard the operator from slipping and falling.
- Do not attempt to oil the machine or make any adjustments to the work while the drill press is in motion.
- Run the drill only at the proper speed. Forcing or feeding material too fast may result in broken or splintered drills and injury.
- Use only properly sharpened drills, sockets, and chucks that are in good condition.
- Remove chips from the table or workpiece with a brush. Do not use a rag or hands to remove chips or loose debris.
- Use only recommended cutting fluids.

LATHES

A *lathe* is a stationary power tool that rotates a piece of wood or metal to allow shaping and finishing of the material with specialized tools. The major function of the lathe is to change the size, shape, or finish of a cylindrical, rotating workpiece by one cut or a series of cuts into the workpiece with an adjustable cutting tool. With proper attachments and adjustments, a lathe can also drill, ream, tap, and thread.

Certain components are common to all lathes. An understanding of the components, their locations, and their use in machining enables one to do more accurate work with less waste. While lathes may vary with respect to power options and controls, they typically have a bed, headstock, tailstock, and tool rest. **See Figure 15-12.**

Lathes

Figure 15-12

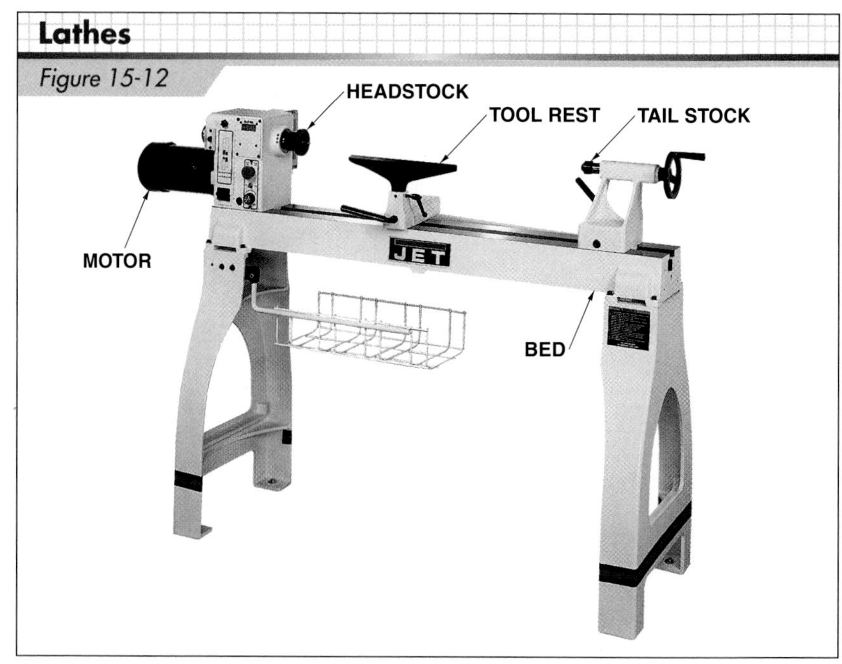

Image Courtesy of Jet

Figure 15-12. *A lathe rotates a piece of wood or metal to allow shaping and finishing of the material with specialized tools.*

To machine materials effectively and accurately with a lathe, the correct size and type of turning tool must be used. **See Figure 15-13.** The turning tool used should be suited for the kind of material being machined and set at the proper position in relation to the work. Turning tools are available in a variety of shapes and sizes. The operation and material involved determines the type of the turning tool to be used.

Turning Tools

Figure 15-13

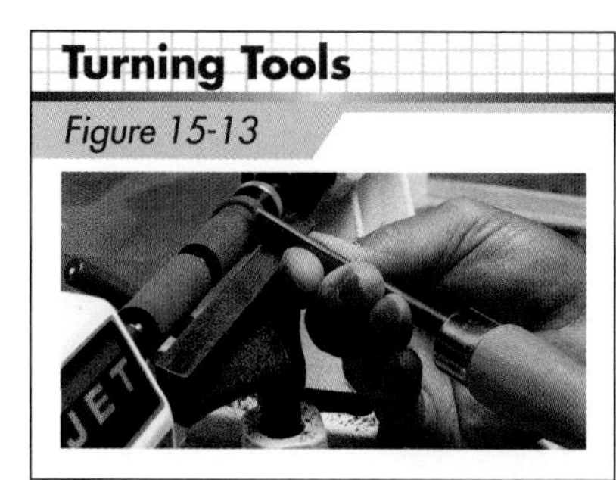

Image Courtesy of Jet

Figure 15-13. *The turning tool used should be suited for the kind of material being machined and set at the proper position in relation to the work.*

Lathe Safety

The hazards of lathe use can be minimized if safety procedures are followed during lathe setup and operation. The following are recommended safety tips for using a lathe:

- The lathe operator should have a thorough knowledge of the lathe components and operation.
- Always wear appropriate eye and hearing protection.
- Do not wear gloves, rings, bracelets, or other jewelry, or aprons, neckties, or loose or torn clothing. Long hair should be secured to prevent entanglement in rotating machinery.
- Remove the chuck wrench immediately after adjusting the chuck.
- Wait until the lathe stops completely before making any adjustments or taking any measurements.

Unit 15 Review and Resources

UNIT 16 Portable Power Drills and Screwdrivers

OBJECTIVES

1. Describe common portable power drills.
2. Identify common applications of variable-speed drills.
3. List and describe common drill bits.
4. Describe common applications of hammer-drills and rotary hammers.
5. Identify common applications of impact wrenches.
6. Identify common applications of power screwdrivers.
7. List common safety precautions of power drills and screwdrivers.

For most applications, portable power drills have replaced hand tools for drilling holes since power drills are faster and more accurate. Power drills with variable-speed controls and adjustable-torque chucks can also be used as power screwdrivers. Specialized power screwdrivers that have greatly increased the efficiency of many fastening operations are also available.

Keep good balance and proper footing at all times when using a portable power drill to maintain good control of the tool, especially in unexpected situations.

PORTABLE POWER DRILLS

Portable power drills are either *cordless* (battery-operated) or *corded* (AC-powered) and are available in D-handle or pistol-grip handle designs. **See Figure 16-1.** Cordless drills receive their power from a removable battery, which is commonly located in the handle. Batteries must be recharged at intervals which depend on the amount and type of usage. Cordless drills can be used for most of the same operations as corded drills. However, cordless drills are most effective for low-torque operations such as drilling thinner wood and driving narrower screws. *Corded drills* are attached to an AC power source through an electrical cord, which is plugged into an electrical outlet, extension cord, or generator outlet. Corded drills are popular for high-torque operations such as drilling into concrete or drilling large holes in lumber.

Milwaukee Electric Tool Corporation

Figure 16-1. *Power drills are available in a variety of sizes and grip configurations and for a variety of power sources. Select the appropriate tool for the operation being performed.*

Power drill sizes range from ¼″ to 1¼″, and are based on the diameter of the largest bit shank that fits into the chuck of the drill. Power drill sizes most often used by construction carpenters are the ¼″, ⅜″, and ½″ models.

A *D-handle,* or spade, design is used for heavy-duty operations, including demolition. A *pistol-grip* design is commonly used for lighter drilling operations. D-handle models are typically available in ⅜″ or ½″ chuck sizes; pistol-grip models in ¼″ or ⅜″ chuck sizes.

Variable-Speed Drills

Drill speed is controlled by pressure on the trigger switch of a *variable-speed drill.* As a result, a variable-speed drill can effectively be used on different materials. Slow drill speeds are more effective with harder materials such as steel. Faster drill speeds work better with softer materials such as wood. Some variable-speed drills have a reversing switch that is useful for withdrawing screws when a screwdriver attachment is used with the drill.

When using a portable power drill, keep the wrist straight to reduce stress on muscles and tendons. In addition, select a drill with vibration-absorbing handles.

Bits

The cutting devices that fit into the jaws of a power drill are referred to as *bits* or *drills.* The most frequently used bit for drilling small holes is the *twist drill.* Twist drills may have a *straight shank* or a *reduced shank.* **See Figure 16-2.** Straight-shank bits are available in 1/64″ to ½″ diameters. Reduced-shank bits are available in diameters up to 1½″. The reduced shank allows a bit that is larger in diameter than the drill capacity to fit a chuck. For example, a ½″ diameter bit will fit a ¼″ diameter chuck if the drill has a ¼″ diameter reduced shank.

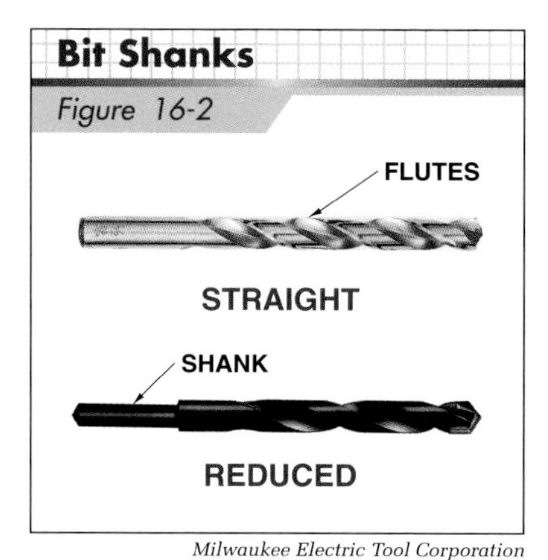

Milwaukee Electric Tool Corporation

Figure 16-2. *Bit shanks are straight or reduced in diameter. Reduced-shank bits allow a bit that is larger in diameter than the drill capacity to fit a chuck.*

Various other types of bits are also used for boring into wood, metal, plastics, masonry, and other materials. **See Figure 16-3.** A *feeler bit* is used to bore deep holes in wood, and is available in 3/16″ to ⅜″ diameters and 12″, 18″, and 24″ lengths.

The *ship auger bit* is also used to bore deep holes in wood. Ship auger bits with screwpoints are used most often in construction. Ship auger bits are available in 9/16″ to 1″ diameters and 12″, 18″, and 24″ lengths. Ship auger bits without screwpoints are designed especially for end-grain boring. Since there is no screwpoint and the heel is backed off, this type of auger has less tendency to drift in the same direction as the grain of the wood. If a ship auger bit without a screwpoint is used, it is necessary to start the hole with a ship auger bit that has a screwpoint.

A carbide-tipped *masonry bit* is used to bore holes in concrete, plaster, slate, stone, brick, and other types of masonry. Masonry bits are available in ⅛″ to 1¼″ diameters. Wear proper personal protective equipment, including appropriate respiratory protection, when drilling concrete, brick, and other masonry products.

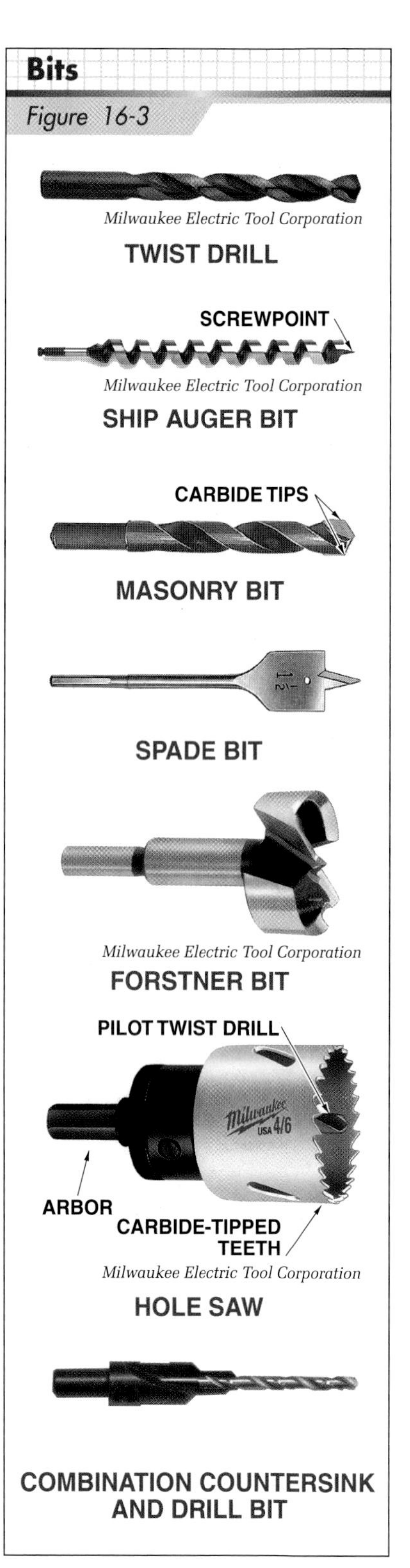

Figure 16-3. *A variety of bits, including twist drills, masonry bits, and hole saws, are commonly used to bore holes in wood and nonwood materials.*

A *spade bit,* or flat bit, is used to bore holes in wood, plastic, and composition materials. Spade bits are available in ¼″ to 1½″ diameters.

A *forstner bit* is used to bore flat-bottom holes in wood. Unlike other drill bits that are guided through the workpiece by a central point, the outside rim of the bit primarily guides forstner bits. Forstner bits are commonly used to drill pocket holes for furniture construction and to drill overlapping holes. Forstner bits range in size from ¼″ to over 3″ in diameter.

Hole saws are used to cut larger holes than can be cut with bits. A pilot twist drill at the center of a hole saw is used as a guide when starting the hole saw. The teeth of a hole saw may be carbide-tipped. Hole saws are used to cut wood, fiberglass, fiberboard, ceramic, tile, and soft metals, and are available in 9⁄16″ to 6″ diameters.

An adjustable *countersink cutter* is slipped over a twist drill and held in place with a socket setscrew. The countersink cutter, available in ⅛″ to ½″ diameters, has cutting edges beveled 82° to match the taper of a flathead screw so it will be flush with the wood surface when installed. An *adjustable counterbore cutter,* available in 3⁄16″ to ½″ diameters, cuts flat-bottomed holes for screws that must be set below the surface of the wood.

Hammer-Drills and Rotary Hammers

Hammer-drills and rotary hammers rotate and drive simultaneously, causing the bit to spin and hammer at the same time. A *hammer-drill* is used to drill holes in concrete using masonry bits or twist drills. **See Figure 16-4.** Hammer-drills should be used when drilling up to ⅝″ holes in concrete. A hammer-drill drills concrete more efficiently than a variable-speed drill because of the rotating and percussive action, which results in faster and cleaner holes. Two factors affect the drilling rate with a hammer-drill—speed and pressure. Speed of bit rotation is affected by the pull of the trigger. The hammering action of a hammer-drill is affected by the pressure applied by an operator. Greater pressure results in better hammering action. The clutches of some hammer-drills can be disengaged so the tool can be used as a conventional drill.

Milwaukee Electric Tool Corporation

Figure 16-4. *A hammer-drill is typically used to drill holes in concrete. Carbide-tipped percussion masonry bits are used to drill concrete.*

A *rotary hammer* is used to drill holes larger than ⅝″ diameter in concrete or masonry. A heavy-duty rotary hammer drills holes in concrete up to 1⅛″ diameter with solid drill bits and holes up to 4″ diameter with core drill bits. Rotary hammers operate at a lower rpm than hammer-drills and should not be used to drill wood or metal. Rotary hammers require slotted drive system or slotted drive shank (SDS) chucks and bits. Pressure applied to a rotary hammer does not affect the hammering action; the weight of the tool properly engages the bit. Some models of rotary hammers also accommodate a cold chisel bit for chipping and edging concrete as shown in **Figure 16-5.**

Milwaukee Electric Tool Corporation

Figure 16-5. *A rotary hammer may be used to chip concrete with a cold chisel bit.*

IMPACT WRENCHES

Impact wrenches are corded or cordless tools that apply a large amount of force to a small area, such as bolt heads, with little effort from the worker. Impact wrenches are used for a variety of applications including installing and removing anchor bolt nuts, assembling metal building frames and panels, and driving lag bolts or screws.

Impact wrenches may be fitted with various size sockets for installing nuts and bolts or a hex adapter for use with hex-shank auger bits or self-feeding bits. **See Figure 16-6.** Socket sizes range from ½″ to 13⁄16″ in 1⁄16″ increments. Impact wrenches used for carpentry typically have a ½″ or ¾″ square drive shank. Some impact wrenches may also be fitted with SDS bits.

POWER SCREWDRIVERS

Power screwdrivers are used for rapid and efficient driving or removal of screws. Power screwdrivers have an adjustable clutch mechanism or an adjustable depth control to prevent the overtightening and stripping of screws. Power screwdrivers are commonly used to install wall and ceiling materials over metal framing members and install gypsum board over wood studs.

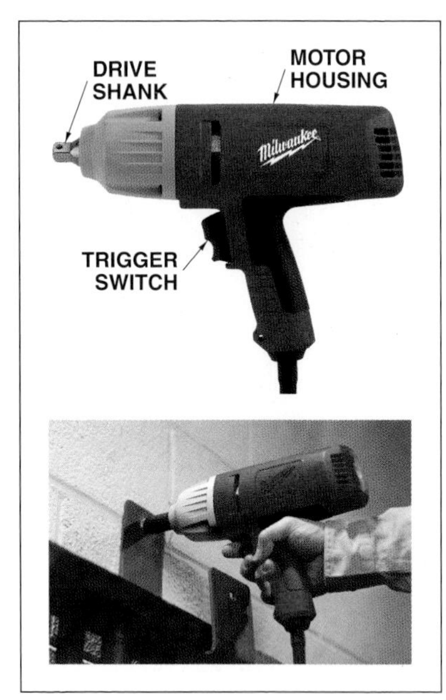

Milwaukee Electric Tool Corporation

Figure 16-6. *Impact wrenches are used to install and remove nuts, bolts, and screws.*

A *drywall screwdriver,* or "screw-shooter," is specifically designed for fastening gypsum board to wall studs and ceiling joists. **See Figure 16-7.** A drywall screwdriver is adjustable so a screw head is driven just below the surface of the gypsum board without cutting through the outside layer of paper.

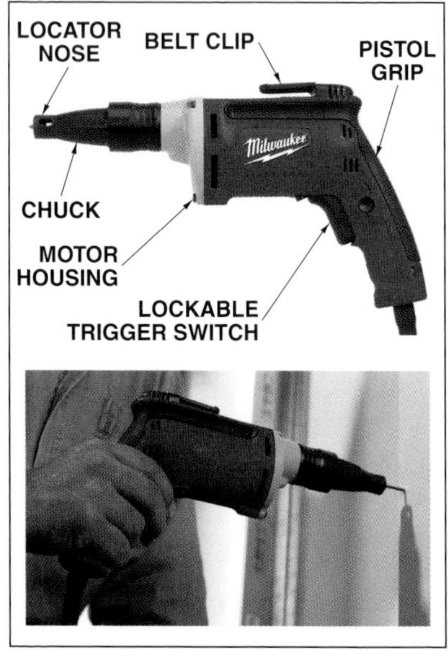

Milwaukee Electric Tool Corporation

Figure 16-7. *A drywall screwdriver is specifically designed for fastening gypsum board to support members.*

Impact Drivers

An *impact driver* is a cordless power screwdriver that uses both rotational force and a hammer mechanism to drive fasteners. **See Figure 16-8.** Impact drivers typically deliver more torque and rpm than cordless drills. Impact drivers usually have a 1/4" hex chuck and accept either Phillips head or Robertson (square) screwdriver bits and hex-shaft drill bits.

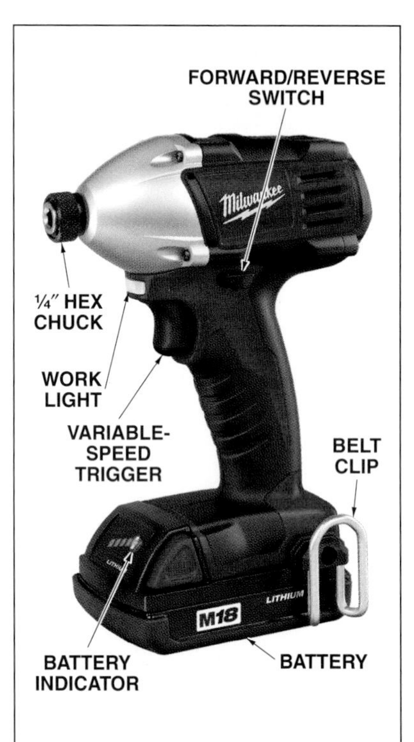

Milwaukee Electric Tool Corporation

Figure 16-8. *Impact drivers use rotational force and a hammer mechanism to drive fasteners.*

The hammer mechanism that produces the torque also creates some forward pressure, which reduces cam-out when using Phillips head bits. *Cam-out* is a condition that occurs when the screwdriver bit slips out of the screw head when exerting pressure to turn the screw. Cam-out often results in damaging the screw head or the screwdriver bit.

POWER DRILL, IMPACT WRENCH, AND SCREWDRIVER SAFETY

Power drills, impact wrenches, and screwdrivers use a rotational motion during operation. Drill and screwdriver bits must be handled carefully to ensure that injuries are not incurred due to cuts from the tips or edges. A power drill, impact wrench, or screwdriver should be allowed to come to a complete stop before the bit is grasped.

Larger and more powerful drills require greater caution on the part of the operator. A large bit that binds in the hole may cause the drill handle to kick back, resulting in hand injuries or knocking the operator off balance. When using a large drill and bit, maintain a firm, well-balanced position and a tight grip on the handle of the drill. Push the drill bit steadily into the work and alternately pull back on the drill to clear the wood chips that may cause binding. Use an extension handle to help prevent binding.

Safety regulations for power drills are included in OSHA 1910.304, *Woodworking Tools.* Power drills are equipped with constant-pressure switches. In addition, a lock-on control is permitted on a power drill provided that it can be turned off by a single motion of the fingers. Additional safety rules to observe when using power drills, impact wrenches, and screwdrivers include the following:

- Always wear appropriate personal protective equipment for the operation being performed, including the proper eye and hearing protection.
- Clamp small objects to a solid surface before drilling.
- Keep drill bits sharp and properly shaped.
- Center a drill bit in the jaws. Tighten the chuck securely before using a drill.
- Disconnect a drill from the power source when inserting or removing bits.
- Do not lock a drill in an ON position while drilling.
- Ensure the switch is in the OFF position before connecting a drill to the power source.

Milwaukee Electric Tool Corporation

Drywall screwdrivers are commonly used to fasten metal framing members together.

Unit 16 Review and Resources

UNIT 17

Portable Power Planes, Routers, and Sanders

OBJECTIVES

1. Describe common portable power planes.
2. List common safety precautions for power planes.
3. Identify common applications of routers.
4. Review common safety precautions for routers.
5. Identify common portable power sanders.
6. Identify common abrasives.
7. Review common safety precautions for power sanders.

Portable power planes, routers, and sanders are used to smooth wood surfaces and prepare the surfaces for finishing. Portable power planes are typically used to replace work performed by bench planes and reduce the amount of labor required to produce smooth surfaces and edges on stock. Routers are used to shape the edges of stock and to create internal openings such as mortises for door hinges. Portable sanders smooth the wood surface using an abrasive paper or belt.

PORTABLE POWER PLANES

Portable power planes consist of a straight or spiral cutter driven by an electric motor. The spiral cutter produces the cutting action of a plane. The depth of cut of a portable power plane is changed with a depth-adjustment lever, which loosens and tightens the front shoe. A portable power plane is used for fitting doors and for smoothing rough surfaces on the sides and edges of boards. **See Figure 17-1.**

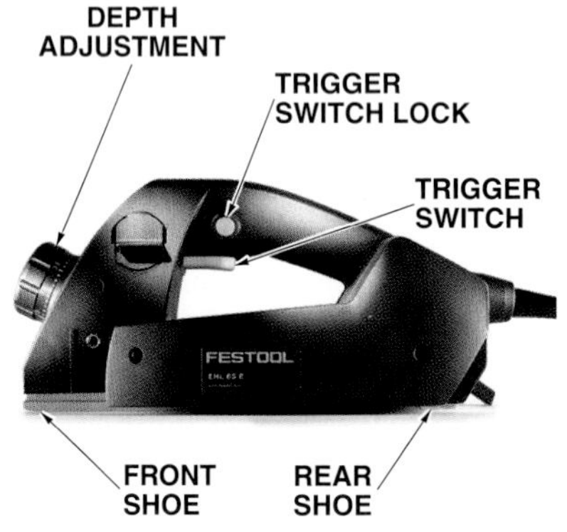

Festool USA

Figure 17-1. *A portable power plane is used to fit doors and smooth rough surfaces on the sides and edges of boards.*

Power plane cutters range from 2¹³⁄₃₂″ to 3¼″ wide. Depth of cut can be adjusted from ¹⁄₃₂″ to ³⁄₁₆″. An adjustable fence allows planing of bevels up to 45°.

A *power block plane,* which is a small portable power plane, is used for planing molding, cabinet doors, and other thin-edged materials. Block planes are also used for planing end grain, cleaning out rough rabbets, and cutting bevels.

OSHA 29 CFR 1926.302, Power-Operated Hand Tools, provides information on safety precautions that must be taken when working with power tools.

Festool USA

A portable power plane is used to smooth wood surfaces.

Festool USA

Stairways with housed stringers are generally prefabricated in a shop. Grooves in the stringers cut by a router to accept the treads and risers, and are secured with glue and wedges.

Power Plane Safety

The sharp and exposed cutter poses the primary danger when using a portable power plane. When a cut is completed, turn off the motor but do not set the plane down until the cutter has stopped rotating. Additional power plane safety rules to observe include the following:

- Wear appropriate personal protective equipment including the proper eye protection and hearing protection.
- Disconnect the plane from the power source before making adjustments to the cutter head or blades.
- Ensure that blade-locking screws are tight.
- Remove all adjustment keys and wrenches prior to turning on the plane.
- Check stock thoroughly for nails, staples, screws, or other foreign objects prior to planing.
- Firmly secure the stock in a comfortable position.
- Keep both hands on the power plane during operation.

When using a router, feed the router against the bit's rotation.

ROUTERS

A *router* is used for on-the-job mortising and shaping operations, such as mortising door hinges or rounding the edges of a countertop. **See Figure 17-2.**

Depending on the operation being performed or the edge shape desired, different types and shapes of bits are available. The bits are inserted into the chuck of a router and secured into position. Router bits and the types of cuts they produce are shown in **Figure 17-3.**

A *laminate trimmer* is a specialized type of router. **See Figure 17-4.** Laminate trimmers are used to trim and shape the edges of plastic laminate countertops.

Router Safety

The sharp and exposed bit poses the primary danger when using a router or laminate trimmer. When a cut is completed, turn off the motor and wait until the bit stops rotating before removing the tool from the work. Between operations, lay the router on its side so as not to damage or dull the bit. Always disconnect the plug from the power source when inserting router bits and making depth adjustments. Additional router safety rules to observe include the following:

- Always wear appropriate personal protective equipment including the proper eye protection and hearing protection.
- Ensure the router bit is securely mounted in the chuck and the base is tight.
- Properly secure stock with a clamp or in a vise.
- Check stock thoroughly for nails, staples, screws, or other foreign objects prior to cutting.
- Keep electrical cords clear of cutting area.
- Keep both hands on router handles until the motor has stopped and the bit has stopped turning.
- Do not overreach; maintain proper footing and balance.

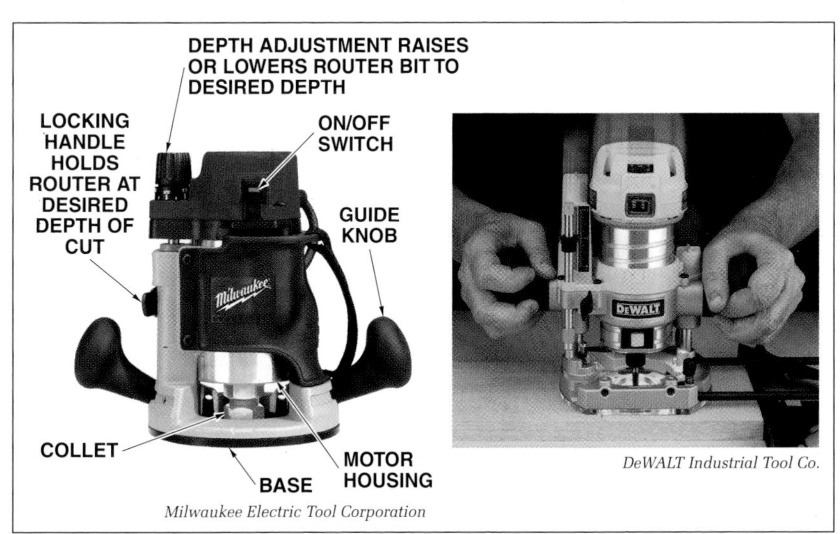

Milwaukee Electric Tool Corporation

DeWALT Industrial Tool Co.

Figure 17-2. *Routers are used for mortising and shaping operations.*

- The sound of a router motor can indicate safe cutting speeds. When a router is fed into material too slowly, the motor makes a high-pitched sound. When the router is pushed too hard into the stock, the motor makes a low-pitched sound.
- When making deep cuts, make two or more passes to prevent kickback.

Most router bits are carbide tipped since solid carbide bits are too expensive. Carbide tips are brazed to the bits.

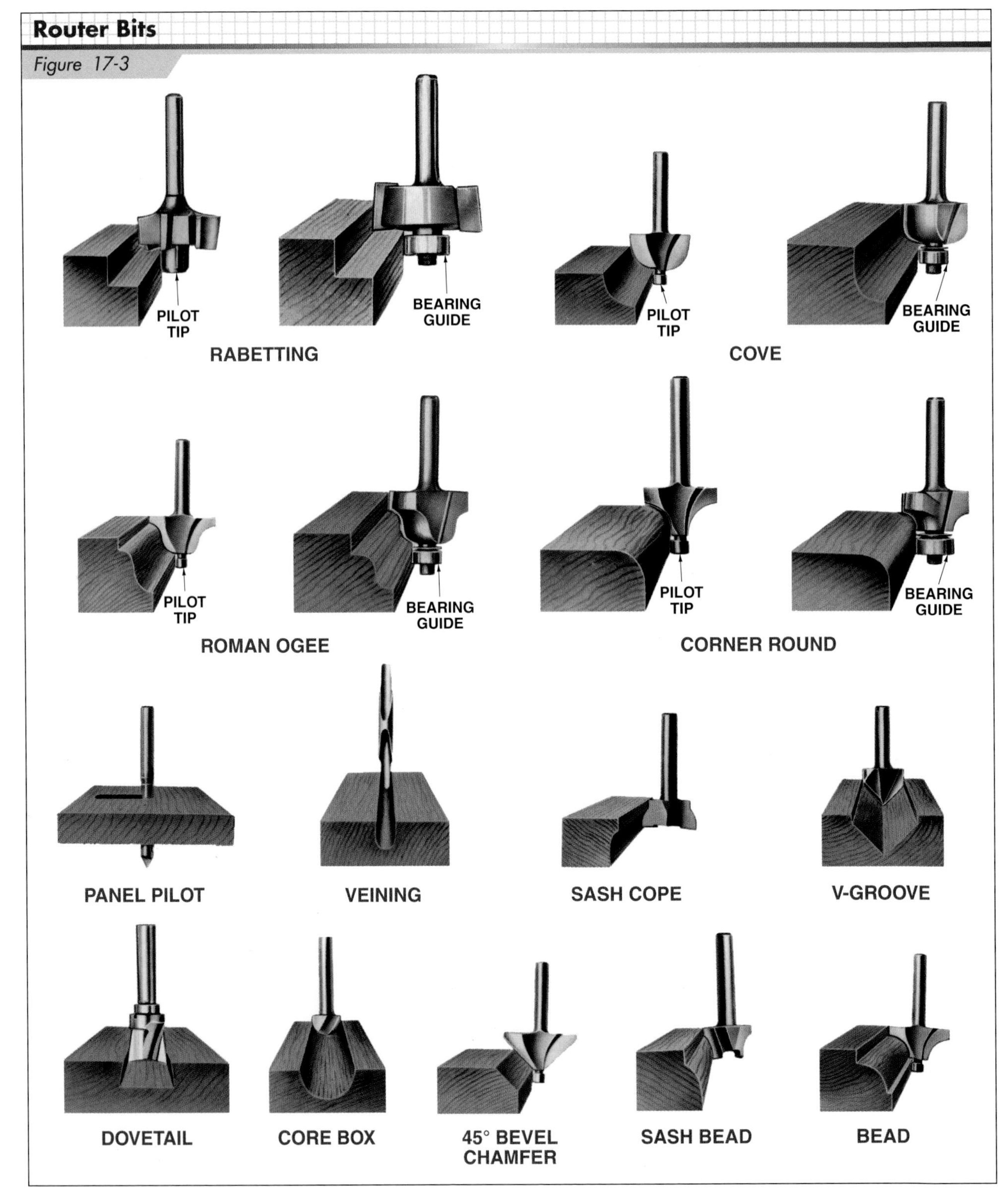

Figure 17-3. *Router bits are inserted into a router chuck to make different cuts.*

DeWALT Idustrial Tool Co.

Figure 17-4. *Laminate trimmers are used to trim and shape the edges of countertops.*

PORTABLE POWER SANDERS

In general, finish materials delivered to a job site require little sanding. However, on occasion the surfaces of materials are scratched or damaged while being transported and require sanding prior to finishing. *Slash grain*, a condition in which the grain changes direction, may occur on a door edge, where it prevents a smooth finish after the door has been planed. Portable power sanders, including belt and finish sanders, are used during the final trim stages of construction.

Belt Sanders

A *belt sander* uses a rotating abrasive belt to produce a smooth finish. **See Figure 17-5.** Abrasive belts are 3″ or 4″ wide, and range in length from 21″ to 27″. Various grades of abrasives are available, which are selected according to the material being sanded and the amount of material to be removed.

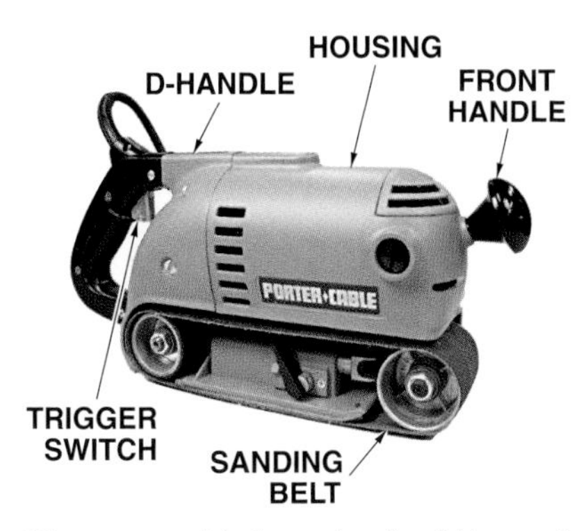

Figure 17-5. *A belt sander should be used with care to avoid gouging the wood.*

A belt sander should be used with care since careless handling and operation can easily gouge a wood surface. When a belt sander is brought into contact with a wood surface, it should remain moving to avoid sanding a groove in the surface. Keep hands on the belt sander handles and away from the rotating belt.

Finish Sanders

A *finish sander* is used for lighter and finer sanding than a belt sander. Three types of finish sanders are available—orbital, random orbital, and detail. **See Figure 17-6.** An *orbital sander* operates with an orbital (circular) motion. A *random orbital sander* has oscillating (back-and-forth) and orbital movements. Detail sanders operate with an orbital motion.

Abrasives

Abrasive belts, sheets, and pads used with portable power sanders are made of four different materials—flint, garnet, silicon carbide, and aluminum oxide. Flint and garnet are natural minerals, which are crushed into fine particles. Silicon carbide and aluminum oxide are synthetic materials that provide excellent service and are recommended for most power sanding operations.

Abrasives are divided into fine, medium, and coarse grades. The grades are further divided into various *grit numbers,* which are stamped on the back of an abrasive belt, sheet, or pad. A chart of grit numbers is shown in **Figure 17-7.** *Grit* refers to the number of abrasive particles per square inch. The lower the grit number, the larger the grit, and the rougher the abrasive paper. The higher the grit number, the smaller the grit, and the smoother the abrasive paper.

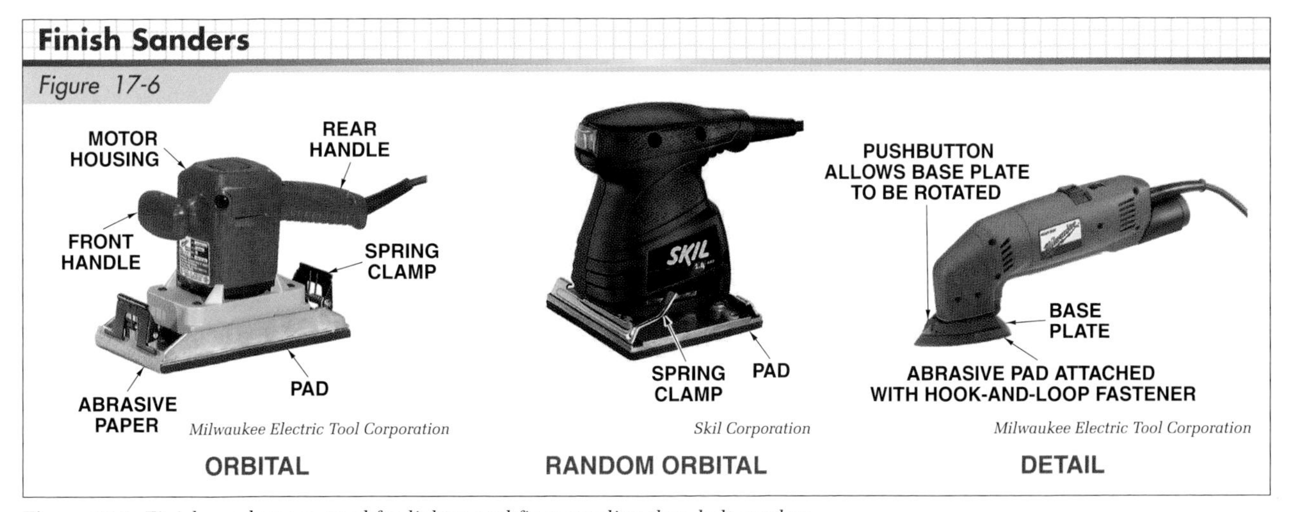

Figure 17-6. *Finish sanders are used for lighter and finer sanding than belt sanders.*

ABRASIVE GRIT SIZES

Grade	Grit Type		
	Synthetic*	Flint	Garnet
Very Fine	400		
	360		
	320	7/0	
	280	6/0	8/0
	240	5/0	7/0
	220	4/0	6/0
Fine		3/0	
	180		5/0
	150		4/0
		2/0	
	120		3/0
		0	
	100		2/0
Medium		½	
	80		0
		1	
	60		½
	50	1½	1
Coarse		2	
	40		1½
		2½	
	36		2
	30	3	2½
Very Coarse	24		3
	20		3½
	16		
	12		

*includes silicon carbide and aluminum oxide

Figure 17-7. *Grit numbers, typically stamped on the back of the belts, sheets, or pads, indicate the coarseness of the abrasive grits.*

Power Sander Safety

Hand and arm abrasions are the primary type of injury that occurs with power sanders. When using belt sanders, injuries may occur if fingers become pinched between the abrasive belt and housing. Use two hands when operating a sander—one on the trigger switch and the other on the front handle knob. Additional power sanding safety rules to observe include the following:

- Wear appropriate personal protective equipment including the proper eye, hearing, and respiratory protection.
- Lift the sander away from the work before turning off the motor. Wait until the movement has completely stopped before setting down the tool.
- Keep the electrical cord away from the area being sanded.
- Ensure the sander is turned off before connecting the cord to the power source.
- Disconnect the power source before changing a sanding belt, making adjustments, or emptying the dust bag.
- Replace sanding belts that are worn or frayed.
- Do not use a sander without an exhaust system or a dust bag present that is in good working condition. Dust created when sanding can be a fire and explosion hazard and proper ventilation is required.
- Do not overreach. Always keep proper footing and balance.
- Do not start a sander while in contact with a workpiece.

Unit 17 Review and Resources

UNIT 18 Pneumatic and Powder-Actuated Tools

OBJECTIVES

1. Describe common pneumatic nailers and staplers.
2. Identify common safety precautions for pneumatic tools.
3. Describe the principles of powder-actuated tools.
4. Describe powder loads and powder-actuated fasteners.
5. Identify factors for installing powder-actuated fasteners.
6. List common safety precautions for powder-actuated tools.

Pneumatic nailers and staplers and powder- actuated tools are commonly used in the construction industry. Greater worker productivity and less worker fatigue result when pneumatic and powder-actuated tools are used. Pneumatic tools are extensively used in framing operations, as well as for attaching sheathing, light metals, and plastic materials. Powder-actuated tools are used to fasten materials to concrete and steel.

PNEUMATIC NAILERS AND STAPLERS

Pneumatic nailers and staplers are used in framing and finish work, as well as for a variety of other structural applications. The basic component of pneumatic nailers and staplers is a *piston head* charged by air that, in turn, pushes a *piston* that drives a nail or staple. **Figure 18-1** illustrates the operation of a pneumatic nailer.

Pneumatic nailers and staplers are designed to fire in a number of ways. With *contact trip* pneumatic nailers and staplers, the trigger is pulled first and the tool fires each time the tip is pressed against a hard surface. *Sequential trip* pneumatic tools fire only one fastener with a single pull of the trigger. Pneumatic nailers and staplers have an integral safety feature that prevents the tool from firing unless the tip is pressed firmly against a surface.

Air Compressors

Most pneumatic tools are powered by compressed air traveling through an air hose, which is connected to an *air compressor.* A compressor has one or more pistons, which pull air into the cylinder on the downstroke and push the air into the hose and pneumatic tool on the upstroke. A constant supply of compressed air flows through the hose and into the tool reservoir. The amount of air pressure required to operate a pneumatic nailer or stapler depends on its size and the type of operation being performed. Air pressure delivered to the pneumatic tool is adjusted by a *regulator valve* on the compressor.

Air compressors are available in different sizes. Large compressors can supply proper pressure for several tools at one time. Portable, lightweight air compressors, however, are more practical for many construction jobs. **See Figure 18-2.** Most large air compressors are gasoline-powered. Smaller, portable air compressors are often powered by electricity.

Pneumatic Nailers

Most pneumatic tool manufacturers produce an assortment of nailers. *Framing* and *brad and finish nailers* are most commonly used by carpenters. Specialized pneumatic nailers are also available for fastening roofing materials, insulation, and house wrap.

Framing Nailers. As their name implies, *framing nailers* are used for framing operations, including nailing studs to top and bottom plates and nailing joists and rafters to top plates. Framing nailers are also used to fasten sheathing and subfloor to framing members. In addition, framing nailers may be used to fasten metal connectors to framing members. However, the framing nailer must be equipped with a hole-locating mechanism for use with metal connectors, or be especially designed for fastening metal connectors.

Framing nailers are available in various power capacities and drive a wide range of sizes and types of nails. A large framing nailer can drive a 16d (3½″) nail into a piece of oak. Special hardened steel nails can be used to fasten wood plates to a concrete slab.

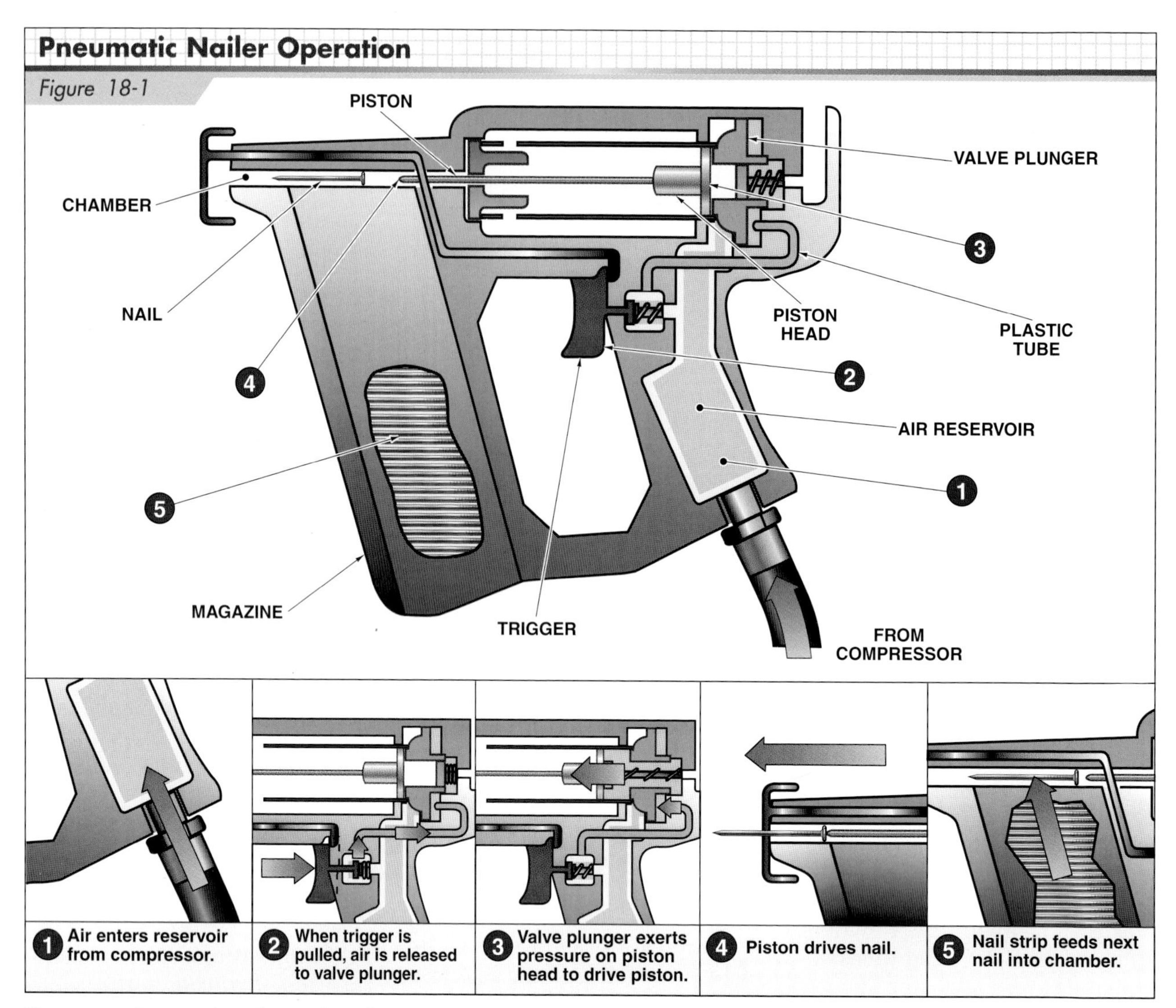

Figure 18-1. *Pneumatic nailers and staplers use compressed air to drive nails and staples.*

Ingersoll-Rand Material Handling

Figure 18-2. *An air compressor provides the power for pneumatic tools. The portable compressor shown here can accommodate two nailers at one time.*

Air pressure for a pneumatic framing nailer should be adjusted so that nail heads are driven flush with or slightly below the material surface. A nail may drive too deeply into the wood if the air pressure is too high, resulting in decreased holding power of the nail.

Nails are fed into the chamber of a pneumatic nailer using a straight strip or coil. **See Figure 18-3.** The magazine of a strip-fed nailer may be horizontal to the work or slanted. A slanted magazine offers an advantage when nailing in tight spaces.

The nails used in framing nailers are identified as *round-head* (full head) and *clipped-head.* Round-head nails have conventional circular heads and are commonly used in the southern United States and areas prone to hurricanes and other severe weather conditions. Clipped-head nails are flat on one side and round on the other, and are commonly used in areas with milder climates. The main advantage of clipped-head nails is that more nails are in a nailing strip, allowing for fewer reloads of the magazine.

Pneumatically driven nails have greater holding power than hammer-driven nails. Most types of nails used in pneumatic nailers are held together in the magazine or coil by a resin-adhesive coating on the shaft of each nail. Friction created when driving a nail produces heat to melt the resin. The resin lubricates the nail, helping the nail to penetrate the wood more easily. The resin returns to a solid state within seconds after penetration, providing additional holding power for the nails.

Figure 18-3. *Framing nailers can be used for most framing applications. The strip-fed pneumatic nailer has an angled magazine that can hold 65 2″ to 3½″ round-head nails. The coil nailer holds 250 2″ to 3¼″ nails.*

Palm Nailers. *Palm nailers,* or hand nailers, are compact pneumatic tools used to drive fasteners for metal connectors. **See Figure 18-4.** Palm nailers are capable of driving common, spiral-shank, and annular nails ranging in size from 5d to 70d. Nails are loaded individually into the palm nailer. The nail tip is placed through the opening of the metal connector, and pressure is applied against the framing member to trigger the nailer.

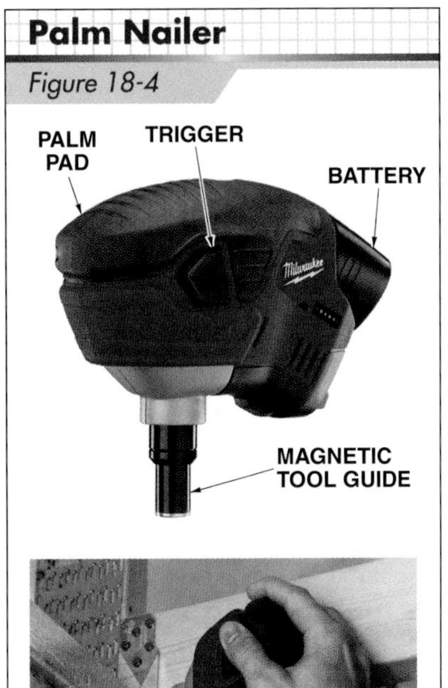

Milwaukee Electric Tool Corporation

Figure 18-4. *Palm nailers are used to drive fasteners for metal connectors.*

Finish and Brad Nailers. *Finish and brad nailers* are used to fasten interior trim materials such as baseboard, wall moldings, paneling, door and window trim, and cabinet assemblies. **See Figure 18-5.** Finish and brad nailers accommodate a variety of finish nail and brad sizes. Finish and brad nailers are adjustable to countersink nails below the material surface or flush with the surface.

ITW Paslode

Figure 18-5. *Finish and brad nailers are used to fasten trim materials such as door and window trim, baseboard, and paneling.*

Cordless Pneumatic Nailers. Although the majority of pneumatic nailers require an air compressor as a power source, *cordless pneumatic nailers* are also available that do not require a compressor. Cordless pneumatic nailers are powered by an internal combustion power source, which includes a battery and replaceable fuel cell. The battery causes a spark to ignite a fuel and air mixture, which produces the power to drive the piston. **See Figure 18-6.**

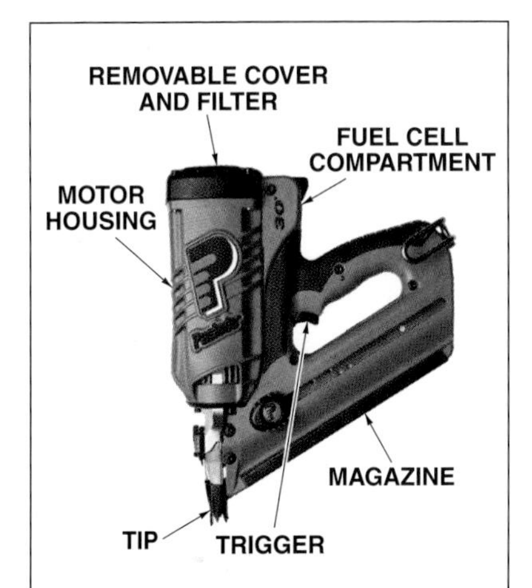

ITW Paslode

Figure 18-6. *A cordless pneumatic nailer is a self-contained tool containing a rechargeable battery and fuel cell. The nailer shown here can drive approximately 4000 nails before recharging its battery. The fuel cell life is approximately 1100 to 1300 nails.*

Pneumatic Staplers

Pneumatic staplers are used for applications similar to pneumatic nailers. Under some conditions, a staple has advantages over a nail as a fastener. A staple will not split the wood as easily as a nail when driven near the end of a board. Since a staple is a two-legged device covering a greater surface area, a staple is an excellent fastener for sheathing, shingles, building paper, and other construction materials.

Pneumatic staplers are available in various sizes ranging from models for heavier operations such as sheathing and subfloor, to light-duty models for trim work and cabinet assembly. **See Figure 18-7.** Pneumatic staplers accommodate different lengths of staples for fastening different types and thicknesses of materials.

Senco Products, Inc.

Figure 18-7. *Pneumatic stapler models range from heavy-duty models for fastening sheathing and subfloor, to light-duty models used for trim work and cabinet assembly.*

Pneumatic Tool Safety

When pneumatic nailers and staplers are used, the safety of not only the operator, but also other people in the immediate area must be considered. For example, before panels are fastened to studs or joists, ensure that no one is on the other side of the wall or below the joists. If the nail or staple misses the stud or joist, it could go through the panel and severely injure someone. Other essential safety precautions that must be observed when using pneumatic nailers and staplers are as follows:

- Follow proper operating procedures as outlined in the operator manual for the particular tool.
- Wear safety glasses or a face shield. Wood chips, concrete, or a deflected nail can cause serious eye injury.
- Ensure that air hoses and connections are in good operating condition. Air hoses that are larger than ½″ diameter must have a safety device at the air compressor or branch line to stop air flow in case of hose failure.
- When using a portable air compressor, ensure that exposed belts have guards on both sides to reduce the possibility of injuries to fingers and hands.
- Use the right nailer or stapler for the job, and use the correct size of nail or staple. Refer to manufacturer recommendations.
- Do not lift or carry a pneumatic nailer or stapler by the air hose.
- Inspect pneumatic nailers and staplers daily to ensure that the firing mechanism and safety features are operating properly. Test-fire the tool into a block of wood before using the nailer or stapler for the desired application.
- Point the tip of a pneumatic nailer or stapler away from your body and other workers.
- Per OSHA 29 CFR 1926.302, *Power-operated Hand Tools,* pneumatic nailers and staplers that operate at more than 100 psi air pressure must have a safety device on the tip to prevent the nailer or stapler from ejecting fasteners unless the tip is in contact with the work surface.
- Disconnect pneumatic tools from the air supply when the tools are not in use. Pneumatic tools should be equipped with a fitting that releases air pressure from the tools when disconnected.
- When toenailing, do not support or back up a workpiece with a foot or knee.
- Always keep your finger off the trigger when a pneumatic tool is not in use.

POWDER-ACTUATED TOOLS

Powder-actuated tools, often referred to as *PATs* or *stud guns,* are used to fasten building materials or other objects to concrete, masonry, and steel without predrilling holes for anchors. Powder-actuated tools are used to fasten wood to concrete, wood to steel, steel to concrete, and metal connectors to steel. **See Figure 18-8.** For example, powder-actuated tools are commonly used to fasten bottom plates of wood-framed walls to concrete slabs and foundation walls.

In concrete buildings, structural components such as top and bottom plates and corner studs of wood interior walls are secured to the floors, walls, and ceilings with powder-actuated tools. Powder-actuated tools are also used to attach the top and bottom channel tracks for metal-framed walls and solid gypsum board partitions.

Only workers who are trained in the proper operation of a powder-actuated tool and have obtained an operator license can operate powder-actuated tools. This requires the successful completion of a test on the safe operation of power-actuated tools.

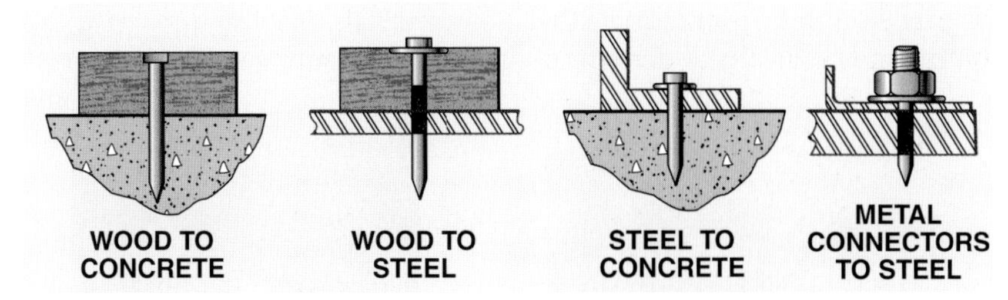

Figure 18-8. *Powder-actuated tools are used to fasten wood to concrete, wood to steel, steel to concrete, or metal connectors to steel.*

Operating Principles

Powder-actuated tools are similar in operating principle to conventional firearms. Powder-actuated tools operate on the indirect-acting principle. A powder charge triggered in the chamber of a powder-actuated tool produces expanding gases. The gases propel a piston, which in turn drives the fastener into the material being attached to the concrete or steel. **See Figure 18-9.**

Powder-actuated tools are designed so they will not operate unless pressed against a work surface with a force at least 5 lb greater than the total weight of the tool. In addition, the tool may not operate if it is tilted more than 8° from the work surface. An integral safety feature does not permit the tool to fire until pressure is exerted to depress the safety mechanism and deactivate the trigger interlock.

Types of Powder-Actuated Tools

Powder-actuated tools are classified as *low-*, *medium-*, and *high-velocity* based on ballistics tests of the different combinations of power loads and fasteners used for softer and harder materials. The average test velocity of low-velocity powder-actuated tools does not exceed 328 feet per second (fps), the average test velocity of medium-velocity tools ranges from 328 fps to 492 fps, and the average test velocity of high-velocity tools exceeds 492 fps.

Powder-actuated tools are produced in a variety of styles and configurations. In general, powder-actuated tools are classified as *single-shot*, *semiautomatic*, and *automatic*. **See Figure 18-10.** Powder-actuated tools must meet the requirements of ANSI A10.3, *Safety Requirements for Explosive-Actuated Fastening Systems.*

Single-Shot Powder-Actuated Tools. The powder load and fastener must be manually loaded into a single-shot powder-actuated tool each time the tool is discharged. Single-shot powder-actuated tools are adequate for construction projects requiring less frequent use of the tool. Most single-shot powder-actuated tools are discharged by pulling a trigger. Other single-shot powder-actuated tools are discharged by striking the end of the tool with a hammer.

Semiautomatic Powder-Actuated Tools. For increased production, a semiautomatic powder-actuated tool may be used. The powder loads used with a semiautomatic tool are attached to a strip or disk which is inserted into the tool. The powder loads advance into firing position after the previous load is discharged. The tool operator must load the fasteners individually for each shot.

Automatic Powder-Actuated Tools. For applications where a large number of fasteners must be installed, an automatic powder-actuated tool is used. Automatic tools feature a magazine that holds 10 nails or fasteners. The powder loads are attached to a strip or disk which is inserted into the tool. The powder loads and fasteners automatically advance after the previous load and fastener are discharged.

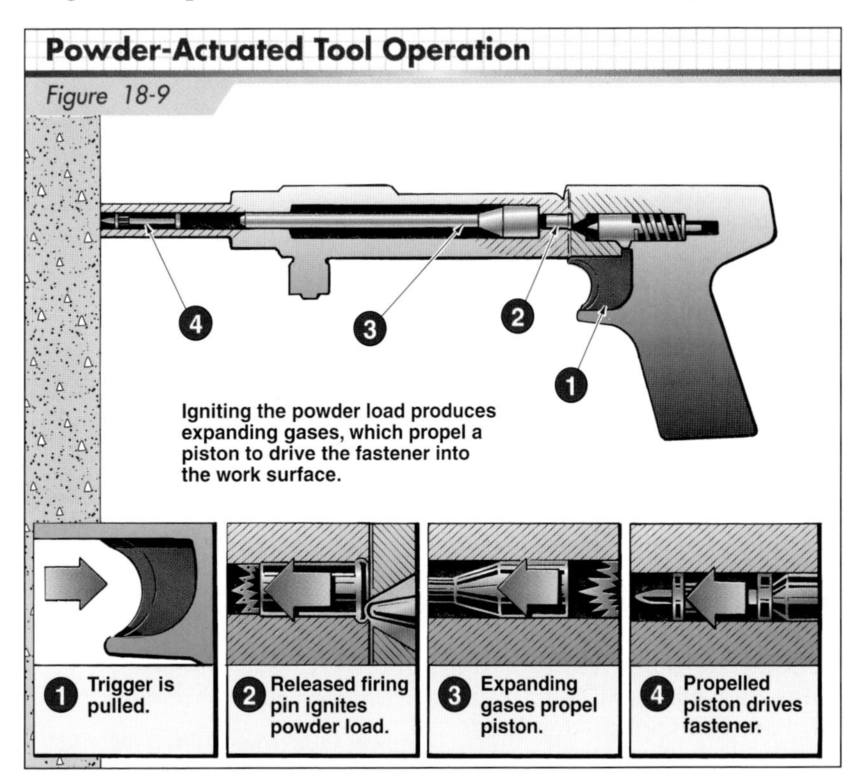

Figure 18-9. *Powder-actuated tools operate on the indirect-acting principle.*

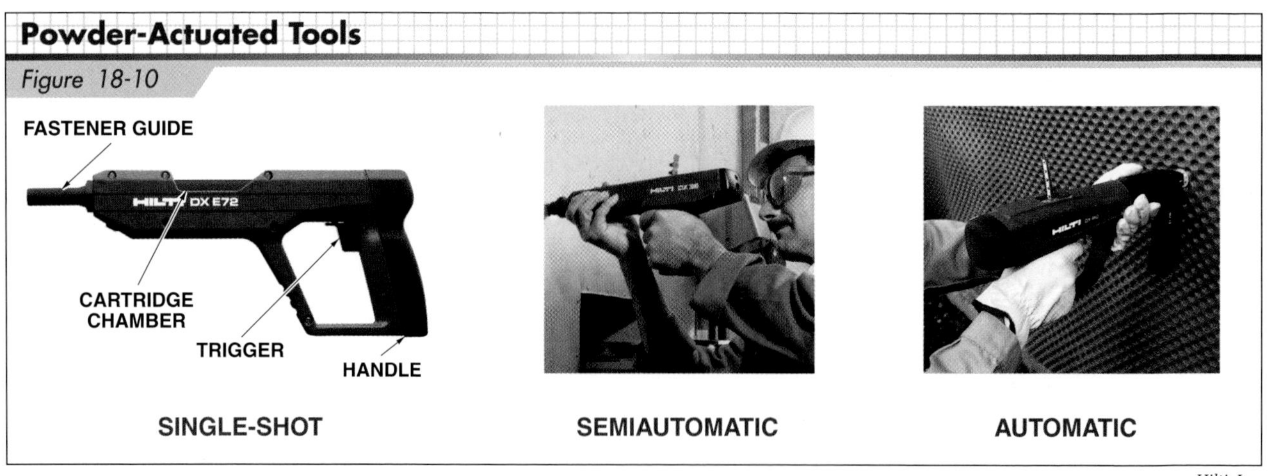

Hilti, Inc.

Figure 18-10. *Powder-actuated tools are classified as single-shot, semiautomatic, and automatic. Single-shot powder-actuated tools are adequate for applications requiring moderate use of the tool. Semiautomatic powder-actuated tools are designed to increase productivity. Automatic powder-actuated tools include a magazine to hold the drive pins.*

Powder Loads

A *powder load* is a metal casing that contains a powder propellant and is crimped closed on the end and sealed. Powder loads resemble a shell casing used in conventional firearms. When the powder propellant is ignited as the powder-actuated tool is discharged, gases from the ignited powder exert pressure against the piston and drive it against the fastener. Most powder-actuated tools have a .22, .25, or .27 caliber bore diameter and require a load of the same caliber. Only powder loads designated by the manufacturer for the particular tool should be used.

Powder loads are available as individual units for single-shot powder-actuated tools. Powder loads are also available in plastic strips and metal disks for use with semiautomatic and automatic powder-actuated tools. **See Figure 18-11.**

Powder loads for powder-actuated tools range from low to higher velocities, with #1 loads being the lowest velocity loads and #6 loads being high velocity loads. Powder loads are also color-coded for ease of identification. **See Figure 18-12.** Each color represents the average velocity of the powder load needed to drive a fastener into a particular material (usually concrete or steel).

Proper powder load charges are needed to drive powder-actuated fasteners to the desired depths.

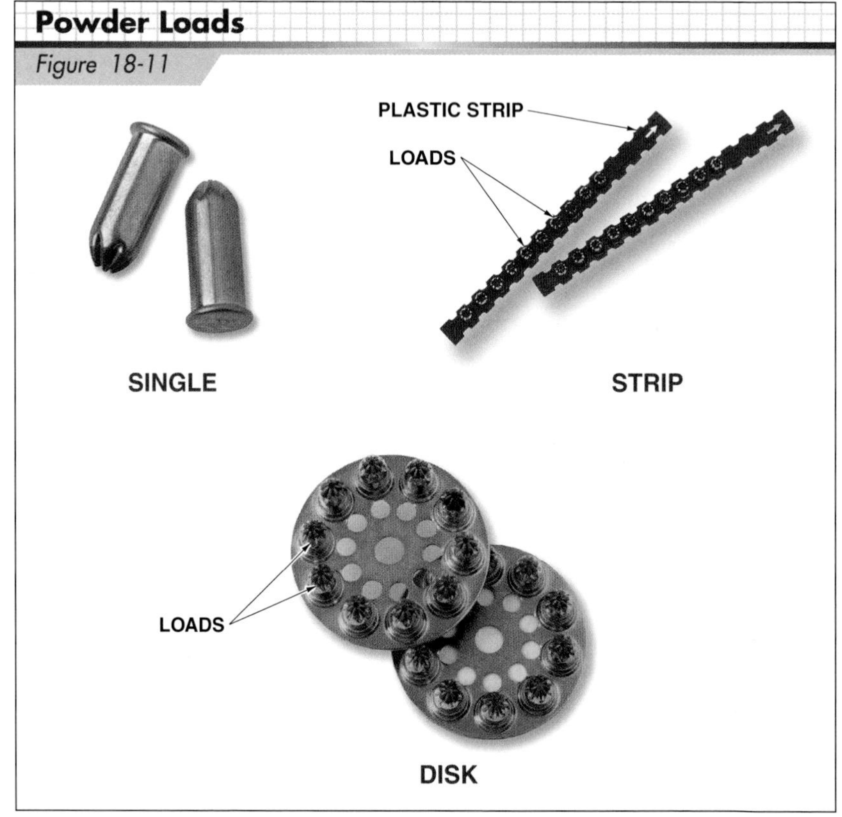

ITW Ramset/Redhead

Figure 18-11. *Powder loads are available as single units, or are collated in groups of 10 into plastic strips and metal disks used with semiautomatic or automatic powder-actuated tools.*

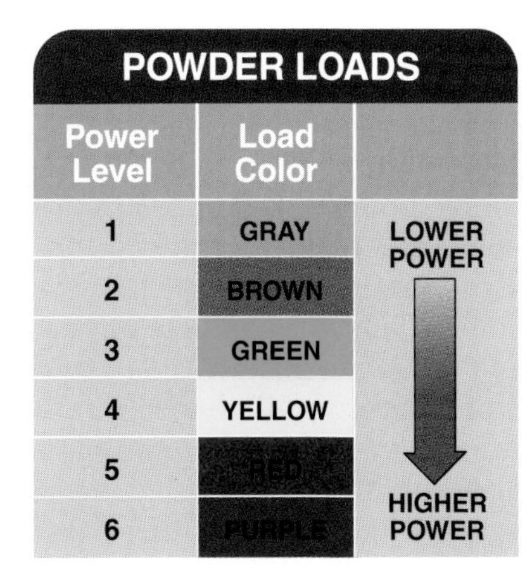

Power Level	Load Color	
1	GRAY	LOWER POWER
2	BROWN	
3	GREEN	
4	YELLOW	
5	RED	
6	PURPLE	HIGHER POWER

Figure 18-12. *Powder loads are available in a wide range of velocities ranging from #1 (gray) loads to #6 (purple) loads.*

Powder-Actuated Fasteners

Powder-actuated fasteners are available in a variety of lengths and diameters. The two basic types of fasteners are *drive pins* and *threaded studs.* Drive pins are used for permanent installation of materials or objects—for example, fastening a sill plate to a concrete slab. Threaded studs are used for applications where the material or object being fastened to the concrete or steel base may need to be moved. Drive pins and threaded studs have a premounted plastic fluting to hold the fastener centered in the tool muzzle prior to driving.

Drive Pins. Drive pins are more commonly used than threaded studs. Drive pins are manufactured from steel that has been treated for extra hardness and must be harder than the material into which they are driven. Drive pins driven into steel often have a knurled shank for extra holding power.

Drive pins range in length from ½″ to 3″. Head and shank dimensions are typically indicated in millimeters and/or inch fractions. The most common head sizes are .300″, 6 mm (¼″), 8 mm (5/16″), and ⅜″. **See Figure 18-13.** Drive pin shank diameters are specified by the diameter of the wire used to make the pin. For example, drive pins with 6 mm and 8 mm heads commonly have .143″ diameter shanks.

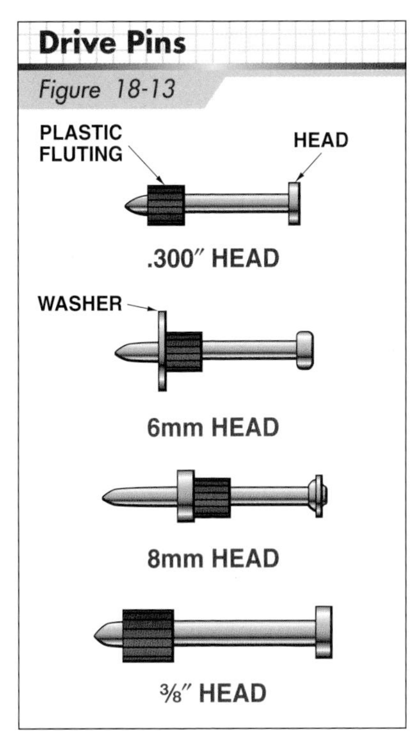

Figure 18-13. *Drive pins range in length from ½″ to 3″ and have a premounted plastic fluting to center the fastener in the muzzle. Drive pins are available with washers that prevent the pins from being driven too deeply into the material being fastened.*

Drive Pin Assemblies. Drive pin assemblies consisting of preassembled clips, drive pins, and plastic fluting are available for a variety of applications. Ceiling clip assemblies are commonly used for attaching suspended ceiling grids. Conduit clip assemblies are used to secure electrical conduit to walls or ceilings. **See Figure 18-14.**

Threaded Studs. Threaded studs are a one-piece fastener with a drive pin on one end and threads at the other end. When a threaded stud has been driven into place, a washer and nut are used to secure the object being fastened, making it possible to remove the object altogether or loosen it when shimming or other adjustment is required. Threaded studs are commonly used for installation of suspended sprinkler and lighting systems and are available in ¼″–20 and ⅜″–16 diameters and threads per inch. **See Figure 18-15.**

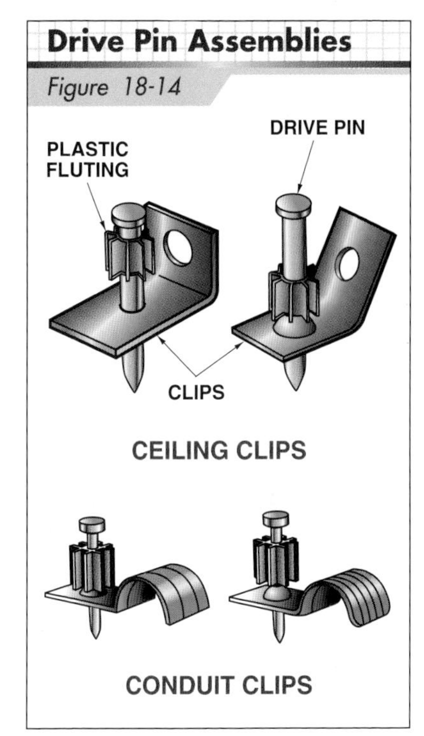

Figure 18-14. *Drive pin assemblies are used for applications such as hanging suspended ceilings or supporting electrical conduit.*

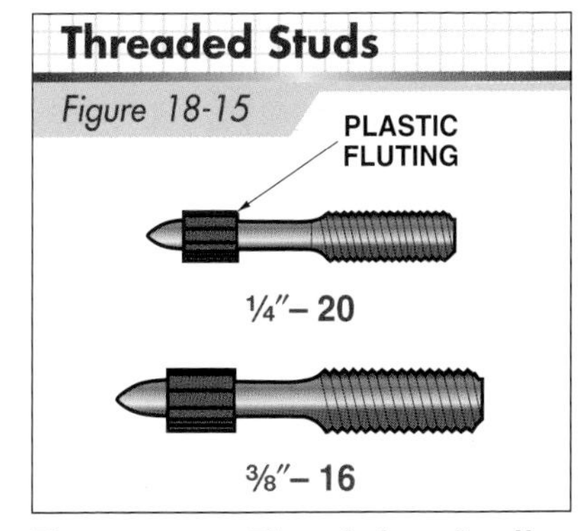

Figure 18-15. *Threaded studs allow objects to be removed or loosened for proper shimming.*

Installing Powder-Actuated Fasteners

When driving a powder-actuated fastener into concrete, a *compression bond* holds the fastener in place. The fastener displaces the concrete, which attempts to return to its original shape, producing a squeezing or compression action against the fastener. **See Figure 18-16.**

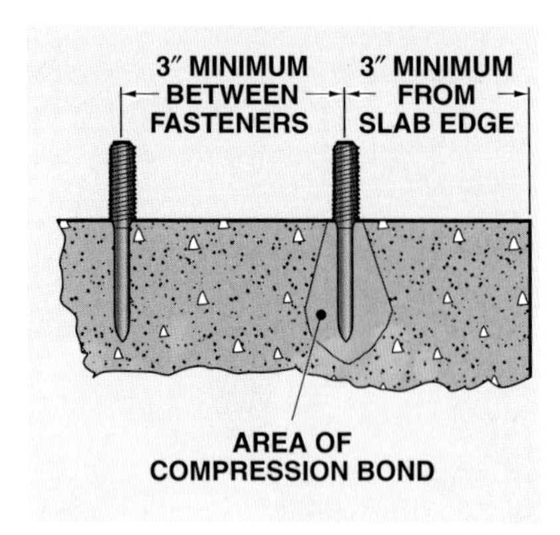

Figure 18-16. *Concrete exerts a compression bond to hold drive pins and threaded studs in place.*

Fasteners should not be driven into concrete unless the concrete thickness is at least three times the fastener shank penetration. In addition, fasteners should not be driven closer than 3″ from the unsupported edge or corner of concrete unless a special guard, fixture, or jig is used. Low-velocity tools should not be used to drive fasteners closer than 2″ from an unsupported edge or corner.

Concrete that has cured less than seven days will not have the compressive strength required to adequately secure a powder-actuated fastener. Concrete should cure for a minimum of 28 days before powder-actuated fasteners are installed. Powder-actuated tool fasteners should not be used in excessively hard concrete or in stone or brittle materials such as glass, tile, or brick.

When powder-actuated fasteners are installed in steel, a clamping force is produced on the fastener because of the natural tendency of the steel to return to its original shape. **See Figure 18-17.** Powder-actuated fasteners driven into steel should not be closer than ½″ from the edge. Low-velocity tools should not be used to drive fasteners closer than ¼″ from the edge.

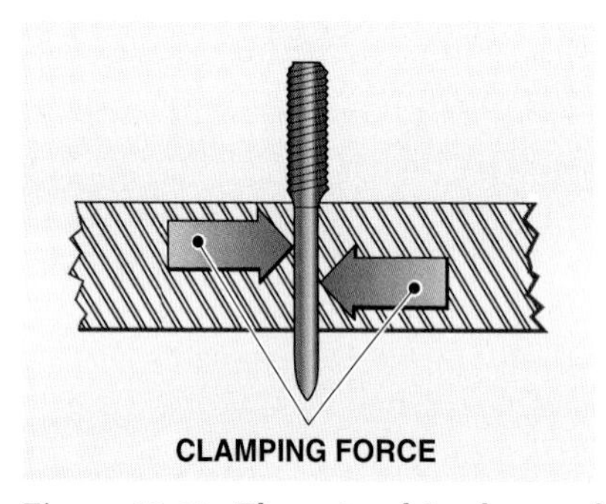

Figure 18-17. *The natural tendency of steel is to return to its original shape, producing a clamping force against a drive pin or threaded stud.*

Center Point Test. A *center point test* is used to determine the suitability of concrete for powder-actuated fasteners. To conduct a center point test, first select the powder-actuated fastener to be used. Place the point of the fastener against the concrete surface and strike the head with a single hammer blow. If the concrete shows a well-defined depression from the impact, the concrete is suitable for powder-actuated fasteners. **See Figure 18-18.**

If the point is blunted, the concrete is too hard for the powder-actuated fastener. If the concrete cracks or shatters, the concrete is too brittle. If the fastener penetrates into the concrete, the concrete may be too soft. If the results of the center point test indicate that powder-actuated fasteners are not suitable for the installation, anchors such as expansion shields should be installed.

Driving Powder-Actuated Fasteners. Fastener shank length and powder load charge must be considered before installing drive pins. The fastener shank must be long enough to go through the material and adequately penetrate the concrete or steel base. If the fastener is driven too deep or too shallow, a weak connection is created. To determine the

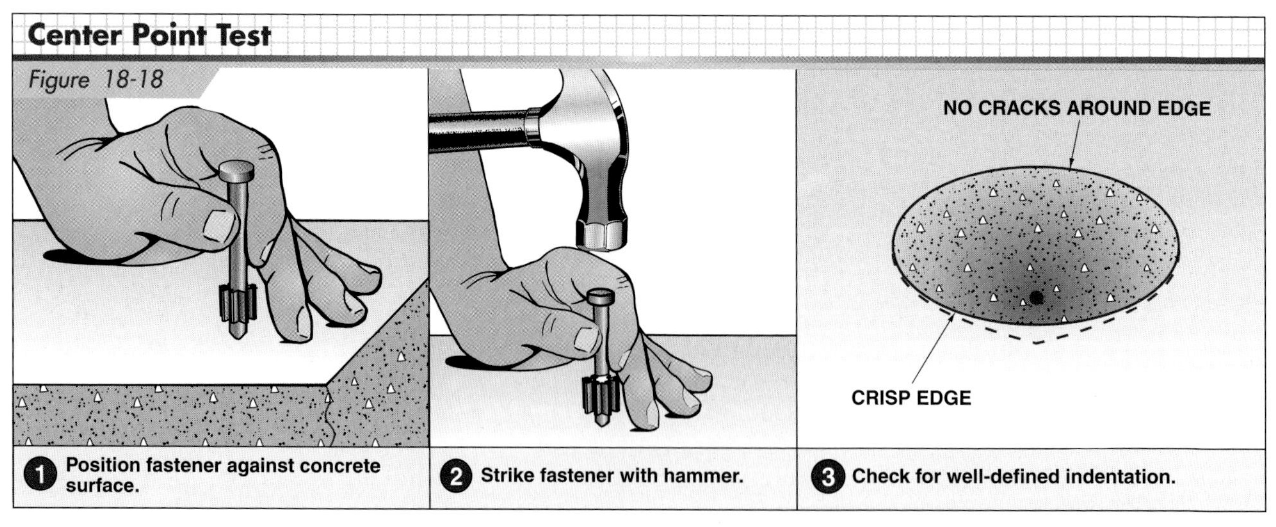

Figure 18-18. *A center point test is used to determine the suitability of concrete for powder-actuated fasteners.*

proper drive pin shaft length for concrete, add the thickness of the material being fastened and the required penetration. To determine the proper drive pin shaft length for steel, add the thickness of the material being fastened, the steel thickness, and ¼″.

The proper powder load charge must be used to drive the pin to the appropriate depth. When fastening a piece of wood to concrete or steel, the top of the drive pin head should be flush with the surface of the wood. If the head is driven below the wood surface, the powder load is too strong. If the head protrudes above the surface, the powder load is too weak.

Threaded studs should be driven so the full threaded portion of the stud is above the concrete or steel surface. To determine the proper thread length for threaded studs driven into concrete or steel, add the thickness of the material being fastened and an allowance for the washer and nut. The allowance is equal to the fastener thread diameter. For example, the allowance for a ¼″–20 threaded stud is ¼″.

A recommended practice to determine the proper powder load for a drive pin or stud is to make a trial installation using a #1 or #2 powder load. If the pin or stud is not driven to the proper depth, increase the load power until the proper installation depth is achieved.

Powder-Actuated Tool Safety

Per OSHA 29 CFR 1926.302, *Power-Operated Hand Tools,* only workers who have been trained in the operation of a powder-actuated tool and have obtained an operator license can operate powder-actuated tools. The operator must take and satisfactorily pass a quiz on the safe operation of powder-actuated tools. A qualified instructor can be located by contacting the regional office of the tool manufacturer or by contacting the Powder Actuated Tool Manufacturers' Institute (PATMI).

Powder-actuated tool operators must wear proper personal protective equipment such as safety goggles and earplugs or earmuffs. Powder-actuated tools should not be used in a flammable or explosive environment since the ignition of the charge produces a spark. Ensure there is adequate ventilation when the tool is used in confined spaces. Other workers should not stand close by or to one side of the operator. If a drive pin hits a reinforcing steel bar in the concrete, the pin can ricochet in a horizontal direction.

Powder-actuated tool operators should not drive fasteners into easily penetrated materials unless the materials are backed by a substance that will prevent the drive pin or threaded stud from passing completely through.

Powder-actuated tools must be inspected and tested each day to ensure that safety devices are in proper operating condition. Tools found not to be operating properly should be immediately removed from service and repaired.

A number of precautions should be taken to prevent a powder-actuated tool from firing accidentally. Powder-actuated tools should not be loaded until immediately prior to firing. Always bring the tool into the intended firing position before pulling the trigger.

If a powder-actuated tool misfires, hold the tool in place for at least 30 seconds and then try to operate the tool a second time. If the tool does not fire again, hold the tool in the operating position for another 30 seconds and then proceed to remove the load following the manufacturer instructions. If the problem is due to a tool defect, remove the tool from service. Additional safety precautions to be observed when using powder-actuated tools include the following:

- Never point any powder-actuated tool at a person, whether the tool is loaded or unloaded.
- Before firing a powder-actuated tool, inspect the unloaded chamber to ensure the barrel is clean and there are no obstructions.
- Keep hands away from the open barrel end.
- Be aware of electrical circuits and use caution when driving fasteners near electrical circuits.
- Do not leave a powder-actuated tool unattended.
- Do not carry a loaded tool from the job.
- Store tools and cartridges in a locked container when they are not in use. Be sure the tool is unloaded before storing it.
- Never carry fasteners or other hard objects in the same container or pocket as powder loads.
- Post warning signs stating the powder-actuated tools are in use within 50′ of the area where the tool is being used.
- Use a spall guard whenever possible.
- Use caution when driving pins into steel beams as the pin may ricochet and exit toward the operator.
- Safely dispose of misfired powder loads by submerging in water.

Unit 18 Review and Resources

UNIT 19

Welding and Metal-Cutting Equipment

OBJECTIVES

1. Describe welding processes.
2. Identify common safety precautions for electric arc welding.
3. Explain oxyacetylene welding and cutting operations.
4. Identify common safety precautions for oxyacetylene welding and cutting.
5. Explain plasma arc cutting operations.
6. Identify plasma arc cutting equipment.

Carpentry work often involves the use of metal materials that must be welded or cut. *Welding* involves the use of high heat which is applied to adjacent metal surfaces to join them. A filler metal may be added to the weld joint to strengthen it. During *cutting* operations, metal is heated and an oxygen jet from a cutting torch is applied, burning through the metal.

Figure 19-1. *Carpenters perform many welding and metal-cutting operations.*

Carpenters perform welding and metal-cutting operations, such as welding anchor bolts, straps, J-bolts, and joist hangers in place. They may also weld reinforcing bars, make repairs on steel forms, and weld structural members of steel-framed walls. **See Figure 19-1.**

In order to perform welding and metal-cutting operations, a carpenter must have the proper training or experience and be certified by passing an examination conducted by the proper authorities. Many apprenticeship programs today include welding courses as part of their required training. Evening courses are often available for journeyman carpenters who want to become proficient and certified in welding.

WELDING PROCESSES

While there are a variety of welding processes, the two basic types of welding processes most used by a carpenter are *electric arc welding* and *oxyacetylene welding.* Each method requires specialized skills and equipment. Carpenters normally use electric arc welding for welding. Oxyacetylene equipment, which can be used for both welding and cutting, is primarily used for cutting rather than for welding in construction work.

Modern welding processes evolved from inventions and discoveries dating back to 2000 BC.

Electric Arc Welding

In electric arc welding, metal surfaces are *fused* (melted together by the heat of an electric arc) together. A large amount of heat is produced by an arc that jumps between a stick electrode or wire electrode and the materials being welded. Arc welding machines, **Figure 19-2,** provide electric current to produce the welding arc. Arc welding machines may be alternating-current (AC) or direct-current (DC) models. Some transformer-rectifier models produce both AC and DC power.

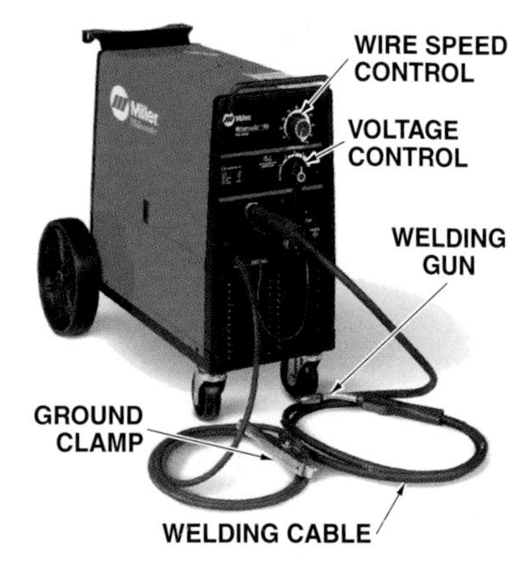

Miller Electric Manufacturing Company

Figure 19-2. *Arc welding machines are available as AC, DC, and transformer-rectifier models that produce both AC and DC power.*

Electric arc welding operates on the principle of a hot arc jumping across an air gap between the electrode and the base metal. The air gap produces a high resistance to current flow. Arc temperature ranges from 6000°F to 10,000°F.

In electric arc welding, a continuous circuit is formed with the arc welding machine and the piece of metal being welded. **See Figure 19-3.** Electrical current from the arc welding machine is transferred to the electrode holder or welding gun and workpiece through an *electrode lead.* An *electrode holder* is an insulated clamp that holds an electrode. A *welding gun* is a device that discharges the wire electrode to the weld joint. A *workpiece connection* is attached to the workpiece and a workpiece lead extends between the connection and the arc welding machine to complete the circuit.

Consumable stick or wire electrodes are available in different types of metals. The electrodes must be compatible with the type of metal being welded. Stick electrodes are approximately 12″ long and are replaced in the electrode holder when the electrode is approximately 1″ to 1½″ long. Wire electrode is available on spools and is fed into a welding gun using a feed control.

Electric Arc Welding Safety. Several potential safety hazards are associated with electric arc welding. Ultraviolet and infrared rays produced in electric arc welding can cause severe eye damage. Droplets of molten metal from an arc can cause burns and ignite flammable materials. OSHA 29 CFR 1926.351, *Arc Welding and Cutting,* includes detailed safety information relating to electric arc welding. Safety rules to observe when using electric arc welding equipment are as follows:

- Wear appropriate eye protection and fireproof clothing and headgear.
- Current-carrying parts of an electrode holder or gun must be fully insulated.
- Cable must be free of cracks or other defects. If such defects exist, immediately remove the cable from service.
- When electrode holders are left unattended, remove the electrode and protect the holders so they cannot make contact with other workers or conducting objects.
- Do not dip electrode holders in water to cool them.
- Whenever possible, shield electric arc welding operations with noncombustible or flameproof screens to protect other workers from the direct rays of the arc.
- Install electric arc welding equipment so it is properly grounded and has a power disconnect switch nearby.
- Shut off the power to an arc welding machine when making repairs.
- Periodically check cable connections to ensure they are tight.
- Do not overload cables or allow them to come in contact with hot metal, water, oil, or grease. Avoid dragging cables over or around sharp corners.
- Do not stand in water or on a wet floor or use wet gloves when welding.
- Dispose of used electrodes properly.
- Do not pick up pieces of metal that have just been welded or heated.
- Weld only in areas that are well-ventilated. When necessary, use an exhaust system to keep toxic gases below the prescribed health limits. Wear a respirator when welding metals that produce toxic fumes.

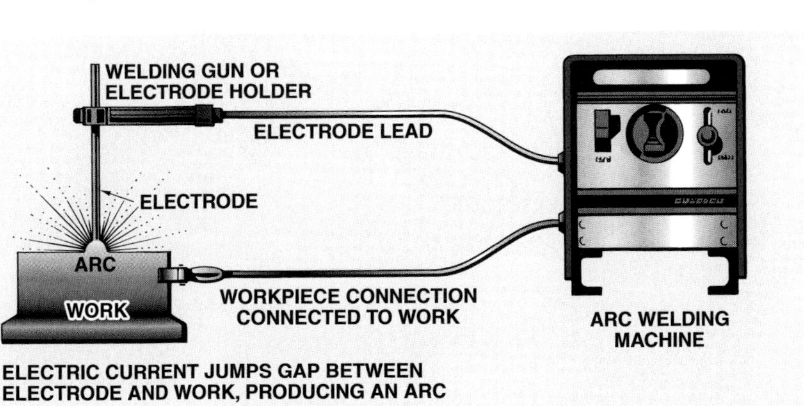

Figure 19-3. *In electric arc welding, metal surfaces are fused by the heat of an electric arc.*

The actual voltage used to provide welding current for electric arc welding is approximately 18 V to 34 V. However, high amperage is needed to produce the heat required for welding.

Oxyacetylene Welding and Cutting

In oxyacetylene welding, metal surfaces are fused together by the heat of a welding flame. The flame temperature may be as high as 6300°F. When the edges of the metal being welded reach the melting point, the two edges flow together to form a solid piece. Some joints require a filler metal to be added. The filler metal is added to the molten pool of the base metal. **See Figure 19-4.**

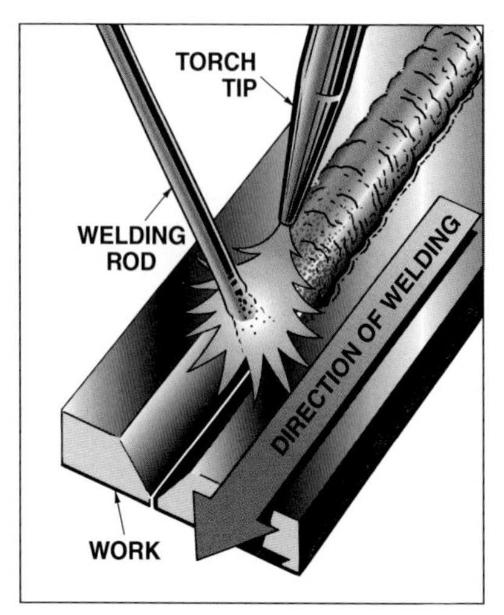

Figure 19-4. *Filler metal is commonly added to a weld joint during an oxyacetylene welding operation.*

Equipment setup for oxyacetylene welding and cutting operations is similar. Two cylinders hold the oxygen and acetylene gases, with oxygen in one cylinder and acetylene in the other cylinder. Hoses lead from the regulators on the tops of the cylinders to the torch. **See Figure 19-5.** Pressure regulators indicate the cylinder pressure and pressure of the gases flowing to the torch. Adjusting (needle) valves are located at the end of the torch to control the gas flow through the torch to the tip or nozzle. The pressures at the cylinders and adjustment of the valves determine the type of flame produced. The gases are mixed within the torch and delivered to the welding tip or nozzle where they are ignited to produce the heat for welding.

Before connecting a cylinder regulator to the cylinder, crack (slightly open) and immediately close the valve to clear dirt or dust in the valve. **See Figure 19-6.** Open a cylinder valve slowly to prevent damage to the regulator. Purge oxygen and acetylene hoses before lighting a torch. Use a friction lighter or other approved device to light torches.

The major difference between the welding and cutting setups is a cutting attachment that replaces the welding torch and tip with a cutting torch. **See Figure 19-7.** The cutting process begins by preheating the metal with a flame fed by the oxygen and acetylene gases. When the required temperature is reached, the oxygen valve lever on the cutting attachment is depressed to release only oxygen. The oxygen causes the hot metal to burn very hot and forces the molten metal from the kerf.

Figure 19-6. *Crack (slightly open) a cylinder valve prior to installing a regulator to discharge dirt or dust from the valve.*

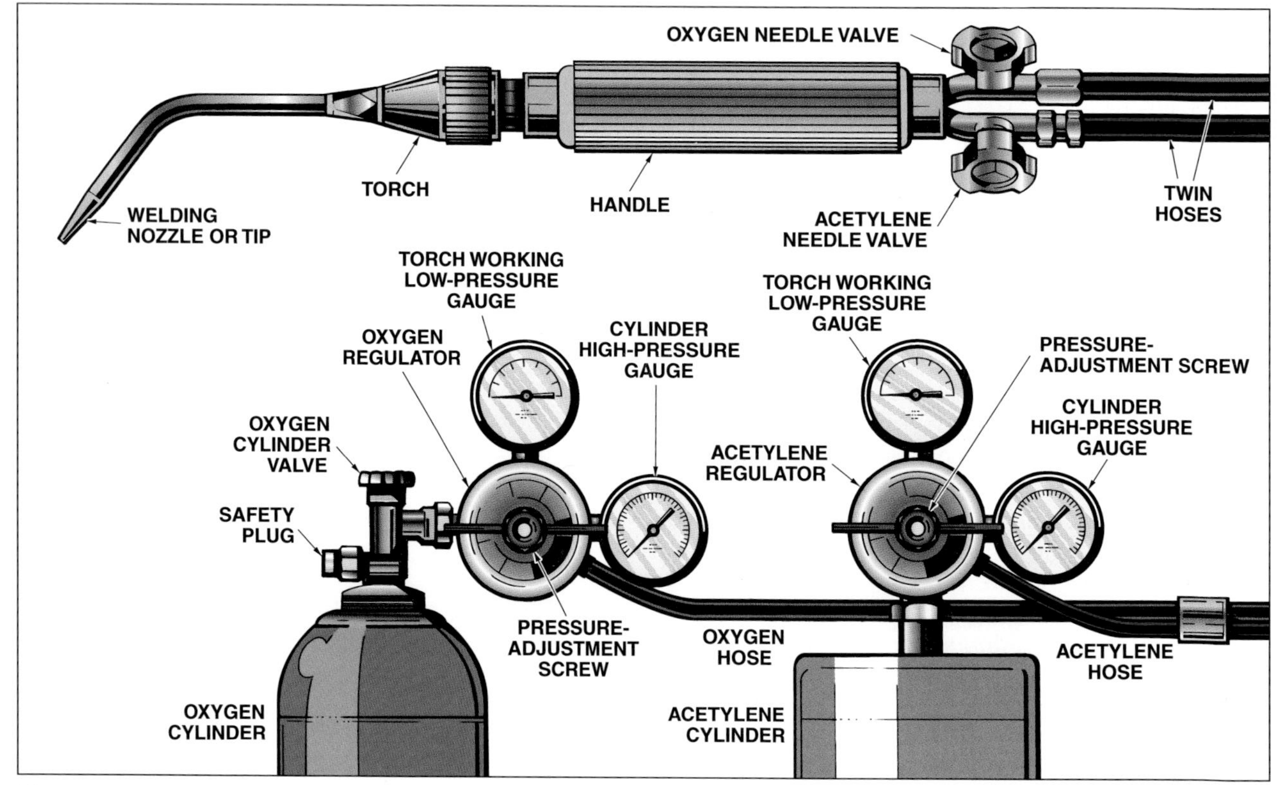

Figure 19-5. *Oxyacetylene equipment setup is similar for welding and cutting operations.*

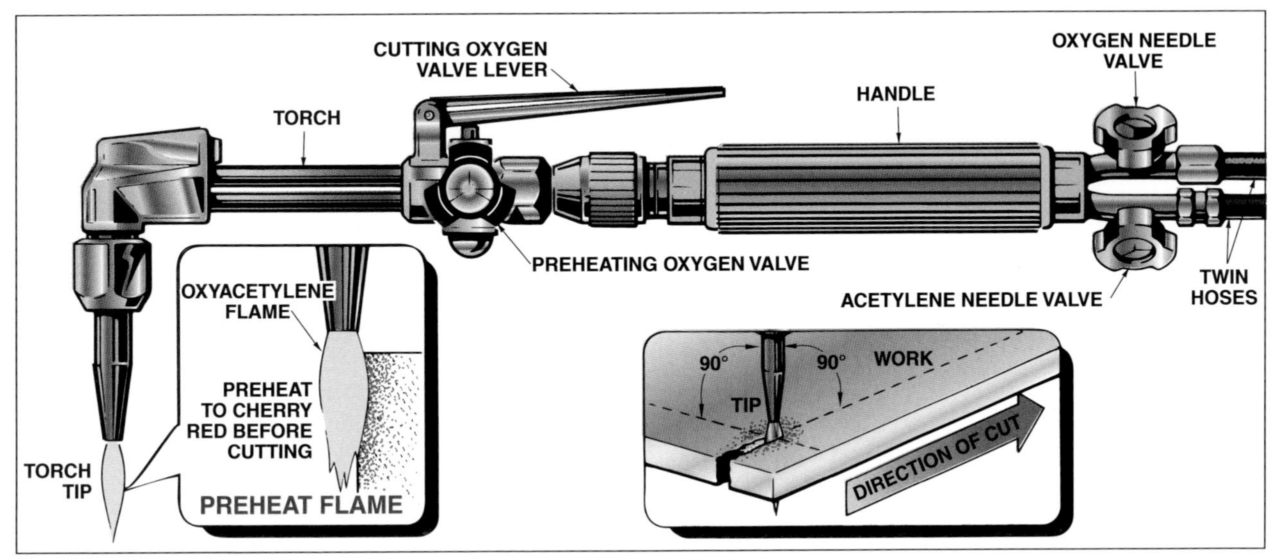

Figure 19-7. *A cutting attachment replaces the welding torch and tip for oxyacetylene cutting operations.*

Oxyacetylene Welding and Cutting Safety. Several potential safety concerns are associated with oxyacetylene welding and cutting processes. OSHA 29 CFR 1926.350, *Gas Welding and Cutting,* details safety precautions that must be observed when using oxyacetylene welding and cutting equipment.

Compressed oxygen and oxyacetylene gases, which are stored in steel cylinders, must be handled carefully. When transporting, moving, and storing cylinders, protective caps must be placed over the valve stems. The cylinder must be secured in a vertical position. In most situations, cylinders are secured with chains. Cylinders are transported by tilting and rolling them on their bottom edges. Cylinders must be kept far enough away from the welding or cutting operation so that sparks, hot slag, or flame will not reach them.

Hoses for oxygen and acetylene cannot be interchanged, and the two hoses should be easily distinguished from one another. Oxygen hoses are usually green and acetylene hoses are red. The oxygen fitting is made with a right-hand thread and the acetylene nut has a left-hand thread. Hoses should be inspected at the beginning of each work shift for cracks or other defects. If defects are identified, the hose must be immediately removed from service.

Torches and cutting attachments must be inspected at the beginning of each work shift. Ensure that shutoff valves, hose couplings, and tip connections do not leak.

Additional safety rules to observe when operating oxyacetylene welding and cutting equipment are as follows:

- Wear appropriate eye protection and fireproof clothing and headgear.
- Weld only in areas that are well-ventilated. When necessary, use an exhaust system to keep toxic gases below the prescribed health limits. Wear a respirator when welding metals that produce toxic fumes.
- Do not weld or cut used drums, barrels, tanks, or other containers unless they have been thoroughly cleaned of all combustible substances.
- Do not weld or cut near flammable materials. If welding or cutting must be done near flammable materials, use fire-resistant guards, partitions, or screens.
- Keep a fire extinguisher nearby any welding or cutting operation.
- Do not weld or cut near ventilators.
- Do not use oxygen to dust off clothing or work.
- Do not use oxygen as a substitute for compressed air.
- Do not allow acetylene gas to come in contact with unalloyed copper except in a torch.

PLASMA ARC CUTTING (PAC)

Plasma arc cutting (PAC) is a cutting process that uses a constricted arc and high-pressure gas to form a high-velocity jet of plasma (ionized gas) for cutting metal. Plasma is a superheated gas capable of conducting an electric arc and is typically considered the fourth state of matter because it does not have the exact properties of a solid, liquid, or gas.

During PAC, a high-velocity jet of plasma emanates from a constricted orifice. Air velocity must increase when passing through the restricted orifice of the nozzle to maintain the same volume of air passing through the nozzle to the workpiece. The orifice directs the superheated plasma stream toward the base metal. As the arc melts the base metal, the high-velocity jet removes the molten metal to form a narrow kerf.

PAC is one of the best processes for the high-speed cutting of metals such as carbon steel, stainless steel, and aluminum. It cuts carbon steel up to 10 times faster than any oxyfuel mixture with equal quality and at a lower cost. A handheld torch may cut materials up to 2″ thick. PAC can be used on dirty, rusty, or painted metal surfaces and does not require precleaning surfaces.

PAC Equipment

PAC equipment includes a welding machine, cutting and shielding gases, a ground clamp attached to a cable, and a cutting torch attached to a power lead. **See Figure 19-8.** PAC produces intense heat and arc radiation. Proper PPE, including a welding helmet with a lens matched to the current, should always be worn when working with a PAC torch. Several metals produce toxic gases when they are cut. For example, stainless steel produces chromium fumes, and galvanized steel produces zinc fumes. Therefore, proper ventilation should always be used. PAC also produces extreme noise levels, which requires hearing protection.

Compressed air is most often used in PAC. The air provides a stream of gas to produce plasma, shields the plasma arc, and aids in pushing the arc through the base metal to be cut. For PAC to operate efficiently, the air must be dry, and there should be no moisture on or around the equipment. To protect the power unit, electrode, and tip from wear and damage, an oil and water separator should be connected between the compressed air supply and the power unit.

PAC Torches. Inside a cutting torch is a set of parts that help produce, shape, and direct a plasma arc. The parts include an electrode, a nozzle, a swirl ring, a retaining cap, and a shield. In a PAC torch, the tip of the electrode is located within the constricting nozzle, which has a small opening to constrict the arc. The electrode conducts the negative charge of electrons from the power supply. The swirl ring controls the flow of high-pressure gas, causing a vortex resulting in a thin stream. The nozzle contacts the electrode and channels the arc.

The retaining cap contains the nozzle, electrode, and swirl ring and attaches to the torch handle. The shield protects the torch components from heat and spatter and cools the tip. Since the gas cannot expand because of the construction of the nozzle, it is forced through the opening and emerges at a high velocity and hotter than any flame (30,000°F to 43,000°F). **See Figure 19-9.** The heat melts any known metal and its velocity blasts the molten metal through the plate creating a kerf. PAC produces a narrower kerf than an oxyacetylene cutting torch.

Plasma Arc Cutting (PAC) Equipment

Figure 19-8

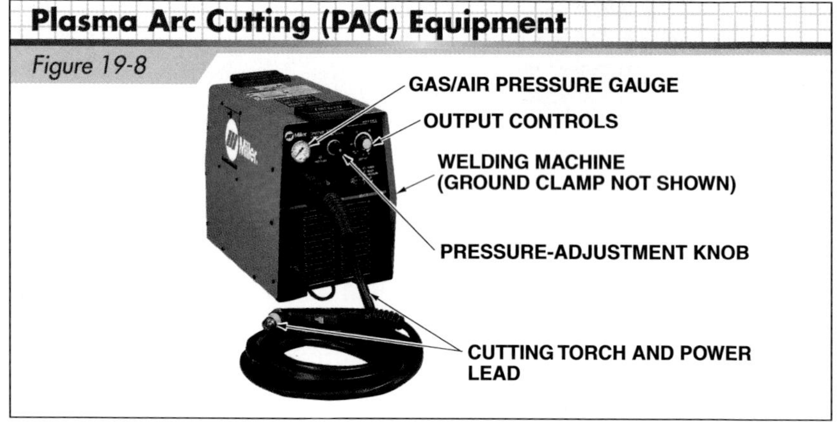

Miller Electric Manufacturing Company

Figure 19-8. *PAC equipment includes a welding machine, cutting and shielding gases, a ground clamp attached to a cable, and a cutting torch attached to a power lead.*

PAC Torches

Figure 19-9

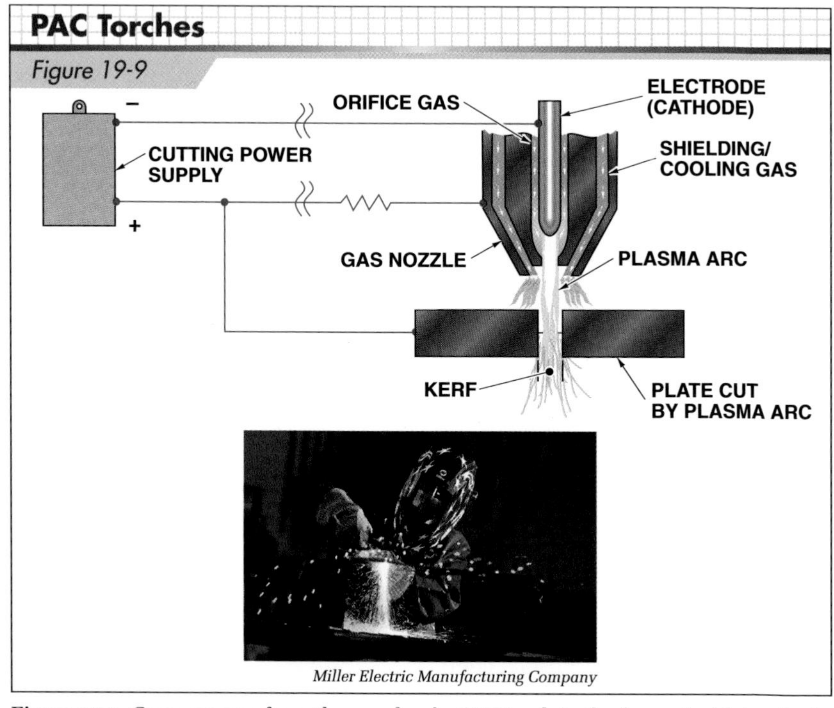

Miller Electric Manufacturing Company

Figure 19-9. *Gases emerge from the nozzle of a PAC torch in the form of a high-velocity jet stream that can blast through the metal, creating a kerf.*

Unit 19 Review and Resources

SECTION

5

Construction Equipment, Job Site Safety, and Working Conditions

CONTENTS

UNIT 20

Scaffolds, Personal Fall-Arrest Systems, and Ladders

OBJECTIVES

1. List and describe basic requirements for scaffold construction.
2. Identify common types of scaffolds.
3. List and describe common components of sectional metal-framed and wood pole scaffolds.
4. Identify other types of scaffolds and support systems.
5. List common safety precautions for scaffolds.
6. Identify common safety precautions when working on scaffolds around electrical hazards.
7. Describe common types of fall-arrest equipment and safety nets.
8. Describe common ladders used in construction.
9. Describe proper ladder climbing techniques.
10. Identify common safety precautions for ladders.

A large number of construction operations require working at heights. These heights may range from a few feet to hundreds of feet above floor or ground level. According to the U.S. Department of Labor, an estimated 2.3 million construction workers, or 65% of the construction industry, frequently work on scaffolds. Although scaffolds are the primary means for performing operations at higher elevations, other climbing equipment, such as aerial lifts and ladders, may be used for lower and more limited operations.

SCAFFOLDS

A *scaffold* is a temporary elevated platform and structure used to support workers and materials. Scaffolds are constructed of steel, wood, aluminum, or fiberglass. **See Figure 20-1.** Steel scaffolds are used most often.

Site Preparation and Design

Prior to scaffold construction, job site conditions should be carefully inspected by a qualified person trained in scaffold erection, or by an engineer. The soil conditions over which the scaffold is to be constructed are of particular concern. Soils vary in their allowable bearing strength per square foot, with rock having the greatest bearing strength. As a rule of thumb, a *heel test* can be performed to determine the general bearing strength of soil. A heel test is performed by firmly pressing the heel of your shoe into the soil while twisting the heel. If the heel penetrates the soil more than 1″, the soil should be tested by an engineer before the scaffold is constructed.

Trus Joist, A Weyerhaeuser Business

Figure 20-1. *Scaffolds are temporary elevated platforms and structures to support workers and/or materials.*

Scaffolds should rest on level surfaces. Ground elevations between scaffold uprights may vary enough to remove soil beneath the uprights to provide a level surface.

Scaffold footing must be stable and must not settle while carrying the *maximum intended load.* The maximum intended load is the total of all loads, including working loads, scaffold weight, and other anticipated loads. Scaffolds and their components must be able to withstand at least four times their maximum intended load.

All scaffolds more than 10′ above the ground must have a *guardrail* system, which includes *toprails* and *midrails.* Guardrails are secured to uprights and erected along the exposed sides and ends of a platform. A midrail is secured to uprights approximately midway between the guardrail and platform. *Toeboards* must be installed if needed to protect workers from the hazards of falling hand tools, material, or debris. Guardrails must be installed with the toprail no less than 38″ or more than 45″ high. Screening, mesh, or solid panel extending from the platform to the toprail may be installed with a midrail.

Various types of scaffolds are used by carpenters during construction or remodeling operations, including *sectional metal-framed, wood pole,* and *suspension scaffolds.* Sectional metal-framed scaffolds are the primary type of scaffold used in construction work. Wood scaffolds, at one time the predominant type of scaffold employed by the building trades, are still used, but far less frequently. Approved scaffolds for specialized operations include outrigger scaffolds, carpenter's and form bracket scaffolds, and pump jack scaffolds.

Sectional Metal-Framed Scaffolds

The framework for sectional metal-framed scaffolds consists primarily of tubular steel or aluminum components. The working platform may be metal or wood. Three basic types of sectional metal-framed scaffolds used on construction projects are *welded-frame, tube-and-clamp,* and *systems scaffolds.* Mobile scaffolds can be constructed with the three basic types of scaffolds.

Welded-Frame Scaffolds. Heavy-gauge tubular steel bearers (transoms) and posts (standards) are welded together to form the frames for welded-frame scaffolds. The frames are joined together with diagonal braces that attach to locking devices on the scaffold frame. **See Figure 20-2.** Frames are joined vertically using coupling pins, which are inserted into the upper ends of the lower frame and lower ends of the upper frame. Holes in the coupling pin are aligned with holes in the frame, and a locking device is inserted through the holes.

Several types of frames are available. *Walk-through* and *open-end frames* are commonly used. Walk-through frames provide openings for workers to move about in the bays between the scaffold sections and also feature an access ladder built into the frames. Open-end frames resemble walk-through frames but do not include a built-in access ladder. **See Figure 20-3.** Both types of frames are available in 3′ to 5′ widths and 5′ to 7′ heights.

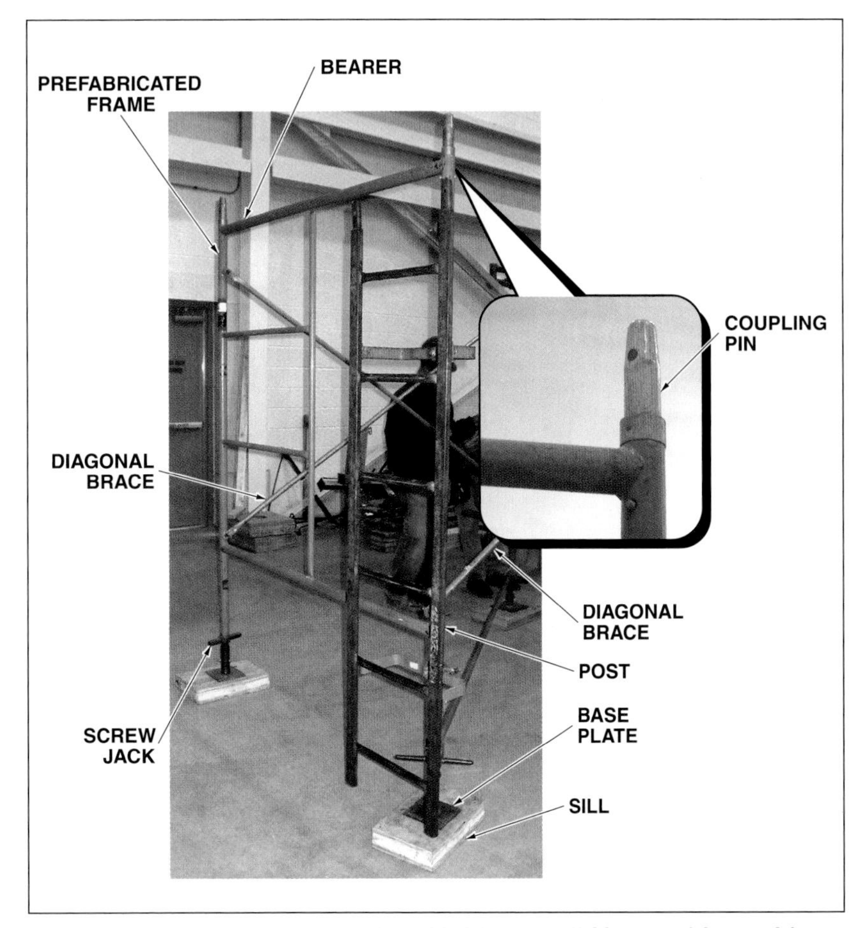

Figure 20-2. *The main component of a welded-frame scaffold is a prefabricated frame.*

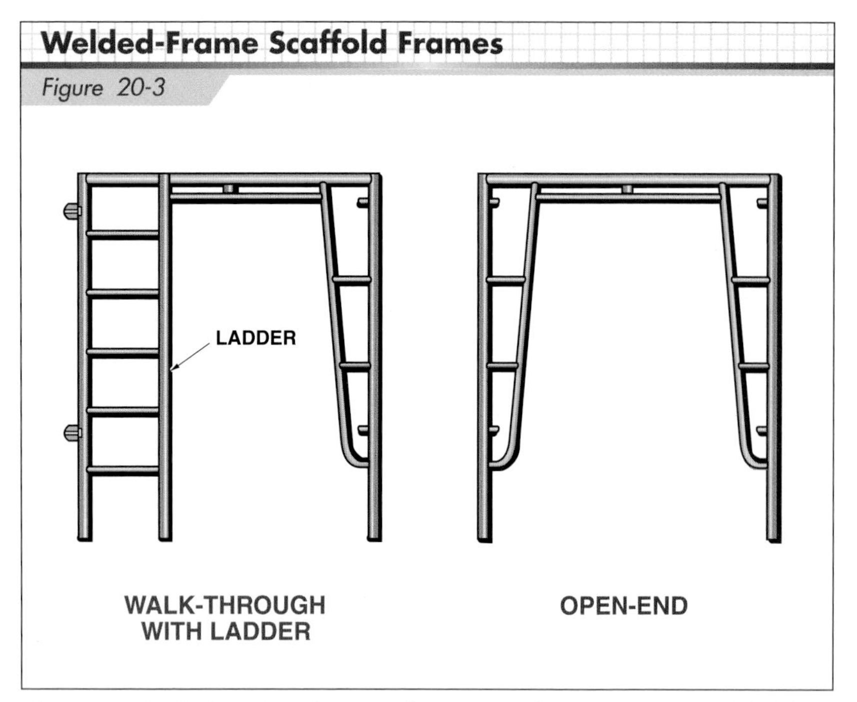

Figure 20-3. *Walk-through and open-end frames are the most common welded-frame scaffold frames.*

Tube-and-Clamp Scaffolds. Tube-and-clamp, or tube-and-coupler, scaffolds consist of heavy-gauge galvanized steel or aluminium tubes that are joined using coupling pins or interlocking ends. Tube-and-clamp scaffolds allow for more flexibility in the shape of the scaffold, particularly when erected around circular structures such as silos and tanks.

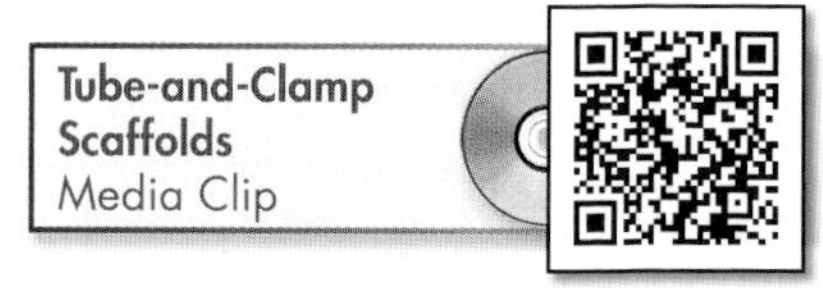

Tubes serve as posts, runners, bearers, and braces. Some tubes have interlocking ends that are twisted together to lock the tubes. Other tubes have open ends and require a coupling pin to secure the tubes together. **See Figure 20-4.**

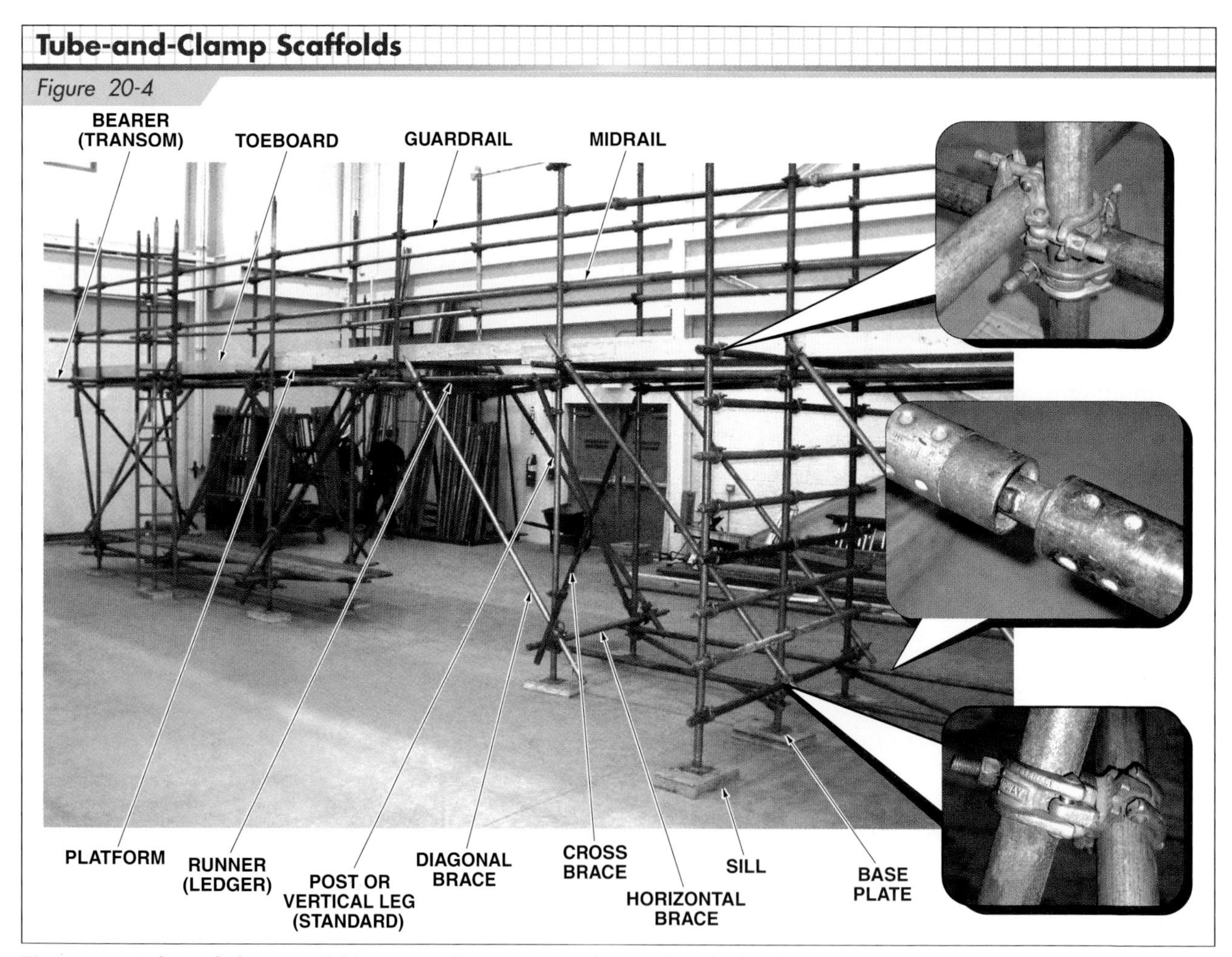

Figure 20-4. *Tube-and-clamp scaffolds consist of heavy-gauge galvanized steel tubes that are joined using coupling pins or interlocking ends.*

The open section of a clamping device should face upward to provide proper support for a horizontal member.

Tubes for runners, bearers, and braces are connected to uprights using right-angle and swivel clamps.

Systems Scaffolds. Systems scaffolds are similar to tube-and-clamp scaffolds. However, systems scaffold *nodes* (rosettes) are welded to the posts and bearers for attachment of braces and railings. **See Figure 20-5.** Less time is required to set up systems scaffolds than tube-and-clamp scaffolds.

Mobile Scaffolds. Mobile scaffolds are equipped with casters (wheels). Mobile scaffolds require a level paved or concrete surface, and are typically used for high interior work. A mobile scaffold may have a power drive system to move it. A *rolling scaffold* is similar to a mobile scaffold, but does not contain a power drive system. *Outriggers* help maintain the balance of taller rolling and mobile scaffolds. **See Figure 20-6.**

Mobile and rolling scaffolds may be moved with a worker on the platform, as long as the worker is advised and aware of each movement in advance. The minimum dimension of the base, when ready for rolling, must be at least one-half the scaffold height. Mobile scaffolds must not be higher than four times the minimum base dimension.

All tools and materials must be removed or secured before a mobile scaffold is moved. After the scaffold has been moved, the casters must be locked to prevent movement while the scaffold is being used.

Putlogs. A *putlog,* also known as a bridge or trestle, is a horizontal truss that extends between two separate scaffolds. A platform is constructed over the putlogs. Putlogs support work platforms but are limited in the amount of weight they can carry. Putlogs are used to span areas where standard scaffolds cannot be easily used, such as over a door opening, thus reducing the number of scaffold parts needed. **See Figure 20-7.**

Figure 20-5. *Systems scaffolds are similar to tube-and-clamp scaffolds but have their nodes welded to bearers and posts.*

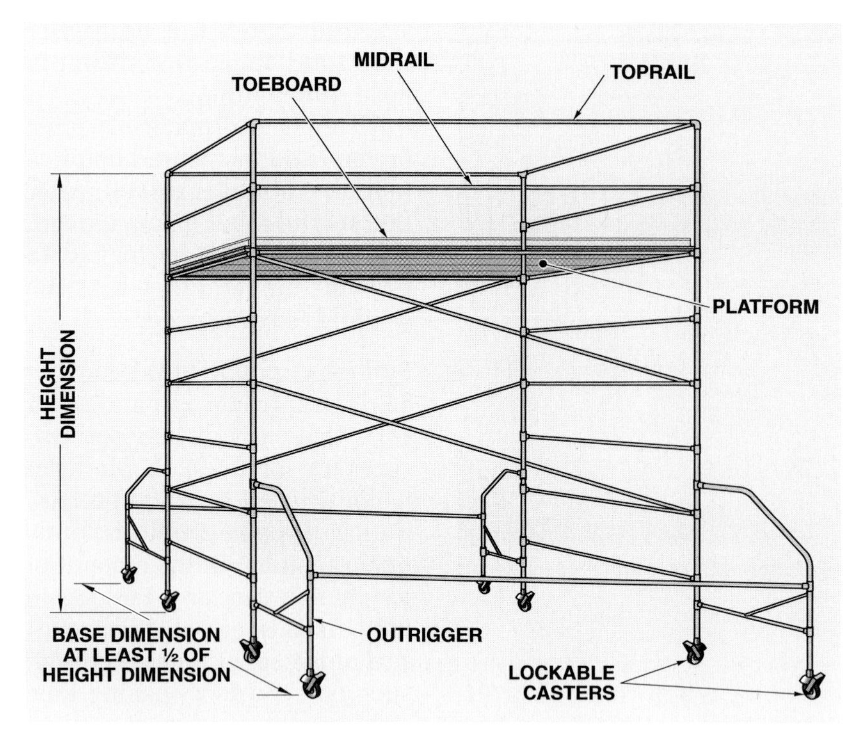

Figure 20-6. *Mobile scaffolds are equipped with casters to allow easy movement around a job site. Outriggers help maintain the balance of taller rolling and mobile scaffolds.*

Figure 20-7. *Putlogs are braced horizontal trusses extending between separate scaffolds and are used to span areas where scaffolds cannot be easily used.*

Sectional Metal-Framed Scaffold Components

The basic components of most sectional metal-framed scaffolds are *base plates* and/or *screw jacks, horizontal braces, diagonal braces, work platforms, toeboards,* and *guardrails.* **See Figure 20-8.** In addition, ladders, stairs, and locking or clamping devices are installed. Only components designed for the particular type of scaffold being erected, and approved by the scaffold manufacturer, should be used when constructing scaffolds.

Base Plates. Typical base plates are 6″ square, with nail holes in the corners for nailing into a wood pad or sill. Single pads for base plates are convenient when positioning a scaffold over uneven ground. Continuous (full-length) sills are commonly used over even ground, smooth floors, and slabs.

Screw Jacks. Screw jacks are used to level scaffold frames, and either are part of the base plate or are slipped onto the base plate and secured with locking pins. Screw jacks are commonly 18″ long, allowing nearly 10″ to 12″ of adjustment.

Horizontal Braces. *Horizontal braces* extend parallel to the ground across the *bay* (space between opposing frames). Horizontal braces are usually installed at the base of the frames. The first row of horizontal braces are placed over the screw jacks, and then rows are placed at every platform level. Horizontal braces supporting the work platform are *bearers.* Horizontal braces are available in 4′ to 10′ lengths in 2′ increments.

Diagonal Braces. *Diagonal braces,* or crossbraces, extend from the bottom of one frame to the top of the opposing frame. Diagonal braces provide lateral stability and rigidity to the scaffold. Diagonal braces are commonly available in pairs that are attached to each other with a pivot pin at the center. Diagonal braces are available to accommodate 2′ to 4′ spacing between attachment points on the frame posts, and 2′ to 10′ spacing of the bays.

Scaffold components from different manufacturers must not be intermixed unless the components fit together without force and the structural integrity of the scaffold is maintained.

Sectional Metal-Framed Scaffold Components

Figure 20-8

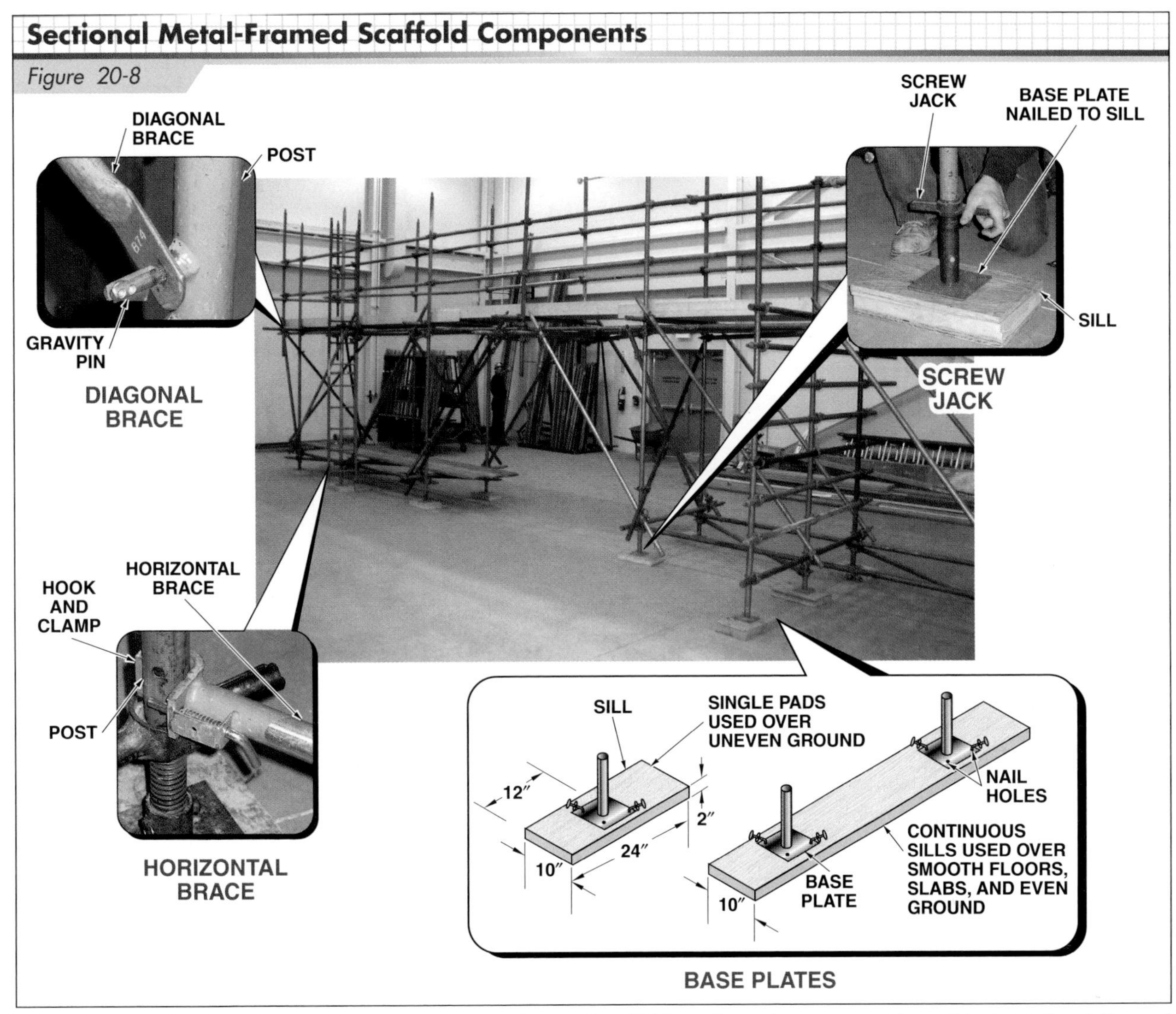

Figure 20-8. *The basic components of most sectional metal-framed scaffolds are base plates, screw jacks, and horizontal and diagonal braces.*

Nuts on scaffold clamping devices must be properly tightened to manufacturer specifications.

Work Platforms. Scaffold work platforms consist of scaffold-grade solid or engineered lumber or prefabricated metal sections. **See Figure 20-9.** Wood and metal scaffold platforms must be at least 18″ wide. When solid or engineered lumber is used as a work platform, the planks must extend 6″ to 12″ past the end supports. Planks must overlap a support a minimum of 12″ and are toenailed together.

Prefabricated metal work platforms are supported with hooks at each end. The hooks are placed over the frame bearers and locked into place.

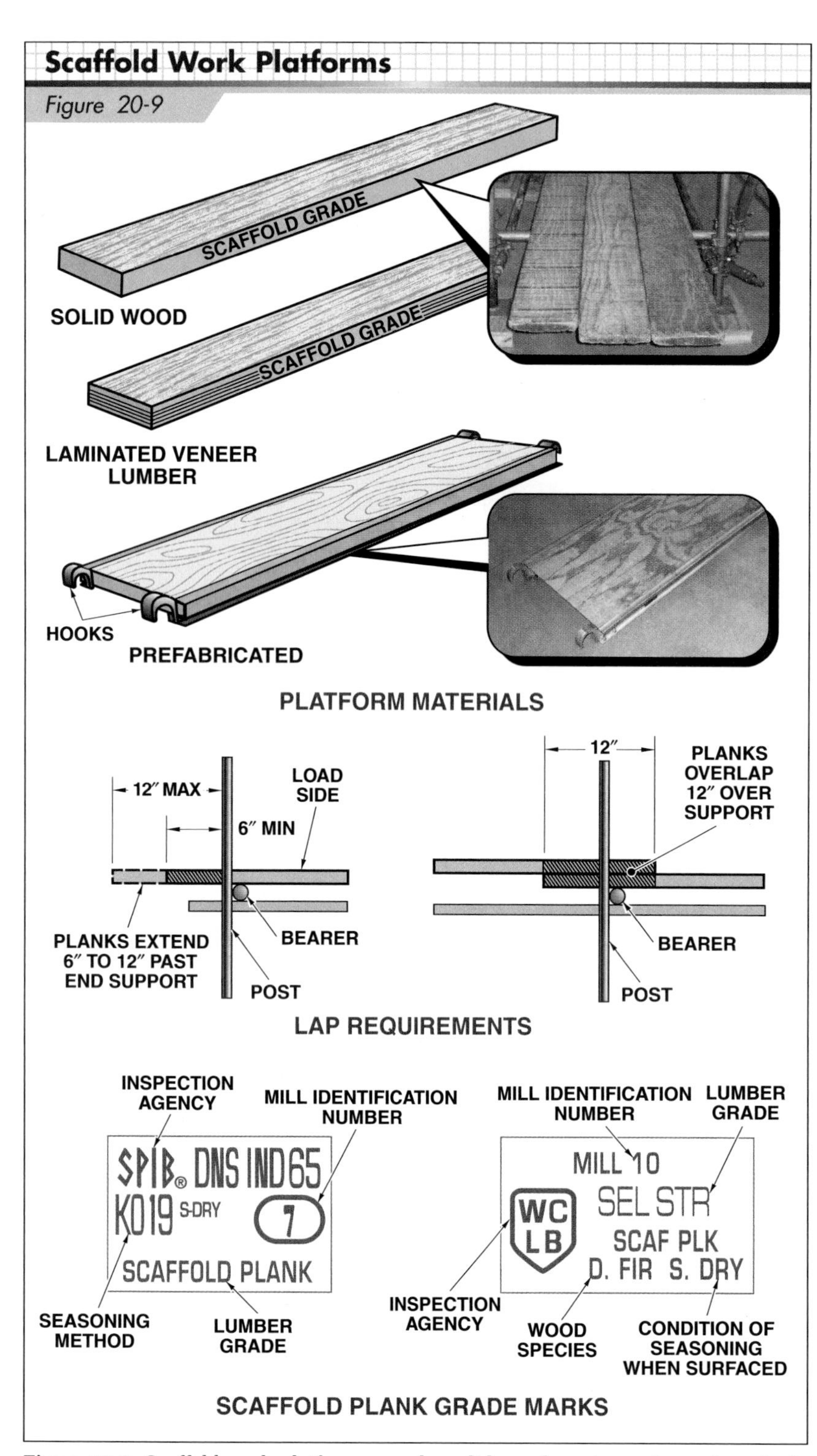

Figure 20-9. *Scaffold work platforms may be solid wood, engineered wood products (such as laminated veneer lumber), or prefabricated metal. Scaffold-grade planks must be used if the work platform is constructed of solid lumber or engineered wood products. Scaffold planks must be properly overlapped.*

Toeboards. *Toeboards* are installed along the work platform edges to prevent objects from falling off the sides and ends. Toeboards may be constructed of 2 × 4s placed on edge and attached to the edges of the work platform. Manufactured toeboards are also available with clips to anchor the toeboards to the scaffold. The gap between toeboards and the work platform should not be greater than ¼″. Panels placed on edge or screens should be used if materials will be piled higher than the toeboard.

Guardrails. Scaffold guardrails are a type of fall protection, and consist of a *toprail* and *midrail.* Guardrails are required at the open sides and ends of scaffold work platforms. The distance from the work platform to the upper surface of the toprail should be 38″ to 45″. A midrail is placed midway between the platform and the toprail. **See Figure 20-10.** Screening, mesh, or solid panels extending from the work platform to the toprail may be used with a midrail.

OSHA's scaffold standard defines a competent person as a person capable of identifying existing and predictable hazards in the surroundings or working conditions which are unsanitary or dangerous and a hazard to employees, and who has authorization to take prompt corrective measures to eliminate them. Among other functions, a competent person is allowed to select and direct employees who erect, dismantle, move, or alter scaffolds.

Attached Stairs and Ladder Units. When scaffold platforms are more than 2′ above or below a point of access, a stairway or ladder must provide safe access to the different levels. **See Figure 20-11.** For large scaffold assemblies, prefabricated internal *stair assemblies* provide safe access to one level from another. Stair assemblies are self-contained units that contain stair treads, handrails, and landings.

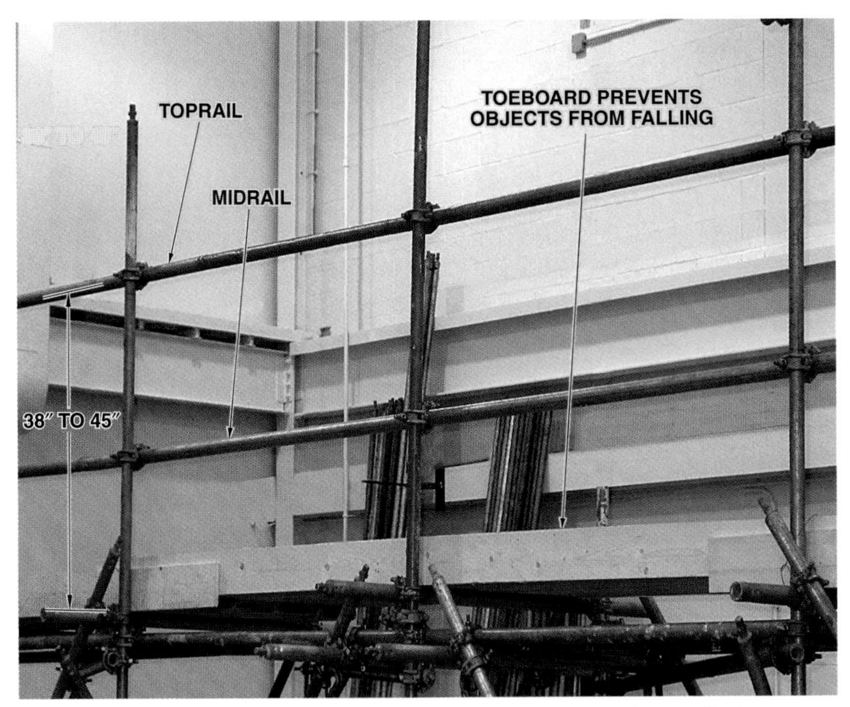

Figure 20-10. *Guardrails, including toprails and midrails, must be installed per OSHA regulations. Toeboards are installed to prevent objects from falling from a scaffold.*

Access must be provided to scaffolds when work platforms are more than 2′ above or below an access point. Diagonal braces cannot be used as a means of access to work platforms. Direct access to the work platform is acceptable when the scaffold is less than 14″ horizontally and less than 24″ vertically from the other surface. Types of access permitted by OSHA include ladders, ramps and walkways, stairways, and prefabricated frames.

Upper-Level Access

Figure 20-11

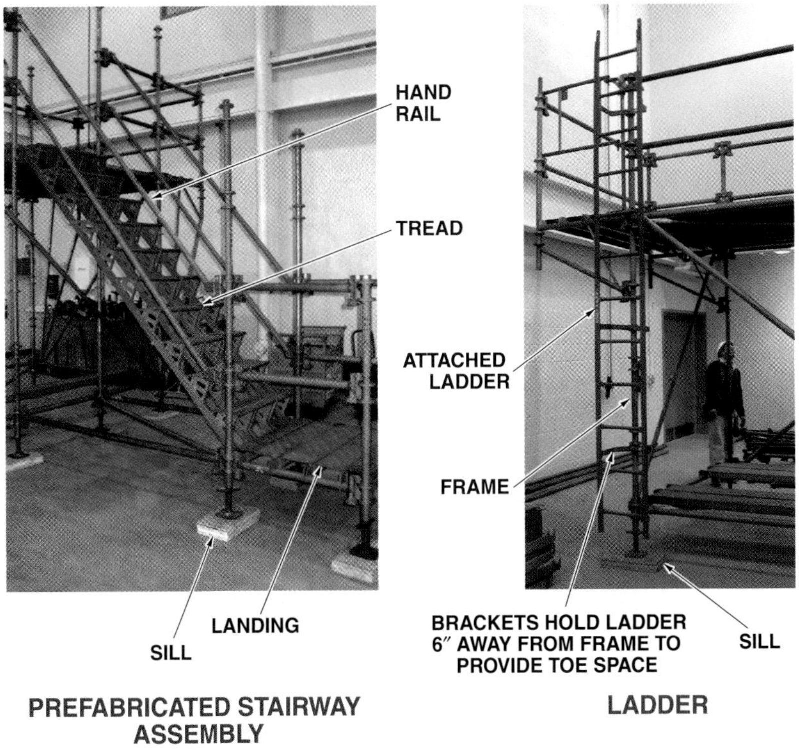

HAND RAIL
TREAD
LANDING
SILL
ATTACHED LADDER
FRAME
BRACKETS HOLD LADDER 6″ AWAY FROM FRAME TO PROVIDE TOE SPACE
SILL

PREFABRICATED STAIRWAY ASSEMBLY

LADDER

Figure 20-11. *Properly installed stair units or ladders are required for access to the different levels of a scaffold.*

Ladders may be used to access upper scaffold levels. Ladders may be built into the frames of welded-frame scaffolds. *Attached ladders* are frequently used and should be placed so as not to tip the scaffold when a worker is climbing. Attached ladders are available in 3′ and 6′ lengths.

Brackets for attached ladders must allow 6″ of toe room when climbing. The bottom rung of ladders should not be more than 24″ above any working level.

Scaffold Clamping and Locking Devices. Clamping and locking devices are a means of tying together various scaffold components. Only clamping and locking devices approved by the scaffold manufacturer must be used when constructing a scaffold.

Clamping devices are used to fasten bearers and braces to scaffold posts. Bolt- and wedge-type clamping devices are commonly used. **See Figure 20-12.** Clamping devices may be rigid (nonadjustable) or adjustable to various angles. Adjustable (swivel) clamping devices are not load bearing and can only be used for supporting braces.

Locking devices are used to connect and retain the position of scaffold components, such as when connecting diagonal braces to scaffold frames. **See Figure 20-13.**

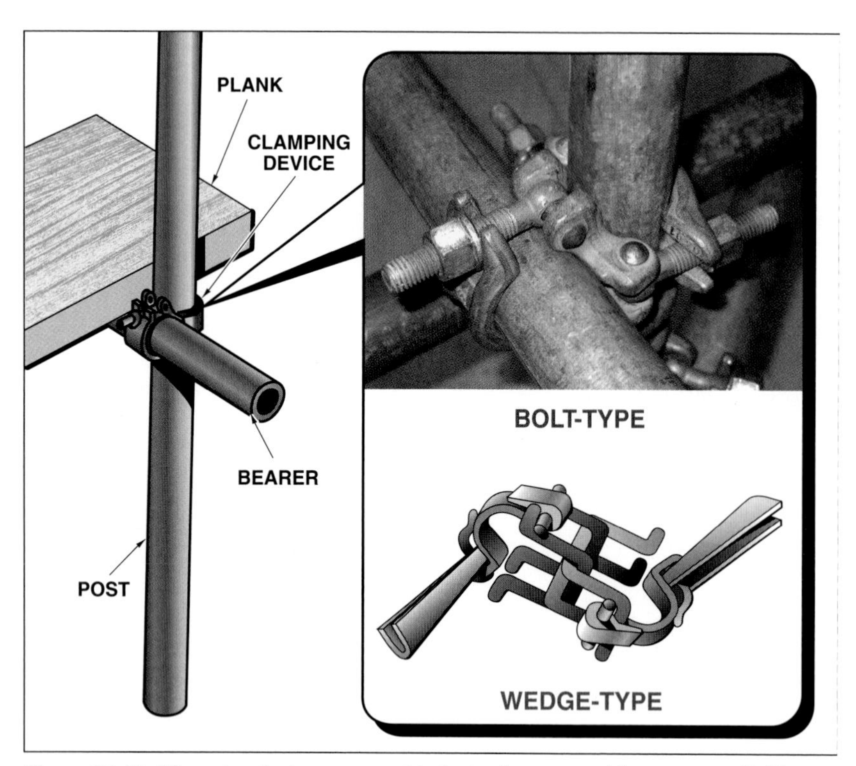

Figure 20-12. *Clamping devices are used to fasten bearers and braces to scaffold posts.*

Scaffold Locking Devices

Figure 20-13

STRAIGHT END INSERTED THROUGH HOLE
POST
GRAVITY PIN
GRAVITY PIN
POST
FRAME
BRACE
BANANA CLIP
CURVED END WRAPS AROUND POST AND LOCKS PIN INTO PLACE
PIGTAIL PIN
COTTER PIN SECURES RIVET PIN
THREADED STUD
NUT
BRACE
BANANA CLIP
FRAME RIVET PIN INSERTED THROUGH HOLE
FRAME RIVET PIN
THUMBSCREW
THREADED STUD
THUMBSCREW

Figure 20-13. *Scaffold locking devices include gravity pins, pigtail pins, frame rivet pins, thumbscrews, and banana clips.*

Wood Pole Scaffolds

Wood pole scaffolds were used before metal scaffolds were introduced to the construction industry. There may still be occasions when a contractor may find it more convenient and economical to erect a wood pole scaffold. *Light-trade pole scaffolds* are designed for operations where heavy tools and heavy materials are not required, such as trim work or painting. *Heavy-trade scaffolds* have the same design as light-trade scaffolds, but are constructed of heavier members. Heavy-trade scaffolds are used to support heavier tools and materials and accommodate trades such as bricklayers and cement masons. Regulations in OSHA 29 CFR 1926 Subpart L, *Scaffolds,* also relate to wood pole scaffolds.

Wood work platforms must not be covered with opaque finishes, such as paint, except that the platform edges may be covered or marked for identification. Work platforms may be periodically coated with wood preservatives, fire-retardant finishes, and slip-resistant finishes. However, the coating may not obscure the top or bottom wood surfaces.

Many basic components of a wood pole scaffold perform the same functions as their metal counterparts. **See Figure 20-14.** Single pads or continuous sills are placed on the ground. Wood *poles* provide vertical support, and are 2 × 4s for heights up to 20′ or 4 × 4s for heights greater than 20′. *Ribbons,* usually 1 × 6s or 2 × 4s, are installed parallel to the ground to provide horizontal stability for the posts. *Ledgers* are installed between two posts to provide support for the work platform. The work platform is commonly constructed of 2 × 10 or 2 × 12 planks, but engineered or prefabricated metal work platforms may also be used. Guardrail requirements, including toprails, midrails, and toeboards, are the same as for metal scaffolds. A *diagonal brace* is attached to the face of the scaffold to provide lateral stability. A procedure for constructing a light-trade wood pole scaffold is shown in **Figure 20-15.**

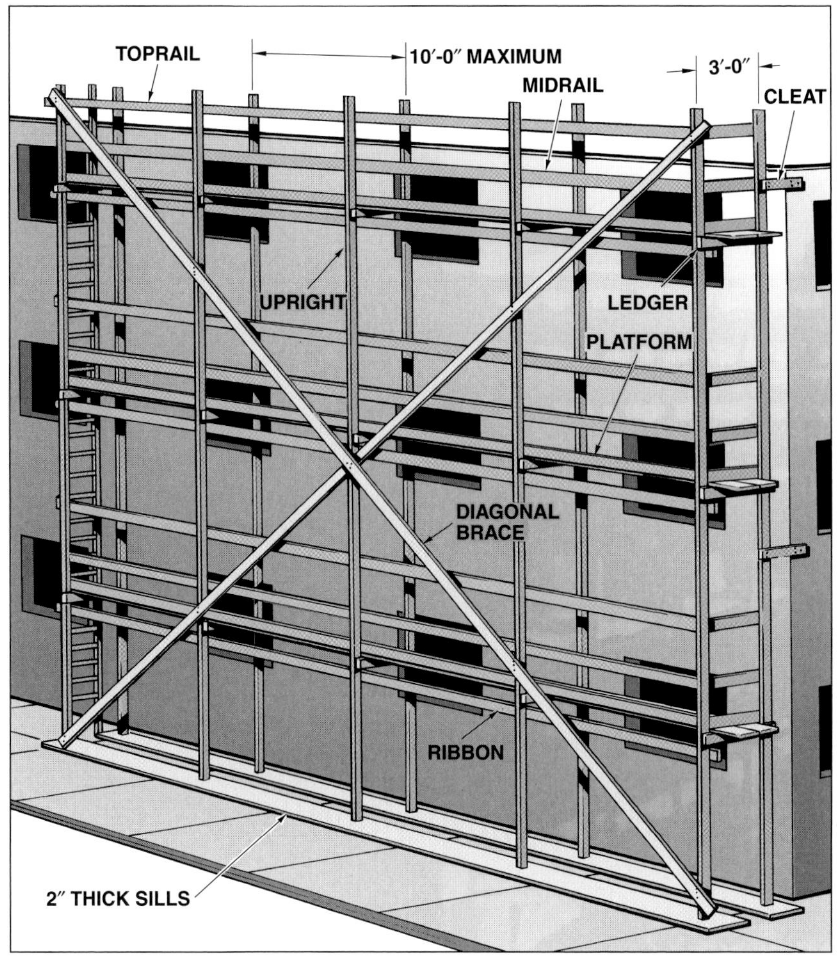

Figure 20-14. *The basic components of a wood pole scaffold are the uprights (poles), ribbons, ledgers, diagonal braces, work platform, and guardrails.*

Gates & Sons, Inc.

Guylines provide support for vertical structures such as formwork and scaffolds.

Tie-Ins and Guylines

Tie-ins and *guylines* provide support for a scaffold and prevent it from tipping forward or backward. Tie-ins are secured directly to a building. Guylines extend from the scaffold to stakes driven into the ground.

Per OSHA 29 CFR 1926.451, *General Requirements,* scaffolds with a height-to-base-width ratio exceeding 4:1 must be restrained from tipping by tie-ins, braces, guylines, or equivalent means. The 4:1 height-to-base-width ratio means that the scaffold must increase 1′ in width for every 4′ of height. For example, a 24′ high scaffold must be 6′ wide at the base. Tie-ins, braces, and guylines should be installed at the closest horizontal member to the 4:1 height and must be installed per manufacturer recommendations. Additional tie-ins, braces, or guylines are installed at 20′ intervals or less for a scaffold that is 3′ wide or less. For wider scaffolds, tie-ins, braces, or guylines are required every 26′ vertically after the initial placement.

The top tie-in, guyline, or brace should be fastened toward the top of the scaffold no more than the scaffold width from the top. For example, if a scaffold is 4′ wide, the top tie-in should be fastened no more than 4′ from the top. Tie-ins, guylines, and braces must be installed at each end of the scaffold and at horizontal intervals not exceeding 30′.

Tie-ins must be able to withstand both tension and compression forces. While tying in with a couple loops of wire may provide some support in tension, it does not support a compressive force. A strut is required between the scaffold and building to counteract the compressive forces.

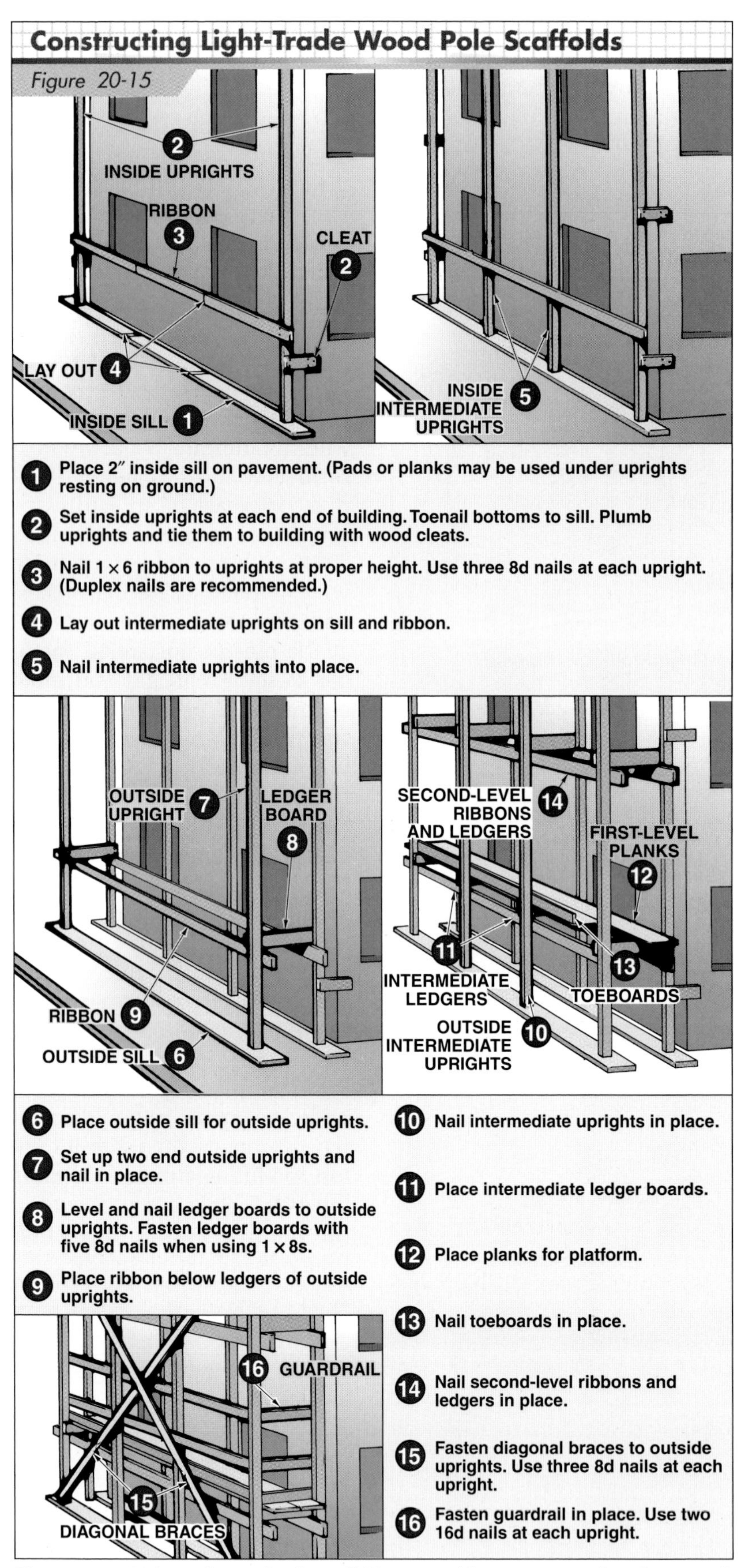

Figure 20-15. *Light-trade wood pole scaffolds are used for operations where heavy tools and heavy materials are not required.*

Tie-ins. Tie-ins may consist of heavy-gauge wire, metal braces, or scaffold tubes extending between the scaffold and the building under construction. **See Figure 20-16.** Tie-ins prevent tension (pull) and compression (push) movement. When heavy-gauge wire is used as tie-ins, 2 × 4 struts (blocks) must be secured between the scaffold and the point of tie-in to prevent the scaffold from pulling in while the wire is tightened.

When tube-and-clamp scaffolds are erected, tie-in tubes are extended from the scaffold through a window opening where the tubes are anchored. The extension tubes may be anchored to horizontal tubes inside the window opening or to a shoring post securely wedged between the floor and ceiling.

Guylines. Guylines are employed when constructing a freestanding scaffold before there is a structure to which it can be attached. Wire rope or cable must be used for guy-lines. Guylines should be attached to a scaffold at the intersections of posts and bearers and should extend to their bottom anchors at an angle to the ground as close to 45° as possible. **See Figure 20-17.** Guy-lines should never be installed at an angle to the ground greater than 60°. More than one guyline may be required at the same post for taller scaffolds.

Other Scaffolds and Support Systems

A variety of other scaffold and support systems may be used on a job site to provide support for workers and materials. Some types of scaffolds are used by different trades for specialized operations that do not require complete tubular or wood pole structures.

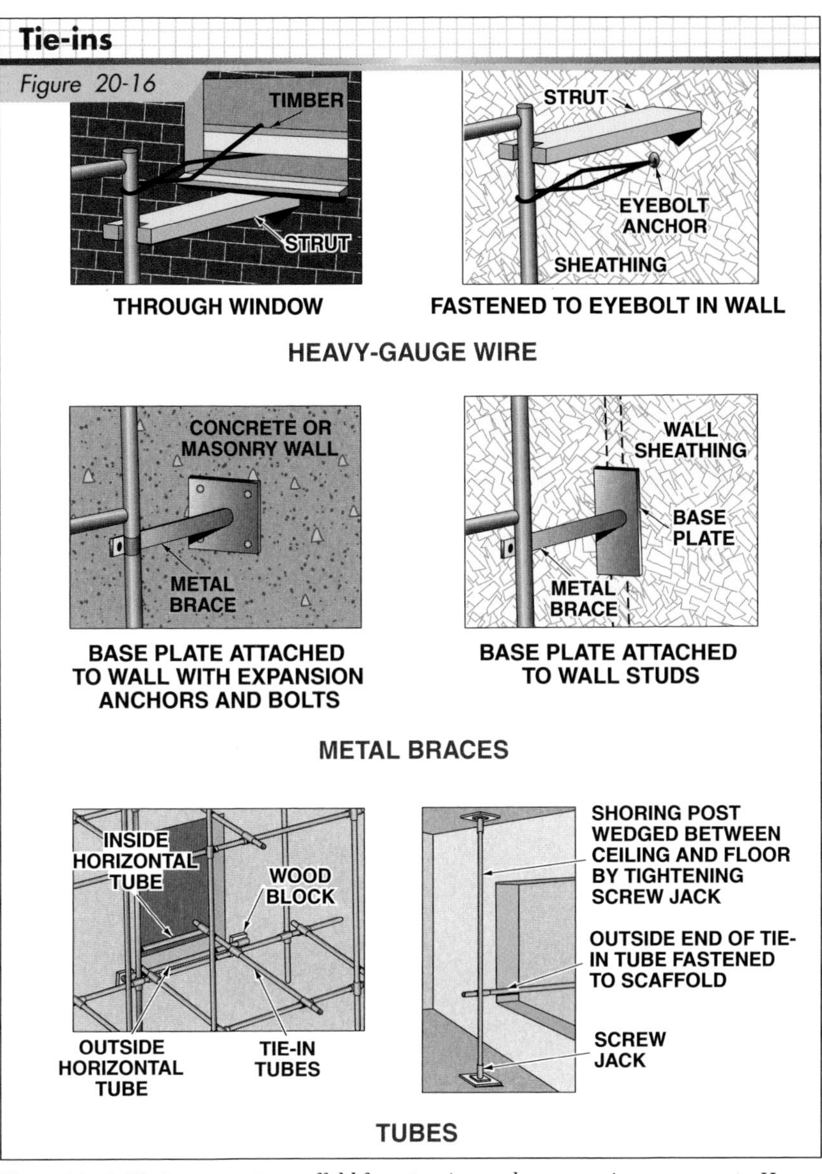

Figure 20-16. *Tie-ins prevent a scaffold from tension and compression movements. Heavy-gauge wire and struts, metal braces, and scaffold tubes are used most often as tie-ins.*

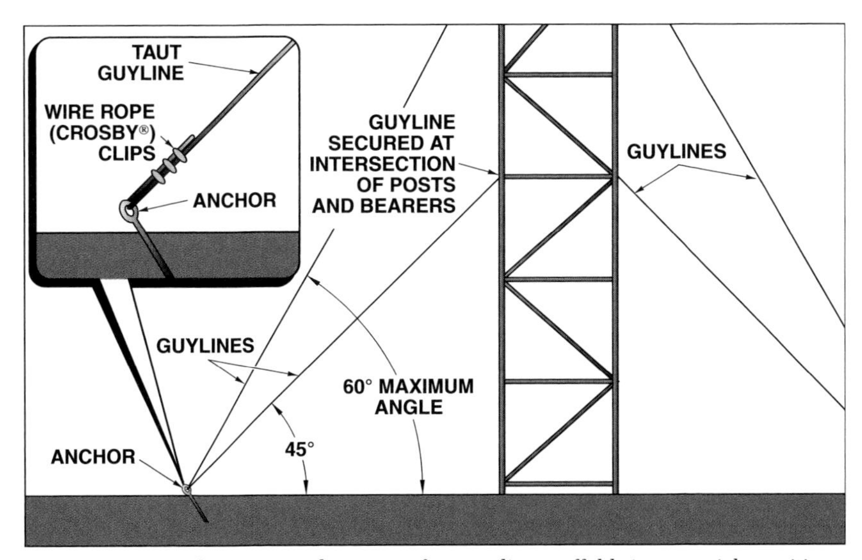

Figure 20-17. *Guylines are used to secure freestanding scaffolds in an upright position.*

Suspension Scaffolds. *Suspension scaffolds* are supported by overhead wire ropes. Per OSHA 29 CFR 1926.451, *General Requirements,* the wire ropes must be capable of supporting at least six times the maximum intended load applied or transmitted to the rope.

Suspension scaffolds include *swinging platform,* and *two-* and *multiple-point suspension scaffolds.* **See Figure 20-18.** A *swinging platform scaffold* consists of a metal grid base supporting a wood platform. The platform is supported at each end by a steel stirrup to which the lower block of a block and tackle is attached. The scaffold is supported by hooks or anchors on the roof of the building.

Suspension scaffolds are heavier than swinging platform scaffolds, and are used for heavier operations and materials. Suspension scaffolds are available in different sizes up to 6′ wide by 12′ long. Suspension scaffolds are supported by outriggers on the roof and are raised or lowered by powered hoists.

Counterweights may be used to balance suspension scaffolds. Only items designed as counterweights may be used for this purpose. Counterweights used for suspension scaffolds must be made of materials that cannot be easily dislocated. Materials such as concrete masonry units, rolls of roofing felt, sand, gravel, and similar materials must not be used as counterweights.

Outrigger Scaffolds. *Outrigger scaffolds* are supported by beams extending out a wall opening (usually a window) and secured inside the building. Planks are placed on top of the beams and guardrails are installed. **See Figure 20-19.**

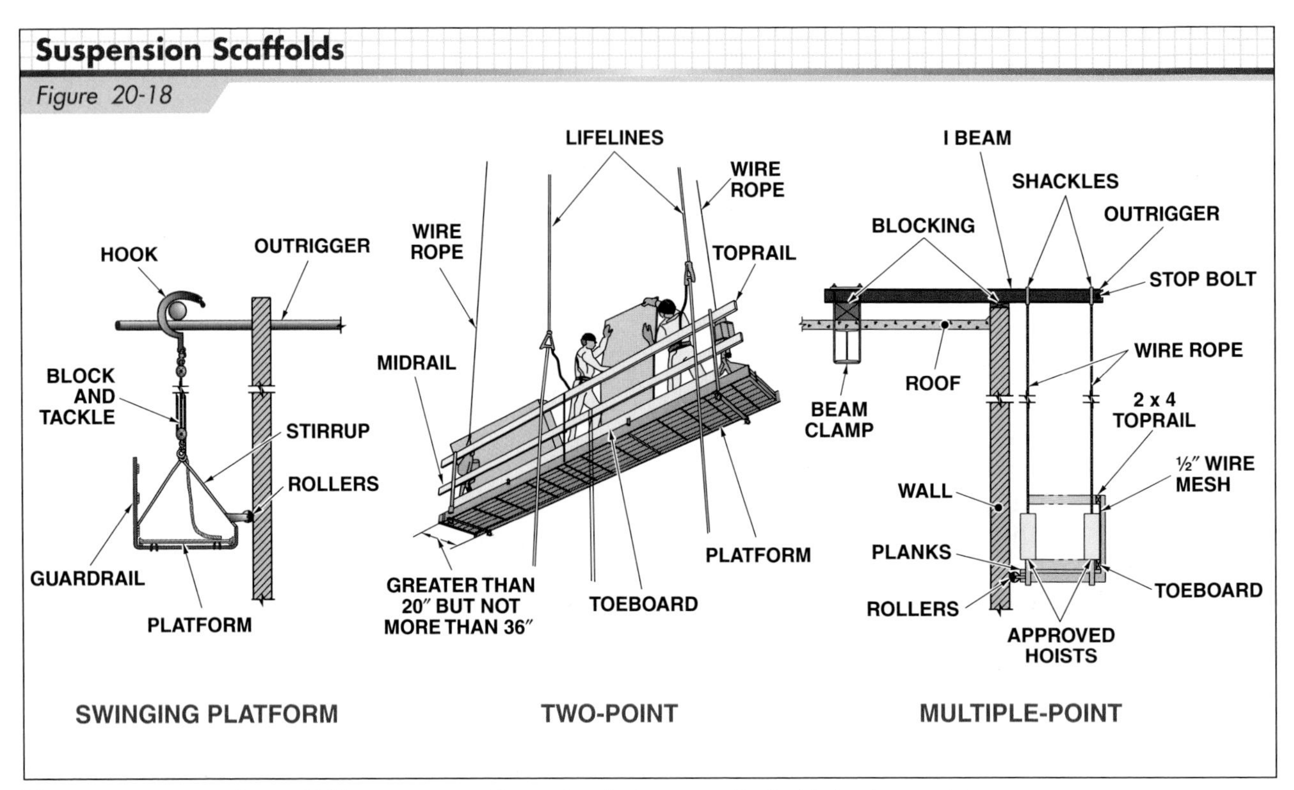

Figure 20-18. *Suspension scaffolds are supported by overhead wire ropes attached to hooks or outriggers.*

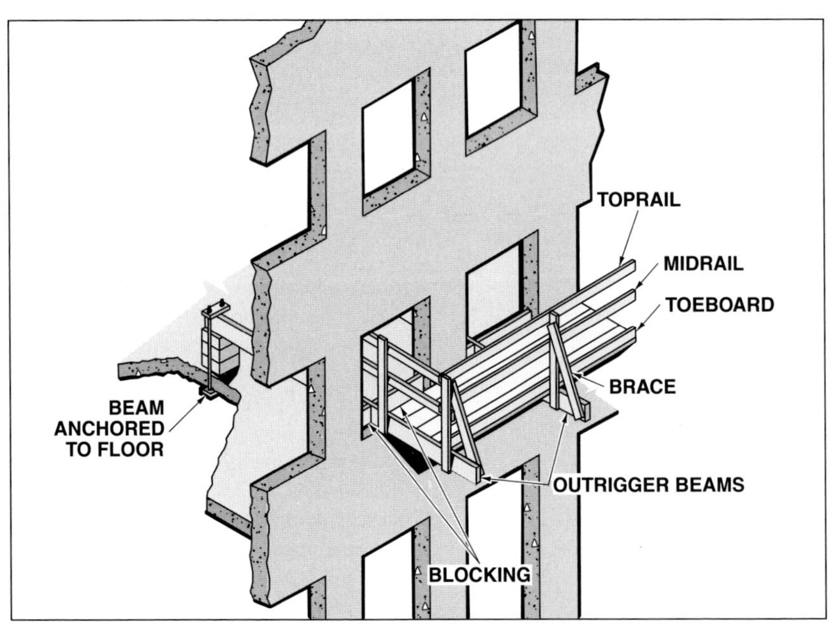

Figure 20-19. *Outrigger scaffolds are supported by beams extending out a wall opening (usually a window) and are secured inside the building.*

Bracket Scaffolds. The more commonly used bracket scaffolds are *carpenter's* and *form bracket scaffolds.* Brackets for carpenter's scaffolds are angle iron or triangular wood frames. The brackets may be bolted to lumber placed behind the wall studs or attached to walls using a hook on the bracket. Form brackets are made of angle iron that attaches to form walers or strongbacks. **See Figure 20-20.**

Pump Jack Scaffolds. A *pump jack scaffold* features an adjustable pump jack that is operated with a foot pedal. Pump jack scaffolds can be used for heights up to 30′. Uprights (4 × 4s) are attached to the building with metal braces or tie-ins. The adjustable brackets are attached to the uprights and provide a 24″ working platform. **See Figure 20-21.**

Per OSHA 29 CFR 1926.451, General Requirements, fall protection must be provided for workers on a scaffold more than 10′ above a lower level. Fall protection includes guardrail systems and personal fall-arrest systems. Personal fall-arrest systems include harnesses, harness components, lifelines, and anchorage points. The fall-protection equipment used for a particular situation depends on the type of scaffold being used.

Platform Support Systems. Metal *trestle jacks* or wood *horse scaffolds* may be used when a low work platform is required. Ledgers, supporting planks at least 2″ thick, are clamped to the tops of the trestle jacks. The planks are fastened to the ledgers with screws or nails. Trestle jack height can be adjusted to accommodate different operations.

A horse scaffold is constructed by supporting planks with a sawhorse. The planks are cleated together along the underside to provide greater stability and the work platform is fastened to the sawhorses with screws or nails. If two levels of sawhorses are used, the upper sawhorse legs are toenailed to the lower work platform. **See Figure 20-22.**

The main function of a sawhorse is to serve as a portable workbench. Sawhorses are constructed in a shop or on the job. The procedure for building a sturdy sawhorse is shown in **Figure 20-23.**

A scaffold work platform must not deflect more than 1/60 of the span when loaded.

Scaffold Regulations and Standards

As of November 1996, OSHA requires any workers using a scaffold to receive training for that type of scaffold and have knowledge of related OSHA standards. Scaffolds must be erected, dismantled, and altered under the supervision of a *competent person* who has been trained and has obtained a *certificate of competence* in scaffold erection, use, and dismantling. A certificate of competence in scaffold erection, use, and dismantling is issued after completion of an authorized training program.

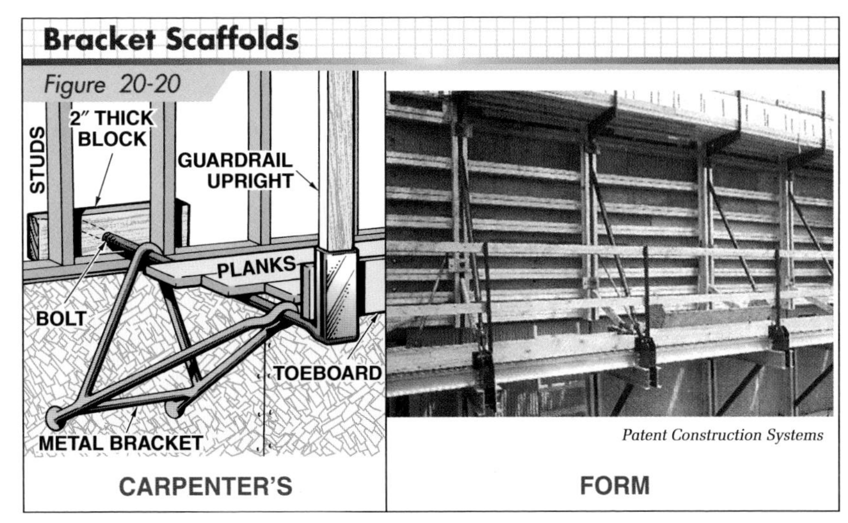

Figure 20-20. *Steel brackets are used for carpenter's and form bracket scaffolds.*

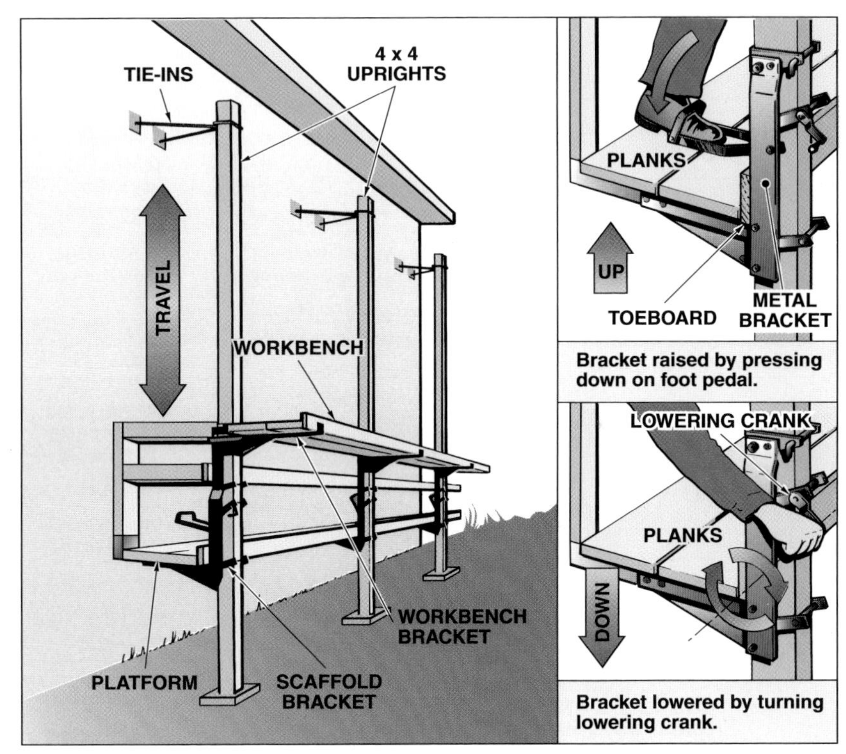

Figure 20-21. *Pump jack scaffolds can be used for heights up to 30′.*

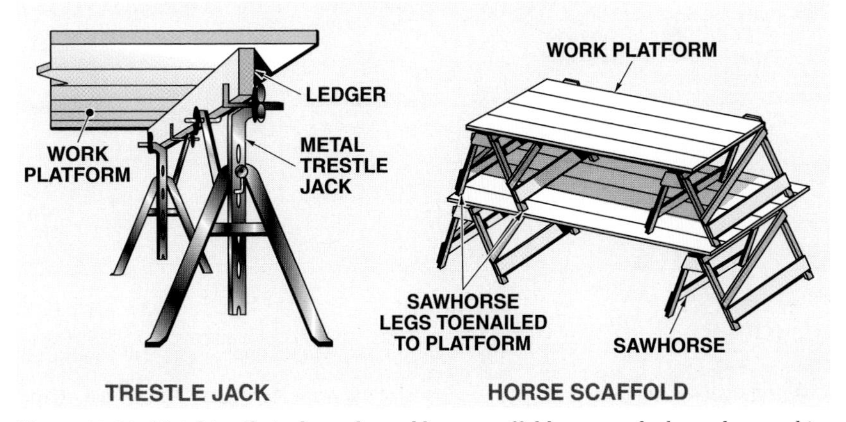

Figure 20-22. *Metal trestle jacks and wood horse scaffolds are used when a low working platform is required.*

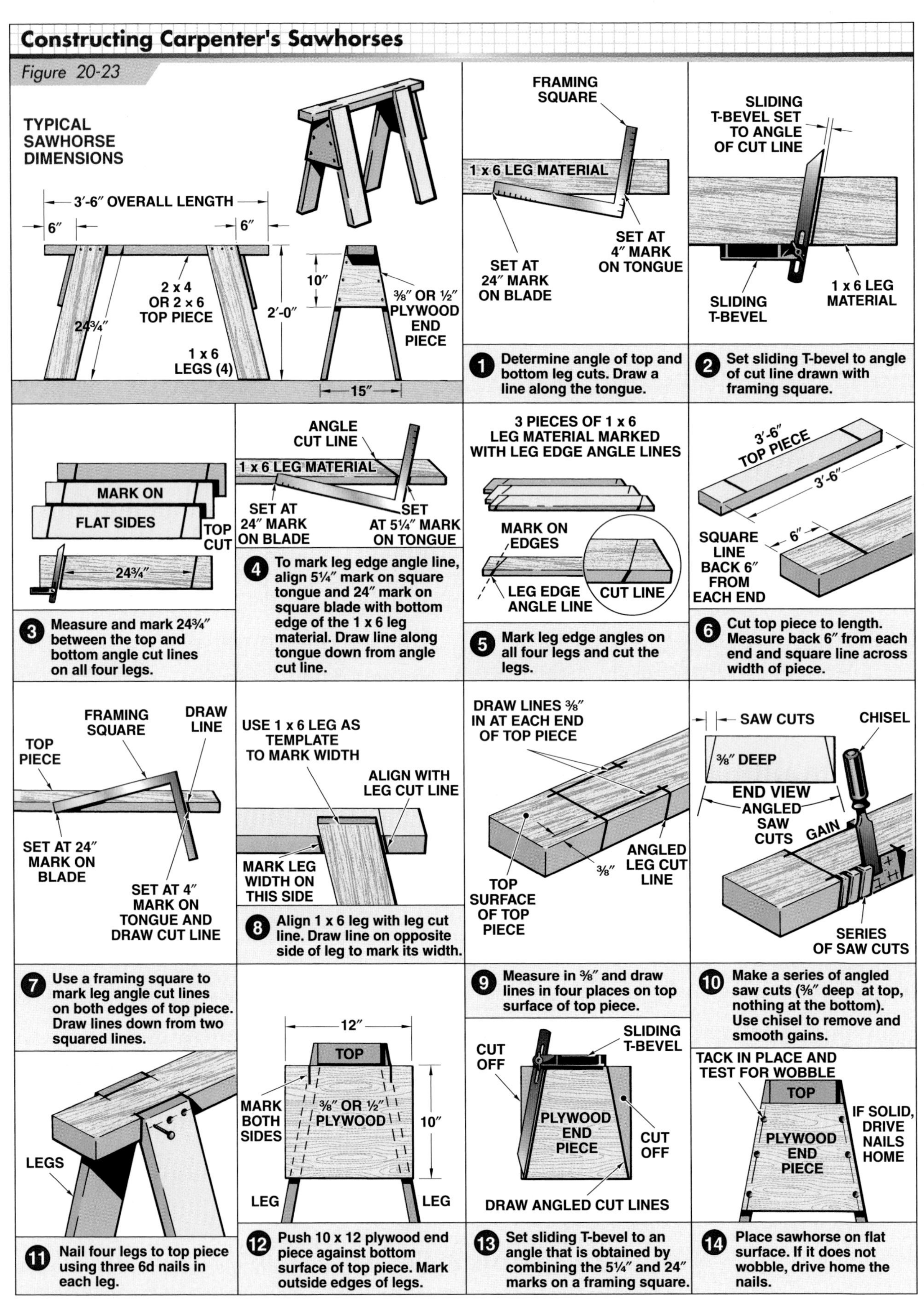

Figure 20-23. *A sturdy sawhorse is an invaluable device for a carpenter.*

OSHA regulations relating to scaffolds are found in OSHA 29 CFR 1926 Subpart L, *Scaffolds;* OSHA 29 CFR 1910.28, *Safety Requirements for Scaffolding;* and OSHA 29 CFR 1910.29, *Manually Propelled Mobile Ladder Stands and Scaffolds (Towers).* ANSI standards are found in ANSI A10.8-1988, *Construction and Demolition Operations—Scaffolding—Safety Requirements.*

Scaffold Safety

OSHA regulations state that scaffolds may be erected, moved, altered, or dismantled only under the supervision of a competent person because the lives of workers depend on the proper construction of scaffolds. Yearly, many scaffold accidents involve workers falling, workers being struck by falling objects, scaffold collapse, or electrocution. General scaffold safety precautions include the following:

- Use only 2″ nominal structural planking that is free from defects, engineered wood, or approved prefabricated metal work platforms.
- Cleat platform end extensions must have a minimum of 6″ overlap and a maximum of 12″.
- Always observe working load limits.
- Install toprails, midrails, and toeboards on all open sides and ends of work platforms more than 10′ above a lower level.
- Lay platform planks with no openings more than 1″ between adjacent planks.
- Provide overhead protection for workers on a scaffold exposed to overhead hazards.
- Do not work on scaffolds during high winds or storms, or when work platforms are ice covered or slippery.
- Lock mobile scaffolds in position when in use.
- Advise all personnel in close proximity of the movement of a mobile scaffold.
- Use fall protection on working heights of more than 10′.
- Use safety nets for workers at any level over 25′ when the workers are not otherwise protected by harnesses or other approved fall-arrest systems.

Electrical Hazards. Overhead power and utility lines pose potential hazards to workers on metal scaffolds. Many electrical accidents occur when erecting or moving metal scaffolds, or when improperly grounded tools are used on the scaffolds. The following safety precautions should be observed when working around electrical hazards:

- Do not erect a metal scaffold closer than 3′ from energized insulated lines or 10′ from noninsulated lines. If the minimum clearances cannot be maintained, contact the utility company to de-energize the power lines or cover them with insulating hoses.
- Wear only approved plastic protective helmets (hard hats) when working around power lines; do not wear metal helmets.
- Use only double-insulated tools when working on scaffolds. Ensure extension cords are attached to a ground fault circuit interrupter (GFCI) receptacle.
- Ensure warning labels are affixed to each scaffold section to warn of electrical hazards.

PERSONAL FALL-ARREST SYSTEMS

A *personal fall-arrest system* is used to arrest (stop) a worker in a fall from a working level. Personal fall-arrest systems may be used, and are many times required for use, with equipment such as aerial lifts. Proper personal fall-arrest systems must be used when working at heights greater than 10′. Personal fall-arrest systems consist of an *anchorage point, connectors,* and a *body harness,* and may include a *lanyard, lifeline, deceleration device,* or combinations of these devices. **See Figure 20-24.**

An *anchorage point* provides a secure point of attachment for lifelines, lanyards, or other deceleration devices. Anchorage points for personal fall-arrest systems must be independent of any anchorage point used to support or suspend platforms and must be capable of supporting 5000 lb minimum per person attached to the anchorage point, or at least twice the anticipated load. Anchorage points must be designed and installed under the supervision of a qualified person.

Connectors, such as D-rings and snap hooks, must have a minimum tensile strength of 5000 lb and must be proof-tested to a minimum tensile load of 3600 lb. Connectors are typically attached to other personal fall-arrest system components such as anchorage points, body harnesses, lifelines, and lanyards. Only locking-type snap hooks are permitted to be used for personal fall-arrest systems. Snap hooks can be attached directly to webbing, rope, or wire rope, to a D-ring, to each other, and to horizontal lifelines.

Body harnesses, when worn properly, protect internal body organs, the spine, and other bones in a fall. The attachment point of a body harness should be located in the center of the back near shoulder level, or above the worker's head. Body harnesses should be inspected before each use to ensure they will properly support a worker in case of a fall. The harness webbing should be inspected for wear, including frayed edges, broken fibers, burns, and chemical damage. D-rings and buckles should be inspected for distortion, cracks, breaks, and sharp edges. Grommets on body harnesses should be inspected to ensure they are tight.

Harnesses must fit snugly and be securely attached to a lanyard. A *lanyard* is a flexible line of rope, wire rope, or strap that generally has a connector at each end for connecting a body harness to a deceleration device, lifeline, or anchorage point. A *deceleration device* dissipates a substantial amount of energy during a fall arrest or limits the energy imposed on a worker during a fall arrest. Common deceleration devices include rope grabs, shock-absorbing lanyards, and self-retracting lifelines or lanyards.

A rope grab is a deceleration device that travels on a lifeline. A rope grab automatically engages a vertical or horizontal lifeline by friction to arrest the fall of a worker. Rope grabs protect workers from falls while

Personal Fall-Arrest Systems

Figure 20-24

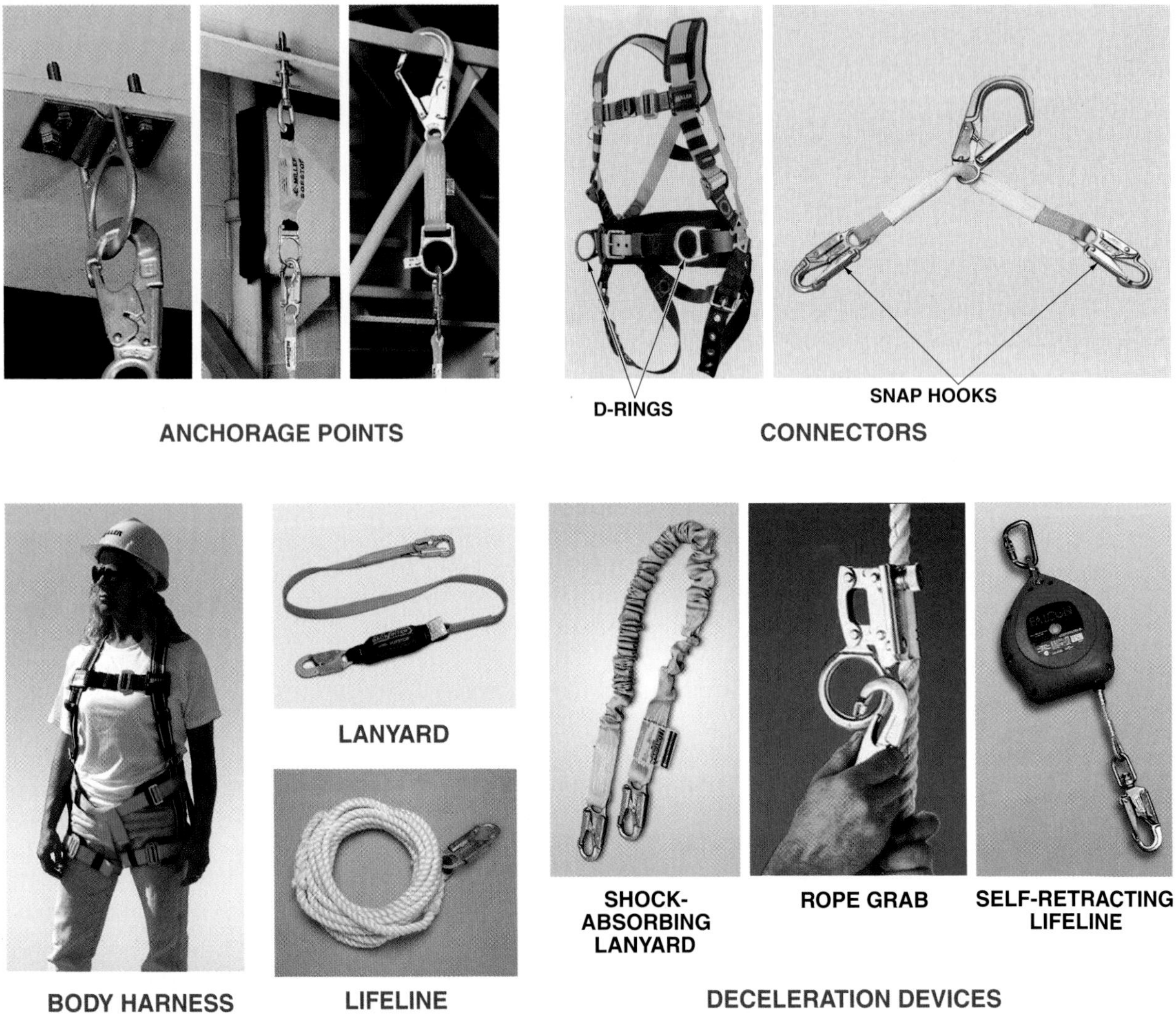

ANCHORAGE POINTS

D-RINGS

SNAP HOOKS

CONNECTORS

LANYARD

SHOCK-ABSORBING LANYARD

ROPE GRAB

SELF-RETRACTING LIFELINE

BODY HARNESS

LIFELINE

DECELERATION DEVICES

Miller Fall Protection

Figure 20-24. *Personal fall-arrest systems must be utilized when working at heights greater than 10′ when other means of fall protection are not provided. Personal fall-arrest systems consist of anchorage points, connectors, body harnesses, lanyards, deceleration devices, and lifelines.*

allowing freedom of movement. A lanyard or lifeline is attached between the body harness and rope grab. A *shock-absorbing lanyard* has a specially woven, shock-absorbing inner core that reduces fall arrest forces. The outer shell of the lanyard serves as the secondary lanyard.

Lifelines are anchored above the work area, offering a free-fall path, and must be strong enough to support the force of a fall. *Vertical lifelines* are connected to a fixed anchor at the upper end that is independent of a work platform such as a scaffold. Vertical lifelines must never have more than one worker attached per line. A *self-retracting lifeline* is a type of vertical lifeline. A self-retracting lifeline contains a line that can be slowly extracted from or retracted onto its drum under slight tension during normal worker movement. When a fall occurs, the drum automatically locks, arresting the fall. *Horizontal lifelines* are connected to fixed anchors at both ends. Workers attach their lanyard to a D-ring on the lifeline, allowing them to freely move horizontally along the lifeline.

A lifeline must be properly terminated (anchored) to prevent the safety sleeve or ring from sliding off its end. The path of a fall must be visualized when anchoring a lifeline. Use an anchored system without any obstructions to the fall. Obstacles below and in the fall path can be deadly.

Personal Fall-Arrest System Requirements

When a personal fall-arrest system is used for fall protection, the system must be able to do the following:

- Limit the maximum arresting force to 1800 lb when used with a body harness.
- Be rigged so the worker can neither fall more than 6′ nor contact any lower level. **See Figure 20-25.**
- Bring a worker to a complete stop and the limit maximum deceleration distance a worker falls to 3′-6″. The *deceleration distance* is the additional vertical distance a falling worker travels, excluding lifeline elongation and free-fall distance, before stopping, from the point at which the deceleration device begins to operate. The deceleration distance is measured as the distance between the location of a worker's body harness attachment point at the moment of activation of the deceleration device during a fall and the location of that attachment point after the employee comes to a full stop.
- Have sufficient strength to withstand twice the potential energy impact of a worker free falling a distance of 6′ or the free-fall distance permitted by the system, whichever is less. The *free-fall distance* is the vertical distance between the fall-arrest attachment point on the body harness before the fall and the attachment point when the personal fall-arrest system applies force to arrest the fall.

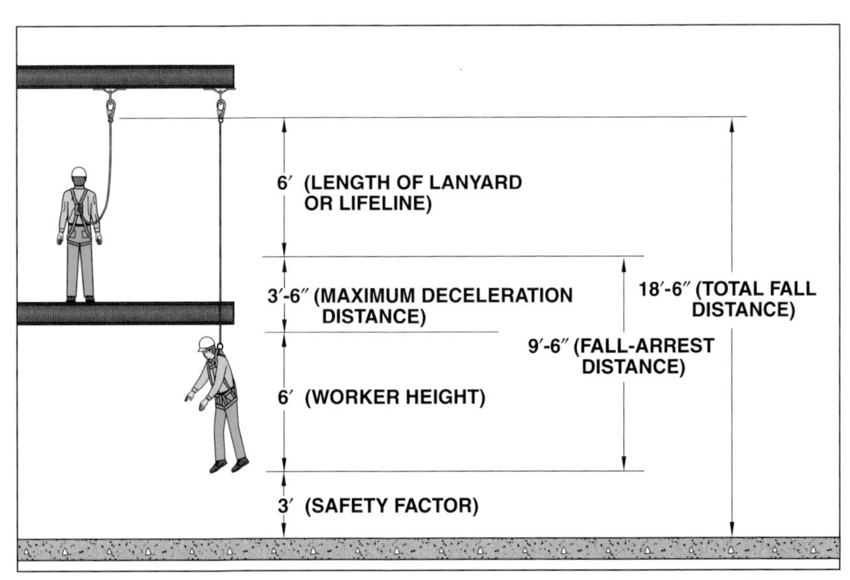

Figure 20-25. *Always determine the estimated fall distance when selecting the proper personal fall-arrest system. Self-retracting lifelines should be used when the estimated fall distance is less than 18′-6″. Self-retracting lifelines or shock-absorbing lanyards can be used when the fall distance is greater than 18′-6″.*

Safety Nets. A *safety net* is a net made of rope or webbing for catching and protecting a falling worker. A safety net must be used anywhere a worker is 25′ or more above the ground, water, machinery, or other solid surface when the worker is not otherwise protected by fall-arrest equipment or scaffold guardrails. **See Figure 20-26.** Safety nets must also be used when public traffic or other workers are permitted underneath a work area that is not otherwise protected from falling objects.

LADDERS

A *ladder* consists of two siderails joined at intervals by rungs or steps for climbing up and down. Ladders are made of wood, aluminum, or fiberglass in lengths of 3′ to 50′. Carpenters and other tradesworkers use ladders to perform light operations and repairs not requiring full scaffolds. Ladders are categorized as *fixed, single, extension,* and *stepladders.* **See Figure 20-27.**

Fixed Ladders

Fixed ladders are permanently attached to a structure, and are commonly used to provide access to a flat roof. Fixed ladders are usually constructed of steel or aluminum.

Single Ladders

A single ladder consists of only one section and is of fixed length. Typical single ladder lengths range from 6′ to 24′. Single ladders are limited in their versatility because a given length ladder may be safely used only within a fixed height range.

The Sinco Group, Inc.

Figure 20-26. *A safety net must be installed whenever a worker is 25′ or more above the ground, water, machinery, or other solid surface when the worker is not otherwise protected by fall-protection equipment or guardrails.*

The rungs and steps of portable metal ladders must be corrugated, knurled, dimpled, coated with skid-resistant material, or otherwise treated to minimize slipping. Rungs must be parallel, level, and uniformly spaced when the ladder is in position for use.

Ladders

Figure 20-27

FIXED — SIDE RAILS, LADDER ATTACHMENTS, ROOF ATTACHMENTS (Ballymore Company, Inc.)

SINGLE — WALL, RUNGS, SIDE RAILS

EXTENSION — TIP, FLY SECTION, BED SECTION, HALYARD, RUNG, FOOT ASSEMBLY, WALL, CENTER SWIVEL PULLEY, PAWL LOCK, SIDE RAILS, BUTT END

STEP — SPREADER, SHELF, SIDE RAILS, STEPS, BRACES (Werner Ladder Co.)

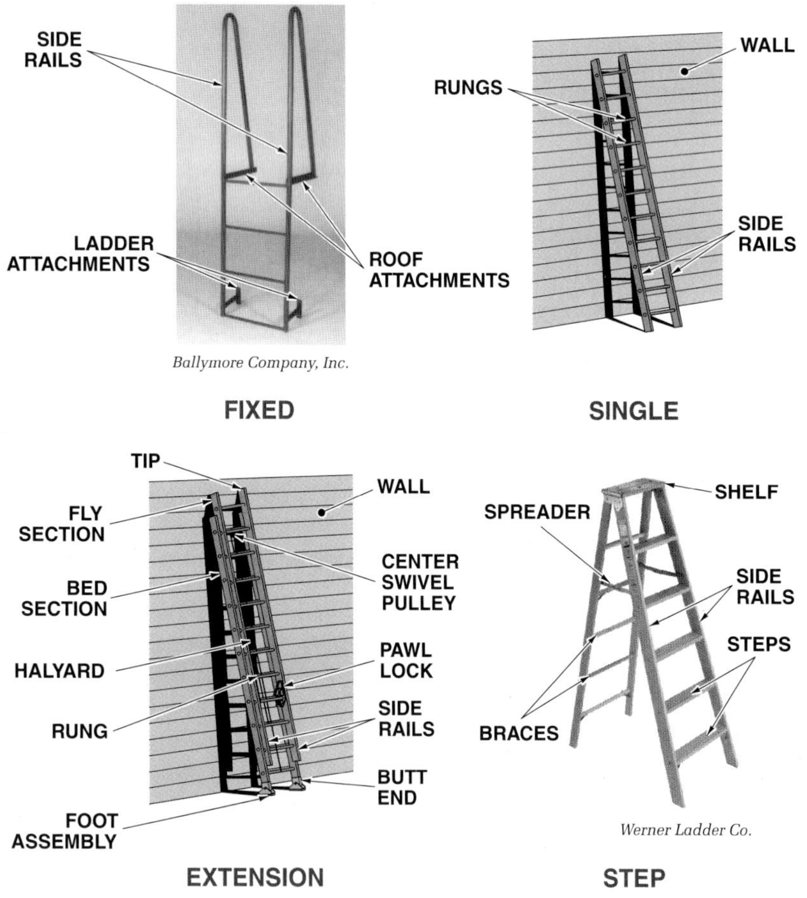

Figure 20-27. *Ladders are categorized as fixed, single, extension, and stepladders. Extension and stepladders are most commonly used on a job site.*

Extension Ladders

An extension ladder is an adjustable-height ladder with a fixed *bed section* and one or more slidable, locking *fly sections.* The bed section is the lower section of an extension ladder. A fly section is an upper section of an extension ladder. A *pawl lock* is attached to a fly section to hold the fly section at the desired height.

Fly sections are raised and lowered with the use of a *halyard.* A halyard must be a minimum of 3/8″ diameter with a minimum breaking strength of 825 lb. The halyard is threaded through the pulley attached to the top rung of the bed section. One end of the rope is attached to the bottom rung of the fly section and the other end is usually tied off at the bottom of the ladder.

Extension and single ladders should be positioned on a 4:1 ratio (75° angle). For every 4′ of *working length,* 1′ of space is required at the base. **See Figure 20-28.** Working length is the distance along the ladder between the foot and top support. The tip of a single or extension ladder should be secured at the top to prevent it from slipping and must be at least 3′ above the roof line or top support. Ladders over 15′ long should also be secured at the bottom. **See Figure 20-29.** Never stand on the top three rungs of a single or extension ladder.

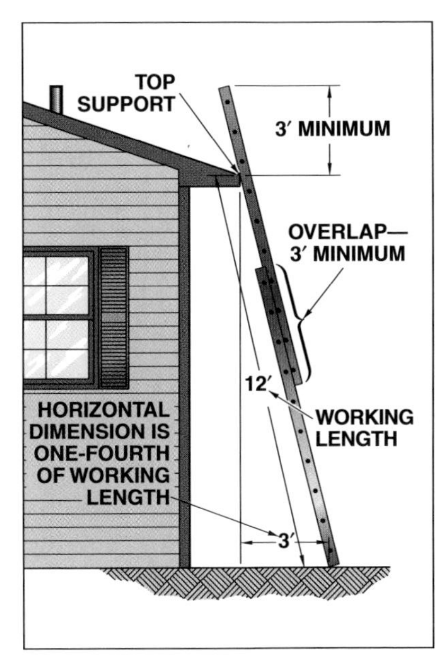

Figure 20-28. *Single and extension ladders should be positioned on a 4:1 ratio, which is approximately a 75° angle.*

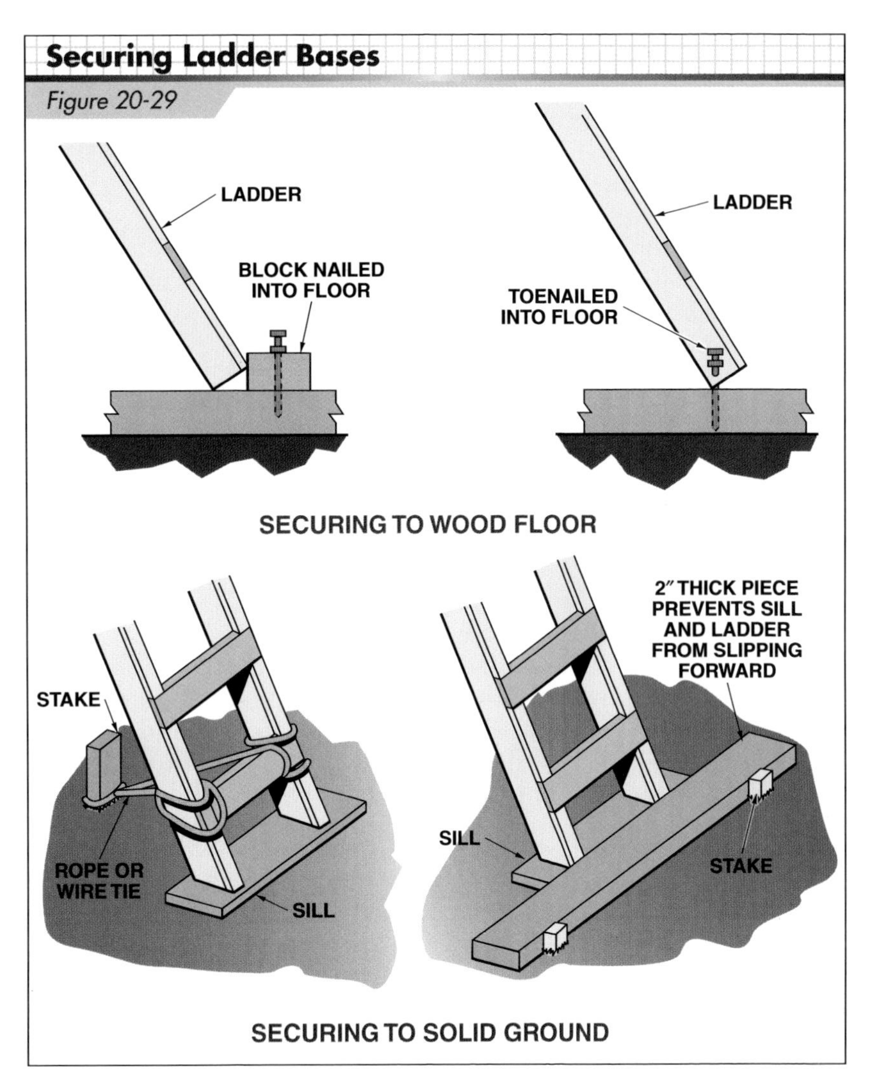

Figure 20-29. *Ladders over 15′ long should be secured at the bottom to prevent ladder movement.*

Stepladders

A stepladder is a folding ladder that stands independent of support. Stepladders are available in 2′ to 12′ lengths, and are used more often than other ladders because they are easily portable and can be used for working at various heights. Never stand on the top two rungs of a stepladder since there is no support for the user.

Ladder Regulations and Standards

Regulations and standards for the design, use, and testing of ladders are published by various federal, state, and standards organizations. OSHA publishes regulations for ladders in OSHA 29 CFR 1926.1053, *Ladders;* OSHA 29 CFR 1910.25, *Portable Wood Ladders;* OSHA 29 CFR 1910.26, *Portable Metal Ladders;* and OSHA 29 CFR 1910.27, *Fixed Ladders.* ANSI publishes standards for ladders in ANSI A14.1, *Ladders—Portable Wood—Safety Requirements for;* ANSI A14.3, *Ladders—Fixed—Safety Requirements;* and ANSI 14.5, *Ladders—Portable Reinforced Plastic—Safety Requirements.*

Ladder Duty Ratings. *Ladder duty rating* is the weight a ladder is designed to support under normal use. The four ladder duty ratings are as follows:

- Type IAA—special duty, industrial, 375 lb capacity
- Type IA—extra heavy-duty, industrial, 300 lb capacity
- Type I—heavy-duty, industrial, 250 lb capacity
- Type II—medium-duty, com-mercial, 225 lb capacity
- Type III—light-duty, household, 200 lb capacity

All ladders must be used only for the purpose for which they are intended.

Ladder Climbing Techniques

Ladder climbing should begin only after a ladder is resting on a firm, level surface and is properly secured. Climbing movements should be smooth and rhythmical to prevent ladder bounce and sway.

Safe climbing employs the three-point contact method. In the three-point contact method, the body is kept erect, the arms are kept straight, and the hands and feet make the three points of contact. Two feet and one hand or two hands and one foot are in contact with the ladder rungs at all times. Each hand should grasp the rungs with the palms down and the thumb on the underside of the rung. Progress up the ladder should be caused by the push of the leg muscles and not the pull of the arm muscles.

Fall-arrest systems should be used when climbing to greater heights, **Figure 20-30.** A fall-arrest system consists of a flexible or rigid rail *carrier* and a *safety sleeve.* The carrier is secured to the ladder or structure, and is the track for the safety sleeve. The ring or a safety harness connects to the safety sleeve. The safety sleeve clamps against the carrier if a fall occurs, protecting the person climbing the ladder from injury.

Safety precautions to observe when using ladders include the following:

- Use ladders only for the purpose for which they were designed.
- Inspect ladders when new and before each use for broken, cracked, or otherwise defective parts.
- Ensure ladder rungs and siderails are free of oil, grease, or other substances that may cause slipping.
- Set the ladder in place, making sure the base is level and firm. Secure the ladder so it will not slip while in use.
- Do not place a ladder against a movable object.
- Use leg muscles when lifting and lowering ladders.
- Do not place a ladder against a window sash. Tack a board over the window opening and place the ladder against the board.
- Always check for the proper angle of inclination before climbing a ladder, keeping the 4:1 ratio in mind.
- Exercise extreme caution when using ladders near electrical conductors or equipment. All ladders conduct electricity when wet.
- Verify that all pawl locks on extension ladders are securely hooked over rungs before climbing.
- Ensure that stepladders are fully open, with spreaders locked, before climbing.
- Face the ladder when ascending and descending.
- Do not carry tools or materials in your hands when climbing up or down the ladder. Carry tools in a work apron or use a rope and bucket to hoist them up.
- Position a ladder so it is clear of doors or passageways.
- Do not stand on the top two rungs of a stepladder or top three steps of an extension or single ladder.
- Do not attempt to reposition an occupied ladder.
- Do not lean or overreach from a ladder.
- Do not use a ladder in high wind conditions.

Miller Fall Protection

Figure 20-30. *A fall-arrest system should be employed on ladders when climbing to greater heights.*

Trestle Ladder and Ladder Jack Scaffolds

Trestle ladder and *ladder jack scaffolds* combine ladders with a platform. **See Figure 20-31.** Trestle ladder and ladder jack scaffolds are designed for light-duty applications such as caulking or painting. Trestle ladders must not be used as part of a scaffold for any other purpose besides access and egress unless they are equipped with guardrails. The use of trestle ladders longer than 20′ is prohibited.

Ladder jack scaffolds are constructed of two extension or single ladders, ladder jacks (brackets), and a platform extending between and resting on the ladder jacks.

Ladder jack platforms cannot be placed higher than 20′ above the floor or ground and must be at least 12″ wide. Personal fall-arrest systems must be used for work 10′ or more above the floor or ground.

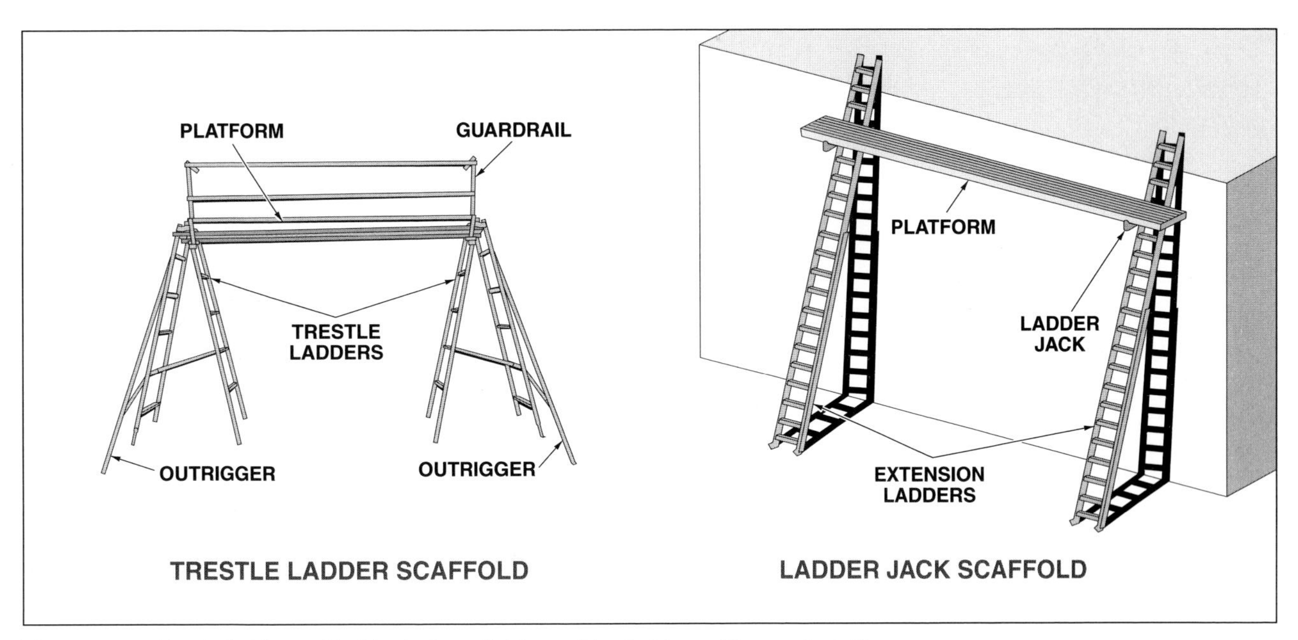

Figure 20-31. *Trestle ladder and ladder jack scaffolds combine ladders with a work platform.*

UNIT 21 Construction Equipment

OBJECTIVES

1. List and describe common earth-moving equipment.
2. List and describe common hoisting, lifting, and rigging equipment.
3. Identify voice and visual crane operating signals.
4. Understand the importance of the center of gravity when lifting a load.
5. Understand rigging calculations.

Earth-moving equipment, such as bulldozers and excavators, is used to remove and transport soil around a job site in preparation for foundation footings, to backfill soil against a foundation, and to prepare a job site for landscaping. Other construction equipment, such as a tower crane, is used on larger construction projects where materials and equipment must be hoisted several floors. The type and number of pieces of construction equipment used on a project depends on the type of construction. Erecting a large concrete building, for example, requires more equipment than building a wood-framed house.

EARTH-MOVING EQUIPMENT

Most new construction projects require a certain amount of excavation or grading to be performed. The amount of excavation or grading depends on the size of the structure being erected and the depth of the foundation. A job site for a residential structure may require a small amount of grading and trenching for foundation footings and perhaps holes drilled for piers. On a large construction project, tons of soil may have to be removed for deep footings and foundations, and pilings may need to be driven into the ground to support the structure.

Carpenters are often required to set up lines and lay out the areas for excavation and trenching. Carpenters may also be called upon to verify the correct depths during excavation and trenching operations.

One type of equipment used for excavation is a *bulldozer,* which is a crawler-tractor with a blade that is mounted perpendicular to the line of travel. Bulldozers are used to start excavations and to strip rocks and topsoil from a job site. Bulldozers are also used to move soil excavated from one part of the site to another part of the site where additional fill is required. **See Figure 21-1.** In addition, bulldozers are commonly used to backfill soil against a foundation.

Courtesy of Deere & Company

Figure 21-1. *A bulldozer is used to start excavations and to strip rocks and topsoil from a job site. It is also used to move soil from one part of the site to another part.*

A *loader* is a wheeled tractor with hydraulic arms that control a bucket mounted in front. Loaders are used to pick up loose soil and rocks and deposit them into trucks for removal, and are also used for stockpiling aggregate materials and backfilling trenches. **See Figure 21-2.**

Courtesy of Deere & Company

Figure 21-2. *A loader is used to pick up loose soil and rocks and deposit them into trucks.*

A *grader*, or motor grader, is used in final grading operations for large construction sites to level the earth surface. Graders are commonly used in road construction to ensure the proper roadbed grade. The blade of a grader can be adjusted to various angles and positions. **See Figure 21-3.**

An *excavator* is used to remove soil and deposit it in trucks for removal or deposit it elsewhere on the job site. Large and small models of excavators are available. **See Figure 21-4.** Different attachments can be installed on an excavator, making it useful for general excavation, trenching, and loading operations. A *power shovel* is a large excavator used to remove large volumes of soil from a job site.

A *backhoe loader* is a piece of construction equipment used for digging and loading operations. The hoe bucket is used to dig trenches for foundation footings and the loader bucket is used for smaller loading jobs. **See Figure 21-5.** Backhoe loaders are equipped with outriggers to provide stability during trenching operations.

Earth-moving equipment, such as backhoe loaders, may become unstable when lifting heavy loads. When transporting a heavy load, travel with the load as close to the ground as possible.

An *earth auger* is a large power-driven drill used to bore holes for deep concrete piers or piles. **See Figure 21-6.** Earth augers are connected to excavators, backhoes, or cranes, which provide power for the augers.

Courtesy of Deere & Company

Figure 21-4. *Excavators remove soil from a job site and are available in large and small models.*

Courtesy of Deere & Company

Figure 21-3. *A grader performs final grading operations. Its blade can be adjusted to various angles and positions.*

Courtesy of Deere & Company

Figure 21-5. *A backhoe loader is used for digging and loading operations. The outriggers behind the cab are used to stabilize the equipment when the backhoe bucket is in use.*

Figure 21-6. *An earth auger is a power-driven drill used to bore holes for deep concrete piers or piles. The belling tool attachment to the right of the earth auger is used to form belled caissons.*

AERIAL LIFTS

An aerial lift is used when work must be performed at elevated or otherwise inaccessible locations. An *aerial lift* is a piece of extendable and/or articulating equipment designed to position personnel and materials in elevated locations. These types of lifts are also known as manlifts or aerial work platforms.

There are several types of aerial lifts. Self-propelled aerial lifts include scissors, articulating Z-boom, and extensible S-boom lifts. The work platform of a scissors lift is raised and lowered by mechanical scissors action using an electrical or hydraulic power source. **See Figure 21-7.** The work platform or bucket of an articulating Z-boom lift is raised and lowered through two or more hinged sections. The work platform or bucket of an extensible S-boom lift is raised and lowered by use of a telescoping arm.

Ensure that the rated load capacity of a hoist or rigging components is not exceeded.

JLG Industries, Inc.

Figure 21-7. *Self-propelled scissors lifts are used to lift personnel to appropriate working heights.*

CONSTRUCTION FORKLIFTS

A *construction forklift* is a piece of equipment used for transporting building materials and equipment around a job site, usually relatively close to the ground. Forklifts use a two-pronged fork to support and lift the load from underneath. Two common types of construction forklifts are rough-terrain forklifts and telehandlers.

A *rough-terrain forklift* is a heavy-duty forklift designed to traverse the rough terrain of construction sites. **See Figure 21-8.** Large, knobby tires allow the forklift to drive over ungraded, muddy, or sloping ground conditions. Some models can lift and deposit materials up to two stories above the ground but must still be driven up close to the destination.

Courtesy of Deere & Company

Figure 21-8. *A rough-terrain forklift is designed for rough-terrain construction work.*

A *telehandler* is a construction forklift designed to place materials up to a few stories in elevation and from a short distance. **See Figure 21-9.** Telehandlers have telescopic arms that extend from the vehicle and can raise and lower materials, providing a greater range of placement. Some telehandlers also feature a rough-terrain chassis for traveling on uneven ground.

JLG Industries, Inc.

Figure 21-9. *A telehandler uses a telescopic arm to place materials in remote locations. The outriggers at the front of the telehandler provide stability to the equipment.*

CRANES

For larger loads and higher lifts, a crane is typically required. A *crane* is a combination of a hoist and a supporting structure designed to move loads aerially. There are many sizes and types of cranes. Cranes used on construction sites are broadly classified into mobile cranes and tower cranes.

Mobile Cranes

A *mobile crane* is a crane that can be moved within and between job sites. Mobile cranes are composed of crane assemblies and vehicle platforms. Many variations are available because the type of platform is relatively independent of the type of crane. **See Figure 21-10.** Therefore, mobile cranes are specified by both the crane type and the platform type.

Mobile Cranes

Figure 21-10

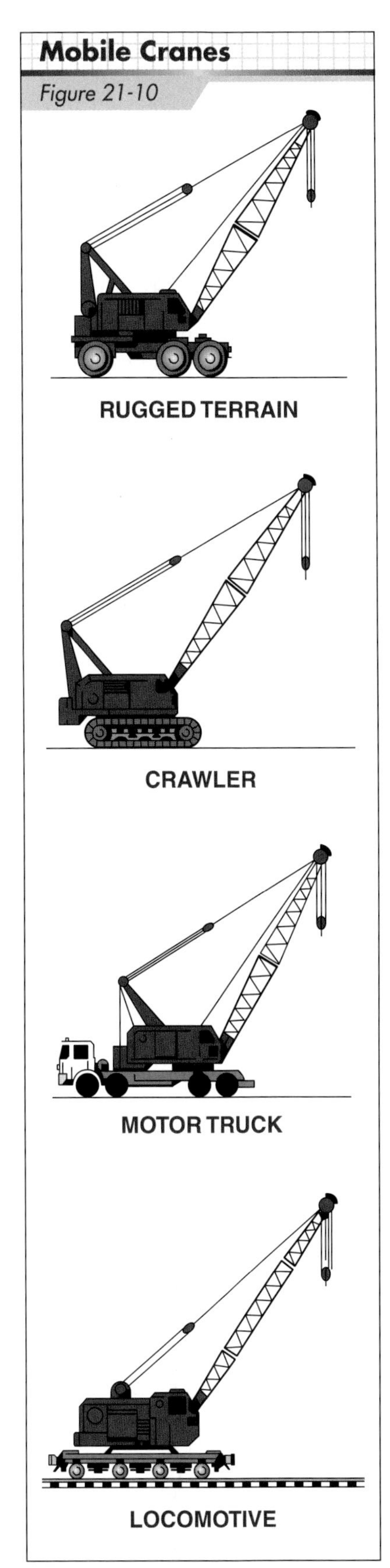

Figure 21-10. *Mobile cranes are crane assemblies that are mounted on mobile platforms. The platform may use a variety of forms of mobility depending on the intended application.*

Telescopic-boom cranes are the most common type of crane on commercial or residential construction job sites. A *telescopic-boom crane* is a mobile crane with an extendable boom composed of nested sections. These booms are similar in design to the telescopic arms of telehandlers. The boom sections are extendable and retractable, allowing a wide range of boom lengths, which changes the distance of the hoist hook from the crane. **See Figure 21-11.**

Telescopic-Boom Cranes

Figure 21-11

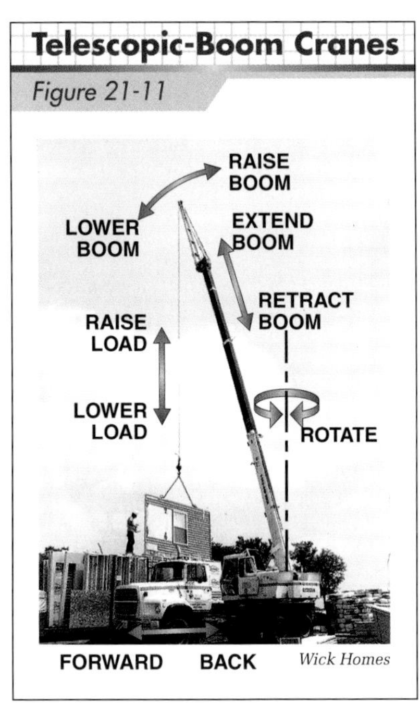

Figure 21-11. *A telescopic-boom crane can extend or retract its boom, providing significant versatility to its reach.*

The boom movement and extension is typically powered by integrated hydraulic systems. When the boom is fully retracted, the crane is easily transported over most regular roads. Telescopic-boom cranes are often mounted onto a motor truck platform.

For larger construction projects, such as high-rise construction, lattice-boom cranes are commonly used. A *lattice-boom crane* is a mobile crane with a boom constructed from one or more gridworks of thin steel members. The lattice structure provides a very strong boom that is light for its size. **See Figure 21-12.**

Lattice-Boom Cranes

Figure 21-12

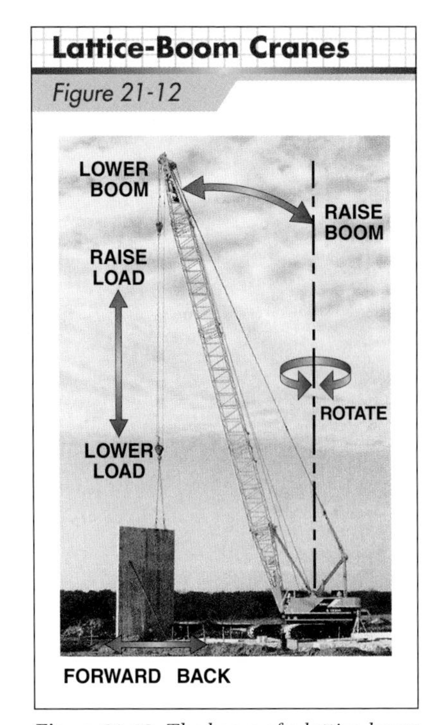

Figure 21-12. *The boom of a lattice-boom crane is constructed of a lightweight, open structure that allows these cranes to have particularly long booms.*

The boom sections must be assembled onto the crane body when on site. This makes a lattice-boom crane less transportable between job sites than a telescopic-boom crane, although its vehicle platform allows it to move around while on site. A lattice-boom crane is typically mounted on a crawler platform, which allows the crane to operate in areas with soft or moderately uneven ground. Crawler platforms are tracked with linked plates, much like military tanks.

Construction equipment must be inspected by a competent person prior to each use, and during use, to ensure safe operation.

Tower Cranes

Tower cranes are commonly used on heavy construction projects such as high-rise structures. A *tower crane* is a fixed crane consisting of a high vertical mast tower topped by a horizontal jib. **See Figure 21-13.** The jib has a long arm, which provides horizontal travel for the hoist trolley, and a short arm, which provides a counterweight. The intersection between the mast and jib contains a slewing unit for jib rotation and an operator's cab.

Tower cranes are transported in sections by truck to a job site where they are assembled. Tower crane types are categorized by the way they are attached at the base. A *climbing tower crane* is a tower crane that is secured to the floor of the high-rise structure being erected and can be periodically raised as new floor levels are added to the structure. A *freestanding tower crane* is a tower crane that is secured to a concrete foundation next to the structure being erected.

One of the most widely used tower cranes is the inside-climbing tower crane. To use this type of crane, openings must be provided in the floor slabs for the tower sections. At the base of the crane, steel collars on a floor slab support the weight of the crane. Additional shoring may be required beneath the floor slab around the opening.

After the completion of six to eight stories of the building, the entire crane is raised with hydraulic jacks. The floor openings are filled when they are no longer required for the lower tower sections. In some situations, the elevator shaft of a building can be used as a floor opening.

When a climbing tower crane is no longer required on a job site, the crane is dismantled in sections and lowered to the ground using a derrick. A *derrick* is a lifting device consisting of a vertical tower with a swinging boom hinged at its base. The derrick is set up on the roof of a building, and after the tower crane is dismantled, the derrick is also dismantled and lowered to the ground.

On small- to medium-size construction projects, small, self-erecting tower cranes are used. **See Figure 21-14.** No additional lifting equipment is needed to assemble or disassemble a self-erecting tower crane.

SECURED TO FLOOR OF STRUCTURE

CLIMBING TOWER CRANE

FREESTANDING NEXT TO STRUCTURE

FREESTANDING TOWER CRANE

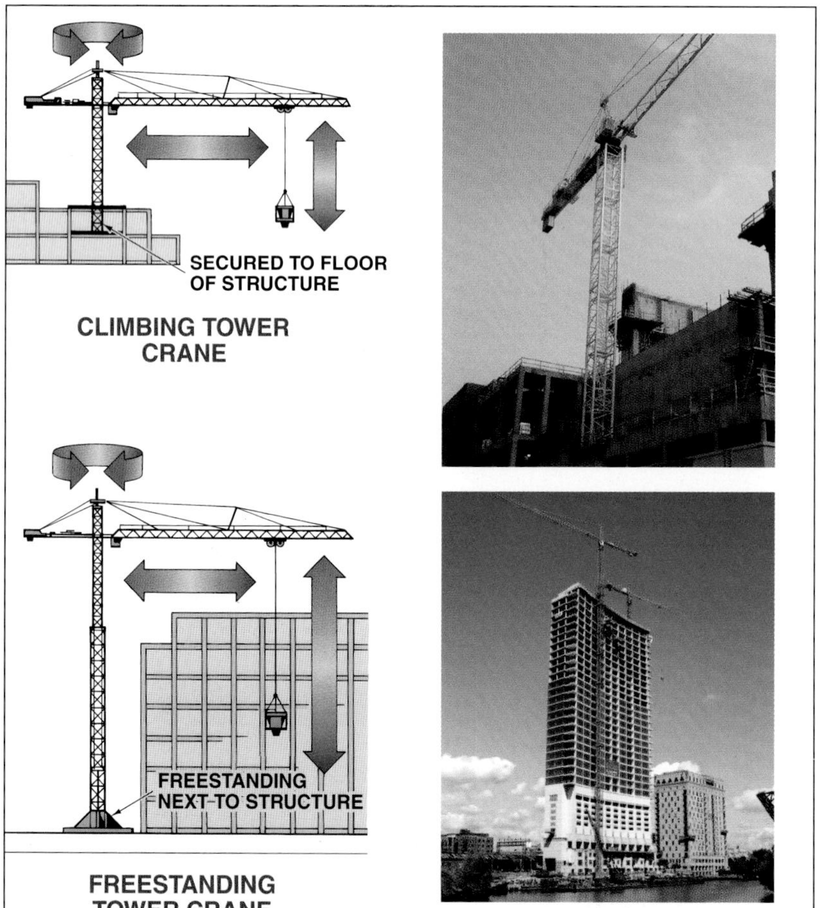

Portland Cement Association

Figure 21-13. *Tower cranes are very tall, semipermanent structures used to construct buildings, particularly high-rises.*

Manitowoc Crane Group

Figure 21-14. *Small, self-erecting tower cranes require no additional lifting equipment for assembly or disassembly.*

Crane Operating Signals

During a lift, a crane operator may not have a clear view of a load's starting position, destination, or other points along the route of travel. This may be because of visual obstructions or because of the distance involved. Also, other activities at a job site may need to be monitored to ensure that they will not interfere with the lift. Therefore, many hoisting operations require ground personnel to properly guide the load by providing crane operating signals. A signalperson must communicate with the crane operator in a clear and effective manner in order to ensure safe operation of the crane.

OHSA 29 CFR 1926 Subpart CC, *Cranes & Derricks in Construction*, provides regulations for the safe preparation and operation of this equipment, including signaling. Allowable types of signals include voice commands, hand signals, sounds, and other signals established between the signalperson and crane operator. Voice commands and hand signals are the two most common types of crane operating signals.

Voice commands are given over two-way radios or similar communications equipment that is free from interference from nearby equipment. The channel must be dedicated to the signalperson-operator team except when other signalpersons and/or crane operators are involved in the same lifting operation. Voice commands must be given in a certain order:

1. Function and direction
2. Distance remaining and/or speed
3. Function and stop command

The distance and/or speed information is repeated throughout the operation in order to provide continual feedback to the crane operator. For example, the commands for hoisting a load given by the signalperson may be, "hoist up, fifty feet... forty feet...thirty feet...twenty feet...ten feet...slow...slow... hoist stop."

Hand signals can provide operating instructions without requiring communications equipment. However, hand signals can only be used when the crane operator can clearly see the signalperson and their hands at all times. The OSHA standard contains a chart of standard hand signals that cover all possible crane motions. **See Figure 21-15.**

Only an experienced and designated individual should give crane signals. In many cases, the signalperson must be certified for the task. The wrong signal may result in damage to materials or serious injury to people working near the crane. The exceptions to this rule are the "stop" and "emergency stop" signals, which can be given by any personnel that become aware of an unsafe situation during a lift operation.

RIGGING

In general, the hoist of a crane cannot be directly connected to a load. Therefore, additional equipment is used as an interface between the crane hoist and lift points on the load. *Rigging* is the equipment required to connect the hoist hook of the crane to attachment points on a load. A *lift point* is any point on a load where rigging can be attached. Different types of components can be used in a variety of ways to assemble any number of rigging arrangements.

Slings

A *sling* is a flexible length of load-bearing material that is used to rig a load. Slings are an intermediate connection for hoist hooks that cannot be directly connected to a load. More importantly, however, the use of slings with other rigging hardware allows a variety of rigging arrangements. For example, flexible slings can often be looped around loads that have no lift points.

Sling Types. The most common types of materials used to construct rigging and hoisting slings are chain, wire rope, synthetic-fiber webbing, and synthetic-fiber round slings. Synthetic rope slings and wire mesh slings are also available but are not as widely used.

The lifting points of precast concrete panels are predetermined before the panel is produced.

Hand Signals for Crane Operators

Figure 21-15

STOP

With arm extended horizontally to the side, palm down, arm is swung back and forth.

EMERGENCY STOP

With both arms extended horizontally to the side, palms down, arms are swung back and forth.

HOIST

With upper arm extended to the side, forearm and index finger pointing straight up, hand and finger make small circles.

LOWER

With arm and index finger pointing down, hand and finger make small circles.

RAISE BOOM

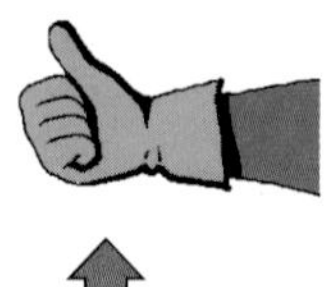

With arm extended horizontally to the side, thumb points up with other fingers closed.

LOWER BOOM

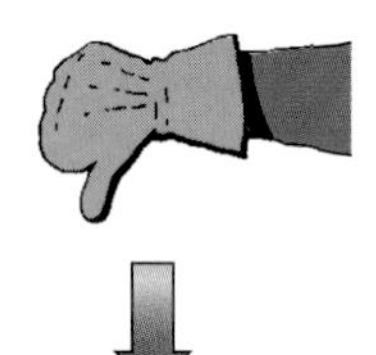

With arm extended horizontally to the side, thumb points down with other fingers closed.

EXTEND TELESCOPING BOOM

With hands to the front at waist level, thumbs point outward with other fingers closed.

RETRACT TELESCOPING BOOM

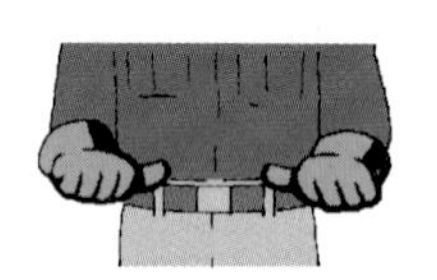

With hands to the front at waist level, thumbs point at each other with other fingers closed.

RAISE THE BOOM AND LOWER THE LOAD

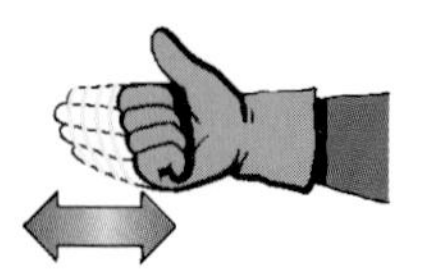

With arm extended horizontally to the side and thumb pointing up, fingers open and close while load movement is desired.

LOWER THE BOOM AND RAISE THE LOAD

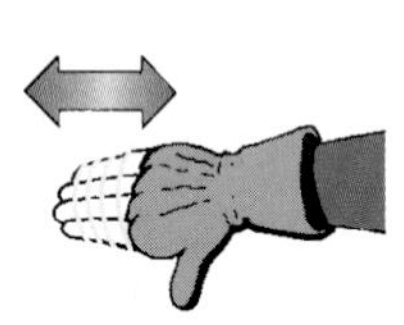

With arm extended horizontally to the side and thumb pointing down, fingers open and close while load movement is desired.

SWING

With arm extended horizontally, index finger points in direction that boom is to swing.

MOVE SLOWLY

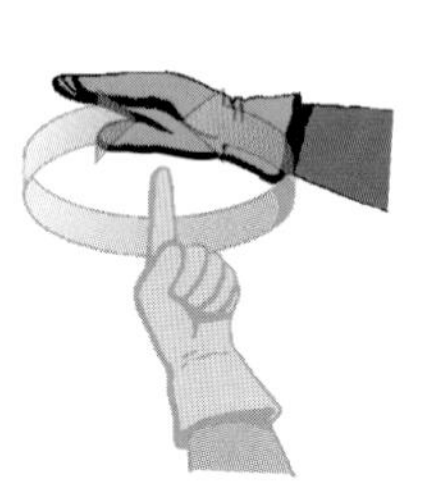

A hand is placed in front of the hand that is giving the action signal.

USE MAIN HOIST

A hand taps on top of the head. Then regular signal is given to indicate desired action.

USE AUXILIARY HOIST

(whipline) - With arm bent at elbow and forearm vertical, elbow is tapped with other hand. Then regular signal is used to indicate desired action.

DOG EVERYTHING

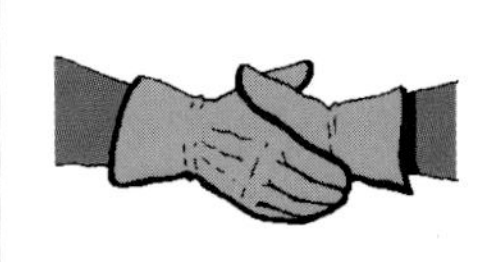

Hands held together at waist level.

TRAVEL/TOWER TRAVEL

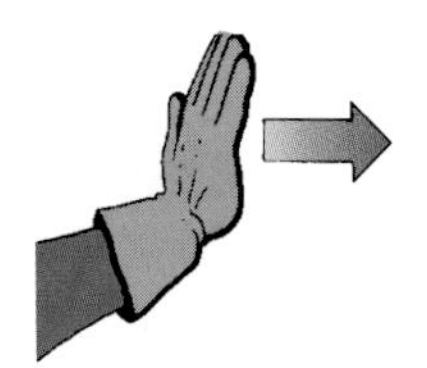

With all fingers pointing up, arm is extended horizontally out and back to make a pushing motion in the direction of travel.

TROLLEY TRAVEL

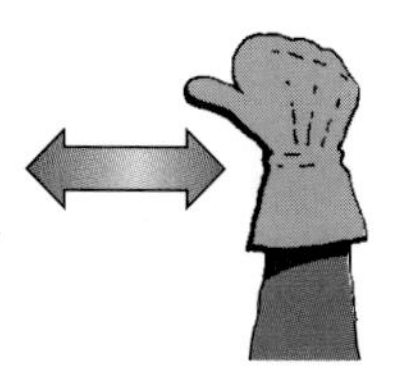

With palm up, fingers closed and thumb pointing in direction of motion, hand is jerked horizontally in direction trolley is to travel.

CRAWLER CRANE TRAVEL, BOTH TRACKS

Rotate fists around each other in front of body; direction of rotation away from body indicates travel forward; rotation towards body indicates travel backward.

CRAWLER CRANE TRAVEL, ONE TRACK

Indicate track to be locked by raising fist on that side. Rotate other fist in front of body in direction that other track is to travel.

Figure 21-15. *Standard hand signals can be used to communicate lifting commands to a crane operator, particularly if the operator cannot see the load and/or destination.*

A *chain* is a series of metal links connected to one another to form a continuous line. **See Figure 21-16.** Chain is recommended for rugged industrial applications where flexibility, abrasion resistance, and long life are required. Chain can often be used in situations in which other materials would be damaged by the load or environment, such as with a load of rough or raw castings or in a high temperature environment.

Figure 21-16. *Chain is a series of interlocking links. Each link is formed from a steel rod.*

Rigging or hoisting chain is designed to deform, up to 15% to 30%, before breaking. However, chain should be removed from service if elongation exceeds approximately 5% or the thickness of any part of a link has decreased by approximately 10%.

Wire rope is constructed in the same general way as fiber rope, except from thin metal wires. Wire rope is made of a specific number of strands wound spirally around a core. **See Figure 21-17.** Each strand is made of a number of metal wires. The core may be another wire strand, a small wire rope, or a strand made from a nonmetallic material, such as fiber or polyvinyl. The strength and flexibility of a wire rope depends on the precise laying of each wire and the way the wires slide against each other as the rope flexes. Wire rope is available in many types of metal, though high-carbon steel and stainless steel are the most common.

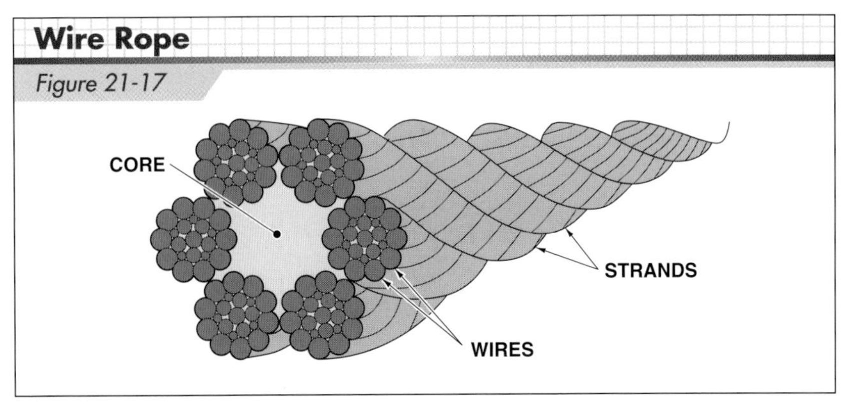

Figure 21-17. *Wire rope is composed of a core surrounded by strands. Each strand is composed of a specific pattern of wires of different sizes.*

Wire rope ends must be fastened to fittings or spliced into loops in order to form a sling. Common wire rope terminations include thimble eyes and sockets. **See Figure 21-18.** A *thimble* is a curved piece of metal that supports a loop of rope and protects it from sharp bends and abrasion. The ends of the loop are secured together with U-bolt clips. A *socket* is a wire rope end attachment that allows connections to other hardware. Closed socket fittings are solid loops. Open socket fittings have a part that can be removed in order to connect to another fitting.

Synthetic fibers, namely polyester and nylon, are used to construct web slings and round slings. Synthetic fiber slings are soft. They do not mar delicate loads and easily conform to the load shape. Also, synthetic fibers are resistant to many chemicals.

A *web sling* is a sling made from flat narrow strapping that is woven from yarns of strong synthetic fibers. **See Figure 21-19.** Webbing for rigging purposes is made of woven nylon or polyester fibers. Colored marker yarns woven into the center of the face of the webbing may be used to identify the material. Colored yarns hidden inside the weave, if they become visible, indicate excessive webbing damage. For strength, the webbing weave along the edge is often drawn tighter. This is known as a selvedge.

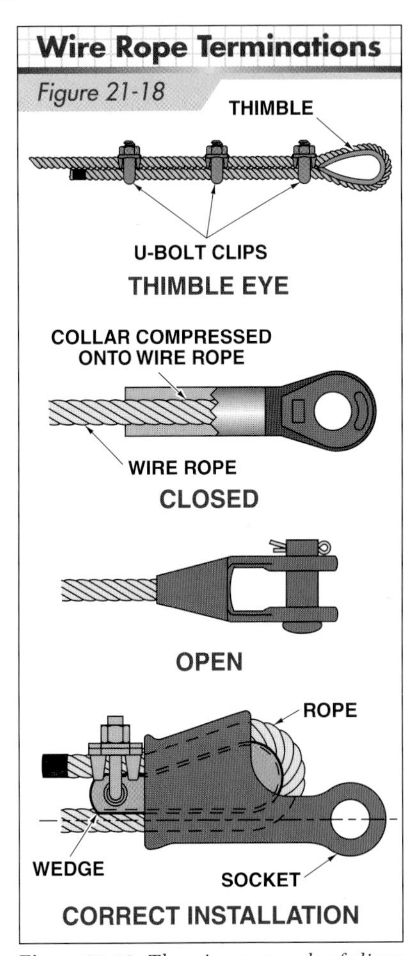

Figure 21-18. *The wire rope ends of slings must be terminated with fittings so that they can be connected to other rigging hardware.*

Web slings are fabricated in at least six standard configurations. **See Figure 21-20.** These provide flexibility in designing appropriate rigging arrangements. Manufacturers may also offer additional configurations for specialized applications.

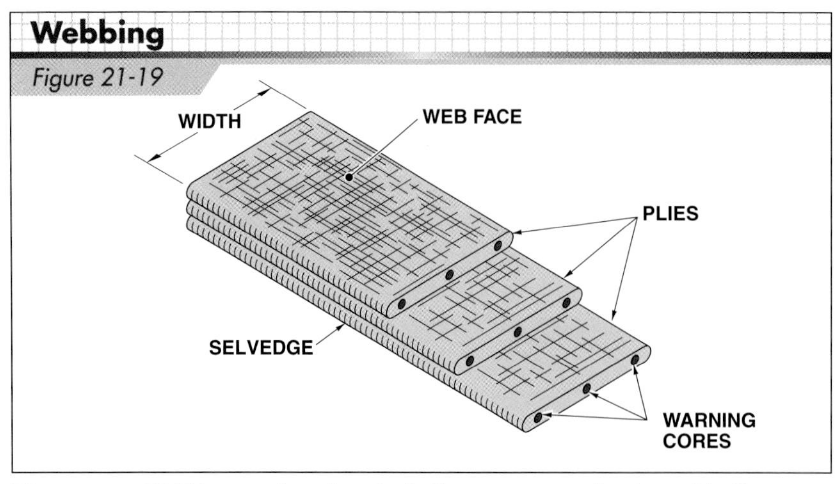

Figure 21-19. *Webbing consists of synthetic fibers woven together into wide, flat straps. Multiple layers, or plies, can be sewn together to make stronger webbing.*

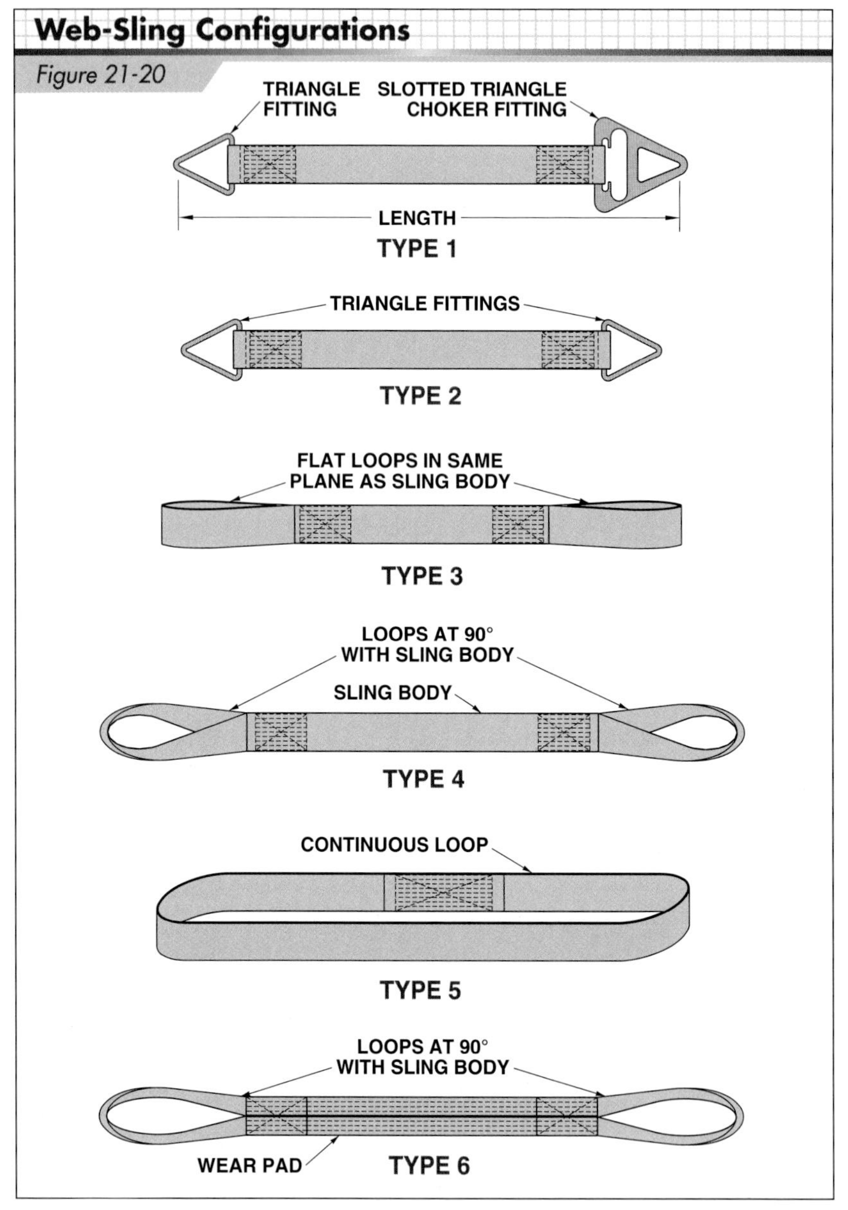

Figure 21-20. *There are six primary web-sling configurations, which are available in various sizes and webbing styles.*

A *round sling* is a continuous loop sling consisting of unwoven synthetic fiber yarns enclosed in a protective cover. The core yarns, typically polyester, are uniformly wound to ensure an even load distribution. **See Figure 21-21.** The cover is made from polyester or nylon fibers woven into a continuous tubular shape. The cover provides protection to the core and is not load bearing. The load rating of a round sling is indicated by the color of its cover.

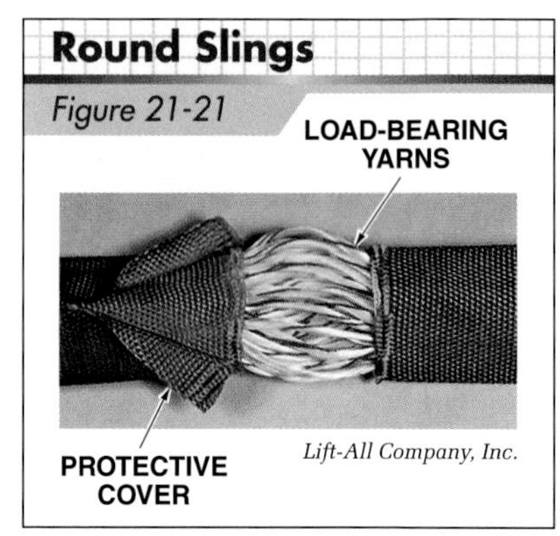

Figure 21-21. *A round sling is composed of a bundle of unwoven yarns of load-bearing synthetic fibers surrounded by a non-load-bearing protective cover.*

Sling Hitches. A *sling hitch* is an arrangement of one or more slings used for connecting a load to a hoist hook. Most slings can be used in a variety of hitches, though this affects their effective load rating and appropriate applications. When deciding on a sling hitch for a load, consideration should be given to how the sling or slings will be attached and how control of the load will be maintained throughout the lift.

There are four basic types of sling hitches. **See Figure 21-22.** A *vertical hitch* is a sling hitch in which one end of the sling connects to the hoist hook and the other end connects to the load. The sling does not wrap around or under the load. A *bridle hitch* is a sling hitch in which two or more slings share a common fitting as a means of attachment to the hoist hook.

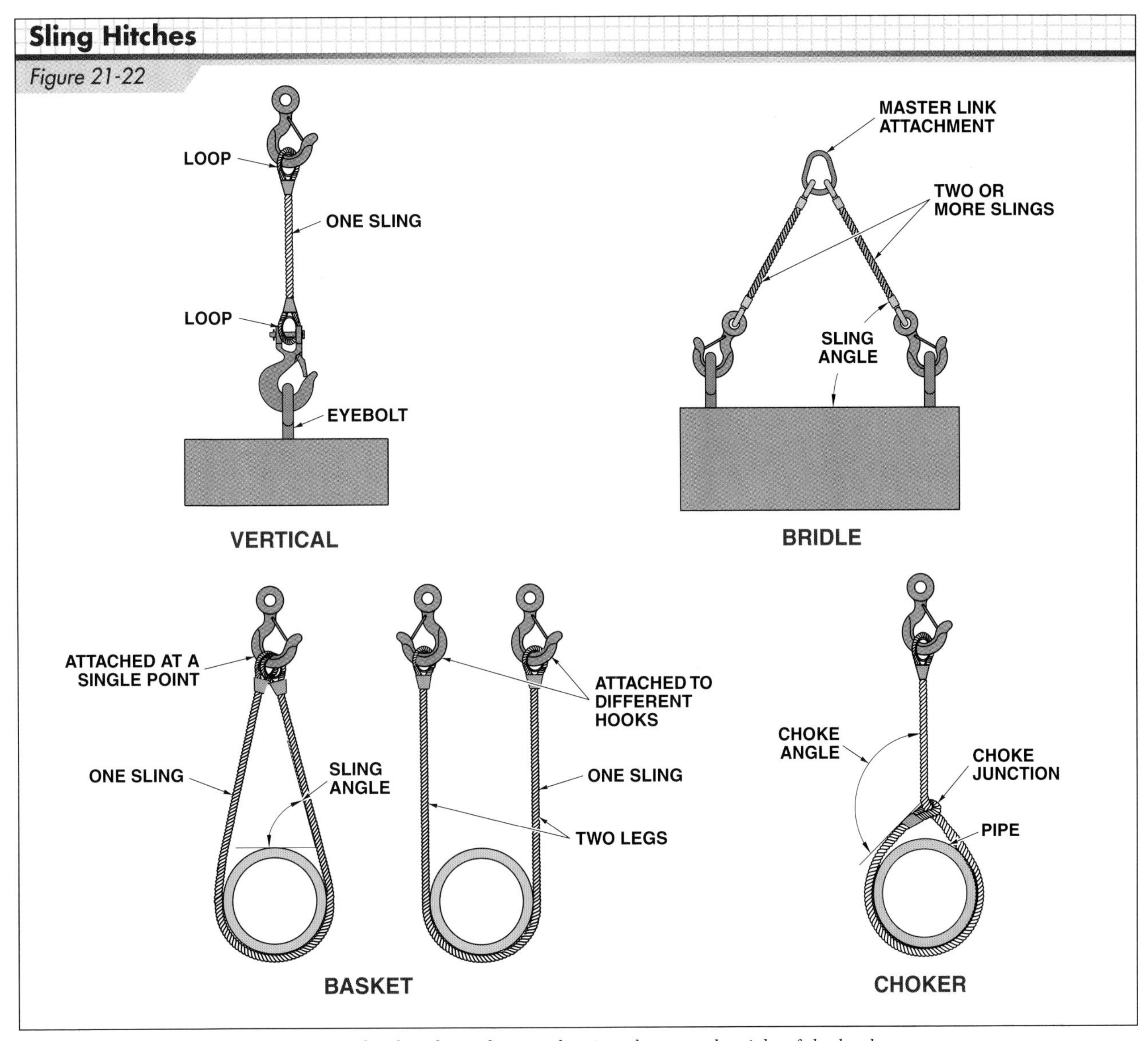

Figure 21-22. *Slings are used in various hitches depending on the size, shape, and weight of the load.*

A *basket hitch* is a sling hitch in which the sling is passed under a load and both ends of the sling are connected to a hoist hook. A *choker hitch* is a sling hitch in which one end of the sling is wrapped under or around the load, passed through the eye at the other end of the sling, and then connected to the hoist hook.

The type of hitch used for a given application is determined by a number of factors. A load with no preexisting lift points usually requires the use of a basket or choker hitch. Choker hitches are also used to bundle a number of pieces of material together for a single lift. Basket hitches are commonly used when multiple hoist hooks are required.

Both choker and basket hitches may also require additional wraps of the sling around the load in order to shorten the effective length of the sling or for additional friction between the sling and the load. Care should be taken so that the wraps of the sling body do not cross and create a pinch point, which would reduce the strength of the sling. For web slings, the additional wraps of the sling may overlap but should not cross.

Hardware

Various pieces of hardware are needed to connect loads and slings, slings and other slings, and slings and hoists. Some hardware is permanently attached to a sling when it is manufactured. However, separate hardware can also be attached individually. The most common rigging hardware includes hooks, shackles, eyebolts, turnbuckles, and links. **See Figure 21-23.**

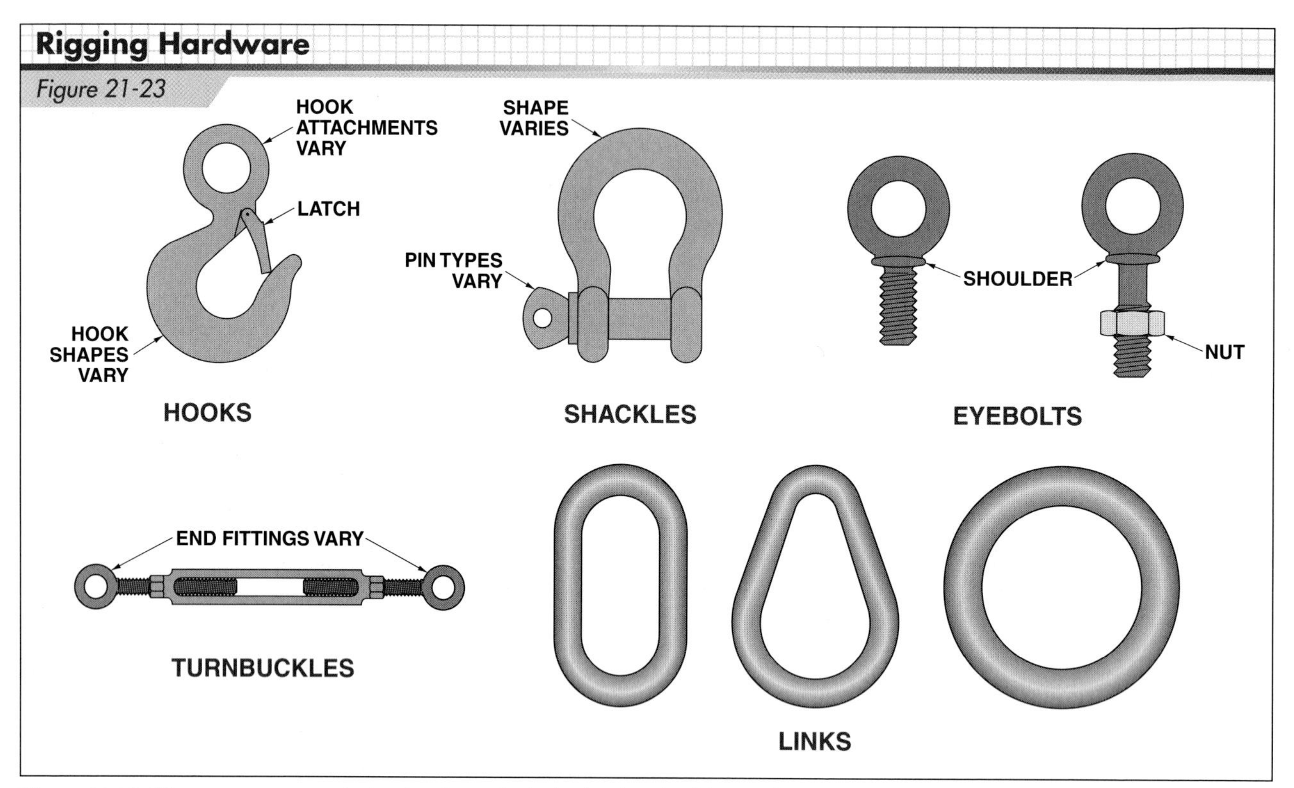

Figure 21-23. *The most common rigging hardware includes hooks, shackles, eyebolts, turnbuckles, and links.*

A *hook* is a curved implement used for quickly and temporarily connecting rigging to loads or lifting equipment. Hooks are made in various designs and sizes. The primary design variations involve the shape of the hook, method of attachment, and latch arrangement.

A *shackle* is a U-shaped metal connector with holes drilled into the ends for receiving a removable pin or bolt. When assembled, the shackle forms a complete loop that is used to securely connect other rigging equipment. The removal of the pin or bolt opens the loop. A shackle is commonly used to make the connection between the rigging assembly and the hoisting hook, though it can also be used for other purposes. The shape of the shackle may determine how many and what type of slings (wire rope, web sling, chain, etc.) can be attached. Shackles are made in various shapes and pin designs.

An *eyebolt* is a bolt with a looped head that is fastened to a load to provide a lift point. Machinery eyebolts have a shoulder and a fully threaded shank. A machinery eyebolt must be screwed in until the shoulder is tight against the load. Nut eyebolts are not fully threaded and require nuts to be secured in place. These eyebolts are often used when the thinness of the load material prevents the use of machinery eyebolts.

A *turnbuckle* is a rigid rigging attachment with an adjustable length. A turnbuckle consists of a pair of threaded fittings, typically eyebolts, screwed into a metal body. One fitting has a right-hand thread and the other has a left-hand thread. Due to this arrangement, when the ends are constrained, the overall length of the turnbuckle can be adjusted by rotating the body. Rotating the body in one direction lengthens the turnbuckle, while rotating it in the other direction shortens it. Turnbuckles are typically used in complex rigging arrangements that require slings of different lengths. Turnbuckles are used to achieve exactly the right length to keep the load level and stable.

A *link* is a plain, rigid, closed loop used to connect multiple rigging components. Links provide a large loop for joining multiple slings or shackles that would not otherwise fit onto a hook. Links are often permanently installed to slings by the sling manufacturer. Links are commonly used as the hoist connections for multiple-leg sling assemblies. However, links can also be used on the load-attachment end of slings. Links are available in a variety of shapes.

Rigging Calculations

A rigging assembly is only as strong as its weakest component, so the strength of each added attachment and any hardware must be considered. Each hoist, sling, piece of hardware, and any other rigging equipment must be carefully selected to be able to withstand the forces induced by the weight of the load and configuration of the rigging.

Working Load Limit. Rigging components and slings are rated by the amount of tension that they can safely tolerate. The *working load limit (WLL)* is the maximum linear force that a rigging component may be safely subjected to. This is also called the load rating, rated capacity, or safe working load (SWL). The designated WLL of a component already includes a significant safety factor above the material's breaking strength.

When selecting rigging equipment for a particular lift, riggers must ensure that the actual tension on a sling is less than its WLL. The WLL of a sling or other rigging hardware is marked on the component or can be looked up in a reference table. **See Figure 21-24.** For slings, multiple WLL ratings may be listed for different sling hitches. However, not every possible configuration can be listed, so calculations may be necessary to determine the tension forces of special configurations.

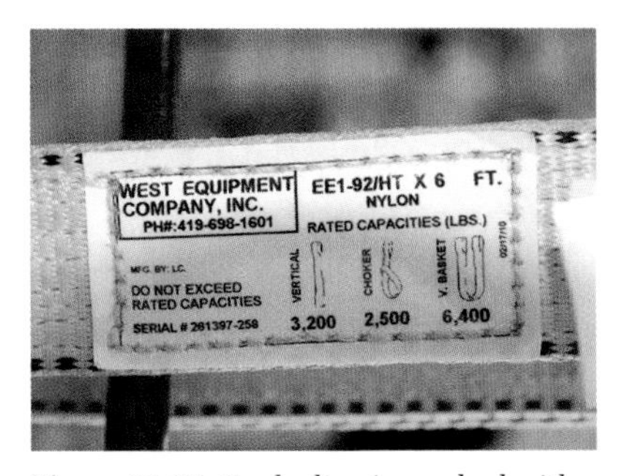

Figure 21-24. *Each sling is marked with a WLL for one or more sling hitches.*

Load Weight. The weight of a load must be determined for all lifting scenarios. Load weight is sometimes found on equipment data plates, shipping documents, or the manufacturer product information. **See Figure 21-25.** However, the load weight may be greater than the weight given in the original data if the load has been modified or is inside a shipping container or on a skid.

Figure 21-25. *Some equipment includes weight information on a nameplate or label.*

If the weight of the load is unknown, it can be calculated using stock material weight tables, or it can be determined using area, volume, and material weight information. The weight of the rigging equipment must also be added in order to calculate the overall load weight.

Riggers must determine the center of gravity of the load. The *center of gravity (CG)*, also known in this context as the *center of mass*, is the point at which an object's total mass can be considered to be concentrated. **See Figure 21-26.** The center of gravity is the balancing point of a load. A load must be lifted from directly over its CG in order to put the least amount of stress on the rigging and provide the safest and most controlled lifting situation.

Equipment manufacturers often mark the CG on their products or include specification sheets with CG information. Otherwise, the CG of the load can be determined by a simple trial and error lifting procedure, but only if lifting an unbalanced load slightly will not create a safety hazard.

If the planned rigging arrangement can be referenced from the rating tag of the sling or from a chart, then no further sling load calculations are needed. As long as the total load weight is less than the WLL of the sling for that configuration, the sling may be used in that application. However, if the necessary rigging arrangement is not specifically listed, then additional calculations are needed.

First, the portion of the total load weight that each lift point must support is determined. These portions are determined by the locations of the lift points in relation to the CG. If each lift point is an equal horizontal distance from the CG, then each shares an equal portion of the load weight with a one-third minimum. That is, if the total weight of a load lifted by four points is 1200 lb, then each point is assumed to support 400 lb (one-third of the total). However, if the lift points are at different

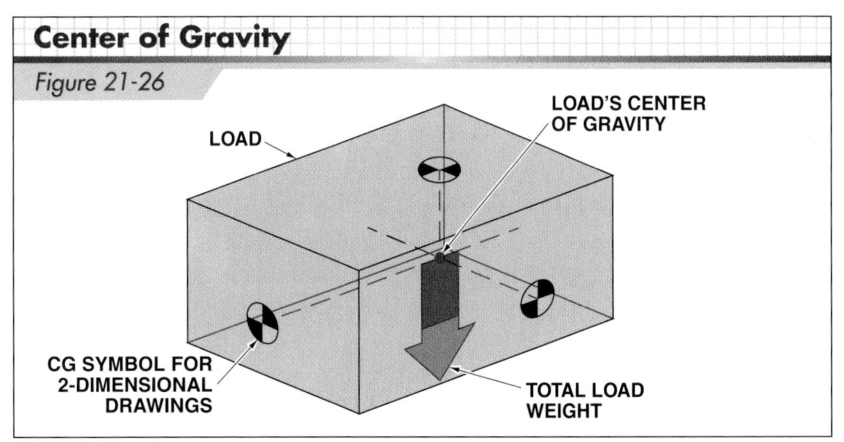

Figure 21-26. *The CG is the point at which the entire load weight can be assumed to be concentrated for calculation and balance purposes. The CG point is designated with a special symbol on two-dimensional drawings.*

distances from the CG, the closer points must support a greater portion of the load weight. Each portion is calculated with the following formula:

$$W_A = W_{total}\frac{D_{CG-B}}{D_{A-B}}$$

where

W_A = portion of load weight supported at lift point A (in lb or t)

W_{total} = total load weight (in lb or t)

$D_{CG\text{-}B}$ = horizontal distance from CG point to lift point B (in ft or in.)

$D_{A\text{-}B}$ = horizontal distance between lift points A and B (in ft or in.)

For example, a load has lift points that are 120″ apart, but the CG is located 40″ horizontally from one of the points. If the load is rigged so that it is hoisted from directly over the CG point, as it should be, the sling at the closer lift point bears a greater portion of the load's weight. **See Figure 21-27.** If the load is 10,000 lb, how is the load weight distributed between the two points?

$$W_A = W_{total}\frac{D_{CG-B}}{D_{A-B}}$$

$$W_A = 10{,}000 \times \frac{80}{120}$$

$$W_A = 10{,}000 \times 0.6667$$

$$W_A = \mathbf{6667\ lb}$$

$$W_B = W_{total}\frac{D_{CG-A}}{D_{A-B}}$$

$$W_B = 10{,}000 \times \frac{40}{120}$$

$$W_B = 10{,}000 \times 0.3333$$

$$W_B = \mathbf{3333\ lb}$$

Two-thirds of the load weight is supported at the closer lift point. The remaining one-third is supported at the other lift point.

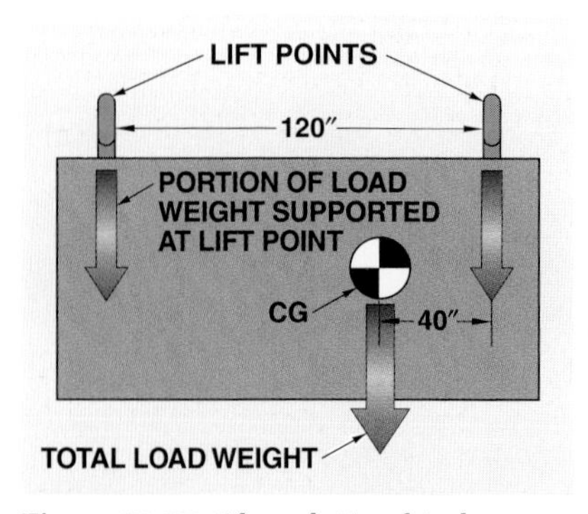

Figure 21-27. *The relationship between the lift points and the CG of a load determines the portion of load weight that each point must support.*

Sling Loads. A vertical sling experiences only the force of the portion of load weight it is supporting. However, a sling loaded at an angle also experiences horizontal forces. The horizontal force is inversely proportional to the sling angle. That is, a smaller angle produces a greater force. The *sling angle* is the acute angle between the horizontal plane and the sling leg. The vertical and horizontal forces combine to result in a total loading on the sling that can be much greater than just its portion of the load weight. **Figure 21-28.** The minimum recommended sling angle is 30°.

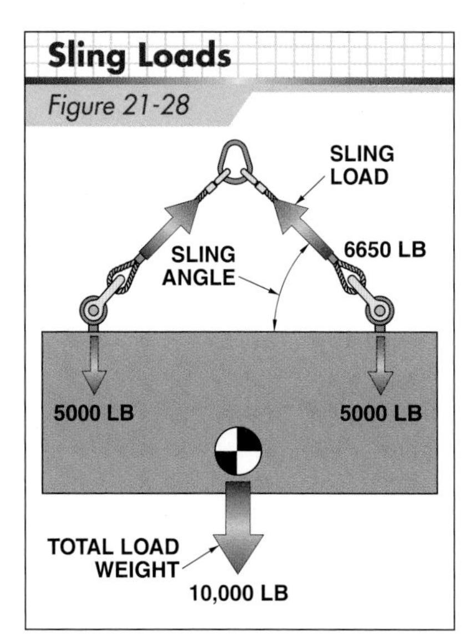

Figure 21-28. *When a sling is used at an angle, the force on the sling is greater than just the portion of load weight the sling is supporting.*

Sling tension can be calculated in two ways. One way uses lengths and the other uses sling angle, though they both produce the same result. Either can be used, depending on which quantities are easier to measure. These relationships are shown in the following formulas:

$$F_{sling} = \frac{L}{H}W_{portion}$$

$$F_{sling} = \csc\alpha \times W_{portion}$$

where

F_{sling} = tension force on sling (in lb or t)

L = length of sling (in ft or in.)

H = vertical distance between lift point and upper point of attachment (in ft or in.)

$W_{portion}$ = portion of load weight supported by sling (in lb or t)

α = sling angle (in °)

For example, a 10,000 lb load is to be rigged with two 8′ long slings. The point of attachment is 6′ above the load. Since the CG is exactly between the two lift points, the load supported by each sling will be 5000 lb (one-half the total load weight).

$$F_{sling} = \frac{L}{H}W_{portion}$$

$$F_{sling} = \frac{8}{6} \times 5000$$

$$F_{sling} = 1.33 \times 5000$$

$$F_{sling} = \mathbf{6650\ lb}$$

Therefore, the slings and any other hardware used for hoisting this load must have a load rating of at least 6650 lb each in order to lift the load safely. This force is compared to the WLL of the rigging equipment to ensure that each piece of equipment has adequate strength.

Knots and Hitches

Securing loads or attaching a tag line with a fiber rope normally requires some form of a knot or hitch. A *knot* is the interlacing of a part of a rope to itself, which is then drawn tight. Knots are designed to form a semipermanent connection that can be untied later. A *hitch* is the binding of rope to another object, usually temporarily. Hitches are designed for quick release.

The portions of rope being worked are identified as working or standing. The *working part* is the portion of the rope involved in making the knot or hitch. The working part of a rope is loose and the end can be passed through loops. The *standing part* is the portion of a rope that is unaltered or not involved in making a knot or hitch. The standing end is often attached to some other item, making it taut and unable to be worked.

There are many different knots and hitches that may be applicable to rigging work, though usually only a few are used. These include the square knot, sheet bend, bowline, clove hitch, and timber hitch. **See Figure 21-29.** Each has features, such as strength, self-tightening ability, and ease of removal, that make it suitable for some applications and not others. Therefore, knot or hitch selection is important. Personnel working with loads or having related material-handling duties should be familiar with several knots and hitches and be able to use them appropriately.

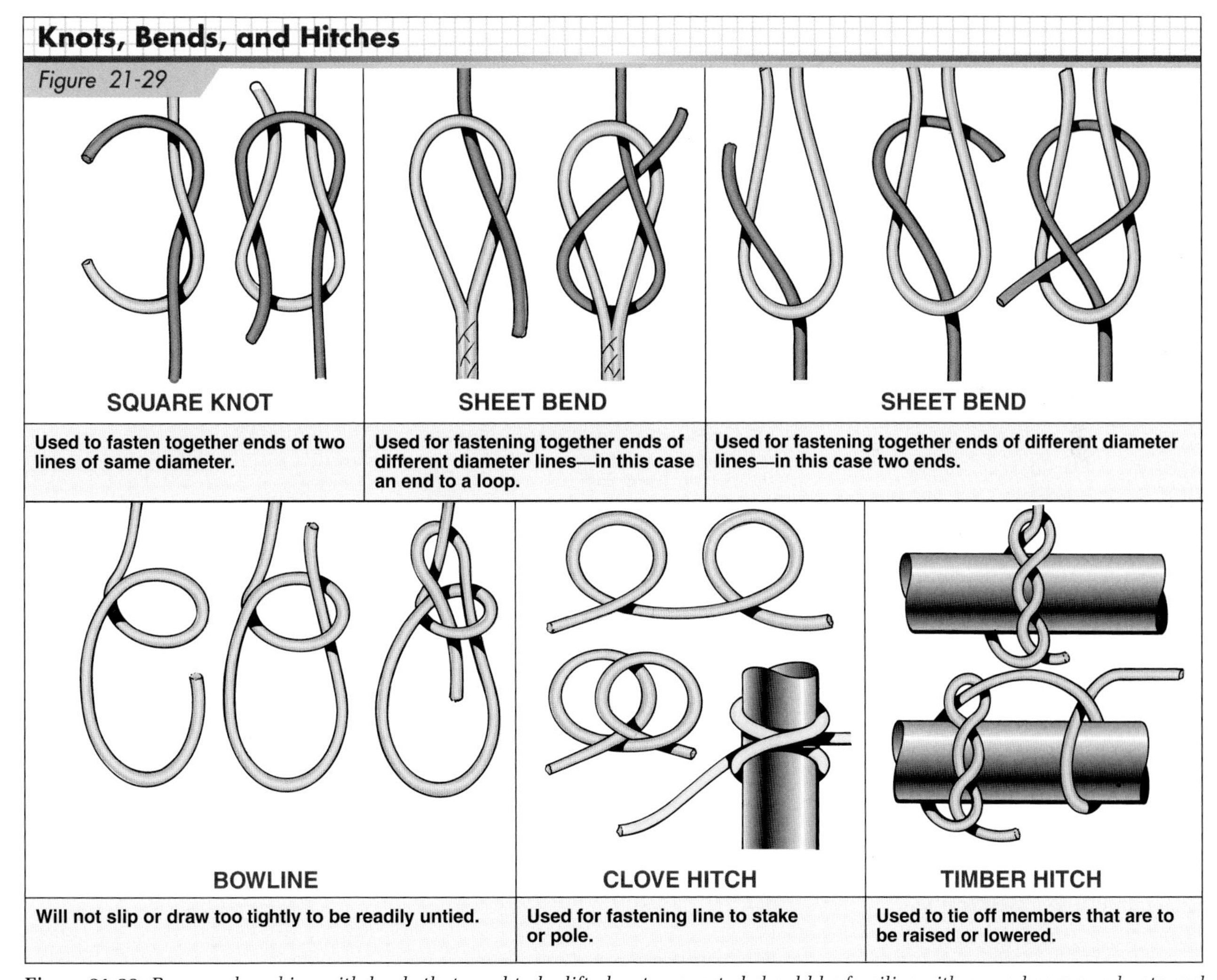

Figure 21-29. *Personnel working with loads that need to be lifted or transported should be familiar with several common knots and hitches.*

Unit 21 Review and Resources

UNIT 22

Job Site Safety and Working Conditions

OBJECTIVES

1. List and describe common personal protective equipment.
2. Review general safety precautions for job sites.
3. Identify common safety precautions for electrical and extension cords.
4. Define material handling and review proper methods for handling materials on a job site.
5. Identify factors to consider before operating a forklift or material lift.
6. Identify common safety precautions for operating a forklift or material lift.
7. List and describe common hazardous materials and their effects on health.
8. Define confined spaces.
9. Describe the procedure for excavations and slopes for different types of soil.
10. Describe procedures for shoring.
11. List and describe types of shielding.
12. Describe barricades and guardrails.
13. Describe uses for ramps, runways, and temporary stairs.
14. List and describe types of fires and appropriate fire extinguishers.
15. Identify common safety guidelines for fire prevention.

Safe work practices are required at a job site to minimize the potential for injury. The Occupational Safety and Health Administration (OSHA) requires all employers to provide a safe environment for employees. A safe environment is free of hazards and has precautions in place to protect employees.

Employers are responsible for safety training and for ensuring that employees follow OSHA regulations. Many large contractors have a comprehensive orientation program to familiarize employees with applicable safety regulations and company standards. Safety meetings and/or toolbox talks are frequently conducted to discuss current safety topics and address employee safety concerns.

In late 2007, OSHA announced a final rule on employer-paid personal protective equipment (PPE). The rule provides that all PPE must be purchased by the employer when used by the worker to comply with one of the OSHA PPE requirements. Personal protective equipment excluded from the rule includes safety toe footwear, prescription safety eyewear, and weather-related work gear.

WORKER RESPONSIBILITIES

Workers must report accidents and safety hazards to the employer or supervisor. Workers must practice job site safety at all times and wear appropriate personal protective equipment. If injuries occur, medical assistance should be immediately obtained.

All companies establish a substance abuse policy that maintains a safe working environment and promotes high work standards. Substance abuse policies prohibit workers from working while under the influence of illegal drugs, alcohol, or other controlled substances. Failure to comply with an established substance abuse policy may result in serious injury and/or employment termination.

Experience has proven that most accidents can be prevented by proper safety practices. Various studies stress the following facts:

- Strain or overexertion is the most common injury suffered by construction workers.
- Slips or falls from elevated work surfaces and ladders account for nearly one-third of construction injuries.
- Worker injuries resulting from use of machines and tools, or from a worker being struck by a tool or machine, account for one-fourth of injuries reported.
- The most common cause of death from job site accidents involves falls from elevations such as upper building levels.

PERSONAL PROTECTIVE EQUIPMENT

Personal protective equipment (PPE) is used to protect against safety hazards on a job site. PPE includes protective clothing, head protection, eye and face protection, hearing protection, hand and foot protection, back protection, knee protection, and respiratory protection.

OSHA regulations for construction are in the Code of Federal Regulations (CFR), Title 29, Part 1926. Individual regulations are identified as 29 CFR 1926.section number.

Hilti, Inc.

Appropriate personal protective equipment (PPE), including head and eye protection, must be worn to protect against safety hazards on a job site.

Protective Clothing

Protective clothing made of durable material, such as denim, provides protection from contact with sharp objects, hot equipment, and harmful materials. Protective clothing should be snug, yet provide ample movement. Pockets should allow convenient access, but should not snag on tools or equipment. Loose-fitting clothing and long hair must be secured, and jewelry must be removed to prevent its getting caught in rotating equipment. Metallic watches and rings should not be worn since serious injury may result if contact is made with an energized electrical circuit.

Head Protection

OSHA 29 CFR 1926.100, *Head Protection,* mandates that workers wear *protective helmets,* or hard hats, in areas where there is the potential for head injury from impact, falling and flying objects, and electrical shock. **See Figure 22-1.** Many contractors require protective helmets to be worn at all times on a job site. Protective helmets resist penetration and absorb impact force. Protective helmet shells are made of durable, lightweight material. A shock-absorbing lining consisting of crown straps and headband keeps the shell away from the head to provide ventilation.

Hilti, Inc.

Figure 22-1. *Protective helmets provide head protection by resisting penetration of falling or flying objects and absorbing the impact force of the objects.*

Eye and Face Protection

Proper eye and face protection must be worn to prevent eye or face injuries caused by flying particles such as wood chips, metal particles, and chemicals. Eye and face protection are detailed in OSHA 29 CFR 1926.102, *Eye and Face Protection.*

Eye and face protection includes *safety glasses, face shields,* and *goggles.* Safety glasses have special impact-resistant glass or plastic lenses, reinforced frames, and side shields. Frames are designed to keep the lenses secured in the frame if an impact occurs. Face shields cover the entire face with a plastic shield, and are used for protection from flying objects or splashing liquids. Goggles have a flexible frame and are secured on the face with an elastic headband. **See Figure 22-2.** Goggles must fit snugly against the face to seal the areas around the eyes, and may be used over prescription glasses. For general operations, goggles with clear lenses are typically worn. For welding and cutting operations, tinted lenses are required to protect against ultraviolet (UV) rays.

Figure 22-2. *Safety glasses and welding helmets provide worker eye protection.*

Hearing Protection

Power tools and equipment can produce excessive noise levels. Carpenters subjected to excessive noise levels may develop hearing loss over a period of time. The severity of hearing loss depends on the noise intensity and the duration of exposure. Noise intensity is expressed in *decibels.* **See Figure 22-3.** Per OSHA 29 CFR 1926.101, *Hearing Protection,* ear-protection devices must be worn when it is not feasible to reduce the noise intensity or duration of exposure. Ear-protection devices inserted in the ears, such as earplugs, must be fitted or determined individually by a competent person. Cotton is not acceptable hearing protection.

Earplugs are made of moldable rubber, foam, or plastic, and are inserted into the ear canal. *Earmuffs* are worn over the ears. A tight seal around an earmuff is required for proper hearing protection.

Hand and Foot Protection

Hand protection, such as gloves, is required to prevent injuries to hands caused by cuts or chemical absorption. The appropriate hand protection is determined by the duration, frequency, and degree of the hazard to hands. Gloves are recommended when installing insulation and when mixing and applying certain adhesives. OSHA 29 CFR 1910.138, *Hand Protection,* details the use of hand protection.

Foot injuries are typically caused by objects falling less than 4′ and having an average weight of 65 lb. Safety shoes with reinforced steel toes protect against injuries caused by compression and impact. **See Figure 22-4.** Some safety shoes have protective metal insoles and metatarsal guards for additional protection. *Metatarsal guards* protect the area of the foot between the toes and ankle. Protective footwear must comply with ANSI Z41, *Personal Protection—Protective Footwear.*

Figure 22-4. *Safety shoes with reinforced steel toes provide foot protection.*

SOUND LEVELS

Noise Intensity*	Loudness	Examples
140	Deafening	Jet airplane taking off, air raid siren, locomotive horn
130	Pain threshold	
120	Feeling threshold	
110	Uncomfortable	
100	Very loud	Chain saw
90	Noise	Shouting, air horn
80	Moderately loud	Vacuum cleaner
70	Loud	Telephone ringing, loud talking
60	Moderate	Normal conversation
50	Quiet	Hair dryer
40	Moderately quiet	Refrigerator running
30	Very quiet	Quiet conversation, broadcast studio
20	Faint	Whispering
10	Barely audible	Rustling leaves, soundproof room, human breathing
0	Hearing threshold	Intolerably quiet

*in dB

Figure 22-3. *The severity of hearing loss depends on the intensity level of the noise (measured in decibels) and the duration of exposure. Protection against the effects of noise exposure is required when average sound levels exceed 90dB over an 8 hr period.*

Low back pain and more serious musculoskeletal back injuries can occur suddenly or develop over a period of time. Sudden quick movements, especially those made while handling heavy objects, can lead to painful muscle strains.

Back Protection

Back injury is one of the most common injuries resulting in lost time. Most back injuries are the result of improper lifting techniques. Back injuries are prevented through proper planning and lifting technique. Assistance should be sought when moving heavy objects. When lifting objects from the ground, ensure the path is clear of obstacles and free of hazards. **See Figure 22-5.** Bend the knees and grasp the object firmly. Next, lift the object, straightening the legs and keeping the back as straight as possible. Finally, move forward after the whole body is in the vertical position. Keep the load close to the body and steady.

Knee Protection

Carpenters who spend considerable time working on their knees, such as floor layers, should wear knee pads. Knee pads are rubber, plastic, or leather pads strapped onto the knees for protection. Buckle straps or Velcro™ closures secure knee pads in position.

Respiratory Protection

Respiratory protection is required to protect against airborne hazards. Harmful vapors, dusts, particles, fumes, mists, or gases may be encountered on a job site due to the use of cleaners and solvent cements or from sawing or drilling operations performed by tradesworkers. The degree of risk from exposure to any given substance depends on the nature and potency of the substance and the duration of exposure.

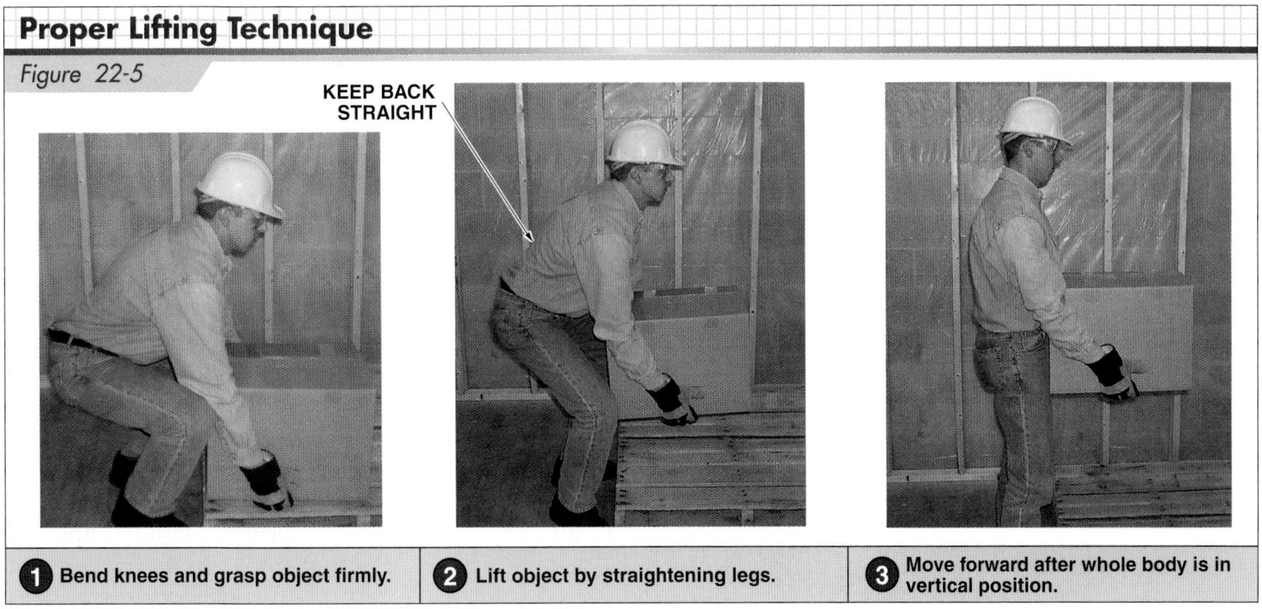

Proper Lifting Technique

Figure 22-5

KEEP BACK STRAIGHT

1. Bend knees and grasp object firmly.
2. Lift object by straightening legs.
3. Move forward after whole body is in vertical position.

Figure 22-5. *Use the proper lifting procedure to avoid back injury.*

OSHA 29 CFR 1926.103, *Respiratory Protection,* details respiratory protection requirements. When harmful airborne substances are present on a job site, engineering control measures, such as enclosure of the operation, ventilation, and substitution of less-toxic substances, should be undertaken. If proper engineering control measures are not feasible, appropriate respiratory protection must be used. The respiratory protection required is determined by the type of airborne hazard. **See Figure 22-6.**

Figure 22-6. *The respiratory protection required is determined by the type of airborne hazard. The particulates mask shown here provides protection against airborne fiberglass particles.*

SAFE WORK HABITS

Many accidents can be prevented by safe work habits including the following:

- When working on an elevated surface, do not place tools where they may fall and injure another worker below.
- Always watch where you step.
- Look out for the safety of other workers as well as your own safety.
- Do not engage in horseplay on the job.

GOOD HOUSEKEEPING

Hazards on a job site may be caused by general sloppiness, poor organization, and careless storage of materials. OSHA 29 CFR 1926.25, *Housekeeping,* details good housekeeping practices on a job site. Rules for good housekeeping are as follows:

- Keep scrap lumber cleared from work areas, passageways, and stairs. Clinch or pull out protruding nails.
- Ensure areas within 6′ of a building under construction are reasonably level. Provide walkways at convenient places to bridge ditches.
- Keep material storage areas free of obstructions and debris.
- Stack materials in such a way that they will not fall, slip, or collapse. Lumber piles should not exceed 8′ in height if the lumber is to be handled manually, or 20′ if it is to be handled with equipment such as a forklift or telehandler.
- Maintain well-defined passageways and walkways on a job site. Keep passageways and walkways well lit and free of tripping hazards.
- Use clean-up crews to periodically remove all waste materials from the job site.
- Store tools and equipment not being used in chests or tool sheds.
- Remove slush or snow from work areas or walkways before it turns into ice. Slipping can be reduced by spreading sand, gravel, cinders, or other gritty material over the work areas.

ELECTRICAL SAFETY

Improper procedures with electrical tools and equipment can result in an electrical shock. *Electrical shock* is a condition that results when a body becomes part of an electrical circuit. Safe work habits and use of proper PPE are required to prevent electrical shock when working with or in proximity to electrical devices.

The severity of electrical shock depends on the amount of electrical current, measured in milliamps (mA), that flows through the body; the length of time the body is exposed to the current flow; the path the current takes through the body; and the physical size and condition of the body through which the current passes. **See Figure 22-7.**

ELECTRICAL SHOCK EFFECTS	
Approximate Current*	Effect on Body†
Over 20	Causes severe muscular contractions, paralysis of breathing, heart convulsions
15–20	Painful shock May be frozen or locked to point of electrical contact until circuit is de-energized
8–15	Painful shock Removal from contact point by natural reflexes
8 or less	Sensation of shock but probably not painful

*in mA

†effects vary depending on time, path, amount of exposure, and condition of body

Figure 22-7. *Effects of electrical shock vary depending on the amount of current, the length of time the body is exposed to the current, the path the current takes through the body, and the physical size and condition of the body.*

Extension Cords

An *extension cord* is used to supply power to portable electric tools and equipment. Heavy-duty, three-wire extension cords must be used for tools and equipment to operate properly. When selecting an extension cord for use, wire (conductor) diameter and length are the key factors to consider. A wire with a small diameter has greater resistance and cannot carry as much power as a wire with a larger diameter. A smaller gauge number indicates larger diameter wire. For example, 12-ga wire has a larger diameter than 16-ga wire.

An extension cord that is too long for an application can create a voltage drop along its length. Voltage drop is another type of resistance, which impedes the flow of power through the extension cord. The farther electricity travels from its source, the larger the voltage drop. To reduce voltage drop, always use the shortest extension cord possible for the application.

When selecting the extension cord to be used for an application, consider the tools or equipment to be connected and their nameplate amperage rating. For example, if a drill that draws 5.5 A and a sander that draws 4 A are to be connected, a 50′ extension cord with 14-ga conductors could be used. **See Figure 22-8.** In most cases, 15 A or less should be drawn through an extension cord.

OSHA 29 CFR 1926.405, *Wiring Methods, Components, and Equipment for General Use,* details safe work practices related to extension cords. Visually inspect extension cords each day for external damage such as deformed or missing prongs, damaged insulation, or indications of possible internal damage. Safety precautions to observe when using extension cords include the following:

- Extension cords should be kept away from high heat areas.
- Always uncoil an extension cord during use to properly dissipate heat from the cord.
- Carefully remove an extension cord plug to avoid breaking the internal wire connection.
- Do not use frayed extension cords and do not use electrical tape to make repairs.
- Extension cords must be of the three-wire type, which have a grounding conductor and ground prong.
- Protect extension cords that pass through doorways or other pinch points from damage.
- Extension cords should not be run through holes in walls, ceilings, or floors.
- Do not conceal extension cords behind walls, ceilings, or floors.
- Extension cords should not be hung from nails or wire, or fastened with staples.

EXTENSION CORDS

Cord Length*	Amperage rating†					
	0–2	2–5	5–7	7–10	10–12	12–15
25	16 ga	16 ga	16 ga	16 ga	14 ga	14 ga
50	16 ga	16 ga	16 ga	14 ga	14 ga	12 ga
100	16 ga	16 ga	14 ga	12 ga	12 ga	
150	16 ga	14 ga	12 ga	12 ga		
200	14 ga	14 ga	12 ga	10 ga		

*in ft

†in A

Figure 22-8. *Select an extension cord with the appropriate length and amperage rating.*

AERIAL LIFT SAFETY

Hard hats and required fall-arrest equipment must be worn while working on elevated work platforms or buckets of aerial lifts. Loose-fitting clothing should not be worn while working in an aerial lift.

Aerial lifts must be inspected and tested daily before use. When using articulating Z-boom or extensible S-boom lifts, a worker qualified in the operation of the ground controls must be stationed at the ground controls. Ground controls must not be operated without the authorization of the elevated worker, except during emergencies.

Elevated work platforms have guardrails, a toeboard, and a means for each person to attach fall-arrest equipment. A guardrail is secured to the uprights and erected along the exposed sides and ends of a platform. Guardrails must be installed with the toprail no less than 36" or more than 42" high, and with a midrail. A midrail is installed approximately midway between the toprail and platform. A toeboard serves a purpose similar to toeboards for scaffolds.

OSHA regulations require that an aerial lift be operated by a person qualified in aerial lifts. OSHA 29 CFR 1926.453, *Aerial Lifts;* OSHA 29 CFR 1910.66, *Powered Platforms for Building Maintenance;* ANSI A92.5, *Boom-Supported Elevating Work Platforms;* and ANSI A92.6, *Self-Propelled Elevating Work Platforms* provide additional information regarding the safe use of aerial lifts. Aerial lift safety precautions include the following:

- Wear appropriate fall-arrest equipment when working in an aerial lift. The lanyard must be attached to the work platform or basket of the aerial lift. **See Figure 22-9.** Workers in aerial lifts must not attach themselves to adjacent poles, structures, or equipment.
- Do not sit or climb on the edge of an aerial lift basket or railing of an elevated work platform.
- Set the brakes on aerial lifts during use. If the lift is equipped with outriggers, they must be positioned on pads or solid surfaces.
- Do not move an occupied aerial lift with an elevated work platform or basket unless the equipment is specifically designed for such work.
- Do not operate aerial lifts with any portion of the lift closer than 10′ from live overhead electrical lines.
- Aerial lift controls must be tested prior to each use to determine if the controls are in safe working condition.
- Boom and basket load limits specified by the manufacturer must not be exceeded.
- Install wheel chocks before using an aerial lift on an incline, provided they can be safely installed.

Genie Industries

Figure 22-9. *Workers in elevated aerial lifts must wear appropriate fall-arrest equipment.*

MATERIAL HANDLING SAFETY

Material handling involves the rigging, lifting, and transporting of a load by mechanical means. **See Figure 22-10.** Materials such as glulam beams and roof trusses must be properly transported around a job site and placed without damaging the materials. Safety procedures must be followed to ensure the safety of operators and other workers moving materials.

Symons Corporation

Figure 22-10. *Large or heavy loads may require specialized rigging equipment.*

Rigging Safety

When the load is properly rigged, the rigging is checked by lifting the load slightly off the ground and ensuring that the load is properly secured and balanced. Additional safety rules for rigging and lifting a load include the following:

- Loads must be well secured.
- A tag line should be used when necessary to help control the load.
- Slings and sling hitches should be selected based on the weight of the load and type of material being lifted.
- Slings should not contain any kinks or knots.

All rigging equipment must be properly stored. Rope and slings must be kept in a clean and dry area, away from harmful fumes or heat. Synthetic webbing and natural fiber rope should be stored out of sunlight and away from areas used for arc welding. Slings should be hung neatly on racks and long rope should be rolled onto spools.

Slings and sling components must not be run over by construction equipment or placed where heavy loads may be set on them. Slings should not be dragged over abrasive surfaces or sharp objects. Rigging equipment must be inspected prior to use and any excessively worn or damaged equipment must be retired. Additional rigging regulations can be found in OSHA 29 CFR 1926.251, *Rigging Equipment for Material Hoisting.*

Hoisting Safety

Safety rules to observe when working around hoisting equipment include the following:

- Do not hoist loads that exceed the rated capacity of the crane.
- Perform a safety inspection of a crane at the start and end of each shift.
- Do not stand in a confined area when loads are being raised or lowered. Ensure there is a safe escape route if lifting equipment should fail.
- Riding on a load or sling attached to the cable of a crane is not permitted.
- Ensure that all tradesworkers are clear of loads about to be lifted and any suspended loads. Loads should never pass over the heads of workers.
- Be alert for material that may fall from suspended loads.
- Areas within the swing radius of a crane must be barricaded to prevent access by unauthorized workers. Workers should also stay clear of the line of travel of a crane.
- Remove any obstructions in the lifting or traveling path prior to the lift.
- Always wear a hard hat when involved in a lift.
- When signaling crane operators, keep signals in one place. If the signaler's gloves and clothing are similar colors, make the signals away from the body where a crane operator can easily see them.
- Wear a harness and attach a lanyard to an approved anchorage point on the work platform when working from an aerial lift. Do not attach the lanyard to an adjacent pole, structure, or equipment.
- Do not sit or climb on the edge of an aerial lift basket or on the rails of an elevated work platform.
- For overhead electrical lines rated 50 kV or less, the minimum clearance between the lines and any part of the aerial lift, crane, or load is 10′. For overhead lines rated over 50 kV, the minimum clearance between the lines and any part of the aerial lift, crane, or load is 10′ plus 0.4″ for each 1 kV over 50 kV, or twice the length of the line insulator, but never less than 10′.

More information about hoisting safety can be found in OSHA 29 CFR 1926.550, *Cranes and Derricks*; OSHA 29 CFR 1926.453, *Aerial Lifts*; and OSHA 29 CFR 1926.600, *Equipment*.

Material Transport Safety

Several factors must be considered when operating a forklift, telehandler, or similar material handling equipment, including the type of surface, type of load and stability, load manipulation required, and equipment and pedestrian traffic.

The surface on which a forklift or telehandler operates must be stable enough to support the lift and load. **See Figure 22-11.** Wet, rough, or uneven surfaces should be carefully crossed at an angle. The safe operating speed of a forklift or telehandler is equivalent to a brisk walking pace. Load size should be kept as small as possible when performing a high lift, when the load is irregularly shaped, or when the travel route includes ramps or uneven surfaces.

JCB Inc.

Figure 22-11. *The surface on which a telehandler operates must be stable enough to support the lift and the load. A load should be kept low when being transported.*

When loads are being moved, the area must be clear of other workers. Turning or traveling should not be done with an elevated load. When transporting a load, the forks should be positioned as close to the ground as possible—1″ to 2″ at the heel of the forks and 4″ to 5″ at the tips, with the load resting against the mast. If clearing an obstacle is necessary, the load should be lifted enough to clear the obstacle and then lowered as soon as possible.

Forklifts and telehandlers must be operated safely to ensure worker safety and prevent property damage. Safety precautions to observe when operating a forklift or telehandler include the following:

- Ensure there is adequate overhead clearance when placing a load on a stack, and that the lift is fully stopped before raising or lowering the load.
- Exercise caution when turning. Check over both shoulders before moving in reverse.
- Stop completely before shifting to reverse; do not use reverse as a brake.
- Do not allow anyone to ride on any part of a forklift.
- Lower forks to ground level when parking.
- Keep arms and legs inside a lift, and keep head, hands, and feet out of uprights when the machine is operating.
- Perform a daily safety inspection of the equipment.
- When climbing into a forklift or telehandler, always maintain three-point contact with the machine—two hands and one foot or two feet and one hand.
- Operators must be properly trained for the equipment and may require periodic performance checks.

More information regarding the safe operation of forklifts or telehandlers can be found in OSHA 29 CFR 1910.178, Powered Industrial Trucks, and ANSI B56.1, *Powered Industrial Trucks*.

HAZARDOUS MATERIALS

A *hazardous material* is any material that poses a risk to health, safety, or property. Materials such as epoxies and solvent cements are classified as hazardous materials. Several years ago, OSHA adopted regulations based on a document developed by the United Nations entitled the *Globally Harmonized System of Classification and Labeling of Chemicals* (GHS), commonly referred to as the Purple Book. This document provides classification criteria for the health, physical, and environmental hazards created by these chemicals. It also assigns standardized labels to these hazard classes and categories. It provides appropriate signal words, pictograms, and hazard and precautionary statements to convey the hazards to users.

The GHS is based on major existing systems around the world, including the chemical classification and labeling systems of many US agencies. Information about the GHS and hazard communication requirements can be found in OSHA 29 CFR 1926.1200, *Hazard Communication.*

Employers must develop, implement, and maintain a written, comprehensive hazard communication program that includes provisions for container labeling, material safety data sheets, chemical inventory, and an employee training program. This hazard communication program should include a list of hazardous materials on the job site. Information must be provided in a language or manner that employees understand.

Container Labeling

Each hazardous material container must have a label, which should be examined before using the product. Specific hazards, precautions, and first-aid information are listed on the label. Hazardous material containers must be labeled, tagged, or marked with appropriate hazard warnings per OSHA 29 CFR 1910.1200(f), *Labels and Other Forms of Warning.* Material stored in a different container than originally supplied from the manufacturer must also be properly labeled. Unlabeled containers pose a safety hazard since users are not provided with content information and warnings.

Material Safety Data Sheets

A *material safety data sheet (MSDS)*, also known as a *safety data sheet (SDS)*, is used to relay hazardous material information from the manufacturer, importer, or distributor to the worker. The information is listed in English, provides precautionary information regarding proper handling, and includes emergency and first-aid procedures. **See Figure 22-12.** All hazardous materials used on a job site must be inventoried and have an MSDS on file. Carpenters should become familiar with the MSDS of any hazardous materials to which they are exposed. An MSDS includes the following information:

- manufacturer information
- product information
- hazardous ingredients and identity information
- physical and chemical characteristics
- fire and explosion hazard data
- reactivity data
- health hazard data
- spill or leak procedures
- safe handling and use information
- special precautions

Hazardous Material Disposal

Construction companies commonly dispose of hazardous materials when their shelf life has expired or if they have become contaminated with other substances. Hazardous materials become *hazardous waste* when the materials are disposed of. In 1986, the Resource Conservation and Recovery Act (RCRA) regulations covering small-quantity generators of hazardous waste went into effect. Disposal options typically include transporting hazardous waste to an approved disposal site or contracting with a firm to pick up and dispose of hazardous material.

Carcinogens

A *carcinogen* is a hazardous material at a job site that can cause cancer. Check the container label or MSDS sheet of a hazardous material to determine if it is carcinogenic. Personal protective equipment is required if control measures at the job site do not reduce exposure to the carcinogen to appropriate levels.

PPE for carcinogenic substances is determined by evalutating the means the substance can enter the body. PPE for airborne carcinogenic substances ranges from disposable particulates masks to respirators, depending on particle size. If the substance is absorbed through the skin, PPE may range from appropriate gloves to a sealed full body suit.

Asbestos

Asbestos is a mineral that has long, silky fibers in a crystal formation, and was a component of many building materials including fireproofing, siding, and tile installed until the late 1980s. Products manufactured with asbestos release asbestos fibers into the air when the product crumbles or is crushed. Airborne asbestos fibers range in size from .1μ to 10μ, and are odorless and tasteless. Asbestos fibers can stay suspended in air for long periods.

Material Safety Data Sheets

Figure 22-12

Fastener Chemical Co.

2929 W. Industrial Dr.
Lincoln, IA 61614

Material Safety Data Sheet

Section I — Product Information

PRODUCT NAME: Fastener Epoxy	EMERGENCY TELEPHONE NUMBER: (555) 434-4414
PRODUCT CLASS: Epoxy	TELEPHONE NUMBER FOR INFORMATION: (555) 344-4144
DOT INFORMATION: Amines, Liquid, Corrosive	DATE PREPARED: 2/22/03

Section II — Hazardous Ingredients/Identity Information

HAZARDOUS COMPONENTS	CAS NUMBER	OSHA PEL	ACHIH TLV	%
Component A: Modified Bisphenol A	25068-38-6	N/E	N/E	>80%
Talc	14807-96-6	5 mg/m^3	5 mg/m^3	<10%
Coated Calcium Carbonate	1317-65-3	5 mg/m^3	10 mg/m^3	<10%
Zinc Oxide	1314-13-2	5 mg/m^3	5 mg/m^3	<10%
Component B: 2,4,6—Triphenol	90-72-2	10 mg/m^3	5 mg/m^3	<5%
Coated Calcium Carbonate	1317-65-3	10 mg/m^3	5 mg/m^3	>20%

Component A contains detectable amounts of a chemical known to the State of California to cause birth defects/cancer or other reproductive harm.

Section III—Physical/Chemical Characteristics

Boiling Point (°F)	N/E	Specific Gravity (H_2O = 1)	A: 1.13 B: 1.64
Vapor Pressure (mm Hg)	N/E		
Vapor Density (Air = 1)	>1	Melting Point	N/E
Solubility in Water	Insoluble	Evaporation Rate (Ether =1)	>1

APPEARANCE AND ODOR:
A: White gel, mild odor. B: Gray gel, characteristic odor

Section IV—Fire and Explosion Hazard Data

Flash Point (Method Used):	>200°F (Cleveland Open Cup)
Flammable Limits:	LEL: N/E UEL: N/E
Extinguishing Media:	Foam, CO_2, Water Fog
Special Fire Fighting Procedures:	None
Unusual Fire and Explosion Hazards:	None. Avoid breathing smoke.

Section V—Reactivity Data

Stability: Stable	Conditions to Avoid: None
Incompatibility (Materials to Avoid):	Strong Oxidizers, Strong Bases
Hazardous Decomposition or Byproducts:	CO, CO_2, NO_x
Hazardous Polymerization: None	Conditions to Avoid: Fires when curing.

Section VI — Health Hazard Data

Carcinoginicity: No NTP? No IARC? No OSHA Regulated? No

Effects and Hazards of Overexposure (Acute and Chronic):
Eyes: May produce irritation, sensitization
Skin: May produce irritation, sensitization
Inhalation: May produce irritation, sensitization
Ingestion: May produce irritation, sensitization

Emergency and First Aid Procedures: Remove any contaminated clothing.
Eyes: Flush immediately with large amounts of water for at least 15 minutes. Contact physician immediately.
Skin: Remove epoxy from skin immediately with a dry cloth or paper towel. Wash area of contact thoroughly with soap and water; solvents should not be used since they carry the irritant into the skin.
Inhalation: If respiratory irritation occurs, go to fresh air. Flood work area with fresh air. If irritation continues, seek medical attention.
Ingestion: Not expected. Contact immediate medical attention. Untrained first aid personnel should not attempt to administer first aid.
Contaminated: Clothing should be washed prior to re-use.

Section VII — Spill or Leak Procedures

Steps to be taken in case material is spilled or leaked: Bind with absorbent material
Waste Disposal Method Per local, state, and federal regulations

Section VIII—Safe Handling and Use Information

Respiratory Protection:	NIOSH/MSHA-approved respirators should be provided and worn if ventilation is inadequate. All workers required to use respiratory protection should be trained in their proper selection, use, and care.
Ventilation (local):	Recommended
Ventilation (mechanical):	Recommended
Ventilation (special):	Recommended when local and mechanical ventilation is inadequate
Protective Gloves:	Recommended
Eye Protection:	Recommended
Other Protective Equipment:	Recommended; splash bib with protective clothing.
Work/Hygienic Practices:	Remove and wash contaminated clothing. Wash hands before eating or smoking.

Section IX—Special Precautions

Handling Precautions:	Store in cool, dry location. Do not allow material to freeze. Store away from sparks and open flames.
Other Precautions:	None

Disclaimer of Warranties: Neither manufacturer or seller have any knowledge or control concerning purchaser's use of product. No expressed warranty is made by manufacturer or seller with respect to the results of any use of product or container that product comes in. No implied warranties including, but not limited to, an implied warranty of merchantability or an implied warranty for fitness for a particular purpose are made with respect to the product. Neither manufacturer or seller assume any liability for personal injury, loss or damage resulting from use of product. In the event that the product shall prove defective, seller or manufacturer shall replace a quantity of the product proved to be defective or refund the purchase price of the product upon return of the product.

Figure 22-12. *All hazardous materials on a job site must be inventoried and have a material safety data sheet. Vital information, such as proper handling and first-aid information, is included on a material safety data sheet.*

Products containing asbestos may be encountered in remodeling jobs. Inhalation of asbestos fibers may cause *asbestosis,* which is a respiratory disease caused by inhaling asbestos fibers and that results in scar tissue forming in the lungs. Workers are not allowed to enter areas where they may be exposed to asbestos or lead unless they have the proper training and are wearing the appropriate PPE.

Lead

Lead is a heavy and dense material with a low melting point, low strength, and high rate of expansion. Lead-based paint is a common source of lead poisoning, and cannot be recognized by the human eye. Only workers trained in lead removal are permitted to remove lead-based paint. Material covered with lead-based paint that is cut, sanded, heated, or burned releases lead into the air. Lead is toxic if swallowed or inhaled. Exposure to lead may cause brain and neurological damage. Guidelines to minimize lead exposure include the following:

- Do not eat or drink in the hazardous work area.
- Wear an appropriate respirator to prevent inhaling lead.
- Cover and seal cabinets and surfaces that cannot be removed from a hazardous work area.
- Clean up solid debris using special vacuum cleaners with high-efficiency particle absorption (HEPA) filters. Wet-mop the area after vacuuming.
- Dispose of protective clothing (jump suits) worn in the hazardous work area after completing the cleanup operation. Soiled protective clothing must not be worn in other areas of a job site or home.

Lead-Based Paint Renovation, Repair, and Painting Program. The 2008 Lead-Based Paint Renovation, Repair, and Painting Program, developed by the Environmental Protection Agency (EPA), was designed to protect against the hazards of lead-based paint associated with renovation, repair, and painting activities. The program is a federal regulatory program affecting contractors, property managers, and others who disturb painted surfaces. It applies to residential buildings and child-occupied facilities, such as schools and day care centers, built before 1978. It includes pre-renovation education, training, certification, and work practice requirements. General work practice requirements must also be followed.

The *Small Entity Compliance Guide to Renovate Right: EPA's Lead-Based Paint Renovation, Repair, and Painting Program* developed by the EPA, provides detailed information on general work practice requirements. For example, renovations must be performed by certified companies using certified renovators. Companies must post signs clearly defining the work area and warn occupants and other persons not involved in renovation activities to remain outside of the work area. These signs should be written in the language of the occupants. Prior to renovation, a company must contain the work area so that dust and debris do not leave the work area while renovation is underway.

Concrete Dust

Concrete sawing and grinding, **Figure 22-13,** abrasive blasting of concrete, demolition of concrete structures, dry sweeping or pressurized air blowing of concrete, and concrete mixing produce dust containing crystalline silica particles. *Crystalline silica,* also known as quartz, is a natural compound found in the earth's crust and is a basic component of sand and granite. *Silicosis* is a lung disease caused by inhaling dust containing crystalline silica particles. As dust containing crystalline silica particles is inhaled, scar tissue forms in the lungs, which reduces the ability of the lungs to extract oxygen from the air. Since there is no cure for silicosis, exposure prevention is the only means of control. Disposable particulates masks are required when crystalline silica particles are present in the air.

Gypsum Board

Gypsum board (drywall) cutting and sanding operations produce airborne crystalline silica particles. The particles may cause irritation to skin, eyes, and/or respiratory tract. Prolonged exposure may cause silicosis. Appropriate respiratory protection is required when crystalline silica particles are present in the air.

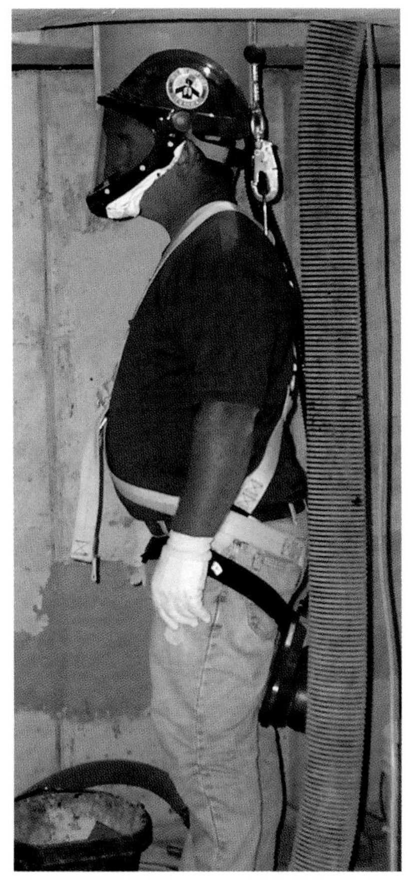

Figure 22-13. *Proper respiratory protection must be worn when working around concrete sawing or grinding operations.*

Insulation

Cutting and installation of cellulose, foam, or fiberglass insulation may produce particles and result in potential health risks. Cellulose insulation may cause irritation or abrasion of the mouth and throat, headaches, nausea, dizziness, and difficulty in breathing if inhaled. Contact with eyes causes irritation and tearing.

Foam insulation may cause irritation to the respiratory tract if inhaled. Repeated exposure may cause lung damage. Contact with eyes causes irritation or corneal injury from abrasion.

Fiberglass insulation may cause headaches, nausea, dizziness, and difficulty in breathing if inhaled. Contact with eyes causes irritation and inflammation of the mucous membranes, tearing, and sensitivity to light. Repeated exposure causes inflammation of the eyelids, digestive disturbances, weight loss, and general weakness. Extended exposure may result in respiratory and lung disease, bronchitis, asthma, or pulmonary heart disease.

Carpenters working on a job site where insulation particles exist should wear appropriate personal protective equipment to avoid health hazards. **See Figure 22-14.**

Figure 22-14. *Fiberglass insulation may cause headaches, nausea, dizziness, and difficulty in breathing if inhaled. Appropriate personal protective equipment must be worn to prevent inhalation and contact with the insulation.*

Bloodborne Pathogens

A *bloodborne pathogen* is a virus or bacteria of a disease in the blood that may be transmitted by another worker coming into contact with the infected worker's blood. Carpenters may be exposed to bloodborne pathogens if someone on the job site infected with a bloodborne pathogen receives a cut or injury.

If it is reasonably anticipated that workers will be exposed to blood or other potentially infectious materials while using first-aid supplies, employers should provide PPE.

OSHA 29 CFR 1910.1030, *Bloodborne Pathogens,* specifies that bloodborne pathogen personal protective equipment must be readily accessible and available in appropriate sizes. Disposable gloves, body gowns, and other first-aid equipment that is required to stop exposure to blood or other infectious materials must be available in the job site first-aid kit. A first-aid kit must be checked at least once a week to ensure supplies are replaced as used.

CONFINED SPACES

A *confined space* is a space large enough and so configured that a carpenter can enter and perform assigned work, has limited or restricted means for entry and exit, and is not designed for prolonged occupancy. Confined spaces are subject to the accumulation of toxic or flammable contaminants or an oxygen-deficient atmosphere. Confined spaces include storage tanks, underground utility vaults, and pits more than 4′ in depth.

Miller Fall Protection

A safety harness with an attached lifeline must be worn when entering a confined space if a safe atmosphere cannot be assured. The safety harness and lifeline will be invaluable in case a rescue is required.

All employees entering a confined space must be instructed as to the nature of the hazards, necessary precautions, and protective and emergency equipment required. A safety harness with an attached lifeline must be worn when entering a confined space if a safe atmosphere cannot be assured in the confined space. Another worker, also wearing a safety harness and lifeline, must constantly observe the carpenter in the confined space.

EXCAVATIONS

An *excavation* is a cut, depression, or trench in the earth's surface formed manually or by earth-moving equipment. **See Figure 22-15.** A multistory structure often requires a deep excavation. Smaller construction projects may require narrow and deep trenches.

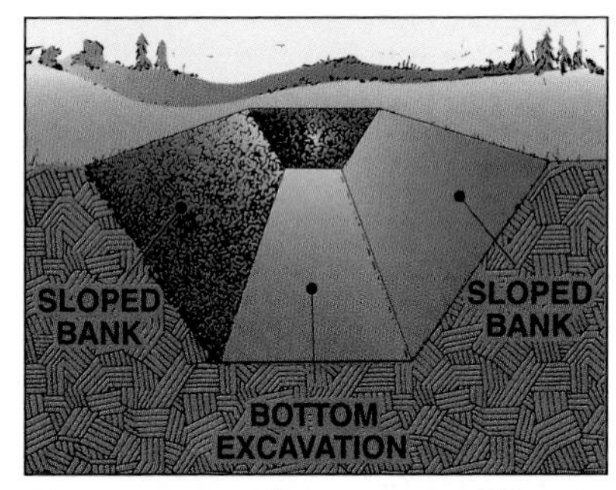

Figure 22-15. *Whenever possible, the banks of a deep excavation should be sloped outward.*

Carpenters and other tradesworkers work in excavations while constructing the footing and concrete wall forms. A potential safety hazard when working in excavations is the collapse of the earth banks. Earth bank collapse can be prevented by sloping or shoring the earth banks of an excavation. The method used to protect against collapsing earth banks depends on the soil type, excavation depth, water table level, type of foundation, and the space around the excavation.

OSHA 29 CFR 1926.652, *Requirements for Protective Systems,* states that workers in excavations must be protected from cave-ins by adequate protective systems except when the excavation is made entirely in stable rock or when excavations are less than 5′ deep and inspection of the ground by a competent person provides no indication of potential cave-ins. Workers in excavations not made in stable rock or excavations more than 5′ deep must be protected by sloping, shoring, or an equivalent means of protection.

When working in or around excavations, the following safety rules must be observed:

- Ensure that excavated soil and other materials and construction equipment are at least 2′ away from the edge of the excavation.
- A ramp, runway, ladder, or stairway must be located within 25′ of workers if the excavation is 4′ or more in depth to provide a means of access and egress in case of an emergency.
- Watch for vibration and increased lateral pressure on sides of excavations such as soil vibrating loose when vehicles travel close to excavations.
- Ensure there is adequate removal of engine exhaust from trenches if gasoline- or diesel-powered construction equipment is being used.
- Where an oxygen-deficient atmosphere (i.e., an atmosphere with less than 19.5% oxygen) or otherwise hazardous atmosphere exists or could reasonably be expected to exist, such as in landfill areas, the atmosphere must be tested before workers enter excavations greater than 4′ deep.
- Do not work in excavations containing water or in which water is accumulating unless adequate precautions have been taken.
- Provide support systems such as shoring and bracing when the stability of adjoining buildings, walls, or other structures is endangered by excavation operations.

Sloping and Benching

The sides of an excavation can be sloped to minimize the possibility of collapse by inclining the sides away from the excavation. Sloping is typically used when there is room around the construction area to slope the sides of an excavation. Sloping cannot be used if buildings or streets are immediately adjacent to the excavation.

Slope angle is based on the soil type of the excavation. OSHA 1926 Subpart P Appendix A, *Soil Classification,* defines soil types as *stable rock, Type A, Type B,* and *Type C* (in decreasing order of stability). Stable rock includes solid rock, shale, and cemented sand and gravels. Type A soils are highly cohesive soils and include clay, silty clay, sandy clay, and clay loam. Type B soils are cohesive soils and include angular gravel, silt, silt loam, and sandy loam. Type C soils are granular soils and include gravel, sand, loamy sand, and soil from which water is freely seeping.

OSHA 29 CFR 1926 Subpart P Appendix B, *Sloping and Benching,* specifies slope angles for various soil types. Excavations occurring in average soil conditions (Type B) require a slope of 45°. **See Figure 22-16.** Less slope is permitted for stable rock and Type A soils. More slope is required for Type C soils including compacted sharp sand or well-rounded loose sand.

Benching is a method of protecting workers in an excavation by excavating the sides to form one or a series of steps, with vertical surfaces between the levels. Benching typically requires more room around the excavation than sloping.

Shoring

In large and deep excavations, several methods of shoring are effective. *Interlocking sheet piling* consists of steel sections that are driven into the earth and provides a fairly watertight system. Sheet piling is lowered by crane into templates that hold it in position and is driven with a pile-driving rig. **See Figure 22-17.** If necessary, braces are installed to support the sheet piling. When construction is complete, sheet piling is removed and reused.

Soldier piles are driven into the ground with a pile-driving rig. *Lagging,* consisting of 3″ thick wood planks, is placed between the flanges of the soldier piles. **See Figure 22-18.** A horizontal steel waler may extend across the fronts of the soldier piles to provide additional stability. The waler is secured into position by *tie-backs,* which are heavy-gauge steel strands secured into grouted holes drilled in the sides of the excavation. The holes may extend as deep as 50′.

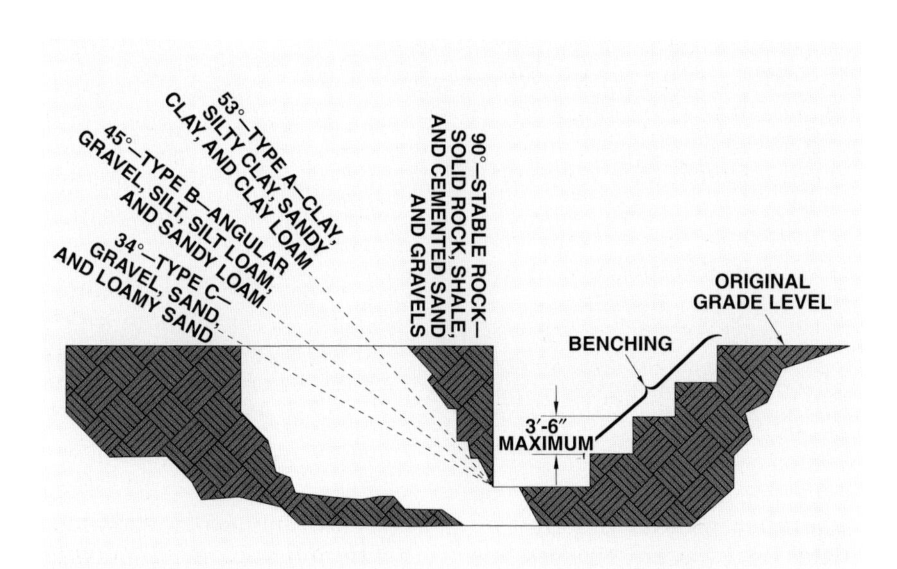

Figure 22-16. *Soil type and conditions determine the amount of slope required for earth banks around excavations.*

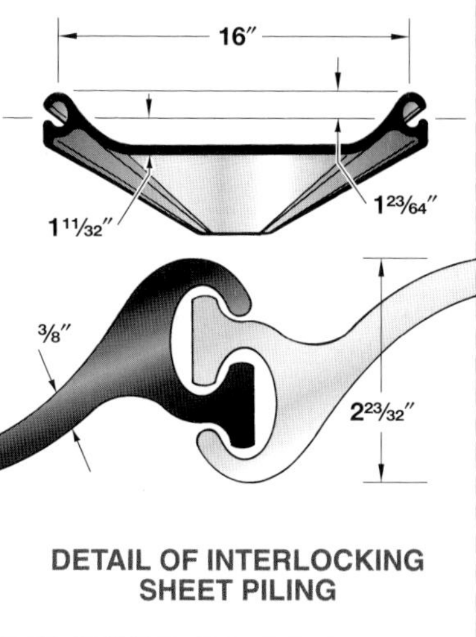

Figure 22-17. *One shoring method is to place steel interlocking sheet piling around the area to be excavated. At top, a section of sheet piling is being lowered by crane.*

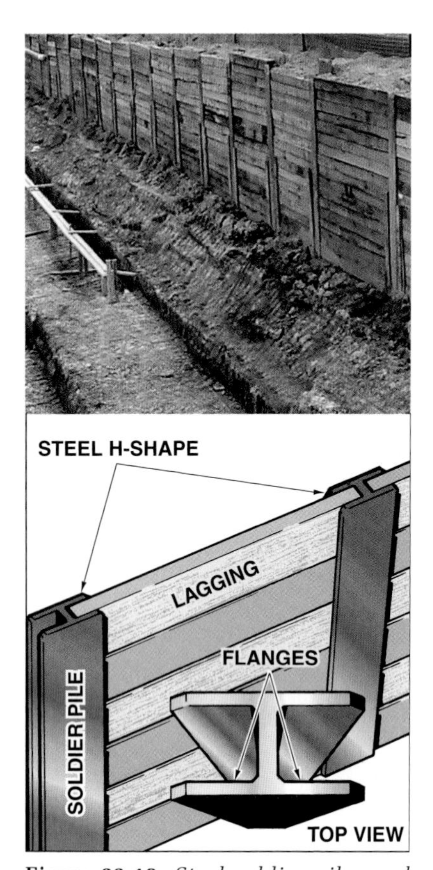

Figure 22-18. *Steel soldier piles and wood lagging are often used for shoring excavations. Here the soldier piles have been driven into the ground and the lagging has been placed between the flanges of the soldier piles.*

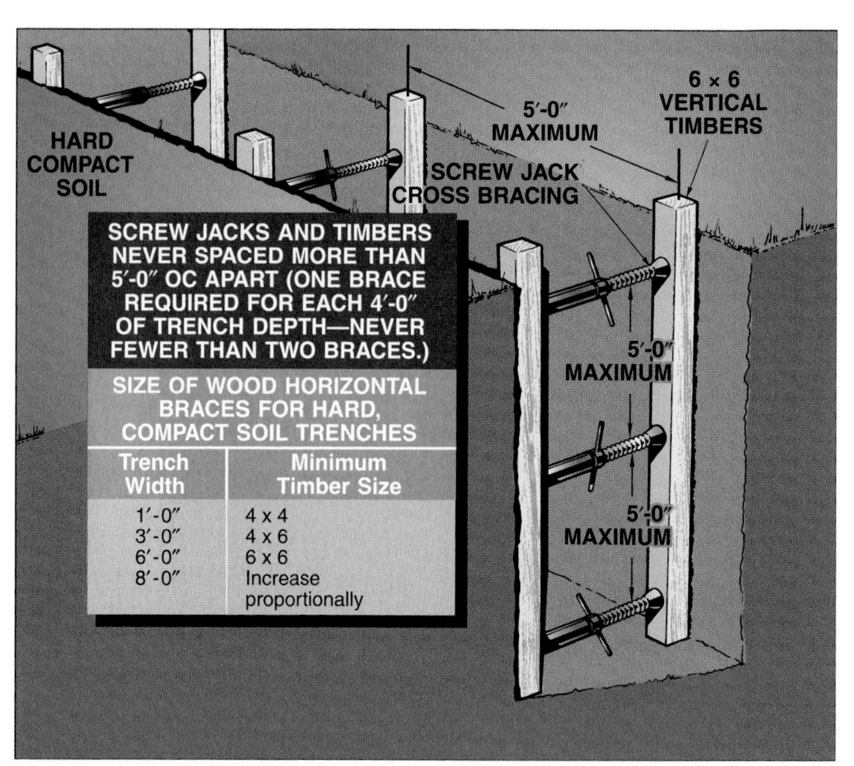

SIZE OF WOOD HORIZONTAL BRACES FOR HARD, COMPACT SOIL TRENCHES

Trench Width	Minimum Timber Size
1′-0″	4 x 4
3′-0″	4 x 6
6′-0″	6 x 6
8′-0″	Increase proportionally

Figure 22-19. *Trenches in hard, compact soil may require no more than spaced vertical timbers held in place by cross braces or screw jacks.*

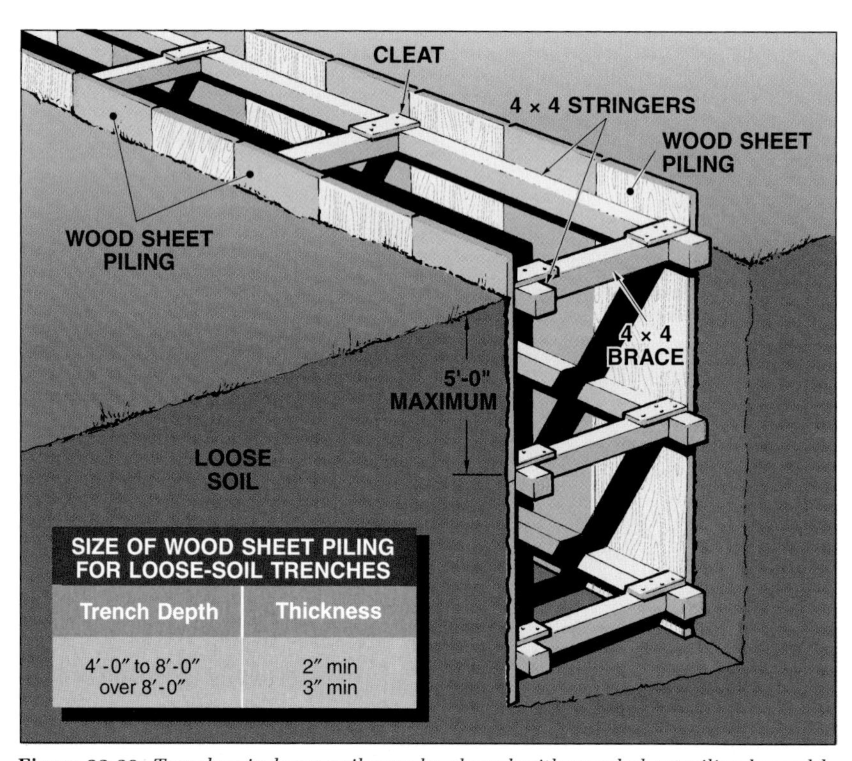

SIZE OF WOOD SHEET PILING FOR LOOSE-SOIL TRENCHES

Trench Depth	Thickness
4′-0″ to 8′-0″	2″ min
over 8′-0″	3″ min

Figure 22-20. *Trenches in loose soil may be shored with wood sheet piling braced by stringers and braces.*

Various shoring methods are recommended for trenching operations. Trenches 5′ or more in depth in hard, compact soil can be shored by placing vertical timbers on opposite sides of the trench. The timbers are held in place by cross braces or screw jacks. **See Figure 22-19.** The uprights should be no more than 5′ apart horizontally and vertically. For trenching operations in loose soil, carpenters may install wood sheet piling, which is supported by stringers and braces. **See Figure 22-20.**

Shielding

Shielding is the use of a portable protective device capable of withstanding cave-in forces in trenches. A *trench shield,* or trench box, is a reinforced assembly consisting of two plates held apart by spacers. **See Figure 22-21.** Trench shields are made from steel, concrete, or plastic and are moved along in a trench as work progresses. Trench shields are used in stable or unstable soil.

LiteGuard

Figure 22-21. *A trench shield is used to withstand cave-in forces in trenches.*

BARRICADES AND GUARDRAILS

On construction jobs where there is pedestrian or vehicular traffic in adjoining areas, carpenters erect *barricades* around the job site to prevent unauthorized persons from entering the construction area. In many cases, barricades provide overhead protection to prevent objects from falling in the traffic area.

Guardrails protect the safety of workers on the job. Guardrail requirements are detailed in OSHA 29 CFR 1926.502, *Fall Protection Systems Criteria and Practices.* Working and walking surfaces 6′-0″ or more above a lower surface must be protected by guardrails or personal fall-arrest systems. Guardrails are placed across openings for exterior doors if there is a drop of more than 4′. Guardrails are also required if the bottom of a window opening is less than 39″ above the working surface. Guardrails must be constructed on all unprotected sides or edges of floor openings such as openings for stairways and skylights. If the openings are used for worker access, such as ladderways, the openings must be protected with a gate so workers cannot walk directly into the opening. Toeboards are installed if there is a possibility that tools or materials may fall into the opening.

Guardrails consist of a *top rail* and *midrail.* Top rails are typically constructed of smooth 2 × 4s and placed 39″ to 45″ above the working surface. **See Figure 22-22.** Midrails are commonly constructed with 1 × 6s or 2 × 4s and are placed midway between the top rail and the working surface. Screens or meshes may be used instead of midrails. If screens or meshes are installed, they must extend from the top rail to the working or walking surface.

Wire rope or cable may be used for top and midrails. Wire rope or cable must be at least 1/4″ diameter. If wire rope or cable is used as a top rail, it must be flagged at not more than 6′ intervals with high-visibility material.

RAMPS, RUNWAYS, AND TEMPORARY STAIRS

Heavy construction jobs may require ramps, runways, and temporary stairs to provide a means for workers to move about on the job and for materials to be transported. Ramps and runways may also be constructed for the movement of wheelbarrows and power buggies used to transport concrete and other materials. **See Figure 22-23.**

FIRE PREVENTION

A serious concern for all construction workers is the ever-present danger of fire on a job site. Workers must be aware of potential fire hazards and understand what creates a fire hazard and how this danger can be reduced.

Structural ramps that are used solely by employees as a means of access or egress from excavations must be designed by a competent person. If there is a risk of falling 6″ or more, the ramp must have the proper guardrail system.

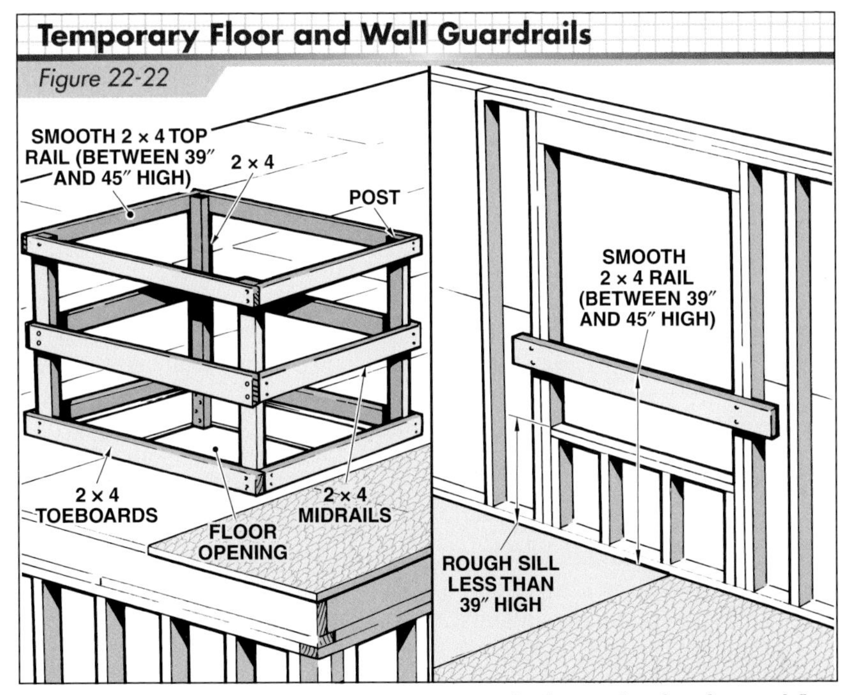

Figure 22-22. *During construction, temporary guardrails must be placed around floor openings. In addition, guardrails must be placed across wall openings for exterior doors if there is a drop of more than 4′ and across window openings if the rough sill is less than 39″ above the working surface.*

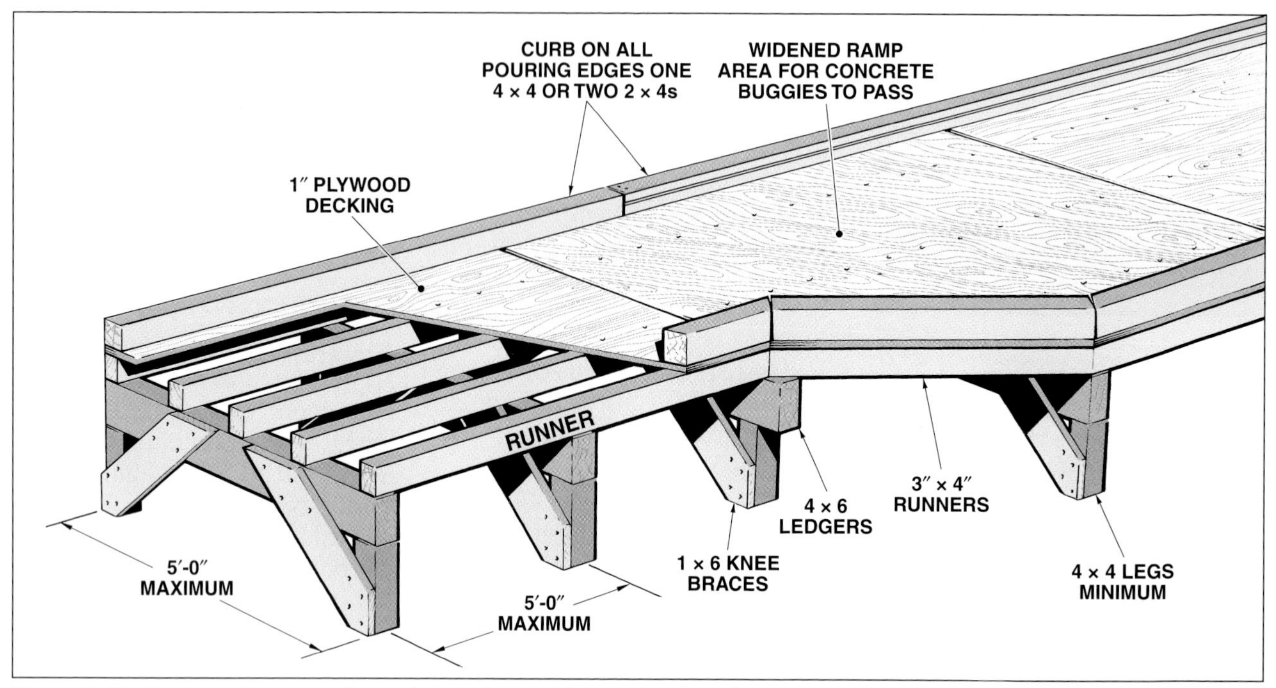

Figure 22-23. *Ramps and runways for workers and materials must be properly constructed to ensure worker safety. The ramp shown here is used for power buggies, which are used to deliver concrete for placement.*

Fire Extinguishers

The National Fire Protection Association (NFPA) classifies fires as Class A, B, C, D, and K, based upon the combustible material. The appropriate fire extinguisher must be used on a fire to safely and quickly extinguish the fire. **See Figure 22-24.** The types of fires and appropriate fire extinguishers are as follows:

- *Class A fires* occur with wood, paper, textiles, and similar materials. Class A fires are extinguished with water and other water-based agents.
- *Class B fires* occur with flammable liquids such as grease or solvent cements. Class B fires are extinguished with smothering agents such as carbon dioxide and chemical foams.
- *Class C fires* occur with live electrical equipment and are extinguished with nonconductive dry chemical agents.
- *Class D fires* occur with combustible metals such as magnesium, sodium, and potassium. Class D fires are extinguished with a coarse powder agent that seals the burning surface and smothers the fire.
- *Class K fires* occur with grease in commercial cooking equipment. Class K extinguishers coat the fuel with wet- or dry-base chemicals.

OSHA 29 CFR 1926.150, *Fire Protection,* details the fire protection requirements for a construction job site. Fire extinguishers must be periodically inspected to ensure proper operation. Fire extinguishers are placed around the job site so the travel distance from any point to the nearest fire extinguisher does not exceed 100′. At least one fire extinguisher must be provided on each floor of a building. In multistory buildings, at least one fire extinguisher shall be located adjacent to a stairway. Fire extinguishers should be located in clear view and should not be obstructed by building materials. **See Figure 22-25.**

If a fire occurs on the job site, an alarm should be sounded and the fire department called. However, small fires can be quickly extinguished if the proper fire extinguisher is present on the job.

Preventive Measures

The following preventive measures should be observed to reduce the threat of fire on a job site:

- Do not allow rubbish and combustible material to accumulate on a job site. Periodically clean the job site and place wood scraps and other rubbish in appropriate disposal containers.
- Keep volatile and flammable materials stored away from the immediate job site.
- Do not smoke near volatile materials such as solvent cements, as these materials readily evaporate at normal temperatures and pressures and produce flammable vapors.
- Keep all flammable liquids such as gasoline, paint thinner, oil, grease, and paint in tightly plugged or capped containers.
- A fire extinguisher must be provided within 50′ of the point where 5 gal. of flammable or combustible liquid or 5 lb of flammable gas are being used on the job site.

Figure 22-24. *Fire extinguisher classes are based on the combustible material.*

Figure 22-25. *Fire extinguishers must be properly placed around a job site for convenient access by workers. Panels can be constructed with 2 × 4s and plywood.*

Fire extinguisher panels can be easily constructed using 2 × 4s and plywood. The tops of the panels can be painted a bright red color for ease in locating the fire extinguisher across a job site.

Unit 22 Review and Resources

SECTION

6

Building Design and Printreading

CONTENTS

UNIT

23 Building Design, Plans, and Specifications

OBJECTIVES

1. Identify common building designs.
2. List and describe professions involved in creating building designs and plans.
3. List and describe the factors influencing building design.
4. Identify the basic shapes and types of buildings.
5. Identify common types of house plans.
6. List and describe the processes for producing plans.
7. Explain the advantages of using building information modeling (BIM).
8. Discuss specifications commonly used with plans.

The design of a building determines its appearance and dictates the methods to be used in its construction. Most buildings are *traditional* or *contemporary* in appearance. Some traditional styles date back to colonial times in the United States, yet are still popular in new construction. **See Figure 23-1.** Contemporary designs are characterized by their simplicity and straight lines. **See Figure 23-2.**

California Redwood Association

Figure 23-2. *The contemporary design of this residence is characterized by its simplicity and straight lines. Vertical redwood siding allows the structure to blend well with its surroundings.*

James Hardie Building Products

Figure 23-1. *Some building designs are based on traditional styles of architecture.*

A design known as *pueblo revival architecture* is popular in the southwestern United States. Pueblo revival architecture is a blend of Native American, Hispanic, and Anglo building traditions. These houses were originally constructed out of adobe brick, a process that is expensive both in terms of labor and material. Currently, most buildings of pueblo revival architecture design are wood-framed with a stucco finish. **See Figure 23-3.**

Figure 23-3. *Pueblo revival architecture is popular in the southwestern United States.*

Architects are licensed professionals qualified to design buildings. In some areas, prospective homeowners, designers, or drafting services can perform residential design services provided the plans conform to code requirements and are approved by the local building authorities. A competent architect has a good understanding of how the building is constructed as well as how it should look.

A set of construction prints for larger buildings typically includes drawings developed by *structural engineers*. Structural drawings include the structural components of a building such as beams, columns, floors, and walls that support and hold together the entire structure. Architects frequently employ the services of *soil engineers* to analyze the ground conditions at the building site.

FACTORS INFLUENCING BUILDING DESIGN

Many factors must be considered in the design of a building. One factor that is considered in building design is the size and shape of the lot. A building must adhere to local building code requirements for setbacks. A *setback* is the distance required between a structure on a piece of property and the front property line. Another factor is the soil condition of the lot, which determines the type of foundation that can be used. Yet another factor in building design is the appearance of the building in relationship to other buildings in the area. If possible, a newly constructed building should conform to the sizes and styles of buildings already in the neighborhood. A newly constructed building must also conform to the building code and zoning regulations in effect in that jurisdiction.

Orientation is the position of a building on the lot and the direction in which different walls will face. Wind conditions and sun position during the day help determine the ideal orientation for a building on the lot. Privacy is another concern in determining building orientation. The location of doors and windows, and even the type of windows, are largely determined by the orientation of a building.

BASIC SHAPES AND TYPES OF BUILDINGS

Whether traditional or contemporary, the design of a one-family dwelling is usually derived from one of several basic shapes—rectangular, T-shape, L-shape, or U-shape. **See Figure 23-4.** Basic house types are one-story, one-and-one-half-story, two-story, and split-level. Most house designs can be built over a concrete slab, crawl space, or full basement foundation.

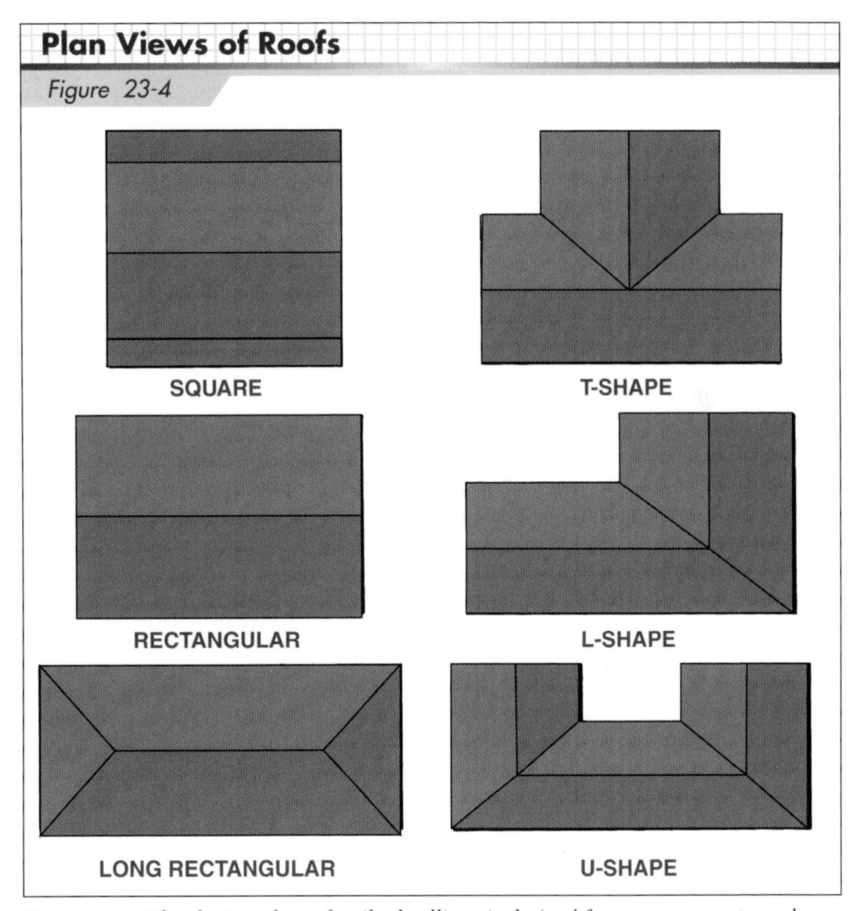

Figure 23-4. *The design of one-family dwellings is derived from square, rectangular or T-, L-, or U-shapes.*

APA—The Engineered Wood Association

Advanced carpentry skills are required to construct complex roofs.

One-Story House

In a one-story house, all habitable rooms are on the same level. A *habitable room* is a room used for living purposes, such as a bedroom, living room, dining room, family room, or bathroom. **See Figure 23-5.** The attic of a one-story house is usually too small for living purposes but may be used for storage.

One-and-One-Half-Story House

A one-and-one-half-story house has a high-pitched roof, which allows some of the attic space to be used for living purposes. **See Figure 23-6.** Dormers, or roof windows, provide light and ventilation. Rooms in the attic are usually additional bedrooms and a second-floor bathroom.

Heating costs for a one-and-one-half-story house are minimized due to the small outside wall area when compared to the amount of interior space.

Figure 23-5. *One-story house designs can be built over a concrete slab, crawl space, or full basement. All habitable space is located on one floor.*

Figure 23-6. *A one-and-one-half-story house has a high-pitched roof, allowing some of the attic space to be used for living purposes. Note the dormers in the roof.*

Two-Story House

A two-story house consists of two full stories under a pitched roof. **See Figure 23-7.** The second story usually consists of bedrooms and a second bathroom. The laundry room may also be located on the second floor.

Split-Level House

A split-level house is more complicated in design. A split-level house is actually a one-story house with a basement under one section of the house that is partially below and partially above ground level. The other section of the house is constructed on a concrete slab or crawl space. There are a number of variations on the split-level design, as shown in **Figure 23-8.** Split-level houses are practical for sloping lots, but can also be constructed on level lots.

Little or no hallway space is required in a split-level house.

DEVELOPING AND DRAWING BUILDING PLANS

Prior to construction starting on a building, a set of building plans must be produced. Buildings may be produced on speculation, from stock plans, or may be custom-built.

Plans may be prepared for general contractors who construct individual homes or apartments on *speculation*. These structures are financed and built by the contractors or real estate developers and are sold during or after completion of the work. Housing tracts and condominiums are examples of buildings constructed on speculation.

Stock plans are working drawings developed by architects and then purchased from design companies that produce a wide variety of working drawings for home construction.

Custom-built plans are a set of plans designed by an architect to serve the needs of the potential owner. Custom-built plans usually begin with preliminary sketches based on discussions between the architect and client. Factors such as the number of bedrooms and bathrooms, living space, garage size, lot size, and approximate budget are determined in the discussions.

All states require architects to take an examination to be licensed.

Working drawings are developed either manually or using a *computer-aided design (CAD)* program. CAD, also known as computer-aided design and drafting (CADD), programs allow an architect to develop the working drawings more quickly than using manual drafting techniques. In addition, CAD programs generate clear and concise drawings, and allow an architect to make design changes more easily than using manual drafting techniques. **See Figure 23-9.** The working drawings are reviewed and approved by the client.

Figure 23-7. *A two-story house typically has a pitched roof and may be constructed on a full basement or slab foundation.*

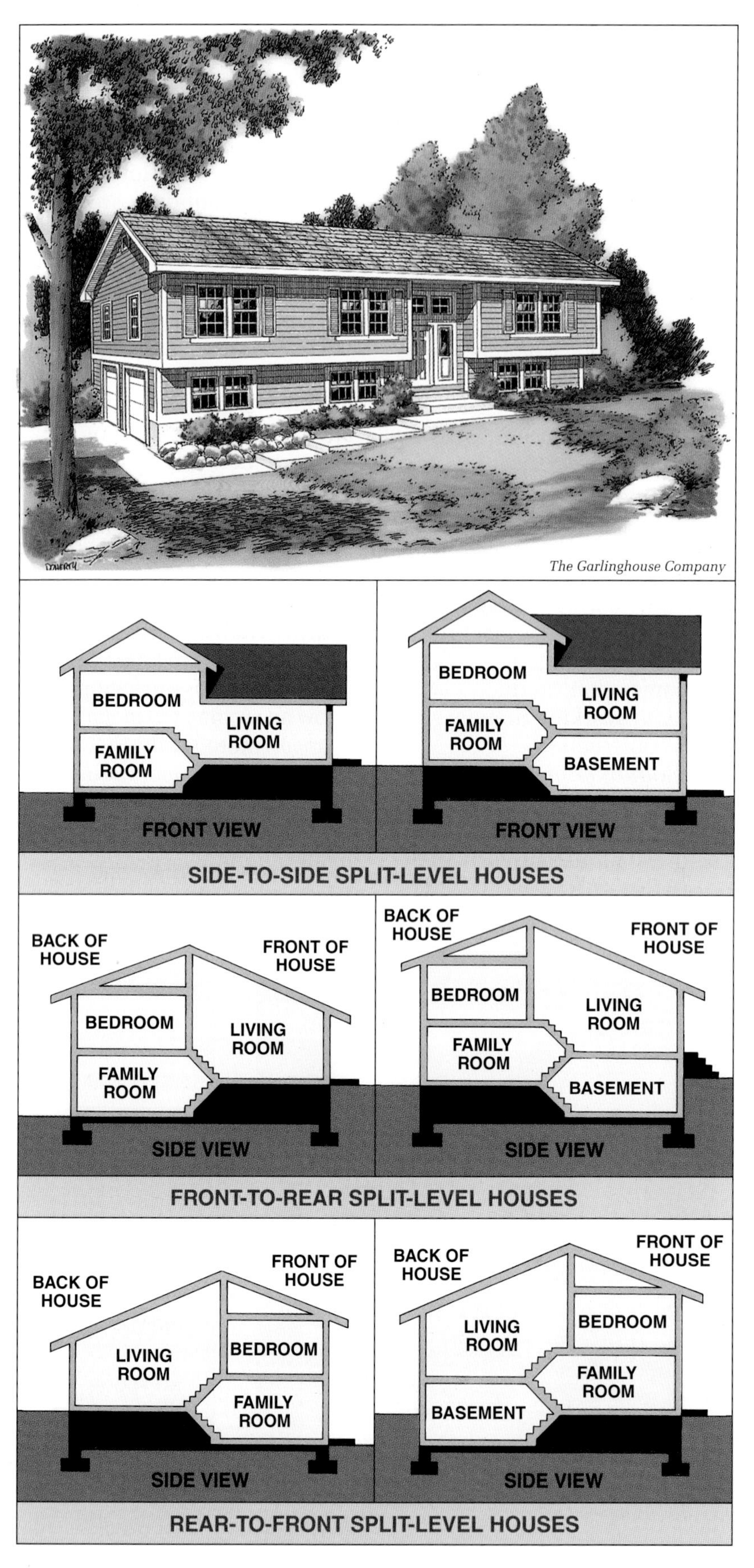

Figure 23-8. *The basement of a split-level house is partly below and partly above the ground level.*

The original working drawings are reproduced for distribution to contractors and subcontractors who are awarded the contracts for the construction project. In the past, original working drawings were reproduced on an off-white, chemically coated paper. Ammonia vapor was used to process the paper, creating white lines against a dark blue background. Plans produced using the ammonia vapor process were referred to as *blueprints.*

The blueprint process was eventually replaced by the *Ozalid® transfer* method that produced blue or black lines against a white background. Even though the term "blueprint" is commonly used in the trade, it is no longer accurate since most drawings are copied using wide-format engineering copiers.

Both blueprints and diazo prints were replaced with the development of electrostatic printing. Electrostatic prints are produced by the same process office photocopiers use. Full-size working drawings are exposed to light and projected directly through a lens onto a negatively charged drum that transfers the image to positively charged copy paper.

While some electrostatic prints are still produced today, the majority of construction drawings are produced from digital files using wide-format digital printers. These wide-format digital printers, sometimes referred to as plotters, use light-emitting diodes (LEDs), laser, or inkjets to produce clear, accurate construction documents quickly and cleanly. **See Figure 23-10.**

The fastest wide-format digital printers are capable of printing over 20 D-sized (22″ × 34″) prints per minute.

Figure 23-9. *A CAD program allows an architect to develop working drawings more proficiently than using manual drafting techniques.*

Wide-Format Digital Printers

Figure 23-10

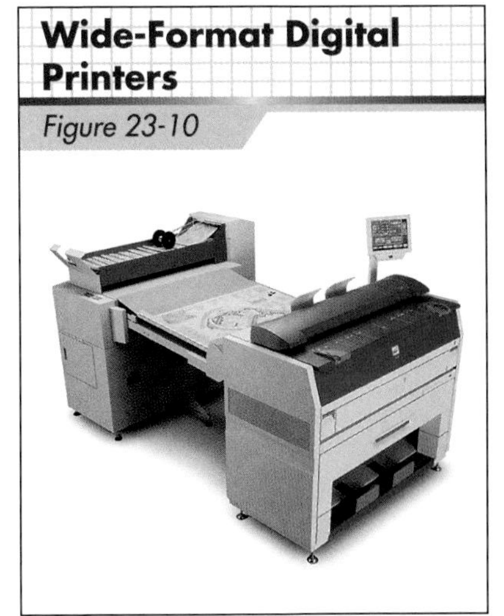

KIP America

Figure 23-10. *Wide-format digital printers are used to produce clear, accurate copies of construction drawings quickly.*

Building Information Modeling (BIM)

Building information modeling (BIM) is an integrated, electronically managed system that aligns all working drawings, structural drawings, and shop drawings into a consistent system. BIM modeling, also known as virtual design and construction (VDC), allows for integration of CAD-generated documents or building models from various sources to create a complete set of construction documents.

With BIM, contractors, subcontractors, and suppliers can all import the electronically produced construction documents or models into a common system that recognizes all the individual construction components in a fully integrated platform. Prints are then generated from the integrated model. BIM technology also allows for the integration of estimating and scheduling software. **See Figure 23-11.**

BIM has several advantages over conventional CAD drawing development. Some BIM software can identify areas of inconsistency or clashes between various building components. These clashes are automatically identified by the software and can be reviewed and resolved prior to construction.

Another advantage of BIM is the automatic updating of views on the model when a change is made. For example, if a door is added to a floor plan, the door will automatically appear on the elevation views and the door schedule. This eliminates the errors associated with having to redraw multiple drawings. BIM information can also be exported to different analytical programs for use in energy efficiency and dynamic thermal analysis, load calculations, or daylighting assessments.

BIM is useful in estimating, scheduling, and project management processes.

Building Information Modeling (BIM)

Figure 23-11

Vico Software, Inc.

Figure 23-11. *BIM allows for integration of CAD-generated documents or building models from various sources to create a complete set of construction documents and for the integration of estimating and scheduling software.*

SPECIFICATIONS

A complete set of building plans typically includes a legal document called *specifications* (specs). Specifications are written information from the architect and engineers that supplement the plans and provide details that could not be shown on the plans or that require additional description. Depending on the size of the construction project, specifications may consist of a few notations on the prints, a few sheets of paper, or possibly a detailed book with hundreds of pages.

Specification formats vary from architect to architect. However, a common practice is to divide the specifications into *divisions* that pertain to different work areas on the construction project. The Construction Specifications Institute (CSI) publishes the *CSI MasterFormat*™, which is a master list of the divisions and titles for organizing information about construction requirements and activities into a standard sequence. The CSI MasterFormat contains 50 divisions that define the broad areas of construction. **See Figure 23-12.** Each division is divided into subclassifications, which provide greater detail. A list of the CSI MasterFormat divisions and complete list of the subclassifications is available from the Construction Specifications Institute.

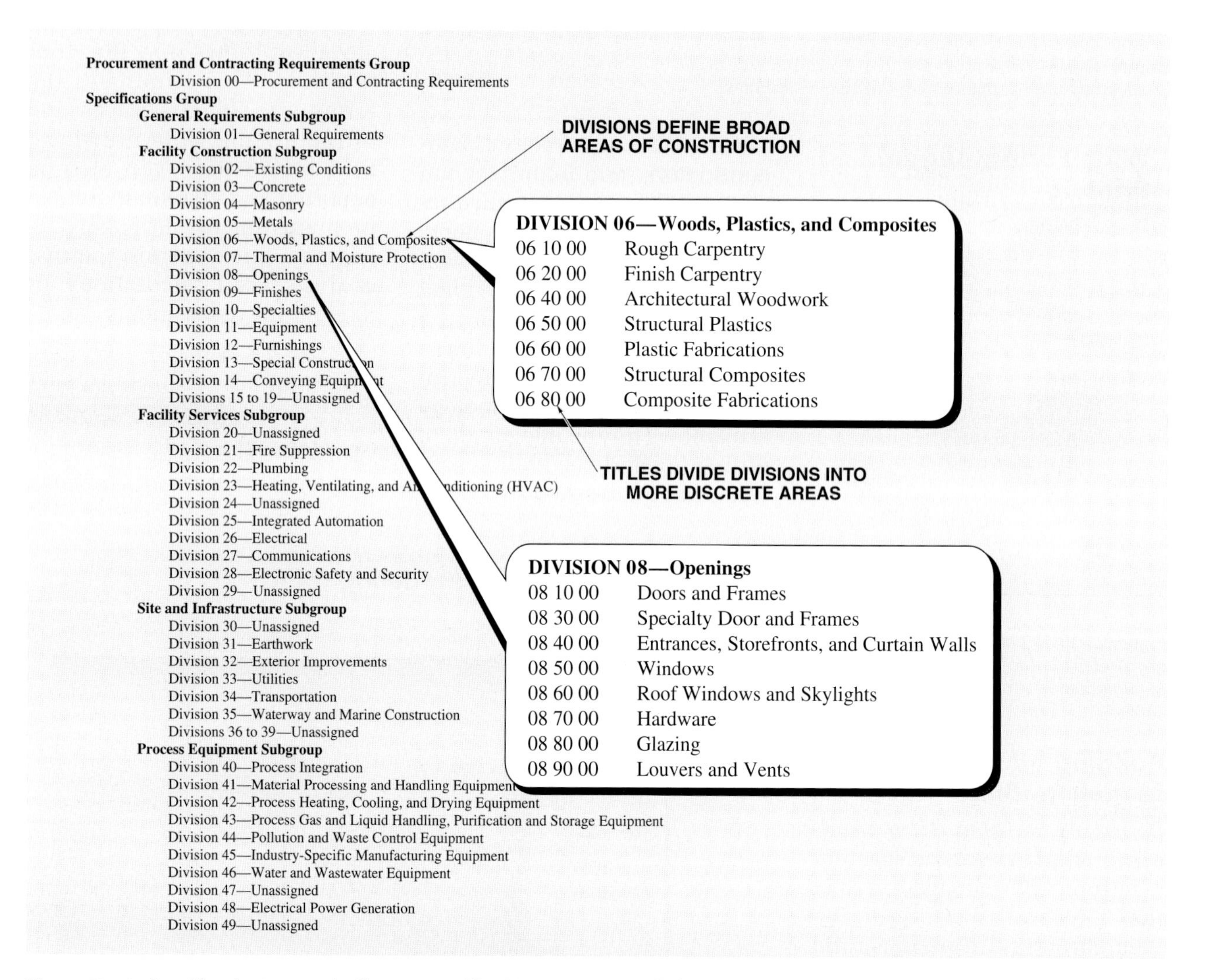

Figure 23-12. *Specifications are typically organized by CSI MasterFormat™ divisions.*

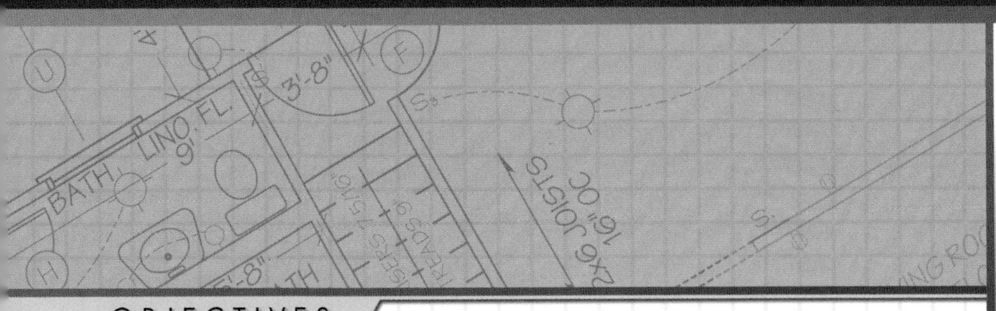

UNIT 24 Understanding the Language of Prints

OBJECTIVES

1. Describe information provided by prints.
2. Identify linetypes used on a print.
3. Explain the function of dimensions on a print.
4. Identify the tools used for scaling dimensions on prints.
5. List and describe the functions of symbols and abbreviations used on prints.

A set of *prints*, also referred to as working drawings or plans, acts as a step-by-step guide to the construction of a building. Some prints provide more detail than others, but all prints include certain basic information. The three-bedroom house plan used as an example in this section consists of the following:

- specifications
- plot plan
- foundation plan
- exterior elevations
- section views
- framing plans
- details
- door and window schedules
- finish schedule

In its original state, the three-bedroom house plan fills five 22″ × 36″ pages, which is typical for this type of building. The drawings reproduced in this text have been reduced in size, and in some cases rearranged, in order to fit on the page.

Prints provide an *orthographic,* rather than a *pictorial,* view of each part of the building from above or from the sides. A comparison of orthographic and pictorial drawings is shown in **Figure 24-1.** While pictorial drawings provide a more realistic depiction of a building to an owner, the drawings do not contain the details necessary to construct the building.

Although a complete set of prints consists of many different plans, all plans relate to each other. During the various stages of construction, carpenters may have to refer to several plans to get a complete understanding of the work to be done.

LINES, DIMENSIONS, AND SCALE

Different *linetypes* have different meanings in a print. **Figure 24-2** shows a section of a foundation plan and identifies the lines used. A solid line indicates the visible outline of an object, while a dashed line indicates an edge that is hidden from view. A *centerline* establishes the center of an area. A *cutting plane line* indicates where an object is "cut" so interior features may be seen. A *break line* indicates a shortened view of a part that has a uniform shape. A *leader line* points from a note or measurement to a part of the building.

Linetypes on construction prints are established in ASME Y14.2M, Line Conventions and Lettering.

Southern Forest Products Association

Carpenters must understand the details of individual drawings and how the drawings relate to other drawings in a set of prints.

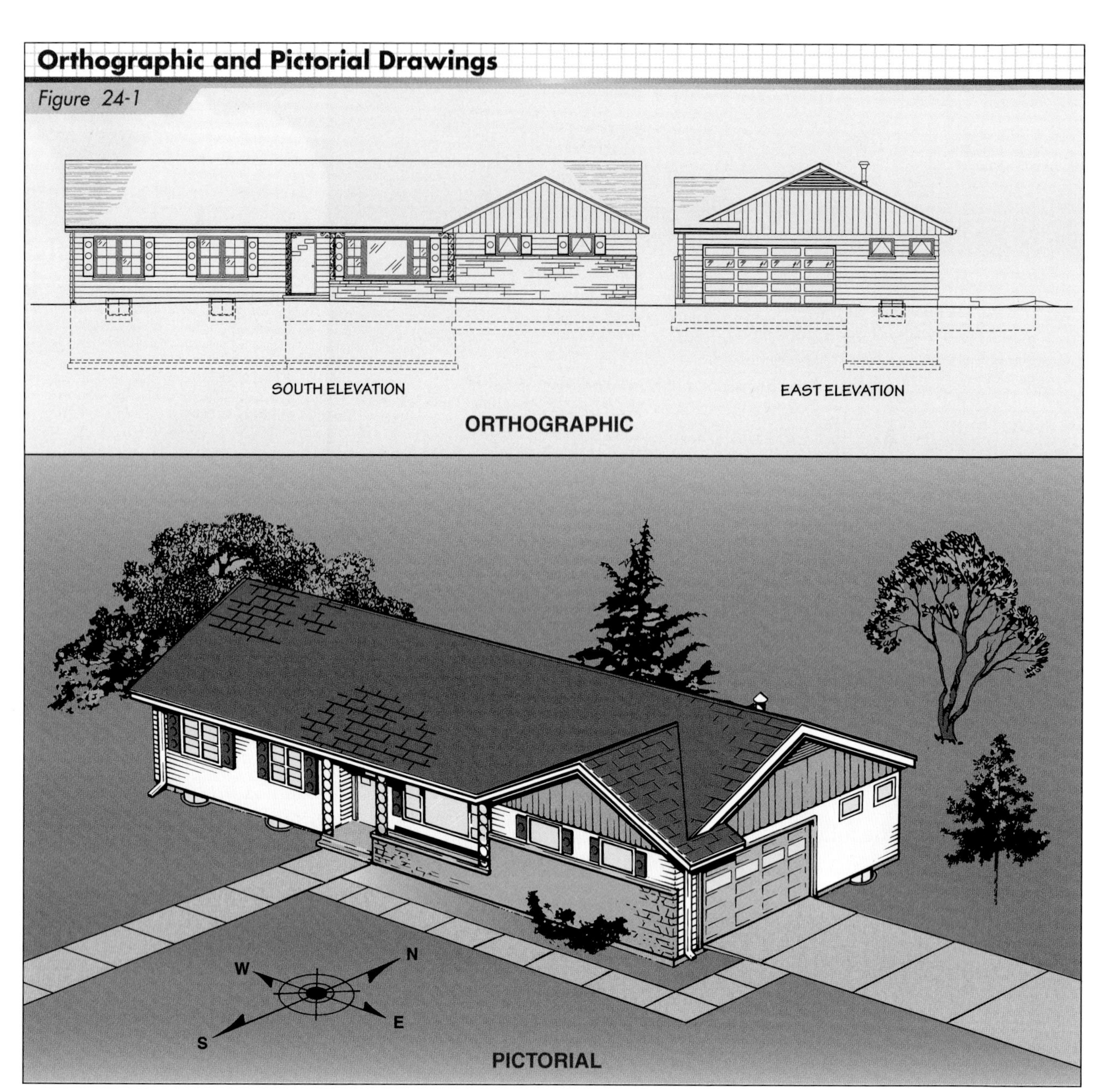

Figure 24-1. *Prints are typically composed of several orthographic drawings. Pictorial drawings provide a more realistic depiction of a building to an owner, but do not provide the details necessary to construct the building.*

Dimensions are measurements that give the distances between different points such as walls, columns, beams, and other structural parts. Dimensions also show the heights of different building components such as walls, window openings, and door openings. Figure 24-2 includes examples of how dimensions are used with dimension lines.

Obviously, a set of plans cannot have drawings that are as large as the actual size of a building. The drawings must be made to *scale.* Inches or fractions of an inch are used to represent feet of the actual measurement of the building. For example, in a plan drawn to 1/4″ scale, 1/4″ on the drawing usually represents 1′-0″ of the building. The scale for a drawing is usually identified directly below the drawing. The same scale is not usually used for all drawings in a set of plans. The following scales may be found in a set of plans:

- 1/16″ = 1′-0″
- 3/32″ = 1′-0″
- 1/8″ = 1′-0″
- 3/16″ = 1′-0″
- 1/4″ = 1′-0″
- 3/8″= 1′-0″
- 1/2″ = 1′-0″
- 3/4″ = 1′-0″
- 1″ = 1′-0″
- 1 1/2″ = 1′-0″
- 3″ = 1′-0″

Smaller scales are used for buildings with larger dimensions; larger scales are used for details.

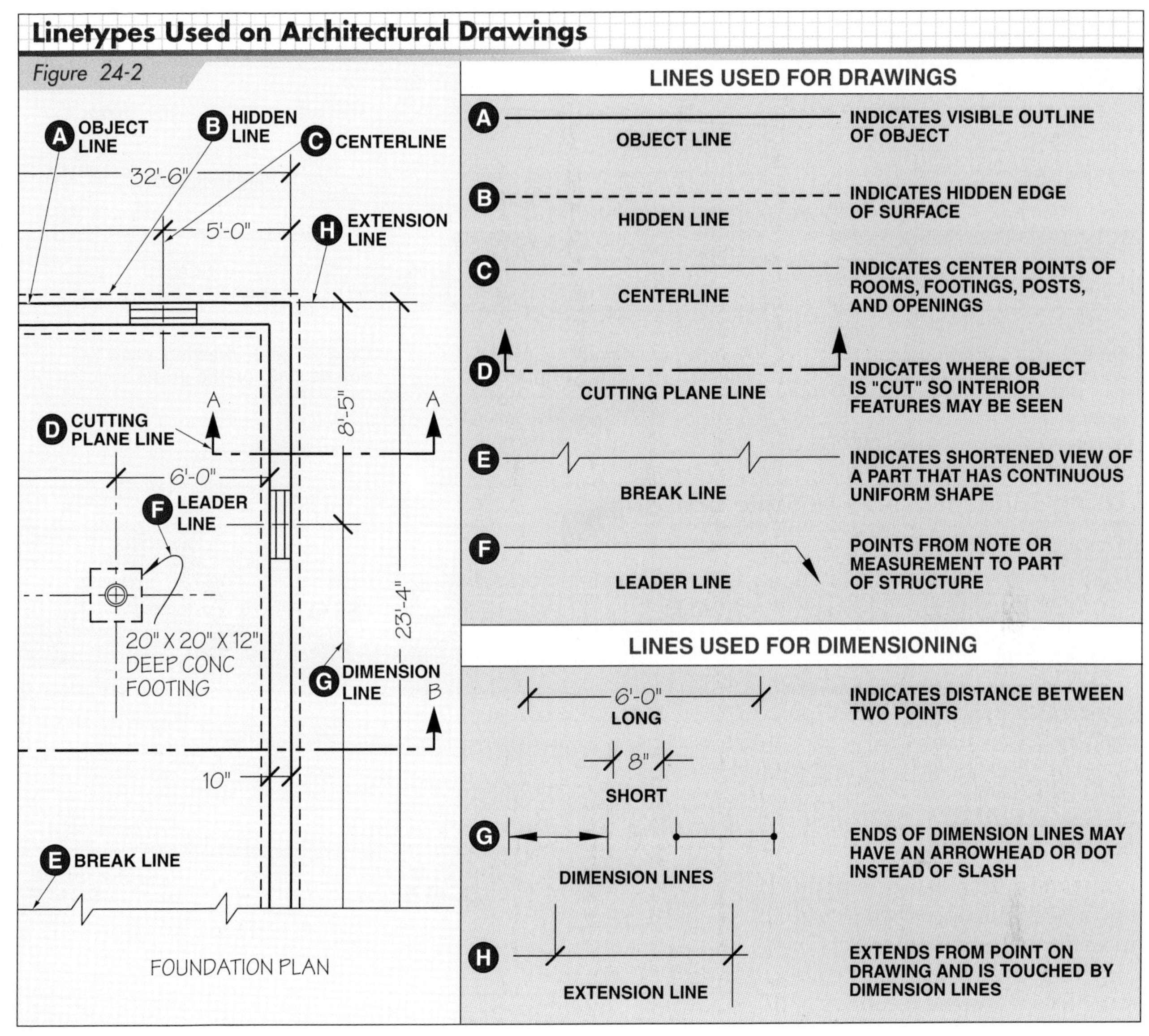

Figure 24-2. *Different linetypes have different meanings in a print.*

Architect's scales are based on 12 units to the foot while engineer's scales are based on 10 units to the foot. When using a scale to determine dimensions on a print, ensure the appropriate type of scale is used.

The scale used for a particular drawing should be large enough to present a clear drawing. However, the scale used is also determined by the size or area being presented in the drawing. A ¼″ scale is used most often, although 3⁄32″, ½″, and 3″ scales are also typical for a set of residential prints.

An *architect's scale* is used for scaling dimensions on prints. An architect's scale contains several different scales. Examples of how an architect's scale is used are shown in **Figure 24-3.** Although not as accurate or convenient, a tape measure can also be used to scale dimensions on prints. **See Figure 24-4.** However, all measurements scaled with a tape measure should be verified with the architect or engineer to ensure accuracy.

An *engineer's scale* is used to draw large areas such as building lots on plot plans. Scales on an engineer's scale are 1″ = 10′, 1″ = 20′, 1″ = 30′, 1″= 40′, 1″ = 50′, and 1″ = 60′. **See Figure 24-5.** Since engineer's scales are used for drawing large areas, the inch divisions of a foot on an engineer's scale are omitted.

On a metric engineer's scale, common scales are 1:100, 1:200, 1:250, 1:300, 1:400, and 1:500.

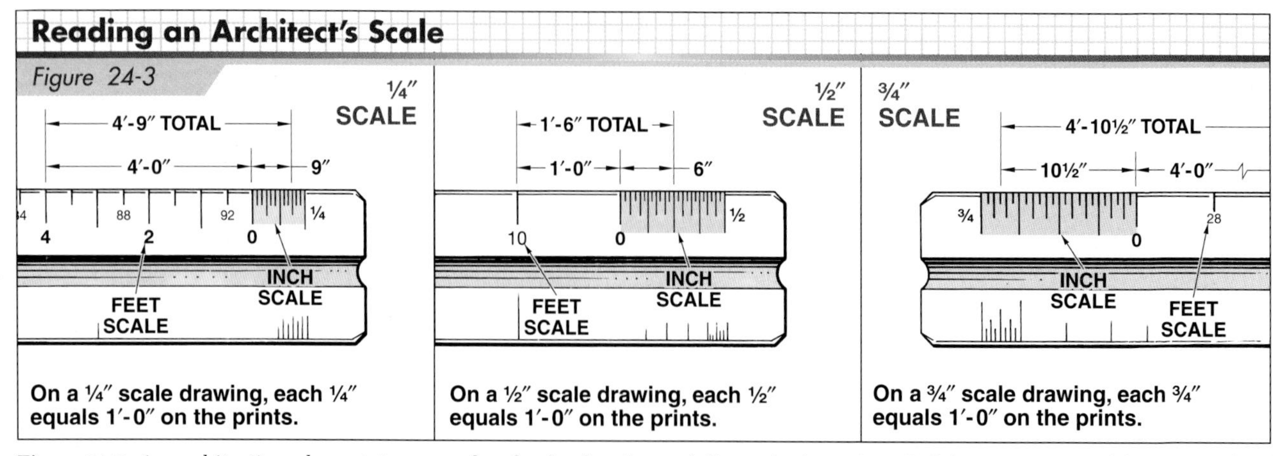

Figure 24-3. *An architect's scale contains several scales for drawing and dimensioning prints. Full feet are measured from one side of the zero point and inches and fractions of an inch are measured from the opposite side of the zero point.*

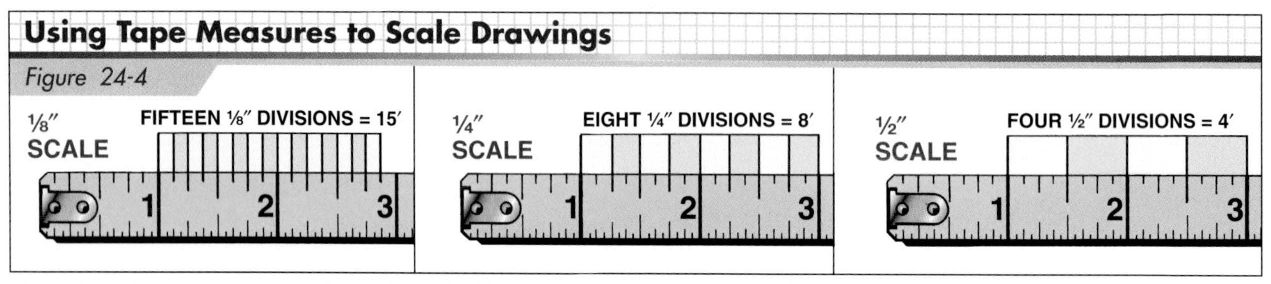

Figure 24-4. *A tape measure may be used to scale dimensions on prints. However, all measurements scaled with a tape measure should be verified with the architect or engineer to ensure accuracy.*

SYMBOLS AND ABBREVIATIONS

Prints must graphically explain the different materials being used in a building and provide details regarding structural parts. Prints must also show the locations of electrical switches, plugs, and outlets. Plumbing fixtures such as lavatories (sinks), bathtubs, shower enclosures, and water closets (toilets) are identified with symbols and abbreviations on prints.

Symbols on prints are a pictorial representation of a structural or material component. **Figures 24-6** through **24-10** show symbols commonly used on prints.

Little space is allowed for writing on the sheets that make up a set of plans. Therefore, abbreviations are used whenever possible. The table in **Figure 24-11** shows common abbreviations used in the construction trade. Abbreviations may vary from print to print.

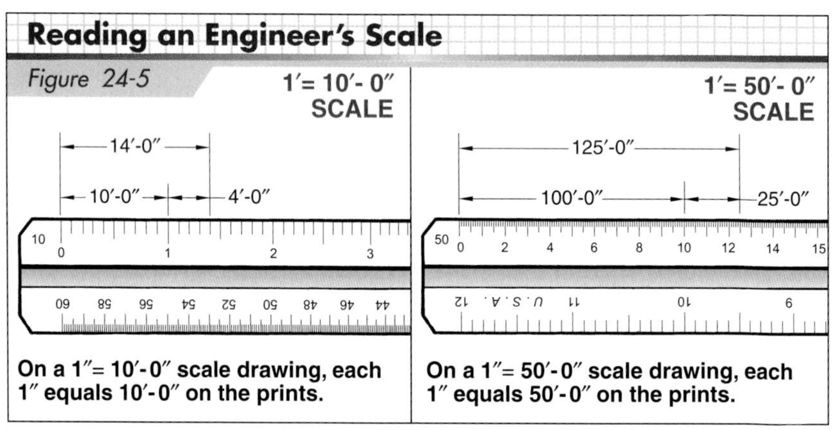

Figure 24-5. *An engineer's scale is typically used to draw plot plans.*

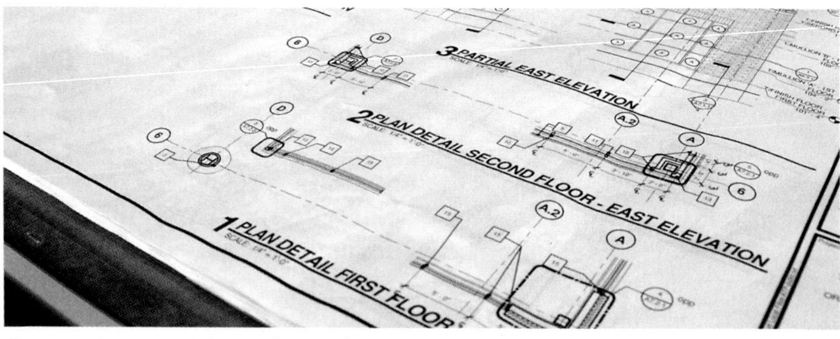

Construction materials are depicted on prints using symbols and abbreviations.

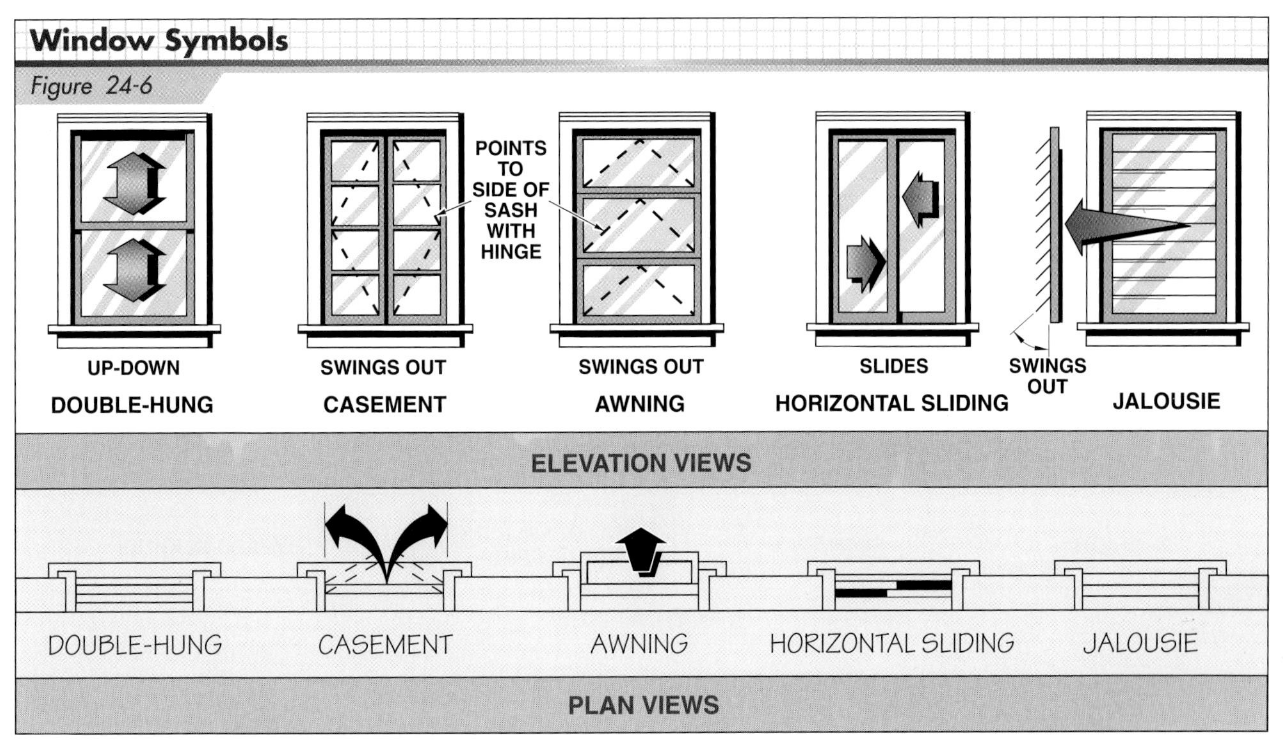

Figure 24-6. *Standard symbols are used to identify windows on a print. An elevation view of a window or other building component is created by looking directly at the front of the window and a plan view is created by looking down at the window from the top of a wall.*

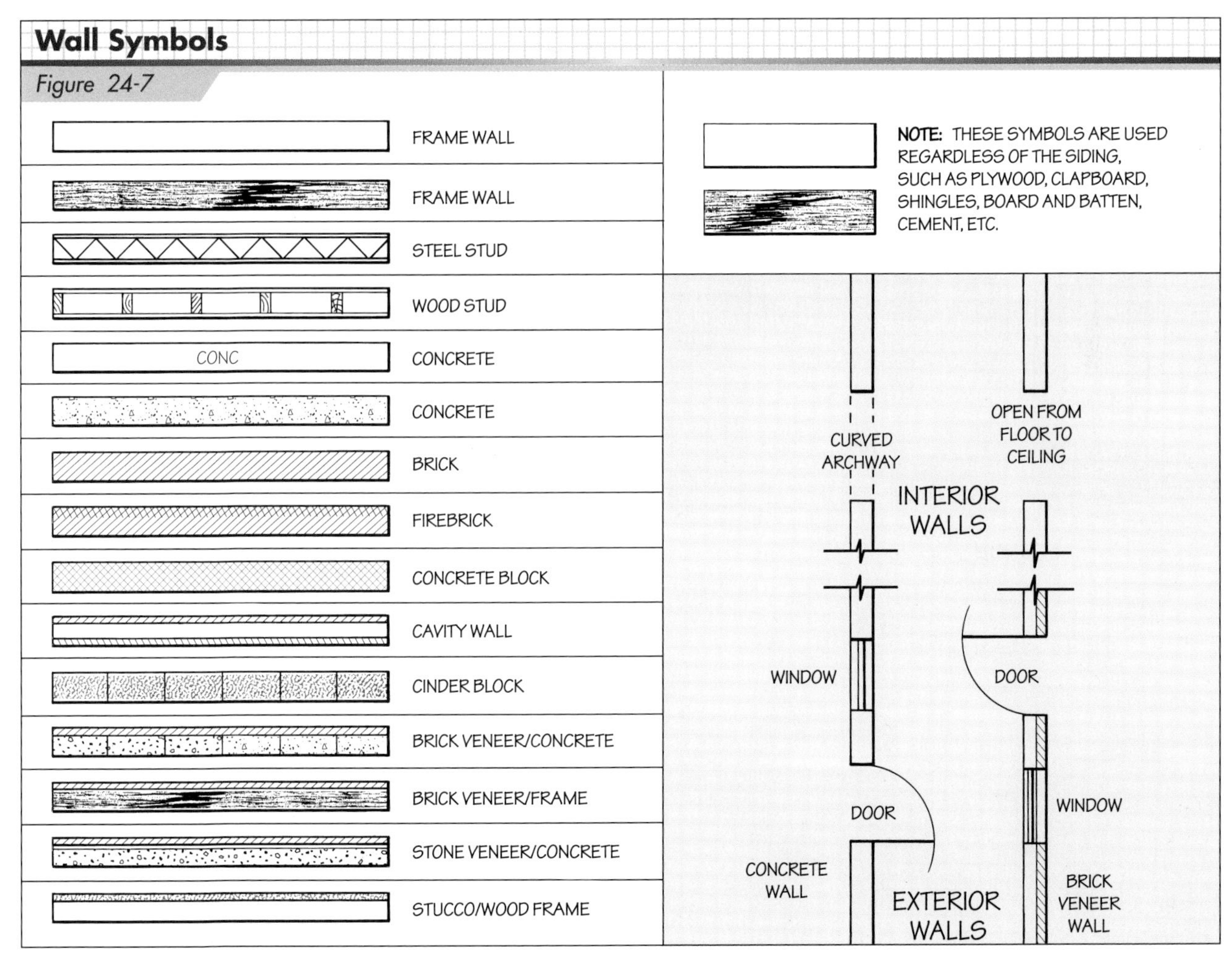

Figure 24-7. *Building materials are each assigned a distinctive symbol.*

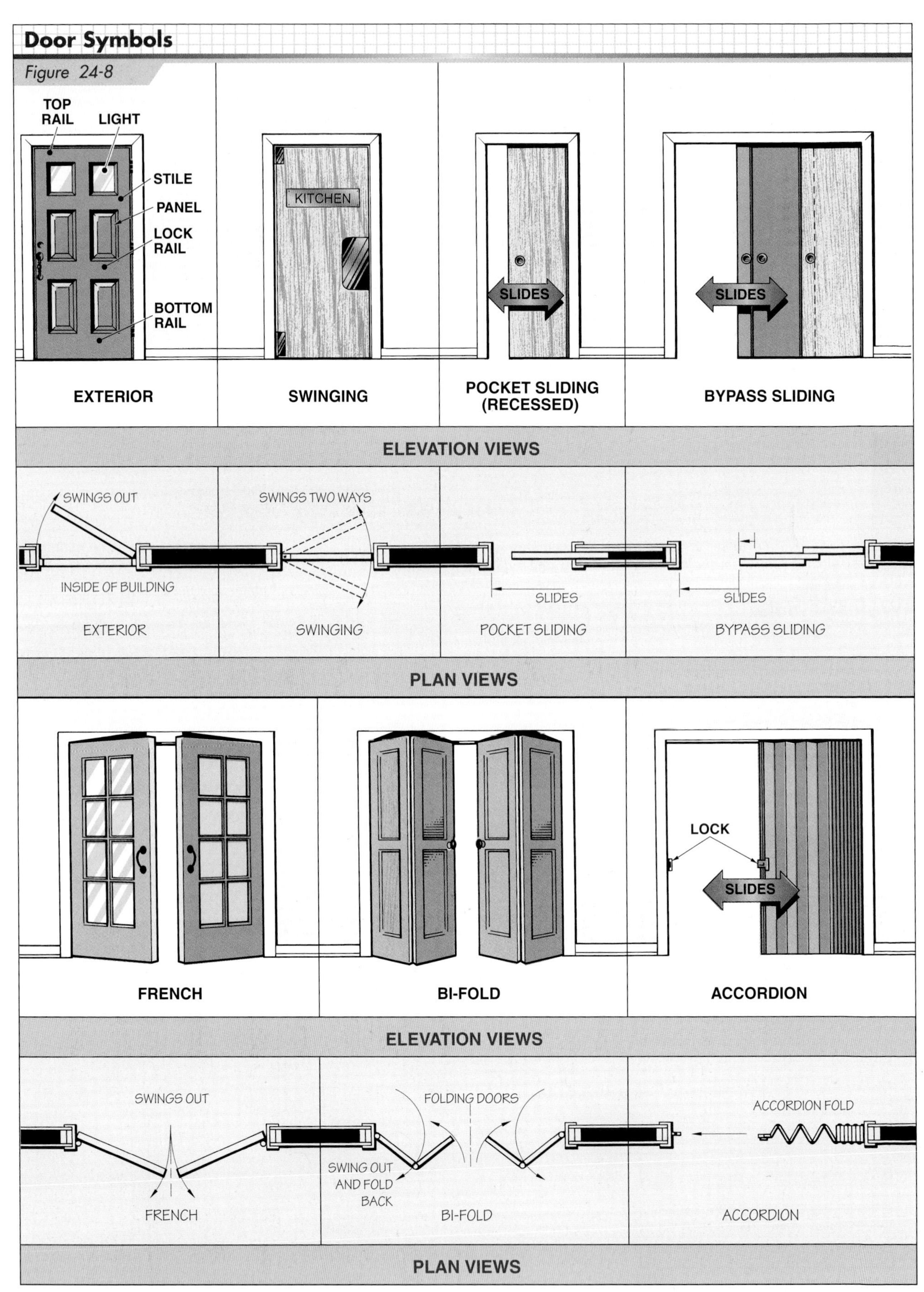

Figure 24-8. *Elevation views of doors are typically shown on elevation drawings, while plan views are shown on floor plans.*

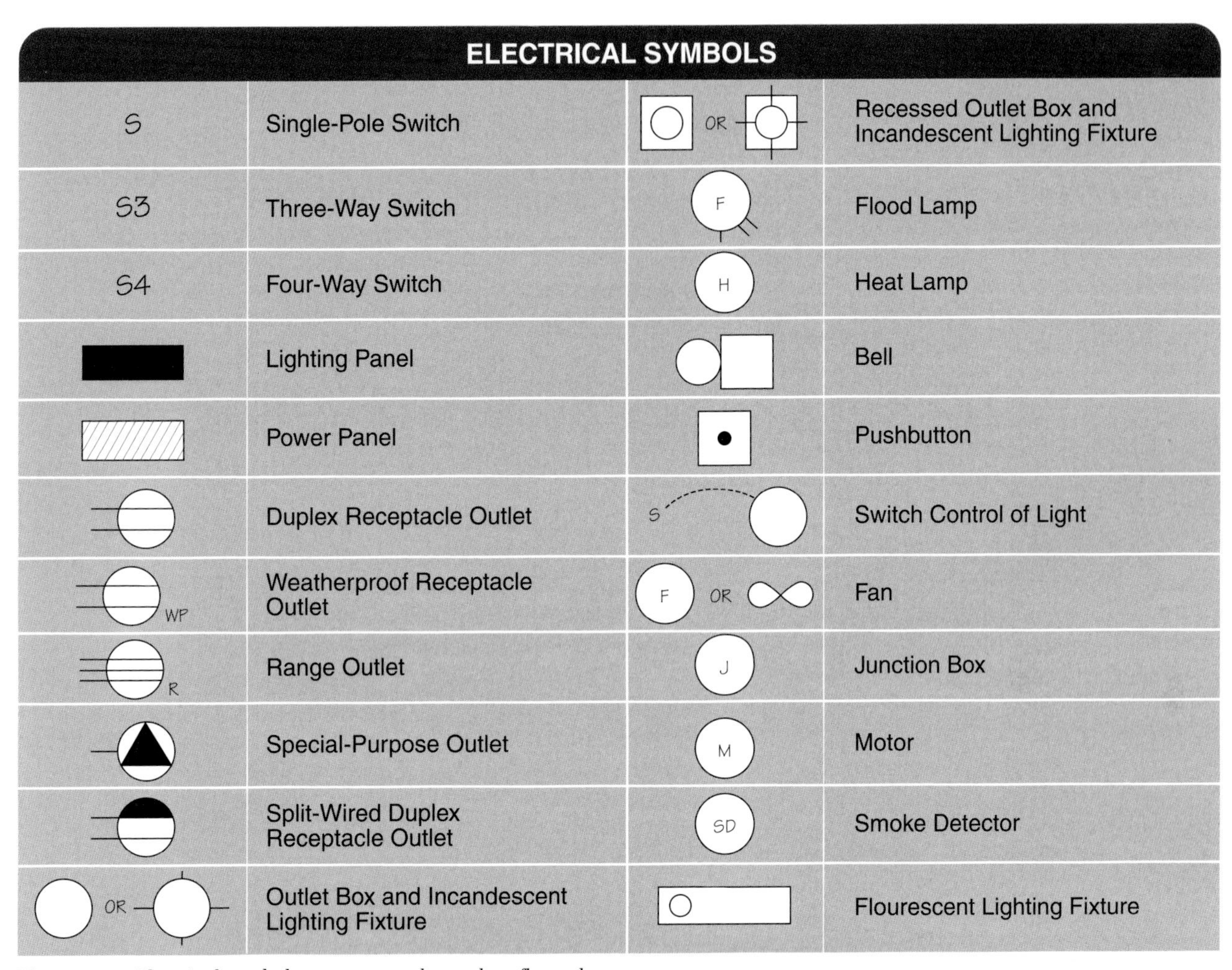

Figure 24-9. *Electrical symbols are commonly used on floor plans.*

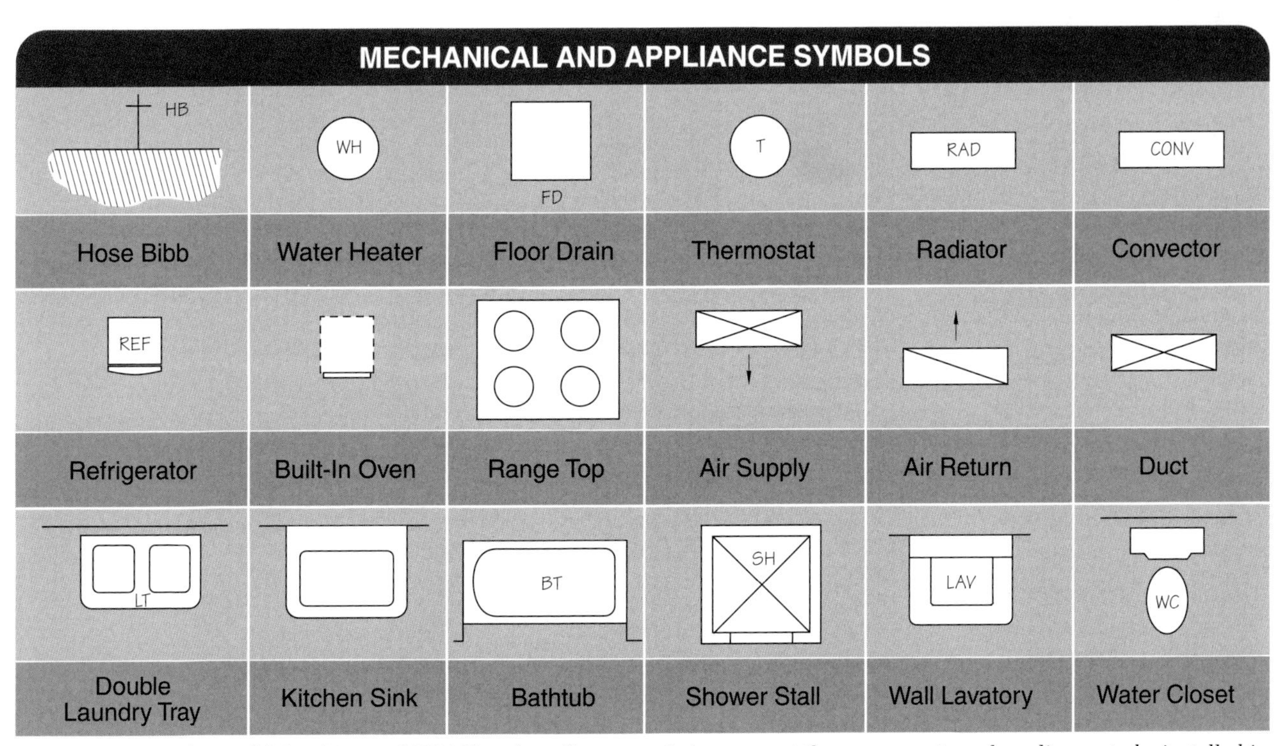

Figure 24-10. *Mechanical (plumbing and HVAC) and appliance symbols represent the components and appliances to be installed in a structure.*

ABBREVIATIONS

Term	Abbreviation	Term	Abbreviation	Term	Abbreviation
Aluminum	AL	Flashing	FL	Roof Drain	RD
Anchor Bolt	AB	Floor	FL	Roofing	RFG
Asphalt Tile	AT.	Footing	FTG	Room	RM
At Finished Face	AFF	Foundation	FDN	Rough Opening	RO
Basement	BSMT	Furnace	FURN	Screen	SCR
Bathroom	B	Gauge	GA	Sewer	SEW.
Bathtub	BT	Galvanized Iron	GI	Shake	SHK
Beam	BM	Girder	GDR	Sheathing	SHTHG
Bedroom	BR	Glass	GL or GLS	Sheet Metal	SM
Benchmark	BM	Grade	GR	Shingle	SHGL
Block	BLK	Ground	GND or GRD	Shower	SH
Board	BD	Gypsum Board	GYP BD	Siding	SDG
Brick	BRK	Hardboard	HBD	Sill	SL
Building	BL or BLDG	Hardwood	HDWD	Sink	S or SK
Building Line	BL	Head	HD	Skylight	SLT
Cabinet	CAB.	Heat	H or HT	Sliding Door	SL DR
Casement	CSMT	Hose Bibb	HB	Soffit	SF or SOF
Cedar	CDR	Insulation	INS or INSUL	Soil Pipe	SP
Ceiling	CLG	Interior	INT	Solar Panel	SLR PAN.
Cement	CEM	Jamb	JB or JMB	South	S
Center	CTR	Joist	J or JST	Stack Vent	SV
Centerline	CL	Kitchen	K or KIT.	Stairs	ST
Chimney	CHM	Laundry	LAU	Stairway	STWY
Closet	CLO	Lavatory	LAV	Steel	STL
Column	COL	Light	LT	Stone	STN
Concrete	CONC	Linen Closet	LC or LCL	Street	ST
Concrete Block	CONC BLK	Linoleum	LINO	Tongue-and-Groove	T&G
Cornice	COR	Living Room	LR	Top of Concrete	TOC
Corrugated	CORR	Louver	LV or LVR	Top Hinged	TH
Detail	DET or DTL	Medicine Cabinet	MC	Top of Slab	TOS
Diameter	D or DIA	Metal	MET. or MTL	Top of Steel	TOS
Dining Room	DR	Noncombustible	NCOMBL	Tread	TR
Dishwasher	DW	North	N	Typical	TYP
Door	DR	On Center	OC	Unexcavated	UNEXC
Dormer	DRM	Opening	OPNG	Utility Room	URM
Double-Hung Window	DHW	Overhang	OH.	Vent	V
Douglas Fir	DF	Overhead Door	OH. DR	Ventilation	VENT.
Downspout	DS	Panel	PNL	Vent Stack	VS
Drain	DR	Partition	PTN	Vinyl Tile	VA TILE
Drywall	DW	Plate	PL	Washing Machine	WM
East	E	Plywood	PLYWD	Water	W
Electric	ELEC	Porch	P	Waterproof	WP
Elevation	EL	Pressure-Treated	PT	Water Closet	WC
Excavate	EXC	Rafter	RFTR	Water Heater	WH
Exterior	EXT	Redwood	RWD	Welded Wire Reinforcement	WWR
Face Brick	FB	Refrigerator	REF	West	W
Fill	F	Reinforced	REINF	White Pine	WP
Finish	FNSH	Reinforcement Bar	REBAR	Wide Flange	WF
Finish Floor	FNSH FL	Retaining Wall	RW	Window	WDW
Fireplace	FP	Ridge	RDG	Wood	WD
Fireproof	FPRF	Riser	R	Wood Blocking	WBL
Fixture	FXTR	Roof	RF	Yellow Pine	YP

Figure 24-11. *Abbreviations are used on prints to avoid cluttering drawings.*

Unit 24 Review and Resources

UNIT 25

Plot Plans

OBJECTIVES

1. List and describe the features of a plot plan.
2. Identify other information shown on a plot plan.
3. Explain information about finish grades and elevations as they are shown on a plot plan.
4. Explain the significance of a benchmark.
5. Discuss the differences between existing (natural) grades, finish grades, and elevations.
6. Explain the procedure for converting decimal foot measurements to inches and fractions.

A *plot plan*, or *site plan*, is one of the first drawings to be considered in a construction project. Before construction can begin, the exact location of the building on the property must be known. High and low points of the property must be determined so the lot can be graded to provide proper water drainage away from the building. The plot plan provides location and elevation information, and other information typically used by an operating engineer who clears the property.

BASIC PLOT PLAN INFORMATION

The plot plan for the three-bedroom house plan is shown in **Figure 25-1.** When reviewing the plot plan, compare the features with the pictorial drawing shown in **Figure 25-2.** Various items of note shown on the plot plan are as follows:

A. *Terrace:* The outline of a terrace is indicated. Measurements for the terrace are provided on other drawings.
B. *Electrical utilities:* A power pole is located along the east side of the lot. Electric and telephone lines extend from the power pole to the house. In some communities, electrical utilities are placed underground.
C. *Side yard:* The distance (18′-0″) from the side property line to the east wall of the house is the side yard.
D. *Driveway:* A 15′-0″ wide concrete driveway extends from the two-car garage to the street.
E. *Roads:* Roads are south and east of the property. Rose Street borders the property on the east and Virginia Avenue borders the property on the south.
F. *Benchmark:* A point designated as 100.0′ has been established at the street curb as a benchmark. Benchmarks and how they relate to finished grades and elevations are discussed later in this unit.
G. *Finish grades:* A finish grade height of 100.2′ is noted at the southeast corner of the property. Finish grade references are also noted at other points on the property, including the other lot corners and building corners.
H. *Sidewalks:* A 6′-0″ wide sidewalk borders the property on the south and east side.
I. *Finish floor elevations:* A finish floor elevation of 105.0′ is noted within the building outline. This is the height of the finish floor in relation to the benchmark.
J. *Walk:* A walk extends from the sidewalk to the front porch and branches off toward the driveway.
K. *Planter strip:* A 4′-0″ wide planter strip, or parkway, is located between the street curb and the sidewalk.
L. *Front setback:* A 20′-0″ front setback is shown between the southwest property line and the southwest wall of the building.
M. *Plumbing utilities:* Public utilities such as gas, water, storm drainage, and sewer drainage are below the street surface.
N. *Building lines:* The outline of the building to be constructed on the lot is shown by building lines. Dimension lines show the width and depth of the building.
O. *Trees:* Before construction can begin, some trees on the property may need to

be removed. Trees that will remain on the lot are shown on the plot plan.

P. *Property lines:* The property lines, or lot lines, show the shape of the lot. The lot in the example is rectangular. The width and length of the property are also shown.

Q. *Compass direction:* A plot plan typically includes an arrow pointing north, which is used to properly position the building on the lot. The different sides of the building are referred to by compass directions (north, south, east, west) in other drawings of the prints.

R. *Swale:* A swale is a gradual grade or depression designed for channeling surface water away from a building.

Additional Plot Plan Information

In addition to the information shown on the example plot plan, *retaining walls* and *easements* may also be shown. Retaining walls are structures made of concrete or concrete blocks and keep earth from sliding. Retaining walls are commonly constructed on sloping lots or where excavations are made along points of access to a structure, such as along a driveway. Easements are right-of-way provisions on the property. Easements may indicate, for example, the right of a neighbor to build an access road or a public utility to install water and gas lines on the property. A property owner cannot build on an area where an easement has been provided.

FINISH GRADES AND ELEVATIONS

A plot plan provides information about the shape of a lot. The lot may be flat, or it may have a steep or gradual slope. If a lot is sloped, it will be higher at some points and lower at others. A sloped lot may have to be *graded* by removing or adding soil so surface water caused by rain or melting snow will be directed away from the building and into the street.

The finish grades and elevations shown on a plot plan are based on data provided by a professional surveyor or engineer. Finish grades and elevations are recorded on a plot plan in feet and tenths of a foot rather than in feet and inches.

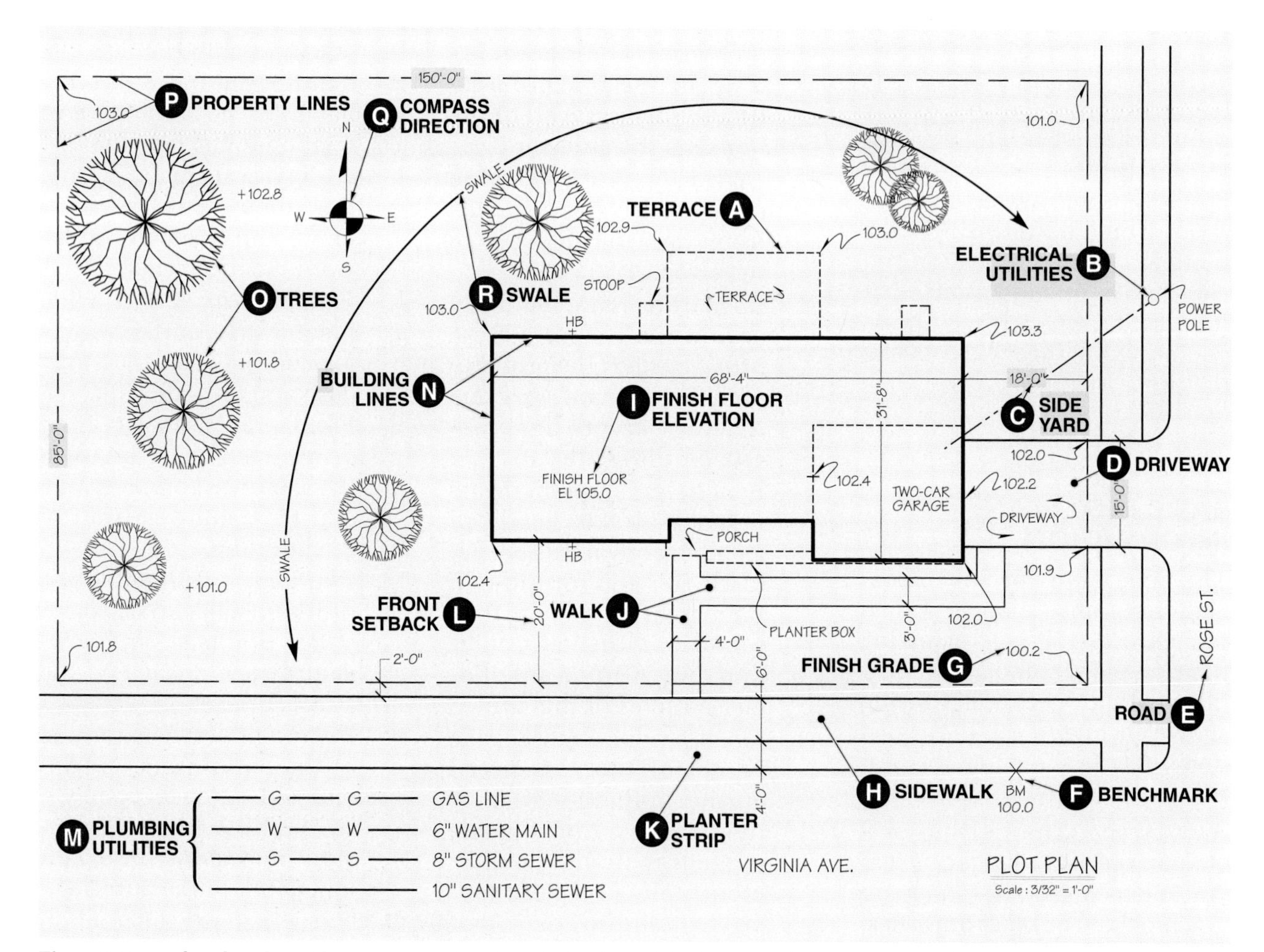

Figure 25-1. *A plot plan shows the shape and size of a building lot and the location, shape, and size of the building on the lot. Compare this plan with the pictorial drawings in Figures 25-2 and 25-3.*

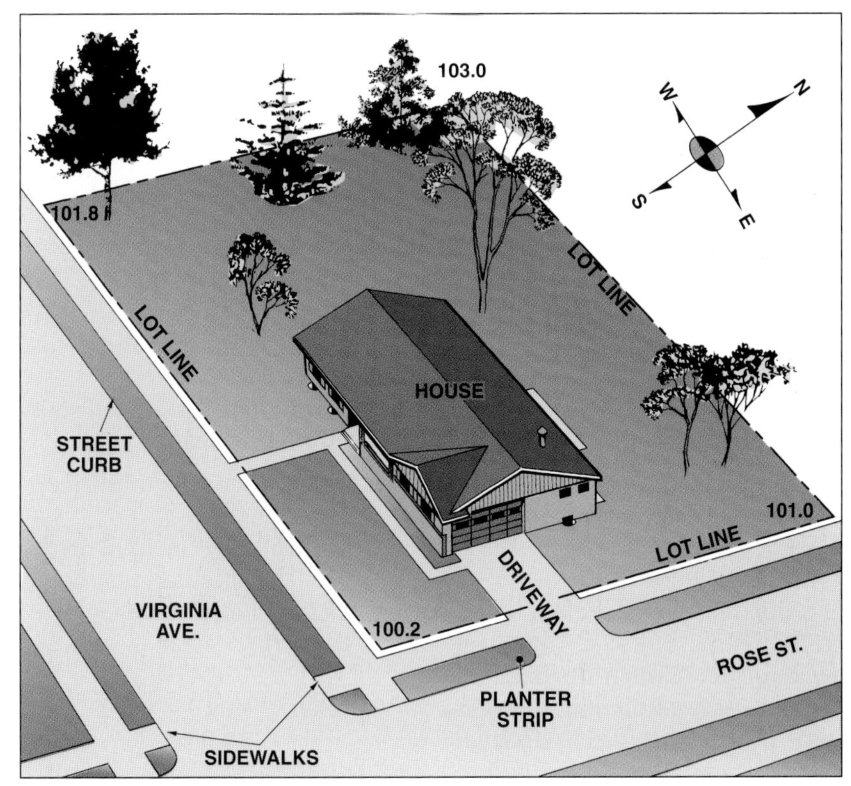

Figure 25-2. *This pictorial drawing is based on the plot plan in Figure 25-1.*

Benchmarks

A *benchmark*, or job datum, is a point established by a surveyor on or close to the property and is often placed at one corner of the lot. A benchmark may also be placed at a point along the street curb next to the property. Benchmarks may be identified by a plugged pipe driven into the ground, brass marker, or wood stake. Benchmarks are also identified by scratching a mark into existing concrete.

The location of a benchmark is shown on the plot plan with a grade figure next to it. The grade figure may be the number of feet above sea level at that point, or it may be the number 100.0′. In the example plot plan, a 100.0′ benchmark is shown at the street curb near the southeast corner of the lot.

Finish Grades

The plot plan in Figure 25-1 shows finish grades at all corners of the lot and at various other points on the lot, including the building corners, in the garage, and at the driveway.

All finish grades are based upon their relation to the 100.0′ benchmark. For example, the grade at the southeast corner of the lot is 100.2′, meaning that the ground is .2′ (2⁄10′) higher at this point than at the benchmark (100.2′ – 100.0′ = .2′). The finish grade at the northwest corner of the lot is 103.0′, meaning that the ground is 3′ higher at this point than at the benchmark.

The lot corner grades show that the lot slopes down from the northwest to the southwest corner. Notice that the finish grades closest to the building are higher than those farther out on the lot. The surface of the lot is graded so surface water will drain away from the house. **See Figure 25-3.**

As shown in Figure 25-1, finish grades at the front and back of the garage indicate that the concrete slab in the garage will slope .2′ toward the door opening (102.4′ – 102.2′ = .2′). The slope allows rainwater blown into the garage or melting snow on the vehicles in the garage to drain out. In the driveway, finish grade points show a slope away from the garage and from north to south.

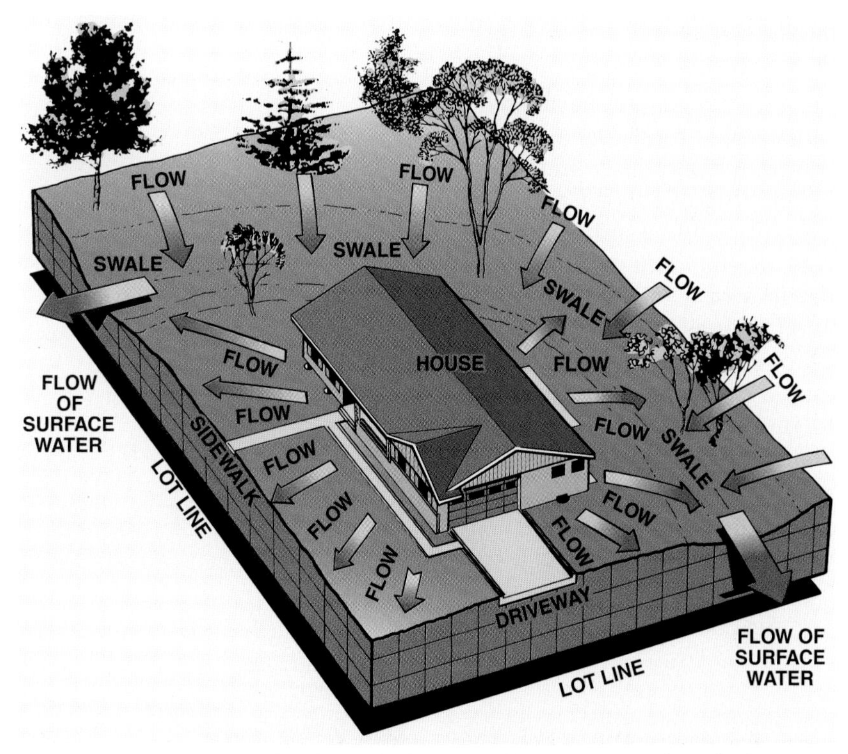

Figure 25-3. *A lot is graded to direct the flow of surface water away from the house.*

Natural Grades and Contours

Some plot plans include the existing (natural) grades as well as the finished grades. The existing grade refers to the condition of the lot before grading. *Contour lines* are sometimes drawn on the plot plan to show the existing and finished surface shapes of the lot.

Elevations

The term "elevation" is often used interchangeably with the term "grade." More precisely, however, *elevations* are the heights established for different levels of the building. A plot plan usually shows the finish floor elevation of a building, which is the level of the first floor of the building in relation to the benchmark. The plot plan in Figure 25-1 shows the finish floor elevation as 105.0′ (5′ higher than the benchmark grade). During the construction of the building, many measurements are taken from the finish floor elevation.

Converting Decimal Foot Measurements to Inches and Fractions

The grades on a plot plan are usually indicated in feet and tenths of a foot. Occasionally, the grades are shown in feet and hundredths of a foot. Conversion charts provide a quick and easy way to change tenths and hundredths of a foot to the inches and fractions typically shown on a tape measure. The following example shows how to convert .86′ to inches using the conversion chart shown in **Figure 25-4:**

1. Locate .86 in the conversion chart.
2. Record the number of inches (10″) shown at the top of the column.
3. Record the number "3" shown at the far left of the horizontal row. The number "3" represents the number of eighths of an inch. Therefore, the recorded number "3" represents 3⁄8″.
4. Combine the results of steps 2 and 3: .86′ = **10 3⁄8″**.

A carpenter should be able to make decimal foot conversions mathematically in case a conversion chart is not available. Decimal foot values are converted to their inch equivalents by multiplying the decimal foot value by 12. For example, the inch equivalent of .75′ is 9″ (.75 × 12 = 9). If a decimal value remains after multiplying the decimal foot value by 12, convert the decimal value to its fractional inch equivalent by multiplying the decimal by a common fraction denominator (lower number) such as 16, 8, or 4. For example, convert .7′ to inches using the following procedure:

1. Multiply the decimal .7′ by 12. The answer results in inches and decimal part of an inch.

$$\begin{array}{r} .7' \\ \times\ \underline{12} \\ 8.4'' \end{array}$$

2. Multiply the decimal (.4) by a common fraction denominator such as 16, 8, or 4. In this example, the 16 denominator is used.

$$\begin{array}{r} .4 \\ \times\ \underline{16} \\ 6.4 \end{array}$$

3. Round the answer to the nearest sixteenth. In this example, 6.4 is rounded to 6 and 6⁄16″ = 3⁄8″.
4. Combine the results of steps 1 and 3: .7′ = **8 3⁄8″**.

In the following example, convert .84′ to inches.

1. Multiply the decimal .84 by 12. The answer results in inches and decimal part of an inch.

$$\begin{array}{r} .84' \\ \times\ \underline{12} \\ 10.08'' \end{array}$$

2. Multiply the decimal (.08) by a common fraction denominator; in this example, 16.

$$\begin{array}{r} .08'' \\ \times\ \underline{16} \\ 1.28 \end{array}$$

3. Round the answer to the nearest sixteenth. In this example, 1.28 is rounded to 1, or 1⁄16.
4. Combine the results of steps 1 and 3: .84′ = **10 1⁄16″**.

DECIMAL FOOT TO INCH AND FRACTIONAL INCH CONVERSION

8TH	0″	1″	2″	3″	4″	5″	6″	7″	8″	9″	10″	11″
0	.00	.08	.17	.25	.33	.42	.50	.58	.67	.75	.83	.92
1	.01	.09	.18	.26	.34	.43	.51	.59	.68	.76	.84	.93
2	.02	.10	.19	.27	.35	.44	.52	.60	.69	.77	.85	.94
3	.03	.11	.20	.28	.36	.45	.53	.61	.70	.78	.86	.95
4	.04	.13	.21	.29	.38	.46	.54	.63	.71	.79	.88	.96
5	.05	.14	.22	.30	.39	.47	.55	.64	.72	.80	.89	.97
6	.06	.15	.23	.31	.40	.48	.56	.65	.73	.81	.90	.98
7	.07	.16	.24	.32	.41	.49	.57	.66	.74	.82	.91	.99

Figure 25-4. *A conversion chart can be used to convert decimal foot measurements to their inch and fractional inch equivalents.*

Unit 25 Review and Resources

UNIT 26

Foundation Plans

OBJECTIVES

1. Explain the purpose of foundation plans.
2. Describe types of foundations.

The first stage in the construction of a building involves a foundation. The foundation must be designed to support its own weight as well as the rest of the structure. Footings, foundation walls, and piers are the basic features of a foundation. A *foundation plan* provides information regarding the foundation and also includes information about posts and beams that help support the floor unit above the foundation. If joists are used to support a floor unit, the size and spacing of the joists are noted in the foundation plan.

One-story buildings, such as the example three-bedroom house, can be constructed over a full-basement, crawl-space, or concrete slab foundation. Foundation plans for the full-basement foundation and the crawl-space foundation are discussed in this unit.

FULL-BASEMENT FOUNDATION

The foundation plan for the full-basement foundation of the three-bedroom house is shown in **Figure 26-1.** The foundation plan shows the foundation walls resting on footings. The footings extend from each side of the walls, making a T-shape. A *T-foundation* is commonly used in a full-basement foundation to distribute the load of the house to be constructed above. Carpenters construct the forms according to the foundation plan and concrete is placed in the completed forms.

The following is an explanation of the items identified on the foundation plan in Figure 26-1. When reviewing the foundation plan, compare the features with the pictorial drawing shown in **Figure 26-2.**

A. *Foundation footings:* Hidden (dashed) lines on each side of the foundation walls indicate the foundation footings. Hidden lines are used to represent footings because the outside edge of the footings is covered by soil and the inside edge is covered by the basement concrete slab.
B. *Rear stoop:* Rear stoops are located at the two rear entrances to the house. For each stoop, a low concrete foundation wall around the outside edges supports a concrete slab. The stoops are filled with soil and tamped (pressed down) before the slab is placed.
C. *Garage area:* Note the "unexcavated, fill and tamp" instructions. Soil will not be removed in the garage area. If necessary, soil will be added and tamped before concrete for a slab is placed (poured).
D. *Front porch:* For the front porch, a concrete slab is supported at the front edge by a low foundation wall without a footing. A footing is not required in this area since heavy loads are not anticipated for the front porch.
E. *Column footing:* Column footings measuring 24″ × 24″ × 12″ are placed below the pipe columns.
F. *Pipe columns:* Hollow steel columns, 4″ in diameter, rest on the column footings. Dimensions from the foundation walls to the centers of the columns, and from the center of one column to the next one, are provided.
G. *Foundation wall:* Foundation walls are shown with object (solid) lines on a foundation plan. Foundation walls extend around the outside of the house and garage.

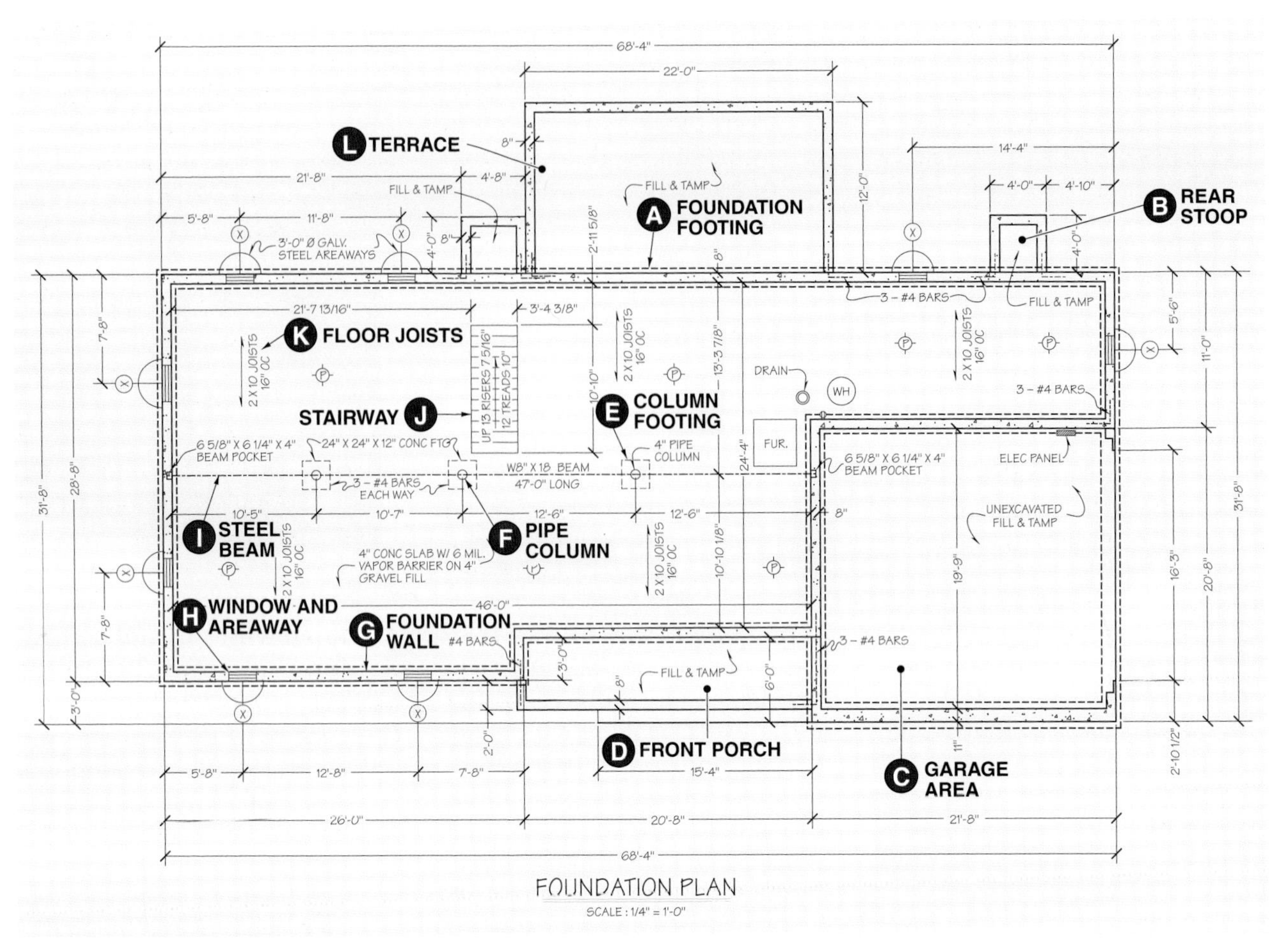

Figure 26-1. *The example three-bedroom house can be built over a full-basement foundation. Compare this foundation plan with the pictorial drawing in Figure 26-2.*

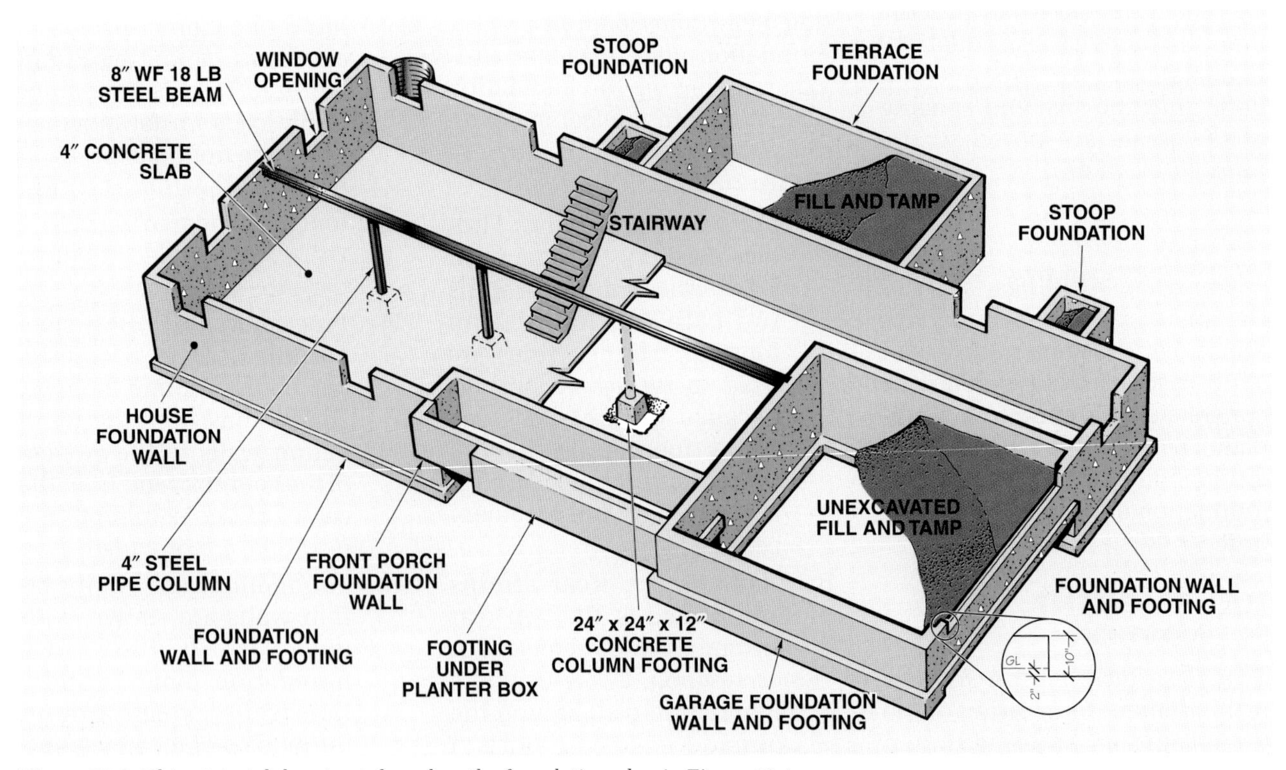

Figure 26-2. *This pictorial drawing is based on the foundation plan in Figure 26-1.*

H. *Window and areaway:* Basement windows extend below grade level. The open space around each basement window is an *areaway* (window well), and is used for light and ventilation. Each areaway is surrounded by 3′-0″ diameter galvanized steel.

I. *Steel beam:* An 8″ wide-flange steel beam that weighs 18 lb/ft is used to support the loads across the center of the house.

J. *Stairway:* The stairway leading from the basement to the main floor has 13 risers and 12 treads.

K. *Floor joists:* The direction in which the floor joists run is shown by an arrow. One end of the joists rests on the outside foundation walls and the other end rests on the steel beam. The joists are spaced 16″ OC (on center).

L. *Terrace:* A low concrete foundation wall, measuring 8″ thick, extends around the outside edges of the terrace. The terrace should be filled and tamped.

CRAWL-SPACE FOUNDATION

Although the example three-bedroom house plan specifies a full-basement foundation, the same building could also be constructed over a crawl-space foundation. A crawl-space foundation does not have a basement area. *Crawl space* refers to the distance (typically 18″ or more) between the bottoms of the floor joists and the ground. The main difference between the full-basement and crawl-space foundation is the height of the walls. In addition, stairways and areaways are not required in a crawl-space foundation.

An example of a crawl-space foundation plan for the three-bedroom house is shown in **Figure 26-3.** When reviewing the crawl-space foundation plan, compare the features with the pictorial drawing shown in **Figure 26-4.** Wood posts (6″ × 6″) are used to support a 6″ × 10″ glulam beam. The wood posts rest on 24″ × 24″ × 12″ concrete footings. Dashed lines on the crawl-space foundation plan indicate that the joists are doubled where the walls above run in the same direction as the joists.

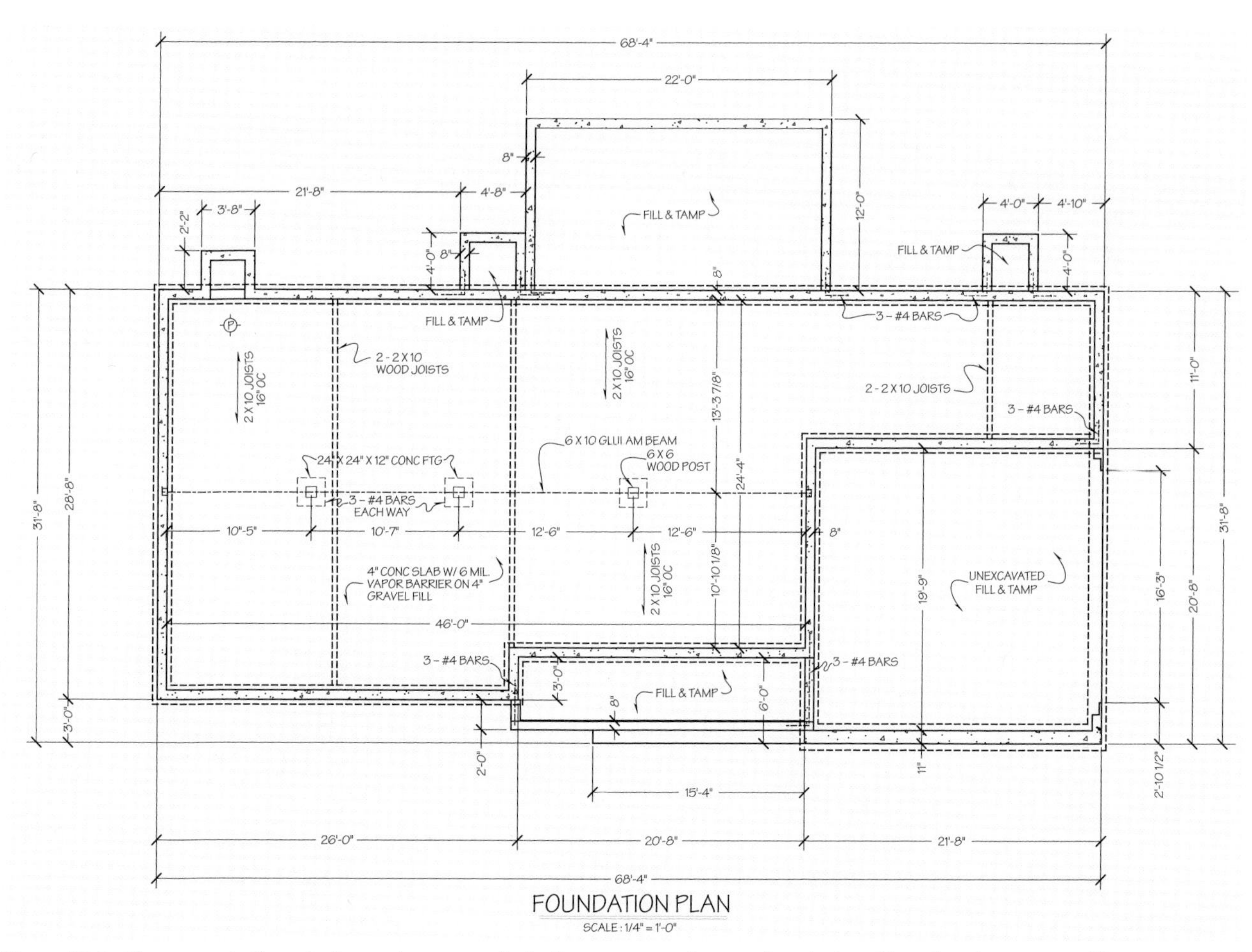

Figure 26-3. *The example three-bedroom house can be built over a crawl-space foundation. Note that the posts and beam are wood products. Double joists, indicated with dashed lines, are used where the walls above run in the same direction as the joists. Compare this foundation plan with the pictorial drawing in Figure 26-4.*

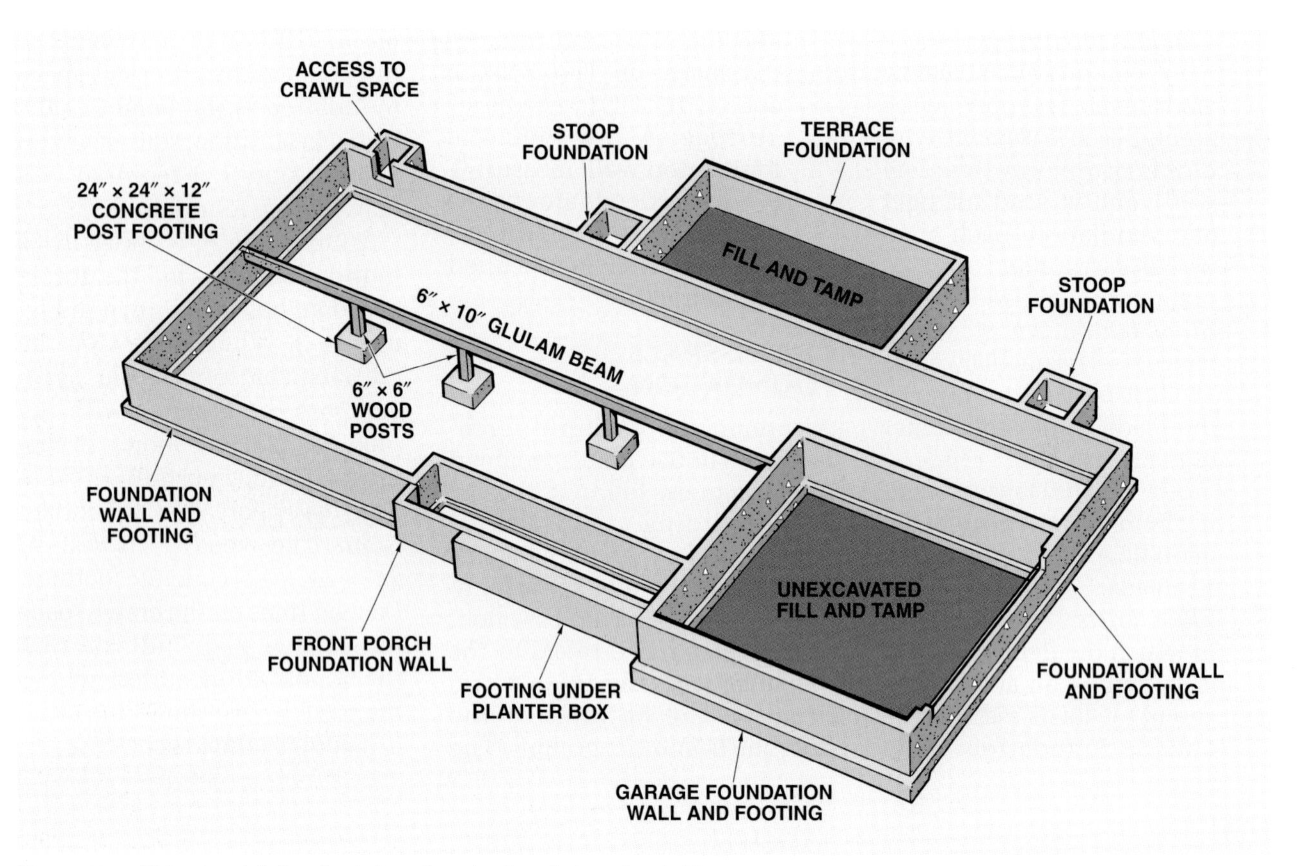

Figure 26-4. *This pictorial drawing is based on the foundation plan in Figure 26-3.*

UNIT 27

Floor Plans

OBJECTIVES

1. Identify information provided on a floor plan.

Floor and wall construction begins after the foundation has been completed. A *floor plan* provides most of the information for floor and wall construction and gives a plan view (view looking down) of the floor level above the foundation. A one-story house requires only one floor plan. Most multistory buildings require a separate floor plan for each level. However, multistory buildings where the same room layout is desired on each level, such as office complexes and hotels, often provide one floor plan for each level with identical layouts.

Floor plans indicate the positions of exterior walls and partitions (interior walls). The shape and arrangement of the rooms can be envisioned by studying the floor plan. The floor plan also shows all the door and window openings. Electrical items such as receptacles, light outlets, and switches are noted. Locations of plumbing fixtures and appliances such as lavatories, water closets, bathtubs, shower stalls, and stoves are also provided.

Many floor plans also provide information about the heating and cooling systems used in the building. Locations of baseboard heating units or wall or floor openings for supply and return registers that connect with a central heating system are shown.

Figure 27-1 is the floor plan for the example three-bedroom house. When reviewing the floor plan, compare the features with the pictorial drawing shown in **Figure 27-2.** The following is an explanation of the items identified on the floor plan in Figure 27-1:

A. *Lighting outlet:* The symbol indicates an overhead (ceiling) light. The overhead light symbol appears in most rooms of the house (except the living room). The symbol for a fluorescent ceiling light is shown in the kitchen.

B. *Pocket door:* When opened, a pocket door slides into a pocket inside the wall.

C. *Sliding glass doors:* Two sliding glass doors open from the family room to the terrace.

D. *Terrace:* A 3″ thick concrete slab is covered by 2″ thick flagstone to form the terrace.

E. *Kitchen cabinets:* Cabinets are shown along the north and south walls of the kitchen. Cabinet details are provided in other drawings.

F. *Garage area:* A 4″ thick concrete slab is identified. The slab, which measures 20′-8″ long, slopes 2″ toward the door so rainwater and melting snow do not accumulate in the garage.

Southern Forest Products Association

A multistory structure requires a floor plan for each level.

G. *Attic access:* Also known as a *scuttle,* the attic access is a ceiling opening covered by a removable panel.
H. *Planter box:* The width and length of the planter box are 1′-8″ × 15′-4″. Details regarding the structure and material used are provided in section drawings in the plans.
I. *Wall opening:* Hidden lines indicate a wall opening without a door. The opening is between the living room and the family room.
J. *Ceiling joists:* The arrow indicates the direction in which the ceiling joists run. The size and spacing of the joists are 2 × 6–16″ OC.
K. *Front entrance door:* The front entrance door swings into the living room. The circled "A" on the print refers to a door schedule, which provides the size of the door and other information. There is a step from the front porch to the living room level.
L. *Front porch:* The porch area consists of a concrete slab 4″ thick. A step from the walk to the porch is indicated.
M. *Hose bibb:* Hose bibbs, or sillcocks, are threaded water faucets to which hoses are attached. Hose bibbs must be fitted with a vacuum breaker to protect against back-siphonage and prevent contaminated water from entering the water supply system.
N. *Exterior wall:* Lengths of exterior walls are shown by dimension lines. Note that the exterior walls are not aligned along the south side (front) of the house.
O. *Electrical receptacle:* The symbol indicates a duplex (double) receptacle outlet. Duplex receptacle outlet symbols are shown in all rooms of the house. In the kitchen and bathrooms, a ground fault circuit interrupter (GFCI) outlet is indicated.

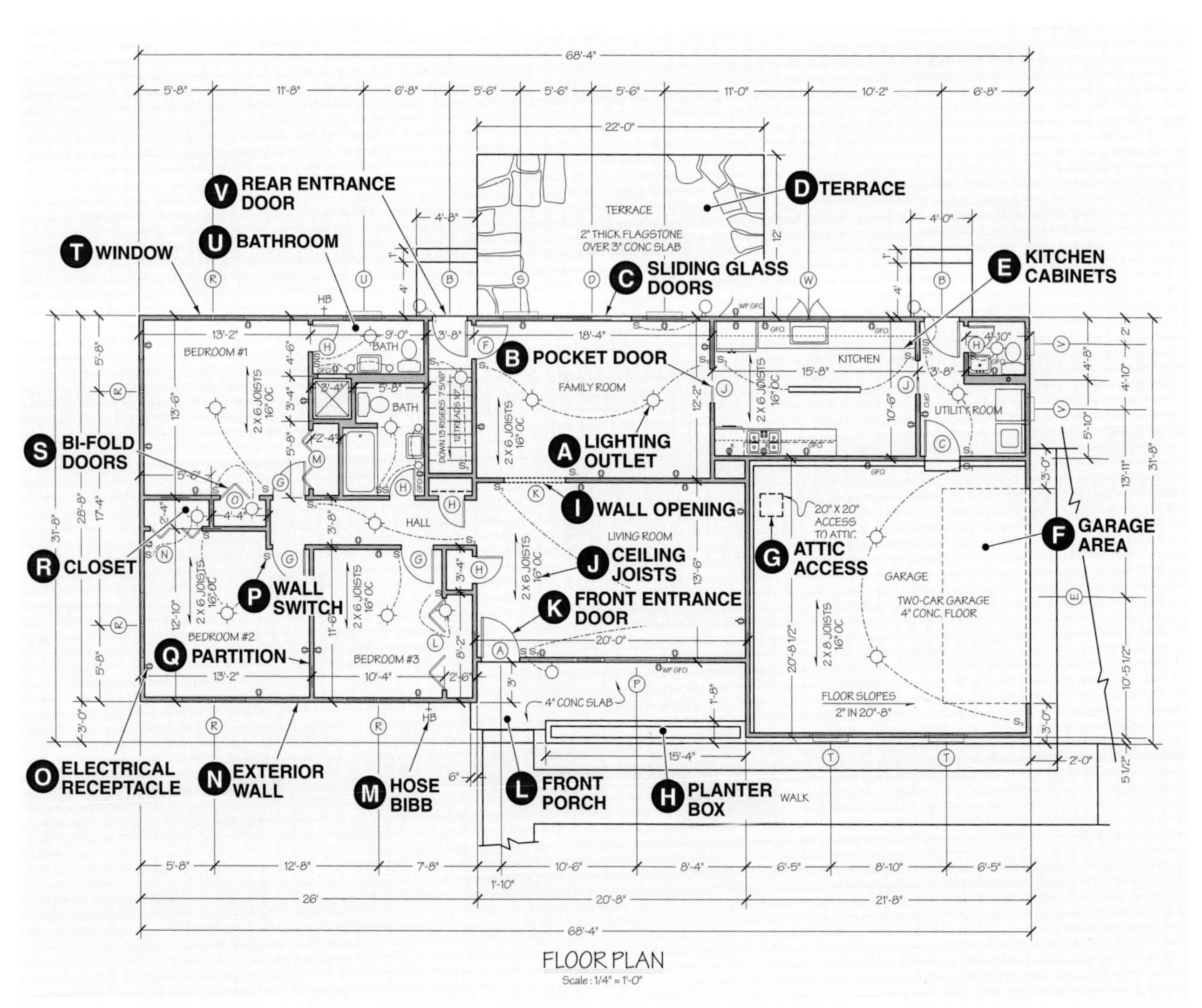

Figure 27-1. *The three-bedroom house includes three bathrooms, amily room, living room, kitchen, and an attached two-car garage. Compare the floor plan with the pictorial drawing in Figure 27-2.*

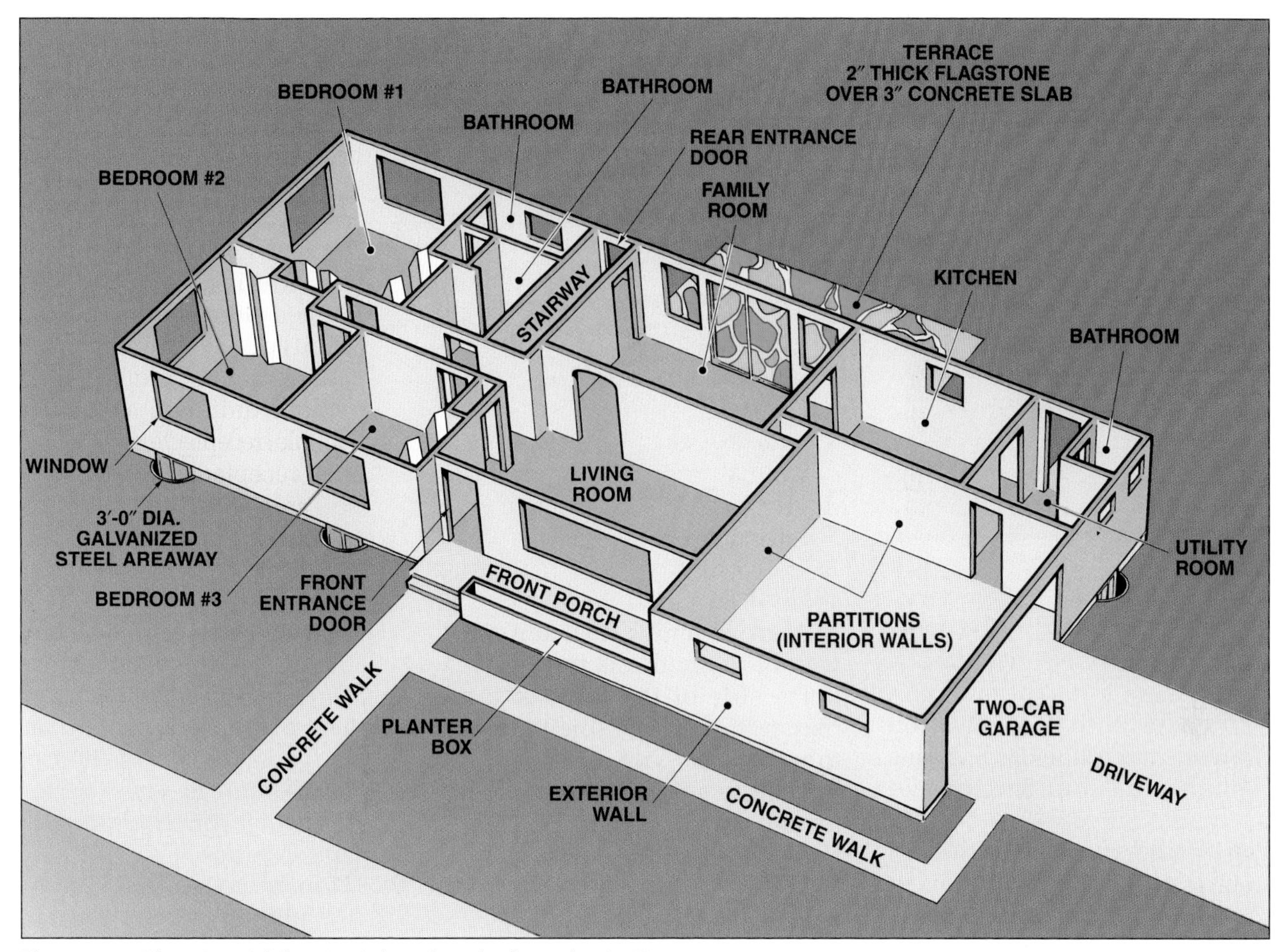

Figure 27-2. *This pictorial drawing is based on the floor plan in Figure 27-1.*

P. *Wall switch:* A wall switch is located next to the door opening. A hidden line extends from the switch to the light it controls.

Q. *Partition:* Interior, non-load-bearing walls are typically referred to as *partitions.* Measurements to the partitions of the three-bedroom house plan are to the centers of the walls. In this example, the partition is laid out by measuring 13′-2″ from the outside face of the west exterior wall to the center of the partition.

R. *Closet:* Bedroom closets are equipped with bi-fold doors. An overhead light is shown in the closet, with a wall switch immediately outside the closet.

S. *Bi-fold doors:* Bi-fold doors are attached to the closet opening at the sides and have a hinge in the middle. Bi-fold doors are available in a variety of sizes.

T. *Window:* A pair of double-hung windows are shown here. The circled "R" above the window refers to a window schedule, which provides window size and other pertinent information. Window location is established by a dimension line that identifies the center of the window unit. In this example, the "R" type window is laid out by measuring 5′-8″ from the outside face of the west wall to the center of the window.

U. *Bathroom:* Three bathrooms are shown on the floor plan. One bathroom (master bathroom) is entered from Bedroom #1, a second bathroom is entered from the hallway, and the third bathroom is entered from the utility room. Plumbing fixtures in the master bathroom include a lavatory, water closet, and shower stall.

V. *Rear entrance door:* The identified stoop, which leads to a rear entrance door, is 16″ above ground level. (The grade elevation difference is shown in the west elevation in Unit 28.) Another stoop and rear entrance door lead into the utility room.

Unit 27 Review and Resources

UNIT 28 Exterior Elevations

OBJECTIVES

1. Identify information provided on an elevation drawing.
2. Describe how views are identified on elevation drawings.

An *exterior elevation* is a view from the side of a structure. An elevation drawing of the side of a building includes the wall surface and the roof. Elevation drawings clarify and provide additional details regarding information shown on a floor plan. For example, a floor plan indicates where the doors and windows are located in the exterior walls. An elevation view of the same wall shows the appearance of the doors and windows.

Elevation drawings typically identify the materials used to finish the outside surfaces of the walls and roof. The height from the finished floor to the finished ceiling and the height from the floor to the top of the door and window openings are also provided.

Downspouts leading from roof gutters and the roof vents also appear on elevation drawings. Flashing required over doors and windows and on the roof is described. The locations of diagonal bracing may be indicated by dashed lines.

Elevation drawings are commonly identified by compass direction, such as north, south, east, or west. The plot plan included in Unit 25 shows that the rear wall of the house is toward the north side of the lot. For this reason, the rear wall is referred to as the north wall. Another way to identify elevation drawings is to refer to them as the front, rear, left, and right elevations.

SOUTH AND EAST ELEVATIONS

The south and east elevations show the front and right side of the example three-bedroom house. **See Figure 28-1.** The exterior wall in the south elevation extends across Bedroom #2, Bedroom #3, living room, and one side of the garage. The exterior wall of the east elevation extends along the front of the garage, one side of the utility room, and the bathroom located off the utility room. When reviewing the elevations, compare the features with the pictorial drawing shown in **Figure 28-2.** The following is an explanation of the items identified on the elevations in Figure 28-1:

A. *Cutting plane line B:* Cutting plane line B refers to the Garage Section B-B drawing (shown in Unit 29).
B. *Stone veneer:* Cut stone, which is 4″ thick, finishes off the exterior wall in this area. The cut stone extends along the front of the planter box.
C. *Garage foundation wall:* There is no basement under the garage area of the house. Less depth is required for the garage foundation walls and footings.
D. *Planter box:* A front view of the planter box is shown.
E. *Front entrance door:* A flush door with three glass lights (panes) is shown at the front porch.
F. *Front porch foundation wall:* Hidden lines identify the low concrete foundation that supports the front edge of the concrete floor slab. No footing is required since heavy loads are not anticipated in this area.
G. *Basement floor slab:* Hidden lines indicate the top of the concrete basement floor slab.
H. *Shutters:* Wood shutters are placed at each side of the window units.
I. *Foundation wall:* The outlines for the foundation walls are shown with vertical hidden lines.
J. *Double-hung window units:* The windows shown in the south elevation are the same "R" type window units shown in the south bedroom walls on the floor plan.

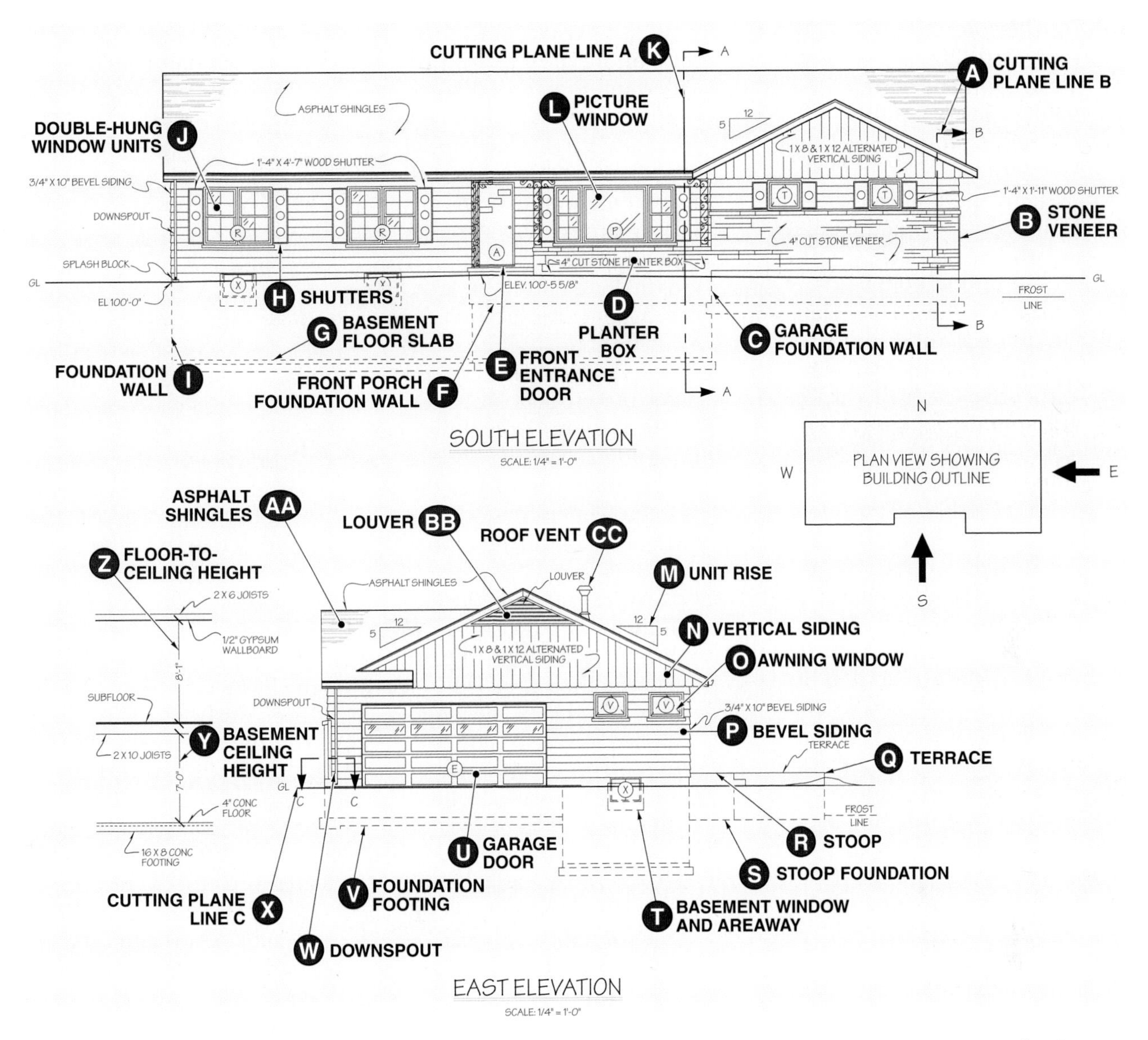

Figure 28-1. *Exterior elevations, such as the south and east elevations, are views from the side of a building. Compare the elevations with the pictorial drawing in Figure 28-2.*

Figure 28-2. *This pictorial drawing is based on the south and east elevations in Figure 28-1.*

K. *Cutting plane line A:* Cutting plane line A refers to Thru-House Section A-A (shown in Unit 29).
L. *Picture window:* The picture window, or *fixed-sash window,* cannot be opened. Double-hung window units are shown at each side of the picture window.
M. *Unit rise:* The small triangle indicates a 5″ unit rise, meaning that the roof rises vertically 5″ for every 12″ of horizontal run. Unit rise is also referred to as pitch or slope (in Canada).
N. *Vertical siding:* The gable end of the intersecting roof is finished with 1 × 8 and 1 × 12 vertical redwood or cedar boards.
O. *Awning window:* Awning windows are installed in the south garage wall. The hidden lines indicate the windows are hinged at the top.
P. *Bevel siding:* Redwood ¾″ × 10″ boards are used for exterior finish in this portion of the wall.
Q. *Terrace:* A side view of the terrace is shown.
R. *Stoop:* A side view of one of the rear stoops is shown. Note the step up to the stoop.
S. *Stoop foundation:* The dashed lines show a small foundation that provides support around the perimeter of the stoop.
T. *Basement window and areaway:* Steel-framed basement windows extend below the surface of the ground.
U. *Garage door:* A panel overhead garage door with four lights is shown.
V. *Foundation footing:* The top and bottom of the footings are identified with hidden lines. The width and thickness of the footing are provided in the view to the left of the east elevation.
W. *Downspout:* A *downspout* is a rectangular channel, usually aluminum or plastic, which carries rainwater from roof gutters to the ground. A splashblock at the base of the downspout directs rainwater away from the building.
X. *Cutting plane line C:* Cutting plane line C refers to Section C-C (shown in Unit 29).
Y. *Basement ceiling height:* The distance from the top of the basement floor slab to the bottom of the floor joists is 7′-0″.
Z. *Floor-to-ceiling height:* The distance from the top of the subfloor to the bottom of the ceiling joists is 8′-1″.
AA. *Asphalt shingles:* The material used as finish roof covering is asphalt shingles.
BB. *Louver:* A louver is located under the roof ridge to provide attic ventilation.
CC. *Roof vent:* Exhaust from gas appliances, such as the stove, water heater, and furnace, is expelled through the roof vent.

NORTH AND WEST ELEVATIONS

The north and west elevations in **Figure 28-3** show the back and left sides of the house. For consistency in appearance, the same type of bevel siding specified for the south and east elevations is used. Asphalt shingles are again identified on the roof. When reviewing the elevations, compare the features with the pictorial drawing shown in **Figure 28-4.** The following is an explanation of the items identified on the elevations in Figure 28-3:

A. *Roof vent:* Another view of the roof vent shown on the east elevation is shown in the north elevation.
B. *Casement window:* Casement windows are hinged on the side and swing outward (similar to doors). Hidden lines indicate side hinges. The only casement windows used in the building are located in the wall area above the kitchen sink.
C. *Sliding glass doors:* Sliding glass doors lead from the family room to the terrace. One door is usually stationary, and the other door slides to provide access to the house.
D. *Rear stoop:* A front view of the rear stoop is shown in the north elevation. Hidden lines indicate the stoop foundation. Another rear stoop is located directly to the right of the terrace.
E. *Rear entrance door:* A panel door with three lights leads into the utility room. An identical door over the other stoop leads into the stair landing off the family room.
F. *Barge rafter and trim:* The gable end of the roof is finished with barge rafters, or fascia rafters.
G. *Stairway:* Hidden lines show the outline of the stairway that leads from the basement to the main floor.
H. *Front porch:* A side view of the porch is shown in the west elevation. Hidden lines indicate the porch foundation.
I. *Planter box:* A side view of the planter box is shown in the west elevation. Hidden lines indicate the planter box foundation.

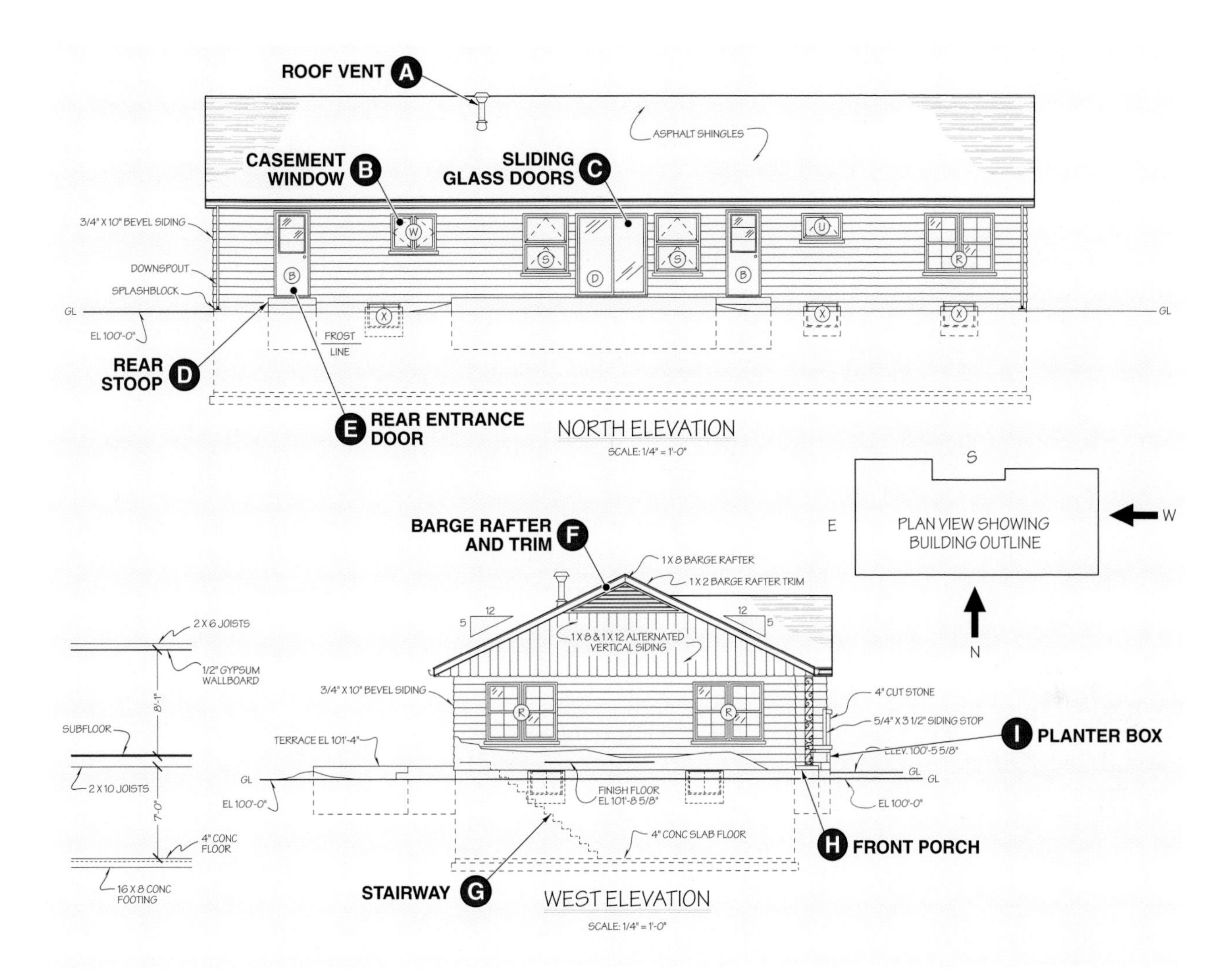

Figure 28-3. *The north and west elevations of the three-bedroom house show the rear and left side of the building. Compare the elevations with the pictorial drawing in Figure 28-4. Note that the plan view showing the building outline has been reversed so that the north (back) wall is at the bottom of the drawing and the west (side) wall is at the right.*

Figure 28-4. *This pictorial drawing is based on the north and west elevations in Figure 28-3.*

UNIT 29

Section Views

OBJECTIVES

1. Define and identify section views as they relate to cutting plane lines.
2. Describe section view examples based on cutting plane lines shown in Unit 28.

A *section view* is a drawing showing the part of a building that would be revealed if a vertical or horizontal cut was made through the building. Section views provide information that cannot be obtained from other drawings in a set of prints. For example, section views show the structural members and materials used inside the walls and on exterior surfaces. The height, thickness, and shape of the walls are shown and window and door heights are often specified. The prints for the three-bedroom house contain several section views.

Section views must be related to the other drawings of the prints for clarity. Cutting plane lines are usually found on foundation plans, floor plans, and exterior elevations. Section views shown in this unit are based on the position of cutting plane lines appearing in plans and elevations shown in Units 26, 27, and 28.

THRU-HOUSE SECTION A-A

To the right of the picture window in the south elevation shown in Unit 28, a cutting plane line extends through the roof, wall, and foundation. At each end of the line, an arrow points to the right with an "A" next to it. The arrows point in the direction of what would be seen if a transverse section cut were made across the house. A *transverse section* presents a view across the width of the house. (A *longitudinal section* presents a view across the length of the house.) The "A" refers to a drawing called Thru-House Section A-A, which is shown in **Figure 29-1.** When reviewing the section view, compare the features with the pictorial drawing shown in **Figure 29-2.**

The following is an explanation of the items identified in Thru-House Section A-A in Figure 29-1:

A. *Roof rafter:* The size and spacing of the roof rafters are indicated on the left side of the section view.

B. *Roof slope:* Similar to the elevations, the unit rise of the roof is provided on the section view.

C. *Cornice construction:* The area under the roof overhang is the *cornice.* When a cornice is closed in, as shown in the drawing, nailing blocks are placed between the ends of the rafters and the wall. The material used to close in the cornice is 3/8" plywood.

D. *Window and door heights:* Typically, the tops of all doors and windows align horizontally. A dimension (6'-8½") is given from the subfloor to the top of the doors and windows.

E. *Front porch:* A 4" thick concrete slab with steel reinforcement is shown for the front porch.

F. *Planter box:* The inside wall of the planter box rests on the front porch. The outside wall is supported by a footing. The height of the planter box (2'-0") is measured from the porch slab. A dimension line at the top of the planter box indicates the overall width of the planter box is 1'-8". Material used for construction of the planter box is also specified.

G. *Grade line:* The grade line is the level of the ground at the outside of the foundation walls, and is indicated with a horizontal line marked "GL." The grade elevation is 103.3', which is .3' higher than the benchmark identified in the plot plan.

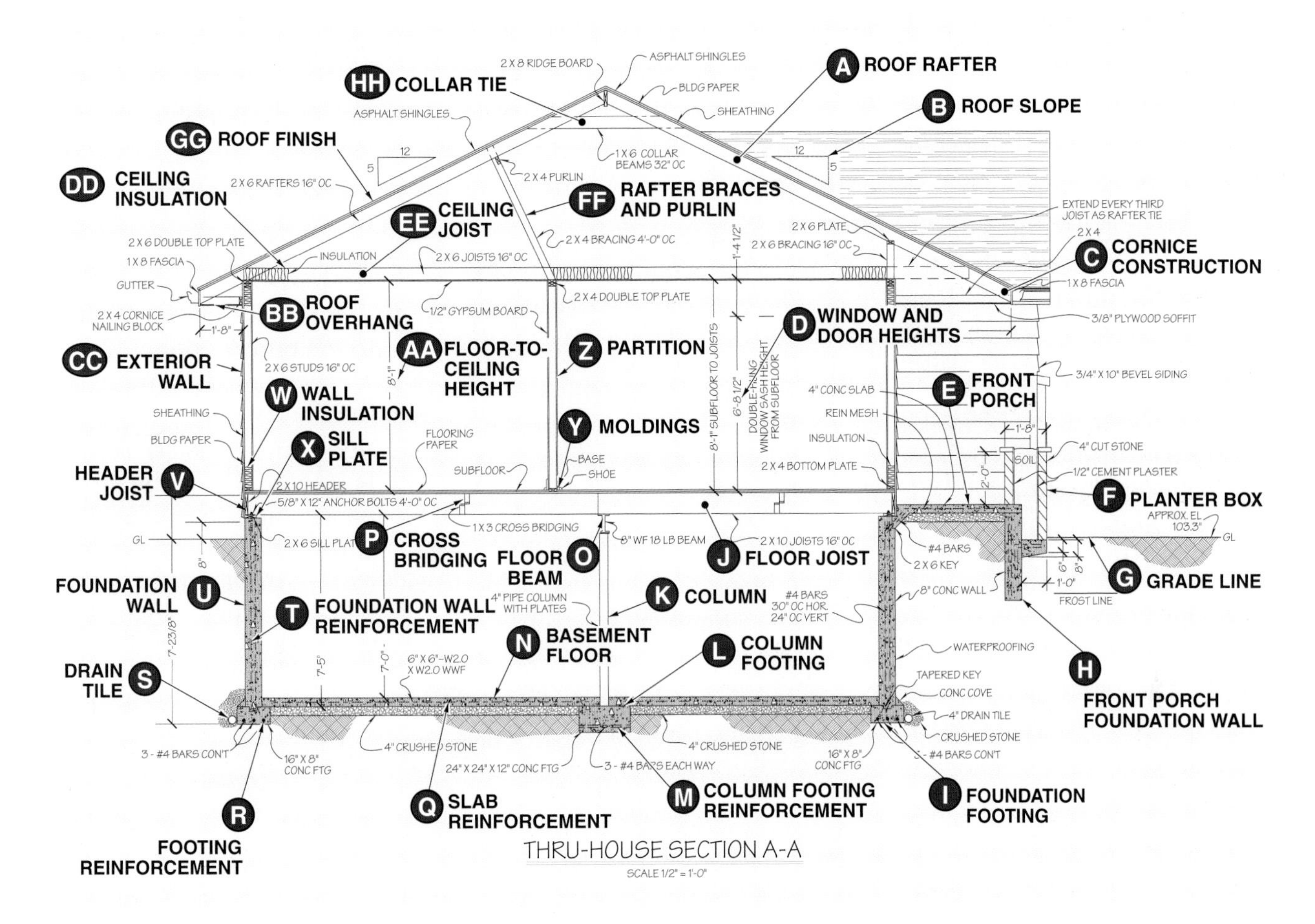

Figure 29-1. *Thru-House Section A-A represents a transverse section across the width of the house from north to south and shows the foundation, first floor, walls, ceiling, and roof. Compare Thru-House Section A-A with the pictorial drawing in Figure 29-2.*

H. *Front porch foundation wall:* The thickness of the foundation wall is 8″. The bottom of the foundation wall is below the frost line.

I. *Foundation footing:* The width and thickness of the foundation footing are 16″ × 8″.

J. *Floor joist:* The size and spacing of the lapped floor joists are 2 × 10–16″ OC. One end of the joists rests on the foundation wall. The lapped end rests on top of the steel beam running the length of the house.

K. *Column:* Steel pipe columns (4″ diameter) support the steel beam. The spacing of the columns is shown in the foundation plan.

L. *Column footing:* The square concrete base for each column is the *column footing.* The dimensions of the footing are 24″ × 24″ × 12″.

M. *Column footing reinforcement:* Column footings require reinforcement for additional strength. The spacing and number of reinforcing bars are three #4 rebar each way.

N. *Basement floor:* The concrete slab thickness for the basement floor is given. The slab is placed over a 4″ layer of crushed stone.

O. *Floor beam:* A wide-flange steel beam is used. The vertical dimension of the beam is 8″ and it weighs 18 lb/ft.

P. *Cross bridging:* Cross bridging is placed between the joists at the center of their spans. Cross bridging is used to prevent the joists from tipping or rolling.

Q. *Slab reinforcement: Welded wire reinforcement* (wire fabric) used for slab reinforcement is shown by the hidden lines inside the slab. The welded wire reinforcement (WWR) used for slab reinforcement is 6 × 6–W2.0 × W2.0 WWR.

R. *Footing reinforcement:* Footings are reinforced with three #4 reinforcing bars (rebar). A #4 bar is 4/8″, or 1/2″ in diameter.

S. *Drain tile:* Drain tile is placed in a layer of crushed rock next to the footing to move water away from the foundation.

T. *Foundation wall reinforcement:* Rebar size and

horizontal and vertical spacing are indicated toward the right foundation wall.

U. *Foundation wall:* The height of the foundation walls from the top of the footing to the bottom of the first floor joist is shown. The thickness of the foundation walls is 8″.

V. *Header joist:* The pieces that nail into the ends of the regular joists are the *header joists.*

W. *Wall insulation:* Blanket or batt insulation is placed between the studs.

X. *Sill plate:* A sill plate, or mudsill, is installed on top of the foundation wall. The anchor bolts are 5/8″ diameter by 12″ long and are spaced 4′-0″ OC.

Y. *Moldings:* The finish pieces placed at the bottom of a wall are the base and shoe moldings.

Z. *Partition:* A partition, or interior wall, is shown.

AA. *Floor-to-ceiling height:* The distance from the top of the subfloor to the bottom of the ceiling joists is 8′-1″.

BB. *Roof overhang:* Roof overhang (1′-8″) is the horizontal distance from the side of the house to the end of the rafters.

CC. *Exterior wall:* The size and spacing of the studs for the exterior wall are 2 × 6–16″ OC.

DD. *Ceiling insulation:* Blanket or batt insulation is placed between the ceiling joists.

EE. *Ceiling joist:* The ceiling joists are 2 × 6–16″ OC.

FF. *Rafter braces and purlin:* Rafter braces (2 × 4s) support a 2 × 4 purlin, which is a horizontal member that provides additional support to the rafters. The purlin is notched into the top ends of the braces.

GG. *Roof finish:* Wood sheathing is covered with building paper and then with asphalt shingles.

HH. *Collar tie:* Collar ties tie opposite rafters together.

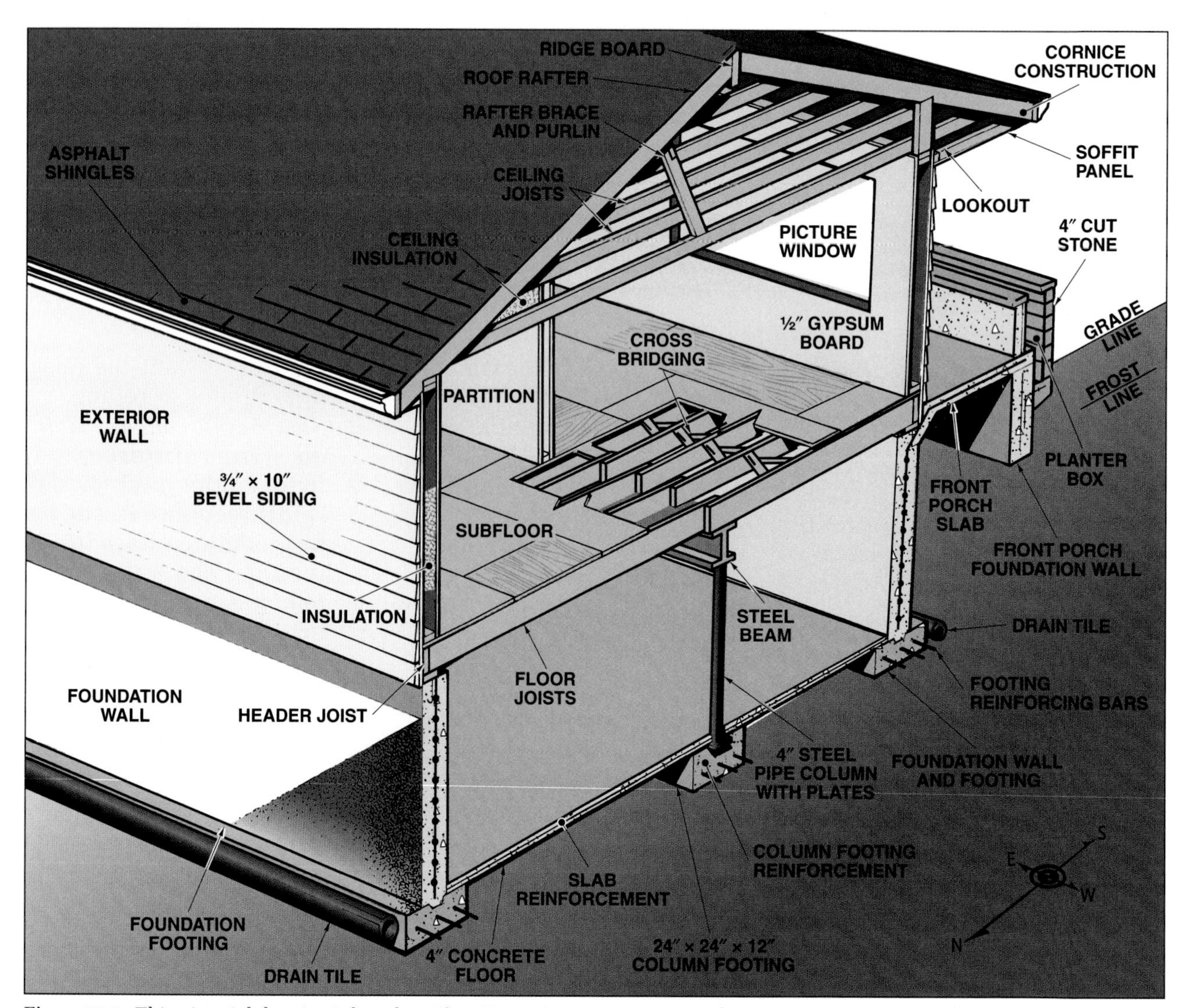

Figure 29-2. *This pictorial drawing is based on Thru-House Section A-A in Figure 29-1.*

SECTION B-B

A section view of the garage wall is also included in the three-bedroom house plan. At the right side of the south elevation shown in Unit 28, a cutting plane line with arrows labeled with a "B" extends through the garage. The "B" refers to a drawing called Section B-B, which is shown in **Figure 29-3.** When reviewing the section view, compare the features with the pictorial drawing shown on the right side of Figure 29-3. The following is an explanation of the items identified in Section B-B:

A. *Roof rafters:* The roof rafters are 2 × 6–16″ OC.

B. *Roof finish material:* Asphalt shingles are identified for the roof.

C. *Cornice trim:* Trim (1″ × 2″) is nailed at the top of the 1 × 8 barge rafter.

D. *Ceiling joist:* The ceiling joists are 2 × 8–16″ OC.

E. *Exterior finish:* The lower portion of the exterior walls in the garage area has a stone veneer. The upper part of the walls has bevel siding. Sheathing covered with building paper is placed against the stud wall before the stone and bevel siding are applied. A 1″ air space is behind the stone veneer.

F. *Ceiling height:* The distance from the top of the concrete slab to the bottom of the ceiling joists is 9′-1″.

G. *Garage wall:* The size and spacing of the studs for the garage are 2 × 6–16″ OC.

H. *Garage floor:* The concrete slab for the garage floor is 4″ thick. A 4″ layer of crushed stone is shown beneath the slab.

I. *Foundation wall:* The thickness of the garage foundation wall is 11″ and the minimum distance that the wall must extend above the ground is 8″.

J. *Foundation footing:* The width and depth of the footings are 22″ × 11″. Steel reinforcement consists of three #4 rebar running the length of the footing.

K. *Frost line:* The top of the footing must be below the frost line in the area.

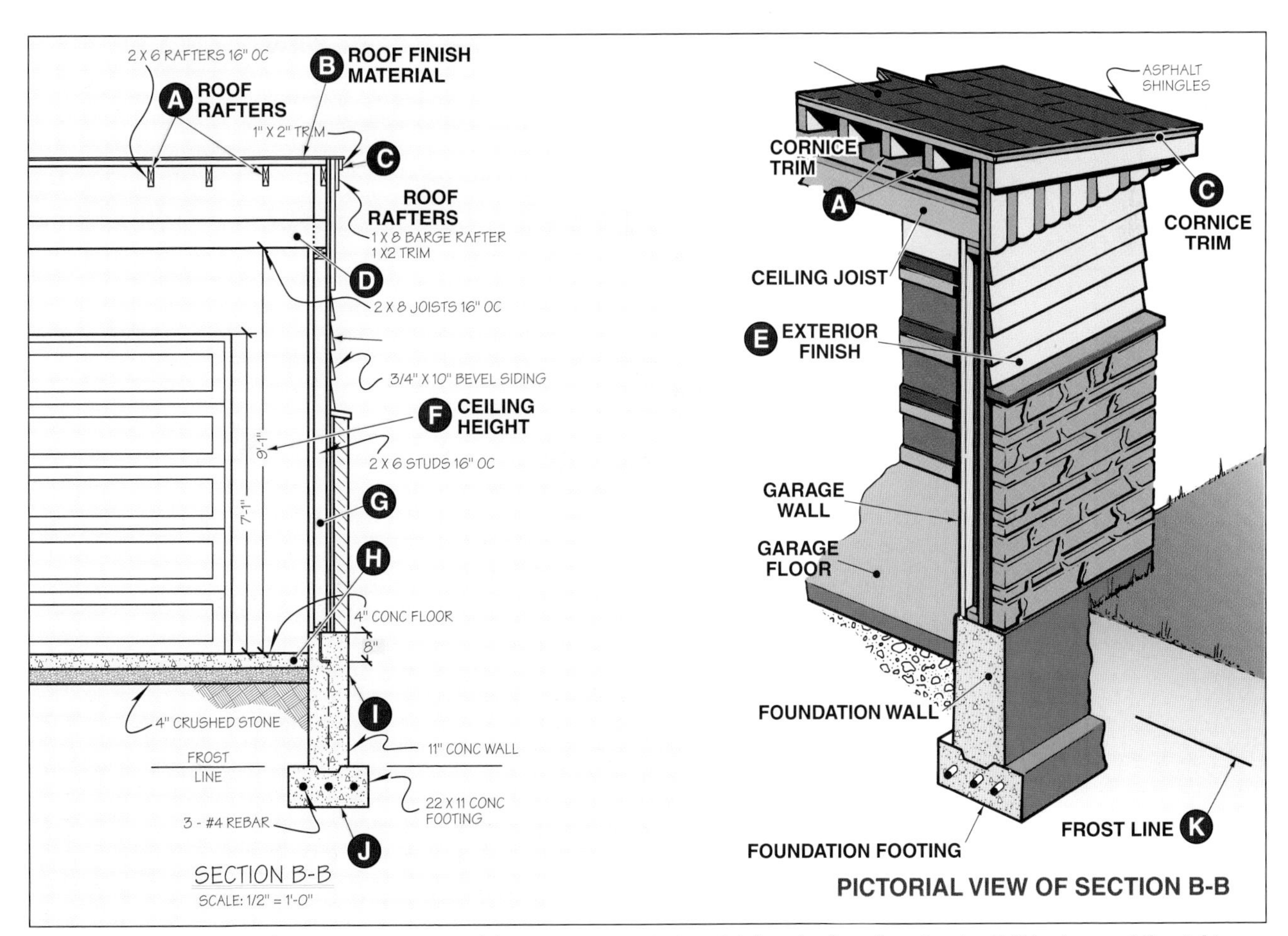

Figure 29-3. *Section B-B shows a section view of the garage area. A pictorial drawing based on Section B-B is shown at the right.*

SECTION C-C

The three-bedroom house plan includes a section view to clarify the corner construction at the southeast corner of the garage. The east elevation shown in Unit 28 includes a cutting plane line with arrows at the ends labeled with a "C." Note that the arrows are pointing down, which indicates that the section will be a view looking down. When reviewing the section view, compare the features with the pictorial drawing shown on the right side of **Figure 29-4.** The following is an explanation of the items identified in Section C-C:

A. *Masonry veneer:* The edge of the cut stone veneer is shown.

B. *Corner framing:* The corner is framed with three 2 × 6 studs.

C. *Wall sheathing:* Panel sheathing is nailed against the stud wall.

D. *Siding:* A top view of the bevel siding is shown.

E. *Corner trim:* A 1⅛″ × 3⅝″ board fits against the siding to cover the gap between the sheathing and the cut stone.

Masonry veneers are generally between 1″ and 1¾″ thick.

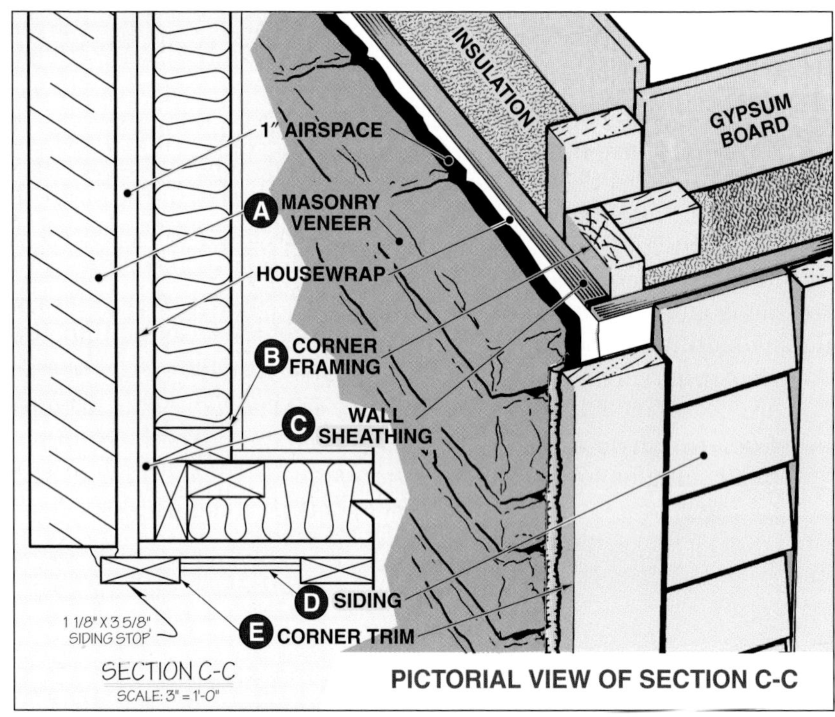

Figure 29-4. *Section C-C shows a section view of the southeast corner of the three-bedroom house. A pictorial drawing based on Section C-C is shown at the right.*

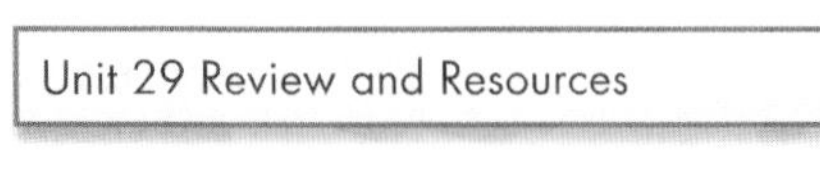

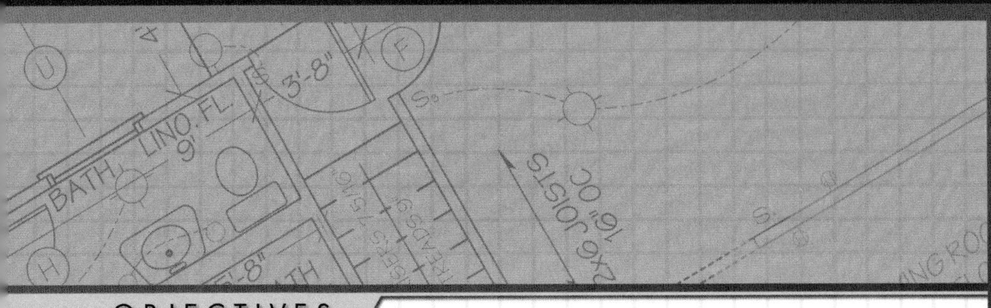

UNIT 30 Details and Framing Plans

OBJECTIVES

1. Describe the purposes of detail and framing plans.
2. Identify common components of window details.
3. Recognize common components of door details.
4. Identify common components of stair details.
5. Identify common components of kitchen cabinet details.
6. Describe common components of framing plans.

A *detail* provides an enlarged view of a part of a building that cannot be fully shown or explained in other drawings of the prints. Details may appear on the same sheet as a related drawing, or may be on a separate sheet consisting of detail drawings and/or section views. Examples of construction features that often require details are door and window units, kitchen cabinets, stairways, fireplaces, roof cornices, and trim materials.

A set of prints for a wood-framed house may also include *framing plans.* Framing plans provide information on constructing the framework of the building.

DETAILS

The three-bedroom house plan includes elevation and plan view detail drawings for windows, doors, stairs, and kitchen cabinets. The detail drawings clarify and provide additional information for these house components, which are shown on the floor plan and exterior elevations.

Window Details

The floor plan and elevation drawings of the three-bedroom house plan show three types of windows. *Double-hung windows* are located in the bedrooms and on each side of the picture window in the living room. Double-hung window details are shown in **Figure 30-1.** Awning windows are to be installed in the garage, utility room, family room, and bathrooms. Awning window details are shown in **Figure 30-2.** Casement windows are located over the sink in the kitchen and fixed windows are installed in the foundation wall; however, detail drawings are not provided for these windows in the plans.

Trus Joist, A Weyerhaeuser Business

Windows must be properly framed to ensure that roof loads are transferred to the foundation. Temporary braces are installed until structural sheathing is applied to the walls.

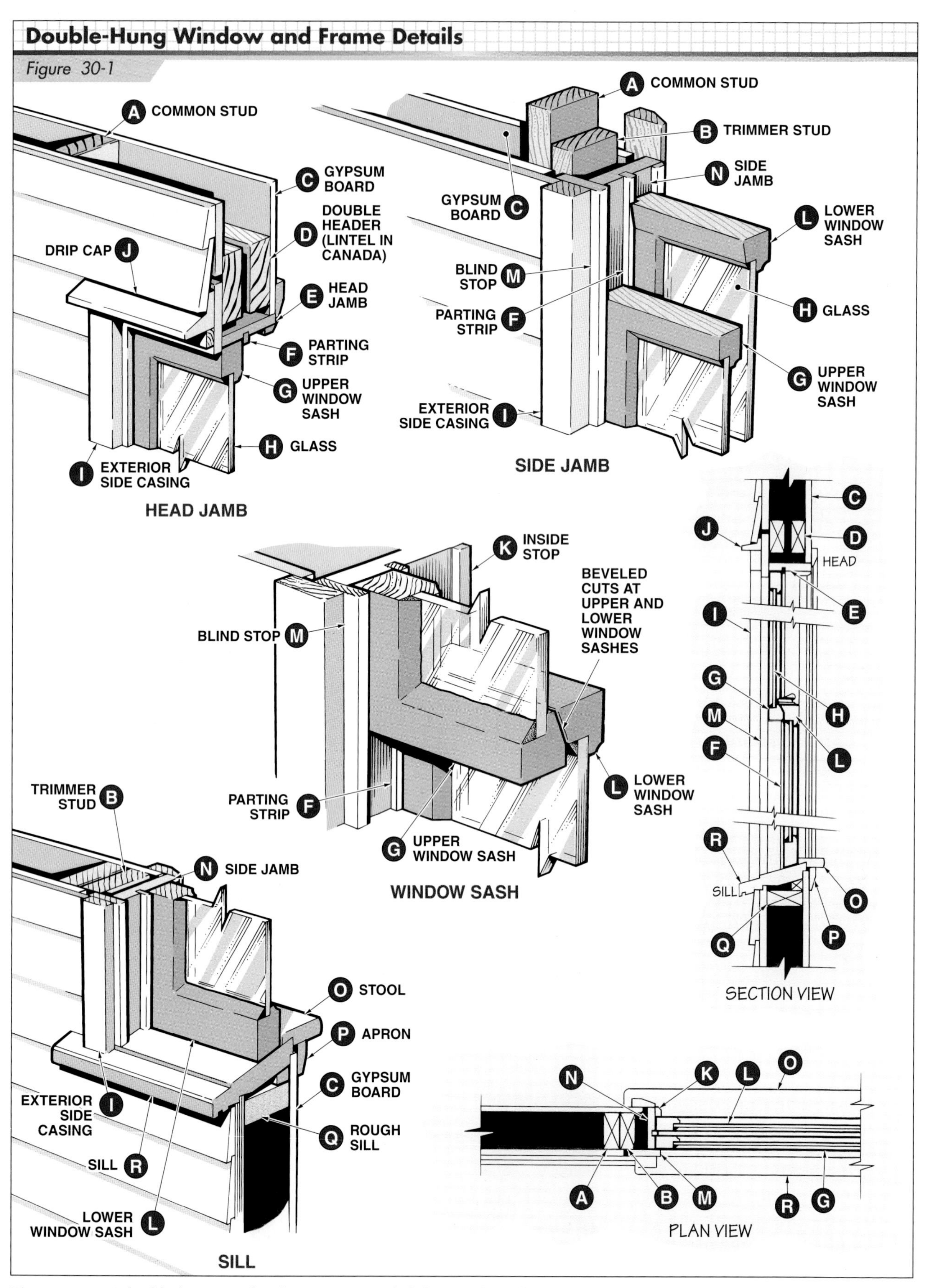

Figure 30-1. *A double-hung window has two vertical sliding sashes placed in adjoining tracks. Both sashes can be opened at the same time and slide past one another.*

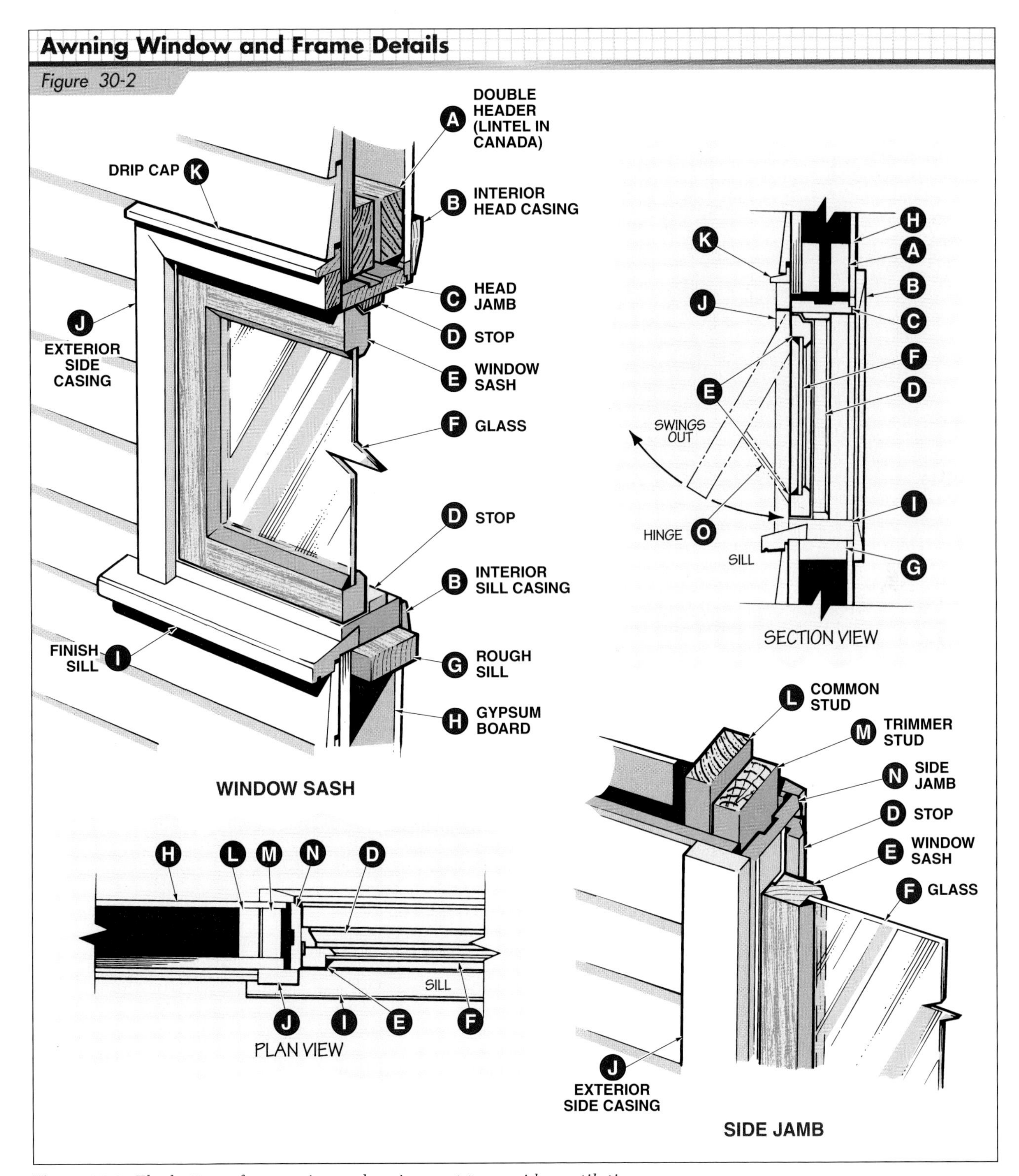

Figure 30-2. *The bottom of an awning sash swings out to provide ventilation.*

Door Details

The floor plan and elevation drawings show three types of doors. *Hinged doors* are located at all the outside entrances to the house. Hinged doors swing back and forth on hinges installed on the side jamb. *Exterior doors* provide access to the outside of the house. **See Figure 30-3.** *Interior doors* are located at many of the openings between rooms inside the house. **See Figure 30-4.**

Main entrance doors are made of solid wood, fiberglass, or steel with an insulated foam core. Many main entrance doors have a fixed window.

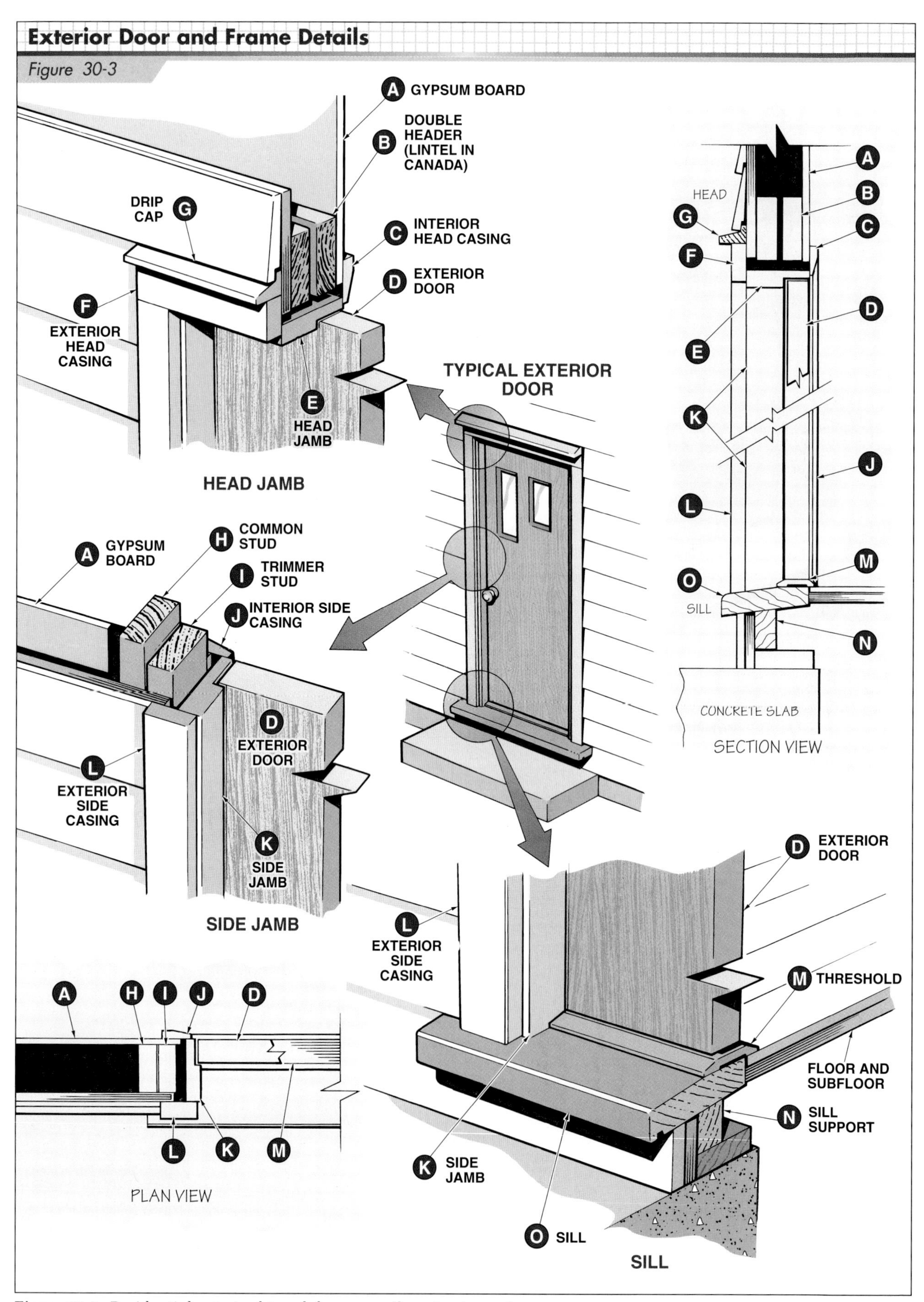

Figure 30-3. *Residential exterior hinged doors usually swing inward.*

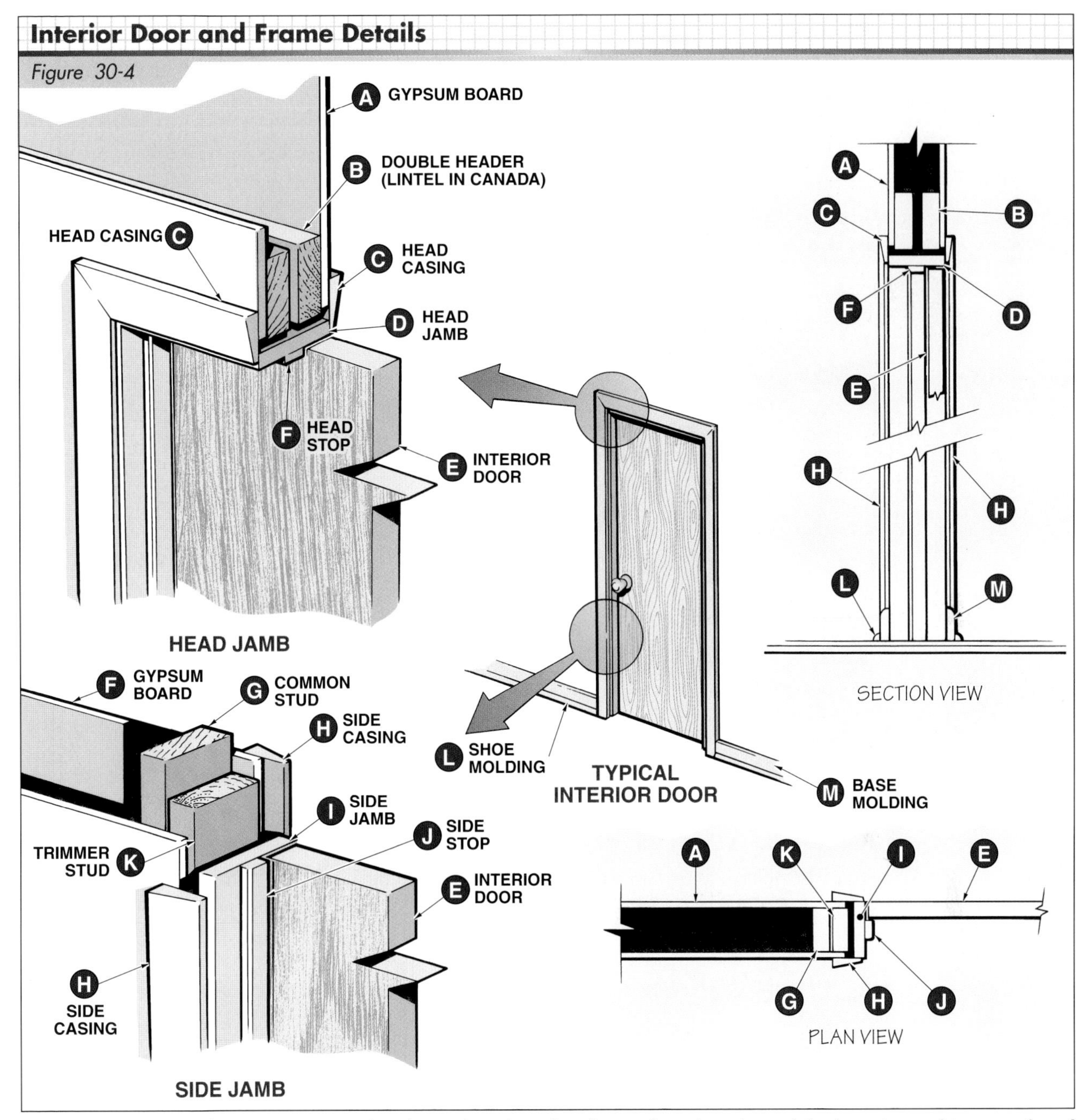

Figure 30-4. *Interior door construction details are found on the plan and section view, while door size and type are found on the door schedule.*

Pocket sliding (recessed) doors are located in two walls of the kitchen. Pocket sliding doors slide into a space (pocket) framed in the wall. **See Figure 30-5.**

Bi-fold doors are located at the entrances of all bedroom closets. **See Figure 30-6.** The closets in Bedrooms #1, #2, and #3 have double bi-fold doors. The closet in Bedroom #1 also has a single bi-fold door.

A bead of waterproof caulk is applied between the sill and threshold of an exterior door to reduce air infiltration into the building.

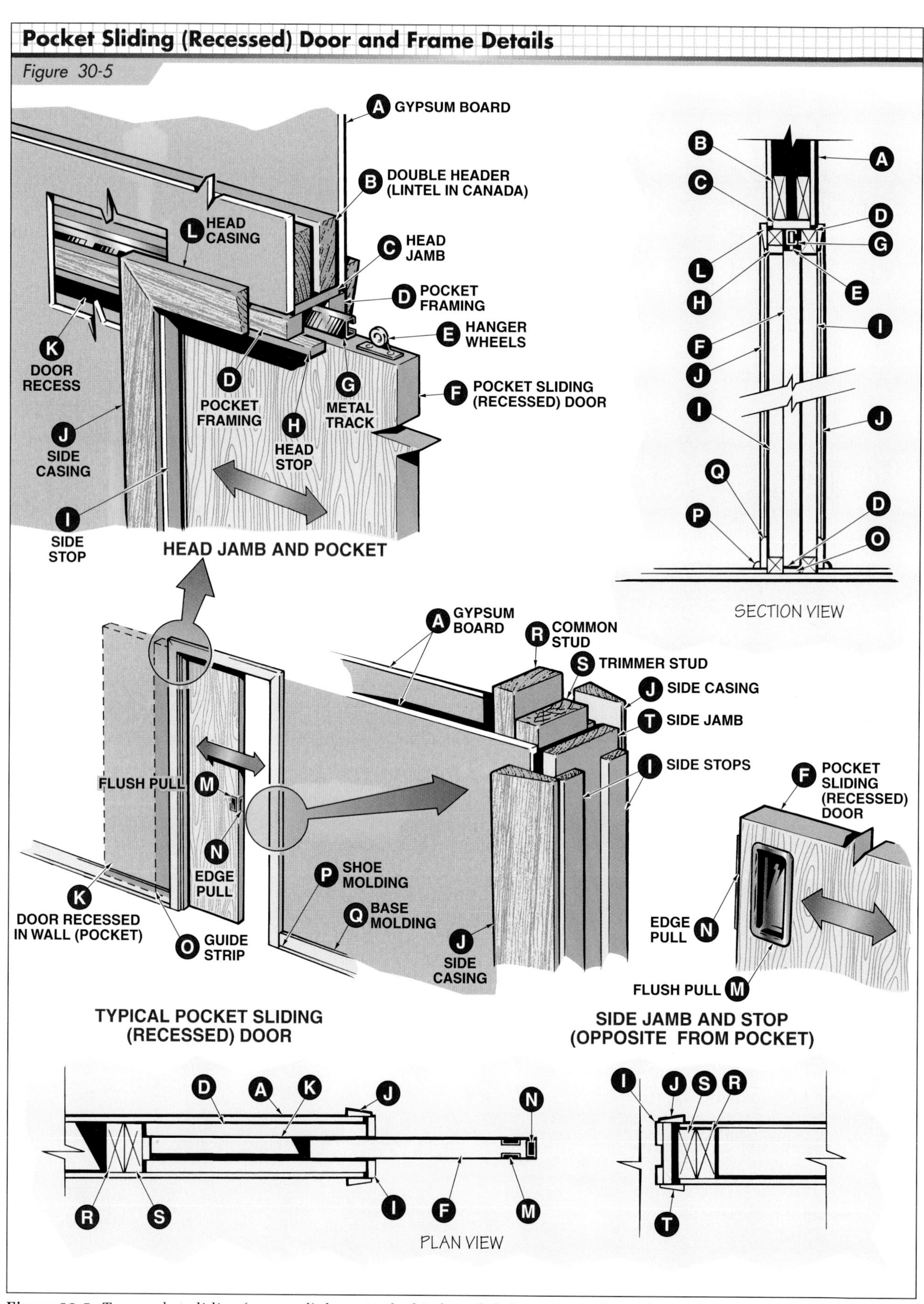

Figure 30-5. *Two pocket sliding (recessed) doors in the kitchen slide into spaces framed into the wall. The details show how to frame the pocket and hang the door.*

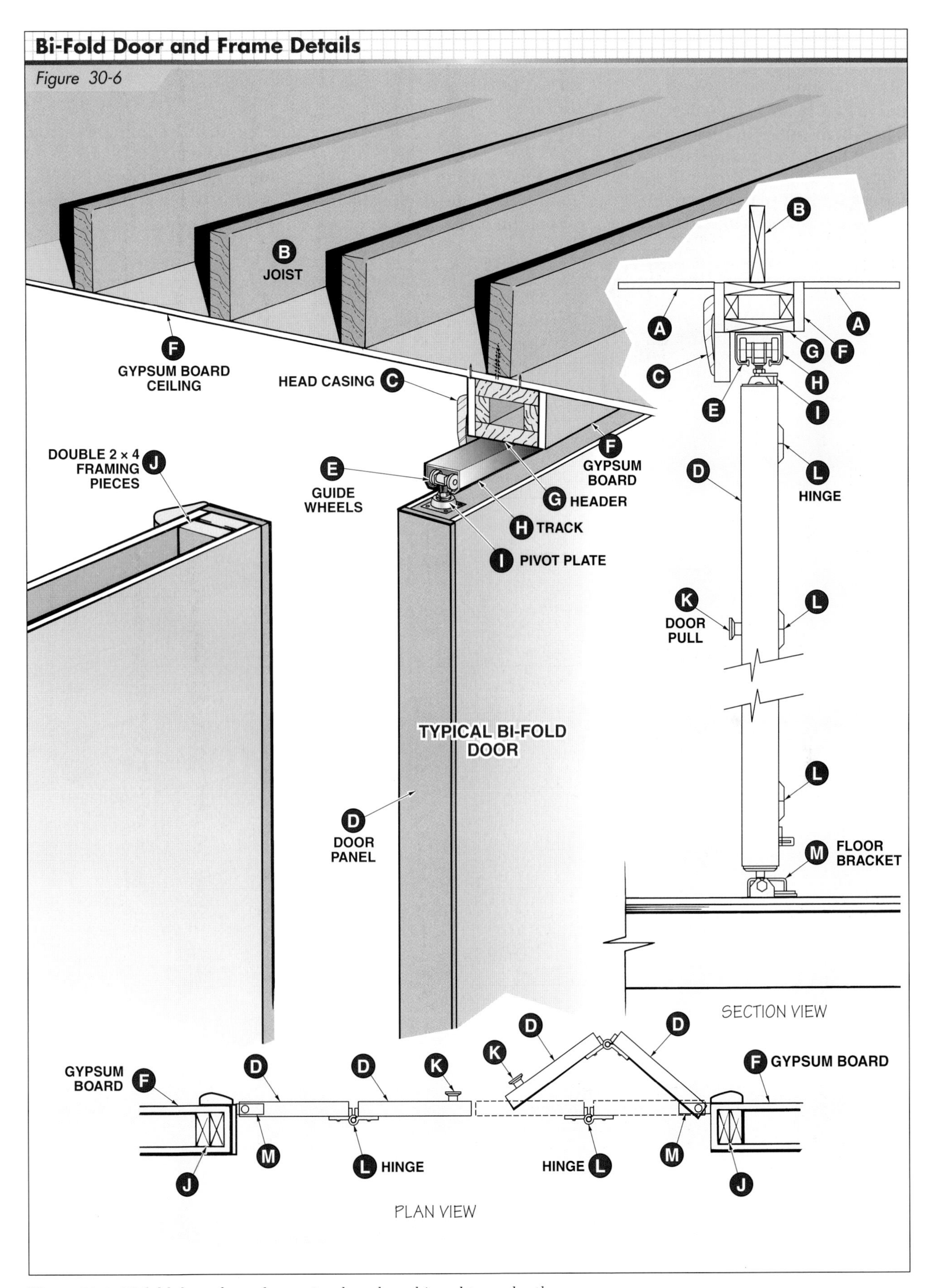

Figure 30-6. *Bi-fold doors hang from a track and are hinged to each other.*

Stair Details

A good example of a structural detail drawing is the stairway detail in **Figure 30-7.** The stairway (shown in both the floor plan and foundation plan) extends from the stair landing off the family room into the basement. A plan view of the framing around the stairwell is included in the detail. The following is an explanation of the items identified on the elevations in Figure 30-7:

A. *Double header:* The header is the framing member at the ends of the stairway floor opening. Framing members around a floor opening are doubled to provide additional support.

B. *Handrail:* The thickness and width of the railing are given.

C. *Double trimmer:* Double 2 × 10 trimmers are used along the sides of the stairway opening for added support.

Basement Stairway and Frame Sections

Figure 30-7

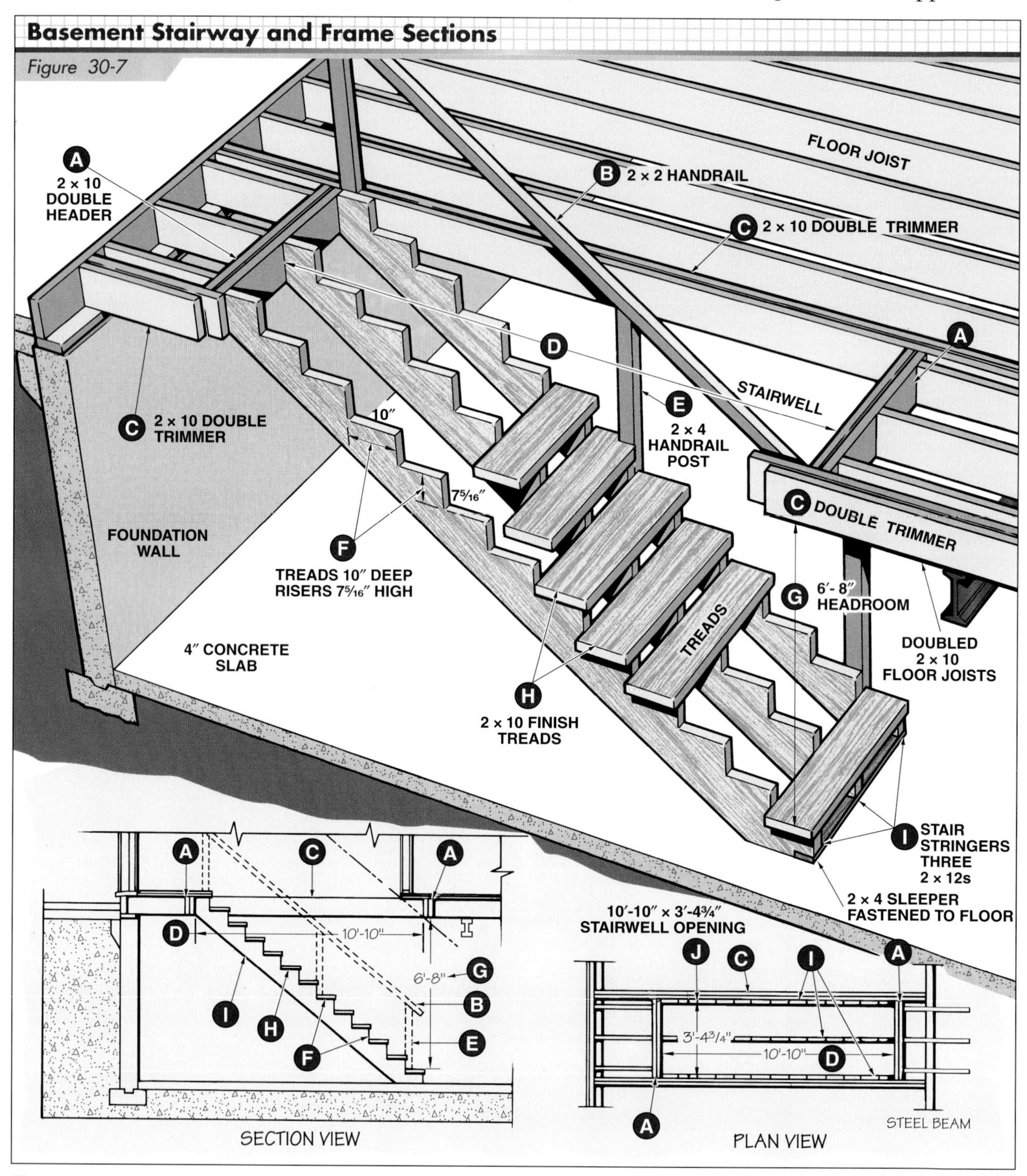

Figure 30-7. *A section view and details provide framing information for the stairway and stairwell opening. The framing information includes proper dimensions for the stairwell opening and the headroom for the stairway.*

D. *Stairwell:* The floor opening for the stairway is the stairwell. The stairwell length (10′-10″) is given in the elevation section view and plan view.

E. *Handrail post:* The thickness and width of the railing posts are given.

F. *Treads and risers:* The depth of the treads (10″) and height of the risers (7 5/16″) are given.

G. *Headroom:* Headroom is measured from the bottom of the floor opening header to a line running diagonally through the front edge of each tread.

H. *Finish treads:* Finish treads are the surface that is walked upon when ascending and descending the stairway.

I. *Stair stringers:* Stair stringers are the main support for the stairway, and are located at the center and on both sides of the stairway.

J. *Stairwell width:* The stairwell width is shown in the plan section view.

Kitchen Cabinet Details

The floor plan for the three-bedroom house plan indicates that kitchen cabinets are to be installed along the north and south kitchen walls. Details for the kitchen cabinets are shown in **Figure 30-8.** When reviewing the details for the kitchen cabinets, compare the features with the pictorial drawing shown in **Figure 30-9.** The following is an explanation of the items identified on the details in Figure 30-8:

A. *Soffit:* The space between the top of the wall cabinets and the ceiling will be closed off flush to the face of the cabinets.

B. *Cabinet doors:* All cabinet sections have double doors, allowing access from the left and right.

C. *Countertop range:* The range fits into the countertop.

D. *Oven:* The proper amount of space must be provided for the oven.

E. *Base cabinet shelves:* Hidden lines indicate shelves in each section.

F. *Width of base sections:* The base cabinets for the three-bedroom house consist of a number of sections that are fastened together. The width is provided for each section.

G. *Length of base cabinets:* The total length of the base cabinets is given. A space to the left of the north wall cabinets is provided for a refrigerator.

H. *Door handles:* Short lines toward the top of the base cabinet doors indicate handles. Handles are also installed at the bottoms of the wall cabinets.

I. *Window space:* The north kitchen wall plan indicates the location of a window. The window schedule gives the width and height of the window.

J. *Depth of wall cabinets:* The section view shows the depth of the wall cabinets.

K. *Height of wall cabinets:* The distance from the bottom to the top of the wall cabinets is shown.

L. *Distance between base and wall cabinets:* The distance from the countertop to the bottom of the wall cabinet is given.

M. *Depth of base cabinet:* The distance from the front to the back of the base cabinet is given.

N. *Sink area:* A stainless steel self-rimming kitchen sink is installed in an opening in the base cabinet countertop. Part of the faucet fixture is visible above the counter.

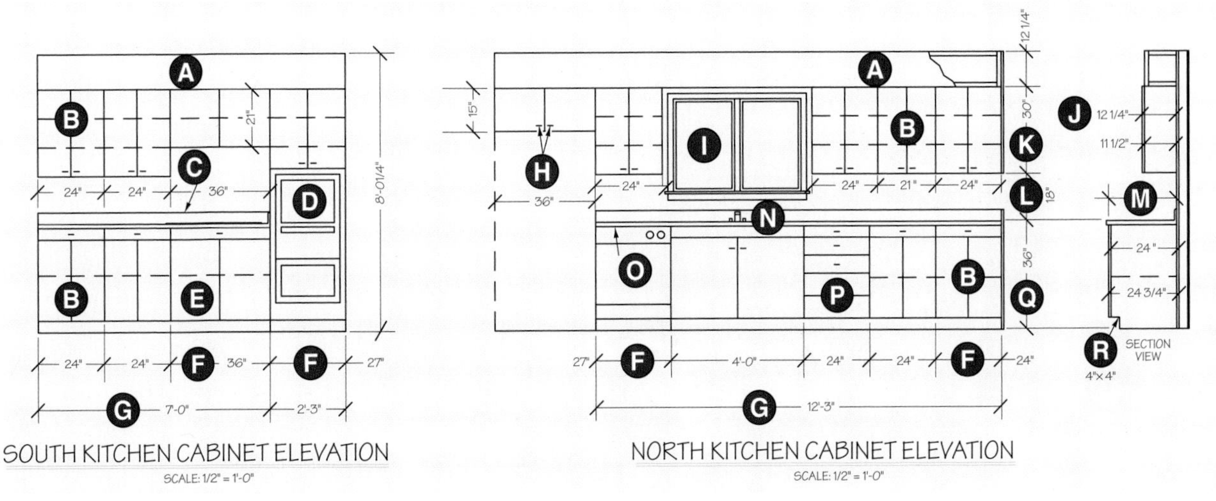

Figure 30-8. *Kitchen elevations provide detailed information to ensure proper spacing of cabinets and appliances. A pictorial drawing based on the detail of the north wall cabinets is shown in Figure 30-9.*

O. *Dishwasher:* An automatic dishwasher is located under the counter to the left of the sink.

P. *Drawers:* The section to the right of the sink consists of five drawers. Drawers are at the top of other sections. There are no drawers in the sink area (north wall) and range area (south wall).

Q. *Height of base cabinet:* The distance from the floor to the countertop is given.

R. *Toe space:* A toe space is provided at the bottom of the base cabinets, allowing kitchen occupants to comfortably rest against the cabinets.

Trim Material Details

A set of prints often includes details showing the size and shape of the molding for the interior finish of the house. **See Figure 30-10.**

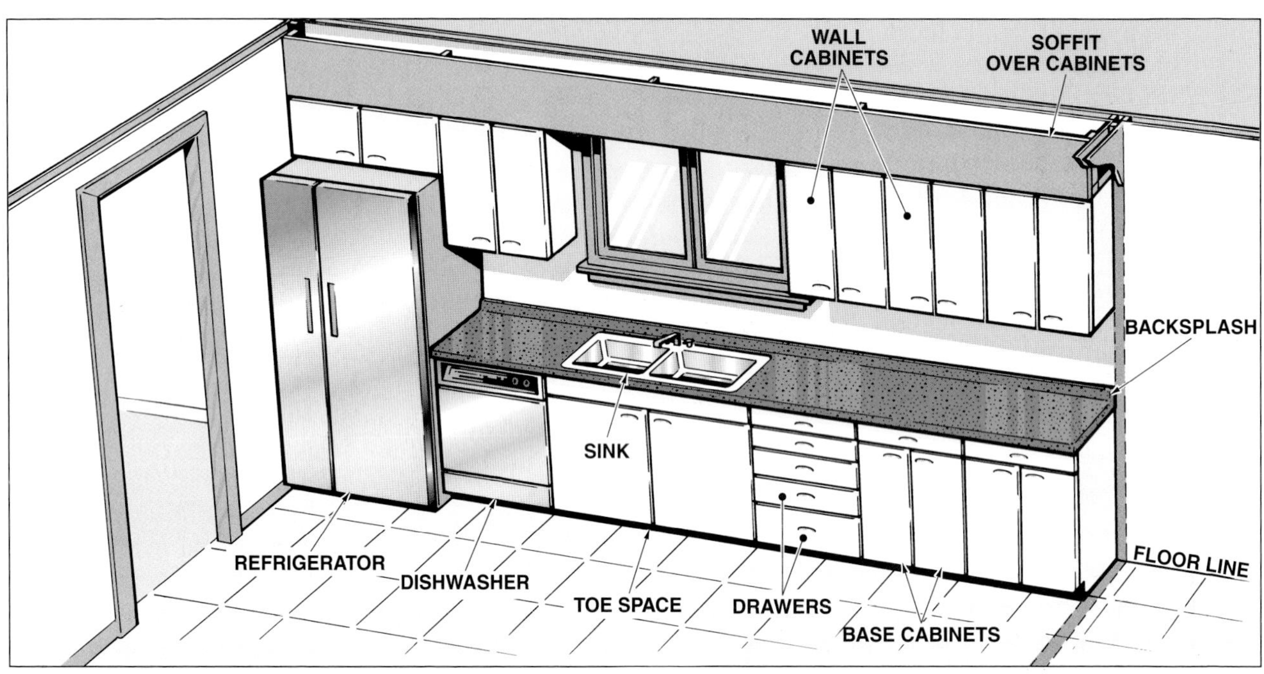

Figure 30-9. *Wall cabinets must be properly spaced around the window. The kitchen sink is installed directly in front of the window.*

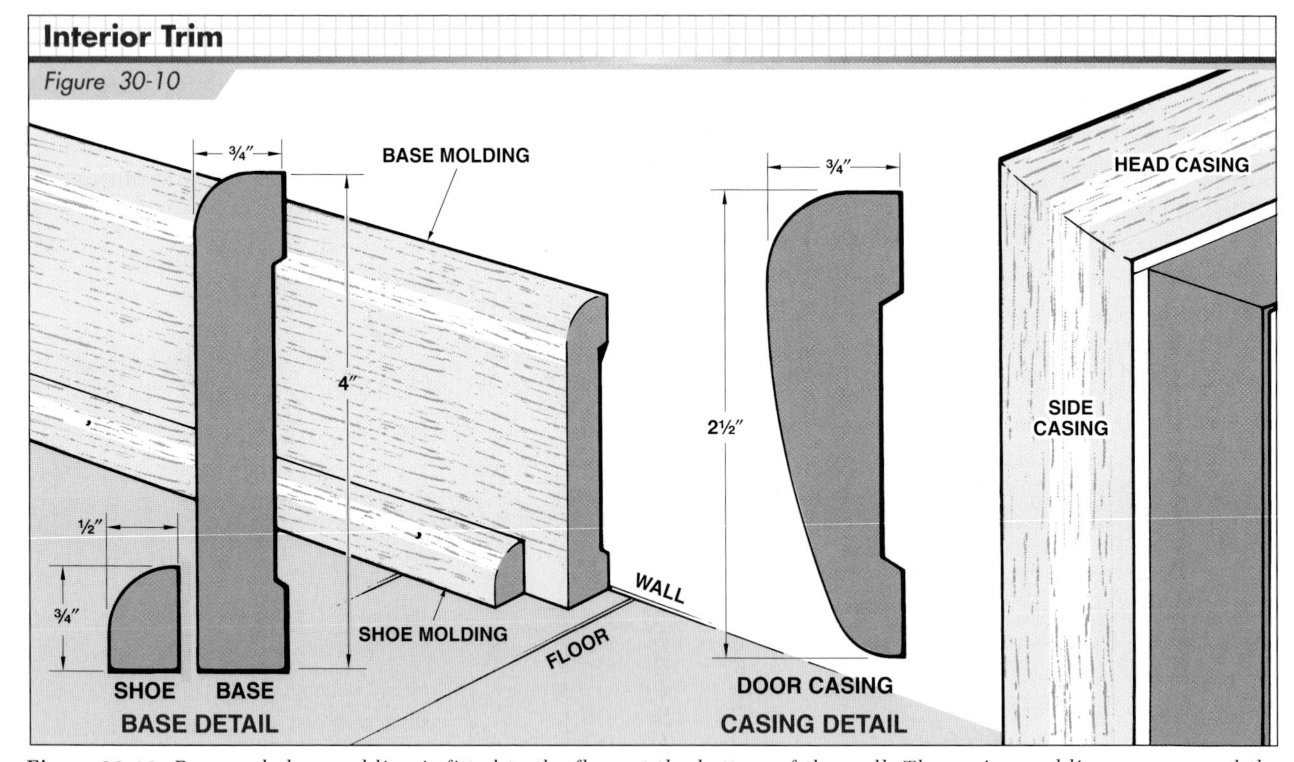

Figure 30-10. *Base and shoe molding is fitted to the floor at the bottom of the wall. The casing molding goes around the door and window openings.*

FRAMING PLANS

The three-bedroom house plan includes framing plans for exterior walls, floor and ceiling units, and roof. **Figure 30-11** shows the framing plan for the south and east exterior walls. Note the header size for the window header and garage door header. A 2 × 14 is used as a garage door header to prevent sagging due to the long unsupported span and the weight of the door on the header.

Figure 30-12 shows the framing plan for the north and west exterior walls. OSB or another performance rated panel is commonly used to provide lateral support, eliminating the need for wood or metal let-in bracing.

Door and window headers are sometimes referred to as lintels.

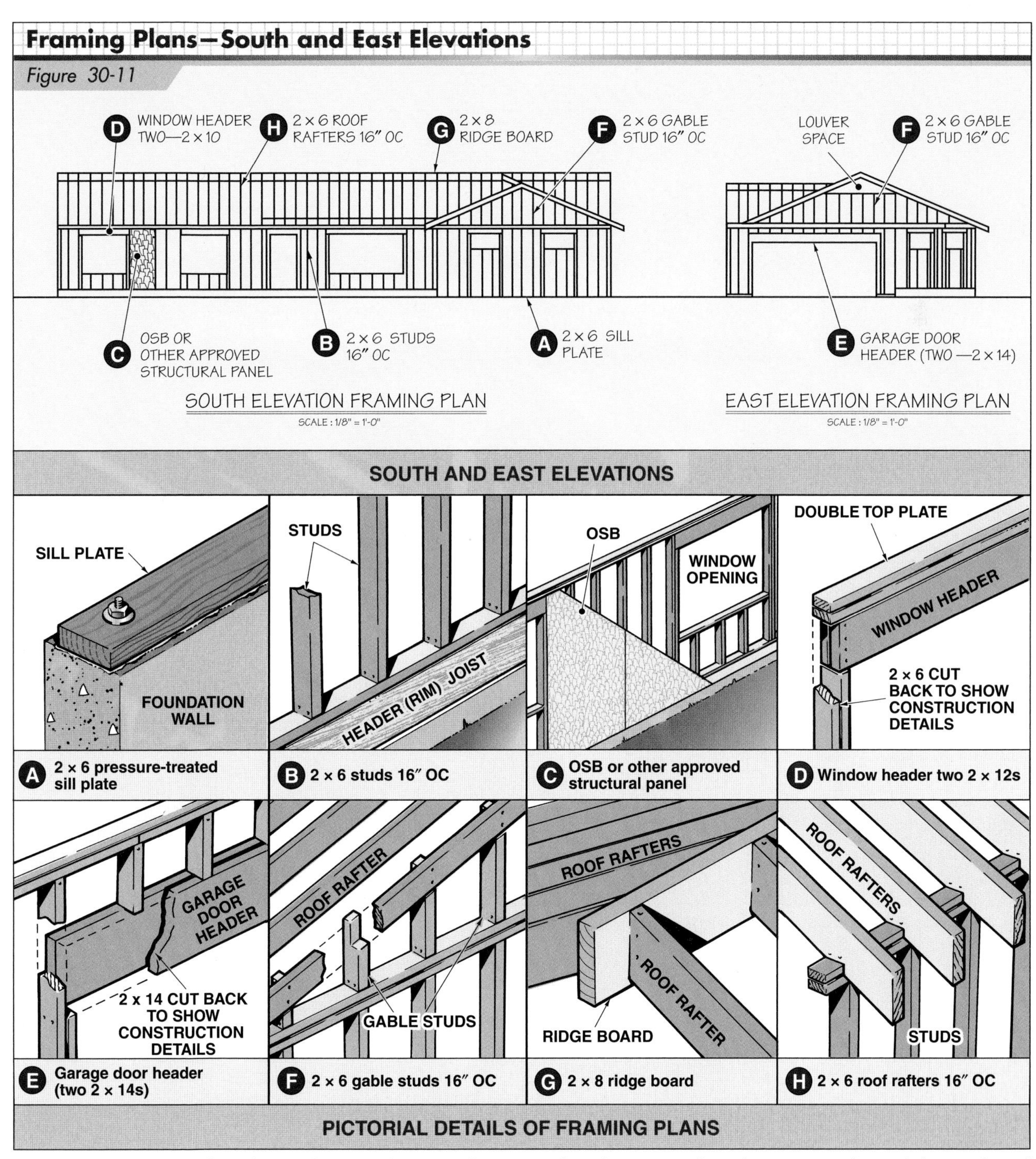

Figure 30-11. *The south and east elevation framing plans provide information about framing members of the south and east exterior walls.*

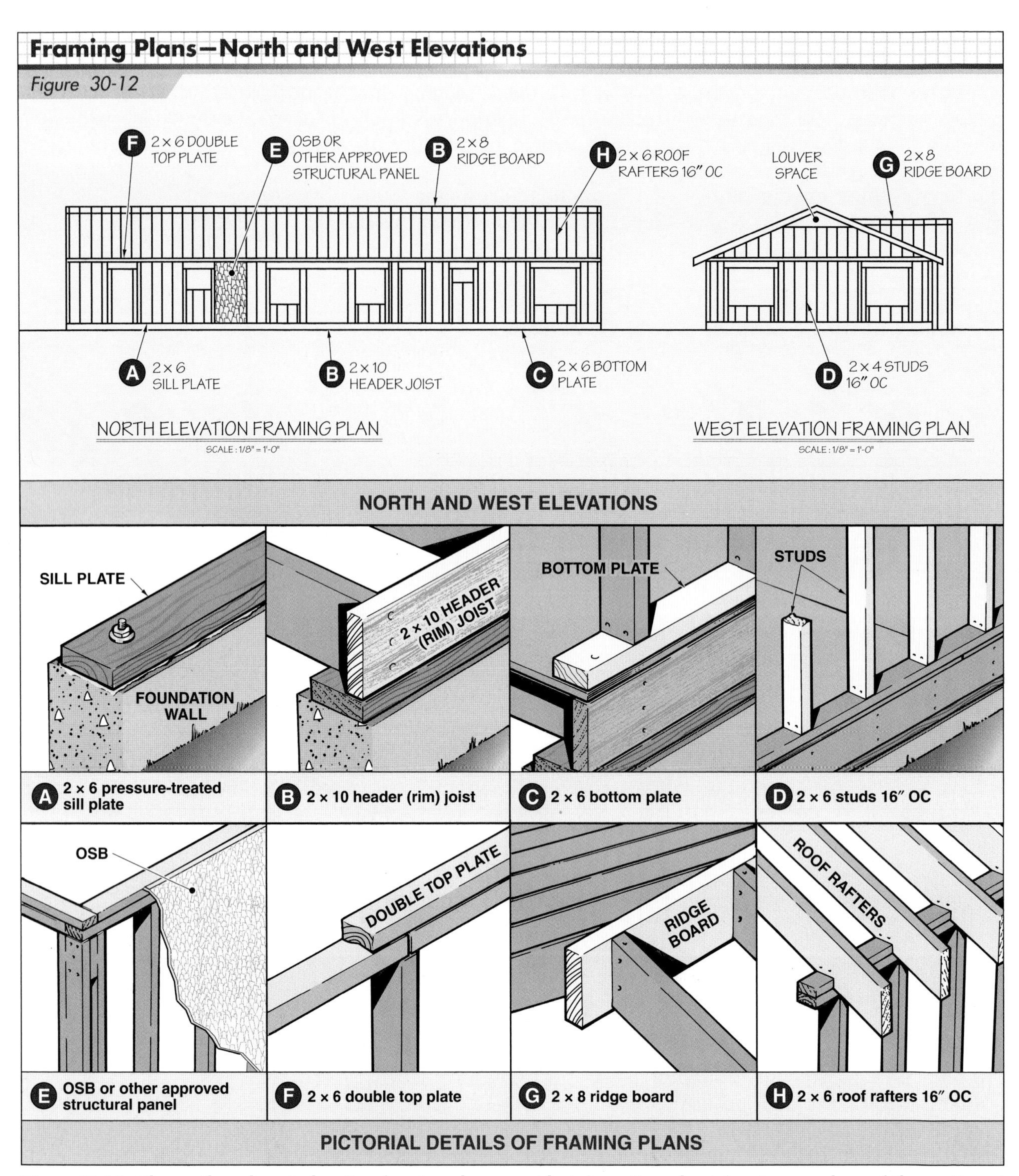

Figure 30-12. *The north and west elevation framing plans provide information about framing members of the north and west exterior walls.*

Figure 30-13 shows the floor joist framing plan for the three-bedroom house. Joists are doubled under all partitions running parallel to the joists. Metal cross bridging is installed to prevent joists from twisting.

The framing plan for the ceiling joists is shown in **Figure 30-14.** Ceiling joists and rafters should be aligned with the wall studs as close as possible to provide maximum support for the roof loads. Ceiling joists should be properly lapped at wall studs where required.

Figure 30-15 shows the roof rafter framing plan. A ridge board spans the length of the roof with roof rafters extending down to the exterior wall studs.

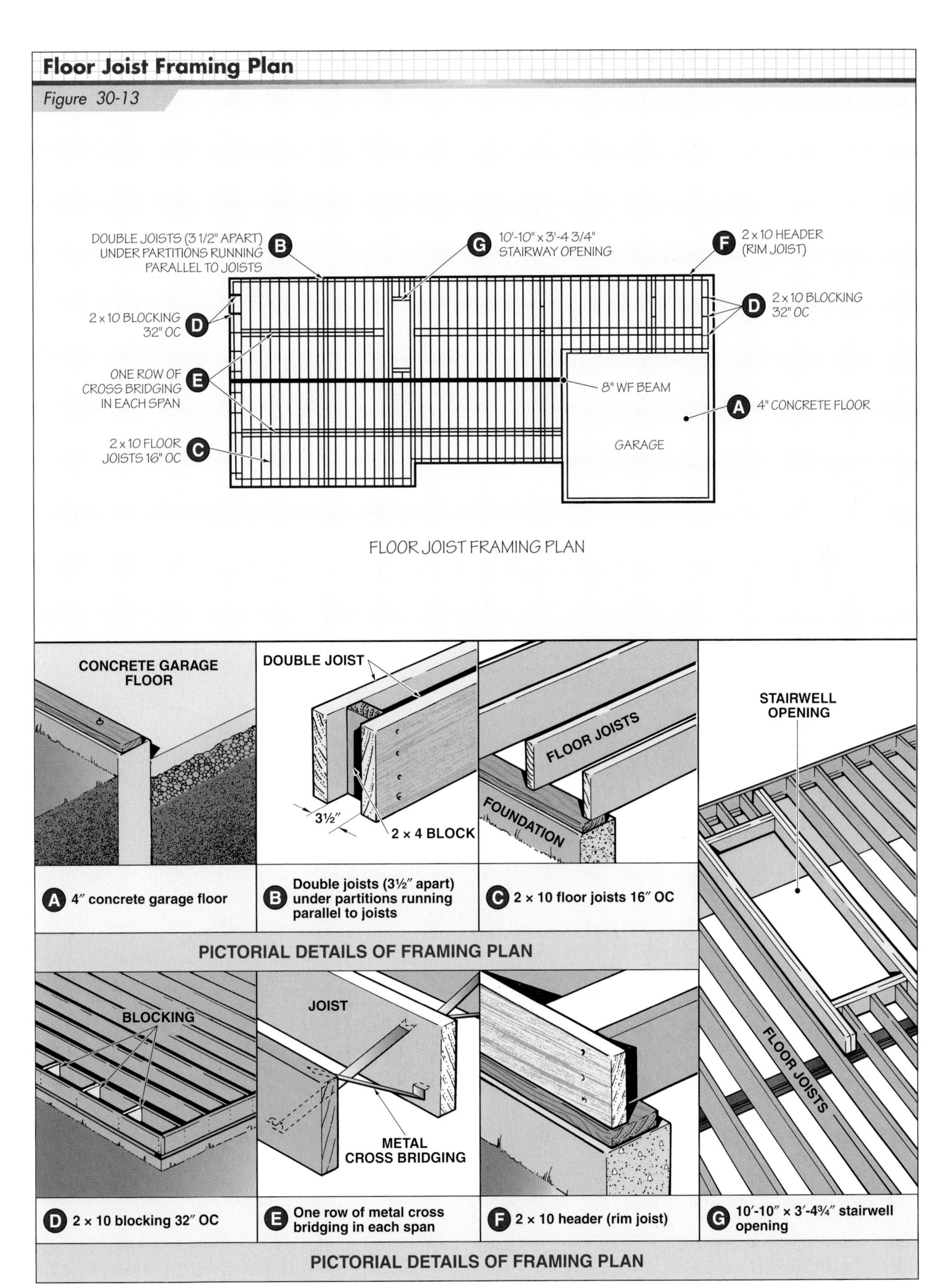

Figure 30-13. *The floor joist framing plan includes details about the stairway floor opening. Double joists are installed under partitions running parallel to joists.*

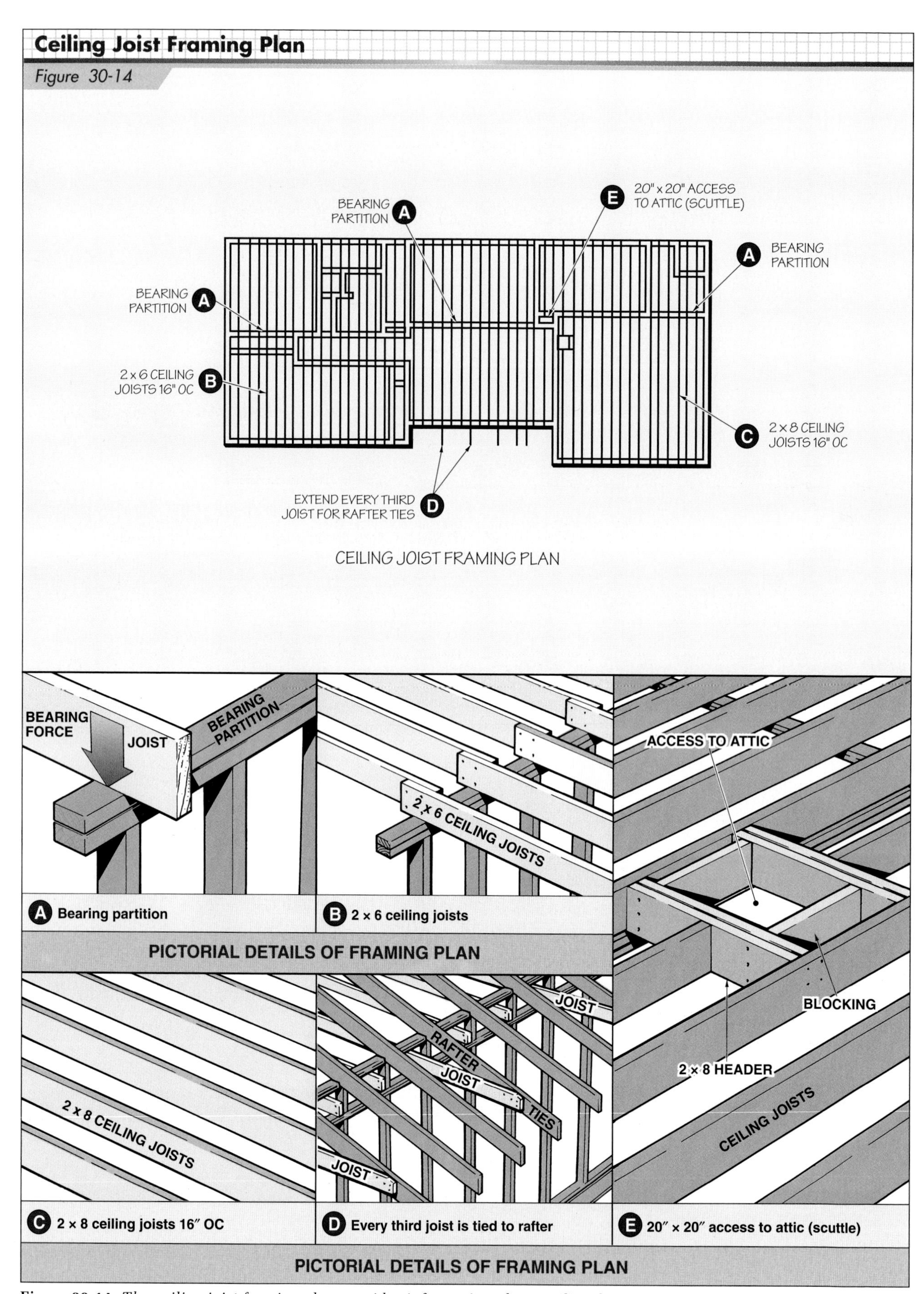

Figure 30-14. *The ceiling joist framing plan provides information about ceiling framing.*

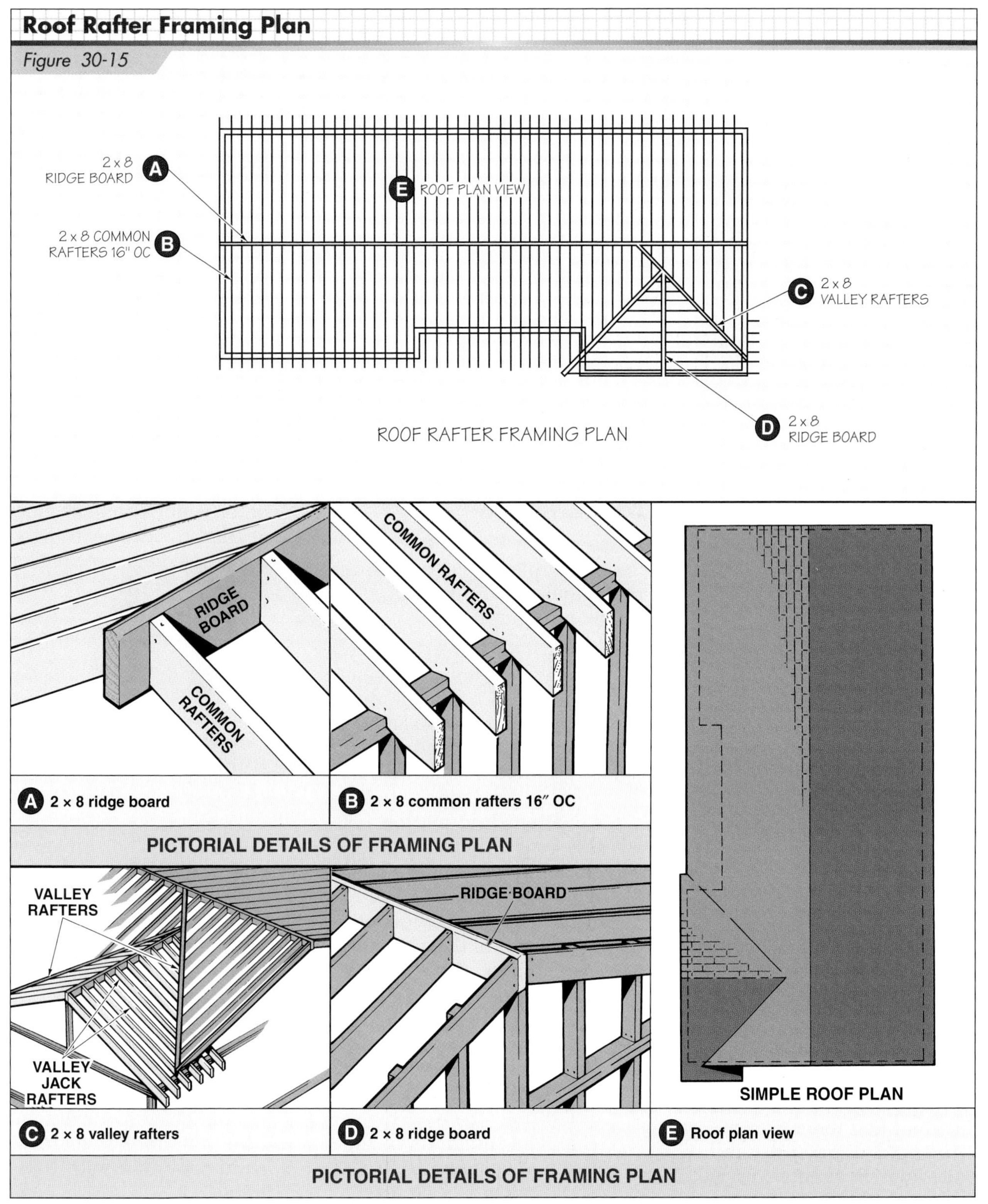

Figure 30-15. *A roof rafter framing plan indicates ridge board and rafter size, spacing, and direction. Many sets of prints provide only the simpler type of roof plan shown at lower right.*

UNIT 31

Door, Window, and Finish Schedules

OBJECTIVES

1. Define a schedule as it relates to construction prints.
2. Identify typical information included in a door schedule.
3. List typical information included in a window schedule.
4. Identify typical information included in a room finish schedule.

A *schedule* is detailed list that provides information about building components such as doors and windows. A set of construction prints typically includes a *door schedule* and *window schedule*. In many prints, a *room finish schedule* is also included. Letters and/or numbers on the prints refer to the schedule.

DOOR AND WINDOW SCHEDULES

The locations of all door and window openings except the foundation windows are shown on the floor plan. The foundation windows appear on the foundation plan. In some cases, the sizes (widths and heights) of the doors and windows are noted on the floor plans next to the openings. More typically, the door and window sizes are shown in door and window schedules. Schedules vary with different plans; some provide more information than others.

Additional schedules, such as lighting and plumbing fixture schedules, are commonly included in a set of construction prints.

Door Schedule

As shown in **Figure 31-1**, the door schedule for the three-bedroom house is divided into the following columns:

- Code: The letter used to identify the door is given; some plans use a number and/or letter. For example, a circled letter "A" is shown by the front entrance door in the floor plan and the south elevation. Information for this type of door is given on the first line of the door schedule.
- Quan: The number (quantity) of doors of this type and size is given. For example, there is one "A" door and one "E" door (garage door).
- Size: The width and height of doors are given (width always first). The "A" door is 3′-0″ wide by 6′-8″ high.
- Thk: The thickness of the door is shown; the "A" door is 1¾″ thick.
- Rough Opening: The width and height of the opening that must be provided in the wall to accommodate the door and the door frame (jamb) is the *rough opening.* A ½″ clearance is usually allowed at the sides and top of the jamb. Clearance must also be provided beneath the door. The rough opening for the "A" door is 3′-3″ wide by 6′-10¼″ high.

Some door and window schedules do not include rough openings. When rough openings are not provided, a carpenter must know how to calculate the rough opening from information found in different parts of the plans.

- Jamb Size: The thickness and width of the door frame are given. The jamb for the "A" door is 1³⁄₁₆″ thick by 4⅞″ wide.
- Type: The door action is indicated by its type. The "A" door is a hinged door which swings on a set of hinges.
- Design: The appearance and construction of the door are indicated by its design. The "A" door is a flush, solid-core door with three lights.
- Remarks: Special information such as the location of a door, special operating instructions, and other pertinent information may be given. The "A" door is identified as a front entrance door.

DOOR SCHEDULE									
CODE	QUAN	SIZE	THK	ROUGH OPENING	MASONRY OP'G	JAMB SIZE	TYPE	DESIGN	REMARKS
A	1	3′-0″ × 6′-8″	1 3/4″	3′-2″ × 6′-10 1/4″		1 3/16″ × 4 7/8″	HINGED	3 LIGHTS; FLUSH SOLID-CORE	FRONT ENTRANCE DOOR
B	2	2′-8″ × 6′-8″	1 3/4″	2′-10″ × 6′-10 1/4″		1 5/16″ × 4 7/8″	" "	3 LIGHTS; 1 PANEL	REAR ENTRANCE DOORS
C	1	2′-8″ × 6′-8″	1 3/8″	2′-10″ × 6′-10 1/4″		1 5/16″ × 4 5/8″	" "	FLUSH HOLLOW-CORE	DOOR BETWEEN UTILITY ROOM & GARAGE
D	1	6′-0 1/8″ × 6′-10″		6′-0 1/2″ × 6′-10 3/8″			SLIDING	GLASS	SLIDING GLASS DOOR
E	1	16′-0″ × 7′-0″	1 3/8″	16′-3″ × 7′-1 1/2″		3/4″ × 6″	OVERHEAD	4 LIGHTS; 16 PANELS	GARAGE DOOR
F	1	2′-8″ × 6′-8″	1 3/8″	2′-10″ × 6′-10 1/4″		3/4″ × 4 5/8″	HINGED	FLUSH HOLLOW-CORE	INTERIOR DOOR
G	3	2′-6″ × 6′-8″	1 3/8″	2′-8″ × 6′-10 1/4″		3/4″ × 4 5/8″	" "	" "	" "
H	5	2′-0″ × 6′-8″	1 3/8″	2′-2″ × 6′-10 1/4″		3/4″ × 4 5/8″	" "	" "	" "
J	2	2′-6″ × 6′-8″	1 3/8″	5′-2″ × 7′-0″			SLIDING	" "	POCKET DOOR
K	1	3′-0″ × 6′-8″		3′-2 1/2″ × 6′-10 1/2″		3/4″ × 4 5/8″			CASED OPENING
L	1	7′-7 9/16″ × 7′-11″		7′-8 9/16″ × 7′-11 3/4″		3/4″ × 4 5/8″	BI-FOLD	FLUSH	BEDROOM #3
M	1	5′-3 3/8″ × 7′-11″		5′-4 3/8″ × 7′-11 3/4″		3/4″ × 4 5/8″	" "	" "	BEDROOM #1
N	1	4′-11 9/16″ × 7′-11″		5′-0 9/16″ × 7′-11 3/4″		3/4″ × 4 5/8″	" "	" "	BEDROOM #2
O	1	3′-11 3/8″ × 7′-11″		4′-0 3/8″ × 7′-11 3/4″		3/4″ × 4 5/8″	" "	" "	BEDROOM #1

Figure 31-1. *A door schedule provides the information necessary to lay out the door openings. In commercial and industrial construction, many door schedules also make reference to print details and sections.*

Window Schedule

A window schedule provides information about windows similar to the information that a door schedule provides about doors. The window schedule for the three-bedroom house, shown in **Figure 31-2,** is divided into the following columns:

- Code: The letter used to identify the window is given; some plans use a number and/or letter. For example, a circled letter "R" appears next to each bedroom window unit on the floor plan and elevation drawings. Information for this type of window appears on the second line of the window schedule.
- Quan: The number (quantity) of windows of this type and size is given. For example, there are five "R" windows in this house.
- No. Lts.: The number of *lights* (panes of glass) in each window is given. The "R" window is a double-hung window and has four lights in the top section and four in the bottom section.
- Glass Size: The dimensions of the entire glass area in each window section are given. In an "R" window, the glass is 28″ wide by 24″ high in both the top and bottom sashes.
- Sash Size: The dimensions of the entire window unit after the glass has been set in its frame give the window sash size. The combined sash size of an "R" window is 2′-8″ wide by 4′-6″ high.
- Rough Opening: The width and height are given for the opening that must be provided in the wall to accommodate the window, frame, and clearance around the frame. The "R" window, which is a double-hung window, requires a rough opening 5′-10″ wide by 4′-10″ high.
- Remarks: Information such as the location of the window, operating instructions, and window manufacturer may be included. In this example, the "R" windows are identified as double-hung windows installed as a double unit.

ROOM FINISH SCHEDULE

Some prints also have a room finish schedule, which specifies the interior finish materials for each room in the house. The room finish schedule for the three-bedroom house provides information for the floor, walls, ceiling, base, and trim in each room. **See Figure 31-3.** The base molding is nailed at the bottom of the walls. Trim refers to the material around the door and window openings. Some of the information contained in the room finish schedule is intended for other tradesworkers.

Schedules
Media Clip

WINDOW SCHEDULE							
CODE	QUAN	NO. LTS	GLASS SIZE	SASH SIZE	ROUGH OPENING	MASONRY OP'G	REMARKS
P	1	1 8	56" × 49" 20" × 24"	(1)5'-0" × 4'-6" (2)2'-0" × 4'-6"	9'-8" × 4'-10"		PICTURE WINDOW FLANKED EACH SIDE BY ONE (1) DH WINDOW
R	5	8	28" × 24"	(2)2'-8" × 4'-6"	5'-10" × 4'-10"		DOUBLE-HUNG WINDOW
S	2	2	39" × 22"	3'-8" × 4'-6"	3'-9" × 4'-7$\frac{3}{16}$"		NO. A-12-84 CURTIS CONVERTIBLE AWNING WINDOW UNIT
T	2	1	39" × 17"	3'-8" × 1'-10"	3'-9" × 1'-11$\frac{3}{8}$"		NO. A-11-83 " " "
U	1	1	33" × 17"	3'-2" × 1'-10"	3'-3" × 1'-11$\frac{3}{8}$"		NO. A-11-73 " " "
V	2	1	27" × 17"	2'-8" × 1'-10"	2'-9" × 1'-11$\frac{3}{8}$"		NO. A-11-63 " " "
W	1	3	17" × 27"	3'-8" × 2'-8"	3'-9" × 2'-9$\frac{3}{8}$"		NO. A-21-56 CURTIS CONVERTIBLE CASEMENT WINDOW UNIT
X	8	2	15" × 20"	2'-8$\frac{1}{2}$" × 1'-10$\frac{3}{4}$"		2'-0" × 1'-11"	STEEL BASEMENT WINDOWS

Figure 31-2. *A window schedule provides the information necessary to lay out the window openings.*

ROOM FINISH SCHEDULE					
ROOM	FLOOR	WALLS	CEILING	BASE	TRIM
LIVING ROOM	1" × 3" OAK	$\frac{1}{4}$" PANELING OVER $\frac{1}{2}$" DRYWALL	$\frac{1}{2}$" GYP BOARD	WOOD	WOOD
FAMILY ROOM	1" × 3" OAK	$\frac{1}{2}$" DRYWALL-PAPERED	$\frac{1}{2}$" GYP BOARD	WOOD	WOOD
BEDROOMS	1" × 3" OAK	$\frac{1}{2}$" DRYWALL-PAPERED	$\frac{1}{2}$" GYP BOARD	WOOD	WOOD
KITCHEN	LINOLEUM	$\frac{1}{2}$" DRYWALL-PAINT	$\frac{1}{2}$" GYP BOARD	LINO. COVE	WOOD
UTILITY ROOM	LINOLEUM	$\frac{1}{2}$" DRYWALL-PAINT	$\frac{1}{2}$" GYP BOARD	LINO. COVE	WOOD
HALL	1" × 3" OAK	$\frac{1}{2}$" DRYWALL-PAINT	$\frac{1}{2}$" GYP BOARD	WOOD	WOOD
BATHROOM	LINOLEUM	$\frac{1}{2}$" DRYWALL-PAINT	$\frac{1}{2}$" GYP BOARD	LINO. COVE	WOOD
GARAGE	CONCRETE	$\frac{1}{2}$" DRYWALL-PAINT	$\frac{1}{2}$" GYP BOARD	WOOD	WOOD

Figure 31-3. *A room finish schedule identifies the finish materials for the walls, floors, and ceilings. Room finish schedules may also reference room elevations.*

Unit 31 Review and Resources

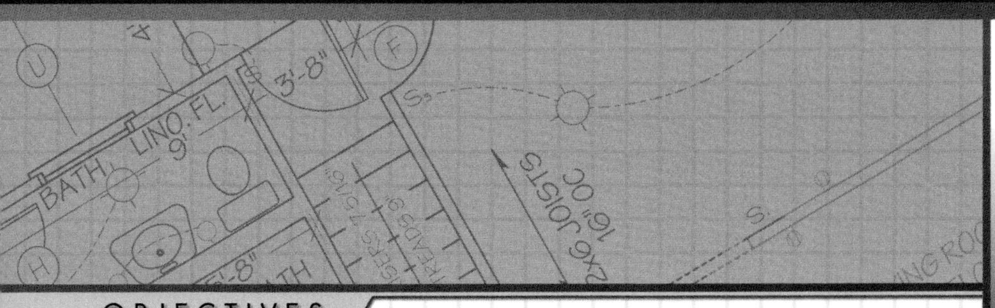

UNIT 32

Building Codes, Zoning, Permits, and Inspections

OBJECTIVES

1. List and describe the purpose of building codes.
2. Describe the purpose of zoning regulations.
3. Describe types of permits required before starting a construction project.
4. List and describe the proper sequence of inspections.
5. Identify the two major green building rating systems.
6. List and describe the major credit categories for the LEED® green building system.

In most sections of the country, particularly in and near towns and cities, strict rules exist for the construction of new buildings. These rules can be used to enforce minimum building and safety standards, ensure the use of proper materials and sound construction methods, and are written in pamphlets or books known as *building codes*. No single factor has a more direct effect on building design and safety than building codes. Architects must have a thorough knowledge of building codes in effect in their jurisdiction before developing a set of plans. Building contractors and tradesworkers must be familiar with the local code regulations as they apply to construction procedures.

BUILDING CODES

Building codes establish the minimum standards required to protect the health, safety, and welfare of persons who will be living or working in the building. Fire-resistant materials, adequate lighting, ventilation, and insulation are among the areas covered in code regulations. Often separate code books exist for electrical and plumbing operations.

State and Local Codes

Most states in the United States have a building code that can be applied in all areas of the state. In addition, larger cities, such as Los Angeles or Chicago, may have their own set of code regulations. City codes are typically based on the state code, but are usually more detailed and often have stricter requirements than statewide codes. Any officially adopted code is enforceable by law.

Model Codes

Model codes date back to the early 1900s when they were initially developed during periodic meetings of building officials and industry representatives. Model codes serve as models that can be adopted by states or local communities. Model codes are periodically revised to keep up with the changing conditions and new materials being developed for the construction industry.

Several model building codes existed at one point in time in the United States. The *Uniform Building Code* was a regional model code that strongly influenced construction procedures in the western states. The *Southern Standard Building Code* was a regional model building code that influenced construction procedures in the southeastern states. The *Basic National Building Code* was a national model building code that dictated construction procedures across the United States. Still in effect today, the *National Building Code* establishes minimum construction standards in Canada.

Since the early 1900s, the United States has operated under a model code system in which individual states adopt regional or national model building codes. The model codes were developed by the Building Officials and Code Administrators (BOCA), International Council of Building Officials (ICBO), and Southern Building Code Congress International (SBCCI). In 1994, BOCA, ICBO, and SBCCI formed the *International Code Council® (ICC)* to develop recognized national model building codes for commercial and residential construction. Members of the ICC developed the *International Building Code® (IBC)*, which pertains to commercial construction, and the *International Residential Code® (IRC)*, which applies to one- and two-family houses, townhouses, and condominiums. The goal of the IBC and IRC is to standardize construction procedures across the country and to facilitate the work of architects, engineers, and building officials. Similar to their predecessors, the IBC and IRC are periodically revised to address changing conditions and new materials in the construction industry. In the event that more than one model code applies to a specific construction project, the code with the stricter requirements applies.

Another model code, the *International Energy Conservation Code®*, is based on the 1995 edition of the *Model Energy Code,* which was developed by the Council of American Building Officials (CABO). The *International Energy Conservation Code* provides guidance to permit the use of innovative approaches and procedures to effectively use energy, and applies to both residential and commercial construction.

Minimum Property Standards

An important guide to code requirements in all parts of the country is the *Minimum Property Standards (MPS),* published by the U.S. Department of Housing and Urban Development (HUD). The Minimum Property Standards establish minimum requirements for buildings constructed under HUD housing programs, including new one-family houses, multifamily dwellings, and health care facilities. Until the mid-1980s, HUD maintained separate MPSs for different types of buildings. However, since that time, HUD has adopted model building codes and local building codes.

ZONING REGULATIONS

Like local building codes, *zoning regulations* strongly influence building design. Most cities, counties, and provinces have specific laws concerning the type of buildings that can be constructed in different areas of the community. A city is usually divided into different zones. The boundaries of the zones are shown on special *zoning maps.* When developing a set of prints, an architect must ensure the type of building being designed is permitted in the area where it is to be built.

There are three major types of zones in larger communities—residential, commercial, and manufacturing. Residential zones are districts limited to buildings in which people live and buildings that serve the neighborhood, such as schools, churches, libraries, and playgrounds. Some residential zones may permit only one-family dwellings while others may permit multifamily dwellings such as apartments. Commercial zones permit buildings such as stores, shopping centers, movie theatres, restaurants, bowling alleys, and hospitals. Manufacturing zones are areas set aside for factories, warehouses, and other types of industry.

Figure 32-1 illustrates a number of common residential zoning regulations. Most cities specify the minimum lot size for that zone. For example, a zoning regulation for a residential section of a city might state that any new building must be located on a lot of no less than 5000 sq ft (square feet) (50′ × 100′ = 5000 sq ft).

Another regulation might specify the maximum amount of the lot that the building is allowed to cover. For example, if a building is allowed to cover no more than 40% of a lot, and if the lot is 5000 sq ft, then the building must cover no more than 2000 sq ft (40% × 5000 = 2000). Consequently, this zoning regulation affects the size of the house designed for that lot.

Other regulations may establish the maximum height of a building in a particular zone. For example, a residential zone may allow one-family buildings up to three stories that do not exceed 35′ in height.

Another set of regulations usually contained in zoning ordinances concerns the minimum setback of the building from the property lines. For example, one regulation may specify that the front setback (distance from the front of the house to the front property line) can be no less than 25′, and that the side setback (distance from the side of the house to the side property line) can be no less than 5′.

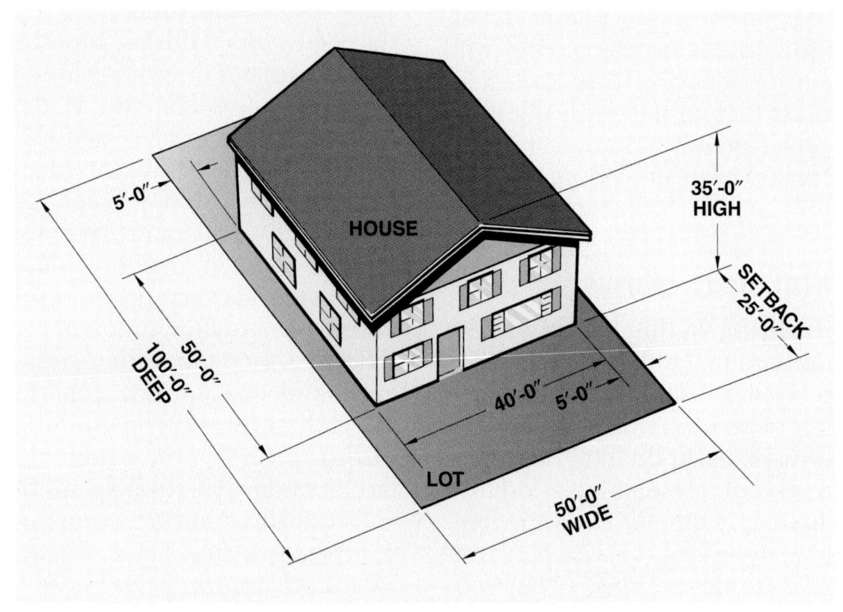

Figure 32-1. *Zoning regulations, such as the minimum lot size, dwelling size, and setbacks, strongly influence building design.*

Zoning regulations are subject to change. Residential zones limited to one-family dwellings may be reclassified to allow multifamily dwellings or commercial structures. Many cities and counties have planning commissions that review requests for changes in zoning regulations and make recommendations to the local government. Since zoning changes may occur at any time, current information should be obtained before the drawings for a building are finalized. Changing a building to conform to a zoning change once construction has begun will not only delay construction but can be very costly.

PERMITS

Before construction begins, the owner or building contractor must apply to local building authorities for a *building permit.* An application must be completed, stating the kind of construction proposed. **See Figure 32-2.** The application also requires a legal description of the land upon which the construction is to take place and the estimated cost of the project. Local building officials also may request other information.

One or two sets of the working drawings and specifications must be submitted with an application. The plans are examined by the proper authorities to ensure they conform to the local building code and zoning regulations. If everything appears to be in order, the plans are approved and a permit is granted to begin construction. **See Figure 32-3.**

In addition to the building permit, separate *electrical* and *plumbing permits* are usually required. The owner or building contractor must pay fees for the permits. The amount is usually based on the total cost of the project.

INSPECTIONS

A building project for which a permit has been granted will be inspected a number of times as the work progresses. The overall structural inspection is conducted by a local *building inspector.* The plumbing and electrical work is usually checked by separate plumbing and electrical inspectors. If inspectors discover a violation of the building code, they have the legal authority to have the work demolished and reconstructed properly, which can be costly to the owner or building contractor.

The International Building Code recommends the following sequence of inspection:

1. Footing and foundation inspection: To be made after the trenches are excavated, the forms constructed, and all the required bolts and reinforcing steel are placed inside the form. The inspection must take place before the concrete is placed in the forms.
2. Concrete slab or underfloor inspection: To be made where a concrete slab floor is to be placed. A concrete slab or underfloor inspection mainly concerns the inspection of electrical conduit, plumbing, ductwork, and other equipment that will be covered by the slab.
3. Lowest floor elevation: In floodplain areas, the elevation of the lowest floor, including the basement, must be certified and submitted to the building inspector prior to additional vertical construction.
4. Framing inspection: To be made after the floors, walls, ceilings, and roof are framed and all blocking and bracing is installed. All the rough electrical work, plumbing, and ductwork must be completed and visible for inspection before the walls can be enclosed.
5. Lath and/or gypsum board inspection: To be made after all exterior and interior lath or gypsum board is in place, and before any plaster is applied or any of the gypsum board joints are taped and finished.
6. Fire-resistant penetration inspection: To be made for protection of joints and penetrations through fire-resistance-rated assemblies prior to being concealed from view.
7. Energy efficiency inspection: To be made for building envelope R- and U-values, fenestration U-value, duct system R-value, and HVAC and water-heating efficiency.
8. Final inspection: To be made after the building has been completed and is ready for occupancy.

A framing inspection is made after the building framing has been completed but before the walls are enclosed. Rough-in electrical and plumbing can also be inspected at this time.

NO. ______________

DO NOT WRITE IN THIS SPACE

FOR VILLAGE CLERK'S USE ONLY

DATE ______________

for use of assessor

VOLUME ______

PAGE ______ LINE ______

BLK. ______ PARCEL ______

APPLICATION FOR BUILDING PERMIT
Village of Emperior, Illinois

SKETCH OF APPLICANT'S LOT SHOWING EXISTING AND PROPOSED IMPROVEMENTS, INCLUDING CONSTRUCTION DETAILS.

FRAME ☐ BRICK ☐ BASEMENT: YES ☐ NO ☐

WIDTH OF LOT ______

REAR YARD

SIDE YARD

SIDE YARD

LENGTH OF LOT

SETBACK LINE

FRONT YARD

do not fold

I HEREBY APPLY FOR A BUILDING PERMIT FOR THE CONSTRUCTION (REPAIR) OF A BUILDING SPECIFIED AS FOLLOWS:

OWNER ______

OWNER'S PRESENT ADDRESS ______

ADDRESS OF PROPOSED STRUCTURE ______

TYPE OF BUILDING ______

GENERAL CONTRACTOR ______

ADDRESS OF GENERAL CONTRACTOR ______

ELECTRICAL CONTRACTOR ______

ADDRESS OF ELECTRICAL CONTRACTOR ______

PLUMBING CONTRACTOR ______

ADDRESS OF PLUMBING CONTRACTOR ______

PLANS PREPARED BY ______

HEATING, AIR COND. CONTRACTOR ______

ADDRESS OF HEATING, AIR COND. CONT. ______

LEGAL DESCRIPTION: LOT ______ BLOCK ______

SUBDIVISION ______

SEC ______ TOWN ______ 36 ______ RANGE ______ 14 ______

COLLECTOR'S TAX BILL VOL. NO. ______ ITEM ______

PERMANENT REAL ESTATE INDEX NO. ______

COST OF BUILDING COMPLETE $
AS THE APPLICATION FOR THIS PERMIT I EXPRESSLY AGREE TO CONFORM TO ALL APPLICABLE ORDINANCES, RULES, AND REGULATIONS OF THE VILLAGE OF EMPERIOR.

______
SIGNATURE OF APPLICANT

BUILDING INSPECTOR'S ANALYSIS AND APPROVAL

ANALYSIS	APPLICATION SATISFACTORY	NOT SATISFACTORY
1. PLOT PLAN RECEIVED	☐	☐
2. DUPLICATE SET OF PLANS RECEIVED	☐	☐
3. WATER CONNECTION LOCATION DESCRIBED IN PLAN	☐	☐
4. SEWAGE DISPOSAL PLAN	☐	☐
5. ZONING ORDINANCE COMPLIANCE	☐	☐
6. ELECTRICAL PLANS, LICENSE, AND BOND	☐	☐
7. PLUMBING PLANS, LICENSE, AND BOND	☐	☐
8. HEATING, AIR COND., & REFRIG. LICENSE & BOND	☐	☐
9. PUBLIC SIDEWALK PLAN	☐	☐
10. IF PUBLIC OR APT. BLDG. FIRE CHIEF APPROVAL	☐	☐
11. IF PAVING TO BE CUT, IS REQUIRED SURETY CO. BOND SUPPLIED AND IN ORDER	☐	☐
12. ______	☐	☐
13. ______	☐	☐
14. ______	☐	☐

APPROVED

DATE: ______ ______
BUILDING INSPECTOR, EMPERIOR, IL

CERTIFICATE OF OCCUPANCY ISSUED

DATE: ______ ______
BUILDING INSPECTOR, EMPERIOR, IL

COMPUTATION OF FEES

BUILDING PERMIT NO. $ ______

ELECTRICAL PERMIT NO. ______

SEWER TAP & PLUMBING PERMIT NO. ______

WATER TAP ______

HEATING, AIR COND., & REFRIGERATION NO. ______

WATER METER ______

WATER USE DURING CONSTRUCTION ______
(THIS DOES NOT AUTHORIZE USE OF WATER FROM ANY FIRE HYDRANT. ANYONE TAMPERING WITH A FIRE HYDRANT WILL BE ARRESTED)

SIDEWALK CONSTRUCTION PERMIT NO. ______

EXCAVATION FOR WATER TAP ______

EXCAVATION FOR SEWER TAP ______

BUILDING INSPECTOR'S FEE;

PLANS AND SPECIFICATIONS ______

OCCUPANCY & FIELD INSPECTIONS ______

STREET & ALLEY ______

TOTAL ______

PENALTY – $5.00 FOR CHANGE OF SUBCONTRACTOR

Figure 32-2. *An owner or building contractor must apply for a building permit before construction can commence. Local building authorities may also request additional information.*

VILLAGE OF EMPERIOR
SLAKE COUNTY, ILLINOIS

NOTICE!

BUILDING PERMIT

Must Be Posted on FRONT of Building

Any person willfully destroying this permit before completion of this building WILL BE PUNISHED TO THE FULL EXTENT OF THE LAW.

BUILDING PERMIT No. ____________

Is issued for the [ERECTION / ALTERATION] **of a** ________ **story**

Description of Building ____________

No. ________ **STREET** ____________

GENERAL CONTRACTOR ____________

Date ____________

Signed ____________
Building Commissioner

Figure 32-3. *A building permit must be obtained and properly posted before beginning construction.*

Inspection Record Card

Many local code authorities require that an inspection record card be posted in a conspicuous place on the job. This card lists all the inspections required for the job. Inspectors sign the card as the different stages of construction are completed.

GREEN BUILDING RATING SYSTEMS

Green building rating systems provide a way to measure the effectiveness of green building design and construction methods and the efficient use materials, energy, water, and other natural resources. Voluntary rating systems provide architects, engineers, construction managers, lenders, and government officials with standards and guidance for green buildings. The majority of green buildings are guided by two voluntary rating systems: Leadership in Energy and Environmental Design® (LEED®) and Green Globes®. Both use a point system to grade the major areas of construction projects leading to certification.

LEED CERTIFICATION

The most widely used system in the U.S. and Canada is Leadership in Energy and Environmental Design® (LEED®), which was created by the United States Green Building Council (USGBC). LEED is a program that provides verification that individual buildings, homes, or entire neighborhoods and communities meet a minimum performance standard. The LEED program provides guidance and standards for architects, engineers, construction managers, lenders, and government officials. There is currently a movement to incorporate such a system into federal and state law, and it has been embraced by some government agencies such as the U.S. General Services Administration (GSA) and the U.S. Navy.

The LEED rating system is separated into nine major categories based on project or building type. **See Figure 32-4.** In all of these categories, projects must meet prerequisites and attain a minimum number of credits. The number of credits attained determines the level of LEED certification. There are four levels of LEED certification: Certified, Silver, Gold, and Platinum. To achieve Platinum certification for all of the categories, except LEED for Homes, 80 points are needed. Due to LEED for Homes using slightly different rating guidelines, 90 points are needed to achieve Platinum certification.

The LEED point system applies to the following five major areas for most of the credit categories:

- sustainable sites
- water efficiency
- energy and atmosphere
- materials and resources
- indoor environmental quality

There are also additional areas where credits can be achieved. The first, innovation in design, grants up to six points for sustainable building expertise as well as design ideas not covered under the other credit categories. Regional priority credits allow four points for meeting environmental requirements that have been deemed particularly important for a project's specific geographical location.

Leadership in Energy and Environmental Design® (LEED®) Major Certification Categories

Figure 32-4

- New Construction and Major Renovations
- Existing Buildings Operations and Maintenance
- Commercial Interiors
- Core and Shell Components
- Retail
- Schools
- Homes
- Healthcare
- Neighborhood Development

Figure 32-4. *The LEED certification category that a project will fall under is based on the type of project or building that will seek certification.*

Sustainable Sites

A *sustainable site* is a building site that has reduced disruption of local plant and animal life, conserves existing natural areas, and restores damaged areas. When possible, green buildings should be oriented to take advantage of natural sunlight and cooling breezes. Sustainable site credits are awarded to building owners who address issues such as site selection, land disturbance, pollution reduction, storm water, and erosion control. One way that credits can be achieved is through the reduction of heat island effect.

The heat island effect describes the higher air and structure temperatures in an urban setting as compared to lower temperatures found in rural areas. Dark roofs and nonreflective surfaces release heat absorbed from sunlight into the surrounding atmosphere. The result can be an increase in outdoor air temperature of 2°F to 10°F. A reduction in the number of heat islands or their intensity limits the amount of extra energy that must be used to cool the excessively hot buildings. The production of the extra energy that is required to cool the buildings causes increased pollution and wastes natural resources. Lighter, more reflective roofing or building material colors, such as white or light grey, are recommended to reduce the amount of heat absorbed.

Green roofs are also an option to help reduce the heat island effect. An engineered green roof system is used to support plants and a growing medium. Green roofs provide natural roof cooling and can also reduce the amount of storm runoff from the roof area. **See Figure 32-5.**

A LEED Green Associate credential can be earned by passing an exam that covers the principles of sustainable design and construction.

Green Roofs

Figure 32-5

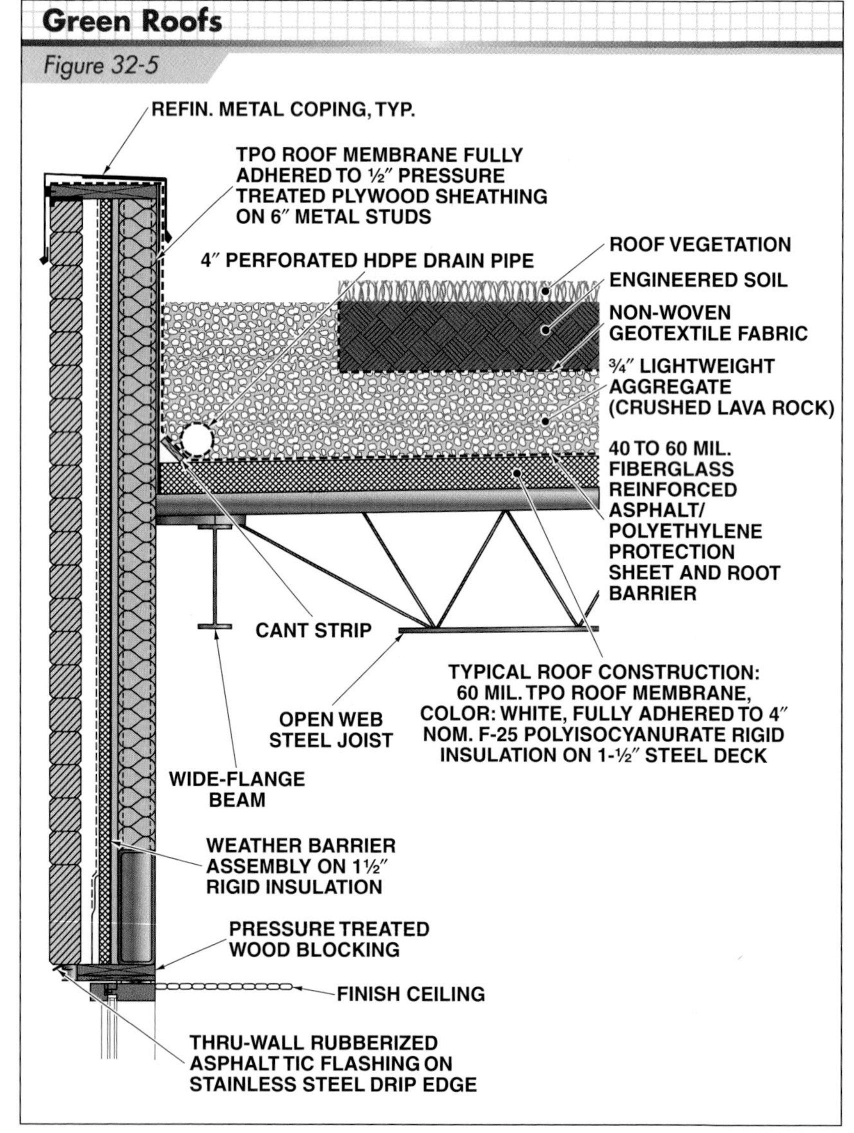

Figure 32-5. *An engineered green roof system is used to support plants and a growing medium.*

Water Efficiency

Indoor and outdoor water conservation has become one of the major concerns in today's environment. Therefore, green building design must include devices and methods that reduce water use. Water efficiency credits are awarded to building owners who limit or eliminate the use of supplied potable water. A reduction of water usage decreases operating costs and lessens the burden on local municipalities of supplying water. One way to reduce water usage is the use of ultra-low fixture devices in faucets, showers heads, and toilets. Dual-supply plumbing systems can be installed to use recycled (gray) water for toilets and outside irrigation. **See Figure 32-6.** *Gray water* is wastewater generated from domestic processes such as bathing, cleaning, and washing laundry. Gray water makes up 50% to 80% of residential wastewater.

A rainwater harvesting system is another way to reduce potable water consumption. A *rainwater harvesting system* is an on-site water collection and holding system for rainwater collected from roofs, impervious surfaces, and parking lots. The collection system can be as simple as a rain barrel under a downspout with a spigot at the bottom attached to a garden hose. It also can be a large, elaborate system that collects rainwater from roofs and pipes it to large cisterns for storage. The stored rainwater can then be used for irrigation in place of potable water.

Energy and Atmosphere

Buildings in the United States consume greater than 30% of the nation's total energy load and about 60% of the nation's electricity. Fossil fuels such as oil and coal account for about three-quarters of national energy production, significantly contributing to global warming. Therefore, green building design requires the reduction of energy use. The use of high-efficiency or renewable energy systems plays a major role in reducing a building's energy use. The more efficient the system and the tighter the building envelope, the greater number of points achieved.

Materials and Resources

LEED emphasizes the efficient use of materials, selection of environmentally friendly materials, and reduced waste during construction. Whenever possible, the selection of local or regional sources is encouraged to save energy from and reduce cost of transportation. Wood products that are certified by the Forest Stewardship Council (FSC) are strongly recommended. This certification indicates that lumber originates from sustainable, well-managed forests and includes a chain of custody. A chain of custody is the path that raw materials harvested from FSC-certified forests take through processing, manufacturing, and distribution. Today more than 170 million acres in the U.S. and Canada are certified under the FSC system.

Gray Water Systems

Figure 32-6

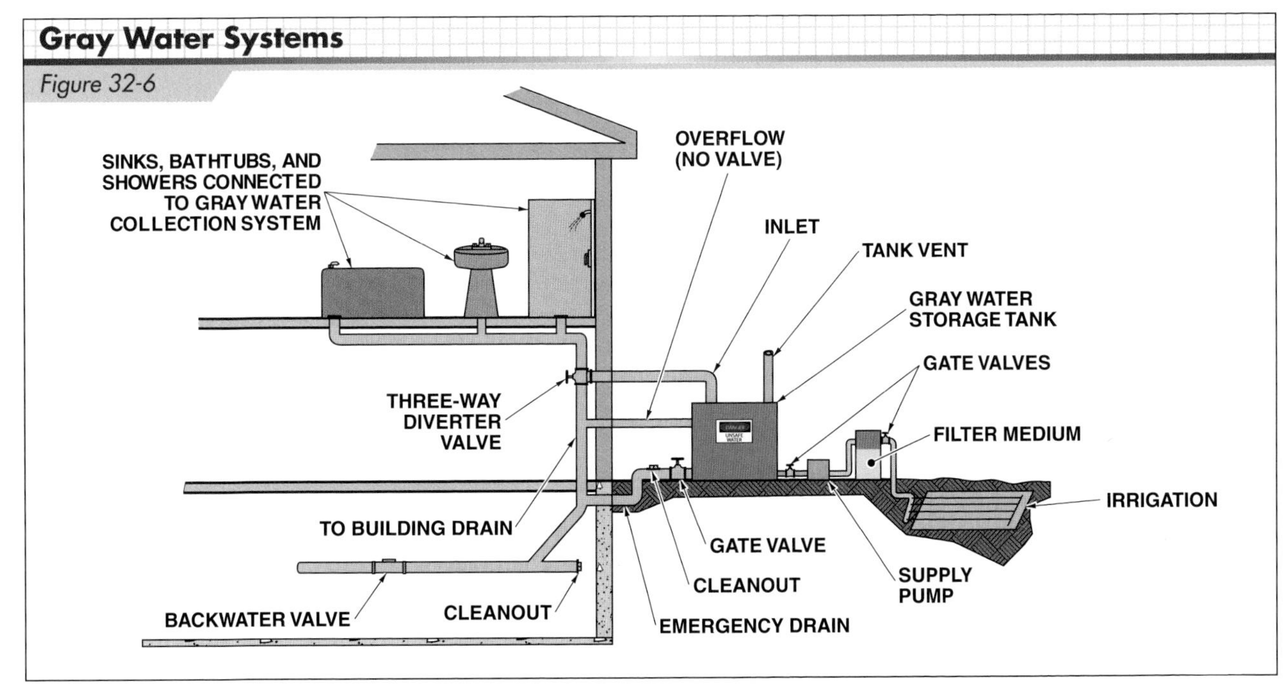

Figure 32-6. *Gray water generated from domestic processes such as bathing, cleaning, and washing laundry can be used for irrigation.*

LEED places great importance on recycling. For example, during the construction of a 2000 sq ft house, approximately 8000 lb of waste is transported and thrown into landfills, contributing to the overburdening of landfill facilities. A large amount of this waste (metal, plastic, etc.) can be deposited into separate dumpsters, taken to recycling facilities, and reused in manufacturing. For example, light-gauge metal framing members often consist of more than 25% recycled steel. In turn, these framing members are entirely recyclable if a building is demolished.

Indoor Environmental Quality

Most Americans spend as much as 90% of their time indoors. Indoor pollutants can be 2 to 5 times higher than outdoor pollutants. This leads to health problems for many Americans. It has been estimated that as many as 17 million Americans suffer from asthma and 40 million from some sort of allergy. Newly constructed buildings are tightly sealed against air infiltration compared to older buildings. In older buildings, it was common for air in a house to be replaced every hour due to normal air infiltration. In new construction, complete air exchanges may only occur every six hours. Therefore, toxins released indoors stay in circulation longer.

LEED certification emphases the protection of indoor air quality. This means that materials and finishes used on the interior of buildings should contain few or no volatile organic compounds (VOCs) that contain carcinogens or other irritants. A *carcinogen* is a cancer causing agent or material. A very common VOC substance is formaldehyde. It is common in many building materials such as solvents, adhesives, paints, insulation, particleboard, plywood paneling, pressed woods, carpeting, and hardwood and vinyl flooring. Using materials that have little or no VOCs can have a big impact on the indoor air quality of a building with no extra cost. **See Figure 32-7.**

Indoor Environmental Quality

Figure 32-7

Figure 32-7. *Millwork, such as shelving comprised of particleboard or plywood, that is used in a LEED building should contain little or no VOC-containing compounds.*

When using mechanical ventilation (HVAC) systems, the exchange rate must be high enough to ensure that the majority of air in the space is coming from outdoors. This reduces the amount of pollution inside. Properly functioning air handling systems that use fans and ductwork will remove indoor air and distribute filtered and conditioned air to strategic spaces throughout the building.

Another important indoor environmental quality credit is daylight and views. **See Figure 32-8.** *Daylighting* is a method of capturing and redirecting natural light for use in the interior of a building using special design or equipment. Increasing the amount of natural light indoors is good for the environment because it reduces the amount of electricity used. Architectural design, such as window size and placement, or special materials or components may be used for solar lighting to bring additional light indoors.

Innovation in Design

The purpose of the innovation in design category is to make it possible for design teams to receive extra credits for exceptional performance beyond the standards set by the green rating system. A few examples are exceeding the requirements for water use reduction, implementing educational outreach programs, keeping 50% to 75% of the site's native vegetation, and using building materials that go beyond required performance requirements.

In addition, a point will be awarded if at least one member of the project team is a LEED Accredited Professional (LEED AP). To obtain a LEED AP designation, a person must demonstrate knowledge of the LEED process and green building by passing an exam and documenting previous LEED experience.

Regional Priority

Regional priority credits are incentive credits that vary by geographic location. An extra point is given in one of six designated credit areas that have significant importance by USGBC regional councils and local chapters. Up to four bonus points may be earned out of the six possible. The specific locations are referenced by the ZIP code the project is registered under.

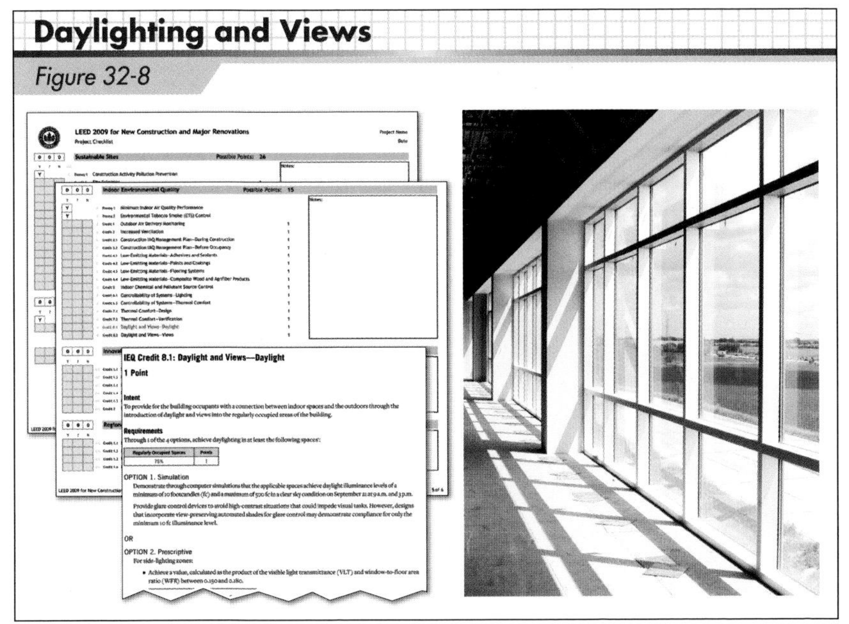

U.S. Green Building Council

Figure 32-8. *A LEED-certified building may have design elements such as floor-to-ceiling windows to reduce energy use and increased occupant comfort.*

GREEN GLOBES

Green Globes® is a green building rating system that originated in Canada and has spread to the U.S. In the U.S., the Green Building Initiative (GBI) operates Green Globes. There is a great deal of similarity and overlap (around 80%) between Green Globes and LEED. Like the LEED system, Green Globes emphasizes the efficient use of energy, water and resource efficiency, site ecology, indoor air quality, and pollution. Supporters of the Green Globes system claim that it is easier to use, even for those with limited design experience.

Green Globes® is a point-based system that offers a certified rating of performance and includes four rating levels similar to LEED. There are a maximum of 1000 points that can be achieved. In order to receive Green Globes certification, a minimum of 35% of the 1000 points must be reached. The Green Globes system certification levels are represented by the number of globes achieved. A project can obtain one to four globes. The more globes attained, the higher the level of certification.

One important difference between LEED and Green Globes is the treatment of wood used in the construction of buildings. At this time, LEED only recognizes timber certified by the FSC. Green Globes also accepts timber certified by other organizations such as the American Tree Farm System® (ATFS), Canadian Standards Association (CSA), and the Sustainable Forestry Initiative (SFI).

Unit 32 Review and Resources

SECTION 7

Survey Instruments and Operations

CONTENTS

UNIT

33 Builder's Levels, Automatic Levels, and Transit-Levels

OBJECTIVES

1. Identify common components of builder's levels.
2. Identify factors involved in setting up a builder's level.
3. Describe common methods for using leveling rods and targets.
4. Describe common operations performed with builder's levels.
5. Describe functions of automatic levels.
6. Identify common components of transit-levels.
7. Review the procedure for setting up transit-levels.
8. Describe common operations performed with transit-levels.

Survey instruments are used by carpenters and other tradesworkers to accurately lay out a building site, to verify grades and elevations, and in other construction operations. Traditional survey instruments are builder's levels, automatic levels, and transit-levels. Experience and knowledge gained using traditional survey equipment provides a solid basis for using advanced survey equipment such as laser levels and total station instruments.

BUILDER'S LEVELS

A *builder's level* is used to establish and verify grades and elevations and to set up level points over long distances. The main parts of a builder's level are the *telescope, leveling vial,* and *leveling screws.* The builder's level assembly is mounted over a circular base. **See Figure 33-1.** The telescope includes a focusing knob and has vertical and horizontal *crosshairs* within the barrel. The sensitive leveling vial is mounted parallel with the telescope and is located above or below the barrel. A builder's level is adjusted to be level with the leveling screws.

Most builder's levels have a *horizontal clamp screw* to hold the instrument in a fixed horizontal position. A *horizontal tangent screw* allows precise adjustment of the telescope in a left or right horizontal direction. A *horizontal graduated circle* and *vernier scale* are situated over the circular base and provide a means for angular measurement.

A chain and plumb bob hook are attached to the bottom of a builder's level so a plumb bob can hang beneath the builder's level. A plumb bob is used with the builder's level when the instrument must be set up over a specific point.

A dust cap to protect the *objective lens* of the telescope when it is not in use is usually included with the instrument. Sunshades are furnished with many models to be slipped over the end of the telescope to reduce or eliminate glare.

Always allow a survey instrument to acclimate to the surrounding temperature before use.

Builder's levels range from the less-expensive 12-power telescopes to highly sensitive 32-power models. The *power* of a telescope determines how much closer an object will appear when viewed through the telescope. A target seen through an 18-power telescope will appear to be 18 times closer than when it is seen with the naked eye.

A builder's level is mounted on a *tripod.* **See Figure 33-2.** The *tripod head* is supported by three legs. Locking levers or wing nuts tighten the legs into position. Some tripods have adjustable extension legs, which accommodate sloping or uneven ground. Tripods are available in hardwood or aluminum.

Setting up Builder's Levels

When setting up a builder's level, place the instrument where it will provide an unobstructed view of the work area. Set the tripod in a stable position and fasten the builder's level to the tripod. Adjust the instrument until it is exactly level. A poorly adjusted instrument is of no value in a layout operation.

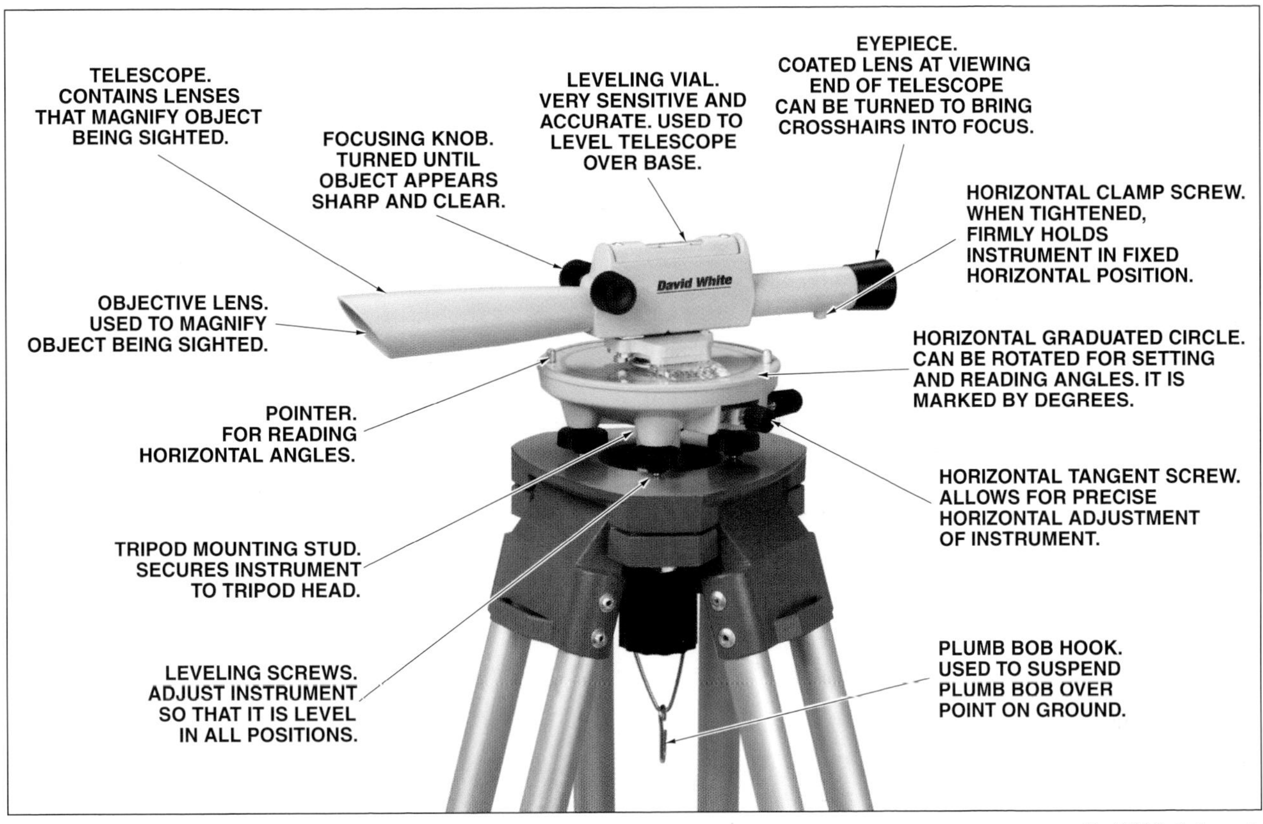

David White Instruments

Figure 33-1. *A builder's level is used to establish and verify grades and elevations and to set up level points over long distances. The instrument is secured to the tripod head by hand tightening the cup assembly on the tripod head.*

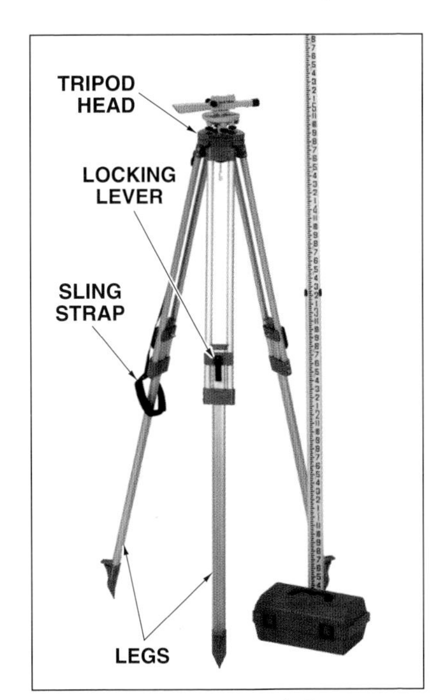

David White Instruments

Figure **33-2.** *A builder's level is mounted on a tripod. When the tripod is not in use, its head should be covered with a protective cap.*

Placing Tripods. Extend the legs and spread them about 3′ apart. Position the legs so the tripod head is level. Push the legs firmly into the ground and tighten the locking levers or wing nuts. On sloping ground, place one leg of the tripod into the slope. **See Figure 33-3.** A tripod set on soft or marshy soil may require a base of three stakes driven into the ground. **See Figure 33-4.** For placing a tripod over concrete or any other smooth surface, a triangular wood base helps to ensure the tripod legs will not shift. **See Figure 33-5.**

Fastening Builder's Levels to Tripods. Builder's levels are typically transported or stored in a carrying case. When lifting the instrument from the case, grasp the instrument by the base plate or *standard.* The standard is the frame in which the telescope is mounted. Hold the instrument directly over the tripod head. If a chain and plumb bob hook are fastened to the bottom of the instrument, they must hang freely through the hole in the tripod head before the base plate is screwed down.

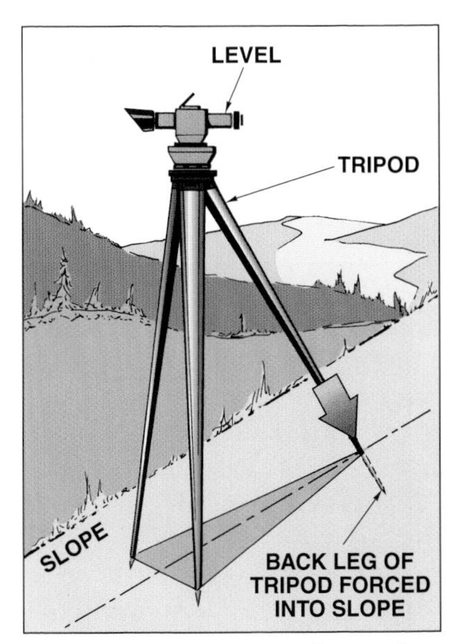

Figure 33-3. *Set up a tripod on a slope so that one leg of the tripod faces into the slope.*

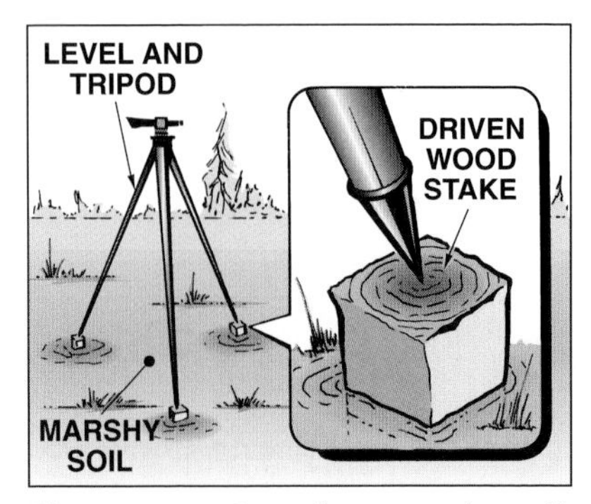

Figure 33-4. *On soft or marshy soil, stakes should be driven into the ground to support the tripod.*

Leica Geosystems Inc.

Laser levels permit one-person operation and provide accurate readings. Basic principles of traditional survey instruments such as builder's levels also apply to advanced instruments such as laser levels.

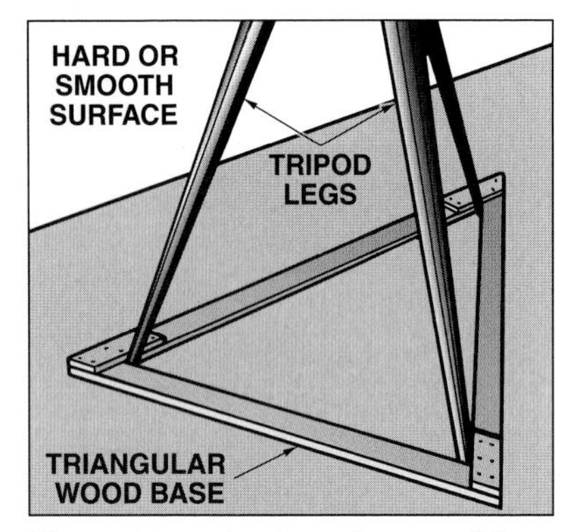

Figure 33-5. *A triangular wood base stabilizes a tripod placed on a hard concrete surface.*

One of two arrangements is generally used to fasten a builder's level to a tripod, depending on the design of the tripod head. **See Figure 33-6.** If the tripod head is the threaded type, the base of the builder's level is screwed directly onto it. If the tripod head has a cup assembly, a threaded mounting stud at the base of the builder's level is screwed into the cup assembly.

Adjusting Builder's Levels. The telescope of a builder's level rotates on top of the circular base. To function properly, the telescope must be level in all positions over the base. Leveling a telescope is accomplished by adjusting the leveling screws. The procedure for this operation is shown in **Figure 33-7.**

Although there should be firm contact between the leveling screws and base, overtightening the screws can damage the instrument. The leveling procedure becomes quick and simple with practice. When leveling a builder's level, opposite leveling screws must be turned equally, at the same time, and in opposite directions. The direction that the left thumb moves is the direction that the bubble moves. **See Figure 33-8.**

When a builder's level is used to lay out horizontal angles, it must be positioned directly over a specific point while it is being adjusted. However, a builder's level is not typically used for this application because it is easier to lay out horizontal angles with a transit-level. The procedure for setting up over a point is discussed in the transit-level section of this unit.

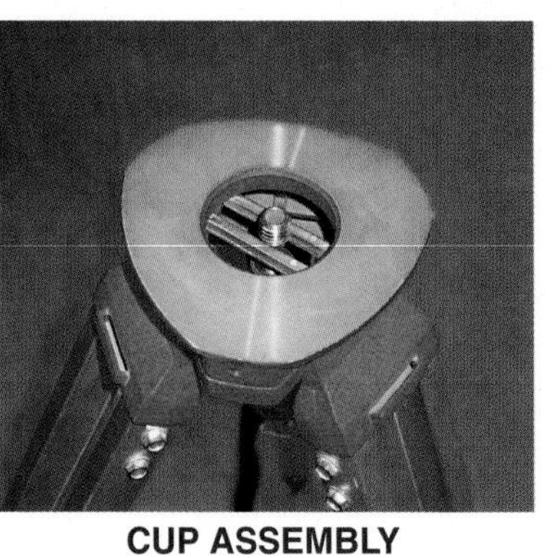

Figure 33-6. *Two types of tripod heads are the threaded and the cup assembly types.*

Focusing and Sighting. To view an object through a builder's level, focus the telescope by turning the focusing knob until the object being sighted is sharp and clear. The focusing knob adjusts the lenses inside the telescope barrel.

A builder's level is used to focus on a very small target and the *field of vision* is very small. **See Figure 33-9.** The field of vision is the total magnified area seen through the telescope. Some builder's levels have an eyepiece focusing ring for focusing the crosshairs within the telescope. The *crosshairs* are fine horizontal and vertical lines in the telescope that permit an object being sighted to be placed exactly in the center of the field of vision.

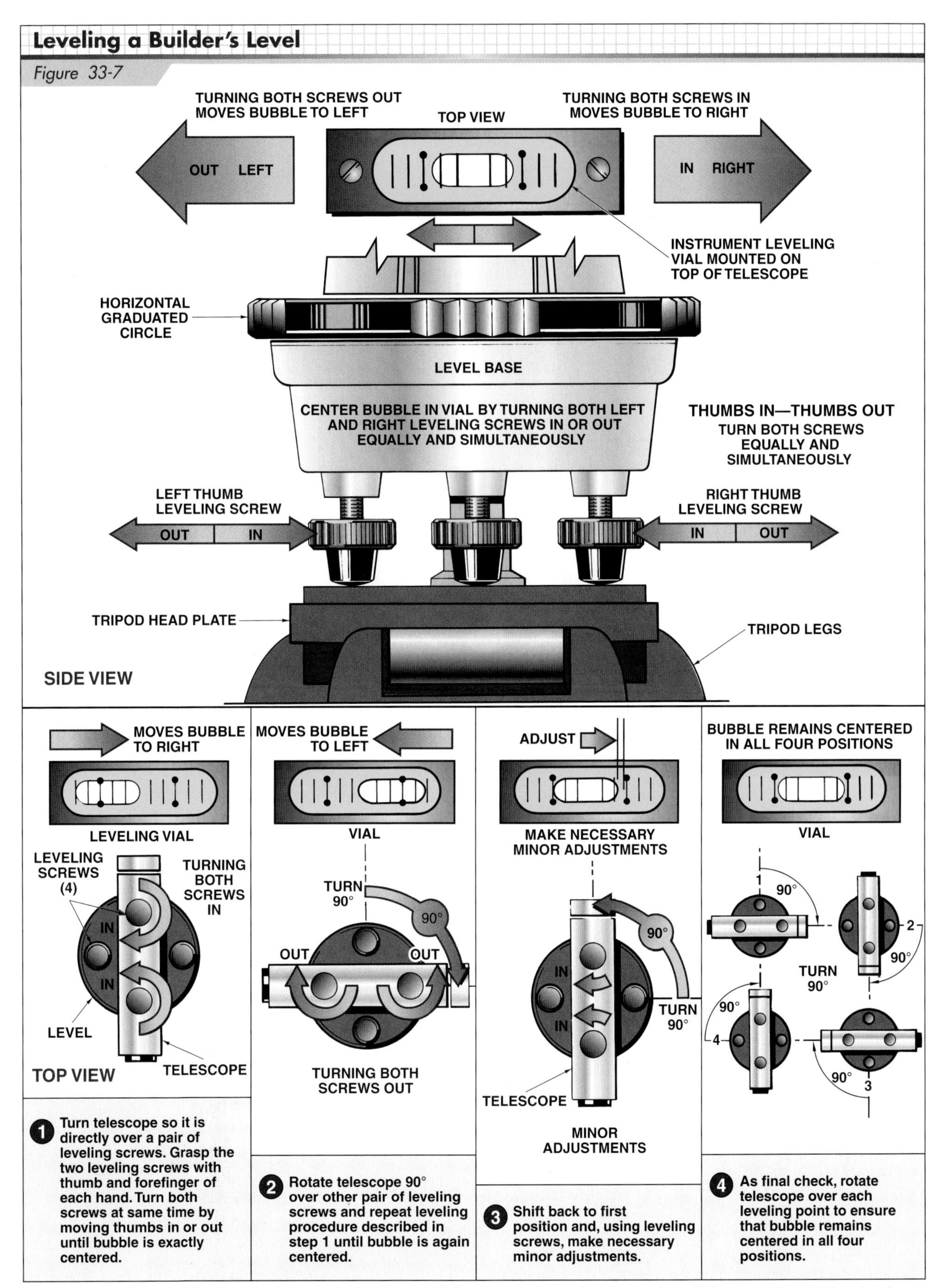

Figure 33-7. *A builder's level is level when the bubble in the leveling vial remains centered in all four positions. When leveling a builder's level, do not overtighten the leveling screws.*

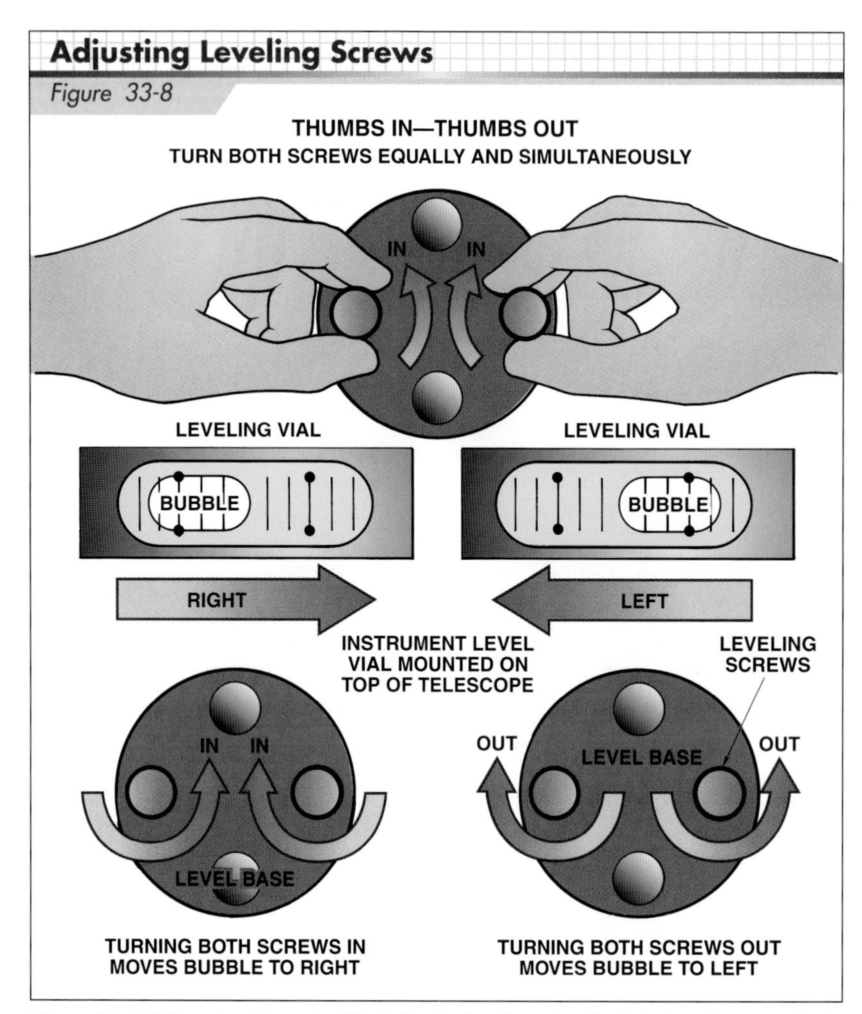

Figure 33-8. *When leveling a builder's level, the direction the left thumb moves is the direction that the bubble moves.*

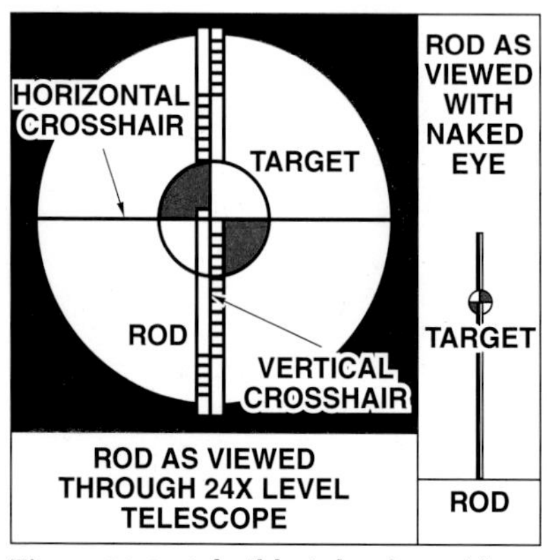

Figure 33-9. *A builder's level provides a very small field of vision.*

The procedure for locating a target and focusing on it with the builder's level is as follows:

1. Aim the telescope at the target by looking across the top of the barrel. Some instruments have devices similar to gun sights at the top of the barrel.
2. Focus the telescope until the target is clear.
3. When the target is sighted as closely as possible, tighten the horizontal clamp screw. Make final adjustments by using the horizontal tangent screw to move the telescope into position.

The dashed line in **Figure 33-10** shows the principle of the *line of sight,* or height of instrument. The line of sight is an imaginary, perfectly level line extending from the horizontal crosshair at the center of the telescope barrel to the target.

Leveling Rods and Targets

When using a builder's level, a second worker must hold a vertical measuring device, such as a rod, in the area where the grade or elevation is being established or verified. **See Figure 33-11.** The horizontal crosshair is then aligned with a measurement or mark on the rod.

Manufactured Leveling Rods. Wood, plastic, or aluminum rods have been specially designed for use with leveling instruments. Most manufactured leveling rods have adjustable sections. Typical two-section rods extend from 8′ to 9½′; three-section rods extend from 12′ to 14′. The numbers and graduation marks on a rod are large so that they can be easily read from a distance. The numbers indicating feet are the largest and are usually printed in red. The numbers and graduations between the foot numbers are usually printed in black. Movable metal targets are fitted to rods to make sighting easier at longer distances. **See Figure 33-12.**

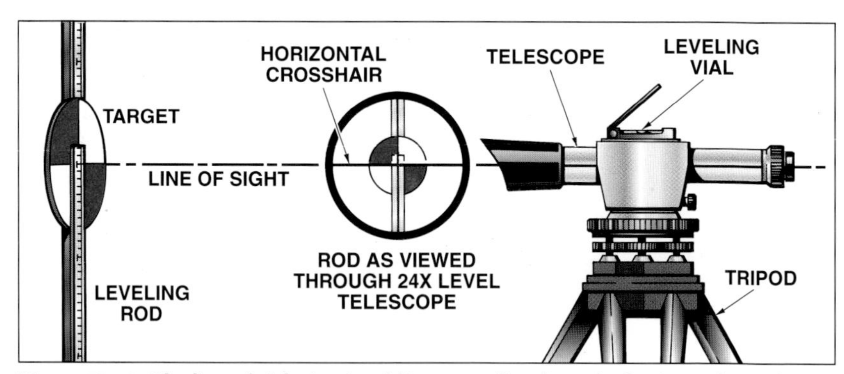

Figure 33-10. *The line of sight is a level line extending from the horizontal crosshair at the center of the telescope barrel to the target.*

David White Instruments

Figure 33-11. *A leveling rod must be held in a vertical position as the target is being sighted with a builder's level. In this photo, the elevation of the top of a cast-in-place concrete footing is being determined.*

Leveling rods are available with three types of graduations; two types are U.S. customary (English) and the third type is SI metric graduations. U.S. customary measurements are shown on the *architect's rod* and *engineer's rod* in **Figure 33-13.** An architect's rod is graduated in feet, inches, and eighths of an inch, and is typically used by carpenters and other tradesworkers. An engineer's rod is graduated in feet, tenths of a foot, and hundredths of a foot, and is used by surveyors and engineers.

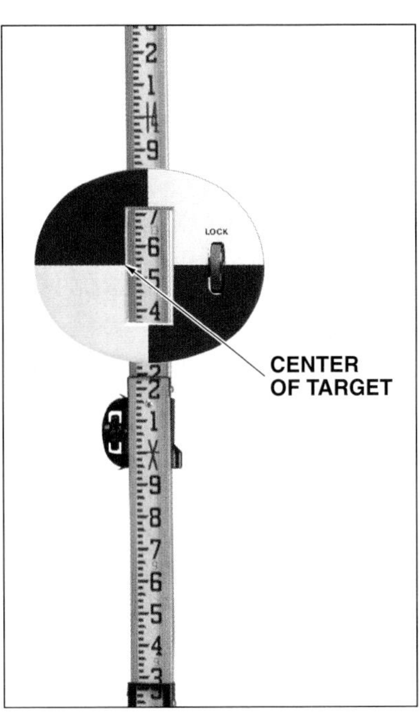

Figure 33-12. *An adjustable direct-reading leveling rod with a movable metal target is commonly used for construction applications. The center of the target is aligned with the reading being taken on the rod.*

Stick-and-Rule Method. When sighting short distances, carpenters frequently use a standard tape measure held against a straight piece of wood. **See Figure 33-14.**

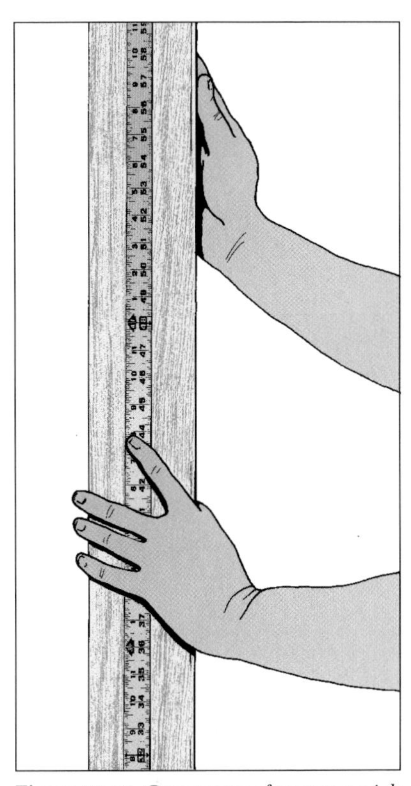

Figure 33-14. *Carpenters often use a stick and tape measure as a leveling rod.*

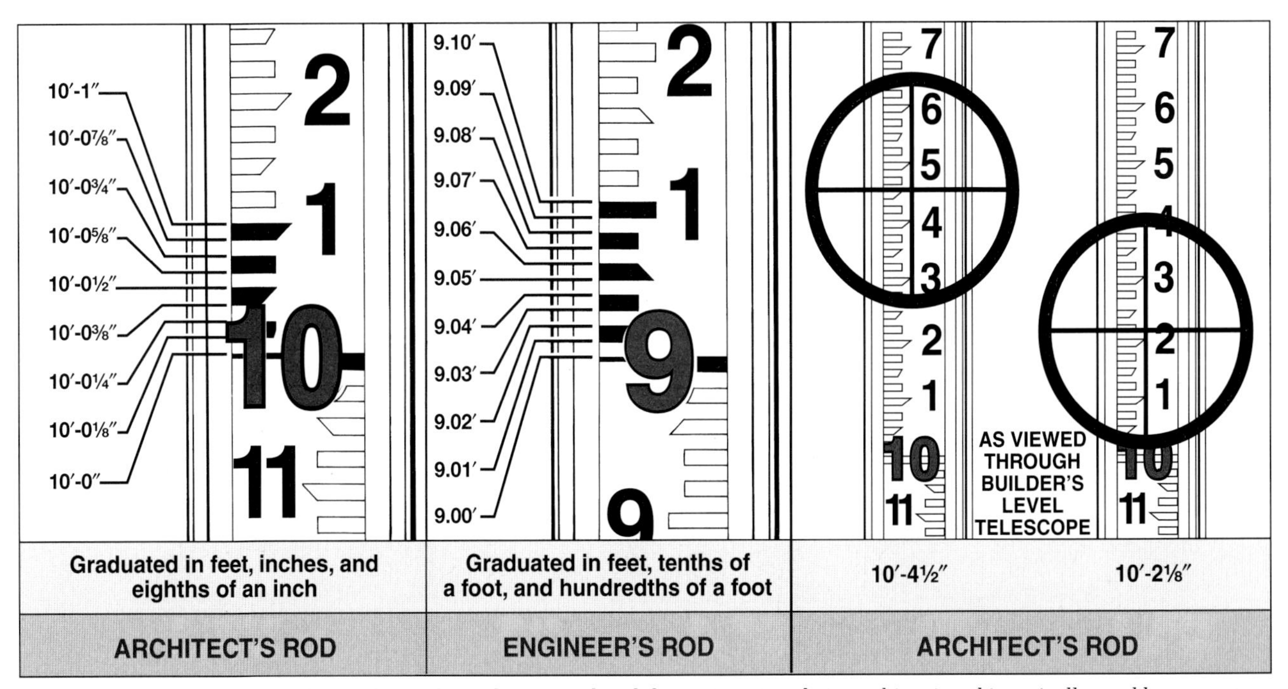

Figure 33-13. *Two types of leveling rods are the architect's rod and the engineer's rod. An architect's rod is typically used by carpenters.*

Plain Stick Method. In some situations, the most convenient way to perform layout operations with a builder's level is to use an unmarked, straight piece of wood. The line of sight is marked on the rod as it is held over an established point. The rod can be moved to other locations to establish grades.

Arm Signals. For accurate readings with the builder's level, the person holding the leveling rod must hold it in a plumb (perfectly vertical) position. When the rod is a great distance from the builder's level, the operator of the level may use a two-way radio or arm signals to instruct the person holding the rod to bring the rod into a plumb position or otherwise move it. **Figure 33-15** shows standard arm signals for bringing the rod into plumb position. **Figure 33-16** shows standard arm signals for bringing the target on the rod into an on-grade position.

Operations with Builder's Levels

Common operations performed with a builder's level include verifying grade differences and establishing elevations and level points.

Verifying Grade Differences. All lots have low and high grade points. The grade points should be known before the first corner height of the foundation wall is established. If excavation is required on the lot, a builder's level and rod can be used to verify the heights while the excavation work is performed. **Figure 33-17** shows how a builder's level is used to verify grade differences. When a rod is used in this manner, the higher the rod reading, the lower the grade. When a lot has a steep slope, the procedure is more complicated. **See Figure 33-18.**

Establishing Elevations. Job site elevations determine the major structural levels of the building, such as the top of the foundation walls and the finished floor height of the first floor. All job site elevations relate to a *benchmark* or *datum* (sometimes called *point of beginning*) that has been established for the job. On many jobs, the top of a stake driven at one corner of the lot or a chiseled mark at the top of a concrete curb identifies the benchmark. The location of a benchmark is also marked on the plot plan of a set of prints.

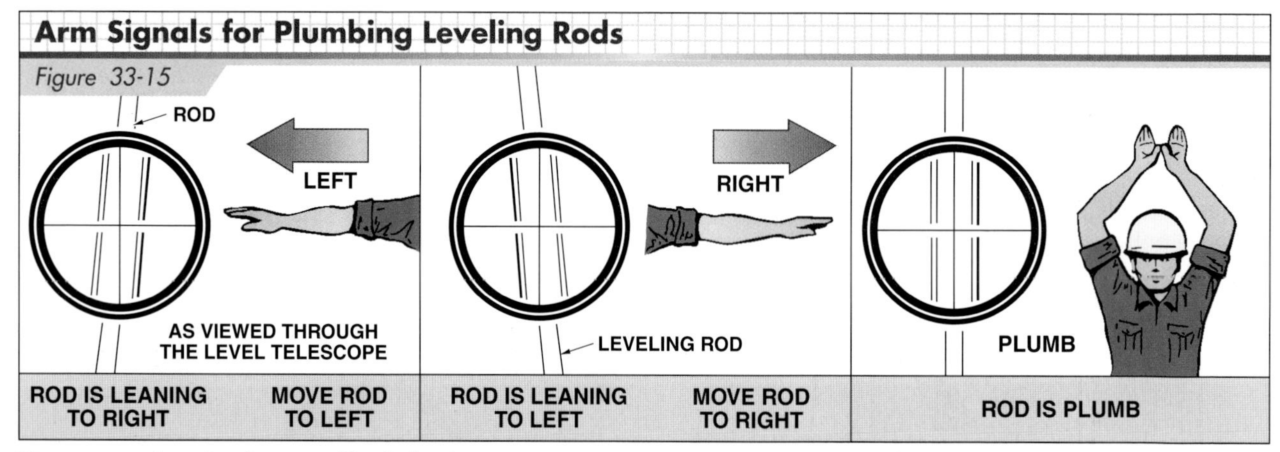

Figure 33-15. *Arm signals are used by the leveling instrument operator to communicate with the worker holding the leveling rod when bringing the rod into plumb position.*

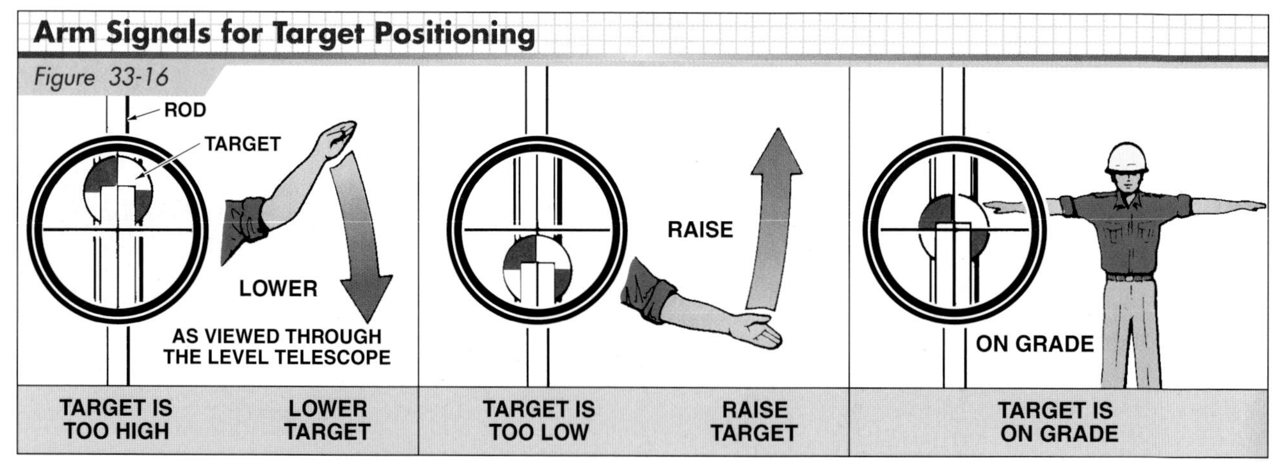

Figure 33-16. *Arm signals are used to bring the target on a leveling rod into an on-grade position.*

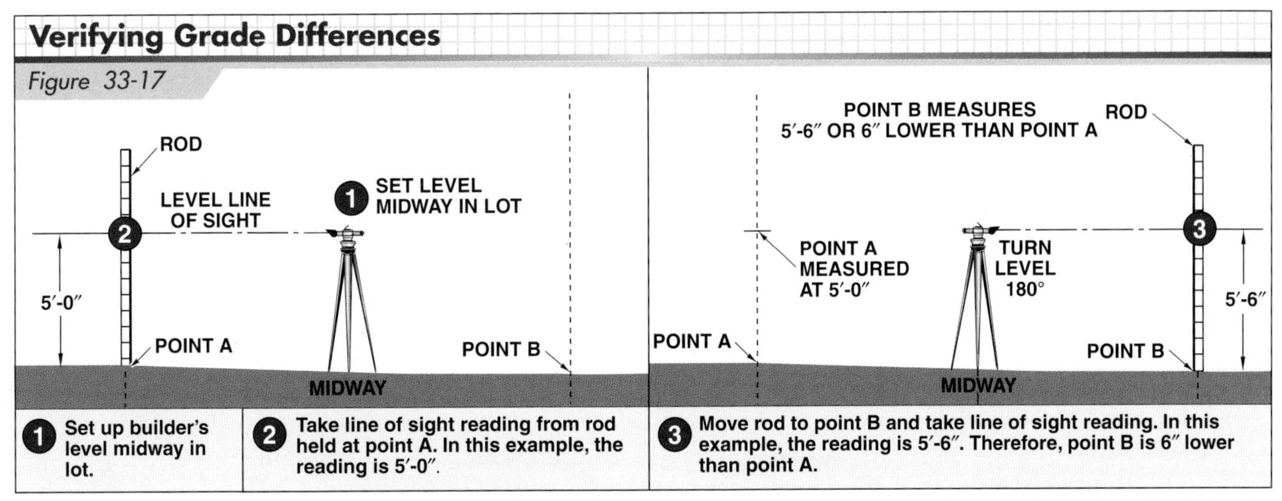

Figure 33-17. *The grade points should be known before the first corner height of a foundation wall is established. A builder's level is used to verify the grade differences of the building lot.*

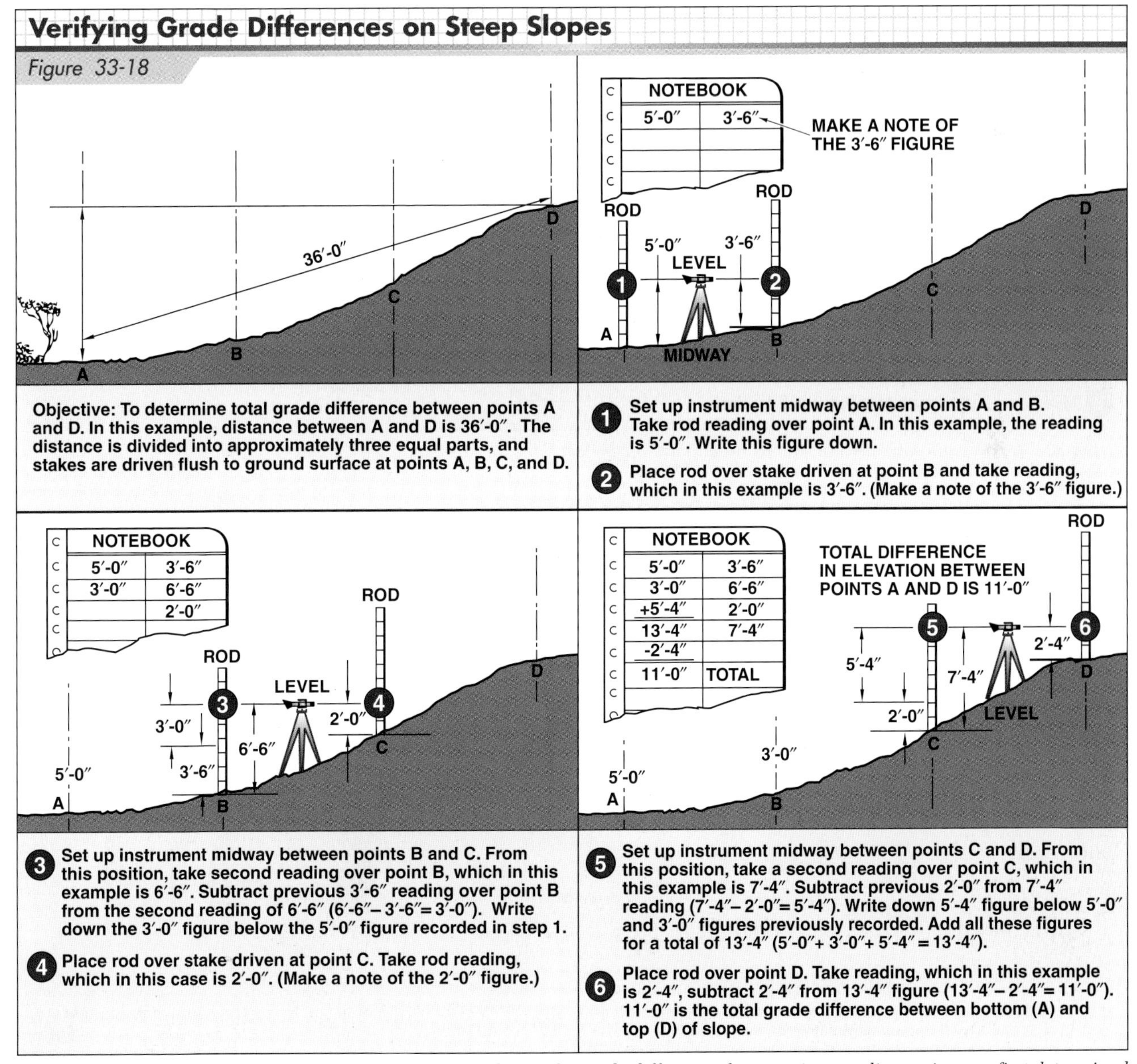

Figure 33-18. *When verifying grade differences on steep slopes, the grade differences between intermediate points are first determined and then added together.*

Figures 33-19 and **33-20** show how the builder's level is used to establish elevations in relation to a benchmark. The plain stick method is used in Figure 33-19 and the stick-and-rule method is used in Figure 33-20.

A method of measuring the differences in elevations of a building under construction is shown in **Figure 33-21.**

Leica Geosystems Inc.

Graders and other earth-moving equipment may be equipped with computer-assisted survey equipment and receivers to control the blade height, shift, and rotation when excavating a large job site.

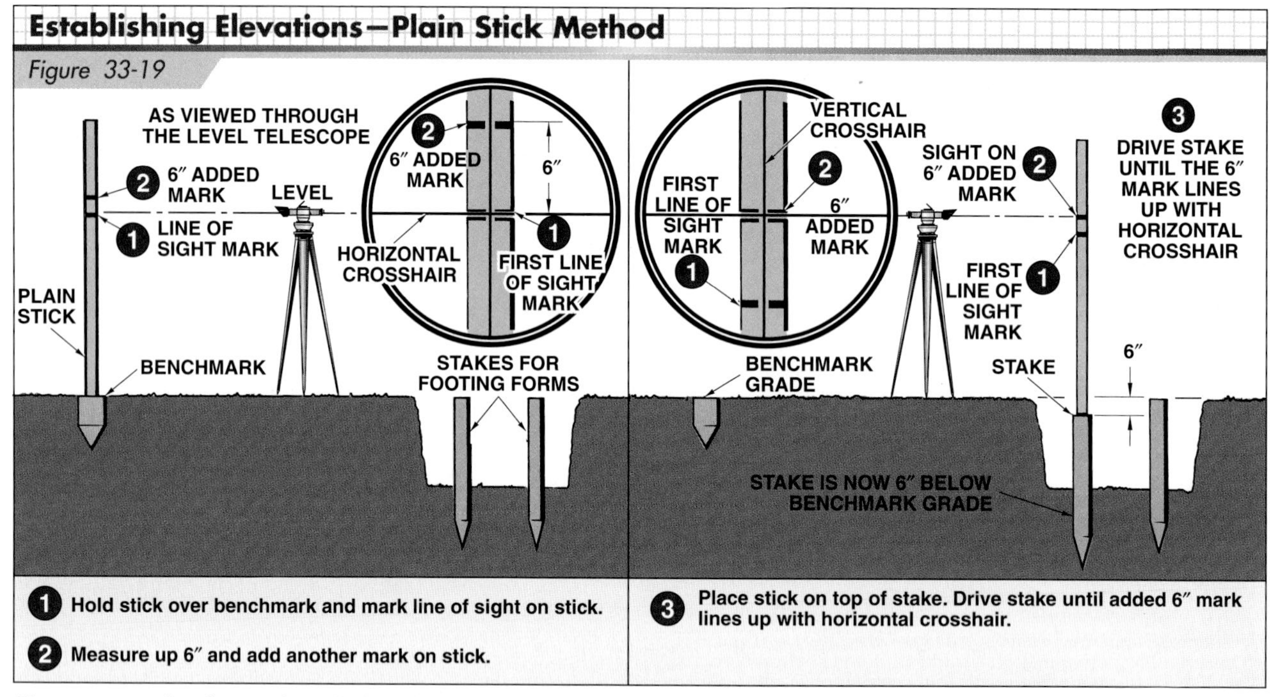

Figure 33-19. *The plain stick method can be used to establish elevations. In this example, the height of footing form stakes must be 6″ below the benchmark.*

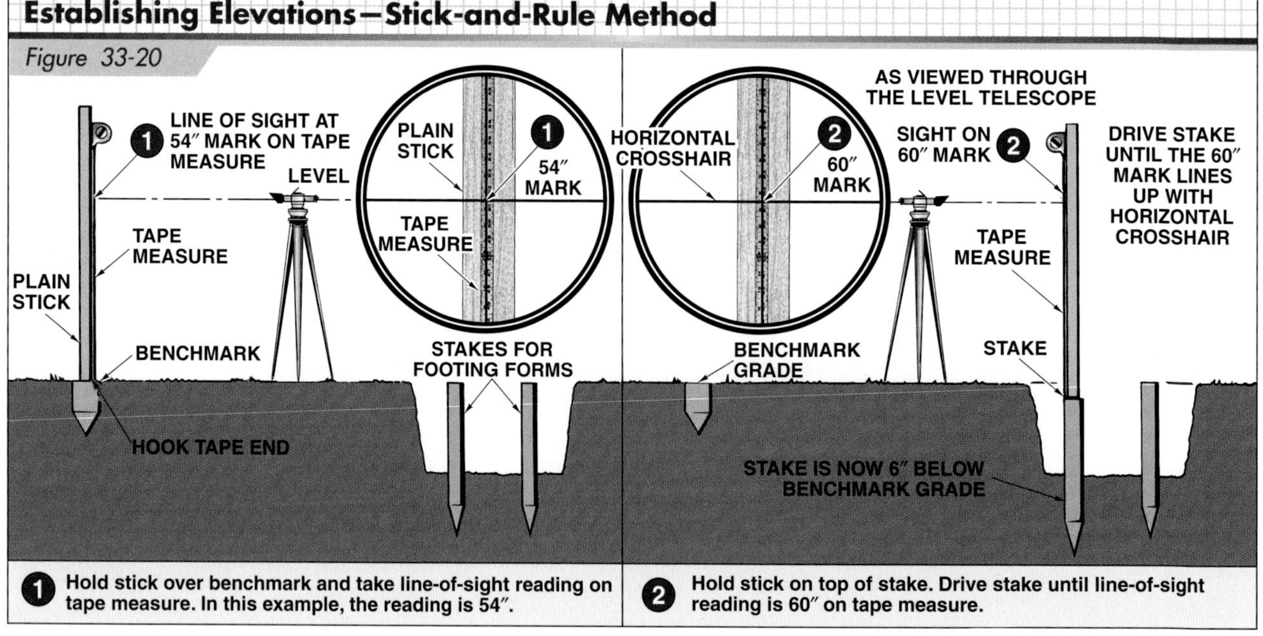

Figure 33-20. *The stick-and-rule method can be used to establish elevations. In this example, the height of footing form stakes must be 6″ below the benchmark.*

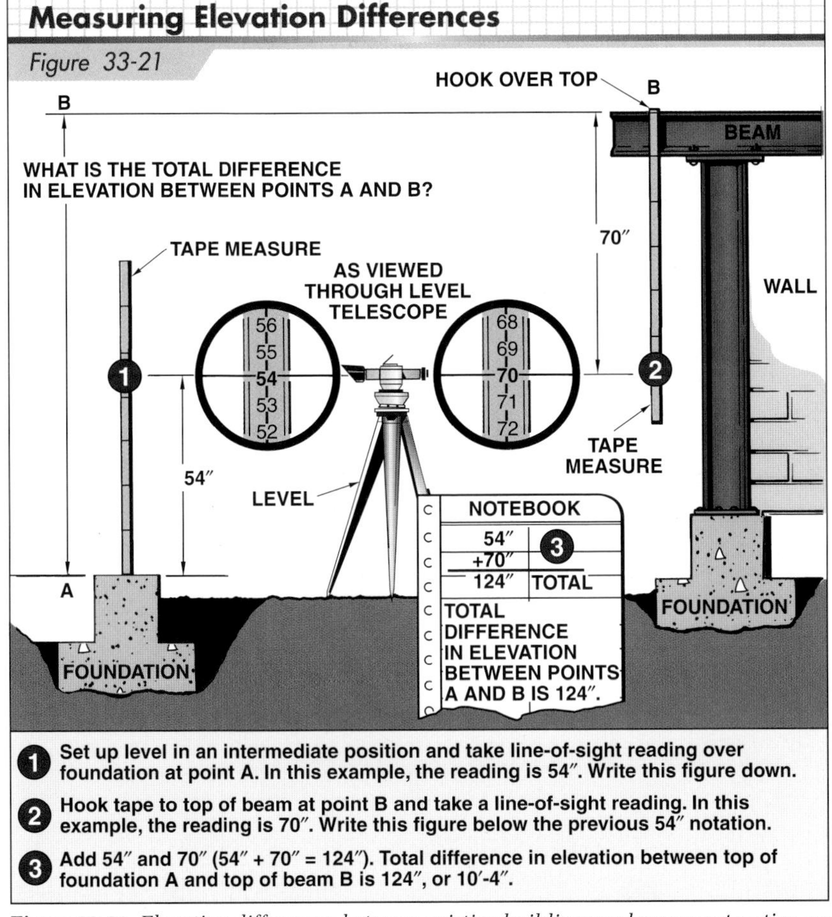

Figure 33-21. *Elevation differences between existing buildings and new construction are determined using a builder's level.*

Establishing Level Points. After an elevation has been established in relation to a benchmark, the elevation often must be transferred to other points. For example, level points must be established at the corner stakes when forms are constructed for the concrete foundation of a building. **Figure 33-22** shows how level points are established with a builder's level.

Clean the objective and eyepiece lenses using a lens tissue.

AUTOMATIC LEVELS

Automatic levels are popular instruments on construction jobs since the levels can be set up and adjusted more quickly than builder's levels. In addition, automatic levels maintain a high degree of accuracy, which is important when sighting across long distances. **See Figure 33-23.** Automatic levels are used for many of the operations for which a builder's level is used.

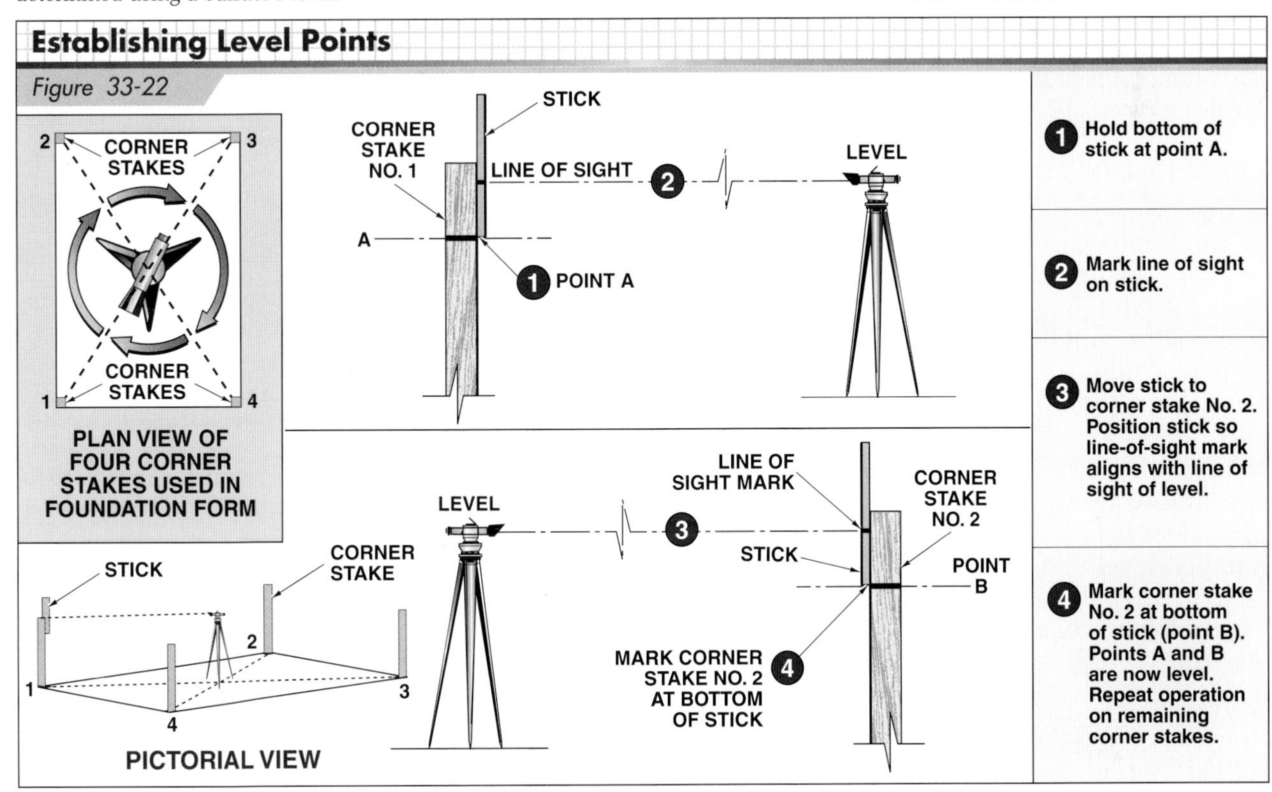

Figure 33-22. *Level points on corner stakes for a foundation form are established using a builder's level. The height at point A has been established from a benchmark.*

David White Instruments

Figure 33-23. *An automatic level can be adjusted very quickly and it maintains a high degree of accuracy.*

An automatic level is attached to a tripod using a threaded base or a tripod draw screw. Three leveling screws are used to adjust an automatic level to be roughly (approximately) level. **See Figure 33-24.** An automatic level is in rough adjustment when the bubble in the circular leveling vial is centered. An automatic level is leveled without rotating the instrument. After the instrument is in rough adjustment, an *internal compensator* aligns the instrument into a level position. Some models of automatic levels include an electronic safety system that alerts the user if the instrument has shifted out of its level position.

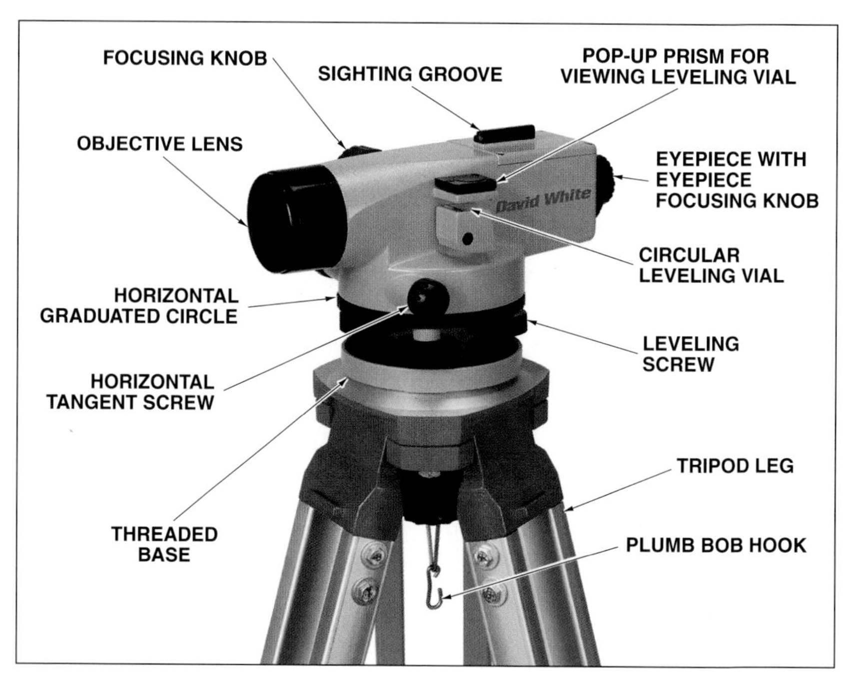

David White Instruments

Figure 33-24. *After an automatic level is roughly leveled using the leveling screws, an internal compensator aligns the instrument into a level position.*

TRANSIT-LEVELS

A *transit-level*, or level-transit, is similar in appearance to a builder's level. **See Figure 33-25.** The main difference between a transit-level and builder's level is that the telescope of a transit-level can be tilted vertically (up and down) as well as rotated horizontally (side to side). Vertical tilting allows a number of operations to be performed with a transit-level that are not possible with a builder's level such as plumbing a column or setting stakes in a straight line.

David White Instruments

Figure 33-25. *The telescope of a transit-level can be moved vertically and horizontally, making more operations possible than with a builder's level.*

Keep a survey instrument in its carrying case when not in use.

Transit-levels have a *lock lever* or clamp to hold the telescope in a fixed, level position. **See Figure 33-26.** When the telescope is locked in a level position, the transit-level can perform the functions of the builder's level.

When the lock lever or clamp is released, which allows vertical movement of the telescope, a *vertical clamp screw* is tightened to hold the telescope in the desired vertical position. A *vertical tangent screw* is used to make vertical fine adjustments.

Horizontal Circle and Horizontal Vernier Scales

A transit-level has a *horizontal circle* and *horizontal vernier scale*. **See Figure 33-27.** The horizontal circle and horizontal vernier scale are intersecting scales used to measure horizontal angles. The horizontal circle is divided into four quadrants, each reading from 0° to 90°. The vernier scale has 12 graduations (labeled 0 to 60 minutes) at each side of the 0 index.

The horizontal circle is turned by hand and it does not move when the telescope rotates. The horizontal vernier scale is attached to the instrument frame, and it rotates around the inside of the horizontal circle as the telescope is turned to the right or left.

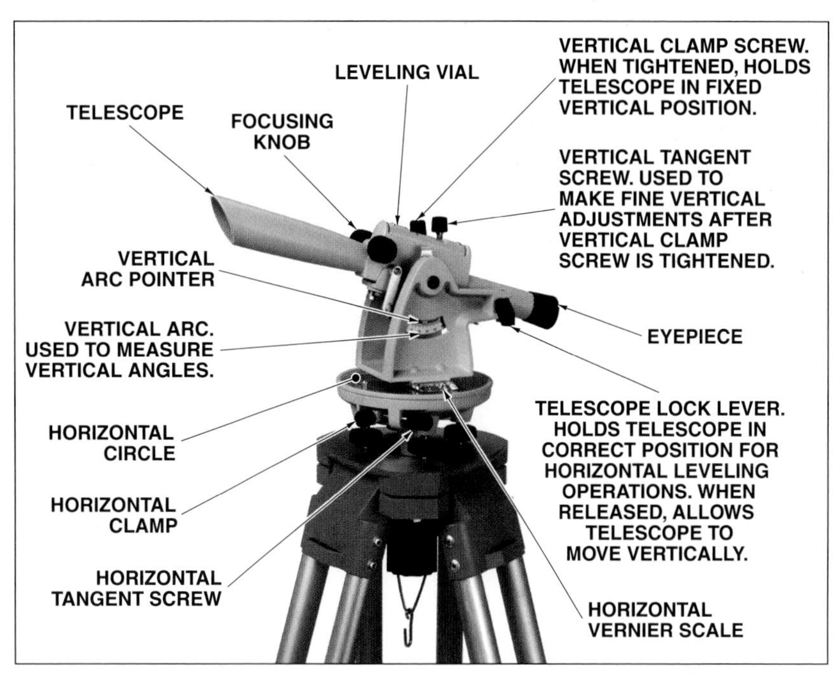

David White Instruments

Figure 33-26. *Vertical and horizontal angles can be determined with a transit-level.*

Vertical Arc and Vertical Vernier Scales

Transit-levels are equipped with a *vertical arc,* which is similar to the horizontal circle. A vertical arc is used to measure vertical angles and is graduated from 0° to 45° in two directions. A vertical arc moves with the vertical motion of the telescope. Some transit-level models have a vertical vernier scale, which is attached to the frame and does not move. Similar to the horizontal vernier scale, a vertical vernier scale indicates smaller increments than a vertical arc.

Some transit-levels have a vertical arc pointer to indicate the vertical angle to which the telescope has been set. **See Figure 33-28.**

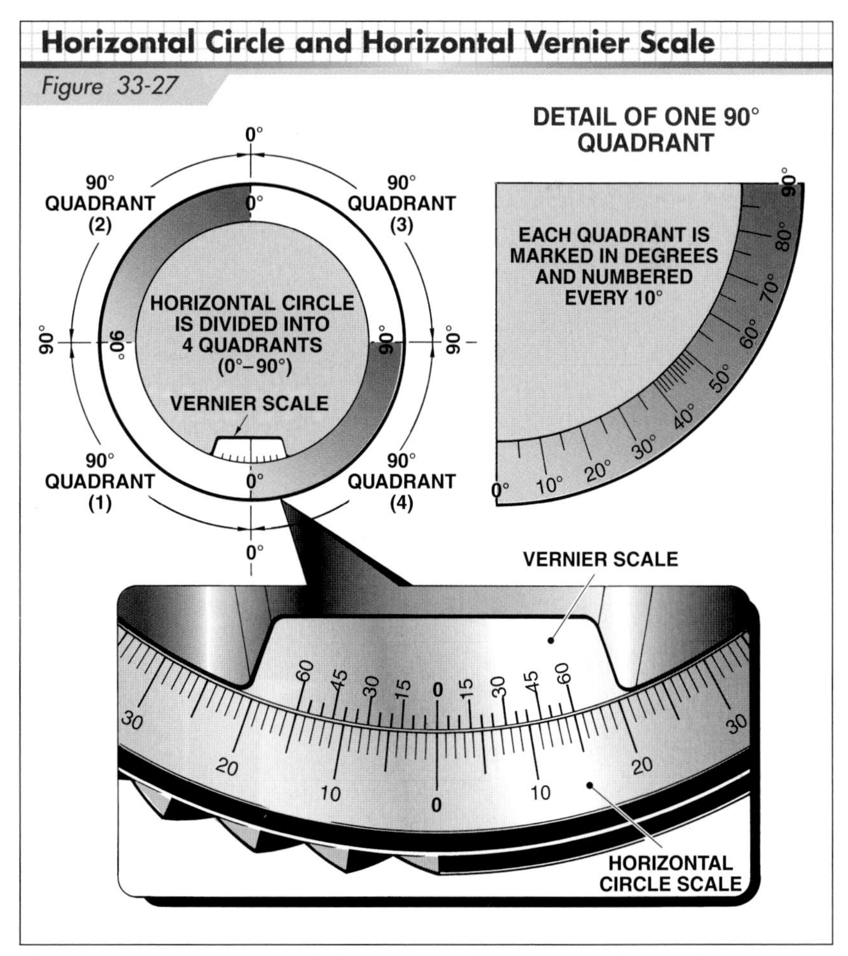

Figure 33-27. *Each quadrant of a horizontal circle scale is divided into degrees ranging from 0° to 90°. The vernier scale is used for more accurate readings in minutes.*

Setting up Transit-Levels

The procedure for setting up transit-levels is similar to setting up builder's levels. The telescope must be locked into its horizontal position before a transit-level is adjusted with the leveling screws.

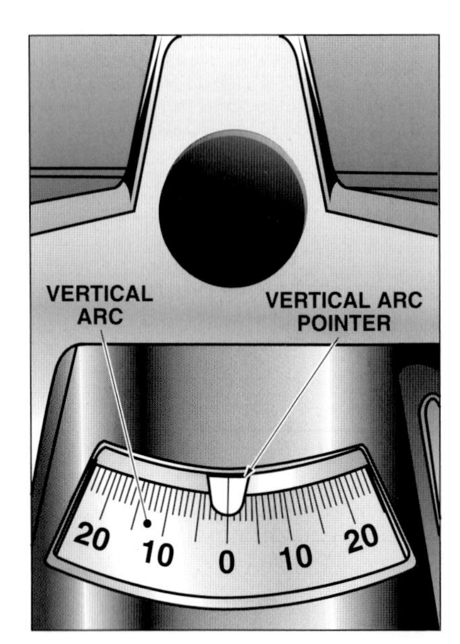

Figure 33-28. *A transit-level may have a pointer instead of a vernier scale. A transit-level with a pointer can measure vertical angles only to whole degrees.*

Setting Up over a Point. For measuring and laying out horizontal angles, transit-levels must be set up over a specific point on the ground. As shown in **Figure 33-29,** a plumb bob is used in this procedure as follows:

1. Spread the legs of the tripod and place the tripod head directly over the desired point.
2. Secure the transit-level to the tripod. Form a slip knot in the plumb bob line and attach the line to the plumb bob hook. A slip knot allows a plumb bob to be raised or lowered by sliding the knot along the cord. Adjust the line length so the plumb bob is about ¼″ above the point.
3. If necessary, shift the tripod so the plumb bob is directly over the point.
4. Roughly level the transit-level with the leveling screws. Do not tighten the screws.
5. Most transit-levels have a *shifting center,* which allows for slight movement of the instrument on the leveling plate. When the plumb bob is close to the point, make the final adjustment by shifting the transit-level on the leveling plate.
6. Adjust and tighten the leveling screws. Ensure the transit-level is level in all positions.

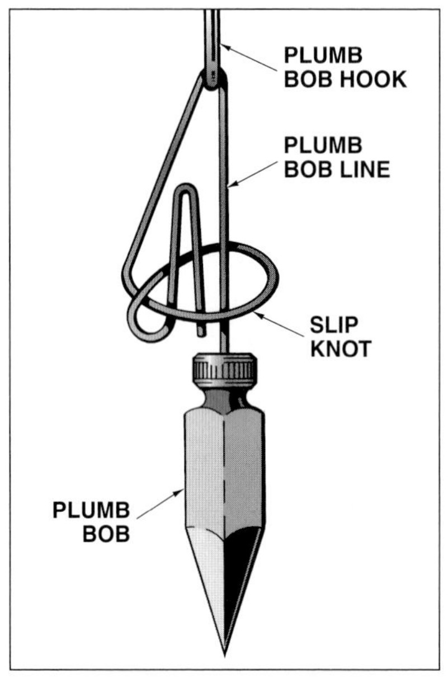

Figure 33-29. *A plumb bob is used to locate a transit-level over a specified point on the ground. A slip knot permits the plumb bob to be raised or lowered as needed.*

Transit-Level Operations

The right angle (90°) is the angle used most often in construction. Most buildings consist of one or more rectangular parts. Each line in a rectangular shape forms a 90° corner at the point where it meets another line. Walls constructed according to these lines are referred to as being *square* to each other. **See Figure 33-30.**

One of the first stages of a construction project is the construction of a foundation. Before any foundation work can occur, building lines must be set up indicating the exact outline of the foundation. Building lines are usually at a 90° angle to each other. One of the fastest ways to lay out building lines is to use a transit-level. The transit-level can also be used to lay out angles other than 90°, to plumb, and to establish points in a straight line.

Survey instruments may have an optical plummet, allowing a level to be positioned over a specific ground point without a plumb bob.

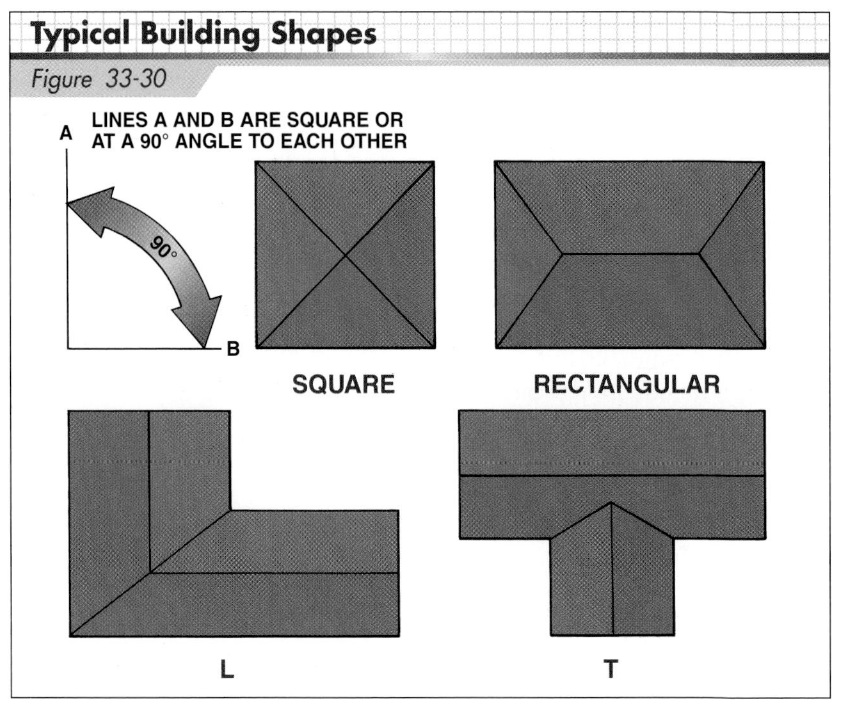

Figure 33-30. *Buildings usually consist of lines that are square to each other.*

Measuring 90° Horizontal Angles. The points at both ends of a line must be established before a 90° horizontal angle can be measured. The transit-level is set up over one of the points. A sight is taken on the second point. The transit-level is then rotated 90° to establish a second line that is square to the first line. This procedure is shown in **Figure 33-31.**

Measuring Horizontal Angles Other than 90°. The horizontal circle on a transit-level is divided into four quadrants, each reading from 0° to 90°. Angles greater than 90° angles can be measured by rotating the telescope to the right or left and then performing addition or subtraction as necessary. Angles of 90° or less can be measured directly on the horizontal circle scale without any addition or subtraction. Examples of procedures for measuring various angles are shown in **Figure 33-32.** Keep in mind that the vernier scale moves with the swing of the instrument.

Measuring Horizontal Angles in Degrees and Minutes. For some applications, measurements must be expressed in degrees and minutes (and seconds in some situations). *Minutes* and *seconds* are used to express fractions of a degree. The breakdown of degrees is as follows:

Full circle = 360°
1° = 60 minutes (60′)
1′ = 60 seconds (60″)

When removing a survey instrument from a tripod, loosen two adjacent leveling screws and unscrew the tripod-mounting stud while holding on to the instrument. Carefully remove the instrument from the tripod and secure it in its carrying case. When closing the cover, ensure that all components are properly protected.

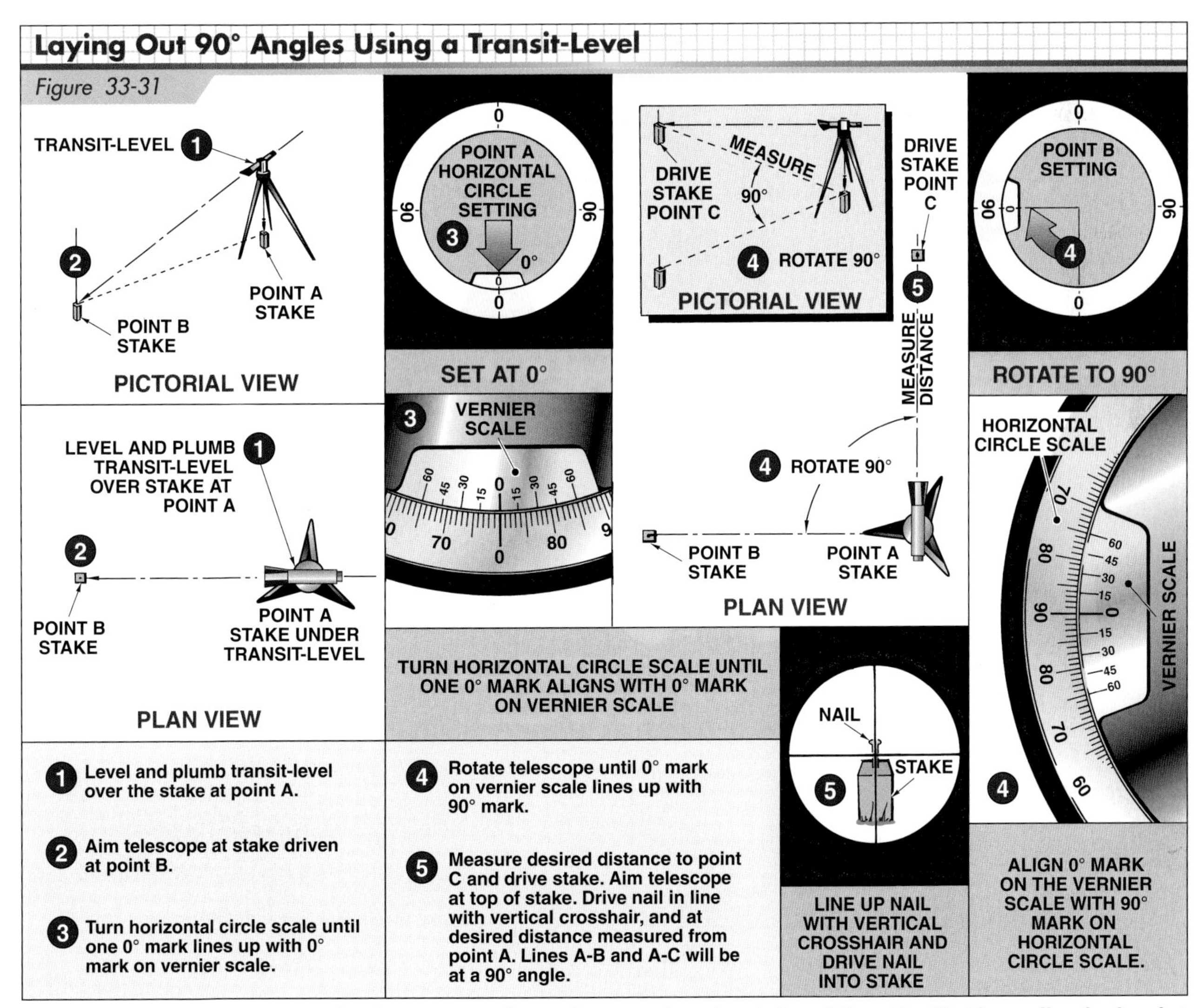

Figure 33-31. *Transit-levels are used to lay out 90° angles. In this example, points A, B, and C are identified by small nails placed at the top of wood stakes.*

In **Figure 33-33,** an explanation and comparison are given between a degree-only reading and a degree-and-minute reading. Additional examples of degree-and-minute readings are shown in **Figure 33-34.**

A horizontal circle is divided into degrees, which are numbered every 10 degrees. A vernier scale is used to divide each whole degree into minutes, allowing precise measurements to be made.

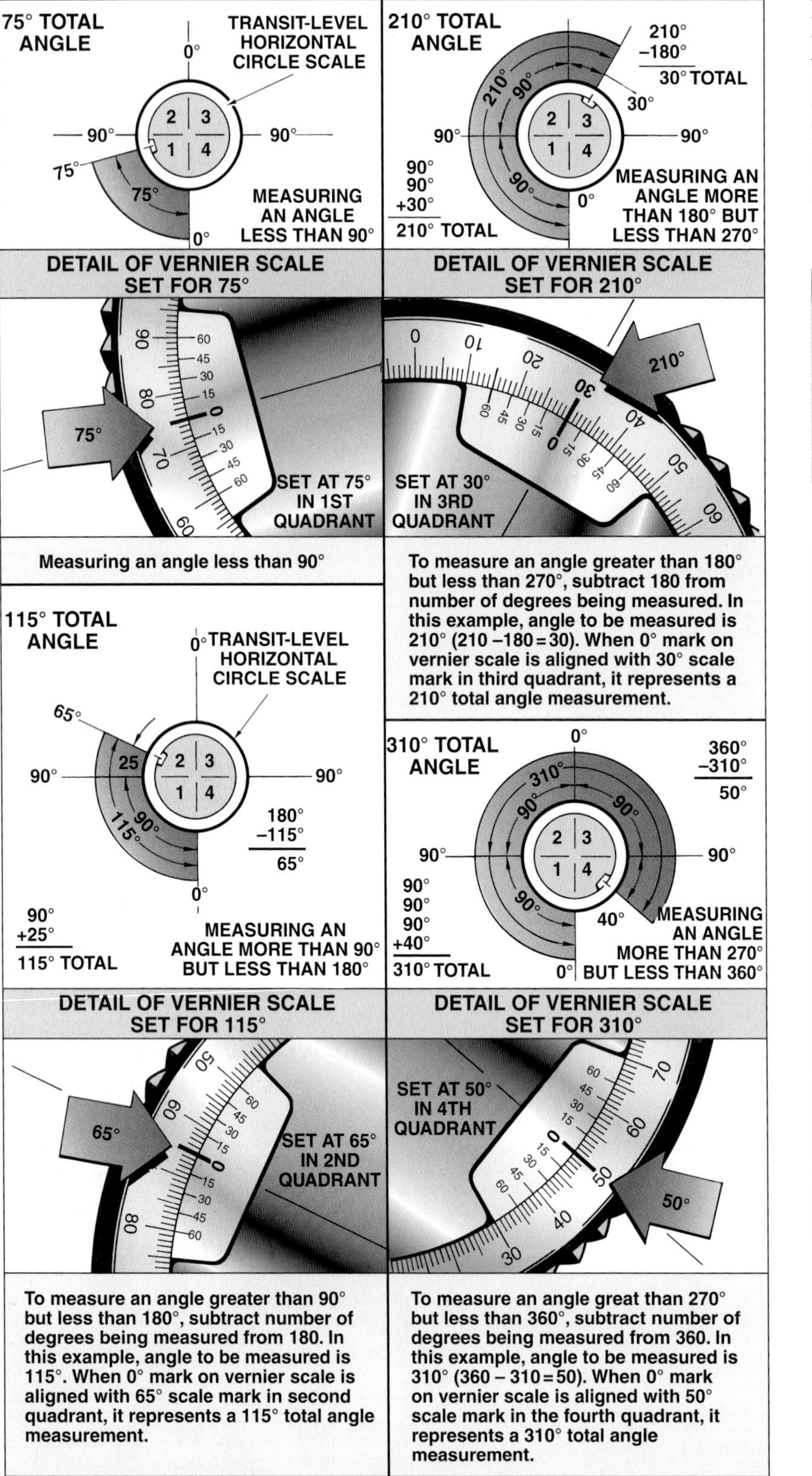

Figure 33-32. *Any angle can be measured by rotating the telescope to the right or left and then performing addition or subtraction as necessary.*

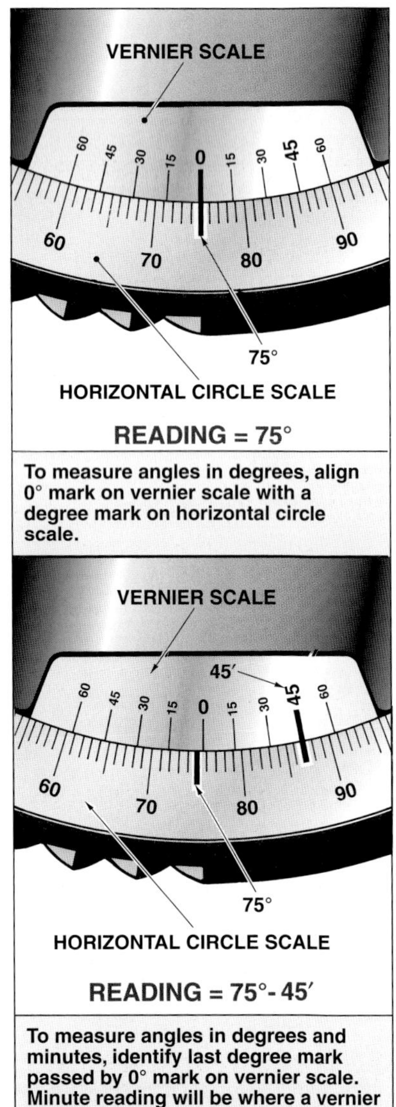

Figure 33-33. *The horizontal circle and vernier scales are used to measure angles in degrees and minutes.*

Transit-levels and builder's levels measure angles in degrees and minutes by correlating the degree graduations on the horizontal circle to the minute graduations on the vernier scale. The numbers on the vernier scale run from 0° to 60° to the right and left of the 0 index. Each small mark on the vernier scale represents 5′ (minutes).

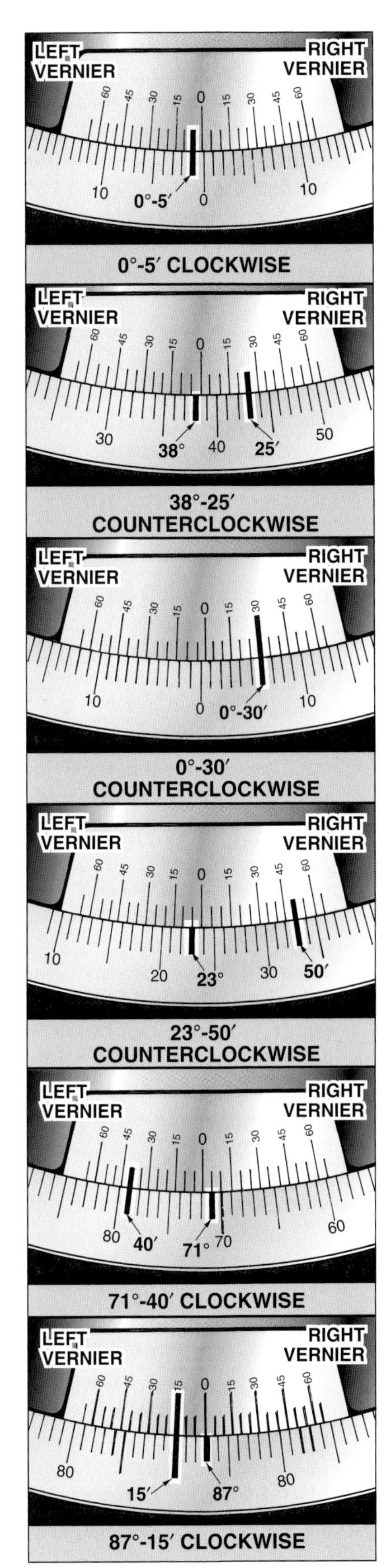

Figure 33-34. *When degree numbers increase to the left, clockwise angles are being read. When degree numbers increase to the right, counterclockwise angles are being read.*

Plumbing. A column, building corner, or other vertical member can be plumbed accurately with a transit-level. When using a transit-level for plumbing, place the instrument at a convenient distance from the object being plumbed. If possible, the distance should be greater than the height of the object. The procedure for plumbing a column is shown in **Figure 33-35.**

Establishing Points in a Straight Line. A transit-level simplifies the task of establishing points in a straight line and can be used for applications such as setting stakes for a form wall and establishing centers of piers or footings. The procedure for establishing points in a straight line with a transit-level is shown in **Figure 33-36.**

CARE OF LEVELING INSTRUMENTS

Leveling instruments, such as builder's levels, automatic levels, and transit-levels, are expensive precision instruments. They can withstand many years of normal use on a construction job if handled carefully.

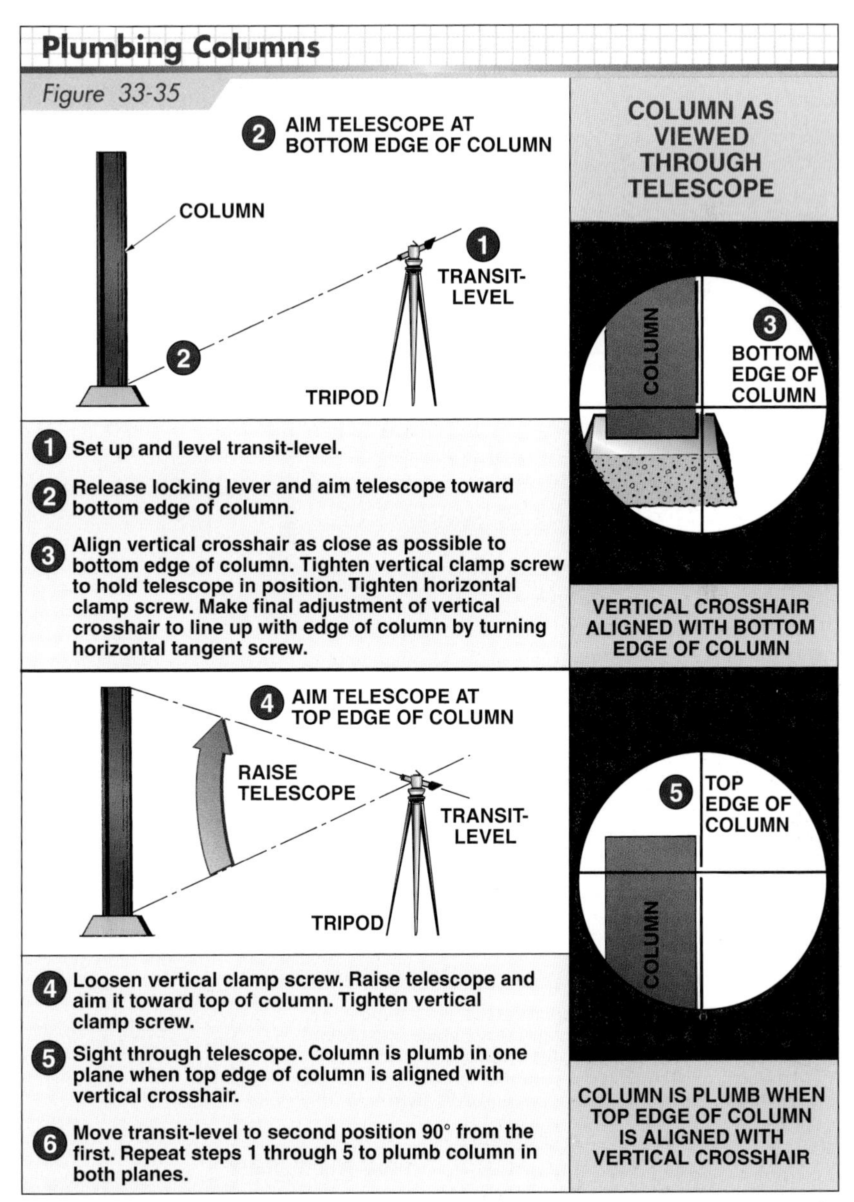

Figure 33-35. *When plumbing a column with a transit-level, the distance between the transit-level and object being plumbed should be greater than the height of the object, if possible.*

When tradesworkers are unfamiliar with a particular brand or model of leveling instrument, they should refer to the instruction or operator manual before using the instrument.

Most problems that occur with leveling instruments are caused by improper care or use. For example, continued exposure to rain and dust may cloud the telescope lenses. A slight blow or continual vibration may cause the instrument to become misaligned and reduce its accuracy. A hard blow or a fall can ruin the instrument.

The following precautions should be observed with all leveling instruments:

- Keep leveling instruments in their carrying case when they are not in use.
- Adjust the leveling and clamp screws so instrument parts will not move about freely when the instrument is in the case. However, do not tighten the screws enough to put the instrument in a rigid position.
- When transporting leveling instruments in a vehicle, do not place the carrying case on a hard floor where it will be subjected to vibration.
- Never set up a leveling instrument without fully spreading the tripod legs. When possible, press the legs firmly into the ground.
- Do not overtighten the leveling screws. Overtightening can cause the plate to become pitted and warped. A firm adjustment of the screws is adequate.
- Protect a leveling instrument with a waterproof cover if the instrument must be in rain or heavy dust for an extended period.
- Use a sunshade to help protect the objective lens from dust.
- Never rub dust or dirt off a lens; blow the dust or dirt off or use a camel's hair brush.
- If leveling screws or other movable parts require cleaning, wipe them with an oiled cloth or brush. Use a light instrument oil.
- Do not allow unauthorized persons to use leveling instruments.
- When moving a leveling instrument to another location on a job site while mounted on a tripod, slightly loosen the leveling screws to take pressure off the tripod plate. Hold the instrument in front of you with the tripod under your arm.

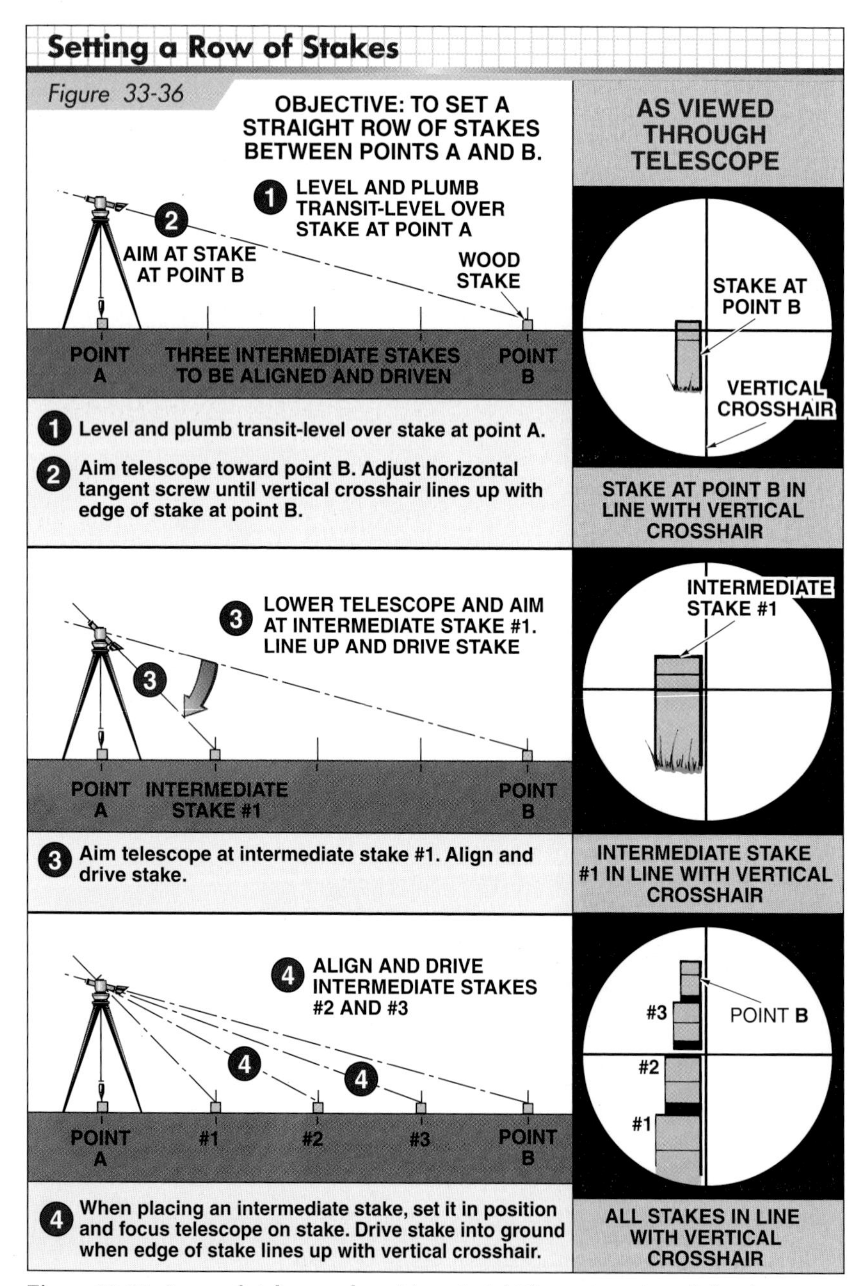

Figure 33-36. *A row of stakes can be set in a straight line using a transit-level.*

Unit 33 Review and Resources

OBJECTIVES

1. Describe functions of laser levels.
2. Describe the principles of laser beams.
3. List functions of laser beam detectors.
4. Describe common types of laser levels.
5. Review the general procedure for setting up and operating laser levels.
6. Describe laser safety classifications.
7. Identify common safety precautions for laser levels.
8. List and describe functions of total station instruments.
9. List and describe common types of total stations.
10. Explain the process and advantages of high-definition surveying.

UNIT 34 Laser Levels and Total Station Instruments

Laser levels and total station instruments are two of the more technologically advanced layout and survey instruments used on construction projects. Laser levels offer several advantages over traditional surveying equipment: they are easy to set up and adjust, and they allow a single worker to perform a variety of operations accurately and quickly.

Total station instruments combine digital data processing and survey technology. Total stations are used to electronically measure distances, record and store data, and perform mathematical calculations.

David White Instruments

Figure 34-1. *Only one worker is needed to perform layout work with a laser level. After a laser level is properly set up, the worker uses a leveling rod and detector to establish or verify elevations.*

LASER LEVELS

Laser levels, also referred to as construction lasers, perform the functions of traditional surveying equipment more efficiently. When using traditional surveying equipment, one worker sights the instrument and a second worker holds a target rod. When using a laser level to perform the same operations, only one worker is required. **See Figure 34-1.**

A variety of laser levels are available. Some laser levels are used for general construction operations while others are designed for specialized operations such as laying and sloping pipe. Compact, less-expensive laser levels are also available. **See Figure 34-2.** These laser levels provide many of the functions included in larger laser levels.

A laser level is used to establish and verify grades and elevations and to set up level points over long distances. The main parts of a laser level are the *rotating beacon, out-of-level indicator,* and *leveling screws.* A laser level assembly is attached to a base, which may be mounted to a tripod. **See Figure 34-3.** The rotating beacon projects a laser beam. The out-of-level indicator indicates when the laser level is out of level.

PLS • Pacific Laser Systems

Figure 34-2. *Compact laser levels provide many of the functions of larger laser levels.*

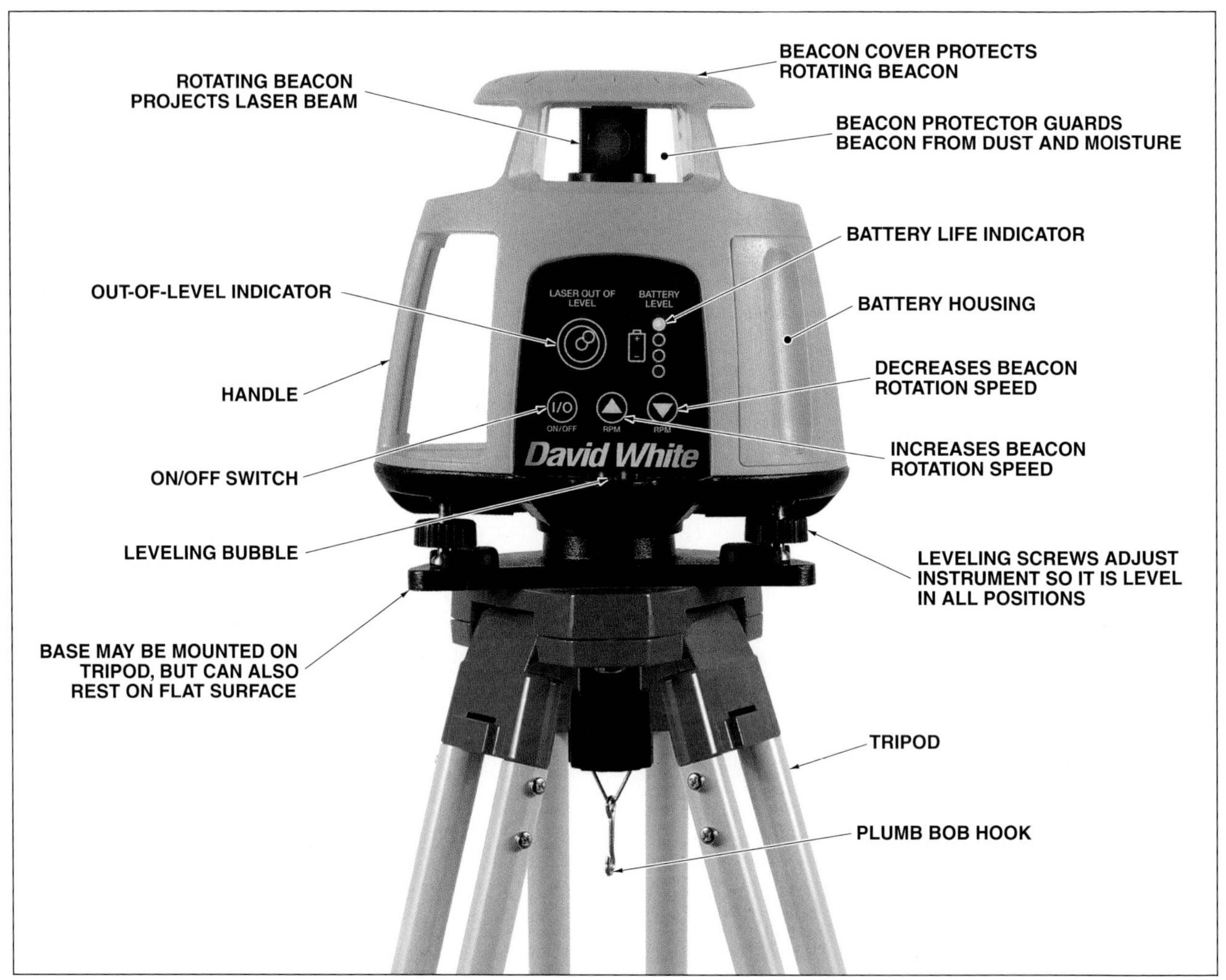

David White Instruments

Figure 34-3. *Most laser levels perform similar functions, but the locations of the parts may differ, depending on the manufacturer.*

A laser level must be level for proper operation and accurate readings. Leveling screws permit a laser level to be adjusted so that it is level. A laser level is level when the bubble in the circular leveling vial is centered. Fully self-leveling laser levels do not include leveling screws.

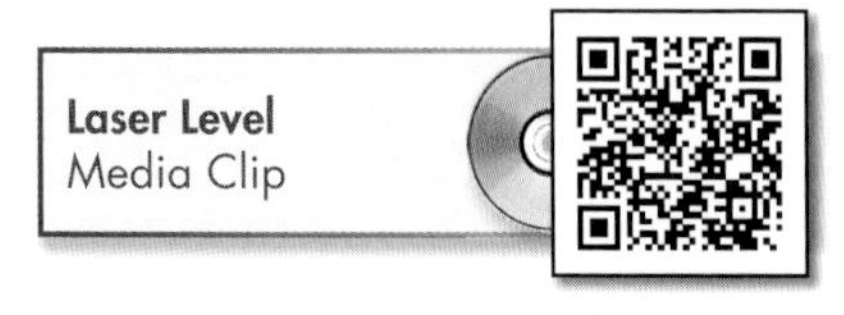

Laser Beams

The term *laser* is formed from the initial letters of the words *light amplification by stimulated emission of radiation.* When a laser level is turned on, a *laser diode,* which is a solid-state device, produces a highly concentrated laser beam. The laser beam is then projected from the rotating beacon.

A small-diameter (3⁄8″) laser beam is produced in a laser level, which can extend the horizontal operating distance up to 1500′. Typical accuracy for high-quality laser levels is within 1⁄16″ per 100′. Accuracy will decrease significantly at distances over 1000′ due to the curvature of the earth and distance from the laser level. Strong air disturbances may also bend the laser beam and decrease its accuracy.

A laser beam may be visible or infrared (invisible). Laser levels with *visible beams* are generally used for indoor applications where they replace less-accurate tools and supplies such as hand levels and chalk lines. Visible beam laser levels are commonly used to hang suspended ceiling grids, level and hang cabinets, and lay out and align wall studs. A detector may be used with a visible beam laser level.

Laser levels with *infrared beams* are used in general construction work including outdoor applications. Infrared laser beams have the ability to travel through, and be less affected by, atmospheric conditions at greater distances than visible beams. A detector is used to locate the infrared beam position. Infrared beam laser levels are commonly used to set concrete forms and check elevations.

Types of Laser Levels

Three basic types of laser levels are *manual-leveling, compensated self-leveling,* and *fully self-leveling* instruments.

Manual-Leveling Laser Levels. Manual-leveling laser levels have tubular or circular vials with air bubbles for leveling the instrument. An operator must manually level the laser level using the leveling screws in a manner similar to traditional surveying equipment. Temperature or atmospheric changes during the day may cause the air bubbles to drift, therefore requiring releveling of the instrument. Manual-leveling laser levels are the least expensive and the least accurate type of laser level.

Compensated Self-Leveling Laser Levels. A compensated self-leveling laser level requires that the laser first be manually leveled using the leveling screws. A compensated self-leveling laser level is equipped with a *compensator* that automatically maintains the levelness of the laser level while it is being used. If the level is jarred out of its self-leveling range, the level will shut down and emit a visual or audible signal notifying the operator that the instrument needs to be releveled.

Fully Self-Leveling Laser Levels. Fully self-leveling laser levels are the most accurate and easiest type of laser level to operate. Fully self-leveling lasers have motor-driven electronic leveling bubbles and require no manual adjustment. If the laser level is jarred, the level will shut off the beam, relevel itself, and turn on the beam again. Many fully self-leveling instruments are equipped with an audible alert that notifies the operator when the level is out of level.

Laser Beam Detectors

Laser beam *detectors,* also known as sensors or receivers, serve as a target for laser beams. **See Figure 34-4.** Detectors are battery-powered and are clamped to a leveling rod or a plain wood rod. Some detectors can be hand held or suspended from a metal surface with a magnetic mount.

The *capture window* of a detector receives the laser beam from a laser level. The detector is moved vertically on the rod until it is directly aligned with the laser beam. An indicator shown in the liquid crystal display (LCD) on the detector indicates whether the detector is above center, below center, or on center. If the detector is above or below center, it must be moved down or up, respectively. When the detector is on center, a level line of sight extending from the laser level to the detector is established. Many detectors emit audible tones to indicate the sensor position.

Direct sunlight and highly reflective surfaces can affect the ability of a detector to accurately sense a laser beam. If possible, a laser level and/or detector should be repositioned so they are shaded from direct exposure to the sun.

Remote Controls. Many laser levels can be adjusted with a remote control device, making it possible to move the beam vertically or focus the laser beam while the operator stands next to the target. Effective ranges for laser level remote controls are up to 160′ indoors and 65′ to 160′ outdoors.

Positioning Laser Levels. Tripods are commonly used to position and mount laser levels. For certain applications, laser levels are attached to metal brackets or other devices fastened to walls or ceilings. Laser levels used for plumbing and squaring operations can be placed directly on the floor and manually leveled. **See Figure 34-5.**

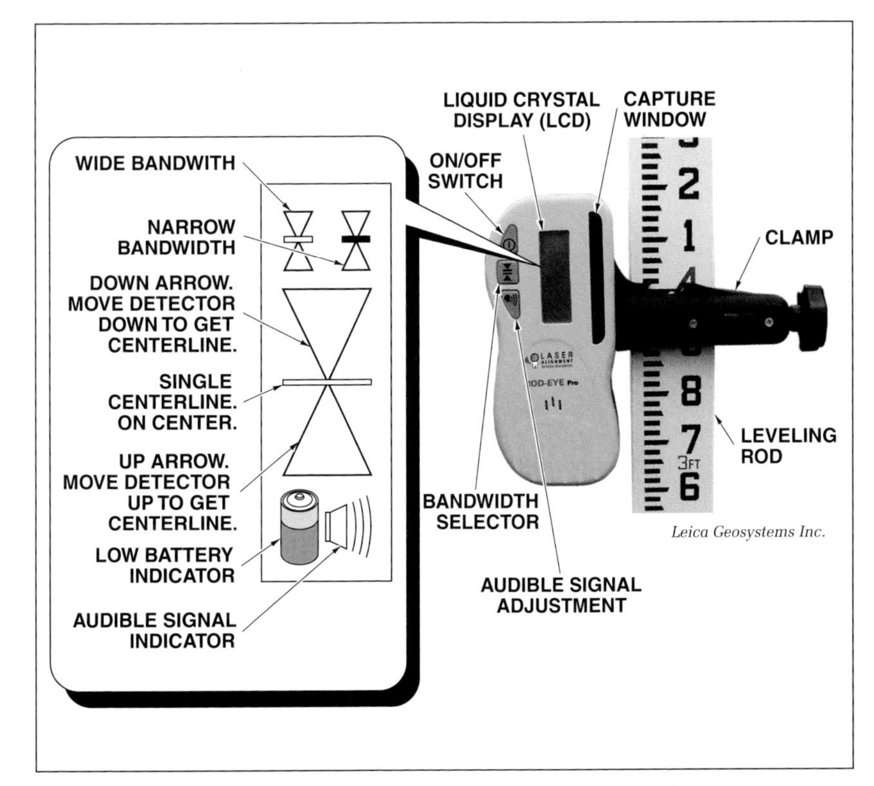

Figure 34-4. *Laser beam detectors function as a target for the laser beam.*

Figure 34-5. *Laser instruments can be mounted on tripods or brackets or placed directly on the floor and manually leveled.*

Selecting Laser Levels

A variety of laser levels are available for various construction operations. When selecting a laser level, the following factors should be considered:

- size of the job site and required range of the instrument
- degree of accuracy required
- primary laser level application (interior or exterior work). Are vertical layout capabilities required? Will sloping grades need to be established and verified?

Laser levels may be able to fulfill one or a combination of these operations. The two types of laser levels most frequently used by carpenters are *rotating laser levels* and *split-beam laser levels.*

Rotating Laser Levels. The laser beam of a rotating laser level moves 360° in a continuous level motion. **See Figure 34-6.** The rotation speed of most rotating laser levels can be adjusted, allowing the laser beam to appear stationary or to sweep across a surface. Rotating levels are commonly used for establishing grades and elevations such as when setting foundation form heights, aligning tops of window frames, and aligning tops of kitchen cabinets.

Split-Beam Laser Levels. A split-beam laser level generates two laser beams which are projected 90° to one another. A single beam is directed into a beam-splitting prism which accurately divides the beam into two laser beams. **See Figure 34-7.** Both beams can be projected simultaneously, or a single horizontal or vertical beam can be projected individually.

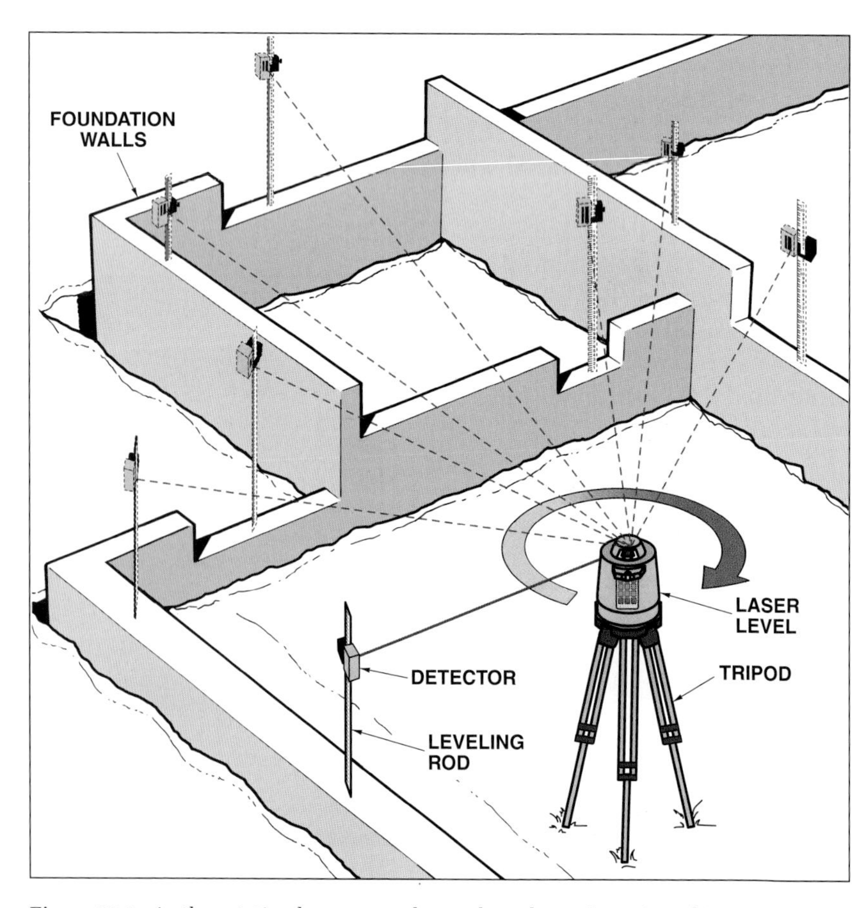

Figure 34-6. *As the rotating beacon revolves, a laser beam is projected in a continuous level, circular motion.*

PLS • Pacific Laser Systems

Figure 34-7. *Split-beam lasers generate two laser beams at 90° to one another. Note the optical laser plummet directly below the laser level.*

Laser levels are commonly used to assist in leveling a suspended ceiling. The laser level shown here is mounted on an elevating tripod. The laser level is adjusted in elevation by using a crank handle on the tripod.

Split-beam laser levels are primarily designed for plumbing and leveling operations, even though they can be used for other laser level applications. The vertical beam can be used to plumb vertically oriented items such as high wall forms and column forms. The horizontal beam can be used for operations such as leveling and setting cabinets and countertops, squaring lines for floor and ceiling tiles, and leveling suspended ceiling grids. Carpenters also use split-beam laser levels to plumb and align window and door frames, as well as to place trim and millwork.

Setting Up and Operating Laser Levels

Features and operation of laser levels vary with the model and manufacturer. If you are not familiar with the instrument, always refer to the manufacturer instructions before operating a laser level. A general procedure for setting up and operating a compensated self-leveling laser level is as follows:

1. Position the laser level so there is an unobstructed path from the laser level to the detector. Mount the laser level on a tripod, wall or ceiling mount, or a flat level surface. The tripod table or other surface on which the laser level is to be placed should be approximately level, A typical laser level requires the table to be ± 3° for the compensator to operate properly.
2. Turn on the laser level. If the out-of-level indicator illuminates, use the leveling screws to level the laser level to within its self-leveling range. **See Figure 34-8.** When the level is within its self-leveling range, the compensator will make the final level adjustments and the out-of-level indicator will turn off.

3. Attach the detector to a leveling rod or a plain 2 × 2 rod. Lightly secure the bracket to the rod so the detector can be moved up and down the rod.
4. Turn on the detector and select the appropriate sensitivity setting and audible signal.
5. Slide the detector up or down the rod until the beam registers in the LCD.
6. Adjust the detector until the on-center display and/or audible signal is obtained. Tighten the bracket lock knob to secure the detector in position.

Laser Level Safety

A laser level produces an intense and highly focused laser beam. Low-power laser levels emit levels of laser beams that are not hazardous. However, high-power laser levels can cause serious eye and skin injuries if improperly used.

Lasers are generally divided into four basic classifications, based on potential risk. The higher the class, the greater the potential risk for the operator and other workers in the area. Brief descriptions of the laser classes are as follows:

- *Class I:* Under normal operating conditions, Class I lasers pose no potential risks. Infrared lasers are classified as Class I.
- *Class II:* Low-power visible laser beams, which normally do not present a potential hazard if viewed for short periods of time. Many laser levels used for construction are Class II lasers.
- *Class III:* Class III lasers are further classified into *Class IIIA* and *Class IIIB.* Class IIIA lasers do not pose a hazard if viewed only momentarily without protection. Class IIIB lasers can be hazardous if viewed directly.
- *Class IV:* Class IV lasers are hazardous to view under all conditions.

Laser level classification and maximum output are listed on a label found on the body of the laser level. Be sure to determine the classification of the laser level you are working with or near to determine the proper safety precautions. If the label is missing from the laser level, determine its classification and the safety precautions that should be taken before turning on the level.

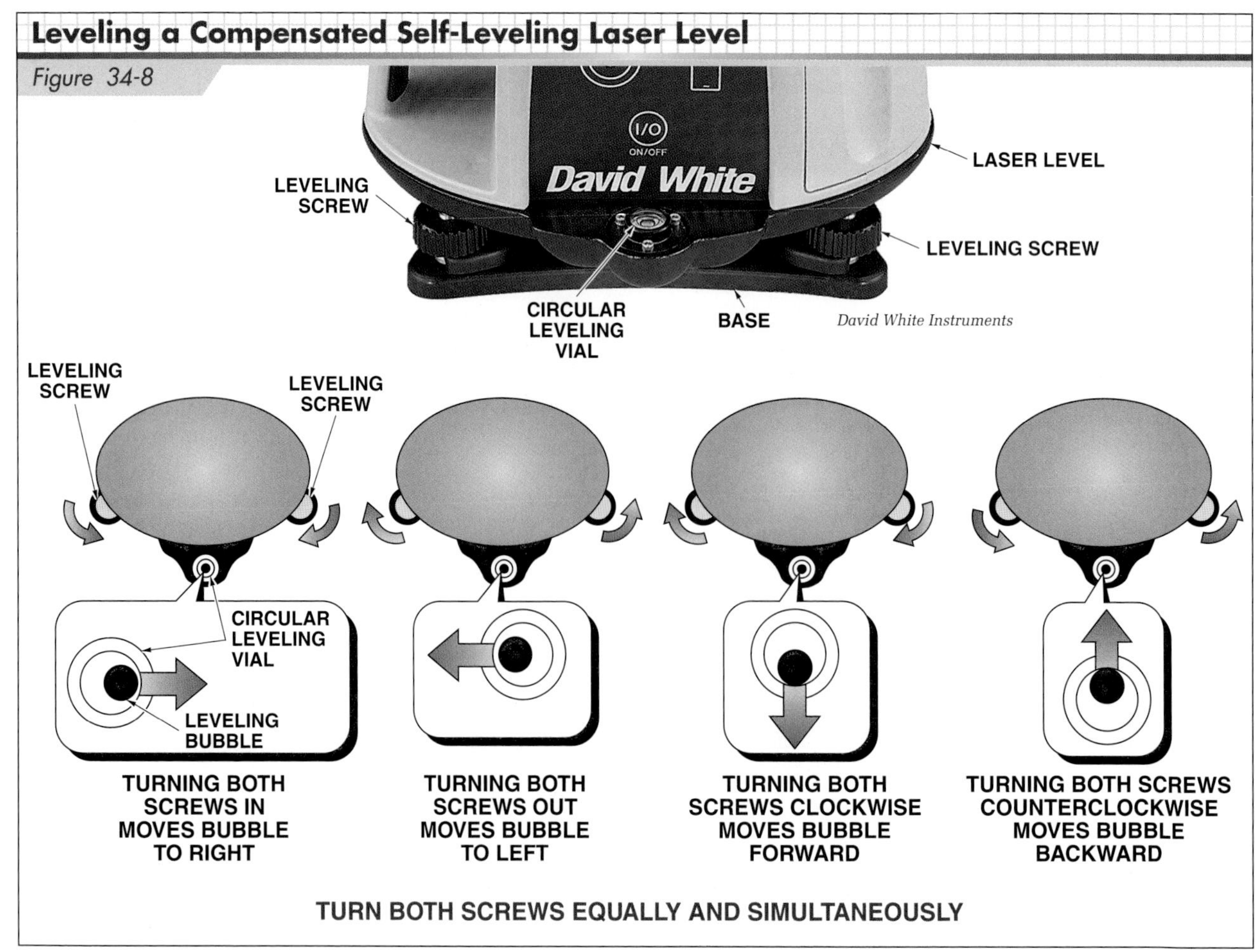

Figure 34-8. *Leveling screws are used to adjust a laser level to within its self-leveling range.*

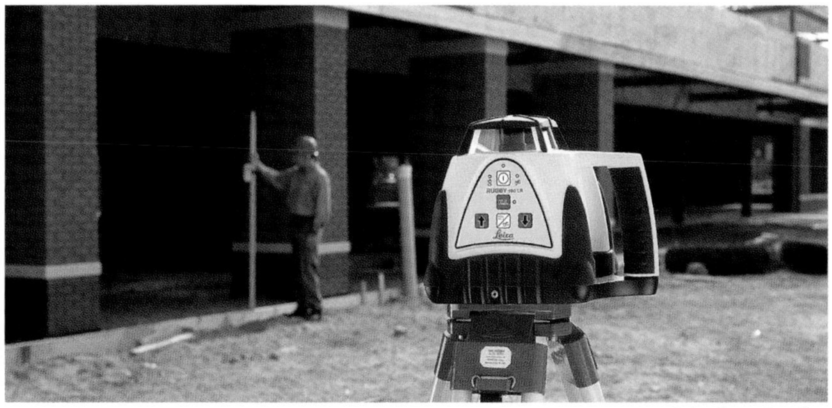

Leica Geosystems Inc.

Leveling forms can be accomplished by one worker using a laser level.

Laser levels must be operated only by workers trained in the proper installation, adjustment, and operation of laser instruments. Proof of qualification of the laser level operator must be available and in the possession of the operator at all times. Manufacturer recommendations must always be observed when using a laser level. Employees on a job site must be aware of potential hazards posed by laser levels. OSHA 29 CFR 1926.54, *Nonionizing Radiation,* provides information regarding the safe use of laser levels. Safety precautions to observe when using laser levels include the following:

- Depending on the laser classification, place the proper laser warning placards in areas where lasers are being used. **See Figure 34-9.**
- Set up laser levels above or below the heads of employees, when possible. A low laser level position provides greater stability and minimizes the chance that the instruments may be knocked over due to wind.
- Use beam shutters or caps, or turn off the laser when laser transmission is not required.
- Turn off the laser when left unattended for a period of time, such as during lunch hour or a change of shifts, or overnight.
- Use only mechanical or electronic means as a detector for receiving the laser beam.
- Do not direct the laser beam at other workers.
- When practicable, discontinue laser operation when it is raining or snowing, or when there is a large dust concentration at the job site.
- Per OSHA 29 CFR 1926.102, *Eye and Face Protection,* workers must wear appropriate antilaser eye protection when working in areas where there is a potential exposure to direct or reflected laser light greater than 5 milliwatts.

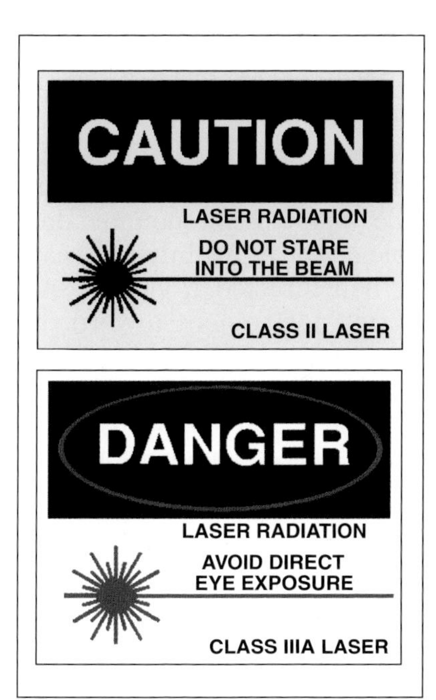

Figure 34-9. *Warning placards are used in areas where laser levels are being used. Examples of warning placards are shown.*

TOTAL STATION INSTRUMENTS

Total station instruments have evolved considerably in the years since they were first introduced as survey instruments. Initially, total station instruments were primarily operated by surveyors for a variety of activities such as land terrain surveys, cartography (mapmaking), bridge building, and mining operations. Recently, total station instruments have seen increasing use by carpenters and other building tradesworkers for job site layout.

Total stations combine digital data processing and survey technology. Total stations perform the functions of traditional transit-levels including leveling, plumbing, and horizontal and vertical measurements. In addition, total stations electronically measure distances, record and store data using an integral computer, and perform mathematical calculations.

Total station instruments are available from many manufacturers. Each manufacturer offers several models of total station instruments falling within different price ranges. However, the basic features and operating procedures are similar. **Figure 34-10** shows a total station instrument and identifies its primary operating and adjustment devices.

Setting up Total Station Instruments

Features and operation of total stations vary with the model and manufacturer. If you are not familiar with the instrument, always refer to the manufacturer instructions before operating a total station instrument. A general procedure for setting up and operating a total station instrument is as follows:

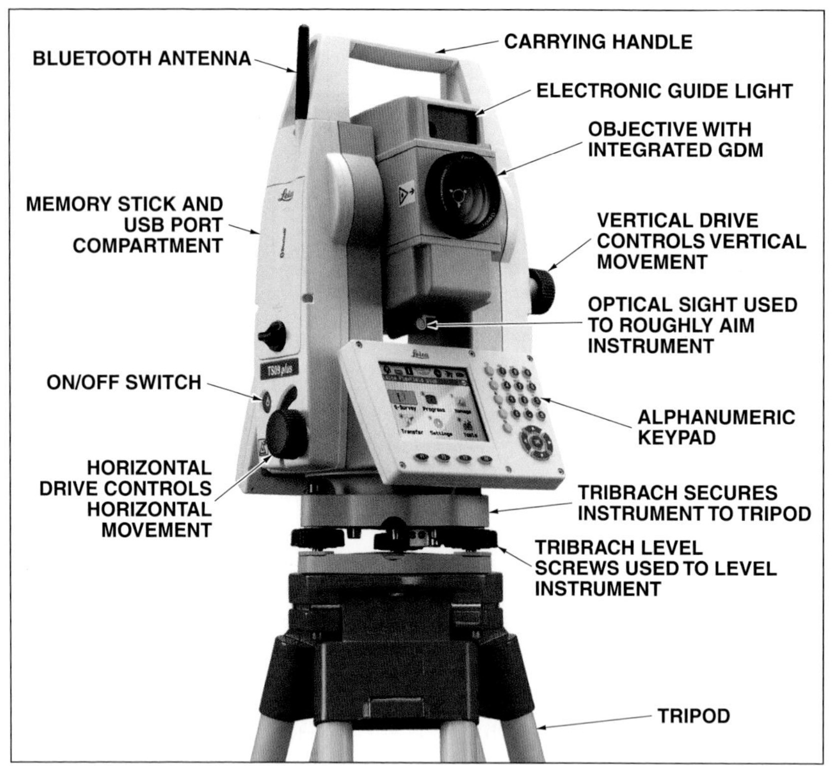

Leica Geosystems Inc.

Figure 34-10. *Total station instruments include many operating and adjustment devices and display data on a liquid crystal display.*

1. Place the tripod directly over an established *ground mark* (station point) so that the tripod head is roughly level. Press the tripod legs firmly into the ground so that the tripod is in a stable position. **See Figure 34-11.**
2. Mount the total station instrument on the tripod.
3. Level the total station instrument using the tribrach leveling screws.
4. Ensure the total station instrument is directly over the ground mark by turning on the optical laser plummet and shifting the tribrach until the laser plummet aligns with the ground mark. The optical laser plummet projects a small-diameter laser beam downward. For older models of total station instruments, use a plumb bob to align the instrument with the ground mark.
5. Complete the final leveling of the total station instrument.

Electronic Distance Measurement

An integral feature of a total station instrument is *electronic distance measurement (EDM).* EDM allows distances to be measured between points with a high degree of accuracy (within .001′) without the use of a measuring tape.

The total station is focused on reflective tape adhered to a surface or on a prism mounted on a pole or stake. An infrared beam is then directed from the total station instrument toward the tape or prism. The beam bounces back to the total station and the distance is calculated from the time elapsed. The measurement is recorded by an integral data collector or by a data-collection unit attached to the total station. The information can also be transmitted to a computer, often located in the contractor's office, where computer-aided design and drafting (CADD) software can develop precise field drawings and contour maps.

Prisms. Traditional survey equipment and laser levels use a leveling rod or plain rod with an attached target or detector. The target for a total station instrument is a reflective device called a *prism* that receives the infrared beam from the instrument. The prism is commonly mounted on a metal prism pole and can be moved up or down as required. **See Figure 34-12.** Prisms can also be held by hand, but holding by hand is not as accurate.

Data Recording

Total station instruments are equipped with an integral computer that not only records survey data, but can also run other software programs. The computer can be used to calculate sines, cosines, and tangents, as well as to determine angles and measure distances. The data can electronically transferred to laptop and desktop computers for other engineering and construction applications.

Types of Total Stations

Additional total station capabilities are obtained with advanced types of total stations. *Robotic, motorized,* and *reflectorless* total station instruments use more powerful software and electronics to provide additional capabilities.

Robotic Total Stations. Only one person is needed to perform survey functions when using a robotic total station instrument. Data is entered into a *remote keyboard,* which is contained in the prism. The prism sends signals back to the total station. Motors within the total station adjust the instrument to point in the direction of the prism pole. Data previously input to the total station computer directs the operator as to which direction to move to establish a point or coordinate. The primary advantages of a robotic total station are that only one person is needed to operate the instrument, and the speed of operation. However, a robotic total station is accurate only up to 875 yards (800 meters).

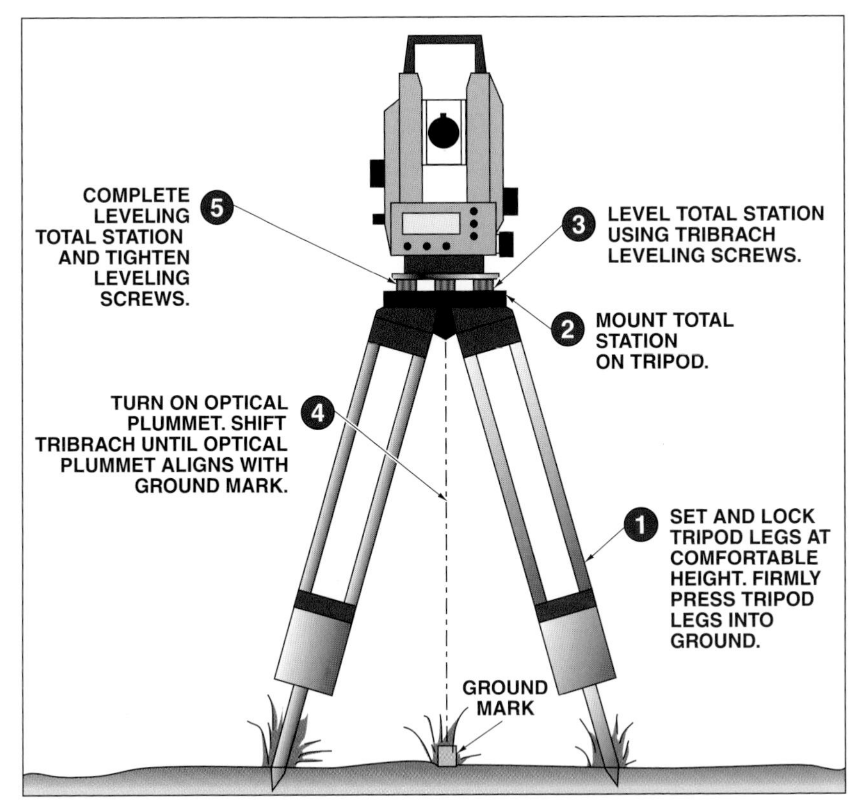

Figure 34-11. *A total station must be leveled and plumbed over a ground mark with great accuracy.*

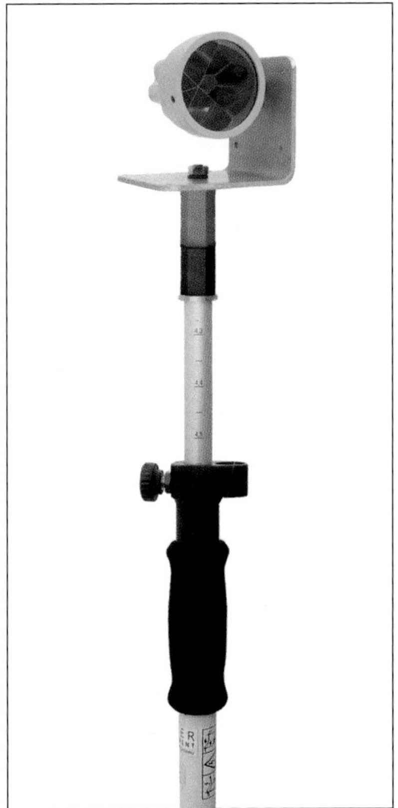

Leica Geosystems Inc.

Figure 34-12. *Total station prisms are frequently mounted on metal poles. Handheld prisms may also be used.*

Motorized Total Stations. A motorized total station requires two workers to establish a point or coordinate. After the total station is properly set up and data is input, the instrument will automatically rotate and point at the location where the prism should be set up. The operator then verbally directs the other worker to position the prism.

Reflectorless Total Stations. A reflectorless total station is equipped with two types of EDM. In addition to an infrared beam that requires a prism, the instrument can also project a visible red beam toward a target such as an inaccessible location on the job site. A red dot appears on the target and the total station measures the distance without the use of a prism. The primary advantage of a reflectorless total station is that the instrument can be used at locations where it is not convenient to place a prism directly on the target point. Reflectorless total stations, however, are not as accurate as the infrared EDM and prism.

Total Station Instrument Care

Total station instruments are precise survey instruments which must be handled carefully to ensure accurate measurements. The following items should be considered when setting up and using a total station instrument:

- After removing the instrument from its carrying case, immediately attach the instrument to the tripod, which has been firmly pressed into the ground. Do not leave the instrument unfastened on the tripod mounting plate.
- Verify that the instrument has been correctly set up and that accessories are properly attached.
- Allow the instrument to adjust to the ambient temperature prior to turning on the instrument.
- Limit the amount of time the instrument is used during inclement weather conditions. Do not use a total station instrument when lightning is present.
- Cover the instrument with a protective hood between operations.
- Remove batteries from an instrument if it will not be used for extended periods.
- After removing the instrument from the tripod, immediately place it back in its carrying case.

Total Station Operation
Media Clip

HIGH-DEFINITION SURVEYING

High-definition surveying (HDS) is a surveying process that utilizes a 3D laser scanner to obtain three-dimensional data for a variety of civil, industrial, and construction applications. **See Figure 34-13.** Conventional surveying equipment captures specific individual points one at a time. An HDS scanner captures up to a million points per second.

There are two steps to producing HDS scans. First, the construction site is scanned using an HDS scanner to create the point cloud. Second, the point cloud is converted using specialized software to produce 2D plans and elevations, BIM models, panoramic images with measurement information from each point, clearances, point-to-point measurements, and sections and profiles. **See Figure 34-14.**

High-Definition Surveying (HDS)

Leica Geosystems Inc.

Figure 34-13. *High-definition surveying uses a 3D laser scanner to obtain three-dimensional data for a variety of civil, industrial, and construction applications.*

The scanner obtains details of an entire job site, similar to a 360° camera, but with an accurate and measureable X, Y, and Z position for every point. The scanner rotates horizontally while emitting a rapidly pulsating laser beam. A rotating mirror directs the laser beam up and down vertically. The combination of the vertical and horizontal rotation allows the scanner to obtain information from the entire site. When the laser hits an object, part of the beam is reflected back at the scanner. These reflections are then used to create the 3D image, called a point cloud.

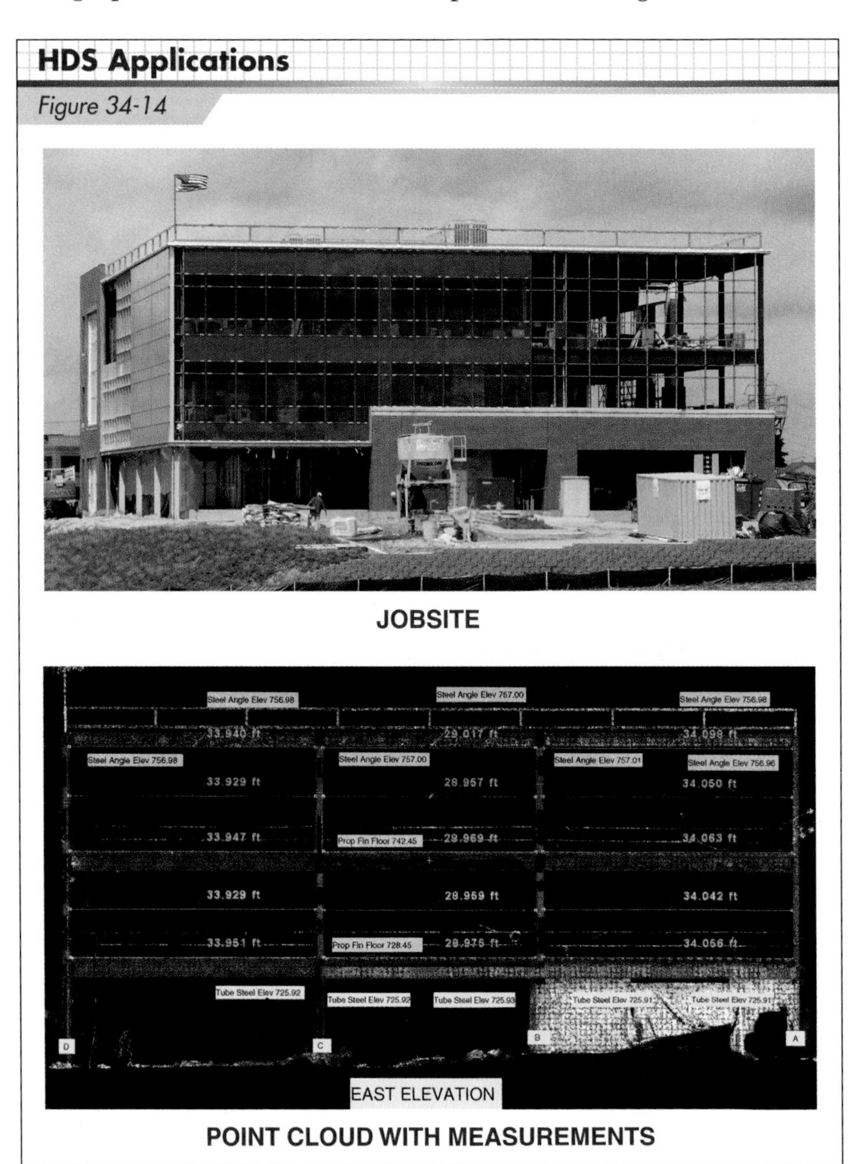

Figure 34-14. *Once the point cloud has been produced using HDS, measurements may be obtained from points in the scan. In this example, as-built elevations of the steel were used to fabricate the building's glass and aluminum curtain wall.*

Unit 34 Review and Resources

SECTION

8

Foundation and Outdoor Slab Construction

CONTENTS

UNIT 35

Building Site and Foundation Layout

OBJECTIVES

1. Identify factors to consider before starting a new construction project.
2. List and describe procedures for laying out foundation walls.

The exact location of a building on a building site must be known before construction can begin. A plot plan provides location and elevation information for the building site, and other information commonly used by an operating engineer who clears the property.

BUILDING SITES

In residential areas where streets have been established, a piece of property is referred to as a *lot.* When a building is to be constructed on a lot, the lot becomes a *building site.* Features of a building site help determine the type of foundation best suited to the building. For example, the shape and size of the lot and ground slope must be considered in designing the foundation. Other factors to be considered are weather conditions of the area and soil conditions of the site. Soil conditions of one building site may be different from conditions of other sites in the same general area.

Soil Conditions

Since the full load of a building rests on the ground below it, soil conditions are important to the integrity of the building foundation. The American Society for Testing and Materials (ASTM) classifies soils as *gravel, sand, silt,* or *clay.* The main difference among the four soil types is the size of the *grains* (particles) that make up the soil. Clay is composed of the smallest particles. Gravel is composed of the largest particles. Silt and sand particles are larger than clay but smaller than gravel.

Since gravel particles are larger, they compress (press together) less than clay particles when subjected to heavy pressure. **See Figure 35-1**. Consequently, there is less *settlement* (downward movement) of a building constructed on sand than of a building constructed on clay. In addition, since larger voids exist between sand particles, sandy soil drains water away more quickly than clayey soils.

A certain amount of settlement can be expected with any newly constructed building, unless it has been built on *bedrock* (solid rock). No problems arise if settlement is minimal and takes place evenly. A large amount of uneven settlement, however, can cause cracks in the foundation and structural damage to the building.

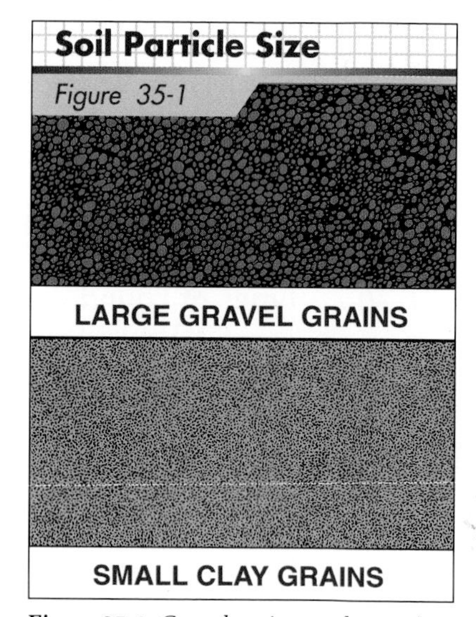

Figure 35-1. *Gravel grains are larger than clay grains and more space exists between the grains. Therefore, gravel compresses less than clayey soil.*

When drawing foundation plans, an architect must consider the soil conditions of the building site. Sometimes the architect consults with a soil engineer, who makes test bores and analyzes soil samples taken from the lot.

Earthquake and Weather Conditions

A foundation constructed in an area where there is danger of earthquakes *(seismic risk zone)* must be designed to withstand great stress. Reinforcing steel bars are normally required in all concrete or masonry foundation walls constructed in these zones. In addition, special ties or other structural reinforcement may be required to fasten the foundation to the structure.

The depth of the *frost line* also affects foundation design. The frost line is the depth to which the soil freezes. In colder regions, moisture from rain and snow penetrates the ground and freezes during the winter season. The freezing and thawing cycles of soil cause it to expand and contract. If a foundation footing is placed below the frost line, there will be no movement of the foundation during freezing and thawing cycles. **See Figure 35-2.** The local building code usually specifies how deep foundation footings should be placed below the frost line.

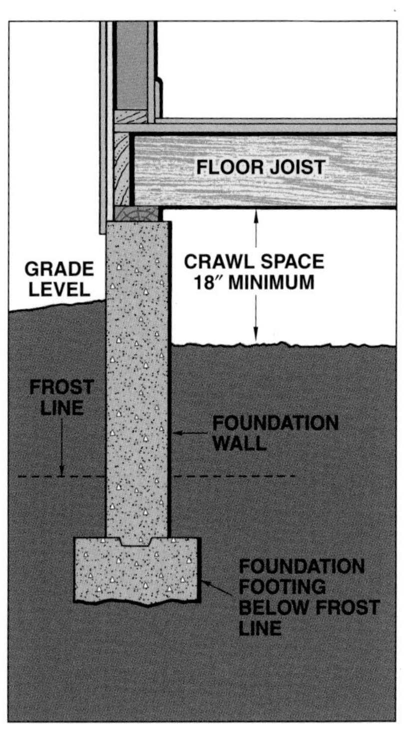

Figure 35-2. *The bottom of a footing must be below the frost line.*

LAYING OUT FOUNDATION WALLS

Before foundation construction can begin, the precise boundaries of a building site, called *property lines (PL),* must be verified with a lot survey. Next, building lines are set up to show the exact location of the building. Information required to lay out foundation walls is found on the plot plan. A plot plan not only shows the building location on the lot, but also identifies different grade levels of the lot. **See Figure 35-3.**

Establishing Lot Boundaries by Lot Surveys

Lots are usually mapped, and the maps are recorded by local building or zoning authorities. A lot surveyor studies the zoning maps

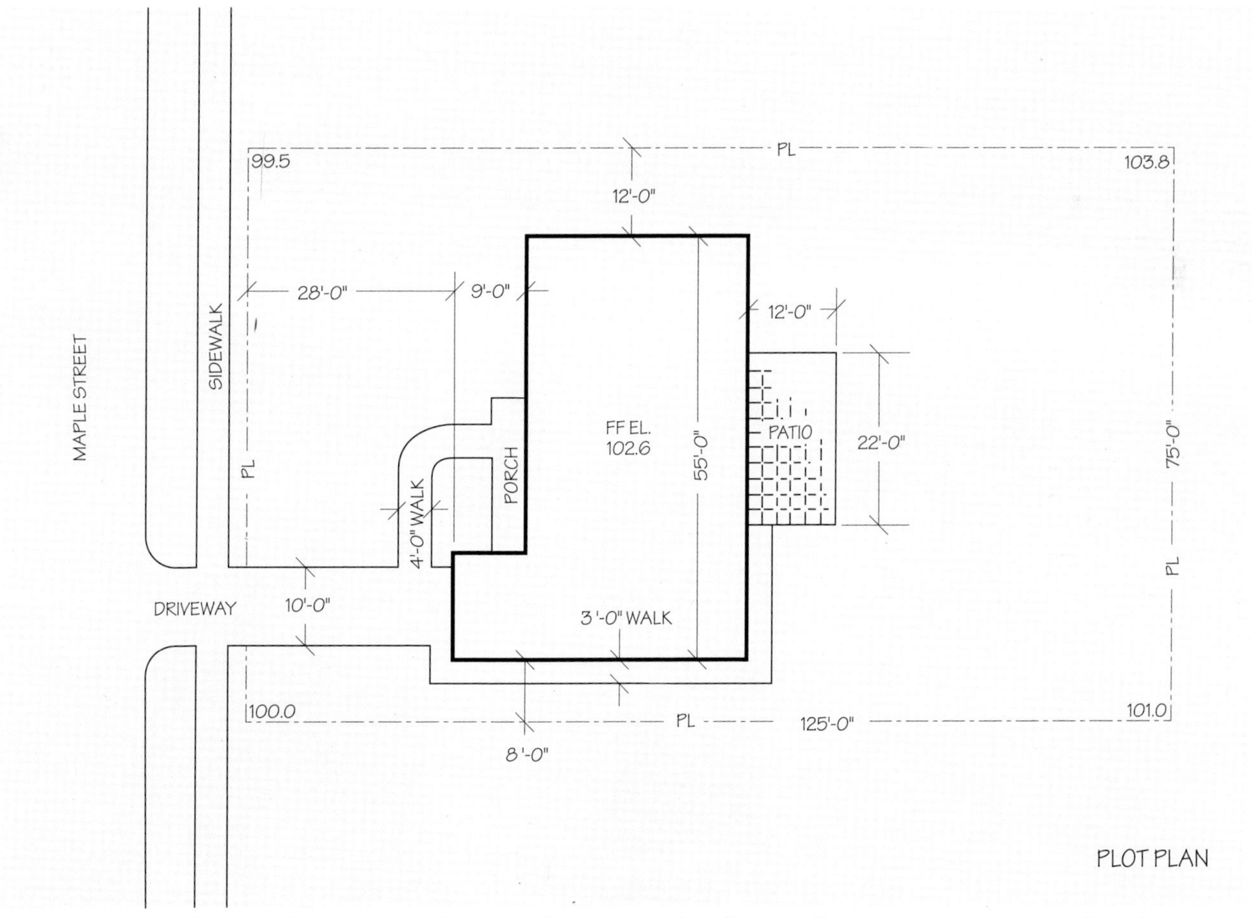

Figure 35-3. *A plot plan provides necessary information for laying out foundation walls. Measurements are given from the property lines to the sides of the building. Grade levels indicate the high and low points of the ground.*

and records to determine the precise boundaries of a lot. By measuring from the proper reference points, a surveyor can establish the two front corners of a lot. Often, street curbs are used as reference points. In some areas of North America, however, the front corners are established by measuring from the center of a road or from special monuments placed in sidewalks. A transit-level is commonly used to establish the two rear corners of the lot.

Corner Hubs and Property Points. A surveyor places a marker, or *hub,* at the lot corners. In the past, a hub was often a 2 × 2 wood stake driven into the ground with the top flush with the surface. A small nail was driven into the top of the stake to identify the exact property corner. Today, a hub usually consists of a rebar driven into the ground with a colored plastic cap. A pipe with a lead or plastic plug may also be used. This is called a *property point* or *survey point.* The property lines drawn on a plot plan represent lines extending from the four corners of a lot.

Locating Buildings on Lots

Local zoning regulations or building codes identify the minimum setbacks permitted for buildings. Setbacks vary from one city or county to another and often vary among different sections of the same city.

A common setback for buildings located in a one-family residential zone is 20′ minimum from the front property line to the front of the house and 4′ minimum from the side property lines to the sides of the house. In contrast, buildings in commercial zones are sometimes built within inches of property lines.

Establishing Building Corners

The most efficient way to establish building corners is to use a transit-level. **Figure 35-4** shows how to use a transit-level to establish building lines.

To establish building corners without a transit-level, use the 3-4-5 method of determining 90° corners. In the 3-4-5 method, one stake is driven in line 3′ from the corner stake and another stake is driven in line 4′ from the corner stake on the adjoining wall. The diagonal distance between these two stakes should be exactly 5′. A greater degree of accuracy can be obtained by multiplying the distances that the auxiliary stakes are driven from the corner stake by any number. For example, the first stake is driven 6′ from the corner stake, and the second stake is driven 8′ from the corner stake on the adjacent wall. The diagonal distance across these two stakes is 10′.

Building Lines and Batterboards

After building corners have been established, building lines must be set up to mark building boundaries. Foundation wall forms are then set to the building lines. Since it is impractical to attach the strings to low stakes that mark building corners, *batterboards* are constructed to hold the building lines while forms are being constructed. **See Figure 35-5.**

Laser levels are commonly used to establish elevations on a job site.

Geotextiles stabilize and retain soil or earth in position on slopes or in other unstable conditions after excavation.

Batterboards are level boards nailed to stakes driven firmly into the ground. When stakes cannot be driven firmly, braces may be required to support batterboards. When forms are being constructed for foundations on fairly level lots, batterboards at all corners of the building should be level to each other. Batterboards are usually placed 4′ to 6′ behind each building corner to provide working room between the batterboards and form construction.

Using Transit-Levels to Establish Building Lines...

Figure 35-4

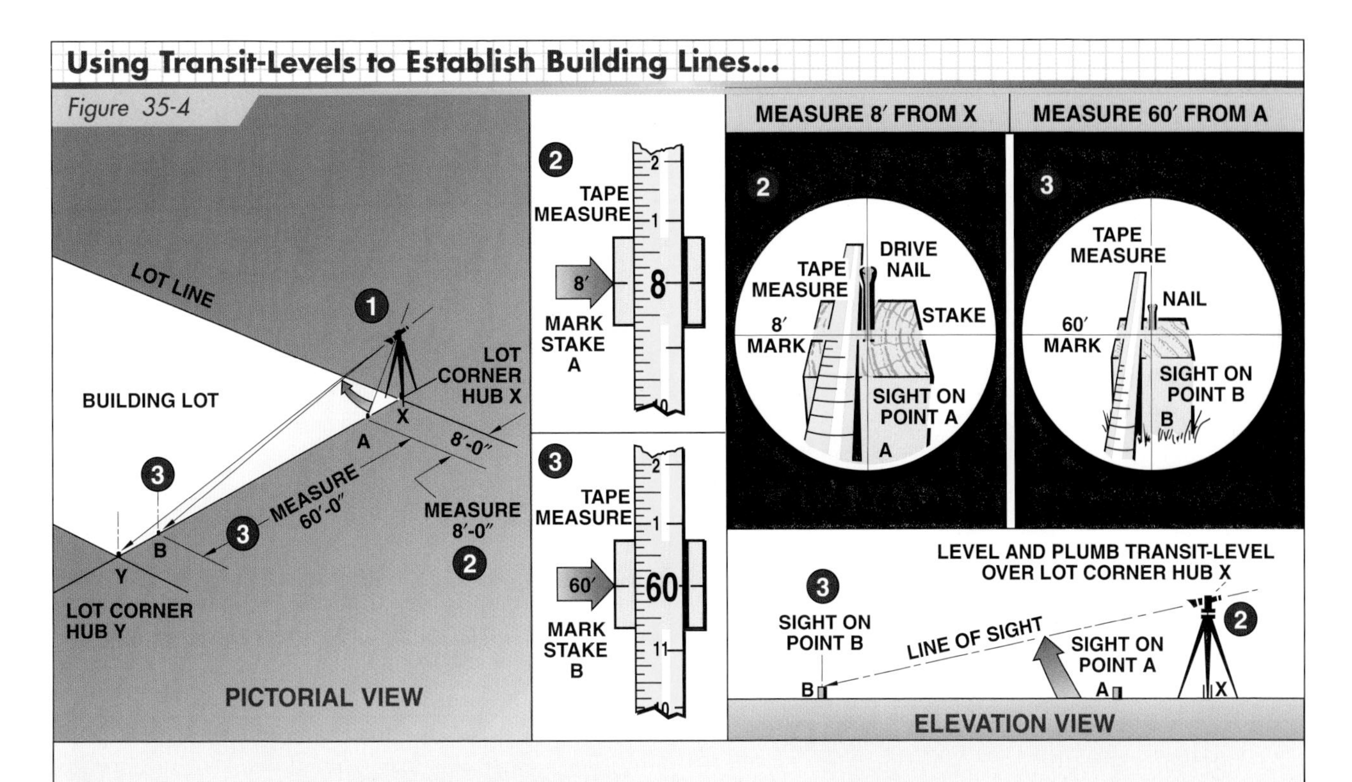

1. Level and plumb transit-level over lot corner hub X. Sight down to opposite lot corner hub Y.

2. Distance from property line to one side of building is 8′-0″. Drive a stake 8′-0″ from corner hub X and align with lot corner hubs X and Y. This is done by lowering telescope until vertical and horizontal crosshairs are close to center of stake top at the 8′-0″ measurement. Using fine adjustment tangent screws of the transit-level, sight through the instrument and drive nail into stake top exactly in line with vertical and horizontal crosshairs at 8′-0″ reading of tape measure.

3. Measure 60′-0″ from point A, which is width of building. Drive another stake in line with lot corner hubs X, Y, and point A. Holding tape measure in position over this hub, raise telescope until the horizontal crosshair coincides with 60′-0″ mark on tape. Now, using vertical crosshair, align and drive nail into top of stake, establishing point B.

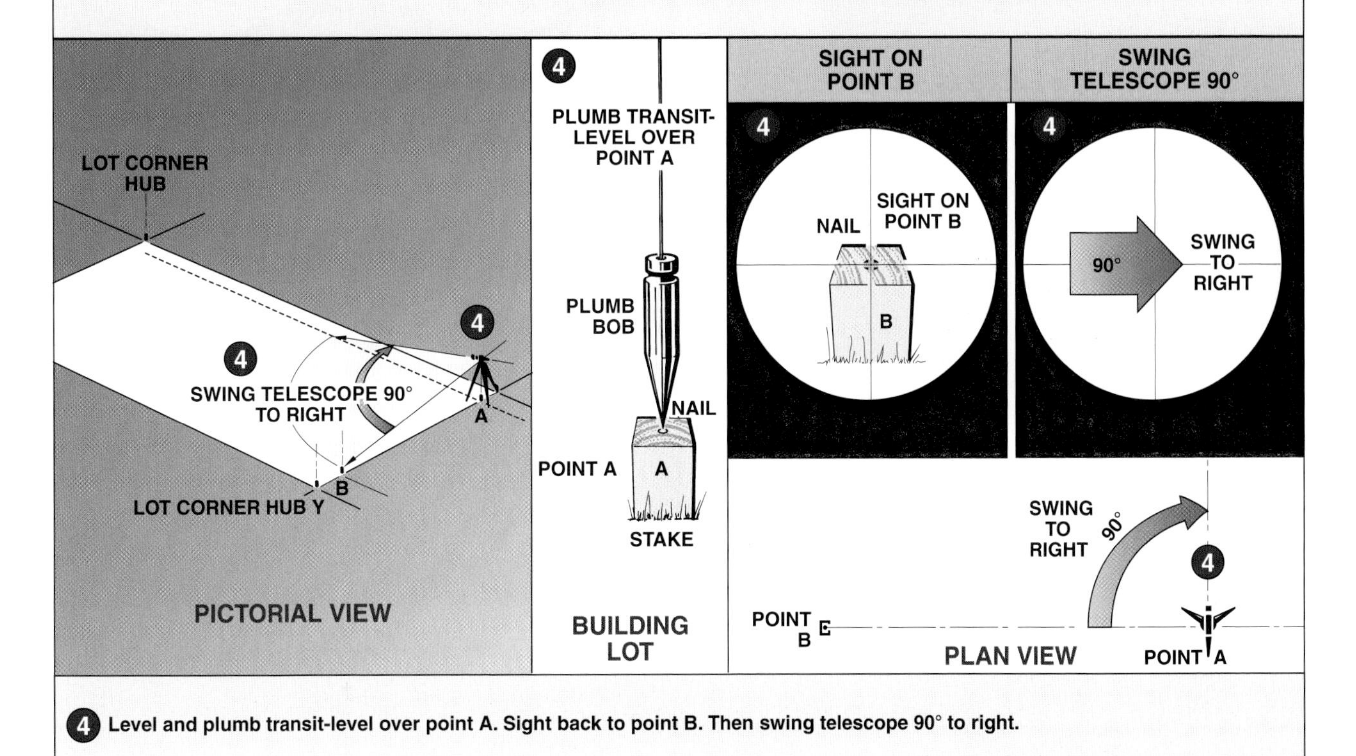

4. Level and plumb transit-level over point A. Sight back to point B. Then swing telescope 90° to right.

Figure 35-4 . . .

...Using Transit-Levels to Establish Building Lines

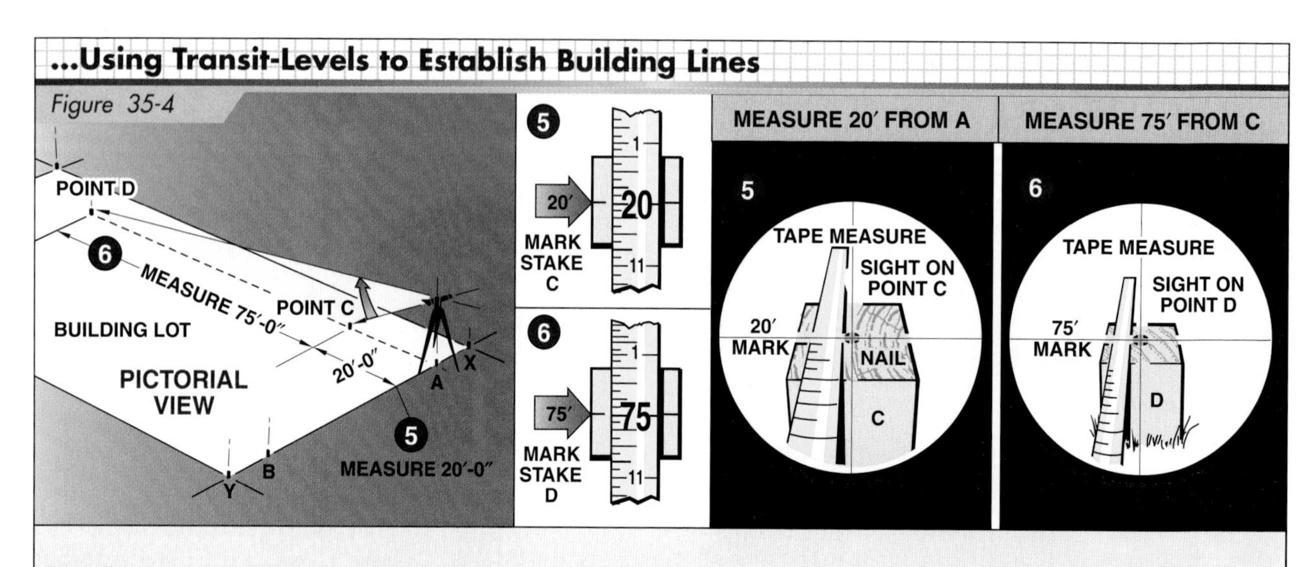

5 Measure 20′-0″ from point A, which is distance from front property line to front of building. Lower telescope until horizontal crosshair coincides with 20′-0″ mark on tape measure. Drive stake and nail, establishing point C, which is first corner of building.

6 Measure 75′-0″ from point C, which is length of building. Raise telescope until horizontal crosshair coincides with 75′-0″ mark on tape measure. Drive stake and nail, establishing point D, which is second corner of building.

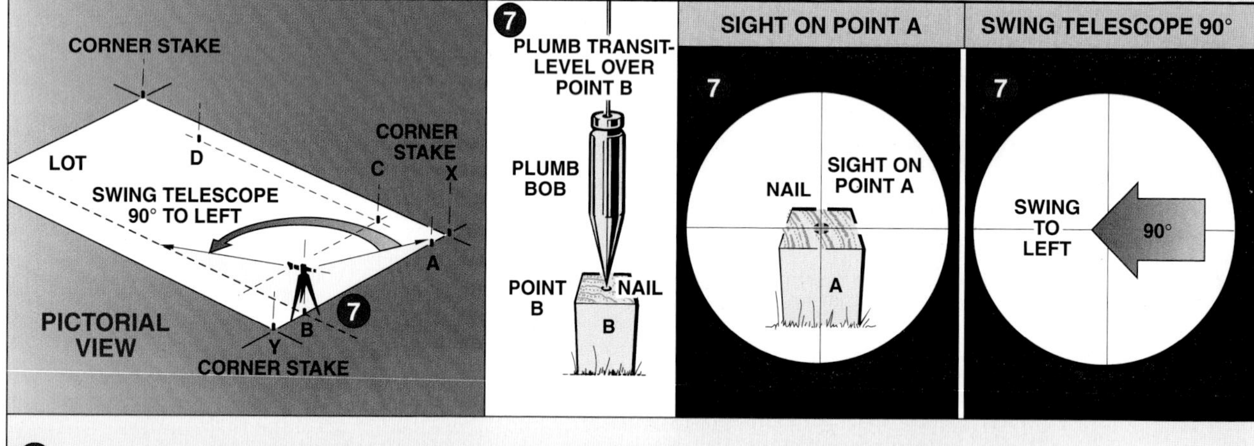

7 Level and plumb transit-level over point B. Sight back to point A, then swing telescope 90° to left.

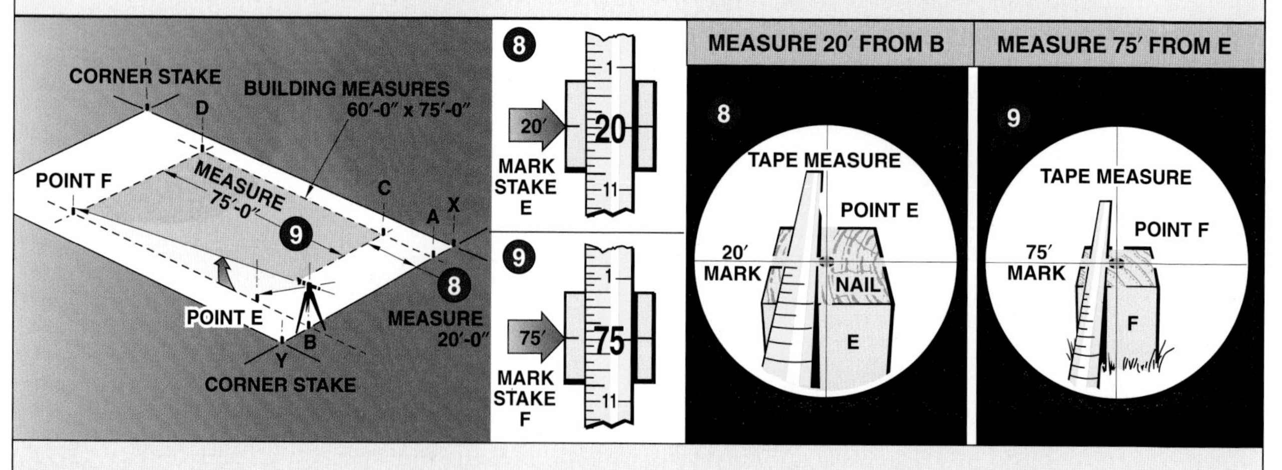

8 Measure 20′-0″ from point B. Lower telescope until horizontal crosshair coincides with 20′-0″ mark on tape measure. Drive stake and nail, establishing point E, which is third corner of building.

9 Measure 75′-0″ from point E. Raise telescope until horizontal crosshair coincides with 75′-0″ mark on tape measure. Drive stake and nail, establishing point F, which is fourth corner of building.

. . . **Figure 35-4.** *Building corners are established with a transit-level.*

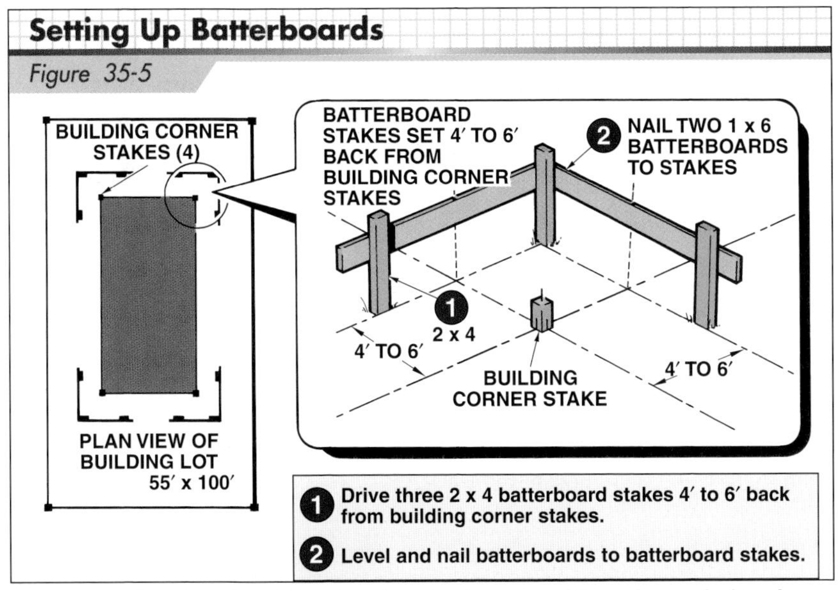

Figure 35-5. *Batterboards hold building lines in place to mark boundaries of a foundation.*

Groundwork

Before foundation forms can be constructed, preliminary groundwork must be completed. Excavation and finish grading may be needed.

Depending on the soil conditions and the foundation type, trenches may have to be dug for foundation footings.

Most groundwork is performed by operating engineers running earth-moving equipment. **See Figure 35-7.** Carpenters often work with the operating engineers by setting up lines or placing stakes to guide the excavation and trenching operations.

Building lines may be set up by stretching each line tightly from one batterboard to the opposite batterboard. Building lines are positioned directly over a building corner stake using a straightedge and level or plumb bob as shown in **Figure 35-6.** The same operation can be performed using a transit-level to transfer the building corners from the stakes to the batterboards. Regardless of the method used to set up the building lines, nails should be driven or shallow kerfs should be cut in the upper edge of the batterboards to prevent the lines from moving out of position.

After building lines are secured, carpenters should verify the measurements between the lines to ensure they are accurate. The diagonals across the lines should be measured to ensure the building lines are square. Any errors in layout can be easily corrected at this time. A foundation built out-of-square or to a wrong dimension will create many problems for the construction that follows.

Nails may also be used to secure building lines to batterboards.

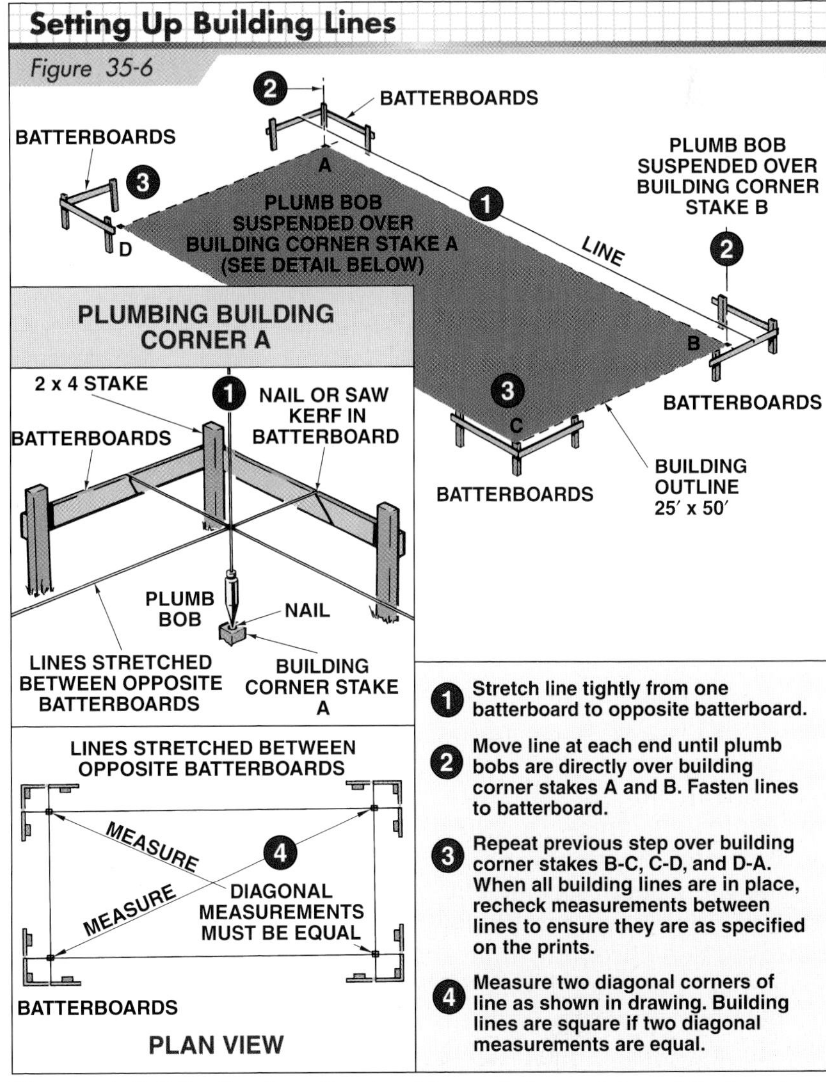

Figure 35-6. *Building lines (usually nylon string or wire) accurately mark the boundaries of a foundation.*

© Case Foundation

Heavy construction foundations may require piles to be driven deep into the earth to support the foundation.

As digging proceeds, depths are checked with a laser transit-level or builder's level and rod. **See Figure 35-8.** If batterboards have been set close to the foundation wall height, a line can be stretched and the depth measured from the line. A rod cut to the required measurement is more convenient to use than a tape measure. **See Figure 35-9.** Trenches for stepped footings must also be level.

Photo Courtesy of Deere Construction

Figure 35-7. *Most groundwork is performed by an operating engineer operating earth-moving equipment such as this excavator.*

Footing depth is usually indicated in section views of a foundation plan. Trenches must be wide enough to allow room for foundation form construction. Trench bottoms must be level. Most lots have some slope; therefore, the starting point from which the trenches are leveled should be at the lowest point of the excavation.

Angle of repose is the slope a material maintains without sliding.

Figure 35-8. *Footing depth is verified using a laser level (shown) or builder's level and leveling rod.*

Excavation. An *excavation* is a cut, depression, or trench in the earth surface formed manually or by earth-moving equipment. The excavating process may involve leveling a slope or digging the ground to the proper depth for a full-basement foundation.

Trenching. Fairly level ground beneath a crawl-space foundation generally requires little excavation but may require *trenching* for footings. Trenches must be dug so footings can be placed at the depth specified by the local building code. Some building codes require that a footing extend an additional 6″ deeper into natural, undisturbed soil. In areas where the ground will freeze, the trench bottom must be below the frost line.

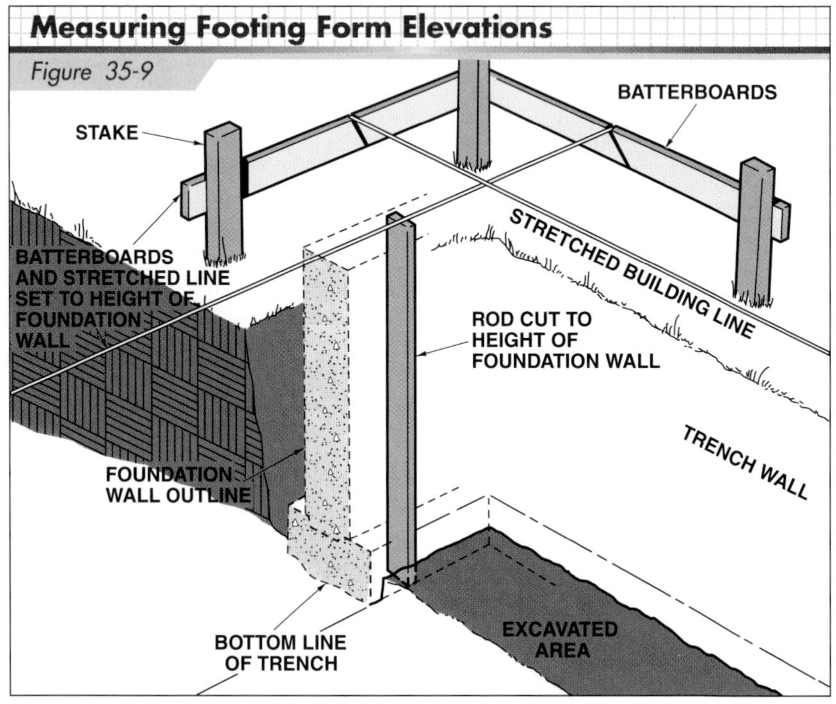

Figure 35-9. *When batterboards have been set to the foundation wall height, the depth of the trench can be measured from the building line. A rod cut to the required length is a convenient method of verifying trench depth.*

Grading. Many lots require *grading,* which is reshaping the surface of a lot. Soil is removed from some sections and added to other sections. Lots are graded so surface water (rain or melting snow) flows away from the building. Recommended minimum slope is 6″ in 10′, or 5%. However, if the area surrounding the foundation is to be paved, a 1/8″ in 1′, or 1%, slope is adequate.

The plot plan in **Figure 35-10** shows *natural grades (NG)* and *finish grades (FG)* at all four corners of the lot. Natural grade is the level of the ground before groundwork begins. The finish grade is the level of the ground after soil has been added or removed. Soil on the lot is added or removed to correspond to the finish grades on the plot plan.

Plot plans also include contour lines, which show the shape produced by the varying grades of the lot. Dashed contour lines represent the natural contour (shape) of the ground; solid lines show the finished contour. Lot grades are checked with a laser level or builder's level and leveling rod.

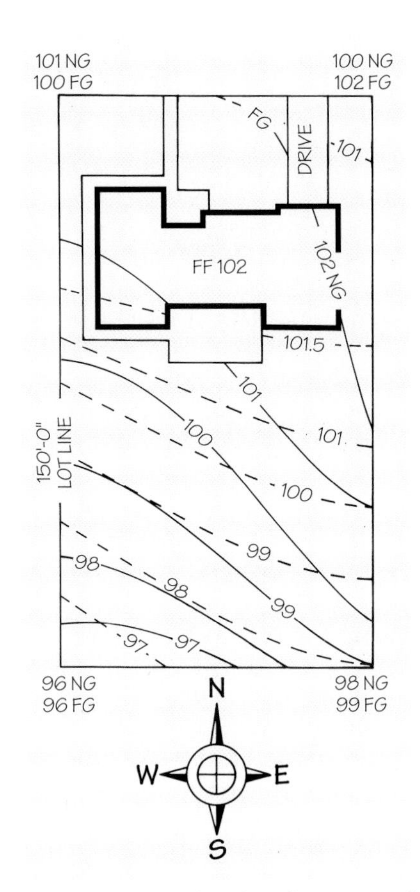

Figure 35-10. *A plot plan shows natural and finish grade levels. In this plot plan, the lot is to be graded so that it slopes toward the southwest and northeast corners. Soil will be removed from the northwest corner and added to the southeast and northeast corners. The southwest corner will remain at the same grade.*

Backfilling. Backfilling is done after forms have been removed and work such as waterproofing, installing exterior insulation, and placing drain tiles has been completed. Soil used for backfilling should be free of wood scraps and other waste material. Backfill is placed carefully against the foundation walls and compacted (pressed down). **See Figure 35-11.** On many jobs, gravel is required for backfill since gravel allows better water drainage away from the building.

Figure 35-11. *Backfill is placed against completed foundation walls. Curved steel areaways prevent backfill from entering the basement and allow light to enter through a basement window.*

Unit 35 Review and Resources

UNIT 36

Types of Foundations

OBJECTIVES

1. Describe the common types and functions of foundation walls and footings.
2. Define sill plates and explain how sill plates are fastened to foundation walls.
3. List and describe common foundation systems.
4. Describe the use of concrete masonry unit foundations.
5. Explain the procedure for preparing a site for a wood foundation.

Most building foundations are constructed of concrete or concrete masonry units (CMUs). Wood foundations are permitted in some parts of the country. Major factors that determine foundation design are the live and dead loads the foundation must support and the soil conditions beneath the foundation. Local environmental factors such as climate and seismic conditions also play a large role in foundation design.

WALLS AND FOOTINGS

Most foundation designs consist of footings and walls. *Footings* rest directly on the soil and act as a base for foundation walls. In most foundation designs, footings serve to distribute the weight of the building over a wider soil area.

The most visible parts of foundations are the walls around the perimeter (outside) of the building. Additional foundation walls may be located under interior sections of the building to support loads from the framed units above. Concrete *piers* (square, round, or battered structures that support posts or beams) and steel columns or wood posts are frequently used to support interior sections of a building.

Common foundation shapes are *inverted T-shaped, battered, L-shaped,* and *rectangular.* **See Figure 36-1.** An inverted T-shaped foundation is constructed with a stem wall supported by a spread footing. The sides of a battered foundation are wide at the base and taper toward the top. An L-shaped foundation is similar to an inverted T-shaped foundation,

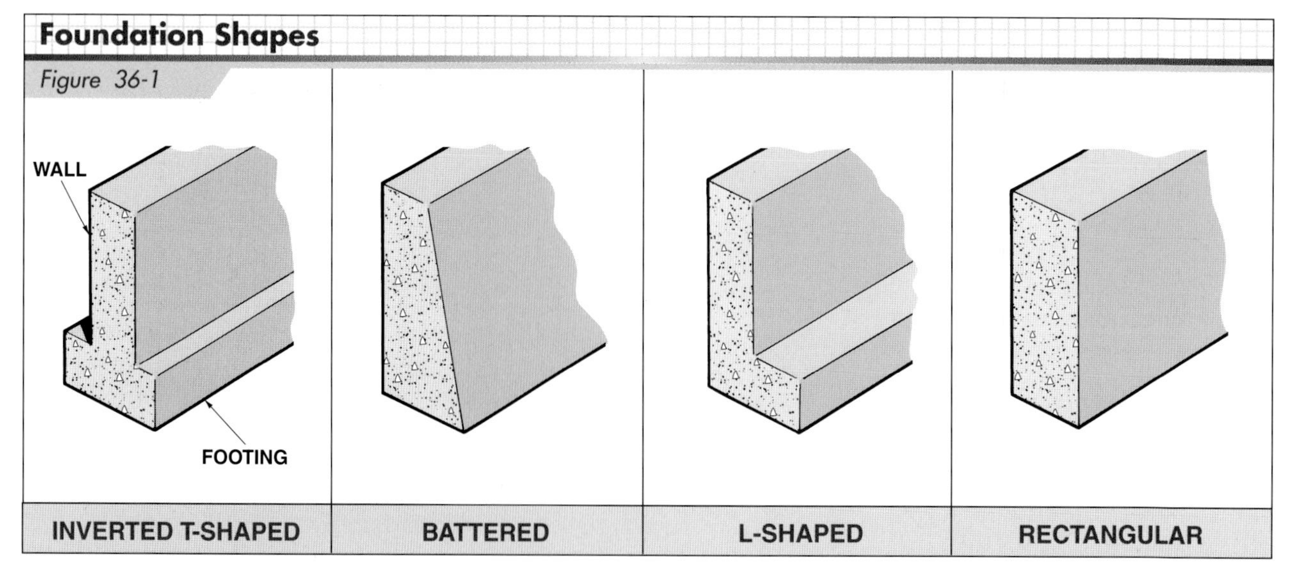

Figure 36-1. *Foundation design for a particular building site depends on soil conditions and anticipated loads. The rectangular foundation can be used only for light wall loads and firm soil conditions.*

but the footing extends from only one side of the foundation wall. A rectangular foundation, designed for light wall loads and firm soil conditions, has uniform wall thickness over its entire height.

The dimensions of foundation walls and footings are normally provided in section views of the foundation plan. The dimensions may also appear in a full-house section found in another section of the prints. **Figure 36-2** shows formulas that may be used to determine inverted T-shaped foundation dimensions over normal soil.

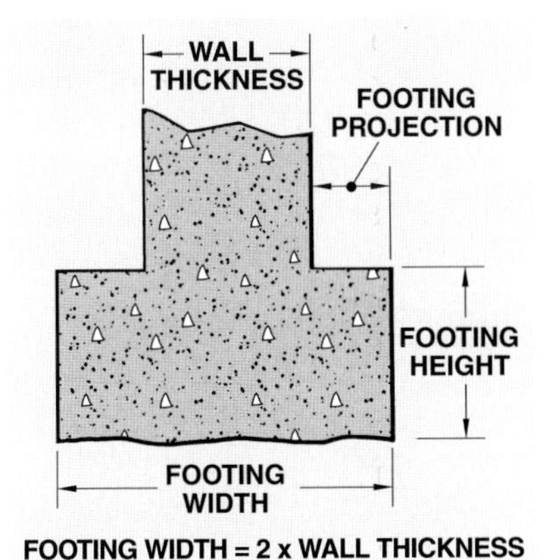

Figure 36-2. *The dimensions of a footing for normal soil conditions can be determined if the foundation wall thickness is known.*

Concrete *piers* support and anchor the bottoms of steel columns and wood posts which support beams that are part of the framing system of a building. A square footing is usually placed beneath a steel column. Other shapes, such as round or battered concrete piers, are often placed beneath wood posts. **See Figure 36-3.**

FOUNDATION SILLS

Many parts of North America experience extreme weather conditions, such as hurricanes, tornadoes, and earthquakes, which can have devastating effects on buildings and human life. One result of these conditions is the building separating from its foundation. Extreme wind conditions can cause the building to rack, slide, or overturn from its foundation. Racking and sliding occur when the wind exerts an extreme horizontal pressure on the building. Overturning occurs when the building is unable to either rack or slide, which causes the building to rotate off its foundation. **See Figure 36-4.**

Earthquakes are caused by the rupture of geological fault lines deep below the Earth's surface. Earthquakes occur suddenly with little or no warning. Two types of forces—lateral and uplift—are involved in earthquakes. *Lateral forces* produce longitudinal waves, resulting in a forward-and-backward motion. *Uplift forces* produce transverse waves, resulting in an up-and-down motion.

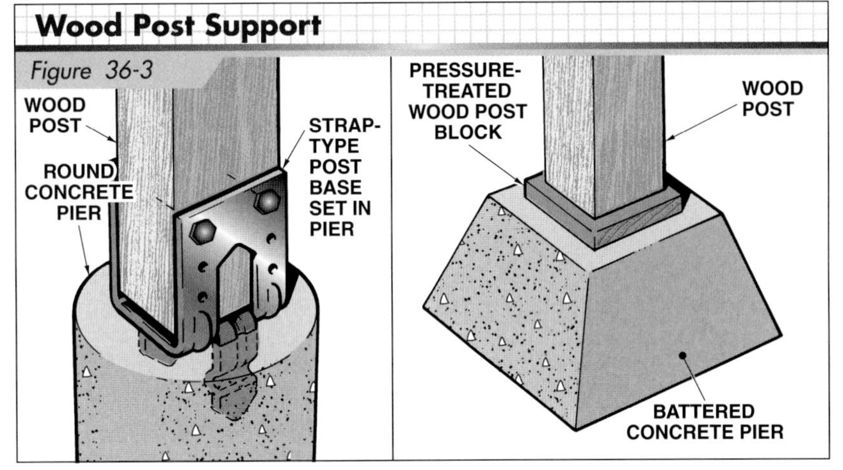

Figure 36-3. *Round or battered piers are often used to support wood posts.*

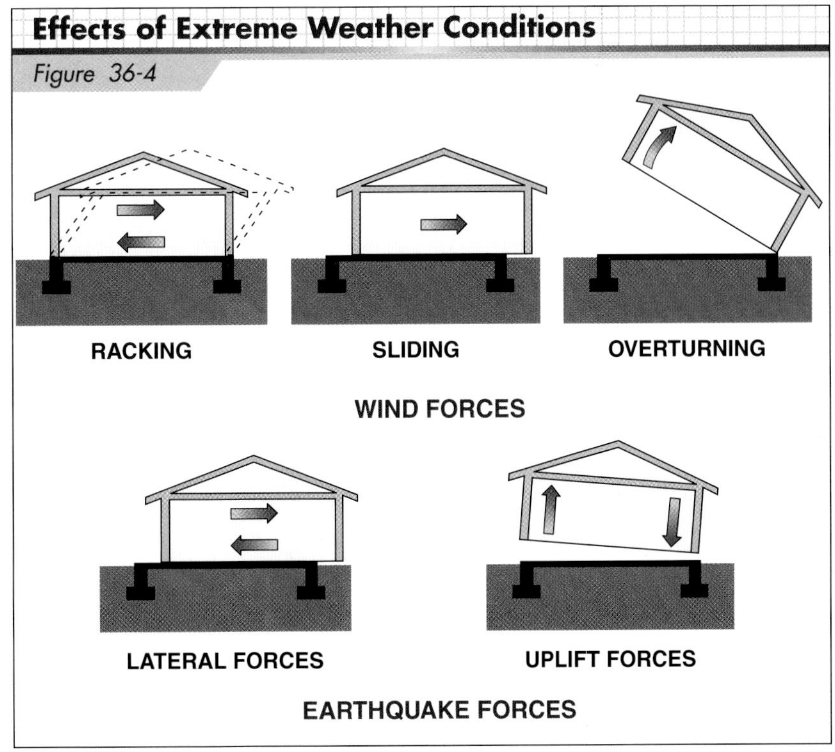

Figure 36-4. *A building can be separated from its foundation by racking, sliding, or overturning. Lateral and uplift forces from earthquakes also affect buildings and their foundations.*

To minimize damage due to extreme weather conditions, a building must be properly fastened to its foundation. Wood *sill plates,* also called mudsills, are fastened to the tops of foundation walls. Sill plates provide a nailing base for joists or studs resting directly over a foundation. Pressure-treated wood is recommended for foundation sills because of its superior resistance to decay and insect attack. Other decay- or insect-resistant lumber species, such as redwood, may be used. In residential and other light construction, foundation sills are usually 2 × 4s or 2 × 6s. Heavier construction may require 4″ thick or larger pieces.

Sill plates are fastened to foundations using anchor bolts or metal connectors such as mudsill anchors. Sill plates may also be fastened to foundations using heavy-duty concrete screws. Always refer to the construction specifications and local building code to ensure adherence to the building code.

Fastening with Anchor Bolts

A common method of fastening sill plates to a foundation is using *anchor bolts,* which are also known as J-bolts because of their shape. **See Figure 36-5.** Anchor bolts should have a ½″ minimum diameter. Information regarding anchor bolt size and spacing is usually found in section views of a foundation plan.

Local building codes specify the minimum spacing and depth of embedment of the anchor bolts into the fresh concrete. A common code requirement is that the bent end of an anchor bolt be embedded a minimum of 7″ into reinforced concrete. In unreinforced masonry, anchor bolts should penetrate 15″. The code further states that anchor bolts must be spaced no farther apart than 6′ OC, and that there must be a bolt within 12″ of the ends of a sill plate. A short sill plate must have at least two anchor bolts spaced no more than 6′ apart.

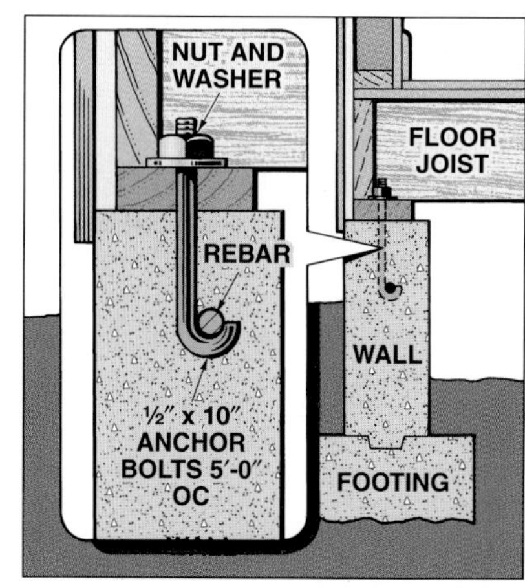

Figure 36-5. *Anchor bolts are commonly used to fasten sill plates to foundation walls. In this example, floor joists rest on the sill plates.*

Sill plates may be secured to foundation walls immediately after concrete is placed in the wall forms. **See Figure 36-6.** The plates are cut to length ahead of time and the holes are drilled. Anchor bolts are placed in the holes, and the nuts are started on the threads. Washers should always be used under the nuts. When concrete placed in the forms reaches the required height and is properly consolidated and leveled, the sill plates are pressed into the fresh concrete. At the same time, the bolts are tapped down using a wood block to prevent damage to the threads. The sill plates are then leveled and cleats are nailed across the form to hold the plates in place. When the concrete has hardened for several days, the nuts can be tightened.

Another method used to fasten sill plates to foundations is to set the bolts in position before concrete is placed in the forms. An advantage of this method is that the anchor bolts can be hooked beneath the upper horizontal rebars, providing greater holding power. The anchor bolts are placed in templates, which allows the bolts to remain in a plumb position while concrete is being placed. The templates also keep the anchor bolts extending the proper amount above the concrete. The templates are removed when the concrete has sufficiently set. **See Figure 36-7.**

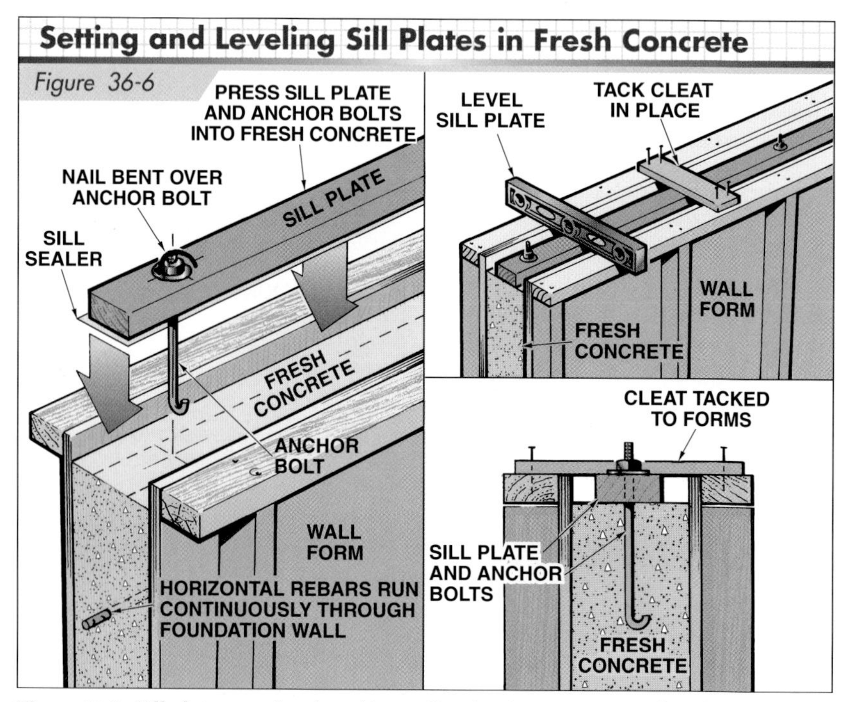

Figure 36-6. *Sill plates may be placed immediately after concrete is placed and the fresh concrete meets the required level in the wall form. In this example, the foundation wall forms consist of panels reinforced by studs and a top plate.*

Figure 36-7. *Anchor bolts extend from the top of a foundation wall. Holes are drilled in sill plates and the plates are placed over the anchor bolts and fastened into position with washers and nuts.*

After the concrete has set, holes are laid out and drilled in the sill plate. A layer of *grout* (mixture of sand, cement, and water) may be placed below the sill plates to provide a level and even base for the plates. In most situations, a foam *sill sealer* (sill gasket) may be installed to minimize air infiltration along the sill plate.

Fastening with Mudsill Anchors

Mudsill anchors are metal connectors used to fasten the sill plate to a foundation. Mudsill anchors are available in a variety of designs. **See Figure 36-8.** Always refer to the manufacturer instructions for proper installation of mudsill anchors. When properly installed, some mudsill anchors provide greater strength than conventional bolts.

Mudsill anchors must be spaced at the correct intervals and placed along the center of the sill plate for optimum anchorage. Typically, mudsill anchors have less resistance value than anchor bolts and are installed at closer intervals than anchor bolts to compensate for the difference.

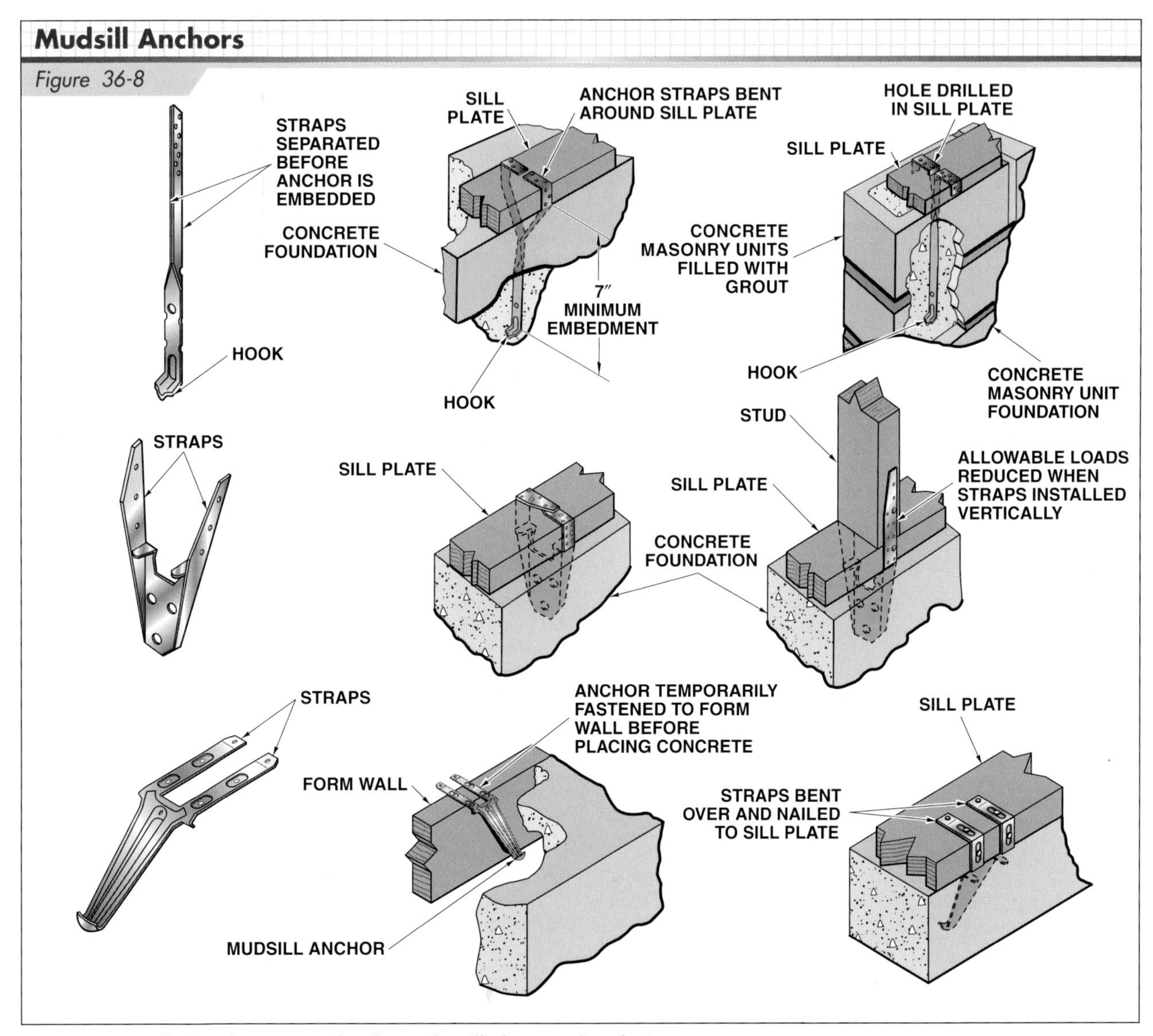

Figure 36-8. *Mudsill anchors are used to fasten the sill plate to a foundation.*

The material in which a mudsill connector is installed commonly determines the installation method. When installing a mudsill anchor in a concrete foundation, the straps of the anchor are spread to accommodate the width of the sill plate, and the hook is embedded in the concrete to a minimum depth of 7″. When the mudsill anchor is installed in a concrete masonry unit (concrete block) foundation, the hook is embedded at least two courses deep. Holes are drilled in the sill plate to align with the straps, and the sill plate is placed in position after the concrete has properly set. The straps of the mudsill anchor are then spread and fastened to the sill plate.

Foundation anchors are metal connectors that provide a solid connection between the foundation and structural members to resist uplift and lateral forces. **See Figure 36-9.** Foundation anchors are fastened to foundation walls using anchor bolts. The other ends of the anchors are attached to studs or joists using bolts, screws, or nails. Always refer to manufacturer instructions for proper installation procedures.

Retrofitting Existing Structures

Older existing structures, including residential structures such as homes, may not be properly anchored to foundation walls and may be subject to damage due to extreme weather conditions. The structures should be *retrofit* to strengthen the foundation-to-structure connection to provide greater resistance to wind and earthquake forces.

If there is sufficient room in a basement or crawl space area to operate a drill, bolting the sill plate is the easiest and most economical means of fastening the sill plate to the foundation wall. A rotary hammer is used to drill holes through the sill plate and into the top of the foundation wall. The holes should be 1⁄16″ larger than the diameter of the anchor bolt, spaced 4′ to 6′ OC, and be at least 12″ from the ends of the sill plates. Wedge-type stud-bolt anchors are then inserted into the holes and tightened in place. A nut and washer is placed over the threaded end of the anchor and sill plate and tightened. Threaded anchor bolts can also be secured into position using an epoxy adhesive.

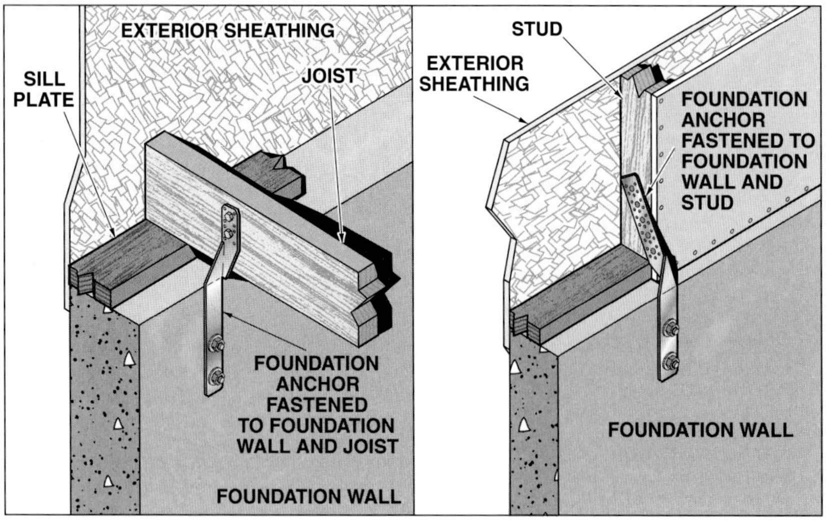

Figure 36-9. *Foundation anchors provide a solid connection between the foundation and structural members to resist uplift and lateral forces.*

Bearing plates may be used instead of round washers when retrofitting existing foundation connections. Bearing plates are 3″ square steel plates with an oblong hole in the middle to allow the plates to be adjustable in case the anchor bolt is not centered on the sill plate. Bearing plates significantly increase the strength of anchor bolt connections.

If there is minimal room between the foundation and top of the sill plate, *foundation plates* can be used to fasten the foundation to the side of the sill plate. A foundation plate is fastened to the foundation wall using anchor bolts and is fastened to the side of the sill plate using the proper screws. **See Figure 36-10.**

In hurricane-prone areas, framing connections should be secured with straps, ties, or clips.

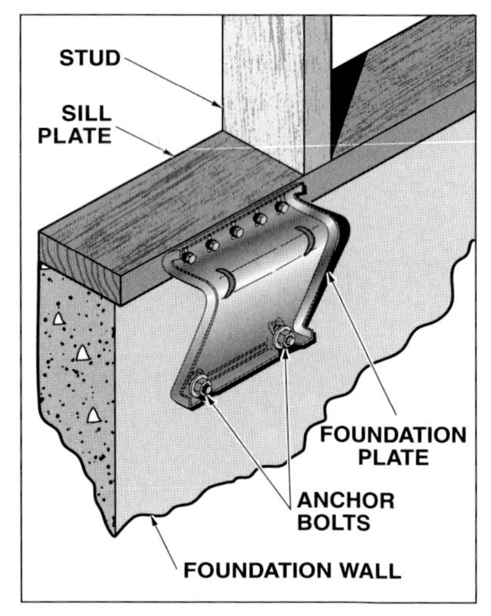

Figure 36-10. *Foundation plates are used to fasten the foundation to the side of a sill plate.*

Fastening to Concrete Slabs

When anchor bolts are used to fasten sill plates to concrete slabs, they are set in position as concrete is placed. The sill plate is set into position and the nuts on the anchor bolts are tightened after the concrete sets.

Sill plates for interior framed walls are often fastened into position with a powder-actuated fastener. Concrete must be allowed to cure the appropriate amount of time prior to driving fasteners into fresh concrete.

FOUNDATION SYSTEMS

The main foundation systems are the *crawl-space foundation, full-basement foundation,* and *slab-at-grade foundation.* The primary difference between a crawl-space and full-basement foundation is the wall height. The slab-at-grade foundation features a concrete slab floor instead of a framed floor unit.

Crawl-Space Foundations

A crawl-space is a narrow space between the bottom of a floor unit and ground. **See Figure 36-11.** Crawl-space foundations are also known as *basementless* foundations. Local building codes usually specify a minimum crawl-space distance for buildings. A common minimum distance is 18″ from the bottom of the floor joists to the ground and 12″ between the ground and beams that support joists. Crawl-space foundations are often constructed with 2′ clearance beneath the joists to provide easier access for plumbing, electrical, or other mechanical repairs.

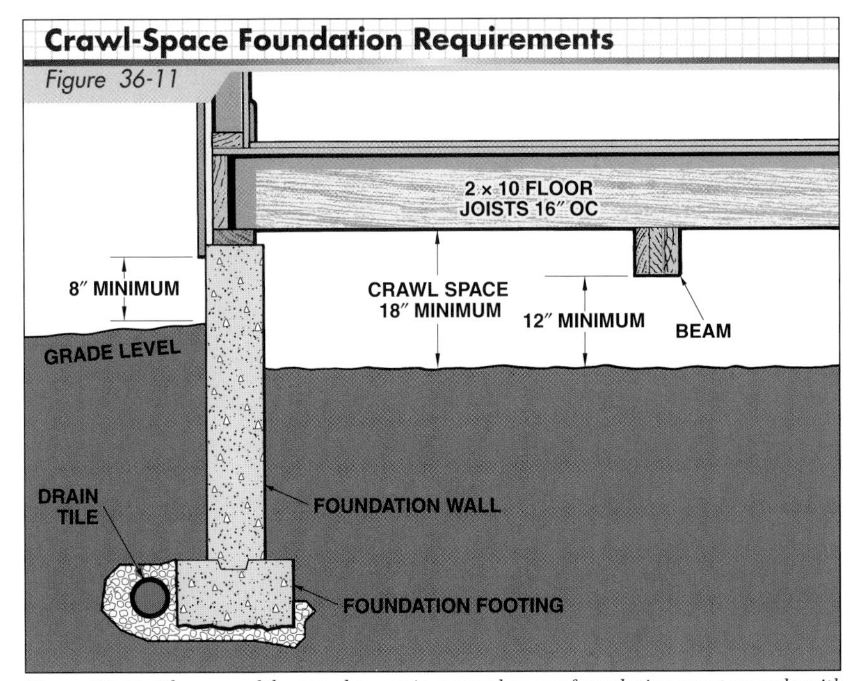

Figure 36-11. *The area of the crawl space in a crawl-space foundation must comply with local building code requirements.*

Full-Basement Foundations

In a full-basement foundation, the foundation wall also serves as a basement wall. **See Figure 36-12.** The top of the foundation wall should extend at least 8″ above finish grade.

Foundation walls are usually 7′-0″ to 8′-0″ from the floor slab to the bottoms of the joists. The walls must withstand the lateral pressure of the soil and be fully waterproof.

Areaways. When basement windows are located below the finished grade level, areaways (window wells) are required. Areaway walls are constructed of sheet metal, concrete, hollow concrete blocks, or brick. Areaways must project above the finished grade and below the bottom of the window. **See Figure 36-13.**

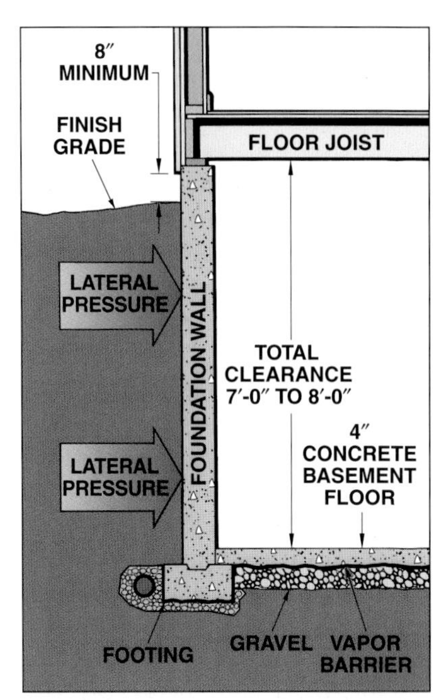

Figure 36-12. *A full-basement foundation must have adequate clearance between the basement floor and ceiling.*

Figure 36-13. *Areaways are installed when window openings are located below finished grade level.*

Slab-at-Grade Foundations

In a slab-at-grade foundation, the foundation walls are combined with a concrete floor slab. The slab surface is level with the top of the foundation wall. In a slab-at-grade foundation, the floor slab receives its main support from the ground. Slab-at-grade foundations are not practical over steeply sloped lots or where the water table is near the ground surface.

Concrete for slab-at-grade floors can be placed as concrete for the foundation walls is placed or the floors can be placed separately. Slab-at-grade floors can also be combined with grade beams and inverted T-shaped, rectangular, or battered foundation walls.

Foundations for Sloped Lots

Stepped foundations are used on a steeply sloped lot. Sections of the footings are shaped like steps and a foundation wall with a level top is constructed over the footings. A level foundation for a sloped lot would require more concrete and more excavation of the hillside than a stepped foundation. A stepped foundation may provide both a crawl space and a full basement.

Stepped foundations require vertical (plumb) and horizontal (level) footings. **See Figure 36-14.** Many building codes require a 2′ minimum distance between one horizontal step and another. The vertical footing must be at least 6″ thick and no higher than three-quarters of the distance between one horizontal step and another.

Grade Beams. Grade beams are foundation walls that receive their main support from piers extending deep into the ground. **See Figure 36-15.** Grade beams are often used with stepped foundations erected on steeply sloped lots. Grade beams are also used on level lots where unstable soil conditions exist.

CONCRETE MASONRY UNIT FOUNDATIONS

Foundations constructed of *concrete masonry units (CMUs)*, also referred to as concrete blocks, are a common alternative to solid concrete foundations. CMU foundations are erected over cast-in-place concrete footings. **See Figure 36-16.** Carpenters usually construct the footing forms. Wood- or light steel-framed walls are usually built over CMU foundations.

CMU foundation walls must extend above the finished grade a minimum of 4″ where masonry veneer is used.

Buildings may be designed with CMU walls that extend above the foundation level and are used as the exterior walls of a building. Construction of the CMU walls is performed by bricklayers, while carpenters set, plumb, and brace wood or metal window frames and door jambs. **See Figure 36-17.** In general, metal frames and jambs are preferred in masonry walls.

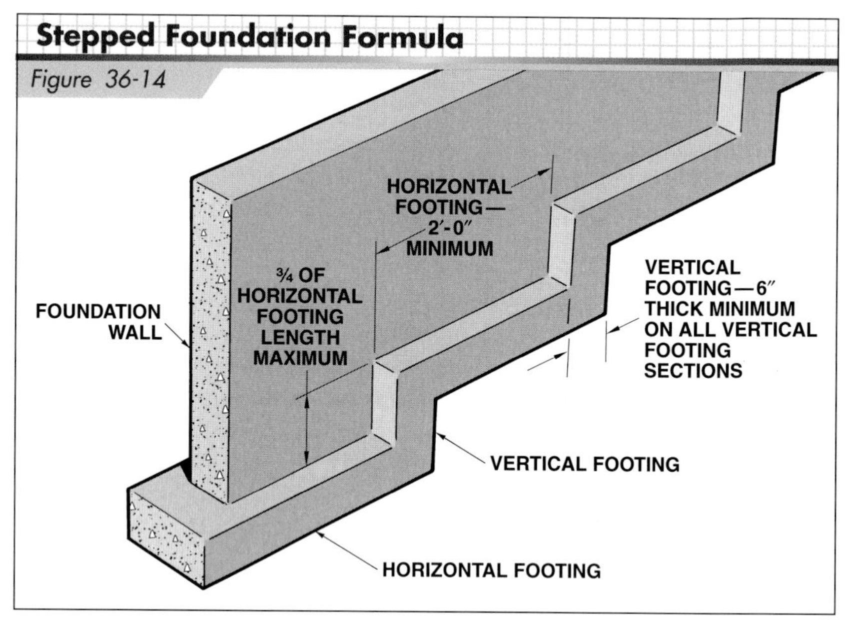

Figure 36-14. *Building codes typically specify a formula for determining dimensions of stepped foundations.*

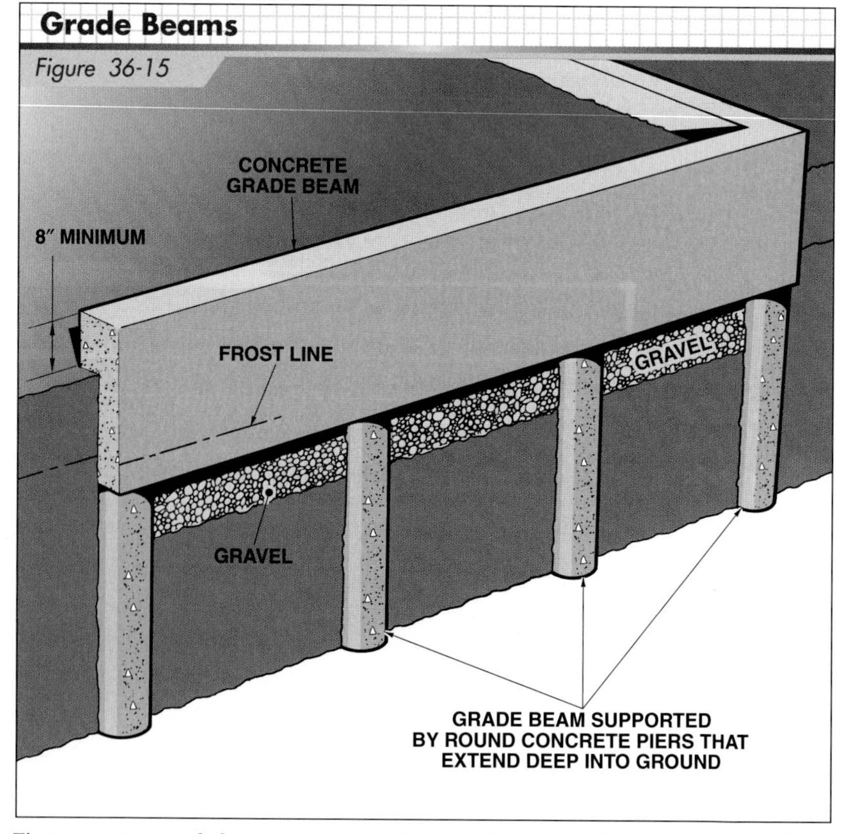

Figure 36-15. *A grade beam is supported by piers that extend deep into the ground.*

Figure 36-16. *Bricklayers use hollow concrete masonry units to construct masonry foundation walls.*

Figure 36-17. *Carpenters plumb and brace window frames in masonry foundation walls.*

CMUs are made with a combination of cement and lightweight aggregate, making them lighter in weight. CMUs are available in a modular shapes and sizes. CMUs used for foundations are commonly 7⅝″ high and 15⅝″ long. However, block size and web thickness vary, depending on the foundation wall requirements. **See Figure 36-18.**

CMUs are joined using mortar along the top, bottom, and sides. A ⅜″ joint is typical for 7⅝″ × 15⅝″ blocks. Steel reinforcement may be required for high foundation walls. Vertical rebar is placed in the cores of the CMUs and the cores are filled with concrete. Horizontal reinforcement is achieved by notching the CMUs and placing rebar horizontally, or by placing metal ties between the courses.

WOOD FOUNDATIONS

In some sections of North America, wood foundations, including *all-weather wood foundations* and *permanent wood foundations,* are considered acceptable alternatives as foundation systems for residential construction. Wood foundations were researched and developed through a joint effort by the National Forest Products Association and the American Wood Preservers Institute, and are approved by the U.S. Department of Housing and Urban Development (HUD) and a number of national and regional building code agencies.

Wood foundations can be used to construct a full-basement or a crawl-space foundation. Load-bearing walls are framed with pressure-treated lumber. The walls are sheathed with pressure-treated APA plywood. **See Figure 36-19.** The lumber moisture content must be no more than 19% for the framing lumber and 18% for the plywood. Nails and other fasteners used must be corrosion-resistant.

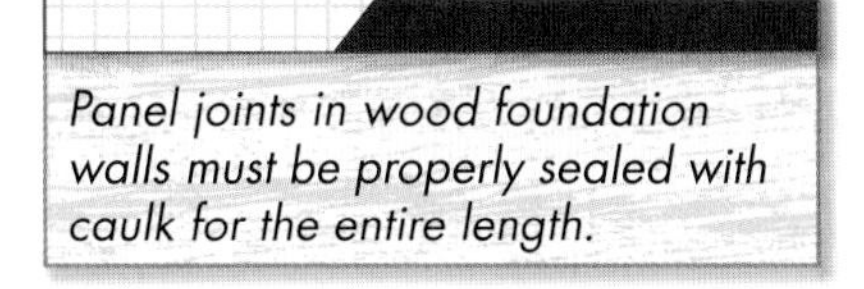

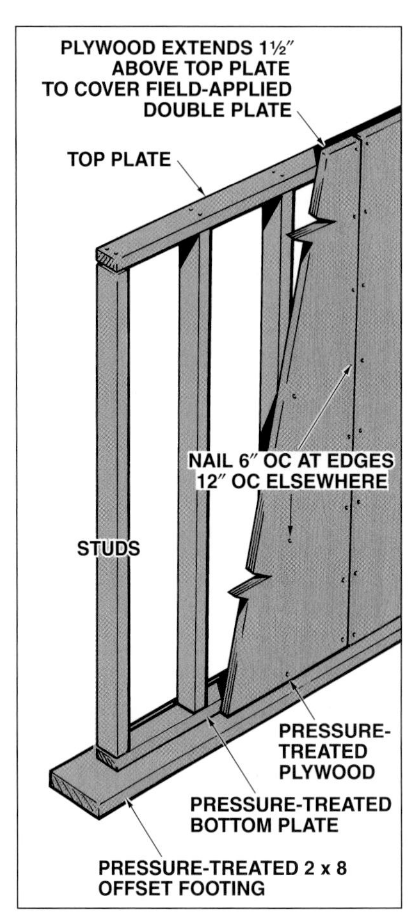

Figure 36-19. *A wood foundation wall is basically a stud wall with a top and bottom plate. Plywood sheathing is nailed to the exterior side of the wall, and the wall rests on 2″ thick wood footing.*

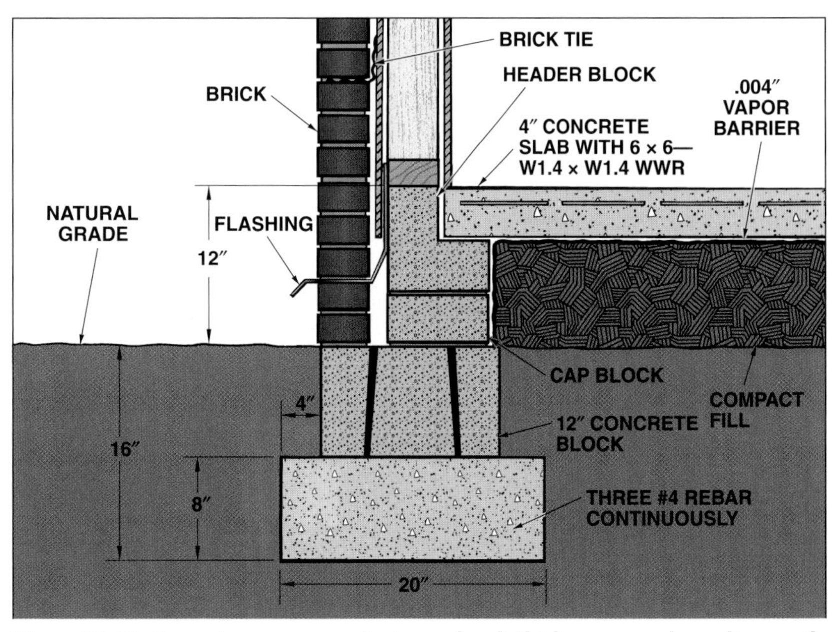

Figure 36-18. *Concrete masonry unit size and web thickness vary depending on the foundation wall requirements.*

Wood foundation walls are designed to withstand backfilling and vertical structural loads. The walls are basically a stud wall with a top and bottom plate. Studs are normally 2 × 6s spaced 16″ OC. Under conditions where greater soil pressures exist, 2 × 6s are spaced 12″ OC. Plywood sheathing is nailed to the exterior side of the stud wall. Wood foundation walls rest on a 2″ thick pressure-treated wood footing or cast-in-place concrete slab.

Preparing Sites for Wood Foundations

Footings for a wood foundation must rest on firm soil and be below the frost line, as required for concrete footings. For a crawl-space wood foundation, gravel is placed where the footings will rest. The thickness of the gravel bed must be at least three-quarters the width of the footing plate. The gravel helps distribute the load from the foundation wall to the underlying soil. **See Figure 36-20.**

Southern Forest Products Association

Figure 36-20. *A section of a wood foundation is placed into position on a wood footing. Gravel has been spread below the footing.*

For a full-basement wood foundation, a layer of gravel is also spread over the area where the concrete slab will be poured after the foundation walls have been completed. **See Figure 36-21.**

Constructing Wood Foundations

Wood foundation walls are built in sections. The sections may be constructed in a shop and delivered to the job site or constructed on the site. The footing plates are placed on the gravel and leveled. The panels are set in the proper positions, aligned with a string to ensure plumb and level, and temporarily braced. Where wall sections are joined, caulk is applied to the butting edges of the plywood. A waterproof 6 mil polyethylene film is placed on the outside of the wall areas that will be below grade.

Moisture control is extremely important for wood foundations. All procedures described in Unit 41 relating to concrete foundations should be referred to for wood foundations. The ground surrounding wood foundations should be sloped to allow surface water to drain away from the foundation. In some situations, a sump may also be required to collect water, which can then flow through a drain and away from the foundation. Backfilling should not occur until concrete for the basement floor has been placed and floor joists have been installed.

APA—The Engineered Wood Association

Figure 36-21. *The sections of a wood foundation wall are braced into proper position. The gravel bed will support a concrete slab and provides proper drainage of water that collects below the slab.*

Unit 36 Review and Resources

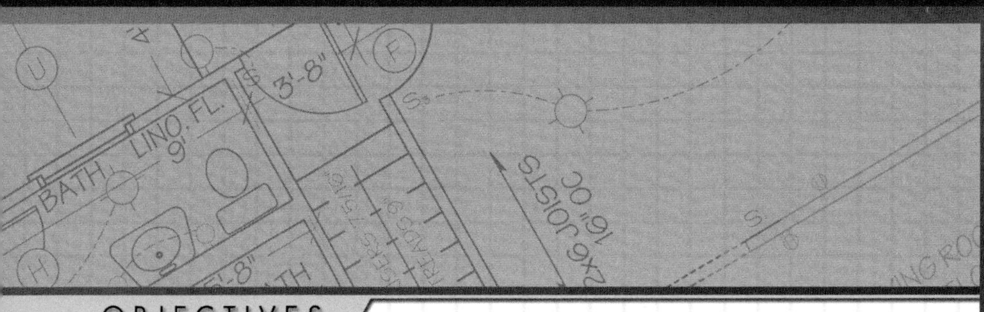

UNIT 37 Concrete

OBJECTIVES

1. Describe the basic composition of concrete.
2. Explain how the proportions of ingredients in concrete mixtures affect concrete strength.
3. Describe how concrete is prepared and delivered to a job site.
4. Describe the proper method for placing concrete.
5. Define curing and its effect on concrete strength.
6. Explain the purpose of reinforcement in concrete.

Concrete is the strongest and most durable material for a foundation. It is placed (poured) in wood forms built to the shape of the foundation walls. **See Figure 37-1.** After the concrete has hardened and set, the forms are stripped to expose the concrete. **See Figure 37-2.** A large amount of carpentry work concerns the construction of concrete forms. Therefore, carpenters should understand the composition and placement of concrete.

CONCRETE COMPOSITION

Ancient Romans were the first to use a form of concrete in 27 B.C. in the construction of buildings, roads, and aqueducts. Many of these structures are still standing. Concrete technology continued to evolve throughout the Middle Ages, when the Spanish introduced a type of concrete consisting of lime, sand, and aggregate (gravel, shells, and stones) mixed with water. Major developments in concrete occurred in the 1880s, when portland cement was developed and refined, and steel reinforcement began to be used. Currently, approximately 15,000,000 tons of portland cement are used annually for residential applications in the United States.

Concrete is often used to describe many building materials used in the construction industry. A common misconception is that concrete is the same as cement. Concrete is a mixture of *portland cement, fine* and *coarse aggregate,* and *water.* When water is added to portland cement and aggregate, a chemical reaction called *hydration* takes place, resulting in the hardening of the concrete.

Figure 37-1. *Foundation forms are constructed by carpenters. The forms are stripped after the concrete has hardened and set.*

Figure 37-2. *After the forms have been removed from the foundation, areaways are installed. Note the sill plate bolts extending from the top of the foundation wall.*

Portland Cement

Portland cement is a mixture of limestone, shells, clay, silica, marble, shale, sand, bauxite, and iron ore that is ground, blended, fused, and crushed to a powder. Portland cement acts as the bonding agent in concrete when mixed with water. Joseph Aspdin, an English builder, developed the modern process of manufacturing portland cement. The material was named portland cement because, after hardening, it resembled the natural limestone on the island of Portland near the coast of England.

Limestone, which is the principal ingredient in cement, is obtained by aboveground or underground mining operations; it may also be dredged from deposits covered by water. After limestone has been extracted from the ground, it is broken down, transported to a cement mill, and pulverized into a hard powder. The powder is mixed with other chemicals such as silica, iron oxide, and alumina. The combined ingredients are heated to 2600°F to 3000°F in a rotary kiln until they form small pellets called *clinkers.* The clinkers are ground to a fine powder. Gypsum, which affects the setting time (hardening) of the cement, is then added.

Aggregate

Aggregate is hard, granular material, such as gravel and sand, which is mixed with cement to provide structure and strength in concrete. Aggregate makes up the largest volume of material in concrete. Fine and coarse aggregate are used in concrete. Fine aggregate is sand; coarse aggregate is gravel or crushed stone ranging in size from 1/4″ to 1 1/2″ in diameter. Aggregate must be clean, hard, strong, durable, and be free from chemicals or coatings that may inhibit the bond between cement and aggregate.

Water

The amount of water in a concrete mixture is a primary factor in determining concrete quality and strength. Too much water dilutes the cement paste, causing the cement to separate from the aggregate and rise to the surface of the mixture. Separation of the ingredients results in a weak concrete. Too little water results in poor mixing action of the cement and aggregate, again producing a weakened concrete. Water used for a concrete mixture should be clean and free of oil, alkali, or acid. Water with a *pH* of 6 to 8 is suitable for a concrete mixture. The pH scale represents the pH level from 0 to 14, based on whether a solution is acidic, alkaline, or neutral. Most drinking water is satisfactory for mixing concrete. However, drinking water with a small amount of sugar or citrates is not recommended.

CONCRETE MIXTURE VARIATIONS

The proportions of cement, aggregate, and water vary according to the type of concrete mixture, or mix, needed for a particular job. **Figure 37-3** gives formulas for several concrete mixtures. The table also shows the maximum diameter of coarse aggregate for each concrete mix. The size of coarse aggregate used in a concrete mixture primarily depends on the rebar spacing and the wall thickness. Thin walls and walls with a large amount of rebar may require a concrete mixture with smaller aggregate.

Although engineers or concrete field specialists usually calculate concrete mixture proportions, print specifications often stipulate the minimum cement content in relation to the lowest *water-cement ratio.* The water-cement ratio largely determines the *compressive strength* of concrete. In addition, the size and amount of coarse and fine aggregate and types of admixtures (added ingredients) are often indicated in print specifications. For smaller concrete structures, such as residential foundations, local building codes may also furnish compressive strength and aggregate size information. Another source of information is local batch plants that produce ready-mixed concrete.

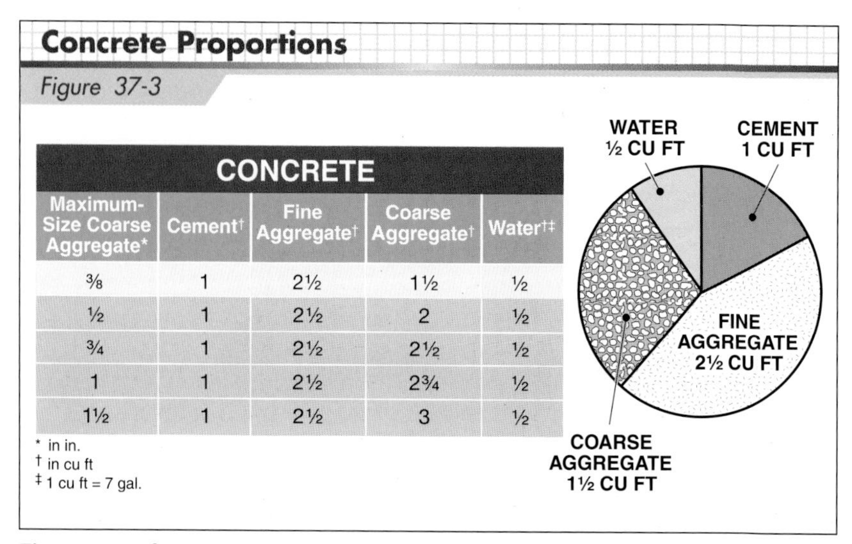

CONCRETE				
Maximum-Size Coarse Aggregate*	Cement†	Fine Aggregate†	Coarse Aggregate†	Water†‡
3/8	1	2 1/2	1 1/2	1/2
1/2	1	2 1/2	2	1/2
3/4	1	2 1/2	2 1/2	1/2
1	1	2 1/2	2 3/4	1/2
1 1/2	1	2 1/2	3	1/2

* in in.
† in cu ft
‡ 1 cu ft = 7 gal.

Figure 37-3. *The proportions of the ingredients determine the compressive strength and other properties of concrete. The pie chart illustrates the cubic foot quantities of the first concrete mixture listed in the table.*

DELIVERING CONCRETE

Small amounts of concrete can be mixed by hand or with small mechanical mixers. For the large amounts required for most construction projects it is standard practice to use *ready-mixed concrete.* Ready-mixed concrete is prepared to specification at a *batch plant* and delivered to a job site by truck.

The estimated amount of concrete required for a job is ordered from a batch plant by the cubic yard. (One cubic yard equals 27 cubic feet.) In Canada, concrete may be ordered by the cubic meter. Automatic controls ensure the proper proportions of cement, aggregate, and water. If a job site is a short distance away, water is added at the plant. If the job site is a long distance away, water is added en route or upon arrival at the site. Under normal conditions, concrete must be discharged from a truck within 1½ hours after water has been added.

Various kinds of trucks are used for delivering concrete, but a *transit-mix truck* is most often used. **See Figure 37-4.** A transit-mix truck is equipped with a large revolving drum operated by an auxiliary engine. The drum rotates to mix the concrete en route to the job site. Drum capacities range from 1 to 12 cu yd. A transit-mix truck also has a water tank so water can be added to the concrete mixture if necessary.

PLACING CONCRETE

Concrete is not a liquid; therefore, technically, it is *placed* rather than poured. However, concrete placement is commonly referred to as the concrete pour.

Concrete is placed at or below ground level by chuting it directly from a transit-mix truck if the truck is able to maneuver close enough to the forms. Another method of placing concrete is using a concrete pumping apparatus. A transit-mix truck deposits the concrete into the pumping apparatus, which in turn pumps the concrete through hoses supported by a boom. **See Figure 37-5.** Concrete can easily be placed in remote locations using a pumping apparatus.

Tierra de Zia Construction

Figure 37-4. *Concrete is usually delivered to a job site in a transit-mix truck. The chute extending from the back of the truck directs the concrete to the proper location.*

Figure 37-5. *A concrete pump is commonly used to place concrete in hard-to-reach locations.*

Concrete should be deposited as close as possible to its final location to avoid *mixture segregation.* Mixture segregation occurs when coarse aggregate is separated from the remaining concrete mixture as the concrete is moved from one location to another. Concrete should not be discharged in one corner or placed in high piles and then moved to another location and leveled. Rather, concrete should be discharged from different positions until an even 12″ to 20″ thick lift (layer) has been placed in the entire form. The procedure is repeated until enough concrete has been deposited to fill the form.

As concrete is placed, it should be consolidated using a tamping rod or mechanical vibrator to remove air pockets and create a close arrangement of solid particles in the concrete. Concrete that is not properly consolidated may have honeycombing or other defects. Honeycombs are open spaces in the concrete caused by trapped air pockets during concrete placement.

A *tamping rod* is a straight steel rod with a circular cross section and rounded ends that is moved up and down by hand as concrete is being placed in a wall form. Mechanical vibrators are used to consolidate concrete in wall forms and slab forms on larger construction projects. Mechanical vibrators generate high-frequency vibrations which are transmitted through the concrete. **See Figure 37-6.** A *jitterbug,* or tamper, is a long-handled tool device with a steel grid base. A jitterbug, commonly used for concrete slabs, forces coarse aggregate below the surface, allowing the cement paste to rise to the surface for finishing.

Figure 37-6. *Mechanical vibrators are used to consolidate concrete after it is placed.*

CURING CONCRETE

Concrete hardens because of a chemical reaction called *hydration* that occurs between the water and cement in a concrete mixture. Hydration begins as soon as water and cement are combined. If water in a mixture evaporates too quickly, the hydration process will end before the concrete attains its full strength. Rapid water loss may also cause the concrete to shrink and crack.

Curing is the process of maintaining concrete moisture long enough to allow proper hydration to occur. Under normal conditions, concrete walls are cured by allowing wall forms to remain in place for a sufficient period of time (usually three days to a week) after the concrete has been placed. During hot and dry weather, wall forms should be dampened by sprinkling them with water. Floor slabs are more difficult to cure than walls. Floor slabs may need to be misted or flooded with water, covered with water-soaked burlap or polyethylene film, or sprayed with chemical sealing compounds. **See Figure 37-7.**

Portland Cement Association

Figure 37-7. *A concrete slab-at-grade floor is covered with water-soaked burlap to keep concrete moist during curing.*

The first three days after concrete is placed are the most critical to concrete quality since concrete is most vulnerable to damage during this time. Concrete reaches about 70% of its strength after being in place for 7 days. At 14 days, concrete achieves approximately 85% of its strength. Under normal conditions, maximum strength is reached at 28 days. Special considerations such as unusual weather conditions, size of the structural members, and mixture proportions may make it necessary to allow a longer curing period.

REINFORCED CONCRETE

Two different kinds of pressure are exerted on a foundation wall—vertical and lateral. **See Figure 37-8.** Vertical pressure is exerted on a foundation by live and dead loads, including the weight of the structure, furniture, and appliances. Lateral pressure is exerted on a foundation by the soil. A wall made of concrete has a great deal of *compressive strength,* which is its ability to hold up under vertical pressure. However, concrete has far less resistance to lateral forces, which push against the wall sides.

Reinforced concrete is used to help resist lateral forces. Reinforced concrete is a concrete mixture in which tensile members, such as rebar, welded wire reinforcement, or plastic or steel fibers, are placed in the concrete to improve resistance to lateral pressure.

Rebar

Rebar is a deformed steel bar with ridges on the surface to allow the bars to interlock with concrete. Rebar is positioned inside the form walls before concrete is placed, and may be positioned vertically, horizontally, or a combination of both positions. Rebar is commonly tied to metal form ties or other rebar using lightweight wire. **See Figure 37-9.** Solid fiberglass reinforcing bar may be used instead of steel rebar. Fiberglass reinforcing bar has the same tensile strength as steel rebar, but will not rust.

Rebar is identified by numbers from #3 through #18. The bar diameter is determined by multiplying the number designation by ⅛″. For example, a #4 bar is 4/8″, or ½″ (4 × ⅛″ = 4/8″). Rebar size, placement, and spacing are shown on the foundation plan.

Masonry walls may be reinforced with rebar. Vertical rebar is placed in the cores of concrete masonry units (CMUs) and the cores are filled with concrete. In seismic risk zones, vertical rebar extends from the concrete footing and is tied to horizontal rebar. The CMUs' cores are filled with concrete as the wall is being constructed. Horizontal reinforcement is achieved by notching the CMUs and placing rebar horizontally, or by placing metal ties between the courses.

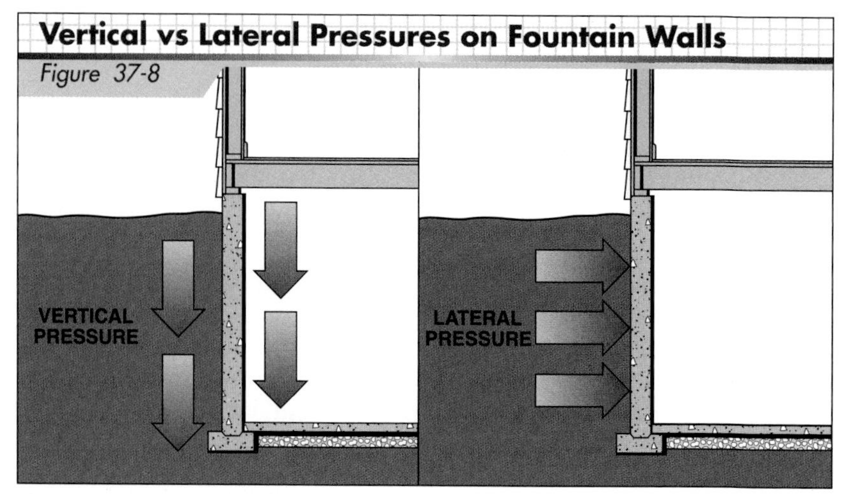

Figure 37-8. *Concrete has greater resistance to vertical pressure than lateral pressure.*

Figure 37-9. *Rebar is placed in foundation wall forms to provide resistance against lateral pressure.*

Welded Wire Reinforcement

Welded wire reinforcement (WWR), or welded wire fabric, is heavy-gauge wire joined in a grid and commonly used to reinforce concrete slab-at-grade floors, sidewalks, and driveways. WWR helps prevent cracks in the concrete from occurring later due to settlement. WWR is identified by the wire spacing and cross-sectional area of the wires. **See Figure 37-10.** For example, 6 × 6—W1.4 × W1.4 refers to WWR with 6″ wire spacing that is constructed with wire having a cross-sectional area of .014 sq in.

WWR is available in sheets and rolls. The WWR is laid in position before the concrete is placed. **See Figure 37-11.**

Figure 37-11. *Welded wire reinforcement is positioned prior to concrete being placed.*

Fiber-Reinforced Concrete

Fiber-reinforced concrete (FRC) is concrete reinforced with plastic or steel fibers, which minimize shrinkage cracking and increase tensile strength. **See Figure 37-12.** Fibers can be used as the sole reinforcement or they can be used in combination with rebar and WWR.

Plastic and steel fibers are available in various lengths and diameters, ranging in length from ¼″ to 2½″. Short fibers are commonly used for thin-wall applications. Longer fibers are used for thick slab-at-grade floors. Orientation of the fibers in the mixture is generally random.

Figure 37-12. *Plastic fibers may be added to a concrete mixture to provide greater tensile strength.*

COMMON STOCK SIZES OF WELDED WIRE REINFORCEMENT

New Designation (W-Number)	Old Designation (Wire Gauge)	Diameter*		Steel Area†		Weight‡
				Longitudinal	Transverse	
6 × 6 – W1.4 × W1.4	6 × 6 – 10 × 10	.134	⅛	.028	.028	21
6 × 6 – W2.0 × W2.0	6 × 6 – 8 × 8	.160	5⁄32	.040	.040	29
6 × 6 – W2.9 × W2.9	6 × 6 – 6 × 6	.192	3⁄16	.058	.058	42
6 × 6 – W4.0 × W4.0	6 × 6 – 4 × 4	.226	¼	.080	.080	58
4 × 4 – W1.4 × W1.4	4 × 4 – 10 × 10	.134	⅛	.042	.042	31
4 × 4 – W2.0 × W2.0	4 × 4 – 8 × 8	.160	5⁄32	.060	.060	43
4 × 4 – W2.9 × W2.9	4 × 4 – 6 × 6	.192	3⁄16	.087	.087	62
4 × 4 – W4.0 × W4.0	4 × 4 – 4 × 4	.226	¼	.120	.120	85

* in In.
† in sq in./ft
‡ in lb per 100 sq ft

Figure 37-10. *Welded wire reinforcement is available in various wire diameters and spacings.*

Unit 37 Review and Resources

UNIT 38

Forming Methods and Materials

OBJECTIVES

1. Identify the principles of form construction.
2. List materials that can be used for wall sheathing.
3. Describe the purpose of framing and bracing materials.
4. List common types of ties.
5. Describe how job-built forms are constructed.
6. Describe how panel forms are constructed.
7. Explain how door and window openings are formed in concrete walls.

Various methods are used for foundation form construction. All methods require sheathing, studs and/or walers, bracing, and a means of tying the form walls together. **See Figure 38-1.**

Since forms are temporary structures, the forms should be constructed for easy dismantling. Duplex nails are used wherever practical since they can be quickly removed. Sheathing is fastened to stakes or studs with just enough nails to hold it in place. An adequate number of braces and ties should be used to keep the walls aligned and in place.

Stud and Double Waler Foundation Wall Form

Figure 38-1

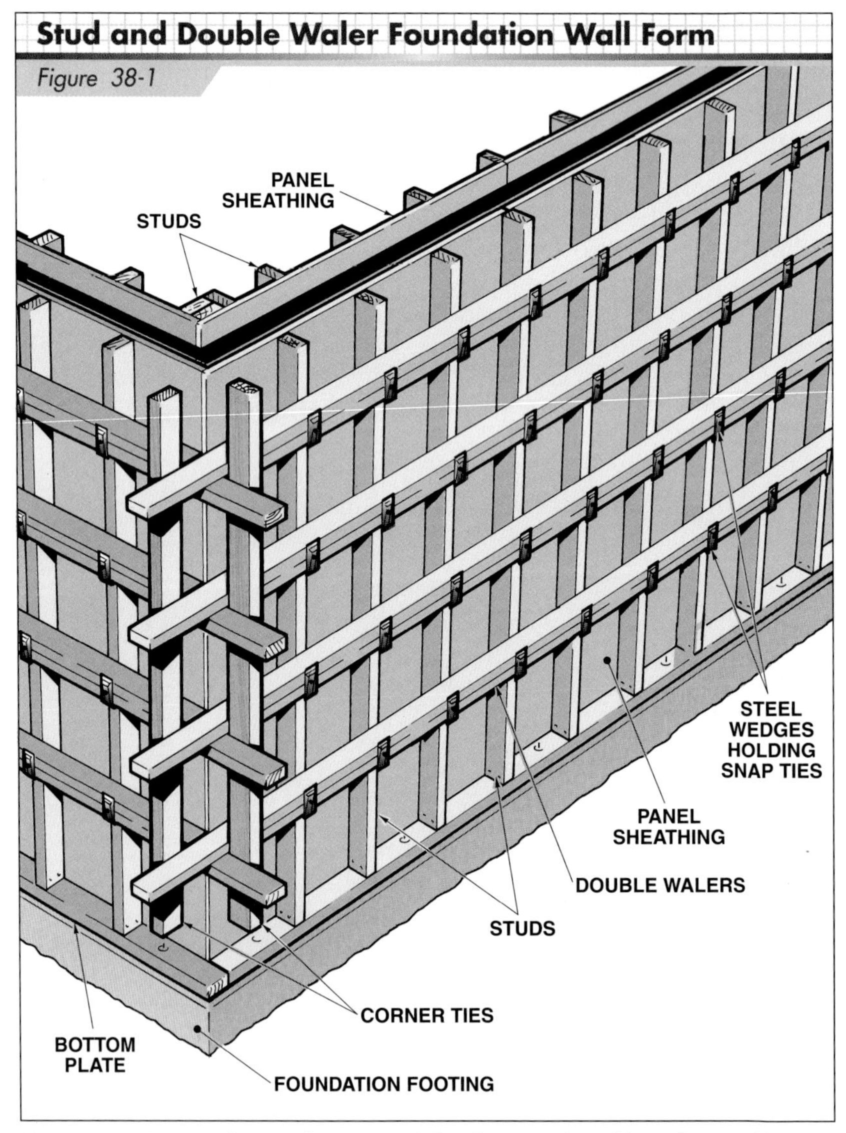

Figure 38-1. *Concrete formwork consists of sheathing, studs and/or walers, and bracing. Note the use of corner ties to lace the walers together.*

SHEATHING

Various panel products may be used for form wall sheathing, including high density overlay (HDO) plywood, Plyform®, and fiber reinforced plastic (FRP) plywood. Plyform is manufactured specifically for concrete form construction. Plyform panels are available in 4′ × 8′ sheets and ⅝″, ¾″, 1⅛″, and 1¼″ thicknesses. HDO plywood, Plyform, and FRP plywood can be reused many times.

Spread footings for inverted T-shaped foundations are usually formed with 2″ thick planks. **See Figure 38-2.** Planks are also used to form foundation walls in some forming systems.

Figure 38-2. *Spread footings are commonly formed with 2″ thick planks.*

FRAMING AND BRACING MATERIALS

A wall form is subjected to great pressure when concrete is placed. The pressure increases as the wall height increases. A fast concrete placement rate also places a greater strain on the forms.

Walers, or wales, reinforce and stiffen foundation wall forms. Walers run horizontally and are toenailed to *studs,* which run vertically. For some applications, walers are fastened directly to the sheathing. The distance between walers depends on the thickness and height of the wall.

When ¾″ panels are used to form low walls, studs or stakes are usually spaced 2′ apart. Higher walls may require a spacing of 16″ or 12″.

Proper *bracing* is required to hold wall forms in position while concrete is being placed. One end of a brace is fastened to studs or walers and the other end is usually nailed to a stake driven into the ground.

Lumber used for walers, studs, and bracing is usually cut from 2 × 4s. Structural light framing lumber or another good grade of softwood lumber should be used. Metal stakes may be used instead of wooden stakes to hold the forms in place. **See Figure 38-3.**

The required plywood form thickness and size and spacing of framing depend on the maximum load.

An area must be properly excavated before foundation forms are constructed.

Figure 38-3. *Reusable metal stakes may be used to hold planks in position. Duplex nails are driven through the holes provided in the stakes and into the planks.*

TIES

Form walls must be tied together so they will not shift during concrete placement. Small form walls may be tied together by braces and wood cleats. **See Figure 38-4.** Larger walls require metal ties to hold the form walls together and maintain the proper spacing between the walls.

Figure 38-4. *Braces and wood cleats are sufficient to tie low foundation form walls together.*

Figure 38-5 shows plank form walls tied together with steel wedge form ties. **Figure 38-6** shows a system of snap ties and steel wedges used with single and double walers. *Spreader cones* maintain the correct spacing between the form walls. Buttons at the ends of the snap ties hold the steel wedges that are driven behind the walers. *Breakbacks* are grooves behind the spreader cones. After forms are stripped from foundation walls, the portion of the snap tie that protrudes from the wall is snapped off at the breakbacks.

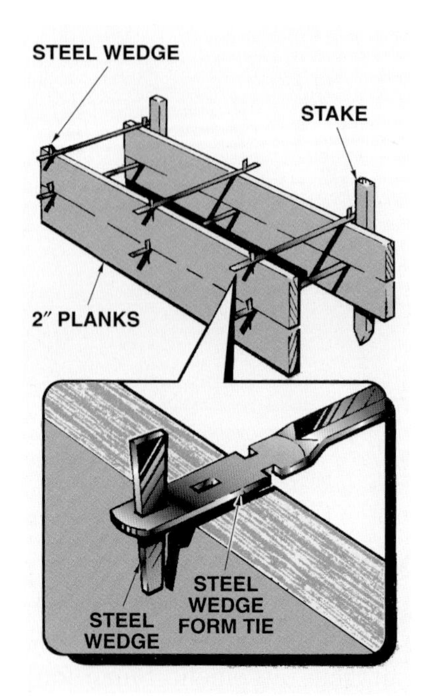

Figure 38-5. *Wedge ties are used to tie low form walls together.*

JOB-BUILT FORMS

The oldest method of form construction is the job-built method. **See Figure 38-7.** Form walls may be sheathed with panels or planks. When 2″ thick planks are used as sheathing, walers are not required and studs and stakes may be placed farther apart.

PANEL FORMS

Many carpenters and contractors consider panel forms a more efficient method of form construction than built-in-place forms. Panel forms consist of studs and top and bottom plates nailed to a 4′ × 8′ panel. **See Figure 38-8.** When the sections are set in place, the end studs are fastened to each other with duplex nails.

Snap ties are laid out horizontally at 2′-0″ OC. Vertical tie spacing depends on wall height and concrete placement rate. The first tie of the horizontal layout must clear the adjoining wall when doubled and then continue 2′-0″ OC.

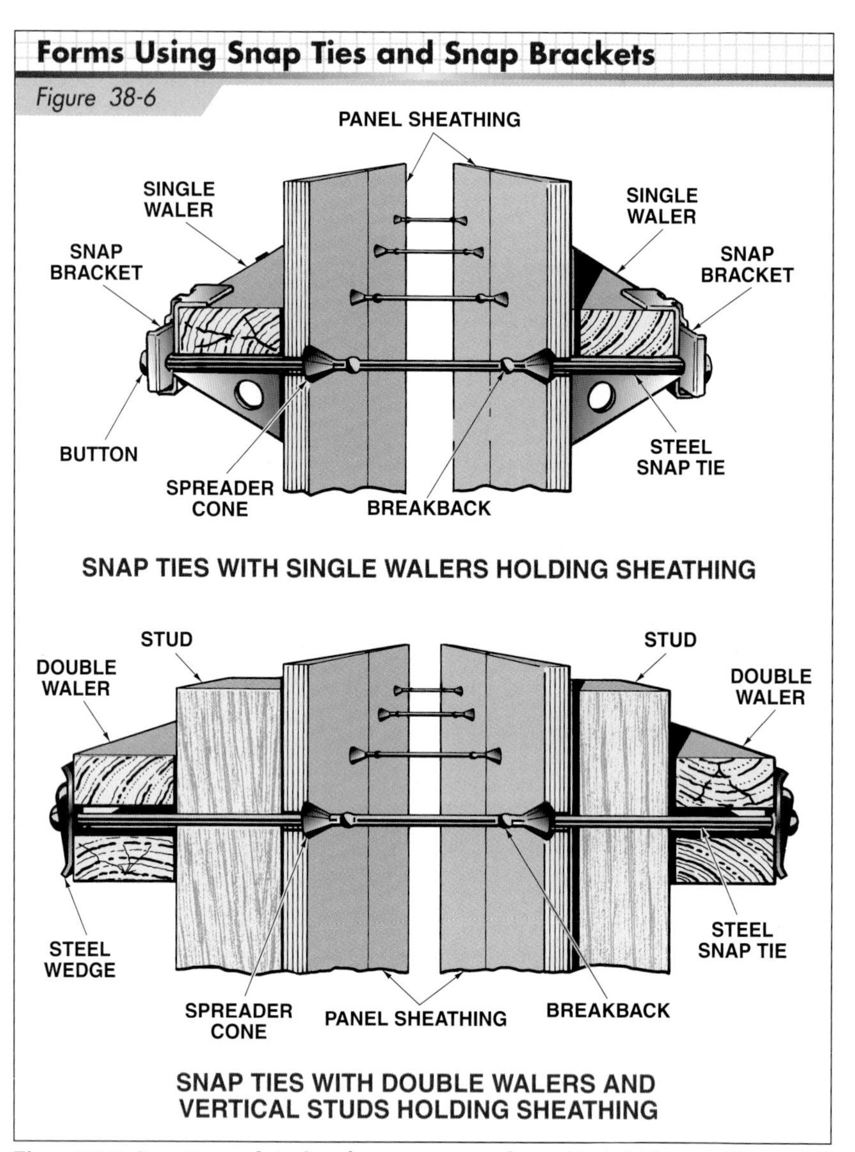

Figure 38-6. *Snap ties and steel wedges are commonly used to hold form walls together. Spreader cones set the walls to the correct width. The buttons at the ends of the ties hold steel wedges that are driven behind the walers. Snap ties are designed for both single-waler and double-waler systems. Breakbacks are grooves behind the spreader cones. After the forms are stripped from the foundation wall, sections of snap ties protruding from the wall are snapped off at the breakback points when bent back and forth.*

ECO-Block, LLC

Insulating concrete forms (ICFs) may be used to form residential and commercial foundations.

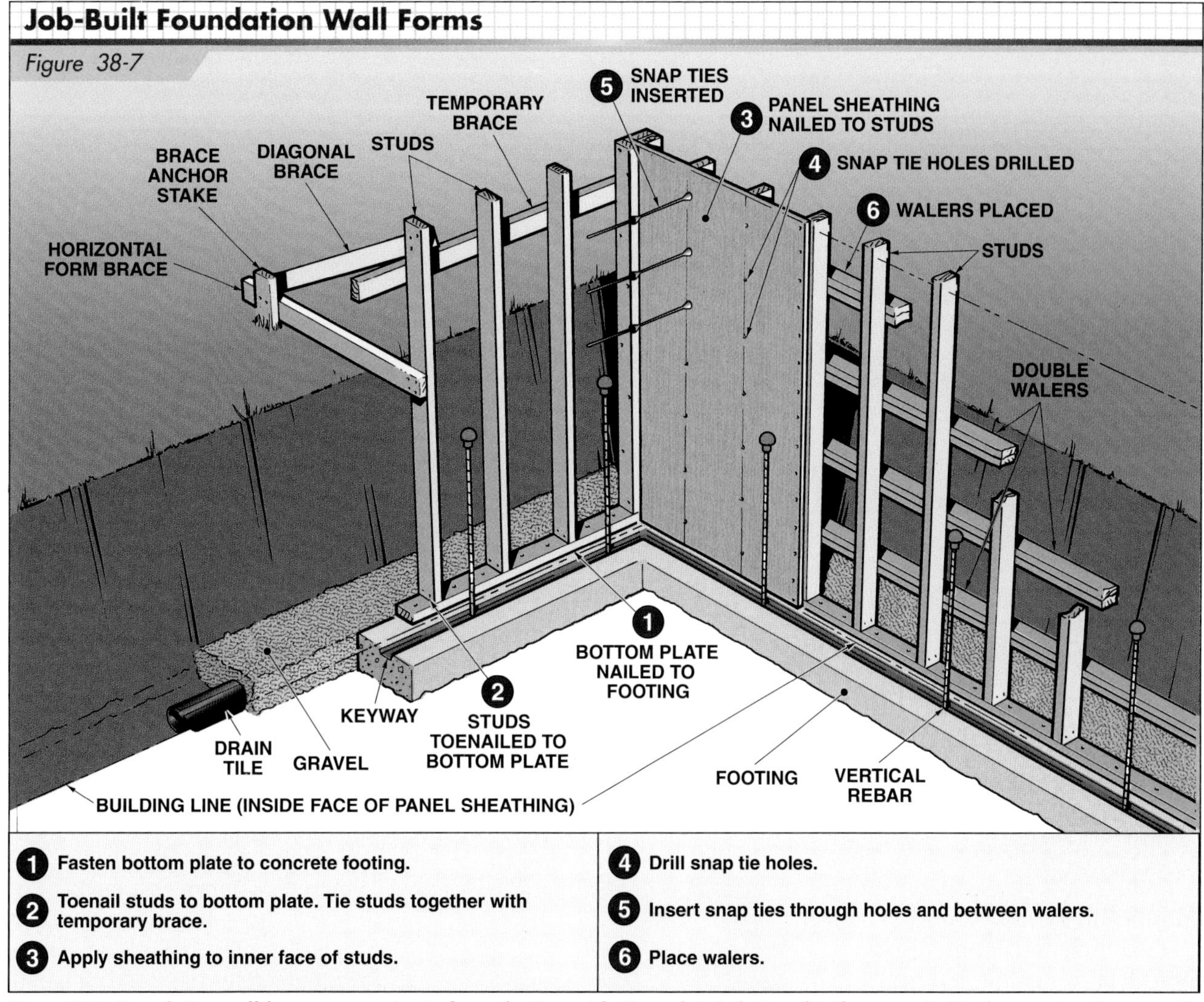

Figure 38-7. *Foundation wall forms are constructed over footings. A bottom plate is fastened to the concrete. Studs are set up, temporarily tied together, and braced. Sheathing is then applied to the inner face of the studs. The walers are then placed.*

Footings for slab-at-grade foundations may be formed by excavating a trench. Rigid insulation is placed along the outer edge of the footing and beneath the outer edge of the slab. Plastic tubing is installed in this slab to provide radiant heat for the structure.

Panel forms can be constructed in the shop or on the job. Panel forms are convenient for use in housing developments where one foundation design is repeated. The panel form sections can be reused after being stripped from the foundation walls. Tie holes are patched with small pieces of sheet metal. Patented wood and metal panel forms can be rented or purchased.

Concrete pressure on the forms is affected by several factors, including concrete temperature, placement rate, concrete slump (consistency), type of cement, concrete density, method of vibration, and height of the forms.

Foundation Wall Form Using Panel Form Systems

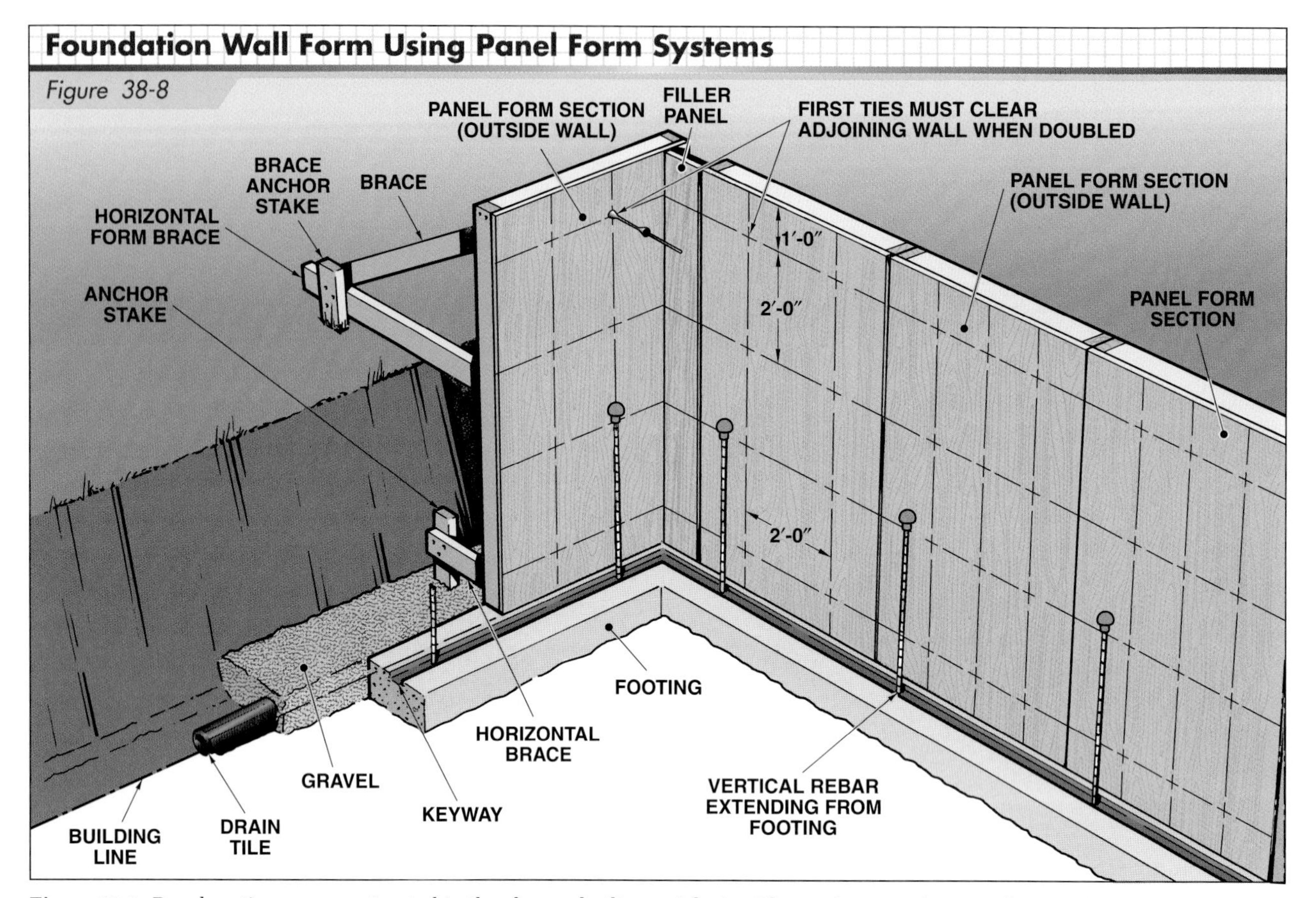

Figure 38-8. *Panel sections are constructed in the shop or built on a job site. The sections can be reused many times.*

Figure 38-9 shows lightweight aluminum forms. Round stakes driven through holes in the aluminum sections hold them in place. Wedge locks at the stakes allow each section to be positioned at the proper height. **Figure 38-10** shows a lightweight aluminum panel system for foundation walls.

Western Forms, Inc.

Figure 38-9. *Reusable aluminum forms are secured in position using round stakes driven through holes along the edges. Wedge locks at the stakes allow the height of each form section to be adjusted.*

Figure 38-10. *Lightweight aluminum form panels were used to form this basement foundation wall. Some of the panels are still in place on the wall at right.*

DOOR AND WINDOW OPENINGS

Full-basement concrete foundations often have windows and may also have a door to provide access from the outside of the building. Preparations must be made for door and window openings when the forms are built. In most cases, finish metal window frames and door jambs are installed in the form walls and secured in place with temporary fasteners. Brackets attached to the outside of the frames extend into the concrete when it is placed. The brackets hold the frame in a permanent position in the wall when the concrete has set and hardened.

Per the International Residential Code, habitable spaces and bedrooms in a basement must have at least one openable emergency escape and rescue opening.

Another method to allow for door or window openings is to construct a well-braced frame called a *buck,* which is set in place and fastened to the outside form wall with duplex nails. **See Figure 38-11.** After the inside form walls are placed, duplex nails are driven through the inside form walls into the buck frame. Bucks may be removed when the concrete has set and the form walls have been stripped from the wall.

CertainTeed Corporation

Figure 38-11. *Bucks are installed between wall forms to create door and window openings in the concrete.*

Two types of bucks may be used to frame door or window openings. One type, used for stucco finish, is recessed between the forms flush with the edges of the rough opening. Another type is a flanged buck that allows window or door trim to be attached directly to the outside buck flange.

There are several ways to fasten finish frames into a door or window opening. Finish frames may be attached to the openings with powder-actuated fasteners. Another method is to drive bolts or screws into expansion anchors installed in the concrete.

Unit 38 Review and Resources

UNIT 39

Foundation Designs—Form Construction

OBJECTIVES

1. Describe the use of inverted T-shaped foundation forms.
2. Review the procedure for constructing footing forms.
3. Describe the construction of wall forms.
4. Describe the construction of monolithic inverted T-shaped foundation forms.
5. Explain how the height of pour is established.
6. Describe the use of rectangular and battered forms.
7. Explain how pier forms are constructed.
8. Describe the construction of grade beam forms.
9. Review the construction of slab-at-grade forms.
10. Define insulating concrete forms and describe their use in residential and light commercial construction.
11. List and describe common tools used when working with insulated concrete forms.
12. List the devices and methods used to attach finish materials to insulating concrete forms.

Foundation forms and forms for slab-at-grade foundations are constructed by carpenters. The foundation location is transferred from the building lines, which were established when the building site was laid out. Forms must be properly constructed and braced to ensure the forms do not move when concrete is placed in them.

INVERTED T-SHAPED FOUNDATION FORMS

Inverted T-shaped foundations are used for full basements or crawl spaces. Spread footings of inverted T-shaped foundations provide good load-bearing on all types of soil. The forming procedure used for inverted T-shaped foundations depends on the climate, soil condition, and height of the foundation walls. For low inverted T-shaped foundations, concrete for footings and walls is placed at the same time, or *monolithically.* For high inverted T-shaped foundations, concrete for footings and walls is placed separately.

Footings

Where the soil condition is firm and stable, footing forms may not be needed. Instead, a trench is dug to the width and depth of the footing and concrete placed directly in the trench. **See Figure 39-1.** Where the soil condition is unstable, a footing form is built using 1″ or 2″ lumber. **See Figure 39-2.** With 2″ lumber, stakes are placed farther apart and less bracing is required. A procedure for constructing footing forms is shown in **Figure 39-3.**

Figure 39-1. *Concrete for footings may be placed directly in the trench when firm and stable soil conditions exist.*

Figure 39-2. *Footing forms may be required on one or both sides to contain concrete when the soil condition is unstable.*

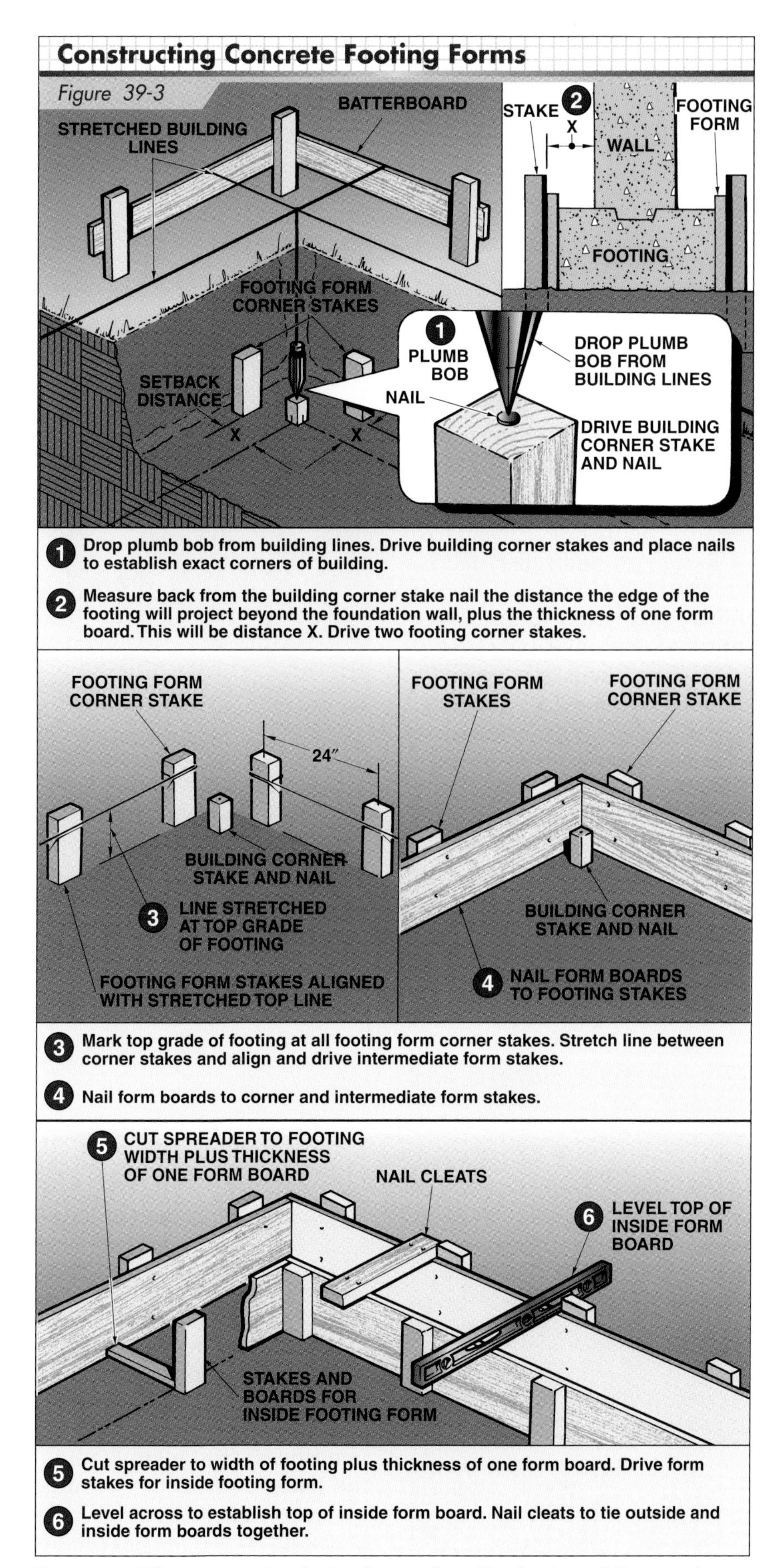

Figure 39-3. *When constructing footing forms, 1″ or 2″ lumber is used. Fewer stakes are required when 2″ lumber is used.*

Keyways. Immediately after the concrete has been placed, chamfered pieces of 2 × 4s called *key strips* may be pressed into the concrete toward the center of the footing. When the pieces are removed after the concrete has set, a groove called a *keyway* has been formed in the concrete. See **Figure 39-4.** A keyway helps to secure the bottom of a foundation wall to the footing.

Reinforcing Steel. In seismic risk areas, rebar (reinforcing steel) are positioned in footing forms before concrete is placed. Vertical rebar project out of the footing forms and are later tied to rebar placed in the foundation wall. **See Figure 39-5.** Concrete is placed after footing form construction is completed, rebar positioned, and provisions made for keyways.

Rebar consists of steel rods containing lugs (protrusions) that allow the rebar to interlock with the concrete. Refer to floor plans and elevations when placing vertical rebar in footings so rebar is eliminated at door openings and are appropriate height at window openings.

Walls

Form walls are erected after concrete for the footings has hardened and set. Bottom plates are fastened to the fresh (green) concrete as a base for the outside form walls of either job-built or panel forming systems.

Except in the case of low form walls, rebar is usually installed after the outside form walls are set in place. **See Figure 39-6.** When the rebar has been installed, the inside form walls are constructed. When a large amount of reinforcing steel is required, the steel is typically placed by reinforcing-metal workers.

Figure 39-4. *A keyway helps to secure the bottom of a foundation wall to the footing. Note the drain tile placed around the outside of the footing.*

Figure 39-5. *Rebar projecting from the footing will be tied to reinforcing steel in the foundation wall. The rebar are temporarily capped to ensure worker safety.*

Rebar should be clean and free of loose rust and other debris, form oil, and form-release oil when installed in forms. Rebar must be positioned and secured in place so it will be covered by an adequate layer of concrete. Rebar is typically fastened to adjacent rebar using wire ties.

Preparations must be made for door and window openings when the form walls are built. Finish window frames and door jambs may be attached to the form walls. Door or window bucks are attached to the form walls if finish frames and jambs are not installed.

Figure 39-6. *Outside form panel sections are fastened to the top of the footing. Note the snap ties projecting from the panels and the rebar at the back wall.*

A traditional double-waler outside wall is shown in **Figure 39-7.** A procedure for building a single-waler panel wall form for an inverted T-shaped foundation is shown in **Figure 39-8.**

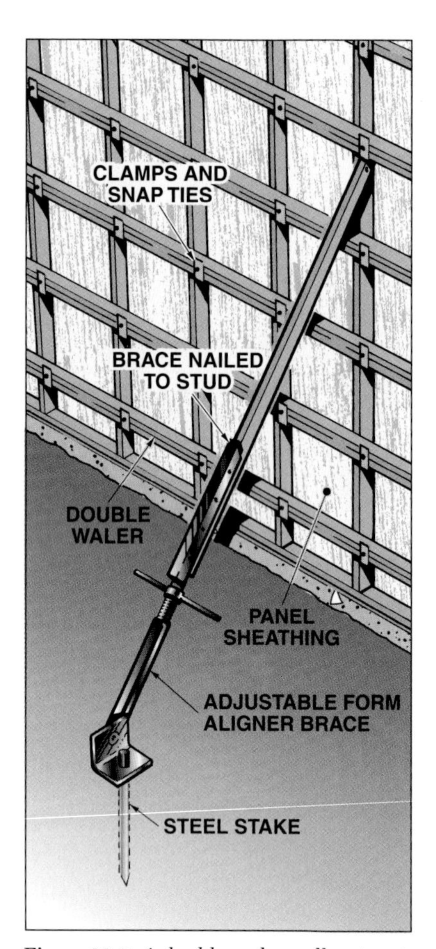

Figure 39-7. *A double-waler wall system is commonly used to form high foundation walls.*

Patented panel systems are widely used in the construction of foundation walls. Patented panel systems generally consist of panel sheathing set in metal frames. **See Figure 39-9.** Depending on the manufacturer, the panel sections are secured to each other with wedge bolts or clamps. When wedge bolts are used to secure the panel sections together, one wedge bolt is inserted in a slot provided in the side rails and the other wedge bolt is inserted in a slot in the first wedge bolt. **See Figure 39-10.**

Many patented panel systems are aligned and secured with wood braces, which are equipped with metal turnbuckles on their upper ends. The turnbuckle is secured to the panel with wedge bolts or duplex nails. **See Figure 39-11.** The lower end of each brace is nailed to a wood or metal stake driven into the ground. Panels are then adjusted to the proper position by turning the turnbuckle.

High density overlay (HDO) plywood panels have resin-treated fiber overlays bonded to plywood with waterproof glue under heat and pressure. HDO plywood concrete forms can be reused 20 to 50 times if properly maintained.

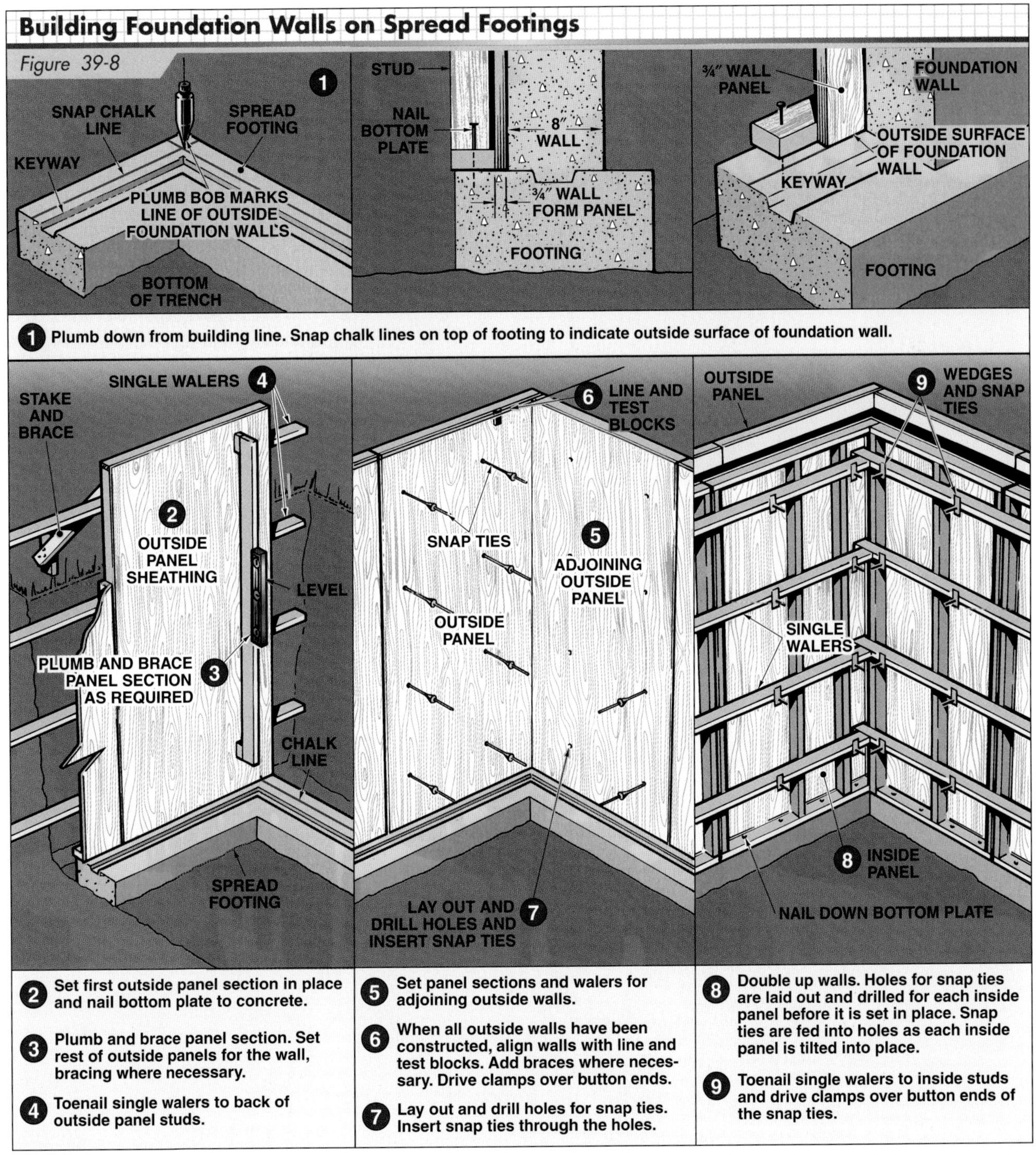

Figure 39-8. *Foundation wall forms are fastened to a spread footing using concrete nails. The outside form wall is constructed first and properly braced. Snap ties are then inserted through the panels, and the inside form wall is constructed and braced. When laying out snap tie holes (step 7), ensure the snap ties clear the adjoining doubled wall.*

Figure 39-9. *Patented panel systems consist of panel sheathing set in metal frames.*

Figure 39-12. *Wire ties space and hold together opposite wall forms.*

Figure 39-10. *Wedge bolts secure panel sections together. One wedge bolt is inserted in a slot in the side rails and another wedge bolt is inserted in a slot in the first wedge bolt.*

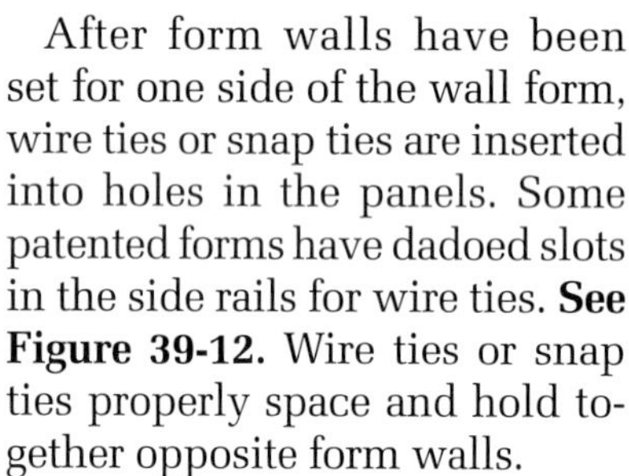

After form walls have been set for one side of the wall form, wire ties or snap ties are inserted into holes in the panels. Some patented forms have dadoed slots in the side rails for wire ties. **See Figure 39-12.** Wire ties or snap ties properly space and hold together opposite form walls.

For many patented panel systems, walers are required only at the upper section of the form walls. The lower ends of the form walls are secured to the footing. **Figure 39-13** shows a double-waler snap tie-and-clamp system.

Monolithic Inverted T-Shaped Foundations

When low walls are required for a crawl-space foundation, the wall and footing forms may be built as one unit. Since the concrete for the walls and footings is placed at the same time, the foundation is called a *monolithic* foundation. An advantage of monolithic foundations is that there is no possibility of moisture seeping through a *cold joint* afterwards. A cold joint occurs where the concrete for a foundation wall is placed over the hardened concrete of a footing.

Figure 39-11. *This patented panel system uses wood braces with metal turnbuckles that are secured to walers with duplex nails.*

Figure 39-13. *Walers are placed toward the top of a patented panel wall form with a snap tie-and-clamp system.*

Figure 39-14 shows a monolithic inverted T-shaped foundation form built over a trench. Forms are not required for the footings since the trench sides form the footing. **Figure 39-15** shows the forming procedure used when soil conditions require footings to be formed along with the wall.

Establishing Height of Pour

Concrete is placed in a form until it reaches the level required for the top of the foundation wall. When forms are constructed to the actual height of the finished foundation wall, the concrete is struck off when it reaches the top of the forms. However, form walls are often higher than the level required for the foundation. When this occurs, a builder's level is used to establish elevation points at intervals and at all corners of the form. Lines are snapped and a narrow strip of wood called a *pour strip* is tacked above the line. **See Figure 39-16.** When the concrete is placed to the proper level, the pour strip is removed.

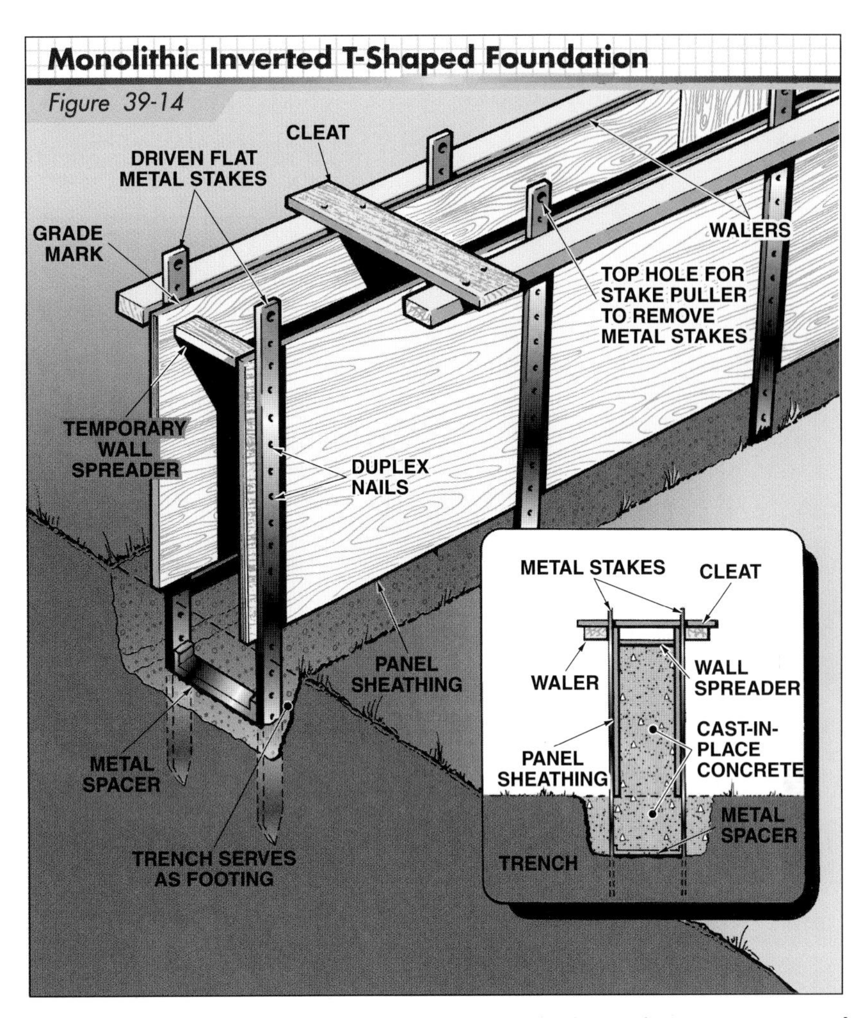

Figure 39-14. *When firm and stable soil conditions exist, footing forms are not required for monolithic inverted T-shaped foundations. The flat metal stakes are pulled out when the concrete begins to set.*

Double walers are secured in place with waler brackets, which are attached to siderail holes.

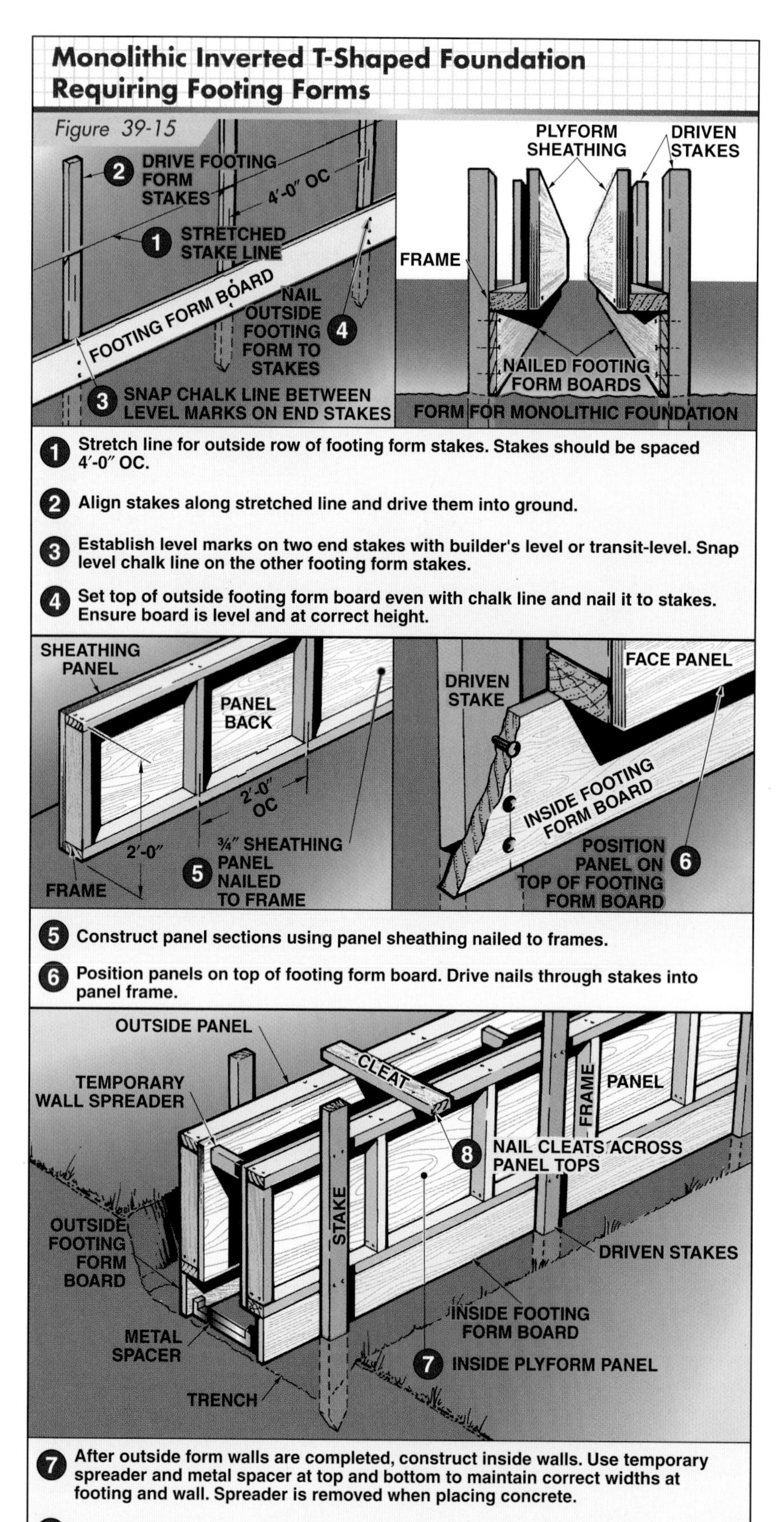

Figure 39-15. *When unstable soil conditions exist, footing forms are required for monolithic inverted T-shaped foundations. Different wall thicknesses require other sizes of framing materials.*

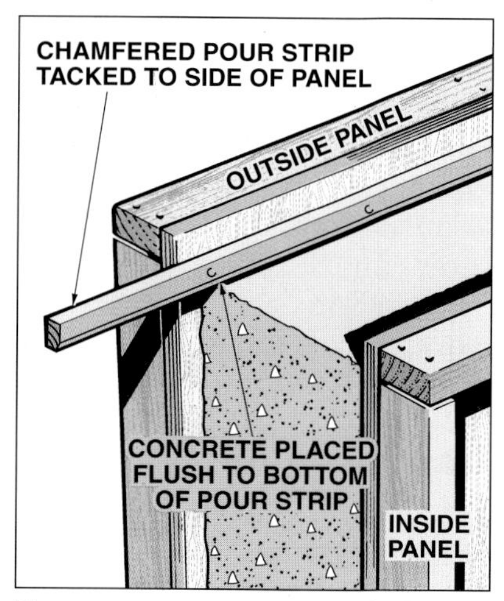

Figure 39-16. *A pour strip tacked to one side of a form indicates the height to which the concrete will be placed. The chamfer prevents concrete from breaking along the edge. Grade nails may be installed instead of a pour strip.*

RECTANGULAR AND BATTERED FORMS

Rectangular or battered foundation walls may also be used to provide support for a building. Rectangular foundations are often used as grade beams. Battered foundations are sometimes used with crawl-space and slab-at-grade foundations.

Rectangular and battered forms can be constructed using job-built forms or patented panel systems. Components used to construct rectangular and battered forms, such as stakes, sheathing, walers, and bracing, are the same as the components used for inverted T-shaped foundations. The wood or metal stakes holding the sheathing are driven directly into the ground. Where hard soil makes stake driving difficult, bottom plates are set on the ground and held in position with steel dowels driven through holes bored in the plates. **See Figure 39-17.** If a footing is not required, a concrete skimcoat, or *mudslab*, may be placed to maintain exposed excavation surface conditions and to provide a clean surface for formwork layout. A forming method for battered foundations is shown in **Figure 39-18.**

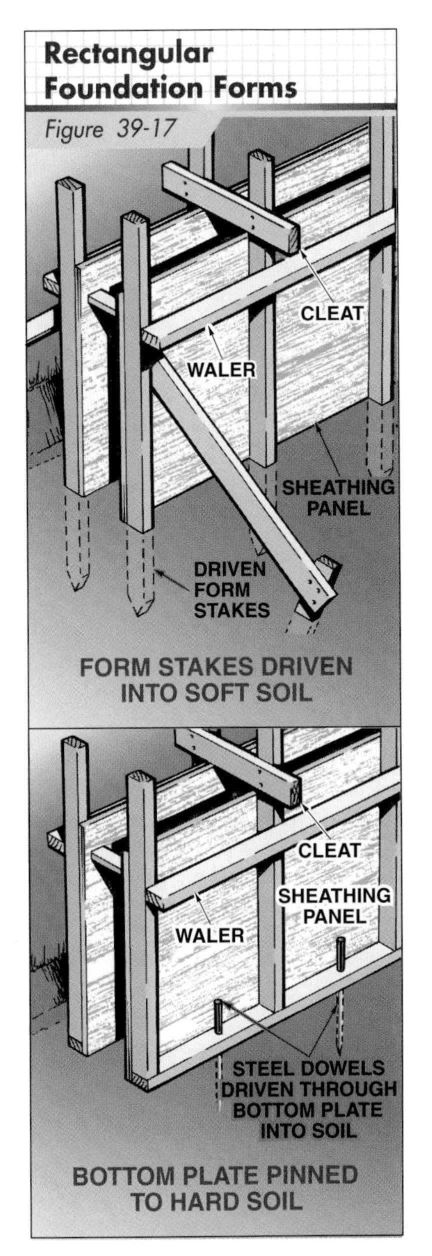

Figure 39-17. *Where suitable soil conditions exist, rectangular forms can be constructed with stakes driven directly into the ground. In hard soil, bottom plates are pinned to the ground with steel dowels.*

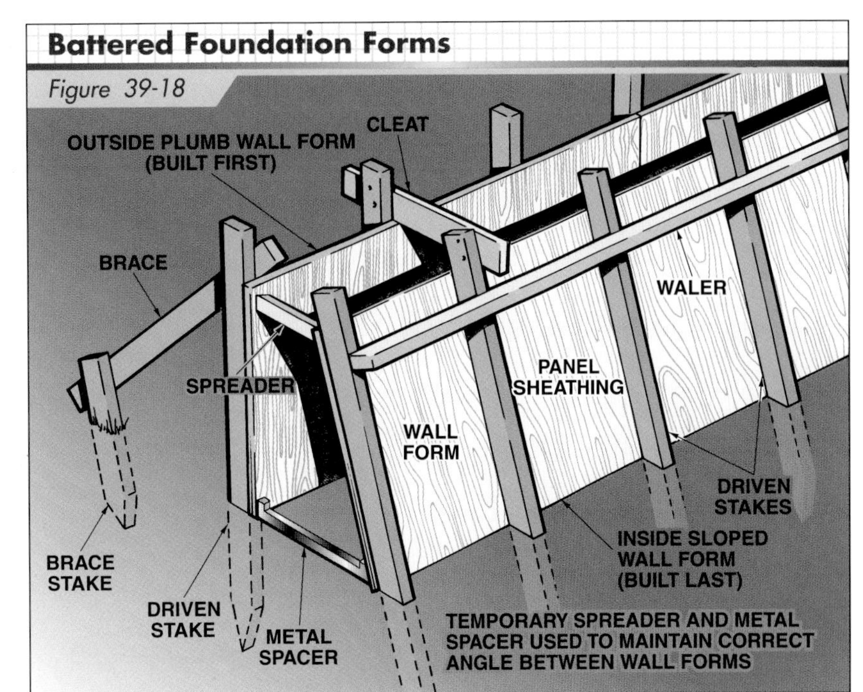

Figure 39-18. *This battered foundation has a plumb outside wall and the inside wall slopes to provide a wider base against the ground. The plumb wall is built first. Upper and lower spreaders position the sloped wall so stakes can be driven at the proper angle.*

PIER FORMS

Piers are circular, square, or battered footings used as a base for wood posts or steel columns, which are placed under beams to support the floor and wall units of a building. Regardless of the pier shape design, pier bottoms must rest on firm soil and be below the frost line. The lower end of a post or column must be fastened securely to the pier.

Circular piers are popular in most areas. The pier form is cut from tubing made from treated cardboard or other fibrous material. **See Figure 39-19.** When circular pier forms are set in position and plumbed, soil is placed around the forms to hold them in place while being filled with concrete.

Figure 39-19. *Circular pier forms must be set in position and properly plumbed before concrete is placed. For this application, a metal post base connector is placed in fresh concrete.*

Gates & Sons, Inc.

Circular footings may be required to support a circular wall form.

Other common pier shapes include square, tapered, and stepped. **Figure 39-20** shows the construction of a square pier used to support a post or column. **Figure 39-21** shows a procedure for setting bolts that secure the base of a steel column to a square pier. The bolt holes are laid out on a template that is centered on the pier form.

Figure 39-22 shows the construction of a tapered pier form. A pier block may be used to provide a nailing base for the bottom of a wood post. Common or duplex nails are driven at angles into the pier block, which is then pressed into the fresh concrete. The angled nails hold the block securely after the concrete has hardened and set.

Figure 39-23 shows the construction of a stepped pier form. Cleats secure the top section of the form to the bottom section. A metal dowel may be used to hold a wood post in position. Stepped piers are also used to support steel columns.

Pier forms must be level and positioned according to the dimensions provided on a foundation plan. A typical pier layout procedure is shown in **Figure 39-24.**

Anchor bolts must be accurately located to ensure proper positioning of posts and columns. Anchor bolts may be embedded in fresh concrete or installed using adhesive anchors after the concrete has hardened.

Square Pier Form Construction

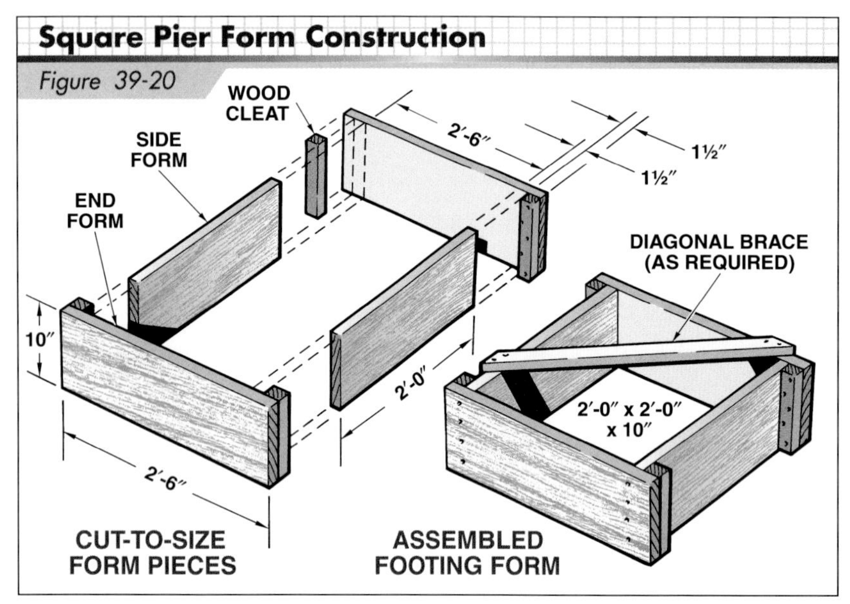

Figure 39-20. *A square pier supports a post or column. In this example, the pier dimensions are 2′-0″ × 2′-0″ × 10″.*

Steel Column Bolt Hole Template

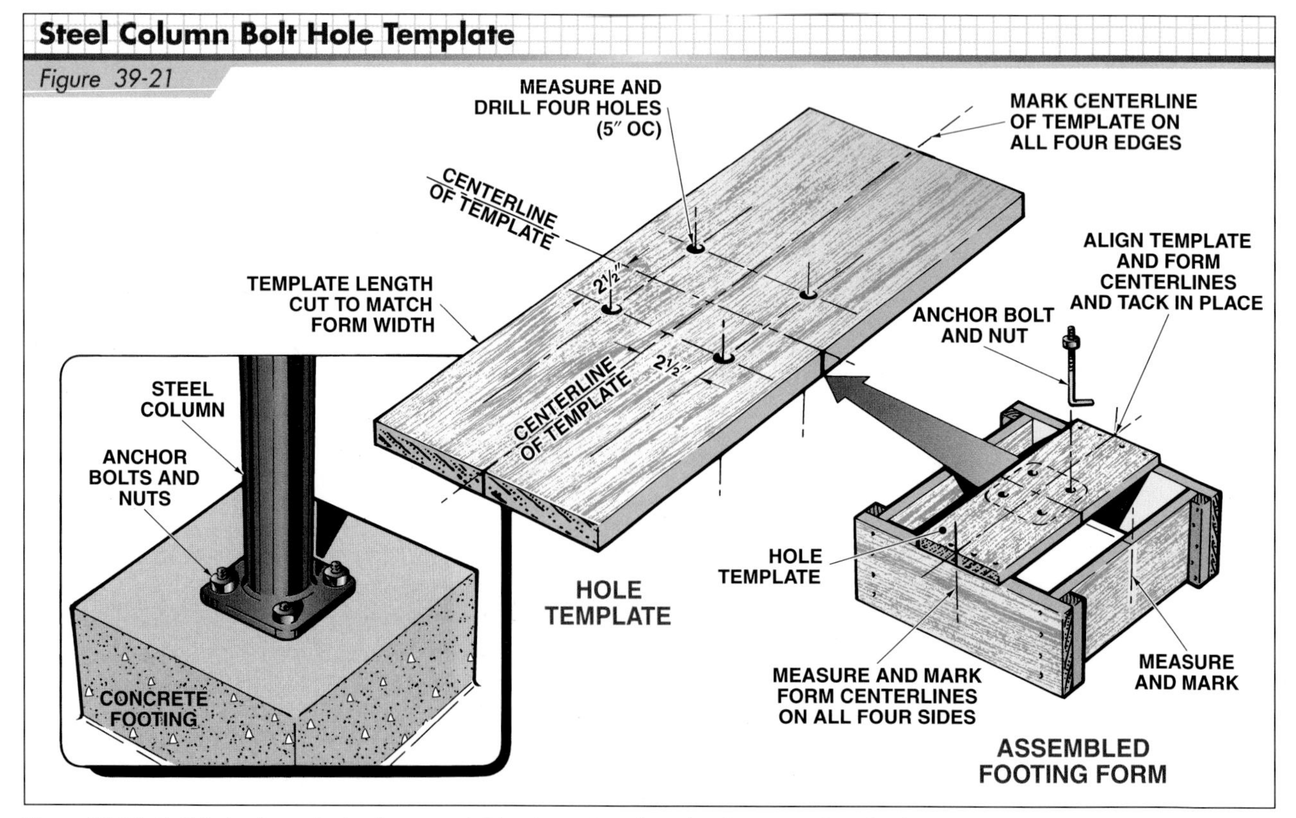

Figure 39-21. *Bolt holes for a steel column are laid out on a template that is centered on the form.*

Tapered Pier Form Construction

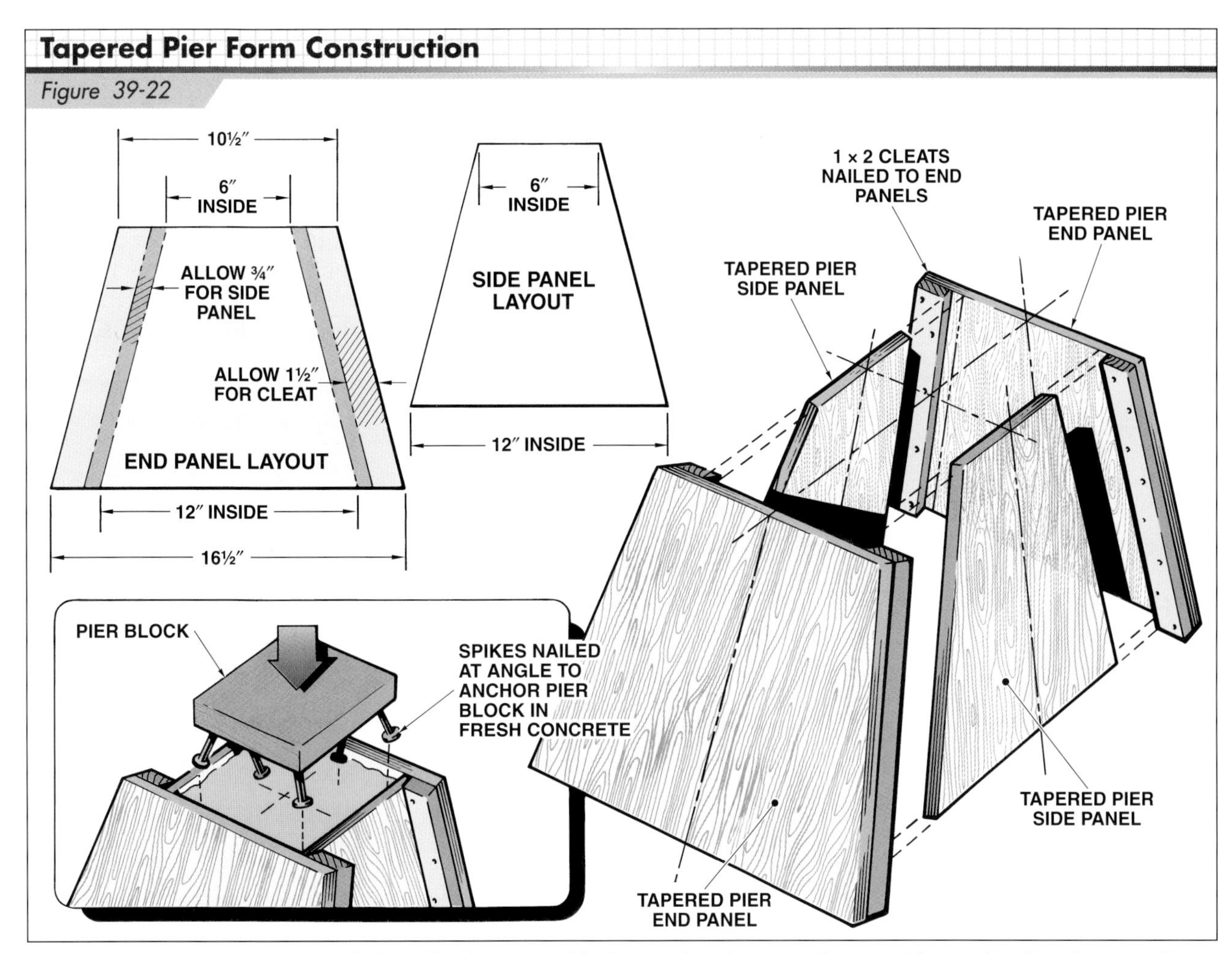

Figure 39-22. *Tapered piers are angled on all sides. A pier block is used in this example to provide a nailing base for a wood post. Nails are driven at angles into the pier block that is pressed into the fresh concrete. The angled nails hold the pier block securely after the concrete has hardened and set.*

A circular pier form is cut from tubing made with treated cardboard, braced, and filled with concrete. Electrical connections for a future sign protrude from the top of the footing. Long anchor bolts, shown in a template before being inserted into the concrete, will provide anchorage for the signpost.

GRADE BEAM FORMS

A *grade beam* is a reinforced concrete beam placed at ground level and supported by piles or piers at the ends and at intermediate positions. A grade beam should extend at least 8″ above the finished grade of the lot and the bottom must be below the frost line. Many building codes require the soil directly beneath grade beams be removed and replaced with a layer of coarse rock or gravel. The coarse rock or gravel provides drainage and reduces the chance of freezing action causing movement in the foundation walls.

Foam insulation under a grade beam provides thermal resistance.

Stepped Pier Form Construction

Figure 39-23

Figure 39-23. *Stepped pier forms are essentially two stacked square pier forms. Cleats hold the top form in place over the bottom form. A metal dowel may be used to hold a wood post in position. Stepped piers may also be used for steel columns.*

Cast-in-Place Grade Beams

Pier holes must be dug before grade beam forms can be constructed. Pier holes are dug using manual means such as shovel or earth auger or using a mechanical drilling rig that drives an earth auger into the ground. **See Figure 39-25.**

Figure 39-25. *Pier holes are dug with a mechanical drilling rig that drives an earth auger.*

Rebar are placed in the pier holes and extend above the top of the concrete to tie into rebar in the grade beam. In firm and stable soil, concrete can be placed directly into the open hole. In soft and unstable soil, a circular fiber form should be properly positioned in the pier hole before concrete is placed. **See Figure 39-26.**

The grade beam wall form is built directly over the piers. The rebar extending from the piers are tied to rebar placed in the grade beam forms. **See Figure 39-27.**

Flare the bottoms of pier holes to increase the surface area of the pier and help distribute weight, and to resist forces of frost heaving the piers.

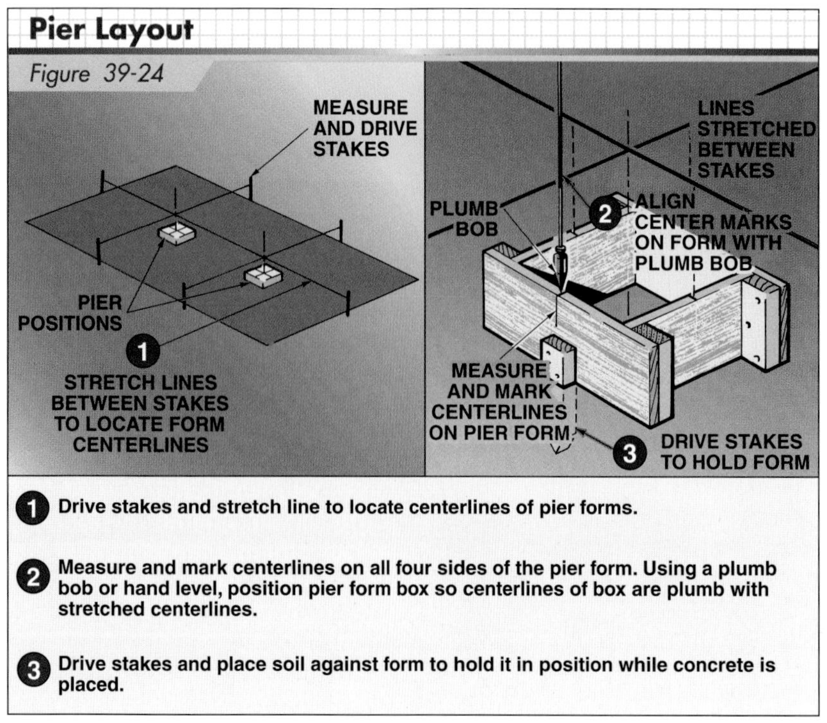

Figure 39-24. *Pier forms must be accurately placed.*

Figure 39-26. *Circular piers are used to support grade beams. Reinforcing steel extending from the top of the pier ties into the reinforcing steel of the grade beam.*

Precast Grade Beams

In some areas, precast grade beams are used for residential foundations. Precast grade beams are formed at a precast manufacturing facility and transported by truck to the job site, where they are lifted and set in place by a crane. Precast grade beams allow fast and efficient construction of crawl-space foundations that are placed on fairly level ground.

The precast system shown in **Figure 39-28** uses concrete grade beams that are 6″ wide by 12″ high. The beams are temporarily supported by wood blocks when they are placed over the pier forms. A ½″ rebar is inserted through a predrilled hole in the beam and is tied to another rebar that has been placed in the pier. After the grade beams have been properly leveled and aligned, concrete is pumped into the pier forms up to the bottom of the beams. The pier forms must be properly secured in position during concrete placement.

Plyform® is a performance rated plywood panel used for concrete forms. Plyform panels have an overlay which provides additional water and abrasion resistance.

SLAB-AT-GRADE FORMS

In slab-at-grade construction, a concrete slab rests directly on a bed of gravel that has been placed on the ground. Forms are constructed around the perimeter of the slab. **See Figure 39-29.**

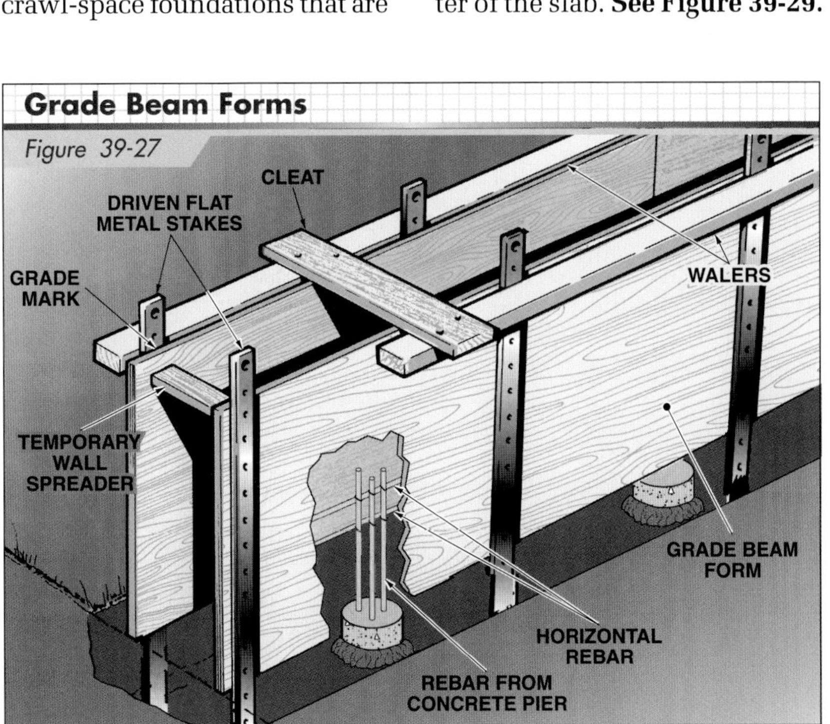

Figure 39-27. *Grade beam forms may be constructed over concrete piers.*

Figure 39-28. *Grade beams are placed by crane over the pier holes. The grade beams are temporarily supported by blocks and wedges.*

Figure 39-29. *Slab-at-grade forms are constructed around the perimeter of a slab.*

Two methods are used in slab-at-grade foundation construction. In one method, concrete for foundation walls and footings is placed first and concrete for the slab is placed separately, after the walls and footings have hardened and set. **See Figure 39-30.** Placing the foundation and slab separately is recommended for colder climates. Concrete floors tend to be cold because of heat loss occurring around the edges of the slab and exterior foundation walls. To help prevent heat loss, rigid insulation is placed around the perimeter of the slab and extends 2′-0″ beneath the concrete slab.

In warmer climates, concrete for the walls and floor is placed at the same time (monolithic construction). **See Figure 39-31.** The foundation and floor slab are one unit.

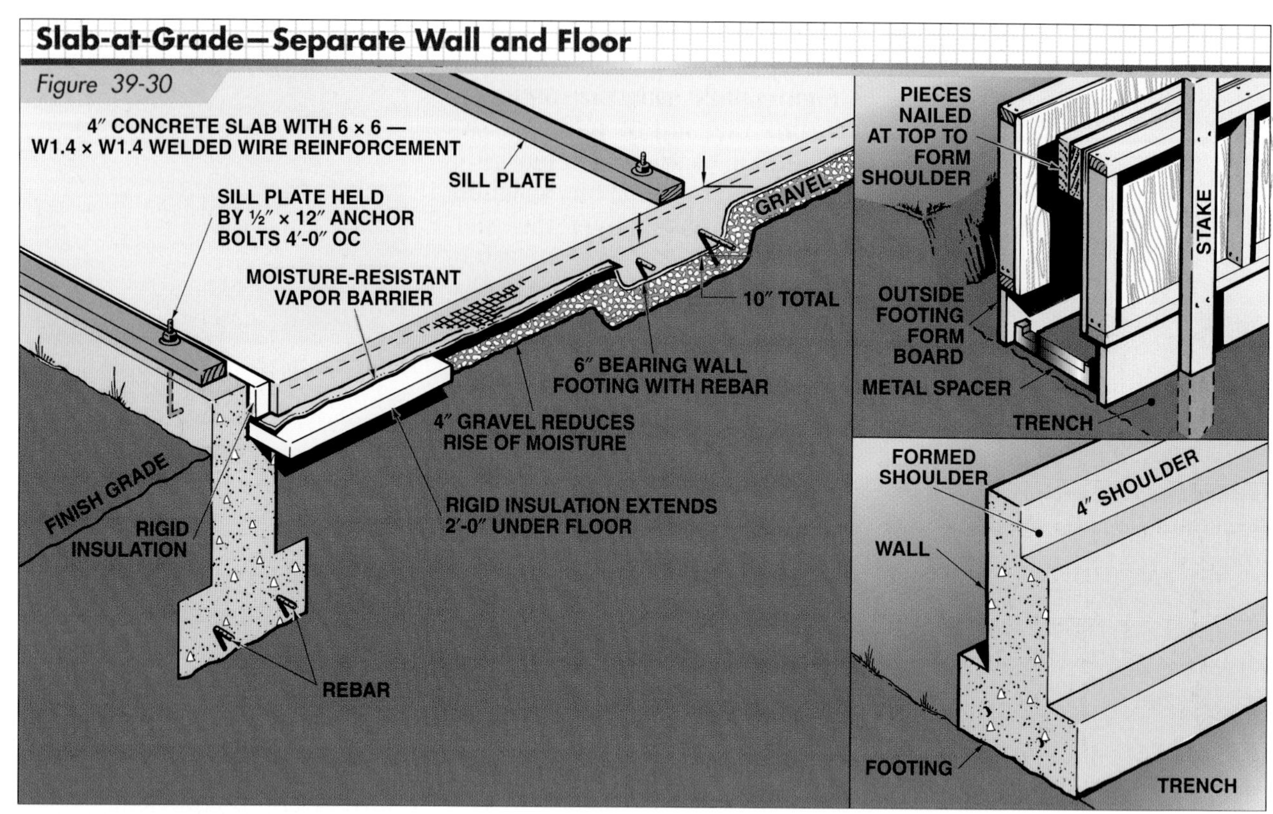

Figure 39-30. *In this slab-at-grade system, concrete for the foundation walls and slab is placed separately. The shoulder at the top of the wall supports the edges of the slab.*

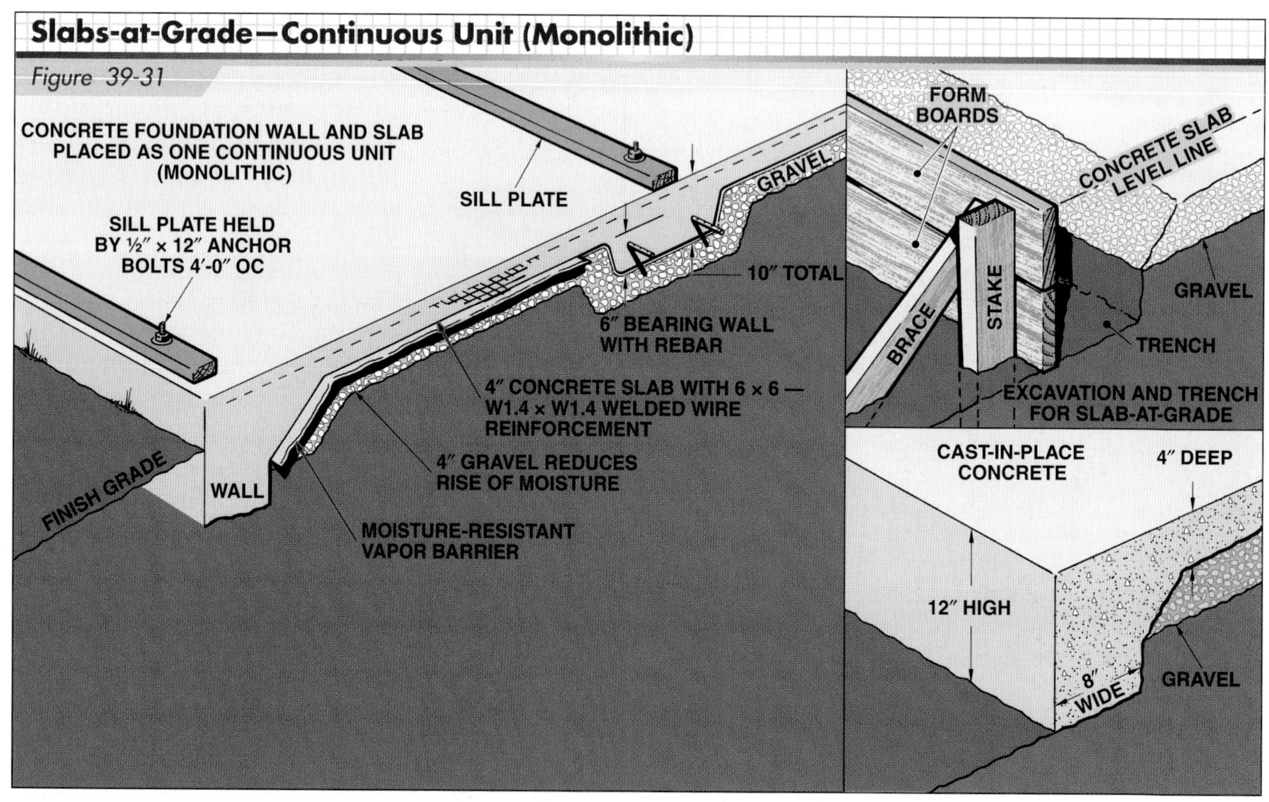

Figure 39-31. *The walls and slab form a continuous unit in a monolithic slab-at-grade.*

General rules to observe when constructing slab-at-grade foundations include the following:

1. Remove topsoil in the slab area.
2. Install required pipes and ducts and place a 4″ to 6″ layer of gravel to prevent groundwater from collecting beneath the floor. **See Figure 39-32.**
3. Place a moisture-resistant vapor barrier, such as 6 mil polyethylene plastic, over the gravel. Joints should be lapped at least 4″. Ensure that the vapor barrier is not punctured while it is being spread over the gravel.
4. Slabs-at-grade should be at least 4″ thick and reinforced with 6 × 6—W1.4 × W1.4 welded wire reinforcement.

Post-Tensioned Slabs

Post-tensioned slabs may be used in conjunction with slab-at-grade floor systems in residential and industrial construction. Post-tensioned slabs improve crack control with fewer joints in the slab and also reduce slab deflection. Less excavation is required and installation of the slab is faster because less reinforcing steel is required.

In post-tensioned slabs, high-strength tendons replace the welded wire reinforcement and rebar used in conventional construction. In a typical 4″ slab, tendons are positioned at the center of the slab thickness and run in both directions at 2′-0″ to 5′-0″ intervals. **See Figure 39-33.** One end of the tendons is securely anchored. The other end of the tendons is inserted through an anchoring device and is attached to a hydraulic jack. Slabs constructed over less stable soil conditions may require post-tensioned grade beams spaced 10′ to 20′ apart.

Concrete is placed after the tendons have been positioned and properly supported. When the concrete has reached sufficient strength, the tendons are stressed by hydraulic jacks to an effective force of about 25,000 lb. Anchoring devices at the ends of the tendons transfer the force to the concrete slab. **See Figure 39-34.**

Care must be taken to prevent damage to the plastic sheath covering post-tensioning tendons. Tendons within the sheath are coated their entire length with corrosion-inhibiting grease to protect the tendons from corrosion, weakening, and potential breakage. Also, do not make sharp angles with the tendons.

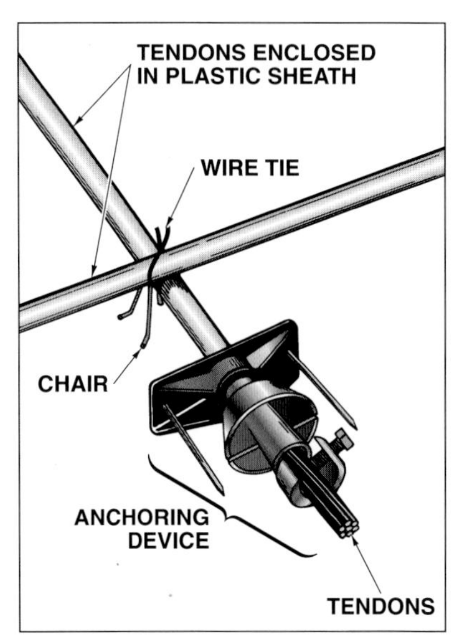

Figure 39-34. *Anchoring devices hold the ends of the stressed tendons.*

Figure 39-32. *Pipes, ducts, and welded wire reinforcement are installed in a slab-at-grade before concrete is placed. The orange tubing in this photo will provide radiant heat for the home.*

Portland Cement Association

Figure 39-33. *Tendons (encased in green plastic sheath) run in both directions at 2′-0″ to 5′-0″ intervals in a typical 4″ thick post-tensioned slab-at-grade.*

INSULATING CONCRETE FORMS

Insulating concrete forms (ICFs) are a type of concrete-forming system that consists of a layer of concrete sandwiched between expanded polystyrene (EPS) foam forms on each side. The forms remain in place after the concrete has been placed and become a permanent part of the walls or floors. **See Figure 39-35.** Interior walls, floors, and ceilings are then constructed using standard wood or metal framing members. ICF systems are increasingly used for above- and below-grade residential and commercial construction, including walls and floor and roof decks. In addition, ICF construction techniques and materials can be used for tilt-up construction. While forms and components for ICF construction from different manufacturers are similar, they are typically not interchangeable. Always consult the manufacturer instructions regarding the proper construction of insulating concrete forms.

ECO-Block, LLC

Insulating concrete forms can be used to form a short foundation wall, which will be tied into a slab.

After placing the concrete, ICFs remain permanently attached to the concrete and are not removed like traditional concrete-forming systems. The insulating forms combined with the concrete provide a continuous insulation system and an excellent sound barrier. The forms also serve as backing for gypsum board for the building interior and exterior finish such as wood siding, brick, and stucco.

Some building codes, particularly in the South, require that below-grade ICF wall sections be treated to resist the possible infestation of termites and carpenter ants.

A completed ICF building looks no different from a framed structure. ICF construction offers many advantages over traditional wood- or metal-framed buildings, including minimal air infiltration, reduced heating and cooling loads, and better fire resistance. In addition, ICF construction can contribute to LEED® certification. ICFs do not contain any formaldehyde and the waste form materials are 100% recyclable.

Quad-Lock Building Systems Ltd.

Figure 39-35. *Insulating concrete form (ICF) systems are used for above- and below-grade residential and commercial systems. When the interior and exterior finish materials are applied, ICF buildings have thicker walls than buildings constructed using traditional methods.*

ICF Systems

The three main ICF systems are *block, panel,* and *plank* systems. **See Figure 39-36.** The ICF units typically fit together with tongue-and-groove or serrated joints. Foam adhesives may be used to reinforce the bond between joints.

Insulating Concrete Form (ICF) Systems

Figure 39-36

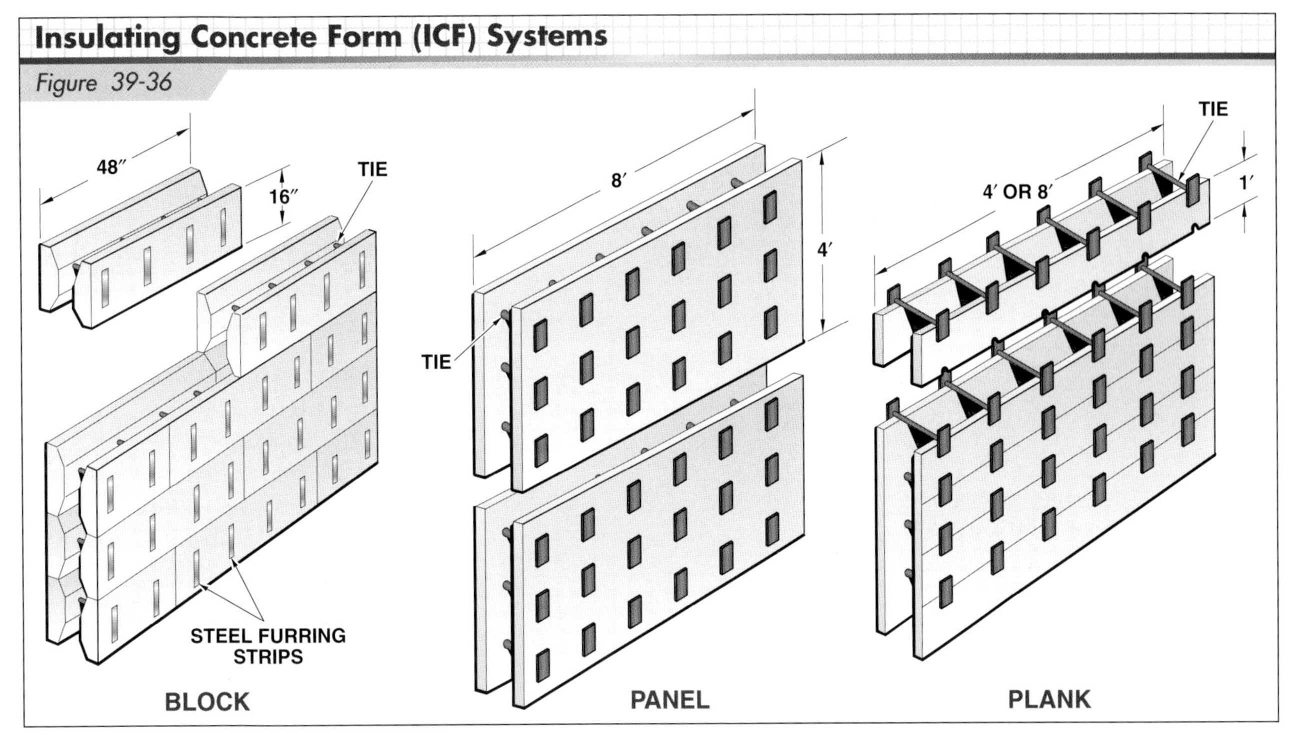

Figure 39-36. *The main types of insulating concrete forms are blocks, panels, and planks.*

ICF Block Systems. Block forms resemble concrete masonry units (CMUs), although their dimensions may vary from a typical CMU. Block units arrive on the job site ready for wall construction.

An ICF block consists of EPS face shells that are attached to each other using plastic or steel ties. **See Figure 39-37.** The face shells are typically 2″ to 2¾″ thick depending on the manufacturer and desired R value of the completed walls. The ties are crosspieces whose ends are molded into the face shells. In addition to tying the face shells together, the ties also maintain a consistent space between the interior surfaces of the face shells and support rebar. Wall thicknesses range from 4″ for above-grade applications to 12″ or more for foundations and below-grade applications. The length of the ties determines the wall thicknesses. While a common block size is 16″ high by 48″ long, smaller or larger units are also available. In addition to straight blocks, 45°and 90° corner blocks are also available.

ICF Panel and Plank Systems. Panel and plank systems differ from block forms in that they are shipped flat to the job site without the ties installed. Panel forms can be up to 4′ wide by 8′ long, while plank forms are typically 1′ wide by 4′ or 8′ long. **See Figure 39-38.** Ties for panel systems are installed between the opposing form sides before the units are placed in position. Ties for plank systems are installed between the form sides as the ICF planks are placed in position.

Quad-Lock Building Systems Ltd.

Insulating concrete form plank systems are shipped flat to a job site and erected in courses.

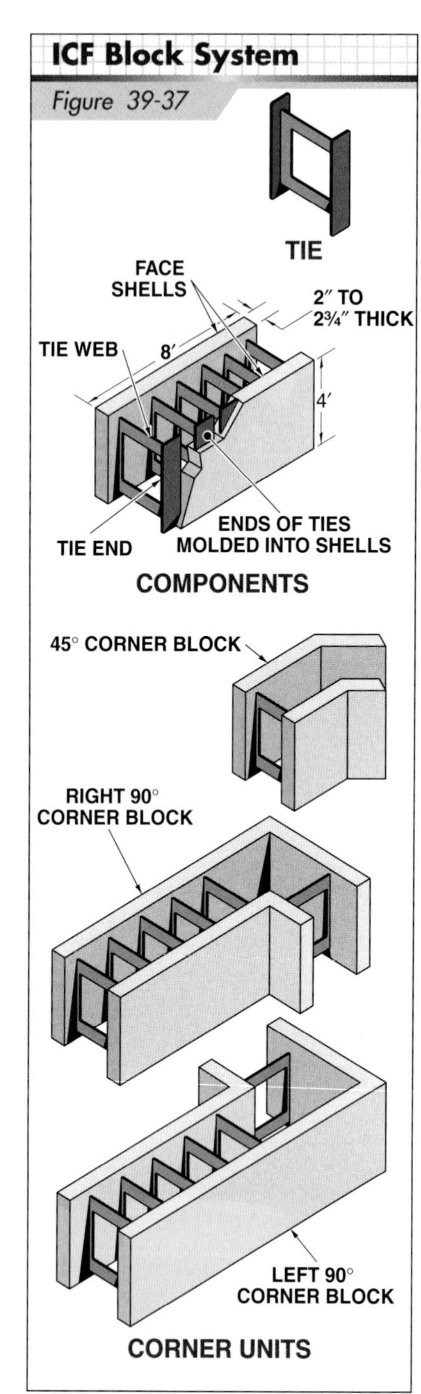

Figure 39-37. *Insulating concrete form (ICF) block systems consist of face shells and ties whose ends are molded into the shells.*

ICF Wall Designs

Three basic ICF wall designs are *flat core, waffle grid,* and *screen grid.* **See Figure 39-39.** The wall contours are formed by the shape of the interior of the ICFs.

Flat Core Walls. Flat core walls are similar to traditional cast-in-place concrete foundation walls with a layer of expanded polystyrene insulation on each side. Flat core walls vary in thickness, and commonly are 4″, 6″, 8″ or 10″.

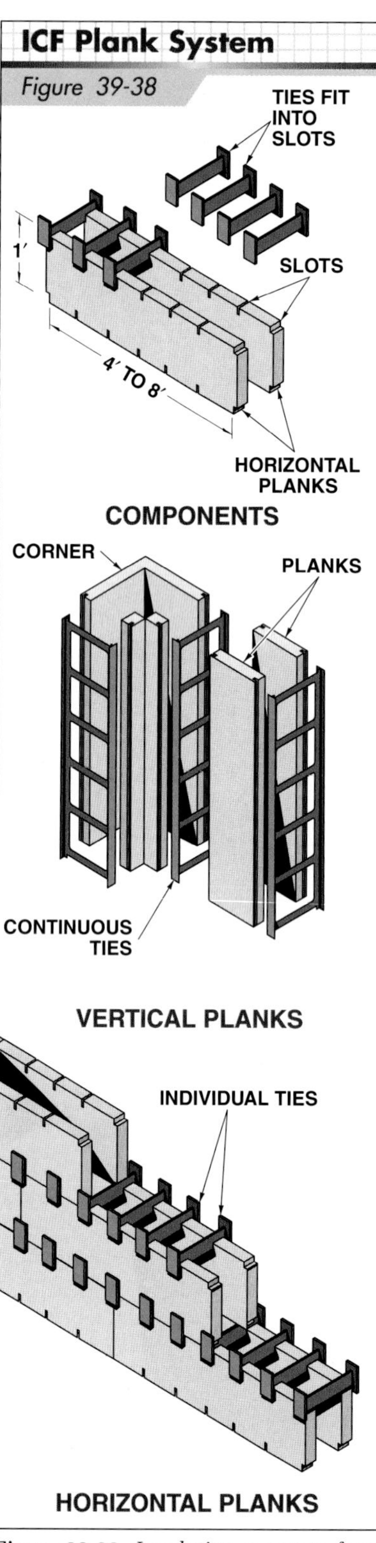

Figure 39-38. *Insulating concrete form (ICF) planks have slots to receive plastic or metal ties. Vertical ICF planks extend the entire height of a wall.*

Waffle Grid Walls. Less concrete is used in a waffle grid design than a flat core design. The horizontal and vertical core thicknesses are usually 6″ or 8″. Web thickness between the cores should be a minimum of 2″. The maximum spacing of horizontal and vertical cores is 12″ OC.

Screen Grid Walls. Screen grid walls, also known as post and beam forms, have columns spaced approximately 48″ OC and horizontal beams spaced 4′ or 8′ OC. Column and beam thicknesses are usually 6″ or 8″. Unlike the waffle grid design, screen grid systems do not have webs between the columns and beams.

ICF Tools and Handling

Insulating concrete forms are significantly lighter in weight than traditional concrete forms. ICFs weigh approximately 1 lb to 2 lb per square foot. Basic layout tools such as tape measures, framing squares, a chalk line, and a builder's level or laser transit level are used to lay out the forms and ensure walls are plumb and square and that floors are level. Fastening tools, including a hammer and drywall screwgun, are used to fasten materials to the ICF faces. In addition, a caulking gun or foam applicator is needed to apply foam adhesives to the forms.

Standard powered cutting tools are typically used to cut ICFs. A table saw with a fine-toothed blade is used for long, straight cuts. A circular saw or reciprocating saw can be used for cutouts when the walls are in place. Curved pruning saws work well for cutting most ICF materials. Electric hot knives can also be used to cut grooves or other recesses in ICFs. **See Figure 39-40.**

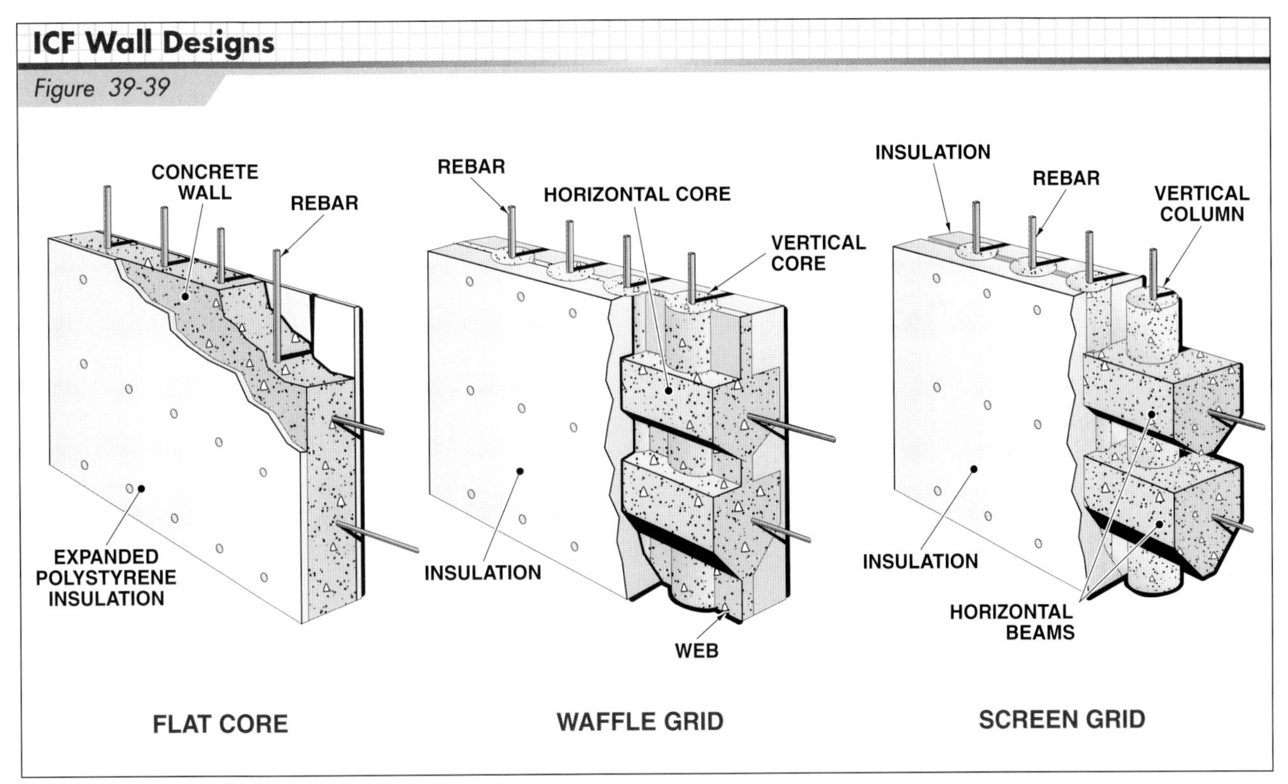

Figure 39-39. *The basic insulating concrete form wall designs are flat core, waffle grid, and screen grid.*

Quad-Lock Building Systems Ltd.

Figure 39-40. *Cutouts can be created in insulating concrete forms (ICFs) using an electric hot knife.*

Door and Window Bucks

When erecting ICF walls, openings must be created for doors, windows, or other openings such as those required for ventilation. Door and window *bucks* are required to frame door and window openings in ICF walls and prevent concrete from escaping from the forms during concrete placement. **See Figure 39-41.** Door and window bucks remain in place after the concrete is placed and serve as door and window frames. Metal *sleeves* may be used to create openings for ventilation or utility runs. Bucks and sleeves must be properly braced and supported to prevent movement during concrete placement.

Door and window bucks are constructed from plastic, vinyl, or pressure-treated wood. Plastic or vinyl bucks are preferred by many contractors. Unlike wood bucks, which will deteriorate over time, plastic and vinyl bucks remain stable and are not subject to water or mold damage. Integral flanges extending out from the frames anchor plastic or vinyl bucks to the concrete. Wood bucks can be prebuilt and delivered to the job site or constructed on the job site. The sides and top of door and window bucks should be the width of the wall. The bottom of a wood buck is commonly constructed of two or more 2 × 4s. One of the 2 × 4s is removed during concrete placement to provide an opening for the concrete. When concrete fills the wall cavity below the buck, the 2 × 4 is fastened into position and concrete placement continues. Untreated lumber can also be used to construct a buck, but a waterproof barrier must be installed between the buck and concrete.

Quad-Lock Building Systems Ltd.

Figure 39-41. *Window and door bucks frame door and window openings in insulating concrete form (ICF) walls.*

Attaching to Insulating Concrete Forms

A variety of devices and methods are used to attach finish materials to insulating concrete forms. Exterior materials such as wood siding, brick, stone, or stucco can be applied to exterior surfaces. Gypsum board, solid wood boards, or wall paneling can be fastened to interior ICF surfaces. Metal hangers and other devices that support wood or metal ceiling joists, ledgers, and beams may also need to be installed.

Ledger Connectors. Ledger connectors can be installed before or after concrete is placed in the forms. Ledger connectors support the wood or metal ledger boards to which joist hangers or other metal framing connectors are fastened. **See Figure 39-42.** When installing ledger connectors before concrete is placed, a chalk line is snapped for the bottom of the ledger board, and the on-center spacing of the connectors is marked along the line. Some ledger connectors can be pushed through the forms, while other ledger connectors require kerfs to be cut for the legs. The ledger connector is then inserted into the kerfs until it is flush with the surface of the ICF. Concrete can then be placed for the walls.

Wood ledger boards can be fastened to ICFs after the concrete has properly hardened using concrete screws. A chalk line is snapped for the bottom of the ledger board. The concrete screws are driven through the connector and ledger board. The ledger board is then aligned with the chalk line and the screws are driven through forms and into a metal connector plate embedded in the concrete.

Insulating concrete forms are constructed around horizontal and vertical rebar that are tied together.

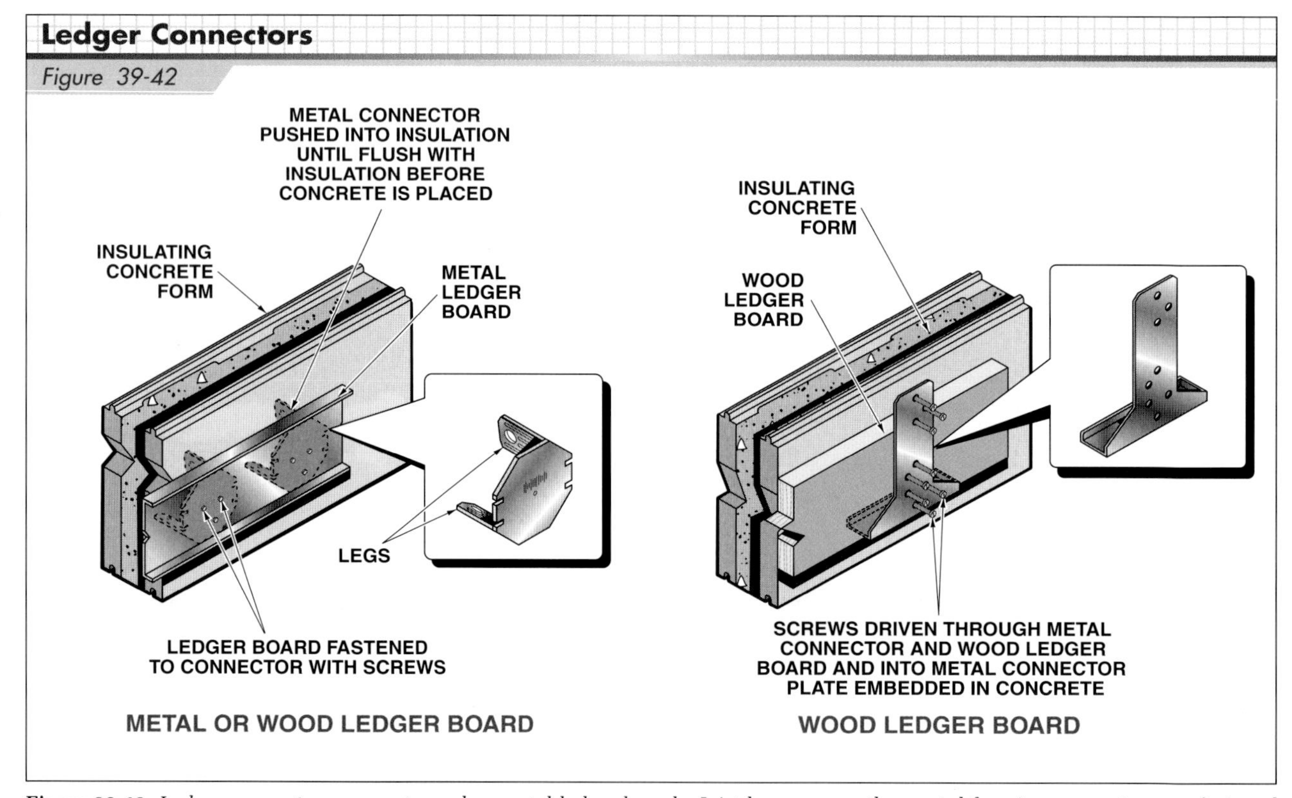

Figure 39-42. *Ledger connectors support wood or metal ledger boards. Joist hangers or other metal framing connectors are fastened to the ledger boards.*

Wall Finish. Interior finish materials may be applied directly to insulating concrete forms using adhesives, or they may be attached to plastic or metal furring strips embedded in the forms. When using adhesives, always consult the ICF manufacturer recommendations for the proper type of adhesive to be used. **See Figure 39-43.**

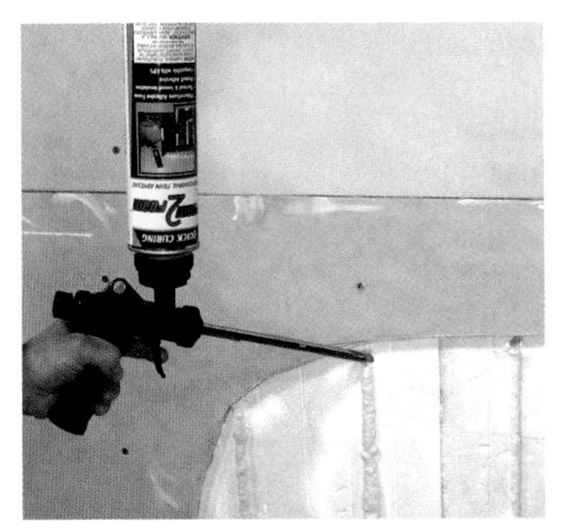

Wind-Lock Corporation

Figure 39-43. *Foam adhesive may be used to attach gypsum board panels to ICF walls.*

Plastic or metal furring strips may be embedded in ICFs, providing a good base for attaching gypsum board or finish products. Self-tapping screws should be used when attaching finish products to plastic or metal furring strips. Ring-shank nails should be used when nailing to wood furring strips.

For certain applications, sheets of 24-, 25-, or 26-ga galvanized sheet metal are installed under the gypsum board to provide proper support for heavy items. The sheet metal is fastened to furring strips using self-tapping screws or pop rivets. The gypsum board is then attached to the sheet metal. The heavy items, such as cabinets, are secured in place using self-tapping screws with sufficient length to penetrate the sheet metal backing.

Constructing ICF Walls

When constructing ICF walls, the first row of blocks, panels, or planks are set in position on a concrete footing or slab with vertical rebar extending vertically. The footing or slab should be ±¼″ from level to avoid adjusting the forms later in the wall construction. The general procedure for constructing an ICF wall is as follows:

1. Snap chalk lines on the footing or slab to position the walls. Fasten wood or metal bottom plates to the footing or slab to prevent the base of the ICF wall from moving. **See Figure 39-44.** Dabs of foam adhesive can be used to prevent the base from moving in lieu of the plates.

REBAR
INSULATING CONCRETE FORM
TIE
FOOTING
BOTTOM PLATES FASTENED TO FOOTING
WOOD BOTTOM PLATES
REBAR
INSULATING CONCRETE FORM
TIE
FOOTING
METAL C-SHAPE BOTTOM PLATE FASTENED TO FOOTING
METAL BOTTOM PLATES

Figure 39-44. *Wood or metal bottom plates are fastened to the footing or slab to prevent the base of the insulating concrete wall form from moving.*

2. As the ICF wall is constructed, place the door and window bucks where required. Also, install sleeves for other wall openings such as holes for utilities and ventilation. Bucks and sleeves must be properly braced and supported to prevent movement as concrete is being placed.
3. Place horizontal rebar as the wall is constructed. Depending on the type of ICF system and the manufacturer recommendation, the horizontal rebar may be supported by and tied to the ICF ties.
4. Align and brace forms to keep the walls and openings plumb and square during concrete placement. For higher walls, one method of aligning and bracing the forms is to construct a scaffold with uprights fastened to one side of the wall. Braces attached to the uprights of the scaffold are secured to stakes in the ground at their lower ends. The scaffold also serves as a working platform. **See Figure 39-45.** Typical corner braces consist of two 2 × 6s held in place by diagonal 2 × 4s fastened to stakes in the ground or cleats nailed to the top of the footing or slab. **See Figure 39-46.**

Quad-Lock Building Systems Ltd.

Figure 39-45. *Braces are required to support the insulating concrete wall forms during concrete placement. Note the use of ladder frames and work platforms for easy access to the top of the forms.*

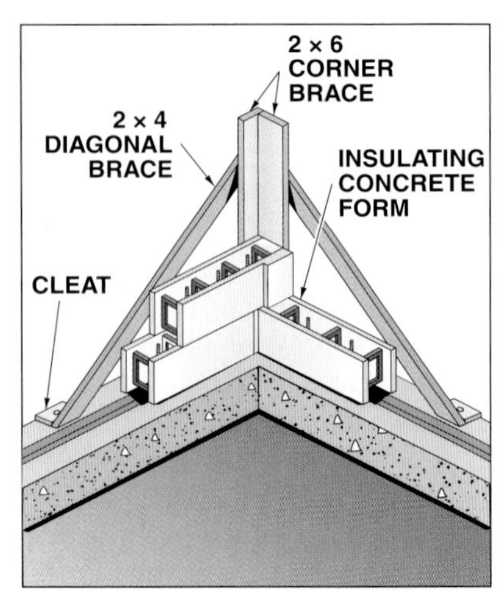

Figure 39-46. *Corners of insulating concrete forms must be properly supported and braced.*

5. Install miscellaneous metal connectors such as ledger connectors and anchor bolts as required.
6. When required by the manufacturer, install wire ties along the top edge of the wall to prevent the upper form edges from spreading. **See Figure 39-47.**
7. Place the concrete as shown in **Figure 39-48.**

Quad-Lock Building Systems Ltd.

Figure 39-47. *Wire ties may be installed along the top edge of insulating concrete forms (ICFs) to prevent upper form edges from spreading. A metal cap may also be installed to protect the tops of ICF wall forms from concrete that may be splattered during placement.*

After the concrete hardens and cures properly, the plumbing and electrical installation can begin. The foam can be cut to create channels for electrical lines and water pipes using an electric hot knife.

Quad-Lock Building Systems Ltd.

Figure 39-48. *Concrete should be slowly introduced into the forms in 2′ to 4′ lifts.*

Placing Rebar and Concrete

The procedure for positioning vertical and horizontal rebar inside ICFs depends on the form system being used. Rebar placement in plank and panel forms is similar to placing rebar in traditional panel forms. Vertical and horizontal rebar are joined by tying the rebar together with wire. Rebar placement in block forms is similar to placing rebar in concrete masonry units. Block forms may be designed with cradles, making it more convenient to place the horizontal rebar.

Insulating concrete forms are not braced with a system that includes walers. Therefore, there is an increased possibility of *pillowing* (bulges) occurring in the walls. To prevent pillowing, concrete should be introduced into the forms slowly in 2′ to 4′ lifts using a concrete pump and hose. A reducer is installed on the hose to narrow the discharge to 2″. A concrete mix design that includes a plasticizer agent will ensure proper flow of the concrete into all areas of the forms. When permitted by the ICF manufacturer, mechanical vibration may be used to help consolidate the concrete.

ICF Floor and Roof Decks

Similar to ICFs for wall construction, the ICFs for floor and roof deck systems remain in place after concrete has hardened, adding greater insulation value to the building. Form shapes and installation methods vary with different manufacturers. However, ICF floor and roof deck systems typically consist of a thick layer of foam, integral steel joists, and a 2″ to 4″ thick layer of concrete placed on top of the foam. **See Figure 39-49.** Rebar are placed in the concrete beam pocket between the ICF units. Hollow cores are provided in the forms for plumbing and electrical installation. Most ICF floor and roof systems require temporary shoring until the concrete has achieved design strength. **See Figure 39-50.**

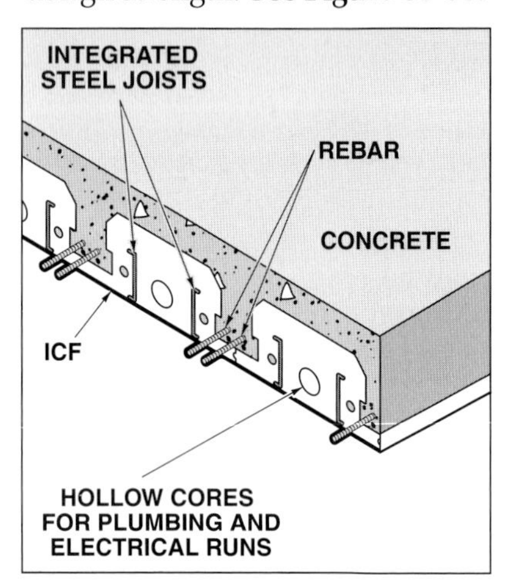

Figure 39-49. *Insulating concrete form (ICF) floor and roof deck systems consist of a layer of foam, steel joists embedded in the foam, and a layer of reinforced concrete.*

Quad-Lock Building Systems Ltd.

Figure 39-50. *Insulating concrete forms (ICFs) for floor and roof decks must be properly shored during concrete placement and must remain in place for the amount of time specified in the construction specifications.*

UNIT 40

Stairway and Outdoor Slab Forms

OBJECTIVES

1. Describe the procedure for constructing stairway forms.
2. Explain the procedure for forming and placing concrete for outdoor slab forms.

Stairway and outdoor slab forms are constructed by carpenters, usually after most of the building construction is complete to minimize the wear on the surfaces. The local building code should be consulted to determine the permitted stairway riser height and tread width, as well as thicknesses and widths of walkways and driveways.

STAIRWAY FORMS

Due to the finished floor height of a building, an entrance platform and stairway may be required. Concrete is a practical material for entrance platforms and exterior stairways since it is durable and will not decay under damp conditions.

Figure 40-1 shows a procedure for constructing entrance platform and stairway forms. In this example, the riser height is 7½″ and the tread width is 10″. A center brace and cleats are required on wide stairways to support the riser boards. The bottom edges of the riser boards should be beveled to allow a cement mason to trowel into the corner. **See Figure 40-2.** Riser boards should be removed as soon as possible in order to finish the vertical concrete surface.

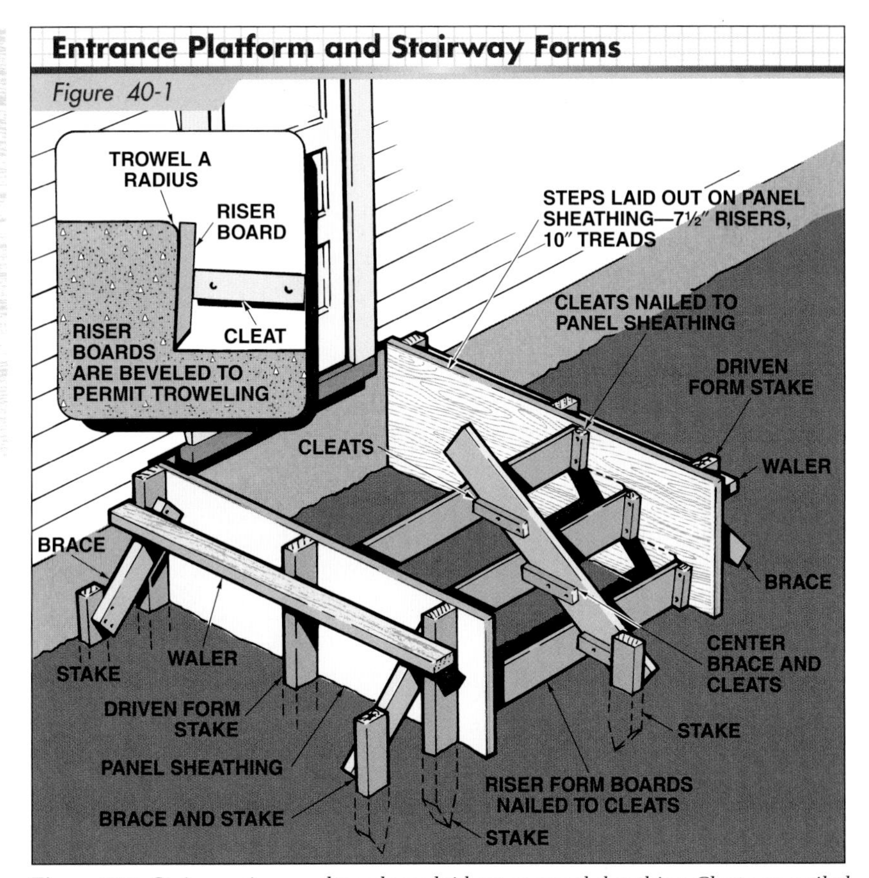

Figure 40-1. *Stairway risers and treads are laid out on panel sheathing. Cleats are nailed to the sheathing and the riser form boards are fastened to the cleats. For wide stairways, a brace is placed at the center of the riser boards.*

OUTDOOR SLAB FORMS

Outdoor concrete slabs include *sidewalks, patios,* and *driveways.* Information regarding the widths and locations of sidewalks, patios, and driveways is provided on the plot plan. Sidewalk, patio, and driveway forms are usually built as construction work on the building nears completion. Final grading of the soil should be completed and the ground settled and compacted before the forms are built.

Figure 40-2. *Riser boards are beveled to allow a cement mason to trowel into the corner.*

Sidewalk, patio, and driveway slabs are usually placed directly on the soil. For many applications, soil is removed from the slab area so the top of the slab is flush with the surrounding area. If the building site has significant surface water or if problems could result from frost conditions, a layer of gravel is laid before the concrete is placed.

Carpenters or cement masons typically set the forms for outdoor slabs. Outdoor slab forms usually are 2 × 4s or 2 × 6s placed on edge and held in place by stakes. *Spreader boards* are placed at intervals to ensure proper spacing between the outside form boards and to retain the concrete while it is being placed in different sections of the form. The spreader boards are removed as concrete is placed. A procedure for setting sidewalk forms is shown in **Figure 40-3.**

When concrete is placed for an outdoor slab, provisions must be made to strike off (screed) the concrete to the required level. For a sidewalk form, a straightedge known as a *strike board,* or screed board, is laid across the top of the outer form boards. The strike board rests on top of the form boards and is moved back and forth using a saw-like motion to level the concrete surface.

For larger slabs, such as patios or terraces, temporary wood or metal pieces called *screed rails* are placed at intervals in the slab area. The tops of the screed rails are set to the desired finished surface of the concrete. A strike board is moved back and forth along the screeds after the concrete has been placed. **See Figure 40-4.** Large slabs can also be struck off using power-driven screeds that make the job easier. **See Figure 40-5.** After the concrete has been placed and struck off, cement masons work the concrete to the desired finish.

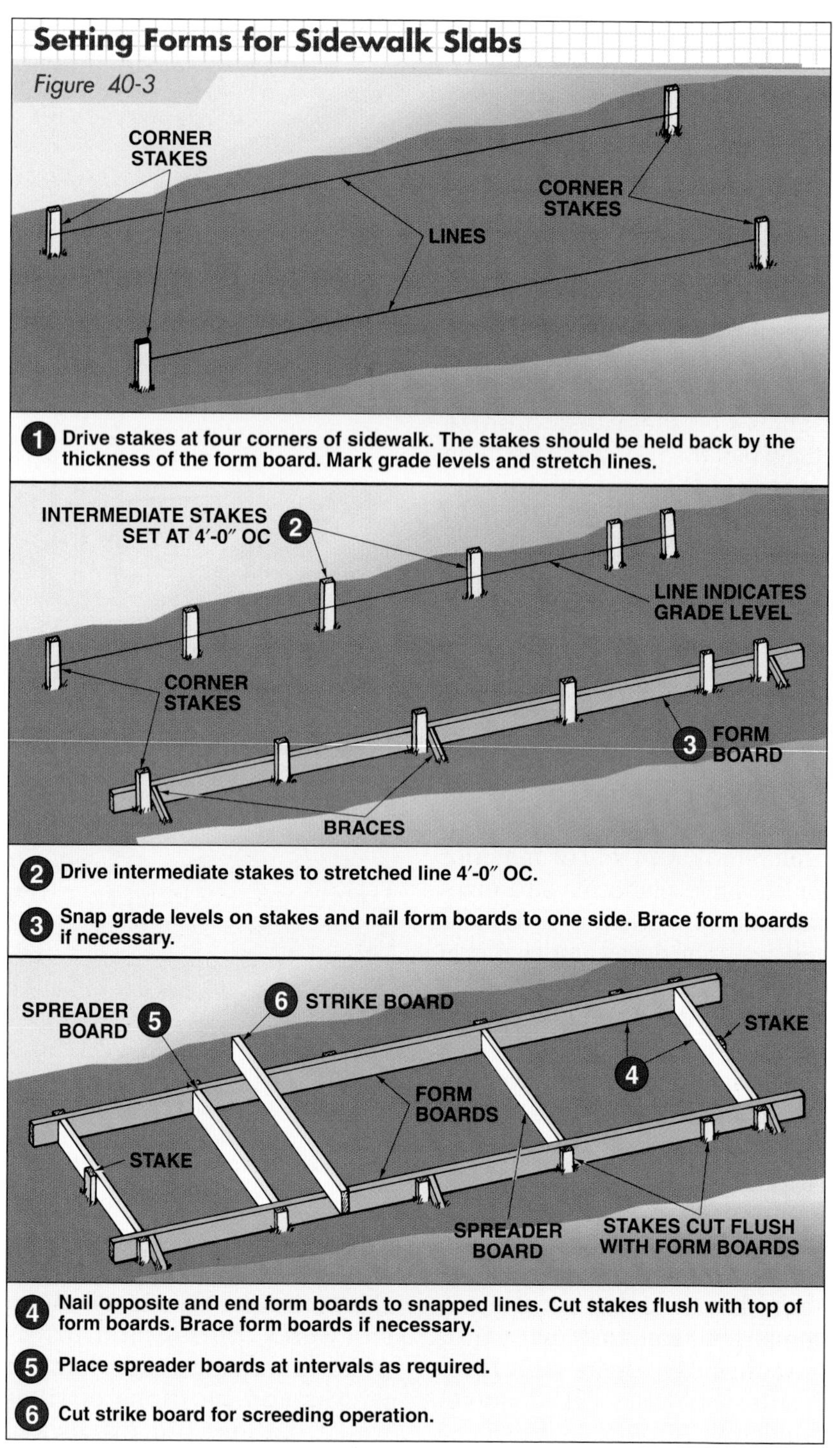

Figure 40-3. *Sidewalk slabs are formed with 2 × 4s or 2 × 6s.*

Figure 40-4. *When striking off concrete, the strike board is moved back and forth using a saw-like motion.*

Figure 40-5. *Power-driven screeds, such as a vibratory wet screed, are commonly used for large slabs-at-grade.*

Walkways and Patios

The width required for a walkway depends on whether it is a *sidewalk, front walk,* or *service walk.* Sidewalks run along a street and border the building lot. Sidewalks are typically 4′ to 5′ wide. Front walks extend from the sidewalk or driveway to the front entrance of a building. Front walks are usually 3′ wide. A service walk extends from a driveway or sidewalk to the rear entrance of a building. Service walks are commonly 2′ to 6′ wide and are set approximately 2′ from a building. Walkways and patios are 4″ thick, although walks should be thicker if trucks carrying heavy loads frequently cross over them.

The slope across the width of a walkway, known as the *cross slope,* should be 1/8″ to 1/4″ per foot to allow proper water drainage. A slope of 1″ in 12′ is adequate for patios.

Driveways

One-car driveways are usually 10′ wide. Two-car driveways range from 18′ to 20′ wide. Driveways used for passenger cars should be at least 4″ thick while driveways used for trucks or other heavy vehicles should be 6″ thick. Driveway slabs that abut garage floors should have a finished surface that is 1/2″ lower than the garage floor to prevent water from entering the garage.

Driveways are reinforced with welded wire reinforcement to minimize cracking. Driveways for heavy vehicles are reinforced with rebar. Minimum cross slope of a driveway should be 1/8″ to 1/4″ per foot for proper water drainage.

Control and Expansion Joints

A *control joint,* or contraction joint, is a groove made in a concrete slab to create a weakened plane and control the location of cracking in the slab. Control joints are placed at intervals to a depth of one-fourth the slab thickness. Control joints are tooled into the concrete surface using a hand groover as the concrete is finished, or they can be cut into the concrete using a concrete saw after the concrete has set. **See Figure 40-6.**

Figure 40-6. *A control joint is tooled using a hand groover and straightedge.*

Spacing of control joints is determined by the type of slab. In general, control joints should be placed approximately 2 1/2 times (in feet) the slab depth (in inches). For example, control joint spacing for a 4″ thick slab is approximately 10′. In driveways, control joints running across the width should be no more than 10′ apart. For driveways 12′ wide or more, a control joint is also recommended along the length of the driveway.

An *expansion joint,* or isolation joint, is provided to separate dissimilar construction materials and to separate adjoining sections of concrete to allow movement caused by expansion and contraction of the slabs. For example, an expansion joint is used where a sidewalk butts against a foundation wall, stairway, or driveway. An expansion joint passes through the entire slab thickness. An asphalt-impregnated strip or other type of material is normally placed at the juncture of the isolation joint.

Unit 40 Review and Resources

UNIT 41

Foundation Moisture Control and Insect Prevention

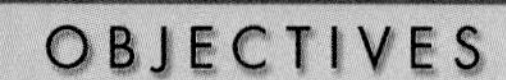

1. Describe the principles of moisture control in buildings.
2. List and describe the various materials used for waterproofing foundation walls.
3. Review the procedure for installing footing drains.
4. Describe the procedure for installing a sump well.
5. Describe factors to consider for mold prevention.
6. Identify the prevention methods used to treat common insect infestations.

The foundation of a building is critical to the integrity and soundness of the entire building. Foundations must be properly designed and constructed according to local building code regulations. In addition, moisture control and insect prevention are key factors in preventing serious foundation damage as the building ages.

MOISTURE CONTROL

Water conditions on and below the ground surface must be considered in foundation construction. Precautions must be taken to ensure that water does not enter the living area of a full-basement foundation. In addition, water collecting on the ground inside and surrounding a crawl-space foundation creates dampness that causes wood decay, unpleasant odors, mold, and rust.

Water from rain and melting snow that stays on the ground surface is called *surface water*. Finish grades around a foundation should be sloped away from the building to divert water away from the building.

Most problems related to water conditions of the ground concern the water table, which is located beneath the ground surface. The *water table* is the highest point below the surface of the ground that is normally saturated with water in a given area. Water tables tend to rise during wet seasons and subside during dry seasons.

Capillary Action

A certain amount of water and water vapor rise from the water table through a process known as *capillary action*. Capillary action occurs in all types of soil, although water will rise higher in more porous soils, such as silt and clay, than in less porous soils. **See Figure 41-1.** Even when the water table is well below the finish grade, capillary action can cause damp conditions at the ground surface.

Capillary action does not occur in coarse material such as gravel. Therefore, one method of water control is to excavate the porous soil around a foundation and replace it with gravel.

Moisture-Resistant Vapor Barriers

One approach to solving the problem of water collecting under a crawl-space foundation is to cover the ground with a moisture-resistant vapor barrier. The material must also be decay- and insect-resistant. Four or six mil polyethylene film is commonly used. After the film is placed, it should be covered with a layer of pea gravel, sand, or a skim coat of concrete. **See Figure 41-2.**

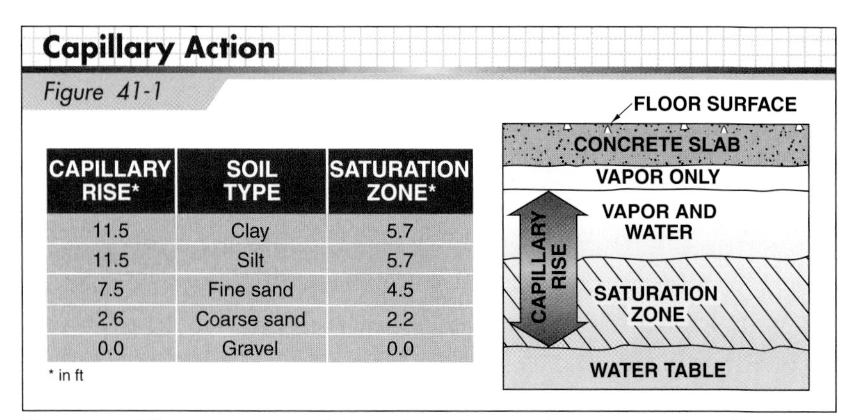
Capillary Action

Figure 41-1

CAPILLARY RISE*	SOIL TYPE	SATURATION ZONE*
11.5	Clay	5.7
11.5	Silt	5.7
7.5	Fine sand	4.5
2.6	Coarse sand	2.2
0.0	Gravel	0.0

* in ft

Figure 41-1. *Water moves toward the surface of most soils as a result of capillary action. Water rises higher in fine-grained soils such as clay or silt.*

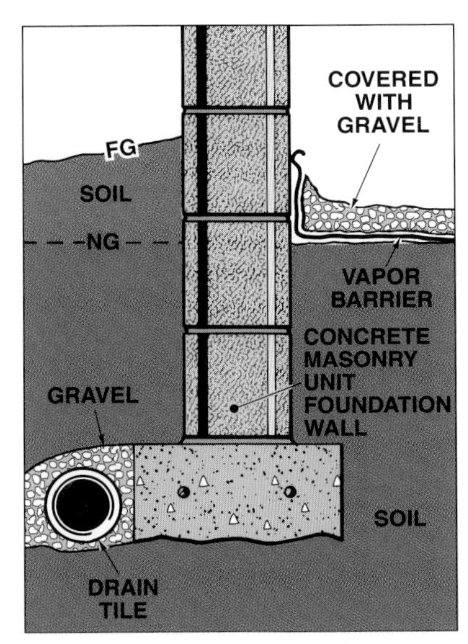

Figure 41-2. *A vapor barrier helps prevent moisture from accumulating underneath a house with a crawl-space foundation.*

Waterproofing Foundation Walls

Although solid concrete walls are highly resistant to water penetration, they can still absorb moisture if not properly waterproofed. Concrete masonry unit walls are less resistant to water penetration since the walls are constructed with individual units. Water penetration through foundation walls can damage interior wall finishes, equipment, and furnishings and produce mold and decay. Foundation walls should be treated from the bottom of the wall to the finish grade.

Most moisture penetration occurs through *leaks, seepage,* or *condensation.* Leaks occur in cast-in-place concrete and CMU foundations through cracks that may develop over a period of time. Seepage results from groundwater hydrostatic pressure. Hydrostatic pressure is the water pressure at a given depth. Rainfall, melting snow, or lawn irrigation can increase the hydrostatic pressure exerted on a building. Condensation occurs on the inside surface of a foundation wall and occurs when the interior of a wall is warmer than the exterior surface.

Effective methods and materials for waterproofing foundation walls include *asphalt sheet membranes, liquid membranes, cementitious products, clay materials,* and *rigid foundation insulation panels.*

Asphalt Sheet Membranes. Asphalt sheet membranes are self-adhesive 60 mil sheets of rubberized asphalt laminated to polyethylene film. The adhesive side of the sheet is covered with a release paper, which is removed at the time of application. **See Figure 41-3.** The primary advantage of asphalt sheet membranes is their uniform thickness. Asphalt sheet membranes provide good waterproofing capabilities when properly applied. Asphalt sheet membranes must be lapped a minimum of 6″. Some manufacturers recommend a bead of mastic be applied to lap joints within 12″ of a corner.

Figure 41-3. *Asphalt sheet membranes are self-adhesive waterproofing material applied directly to foundation walls.*

Liquid Membranes. Liquid membranes are applied by spraying, roller, or trowel. The liquid cures into a rubbery coating after being applied to foundation walls. **See Figure 41-4.** Liquid membranes are commonly composed of *polyurethane* or *polymer-modified asphalt.* Precautions must be taken by the applicator to ensure a minimum membrane thickness of 60 mil.

WATCHDOG WATERPROOFING®

Figure 41-4. *Liquid membranes cure to a rubbery texture after being applied to foundation walls.*

Cementitious Products. Cement combined with various admixtures produces a cementitious product that can be easily applied with a brush, trowel, or low-pressure spray pump. Cementitious products are rigid materials and will not stand up well to joint or crack movement.

Clay Material. *Sodium bentonite,* a natural mineral primarily consisting of aluminum silicate, is a highly expansive clay material (combined with admixtures) used for foundation waterproofing. Sodium bentonite is manufactured as panels and is fastened to foundation walls with nails or powder-actuated fasteners. As sodium bentonite absorbs water, the material swells to 15 times its original volume and pushes itself into voids and cracks. When the material reaches its maximum volume, it remains in the voids and cracks to seal against water entering the building.

Rigid Panels. Rigid foundation insulation panels provide a means for groundwater to flow, protect the liquid membrane under the panels, and provide good insulation capabilities. Rigid panels are placed over a liquid membrane which holds the panels in position until the foundation is backfilled. **See Figure 41-5.**

Footing Drains

Footing drains, commonly referred to as *drain tile,* are perforated pipes laid along the outside of foundation footings to collect rainwater and water from melting snow percolating down through the backfill and move it away from the foundation. In addition, footing drains relieve hydrostatic pressure from rising groundwater. Four-inch diameter flexible or rigid plastic drain tile are commonly used as footing drains. **See Figure 41-6.**

Drain Tile. In the past, footing drains were often connected to the sanitary sewer. However, this practice caused problems since sanitary sewer systems were not designed to handle large volumes of surface water from rain and melting snow. Modern drain tile design allows groundwater to enter the tile and flow away from the building or flow into a sump well where it collects and is directed to the outside of the building where it can drain away.

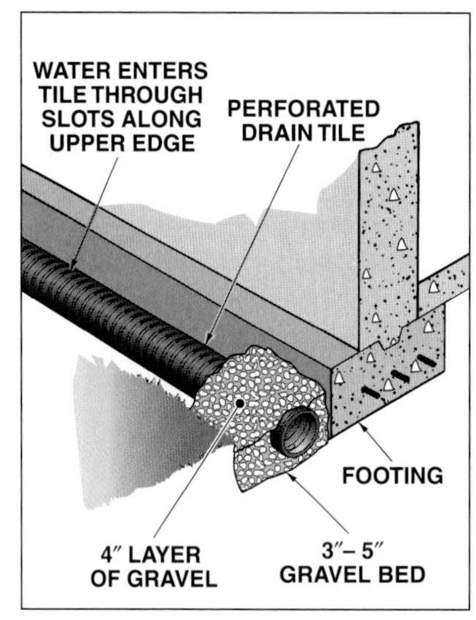

Figure 41-6. *Drain tile collects and moves water away from a foundation.*

Various types of drain tile are available. Flexible high-density polyethylene (HDPE) drain tile are slotted all the way around. Surface water enters drain tile slots along the upper edge and is directed away from the foundation through slots along the bottom edge. One type of rigid polyvinyl chloride (PVC) drain tile has holes around the entire circumference and directs surface water away from the foundation in a similar manner to HDPE drain tile. Drain tile used in fine soils are available with a cloth sock that surrounds the tile to prevent small particles from clogging the holes.

Installing Drain Tile. When installing drain tile, a 3″ to 5″ layer of gravel is placed at the bottom of the footing trench. The drain tile is placed on top of the gravel base. For best results, the bottom of the drain tile should be flush with the bottom of the footing. Drain tile is covered with a 4″ layer of gravel.

Sump Well

A *sump well* may be required for full-basement foundations when soil conditions do not provide adequate drainage of surface water away from a building. A sump well is a pit that collects surface water, which is then pumped to the outside of the building where it can drain properly. **See Figure 41-7.** A sump well is commonly lined with an HDPE liner, which has molded outlets for drain tile connections. Drain tile divert water to the sump well, where it collects and is then drained to the outside.

Based on the local building code, the liner is placed with the top flush with the foundation slab or slightly above the slab. A 4″ gravel base is placed around and below the sump well.

TUFF-N-DRI®

Figure 41-5. *Rigid foundation insulation panels are applied over liquid membranes.*

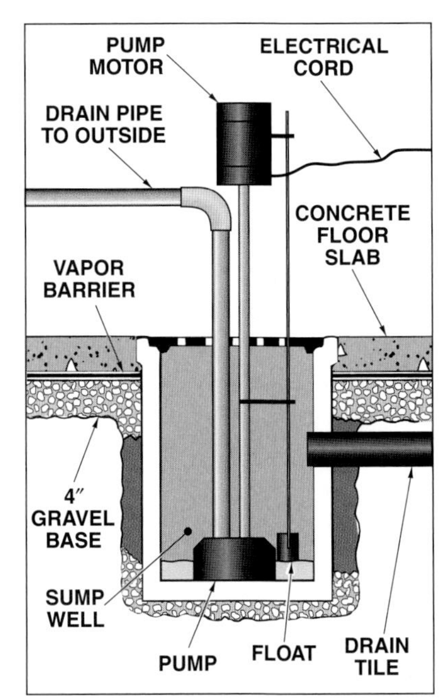

Figure 41-7. *A sump well is commonly installed in the floor of a full-basement foundation system. Water flows through drain tile and collects in the sump well. A sump pump is installed in the well to expel the water to the outside of the building where it can properly drain.*

MOLD PREVENTION

Mold is becoming an increasing concern in building construction. Many construction materials are exposed to adverse weather conditions, such as rain and snow, and take on moisture during exposure. Some construction materials, such as framing members, are enclosed behind finish materials, resulting in conditions prone to mold growth. Using proper construction materials and techniques can help reduce susceptibility to mold.

In order for mold to grow, it requires a nutrient source, the appropriate temperature, and moisture. Materials containing cellulose, including paper, wood paneling, plywood, and OSB, are a good source of nutrition for mold. Traditional construction materials contain natural chemicals that resist mold growth. Synthetic construction materials, including pressed-wood products, do not contain these natural chemicals and are more susceptible to mold growth.

Mold can grow at a wide range of temperatures, depending on the type of mold. Mold can survive and grow at temperatures as low as 50°F or as high as 122°F. Since buildings are typically maintained at 65°F to 75°F, building temperatures are appropriate for mold growth.

Of the three requirements for mold growth, the only one that can be controlled is moisture. Sufficient moisture must be maintained for a period of time for mold to grow. Building moisture is affected by three factors:

- water infiltration
- building tightness
- moisture condensation

The building envelope must be properly enclosed to minimize water infiltration into a building and water accumulation on mold-susceptible materials. Water may penetrate a building envelope at the following locations:

- roof leaks
- foundation leaks
- poorly designed or missing flashing
- door and window leaks

Regardless of the location or the cause, water infiltration must be prevented.

New and improved materials and techniques have contributed to the increasing tightness of buildings. In response to the energy crisis in the 1970s and the green building movement's growing influence, energy conservation and efficiency are vital to today's buildings. Therefore, new materials and techniques have been developed to minimize air infiltration into a building. In using these materials and techniques, less air is exchanged between the indoor and outdoor space. Since less air moves in and out of a building, mold spores have a greater opportunity to grow.

In cold climates, moisture can accumulate in and on internal surfaces of external walls. When the temperature is at or below the dewpoint temperature of the room, condensation may occur. This condensation provides the moisture required for mold to grow.

In warm climates, moisture may accumulate on the back side of internal surfaces of external walls. The wall surface is cooled by air conditioning, moisture is not allowed to escape due to low-permeable wall covering, and the moisture will condense on the back side of the wall material. This condensation also provides moisture for mold growth.

INSECT PREVENTION

Insects, such as termites, carpenter ants, wood wasps, and wood-boring beetles, can cause serious structural damage to wood buildings. Although most insect infestation occurs after construction is completed, precautions can be taken during excavation and foundation construction to minimize insect damage later.

Termites

Termites are small insects that can cause serious structural damage to wood buildings. Termite species frequently encountered include eastern subterranean, Formosan, and drywood termites.

Two general categories of termites are *subterranean* and *nonsubterranean*. Subterranean termites account for 95% of termite damage in the United States and require warm and moist conditions to survive. In the past, subterranean termites were a serious problem in the southern and western United States. In recent years, however, subterranean termites have become more active in northern regions. Improved methods of heating and insulating buildings have provided warm and favorable conditions for year-round termite activity.

Subterranean termites live in underground nests and build tunnels to travel to the earth surface. Termites attack the wood portion of a building in several ways. If the wood is in direct contact with the ground, termites can simply penetrate the wood. Termites also work their way up through small cracks in foundation walls and often build small mud tunnels up the outside of a foundation wall until they reach wood. Termites consume the interior of wood members but leave the outside shell to protect themselves. For this reason, termite attack may go unnoticed until very serious damage has occurred. **See Figure 41-8.**

Nonsubterranean, or drywood, termites have the ability to live without moisture or contact with the ground. Nonsubterranean termites can also cause damage, but they are responsible for much less damage than subterranean termites.

Termite Prevention. Termite prevention begins with proper foundation construction. Wood or wood-based materials placed too close to the ground increase the probability of termite attack. Local building codes usually specify the clearances required between wood members and the finished grades around a foundation.

Since subterranean termites require moist conditions to survive, moisture control is an effective preventive measure, especially for buildings with crawl-space foundations. The ground should be graded to prevent water accumulation inside foundation walls. A properly installed vapor barrier is also helpful to control moisture. An adequate number of vents should be provided to ensure good circulation of air beneath the building.

Pressure-treated lumber should be used for wood members close to or in contact with the ground. Chemicals forced into the wood fibers increase resistance to termites.

Local building codes may require metal *termite shields.* Termite shields are placed between the sill plate and top of a foundation wall. The shield should be made of 24 ga galvanized iron, extend 2″ on each side of the wall, and be bent down at a 45° angle. Holes drilled in the termite shield for foundation bolts must be properly sealed.

The use of soil treatment to prevent termites has increased in recent years. Special soil poisons are applied to the ground during construction. The poisons form a chemical barrier that is toxic to termites and other harmful soil insects. Refer to the local building code to determine whether soil poisons are permitted. The most effective termite prevention combines a number of the previous recommendations. **See Figure 41-9.**

Orkin

Figure 41-8. *Termite damage may go unnoticed until the wood begins to crumble.*

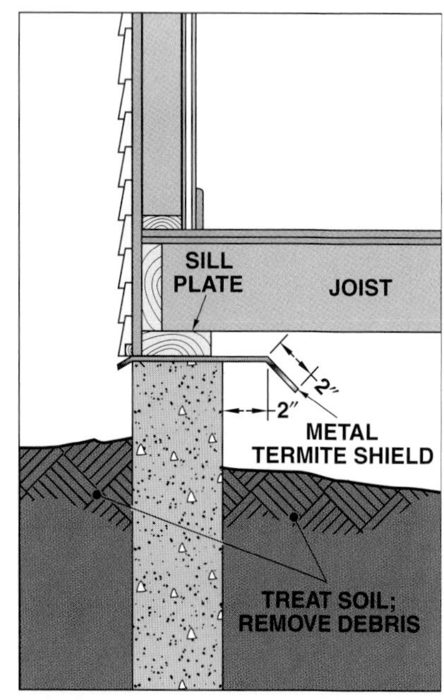

Figure 41-9. *The most effective termite prevention includes treated lumber close to the soil, a metal termite shield, and chemical treatment of the soil surrounding the upper part of the foundation wall.*

Wood-Boring Insects

Although not as destructive as termites, a variety of other wood-boring insects can cause structural and cosmetic damage to a building. In their natural habitat, most wood-boring insects feed on the wood and bark of trees and shrubs. A building may be infested by wood-boring insects that have survived in lumber or may have been brought into a building with firewood. Other wood-boring insects are attracted to moist conditions. Examples of common wood-boring insects are *carpenter ants, wood wasps,* and *wood-boring beetles*.

Southern Forest Products Association

All lumber and plywood components of a building with a permanent wood foundation (PWF) must be pressure-treated to withstand decay from moisture and insect damage.

Carpenter Ants. Carpenter ants are black or red and black in color and range in size from ¼″ to ¾″ long. Winged carpenter ants are commonly mistaken for termites. **Figure 41-10** provides a comparison between carpenter ants and termites. Buildings near wooded areas or brush-covered vacant lots are most vulnerable to carpenter ant infestation. Carpenter ants may migrate to buildings if there is a nearby ant colony and enter buildings through holes used for utility wiring and plumbing. Carpenter ants use wood as a nesting site, but do not eat the wood. Instead, they feed on other insects, plant juices, and food scraps.

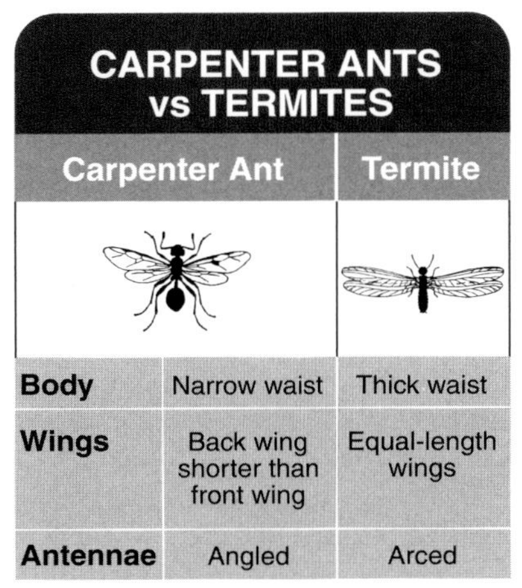

CARPENTER ANTS vs TERMITES

	Carpenter Ant	Termite
Body	Narrow waist	Thick waist
Wings	Back wing shorter than front wing	Equal-length wings
Antennae	Angled	Arced

Figure 41-10. *Carpenter ants and termites are similar in appearance, but have a few distinguishing features.*

The first step in controlling carpenter ant infestation is to locate the nests. Moist locations such as eaves, columns, porch or deck floors, and wood in contact with the ground are common nesting sites. Insecticides are applied to baseboards, moldings, and sill plates to treat carpenter ant infestation. To prevent future infestation, leaks should be repaired and good ventilation should be provided for roofs and crawl-space foundations. Tree branches touching a building and stumps near a building should be removed. Firewood should be stored away from the building and off the ground.

Wood Wasps. Wood wasps, commonly referred to as horntail wasps, are ½″ to 1⅝″ long. **See Figure 41-11.** Although wood wasps are similar in appearance to other varieties of wasps, they are not aggressive and do not sting humans. Adult female wasps deposit eggs into dead or dying trees. Larvae develop from the eggs and feed on fungus within the wood, creating cylindrical tunnels within the wood. When the larva develops into an adult wasp, it chews its way out of the tunnel and will not return to deposit more eggs in the wood.

Figure 41-11. *Wood wasps cause cosmetic damage to structures.*

Wood wasps leave ⅛″ to ¼″ diameter holes when they emerge from wood. Typically, there are not enough holes to cause structural damage to the building. However, the holes may appear on the interior of a building. Wood wasps are known to bore through painted walls, linoleum, carpet, gypsum board, plaster, and nonceramic tile. Since wood wasps do not return to the wood from which they emerge, controlling existing infestation is not an issue. In new construction, the use of kiln-dried lumber will ensure no infestation in the lumber.

Wood-Boring Beetles. Wood-boring beetles feed on old and weakened trees, as well as fire- and insect-damaged trees. Typically, wood-boring beetles do not attack harvested lumber. However, the presence of wood-boring beetles in structural lumber comes from harvesting lumber from infested logs. Wood-boring beetles are difficult to control once they infest the lumber.

Common wood-boring beetles include *powderpost beetles, false powderpost beetles,* and *deathwatch beetles.* **See Figure 41-12.** Powderpost beetles are reddish-brown to black in color and range from ⅛″ to ¼″ long. Powderpost beetles prefer hardwoods over softwoods since hardwoods have larger pores into which female beetles can lay their eggs. Since softwoods have smaller pores and have a low starch content, powderpost beetles typically do not infest softwoods. Tunnels produced by powderpost beetles run parallel to the wood grain and exit the wood as 1⁄16″ diameter holes.

False powderpost beetles are dark brown in color, and may have red mouths, legs, and antennae. False powderpost beetles range in length from ¼″ to 2″. False powderpost beetles infest hardwood floors, paneling, and furniture. Adults of some species of false powderpost beetles are known to bore through soft metals such as lead.

Deathwatch beetles are dark brown with patches of yellow hair and range in length from ¼″ to ⅜″. Deathwatch beetles prefer wood with a moisture content greater than 14%. They are primarily found in softwood lumber including studs, beams, and floor systems.

Preventive measures to minimize infestation should be taken at all stages of lumber manufacture and handling. While kiln drying will destroy any infestation, it will not prevent a reoccurrence. Dead tree limbs and scrap lumber around buildings or near lumber storage areas should be removed. Wood can be protected by painting or varnishing to seal pores, cracks, or holes where eggs could be laid.

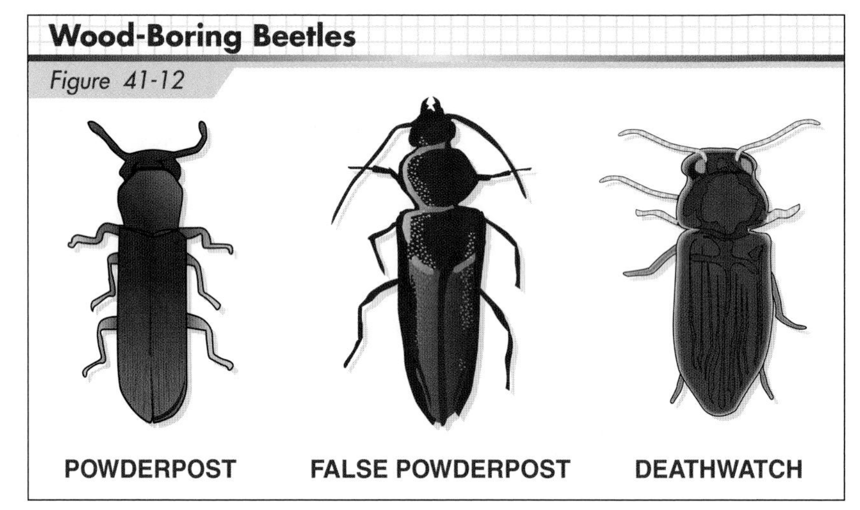

Figure 41-12. *Wood-boring beetles feed on old and weakened trees, as well as fire- and insect-damaged trees.*

Unit 41 Review and Resources

SECTION

Floor, Wall, and Ceiling Frame Construction

CONTENTS

OBJECTIVES

1. Identify basic parts of floor units.
2. Define cripple walls and describe their uses.
3. Describe how wood posts are installed to support floor units.
4. Explain the purpose of wood beams to support floor units.
5. Describe the use of steel pipe columns and steel beams to support floor units.
6. Explain the installation of wood posts and beams.
7. Explain the purpose and installation of floor joists.
8. Describe how joists are supported.
9. Explain the use of bridging.
10. Describe construction of floor, fireplace, and chimney openings in floor units.
11. Review the placement of floor joists.
12. Explain the purpose of a subfloor.
13. Describe how subfloor panels are installed.
14. Explain the purpose of floor underlayment.
15. Describe the use of floor trusses.
16. List the advantages of wood I-joists.
17. Explain installation methods for wood I-joists.

Floor framing begins after the foundation work has been completed. In platform construction, the floor unit is framed directly over the foundation walls or short stud walls used as cripple walls. Most floor units include *posts, beams, joists, bridging* or *blocking,* and a *subfloor.* **See Figure 42-1.** Posts and beams support the lapped or butted ends of the joists, or they may provide central support for long joist spans. Bridging or blocking keeps the joists aligned and helps to distribute the load carried by the floor unit. The subfloor is the wood deck that rests on top of the joists.

FLOOR UNIT RESTING ON SILL PLATES

The floor unit may be framed directly on the sill plates of a building with a crawl-space foundation and low foundation walls. **See Figure 42-2.** For a crawl-space foundation, one end of the joists rests on the outside foundation walls. The lapped ends rest on top of an interior foundation wall running down the center of the building. Posts and beams provide midspan support to the long span of the joists. There should be at least 18″ clearance between the bottoms of the floor joists and the soil, and at least 12″ between the bottom of the beam and the soil.

The floor unit may also be framed directly on the sill plates of a building with a full-basement foundation and high foundation walls. In this case, one end of the joists laps over a beam.

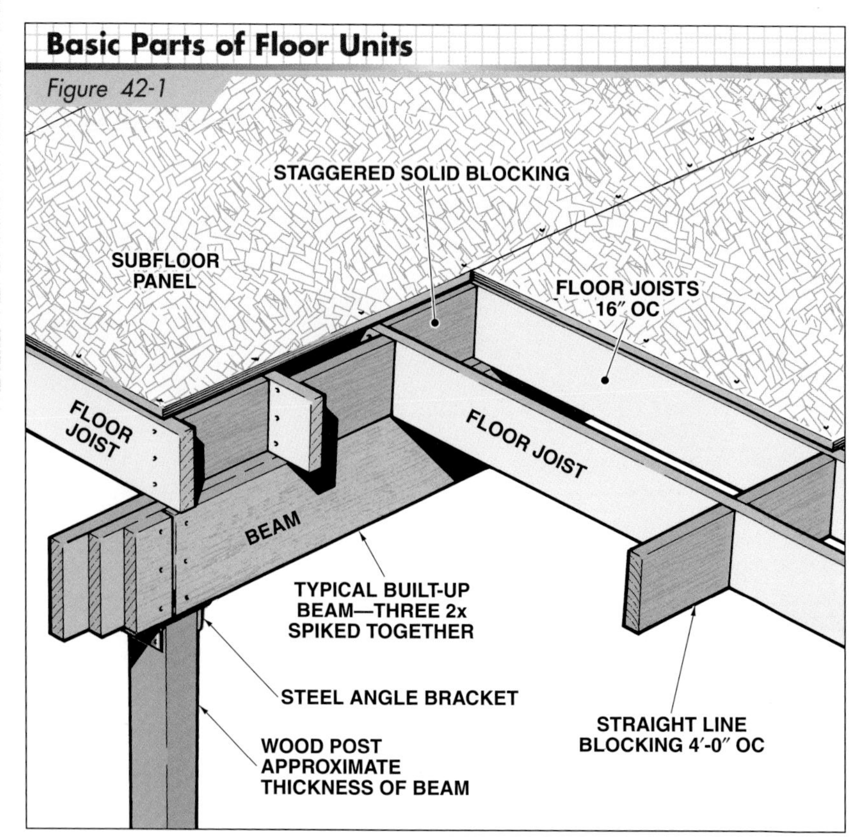

Figure 42-1. *Basic parts of a floor unit include posts, beams, joists, bridging or blocking, and a subfloor. Floor framing begins after foundation work is completed.*

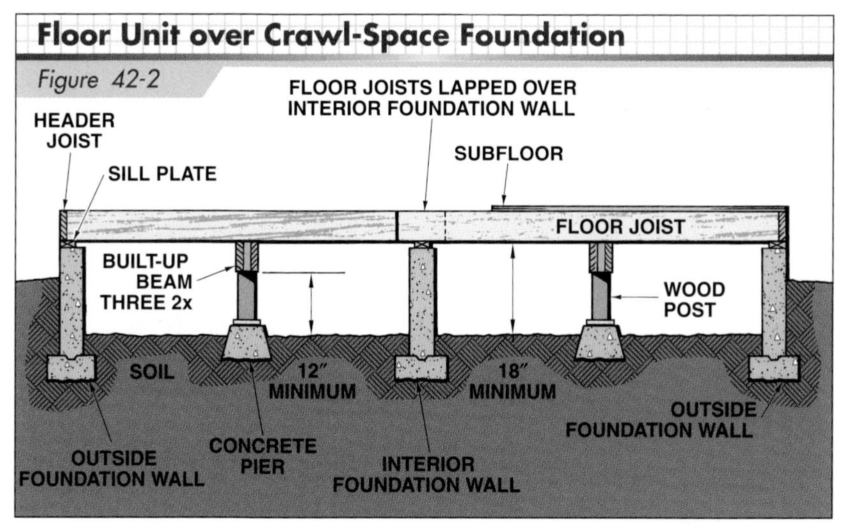

Figure 42-2. *When constructing a floor unit resting on sill plates with a crawl-space foundation underneath, an 18″ minimum clearance should be maintained between the bottoms of floor joists and the soil, and 12″ between the bottom of the beam and the soil.*

FLOOR UNIT RESTING ON CRIPPLE WALLS

Cripple walls are used to extend the height of low foundation walls, usually for a full-basement house. **See Figure 42-3.** Under some conditions, cripple walls are less costly than the additional concrete and formwork required for higher foundation walls. Cripple wall studs are toe-nailed to the sill plates and are commonly spaced 16″ OC. Cripple wall studs should be no smaller in cross section than wall studs above the floor. Floor joists placed on top of the wall should be directly above the cripple wall stud (stacked).

A stepped foundation is another example of where cripple walls are used to reduce construction costs. Stepped foundations are needed on hillsides and steeply sloping lots. In a typical stepped foundation, the floor unit rests directly on the sill plate at the highest area of the foundation. Cripple walls are placed over the lower walls, and in that area the floor unit rests on the cripple walls.

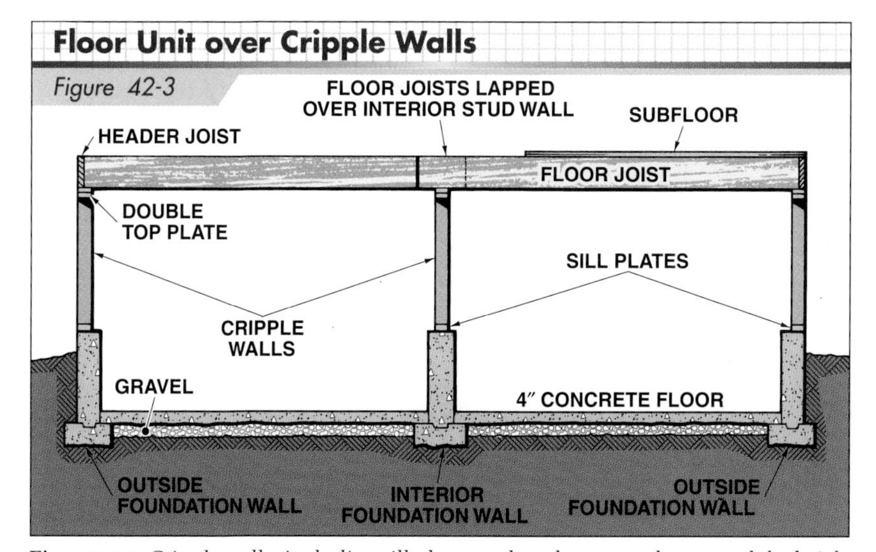

Figure 42-3. *Cripple walls, including sill plates and studs, are used to extend the height of low foundation walls.*

POSTS AND BEAMS

Posts and beams help support the floor joists and subfloor. Posts and beams may be steel or wood, and their size depends upon the load they are required to carry. The dimensions and placement of beams are shown on a foundation plan.

Wood Posts

Wood posts are placed directly below wood beams. In general, the width of the wood post should be approximately equal to the width of the beam it supports. A 4″ wide beam requires 4 × 4 or 4 × 6 posts. An 8″ wide beam requires 6 × 8 or 8 × 8 posts.

The bottom of each post may be nailed to a pier block that is secured to the top of a concrete pier. Another method is to place a ½″ steel dowel in the concrete pier at the time concrete is placed. The dowel fits into a hole drilled at the bottom of the post and holds the post in position. **See Figure 42-4.** The dowel should extend at least 3″ into the concrete and 3″ into the post.

Fixed metal post bases are placed when concrete is placed. An adjustable post base is placed after the concrete has set. A J-bolt may be set into the concrete as it is placed, or a hole can be drilled and a threaded rod fixed into place with epoxy. **Figure 42-5** shows two types of metal bases that are set into a concrete pier when concrete is placed. A wood post fits into the base and is secured to the base with lag bolts. No bolts or other fasteners are required to attach the base to the concrete. These bases must be positioned in the concrete very accurately.

Lumber should not be in direct contact with the ground while being stored. Place lumber on stringers to allow air circulation.

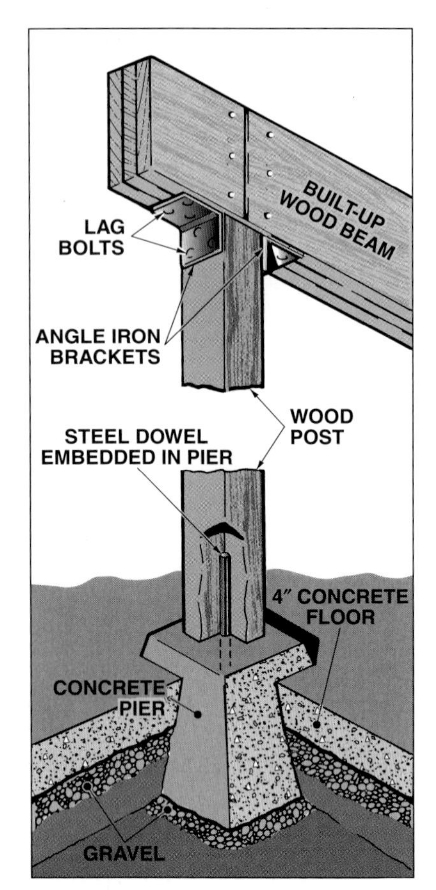

Figure 42-4. *Posts are fastened at the top of the pier and at the bottom of the beam.*

An adjustable, or elevated, metal base may also be used to support a wood post and provide moisture protection for the post. An adjustable base is secured to concrete using a ½″ J-bolt set in the concrete when concrete is placed. A standoff plate provides a flat bearing area for the post and keeps the post 1³⁄₁₆″ above the concrete surface to guard against wood rot and termite damage at the bottom of the post. A slotted adjustment plate permits movement for plumbing the post.

The top of each post is fastened to the bottom of the beam. Angle iron or a metal connector is typically used to create a strong tie between the beam and post.

Figure 42-6 shows two types of metal post caps used to tie together wood posts and beams. Post caps are nailed or bolted to the posts and beams. The twin-design post cap can be installed after the beam has been placed on top of the post since it is available in two pieces.

Wood Beams

Wood beams, also referred to as *girders,* may consist of a solid timber, built-up lumber, or engineered products such as glulam or laminated veneer lumber. A *built-up beam,* for example, may be fabricated with two or three or more planks. Joints between the planks are staggered. When assembling a built-up beam, the joints of the planks should fall directly over a post. **See Figure 42-7.** Three 16d nails are driven at the ends of the planks, and other nails are staggered 32″ OC.

Western Wood Products Association

Heavy timber posts and beams require similar, but heavier gauge, post bases and caps.

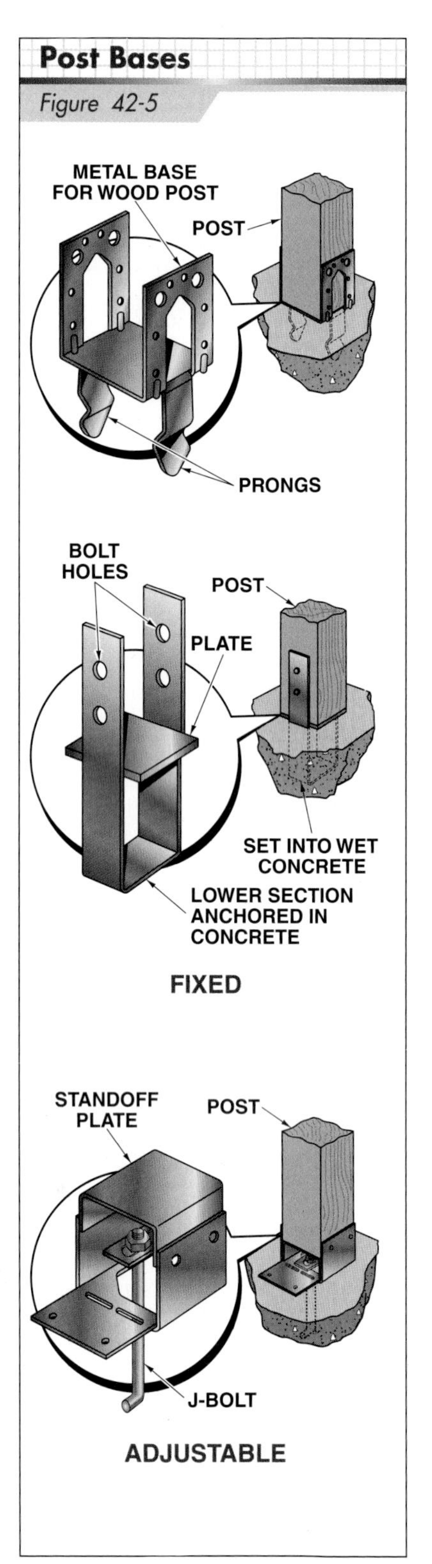

Figure 42-5. *No bolts or other types of fasteners are necessary to secure the fixed post bases to concrete; the lower section of the base is set into a concrete pier when concrete is placed. An adjustable post base should have 1″ to 2½″ standoff height above the concrete. An adjustable base is secured in position by drilling a hole into hardened concrete and fixing a threaded rod into place with epoxy, or by setting a J-bolt into fresh concrete.*

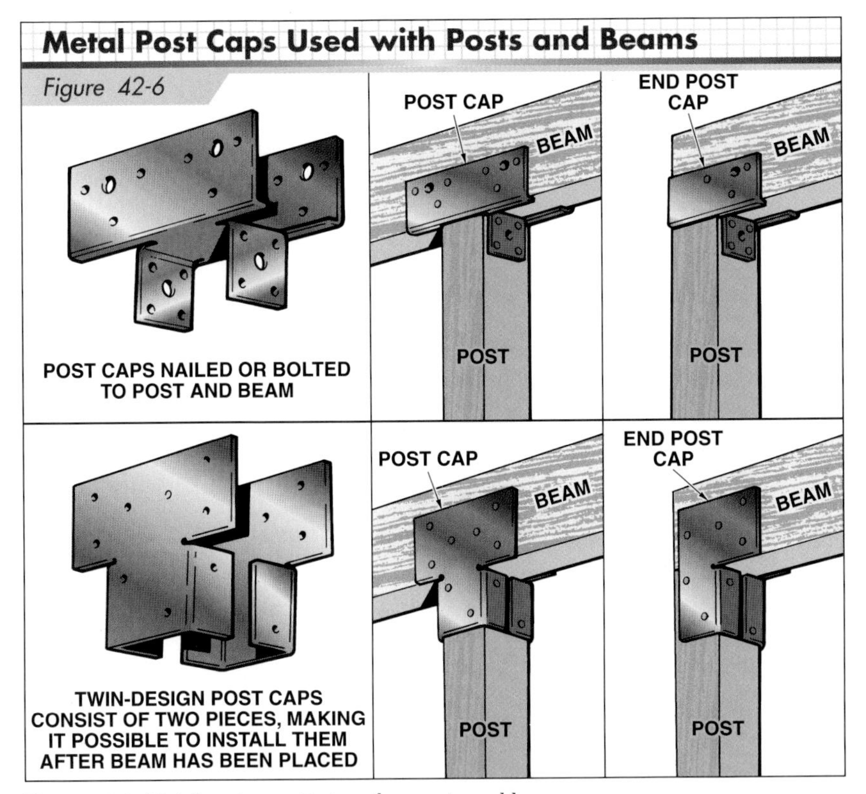

Figure 42-6. *Metal post caps tie together posts and beams.*

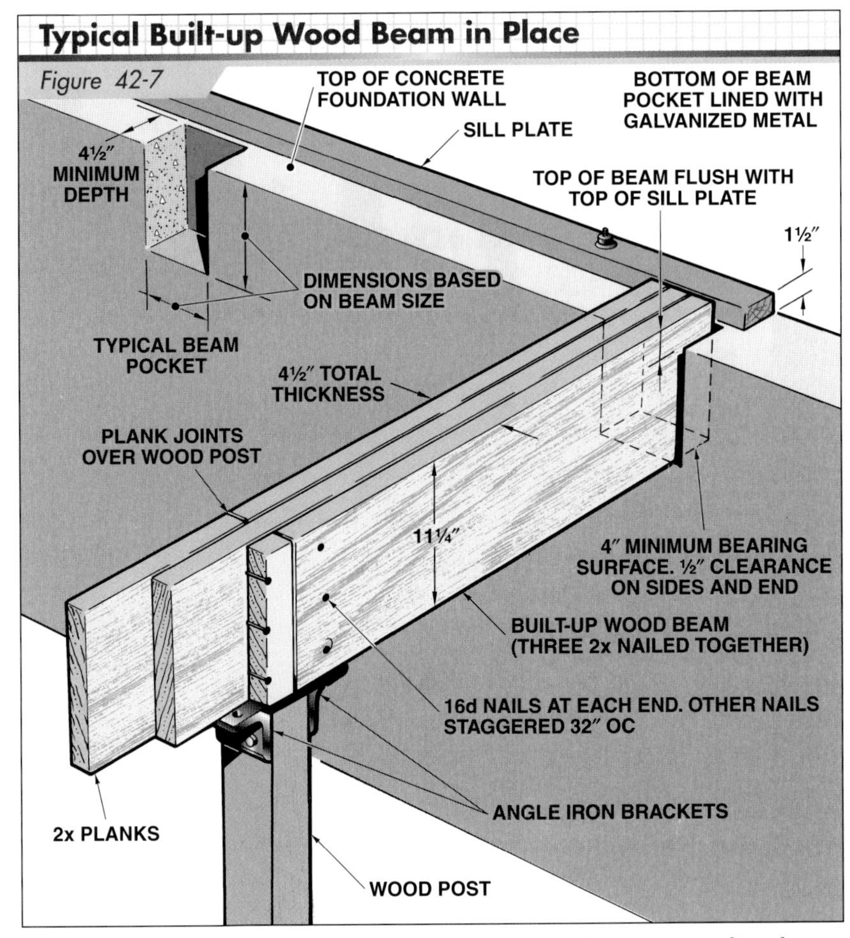

Figure 42-7. *Joints on a built-up beam are staggered. Plank joints occur directly over a post.*

Glulam beams and laminated veneer lumber (LVL) beams are used for residential and heavy timber construction. Glulam beams and LVL beams consist of flat pieces that are glued and pressed together under intense pressure.

The ends of a beam often rest in pockets formed in a concrete wall. When beam pockets are used, the beam ends must bear at least 4″ on the wall, and the pocket should be large enough to provide ½″ clearance around sides and end of the beam. The ends of the beam should be treated with a preservative to avoid termite damage. As a further precaution, the beam pocket should be lined with galvanized metal.

Beams are classified as *non-load-bearing* and *load-bearing* according to the amount and type of load they support. Non-load-bearing beams are for cosmetic purposes only and do not support loads. Load-bearing beams must support a wall framed directly above, as well as the live load and dead load of the floor. The *dead load* is the weight of the material used for the floor unit itself. The *live load* is the weight of people, furniture, appliances, and other items placed on the floor after its construction.

Allowable Spans for Beams. The *allowable span,* also referred to as the *clear opening,* is the distance between supporting posts permitted for different size beams. Factors such as beam size, grade of lumber, and weight of live and dead loads determine allowable beam spans. Tables are available for determining the allowable beam size for different species of lumber (and steel beams) and different live loads and dead loads. The table in **Figure 42-8** is used to determine allowable beam size for Southern

Pine beams supporting a 40 pound per square foot (psf) live load and 10 psf dead load. The *load duration factor* shown at the top of the table is an engineering calculation relating to the capability of the wood to carry maximum loads for longer or shorter durations.

Beams should not be cut, drilled, or notched.

Determining Allowable Beam Sizes

Figure 42-8

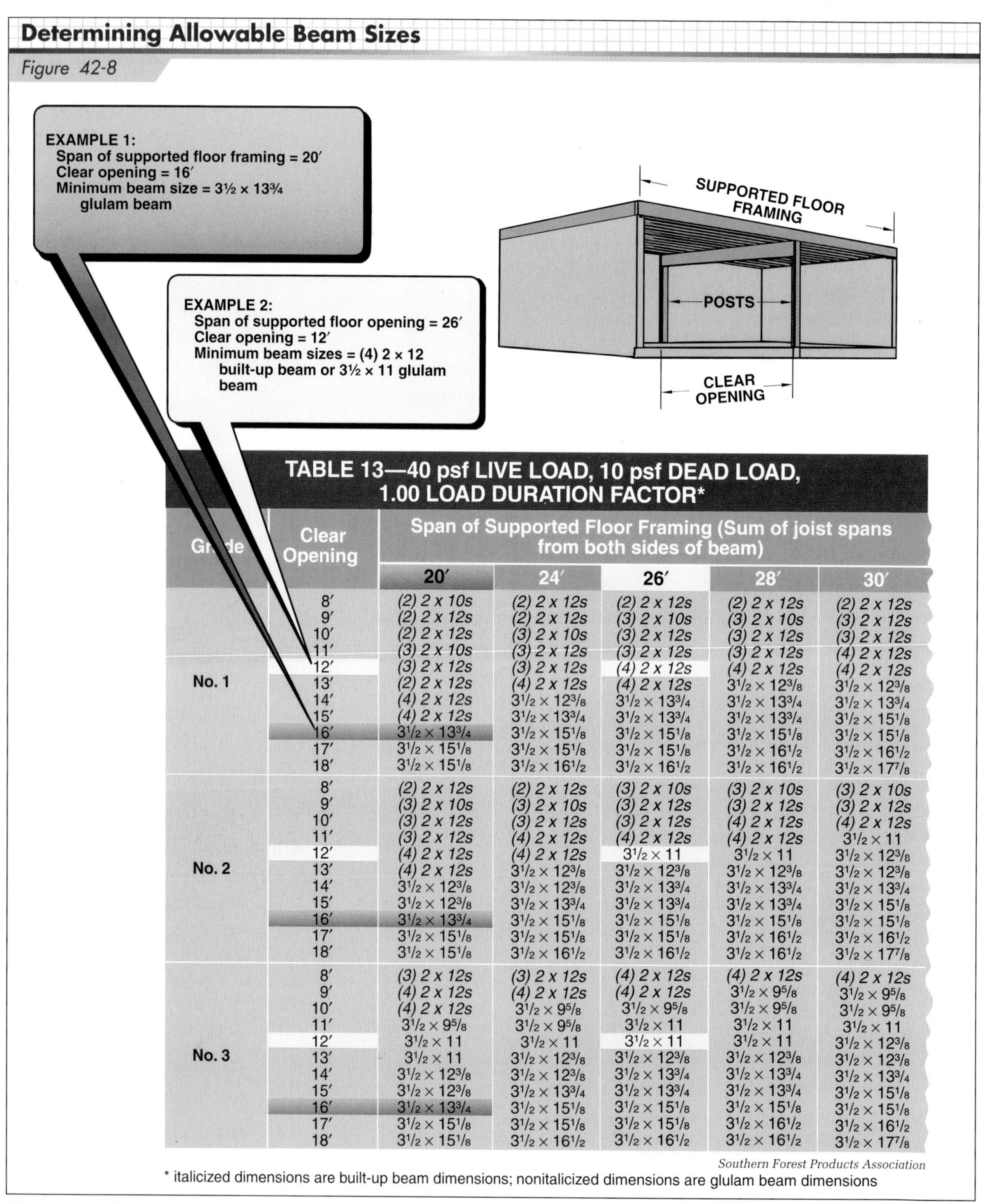

TABLE 13—40 psf LIVE LOAD, 10 psf DEAD LOAD, 1.00 LOAD DURATION FACTOR*

Grade	Clear Opening	Span of Supported Floor Framing (Sum of joist spans from both sides of beam)				
		20′	24′	26′	28′	30′
No. 1	8′	*(2) 2 x 10s*	*(2) 2 x 12s*	*(2) 2 x 12s*	*(2) 2 x 12s*	*(2) 2 x 12s*
	9′	*(2) 2 x 12s*	*(2) 2 x 12s*	*(3) 2 x 10s*	*(3) 2 x 10s*	*(3) 2 x 12s*
	10′	*(2) 2 x 12s*	*(3) 2 x 10s*	*(3) 2 x 12s*	*(3) 2 x 12s*	*(3) 2 x 12s*
	11′	*(3) 2 x 10s*	*(3) 2 x 12s*	*(3) 2 x 12s*	*(3) 2 x 12s*	*(4) 2 x 12s*
	12′	*(3) 2 x 12s*	*(3) 2 x 12s*	*(4) 2 x 12s*	*(4) 2 x 12s*	*(4) 2 x 12s*
	13′	*(2) 2 x 12s*	*(4) 2 x 12s*	*(4) 2 x 12s*	3½ × 12⅜	3½ × 12⅜
	14′	*(4) 2 x 12s*	3½ × 12⅜	3½ × 13¾	3½ × 13¾	3½ × 13¾
	15′	*(4) 2 x 12s*	3½ × 13¾	3½ × 13¾	3½ × 13¾	3½ × 15⅛
	16′	3½ × 13¾	3½ × 15⅛	3½ × 15⅛	3½ × 15⅛	3½ × 15⅛
	17′	3½ × 15⅛	3½ × 15⅛	3½ × 15⅛	3½ × 16½	3½ × 16½
	18′	3½ × 15⅛	3½ × 16½	3½ × 16½	3½ × 16½	3½ × 17⅞
No. 2	8′	*(2) 2 x 12s*	*(2) 2 x 12s*	*(3) 2 x 10s*	*(3) 2 x 10s*	*(3) 2 x 10s*
	9′	*(3) 2 x 10s*	*(3) 2 x 10s*	*(3) 2 x 12s*	*(3) 2 x 12s*	*(3) 2 x 12s*
	10′	*(3) 2 x 12s*	*(3) 2 x 12s*	*(3) 2 x 12s*	*(4) 2 x 12s*	*(4) 2 x 12s*
	11′	*(3) 2 x 12s*	*(4) 2 x 12s*	*(4) 2 x 12s*	*(4) 2 x 12s*	3½ × 11
	12′	*(4) 2 x 12s*	*(4) 2 x 12s*	3½ × 11	3½ × 11	3½ × 12⅜
	13′	*(4) 2 x 12s*	3½ × 12⅜	3½ × 12⅜	3½ × 12⅜	3½ × 12⅜
	14′	3½ × 12⅜	3½ × 12⅜	3½ × 13¾	3½ × 13¾	3½ × 13¾
	15′	3½ × 12⅜	3½ × 13¾	3½ × 13¾	3½ × 13¾	3½ × 15⅛
	16′	3½ × 13¾	3½ × 15⅛	3½ × 15⅛	3½ × 15⅛	3½ × 15⅛
	17′	3½ × 15⅛	3½ × 15⅛	3½ × 15⅛	3½ × 16½	3½ × 16½
	18′	3½ × 15⅛	3½ × 16½	3½ × 16½	3½ × 16½	3½ × 17⅞
No. 3	8′	*(3) 2 x 12s*	*(3) 2 x 12s*	*(4) 2 x 12s*	*(4) 2 x 12s*	*(4) 2 x 12s*
	9′	*(4) 2 x 12s*	*(4) 2 x 12s*	*(4) 2 x 12s*	3½ × 9⅝	3½ × 9⅝
	10′	*(4) 2 x 12s*	3½ × 9⅝	3½ × 9⅝	3½ × 9⅝	3½ × 9⅝
	11′	3½ × 9⅝	3½ × 9⅝	3½ × 11	3½ × 11	3½ × 11
	12′	3½ × 11	3½ × 11	3½ × 11	3½ × 11	3½ × 12⅜
	13′	3½ × 11	3½ × 12⅜	3½ × 12⅜	3½ × 12⅜	3½ × 12⅜
	14′	3½ × 12⅜	3½ × 12⅜	3½ × 13¾	3½ × 13¾	3½ × 13¾
	15′	3½ × 12⅜	3½ × 13¾	3½ × 13¾	3½ × 13¾	3½ × 15⅛
	16′	3½ × 13¾	3½ × 15⅛	3½ × 15⅛	3½ × 15⅛	3½ × 15⅛
	17′	3½ × 15⅛	3½ × 15⅛	3½ × 15⅛	3½ × 16½	3½ × 16½
	18′	3½ × 15⅛	3½ × 16½	3½ × 16½	3½ × 16½	3½ × 17⅞

Southern Forest Products Association

* italicized dimensions are built-up beam dimensions; nonitalicized dimensions are glulam beam dimensions

Figure 42-8. *Allowable beam sizes are based on the span of supported flooring, clear opening, lumber grade, and live and dead loads. This table is for a floor beam supporting a 40 psf live load and 10 psf dead load.*

Steel Pipe Columns

Steel pipe *(Lally)* columns are often used as posts in wood-framed buildings, and can be placed beneath either steel or wood beams. **See Figure 42-9.** The base of a steel pipe column is bolted to the top of a pier or floor slab. The bolts holding the base must be accurately set into the concrete during concrete placement. The cap at the top of a steel pipe column is secured with machine bolts and nuts or by welding when it is attached to a steel beam. Lag bolts are used when the columns are attached to a wood beam. **Figure 42-10** shows a hollow steel section (HSS) column supporting an LVL beam.

Steel Beams

A standard steel beam, called a *wide-flange beam,* is commonly used with wood framing. Wood joists either rest on top of the beam or butt against the sides of the beam. **Figure 42-11** shows a wood plate attached to the top of a steel beam to provide a nailing base for joists.

A wide-flange beam is a structural steel member with parallel upper and lower flanges that are joined with a web. On construction prints, a wide-flange beam is designated with the letter "W" followed by the measurement to the outside edges of the flanges and the weight per running foot. For example, a W14 × 34 beam measures 14″ outside the flanges and weighs 34 lb/ft.

Placing Wood Posts and Beams

Posts must be cut to length and set up before the beams can be installed. The upper surface of the beam may be aligned with the upper surface of the foundation sill plate, or the beam ends may rest on top of the walls. Long beams must be placed in sections. Solid beams must be measured and cut so that the ends will fall over the center of a post. Built-up beams should be placed so that their joints fall over the posts.

A ½″ clearance must be provided at the ends of wood beams to allow for expansion and contraction of the beams and to allow air circulation.

Trus Joist, A Weyerhaeuser Business

Parallel strand lumber (PSL) is commonly used as load-bearing beams that span longer areas.

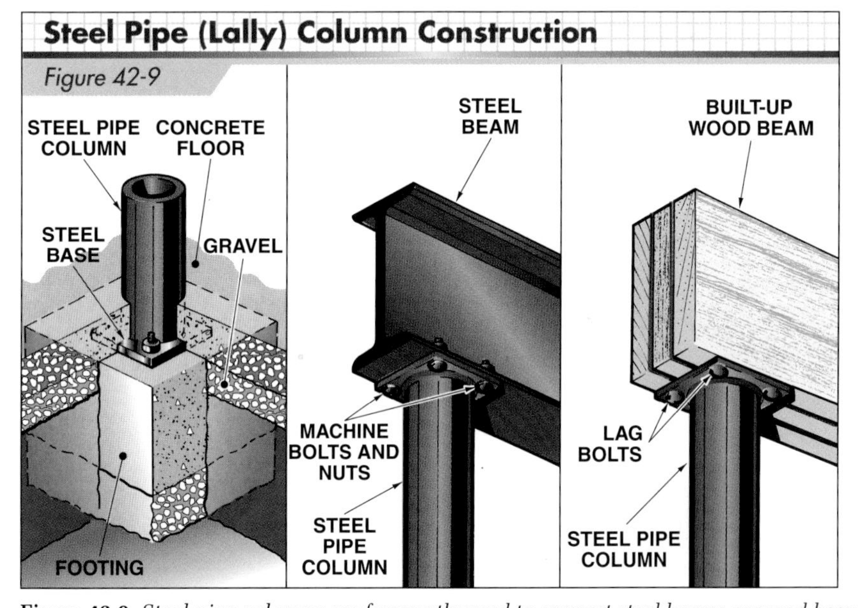

Figure 42-9. *Steel pipe columns are frequently used to support steel beams or wood beams.*

Figure 42-10. *A hollow steel section column may be used to support the end of an LVL beam.*

Figure 42-11. *Wood plates may be attached to the top of wide-flange steel beams with a powder-actuated tool.*

One procedure for placing posts and beams using a line stretched from opposite walls is shown in **Figure 42-12**; in this procedure, the upper surface of the beam is aligned with the upper surface of the sill plate. The same procedure can also be performed using a laser level or other surveying equipment.

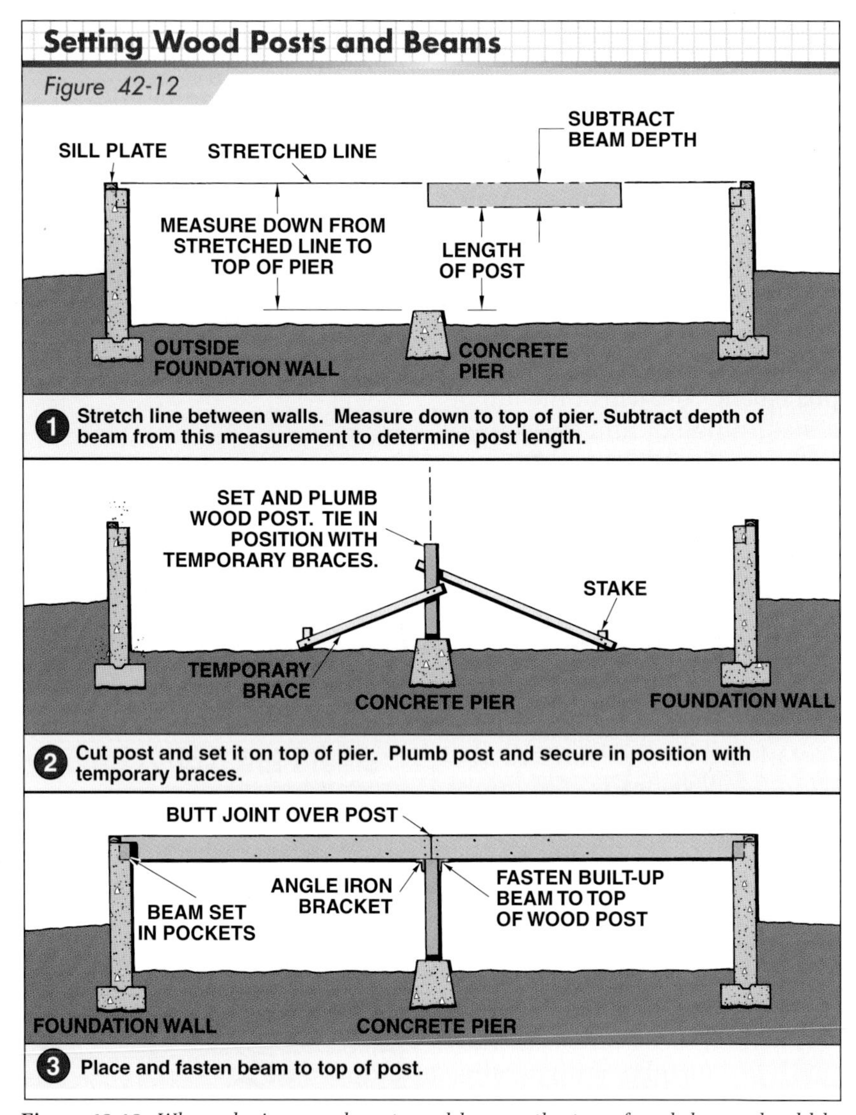

Figure 42-12. *When placing wood posts and beams, the top of each beam should be aligned with the upper surface of the sill plate.*

FLOOR JOISTS

In platform framing, one end of the floor joist rests directly on the sill plate of the exterior foundation wall or on the top plate of a framed outside wall. The bearing surface should be at least 1½″. The opposite end of the joist laps over or butts into an interior beam or wall. The size of joist material (2 × 6, 2 × 10, 2 × 12, etc.) is determined by considering the span and weight of the load to be carried. The foundation plan usually specifies joist size, joist spacing, and direction the joists should travel.

Floor joists are typically spaced 16″ or 19.2″ OC. However, some systems allow wider spacing (24″ to 32″) between floor joists. Narrower joist spacing may be specified based on joist span and anticipated loads.

One end of the floor joists rests directly on the sill plate of an exterior foundation wall.

Allowable Joist Spans

Tables are available for finding the allowable joist spans for different species of lumber as well as different live loads and dead loads. **Figure 42-13** shows a table providing allowable joist spans between exterior walls or posts for a variety of framing lumber species supporting a 40 psf live load and 10 psf dead load. Species, lumber grade, and loads are key factors in determining joist spans.

A ½″ expansion gap must be provided at the end of a wood beam in a beam packet.

Determining Allowable Joist Spans

Figure 42-13

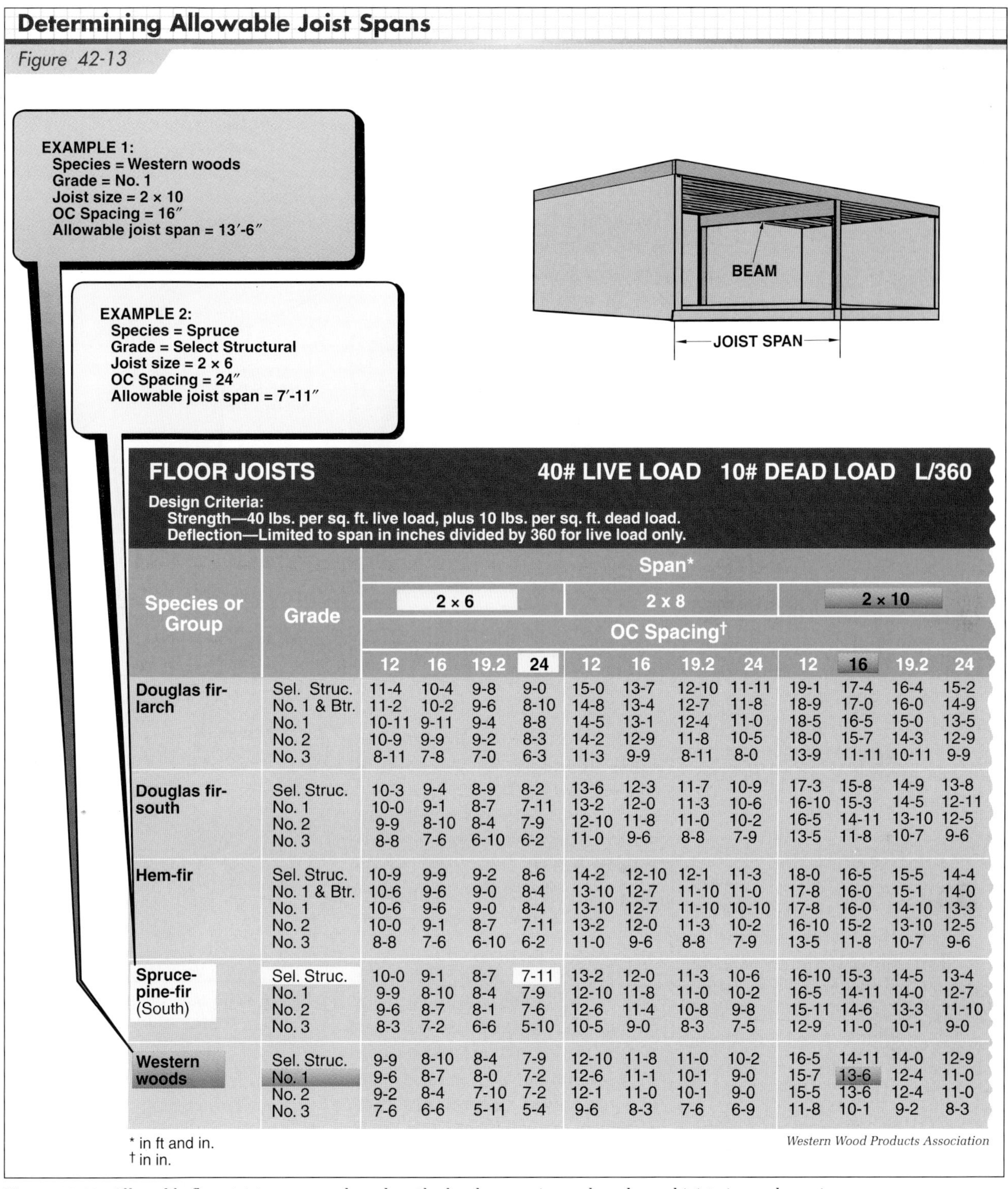

FLOOR JOISTS — **40# LIVE LOAD 10# DEAD LOAD L/360**

Design Criteria:
Strength—40 lbs. per sq. ft. live load, plus 10 lbs. per sq. ft. dead load.
Deflection—Limited to span in inches divided by 360 for live load only.

Species or Group	Grade	Span*											
		2 × 6				2 × 8				2 × 10			
		OC Spacing†											
		12	16	19.2	24	12	16	19.2	24	12	16	19.2	24
Douglas fir-larch	Sel. Struc.	11-4	10-4	9-8	9-0	15-0	13-7	12-10	11-11	19-1	17-4	16-4	15-2
	No. 1 & Btr.	11-2	10-2	9-6	8-10	14-8	13-4	12-7	11-8	18-9	17-0	16-0	14-9
	No. 1	10-11	9-11	9-4	8-8	14-5	13-1	12-4	11-0	18-5	16-5	15-0	13-5
	No. 2	10-9	9-9	9-2	8-3	14-2	12-9	11-8	10-5	18-0	15-7	14-3	12-9
	No. 3	8-11	7-8	7-0	6-3	11-3	9-9	8-11	8-0	13-9	11-11	10-11	9-9
Douglas fir-south	Sel. Struc.	10-3	9-4	8-9	8-2	13-6	12-3	11-7	10-9	17-3	15-8	14-9	13-8
	No. 1	10-0	9-1	8-7	7-11	13-2	12-0	11-3	10-6	16-10	15-3	14-5	12-11
	No. 2	9-9	8-10	8-4	7-9	12-10	11-8	11-0	10-2	16-5	14-11	13-10	12-5
	No. 3	8-8	7-6	6-10	6-2	11-0	9-6	8-8	7-9	13-5	11-8	10-7	9-6
Hem-fir	Sel. Struc.	10-9	9-9	9-2	8-6	14-2	12-10	12-1	11-3	18-0	16-5	15-5	14-4
	No. 1 & Btr.	10-6	9-6	9-0	8-4	13-10	12-7	11-10	11-0	17-8	16-0	15-1	14-0
	No. 1	10-6	9-6	9-0	8-4	13-10	12-7	11-10	10-10	17-8	16-0	14-10	13-3
	No. 2	10-0	9-1	8-7	7-11	13-2	12-0	11-3	10-2	16-10	15-2	13-10	12-5
	No. 3	8-8	7-6	6-10	6-2	11-0	9-6	8-8	7-9	13-5	11-8	10-7	9-6
Spruce-pine-fir (South)	Sel. Struc.	10-0	9-1	8-7	7-11	13-2	12-0	11-3	10-6	16-10	15-3	14-5	13-4
	No. 1	9-9	8-10	8-4	7-9	12-10	11-8	11-0	10-2	16-5	14-11	14-0	12-7
	No. 2	9-6	8-7	8-1	7-6	12-6	11-4	10-8	9-8	15-11	14-6	13-3	11-10
	No. 3	8-3	7-2	6-6	5-10	10-5	9-0	8-3	7-5	12-9	11-0	10-1	9-0
Western woods	Sel. Struc.	9-9	8-10	8-4	7-9	12-10	11-8	11-0	10-2	16-5	14-11	14-0	12-9
	No. 1	9-6	8-7	8-0	7-2	12-6	11-1	10-1	9-0	15-7	13-6	12-4	11-0
	No. 2	9-2	8-4	7-10	7-2	12-1	11-0	10-1	9-0	15-5	13-6	12-4	11-0
	No. 3	7-6	6-6	5-11	5-4	9-6	8-3	7-6	6-9	11-8	10-1	9-2	8-3

* in ft and in.
† in in.

Western Wood Products Association

Figure 42-13. *Allowable floor joist spans are based on the lumber species and grade, and joist size and spacing.*

Supporting Joists over Exterior Walls

Floor joists are supported and held in position over exterior walls by header joists or by *solid blocking* between the joists. Header joists are used more often.

Header Joists. Header joists, also known as *rim* or *band joists*, run along the outside walls. Three 16d nails are driven through the header joists into the ends of the regular joists. **See Figure 42-14.** Header joists prevent the regular joists from rolling or tipping, support the wall above, and fill in the spaces between the regular joists.

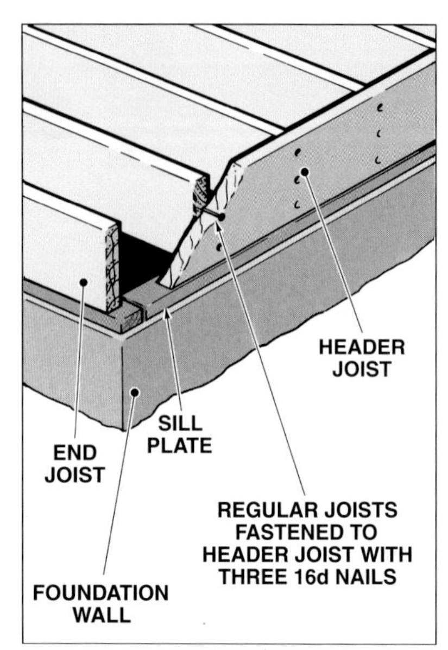

Figure 42-14. *Header joists prevent regular joists from rolling or tipping, support the wall above, and fill spaces between regular joists.*

Blocking between Joists. Another system of providing exterior support to joists is to place solid blocking between the outside ends of the joists. This allows the ends of the joists to have greater bearing on the outside walls.

Interior Support of Joists

Floor joists typically extend across the full width of the building. However, extremely long joists are expensive and difficult to handle. Therefore, two or more shorter joists are used instead. The ends of these joists are supported by lapping or butting them over a beam, butting them against a beam, lapping them over a wall, attaching them over a steel beam, or butting them against a steel beam.

Joists Lapped over Beams or Walls. Often, joists are lapped over a beam running down the center of a building. The lapped ends of the joists may also be supported by an interior foundation or framed wall. It is standard procedure to lap joists the full width of the beam or wall supporting them. **See Figure 42-15.** The minimum lap should be 4″. Joists should not extend more than 4″ past the beam. Solid blocking is installed between the lapped ends as the joists are placed or after all the joists have been nailed down.

Southern Forest Products Association

Figure 42-15. *Joists should be lapped the full width of the beam or wall supporting them.*

Joists Butted over Beams. The ends of the joists can also be butted over a beam. The joists should then be scabbed together with a metal or wood tie. **See Figure 42-16.** The ties can be omitted if the plywood subfloor panels straddle the butt joints.

Joists Butted against Wood Beams

Metal *joist hangers* are required when joists butt against a beam or ledger board. Joist hangers are metal connectors that, when properly installed, provide a stronger joint than nails alone. Joist hangers may also be installed at the connection between wood girders or other joists. Joist hangers are available in many sizes and configurations to support different lumber sizes and engineered wood products, such as wood I-joists. Joist hangers are available with galvanized finishes or as stainless steel.

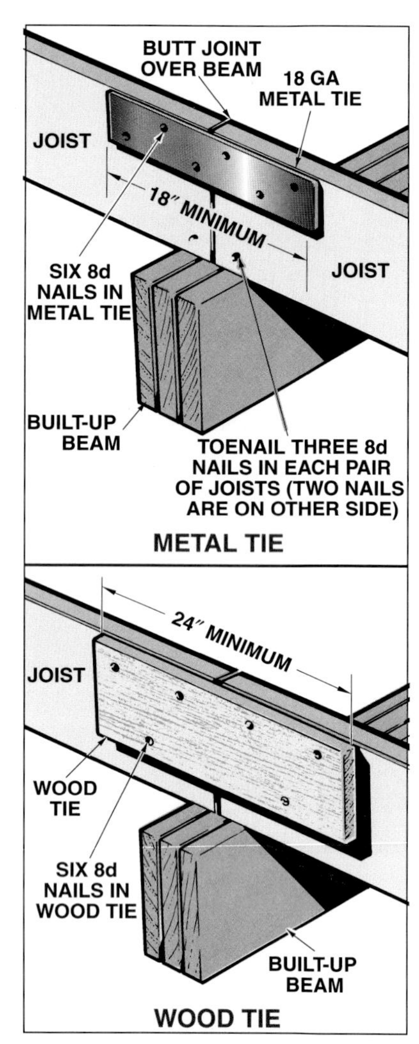

Figure 42-16. *Joists butting over a beam should have the ends scabbed together with a metal or wood tie.*

Face-mount and *top-flange hangers* are the most common types of joist hangers. **See Figure 42-17.** The flanges of face-mount hangers are fastened to the face of a wood beam or ledger board using galvanized nails. Heavy-duty face-mount joist hangers may be fastened to the beams or ledger boards with nails or screws. The flanges of top-flange hangers are fastened to the top of wood beams using galvanized nails. Top-flange hangers are faster and easier to install and support heavier loads than face-mount hangers. Joist hangers are available for solid lumber or engineered lumber for skewed (angled) and adjustable applications.

Double-shear joist hangers provide greater strength than standard joist hangers of the same size and are easier and more economical to install. Joist fastener holes are capped with a dome or tab to guide the nails through the joist and supporting member at a 45° angle, resulting in a strong connection. **See Figure 42-18.**

When installing joist hangers, select the deepest hanger that fits the joist being installed. Galvanized nails should be used to install the joist hangers per the manufacturer's instructions. In some cases, nails may have a mark on the head for ease of identification for inspection after installation. Tables that provide general load and installation information are provided by manufacturers. **See Figure 42-19.** Improperly installed joists may not be able to support anticipated loads and can result in floor squeaks.

A floor framing plan or floor joist framing plan includes information about the spacing and direction of floor framing members including beams, joists, and bridging. Information about framing for floor openings, such as those required for stairways, is also included on framing plans. In addition, information regarding metal connectors and their nailing schedule is often included on a floor framing plan. Unfortunately, framing plans are not always included in a set of plans.

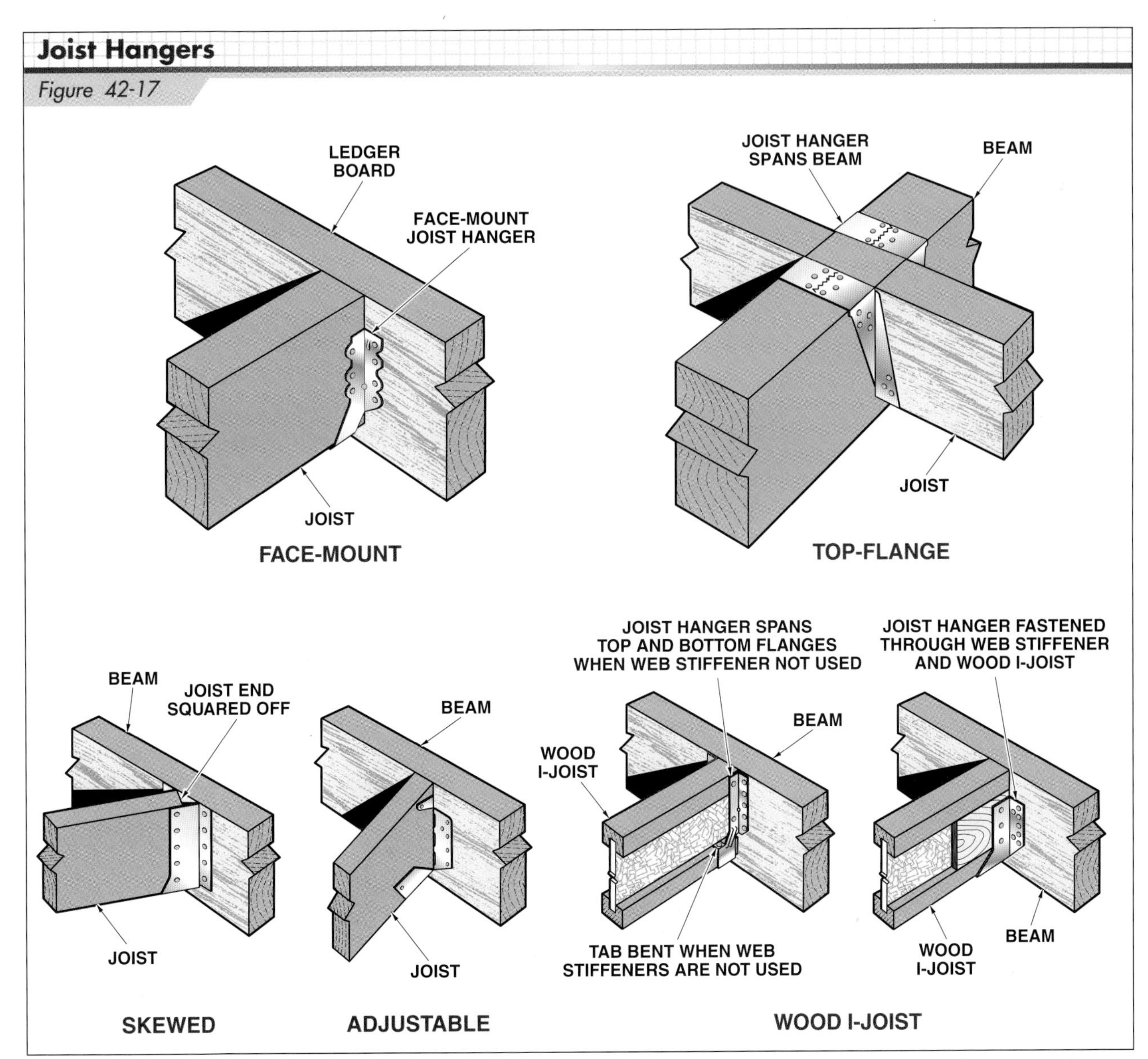

Figure 42-17. *Joist hangers are available in a variety of standard configurations. Face-mount and top-flange joist hangers are the most common types.*

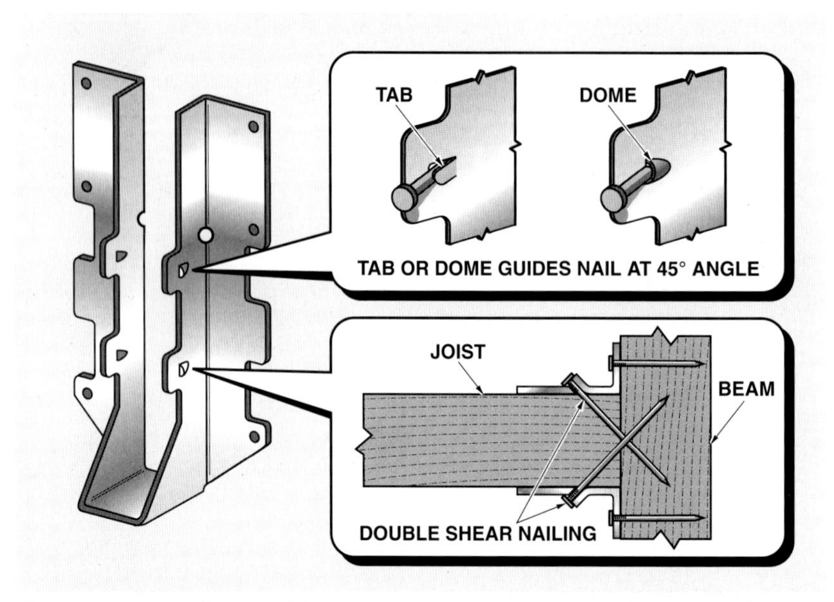

Figure 42-18. *Double-shear joist hangers provide greater strength than standard joist hangers.*

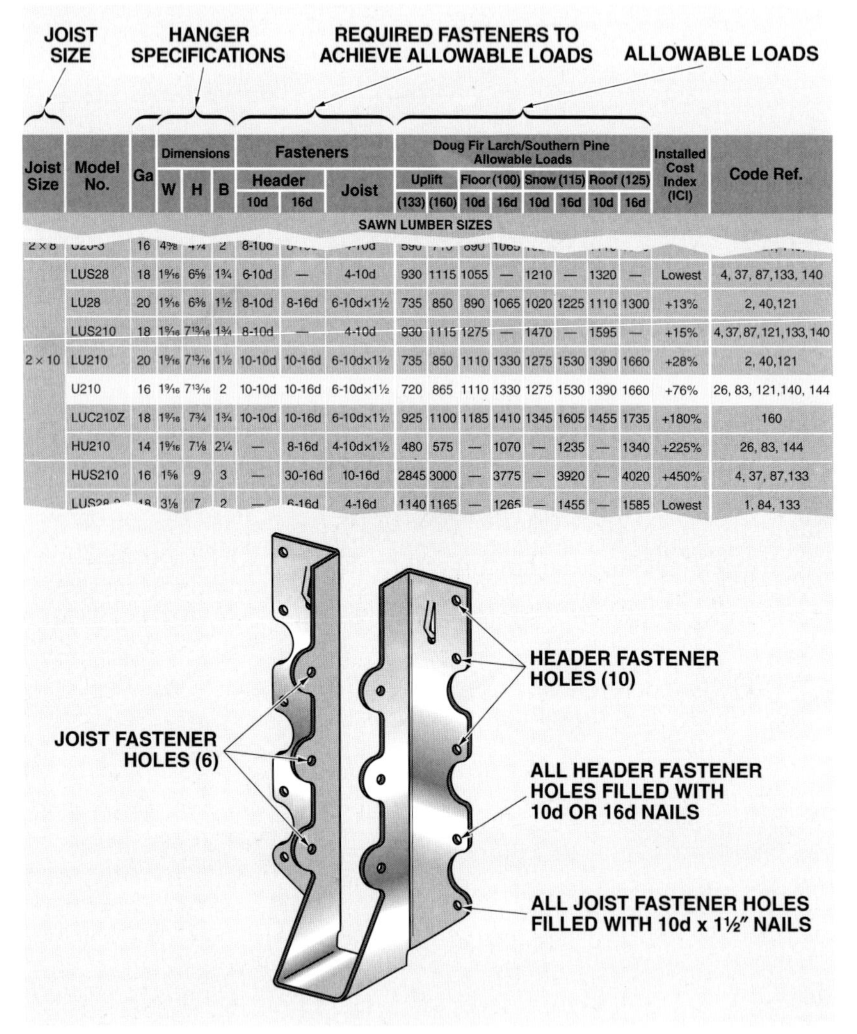

Joist Size	Model No.	Ga	Dimensions W	Dimensions H	Dimensions B	Fasteners Header 10d	Fasteners Header 16d	Fasteners Joist	Uplift (133)	Uplift (160)	Floor (100) 10d	Floor (100) 16d	Snow (115) 10d	Snow (115) 16d	Roof (125) 10d	Roof (125) 16d	Installed Cost Index (ICI)	Code Ref.
SAWN LUMBER SIZES																		
2 × 8	[illegible]	16	4⅝	[illegible]	2	8-10d	[illegible]	[illegible]	590	[illegible]	890	1065	[illegible]	[illegible]	[illegible]	[illegible]	[illegible]	[illegible]
2 × 10	LUS28	18	1 9/16	6⅝	1¾	6-10d	—	4-10d	930	1115	1055	—	1210	—	1320	—	Lowest	4, 37, 87, 133, 140
	LU28	20	1 9/16	6⅜	1½	8-10d	8-16d	6-10dx1½	735	850	890	1065	1020	1225	1110	1300	+13%	2, 40, 121
	LUS210	18	1 9/16	7 13/16	1¾	8-10d	—	4-10d	930	1115	1275	—	1470	—	1595	—	+15%	4, 37, 87, 121, 133, 140
	LU210	20	1 9/16	7 13/16	1½	10-10d	10-16d	6-10dx1½	735	850	1110	1330	1275	1530	1390	1660	+28%	2, 40, 121
	U210	16	1 9/16	7 13/16	2	10-10d	10-16d	6-10dx1½	720	865	1110	1330	1275	1530	1390	1660	+76%	26, 83, 121, 140, 144
	LUC210Z	18	1 9/16	7¾	1¾	10-10d	10-16d	6-10dx1½	925	1100	1185	1410	1345	1605	1455	1735	+180%	160
	HU210	14	1 9/16	7⅛	2¼	—	8-16d	4-10dx1½	480	575	—	1070	—	1235	—	1340	+225%	26, 83, 144
	HUS210	16	1⅝	9	3	—	30-16d	10-16d	2845	3000	—	3775	—	3920	—	4020	+450%	4, 37, 87, 133
	LUS2[illegible]	18	3⅛	7	2	—	6-16d	4-16d	1140	1165	—	1265	—	1455	—	1585	Lowest	1, 84, 133

Figure 42-19. *Always refer to manufacturer installation information regarding the proper nail size and required number of nails in the header and the joist.*

Joists Supported by Steel Beams. Often, wood floor joists are supported by a steel beam instead of a wood beam. The joists may rest on top of the steel beam and sill plate as shown in **Figure 42-20.** A plate is fastened to the beam and the joists are toenailed into the plate.

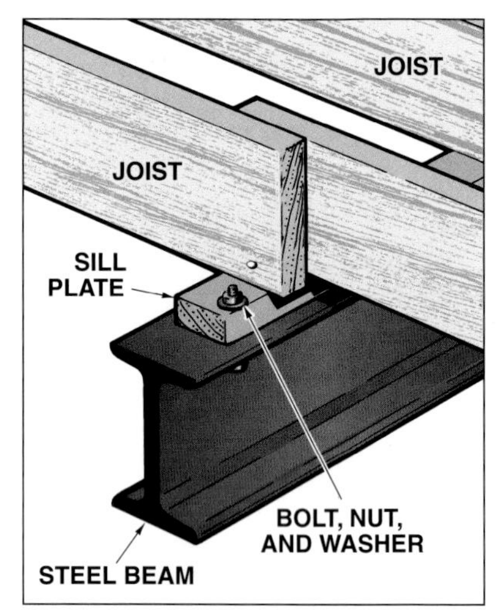

Figure 42-20. *A sill plate can be bolted to the top of a steel beam. Joists are toenailed to the plate. The sill plate may also be attached to the steel beam with a powder-actuated tool.*

Joists Butted against Steel Beams. Joists can be butted to a steel beam by welding joist hangers to the upper surface of the beam. Hot-dipped galvanized nails are driven through the hangers to secure joists in position. **See Figure 42-21.**

The joists can also be notched to fit and butted against the sides of a steel beam. When joists are notched to fit against the sides of the beam, ⅜" clearance should be allowed above the top flange of the beam to allow the wood to shrink without splitting. In some situations, opposite pairs of joists are scabbed across the top with a metal tie. In addition, a wide plate may be welded to the bottom of the beam to provide better support. Wood blocking is placed at the bottom of the joists to help keep them in position.

Doubled Joists

Joists should be doubled under partitions that run in the same direction as the joists. Some walls have water pipes, vent stacks, or heating ducts coming up from the basement or the crawl space below. Blocks are placed between the doubled joists to allow space for these utilities. **See Figure 42-22.**

Cantilevered Joists

Cantilevered joists are used when a floor or balcony of a building projects past the wall below it. **See Figure 42-23.** The permissible amount of cantilever is based on the intended application. For exterior load-bearing walls and roofs, 1′ of cantilever is allowed per each 3′ of backspan (amount of joist on opposite side of load-bearing wall). For exterior balconies, 1′ of cantilever is allowed for each 2′ of backspan. A header joist is nailed to the ends of the joists.

When regular floor joists run parallel to the intended overhang of the floor, the inside ends of the cantilevered joists are fastened to doubled joists. **See Figure 42-24.** Nailing should be through the first regular joist into the ends of the cantilevered joists. Metal connectors are strongly recommended and often required by the building code. A header joist is also nailed to the outside ends of the cantilevered joists.

Bridging between Joists

Many local building codes require bridging between the joists. The bridging keeps the joists aligned and helps distribute the load carried by the floor unit. Bridging is typically required when the joist spans are greater than 8′. Joists with a 15′ span need one row of bridging down the center of the span.

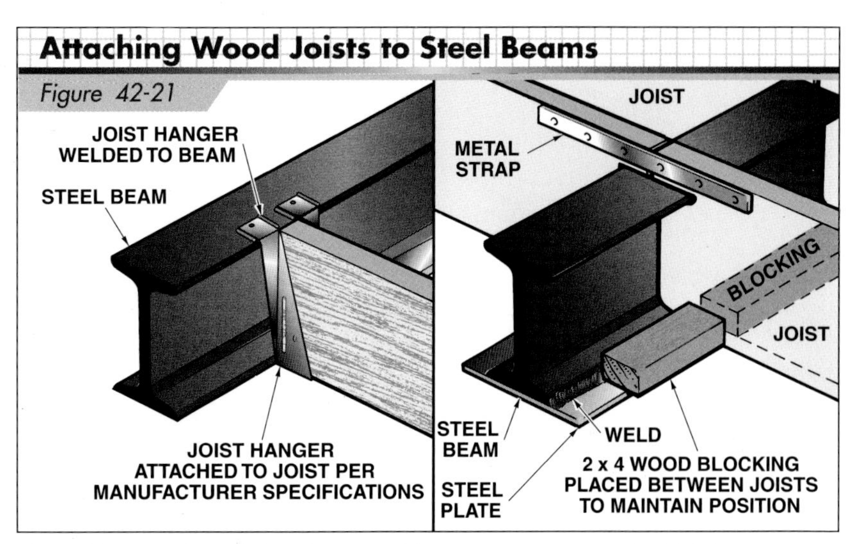

Figure 42-21. *Joists butted against a steel beam may be supported with joist hangers or a steel plate welded to the bottom of the beam.*

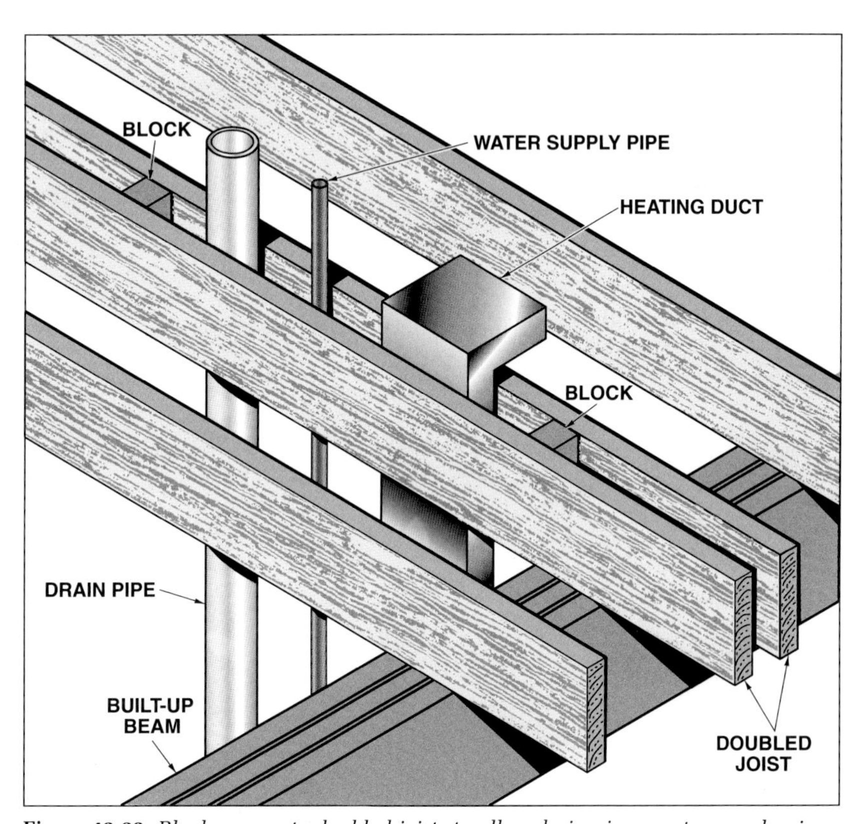

Figure 42-22. *Blocks separate doubled joists to allow drain pipes, water supply pipes, and heating ducts to extend into a wall cavity above.*

For longer spans, such as an 18′ span, two rows of bridging spaced 6′ apart are required.

Cross Bridging. Also known as *herringbone bridging,* cross bridging usually consists of 1″ × 3″ or 2″ × 2″ wood members installed as shown in **Figure 42-25.**

Per the International Residential Code, 2 × 12 or larger joists must be supported laterally by solid blocking, diagonal wood or metal bridging, or a continuous 1 × 3 strip nailed across the bottom of and perpendicular to the joists at intervals not exceeding 8′.

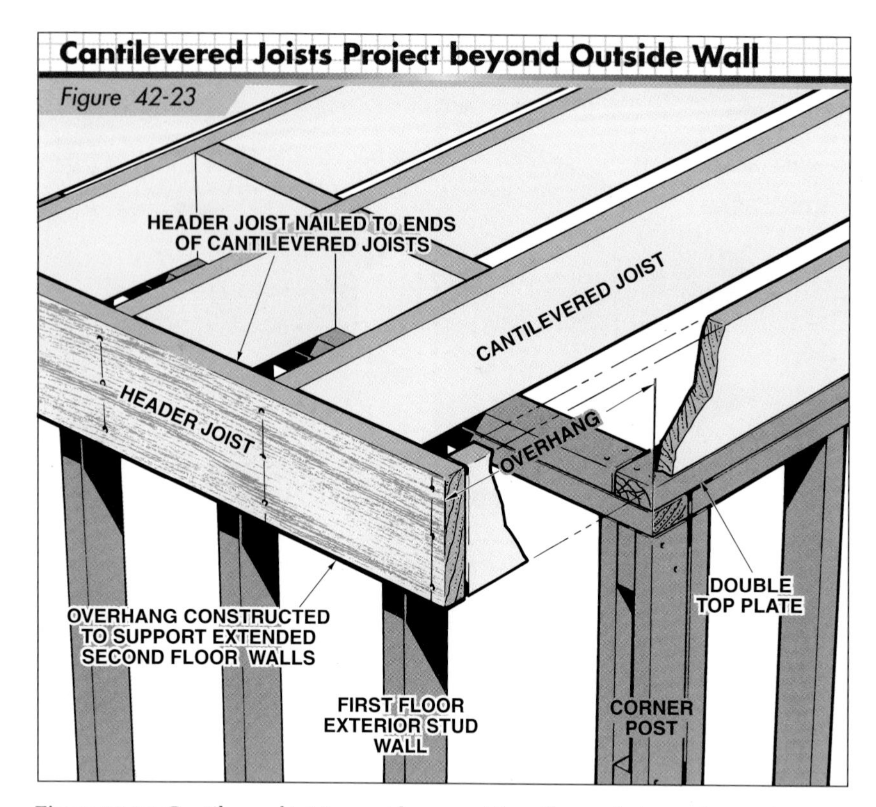

Figure 42-23. *Cantilevered joists provide support for a floor or balcony that projects past the wall below it.*

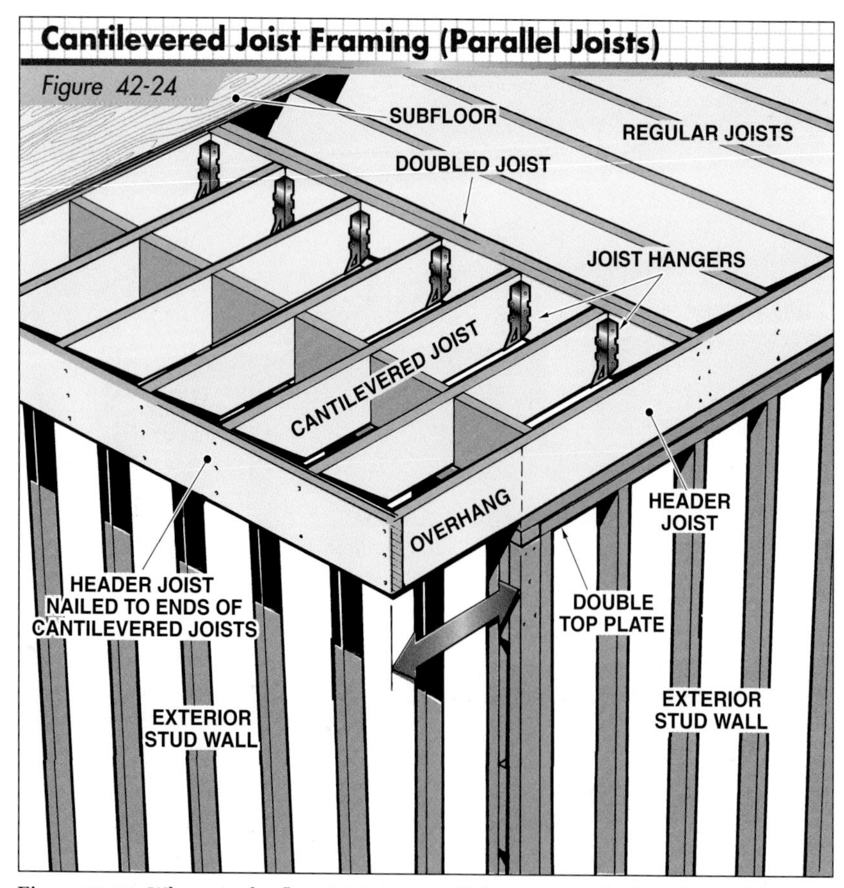

Figure 42-24. *When regular floor joists run parallel to the intended overhang of the floor, the inside ends of the cantilevered joists are fastened to doubled joists.*

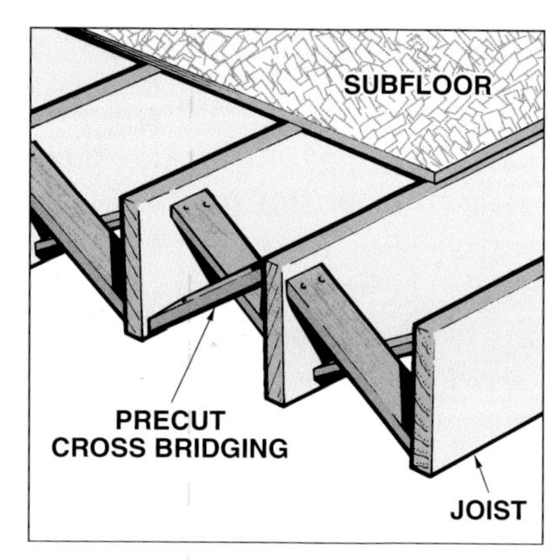

Figure 42-25. *Cross bridging is precut to fit between the joist spans.*

Cross bridging is toenailed at each end with 6d or 8d nails. The pieces are usually precut on a power miter box or radial arm saw. The nails are started at each end before the cross bridging is placed between the joists. The upper ends of the cross bridging are nailed into the joists and the nails at the lower end are not driven in until the subfloor has been placed. If both ends are nailed in place before the subfloor is placed, the joist could be pushed out of line when the bridging is nailed in. An efficient method for placing wood cross bridging is shown in **Figure 42-26.**

Another approved system of cross bridging uses metal pieces instead of wood and requires no nails. **See Figure 42-27.** Metal cross bridging is available for 12″, 16″, and 24″ joist spacing.

Cross bridging reduces the amount of floor deflection by spreading an applied load. The effectiveness of cross bridging is limited by the fasteners used to attach the cross bridging and the quality of the connection. If wood cross bridging splits while being installed or if the nail hits a knot and bends, the connection is not as effective.

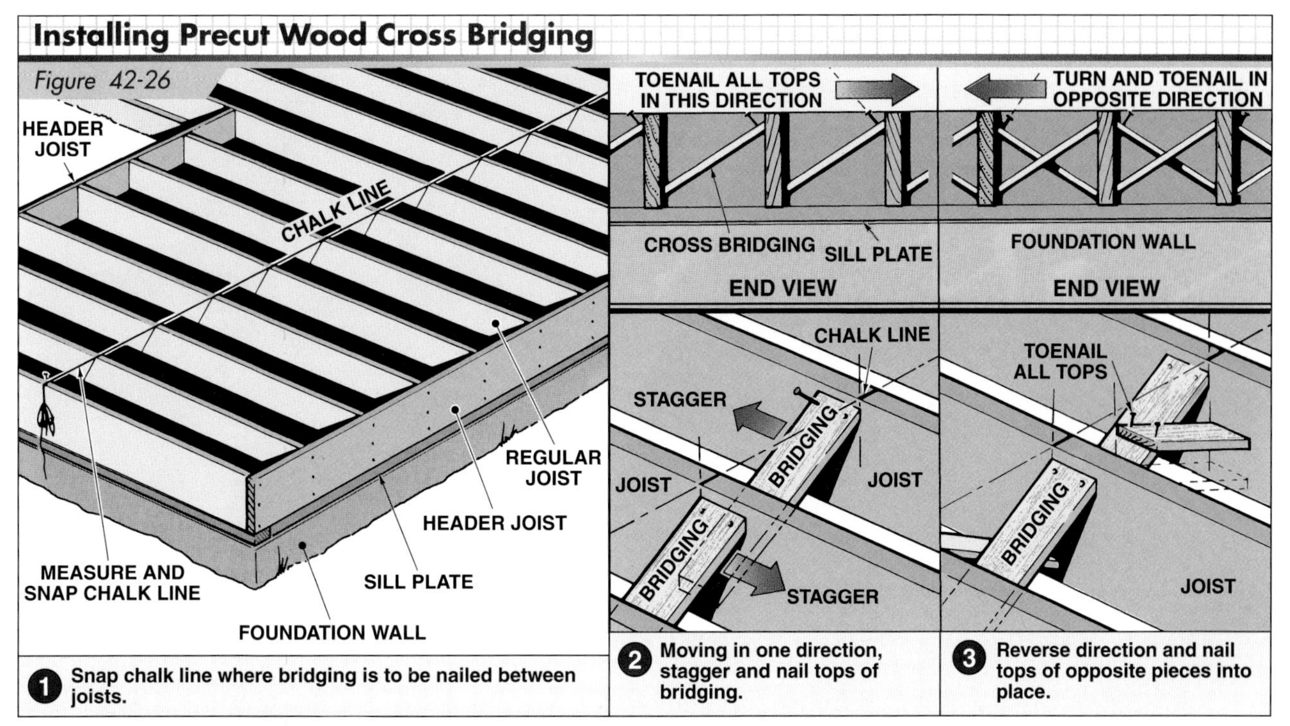

Figure 42-26. *Wood cross bridging is installed by toenailing each end with 6d or 8d nails. Allow ½″ space between adjacent bridging members to prevent them from rubbing together and producing squeaks.*

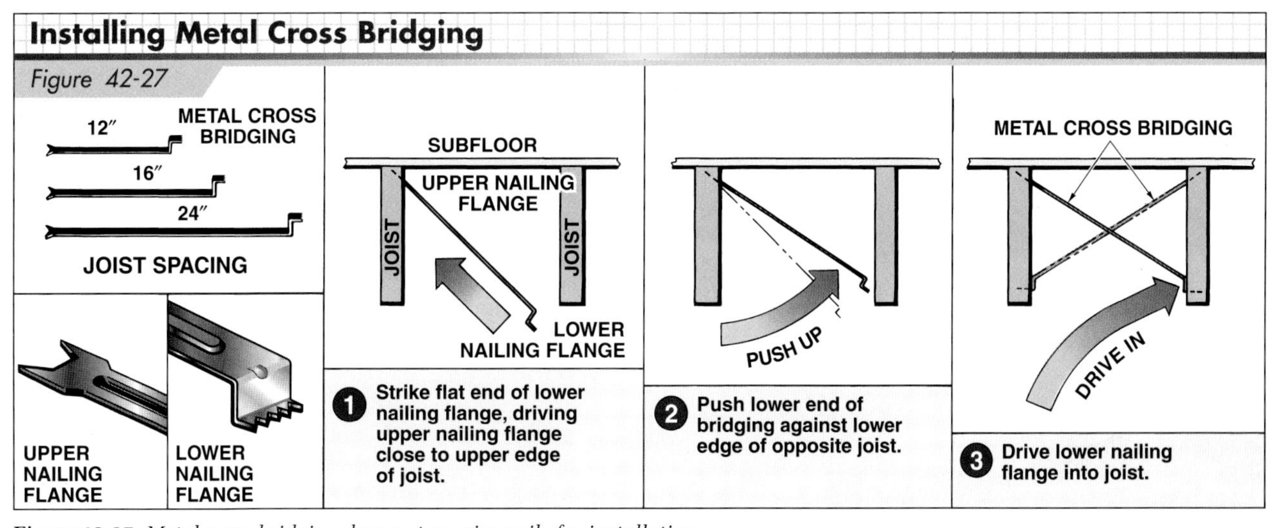

Figure 42-27. *Metal cross bridging does not require nails for installation.*

Solid Bridging. Solid bridging, also known as *solid blocking*, serves the same purpose as cross bridging. Solid bridging is commonly used in conjunction with cross bridging where non-standard joist spacing occurs. Solid bridging members are cut from lumber that is the same width as the joist material and are installed in a straight line or staggered. Straight-line bridging may be required every 4′ OC to provide a nailing base for the panel subfloor. If solid bridging is staggered, the members can be spiked from both ends, resulting in a faster nailing operation. **See Figure 42-28.**

Floor frames are bridged to prevent unequal deflection of the joists and to help solidify the frames. Bridging allows an overloaded joist to receive support from adjacent joists in the floor frame. Solid bridging provides maximum rigidity to the floor frame.

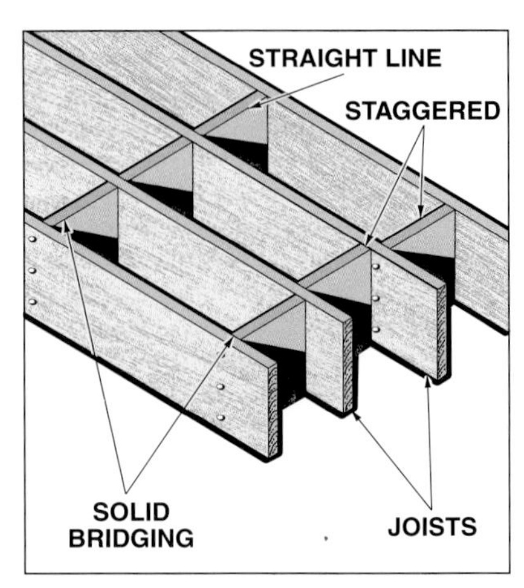

Figure 42-28. *Solid pieces are often used for bridging between joists.*

Cross bridging keeps the joists aligned and helps distribute the load carried by the floor unit.

Floor, Fireplace, and Chimney Openings in Joists

A floor opening must be framed where stairs rise to the floor. Fireplaces and chimneys also require specially framed floor openings. When joists are cut for floor openings, there is a loss of strength in the area of the opening. The opening must be properly framed to restore the lost strength as shown in **Figure 42-29.**

Figure 42-30 shows the procedure for framing a floor opening. A pair of joists called *trimmers* are placed at each side of the opening to support the headers. The headers should be doubled if the span is more than 4′. Nails supporting the ends of the headers are driven through the trimmers into the ends of the headers. *Tail joists* run from the header to a supporting wall or beam. Nails are driven through the header into the ends of the tail joist.

Beams must be properly braced to stabilize them during floor framing.

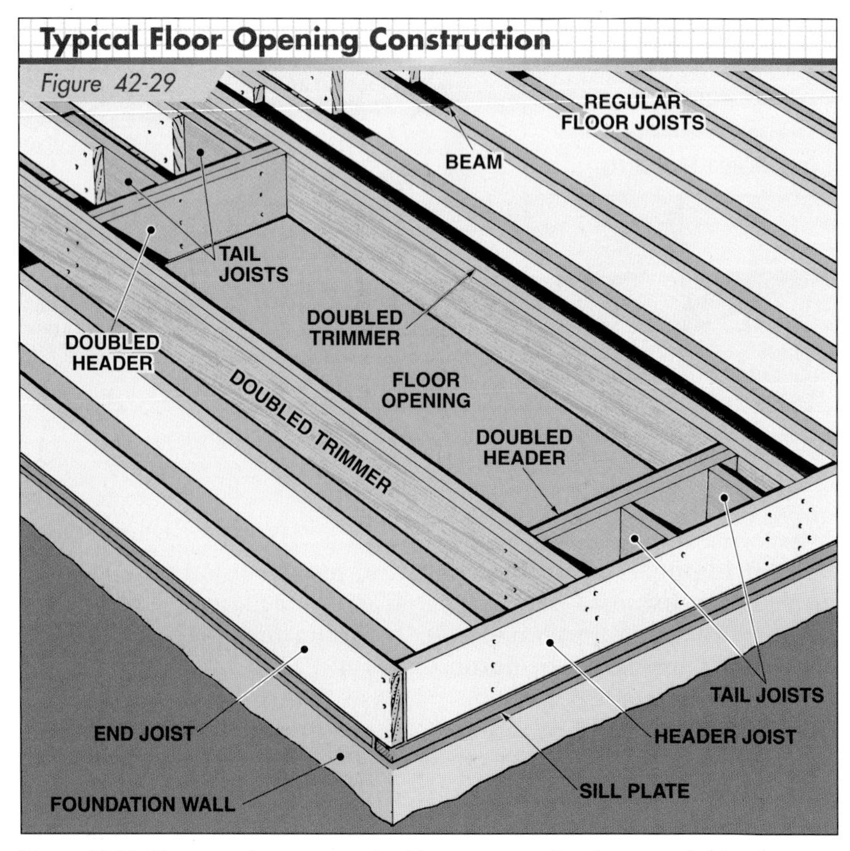

Figure 42-29. *Floor openings are required for stairways, fireplaces, and chimneys.*

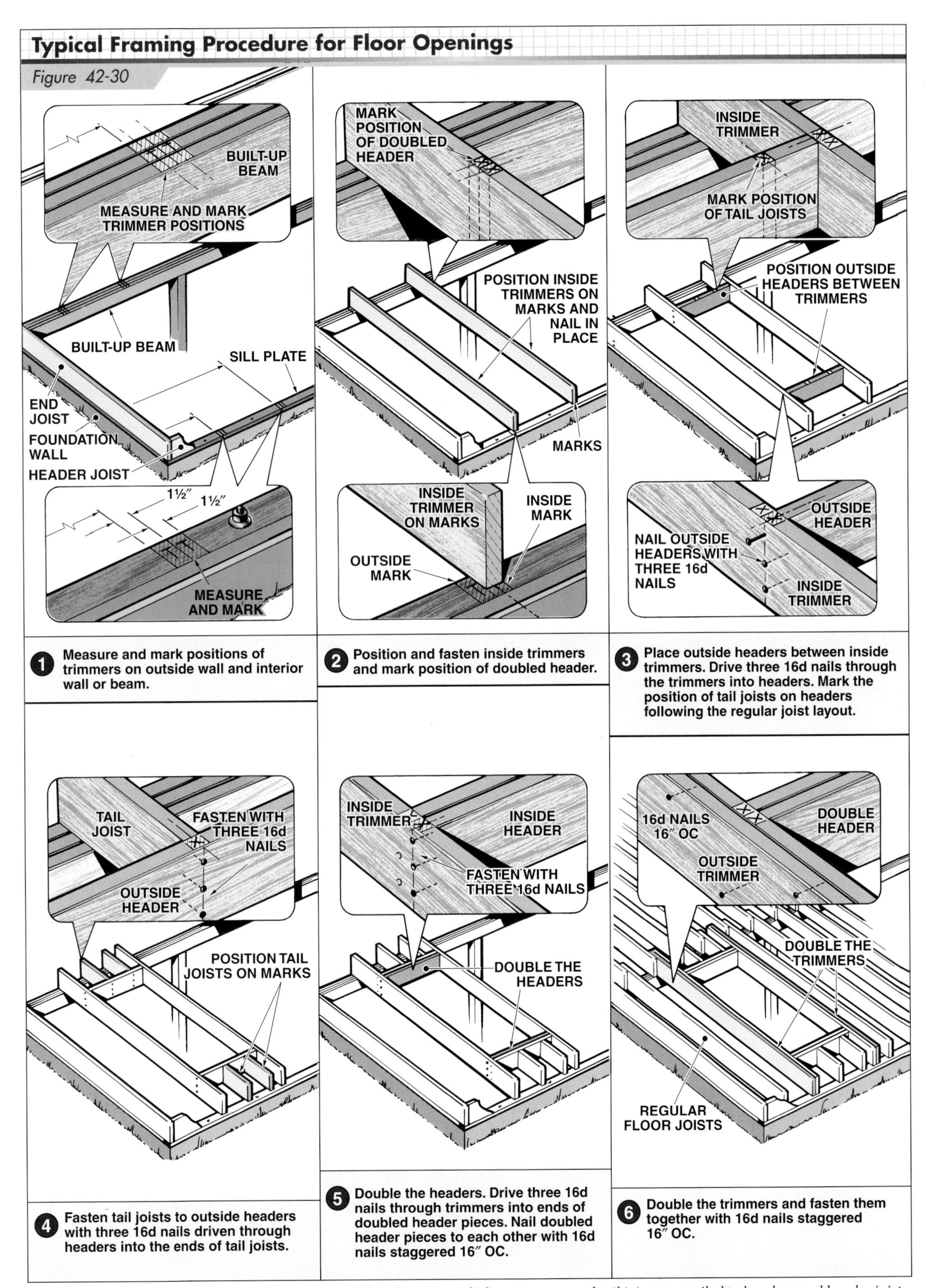

Figure 42-30. *When framing a typical floor opening, headers are nailed to trimmers and tail joists are nailed to headers and header joists.*

Senco Products, Inc.

Subfloor panels can be fastened to floor joists using pneumatically driven nails.

Various metal connectors may also be used to strengthen a framed floor opening. **See Figure 42-31.**

Placing Floor Joists

Before floor joists are placed, the sill plates and beams must be marked to show where the joists are to be nailed. Floor joists are usually placed 16″ OC. Most carpenters use a 20′ or longer steel tape for layout purposes, and these tapes have special markings every 16″ and 19.2″ to aid in joist layout. Some framing systems permit 24″ OC joist spacing. **Figure 42-32** shows on-center layout for joists spaced 16″ OC.

For joists that rest directly on the foundation walls, layout marks may be placed on the sill plates or the header joists. Lines must also be marked on top of the central beams, beams, or walls over which the joists will lap. If framed walls are below the floor unit, the joists are laid out on top of the double plate. The floor layout should also show where any joists are to be doubled because of partitions resting on the floor that run in the same direction as the floor joists. Floor openings for stairwells must also be marked.

Joists should be laid out so that the edges of standard subfloor panels fall directly over the centers of the joists. This layout eliminates additional cutting of panels when they are being fitted and nailed into place. One method of laying out joists is to mark the first joist 15¼″ OC from the edge of the building with subsequent joist spacing of 16″ OC. **See Figure 42-33.** A procedure for laying out the entire floor is shown in **Figure 42-34.**

Most framing members should be precut before construction begins. The joists should all be trimmed to their proper lengths. Cross bridging and solid blocks should be cut to fit between the joists that have a common spacing. The distance between joists is usually 14½″ for joists spaced 16″ OC. Blocking for the odd spaces is cut afterward. A typical procedure for framing is shown in **Figure 42-35.**

Crowns. Most joists have a *crown* (a bow shape) on one edge. The edge of each joist should be sighted before it is nailed in place to ensure that the crown is turned up. The joist will later settle from the weight of the floor and straighten out.

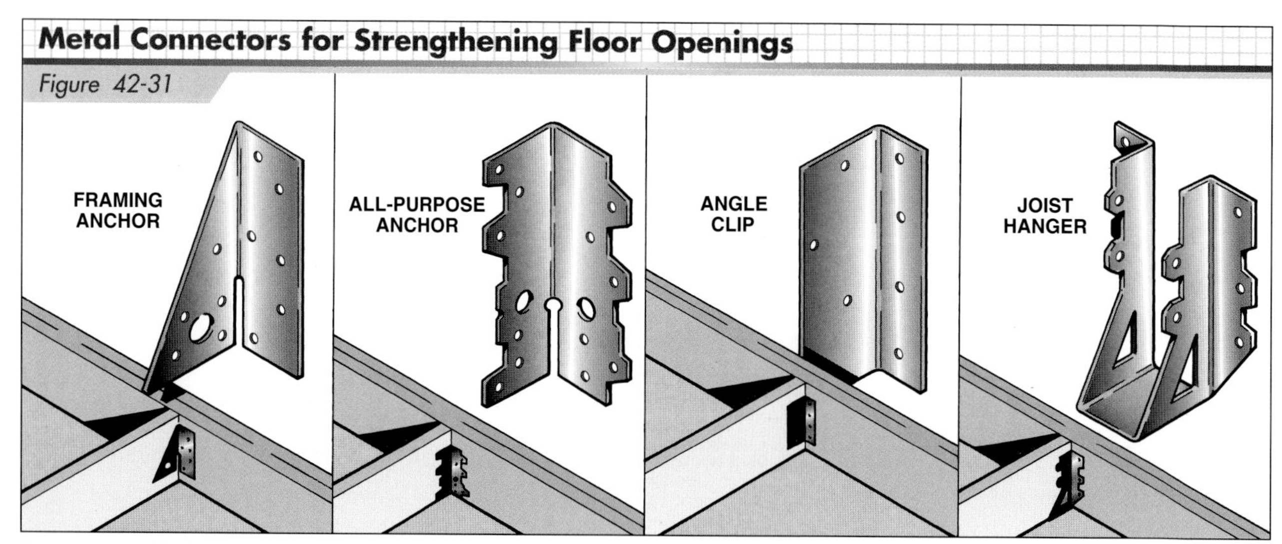

Figure 42-31. *Various metal connectors are used to strengthen framed floor openings.*

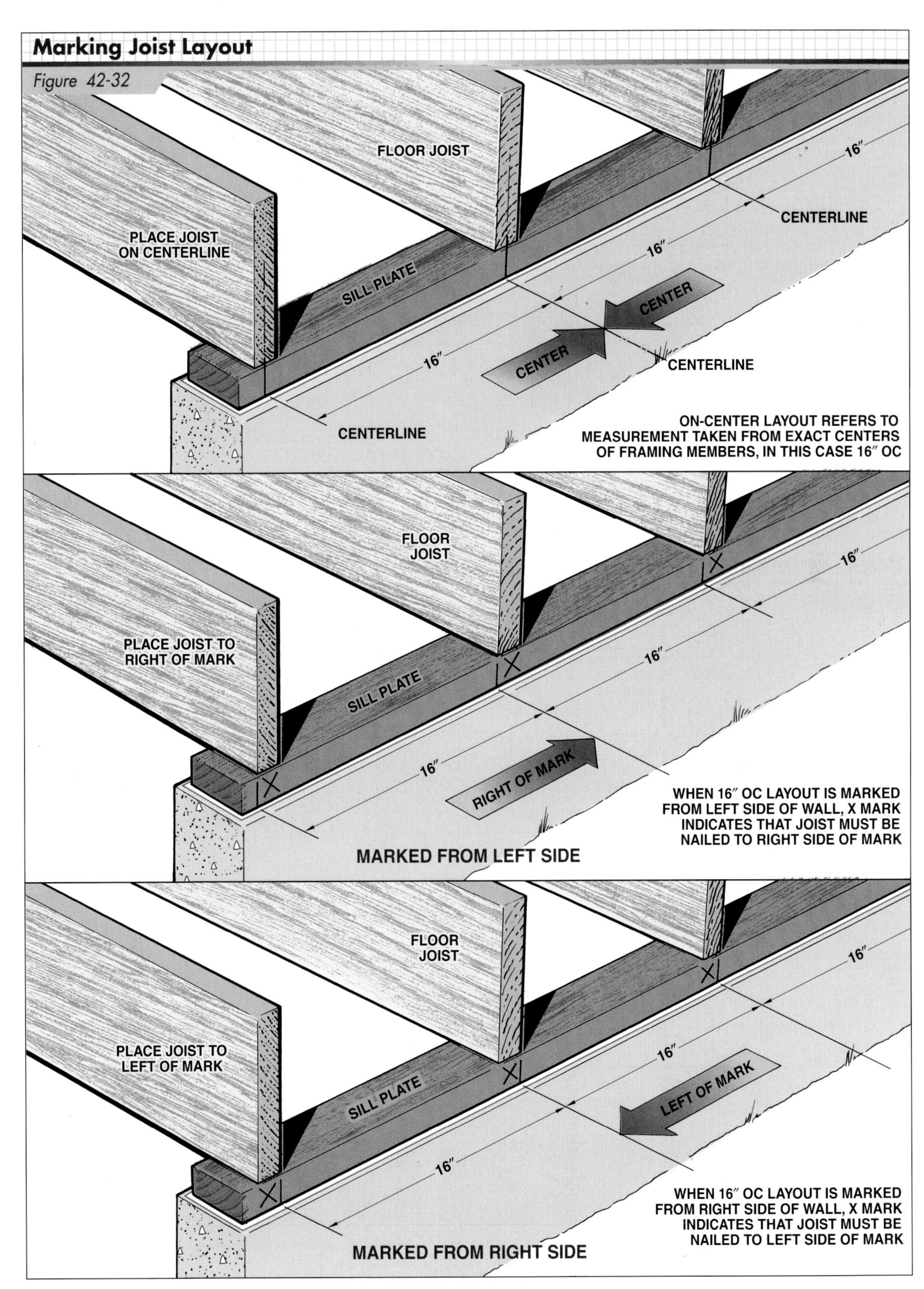

Figure 42-32. *When laying out joists, an "X" mark indicates the side where the joists are to be nailed.*

Floor Joists 16″ OC with Panel Subfloor

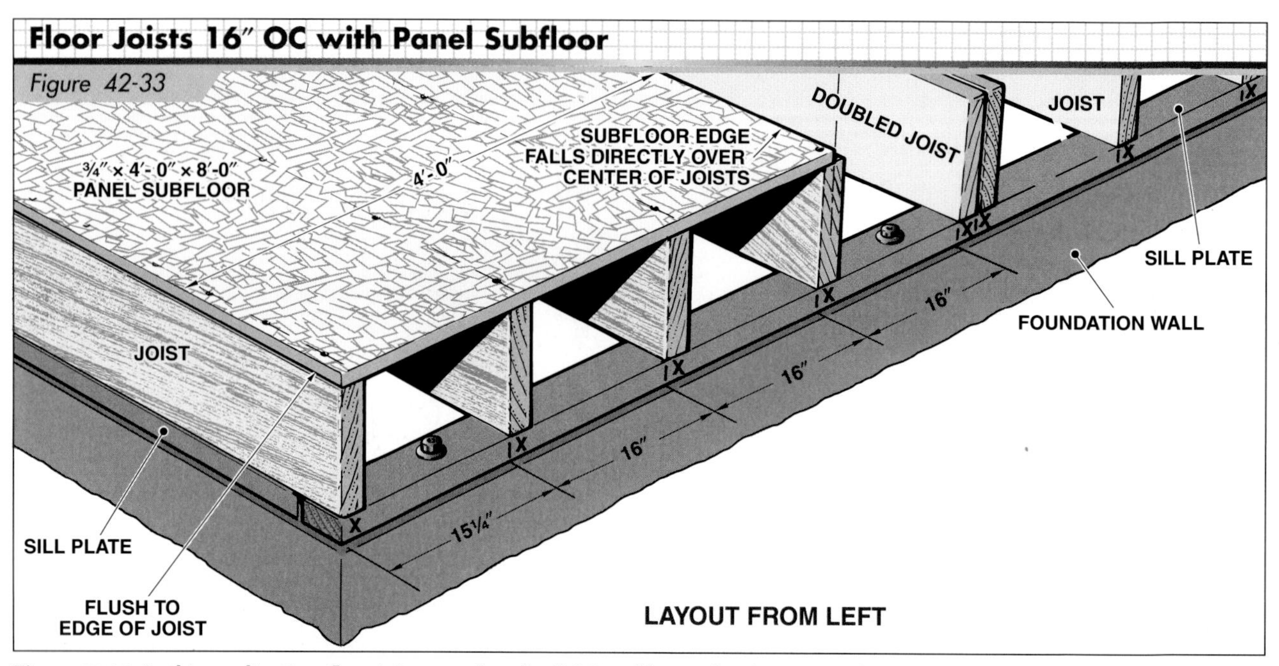

Figure 42-33. *In this application, floor joists are placed 16″ OC and layout has been started from the left side. The first joist is marked at 15¼″ to ensure that the 4′ or 8′ edge of a panel will fall on the center of a joist. The layout for the following joists is then 16″ OC. A doubled joist is laid out at the right.*

Marking Joist Layout on Sill Plates

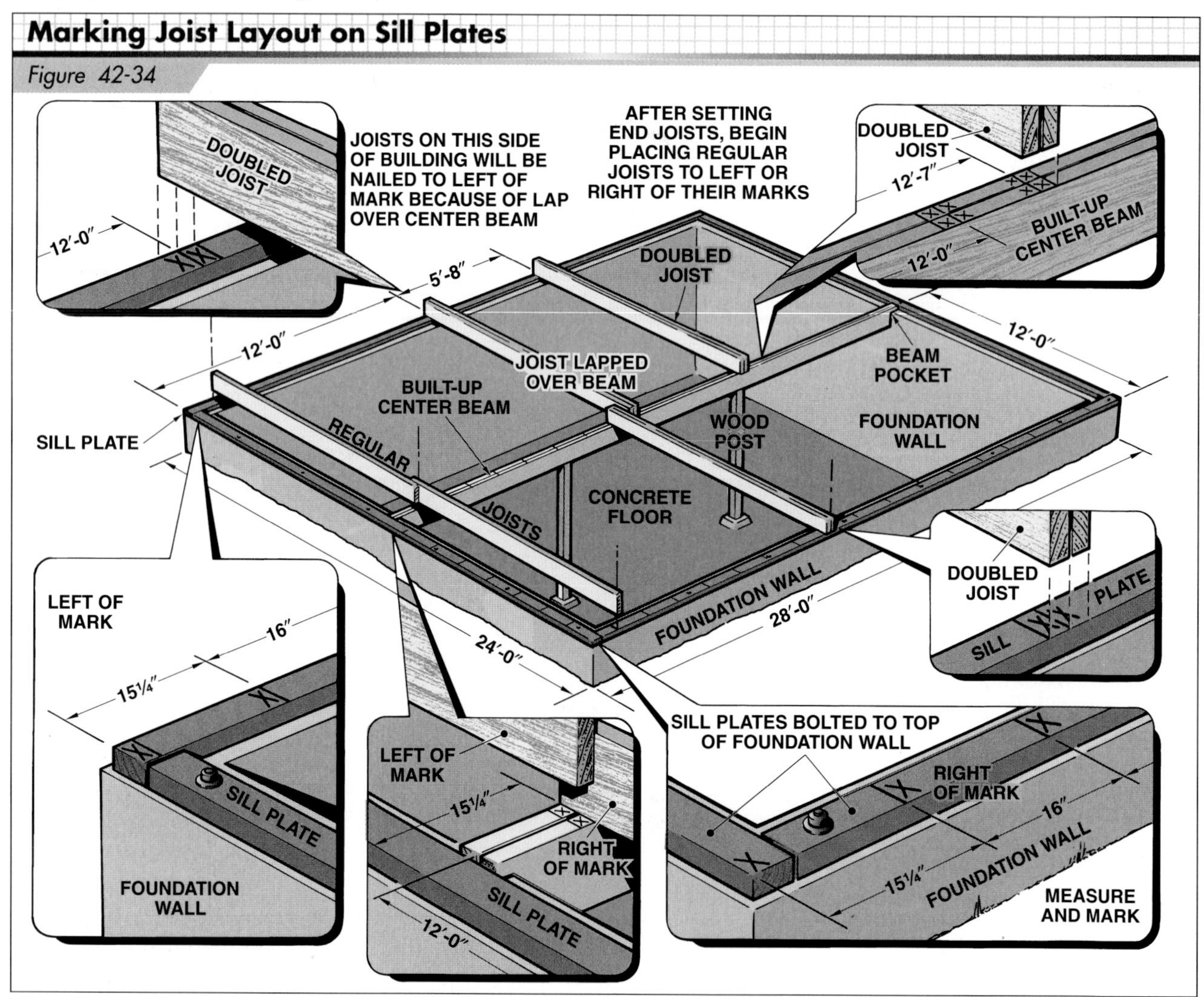

Figure 42-34. *When laying out a floor unit, joist placement is marked on the sill plates.*

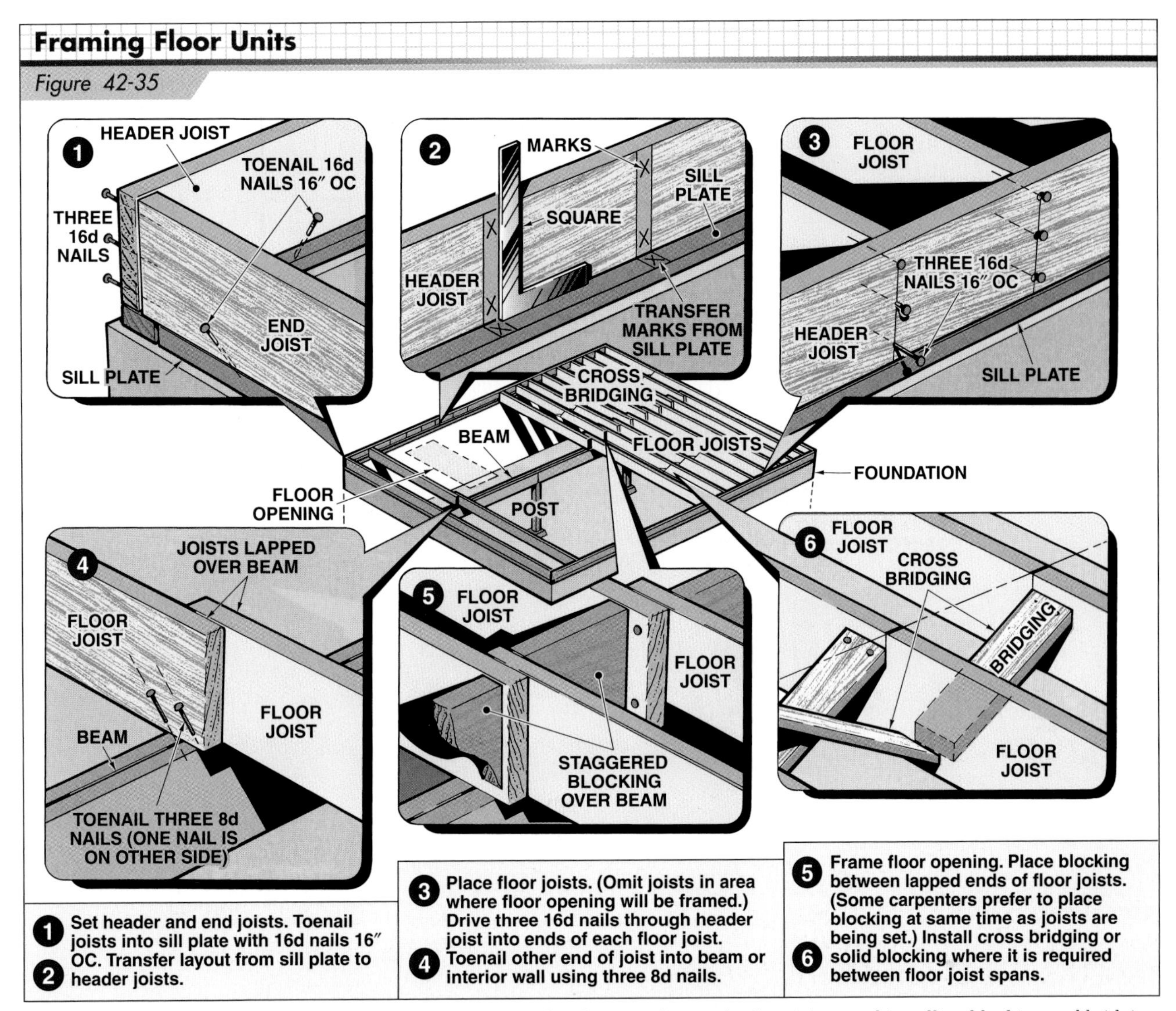

Figure 42-35. *A floor unit is framed by setting the header and end joists, placing the floor joists, and installing blocking and bridging. The floor opening is framed as described in Figure 42-30. Most framing members should be precut before construction begins.*

Installing Joist Hangers. Joist hangers may be specified and/or required by the local building code. After laying out the joist location on the beam, girder, or other structural member using the methods previously described, draw plumb lines on the member and mark an "X" to indicate the side of the line where the joist will be installed.

Two methods may be used to install joist hangers. One method involves toenailing the joist into position and then installing the joist hangers. The other method involves preinstalling the joist hangers and then installing the joists as follows:

1. Insert a small block of joist scrap material into the joist hanger. **See Figure 42-36.**
2. Carefully position the joist hanger on the structural member along the layout line. In some cases, the joist hanger may be flush with the top of the structural member while in other cases it may be placed lower along the line.
3. Drive two nails through each side of the hanger and into the structural member. Some joist hangers have nailing tabs that can be driven into the structural member to hold it in position before nailing.
4. Remove the joist block and finish nailing the hanger on each side.
5. Cut all joists to length and place the ends in the hangers.
6. Drive the nails through the joist hangers and into the joist on each side.

In most cases, all nail holes in the joist hangers must be filled with nails to support the design loads. Always refer to the manufacturer instructions regarding the size, type, and number of nails to be used in each joist hanger.

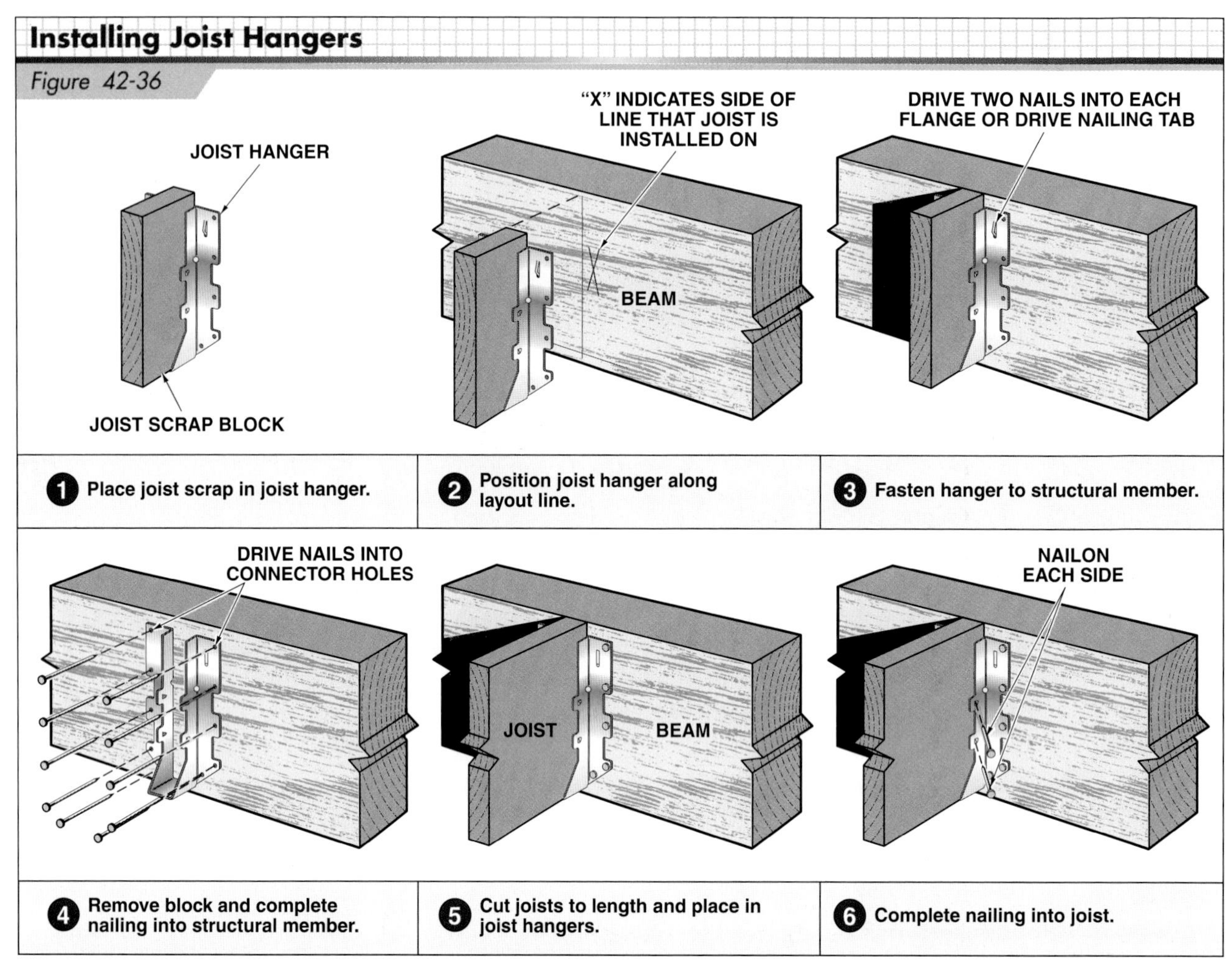

Figure 42-36. *Joist hangers can be installed prior to installing the joists. Joist hangers may be installed using a hammer, palm nailer, or pneumatic nailer specifically designed for metal connectors.*

SUBFLOOR

The *subfloor*, also known as rough flooring, is nailed to the top of the floor frame. **See Figure 42-37.** The subfloor strengthens the entire floor unit and serves as a base for the finish floor material. The walls of a building are laid out, framed, and raised into place on top of the subfloor. Panel products are typically used for the subfloor.

Plywood is the oldest type of panel product used for the subfloor in residential and other light-framed construction. However, the use of non-veneered panels such as oriented strand board and composite board continues to grow. Manufacture and composition of nonveneered panel products is discussed in Section 2.

APA—The Engineered Wood Association

Figure 42-37. *Subfloor panels are placed over joists so the long sides run at a right angle to the joists. Note that the panel joints are staggered.*

Panels used for subfloors must be the proper grade and thickness for the floor system of the building. APA performance-rated panels meet the code requirements for subfloors in all parts of the United States. Subfloor panels may have square or tongue-and-groove (T&G) edges.

Applying Subfloor Panels

Subfloor panels are applied with the longer edge at a right angle to the joists. Blocking should be nailed under the long edges of panels if tongue-and-groove panels are not used. A 1/8″ space should be provided between the end joints, with a 1/8″ space between the edge joints for square and T&G edges, except where otherwise indicated by the manufacturer. **See Figure 42-38.**

Panel Subfloor Blocking and Nailing Patterns

Figure 42-38

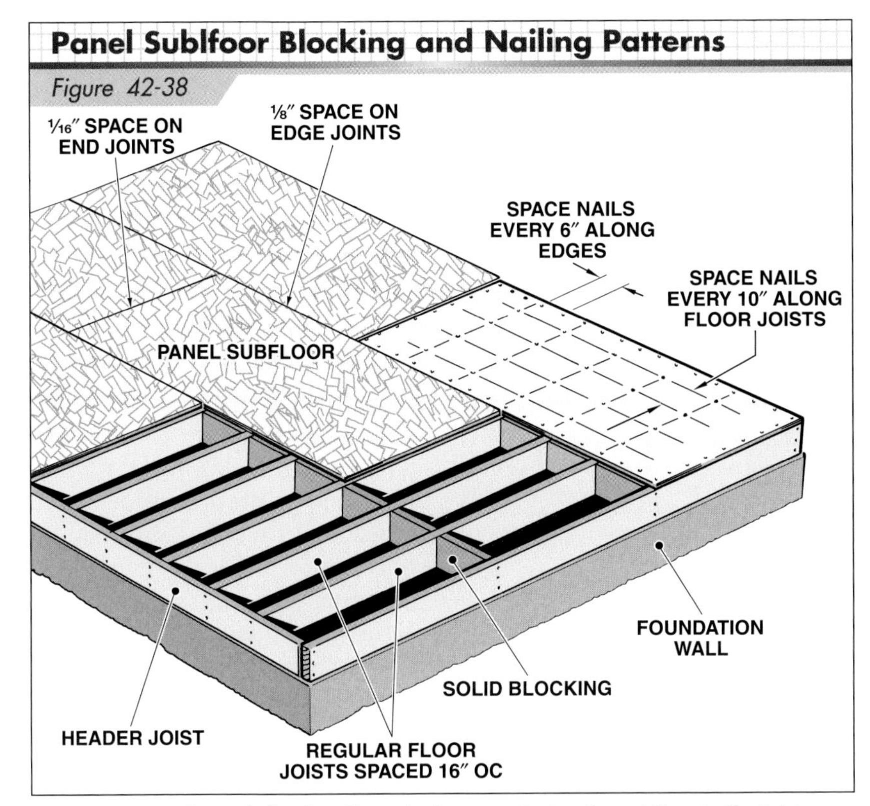

Figure 42-38. *Deformed-shank nails or deck screws fasten the subfloor to the joists.*

The nailing schedule for most types of subfloor panels requires 6d common nails for materials up to 7⁄8″ thick and 8d nails for panels up to 1 1⁄8″ thick. Deformed-shank nails are strongly recommended. Nails are usually spaced 6″ OC along the panel edges and 10″ OC along the floor joists.

No. 8 gauge deck screws are also used for fastening subfloor panels to the floor unit. Screw lengths range from 1 3⁄4″ to 2 1⁄2″, depending on the thickness of the panel. A typical layout of the screws is 6″ OC along the panel edges and 6″ OC along the floor joists. In some cases, 12″ OC is considered adequate spacing for screws along the floor joists.

When installing subfloor panels, carpenters initially use just enough fasteners to secure the panels in position. When all the panels are placed, chalk lines are snapped to locate the centers of the joists below. The entire subfloor is then nailed or screwed down in one operation. Heavy-duty pneumatic nailers are used to drive nails. **See Figure 42-39.** Self-feeding electric screwguns are used to drive screws. A procedure for placing subfloor panels is shown in **Figure 42-40.**

The weight of the joists on the sill plates may expose a few more threads on the anchor bolts. After installing the floor joists, tighten down the anchor bolt nuts once again to ensure a strong connection.

Glue-Nailing Panels. For some construction projects, a construction adhesive is used in addition to nails or screws for fastening the panels to the joists. Although this system is commonly referred to as glue-nailing, a mastic adhesive rather than a glue is used. The mastic is applied to the upper edges of the joists, and the panel is then set in place and fastened before the mastic sets. **See Figure 42-41.** The combination of an adhesive with nails or screws helps to further stiffen the entire floor system. In addition, there is less squeak to the floor, and the nails or screws are less likely to pop over a period of time. The nailing schedule for a glue-nailed panel requires nails or screws only 12″ OC along the panel edges and along the joists.

Installing Subfloor Panels

Figure 42-39

Figure 42-39. *Subfloor panels are attached to joists using a heavy-duty pneumatic nailer or self-feeding screwgun.*

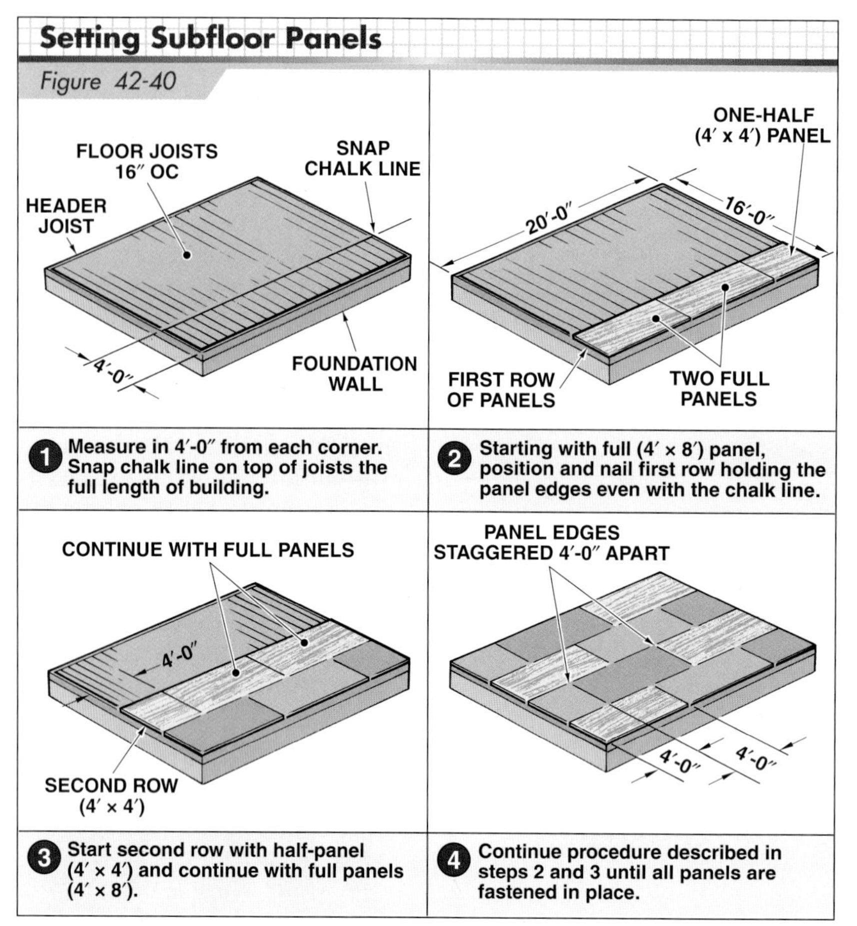

Figure 42-40. *Aligning the first row of panels with a chalk line 4′-0″ from the edge of the building ensures that the first row of panels will be perfectly straight even if the edge of the building is not straight. When placing the remaining panels, drive a few fasteners to secure the panels in place. Then, snap chalk lines across the panels in line with the joist centers below and complete nailing or screwing down the panels.*

APA—The Engineered Wood Association

Figure 42-41. *In the glue-nailed panel method, a construction adhesive is applied to the top of the joists before subfloor panels are placed.*

Post-and-Beam Subfloor System

In a post-and-beam subfloor system, the floor unit receives its main support from floor beams rather than from floor joists. **See Figure 42-42.** The beams are usually spaced 4′ OC and are supported by wood posts resting on concrete piers. The subfloor panels are 1⅛″ or 1¼″ thick and have tongue-and-groove edges. Nails are 10d common, spaced 6″ apart.

FLOOR UNDERLAYMENT

Floor underlayment consists of thin panels that are placed directly over the subfloor. **See Figure 42-43.** Underlayment provides a smooth, even surface for finish floor materials such as tile, linoleum, and carpet. Underlayment panels may be oriented strand board, plywood, particleboard, or hardboard. Underlayment panels must be a high grade of material that has good resistance to dents and punctures from concentrated loads.

Some types of subfloor panels, such as APA performance-rated Sturd-I-Floor® panels, have surfaces that are smooth enough for the direct application of the finish floor materials. However, the surface of any subfloor becomes scratched, dented, and roughened during construction. For this reason it may be necessary to later place underlayment panels over the Sturd-I-Floor panels in the kitchen and bathroom areas, or anywhere else that resilient materials such as tile or linoleum will be used for the finish floor. Underlayment is also placed under tile and linoleum to bring the finish floor up to the level of carpeted or hardwood floors in other parts of the building.

Fastening Methods for Underlayment

The APA nailing schedule for ¼″ thick plywood under-layment recommends 3d ring-shank nails or 16 ga to 18 ga staples. Either type of fastener should be spaced 3″ apart along panel edges and 6″ each way at the intermediate sections.

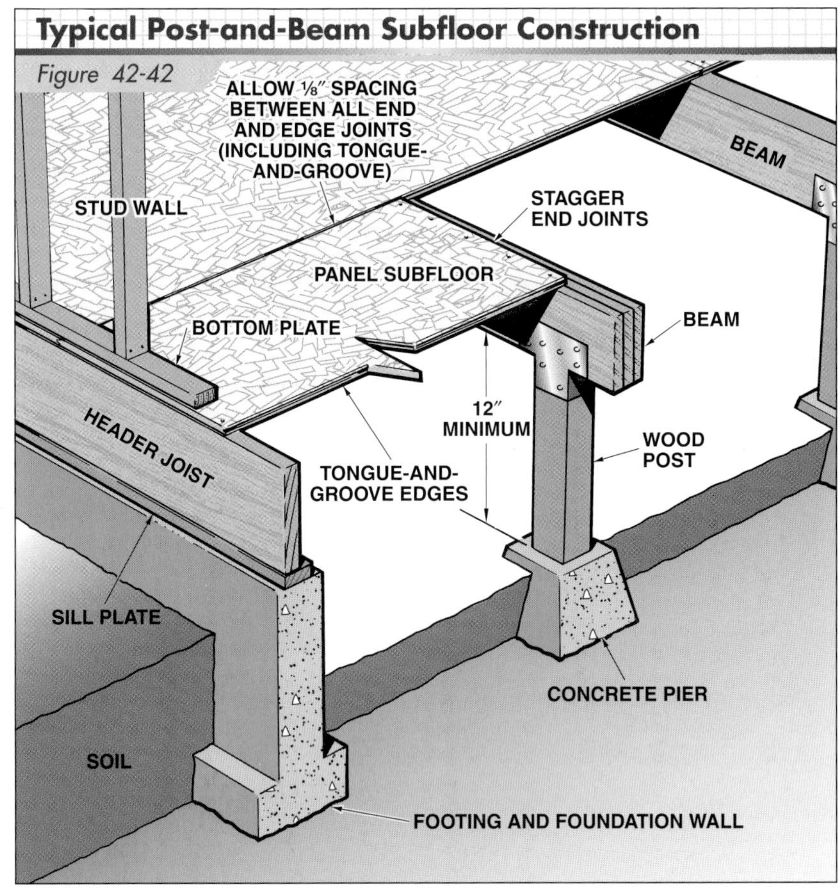

Figure 42-42. *In a post-and-beam subfloor, the floor unit receives its main support from beams rather than from floor joists.*

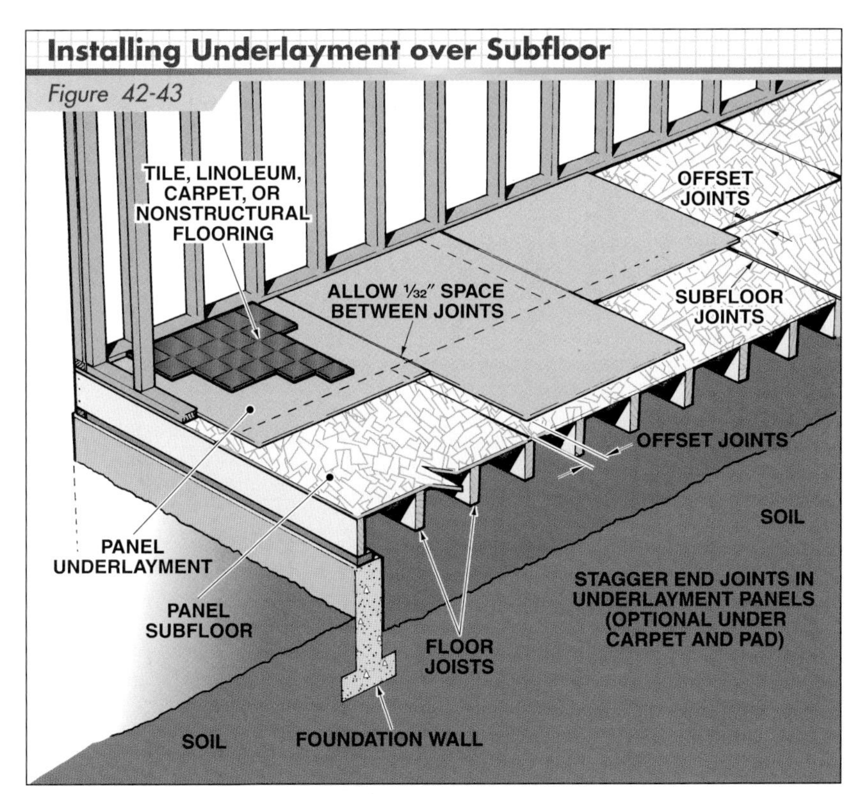

Figure 42-43. *Underlayment is placed over the subfloor to provide a smooth and even surface for finish floor materials.*

The Composite Panel Association suggests 4d nails for fastening particleboard panels thinner than 3/8″, spaced 3″ apart around the edges, and 6″ OC each way throughout the body of the panel. For panels 3/8″ to 5/8″ thick, the Composite Panel Association recommends 6d nails spaced 6″ OC around the edges, and 10″ OC each way throughout the body of the panel. If staples are used with particleboard under-layment, they should be a minimum of 7/8″ long, 18 ga, with a 3/16″ crown for 1/4″ thick panels. Staples should be 1 1/8″ long, 16 ga, with a 3/8″ crown for 1/2″ or 5/8″ thick panels.

Glue-nailing is also recommended for underlayment. When mastic adhesive is used in addition to nails, the spacing of the nails can be increased to 16″ OC along the edges and throughout the body of the panel.

The subfloor must be dry before underlayment is installed. The structure must be closed in and other building materials, such as concrete, plaster, and paint, should have dried to the moisture conditions that will occur in the building when it is occupied. Subfloor joints may require sanding due to swelling during construction.

Cement Board Underlayment

Cement board may be required as underlayment for ceramic tile floors. Cement board is a reinforced panel composed of concrete and aggregate. Cement board underlayment is available in various thicknesses ranging from 1/4″ to 1/2″ and in 4′ × 4′, 3′ × 5′, and 4′ × 8′ panels. Cement board is smooth on one side for applications involving adhesives, and textured on the other side for thin-set mortar applications.

Cement board can be cut using a utility knife by scoring both sides of the panel through the fiber-reinforcing mesh and snapping the panel to size. A circular saw with a carbide-tipped blade can also be used to cut cement board.

Cement board is secured to the subfloor using adhesives or mortar and nails or screws. Use a ⅝″ V-notched trowel when applying an adhesive or a ¼″ square-notched trowel when applying mortar. Place the cement board in position, staggering the joints with the subfloor joints. Allow a ⅛″ gap at the ends and along the sides of the panels. Fasten the cement board to the subfloor with 1½″ hot-dipped galvanized roofing nails or 1 ¼″ wood screws spaced 8″ OC in both directions.

FLOOR TRUSSES

Floor truss systems are frequently used in residential and commercial wood-framed construction. A typical floor truss is composed of top and bottom chords tied together by webs. **See Figure 42-44.** The webs are fastened to the top and bottom chords with metal connector plates. Webs may also be connected to the chords by finger-jointing and gluing.

Another type of truss design uses tubular steel webs. **See Figure 42-45.** Although the top and bottom chords in this type of truss are 2 × 4s, the webs are tubular steel. A 2 × 8 is installed below the top chords at standard intervals to provide lateral support for the truss.

Floor trusses are usually built in a manufacturing facility by the truss manufacturer. Truss design and length are determined by information on the construction prints. The finished trusses are delivered to the job site and installed by carpenters.

Floor trusses offer the advantage of covering long spans without requiring intermediate support of a wall or beam. Also, the space between the top and bottom chords makes it convenient to run pipes, electrical wires, and ducts through the flooring system. Floor trusses are usually spaced 24″ OC. They provide a 3½″ nailing surface for the subfloor.

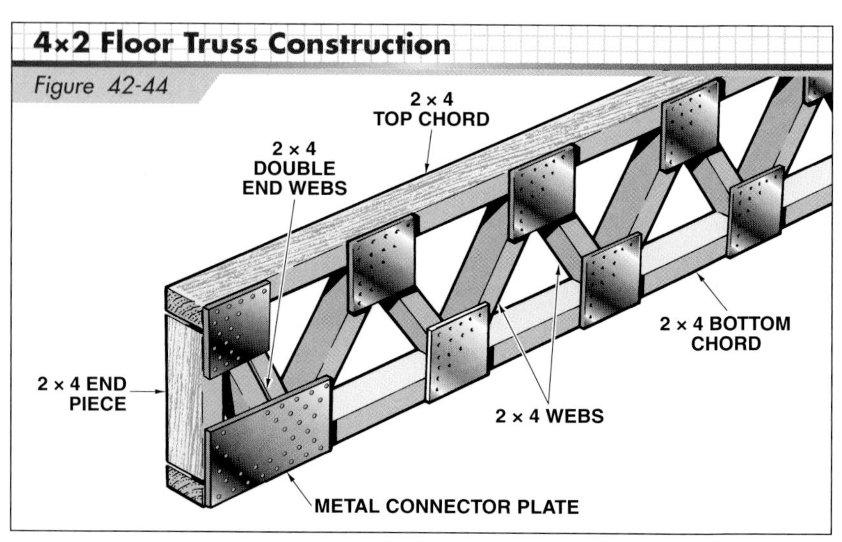

Figure 42-44. *The top and bottom chords of floor trusses are tied together by webs. The chords and webs are typically 2 × 4s and are joined together with metal connector plates.*

Southern Forest Products Association

Figure 42-45. *Floor trusses may be constructed with 2 × 4 wood top and bottom chords and tubular steel webs. The 2 × 4 shown along the bottom chord toward the left side stabilizes and ties the trusses together.*

WOOD I-JOISTS

Wood I-joists are an engineered wood product that are widely used in the construction of wood-framed floor units and flat roofs, and are a high-performance alternative to solid dimension lumber. Wood I-joists are structural, load-carrying products that provide high bending strength and good stiffness characteristics.

Wood I-joists consist of top and bottom flanges and webs that fit into grooves machined in the flanges. Flanges are dimension lumber or LVL and webs are plywood, LVL, or OSB. **See Figure 42-46.** The webs are bonded to the flanges with exterior-type adhesives.

Trus Joist, A Weyerhaeuser Business

Figure 42-46. *Wood I-joists provide a wide nailing surface for a subfloor.*

Wood I-joists are available in 9½″, 11⅞″, 14″, and 16″ depths. The flanges are 1½″, 1¾″, 2⁵⁄₁₆″, 2½″, or 3½″ wide, depending on the design. Wood I-joists are manufactured in lengths up to 60′, eliminating joist overlapping at central beams or walls. Wood I-joists are easy to grip and are lightweight; a 26′ wood I-joist weighs approximately 50 lb.

APA *Performance Rated I-joists (PRI)* are wood I-joists designed for use in residential floor construction. APA PRIs are manufactured per APA Standard PRI-400, *Performance Standard for APA EWS I-Joists.* Wood I-joists are identified by their depth (overall measurement from top to bottom of I-joist) and a joist designation ranging from PRI-20 to PRI-80. A typical I-joist trademark is shown in **Figure 42-47.**

In general, wood I-joists do not require intermediate blocking between spans if the I-joists conform to specifications in wood I-joist span tables. **See Figure 42-48.** Blocking is required, however, if header joists or rim boards are not installed between wood I-joists supporting load-bearing floors, and between cantilevered wood I-joists.

Workers should not walk on I-joists until the joists are completely installed and braced.

WOOD I-JOIST SPAN TABLE

Depth*	Joist Designation	Simple Spans			
		On-Center Spacing*			
		12	16	19.2	24
9½	PRI-20	16′-7″	15′-2″	14′-4″	13′-4″
	PRI-30	17′-1″	15′-8″	14′-10″	13′-11″
	PRI-40	18′-0″	16′-6″	15′-7″	14′-1″
	PRI-50	17′-10″	16′-4″	15′-5″	14′-5″
	PRI-60	19′-0″	17′-4″	16′-4″	15′-4″
11⅞	PRI-20	19′-11″	18′-2″	17′-2″	15′-5″
	PRI-30	20′-6″	18′-9″	17′-9″	16′-7″
	PRI-40	21′-6″	19′-7″	18′-2″	16′-3″
	PRI-50	21′-4″	19′-6″	18′-5″	17′-3″
	PRI-60	22′-8″	20′-8″	19′-6″	18′-3″
	PRI-70	23′-0″	21′-0″	19′-10″	18′-7″
	PRI-80	24′-11″	22′-8″	21′-4″	19′-11″
	PRI-90	25′-8″	23′-4″	22′-0″	20′-6″
14	PRI-40	24′-4″	22′-1″	20′-2″	18′-0″
	PRI-50	24′-4″	22′-3″	21′-0″	19′-8″
	PRI-60	25′-9″	23′-6″	22′-2″	20′-9″
	PRI-70	26′-1″	23′-10″	22′-6″	21′-0″
	PRI-80	28′-3″	25′-9″	24′-3″	22′-8″
	PRI-90	29′-1″	26′-6″	24′-11″	23′-3″
16	PRI-40	27′-0″	24′-0″	21′-11″	19′-7″
	PRI-50	27′-0″	24′-8″	23′-4″	20′-2″
	PRI-60	28′-7″	26′-1″	24′-7″	23′-0″
	PRI-70	29′-0″	26′-5″	24′-11″	23′-1″
	PRI-80	31′-4″	28′-6″	26′-11″	25′-1″
	PRI-90	32′-2″	29′-3″	27′-7″	25′-9″

* in in.

APA—The Engineered Wood Association

Figure 42-48. *Span tables specify on-center spacing of various sizes of wood I-joists.*

One benefit of using wood I-joists in residential construction is that holes may be drilled or cut in the webs to accommodate plumbing, electrical, and other mechanical systems. Top and bottom flanges must never be cut, notched, or otherwise modified. *Knockouts* are prescored holes that are commonly provided by I-joist manufacturers to allow electrical wiring, plumbing lines, or other mechanical systems to be easily installed. Knockouts are typically 1⅜″ to 1¾″ diameter and are spaced 12″ to 24″ OC along the I-joist length. If knockouts are not provided, holes should be cut using a sharp saw. Holes should be cut only in areas designated by the manufacturer or engineer and should not exceed stated specifications.

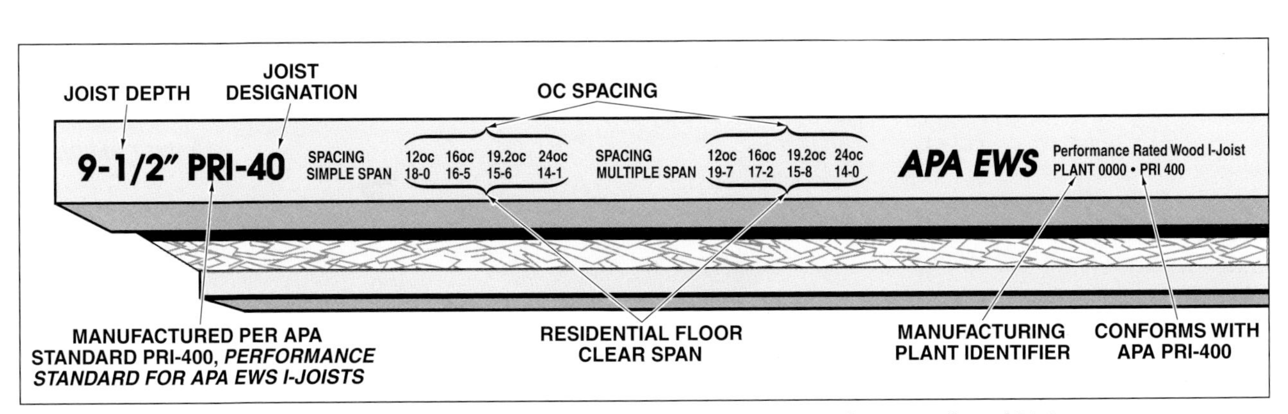

Figure 42-47. *Trademarks provide information regarding recommended clear spans and spacing of wood I-joists*

Construction Methods

The spacing and layout of wood I-joists is the same as for solid dimension lumber. APA Performance Rated *rim boards* are used as header joists for wood I-joists. **See Figure 42-49.** A rim board ties together the wood I-joists, and also transfers lateral and vertical bearing forces. Rim boards manufactured from plywood, OSB, glulam timber, or LVL are used to tie together wood I-joists since they shrink less than dimension lumber and correspond to the depth of the wood I-joist and other engineered wood products.

Wood I-joists are fastened to rim boards or other flat surfaces using top-flange or face-mount joist hangers. Hangers should be deep and wide enough to allow the bottom flange of the I-joist to fully seat in the hanger and support the top flange. **See Figure 42-50.** If joist hangers are not deep enough, bearing stiffeners must be used to provide proper support. Some joist hangers have a tab that is bent over the upper surface of the bottom flange and fastened with a 10d nail.

Due to the "I" shape of wood I-joists, *backer blocks, web stiffeners, filler blocks, squash blocks,* and *blocking* may be installed to prevent twisting or to provide additional support and reinforcement. Blocks and stiffeners are typically made from OSB, dimension lumber, or structural panel material.

Trus Joist, A Weyerhaeuser Business

Figure 42-50. *Joist hangers for wood I-joists should be deep enough to support the top flange.*

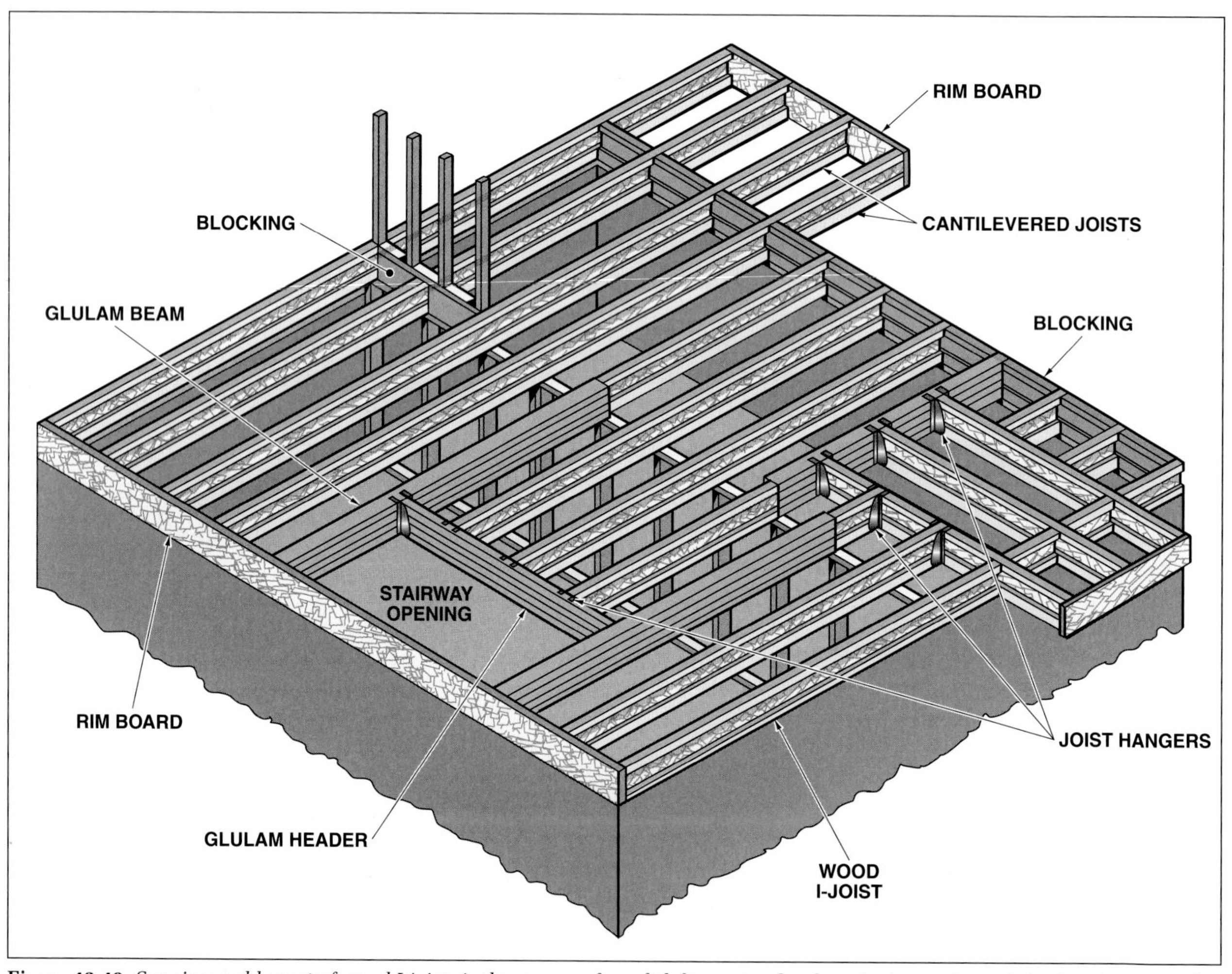

Figure 42-49. *Spacing and layout of wood I-joists is the same as for solid dimension lumber. Engineered wood rim boards are used to support the ends of I-joists.*

Backer Blocks. Backer blocks provide a flat, flush surface for attachment of top- or face-mounted joist hangers or other structural elements by filling the space between the outside edge of the I-joist flange and the web of the adjoining I-joist. Backer blocks must be long enough to permit nailing without splitting. The depth of backer blocks should be approximately 1/8″ less than the distance between the flanges.

Backer blocks should fit tight against the top flange of an I-joist and are fastened into position with three 10d nails, which are clinched if possible. **See Figure 42-51.** For top-flange hangers, a backer block is required only on the nailing side. For face-mount hangers, backer blocks are required on both sides of the adjoining I-joist to provide good anchorage for joist hangers.

Web Stiffeners. Web stiffeners are dimension lumber, OSB, or structural panel materials used to reinforce an I-joist web. Wood I-joist webs may be reinforced in the following situations:

- when the webs, especially in deeper I-joists, are in danger of buckling out of plane
- when the webs may break through the grooved portion of the flange due to increased vertical loads
- when the I-joist is supported in a joist hanger that does not extend to the top flange, allowing the I-joist to deflect laterally

Bearing stiffeners and *load stiffeners* are two types of web stiffeners and are differentiated by the applied load and the location of the gap between the stiffener and top or bottom flange. **See Figure 42-52.** Bearing stiffeners are installed when increased load-bearing capacity is required. Load stiffeners are installed between supports when significant load points are anticipated. The depth of web stiffeners is 1/8″ to 1/4″ less than the distance between flanges. Bearing stiffeners are fitted tightly against the upper surface of the bottom flange; load stiffeners are fitted tightly against the lower surface of the top flange. Most web stiffeners are fastened into position with four staggered 8d nails, which are clinched if possible. Web stiffeners for wood I-joists with 3½″ flanges are fastened into position with four staggered 10d nails.

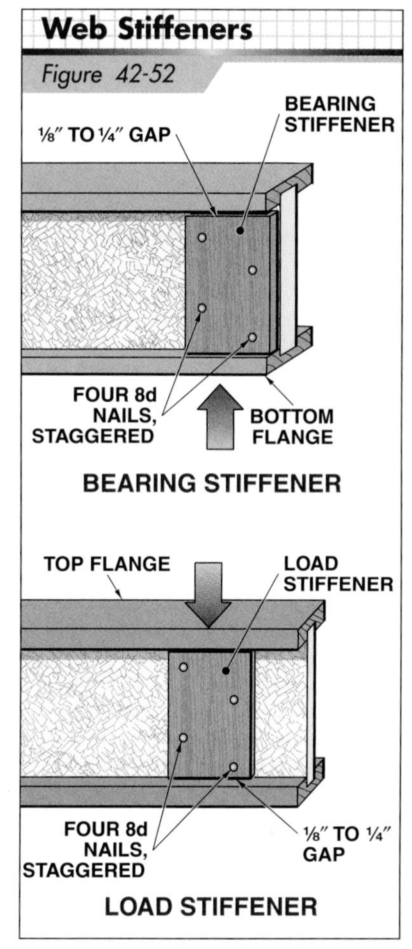

Figure 42-52. *Web stiffeners reinforce I-joist webs. The direction of an applied load determines the location of the gap between the stiffener and top or bottom flange.*

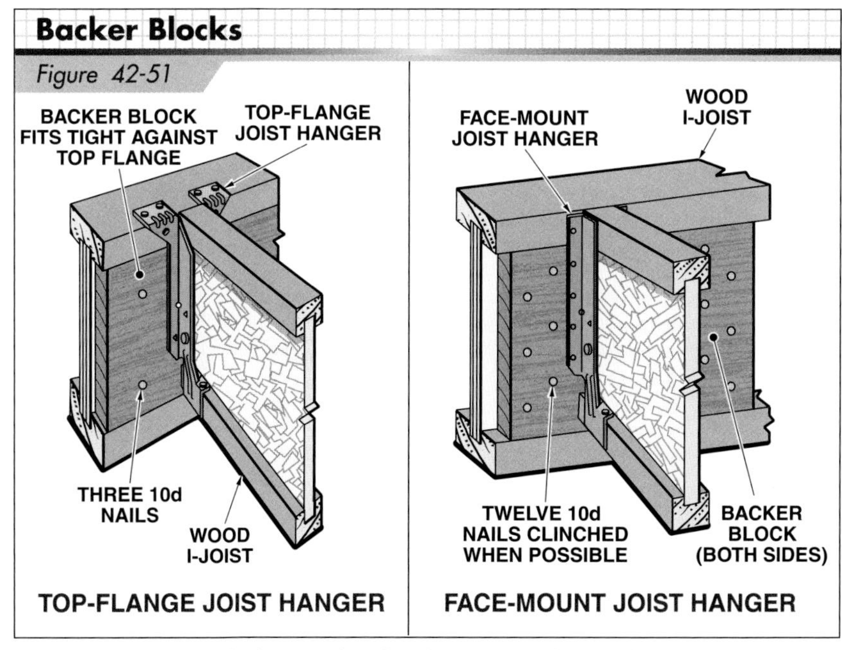

Figure 42-51. *Backer blocks provide a flat, flush surface for attachment of top-flange or face-mount joist hangers or other structural elements.*

Filler Blocks. Filler blocks fill the rectangular space between a pair of I-joists that are used as a single member, such as when I-joists are doubled at door or window openings. Filler blocks permit the vertical load to be shared between the two I-joists and force each joist to absorb equal amounts of the load and bend the same amount under the applied load.

Filler blocks are installed the full length of the double I-joists, either as a single member or as shorter pieces. The depth of filler blocks is $\frac{1}{8}$″ less than the distance between the flanges. Filler block thickness must also be appropriate for the size of wood I-joist being used. Filler blocks slightly thicker than the recommended size shown in **Figure 42-53** can be used without affecting performance of the structural unit. Thinner filler blocks leave a void between the webs and filler blocks, permitting twisting or rotation of the webs during nailing.

Squash Blocks. Squash blocks carry a *point load* that would otherwise bear directly on a wood I-joist, such as when a post is installed directly over another post on a lower level. Squash blocks are pieces of dimension lumber or rim board, and are typically installed in pairs to allow even distribution of a load. If pieces of dimension lumber, such as 2 × 4s or 2 × 6s, are used as squash blocks, they should be cut $\frac{1}{16}$″ longer than I-joist depth to ensure the squash blocks provide support for the vertical loads. **See Figure 42-54.** If rim board is used as squash blocks, select rim board of compatible depth with the I-joists being installed.

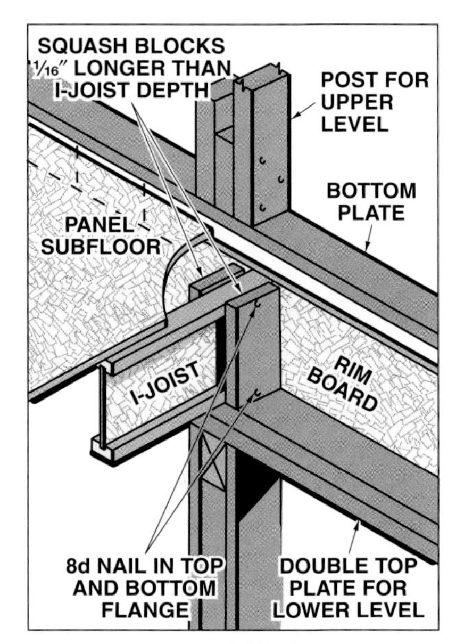

Figure 42-54. *Squash blocks carry a point load that would otherwise bear directly on a wood I-joist.*

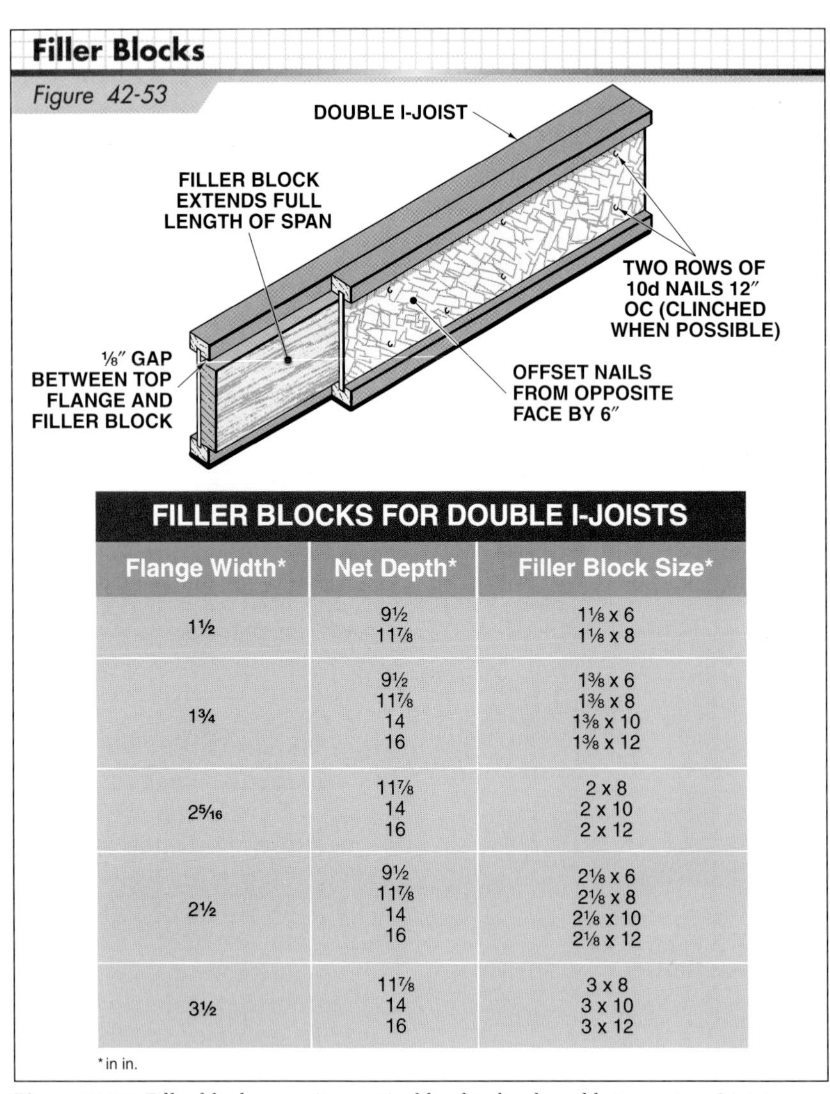

FILLER BLOCKS FOR DOUBLE I-JOISTS

Flange Width*	Net Depth*	Filler Block Size*
1½	9½ 11⅞	1⅛ x 6 1⅛ x 8
1¾	9½ 11⅞ 14 16	1⅜ x 6 1⅜ x 8 1⅜ x 10 1⅜ x 12
2$\frac{5}{16}$	11⅞ 14 16	2 x 8 2 x 10 2 x 12
2½	9½ 11⅞ 14 16	2⅛ x 6 2⅛ x 8 2⅛ x 10 2⅛ x 12
3½	11⅞ 14 16	3 x 8 3 x 10 3 x 12

* in in.

Figure 42-53. *Filler blocks permit a vertical load to be shared between two I-joists.*

Blocking. Wood I-joists are commonly installed with engineered wood or sections of I-joist placed between joists at various locations. Blocking provides lateral support for floor I-joists, transfers shear loads from walls above to floor or foundation below, and transfers vertical loads from the walls above to the floor or foundation below.

Job-built blocking is fabricated from available engineered wood products, including wood I-joists, rim board, or LVL. Engineered wood products must be used for blocking because shrinkage encountered in dimension lumber would affect the stability and ability to transfer loads. Blocking must fit tightly between I-joists. Blocking is recommended for the following situations:

- at each end of floor joists that are not restrained by a header joist or rim board. **See Figure 42-55.**
- between floor joists supporting load-bearing walls extending perpendicular to joists
- between cantilevered wood I-joists over an adjacent supporting wall

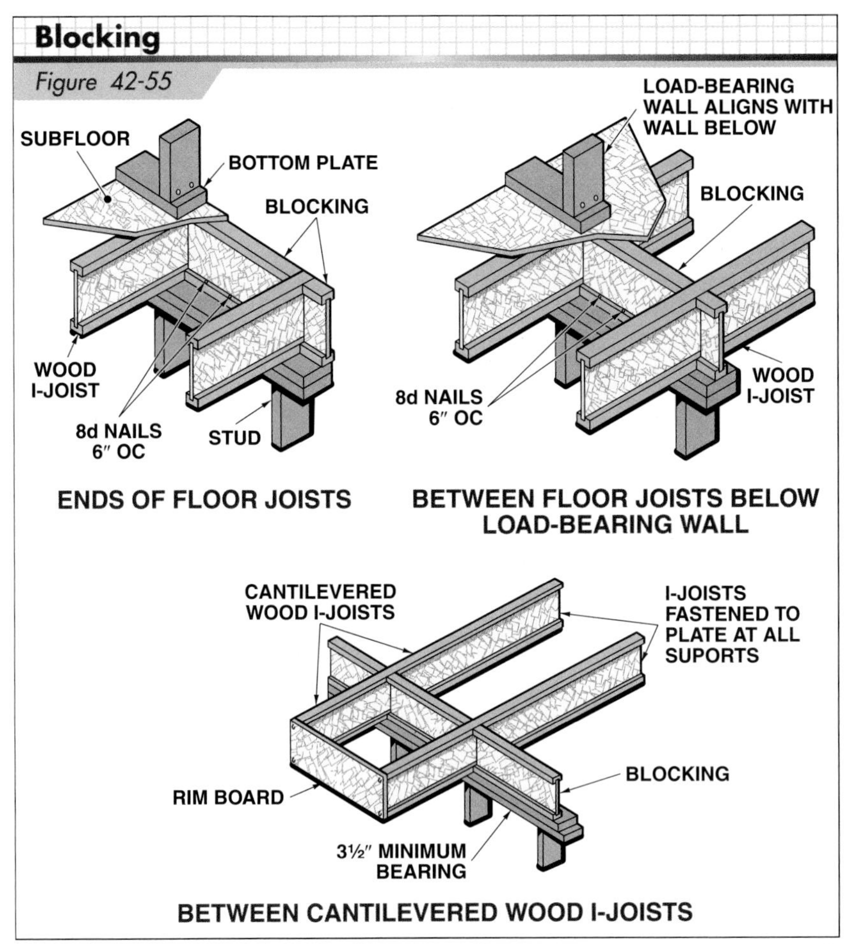

Figure 42-55. *Blocking provides lateral support for floor I-joists, transfers shear loads from walls above to floor or foundation below, and transfers vertical loads from the walls above to the floor or foundation below.*

UNIT 43 Wall Framing

OBJECTIVES

1. Compare the construction of exterior walls and interior partitions.
2. List and describe wood-framed wall components.
3. Describe the construction of door and window openings.
4. Describe the purpose of fireblocking and fireblocking caulk.
5. Describe the construction of wood-framed walls.
6. Describe the procedure for laying out walls according to a floor plan.
7. Explain how vertical layout is performed.
8. Describe the procedure for constructing corner posts.
9. Discuss the procedure for framing door and window openings.
10. Describe the procedure for assembling wood-framed walls.
11. Review the procedure for raising, plumbing, and aligning walls.
12. Describe the procedure for framing over concrete slabs.
13. List the procedures for sheathing exterior walls.
14. Describe the use of weather barriers and rainscreen walls.
15. Describe the purpose and construction of shear walls.
16. Describe the use of anchor tiedown systems.
17. Describe the advantages of advanced framing.

Wall construction begins after the subfloor has been fastened in place. The wall system of a wood-framed building consists of exterior (outside) and interior (inside) walls. Exterior walls have door and window openings. Interior walls, known as *partitions,* commonly have door openings. Partitions divide the living area of the house into separate rooms. Door openings and archways provide access from room to room.

Partitions are either *load-bearing* or *non-load-bearing.* Load-bearing partitions support the ends of the floor or ceiling joists. Non-load-bearing partitions usually run in the same direction as the joists and carry little weight from the floor or ceiling above.

Traditionally, 2 × 4 lumber was used for the exterior wood-framed walls of one-story buildings. Today, 2 × 6s are commonly used for exterior wall framing since the additional width of the studs and plates provides room for thicker insulation materials.

WOOD-FRAMED WALL COMPONENTS

Wood-framed walls consist of structural parts referred to as *wall components* or *framing members.* Components of a wood-framed wall include *studs, plates, headers, trimmer studs, sills, cripple studs,* and *corner posts.* **See Figure 43-1.** For some applications, *diagonal braces* may also be required. Each component has a special function within the total wall structure.

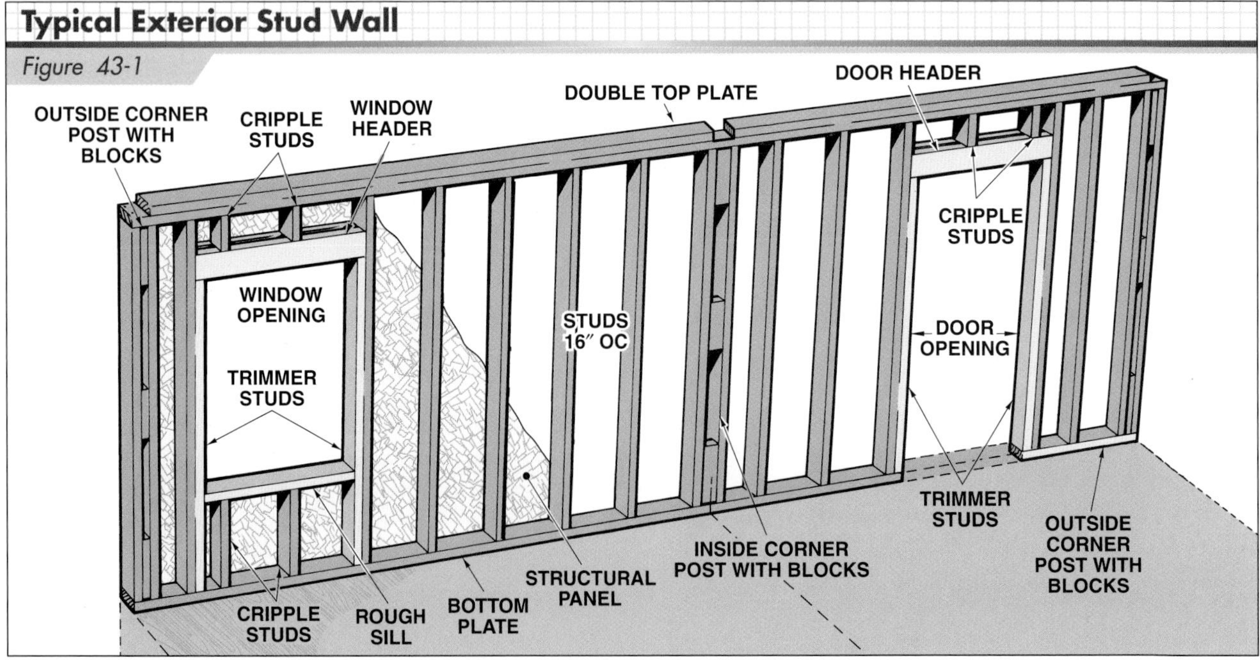

Figure 43-1. *Wood-framed wall components include studs, plates, headers, trimmer studs, sills, cripple studs, and corner posts. Exterior walls are commonly covered with structural panels such as oriented strand board or plywood.*

Studs, Plates, and Corner Posts

Studs are vertical framing members that run between the wall plates. Studs are usually spaced 16″ OC. In some areas of the country, advanced framing methods can be used that allow 24″ OC stud spacing in certain buildings.

The *plate* at the bottom of a wall is the *bottom plate,* or sole plate. The plate at the top of the wall is the *top plate.* The top plate is typically doubled for additional stability. The topmost plate of a double top plate is sometimes called a *doubler.* A doubler strengthens the upper section of a wall and helps carry the weight of floor or ceiling joists. Wall plates are nailed to the wall studs, including trimmer studs and cripple studs, and tie the entire wall together.

Corner posts, also called corner assemblies, are constructed wherever a wall ties into another wall. Outside corners are at the ends of a wall. Inside corners occur where a partition ties into a wall at some point between the ends of the wall. Three typical designs for corner posts are shown in **Figure 43-2.** All corner posts should be constructed from straight studs and should be well-nailed.

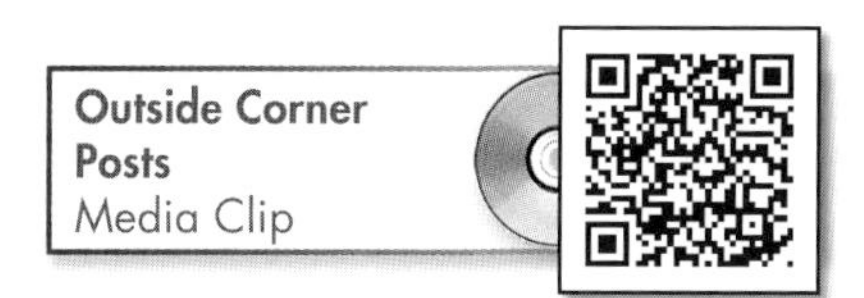

When large quantities of inside and outside corner posts of the same dimension are required for a project, a bench can be constructed using 2 × 6s to hold the corner post studs and blocks in position while nailing.

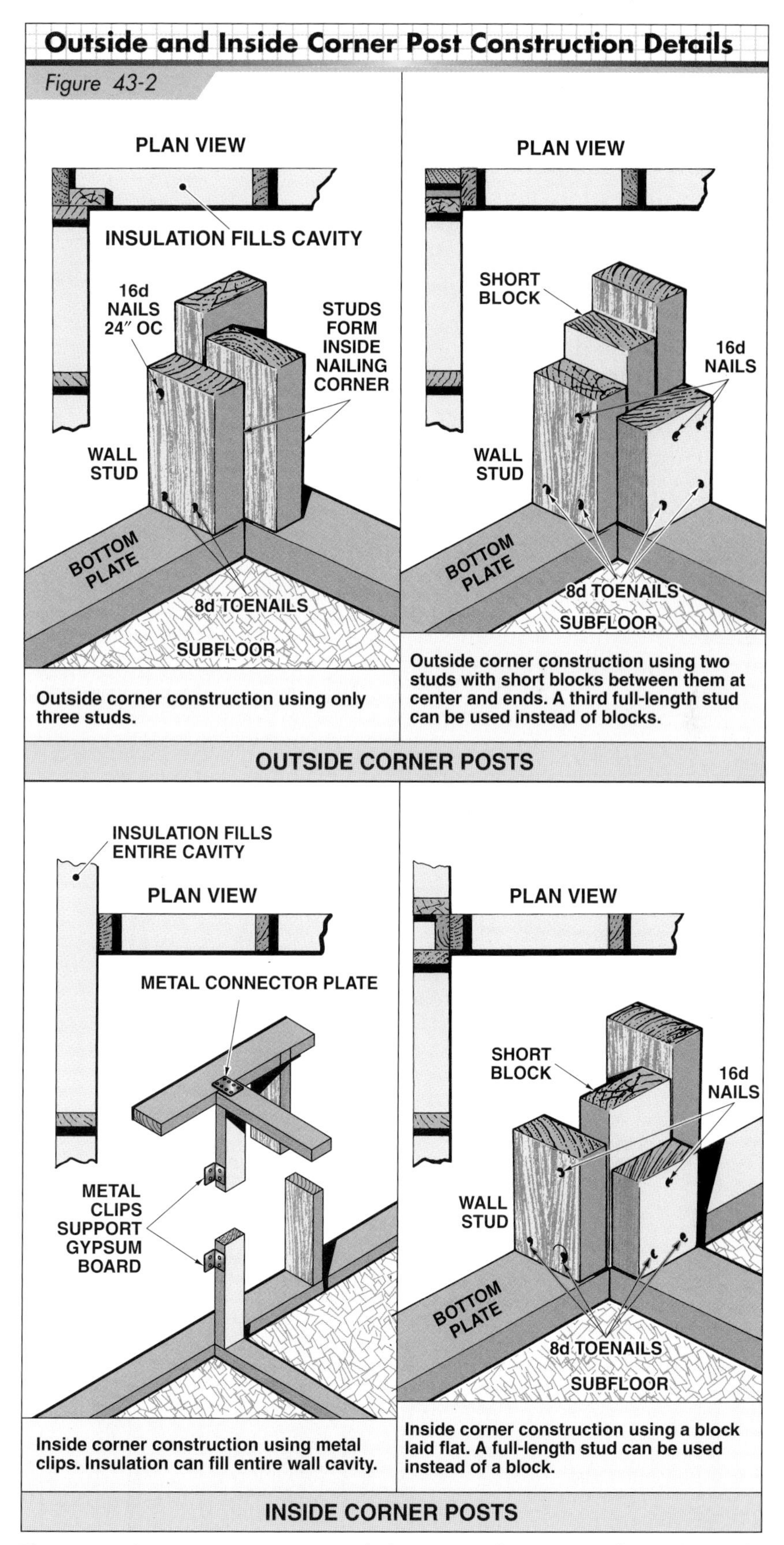

Figure 43-2. *Corner posts are constructed wherever a wall ties into another wall. Straight studs are used to frame corner posts. When an inside corner is constructed using a block laid flat, a carpenter must insulate the U-shaped cavity before exterior wall sheathing is applied. When metal clips are used, the wall cavity can be insulated after exterior wall sheathing is applied.*

Door and Window Openings

A *rough opening* must be framed in a wall wherever a door or window is planned. Rough opening dimensions must allow for the finish frame into which the door or window will fit and for the required clearance around the frame. Methods of calculating the dimensions are explained later in this unit.

Rough Door Openings. The rough opening for a door is framed with a *header, trimmer studs*, *wall studs*, and, in some cases, *cripple studs.* The rough opening for a typical window includes the same members as for a door, plus a rough sill and bottom cripple studs. **See Figure 43-3.**

Glulam timbers may be used as door and window headers.

Wood-Framed Rough Openings Media Clip

Southern Forest Products Association

Rough sills may be doubled to provide support for heavy window units.

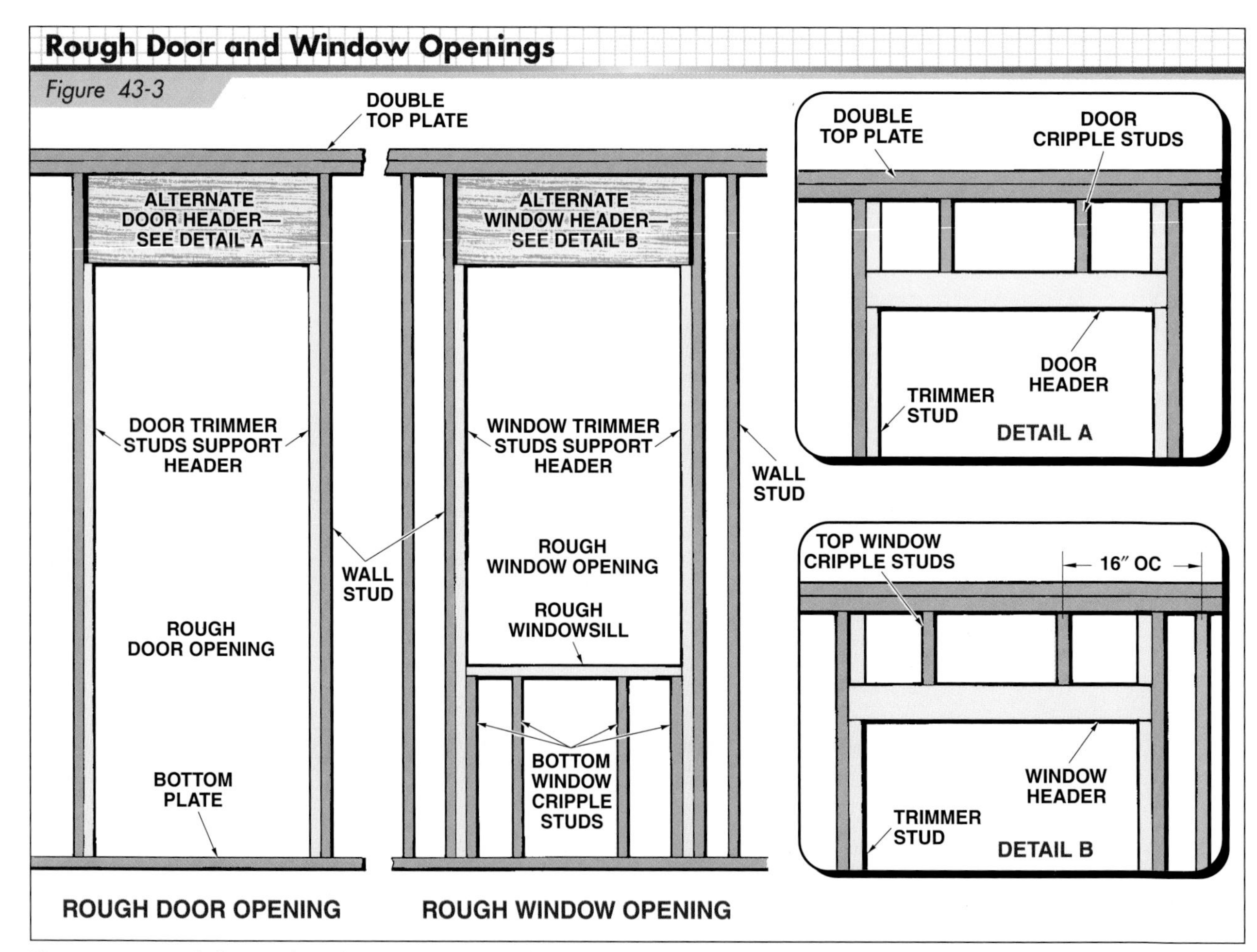

Figure 43-3. *Rough door and window openings must allow for the finish frame and a required clearance around the frame.*

A header is placed at the top of a rough opening and must be strong enough to carry the weight bearing down on that section of the wall. In some areas of North America, headers are commonly referred to as *lintels*. A header is supported by *trimmer studs* that fit between the bottom plate and bottom of the header. Trimmer studs are nailed into the wall studs at each side of the header. Nails are also driven through the wall studs into the ends of the header.

Headers may be solid pieces or may be built up of two pieces with ½″ spacer blocks between them. Spacer blocks are needed to bring the width of the header to 3½″, which is the actual width of a 2 × 4 stud wall. A built-up header is as strong as a solid piece. However, it involves extra labor to construct a built-up header. Another type of built-up header is made of two planks, such as 2 × 10s, with a 2 × 4 piece nailed at the bottom. For 2 × 6 walls, a solid piece of lumber (for example, a 6 × 12) may be used as a header. Another common header for 2 × 6 walls consists of two 2 × 10s with a 2 × 6 nailed to the bottom. **See Figure 43-4.**

Another type of header is similar to a wood I-joist. However, the web is constructed of expanded polystyrene (EPS) insulation sandwiched between two pieces of OSB. This type of header is available with 3½″ and 5½″ flanges and in depths of 9¼″ and 11¼″ or greater.

The required header type and size are shown on the prints. Header size is determined by the width of the opening and the load bearing down from the floor above. **See Figure 43-5.**

The tops of door and window openings in all walls are usually aligned with each other. Therefore, all headers are the same distance from the floor. The standard height of walls in most residential wood-framed buildings is 8′-0¾″ or 8′-1″ from the subfloor to the ceiling joists. Standard door height is 6′-8″.

For standard-height door openings in standard-height walls, cripple studs are not necessary if a 12″ wide header is placed directly below the top plate. The distance between the bottom of the 12″ wide header and the subfloor allows for the 6′-8″ door height, thickness of the head jamb, clearance above the head jamb, and required clearance below the door. **See Figure 43-6.** The clearance below the door varies with the thickness of the finish floor material.

If a header less than 12″ wide is used, cripple studs are necessary. Cripple studs are nailed between the header and the double top plate of a door or window opening to carry the weight from the top plate to the header. The cripple studs are generally spaced on regular stud layout.

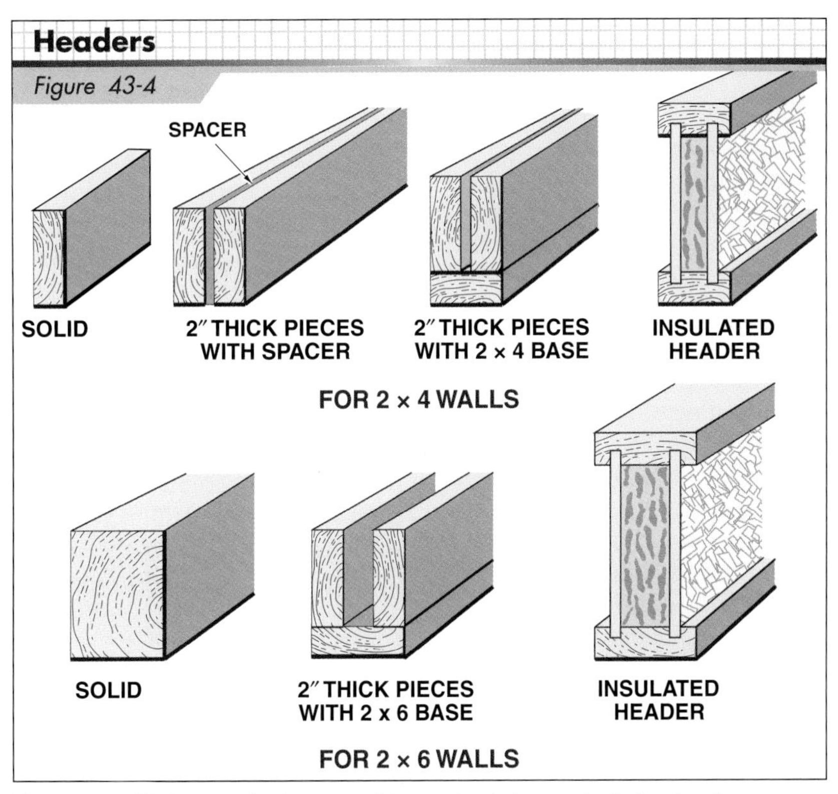

Figure 43-4. *Various methods are used to construct door and window headers.*

HEADER SIZES FOR DOOR AND WINDOW OPENINGS

Lumber Size	Species	Minimum Grade	Maximum Allowable Span*
4 × 4	Douglas Fir	No. 2	4
4 × 6	Douglas Fir	No. 2	6
4 × 8	Douglas Fir	No. 2	8
4 × 10	Douglas Fir	No. 2	10
4 × 12	Douglas Fir	No. 2	12
4 × 14	Douglas Fir	No. 1	16 (commonly used for garage door)

* in ft

Figure 43-5. *Header size is determined by the width of the opening and load bearing down from above. For example, a 4 × 6 Douglas fir header can be used for an opening up to 6′-0″ wide.*

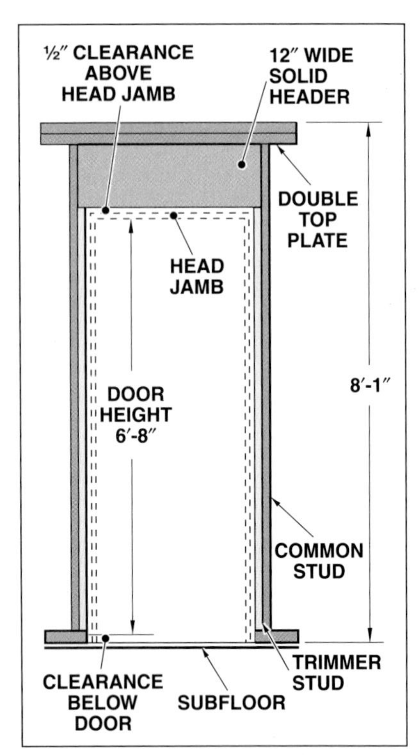

Figure 43-6. *A 12″ wide solid header can be used to accommodate a 6′-8″ door in an 8′-1″ standard-height wall.*

Header-cripple stud construction is usually used only where walls are more than 8′-1″ high. For walls higher than 8′-1″, a 12″ wide header placed directly under the top plate would result in a rough opening that would be too tall for standard doors. Cripple studs must be used to provide a rough opening of the correct height. **See Figure 43-7.**

Trus Joist, A Weyerhaeuser Business

Figure 43-7. *When a header less than 12″ wide is installed, cripple studs are installed between the top of the header and top plate.*

Rough Window Openings. A rough sill is installed at the bottom of a rough window opening. The *rough sill* provides support for the finish window and frame that will later be placed in the wall. The distance between the sill and header is determined by the dimensions of the window, window frame, and necessary clearances at the top and bottom of the frame. Cripple studs are nailed between the sill and bottom plate and spaced 16″ OC like the rest of the wall. Additional cripple studs may be placed under each end of the sill. Some building codes require that sills be doubled if the opening is more than 4′ wide.

The framing for the rough openings of oval or round windows may require corner bracing so that the finish window frames can be secured from all sides. **See Figure 43-8.** Some oval, round, or irregularly shaped windows have bracing and other framing members in place when they are delivered. The window units are installed in the rough opening and fastened in place.

Figure 43-8. *Corner braces may need to be installed in rough openings for installation of oval or round windows.*

Diagonal Braces

When exterior walls are covered with structural sheathing, such as OSB or plywood, diagonal braces may not be required. Diagonal braces, however, are required when improved lateral wall strength is needed. When diagonal braces are required, braces should be placed at both ends (where possible) and at 25′ intervals in exterior walls and main interior partitions.

Diagonal braces are most effective when installed at a 45° to 60° angle. The braces are installed in the wall after the wall has been squared and is still lying on the subfloor.

Two diagonal bracing systems are metal wall braces and 1 × 4 let-in braces. Metal wall braces are 16 ga to 22 ga L- or T-shaped metal pieces that are installed in a saw kerf running diagonally across a wall section. **See Figure 43-9.** A diagonal line is snapped diagonally across the wall section and a saw kerf is made along the line. A metal wall brace is placed in the kerf and secured to the top plate. The wall is squared and the brace is secured to the studs it crosses.

Figure 43-9. *T-shaped metal braces fit into saw kerfs in the studs.*

For a 1 × 4 wood let-in brace, the studs are notched so a 1 × 4 brace will fit flush with the surface of the studs. Metal wall braces are more commonly used than wood diagonal bracing because they are easier and faster to install than wood let-in bracing.

Wood Let-In Braces
Media Clip

Cripple studs support the rough sill at the bottom of a window opening.

Fireblocking

Fireblocking is required in concealed spaces to cut off vertical and horizontal draft openings and to form a barrier between stories, and between the top story and attic or roof space. Fireblocking creates a barrier to prevent smoke, toxic fumes and gases, and fire from spreading from one room to another, or from one floor to another, allowing inhabitants more time to escape from the building.

Fireblocking requirements are different for residential and commercial construction, and vary from jurisdiction to jurisdiction. Always consult the applicable local building code for specific fireblocking requirements. In general, fireblocking is required in the following locations for residential construction:

- stud walls and interior partitions at the floor and ceiling level
- junction between concealed wall spaces, such as those at soffits, coves, or drop ceilings. **See Figure 43-10.**
- concealed spaces between stair stringers at the top and bottom of the stringers
- concealed spaces around fireplace and chimney openings
- openings around ducts, pipes, and vents at the floor and ceiling level

A variety of materials can be used for fireblocking, including the following:

- 2″ nominal lumber
- two thicknesses of 1″ nominal lumber with reinforced joints
- one thickness of 23/32″ wood structural panels with reinforced joints
- one thickness of 3/4″ particleboard with reinforced joints
- batts or blankets of mineral wool or glass wool insulation
- 1/2″ gypsum board
- 1/4″ cement board
- approved nonrigid materials, such as caulk and foam

For platform-framed construction, the top and bottom plates serve as vertical fireblocking.

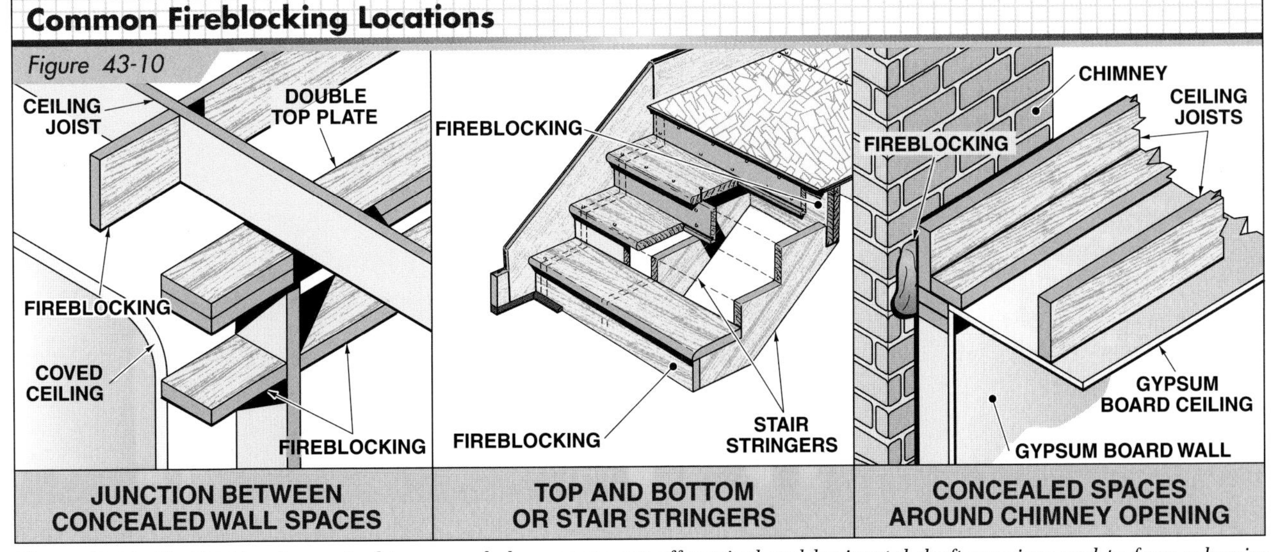

Figure 43-10. *Fireblocking is required in concealed spaces to cut off vertical and horizontal draft openings and to form a barrier between stories.*

Fireblocking Caulk. *Fireblocking caulk* is commonly used to seal holes drilled through walls or structural members. **See Figure 43-11.**

Figure 43-11. *Fireblocking caulk is applied to joints or openings in walls or structural members to inhibit the spread of fire, smoke, and fumes.*

Two types of fireblocking caulk are available—*intumescent* and *endothermic.* Intumescent caulk quickly expands in volume when exposed to high temperatures and closes off voids left by burning or melting construction materials. Endothermic caulk releases water vapors when exposed to high temperatures, slowing the spread of fire in the building.

Fireblocking caulk is dispensed from a tube which is placed in a caulking gun, or from a refillable caulking gun. The caulk is applied to joints or openings to the proper thickness. If a smooth, clean surface is desired, soapy water is applied from a spray bottle and the caulk is smoothed with a putty knife, paint brush, or similar tool.

CONSTRUCTING WOOD-FRAMED WALLS

All components of a wall should be cut before any components are assembled. By studying the prints, a carpenter can determine the number and lengths of studs, trimmer studs, cripple studs, headers, and rough sills required for the building. If necessary, the components are cut to length with a circular saw or radial-arm saw on the job site. For buildings on large housing tracts, framing members are often cut to length at the mill or lumberyard and then delivered to the job.

Wall components are assembled on the subfloor. The components are nailed together and the completed wall is raised into place. **Figure 43-12** shows a partially assembled wall lying on the subfloor of a building under construction.

Laying Out Walls

Carpenters must lay out where each wall is to be placed before construction begins. Wall layout requires printreading abilities and a thorough understanding of wall construction. Wall layout is usually performed by a lead carpenter or job supervisor. Two types of procedures are involved in wall layout—horizontal plate layout and vertical layout.

Fireblocking Caulk
Media Clip

Figure 43-12. *Wood-framed walls may be assembled while they are lying on the subfloor.*

Trus Joist, A Weyerhaeuser Business

Laminated strand lumber (LSL) headers are 3½" thick and are used for short-span window and door headers.

Horizontal Plate Layout. The first step in wall layout is to snap chalk lines on the subfloor to indicate the exact locations of walls. The locations are determined from the measurements provided on the floor plan. **See Figure 43-13.**

After the lines are snapped, the top and bottom plates are cut to length and tacked next to the chalk lines. **See Figure 43-14.** The plates are then marked for corner posts and regular studs, as well as for studs, trimmer studs, and cripple studs for rough openings. All framing members must be clearly marked on the plates for carpenters to frame the wall efficiently. **Figure 43-15** shows a wall with framing members nailed in position according to layout markings.

A procedure for marking outside and inside corners for stud-and-block corner posts is shown in **Figure 43-16.** A procedure for laying out studs for the first exterior wall is shown in **Figure 43-17.** In this common layout method, plates are marked for the first stud from a corner post by measuring 15¼″ from the end. Subsequent studs follow 16″ OC layout. This layout method ensures that the edges of standard wall sheathing or gypsum board panels fall on the centers of the studs. Cripple studs are laid out to follow the stud layout.

Figure 43-14. *The top and bottom plates are tacked next to the snapped chalk lines before laying out the wall components.*

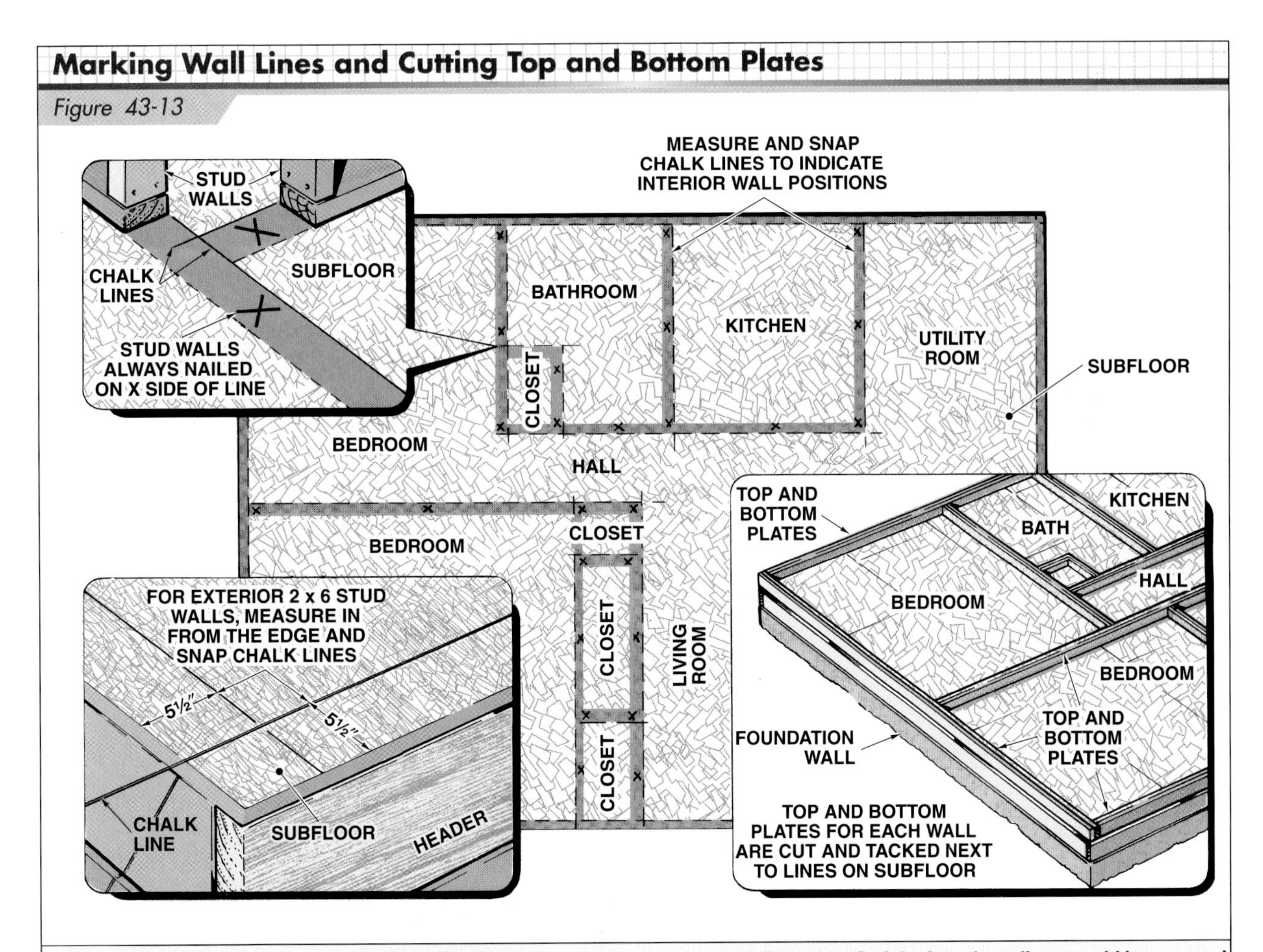

Figure 43-13. *Lines are snapped to indicate the exact locations of walls, and top and bottom plates are cut to length and tacked next to the chalk lines.*

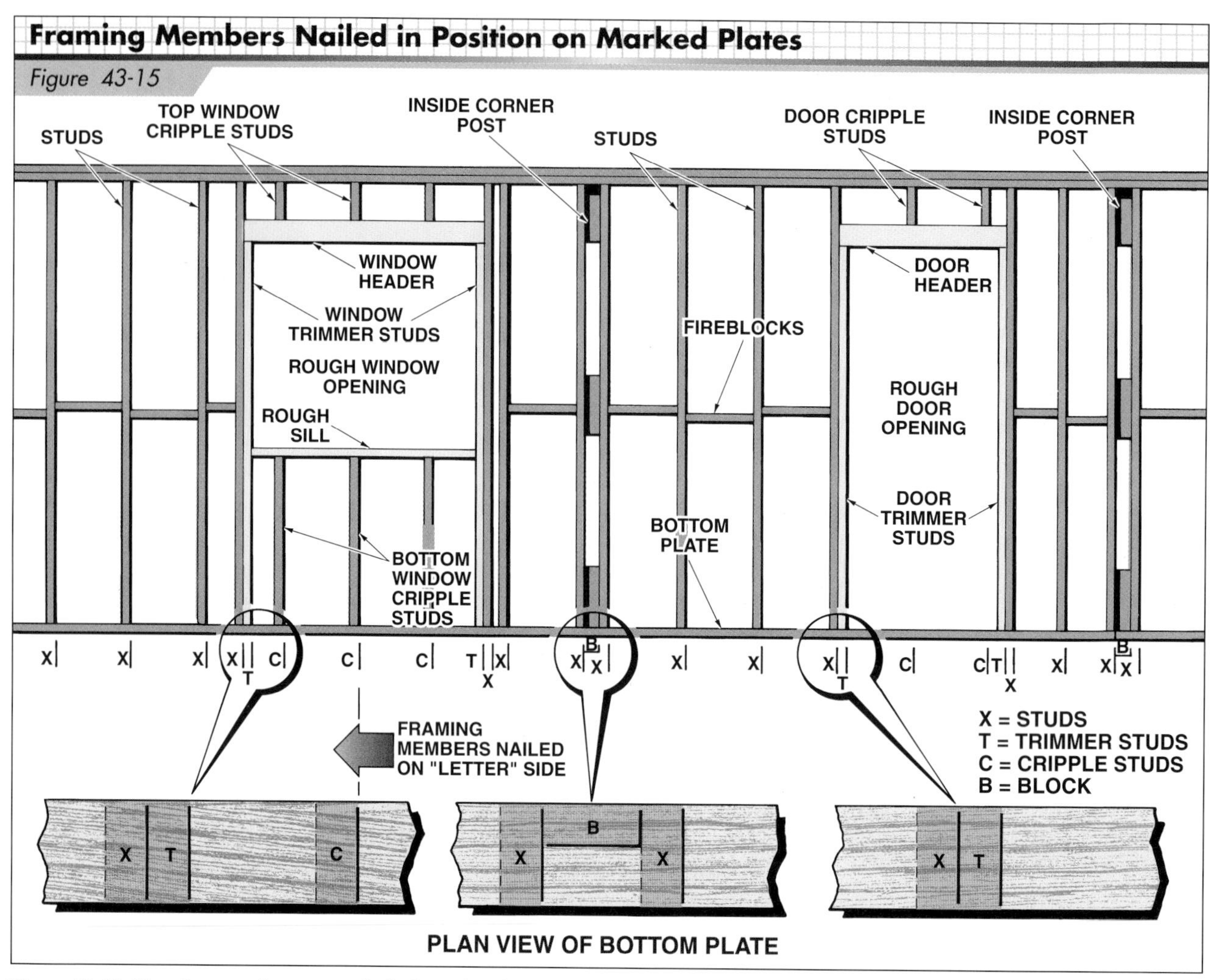

Figure 43-15. *Framing members are nailed where the plates are marked.*

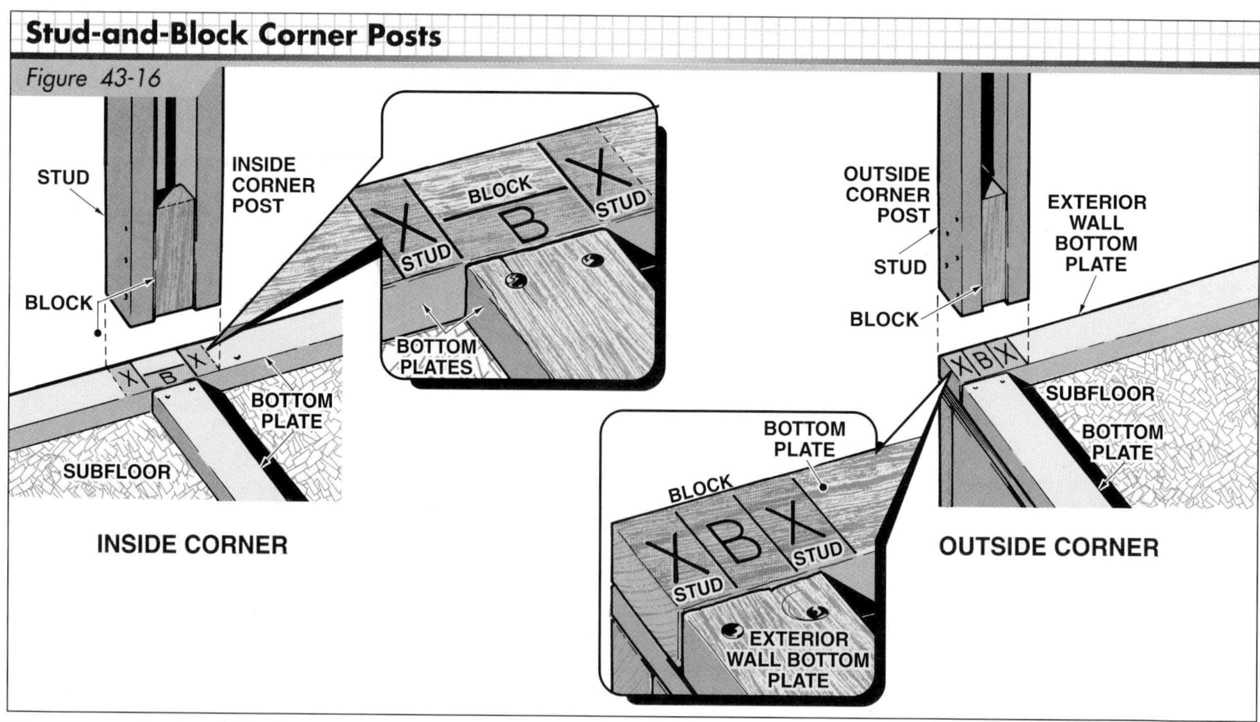

Figure 43-16. *Stud-and-block corner posts are constructed with full-length studs and blocks.*

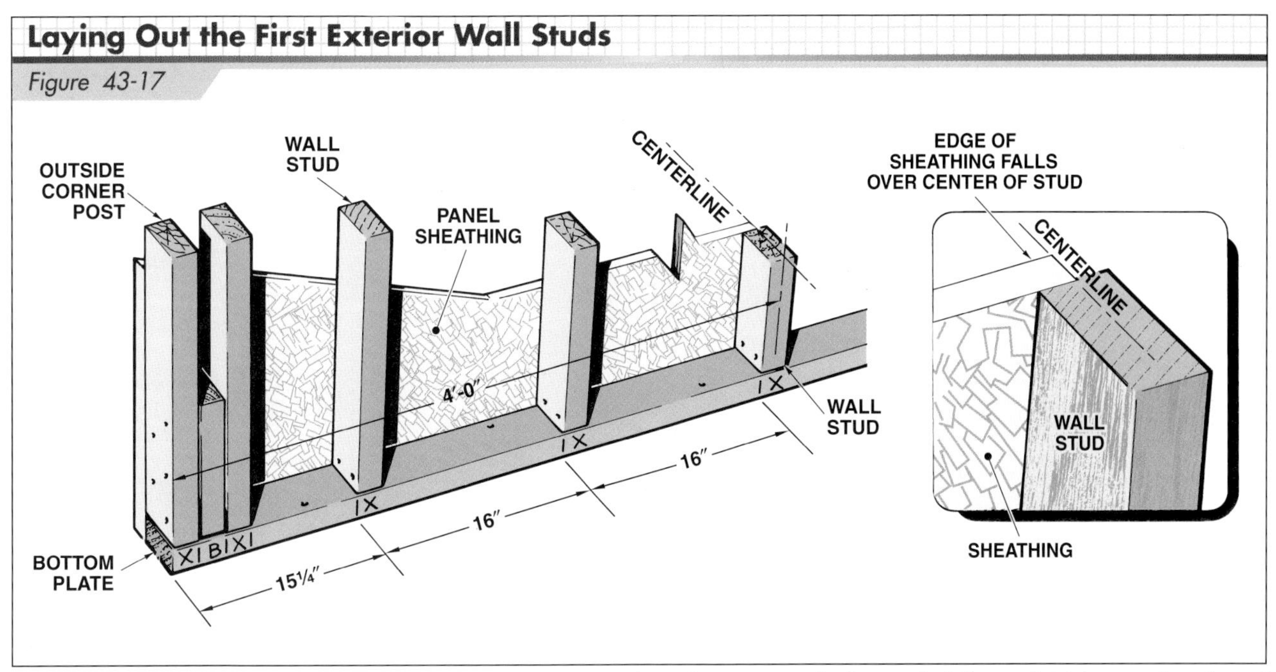

Figure 43-17. *If the first stud of an exterior wall is placed 15¼" from the corner and other studs follow 16" OC layout, the edges of standard-size panels will fall over the stud centers.*

A procedure for laying out studs for the second exterior wall is shown in **Figure 43-18.** The plates are marked for the first stud to be placed 15¼" from the outside edge of the panel on the first wall. This layout allows the corner of the first panel on the second wall to align with the edge of the first panel on the first wall. In addition, the opposite edge of the panel on the second wall will break on the center of a stud.

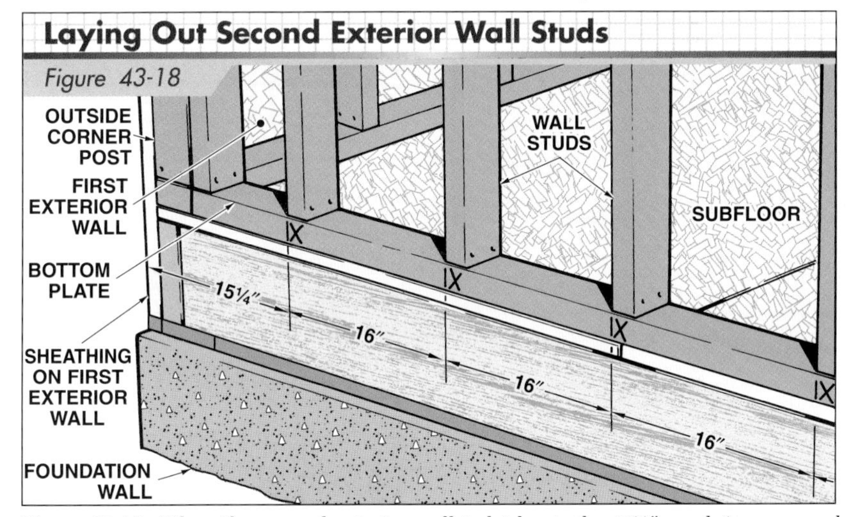

Figure 43-18. *When the second exterior wall is laid out, the 15¼" mark is measured from the outside edge of the panel on the first wall. The corner of the first panel on the second exterior wall will align with the edge of the first wall panel. The opposite edge of the panel will fall on the center of a stud.*

A procedure for laying out studs for partitions is shown in **Figure 43-19.** If wall panels are placed on the exterior wall first, followed by the partitions, wall plates for the partition are marked for the first stud to be placed 15¼" from the edge of the panel on the exterior wall. If panels are to be placed on the partitions before they are placed on the exterior wall, then the wall plates of the interior wall are marked for the first stud to be placed 15¼" from the unpaneled exterior wall.

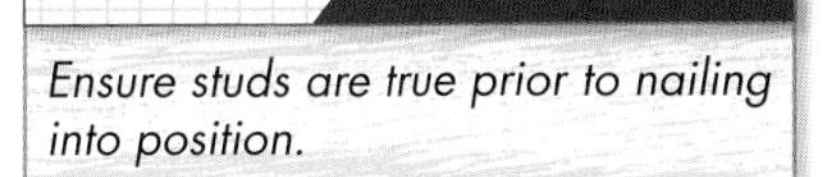

Ensure studs are true prior to nailing into position.

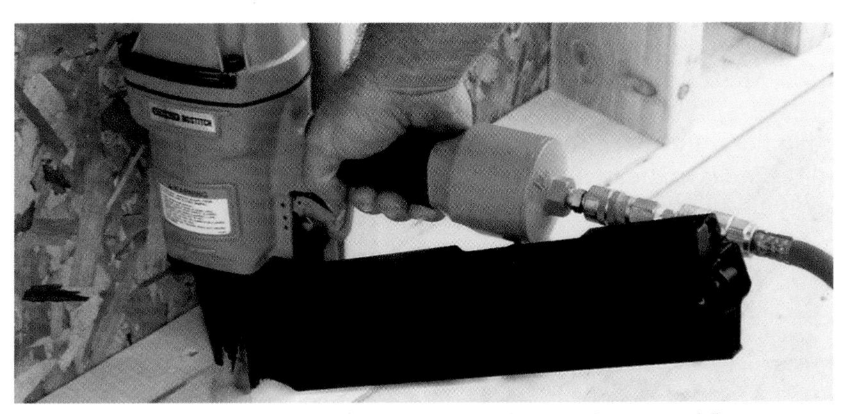

A pneumatic nailer is commonly used to fasten the bottom plate to a subfloor.

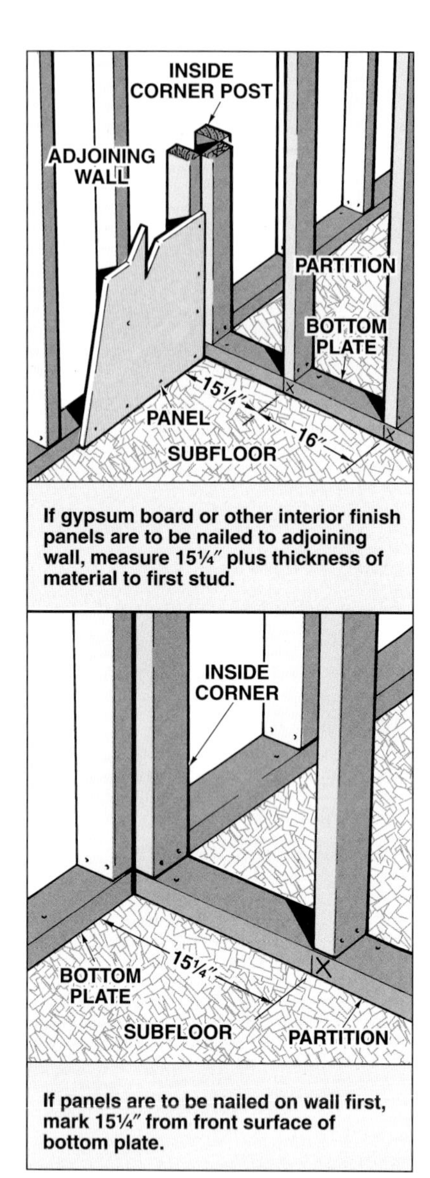

Figure 43-19. *On partitions, the 15¼″ measurement ensures that standard-size gypsum board or interior finish panels will fall over the center of a stud.*

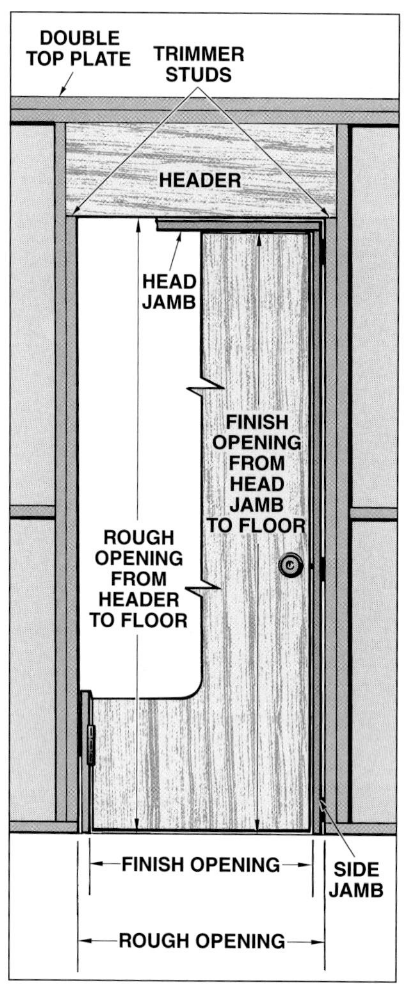

Figure 43-20. *The finish door opening is the width of the door and the distance from the head jamb to the floor. The rough door opening is the distance between the trimmer studs and the height from the floor to the header.*

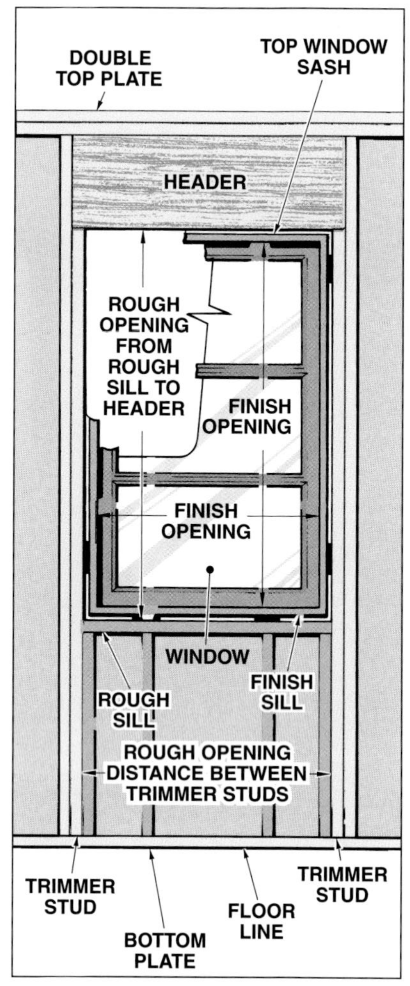

Figure 43-21. *The finish window opening width and length is the frame-to-frame dimension. The rough window opening is the distance between the trimmer studs and the height from the rough sill to the header.*

Dimensions for the widths of rough door and window openings must also be marked on the wall plates. Rough openings are calculated based on door or window width, finish frame thickness, and ½″ shim clearance at the sides of the frame. **See Figures 43-20 and 43-21.**

A rough opening is the distance between the trimmer studs and the height from the rough sill to the header.

Door and window schedules found on prints may provide rough opening dimensions. When rough openings are not included on schedules, carpenters must know how to calculate the rough openings. The procedure for calculating rough openings is shown in **Figures 43-22** and **43-23.**

A rough opening for a metal window often requires a ½″ clearance around the entire frame. If the measurements are not given in the window schedule, they can be obtained from the manufacturer instructions supplied with the windows.

A completely laid out wall plate includes markings for corner posts, rough openings, studs, and cripple studs. A procedure for laying out walls according to the floor plan shown in **Figure 43-24** is shown in **Figures 43-25** through **43-28.** First, wall plates are laid out and chalk lines are snapped. The corner posts are then laid out and properly marked to indicate studs and blocking. Next, the 16″ marks for the studs and cripple studs are marked. Finally, rough openings for doors and windows are laid out and marked.

Carpenters may prefer to lay out rough openings before studs and cripple studs are laid out. However, there is an advantage to laying out the 16″ OC marks for the studs first. Often, studs and trimmer studs for a door or window fall close to a 16″ OC stud mark. Slightly shifting the position of a rough opening may eliminate a stud from the wall frame.

Trimmer and cripple studs support the rough sill. Clips along the sides and top help to tie the window unit to the framing members.

Measuring for Rough Openings in Corners

Figure 43-22

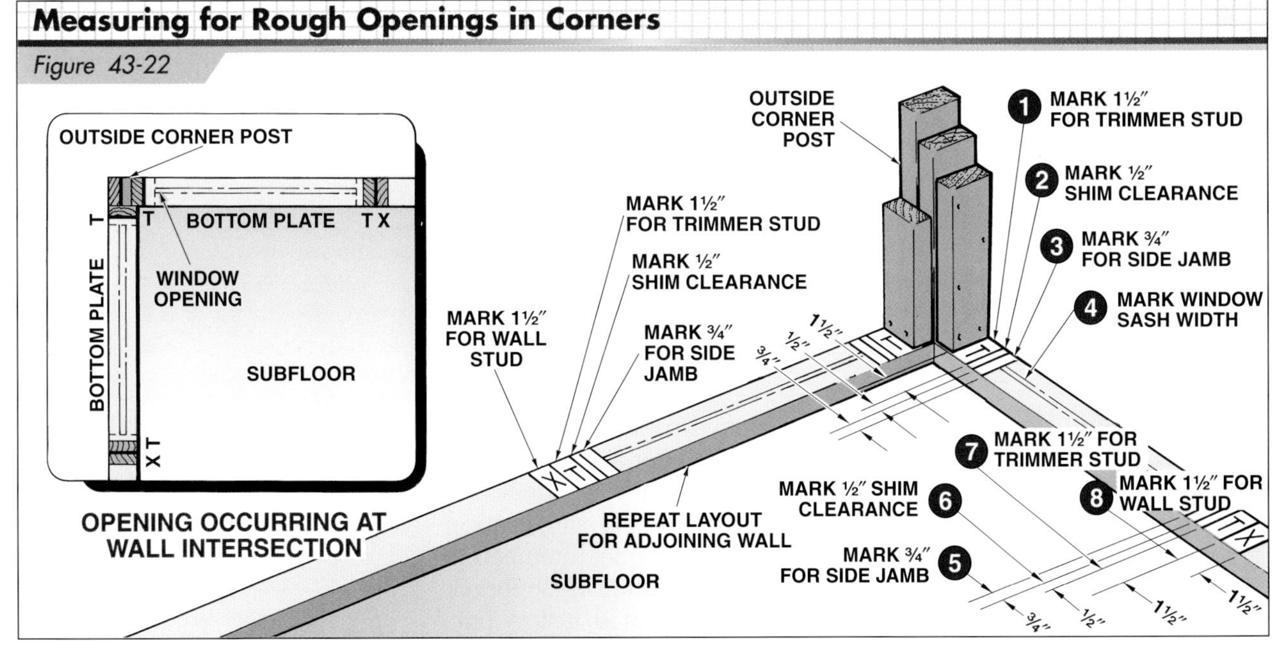

Figure 43-22. *When laying out the width of a door or window rough opening at the corners of intersecting walls, allowances must be made along the corner assembly for the trimmer stud, side jamb thickness, and shim clearance.*

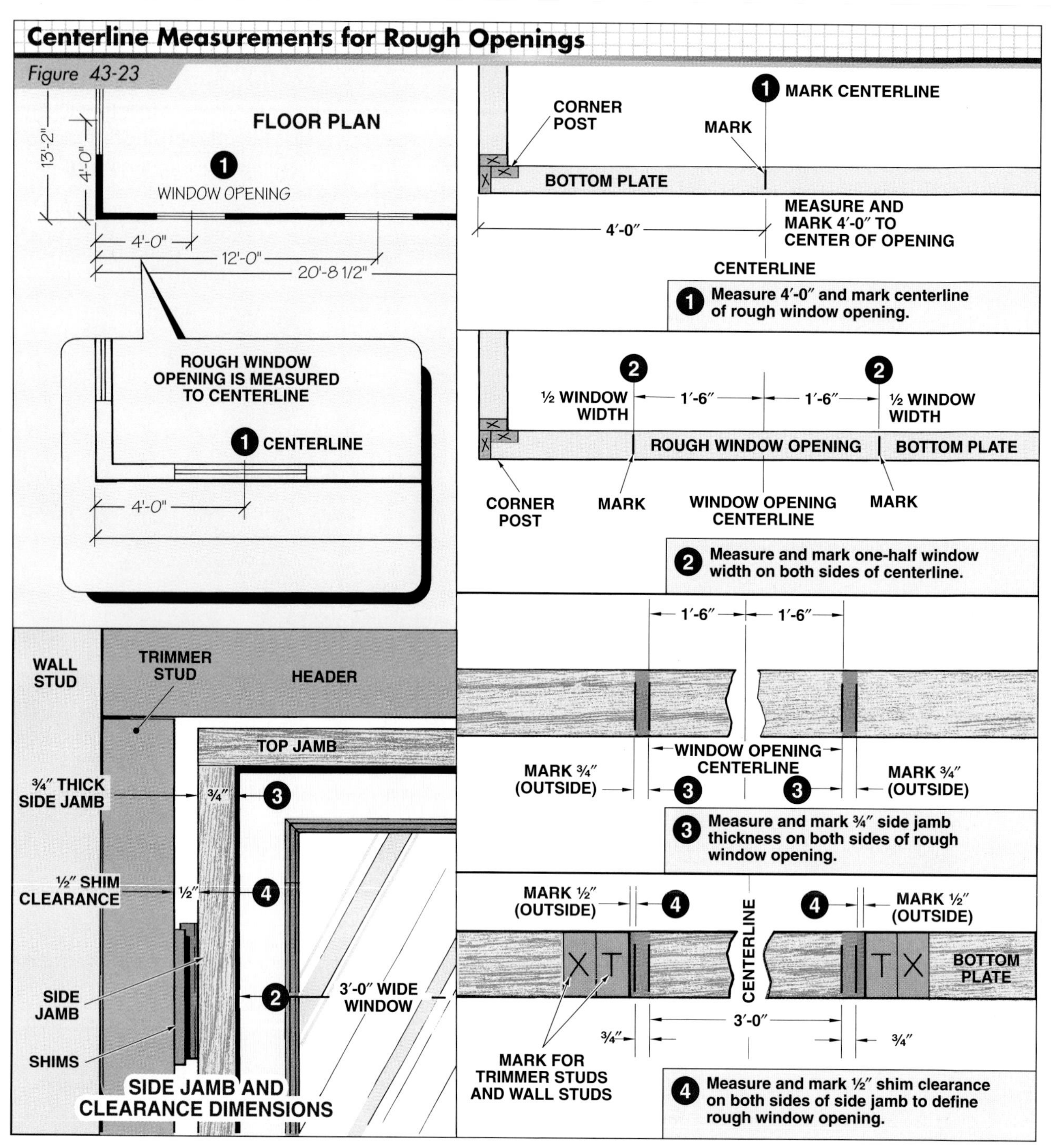

Figure 43-23. *When laying out the width of a door or window rough opening, the center of the opening is first laid out. One-half of the rough opening width, side jamb thickness, and shim clearance is laid out on each side of the centerline.*

Vertical Layout. Vertical layout is a procedure for calculating the lengths of vertical framing members, making it possible to precut the studs, trimmer studs, and cripple studs required for a building.

The most efficient vertical layout procedure involves a wall framing *story pole.* A story pole is a 1 × 2 or 1 × 4 marked to indicate the lengths of studs, trimmer studs, and cripple studs. A story pole reduces the possibility of making mathematical errors and allows measurements to be retained for future use. Information required to mark a story pole is taken from wall section views, door and window schedules, and door and window details shown on the prints.

The rough openings for windows and doors are dimensioned to the centerlines on prints.

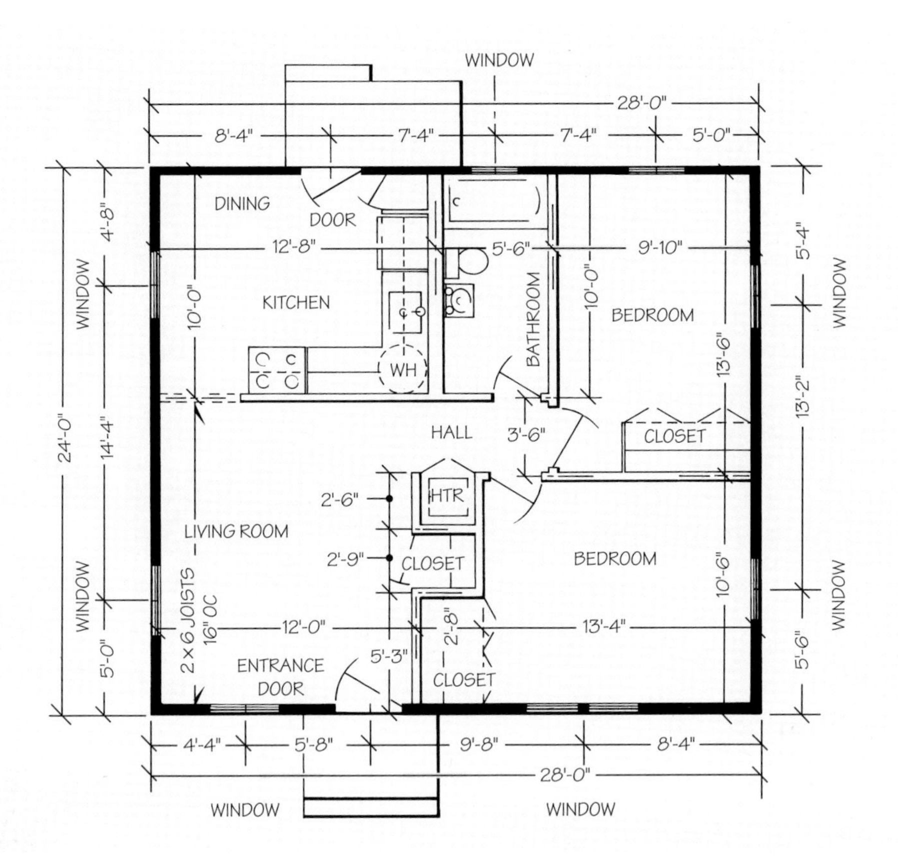

Figure 43-24. *Floor plans provide information to properly lay out walls. Information on this floor plan is used in the layout calculations for Figures 43-25 through 43-28.*

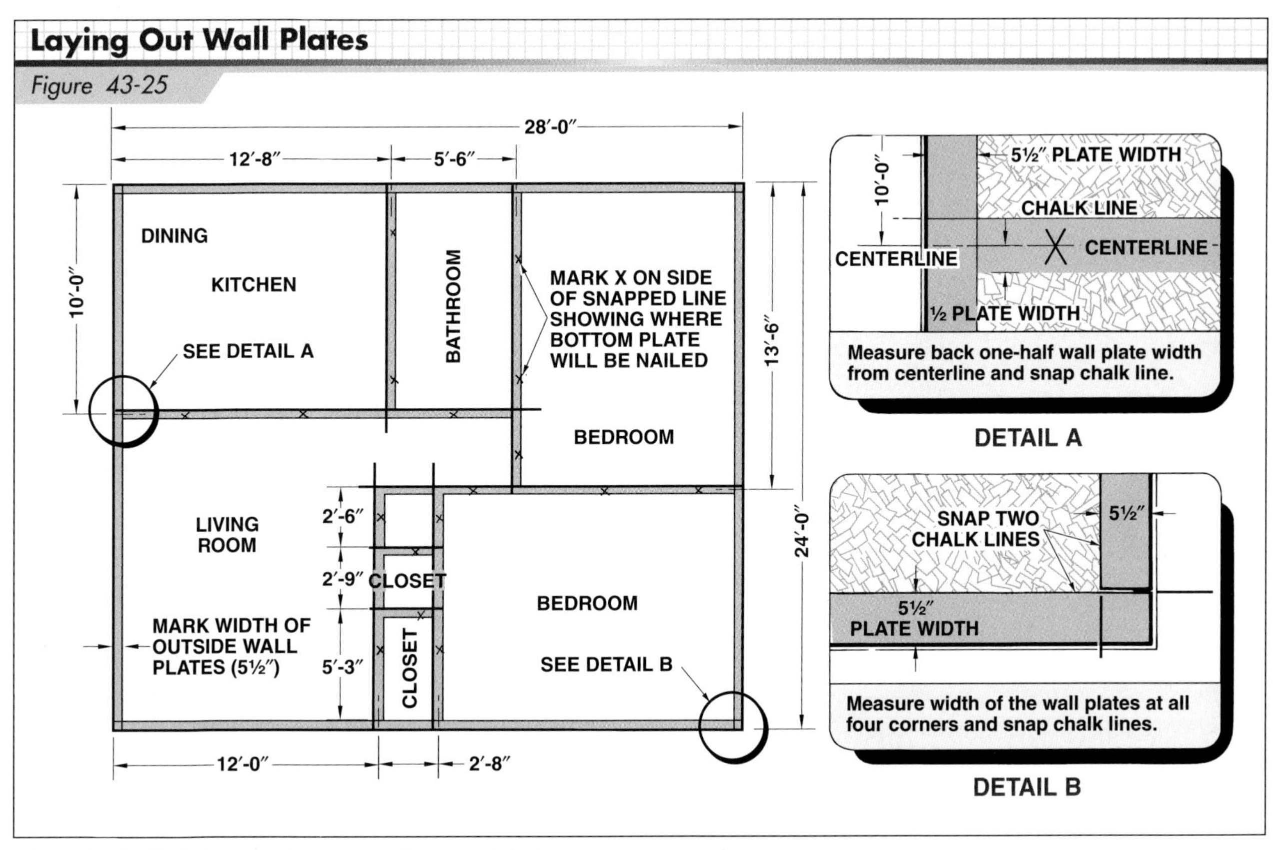

Figure 43-25. *Wall plate locations are laid out and chalk lines are snapped.*

Laying Out Outside and Inside Corner Posts

Figure 43-26

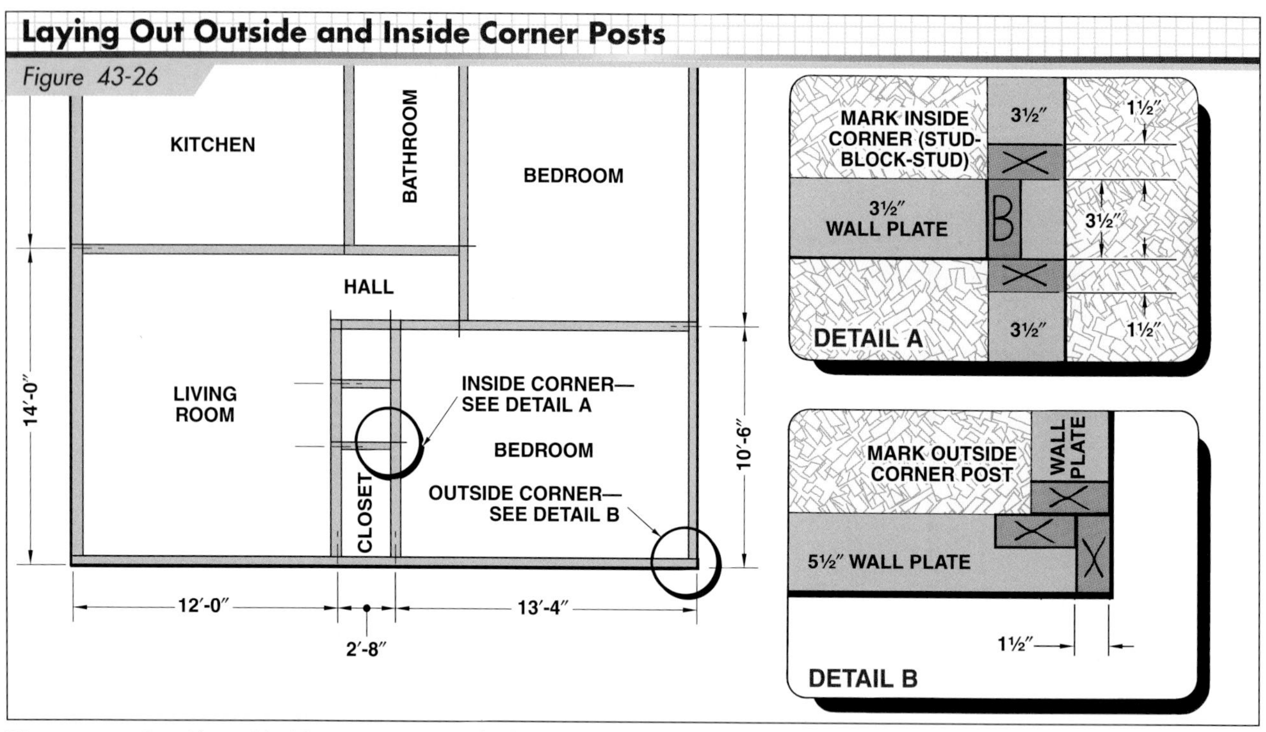

Figure 43-26. *Outside and inside corner posts are laid out and studs are laid out 16″ OC.*

Laying Out Wall Studs 16" OC

Figure 43-27

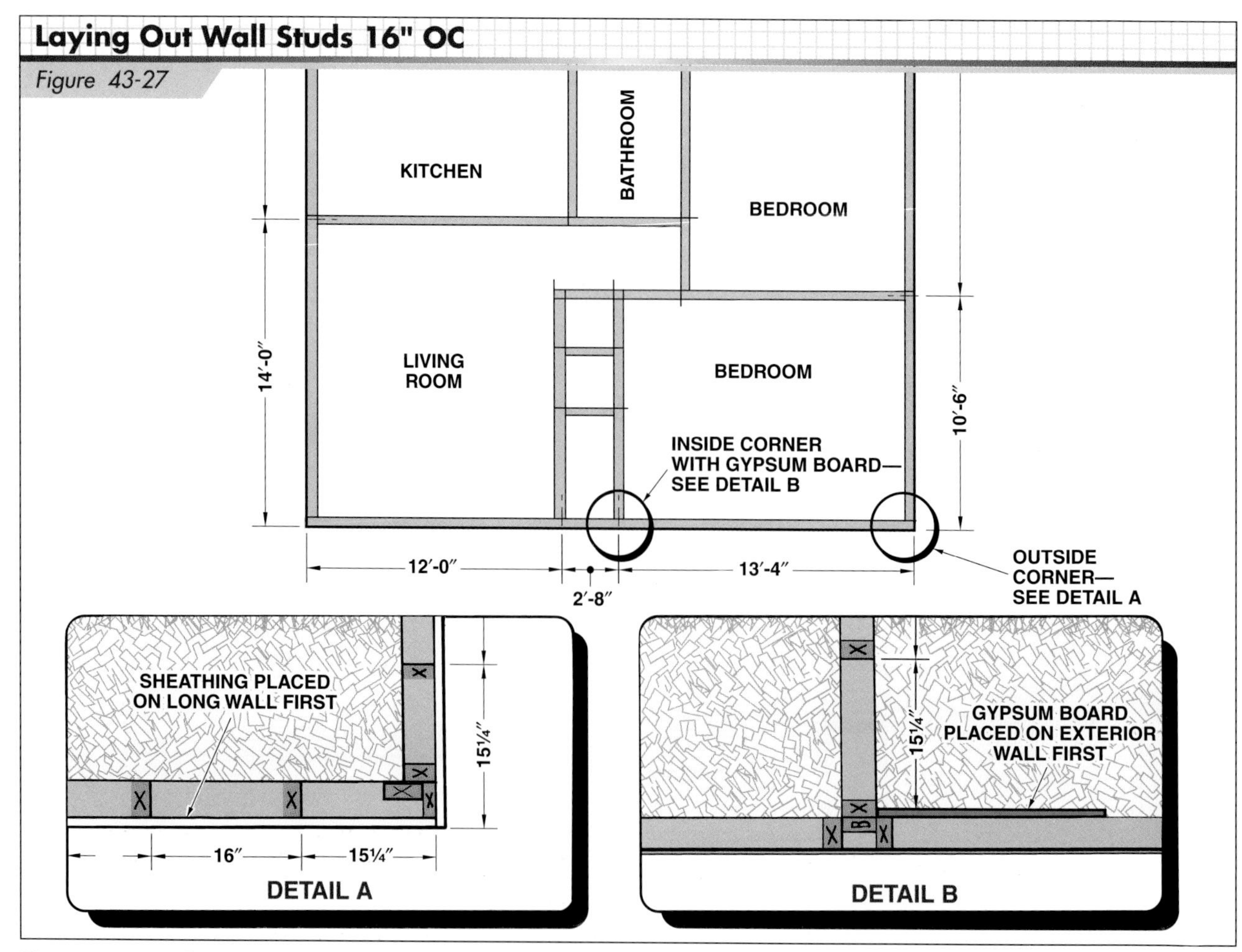

Figure 43-27. *Studs are commonly laid out 16″ OC.*

Most sets of prints contain section views to indicate the rough heights of walls. The rough wall height is the distance from the subfloor to the bottom of ceiling joists. The rough door height (distance from the subfloor to the bottom of the door header) may also be noted on a section drawing or door schedule. The door rough height establishes the dimension for the rough window heights since window headers are usually aligned with door headers.

The rough window length is the distance from the top of a sill to the bottom of a window header. The rough window length is laid out by measuring down from the bottom of the window header using dimensions provided in the rough opening column of a window schedule. Although a building usually contains windows of more than one length, only the dimensions of the window type most commonly installed in the building are marked on the story pole.

Figure 43-29 shows a typical wall section view and door and window schedules. The wall section and door and window schedules provide information needed to lay out a story pole. **Figure 43-30** shows a procedure for laying out a story pole using information obtained from the wall section view and door and window schedules.

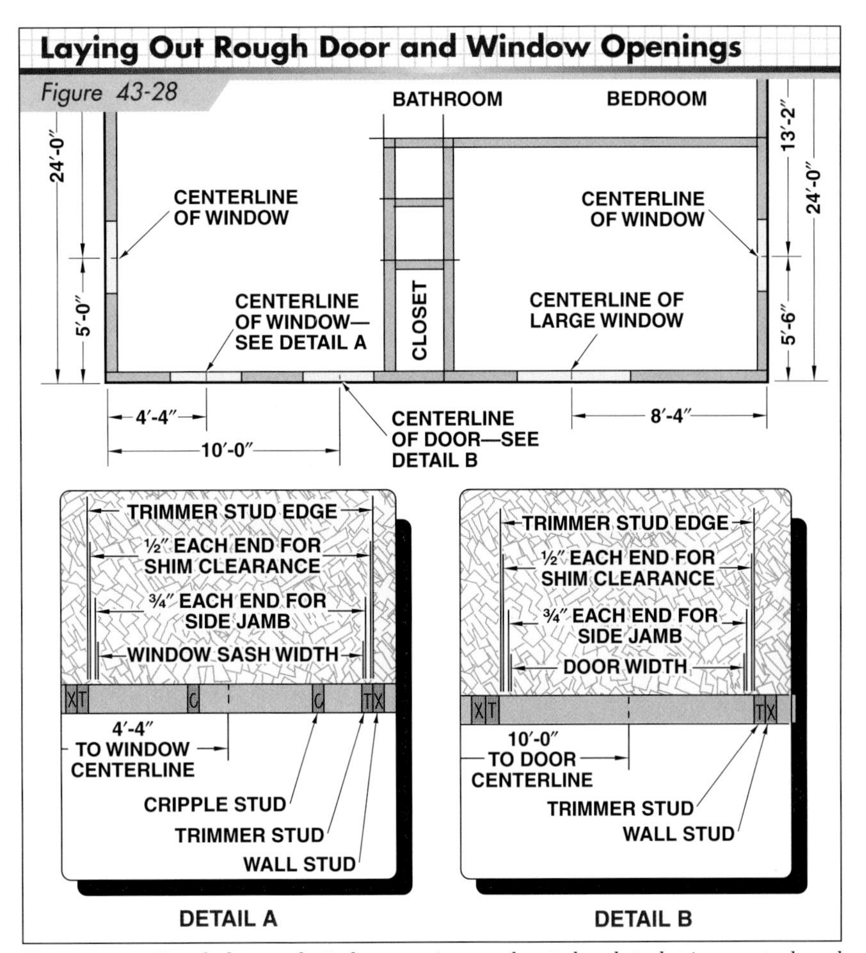

Figure 43-28. *Rough door and window openings are located and stud, trimmer stud, and cripple stud locations are marked.*

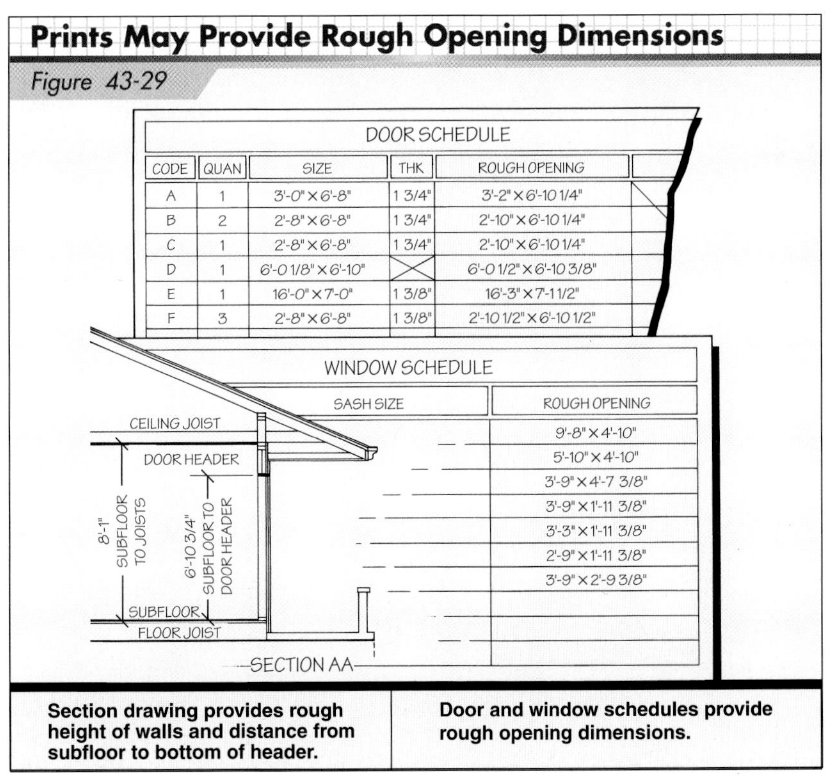

DOOR SCHEDULE

CODE	QUAN	SIZE	THK	ROUGH OPENING
A	1	3'-0" × 6'-8"	1 3/4"	3'-2" × 6'-10 1/4"
B	2	2'-8" × 6'-8"	1 3/4"	2'-10" × 6'-10 1/4"
C	1	2'-8" × 6'-8"	1 3/4"	2'-10" × 6'-10 1/4"
D	1	6'-0 1/8" × 6'-10"		6'-0 1/2" × 6'-10 3/8"
E	1	16'-0" × 7'-0"	1 3/8"	16'-3" × 7'-1 1/2"
F	3	2'-8" × 6'-8"	1 3/8"	2'-10 1/2" × 6'-10 1/2"

WINDOW SCHEDULE

SASH SIZE	ROUGH OPENING
	9'-8" × 4'-10"
	5'-10" × 4'-10"
	3'-9" × 4'-7 3/8"
	3'-9" × 1'-11 3/8"
	3'-3" × 1'-11 3/8"
	2'-9" × 1'-11 3/8"
	3'-9" × 2'-9 3/8"

Figure 43-29. *Rough door and window opening dimensions may be obtained from the door and window schedules on the prints.*

A story pole is a straight 1 × 2 or 1 × 3 which extends from the subfloor to the bottom of the ceiling joists above. A story pole is a convenient means of laying out rough wall openings to determine lengths of studs, trimmer studs, and cripple studs. In addition, the thickness and height of rough sills and locations of headers are indicated on story poles.

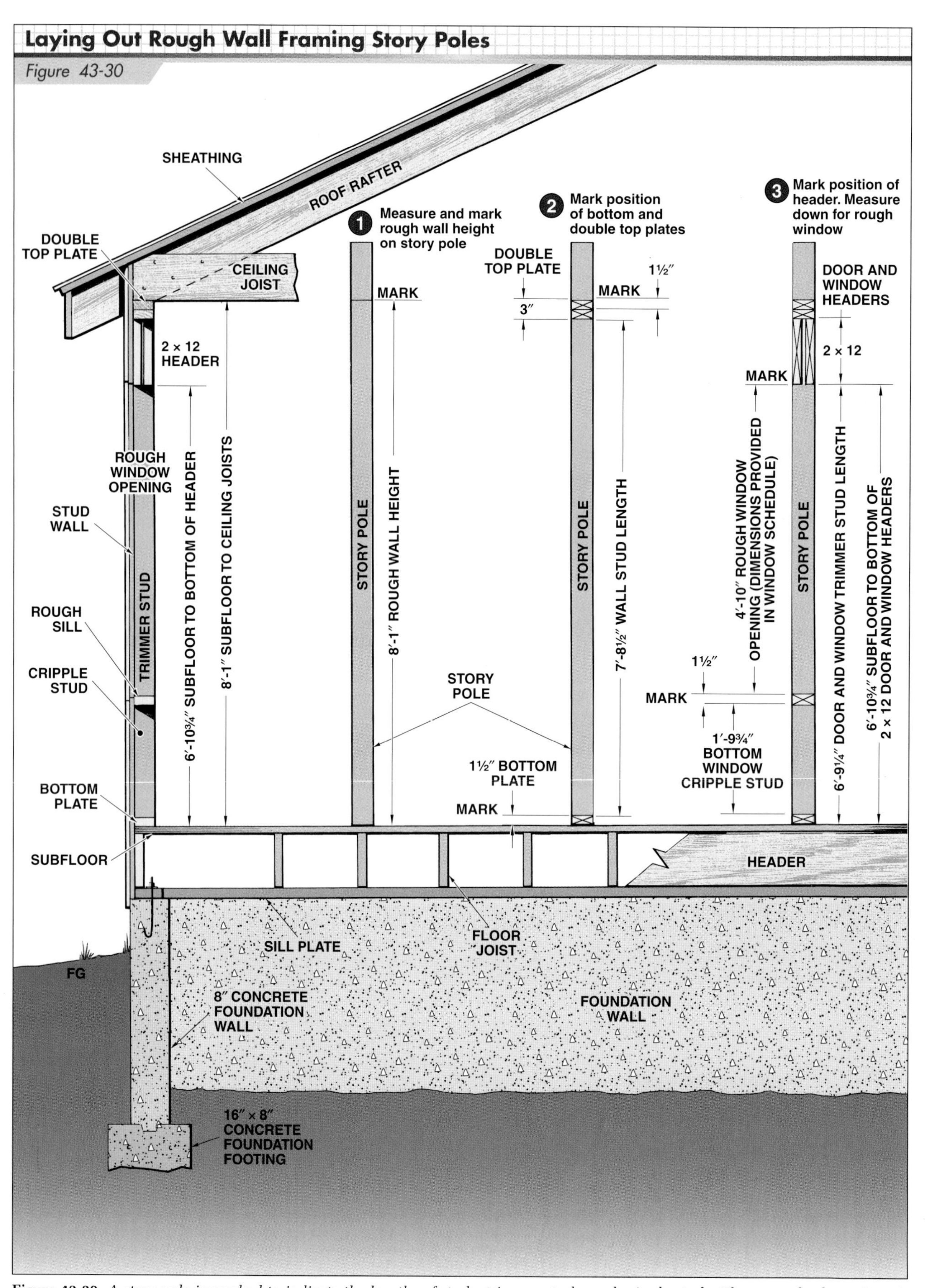

Figure 43-30. *A story pole is marked to indicate the lengths of studs, trimmer studs, and cripple studs. The example shown here is based on information in Figure 43-29.*

Prints may only provide finish opening dimensions in the wall section views and door and window schedules. If only finish opening dimensions are provided, additional information must be obtained from detail drawings and the finish schedules to lay out the story pole. **Figure 43-31** shows a wall section view that provides finish dimensions, parts of door and window schedules, and detail drawings. **Figure 43-32** shows a procedure for marking a story pole based on the information provided in Figure 43-31.

Southern Forest Products Association

Ceiling joists, which are cut to the slope of the roof at the exterior wall end, rest on top of and are toenailed to the double top plate.

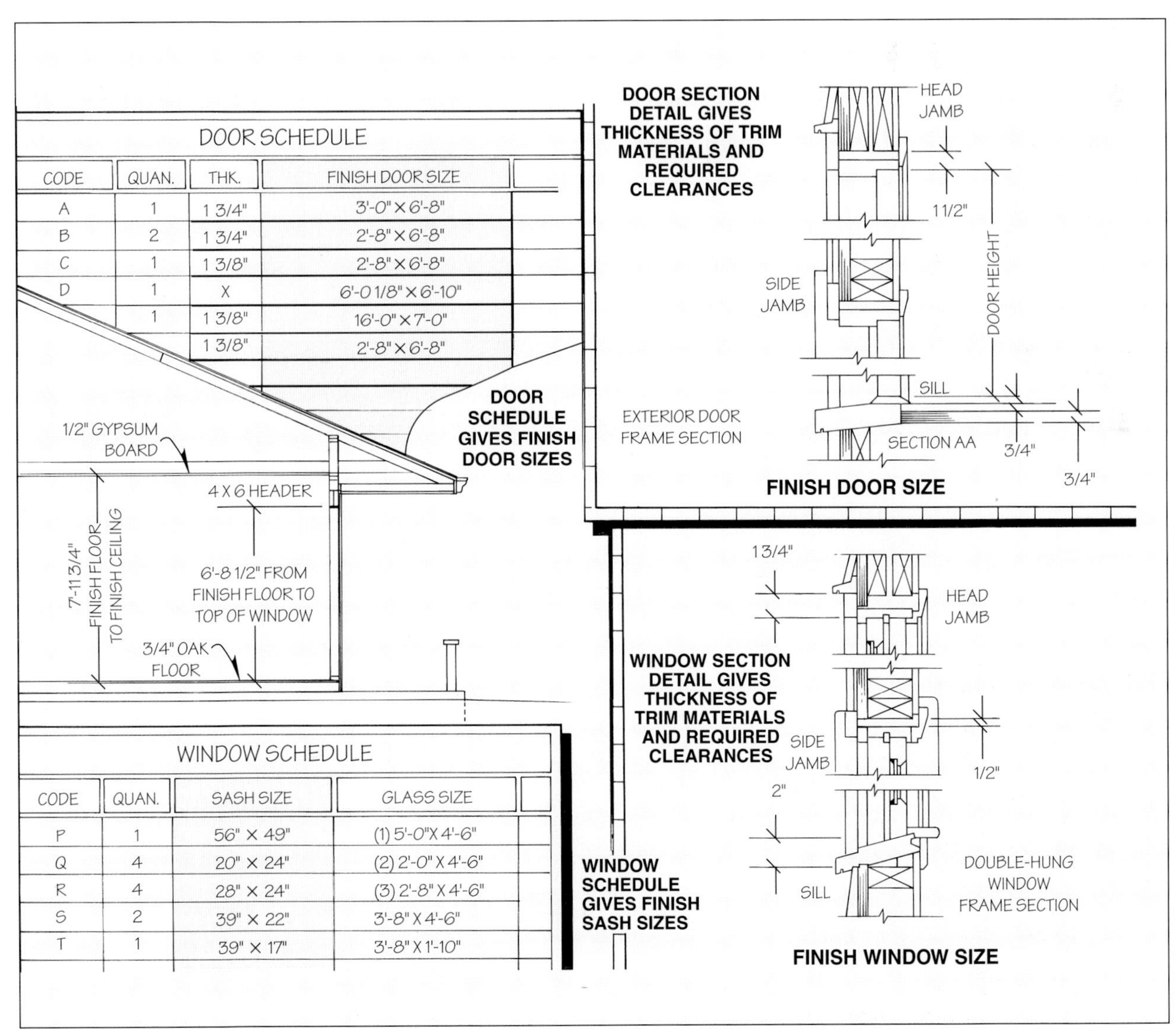

Figure 43-31. *Some prints provide only finish opening dimensions.*

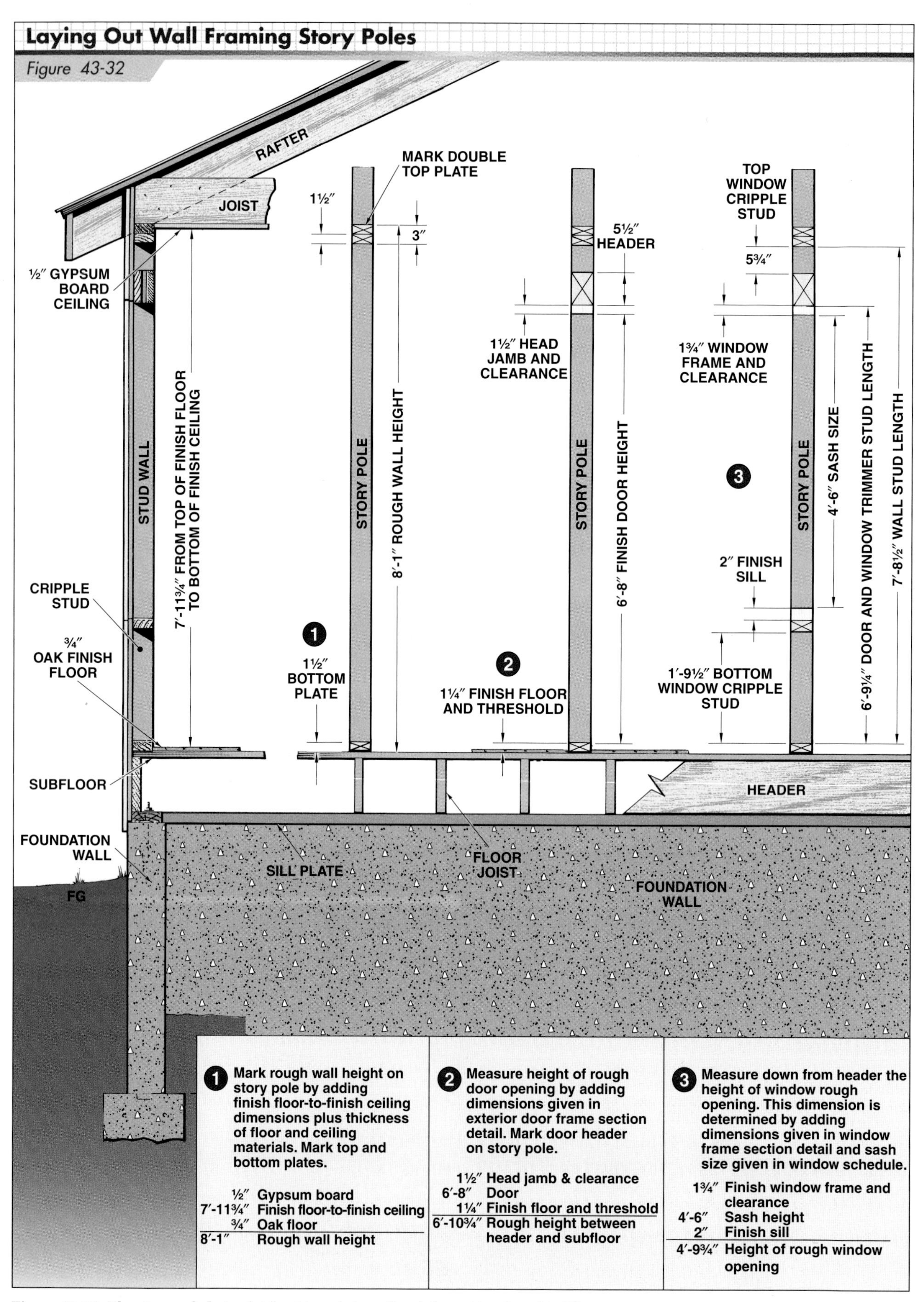

1 Mark rough wall height on story pole by adding finish floor-to-finish ceiling dimensions plus thickness of floor and ceiling materials. Mark top and bottom plates.

½″	Gypsum board
7′-11¾″	Finish floor-to-finish ceiling
¾″	Oak floor
8′-1″	Rough wall height

2 Measure height of rough door opening by adding dimensions given in exterior door frame section detail. Mark door header on story pole.

1½″	Head jamb & clearance
6′-8″	Door
1¼″	Finish floor and threshold
6′-10¾″	Rough height between header and subfloor

3 Measure down from header the height of window rough opening. This dimension is determined by adding dimensions given in window frame section detail and sash size given in window schedule.

1¾″	Finish window frame and clearance
4′-6″	Sash height
2″	Finish sill
4′-9¾″	Height of rough window opening

Figure 43-32. *The story pole being laid out here is based on information found in Figure 43-31.*

Constructing Corner Posts

For improved productivity, corner posts for a building can be constructed at one time using a bench as shown in **Figure 43-33.** **Figure 43-34** shows a procedure for assembling inside and outside corner posts. Blocks should be held slightly back from the ends of the studs. Before nailing, ensure the ends of the studs align with each other.

Outside corner posts constructed of three full-length studs provide room for insulation to be placed between the corner studs and exterior wall sheathing.

Framing Door and Window Openings

Many carpenters prefer to frame door and window openings before assembling the rest of the wall. **Figure 43-35** shows a typical procedure for framing a door opening. In this example, cripple studs are required over the header. To frame a door opening with a 4 × 12 header in a wall no more than 8′-1″ high, the procedure is the same except that cripple studs are not needed.

For window openings, a rough sill is added to the bottom of a window opening. Cripple studs that follow the stud layout (usually 16″ OC) are nailed between the sill and bottom plate. Cripple studs are also placed under each end of the sill. The procedure for framing window openings is the same as the procedure for framing door openings, and then a rough sill and bottom cripple studs are added. **See Figure 43-36.**

Arched Openings
Media Clip

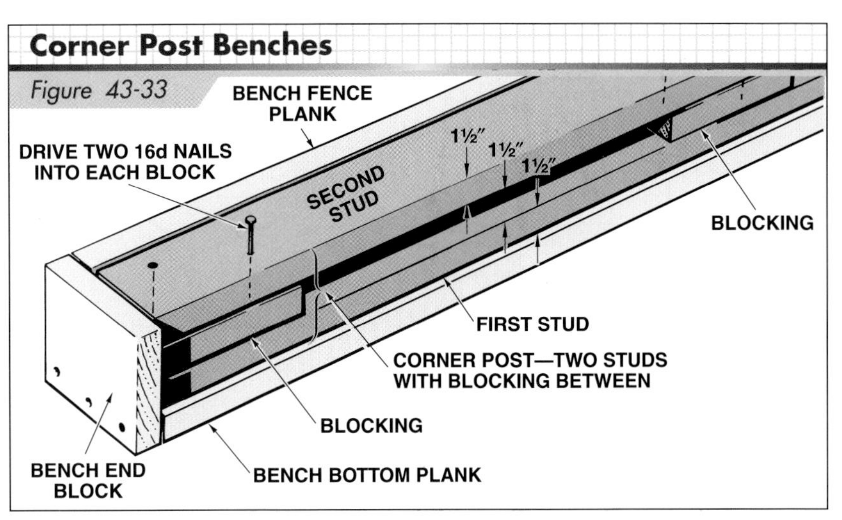

Figure 43-33. *A bench is useful for making up large quantities of corner posts. The material should be held tightly against the fence and end block when pieces are nailed together.*

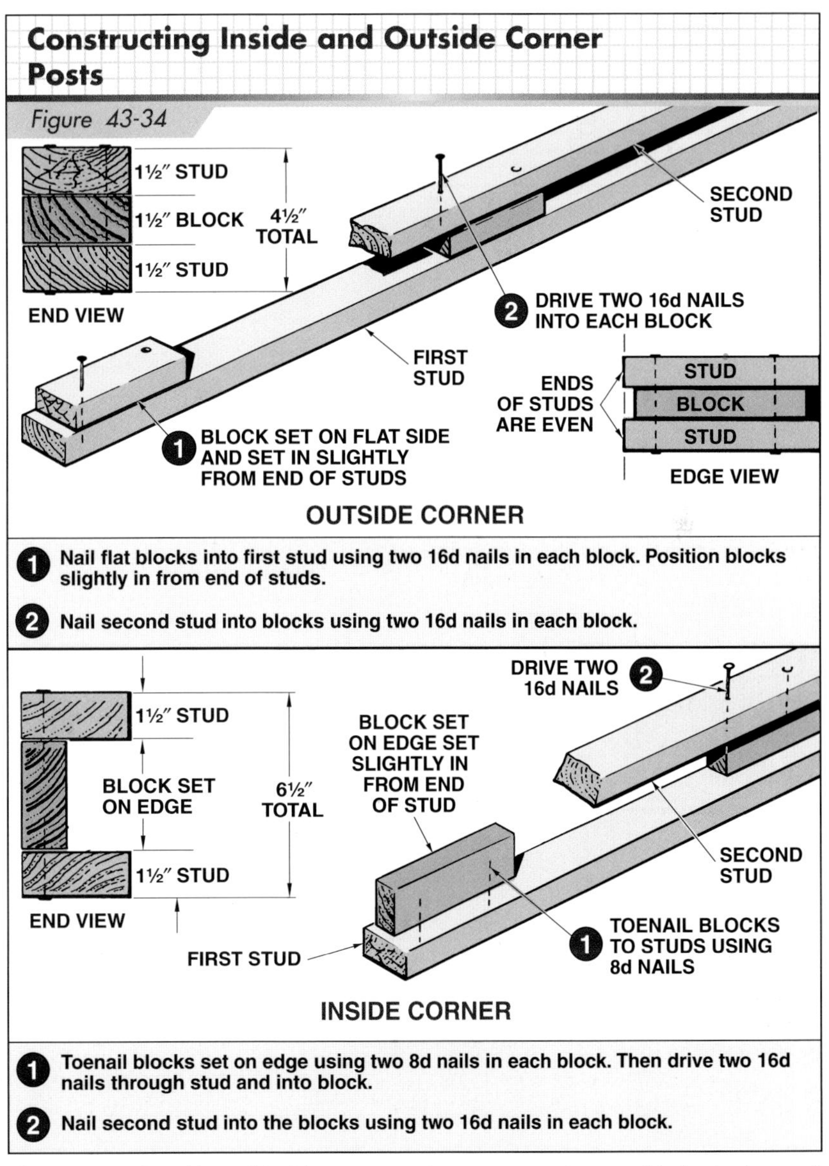

Figure 43-34. *Outside and inside corner posts are commonly assembled prior to wood-framed walls being assembled. Before nailing, align the ends of the studs.*

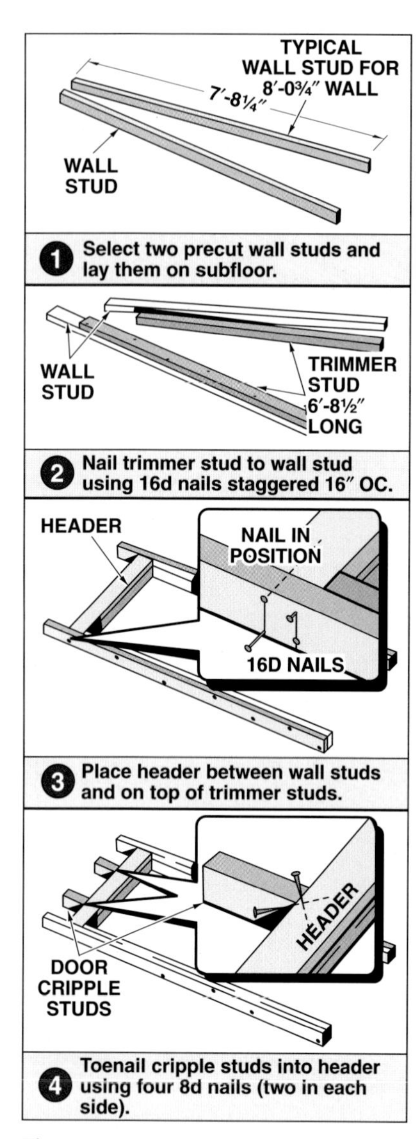

Figure 43-35. *Door openings are assembled before assembling the rest of the wall. In this example, cripple studs are required over the header.*

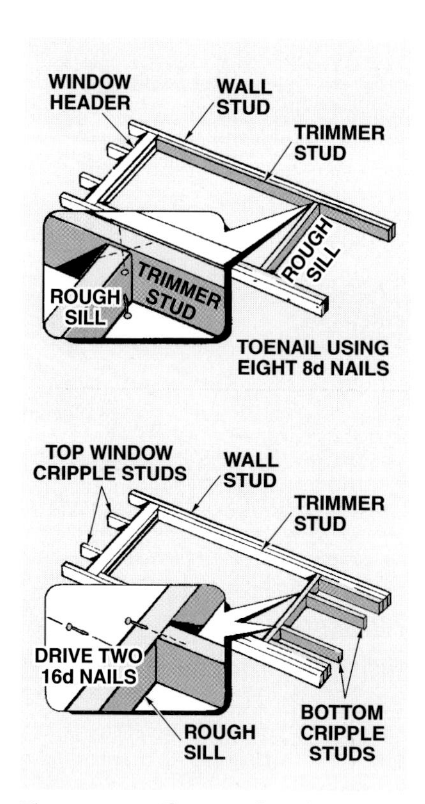

Figure 43-36. *The procedure for framing window openings is the same as framing door openings. A rough sill and bottom cripple studs are then added. Cripple studs follow the stud layout (usually 16″ OC).*

Assembling Wood-Framed Walls

After corner posts and door and window openings have been constructed, the entire wall is nailed together on the subfloor. Position top and bottom plates on the subfloor at a distance slightly greater than the length of the studs. **See Figure 43-37.** Position corner posts and rough openings between the plates according to the plate layout. Place studs in position with crown side up. Nail the plates into the studs, cripple studs, and trimmer studs.

On long walls, breaks in the plates should occur over a stud or cripple stud. If a 4 × 12 header without cripple studs is used, breaks in the plates should occur over a stud or over the header. **See Figure 43-38.**

Proper lumber storage techniques are essential to the efficient use of lumber. Proper storage protects lumber from fungi and insects, prevents defects due to varying moisture levels, and helps to maintain dimensional stability of the lumber. Lumber should be unloaded in a dry place and not in water or muddy areas. Lumber should not be allowed to come into direct contact with the ground. Lumber stored in open areas should be covered with a material porous enough to allow moisture to escape.

After the building is enclosed, utilities and insulation are installed in the walls, floor, and ceiling. Note the framing for the pocket sliding (recessed) door at the left.

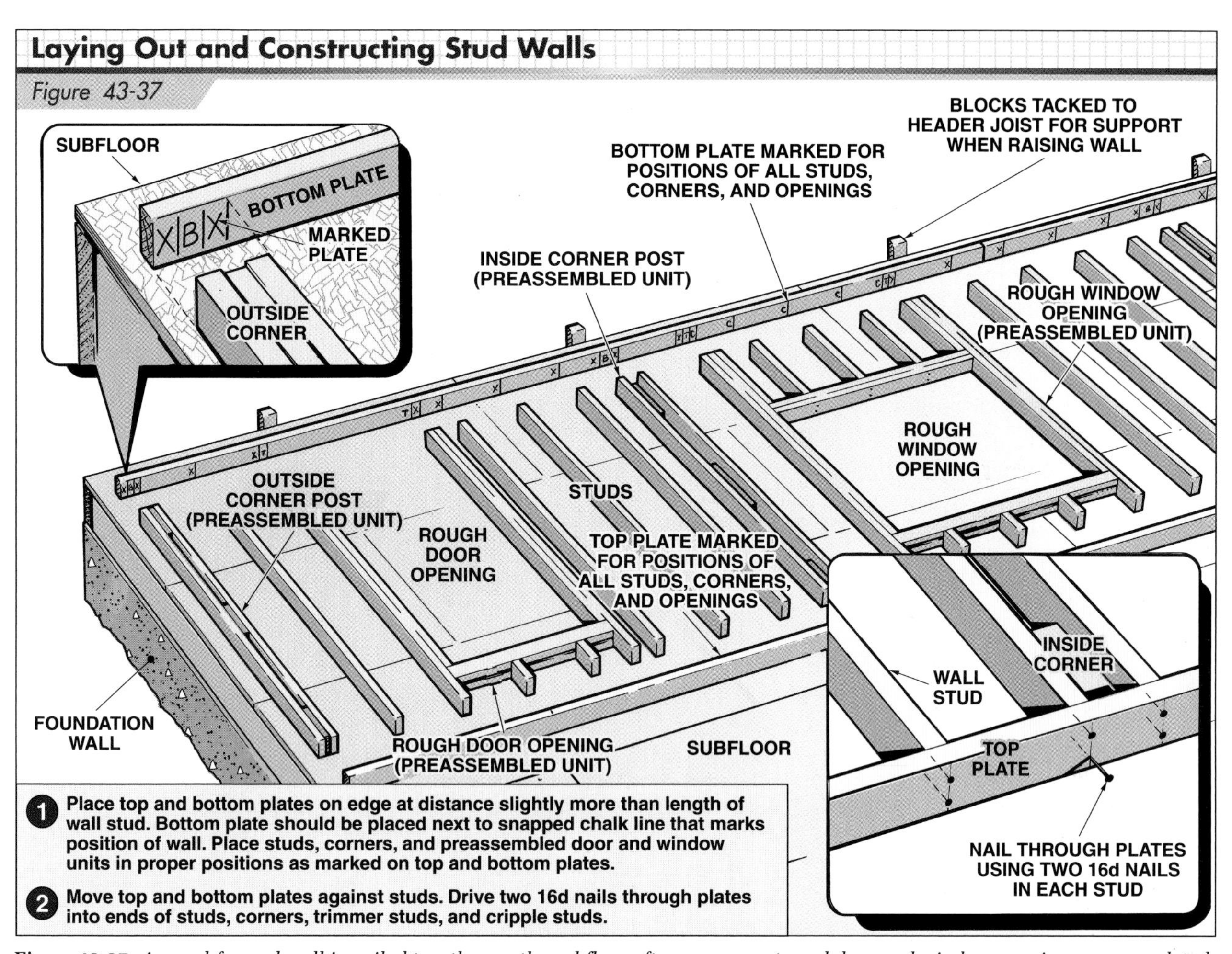

Figure 43-37. *A wood-framed wall is nailed together on the subfloor after corner posts and door and window openings are completed.*

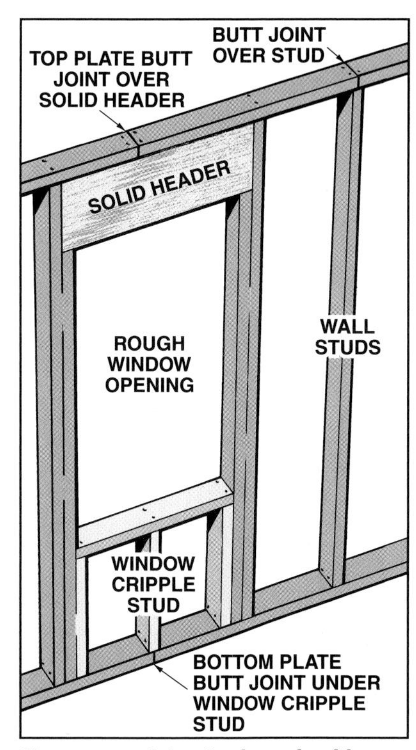

Figure 43-38. *Joints in plates should occur over a full header or at the center of a stud or cripple stud.*

Door headers in non-load-bearing partitions can be formed with two 2 × 4s and supported by trimmer studs.

Placing Double Top Plates

A top plate may be doubled while the wall is lying flat on the subfloor or after all walls have been raised. The topmost plates are nailed so they overlap the plates below them at all corners to tie the walls together. **See Figure 43-39.** All top plate ends are fastened with two 16d nails. Between the ends, 16d nails are placed every 16″ OC (over the studs) so that it is easier to identify stud locations when attaching sheathing. Butt joints between the topmost plates should be at least 4′ from any butt joint between the plates below them.

Squaring Walls and Placing Braces

A framed wall is often squared while it is lying on the subfloor. Walls are squared and braced. Braces and structural panels or other exterior wall sheathing is nailed in place to keep the walls square while being raised. **See Figure 43-40.**

Diagonal braces, such as metal braces or wood let-in braces, may be used as diagonal bracing. Three types of metal braces are flat, T-, and L-shaped. Nail holes in the braces allow them to be easily fastened to plates and studs. When installing metal braces to exterior walls, the braces are nailed to the outside surface. T- and L-shaped metal braces provide better rigidity, but require saw kerfs cut into the plates and studs.

Extreme care should be exercised when raising walls, especially when windy conditions prevail. Wall panels can be easily caught and damaged by the wind. Before raising a wall, ensure that all braces and tools required to fasten the braces to the subfloor are readily available.

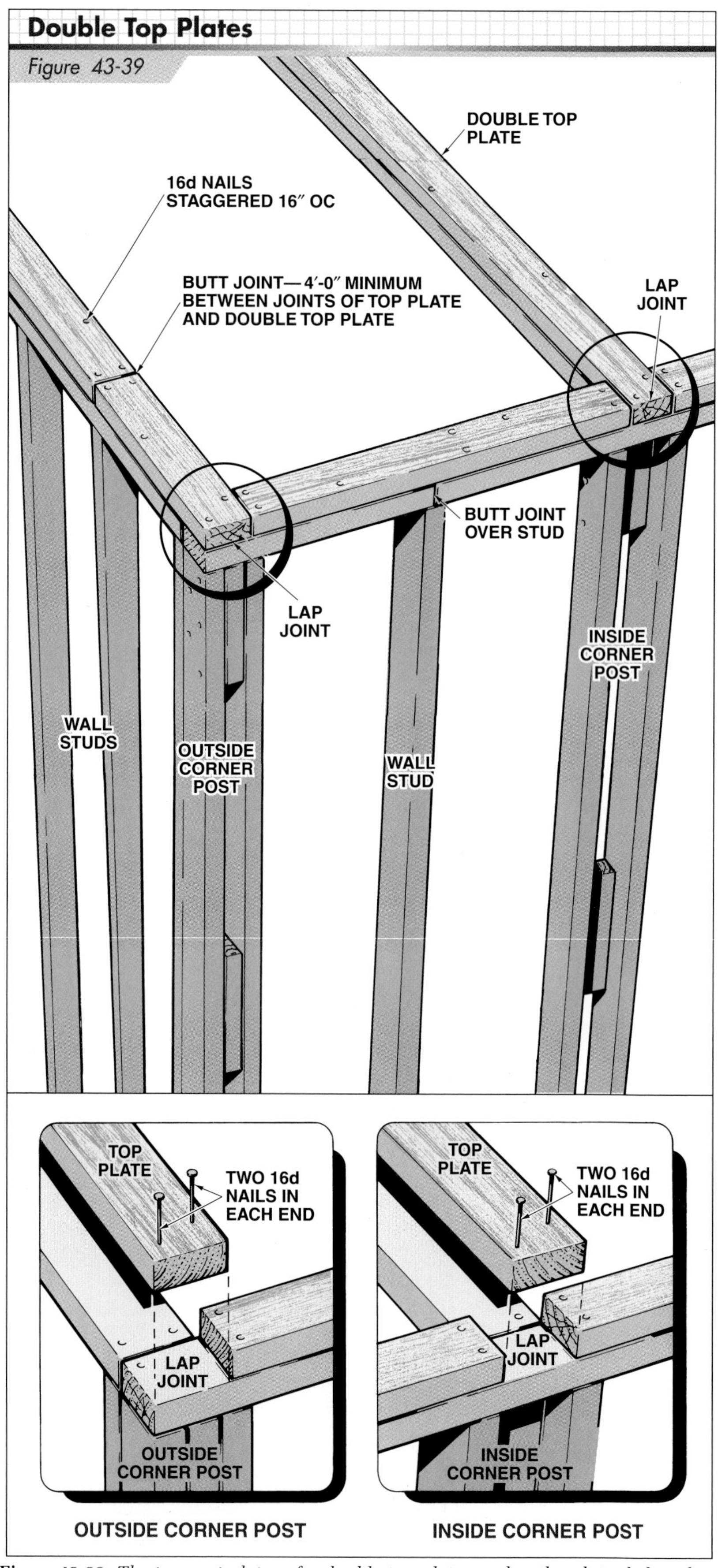

Figure 43-39. *The topmost plates of a double top plate overlap the plates below them at all inside corners.*

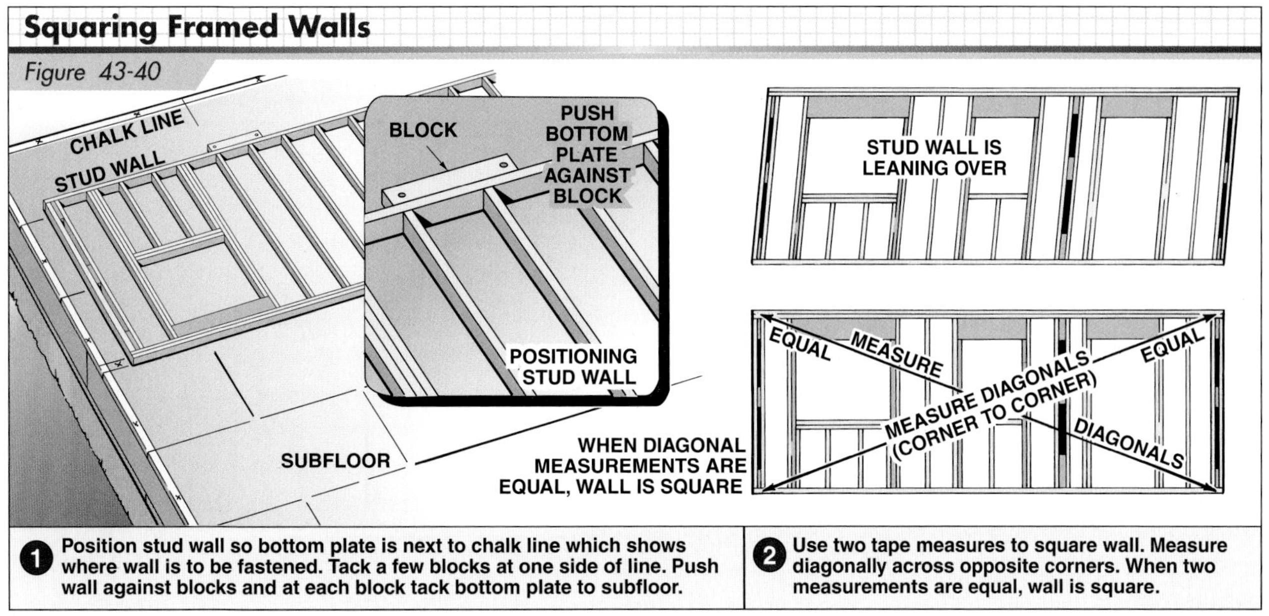

Figure 43-40. *A framed wall is often squared while it is lying on the subfloor. Temporary braces must be attached to keep the wall square while it is being raised. Wall sheathing may also be applied while the wall is lying on the subfloor to keep the wall square when raising it.*

When installing metal braces, lay out and snap lines on the studs to show the brace location. **See Figure 43-41.** Shallow kerfs are cut in the edges of the studs to allow for metal braces. A brace is installed in the kerfs and fastened in place. Metal braces should be placed at the corners of walls and approximately every 25′ in longer walls.

When installing wood let-in braces, the studs are marked and notched for the brace. **See Figure 43-42.** Tack the brace to the studs while the wall is lying on the subfloor. Tacking instead of nailing allows for some adjustment after the wall is raised. After any necessary adjustment is made, the nails can be securely driven in.

Placing Fireblocking

Fireblocking may be placed in higher walls to slow the rate of a fire that may occur inside the walls. In addition, fireblocking can also be used as a nailing base for the edges of plywood or gypsum board panels.

Raising Walls

Most walls can be raised manually if enough carpenters are available on the job. As a general rule, one person is required for every 10′ of wall for the lifting operation. The order in which walls are framed and raised may vary from job to job. Generally, longer exterior walls are raised first. Shorter exterior walls are then raised and the corner posts are nailed together. The order of framing and raising interior partitions depends on the floor layout.

Walls must be properly braced after being raised into position.

Diagonal Metal Brace Layout

Figure 43-41

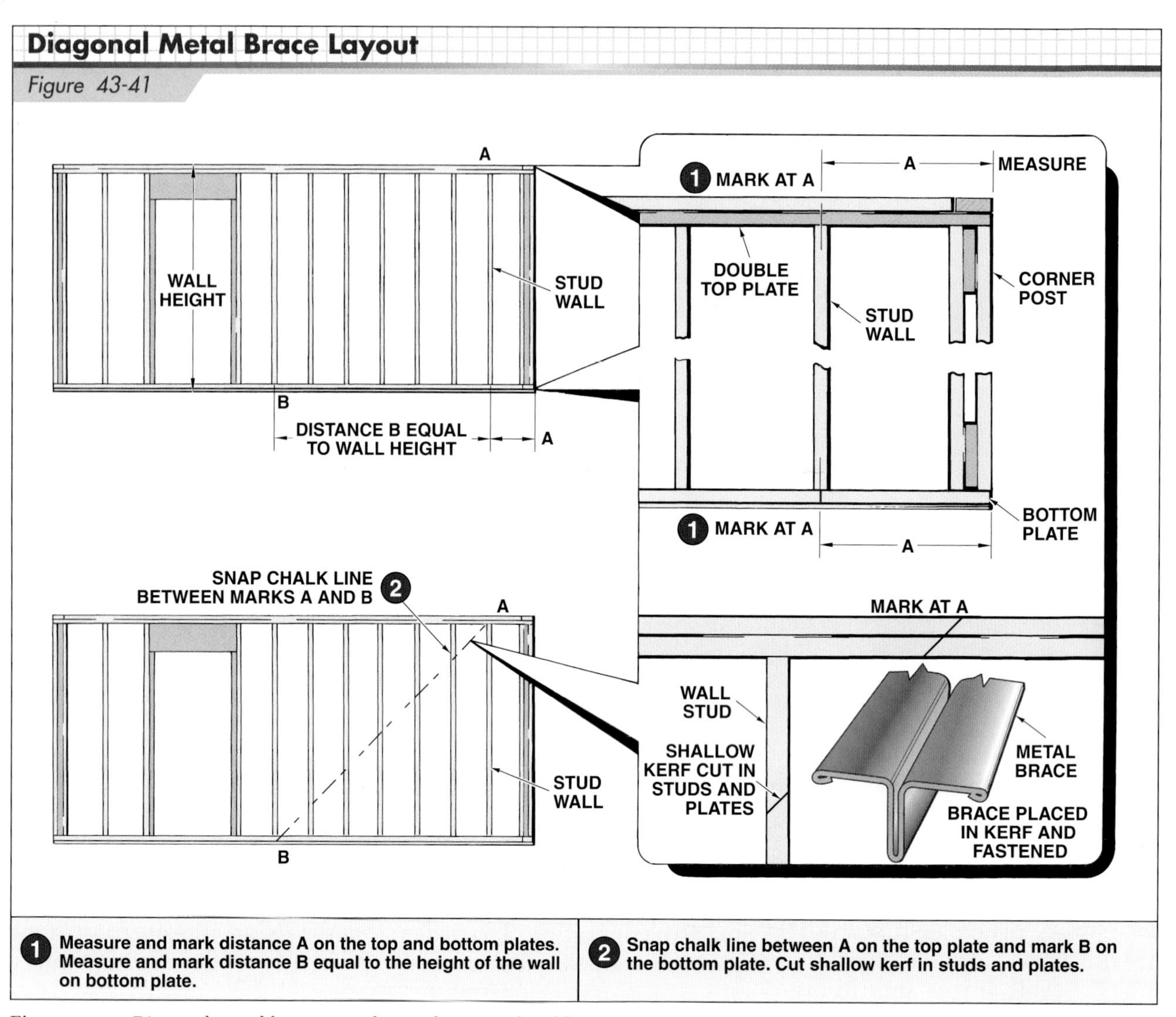

1 Measure and mark distance A on the top and bottom plates. Measure and mark distance B equal to the height of the wall on bottom plate.

2 Snap chalk line between A on the top plate and mark B on the bottom plate. Cut shallow kerf in studs and plates.

Figure 43-41. *Diagonal metal braces may be used to provide additional wall stability. A shallow kerf is required for T- and L-shaped braces.*

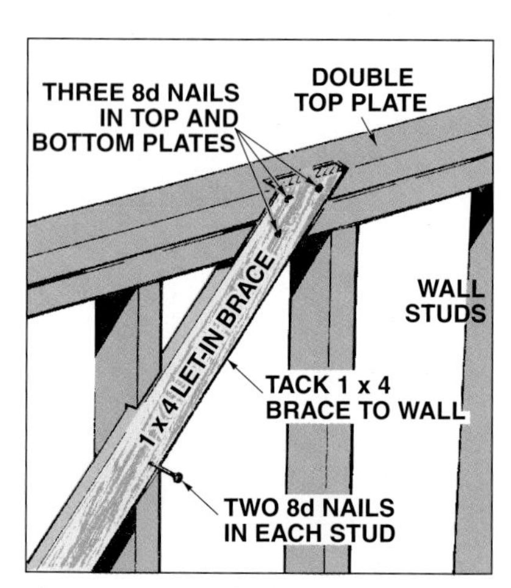

Figure 43-42. *Wall plates and studs are notched for wood let-in braces. The braces are tacked in place when the wall section is lying on the subfloor. After the wall is raised and the brace adjusted to its final position, the nails are driven in.*

After a wall has been raised, its bottom plates must be securely fastened to the floor. For walls resting on a wood subfloor and joists, 16d nails should be driven through the bottom plate and subfloor and into the floor joists.

Raising Walls with Wall Jacks. *Wall jacks* may also be used to raise walls. Wall jacks are lightweight devices commonly use to raise walls when small carpentry crews are working on a job site. A wall jack consists of an adjustable cylindrical tube with a hinged foot at the bottom and a pulley at the top with a cable that extends to a winch mounted on the tube. An adjustable wall stop is secured to the wall jack for proper positioning of the wall.

A minimum of two wall jacks must be used when raising a wall. The number of wall jacks used depends on the wall length, height, and overall wall structure. The general procedure for raising walls using walls jacks is shown in **Figure 43-43.** Always follow manufacturer instructions when raising walls using wall jacks. When walls are being raised, no workers should be allowed directly below the wall or on the opposite side of the wall.

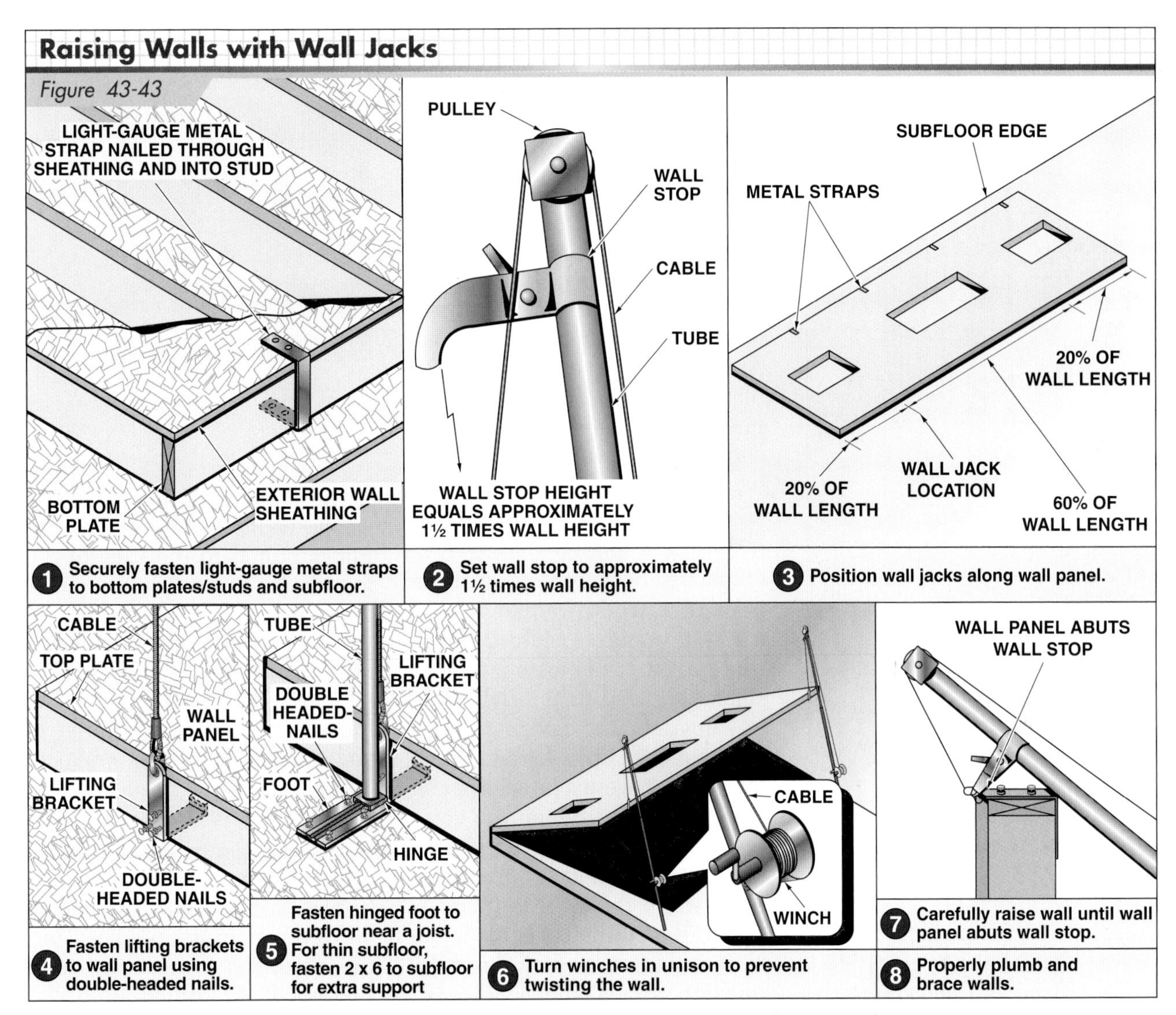

Figure 43-43. *Wall jacks can be used to raise walls when small carpentry crews are working on a job site.*

Plumbing and Aligning Walls

Plumbing of corners is performed after all walls are raised. Plumb the outside corners first and then securely brace for stability. Most framing material is not perfectly straight; therefore, a plate level or straightedge and a hand level should be used to plumb walls. **See Figure 43-44.** A straightedge is ripped out of plywood or a straight 2 × 4. Blocks ¾″ thick are nailed to each end of the straightedge to accurately plumb the wall from the bottom plate to the top plate. Walls should not be plumbed by placing a short hand level directly against the end stud.

Plumbing corners requires two carpenters working together. One carpenter releases the nails at the top end of the temporary corner brace so the top of the wall can be moved. The other carpenter places a straightedge and level along the corner post and watches the level. The bottom end of the brace is renailed when the wall is perfectly plumb.

Wall tops are straightened (aligned) after all corners have been plumbed, but before floor or ceiling joists are nailed to the tops of the walls. A string is attached to the top plate at one corner of the wall. The string extends to, and is fastened to, the top plate at the opposite end of the wall. Three small blocks are cut from a 1 × 2. One block is placed under each end of the string so the line is clear the entire length of the wall. The third block is used as a gauge to check the wall at 6′ or 8′ intervals. **See Figure 43-45.** At each check point, a temporary brace is fastened to a wall stud.

When fastening a temporary brace to the wall stud, adjust the wall so the string is barely touching the gauge block. Nail the other end of the brace to a short 2 × 4 block fastened to the subfloor. Temporary braces are not removed until the framing and sheathing for the entire building have been completed.

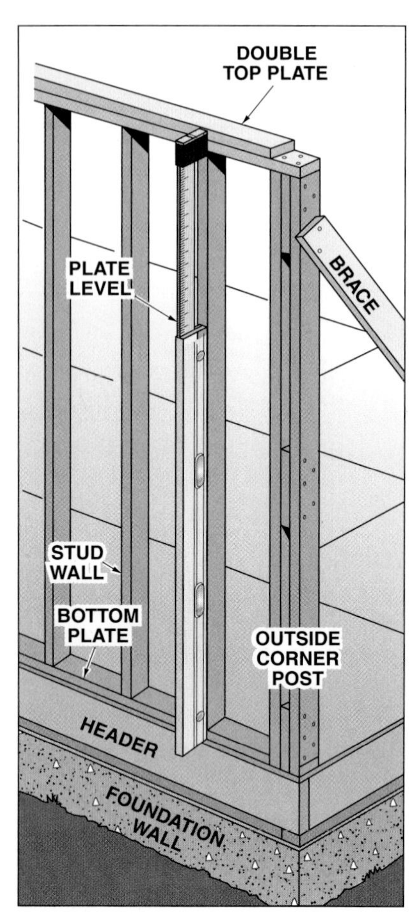

Figure 43-44. *A plate level is used to plumb a wall corner.*

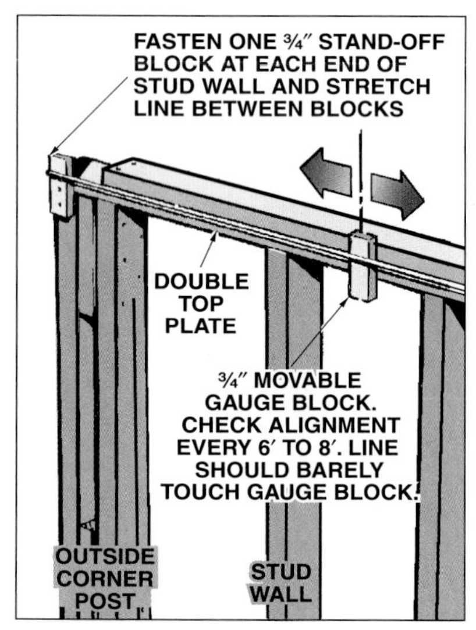

Figure 43-45. *After the corners are plumbed, the tops of the walls are straightened using a line and ¾″ blocks.*

If a laser level or other survey instrument is available, it may be used for plumbing and aligning walls.

Pick-up Framing Operations

Certain smaller framing operations, often referred to as *pick-up operations or back framing,* are performed after the walls have been raised. Pick-up operations include tying corners together, cutting and removing bottom plates for door openings, straightening studs, and cutting openings in walls for heating vents. Another pick-up operation is the placement of wall backing where plumbing fixtures, such as sinks, bathtubs, and water closets, will later be fastened to the wall.

Framing over Concrete Slabs

Often the basement or ground floor of a wood-framed building is a concrete slab. Bottom plates must be bolted to the slab or secured to the slab with powder-actuated fasteners. **See Figure 43-46.** Bolts for attaching bottom plates must be accurately set into the slab when the concrete is placed. Bolt holes are laid out and drilled in the bottom plate when the wall is framed. The wall is slipped over the bolts as it is raised and secured in position with washers and nuts.

Simpson Strong-Tie Company, Inc.

Figure 43-46. *Powder-actuated fasteners are often used to fasten bottom plates to concrete slabs.*

The bottoms of basement and garage walls usually rest on sill plates bolted to the tops of the foundation walls. The top plates are nailed to wall studs and the walls are raised. The bottom ends of the studs are toenailed to the sill plate. The remainder of the framing procedure is similar to walls nailed on top of a subfloor.

Metal Connectors

Metal connectors strengthen wood-framed wall connections by providing additional support between framing members and the concrete foundation. **See Figures 43-47** and **43-48.**

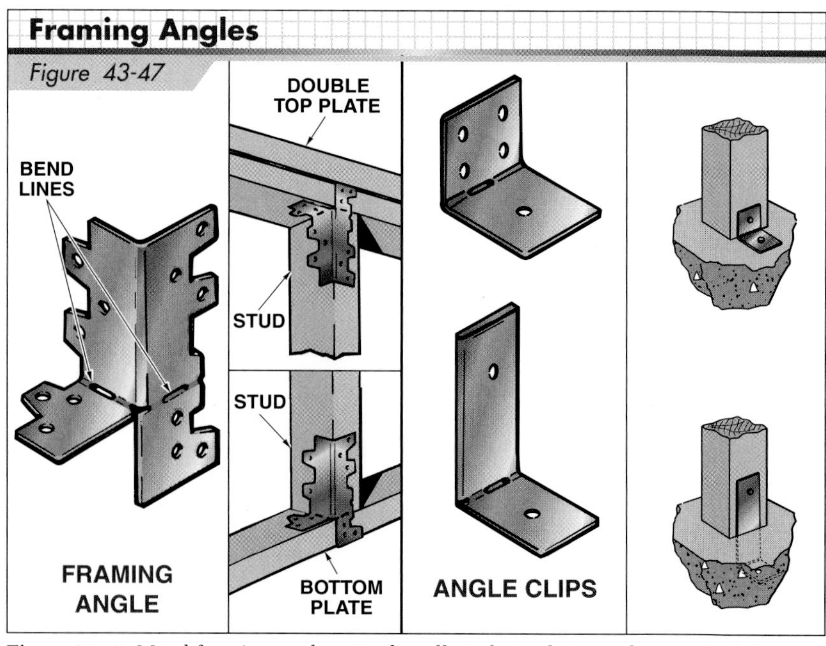

Figure 43-47. *Metal framing angles attach wall studs to plates and concrete slabs.*

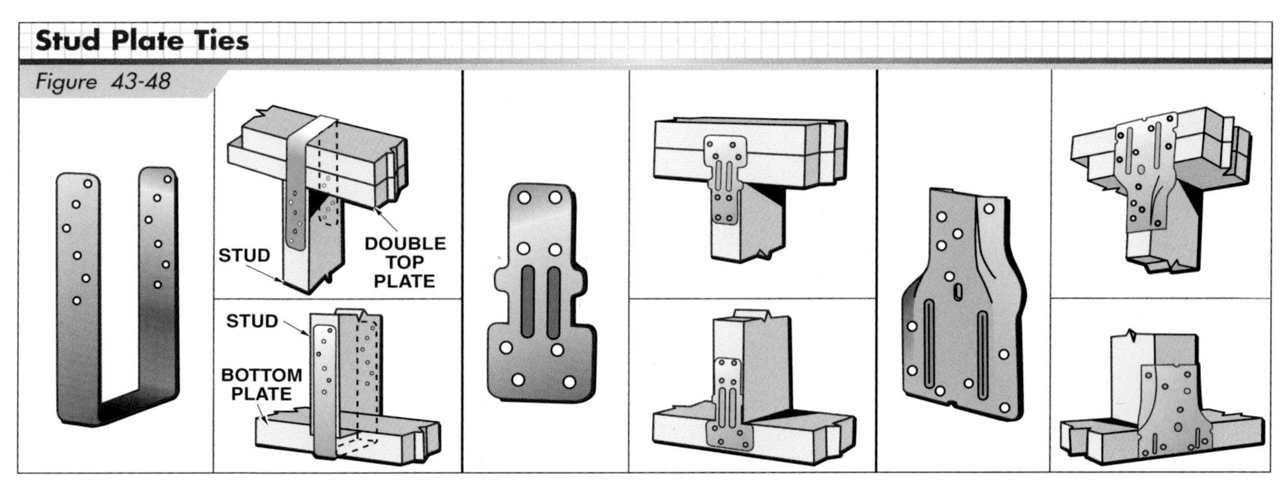

Figure 43-48. *Stud plate ties secure wall studs to plates.*

Seismic or hurricane ties are required in areas that experience earthquakes and high winds. **See Figure 43-49.** Wood-to-wood ties, such as floor ties and plate ties, are attached to the framing members with galvanized nails. Wood-to-masonry (or concrete) ties are attached to the masonry or concrete with self-tapping screw anchors. Refer to the local building code to determine metal connector and fastener requirements.

SHEATHING EXTERIOR WALLS

Exterior wall sheathing is the panels fastened to the outside of exterior wall studs around the perimeter of a building. Oriented strand board (OSB) is commonly used as wall sheathing for residential and light-frame structures. Plywood may also be used as wall sheathing. Structural panel sheathing adds enough lateral shear strength to wood-framed walls to eliminate the need for diagonal braces.

Wall sheathing does not include the finish surface of a wall. Siding, shingles, stucco, or brick veneer is placed over the sheathing to finish the wall.

Wall Sheathing

Wall sheathing, such as APA Rated Sheathing and APA Rated Wall Bracing panels, must conform to PS 2-92, *Performance Standard for Wood-Based Structural-Use Panels.* APA Rated Sheathing panels are commonly used for wall sheathing when diagonal braces are not required. APA Rated Wall Bracing panels can also be used over walls requiring diagonal braces.

Wall sheathing panels range in thickness from 3⁄8″ to 3⁄4″ and are commonly available in 4′ × 8′ to 4′ × 12′ sizes. Larger panel sizes can be special-ordered.

Wall sheathing panels are typically placed with the long edges in a vertical position, although the long edges may also be placed horizontally. If panels are placed horizontally, nailing blocks should be installed between the studs. Building codes may require nailing blocks along the long edges of horizontal panels. Joints at panel edges should be staggered so no joints are aligned. A 1⁄8″ space should be allowed between the panels to prevent buckling, which may result from panel expansion. **See Figure 43-50.**

Fasteners. Typical code regulations require 6d nails for wall sheathing panels 1⁄2″ thick or less, and 8d nails for panels more than 1⁄2″ thick. Nail spacing is commonly 6″ OC along the edges and 12″ OC at intermediate studs.

When using screws to attach wall sheathing panels, 2″ screws should be used for panels 1⁄2″ thick or less, and 2 1⁄2″ screws for panels more than 1⁄2″ thick. Screws should be spaced 6″ OC along the edges and 12″ OC at intermediate studs.

When staples are used to attach wall sheathing panels, a minimum of 1 1⁄4″ long staples should be used. Staples are spaced 4″ OC along the edges and 8″ OC at intermediate studs. Staples should not be used in shear wall construction.

Placing Panels. When placing wall sheathing panels, ensure the first panel is plumb along its vertical edge and level along the horizontal edge. Secure the panel with nails, screws, or staples at the corners. Snap vertical chalk lines locating the intermediate studs before installing nails or screws. **See Figure 43-51.** Some wall sheathing panels are available with factory-painted lines at 16″ and 24″ OC.

Panel sheathing may be applied when a wall has been squared and is lying on the subfloor. However, problems can occur after the wall is raised if the floor is not level. Builders often prefer to place all panels after the building has been framed.

Simpson Strong-Tie Company, Inc.

Figure 43-49. *Seismic or hurricane ties are required in areas that experience earthquakes and high winds.*

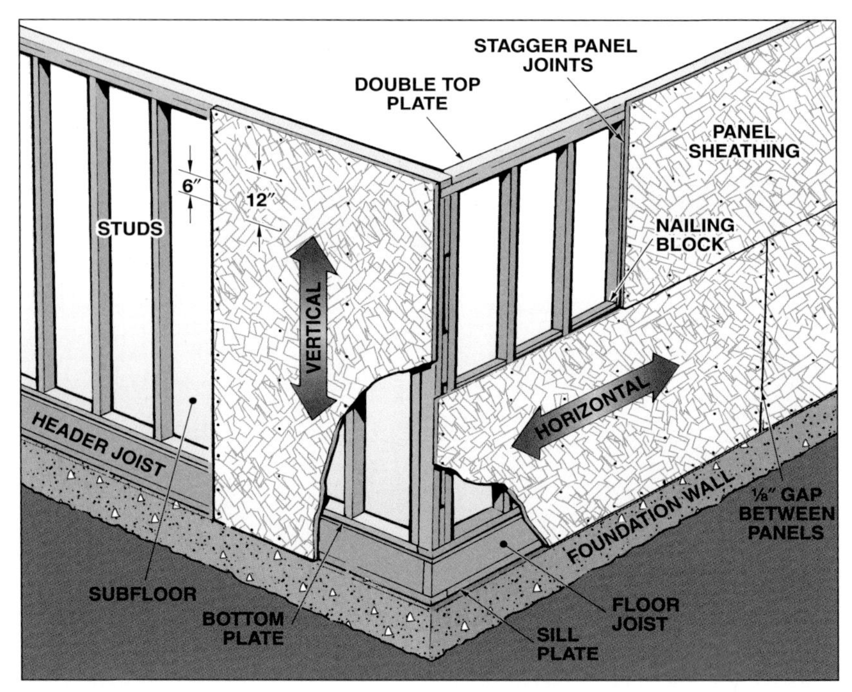

Figure 43-50. *Wall sheathing panels can be placed with the grain running vertically or horizontally.*

Figure 43-51. *After placing wall sheathing panels and securing them at the corners, snap vertical chalk lines locating the intermediate studs before installing nails or screws.*

Structural Insulated Sheathing (SIS)

Structural insulated sheathing (SIS) is wall panel sheathing that adds structural strength and insulation value to an exterior wall. A typical SIS panel consists of a core of rigid polyisocyanurate foam sandwiched between a water-resistant exterior layer and a high-pressure laminated structural member. **See Figure 43-52.** The panels come in ½″ and 1″ nominal thicknesses and are available as 4′ × 8′, 4′ × 9′, or 4′ × 10′ panels. SIS can add an R-value of 3.0 for the ½″ thickness and 5.5 for the 1″ thickness.

Structural Insulated Sheathing (SIS)

Figure 43-52

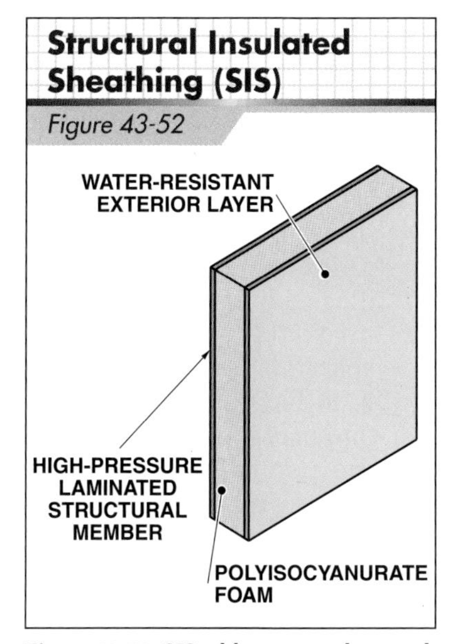

Figure 43-52. *SIS adds structural strength and insulation value to an exterior wall.*

The use of SIS may eliminate the need for separate layers of building wrap or felt to prevent water and air infiltration. **See Figure 43-53.** Polypropylene construction tape is used to seal the joints between panels, seal around penetrations, or patch any damage to the panels that occurred during installation. Construction tape has a UV-resistant film to protect it from degrading in sunlight.

SIS Installation

Figure 43-53

OX Engineered Products

Figure 43-53. *The use of SIS may eliminate the need for separate layers of building wrap or felt to prevent water and air infiltration.*

WEATHER BARRIERS

Weather barriers are attached to the wall sheathing to help prevent air and water infiltration into the building. Building paper and housewrap are two commonly used weather barriers. A traditional weather barrier is 15 lb *building paper*, which is available in 3′ wide rolls. Building paper must be sufficiently waterproof to resist water penetration from the outside, yet not so resistant that it prevents the escape of water vapor from the inside of the building during cold weather. If water vapor is prevented from escaping, condensation could occur inside the walls.

Housewrap

Housewrap is a common weather barrier used over panel sheathing. Housewrap is a translucent spun plastic sheet material that is tightly wrapped around a building to prevent air and water infiltration. **See Figure 43-54.** Housewrap allows a building to "breathe" by allowing water vapor and gases from a building interior to move outward without trapping moisture in the wall cavity. Housewrap is available in 3′ to 9′ rolls. The 9′ wide rolls, which are the most frequently used, should be installed by two workers.

Housewrap is installed before window and door frames are installed. Housewrap is available from various manufacturers and each manufacturer has specific instruction for proper installation of their product. Always consult the manufacturer's instructions before installing housewrap. The general procedure for installing housewrap is as follows:

1. Start the lowest course of housewrap at a building corner. Place a roll of housewrap vertically and extend the housewrap 2′ to 3′ past the corner. **See Figure 43-55.** Position the housewrap so the bottom edge sill plate and approximately 1″ of the foundation wall are covered. Wrap the housewrap entirely around the house, keeping it level and tight. Make tight inside corners by holding a 2 × 4 in the corner. Overlap the beginning of the housewrap approximately 6″. Each subsequent row of housewrap should overlap the row below by approximately 6″ to 12″.
2. Secure housewrap firmly in place by fastening it every 6″ along the perimeters and along door and window openings, and every 12″ to 16″ through the sheathing into the intermediate studs. Use large-headed or plastic cap nails or 1″ crown staples. Fasteners should penetrate a minimum of ½″ into the studs.
3. After fastening the housewrap for the entire building, attach the bottom of the first row to the foundation wall using tape. Use only tape approved by the housewrap manufacturer. Tape all vertical seams, holes, and any other openings in the housewrap. Tape around plumbing and electrical extrusions as well as vent openings.

Figure 43-54. *Weather barriers, such as housewrap, prevent water and air penetration into a building while allowing moisture vapor and gases to escape. Housewrap for the prefabricated panels forming the exterior walls of this house is applied at the manufacturing facility. Housewrap that is folded back at the corners and along the joists will be laid into position and fastened.*

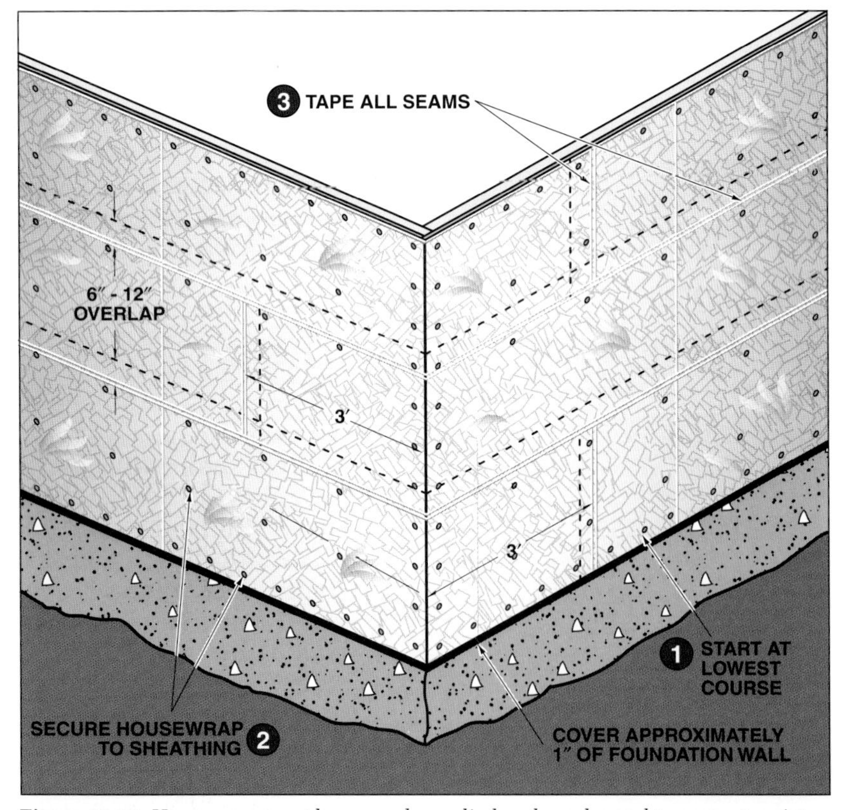

Figure 43-55. *Housewrap must be properly applied and overlapped to prevent moisture from infiltrating the building envelope.*

Window and Door Openings in Housewrap. Window and door openings are particularly prone to leakage unless they are properly prepared prior to installing the frames. Always consult the manufacturer's instructions for installation procedures for their products. The general procedure for cutting window and door openings in housewrap is as follows:

1. Make an inverted Y cut in the middle of the window opening. **See Figure 43-56.**
2. Fold back the flaps at the two sides and bottom of rough opening and fasten to the inside of the rough opening using nails or staples spaced 6″ apart. If a sill pan is to be installed, the side flap may remain loose until the pan is installed. The flaps are then folded into the rough opening and secured.
3. Make angled cuts at the top of the rough opening and fold the flap up. When the window frame is in place, the flap will be folded down and secured over the frame.

Preparation for door openings will be the same for the top and sides of the opening.

RAINSCREEN WALLS

Oriented strand board or structural plywood sheathing provides lateral strength against high winds, earthquakes, and other natural forces. Moisture trapped between the exterior finish material and sheathing will cause fungal growth, physical deterioration of the sheathing, and possible loss of structural capabilities of the sheathing. One method for preventing moisture vapor from accumulating between the exterior finish material and weather barrier is using a rainscreen wall. A *rainscreen wall* is a moisture-management system that incorporates a vented or porous exterior finish, an air cavity, a drainage plane, and an airtight interior support wall. **See Figure 43-57.** Moisture vapor or moisture entering the wall cavity is properly drained away and air is allowed circulate within the cavity to maintain dryness within the wall cavity.

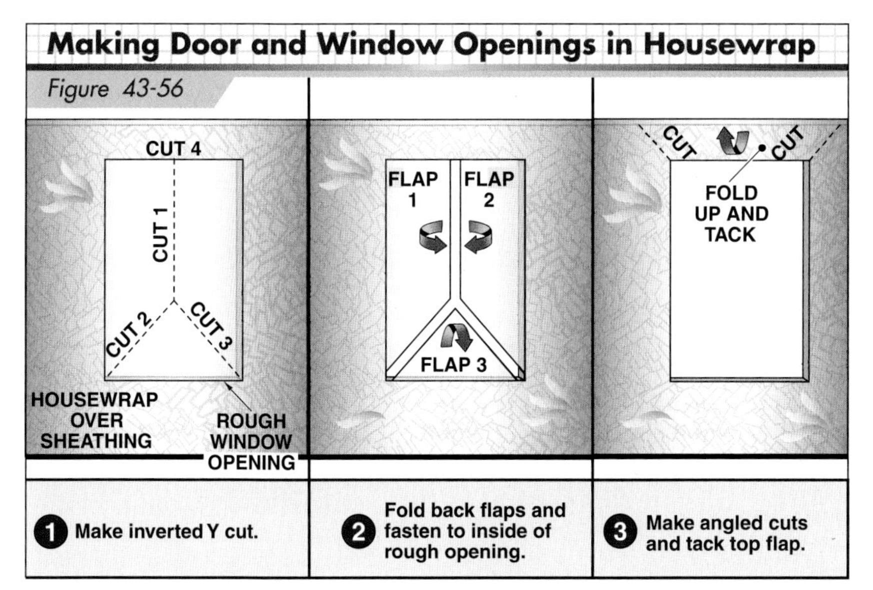

Figure 43-56. *An inverted Y cut is made in the housewrap, and the flaps are folded back and fastened to the inside of the rough opening. The top flap is folded up and tacked to the exterior side of the wall. The flap will be folded down and secured after the window or door is installed in the opening.*

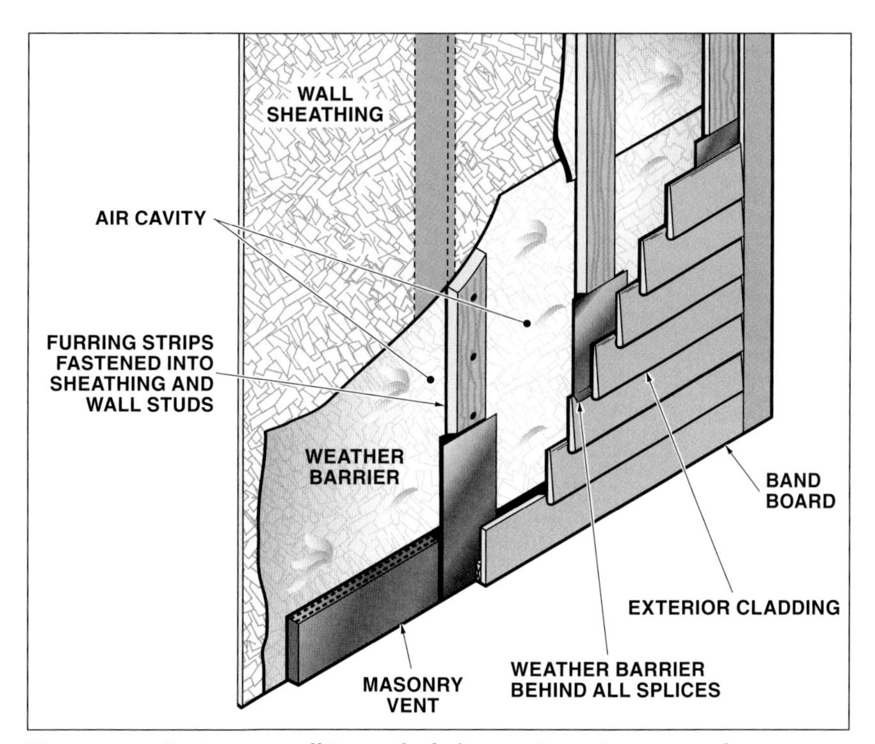

Figure 43-57. *A rainscreen wall is a method of preventing moisture vapor from accumulating between the exterior finish material and weather barrier.*

A layer of building paper or housewrap is applied to the exterior face of the support wall to act as a drainage plane. When rainscreen walls are installed in areas subjected to large rainfall amounts, water and ice guard may be applied directly to the sheathing extending 3′ to 4′ from the bottom edge, and the building paper or housewrap is applied over the guard. Ensure that the building paper or housewrap is lapped properly.

Furring strips are fastened through the wall sheathing and into studs where possible. The thickness of the furring strips creates the air cavity for the rainscreen wall. The air cavity separates the exterior finish material from interior support wall and reduces any capillary action to transfer moisture from the exterior finish material to the support wall.

Large, protected vents at the top and bottom of the wall promote airflow and allow moisture to drain and/or evaporate. Some types of vent openings may be covered with window screens to prevent insects from entering the air cavity.

The exterior finish material helps to shed rain and prevent moisture from entering the air cavity. Moisture may be forced through the exterior wall finish material itself or through joints and splices between the finish material. When installing wall finish material, place a piece of weather barrier at all splices to direct moisture into the air cavity.

STRUCTURAL INSULATED PANELS (SIPs)

A structural insulated panel (SIP) consists of two outer structural panels (skins) with a thick inner foam core. **See Figure 43-58.** The outer panels are commonly OSB, but plywood may also be used. SIPs are available in many standard sizes ranging from 4′ × 6′ to 8′ × 24′. The foam core is usually expanded polystyrene (EPS) or polyisocyanurate rigid foam, although other foam products such as extruded polystyrene and polyurethane are also used. Foam for the core has high insulation values and keeps its shape permanently. SIP construction is an environmentally responsible alternative to standard platform framing.

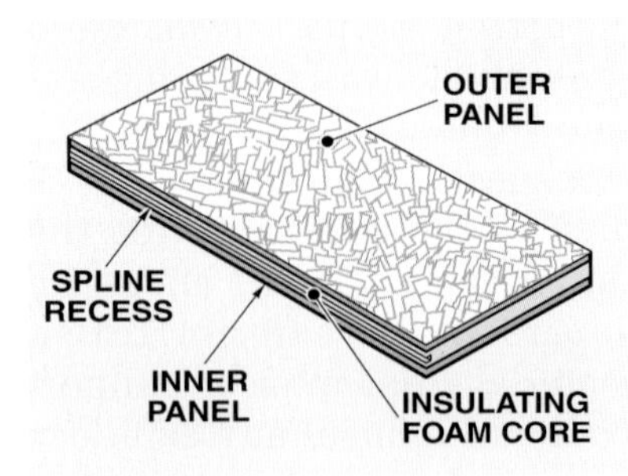

Figure 43-58. *Structural insulated panels (SIPs) are composed of two outer panels and an insulating foam core.*

The outer SIP panels are usually oriented strand board (OSB), but other surfaces are also available. SIPs are structurally rated panels, which provide lateral stiffness and strength and uniform load capacity. SIPs are also a good base for fasteners. Door and window openings may be cut out before or after panels have been installed.

Joining and assembly procedures for SIPs vary with the manufacturer. However, glue can be used in addition to nails and/or panel screws to fasten SIPs to the framework. Liquid expanding foam is usually applied at the joints to ensure airtight seals. Many manufacturers use spline connections at the joints. **See Figure 43-59.**

An SIP building must be constructed on and anchored to a solid foundation. A pressure-treated sill plate equal in width to the space between the SIP panels is secured to the foundation using anchors at a spacing specified by the manufacturer. **See Figure 43-60.** The panels are then hoisted and lowered over the sill plates using a crane or other hoisting equipment, depending on the panel size. Smaller panels may be fitted into position by hand. Sealant should be applied as specified by the manufacturer. Nails are then driven through the lower edges of the panels and into the sill plates. The panels should be properly braced until the roof or floor framing for the next floor is complete.

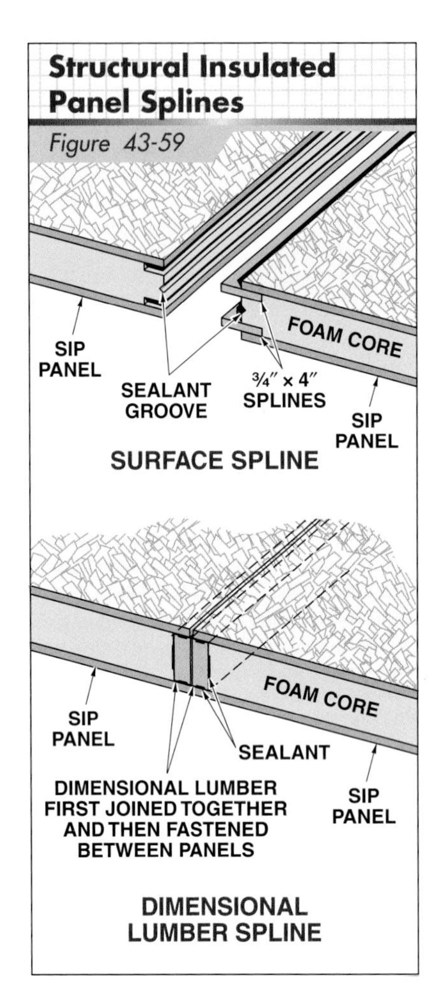

Figure 43-59. *A surface or dimensional lumber spline may be used to join SIPs together.*

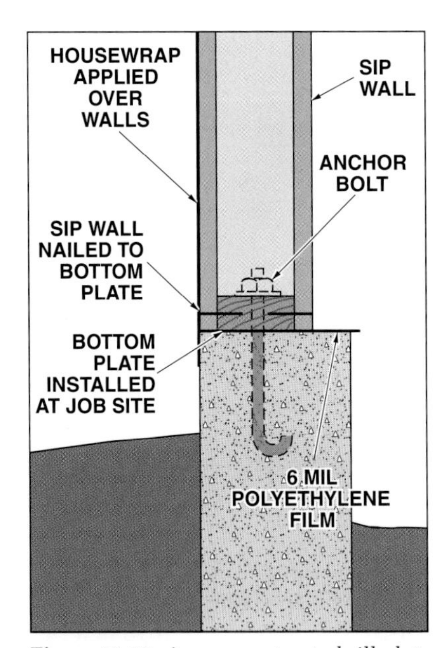

Figure 43-60. *A pressure-treated sill plate is secured to the foundation wall, and SIP panels are lowered over and fastened to the sill plate.*

Metal connectors are used to support floor joists for an upper floor of an SIP building. Top-flange joist hangers are fastened to the top plate of the SIP wall panels before the panels for the upper floor are placed and fastened into position. **See Figure 43-61.** The joists can then be installed and fastened into position. Metal straps are fastened to the exterior panel faces to tie the wall panels together.

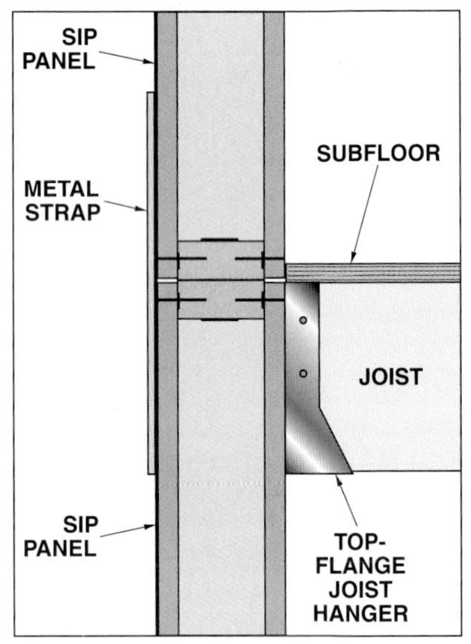

Figure 43-61. *Top-flange joist hangers are used to support upper floors in SIP construction.*

SHEAR WALLS

Shear walls are panel-sheathed walls designed to withstand severe seismic activity or high wind loads. Shear walls are interior or exterior, load-bearing or non-load-bearing walls that are designed to meet the design load requirements. Shear walls are required by building codes in many areas of the country including areas subject to strong seismic activity such as the West Coast and areas subject to excessive wind loads caused by hurricanes and tornadoes such as the East Coast and southern states.

The floors, roof, walls, and foundation of shear walls are tied together using specialized fasteners and anchors to provide the building with greater *shear load* and *uplift* resistance. Shear load is the lateral pressure against a wall. Two major contributing factors of shear load are high winds, such as winds generated in hurricanes, and seismic forces. Uplift occurs at the side of the lateral force, and is generally caused by forces generated during earthquakes. A downward force occurs at the opposite end of a wall experiencing uplift. **See Figure 43-62.**

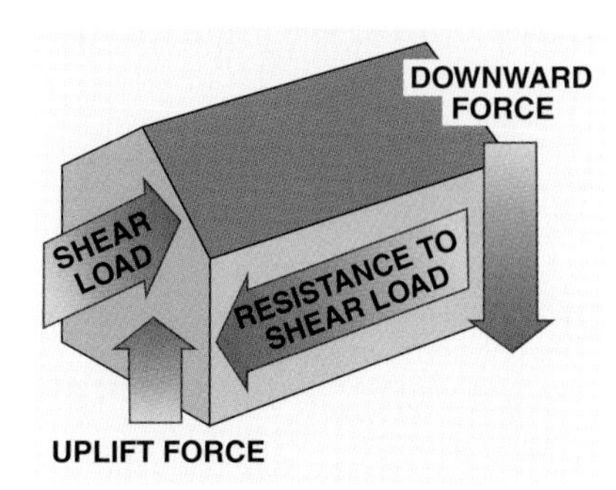

Figure 43-62. *The lateral direction of the shear load against the wall may cause uplift at the bottom of the wall, resulting in a downward force at the opposite end of the wall.*

Shear walls are most effective at the perimeter of a building since they can easily tie into the foundation. Shear walls are efficient for use as load-bearing walls. A building design may also include interior shear walls. If interior shear walls are required over crawl-space or full-basement foundations, they must be supported by the floor unit with additional support below the floor unit.

Constructing Shear Walls

As shown in **Figure 43-63**, shear wall construction is similar to conventional wood-framed walls, but there are several important differences, including the following:

- additional shear anchor bolts in bottom plate
- holddown anchors at the ends of shear walls
- tighter sheathing nailing patterns
- thicker sheathing
- special fastening requirements at the top of shear walls
- different lumber framing grades, species, and sizes

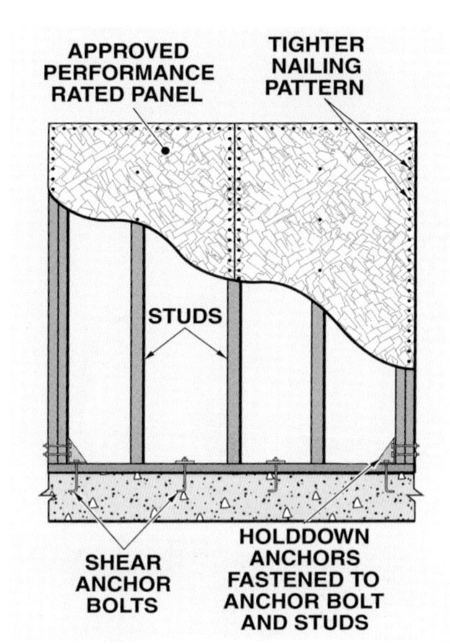

Figure 43-63. *Shear wall construction is similar to conventional wood-framed walls. Additional shear anchor bolts and holddown anchors are installed at the ends of shear walls.*

Shear anchor bolts or holddown straps are embedded in the concrete or masonry foundation wall or footing. Holddown straps may also be bolted to the sides of foundation walls, and are commonly used when retrofitting a building. **See Figure 43-64.**

When a shear wall panel is raised, the sill plate is properly positioned and fastened into position. Metal washers and nuts are used on intermediate anchor bolts, while holddown anchors are secured in position at the ends of shear walls. The holddowns are also fastened to the adjoining wall studs using lag or carriage bolts. **See Figure 43-65.**

Panel installation should begin at the corners of a building, leaving a 1/8″ space at the ends and edges of the panels. For multistory buildings, the tops of the panels are fastened to the top plate, as well as to the rim joists of the upper story to create a strong connection between different floor levels. The same procedure can be used for shear walls of a one-story building with a flat roof. For a one-story building with a pitched roof, the tops of the panels are nailed to the top plates. Seismic or hurricane roof ties strengthen the connection between the rafters and top plates. In multistory buildings, shear wall panels must align vertically to obtain the rated shear resistance.

Prefabricated shear walls provide high-load ratings and can be used in place of site-built shear walls.

Nailing patterns and penetration are determined by panel grades and thickness, and anticipated shear load. For APA Structural I and APA Rated Sheathing panels ranging in thickness from 1/4″ to 3/8″, 6d nails are used to fasten the sheathing to the studs and plates. For 7/16″ and 15/32″ panels, 8d nails are used, while on 19/32″ panels, 10d nails are used. Nails along the edges are spaced 2″ OC to 6″ OC, depending on the anticipated shear load. (When nails are spaced closer together, the permitted load increases considerably.) Nails are spaced 12″ OC for intermediate studs. Refer to print specifications for nailing patterns and anticipated shear loads.

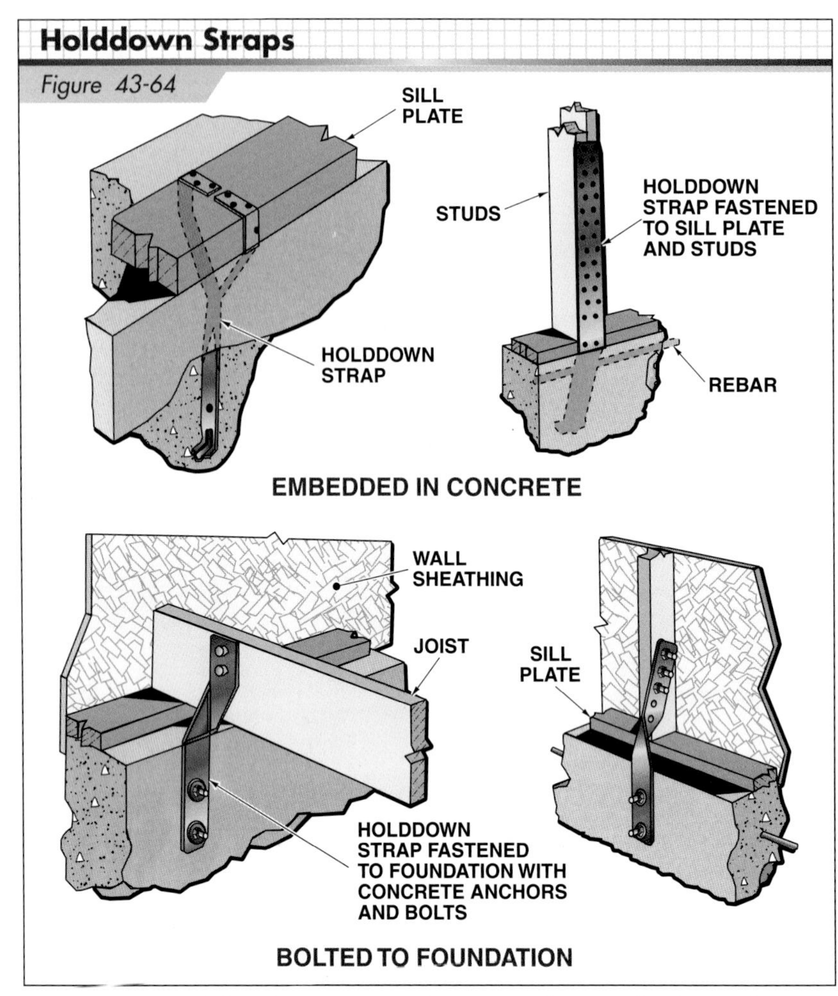

Figure 43-64. *Special anchors are required to secure sill plates of a shear wall system. Some anchors are embedded in concrete as it is being placed in the forms or CMUs. Other anchors are bolted to the foundation.*

Figure 43-65. *Holddowns are secured to anchor bolts embedded in the foundation. Lag or carriage bolts are driven into the adjoining wall studs.*

The bottoms of the wall sheathing panels are also nailed to the sill plates. The sill plates have, in turn, been secured at regular intervals to the foundation wall or footing using holddowns.

Anchor Tiedown Systems. *Anchor tiedown systems* are used in areas where the most extreme uplift due to seismic conditions or high winds is anticipated. Anchor tiedown systems are more effective than other shear wall anchor and fastening methods, and local building codes may require use of the systems.

In one-story anchor tiedown systems, a long threaded rod extends from the double top plate and is coupled to a bolt embedded in the concrete foundation. **See Figure 43-66.** The tension on the rod is adjusted by turning a nut positioned over a metal bearing plate, which rests on the double top plate. In multistory buildings, sections of threaded rod extend the height of each wall and are connected to one another using a *take-up device*. The take-up device is a heavy spring mechanism that compensates for wood shrinkage and settlement caused by dead loads. Connected anchor tiedown systems can extend up to five stories. **See Figure 43-67.**

Many jurisdictions in hurricane- and earthquake-prone areas have strict requirements for holddown and anchor tiedown systems.

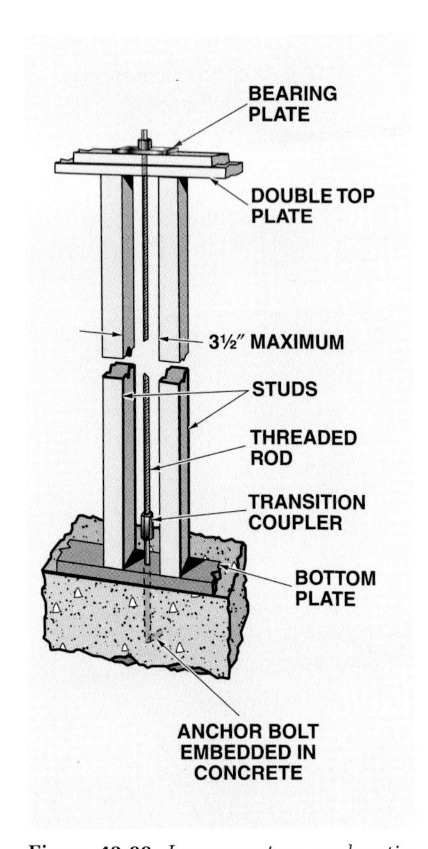

Figure 43-66. *In a one-story anchor tiedown system, a long threaded rod extends from the top double plates and is coupled with an anchor bolt embedded in the concrete.*

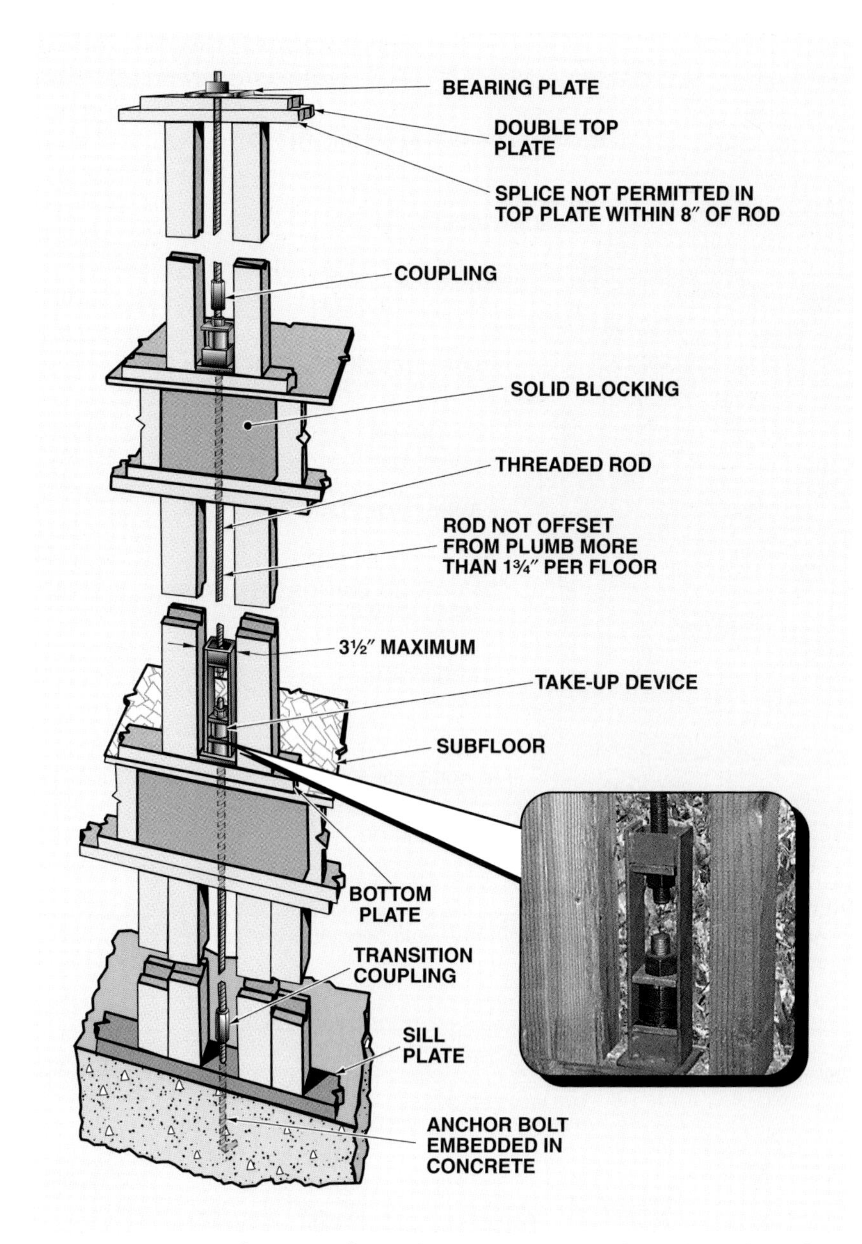

Figure 43-67. *In a multistory anchor tiedown system, a threaded rod is tightened and secured to the next story with a take-up device.*

ADVANCED FRAMING

Advanced framing is an engineered framing system that reduces material costs and construction waste and allows for increased insulation use. The use of advanced framing for the construction of wood-framed buildings has been increasing in recent years. The main purpose of advanced framing is to reduce lumber use and waste. This results in significantly lower material and labor costs.

Advanced framing methods improve energy efficiency by allowing more insulation to be added to the walls, floors, and ceilings. Many of the methods used in advanced framing are also accepted in green building construction. Advanced framing must be done without weakening the structure and must conform to local building codes. Because advanced framing is relatively new, it requires greater teamwork, coordination, and collaboration between the architects, the engineers, the builder, and all other parties involved in the project.

The main use of advanced framing is in walls. Advanced wall framing involves using stack framing and alternative stud spacing, using single top plates instead of double, framing corners with two studs instead of three, building and installing alternate door and window headers, and using SIS for increased R-value and strength.

Stack Framing

Stack framing, also known as in-line framing, is a key element in the advanced framing method. *Stack framing* is a method of construction where roof trusses, wall studs, and floor joists are constructed in a vertical line so that the load transfers directly to the structural member below it. **See Figure 43-68.**

Studs should be 2 × 6s spaced 24″ on center (OC). The use of 24″ OC stud spacing rather than the conventional 16″ OC reduces the amount of framing lumber used and makes it possible to install more insulation with less thermal bridging. *Thermal bridging* is the interruption of insulation by a framing member that has higher thermal conductivity than the insulation.

Single Top Plates

Stacked framing makes it possible to use only a single top plate for load-bearing and non-load-bearing walls and partitions. This is possible because there is no load bearing down between the studs below. The load placed on a stud is transferred to the framing member directly below it. The joints of single top plates must be reinforced with metal plates or solid blocking. Metal plates or blocking is also used when tying walls together at corners or intersections. **See Figure 43-69.**

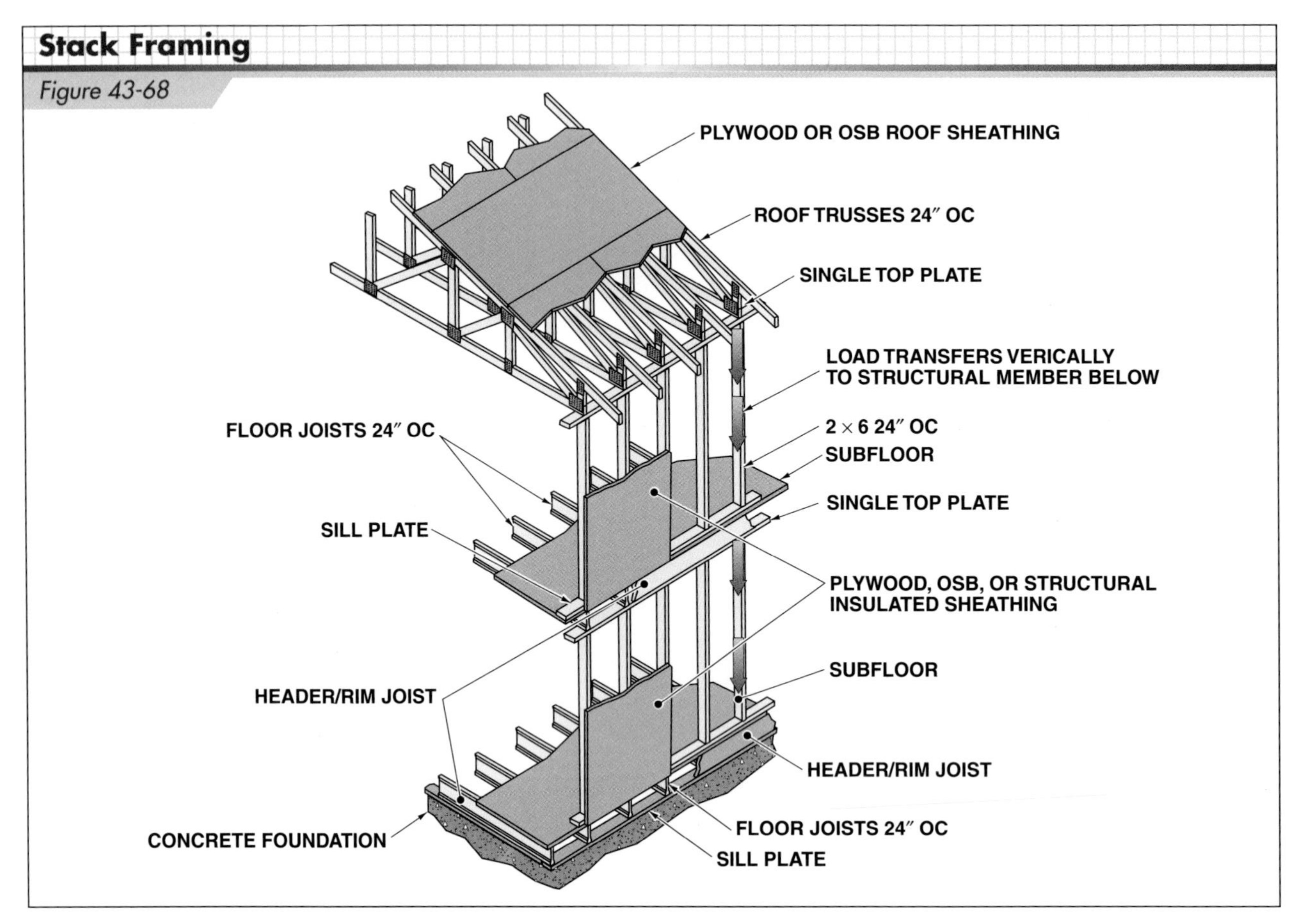

Figure 43-68. *Stack framing is a method of construction where the roof trusses, wall studs, and floor joists are in a vertical line so that the load transfers directly to the structural member below it.*

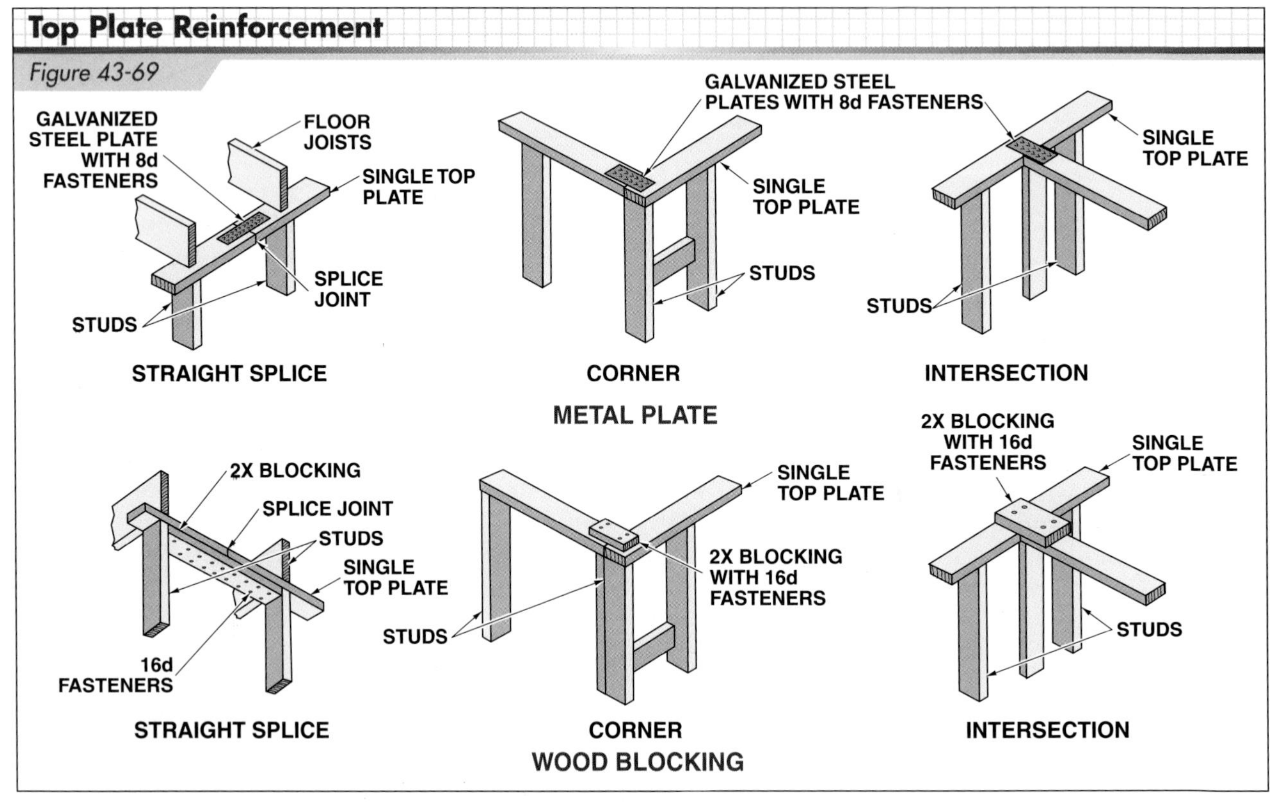

Figure 43-69. *Single top plates must be reinforced with metal ties or solid blocking.*

Corner Framing

Traditional inside and outside corner posts require three studs. This arrangement provides backing when nailing drywall or other types of wall finish but leaves little room for adequate insulation and is a waste of framing material. The advanced framing method eliminates one stud and uses drywall clips or a 1 × 4 backer board. **See Figure 43-70.**

In situations where one wall intersects another, called a T-wall, horizontal blocking may be placed in the receiving wall. Alternatively, either drywall clips or a full length 1 × 6 or 2 × 6 behind the partition stud can be used.

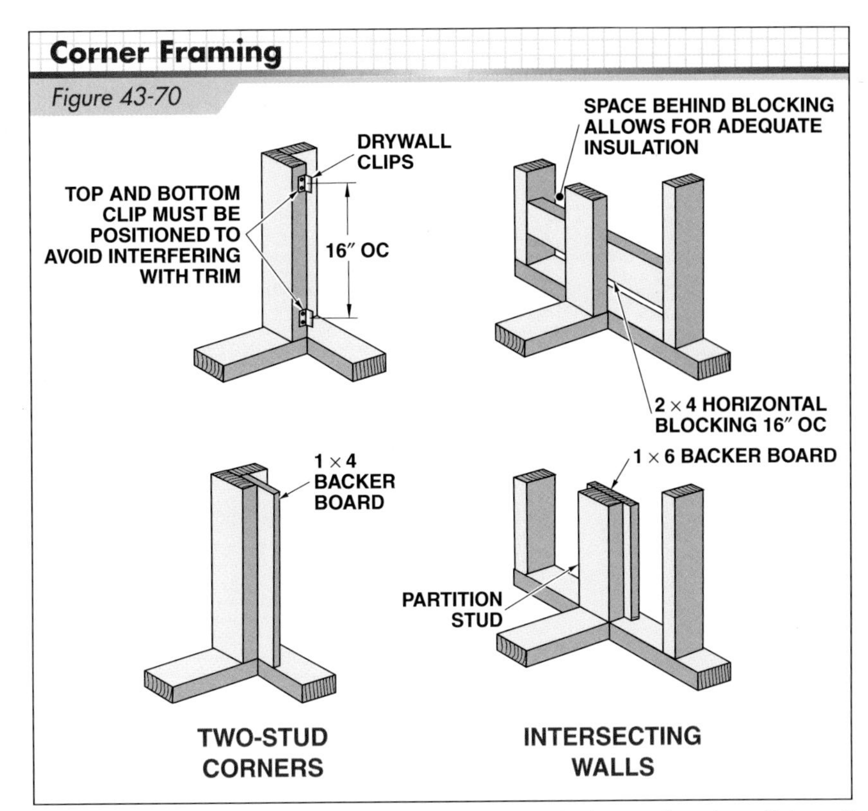

Figure 43-70. *Drywall clips or backer boards are used when nailing drywall in wall corners and intersections.*

Door and Window Headers

Advanced framing headers for load-bearing walls differ from those used in traditional framing. Advanced framing headers are sized for the loads they carry and allow a cavity space for insulation inside of the header. The header should span the entire length of the opening and be flush with the outer edge of the studs. The two ends of the header are supported by trimmer studs or metal header hangers. **See Figure 43-71.**

Box headers can also be used in advanced framing. Box headers are used on load-bearing exterior walls and can be built on the job site. They consist of 15/32″ sheathing on one or both sides. The sheathing is nailed every 3″ to the studs at the side of the opening, the top and bottom plates, and the cripple studs in between the panels. If the size of the rough opening is greater than 48″, trimmer studs are required for extra support. The space in between the sheathing is filled with insulation material.

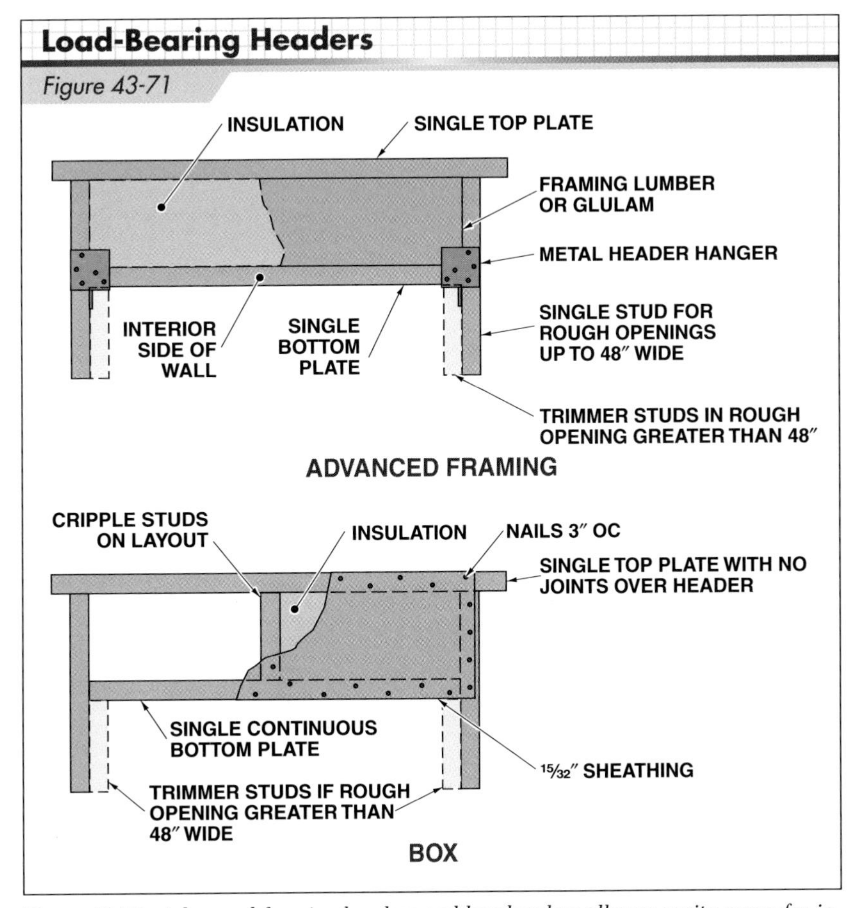

Figure 43-71. *Advanced framing headers and box headers allow a cavity space for insulation in load-bearing walls.*

Conventional headers can be replaced with simple framing at openings in non-load-bearing walls. The top of the opening is a single plate, and cripple studs are installed between the plate over the opening and the top wall plate. **See Figure 43-72.**

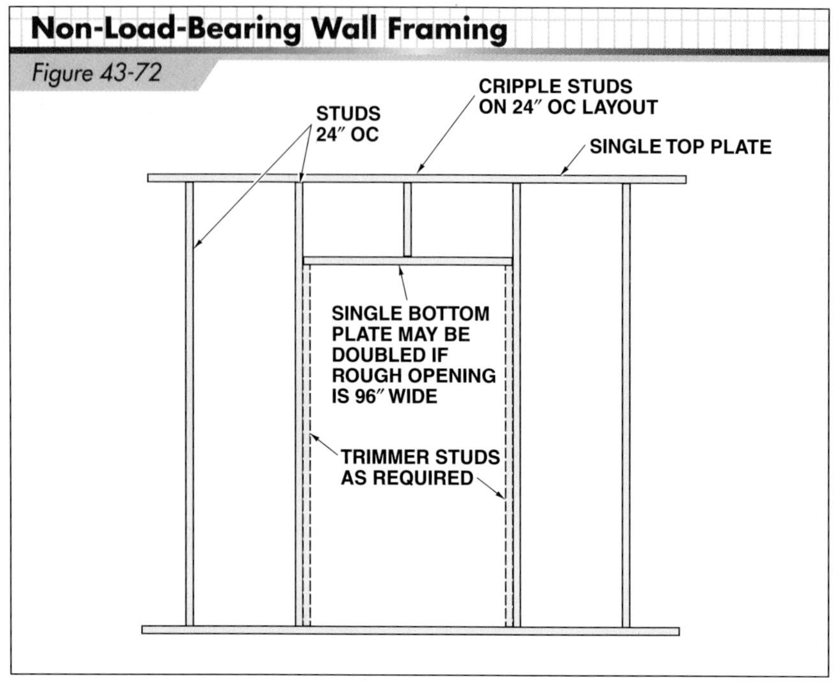

Figure 43-72. *Conventional headers can be replaced with simple framing at openings in non-load-bearing walls.*

UNIT 44
Ceiling Framing

OBJECTIVES

1. Explain the purpose of a ceiling frame.
2. Describe the installation of ceiling joists.
3. Describe the use of stub joists, ribbands, and strongbacks.
4. Explain methods of laying out ceiling frames.
5. Describe the construction of ceiling frames.
6. Explain the use of backing in roof construction.
7. Review the construction of a scuttle.
8. Describe the construction of flat roofs.

Ceiling construction begins after all walls have been set in place and fastened to the subfloor, the corners of the walls have been plumbed, and the tops of the walls have been aligned and supported with temporary bracing. One type of ceiling supports an attic area beneath a pitched (sloping) roof. Another type of ceiling serves as the framework for a flat roof. When a building has two or more floors, the ceiling unit of a lower story is part of the floor unit of the story above.

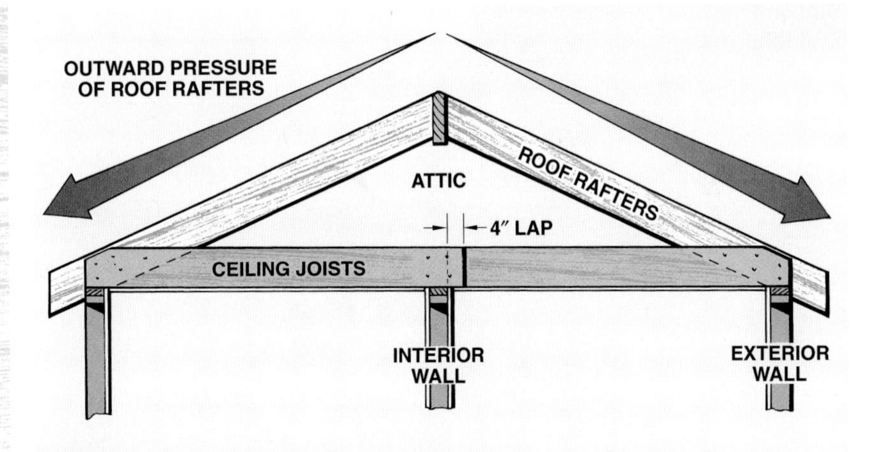

Figure 44-1. *A ceiling frame ties together the exterior walls and resists the outward pressure of roof rafters.*

One of the main structural functions of a ceiling frame is to tie together the exterior walls of a building. When located under a pitched roof, the ceiling frame also resists the outward pressure exerted on the walls by the roof rafters. **See Figure 44-1.** The tops of interior partitions are fastened to the ceiling frame. In addition to supporting the attic area beneath the roof, the ceiling frame supports the weight of the finish ceiling materials, such as gypsum board.

Various materials are used to finish ceilings including gypsum board, wood paneling, and acoustical or decorative ceiling tiles.

CEILING JOISTS

Joists are the most important framing members of a ceiling. Joist size, spacing, and direction of travel are shown on a floor plan. The spacing between ceiling joists is usually 16″ or 24″ OC, although 12″ and 19.2″ OC spacing may be used. Ceiling joist size is determined by the amount of weight the joist must carry and the span the joist covers from one wall to the other. The table in **Figure 44-2** gives allowable ceiling joist spans.

Although it is convenient to have all joists running in the same direction, designs may specify for different sets of joists to run at right angles to each other.

Interior Support of Joists

One end of a ceiling joist rests on an outside wall and the other end often overlaps an interior bearing partition or beam. Ceiling joists should overlap the partition or beam at least 4″. Ceiling joists may be butted over a partition or beam. In this case, the joists must be cleated with a ¾″ × 24″ piece of plywood or an 18 ga × 18″ long metal strap. **See Figure 44-3.** Ceiling joists may also butt against a beam and be supported by a ledger strip or joist hangers in a manner similar to floor joists.

The structural design of a building may specify a beam be placed above the ceiling joists. In this case, the inside ends of the joists are hung from the beam with special joist hangers.

CEILING JOIST SPANS*

Species or Group	Grade	Span† 2 x 4				2 x 6				2 x 8				2 x 10			
		OC Spacing‡															
		12	16	19.2	24	12	16	19.2	24	12	16	19.2	24	12	16	19.2	24
Douglas Fir-Larch	Sel. Struc.	13-2	11-11	11-3	10-5	20-8	18-9	17-8	16-4	27-2	24-8	23-3	21-7	34-8	31-6	29-8	27-6
	No. 1 & Btr.	12-11	11-9	11-0	10-3	20-3	18-5	17-4	16-1	26-9	24-3	22-10	21-2	34-1	31-0	29-2	26-10
	No. 1	12-8	11-6	10-10	10-0	19-11	18-1	17-0	15-9	26-2	23-10	22-5	20-1	33-5	30-0	27-5	24-6
	No. 2	12-5	11-3	10-7	9-10	19-6	17-8	16-8	15-0	25-8	23-4	21-4	19-1	32-9	28-6	26-0	23-3
	No. 3	11-1	9-7	8-9	7-10	16-3	14-1	12-10	11-6	20-7	17-10	16-3	14-7	25-2	21-9	19-10	17-9
Douglas Fir-South	Sel. Struc.	11-10	10-9	10-2	9-5	18-8	16-11	15-11	14-9	24-7	22-4	21-0	19-6	31-4	28-6	26-10	24-10
	No. 1	11-7	10-6	9-11	9-2	18-2	16-6	15-7	14-5	24-0	21-9	20-6	19-0	30-7	27-9	26-2	23-7
	No. 2	11-3	10-3	9-8	8-11	17-8	16-1	15-2	14-1	23-4	21-2	19-11	18-6	29-9	27-1	25-3	22-7
	No. 3	10-10	9-5	8-7	7-8	15-10	13-9	12-6	11-2	20-1	17-5	15-10	14-2	24-6	21-3	19-5	17-4
Hem-Fir	Sel. Struc.	12-5	11-3	10-7	9-10	19-6	17-8	16-8	15-6	25-8	23-4	21-11	20-5	32-9	29-9	28-0	26-0
	No. 1 & Btr.	12-2	11-0	10-4	9-8	19-1	17-4	16-4	15-2	25-2	22-10	21-6	19-11	32-1	29-2	27-5	25-5
	No. 1	12-2	11-0	10-4	9-8	19-1	17-4	16-4	15-2	25-2	22-10	21-6	19-10	32-1	29-2	27-1	24-3
	No. 2	11-7	10-6	9-11	9-2	18-2	16-6	15-7	14-5	24-0	21-9	20-6	18-6	30-7	27-8	25-3	22-7
	No. 3	10-10	9-5	8-7	7-8	15-10	13-9	12-6	11-2	20-1	17-5	15-10	14-2	24-6	21-3	19-5	17-4
Spruce-Pine-Fir-South	Sel. Struc.	11-7	10-6	9-11	9-2	18-2	16-6	15-7	14-5	24-0	21-9	20-6	19-0	30-7	27-9	26-2	24-3
	No. 1	11-3	10-3	9-8	8-11	17-8	16-1	15-2	14-1	23-4	21-2	19-11	18-6	29-9	27-1	25-5	22-11
	No. 2	10-11	9-11	9-4	8-8	17-2	15-7	14-8	13-8	22-8	20-7	19-5	17-8	28-11	26-3	24-2	21-7
	No. 3	10-3	8-11	8-2	7-3	15-0	13-0	11-11	10-8	19-1	16-6	15-1	13-6	23-3	20-2	18-5	16-5
Western Woods	Sel. Struc.	11-3	10-3	9-8	8-11	17-8	16-1	15-2	14-1	23-4	21-2	19-11	18-6	29-9	27-1	25-5	23-3
	No. 1	10-11	9-11	9-4	8-8	17-2	15-7	14-7	13-0	22-8	20-2	18-5	16-6	28-6	24-8	22-6	20-2
	No. 2	10-7	9-8	9-1	8-5	16-8	15-2	14-3	13-0	21-11	19-11	18-5	16-6	28-0	24-8	22-6	20-2
	No. 3	9-5	8-2	7-5	6-8	13-9	11-11	10-10	9-8	17-5	15-1	13-9	12-4	21-3	18-5	16-9	15-0

* design criteria: strength = 10 lb/sq ft live load plus 5 lb/sq ft dead load
deflection = limited in span (in in.) divided by 240 for live load only

† in ft and in.

‡ in in.

Western Wood Products Association

Figure 44-2. *Ceiling joist size is based on the amount of weight the joist must carry and the joist span. For example, No. 1 2 × 6 Douglas fir joists spaced 24″ OC can span up to 14′-5″.*

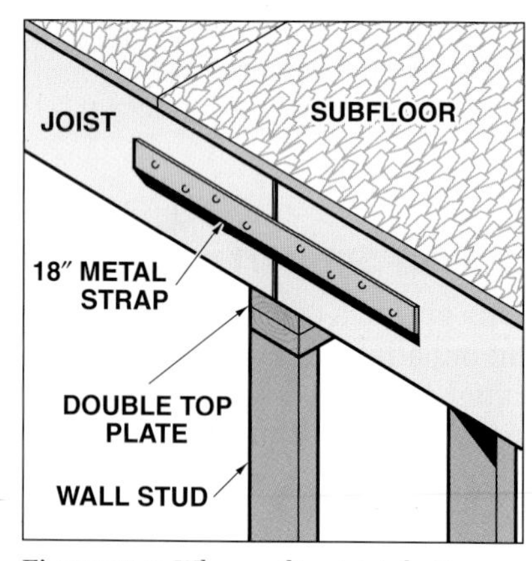

Figure 44-3. *When ceiling joists butt over a partition, an 18 ga × 18″ long metal strap can be used to tie the joists together.*

Ceiling Joists and Roof Rafters

Whenever possible, ceiling joists should run in the same direction as roof rafters. Nailing the outside end of each ceiling joist to the heel of a rafter as well as to the top plate strengthens the tie between the exterior walls of a building. **See Figure 44-4.**

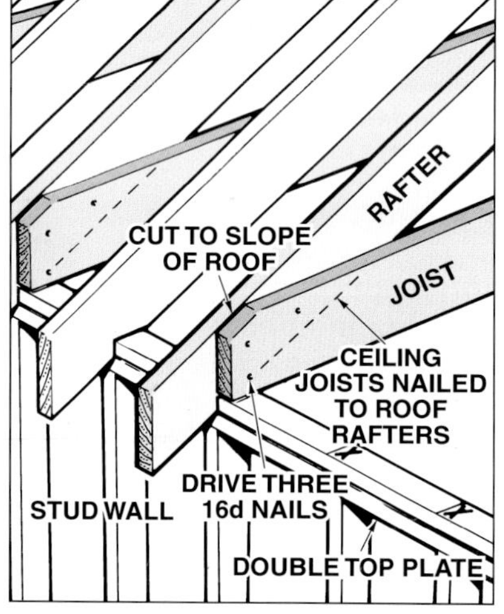

Figure 44-4. *Whenever possible, ceiling joists should be nailed to rafters.*

A building may be designed so ceiling joists do not run parallel to the roof rafters. To prevent rafters from pushing out the walls, ties (2 × 4s or 2 × 6s) are installed running in the same direction as the rafters. The ties are then fastened to the top edge of each ceiling joist with two 16d nails. **See Figure 44-5.** The ties should be spaced no more than 4′ apart. The tie ends should be secured to the rafter heels by nailing through the rafters into the edge of the tie.

Cutting Ends for Roof Slope

When ceiling joists run in the same direction as roof rafters, the outside ends must be cut to the slope of the roof. Ceiling frames may be constructed with *stub joists.* **See Figure 44-6.** Stub joists are necessary when, in certain sections of the roof, rafters and ceiling joists do not run in the same direction. For example, a low-pitched hip roof requires stub joists in the hip section of the roof.

Ribbands and Strongbacks

Ceiling joists that do not support a floor above do not require header (rim) joists or blocking.

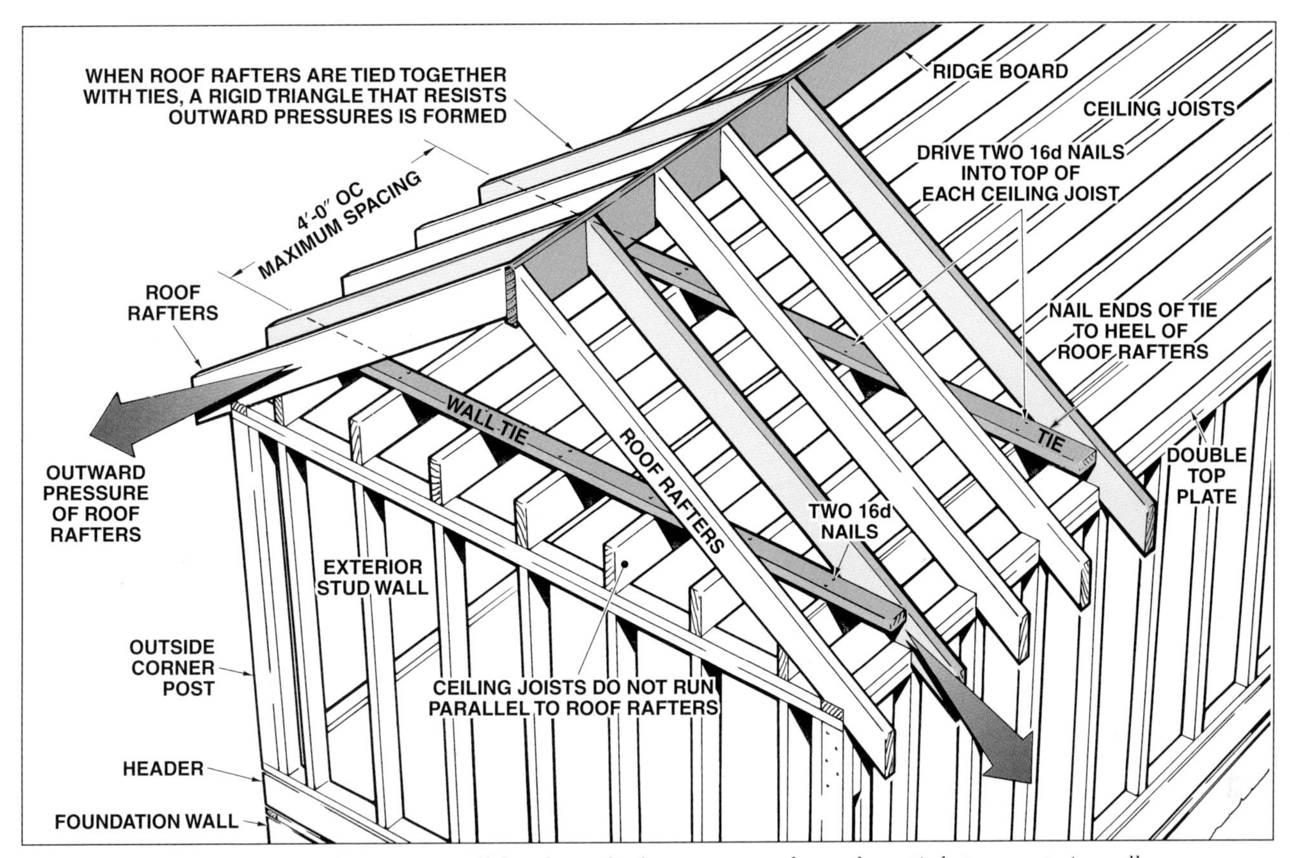

Figure 44-5. *When ceiling joists do not run parallel to the roof rafters, 2 × 4s can be used as a tie between exterior walls.*

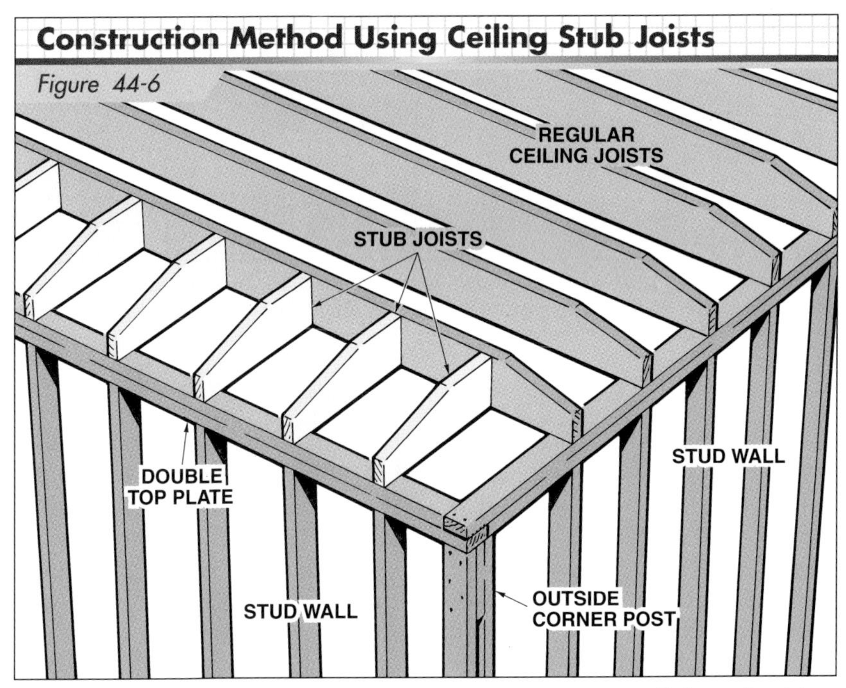

Figure 44-6. *Stub joists may be required where rafters do not run parallel to ceiling joists.*

Without additional header joists, however, ceiling joists may twist or bow at the centers of their span. To prevent twisting or bowing, a *ribband* is nailed at the center of the spans. **See Figure 44-7.** The ribband is laid flat and fastened to the top of each joist with two 8d nails. The end of each ribband is secured to the exterior walls of the building.

A more effective method of preventing twisting, bowing, or sagging of ceiling joists is to use a *strongback.* A strongback is built with 2 × 6s or 2 × 8s nailed to the side of a 2 × 4. The 2 × 4 is fastened with two 16d nails to the top of each ceiling joist as shown in **Figure 44-8.** Strongbacks are blocked up and supported over the exterior walls and interior partitions. Strongbacks hold the ceiling joists in alignment and also help support joists at the centers of their spans.

LAYING OUT CEILING FRAMES

Ceiling joists should be placed directly above wall studs when joist spacing is the same as stud spacing. Aligning ceiling joists and wall studs makes it easier to install pipes, flues, or ducts that have to run up the wall and through the roof. However, for buildings with walls with double top plates, most building codes do not require ceiling joists to align with the studs below.

1 × 4 Ribbands Prevent Joists from Twisting

Figure 44-7

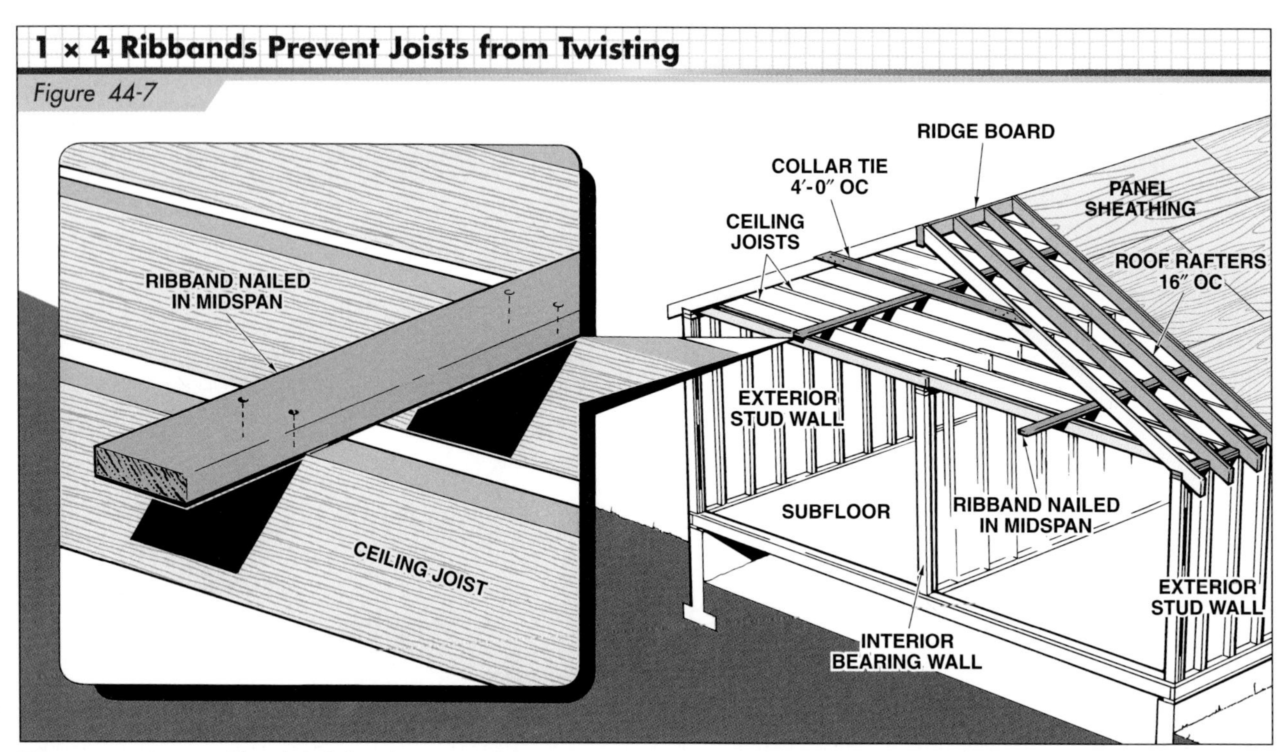

Figure 44-7. *A 1 × 4 ribband nailed at the center of the joist spans prevents twisting and bowing of the joists.*

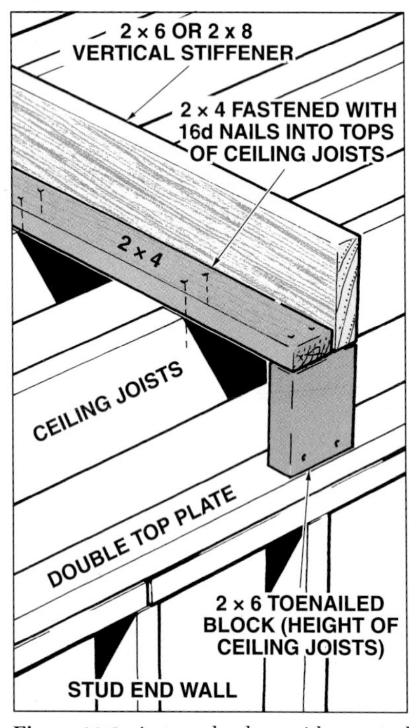

Figure 44-8. *A strongback provides central support for the joist span.*

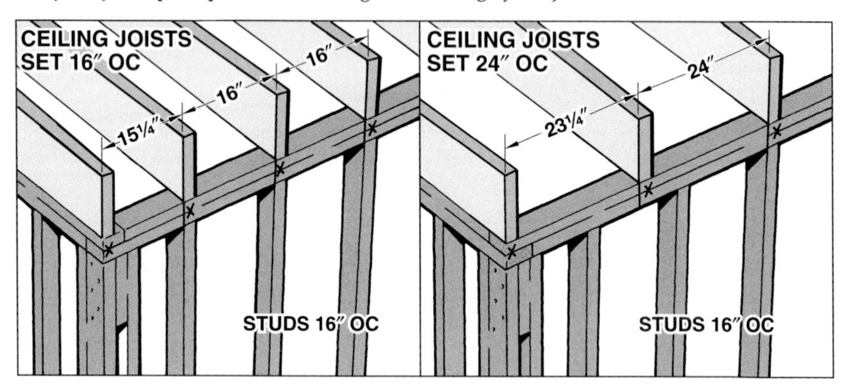

Figure 44-9. *When studs and ceiling joists are spaced 16″ OC, the joists rest directly over the studs. When studs are spaced 16″ OC and ceiling joists are spaced 24″ OC, every other joist will align with the studs below.*

If joists are placed directly above the studs, the joists will follow the same layout as the studs below. **See Figure 44-9.** If joist spacing is different from that of the studs below (for example, if joists are spaced 24″ OC over 16″ OC studs), lay out the first joist at 23¼″ and then at every 24″ OC.

The positions of the roof rafters should be marked when the ceiling joists are being laid out. If ceiling joist spacing is the same as roof rafter spacing, there will be a rafter next to every joist. Often, ceiling joists are spaced 16″ OC and roof rafters are spaced 24″ OC. In this case, only every other rafter can be placed next to a ceiling joist.

CONSTRUCTING CEILING FRAMES

Joists for a ceiling frame should be cut to length before they are placed on top of the walls. On houses with pitched roofs, the outside ends of the joists should also be cut at an angle on the crown edge of the joist to accommodate the roof slope. The prepared joists can then be handed up to the carpenters working on top of the walls. The joists are laid flat along the walls, close to where they will be nailed. **Figure 44-10** shows a procedure for constructing the ceiling frame. In this example, the joists will lap over an interior bearing partition.

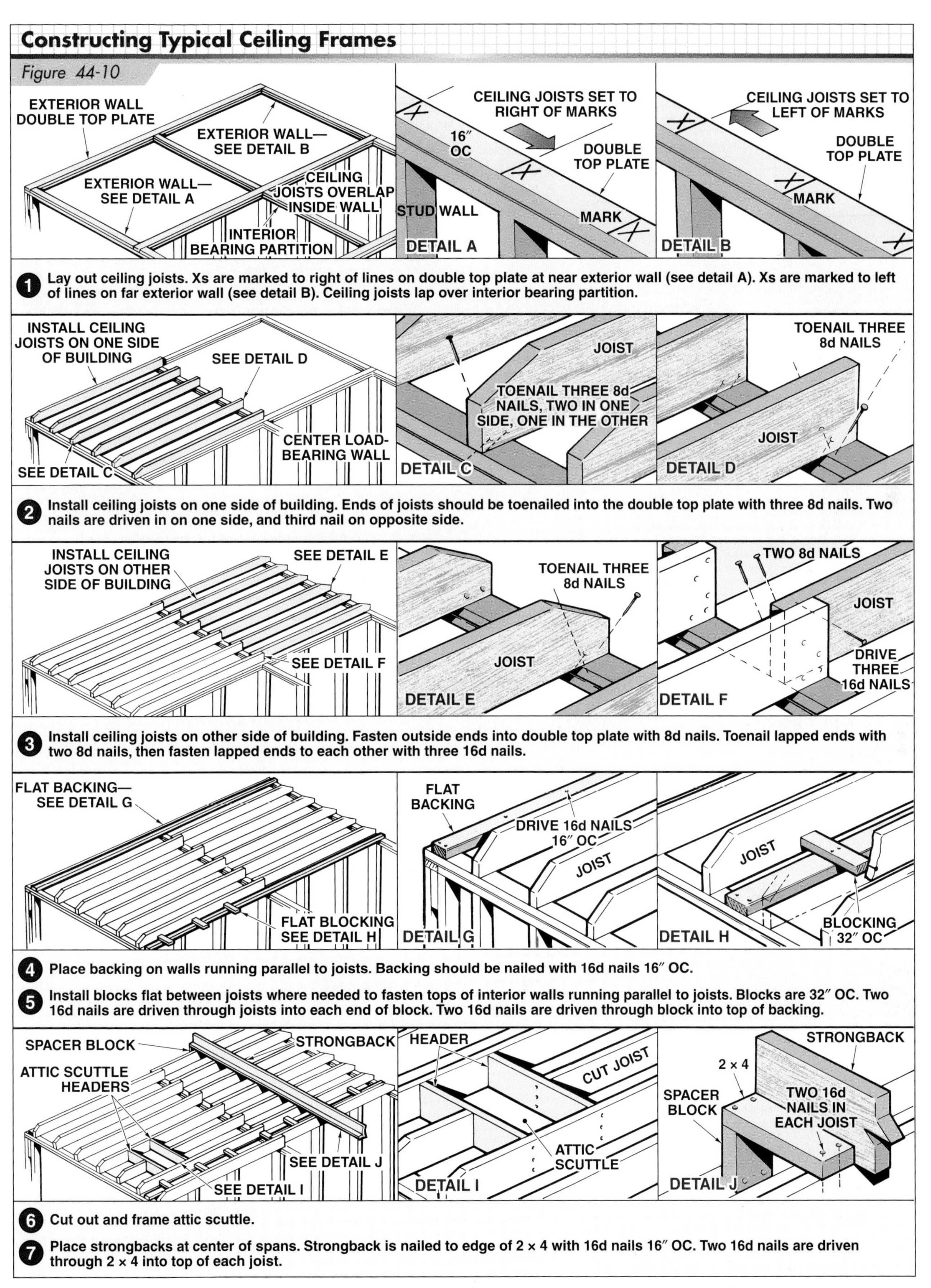

Figure 44-10. *A ceiling frame is constructed over walls that have been straightened and properly braced. In this example, joists lap over an interior bearing partition.*

Ceilings may be *furred* down for insulation or sound-deadening purposes. **Figure 44-11** shows an example where a ledger board the same width as the joist material is nailed against the wall studs. The ends of the joists are then toenailed or fastened with metal joist hangers to the ledger board.

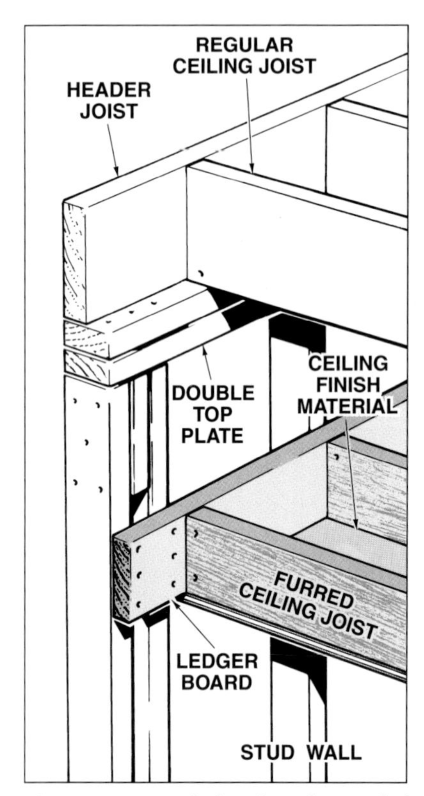

Figure 44-11. *A ledger board is nailed against the wall to provide a nailing surface for lower ceiling joists.*

Applying Backing to Walls

Walls running in the same direction as ceiling joists require *backing* (sometimes called deadwood) to provide a nailing surface for the edges of the finish ceiling material. Backing is usually 2″ lumber, although 1″ lumber may be used.

Figure 44-12 shows backing placed on top of walls. The 2 × 6 backing nailed to the exterior wall projects to the inside of the building. The interior wall requires a 2 × 6 or 2 × 8 centered on the double top plate and projecting on both sides of the wall. Backing is fastened to the top plates with 16d nails spaced 16″ OC.

Backing may also be used where ceiling joists run at a right angle to an interior bearing partition. **See Figure 44-13.** The backing is cut to fit between the lapped ceiling joists and nailed to the double top plate.

Noises such as footsteps and vibrations from the floor above can be reduced by installing resilient channels along the bottoms of floor joists.

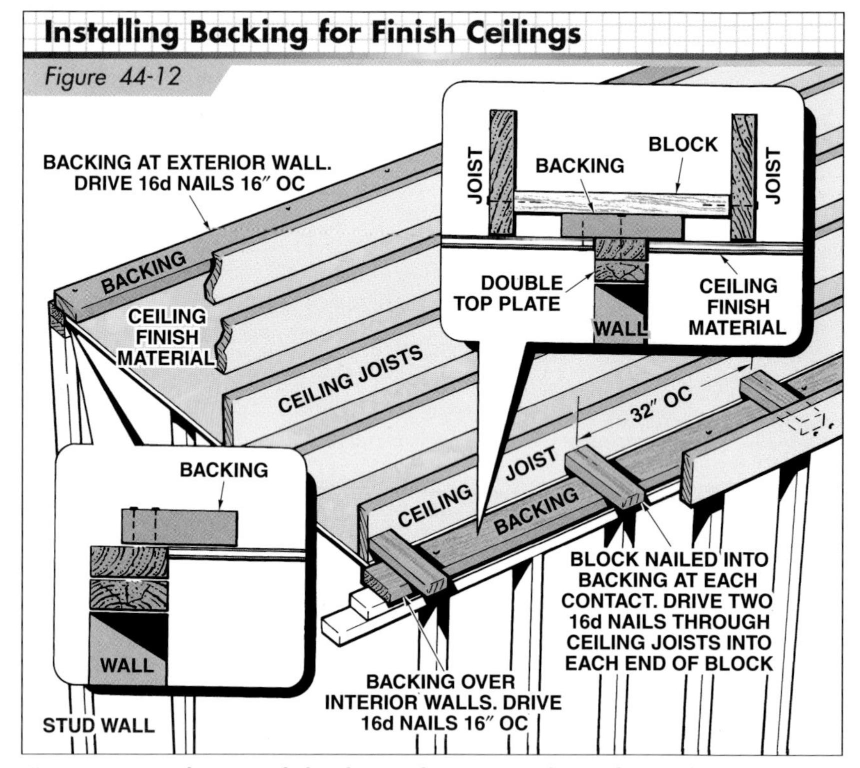

Figure 44-12. *Backing is nailed to the top plates to provide a nailing surface for the edges of the finish ceiling material. Backing may be 2 × 6s or 2 × 8s.*

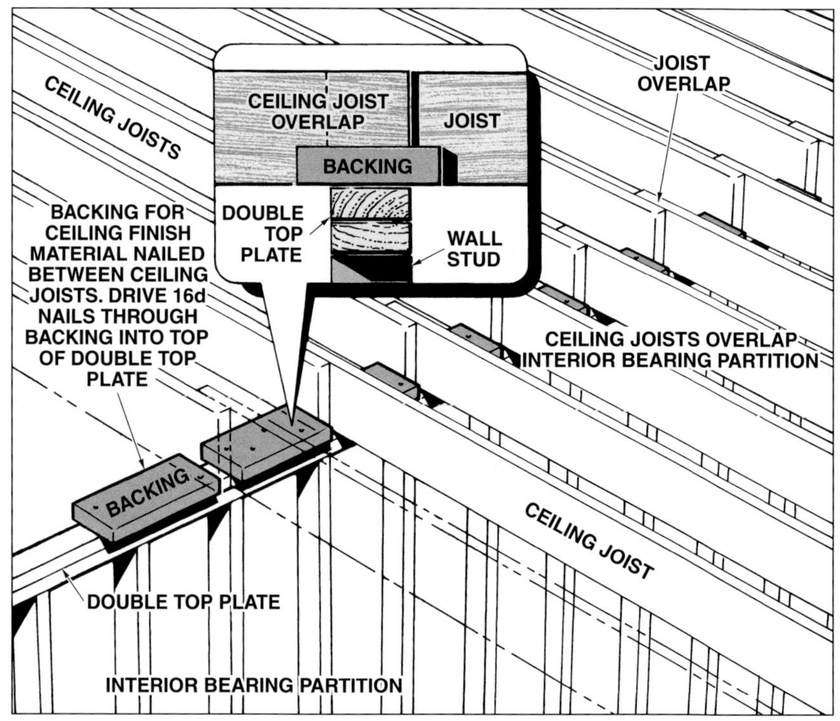

Figure 44-13. *Backing is sometimes nailed on top of bearing partitions.*

Fastening Partitions to Ceiling Frames

The tops of partitions running in the same direction as the ceiling joists must be securely fastened to the ceiling frame. **See Figure 44-14.** Blocks, 2 × 4s spaced 32″ OC, are placed flat over the backing, which is fastened to the top of the partition. The ends of each block are fastened to the joists with two 16d nails. Two 16d nails are also driven through each block into the backing and top of the partition.

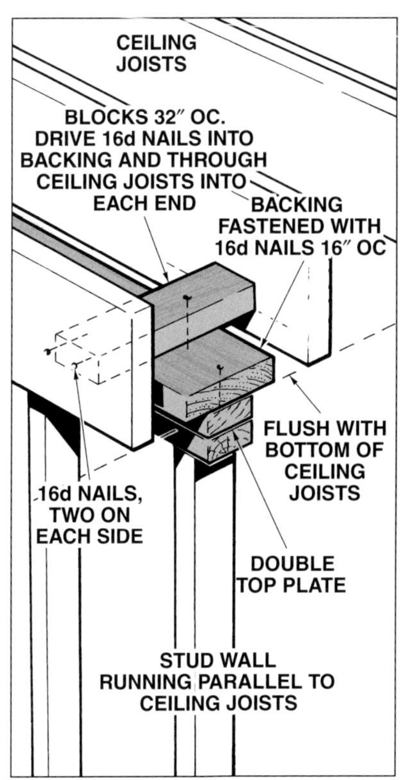

Figure 44-14. *Blocks and backing are installed between joists to secure tops of walls running parallel to ceiling joists.*

ATTIC SCUTTLES

A *scuttle* is an opening framed in the ceiling to provide an entrance into the attic. In buildings with a combustible ceiling or roof, an attic scuttle must be provided if the attic is larger than 30 sq ft in size and has a height greater than 30″. The rough framed opening must be at least 22″ × 30″ and be located in a hallway or other accessible location.

A scuttle is framed in a similar manner as a floor opening. If a scuttle is no more than 3′ square, it is not necessary to double the joists and headers. Scuttles must be placed away from the lower areas of a sloping roof. The opening may be covered by a piece of plywood resting on stops. The scuttle opening can be cut out after all the regular ceiling joists have been nailed in place.

CONSTRUCTING FLAT ROOF CEILINGS

Flat roofs are economical to build. The joists used to frame a flat roof are also the ceiling joists for the living area below. A flat roof should have a small pitch (slope), no less than ¼″ per foot, for proper water drainage. To strengthen a flat roof in areas of heavy snowfall, heavier joists spaced closely together may be used. Since the rafters for a flat roof must support the combined load of the ceiling and roofing materials, 2 × 10 or larger pieces spaced 16″ OC are usually used. In northern states, flat roofs are usually required by local building codes to withstand a live load (snow and wind load) of 40 lb/sq ft. In central states, flat roofs are required to withstand a live load of 30 lb/sq ft. In southern and western states, flat roofs must withstand a live load of 20 lb/sq ft.

Figure 44-15 shows a section of the framework for a flat roof. Since the joists also serve as a base for the roof, the joists are typically referred to as rafters. The rafters are tied together with a header where they extend past the exterior walls of a building. Shorter, cantilevered joists are called *lookout rafters.* One end of each lookout rafter is fastened to a regular roof rafter using a metal framing anchor. The roof rafter will be doubled after the lookouts have been installed.

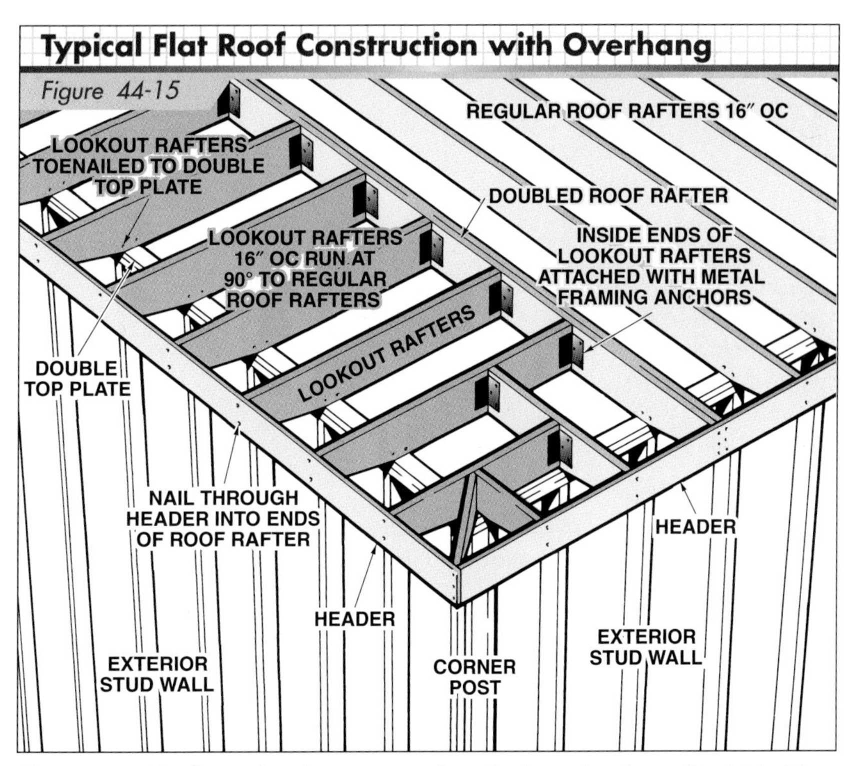

Figure 44-15. *For flat roofs, rafters serve as a base for the roof and as ceiling joists. Since joists for a flat roof must support the combined load of the ceiling and roof materials, 2 × 10s or larger framing members are used.*

Figure 44-16 shows a flat roof overhang. The ceiling material is gypsum board and the roof is panel sheathing over insulation board. Flat roofs may also be constructed with *parapets* (short walls) above the roof deck. **See Figures 44-17** and **44-18.**

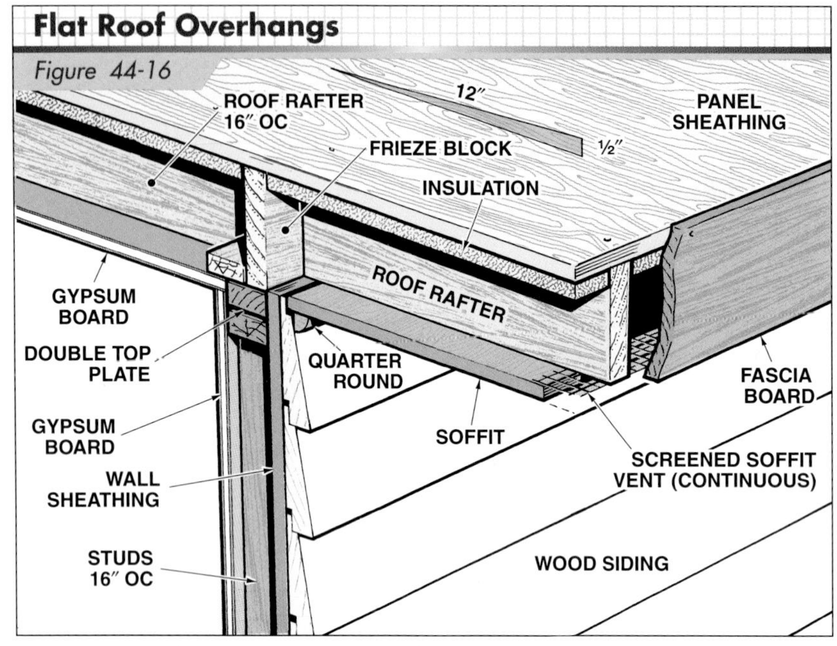

Figure 44-16. *A flat roof must have a small amount of pitch for proper water drainage.*

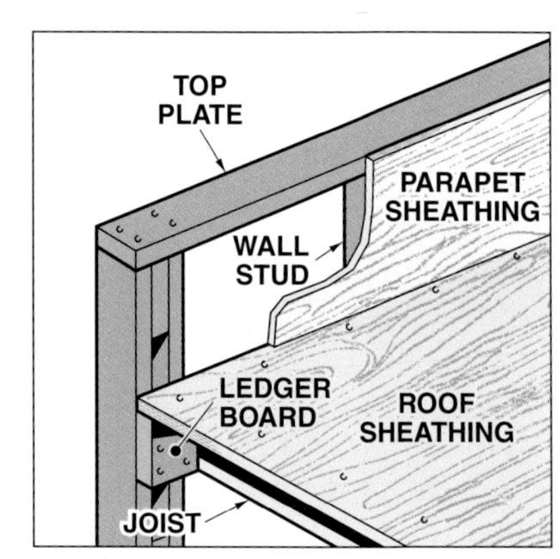

Figure 44-18. *Parapets can be framed into exterior walls.*

Figure 44-17. *Parapets can be framed over top plates.*

OBJECTIVES

1. List and describe common types of metal framing members.
2. Describe how metal framing members are fastened together.
3. Identify tools used to work with metal framing members.
4. Discuss metal framing safety precautions.
5. Describe the construction of floor units when using metal framing.
6. Describe wall construction methods using steel framing.
7. Discuss the procedure for framing walls.
8. Review the procedure for raising walls.
9. Describe the placement of ceiling-floor joists.
10. Describe metal roof rafter installation.
11. Discuss the use of metal roof trusses.
12. Define panelization.
13. Describe structural steel framing.
14. Describe the installation of building envelopes and insulation.
15. Discuss methods of attaching interior finish to metal-framed walls.

Metal framing is a method of constructing the framework of a building using light-gauge and heavy-gauge steel framing members. *Light-gauge steel (LGS) framing* was first used in commercial structures with limited heights. By the late 1950s, light-gauge steel framing methods were also used in high-rise structures. Today, light-gauge steel is specified for residential, light commercial, and heavy commercial structures. Carpenters skilled in wood framing methods should be readily able to adapt to light-gauge framing methods.

Metal framing offers many advantages over conventional wood framing. Light-gauge steel framing members are lightweight and straight, allowing a carpenter to easily cut and install the members. Metal framing members are not affected by insects or decay, and do not shrink or otherwise deform like similar wood framing members. Metal framing offers additional LEED® points since 25% to 35% of metal framing members are manufactured from recycled steel. In addition, metal framing has more fire resistance than comparable wood framing members.

Many metal framing methods are employed today. Light-gauge steel construction is often referred to as "steel for stick" construction since each wood framing member in a typical assembly is replaced with a metal framing member. **See Figure 45-1.** Another metal framing method, similar to light-gauge steel construction, involves the use of heavier-gauge light-gauge steel studs spaced 4′ OC, with lighter gauge members installed between the studs.

The use of structural steel framing members for residential and light-commercial structures is increasing throughout North America. *Stuctural steel framing,* also referred to as red iron steel framing because of the characteristic color of the framing members, involves the use of bolted or welded connections to form the framework of a structure. Structural steel framing involves the use of steel columns, beams, and girders placed in a manner similar to wood post-and-beam construction. Metal-framed buildings can be constructed on the job site or prefabricated and transported to a job site for erection.

TrusSteel Divisions of Alpine Engineered Products, Inc.

Figure 45-1. *The use of light-gauge steel framing continues to increase in commercial and residential construction.*

INDUSTRY AND CODE REGULATIONS

The *Steel Framing Alliance (SFA)*, formerly the North American Steel Framing Alliance, is an organization established by the American Iron and Steel Institute to promote the use of light-gauge steel framing in construction. *The Prescriptive Method for Residential Cold-Formed Steel Framing* is a metal framing standard that was codeveloped by the American Iron and Steel Institute (AISI), the National Association of Home Builders (NAHB), and the Department of Housing and Urban Development (HUD), with the assistance of steel manufacturers and producers, code officials, engineers, and builders experienced in cold-formed steel framing. The standard provides construction details and other information for the construction of one- and two-family residential dwellings using light-gauge steel framing members. Information such as floor and ceiling joist span tables, rafter span tables, wall stud tables, wall bracing requirements, header span tables, and connection requirements is included in the standard.

LIGHT-GAUGE STEEL FRAMING MEMBERS

Light-gauge steel framing members are fabricated from structural-quality sheet steel by *cold-forming*. The desired shapes of the framing members are achieved through rolling, hammering, or stretching the steel at a low temperature, which may often be room temperature. No additional heat is applied in the manufacture of light-gauge steel framing members.

Light-gauge steel framing members may have a dimpled texture applied during manufacturing. The dimples provide additional strength, grabs screws better (and resists "walking"), and holds the screws tighter so wall sheathing does not pull away. Lighter-gauge members with the dimples can be used in lieu of heavier-gauge metal.

Light-gauge steel framing members receive a protective coating to prevent corrosion during transportation, storage, and final placement. The protective coating is applied using a hot-dipped steel galvanizing treatment.

Shapes and Dimensions

Light-gauge steel framing members are available in various shapes, thicknesses, and strengths. Common light-gauge steel framing member shapes include the *C-shape, track, U-channel,* and *furring channel.* **See Figure 45-2.** The C-shape is the most common shape produced for light-gauge steel framing, and is used for studs and joists. C-shape members consist of a *web, flange,* and *lip*. Web depths range from 1⅝″ (1.625″) to 12″. The flanges stiffen the web and provide surfaces for attaching sheathing and gypsum board. The lips extend from the flanges on the open side and stiffen the flanges.

The term *gauge* traditionally has been the unit of measurement for identifying sheet steel thickness. The higher the gauge number, the thinner the steel. Steel thicknesses also have traditionally been given on prints as *mils*. A mil is equal to one thousandth of an inch (1 mil = .001″). Metric dimensions (in millimeters) are increasingly being used to express steel thickness. One millimeter is equal to .001 meter (1 mm = .001 m). The table in **Figure 45-3** provides gauge dimensions and their decimal inch, mil, and SI metric equivalents.

Identifying Light-Gauge Framing Members. The *Steel Stud Manufacturers Association (SSMA)* has developed a four-part identification code, which identifies the web depth, shape, flange width, and steel thickness. This identification system, known as *The Right S-T-U-F,* eliminates the confusion stemming from the widely varying properties and loads published by various manufacturers. The S-T-U-F shape designators represent the following:

- S—studs or joist sections with lips (C-shape)
- T—track sections
- U—U-channels or channel studs (without lips)
- F furring channels

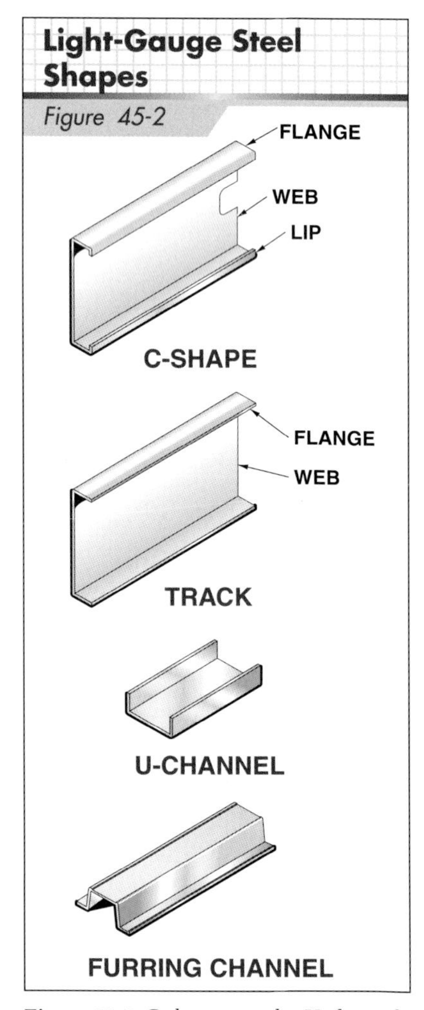

Figure 45-2. *C-shapes, tracks, U-channels, and furring channels are the most common light-gauge steel framing member shapes.*

Gauge	Decimal Inch Thickness*	Mils	SI Metric Thickness†
20 (light-duty)	.033	33	.85
20 (standard)	.036	36	.91
18	.048	48	1.22
16	.060	60	1.52
14	.075	75	1.91
12	.105	105	2.67

STEEL GAUGE EQUIVALENTS

* in in.
† in mm

Figure 45-3. *Light-gauge steel thickness is expressed as gauge, mils, and millimeters.*

When using The Right S-T-U-F system, the web depth and flange width are expressed in hundredths (1⁄100) of an inch. For example, the web depth for a stud with a 6″ depth is expressed as 600 (6″ = 1⁄100 × 600). The flange width for a stud with 1⅝″ flanges is expressed as 162 (1⅝″ = 1.625″; 1.62 = 1⁄100 × 162). Steel thickness of light-gauge steel framing members is the thickness of the base metal (in mils) before being galvanized. **See Figure 45-4.**

FASTENERS

Proper fastening systems are crucial for establishing proper connections between steel framing members. Fastening systems include self-tapping screws (including self-drilling and self-piercing screws), welds, and spot clinching. Pneumatically driven pins are commonly used to fasten subfloor and sheathing to steel joists and studs. Powder-actuated fasteners may be used to fasten tracks for non-load-bearing walls to concrete and masonry.

Self-Tapping Screws

Self-tapping screws are used to fasten metal framing members to each other and to fasten other materials to metal framing members. As the name implies, self-tapping screws cut their own threads as they are being driven into the metal framing members.

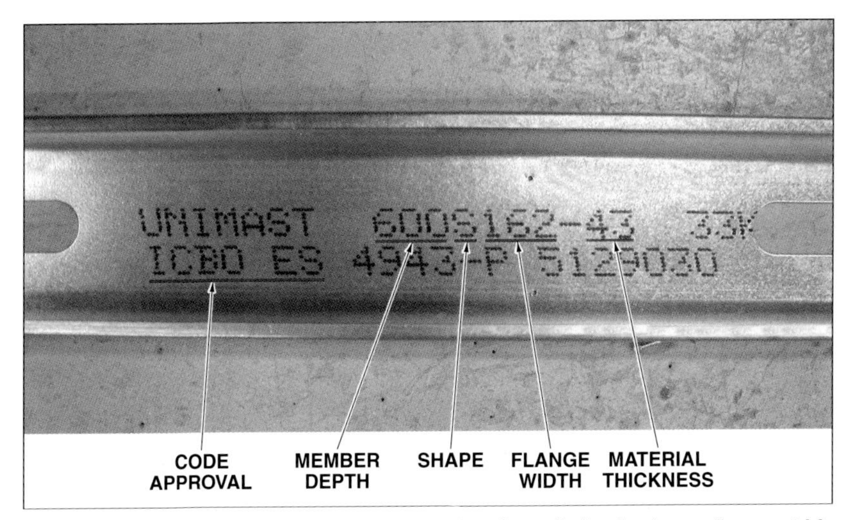

Figure 45-4. *The Right S-T-U-F system identifies the web depth, shape, flange width, and steel thickness.*

Self-tapping screws include *self-drilling* and *self-piercing screws.* **See Figure 45-5.**

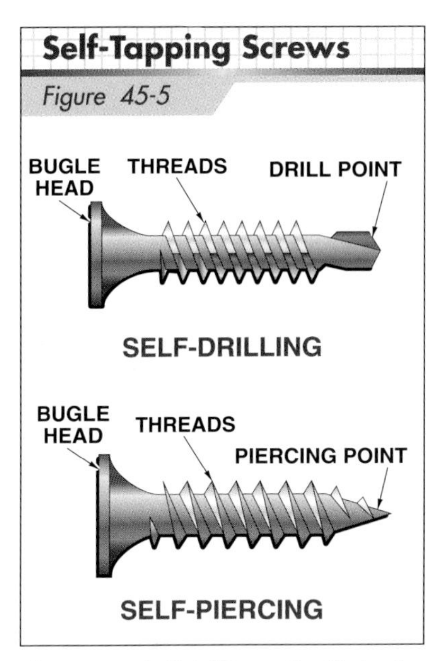

Figure 45-5. *Self-drilling and self-piercing screws are used to connect light-gauge steel members.*

Self-drilling screws are the most frequently used steel-to-steel fasteners. Most screws used in light-gauge framing have a Phillips recess or hex head and a galvanized finish. The points of self-drilling screws drill through the steel layers before the screw threads engage. The drill point of the screw must be sharp and long enough to penetrate the steel being fastened together. Self-drilling screws should be ⅜″ to ½″ longer than the materials being fastened together to ensure the screw threads do not engage the steel before the hole is drilled through subsequent layers of material and to ensure that at least three threads will extend past the fastened material. Coarse screw threads are commonly used for light-gauge steel framing operations. Finer threads may be used when tapping into thicker steel material.

Self-piercing screws have a sharp point capable of penetrating and tapping thin metal. Screw diameters are identified by gauge numbers, which range from #6 to #14. The most frequently used self-piercing screw diameters are #6, #8, and #10, which are .138″, .164″, and .190″, respectively.

A variety of screw head types are available for different metal framing operations. **See Figure 45-6.** Screws with pan, hex washer, or pancake heads are recommended for fastening steel to steel. **Figure 45-7** provides a chart of fasteners and their applications.

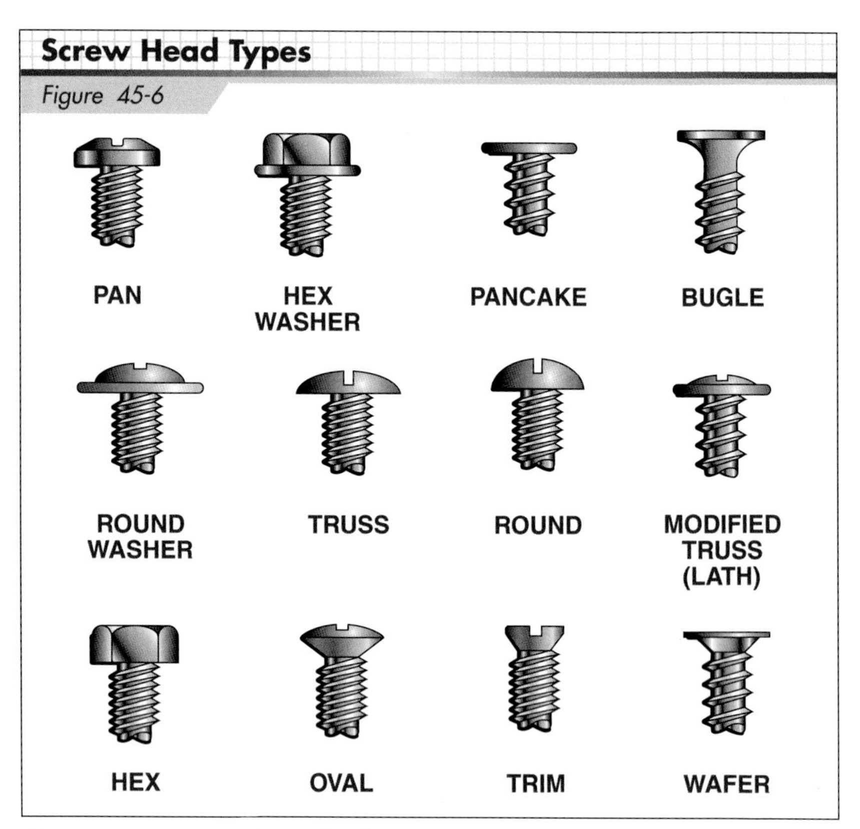

Figure 45-6. *A variety of screw head types are available for metal framing operations. Pan, hex washer, and pancake heads are commonly used for light-gauge steel framing operations.*

Drive Pins

Drive pins may be used to fasten panel sheathing, such as OSB or plywood, to metal framing members for walls, floors, and roofs. Drive pins are driven by pneumatic nailers specifically designed to fasten panel products to metal framing members.

Drive pins are made of heat-treated high-carbon steel that produces a very hard fastener. Drive pins have spiral grooves or knurls on the shanks to draw the pins tight when driven. As a pin is driven, the point penetrates the sheathing and bores through the steel. This action pushes the steel outward away from the pin. The steel then compresses and grips the pin. **See Figure 45-8.**

Pin length is measured from the underside of the head to the tip of the point. The shank diameter is identified by the diameter of the steel wire used to make the pin. Common pin shank diameters are 0.100″ (3/32″) and 0.120″ (1/8″). Head diameters are usually 1/4″ for thinner shank pins.

Since drive pins may be exposed to damp conditions, a corrosion-resistant coating is required. Drive pins may be galvanized or another type of coating may be applied.

Always refer to the manufacturer specifications before driving drive pins.

METAL FRAMING FASTENERS

Application	Fastener*
Brick ties	#8 (minimum) self-drilling hex head screws
Diagonal bracing	#8 (minimum) self-drilling screws with low-profile heads
Gypsum board	#6 (minimum) self-piercing screws for 33 mil and thinner material. #6 self-drilling screws for 33 mil and thicker material.
Interior trim	#6 trim head screws, or finish nails and adhesive
Rigid foam insulation	Roofing nails for structural sheathing, or #6 self-drilling bugle head screws with washer to steel
Siding—hardboard, fiber cement, or panel	#8 (minimum) self-drilling bugle head screws to steel or #8 self-piercing screws to structural sheathing
Steel-to-steel, load-bearing	#8 (minimum) self-drilling screws with low-profile heads for gypsum board and sheathing. Use hex head screws for other applications.
Steel-to-steel, non-load-bearing (less than 33 mil)	#6 (minimum) self-piercing screws with low-profile heads
Structural sheathing	#8 (minimum) self-drilling bugle heads screws or pneumatic pins
Stucco lath	Roofing nails for structural sheathing or #8 self-drilling low-profile screws to steel
Vinyl siding	#8 (minimum) self-piercing screws to structural sheathing or #8 self-drilling screws to steel

* Fastener length varies depending on thickness of material being joined. Screws must penetrate all layers of material and have three exposed threads.

Figure 45-7. *No. 8 screws are used most often as light-gauge steel fasteners.*

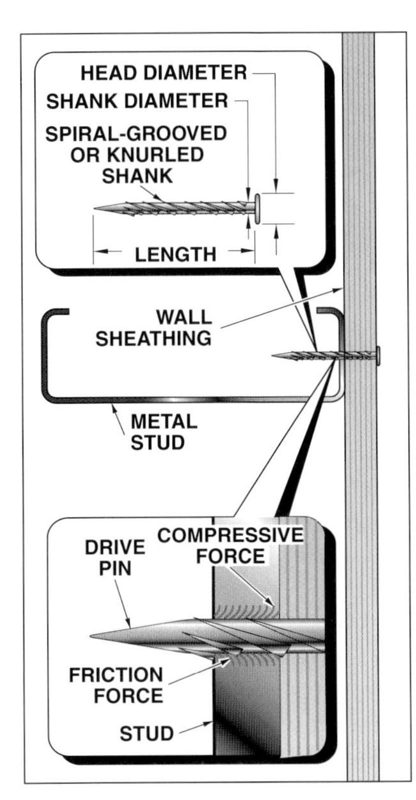

Figure 45-8. *Drive pins are held in place by the compressive force of metal framing members.*

The Lincoln Electric Company

Figure 45-9. *Welding must be performed by a certified welder.*

Welding

Welding can be used to prefabricate wall sections. Most often, wall sections are welded in a shop, rather than on the job site. **See Figure 45-9.** Welding must be performed by certified welders and according to American Iron and Steel Institute (AISI) specifications.

When welding light-gauge steel components, the zinc coating applied during the galvanization process will be destroyed in the area surrounding the weld. A corrosion-resistant coating must be applied to this area.

Metal framing products are 100% recyclable. The amount of metal framing products recycled over the past 10 years has extended the life of landfills by more than three years.

Spot Clinch Connections

Spot clinching is a method for joining two metal framing members and provides a strong connection without the use of mechanical fasteners. In the spot clinching process, a pneumatic *clincher* is used to press a section of one framing member into the adjoining member, leaving a button or stitch indentation. **See Figure 45-10.** Spot clinching is currently approved by the International Code of Building Officials (ICBO), but is not yet approved by other code organizations.

If a clinch is made in a wrong place, the connection has to be drilled or cut out. Therefore, accurate layout is essential when spot clinching metal framing members.

FRAMING TOOLS

A variety of power and hand tools may be used when working with metal framing members, including a screwgun or drywall screwdriver; and pneumatic steel framing tool, chop saw, power shears, plasma arc cutting equipment, aviation snips, and clamps.

Screwguns and Drywall Screwdrivers

An adjustable-clutch screwgun is used to drive self-tapping screws when fastening metal framing members together. Screwguns are available in cordless (battery-operated) and corded (AC-powered) models. Self-feeding screwguns allow rapid installation of screws and eliminate the need for manually placing screws before driving them. Screwguns used for metal framing should include a variable speed control, adjustable depth locator, and adjustable clutch. The adjustable clutch disengages when the screw has reached its proper tightness and depth.

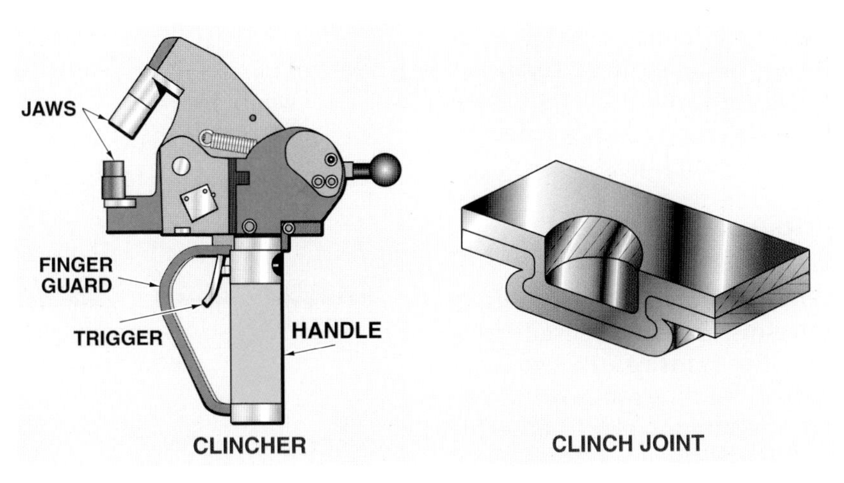

Figure 45-10. *Spot clinching is a method for joining two metal framing members and provides a strong connection.*

A drywall screwdriver is used to fasten metal framing members together and to fasten panel sheathing or gypsum board to metal framing members. **See Figure 45-11.** The nose piece on a drywall screwdriver controls the depth to which the screw is driven and prevents screws from going through the panel or gypsum board.

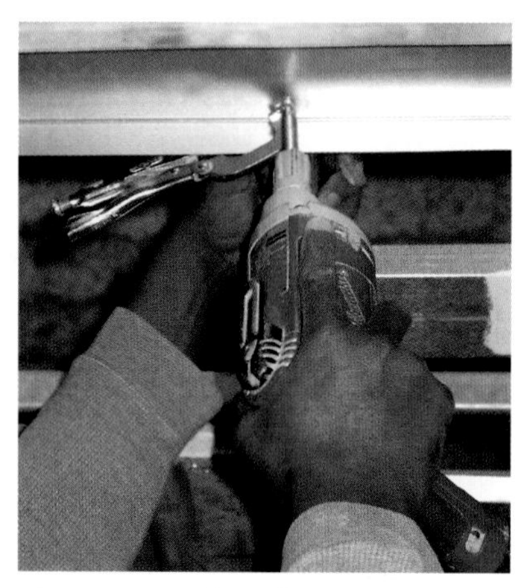

Figure 45-11. *A drywall screwdriver is commonly used in metal framing operations.*

Driving Screws. When connecting metal framing members of different thicknesses, drive screws through the thinner material and into the thicker material. The screw head will bring the thinner and more flexible metal into tight contact with the thicker metal.

When driving screws, start the screw spinning slowly to allow the point to cut through the metal. Increase the speed as the screw penetrates through the metal and maintain the speed until the screw has reached its maximum tightness.

Pneumatic Steel Framing Tools

Pneumatic steel framing tools, or pin guns, are used to install drive pins. **See Figure 45-12.** Pneumatic steel framing tools operate on air pressure ranging from 90 psi to 120 psi. The pressure required to install drive pins depends on the steel thickness. Pneumatic steel framing tools have a safety contact to prevent the gun from firing unless the trigger is pressed and the tool is firmly pressed against a work surface. An *overdrive control mechanism* prevents overdriving of the pins. In addition, the air compressor line pressure must be adjusted to the proper setting for the steel to be fastened.

Driving Pins. A drive pin should be driven so its head is flush with the panel surface. **See Figure 45-13.**

An over-driven pin will rupture the top surface of the wood and weaken the panel where the pin has been driven.

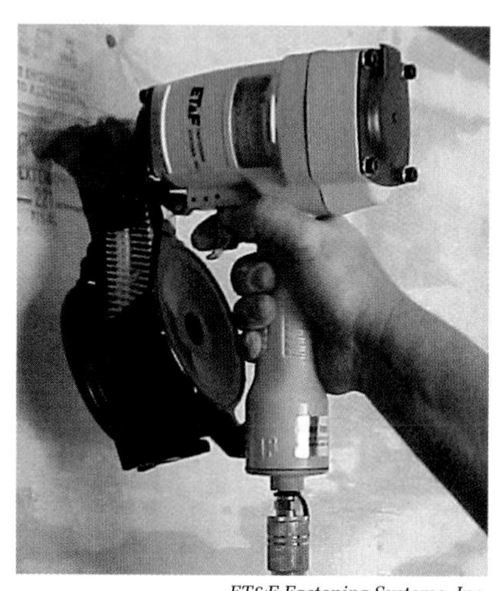

ET&F Fastening Systems, Inc.

Figure 45-12. *Pneumatic steel framing tools are used to install drive pins.*

An underdriven pin will cause a weak, loose connection between the panel and metal framing member and contribute to a squeaky floor. An underdriven pin should be removed using a hammer or prybar. Do not attempt to set an underdriven pin by shooting it again. Do not hit the pin with a hammer as this will only loosen the pin.

When installing drive pins, a tight fit must be maintained between the surfaces being joined. Drive pins will not pull panel sheathing tightly against the metal framing member.

Abrasive Cutoff Saws

An *abrasive cutoff saw,* or *chop saw,* is a power tool used to cut metal framing members. Unlike power miter saws, abrasive cutoff saw blades cannot be rotated from side to side for angled cuts. Rather, the saw fence is adjusted from side to side for angled cuts. An abrasive cutoff saw is equipped with an abrasive blade. **See Figure 45-14.**

Metal framing members are produced to strict strength requirements.

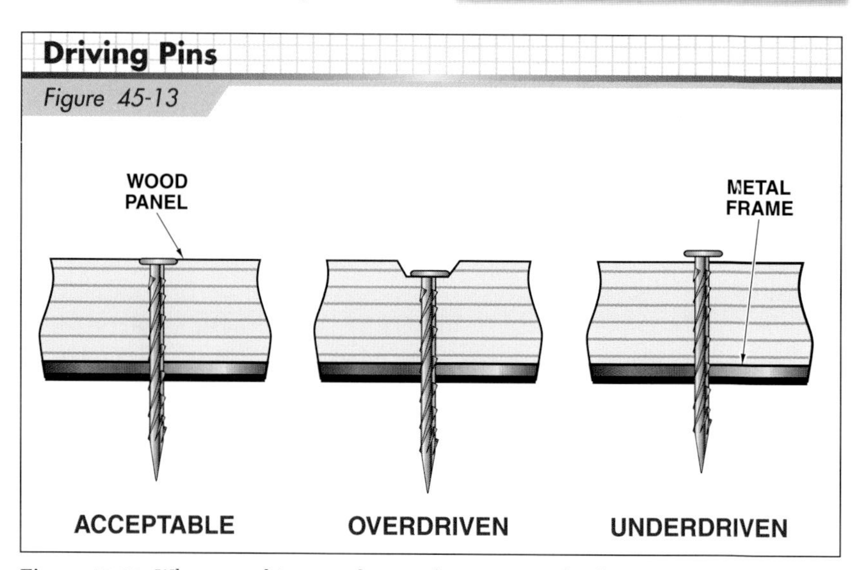

Figure 45-13. *When attaching wood to steel, a pin must be driven so its head is flush with the surface of the wood.*

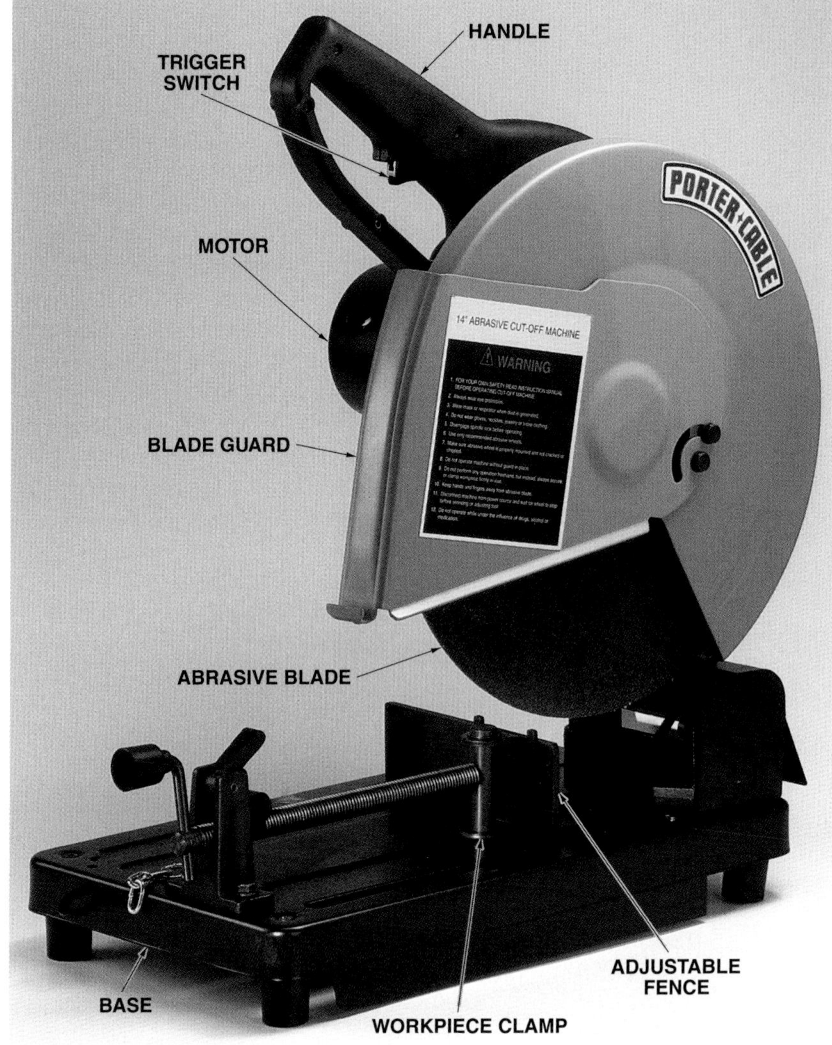

Porter-Cable Corp.

Figure 45-14. *An abrasive cutoff saw is used to cut heavy-gauge framing members.*

Abrasive Cutoff Saw Safety. Hand and eye injuries are typcially assocated with abrasive cutoff saw accidents. While the abrasive blade does not contain teeth like a standard wood-cutting blade, hand or finger injuries may result from them being in the travel path of the blade. In addition, proper eye and face protection should be worn since airborne metal particles are generated when cutting. Abrasive cutoff saw safety rules to observe include the following:

- Ensure the proper type of blade is being used for the material being cut.
- Ensure the metal framing member is properly secured in the workpiece clamp before turning on the saw.
- Check the blade for cracks or chips before use.
- When mounting an abrasive blade, ensure the arrow on the blade corresponds to the direction of spindle travel.
- After starting the saw, allow it to attain full speed. Observe the saw before cutting to ensure the saw is in balance and the blade is running true.
- Ensure the blade guard is in the proper position.
- Do not force the blade through the material being cut; allow the blade and weight of the saw to make the cut for you.
- Do not allow excessive heat to be generated at the cutting edge of the blade.
- Always wear appropriate personal protective equipment, including head, eye, and ear protection.

Abrasive Cutoff Saws
Media Clip

Power Shears

Hand-held *power shears* can cut metal up to a thickness of 68 mils. **See Figure 45-15.** Various models of power shears are available.

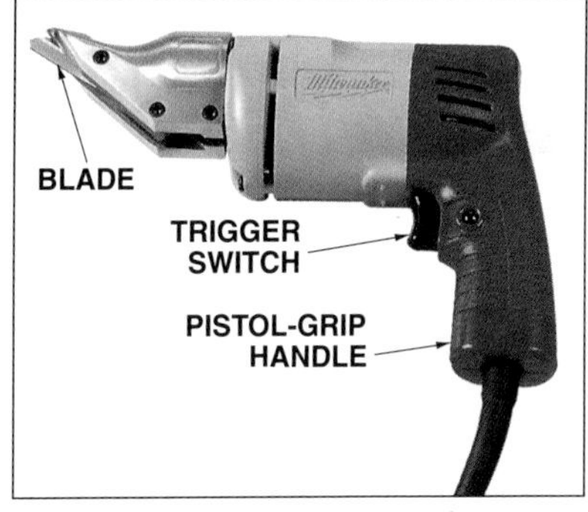

Milwaukee Electric Tool Corporation

Figure 45-15. *Power shears can cut metal up to a thickness of 68 mils.*

Aviation Snips

Metal framing members up to 43 mil thick are typically cut with *aviation snips.* Aviation snips are available in straight-, left-, or right-cutting models. Straight-cutting aviation snips are used for most metal framing operations.

PLS • Pacific Laser Systems

Small, self-leveling lasers are used to lay out metal-framed walls quickly.

Clamps

A *locking C-clamp* is convenient for holding metal framing members tightly together while they are being fastened. **See Figure 45-16.** Small bar clamps may also be used to hold members together, but are used less frequently than locking C-clamps.

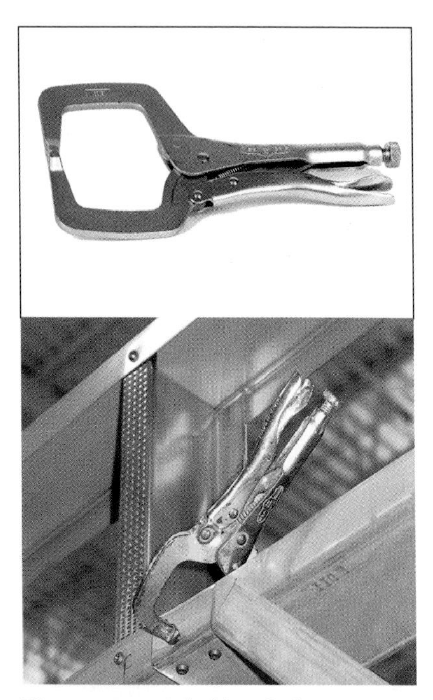

Figure 45-16. *A locking C-clamp secures metal framing members tightly together while they are being fastened.*

METAL FRAMING SAFETY

General safety procedures used for wood framing also apply to steel construction. Additional safety procedures for metal framing are as follows:

- Wear gloves whenever possible. The gloves should be thin enough to hold screws and other fasteners, but thick enough to protect hands from cuts.
- Be aware of sharp edges when handling steel.
- Wear ear protection when subjected to loud noises, especially when cutting metal framing members using a chop saw.
- Wear goggles when using a chop saw to protect from flying steel fragments. Goggles should also be worn when cutting or fastening overhead framing members.
- Use care when handling wet metal framing members as they may be slippery.
- Do not drop metal framing members on electrical cords. The sharp steel edges can cut through the cord insulation and create an electrical hazard.

LIGHT-GAUGE STEEL CONSTRUCTION METHODS

A carpenter experienced in wood framing can easily adapt to steel framing. The layout and components of a metal-framed building are similar to the layout and components of a wood-framed building. The main differences are the fastening devices and tools used to construct metal-framed buildings.

First-Floor Unit

A metal-framed first-floor unit resting on full-basement or crawl-space foundation walls is secured to the foundation with anchor bolts. The main components of a metal-framed floor unit are the joists, rim track, cross bridging, and blocking. A plywood or OSB subfloor is installed after the floor unit has been framed.

Load-bearing C-shape steel joists are available in sizes comparable to wood joists and range from 2 × 6s to 2 × 14s. Joist thickness and size are determined by the live and dead loads the floor must support and its unsupported span. **See Figure 45-17.** Openings are provided in the webs for utilities such as electrical wiring and plumbing. Floor joists should be positioned so the openings are at least 12″ from any bearing points. The outer ends of the joists are attached by metal angles to a *rim track* that serves the same purpose as a header joist in wood-frame construction.

The Steel Truss and Component Association (STCA) promotes and advances the use of cold-formed steel trusses and components. The STCA works with building code officials and governing bodies to develop standards relating to steel trusses.

ALLOWABLE SPANS FOR COLD-FORMED STEEL FLOOR JOISTS SINGLE SPAN WITH WEB STIFFENERS								
Joist Designation	30 psf Live Load Spacing*				40 psf Live Load Spacing*			
	12	16	19.2	24	12	16	19.2	24
550S162-33	11′-7″	10′-7″	9′-11″	9′-1″	10′-7″	9′-7″	9′-0″	8′-1″
550S162-43	12′-8″	11′-6″	10′-10″	10′-0″	11′-6″	10′-5″	9′-10″	9′-1″
550S162-54	13′-7″	12′-4″	11′-7″	10′-9″	12′-4″	11′-2″	10′-6″	9′-9″
550S162-68	14′-7″	13′-3″	12′-5″	11′-6″	13′-3″	12′-0″	11′-4″	10′-6″
550S162-97	16′-2″	14′-9″	13′-10″	12′-10″	14′-9″	13′-4″	12′-7″	11′-8″
800S162-33	15′-8″	13′-5″	12′-3″	11′-0″	14′-0″	12′-0″	11′-0″	9′-2″
800S162-43	17′-1″	15′-6″	14′-7″	13′-7″	15′-6″	14′-1″	13′-3″	12′-3″
800S162-54	18′-4″	16′-8″	15′-9″	14′-7″	16′-8″	15′-2″	14′-3″	13′-3″
800S162-68	19′-8″	17′-11″	16′-10″	15′-7″	17′-11″	16′-3″	15′-4″	14′-2″
800S162-97	22′-0″	20′-0″	18′-10″	17′-5″	20′-0″	18′-2″	17′-1″	15′-10″
1000S162-43	20′-6″	18′-8″	17′-0″	15′-3″	18′-8″	16′-8″	15′-3″	13′-7″
1000S162-54	22′-1″	20′-1″	18′-10″	17′-6″	20′-1″	18′-3″	17′-2″	15′-11″
1000S162-68	23′-8″	21′-6″	20′-3″	18′-10″	21′-6″	19′-7″	18′-5″	17′-1″
1000S162-97	26′-6″	24′-1″	22′-8″	21′-0″	24′-1″	21′-10″	20′-7″	19′-1″
1200S162-43	23′-5″	20′-3″	18′-6″	16′-7″	20′-11″	18′-2″	16′-7″	13′-4″
1200S162-54	25′-9″	23′-4″	23′-8″	19′-7″	23′-4″	21′-3″	21′-6″	17′-6″
1200S162-68	27′-8″	25′-1″	23′-8″	21′-11″	25′-1″	22′-10″	21′-6″	19′-11″
1200S162-97	30′-11″	28′-1″	26′-5″	24′-6″	28′-1″	25′-6″	24′-0″	22′-3″

* in in.

Figure 45-17. *Joist thickness and size are determined by the live and dead loads the floor must support and its unsupported span.*

Light-gauge steel framing members with a dimpled texture provide many benefits including better screw-holding capabilities.

Figure 45-18 shows a framed floor unit over a full-basement foundation. The rim track is directly fastened to the foundation with clip angles. One part of the clip angle is secured to the foundation wall using an anchor bolt and the other part is secured to the rim track with screws. Bearing stiffeners tie together the ends of the joists and the rim track. In hurricane and seismic areas, holddowns may be required to tie the wall studs and track together and fasten the assembly to the foundation. **See Figure 45-19.**

Refer to manufacturer instructions before installing holddowns.

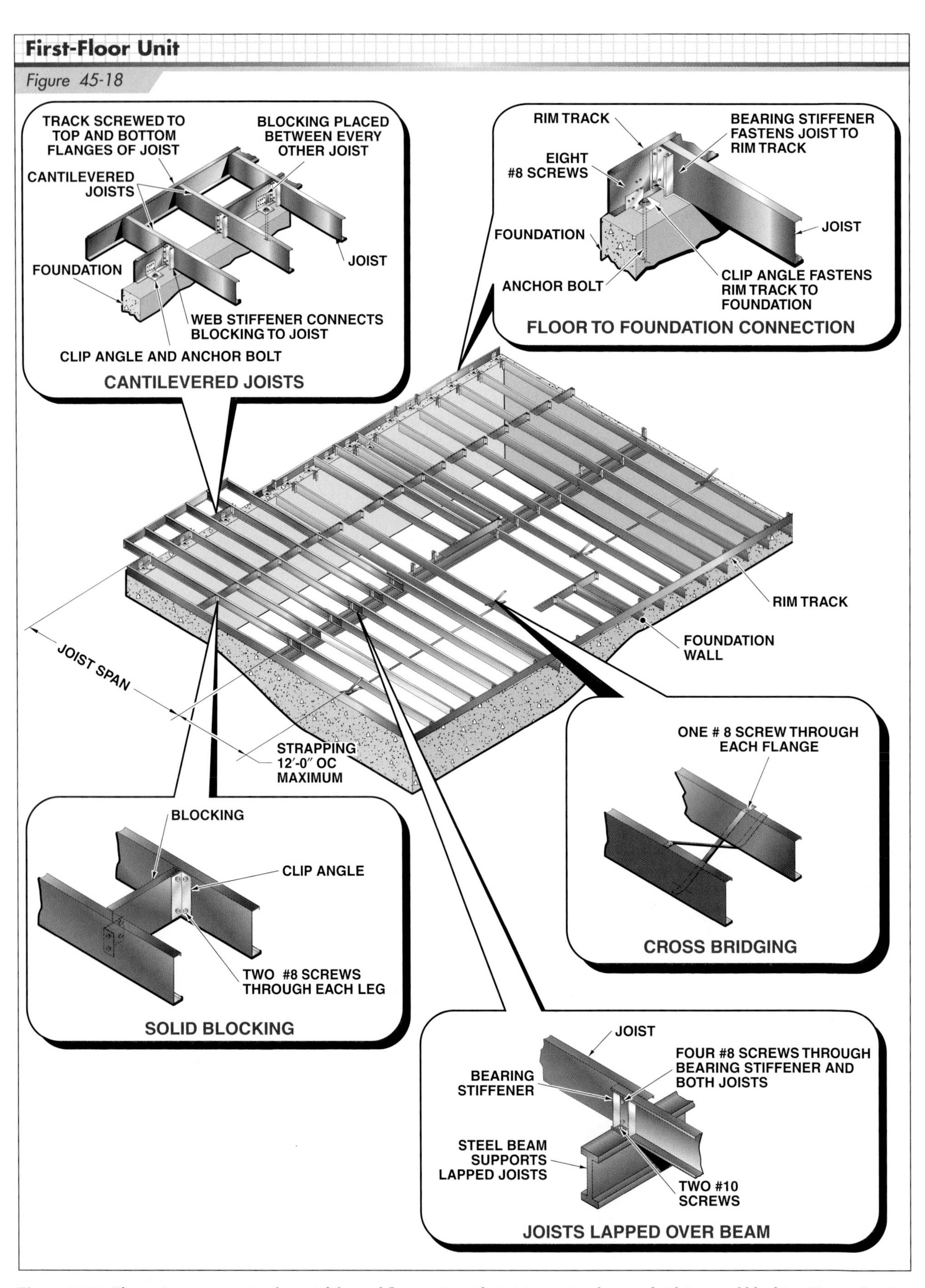

Figure 45-18. *The main components of a metal-framed floor unit are the joists, rim track, cross bridging, and blocking. Ensure framing components are installed in the same direction so cutouts align to allow easy installation of utilities.*

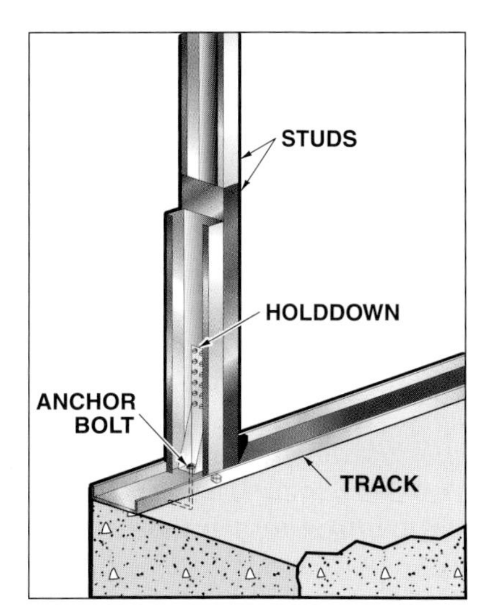

Figure 45-19. *Holddowns tie the wall studs and track together and fasten the assembly to the foundation.*

Where intermediate support is required, the joists are lapped and tied together over a beam or load-bearing wall. For spans over 12′, solid blocking or cross bridging is required at the centers of the spans to prevent the joists from twisting.

Where cantilevered joists are required, blocking must be installed where the joists rest on the foundation wall. Blocking is installed between every other joist.

The floor opening for a stairway is framed with headers tied to trimmer joists. **See Figure 45-20.** A trimmer joist is constructed by fitting a C-shape joist member inside a track member of equal size and securing the two members together with #8 screws 24″ OC. Clip angles tie the headers to the floor joists.

Installing Subfloors

A plywood or OSB subfloor is installed after the floor frame is complete. Subfloor material and thickness are specified based on the local building code. When floor joists are spaced 24″ OC, 23/32″ tongue-and-groove APA-rated sheathing is typically used for the subfloor. Ends of the subfloor panels are staggered in a manner similar to subfloor panels for a wood-framed building. Self-drilling screws should be used to secure the subfloor in place as they will penetrate the panels without lifting them. Spacing between screws is commonly 6″ OC along the edges and 12″ OC at intermediate joists unless the manufacturer recommends closer spacing.

The proper spacing of screws helps prevent squeaky floors. To further reduce squeaking, an adhesive or foam tape specified for wood-to-steel connections can be placed between the joists and subfloor panels. In addition to preventing squeaking, foam tape provides a thermal barrier over basements and crawl spaces. When using an adhesive between the joists and subfloor panels, ensure that the panels are placed and fastened before the adhesive sets. If the adhesive is allowed to set before placing the panel, the adhesive will form a skin and act like a flexible gasket rather than bond with the wood, resulting in potentially squeaky floors.

Walls

In-line framing (stacking) is used for light-gauge steel construction. In-line framing requires that all joists, studs, and roof rafters are in a direct line with one another. **See Figure 45-21.** A maximum variance of 3/4″ on either side of the centerline is allowed. Unlike the double top plates of a wood-framed building, the horizontal tracks of a metal-framed wall cannot adequately support weight between the spans of the studs or joists.

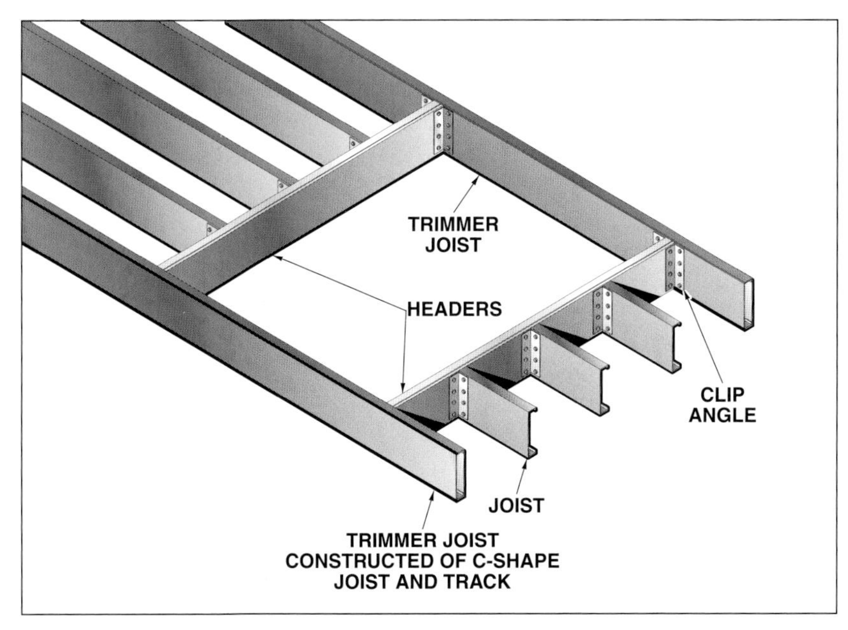

Figure 45-20. *Similar to wood-framed floor openings, additional support must be provided around floor openings in metal-framed buildings.*

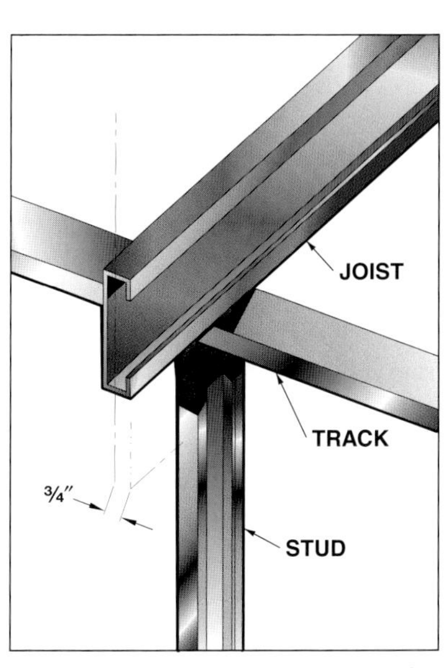

Figure 45-21. *In-line framing requires that the joists, studs, and roof rafters be in a direct line ±3/4″.*

Steel Studs and Tracks

Steel studs are the vertical uprights, which serve the same purpose as wood studs. The steel studs used most often have a 3½″ or 5½″ web dimension, corresponding to wood 2 × 4s and 2 × 6s, respectively. *Knockouts,* or *punchouts,* are provided in the webs for utilities such as electrical wiring and plumbing pipes.

Top and bottom *tracks* are comparable to the top and bottom plates in wood-frame construction. Tracks consist of a web and two flanges. The web width of a track must be equal to the web width of the studs. **See Figure 45-22.** Track flanges may be bent slightly inward to ensure a tight fit between the track and studs. The studs are placed at an angle between the track flanges and turned so they are perpendicular to the track.

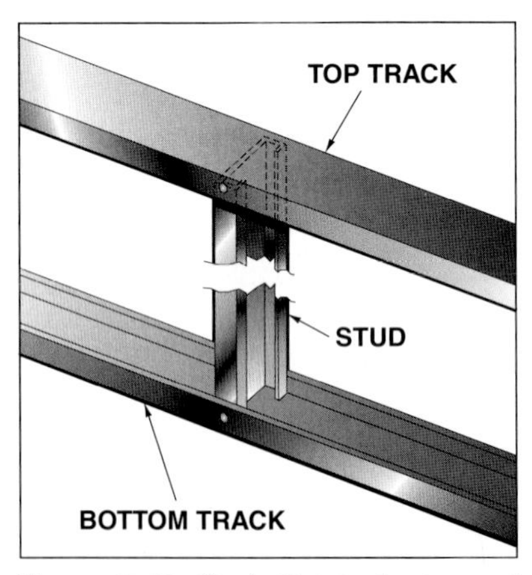

Figure 45-22. *Studs fit into the top and bottom tracks and are secured with one #8 screw in each flange. Studs should be oriented the same way to ensure utility cutouts align properly.*

Load-Bearing and Non-Load-Bearing Studs. Steel studs are categorized as *load-bearing* and *non-load-bearing.* Load-bearing studs, commonly referred to as *heavy-gauge framing,* range in thickness from 20 ga (.835 mm) to 14 ga (1.91 mm), and are used to support heavier vertical loads. The heavier thicknesses may be used in multistory construction.

Non-load-bearing studs, also referred to as *drywall studs,* range in thickness from 25 ga (.454 mm) to 20 ga (.835 mm). Non-load-bearing studs are only used as interior partition studs and do not support loads other than the attached gypsum board or plaster finishes.

Shear Walls. Steel-framed walls must have sufficient *shear strength* to resist lateral loads caused by high winds and seismic conditions (earthquakes). The steel framework of a typical *shear wall* in residential and light commercial construction includes APA-rated panels (OSB or plywood) on the exterior surface and gypsum board on the interior surface, which are attached with self-drilling screws. Steel diagonal bracing and horizontal bracing are also required. Shear walls in commercial construction may consist of steel sheathing on one side and diagonal bracing on the opposite side.

Walls over Concrete Slabs. Anchor bolts are commonly used to secure the bottom tracks of load-bearing exterior walls to concrete slabs or foundations. Bottom tracks must be reinforced with a washer, hold-down bracket, or plate to ensure a proper connection. **See Figure 45-23.** Anchor straps and mushroom spikes may also be used to secure the bottom tracks in position. Powder-actuated fasteners should not be used to permanently anchor bottom plates to slabs or foundation walls.

The bottom tracks of non-load-bearing interior partitions are commonly secured with powder-actuated fasteners.

Wall Construction

A first-floor exterior wall placed over the floor joists and subfloor is shown in **Figure 45-24.** Note the in-line framing of the studs directly above the joists.

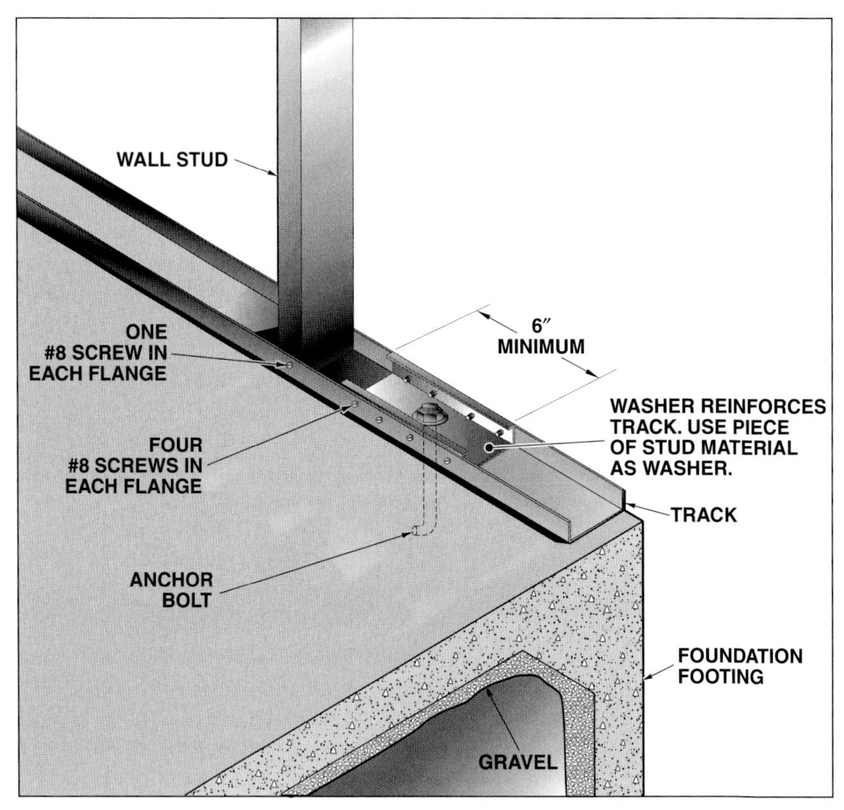

Figure 45-23. *The bottom track of a wall is secured to the foundation with anchor bolts. Bottom tracks must be reinforced to ensure a proper connection.*

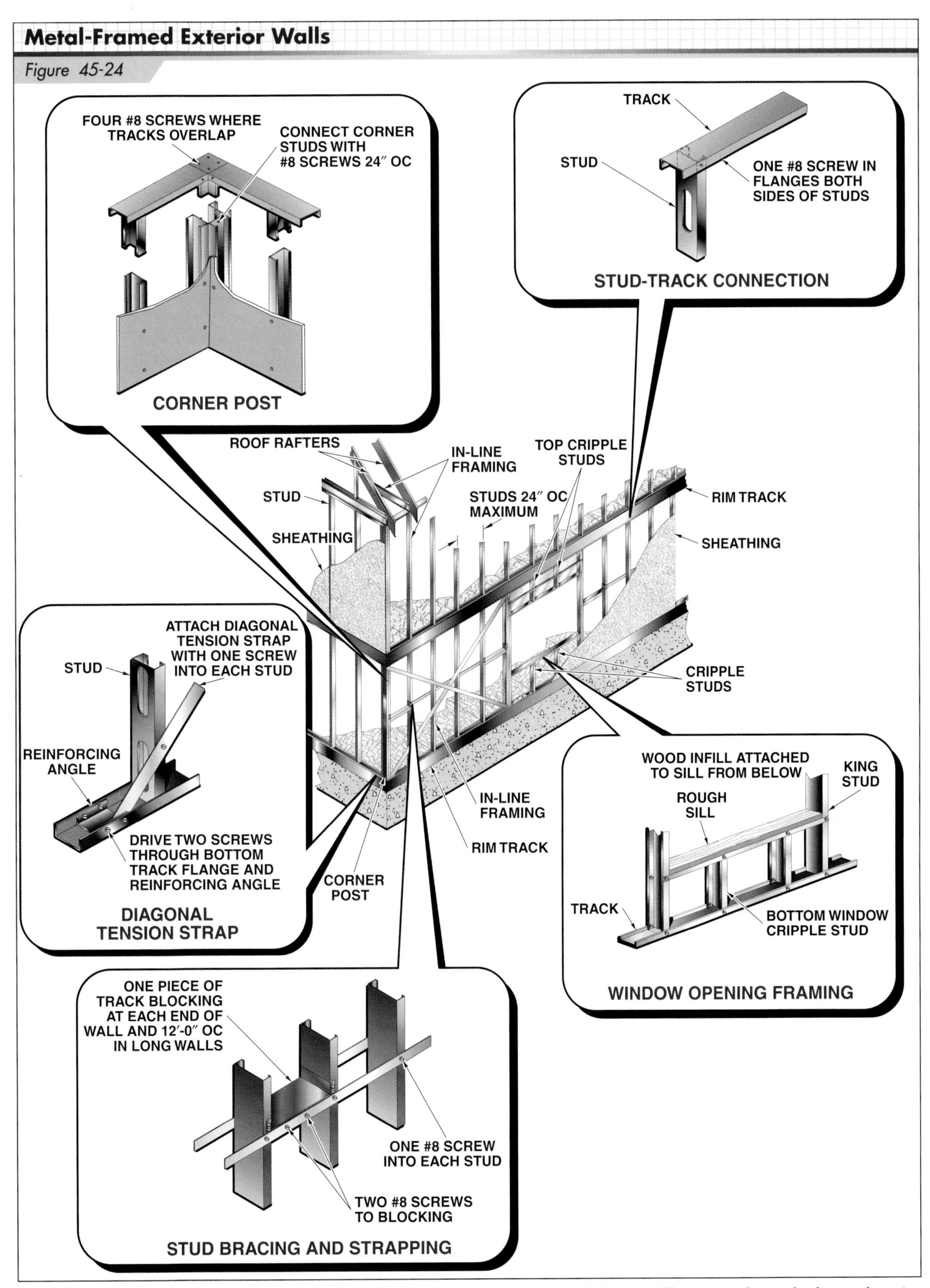

Figure 45-24. *The main components of an exterior wall placed over the floor unit include top and bottom tracks, studs, diagonal tension straps, horizontal bracing, strap stud bracing, and corner posts.*

The wall includes a window opening with a header, sill plate, and top and bottom window cripple studs. Diagonal braces, or *tension straps*, are used to brace studs against lateral movement. An outside corner post and stud-to-track connections are identified. Exterior walls are sheathed with APA-rated structural panels (OSB or plywood).

Headers. Similar to wood-frame construction, headers are installed above wall openings in exterior walls and interior load-bearing walls. Headers are formed with two equal-size C-shape framing members *(box beam header* or *back-to-back header),* or may be constructed with one or two angle pieces that fit over the top track *(L-header).* **See Figure 45-25.**

Box beam headers are commonly used if holddown straps are used to anchor trusses or rafters to the top of the wall. Box beam headers must be insulated before installing them. Back-to-back headers can be insulated after installation.

Inside Corner Posts. An inside corner post is required for proper attachment of interior wall finish material. An interior corner post can be constructed in several ways. **Figure 45-26** shows two methods for constructing inside corner posts. Where a non-load-bearing wall intersects with a load-bearing wall, a 6″ or larger stud can be installed in the exterior wall and a smaller stud is fastened back-to-back with the larger stud. Another method, similar to constructing inside corners for wood-framed buildings, involves fastening two equal-size studs back-to-back.

Unreinforced headers are used to span openings and doorways in non-load-bearing metal-framed walls. Additional bracing may be required for longer spans to ensure a straight and plumb run.

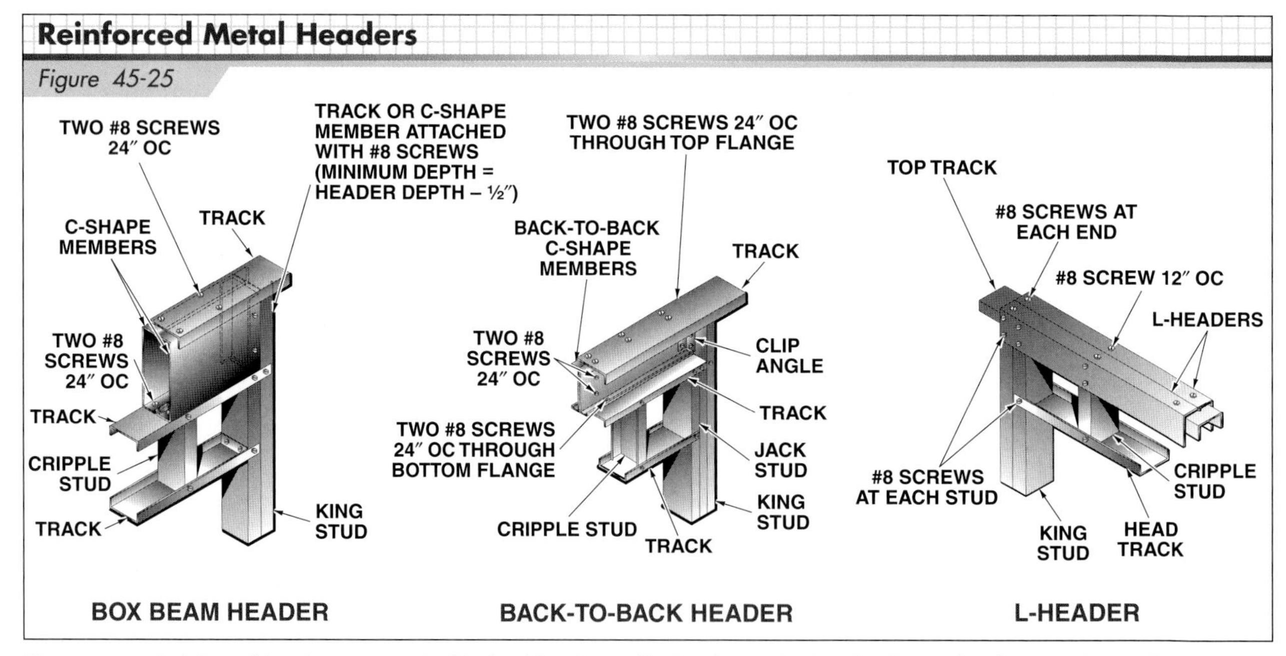

Figure 45-25. *Reinforced headers are required in load-bearing walls. Box beam, back-to-back, or L-headers may be used.*

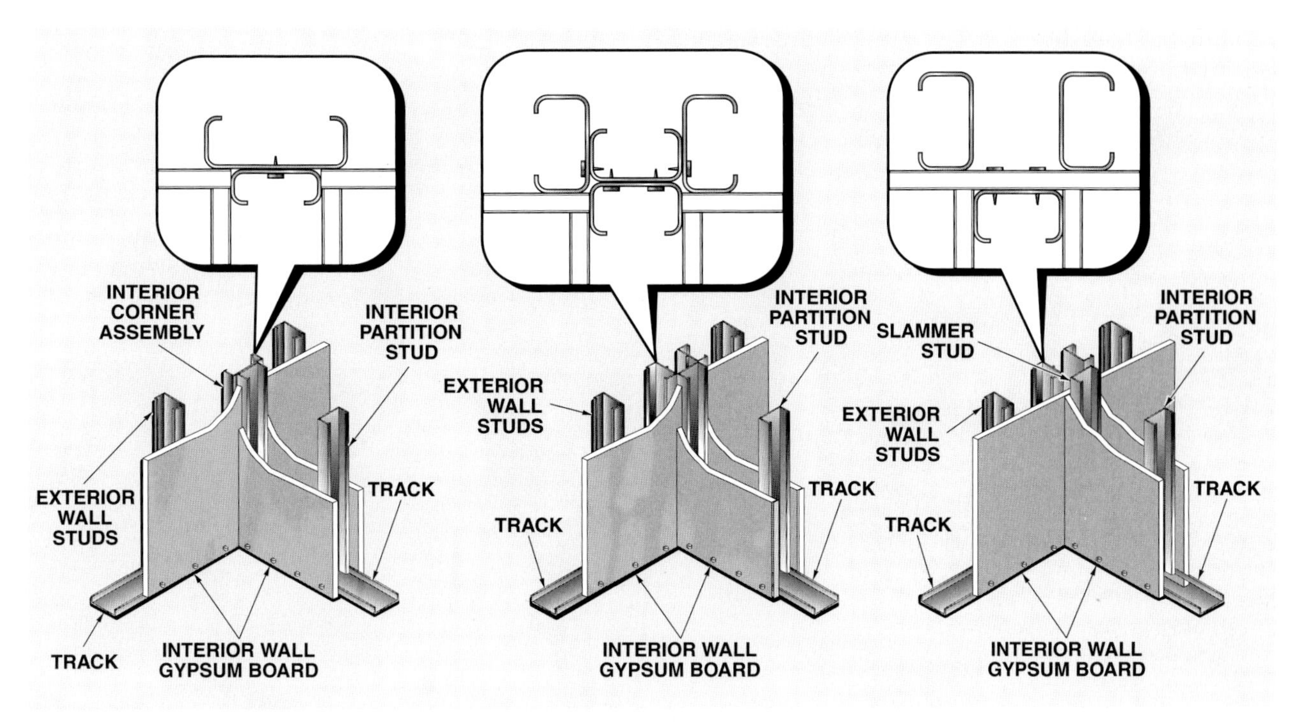

Figure 45-26. *An inside corner post provides support for interior wall finish materials.*

Another method involves the use of a *slammer stud,* which is attached to the top and bottom tracks to provide a bearing surface for interior wall finish material. Blocking is installed between the studs on each side of the slammer stud for additional support.

Framing Walls. Wall studs can be precut to the proper length and delivered to the job site, or they can be cut on the job site with a chop saw. The walls are framed on the subfloor or a separate panel table. A common wall-framing procedure is as follows:

1. Lay out the walls on the subfloor or panel table and snap lines.
2. Cut the bottom and top tracks. Where it is necessary to splice the tracks, insert a short piece of stud material and fasten it to the tracks where they butt together. **See Figure 45-27.**
3. Position the tracks on edge next to each other. Use a black felt-tip marker to lay out the studs, usually 24″ OC. Mark the tracks using standard conventions as shown in Unit 43.

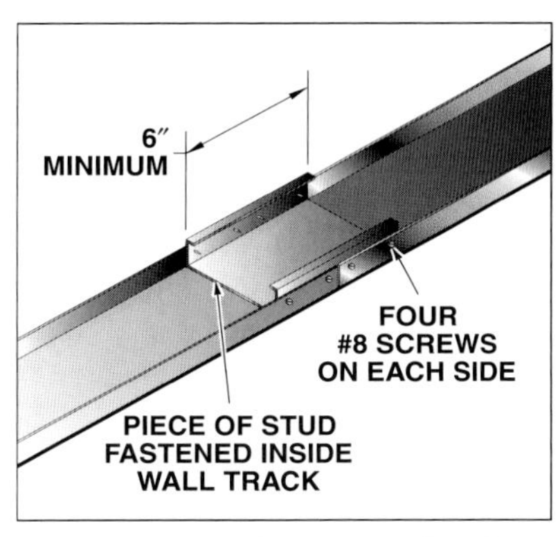

Figure 45-27. *A short piece of stud material is commonly used to splice tracks.*

4. Lay out door and window openings using a red felt-tip marker.
5. Preassemble the door and window framing members and place them as a unit between the tracks.
6. Position the studs between the flanges of the top and bottom tracks. Tap the track flanges against the studs to ensure the studs are tight against the track. Secure the studs and tracks together with a locking C-clamp. Fasten the studs to the track flanges with #8 self-drilling, low-profile screws.
7. Turn the wall over after the studs and tracks have been fastened together. Secure the track flanges to the studs using the previous procedure.
8. Square the wall by measuring diagonally from corner to corner. Adjust the wall squareness so equal diagonal measurements are obtained in each direction.
9. Attach diagonal and horizontal bracing to the wall studs, especially at door openings where the bottom track is weak. Studs are commonly used for diagonal bracing.

Metal-Framed Rough Openings
Media Clip

Raising Walls. A common procedure for raising and placing the wall is as follows:

1. Measure the locations of the foundation anchor bolts. Transfer the measurements to the bottom tracks and drill holes so the tracks fit over the bolts.
2. Raise the wall and position it over the anchor bolts. **See Figure 45-28.**
3. Attach braces to the ends of the wall and approximately every 8′ to 12′ between the ends. Studs about 12′ long work well for braces. Nail wood blocks to the subfloor next to the lower ends of the intermediate braces.
4. Plumb both corners of the wall using a magnetic level. Fasten the lower ends of the braces to the outside joists.
5. Run a string from corner to corner with stand-off blocks at each end. Using a gauge block equal in thickness to the stand-off blocks, align the top of the wall. Fasten the lower ends of the intermediate braces to the blocks nailed to the subfloor.
6. Allow the temporary braces to remain in place until the entire building has been framed and the sheathing or straps have been fastened to walls.

United States Steel Corporation

Figure 45-28. *After a wall has been framed and squared, it is raised into position. Temporary diagonal bracing remains in place until structural sheathing is fastened to the wall.*

Ceiling-Floor Joists

Ceiling joists are placed after the walls below have been plumbed, aligned, and braced. For a multistory building, the ceiling joists also serve as floor joists. The framing procedure, fasteners, and components are the same as those used for the first-floor joists. Ceiling joist bracing is fastened to the bottoms of the joists where spans exceed 12′-0″. **See Figure 45-29.**

Continuous Metal Straps
Media Clip

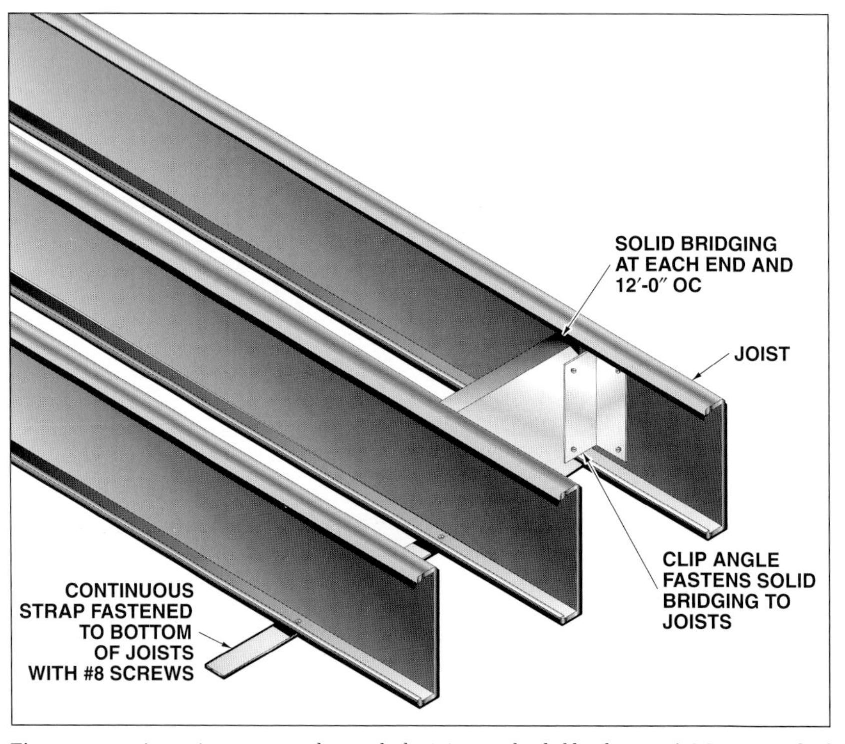

Figure 45-29. *A continuous strap beneath the joists and solid bridging 12′ OC are attached to the ceiling joists to provide rigidity.*

Ceiling joists must be placed directly above and in line with the studs below. When joists are lapped over an interior load-bearing wall, the first-floor wall studs will be directly under one of the lapped ends. **See Figure 45-30.** A bearing stiffener is used to secure the lapped ends in position.

If continuous-span joists supported by a load-bearing wall at the midpoint are used, the stud should be placed directly underneath. Where walls below run parallel with the ceiling joists, blocking is placed between the joists a maximum of 48″ OC to provide anchorage for fastening the top of the walls to the ceiling. **See Figure 45-31.**

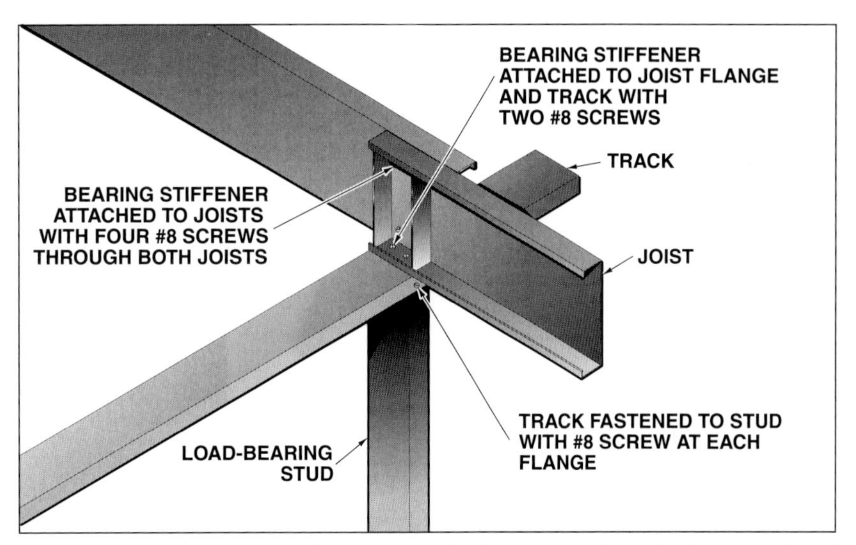

Figure 45-30. *When joists are lapped over a load-bearing wall, studs should be placed directly below the lapped ends. A bearing stiffener provides additional support.*

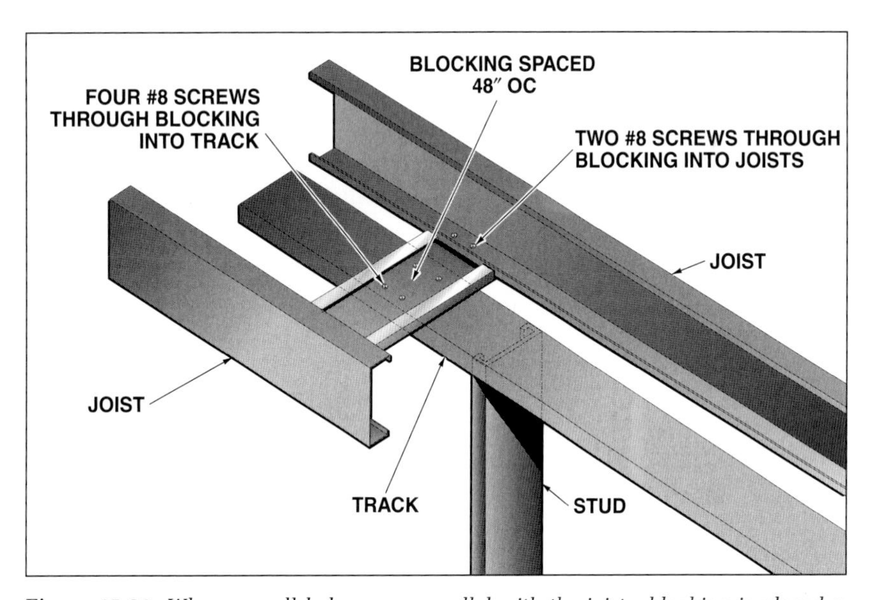

Figure 45-31. *Where a wall below runs parallel with the joists, blocking is placed a maximum of 48″ OC between the joists for fastening the tops of the wall to the ceiling.*

Roof Rafters

The heel (lower end) of a roof rafter rests on the track of a load-bearing wall below. Roof rafters are positioned next to ceiling joists and are securely fastened to the webs of the ceiling joists. The upper ends of the rafters are fastened to the ridge.

For rafters covering longer spans, *rafter support braces* extending from the ceiling joists to each of the rafters are installed. **See Figure 45-32.** Rafter support braces are typically 2 × 4 × 33 mil C-shape members which are fastened to each ceiling joist and rafter with four #10 screws at each end. The rafter support braces and rafters are stiffened by *lateral braces.* Lateral support braces for rafter support bracing are C-shape or track members spaced 4′ OC. Lateral support braces for rafters are fastened to the bottom flange of the rafters, and could be flat straps, C-shape members, or track members.

Roof Trusses

Metal roof trusses are frequently used to construct roofs over metal-framed buildings. **See Figure 45-33.** The design, shapes, and engineering principles of metal roofs are similar to wood trusses (discussed in Unit 50). Temporary and permanent bracing methods for metal trusses are also similar to those for wood trusses.

Metal roof trusses can be constructed on the job site. Most often, however, metal roof trusses are prefabricated in a shop. **See Figure 45-34.** The trusses may be constructed on a level surface, such as the shop floor or subfloor, or on an elevated jig platform. Many manufacturers design the entire roof structure and furnish shop drawings and specifications to a general contractor.

Ensure that metal framing members being installed match the specified size, type, mechanical properties, and spacing indicated on the prints or in the specifications.

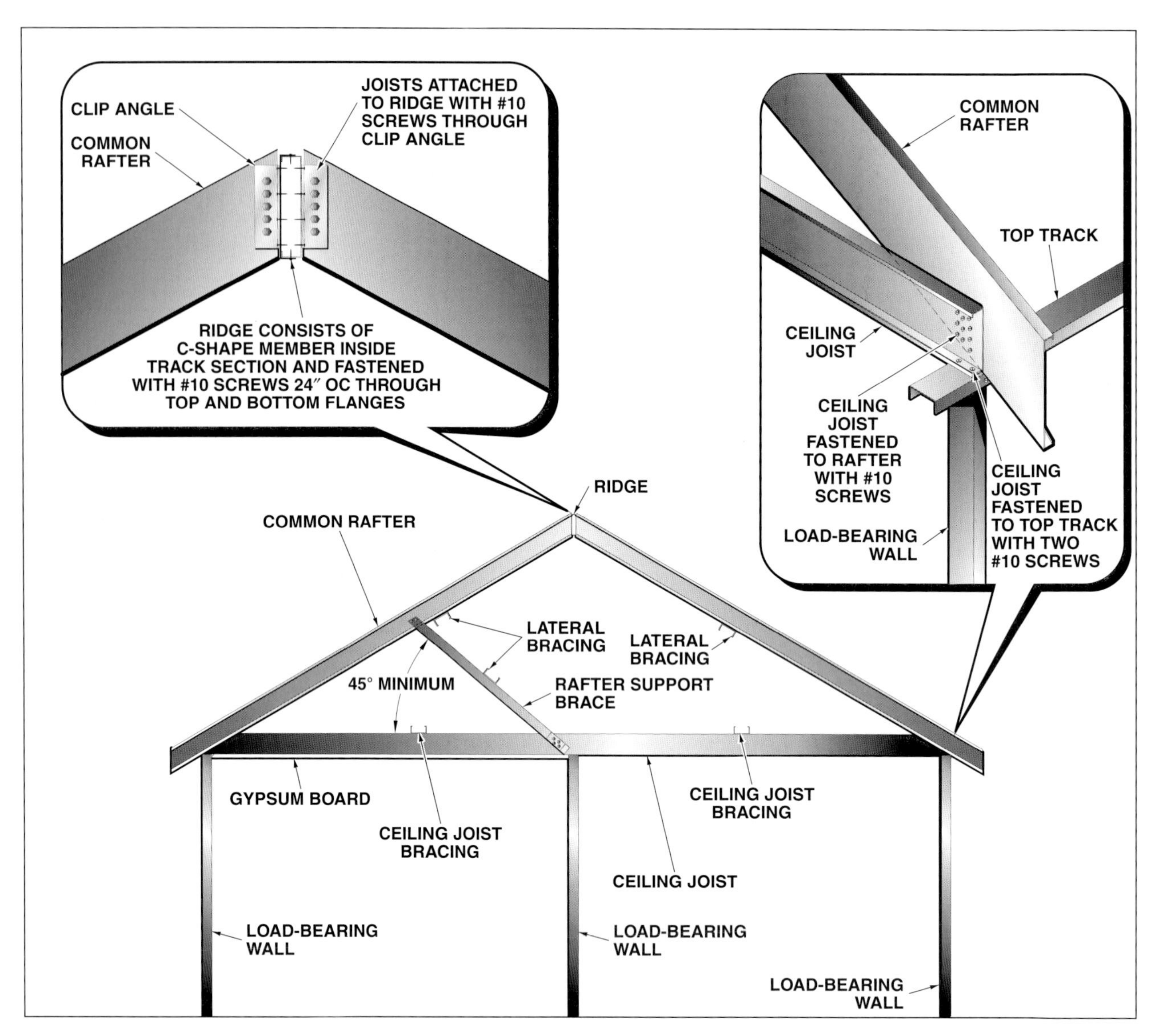

Figure 45-32. *A light-gauge steel roof framework requires considerable bracing.*

Figure 45-33. *Light-gauge steel trusses are very frequently installed over steel-framed walls.*

TrusSteel Divisions of Alpine Engineered Products, Inc.

Figure 45-34. *Light-gauge steel trusses are commonly prefabricated in a shop.*

Light-gauge steel trusses may not be modified or repaired in the field without the consent of the fabricator or design professional.

Seismic and hurricane ties should be used with metal rafters and trusses in areas prone to earthquakes and severe winds. **Figure 45-35** shows metal connectors used to tie metal-framed walls to metal-framed roofs.

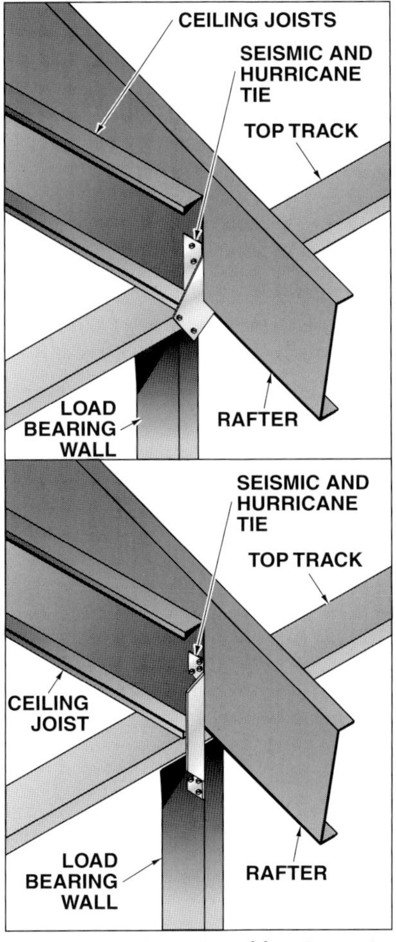

Figure 45-35. *Seismic and hurricane ties should be used to tie metal-framed walls to metal trusses in areas prone to earthquakes and severe winds.*

Panelized Systems

Panelization is a system for prefabricating walls, floors, and roofs in separate sections. Panelization is practical and efficient when many panel types of the same dimensions are required, such as for an apartment complex. The panels can be made on the job site, but are most often prefabricated in a shop. In either case, a jig platform (table) is set up for each panel type. Studs and joists are precut to length and placed into the jig where they are fastened together by screws or welding. Exterior sheathing is then fastened to the walls. Exterior finish may also be applied to the panels. After the panels are completed, they are transported by truck to the job site.

STRUCTURAL STEEL FRAMING

Structural steel framing, also referred to as red iron steel framing because of the characteristic color of the framing members, involves the use of bolted or welded connections to form the framework of a structure. The framing members are painted with a durable red oxide coating that resists corrosion after being manufactured to specified sizes. The use of structural steel framing is on the increase for residential and light-commercial buildings up to two stories in height. **See Figure 45-36.**

Depending on the size of the framing members, structural steel framing members may be joined on the ground at the job site and erected using a crane, or the framing components may be erected individually and joined together. The assemblies or individual structural steel framing members are erected using a crane or other hoisting equipment. Columns are fastened to the foundation or concrete slab using anchor bolts. The assemblies and individual framing members are joined using bolts and/or welding. **See Figure 45-37.**

Figure 45-36. *Structural steel framing is on the increase for residential and light-commercial buildings. Large open spaces can be created within a structure.*

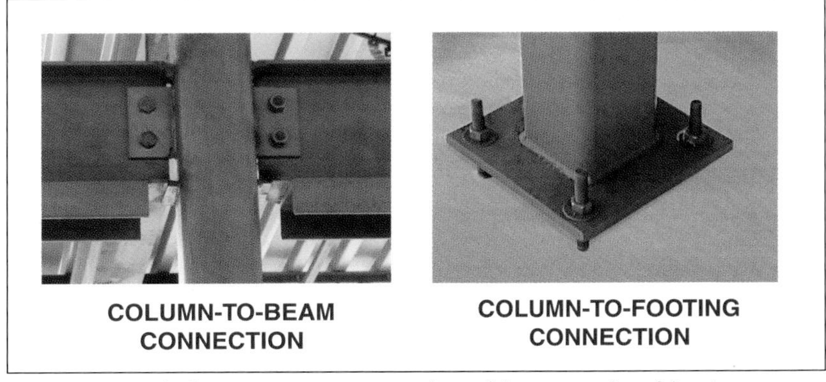

Figure 45-37. *Bolted connections are commonly used for structural steel framing.*

Due to the size of the structural steel framing members, larger open spaces can be created within a building, offering greater flexibility in room layout. Structural steel framing members are commonly spaced 8′-0″ OC, and heavy-gauge steel framing members spaced 16″ to 24″ OC are used to fill the spaces between the structural steel members. **See Figure 45-38.** Door and window openings are framed using wood or heavy-gauge steel framing members. **See Figure 45-39.** Heavy-gauge steel is also used for ceiling joists and roof purlins.

Figure 45-38. *Spaces between structural steel framing members are filled with heavy-gauge steel framing members spaced 16″ to 24″ OC.*

Figure 45-39. *Window and door openings are framed with wood or heavy-gauge steel framing members.*

BUILDING ENVELOPE AND INSULATION

After the building has been completely framed, carpenters begin work on the building envelope, which mainly consists of the wall sheathing, insulation, and exterior finish. Electricians, plumbers, and sheet metal workers also begin roughing in the wiring, plumbing, and ductwork.

Sheathing

Plywood or OSB panels are typically used for roof and metal-framed wall sheathing. The panels, combined with diagonal bracing, produce a shear wall. Panel thickness and screw or drive pin sizes are usually included in the print specifications. A common example for wall and roof sheathing is 7⁄16″ thick OSB fastened with #8 bugle-head self-drilling screws. Specialty screws, such as winged drill point screws, can also be used to attach sheathing to metal framing members. The wings create a pilot hole through the sheathing and snap off when the screw engages in the metal framing member.

Screws or drive pins are installed 3⁄8″ from the panel edge and typically spaced 6″ OC along the edge and 12″ OC at intermediate studs. **See Figure 45-40.** Increased strength is obtained by applying a continuous bead of construction adhesive along the studs and rafters before fastening the panels in place.

Exterior Finish

Metal-framed buildings can be finished with the same types of exterior wall and roof materials used for wood-framed buildings. A weather barrier, such as asphalt-saturated felt or housewrap, is first applied against the sheathing or framework (if sheathing is not required).

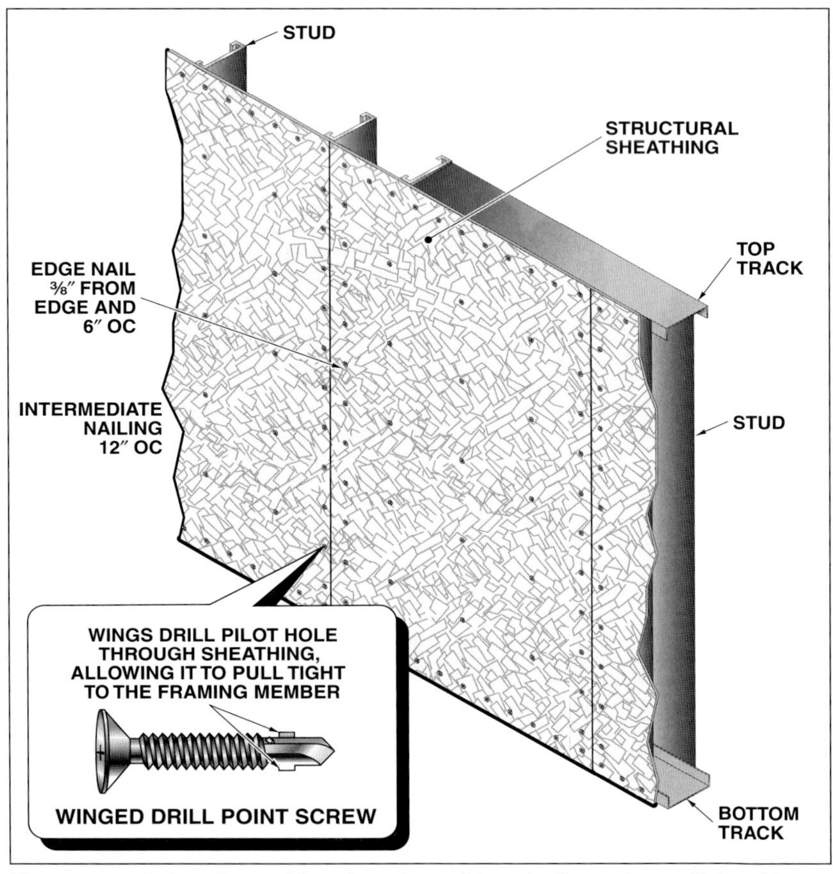

Figure 45-40. *Oriented strand board or plywood is typically used as wall sheathing on steel-framed buildings.*

Where sheathing has been applied to the walls, the exterior finish materials are fastened to the sheathing using screws or nails. If shear walls are not required and no sheathing is placed first, the exterior finish materials are attached directly to the metal studs using screws.

Masonry Veneer. When masonry is used as an exterior finish, wall ties fastened to the studs are embedded between the masonry courses. Several wall tie designs are available. One example is shown in **Figure 45-41.**

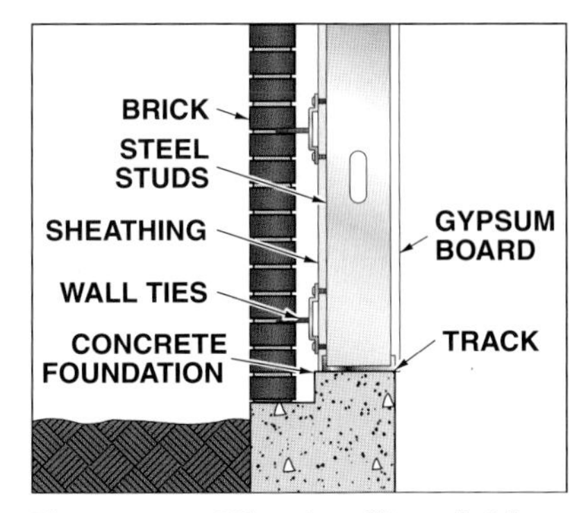

Figure 45-41. *When installing a brick veneer wall, the wall ties are fastened to the sheathing and metal studs and embedded in the mortar between bricks.*

Exterior insulation and finish systems (EIFS) are common as exterior finish on metal-framed buildings. **See Figure 45-42.** One of the biggest advantages of using EIFS with metal-framed buildings is the elimination of *cold bridging*. Cold bridging occurs when construction components (usually structural members or door or window frames), which have greater thermal conductivity than the rest of the building, extend from the interior face to the exterior face of the building. Since structural members do not extend across the faces, cold bridging in EIFS construction is eliminated. In addition, there are no fasteners in contact with the exterior to transfer the cold to the interior of the building.

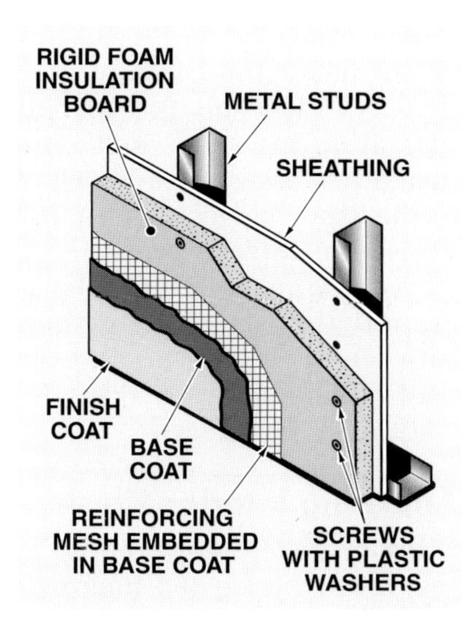

Figure 45-42. *Exterior insulation and finish systems are commonly installed over metal-framed buildings.*

Insulation

Metal studs conduct heat flow faster than wood; therefore, adequate insulation takes on added importance for metal-framed structures. Types of insulation similar to those used for wood-framed buildings are used for metal-framed buildings. **See Figure 45-43.**

R-value is the measurement of a material's ability to resist the flow of heat. The higher the R-value, the greater the resistance. Fiberglass batt insulation ranges from R-2.9 to R-4.3 per inch of thickness.

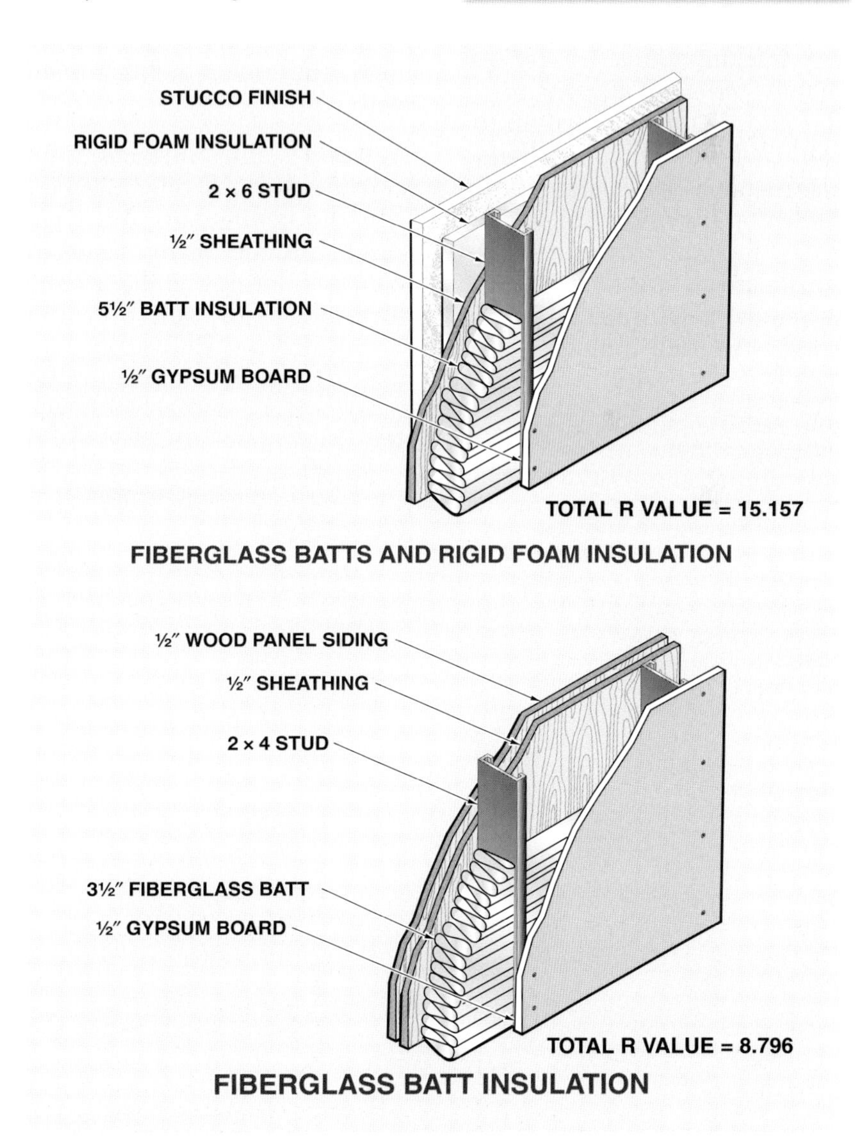

Figure 45-43. *Insulation for steel-framed buildings can be rigid foam insulation, fiberglass blankets or batts, or a combination of both.*

Rigid Foam Insulation Panels. A layer of rigid insulation over wall sheathing is considered the most effective insulation. When installing rigid foam insulation panels, self-drilling screws with washers should be used to prevent the screws from pulling through the insulation. Construction adhesive is also recommended.

Blanket and Batt Insulation. Metal-framed walls require full 16″ or 24″ faced or unfaced blankets or batts, depending on the stud spacing. The insulation is placed between the webs of adjoining studs in a manner similar to wood-framed construction. Friction holds the insulation in place. For a proper friction fit, blanket or batt thickness must be approximately the same dimension as the stud width and must fill the stud cavity. Friction installation also requires that sheathing be installed on the outside of exterior walls, and wall paneling or gypsum board be directly fastened to the studs. The insulation is then enclosed on all four sides and cannot settle or shift.

If the steel stud width exceeds the insulation thickness by more than ½″, faced blanket or batt insulation must be used. The insulation flanges are fastened to the stud flanges using tape. When the interior wall finish material is installed, the flanges will then be sandwiched between the stud faces and the wall finish and secured by the wall finish fasteners.

Attaching Interior Finish

Different fastening methods are required when fastening finish materials to metal-framed walls than when fastening the materials to wood-framed walls. Adhesives and screws are the primary means of attaching finish material to metal-framed walls.

Wall and Ceiling Surfaces. Gypsum board is most often used for wall and ceiling surfaces. Gypsum board is typically fastened to metal framing members using #6 bugle-head, self-piercing screws (drywall screws). Recommended screw lengths are 1⅛″ for securing ½″ gypsum board and 1¼″ for ⅝″ gypsum board. When attaching gypsum board to 18 ga studs, self-drilling screws should be used. Drywall screwguns with a depth-setting nose piece prevent screws from being driven too deeply or through the gypsum board.

Although used less often, plaster may also be used as a surface finish. Lath must be fastened to the studs as a base for the plaster using low-profile screws with washers.

Molding. Trim head screws can be used to fasten molding to metal studs. When using hardwood molding, a pilot hole should first be drilled through the molding. Molding can also be fastened in place using adhesives, with finish nails driven at an angle through the molding and gypsum board.

Wall Cabinets. Wood or metal backing must be placed inside of the walls before they are covered with gypsum board or plaster. The cabinet backs are screwed into the backing.

Unit 45 Review and Resources

SECTION

10

Roof Frame Construction

CONTENTS

UNIT 46

Basic Roof Types and Roof Theory

OBJECTIVES

1. List and describe basic roof types.
2. Discuss basic roof layout principles.
3. List and describe the structural factors in roof design.
4. Describe the installation of roof sheathing.

The design of a roof affects the appearance of a building and its construction. Roof appearance must be in harmony with the remainder of the building and the surrounding environment. Proper water drainage from rain and melting snow is an important aspect of roof design.

BASIC ROOF TYPES

Basic types of pitched roofs are *shed, gable,* and *hip roofs*. **See Figure 46-1.** Shed, gable, and hip roofs are traditional roof designs that have proven practical over hundreds of years and are still used in all types of construction. Flat roofs are most often used for buildings of modern design.

The shed roof, or lean-to, is the simplest type of roof to construct as it slopes in only one direction. Shed roofs may be used as the main roof of a building, but they are more often employed for small sheds, covered porches, and additions to a building.

A gable roof has a ridge at the center and slopes in two directions. Gable roofs are used more often than any other type of roof.

Roof Types
Media Clip

A hip roof has four sloping sides. Hip roofs are the strongest type of roof because they are braced by four hip rafters. Hip rafters run at a 45° angle from the corners of the building to the ridge. A hip roof is more difficult to construct than a gable roof.

Two other types of roofs are *gambrel* and *mansard roofs*. **See Figure 46-2.** Both roof types provide additional living space directly underneath. A gambrel roof is similar to a gable roof except that the slope of a gambrel roof is broken near the center of the roof, making a double slope on each side. A mansard roof is similar to a hip roof except that it has a double slope on each of its four sides.

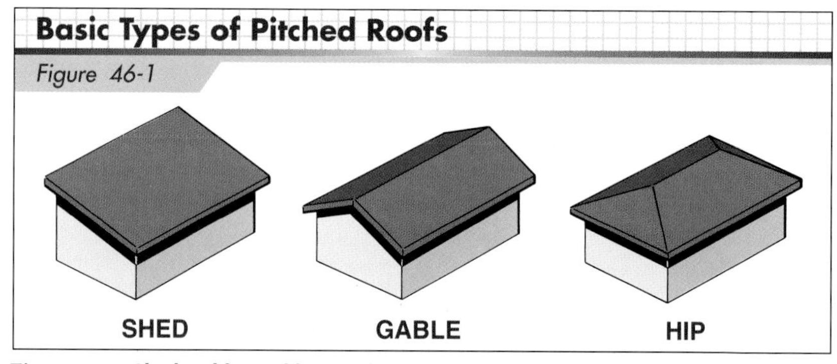

Figure 46-1. *Shed, gable, and hip roofs are the basic types of pitched roofs.*

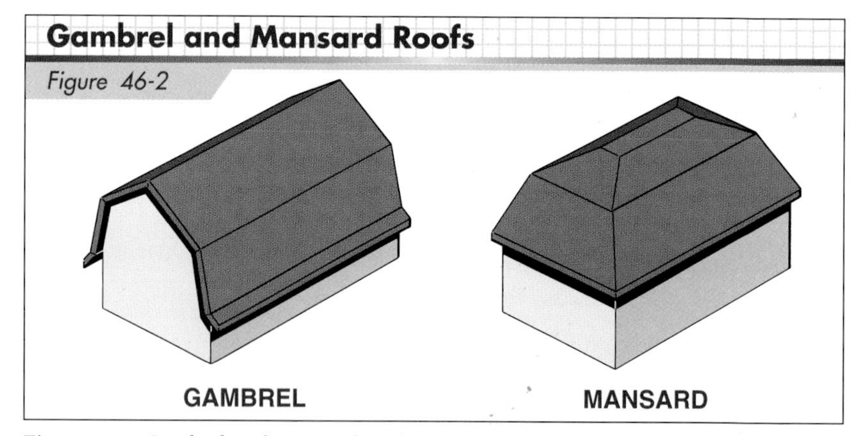

Figure 46-2. *Gambrel and mansard roofs are similar to gable and hip roofs except that they have double slopes on each side.*

Other roof styles, such as the butterfly, monitor, and continuous-slope gable roofs, are variations of the shed, gable, and hip roofs. **See Figure 46-3.** The basic types can also be combined in various ways, producing *intersecting roofs.* **See Figure 46-4.**

Shed and Gable Roof Variations

Figure 46-3

BUTTERFLY SHED ROOF

MONITOR SHED ROOF

SHED ROOF VARIATIONS

CONTINUOUS-LOW-SLOPE GABLE ROOF

CONTINUOUS-SLOPE GABLE ROOF

GABLE ROOF VARIATIONS

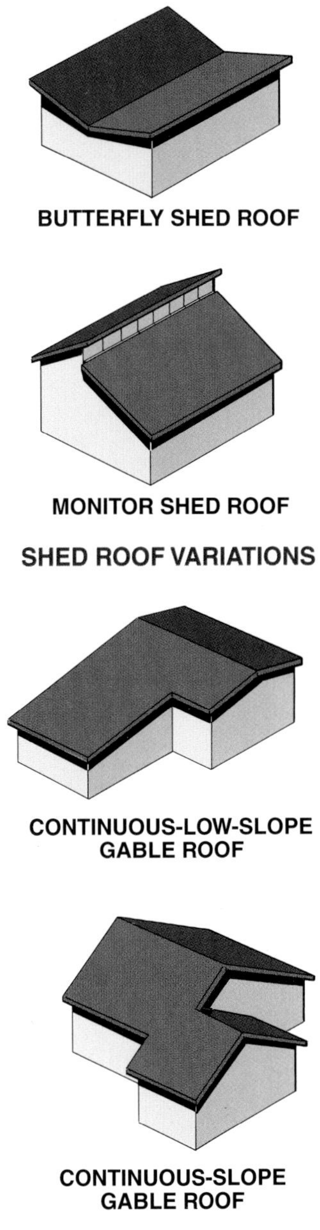

Figure 46-3. *Butterfly and monitor roofs are variations of the shed roof design. Continuous-slope gable roofs are a variation of the gable roof design.*

ROOF LAYOUT PRINCIPLES

A roof that slopes in two or more directions is based on the shape of two or more right triangles. A shed roof, which slopes in one direction, is based on the shape of one right triangle. A gable roof slopes in two directions and is similar to the shape of two right triangles placed together. **See Figure 46-5.**

Carpenters must have the following information to lay out a sloping roof:

- *Total span.* The overall width of a building is the total span. The total span dimension is shown on the floor or roof plan of the prints.
- *Total run.* One-half the total span is the total run.
- *Total rise.* The actual height of a roof is the total rise, or *true rise.* Total rise is measured from the top wall plate to the ridge of the roof.

Roof Pitch and Unit Rise

Pitch refers to the angle (slope) of a roof. The amount of pitch is determined by the *unit rise.* On the prints (usually on elevation or section drawings), a small triangle, or *slope diagram,* is shown with the unit rise and the *unit run* of the roof. The *unit run,* indicated along the base of the slope diagram, is always 12″. The unit rise is indicated along the side of the slope diagram.

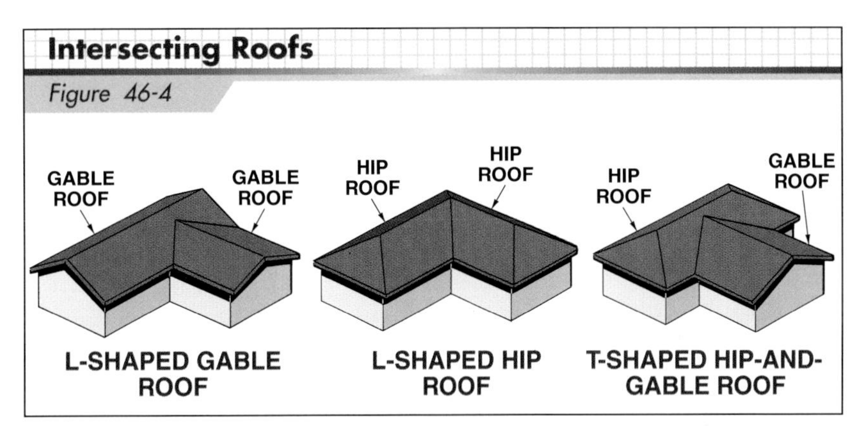

Figure 46-4. *An intersecting roof is formed when two roofs are combined. Among the many possible variations are the L-shaped gable and L-shaped hip intersecting roofs, and the T-shaped hip-and-gable intersecting roof.*

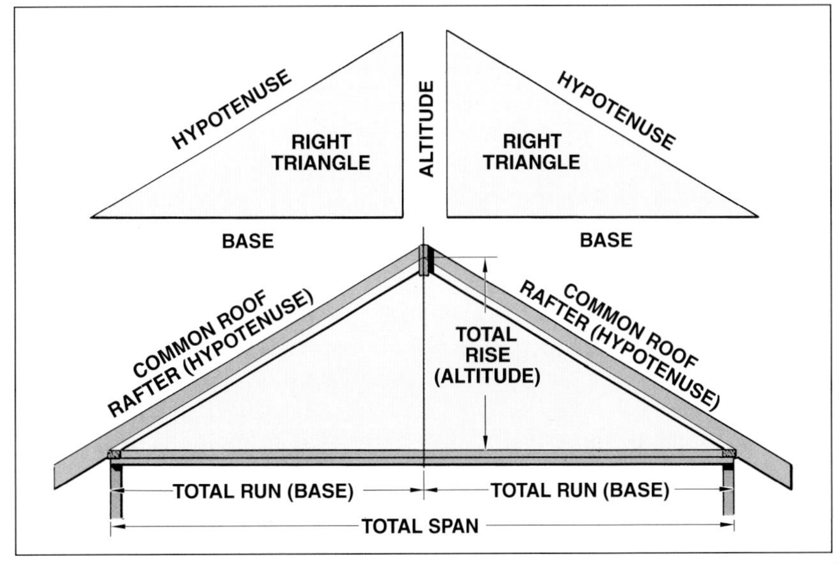

Figure 46-5. *The shape of a gable roof is based on two right triangles. A common roof rafter forms the third side (hypotenuse) of each triangle.*

The unit rise is the number of inches that the rafter rises vertically for every foot of unit run. **See Figure 46-6.** As the unit rise of the roof increases, the slope of the roof becomes steeper. Unit rise is specified on the vertical leg of the slope diagram.

Depending on the house and roof design, a moderate roof pitch of 3:12 to 6:12 is an optimal pitch to reduce wind loads. Gable roofs are more susceptible to wind loads than hip roofs. Roof overhangs wider than 2′ produce higher uplift forces than narrower overhangs.

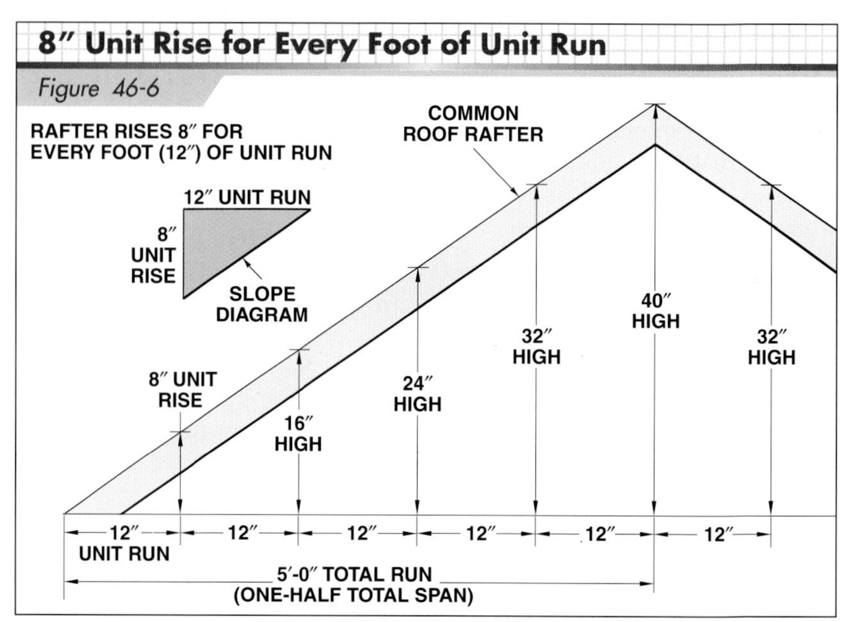

Figure 46-6. *The unit rise of a roof is the number of inches the rafter will rise vertically for every foot of run.*

Total Rise

The total rise of a roof must be known before setting the roof ridge to its correct height and attaching the rafters. The total rise is calculated by multiplying the total run (in feet) by the unit rise (in inches).

Figure 46-7 shows two examples of calculating total rise of a roof. The examples shown and explained in this unit demonstrate theoretical roof dimensions that do not include roof overhangs and ridge boards. The inclusion of an overhang and ridge board will affect the actual total rise, and is explained in detail in Unit 47.

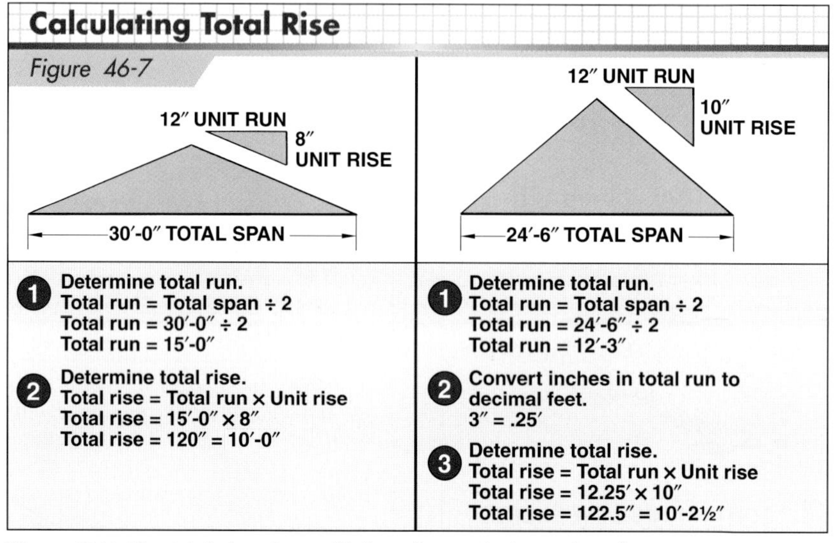

Figure 46-7. *The total rise of a roof is based on unit rise and total run.*

STRUCTURAL FACTORS IN ROOF DESIGN

A roof must withstand a great deal of weight and stress. To guarantee structural strength, the *dead load* and *live load* that a roof will bear must be considered in roof design and construction. Roof rafters, ceiling joists, and bracing such as *collar ties* and *purlins* are factors in roof strength. **See Figure 46-8.**

Dead Load

The dead load is the weight of the materials used to construct a roof. Roof rafters, sheathing, insulation, and finish covering (such as shingles or built-up roofing) are included in the dead load.

Live Load

The live load is the weight and pressure of wind and snow to which the roof will be subjected. In most parts of the United States, the combined wind and snow load for a pitched roof will not exceed 30 lb/sq ft. However, wind and snow loads differ from one part of the country to another, and local building codes reflect this difference.

Flat roofs do not shed snow as easily as pitched roofs. Therefore, in cold climates, flat roofs carry heavier snow loads. In the northern states, a flat roof is usually required to withstand a live load of 40 lb/sq ft. Unit 44 provides information regarding flat roofs.

Allowable Rafter Spans

Dead and live loads have a direct effect on the allowable span of rafters used in a roof. The allowable span is the distance from the ridge to the outside wall plates. Examples of allowable spans for roof rafters are given in **Figure 46-9.** Rafters for low-pitched roofs must be able to support greater live loads than rafters for steeper roofs.

Ceiling Joists

Ceiling joists are an important structural factor in roof systems. Ceiling joists secure the tops of the walls in place and prevent the weight of the roof from pushing walls apart. Ceiling joists usually run in the same direction as roof rafters. Whenever possible, seat ends of rafters should be nailed to the sides of ceiling joists and top wall plates. Ceiling joists are usually spaced 16″ OC and roof rafters are usually spaced 24″ OC. Under these conditions, rafters can be nailed only into every third ceiling joist.

The following nailing schedule is recommended for ceiling joists and roof rafters at exterior wall plates:

- ceiling joists to wall plates, toenail: three 8d nails
- ceiling joists to parallel roof rafters, face nail: three 16d nails
- rafters to wall plates, toenail: three 8d nails

Refer to the local building code for nailing schedules permitted in your area.

Lumber used for ceiling joists is often wider than lumber used for roof rafters. In this case, a slope must be cut at the end of the joist. A framing square can be used to mark the angle of the slope, as shown in **Figure 46-10.**

Collar Ties and Purlins

One method of strengthening a roof is to install collar ties (collar beams) at every second or third pair of rafters. Collar ties should be installed in the upper third area of the attic space and fastened at each end with four 8d common nails.

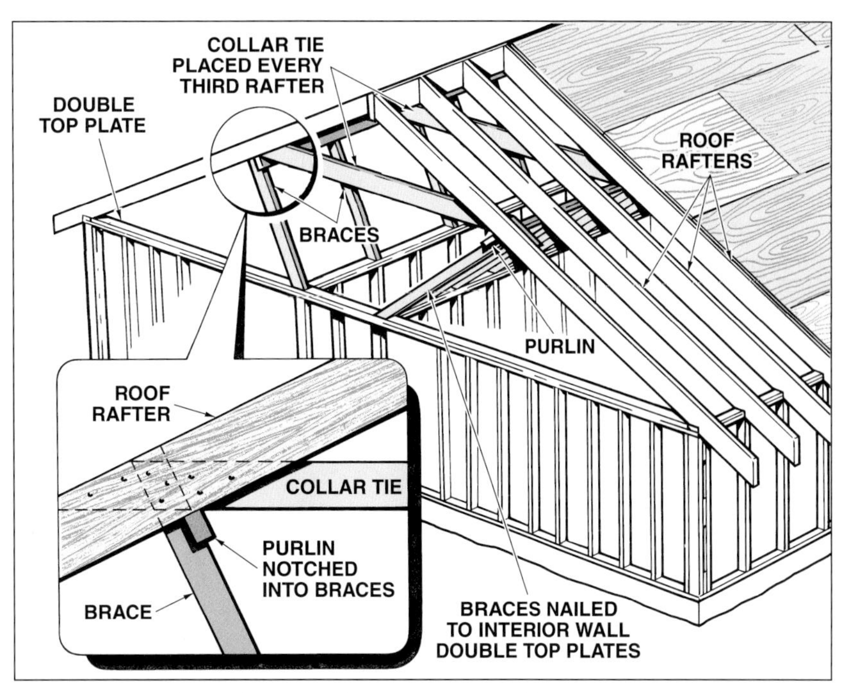

Figure 46-8. *Purlins and braces help support spans of longer rafters.*

ALLOWABLE SPANS

Rafter Size	Rafter OC Spacing†	Maximum Allowable Span*	
		Slope Less than 4 in 12	Slope 4 in 12 to 12 in 12
2 × 4	12	9′-6″	10′-0″
	16	8′-0″	9′-0″
	24	6′-6″	7′-6″
	32	6′-0″	6′-6″
2 × 6	12	16′-6″	17′-6″
	16	14′-6″	15′-6″
	24	12′-0″	12-6″
	32	10′-6″	11′-0″

* in ft and in.
† in in.

Figure 46-9. *The allowable span is the distance from the ridge to the outside wall plates. Rafter spacing and roof slope determine the allowable span. For example, 2 × 4 rafters spaced 16″ OC have an allowable span of 8′-0″ if the unit rise of the roof is less than 4″. The allowable span is 9′-0″ if the unit rise is 4″ or more.*

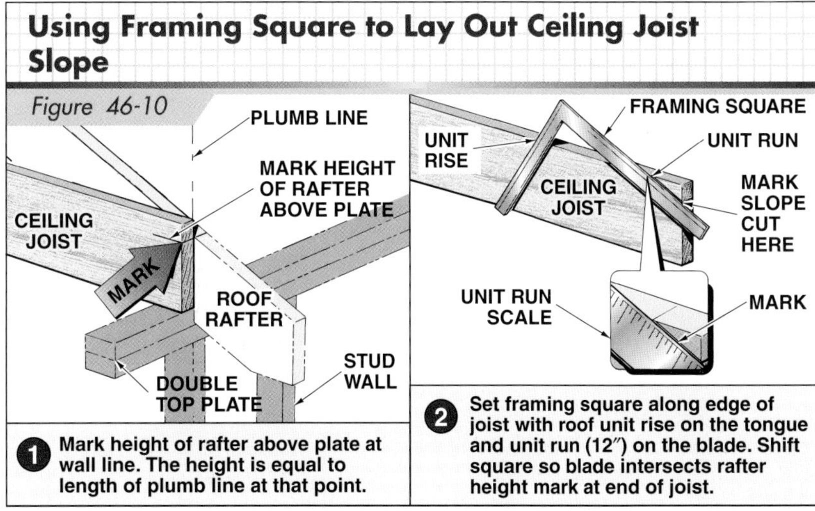

Figure 46-10. *A slope cut must be made on the ends of ceiling joists.*

Long rafters can be supported by purlins, which are horizontal members placed beneath and perpendicular to the rafters at an intermediate point between the roof ridge and exterior wall. Purlins are supported by braces that extend to the nearest interior partition.

Rafter and Truss Anchors

Roof and other structural damage may occur as a result of a roof separating from its supporting walls. Extreme weather conditions, such as hurricanes and tornados, may generate enough force to significantly damage a roof or possibly remove the roof from the building. As high-velocity winds pass over a pitched roof, an uplift force exerts pressure on the roof surface. **See Figure 46-11.** If roof sheathing is not adequately fastened to the rafters or trusses, the sheathing panels may become dislodged. If the roof sheathing performs as designed and stays attached, the uplift force is transferred to the rafters or trusses and their connections to the building walls. If the rafter-to-wall or truss-to-wall connection is successful, the uplift force is transferred to the wall-to-foundation connections.

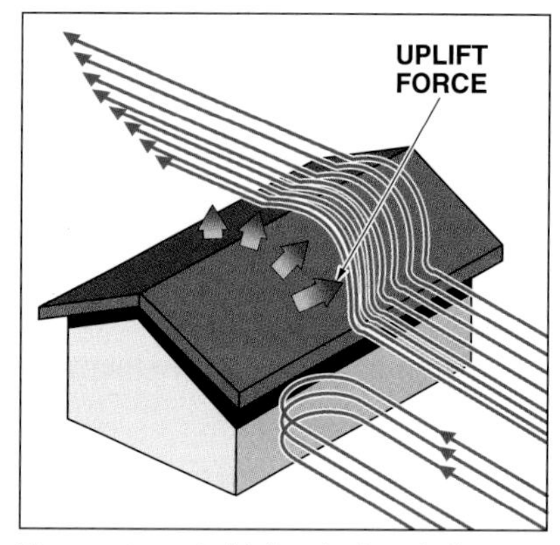

Figure 46-11. *As high-velocity wind passes over a pitched roof an uplift force is created, pulling on the roof surface.*

Metal connectors, such as rafter and truss anchors, are vital in connecting the wall frame to the roof rafters or trusses. Rafter and truss anchors are available in a variety of designs. Rafter anchors are nailed to the rafter and into the plates or the studs below. **See Figure 46-12.** Truss anchors are embedded in grout in the lintel course of a CMU wall, fastened to a CMU or concrete wall with screws, or fastened to a wood-framed wall with nails. Always refer to the manufacturer instructions for proper installation and to the local building code for wall-to-roof connection requirements.

ROOF SHEATHING

Sheathing should be nailed to a roof as soon as framing is completed. Roof sheathing serves as a base for the finish roof material and also strengthens the roof structure. **See Figure 46-13.**

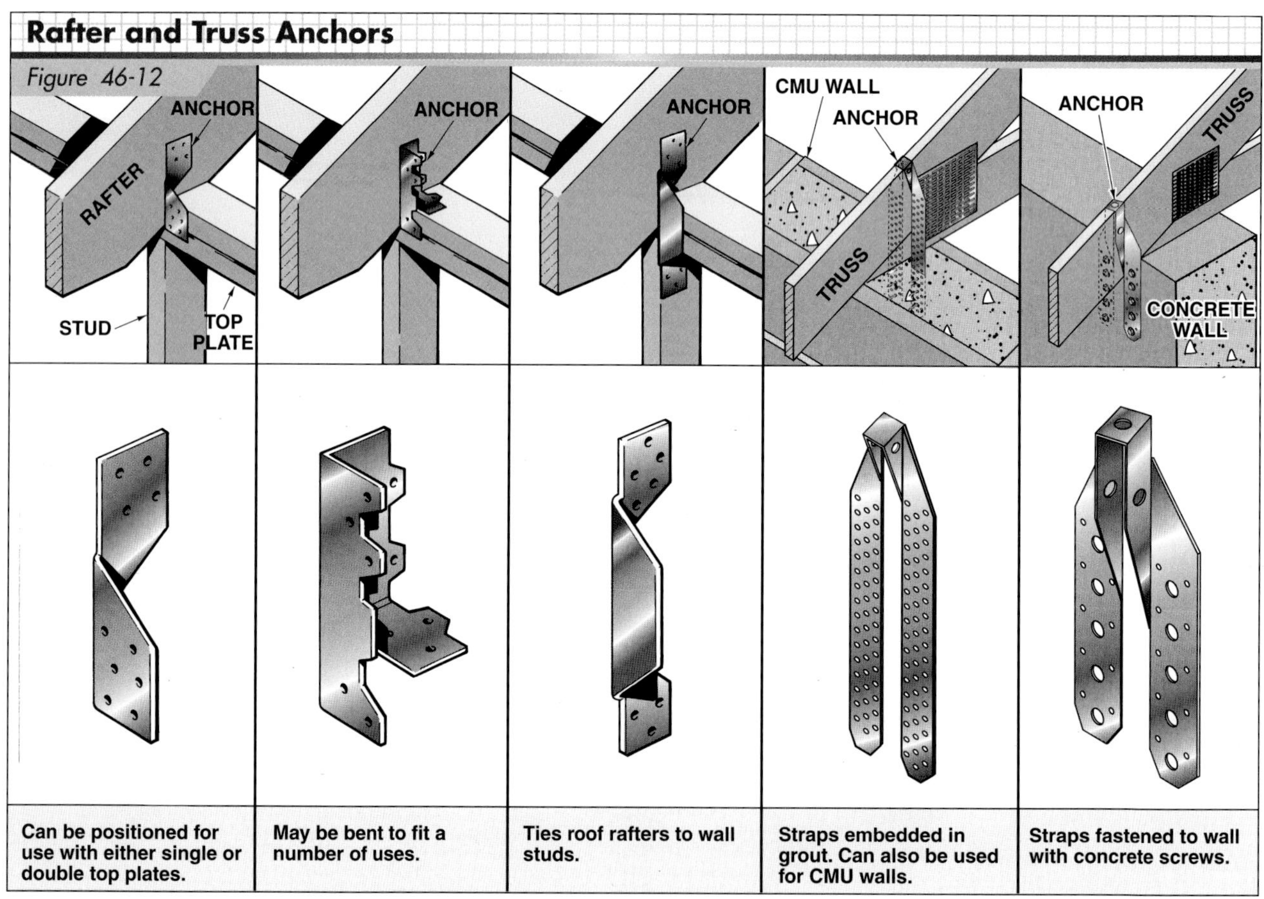

Figure 46-12. *Rafter and truss anchors tie rafters and trusses to the building walls.*

Plywood and nonveneered panel products, such as oriented strand board (OSB) and composite panels, are used for roof sheathing. When OSB is used for roof sheathing, the stamped face should be placed down with the rough side facing up. *Structural insulated panels (SIPs)* may also be used for roof sheathing. Structural insulated panels consist of a thick layer of rigid foam insulation pressed between two OSB or plywood panels. SIPs are commonly used with exposed-beam roofs, but may be used with most construction methods. *Spaced-board sheathing* may also be used for roof sheathing. Typically, spaced-board sheathing is used as a base for certain types of roof shingles.

Panels are placed with the long dimension parallel to the roof ridge and at a right angle to roof rafters. When nailing panels to the roof, ensure the first row of panels is in a straight line. Since the end joints of panels should be staggered, start the second row with a half panel. Roof sheathing panels should be continuous over two or more rafters.

Roof sheathing panels are usually ⅝″ or more, depending on rafter spacing. Plywood and OSB panels are usually fastened with 8d smooth- or ring-shank common nails spaced 6″ OC along the edges and 12″ OC at intermediate rafters. However, in areas where higher wind loads are anticipated, ring-shank or specialized hurricane nails should be used and spaced 3″ to 6″ apart. Always refer to the local building code to determine approved fasteners and spacing. Various sections of the roof, such as along the eaves, corners, ridges, and gable ends, may experience greater wind uplift pressures and require more closely spaced nailing patterns than interior sections of the roof.

Screws or staples may also be used to fasten roof sheathing to rafters. Screws (#8 × 1¾″) are spaced in the same pattern as nails. Staples may be used with thinner roof sheathing panels if permitted by the local building code. Staples should be 1⅜″ long for ⅜″ panels and 1½″ long for ½″ panels. Staples are driven 4″ apart at panel edges and 8″ apart at intermediate rafters.

For certain applications, rafters may be spaced farther apart than the allowable distance between rafters for the thickness of the panels. In these situations, blocking is nailed between the rafters, or panels with tongue-and-groove edges are used. Panel clips may be used instead of blocking or tongue-and-groove panels. **See Figure 46-14.**

When placing sheathing on a pitched roof, carpenters should fasten roofing jacks and planks to the roof sheathing as they work toward the top of the ridge. In addition, appropriate personal fall-protection equipment and techniques should be employed to prevent falls.

Southern Forest Products Association

Senco Products, Inc.

Figure 46-13. *Panel products such as plywood and OSB are typically used for roof sheathing. Panels are attached to roof rafters using 8d common nails or screws.*

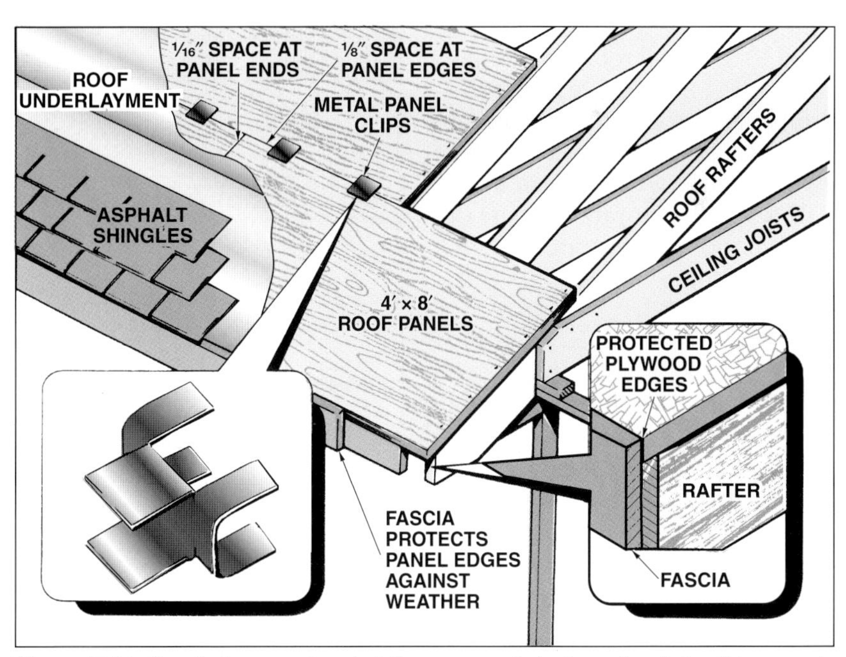

Figure 46-14. *Panel clips are placed at the unsupported edges of roof sheathing panels and eliminate the need for blocking between rafters. Panel clips are made of galvanized steel and accommodate all panel thicknesses used for roof sheathing.*

Unit 46 Review and Resources

UNIT 47

Gable, Gambrel, and Shed Roofs

OBJECTIVES

1. List the characteristics of gable roofs.
2. Identify the cuts that must be made for common rafters.
3. Describe the procedure for calculating length of common rafters.
4. Explain the procedure for constructing gable roofs.
5. Review the procedure for constructing gambrel roofs.
6. Describe the procedure for laying out a shed roof.
7. Explain the construction of dormers.

Gable roofs are the most common type of roof for small residential structures. Gable and gambrel roofs slope in two directions, while shed roofs slope only in one direction.

GABLE ROOFS

Next to the shed roof, which has only one slope, a gable roof is the simplest type of pitched roof to build because it slopes in only two directions. **See Figure 47-1.** The basic structural members of a gable roof are the *ridge board, common rafters,* and *gable studs.* **See Figure 47-2.**

A ridge board is placed at the peak of a gable roof to provide a nailing surface for the upper ends of the common rafters. Common rafters extend from the top wall plates to the ridge. Gable studs are upright framing members that provide a nailing surface for siding and sheathing at the gable ends of the roof.

Traditionally, the framing square has been used to lay out angled cuts and lengths on common rafters. The Speed® Square is now also being used to lay out common rafters. Common rafter layout using the framing square and Speed Square are both discussed in this unit.

APA—The Engineered Wood Association

Figure 47-1. *A gable roof slopes in only two directions.*

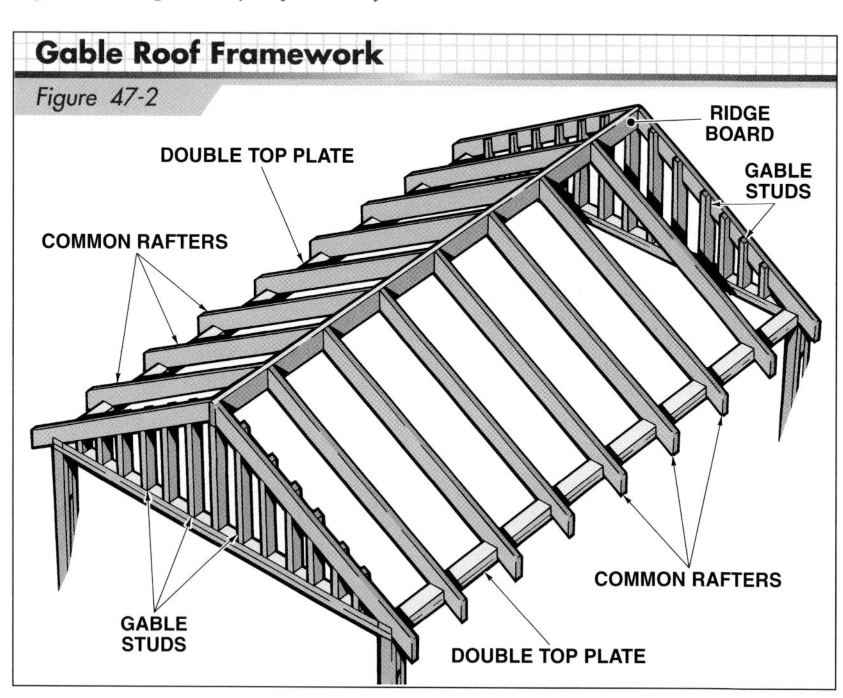

Figure 47-2. *Framework for a gable roof includes ridge board, common rafters, and gable studs.*

Common Rafters

All common rafters for a gable roof are the same length and they can be precut before the roof is assembled. Most common rafters include an *overhang.* An overhang is the part of a rafter that extends past the building line. The *run of the overhang* is the horizontal distance from the building line to the tail cut on the rafter.

Plumb cuts are made at the ridge, heel, and tail of a common rafter. A *seat cut,* or level cut, is made where the rafter rests on the top wall plates. The notch formed by the seat and heel plumb cut lines is often referred to as a *bird's mouth.* **See Figure 47-3.**

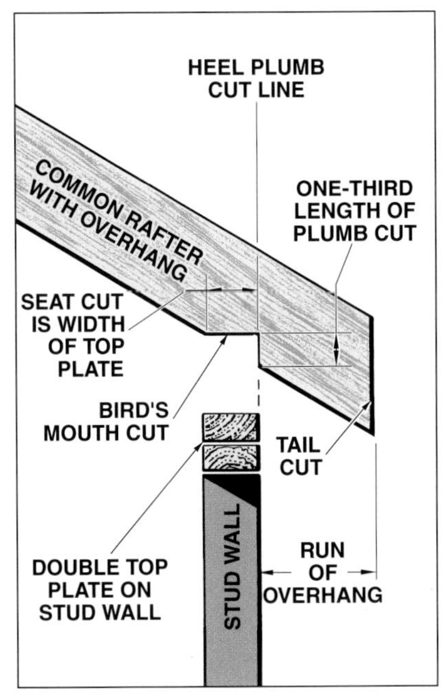

Figure 47-3. *A bird's mouth is formed by the heel plumb cut line and seat line.*

The overall appearance of a building is affected by the roof lines and the type of roof finish material. The roof design should complement the building design and the roof finish material should be appropriate for the geographic area.

The length of a seat cut should be the width of the wall plates. (For 2 × 6 wall plates, the seat cut would be 5½"). In addition, the seat cut should not be deeper than one-third the length of the plumb cut. The procedure for marking these cuts is explained later in this unit.

Seat Cut and True Rise. The amount of stock remaining above a seat cut, referred to as the *stand* or *HAP* (height above plate), is added to the *total rise.* The true rise is the total rise if there is no seat cut. The stand should be two-thirds the length of the seat cut. However, this distance will vary depending on the width of the rafter. When calculating true rise, an allowance is made for the drop at the ridge. **See Figure 47-4.**

Calculating Length of Common Rafters. The length of common rafters is based on the unit rise and total run of the roof. The unit rise and total run are obtained from the prints. Four different procedures may be followed to calculate common rafter length. One procedure uses a framing square rafter table. Another procedure utilizes a book of rafter tables. In another procedure—the step-off method—rafter layout is combined with calculating length. Common rafter length can also be calculated mathematically.

Calculating True Rise

Figure 47-4

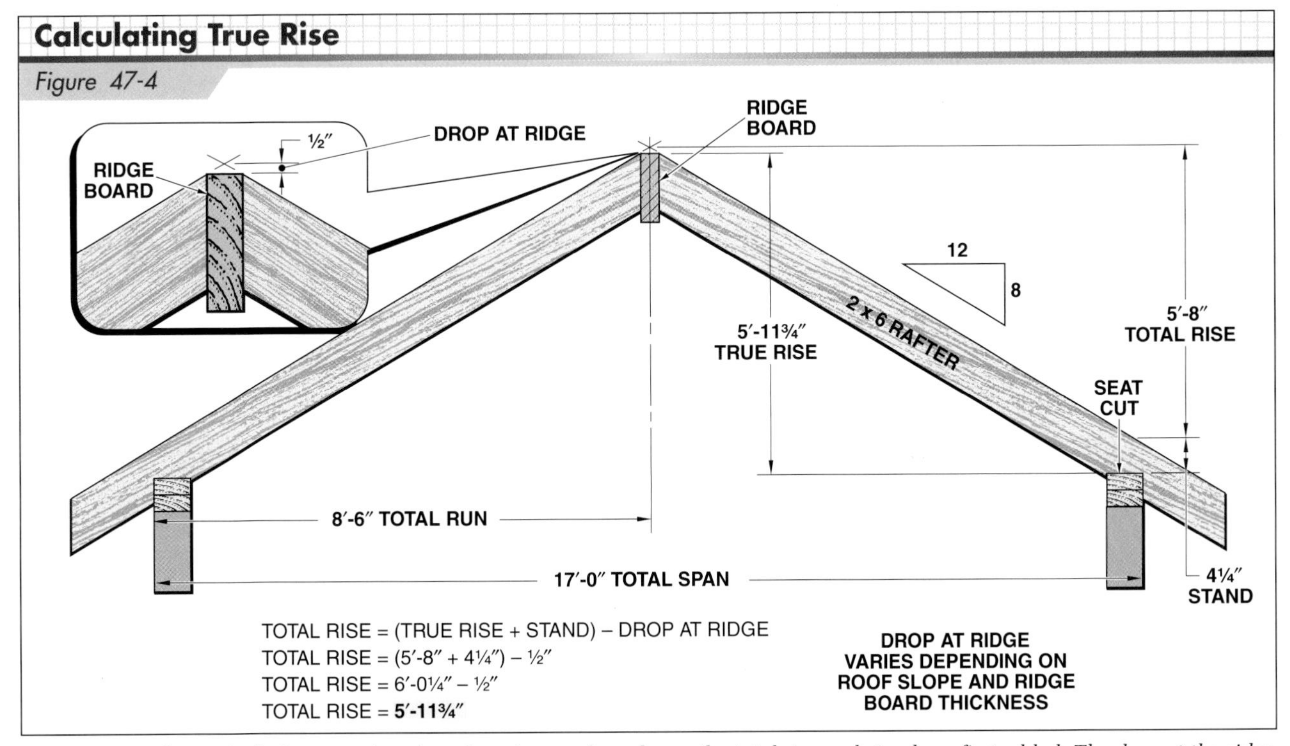

Figure 47-4. *When calculating true rise when there is a roof overhang, the total rise and stand are first added. The drop at the ridge is then subtracted.*

Framing Square Rafter Table. Framing squares typically have a rafter table printed on the face side of the blade. **See Figure 47-5.** The rafter table makes it possible to determine the lengths of all types of rafters for pitched roofs with 2″ to 18″ unit rises. The method for using the framing square rafter tables is given in the following two examples:

Example 1. The roof has a 7″ unit rise and a 16′ total span. **See Figure 47-6.** Look at the first line of the rafter table on the framing square to find the "length of common rafters per foot of run." Since the roof in this example has a 7″ unit rise, locate the number "7" at the top of the square. Directly beneath the number "7" is the figure "13.89," which means that a common rafter with a 7″ unit rise will be 13.89″ long for every foot of run.

To determine the rafter length, multiply 13.89″ by the total run in feet. (Total run equals one-half the total span.) A roof with a 16′ total span has an 8′ total run; therefore, multiply 13.89″ by 8′ to calculate the rafter length. The mathematical steps in this procedure are as follows:

1. Multiply the length of the common rafter per foot of run (13.89″) by the total run (8′).

   $13.89'' \times 8' = 111.12''$

2. Convert the decimal inch portion of the previous calculation to fractional inches by multiplying it by 16.

   $.12 \times 16 = 1.92$ or $\frac{1}{8}''$

   The number "1" to the left of the decimal point represents $\frac{1}{16}''$. The number ".92" to the right of the decimal points represents $\frac{92}{100}$ of $\frac{1}{16}''$. For practical purposes, 1.92 is calculated as being equal to $\frac{1}{8}''$. (As a general rule in this kind of calculation, if the number to the right of the decimal point is more than 5, add $\frac{1}{16}''$ to the figure on the left side of the decimal.) The decimal can also be converted to its fraction equivalent using the hundredth scale located on the heel on the back side of a framing square.

3. Determine the total common rafter length by combining the results of steps 1 and 2 and converting the results to feet and inches.

   $111'' + \frac{1}{8}'' = 111\frac{1}{8}''$

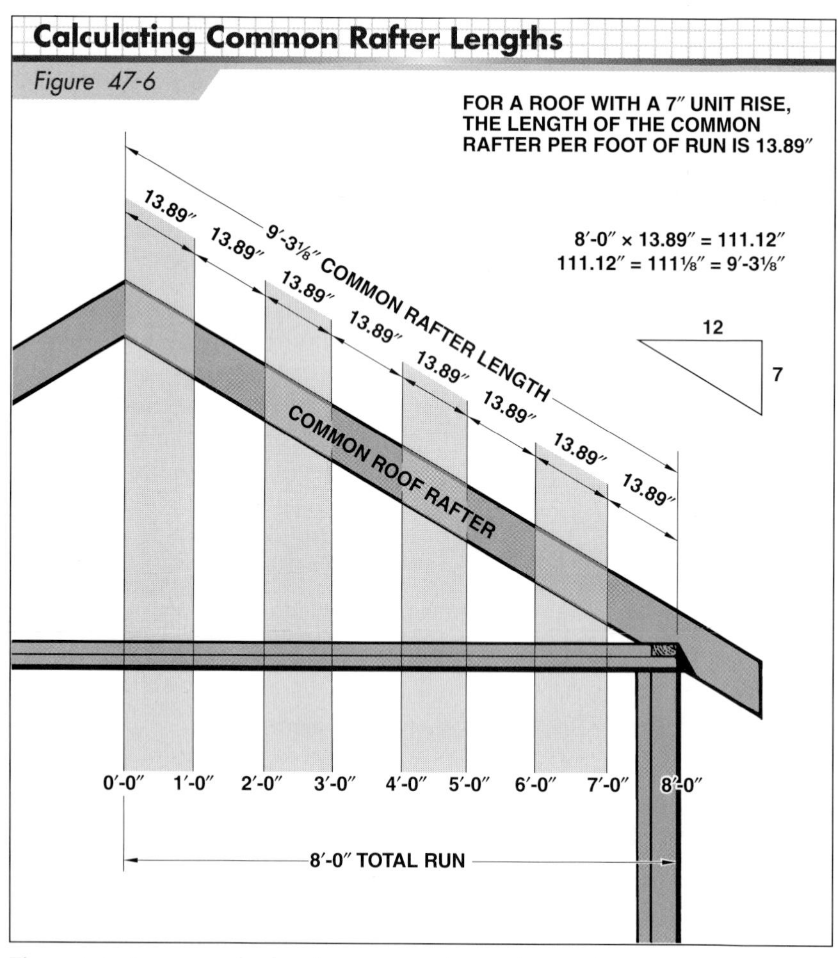

Figure 47-6. *Common rafter length is calculated by multiplying the total run by the length of the common rafter per foot of run.*

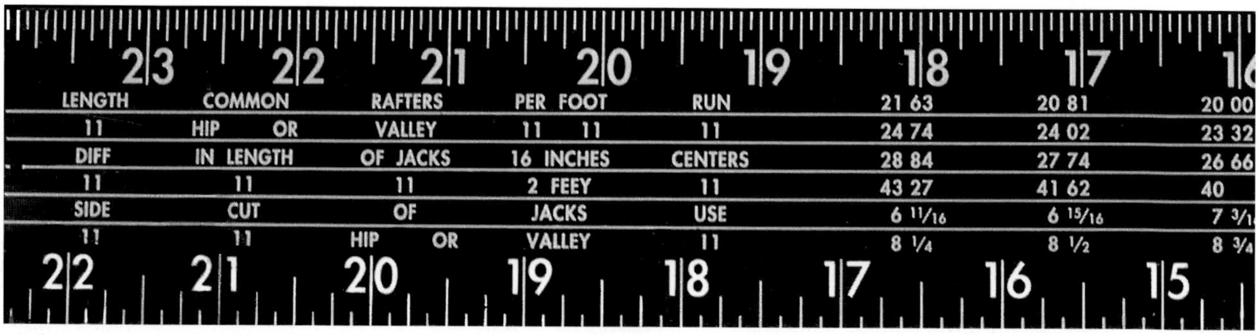

23 22 21 20 19 18 17 16
LENGTH COMMON RAFTERS PER FOOT RUN 21 63 20 81 20 00
11 HIP OR VALLEY 11 11 11 24 74 24 02 23 32
DIFF IN LENGTH OF JACKS 16 INCHES CENTERS 28 84 27 74 26 66
11 11 11 2 FEEY 11 43 27 41 62 40
SIDE CUT OF JACKS USE 6 11/16 6 15/16 7 3/1
11 11 HIP OR VALLEY 11 8 1/4 8 1/2 8 3/4
22 21 20 19 18 17 16 15

Figure 47-5. *The rafter table on a framing square can be used to calculate rafter lengths for pitched roofs with 2″ to 18″ unit rises.*

4. Divide the whole number (111″) by 12 to convert to feet and inches and add the fraction (⅛″).

   111″ ÷ 12 = 9′-3″

   9′-3″ + ⅛″ = **9′-3⅛″**

Example 2. A roof has a 6″ unit rise and a 25′ total span. The total run of the roof is 12′-6″. The mathematical steps in this procedure are as follows:

1. Convert the inch portion of the total run (6″) to decimal feet by dividing by 12.

   6″ ÷ 12 = .5′

2. Multiply the length of the common rafter per foot of run (13.42″) by the total run (12.5′). The number "13.42" is obtained from the framing square.

   13.42″ × 12.5′ = 167.75″

3. Convert the decimal inch portion of the previous calculation to fractional inches by multiplying it by 16.

   .75 × 16 = 12

   ¹²⁄₁₆″ = ¾″

4. Determine the total common rafter length by combining the results of steps 2 and 3 and convert the results to feet and inches.

   167″ + ¾″ = 167¾″

5. Divide the whole number (167″) by 12 to convert to feet and inches and add the fraction (¾″).

   167″ ÷ 12 = 13′-11″

   13′-11″ + ¾″ = **13′-11¾″**

Book of Rafter Tables. Most carpenters prefer to use a book of rafter tables for calculating common rafter length. A book of rafter tables contains rafter measurements for a wider range of roof spans and unit rises than the framing square rafter table.

Shortening Common Rafters. Rafter length found by any of the methods previously discussed is the measurement from the heel plumb cut line to the center of the ridge. This is known as the *theoretical length of the rafter.* Since a ridge board, usually 1½″ thick, is placed between the rafters, one-half of the ridge board thickness (¾″) must be deducted from each rafter. This calculation is known as *shortening the rafter,* and is done when the rafters are laid out. The *actual length of a rafter* is the distance from the heel plumb line to the shortened ridge plumb line. **See Figure 47-7.**

THEORETICAL LENGTH OF RAFTER
ACTUAL LENGTH OF RAFTER
½ THICKNESS OF RIDGE BOARD
THEORETICAL LENGTH
ACTUAL LENGTH
RAFTER
RAFTER

Theoretical length of rafter is the distance from heel plumb cut line to centerline of ridge board.

Actual length of rafter is the distance from heel plumb cut line to centerline of the ridge board minus one-half ridge board thickness.

Figure 47-7. *The actual length of a common rafter is calculated by deducting one-half the ridge thickness perpendicular to the plumb line.*

Laying Out Common Rafters. Before rafters can be cut, the angles of the cuts must be laid out. Layout consists of marking the plumb cuts at the ridge, heel, and tail of the rafter, and the seat cut where the rafter will rest on the top wall plates. The angles are laid out with a framing square or Speed® Square.

A pair of square gauges is useful when using a framing square. One square gauge is secured to the tongue of the square next to the number that is the same as the unit rise. The other square gauge is secured to the blade of the square next to the number that is the same as the unit run, which is always 12″. When the framing square is placed on the rafter stock, the plumb cut is marked along the tongue (unit rise) side of the square. The seat cut is marked along the blade (unit run) side of the square. **See Figure 47-8.**

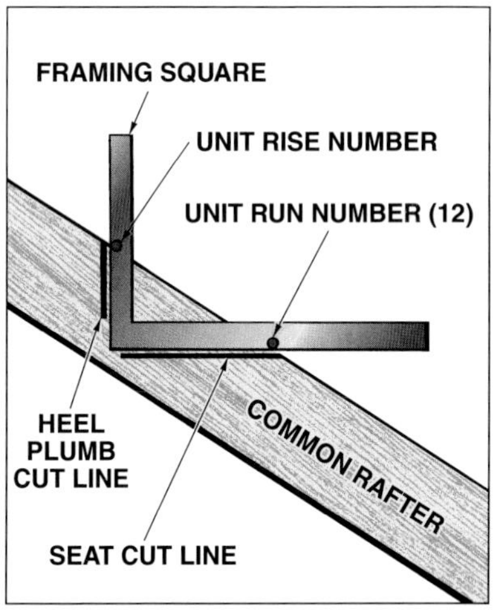

Figure 47-8. *A framing square is used to lay out the plumb and seat cuts.*

Rafter layout also includes marking off the required overhang and making the shortening calculation. A procedure for laying out a common rafter is shown in **Figure 47-9**. A Speed Square can be used to lay out the angled cuts on a common rafter. Rafter length tables are typically included in a book of rafter tables. The procedure for laying out common rafters using a Speed Square is similar to the procedure using a framing square.

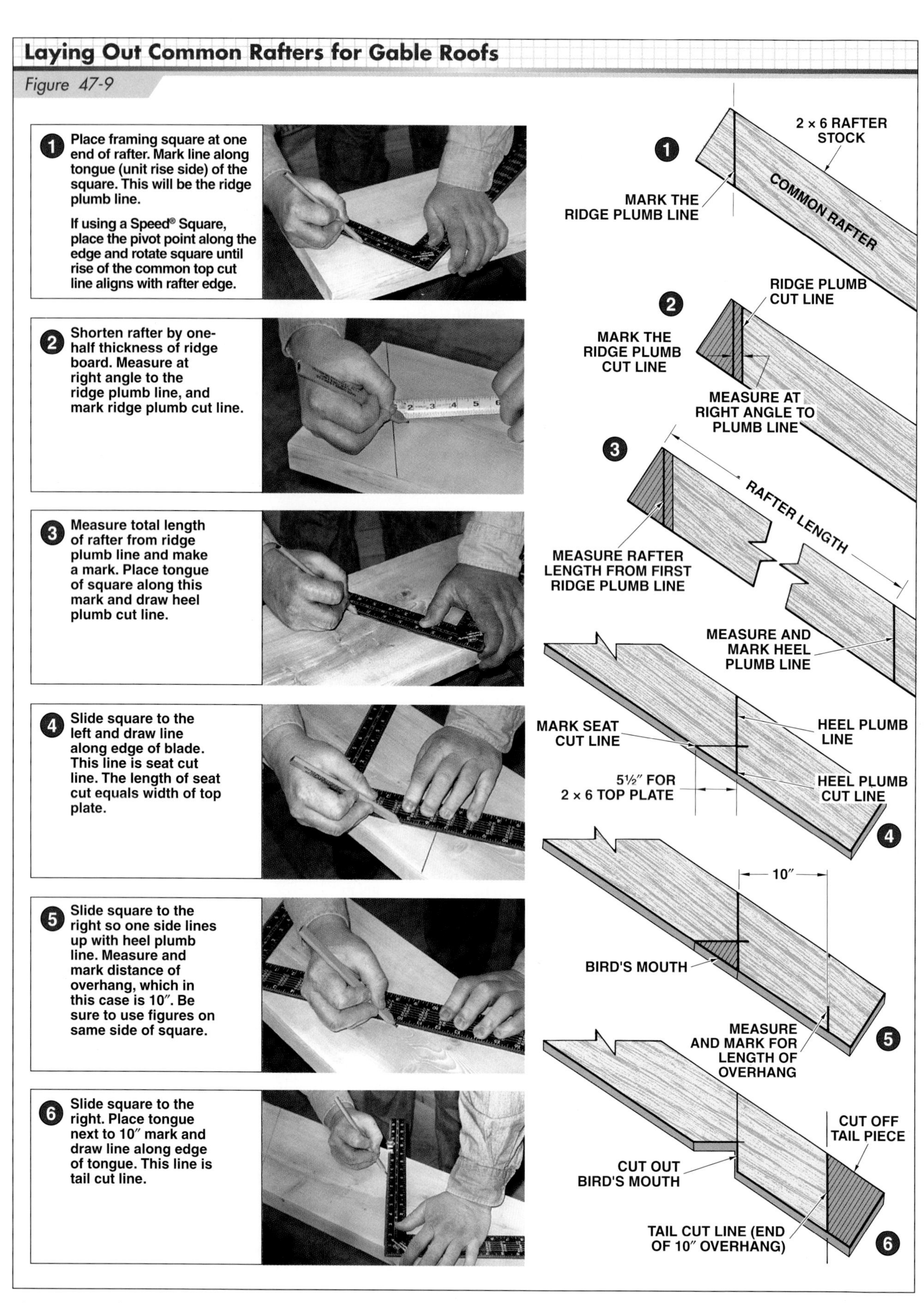

Figure 47-9. *A framing square or a Speed® Square can be used to lay out a common rafter for a gable roof. The roof in this example has an 8″ unit rise and a 10″ overhang. The same procedure is used regardless of the layout tool being used.*

Step-off Method for Calculating Length and Laying Out Common Rafters. The step-off method for rafter layout is an older, but still practiced, method. The step-off method combines the procedure for laying out the rafters with a procedure for stepping off the length of the rafter. **Figure 47-10** shows the step-off method. Extreme care must be used using the step-off method for calculating after length.

Constructing Gable Roofs

The major part of gable roof construction is setting the common rafters in place. **See Figure 47-11.** The most efficient method of constructing gable roofs is to precut all common rafters and then fasten them to the ridge board and the wall plates in a continuous operation.

Marking Wall Plates. Rafter locations should be marked on the top wall plates when the positions of ceiling joists are laid out. Proper roof layout will ensure the rafters and joists tie into each other wherever possible. Roof rafters are often spaced 24″ OC and ceiling joists are commonly spaced 16″ OC. An example of this type of layout is shown in **Figure 47-12.**

A gable roof is the most common type of roof used for a residential garage.

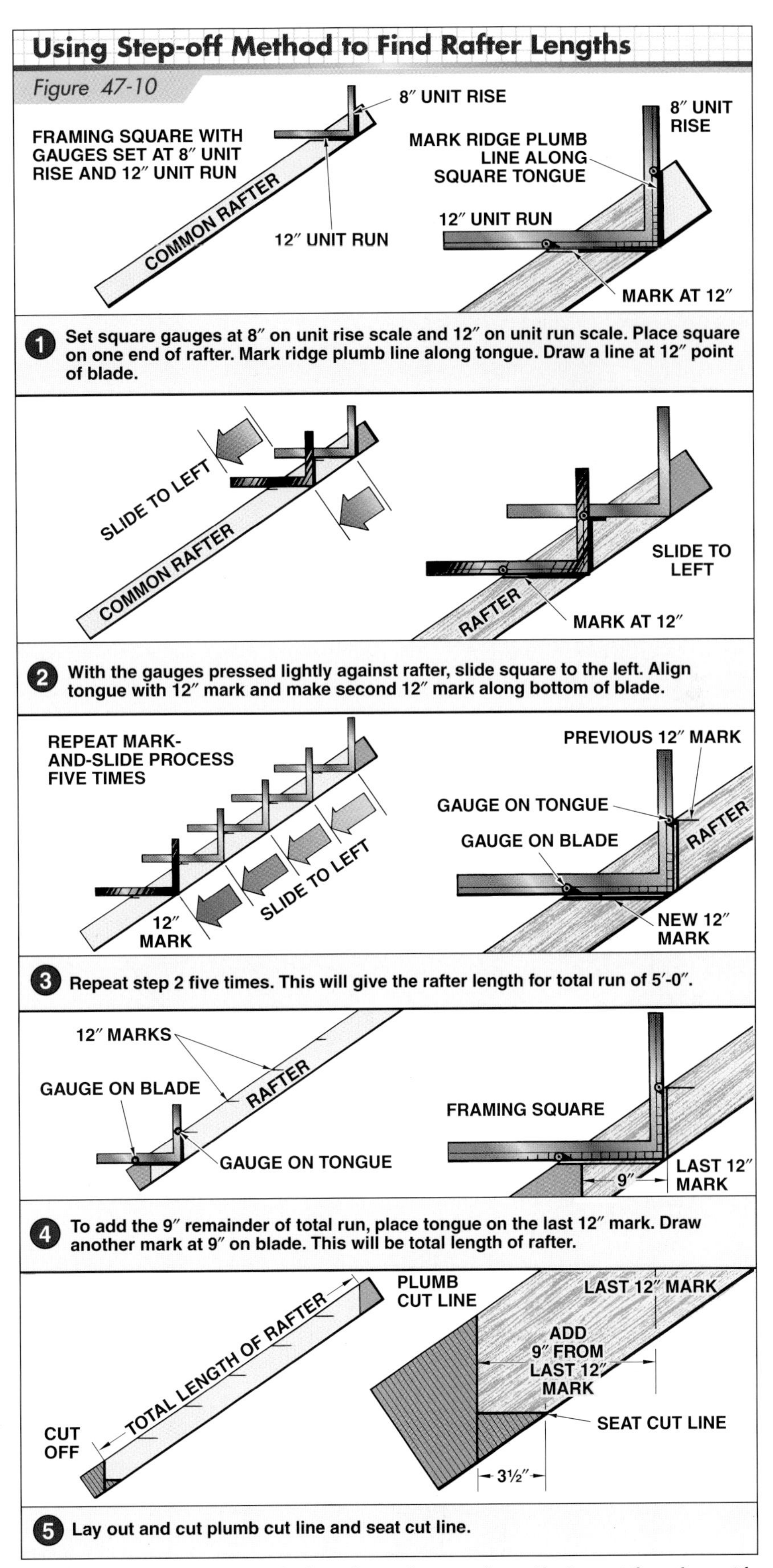

Figure 47-10. *The step-off method combines the procedure of laying out the rafters with a procedure of stepping off the length of the rafter. In this example, the roof has an 8″ unit rise and a total run of 5′-9″.*

Figure 47-11. *When constructing a gable roof, common rafters are precut and then fastened to the ridge board and wall plates in a continuous operation.*

Cutting and Marking a Ridge Board. The ridge board, like the common rafters, should be precut. Rafter locations are then copied onto the ridge board from the markings on the wall plates. **See Figure 47-13.** The ridge board should be the length of the building plus the overhang at the gable ends.

Lumber used for ridge boards is wider than rafter stock. For example, a 2 × 8 ridge board would be used with 2 × 6 rafters. Ridge boards must be at least 1″ thick (nominal) and must be wider than the cut end of the rafter. Some buildings are long enough to require more than one piece of ridge material. The breaks between the ridge pieces should occur at the center of a rafter.

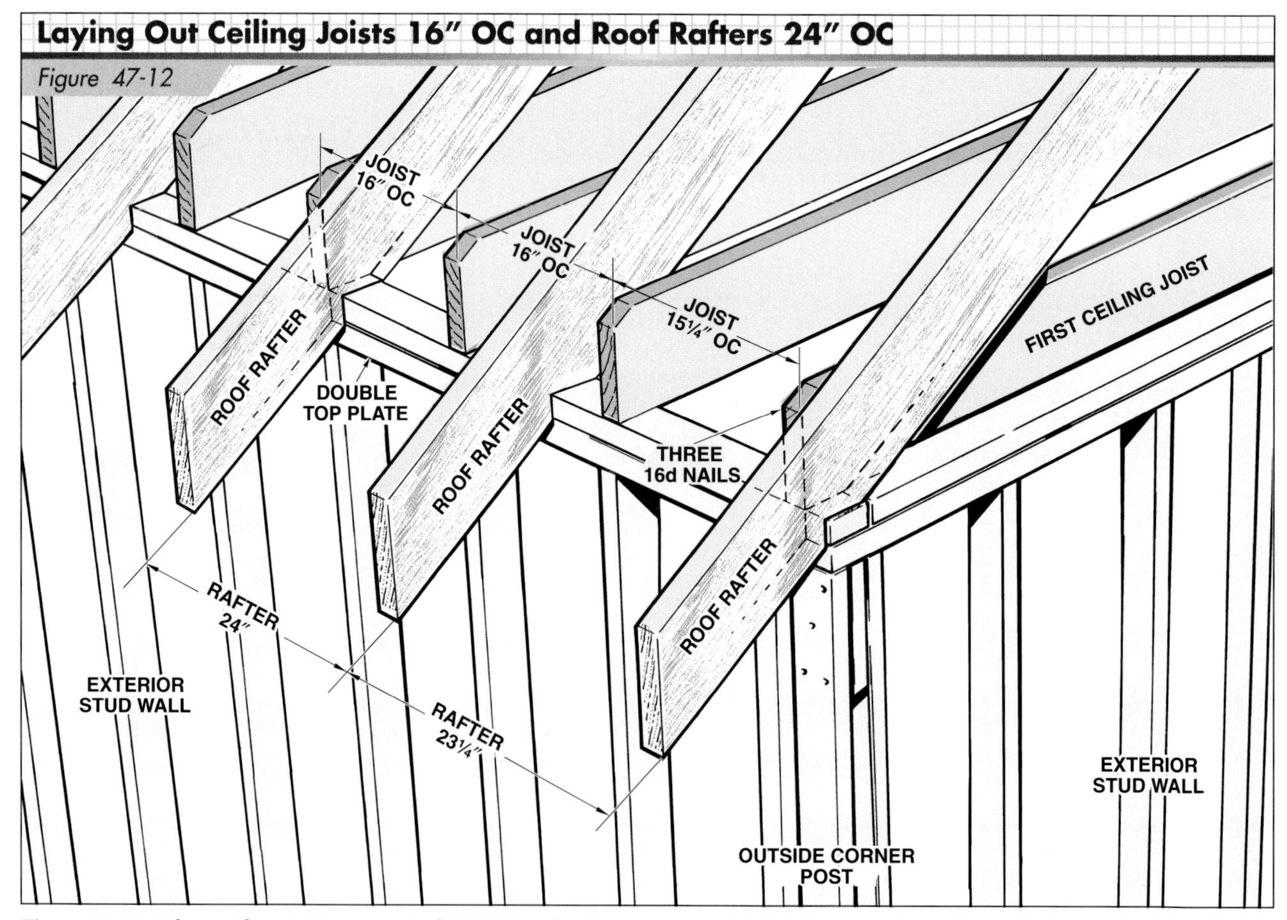

Figure 47-12. *When ceiling joists are spaced 16″ OC and rafters are spaced 24″ OC, every other rafter will tie into the joists. Proper layout ensures that rafters and joists will tie into each other wherever possible.*

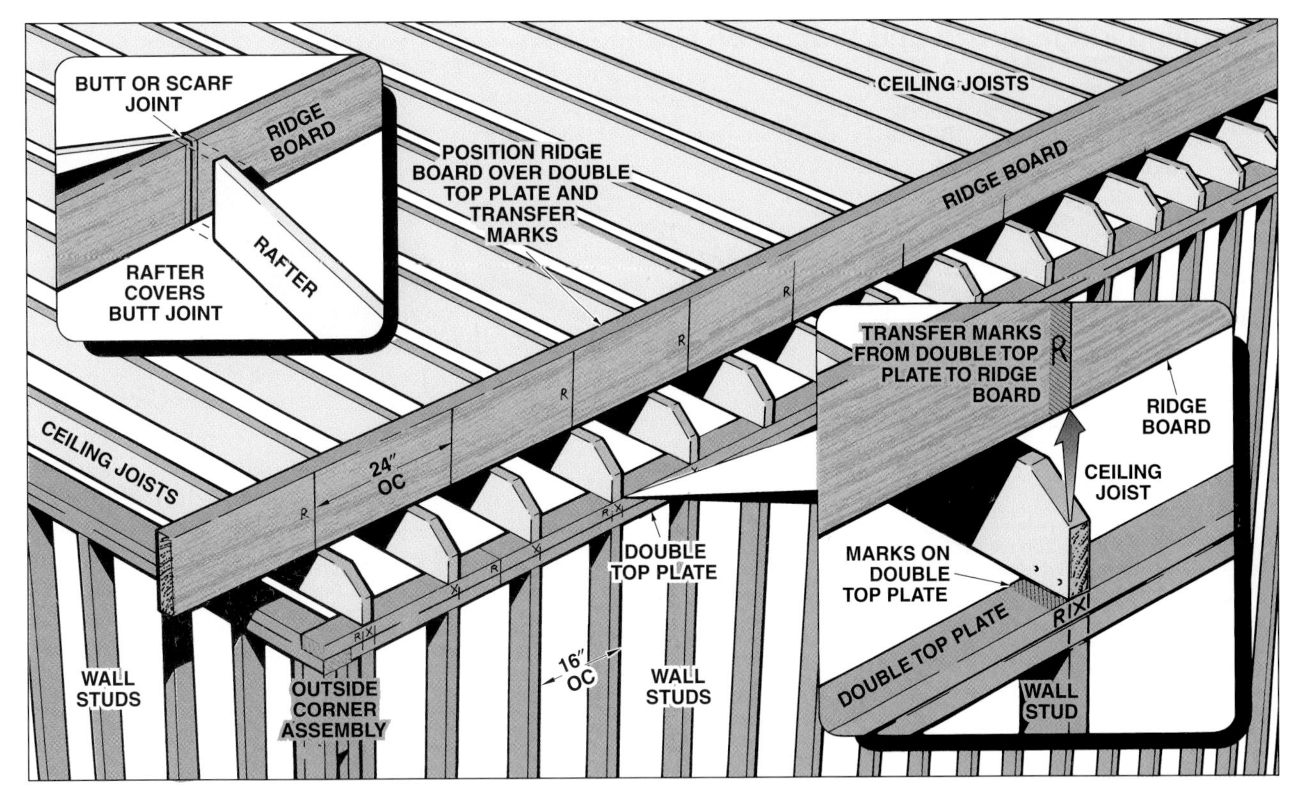

Figure 47-13. *Ridge board layout is copied from markings on wall plates. Butt joints for ridge boards must occur at the center of a rafter.*

Cutting Common Rafters. One pair of rafters should be cut and checked for accuracy before the other rafters are cut. To check the first pair for accuracy, set them in position with a 1½″ piece of wood fitted between them. If the rafters are the correct length, they should fit the building. If, however, the building walls are out of line, adjustments will have to be made to the rafters.

After the first pair of rafters fits the building satisfactorily, one of the pair can be used as a template for marking the other rafters. A circular saw or a radial-arm saw is then used to cut the rafters.

Placing a Ridge Board and Common Rafters. Various methods may be used to set up the ridge board and attach rafters to it. Plywood or OSB panels should be laid on top of the ceiling joists where the framing will take place. The panels provide safe and comfortable footing for carpenters and also provide a place to put tools and materials.

When only a few carpenters are present on a job site, the most convenient procedure is to set the ridge board to its required height (total rise) and hold it in place with temporary props and braces. **See Figure 47-14.** The rafters can then be nailed to the ridge board and the top wall plates. A faster system that can be used when a larger crew is present is shown in **Figure 47-15.**

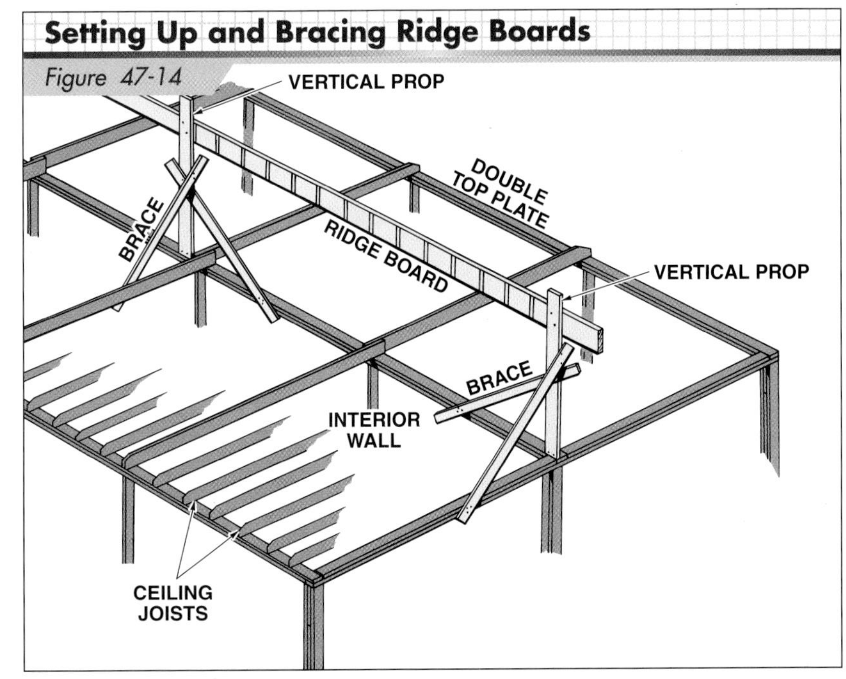

Figure 47-14. *When only a few carpenters are present on a job site to set the ridge board, the ridge board is secured in position with props and braces.*

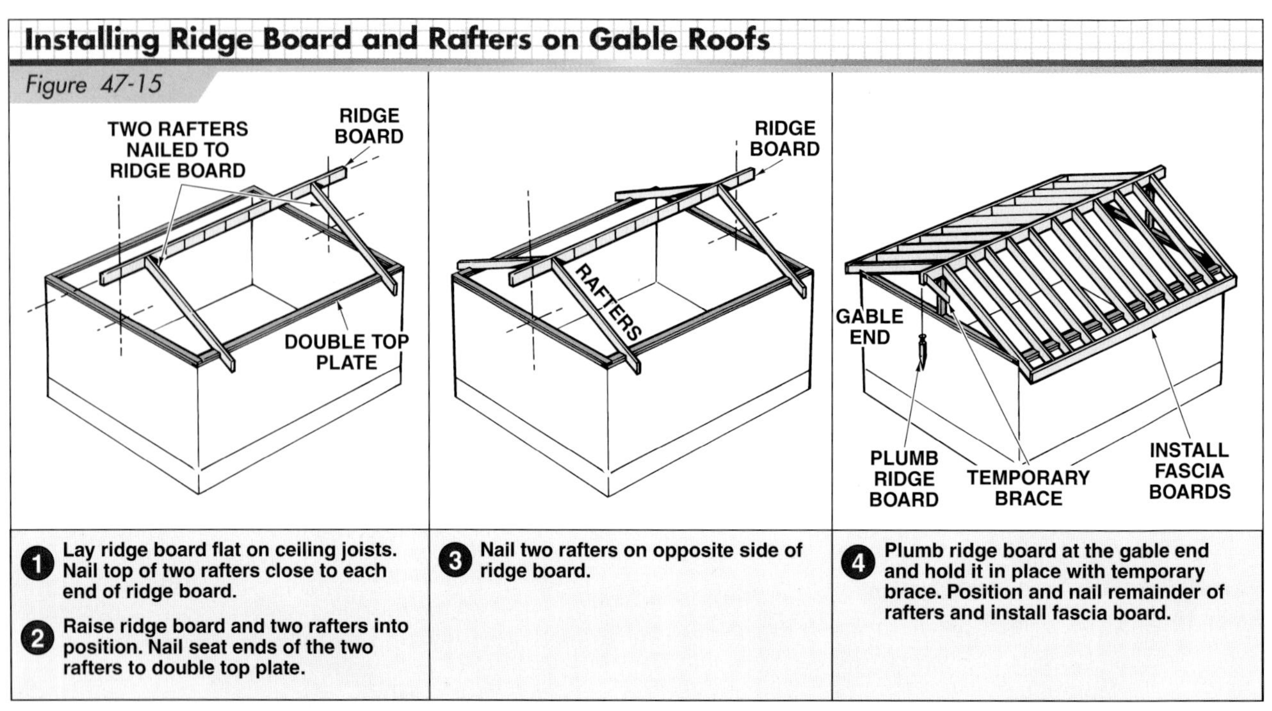

Figure 47-15. *When several carpenters are present to raise a gable roof, the method shown is used to set the ridge board and nail rafters for a gable roof.*

Cutting Common Rafter Overhang. Common rafter overhang can be laid out and cut before the rafters are set in place. However, carpenters may prefer to cut the overhang after rafters are fastened to the ridge board and wall plates. A chalk line is snapped from one end of the building to the other and the tail plumb lines are marked with a sliding T-bevel. **See Figure 47-16.** The rafters are then cut with a circular saw. Cutting common rafter overhang after the rafters are set in place ensures the line of the overhang will be perfectly straight, even if the building is not.

Framing Gable End Overhang. Another overhang is formed over each gable end of the building. The main framing members of the gable end overhang are *fascia rafters,* or barge rafters. Fascia rafters are fastened to the ridge board at one end and to the fascia board at the lower end. Fascia boards are often nailed to the tail ends of the common rafters to serve as a finish piece at the edge of a roof. By extending past the gable ends of the house, fascia boards also help to support the fascia rafters. **Figures 47-17** and **47-18** show different methods used to frame gable end overhangs.

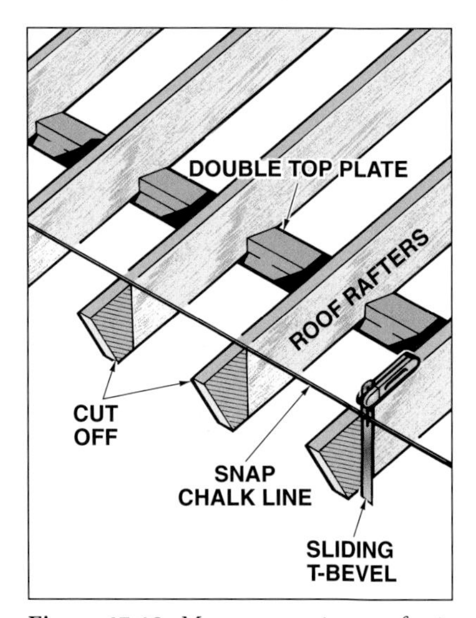

Figure 47-16. *Many carpenters prefer to cut the overhang after rafters are fastened to the ridge board and wall plates. A chalk line is snapped along the rafters and a sliding T-bevel is used to lay out the tail plumb lines.*

James Hardie Building Products

Gable dormers add space and provide light and ventilation to a one-and-one-half-story house.

Gable End Overhang Ladder-Framed Lookouts over End Walls

Figure 47-17

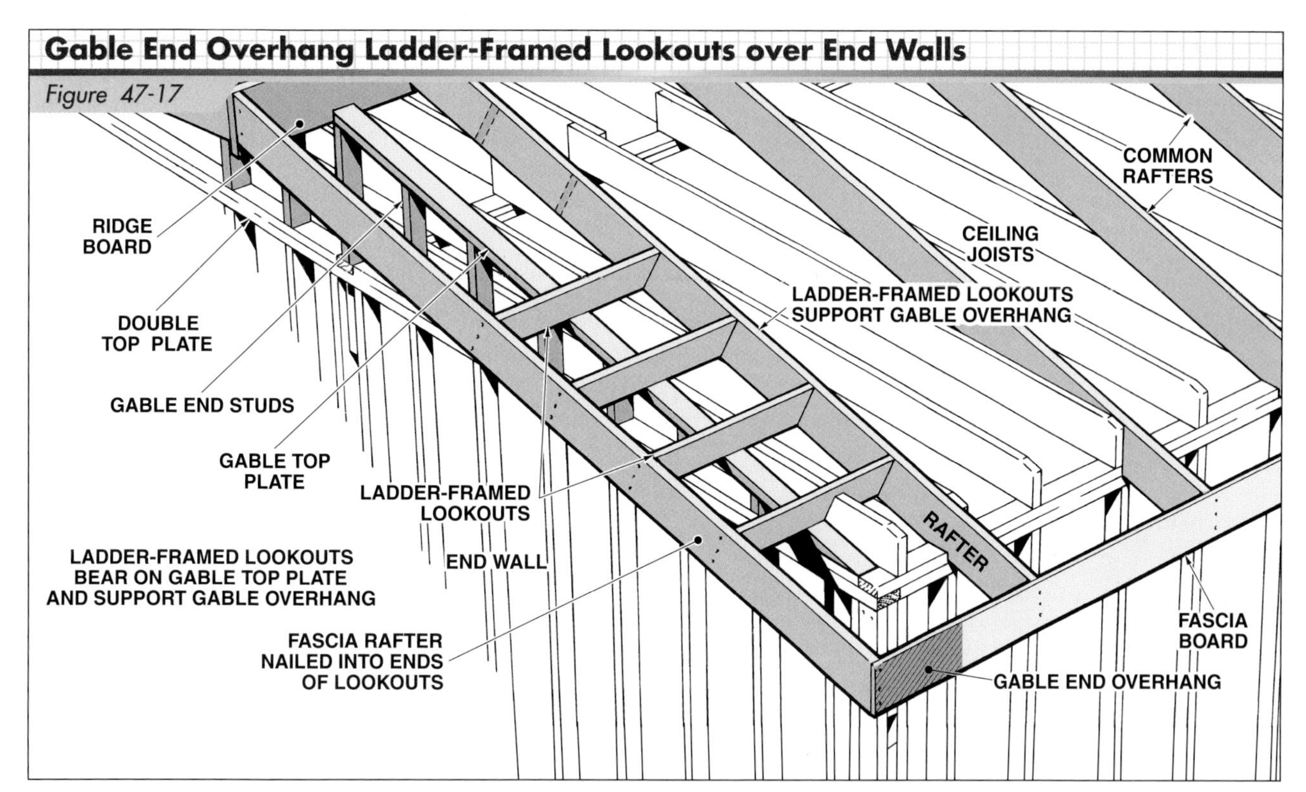

Figure 47-17. *A gable end overhang can be framed with lookouts and a fascia rafter. The fascia rafter is nailed to the ridge board and fascia board. Blocking rests on the end wall and is nailed between the fascia rafter and the rafter next to it. An overhang is further strengthened when the roof sheathing is nailed to it.*

Gable End Overhang Framed Directly over End Walls

Figure 47-18

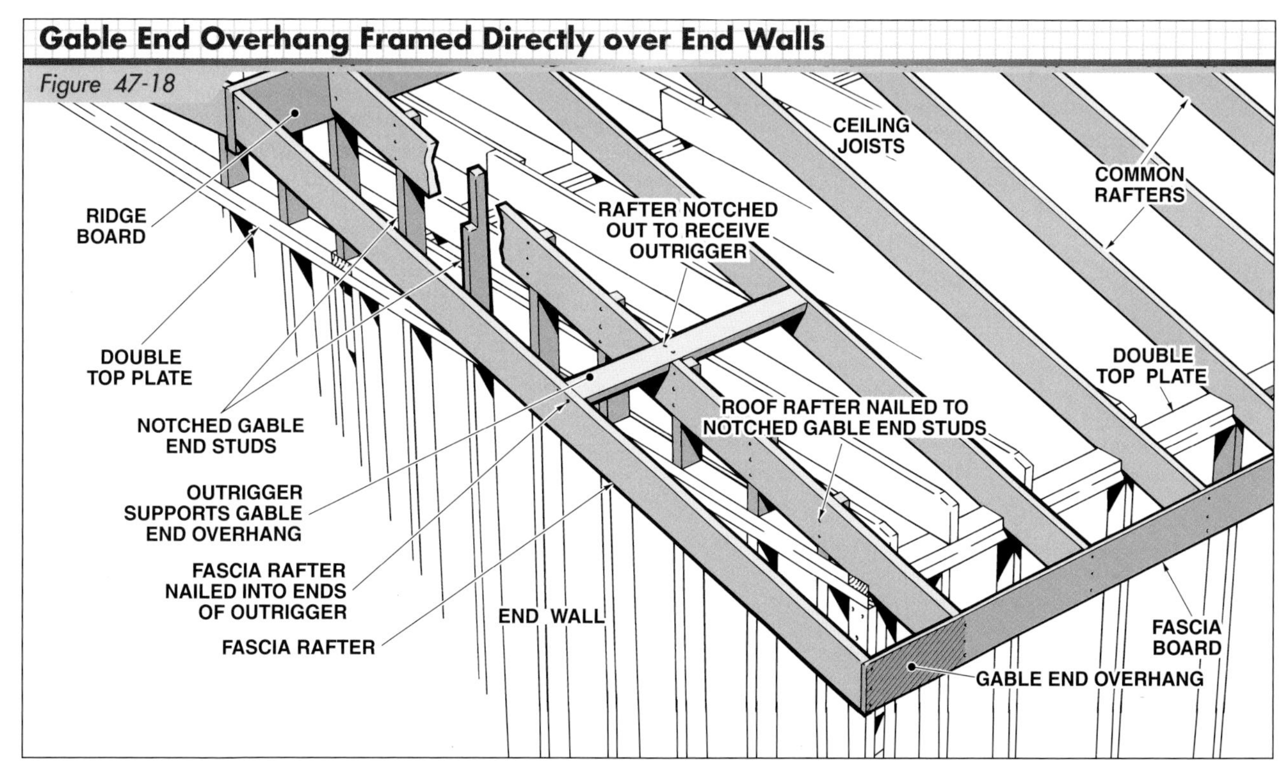

Figure 47-18. *A gable end overhang can be framed with a fascia rafter supported by the ridge board, outrigger, and fascia board. Two common rafters are placed directly over the gable ends of the building. The fascia rafters are installed between the ridge board and the fascia boards. The gable studs are notched to fit against the rafter above.*

Cutting and Placing Gable Studs. At each gable end, vertical members called gable studs are placed. Gable studs decrease in length from the ridge section toward the exterior side walls. A method for finding the common length difference for gable studs is shown in **Figure 47-19.** Another method, using a framing square, is shown in **Figure 47-20.** The common length difference can also be calculated mathematically by dividing the gable stud spacing by the unit run and then multiplying by the unit rise ($[16 \div 12] \times 8 = 10.67''$ or $10\frac{11}{16}''$). Gable studs also require an angle cut where they fit beneath a top plate or rafter. The framing square can be used to lay out the angle. **See Figure 47-21.** A finished gable stud will appear similar to the one shown in **Figure 47-22.**

Placing Purlins and Collar Ties. If purlins are required, they are nailed beneath the rafters after the roof framing is completed. The rafters should be aligned with a string while the posts are being fitted between the purlin and the supporting partition. Collar ties are also installed at this time.

GAMBREL ROOFS

A gambrel roof is basically a gable roof with a double slope on each side. The double slope is created by two common rafters that meet between the top of the wall and ridge. The rafters must be supported either by a wall or by purlins and posts where the rafters meet.

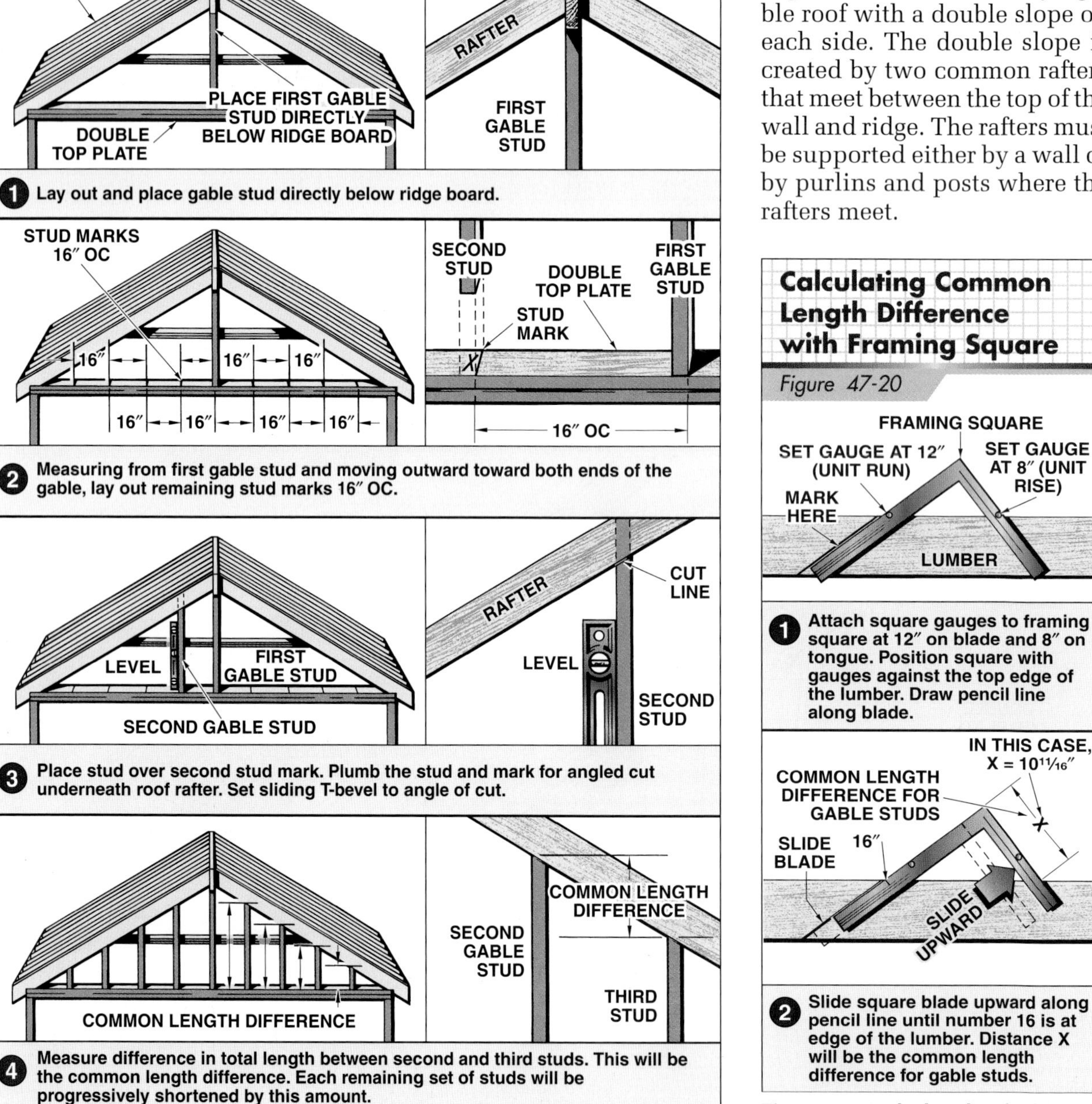

Figure 47-19. *Gable studs decrease in length from the ridge section toward the walls.*

Figure 47-20. *The lengths of gable studs can be calculated with a framing square.*

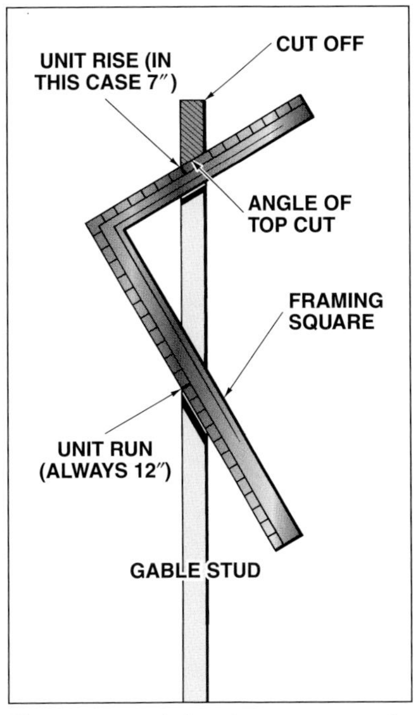

Figure 47-21. *A framing square can be used to find the angle of the top cut on gable studs. Combine the unit rise and 12″ and mark the cut along the unit rise side.*

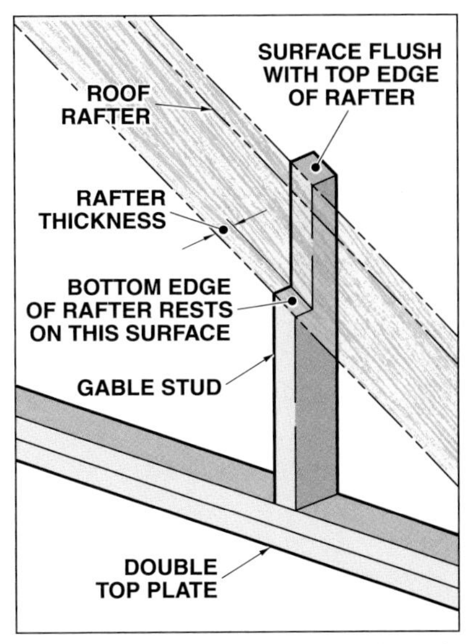

Figure 47-22. *Gable studs provide gable wall structure and support rafters along their bottom edges.*

Figure 47-23 shows a section view of a gambrel roof. The meeting points of the upper and lower rafters in this gambrel roof are supported by walls.

Calculating Length of Common Rafters

The procedures used to calculate the length of common rafters for gable roofs also apply to common rafters for gambrel roofs. The length of rafters for the upper portion of the roof is based on the total run for the upper portion, and the length of the rafters for the lower portion of the roof is based on the total run for the lower portion.

Laying Out Common Rafters

A procedure is shown in **Figure 47-24** for laying out common rafters for the gambrel roof shown in Figure 47-23. Angles for plumb cuts are marked at the ridge, tail, and heel of the common rafters. A seat cut is marked where the rafter rests on the wall. Finally, the overhang cut is marked.

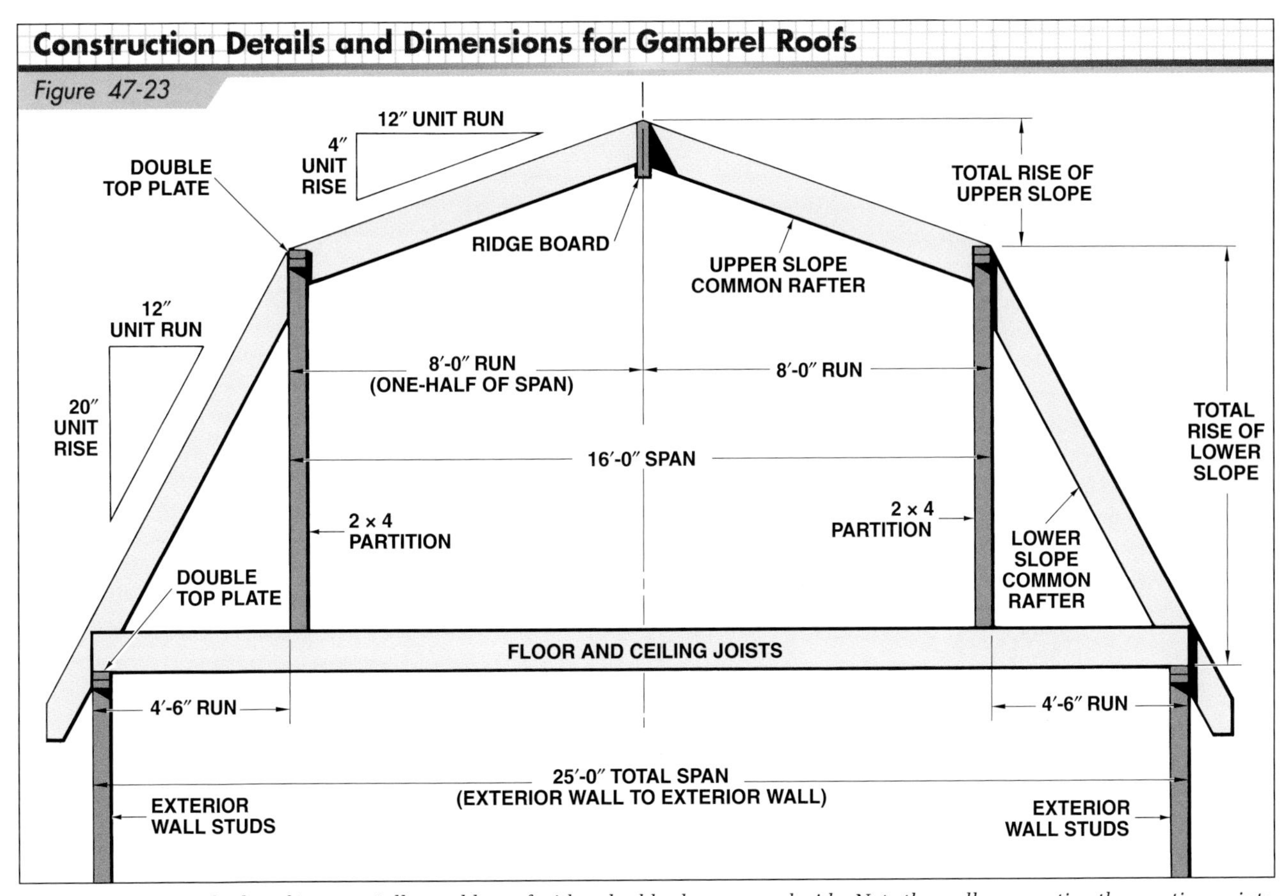

Figure 47-23. *A gambrel roof is essentially a gable roof with a double slope on each side. Note the walls supporting the meeting points of the upper and lower rafters. The length of the rafters for the lower portion of the roof is based on a 20″ unit rise and a 4′-6″ run. The length of the rafters for the upper section is based on a 4″ unit rise and an 8′-0″ run.*

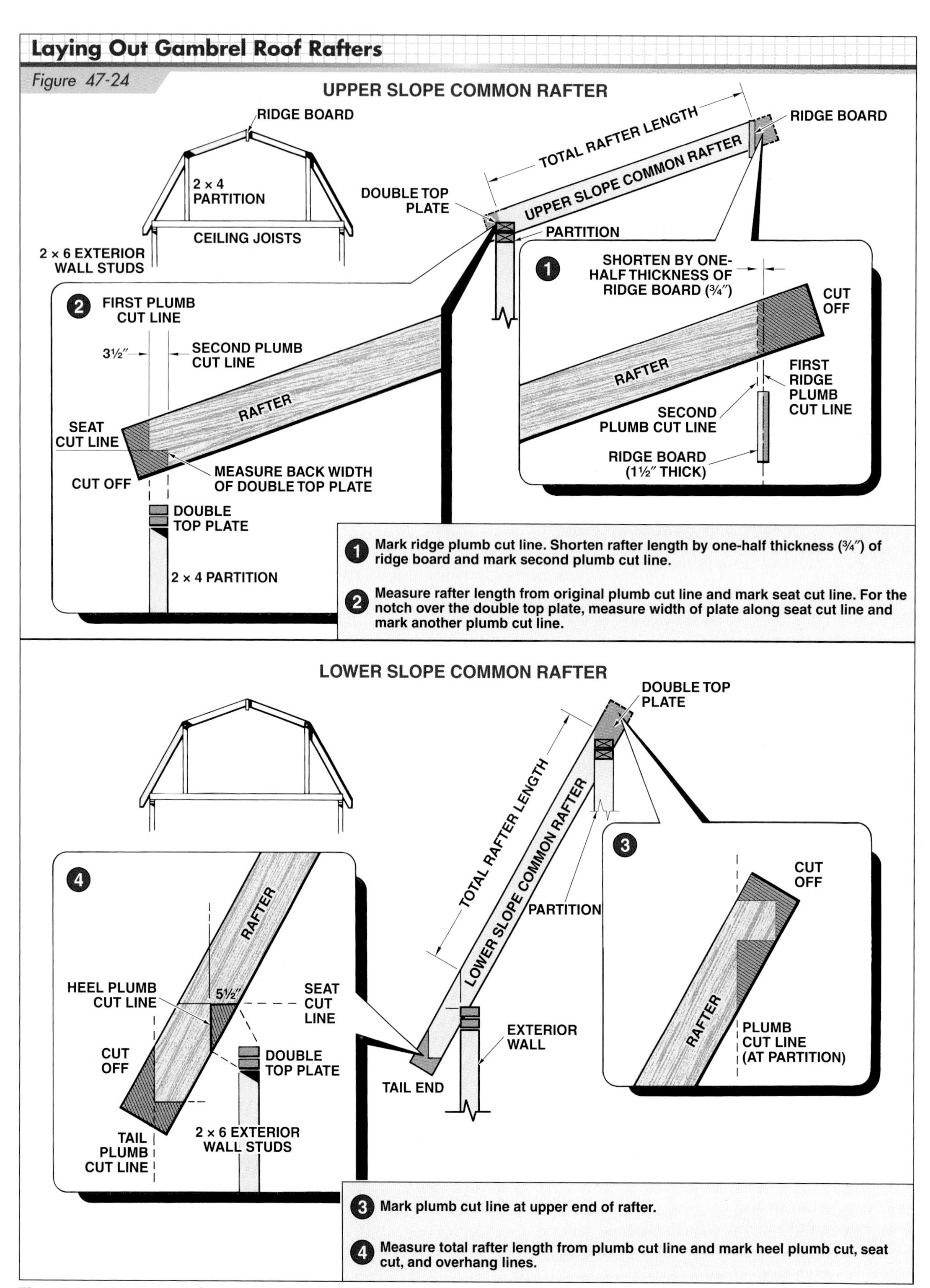

Figure 47-24. *Common rafters for the upper and lower portion of a gambrel roof are laid out differently. The cuts laid out in this illustration are based on the unit rises and runs shown in Figure 47-23.*

Constructing Gambrel Roofs

If the attic space provided by a gambrel roof is to be used for a living area, the subfloor should be installed before roof framing begins. The two outside rows of subfloor panels should be held back or cut so the seat of the rafters can be easily nailed to the top of the wall plates. One recommended procedure for framing a gambrel roof is shown in **Figure 47-25.**

SHED ROOFS

A shed roof has only one slope. **See Figure 47-26.** Common rafters for a shed roof are marked on each end for seat cuts where the rafters will rest on the two opposite walls of the building. The rafters are also marked for overhang cuts on each end. The length of the common rafters is based on the unit rise of the roof and the total run. The total run for a shed roof is the width of the building minus one wall thickness. A procedure for laying out a shed roof is shown in **Figure 47-27.**

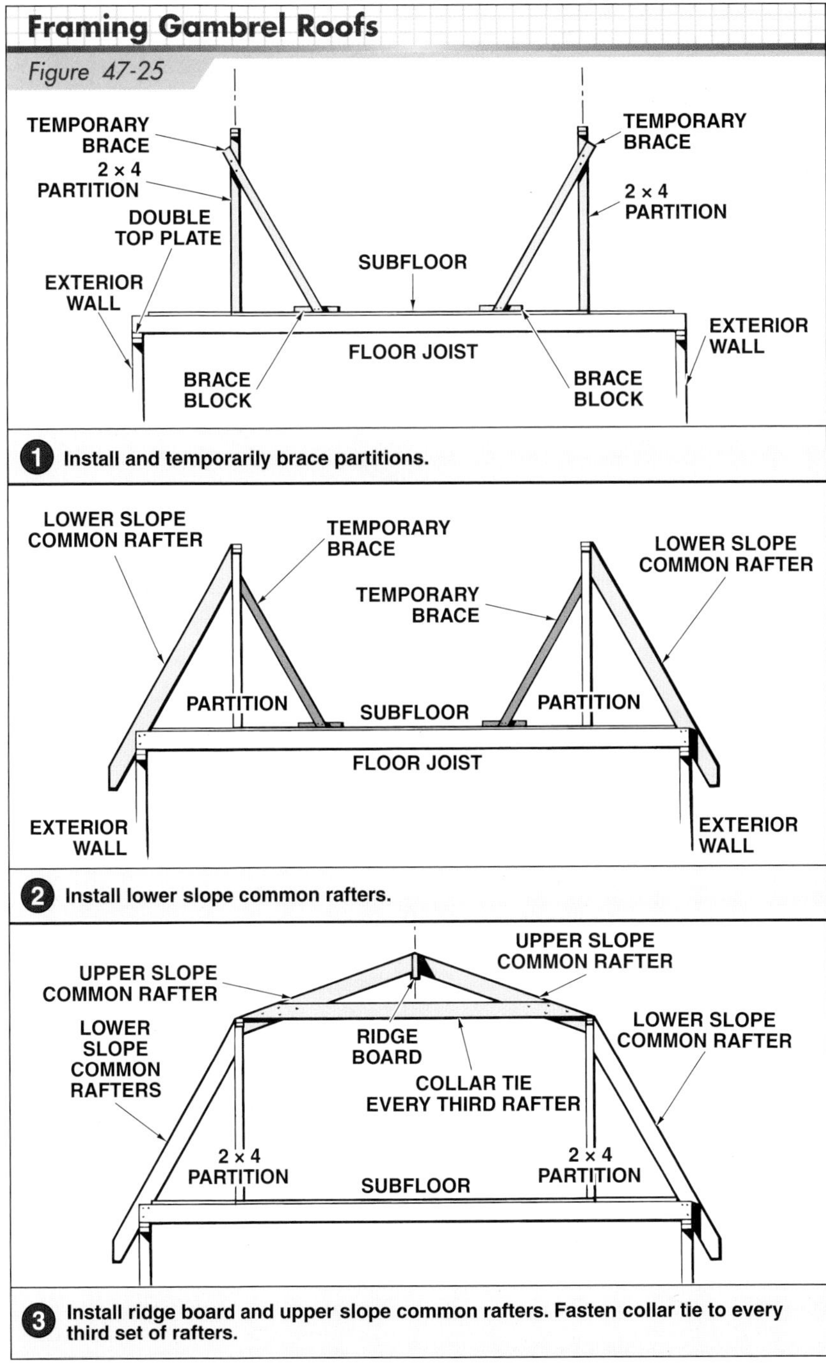

Figure 47-25. *When constructing a gambrel roof, the lower portion of the roof is constructed first. If the attic space is to be used for a living area, the subfloor should be installed before the roof is constructed.*

DORMERS

Dormers add space, and provide light and ventilation to a one-and-one-half-story house or to an attic area. Dormers require a roof with a steep slope and a high ridge. Gambrel roofs are particularly well suited for dormer construction.

Most dormers are of gable or shed design. The construction of a gable dormer is similar to the construction of a gable intersecting roof. The gable dormer consists of a ridge board, common rafters, valley rafters, and valley jack rafters. **See Figure 47-28.**

The front wall of a shed dormer is usually directly over the exterior wall of a building. The dormer rafters extend from the main ridge. **See Figure 47-29.** They must be pitched enough to shed water and snow.

When an opening is framed for a dormer roof, the rafters on both sides of the opening are doubled. Also, double headers are placed at the top and bottom of the opening.

The original purpose of a gambrel roof was economy; the less-pitched upper section required less lumber than would be needed if the steeper section was extended to a peak. Dutch gambrel roofs use shorter upper slope common rafters. Queen Anne gambrel roof rafters extend close to the ground.

Typical Shed Roof Rafter with Overhang

Figure 47-26

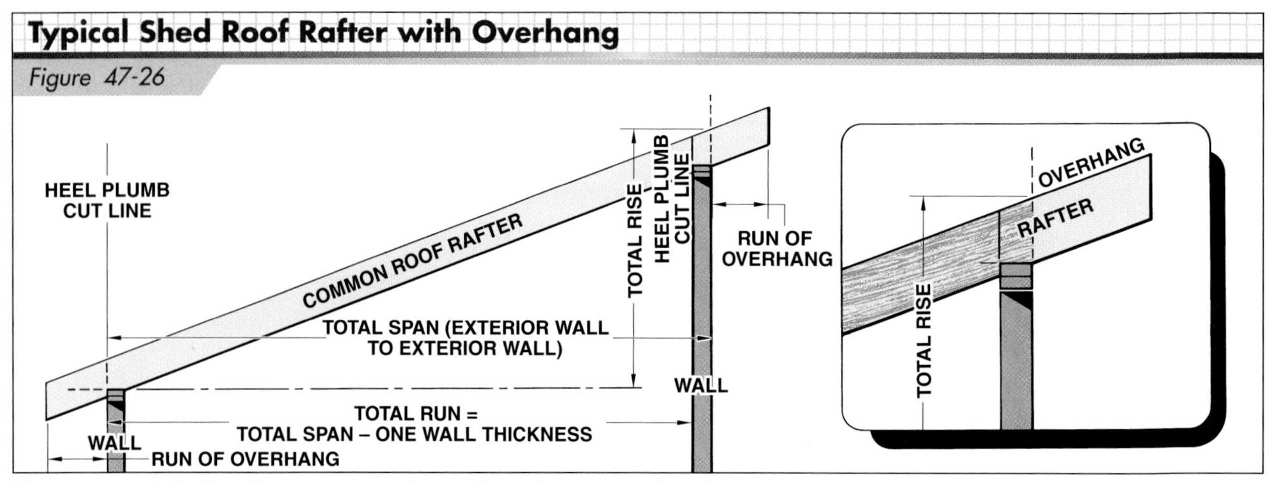

Figure 47-26. *A shed roof has common rafters with overhangs at each end.*

Laying Out Shed Roof Rafters

Figure 47-27

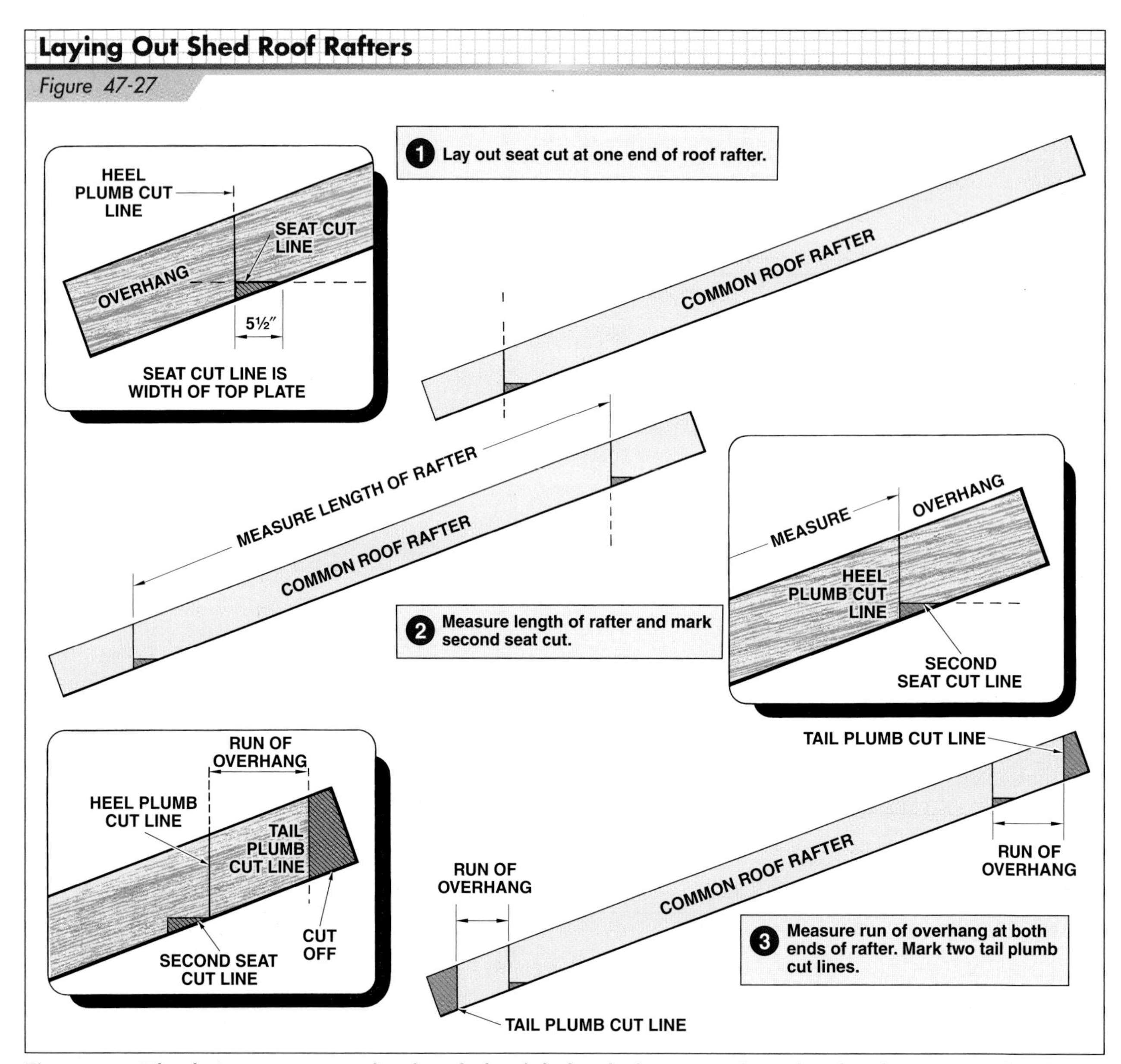

Figure 47-27. *When laying out common rafters for a shed roof, the length of common rafters is based on the unit rise and total run of the roof.*

Constructing Gable Dormers

Figure 47-28

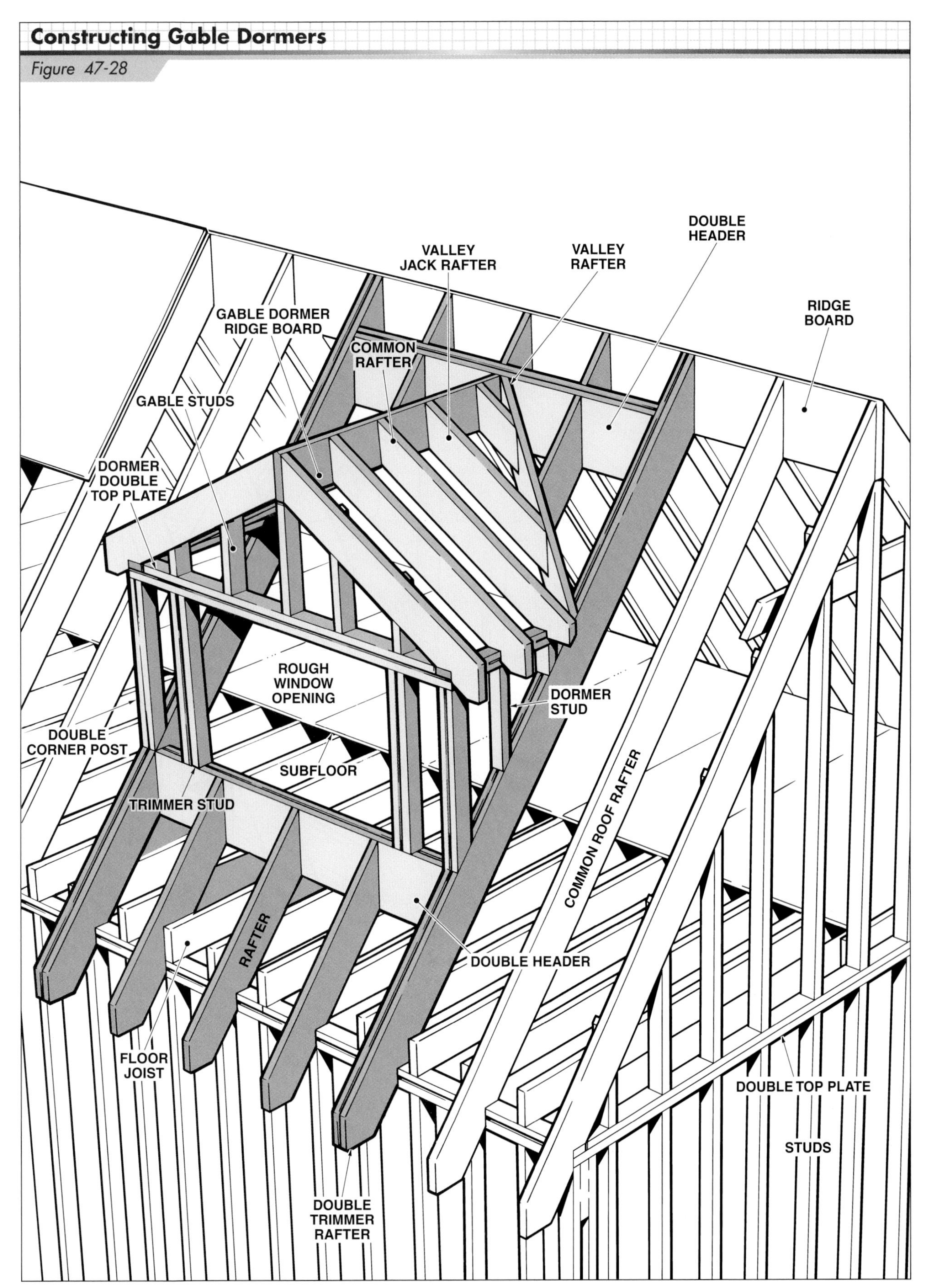

Figure 47-28. *The construction of a gable dormer is similar to that of a gable intersecting roof. A gable dormer has a level ridge board.*

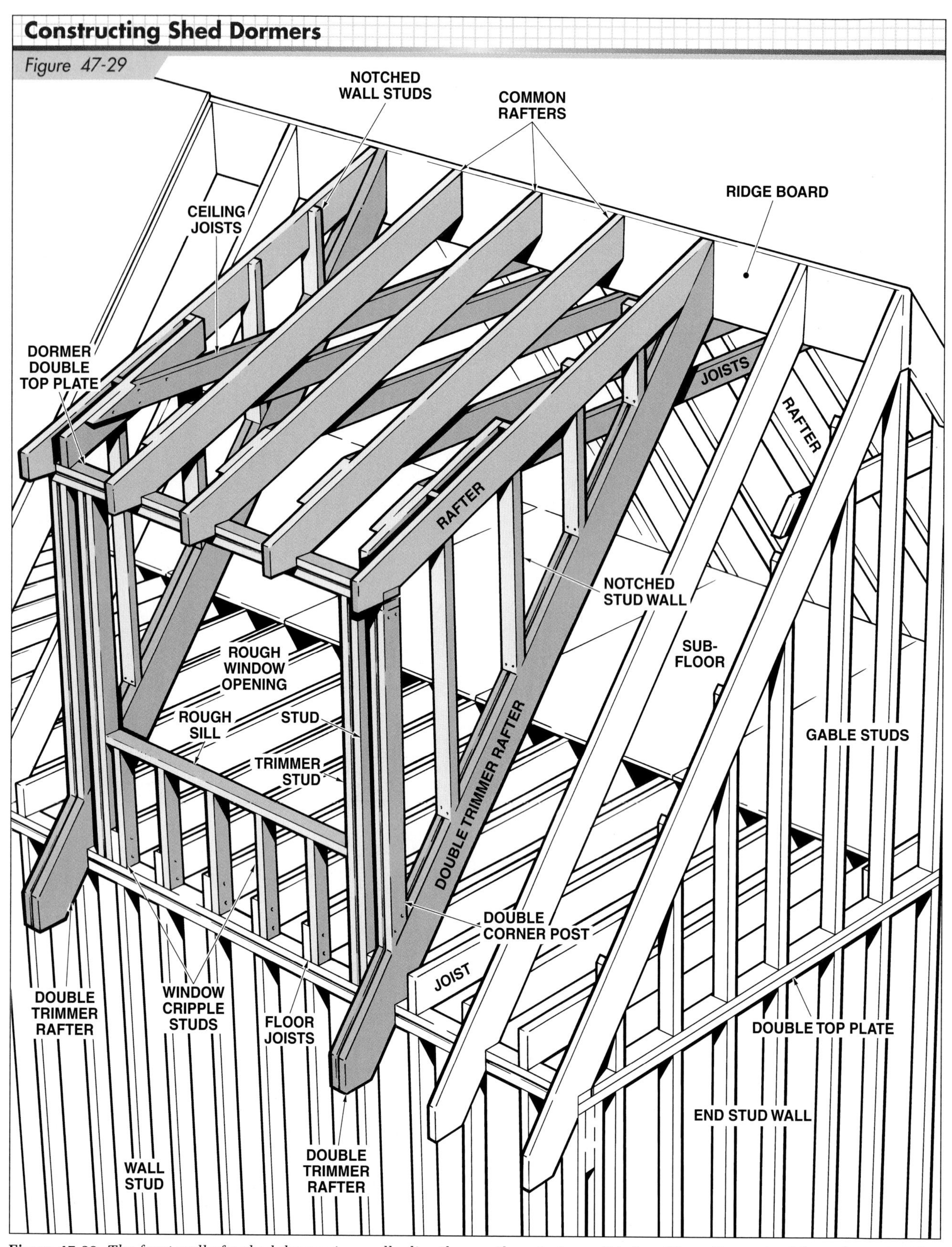

Figure 47-29. *The front wall of a shed dormer is usually directly over the exterior wall below. The rafters extend from the main ridge board. The rafters must be pitched enough to shed water and snow.*

Unit 47 Review and Resources

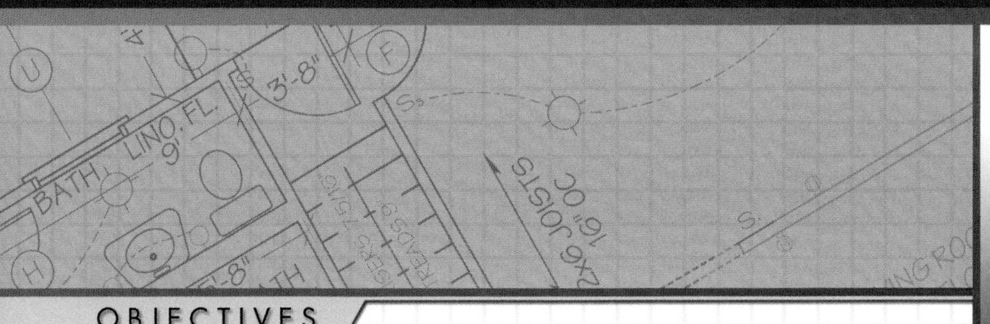

UNIT 48 Hip Roofs

OBJECTIVES

1. Identify general features of hip roofs and hip rafters.
2. Explain the procedure for calculating hip rafter lengths.
3. Describe the procedure for laying out hip rafters.
4. Explain the procedure for calculating hip jack rafter lengths.
5. Describe the procedure for laying out hip jack rafters.
6. Discuss the procedure for framing hip roofs.

A hip roof has four sloping sides. Four *hip rafters* run at a 45° angle from the corners of the building to the ridge board. *Hip jack rafters* frame the space between the hip rafters and the tops of the exterior walls. **See Figure 48-1.**

Common rafters for hip roofs extend from the ridge board to the wall plates, similar to gable roofs. The *king common rafter* extends from the ends of the ridge board to the top plate. Another common rafter, commonly called the *side king common rafter*, extends from the end of the ridge board at a 90° angle.

HIP RAFTERS

A hip rafter travels at a diagonal (45° angle on a plan view) to reach the ridge board and is longer than a common rafter. Hip rafters differ from common rafters in other ways. In addition to plumb cuts at the ridge, heel, and tail, a hip rafter requires *side cuts* where it meets the ridge. Side cuts are also necessary at the tail in order for the overhang of the hip rafters to align with the overhang of the common rafters. **See Figure 48-2.** The procedure for marking rafter side cuts is explained later in this unit. Layout is usually not done until the rafter lengths are calculated.

The unit run of a hip rafter is 17″, compared with 12″ for a common rafter. Since a hip rafter runs at a 45° angle to the common rafter, its unit run is calculated using the diagonal of a 12″ square. The diagonal of a 12″ square is 16.97″, which rounds up to 17″. A hip rafter must run 17″ in order to reach the same height that a common rafter reaches in 12″. **See Figure 48-3.**

Over 90% of structural lumber produced in North America comes from the Douglas Fir/Larch, Hemlock/Fir, Southern Pine, and Spruce/Pine/Fir species groups.

Hip Roof Framework Construction

Figure 48-1

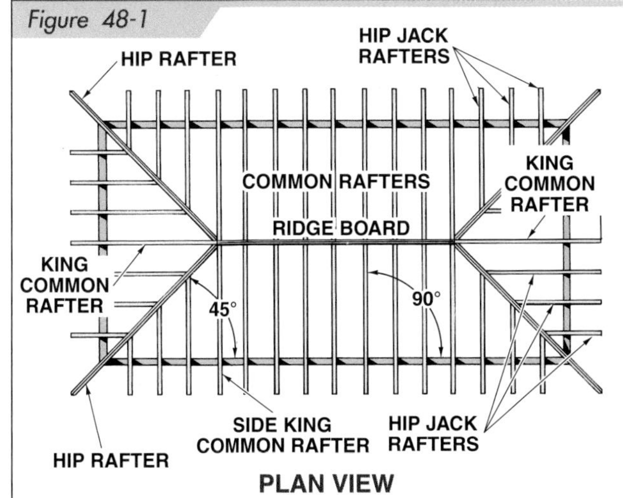

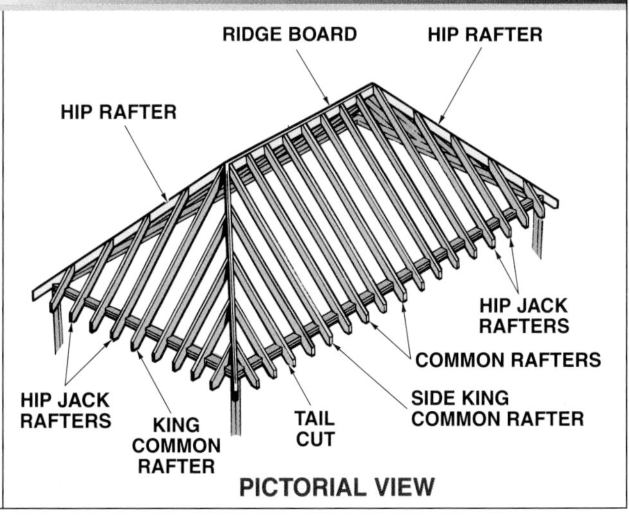

Figure 48-1. *A hip roof has four sloping sides. Four hip rafters run at a 45° angle from the exterior walls. Hip jack rafters frame the space between the hip rafters and the top of the exterior walls.*

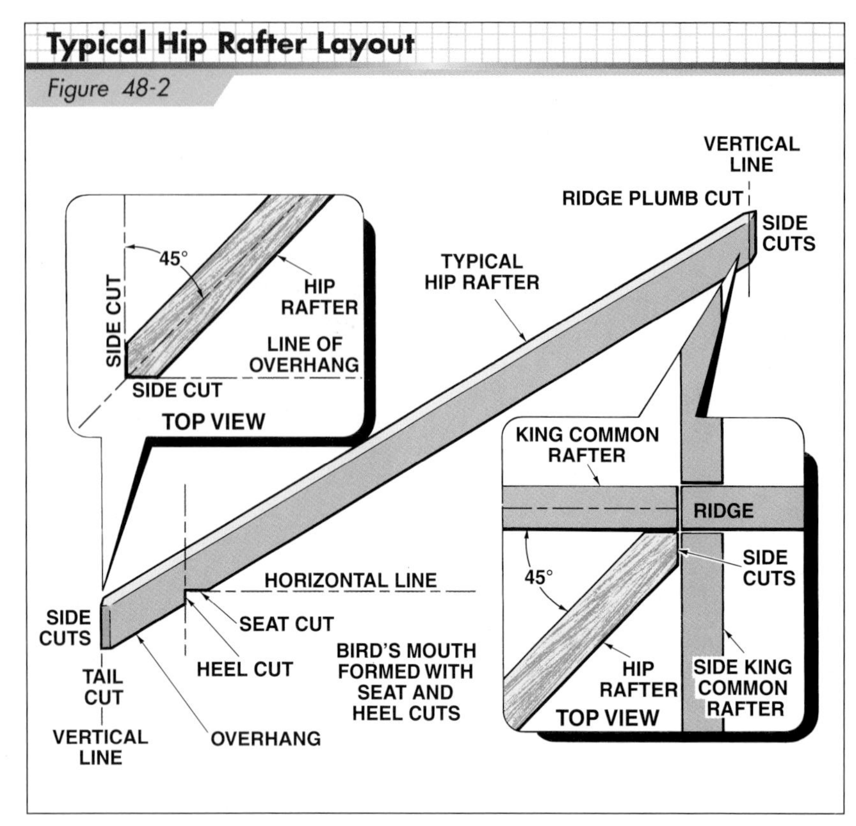

Figure 48-2. *Side cuts are required at the ridge and tail of a hip rafter.*

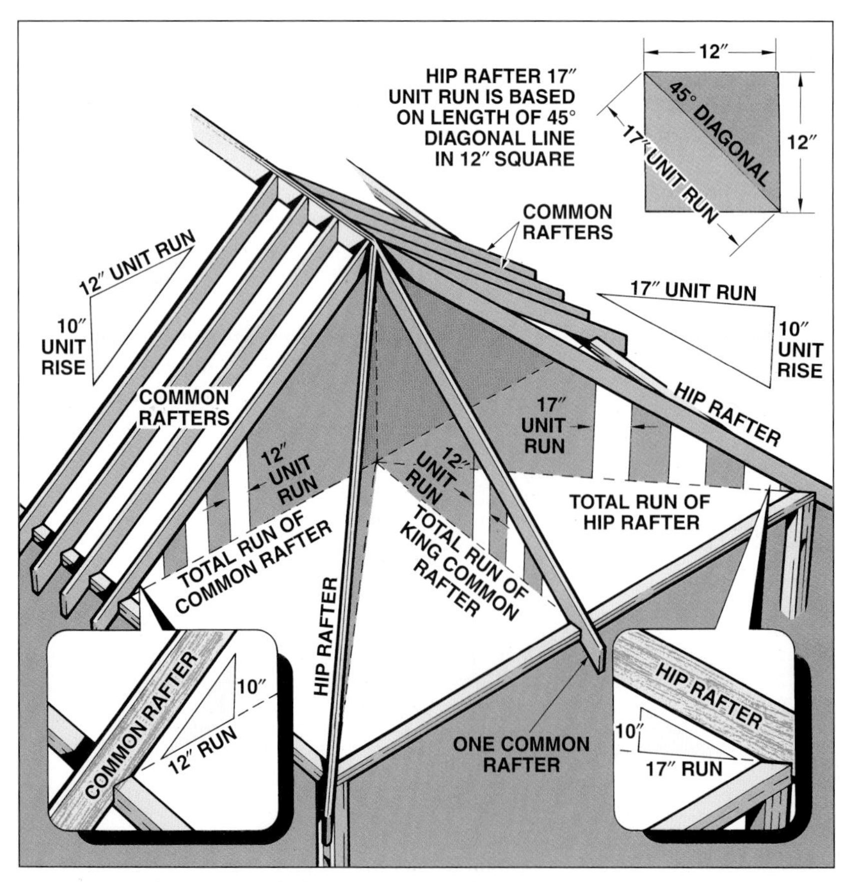

Figure 48-3. *The unit run of a hip rafter is 17″, which is the 45° diagonal of a 12″ square.*

Traditionally, the framing square has been used to lay out angled cuts and lengths on hip and hip jack rafters. The Speed® Square is now also being used to lay out these rafters. Hip and hip jack rafter layout using the framing square and Speed® Square are the same and are both discussed in this unit.

Calculating Hip Rafter Lengths

Hip rafter lengths, like common rafter lengths, can be calculated by four methods—the framing square rafter table, a book of rafter tables, the step-off method that combines laying out with calculating length, and through mathematical calculations.

Framing Square Rafter Table. A framing square imprinted with a rafter table can be used for calculating hip rafter lengths using the same procedure as used for common rafters, except that the second line of the table is used rather than the first line. The second line shows the *length of hip or valley per foot run.* The mathematical procedure is shown in the following example.

Example. The roof has a 6″ unit rise and a 28′ total span.

1. Divide the total span (28′) by 2 to determine the total run.

   28′ ÷ 2 = 14′ total run

2. Look below the number "6" at the top of the framing square. On the second line, the number "18" is given as the length of hip or valley per foot of run.
3. Multiply the length of hip per foot of run (18″) by the total run (14′).

   18′ × 14″ = **252″** or **21′-0″**

Book of Rafter Tables. Hip rafter lengths can also be calculated using a book of rafter tables as described in Unit 47.

Shortening Hip Rafters. All methods for calculating hip rafter length discussed in this text use the theoretical length rather than the actual length. The theoretical length is the distance from the heel plumb cut line to the center of the ridge board. To determine the actual length, one-half the diagonal thickness of the ridge board that fits between the rafters must be subtracted from the theoretical length. For example, the diagonal thickness of a 1½″ thick ridge board is 2⅛″. Therefore, 1¹⁄₁₆″ is subtracted from the theoretical length (2⅛″ ÷ 2 = 1¹⁄₁₆″) at a right angle to the plumb line. Hip roofs may or may not be framed with king common rafters. If king common rafters are used, one-half the diagonal thickness of the common rafter must be deducted from the hip rafter. The procedure for shortening hip rafters is shown in **Figure 48-4.**

Laying Out Hip Rafters

The layout procedure for hip rafters begins with marking the ridge plumb and side cuts. Next, the seat and heel plumb cuts are marked, followed by the lines showing the overhang. A procedure for laying out hip rafters using a framing square or Speed® Square is shown in **Figure 48-5.**

Plumb and Seat Cuts. The 17″ unit run of the hip rafter makes the angles of the plumb and seat cuts different from those of a common rafter. To mark the plumb and seat cuts, use the unit rise measurement on the tongue of the framing square and the 17″ measurement on the blade. Mark the plumb cut along the tongue and the seat cut along the blade.

When positioning the square to mark the seat cut of a hip rafter, first check the amount of stock left above the seat cut at the heel plumb cut line of the common rafter for the roof. Lay out the same distance at the heel plumb line of the hip rafter and draw the seat cut line. The length of the hip rafter seat cut is greater than the seat cut of a common rafter because a hip rafter rests diagonally at the corner of a building and has a lower angle of slope.

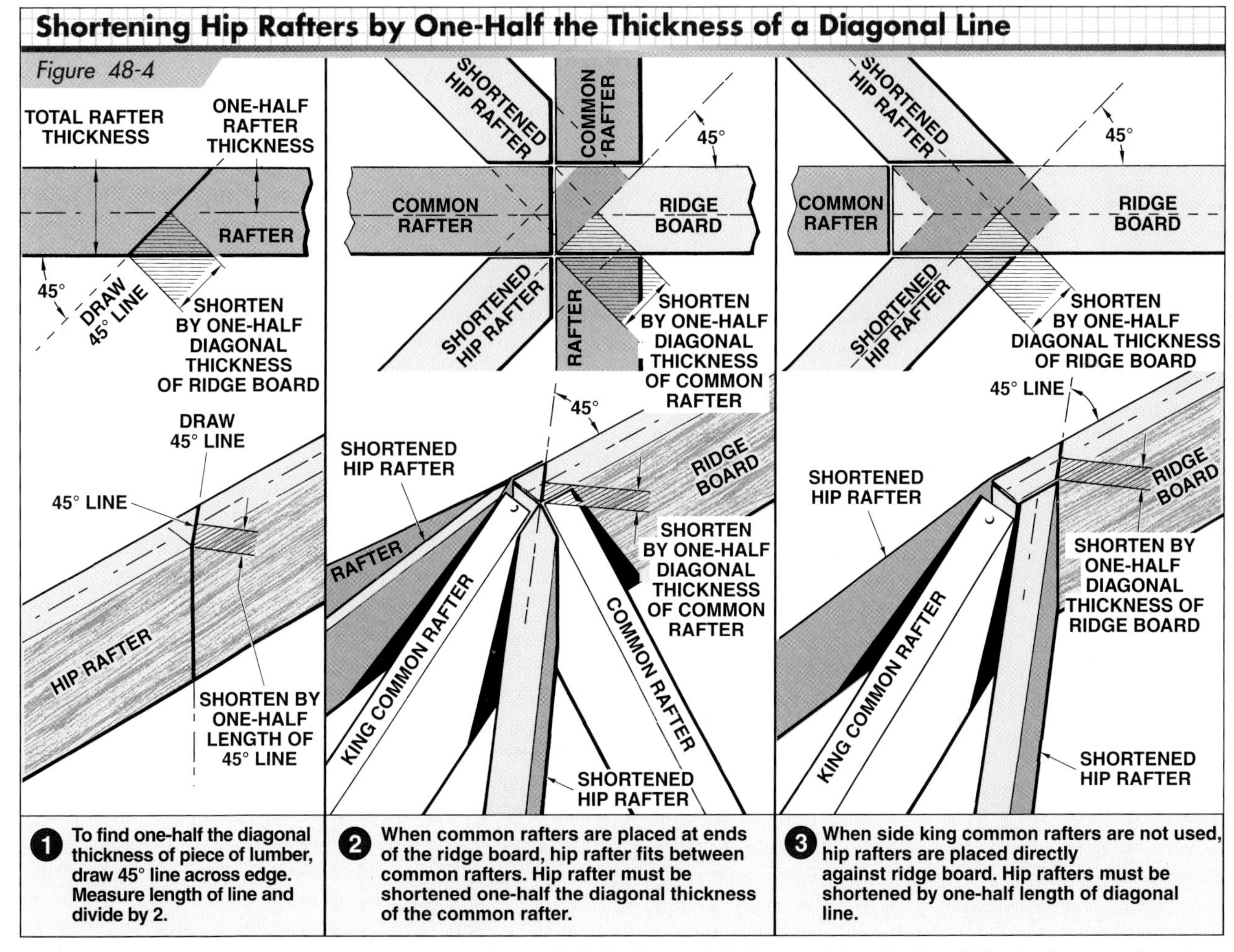

Figure 48-4. *To calculate the actual hip rafter length, one-half the diagonal thickness of the ridge board that fits between the rafters is deducted from the theoretical length. When shortening the hip rafters, make the measurement perpendicular to the plumb line, not along the length of the roof member.*

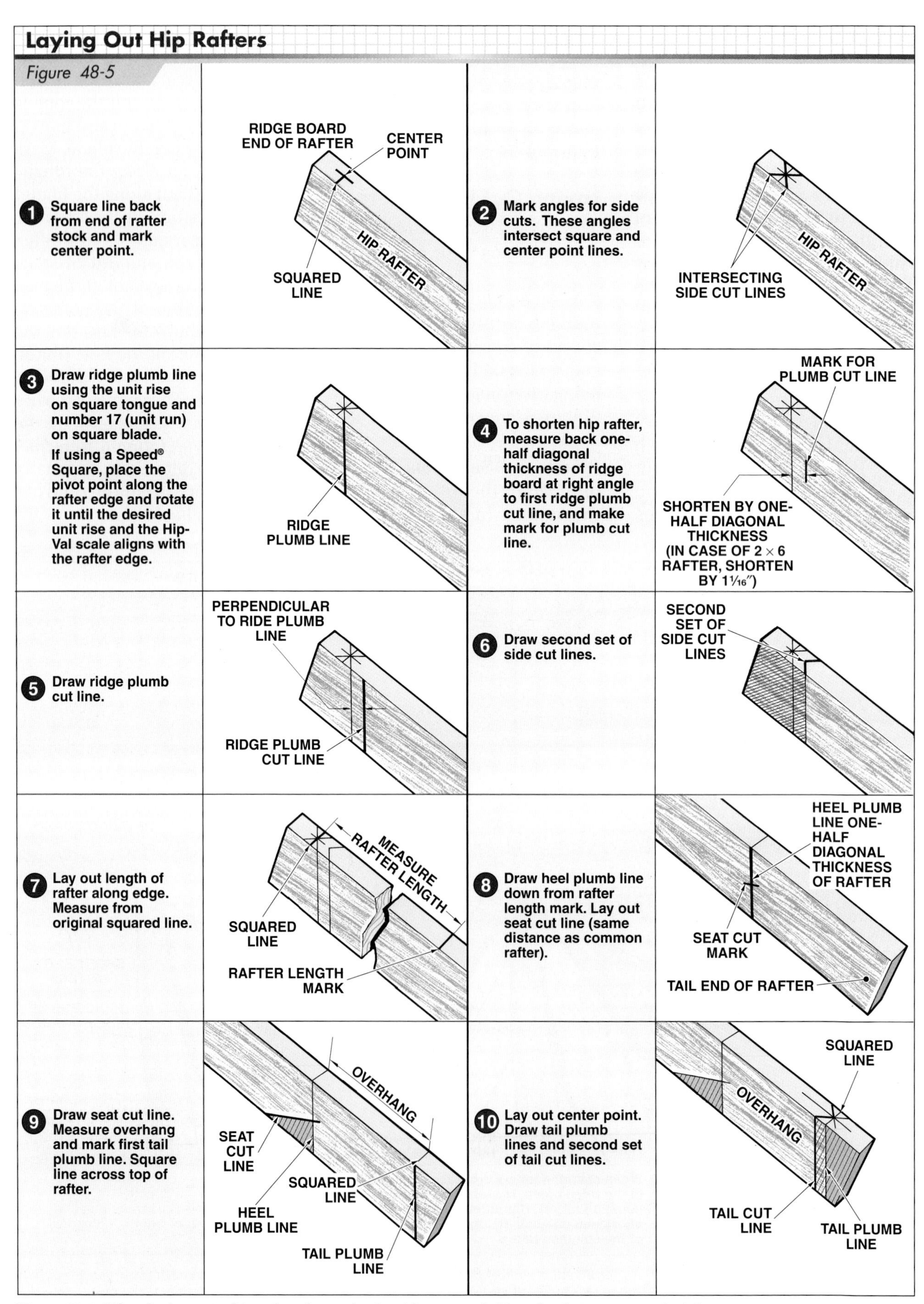

Figure 48-5. *When laying out a hip rafter, the angles for side cuts and ridge plumb cuts are marked first.*

Side Cuts. For a 45° hip roof, a circular saw is always set to 45°. This blade angle will produce the proper 45° angle cut in plan view. When viewed perpendicular to the edge of the rafter, the angle will lay out at less than 45°. The rafter table on a framing square can be used to determine the exact angle for the side cut of a hip rafter when laid out on the top edge of the rafter. The procedure for determining the proper angle is as follows:

1. Locate the unit rise (7″) along the blade above the rafter tables. **See Figure 48-6.**
2. Follow the column down to the last line, which indicates the "Side Cut Hip or Valley." The number is 11¹⁄₁₆″.
3. Using the 11¹⁄₁₆″ on the tongue and 12″ on the blade, place the framing square along the edge of a piece of the rafter.
4. Mark the side cut along the 12″ side.

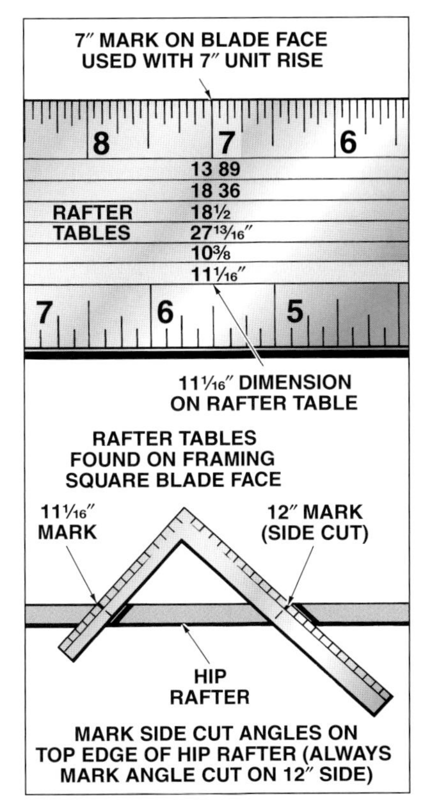

Figure 48-6. *The rafter table on a framing square can be used to determine the exact angle for the side cut of a hip rafter. The roof in this example has a 7″ unit rise.*

Overhang. Since a hip rafter runs at a diagonal, its overhang is longer than the common rafter overhang. The run of a hip rafter overhang is 1.42″ for every 1″ of common rafter overhang. Therefore, to find the run of the hip rafter overhang, multiply 1.42″ by the run of the common rafter overhang. **See Figure 48-7.**

A framing square can also be used to calculate the run of the hip rafter overhang. Take a diagonal measurement from two points that are equal to the sides of a square formed by the common rafter overhang. **See Figure 48-8.**

Step-off Method for Calculating Length and Laying Out Hip Rafters

The step-off method described in Unit 47 for common rafters can also be used for hip rafters. The framing square is set up with the unit rise measurement on the tongue and the 17″ (rather than 12″) measurement on the blade. Use extreme care when using the step-off method to ensure accuracy.

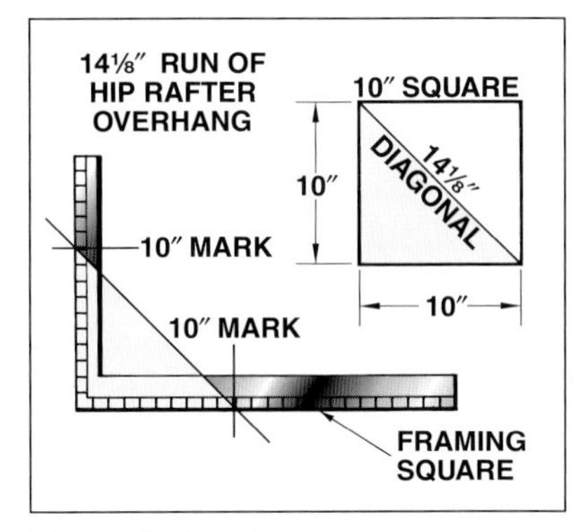

Figure 48-8. *A framing square can be used to calculate the run of the hip rafter overhang. In this example, the run of the roof overhang is 10″.*

Backing or Dropping Hip Rafters

Chamfering the top edges of a hip rafter is called *backing* the rafters. **See Figure 48-9.** Backing prevents roof sheathing from being higher where it covers hip rafters than where it covers common and jack rafters.

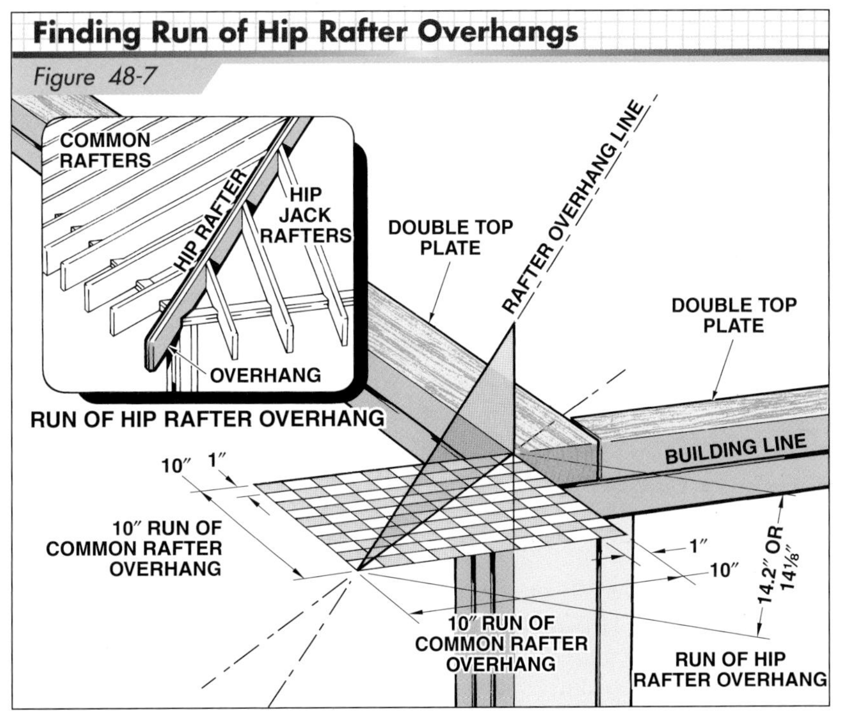

Figure 48-7. *A hip rafter overhang runs 1.42″ for every 1″ of common rafter run. In this example, the run of the common rafter is 10″. To find the run of the hip overhang, multiply 1.42″ by 10. The result is 14.2″, or 14⅛″.*

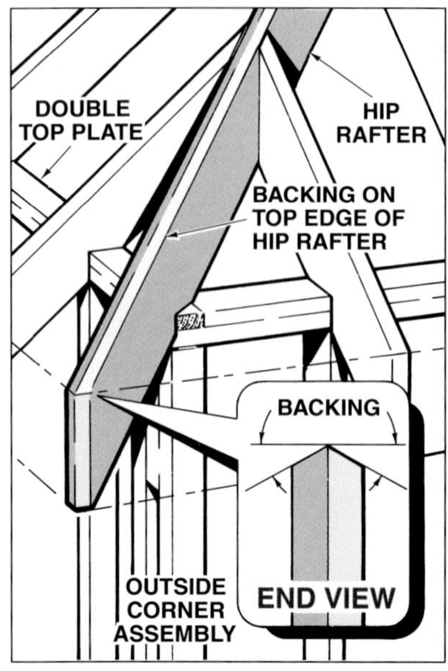

Figure 48-9. *Backing a hip rafter allows roof sheathing to lie in the same plane as common and jack rafters.*

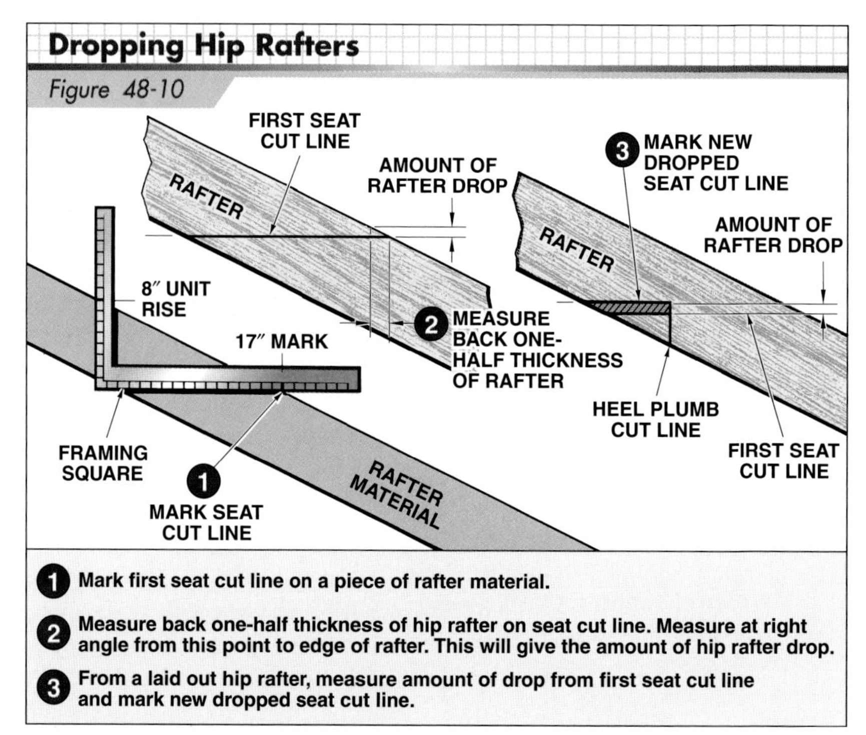

Figure 48-10. *Dropping a hip rafter is faster than backing a rafter and accomplishes the same purpose.*

Another method to prevent roof sheathing from being higher over hip rafters is *dropping* the hip rafters. **See Figure 48-10.** The seat cut is enlarged, causing the rafter to drop. Consequently, the sheathing rests on the top corners of the rafter and is in line with the roof. Most carpenters use the dropping method because it is faster than the backing method.

HIP JACK RAFTERS

Hip jack rafters frame the space between hip rafters and wall plates. Hip jack rafters run in pairs and are spaced the same distance apart as common rafters. If common rafters are spaced 24″ OC, the hip jack rafters are spaced 24″ OC.

Common Length Difference

Hip jack rafters decrease in length as they get closer to the end of a building. Hip jack rafters have a *common length difference* as long as they are equally spaced. When the length of one hip jack rafter is known, the length of other hip jack rafters can be found by subtracting or adding the common length difference.

The common length difference of hip jack rafters can be easily calculated using the framing square rafter table. The common length difference is expressed as the "difference in length of jacks." Measurements on the third line of a rafter table are for jack rafters spaced 16″ OC. Measurements on the fourth line are for jack rafters spaced 2′ (24″) OC.

Laying Out Hip Jack Rafters

As shown in **Figure 48-11,** layout for hip jack rafter placement may begin from a common rafter at the end of the ridge board. Layout may also begin from a common rafter located at some point other than the end of the ridge board. **See Figure 48-12.** In a third hip jack rafter layout method, layout begins at the corner of the building. **See Figure 48-13.**

Diagonal bracing is used to keep the exterior walls straight and plumb until the rafters or trusses are installed.

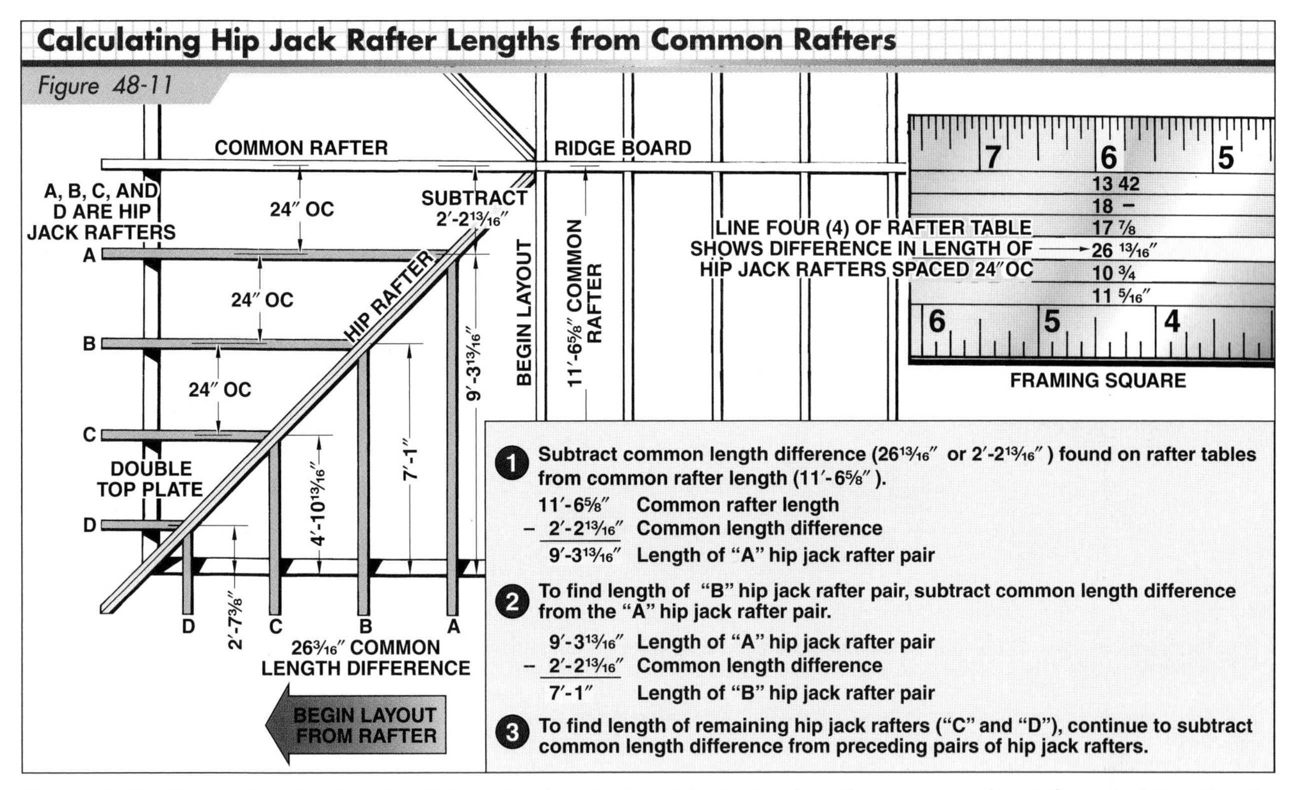

Figure 48-11. *When calculating lengths of hip jack rafters for layout beginning from the common rafter at the end of the ridge, the common length difference must first be determined. In this example, the roof has a 6" unit rise and a 20'-8" span. The length of common rafters is 11'-6⅝". The hip jack rafters are spaced 24" OC. The common length difference is 26¹³⁄₁₆".*

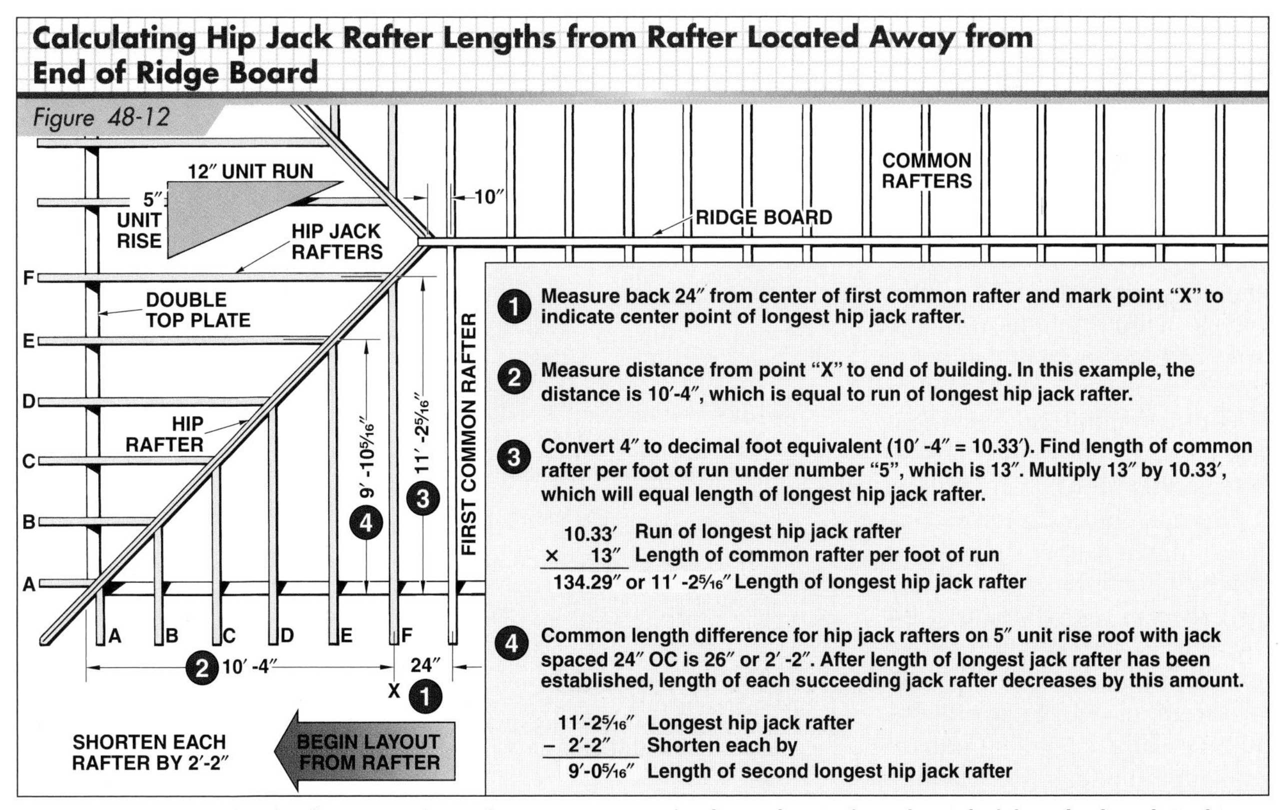

Figure 48-12. *Hip jack rafter layout can begin from a common rafter located away from the end of the ridge board. In this example, the roof has a 5 unit rise and 23'-0" span. The hip jack rafters are spaced 24" OC. The common length difference is 2'-2".*

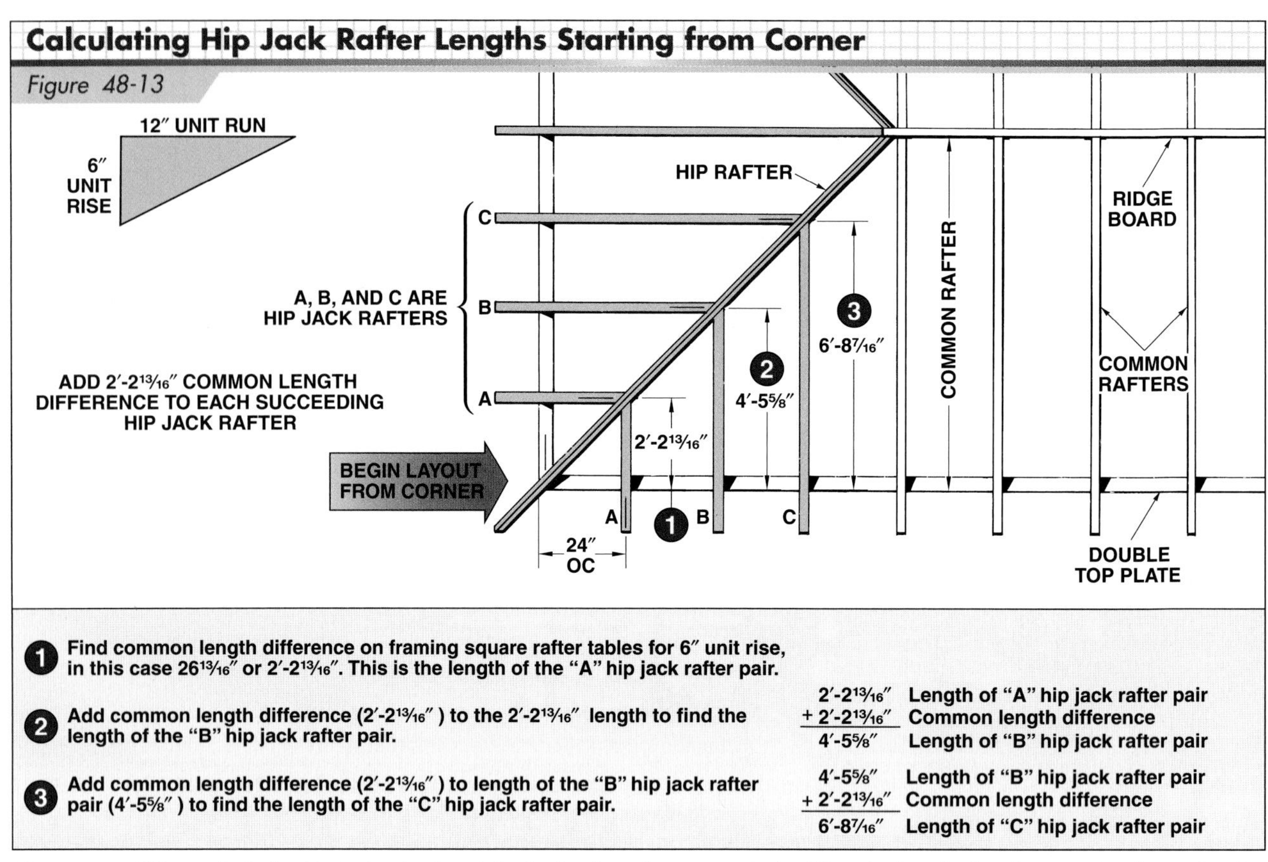

1. Find common length difference on framing square rafter tables for 6″ unit rise, in this case 26¹³⁄₁₆″ or 2′-2¹³⁄₁₆″. This is the length of the "A" hip jack rafter pair.	2′-2¹³⁄₁₆″	Length of "A" hip jack rafter pair
2. Add common length difference (2′-2¹³⁄₁₆″) to the 2′-2¹³⁄₁₆″ length to find the length of the "B" hip jack rafter pair.	+ 2′-2¹³⁄₁₆″	Common length difference
	4′-5⅝″	Length of "B" hip jack rafter pair
3. Add common length difference (2′-2¹³⁄₁₆″) to length of the "B" hip jack rafter pair (4′-5⅝″) to find the length of the "C" hip jack rafter pair.	4′-5⅝″	Length of "B" hip jack rafter pair
	+ 2′-2¹³⁄₁₆″	Common length difference
	6′-8⁷⁄₁₆″	Length of "C" hip jack rafter pair

Figure 48-13. *When calculating hip jack rafter length beginning from the corner of a building, the length of each succeeding hip jack rafter length is increased by the common length difference. In this example, the roof has a 6″ unit rise. The hip jack rafters are spaced 24″ OC. The 26¹³⁄₁₆″ or 2′-2¹³⁄₁₆″ common length difference shown in the rafter table is the length of the first hip jack rafter placed 24″ OC from the corner of the building.*

The hip jack rafter has plumb cuts where it fastens to the hip, as well as at the heel and tail. A seat cut is made where it rests on the plate. The plumb and seat cuts can be laid out using the unit rise and 12″ on the framing square, as is done for the common rafter. The plumb cut is marked on the unit rise side of the square and the seat cut is marked on the 12″ side. The bird's mouth and overhang are marked the same way they are marked for common rafters.

Hip jack rafters require a single side cut where they fasten to the hip rafter. A framing square can be used to find the angle of the side cut as viewed and laid out perpendicular to the top edge of the rafter. **See Figure 48-14.** A procedure for laying out the cuts on a hip jack rafter using a framing square or Speed® Square is shown in **Figure 48-15.**

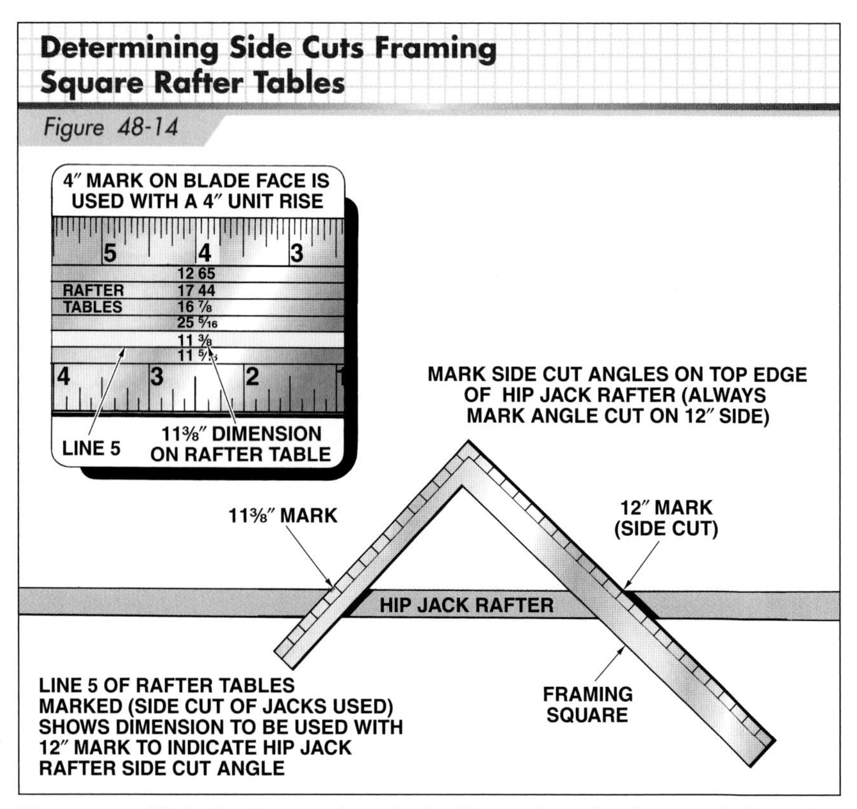

Figure 48-14. *Hip jack rafters require a single side cut where they fasten to the hip rafter. In this example, the roof has a 4″ unit rise.*

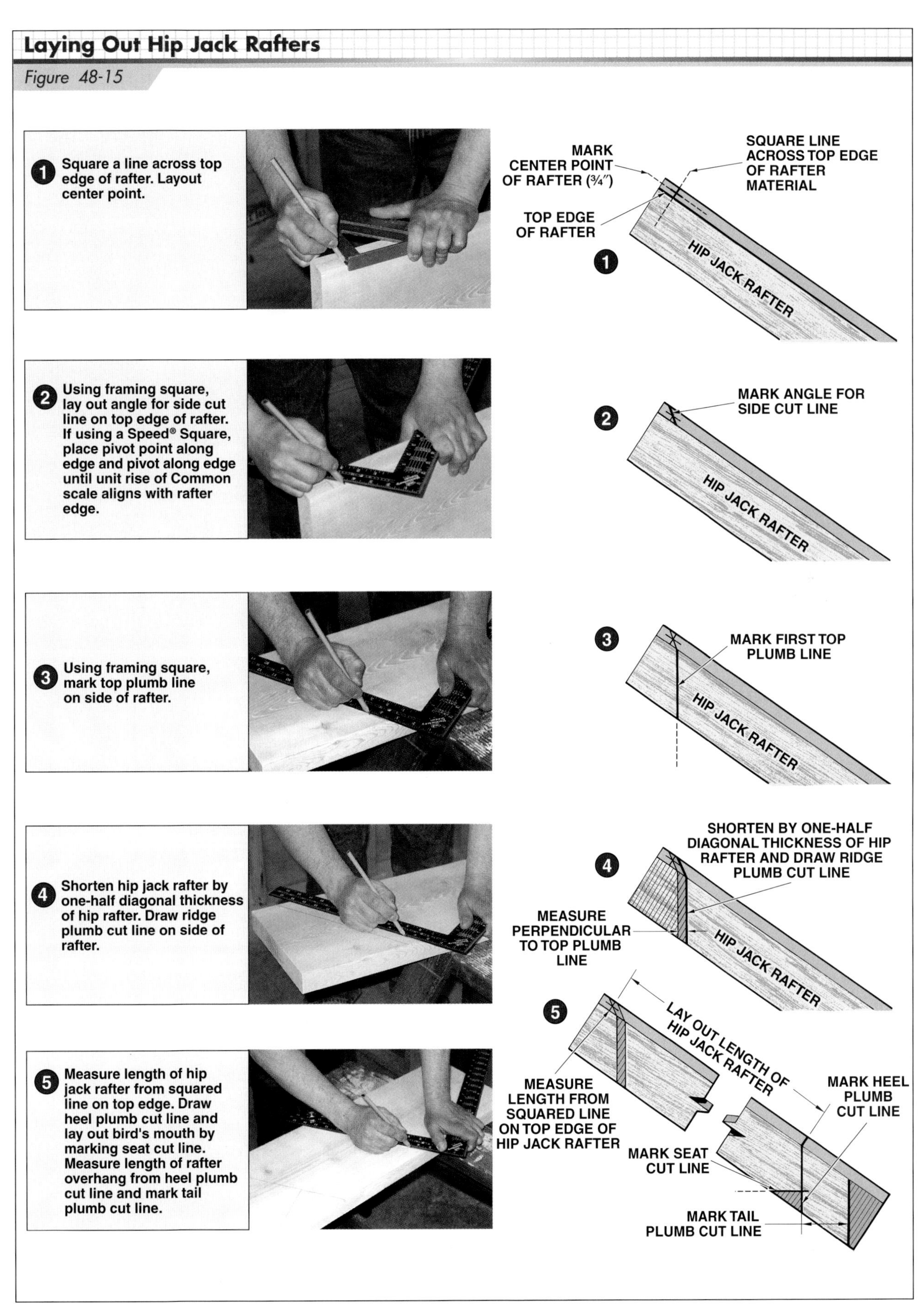

Figure 48-15. *A hip jack rafter has plumb cuts where it fastens to the hip rafter as well as cuts at the heel and tail.*

CONSTRUCTING HIP ROOFS

The main framing members for hip roofs should be precut prior to constructing the roofs. The ridge board for a hip roof must be precut to its exact length before it can be set in place. A procedure for finding the theoretical length of the ridge board is shown in **Figure 48-16.** The actual length is affected by the framing method used at the end of the ridge. As previously mentioned, one method places common rafters at the end of the ridge board, whereas a second method does not. **See Figure 48-17.**

Hip Roof Framing Procedure
Media Clip

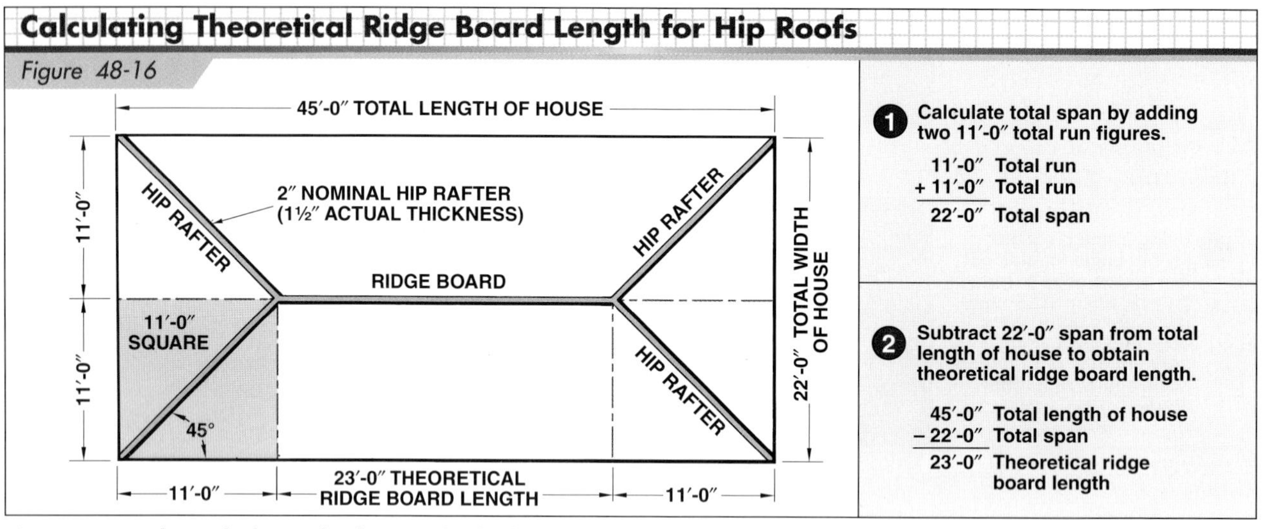

Figure 48-16. *When calculating the theoretical ridge board length of a hip roof, subtract the total span from the total length of the building. In this example, the total span of the roof is 22′ and the total length of the building is 45′.*

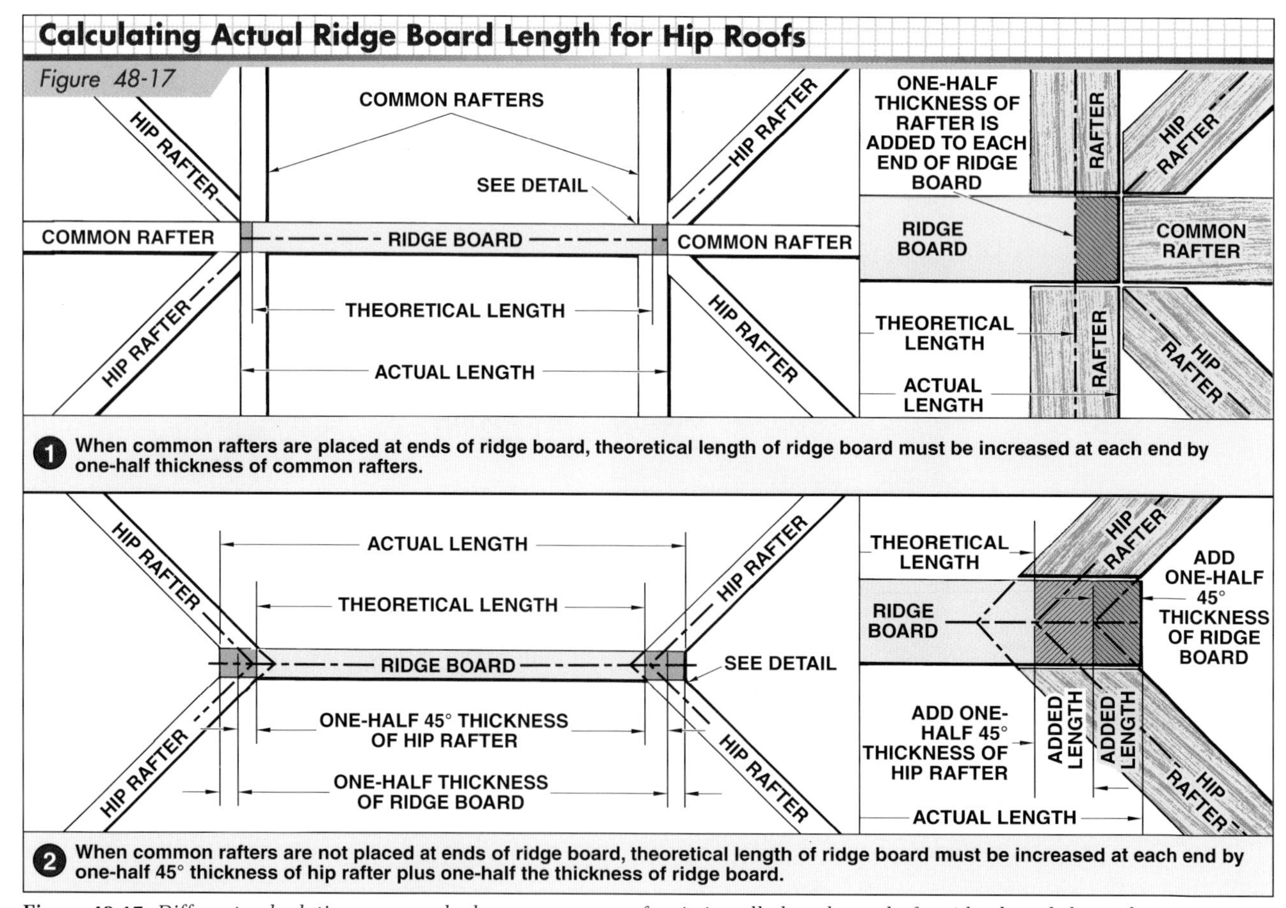

Figure 48-17. *Different calculations are used when a common rafter is installed at the end of a ridge board than when a common rafter is not installed at the end.*

After the ridge board is cut to length, the layout markings for rafter placement can be transferred from the top wall plates to the ridge board.

Conditions that require the use of purlins, braces, and collar ties are described in Unit 46. Collar ties may be placed at every second or third pair of rafters. Purlins may be used to support longer rafters and are placed beneath the rafters at an intermediate point between the roof ridge and the exterior wall. Braces extending to the nearest interior partition support the purlins. General construction procedures for erecting hip roofs include the following and are shown in **Figure 48-18.**

1. Install common rafters at two ends of ridge board.
2. Install hip rafters at corners.
3. Install hip jack rafters and remaining common rafters.

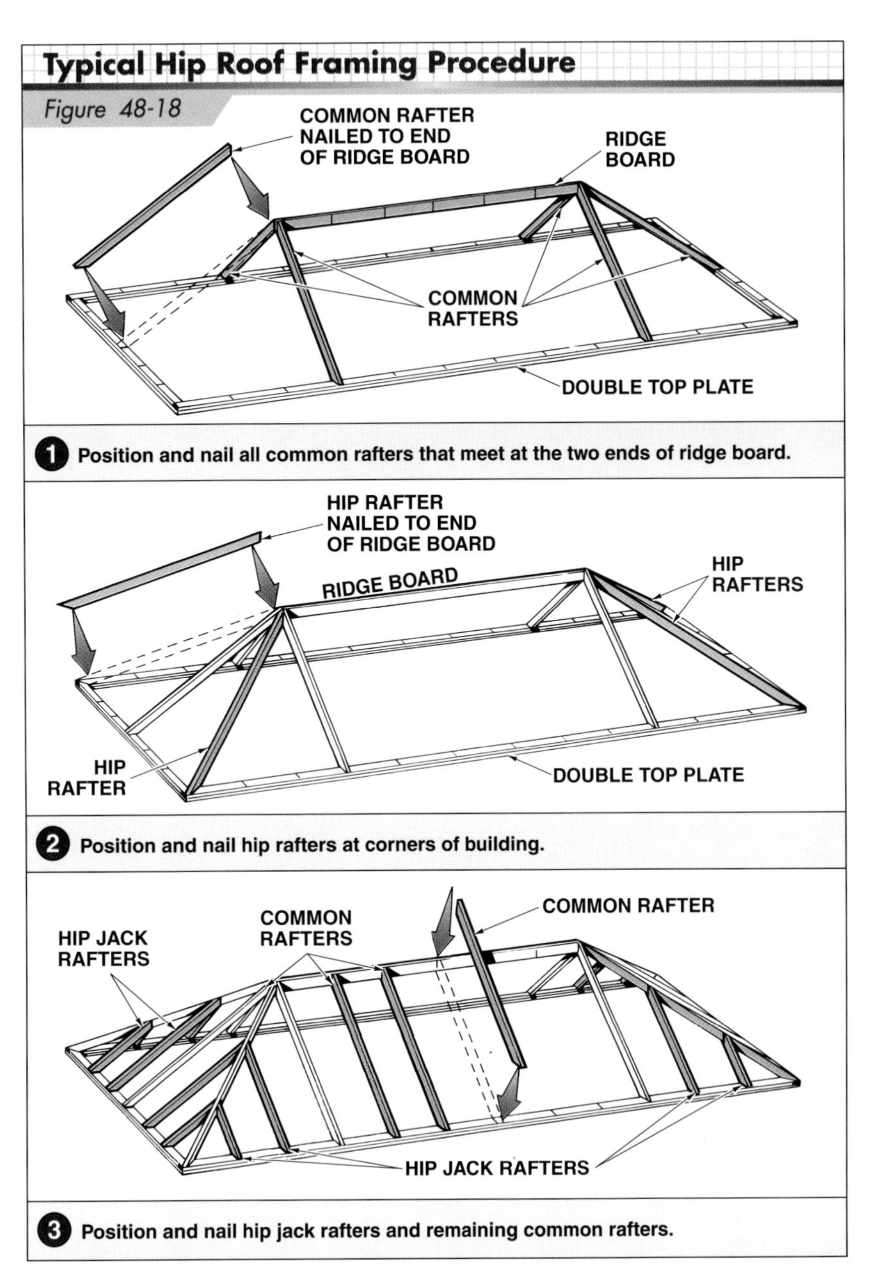

Figure 48-18. *When constructing a hip roof, the main framing members should be precut before construction begins.*

Unit 48 Review and Resources

UNIT 49 Intersecting Roofs

OBJECTIVES

1. Define intersecting roof.
2. Describe the construction of intersecting roofs with equal spans.
3. Describe the construction of intersecting roofs with unequal spans.
4. Discuss the layout of valley rafters.
5. Explain the procedure for laying out valley rafters with a Speed Square®.
6. Explain the procedure for calculating valley jack rafter lengths.
7. Review the procedure for calculating hip-valley cripple jack rafter lengths.
8. Discuss the use of valley cripple jack rafters.
9. Explain the procedure for constructing intersecting roofs.
10. Discuss the procedure for framing intersecting roofs with equal spans.
11. Explain the procedure for framing intersecting roofs with unequal spans.
12. Review blind valley construction on intersecting roofs.

An *intersecting roof,* also known as a combination roof, consists of two or more roof sections sloping in different directions. A *valley* is formed where the different sloping sections are joined. **See Figure 49-1.**

The two sections of an intersecting roof may or may not be the same width. If the two sections are the same width, the roof is said to have *equal spans.* If the two sections are not the same width, the roof is said to have *unequal spans.*

Traditionally, the framing square has been used to lay out angled cuts and lengths on rafters for intersecting roofs. The Speed® Square is now also being used to lay out these rafters. Valley and valley jack rafter layout using the framing square and Speed Square are both discussed in this unit.

INTERSECTING ROOFS WITH EQUAL SPANS

In a roof with equal spans, the total rise (height) is the same for the two ridges. **See Figure 49-2.** Where the slopes of the roof meet to form a valley between the two sections, a pair of *valley rafters* are placed. Valley rafters extend from the inside corners formed by the two sections of the building to the corners formed by the intersecting ridges. *Valley jack rafters* run from the valley rafters to the ridges. *Hip-valley cripple jack rafters* are placed between the valley rafter and hip rafter.

INTERSECTING ROOFS WITH UNEQUAL SPANS

An intersecting roof with unequal spans requires a *supporting valley rafter* to run from the inside corner formed by the two sections of the building to the main ridge. **See Figure 49-3.** A *shortened valley rafter* runs from the other inside corner of the building to the supporting valley rafter. Similar to an intersecting roof with equal spans, an intersecting roof with unequal spans also requires valley jack rafters and hip-valley cripple jack rafters. In addition, a *valley cripple jack rafter* is placed between the supporting valley rafter and shortened valley rafter.

VALLEY RAFTERS

In intersecting roofs with equal spans, valley rafters run at a 45° angle to the exterior walls of a building and are parallel to hip rafters. Therefore, valley rafters are the same length as hip rafters.

APA—The Engineered Wood Association

Figure 49-1. *In this intersecting roof, the hip section over the garage at the right intersects with the main gable roof.*

Intersecting Roofs with Equal Spans

Figure 49-2

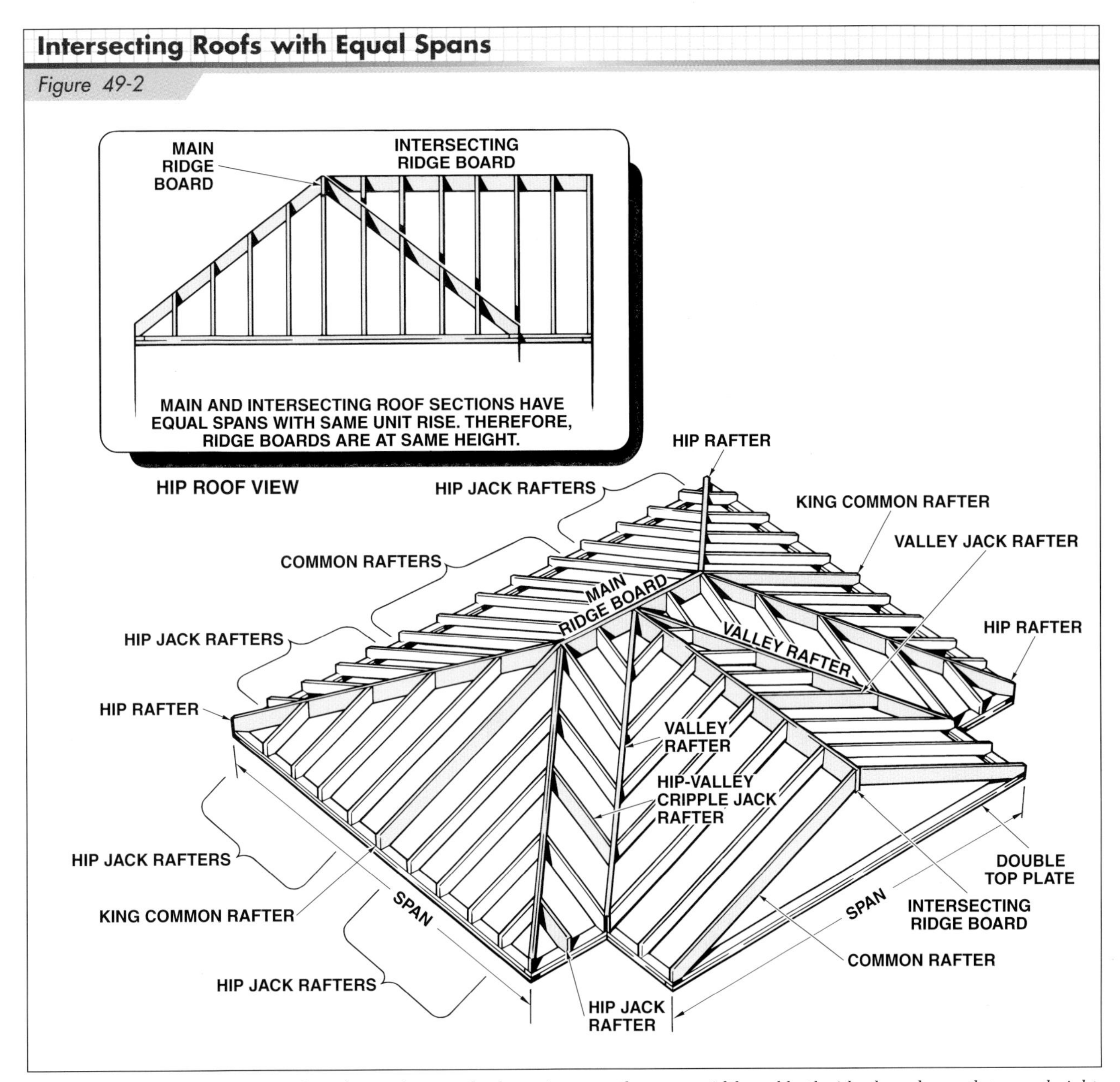

Figure 49-2. *For intersecting roofs with equal spans, both sections are the same width and both ridge boards are the same height.*

Structural Insulated Panel Association

An intersecting roof consists of two or more roof sections sloping in different directions.

Laying Out Valley Rafters

The layout of a valley rafter is almost identical to the layout of a hip rafter. The unit rise measurement and 17″ measurement on a framing square provide the angles for the plumb and seat cuts. The side cut angles for valley rafters are the same as the angles for hip rafters. The only difference in layout occurs at the seat and tail of the valley rafter. Side cuts must be angled back at the heel plumb cut line to allow the valley rafter to drop down into the inside corner of the building. **See Figure 49-4.** Side cuts are also required at the tail of the overhang so that the corner formed by the valley will align with the rest of the roof overhang. **See Figure 49-5.**

The angle of the side cuts at the heel and tail is the same as the angle where the rafter connects with the ridge board. The layout procedure for a valley rafter is shown in **Figure 49-6.** Note that, unlike hip rafters, valley rafters do not require backing or dropping in an equal span roof, but they may require dropping in an unequal span roof.

James Hardie Building Products

The intersecting ridge board of an intersecting roof with unequal spans is lower than the main ridge board.

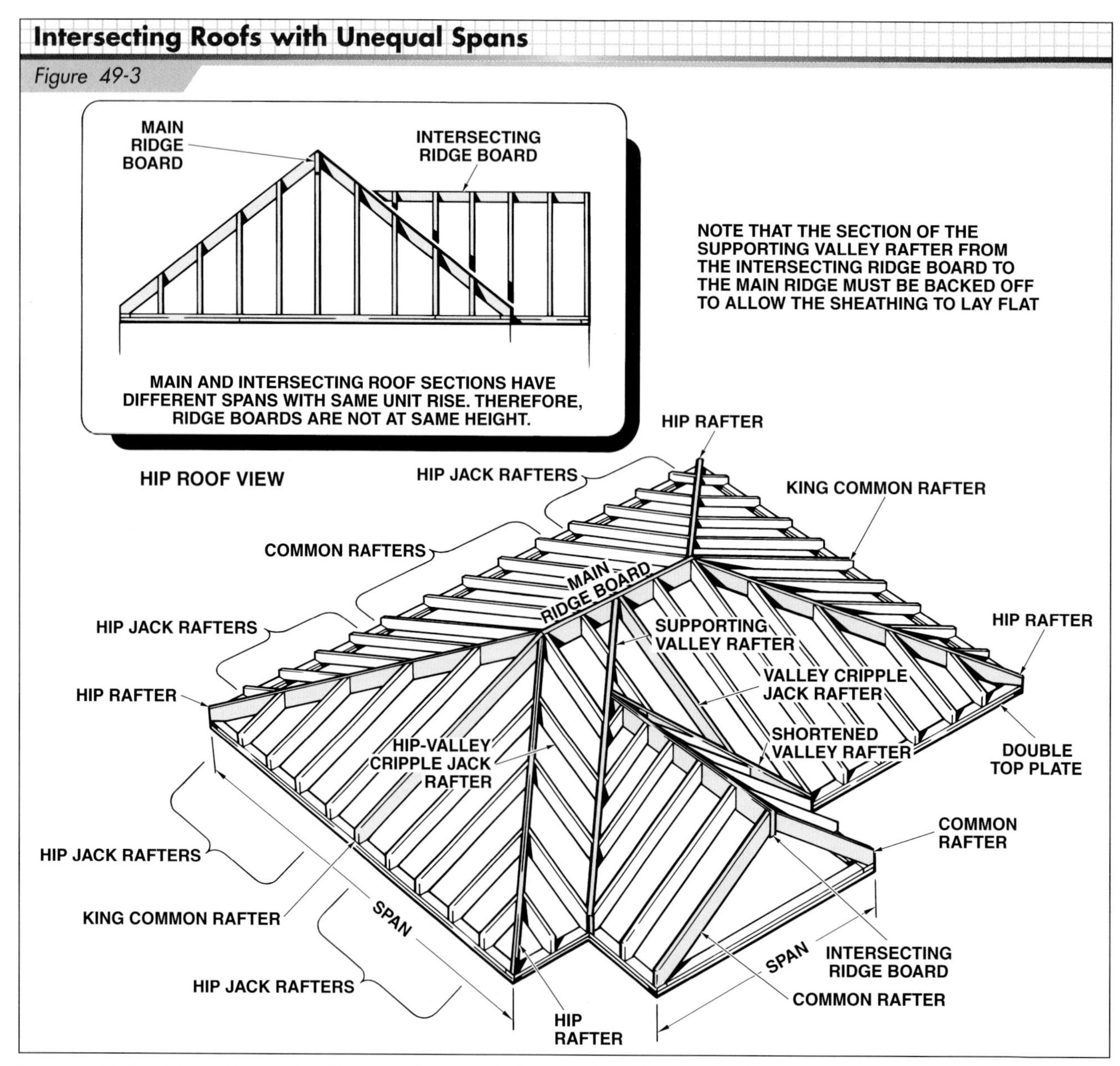

Figure 49-3. *For intersecting roofs with unequal spans, the intersecting ridge board is lower on the section with the smaller span.*

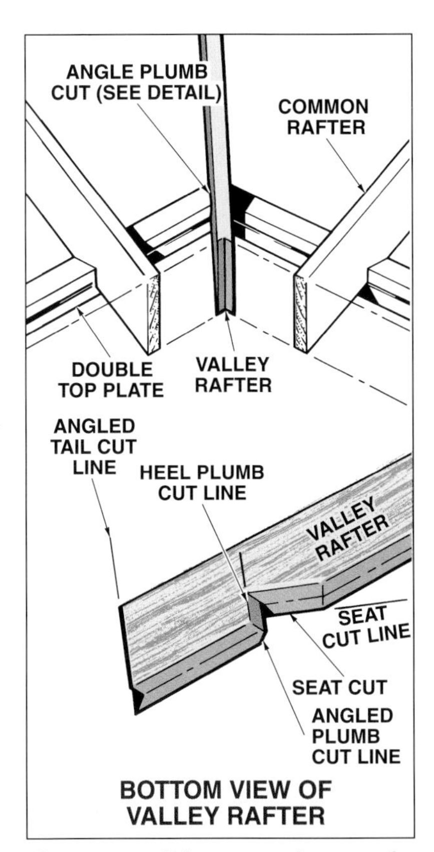

Figure 49-4. *Side cuts at the seat of a valley rafter must be angled back at the heel plumb cut line to allow the rafter to drop down into the inside corner of the building.*

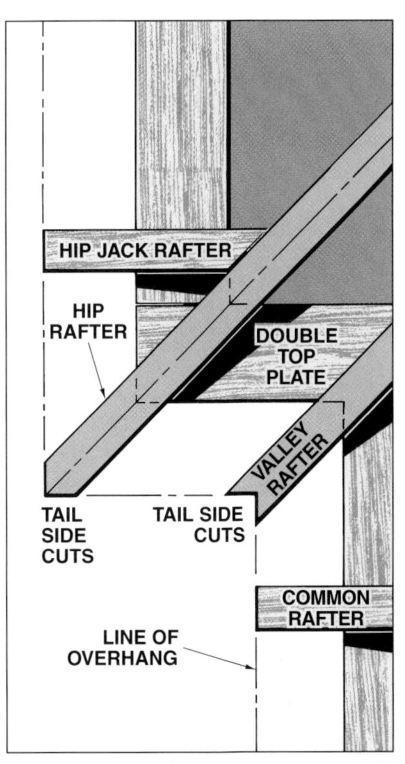

Figure 49-5. *Valley rafters require side cuts at the seat and tail of the overhang.*

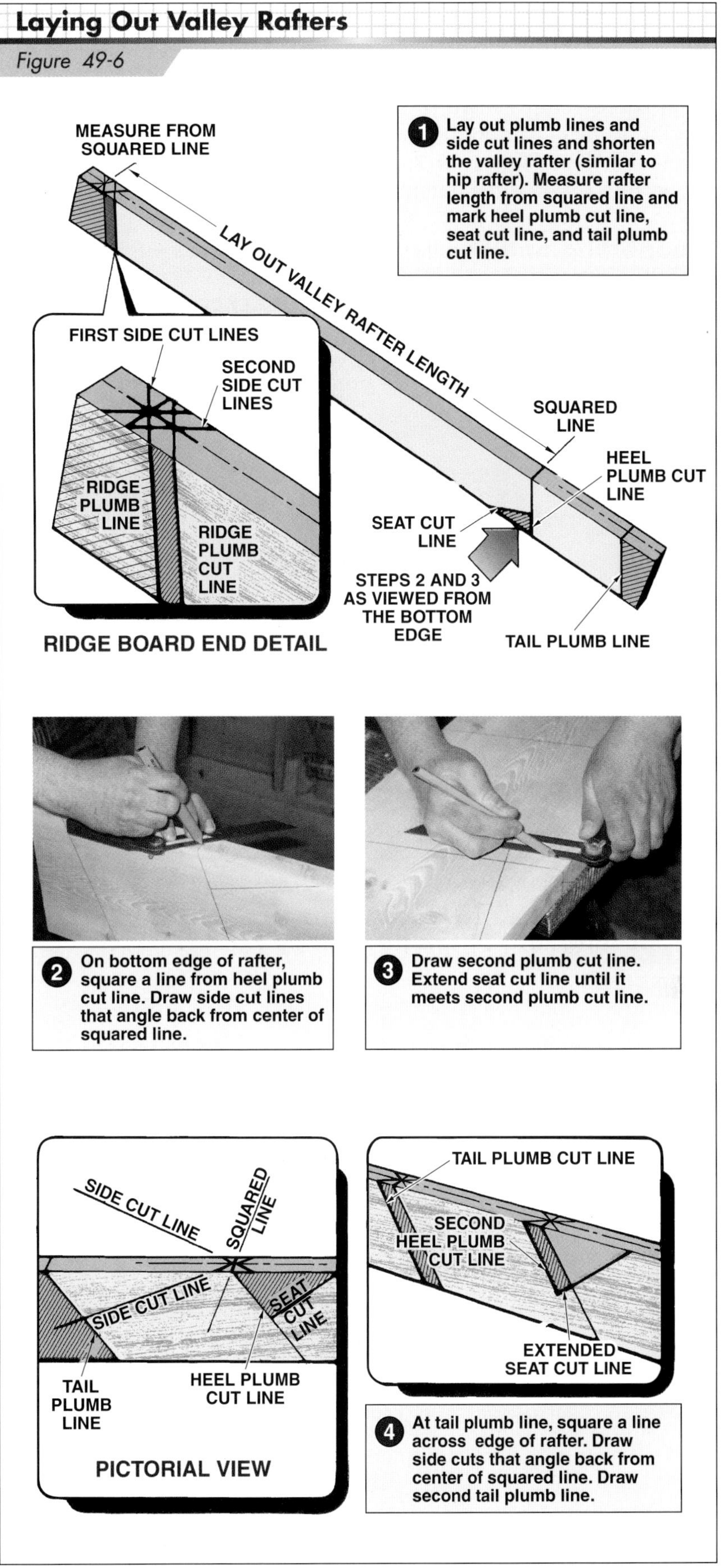

Figure 49-6. *Valley rafters require angled side cuts at the ridge, heel, and tail.*

Supporting and Shortened Valley Rafters

An intersecting roof with unequal spans requires two types of valley rafters—supporting and shortened. **See Figure 49-7.** A *supporting valley rafter* extends from the wall plate to the main ridge board and has a single side cut where it fits against the ridge board. A *shortened valley rafter* runs at a 90° angle to the supporting valley rafter. Shortened valley rafters have a square cut where they butt against the supporting valley rafter. The length of a shortened valley rafter is based on the run of the narrower roof. **See Figure 49-8.** The layout procedure for a shortened valley rafter is shown in **Figure 49-9.**

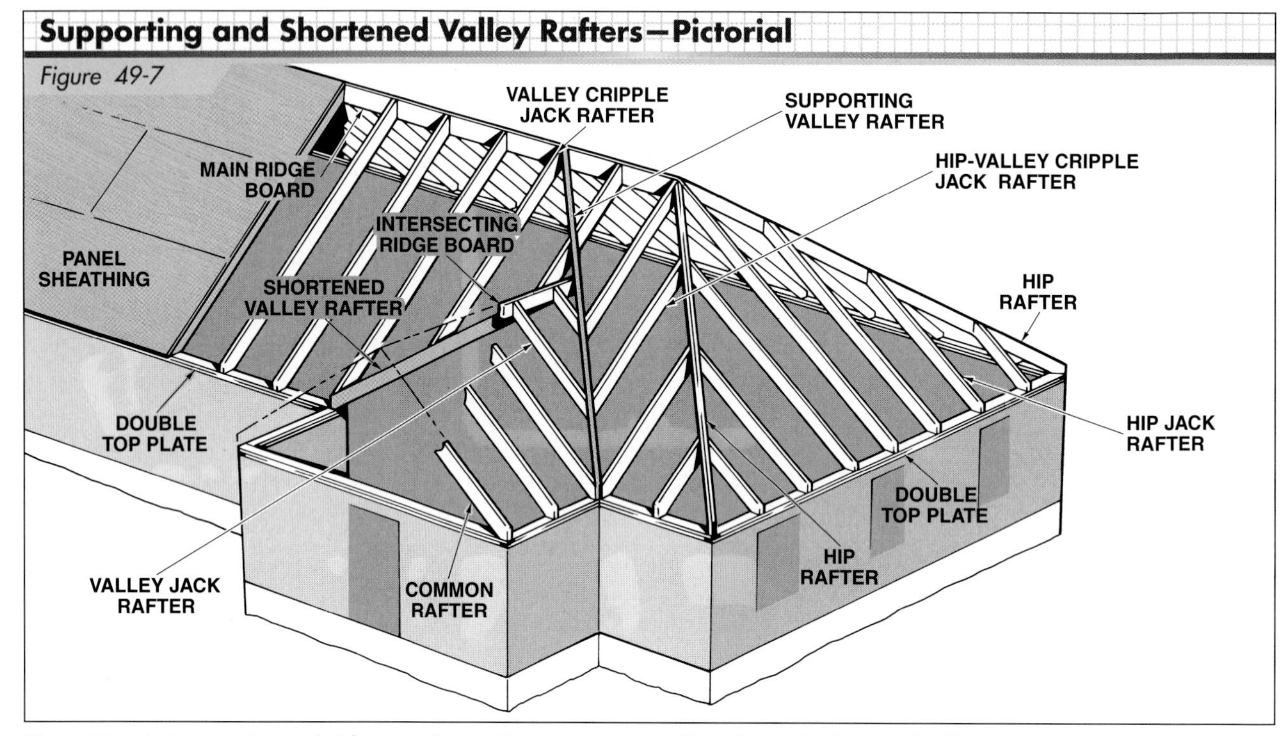

Figure 49-7. *An intersecting roof with unequal spans has a supporting valley rafter and a shortened valley rafter.*

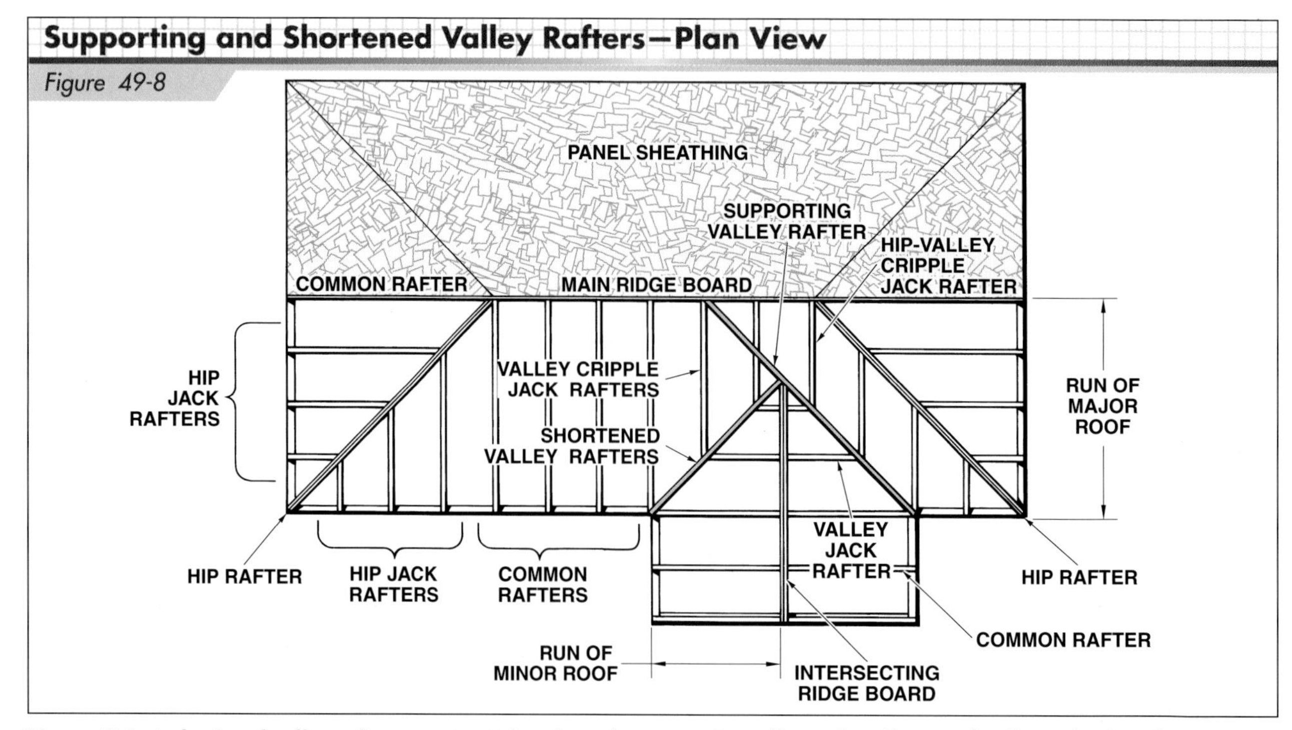

Figure 49-8. *A shortened valley rafter runs at a 90° angle to the supporting valley rafter. Shortened valley rafter length is based on the run of the minor roof section.*

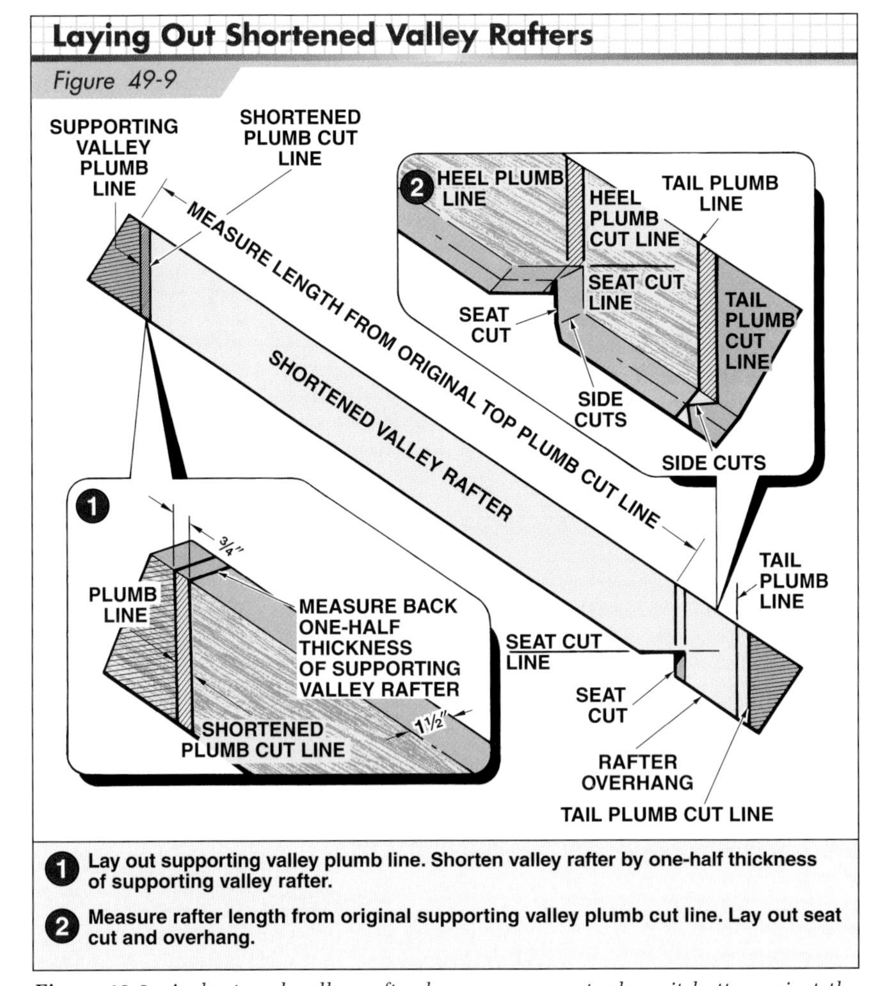

Figure 49-9. *A shortened valley rafter has a square cut where it butts against the supporting valley rafter.*

Roof valleys must have the proper underlayment and waterproofing membrane to prevent ice buildup damage in colder climates.

Laying Out Valley Rafters with a Speed Square

Layout of valley rafters is almost identical to the layout of hip rafters. The Hip-Val scale on the Speed Square is used to lay out the proper angle based on the unit rise. The side cut angles for valley rafters are the same as the angles for hip rafters. The only difference in layout occurs at the seat and tail cuts of the valley rafters. Side cuts are angled back at the second heel plumb cut line to allow the valley rafter to drop down into the inside corners of the building. **See Figure 49-10.**

Side cuts may also be required at the tail of the overhang so the corner formed by the valley will align with the rest of the roof overhang. If side cuts are used, one-half the rafter thickness will need to be added to the overhang length to allow for the side cuts.

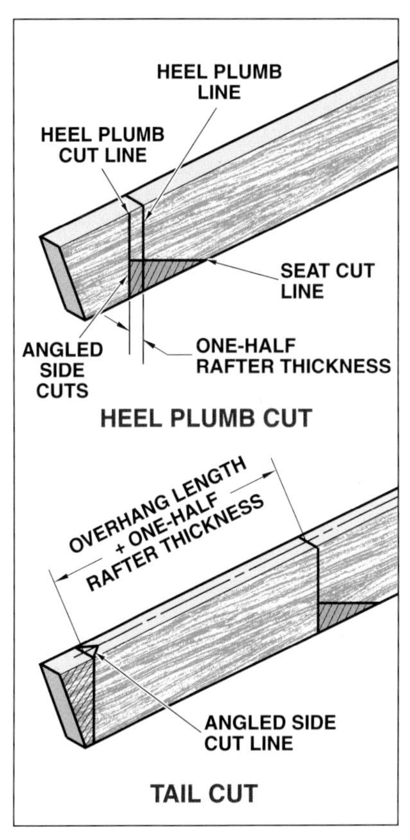

Figure 49-10. *Angled side cuts are required at the heel plumb line. Angled tail cuts may also be needed.*

VALLEY JACK RAFTERS

Valley jack rafters bridge the area between valley rafters and the ridges of an intersecting roof. Spacing of valley jack rafters is the same as the spacing of common roof rafters.

Calculating Valley Jack Rafter Lengths

Valley jack rafters decrease in length as they get closer to the top of the roof. Valley jack rafters have a common length difference if they are spaced the same distance apart. The common length differences are the same as those for hip jack rafters. The third line of a framing square rafter table provides the common length difference for jack rafters spaced 16″ OC. The fourth line provides the common length difference for jack rafters spaced 24″ OC.

The chosen procedure for calculating valley jack rafter lengths depends on how the rafters are positioned on the roof. **Figure 49-11** shows the procedure to use when the valley jack rafter spacing begins from the inside corner of a building. **Figure 49-12** shows the procedure to use when valley jack rafter spacing begins from a common rafter positioned away from the inside corner. **Figure 49-13** shows the procedure to use when valley jack rafter spacing begins from the center point of the intersecting ridges.

The quality of wood products and the design of wood load-bearing members must conform to the National Design Specification® (NDS) for Wood Construction.

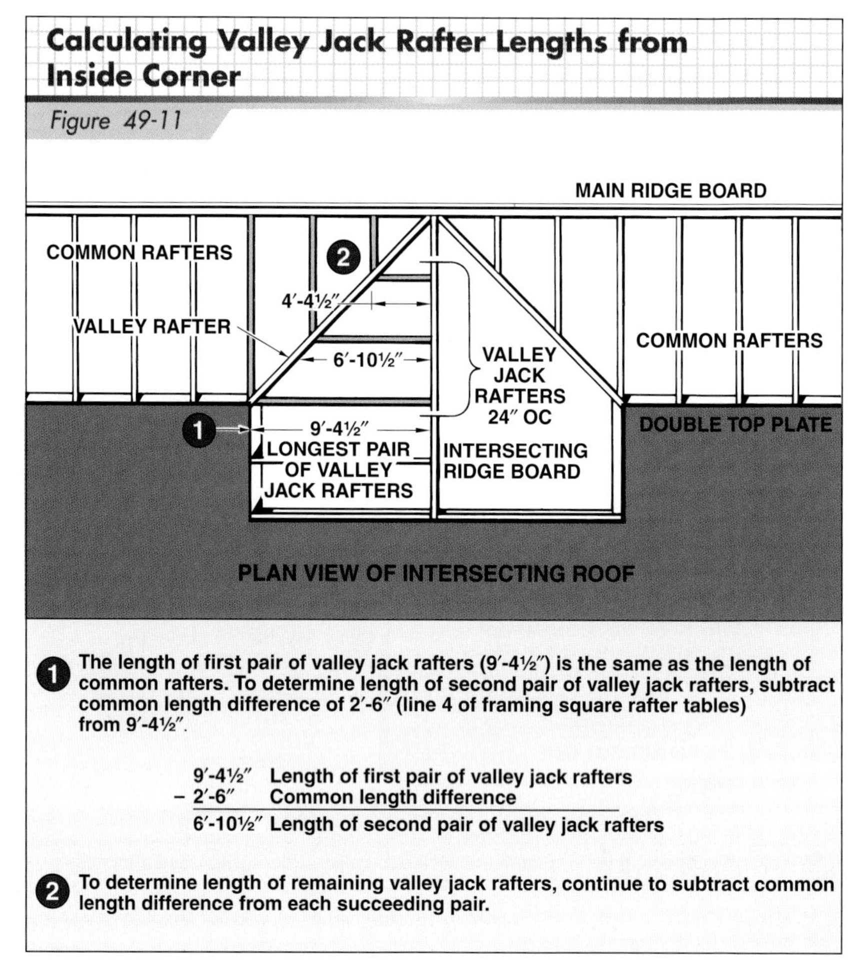

Figure 49-11. *Valley jack rafter lengths can be calculated from the longest jack rafter at the inside corner of the building. In this example, the roof has a 9″ unit rise and the rafter spacing is 24″ OC.*

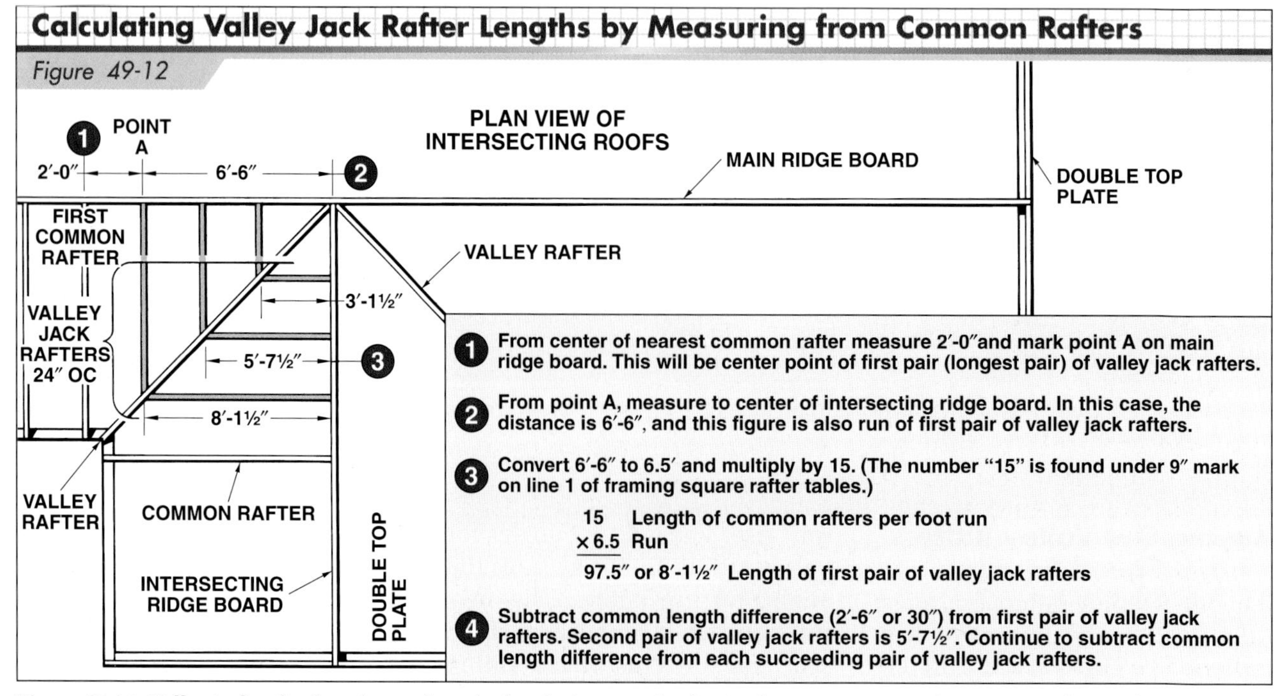

Figure 49-12. *Valley jack rafter lengths can be calculated when spacing begins from a common rafter positioned away from the inside corner. In this example, the roof has a 9″ unit rise and the rafter spacing is 24″ OC.*

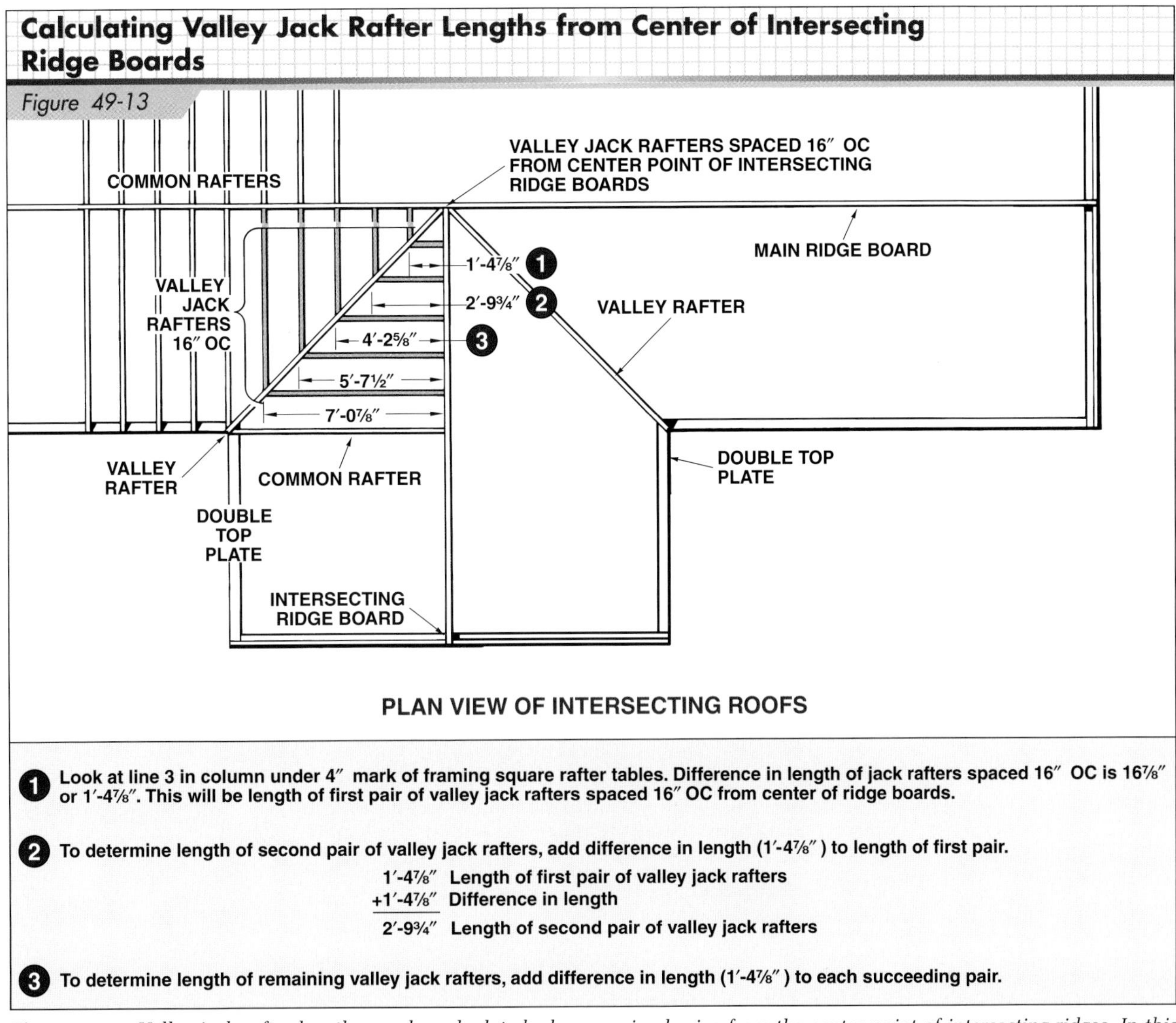

Figure 49-13. *Valley jack rafter lengths can be calculated when spacing begins from the center point of intersecting ridges. In this example, the roof has a 4″ unit rise and the rafter spacing is 16″ OC.*

Laying Out Valley Jack Rafters

For the plumb cut of a valley jack rafter, use the unit rise measurement and the 12″ measurement on a framing square, and place a mark on the unit rise side. For the side cut, use the number found on the fifth line of the rafter table and the number "12." Mark on the side of the 12.

Valley jack rafters require a square cut where they are nailed against the ridge and a side cut where they meet the valley rafter. **See Figure 49-14** for the procedure for laying out a valley jack rafter.

APA—The Engineered Wood Association

Complex roofs may have multiple ridges, hips, and valleys.

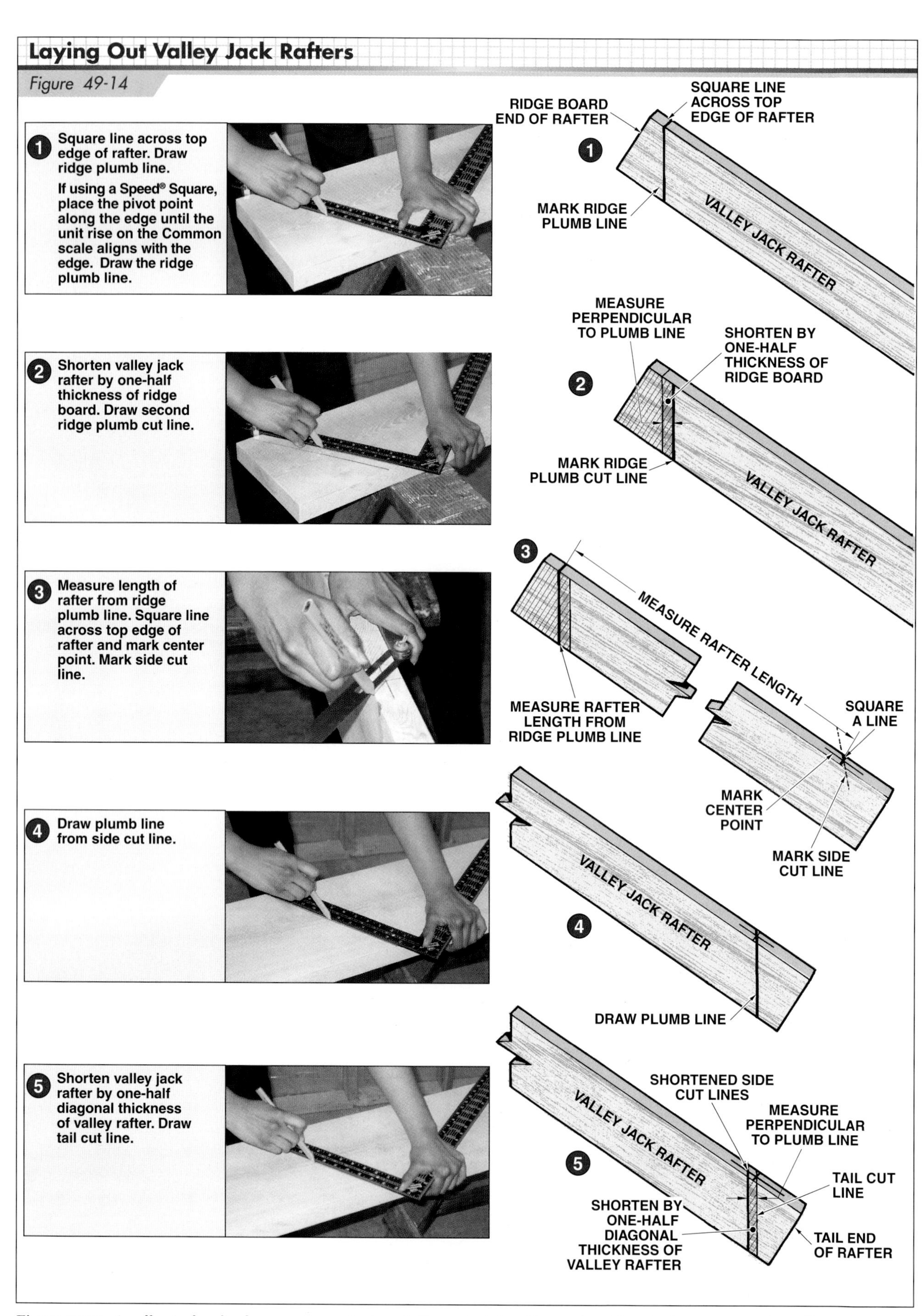

Figure 49-14. *A valley jack rafter has a square cut where it fits against the ridge and a side cut where it meets the valley rafter.*

HIP-VALLEY CRIPPLE JACK RAFTERS

When hip and valley rafters are placed close together, the space between them is framed with hip-valley cripple jack rafters. All hip-valley cripple jack rafters are the same length.

Calculating Hip-Valley Cripple Jack Rafter Lengths

A framing square rafter table can be used to find the lengths of hip-valley cripple jack rafters. First, find the distance from the end of the main roof section to the intersecting roof section. Multiply the number of feet in this distance by the number on the common rafter line under the unit rise number at the top of the square. This procedure is shown in **Figure 49-15.**

APA—The Engineered Wood Association

Hip roofs have four sloping sides and are the strongest type of roof since they are braced by four hip rafters.

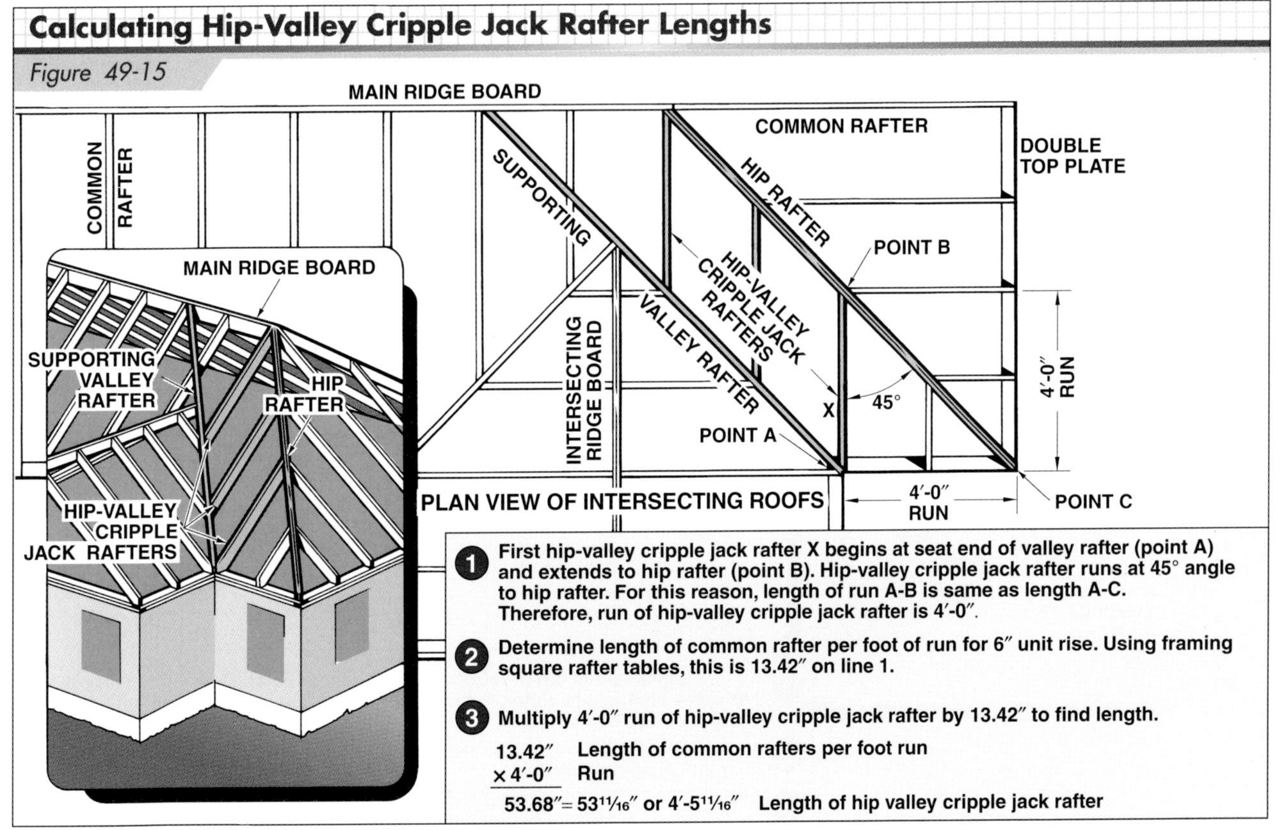

Figure 49-15. *Framing square rafter tables may be used to calculate the length of hip-valley cripple jack rafters. In this example, the roof has a 6″ unit rise. Additional information required is the distance from the end of the main section to the intersecting roof section. This distance, shown as length A-C on the drawing, is 4′-0″.*

A book of rafter tables can also be used to find the lengths of hip-valley cripple jack rafters.

Laying Out Hip-Valley Cripple Jack Rafters

Since hip-valley cripple jack rafters fit between the hip and the valley rafter, they require a plumb cut and side cut at each end. A layout procedure is shown in **Figure 49-16.**

VALLEY CRIPPLE JACK RAFTERS

Valley cripple jack rafters are used only on intersecting roofs with unequal spans. Valley cripple jack rafters are placed between the shortened and the supporting valley rafters to bridge the space in the main roof section between the supporting and shortened valley rafters.

Calculating Valley Cripple Jack Rafter Lengths

The run of a valley cripple jack rafter is always twice the run of the valley jack rafter that it meets at the shortened valley rafter. For this reason, the length of a valley cripple jack rafter is also twice the length of that valley jack rafter. **See Figure 49-17.** Spacing of valley cripple jack rafters is the same as common rafters.

Laying Out Valley Cripple Jack Rafters

The angles for plumb cuts and side cuts on valley cripple jack rafters are found by using the same framing square method described for laying out other types of jack rafters. A procedure for laying out valley cripple jack rafters is shown in **Figure 49-18.**

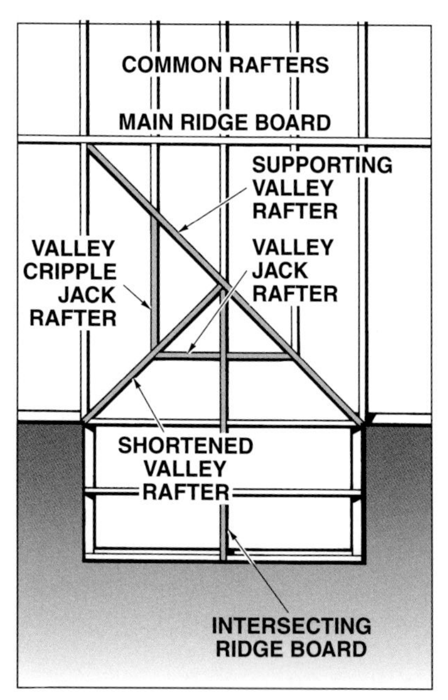

Figure 49-17. *A valley cripple jack rafter is always twice the length of the valley jack rafter. Note that the valley cripple jack rafter and the valley jack rafter meet at the same point on the shortened valley rafter.*

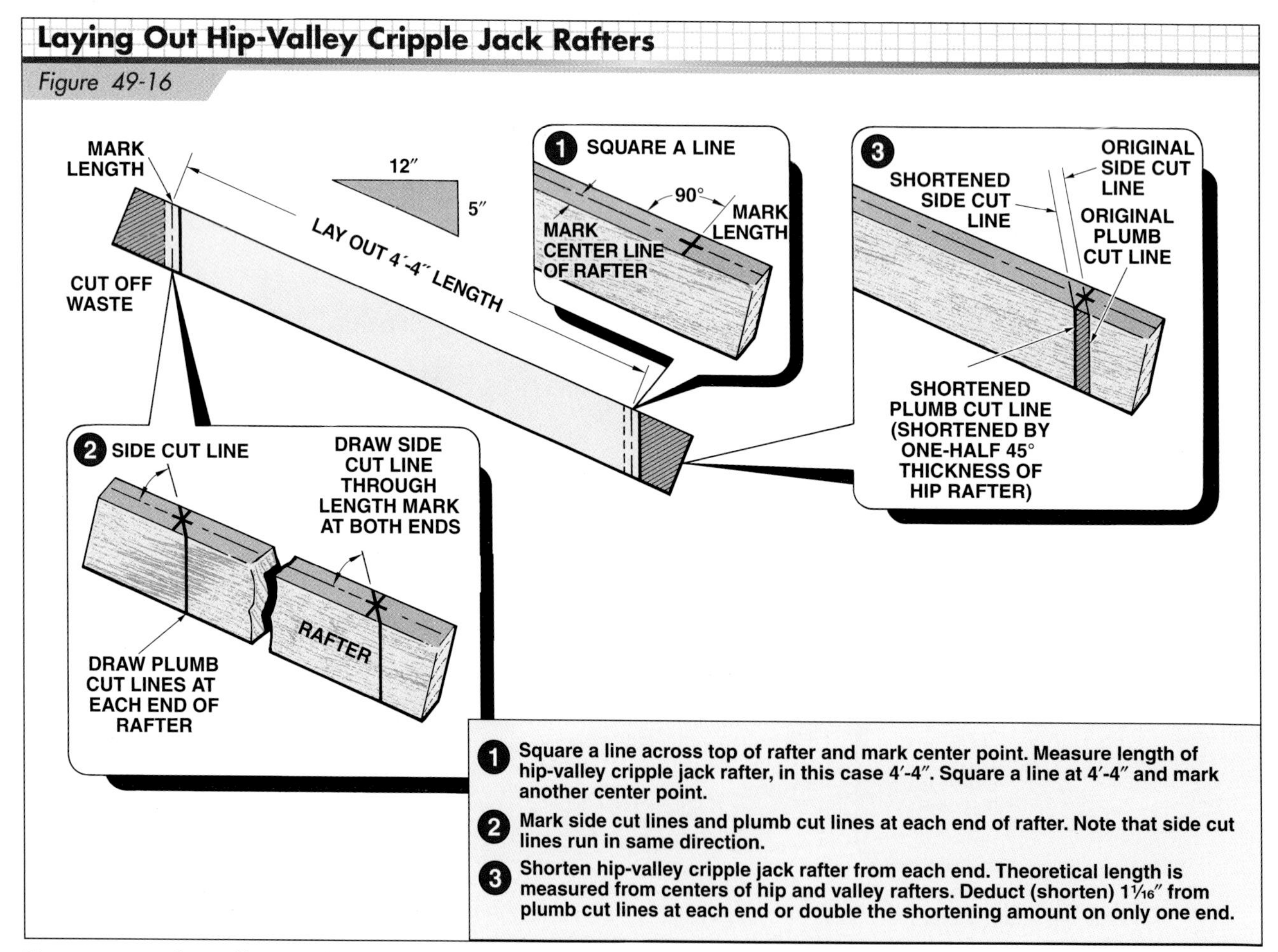

Figure 49-16. *A hip-valley cripple jack rafter has a plumb cut and side cut at each end.*

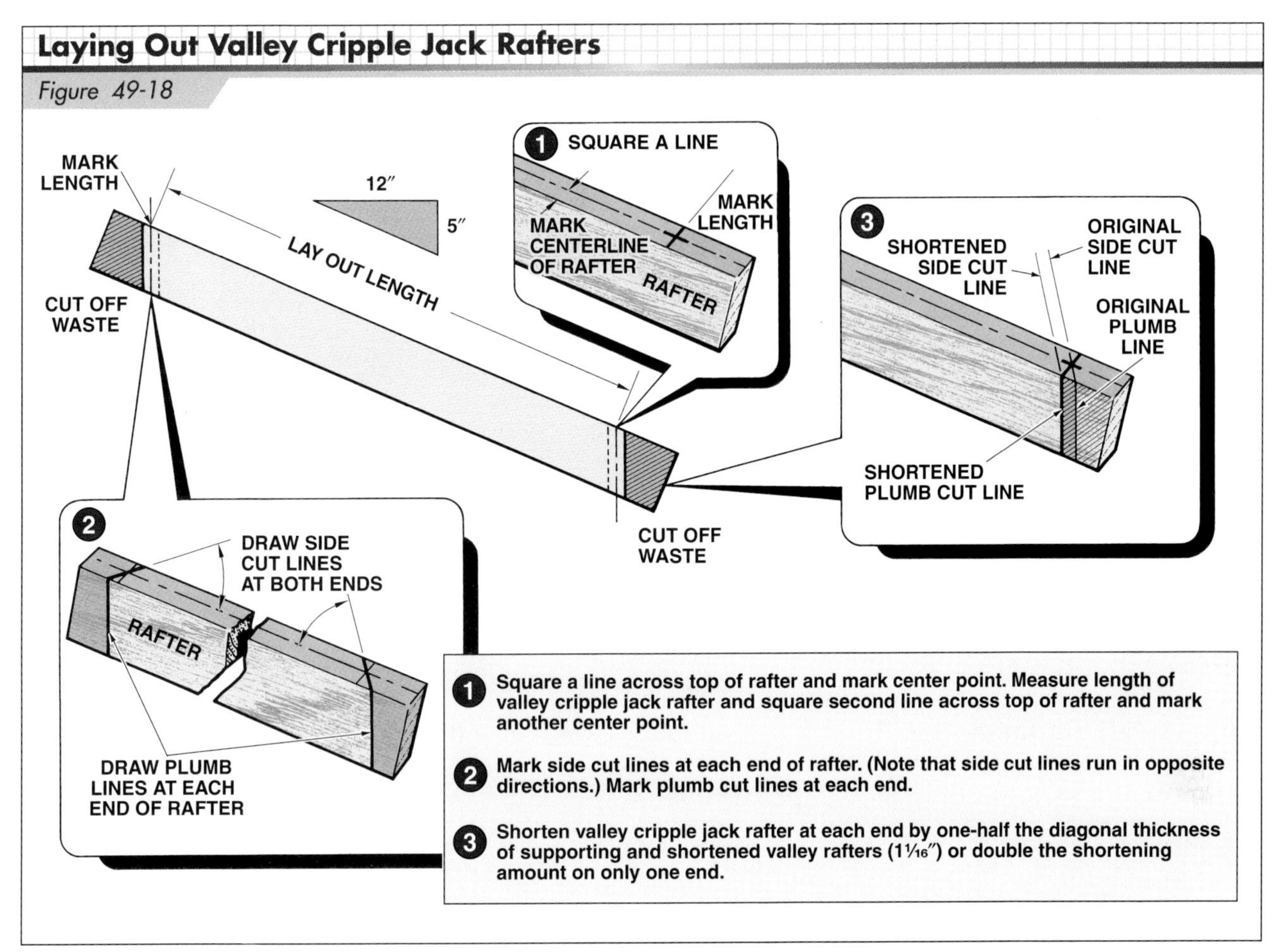

Figure 49-18. *The run of a valley cripple jack rafter is twice the run of the valley jack rafter that it meets at the shortened valley rafter. Side cuts on valley cripple jack rafters run in opposite directions.*

CONSTRUCTING INTERSECTING ROOFS

Construction of an intersecting roof usually begins with setting up the ridge board of the main roof. The main ridge board is supported by a pair of common or hip rafters at each end, as described in earlier units for gable and hip roofs.

Locating Points of Intersection

After the main ridge board has been set in place, the ridge board of the intersecting portion of the roof can be erected. First, however, the correct point of intersection must be marked on the main ridge board. Methods to locate the point of intersection for three different types of intersecting roofs are shown in **Figures 49-19** and **49-20.**

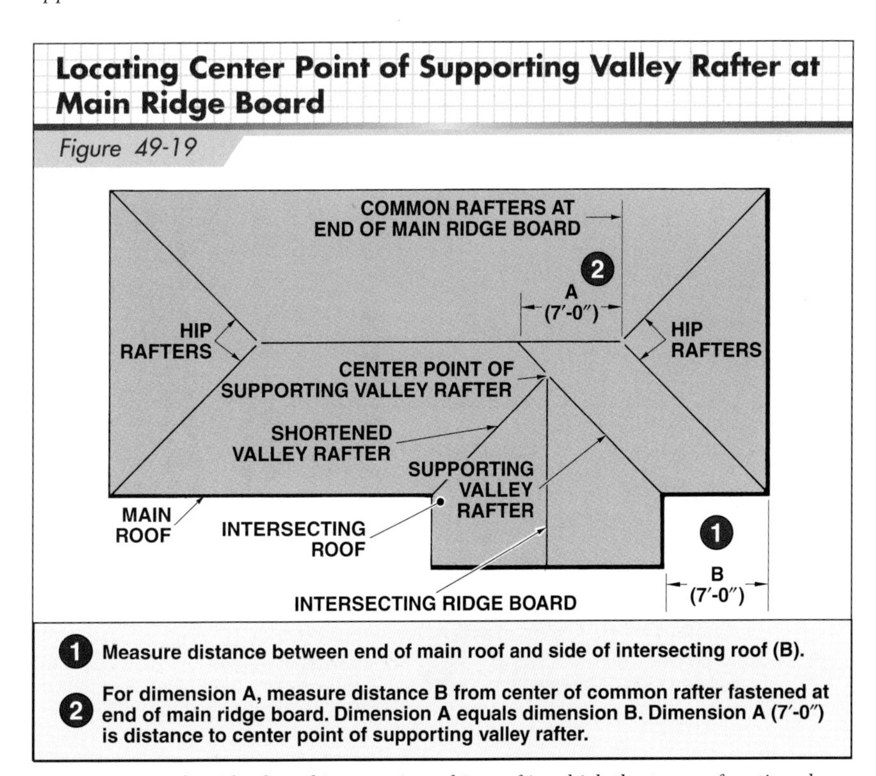

Figure 49-19. *The ridge board intersecting a hip roof in which the two roof sections have unequal spans fits into the corner formed by the supporting and shortened valley rafters.*

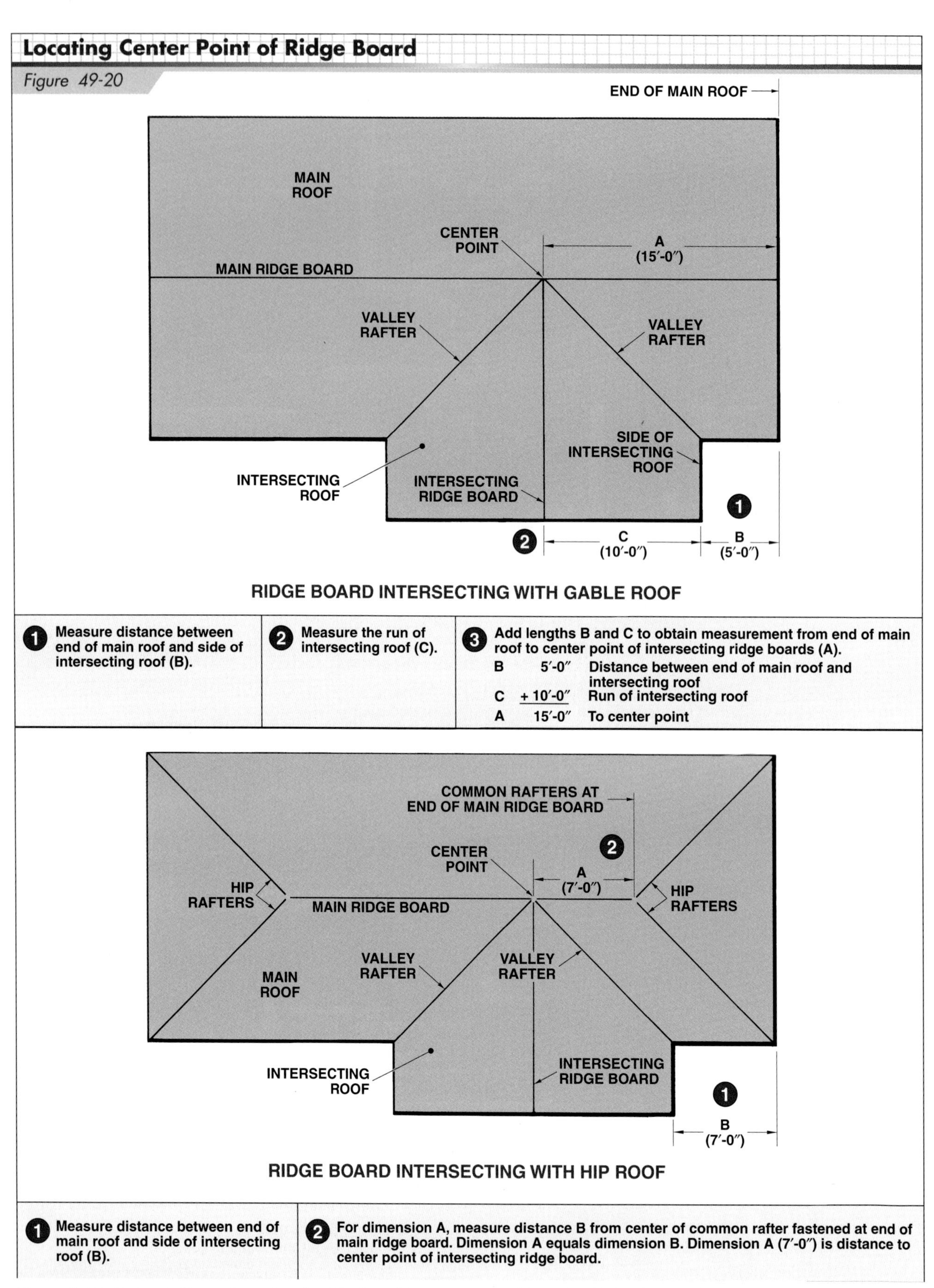

Figure 49-20. *To locate the center point of a ridge board intersecting a gable roof, add the distance between the end of the main roof and side of the intersecting roof to the run of the intersecting roof. The center point of a ridge board intersecting a hip roof in which the two roof sections have equal spans must be located accurately.*

Calculating Intersecting Ridge Board Lengths

Whenever possible, intersecting ridge boards should be cut to their exact lengths before they are set in place. Methods to calculate ridge board lengths for three different types of intersecting roofs are shown in **Figures 49-21**, **49-22**, and **49-23.**

Square gauges are used with a framing square when laying out several rafters that are to be cut at the same angle. Square gauges are clamped to the blade and tongue of the square. If square gauges are not available, make a straightedge of a short 1 × 4 fastened to the blade and tongue of the square with small C-clamps. The straightedge is placed flush against the edge of the rafter stock at the proper position and lines are drawn along the blade and tongue.

Simpson Strong-Tie Company, Inc.

Seismic and hurricane rafter ties may be used to attach rafters to the double top plate. Five 8d nails are driven into the rafter and double top plate.

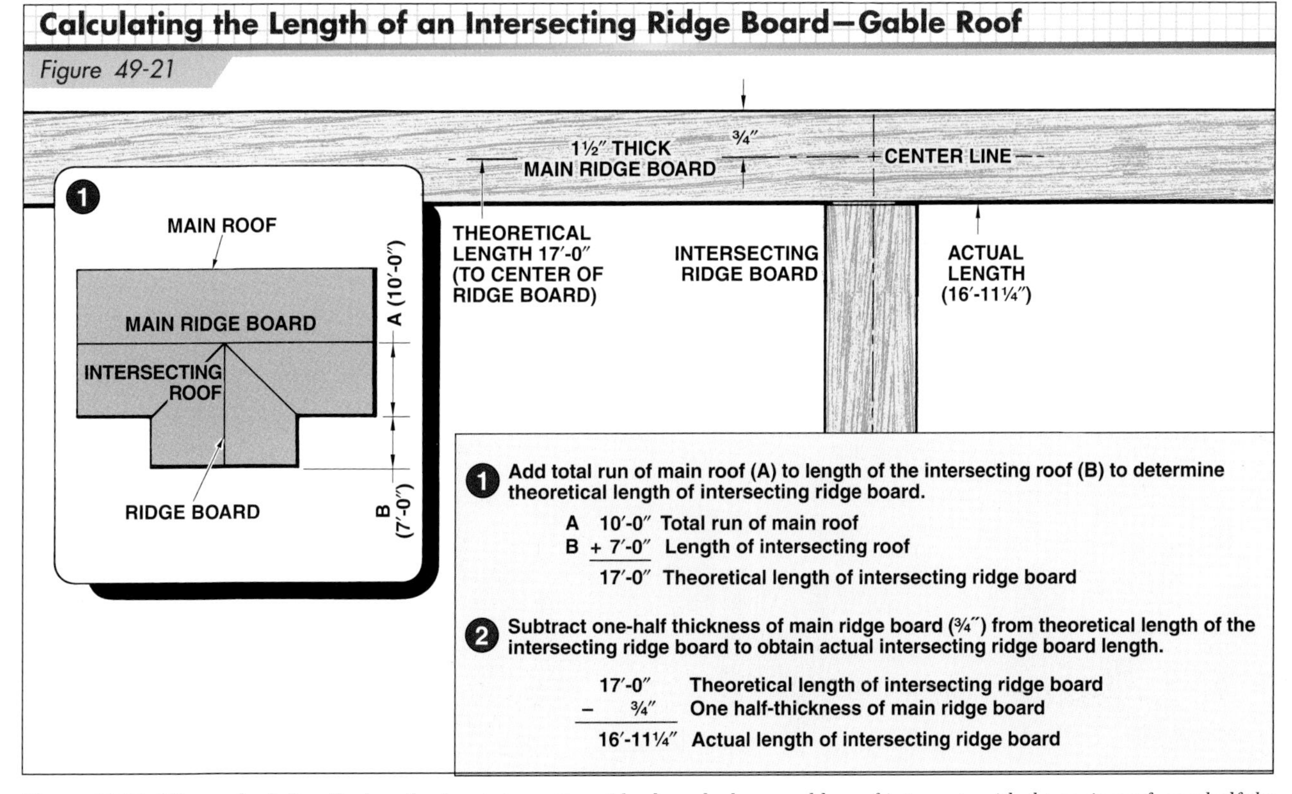

Figure 49-21. *When calculating the length of an intersecting ridge board when a gable roof intersects with the main roof, one-half the thickness of the main ridge board is subtracted from the theoretical length.*

Calculating the Length of an Intersecting Ridge Board—Hip Roof

Figure 49-22

MAIN ROOF
MAIN RIDGE BOARD
INTERSECTING ROOF
INTERSECTING RIDGE BOARD
COMMON RAFTER
HIP RAFTER
A 13′-0″
E 21′-0″
B 8′-0″
C 13′-0″
D

1½″ THICK MAIN RIDGE BOARD ¾″
THEORETICAL LENGTH
ACTUAL LENGTH
INTERSECTING RIDGE BOARD
SUBTRACT ONE-HALF RIDGE BOARD THICKNESS
ADD ONE-HALF COMMON RAFTER THICKNESS
¾″
COMMON RAFTER
HIP RAFTER
COMMON RAFTER
HIP RAFTER
COMMON RAFTER

1. Measure distance A (13′-0″), which is total run of main roof.

2. Measure C (13′-0″), which is distance from end of intersecting hip roof to center point of common rafter fastened at end of intersecting ridge board. Dimension C is equal to total run (D) of hip roof.

3. Measure B (8′-0″), which is distance from one side of main roof to center point of common rafters fastened at end of intersecting ridge board. This dimension is found by subtracting dimension C from length between side of main roof and end of hip roof.

E	21′-0″	Distance from side of main roof to end of hip roof
C	−15′-0″	Distance from end of hip roof to center point of common rafter at end of intersecting ridge
B	8′-0″	Distance from side of main roof to center point of common rafters at end of intersecting ridge board

4. Add dimensions B (8′-0″) and A to obtain theoretical length of intersecting ridge board (21′-0″).

B	8′-0″	
A	+13′-0″	
	21′-0″	Theoretical length of ridge board

5. To determine actual intersecting ridge board length, subtract one-half thickness of main ridge board (¾″) and add one-half thickness of common rafter at opposite end of intersecting ridge board (¾″).

21′-0″	Theoretical length
− ¾″	One-half common rafter thickness
20′-11¼″	
+ ¾″	One-half common rafter thickness
21′-0″	Actual length of ridge board

In the above example, ridge board and common rafter thicknesses are equal (1½″). Therefore, if one-half ridge board thickness is subtracted and one-half common rafter thickness is added, actual length is same as theoretical length.

Figure 49-22. *The procedure is shown for calculating the length of the intersecting ridge board when a hip roof intersects with the main roof and when the two sections have equal spans.*

James Hardie Building Products

A unique roof line, such as the one shown here, may require advanced layout and framing skills.

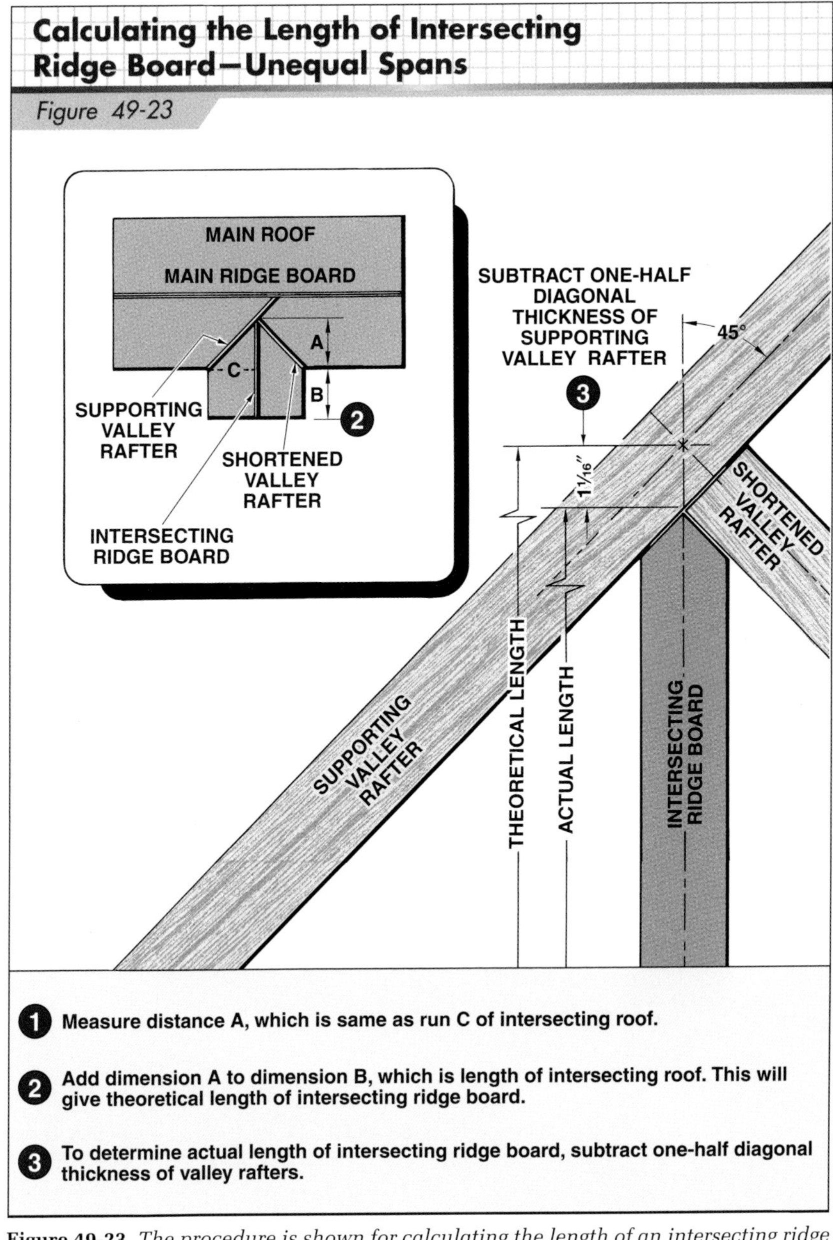

Figure 49-23. *The procedure is shown for calculating the length of an intersecting ridge board when a gable roof intersects with the main roof and when the two sections have unequal spans.*

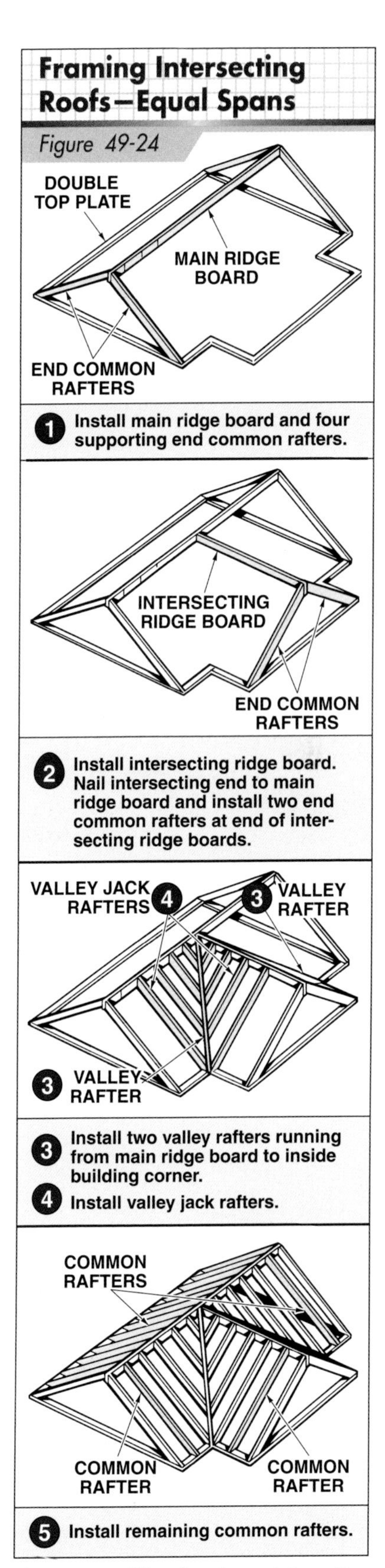

Figure 49-24. *When framing an intersecting roof with equal spans, the main ridge board is set at the proper height and supported with the end common rafters. In this example, both sections of the intersecting roof are gable roofs.*

Framing Intersecting Roofs with Equal Spans

A procedure for framing an intersecting roof with equal spans is shown in **Figure 49-24.** In this example, both sections of the intersecting roof are gable roofs.

Framing Intersecting Roofs with Unequal Spans

The framing procedure for an intersecting roof with unequal spans differs somewhat from the procedure for a roof with equal spans. In a roof with unequal spans, the ridge board of the smaller roof section is lower than the main ridge board. The ridge board of the smaller roof section is fastened to the intersecting point of the shortened valley rafter and the supporting valley rafter. One method for framing an intersecting roof with unequal spans is shown in **Figure 49-25.**

Blind Valley Construction of Intersecting Roofs

Blind valley construction is a method of building intersecting roofs without valley rafters. **See Figure 49-26.** The main roof is sheathed and the intersecting section is built on top of the sheathing.

Boards (1 × 6s) are fastened to the top of the sheathing as a base for nailing the valley jacks. The roof section consists of common rafters and valley jack rafters. The length of the longest set of valley jack rafters is determined by subtracting the common length difference from the common rafter. The valley jack rafters require a seat cut combined with a side cut where they fasten to the 1 × 6. The layout for the valley jack rafter cuts is shown in **Figure 49-27.**

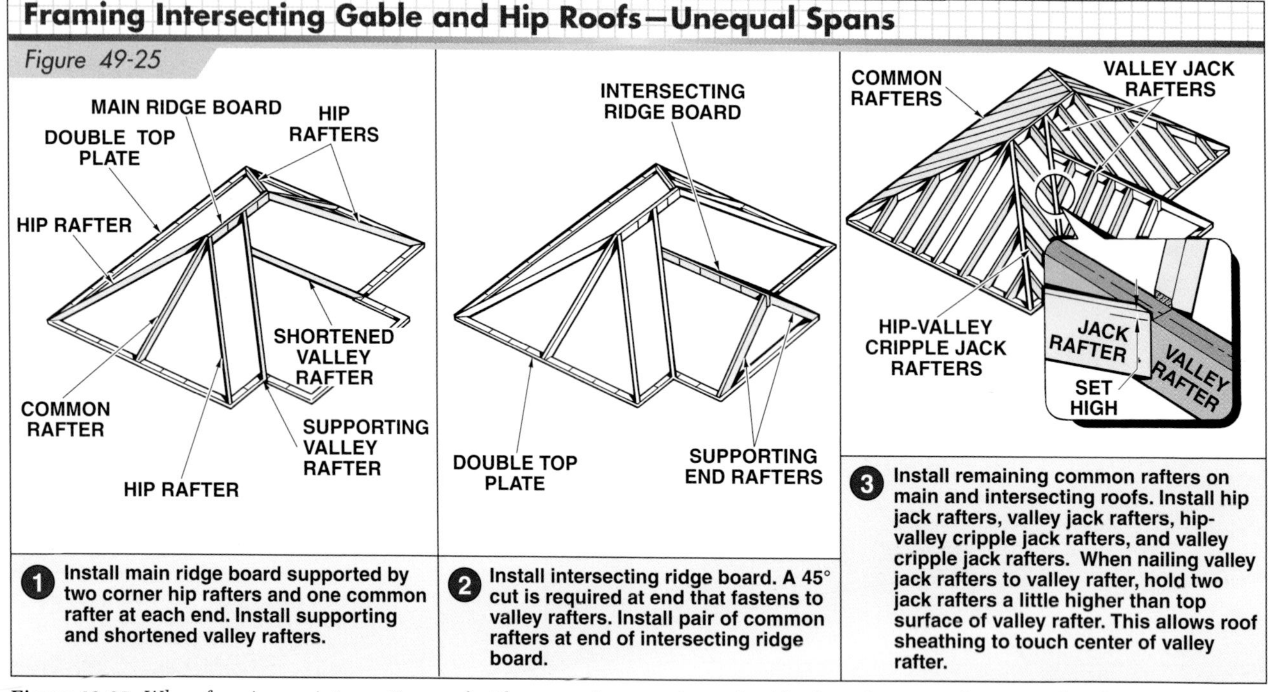

Figure 49-25. *When framing an intersecting roof with unequal spans, the main ridge board is set at the proper height and supported with the hip rafters and common rafters at each end. In this example, one section of the intersecting roof is a gable roof and the other section is a hip roof.*

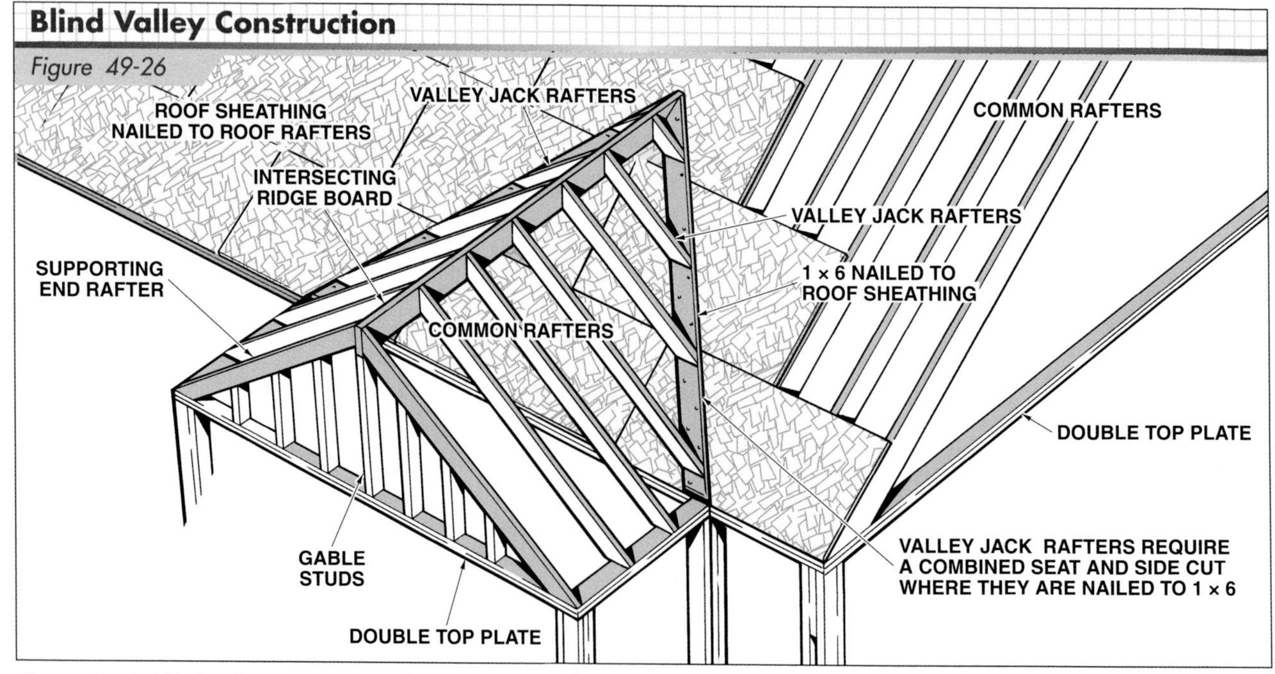

Figure 49-26. *Blind valley construction does not require valley rafters.*

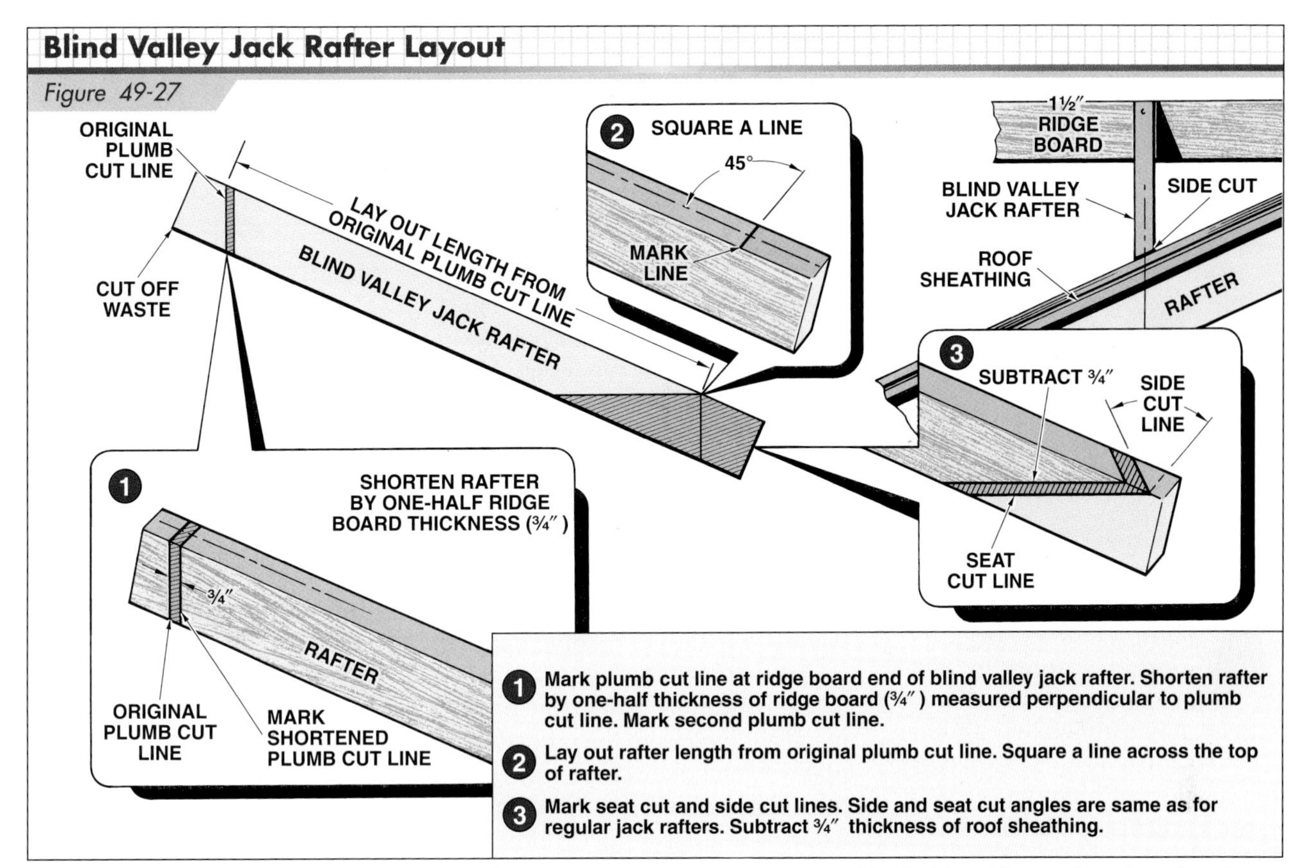

Figure 49-27. *The valley jack rafters for blind valley construction require a seat cut combined with a side cut where the rafters fasten to the 1 × 6.*

UNIT 50 Roof Trusses

OBJECTIVES

1. List and describe advantages of roof trusses.
2. Identify truss types and describe their uses.
3. Discuss the principles of truss design.
4. Describe roof truss installation.
5. List truss safety precautions.

A *roof truss* is an engineered combination of structural members arranged and fastened in triangular units to form a rigid framework for support of loads over a long span. Ends of the trusses bear directly on the opposing exterior walls. **See Figure 50-1.**

Since trusses do not require intermediate support, greater flexibility in room layout is provided. Truss roofs require less labor to construct than traditional framed roofs. Today, over 75% of new homes in the United States and 90% of new homes in Canada are constructed with roof trusses.

Trussed roofs in North America date back to colonial times, and were primarily used with timber-frame buildings. World War II brought about the demand for military housing that could be constructed quickly, and trussed roofs were one of the solutions. The trusses were constructed of dimension lumber and connected with glued and nailed plywood gussets, or the components were simply nailed together. After World War II, an improved fastening method using light-gauge steel plates began to replace the use of plywood gussets. The plates were predrilled to receive nails.

Metal plate connectors with teeth punched from the base metal evolved from the predrilled plates. Metal plate connectors are the most commonly used connector for trusses in residential and light commercial construction. Most trusses used today in residential and light commercial construction are called *metal plate-connected wood trusses.*

TRUSS TYPES AND COMPONENTS

The basic components of a roof truss are the *top* and *bottom chords* and the *web members*. Web members extend between the top and bottom chords and are tied together with metal plate connectors. The top chords serve as the roof rafters and the bottom chords act as ceiling joists. **See Figure 50-2.**

Wood Roof Trusses
Media Clip

Trusses may be placed over most types of walls, including wood-framed, precast concrete, and masonry walls. Trusses can be constructed to form various roof types including gable, hip, mansard, and gambrel. Trusses are also available for intersecting roofs. **See Figure 50-3.**

Figure 50-1. *Roof trusses are commonly used in residential construction. Metal plate connectors fasten the truss components together.*

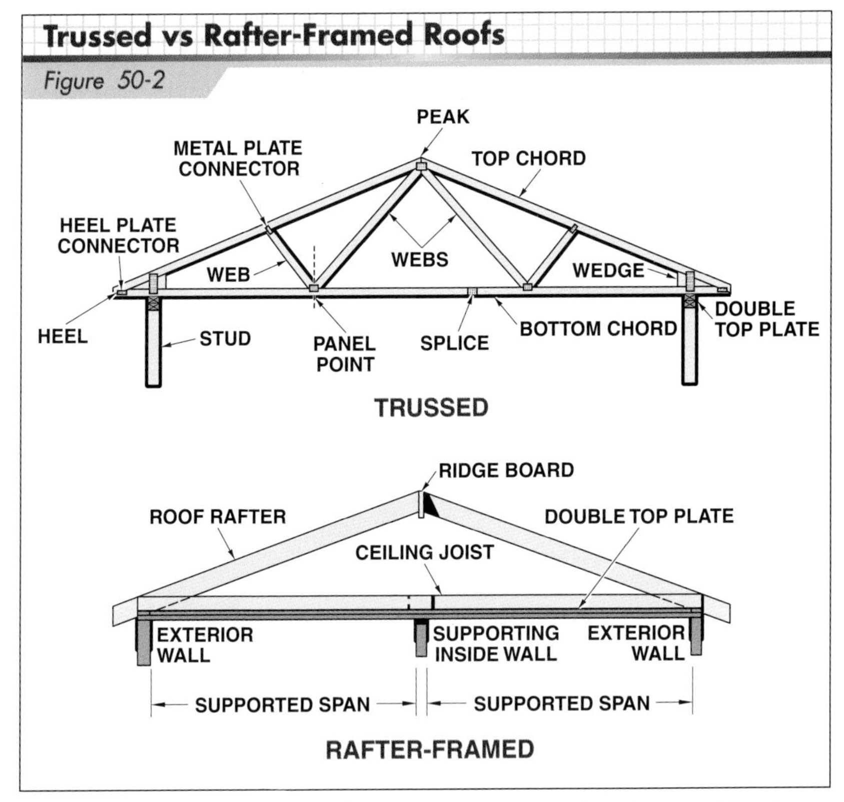

Figure 50-2. *Roof trusses serve the same purpose as rafters but provide a larger unsupported span.*

A variety of truss designs are available. Common truss designs include the king post, W-type (fink), queen post (fan), K-type (howe), room-in-attic, scissors, piggyback, hip (girder), and vault. **See Figure 50-4.**

Residential roof trusses range from 15′ to 50′ long. Roof pitch and span determine the truss height, which commonly ranges from 5′ to 15′.

Lumber

The minimum sizes of lumber used in truss fabrication are 2 × 4s for chords and 2 × 3s for webs. The sizes of chord and web members are increased for greater loads, spans, and stud spacing. The lumber should be identified by the trademark of a lumber inspection bureau or agency. Acceptable grades are Select Structural, No. 1, No. 2, or Machine Stress-Rated (MSR) lumber.

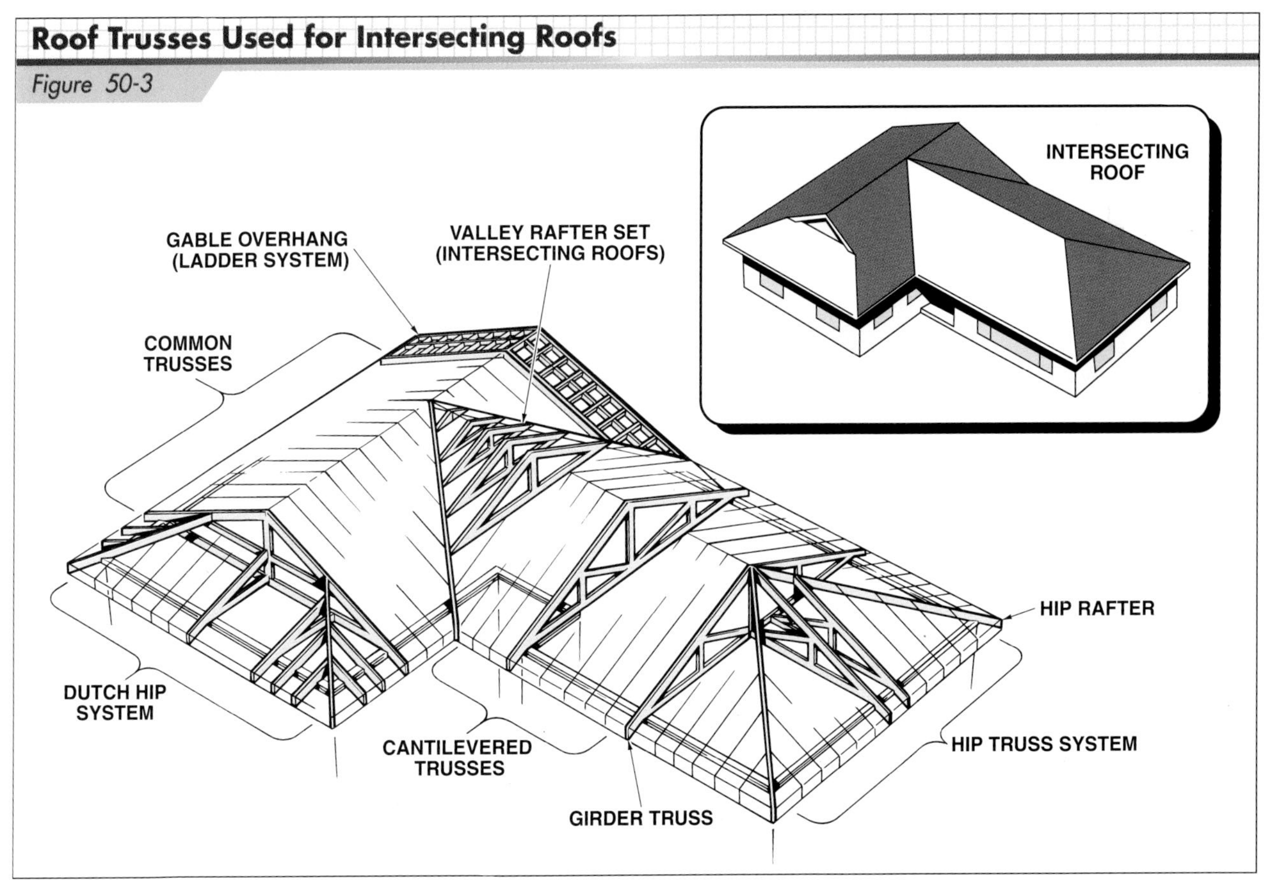

Figure 50-3. *Various roof truss designs can be combined to form intersecting roofs.*

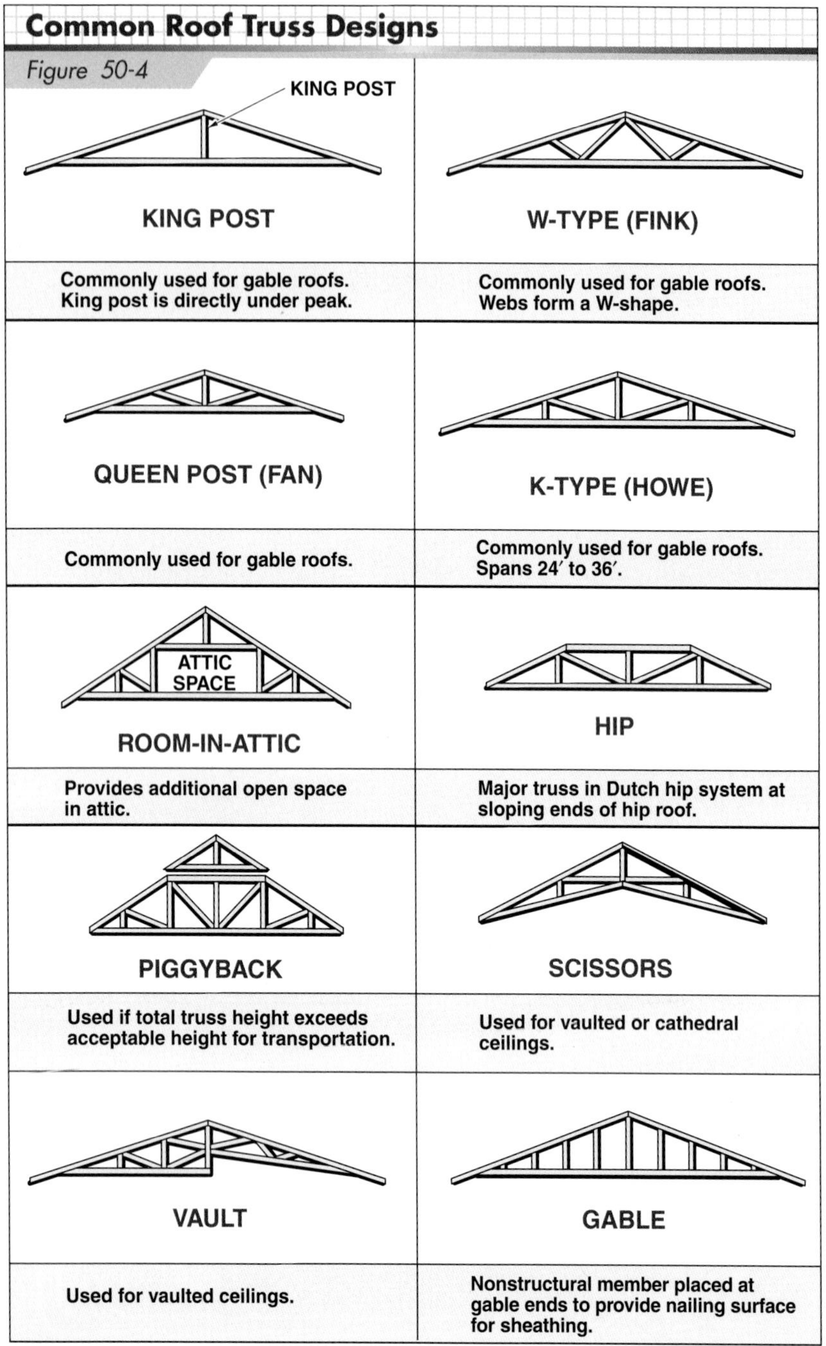

Figure 50-4. *A variety of standard roof truss designs are available. An endless number of variations of the standard designs are possible.*

Douglas fir and southern pine are two wood species used most often in truss fabrication. Other acceptable species are sitka spruce, lodgepole pine, ponderosa pine, and western balsam fir.

Metal Connector Plates

Metal connector plates tie together the chords and web members and distribute and transfer loads between adjacent members. Metal connector plates are manufactured from 16-, 18-, and 20-ga galvanized structural steel that has been machine-stamped to produce small teeth that protrude from the face of the plate. Tooth length ranges from $\frac{1}{4}$″ to 1″. **See Figure 50-5.**

Plates lose some of their effectiveness if the teeth are driven into holes, joint gaps, knots, bark, or pitch pockets. Joint gaps must be held to a minimum in order for the teeth to be an effective connector. The teeth are totally effective if the joint gap is less than $\frac{1}{32}$. As gaps widen, the effectiveness of the teeth decreases. A $\frac{3}{32}$″ gap renders the plate ineffective.

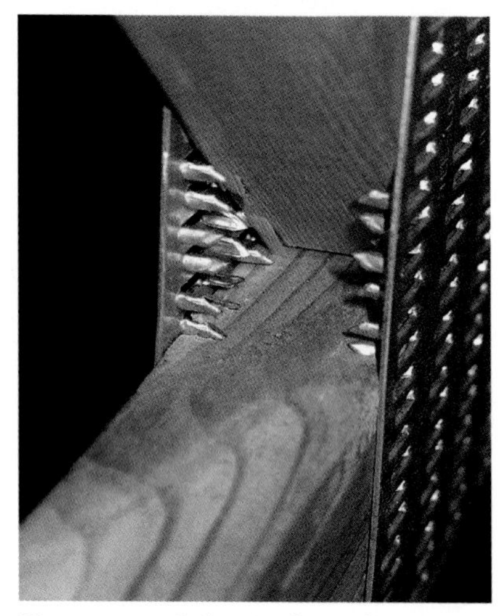

Figure 50-5. *Galvanized metal connector plates tie together chords and web members and distribute and transfer loads between adjacent members. The teeth are very sharp and should be handled carefully.*

Hinge plates are hinged, two-piece metal connectors used as an alternative to piggyback trusses. Hinge plates can be attached to the lower ends of top chords, allowing the chords to fold flat for shipping. At the job site, the top chords are raised into position and another hinge plate is fastened at the peak to join the adjoining top chords together. **See Figure 50-6.**

PRINCIPLES OF TRUSS DESIGN

The members of a roof truss form a series of adjoining triangular shapes. A triangular structure is a rigid geometric shape that resists distortion when force is applied. A roof truss is an engineered structural frame that rests on the two exterior walls of a building. The load carried by the truss is transferred to the exterior walls.

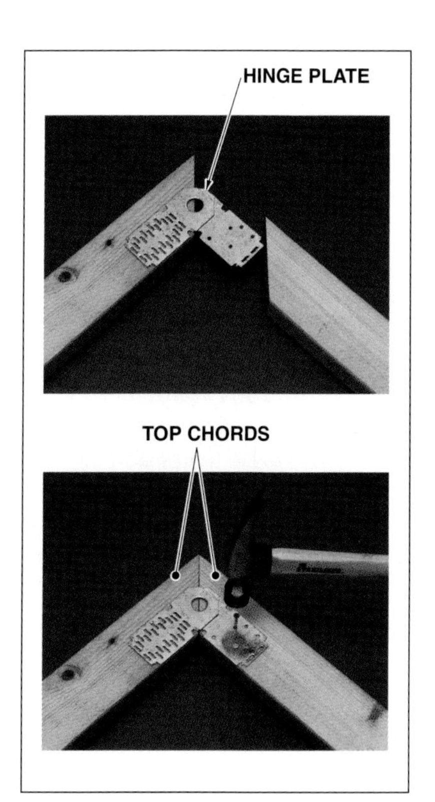

Gerhard Heidersperger, MiTek Industries, Inc.

Figure 50-6. *Hinge plates may be used as an alternative to piggyback trusses. The top chords are aligned at the peak and nails are driven through the prepunched holes.*

Weight and Stress

Truss design must accommodate the weight of the materials used for the truss framework, sheathing, finish roofing materials, and the finish ceiling materials. In addition, local snow and wind conditions must be considered. The less the slope of a truss, for example, the heavier the snow load it may have to support.

Tension and Compression

Each truss component is in a state of tension or compression. Components in a state of tension are subjected to a pulling force. Components in a state of compression are subjected to a pushing force. The balance of tension and compression gives the truss its ability to carry heavy loads and cover wide spans. **See Figure 50-7.**

TRUSS FABRICATION

Roof trusses are typically ordered from a truss manufacturer and delivered by truck to a job site.

A contractor generally furnishes a truss manufacturer with the roof framing plans for the building. The manufacturer then designs the trusses according to local building code and industry specifications. The trusses are then manufactured. Modern truss plants are highly automated facilities which utilize computer-aided controls to set the saws at the proper angles for cuts and lengths. Most truss manufacturers have a supply of stock types and sizes that are used in their surrounding areas, thus assuring quick delivery of the finished trusses to the job site.

Assembly Methods

Trusses are assembled on large tables at a manufacturing plant using various jigs, fixtures, and holddowns. As the truss components are precut, they are laid out on the assembly table and secured in position. A tight fit between truss members is required for structural integrity of the truss. To ensure tight fits, the cuts must be accurate. **See Figure 50-8.**

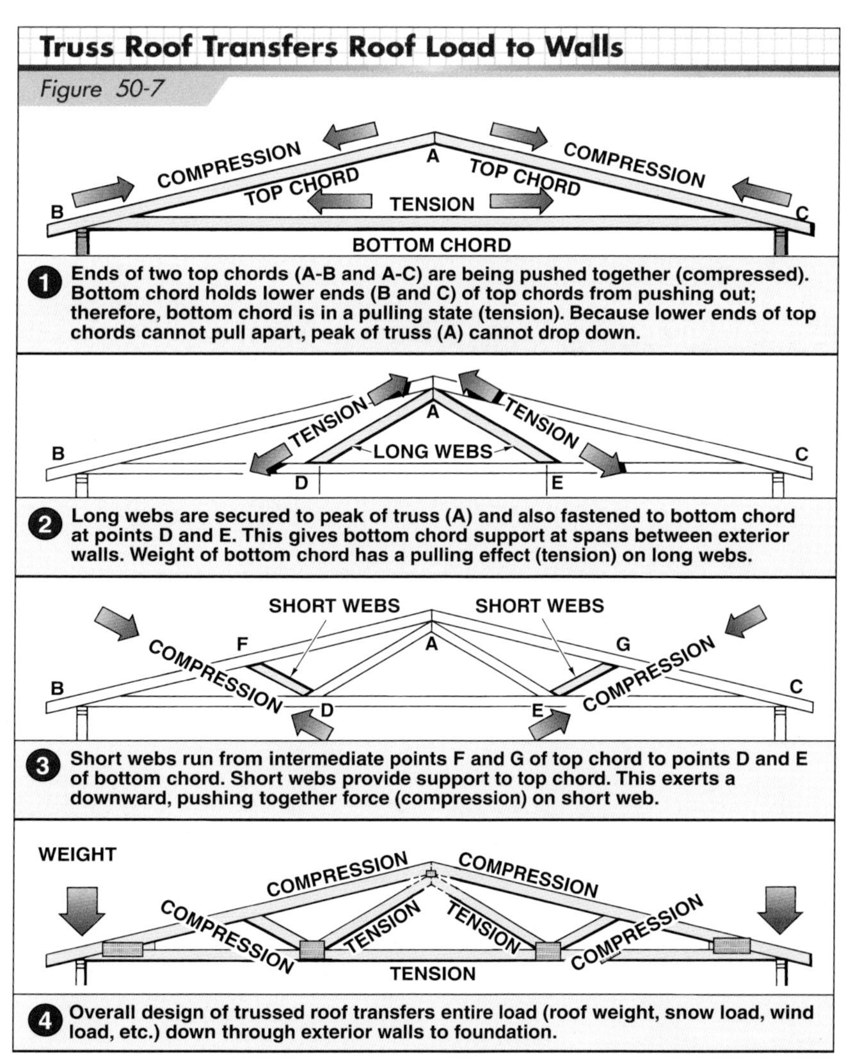

Figure 50-7. *The balance of tension and compression gives a truss its ability to carry heavy loads and cover wide spans.*

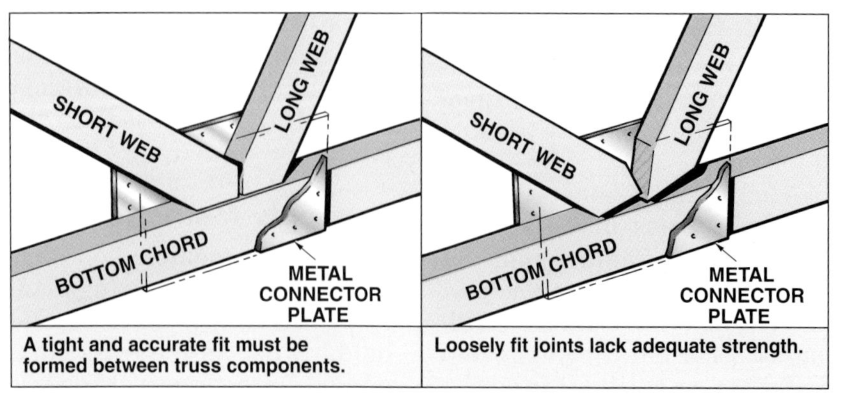

Figure 50-8. *A tight fit between truss members is essential to the structural integrity of the truss.*

The bottom chord of a truss tends to sag at the center after it has been set in place. To prevent sagging, the bottom chord is arched a small amount when the truss is being constructed to produce the desired camber (upward curvature).

Metal plate connectors are then properly positioned and pressed into the truss chords and web members. The metal connector plates are carefully positioned. **See Figure 50-9.** A large gantry press runs along rails on the sides of the assembly table, pressing the plates into the truss components. The truss then passes through a set of rollers that compress the connector plates a final time before shipment.

INSTALLING ROOF TRUSSES

Gable roofs require only one type of truss and are the easiest type of trussed roof to construct. Hip roofs are framed using a combination of trusses and conventional framing or by using a Dutch hip system. When framing intersecting roofs, the most efficient method is to place the valley trusses on top of the sheathing. **See Figure 50-10.** Before trusses are installed, all walls must be aligned and properly braced.

Trusses are commonly placed 24″ OC. Where possible, roof trusses should be placed directly over wall studs.

Wood Truss Council of America

Figure 50-9. *During manufactured truss assembly, metal plate connectors are pressed into truss components.*

Several options exist for safe truss erection including preassembling a roof section on the ground and raising it into place; using an OSHA-approved scaffold, ladder, or aerial lift; or using a roof anchor and fall-arrest system.

When preassembling a roof section, the trusses are positioned on a level surface, such as a subfloor or concrete slab, and properly braced. Sheathing is then attached to the top chords and the roof section is raised and fastened to the double top plate. The preassembled section can then be used as an anchorage point for subsequent sections of the roof.

Smaller and lighter trusses can be placed by hand on one-story buildings. The trusses are placed upside down between the walls. Using lifting poles with a V-shape at one end, one or two carpenters upright the truss and position it. Other carpenters on ladders along the exterior walls toenail the ends of the bottom truss chord to the top plate. If two poles are used to upright the truss, the poles should be positioned close to the quarter points of the span. If one pole is used, it should be placed at the peak when raising trusses. **See Figure 50-11.**

Longer and heavier trusses are hoisted by crane into the upright position. **See Figure 50-12.** If a truss is laid flat while being placed, lateral bending and jarring can place a strain on the connections and cause plates to loosen and pull away from the wood members. Bracket scaffolds may be attached to the inside of exterior walls so carpenters can fasten the trusses to the double top plate. For large commercial trusses, an aerial lift may be used along the inside of opposite exterior walls.

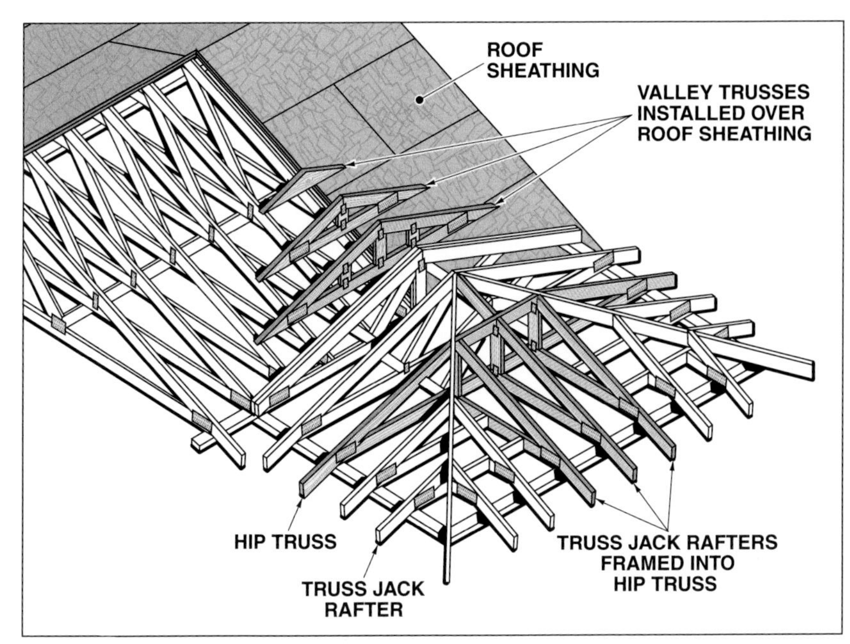

Figure 50-10. *When framing intersecting roofs, valley trusses are installed over the roof sheathing. Hip roofs can be framed using a Dutch hip system where truss jack rafters are attached to a hip truss.*

Some types of truss spacers are not designed to brace roof trusses; they are a convenient means of accurately laying out trusses. Always refer to the manufacturer instructions for proper use of truss spacers.

Metal Anchors and Hangers

Metal anchors, hangers, and angles are used to attach trusses to the tops of framed walls and to one another. **See Figure 50-14.** Metal anchors are used to attach the heels of a truss to the top plate of a framed wall. Truss hangers are used to attach trusses to other trusses such as when attaching hip jack rafters to hip rafters.

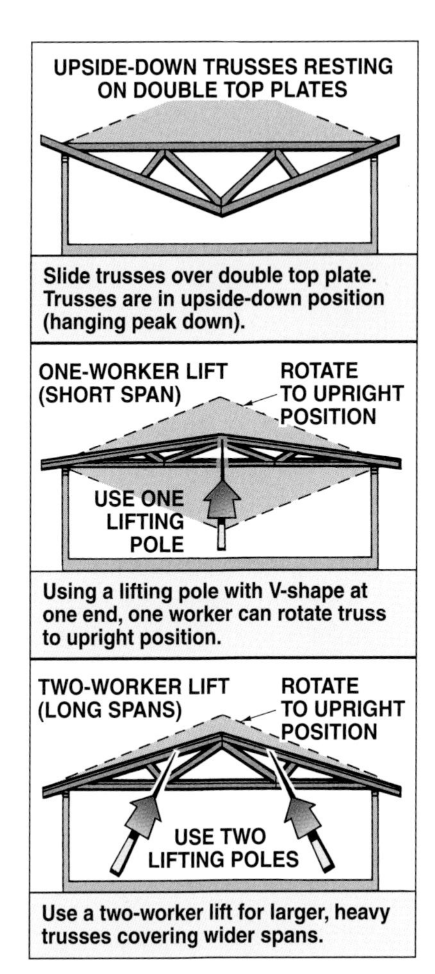

Figure 50-11. *Light trusses can be positioned by hand after they are placed upside down between the exterior walls of a building.*

Metal Truss Spacers

Metal *truss spacers* are a fast and accurate method for spacing trusses and eliminate the need to mark the top plate before placing trusses. **See Figure 50-13.** Truss spacers can remain in place under the roof sheathing.

Occasionally, trusses are damaged when they are unloaded or repositioned on a job site. The truss manufacturer should be immediately informed of damaged or field-modified trusses so the truss designer can produce repair detail drawings.

Southern Forest Products Association

Figure 50-12. *Longer and heavier trusses should always be hoisted by crane in an upright position.*

Metal Truss Spacers

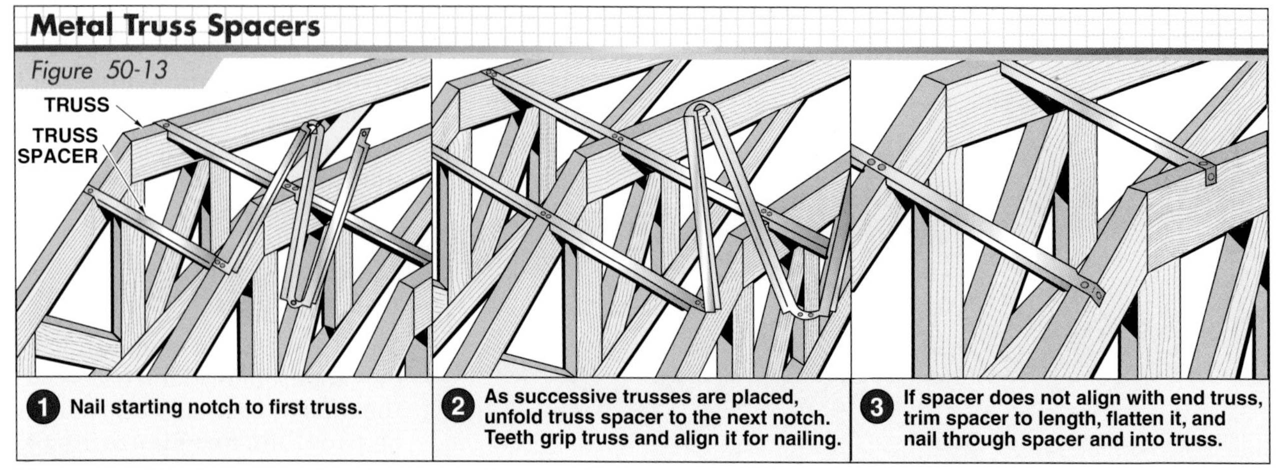

Figure 50-13. *Metal spacers such as these are used to properly space trusses and are not designed to be used as bracing. Sheathing is applied directly over the spacers.*

Truss Anchors and Hangers

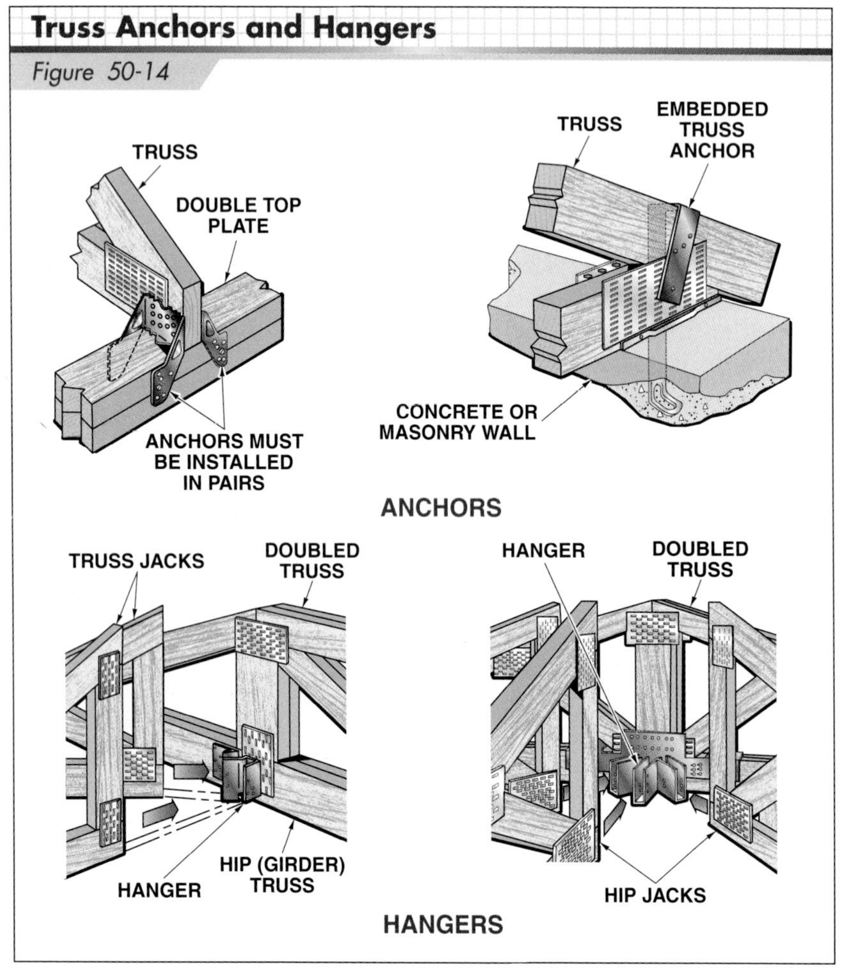

Figure 50-14. *Truss anchors and hangers are used to attach trusses to other structural members.*

Bracing Trusses

Proper temporary bracing while placing trusses is essential to prevent collapse of the trusses, with possible damage to the trusses and worker injury. Inadequate temporary bracing is the main cause of truss collapse during truss placement. Metal or wood braces may be used for bracing roof trusses.

HIB-91, *Commentary and Recommendations for Handling, Installing, and Bracing Metal Plate Connected Wood Trusses,* published by the Truss Plate Institute (TPI), has been developed for handling, installing, and bracing wood trusses spaced 2′-0″ OC or less. While there are alternate safe practices that can be employed when installing trusses, the information provided herein has proven to be an effective and safe means of installing trusses.

Trusses are erected starting at one end of a building and moving toward the other end. When the first truss is placed in a one-story building, the truss must be securely braced to the ground. In multistory buildings, truss braces are fastened to the subfloor. Vertical and diagonal braces should be installed every 8′, with one end nailed to the top chord and the other end nailed to a horizontal tie or stake driven into the ground. **See Figure 50-15.** Lateral braces are installed for additional stability.

Repairs to damaged trusses should not be attempted without proper repair detail drawings.

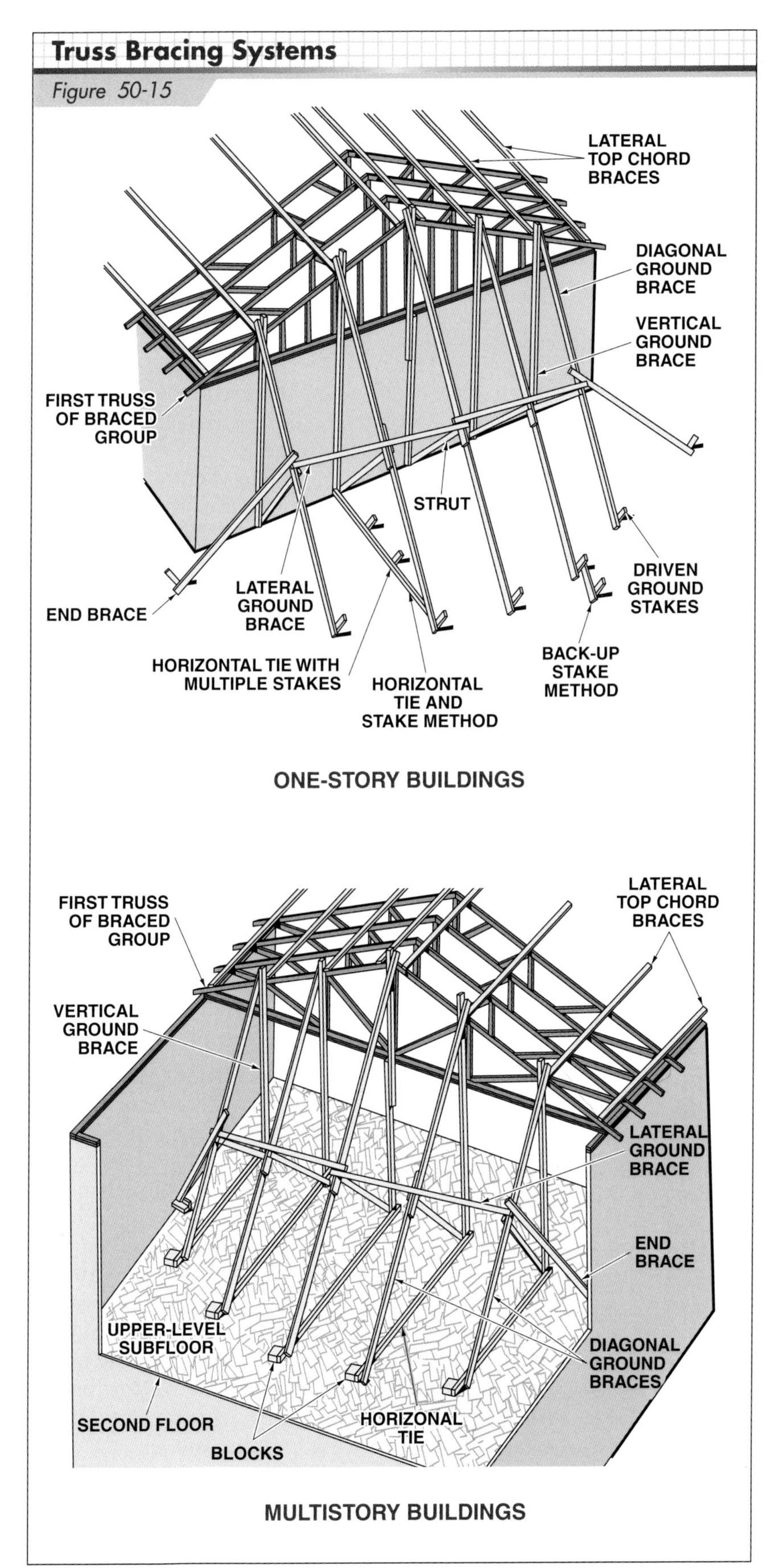

Figure 50-15. *For one-story buildings, the first truss is placed along the double top plate and is secured with vertical, horizontal, and lateral ground braces. For multistory buildings, ground braces are attached to the subfloor.*

One method of installing roof trusses is to first erect and properly brace the gable ends and run a string between them. Other roof trusses can then be aligned to the string.

After the first truss is properly braced, the remaining trusses are placed and properly braced. A 2 × 4 cut to 22⅜″ long can be used to space the trusses at 24″ OC.Starting at the heel end, 2 × 4 lateral braces are installed at 8′ intervals along the top chords. **See Figure 50-16.** Each lateral brace should extend over four or five trusses and be fastened to the top chord of each truss with two 16d nails. The ends of the lateral braces should overlap at least two trusses. Bottom chords should also be braced with 2 × 4s nailed to the chords at 15′ intervals. Diagonal braces can also be installed across several trusses every 30′ starting at one end of the building. Diagonal braces extending from each corner of the building and nailed to the bottom chords are also recommended.

Where permitted by the local building code, permanent metal lateral braces may be installed between roof trusses rather than lateral top chord braces. **See Figure 50-17.** Teeth in the flanges at both ends of the brace are driven into the sides of the truss using a hammer. Permanent metal braces must be supplemented with the same diagonal bracing as used with wood top chord braces. When the braces are in place, roof sheathing can be installed directly over the braces and top chords.

Webs and chords must not be drilled or notched without the approval of the truss designer.

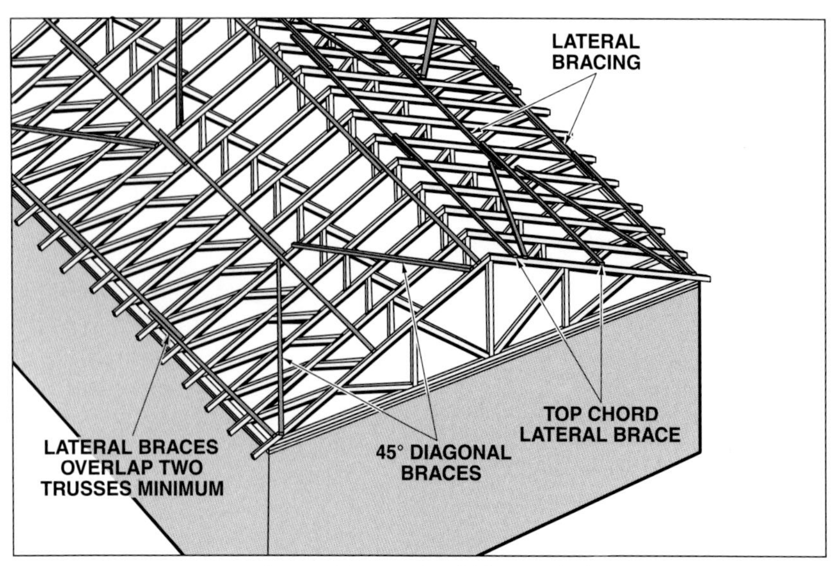

Figure 50-16. *Temporary diagonal braces should be installed across the top chords of the trusses at least every 30′ starting from one end of the building. Temporary lateral braces are installed at 8′-0″ intervals.*

Gerhard Heidersperger, MiTek Industries, Inc.

Figure 50-17. *Permanent metal braces may be installed between roof trusses as lateral bracing.*

Sheathing

Oriented strand board (OSB) or plywood panels are used as roof sheathing. When temporary wood or metal braces are used, the braces are removed as each panel is fastened in place. All braces should not be removed at one time.

TRUSS SAFETY

Fall protection requirements for residential construction are established in OSHA 29 CFR 1926.501, *Duty to Have Fall Protection.* Conventional fall protection is required when performing work at or over 6′. However, OSHA Instruction STD 3.1, *Interim Fall Protection Compliance Guidelines for Residential Construction,* modifies those requirements, permitting employers engaged in certain aspects of residential construction to routinely use alternative procedures instead of conventional fall protection.

Special care must be taken when setting roof trusses in place. Prior to placing the first truss, required scaffolds or ladders should be properly positioned and the necessary guardrails or fall-arrest system must be set up. For walls up to 8′ tall, sawhorse scaffolds can be erected inside the building below where the trusses will be located.

Safety precautions for walls over 8′ tall are as follows:

- Once truss installation begins, workers not involved with the installation must not stand or walk below the roof opening or exterior walls where they may be struck by falling objects.
- Trusses must be properly braced before they may be used as a support.
- Carpenters should be trained to make proper connections at the top plate or peak.

Carpenters should have no other duties during truss/rafter erection. The first two rafters are set from ladders or sawhorse scaffolds. When the heels have been fastened to the top plate, a carpenter will climb a ladder onto an interior partition top plate to secure the truss peaks. Carpenters working along the top plates secure the heels of the trusses to the top plate. Another carpenter detaches trusses from the cranes and secures the trusses at the peaks. Carpenters should work inside the truss webs in a stable position by sitting on a ridge seat or positioning themselves inside previously braced trusses and leaning through the trusses to the next truss.

The size and weight of a truss must be considered when installing trusses. Ensure there are an adequate number of carpenters available to properly support the trusses, especially the bottom chord. When lifting trusses using a crane, proper rigging such as spreader bars and taglines must be used to control the trusses.

When a portion of a trussed roof is properly braced and sheathed, a *roof anchor* and personal fall-arrest system may be used when installing the remaining roof trusses. Several designs of roof anchors are available. **See Figure 50-18.** Some roof anchors are made of metal and are attached to braced trusses or framing members with nails or screws. A D-ring provides an attachment point for a lifeline.

When manufacturing wood trusses, the moisture content of the lumber should not be less than 7%.

Miller Fall Protection

Figure 50-18. *Roof anchors are installed to provide an anchorage point for lifelines of personal fall-arrest systems.*

Unit 50 Review and Resources

SECTION

11

Energy Conservation: Energy Auditing and Construction Methods

CONTENTS

UNIT 51

Energy Auditing and Building Science

OBJECTIVES

1. Describe the energy auditing process and the necessary inspections.
2. Explain heat transfer.
3. List and describe the methods of heat transfer.
4. List and describe ways of measuring heat transfer and resistance.
5. Define condensation and describe how it can be prevented.
6. Describe the pressure difference between the outside and inside of a building.
7. Describe vapor barriers and ventilation.
8. Describe attic rafter vents and crawl space ventilation.

Energy auditing is a survey of a building's current energy use and efficiency. Both residential and commercial buildings may benefit from an energy audit. An energy audit identifies areas of energy waste and provides recommendations for improving energy efficiency. A proper comprehension of building science is important when conducting an energy audit and when analyzing and interpreting the data collected by the audit.

ENERGY AUDITING

An *energy audit* is a comprehensive review of the energy use of a building. Energy audits are performed on residential homes, small commercial buildings, and even large industrial structures. The audit involves collecting detailed information about every use of a particular utility resource, such as electricity, and prioritizing strategies to reduce its consumption. **See Figure 51-1.** When the audit is complete, a report is issued that addresses methods of reducing energy use through changes to the building envelope, equipment, and any procedures used.

Energy Auditing

Figure 51-1

Figure 51-1. *An energy audit involves collecting detailed information about every use of a particular utility resource and prioritizing strategies to reduce its consumption.*

For existing structures or structures in the process of being built, energy auditing is a way to document where a building's level of efficiency is at and identify areas where efficiency can be improved.

Energy audits are conducted by carpenters who have been certified as energy auditors. Almost all energy auditing certifications require a person to receive formal training, pass a certification exam, and demonstrate working knowledge of the subject by performing field exams. Qualified energy auditors must also be highly trained in a number of areas in order to assist homeowners and facility team members with energy-use management. An auditor must conduct, document, and evaluate an energy audit for a variety of building types and settings.

The process of auditing demands training in all types of mechanical and electrical systems and equipment and the use of various test instruments. Auditors must also be familiar with recent energy-efficiency programs, technological advances in building systems, and applicable legislation and regulations.

Energy-Efficiency Benefits

The primary goal of performing an energy audit is to identify ways to improve energy efficiency (using less energy to provide the same or better results). Energy efficiency projects often include improving home or building insulation, correcting any deficiencies in the building envelope, replacing equipment with more energy-efficient model, implementing more sophisticated system controls, adding new technologies, changing maintenance procedures, repairing leaks, and changing occupant activities.

The most obvious benefit of improved energy efficiency is reduced utility costs. However, energy conservation decisions are typically not made for financial reasons alone. There are often several other benefits, such as personal comfort and safety, environmental, or public relations advantages.

Reduced Utility Costs. Improved energy efficiency reduces overall energy use, which results in lower utility costs, especially for electricity and natural gas. These results should be directly reflected in the utility bills for the months following the implementation of efficiency-related changes.

Improved Occupant Safety and Comfort. Lighting and HVAC systems are typically the largest energy consumers and are also important to the safety and comfort of building occupants. Retrofits to these systems often improve both energy efficiency and the working and living environment. Daylighting, weatherization, and ventilation optimization are techniques for addressing both energy use and the health and safety of the building occupants.

Green Building Recognition. The achievement of particular levels of energy efficiency may be used to enhance public relations, particularly by businesses. Standardized levels of efficiency are recognized by independent organizations, and businesses may also publicize other measures of efficiency.

Energy Auditing and Project Costs

Total audit and project costs can vary depending upon the size of the home or business. These costs must be included as part of the energy-efficiency financial strategy of the homeowner or business. In order to choose the projects that are the most cost effective to implement, the potential savings and estimated total costs of each project must be compared.

An energy audit is intended to identify conservation measures that result in savings that outweigh, within a reasonable period of time, the cost of the implementation. Two methods for determining cost-effectiveness are to find the return on investment (ROI) and the payback period. **See Figure 51-2.**

Return on investment (ROI) is the ratio of a project's financial gain to its cost after a certain length of time. The elapsed time is important because savings, and sometimes costs, are often ongoing, which continually changes the ROI ratio. An ROI of less than 100% is not cost-effective because the costs are greater than the savings. An ROI of greater than 100% indicates cost-effectiveness. The *payback period* is the time elapsed until a project's resulting savings equal its costs. The end of the payback period is the point when the ROI equals 100%. At this point, the retrofit has paid for itself and further savings are a net gain.

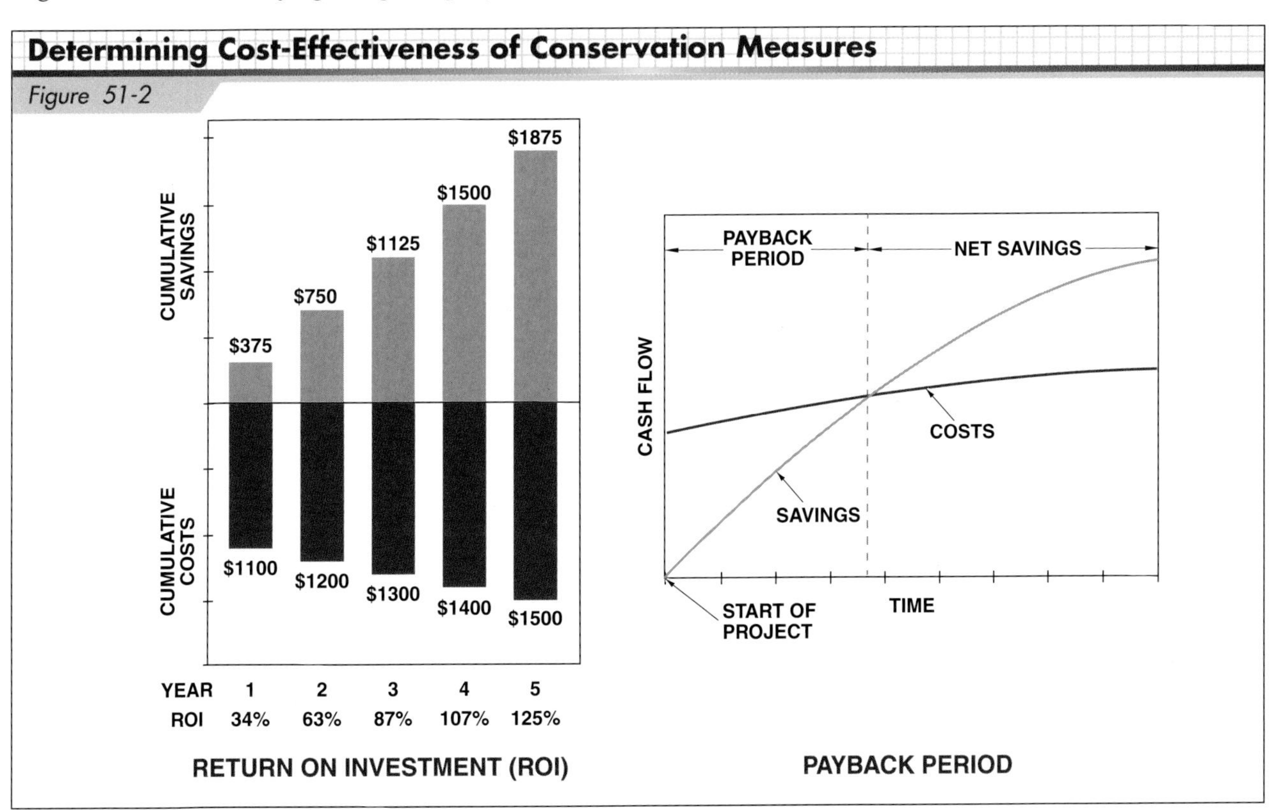

Figure 51-2. *Two methods for determining cost-effectiveness are to find the return on investment (ROI) and the payback period.*

These projections of cost-effectiveness are best guesses. Their accuracy improves with greater audit detail but does not guarantee future results in costs or savings. There are always unforeseeable variables that may affect cost-effectiveness.

CONDUCTING ENERGY AUDITS

The energy audit consists of several phases. The auditor first conducts an investigation, then analyzes the data, and finally, prepares an audit report. There are several data-collection activities and documentation tasks within these phases.

Audit Investigations

An audit investigation involves the inspection of each system within the scope of the audit. It starts at the source of energy and ends at the point of use. The audit investigation is used to assess the efficiency, physical condition, and operating profile of the building and equipment. Multiple test instruments may be used during the investigation to identify energy waste and other abnormal conditions. **See Figure 51-3.**

Data Analysis

The data gathered from energy audit inspections, measurements, and data logging is analyzed to identify and quantify savings opportunities. Some efficiency problems may have more than one possible solution. For each solution, the potential savings and implementation costs are estimated. Accurate data and realistic estimates are critical when calculating ROI.

Equipment specifications and measurements of energy waste are used to calculate potential savings. The exact method of calculation will vary, depending upon the system and energy source. However, the basic method for calculating potential savings involves proposing a change to the system, estimating the energy use of the modified system, calculating the cost of that energy use, and subtracting the new, lower cost from the present cost.

Almost all energy-conservation measures involve financial costs. Some costs may be very small, such as when implementing new procedures or caulking the gaps around a door, while others may be significant, such as when replacing large equipment.

Energy Auditing Test Tools

Figure 51-3

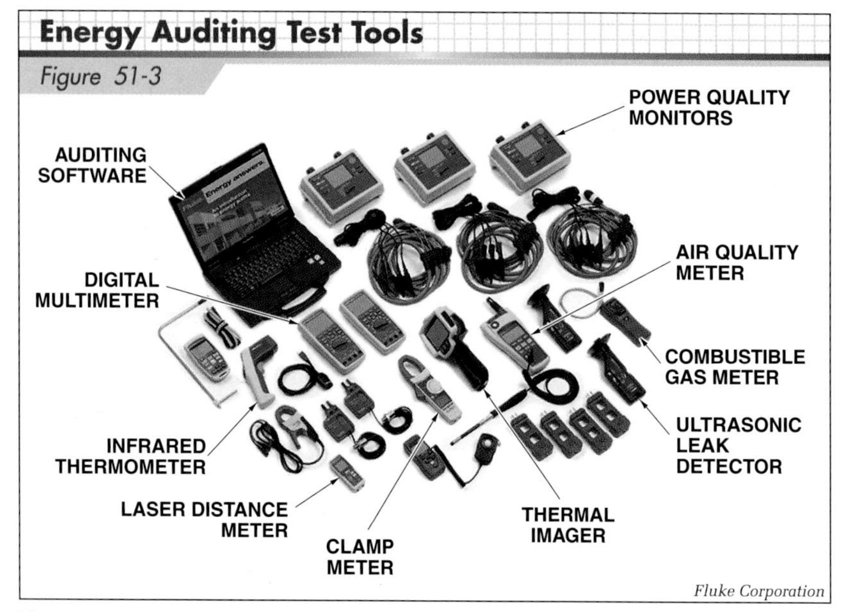

Figure 51-3. *Test instruments used during an energy audit include thermal imagers, digital multimeters, and air quality meters.*

Project costs are figured like any other estimating job. For equipment and materials, it is relatively easy to get purchase prices. Determining the necessary labor for each task is more difficult. This is generally calculated using the contractor's past experience with similar tasks. Some significant measures, such as replacing large equipment, may require soliciting quotes from specialized contractors.

With the potential savings and implementation costs calculated, the rest of the data analysis is relatively simple. Cost-effectiveness is based on the ratio of savings to implementation cost. If the savings are greater than the costs to implement the project, then it makes sense to recommend the project. If the savings are small compared to the implementation costs, then it may not be cost-effective to do the project.

Energy Audit Reports

The energy audit report consists of the complete documentation of the energy audit and the results of the data analysis. The report is necessary for presenting the findings to those who must decide which projects to pursue and when. The goal of the report is to summarize all of the energy savings opportunities and produce a prioritized action plan and implementation schedule for addressing those opportunities. The format of the report may vary, but it must be easy to read and understand by the target audience. Visual components, such as photos, charts, and graphs, are especially helpful at conveying meaning. The critical summary information is often presented first, followed by detailed information. **See Figure 51-4.**

The Building Performance Institute is one of the organizations that provides energy auditing certification.

Energy Audit Reports

Figure 51-4

Priority	Project	Monthly Savings	Project Costs	Project Duration	Payback Period (Months)	1 Year ROI	2 Year ROI	3 Year ROI
1	Turn off all office computers during nonworking hours	$75	$100	1 week	1.5	900%	1800%	2700%
1	Raise summer temperature setpoint to 75°F	$750	$2000	1 week	3	450%	900%	1350%
2	Repair compressed air system leaks	$230	$1200	1 month	6	230%	460%	690%
2	Seal building air leaks	$1000	$7500	2 months	8	160%	320%	480%
3	Replace exhaust fan motor (resizing)	$100	$1900	2 weeks	19	63%	126%	189%
3	Add VFDs to air handler fans, recalibrate controls for variable-volume	$1400	$28,000	2 months	20	60%	120%	180%
4	Retrofit light fixtures for fluorescent lamps	$1950	$39,500	2 months	21	59%	118%	178%

PROJECT SUMMARY

Seal Building Air Leaks

An inspection of the entire building envelope with a thermal imager revealed areas around doors and windows where cold air was entering the building. **See included thermal image.** (At the time of the inspection, the indoor temperature was 71°F and the outdoor temperature was 42°F.) When inspected visually up close, the leaks were found to be due to absent or damaged caulking and absent weather stripping.

The recommended repair is to remove all existing

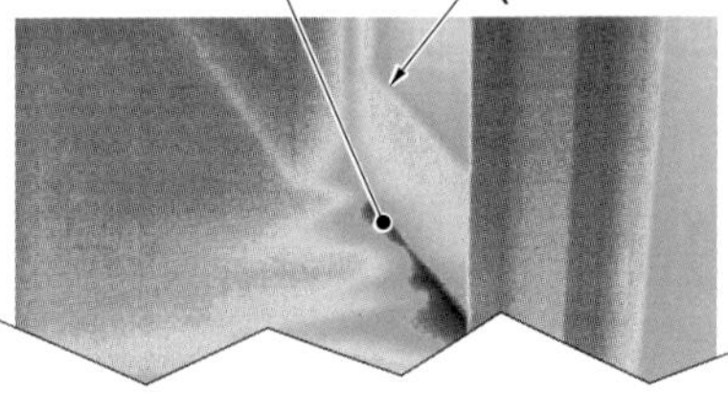

COLD OUTDOOR AIR

EXTERIOR DOOR (NW CORNER)

PROJECT DETAILS

Figure 51-4. *The energy audit report consists of the complete documentation of the energy audit and the results of the data analysis.*

Based on the results outlined in the audit report, the recommended projects, budget, and constraints can be evaluated. The audit report should prioritize savings opportunities because not all will be implemented, at least not initially. The projects with the highest ROI are typically chosen, while the others are kept for later consideration. Therefore, while the audit report should include estimated schedules, the final implementation plan cannot be compiled until the project list is determined.

VISUAL INSPECTION PROCESS

A residential energy audit begins by inspecting the overall size, shape, orientation, and features of a home or business and its surroundings, which can be done visually. This information, along with a sketch of the site, is documented on the appropriate audit forms. Photographs should be taken of the site to document the overall condition of the structure's features.

The visual inspection begins by looking at any protrusions or penetrations through the exterior walls and roofs. Any penetrations of plumbing piping and vents should be examined carefully. Areas where walls meet the roof or breaks in the continuity of a wall are places where the infiltration or exfiltration of air is possible. Cupolas, porches, cantilevers, bump outs, and similar features are potential weak links in the building envelope. Signs of problems include siding deterioration, large gaps where two types of building materials meet, masonry cracks or efflorescence, peeling paint or rotting trim, window fogging, and wet spots. **See Figure 51-5.** Any suspected leaks also should be noted for further investigation.

Photographs should be taken of any potential problem areas and should be included with the audit report. These photographs also allow for comparison should any repair work be needed after the energy audit.

Visual Inspections

Figure 51-5

Figure 51-5. *Potential leaks in the building envelope may be found through visual inspection.*

Fenestrations

Figure 51-6

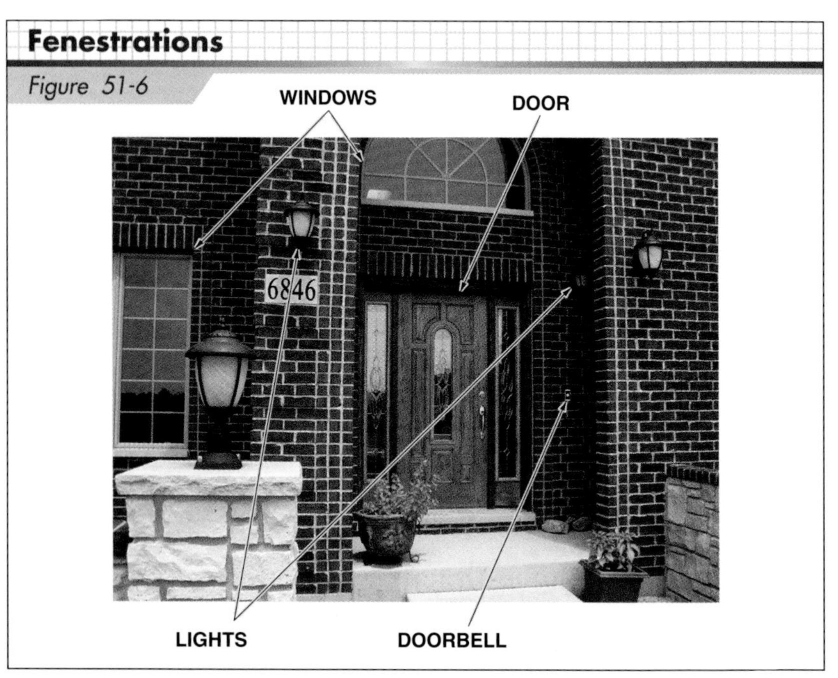

Figure 51-6. *Some examples of fenestrations include windows, doors, outdoor lights, and doorbells.*

A person performing an energy audit must be able to determine the construction method of the building by visual inspection and previous experience or by examining the construction prints, if available. Knowing the building construction method is important for two reasons. First, it helps to understand the nature of building envelope leaks if they are discovered during later audit testing. For example, since the inside of a wall is hidden, the cause of a problem may be located somewhere other than where it is discovered. Understanding wall construction can help in discovering the cause. Second, knowing the construction method is also necessary when it is time to develop, recommend, and implement energy-saving retrofits that help increase the building's energy efficiency.

All fenestrations need to be examined in great detail. A *fenestration* is an intentional opening in a building envelope. The most common fenestrations are doors and windows. Other types of fenestrations include doorbells, outdoor lights and outlets, bathroom or exhaust fan vents, and chimney flues. **See Figure 51-6.**

Because they are potential avenues of heat loss or gain through the building envelope, it is important to consider fenestrations during an energy audit. Any loose or missing caulk or weather stripping should be noted and any frame or glazing that might be missing, damaged, or cracked should be noted as well. It is also necessary to determine and document the types and insulation value of doors and windows.

Water Infiltration

The auditor should be aware of any sign of water infiltration in the building. Signs of water infiltration include stained foundation walls or interior walls or ceilings, bubbling or peeling drywall or plaster, visible mold or mildew, or a musty smell. Water infiltration can occur from exterior sources such as roof or foundation leaks or from interior sources such as inadequate ventilation, plumbing piping or fixture leaks, or cracks in tubs or shower walls or bases. **See Figure 51-7.**

Water Infiltration

Figure 51-7

Figure 51-7. *Water infiltration can cause both building damage and health problems.*

The damaging power of moisture is significant. Water destroys a building faster and more effectively than any other force, aside from a natural disaster. Besides physical damage, moisture can also lead to mold growth, which can be a serious health problem. From an energy-use perspective, moisture can lower the thermal resistance of building materials such as insulation, allowing heat to be conducted more easily into or out of a home. As a rule of thumb, for every 1% of water absorption, there is a reduction of about 5% in the R-value of a material.

BUILDING ENVELOPE INSPECTION

The *building envelope* is a continuous thermal and air boundary separating the conditioned space from any unconditioned space or from the outside. Leaks in the building envelope are a major cause of energy loss due to the infiltration of outdoor air or exfiltration of conditioned (heated or cooled) air. Infiltration and exfiltration of air can usually be detected with a thermal imager.

A *thermal imager* is a device that detects heat patterns in the infrared wavelength spectrum without making direct contact with the targeted area. **See Figure 51-8.** When inspecting the building envelope, a thermal imager can immediately detect the temperature differences that can indicate air leaks or moisture intrusion.

Blower Door Tests

The most popular and convenient way to test the integrity of a building envelope is with a blower door test. A blower door exhausts air out of the sealed space (the inside of the building envelope) to create a pressure difference of 50 Pa in relation to the outside. **See Figure 51-9.** The exhausted air creates negative pressure in the building, which causes the outside air to leak into the building through cracks and holes. A pressure gauge is then used to calculate the amount of air leakage into the building.

Retrotech Inc.

The smoke generated from an air current tester is used to determine the location of air leaks in a building.

The basic system for a blower door test includes three components: a variable-speed calibrated fan, a door panel system, and instruments to measure fan airflow and building pressure. The fan is located in an exterior doorway opening using a door panel system, which seals the opening against any other airflow. Initially, the fan is sealed to allow the instruments to determine the standard pressure difference between the inside and outside of the building envelope. **See Figure 5-10.** The fan is then uncovered and used to blow air into or out of the building, which creates a small pressure differential between the inside and outside.

Pressure differential is the difference between the pressures on either side of a barrier. This pressure differential forces air through all holes and penetrations in the building envelope. A tighter building (one with fewer leaks) requires less airflow from the fan to maintain a predetermined and constant pressure differential than a building with more leaks.

Thermal Imagers

Figure 51-8

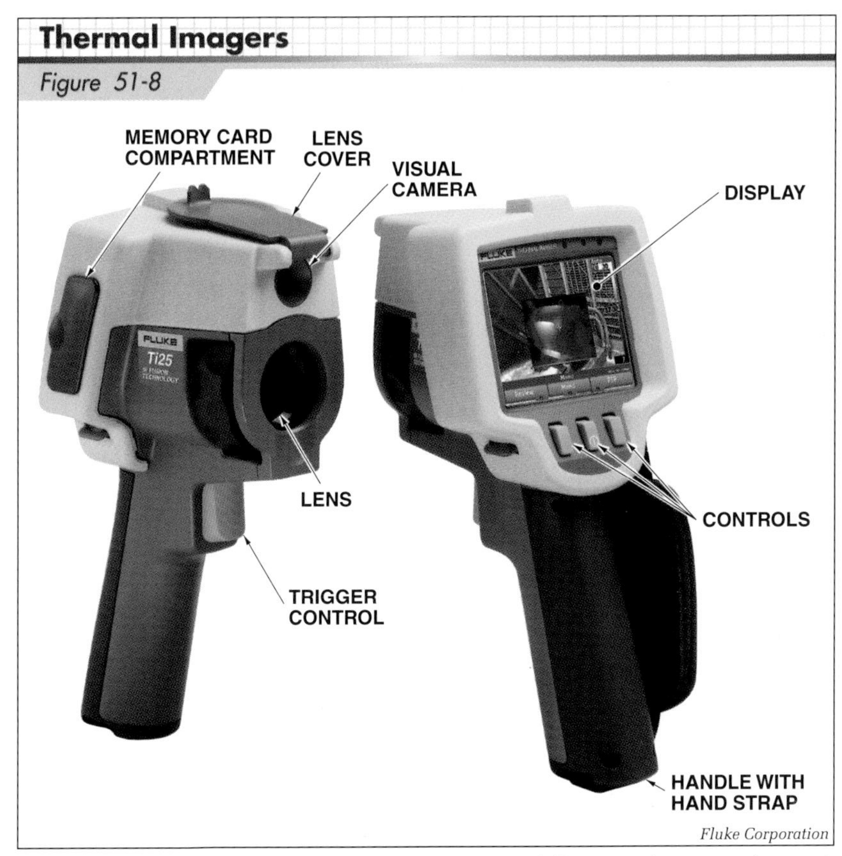

Fluke Corporation

Figure 51-8. *Thermal imagers detect the temperature differences that can indicate air leaks or moisture intrusion into the building envelope.*

Blower Door Tests

Figure 51-9

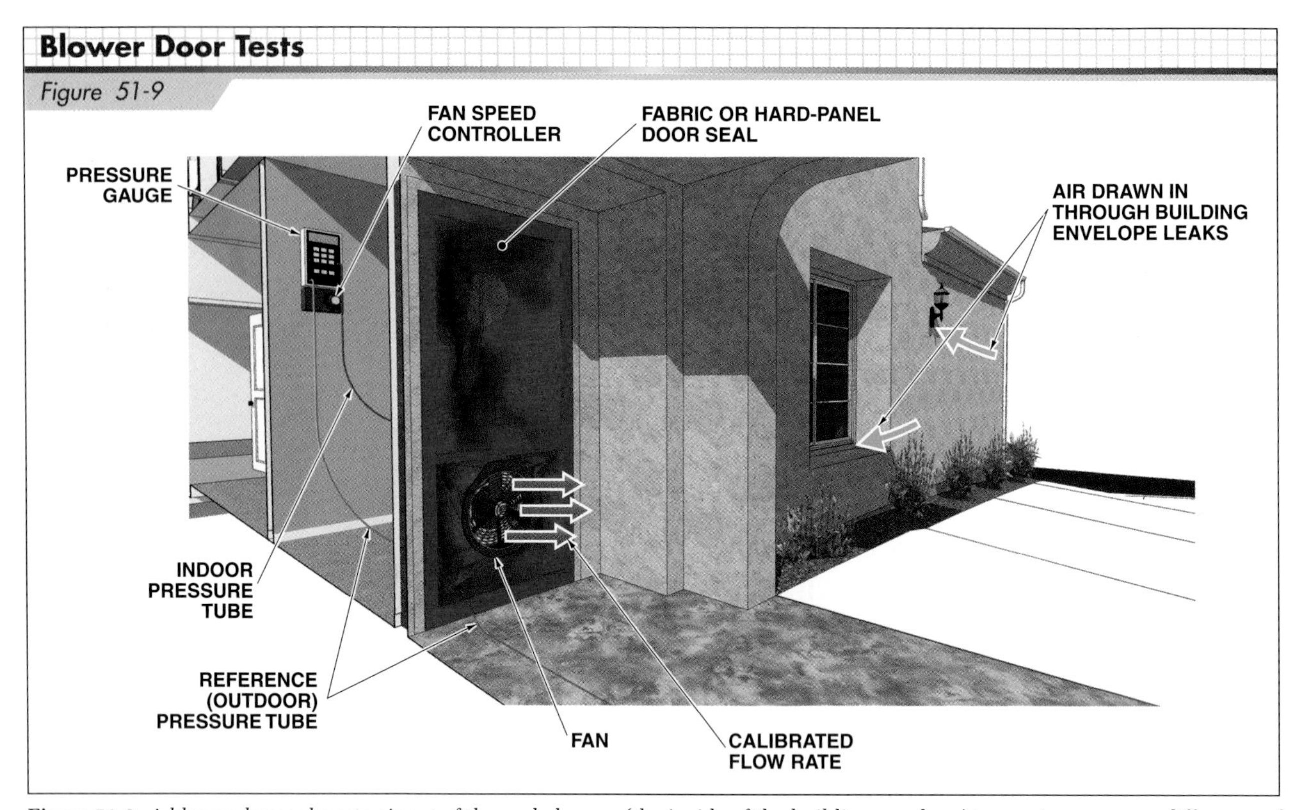

Figure 51-9. *A blower door exhausts air out of the sealed space (the inside of the building envelope) to create a pressure difference of 50 Pa in relation to the outside.*

Blower Door Test Equipment

Figure 51-10

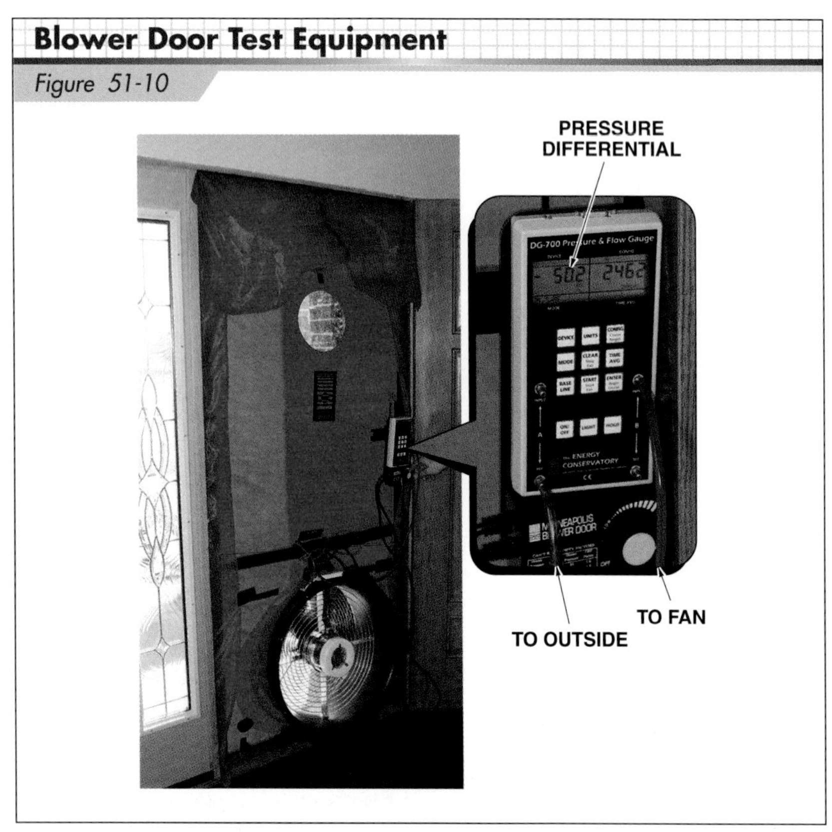

Figure 51-10. *For a blower door test, a fan is temporarily sealed into an exterior doorway using the door panel system and is used to create a small pressure differential between the inside and outside of the structure.*

Before conducting the blower door test, the following conditions must be met:

- Windows and exterior doors must be closed.
- Combustion devices, including pilot lights for furnaces or water heaters, must be turned off.
- Fireplaces must be clean of ashes, and chimney dampers must be closed. If ashes remain in the fireplace, they should be covered with wet newspaper to prevent them from being blown into the house. Other loose household materials must also be secured. These materials can be moved around by airflow, especially if located close to a major leak or the fan itself.
- House thermostats must be off.
- All air movers (attic exhausts, ceiling fans, exhaust fans, appliance fans, etc.) must be off.

- Window air conditioner vents to the outside must be closed.
- Interior room doors must be open (except closet doors, which can remain closed) and secured in the open position. If a door suddenly shuts while using a door fan, the sudden change in pressure can be enough to damage an enclosure or pop the fan out of the panel. Also, the accuracy of the test may be affected since the area behind the closed door will not be subject to the test.
- Duct registers must be open.
- Return air filters must be removed if they are dirty.

Failure to address these conditions can result in both inaccurate readings and damage to the home. When these conditions are met, the test can begin. First, the blower door fan is turned on and the speed is slowly increased until the differential pressure reading reaches 50 Pa. Once the pressure is steady, the instruments indicate the amount of airflow being created by the fan. This value is referred to as CFM50 and is used later in the audit to calculate air infiltration.

Next, the fan is left on while the auditor checks all of the rooms for signs of infiltration. There are many areas where air leakage can occur. **See Figure 51-11.** Some air infiltration can be easily felt with the hand, especially if the outdoor air temperature is low. However, a smoke generator or thermal imager provides a more accurate way to detect the presence of airflow from suspected leaks. **See Figure 51-12.** Both are good tools for identifying air leaks because the blower door test forces air from the exterior through leaks in the building envelope.

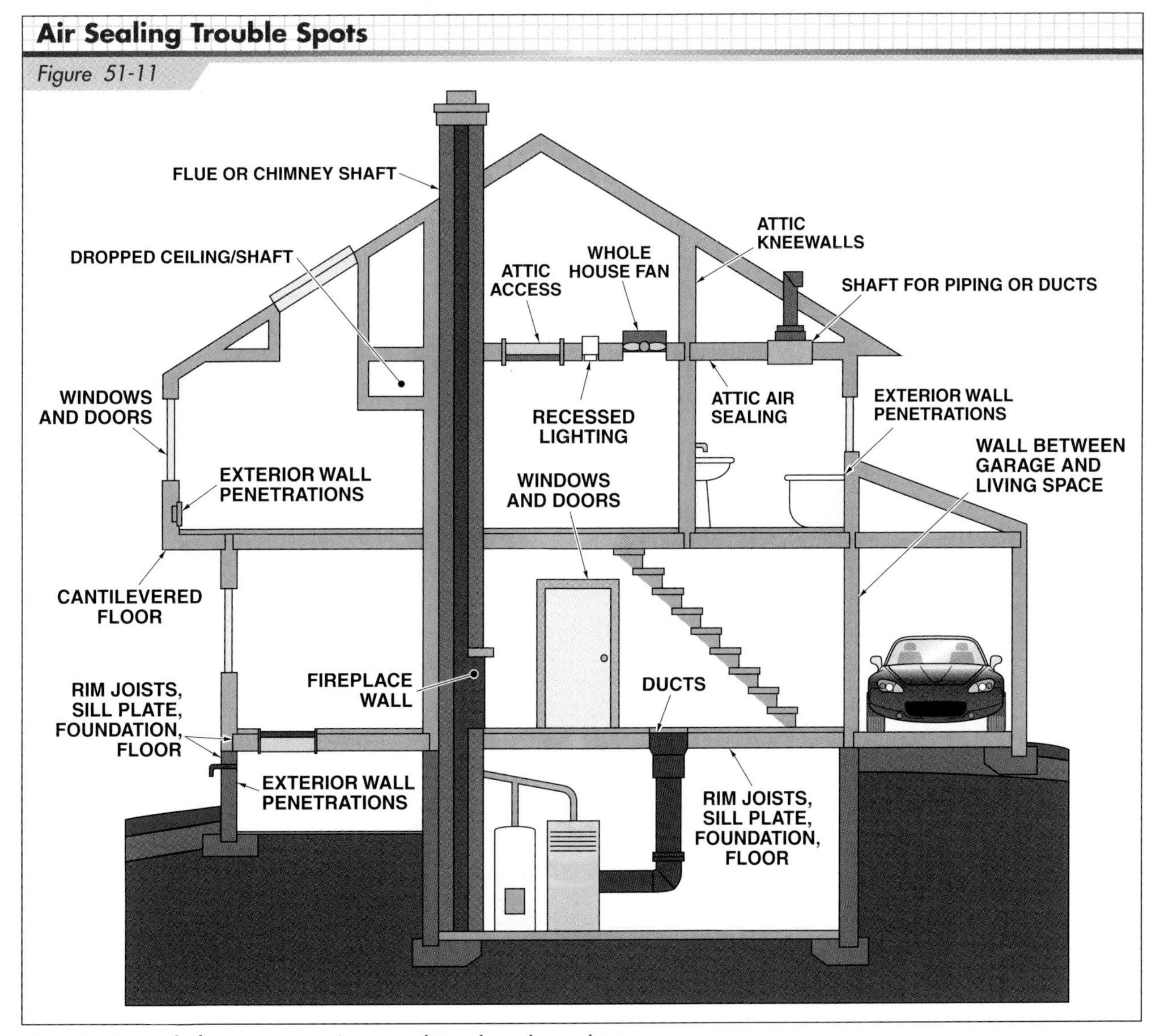

Figure 51-11. *Air leakage may occur in many places throughout a home.*

Air Infiltration Detection

Figure 51-12

SMOKE GENERATOR

THERMAL IMAGER

Figure 51-12. *A smoke generator or thermal imager is used to detect air leaks.*

With a thermal imager, even quick scans of an area from the middle of the room can reveal suspected leaks. A thermal imager can also detect cold areas in a wall that may not be the result of air infiltration. These thermal cold spots may be the result of sections of missing insulation, wet insulation, or a missing air barrier that allows the exterior environment to affect the interior of the structure. The images of any air infiltration or thermal cold spots should be included with the energy audit report with a description of where the problem occurred and an explanation of what is being shown in the image.

After the blower door test is complete, the house is returned to its condition prior to the test. The auditor must also remember to relight any pilot lights on any of the combustion appliances.

Commercial Blower Door Tests

Blower door testing of large commercial or industrial facilities may require a blower door with a more powerful fan or multiple fans. **See Figure 51-13.** Also, the structure may need to be partitioned into smaller test sections if the area to be tested is too large or complex. This process allows each section to be tested separately, but requires more time for site preparation and equipment setup, extra coordination of building occupant movement, and more complex calculations to determine overall building efficiency.

Retrotec Inc.

Figure 51-13. *Blower door testing of large commercial or industrial facilities may require a blower door with a more powerful fan or multiple fans.*

HVAC EQUIPMENT TESTS AND OTHER INSPECTIONS

Another major portion of an energy audit consists of tests to evaluate the performance of a building's HVAC system, which is usually the largest consumer of energy. Equipment efficiency is not easily tested in the field, so the ratings for each major unit are collected from the manufacturer. However, since it cannot be assumed that the equipment remains as efficient as it was when new, other information, such as its age, condition, and overall energy consumption, is gathered to help an auditor estimate equipment efficiency.

While building envelope integrity and HVAC system operation typically have the most influence on energy use in a building, noticeable improvements can be made to other systems as well. Infrared thermometers or thermal imagers can be used to inspect fuses, circuit breakers, and many other types of electrical equipment, loads, and panels. Thermal inspections may reveal both electrical problems, such as poor power quality, and mechanical problems, such as bad bearings in motors.

BUILDING SCIENCE

In order to properly conduct an energy audit and interpret the audit's findings, there must be an understanding of the building science associated with a residential or commercial structure and how the mechanical and electrical systems of a building interact with the building envelope. It is important to understand the physics of heat transfer, moisture control, and air infiltration and how these principles affect and interact with one another.

Building science, also known as building dynamics or building physics, is the study of interaction between a building (including its HVAC, mechanical, and electrical systems), its inhabitants, and the surrounding environment.

HEAT TRANSFER

Heat always moves from warm areas toward cold areas. In the winter, warm air inside a building escapes through the framework of the building, moving to the cooler air outside. In the summer, warm air outside a building flows toward the cooler air inside. **See Figure 51-14.** The flow of heat is called *heat transfer.*

Construction materials used for the exterior walls of buildings (including wood, brick, concrete, and masonry) prevent a certain amount of heat flow through the walls. However, a larger amount of heat flow can be prevented by adding *thermal insulation.* The term "thermal" describes materials that have high heat flow resistance. Insulation is installed by carpenters or insulation specialists, and is placed in the walls, floors, and ceilings that surround the living areas of a building. Insulation should be placed in any section of a wall, ceiling, or floor that is adjacent to the exterior of a building or an unconditioned space. **See Figure 51-15.**

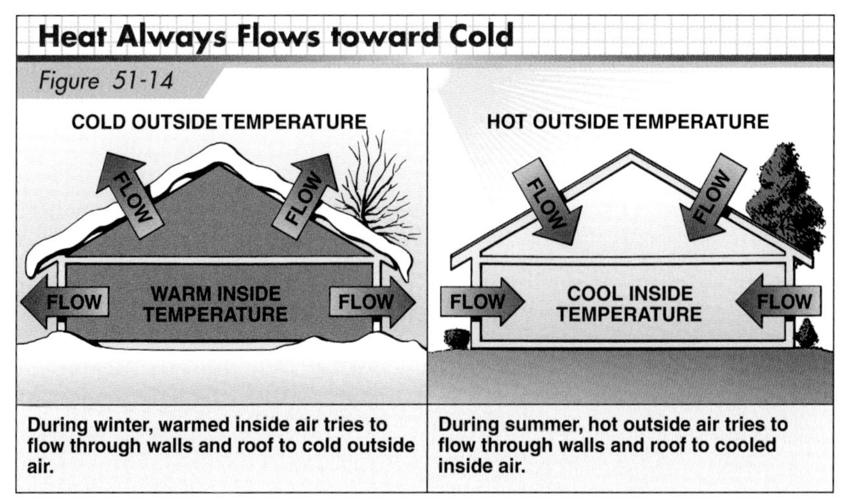

Figure 51-14. *Heat always moves from warm areas toward cold areas.*

Heat Transfer Methods

The three methods by which heat transfer occurs are *conduction, convection,* and *radiation.*

Conduction. Conduction is the movement of heat through a solid or liquid. Heat passes from one molecule to another. When heat is applied to one end of a metal rod, for example, the other end of the rod eventually becomes hot by means of conduction. **See Figure 51-16.** Dense materials, such as metals, are called *conductors* because they transfer large quantities of heat quickly. Less dense materials, such as wood and plastic, are called *insulators* because they do not transfer heat quickly. Therefore, insulators and insulation are used to retard heat flow through a solid, such as a door.

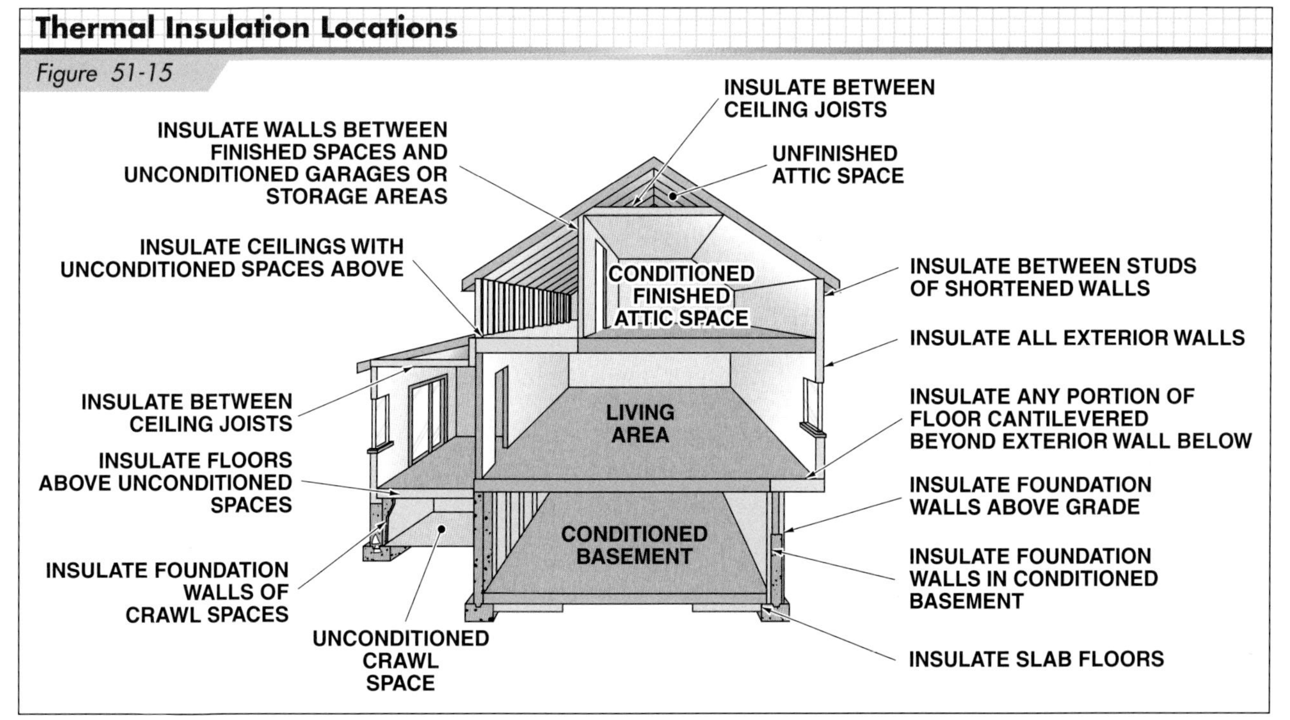

Figure 51-15. *Insulation is typically placed in the walls, floors, and ceilings that surround the working areas of an enclosed structure.*

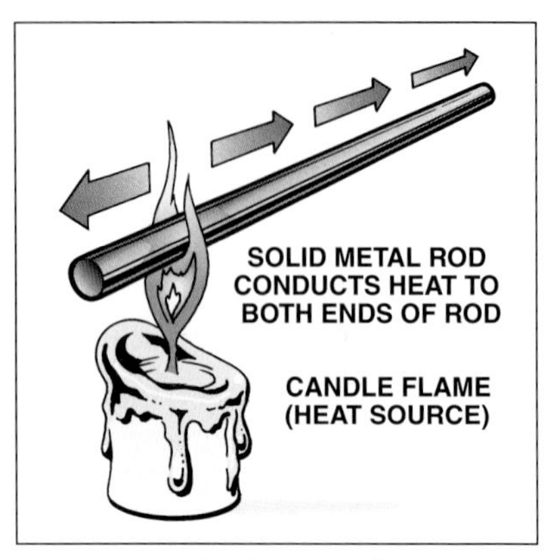

Figure 51-16. *Conduction is the movement of heat through a solid or liquid. For example, the heat from a flame is conducted from one end of a metal rod to the other.*

Convection. Convection is the movement of heat through the circulatory motion of air or liquid. An example of heat transfer by convection is air heated in a furnace and carried to different areas of a house. **See Figure 51-17.** When air is heated, it expands, becomes lighter, and rises. Colder air then replaces the warmer air. As a result, a convection loop is established that circulates the heat throughout the house.

Another example of convection is the way heat is circulated in a pipe by hot water moving through it. In convection, heat is transferred by the movement of a fluid (gas or liquid) substance.

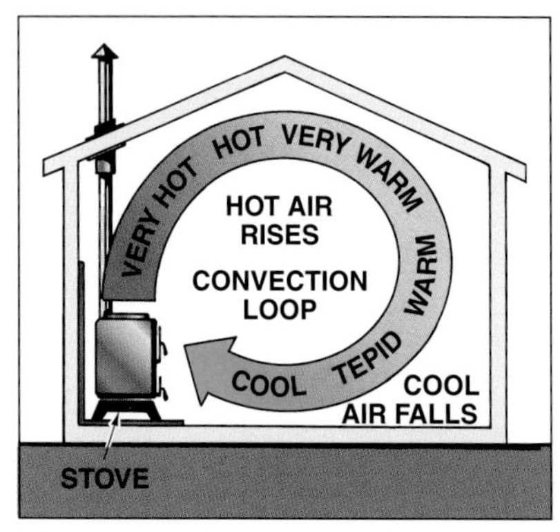

Figure 51-17. *Convection is the movement of heat through a fluid substance such as air or liquid. For example, air warmed by a source such as a furnace or wall heater rises. It is then replaced by colder air, which in turn is heated and also rises. A convection loop circulates the heat throughout the area.*

Radiation. Radiation is the direct transmission of heat by invisible waves similar to light waves. An example of radiation is the way radiant heat waves travel through the vacuum of space and warm the surface of the earth. Another example of radiation is heat that a person feels when standing near a fire. **Figure 51-18** shows the principle of an underfloor radiant heating system. Heat waves from heated water flowing through tubes installed under the floor strike and warm objects such as the walls, ceilings, and furniture. In addition, some of this heat transferred by radiation is reflected and warms the air by convection.

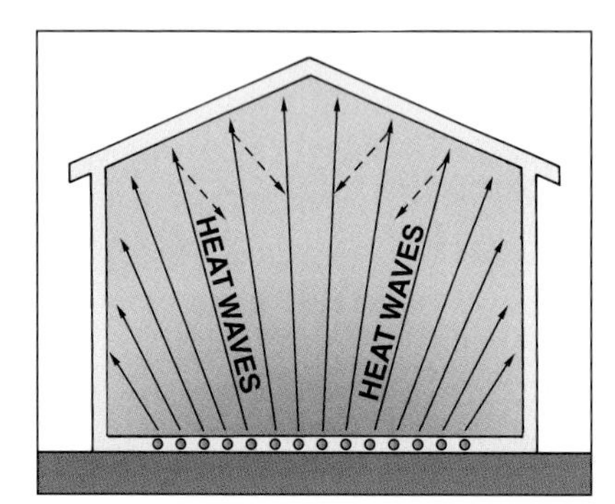

Figure 51-18. *Radiation is the direct transmission of heat waves from a heat source. For example, heat waves from an underfloor radiant heating system strike and warm objects such as the walls, ceilings, and furniture. Some of the heat is reflected and warms the air through convection.*

Insulation as Protection Against Heat Transfer

Materials used for insulation protect a building from losing heat or coolness by any method of heat transfer. Insulation material should be a poor conductor of heat to prevent heat transfer by conduction. The insulation material should stop or slow the airflow from warm to cold areas of the building to prevent heat transfer by convection. Finally, the insulation material should reflect heat rather than absorb it in order to prevent heat transfer by radiation.

Measuring Heat Transfer and Resistance

Many different kinds of insulation are available. Knowledge of how various materials used for insulation resist the passage of heat allows an architect or builder to select the materials best suited to insulating a particular building.

Technical information about each type of insulation is expressed in letters that represent certain *factors*—conductivity (k), conductance (C), and transmittance (U). A *value* of resistance (R) is based on these factors.

Heat is measured in *British thermal units (Btu).* One Btu is the amount of heat required to raise the temperature of 1 lb of water 1°F (Fahrenheit). **See Figure 51-19.** Heating equipment used in buildings, such as furnaces and boilers, is rated by its output in Btu. When specifying heating equipment, engineers must calculate the amount of space to be heated in addition to the expected heat loss in the building.

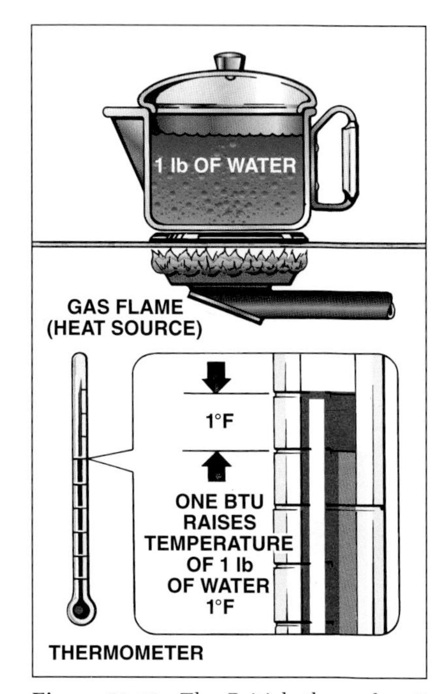

Figure 51-19. *The British thermal unit (Btu) is the unit measurement of heat transfer. One Btu is the amount of heat needed to raise the temperature of 1 lb of water 1°F.*

Conductivity (k) and Conductance (C) Factors. The best materials for insulation have a low k- or C-factor. The *k-factor* (conductivity) measures the amount of heat that travels through homogenous material. Concrete, wood, and polyurethane foam insulation are examples of homogenous materials—they have the same composition throughout. The k-factor is a decimal measurement of how many Btu per hour (Btu/hr) pass through 1 sq ft of material that is 1″ thick and has a temperature difference of 1°F between its inside and outside surfaces. Wood has a lower k-factor than concrete, and polyurethane foam insulation has a lower k-factor than wood. **See Figure 51-20.**

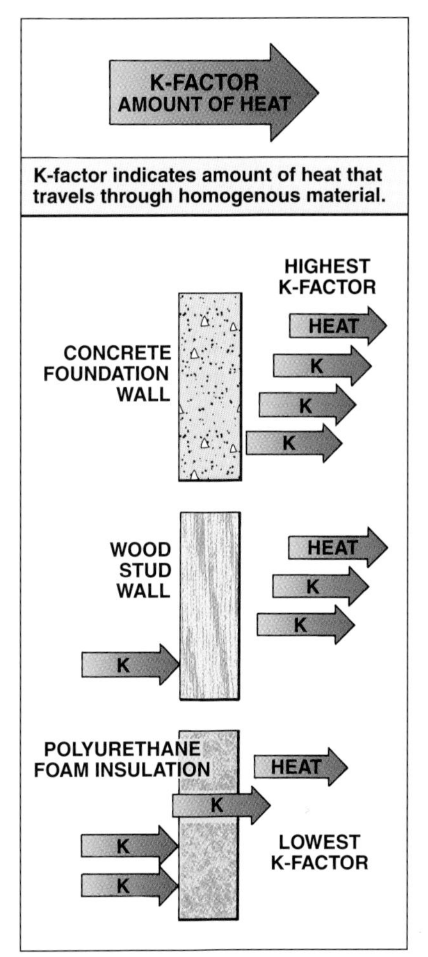

Figure 51-20. *The k-factor represents the amount of heat that travels through a homogenous material.*

The *C-factor* (conductance) is based on the same principle as the k-factor; however, the C-factor applies to materials of any thickness that are not homogenous or that have air cavities, such as hollow concrete masonry units.

Transmittance (U) Factor. The *U-factor* (transmittance) is a decimal measurement of how many Btu/hr pass through 1 sq ft of the combination of materials that make up the floor, ceiling, wall, or roof area of a building when a 1°F difference in temperature exists between the inside and outside of the building. **See Figure 51-21.** The U-factor differs from k- and C-factors in that it expresses heat transfer through combined materials plus any air space that may exist rather than through a single material.

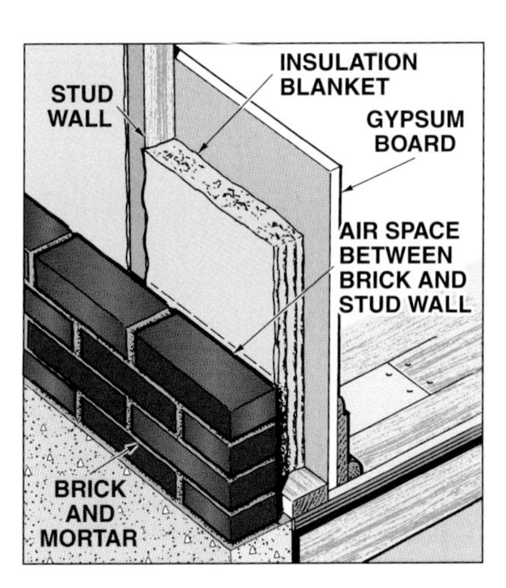

Figure 51-21. *The U-factor for a brick-veneered wood stud wall with insulation expresses heat transfer through the combined materials plus the air space.*

The R-value of a material can be affected by the quality of the installation. For example, if fiberglass batt insulation installed with a wall cavity contains any gaps or voids between the insulation and framing, the effective R-value of the insulation can be reduced by over 30%.

Resistance (R) Value. The *R-value* (resistance) represents the ability of a material to resist heat flow. Most insulation products are labeled with an R-value on the wrapper. The total heat flow resistance of a wall includes the R-value of the material used to construct the wall, such as wood, brick, concrete, or hollow concrete masonry units, plus the R-value of the insulation material placed in the wall. An air space, such as the kind found between a wood frame and brick veneer wall, adds to the total R-value of the wall.

The total R-value needed for a building depends on regional weather conditions, since buildings in colder climates require more insulation. Local building codes often specify insulation requirements for the area. **Figure 51-22** gives examples of the R-values (based on k- and C-factors) of a number of different building materials.

An air space between the brick and insulation allows for proper water drainage and adds to the total R-value of the wall.

R-VALUES OF COMMON BUILDING MATERIALS				
Material or Product	Conductivity (k)	Conductance (C)	Resistance (R)	
			Per 1″ Thickness (1/k)	Per Thickness Shown (1/C)
Concrete	12.0		0.08	
Face brick	9.0		0.11	
Hollow concrete masonry units, 8″		0.90		1.11
Stucco	5.0		0.20	
Metal lathe & plaster, ¾″		7.70		0.13
Gypsum board, ½″		2.22		0.45
Plywood, ½″		1.60		0.62
Pine, fir, other softwoods	0.80		1.25	
Oak, maple, other hardwoods	1.10		0.91	
Asphalt shingles		2.27		0.44
Built-up roofing, ⅜″		3.00		0.33
Wood shingles		1.06		0.94
Structural insulation board, ½″		0.76		1.32
Mineral wool batts, 3″– 4″		0.09		11.00 INSULATION MATERIALS
5″– 6″		0.05		19.00 INSULATION MATERIALS
6½″– 7″		0.05		22.00 INSULATION MATERIALS
8½″– 9″		0.03		30.00 INSULATION MATERIALS
Expanded polystyrene, 1″		0.28		3.57
Inside surface air film*				0.68
Outside surface air film†				0.17
Air space ¾″, nonreflective‡		1.0		1.01
Air space ¾″, reflective‡				3.48

* heat flow horizontal, still air
† heat flow any direction, 15 mph wind
‡ heat flow horizontal

Figure 51-22. *The R-value represents the ability of a material to resist heat flow. Shown are the R-values of common building materials based on their conductivity (k) or conductance (C). Note how much higher the R-values are for insulation materials than for other materials.*

CONDENSATION

Moisture is always present in the air; usually it is invisible. When air temperature increases, the moisture-holding capacity of the air also increases.

Moisture in the form of water vapor is produced from many sources within a building. Activities of an average family of four in a heated building generate approximately 22.5 lb of water vapor within a 24-hr period. **See Figure 51-23.** As much as 30 lb or more of water vapor can be generated by clothes washing and drying inside a building.

WATER VAPOR	
Source or Function	Amount Generated*
Breathing and perspiring	13
Cooking	5
Bathing	1
Dishwashing	3.5

* in lb

Figure 51-23. *In a heated building, an average family of four generates about 22.5 lb of water vapor in a 24-hr period.*

Indoor humidity levels should be between 25% and 40% in the winter and below 60% in the summer.

When the warm, moisture-laden air within a building comes into contact with the cooler outside surfaces of the building, the air is unable to retain all of its moisture. Some moisture comes out of the air and forms water droplets on cooler surfaces. This process is known as *condensation.*

The temperature at which condensation occurs is the *dew point.* Moisture problems because of condensation are greater in colder climates, where houses require more heating during the winter. **See Figure 51-24.**

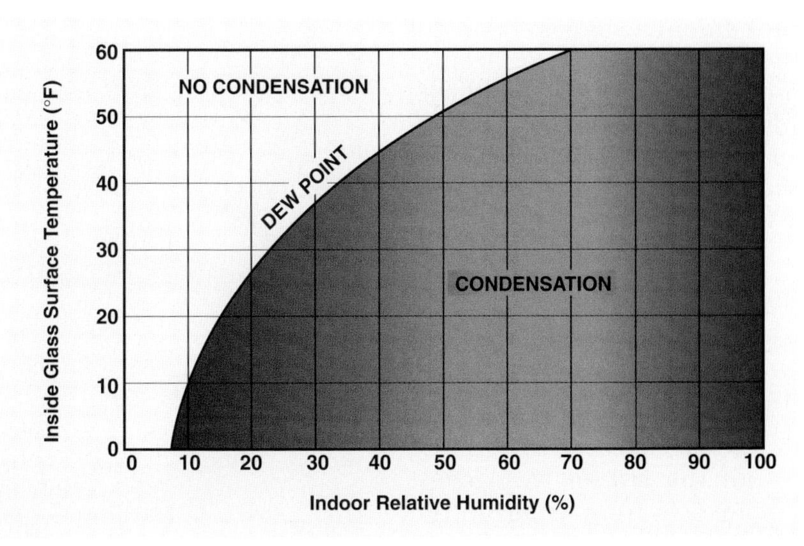

Figure 51-24. *With an indoor air temperature of 70°F, cooler glass temperatures cannot hold as much moisture as higher glass temperatures. This results in condensation at lower temperatures.*

Condensation that occurs on the face of a wall is called *surface condensation.* Surface condensation is easily visible. Often, however, condensation occurs inside a wall. This *concealed condensation* can cause many problems. Over a period of time, concealed condensation can cause serious decay in framing members. Blistering may occur on the outside painted surfaces of a building. Plaster or gypsum board inside a building may begin to crack and crumble.

In the past, houses were not built to resist the effects of weather as effectively as they are today. Construction allowed more air leakage, which in turn permitted more of the interior moisture to move out of the building. Modern houses are generally more tightly constructed and better insulated than older ones. Little air leakage occurs in modern houses, resulting in more air moisture and condensation problems than in older buildings.

Airflow

Uncontrolled airflow through the building envelope can result in major heat loss. The air can carry water vapor with it into the building envelope. For air to move through the building envelope, a hole must be in the envelope or there is a pressure difference between the outside and inside of the house. Holes in the building envelope may be a result of poor design or craftsmanship. A pressure difference between the outside and inside of a building may be the result of *wind effect, stack effect,* or *combustion-and-ventilation effect.*

The wind effect is created when the wind blows against a building. A high-pressure area is created on the windward side and air is forced into the building. A low-pressure area is created on the opposite side of the building allowing air within the building to escape.

The stack effect is based on the principle that warm air rises. As the warm air within a heated building rises, a high-pressure area is created toward the top of the building. Air escapes from the building through ceiling holes or cracks, or through spaces around upper-level windows and exterior doors. The force of the rising warm air creates a low-pressure area on the lower level, allowing outside air to enter the building through cracks and openings around the lower floors.

The combustion-and-ventilation effect is created by wood-, oil-, or natural gas-burning appliances that require makeup air to support combustion. The draft created by the appliances draws air up the chimney. The air must be replaced by air drawn into the building envelope.

Moisture Control

Condensation in the floors, walls, ceiling, and roof of a building can be controlled by means of vapor barriers and ventilation.

Vapor Barriers. A vapor barrier is a thin material through which water cannot easily pass. Many insulation products have a vapor barrier on one surface. These products should be placed with the vapor barrier facing the warm side (inside surface) of the wall. **See Figure 51-25.** The vapor barrier prevents damage to the insulation material from the moisture that collects inside a wall. If the insulation product does not have a vapor barrier, a separate vapor barrier should be attached to the warm side.

Figure 51-25. *Insulation should be installed with the vapor barrier facing the warm side of the wall.*

Depending on thickness, aluminum foil has a perm rating of 0.01 to 0.05, while polyethylene film has a perm rating of 0.06 to 0.16.

Information about a vapor barrier material often refers to its perm rating. The *perm rating* indicates *permeability* (or permeance) and is based on a formula that measures the water vapor flow through the thickness of single material or combined materials. Most building materials are permeable to some degree. A few materials, such as metals and glass, are completely *impermeable,* meaning that they allow no vapor to pass through. An impermeable material has a perm rating of 0.00. Materials with a perm rating of 1 or less qualify as vapor barriers. However, construction conditions today often require a vapor barrier with a perm rating of 0.5 or less. Polyethylene film and aluminum foil have very low perm ratings and are often used as vapor barriers.

Ventilation. Vapor barriers and insulation can effectively control condensation in the walls, floor, and ceiling of inhabited parts of a house. However, ventilation is needed in uninhabited areas, such as the attic space above the ceiling and below the roof, and the area between the floor and ground in a house with a crawl-space foundation. Ventilation is provided by openings that permit warm air to escape, as well as allow the circulation of air in the enclosed areas.

Attic Ventilation. Condensation occurs beneath the roof when water vapor is unable to escape through the roofing material. Wood shingles or shakes do not prevent vapor movement. Asphalt shingles or built-up roofs are highly resistant to vapor movement. Several venting methods are possible for attic areas of wood-framed buildings. Most methods have shortcomings when used by themselves but are effective when combined with other methods.

- *Gable end louvers* are probably the oldest ventilation system for gable roofs. **See Figure 51-26.** Slanted boards in the louver prevent rainwater from entering. A screen covers the inside of the louver to prevent insects from entering the building. Alone, gable end louvers are considered less efficient than other systems as they only work well when a breeze is blowing at a right angle to the louvers.

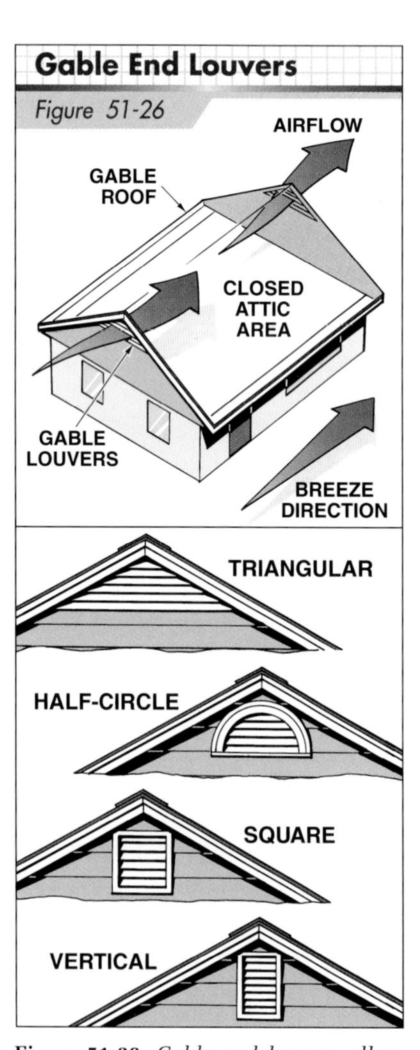

Figure 51-26. *Gable end louvers allow airflow through an attic.*

Gable end louvers are used with soffit vents to ventilate unconditioned attics.

- *Soffit vents* are located beneath a roof cornice. They may consist of a series of small openings or one continuous slot. **See Figure 51-27.** With soffit vents, most of the ventilation occurs over the attic floor and there is little air movement beneath the roof sheathing.
- *Roof vents* are located on top of a roof. **See Figure 51-28.** They allow warm air to escape, but do not allow much airflow in the attic space.
- *Continuous ridge vents* run along the ridge of a roof. **See Figure 51-29.** Continuous ridge vents are unique in that their design provides some airflow resulting from temperature differences alone. General air circulation to the attic space is limited, however.

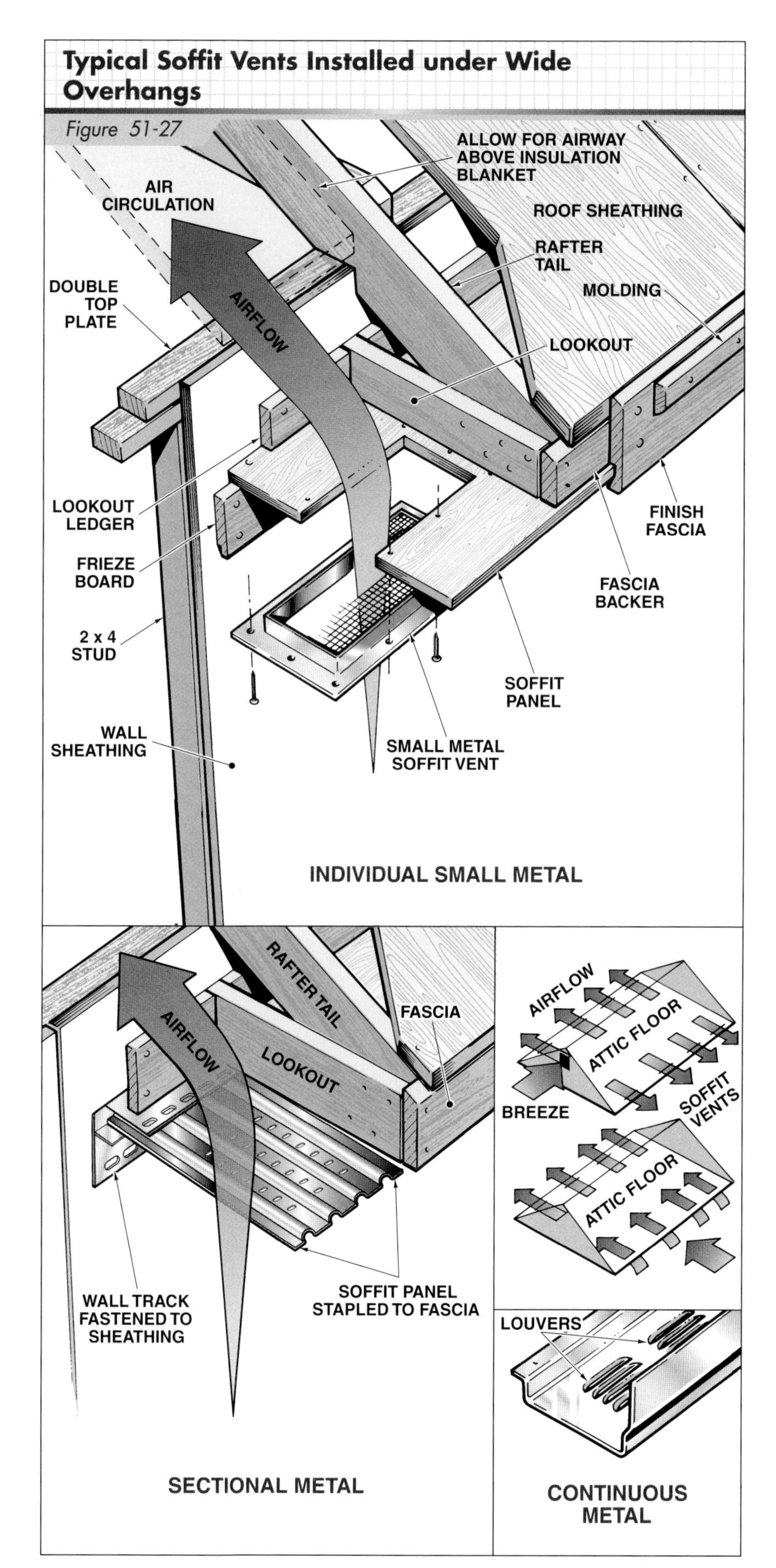

Figure 51-27. *Most ventilation through soffit vents occurs over the attic floor.*

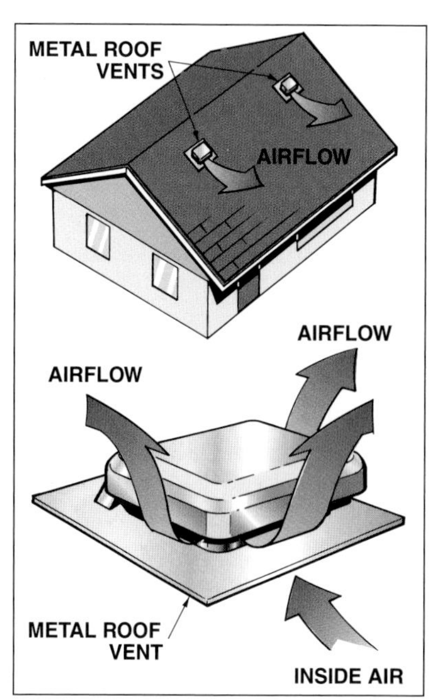

Figure 51-28. *Roof vents are effective in allowing warm air to escape but do not allow much airflow into the attic space.*

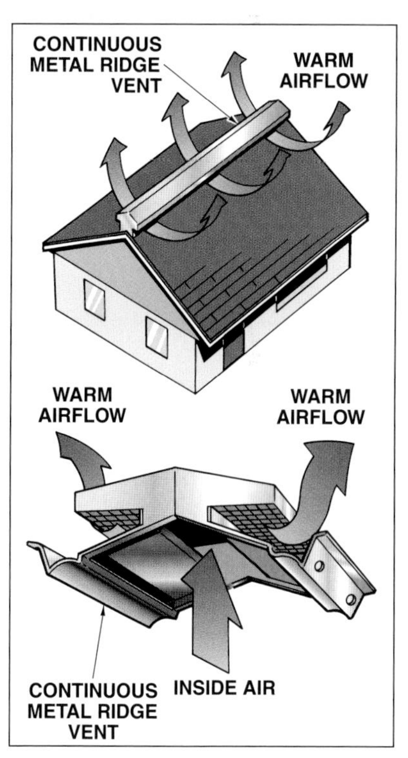

Figure 51-29. *Continuous ridge vents run along the entire ridge of a roof.*

To comply with most manufacturer warranties, attics must be ventilated when asphalt shingles are applied as roofing.

Common combinations of these venting systems are *gable louver-and-soffit, roof-and-soffit,* and *ridge-and-soffit.* **See Figure 51-30.**

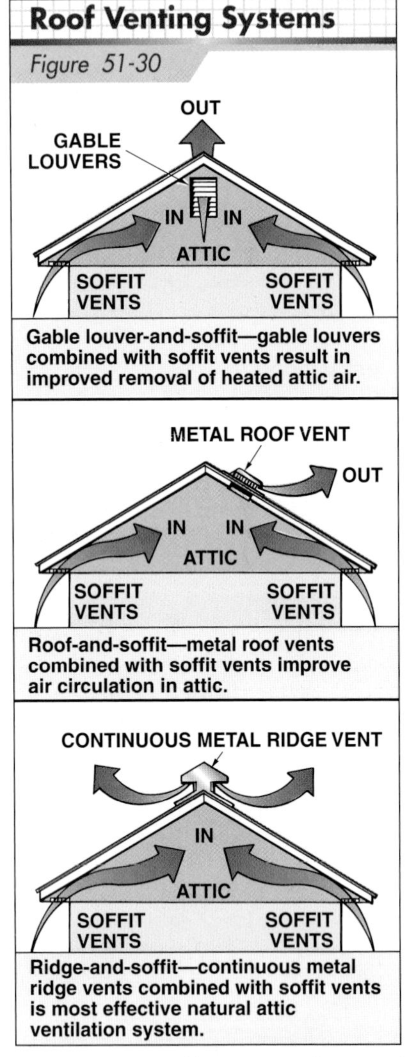

Figure 51-30. *Efficient ventilation of an attic is gained by combining venting systems.*

Attic Rafter Vents

Attic rafter vents, also known positive-ventilation chutes, fit between the rafters or trusses in an attic to prevent insulation from blocking airflow from soffit vents. **See Figure 51-31.** Attic rafter vents should be placed between every rafter to ensure proper airflow. Ensure the attic rafter vent does not obstruct the soffit vent. During warm temperatures, air flows through the soffit vents and forces warmer air from the attic through ridge vents. During cold temperatures, ice dams are eliminated from the eaves.

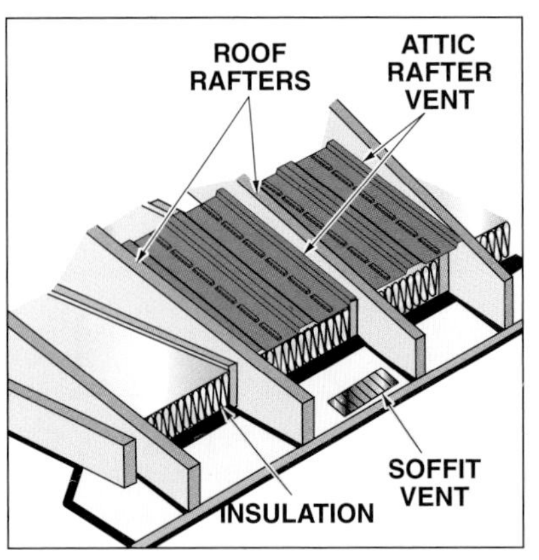

Figure 51-31. *Attic rafter vents prevent insulation from blocking the airflow from soffit vents.*

Crawl Space Ventilation. Moisture rising by capillary action from the ground can cause condensation that results in stain and decay beneath the floor unit of a building with a crawl-space foundation.

Vapor barriers placed over the ground help reduce moisture. In addition, vent openings should be provided to provide airflow in the crawl space. **See Figure 51-32.** In general, 1 sq ft of vent is required for every 150 sq ft of crawl space area. However, more vents may be needed for a crawl space with a dirt floor in a shaded site or damp climate. Fewer vents may be required for a crawl space with a concrete floor in a dry, windy climate. Foundation vents should be placed as high as possible and in opposite walls. Various styles of vent coverings are available.

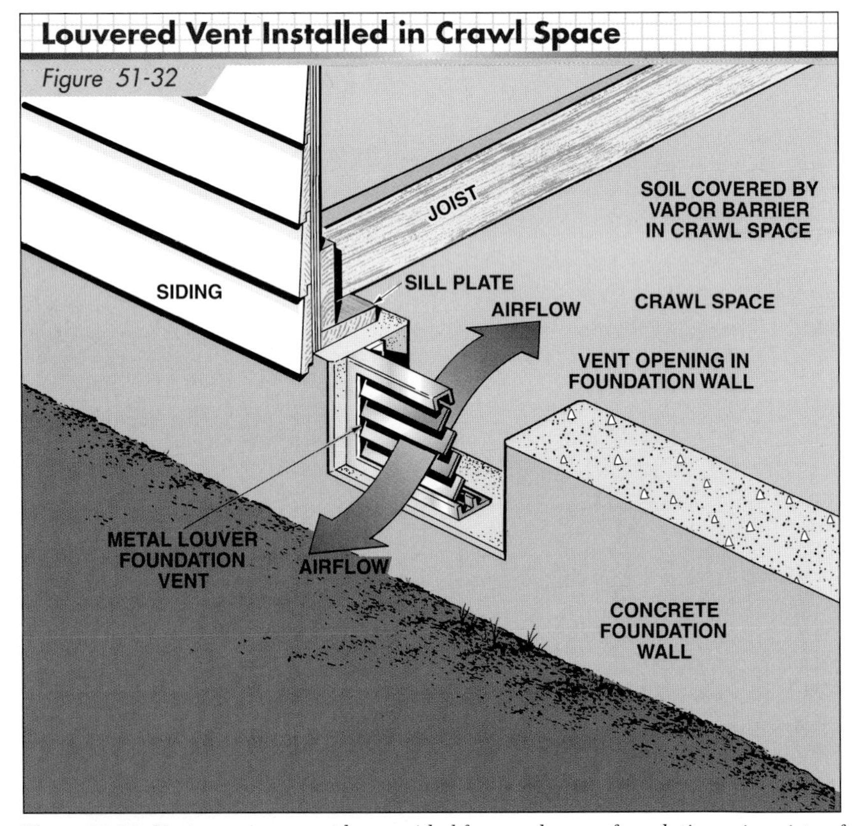

Figure 51-32. *Vent openings must be provided for crawl-space foundations. A variety of vent designs are available.*

Unit 51 Review and Resources

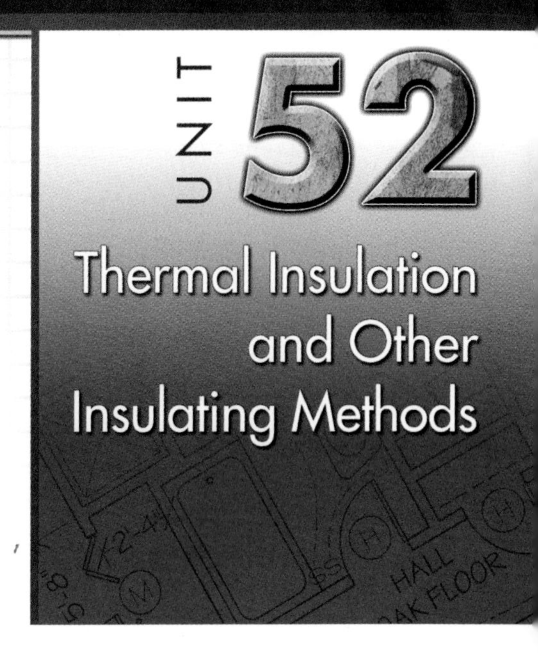

OBJECTIVES

1. Discuss the use of insulating materials and how they are rated.
2. List and describe the types of loose fill insulation.
3. Describe how blanket and batt insulation is installed.
4. Discuss how foamed-in-place insulation is applied.
5. Describe rigid foam insulation.
6. Discuss how rigid foam insulation is attached to foundation walls.
7. Explain how rigid foam insulation is attached to wood-framed walls.
8. Describe the procedure for installing exterior insulation and finish systems (EIFS).
9. List and describe materials used as rigid roof insulation.
10. Describe roofing systems for pitched roofs.
11. Describe roofing methods for flat roofs.
12. Explain heat loss through doors and windows.

Studies by government agencies have shown that 20% of energy used in an average home is used for heating water. An additional 10% is used for lighting rooms, cooking food, and powering appliances such as dishwashers, clothes washers, and clothes dryers. The remaining 70% of the energy used in an average home is for heating and cooling the building. Today, the amount of energy required to heat and cool a building can be greatly reduced by using better insulation and improved construction methods.

THERMAL INSULATION

Thermal insulation is placed inside and on the surfaces of framed floors, walls, ceilings, and roofs of wood- and metal-framed buildings. Thermal insulation is also applied to concrete and masonry surfaces. **See Figure 52-1.** Thermal insulation resists heat flow out of a building during cold weather and heat flow into a building during hot weather. Most insulation also reduces sound transmission, and many types of insulation have high fire-resistive ratings.

Insulation is rated by its *R-value,* which indicates its resistance to heat flow. Buildings in colder climates require insulation with higher R-values. The map in **Figure 52-2** shows R-values required for different areas of the United States. The chart in **Figure 52-3** lists some of the more widely used insulation materials and their R-values. **Figure 52-4** lists the thickness of each insulation material required to achieve different R-values.

Most homes should have between R-22 and R-49 insulation in the attic.

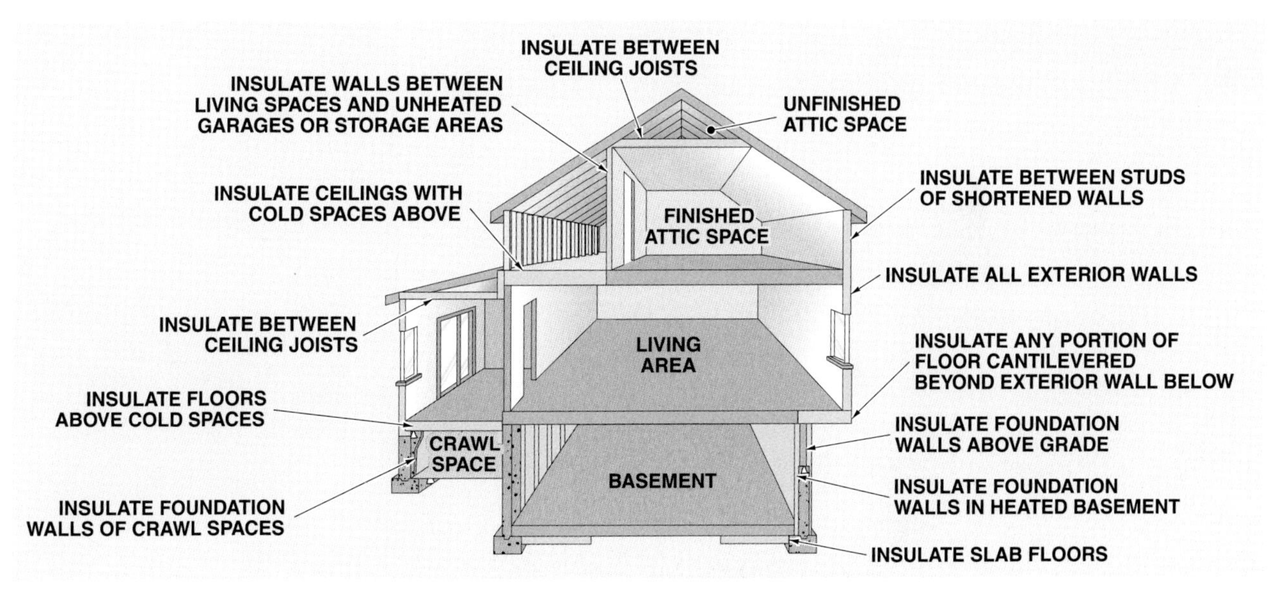

Figure 52-1. *Insulation is installed inside and on the surfaces of framed floors, walls, ceilings, and roofs.*

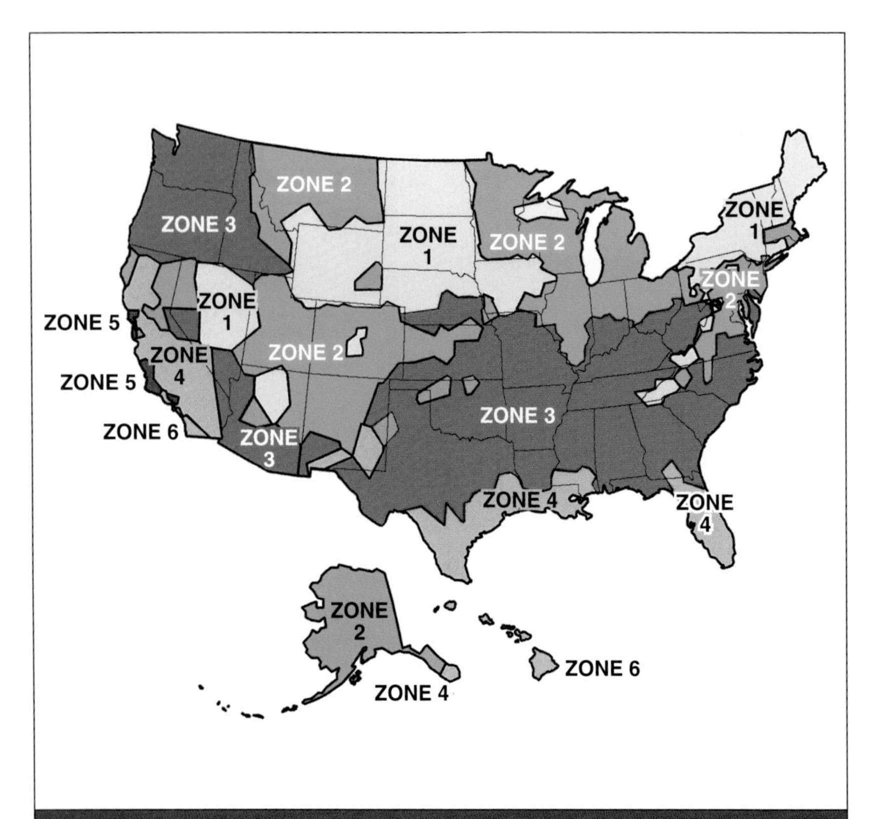

RECOMMENDED INSULATION R-VALUES

ZONE	HEAT SOURCE				CEILING						BASEMENT	
	Gas	Heat Pump	Fuel Oil	Electric Furnace	Attic	Cathedral	Wall*	Floor	Crawl Space†	Slab Edge	Interior	Exterior
1	✓	✓	✓		R-49	R-38	R-18	R-25	R-19	R-8	R-11	R-10
				✓	R-49	R-60	R-28	R-25	R-19	R-8	R-19	R-15
2	✓	✓	✓		R-49	R-38	R-18	R-25	R-19	R-8	R-11	R-10
				✓	R-49	R-38	R-22	R-25	R-19	R-8	R-19	R-15
3	✓	✓	✓	✓	R-49	R-38	R-18	R-25	R-19	R-8	R-11	R-10
4	✓	✓	✓		R-38	R-38	R-13	R-13	R-19	R-4	R-11	R-4
				✓	R-49	R-38	R-18	R-25	R-19	R-8	R-11	R-10
5	✓				R-38	R-30	R-13	R-11	R-13	R-4	R-11	R-4
		✓	✓		R-38	R-38	R-13	R-13	R-19	R-4	R-11	R-4
				✓	R-49	R-38	R-18	R-25	R-19	R-8	R-11	R-10
6	✓				R-22	R-22	R-11	R-11	R-11	‡	R-11	R-4
		✓	✓		R-38	R-30	R-13	R-11	R-13	R-4	R-11	R-4
				✓	R-49	R-38	R-18	R-25	R-19	R-8	R-11	R-10

* R-18, R-22, and R-28 exterior wall systems can be achieved by either cavity insulation or cavity insulation with insulating sheathing. For 2 × 4 walls, use 3½″ thick R-15 or 3½″ thick R-13 fiberglass insulation with insulating sheathing. For 2 × 6 walls, use 5½″ thick R-21 or 6¼″ thick R-19 fiberglass insulation.

† Insulate crawl space wall only if the crawl space is dry all year, the floor above is not insulated, and all ventilation to the crawl space is blocked. A vapor retarder (e.g., 4 or 6 mil polyethylene film) should be installed on the ground to reduce moisture migration into the crawl space.

‡ No slab edge insulation is recommended.

Figure 52-2. *Insulation R-values are based on the climatic conditions of the area. Note the difference between the recommended R-values for western and northeastern areas.*

Mudsill anchors must be spaced at the correct intervals and placed along the center of the sill plate for optimum anchorage. Typically, mudsill anchors have less resistance value than anchor bolts and are installed at closer intervals than anchor bolts to compensate for the difference.

Insulation is manufactured from a variety of materials. The most widely used insulation—fiberglass insulation—is manufactured from melted glass that is spun into fibers. Foam plastic insulation is manufactured from molten plastic that is infused with air. When cooled, the foam plastic insulation contains millions of tiny air bubbles that reduce heat flow through the material. The more air bubbles per inch of thickness, the higher its resistance to heat flow.

The four basic forms of insulation are *loose fill, blanket and batt, foamed-in-place,* and *rigid foam insulation.*

Lightly press blanket or batt insulation into place, allowing friction to hold it in place.

R-VALUES OF COMMON INSULATION MATERIALS	
Product	**R Per 1″ Thickness***
Loose Fill	
Mineral fiber (rock, slag, or glass)	2.20–3.00
Cellulose	3.70
Perlite	2.70
Vermiculite	2.13
Flexible	
Mineral wool batts	3.10–3.70
Fiberglass batts	3.10–3.70
Rigid	
Cellular glass	2.63
Expanded polystyrene (extruded)	5.00
Expanded polystyrene (molded)	3.57
Expanded polyurethane	6.25
Mineral fiberboard	3.45
Polyisocyanurate	7.20
Reflective	
Aluminum foil†	3.48
Foamed-in-Place	
Urea formaldehyde	4.20
Polyurethane	6.25

* all values are for 75° F mean temperature
† thickness of foil not a factor; ¾″ airspace on room side

Figure 52-3. *Various insulation materials provide differing R-values.*

THICKNESS* OF INSULATION TO ACHIEVE VARIOUS R-VALUES					
Material	**R-Values**				
	11	**13**	**19**	**22**	**30**
Loose Fill					
Fiberglass	5.0	5.5	8.5	8.5	13.0
Rock wool	3.5	4.0	6.0	6.0	9.0
Cellulose	3.0	3.5	5.5	5.5	8.5
Vermiculite	5.0	6.0	10.5	10.5	14.5
Batts/Blankets					
Fiberglass	3.5	4.0	7.0	7.0	8.5
Rock wool	3.5	4.0	7.0	7.0	8.5
Rigid Foam					
Polystyrene	3.0	3.5	3.5	5.5	7.5
Urethane	2.0	2.0	2.0	3.5	5.0
Fiberglass	3.0	3.5	3.5	5.5	7.5

* in in.

Figure 52-4. *Increased insulation thicknesses provide greater R-values. For example, 3.5″ of fiberglass insulation has an R-value of 11.*

Loose Fill Insulation

Loose fill insulation is poured directly from a bag or blown in place with a pressurized hose. The more frequently used loose fill materials are rock wool, fiberglass, and cellulose. Some types of loose fill insulation are produced from recycled materials.

Loose fill insulation is often placed in attics by depositing it between ceiling joists. **See Figure 52-5.** The insulation is placed directly on top of the ceiling below in a uniformly thick layer. A rake constructed from plywood is useful in spreading an even layer of insulation. **See Figure 52-6.** Clearances around electrical fixtures in the attic ceiling must be maintained.

ASTM C739, Standard Specification for Cellulosic Fiber (Wood-Base) Loose-Fill Thermal Insulation, covers the composition and physical requirements of chemically treated recycled cellulose loose fill insulation for use in attics or enclosed spaces in housing and other framed buildings. The specification pertains to structures within the ambient temperature range of –50°F to 180°F in which the insulation is blown or poured in.

Loose fill insulation can also be blown or packed into exterior walls of older buildings that were not insulated when originally constructed. Holes are cut between studs at the top of the exterior walls using a hole saw. Insulation is blown into the wall cavity until the insulation begins to back up out of the hole. **See Figure 52-7.** The holes are then sealed with the plugs that were removed when cutting the holes. If the building was built before the mid-1900s or if the walls are over 8′ tall, fire blocking may be installed between the studs. In this situation, a second hole is cut beneath the blocking and the process is repeated.

Blown-in-blanket insulation is a type of loose fill insulation used for new construction. A retention fabric made of spun polypropylene is fastened to wall studs using adhesive or staples before gypsum board or other interior wall covering is applied. A small opening is made in the fabric and loose fill fiberglass insulation is blown into the wall cavity. **See Figure 52-8.** The insulation should bulge approximately ½″ to 1″ past the studs to ensure proper insulation density (1.8 to 2.5 lb/cu ft minimum). Where required, a vapor barrier is installed over the fabric and interior wall covering is applied.

Blanket and Batt Insulation

Blanket and batt insulation is the material most widely used to insulate walls, floors, ceilings, and attics. Blanket and batt insulation is usually made from fiberglass or rock wool that is fire-, moisture-, and vermin-resistant.

Blankets and batts are similar in composition and appearance but differ in length. Batts are usually 48″ long, whereas blankets are longer and are available in rolls. **See Figure 52-9.** Blankets and batts are available in widths suitable for 16″ OC and 24″ OC stud and joist spacing. Insulation thicknesses used in walls, floors, and ceilings range from 3″ to 7″ or more. More than one layer of blanket insulation may be used to increase the R-value.

Blanket and batt insulation is available with *faced* or *unfaced* surfaces. Faced insulation has a specially treated paper or aluminum-foil vapor barrier on one side. Flanges extending from both edges of the vapor barrier are stapled to joists and studs. Unfaced blanket and batt insulation does not have a vapor barrier and relies on friction to hold it in place. Some types of blanket and batt insulation are available with a covering to protect the hands of an installer from irritation when handling the material.

CertainTeed Corporation

Figure 52-5. *A pressurized hose is used to blow loose fill insulation between the ceiling joists.*

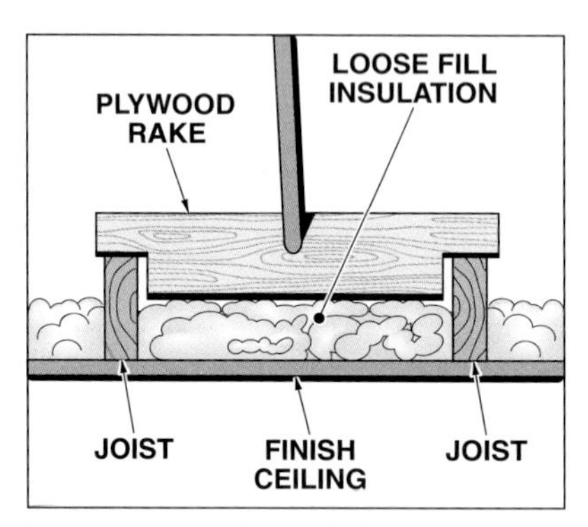

Figure 52-6. *A plywood rake is convenient to spread and level off loose fill insulation between ceiling joists.*

Compressing thick fiberglass insulation into a smaller wall space will not increase the insulation efficiency. Fiberglass insulation works on the principle of trapped air pockets. By compressing fiberglass insulation, the amount of air trapped in the material is reduced. R-15 fiberglass insulation provides better performance in 2 × 4 walls. R-21 fiberglass insulation works best in 2 × 6 walls.

CertainTeed Corporation

Figure 52-7. *Loose fill insulation can be blown between the studs into wall cavities.*

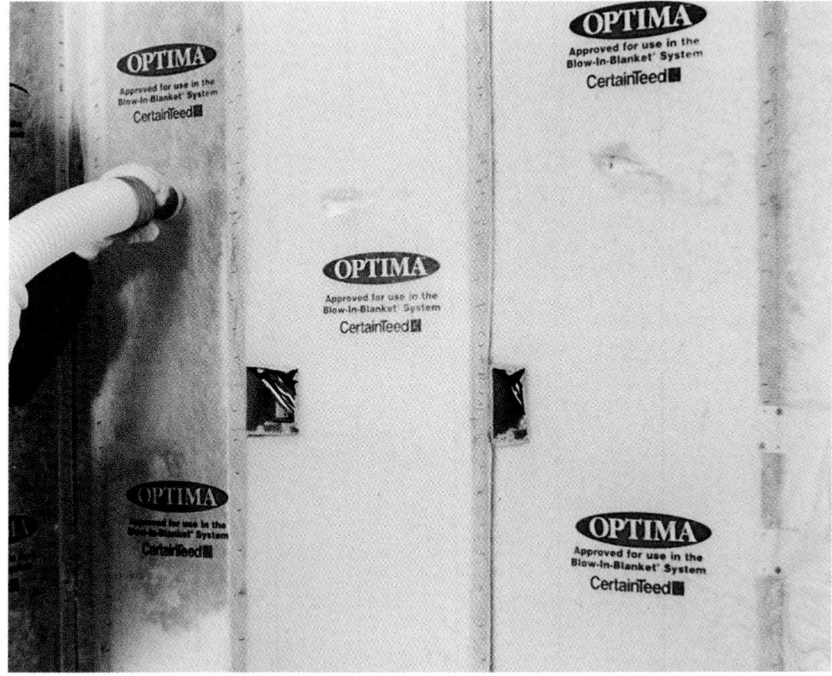

CertainTeed Corporation

Figure 52-8. *In the blown-in-blanket system, loose fill insulation is blown into a bag made of retention fabric, which has been fastened to the studs.*

Figure 52-9. *Flexible batt and blanket insulation differ only in length.*

Applying Blanket and Batt Insulation. General techniques for installing blanket and batt insulation are as follows, and are shown in **Figure 52-10:**

- Apply faced insulation with the vapor barrier toward the interior of the building.
- Always place insulation on the exterior side of pipes or ducts to protect them from freezing.
- Pack insulation into voids or other open spaces around window and door openings.
- Do not compress blankets and batts when installing them. Rather, lightly press the insulation into place, allowing friction to hold it in place.
- Secure faced insulation in place by stapling the flanges to the edges of the studs, joists, or rafters.
- Unfaced insulation should be wedged between studs or joists.
- For maximum efficiency, use double back insulation at the ends. Place insulation in the attic so it does not interfere with airflow through the vents and prevent proper air circulation in the attic.
- Do not place a vapor barrier between two pieces of insulation as this can lead to condensation and lower the effectiveness of the insulation.
- Tears or punctures in the vapor barrier or insulation should be repaired.

CertainTeed Corporation

Blown-in blanket insulation may be used to insulate ceilings.

Foamed-in-Place Insulation

Foamed-in-place insulation is a chemical foam that is poured or sprayed into wall cavities. The foam then expands and completely fills the cavity. Some foamed-in-place products are sprayed on the inside surface of an open wall. **See Figure 52-11.** The foam expands and the excess is trimmed away with long blades. Chemical foam has a very high R-value and is particularly well suited for insulating older buildings that were not insulated when originally constructed.

Foamed-in-place insulation has an R-value of approximately 3.6 per inch.

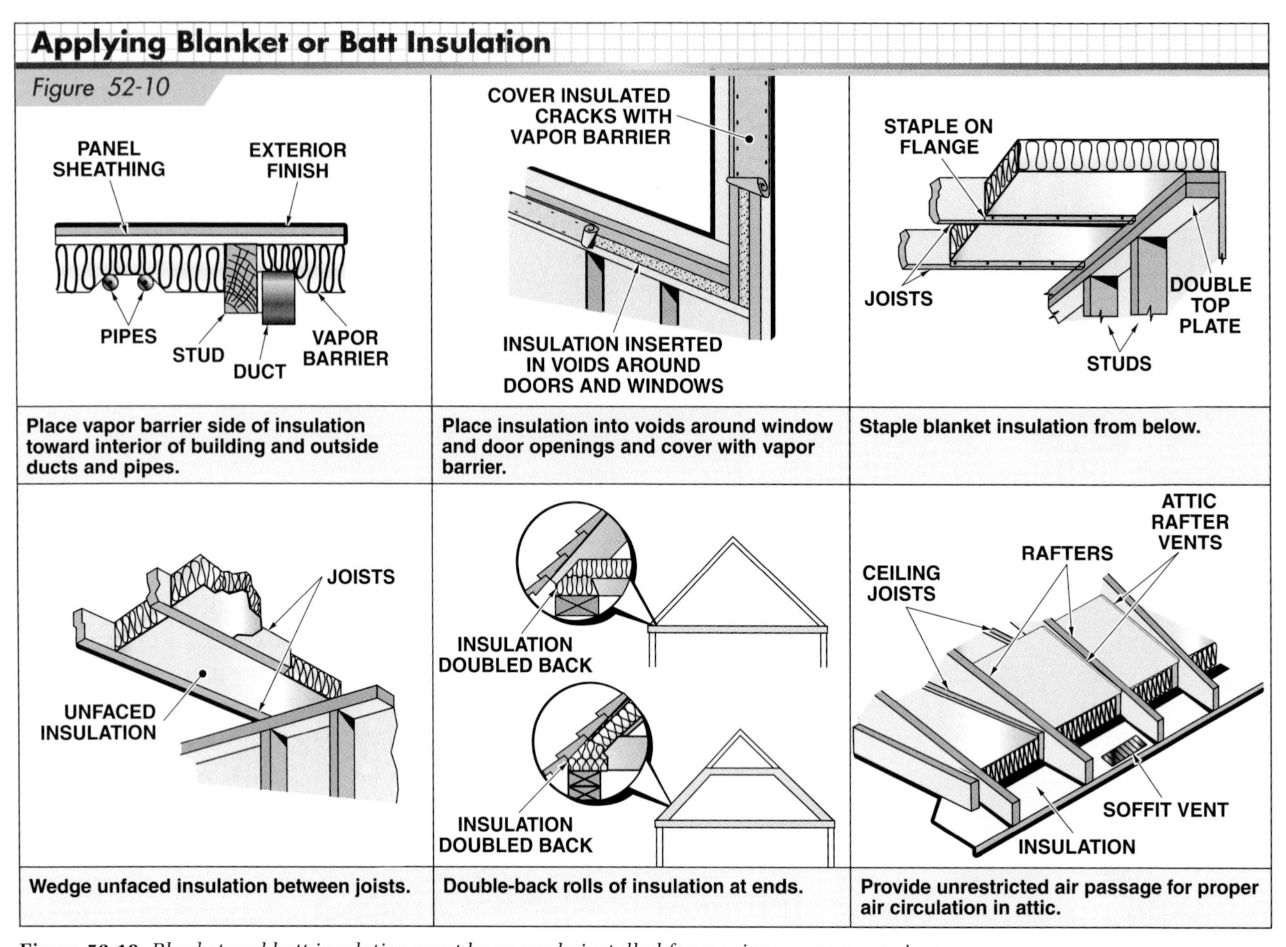

Figure 52-10. *Blanket and batt insulation must be properly installed for maximum energy savings.*

Icynene, Inc.

Figure 52-11. *A thin layer of foamed-in-place insulation is sprayed on a surface. The foam then expands to fill the wall cavity.*

Before foam is sprayed into a wood-framed wall cavity of an older building, a course of shingles or siding near the top and bottom of the wall is removed. Small holes are then drilled through the sheathing and between the studs. The foam is sprayed in with a pressurized air hose. The holes are plugged after the foam is in place. Foam can also be sprayed into the cavities of masonry walls if individual blocks are removed and then replaced.

When spraying foamed-in-place insulation, proper PPE such as a supplied-air respirator, eye protection, and chemical-resistant clothing and gloves must be worn. Failure to do so could cause lung damage, breathing problems, or skin and eye irritation.

CertainTeed Corporation

Fiberglass metal building insulation is flexible, unfaced, blanket insulation, which is primarily used in exterior walls and under standing-seam roofs. A vapor barrier appropriate for the particular job is laminated to the insulation.

Rigid Foam Insulation

Rigid foam insulation consists of polyurethane and polystyrene foam or fiberglass strands pressed into panels. Rigid foam insulation has a higher R-value per inch of thickness than any other type of insulation. Rigid foam insulation panels contribute to the structural support of a building despite their light weight. Rigid foam insulation is usually applied on the exterior of a building, and is widely used on new construction and when retrofitting existing residential and commercial buildings.

Rigid foam insulation is available in faced and unfaced panels. Faced rigid insulation panels have a reflective material covering that acts as a vapor barrier. Unfaced rigid insulation requires the use of a separate vapor barrier when installed. Common panel dimensions are 4′ × 4′ and 4′ × 8′, with thicknesses ranging from 1″ to 4″. Other panel sizes and thicknesses are also available from manufacturers.

Rigid Foam Insulation for Foundations. Most heat loss through concrete and masonry basement walls occurs where the walls are exposed to cold air above grade, and where frozen ground may contact the walls below grade. In full-basement and crawl-space foundations, heat loss begins through the floor joists and continues through the foundation walls. Slab-at-grade floors lose a considerable amount of heat through the exposed slab edge and through the soil beneath the slab. **See Figure 52-12.** Heat loss can be substantially reduced by installing 2″ to 4″ of rigid foam insulation.

Some types of rigid foam insulation have integral grooves in their faces to direct rainwater away from the foundation.

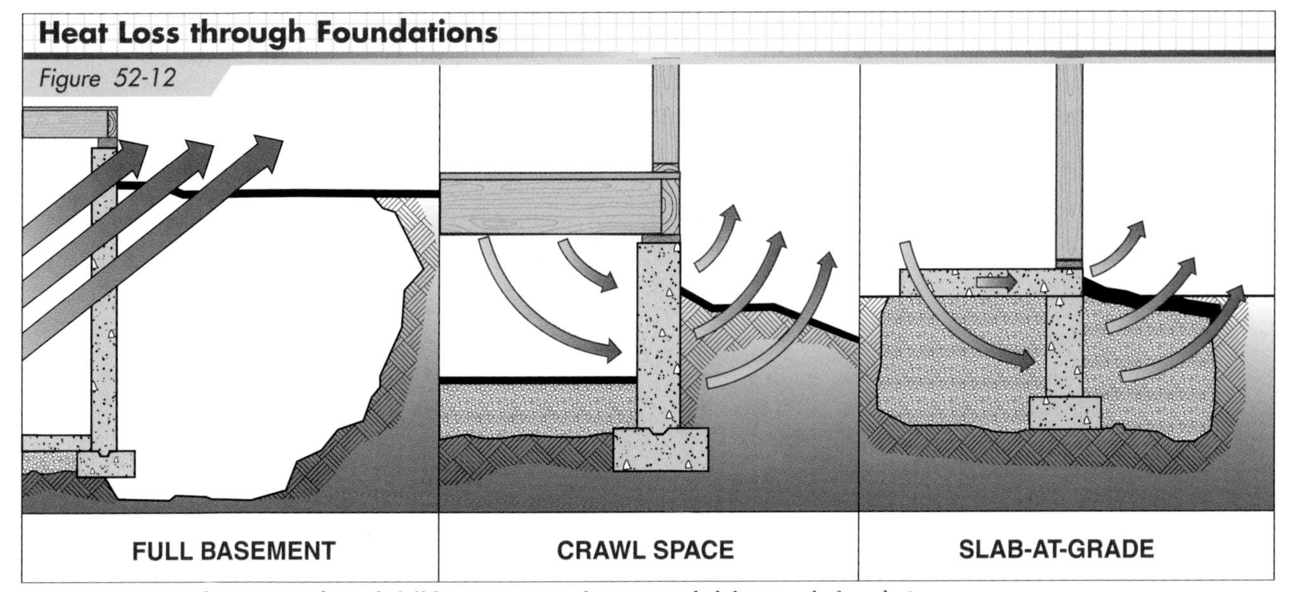

Figure 52-12. *Heat loss occurs through full-basement, crawl-space, and slab-at-grade foundations.*

Rigid foam insulation can be applied to the exterior or interior of full-basement and crawl-space foundations. **See Figure 52-13.** Both application methods are equally effective in reducing heat flow to the outside of the building. However, the exterior application is used more often. In addition to insulating the walls, exterior application of rigid foam insulation panels offers the following advantages:

- ease of insulating and sealing areas where the foundation wall and floor meet
- less possibility for soil freezing to the surface of the foundation and lifting it
- protection of the foundation wall, reducing the risk of condensation inside the wall
- fewer problems with water seepage

CertainTeed Corporation

Rigid insulation panels are available unfaced, or with a white kraft-scrim-foil or foil-scrim-kraft facing.

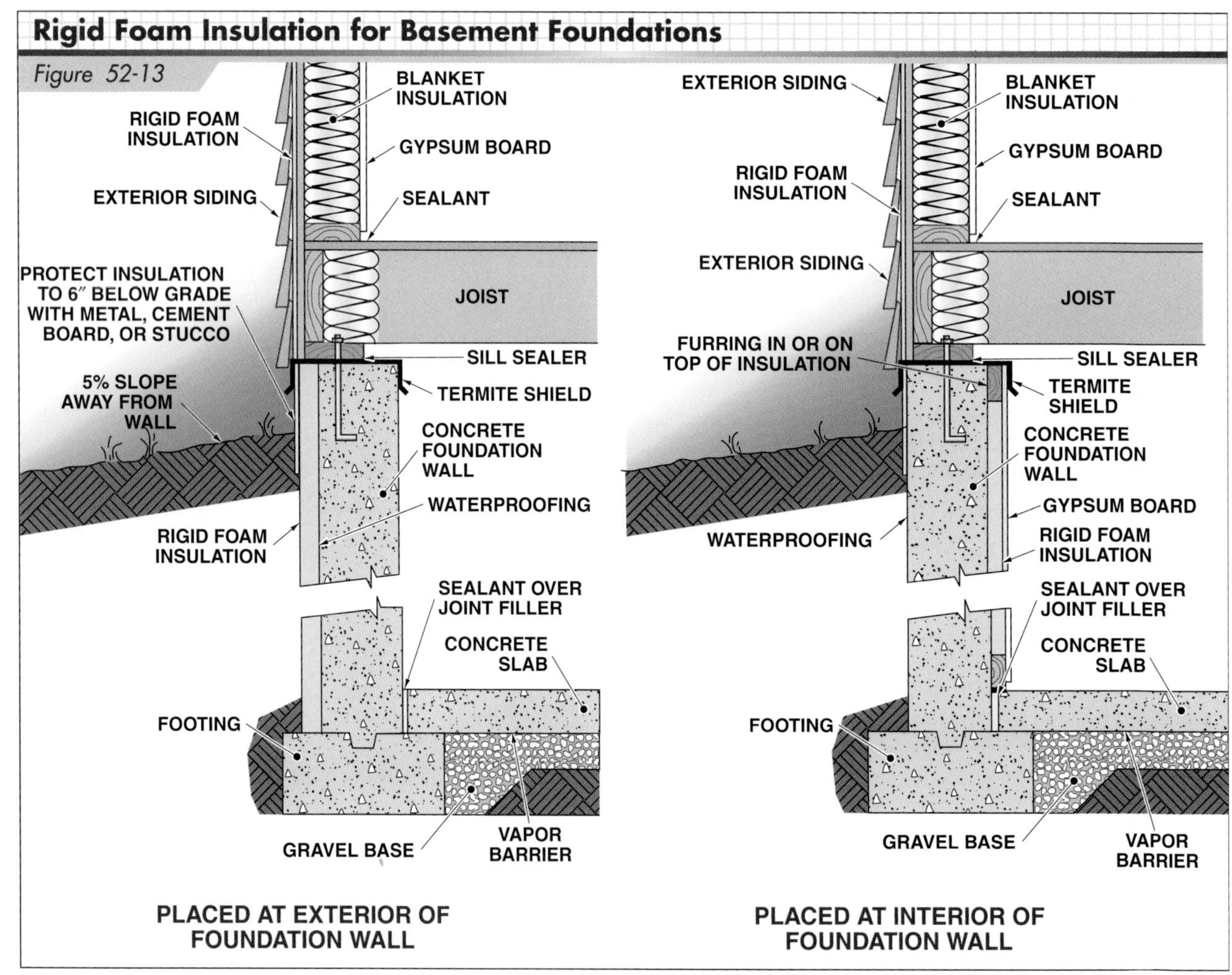

Figure 52-13. *Rigid foam insulation can be applied to the exterior or interior of concrete foundation walls.*

Attaching Rigid Foam Insulation to Foundations. Mechanical fasteners, powder-actuated fasteners, and/or adhesives are used to fasten rigid foam insulation panels to foundation walls. Panels are butted together along the walls and the joint is sealed with fiberglass tape. Panels at the corners are overlapped and butted together. The corner joints are also sealed with fiberglass tape.

Self-tapping screws or bolts with washers are spaced 24″ OC vertically and horizontally. Adhesive is applied to the back of the insulation panel to secure the panel in position while the mechanical fastener is driven in. Screw or bolt heads must be recessed into the panel when driven to allow the exterior finish material to be properly applied. **See Figure 52-14.** Fiberglass tape should be applied over fastener heads to seal the insulation.

Adhesive should be applied per manufacturer recommendations. A few mechanical fasteners may be used to secure the panels to the wall while the adhesive sets.

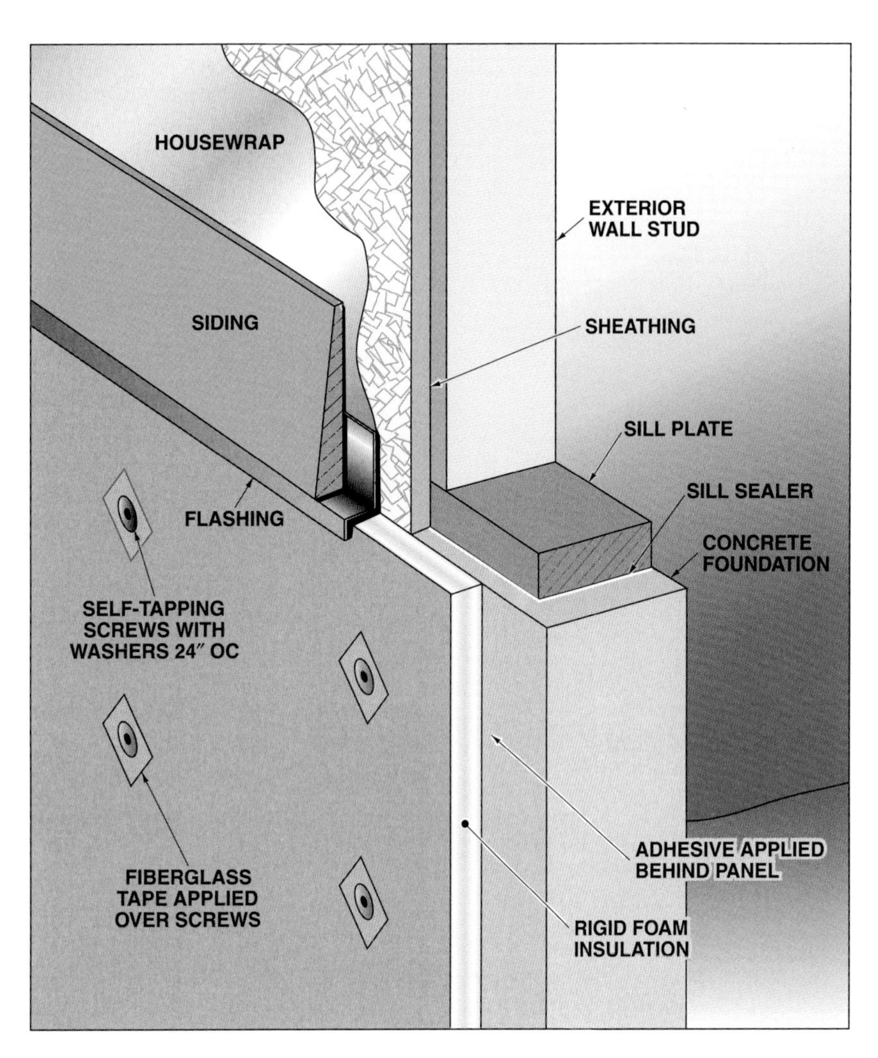

Figure 52-14. *Metal fasteners may be used to attach rigid foam insulation to concrete walls. Note the metal channel cap at the top of the insulation to prevent water from getting behind the insulation.*

For this slab-at-grade foundation, rigid foam insulation is placed at the exterior of the foundation and below the edges of the slab.

Rigid foam insulation must be protected above ground from the sun and physical damage. Ultraviolet light from the sun degrades the insulation value of rigid foam insulation. Contact with sharp objects can tear the surface of insulation panels. Various coverings can be used to protect the surface, including stucco or other cementitious products, cement board, and fiberglass panels.

Slab-at-grade foundations are insulated by placing rigid foam insulation panels along the outside of the foundation wall or between the footing and the slab. **See Figure 52-15.**

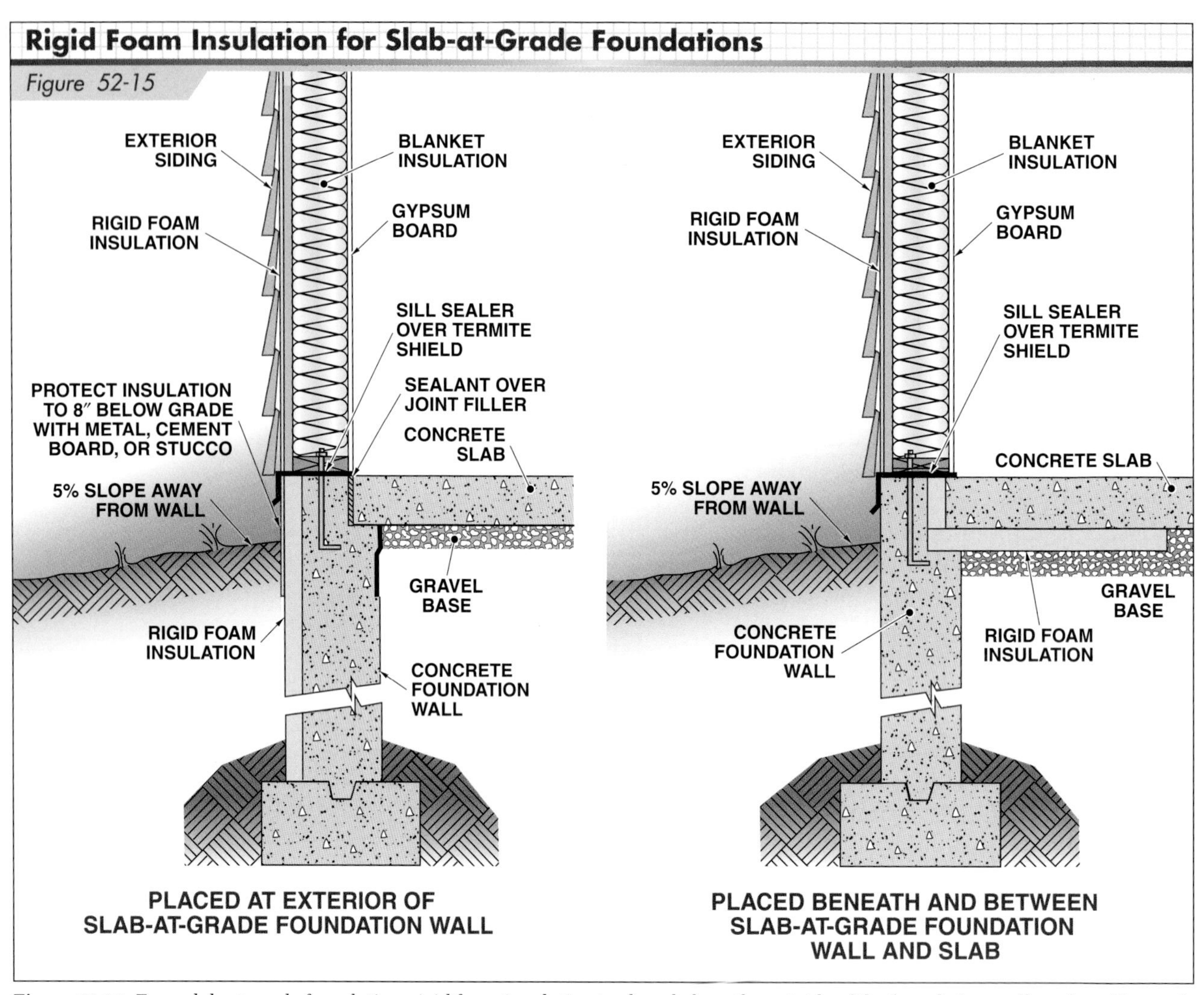

Figure 52-15. *For a slab-at-grade foundation, rigid foam insulation is placed along the outside of the foundation wall or placed beneath and between the wall and slab.*

Rigid Foam Insulation for Wood-Framed Walls. In residential construction, ½″ to 1″ rigid foam insulation panels are attached to the outside of exterior walls. Since blanket and batt insulation is installed between wall studs, rigid foam insulation increases the total R-value of the walls. Lateral structural stability is provided by metal diagonal or wood let-in braces; therefore, panel sheathing is not required. Finish board or panel siding is installed over the rigid insulation. If brick or other masonry veneer is used to finish the walls, additional insulation is provided by a 1″ air space between the masonry and insulation. **See Figure 52-16.**

ECO-Block, LLC

Insulating concrete forms (ICFs) provide insulation along the exterior and interior faces of this foundation and slab form.

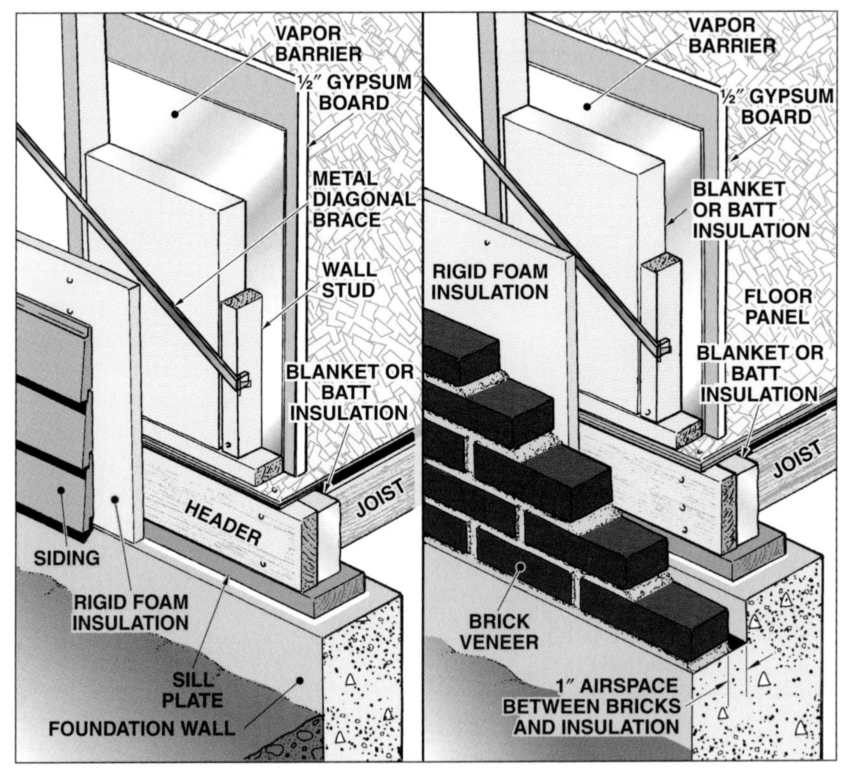

Figure 52-16. *Rigid foam insulation is often placed behind siding or masonry veneer.*

EXTERIOR INSULATION AND FINISH SYSTEMS (EIFS)

Exterior insulation and finish systems (EIFS) are a highly effective multilayer insulating wall system that continues to gain wider acceptance in commercial and residential construction. EIFS require an underlying wood panel sheathing. Polystyrene or polyisocyanurate foam insulation panels are fastened to the substrate using an approved adhesive and/or mechanical fasteners. A base and finish coat of an acrylic copolymer is then applied to the surface, providing a waterproof, colorfast, and crack-resistant finish. **See Figure 52-17.** The general application of EIFS is as follows:

1. Secure rigid foam insulation to the substrate with an approved adhesive. Screws and washers, or nails with wide, flat heads are also used. For best results when nailing, use annular ring shank nails.
2. Install metal flashing and sealant where required around all door and window openings and at the junction of the top of the wall and the roof soffits.
3. Apply a base coat of the acrylic copolymer. Embed reinforcing mesh in the base coat.
4. Apply a finish coat after the base coat has properly set.

ROOF INSULATION

Approximately 40% of the heat loss from a building can occur through the roof. Roof insulation can reduce the heat loss. When there is adequate working space in an attic, blanket insulation can be installed between the rafters, or rigid foam insulation can be fastened to the underside of the rafters. For low-pitch and flat roofs, rigid foam insulation is usually applied to the upper surface of the rafters.

Rigid Roof Insulation Materials

Rigid foam insulation for roofs is available in 4′ × 4′ or 4′ × 8′ panels with thicknesses ranging from 1″ to 3″. A variety of materials are available as rigid roof insulation, including the following:

- *Cellular glass* consists of heat-fused closed glass cells. Cellular glass has an R-value of 2.86 per inch of thickness.
- *Mineral fiber* is manufactured using rock, glass, or slag. Mineral fiber has an R-value of 4 per inch of thickness.
- *Perlite* is composed of expanded volcanic minerals combined with organic fibers and binders. Perlite has an R-value of 2.78 per inch of thickness.
- *Polyisocyanurate* is a closed-cell material typically sandwiched between asphalt-saturated felt or glass fiber facing. R-values vary depending on the material used for the facing.
- *Polystyrene* insulation used in its expanded (EPS) or extruded (XPX) form. R-values vary.
- Wood fiber insulation is manufactured from wood or cane fibers, and is combined with various binders. Its common R-value is 2.78 per inch of thickness.

Rigid Roof Insulation Systems

Rigid roof insulation can be used for low-pitched roofs or flat roofs. Rigid insulation for pitched roofs is finished with shingles or tile. Rigid insulation for flat roofs is finished with single-ply membranes or roofing felt and bitumen.

Fiberglass rigid roof insulation is composed of glass fibers bonded with a resinous binder.

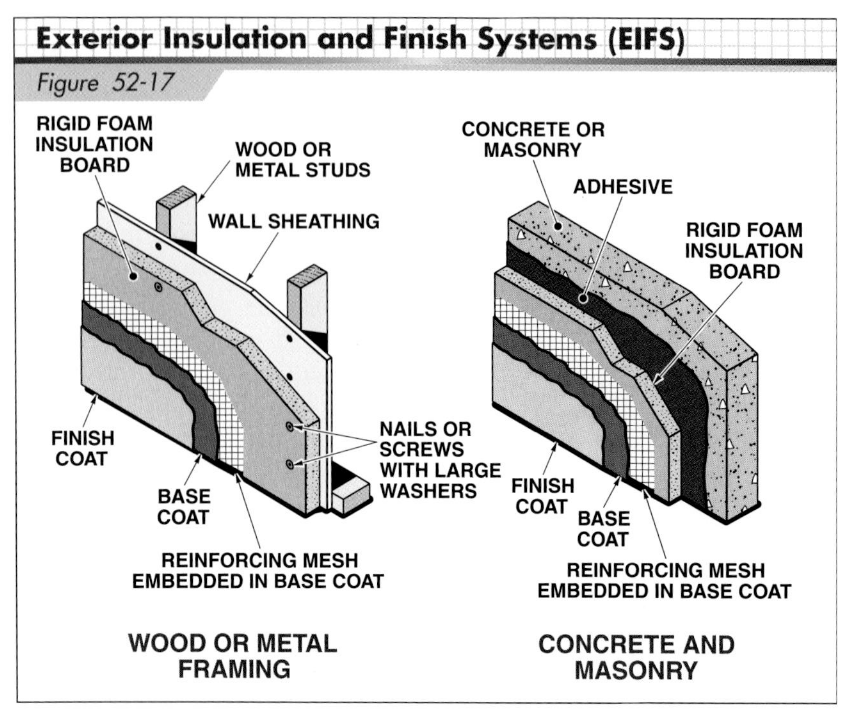

Figure 52-17. *Exterior insulation and finish systems (EIFS) provide an excellent waterproof insulation system.*

Even though the same type of rigid roof insulation panels are applied to pitched and flat roofs, there may be variations in application and installation procedures. Therefore, be sure to follow manufacturer instructions when installing the insulation.

Pitched Roofs. In one method, rigid foam insulation is fastened to the upper edges of the rafters. **See Figure 52-18.** (The insulation may also be cut to fit between the upper edges of the rafters.) When applying roofing tiles, counter battens and tiling battens are installed and roofing felt is applied.

In another system, the rigid foam insulation is attached with metal fasteners or adhesive to the roof sheathing. Oriented strand board or plywood cover boards are installed on top of the rigid foam insulation. Roofing felt is then placed over the cover boards and the shingles or shakes are nailed in place. **See Figure 52-19.**

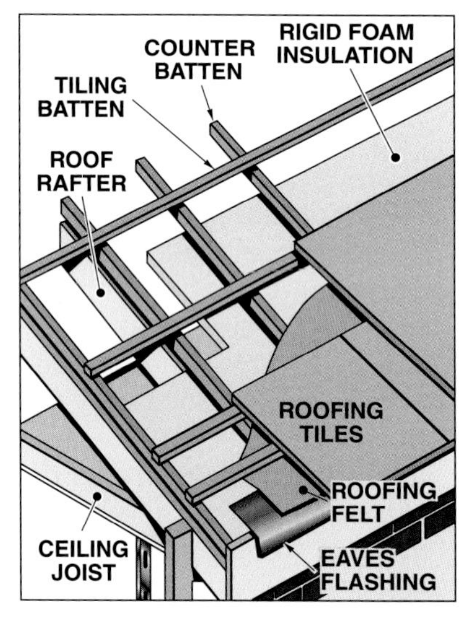

Figure 52-18. *Rigid foam insulation is applied to the upper surfaces of the rafters. Tiling and counter battens and roofing felt are installed prior to installing roofing tile.*

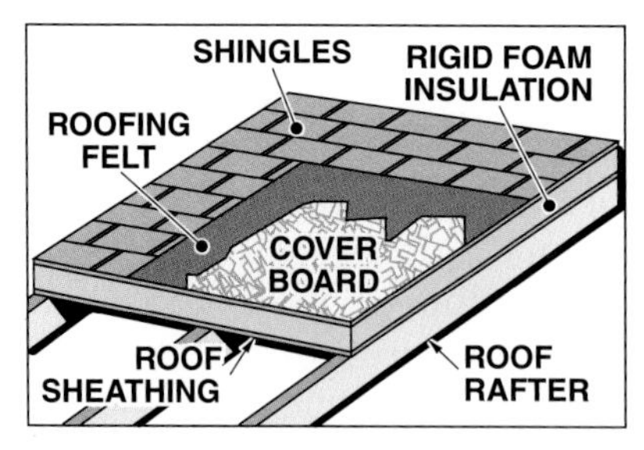

Figure 52-19. *Shingles or shakes can be directly fastened to the cover board.*

Single-Ply Flat Roofs. Single-ply and built-up roofing systems are the two most frequently used flat roof covering systems. For both systems, vapor barriers are laid down and rigid foam insulation is attached to the OSB or plywood sheathing and then protected by cover boards.

Single-ply membranes are made of flexible sheets of combined synthetic plastic materials. They are attached to wood and steel decks with fasteners. Specified adhesives may also be used to attach the membranes to wood decks.

Built-up roofing consists of layers (plies) of roofing felt bonded together on the job site with hot bitumen. The final layer is covered with a layer of light-colored (commonly white) pebbles to help reflect heat from the sun. Roofing felt consists of organic or inorganic fibers saturated with bitumen. Bitumen is derived from asphalt or coal-tar pitch combined with other elements.

Flat roofs must always have some slope to prevent the accumulation of water (ponding) and direct water to the gutters and downspouts. This can be accomplished by installing sloped insulation on the roof sheathing. The recommended slope is 1/4″ per foot. **See Figure 52-20.**

HEAT LOSS THROUGH DOORS AND WINDOWS

In a typical house, more heat is lost through windows and exterior doors than through any other part of the building. Studies conducted by the U.S. Department of Housing and Urban Development show that, in many areas of the country, 70% of the total heating load is used to replace heat lost through doors and windows.

Most heat loss occurs as a result of *infiltration,* which is air leakage through cracks around window and door frames. Heat loss is also due to heat *transmission* (transfer) through the door or window material.

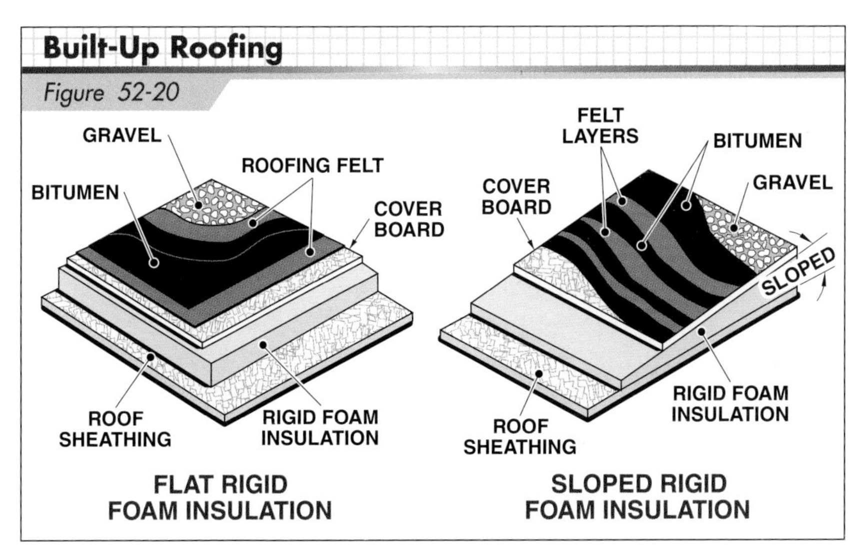

Figure 52-20. *Rigid foam insulation is installed below the roofing felt plies of a built-up roof. The insulation may be tapered to provide a sloped roof.*

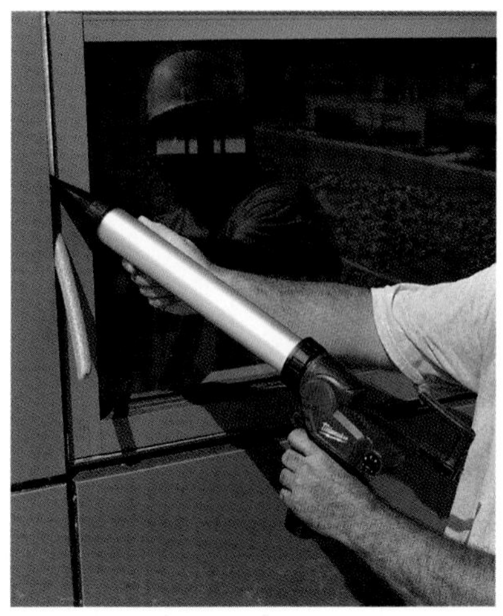

Milwaukee Electric Tool Corporation

Figure 52-21. *A caulking gun with cartridge is used to apply caulk between a window frame and concrete block wall.*

Fixed windows have the smallest amount of infiltration since the frames are sealed. Casement windows that can be tightly shut with compressive weatherstripping allow less heat infiltration than other types of operable windows.

In older homes, *storm windows* can be installed to reduce air infiltration. Storm windows are fastened to the outside of the window frame.

Energy Star® is a program designed to help businesses and individuals protect the environment through better energy efficiency.

Preventing Infiltration

Most heat loss caused by infiltration can be eliminated by sealing cracks around door jambs and window frames. Caulking, weatherstripping, and installing thresholds under the doors are three ways of preventing air infiltration.

Caulking. The best method for sealing cracks between a door jamb or window frame and the exterior wall is caulking. *Caulk* is a composite material that is available in cartridge-type tubes and is applied with a caulking gun. **See Figure 52-21.** Be sure to select the appropriate type of caulk for the job. Silicone caulk cannot be painted; latex and acrylic caulks readily accept paint.

Weatherstripping. Weatherstripping materials are used to prevent air infiltration between a door or window and its frame. Common types of weatherstripping are adhesive-backed foam rubber, wood-backed foam rubber, rolled vinyl, and V-strip. **Figure 52-22** shows each type and explains its applications.

Sealing Door Bottoms. The space at the bottom of a door is a source of air infiltration. **Figure 52-23** shows how the space can be sealed with a sweep, interlocking threshold, vinyl bulb threshold, or automatic sweep.

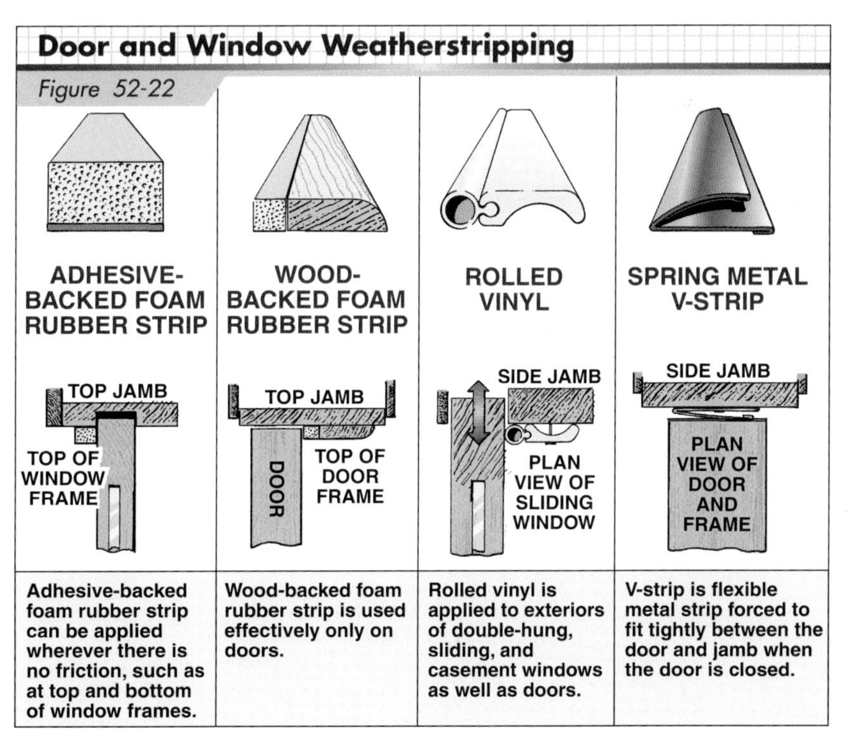

Figure 52-22. *Different types of weatherstripping may be used around doors and windows.*

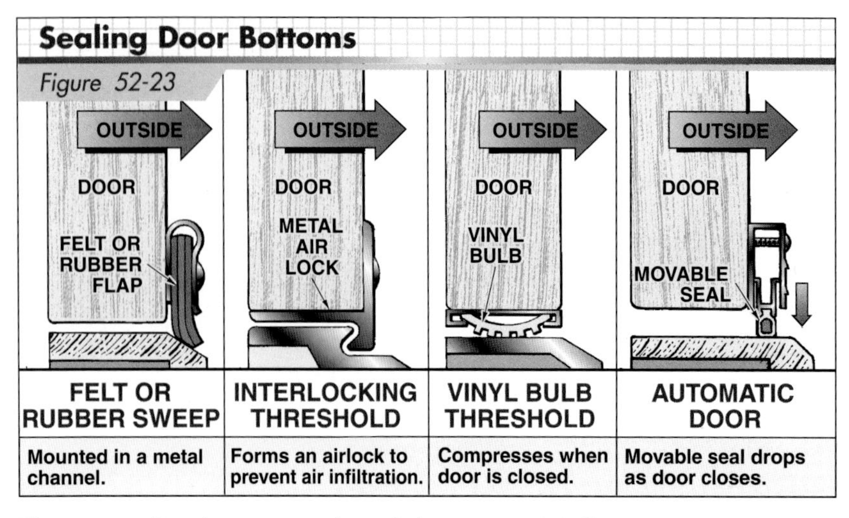

Figure 52-23. *Door bottoms must be sealed to prevent air infiltration.*

Insulated Doors and Storm Doors

Both wood and metal-clad entry doors are available with plastic foam cores that allow very little heat transmission. **See Figure 52-24.**

Composite storm doors have color molded through the entire door thickness. If the door is scratched, the color remains. In addition, the composite material is impervious to water.

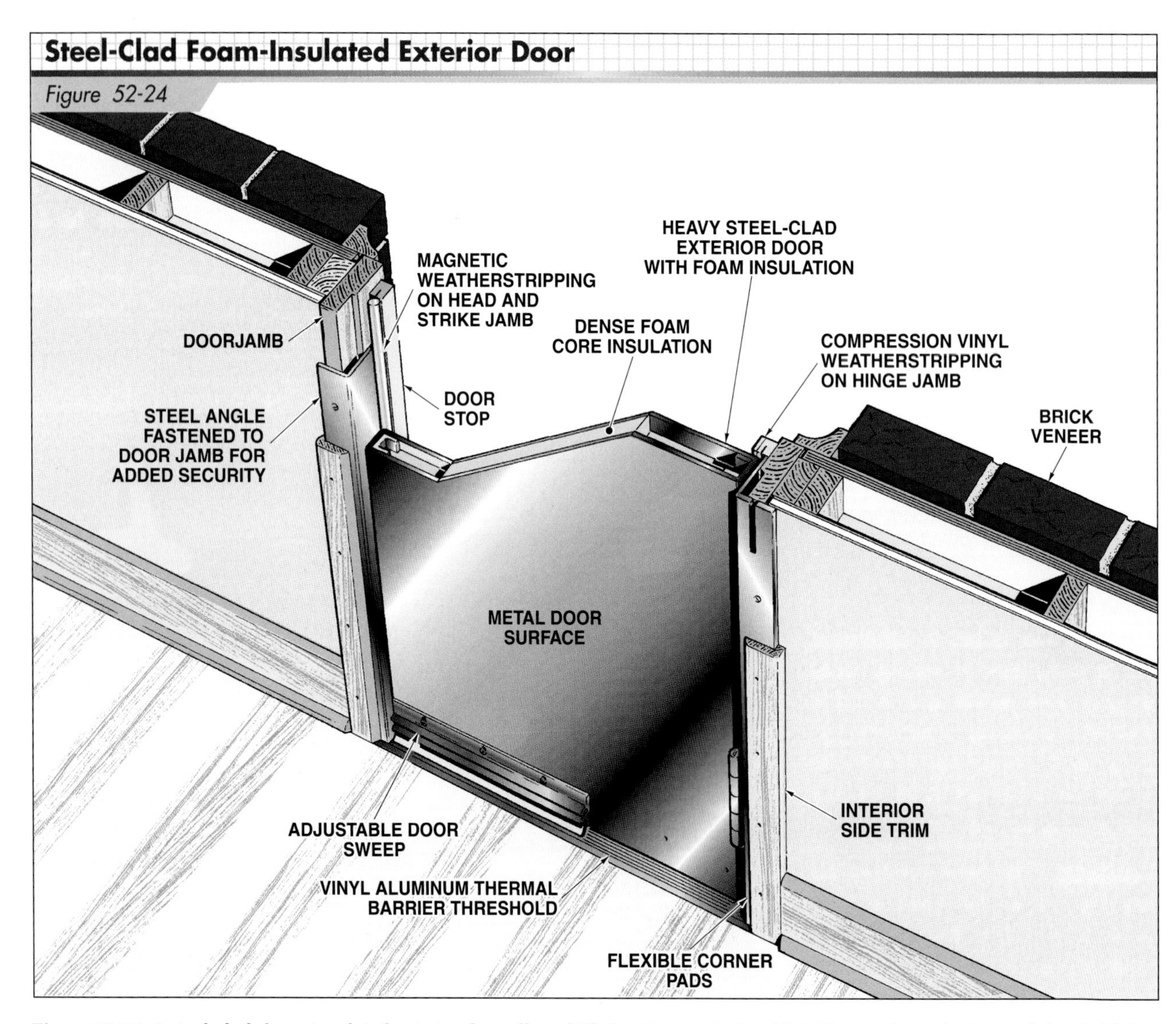

Figure 52-24. *A steel-clad, foam-insulated exterior door allows little heat transmission. Note the weatherstripping and threshold that have been installed.*

Heat transmission is also reduced by a *storm door.* A storm door is an additional wood or metal door hung on the outside of the door frame. Many storm doors have interchangeable glass and screen panels. **See Figure 52-25.** The glass panel is installed during colder temperatures and the screen panel is installed during warmer temperatures for increased airflow.

Multiple-Pane Windows

Glass is a poor insulator. Therefore, windows can be the greatest source of heat loss during the winter and heat gain during the summer. Studies have shown that 1 sq ft of ¼″ clear glass conducts 6 to 10 times more Btu per hour than 1 sq ft of a wood-framed wall.

Figure 52-25. *Storm doors are hung on the outside of the door frame. The glass pane can be replaced with a screen for airflow in warmer months.*

An effective way to increase U-value is to use multiple panes of glass. A single-pane window has a thermal resistance (U-value) of 1. Double-glazed windows, constructed with two panes of glass separated by a half-inch of air space, have an U-value of 2 since they are twice as resistant to heat loss. Windows installed in colder climates should be at least double-glazed. **See Figure 52-26.** Some window styles are available as triple-glazed.

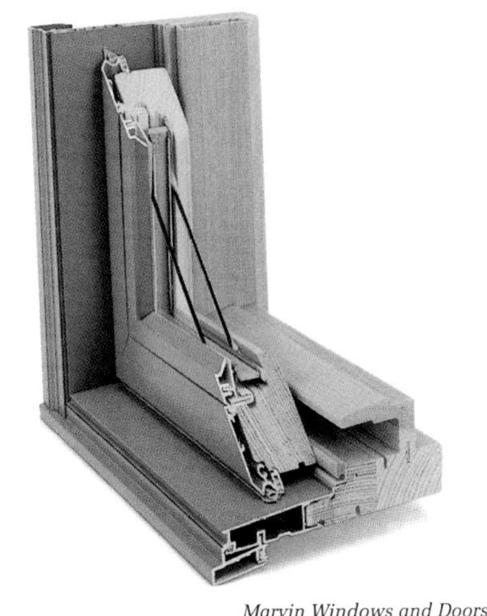

Marvin Windows and Doors

Figure 52-26. *This double-glazed window has special ½″ sealed insulated glass and an air space to provide a highly insulated unit.*

Gas-Filled Windows. Gas-filled windows further increase the U-value of multiple-pane windows. Gas-filled windows are three to four times more energy efficient than standard multiple-pane windows. **See Figure 52-27.** The key components of gas-filled windows are as follows:

- Double- or triple-glazed windows provide a sealed space between the panes.
- A *low-emittance (low-E) coating* is applied to the glass to reduce the passage of heat and ultraviolet rays through the windows.
- Argon, krypton, or xenon gas is introduced between the panes of glass and then sealed off. The gas minimizes convection currents and reduces overall heat transfer through the window unit. In addition, less solar radiation is allowed through the windows in hot climates, while a higher interior temperature is maintained in cold climates. The higher interior temperature of the glass surface also helps to reduce condensation.
- Vinyl-clad or fiberglass window frames and dividers provide good insulation qualities around the glass. Metal frames should not be used with gas-filled windows.

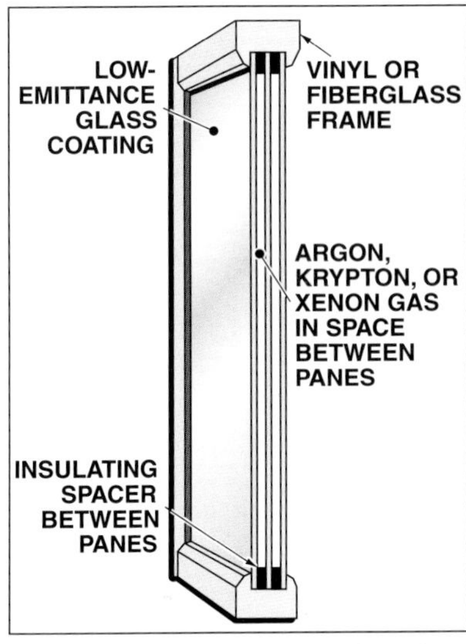

Figure 52-27. *Gas-filled windows have greater U-values than traditional multiple-pane windows.*

BUILDING SYSTEMS FOR IMPROVED INSULATION

Building design and construction methods are continually being modified for energy efficiency. One significant improvement is the use of 2 × 6s instead of 2 × 4s for exterior wall framing. In the past, 2 × 4s were used for framing exterior walls in most one-story residential construction. The 2 × 6 lumber allows the use of thicker insulation materials with higher R-values inside the walls. **Figure 52-28** shows a cross-section of an energy-efficient house utilizing thicker exterior walls and other energy-saving methods.

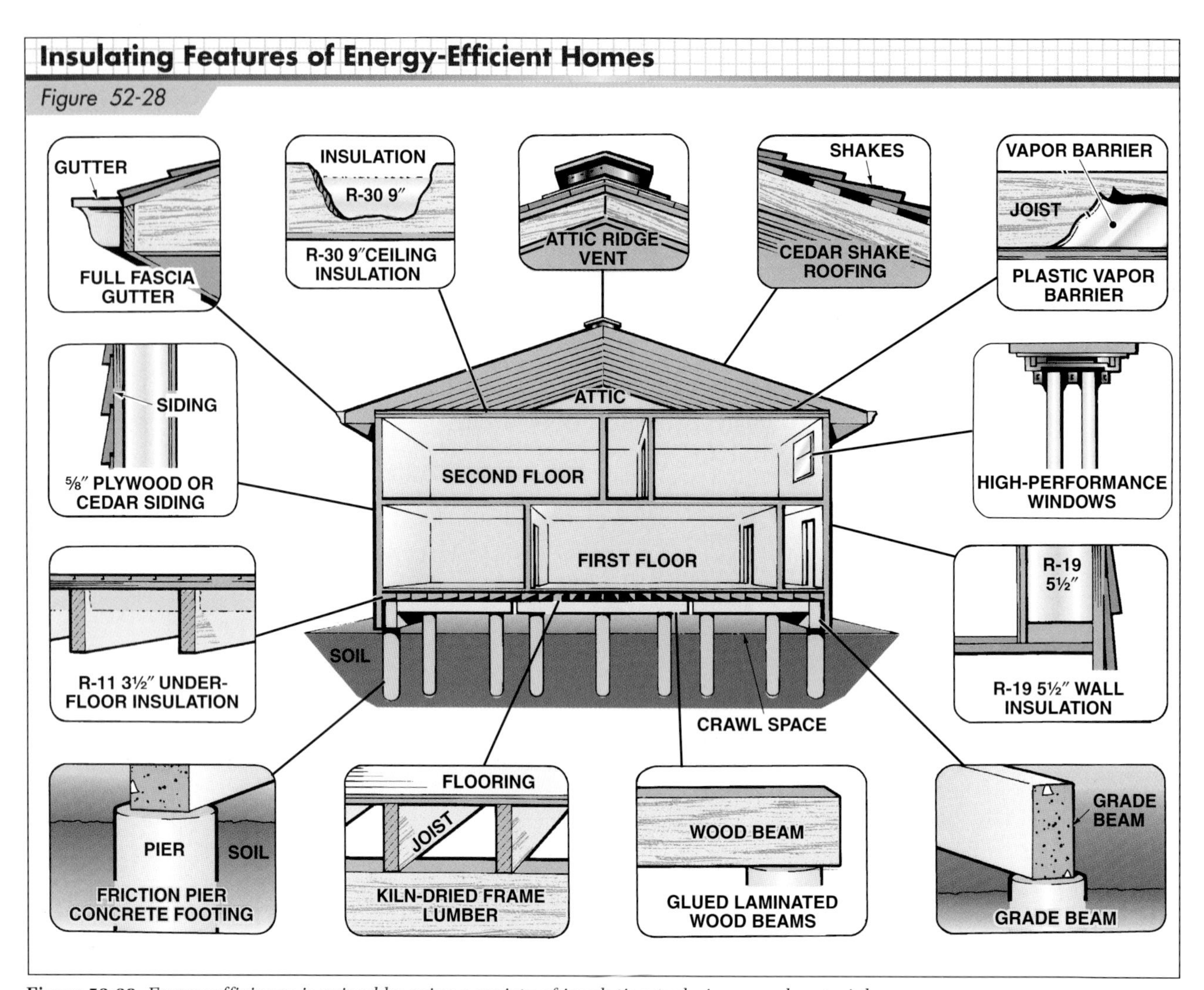

Figure 52-28. *Energy efficiency is gained by using a variety of insulation techniques and materials.*

UNIT 53 Sound Control

OBJECTIVES

1. Describe methods of sound transmission.
2. Discuss how sound transmission can be controlled.
3. Explain methods of floor/ceiling sound control.
4. Explain how sound control is affected by building design and construction practices.
5. List and describe principles of room acoustics.

Noise disturbance can be a serious problem in homes and offices. High noise levels interfere with rest and relaxation at home and contribute to inefficiency and fatigue at work. Constant exposure to extremely high noise levels can cause ear injury and hearing loss.

Noise problems can be reduced through the use of insulation materials and special sound-reducing construction methods for walls and ceilings. Many materials used for sound insulation are similar to materials used for thermal insulation.

SOUND TRANSMISSION AND SOUND LEVELS

Sound travels in different ways and at various levels. These factors must be considered in designing an effective sound control system.

Airborne Sound Transmission

Airborne noise, such as speech or music, is transmitted through the air as pressure waves that strike against a wall surface. The sound is then conducted through the air in the wall cavity to the opposite wall surface, causing it to vibrate and transmit the sound to the adjoining room. **See Figure 53-1.** Airborne noise travels through floors and ceilings in the same way that it travels through walls.

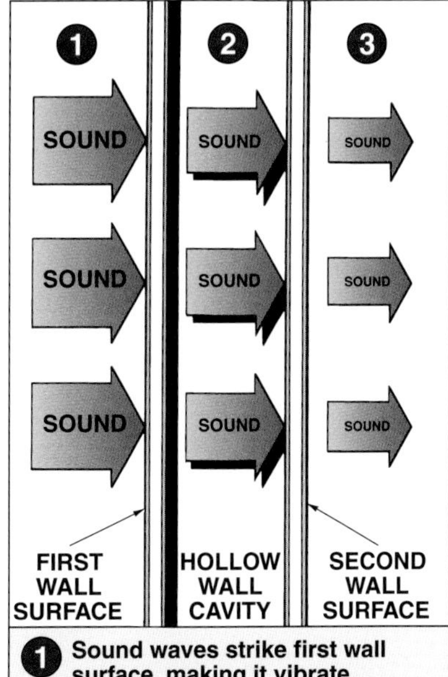

Figure 53-1. *Airborne sound transmission is conducted through the air as pressure waves.*

Resistance of a wall, floor, or ceiling system to airborne sound transmission is indicated by its *sound transmission coefficient (STC)*. **See Figure 53-2.** A high STC number indicates more resistance than a low number. Model building codes specify the minimum acceptable STC rating for multifamily dwellings and public buildings. Acceptable STC ratings range from approximately 39 to 65.

STC RATINGS AND DESCRIPTIONS

Rating	Description
25	Normal speech understood quite easily
30	Loud speech understood fairly well
35	Loud speech audible, but not intelligible
42	Loud speech audible as a murmur
45	Must strain to hear loud speech
48	Some loud speech barely audible
50	Loud speech not audible

Figure 53-2. *In sound transmission coefficient (STC) ratings, the higher the number, the better the sound resistance.*

Structure-Borne Sound Transmission

Structure-borne sound transmission is produced when a part of a building, such as the floor, is set into vibration by a direct impact. Walking on the floor or dropping objects on it, and striking the surface of a wall or ceiling are examples of the way structure-borne sound transmission is produced by impact. **See Figure 53-3.**

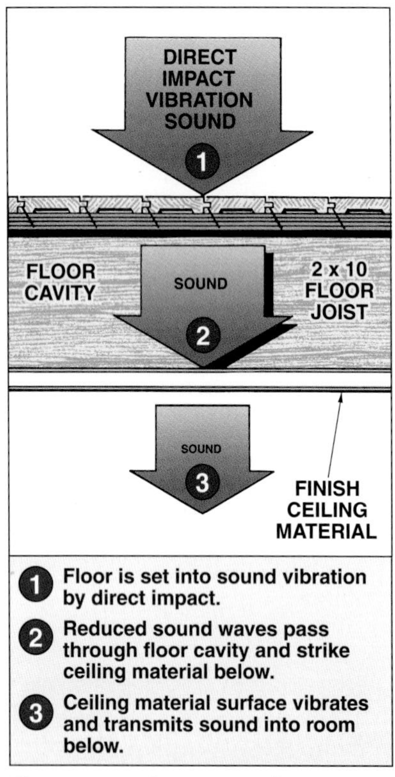

Figure 53-3. *Impact produces structure-borne sound transmission.*

Resistance of any part of the building structure to structure-borne sound transmission is indicated by its *impact insulation class (IIC) rating.* A high number indicates more resistance to sound transmission than a low number. An IIC rating of 55 is required for good control of impact noise. A rating of 60 or more is required for maximum privacy. A rating of less than 50 is considered unacceptable for sound control.

Structure-borne transmission is also caused by appliances such as vacuum cleaners, clothes washers and dryers, food waste disposers, fans, and compressors, and by pipes and ducts.

Sound Levels

Sound intensity is expressed in *decibels (dB).* One decibel is equal to the smallest change in sound intensity that can be detected by the average human ear. **Figure 53-4** shows a decibel scale ranging from 0 to 140.

SOUND LEVELS		
Noise Intensity*	Loudness	Examples
140	Deafening	Jet airplane taking off, air raid siren, locomotive horn
130	Pain threshold	
120	Feeling threshold	
110	Uncomfortable	
100	Very loud	Chain saw
90	Noise	Shouting, air horn
80	Moderately loud	Vacuum cleaner
70	Loud	Telephone ringing, loud talking
60	Moderate	Normal conversation
50	Quiet	Hair dryer
40	Moderately quiet	Refrigerator running
30	Very quiet	Quiet conversation, broadcast studio
20	Faint	Whispering
10	Barely audible	Rustling leaves, soundproof room, human breathing
0	Hearing threshold	Intolerably quiet

*in dB

Figure 53-4. *Sound intensity is expressed in decibels.*

SOUND CONTROL FOR WALLS

A typical interior wood stud wall, with a single layer of ½″ gypsum board (drywall) on each side, has an STC rating of 34. **See Figure 53-5.** According to the chart in Figure 53-2, this type of wall is effective against quiet conversation on the other side of the wall, but not against loud speech.

Three methods of reducing sound transmission are to provide cavity absorption, to increase mass, and to break the sound vibration path. Best results are obtained when the methods are combined. Staggered-stud or double-stud walls also reduce sound transmission.

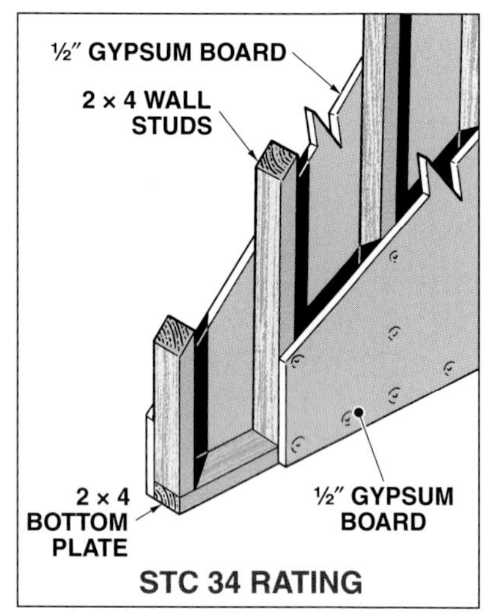

Figure 53-5. *A single wood-framed wall with a single layer of ½″ gypsum board on each side and no insulation between the studs has an STC rating of 34.*

CertainTeed Corporation

Insulation is placed in the walls between rooms to reduce sound transmission through the walls. The insulation may be friction-fit or stapled in place.

Cavity Absorption

One method to reduce sound transmission through a wall is to fill the wall cavity with sound-absorbing materials such as fiberglass insulation. The STC rating of a single wood-framed wall with ½″ gypsum board on each side can be increased from 34 to 39 by placing one thickness of insulation material between the studs. **See Figure 53-6.**

Increasing Mass

Heavy materials block sound better than light materials. Adding another layer of gypsum board to one or both sides of a partition increases its mass, thus increasing the STC rating. Doubling gypsum board on one side increases the STC rating of a single wood-framed wall with one thickness of insulation material from 39 to 40; doubling gypsum board on both sides provides an STC rating of 45. **See Figure 53-7.**

Breaking Vibration Path

Most sounds traveling through walls are a result of vibrations transmitted from one face of the wall to the other through structural members such as wall studs. Metal is more resilient in relation to sound transmission than wood. Therefore, walls constructed with metal studs transmit more sound vibration than wood-framed walls.

An effective way of improving the STC rating of wood stud walls is to place a metal resilient channel between the gypsum board and the studs to break the vibration path. A single wood-framed wall with one thickness of R-11 insulation material between the studs, double layers of gypsum board on both sides, and a metal resilient channel has an STC rating of 56. **See Figure 53-8.**

Placing walls and ceilings at different (nonparallel) angles helps to reduce sound deflection.

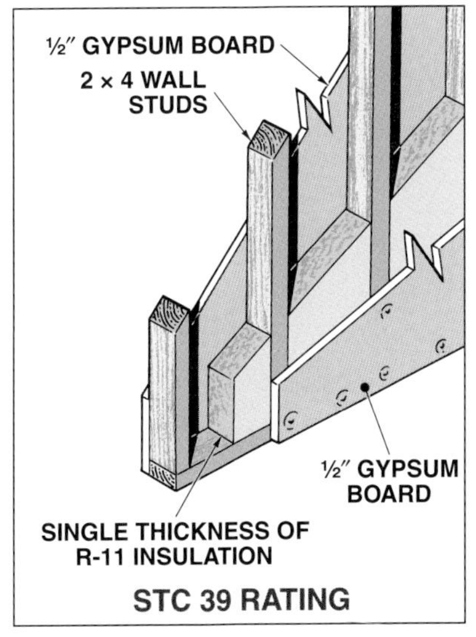

Figure 53-6. *The STC rating of a single wood-framed wall with ½″ gypsum board on each side is increased from 34 to 39 by placing a single thickness of R-11 insulation between the studs.*

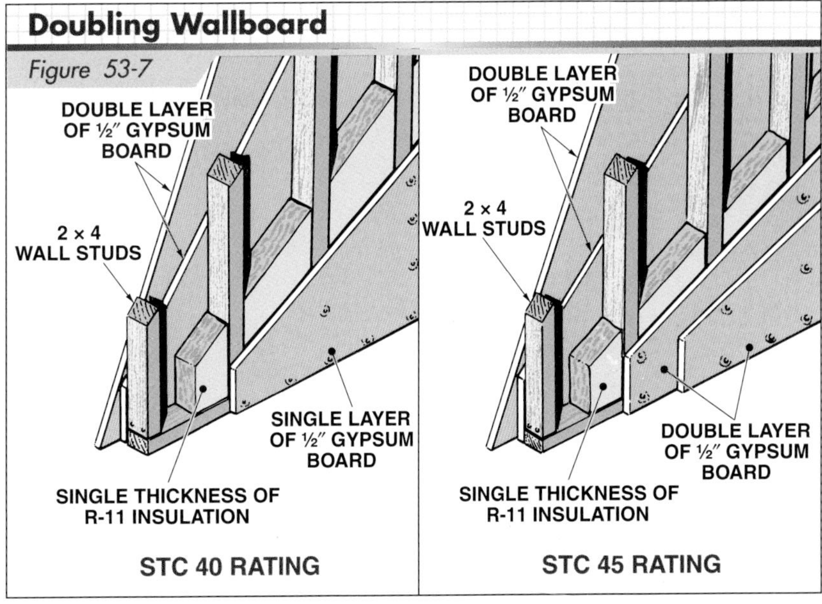

Figure 53-7. *Doubling the layer of gypsum board on one side increases the STC rating of a single wood-framed wall with a single thickness of R-11 insulation from 39 to 40. Doubling the gypsum board on both sides provides an STC rating of 45.*

Installing Resilient Metal Channel

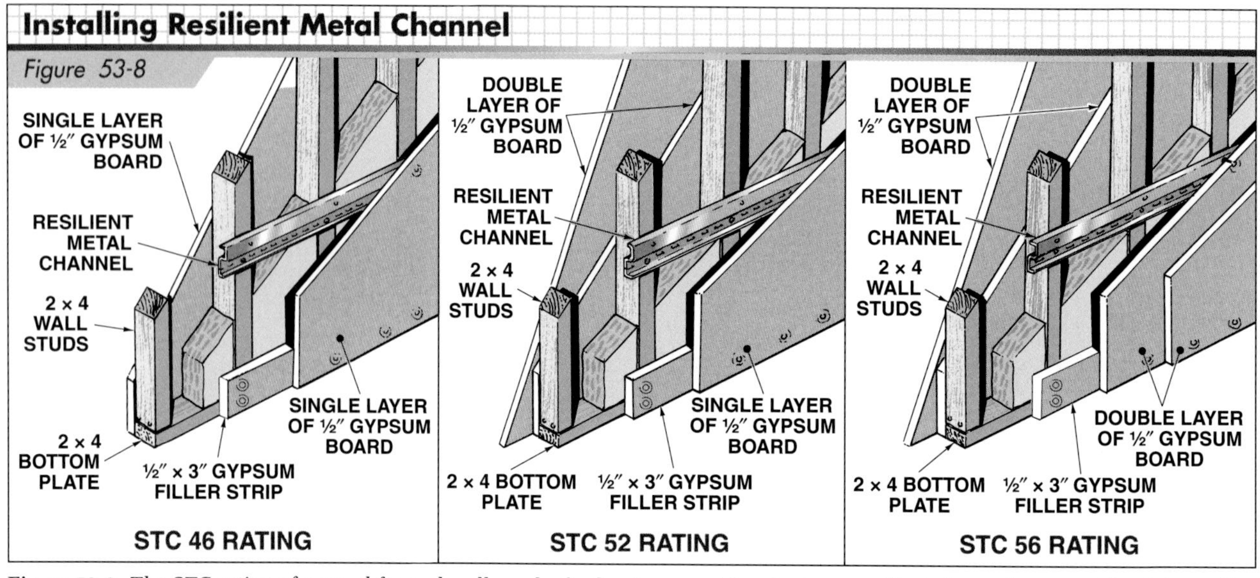

Figure 53-8. *The STC rating of a wood-framed wall can be further increased by placing a resilient channel on one side of the wall.*

Staggered-Stud and Double-Stud Walls

In a staggered-stud system, 2 × 6 bottom and top plates are used with 2 × 4 wall studs. **See Figure 53-9.** In the double-stud system, two separate wood-framed walls are built with a space between them for insulation. **See Figure 53-10.** With insulation material between the studs and double layers of gypsum board on both sides of the wall, a staggered-stud wall has an STC rating of 55 and a double-stud wall has an STC rating of 63.

FLOOR/CEILING SOUND CONTROL

One of the most disturbing household noises is footsteps and vibrations from the floor above. Overhead sounds are also more difficult than other sounds to eliminate. Several floor/ceiling sound control methods have been developed. **See Figure 53-11.** They all require carpets and pads over the subfloor. Mineral insulation is placed between the joists. The ceiling below consists of gypsum board fastened to resilient channels.

Floors with at least 6″ of sound control insulation improve to about 35 dB of noise reduction.

Installing Insulation in Staggered Wood Stud Walls

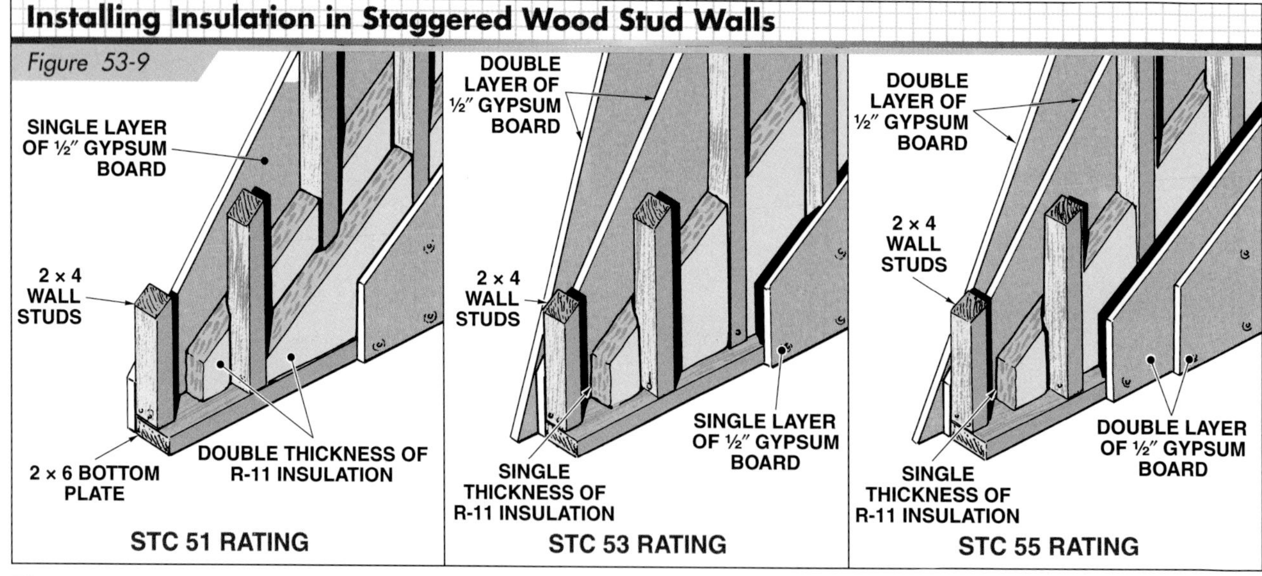

Figure 53-9. *A staggered-stud wall combined with insulation is another effective method of sound control.*

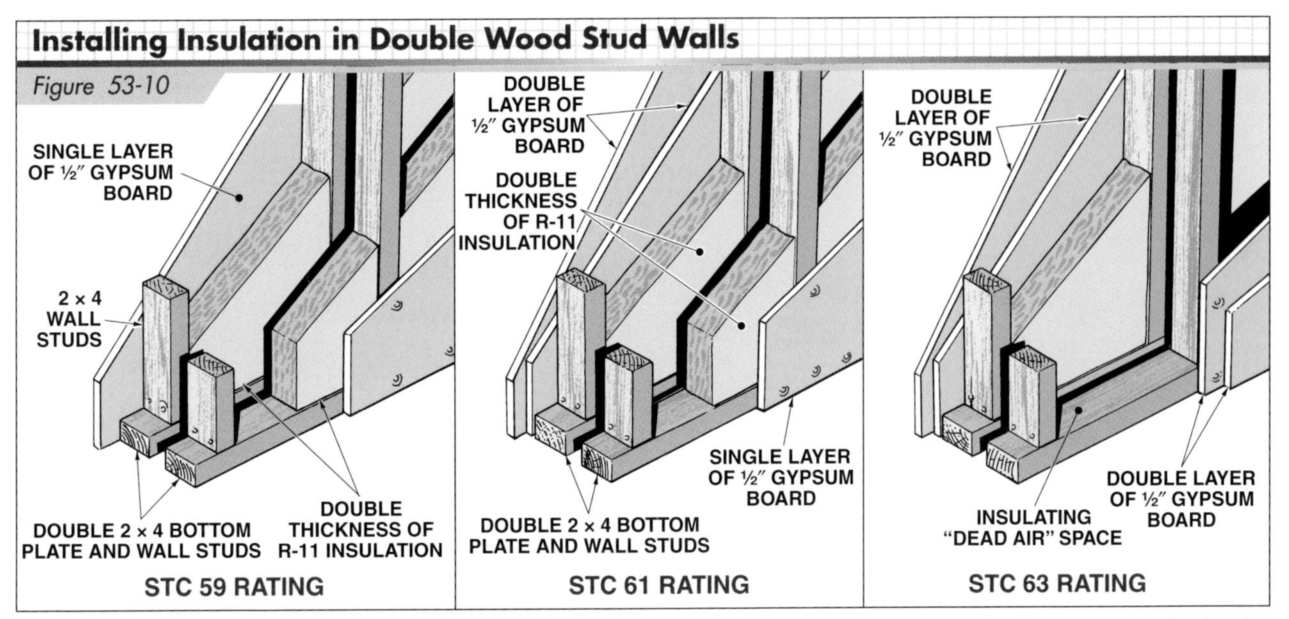

Figure 53-10. *A double wood-framed wall with R-11 insulation between the studs and a double layer of gypsum board on both sides provides extremely effective sound control (STC rating of 63).*

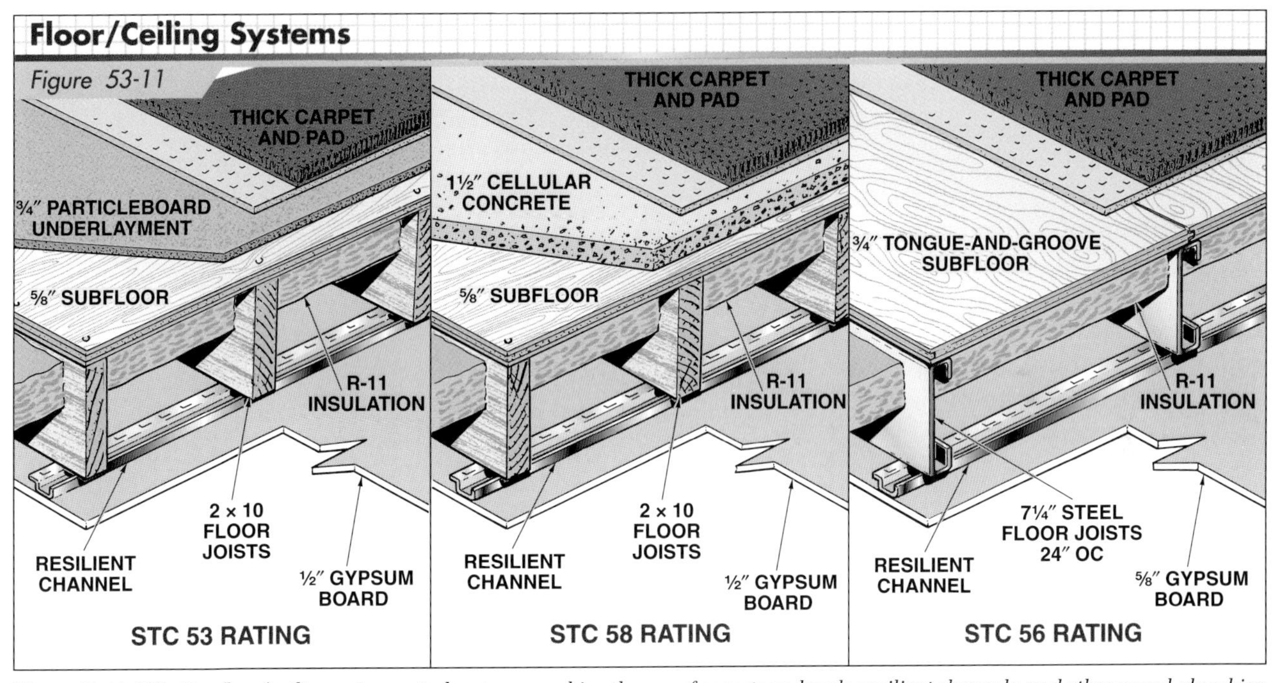

Figure 53-11. *Effective floor/ceiling noise control systems combine the use of carpets and pads, resilient channels, and other sound-absorbing materials.*

SOUND CONTROL BY BUILDING DESIGN AND CONSTRUCTION PRACTICES

The locations of doors and windows in a building are important to sound control. When possible, windows should not face noisy areas and should be separated to reduce cross talk. Door openings on opposite sides of a hallway should be staggered. **See Figure 53-12.**

Acoustic windows may be installed to minimize the penetration of exterior noise into a building. Acoustic windows are triple-pane windows with the inner layer made of laminated glass. *Laminated glass* is a specialty glass produced by placing a sheet of polyvinyl butyral (PVB) between two thin sheets of glass and subjecting the composition to intense heat and pressure to become a single pane of glass. The PVB layer provides a greater sound insulation rating than traditional glass.

The noise reduction gained by an acoustic window depends on the window's mass, area, and air tightness. In general, thicker glass and more panes of glass in a window unit will reduce the amount of sound transmission. The amount of space between the individual glass panes in a window unit also affects the sound transmission; greater space equals greater noise reduction.

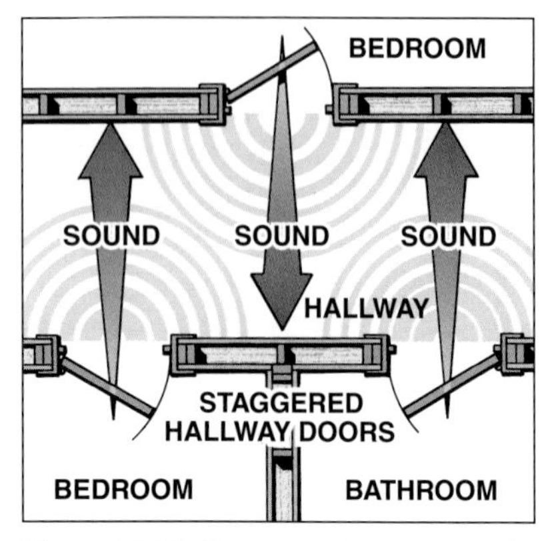

Figure 53-12. *Doors opening on opposite sides of a hallway should be staggered whenever possible.*

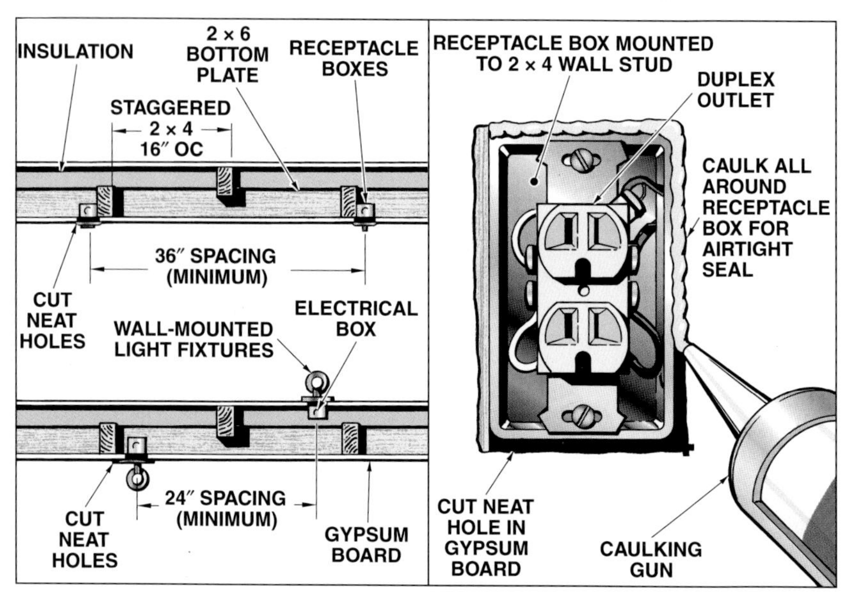

Figure 53-13. *Cut holes for electrical outlets neatly to reduce noise leaks. Place resilient caulk around the outlets before installing cover plates.*

Precautions should be taken in construction to reduce sound transmission. For example, walls that butt against each other or against floors and ceilings should be properly sealed. A nonhardening, permanent resilient caulk, such as butyl-rubber-based compound, is recommended for both sides of a partition at the top and bottom plates. Joint compound and tape applied on multiple-layered and staggered wallboard is also an effective seal.

The use of insulated glass in windows and proper weatherstripping around windows and doors helps reduce transmission of outside sound into the building. Solid wood or foam-core metal doors should be used. Sliding doors should be avoided.

Proper sealing around electrical and plumbing installations is also important to sound control. Cut tight-fitting holes for electrical boxes, and apply resilient caulk around the outlets before the plates are installed. **See Figure 53-13.** Do not place light switches or outlets back to back. Seal around surface-mounted ceiling fixtures in gypsum board to make them airtight. **See Figure 53-14.**

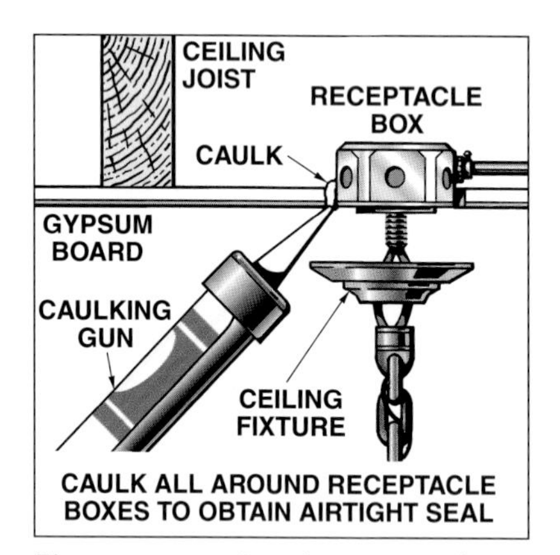

Figure 53-14. *Openings around surface-mounted ceiling fixtures should be sealed to make them airtight.*

Apply resilient caulk or other resilient material around all openings made for pipes in the wall plates as well as in the wallcovering. **See Figure 53-15.** Seal all joints between subfloor panels. **See Figure 53-16.**

Noise can also be controlled by enclosing noisy equipment when possible. Also, if cabinets are placed on opposite sides of a wall, they should not be placed back to back unless they are surface-mounted. **See Figure 53-17.**

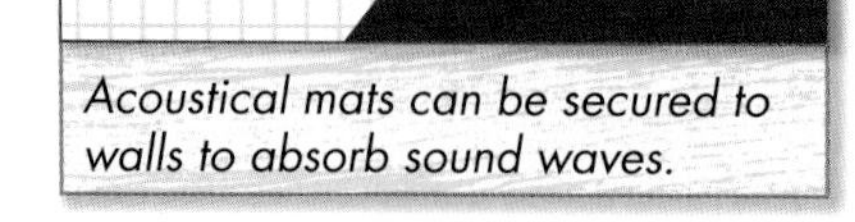

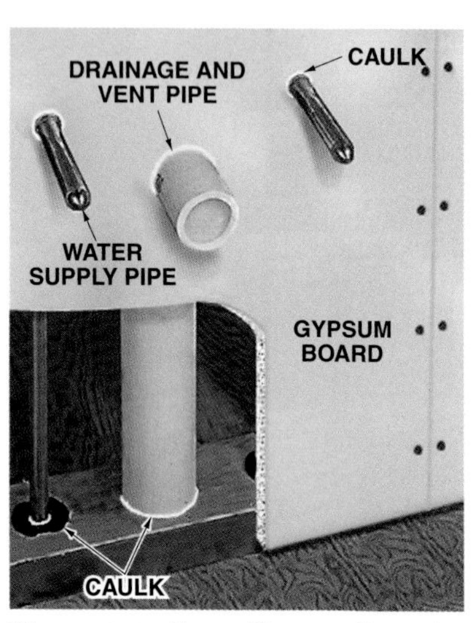

Figure 53-15. *Use resilient caulk or other resilient materials around all openings made for pipes in bottom plates or the wallcovering.*

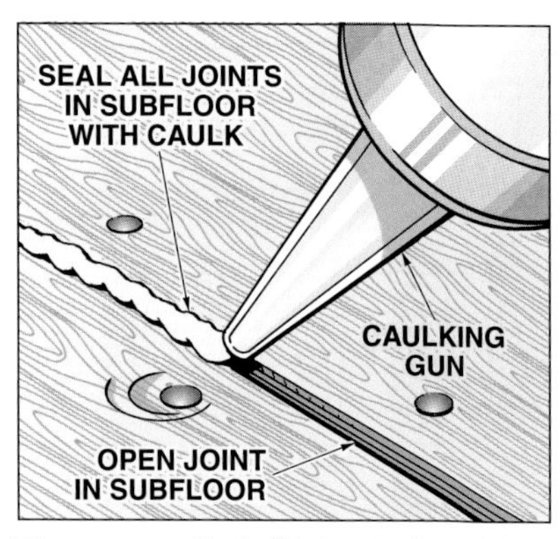

Figure 53-16. *Seal all joints in the subfloor with caulk to make them airtight.*

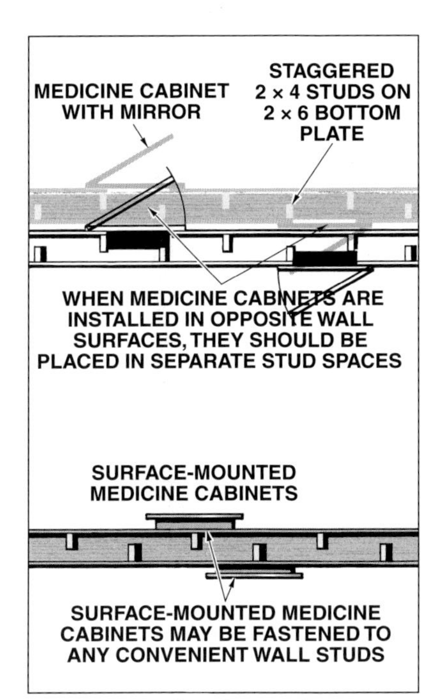

Figure 53-17. *When placing cabinets on opposite sides of a wall, separate the cabinets or surface-mount them as shown.*

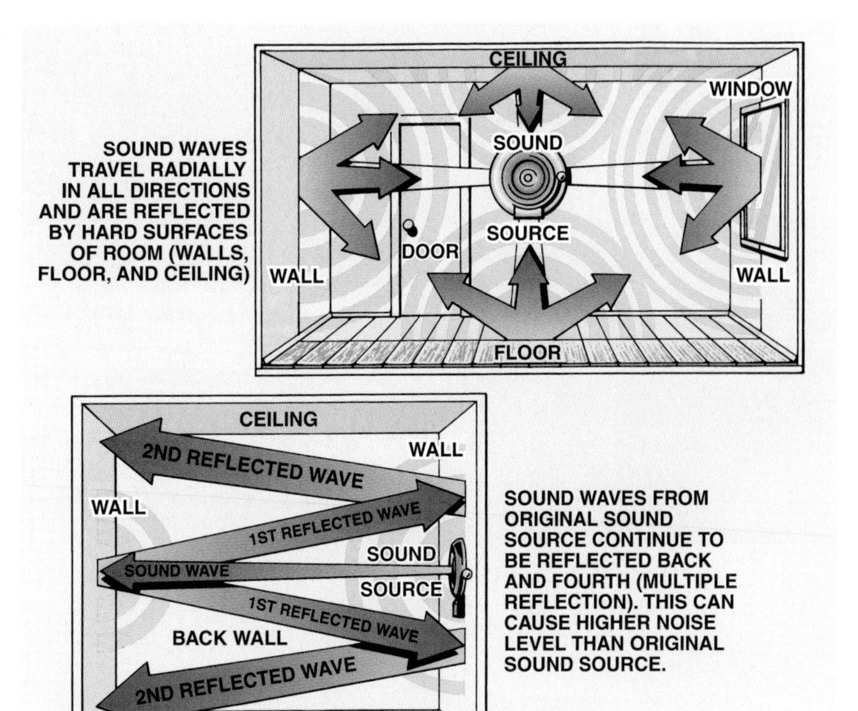

Figure 53-18. *Reflection creates high sound levels within an enclosed area.*

ROOM ACOUSTICS

Airborne and structure-borne noises enter a room through the walls, floors, and ceilings. However, many other noises are also created within the room such as voices, the sound of operating appliances, and the sound produced by walking on uncarpeted floors. In offices, shops, and factories, high noise levels are created by machinery and other equipment.

High sound levels created within an enclosed space, such as a room, are caused by *reflection*. For example, the sound waves created by a reciprocating saw travel in all directions until they strike against an obstacle such as a wall or ceiling. The sound waves bounce off these reflective surfaces, creating a greater noise disturbance than the original sound. **See Figure 53-18.**

Absorbing Sound

Smooth and hard building surfaces reflect up to 98% of the sound that strikes them. The amount of reflection can be greatly reduced by covering some of the room surfaces with acoustical materials that are designed to absorb sound. Sound-absorbing materials such as acoustical tile or acoustical plaster can reduce sound reflection as much as 50% when properly applied.

Acoustical tiles are manufactured from wood fiber or similar materials and are available in different sizes of square or rectangular shapes. They are attached directly to a drywall substrate or are used in suspended ceilings. Acoustical tile is designed with numerous tiny sound traps, comprised of holes or fissures, in the tile surface. When sound strikes the tile, most of its energy is trapped in the holes or fissures. **See Figure 53-19.** Acoustical plaster is a mixture of ground gypsum combined with chemical ingredients that cause tiny air bubbles to form in the material. The air bubbles trap and absorb sound.

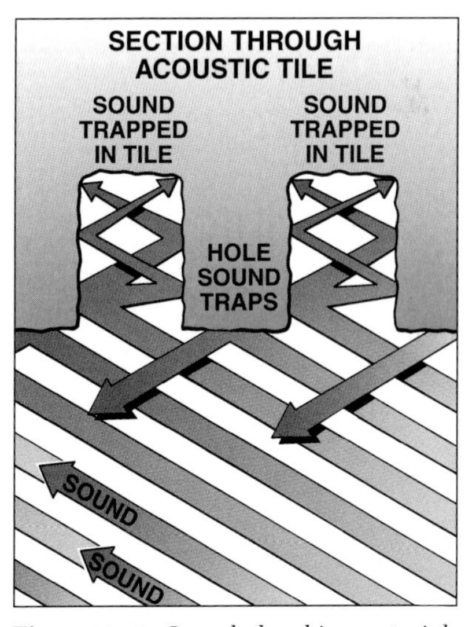

Figure 53-19. *Sound-absorbing materials such as acoustical tile have holes and/or fissures to trap sound.*

Unit 53 Review and Resources

UNIT 54

Solar Energy

OBJECTIVES

1. List and describe two basic methods of solar heating.
2. Identify the elements of passive solar heating systems.
3. Describe how solar heat is distributed and controlled.
4. List and describe three types of passive solar heating systems.
5. Explain the operation of a passive solar cooling system.
6. Identify the components of an active solar heating system.
7. Identify the components of an active solar liquid heating system.

Solar energy is an important alternative to generated electricity. Many buildings are being constructed to accommodate solar heating. Solar heating systems are also being installed in many older structures.

Two basic types of solar heating are used. One method, the passive method, does not rely on any mechanical means. The other, the active method, requires a complex system of collection, transport, and storage. Both solar heating methods rely on direct energy from the sun's rays, which produce the energy that is collected and distributed within a building.

Solar heating systems cannot entirely replace other types of heating systems within a building. During prolonged cloudy or inclement weather, conventional gas, electric, or wood-burning methods of heating are required. However, a solar heating system can significantly reduce heating costs in a building and reduce the use of fossil fuels.

PASSIVE SOLAR HEATING

Passive solar heating principles date back 2500 years. Using these principles, the Greeks built entire cities facing south to take full advantage of the sun's rays and reduce their dependence on wood, which was used as fuel at the time.

In a modern passive solar heating system, the building structure is designed to collect and store solar energy. The south-facing side of the building must have large areas of glass or plastic glazing through which sunlight can pass into the building. **See Figure 54-1.** Once collected, the heat is absorbed and stored by thick masonry materials or by water.

The following elements are necessary for a passive solar heating system:

- *collectors* through which sunlight enters the building
- an *absorber* to take in the heat
- *storage* medium to retain the heat
- a method of *distribution* of the heat
- a *control* element to prevent heat loss back through the collector

L&S Technical Associates

Figure 54-1. *A passive solar-heated building requires a broad expanse of south-facing windows to collect sunlight for heating the structure.*

Collectors

Collectors are the large glass or plastic areas through which sunlight enters the building. Collectors should face true south (±15°) and should not be shaded by trees or other buildings. **See Figure 54-2.**

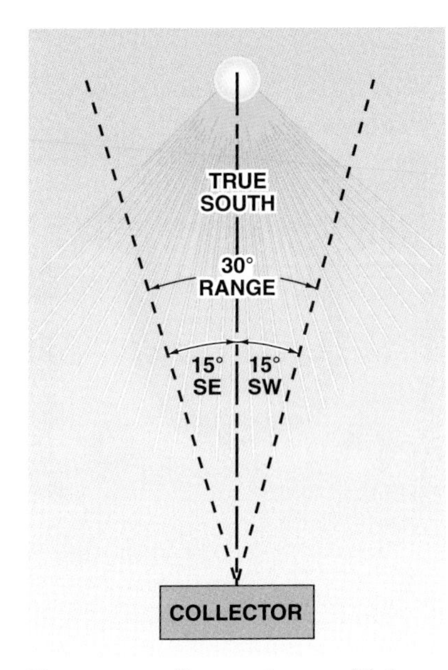

Figure 54-2. *For maximum efficiency, solar collectors must face as closely as possible to true south (±15°).*

Absorber

The absorber is the hard, darkened surface of the storage material. The surface of an absorber in a passive solar heating system should be in the direct path of the sunlight that enters through the collectors. The sunlight is absorbed as heat.

Storage

The storage medium for a passive solar heating system is usually masonry (concrete, concrete masonry units, or terra cotta tile) or water. The storage medium is positioned directly behind or below the absorber. The absorber for masonry storage material is simply the exposed surface of the storage material. The heat absorbed from sunlight by the absorber is retained in the storage medium.

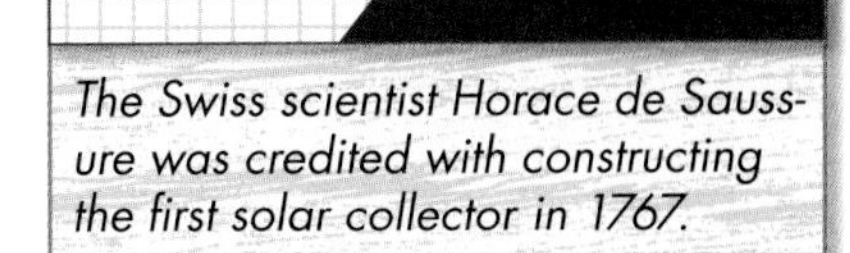

The Swiss scientist Horace de Saussure was credited with constructing the first solar collector in 1767.

Distribution

The method of distribution of solar heat is conduction, convection, and radiation. Sometimes mechanical devices such as fans, ducts, and blowers are used to help circulate heat throughout the building.

Control

The control element is movable insulation such as screens, drapes, or aluminized shades that can be applied at nighttime to the collectors to prevent heat loss back through the collectors. The movable insulation is also applied to the collectors during summer to keep the building cool.

The three main types of passive solar heating system are *indirect gain, direct gain,* and *isolated gain.* **See Figure 54-3.**

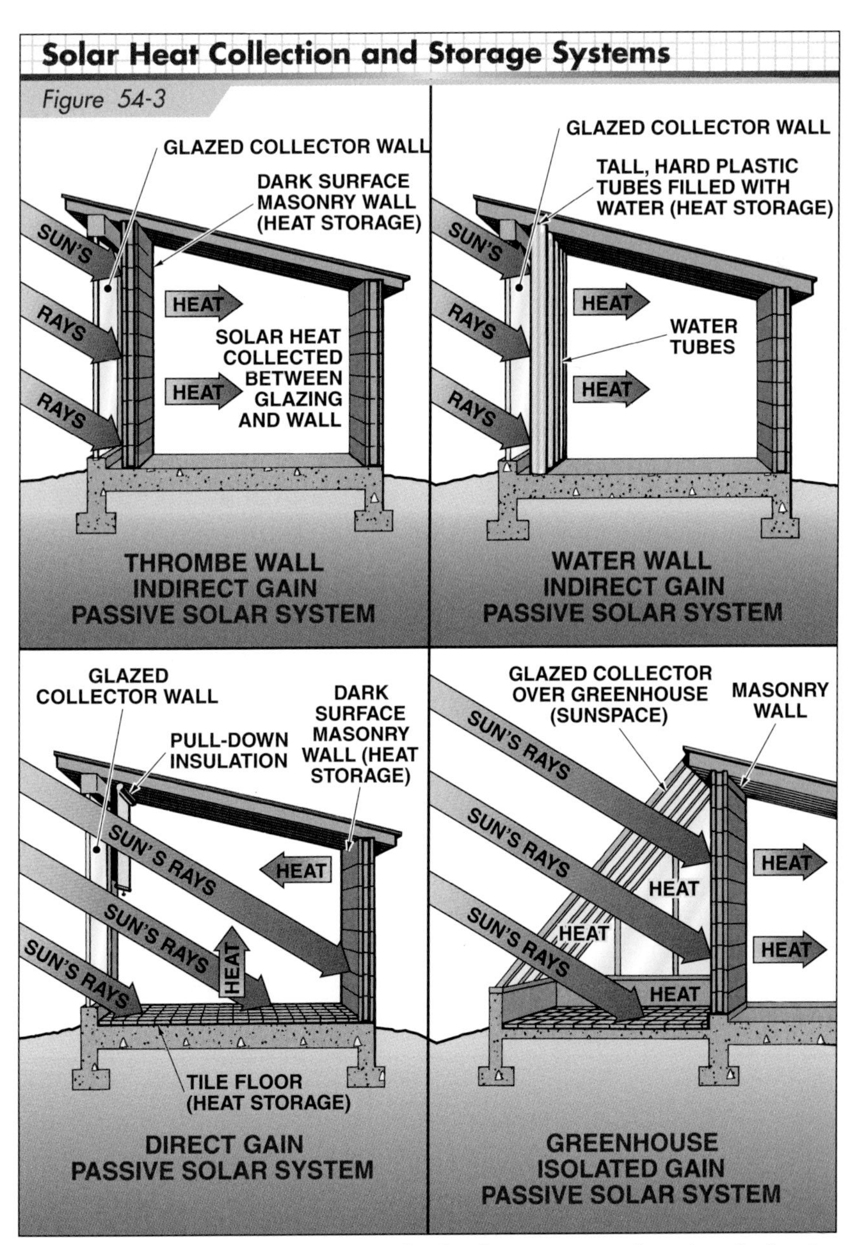

Figure 54-3. *Indirect gain, direct gain, and isolated gain are types of passive solar heating systems. One indirect gain system uses a Thrombe wall and the other uses a water wall.*

Indirect Gain

Two common passive solar heating systems are indirect gain systems—the *Thrombe wall* and the *water wall.* In indirect gain systems, the solar radiation is intercepted by an absorber-and-storage unit that separates the south-facing glass from the room.

A Thrombe wall system uses an 8″ to 16″ masonry wall. A single or double layer of glass or plastic glazing is mounted approximately 4″ in front of the masonry wall surface. The dark outside surface of the wall absorbs the heat, which is stored in the wall mass.

The water wall system uses water-filled containers instead of a masonry wall. Water walls can be built in a number of ways. One way to construct a water wall is to use tall, hard plastic tubes. A water wall absorbs and stores more heat than a masonry wall of equal volume.

Direct Gain

The simplest passive solar heating method is the direct gain method. Sunlight enters through large, south-facing windows (collectors) and strikes the walls and floors. The walls and floors should be 4″ to 8″ thick masonry material such as concrete, concrete masonry units, or brick. The surfaces should be a dark color in order to absorb the sun's heat, which is then stored in the masonry.

At night, as the room cools, the heat stored in the masonry radiates into the room. This is called a *time-lag heating process.* To control heat loss, movable insulation is pulled down to cover the collector at night. During the summer, the movable insulation covers the collector during the day to prevent the sun from overheating the building.

Isolated Gain

The isolated gain passive solar heating method requires a separate space such as a solarium, atrium, or greenhouse to collect the solar radiation.

A solarium can be built as part of a new building or added to an existing building. Solar heat is collected through the solarium glazing and can be absorbed and stored in various ways. A masonry wall or water-filled containers can be used for storage.

Several methods can be used to transfer the heat collected in the solarium to the living space. If a masonry wall is used for heat storage, a time-lag heating process can take place. Ceiling and floor-level vents may allow for a natural convective loop of warm and cold air. A low-horsepower fan or blower can also be used.

Passive Solar Cooling

Solar cooling may seem like a contradiction, but there are a number of ways in which the sun's energy can be used indirectly to cool the interior of buildings. Proper tree placement and natural wind ventilation are two common methods of passive solar cooling. Deciduous (broad-leaved) trees located around the perimeter of a building will shade the building. Tree shade can also lower the temperature several degrees around a building through a process called *transpiration,* an evaporative process by which plants give off water to the atmosphere.

Increasing the natural flow of air through a building by capturing the prevailing winds is another effective cooling method. Capturing the wind is accomplished by designing a building so operable windows (casements offer the best airflow) and interior partitions are placed where they can best direct the natural breezes through the house. *Wing walls* are another method to increase airflow through a building. Wing walls are vertical solid panels placed along the windows and perpendicular to the windward side of the house. **See Figure 54-4.** Wing walls increase the natural wind speed because of pressure differences created by the wings.

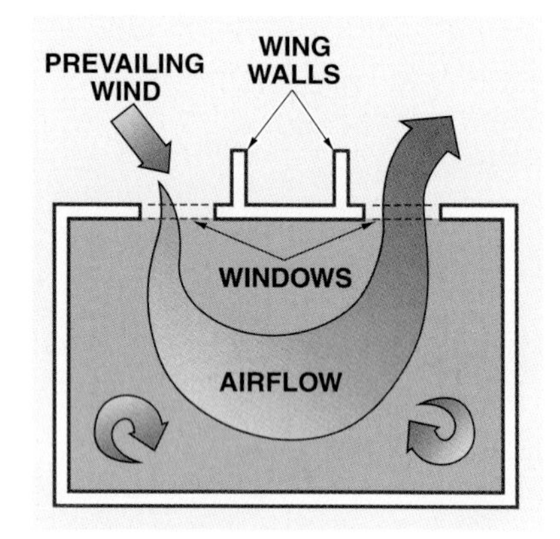

Figure 54-4. *Wing walls increase airflow through the living space.*

Thermal chimneys use air currents to convey air out of a building. Air is conveyed through the building by creating a warm or hot zone with an outside exhaust outlet. As air is drawn out of the building, replacement air is drawn into the house, creating air currents. **See Figure 54-5.**

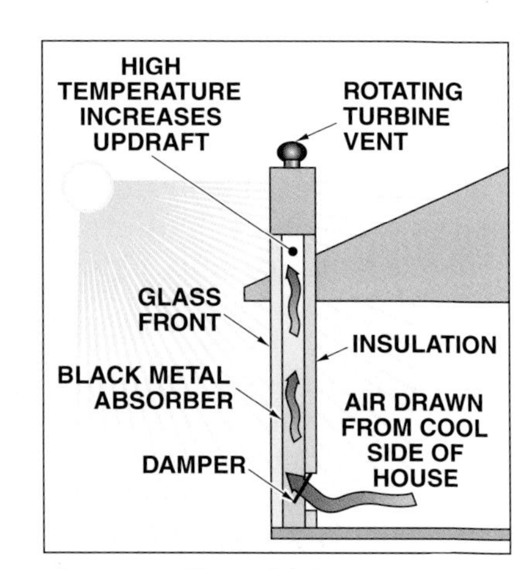

Figure 54-5. *Thermal chimneys use convective currents to draw air out of a building.*

Sunrooms and solariums can also be designed to convey air out of a building. Vents are placed on the roof of the heated south-facing room. Vents are placed in the lower part of interior partitions between the sunroom and adjacent living space. Windows on the north side of the building must be open. All vents in upper sections of interior partitions and operable windows in side walls should be closed. Airflow created through the living space has a cooling effect and is exhausted though sunroom roof vents. **See Figure 54-6.**

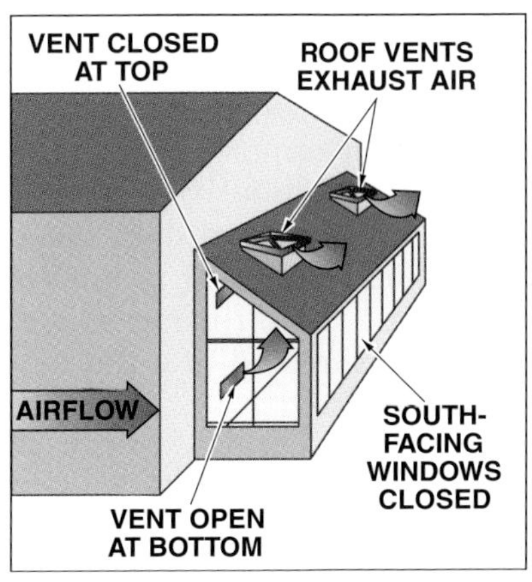

Figure 54-6. *Sunrooms and solariums can be designed to induce airflow through a building.*

ACTIVE SOLAR HEATING

Unlike a passive system, an *active solar heating system* or *photovoltaic system*, does not require a specially designed building. An active solar heating system is a mechanical method that can be installed in most new or existing buildings. Active solar heating systems heat air and/or water.

In active solar heating systems, heated air or liquid is used to move heat into the living space. Heated air and liquid systems require the following components:

- *collectors* to convert the sun's rays into heat
- a means of *moving* the heated air or water
- a method of *storing* heat when it is not being used
- an *auxiliary heating system* for times when direct sunlight is not available over longer periods
- automatic *controls* to operate dampers and/or control valves

Active Solar Air Heating Systems

Active solar air heating systems can produce 40% to 80% of the heating needs for a one-family residence. However, recent studies show that active solar air heating systems are most economical when they are designed to handle about 50% of home heating. A schematic of an active solar air heating system is shown in **Figure 54-7.**

Collectors. Active solar air heating system collectors are usually rectangular containers with transparent glass or plastic glazing to allow sunlight to enter, an absorber plate to absorb the heat from sunlight, and insulation to reduce heat loss from the collector. **See Figure 54-8.**

Solar collectors are usually mounted singly or in rows on the roof or the south wall of a building. **See Figure 54-9.** Collectors should face true south (±15°) and should not be shaded by trees, hills, buildings, or other obstructions. The tilt angle is different in different geographic areas. Generally, collectors should be tilted at an angle equal to the local latitude plus 15°. Collectors receive most solar radiation between 9 AM and 3 PM.

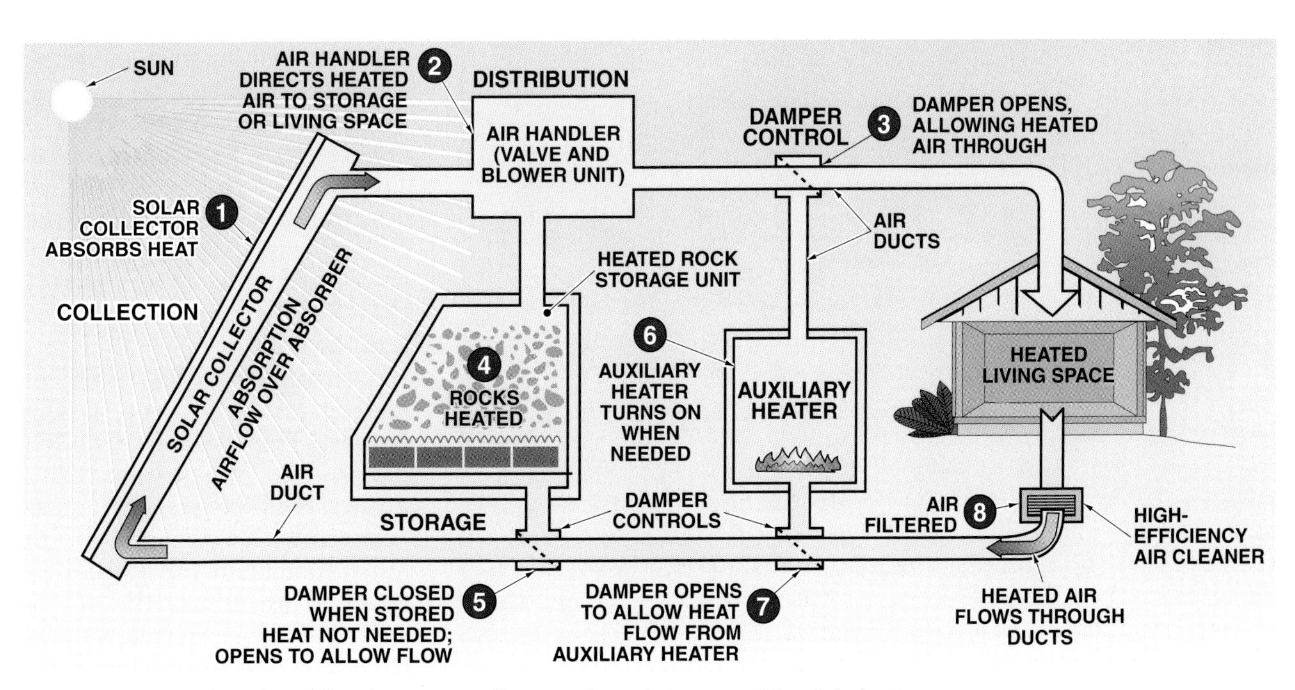

Figure 54-7. *An active solar air heating system relies on a thermal storage unit to retain heat.*

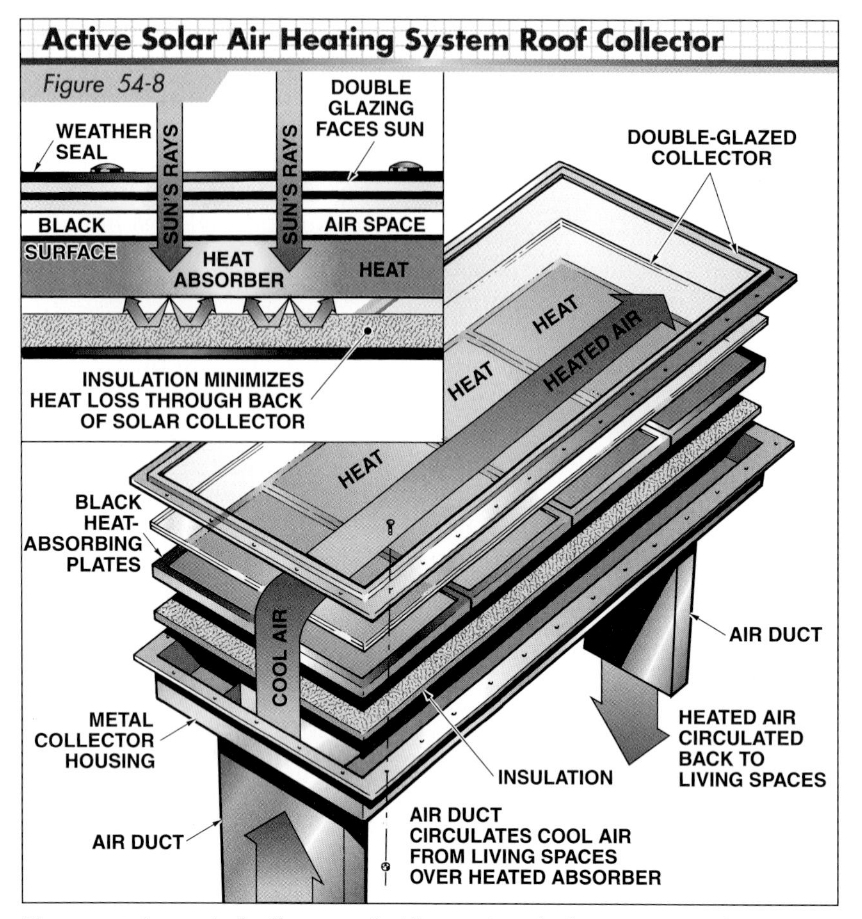

Figure 54-8. *In a typical collector used with an active solar heating system, the sun's rays enter through the double glazing. The black absorber plate absorbs the heat transmitted through the glass and the insulation behind the absorber minimizes heat loss.*

Figure 54-9. *Solar collectors for active solar systems face true south for maximum exposure to the sun.*

In new construction, collectors are installed between the roof rafters. When placed on older homes, collectors are usually mounted on wood strips or metal brackets fastened to the top of an existing roof.

Collectors may also be installed on a structure remote from the building to be heated. A remote collector is used when the roof of a building does not lend itself to southern exposure.

Air Movement after Entering Collectors. Air is introduced into one end of the collectors through a pipe or duct. The air is heated as it passes along the surface of the absorber plate within the collector. The heated air exits the opposite end of the collectors, where it is carried directly into the building space to be heated. If heat is not required in the building space, it can be directed to a thermal storage unit.

Thermal Storage Units. A method for storing the heat is necessary since much of the heating needs to occur in the evening after the sun has set and solar radiation is no longer available. A thermal storage unit for active solar air heating systems often consists of small rocks (1″ to 1½″ diameter) housed in a wood or concrete box. **See Figure 54-10.** The size of the rock storage area depends on the size of the collector area. Most manufacturers recommend approximately ½ cu ft to 1 cu ft of rock per 1 sq ft of the collector.

A thermal storage unit requires an open area at the bottom. This open area, called a *plenum,* can be created with hollow concrete masonry units covered with a steel grate, with allowance made for another air space at the top of the enclosure. The entire thermal storage unit must be well sealed and insulated.

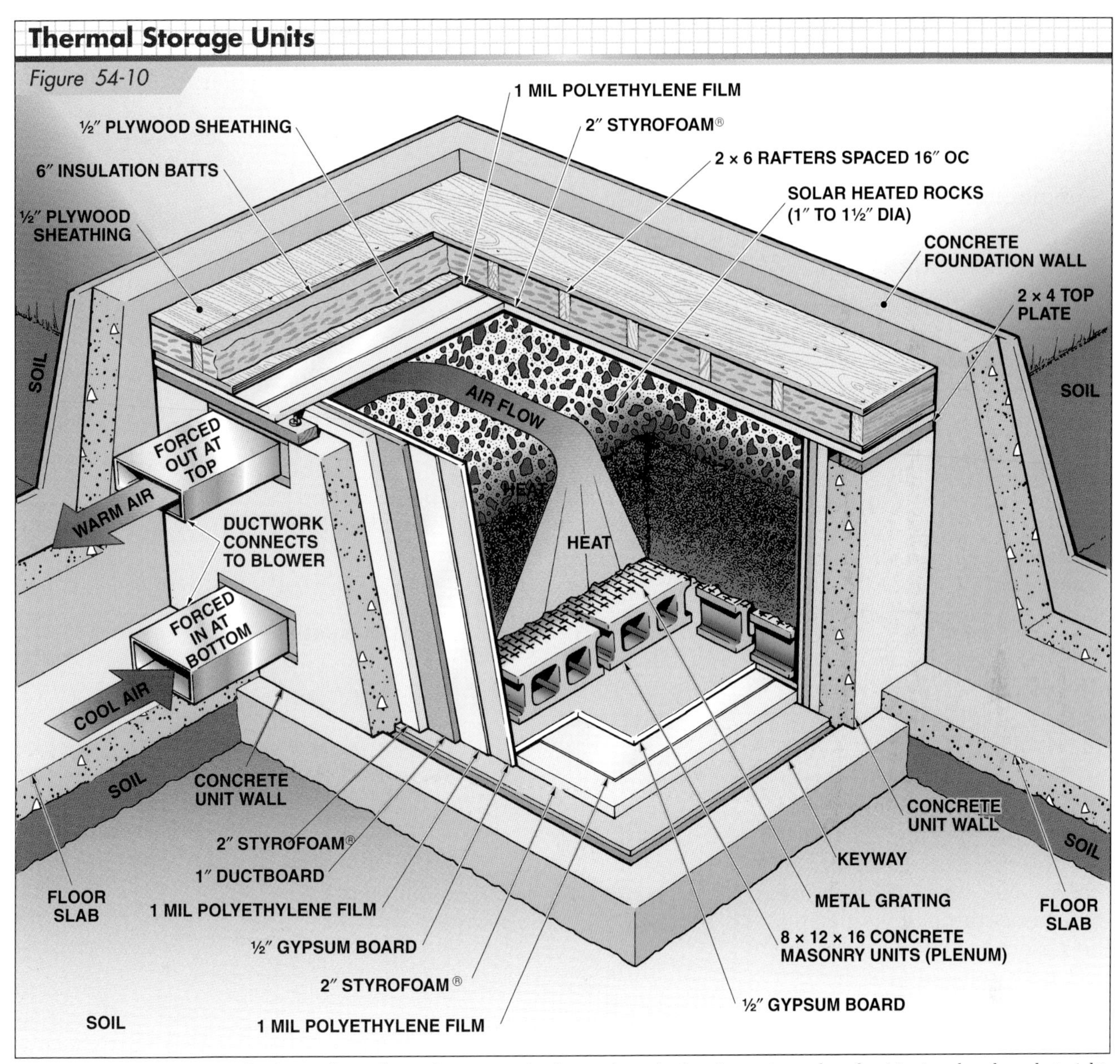

Figure 54-10. *Heated air can be directed to a thermal storage unit of an active solar heating system where heat is stored and used at night.*

Air Handlers. An air handler includes a blower assembly and automatic dampers and is usually positioned next to the thermal storage unit. Through control of the automatic dampers, the heated air is directed to the living space or to the thermal storage unit. Later, the air handler can be used to direct the heated air from storage into the space to be heated.

Auxiliary Heating Systems. Buildings with an active solar air heating system should have an auxiliary heating system for use on days when there is not adequate radiant energy to operate the solar system. Auxiliary systems are powered by natural gas, propane, fuel oil, or electricity.

Automatic Controls. Sophisticated active solar air heating systems feature a fully automatic control system, which coordinates the solar heating operation with thermostats in the living area of a building. As a result, the following operating modes are possible:

- Collectors-to-house mode. In this mode, heated air is conveyed directly from the collectors to the living area.
- Collectors-to-storage mode. In this mode, when the proper living area temperature is reached, heated air from the collector is directed to the thermal storage unit.
- Storage-to-house mode. In this mode, when solar radiation is not available to produce heated air in the collectors, heated air is forced up through the

thermal storage unit and conveyed to the living area.

- Auxiliary-to-house mode. In this mode, mechanical heating equipment is used to provide heated air to the building.

Active Solar Liquid Heating Systems

Active solar liquid heating systems operate in a similar manner to an air heating system except that liquid, rather than air, is heated in the collector. An active solar liquid heating system includes collectors, pumps (instead of fans), storage tanks (instead of thermal storage units), pipes (instead of ducts), a heat exchanger, and controls.

Collectors. A typical liquid heating collector panel is an insulated weatherproof box which encloses a dark solar absorber plate and is sealed with at least one layer of glazing. The box is usually made of aluminum or other metal. As heat from sunlight passes through the glazing, the absorber plate absorbs the heat and transmits the heat to water flowing through pipes in the collector. **See Figure 54-11.**

Heat Transfer and Storage. In one commonly used system, heated water runs from the collector to a storage tank. The heated water can then be directed to baseboard radiators or an underfloor radiant system. When sufficient heated water is not available through solar means, an auxiliary boiler tank controls the distribution of heated water through the system. **See Figure 54-12.**

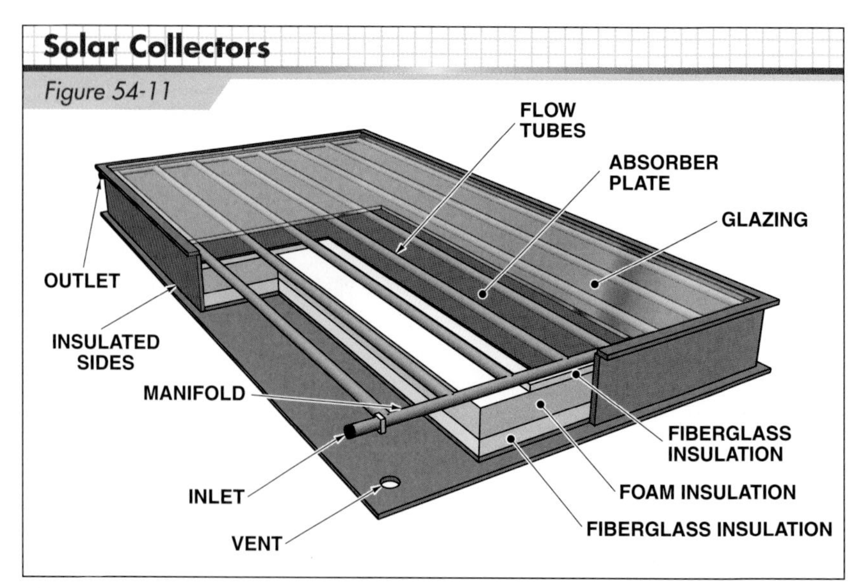

Figure 54-11. *Water or other liquid flows through tubes in the solar collector where it is heated.*

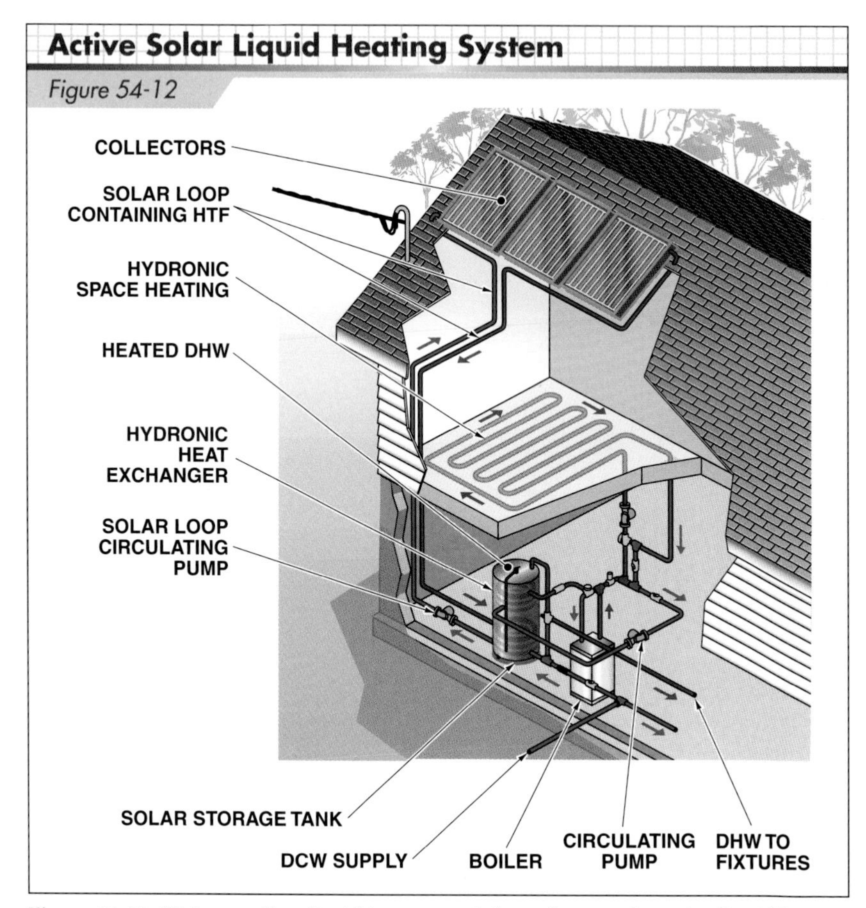

Figure 54-12. *Water or other liquid is conveyed through an active solar liquid heating system to transfer heat to the living space.*

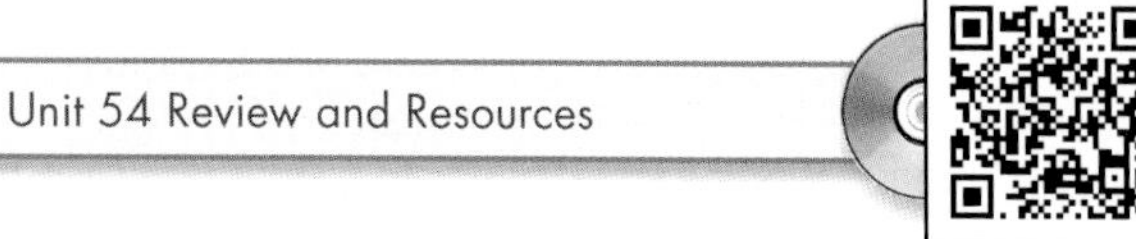

SECTION

12

Exterior Finish

CONTENTS

UNIT 55 Roof Finish

OBJECTIVES

1. Define roof overhangs and cornices.
2. Describe open cornice construction.
3. Describe closed cornice construction.
4. Identify the materials used for roof overhangs and cornices.
5. List and define key terms related to shingle application.
6. Identify the materials needed to prepare a roof for finishing.
7. Describe the characteristics and application of asphalt shingles.
8. Describe the characteristics and application of wood shingles and shakes.
9. Describe the characteristics and application of tile roofing.
10. List advantages of using a metal roof covering.
11. Describe the characteristics and installation of metal roof coverings.
12. Describe the characteristics and application of built-up roof coverings.

Roof finish includes overhang and cornice work as well as applying finish materials to the roof sheathing. Finish work can begin immediately after the roof has been sheathed and after all roof projections, such as vent pipes and chimneys, have been placed. Completing the overhangs and cornices is the first step in finishing a roof.

ROOF OVERHANGS AND CORNICES

Roof *overhangs,* or *eaves,* are the portions of a pitched roof that project past the side walls of the building. The *cornice* is the area beneath the overhangs. Several basic designs are used to finish roof overhangs and cornices. Most of these designs can be categorized as *open cornice* or *closed cornice.* Cornice designs not only add to the attractiveness of a building but also perform a practical function by protecting the side walls of a building from rain and snow. Wider overhangs also shade windows from the sun.

Open Cornices

In open cornice construction, the undersides of the rafters and roof sheathing are exposed. A 1″ or 2″ thick *fascia board* is usually nailed to the tail ends of the rafters. **See Figure 55-1.** Most spaces between rafters are blocked; some spaces are left open (and screened) to allow attic ventilation.

Usually a *frieze board* is nailed to the wall below the rafters. The frieze board is a horizontal piece that provides a watertight connection between the top of the siding and the *cornice.* In some cases, the frieze board is notched between the rafters and bed molding is nailed over it. **See Figure 55-2.**

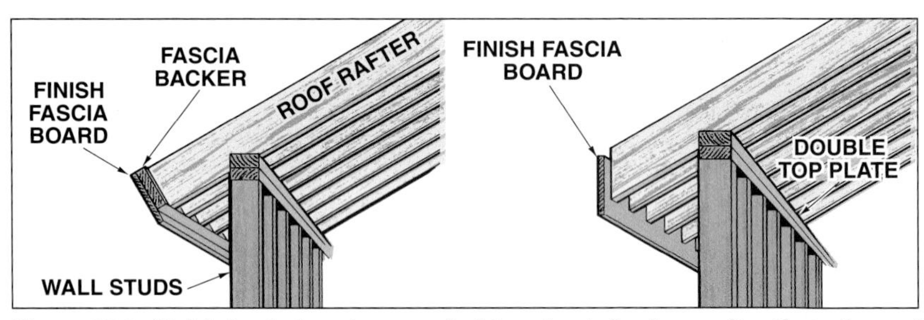

Figure 55-1. *Finish fascia boards are nailed to a fascia backer or directly to the roof rafter tails.*

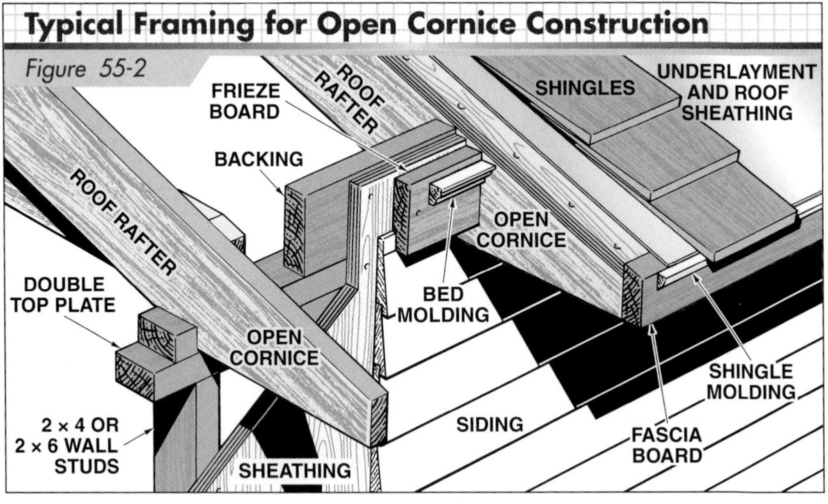

Figure 55-2. *In open cornice construction, the frieze board is cut between the rafters and molding is nailed over the frieze board.*

Closed Cornices

In closed cornice construction, the bottom of the roof overhang is enclosed. The two most common types of closed cornices are the *flat box cornice* and *sloped box cornice.*

A flat box cornice requires framing pieces called *lookouts* that are toenailed to the wall and facenailed to the ends of the rafters. Lookouts provide a nailing base for the *soffit,* which is the material that is fastened to the underside of the cornice. A typical flat box cornice is shown in **Figure 55-3.**

For a sloped box cornice, the soffit material is nailed directly to the underside of the rafters. **See Figure 55-4.** Sloped box cornice designs are common on houses with wide overhangs.

The construction of a *gable end overhang,* also known as the *rake section,* is explained earlier in this textbook. The basic trim pieces in the gable end overhang are the fascia board and soffit material. **Figure 55-5** shows the finished gable end overhang for a flat box cornice. **Figure 55-6** shows the finished gable end overhang for a sloped box cornice.

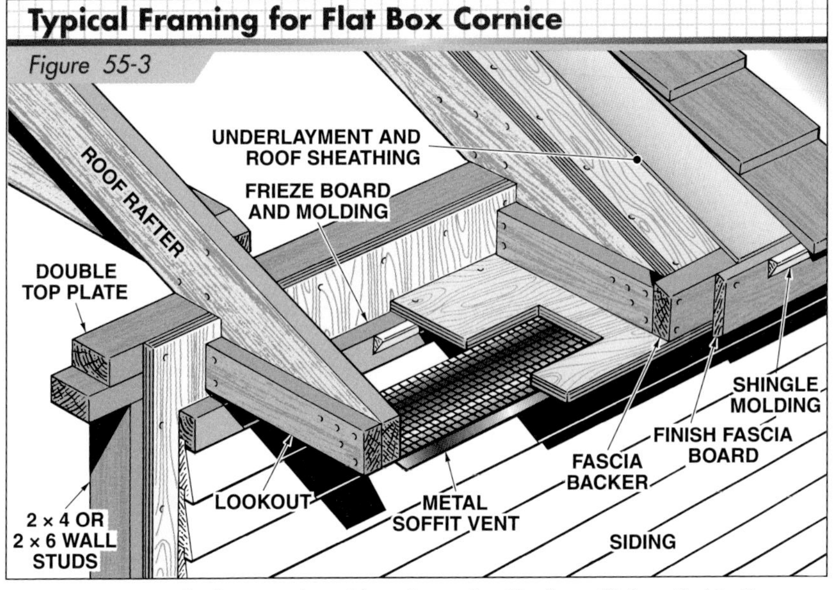

Figure 55-3. *For a flat box cornice with a plywood soffit, the soffit is nailed to the upper edge of the frieze board and to the underside of the lookouts. Note the metal soffit vent installed in the soffit for ventilation.*

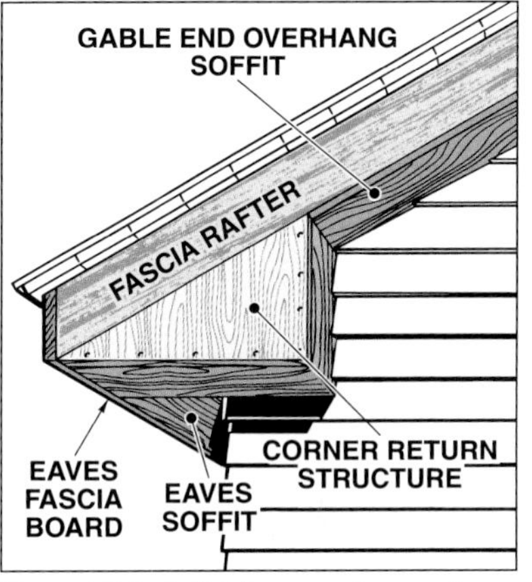

Figure 55-5. *A flat box cornice requires a cornice return at the corner intersection of the gable end overhang and eaves soffits.*

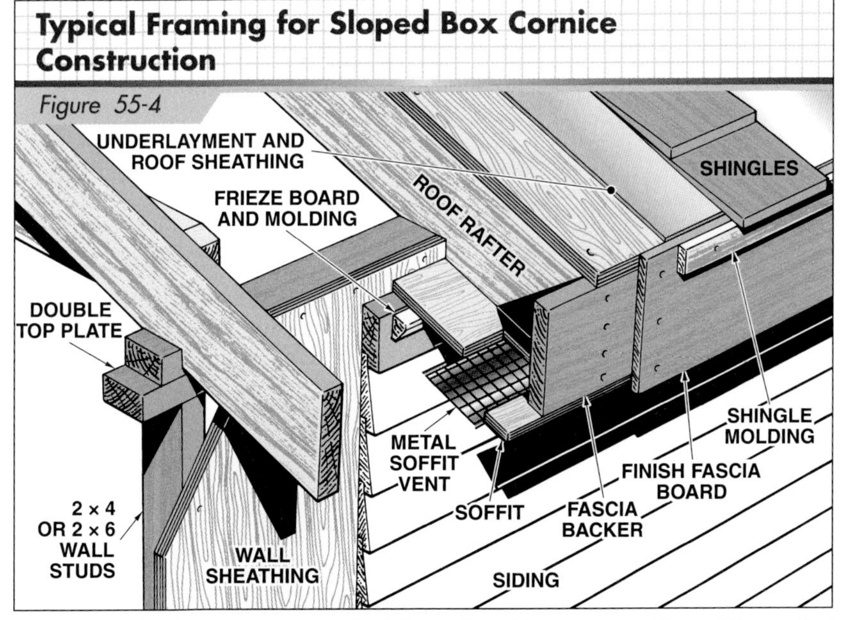

Figure 55-4. *Lookouts are not required for a sloped box cornice. The soffit is nailed directly to the underside of the rafters. Note the soffit angle when compared to the flat box cornice shown in Figure 55-3.*

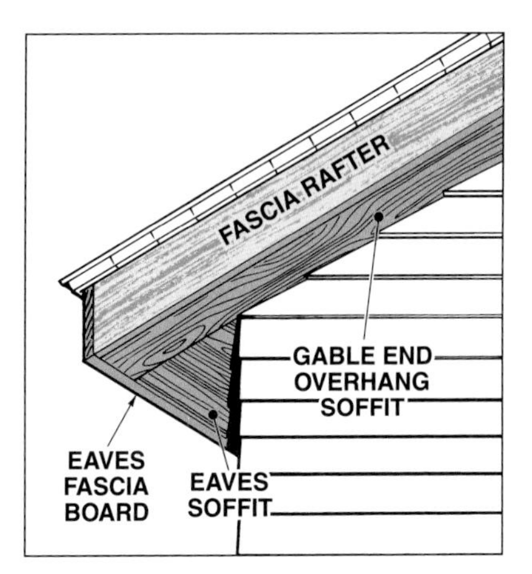

Figure 55-6. *A sloped box cornice does not require a cornice return at the gable end overhang.*

Cornice Soffit Systems

Plywood, hardboard, OSB, and fiberboard panels may be used for cornice soffits. For a more rustic effect, different types of siding patterns may be applied. Cornice trim and soffit systems are also available in aluminum and vinyl, in a variety of colors and designs. **See Figure 55-7.** Manufacturer instructions should be carefully followed when installing aluminum and vinyl systems.

Gutters and Downspouts

Gutters fastened to fascia boards collect water runoff from the roof and channel the water to downspouts. Gutters and downspouts are made of aluminum, plastic, or galvanized steel. Sectional gutters and downspouts are commonly installed on residential structures. **See Figure 55-8.** Seamless aluminum gutters generally are installed on light commercial and commercial buildings.

COVERING ROOFS

A finish roof covering is used to shed water and protect a building from sun, wind, and dust infiltration. Many types of roofing materials are available. The type best suited for a particular building depends on the roof slope, overall design of the structure, and local building code regulations. The color, texture, and pattern of a roofing material can add greatly to the general attractiveness of a building.

Figure 55-8. *Gutters and downspouts drain water from a roof.*

Some types of roofing materials may be applied by carpenters, while other types are applied by specialists called *roofers.* For example, built-up roof coverings are applied by roofers.

Within the carpentry trade, as in many other building trades, there has been a trend toward specialization of roof application. In many areas of the country, wood shingle and shake installation is performed by specialized roofing carpenters. Composition shingles, such as asphalt, mineral fiber, and

CertainTeed Corporation

Figure 55-7. *Vinyl cornice soffits and trim finish the overhang of this building. Note the vent openings in the soffit panels of the upper level.*

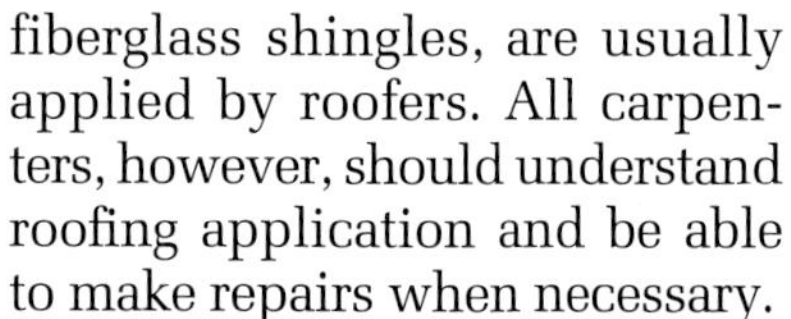

fiberglass shingles, are usually applied by roofers. All carpenters, however, should understand roofing application and be able to make repairs when necessary.

Wood shingles and shakes are produced from cedar logs. *Wood shingles* are cut from the logs. *Shakes* are similar to wood shingles but are split rather than sawn from logs. Shakes are applied when a more rustic effect is desired. Regardless of the shingle material, the following key terms are related to shingle application:

- *Shingle width:* total distance across the top of either a strip or individual shingle. **See Figure 55-9.**
- *Toplap:* distance that one shingle overlaps a shingle in the course (row) below it
- *Sidelap:* distance that one shingle overlaps a shingle next to it in the same course
- *Headlap:* distance that one shingle overlaps a shingle two courses below it. Headlap is measured from the bottom edge of an overlapping shingle to the nearest top edge of an overlapped shingle
- *Exposure:* distance between the exposed edges of overlapping shingles

For the best protection against leakage, shingles and shakes should be applied only on roofs with a unit rise of 4″ or more. Less slope creates slower water runoff, which increases the possibility of leakage as a result of wind-blown rain or snow being driven under the butt ends of the shingles.

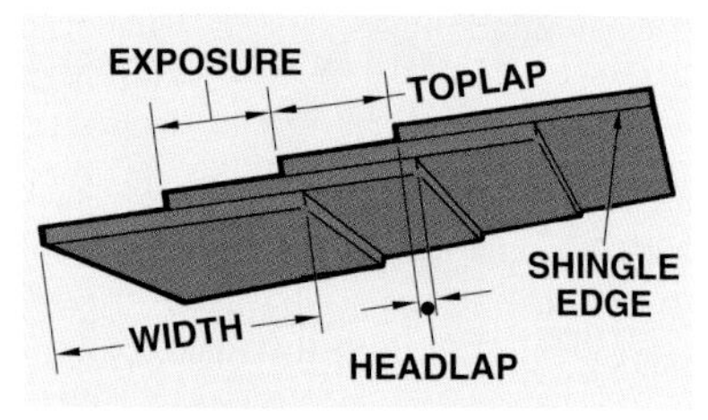

Figure 55-9. *Shingles and shakes must be properly lapped to ensure the roof is watertight.*

Preparing Roofs for Finish Covering

Before roof finish materials are applied, the roof must be sheathed, underlayment installed (except under wood shingles), and flashing placed wherever conditions require it.

Sheathing. Roof sheathing is usually plywood, OSB, or other nonveneered panel product. Post-and-beam roofs often have 2″ tongue-and-groove planks. Spaced 1 × 3s, 1 × 4s, or 1 × 6s are frequently used as sheathing under wood shakes or shingles. Other sheathing materials include various rigid roof insulation products.

Ice and Water Guard. In cold-weather areas, flashing is also recommended along the roof eaves to prevent an ice dam. *Ice dams* are created when warm air rising through the attic is combined with melting snow, water, and cold outside temperature. Snow and ice accumulate on a roof and slowly melt. The melting snow and ice then contact the cold roof surface directly above the uninsulated eaves as the moisture flows and forms an ice dam when it refreezes.

Eaves may be flashed using asphalt-saturated felt or an ice and water guard membrane. A strip of 50-lb asphalt-saturated felt is applied over the regular layer of underlayment at the roof overhang. **See Figure 55-10.** Flashing for the eaves should extend 12″ to 24″ inside the building. A *drip edge* at the top of the fascia is a metal piece that protects the edges of the roof deck and also helps to prevent leakage.

Ice and water guard is a flexible adhesive membrane used to prevent moisture penetration and interior damage from water backup due to an ice dam or wind-driven rain. Ice and water guard membranes are available in 12″, 30″, and 60″ widths and in 75′ and 100′ lengths. The membranes are adhesive on both sides to adhere to the roof sheathing and overlayment.

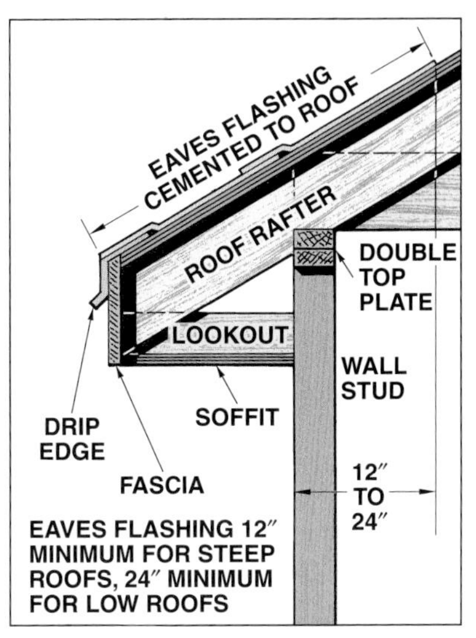

Figure 55-10. *In cold climates, eaves flashing is installed along the roof eaves and extends 12″ to 24″ inside the building wall line.*

Ice and water guards are installed along eaves and rake edges, in valleys, and on low-sloping roofs. Ice and water guard is applied directly to the roof sheathing and then rolled to ensure maximum adhesion and contact. **See Figure 55-11.** Side laps should be a minimum of 3½″, while end laps should be a minimum of 6″. Always work from the low point of a roof to its high point. Apply ice and water dam membranes in valleys prior to applying the membrane to the eaves. When applying the membrane in a valley, start at the bottom and work toward the top of the roof and roll the membrane from the center outward. When applying the membrane for a ridge, center the membrane over the peak and roll from the center outward in both directions. When the underlayment is installed, the release paper on the upper side of the membrane is removed, and the underlayment is positioned over it and rolled to provide proper adhesion. Always refer to manufacturer installation instructions for ice and water guard.

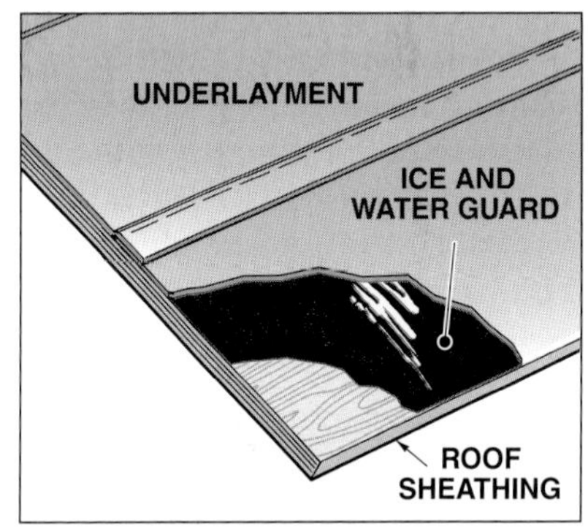

Figure 55-11. *The adhesive ice and water guard membrane is applied directly to the roof sheathing.*

Roofing sheathing platforms can be constructed using 2 × 4s and scrap roof sheathing to reinforce the sides

Underlayment. Roof *underlayment* has several important functions. Underlayment protects sheathing from moisture until shingles are placed. Underlayment also provides additional weather protection from wind-driven rainwater that may penetrate under shingles. In addition, underlayment prevents direct contact between the shingles and sheathing, which is important when asphalt shingles are used. Wood resins in sheathing can cause chemical reactions that are damaging to asphalt shingles.

Underlayment is always used under asphalt and fiberglass shingles, but is not always required under wood shingles and shakes. *Asphalt-saturated felt* (tar paper) is used for underlayment. Underlayment should be applied over the entire roof surface as soon as the sheathing installation has been completed. A 2″ toplap is required at horizontal joints and a 4″ sidelap is required at end joints of the underlayment. **See Figure 55-12.**

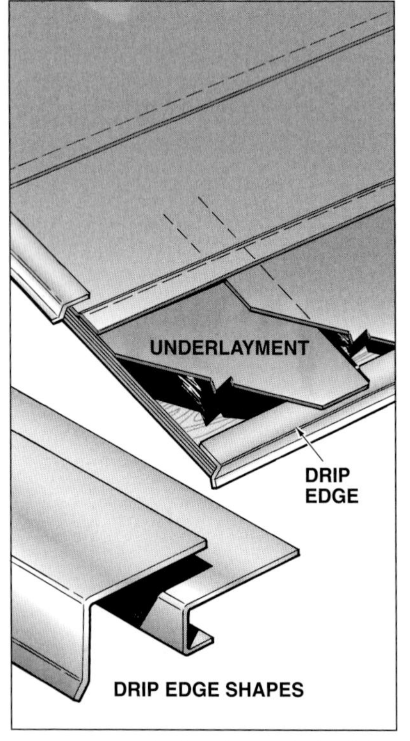

Figure 55-12. *A 2″ toplap and 4″ sidelap are required for shingle underlayment.*

Flashing. Roof shape and construction may create water leakage problems if *flashing* is not properly placed around vulnerable areas of the roof. Common vulnerable roof areas occur around a chimney or other roof projections.

Flashing materials used on roofs may be metal, asphalt-saturated felt, or plastic. Metal flashing is made of aluminum, galvanized steel, or copper, and is installed around roof projections and wall intersections. Asphalt-saturated felt flashing is generally used at the ridges, hips, and valleys since it easily forms to the intersections.

Asphalt Shingles

Most residential roofs in North America are covered with *asphalt shingles.* Asphalt shingles are less expensive than other types of shingles, are easy to apply, and have a life expectancy of 15 to 25 years.

Asphalt shingles are made of asphalt-saturated felt coated with mineral granules. Asphalt shingles are available in a wide range of styles, textures, and colors. Light-colored shingles reflect heat and keep an attic cooler. Dark-colored shingles absorb heat more quickly than light-colored shingles.

Asphalt strip shingles are available in several different designs. **See Figure 55-13.** Strip shingles are commonly used for new construction. *Individual shingles* are primarily used for restoration work. **See Figure 55-14.**

CertainTeed Corporation

Asphalt shingles are available in a variety of colors to complement exterior finish material.

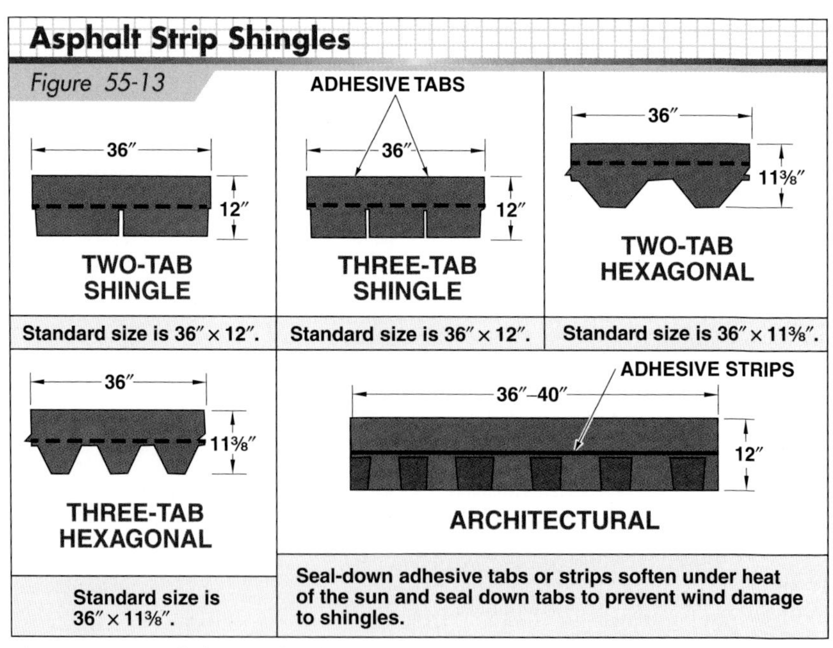

Figure 55-13. *Asphalt strip shingles are available in many designs. Seal-down adhesive tabs help to prevent wind damage to the shingles.*

Preparing Roofs for Asphalt Shingles. An asphalt-shingle roof requires underlayment and usually some type of flashing. *Open valley* or *closed valley flashing* may be used. **See Figure 55-15.** In open valley flashing, 18″ strips of mineral-surfaced roofing material are placed face down at the valley. A second piece (36″ strip) is cemented face up over the first strip. The shingles are applied over the flashing. Shingle edges along the valley should be marked and trimmed parallel to the valley. To avoid penetrating the flashing, shingles should not be cut with a utility knife while they are laying in position.

In closed valley flashing, a piece of mineral-surfaced roofing material is placed face up in the valley. The shingle strips are then placed to cross the valley.

Flashing is required where a pitched roof meets a vertical wall, **Figure 55-16.** Metal stepped flashing extends 3″ under the shingles and at least 4″ vertically on the wall. The flashing must overlap a minimum of 2″.

Flashing must also be installed around roof projections such as ventilation vent pipes. One method of installing flashing around roof projections is shown in **Figure 55-17.**

Applying Asphalt Shingles. Special noncorrosive, hot-dipped galvanized steel or aluminum nails are manufactured for asphalt shingles. **See Figure 55-18.** These nails have flat heads 3/8″ to 7/16″ in diameter and sharp points. Staples are also commonly used to fasten asphalt shingles. Pneumatic nailers and staplers are commonly used to apply underlayment and asphalt shingles.

Release tape on the back of asphalt shingles should remain in place for the life of the shingles.

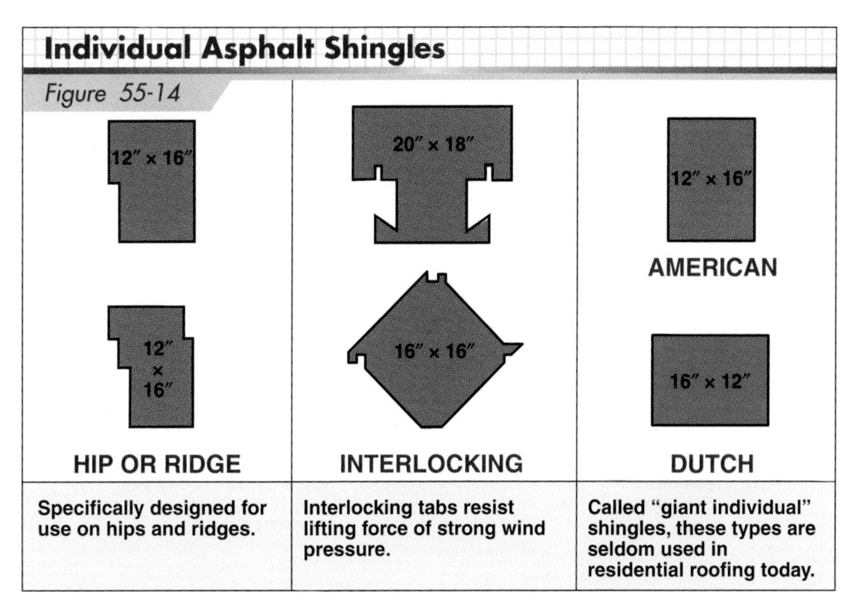

Figure 55-14. *Individual asphalt shingles are primarily used for restoration work.*

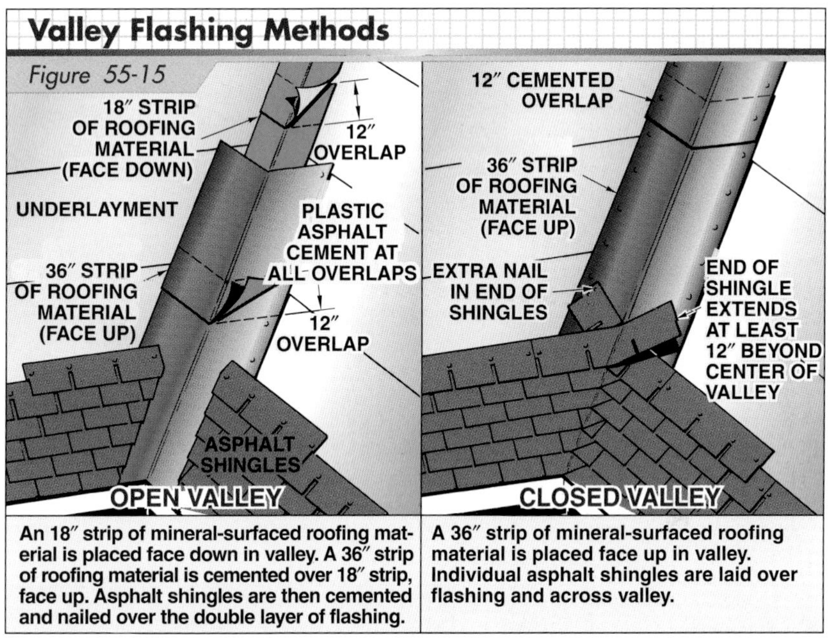

Figure 55-15. *Open valley or closed valley flashing may be used on asphalt-shingled roofs.*

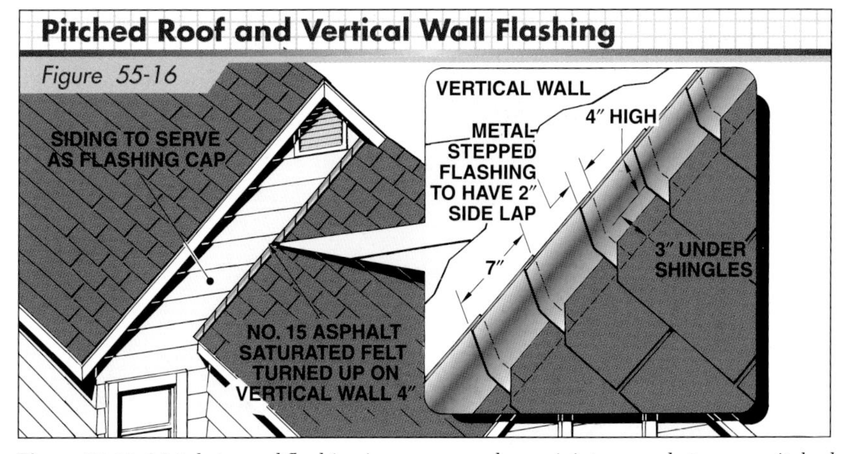

Figure 55-16. *Metal stepped flashing is necessary where a joint occurs between a pitched roof and vertical wall.*

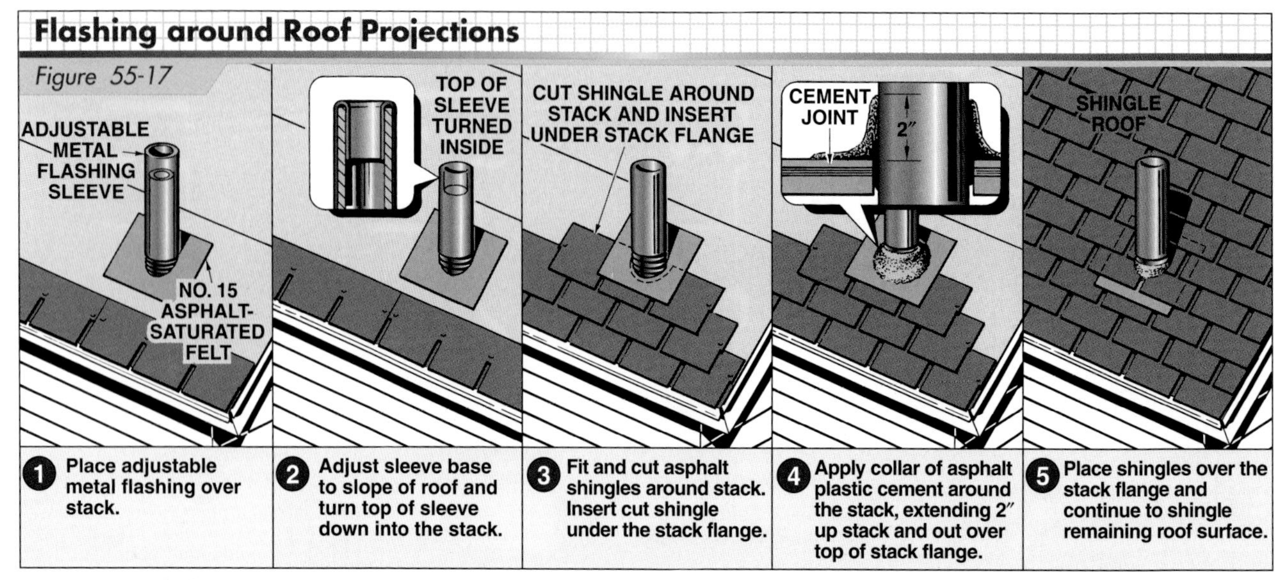

Figure 55-17. *Flashing is required where stacks project above the roof.*

RECOMMENDED NAIL LENGTH*		
Application	1″ Sheathing	⅜″ Plywood
Strip or individual shingle (new construction)	1¼	⅞
Over asphalt roofing (reroofing)	1½	1
Over wood shingles (reroofing)	1¾	—

* in in.

Figure 55-18. *Smooth, annular, or screw-shank nails are used to fasten asphalt shingles to roof sheathing.*

CertainTeed Corporation

A double layer of mineral-surfaced roofing material is used to form an open valley for an asphalt-shingled roof.

Figure 55-19 shows a procedure for laying down asphalt strip shingles. A starter strip is placed beneath the first course of shingles. The first course starts with a full shingle. Joints between subsequent courses must be staggered to provide a tight seal. To stagger the shingle cutouts on the next course, the first tab of the first shingle must be cut. For example, the first shingle of the second course is 30″ wide when the first tab (6″) is cut off (36″ – 6″ = 30″). An additional 6″ is cut off of the first shingle of the next course, resulting in a width of 24″ (36″ – 12″ = 24″). When the width of the first shingle of a course has been reduced to 6″, the next course begins with a full shingle.

A common procedure for finishing off the ridge and hips is the *Boston method.* **See Figure 55-20.** In this method, shingles specially formed for the ridge and hips are overlapped with a 5″ exposure. Nails are driven into the areas of the shingles covered by the overlapping shingle.

In cold weather, hip and ridge shingles may need to be warmed to prevent cracking.

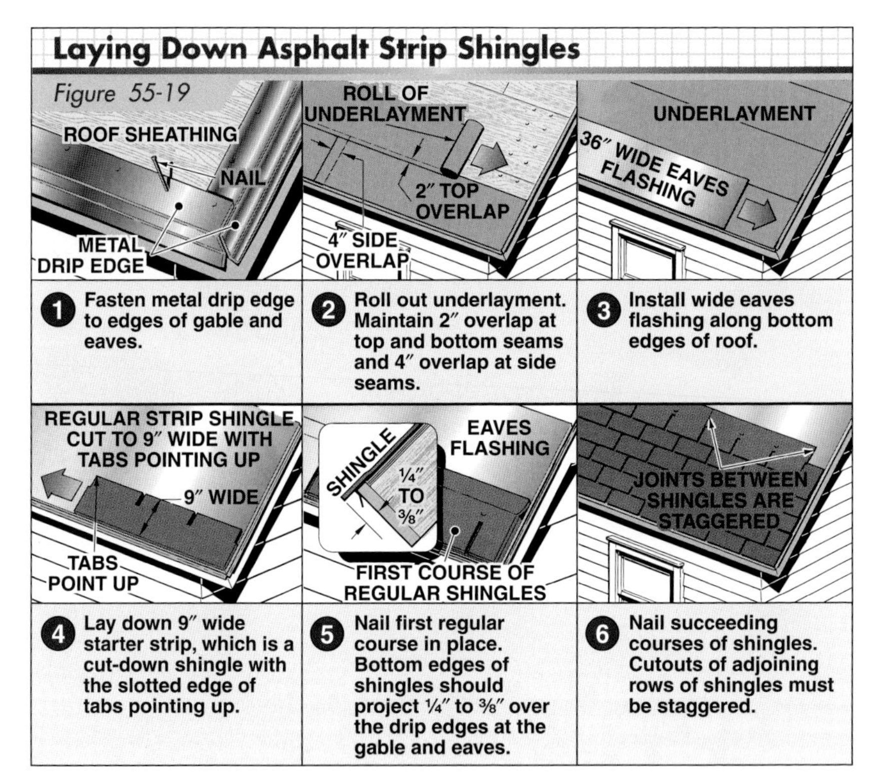

Figure 55-19. *Drip edges, underlayment, and eaves flashing must be installed prior to installing asphalt strip shingles. In this example, three-tab strip shingles are used.*

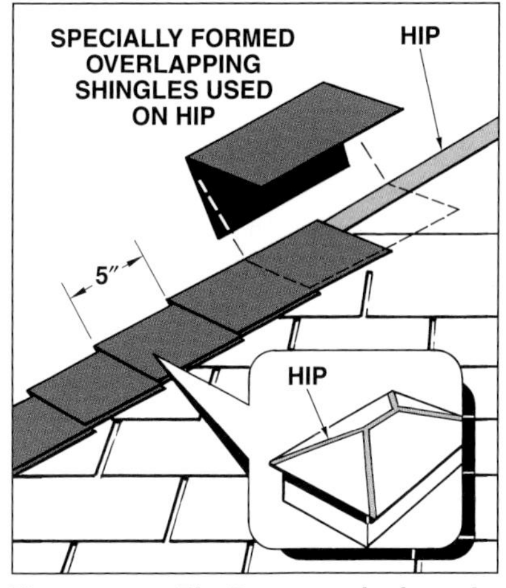

Figure 55-20. *The Boston method may be used for finishing the ridge and hips of an asphalt shingle roof.*

Wood Shingles and Shakes

Wood shingles and *shakes* are among the oldest types of roof coverings and are commonly applied when a rustic architectural effect is desired. In some areas, however, local fire codes prohibit wood roof coverings because of the fire potential.

Most shingles and shakes are produced from western red cedar trees, which are slow-growing coniferous trees found in the Pacific Northwest. Western red cedar has exceptional strength in proportion to its weight and has little expansion and contraction with changes in its moisture content. These factors, along with its decay resistance, make western red cedar a superior wood for roofing or any other exterior use.

Shingles have a smoother finish than shakes. **See Figure 55-21.** They are sawn from cedar blocks by a shingle-cutting machine. Most wood shingles are produced in random widths ranging from 3″ to 14″ and in standard lengths of 16″, 18″, and 24″. Wood shingles are tapered to be thicker at the exposed butt end than at the concealed end.

Different grades of wood shingles are available. No. 1 grade shingles, which are recommended for roofs with a 3:12 slope or greater and exterior walls where an excellent appearance is desired, are cut from clear heartwood. No. 1 grade shingles are 100% *edge-grained,* meaning that the grain runs in the direction of the long dimension of the shingle. No. 2 grade shingles have limited knots and defects above the clear portion, and are used on roofs with a 3:12 slope or greater and exterior walls where a good appearance is desired. No. 3 grade shingles are recommended for roofs with a 3:12 slope or greater and exterior walls where an economical product is acceptable. No. 3 grade shingles have unlimited sapwood and flat grain. No. 4 Undercoursing grade is used for starter courses at the eaves and has unlimited defects, sapwood, and flat grain.

Figure 55-21. *Shingles have a smoother finish than shakes and are tapered from the butt to the concealed end.*

Shakes are similar to shingles, but are typically split rather than sawed from cedar logs. Splitting produces a rougher and more rustic appearance than shingles. Three types of shakes are produced for exterior finish work. *Tapersplit shakes* are cut at a taper. *Straightsplit shakes* do not have a taper. *Hand-split and resawn shakes* are split and then resawn at a taper. Shakes are available in random widths starting from 4″ and in standard lengths of 18″, 24″, and 32″.

Hand seamers are used to form metal flashing while working on a rooftop.

Hand-split and resawn shakes feature a rough split face and sawn back and are available in Premium, No. 1, and Standard grades. Premium grade shakes are 100% edge grain and are used on walls and roofs with a 4:12 slope or steeper where excellent appearance is desired. No. 1 grade shakes are recommended for walls and roofs with a 4:12 slope or steeper where a good appearance is desired. No. 1 grade shakes are manufactured from clear heartwood. Standard grade wood shakes are edge grain, flat grain, or a combination of edge and flat grain, and are graded from the split or best face.

Tapersawn shakes are a combination of a split shake and a sawn shingle and are used for roofs and vertical wall applications. Tapersawn shakes are available in Premium, No. 1, No. 2, and No. 3 grades. Premium and No. 1 grades are installed where a high-quality appearance is desired, and are used on walls and roofs with a 4:12 slope or steeper. The face of No. 2 grade tapersawn shakes is clear along the exposed surface. Tight knots and other small defects are allowed in the top half of the shakes. No. 3 grade tapersawn shakes allow limited knots and defects on the entire face and are commonly used for walls where a rustic appearance is desired.

Preparing Roofs for Wood Shingles or Shakes. Spaced or solid sheathing may be used as a base for nailing shingles and shakes. Spaced sheathing, except over the eaves, is recommended in most climates. **See Figure 55-22.** Wood shingles and shakes tend to absorb rainwater. For this reason, air circulation underneath the shakes or shingles prevents uneven drying, which can eventually cause splits to develop.

Boards used for spaced sheathing are 1 × 4s or 1 × 6s and are spaced on centers equal to the exposure of the shingles or shakes that will be applied. (The exposure is the distance between the exposed edges of overlapping shakes or shingles.)

Asphalt-saturated felt underlayment should be installed in strips, with the felt overlapping the upper portion of each course of shakes. Additional underlayment is required along the eaves.

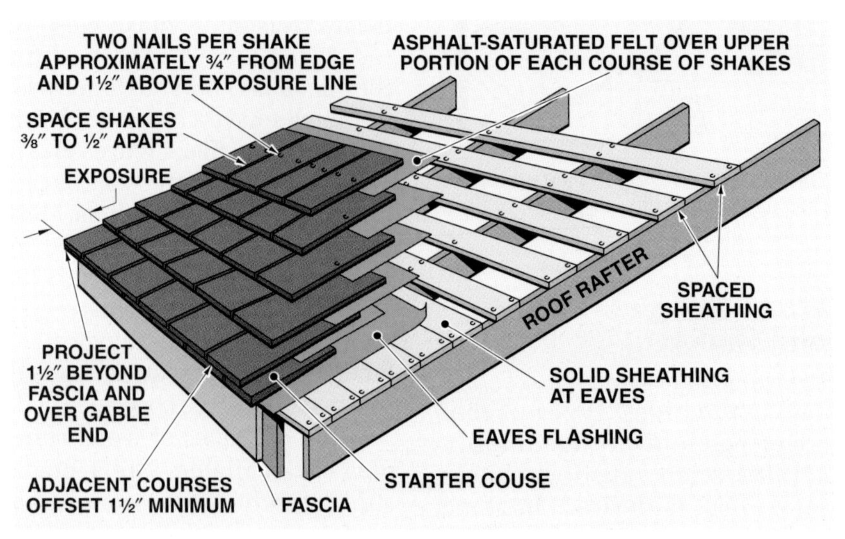

Figure 55-22. *Spaced sheathing is commonly used with wood shake or shingle roofs to provide air circulation beneath the roof. Note that solid sheathing is installed over the eaves.*

Installing Wood Shingles. Hot-dipped, zinc-coated nails are recommended for wood shingles and shakes. Aluminum and stainless steel nails are also acceptable. The nails must be long enough to pass through the shingle or shake and penetrate the sheathing at least ½″. A proper nailing procedure is shown in **Figure 55-23.** The nail heads should rest on the surface. If the nail is driven further, it will have less holding power. Two nails are used per shingle, each ¾″ from the edge and approximately 1½″ above the exposure line.

The recommended exposure for wood shingles depends on the size of the shingle and the roof pitch. **See Figure 55-24.** A procedure for installing wood shingles is shown in **Figure 55-25.**

Installing Wood Shakes. The procedure for installing wood shakes is similar to the procedure for shingles. Shakes are much thicker than shingles; therefore, longer nails should be used. Shakes are also longer than shingles; therefore, they have a greater exposure. Common exposures are 7½″ for 18″ shakes, 10″ for 24″ shakes, and 13″ for 32″ shakes.

The rough and uneven surfaces of shakes increase the possibility of infiltration by wind-driven rain and snow. For this reason, it is necessary to place a strip of underlayment between each course as the shakes are being applied. **See Figure 55-26.** A 36″ wide strip of 15-lb (minimum) asphalt-saturated felt is laid over the eave line. The bottom course of shakes is doubled. After each course of shakes, an 18″ wide strip of 15-lb felt is placed over the top portion of the shake, extending onto the sheathing. The bottom edge of the felt placed on top of the shake is positioned from the butt a distance equal to twice the exposure of the shake. To allow for possible expansion, the shakes are spaced about ½″ apart and joints between the shakes are offset at least 1½″ from adjacent courses.

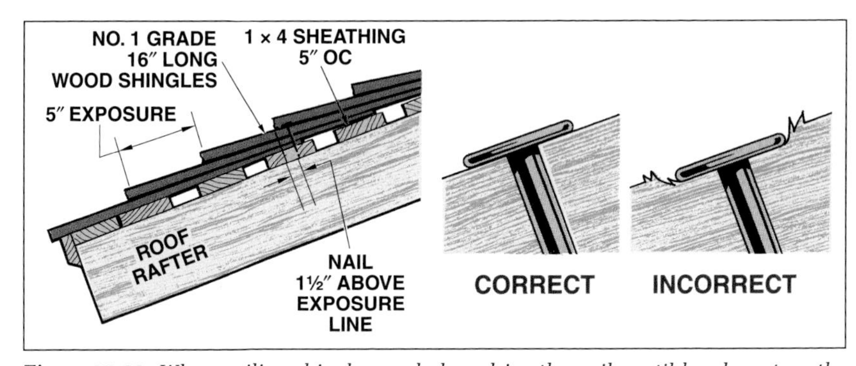

Figure 55-23. *When nailing shingles or shakes, drive the nails until heads rest on the surface. If nails are driven into the surface, they will have less holding power. Use only two nails to a shake, placed ¾″ from each edge and 1½″ above the exposure line. Hot-dipped, zinc-coated nails are recommended when applying wood shingles. Aluminum and stainless steel nails are also acceptable.*

RECOMMENDED WOOD SHINGLE EXPOSURE

Shingle Length*	Shingle Thickness (Green)	Maximum Exposure*	
		Slope Less than 4 in 12	Slope 5 in 12 and Over
16	5 butts in 2″	3¾	5
18	5 butts in 2¼″	4¼	5½
24	4 butts in 2″	5¾	7½

* in in.

Figure 55-24. *Greater shingle exposure is allowed for roofs with steeper slopes.*

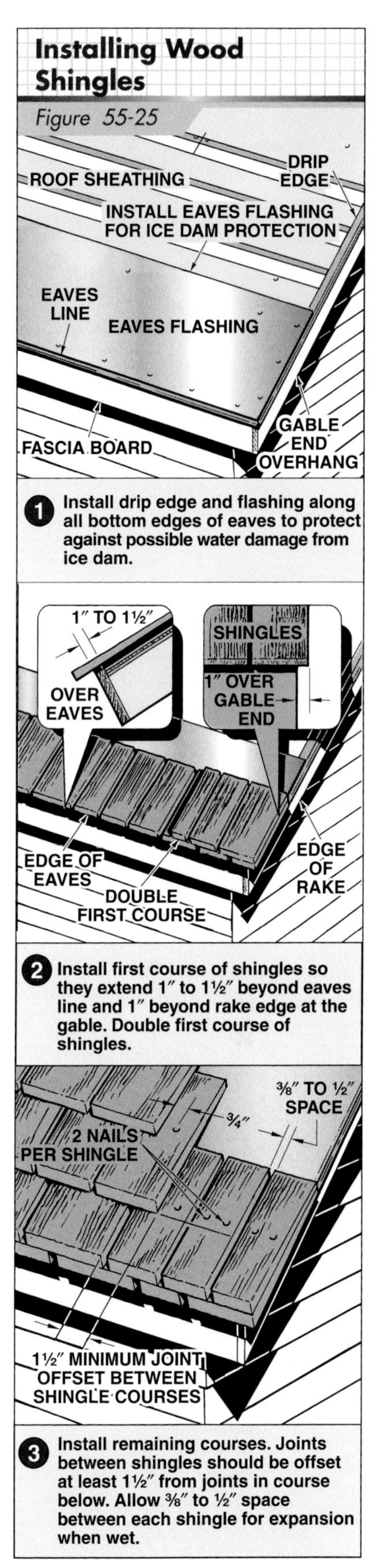

Figure 55-25. *The first course of wood shingles should extend 1″ to 1½″ beyond the eaves and 1″ beyond the rake edge of the gable.*

DC Roofing/Cedar Shake & Shingle Bureau

Figure 55-26. *Since wood shakes have rough and uneven surfaces, underlayment must be placed between each course.*

After installing the first course of shingles, lay out succeeding courses across the roof several courses at a time. Ensure joints between adjacent courses are offset at least 1½″ and that no joint in three adjacent courses is aligned. In addition, joints should not be aligned with shingle defects, such as knots.

Ridges, Hips, and Valleys. The same methods are used with wood shakes as with wood shingles for finishing ridges, hips, and valleys.

Hip sections of the roof are finished with a *hip cap.* **See Figure 55-27.** A strip of 15-lb asphalt-saturated felt, at least 8″ wide, must first be applied, and the starter course area of the hip cap is doubled. **See Figure 55-28.** The joints between the shakes are beveled. The directions of the bevels are alternated with each course.

DC Roofing/Cedar Shake & Shingle Bureau

Figure 55-27. *Hip sections of a roof are finished off with a hip cap.*

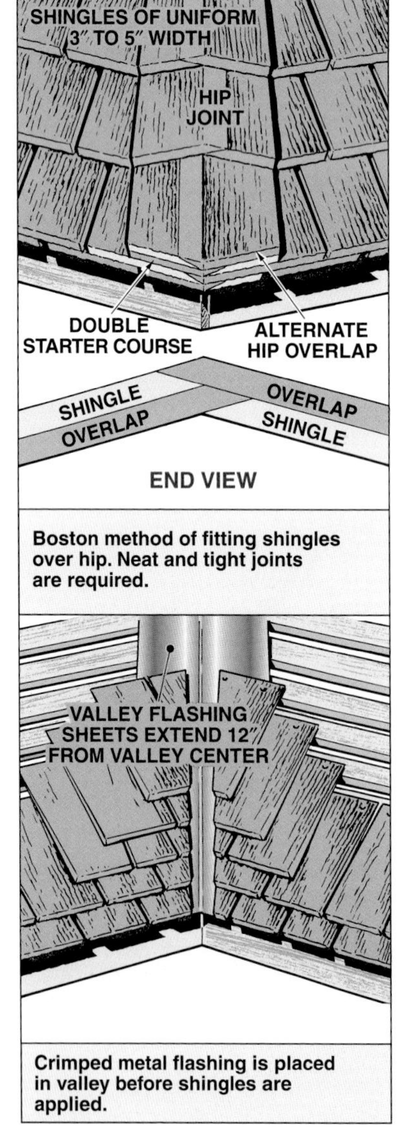

Figure 55-28. *When finishing a hip roof, shingles may be overlapped at the hip and flashing is installed in valleys.*

Valleys must be flashed with metal flashing before shingles or shakes are applied. Valley flashing should extend 12″ from valley center in each direction.

The ridge of a roof is finished off with a ridge cap, which is similar to a hip cap. **See Figure 55-29.** The general procedure for placing a ridge cap is the same as for placing a hip cap.

Figure 55-29. *Ridge and hip caps are also available as factory-manufactured units.*

The potential for roof leaks occurs in areas where water is channeled for running off the roof, such as valleys between intersecting roofs. The open valley method is considered the most practical method of finishing a valley. **See Figure 55-30.** The first step is to apply 15-lb felt directly over the sheathing. Sheets of metal valley flashing at least 24″ wide, and with a 4″ to 6″ headlap, are then nailed down. The metal flashing should be 26-ga or heavier galvanized iron. Shingles or shakes placed in the valley should be fitted so they run parallel to the valley and form a 6″ gutter.

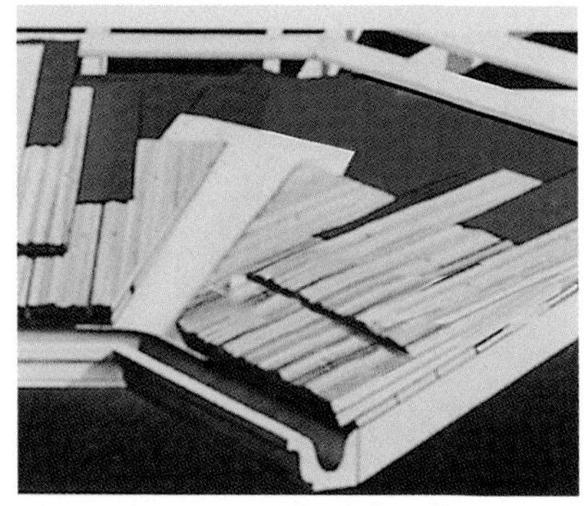

Cedar Shake & Shingle Bureau

Figure 55-30. *The open valley method is considered the most practical for finishing a valley.*

Other areas vulnerable to leakage are where a roof butts up against a vertical wall or around a chimney. *Base flashing* and *counterflashing* should be installed to prevent leakage. The bottom leg of the L-shaped base flashing rests on the roof and extends 6″ under the shingles. The counterflashing laps a minimum of 3″ over the base flashing. When flashing a brick chimney, the upper edge of the counterflashing is inserted at least ¾″ into a mortar joint. Caulking is placed between the counterflashing and the chimney. **See Figure 55-31.**

Tile Roofing

Tile is one of the oldest types of finish covering used on pitched roofs. Tile roofing has grown increasingly popular because of its fireproofing qualities. **See Figure 55-32.** Clay or concrete tile roofing is available in a variety of styles. Clay tile is manufactured by baking plates of molded clay into tile. Clay tile is lighter than concrete tile; however, concrete tile is more durable and is used more often than clay tile. Concrete tile is composed of portland cement, sand, and water.

Field Tile. Field tile is generally classified as *flat tile* or *roll tile.* Flat tile is flat in cross section. Roll tile is curved in cross section. *Accessory tiles* are specially designed tiles used for different intersecting points on a roof. **See Figure 55-33.**

Standard dimensions of a field tile are 13″ wide by 16½″ long (dimensions may vary slightly among different manufacturers). A typical field tile has *head lugs* (anchors) on its underside toward the top of the tile. *Transverse bars* at the lower end of the underside act as weather checks to inhibit the movement of rainwater or snow under the tile. *Interlocking ribs* on the long edges of tiles act as a waterlock and control latitudinal movement of the tile. One nail hole is located near the top of flat tile. Two nail holes are located near the top of roll tile.

Accessory Tiles. *Ridge* and *hip tiles* are used to cover ridges and normal hips. *Hip starter tiles* are used to start tile at the eaves of hip roofs. *Rake tiles* are used to finish the gable ends of a roof. *Three-way apex tiles* are used to cap the intersection of the end of the roof ridge and hip. *Four-way apex tiles* are used to cap the peak of a pyramid-shaped roof.

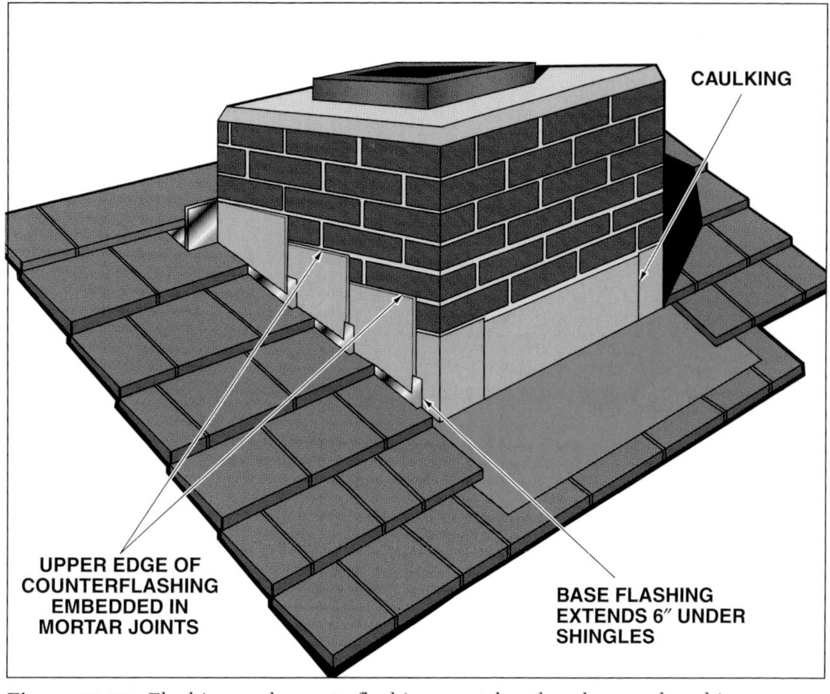

Figure 55-31. *Flashing and counterflashing must be placed around a chimney on a wood shake roof.*

Owens-Corning Fiberglass Corp.

Figure 55-32. *Tile roofing provides an attractive and fireproof finish roof covering.*

Figure 55-33. *Tile roofing is available in a variety of field tiles and accessories.*

Preparation for Tiles. Tiles are placed on spaced or solid sheathing that conforms to building code requirements for the anticipated loads. Spaced sheathing normally requires minimum 1 × 6s spanning a maximum of 24″ between rafters. Underlayment material must be at least No. 30 asphalt-saturated felt installed with a minimum 2″ headlap and 6″ side lap. Eaves flashing may be required in cold climates.

Battens. *Battens* are frequently required under roofing tiles for installation over solid sheathing to provide positive anchorage for the tiles. Battens are1 × 2s nailed to the solid sheathing after snapping chalk lines on the underlayment. The head lugs of each tile lap over and rest on the batten. A ½″ to 1″ break every 4′-0″ in each row of battens provides for water drainage. An alternate water drainage method is to install ¼″ shims under the battens to create a continuous space beneath them. The layout and placement of battens governs the final position of the tiles.

Battens are spaced 13″ to 13½″ OC for 16½″ long tiles. This layout provides a 3″ to 3½″ minimum headlap of the tiles. A line snapped for the upper edge of the top batten is 2″ from the ridge to allow room for a 2 × 3 nailer piece for the ridge tiles.

The line snapped for the upper edge of the first batten measured from the eaves is 15″ to allow for a 1″ tile overhang past the eaves. To determine the number of courses required, the distance between the chalk lines is measured and divided by 13. The exact spacing of the battens is then found by dividing the number of required courses into the distance between the top and bottom chalk lines. **See Figure 55-34.**

Flashing. Corrosion-resistant, 28-ga metal flashing is required at valleys, chimneys, skylights, and where the roof butts against a vertical wall. Valley flashing must extend at least 11″ in both directions from the centerline, with a splash diverter rib not less than 1″ high at the flow line. In open valley construction, tiles are cut to the angle of the valley and held back 2″ from the diverter strip. In closed valley construction, the tiles butt against the diverter strip.

Pan flashing is used where tiles butt against a vertical wall. Counterflashing is also recommended. Flashing placed around flues and vents is copper or other approved material that can be formed to follow the shape of the tile. **See Figure 55-35.** The sides of chimneys and skylights are flashed in the same manner as straight walls. The top and bottom areas require flexible flashing material.

Placing Tiles. Field tiles are placed in vertical and horizontal alignment. Vertical interlocking joints must be free of any foreign matter to ensure a proper fit and interlock of the tiles. Tiles are placed from left to right because of the vertical keyway between the tiles. The number of tiles to be fastened with nails depends on the roof slope, whether or not battens are used, and the anticipated wind velocity in the area. Local building codes should be consulted for the tile nailing schedule.

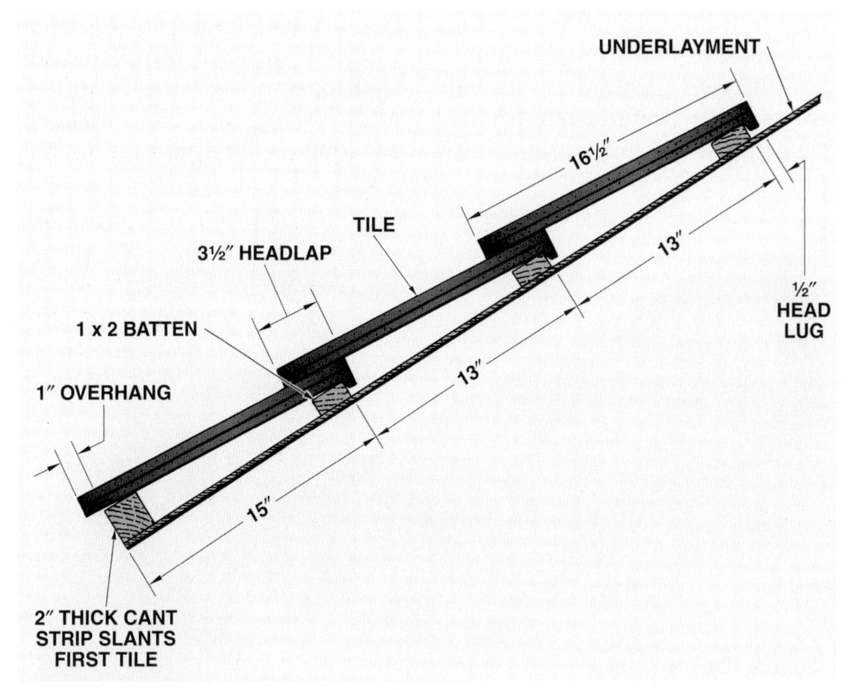

Figure 55-34. *Battens are laid out to provide proper headlap for tiles.*

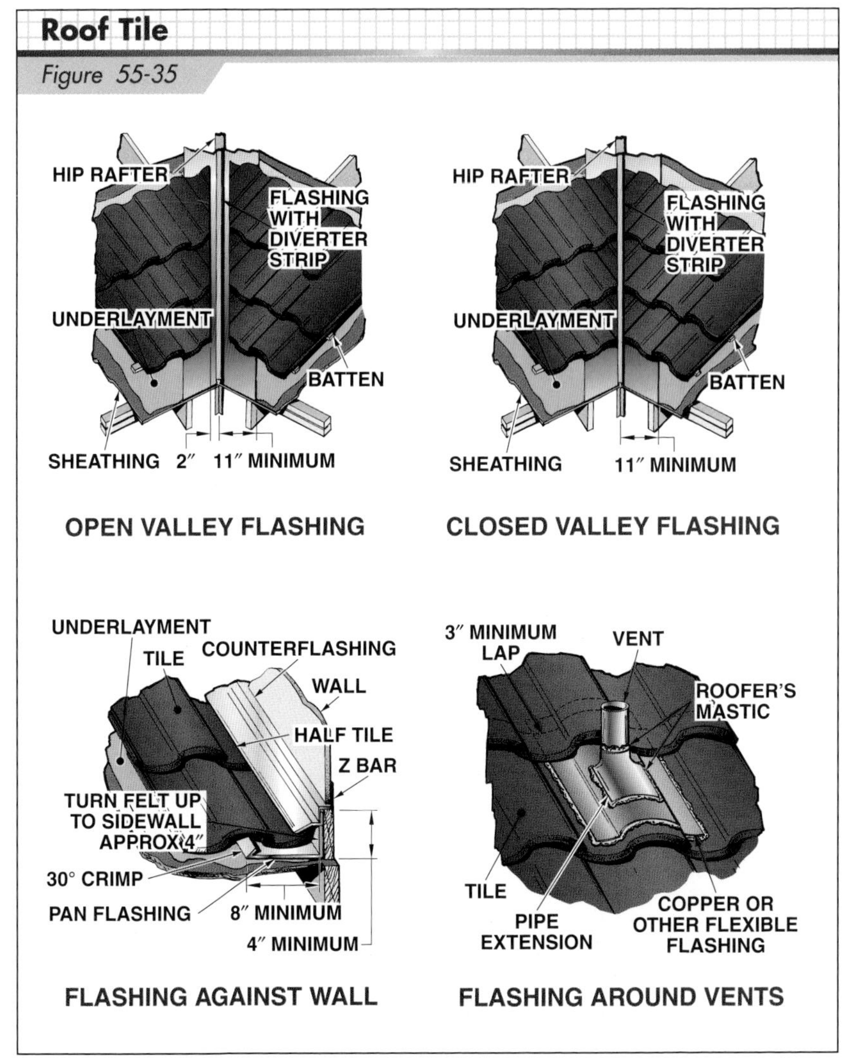

Figure 55-35. *Metal flashing is placed in all roof areas where leaks may occur.*

A cant strip used at the eaves must be twice the thickness of the battens to properly slant the first course of tiles. For rake tiles used to finish gable ends, the field tiles are cut and held back 1″ to 2″ from the outside edge of the sheathing.

Ridge and hip tiles are nailed or wired to a 2 × 3 nailer strip fastened to the roof. A bead of roofer's mastic is applied where the ridge tiles overlap. The juncture where the ridge and hip tiles rest on the field tiles is weatherproofed with mortar or an approved dry ridge/hip system. Ends of a ridge can be finished with pieces cut from rake tiles. A starter tile is placed at the lower end of the hip section when starting hip tiles.

Metal Roof Covering

Metal roof coverings have been in use for a long time. Contemporary metal roof coverings include aluminum and steel shingles and standing-seam metal panels. Metal roof covering provides the following advantages over other roofing materials:

- Can be placed directly over old roofs when reroofing, thereby saving money and eliminating a great deal of mess.
- Reflects most of the sun's rays, and therefore doesn't retain as much heat as shingles made from other materials. Reflecting the rays keeps the attic and floor below cooler, resulting in reduced energy bills.
- Under normal conditions, metal roof covering will never need to be replaced.
- Will not dry out, rust, split, curl, peel, or flake.
- Can be walked on without crushing or damaging the metal roof covering.
- Is much lighter than many other materials used for shingle roofs.
- Is fire resistant.

Shingles. Metal shingles are made of aluminum or galvanized steel and are available in many solid colors as well as designs that give the appearance of wood shingles or shakes. **See Figure 55-36.**

Metal shingles may be applied individually or in panels. Battens should be installed when installing metal shingles. **See Figure 55-37.** Metal shingle panels may have prefabricated battens attached to the backs of the panels.

Standing-Seam Roll-Formed Panels. Roll-formed panels are manufactured from aluminum, copper, zinc, and galvanized steel. Galvanized steel is the most common panel material. The panels can be preformed and/or textured to provide the panels with the appearance of shingles or tiles. Standing-seam pitched metal roofs provide good drainage for rainwater and melting snow. **See Figure 55-38.**

Standing-seam roll-formed panels are secured in place using clips that are concealed inside the seams. The clips are attached to the roof sheathing using lag bolts. While securing the roof in place, the clips also allow the panels to expand and contract with temperature changes, reducing the chance of roof leakage. **See Figure 55-39.**

Built-up Roof Covering

A *built-up roof covering* consists of three, four, or five individual layers of roofing felt. Built-up roofs are usually used on flat decks. Each layer is mopped down with hot tar or asphalt. The final layer is coated with gravel, which is embedded in the tar or asphalt. **See Figure 55-40.**

Built-up roofs are also used with flat roof decks constructed with *parapets.* Parapets are low walls along the edges of a roof. Flashing and counterflashing are installed along the edges of the roof deck where it intersects the parapet wall. Asphalt-saturated felt is then installed over the flashing and counterflashing. **See Figure 55-41.** Flat roofs are pitched to direct rainwater and melting snow toward the gutters and downspouts. Nail holes are avoided in the upper felt layers.

AHI Roofing

Figure 55-36. *Metal shingles are textured to provide the appearance of wood shakes or shingles.*

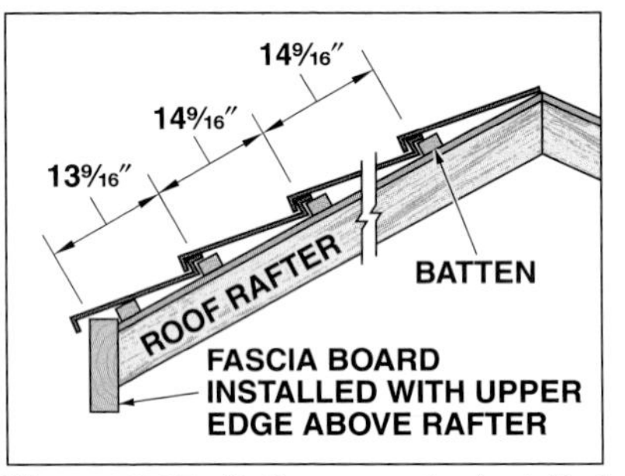

Figure 55-37. *Battens may be installed beneath metal shingle panels.*

Wood and metal roof trusses can be designed for on-center spacings greater than the traditional 16″ OC or 24″ OC by upsizing chord members, or specifying a better grade of wood or thicker gauge of steel for the truss members. Some metal roofing covering products can span distances up to 5′ OC.

Classic Products, Inc.

Figure 55-38. *Standing-seam metal roofs provide excellent water drainage.*

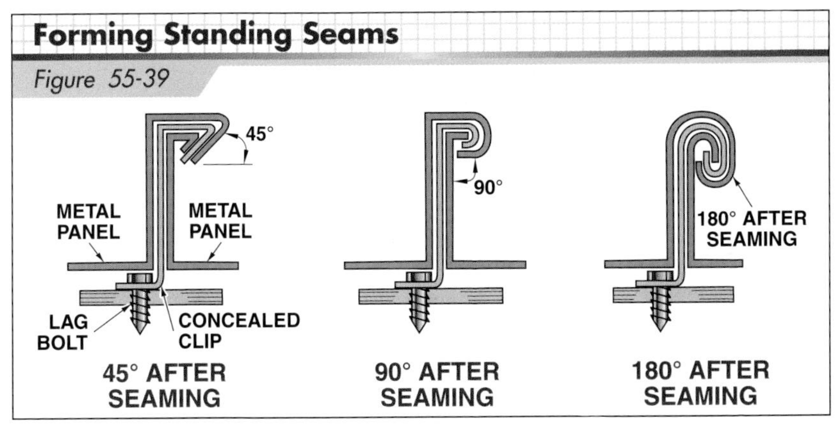

Figure 55-39. *Standing-seam roll-formed panels are secured in place with clips concealed inside the seams. A factory-applied sealant inside the seam produces a watertight connection when the seams are properly installed.*

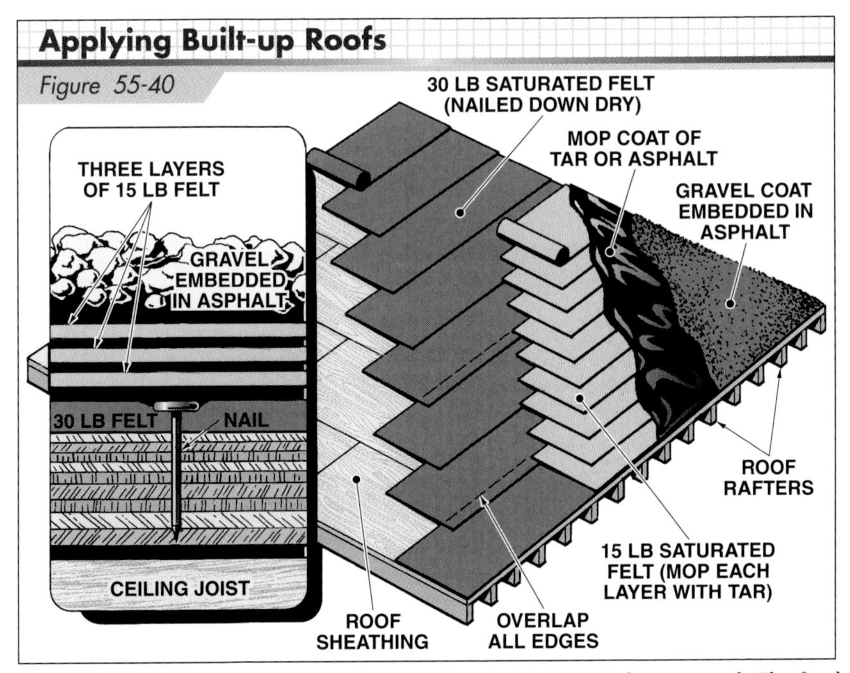

Figure 55-40. *When applying a built-up roof, each felt layer is hot-mopped. The final surface is covered with gravel embedded in asphalt or tar.*

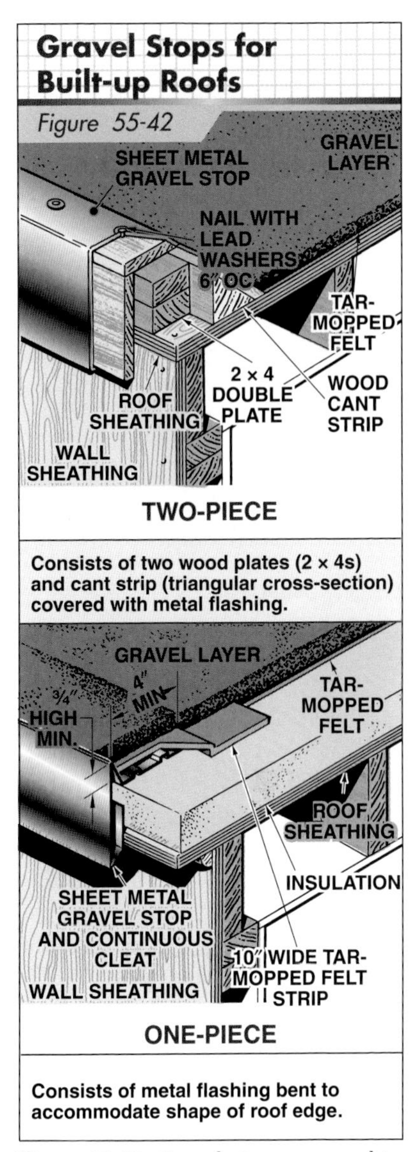

Figure 55-42. *Gravel stops are used to finish the edges of built-up roofs.*

Figure 55-41. *Built-up roof coverings are commonly used with flat roof decks constructed with parapets. Flashing, counterflashing, and asphalt-saturated felt are installed around the edges of the roof deck and other roof projections to provide a watertight seal.*

Built-up roof coverings are installed by roofing contractors who specialize in built-up roofing. Carpenters are not involved in the application of the built-up roof; however, they may perform certain preparatory work such as installing gravel stops, cant strips, and flashing. **See Figure 55-42.**

Gravel stops may consist of one or two pieces. A two-piece gravel stop is made of two wood plates (2 × 4s) and a cant strip covered with metal flashing material. A one-piece gravel stop is made of galvanized steel or copper that is bent to accommodate the shape of the roof edge.

Unit 55 Review and Resources

UNIT 56 Exterior Window and Door Frames

OBJECTIVES

1. Describe window units.
2. List common factors of proper window selection.
3. List and describe common types of windows.
4. Describe the main steps for installing window units.
5. Explain the applications for skylights.
6. List and describe common types of exterior doors.
7. Describe procedures for installing exterior door frames.
8. Identify common types of sliding glass doors.
9. Describe overhead garage doors.
10. Explain the installation of overhead garage doors.

Window units and exterior door units are usually assembled at a factory or mill-cabinet shop. The window and door units are delivered to a job site and installed by carpenters. Frames can be obtained unassembled. However, preassembled frames are usually ordered with window sashes and doors already fitted in them. **See Figure 56-1.** Weatherstripping is often included and the exterior casing may also be attached.

Information regarding the door and window units is provided on the floor plans, elevation plans, and details of the prints. Most prints also include a door and window schedule that specifies the types of doors and windows to be installed.

WINDOW UNITS

Windows allow air and light into a building. Local building codes often specify the minimum glass and venting area required for inhabitable rooms. One building code, for example, recommends that inhabitable rooms be provided with natural light with an area not less than one-tenth of the floor area, with a minimum of 10 sq ft. The code further states that inhabitable rooms must be provided natural ventilation. The area required for ventilation must be not less than one-twentieth of the floor area or a minimum of 5 sq ft, which is greater. **See Figure 56-2.** Openings for natural ventilation are not required if a complete mechanical ventilation system exists in the structure.

Although windows are available in a variety of sizes, the window tops usually align with the door tops. **See Figure 56-3.** Door heights are usually 6′-8″ in residences and 7′-0″ in public buildings.

Various materials are used to construct window frames, sashes, and other window components. **See Figure 56-4.** Wood is the oldest type of material used for window units. Aluminum, steel, and vinyl are also widely used. Aluminum windows are available in natural color or anodized with a colored finish. Vinyl- and aluminum-clad windows are also available. Vinyl- and aluminum-clad windows have a wood core and are wrapped with vinyl or aluminum, respectively.

Figure 56-1. *Preassembled window and door units are set in place by carpenters.*

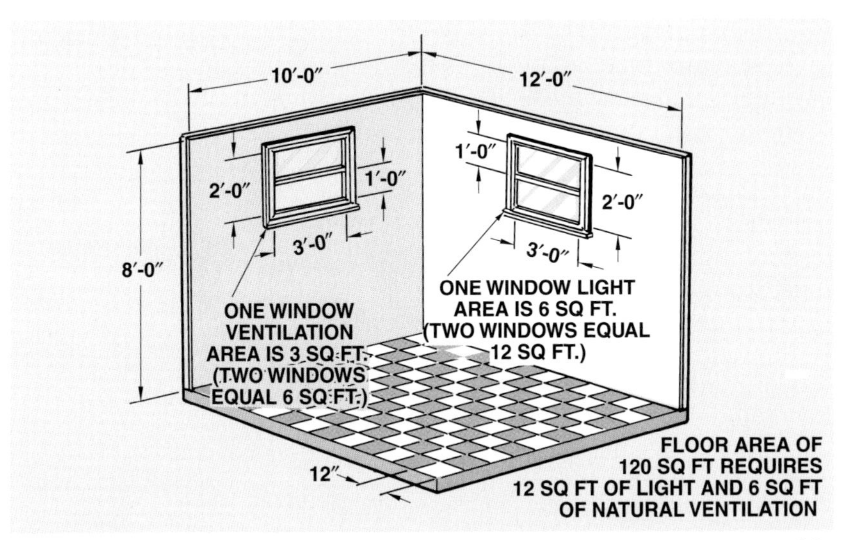

Figure 56-2. *Building codes often specify the minimum glass and venting area required for inhabitable rooms. In this example, the two windows provide 12 sq ft of glass area, which is 10% of the 120 sq ft floor area, and 6 sq ft of natural ventilation, which is 5% of the floor area.*

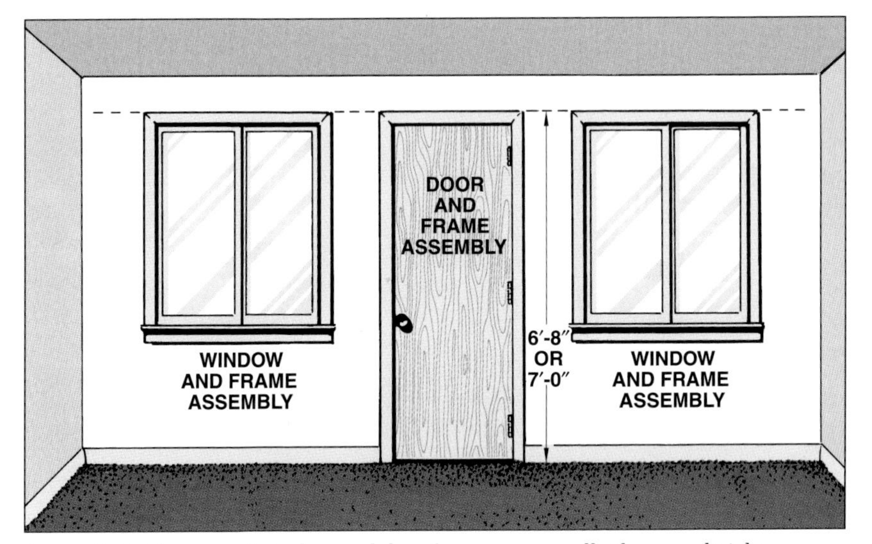

Figure 56-3. *The tops of window and door frames are usually the same height.*

Window Selection

Solar radiation is one of the most efficient methods used to warm a building, particularly a house. Unfortunately, in the summer months or in warm climates, solar radiation can also be the source of unwanted heat. Since windows transmit more solar radiation than the walls or roof, they are key in controlling the heat gain or loss for a building.

Building orientation on the property affects the time and level of heat gain from solar radiation. South-facing windows are exposed to direct sunlight for most of the day. East- and west-facing windows are exposed to direct sunlight during the morning and evenings, respectively. North-facing windows are exposed to indirect sunlight during the summer months and have minimal direct sunlight exposure in the winter months.

Many organic materials, such as carpet and fabric, may fade due to prolonged exposure to sunlight. Selecting the proper windows can impact the type and intensity of transmitted radiation from the sunlight. The most harmful radiation is ultraviolet (UV) rays.

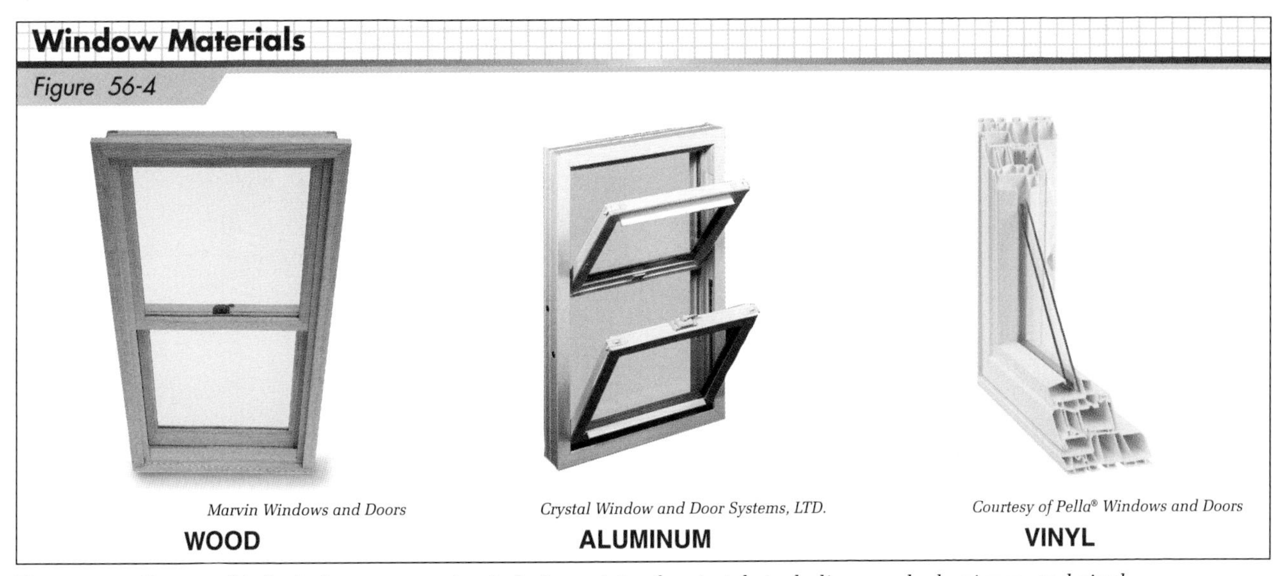

Figure 56-4. *Preassembled windows are constructed of a variety of materials including wood, aluminum, and vinyl.*

Airtightness of a window unit depends on the construction and fit of the sash and frame and installation procedures in the rough opening. Air infiltration occurs through cracks and joints around the window members, and often transmits moisture vapor to the interior of the building. When moisture vapor is transmitted, condensation on the inside of the window will occur when certain air temperatures and relative humidities are achieved. **See Figure 56-5.**

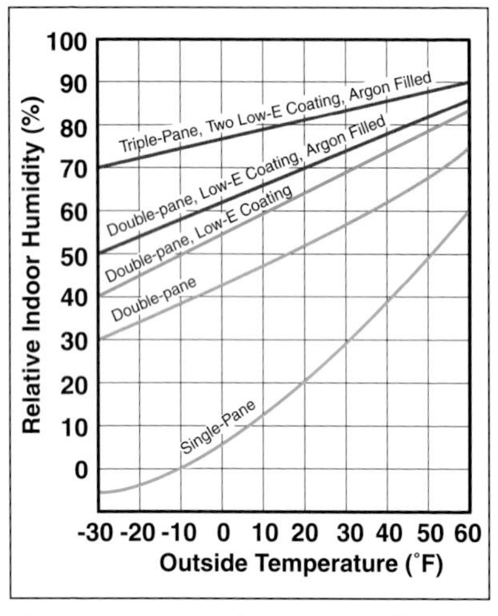

Figure 56-5. *Condensation will occur when certain air temperatures and relative humidity is achieved.*

The National Fenestration Rating Council (NFRC) is the only organization recognized by the U.S. Department of Energy (DOE) for determining the energy performance ratings of windows, skylights, doors, and curtain wall systems. The NFRC has developed several industry standards and administers a rating program that informs consumers about the energy ratings for windows, doors, skylights and curtain wall systems.

NFRC window labels provide information about the U-factor, solar heat gain coefficient, visible transmittance, air leakage, and condensation resistance of a window unit. **See Figure 56-6.** The *U-factor* indicates how well a window unit prevents heat from escaping from the building. A lower U-factor indicates a better insulating value. The *solar heat gain coefficient (SHGC)* indicates how well a window unit blocks heat caused by sunlight. A lower value indicates less solar heat gain in the building through the window unit. The *visible transmittance* value indicates the amount of light transmitted through a window. Even though windows are transparent, varying amounts of light are transmitted due to manufacturing processes, coatings, and type of glass. A higher value indicates more light transmitted by the window unit. *Air leakage* is expressed as a value indicating the equivalent cubic feet of air passing through a square foot of window area (cfm/sq ft). Heat loss and gain occur by air infiltration through cracks or voids in a window unit. The lower the air leakage value, the less air infiltration through the window unit. *Condensation resistance* indicates the ability of a window unit to resist the formation of condensation on the interior surface of that product. The higher the condensation-resistance rating, the better the window unit is at resisting condensation formation.

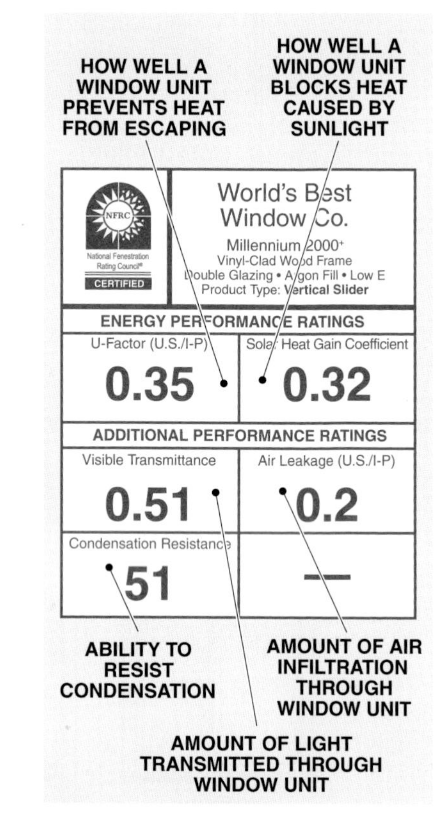

Figure 56-6. *National Fenestration Rating Council (NFRC) labels provide information pertinent to window performance.*

Pella Corporation

National Fenestration Rating Council (NFRC) energy performance ratings apply to skylights and other types of windows.

Window Unit Components

A traditional double-hung window unit is shown in **Figure 56-7.** The *frame* is composed of a top piece and two side pieces made of wood or metal. The width of these pieces is equal to the finished wall thickness. The *sill* is a slanted piece at the bottom of the frame. A sill permits moisture to properly drain away from the building. The space between the frame and the interior and exterior wall is covered by a *casing*. *Aprons* are placed below the finished sill and stool at the bottom of the frame. The window *sash* is the frame that holds the glass that fits into the window frame.

A complete window unit includes one or more sashes. The upper sash in Figure 56-7 is separated into several *lights* (panes of glass). The lights are separated by *muntins.* The complete wood or metal frames including sash units are installed in the rough openings after the walls have been framed and sheathed.

Window Types

The types of windows selected for a building should harmonize with the general design of the building. Some windows look better with traditional building designs, while others blend well with more modern structures. Types of windows include *fixed-sash, double-hung, casement, horizontal sliding, awning, hopper,* and *jalousie windows.*

Fixed-Sash Windows. Also called stationary windows, fixed-sash windows do not open or close. Since fixed-sash windows do not allow air movement through them, they are often used in combination with operable windows. **See Figure 56-8.**

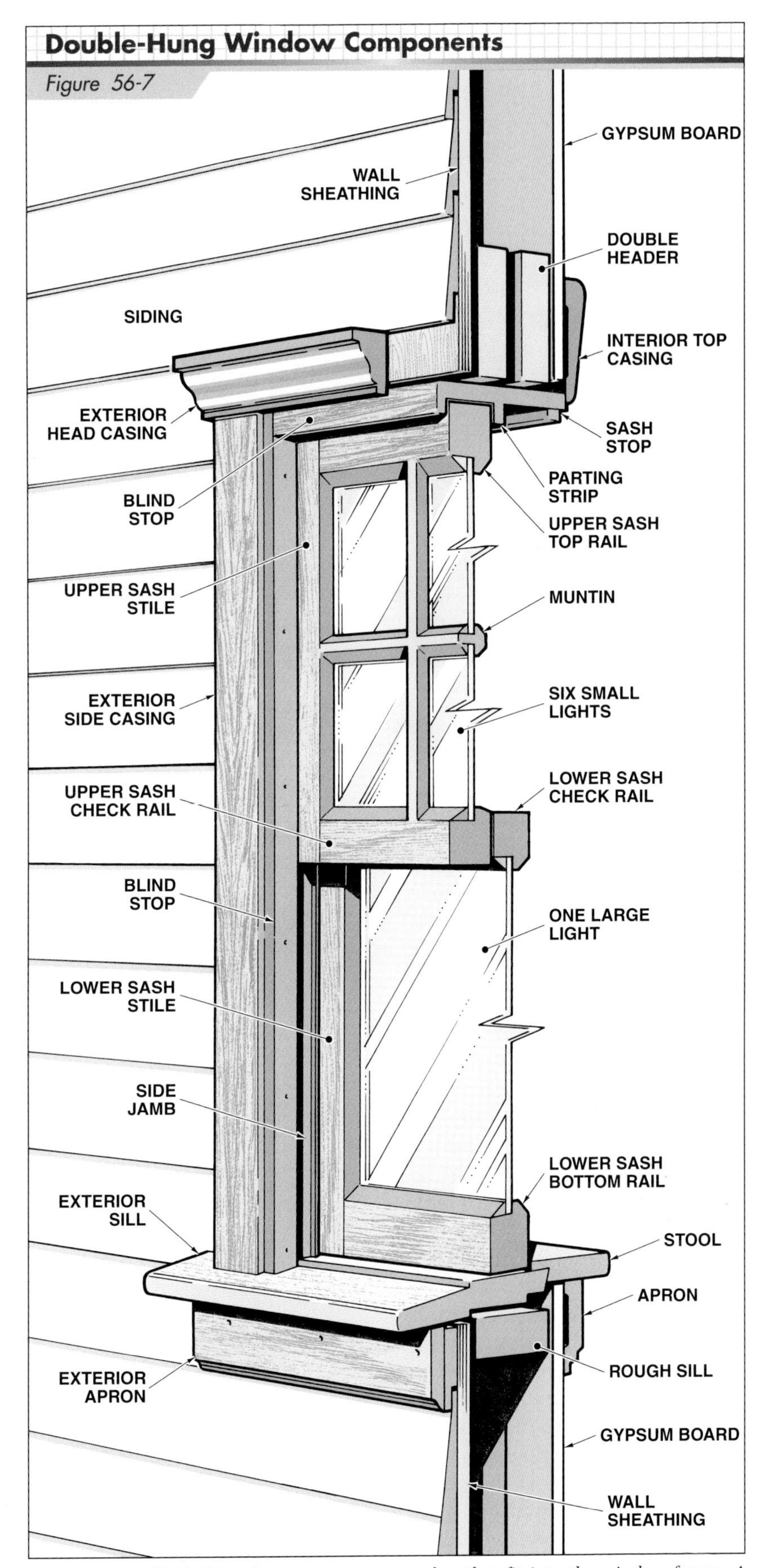

Figure 56-7. *All windows have one or more sashes that fit into the window frame. A double-hung unit with two sashes is shown here.*

CertainTeed Corporation

Figure 56-8. *Fixed-sash windows are often used in combination with operable windows. Fixed-sash windows are shown in the middle and at the top of this assembly, and double-hung windows flank the large window at the bottom.*

Double-Hung Windows. Double-hung windows have upper and lower sash sections that move vertically in tracks provided in the window frame. **See Figure 56-9.** Contemporary double-hung windows commonly have tilting sashes that allow cleaning the exterior side of the window from inside the building. Ventilation offered by double-hung windows is limited to 50% of the window opening.

Contemporary double-hung windows are fitted with balancing devices to hold the sashes in open positions. **See Figure 56-10.** Compressible weatherstripping is sometimes used as a balancing device with lightweight windows. A sash weight and rope were used as a balancing device in the past.

Typical hardware used with double-hung windows includes a sash lift and a sash lock. For some types of double-hung windows, the sash lift is milled into the lower sash bottom rail.

Andersen Corporation

Figure 56-9. *A double-hung window unit has upper and lower sash sections that move vertically in tracks provided in the window frame. The sashes for some double-hung windows can be tilted inward to allow cleaning of the exterior side.*

Casement Windows. Casement windows are hinged on one side and have the same swing action as hinged doors. **See Figure 56-11.** Ventilation from a casement window is 100% of the window opening. Most casement windows are installed to swing out from the building. However, they can also be installed to swing into the building. In addition to hinges, a crank operator and sash lock are installed on casement windows. Double or single casement windows are available.

Horizontal Sliding Windows. Horizontal sliding windows move along tracks (or guides) located above and below the window. Sliding window units often consist of one movable and one stationary window. **See Figure 56-12.** A three-sash design consists of a fixed middle window and sliding windows on both sides. A locking device is the only hardware required.

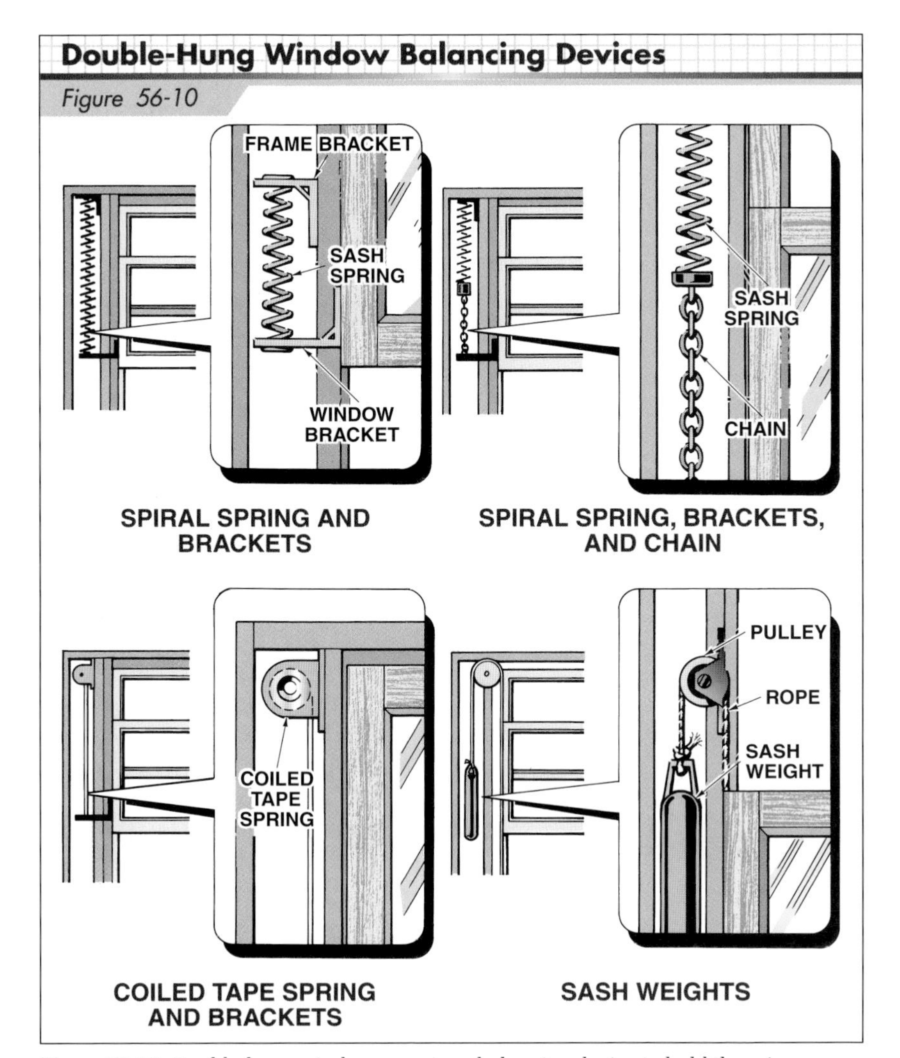

Figure 56-10. *Double-hung windows require a balancing device to hold them in an open position.*

Marvin Windows and Doors

Figure 56-11. *Casement windows are hinged on one side. The casement window shown here swings outward as the crank operator at the bottom of the window is turned. Sash locks are located along the inner edges of the window.*

CertainTeed Corporation

Figure 56-12. *Sliding windows often feature one movable and one stationary window.*

Awning Windows. Awning windows are hinged at the top and swing out at the bottom. **See Figure 56-13.** Awning windows are often combined with a fixed-sash window. Sometimes a series of awning windows are placed in the same opening. Awning windows require hinges and a crank operator or push bar.

Hopper Windows. Hopper windows are similar to awning windows, but are hinged at the bottom instead of at the top. Hopper windows swing in or out and require a crank operator or push bar and a locking device.

Jalousie Windows. Jalousie windows are composed of a series of small glass lights set into metal clips on both sides. A crank operator pivots the glass lights up to 90°.

Tempered glass is glass that is strengthened by reheating it to just below the melting point and then suddenly cooling it. Tempered glass is required in patio doors and entrance doors.

Installing Window Units

Wood, metal, and clad window units are usually delivered to a job site with the sash in the frame. Manufacturer instructions for installation should be closely followed. Prior to installation, the rough openings must be laid out and properly constructed. If the opening has been properly laid out and framed for each window unit, there should be no problem with installation.

Preparing the Rough Opening. The rough opening for window units must be properly prepared to minimize air and moisture infiltration and ensure the integrity of the building envelope. After housewrap is attached to the building sheathing and openings in the housewrap are created, the windows must be properly flashed using metal, plastic, or flexible membrane flashing, and in certain cases, sill pans must be installed.

Pella Corporation

Figure 56-13. *Awning windows are hinged at the top and swing out at the bottom.*

Flashing Window Openings. Window openings are one of the most vulnerable portions of the building envelope to air and water infiltration. A good seal must be created at window openings to protect the building interior and direct water around the opening for proper drainage. Three basic types of flashing—metal, plastic, and flexible membrane tape—can be used to minimize air and water infiltration around windows and doors.

Metal flashing is one of the oldest types of flashing materials used on buildings. Noncorrosive metal flashing, such as copper, aluminum, and galvanized flashing, is available in common shapes and sizes as sheet or preformed shapes. Metal flashing is most commonly used for commercial construction.

Plastic flashing is commonly substituted for preformed metal flashing. Plastic flashing, such as sill pans, is less expensive than metal flashing, and is flexible and easily cut into various shapes. *Sill pans* are preformed polyvinyl chloride (PVC) plastic flashings used at the base of window and door openings to direct water that may have penetrated in or around the frames to the outside. **See Figure 56-14.** Sill pans are available in multiple pieces to allow them to be adjusted to the desired rough opening width. For a two-piece sill pan, one side of the pan is positioned at the base of the rough window opening over the weather barrier and sheathing. Nails or staples are driven through the flange to secure the first piece in position. The same procedure is used for the second piece of the sill pan. A bead of sealant may be applied between the pieces at the lap joint. PVC cement can also be used to solvent-cement the two pieces together. The seam and nails or staples are then covered with membrane tape flashing before installing the window.

When installing sill pans, ensure they are slightly angled to the exterior of the building to direct water away from the interior. In many cases, a beveled wood strip or dam strip may need to be installed along the inner edge of the rough sill to provide the proper angle for the sill pan. **See Figure 56-15.** The strips may also be used without a sill pan directly underneath flexible membrane flashing.

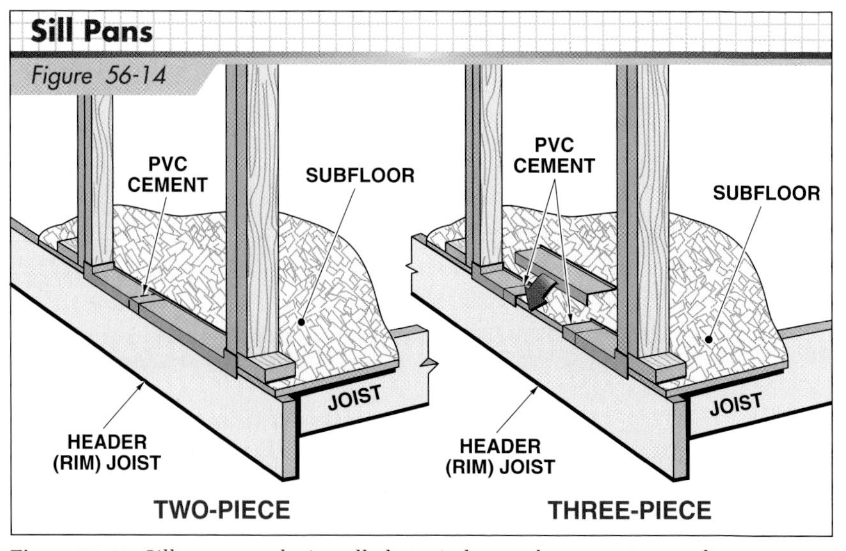

Figure 56-14. *Sill pans may be installed at window or door openings to divert water to the outside.*

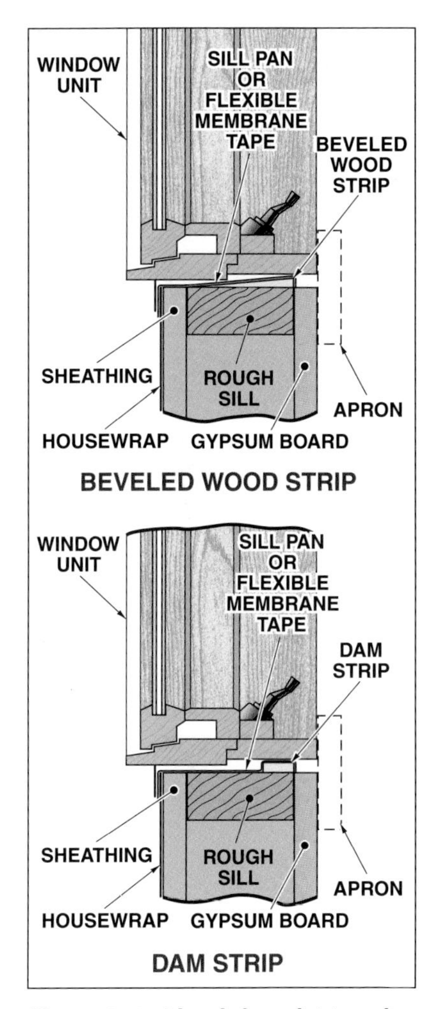

Figure 56-15. *A beveled wood strip or dam strip may be installed below a sill pan to provide a slight angle.*

Flexible membrane flashing, commonly known as *peel-and-stick flashing,* is commonly used as flashing for door and window openings. Flexible membrane flashing is available in 4″ to 12″ wide rolls that can be cut to the desired size. Flexible membrane flashing consists a bituminous or butyl rubber adhesive core with release paper on one side and a metal foil or flexible plastic backing on the other side. The release paper is removed immediately before the flashing is applied to the rough opening. The foil or plastic backing protects the membrane flashing. A hand roller can be used to apply pressure to the membrane flashing for optimal cohesion to the framing members.

Flexible membrane flashing is applied directly over the housewrap. Cut a piece of flashing approximately 1′ longer than the width of the rough opening. If necessary, install a beveled wood strip or dam strip along the inner edge of the rough opening. Peel the release paper from the flashing and position the flashing along the bottom and approximately 6″ up the sides of the rough opening. **See Figure 56-16.** Do not stretch the flashing while applying it. When the flashing is in the proper position, apply firm pressure to the flashing working from the center to the corners and up the jamb to remove air bubbles. Secure the outer edges of the flashing with plastic cap nails.

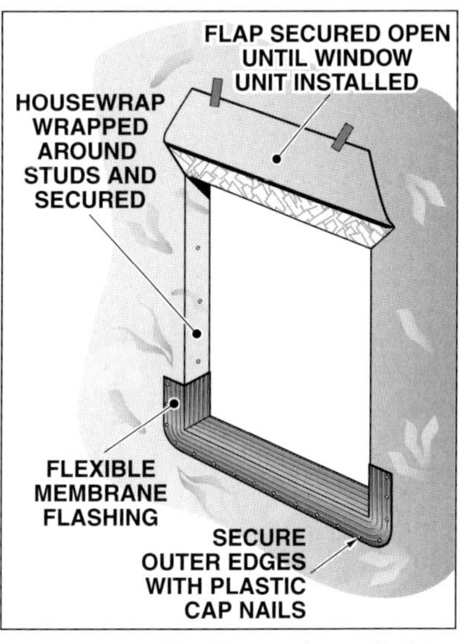

Figure 56-16. *Flexible membrane flashing is commonly used as window and door flashing.*

Building paper may also be used as a weather barrier. When using building paper around window openings, a sill pan is created by cutting a piece of the paper and folding it to fit the corner of the rough opening. **See Figure 56-17.** The corner piece is stapled along the front edge of the rough opening. Another piece of building paper is slid into position over the corner piece and fastened along the side jamb using staples.

Sealing Window Openings. When the window units are ready to be installed, a ¼″ to ⅜″ bead of sealant must be applied to the housewrap along the top and sides of the rough opening. **See Figure 56-18.** Do not apply sealant along the bottom of the rough opening. The window unit can then be installed per the manufacturer instructions. If a drip cap is required, the drip cap should be installed and properly sealed.

With the window unit installed, install the flexible membrane flashing vertically along the side jambs. The flashing should extend over the window nailing fin to cover the window fasteners and extend over the housewrap. After installing the side jamb flashing, install the horizontal flashing at the head jamb to cover drip edge and side jamb flashing.

Release the housewrap at the head jamb and position over the head jamb flashing. Apply seam tape over the diagonal cuts in the housewrap at the corners of the head jamb, lapping the tape onto the window unit or casing.

Flexible membrane flashing is applied directly over the housewrap.

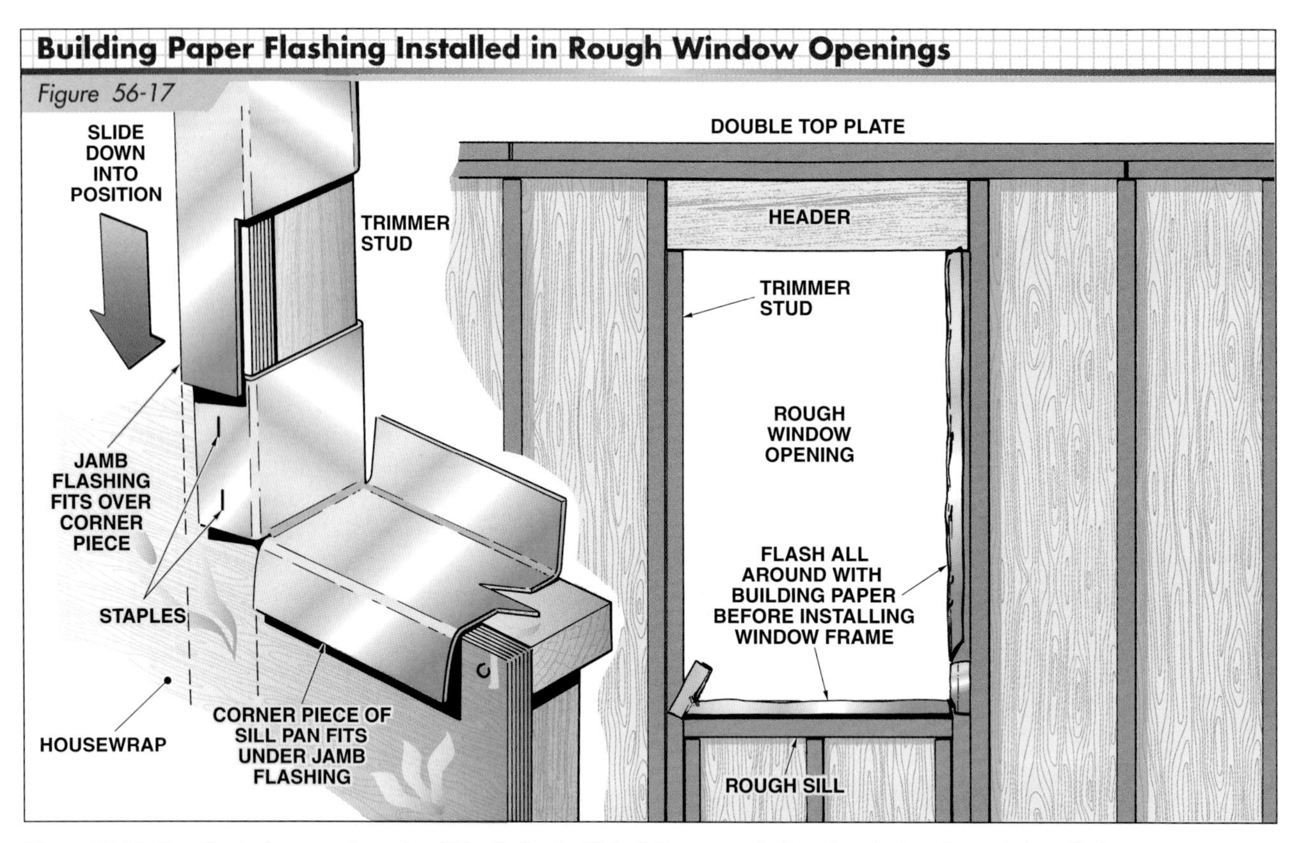

Figure 56-17. *Rough window openings should be flashed with building paper before the window frame is installed.*

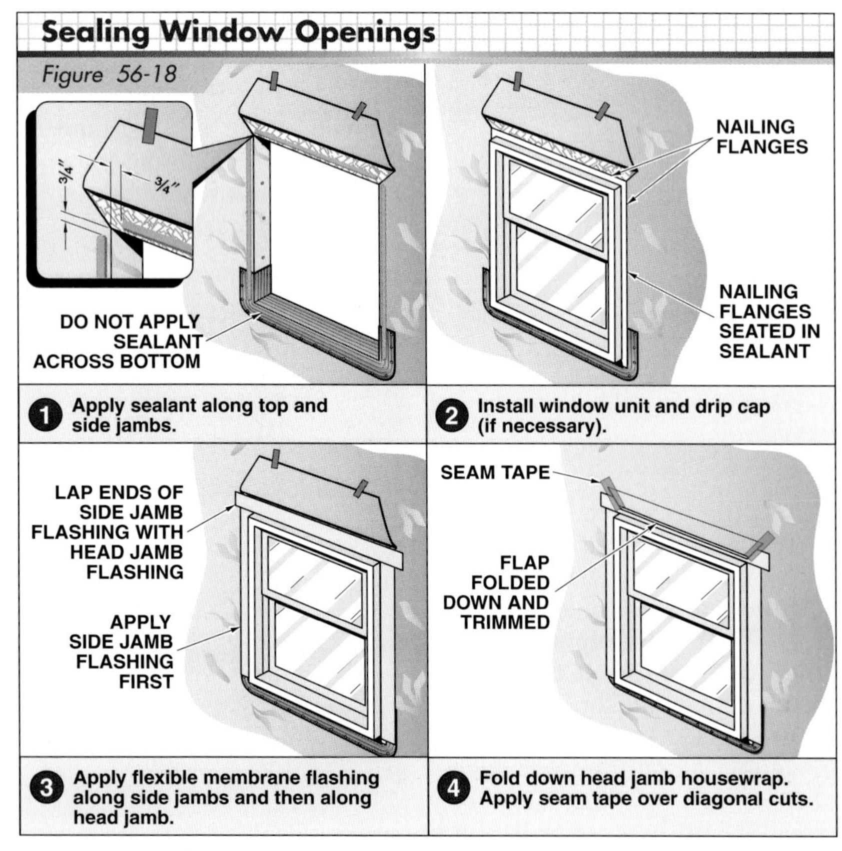

Figure 56-18. *Windows must be properly sealed to prevent air and moisture infiltration.*

Some manufacturers recommend that the window sash be removed before setting the frame to avoid damaging the sash. Others suggest that the sash not be removed, enabling faster installation of the unit. **Figure 56-19** describes one method of installing traditional window units. The procedure for trimming out (finishing) the interior of the window frame is discussed in Section 13.

Metal and Clad Window Units. Most metal and wood- and metal-clad windows have a nailing flange with nail holes on all four sides. Nails or screws are driven through the holes into the sheathing and framework around the opening. Metal and clad window units are easier to install than solid wood units; however, care must be taken to level the bottoms and plumb the sides.

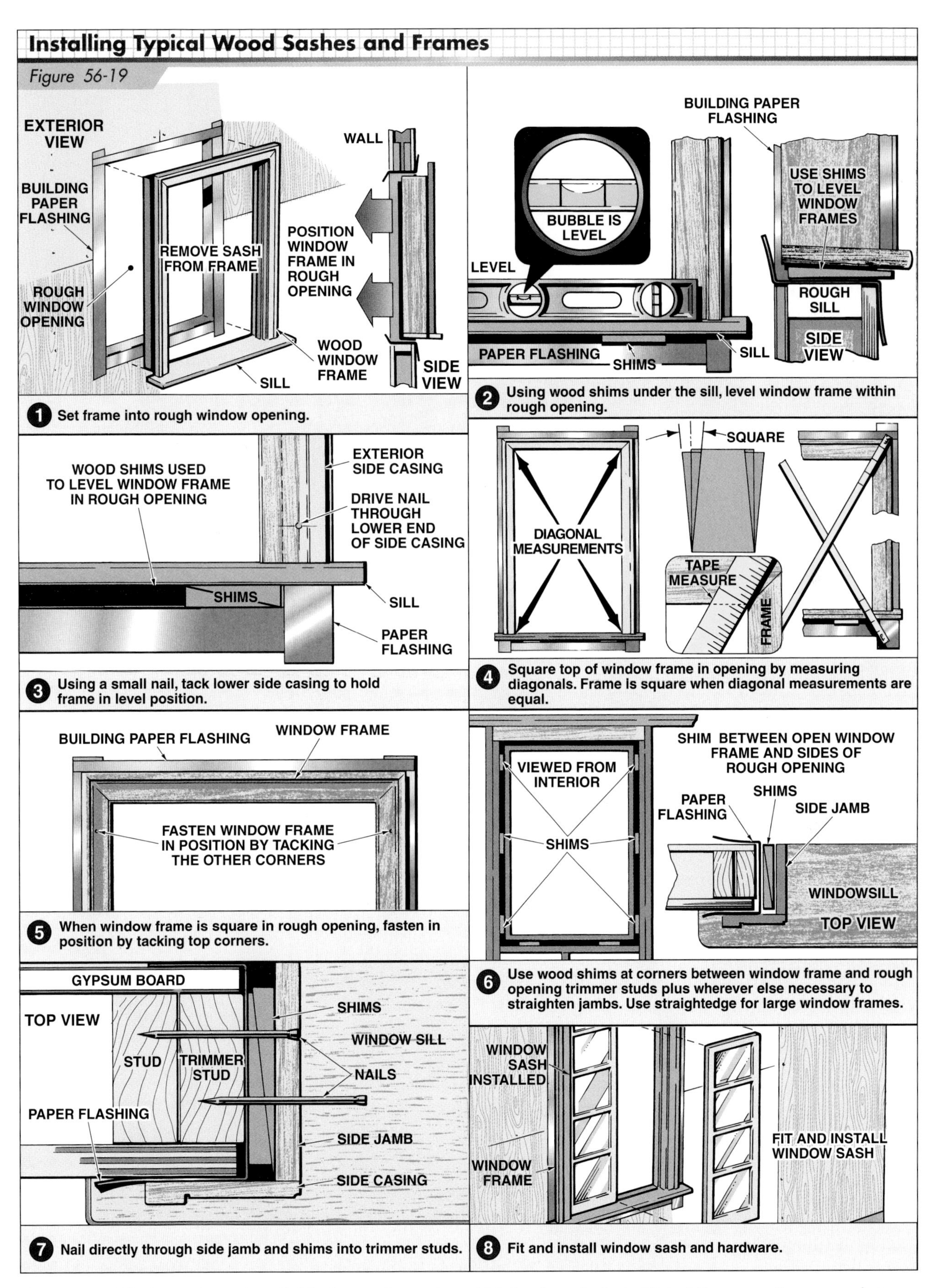

Figure 56-19. *Window units are installed in the rough opening, which has been flashed with building paper. Some manufacturers recommend that the sash be removed before installation.*

Another type of metal unit is designed to fit into an existing wood frame. These units are often used to replace worn wood sashes in older homes. These metal units are attached by driving screws through holes provided in the nailing flange and into the wood frame. **See Figure 56-20.**

Skylights. Skylights are, in effect, windows placed in the roof of a building. Skylights provide additional natural light and are also an additional means of passive heat collection for the interior of the building.

Skylights are available in various shapes and sizes. Smaller skylights are designed to fit between 16″ and 24″ OC rafter spacing. Large skylights require the same framing procedures used for chimneys and flue openings. One or more rafters may have to be cut, and headers must be installed.

Most skylights are made of clear or translucent plastic, which is secured in aluminum or wood frames. Insulating glass may also be used for skylights. Skylights are usually set on curbs constructed over the roof opening. **See Figure 56-21.**

Skylights can be used instead of dormer windows in gambrel or steeply pitched gable roofs to provide ventilation and light. **See Figure 56-22.** Skylights can also be installed over a light shaft that extends through the attic. **See Figure 56-23.**

Pieces of flashing overlap one another beginning at the bottom of the skylight where it penetrates the roof. Each course of roofing material has a piece of flashing (step flashing) on top of the last piece of roofing, which abuts the skylight.

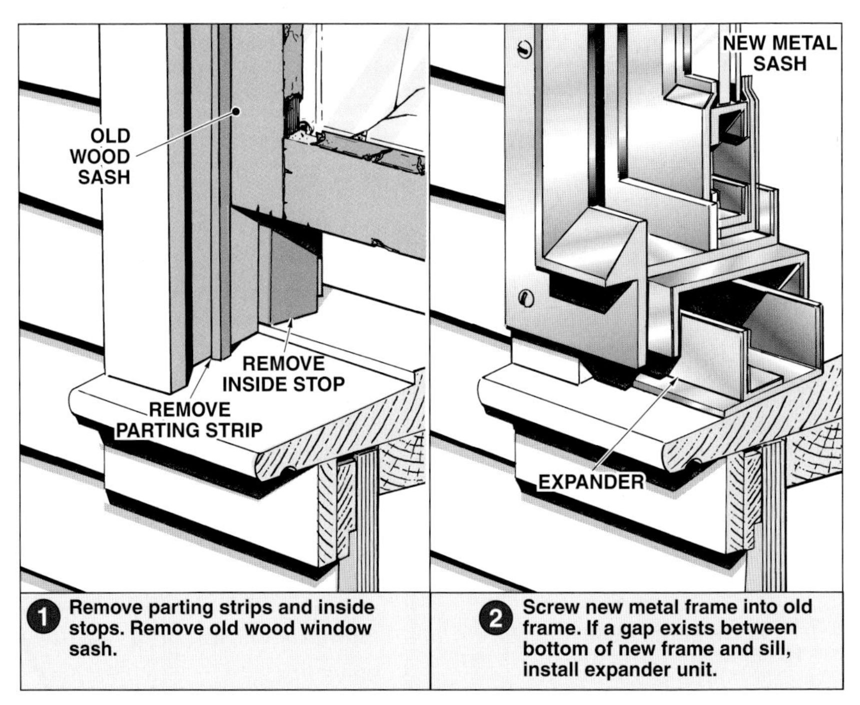

Figure 56-20. *When replacing a wood window unit with a metal unit, the sashes, stops, and parting strips are removed. The metal unit is designed to fit into the existing wood frame in the building.*

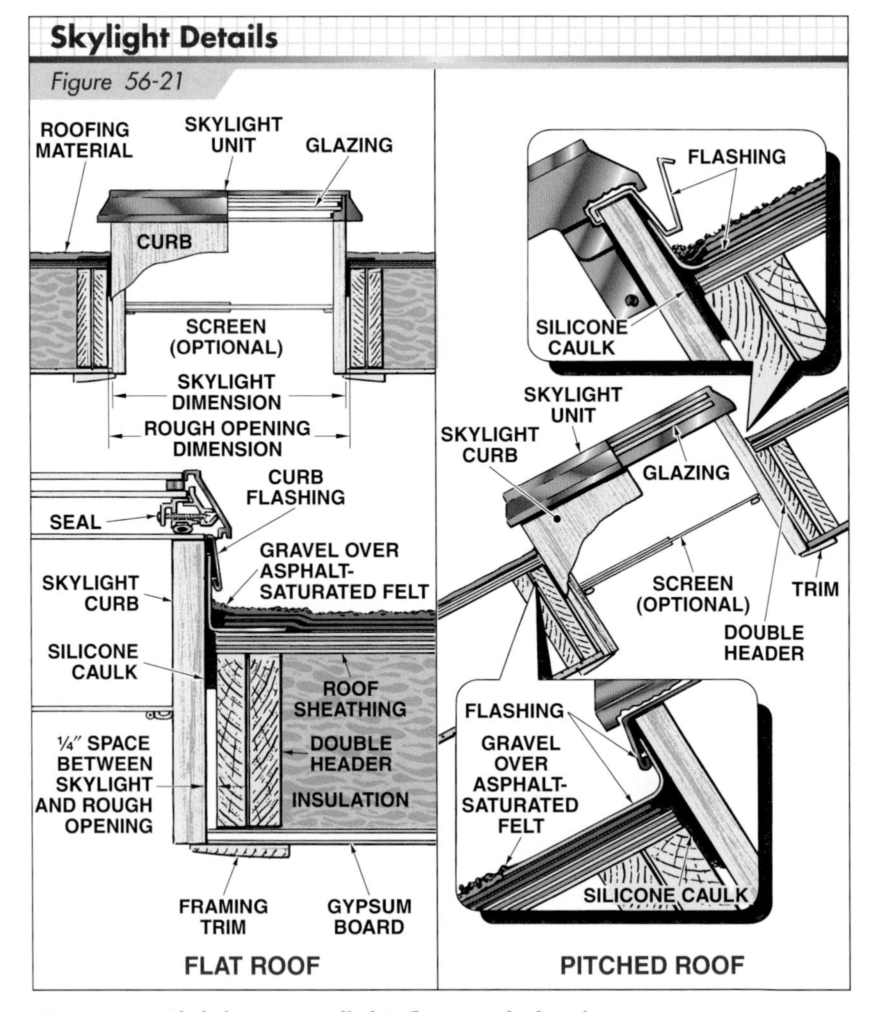

Figure 56-21. *Skylights are installed in flat or pitched roofs.*

Andersen Corporation

Figure 56-22. *Skylights provide additional light and may provide ventilation.*

Figure 56-23. *Skylights may be installed over a light shaft in an attic roof. The light shaft is framed between the ceiling joist and roof rafters and is surfaced with gypsum board or wood paneling.*

EXTERIOR DOOR UNITS

Exterior doors, also known as entry doors, provide passage between the inside and outside of a building. The main entrance door unit is usually at the front or street side of the building. Main entrance doors may be a single door, double doors, or a single door flanked by windows. **See Figure 56-24.** Main entrance doors are made of solid wood, or fiberglass or steel with an insulated foam core. Many main entrance doors have a fixed window. Examples of exterior door styles are shown in **Figure 56-25.**

Passage into a building is also provided by one or more rear entrance (service) doors. Sliding glass or inswing doors are often used for service doors leading to porches, patios, or terraces. **See Figure 56-26.**

Preassembled door units include the door and door frame consisting of a *doorjamb, outside casing, head,* and *threshold* (door sill). A doorjamb is the main vertical member on each side of a door. Doorjambs are either wood or metal. The casing is molding that covers the space between the jamb and wall. The head is the main horizontal member forming the top of a door frame. The threshold is a piece of wood, metal, or stone that covers the space between the jamb and

Main Entrance Doors

Figure 56-24

SINGLE

Therma-Tru

DOUBLE WITH SIDELIGHTS

Pella Corporation

SINGLE WITH SIDELIGHTS

Figure 56-24. *Main entrance doors may be single or double doors. Main entrance doors may be flanked by sidelights or a transom.*

the bottom of the door opening. Often the doors are *prehung,* meaning that they have already been fitted and hinged to the jambs. Prehung exterior door units are installed into the rough door opening.

Solid-Core Doors

A *solid-core door* has wood surface veneers and a solid inner core, which may consist of wood blocks, engineered wood product, or high-density foam. A high-density foam core increases the insulation characteristics of the door. A *fire-rated door* has a mineral-based material core. Fire-rated wood doors are designed to resist fire, and are rated as 20-, 30-, and 45-minute doors. To obtain a fire rating, the doors must be tested by a qualified testing agency and the doors must carry an identifying label.

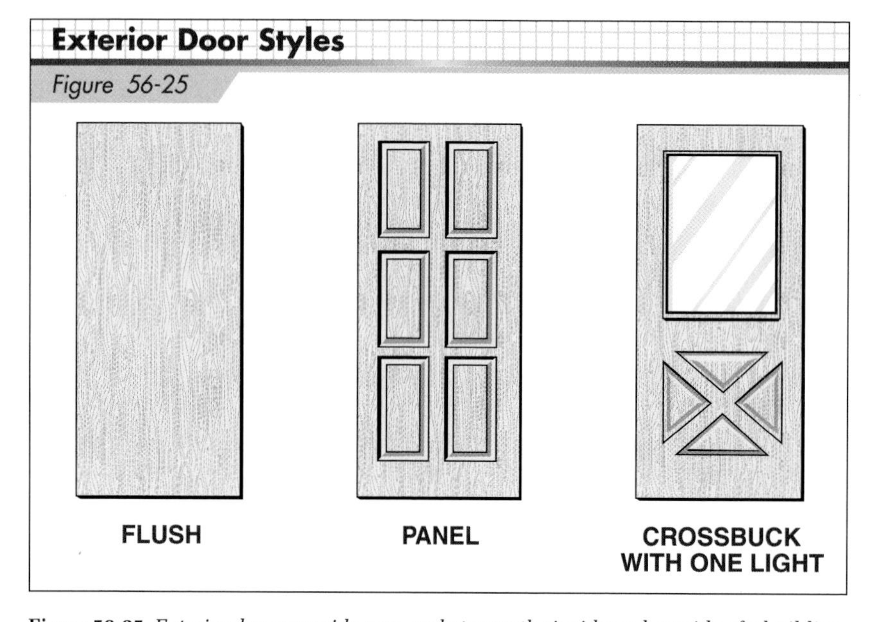

Figure 56-25. *Exterior doors provide passage between the inside and outside of a building. Exterior doors may be flush, panel, or crossbuck doors, with or without lights.*

Andersen Corporation

Figure 56-26. *Exterior glass sliding doors are frequently used for passage to porches, patios, and terraces.*

Fiberglass Composite Doors

Exterior *fiberglass composite doors* have gained wider acceptance in recent years as an alternative to wood doors. Fiberglass composite doors consist of compression-molded fiberglass surfaces, wood stiles and rails, and a polyurethane foam core. The doors have good insulation characteristics and are very durable, even under harsh weather conditions. Fiberglass composite doors may have embossed woodgrain- pattern surfaces which can readily receive wood stains.

Steel Doors

Steel doors have traditionally been associated with commercial construction such as office and industrial buildings. While steel doors are still frequently used in commercial construction, they are gaining wider acceptance as exterior doors in residential construction. Steel doors are available in most standard sizes and in 1⅜″ and 1¾″ thicknesses. Other sizes and thicknesses can be ordered.

Most steel doors are constructed of heavy-gauge galvanized surfaces enclosing a steel or wood frame. A high-density foam core, such as polyurethane or polystyrene, ensures high insulation values. Surface finishes are applied at the factory. Steel doors increase the security of a house.

Installing Exterior Door Units

Before a door frame is installed, the rough door opening should be flashed with building paper. A procedure for installing an exterior wood door frame is described in **Figure 56-27.** The manner in which the doors are fitted and hung in the frame is discussed in Section 13.

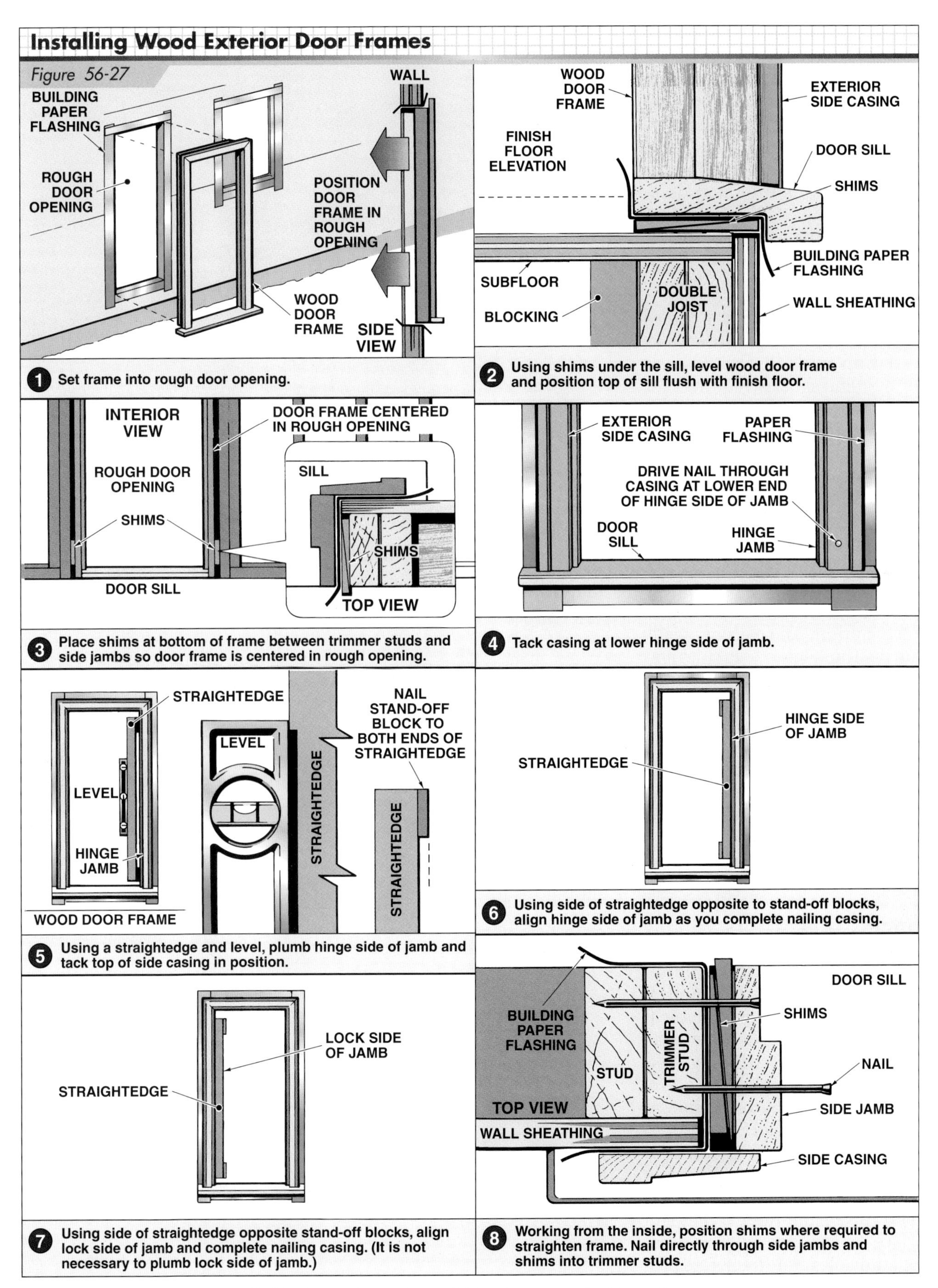

Figure 56-27. *Door units are installed in the rough opening, which has been flashed with building paper.*

Metal Door Frames. Metal door frames are delivered to a job site as packaged units and are assembled on the job according to the manufacturer instructions. Metal door frames must have a fire rating equal to or greater than its fire-rated door. When installed, the units are plumbed and aligned in the rough opening and fastened to the framework with screws.

Wood or Metal Door Frames in Masonry Walls. Close coordination with the bricklayers or stonemasons is required for carpenters to install wood or metal frames in masonry (brick or hollow concrete masonry unit) walls. The frames are set in position when the exterior walls have reached the proper level below the door opening.

Wood or metal frames are plumbed, aligned, and securely braced according to dimensions on the prints. The masonry walls are then built up on both sides and the doorjamb is anchored at the masonry joints. **See Figure 56-28.** The procedure for installing a wood door frame in masonry walls is shown in **Figure 56-29.**

Ensure metal door frames are internally braced to maintain the proper door opening dimensions.

Sliding Glass Doors. *Sliding glass doors,* also known as patio doors, are constructed with wood, vinyl, or aluminum frames and glass lights. A typical sliding glass door unit includes a frame, weatherstripping, and hardware. Sliding glass doors generally have nylon rollers at the bottom to enable the door to move smoothly over tracks set in the sill. The sill is usually made of heavy extruded aluminum. The top of the door is held in place and guided by a channel or guide. Separate tracks are provided for sliding screen sections used during warm weather.

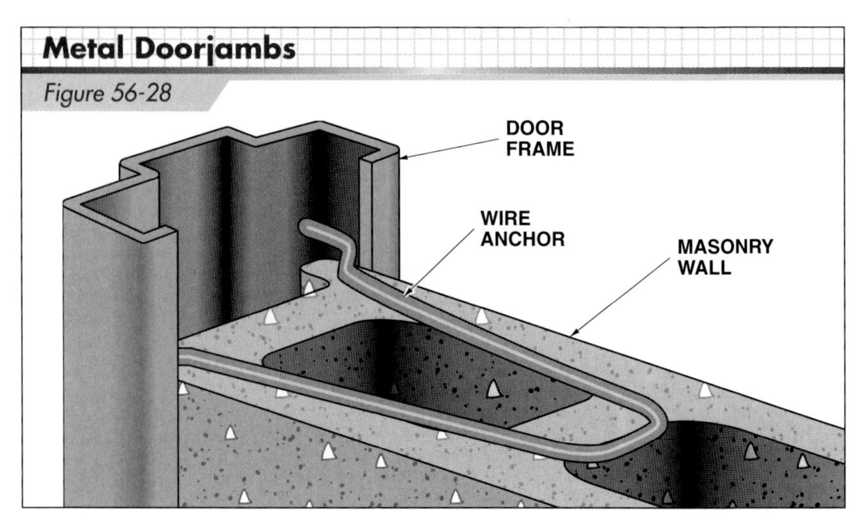

Figure 56-28. *Metal doorjambs are anchored to a masonry wall. One end of the anchors is welded to the doorjamb and the other end is embedded between the masonry courses.*

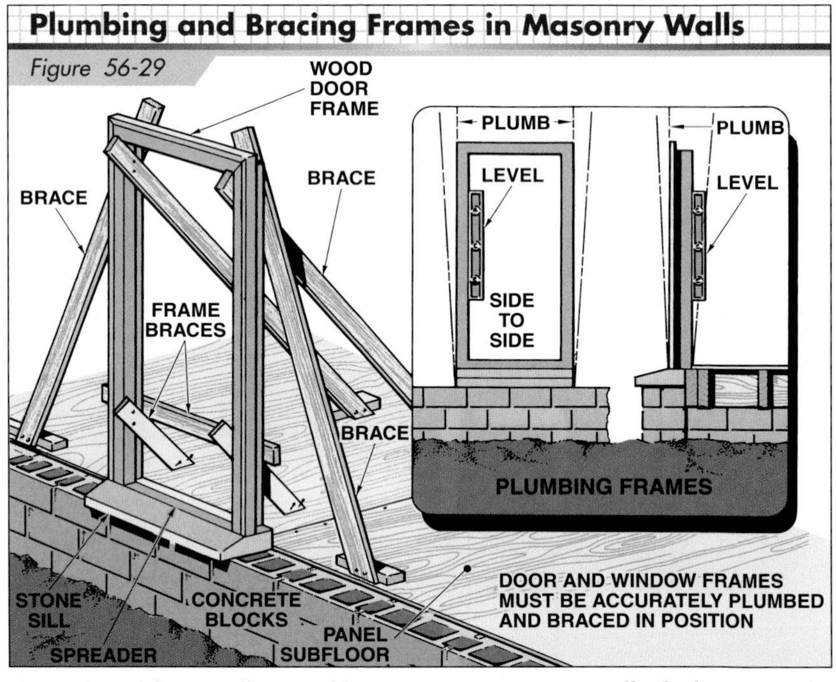

Figure 56-29. *When installing wood frames in exterior masonry walls, the frames must be accurately plumbed, leveled, and braced. Masonry walls are then constructed on both sides.*

Genie Industries

An aerial lift may be used when installing and trimming windows installed in an upper level. Note the proper use of personal fall-arrest equipment.

Some sliding glass doors hang by rollers at the top that enable the door to move along tracks at the top of the frame. The door bottom fits into a channel in the sill that holds the door in place as it moves.

Two- and three-door units are common sliding glass door configurations. In two-door units, both doors may slide or one door may be stationary. In three-door units, one or two of the doors may slide while the other door remains stationary. **See Figure 56-30.** Sliding glass doors typically include double or triple glazing to provide proper insulation.

Figure 56-31. *Overhead garage doors are the most common type of garage door.*

CertainTeed Corporation

Figure 56-30. *Sliding glass doors may be two- or three-door configurations. The two doors on the right of this sliding glass door unit slide while the door on the left is stationary.*

OVERHEAD GARAGE DOORS

Garage doors must be functional and attractive and must complement the building design. **See Figure 56-31.** Overhead garage doors are the most common type of garage door. Older garage doors are hinged and open to the sides.

Overhead garage doors are available in a variety of dimensions for one-car or two-car garages. Garage doors can also be custom built to architect or owner specifications. Flush or panel designs are two common garage door designs.

Garage door frames are wood or metal. Garage doors are sheathed with plywood, hardboard, particleboard, aluminum, and fiberglass, and often are filled with foam insulation.

There are three basic types of overhead garage doors—*one-piece swing-up, sectional roll-up,* and *rolling steel garage doors.* Each type requires special hardware.

A one-piece swing-up door depends on springs and counterbalances for its operation. One-piece swing-up doors are fitted with a ⅜″ to ½″ space on each side to provide ventilation for the garage.

A sectional roll-up door is more expensive and operates more efficiently than a one-piece swing-up door. A sectional roll-up door is constructed in sections that are hinged together. Door movement is controlled by a track system fastened to the side jambs and suspended from the ceiling of the garage. **See Figure 56-32.**

A rolling steel door is normally used in industrial and commercial buildings. **See Figure 56-33.** The door rolls up into a housing unit located over the garage door and does not require attachment to the ceiling, thus saving overhead space.

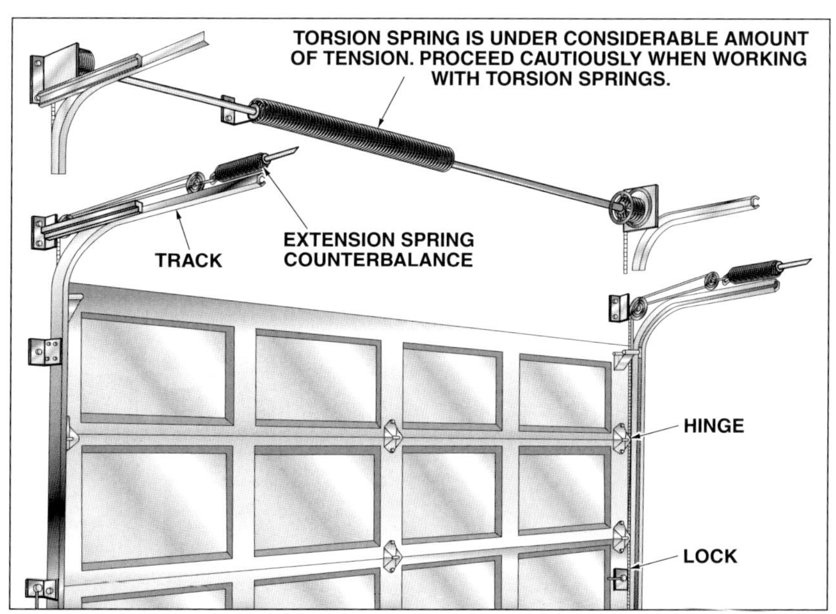

Figure 56-32. *Sectional roll-up garage doors are guided by a track system fastened to the jambs and suspended from the ceiling.*

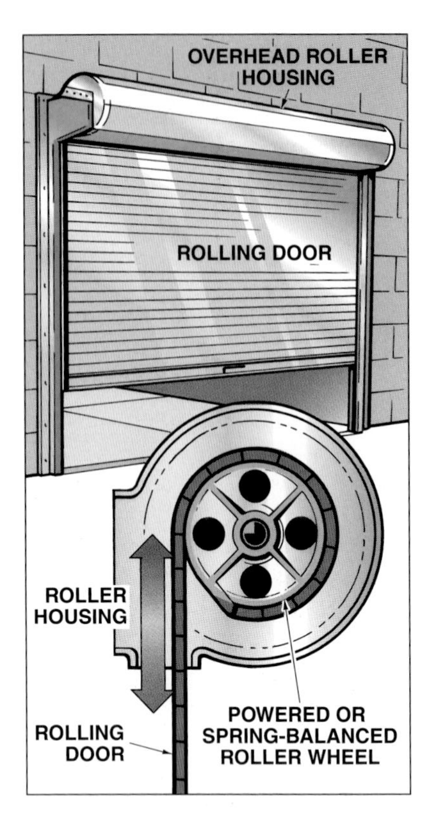

Figure 56-33. *Steel overhead rolling doors are used in industrial and commercial buildings.*

Garage doors are the largest piece of moving equipment in a house. Safety risks are inherent to working with garage doors. Torsion and counterbalance springs are under tension at all times. Only professionally trained workers should relieve the springs of tension.

Automatic Garage Door Openers

For convenience and security, overhead doors may be equipped with automatic opening systems. **See Figure 56-34.** An electric motor operates chains, belts, or cables that lift and lower the door. The electric motor is activated by a radio-controlled unit inside the car or by a switch located in the garage.

Installing Overhead Garage Doors

The rough opening for an overhead garage door must be constructed to the correct dimensions to avoid installation problems. Correct layout for a rough opening is based on the garage door size, jamb thickness, and required clearances. **See Figure 56-35.**

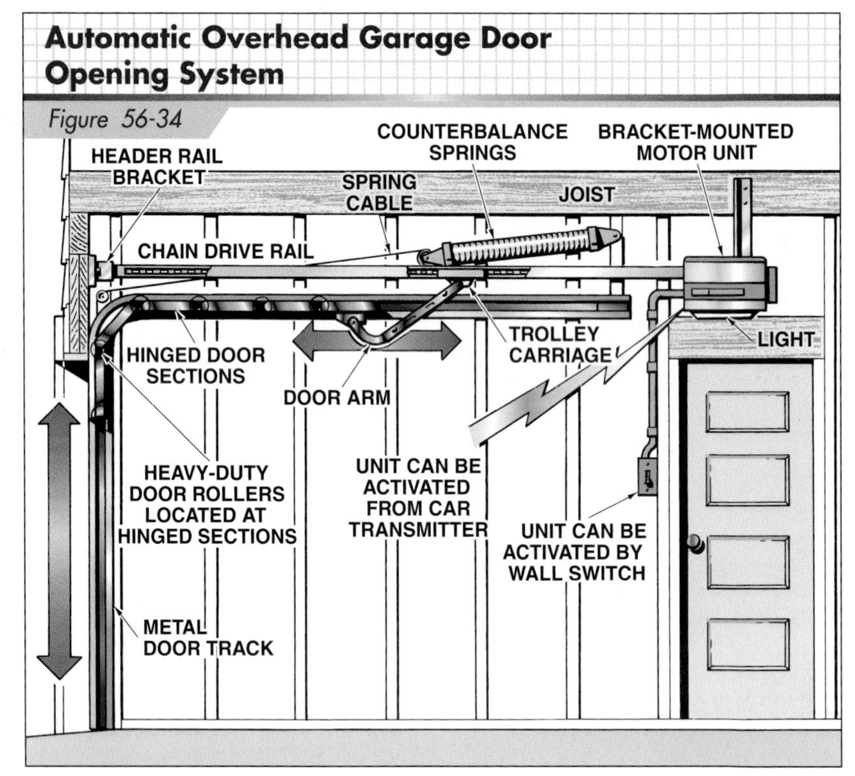

Figure 56-34. *Automatic opening systems are widely used with residential overhead garage doors. The system shown here has a chain-and-cable mechanism operated by an electric motor, which is activated by a pushbutton in the garage or a transmitter carried in an automobile.*

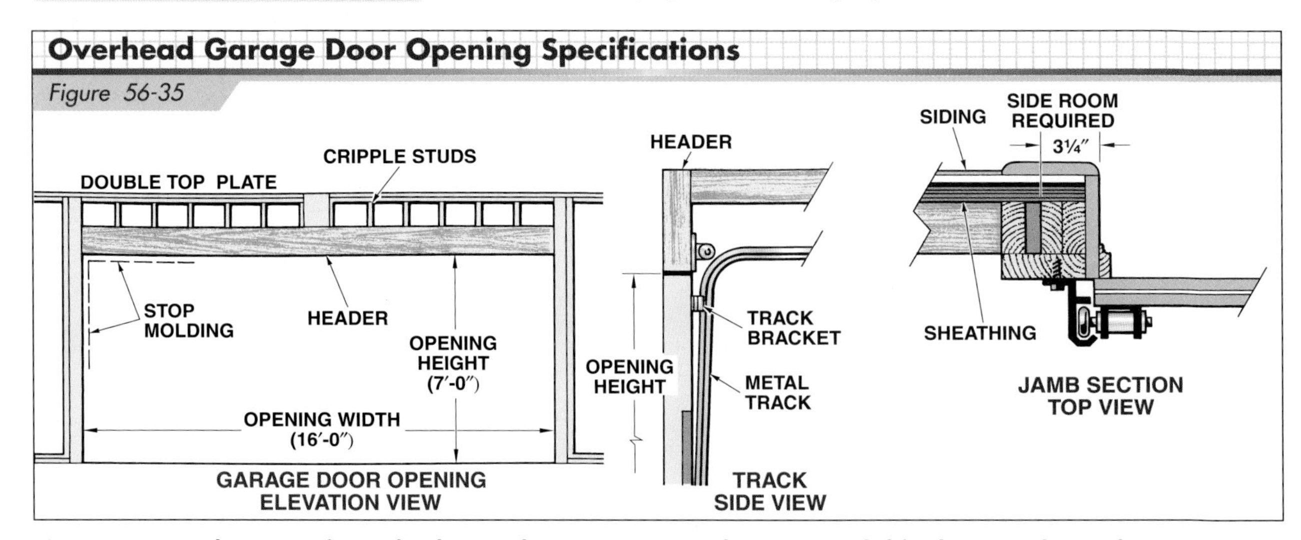

Figure 56-35. *Rough openings for overhead garage doors must accommodate space needed for the garage door in the open position.*

Manufacturer instructions for installing overhead garage door hardware should be carefully followed. Typical hardware is shown in **Figure 56-36.** Angled tracks provide a weathertight seal for a garage door. Torsion springs allow the door to open and close smoothly. An extension spring counterbalance device incorporates a safety cable for use in the event of a broken spring. Other hardware includes heavy-duty rollers, hinges, a bell crank door latch, and reinforcing steel struts.

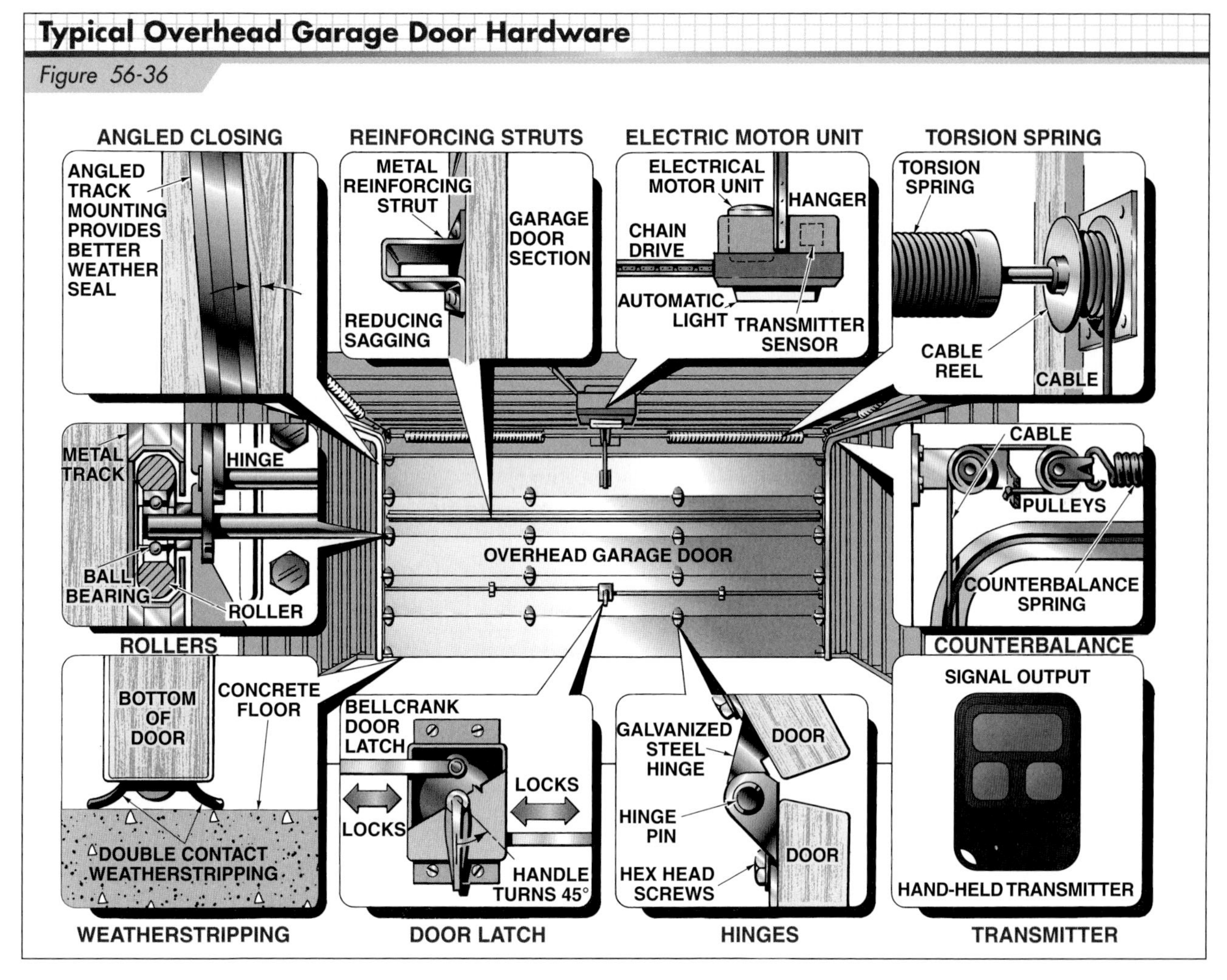

Figure 56-36. *Various hardware components are used for an overhead garage door.*

UNIT 57

Exterior Wall Finish, Porches, and Decks

OBJECTIVES

1. List and describe the components of an exterior building envelope.
2. List and describe common types of wood siding.
3. Describe the procedure for installing board siding.
4. Review the use of panel siding.
5. Explain the procedures for installing panel siding.
6. Describe the application of shingles and shakes.
7. Describe the use of aluminum siding.
8. Describe the use of vinyl siding.
9. Explain the construction of decks.

Exterior wall finish is installed after door and window frames have been placed. A variety of materials are used for exterior wall finish including siding, brick, stone, stucco, and exterior insulation and finish systems (EIFS). In many cases, more than one material is used for the exterior wall finish of a building. **See Figure 57-1.** In addition to exterior wall finish, porches or decks may be constructed.

EXTERIOR BUILDING ENVELOPE

The *exterior building envelope* consists of the exterior wall finish and weather barrier that was applied over the wall sheathing. The first line of protection against air and water infiltration into a building is the exterior wall finish, which generally includes the finish material, trim, windows, and doors. The second line of defense against air and water infiltration is the weather barrier. Weather barriers must prevent air and water infiltration, yet allow a building to "breathe" to prevent moisture from condensing within the exterior walls. Weather barriers and their proper installation is discussed in Unit 43.

Exterior finish material must be installed per the manufacturer instructions to ensure the integrity of the building envelope. Regardless of whether the exterior finish material is wood, metal, vinyl, or composite, it must be properly installed.

WOOD SIDING

Wood is one of the oldest materials used for exterior wall covering and is one of the most attractive and efficient materials used to protect the exterior surface of a building. In addition to solid wood siding, wood panel products, such as plywood and hardboard, are also popular choices for exterior siding. Solid wood and wood panel products used for siding are available in various patterns.

Board Siding

Redwood and cedar are highly recommended for board siding because of their superior decay resistance. Western hemlock, cypress, yellow poplar, spruce, and several pine species are also used for board siding. Wood board siding should be a good grade of lumber, free of warp, twist, knots, and pitch pockets.

Allowable moisture content for wood board siding is approximately 10% to 12% in most areas of the United States. In the southwestern states, 8% to 9% moisture content is acceptable. Board siding should be *back-primed* before nailing. In back-priming, the backs of the boards are painted with a sealer (either paint primer or a water-repellent preservative). As the boards are cut for fitting and nailing to the wall, the freshly cut ends are also painted with sealer.

Hardboard can also be used for siding. Hardboard can be manufactured to appear like many different lumber species. Hardboard siding is made with different textures and is painted or stained.

Board Siding Patterns. Board siding is available in straight boards or a variety of other patterns (shapes). **See Figure 57-2.**

APA—The Engineered Wood Association

Figure 57-1. *Wood panel siding and masonry produce an attractive exterior finish on this traditional-design house.*

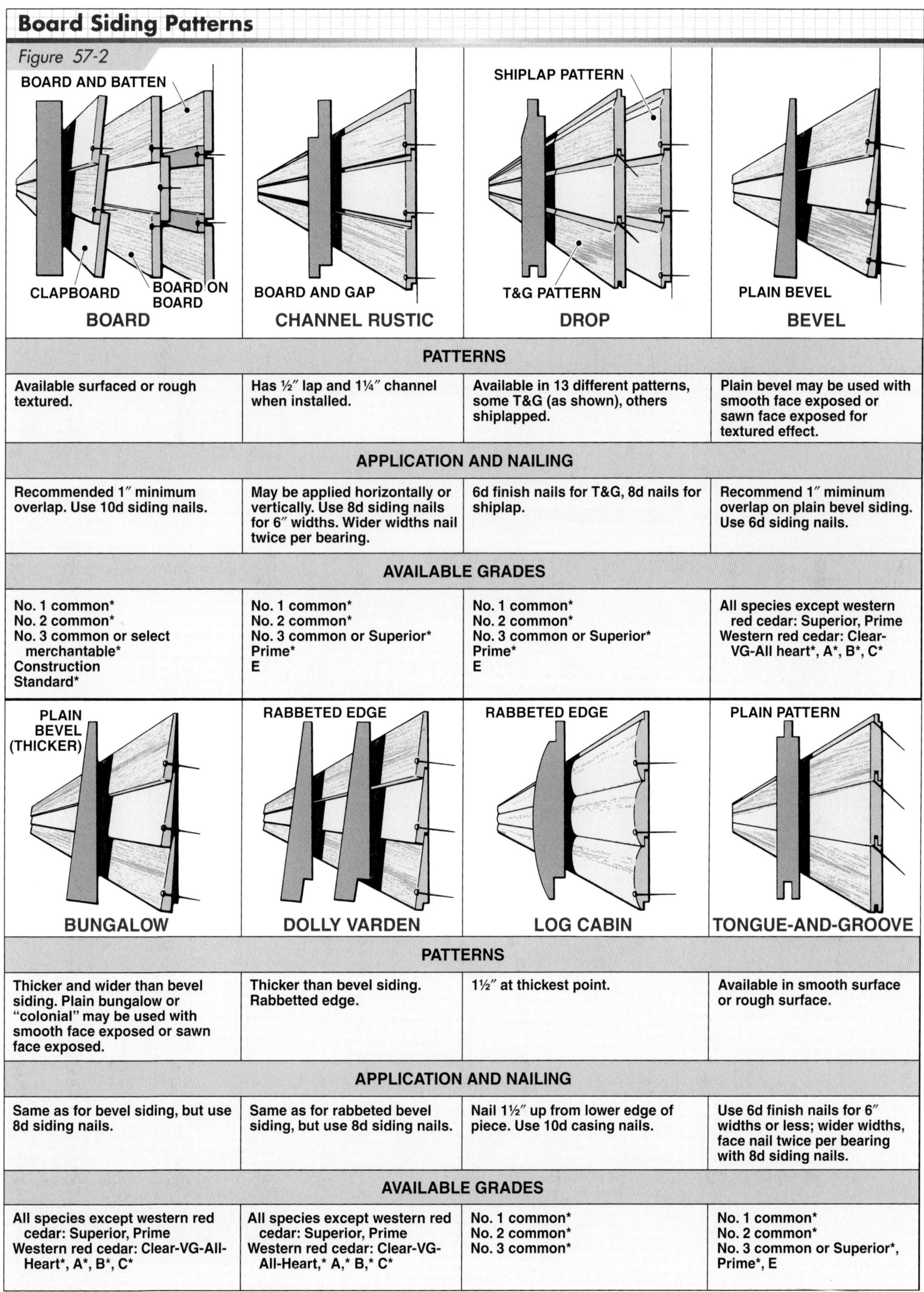

	BOARD	CHANNEL RUSTIC	DROP	BEVEL
PATTERNS	Available surfaced or rough textured.	Has ½″ lap and 1¼″ channel when installed.	Available in 13 different patterns, some T&G (as shown), others shiplapped.	Plain bevel may be used with smooth face exposed or sawn face exposed for textured effect.
APPLICATION AND NAILING	Recommended 1″ minimum overlap. Use 10d siding nails.	May be applied horizontally or vertically. Use 8d siding nails for 6″ widths. Wider widths nail twice per bearing.	6d finish nails for T&G, 8d nails for shiplap.	Recommend 1″ miminum overlap on plain bevel siding. Use 6d siding nails.
AVAILABLE GRADES	No. 1 common* No. 2 common* No. 3 common or select merchantable* Construction Standard*	No. 1 common* No. 2 common* No. 3 common or Superior* Prime* E	No. 1 common* No. 2 common* No. 3 common or Superior* Prime* E	All species except western red cedar: Superior, Prime Western red cedar: Clear-VG-All heart*, A*, B*, C*

	BUNGALOW	DOLLY VARDEN	LOG CABIN	TONGUE-AND-GROOVE
PATTERNS	Thicker and wider than bevel siding. Plain bungalow or "colonial" may be used with smooth face exposed or sawn face exposed.	Thicker than bevel siding. Rabbetted edge.	1½″ at thickest point.	Available in smooth surface or rough surface.
APPLICATION AND NAILING	Same as for bevel siding, but use 8d siding nails.	Same as for rabbeted bevel siding, but use 8d siding nails.	Nail 1½″ up from lower edge of piece. Use 10d casing nails.	Use 6d finish nails for 6″ widths or less; wider widths, face nail twice per bearing with 8d siding nails.
AVAILABLE GRADES	All species except western red cedar: Superior, Prime Western red cedar: Clear-VG-All-Heart*, A*, B*, C*	All species except western red cedar: Superior, Prime Western red cedar: Clear-VG-All-Heart,* A,* B,* C*	No. 1 common* No. 2 common* No. 3 common*	No. 1 common* No. 2 common* No. 3 common or Superior*, Prime*, E

* most commonly used

Figure 57-2. *Board siding patterns are available in various widths, thicknesses, and grades.*

In addition, all patterns are produced in various widths and thicknesses. **Figure 57-3** indicates the nominal and dressed dimensions of some patterns. A triple-lap design is available in hardboard that allows faster installation. **See Figure 57-4.**

Preparing for Board Siding. In most cases, siding is applied over sheathing material and weather barrier. In mild climates, however, sheathing is not always required and the siding may be nailed directly to the wall studs.

Where structural sheathing and rigid insulation board is used as the nailing base, the siding can be nailed into the sheathing material without penetrating the wall studs. When nonstructural sheathing is used, such as gypsum panel products and other types of insulation panels, siding must be nailed through the sheathing and into the wall studs. When no sheathing is used, a weather barrier should be installed on the studs to prevent air infiltration and water leakage. **See Figure 57-5.**

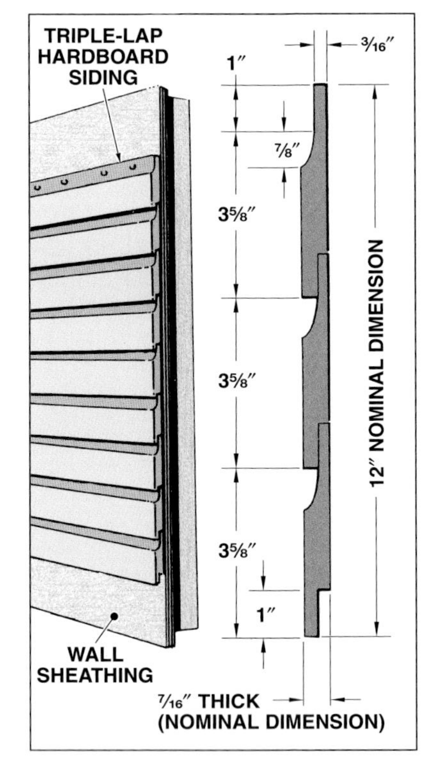

Figure 57-4. *Triple-lap hardboard siding allows fast installation.*

BOARD SIDING SIZES

PRODUCT	NOMINAL SIZE*		DRESSED DIMENSIONS*		
	Thickness	Width	Thickness	Width	
Bevel Siding for WRC Sizes†	½	4 5 6	11/32 butt, 3/16 tip	3½ 4½ 5½	
Wide Bevel Siding (Colonial or Bungalow)	¾	8 10 12	¾ butt, 3/16 tip	7¼ 9¼ 11¼	
				Face	Overall
Rabbeted Bevel Siding (Dolly Varden)	¾ 1	6 8 10 12	⅝ x 3/16 11/16 x 13/32	5 6¾ 8¼ 10¾	5½ 7¼ 9½ 11¼
Rustic and Drop Siding (Dressed and Matched)	1	6 8 10 12	22/32	5⅛ 6⅞ 8⅞ 10⅞	5⅜ 7⅛ 9⅛ 11⅛
Rustic and Drop Siding (Shiplapped, ⅜″ lap)	1	6 8 10 12	22/32	5 6¾ 8¾ 10¾	5⅜ 7⅛ 9⅛ 11⅛
Rustic and Drop Siding (Shiplapped, ½″ lap)	1	6 8 10 12	22/32	4 15/16 6⅝ 8⅝ 10⅝	5 7/16 7⅛ 9⅛ 11⅛
Log Cabin Siding	1½ (6/4)	6 8 10	1½ thickest point	4 15/16 6⅝ 8⅝	5 7/16 7⅛ 9⅛
Tongue-and-Groove (T&G) S2S and CM	1 (4/4)	4 6 8 10 12	¾	3⅛ 5⅛ 6⅞ 8⅞ 10⅞	3⅜ 5⅜ 7⅛ 9⅛ 11⅛

* in in.
† Western red cedar bevel siding available in ½″, ⅝″, and ¾″ nominal thickness. Corresponding surfaced thick edge is 15/32″, 9/16″, and ¾″. Widths 8″ and wider ½″ off.

Figure 57-3. *Nominal and dressed dimensions of board siding vary with the pattern.*

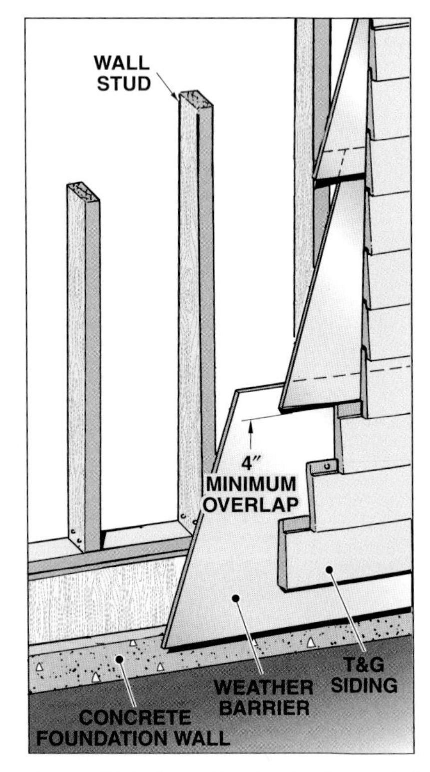

Figure 57-5. *A weather barrier is applied directly to the studs when sheathing is not installed.*

Flashing is recommended over the drip caps of doors and windows to prevent water leakage. A *drip cap* is a piece of molding placed over exterior wood frames to direct water away from the door or window. **See Figure 57-6.** Flashing material on exterior walls is galvanized sheet metal or vinyl.

Other preparations for siding may include metal trim pieces placed around the door and window openings. **See Figure 57-7.**

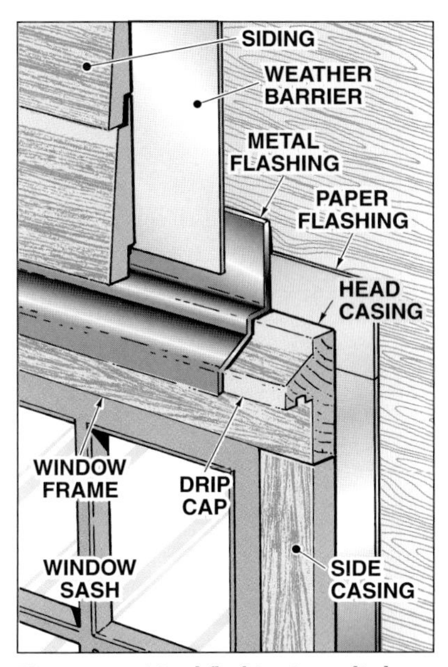

Figure 57-6. *Metal flashing is applied over the drip caps of door and window frames.*

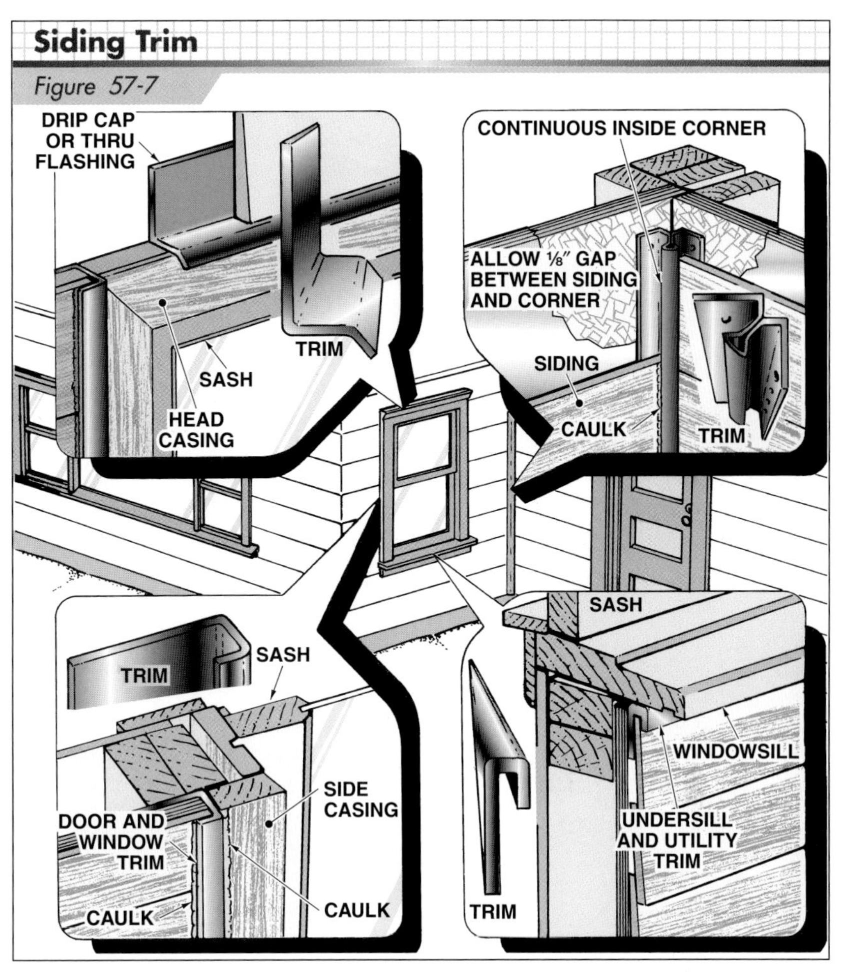

Figure 57-7. *Siding systems may include metal trim around the door and window frames.*

Applying Board Siding. The procedure for applying board siding is similar for the different patterns. Application is usually horizontal. **See Figure 57-8.** However, board, shiplap, channel rustic, and tongue-and-groove patterns can also be applied vertically. **See Figure 57-9.** An attractive effect can be created by combining horizontal and vertical applications. Diagonal siding is also sometimes used, but more waste is created when applying siding diagonally.

Horizontal Application. When applying board siding horizontally, the first row of boards is applied in a level, perfectly straight line. The bottom of the first row of boards should be at least 1″ below the top of the foundation wall to prevent water leakage between the sill plate and foundation wall.

For bevel siding, a spacer strip equal to the thickness of the thin edge of the siding at the exposure line should be nailed to the wall underneath the bottom edge. The spacer strip will slant the first row of siding. **See Figure 57-10.** Bevel siding should overlap at least 1″. To maintain a uniform overlap, a notched spacer block should be used.

Whenever possible, rows of siding should be laid out so they align with the top of the window drip cap and bottom of the windowsill. **See Figure 57-11.** When this layout cannot be followed, the siding boards must be notched around the windows. **See Figure 57-12.**

A carpenter can determine the layout of siding boards by laying out the position and height of the window opening and siding courses on a story pole. One procedure for laying out a siding story pole is shown in **Figure 57-13.** In this case, the story pole shows that the boards can be made to line up with the top and bottom of the window.

Allow wood siding to acclimate to the local humidity after it is delivered to the job site. Boards should be stored horizontally and raised above the ground on wood blocks or scraps.

California Redwood Association

Figure 57-8. *Board siding is usually applied horizontally. California redwood siding is installed on this residence.*

California Redwood Association

Figure 57-9. *Board, shiplap, channel rustic, and tongue-and-groove patterns may be applied vertically.*

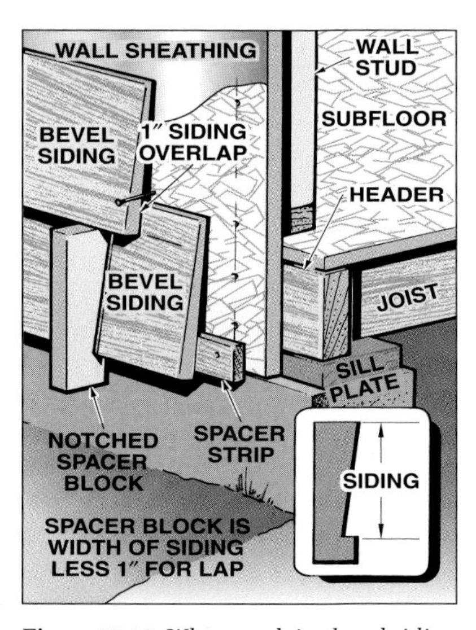

Figure 57-10. *When applying bevel siding, a spacer strip is installed under the bottom edge of the first course.*

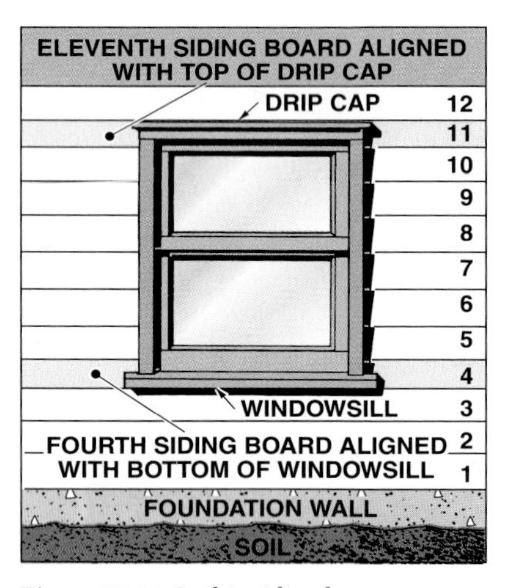

Figure 57-11. *In this siding layout, courses line up with the windowsill and drip cap.*

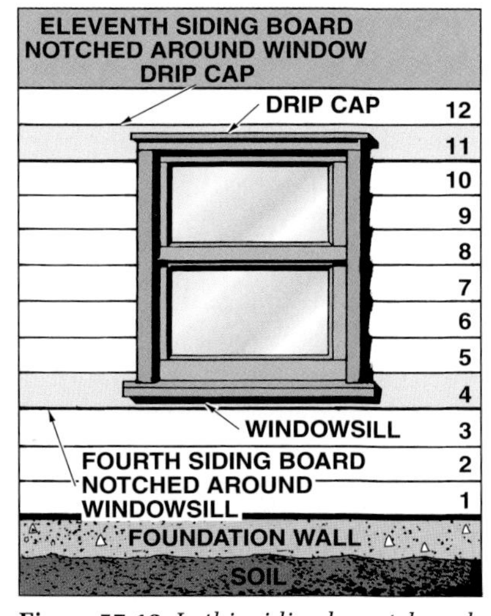

Figure 57-12. *In this siding layout, boards are notched around the windowsill and drip cap.*

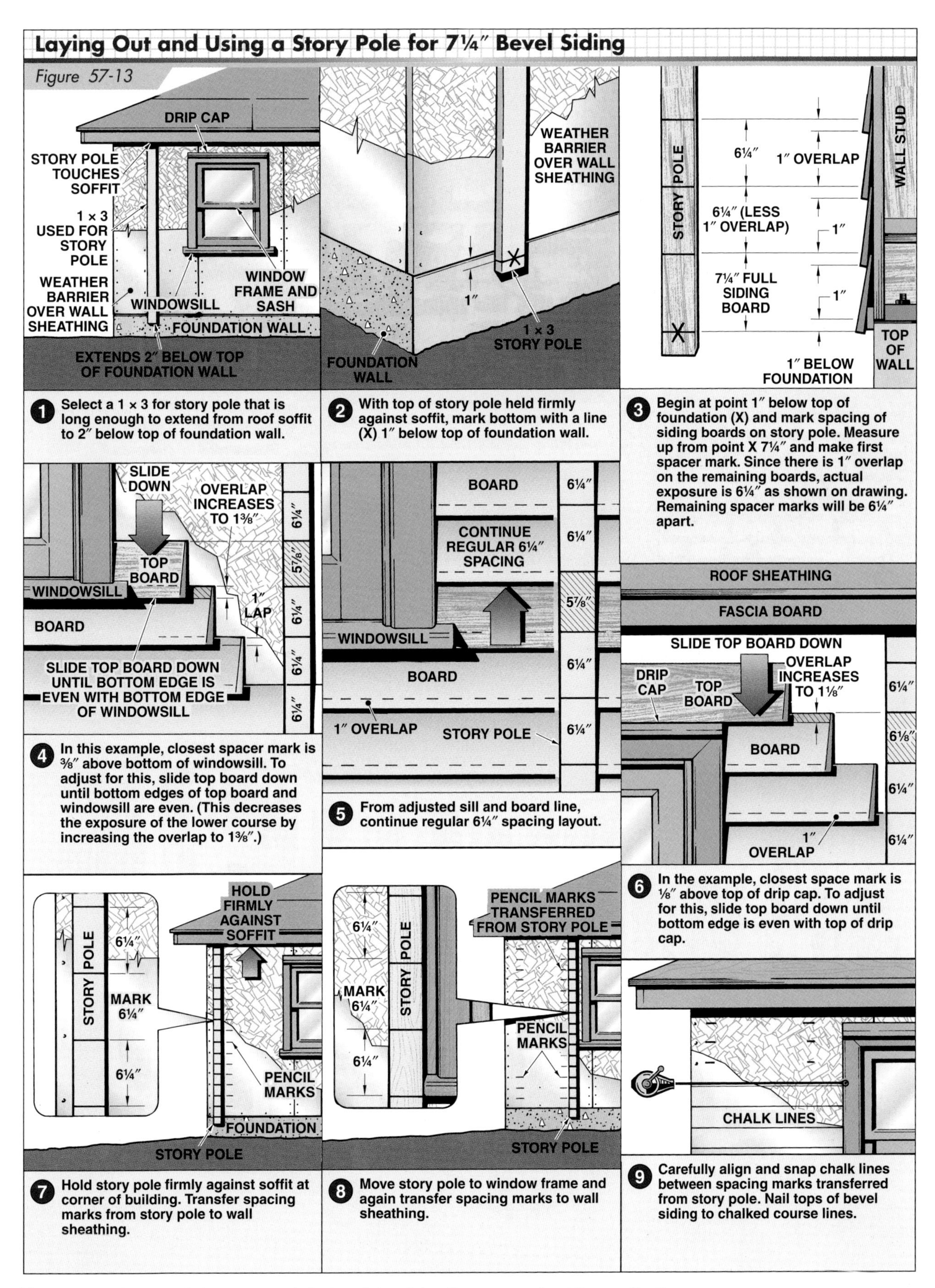

Figure 57-13. *A story pole is recommended when determining the sequence for siding application.*

Vertical Application. The layout procedure is somewhat simpler for vertical siding application than for horizontal application. Position the starting corner board so it is straight and plumb. When vertical siding is placed over walls that are not covered with structural sheathing, blocks for nailing must be placed between the studs 16″ to 24″ apart. **See Figure 57-14.**

Wood is a porous organic material which absorbs and gives off moisture, causing deterioration if the wood is not properly maintained. Horizontal or vertical wood siding must be repainted every five to seven years. Shingle siding provides a rustic appearance, but does not require finishing.

Joints. Butt joints are typically used between the ends of siding boards. The butt joints should be tightly fitted and staggered as shown in **Figure 57-15.** Metal butt joint covers are often used with hardboard siding to allow faster application and provide a weathertight joint. **See Figure 57-16.**

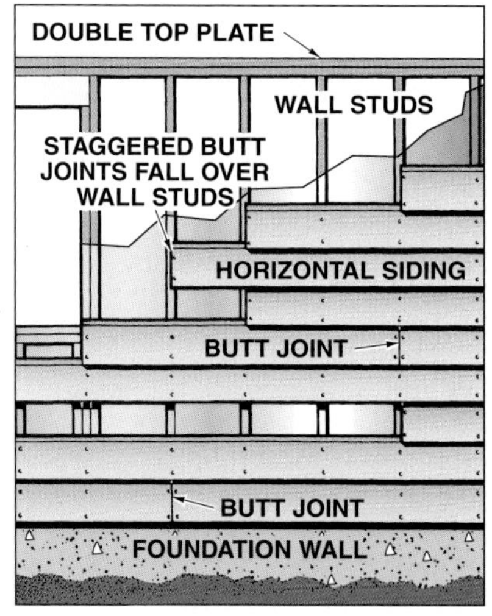

Figure 57-15. *Horizontal siding butt joints should be staggered and fall over a wall stud.*

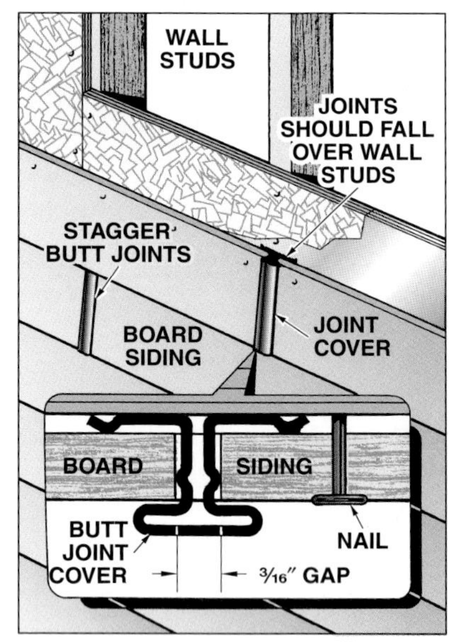

Figure 57-16. *Metal butt joint covers are often used with hardboard siding.*

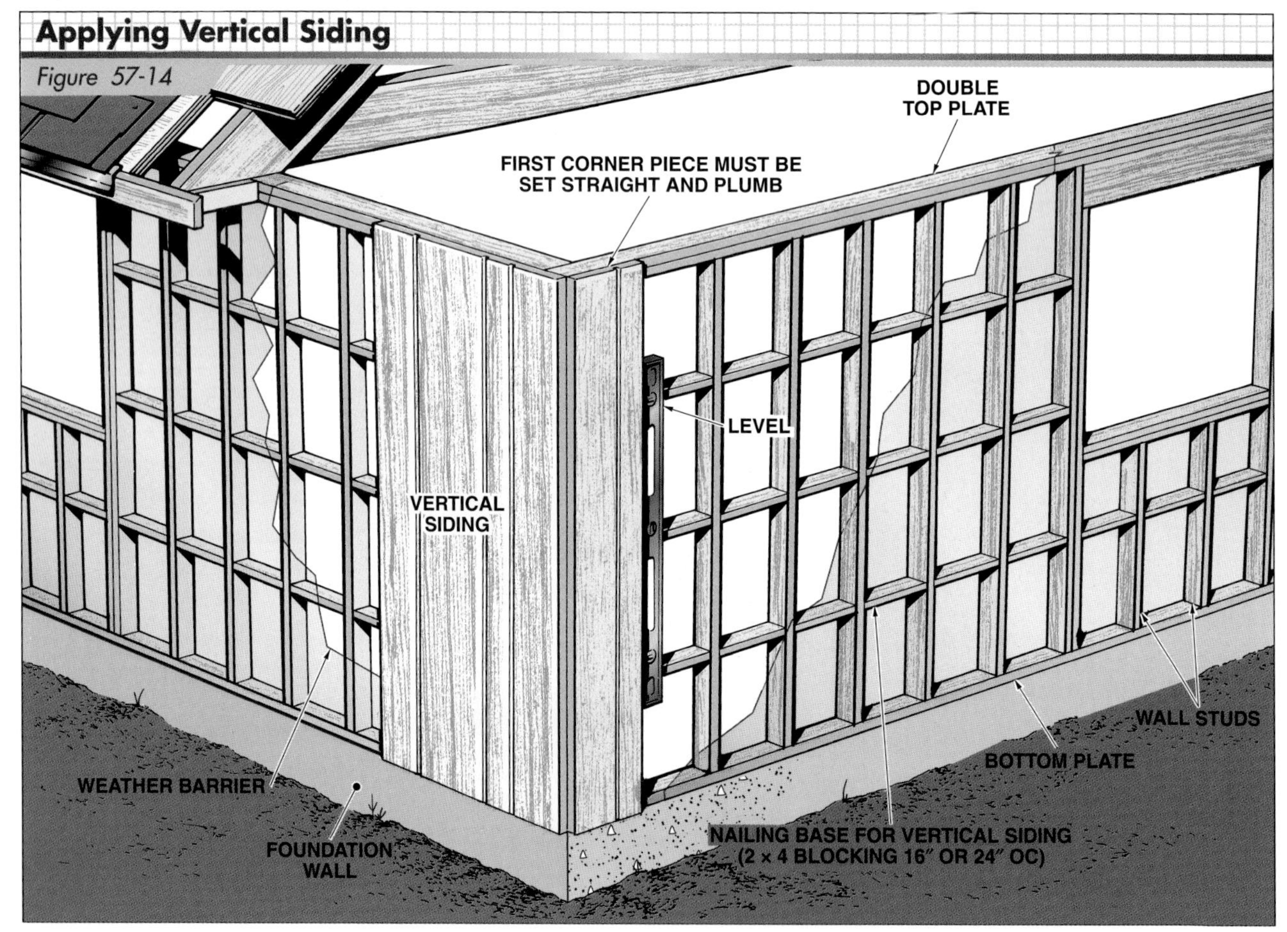

Figure 57-14. *The first corner piece of vertical siding must be set straight and plumb.*

A joint between vertical and horizontal boards occurs on buildings with gable roofs where vertical siding is applied at the gable ends of the roof and horizontal boards are placed on the walls below. When this type of joint occurs, special steps must be taken to prevent water infiltration where the vertical and horizontal siding meet. One method of preventing water infiltration is to offset the upper wall so the vertical siding laps over the top horizontal piece of siding. **See Figure 57-17.** Another method of preventing water infiltration is to install a drip cap. **See Figure 57-18.**

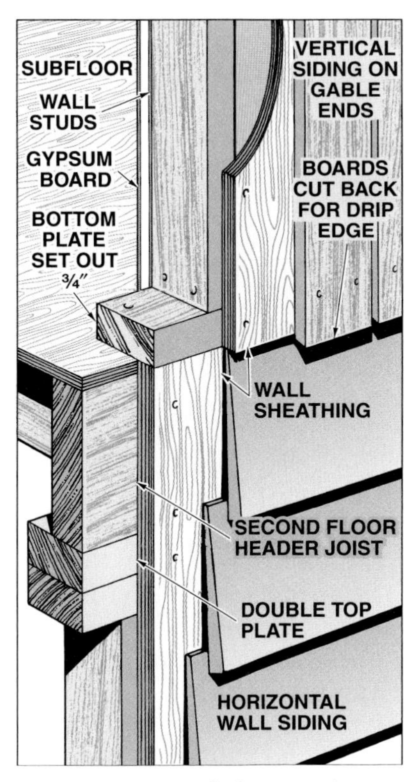

Figure 57-17. *Instead of using a drip cap to prevent water infiltration at the junction between vertical and horizontal siding, the upper wall may be offset so the vertical siding overlaps the top of the horizontal siding.*

Corners. Corners must have a neat appearance and be weathertight. Metal corners provide a mitered effect and are easy to install. **See Figure 57-19.**

Wood outside corner boards and inside corner strips may be used at corners. Outside corners may be mitered, although this is a more difficult procedure since the cut must be extremely accurate to produce a tight joint. **See Figure 57-20.**

Inside and outside corners must be properly flashed prior to installing corner trim pieces.

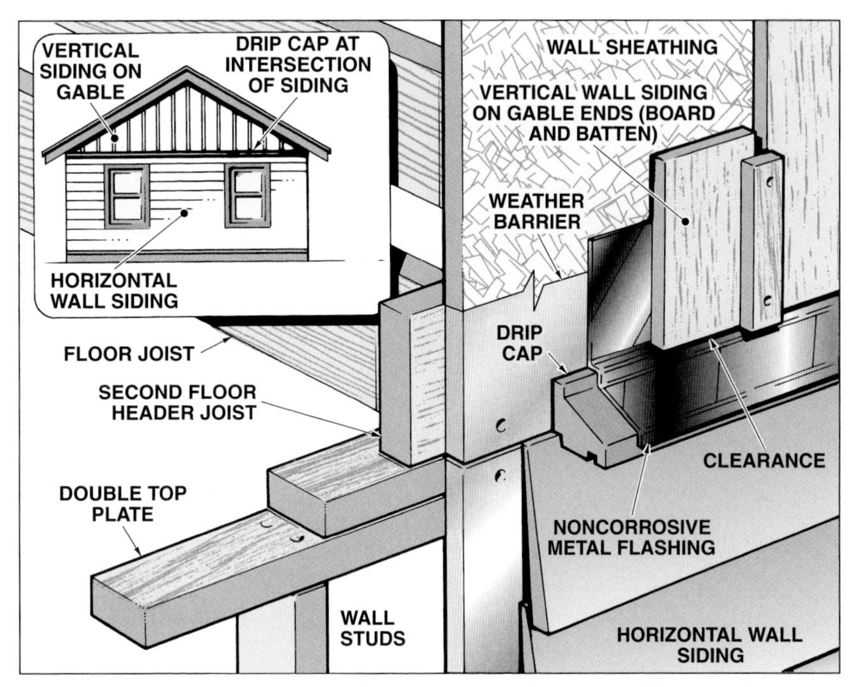

Figure 57-18. *A drip cap is installed at the junction of vertical and horizontal siding to prevent water infiltration.*

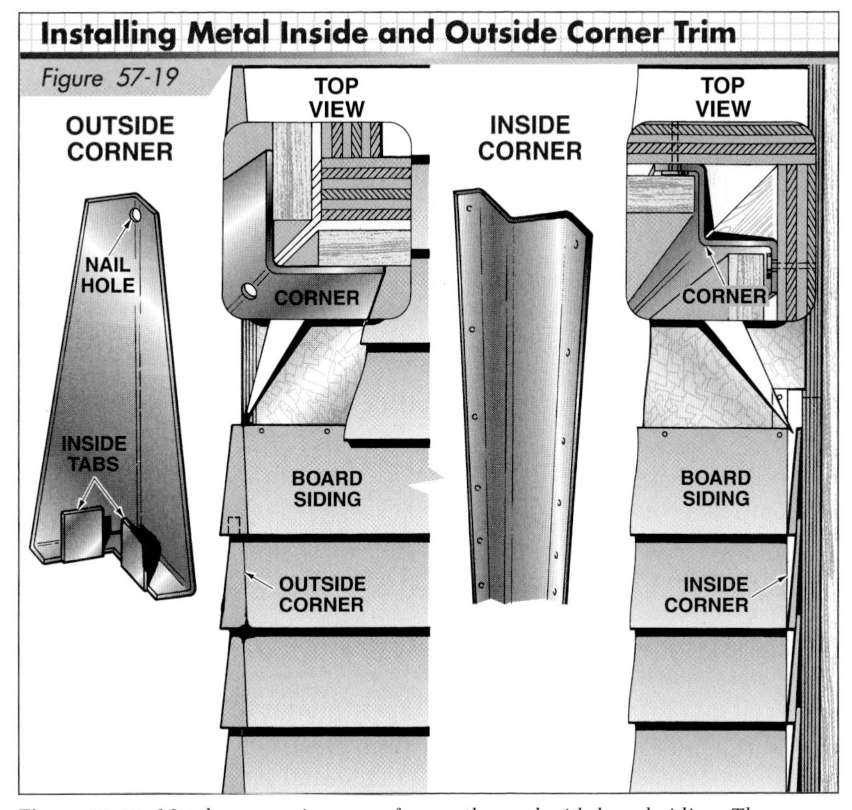

Figure 57-19. *Metal corner pieces are frequently used with board siding. The corner pieces are easy to install and provide a mitered effect to the corners.*

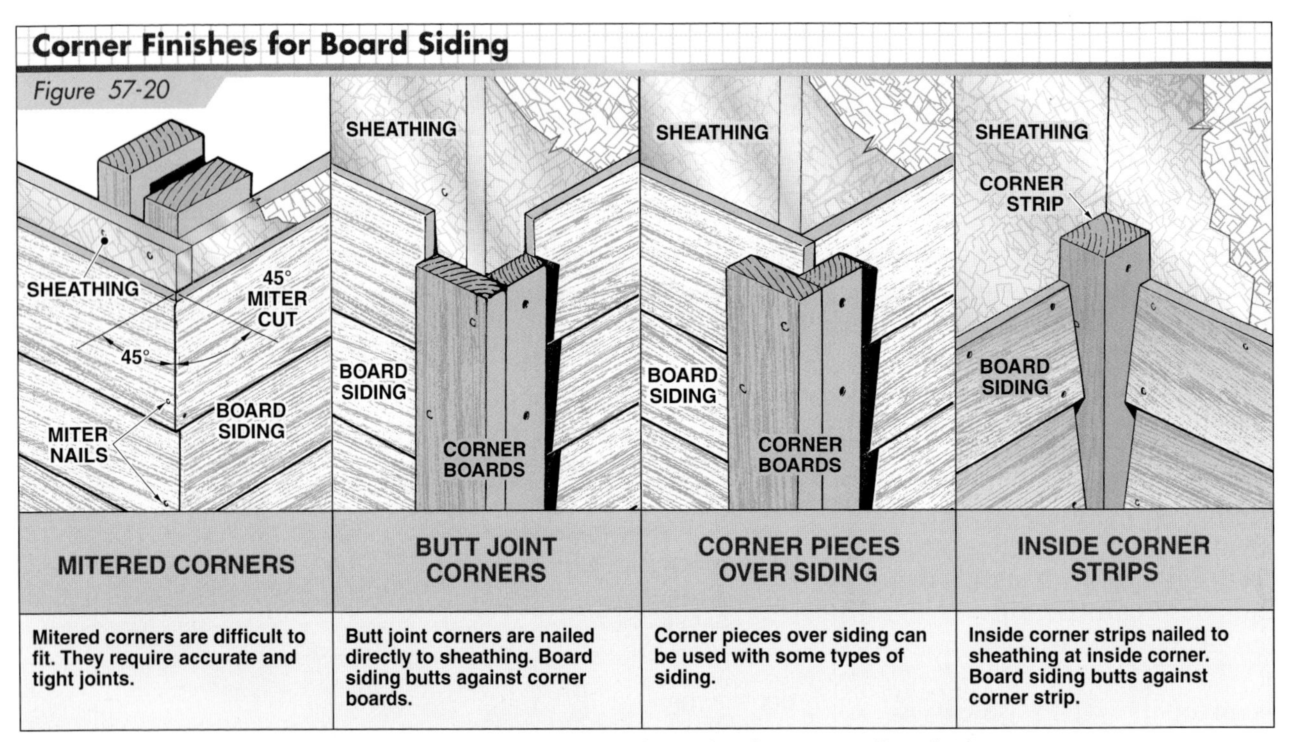

Figure 57-20. *Outside corner boards and inside corner strips may be used at the corners of board siding.*

Cedar is commonly used for board siding because of its resistance to moisture, decay, and insects.

Nailing Methods. Proper nailing is important to the appearance and durability of siding. Noncorrosive nails should be used to avoid unsightly stains or rust streaks. Aluminum alloy, stainless steel, or high-quality hot-dipped galvanized nails are recommended for wood siding. Annular- or spiral-shank nails may be used. **See Figure 57-21.** Although box, casing, and finish nails can be used for siding, *siding nails* are recommended. Siding nails have a slightly tapered head that can be driven flush with the siding or countersunk without crushing surrounding wood. Generally, 6d nails are used for ½″ siding and 8d nails are used for ¾″ siding. Pneumatic nailers and staplers are also used to fasten finish siding. Like nails, corrosion-resistant staples should be used.

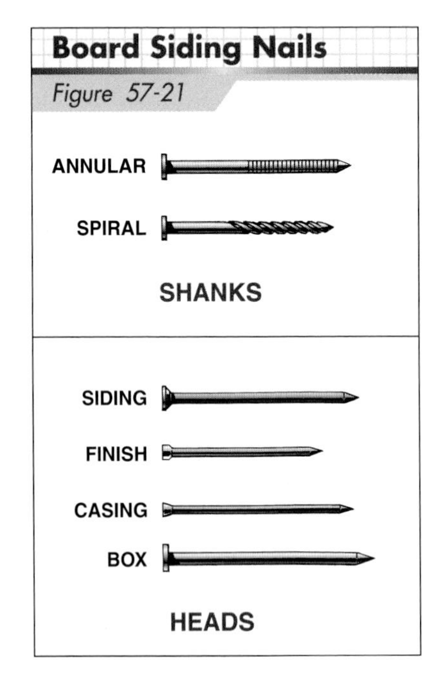

Figure 57-21. *While a variety of nails can be used when installing exterior trim work, siding nails are recommended for board siding. The head is slightly tapered and can be driven flush or countersunk without crushing surrounding wood.*

Although structural sheathing is usually considered an adequate nailing base, better stability is obtained by driving nails through the sheathing and into the wall studs. If the siding is to be painted or a solid-colored stain is to be used, nails should be countersunk and holes are filled with wood putty. If a semitransparent stain is applied, nails should remain flush with the surface.

When installing hardboard siding, drive nails so the backs of the nail heads are in contact with the siding surface. When nailing close to the end of the piece, drive nails into predrilled holes which are approximately three-fourths the nail shank diameter. **Figures 57-22** and **57-23** show nailing methods for commonly used siding patterns.

Use hot-dipped galvanized, aluminum, or stainless steel nails when installing cedar siding. Other types of nails are not recommended since they can corrode and react with the natural preservative oils in cedar, resulting in stains and streaks.

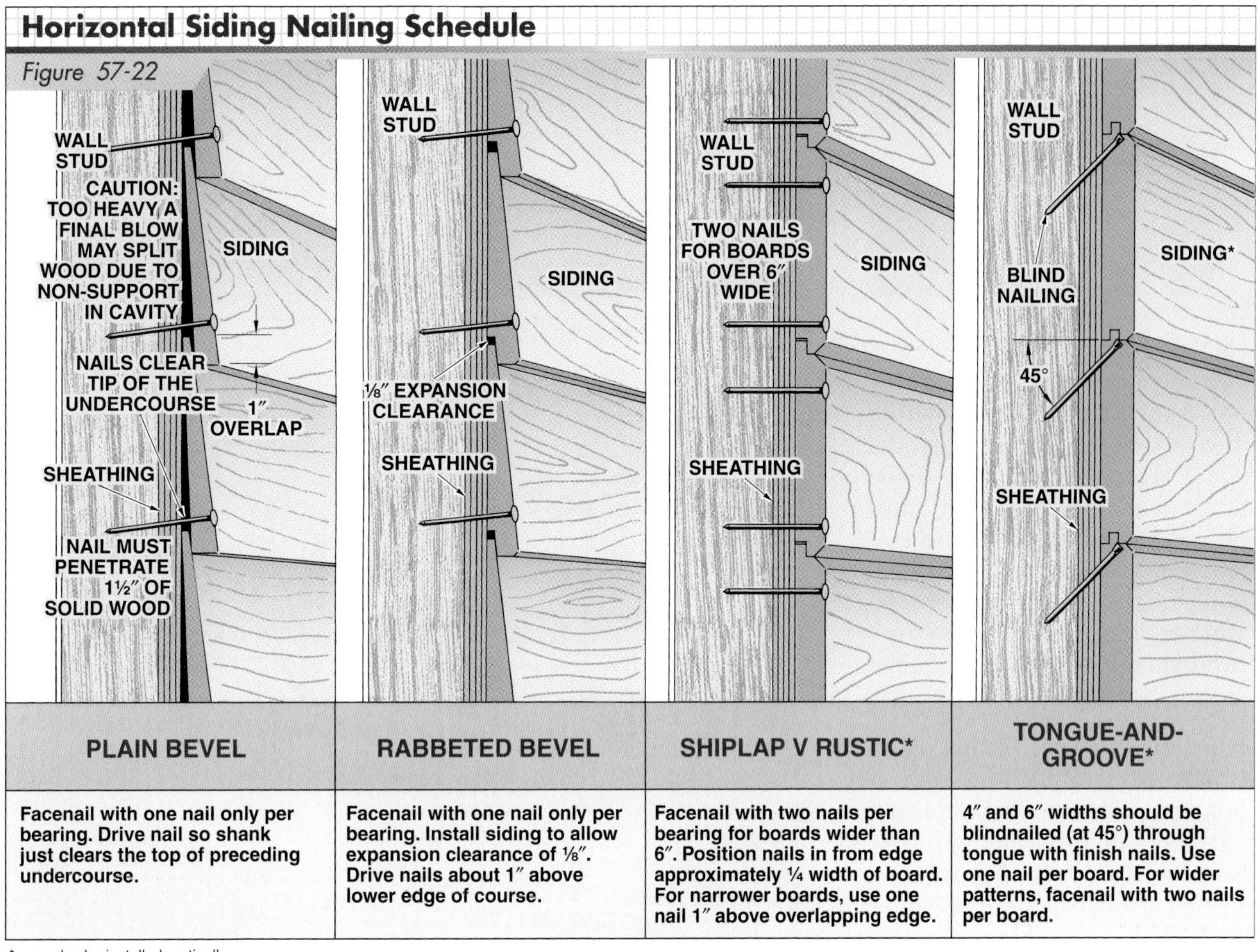

Figure 57-22. *Nailing schedules should be closely followed to provide the best appearance for horizontal siding.*

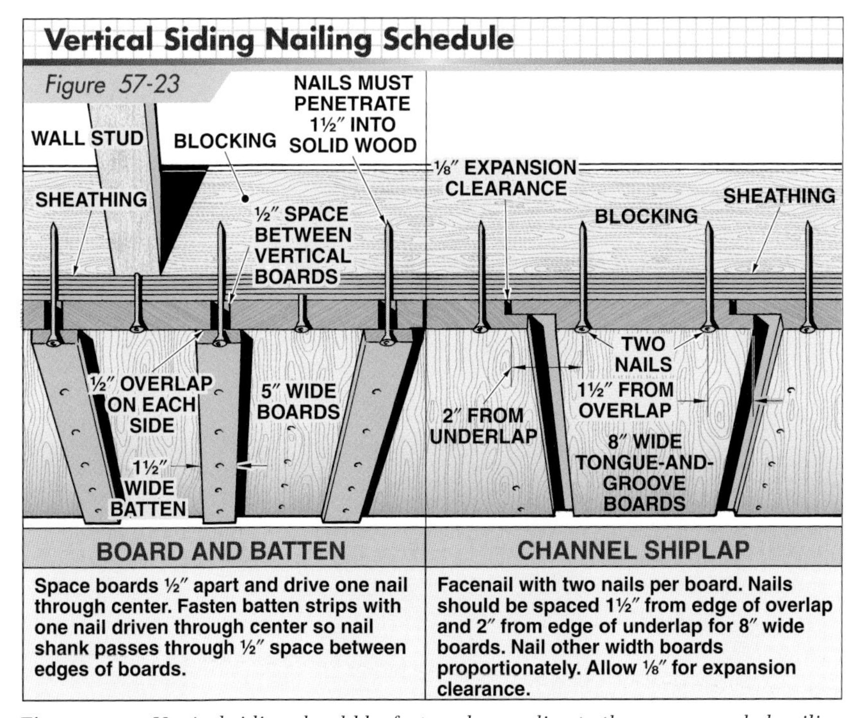

Figure 57-23. *Vertical siding should be fastened according to the recommended nailing schedules.*

Panel Siding

Panel siding can be installed more quickly than board siding. Plywood or hardboard is usually used for panel siding.

Plywood Panel Siding. Plywood panels are available in a variety of textures and patterns. **See Figure 57-24.** Common thicknesses are 11/32″, 3/8″, 15/32″, 1/2″, 19/32″, and 5/8″. Most panels are 4′ in width, and are available in 7′, 8′, 9′, 10′, and 12′ lengths.

Plywood panels may be applied to studs spaced 16″ or 24″ OC or over nonstructural sheathing. The panels are usually applied vertically, but they can be placed horizontally if the horizontal joints are blocked.

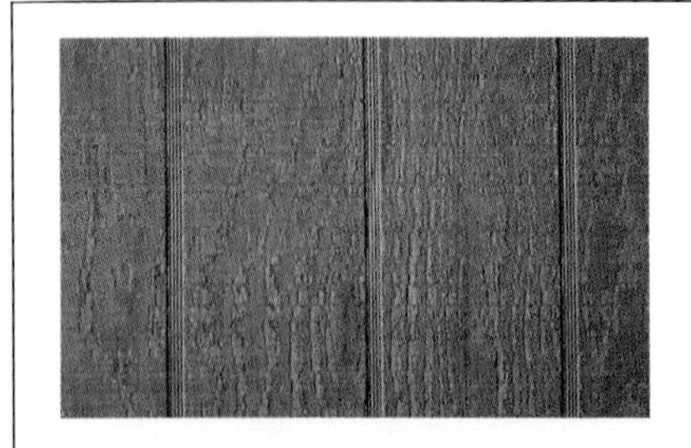

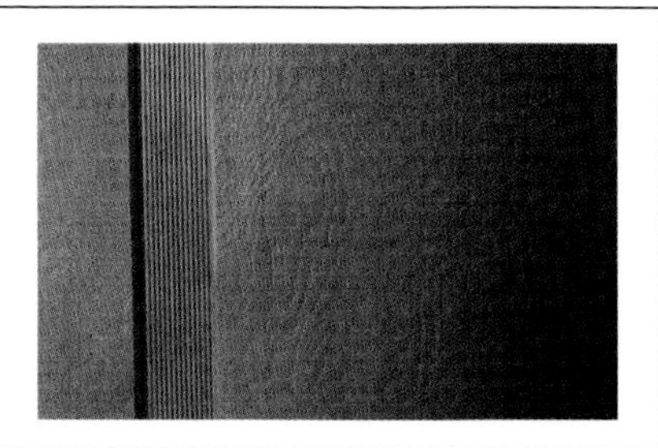

Figure 57-24. *Plywood panel siding is manufactured in a variety of surface textures and patterns.*

Some plywood panel siding has a structural rating, which means that the panels have the same strength and rack resistance characteristics as structural panel sheathing. Therefore, it is not necessary to install diagonal bracing inside the framed wall or another layer of structural sheathing under the panel siding. **See Figure 57-25.**

Hardboard Panel Siding. Similar to plywood panels, hardboard panels are available in a variety of textures and patterns. **See Figure 57-26.** Prefinished panels are also available. Panels are usually 7⁄16″ thick, 4′ wide, and 8′, 9′, or 10′ long.

Hardboard panels do not have the high shear rack resistance of plywood panel siding, and are typically applied over structural sheathing. If hardboard panel siding is applied over nonstructural sheathing, or directly to the studs, diagonal bracing should be installed in the framed wall.

Hardboard panels must be acclimated to the temperature conditions around the building they will be covering. For this reason, they should be stored on the job site at least five days before installation.

Preparing for Panel Siding. Walls are prepared for panel siding in the same manner described for board siding. However, a weather barrier is applied directly against the studs and under the panels. Flashing and drip caps are installed over door and window openings.

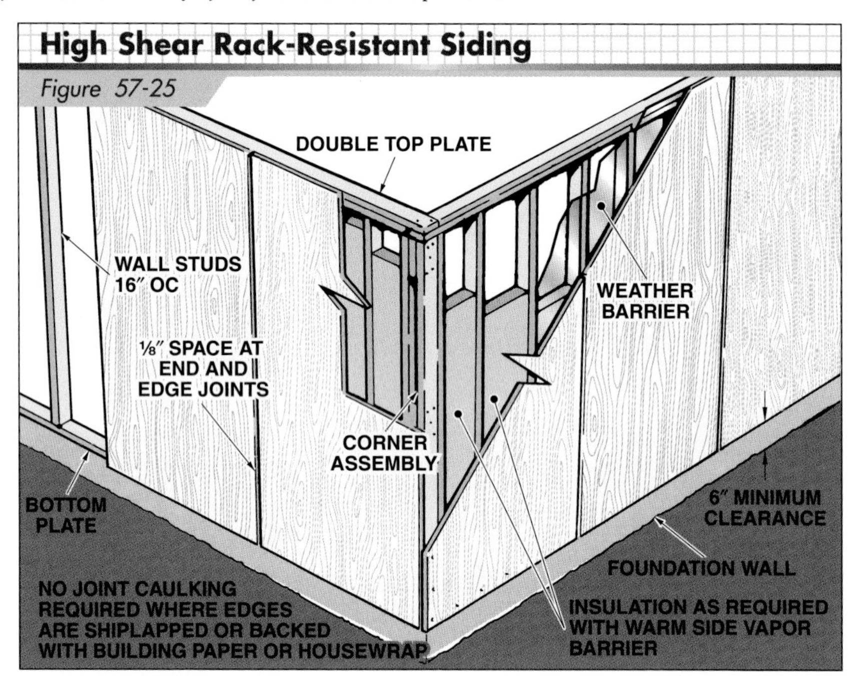

Figure 57-25. *When plywood panel siding with a structural rating is applied to a structure, no structural sheathing is required underneath and diagonal bracing is not necessary in the framed wall.*

James Hardie Building Products

Fiber-cement siding can be embossed to provide a wood grain finish.

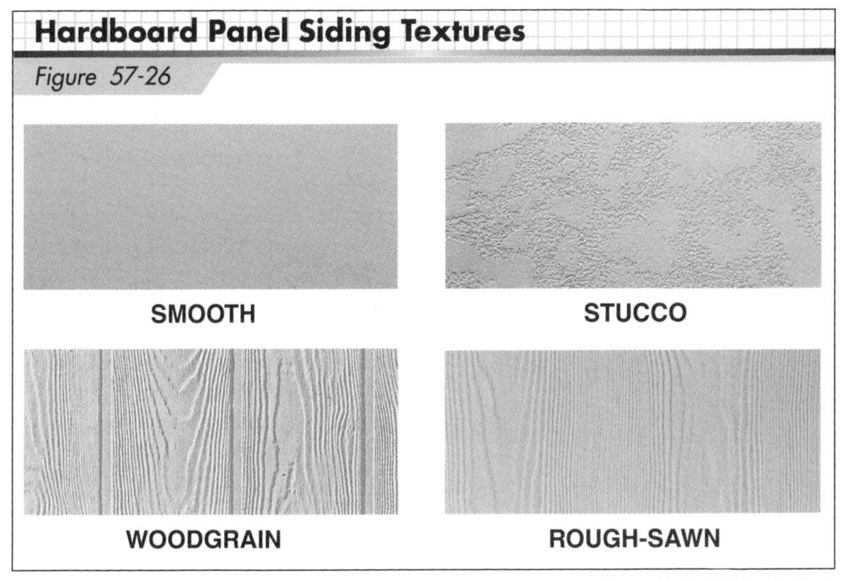

James Hardie Building Products

Figure 57-26. *Hardboard panel siding is available in a variety of textures.*

APA—The Engineered Wood Association

Figure 57-27. *Siding panel edges must be sealed at the time of application.*

Applying Panel Siding. Plywood and hardboard panels are applied the same way. All edges of panels must be sealed at the time of application with either paint primer or a water-repellent preservative. **See Figure 57-27.** Allow a 1/8″ gap along the panel ends and edges to provide for panel expansion from absorption of moisture. (Some manufacturers recommend a 1/16″ gap.)

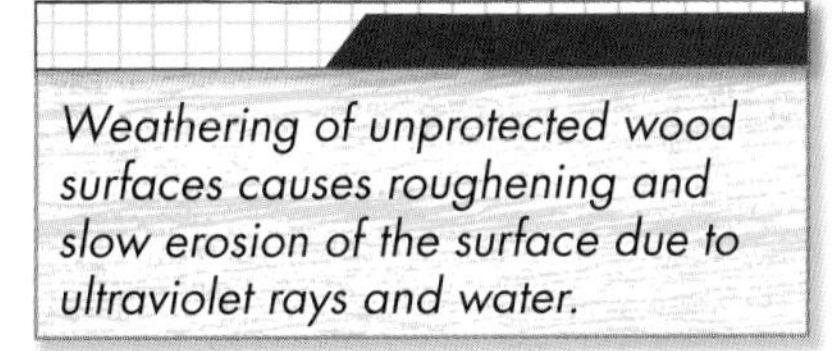

Nail plywood and hardboard panels 6″ OC along the panel edges, 1/2″ from the edge, and 12″ OC at the intermediate studs. Use 6d nonstaining box, casing, or siding nails for panels 1/2″ or less in thickness. Use 8d nails for thicker panels. Panels must be nailed properly for a uniform and flat appearance. A recommended nailing schedule is shown in **Figure 57-28.** A building that has been finished with panel siding is shown in **Figure 57-29.**

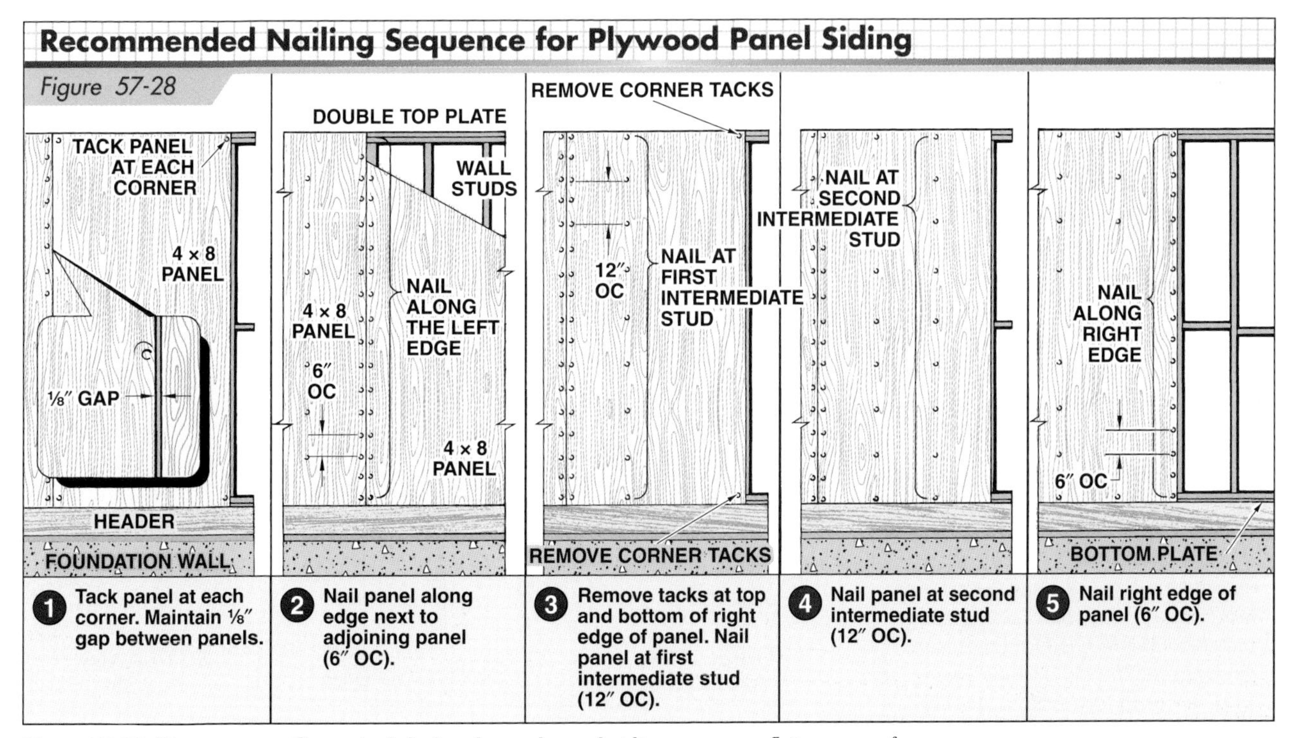

Figure 57-28. *The proper nailing schedule for plywood panel siding ensures a flat, even surface.*

James Hardie Building Products

Figure 57-29. *Panel siding can be combined with other building materials to create an attractive appearance.*

Engineered wood siding may be used as exterior siding for residential and light commercial wood-framed construction. Engineered wood siding is manufactured from wood residue such as planer shavings and sawdust. The residue is processed under heat and pressure, reducing it to fibers. The fibers are then formed into mats and the water is drained. Mats are trimmed and then placed in a press that contains engraving plates to produce surface textures.

Siding panels have square, shiplap, or tongue-and-groove edges along the long sides. Panels are usually placed with their long dimension in a vertical position, with vertical joints falling on studs.

For panels with square edges, a ⅛″ gap should be allowed between panel edges and the gaps should be covered with battens. For shiplap panels, a $\frac{1}{16}$″ gap should be maintained between the edge of the tongue and the groove in the adjacent panel. **See Figure 57-30.** Where horizontal joints occur between the tops of lower panels and the bottoms of panels above, nailing blocks must be installed between the studs. To prevent water leakage, the panels should be lapped or flashing should be applied. **See Figure 57-31.** Corner joints are finished using the same methods described for board siding. The procedure for installing panel siding is shown in **Figure 57-32.**

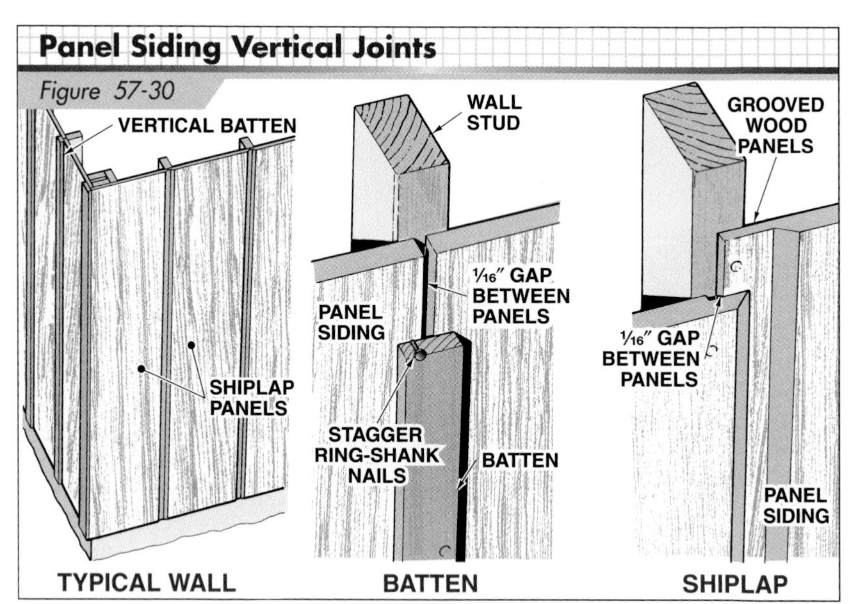

Figure 57-30. *Vertical joints of panel siding should fall over a stud.*

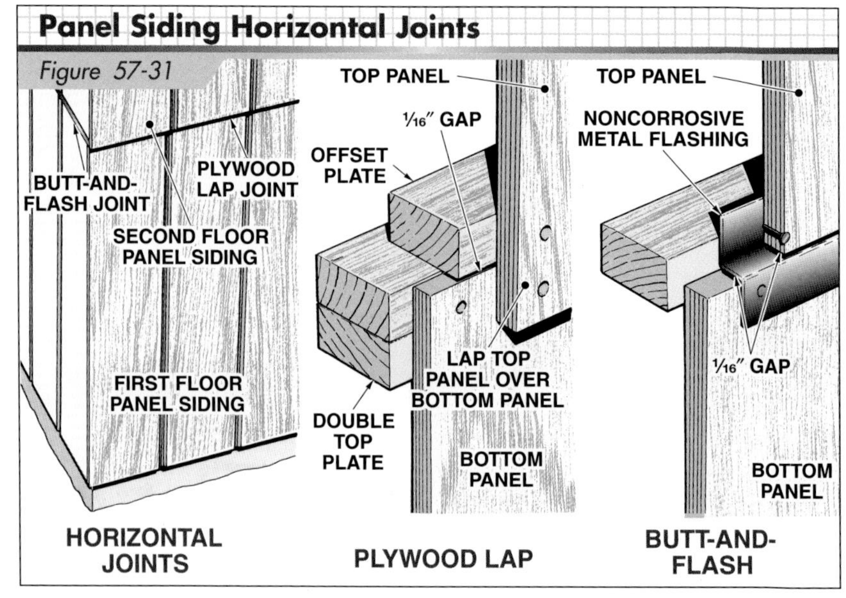

Figure 57-31. *To prevent water leakage, horizontal joints of panel siding should be lapped or flashed.*

Hardboard siding is as strong and durable as most other types of exterior siding. The density and hardness of hardboard siding create a surface that is highly resistant to damage. The uniform thicknesses give hardboard siding superior dimensional stability.

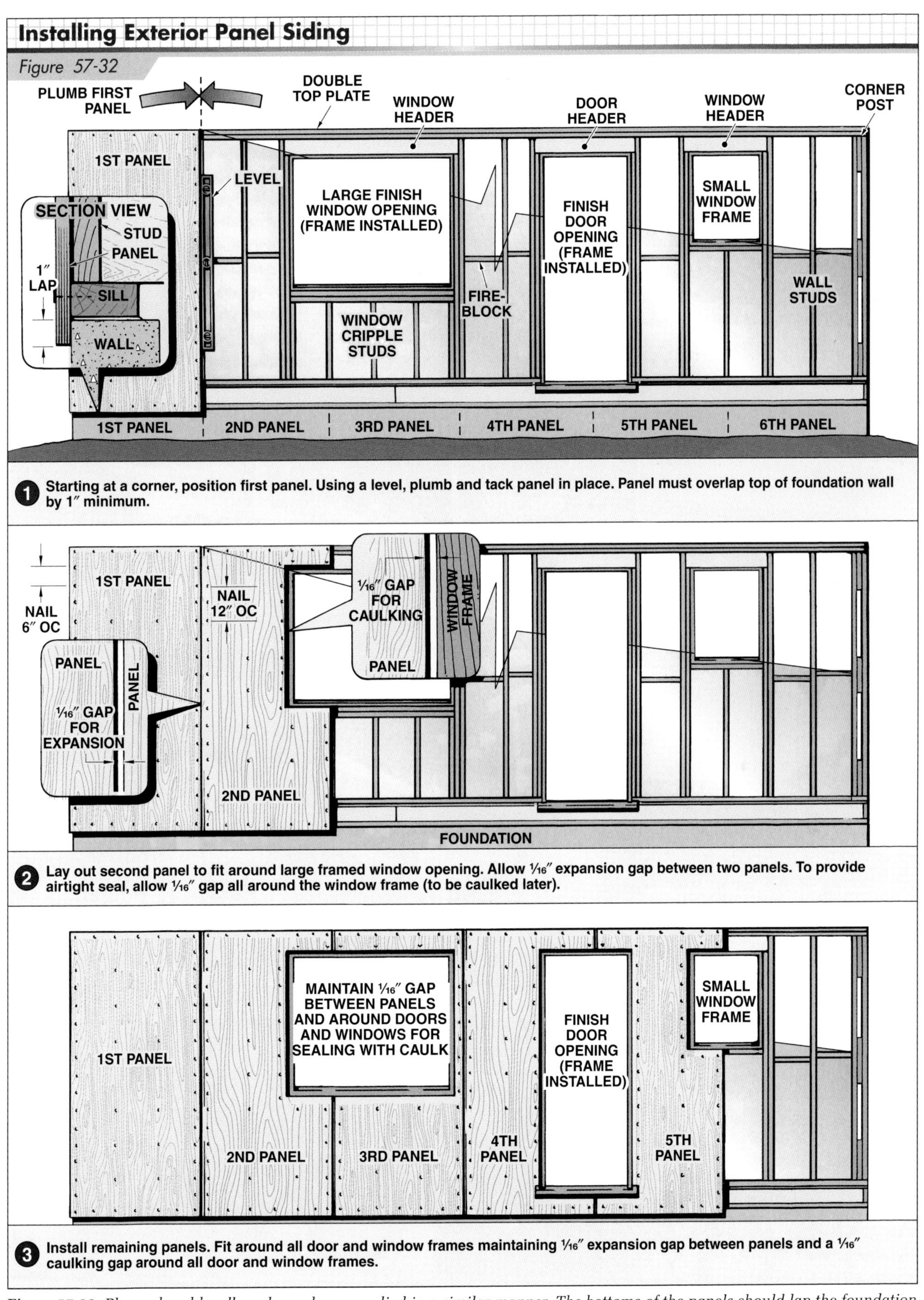

Figure 57-32. *Plywood and hardboard panels are applied in a similar manner. The bottoms of the panels should lap the foundation wall 1″ minimum.*

Wood Shingles or Shakes

Wood shingles or shakes produce an attractive and durable covering for exterior walls. **See Figure 57-33.** The same types of shingles and shakes used for roof covering are used for exterior walls. The traditional method of application is to place one shingle or shake at a time. Shingle or shake panels are also available in which several shingles are glued to a backing, making application more efficient.

Shakertown

Figure 57-33. *Cedar shingles on exterior walls provide an attractive, rustic appearance.*

Applying Individual Shingles or Shakes. A weather barrier is usually placed under the shingles or shakes. Shingles can be nailed directly to structural sheathing. Where nonstructural sheathing is used, horizontal nailing strips are required. Spacing of the nailing strips depends on the shingle or shake length.

Two methods of individual shingle or shake application are *single coursing* and *double coursing.* In single coursing, the wall consists of a single layer of tapered shingles, except for the starting course, which is doubled. **See Figure 57-34.** Shingles or shakes should be applied with about 1/8″ to 1/4″ space between their vertical edges to allow for expansion from water absorption.

In double coursing, two layers of shingles or shakes are applied to the entire wall. **See Figure 57-35.** The first layer (under-course), which is a lower grade shingle or shake, is applied directly to the sheathing or nailing strips. For the second layer, higher grade shingles or shakes are applied with the butt ends extending 1/4″ to 1/2″ below the undercourse.

Recommended exposure for shingles or shakes depends on the shingle or shake length and whether the shingles or shakes are single coursed or double coursed. **See Figure 57-36.** The greater the exposure of the shingle or shake, the more it will tend to curl up.

Shingles or shakes should be applied with noncorrosive nails or staples long enough to penetrate into the sheathing or nailing strips. In single coursing, 3d or 4d galvanized shingle nails are commonly used.

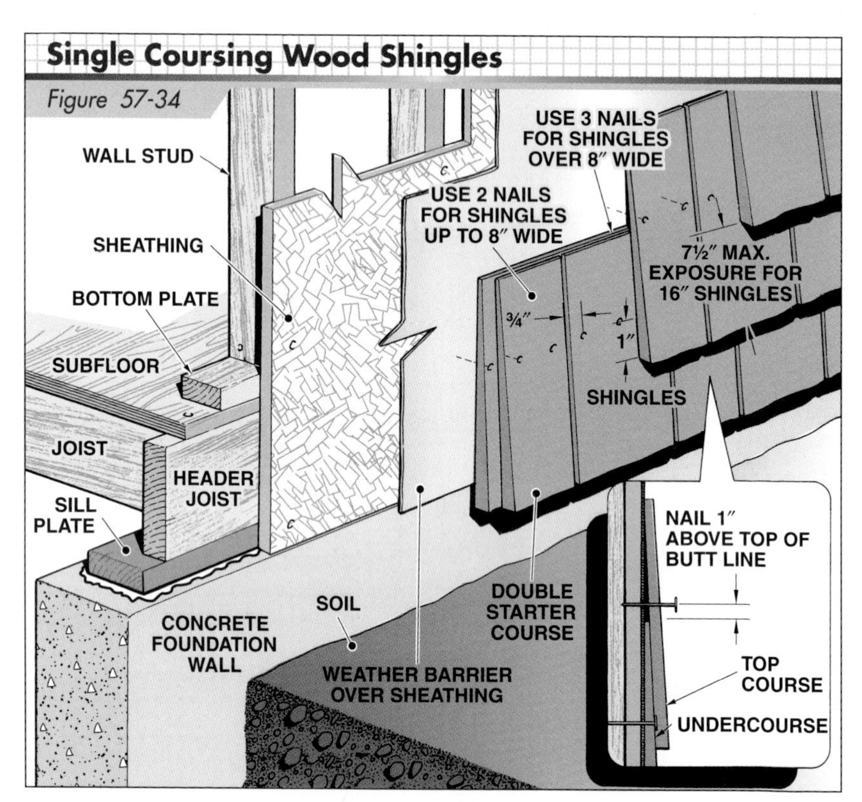

Figure 57-34. *When single coursing shingles, the starting course is doubled.*

Shakertown

Care must be taken when laying out shingle courses to ensure there is minimal waste when trimming around door and window openings.

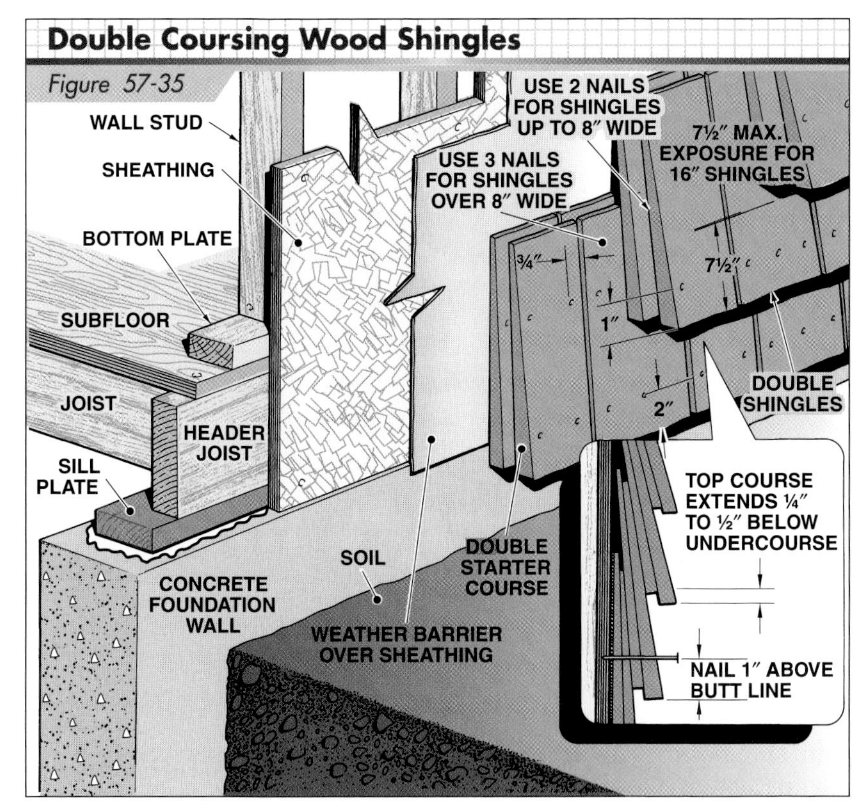

Figure 57-35. *When double coursing shingles, shingles in the first course can be of a lower grade than those in the second course.*

RECOMMENDED SHINGLE AND SHAKE EXPOSURE

MATERIAL	LENGTH*	MAXIMUM EXPOSURE*		
		SINGLE COURSING	DOUBLE COURSING	
			No. 1 Grade	No. 2 Grade
Shingles	16	7½	12	10
	18	8½	14	11
	24	11½	16	14
Shakes (hand split and resawn)	18	8½	14	
	24	11½	20	
	32	15		

* in in.

Figure 57-36. *Recommended exposure for shingles or shakes for walls depends on the shingle or shake length and whether the shingles or shakes are single coursed or double coursed.*

Shakertown

Wood shingle siding and a wood shingle roof covering complement each other.

Nails are placed ¾″ from the edge of the shingle or shake and 1″ above the butt line (thick edge) of the following course to cover the nail by the next course. In double coursing, the heads of the nails driven into the outer layer of shingles are exposed. A 5d nail with a small, flat head is placed 1″ above the butt line of the next higher course.

Shingles at inside corners may be butted to a square strip or mitered. Outside corners may be woven or mitered. **See Figure 57-37.** Boards are also used to finish the corners of shingled or shaked walls. **See Figure 57-38.** Metal corners are also available.

Durability is the primary factor to consider when selecting wood shingles or shakes. The heartwood of western red cedar is rated as very durable because of the amount of natural chemicals in the wood. Sapwood, which is closer to the tree surface, is not as durable since rain and other moisture sources leach natural chemicals from the wood.

Applying Shingle or Shake Panels. Shingle or shake panels consist of several individual shingles or shakes that have been bonded to a wood panel backing. They are available in various designs. Panels with a single row of shingles or shakes are available in 4′ and 8′ widths. **See Figure 57-39.** Panels are also available with two rows of shingles or shakes in an 8′ width.

The procedure for laying out and placing shingle or shake panels is similar to individual shingles. The bottoms of the panels are rabbeted, making them self-aligning. As a result, snapping or stretching lines on the wall for each course is eliminated. Chalk lines are snapped only for the first course.

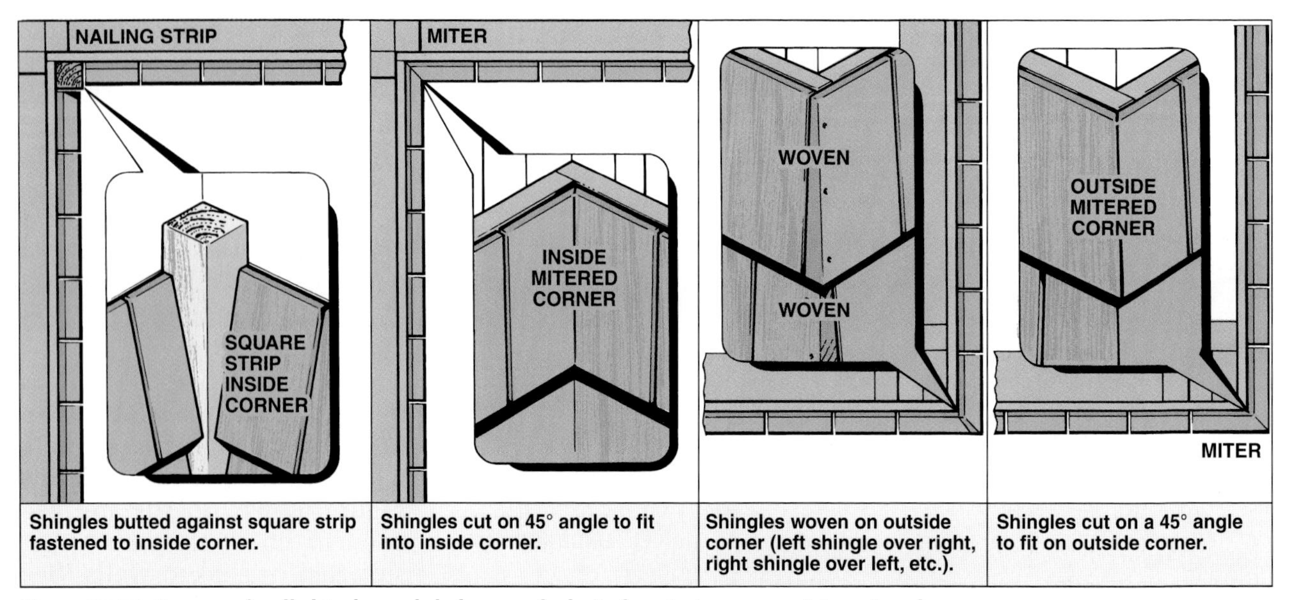

Figure 57-37. *Corners of wall shingles and shakes may be butted against a square strip, mitered, or woven.*

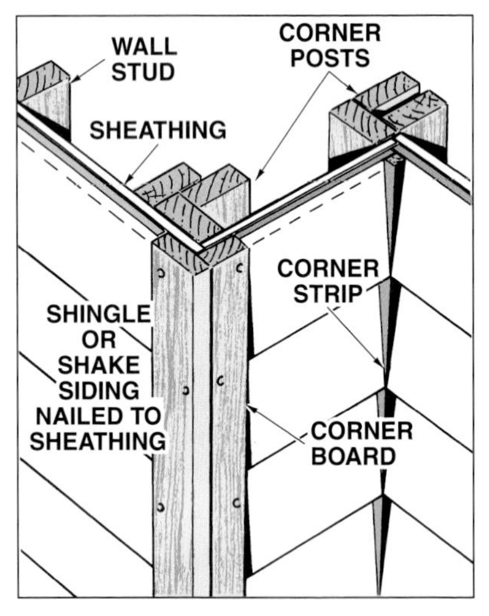

Figure 57-38. *Boards may be used to finish corners of shingled or shaked walls.*

For corners, a special type of single-ply corner trim is used with shingle or shake panels. **See Figure 57-40.**

Water Table

On some framed buildings, a *water table* is placed below the bottom course of the board, panel, or shingle siding. **See Figure 57-41.** A water table may also be used to finish the bottom of a wall that has a stucco finish. The water table diverts water flowing down the siding away from the face of the foundation wall. A water table consists of a board, which is ¾″ or 1¼″ thick and 6″ or 8″ wide, placed beneath a regular drip cap. The *quirk,* a narrow groove beneath the drip cap, prevents water from flowing back to the joint between the board and drip cap.

Shakertown

Figure 57-39. *Single course shingle panels are applied in a manner similar to individual shingles.*

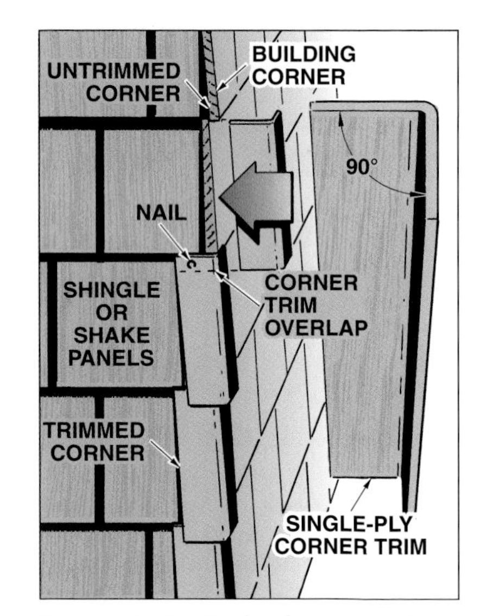

Figure 57-40. *Single-ply corner trim is used for corners when panels of shingles or shakes are applied.*

NONWOOD SIDING

Types of nonwood siding installed by carpenters include aluminum, vinyl, and fiber-cement siding. Aluminum and vinyl siding are lightweight, corrosion-resistant, durable, and require little maintenance.

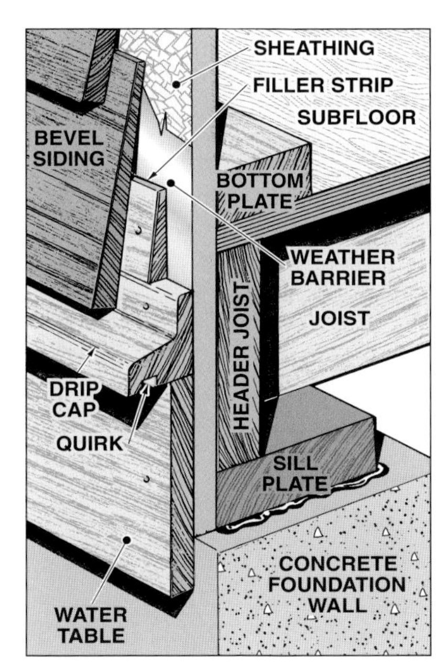

Figure 57-41. *A water table may be placed below the bottom course of siding. A ¾″ or 1¼″ thick by 6″ or 8″ wide board is placed beneath a regular drip cap. The quirk prevents water from flowing back to the joint between the board and drip cap.*

James Hardie Building Products

Figure 57-42. *Fiber-cement siding is manufactured to resemble different styles of wood siding.*

Fiber-Cement Siding

Fiber-cement siding is composed of cement, ground sand, and cellulose fiber that are autoclaved (cured with pressurized steam) to increase its strength and dimensional stability. Fiber-cement siding is durable, insect-, water-, and fire-resistant, and is manufactured to resemble traditional wood siding. **See Figure 57-42.** In addition, fiber-cement siding is environmentally friendly as recycled materials are used in the manufacture of the siding.

Fiber-cement siding is available as planks, panels, and shingles. Planks, which are used for lap siding, are available in 5⁄16″ and 7⁄16″ thicknesses, 5¼″ to 12″ widths, and 10′ to 16′ lengths. Panels are available as 4′ × 8′ panels in 5⁄16″ and 7⁄16″ thicknesses. Shingles are available in planks measuring ¼″ thick by 16″ wide by 48″ long.

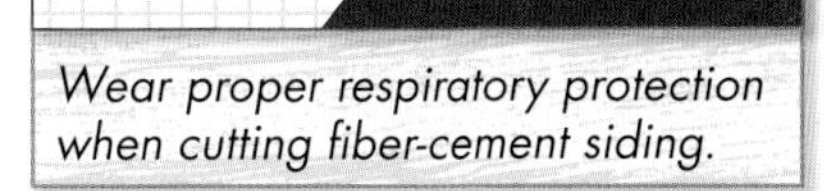

Wear proper respiratory protection when cutting fiber-cement siding.

Specialty tools are required to cut fiber-cement siding due to its content. Hand snips and notchers are used to cut the siding by hand. **See Figure 57-43.** The two lower jaws support the siding and the upper blade shears the siding. Hand snips are primarily used for straight cutting, but can also be used to cut wide arcs. A notcher will cut arcs up to 4″ in diameter to allow for mechanical system components or vents to pass through the siding. Hand tools can be used indoors or outdoors since they do not generate crystalline silica dust.

Power tools, such as shears and saws, can also be used to cut fiber-cement siding. **See Figure 57-44.** Pneumatic power shears are commonly used for productivity applications and can be used indoors or outdoors. Shears are the preferred method of cutting fiber-cement siding since crystalline silica dust is not generated when cutting. Shear attachments are also available that can be mounted to a corded or cordless drill.

A dust-reducing miter saw or circular saw fitted with a special blade can also be used to cut fiber-cement siding. The cutting of fiber-cement siding must only occur outdoors due the crystalline silica dust that is generated. Crystalline silica dust is a known cause of silicosis (discussed in Unit 22). Set up the cutting workstation so that there is good ventilation and where the dust is blown away from other workers. Always wear a properly fitted NIOSH-approved particulate mask or respirator when cutting fiber-cement siding. When possible, the saw should be equipped with a HEPA vacuum system to minimize dust when cutting fiber-cement siding.

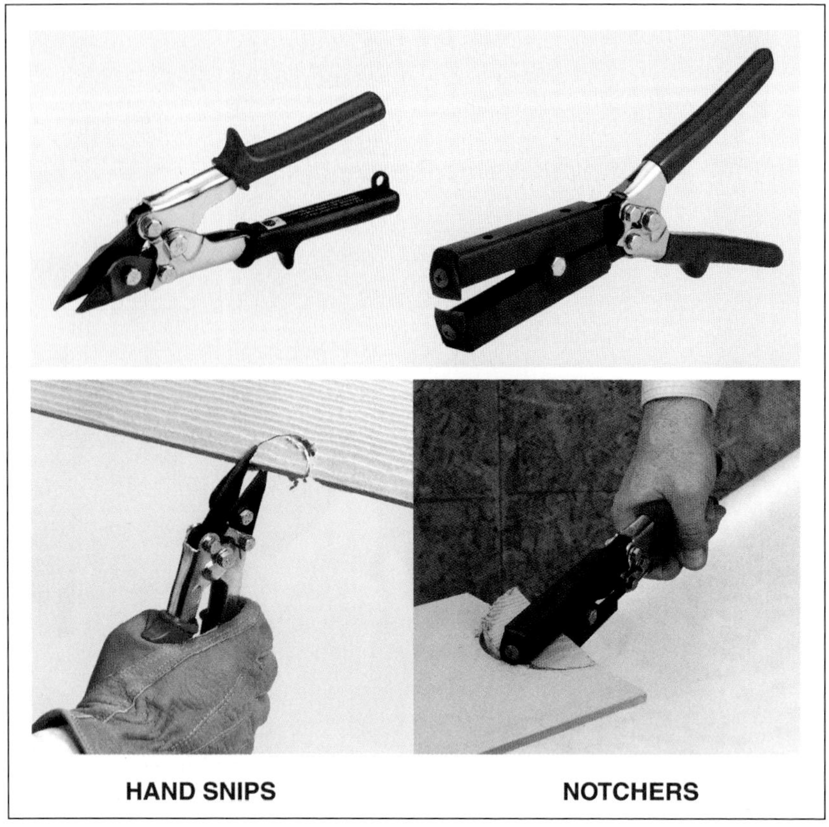

Malco Products, Inc.

Figure 57-43. *Hand snips and notchers can be used to cut fiber-cement siding by hand.*

Malco Products, Inc.

Figure 57-44. *Power shears and saws fitted with special blades can be used to cut fiber-cement siding. Always be sure to wear respiratory protection when cutting fiber-cement siding with a saw.*

Installing Lap Siding. Exterior walls to be covered with fiber-cement lap siding should be covered with sheathing and a weather barrier. If permitted by the local building code, fiber-cement siding can also be installed directly to the studs with a weather barrier in between.

The procedure for installing fiber-cement siding is very similar to the procedure for installing wood board siding (previously discussed). However, always refer to the manufacturer instructions before installing lap siding. Install a 1/4″ thick piece of fiber-cement siding at the base of the wall to ensure a consistent plank angle for the first course. **See Figure 57-45.** Fiber-cement siding is secured to the building using galvanized or stainless steel nails or corrosion-resistant screws. Fasteners should be driven perpendicular to siding and framing with the fastener heads snug against the face of the siding. Do not overdrive the nails. If using a pneumatic nailer, you should drive a nail through a scrap piece of siding to check the driving depth and adjust the nailer accordingly.

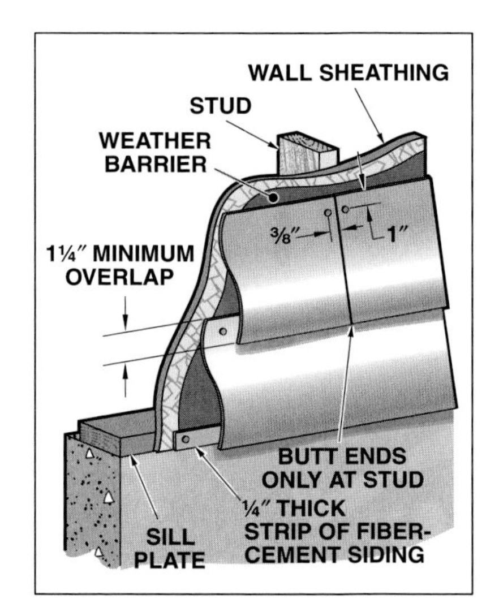

Figure 57-45. *When installing fiber-cement lap siding, a 1/4″ thick piece of fiber-cement siding is fastened along the bottom edge of the wall sheathing. Minimum overlap of lap siding is 1 1/4″.*

Snap a line horizontally along the housewrap or building paper for alignment of the first course of siding. Additional lines for the subsequent courses may also be snapped at this time. Start along the left edge of a wall and align the top edge of the siding to the chalk line. Fasteners are driven into the upper edge of the panel at 12″ to 16″ OC, depending on the panel width and manufacturer recommendations. Allow a small gap (1/8″ maximum) between adjacent panels to allow for expansion. Align the next course to the chalk line and fasten to the building sheathing in a similar manner. A *facing gauge,* also known as an overlap gauge, may be used to help maintain an accurate exposure across the entire plank, **Figure 57-46.** Gaps between the trim pieces and planks, and between adjacent pieces of siding should be caulked.

Handle fiber-cement siding with care to avoid chipping.

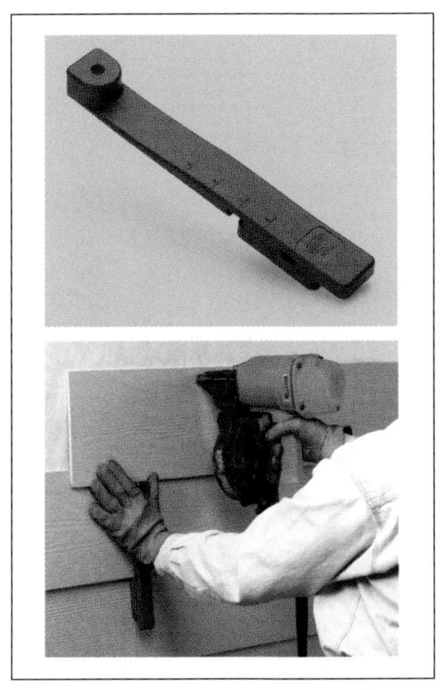

Malco Products, Inc.

Figure 57-46. *A facing gauge is used to help maintain an accurate exposure across the entire plank.*

Installing Panel Siding. Exterior walls to be covered with fiber-cement panel siding should be covered with sheathing and a weather barrier. Installation of fiber-cement panel siding is similar to installing hardboard or wood panel siding (previously discussed). Framing must be provided at vertical and horizontal edges for nailing; vertical panel siding must be joined at studs. Fasteners should be located at least 3/8″ from the panel edges and at least 2″ from the corners.

Vertical joints can remain exposed or a batten joint or H-joint can be used. **See Figure 57-47.** When the joints are to remain exposed, a 1/8″ gap should be left along the edges for caulking. For a batten joint, the vertical joint should be caulked and then the batten is fastened over the joint. An H-joint consists of a metal or PVC H-clip into which the edges of the siding are placed. Z-flashing is required at all horizontal joints to prevent water from infiltrating the building envelope.

Installing Shingle Siding. Fiber-cement shingles are available as individual units or as planks. Individual shingles are available in 18″ lengths and in 6″, 8″, and 12″ widths. The various widths should be used together randomly. Joints between shingles in an upper course should be at least 1½″ away from joints in courses immediately above and below it. Exterior walls to be covered with fiber-cement shingle siding should be covered with sheathing and a weather barrier. Installation of fiber-cement shingle siding is similar to installing wood shingles (previously discussed). Wall sheathing should be a minimum of 7/8″ thick plywood or OSB. The shingles are blind-nailed approximately 1″ from the upper edge. The lower edge of the subsequent course of shingles must overlap at least 1½″. Fiber-cement planks are also manufactured to resemble individual shingles, and install with less effort than individual shingles. **See Figure 57-48.**

Aluminum Siding

Aluminum siding is available with a smooth finish or with an embossed wood grain finish. Whether aluminum siding is smooth or embossed, the siding is prefinished when delivered to the job site.

Aluminum siding is produced with or without an insulating backing board. Different styles are available for horizontal and vertical application. Aluminum panels 12″ × 48″ are also available that provide the appearance of cedar shakes.

Aluminum siding can be applied over sheathing or directly to the stud framing. A weather barrier should be installed beneath the siding. Aluminum siding is popular for residential re-siding projects, in which the siding is attached to furring strips nailed to the existing wall.

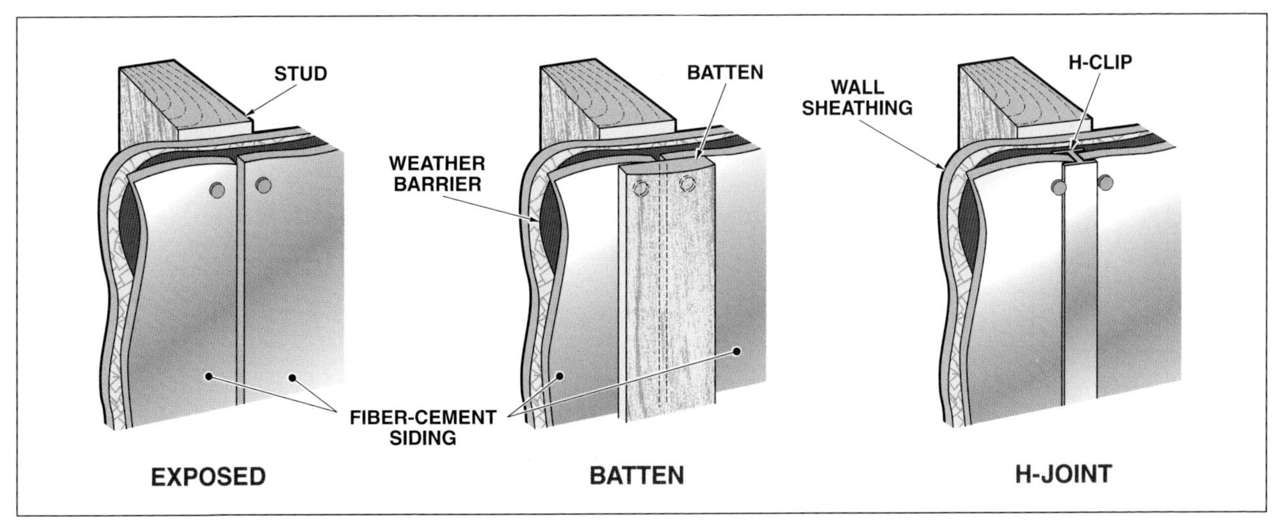

Figure 57-47. *Vertical joints in the siding can be exposed or finished with a batten or H-clip.*

James Hardie Building Products

Figure 57-48. *Fiber-cement siding planks can be manufactured to resemble wood shingles.*

Aluminum Association, Inc.

Figure 57-49. *Aluminum nails are driven every 16″ to 24″ through nailing slots along the nailing flanges of aluminum siding.*

A variety of accessory items and trim are available, such as starter strips, corner pieces, flashing, stiffeners, and trim pieces.

Some building codes require aluminum siding to be grounded against the possibility of electrical shock. Grounding is accomplished by connecting a No. 8 copper wire to a cold water pipe or steel rod embedded in the ground.

Manufacturer instructions for applying aluminum siding should be carefully followed. A circular saw, tin snips, or a utility knife can be used to cut aluminum siding panels. A circular saw with a 10-point aluminum-cutting blade is the quickest way to make precise cuts. A worktable should be set up with a jig that keeps the saw base clear of the siding to avoid scratching its surface.

If siding is cut with tin snips, a duckbill tin snips is recommended for straight cuts. If a utility knife is used, the panel should be heavily scored with the knife, and then bent back and forth until it snaps cleanly along the scored line.

Aluminum siding has nailing slots along the flanges. Aluminum nails are driven every 16″ to 24″ through the slots. **See Figure 57-49.** For straight-shank nails driven into wood studs, the nails should penetrate at least 3/4″. Screw-shank nails should be used to fasten siding to plywood or OSB wall sheathing. Nails should not be driven in tightly. The siding must be able to expand or contract with temperature changes.

A starter strip is installed along the bottom edge of the first course of siding. The first course is securely locked into the starter strip. Joints between siding lengths should be staggered.

Corner caps are used to finish the outside corners of the walls. **See Figure 57-50.** The caps are fitted and installed as each course of siding is placed. The bottom flanges of the corner cap are slipped under the butt of each panel. When the butts of the corner cap and panel are flush, a nail is driven through a prepunched hole at the top of the corner cap.

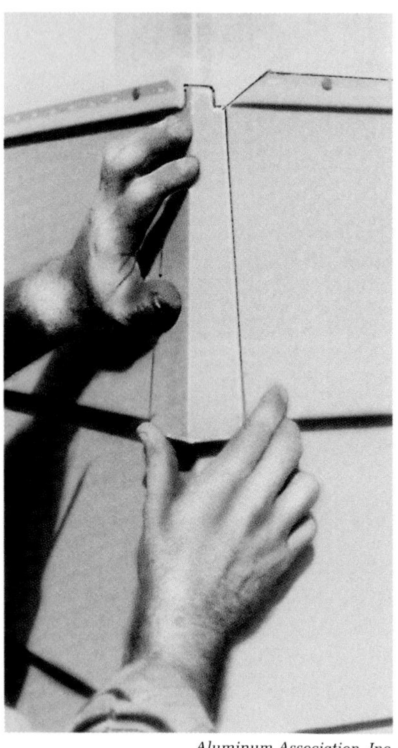

Aluminum Association, Inc.

Figure 57-50. *Corner caps for aluminum siding must be fitted and installed as each course of siding is placed.*

In 1947, Jerome Kaufman invented the first residential baked enamel aluminum siding and formed the Alside Company.

Vinyl Siding

Vinyl siding is a durable, dent- and abrasion-resistant, prefinished plastic product. Many styles of vinyl siding are designed to appear like board siding. Many colors and textures are available, including imitation wood grain patterns. **See Figure 57-51.**

Similar to aluminum siding, vinyl siding is subject to expansion and contraction due to significant temperature changes. Vinyl siding can expand and contract ½″ or more over a 12′-6″ length with temperature changes. Therefore, vinyl siding must be allowed to move freely from side to side. When driving nails or screws into the nailing flanges, allow 1/32″ clearance between the fastener head and siding panel. Drive nails straight and level to prevent panels from buckling and distorting. Fasten nails or screws in the center of the nailing slot to accommodate panel movement. **See Figure 57-52.** Panels should move freely horizontally after nailing.

Aluminum, screws, staples, or galvanized or other corrosion-resistant nails can be used to hang vinyl siding. Nail heads should be 5/16″ diameter minimum and the shank should be ⅛″ diameter. Use 1½″nails for general siding application, 2″ nails for re-siding, 2½″ nails for siding with a backing board, and 1″ to 1½″ nails for trim. When applying vinyl siding, fasteners should be installed 16″ OC for horizontal applications, 12″ OC for vertical applications, and every 8″ to 10″ for accessories and trim.

Screws can be used to hang vinyl siding if they do not restrict siding movement. Corrosion-resistant, No. 8 truss head or pan head, self-tapping sheet metal screws should be used.

If staples are used for hanging siding, the staples must be wide enough to straddle the flange between the nailing slot and edge of siding. Staples must be a minimum of 1″ long.

Alcoa Building Products, Inc.

Figure 57-51. *Vinyl siding is available in a variety of colors and patterns.*

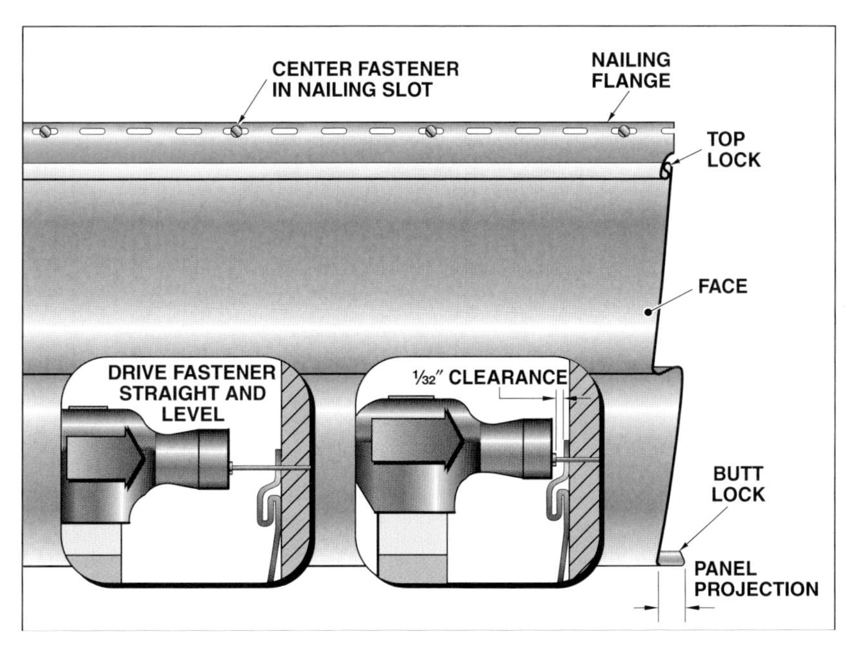

Figure 57-52. *Vinyl siding must be properly installed to allow the siding to move freely from side to side when dimensional changes occur.*

Vinyl siding is cut using common hand and power tools, such as aviator snips, utility knives, circular saws, and radial arm saws. When using aviator snips, avoid closing the blades completely at the end of each stroke to avoid tearing the vinyl. When using a utility knife to cut vinyl siding, score the face of the panel and then snap it into two pieces. Fine-toothed blades should be used in circular saws and radial arm saws. When using a circular saw, mount the blade in the reverse direction and cut slowly. When cutting vinyl siding in cold weather, aviator snips should be used as power saws will chip and crack the siding.

In addition to common hand and power tools, a few specialty tools may be useful when working with vinyl siding. A *snap lock punch* is used to punch lugs in the cut edges of siding to be used for the top or finish course, or underneath a window. **See Figure 57-53**. The lugs allow the siding panel to lock into the upper trim member. A *nail hole slot punch* is used to elongate a nailing slot for expansion and contraction. An *unlocking tool* is used to remove or replace a piece of siding. The curved end is inserted under the panel and hooked into the butt lock. The handle of the unlocking tool is pressed down while the tool is slid along the panel to unlock it. A similar procedure is used to lock a panel back into place.

Before the siding panels are applied, starter strips, corner posts, window and door flashing, trim, and J-channels must be installed. A chalk line is snapped for the starter strip and the starter strip is nailed to the sheathing at 8″ to 10″ intervals. **See Figure 57-54.** A water level can be used to transfer the points to all sides of the building. Allow ¼″ at the ends of starter strips for expansion and provide clearance around the corner posts, J-channels, and trim. For standard corners, install the starter strip approximately 4″ from the corner. For wide corners, allow approximately 6½″ between the end of the starter strip and corner.

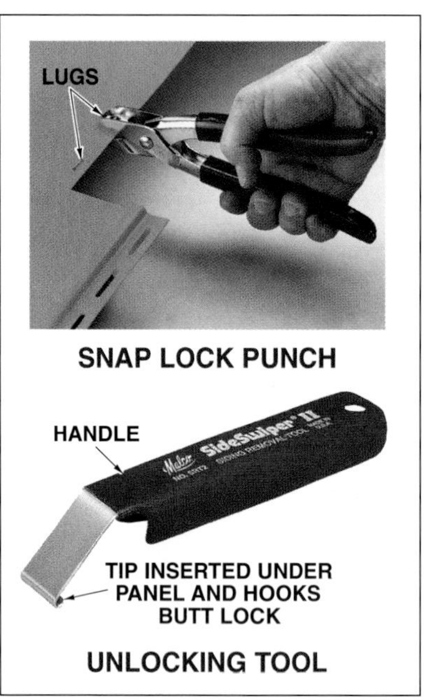

Malco Products, Inc.

Figure 57-53. *A snap lock punch is used to punch lugs in the cut edges of vinyl siding. The lugs allow the siding to lock into trim members.*

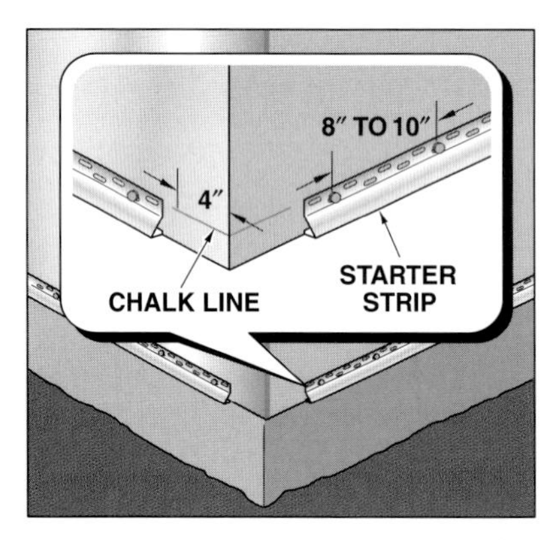

Figure 57-54. *A starter strip is applied along the bottom of the area to be sided, allowing room for the corner posts and other trim member. Approximately ¼″ should be allowed between adjacent starter strips.*

Corner posts receive and conceal the ends of the siding panels. Before installing corner posts, flashing should be applied on inside and outside corners a minimum of 10″ on each side. Remove the bottom ¾″ of the corner post nailing flange so it will not be seen below the installed siding. Place the outside corner post into position, allowing a ¼″ gap between the eaves or soffit and post. The corner post should extend ¾″ below the starter strip. Ensure the outside corner posts are plumb before nailing. The top nail should be positioned at the top of the topmost nailing slot. **See Figure 57-55.** Remaining nails should be centered in the slots at 8″ to 12″ OC. Plugs can be used to seal the ends of the corner posts.

Inside corners can be formed with an inside corner post or with two J-channels. **See Figure 57-56.** Ensure flashing is applied a minimum of 10″ on each side of the corner. Remove the bottom ¾″ of the inside corner post nailing flange so it will not be seen below the installed siding. Place the inside corner post into position, allowing a ¼″ gap between the eaves or soffit and post. Similar to hanging outside corner posts, plumb the inside corner posts and drive a nail at the top of the topmost nailing slot. Center all other nails in their slots. Two J-channels can also be used as an inside corner post. Ensure the corner is properly flashed prior to installing the J-channels.

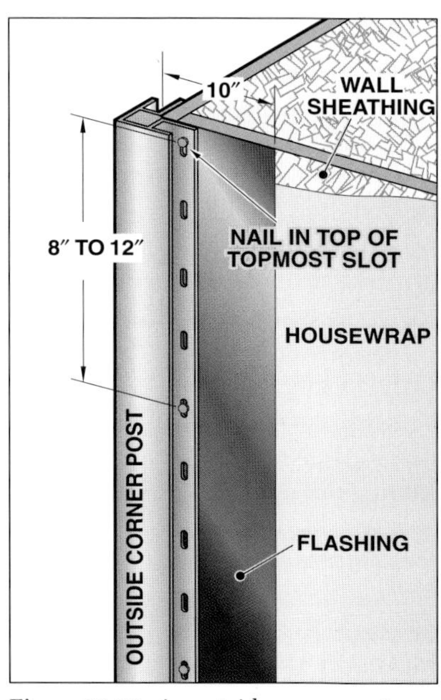

Figure 57-55. *An outside corner post conceals the ends of siding panels.*

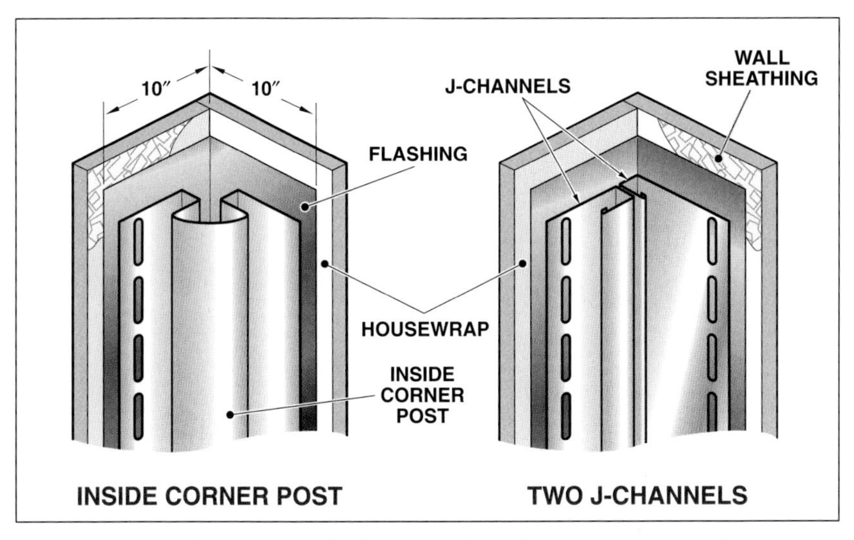

Figure 57-56. *Inside corners can be formed with inside corner posts or with two pieces of J-channel.*

A weather barrier and flashing should be applied around window and door openings. The flashing should extend beyond the nailing flanges to prevent water infiltration. J-channel is installed around doors and windows to receive and conceal the ends of the siding panels. Special attention must be paid when forming the J-channels to ensure that water is properly diverted away from the window. **See Figure 57-57.** Ensure the corners of the J-channels are mitered for a good appearance.

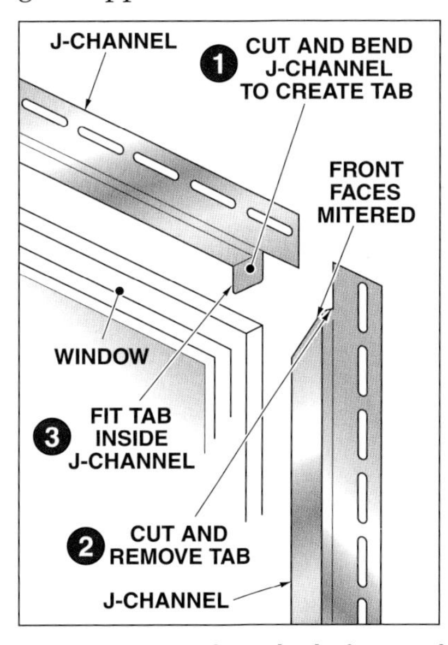

Figure 57-57. *A J-channel to be fit around window and door openings must be modified to properly drain water away from the opening.*

When installing a siding panel, push it up from the bottom until the butt lock fully engages the top lock of the panel below it. While holding the panel in position, reach up and nail it into place. Panel locks should be fully engaged, but the panels should not be under tension or compression when fastened. **See Figure 57-58.** When installing horizontal siding, place the laps away from areas where people typically walk so the lap joints are less easily seen. As additional courses are applied, check the horizontal alignment every fifth or sixth course. Separate joints by at least two courses and avoid joints above and below windows. Short cut-off pieces can be used for narrow openings such as those between window units.

Figure 57-58. *Panel locks must be fully engaged, but the panels should not be under tension or compression when fastened.*

DECKS

Decks provide outdoor living space for recreation, entertaining, and dining. They may have rails, stairways, built-in benches, flower containers, or spas. **See Figure 57-59.**

The deck location should not prevent access to underground utilities such as electrical and telephone lines and plumbing and gas pipes. Always check with the appropriate utility company for location and depth of utilities before digging.

The lot slope and house style are important considerations when planning and building a deck. Decks may be built for all styles of houses and may range from a few inches to several feet above grade level. **See Figure 57-60.**

Deck Materials

In the past, redwood or cypress was used to construct decks because of its insect and decay resistance. Pressure-treated wood is now commonly used for decks because it is decay- and insect-resistant, and is less expensive than redwood or cypress. Most pressure-treated wood is treated with ammoniacal copper arsenate, which provides the lumber with its decay and insect resistance.

Decks are typically constructed with pressure-treated structural members, such as posts, ledger boards, beams, headers, and joists. The decking may be pressure-treated lumber or manufactured building products, which may be a composition of wood and plastic byproducts. Pressure-treated lumber is available in the same nominal thicknesses as untreated lumber, including 1″, 1¼″, 2″, 4″, and 6″. Nominal widths are 4″, 6″, 8″, 10″, and 12″. The 1″, 1¼″ (5⁄4″), and 2″ nominal thicknesses are used for decking.

California Redwood Association

Figure 57-59. *Decks may include benches, stairways, and other features for recreation, entertaining, and dining.*

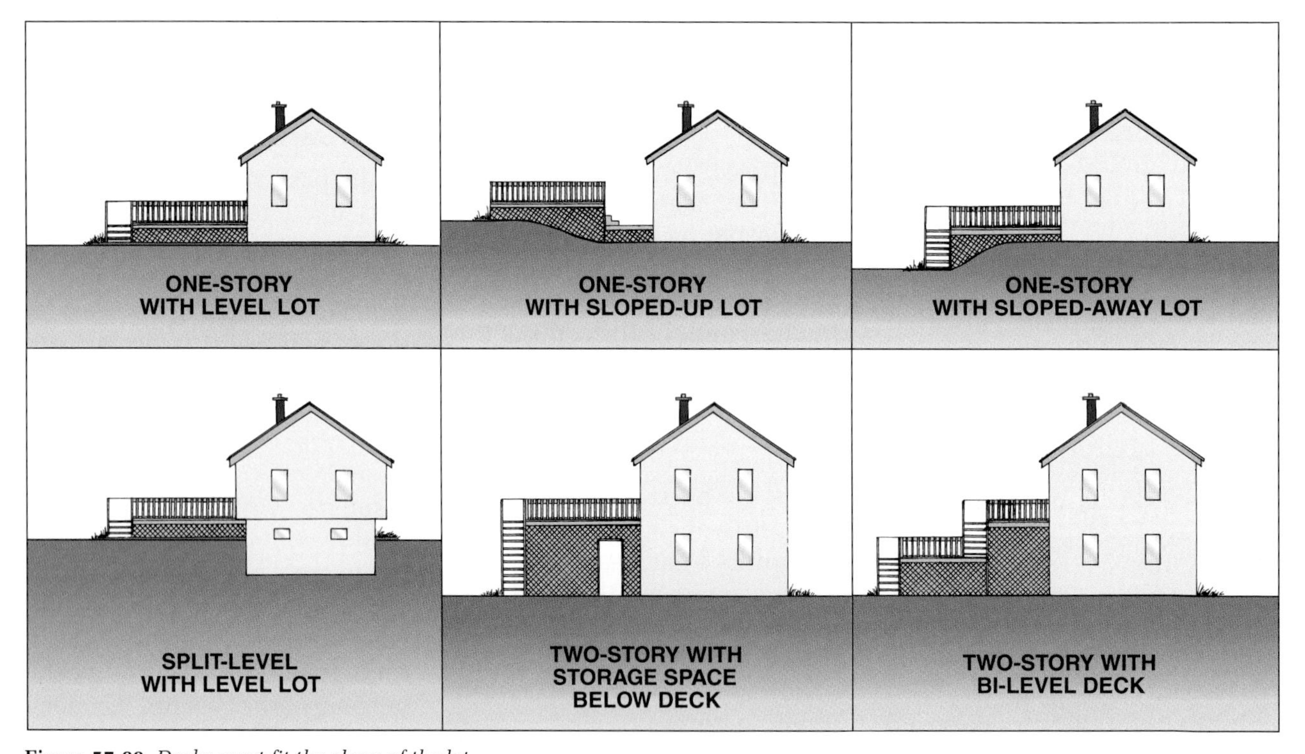

Figure 57-60. *Decks must fit the slope of the lot.*

Fasteners and Connectors

Fasteners and connectors used for deck construction must be hot-dipped galvanized or have another corrosion-resistant coating. Uncoated fasteners and connectors are weakened by corrosion and will stain the wood. Fasteners and connectors are discussed in Unit 8. Composition decking may require special fasteners to prevent unravelling of the material.

Constructing Decks

Before beginning deck construction, the site must be properly prepared to prevent vegetation from growing beneath and up through the deck. The top 2″ to 3″ of soil should be removed and the area covered with black polyethylene plastic, landscaping fabric, or asphalt-saturated felt.

For a deck attached to a building, a ledger board is first fastened to the building. The vertical placement of the ledger board determines the height of the deck floor. When 5/4″ decking is used, position the ledger board with the top edge 1⅜″ below the elevation of the deck floor. When 1½″ decking is used, the ledger board should be 1⅝″ below the finished deck floor. The type of fastener used for the ledger board depends on the material to which it is attached. Use lag bolts when attaching the ledger board to wood; use lag bolts and expansion shields for concrete. **See Figure 57-61.**

Posts are used to support beams, which in turn support the deck joists. Various means may be used as a footing for the posts, **Figure 57-62.** Always consult the local building code to ensure the proper post footing is used and to determine the frost line depth. Footings embedded in the earth must extend below the frost line.

Decks can be constructed using various procedures. One general procedure for constructing an attached deck is described below and illustrated in **Figure 57-63:**

1. Attach the ledger board to the side of the building.
2. Lay out and fasten joist hangers to the ledger board.
3. Construct batterboards with the top of the batterboard at the same elevation as the top of the ledger board.
4. Stretch a line across the batterboards. The line will be used as a guide when setting piers, posts, and beam. For large decks, several batterboards may need to be constructed to ensure the proper elevation and squareness of the deck.
5. Plumb down from the line to position the concrete piers. In this example, 8″ × 8″ precast piers are being used with 4″ × 4″ posts. Therefore, the front of the pier will extend 2″ past the plumb line. Place posts on the piers and mark the tops of the posts at the stretched line. Remove the posts from the piers and cut them to length.
6. Plumb down from line to set posts on top of piers. Secure the post bottoms to the piers.
7. Construct a *sandwich beam* by placing beam 2×s on each side of the posts and securing them in position. The 2×s can be secured to the posts using nails or lag bolts.
8. Plumb and temporarily brace the posts and beam.
9. Set one end of the joists in the joist hangers attached to the ledger board. Secure the joist to the ledger board using fasteners recommended by the manufacturer.
10. Toenail the opposite ends of the joists to the top of the beam.
11. Attach the header (rim) joist to the ends of the floor joist.
12. Secure decking to the joists.
13. Construct the railing.
14. Construct the stairway.

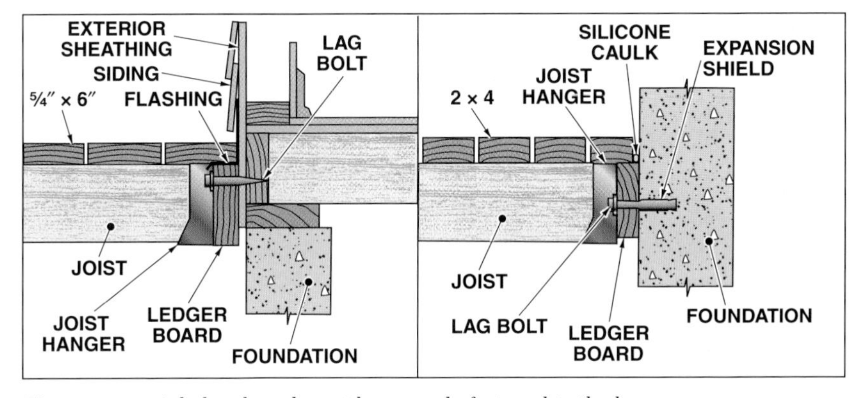

Figure 57-61. *A ledger board must be securely fastened to the house.*

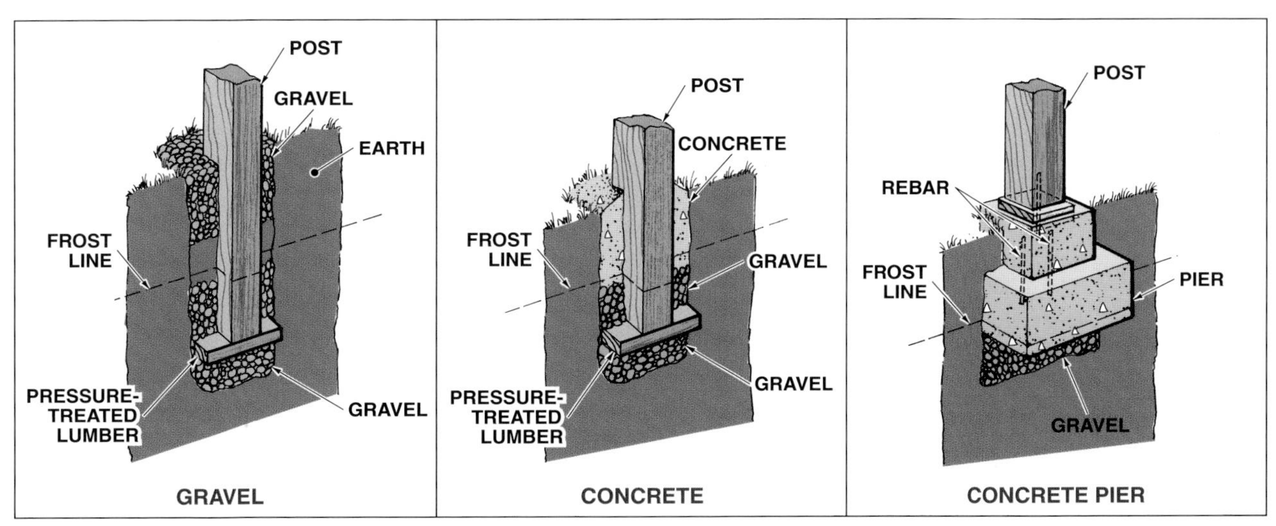

Figure 57-62. *Posts must be set below the frost line.*

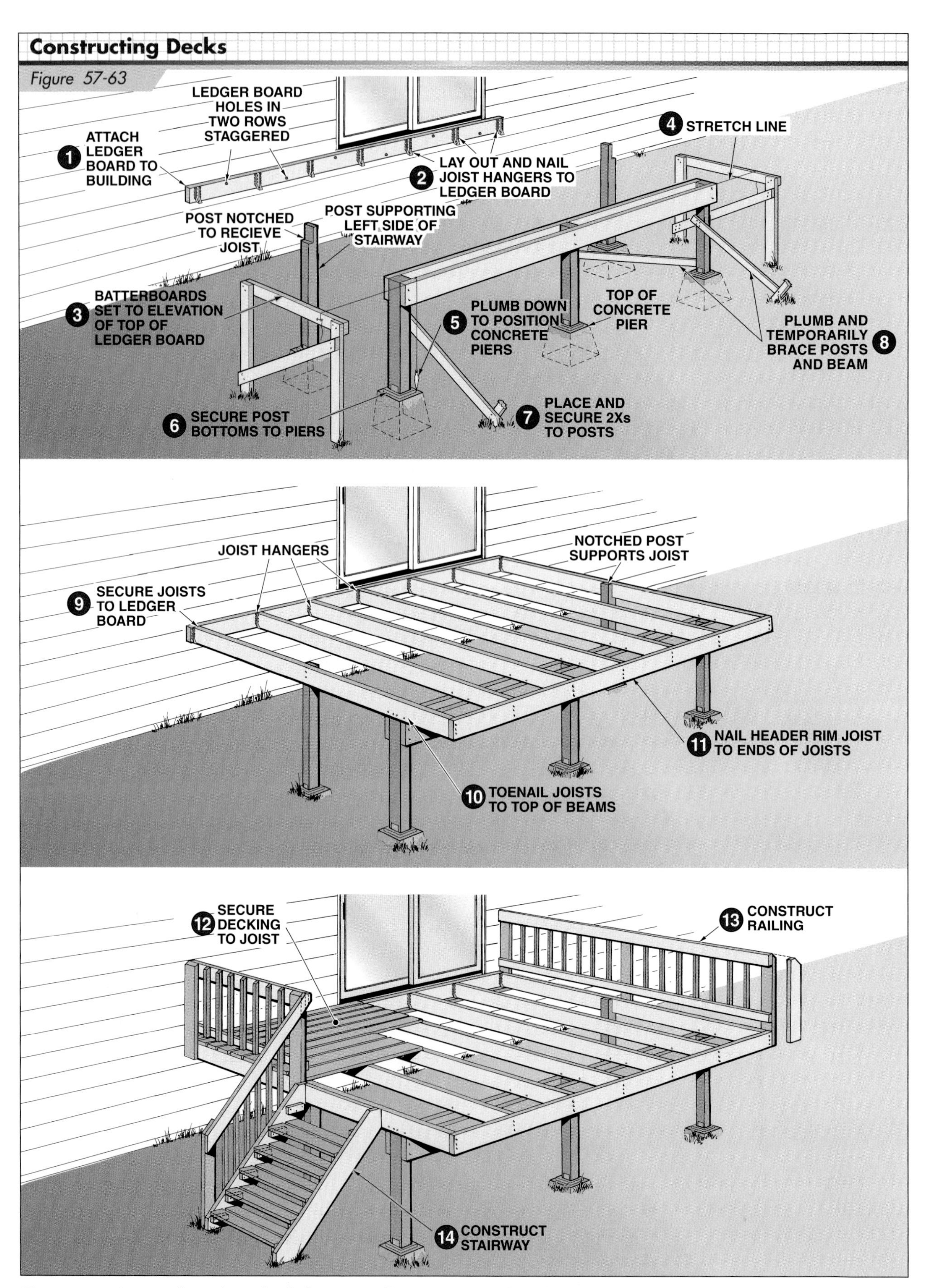

Figure 57-63. *Structural members of a deck are typically pressure-treated lumber. The decking, railings, and stairway are commonly constructed with pressure-treated lumber or composite lumber.*

When using solid wood decking, place the boards with the growth rings facing down. Provide ⅛″ space between decking boards to allow for expansion. (A 10d nail may be used for a spacer.) When using manufactured decking, refer to the manufacturer installation instructions for the recommended spacing.

Decking may be fastened with spiral- or ring-shank nails, or coated screws. Follow the pressure-treated lumber manufacturer recommendations for nail size and spacing.

Decking may be placed in various patterns. For example, a herringbone pattern provides a contemporary appearance, while decking perpendicular to the building provides a more traditional approach. **See Figure 57-64.**

Guardrail posts are fastened to the outside joists or header joists. Balusters are spaced according to the local building code. Stairway stringers are attached to the header joist or outside joist, depending on the stair placement.

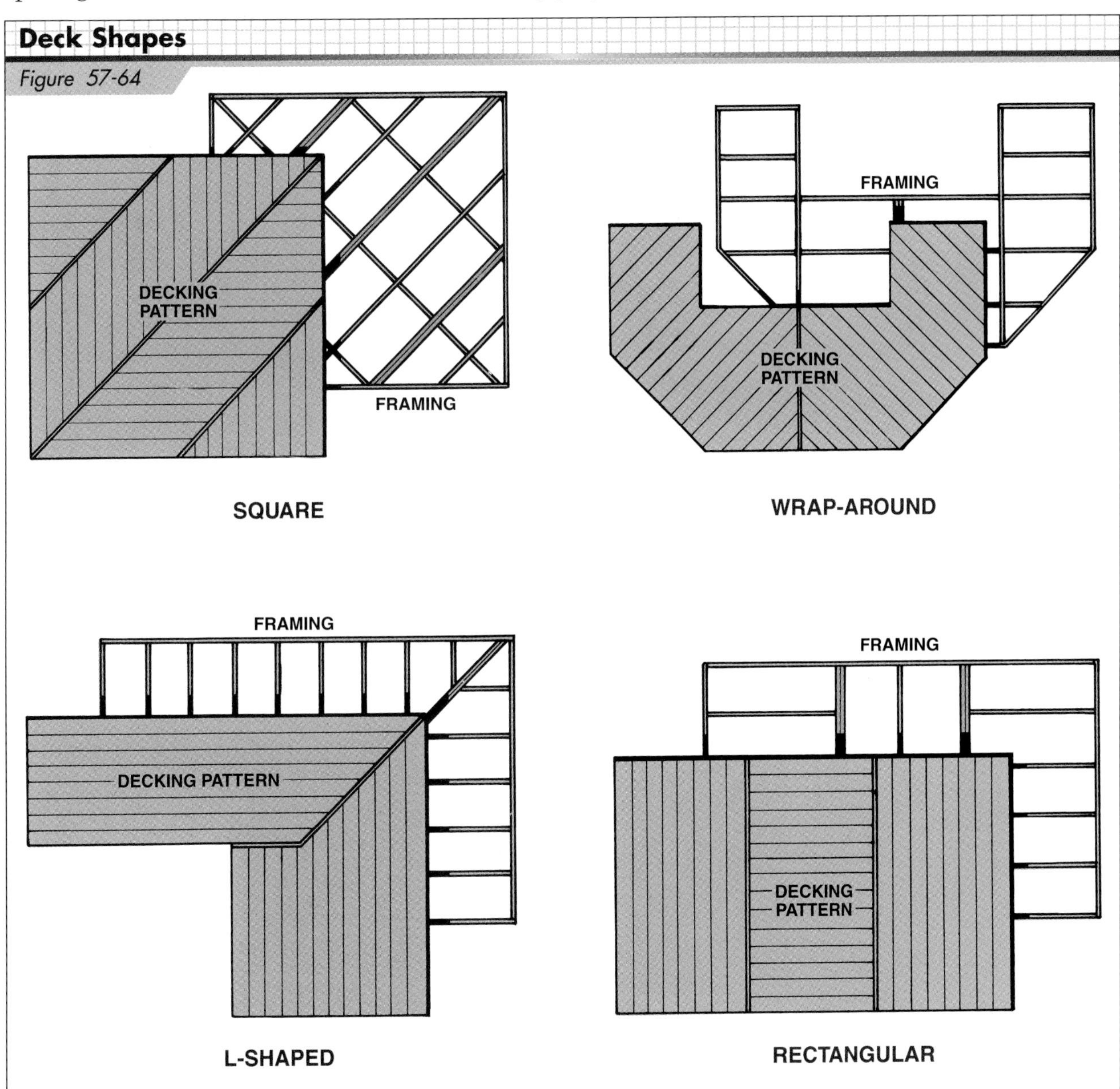

Figure 57-64. *Joists and blocking must be framed to accommodate the deck pattern.*

SECTION

13

Interior Finish

CONTENTS

UNIT

58 Interior Wall and Ceiling Finish

OBJECTIVES

1. List and describe common types of gypsum board.
2. Describe methods of applying gypsum board.
3. Review gypsum board fastening methods.
4. Explain how gypsum board is finished.
5. List and describe common types of wall paneling.
6. Describe how wall paneling is installed.
7. List and describe wall paneling fastening methods.
8. Describe the use of solid board paneling.
9. Review the procedure for applying vertical and horizontal boards.
10. Describe the use of plastic laminate wall covering.
11. Explain the procedure for applying ceiling tiles for nonsuspended ceilings.
12. Explain the procedure for applying ceiling tiles for suspended ceilings.

Interior wall and ceiling finish includes the methods and materials used to finish the wall and ceiling surfaces. Gypsum board, plaster, wood paneling, and plastic laminate are commonly used materials.

GYPSUM BOARD WALLCOVERING

Gypsum board, also known as drywall or wallboard, is commonly used for interior wallcovering. Most new houses and commercial buildings are constructed with gypsum board wall finish. Gypsum board can be fastened to wood, metal, concrete, or masonry. **See Figure 58-1.**

Figure 58-1. *Gypsum board is fastened directly to wood or metal framing members and provides a smooth surface for paint or wallcoverings.*

Gypsum is a substance that is derived from minerals, which are crushed, heated, and combined with other chemicals. The mixture is then sandwiched between two sheets of treated paper to create gypsum board. When properly applied, gypsum board presents a smooth, durable surface suitable for painting or wallcoverings. Gypsum board has good fire-resistance properties and provides a degree of sound insulation. Predecorated gypsum board panels are also available.

Gypsum board is available in a variety of thicknesses, widths, and lengths. Common gypsum board dimensions are 48″ or 54″ wide and 8′, 9′, 10′, 12′, 14′ or 16′ long. Thicknesses range from ¼″ to 1″. For new wall finish, ½″, ⅝″, or ¾″ thicknesses are usually used. Coreboards and liner boards for shafts, firewalls, and stairways are commonly ¾″ or 1″ thick.

Gypsum boards have differently shaped edges along their longer side, depending on their purpose. **See Figure 58-2.** Gypsum board designed to receive a painted finish usually has a tapered edge to provide a recessed joint between the long edges of the boards. The recessed joint receives the paper tape and joint compound that is used to cover the joints between the boards. Predecorated gypsum board has rounded or beveled edges. Boards to be covered with other materials, such as plastic laminate or wood paneling, have tongue-and-groove edges.

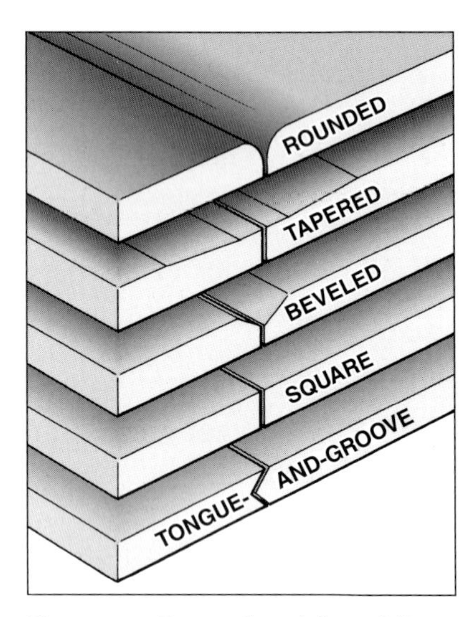

Figure 58-2. *Gypsum boards have different edge shapes depending on their purpose.*

Paperless gypsum board, commonly referred to as *paperless drywall,* is also available for interior and exterior installation. Paperless gypsum board consists of a moisture-resistant gypsum core that may be reinforced with glass fibers, with the panel faces covered with fiberglass mats that are also moisture resistant. Traditional gypsum board panels have a paper facing on both surfaces and hold up well under most circumstances. However, traditional gypsum board panels are vulnerable to mold in damp environments. Some types of mold actually feed on the paper facing of traditional gypsum board panels if the panels become wet and stay moist.

Paperless gypsum board can be used in any part of a building where traditional gypsum board might otherwise be used. Paperless gypsum board is well suited for humid areas, such as bathrooms, if the panels are not in direct contact with water. It can be used for exposed wall surfaces above ceramic tile on bathtubs and showers, and around vanities and water closets. The cutting, placing, and nailing of paperless gypsum board drywall is basically the same as traditional drywall.

Applying Gypsum Board

On large construction projects where large quantities of gypsum board are used, workers specializing in gypsum board application are employed. Where smaller amounts of gypsum board are required, however, carpenters often install gypsum board.

After a gypsum board panel is measured and marked, it is usually cut by scoring the face with a sharp knife, then snapping off the waste piece. **See Figure 58-3.** Jagged edges should be smoothed with a rasp, knife, or serrated-blade forming tool. **See Figure 58-4.** Sometimes cuts can be made more conveniently with a saw.

Figure 58-3. *Gypsum board is cut by scoring it on the face side and then snapping off the waste piece.*

Figure 58-4. *Jagged gypsum board edges can be smoothed with a serrated-blade forming tool.*

The *single-ply system* of application is most often used for gypsum board in residential and other light construction applications. **See Figure 58-5.** A single layer is adequate for fire resistance and sound control. The boards are applied with the long edge in a vertical or horizontal position.

The *double-ply system* provides greater fire resistance and sound control than the single-ply system. The first layer of wallboard is fastened to the studs. The second layer is then applied with an adhesive and tacked in place. **See Figure 58-6.**

Panels are applied to the ceiling first and then to the walls. Wall panels should butt tightly against the ceiling panels. A *foot lift* can be used to raise the wall panels up against the ceiling panels to ensure a tight fit. **See Figure 58-7.**

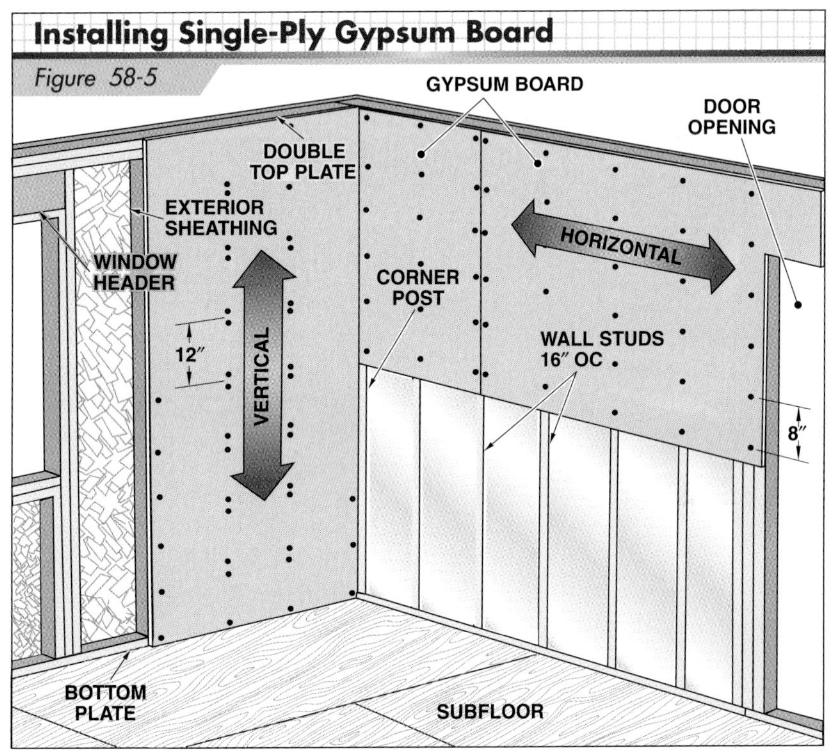

Figure 58-5. *In single-ply application, only one layer of gypsum board is used.*

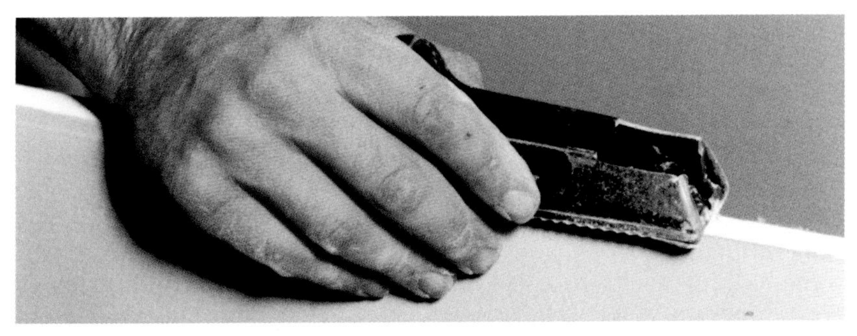

A serrated-blade forming tool, or Surform®, is used to smooth rough gypsum board edges.

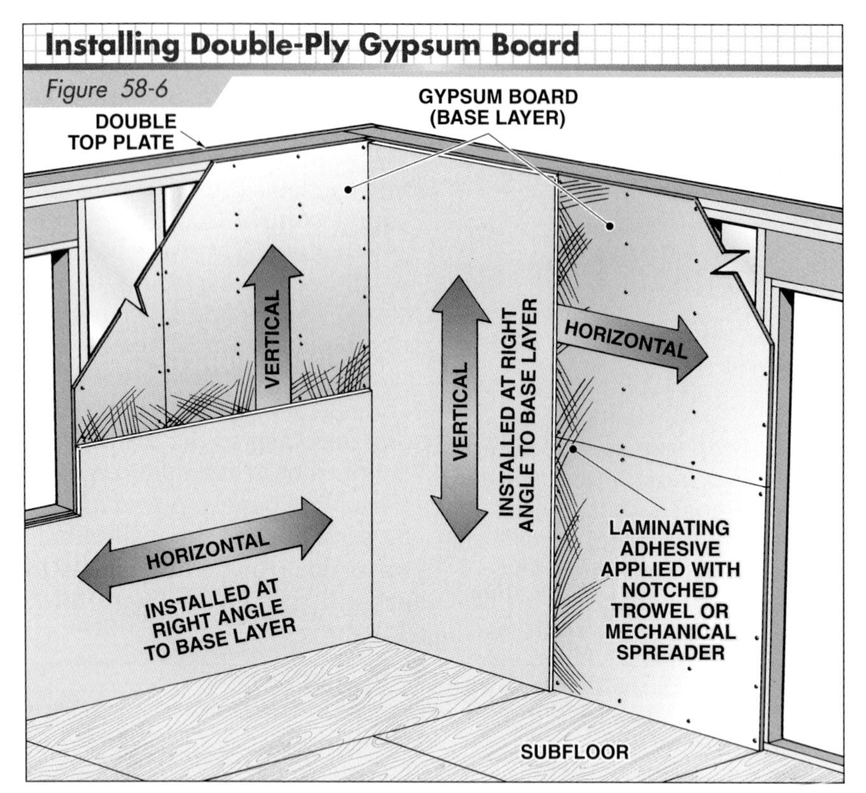

Figure 58-6. *In double-ply application, an adhesive is used to bond the second layer of gypsum board to the base layer. The second layer runs at a right angle to the base layer.*

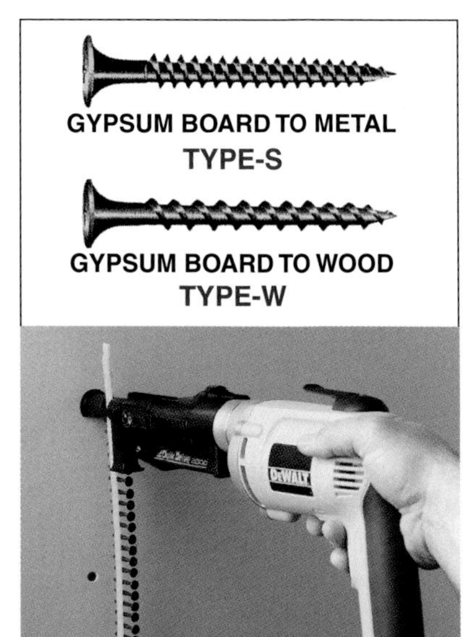

Quik Drive U.S.A., Inc.

Figure 58-8. *Type-S screws are fine-threaded screws and are used for applying drywall to metal studs. Type-W screws are course-threaded screws that are used to fasten drywall to wooden studs and other framing members.*

Figure 58-7. *A metal foot lift is used to raise a gypsum board wall panel up tightly against the ceiling. The space below the panel is covered by base molding.*

In most residential construction, the rough ceiling height is usually 8′-1″. The 8′-1″ height provides clearance between the ceiling and floor for a ½″ or ⅝″ panel on the ceiling and 8′ long panels fastened to the walls. Wall panels may be placed with long edge running vertically or horizontally, whichever results in fewer joints between panels. Joints between the panels should be staggered if possible.

Fastening Methods. Originally, only nails were used to fasten the panels and can still be used. However, drywall screws are preferred by professional drywall installers. Some advantages of drywall screws are speed of installation and superior holding power. Screws go in more easily than nails and resist rust longer. Screws can provide up to 350% more holding power than nails. Drywall screws are applied with a corded or cordless drywall screwdriver. Most drywall screwdrivers have an adjustable nosepiece or tip that ensures consistent fastener depth. Adhesives can also be used in conjunction with nails or screws to help fasten the drywall to the framing.

Screws. There are a wide variety of drywall screws available. Most of these are variations of course-threaded screws for fastening to wood, fine-threaded screws for drilling into metal studs, and self-drilling screws for fastening to thicker metal. Two general categories of drywall screws are type-S and type-W. **See Figure 58-8.**

Type-S screws are fine-threaded screws and are used for applying drywall to metal studs and track. They are self-tapping and very sharp. The screw threads should pass through the studs at least ⅜″. Type-W screws are course-threaded screws that are used to fasten drywall to wooden studs and other framing members. The screws should penetrate the wood at least ⅝″.

The heads of most drywall screws are bugle-shaped. Bugle-shaped heads are designed to catch the gypsum board face paper and pull it in rather than tear it. However, the drywall screw must be driven to the correct depth to properly secure the gypsum board to the framing. **See Figure 58-9.** If a drywall screw is not driven deep enough, it will cause problems with the finishing of the gypsum board. If the drywall screw is driven too deep, the screw may pull through the gypsum board and not hold the drywall tight against the framing.

toward the edges to prevent a sag or bulge at the center of the gypsum board panel. The same nailing patterns are used for walls and ceilings, with the edges of ceiling panels fastened 7″ OC.

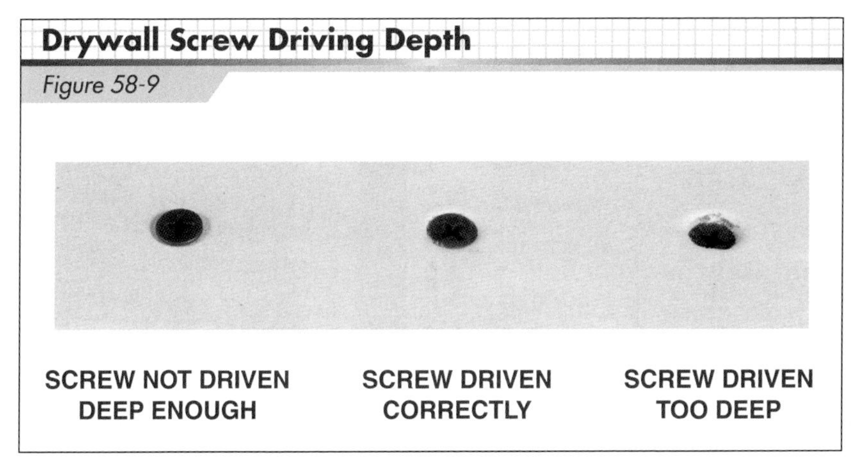

Figure 58-9. *Drywall screws must be driven to the correct depth to properly secure the gypsum board to the framing.*

Nails. Nails used to fasten gypsum board to wood framing members must be long enough to go through the gypsum board and penetrate ¾″ to ⅞″ into the wood. The heads should be at least ¼″ in diameter, flat or concave, and thin at the rim. **See Figure 58-10.**

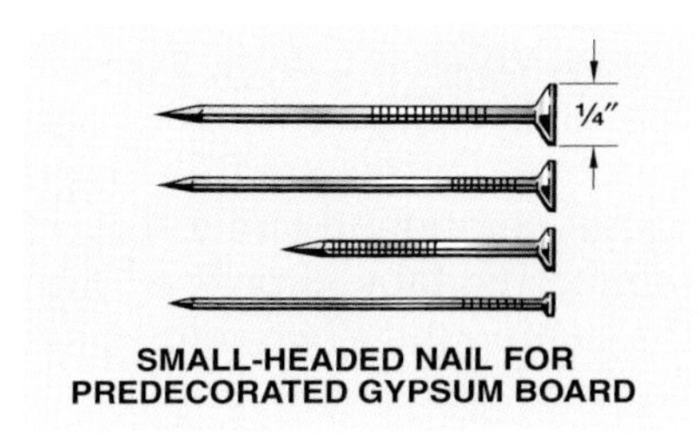

Figure 58-10. *Nails may be used to fasten gypsum board to wood framing members. Small-headed nails are used with predecorated panels.*

Gypsum board should be pressed tightly against wood framing members while nails are driven in. The nail should be driven in far enough to produce a dimple on the surface of the board, but not far enough for the head to cut into the face covering. **See Figure 58-11.** The dimple is later filled with joint compound, hiding the head of the nail.

Improperly driven nails can result in loose gypsum board. As a result, cracks may later appear in the finished wall. Nails that do not catch the wood properly can work loose and eventually pop out. **See Figure 58-12.**

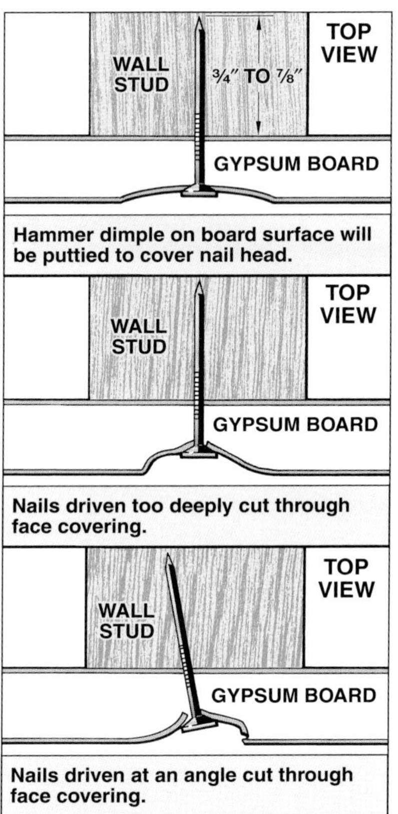

Figure 58-11. *Nails must be driven properly to ensure a solid connection between the gypsum board and wood framing member. Nails driven too deeply, or at an angle, will cut through the face covering.*

The *single-nailing method,* shown in **Figure 58-13**, or *double-nailing method,* as shown in **Figure 58-14**, may be used to fasten gypsum board to framing members. With both methods, first drive the nails in the center of the board and move outward

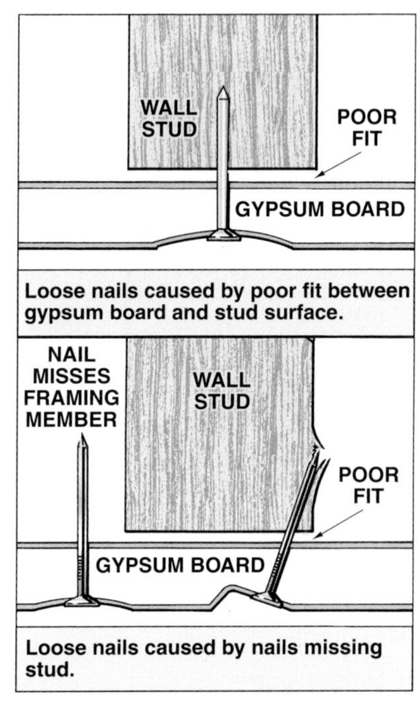

Figure 58-12. *Loose nails are caused by a poor fit of the gypsum board to the framing surface or by nails missing the framing members.*

Adhesives. Adhesives can be used to bond gypsum board directly to wood, metal, masonry, and other wall materials. Manufacturer recommendations must be followed when selecting the proper adhesive for a particular job.

Tube-dispensed *stud adhesive* is used with screws or nails to fasten single-ply gypsum board to wood or metal framing members. A ¼″ to ⅜″ bead is applied to each stud. **See Figure 58-15.**

Laminating adhesives are used for multi-ply application. Some laminating adhesives are available in a dry powder form that must be mixed with water. Contact-type laminating adhesives are also available.

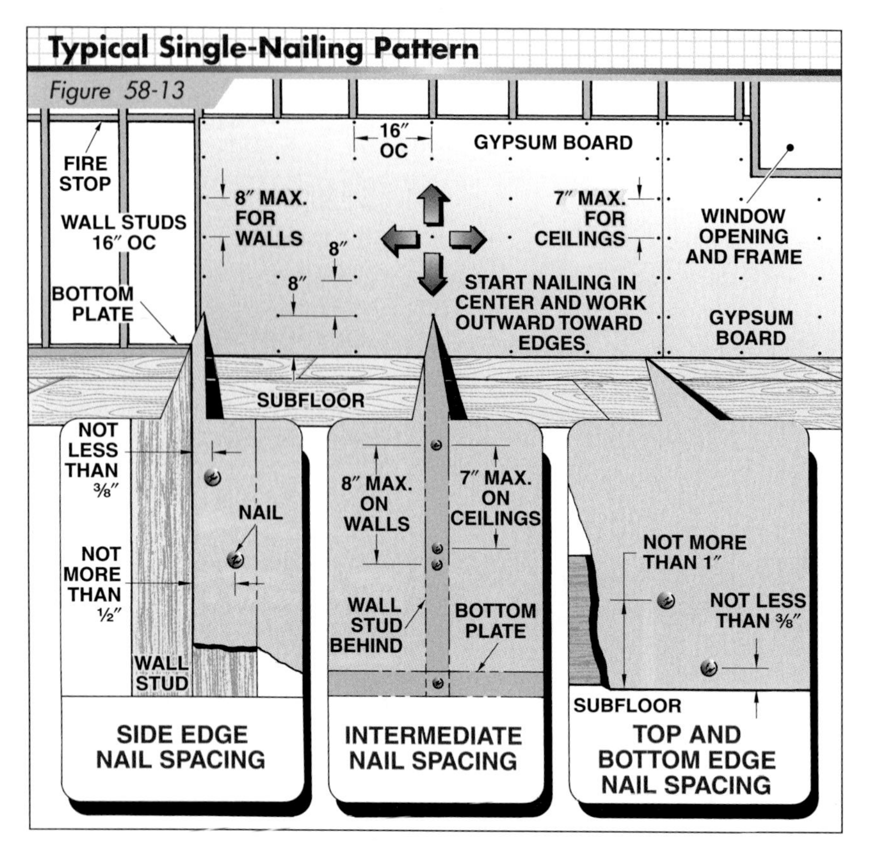

Figure 58-13. *The proper single-nailing pattern ensures a sound connection between the gypsum board and wood framing member. Nailing should begin at the center of the panel and move outward.*

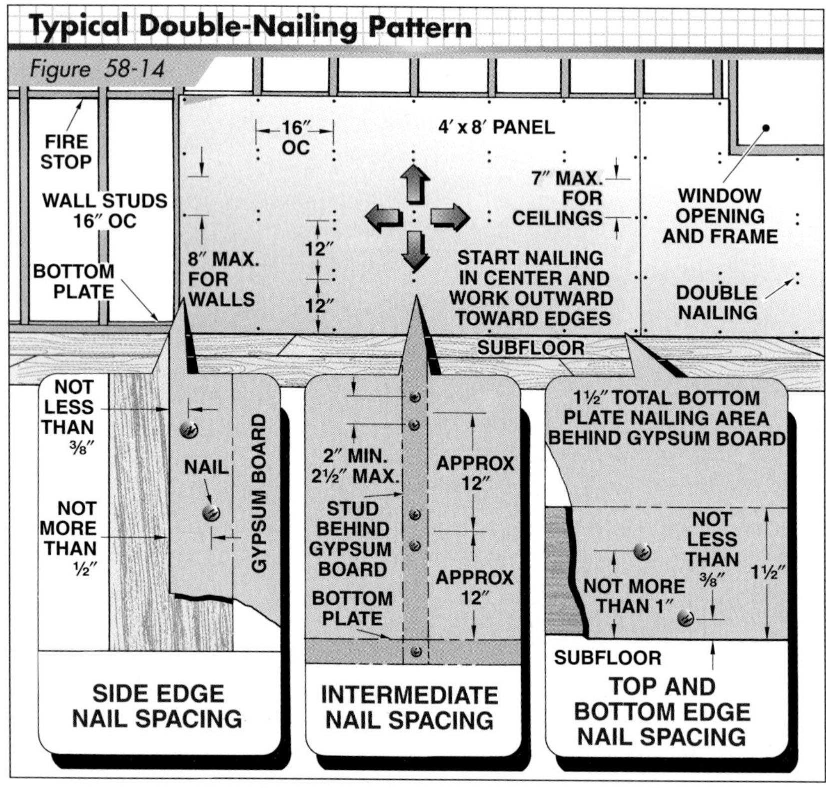

Figure 58-14. *When double-nailing gypsum board, the second set of nails may be driven after the first set is driven across the entire panel, or the second set of nails in each row may be driven after the first set in each row is driven. Nailing should begin at the center of the panel and move outward.*

Figure 58-15. *Stud adhesive is used with screws or nails to fasten single-ply gypsum board to wood or metal framing members.*

Laminating and stud adhesives are used to fasten gypsum board directly to concrete or masonry. The gypsum board panels should be temporarily braced until the bond develops strength.

Finishing After all ceiling and wall panels are nailed in place, joints between panels are finished by painters or other tradesworkers called tapers who specialize in finishing gypsum board. **See Figure 58-16.** Gypsum board joints are filled with a joint compound and a strip of reinforcing tape is pressed into the compound. Two additional layers of joint compound, forming a *topping coat,* are applied. Each layer of joint compound must dry completely and may require light sanding before the next layer is applied. The final coat, called the *finish coat,* of joint compound is applied and feathered out on each side of the joint. Nail or screw dimples are also filled with joint compound and lightly sanded.

Joint compound is also applied to the inside corners. Reinforcing tape is folded and pressed into the compound. Metal corner beads, which reinforce outside corners, are installed and covered with joint compound. **See Figure 58-17.** When all joints, corners, and nail dimples have been treated and sanded, the wall is ready for painting.

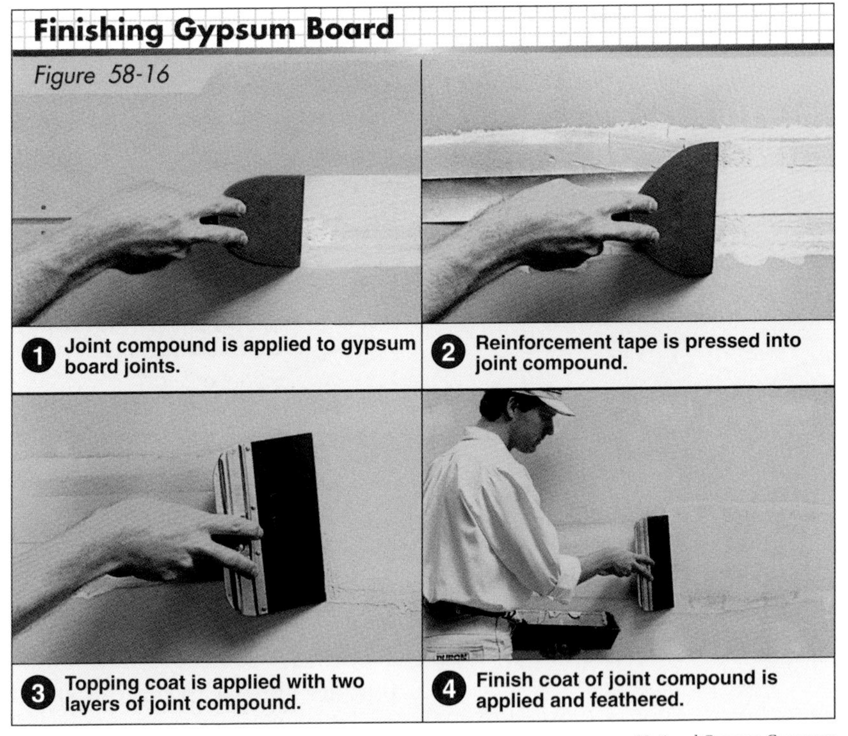

National Gypsum Company

Figure 58-16. *Gypsum board must be properly finished to ensure a smooth and even surface.*

Figure 58-17. *Metal corner beads reinforce outside corners of gypsum board.*

Predecorated Gypsum Board

Predecorated gypsum board has a decorative surface that does not require additional finishing. A popular predecorated gypsum board product is a gypsum panel covered with vinyl. A wide variety of patterns, colors, and textures are available.

Predecorated panels can be applied directly to wall studs or as the outside layer in a multi-ply system. The panels are fastened with adhesives, or with nails with colored heads that match the wallboard.

Predecorated gypsum board panels are usually applied with the long edge in a vertical position. The joints may be exposed or covered with the types of molding shown in **Figure 58-18.**

WALL PANELING

Paneling is an attractive and fast way to finish an interior wall. Wood panel systems include plywood, hardboard, or particleboard panels, as well as solid board applications. Panels surfaced with plastic laminate are also available.

Plywood Paneling

Most plywood wall panels have a hardwood veneer such as oak, ash, beech, walnut, birch, pecan, or mahogany. However, softwood-veneer panels such as redwood, cedar, fir, and southern pine are also available. The surface may be plain or textured. **See Figure 58-19.** Plywood panels are usually ¼″ or ⅜″ thick. Thicker panels, such as ¾″, are also available. Standard sheet sizes are 4′ × 8′ and 4′ × 10′. Longer lengths are available by special order.

Most plywood wall panels are stained, sealed, and varnished at the fabricating plant. Therefore, care must be taken not to scratch or damage the panels during installation.

American Hardwood Export Council

Exotic hardwood inlays complement surrounding hardwood paneling and molding.

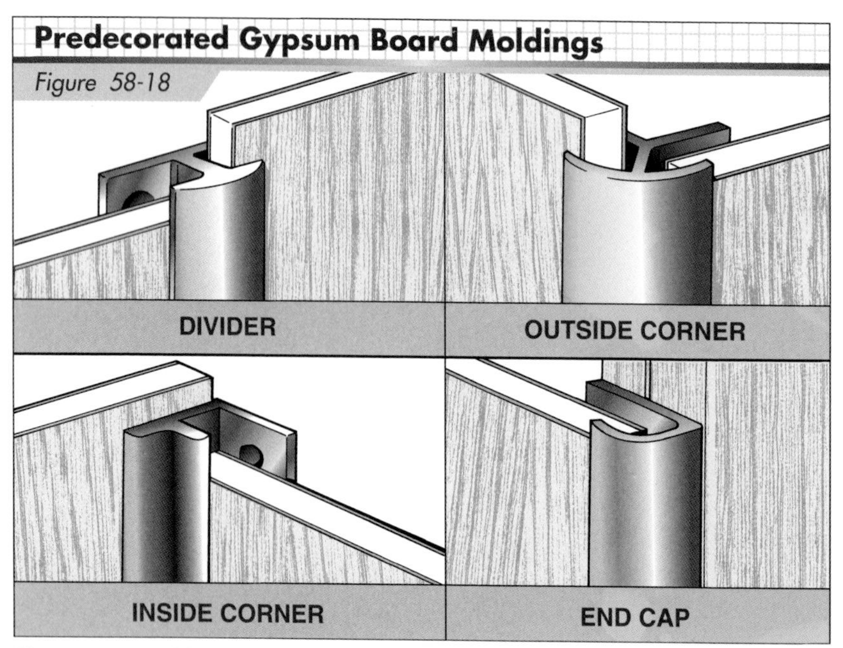

Figure 58-18. *Molding is used to cover joints and corners of predecorated gypsum board.*

Figure 58-19. *Grooved softwood-veneer plywood wall panels create pleasant shadow lines.*

Panels may have a rabbeted or slightly beveled edge. When panels are placed together, rabbeted edges form a *channel joint;* beveled edges form a *V-joint.* **See Figure 58-20.** Both types of joints create an attractive shadow line between the panels.

Preconditioning. Plywood panels should be preconditioned to room temperature for several days before they are applied to a wall. A good procedure is to stack the panels with strips between them. **See Figure 58-21.** This allows air to reach the faces and backs of the sheets, conditioning them to room temperature and humidity. Preconditioning eliminates significant shrinkage after the panels have been nailed in place.

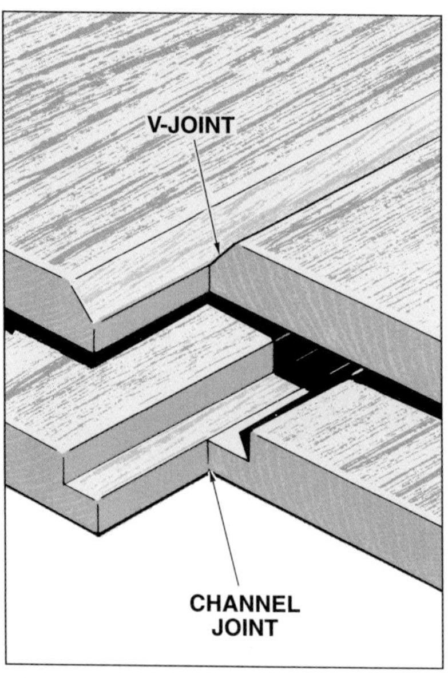

Figure 58-20. *Wall paneling may have rabbeted or beveled edges, forming channel or V-joints respectively.*

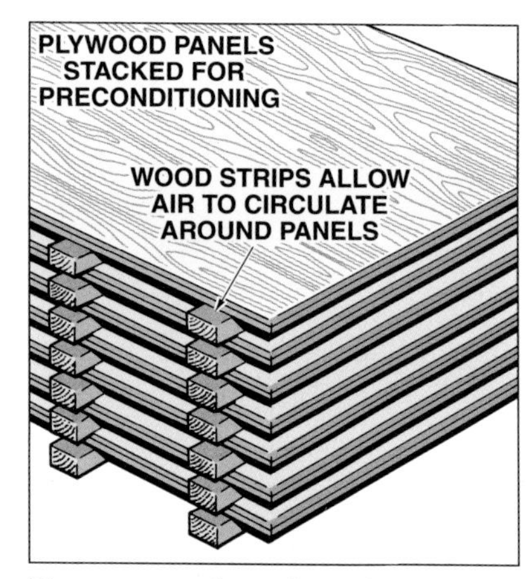

Figure 58-21. *Plywood panels are stacked with wood strips between them for preconditioning to room temperature.*

Hardboard Paneling

Hardboard panels are manufactured with a variety of patterns, including wood grain finishes that are similar to natural wood grain. Panels range from ⅛″ to ¼″ thick. The ⅛″ thickness should be applied only against a solid surface such as plywood, gypsum board, or plaster. Solid backing or furring strips spaced 12″ OC are recommended for ¼″ thick panels. Standard size sheets are 4′ × 8′.

Preconditioning. Hardboard is a wood-based product. Therefore, it will expand and contract with changes in the moisture content of the surrounding air. Panels should be separated and stacked on their edges to allow air to freely flow on both sides of the panels for at least 48 hours.

Applying Wall Paneling

The procedure for applying plywood wall panels is similar to the procedure for applying hardboard panels. The same methods and tools are used to cut and fit both materials. Plywood and hardboard wall panels can be fastened directly to wood studs or furring strips (usually 1 × 2 wood strips) using nails or adhesives. For increased thermal and sound insulation, gypsum board is first applied as a base and the paneling is then glued to the gypsum board.

All studs must be straight and plumb before panels are applied to them. Crooked framing material produces an uneven paneled wall surface. Correct any unevenness before applying panels. Badly warped or twisted studs should be replaced.

Plywood and hardboard panels can be fastened directly to masonry walls with adhesives. This is not advisable, however, if the walls are uneven or subject to damp conditions. If the walls are uneven or if damp conditions exist, furring strips should be nailed or glued to the masonry and the panels fastened to the furring strips.

In remodeling work, panels are often placed directly over the plastered or gypsum board surface of the existing walls. Frequently, these surfaces are very uneven, and furring strips should be used. A procedure for installing furring strips is shown in **Figure 58-22.** Shims tapered from ½″ thick to about ⅛″ thick are used to plumb and straighten the furring strips where necessary.

Before installing panels, determine their arrangement on the walls. **Figure 58-23** shows two possible ways for placing panels on the same wall. Whatever arrangement is used, panels should be set up around the walls so that panels with better matching grain patterns are next to each other. When the final panel arrangement has been determined, number the backs of the panels so they can later be installed in that order.

The first panel must be plumb. Subsequent panels do not have to be checked for plumb if the first panel is plumb. Each panel must be permanently fastened into position before the next panel is placed.

Hardboard paneling moisture content should range from 2% to 9%.

Holes for electrical outlets, switches, or other openings must be laid out on the panel before the panel is set in place. **See Figure 58-24** for the layout procedure. The openings may be cut before the panel is set in place using a keyhole saw or jigsaw, or they can be cut in the panel when the panel is in place using a cutout saw.

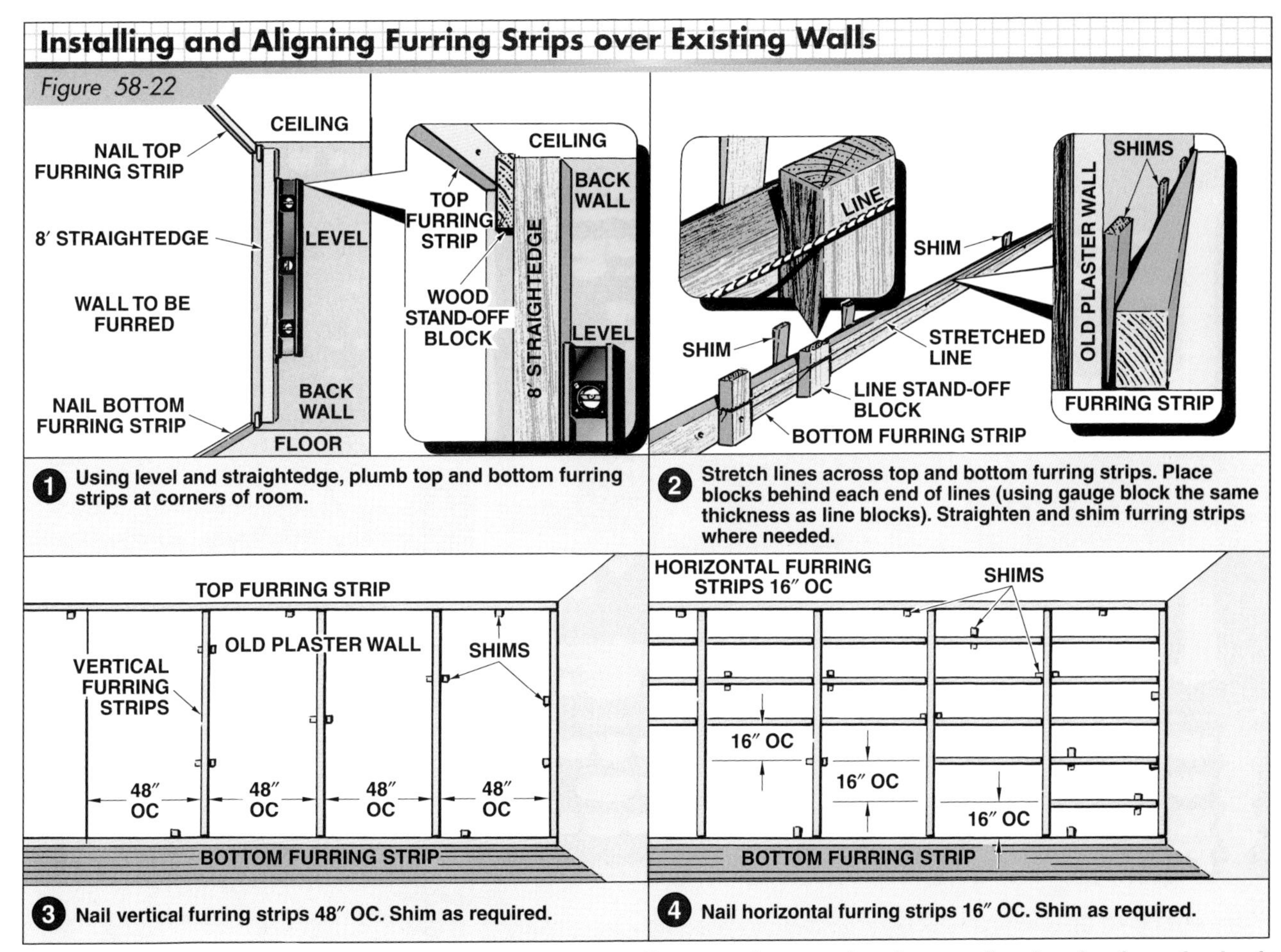

Figure 58-22. *Furring strips are installed over existing uneven walls to provide a smooth and even wall surface for plywood or hardboard panels.*

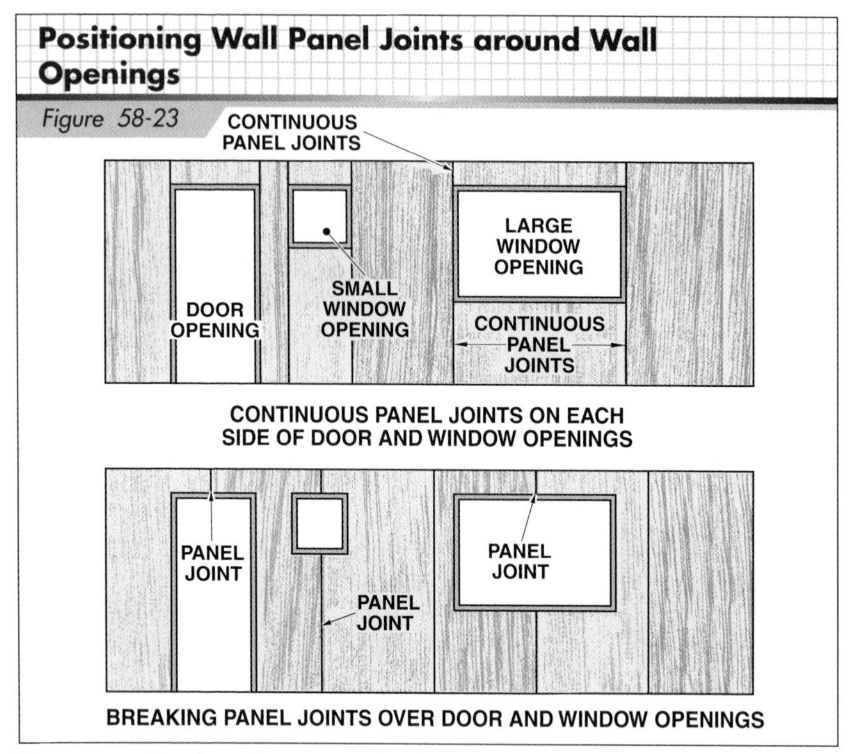

Figure 58-23. *Panels can be positioned so that a continuous joint occurs on each side of the door or so joints break over the openings.*

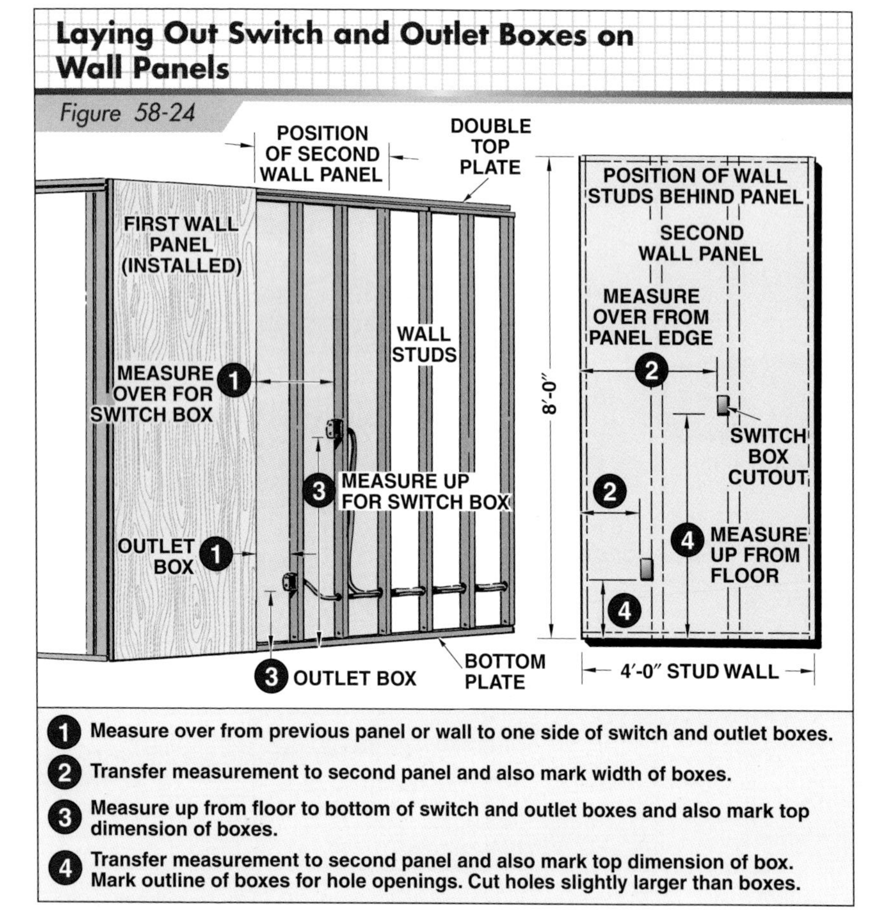

Figure 58-24. *Holes for electrical outlets and other openings must be laid out before the panel is set in place. The openings may be cut before or after the panel is positioned using a cutout saw.*

When cutting the opening before the panel is set in place, the item requiring an opening is used as a pattern for marking the opening. **See Figure 58-25.** The outline is made slightly larger than the item. Starter holes are drilled in the inside corners of the outline. A jigsaw or cutout saw is used to cut along the marked lines. If a jigsaw is not available, a keyhole saw can be used, although it is not as efficient.

When cutting the opening with a cutout saw after the panel is set in place, the center of the opening is marked on the exposed face of the panel. The cut is started at the center, moving toward one of the outer edges of the opening. When the bit contacts the edge of the item requiring an opening, such as an electrical outlet, the bit then follows the edge until the opening is created. The panel is then pushed into place.

A typical method for applying wall panels is shown in **Figure 58-26.**

When using a jigsaw to cut paneling with the exposed face up, a reverse-tooth (down-cut) blade should be used to prevent the edges of the cut from splintering.

Joints and Corners. New panels have a straight "factory" edge. Therefore, joints between panels usually do not require planing in order to fit properly against each other. Slight imperfections are hidden by the shadow line.

When the top and bottom of a panel are covered by moldings, the top and bottom edges do not need to be scribed. **See Figure 58-27.** In contemporary construction, ceiling molding is usually not installed, so the top of the panel must be scribed to the ceiling.

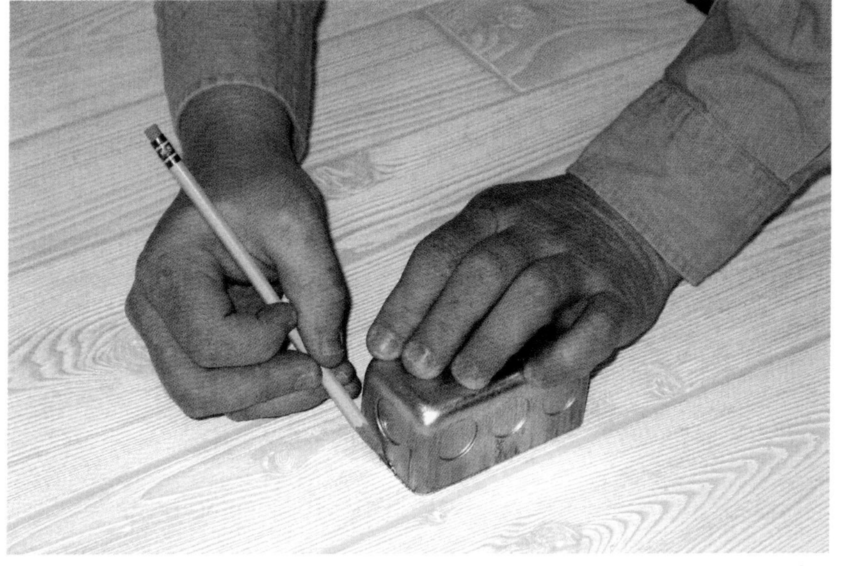

Figure 58-25. *The item requiring an opening—in this case an electrical box—is used as a pattern for marking a panel opening.*

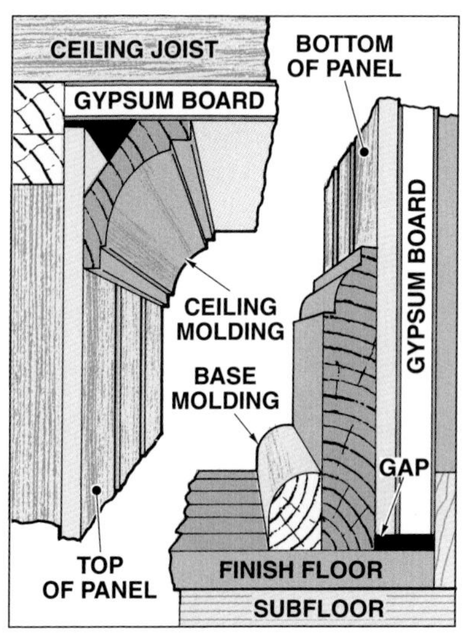

Figure 58-27. *Base and ceiling moldings eliminate the need to scribe the top and bottom panel.*

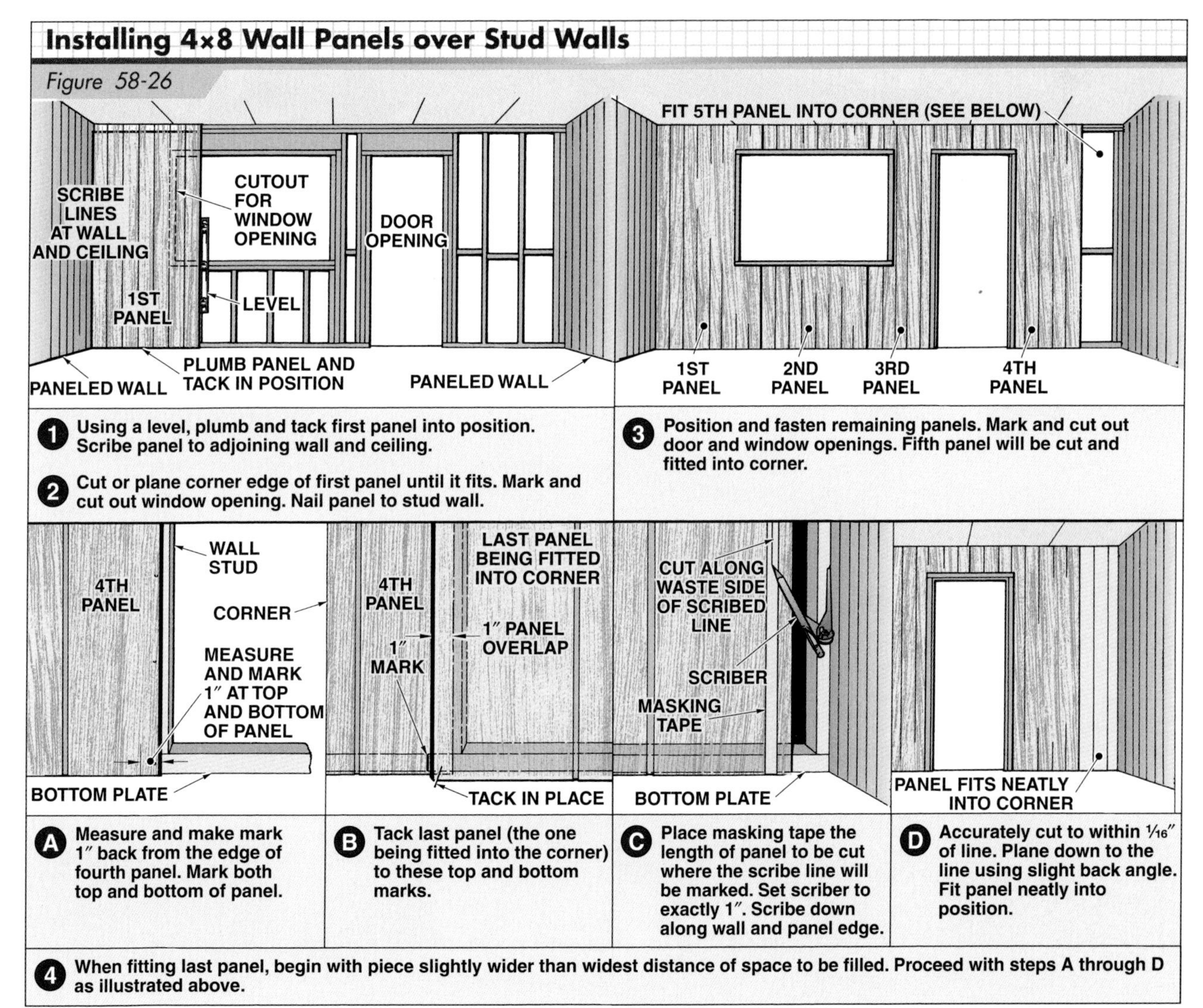

Figure 58-26. *When installing wall panels over stud walls, the edges of the panels must fall over the centers of the studs.*

American Hardwood Export Council

Wall paneling can be applied to irregular or curved walls.

Fastening Methods. Nails or adhesives may be used to fasten wall paneling. Finish nails for paneling should be long enough to penetrate ¾″ into the studs or furring strips. When using grooved panels, drive nails into the grooves located in the body and along the edge of each panel. If panel edges are chamfered, place the nail along the chamfer. **See Figure 58-32.** All nails except those that are colored to match the paneling must be set below the surface with a nail set. The nail heads are later concealed with wood putty that matches the finish of the panel.

Inside and outside corners do not have to be fitted carefully if they are to be covered by moldings. **See Figure 58-28.** If no molding is used on an inside corner, a scriber must be used as shown in **Figure 58-29.** If molding is not used on an outside corner, the corner may be mitered. **See Figure 58-30.** However, these corners are difficult to miter properly and can be easily damaged after mitering. A better method of finishing corners is to fit a corner strip flush with the panel as shown in **Figure 58-31.**

ANSI/AHA A135.5, Prefinished Hardboard Paneling, covers the requirements and methods for testing the dimensions, squareness, edge straightness, and moisture content of prefinished hardboard paneling and for the paneling finish. Hardboard panel finish is specified as Class I or Class II. Class II finishes have limited resistance to heat, humidity, and steam and are not designed to be used where these conditions are excessive, such as around furnaces, showers, and bathtubs.

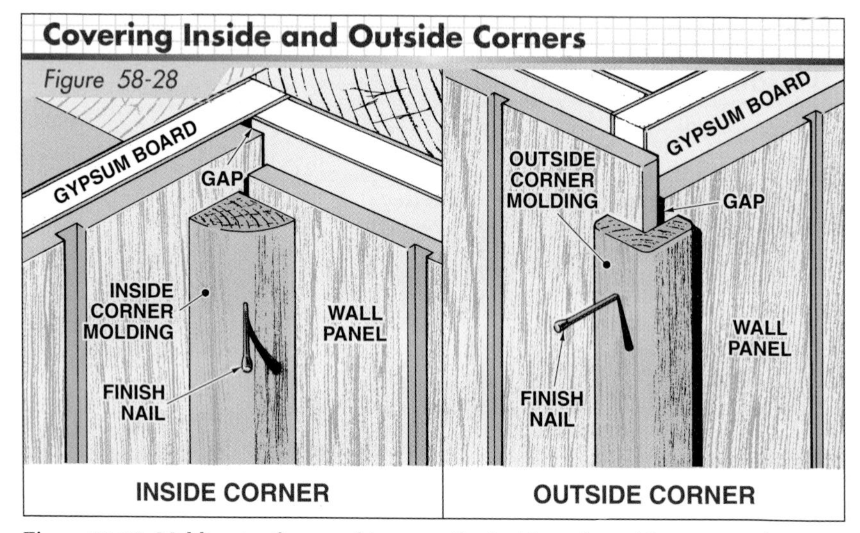

Figure 58-28. *Molding is often used to cover the inside and outside corners of interior paneling.*

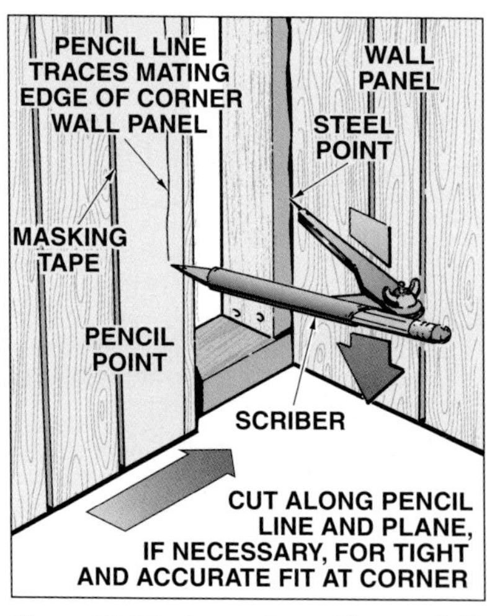

Figure 58-29. *A panel must be carefully scribed to the other wall when no molding covers the inside corner.*

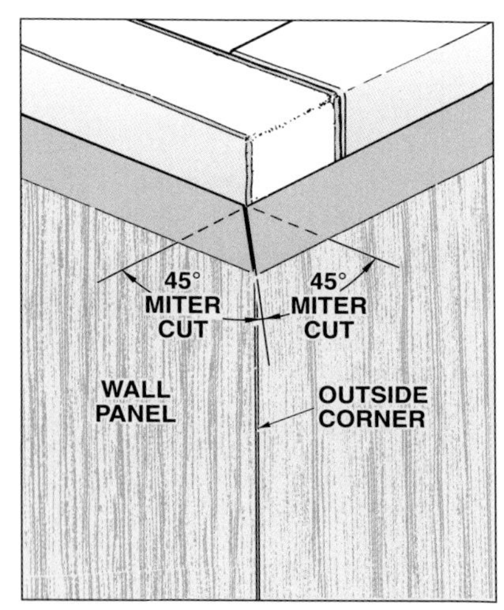

Figure 58-30. *Wall panels may be mitered if molding is not used on an outside corner.*

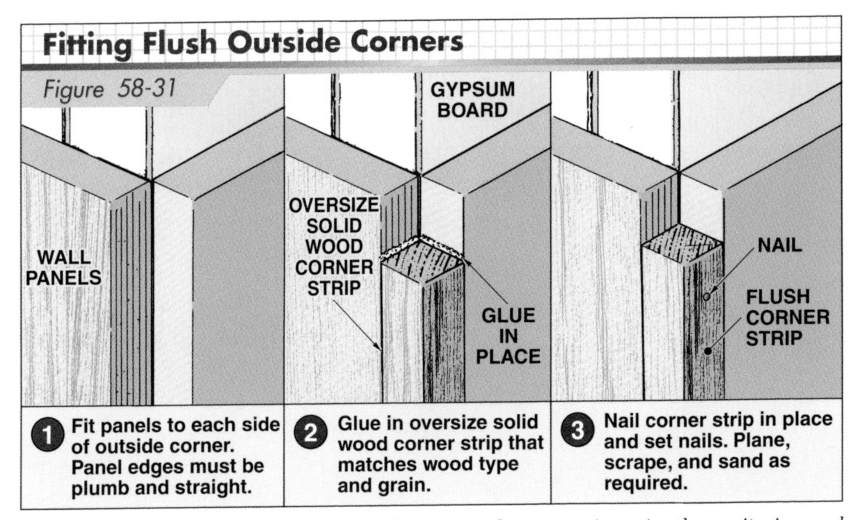

Figure 58-31. *Flush application of panels on outside corners is easier than mitering and panel edges are less susceptible to damage.*

Figure 58-34. *A wall panel is pressed into place after the adhesive has been applied.*

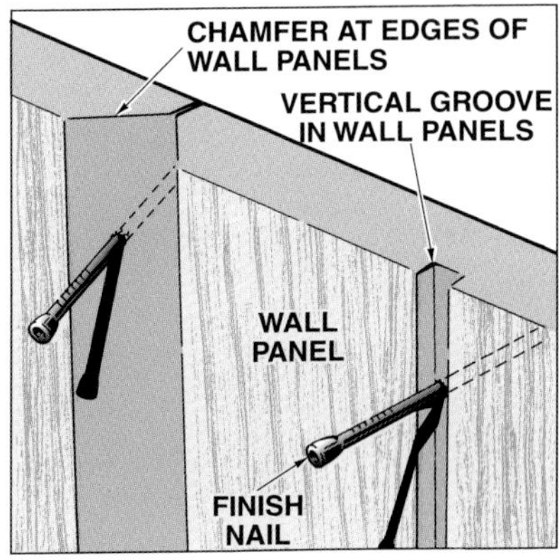

Figure 58-32. *Nails are less noticeable when driven along the bevel or into the grooves of a panel.*

For plywood wall panels that are not fastened with an adhesive, space finish nails 6″ OC along the panel edges and every 12″ OC at the intermediate studs or furring strips. For hardboard panels, space nails 4″ OC along the edges and 8″ OC at the intermediate points.

Most paneling, especially hardboard paneling, is fastened with adhesive rather than nails. A few nails may be required to hold the panels in place until the adhesive sets. Adhesive is applied for the sides and top and bottom edges of the panel in a ⅛″ continuous strip behind the panel joints and at the top and bottom plates. Adhesive is applied in 3″ long beads, 6″ apart at the intermediate studs or furring strips behind the panel. **See Figure 58-33.** The panel is pressed into place with firm, uniform pressure so the adhesive spreads evenly and the panel is tacked at the top. **See Figure 58-34.** The panel is then grasped at the bottom along both edges and slowly pulled away from the stud. **See Figure 58-35.** After two minutes, the panel is pressed back into position.

Hardboard paneling can be bent depending on the board type, thickness, and curvature radius.

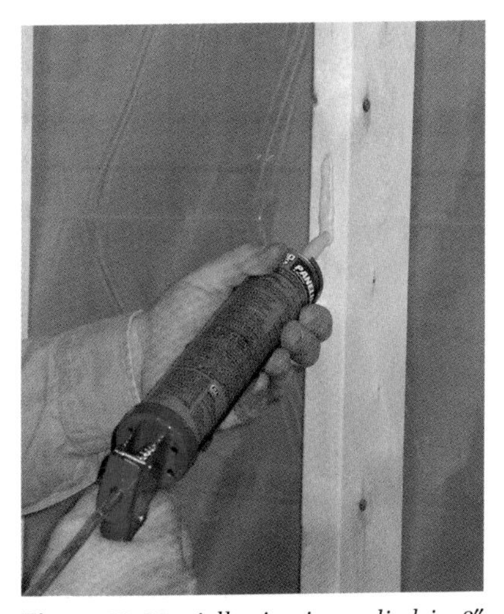

Figure 58-33. *Adhesive is applied in 3″ long beads, 6″ apart, at intermediate studs or at furring strips behind the panel.*

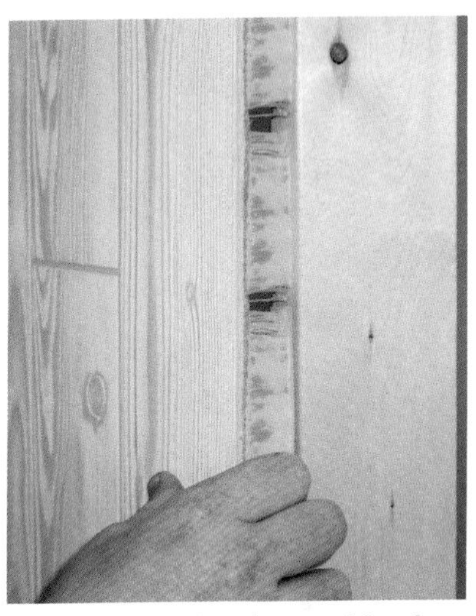

Figure 58-35. *After the panel has been pressed against the adhesive, it is grasped at the bottom along both edges and slowly pulled away from the stud. The panel is then pressed back into position after 2 min.*

Solid Board Paneling

Solid board paneling consists of solid wood boards, usually ¾″ thick and 4″ to 12″ wide. Softwood species such as redwood, fir, pine, hemlock, spruce, and cedar are used. Finishes are smooth, textured, or rough (resawn). Solid board paneling is generally placed horizontally or vertically, although diagonal designs are sometimes used. **See Figure 58-36.** Four common types of solid board panels include the following:

- board-on-board
- board-and-batten
- tongue-and-groove
- channel-rustic

Board-on-board and board-and-batten systems must be applied vertically. Tongue-and-groove and channel-rustic systems can be applied vertically or horizontally. Nails are driven into each stud or every row of blocking or furring strips. Face nails should be set below the surface and covered with wood putty.

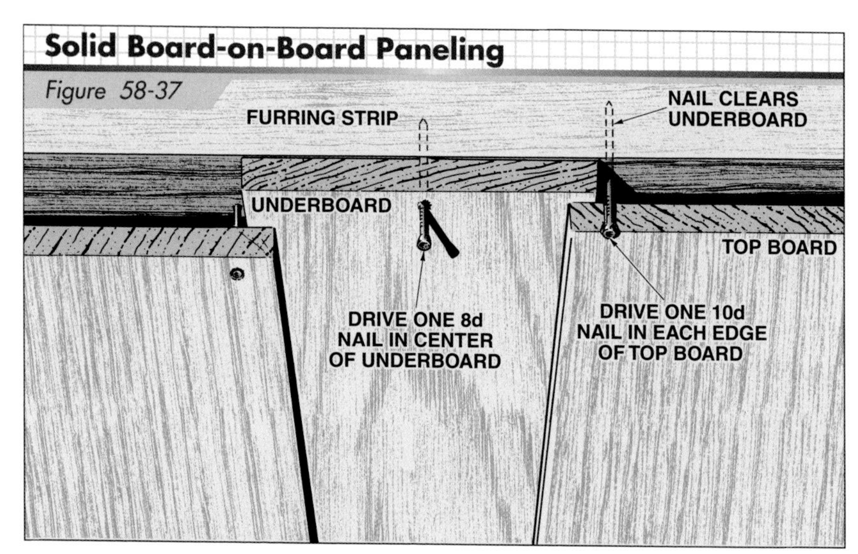

Figure 58-37. *When installing board-on-board wall paneling, drive an 8d nail at the center of the underboard. The top boards should overlap the underboard a minimum of 1″. Fasten the top board with two 10d nails. Make sure the nails clear the underboard to allow for expansion and contraction.*

Figure 58-36. *Solid board paneling may be applied horizontally.*

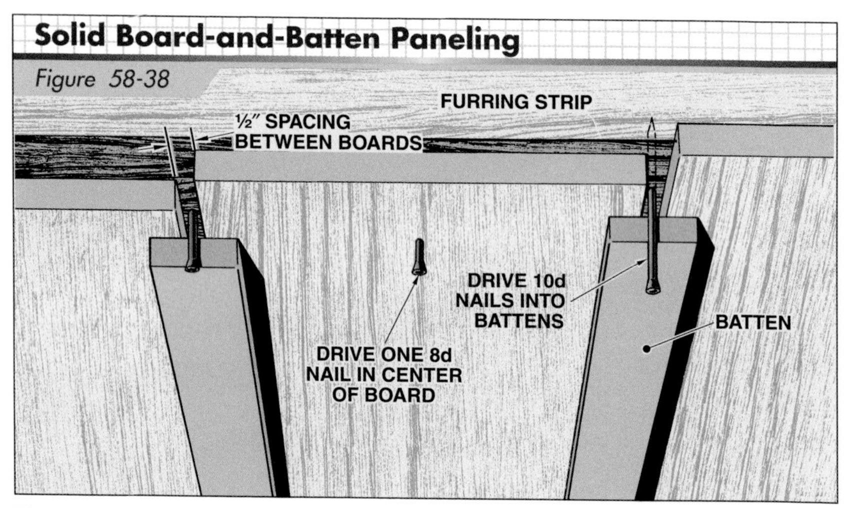

Figure 58-38. *When installing board-and-batten wall paneling, drive an 8d nail at the center of the underboard. The underboards are spaced approximately ½″ apart. Drive a 10d nail through the center of the batten.*

In the board-on-board system, one 8d nail is driven at the center of each underboard. **See Figure 58-37.** Top boards should overlap underboards a minimum of 1″. The top board is fastened with two 10d nails, one in each edge. Ensure the top board nails clear the underboard to allow for expansion and contraction.

In the board-and-batten system, one 8d nail is driven at the center of each underboard. **See Figure 58-38.** The underboards are spaced approximately ½″ apart. One 10d nail is driven at the center of the batten.

In the tongue-and-groove system, boards 4″ to 6″ wide are blind-nailed with 6d finish nails driven at a 45° angle. **See Figure 58-39.** Blind-nailing eliminates the need to countersink and putty the face nails.

In the channel-rustic system, boards up to 6″ require only one face nail. Boards 8″ or wider also require a face nail at the center of each board. **See Figure 58-40.**

Vertical Application. When boards are fastened vertically to a stud wall, blocking must be placed between the studs. **See Figure 58-41.** Masonry walls require horizontal furring strips.

Before solid board paneling is installed, the boards should first be arranged along the wall so the grain and other features can be matched as closely as possible. **See Figure 58-42.** Cut openings for electrical outlets and vents in the boards before they are placed.

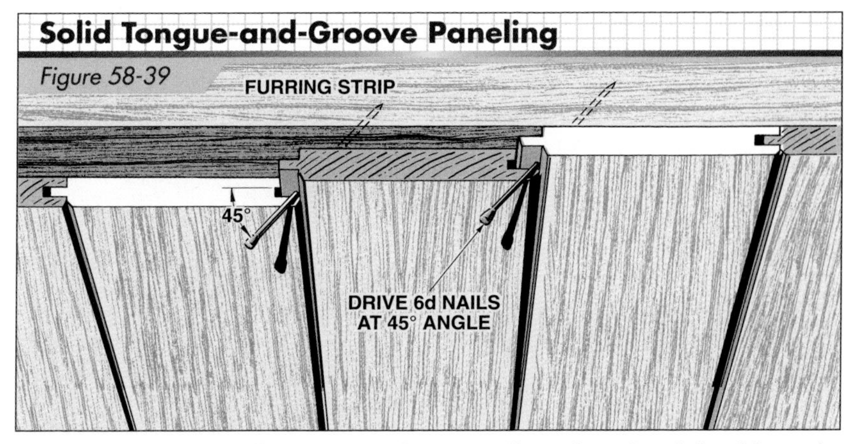

Figure 58-39. *When installing tongue-and-groove wall paneling, 4″ and 6″ widths can be blind-nailed with 6d finish nails driven at a 45° angle. This method eliminates the need to countersink and putty face nails.*

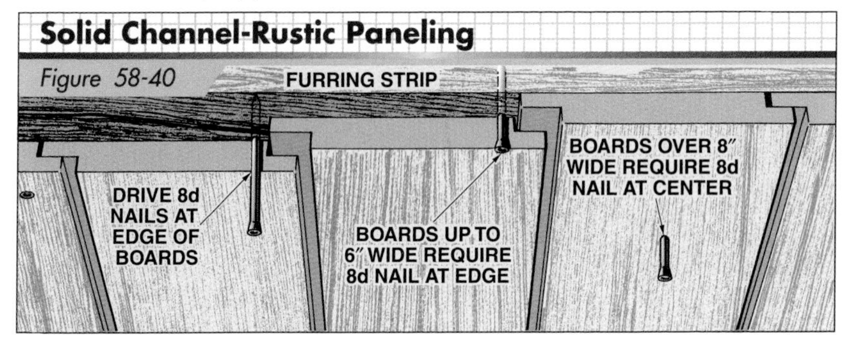

Figure 58-40. *When installing channel-rustic solid board paneling up to 6″ wide, one face nail is adequate to secure the board. Widths over 8″ require a face nail at the center of each board.*

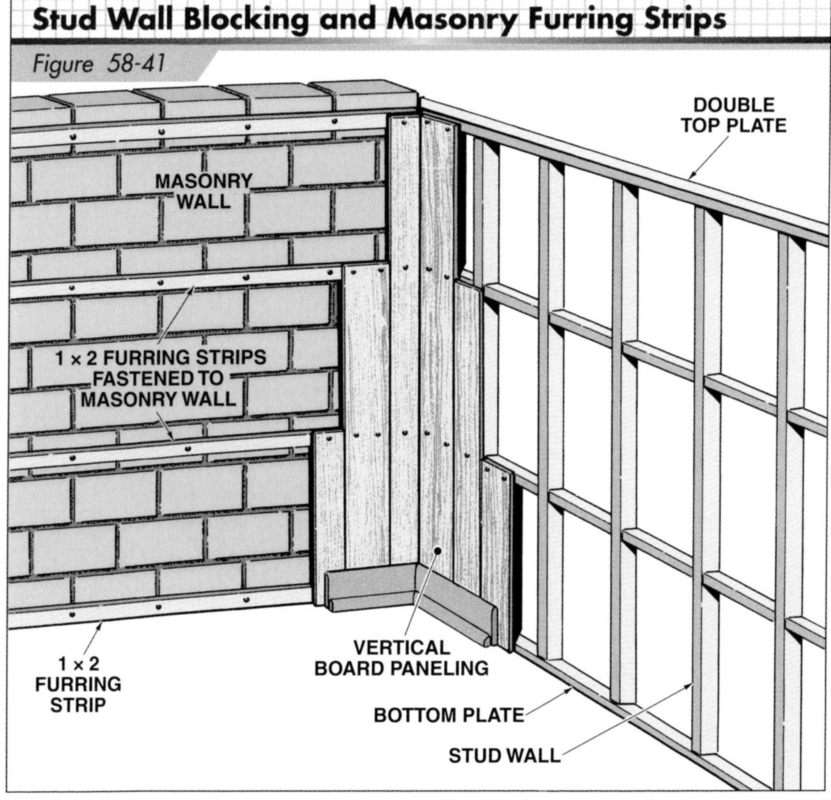

Figure 58-41. *Blocking must be placed in a stud wall to provide a nailing base for vertical board paneling.*

Figure 58-42. *Prior to installation, solid board paneling is arranged along the wall so wood grain and other features can be matched to each other.*

Before installing solid board paneling, precondition the paneling to allow it to acclimate to the moisture conditions of the room and to prevent excessive shrinkage. Approximately 7 to 10 days prior to installation, stack the boards in the room with wood strips between them.

The inside corner board is placed first, in plumb position, and scribed if necessary to match the profile of the adjoining wall. The corner board is fastened securely before the next board is placed. If boards are to be glued, the adhesive is applied to the backs of the prefitted boards. Press the board against the wall so the adhesive spreads evenly. Remove the board, wait a few minutes, then push the board back into place. When tapping a board panel into place, use a piece of material with a grooved edge against the board. **See Figure 58-43.** The grooved edge will protect the board from damage. Face nail the board at the top and bottom with 8d finish nails.

The bottoms of the boards may be covered by, or may rest on top of, a base molding. The tops of the boards may also be covered by a ceiling molding.

Figure 58-43. *When tapping a board panel into place, use a piece of material with a grooved edge against the board to avoid damaging the edge.*

A tongue-and-groove solid board paneling system is installed by blind-nailing.

Horizontal Application. The procedure for placing solid board paneling horizontally is similar to the procedure for placing exterior board panels (siding) horizontally. The first row must be perfectly level. All butt joints between the boards should be staggered and fall over a stud.

PLASTIC LAMINATE WALL COVERING

Originally, plastic laminate was used almost exclusively to surface kitchen and bathroom countertops. Today, plastic laminate is also used to cover shelves, doors, cabinets, and entire wall surfaces.

Plastic laminate is very hard and smooth, and is composed of three or four layers of plastic material bonded under high heat and pressure. Many different patterns are available, including imitation wood grain. **See Figure 58-44.**

Formica Corporation

Figure 58-44. *Plastic laminate wall panels are available with an imitation wood grain finish.*

Plastic laminate is available in 1⁄32″ and 1⁄16″ thick sheets. The 1⁄32″ thick sheets are used for vertical application only. The sheets can be fitted and applied directly to a gypsum board surface with contact cement. However, plastic laminate is often premounted on 3⁄8″ to 1⁄2″ plywood or particleboard panels, 16″ to 48″ wide and 8′ to 10′ long. The panels can be fastened directly to studs or furring strips. Panels are also available with tongue-and-groove edges, which allow blind-nailing of the tongue edge.

Most plastic laminate panel systems require a combination of nails, adhesives, and molding to fasten the panels into place. Panels sometimes used for office partitions are covered with plastic laminate on both sides and are set in tracks secured to the floor and ceiling.

CEILING

Acoustical and decorative ceiling tiles are another method for finishing ceilings. Ceiling tiles are especially practical in remodeling work since the tiles can be directly applied to existing plaster or gypsum board ceilings. Ceiling tile may also be used to form a suspended ceiling. In commercial construction, suspended tile ceiling systems are frequently used. **See Figure 58-45.**

Figure 58-45. *Suspended tile ceiling systems are frequently used in commercial construction.*

Ceiling tiles are fabricated from a variety of materials including fiberboard, mineral fiber, and glass fiber. Various colors and designs are available. Acoustical ceiling tiles absorb sound from the room directly below the ceiling and are an effective means of controlling sound reflection.

Many ceiling tiles have directional arrows on the back. When installing the ceiling tiles, always orient the ceiling tiles in the same direction. Failure to orient the tiles properly will affect the appearance of the ceiling pattern.

Applying Ceiling Tiles for Nonsuspended Ceilings

For direct application to furring strips or to an existing plaster or gypsum board ceiling in an average-size room, 12″ × 12″ tiles, 1⁄2″ or 3⁄4″ thick, with tongue-and-groove edges, are usually the most practical. **See Figure 58-46.**

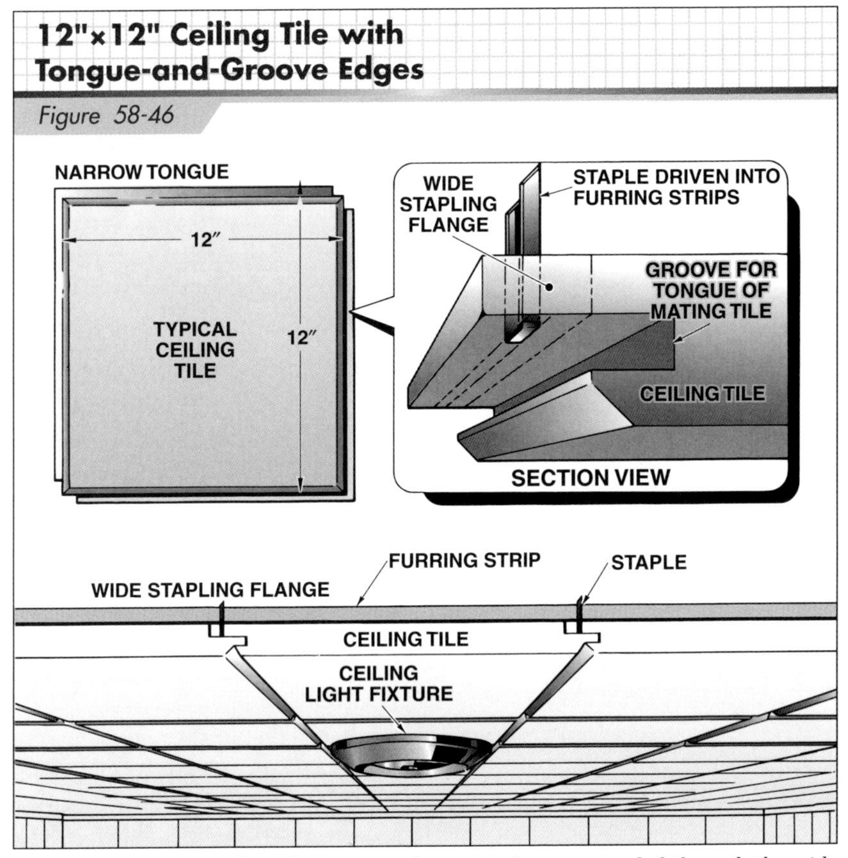

Figure 58-46. *Ceiling tile with tongue-and-groove edges are stapled through the wide flanges into the furring strips.*

Room dimensions are often in feet and inches; therefore, border tiles will not be a full 12″. Border tiles on opposite sides of the ceiling should be the same width, and should be at least 6″ wide. To ensure that the border tiles are the same width and at least 6″ wide, add the width of one tile (12″) to the inch portion of the ceiling length or width dimension and divide by 2. The answer is the proper width for the opposite border tiles. **See Figure 58-49.**

Border tiles can be cut by scoring them with a sharp knife and then snapping the tiles. A handsaw or power saw can also be used. Tiles may be slightly undercut since molding usually covers the joint between the wall and border tiles. The section of tile that is cut off will contain one of the tongued edges. Curved cuts, such as required around light fixtures, can be made with a compass saw or jigsaw.

When installing ceiling tiles directly to an existing plaster or gypsum board ceiling, center-lines are snapped on the ceiling at a right angle (90°) to each other. Ceiling tile placement will begin from the centerlines, ensuring that the first row of tiles will be straight and square to each other.

For ceilings with even-foot dimensions, such as 8′ × 10′ or 10′ × 12′, and using 12″ × 12″ tiles, the center tiles are placed on each side of the centerlines. This placement results in full 12″ border tiles. **See Figure 58-47.** For ceilings with odd-foot dimensions, such as 9′ × 13′, and using 12″ × 12″ tiles, the tiles are centered over the centerlines in order to produce full 12″ wide border tiles. **See Figure 58-48.**

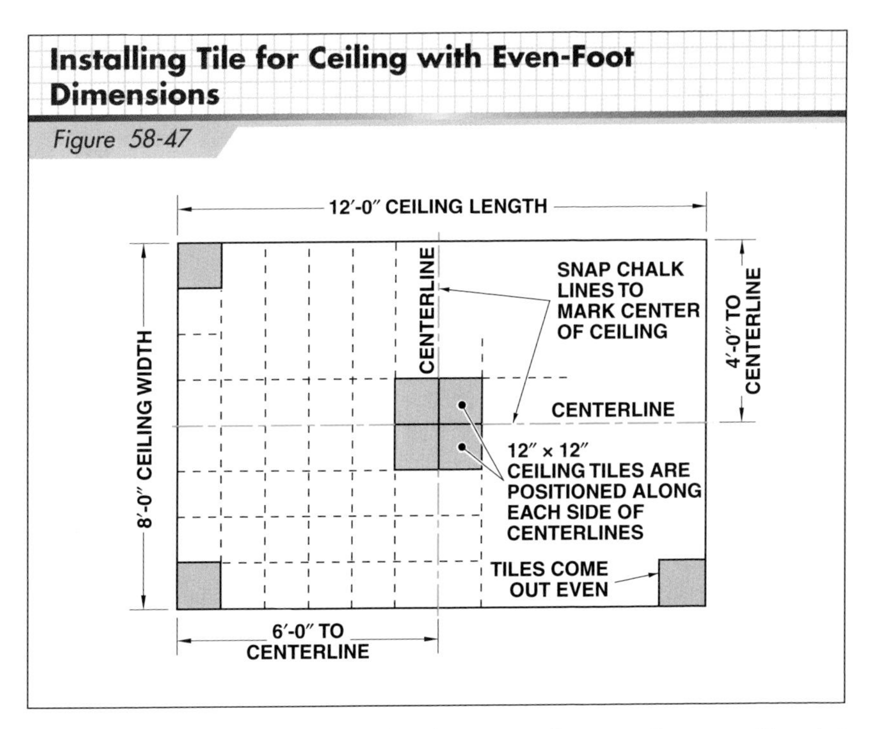

Figure 58-47. *For a ceiling with even-foot dimensions, the center tiles are positioned at each side of the centerlines.*

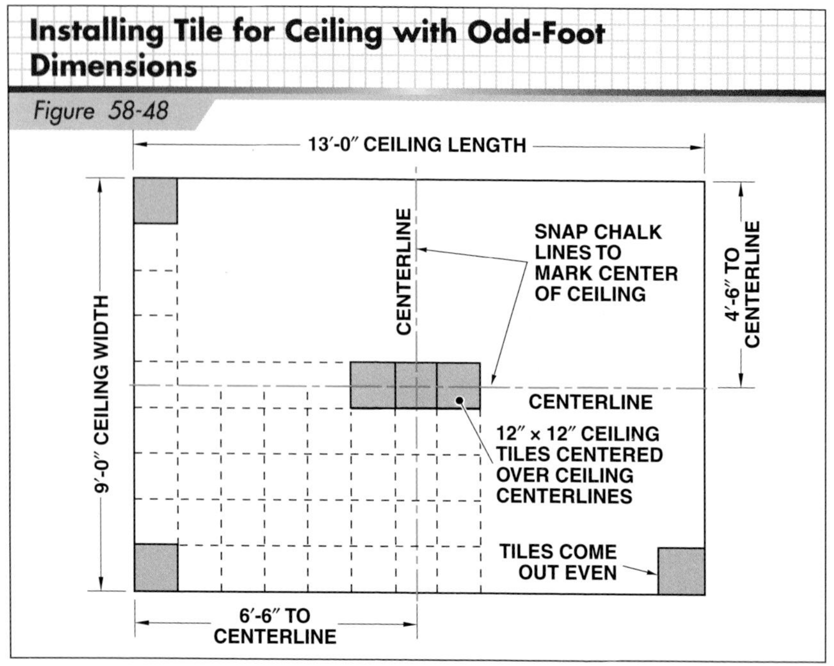

Figure 58-48. *For a ceiling with odd-foot dimensions, the center tiles are positioned so the centerlines align with the middle of the tiles.*

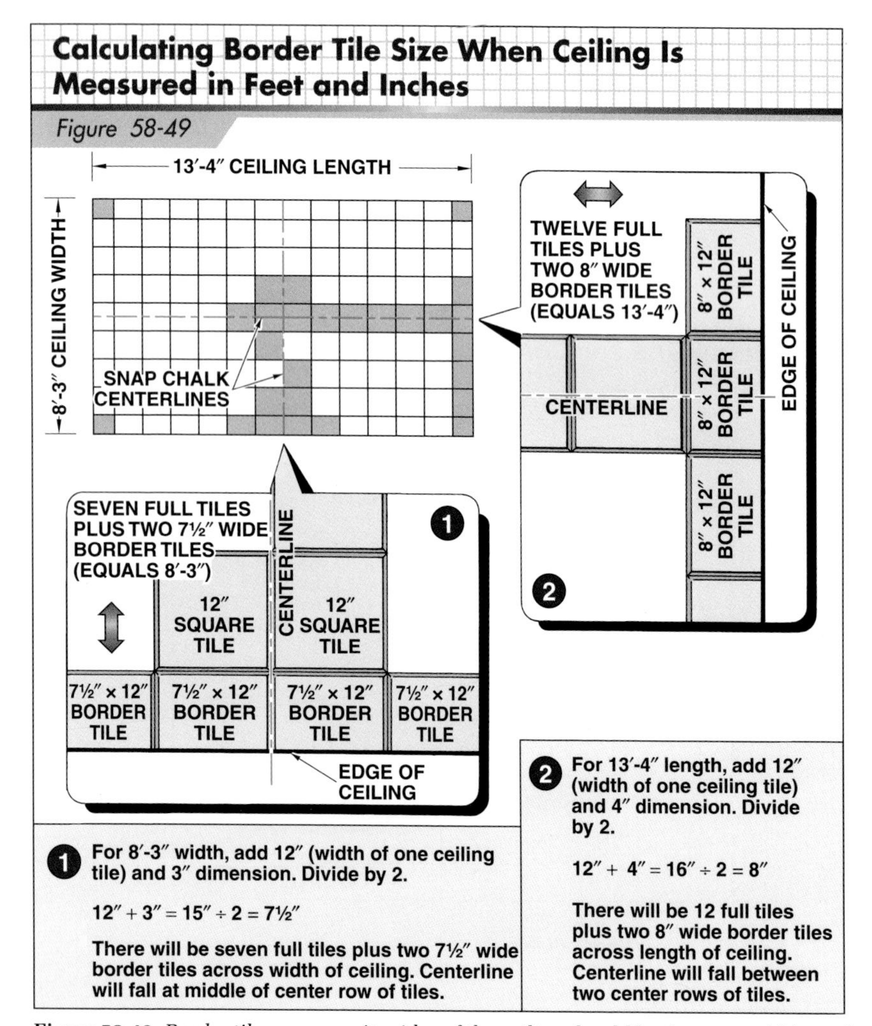

Figure 58-49. *Border tiles on opposite sides of the ceiling should be the same width, and should be at least 6″ wide.*

If the existing plaster or gypsum board ceiling is firm and in good condition, tile can be directly applied to the ceiling using adhesives. Centerlines are snapped on the ceiling. Tiles are applied at the center of the ceiling, working outward toward the walls. A 1½″ to 2″ diameter bead of adhesive is placed about 2″ in from each corner of the tile. The tile is then firmly pressed to the ceiling and against the adjoining tiles.

In new construction, furring strips are fastened directly to the ceiling joists. For remodeling jobs, furring strips are often used when the existing plaster or gypsum board ceiling is wavy, cracked, or flaking. The furring strips are nailed directly over the existing ceiling with nails that are long enough to penetrate into the joists. Shims are used to level low spots. Furring strips are placed 12″ OC according to the layout established for the tile lines. Ceiling tiles are usually fastened to furring strips with staples driven through the wide flange at the grooved edge.

Applying Ceiling Tiles for Suspended Ceilings

For a suspended ceiling, a light metal grid is hung by wire from the original ceiling or from ceiling joists. Tiles (typically 2′ × 2′ or 2′ × 4′) are then placed in the frames of the metal grid. **See Figure 58-50.** Suspended ceilings are often used in commercial construction and in residential structures with high original ceilings.

One advantage of a suspended ceiling is that it reduces the sound traveling from the floor above and increases the insulating capability of the ceiling. Suspended ceilings also allow the use of recessed lighting. **See Figure 58-51.** Pipes, wires, and ductwork can be conveniently run above the suspended ceiling.

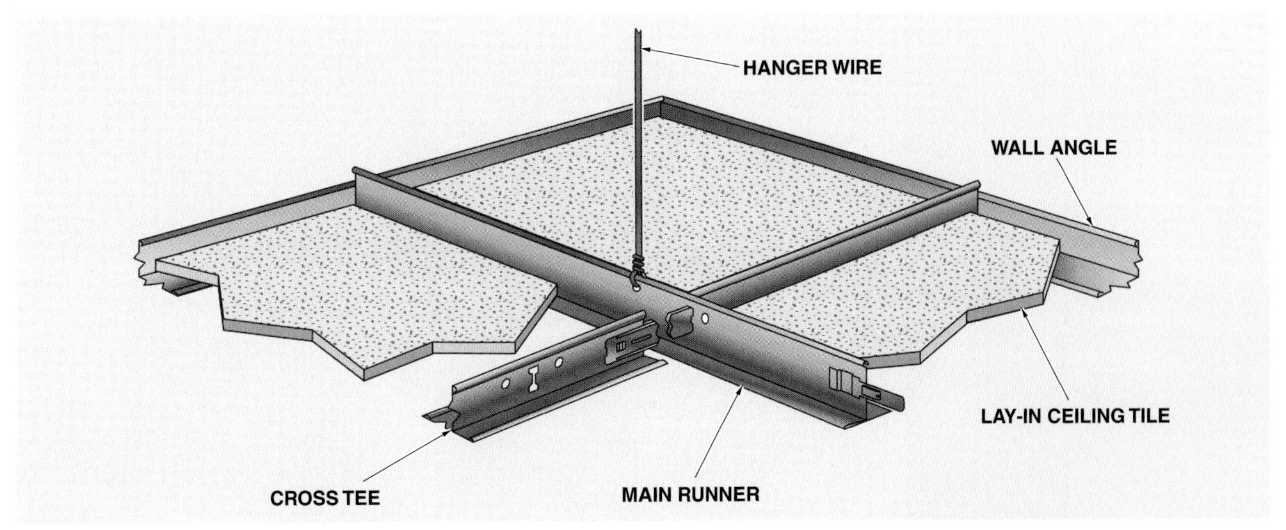

Figure 58-50. *Suspended ceilings are often installed in commercial buildings.*

Figure 58-51. *Recessed lighting can be used with suspended ceilings. Plumbing, lighting, and ventilation fixtures must be properly secured to structural members above a suspended ceiling.*

Many ceiling tiles have directional arrows on the back side. When installing the ceiling tiles in a suspended ceiling, orient all tiles in the same direction.

The first step in installing a suspended ceiling is to snap lines on the wall to establish the correct height for the ceiling. A laser level is commonly used to establish the ceiling height around the entire room. **See Figure 58-52.** *Wall angles* are fastened to the wall. *Main runners* of the metal grid are suspended from the ceiling or other structural elements with hanger wires. **See Figure 58-53.** *Cross tees* are placed between the main runners and secured in position using tabs at the ends of the cross tees, which engage in slots of the runners. Finally, ceiling tiles are placed in the grid flanges.

PLS • Pacific Laser Systems

Figure 58-52. *A laser level is commonly used to establish the suspended ceiling height at several locations along the wall.*

Figure 58-53. *A suspended ceiling framework consists of main runners and cross tees, which are supported by hanger wires. Ceiling tile panels are supported by the runners and cross tees.*

Unit 58 Review and Resources

UNIT 59

Interior Doors and Hardware

OBJECTIVES

1. List and describe components of a door unit.
2. List and describe styles of flush doors.
3. Identify common types of panel doors.
4. Describe the methods of finishing interior door openings.
5. Explain how head and side jambs are sized.
6. Describe the proper procedure for installing doorjambs.
7. Describe the installation of door stops.
8. Explain how doors are fitted.
9. Describe the procedure for hanging doors and installing hinges.
10. Explain how to determine and solve door fit problems.
11. Describe the procedures for fitting and hanging doors using power tools.
12. Describe how prehung door units are installed.
13. List and describe specialty doors.
14. Describe the procedure for installing metal door frames.
15. Review advantages of steel doors.
16. Identify common types of door locks and hinges.
17. List and describe other types of door hardware.

Interior doors are usually installed after the walls have received their finish covering. The basic construction of an interior door differs little from an exterior door. Waterproof adhesive is required for exterior doors; it is not always used for interior doors. Standard exterior doors are 1¾″ thick, while standard interior doors are often 1⅜″ thick. Some interior doors, however, are as thick as exterior doors.

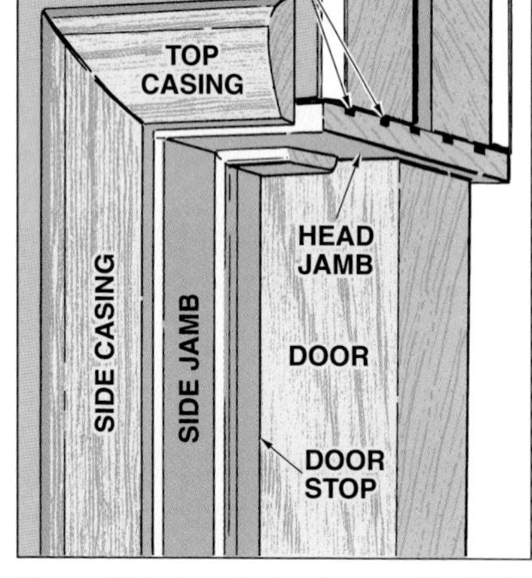

Figure 59-1. *An interior door unit includes a doorjamb, stops, and a casing.*

Stock sizes of interior doors range in width from 1′-6″ to 3′-0″. Common door height is 6′-8″ in residential construction and 7′-0″ in commercial construction.

A traditional interior door unit includes a *doorjamb, stops,* and a *casing*. **See Figure 59-1.** The doorjamb is the finish frame in which the door hangs. Stops hold the closed door in its proper position. Casing is the trim placed around a doorjamb.

Another type of door unit features a doorjamb that is set back from the finish wall. No casing is required. **See Figure 59-2.**

Doorjamb edges are slightly beveled to fit the casing tightly.

Figure 59-2. *Recessed doorjambs are popular with some building designs.*

INTERIOR DOOR STYLES

Two basic styles of interior doors are *flush* and *panel* doors. Flush doors, which have a smooth surface, are often used for interior doors. However, traditional panel doors are very popular. Plywood veneer, hardboard, and vinyl are common finishes for interior doors.

Flush Doors

A flush door consists of a frame covered with plywood or hardboard face panels (skins) about ⅛″ thick. The two main types of flush doors are *hollow-core* and *solid-core*.

Hollow-Core Doors. Hollow-core doors are less expensive than solid-core doors. One type of hollow-core door has a cellular core that provides backing for the face panels. **See Figure 59-3.** Hollow-core doors are not recommended for exterior doors since they do not provide adequate insulation or security capabilities.

Solid-Core Doors. Solid-core doors are heavier than hollow-core doors, provide better acoustical insulation, and have less tendency to warp. The core material is usually particleboard or staggered wood blocks. **See Figure 59-4.** One type of solid-core door has a fire-resistant mineral core and is known as a *fire door.* **See Figure 59-5.**

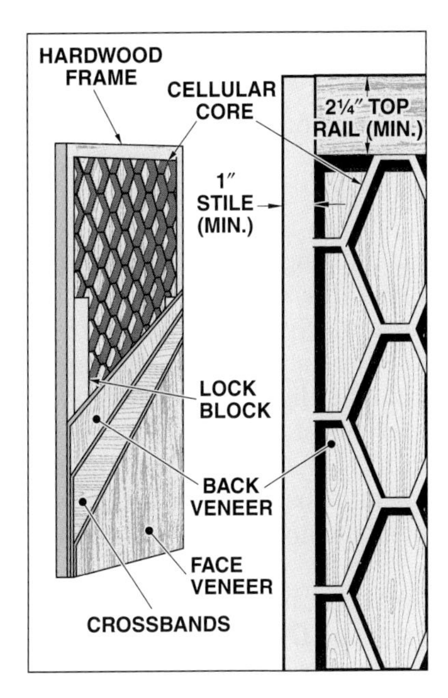

Figure 59-3. *One type of hollow-core door has a hardwood frame and a cellular core, which is covered by veneer, hardboard, or vinyl face panels. A lock block is located where the door lock hole will be drilled.*

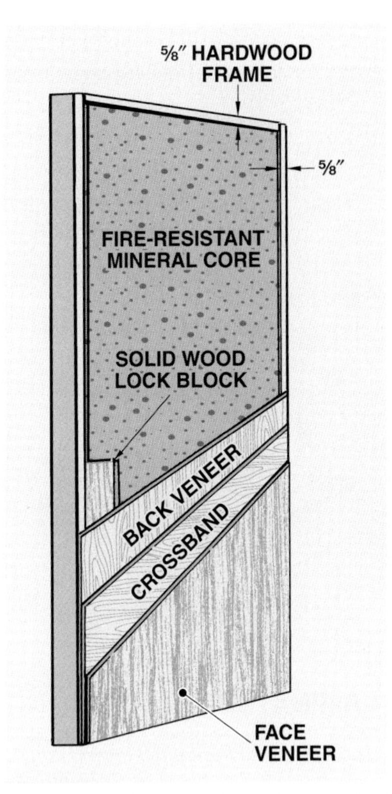

Figure 59-5. *A fire door has a fire-resistant mineral core.*

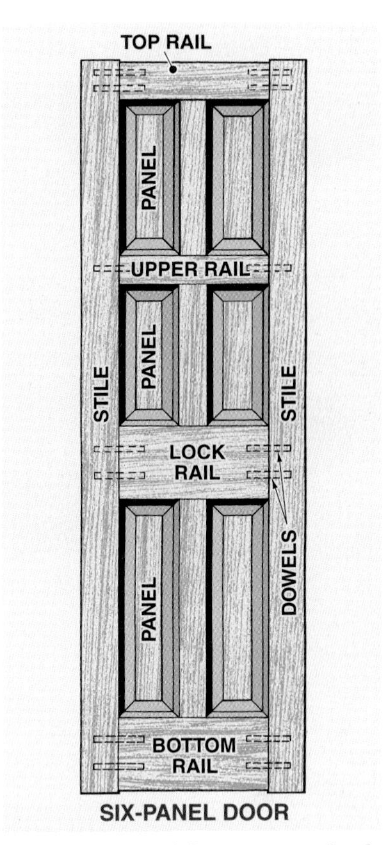

Figure 59-6. *Panel doors consist of rails, stiles, and panels.*

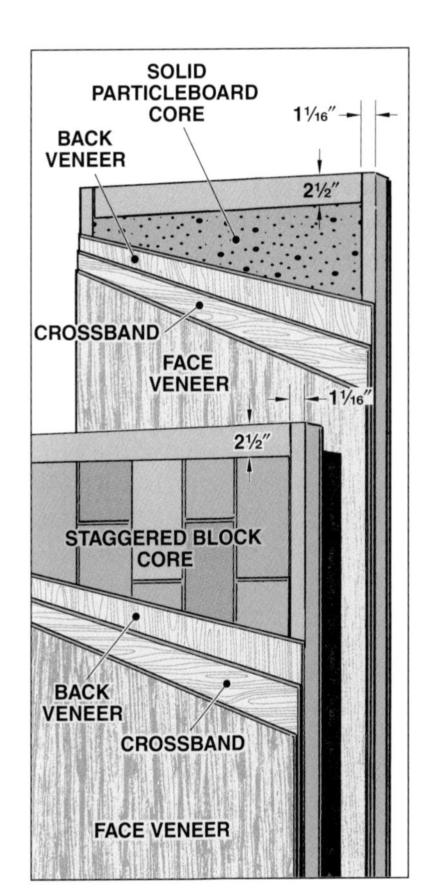

Figure 59-4. *Solid-core doors may have a particleboard or staggered block core.*

Panel Doors

Panel doors are also known as stile-and-rail doors because stiles (vertical members) and rails (cross members) are doweled and glued together to construct the frame. **See Figure 59-6.** Panels fit into the grooved edges of the frame. Panel doors are manufactured in many designs. **See Figure 59-7.**

Material for stiles and rails is usually 1⅜″ or 1¾″ thick. The material is available in solid pieces or in pieces made up of laminated layers and covered with a face veneer. **See Figure 59-8.** Panels may be plain or raised. Raised panels are usually ¾″ thick plywood tapered at the ends to fit into the frame grooves. Plain panels are usually ¼″ thick plywood.

An effect similar to a panel door can be obtained by attaching molding to the surface of a flush door. Panel designs may also be routed on solid-core flush doors. **See Figure 59-9.**

Figure 59-7. *Panels may be arranged in various ways to provide different appearances.*

Panel Door Construction

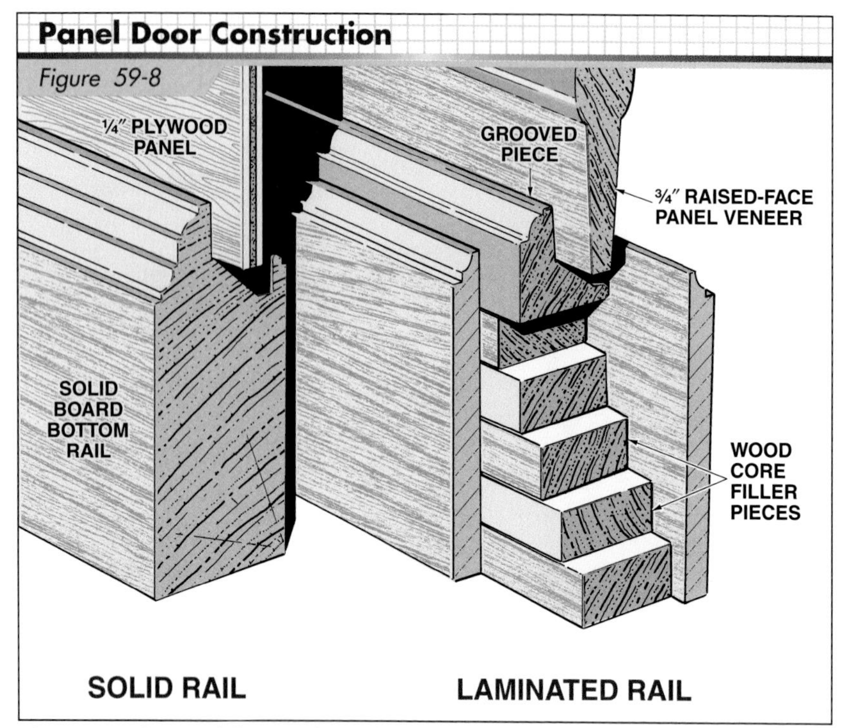

Figure 59-8. *On panel doors, stiles and rails are solid pieces or laminated layers. Panels may be plain or raised. Raised panels are 3/4" plywood tapered at the ends to fit into the grooves of the frame. Plain panels are usually 1/4" thick plywood.*

Figure 59-9. *Solid-core flush doors may be routed out to provide the effect of panel doors.*

Another type of panel door gives the appearance of beveled planks extending from the top to bottom rail with a horizontal piece placed at the lock level. **See Figure 59-10.**

Figure 59-10. *Plank doors give the appearance of beveled planks extending from the top to bottom rail, with a horizontal piece placed at lock level.*

FINISHING DOOR OPENINGS

An interior door opening may be finished in several ways. The traditional method is to set the doorjamb first, attach the casing and stops, and then hang the door. This method is still used, particularly in remodeling and commercial work. In residential and other light construction, various types of *prehung door* systems are most often used. Finish carpenters must be skilled in all the different methods of door installation.

Construction Prints and Interior Doors

Construction prints often include a door schedule, which provides the widths, heights, and thicknesses of doors to be hung in the building. Some prints indicate door sizes on the floor plan. The direction in which a door swings is always shown on the floor plan. Section drawings are usually provided, indicating necessary information about the doorjamb, casing, and stop material to be used.

Doorjambs

A traditional jamb assembly includes a head jamb and two side jambs. **See Figure 59-11.** The two side jambs are dadoed to receive the head jamb. Kerfs (grooves) on the back of the jamb stock prevent later cupping of the material. The back edge of the jamb should be slightly beveled to ensure a tight fit when the casing is nailed against the jamb.

Two traditional doorjamb designs—*rabbeted* and *loose-stop*—are shown in **Figure 59-12.** The rabbeted design includes a one-piece jamb and door stop, and does not require an additional stop to be attached.

The total width of the doorjamb includes the width of the wall stud, plus the width of the wall finish material, plus 1/8" to allow for small variations in the wall.

A standard-width doorjamb for 2 × 4 stud walls covered with ½″ gypsum board on each side is 4⅝″ (3½″ + 1″ + ⅛″ = 4⅝″).

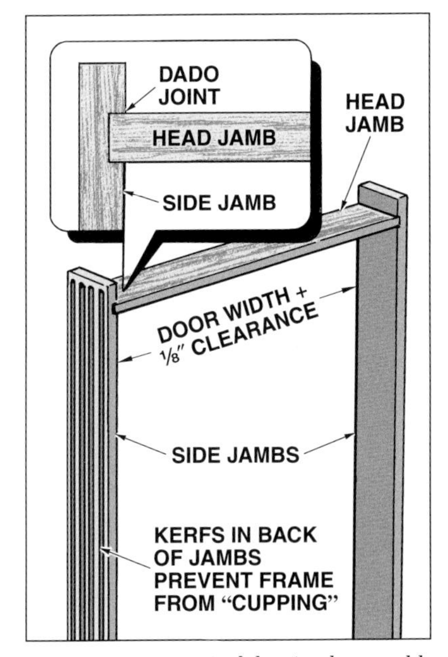

Figure 59-11. *A typical doorjamb assembly includes a head jamb and two side jambs.*

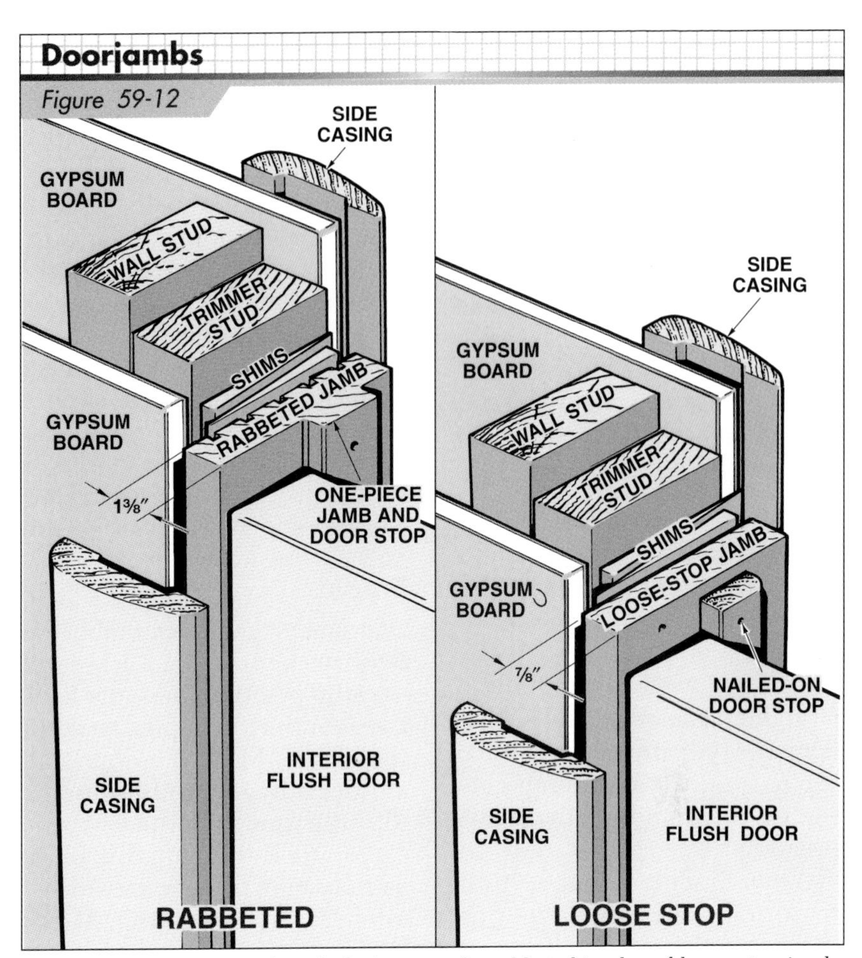

Figure 59-12. *Two typical jamb designs are the rabbeted jamb and loose-stop jamb.*

Assembling Doorjambs. Side jambs are delivered from the supplier a few inches longer than what is needed for the door opening. To figure the exact length needed, the necessary clearance under the door is added to the door height. **See Figure 59-13.** If a thin finish floor material, such as ⅛″ vinyl tile, is to be laid, ½″ clearance between the door and subfloor is adequate. If the floor is to be finished off with thicker finish material, such as carpet or hardwood flooring, additional clearance is necessary. Adequate clearance eliminates the task of cutting the door bottom after installation.

The width of the finish opening should be ⅛″ greater than the door width. This minimizes the amount of planing required to fit the door. The total length of the head jamb includes the door width, the depth of the two dado cuts in the side jambs, plus ⅛″.

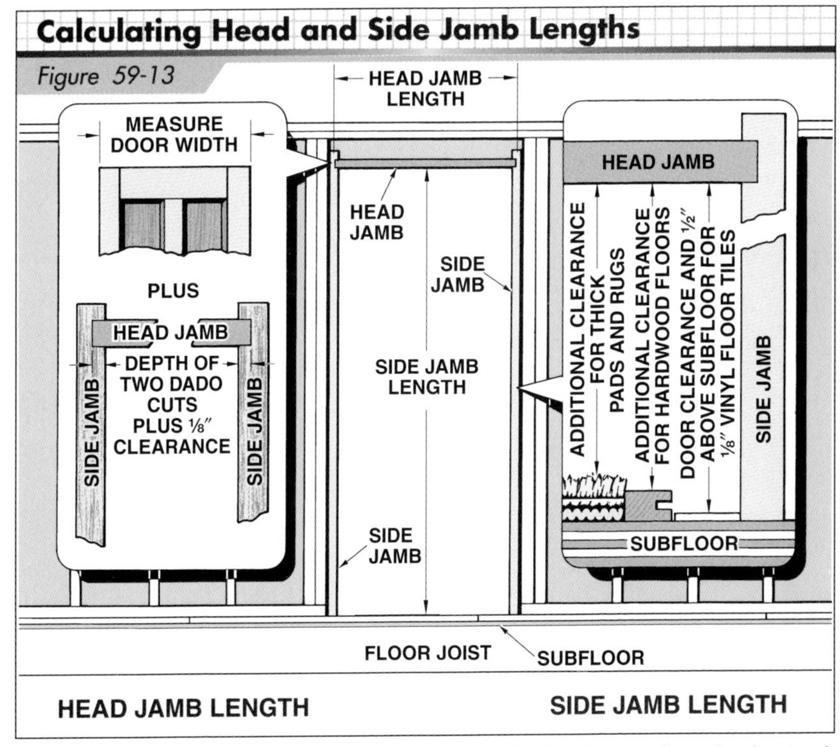

Figure 59-13. *Head and side jamb lengths must be calculated properly so the doorjamb fits in the rough opening.*

When the head jamb and the two side jambs have been cut to their proper lengths, fasten them together with 8d box or casing nails.

If the subfloor is concrete and thin flooring material is to be used, first determine whether the floor is level from one side of the opening to the other. If the subfloor is not level, compensate by cutting the side jamb to account for the amount the floor is out of level.

Installing Doorjambs. Tools needed for installing doorjambs are a carpenter's level, straightedge with stand-off blocks at the ends, and framing square. Shims are used to plumb and straighten the side jambs. Finish nails (8d) are recommended for nailing the doorjamb into the trimmer studs of the rough opening.

When first placing the shims to plumb and align the side pieces of the jamb, some carpenters drive a single nail at the center of the stock above or below (but not through) the shim. Screws may be used to attach the doorjamb. This practice allows the shims to be readjusted, if necessary, without splitting them.

After all the shims have been placed and the frame is perfectly plumb and straight, nails are driven near the two edges of the jamb and through the shims. The shims are then cut flush with the jamb. The saw should be slightly tilted so as not to scar the edge of the jamb. A complete procedure for installing doorjambs is shown in **Figure 59-14.**

Stops

Doorjambs that are not rabbeted require door stops. To lay out the stop position, measure back the door thickness plus an additional 1⁄16″. Mark the doorjamb at intervals. Cut and fasten the top stop and then the two side stops. Tack the stops in place as they may have to be shifted after the door is hung. When the stop on the hinge side is permanently nailed, it should be 1⁄16″ from the door. Many carpenters prefer to wait until the door lock has been fitted before permanently nailing the stop on the lock side. The stop is then adjusted to the door, allowing for a slight amount of movement when the door is in a closed position.

Butt joints are most often used with door stops but miter joints may also be used. **See Figure 59-15.**

FITTING AND INSTALLING DOORS

For commercial construction, doors are commonly delivered to the job site pre-machined. However, in some cases, a door must be fitted before it is hung. A door must be held in a stable position while fitting and hanging it. A *door holder* (or door jack) is used to keep a door in an upright position. Manufactured devices are available, or a job-built door holder can be built with a wedge and a notched 2 × 4. **See Figure 59-16.** Other tools required to fit and install a door are a power plane, butt gauge or butt marker, wide chisel, hammer, and power screwdriver.

Fitting Doors

Place the door in the frame. The door will easily slip into place if the doorjamb is 1⁄8″ wider than the door width. If the door is too wide, plane it to fit.

With the door in the opening, check the sides and top. Using a scriber, mark where the door will have to be fitted. **See Figure 59-17.** Remove the door, place it in the holder, and plane where needed. Do not plane off too much at one time. The strike edge of the door is planed at approximately a 3° bevel so the leading edge does not hit the jamb as the door is closed. Frequently check the door in the opening to ensure proper fit and clearance. **See Figure 59-18.**

Instruction in proper door installation procedures is essential to ensure the correct fit and operation of doors.

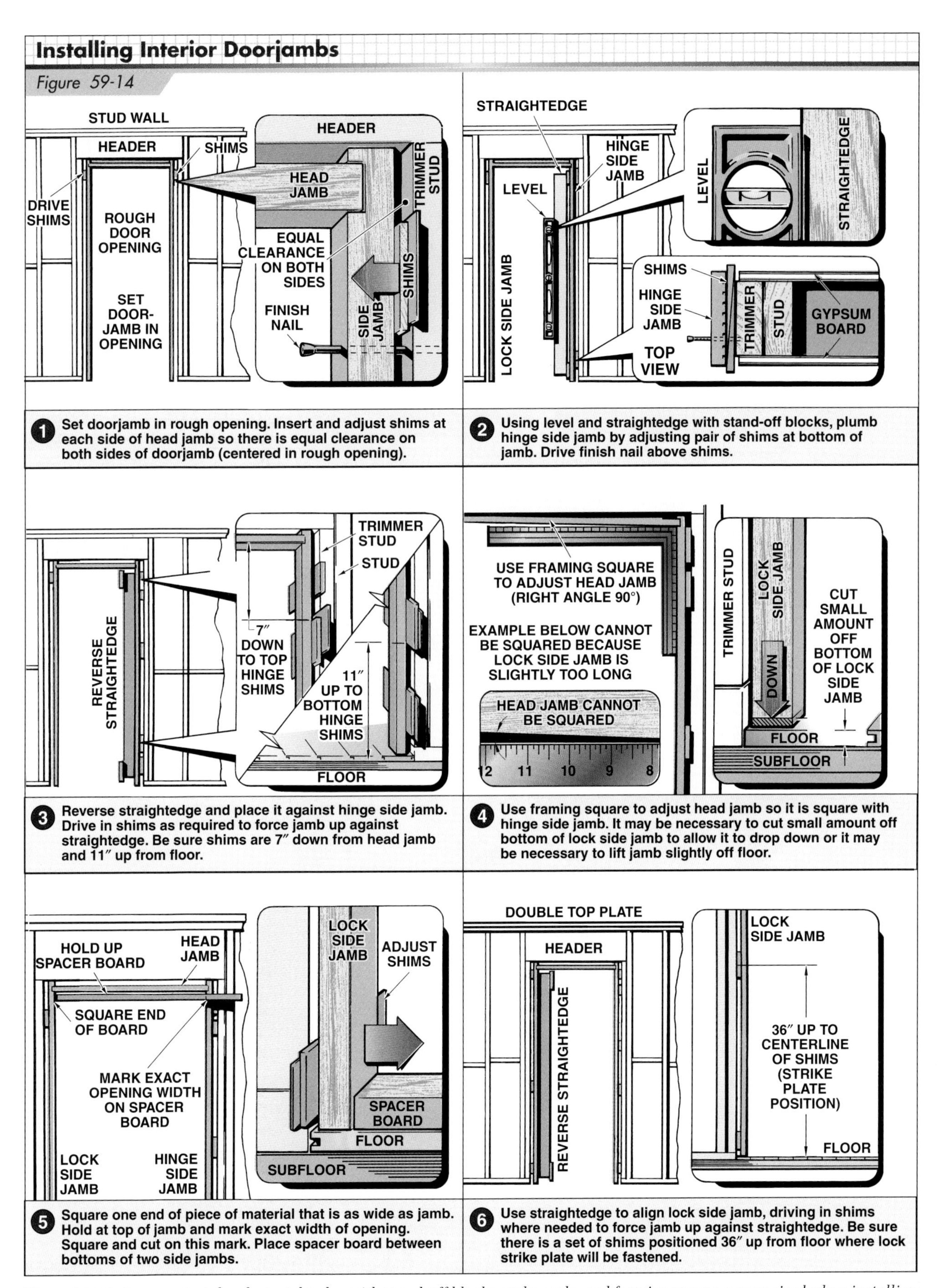

Figure 59-14. *A carpenter's level, straightedge with stand-off blocks at the ends, and framing square are required when installing a doorjamb.*

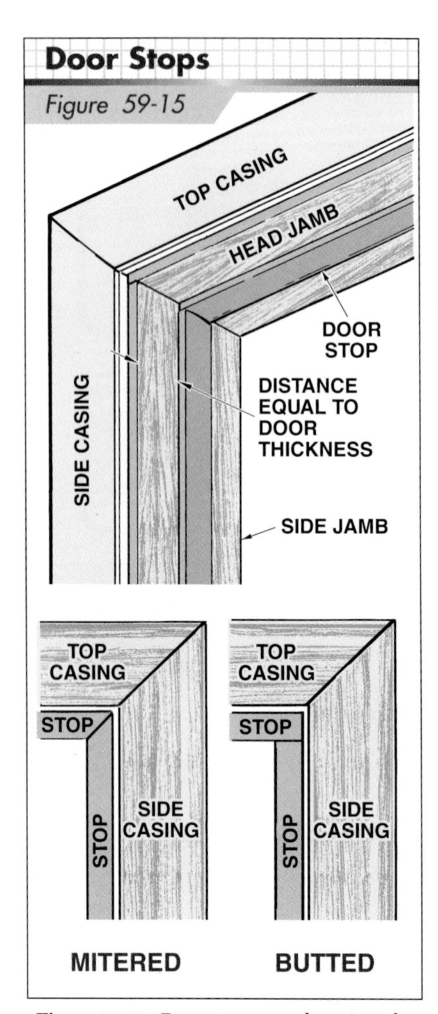

Figure 59-15. *Door stops may be mitered or butted. The top stop is always installed first.*

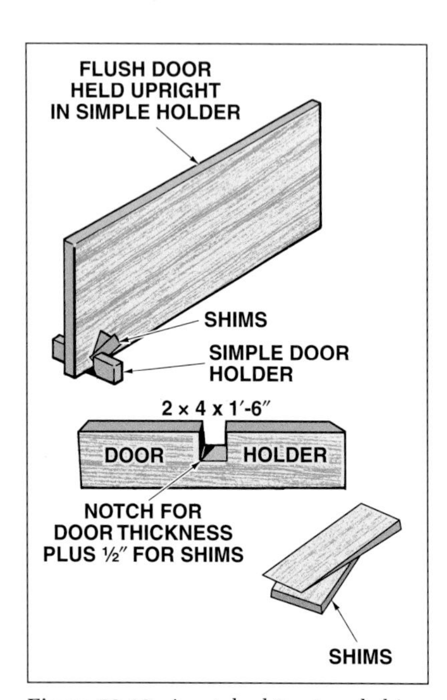

Figure 59-16. *A notched 2 × 4 and shims can serve as a door holder for fitting and installation operations.*

Figure 59-17. *With the door in the opening, check the sides and top to see where the door may need to be fitted.*

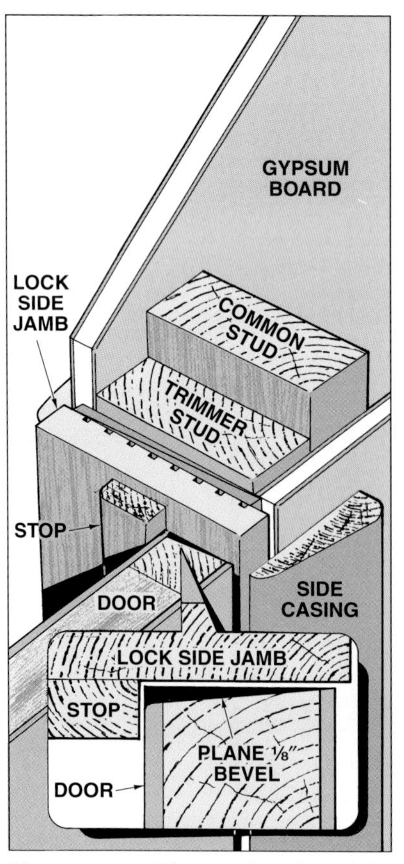

Figure 59-18. *The inside edge of the lock side of the door should be beveled approximately 3° to prevent the inside edge from scraping against the doorjamb as the door is closed.*

Immediately after cutting and fitting a wood door, seal all cut surfaces and ends with a sanding or wood sealer.

Hanging Doors

The hinge (butt) size required to hang a door depends on the thickness, width, and total weight of the door. **See Figure 59-19.** Most doors are hung with *loose-pin hinges.* The pin that fits into the barrel of a loose-pin hinge is removable. As a result, the door can be removed from the frame without unscrewing the hinges. **See Figure 59-20.**

DOOR HINGE SIZES

Door Thickness*	Door Width*	Hinge Size†
1⅛ to 1⅜	Up to 32	3½
1⅛ to 1⅜	32 to 37	4
1⅜ to 1⅞	Up to 32	4½
1⅜ to 1⅞	32 to 37	5
1⅜ to 1⅞	37 to 43	5 Extra Heavy
Over 1⅞	Up to 43	5 Extra Heavy
Over 1⅞	Over 43	6 Extra Heavy

* in in.
† leaf length (in in.)

Figure 59-19. *Hinge size is identified by its leaf length. Hollow-core doors use two hinges. Heavier solid-core exterior doors require three hinges.*

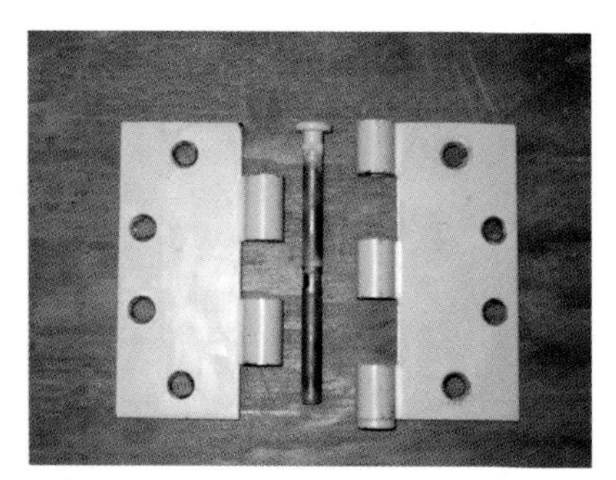

Figure 59-20. *The hinge pin of loose-pin hinges can be removed to allow the door to be taken out of the frame.*

Locating Hinges. After the door has been fitted, the location of the hinges can be laid out. Traditionally, the upper hinge is located 7″ from the top of the door and the lower hinge is located 11″ from the bottom. Another accepted method is to place hinges the same distance from the top of the door as from the bottom of the door—usually 8″ to 10″. Heavier doors and doors over 6′-8″ in height require a third hinge centered between the top and bottom hinges. The hinge leaves must project by at least one-half the casing thickness in the direction that the door swings. **See Figure 59-21.**

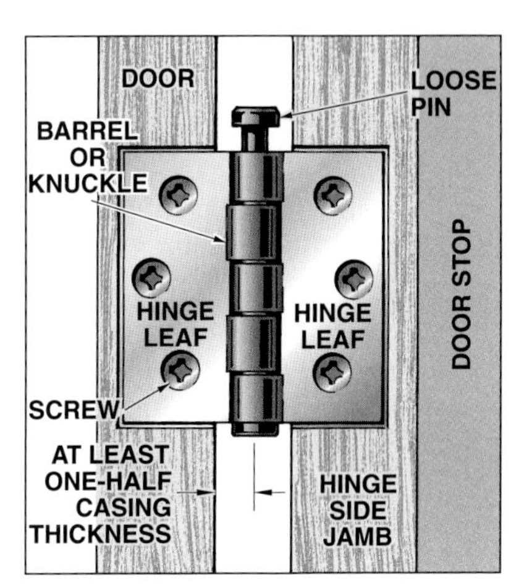

Figure 59-21. *Hinge leaves must project from the door and doorjamb by at least one-half the casing thickness.*

A procedure for marking hinge locations on the door and jamb is shown in **Figure 59-22.** In this example, 7″ and 11″ hinge measurements are used. The door must be placed in the opening at the exact position it will hang, allowing 3/32″ clearance between the top of the door and the head jamb. Lightly draw an arrow on the door edge indicating the top and hinge side of the door. The arrow will ensure proper placement of the door when it is taken out of the opening and placed in the door holder.

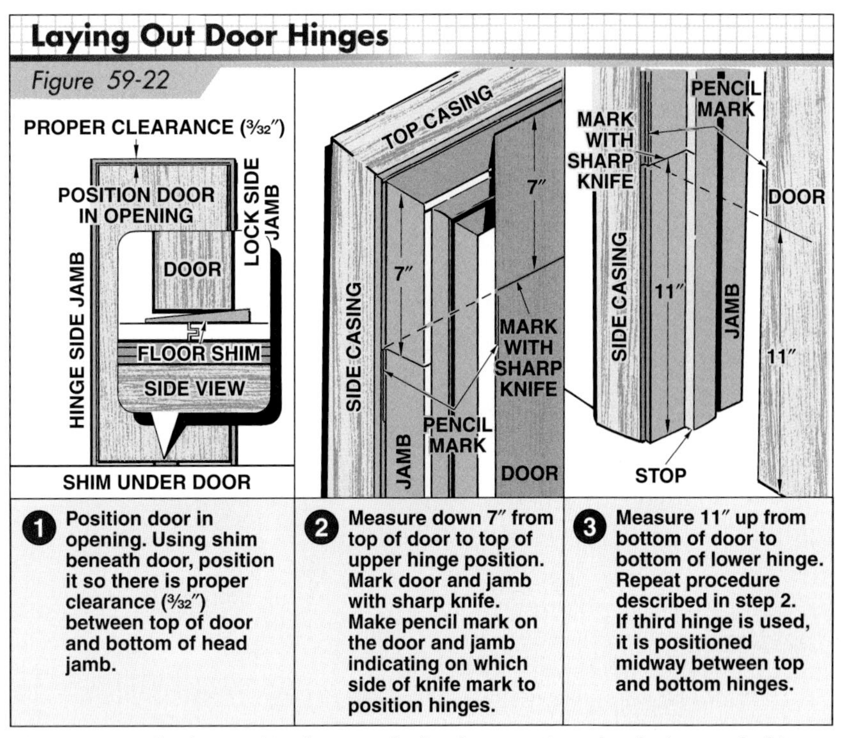

Figure 59-22. *The door and jamb are marked at the same time when laying out the hinges. In this example, the upper hinge is 7″ from the top of the door and the lower hinge is 11″ from the bottom of the door.*

Mortising Gains and Installing Hinges. The door and doorjamb must be *mortised* (notched out) for the hinge to be flush with the surface of the wood. One method of layout is to place the hinge in position and mark around it with a sharp knife. **See Figure 59-23.**

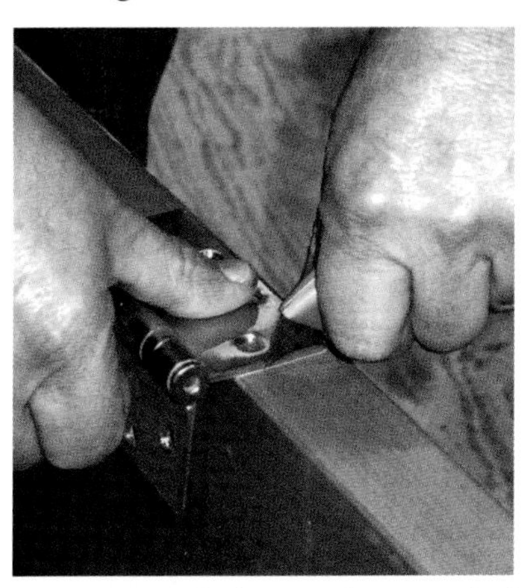

Figure 59-23. *The hinge outline is marked with a sharp knife.*

Cafe doors are short panel or louver doors, and are hung in pairs that swing both directions.

Two tools that make hinge layout and mortising easier are a *butt gauge* and *butt marker.* Both tools quickly mark the door and doorjamb with the outline of the hinge and its depth of gain. The butt gauge has two cutters. One is adjusted to mark the depth and the other is adjusted to accommodate different door thicknesses. **See Figure 59-24.** A butt marker automatically marks the hinge outline and depth of gain when it is struck with a hammer. **See Figure 59-25.**

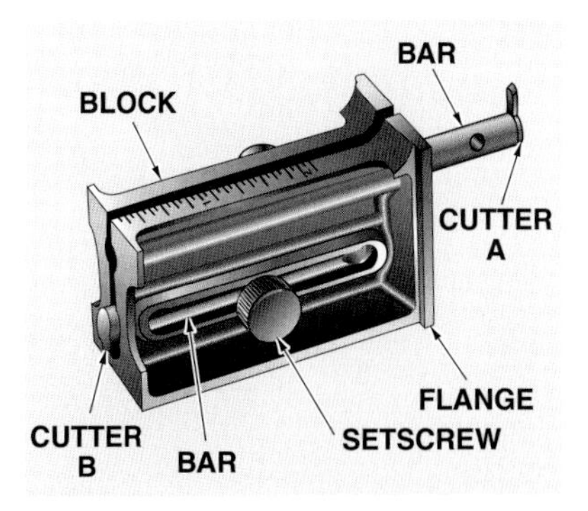

Figure 59-24. *A butt gauge is used to assist in hinge layout. Cutter A is adjusted to the width of the hinge gain. Cutter B is adjusted to the gain depth.*

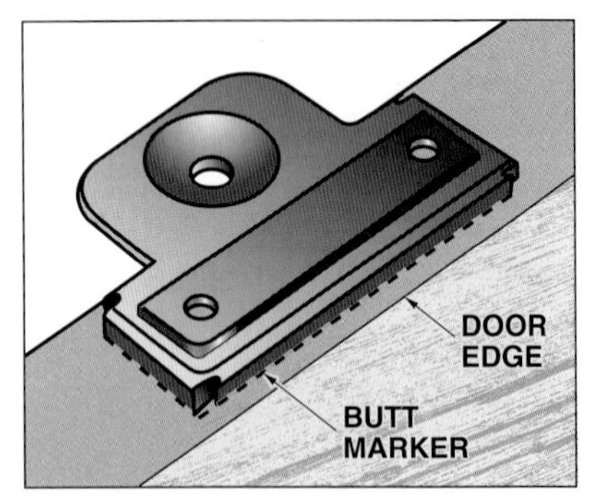

Figure 59-25. *A butt marker is used for hinge layout. The hinge outline and gain depth are marked in the door and doorjamb when the butt marker is hit with a hammer.*

The gain should be removed with a butt chisel. **Figure 59-26** shows a procedure for mortising gains and installing hinges. After the gain is mortised, the hinge leaf is screwed in place. After one leaf of each hinge has been attached to the door, the jamb is mortised and the matching leaves are attached. When the door is hung, the leaves of the hinges are lined up and pushed together, and the loose pins are replaced. **See Figure 59-27.**

Adjusting Hinge Clearance. A properly fitted door has equal clearance along the top and two sides; approximately 1/8″ clearance is recommended. Too little clearance causes the door to bind. Too much clearance looks sloppy. The gap between the bottom of the door and floor should be 5/8″ to 3/4″ unless a threshold is to be installed. Proper clearances should be checked after the door is hung. Some clearances can be adjusted without removing the door. Loosen the screws on the hinge leaf on the jamb or door and place a thin shim such as a cardboard strip toward the front or the back of the leaf, depending on the desired direction of movement. Then retighten the screws. The leaf will move in or out as shown in **Figure 59-28.**

A hinge-mortising template can be used to lay out door hinge locations and guide a router cutting the hinge gains.

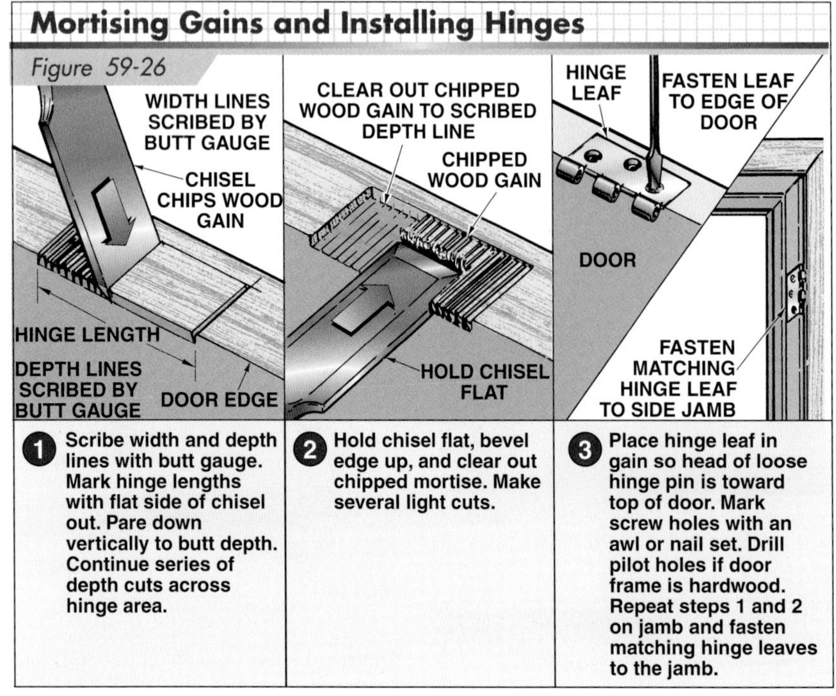

Figure 59-26. *A butt chisel is used to mortise the gain. The hinge should rest flush with the surface.*

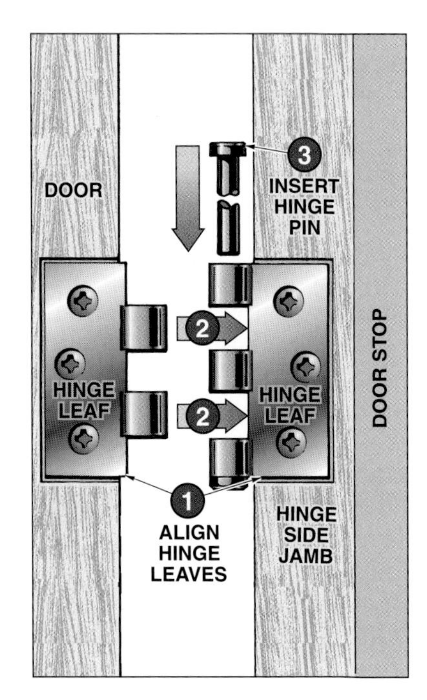

Figure 59-27. *When a door is hung, line up the leaves of the door and jamb. Push the leaves together and replace the pin.*

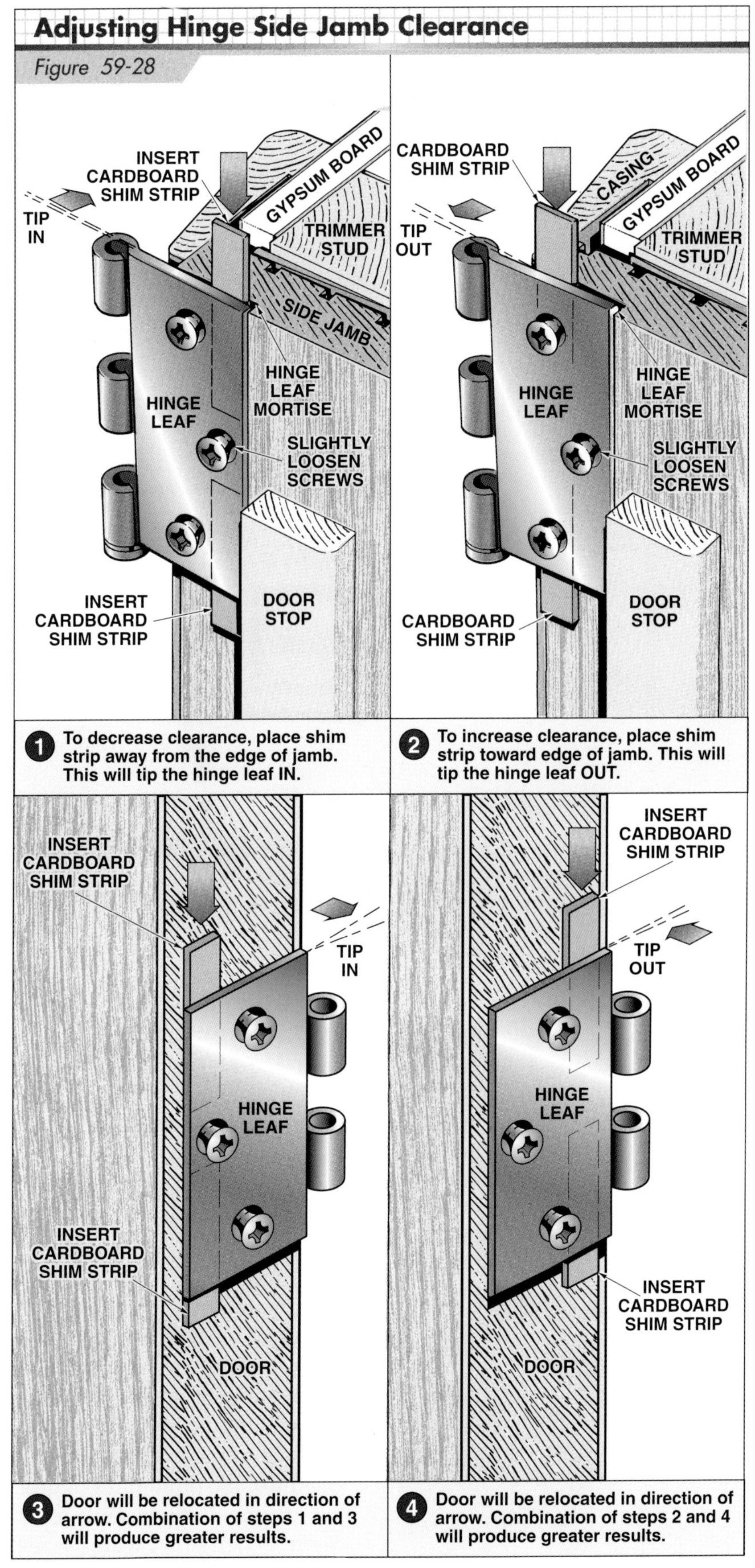

Figure 59-28. *The clearance between the hinge side of the door and doorjamb is adjusted by placing a thin cardboard strip behind the leaf fastened to the doorjamb or door.*

Door Fit Problems. Most door fit problems are the result of improperly assembled or installed frames. Prior to making any adjustments, use the procedure shown in **Figure 59-29** and outlined below to determine door fit problems and possible solutions.

1. With the door closed, measure the clearances at the head and both jambs. Inconsistent clearances on a given side may indicate out-of-square frames, twisted jambs, or improperly positioned hinges.
2. Observe whether the door is flush against the stops of the jamb and the head. Incomplete contact may indicate that the frame is out of plumb, twisted jambs, or the door may be twisted.
3. With the door ajar, observe whether the door remains stationary or moves involuntarily. Involuntary movement indicates out-of-plumb and/or twisted jambs.
4. With the door open, place a framing square in the corner against the jamb and head rabbets to determine the squareness of the head to the jambs. Measure the opening at the head and floor. Out-of-square jambs and heads will cause inconsistent clearances.
5. Measure diagonally across corners in both directions and on both sides of the frame. If the measurements are identical, the frame opening is square.
6. Determine the plumb of both jambs using a 6′ carpenter's level or plumb bob.

In general, rough door openings should be 2″ more than the door width and 2¼″ more than the door height.

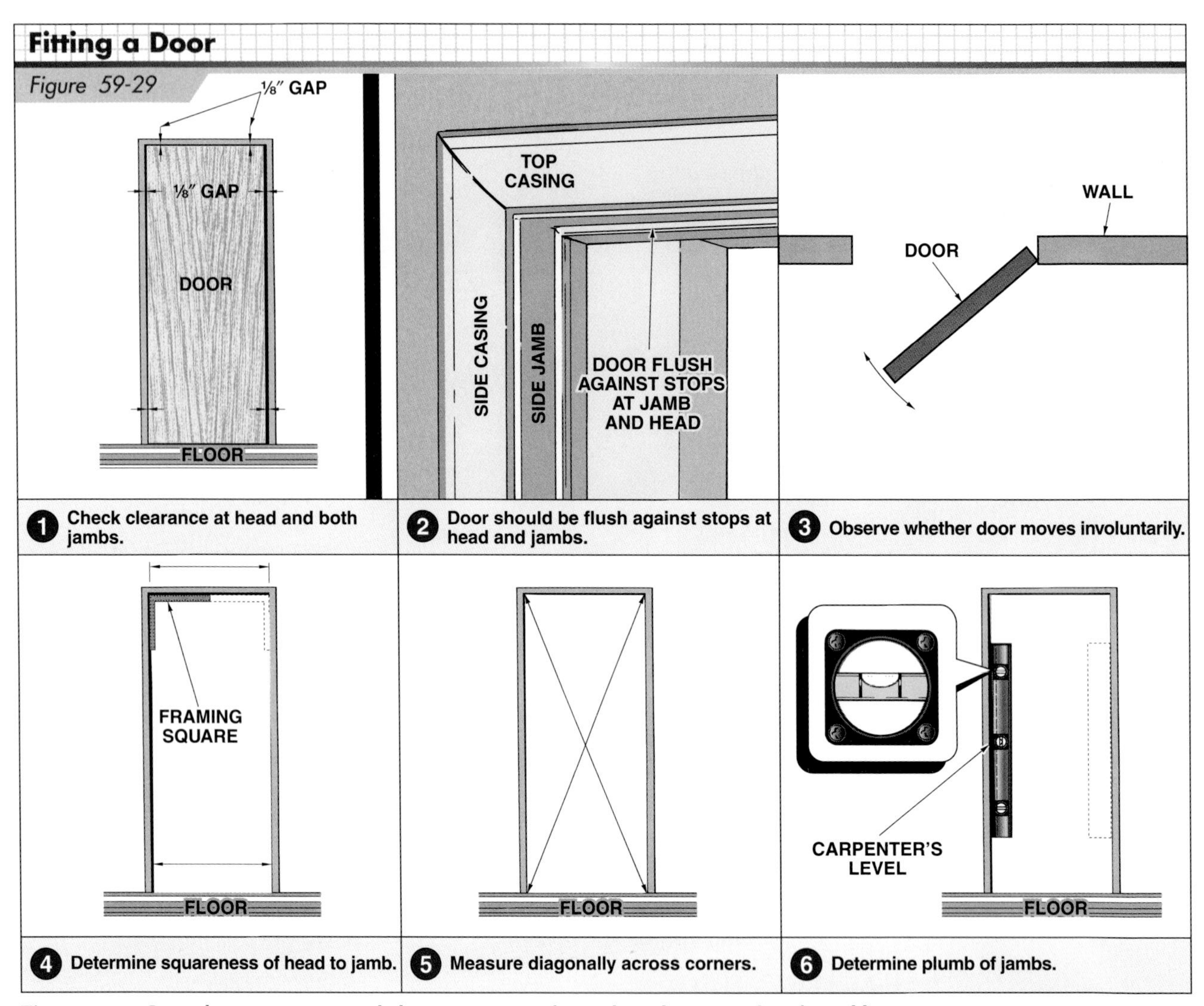

Figure 59-29. *Several measurements and observations must be made to determine door fit problems.*

Fitting and Hanging Doors Using Power Tools

Doors can be fitted and hung more quickly and easily with power tools than with hand tools. A power plane is used to fit the door and quickly bevel the edge. Hinge gains are mortised with a router that is guided by a hinge-mortising template. Templates, which are adjustable for hinge size and spacing, are available from a variety of manufacturers. The template is placed on the door first and the gains are cut with a router that has been set to the proper depth. **See Figure 59-30.** Hinge screws are driven in with a power screwdriver.

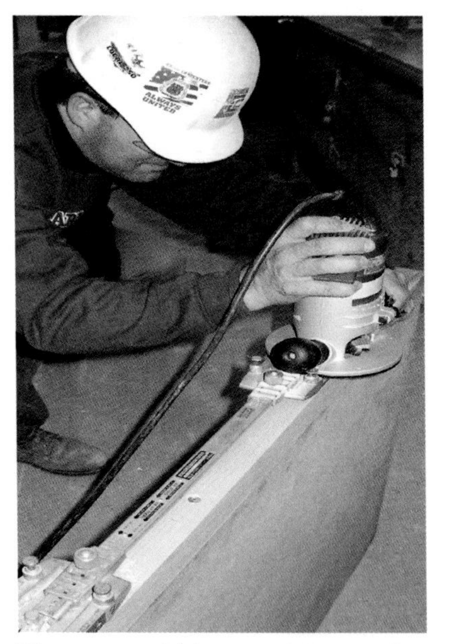

Figure 59-30. *A hinge-mortising template and a router are used to mortise the door edge and doorjamb for hinges.*

Installing Prehung Door Units

A prehung door unit is a door that is already hinged in the frame. In addition, a hole is usually predrilled for a lock and the door stops are tacked to the jamb, unless a rabbeted jamb is used.

Prehung door units may be nonadjustable or adjustable. A section view of a nonadjustable prehung door unit is similar to a conventional jamb assembly. **Figure 59-31** is a pictorial drawing based on a section view of a nonadjustable prehung door unit. The casing, mitered and cut to length, is delivered with the rest of the unit and is nailed around the jamb after it is installed.

Adjustable *(split-jamb)* units can be adapted to different wall widths. They are delivered with the casing already nailed to the jamb section. Pictorial drawings based on section views of three kinds of adjustable units are shown in **Figure 59-32.**

A corner block is a piece of decorative corner trim used primarily on window and door casing. Corner blocks are installed at the intersection of the top and side casings. Casing installation with corner blocks is different from standard mitered casings. The head casing and corner blocks are usually installed first. The side casings are then fitted, marked, and cut to the proper dimensions.

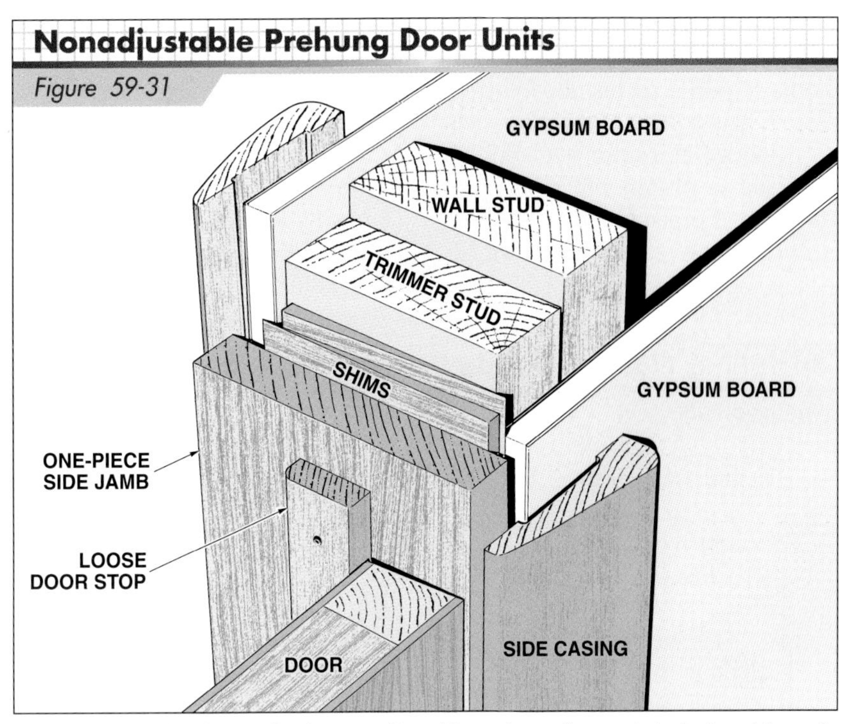

Figure 59-31. *The doorjamb of a nonadjustable prehung door unit is designed for only one wall thickness.*

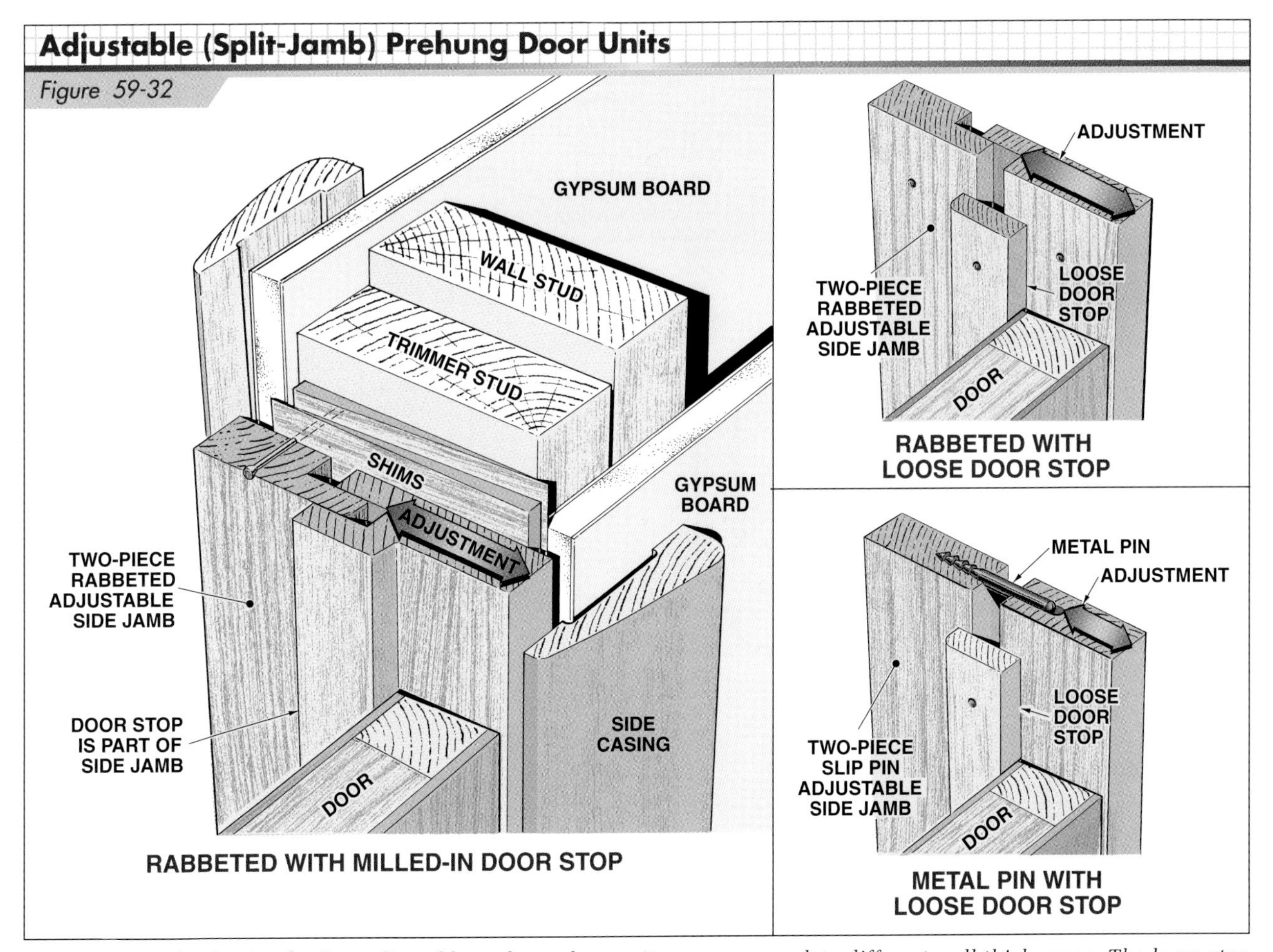

Figure 59-32. *The doorjamb of an adjustable prehung door unit can accommodate different wall thicknesses. The loose stop is installed after the door is installed.*

Prehung units can be installed more quickly than job-hung doors. Tools required to install prehung doors are a hammer, nail set, straightedge, and spirit level. Shims are used to plumb and straighten the jamb. A procedure for installing a nonadjustable unit is shown in **Figure 59-33.** A procedure for installing an adjustable (split-jamb) unit is shown in **Figure 59-34.**

SPECIALTY DOORS

Double doors, sliding doors, and folding doors are commonly used for closets or utility closets to conserve floor space otherwise needed for door swing. Double doors, sliding doors, and folding doors often require specialized hardware, and information on their installation is normally supplied by the manufacturer.

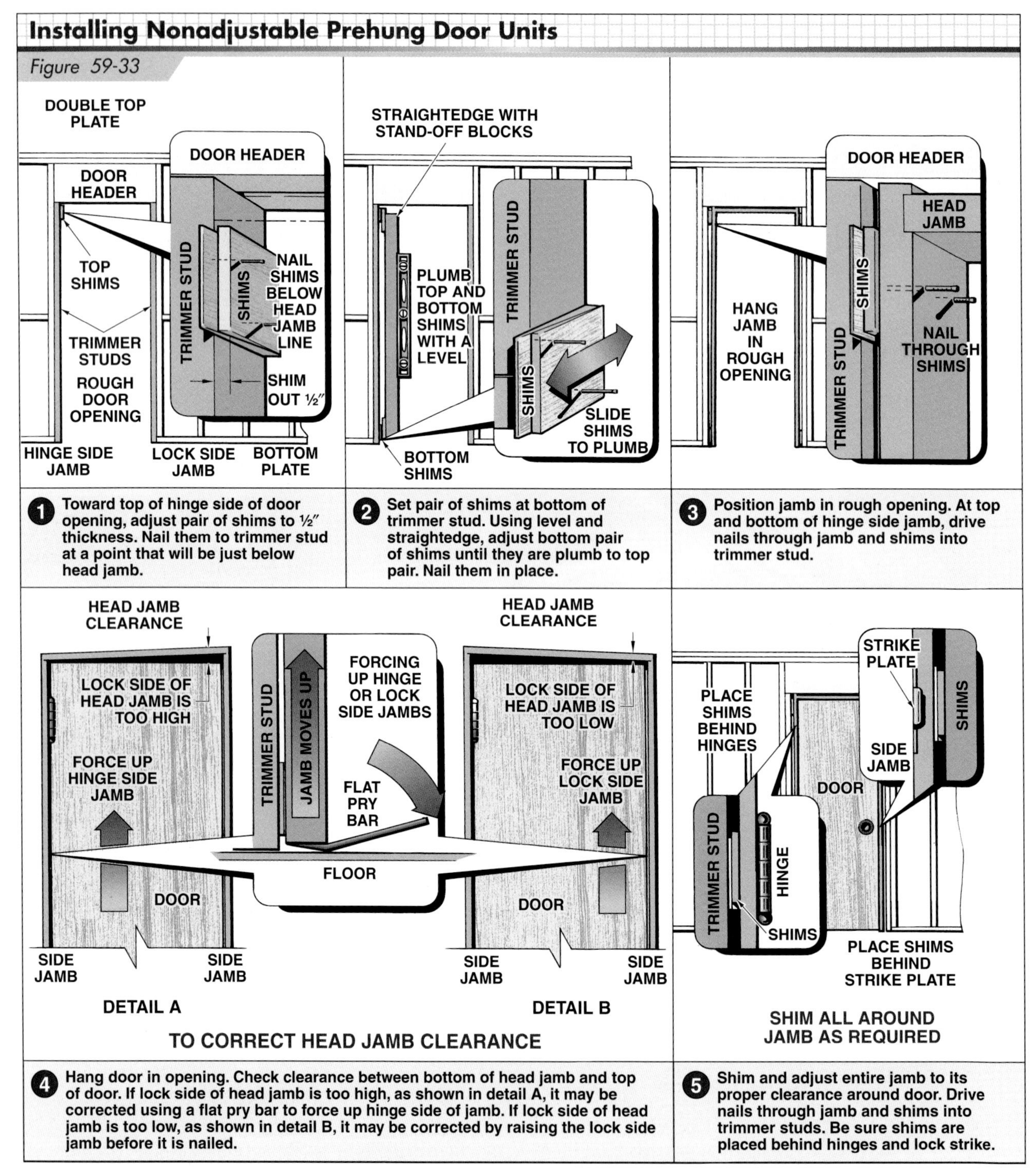

Figure 59-33. *When selecting a nonadjustable prehung door unit, ensure the doorjamb width matches the wall thickness.*

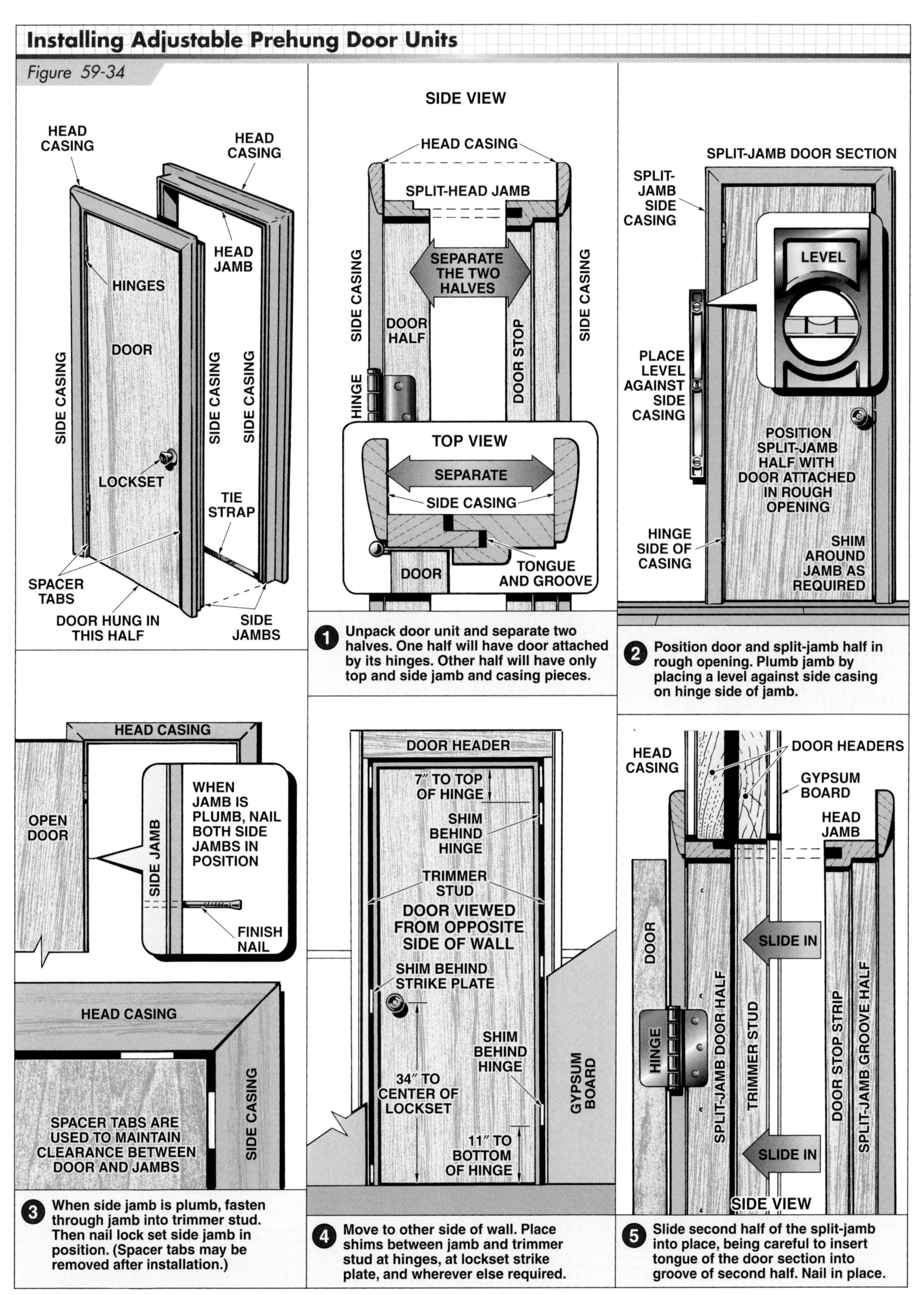

Figure 59-34. *Adjustable (split-jamb) prehung door units are installed in two sections.*

Double Doors

Double doors, consisting of two swinging doors in a single frame, are often found in office and public buildings. Often, one door is *inactive*. Inactive doors are held in place by *flush bolts,* a type of hardware that is mortised into the upper and lower part of the door edge. The inactive door can be opened when necessary by retracting the flush bolts. Usually, however, only the active door is used for passage in and out of the room. The active door generally contains a lock with a knob or handle.

The procedure for installing double doors is similar to installing a single door. The inactive door is fitted and set in place first. The active door is fitted to the inactive door and hung in the frame.

Pocket Sliding (Recessed) Doors

When a swinging door is impractical, a pocket sliding (recessed) door may be used. Hangers with small wheels are fastened to the top of the door, enabling it to slide back and forth on a track mounted in the upper part of the frame. **See Figure 59-35.** When the door is opened, it slides into a pocket frame in the wall.

Pocket sliding doors are produced in standard sizes. Units include the frame, hangers, track, and other required hardware. The frame for a pocket sliding door is set in place when the wall is framed.

Bypass Doors

Bypass doors are commonly used for double closets where two or more doors are used to cover the closet opening. Bypass doors are suspended from tracks with roller hangers. Floor guides hold the bottoms of the doors in position. **See Figure 59-36.**

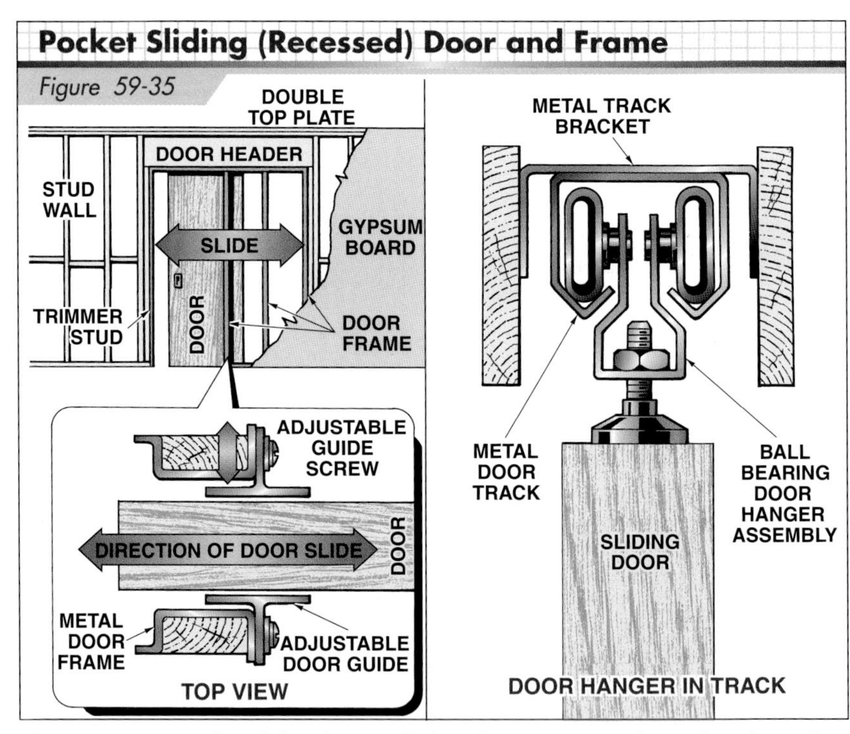

Figure 59-35. *A pocket sliding (recessed) door frame is set in place when the wall is framed. The frames may be wood or metal. The doors are installed later.*

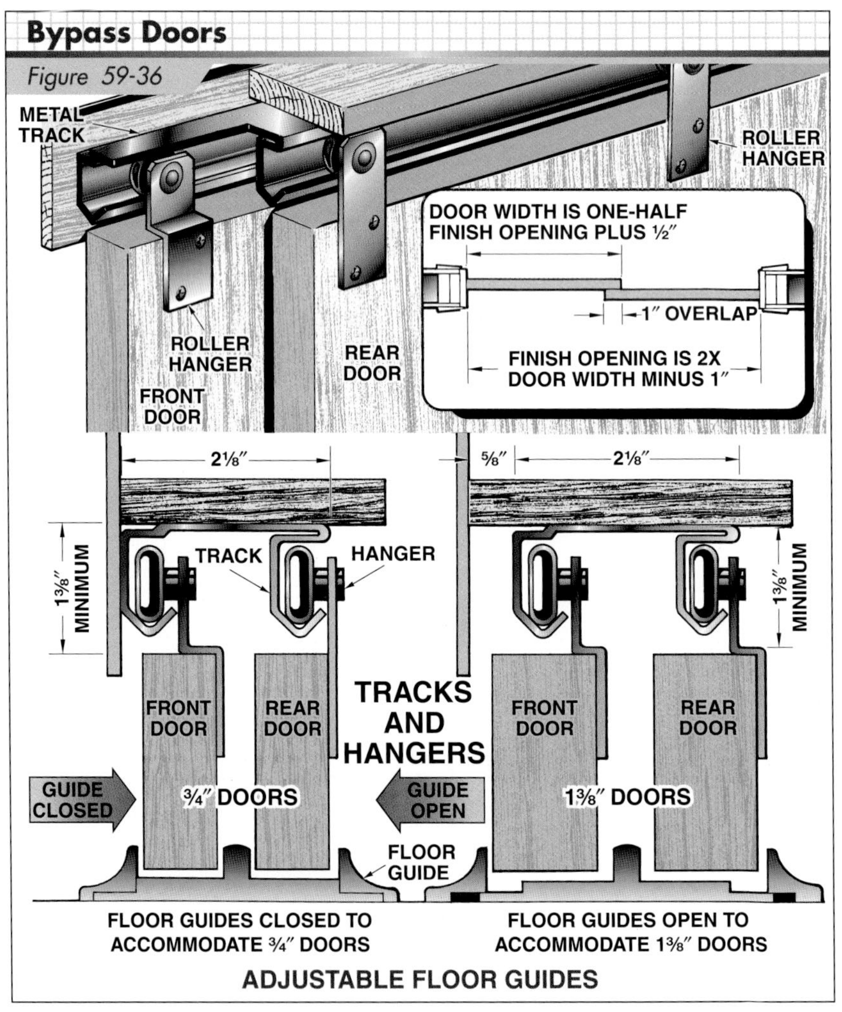

Figure 59-36. *Bypass sliding doors are suspended from tracks with roller hangers.*

Folding Doors

Folding doors, particularly bi-fold doors, are frequently used for closet doors. Folding doors are available in flush, panel, or louver designs. Door thicknesses most often used are 1⅛″ and 1⅜″. Vinyl bi-fold doors are also available.

In bi-fold door installation, the door adjacent to the jamb is secured at the top and bottom with a pivot pin and bracket. The second door is hinged to the first door. The top of the second door has a roller guide that slides along a track mounted to the head jamb. **See Figure 59-37.**

Figure 59-37. *Bi-fold doors are secured next to the jambs with pivot pins and brackets. The second door is hinged to the first door. A roller guide on the second door glides in an overhead track.*

Multifolding doors, also known as accordion doors, operate in a similar manner to bi-fold doors. Multifolding doors are used for closets, room entrances, and room dividers. **See Figure 59-38.**

Multifolding doors are supported by roller hangers that run along an overhead track. The door panels usually have a wood core and are surfaced with wood veneer, vinyl, or plastic laminate. Common door panel thicknesses are ⅜″ and 9⁄16″ and common widths range from 3⅝″ to 5⅝″. Spring hinges run horizontally through the panels and connect to a vertical wood molding that runs the entire length of the door.

Figure 59-38. *Multifolding, or accordion, doors fold against the door jambs.*

METAL DOOR UNITS

Metal door units have been used in commercial construction for many years, and are now commonly used in residential construction for increased security. The metal frames used in commercial construction are fastened to concrete and masonry, as well as to metal and wood studs. Metal frames used in residential and other light construction are designed to be fastened to walls that have a gypsum board (drywall) finish.

Metal frames are available for most standard door sizes and wall thicknesses. A typical steel door frame is made of 16-ga hot-dipped galvanized steel. **See Figure 59-39.** Aluminum frames are also available. Metal door units usually do not have to be trimmed after installation since the casing is molded into the frame. Steel and aluminum frames are available with or without prefinished coatings. Metal frames are fastened to concrete or masonry, as well as to metal and wood studs.

Butt hinge size is determined by the size and weight of the door and frequency of use.

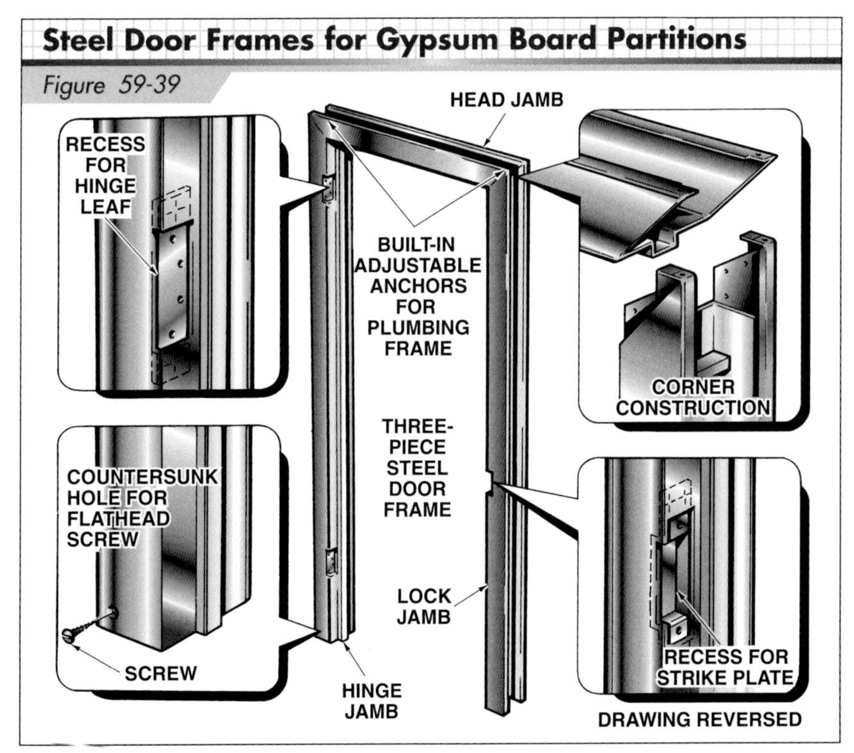

Figure 59-39. *Steel door frames are commonly used in commercial construction. A typical steel door frame is made of 16-ga hot-dipped galvanized steel. Note the recessed areas for the hinges and the lock strike plate.*

Installing Metal Frames

A metal door frame usually consists of three pieces and is installed on the job according to manufacturer instructions. A common installation method is shown in **Figure 59-40.** Hollow metal door frames may also be required to be fastened to the floor. Screws are driven through prefabricated clips inside the jamb and into the floor.

Fastening Systems. The type of door frame installed in a wall opening determines the type of fastener required to secure the frame in place. **See Figure 59-41.** *T-straps* and *wire anchors* are specifically designed for installation in masonry or concrete walls. T-straps and wire anchors must be installed in the masonry or concrete as the walls are being constructed. When installing a door frame in completed masonry or concrete walls, holes for expansion anchors are drilled in the walls and bolts are used to fasten the door frame to the wall surface. *Drywall compression anchors* and *strap anchors* are used to secure metal frames over gypsum board (drywall).

ANSI A250.8, Recommended Specifications for Standard Steel Doors and Frames, covers steel door sizes, types, materials, general construction requirements, and finishing.

Steel Doors

Interior steel doors are widely used in commercial construction such as office and industrial buildings. Interior steel doors are available in most standard sizes and in 1⅜″ and 1¾″ thicknesses. Nonstandard sizes and thicknesses can usually be special ordered.

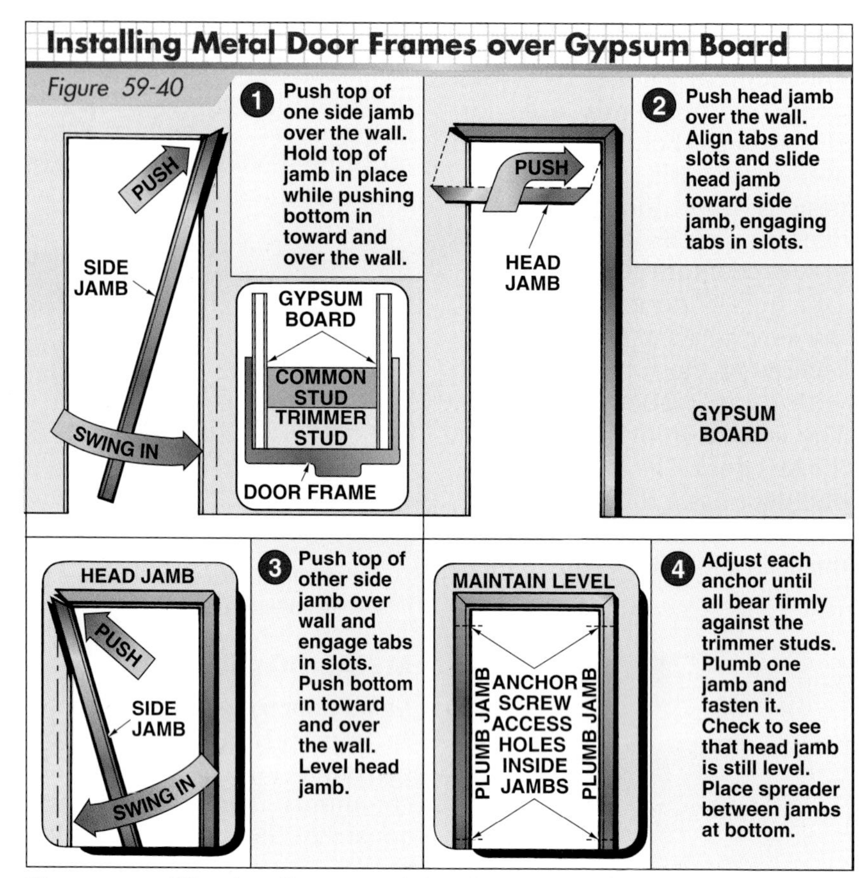

Figure 59-40. *When installing a metal door frame, one side jamb is first installed, followed by the head and other side jamb. The jambs slip over the applied gypsum board.*

A metal door frame consists of three pieces and is installed according to manufacturer instructions.

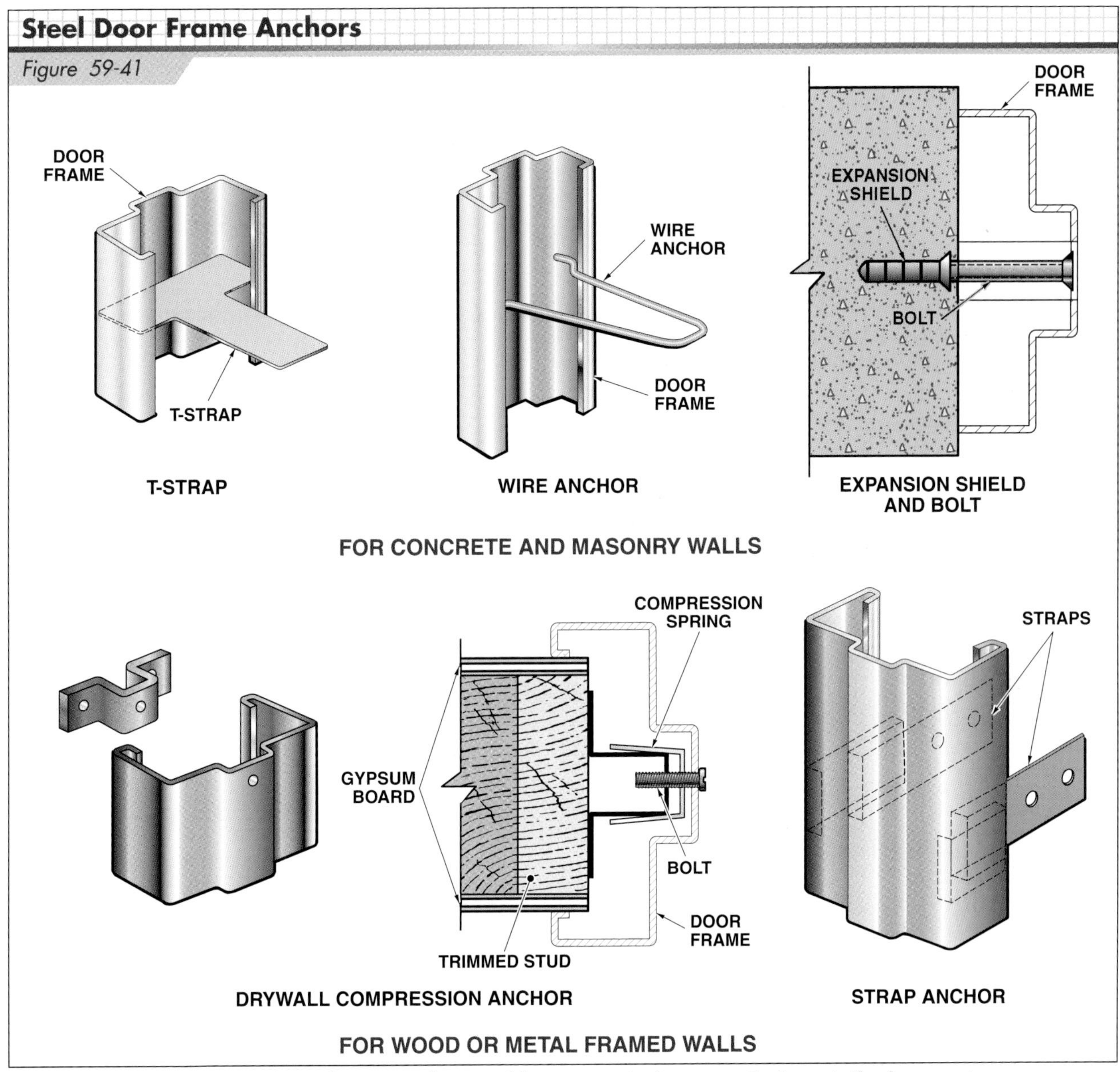

Figure 59-41. *The type of door frame determines the type of fastener required to secure the frame in the door opening.*

Steel doors increase the security of offices. A high-density foam core, such as foamed-in-place polyurethane or polystyrene, ensures high insulation values. Most steel doors are made of heavy-gauge galvanized metal skins enclosing a wood or steel frame. The skins may be plain or embossed with a wood grain texture.

Steel doors have prepared hinge mortises and lock holes. Reinforcement is placed behind the door sections where hinges, locks, door closers, and other types of hardware may be placed. **See Figure 59-42.**

Fire Doors. Steel doors used in office and industrial buildings are usually required to be *fire doors.* Fire doors are constructed with a fire-resistant mineral-material core that retards the passage of fire.

Fire doors are rated by the length of time they will resist the passage of fire. Interior fire doors are rated as *Class A, B,* or *C.* Class A fire doors have a 3-hour rating. Class A fire doors are used in openings in fire division walls and walls that divide a single structure into fire areas. Class B fire doors have a 1- or 1½-hour rating. Class B fire doors are installed in openings in enclosures of vertical communications through structures and in 2-hour partitions, provided there is horizontal fire separation. Class C fire doors have a ¾-hour rating, and are used in wall openings with a fire-spread rating of 1 hour or less.

Fire doors must also be equipped with special hardware to be effective. Fire doors must be self-closing and self-latching, and steel ball bearing hinges must be installed.

DOOR LOCKS

Locks are installed in doors after the doors have been hung. A *lockset* consists of the working components and the trim pieces, which are visible. Trim pieces include the knob or lever, escutcheons, rose, and cylinder. *Escutcheons* are decorative metal plates placed against the door face behind the knob and lock. The *rose* fits against the door and, in some locksets, holds the lock in place. The *cylinder* is the part of the lock that contains the keyhole and tumbler mechanism.

In new construction, a lockset is installed in the unfinished door unit and then removed while the door is painted or stained. After the door is finished, the lockset is reinstalled. Another procedure is to make the first and final installation on the finished door.

The *door hand* is the direction in which a door swings. **See Figure 59-43.** A *left-hand door* hinges on the left and opens inward. A *left-hand reverse door* hinges on the left and opens outward. A *right-hand door* hinges on the right and opens inward. A *right-hand reverse door* hinges on the right and opens outward.

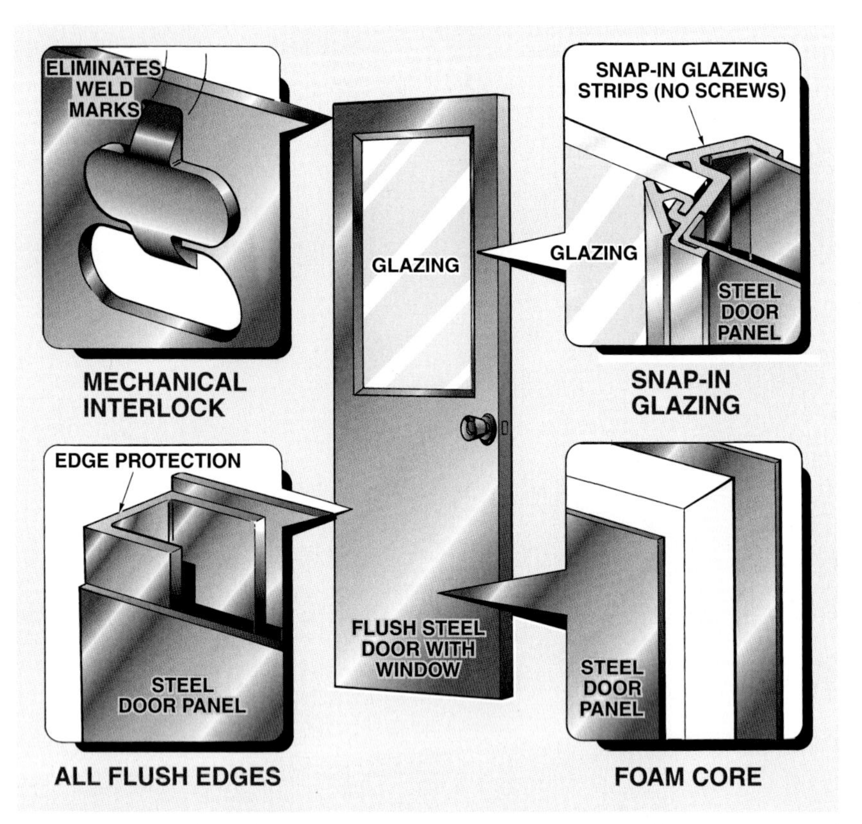

Figure 59-42. *The face of a flush steel door is made of 20-ga steel. Pieces of 12-ga steel reinforce sections where the lock and hinges are to be installed. The holes are drilled and tapped to receive flathead machine screws.*

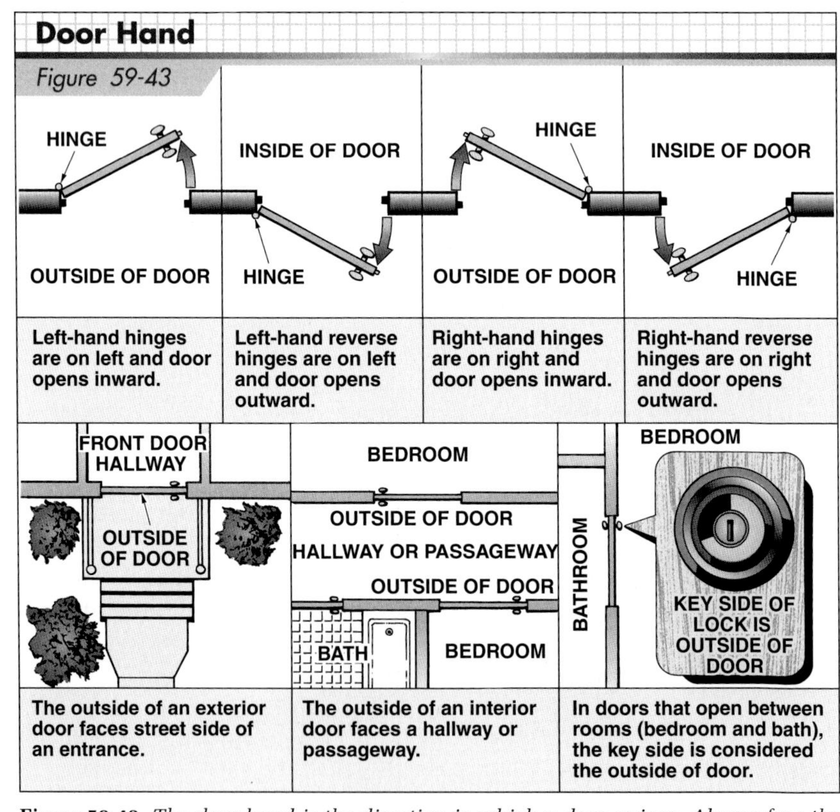

Figure 59-43. *The door hand is the direction in which a door swings. Always face the outside of a door when determining the door hand.*

The panels of panel doors are designed to "float" between the stiles and rails. Before finishing panel doors, ensure the panels are properly aligned with the stiles and rails. If necessary, gently correct the alignment by moving the panel with a soft wood block and rubber mallet.

Baldwin

A lockset consists of the working components and the visible trim pieces.

Many locksets are reversible, meaning that they can be adjusted to operate in doors that swing in either direction. Other locksets, as well as other kinds of door hardware, are designed to work only on doors that swing right or only on doors that swing left. Be sure to identify the door hand before ordering door hardware, including locksets.

To determine door hand, always consider that you are standing on the outside (exterior) of the door. The outside of a door is a hallway or passageway of an interior door or the street side of an exterior door. In doors that open between rooms, the keyed side of a lock is considered the outside of the door.

Types of locks include *cylindrical, tubular, mortised*, and *deadbolt.* For greater security, more than one type of lock may be used on a door. Each type is available in a variety of designs and finishes.

Cylindrical and Tubular Locks

Cylindrical and tubular locks are *bored locks,* meaning that holes must be drilled in a door to install the locks. The locks are operated with a button or key located in the door knob or lever handle. **See Figure 59-44.** The cylinder contains the tumbler mechanism into which the key fits. If the keys are lost or stolen, the tumbler can be replaced and the lock will operate with a new set of keys.

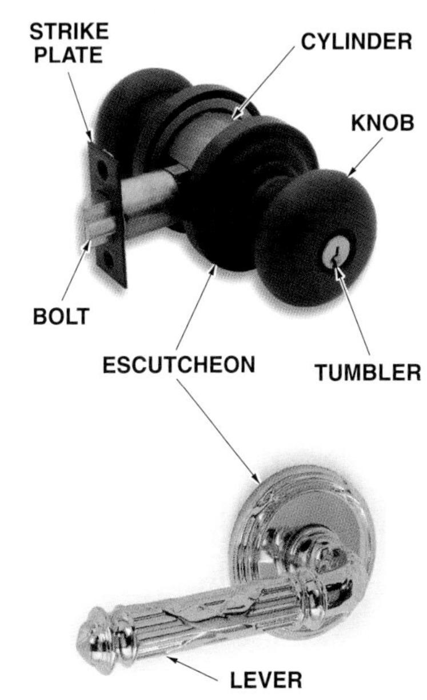

Baldwin

Figure 59-44. *For a cylindrical lockset, the cylinder is located within the knob. The cylinder contains the tumbler mechanism in which the key fits. Locks are operated with a knob or lever.*

Another type of bored lock is a *grip-handle lock.* **See Figure 59-45.** For a grip-handle lock, the latch is retracted by a thumbpiece on the outside of the door or by an inside knob. The cylinder is located above the knob.

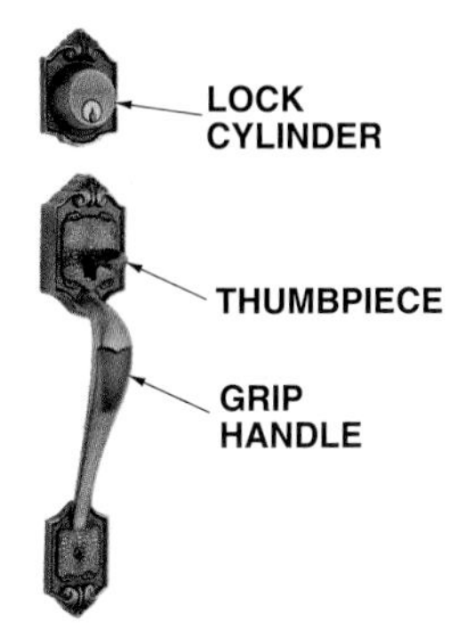

IR Security & Safety

Figure 59-45. *On a grip-handle lock, the cylinder is separate from the knob.*

A cylindrical lock is considered a better-quality lock than the tubular lock and is designed for long life and maintenance-free operation. A tubular lock is less expensive and more simply constructed, and does not have the strength or smooth working action of a cylindrical lock.

Operation. In cylindrical and tubular locks, the latch unit engages with a latch-retractor device in the main body of the lock. When the door knob or key is turned, the latch bolt is pulled back into the door. When the knob is released, the latch bolt springs back to its original position.

Door locks are categorized by "grades." Grade 1 door locks meet commercial building requirements. Grade 2 door locks meet light commercial requirements and exceed residential building requirements. Grade 3 door locks meet residential building requirements.

When the door is in a closed position, the latch bolt is engaged by the strike, which is a metal plate mortised and screwed into the doorjamb. **See Figure 59-46.**

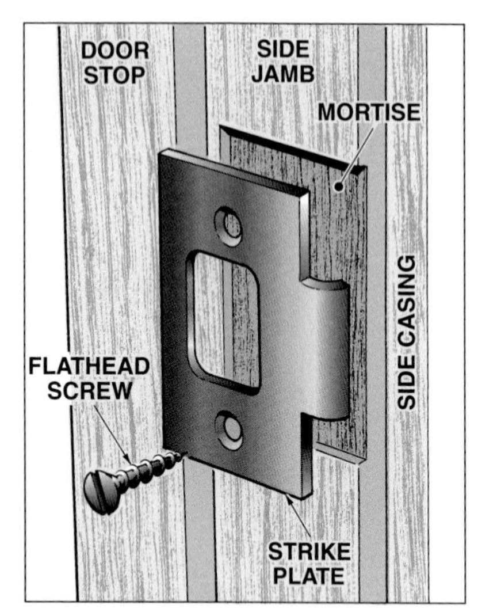

Figure 59-46. *The strike plate is mortised into the side jamb.*

Two types of latch units are the *spring latch* and *deadbolt latch* (or deadlocking latch). **See Figure 59-47.** A spring latch can easily be forced open by a piece of plastic. Therefore, it should not be used on entrance door locks. A deadbolt latch has a small plunger (guard bolt). When the door is closed and the latch bolt has slipped into the strike, the plunger remains in a retracted position. The plunger locks the latch in place so that it cannot be moved.

Keyed and Nonkeyed Locks

For a typical cylindrical or tubular lockset used on entrance doors, the key cylinder is on the exterior side of the door. A *pushbutton* or *turnbutton* is located at the center of the interior knob to lock the door from that side. Pushbuttons and turnbuttons disengage and unlock the door when the interior knob is turned or when the key is inserted and turned. Non-keyed locks may have a pushbutton or turnbutton device on one side, or none on either side.

Installing Locks. In residential construction, the height of a lock from the floor to its center is usually 36″. In public and office buildings, lock height is usually 38″. New locksets are supplied with manufacturer instructions for installation and a template for marking holes to be drilled. When installing cylindrical and tubular locksets, first drill the large hole running from face to face using a hole saw or lockset bit in an electric drill. A standard auger bit inserted in an electric drill is used for the smaller latch unit hole.

In some situations, a carpenter may need to install a lockset without a template or manufacturer instructions. A procedure for laying out and boring the holes and for marking the mortise for the latch unit faceplate is shown in **Figure 59-48.**

Setting Strike Plates. After the lockset is installed, the position of the strike plate is laid out and the doorjamb is mortised for it. If the layout is done accurately, the door will be flush with the edge of the jamb when the latch engages the strike plate. In new door installation, there should be a slight amount of play when the latch bolt engages with the strike plate to prevent binding when the door and jamb are painted or stained. A procedure for establishing the position of the strike plate is shown in **Figure 59-49.**

Installation Kits. Many lock manufacturers offer installation kits that include a boring jig that clamps to the door, drill bits for boring the holes, and special marking chisels for the latch and strike plates. The boring jig can be adapted for a number of different backsets, door thicknesses, and latch unit holes of different diameters, and facilitates rapid and accurate drilling of the lock holes. The marking chisels score the door and jamb for the face plate of the latch unit and strike plate. The strike locator is used to mark the correct position of the strike plate on the jamb.

Figure 59-47. *A deadbolt latch provides greater security than a spring latch.*

The doorjamb must be mortised to accommodate the strike plate and box. The box mortise must be deep enough to allow the latch bolt to fully extend.

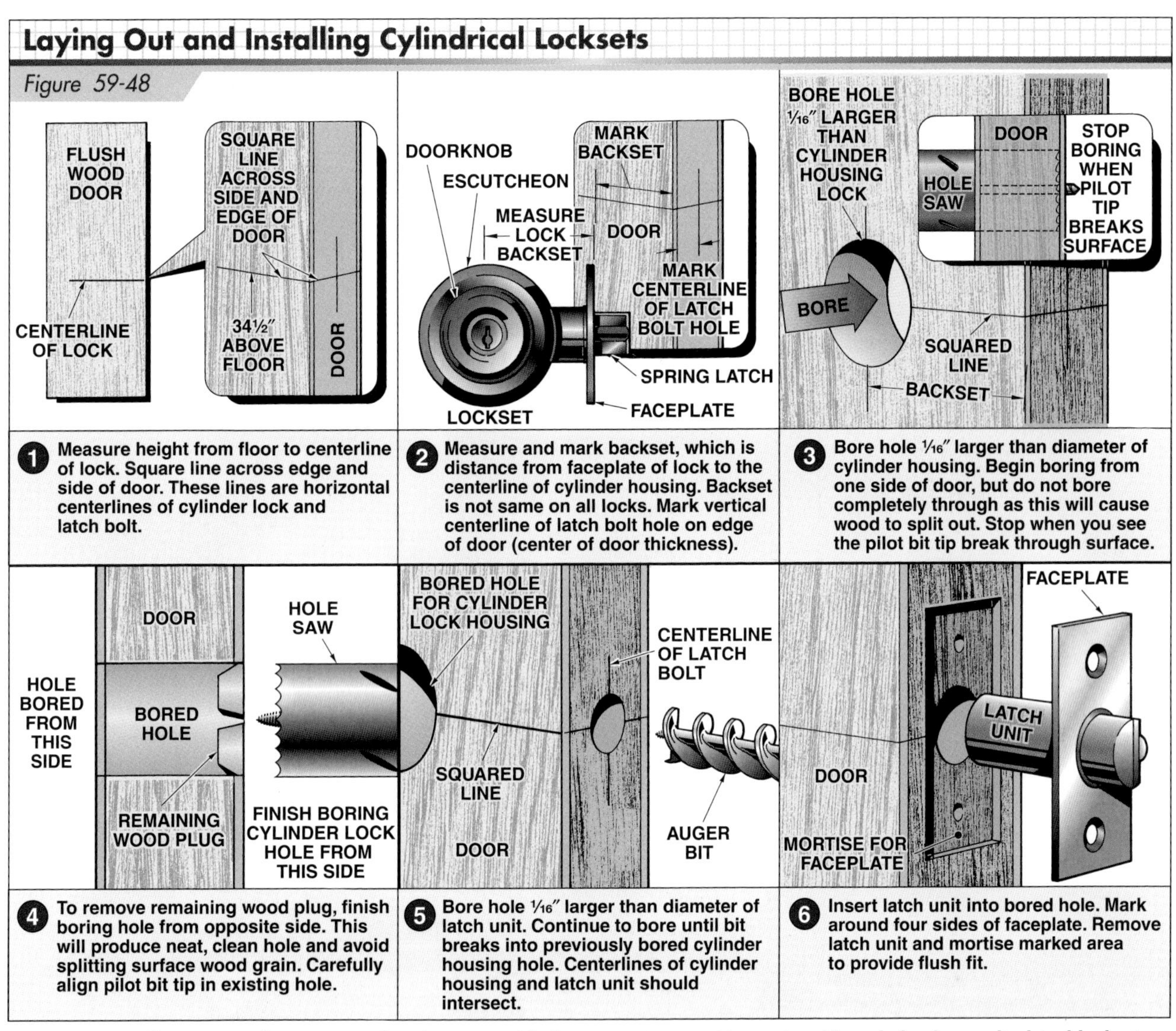

Figure 59-48. *When a manufacturer template is not provided, a carpenter must lay out and bore holes for a cylindrical lockset.*

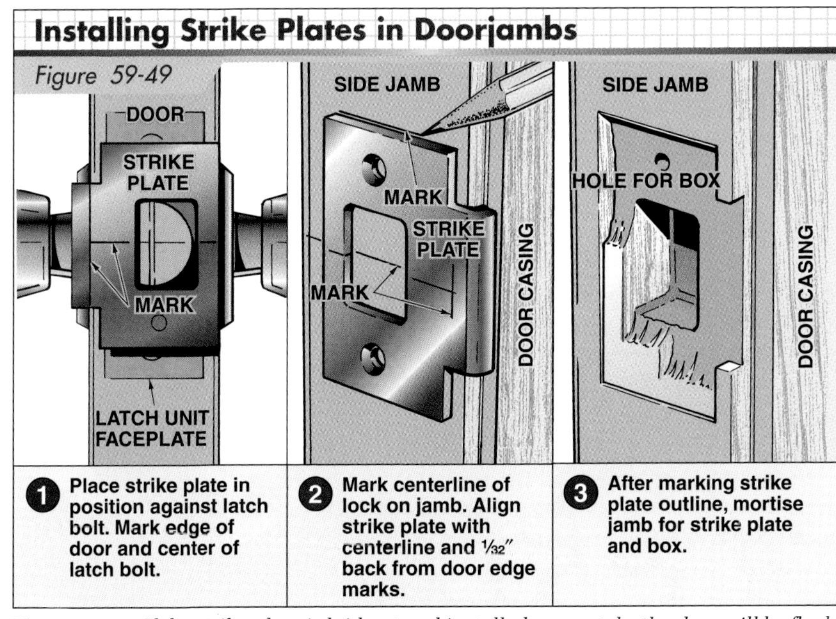

Figure 59-49. *If the strike plate is laid out and installed accurately, the door will be flush with the edge of the jamb when the latch engages the strike plate.*

Doors should be located so they do not interfere with other doors when opened.

Mortised Locks

A mortised lock is a sturdy, durable lock that has been used for many years. **See Figure 59-50.** Mortised locks are more costly and difficult to install than cylindrical or tubular locks and thus are used for specialized installations. However, together with heavy-duty entrance doors, mortised locks provide greater security than cylindrical or tubular locks.

Figure 59-50. *Mortised locks provide greater security than cylindrical or tubular locks.*

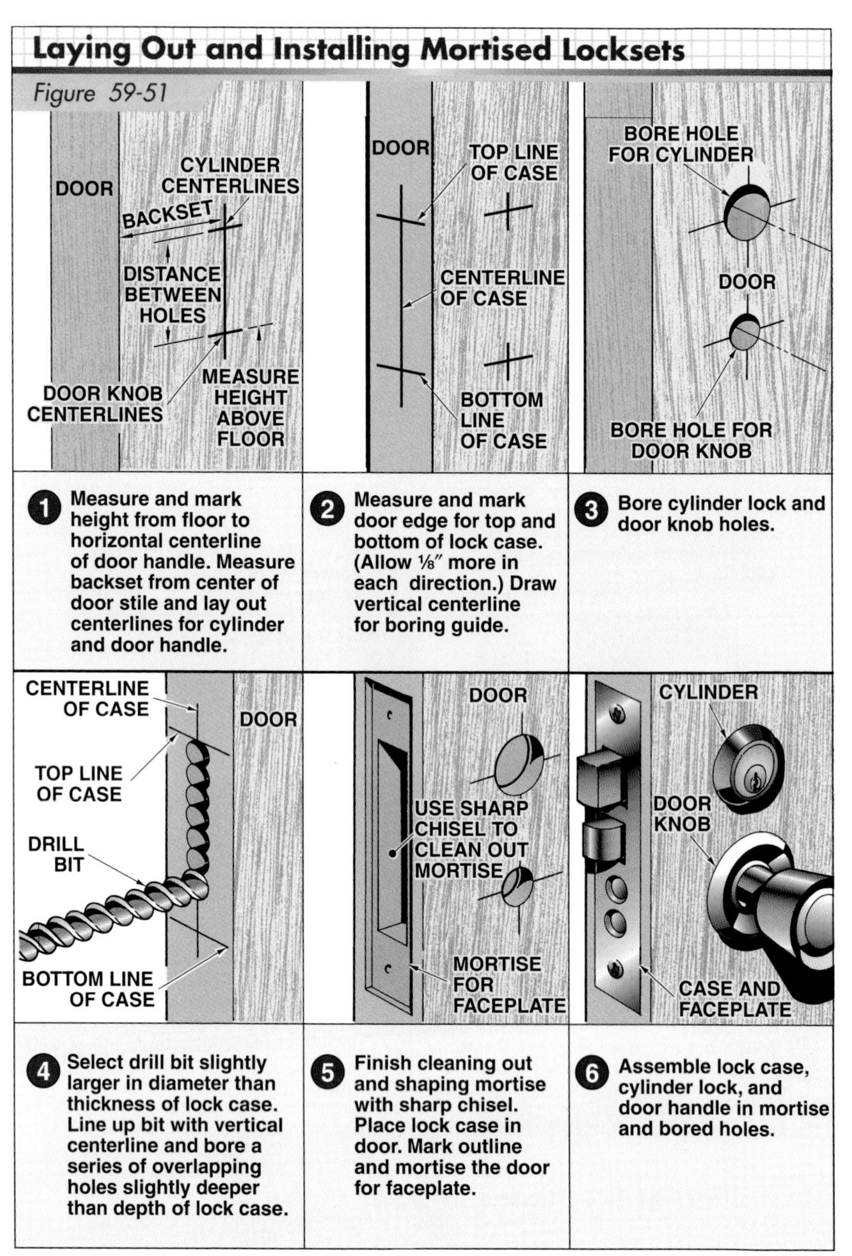

Figure 59-51. *A deep mortise must be created for a mortised lockset.*

Mortised locksets are supplied with a template and manufacturer instructions. In remodeling work, however, an old mortised lock is often removed from one door and placed in another. For this reason, a carpenter must know how to install a mortised lock without a template. **Figure 59-51** shows a procedure for installing a mortised lockset.

When installing mortised locks, some of the holes in the side of the door are not bored completely through the door. Refer to the manufacturer instructions when installing mortised locks.

Deadbolts

A common way to increase security of an exterior door is to install a *deadbolt* in addition to a cylindrical lock. A deadbolt consists of a solid metal bar that must be thrown (locked) and retracted (unlocked) with a knob or key. Deadbolts are usually keyed on the outside of the door and have a knob or handle on the inside. **See Figure 59-52.** Deadbolts are also available with a double cylinder, which is keyed on both sides.

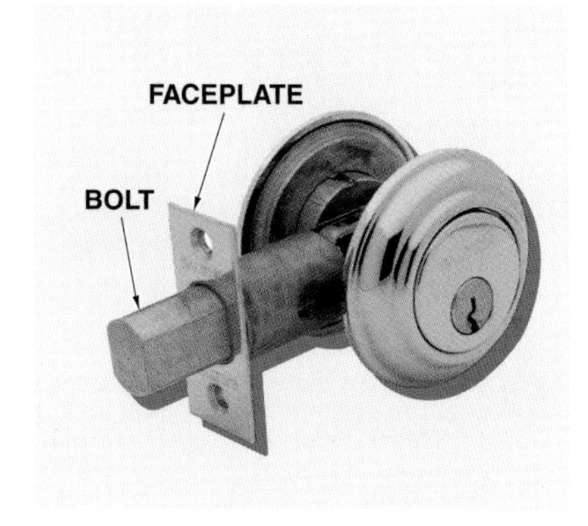

Baldwin

Figure 59-52. *Deadbolts greatly increase the security of an entrance door.*

High-Security Locks

High-security locks combine a cylindrical lockset and a deadbolt mechanism. **See Figure 59-53.** The lock mechanism is protected by an armored plate on both sides of the door. Additional features make forceful entry very difficult.

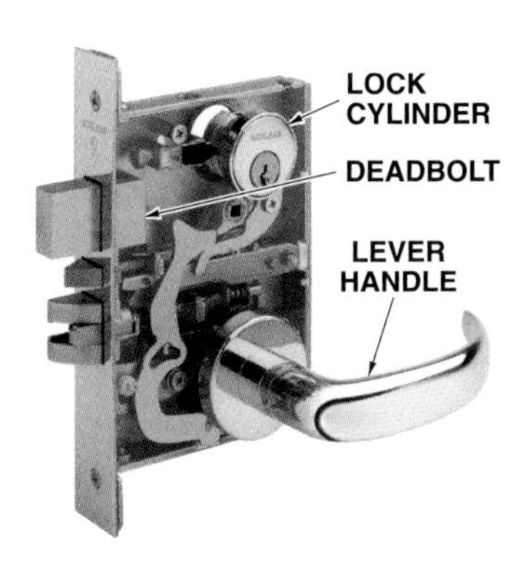

IR Security & Safety

Figure 59-53. *High-security locks combine the features of a cylindrical lockset and a deadbolt.*

HINGES

Hinges are available in a variety of styles, sizes, and finishes. Common hinges include loose-pin, fixed-pin, and ball bearing hinges. Specialized hinges include self-closing, double-acting floor, and concealed hinges.

Standard Hinges

Loose-pin hinges offer the advantage of removing the hinge pin, making door hanging and removal convenient. *Fixed-pin hinges* look identical to loose-pin hinges, but the hinge pin cannot be removed. Under certain conditions, fixed-pin hinges offer more security than loose-pin hinges. *Ball bearing hinges* are installed on heavy doors subject to a great deal of use, such as in schools and public buildings. Ball bearings in the hinges reduce wear at the knuckle joints. Ball bearing hinges may have a loose or fixed pin. **See Figure 59-54.**

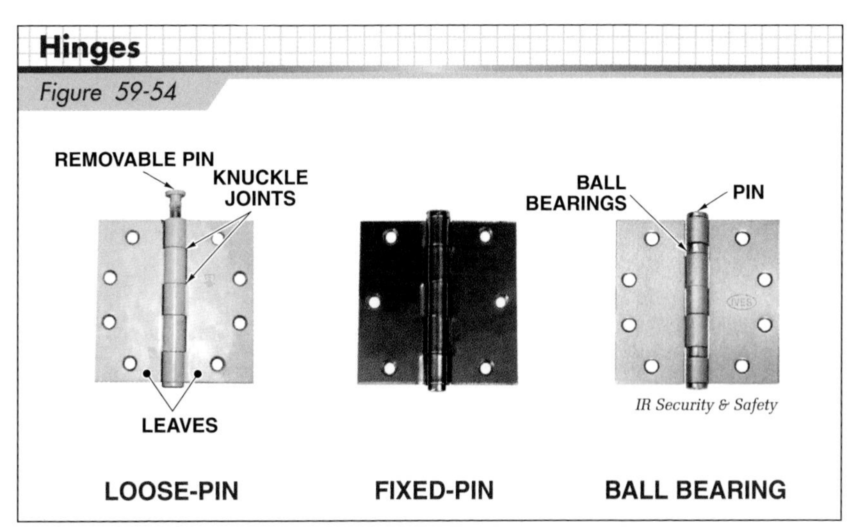

Figure 59-54. *Loose-pin, fixed-pin, and ball bearing hinges are commonly installed on swinging doors.*

A router can be used to create a mortise in the side jamb for the hinge.

Specialized Hinges

Self-closing spring hinges are often used on doors in public buildings to ensure the doors fully close. A door with self-closing spring hinges automatically returns to a closed position after it is opened. A coil spring mechanism in the barrel of the hinge can be adjusted for faster or slower closing action. A hex wrench, provided with the hinge, is inserted into an opening at the top of the hinge and the coil is adjusted for the desired rate of door close. A pin, also provided with the hinge, holds the spring in the desired position. **See Figure 59-55.**

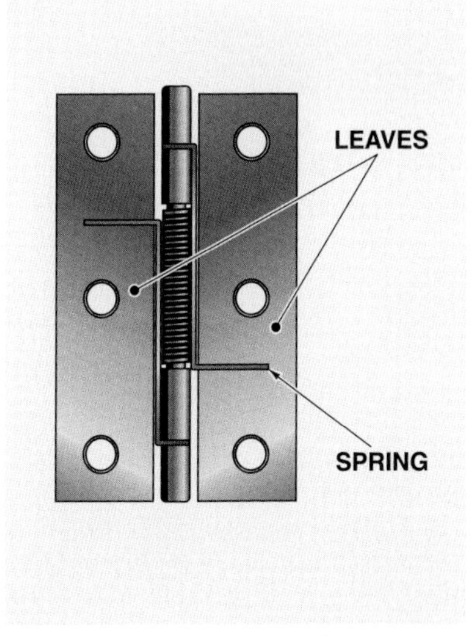

Figure 59-55. *A self-closing spring hinge is often installed on doors in public buildings.*

A *double-acting floor hinge* allows a door to swing in two directions. The bottom of the door is notched to receive the hinge, and the hinge is fastened to the floor with a floor plate. The top of the door is held in place by a pivot piece attached to the door, which fits into a pivot socket mortised into the jamb. **See Figure 59-56.**

Concealed hinges are mortised into the door and doorjamb. No part of a concealed hinge is exposed when the door is closed, making the hinge tamperproof. **See Figure 59-57.**

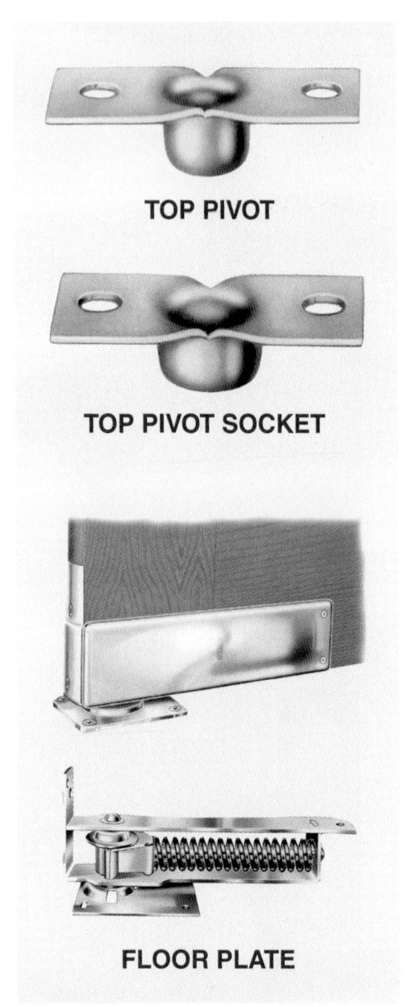

Figure 59-56. *A double-acting floor hinge allows the door to swing in two directions. The top of the door is held in place by a pivot piece attached to the door, which fits into a pivot socket mortised into the doorjamb. The bottom of the door is notched to receive the hinge, which is fastened to the floor with a floor plate.*

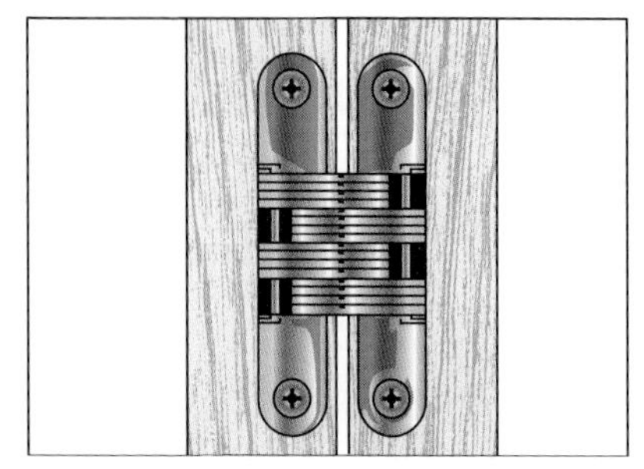

Figure 59-57. *Concealed hinges are mortised into the door and doorjamb. No parts of the hinge can be seen when the door is closed.*

OTHER DOOR HARDWARE

While door locks and hinges are the primary pieces of hardware associated with doors, other door hardware such as door bolts, closers, exit devices, and plates may also be specified on the prints.

Door Bolts

A *flush bolt* holds the inactive door of a double door in place. A handle mechanism is mortised into the edge of the door and a hole is drilled for the door bolt. **See Figure 59-58.** The handle operates the mechanism that moves the bolt up and down. Flush bolts can be installed to project into the floor or upper doorjamb.

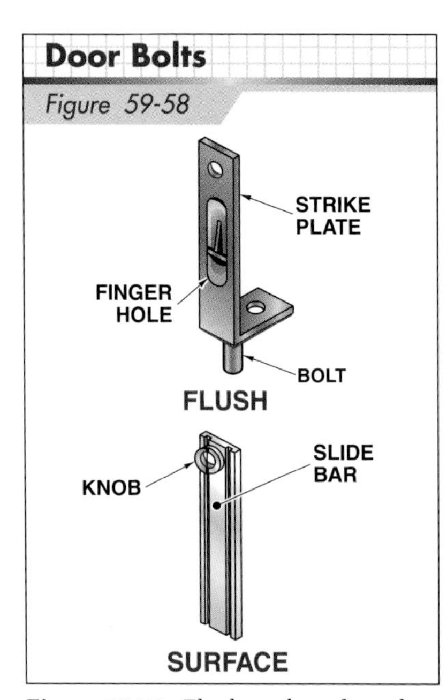

Figure 59-58. *Flush and surface door bolts hold the inactive door of a pair of double doors in a stationary position.*

A *surface bolt* performs the same function as a flush bolt. However, a surface bolt is mounted on the face of a door.

Door Closers

Door closers are frequently used in schools, hospitals, and other public buildings. Door closers offer many benefits including preventing energy loss for air conditioned or heated facilities, greater security, and minimizing sound intrusion through open doors. Several styles of door closers are available. Always refer to the manufacturer recommendations when selecting a door closer, as different door widths and weights require different types of closers.

Door closers can be mounted on either side of a swinging door. However, door closers should always be installed on the inside of a building to prevent damage due to the weather and vandalism. When installed on an interior door, door closers are typically installed on the inside of a room for aesthetics.

A door closer should provide a smooth and controlled closing action. A closer arm transmits motion to a cylinder within the closer housing by means of a rack-and-pinion gear. The force generated by a spring within the cylinder to close the door is controlled by turning an adjusting screw at the end of the cylinder.

If a door opens into a building or room, a regular arm door closer is required. **See Figure 59-59.** If the door opens to the outside of a building or room, a parallel arm, top jamb-mounted, or drop plate-mounted door closer may be used. Parallel arm door closers are commonly used in situations where vandalism may occur, such as schools, since the arm is protected between the closer housing and door frame. Top jamb-mounted door closers are commonly installed on aluminum-framed glass doors. Drop plate-mounted door closers are common in older buildings. Drop plate-mounted door closers should be used with caution, however, as the drop plate reduces headroom clearance. Taller doors may need to be installed with drop plate-mounted door closers to maintain minimum headroom clearance required by the *International Building Code.*

Door Closers

Figure 59-59

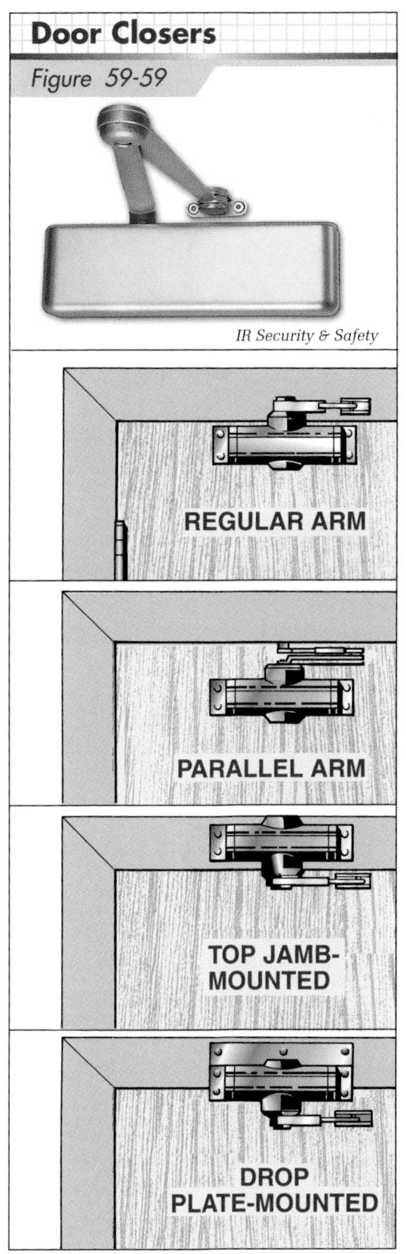

Figure 59-59. *Door closers are operated by hydraulic spring action. Various types of door closers are available.*

Exit Devices

Exit devices, also called fire-exit bolts, are often mounted on doors of public buildings. Pressure on the touch bar of an exit device causes latch bolts at the top and bottom or side of the door to retract, making exit possible. **See Figure 59-60.** In certain applications, exit devices are connected to monitoring devices to indicate entry and exit through the door. Door latches may be retracted or engaged automatically from a remote location.

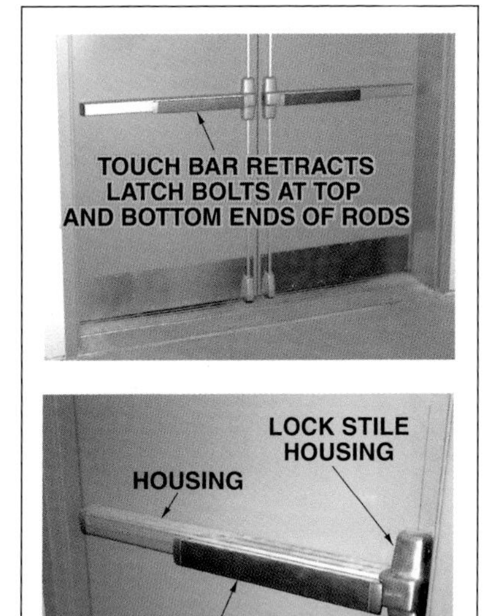

Figure 59-60. *Exit devices are mounted on doors of public buildings. Slight pressure on the touch bar will retract the latch bolts, allowing the doors to be opened.*

Door Holders and Door Stops

Door holders and *door stops* are door hardware typically installed after the doors are hung. Door holders keep the door in an open position. **See Figure 59-61.** A door holder commonly used for lightweight and hollow-core doors is attached to the door with screws and has a lever that drops to the floor to wedge the door open. Another type of holder is mounted on the door and engages with a strike attached to the wall or floor. Magnetic strikes may be installed to allow the door holder to be disengaged from a remote location. A plunger type of door holder can be activated and relieved by foot pressure. A spring in the case retracts the plunger allowing the door to swing.

Door Holders

Figure 59-61

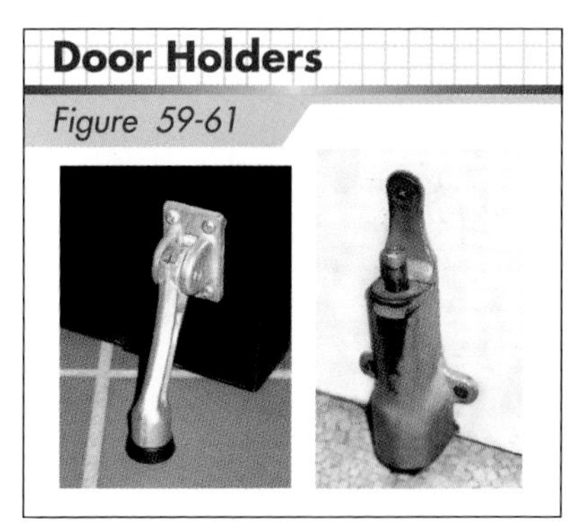

Figure 59-61. *A door holder keeps a door in an open position.*

Door stops prevent doors from bumping against the wall when doors are fully opened. Door stops are mounted on the wall, baseboard, door, or floor. **See Figure 59-62.** Door stops should be installed at the bottom or top of a door.

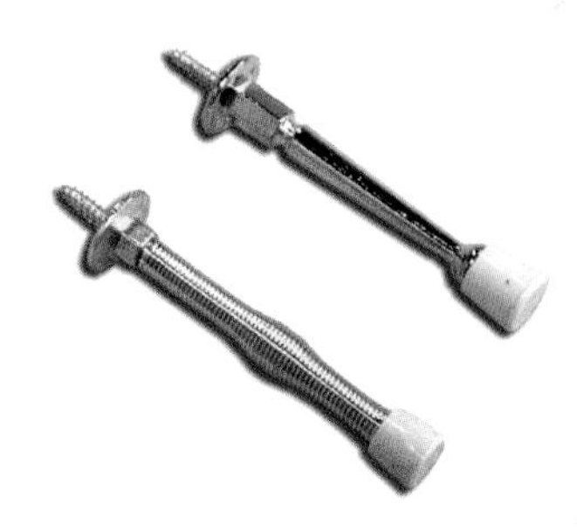

Figure 59-62. *A door stop prevents a door from bumping against a wall.*

The Builders Hardware Manufacturers Association (BHMA) is the trade association for North American manufacturers of commercial builders' hardware. BHMA has developed several standards for builders' hardware, including standards for hinges and locks.

Push Plates, Kick Plates, and Door Pulls

While door stops are designed primarily to protect walls, push plates, kick plates, and door pulls are designed to protect doors. In many applications, push plates, kick plates, and door pulls enhance the appearance of a door.

Push plates and kick plates are frequently installed in schools and other public buildings where doors are in constant use. **See Figure 59-63.** Push plates are mounted at the center of a door, in many cases in lieu of a door knob or door pull. Kick plates are similar to push plates but are mounted at the bottom of the door. Push plates and kick plates are pushed with the hand and feet, respectively.

Door pulls may be flush or may have a handle that projects from the door surface. Flush door pulls are mortised into sliding doors to provide a finger hold for opening and closing a door. **See Figure 59-64.**

Figure 59-63. *Push plates and kick plates are usually installed on commercial doors.*

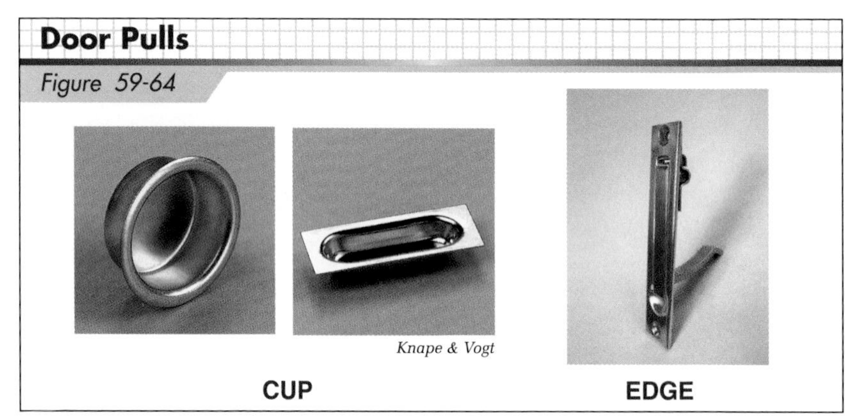

Figure 59-64. *Pulls are mortised into sliding doors to provide a finger hold for opening and closing the doors.*

Unit 59 Review and Resources

OBJECTIVES

1. Describe the construction of cabinets.
2. Describe cabinet drawer construction.
3. Explain the procedure for installing cabinets.
4. Explain the procedure for laying out kitchen wall and base cabinets.
5. Identify common types of cabinet doors and hinges.
6. Describe the procedures for installing built-in units.
7. Describe countertop construction.
8. Discuss installation of plastic laminate countertops.
9. Describe solid-surface countertops.

UNIT 60 Cabinet and Countertop Installation

Cabinets and countertops are usually constructed in a cabinet shop or factory, delivered to the job site, and installed by carpenters. In residential construction, most cabinets and countertops are located in the kitchen and bathroom. **See Figure 60-1.**

Built-in bookcases, storage units, and wardrobes may be required in other parts of a building, and these units are installed in similar fashion to kitchen or bathroom cabinets. In commercial construction, such as hospitals, banks, and retail stores, more cabinets and countertops are required and installed than in residential construction.

CONSTRUCTING CABINETS

Although carpenters on the job site are typically required only to install cabinets, they should have a basic knowledge of cabinet construction. Cabinet parts are fastened together with screws or nails. Nails are usually spaced 6″ OC, and are set below the surface. Surface holes are filled with wood putty. Glue is used at all joints to provide additional rigidity to the cabinet. Clamps are placed on cabinets as the glue sets to ensure a sound joint. A better-quality cabinet is rabbeted along the top, bottom, and sides to accept the back piece and strengthen joints. **See Figure 60-2.** However, butt joints may also be used. If panels are less than ¾″ thick, a reinforcing block should be used with the butt joint. **See Figure 60-3.** Fixed shelves are dadoed into the sides. **See Figure 60-4.**

Figure 60-2. *Rabbet joints make an attractive and sturdy fit for cabinet corners.*

Merillat Industries

Figure 60-1. *Most cabinets and countertops used in residential construction are located in the kitchen.*

Figure 60-3. *Butt joints may be used when constructing cabinets. A reinforcing block should be installed in the corner if the material being used for the cabinet is less than ¾″.*

Figure 60-4. *Dado joints provide good support for cabinet shelves.*

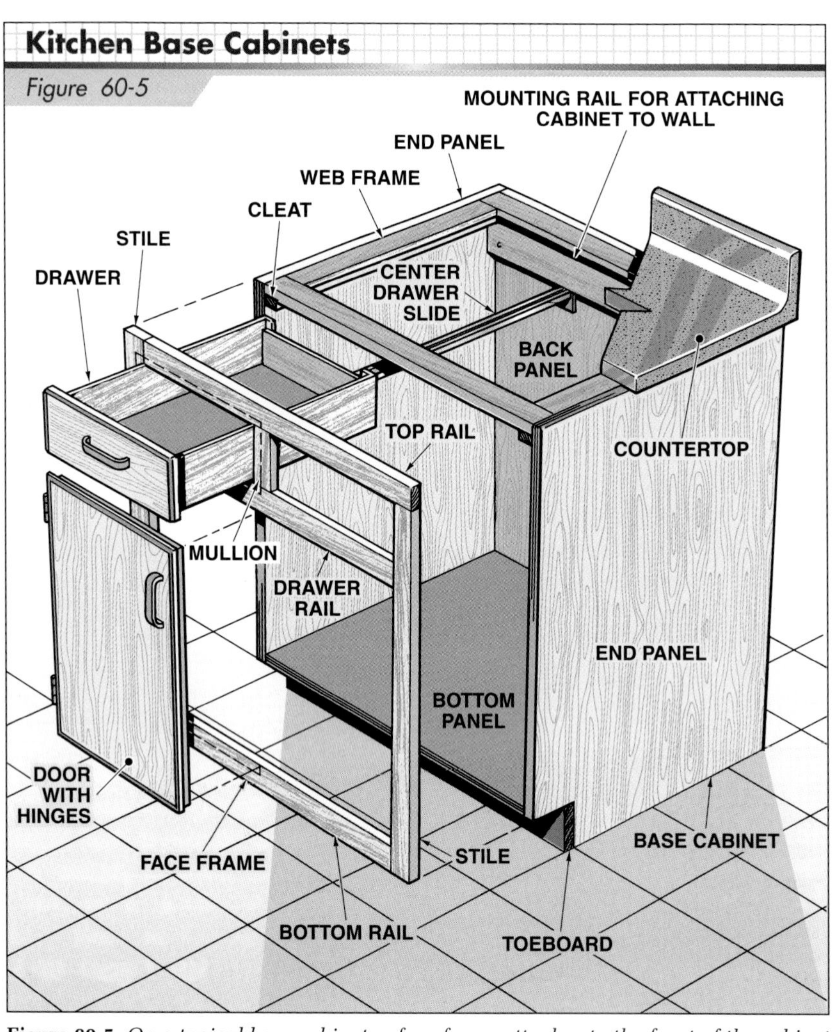

Figure 60-5. *On a typical base cabinet, a face frame attaches to the front of the cabinet and a web frame attaches to the top.*

Cabinets may have a face frame that attaches to the front of the cabinet and a web frame that attaches to the top. **See Figure 60-5.** The rails and stiles of the face frame are joined by mortise-and-tenon, dowel, or plate joints. **See Figure 60-6.**

A Swiss cabinetmaker, Herman Steiner, developed the first plates.

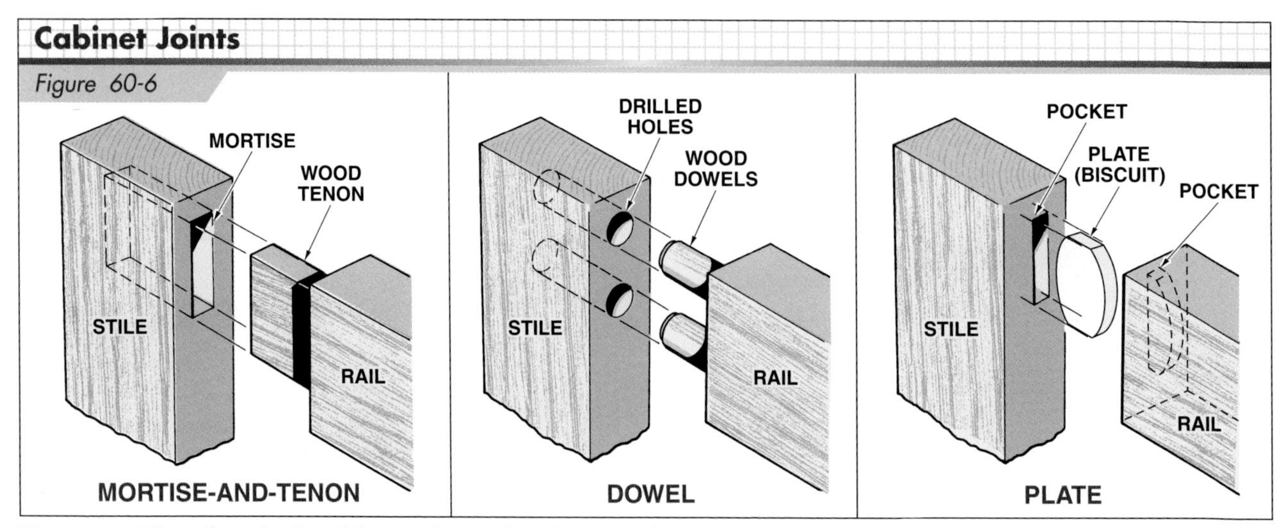

Figure 60-6. *The rails and stiles of the face frame of a cabinet may be joined by mortise-and-tenon, dowel, or plate joints.*

Cabinet backs are best attached by rabbeting the back edge of the cabinet sides. **See Figure 60-7.** A deeper rabbet makes it possible to scribe the side against an uneven or out-of-plumb wall.

Figure 60-7. *Cabinet backs are best attached by rabbeting the cabinet sides. At left, the back is flush with the edge of the sides. At right, the rabbet is deeper and the side projects about ¼″ past the back, making it possible to scribe the side against an uneven or out-of-plumb wall.*

There are many methods of drawer construction. Three common methods are dovetail, lock-shouldered, and square-shouldered. **See Figure 60-8.**

A variety of drawer slides are available. **See Figure 60-9.** Drawer slides are installed prior to cabinet installation. Manufacturer instructions provided with the slides provide allowance dimensions for drawer openings and depth, and installation procedures.

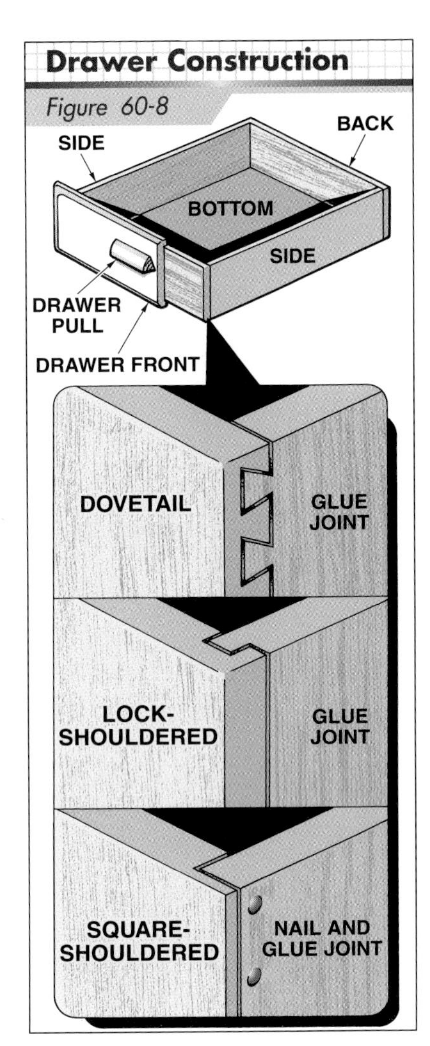

Figure 60-8. *Drawer sides are joined to the front or back with dovetail, lock-shouldered, or square-shouldered joints.*

Drawer slides are designed for loads of various weights. In general, side-mounted drawer slides with ball-bearing rollers can carry more weight than bottom-mounted, single-rail drawer slides.

Wood drawer guides are seldom used because of the construction time required. In addition, wood drawer guides may stick in areas of high humidity. When wood drawer guides are used, the guides are attached to the sides of the drawer or inside openings of the cabinet, depending on drawer design.

The 32 mm System, or System 32, cabinet construction is a type of frameless cabinet construction. The 32 mm System was developed in Germany after World War II due to depleted solid wood resources. The 32 mm System standardized cabinet case construction and hardware mounting.

Shims are placed under the base when leveling base cabinets.

INSTALLING KITCHEN CABINETS

The kitchen cabinet detail drawing found in the prints of the three-bedroom house plan (refer to Unit 30) includes the widths and heights of the cabinets and the height at which wall cabinets are to be placed. **See Figure 60-10.** Spaces for kitchen appliances, such as a refrigerator and dishwasher, are also indicated on the drawing.

The combined widths of the cabinets should be compared to the total distance of the wall on which they are to be mounted before cabinets are installed. This can be easily done by cutting a rod the exact length of the wall and marking on the rod the cabinet widths and widths of any windows or any space provided for appliances or other purposes.

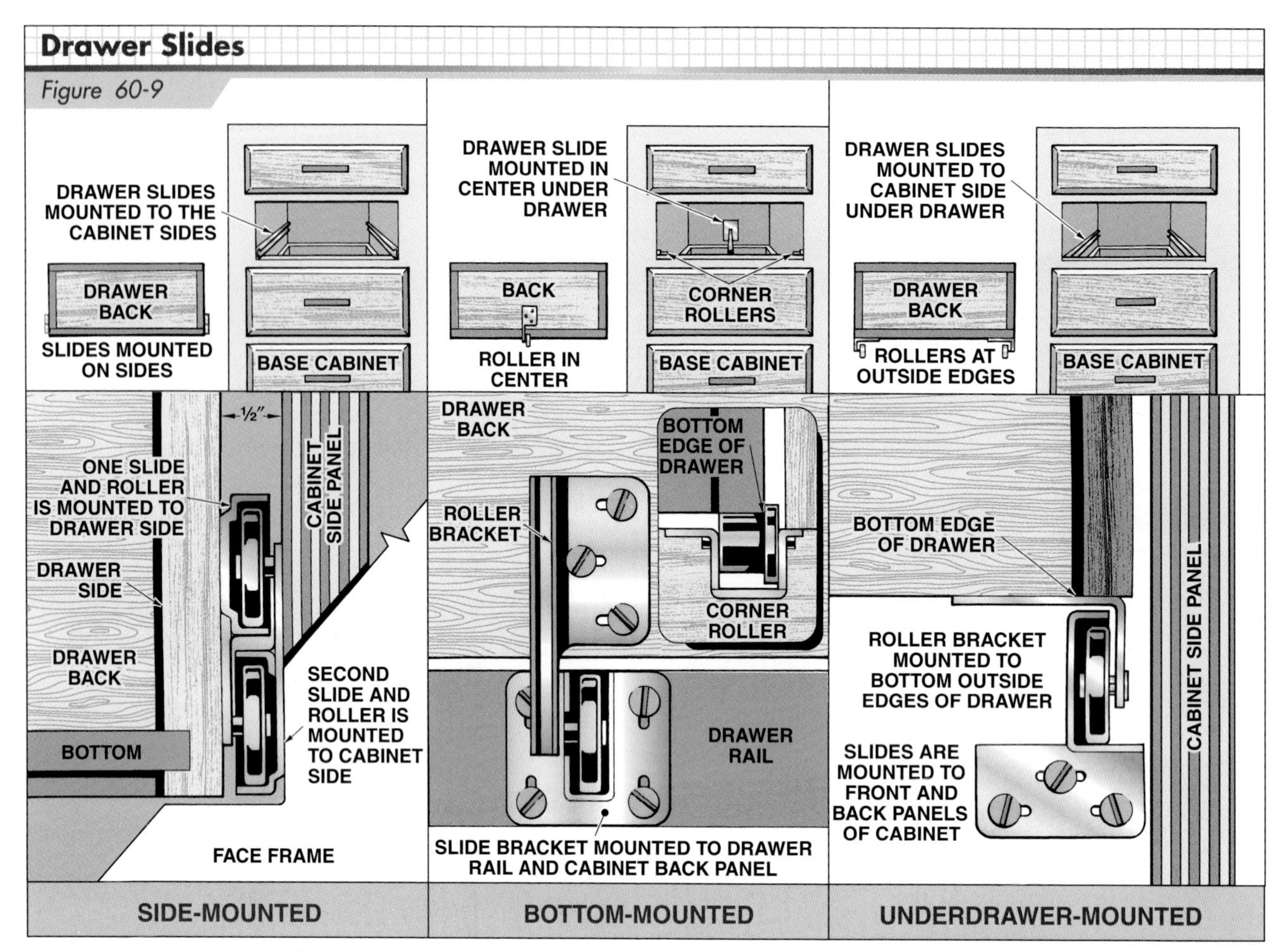

Figure 60-9. *Drawer slides allow drawers to be easily opened and closed. The slides may be mounted on the side, on the bottom, or under the drawer.*

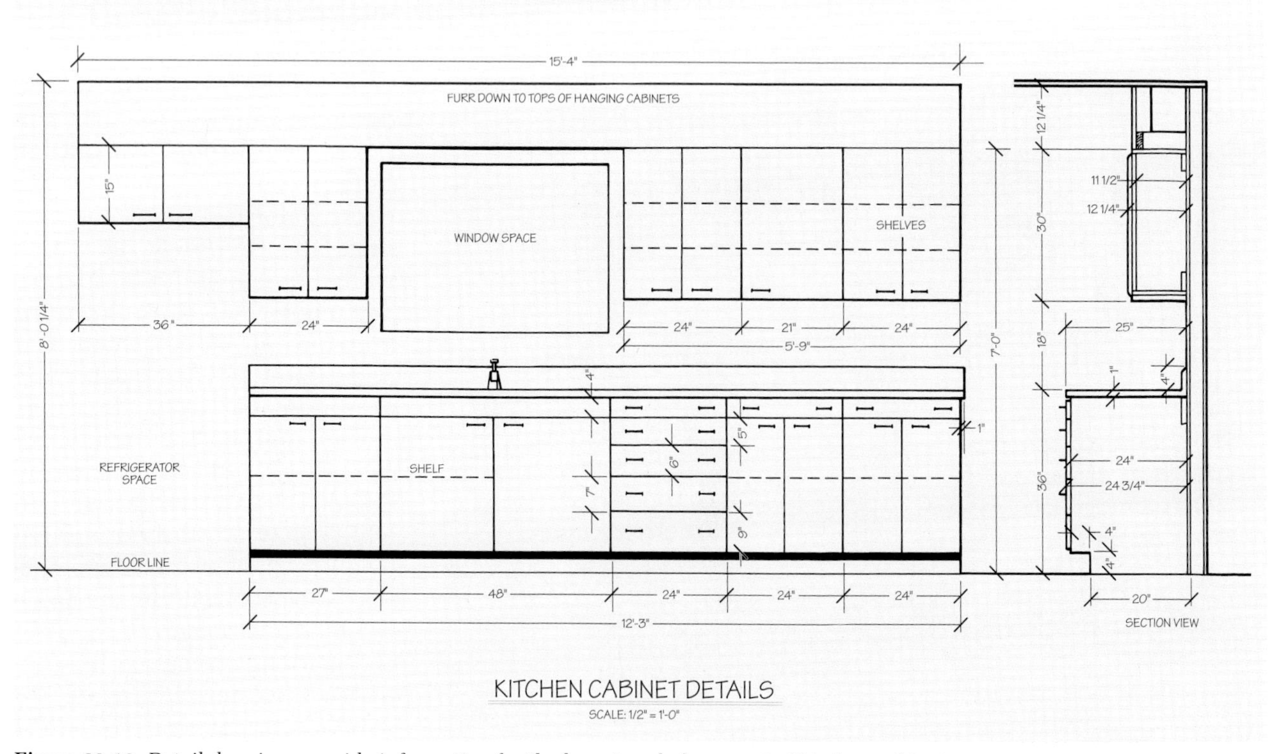

Figure 60-10. *Detail drawings provide information for the layout and placement of kitchen cabinets.*

A typical kitchen consists of a series of base cabinets and a series of wall cabinets along one or more walls of the room. Wall cabinets are more easily installed before base cabinets to avoid awkward lifting over the base cabinets. Care should be taken when installing cabinets as the wall surfaces may already be finished. Basic principles of installation are as follows:

1. Construct T-braces, which are slightly longer than the distance from the floor to the cabinet bottom. A T-brace supports the cabinets while they are being fastened to the wall. **See Figure 60-11.** A cleat may be fastened to unfinished walls to support the cabinets.
2. Place another T-brace along the cabinet front to prevent cabinets from tipping forward while they are being fastened to the wall. **See Figure 60-12.**
3. Using shims where necessary, plumb and level each cabinet while it is being installed. Use wood screws long enough to penetrate into the studs. Better-quality cabinets have *mounting rails,* or hanging rails, at the top and bottom. **See Figure 60-13.**
4. Some carpenters may prefer to fasten a set of wall cabinets together on the floor and hang the set as one unit on the wall. Other carpenters install one cabinet at a time on the wall. Whichever system is used, wood screws are driven through the stiles at the face frames of the cabinets to fasten the cabinets to each other. Small bar clamps or C-clamps are used to temporarily hold the frames tightly together. **See Figure 60-14.**
5. Level and plumb the base cabinets, and fasten them to the wall. **See Figure 60-15.**

Figure 60-16 shows a procedure for installing kitchen cabinets.

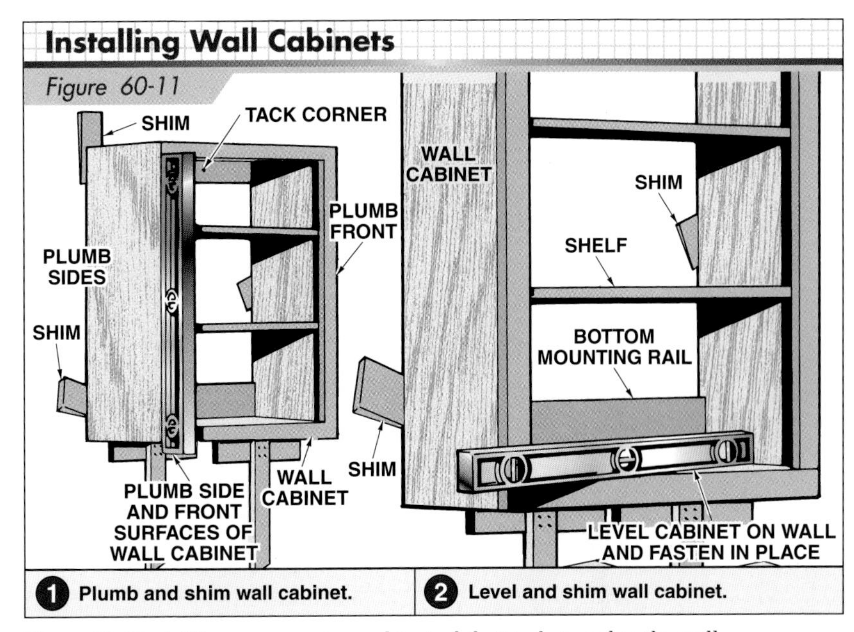

Figure 60-11. *A T-brace supports a cabinet while it is fastened to the wall.*

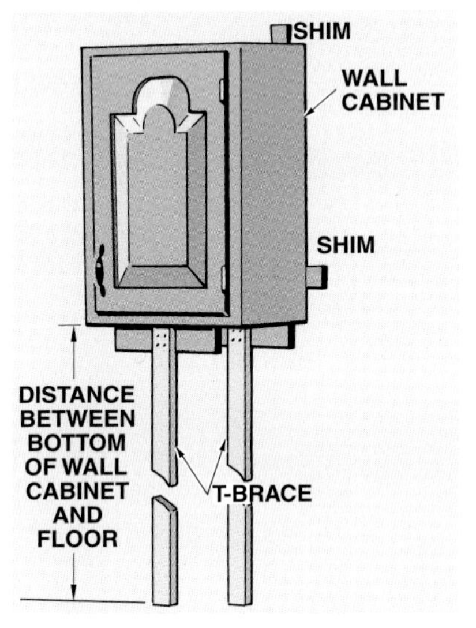

Figure 60-12. *A T-brace prevents a cabinet from falling forward while it rests on the back T-brace.*

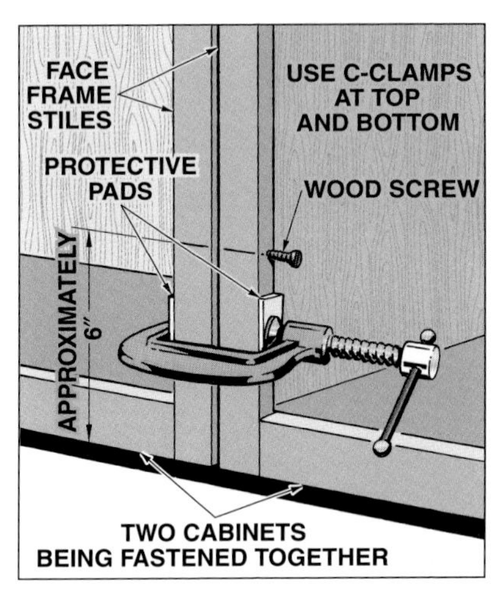

Figure 60-14. *Cabinets are fastened to each other with wood screws driven through the face frame stiles. C-clamps may be used to temporarily hold the stiles tightly together.*

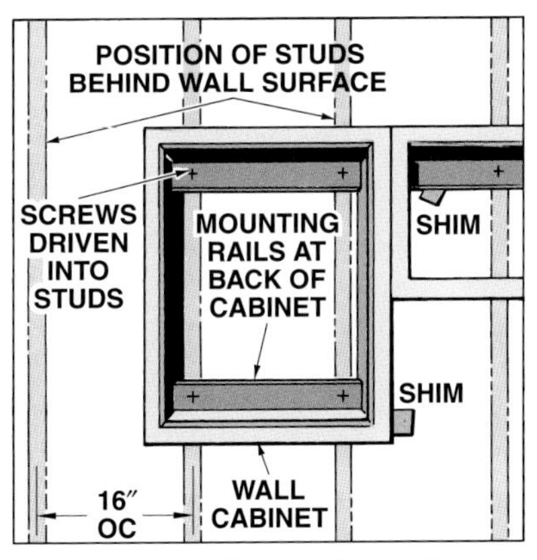

Figure 60-13. *Screws driven through mounting rails and wall surfaces fasten cabinets to wall studs.*

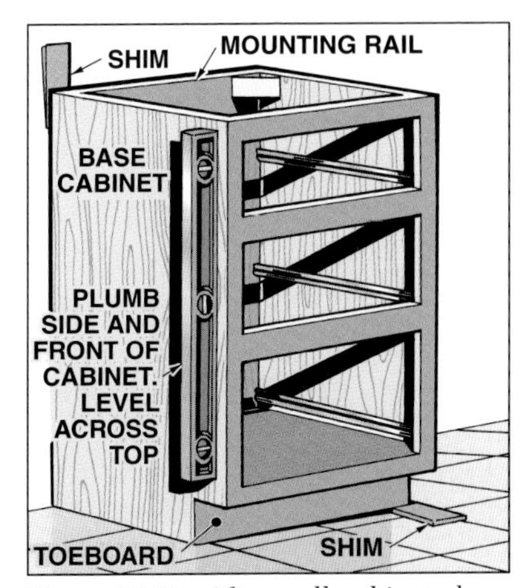

Figure 60-15. *After wall cabinets have been installed, level and plumb base cabinets and fasten them to the wall.*

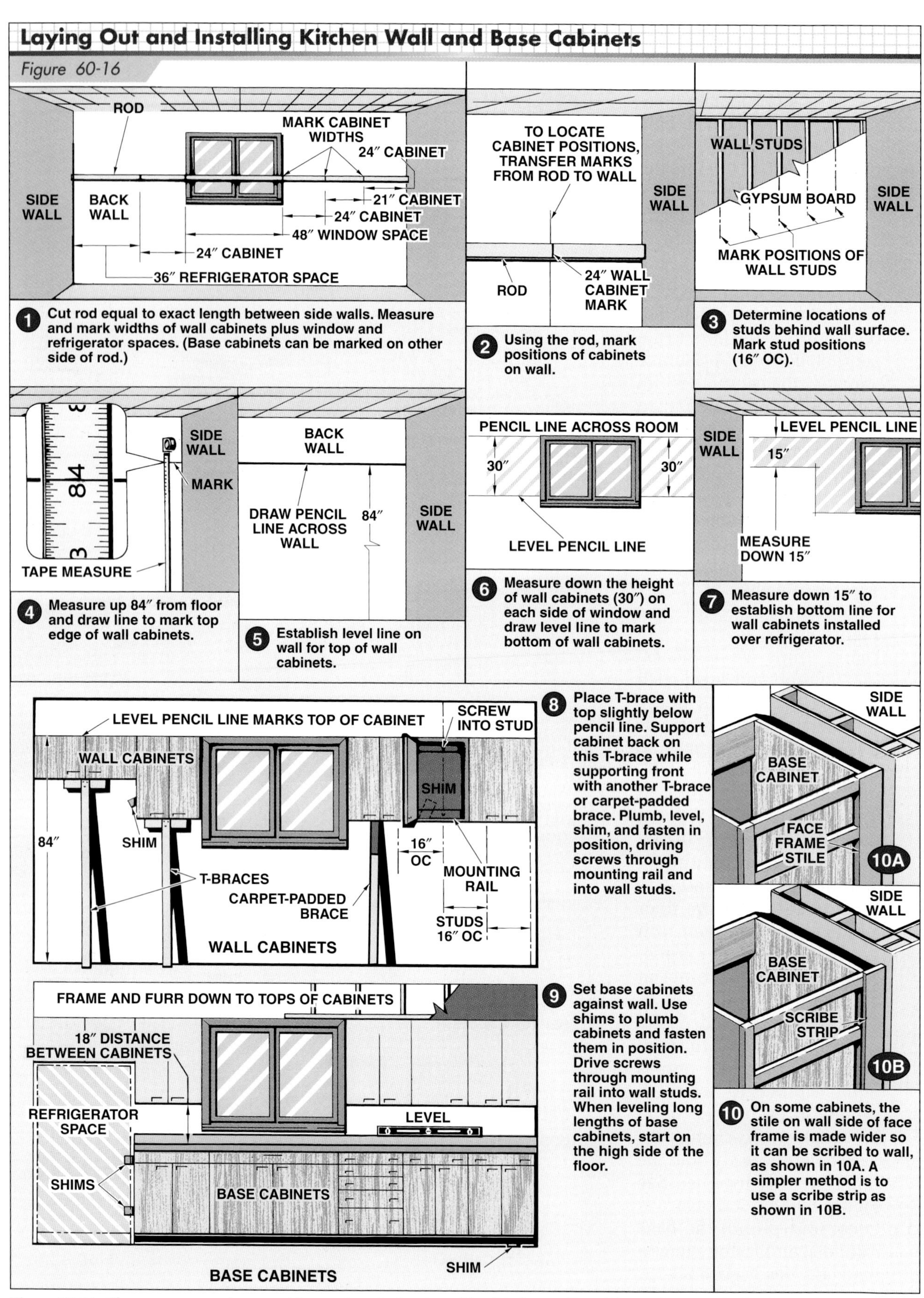

Figure 60-16. *The proper procedure must be employed when installing kitchen wall and base cabinets. The kitchen cabinet details for this installation are shown in Figure 60-10.*

Cabinet Doors and Hinges

Cabinet doors may be hung in a cabinet shop or by carpenters on the job. If cabinets are racked out of square during transportation or installation, prehung doors may be thrown out of alignment. For this reason, many carpenters prefer to hang cabinet doors on the job.

The more commonly installed cabinet doors are *flush, lipped, overlay,* and *sliding doors.* **See Figure 60-17.** Various hinges and other hardware are available for each type of door. **Figure 60-18** shows several types of hinges used to hang cabinet doors. Hinge size depends on the weight and size of the doors. Hinges are available with a variety of finishes in plain and ornamental designs.

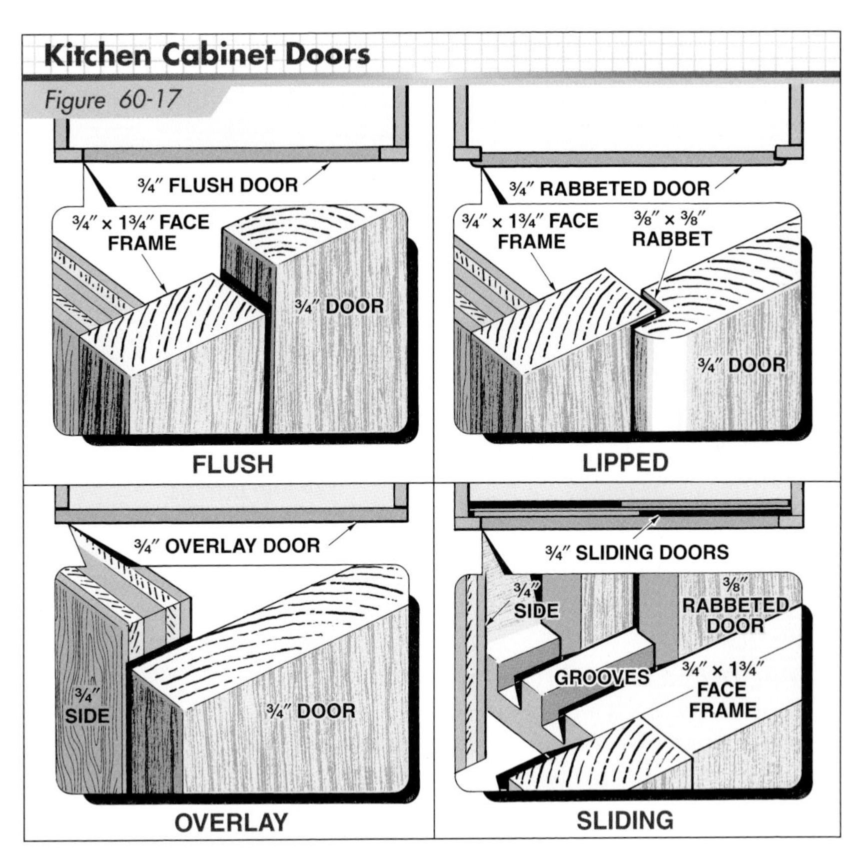

Figure 60-17. *Four types of doors are often used for base and wall cabinets. Various types of hardware are available for each type of door.*

Flush Doors. Flush doors are similar to other swinging doors and are the most difficult to hang. Flush doors must be fitted in the cabinet opening with 1/16″ clearance around all edges. Unequal clearances indicate poor worksmanship.

An ornamental surface hinge, concealed hinge, or semiconcealed loose-pin hinge may be used on a flush cabinet door. A semiconcealed loose-pin hinge looks like a standard loose-pin hinge when the door is closed, but its design allows greater holding power between the hinge leaf and cabinet door back.

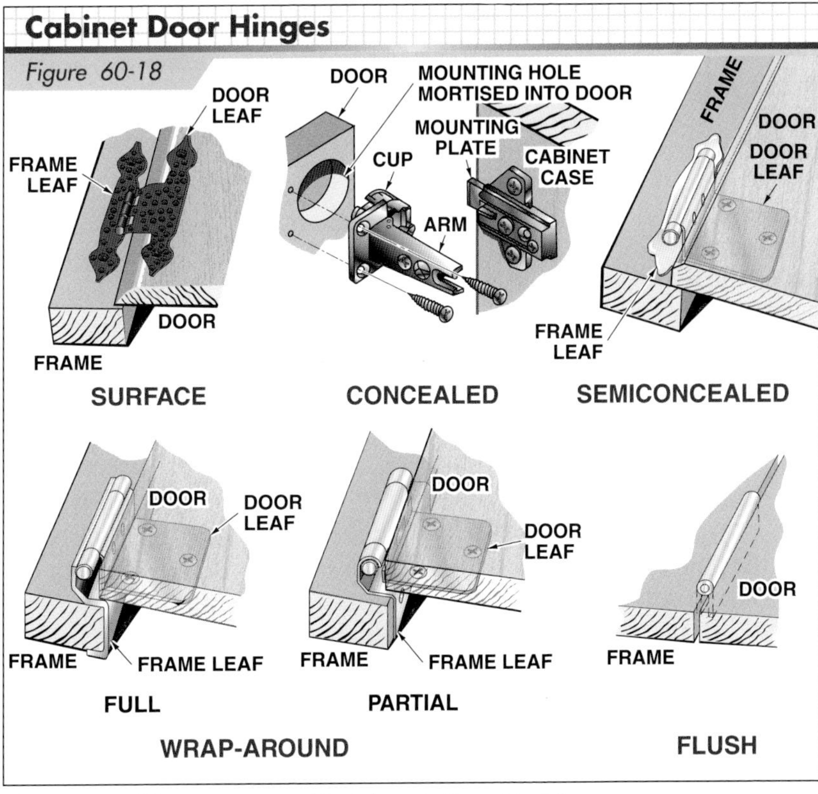

Figure 60-18. *Hinges connect cabinet doors to the cabinet frame or case.*

Lipped Doors. Lipped doors are simpler to install than flush doors since the lip allows for a certain amount of adjustment. A semiconcealed hinge is often used with lipped doors. The hinge may be self-closing or may require a catch to hold the door in a closed position.

A tambour door is a flexible sliding door consisting of narrow wood, veneered medium density fiberboard, or plastic slats bonded to a heavy canvas backing. A tambour door slides in curved tracks in the cabinet case top and bottom or sides. When open, the door is hidden.

Overlay Doors. Overlay doors are common on frameless cabinets. Semiconcealed pivot hinges are often used with overlay doors. A small bevel is required where semiconcealed hinges are placed at the top and bottom edges of the door.

Sliding Doors. Several types of sliding doors are used on cabinets. **See Figure 60-19.** One type of sliding door is rabbeted along its upper and lower edges to fit into grooves at the top and bottom of the cabinet door opening. Thinner doors slide in a plastic track set into the bottom of the cabinet. The top groove is always made deeper to allow the door to be removed by lifting it up and pulling the bottom out.

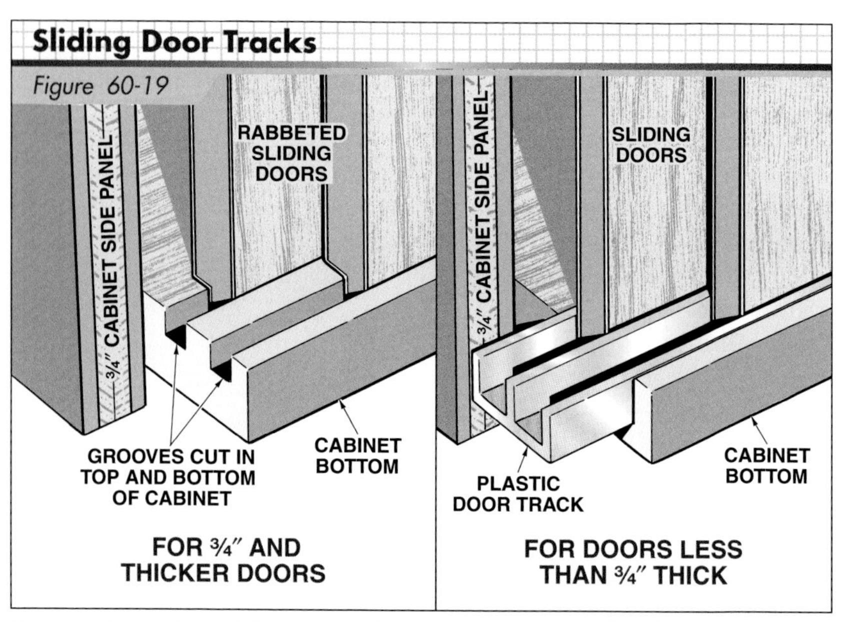

Figure 60-19. *Cabinet sliding doors at least 3/4″ thick may be rabbeted to slide in grooves at the cabinet top and bottom. Cabinet sliding doors thinner than 3/4″ thick are inserted in plastic tracks.*

Pulls and Knobs. Pulls and knobs for cabinets are available in many designs and finishes. **See Figure 60-20.** Holes must be accurately laid out for the machine screws (or bolts) used to fasten the knobs and pulls. When several holes must be laid out on the same size doors or drawers, a jig should be constructed so the holes can be laid out accurately and efficiently. When attaching pulls or knobs to a cabinet door or drawer, drill from the exposed (outside) face to the unexposed (inside) face. To prevent the wood from splitting when using a drill bit, clamp a block to the back of the piece being drilled. **See Figure 60-21.**

Figure 60-20. *Drawer pulls and knobs are available in many designs and finishes.*

Door Catches. A variety of door catches are available to hold doors in place when they are shut. Some of the more common door catches operate by means of friction, magnets, or nylon rollers. **See Figure 60-22.**

Installing Built-in Units

Wardrobe closets, built-in bookcases, and storage cabinets are installed in a similar manner to kitchen cabinets. These permanently attached units often serve as room dividers as well. For example, a long base storage cabinet may divide the work area of the kitchen from the dining room.

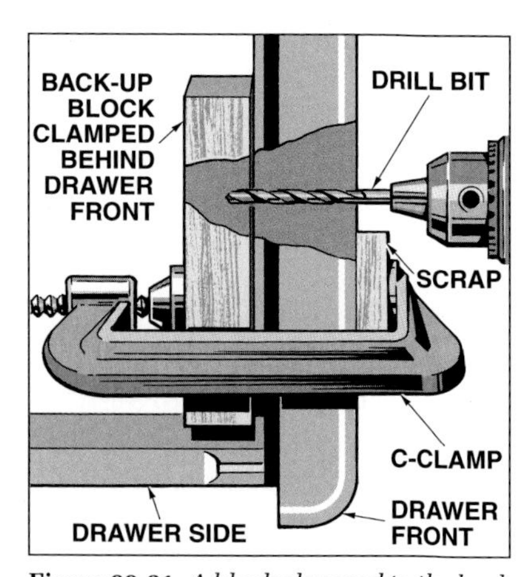

Figure 60-21. *A block clamped to the back of the cabinet piece being drilled prevents the wood from splitting.*

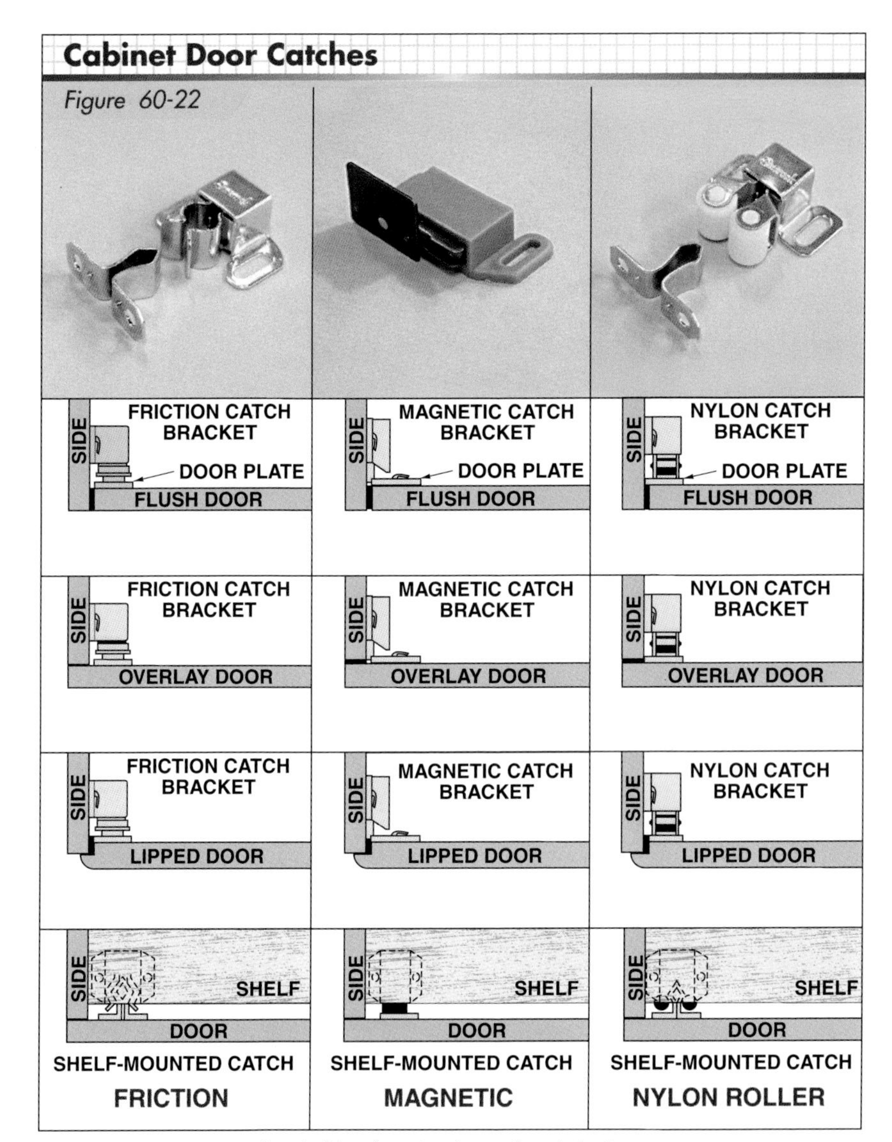

Figure 60-22. *Door catches hold a door in place when it is shut.*

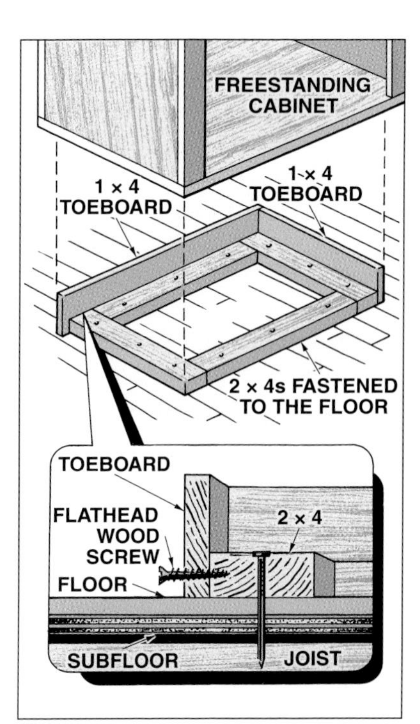

Figure 60-23. *A freestanding cabinet can be secured by driving screws through the toeboards into 2 × 4s fastened to the floor.*

In some cases, built-in units cannot be securely fastened to walls using the same techniques used for kitchen cabinets. Therefore, the units must be secured to the floor. One method of securing built-in units to the floor is to fasten 2 × 4s to the floor inside the unit and then drive screws through the toeboards into the 2 × 4s. **See Figure 60-23.**

COUNTERTOPS

Countertops are constructed in a cabinet shop and delivered to the job site where they are assembled and fastened to base cabinets after the cabinets have been installed. *Plastic laminate* and *solid-surface materials* are popular choices for kitchen and bathroom countertops since the materials are hard, durable, and highly resistant to heat, and resist stains or other damage. Ceramic tile, stone (granite and marble), and wood "butcher block" countertops may also be specified for installation.

Many countertops have a *backsplash,* which is a piece along the back edge of a countertop. A backsplash rests against the wall when the countertop is installed. Backsplashes may be an integral part of the countertop or they may be installed separately.

Plastic Laminate Countertops

Plastic laminate used for countertops is approximately 1/16″ thick and is available in a variety of patterns. Plastic laminate is cut with a table saw, radial-arm saw, circular saw, or jigsaw equipped with a fine-toothed blade.

Contact cement is used to bond plastic laminate to its base core, which is usually 3/4″ plywood or MDO or HDO particleboard. Contact cement is applied to the back of the laminate and to the base core using a brush, roller, or sprayer. The contact cement is allowed to dry, usually 15 min to 20 min under normal conditions. To test for dryness, lightly press a piece of kraft paper against the adhesive. If no glue adheres to the paper, the surfaces are ready for bonding.

Plastic laminate must be positioned carefully when being bonded to the base core. Once the glued surface of the laminate contacts the glued surface of the core, it can no longer be moved. To ensure correct positioning before bonding occurs, place narrow wood strips or dowels on the core at intervals and place the plastic laminate over them. When the laminate is correctly positioned, carefully remove the strips or dowels starting at the center and moving outward. The laminate is then rolled to ensure proper adhesion. When applying plastic laminate for counter-tops, apply the edge strips first, then the top sheet.

Plastic laminate is always cut slightly larger than the base core. After the laminate is bonded to the core, the laminate is trimmed with a laminate trimmer or router equipped with the appropriate laminate trimmer bit to cut and bevel the edges.

Kohler Co.

Figure 60-24. *The sink opening size is determined by the dimensions of the sink installed.*

Cutting Sink Openings. Sink openings may be cut out at the factory, although many carpenters prefer to cut the sink opening at the job site where the exact location and opening size required can be marked on the countertop.

After the opening is cut out, waterproof caulk is placed around the edge of the opening and the sink is set in place. The sink is then secured with hold-down lugs on the underside of the countertop. **See Figure 60-24.**

Attaching Countertops. A base cabinet usually has a 1″ thick frame at the top of the cabinet or triangular pieces in the corners for attaching the countertop. Holes are drilled through the frame or corner pieces and into the countertop. Screws are inserted into the holes and driven into the underside of the countertop. Ensure the screws are short enough so as not to penetrate through the plastic laminate. **See Figure 60-25.**

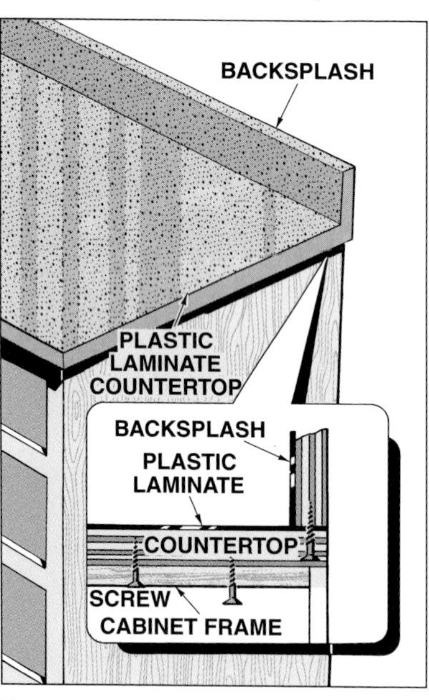

Figure 60-25. *Screws are driven through the web frame and into the bottom of the countertop.*

Most solid-surface countertop manufacturers require that the purchaser of the material have completed training in the techniques and methods of working with their brand of solid-surface material.

Attaching Backsplashes. The pieces making up the backsplash are often fastened to the countertop when it is constructed in the shop. Some carpenters, however, prefer to fasten the backsplash to the countertop on the job so they can make adjustments if the walls are out of square.

If the wall behind the backsplash is irregular, the laminate strip at the top edge of the backsplash may require scribing. For this reason, the strip is sometimes omitted. **Figure 60-26** shows a procedure for scribing and attaching this strip after the backsplash is secured to the countertop.

Solid-Surface Countertops

Solid-surface materials are used for kitchen countertops, bathroom vanity tops, and wall cladding in showers. Because they are nonporous, solid-surface materials are also used in specialized environments such as laboratories and hospitals.

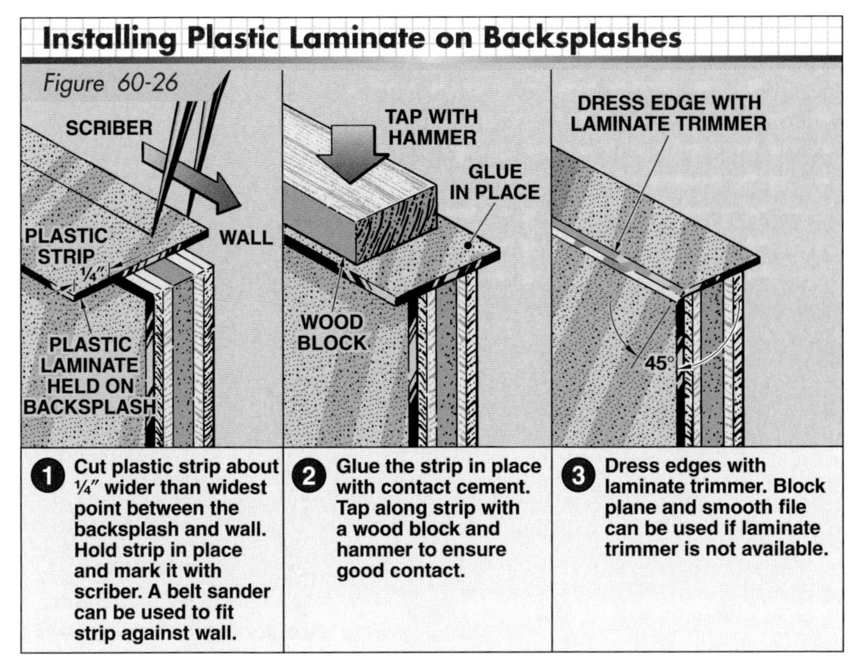

Figure 60-26. *A plastic laminate strip at the top edge of the backsplash may need to be scribed and attached to the top of the backsplash after the backsplash has been secured to the countertop.*

Solid-surface materials are a mixture of an acrylic polymer and alumina trihydrate, which is an inert, flame-retardant mineral. Solid-surface material is cut and shaped with common woodworking tools such as table saws, routers, and jig saws. **See Figure 60-27.** When working with any type of solid countertop material, it is advisable to follow the manufacturer directions for fabrication and installation.

Solid-surface materials are available in ¼″, ½″, and ¾″ thicknesses and in a variety of colors, textures, and patterns. **See Figure 60-28.** The ½″ and ¾″ materials are used for countertops; ¼″ material is used for edge strips and vertical applications. Countertop overhangs can extend past cabinet sides as much as 6″ for ½″ thick countertops and 8″ for thicker materials. Solid-surface countertops should be supported every 18″ for ½″ thicknesses.

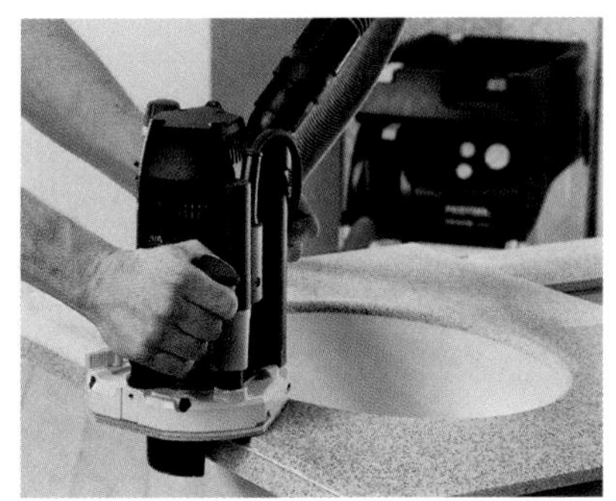

Festool USA

Figure 60-27. *Large routers are used to shape the edges of a solid-surface countertop.*

Similar to other types of countertops, solid-surface countertops are delivered to the job site in sections and are placed in position on the base cabinets. To prevent flexing, solid-surface countertops must be properly supported during delivery and placement. Support (typically 2 × 4s) should be positioned approximately every 18″.

Formica Corporation

Figure 60-28. *Solid-surface materials provide outstanding durability and stain resistance.*

When installing solid-surface countertops, dabs of clear silicone sealant are placed on the base cabinet every 10″ to 12″. The back edge of the countertop is placed into position and the front edge is carefully lowered into position, keeping it as straight as possible. Seams between the sections are filled with a seaming compound, which is allowed to dry. Excess seaming compound and sur-rounding area is then sanded to create an invisible seam.

Unit 60 Review and Resources

Unit 61 Interior Trim

OBJECTIVES

1. Describe the manufacture of wood molding.
2. Identify wood molding profiles.
3. List and describe common types of composite molding.
4. Describe the installation of door casings.
5. Describe the installation of window casings.
6. Describe the movement of miter joints.
7. Explain the procedure for trimming window openings.
8. Explain the procedure for installing base molding.
9. Explain the procedure for installing ceiling molding.

Interior trim is the molding that completes the finish around doors, windows, cabinets, and tops and bottoms of walls. The molding should harmonize with the overall design of the building. For contemporary buildings, less molding, in simpler patterns, is appropriate. Some contemporary building designs completely eliminate molding. For traditional buildings, a greater amount of molding, in more elaborate patterns, is desirable. In commercial structures, plastic or metal trim is commonly installed.

Solid wood molding has been, and continues to be, the predominant material used for interior trim. Installation of composite molding (molding made by combining various materials) has been on the increase in recent years.

WOOD MOLDING

Wood molding is manufactured from softwood and hardwood lumber, as well as composite material such as medium-density fiberboard (MDF). Softwood and composite molding may be stained, but is typically painted. Hardwood lumber is used when a stain finish is preferred.

Wood Molding Manufacture

When manufacturing molding from solid lumber, a strip of lumber is resawn into a *blank,* which is a piece that will produce the desired pattern of molding with the least amount of waste. **See Figure 61-1.** The blank is shaped into molding on a machine called a *molder* (or *sticker*). The molder is equipped with cutter heads and knives that rotate at high speed to create the molding pattern. Computer-controlled molders are used in many mills and shops which produce large quantities of molding.

Medium-density fiberboard is cut to the desired width to create a blank. The blank is shaped into molding on a molder, similar to the process for solid lumber molding. The MDF molding is then sanded and sealed with a paint primer. MDF molding is also available with veneer or paper overlays.

Medium-density fiberboard (MDF) molding with enhanced moisture resistance should be used in bathrooms and for baseboards of slab-at-grade construction.

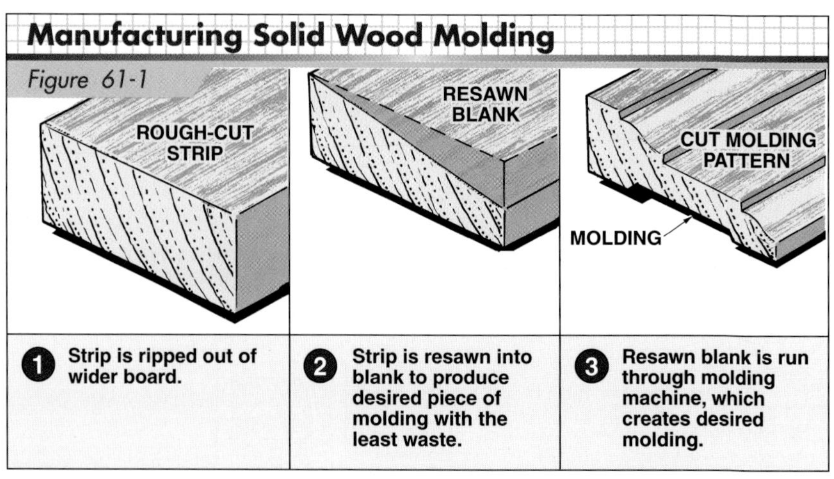

Figure 61-1. *Softwood and hardwood lumber are used in the manufacture of wood molding.*

Wood Molding Profiles

A variety of wood molding profiles (shapes) are available, depending on the purpose for which the molding is used. Molding is usually fastened in place with finish nails. Staples may also be used when installing molding. Nails are usually set below the molding surface. The nail holes are then filled with wood putty that matches the stain or paint finish. When prefinished molding is installed, nails of a matching color may be driven flush with the surface (and not set below the surface).

Typical wood moldings are designed for the intersection of ceilings and walls or walls and floors and for door and window openings, astragals on double doors, and handrails. **See Figure 61-2.** A sequence of interior molding installation is shown in **Figure 61-3.**

Ceiling Molding. Traditional ceiling molding profiles used most often are *cove, bed,* and *crown moldings.* A small rectangular trim piece may be used to cover the joint between finish wall paneling and the ceiling. However, in more contemporary buildings, ceiling molding is typically omitted.

Base Molding. Base molding is installed at the base of a wall. *Base shoes* cover the joint between the base molding and finish floor covering. *Base caps* are nailed to the top of square-edge base molding if a more decorative effect is desired.

Casing. Casing is used to cover the space between the doorjamb and wall on both sides of interior door openings. Casing is also installed along the top and sides of window openings.

Wood molding is a type of architectural millwork.

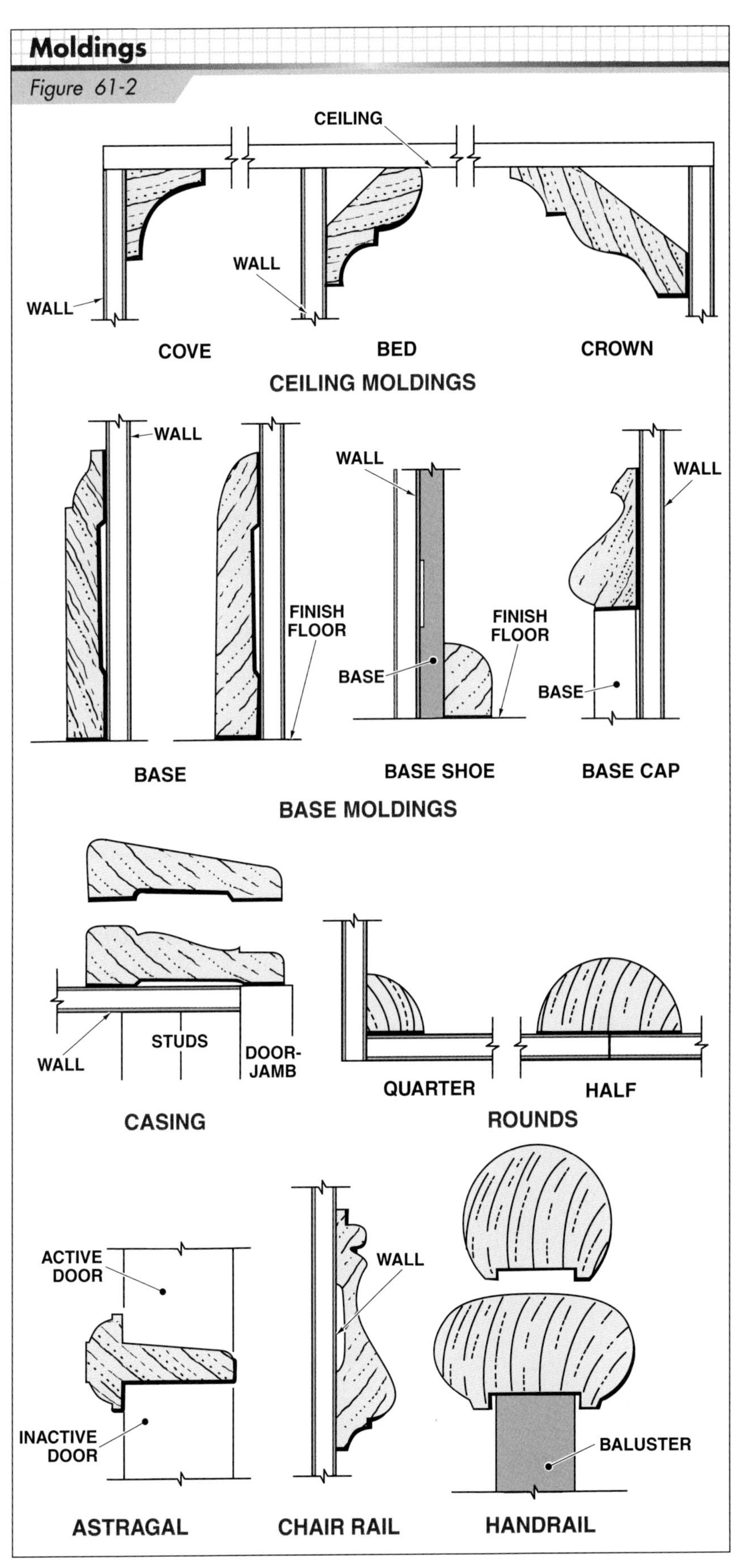

Figure 61-2. *Wood moldings are trim pieces at the juncture of walls and ceilings and floor and walls, and at intersections of other materials.*

Interior Molding Installation

Figure 61-3

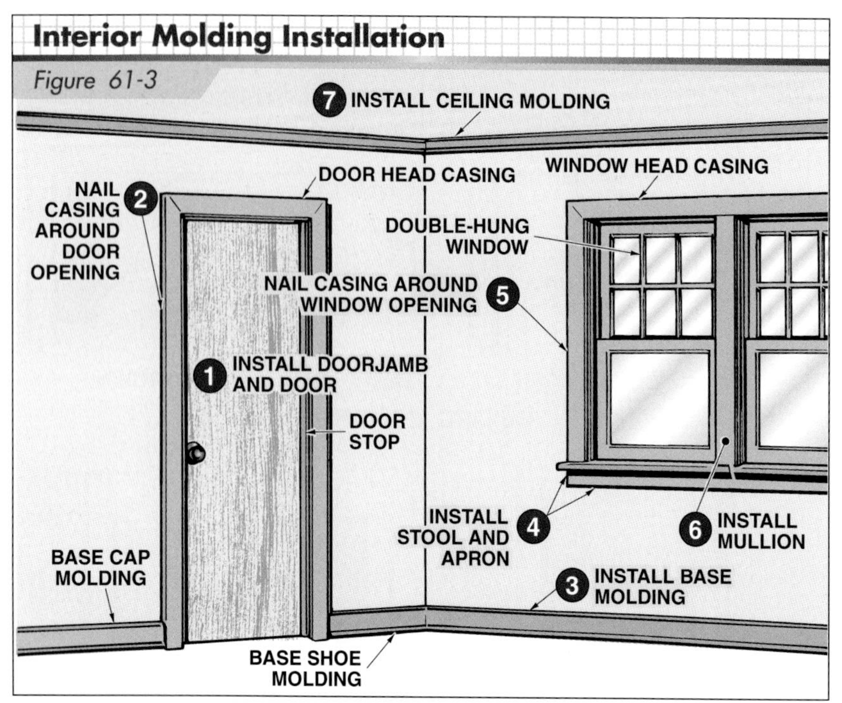

Figure 61-3. *Molding is commonly installed along the floor and around door and window openings. In some cases, ceiling molding is specified.*

Astragals. An *astragal* is fastened to the inactive door of a double door unit. The astragal acts as a stop for the active door.

Rails. *Handrails* and *chair rails* are the two most common types of rails. A handrail is installed along a stairway to provide support for a person ascending or descending the stairway. A chair rail is installed approximately 32″ above the finish floor covering. Originally, chair rails were attached to plaster walls to prevent the back of a chair from damaging the wall surface. Today, the purpose of a chair rail is primarily decorative.

COMPOSITE MOLDING

In recent years, the use of *composite molding* has been increasing. Two examples of composite molding are plastic and gypsum molding.

American Hardwood Export Council

A chair rail is installed approximately 32″ above the finish floor covering and serves a primarily decorative purpose.

Plastic Molding

Plastic molding is made with polyurethane, polystyrene, or polyvinyl chloride foam, which is formed inside molds under intense pressure. Plastic moldings are high-density, strong products. The surfaces may be smooth or a wood grain may be formed in the surface. Plastic molding is manufactured in paint and stain grades. Plastic molding can be shaped to accommodate curved and irregular openings.

Plastic molding is worked with hand and power tools in a similar manner to solid wood molding. Plastic molding is fastened using a hammer and finish nails or a pneumatic nailer. Predrilled holes are not required for attaching plastic molding to walls. A continuous bead of construction adhesive is recommended along the line of the molding before nailing the molding in place.

Gypsum Molding

Gypsum molding is made with gypsum (a natural mineral) combined with glass fibers, burlap, jute, or steel. Gypsum mortar is placed in molds and formed under pressure to produce a wide variety of standard shapes. Custom shapes can also be produced. Gypsum molding is intended only for interior applications, and is finished with paint.

Gypsum molding is cut with a handsaw or power saw. Screw and nail holes must be predrilled when attaching molding, and a gypsum adhesive is required.

TRIMMING DOOR OPENINGS

Door casing is available in many designs. **See Figure 61-4.** Casing material is usually *backed out.* Backing out the material gives the molding flexibility, and produces a tight fit even if there is unevenness between the jamb and the wall.

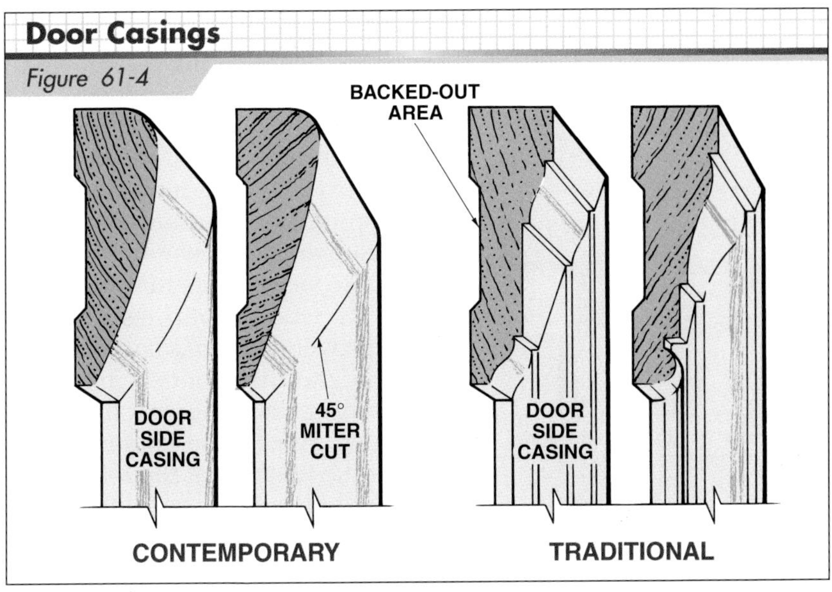

Figure 61-4. *Door casing patterns may be contemporary or traditional in design. Door casing is usually backed out to produce a tight fit.*

When nailed to the jamb, the casing should be held back 1/8″ to 3/16″ from the edge. This space is called a *reveal*. A reveal creates a better appearance and puts the casing out of the way of the door hinges. When tapered casing is used, the narrow edge is nailed to the jamb with 4d or 6d nails and the other edge to the wall studs with 8d nails. **See Figure 61-5.**

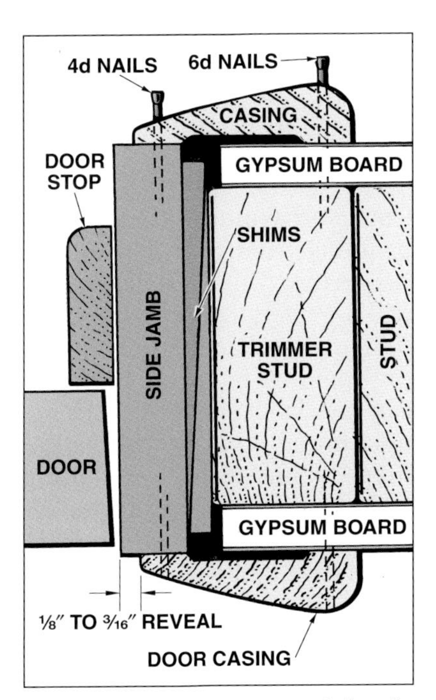

Figure 61-5. *Door casing is nailed to the trimmer stud and side jamb.*

When fastening the side pieces of casing, begin nailing from the top and work down, spacing the nails 16″ OC. If the piece must be straightened, nail into the trimmer stud first, then drive the adjoining nail into the jamb.

A 45° miter is usually cut at the joints between the top piece and the side pieces of casing. **See Figure 61-6.** When freehand cutting with a handsaw, the miter angle is laid out with a combination square. When several openings are to be trimmed, a power miter saw may be used to make accurate miter cuts. A job-built miter box may also be constructed to make accurate miter cuts.

Figure 61-6. *A 45° miter cut is usually required at the joints between the top piece and the side pieces of door casing.*

If a miter joint between the head and side pieces of the casing does not fit properly, use a block plane to make a better fit. When nailing casing to a jamb, drive nails down from the top piece into the side pieces at the miter joint to prevent the joint from opening up later. Applying glue at the joint is also helpful. A procedure for trimming a traditional door opening is shown in **Figure 61-7.**

DeWALT Industrial Tool Co.

A laser level may be used to ensure that a chair rail is level around an entire room.

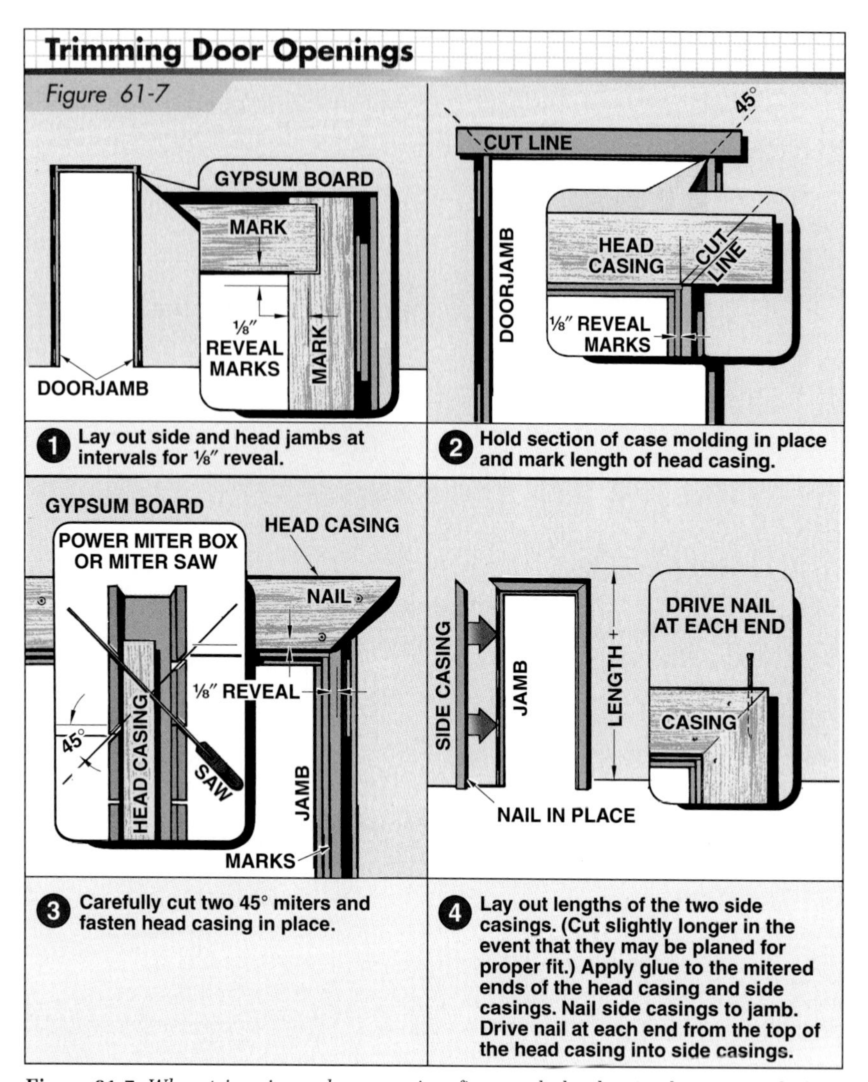

Figure 61-7. *When trimming a door opening, first mark the doorjamb at intervals for a 1/8″ reveal.*

Miter joints may open up at the heel if there is excessive moisture in the building.

Miter Joint Movement

Wood miter joints are known to open up after installation due to seasonal wood movement and moisture content. In addition, today's wood moldings are typically manufactured from younger springwood, which is more susceptible to movement.

As wood moldings dry out, the wood shrinks more across the grain than along the grain, pulling on the face of the miter. Also, since there is more width of the molding at the heel of the miter than at the toe of the miter, the angle of the miter increases and the heel opens up. **See Figure 61-8.** Miter joints for interior applications rarely open up at the heel unless there is excessive moisture in the building.

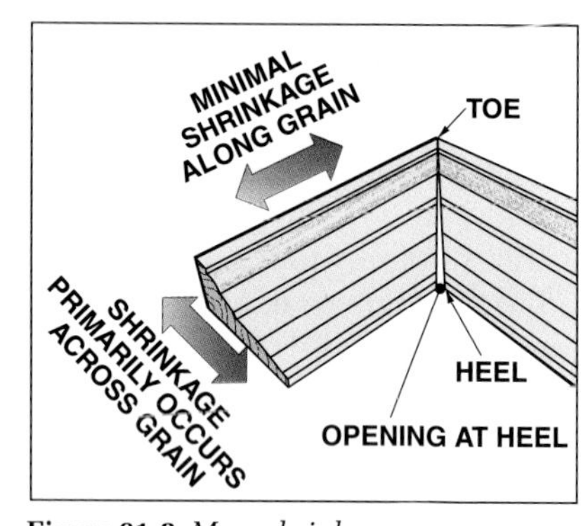

Figure 61-8. *More shrinkage occurs across the grain than along the grain of wood. Since there is more width at the heel than at the toe of the miter, the heel of the miter joint opens up.*

Trim Prefabrication

Door and window trim can be prefabricated and assembled before the trim is fastened to the walls. Prefabrication is typically performed when there are multiple openings requiring trim, and can save labor and ensure strong, tight miter joints. Prefabrication involves measuring and cutting all the trim pieces before assembly on a table or workbench. The miter joints are glued with quick-set cyanoacrylate adhesive or high-quality wood glue. Glue not only helps bind the miter joints together, but also seals the ends against moisture intrusion.

After the glue is applied, the miter joints are clamped tight with specialty clamps before the glue sets. **See Figure 61-9.** A glued joint will not reach its full strength unless pressure is applied to the joint. Nails or counterbored screws are used to secure the joint, and biscuits or splines may be used to reinforce the joint and align the faces of the trim pieces.

Chestnut Tool Co.

Figure 61-9. *Specialty clamps may be used to apply pressure to a miter joint to allow the glued joint to reach its full strength.*

The assembled unit is then set aside to allow the glue to cure. The next unit is then assembled the same as the first, continuing the process. Once the glued miter joints on a trim unit are cured, the clamps are removed and the assembled unit may be nailed into place.

TRIMMING WINDOW OPENINGS

Window casing should be the same material and design as door casing. Most wood-framed windows are trimmed in either *contemporary (picture-frame)* or *traditional* style. **See Figure 61-10.** Some types of windows have only a sill and apron to trim the opening; no casings are installed.

The procedure for placing casing around a contemporary-style window is similar to that for a door. The only difference is that a window requires a fourth piece fitted at the bottom of the frame, which is not required in trimming a door. Drive nails up from the bottom casing piece into the side pieces at the miter joint to prevent the joint from opening up later.

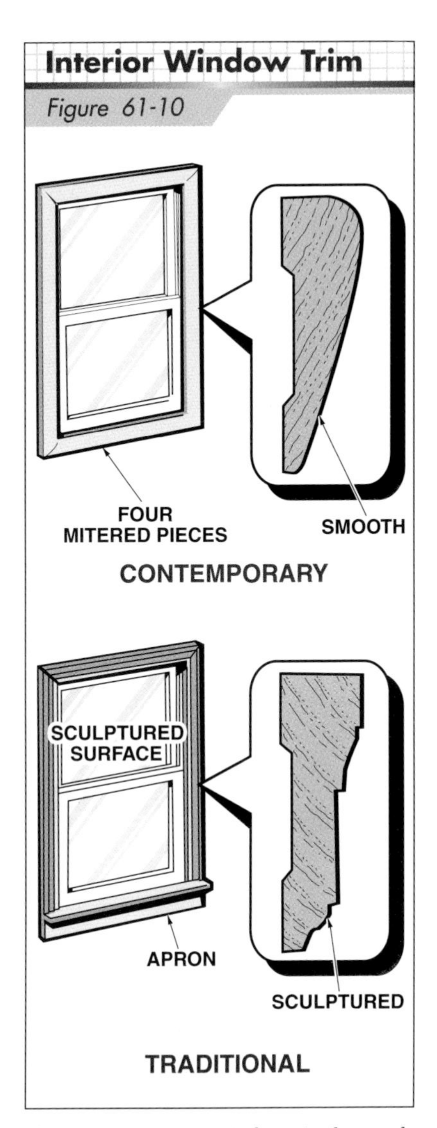

Figure 61-10. *Interior window trim for wood-framed windows may be either contemporary or traditional in design. Window trim should match door trim in the building.*

Traditional-style windows require a rabbeted stool to be installed before the casing can be nailed in place. Some stool designs are shown in **Figure 61-11.** A procedure for trimming a traditional window is shown in **Figure 61-12.** The stool and apron have a 45° miter joint at each end to make a *corner return.* A corner return covers the exposed grain on the end of the stool and apron.

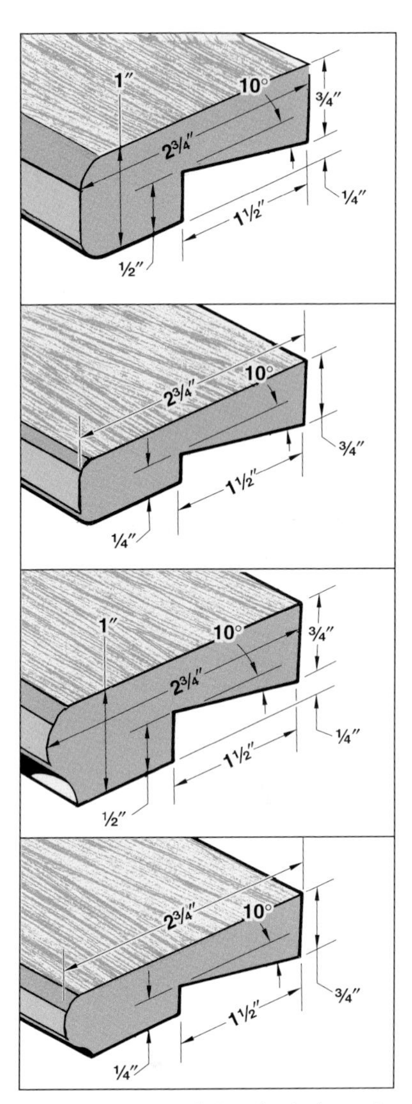

Figure 61-11. *Traditional window trim requires a rabbeted stool. A variety of patterns are available.*

INSTALLING BASE MOLDING

Base molding *(baseboard)* is held tightly to the finished floor and fastened to the wall with 6d or 8d finish nails driven into the bottom plate and studs. Some baseboard patterns are shown in **Figure 61-13.**

> *When installing base molding, nails driven through the molding should hit the studs behind the finish wall material. Locate the position of one stud and then lay out the location of other studs by measuring the stud spacing along the bottom of the wall. Stud positions can also be located using a stud finder.*

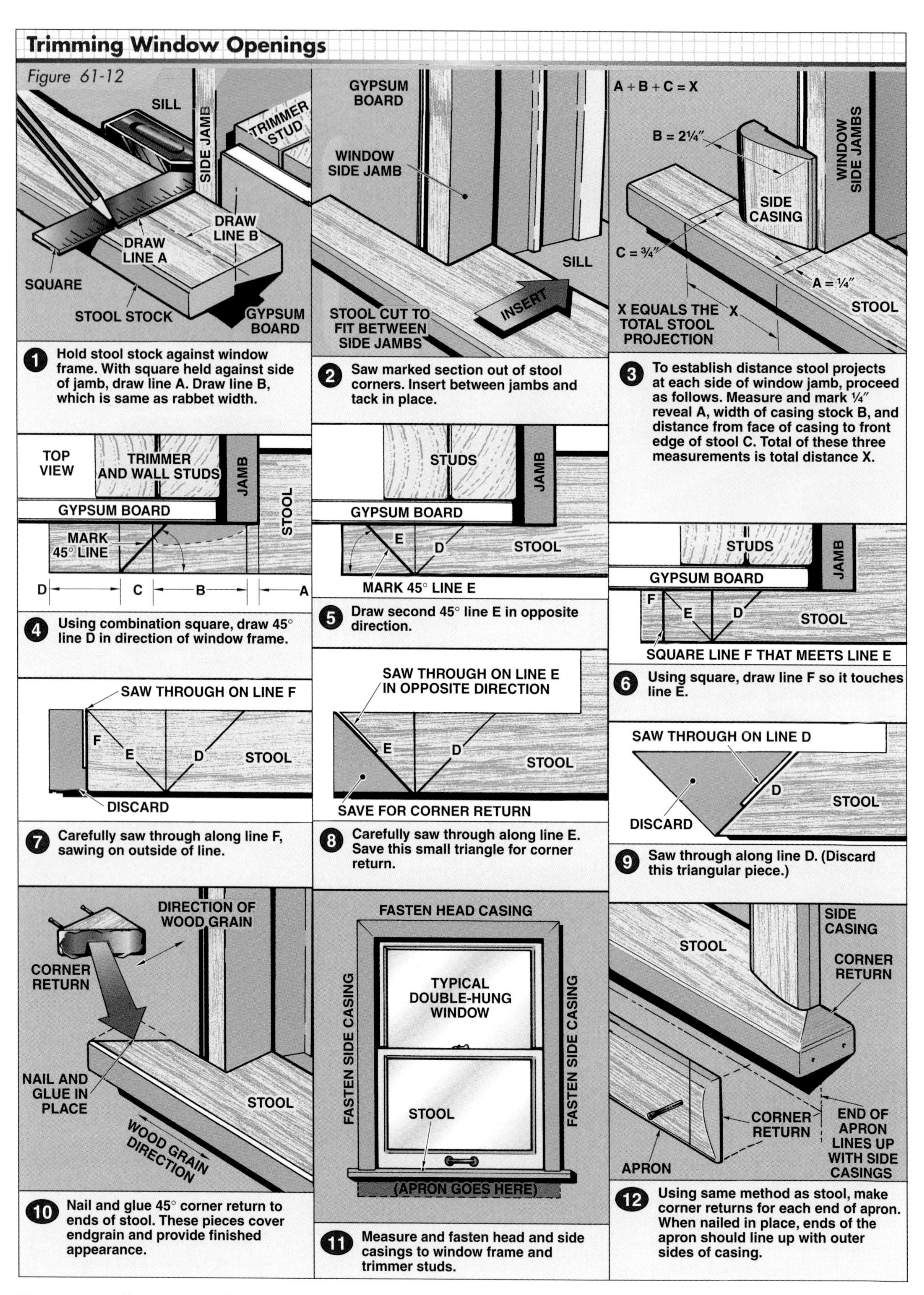

Figure 61-12. *When a stool and apron are installed, a corner return is used to conceal the end grain.*

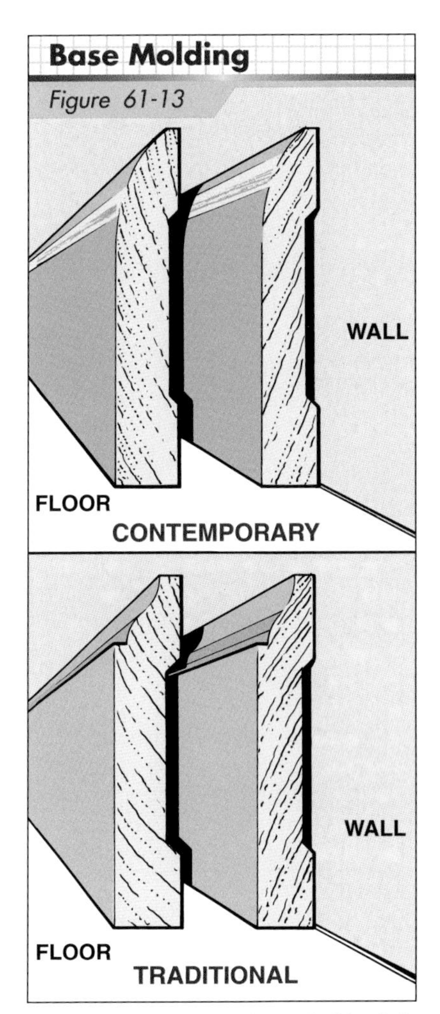

Figure 61-13. *Base molding is held tightly to the finished floor and nailed to the bottom plate and studs.*

Figure 61-14. *Base molding is usually thinner than the door casing it butts against.*

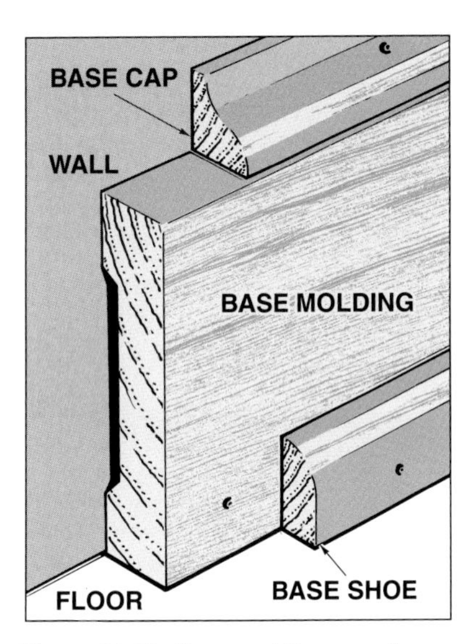

Figure 61-15. *Base molding may be applied with a base shoe and base cap.*

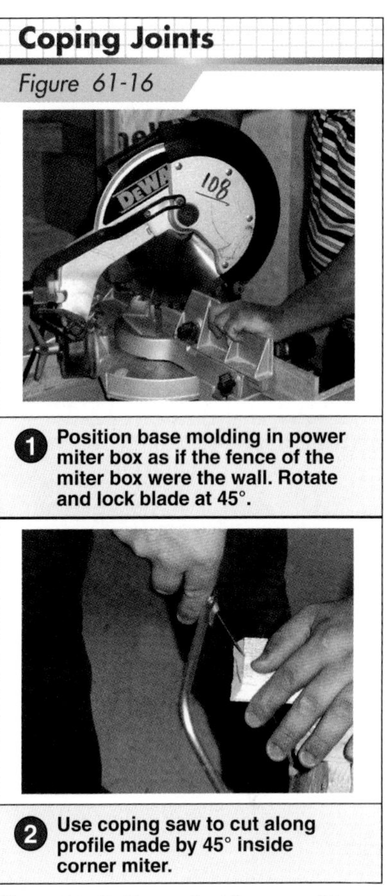

Figure 61-16. *For an inside corner fit, one piece of the base molding must be cut for proper fit.*

Base molding is installed after the door casing has been nailed in place. Base molding is usually thinner than the outside edge of the door casing it butts against. **See Figure 61-14.** A base shoe may be nailed into the floor at the bottom of the base molding to help conceal any unevenness between the base and the floor. A decorative base cap may also be used if molding is traditional style. **See Figure 61-15.**

Miter joints are required at the outside corners of base molding but are not suitable for the inside corners. At the inside corners, miter joints tend to open when being nailed. Later shrinkage of the wood also may cause inside corner miter joints to open. For this reason, a *coped joint* is recommended for inside corners. One piece is coped to the shape of the piece it fits against. **Figure 61-16** shows a procedure for coping.

On long walls where several pieces of baseboard are required, the joints between the pieces should fall over a stud. The best type of fit is a *scarf joint,* which is made by overlapping two 45° angles. **See Figure 61-17.** A procedure for placing baseboard in a room with inside and outside corners is shown in **Figure 61-18.**

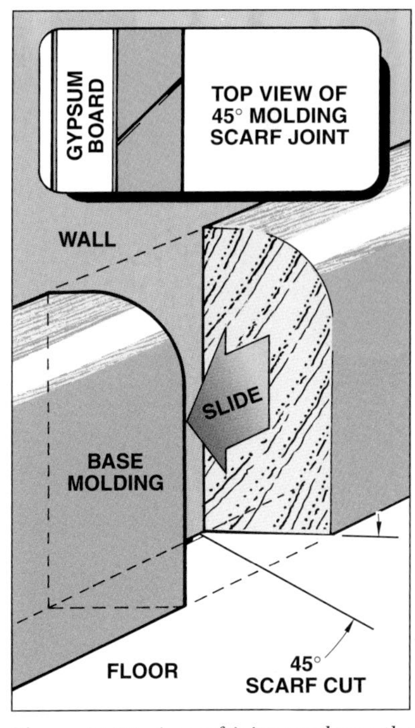

Figure 61-17. *A scarf joint produces the best fit between two pieces of base molding joined along a wall. Scarf joints should always be glued to limit separation.*

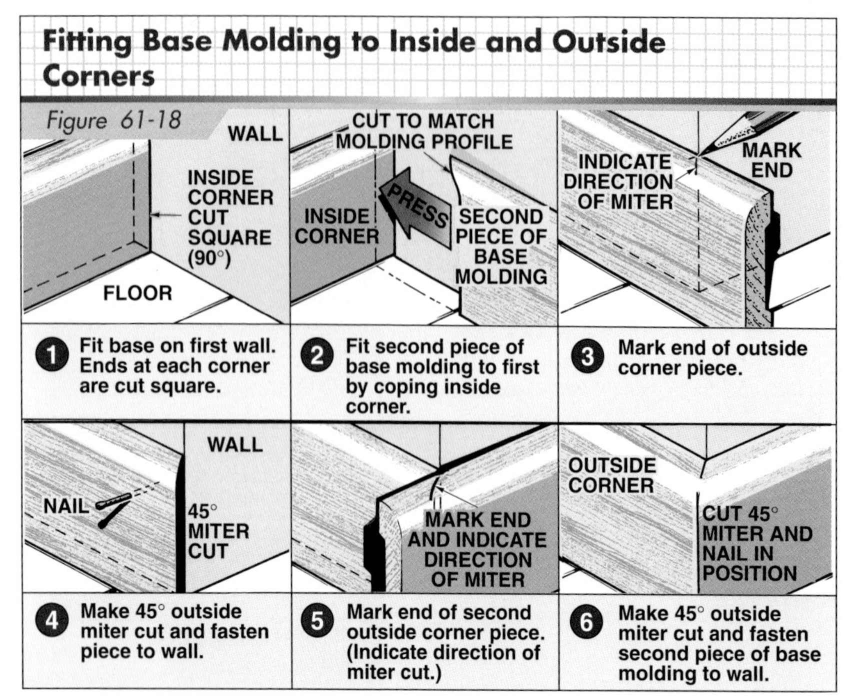

Figure 61-18. *When fitting base molding, first fit and cope the inside corner of one piece, then miter the outside corners.*

American Hardwood Export Council

Various profiles of ceiling molding can be combined to create a dramatic effect.

INSTALLING CEILING MOLDING

Ceiling molding is available in three basic shapes. A contemporary design is rectangular with beveled edges. Traditional profiles include crown and cove moldings. **See Figure 61-19.**

The procedure for placing ceiling molding is similar to that for base molding. One piece of the inside corner is usually coped as shown in **Figure 61-20.** The outside corners are mitered. A scarf joint is recommended for joints between pieces on longer walls. The pieces are attached to the wall with 6d or 8d finish nails driven into the wall studs.

The lower side of crown or cove molding fits against the wall. It has a wider surface than the top side, which rests against the ceiling. To cut the miters (45° angles) for outside corner joints, a different procedure is required. To cut the miter, mark the short side of the 45° angle and position the molding as though the fence of the back side of the miter box is the wall and the bottom of the miter box is the ceiling. **See Figure 61-21.**

Pine and spruce moldings are available as solid and finger-jointed material. Finger-jointed moldings are made from short lengths of wood that are milled and glued end to end. Finger-jointed moldings are less expensive than solid moldings and also are more warp-resistant. If the trim will be painted rather than stained, finger-jointed moldings are recommended.

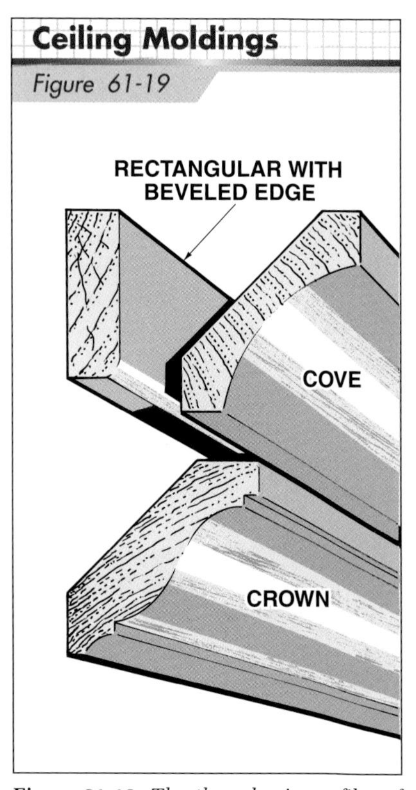

Figure 61-19. *The three basic profiles of ceiling molding are rectangular with a beveled edge, cove, and crown.*

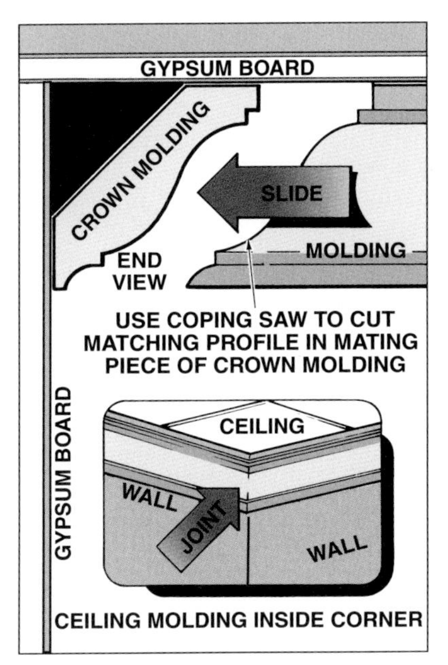

Figure 61-20. *When installing crown molding, one of the inside corner pieces is coped.*

Figure 61-21. *When cutting a 45° angle on crown molding, position the molding with the bottom edge pressed against the fence of the miter box.*

UNIT 62 Finish Flooring

OBJECTIVES

1. Review the types of finish flooring materials.
2. List the types of wood used in wood finish flooring.
3. Identify desirable characteristics of strip flooring.
4. List preparations for the installation of strip flooring.
5. Describe the fastening methods used with strip flooring.
6. List characteristics of plank flooring.
7. Describe the installation of plank flooring.
8. Identify common types of block flooring and describe their installation.
9. Identify common types of laminate flooring and describe their installation.
10. List and describe common types of resilient flooring and their installation.

Finish flooring is applied when the construction of a building is almost completed to minimize wear on the flooring during construction. Many different materials are used as finish flooring, including wood, resilient tile, ceramic tile, and carpeting. Wood flooring is available in strips, planks, and blocks. Resilient tile products, such as linoleum, vinyl, and rubber, are available in sheet form or individual tiles. Ceramic tile, brick, slate, and flagstone may be used to create special effects in different sections of a building. Wall-to-wall carpeting applied directly on top of the subfloor is also considered a finish floor material.

In some parts of the country, carpeting and resilient tile may be placed by carpenters, but this work is usually performed by flooring specialists. Hardwood flooring is installed by carpenters. In some areas of the country, placing hardwood flooring has become a specialty within the carpentry trade.

WOOD FINISH FLOORING

Softwood finish flooring is manufactured primarily from southern pine, Douglas fir, and western hemlock. Hardwood finish flooring, which is more durable, is manufactured from oak, beech, birch, and maple. Wood finish flooring is available in strips, planks, or blocks, each of which produces a different appearance.

Strip Flooring

Strip flooring is available in widths of 1½″ to 3½″ and in random lengths up to 16′. Standard thicknesses are ⁵⁄₁₆″, ⅜″, ½″, ¾″, and ²⁵⁄₃₂″. Thinner strip flooring has square edges and requires face-nailing. Face-nailing requires that the nails be set and the holes later puttied.

The most durable and attractive strip flooring is ²⁵⁄₃₂″ thick by 2¼″ wide strips with tongue- and-groove edges. **See Figure 62-1.** Strip flooring is also *end-matched,* meaning there is a tongue-and-groove end where the pieces butt against each other. With the exception of the wall edge of a starter piece, tongue-and-groove strips are blind-nailed. **See Figure 62-2.** Blind-nailing produces a more attractive appearance than face-nailing. High-quality strip flooring is also *hollow-backed,* which allows the pieces to lie flat and fit snugly against an irregular floor surface.

Figure 62-1. *Strip finish flooring is a durable and attractive finish flooring material.*

Manufacturers recommend that strip flooring be delivered to the job several days before installation and placed in the area of the building in which it will be installed. If the flooring is delivered packaged, open or remove the packaging. The strip flooring should be spread loosely around the floor area so its moisture content can adapt to the building humidity. For larger areas, the strip flooring can be stacked with stickers between the layers to provide proper ventilation. **See Figure 62-3.**

Hardwood flooring is manufactured at 6% to 9% moisture content.

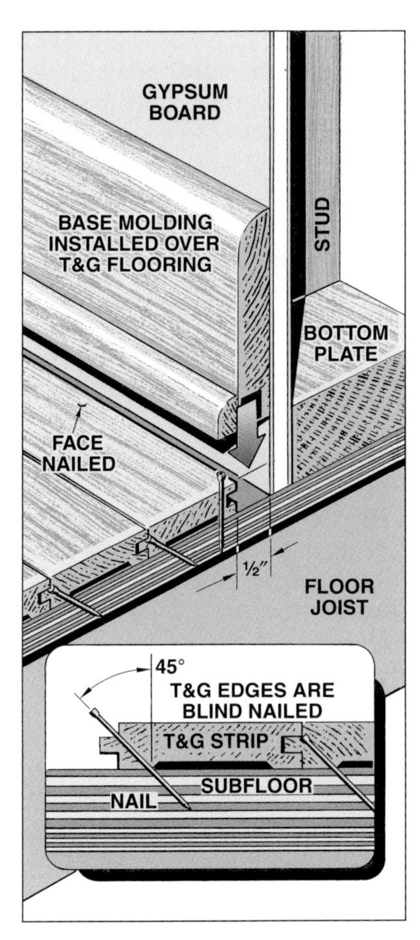

Figure 62-2. *When tongue-and-groove strip flooring is used, only the wall edges of the first and last strip are face-nailed. The remainder of the strips are blind-nailed.*

The subfloor must be cleaned and checked for protruding nails. Plywood, OSB, or other performance-rated panel product can be used as a subfloor. Particleboard should not be used as a subfloor since fasteners do not hold well in the material.

With the proper subfloor in place, a layer of 15 lb asphalt-saturated felt or building paper is applied. The edges of the felt or building paper should overlap 2″ to 4″ to keep out dust, slow moisture movement from the area below the subfloor, and prevent squeaks in dry seasons.

When determining where to begin nailing the strip flooring, consider the floor layout of the building and the number of rooms to receive strip flooring. Strip flooring is laid at right angles to the joists. Snap chalk lines across the felt or building paper to ensure that nails will be driven into joists wherever possible.

Strips nailed along a wall are placed ½″ away from the wall to allow for expansion. The ½″ gap will be covered by the baseboard. Joints between boards are staggered and are no closer than 6″ to joints in adjacent rows.

Fastening Methods. When installing strip flooring, the nails are driven at a 45° to 50° angle so the head is below the edge of the tongue. If necessary, use a nail set to avoid damaging the edge of the flooring material. The nails should be long enough to penetrate through the subfloor and into the joists. Use a short piece of flooring material as a driving block to fit the strips tightly against each other. Proper nailing will help to reduce future squeaks.

Manual or pneumatic nailers are typically used to install strip flooring. Nail strips are inserted into the nailer magazine. As the nailer is operated, individual nails are driven into the edge of the flooring. When using a manual nailer, a floorlayer strikes a plunger at the end of the nailer to actuate the driving mechanism. The driving motion pulls the flooring in tightly, and drives and sets the nail in one motion.

A pneumatic nailer is operated in a similar manner to a manual nailer. However, a pneumatic nailer requires less striking force, which results in less fatigue to the floorlayer and greater productivity. **See Figure 62-4.**

Southern Forest Products Association

Figure 62-3. *The proper moisture content must be maintained in strip flooring before and after installation to prevent cupping. A moisture meter can be used to check the moisture content.*

Senco Products, Inc.

Figure 62-4. *A manual or pneumatic nailer is used to fasten strip flooring to the subfloor.*

Figure 62-5 shows a nailing schedule for strip flooring. The procedure for laying out and installing hardwood strip flooring is shown in **Figure 62-6.** In this example, strip flooring is placed in all rooms of the building except the kitchen and bathrooms.

When two floorlayers work together, one often cuts and fits the pieces while the other nails them in place. An efficient method for a floorlayer working alone is to cut and fit six or eight rows of boards at a time and then nail them in place.

STRIP FLOORING NAILING SCHEDULE			
Flooring Thickness and Edge	Flooring Width*	Fastener Size and Type	Spacing
¾″ T&G	1½, 2¼, 3¼	2 barbed flooring cleat† 7d or 8d flooring nail 2 15-ga staples with ½″ crown†	8″ to 10″
½″ T&G	1½, 2	1½″ barbed flooring cleat 5d screw-shank, cut steel, or wire casing nail	10″
⅜″ T&G	1½, 2	1¼″ barbed flooring cleat 4d bright wire casing nail	8″
5⁄16″ square edge	1½, 2	1″ 5-ga barbed flooring brad	2 nails per 7″

* in in.
† use 1½″ fasteners with ¾″ subfloor on concrete slab

Figure 62-5. *Different strip flooring thicknesses require various fasteners and fastener spacing.*

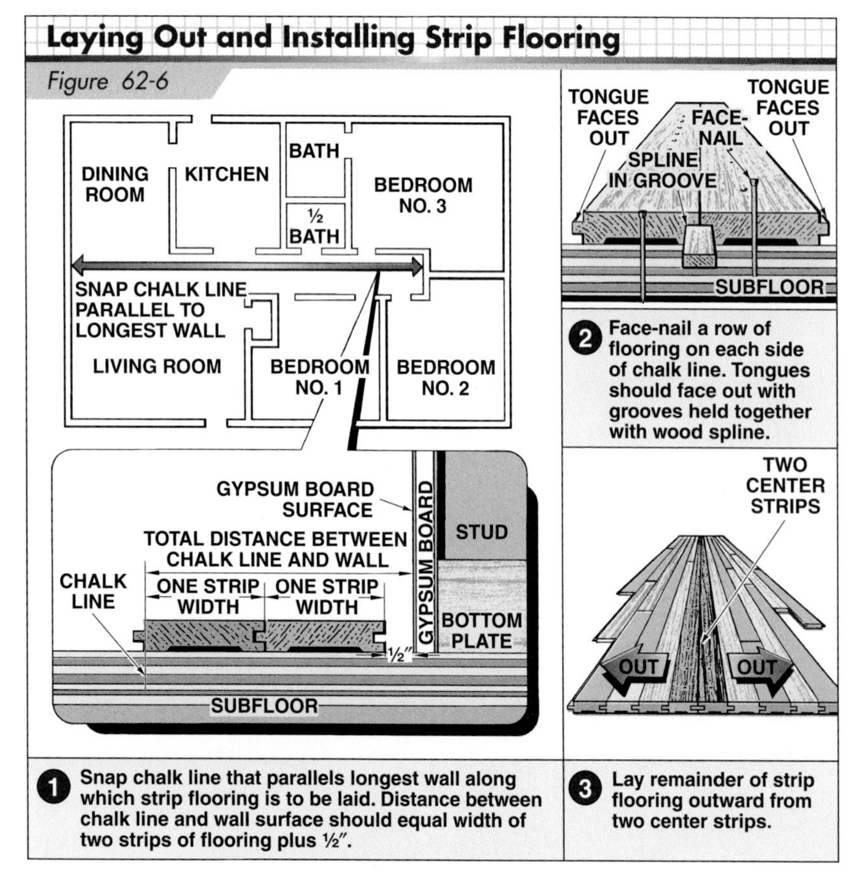

Figure 62-6. *Installation of hardwood strip flooring begins at the center of the area being floored and continues outward toward the walls. Before installation, the strip flooring is spread loosely around the floor area so that its moisture content can adapt to the surrounding air.*

Installing Strip Flooring over Concrete. To install strip flooring over concrete, nailing strips called *sleepers* are used. The sleepers are 1 × 4s or 2 × 4s that are pressure-treated with a wood preservative. Sleepers are fastened to the concrete floor at 16″ OC intervals with concrete nails or mastic. In some cases, the sleepers may need to be shimmed because of irregularities in the floor.

Before sleepers are placed, waterproof mastic must be spread over the concrete floor and allowed to set for approximately 30 min. A layer of polyethylene film is strongly recommended. **Figure 62-7** shows two procedures for preparing concrete to receive strip flooring. One procedure uses nails to fasten the sleepers to the floor, and the other procedure uses mastic to fasten the sleepers over polyethylene film.

Plank Flooring

Plank flooring is normally used in buildings of traditional design. **See Figure 62-8.** Plank flooring was first used in houses built during the colonial period of United States history. Planks are available in widths from 3″ to 8″, and in thicknesses of 5⁄16″ to 25⁄32″. Planks 25⁄32″ thick have tongue-and-groove edges and ends. Since plank flooring is wider than strip flooring, additional time may be required for the planks to acclimate to the moisture content of the building.

Plank flooring is installed in a similar manner to strip flooring. However, screws are installed at the ends of the plank. The screws are counterbored and covered with wood plugs. Often, plank flooring has simulated plugs for a colonial appearance.

Hardwood flooring should be installed 90° to the floor joists.

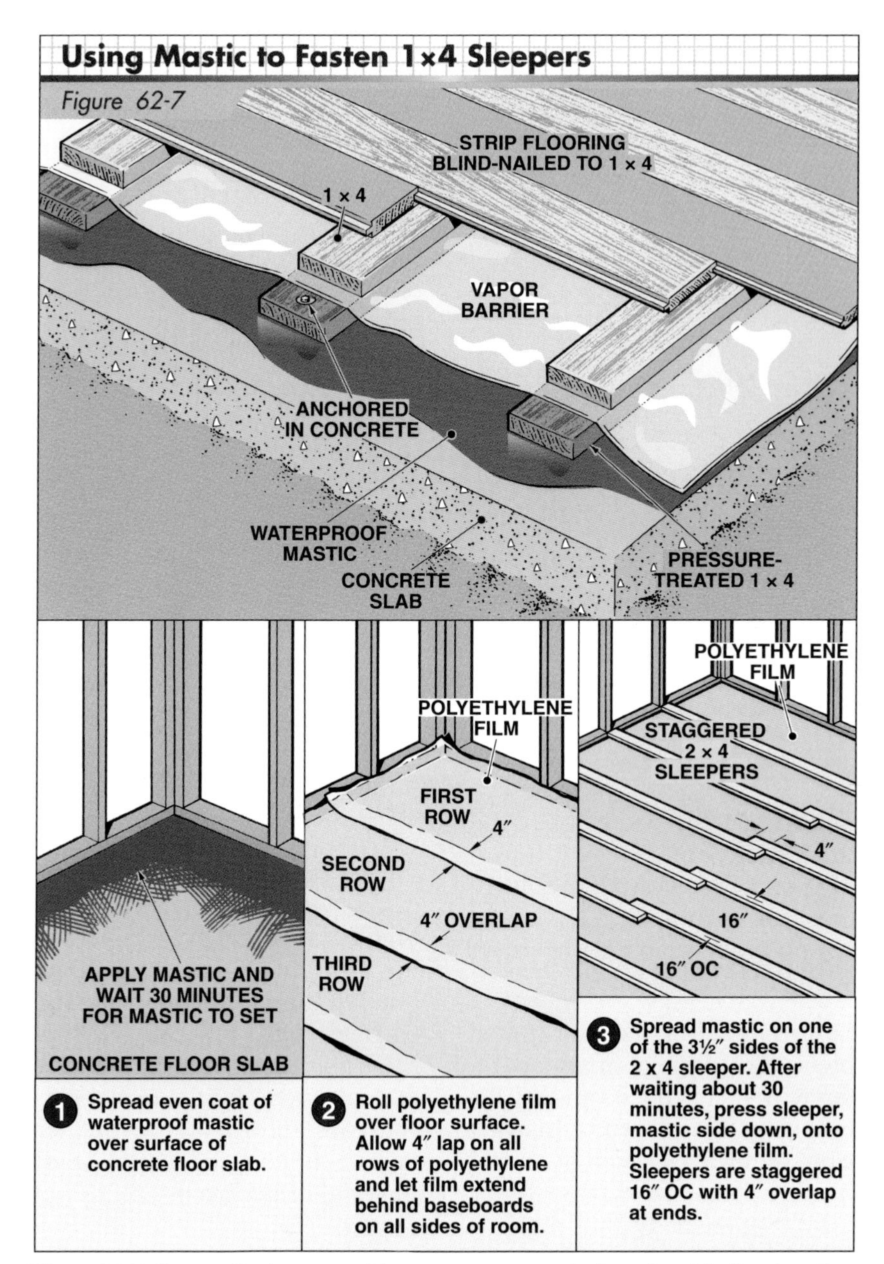

Figure 62-7. *Two methods are used to prepare a concrete floor for strip flooring. One method (shown at the top) uses nails to fasten sleepers to the concrete, and the other method uses mastic to fasten the sleepers.*

Figure 62-8. *In plank flooring, butt joints are used when planks are less than 25/32″ thick. Tongue-and-groove joints are used for 25/32″ thick planks.*

Installing Plank Flooring over Concrete. Plank flooring is installed over concrete in a similar manner as strip flooring is installed, unless special glue-down planks are obtained. Glue-down planks adhere directly to the concrete, making sleepers unnecessary. A 150 lb roller is moved across glue-down planks to ensure proper contact is made between the concrete and the planks.

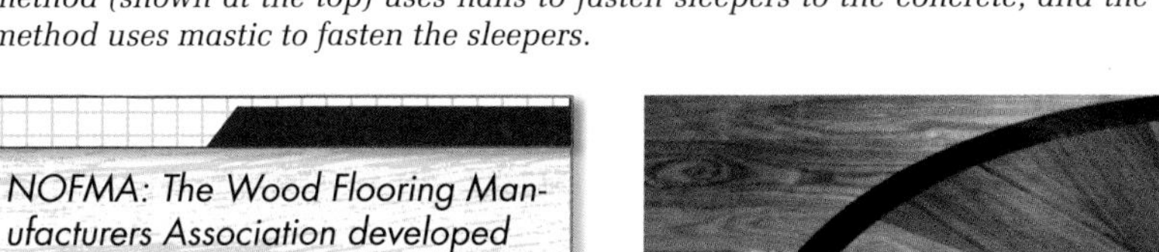

NOFMA: The Wood Flooring Manufacturers Association developed wood flooring standards.

Plank flooring is installed with 2″ barbed flooring cleats, 7d or 8d flooring nails, or 2″ 16 GA staples with ½″ crowns. The fasteners should be spaced approximately 8″ apart. When plank flooring is installed over a ¾″ subfloor on a concrete slab, 1½″ flooring cleats should be used.

National Wood Flooring Association

The skills of an experienced carpenter are required to install an intricate floor inlay.

Block Flooring

Three basic types of block flooring are *solid-unit, laminated,* and *slat.* **See Figure 62-9.** Block flooring is fastened to the floor (wood subfloor or concrete slab) with mastics. Most block flooring has tongue-and-groove edges to ensure alignment between the squares.

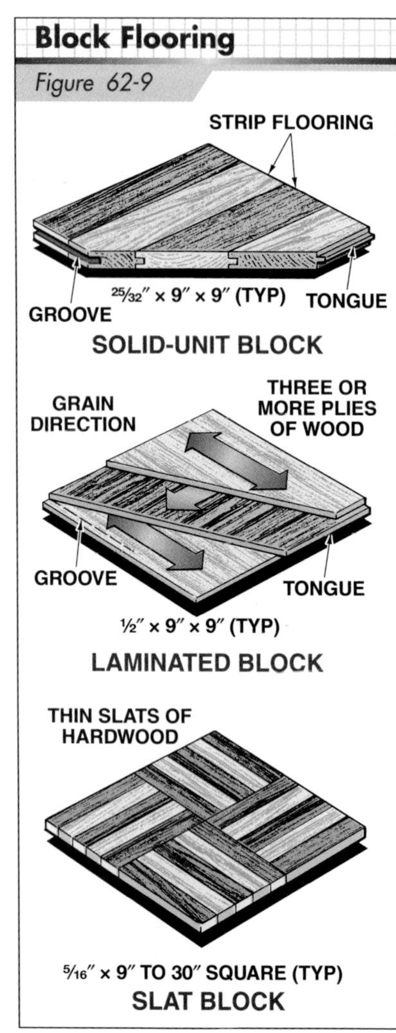

Figure 62-9. *Basic types of block flooring are solid-unit blocks, laminated blocks, and slat blocks. All types are fastened to the floor with mastic.*

Before the mastic is spread, chalk lines must be snapped at right angles to each other and at the center of the room. The method for laying out chalk lines for block flooring is shown in **Figure 62-10.**

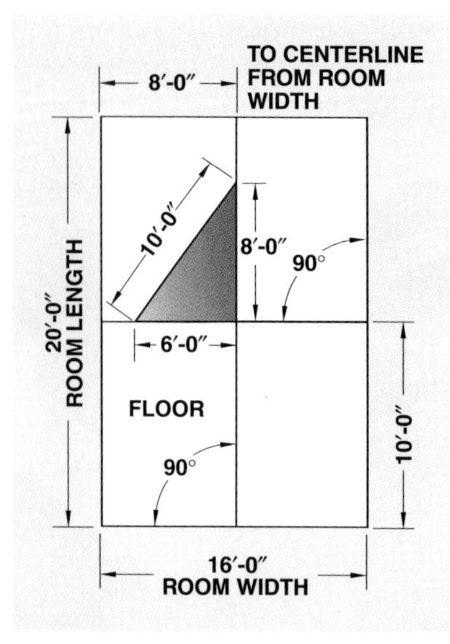

Figure 62-10. *To lay out block flooring, snap chalk lines at the centers of the width and length of the room. The 6-8-10 squaring method can be used to ensure the center lines are perfectly square to one another.*

Solid-Unit Blocks. Strips 25/32″ thick make up most solid-unit blocks. The strips are held together with a wood or metal spline embedded in the lower edge of the material.

Laminated Blocks. Laminated blocks are made of plywood, often with an oak face veneer. Due to its cross-laminated construction, which reduces swelling or shrinkage, laminated block flooring is recommended for damp locations such as concrete slabs resting directly on the ground.

Slat Blocks. Slat blocks are also called *mosaic* or *parquet* flooring. Slat block flooring is composed of wood strips arranged in various patterns. **See Figure 62-11.** In some slat blocks the strips are held together with face paper that is pulled off after the squares have been installed. In other types of slat blocks, the strips are held together by mechanical attachments or a textile web backing.

LAMINATE FLOORING

Laminate flooring is a composite material which consists of a transparent top layer, decorative layer, carrier layer, and bottom layer. **See Figure 62-12.** The top layer of laminate flooring is transparent and provides a great deal of wear resistance. The decorative layer that gives the flooring its grain (or other) pattern is bonded under the top layer. The carrier layer is typically HDF or MDF fiberboard or particleboard, and provides laminate flooring with its stability. The bottom layer may be a stiff paper-based material or laminate.

Figure 62-11. *A parquet floor is made of slat blocks in which wood strips are arranged in various patterns.*

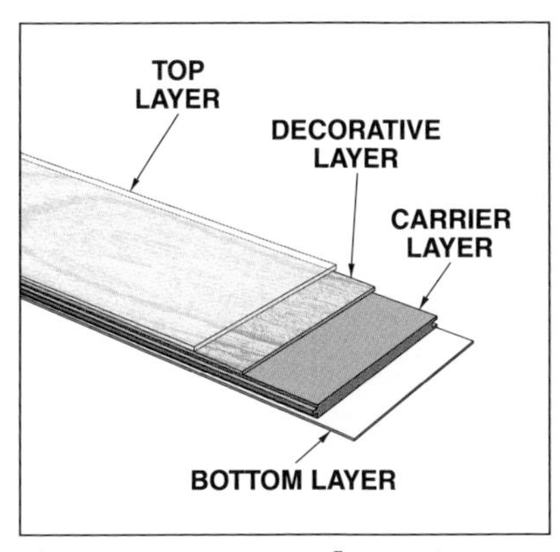

Figure 62-12. *Laminate flooring is a composite material consisting of top, decorative, carrier, and bottom layers.*

National Wood Flooring Association

Wood flooring enhances the beauty of cabinetry and other flooring materials.

Laminate flooring is manufactured using a *high-pressure laminate (HPL)* or *direct-laminate (DL)* process. In the high-pressure laminate process, the components in the top layer are pressed together under intense heat and pressure to form a laminate. The laminate and bottom layer are glued to the carrier to form the finished laminate flooring.

In the direct-laminate process, the components in the top, decorative, carrier, and bottom layers are joined simultaneously under intense heat and pressure. Production costs are lower for direct-laminate flooring than for high-pressure laminate flooring. **See Figure 62-13.**

Laminate flooring is installed by gluing or mechanical joints. Glue-free mechanical joints are preferred since they are a neater and cleaner joint to make than glue joints.

Pergo, Inc.

Figure 62-13. *Laminate flooring provides the appearance of natural wood strip flooring and provides outstanding durability.*

RESILIENT FINISH FLOORING

Resilient, nonwood products are classified according to their basic ingredients as either vinyl, rubber, or cork flooring. Resilient finish flooring is produced in *tile* and *sheet* form. Tile sizes range from 9″ × 9″ to 36″ × 36″. Sheets are generally available in 6′, 9′, or 12′ widths. A wide variety of patterns and colors is available.

The term *resilient* describes the ability of the material to yield when pressure is applied and then return to its original condition. Resilient finish flooring materials hold up well against weight and indentation caused by falling objects, and are effective in reducing sound produced by foot traffic and other types of impact.

Resilient finish flooring is usually laid by flooring specialists. In some parts of the country, however, resilient finish flooring is laid by carpenters.

If the subfloor is uneven, ¼″ underlayment must be nailed or stapled to the subfloor before the resilient flooring is installed. The underlayment surface must be smooth and free from bumps or indentations.

Most resilient flooring products are cut with a scissors or sharp utility knife and are fastened with a mastic to the underlayment or subfloor. Manufacturers usually recommend a specific adhesive.

Resilient Flooring Tiles

The procedure for laying out resilient flooring tiles is similar to laying block flooring. Centerlines are snapped on the floor. The tiles are placed starting from the center of the room and working toward the walls. **See Figure 62-14.** This method ensures that when border tiles are less than full size, they will be of width equal to the border tiles at the opposite side of the room.

When all tiles have been placed, the wall edges are finished with a base molding that is usually the same material as the tile.

Resilient Flooring Sheets

If possible, sheets of resilient flooring are installed as one piece. A template is made from kraft paper to the exact dimensions of the floor by placing the paper on the floor and trimming it. The template is placed on the resilient flooring sheet and taped in position so it does not move while the flooring is being cut. The flooring is carefully cut using a scissors or sharp utility knife. When the flooring is cut to the proper dimensions, it is then placed on the underlayment or subfloor and carefully positioned. When the flooring is in the proper position, half of the flooring is folded back and mastic is applied to the underlayment or subfloor. The flooring is laid back into position and the procedure is repeated for the other half of the flooring. A 150 lb roller is then used to roll the flooring.

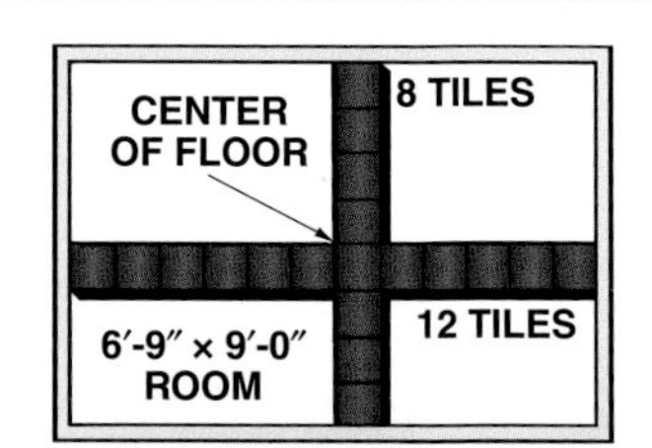

When an even number of tiles is used, edges of beginning rows of tiles will align with centerlines.

EVEN NUMBER OF TILES

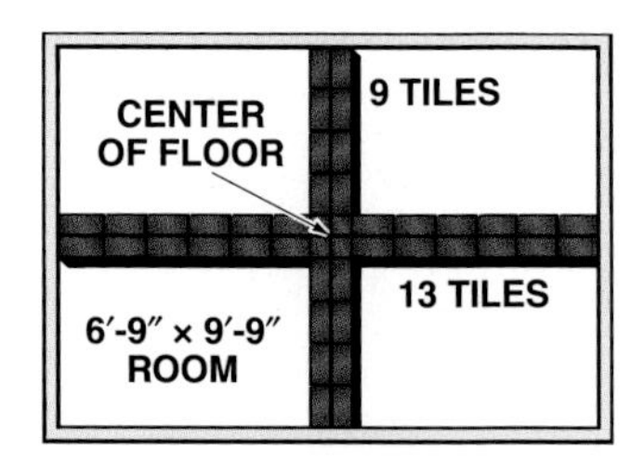

When an odd number of tiles is used, the centerlines will fall at centers of beginning rows of tiles.

ODD NUMBER OF TILES

Figure 62-14. *Tiles are applied beginning at the center of the room and progressing toward the walls.*

Unit 62 Review and Resources

SECTION

14

Stairway Construction

CONTENTS

UNIT 63 Types of Stairways

OBJECTIVES

1. Compare the differences between main and service stairways, and between interior and exterior stairways.
2. List and describe common stairway arrangements.
3. Describe how stairways may be enclosed.
4. Identify common components of an interior stairway with a balustrade.
5. List and describe common types of stringers.
6. Explain the procedure for laying out stringers.
7. Calculate tread width and riser height combinations.
8. Explain the procedure for laying out treads and risers on stair stringers.
9. Discuss the term dropping a stringer.
10. Describe the procedure for attaching stringers.

Main or *primary stairways* serve the inhabited areas of a building. *Service stairways* serve the uninhabited areas such as a basement or attic. *Interior stairways* are located inside a building. **See Figure 63-1.** *Exterior stairways* are located outside a building and lead to entrances or decks. **See Figure 63-2.**

The construction prints for a building provide information on laying out and constructing stairways. Plan and elevation views and detail drawings on the prints provide construction details for stairways.

Figure 63-1. *Interior stairways lead to upper or lower levels within a building.*

BASIC STAIRWAY ARRANGEMENTS

Room design determines the stairway arrangement. **Figure 63-3** shows basic stairway arrangements. The simplest stairway to construct is a *straight-flight (straight-run) stairway* without a landing. A straight-flight stairway runs in a direct line from one floor to another.

In a *straight-flight stairway with a landing,* one section of the stairway runs from the floor to the landing and the other section runs from the landing to the next floor. A straight-flight stairway with a landing is often used in public buildings where there is a long distance (15 steps or more) between stories. Landings are placed at regular intervals to interrupt the climb.

Figure 63-2. *Exterior stairways lead to entrances or decks.*

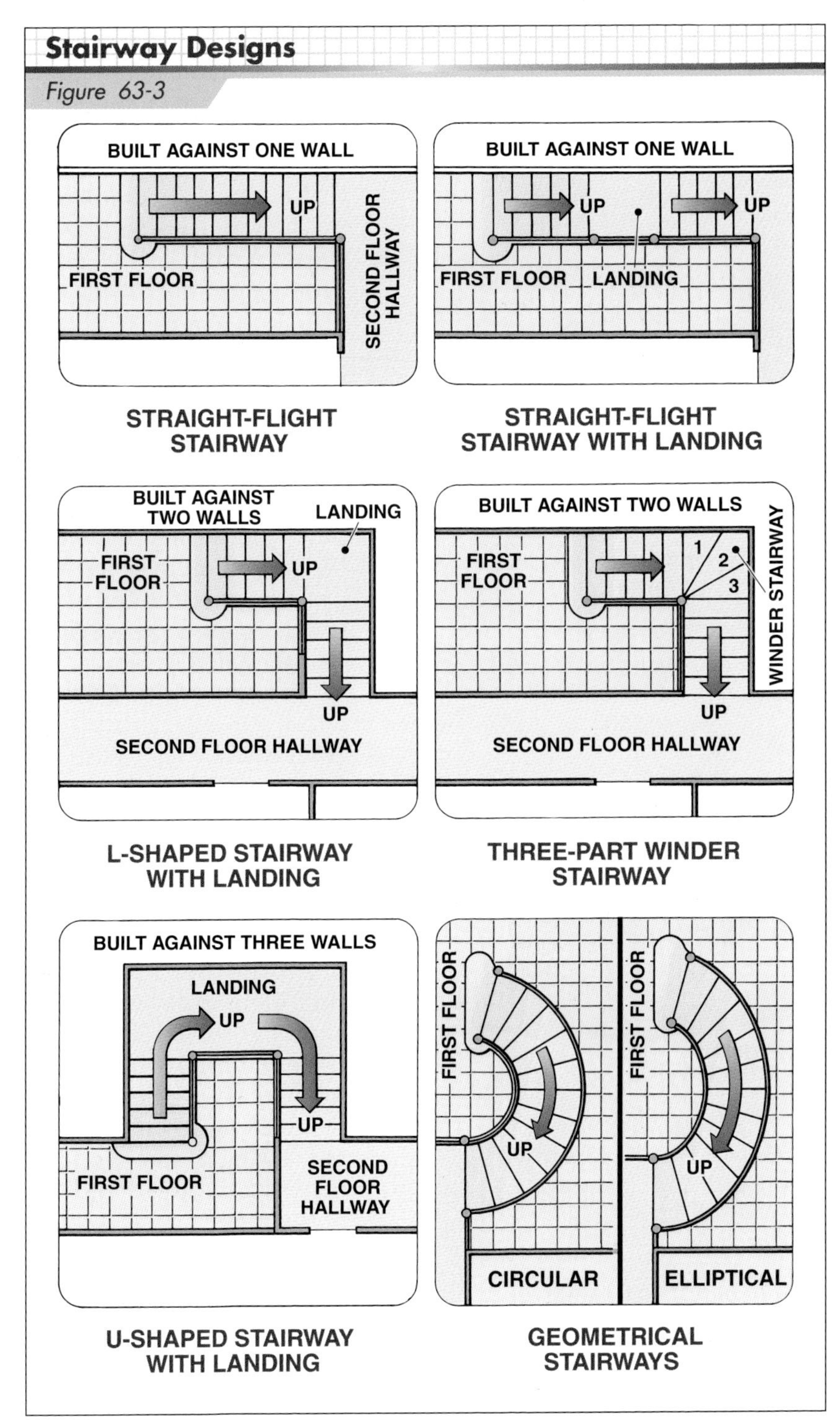

Figure 63-3. *Stairway arrangements are based on room design.*

An *L-shaped stairway* runs along two walls and includes a landing where the stairway changes direction. L-shaped stairways are used when the space is not sufficient for a straight-flight stairway.

U-shaped stairways have two sets of stairways which run parallel to one another. U-shaped stairways may be a wide U or a narrow U. The difference between wide U-shaped and narrow U-shaped stairways is the space between the two flights of stairs. A wide U-shaped stairway has a *stairwell* between the two flights of stairways, while a narrow U-shaped stairway has little or no space between the stairways.

Instead of a landing, a stairway may have a winding section at the turn, which is formed with wedge-shaped steps called *winders.* Stairways with winders are usually L-shaped.

Other types of winding stairways are *geometrical stairways.* Geometrical stairways may be *circular* or *elliptical.* Circular and elliptical stairways are complicated to build and are usually prefabricated in a shop.

Stairways may be enclosed by walls or railings. **See Figure 63-4.** If a stairway is enclosed by walls on both sides, handrails are attached to the walls. In another type of stairway, a short wall may form the railing for one or both sides of the stairway. Still another type of stairway has a balustrade on one or both sides. A balustrade is formed by balusters and a handrail. Balusters (banisters) are upright pieces that extend between the handrail and treads.

STAIRWAY COMPONENTS

Stairways have *treads,* the portion of the stairway that people place their feet on, and *stringers,* the portion that supports the treads. Some stairways also have *landings, handrails,* and *balusters.* The term *stairway* (staircase) includes treads, stringers, landings, handrails, and balusters.

Building codes regulate certain aspects of stair construction including the riser height, tread depth; handrail size, configuration, and height; and baluster spacing.

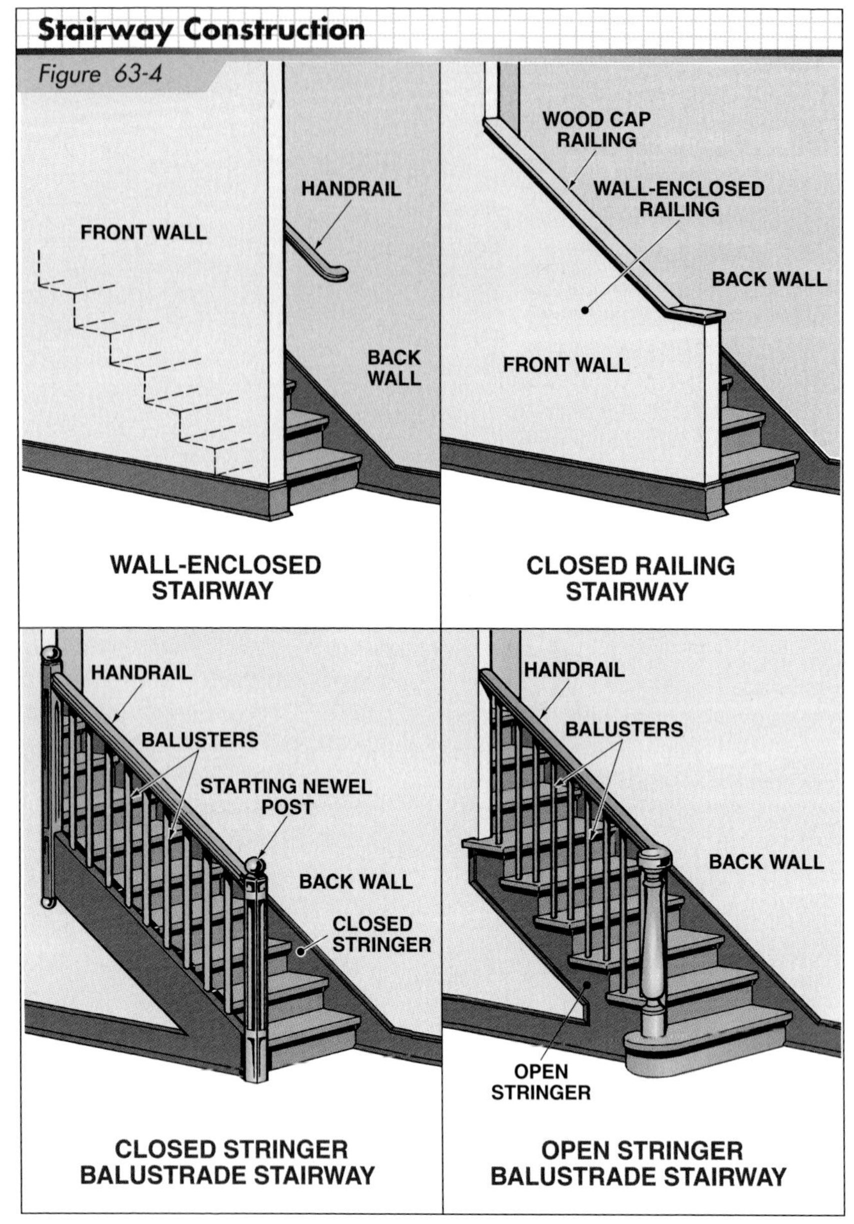

Figure 63-4. *Stairways may be enclosed by walls or railings.*

Figure 63-5 illustrates the components of an interior stairway with a balustrade. An explanation of these components follows:

A. *Landing newel post:* Newel posts are installed at the ends of the stairway and where stairways change direction. The main post supporting the handrail at the landing is the landing newel post.
B. *Gooseneck:* A gooseneck is a curved or bent section at the upper end of a finish handrail. A gooseneck is used to connect the straight section of handrail to a newel post.
C. *Baluster:* The upright pieces that run between the handrail and treads are balusters.
D. *Handrail:* A handrail is grasped by the hand for support when using a stairway.
E. *Starting newel post:* A starting newel post is the main post supporting the handrail at the bottom of a stairway.
F. *Closed finish stringer:* A finish board, known as a closed finish stringer, is nailed against the wall of a stairway and encloses the ends of the treads and risers. The top of a closed finish stringer is parallel to the slope of the stairway.
G. *Tread:* A tread is the step that a person places his or her foot on.
H. *Riser:* A riser is the piece forming the vertical face of a step.
I. *Nosing:* The projection of the tread beyond the face of the riser is the nosing. A nosing usually projects 1″ to 1¼″ beyond the face of a riser.
J. *Nosing return:* A nosing return (end nosing), which is the projection over the face of a stringer at the end of the tread, is used on an open stringer stairway. Similar to a sill or apron return, a nosing return prevents the tread end grain from being exposed.
K. *Open stringer:* An open stringer is cut to support the open side of a stairway. The treads rest on top of the stringer cutouts.
L. *Landing:* A landing is a platform that breaks the stair flight from one floor to another.
M. *Cove molding:* Cove molding may be used to conceal the joint between the tread and stringer and the joint between the tread and top of the riser.

Stringers, Treads, and Risers

The *stringers,* or carriage, provide the main support for a stairway. Finish treads and risers are nailed to the stringers. **See Figure 63-6.**

On wide stairways, at least one additional stringer is installed in the span between the two outside stringers. The tread thickness is a factor influencing the allowable distance between stringers. In general, stringers should be installed 30″ OC when 1¹⁄₁₆″ thick treads are used, and 36″ OC when 1½″ thick treads are used.

Stairway Components

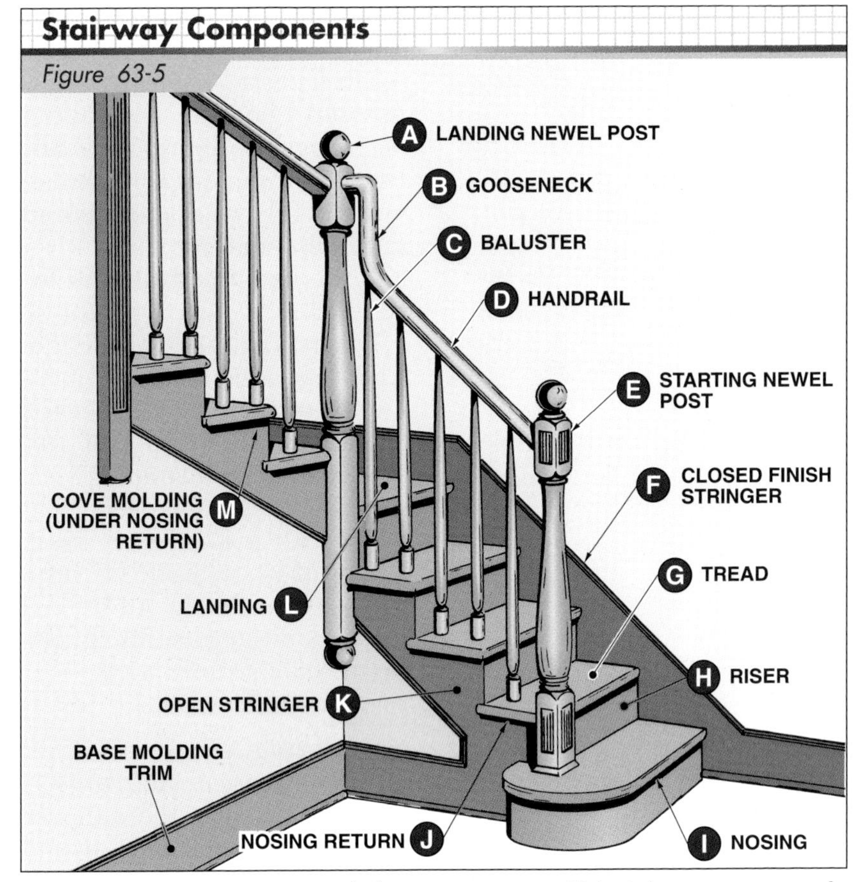

Figure 63-5. *A stairway is composed of many components. Main stairways are commonly outfitted with balusters and newel posts.*

Wood Cut-Out Stringer Stairway

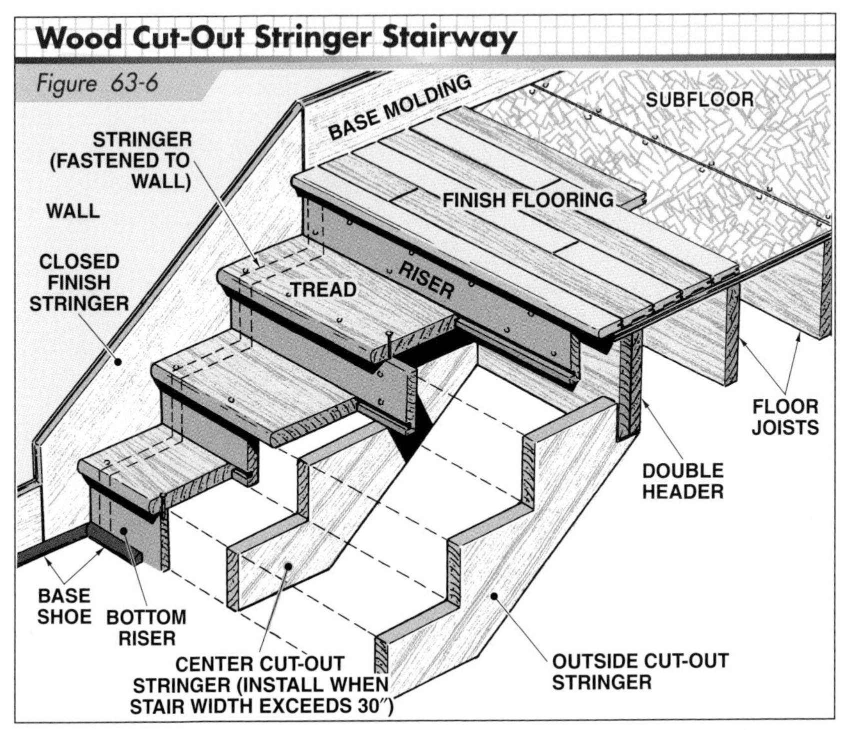

Figure 63-6. *A stringer provides the main support for a stairway. Treads and risers are fastened to the stringers.*

A circular saw blade guard may not properly retract when cutting stair stringers. Do not wedge the guard in an open position.

Stringers may be *cleated, dadoed, cut-out,* or *housed.* **See Figure 63-7.** A cleated stringer is a type of stringer with cleats nailed to it to support the treads. A dadoed stringer is a type of stringer in which treads fit into dadoes cut into the stringer. Cleated and dadoed stringers are typically used for stairways to a basement and other uninhabited areas of a house. However, attractive and modern dadoed stringer designs may be used for main stairways as well.

Cut-out stringers are commonly used for interior and exterior stairways. A cut-out stringer is cut or notched to fit the profile of the stairway. Usually 2 × 10s or 2 × 12s are cut to the tread depth (minus the nosing) and the riser height. At least 3½″ (measured at a right angle from the edge of the stringer) must remain after the cuts have been made.

Housed stringers are also widely used for interior and exterior stairways. Prefabricated stairways usually have housed stringers. Housed stringers are routed to receive the treads and risers, which are wedged and glued into place. When intermediate stringers are installed, housed, cleated, or dadoed stringers are placed on the outside and cut-out stringers are used for intermediate stringers.

Stringer Layout

Stringer layout includes marking off treads and risers. All risers should be the same height and all treads should be the same depth to ensure safe and smooth movement from one level to the next. The maximum deviation allowed between riser heights is ⅜″.

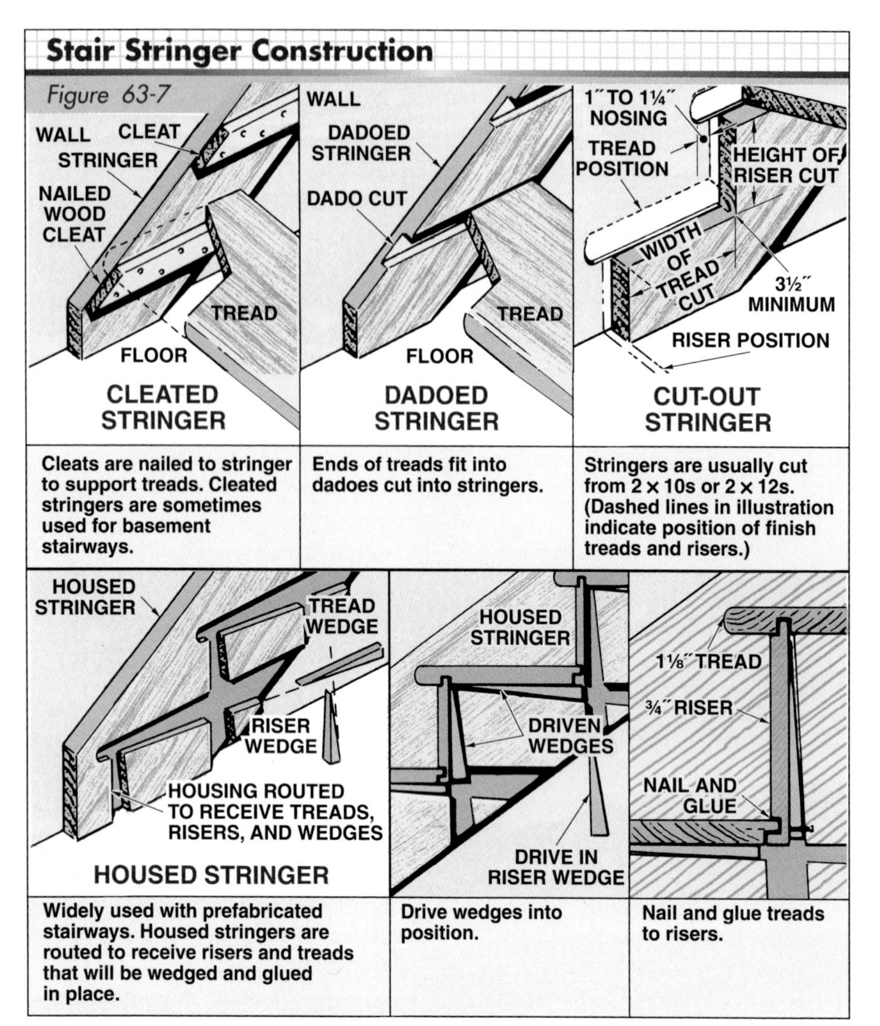

Figure 63-7. *Four types of stringers are cleated, dadoed, cut-out, and housed. Cleated or dadoed stringers are used for stairways to basements and other uninhabited areas of a building. Cut-out or housed stringers are usually used for other stairways.*

The *unit rise* (riser height) is determined by dividing the *total rise* of a stairway by the number of risers. **See Figure 63-8.** The *total rise* is the vertical distance of a stairway from the finished floor to the finished floor above.

The *unit run* (tread depth minus nosing) is determined by dividing the *total run* of a stairway by the number of treads. The *total run* is the horizontal distance measured from the foot of the stairway to a point directly below where the stairway ends at a deck or landing above. There is always one less tread than the total number of risers.

Stringers should be laid out so that a set of stairs will rise at a safe and comfortable angle. **Figure 63-9** shows preferred versus critical angles. *Preferred angles* afford the most comfortable walking angles. *Critical angles* are the minimum and maximum angles for walking safety. The preferred angle for a stairway is 30° to 35°. Critical angles are 20° to 30° and 35° to 50°. For a stairway to rise at a certain angle, the proper combination of tread depth and riser height must be used.

Recommended unit rise for residential stairways is 7″ to 7½″. However, it may not always be possible to divide the total rise of a stairway into equal riser heights that fall between these recommendations. Therefore, a higher or lower unit rise must sometimes be used. The *International Residential Code* (for residential buildings) and *International Building Code* (for commercial buildings) have established minimum and maximum riser heights and tread depths for stairways.

The *International Residential Code* has established a maximum riser height of 7¾″ and a minimum tread depth of 10″. The *International Building Code* requires a minimum riser height of 4″ and a maximum riser height of 7″, with an 11″ minimum tread depth. Additional information about building code requirements for stairways is included in Unit 64.

Calculating Tread and Riser Combinations. A *stair ratio* is the ratio between the unit rise and unit run of a stairway and is expressed as a formula. There are three general stair ratio formulas used in stairway design to determine tread and riser combinations. One of the most widely used stair ratios states that the riser height plus the tread depth should equal 17″ to 18″. Based on this stair ratio, the following riser heights and tread depths would be acceptable:

Riser +	*Tread* =	17″ to 18″
7″	11″	18″
6½″	11¼″	17¾″
6⅜″	11⅝″	18″
6¼″	11″	17¼″

Another stair ratio, which is commonly used for residential stairways, states that two riser heights plus one tread depth should equal 23⅝″ to 25⅝″. Based on this stair ratio, the following riser heights and tread depths would be acceptable:

(2 × *Riser*) +	*Tread* =	23⅝″ to 25⅝″
7″	11″	25″
6¾″	11¼″	24¾″
6½″	11″	24″

Total Rise and Run of Stairways

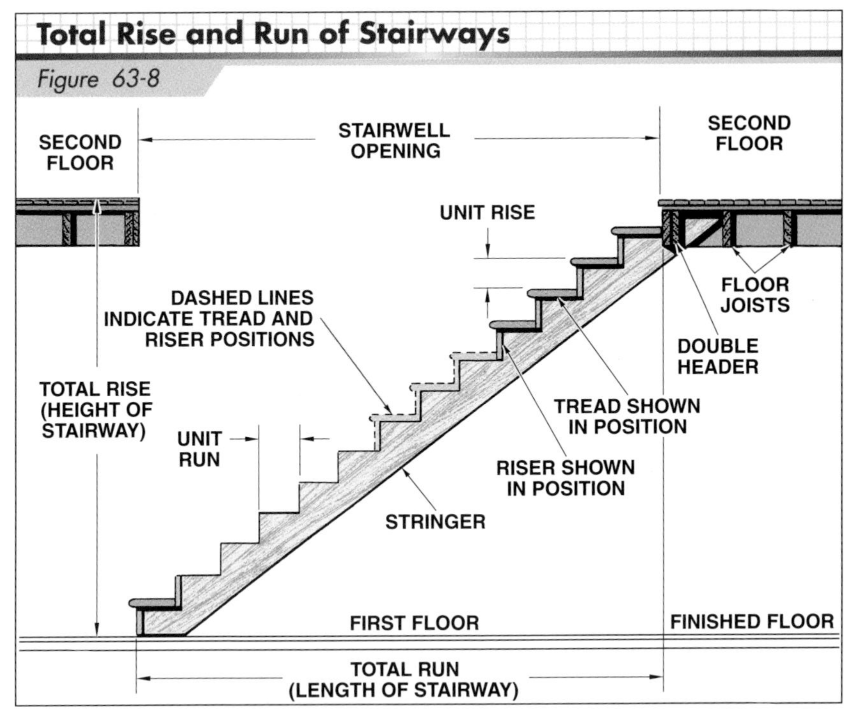

Figure 63-8. *The total rise of a stairway is its total height and the total run is its length. The unit rise is the riser height and the unit run is the tread depth.*

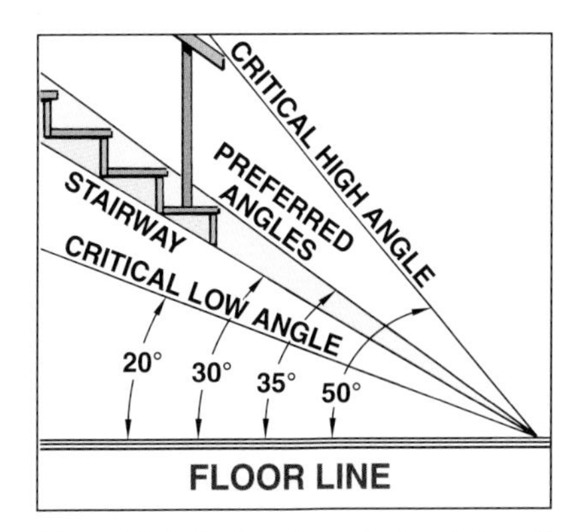

Figure 63-9. *Preferred angles for stairs fall between 30° and 35°.*

The final stair ratio states that the riser height multiplied by the tread depth should equal 72 to 75. Based on this stair ratio, the following riser heights and tread depths would be acceptable:

Riser × *Tread* = 72 to 75

6¾″	11″	74¼″
6⅝″	11¼″	74½″
6½″	11½″	74¾″

For a preferred-angle stairway, a shorter riser must be combined with a deeper tread, or a higher riser must be combined with a narrower tread.

Exact tread and riser sizes are based on the total rise and run of a stairway. **Figure 63-10** shows the procedure for calculating tread and riser sizes for a residential stairway with a total rise of 8′-11″. Total rise is found by adding the rough floor-to-ceiling height, actual header width, and subfloor thickness (8′-1″ + 9¼″ + ¾″ = 8′-11″).

Story Pole. Riser height must be calculated with precision, particularly for long stairways requiring many risers. For example, suppose the riser height is calculated to be 1/16″ less than required for a stringer with 15 risers. On one riser, the error is barely noticeable. However, the top step of the stringer will be 15/16″ less than the desired distance from the landing. To avoid this type of error when calculating riser heights for long stairways, many carpenters check their calculated measurements by marking off a story pole with a pair of dividers. In this manner, an adjustment can be made if necessary.

Marking Treads and Risers. A framing square is used to mark treads and risers on a stringer. Square gauges are positioned on the blade and tongue of the framing square to ensure consistent measurements during layout. The tread depth is indicated on the blade and the riser height is indicated on the tongue. **Figure 63-11** shows a procedure for marking treads and risers on a cut-out stringer with seven 10½″ treads and eight 7¼″ risers.

Walls adjacent to the stairway will be finished with gypsum board. A 1 × 3 is installed between the wall studs and stringer to provide room for the gypsum board.

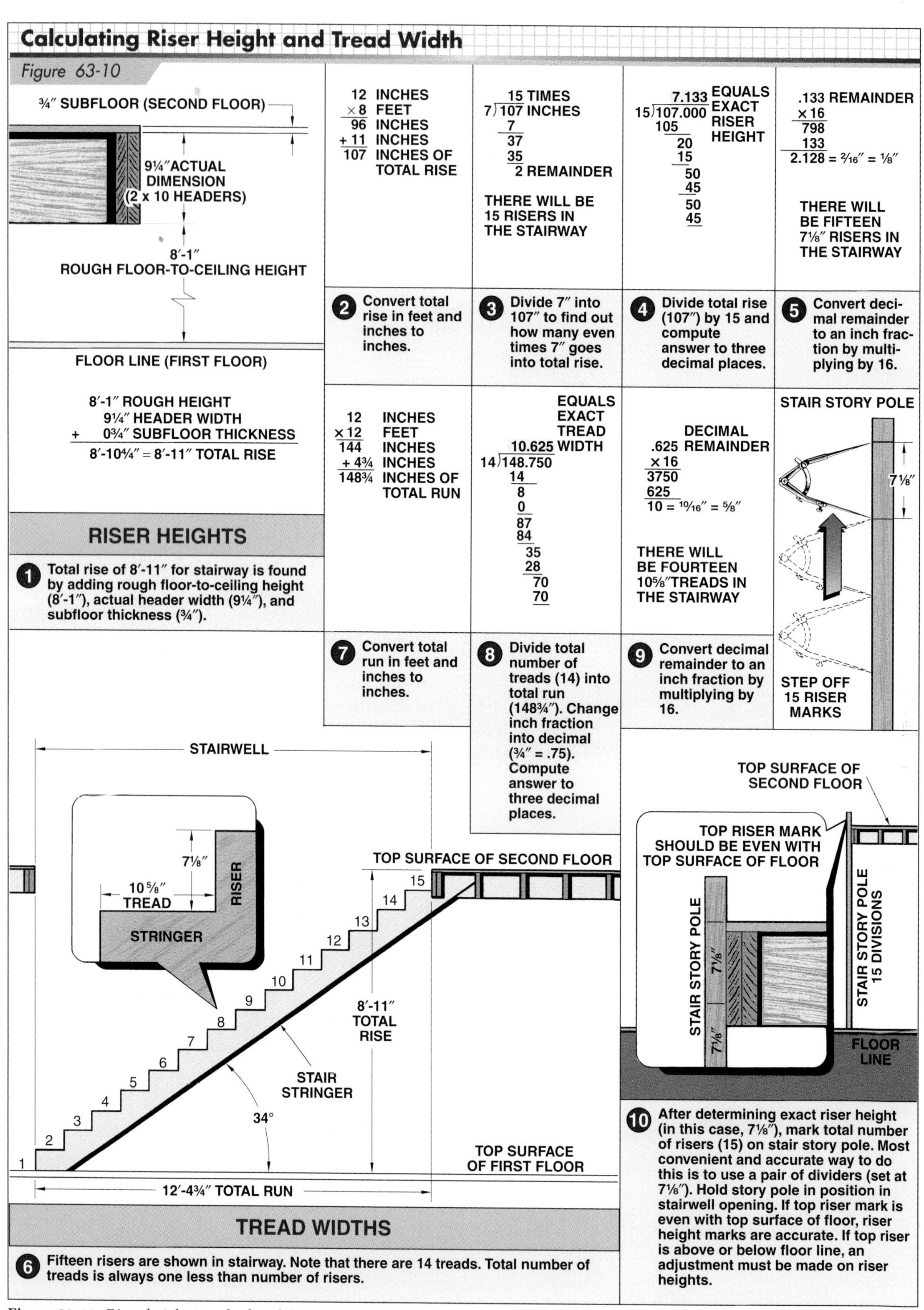

Figure 63-10. *Riser height is calculated first, followed by tread depth calculations. In this example, the total rise of the residential stairway is 8′-11″ and the total run is 12′-4¾″.*

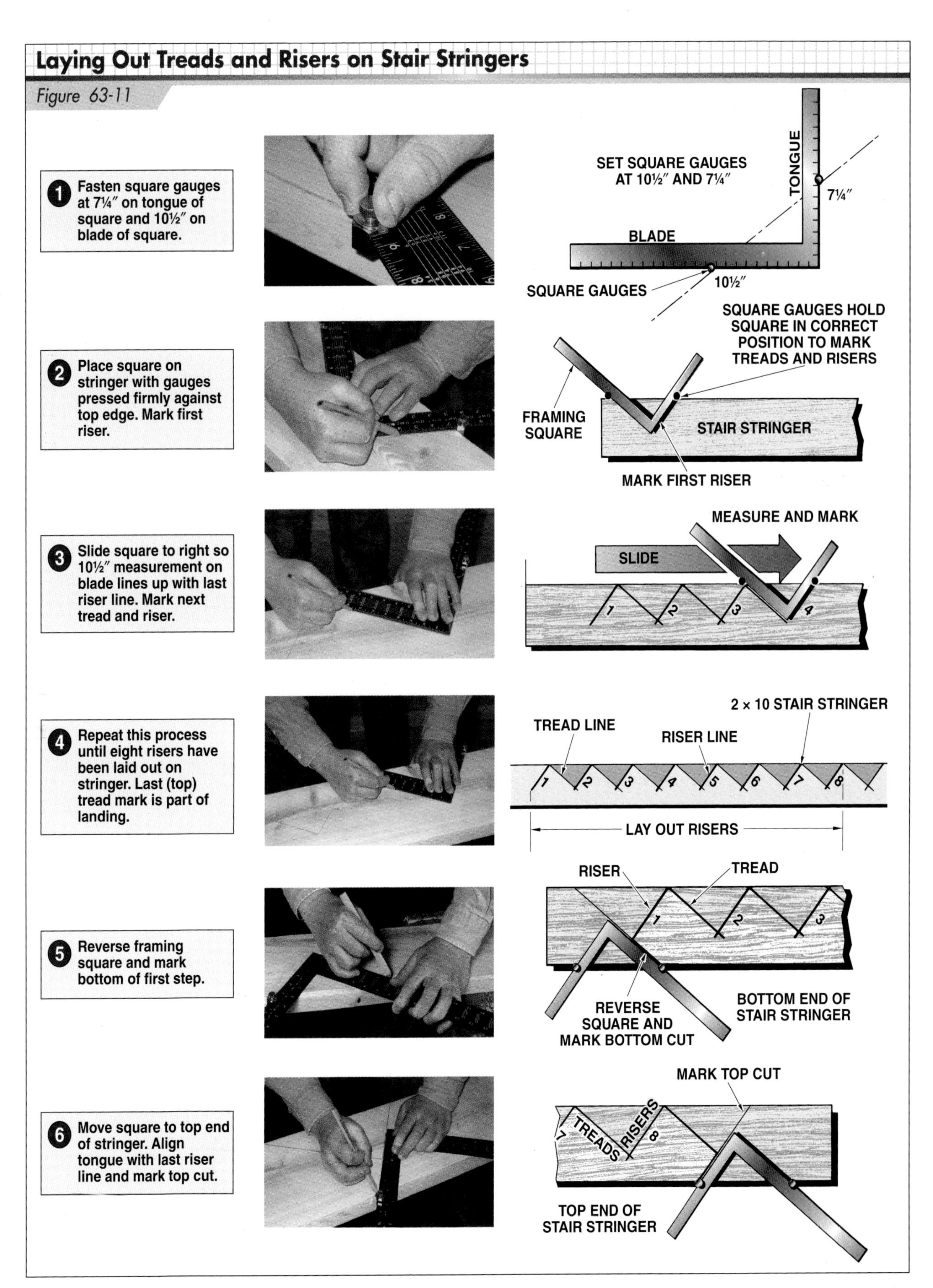

Figure 63-11. *Square gauges are positioned on a framing square to ensure accurate tread and riser measurements. In this example for a residential stairway, the risers are 7¼″ high and the treads are 10½″ deep. The stairway has seven treads and eight risers.*

Dropping Stringers. A small amount of material may need to be deducted from the bottom step so all finished riser heights are equal after the tread material is nailed to the stringer. This calculation is called *dropping the stringer.* **See Figure 63-12.**

Attaching Stringers. After the stringers are cut to the proper size, they must be fastened in place. Stringers carry the main load of a stairway and must be securely fastened for strength. **Figure 63-13** shows common methods for fastening stair stringers at the top and bottom.

Treads may be fastened with their full or partial depth against the stairwell header. The header may act as a backing for the top riser. In some cases, a ledger board is nailed to the header to provide additional support for the stringer. For stair stringers resting on concrete floors, a pressure-treated sleeper is fastened to the concrete and the bottom of the stringer is toenailed to the plate.

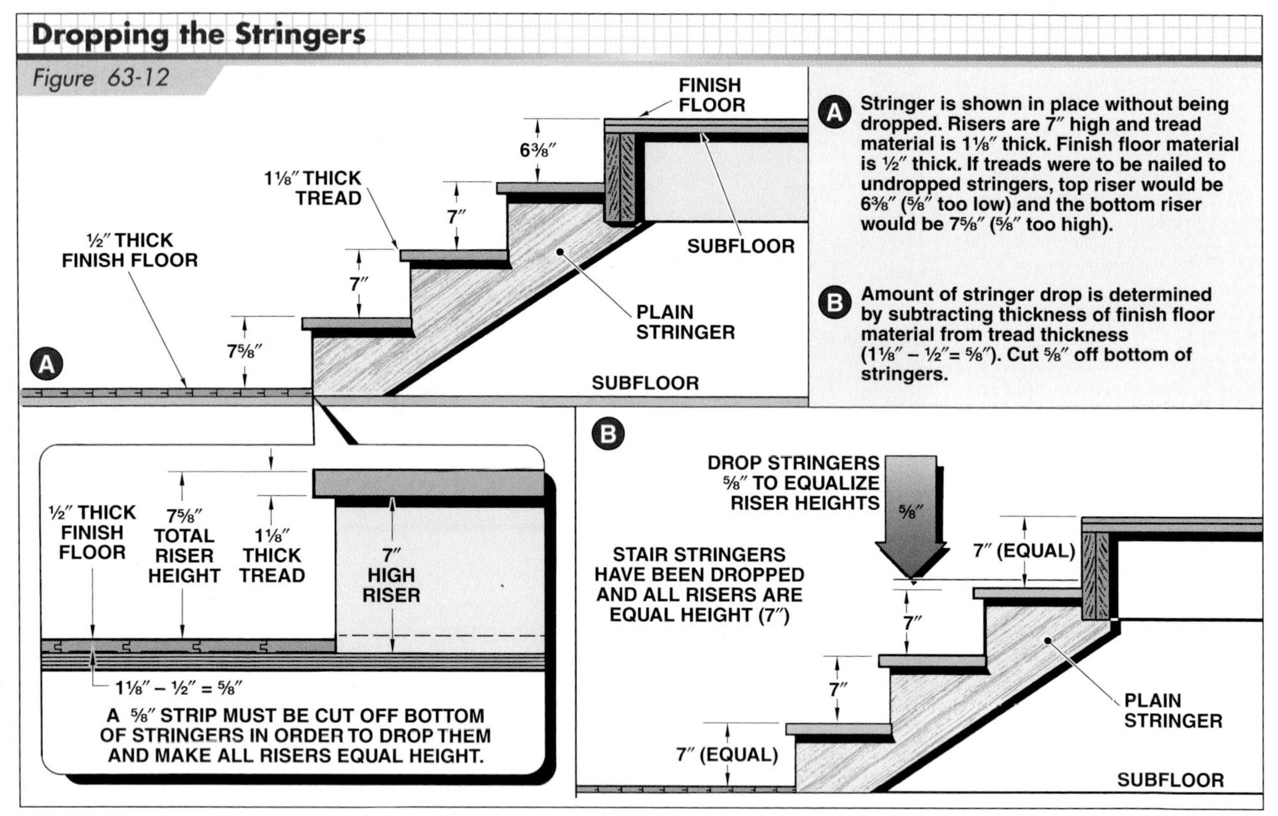

Figure 63-12. *When dropping the stringer, all riser heights will be equal when a finish floor is placed.*

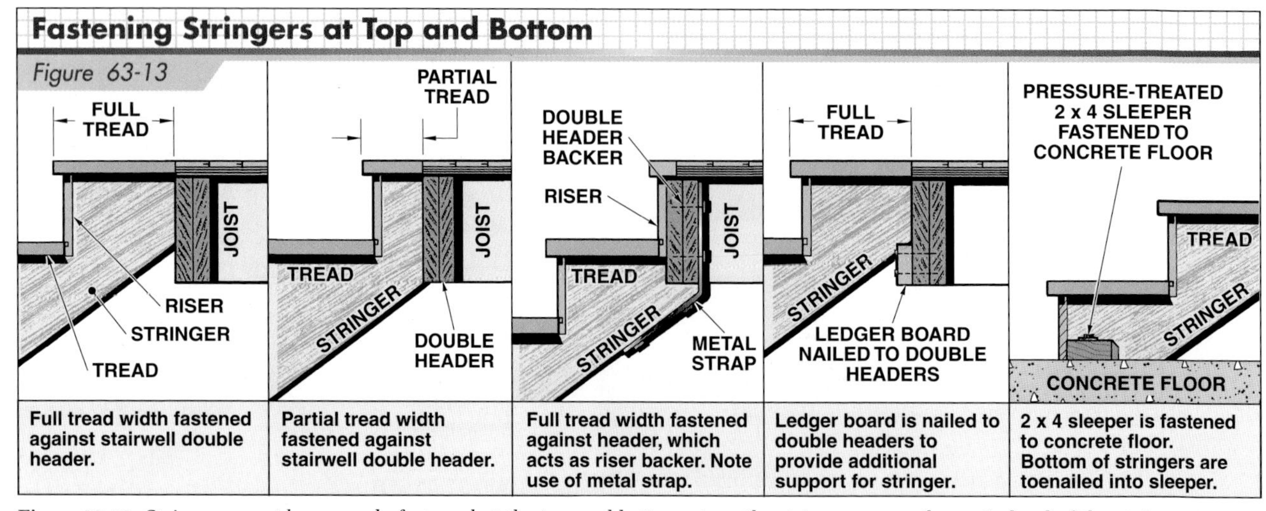

Figure 63-13. *Stringers must be securely fastened at the top and bottom since the stringers carry the main load of the stairway.*

Unit 63 Review and Resources

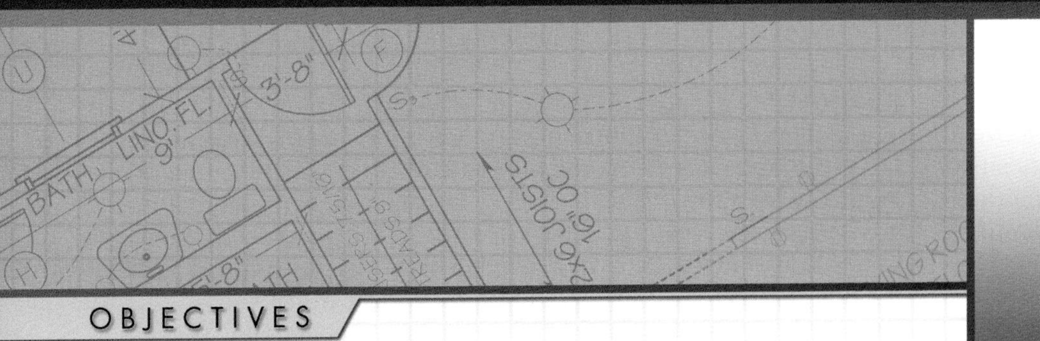

UNIT 64 Stairway Construction

OBJECTIVES

1. List the requirements for residential interior stairways.
2. List the requirements for commercial interior stairways.
3. Describe methods of constructing interior stairways.
4. Describe a method for installing prefabricated stairways.
5. Explain how exterior stairways are constructed.

Safety is a major concern in the design of a stairway. A high percentage of accidents occurring at home take place on stairways. Most building codes include detailed and strict regulations for stairway construction.

Refer to your local building code for information regarding specific stairway requirements for your geographic area.

CONSTRUCTING INTERIOR STAIRWAYS

Stairways must be constructed in accordance with the local building code. Stairway requirements are different for residential and commercial construction. The *International Residential Code* provides stairway requirements for one- and two-family dwellings and townhouses (multiple single-family dwellings) not more than three stories high. *The International Building Code* provides stairway requirements for all buildings except detached one- and two-family dwellings and townhouses not more than three stories high. *The Americans with Disabilities Act (ADA) Standards for Accessible Design* provides stairway requirements for public and commercial buildings.

Residential Stairway Requirements

The *International Residential Code* addresses several residential stairway requirements including stairway width, headroom, treads and risers, profiles, landings, and handrails.

Stairway Width. Stairways for residential structures should not be less than 36″ wide at all points between the top of a handrail and required headroom height. **See Figure 64-1.** Handrails should not project more than 4½″ on either side of a stairway. In addition, the minimum clear stairway width at and below the handrail should not be less than 31½″ with a handrail on one side or 27″ with handrails on both sides.

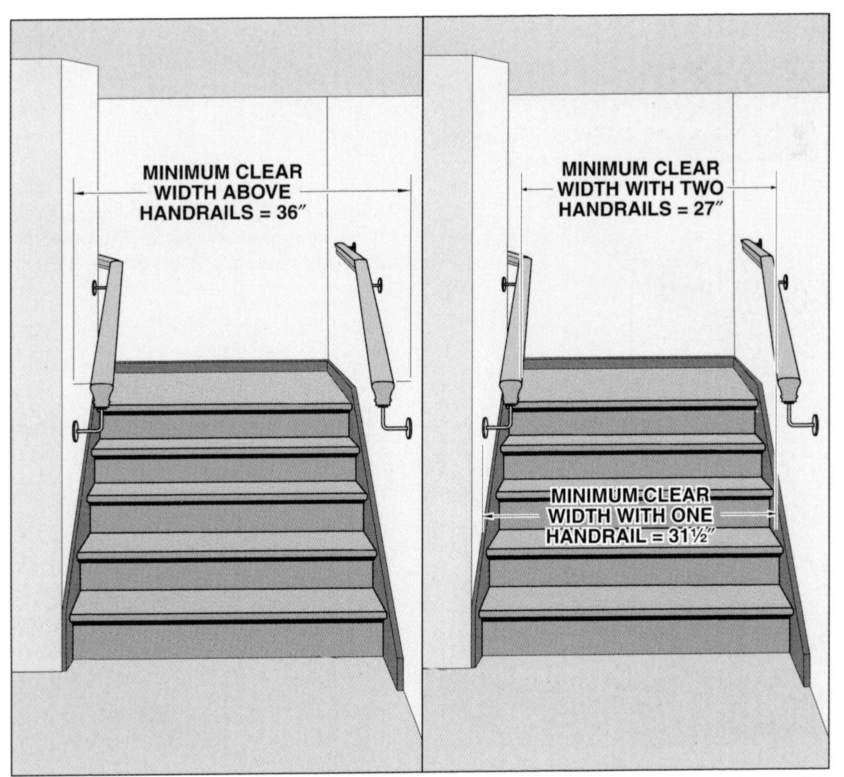

Figure 64-1. *The minimum clear stairway width above handrails is 36″. The clear width at and below the handrail depends on the number of handrails installed.*

Headroom Requirements. *Headroom* is the minimum vertical clearance between tread nosings and the ceiling above. The minimum headroom for residential stairways should not be less than 6′-8″ measured vertically from a line connecting the edges of the nosings. **See Figure 64-2.** The minimum headroom must be maintained between parallel flights of stairs. **See Figure 64-3.**

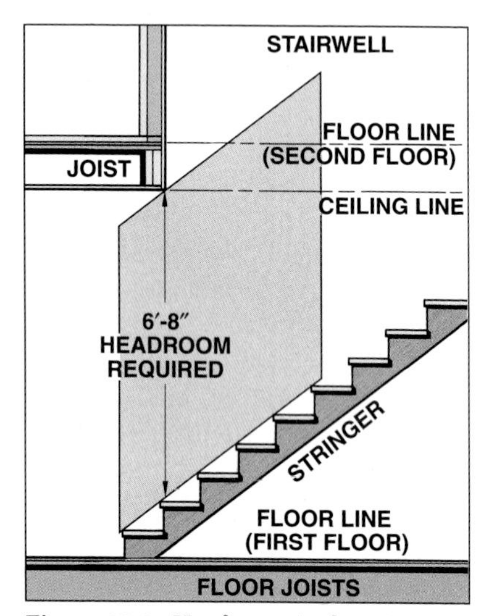

Figure 64-2. *Headroom is the minimum vertical clearance required and is measured from a line connecting the nosings on the stairway to any part of the ceiling above the stairway.*

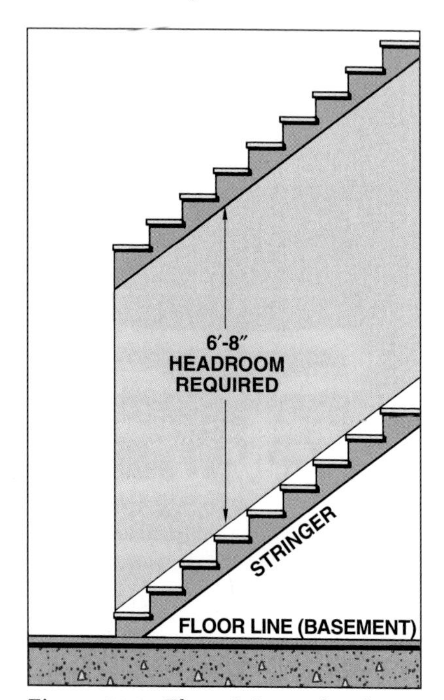

Figure 64-3. *The minimum headroom must be maintained between parallel flights of stairs.*

Riser Heights and Tread Depths. The maximum riser height for residential stairways is 7¾″measured vertically between the nosings of adjacent treads. **See Figure 64-4.** In addition, the variation in riser heights between the largest and smallest riser height should not be more than ⅜″. The minimum tread depth for residential stairways is 10″ with the tread depth measured horizontally from the nosings of adjacent treads. The variation in tread depth between the largest and smallest tread depth should not be more than ⅜″.

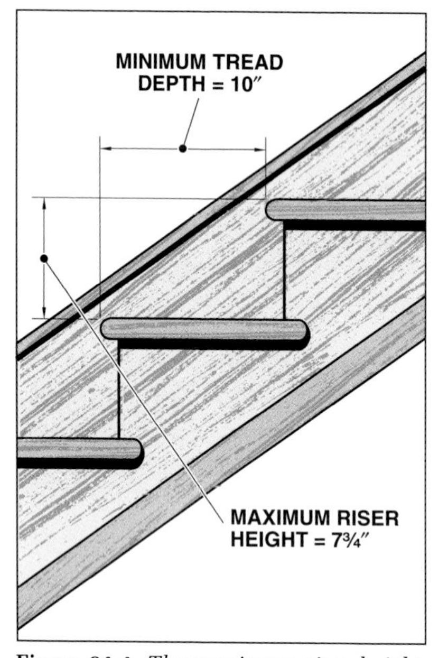

Figure 64-4. *The maximum riser height for residential stairways is 7¾″ and the minimum tread depth is 10″.*

Winder treads must have a 10″ minimum tread depth measured at a right angle to the tread's nosing at a line of travel established 12″ from the narrower edge. **See Figure 64-5.** Minimum tread depth of winder stairways at the narrow end of the tread is 6″. Within a stairway, the largest winder tread depth at a point 12″ from the narrower end should not exceed the smallest tread depth by more than ⅜″.

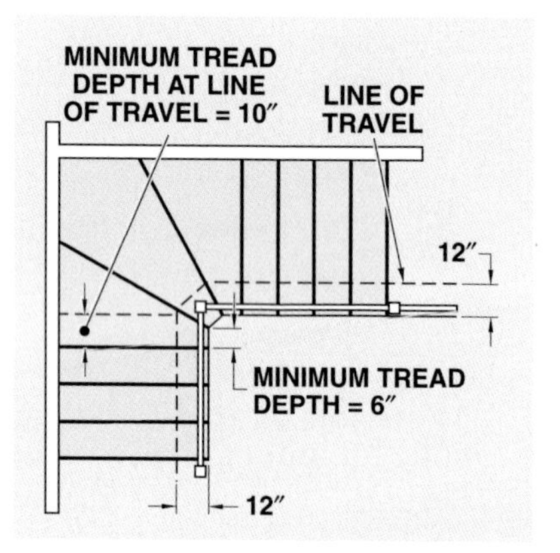

Figure 64-5. *The minimum tread depth at the line of travel is 10″. The narrowest portion of the tread should not be less than 6″ wide.*

The art of stairbuilding dates back to 6000 BC, when stairways were originally developed as exterior additions. In those days, stairways were constructed from notched logs.

Intermediate stringers may be required depending on building code requirements.

Stairway Profiles. Stairway profiles include tread nosings and riser angles. The nosing of a tread is not permitted to have a radius larger than 9⁄16″. Nosings should not be beveled more than ½″. **See Figure 64-6.** For stairways with solid risers, a nosing should not extend from the riser less than ¾″ or more than 1¼″. The largest nosing projection should not be more than ⅜″ greater than the smallest nosing projection. A nosing is not required when the tread depth is at least 11″.

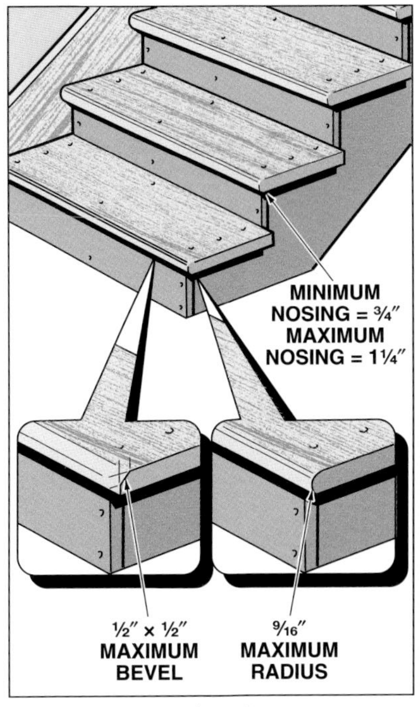

Figure 64-6. *Tread nosings are not permitted to have a radius larger than 9⁄16″. Beveled nosings cannot be greater than ½″ × ½″.*

Ramps may be installed to provide access for individuals with disabilities. Ramps should have a maximum 8.3% (1 in 12) slope.

Risers can be vertical or angled back from the underside of the leading edge of a tread. If angled risers are specified, the angle should be no more than 30° from vertical. Open risers are permitted on residential stairways provided the opening between adjacent treads does not permit the passage of a 4″ diameter sphere. The opening between adjacent treads is not limited to 4″ on stairways with a total rise of 30″ or less.

Landings. A floor or landing must be constructed at the top and bottom of a stairway. However, a floor or landing is not required at the top of an interior stairway when a door does not swing over the stairway. Landings must be at least as wide as the stairways they serve and have a minimum dimension of 36″ measured in the direction of travel. **See Figure 64-7.** When constructing a straight-run stairway, the dimension of the landing in the direction of travel does not need to be greater than 4′-0″. A flight of stairs for a residential stairway should have a 12′-0″ maximum total rise between landings or floors.

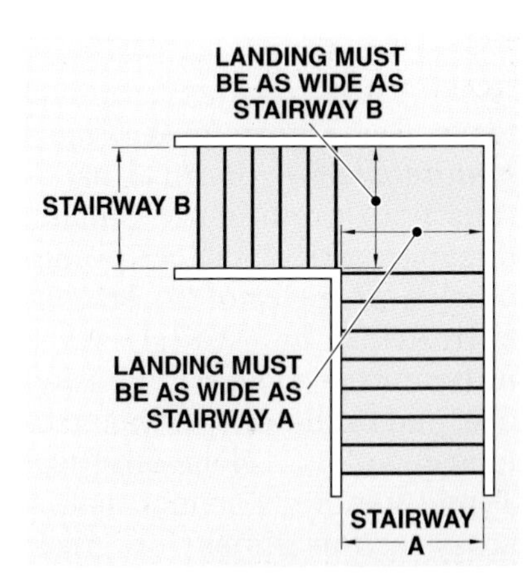

Figure 64-7. *Landings must be as wide as the stairway they serve and must be at least 36″ wide.*

Handrails. Handrails must be provided on at least one side of a stairway with four or more risers. The handrail height must not be less than 34″ or more than 38″, measured from the tread nosings. **See Figure 64-8.** Handrails for residential stairways should be continuous along the entire length of the stairway from a point directly above the top riser to a point directly above the lowest riser. Handrails may be interrupted by a newel post at a turn in the stairway. Handrail ends must be returned or terminate at a newel post or safety terminal. A clearance of at least 1½″ must be provided between the handrail and wall or other surface.

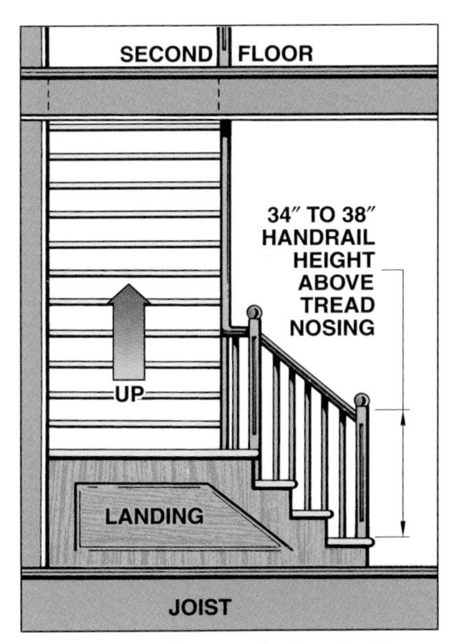

Figure 64-8. *The top of a handrail must be 34″ to 38″ above the tread nosings.*

A volute is a spiral fitting at the end of a handrail that is used to join a section of handrail to a newel post or balusters.

Type I or Type II handrails are required for residential stairways. **See Figure 64-9.** *Type I handrails* are cylindrical handrails with a minimum outside diameter of 1¼″ and maximum outside diameter of 2″, or noncylindrical handrails with a minimum perimeter dimension of 4″ and maximum perimeter dimension of 6¼″. Noncylindrical Type I handrails cannot have a cross-section dimension greater than 2¼″. *Type II handrails* are handrails with a perimeter greater than 6¼″. Due to their perimeter dimension, Type II handrails must have a graspable finger recess on each side.

Commercial Stairway Requirements

The *International Building Code* and *ADA Standards for Accessible Design* address several stairway requirements including stairway width, headroom, treads and risers, stairway profiles, landings, total rise, circular and spiral stairways, and handrails. **See Figure 64-10.**

The International Building Code of Canada establishes minimum requirements to safeguard public health, safety, and welfare through structural strength, means of egress, and adequate light and ventilation. It also establishes minimum requirements to protect the safety of life and property against fire and other hazards.

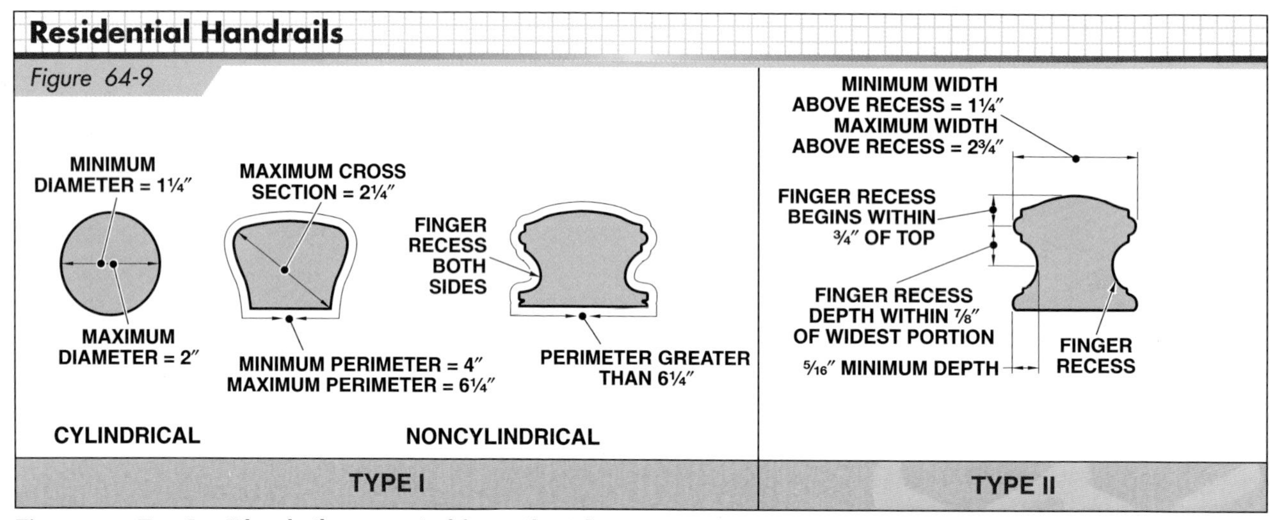

Figure 64-9. *Type I or II handrails are required for residential stairways.*

Type I residential handrails cannot have a cross-section dimension greater than 2¼″.

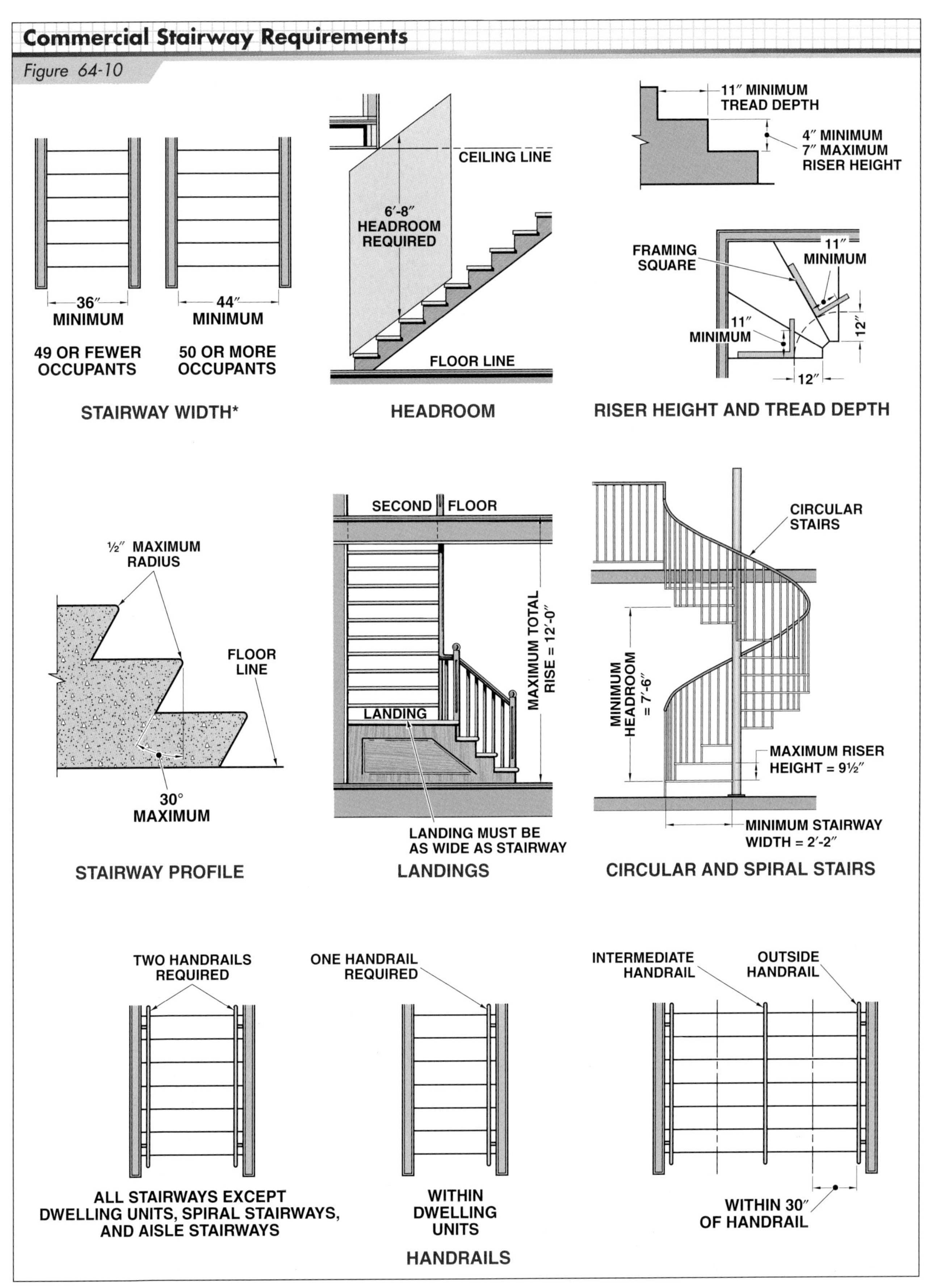

* for stairways in nonhazardous and institutional buildings

Figure 64-10. *Minimum and maximum dimensions for commercial stairways are different than those for residential stairways.*

Stairway Width. The stairway width for commercial buildings is based on building use, the number of occupants for which the building is designed, whether the stairway is equipped with a sprinkler system, and whether the stairway is to be accessible by individuals with disabilities. Buildings involved in the manufacturing, processing, or storage of hazardous materials require wider stairways than typical office buildings for faster egress (exit). For stairways constructed in nonhazardous and institutional buildings, the minimum stairway width is 44″. For occupancy loads of less than 50 people, the minimum stairway width is 36″. However, wider stairways may be required based on the number of occupants and whether a sprinkler system is installed. For stairways with sprinkler systems, the number of occupants is multiplied by 0.2″ to determine the stairway width. For stairways without sprinkler systems, the number of occupants is multiplied by 0.3″. For example, a stairway without a sprinkler system and a design occupancy of 200 people must be at least 5′-0″ wide (200 × 0.3″ = 60″ = 5′-0″).

Headroom Requirements. The minimum headroom for commercial stairways is 6′-8″ measured vertically from a line connecting the edges of the nosings. The minimum headroom must be maintained the full width of the stairway and any associated landings.

Riser Heights and Tread Depths. Riser heights for commercial stairways are 4″ minimum and 7″ maximum with the riser heights measured vertically between the leading edges of adjacent treads. In addition, the variation in riser heights between the largest and smallest riser height should not be more than ⅜″. Tread depths must be at least 11″ with the tread depths measured horizontally from the leading edges of adjacent treads. The variation in tread depth between the largest and smallest tread depth should not be more than ⅜″.

Winder treads must have an 11″ minimum tread depth measured at a right angle to the tread's leading edge at a point 12″ from the narrower edge. For ADA-compliant stairways, all stairs should have uniform riser heights and tread depths. Stair treads must be at least 11″ wide measured from riser to riser. In addition, open risers are not permitted for ADA-compliant stairways.

Stairway Profiles. The leading edge of a tread is not permitted to have a radius larger than ½″. Nosings should not be beveled more than ½″. Risers can be vertical or angled from the underside of the leading edge of a tread. If angled risers are specified, the angle should be no more than 30° from vertical. Nosings are not permitted to project more than 1¼″ and must be of uniform size, including the leading edge of the floor at the top of the stairway. For ADA-compliant stairways, the underside of nosings should not be abrupt and the leading edge of a tread is not permitted to have a radius larger than ½″. In addition, nosing should not project more than 1½″.

Landings. A floor or landing must be constructed at the top and bottom of a stairway. Landings must be at least as wide as the stairways they serve and have a minimum dimension measured in the direction of travel equal to the stairway width. When constructing a straight-run stairway, the dimension of the landing in the direction of travel does not need to be greater than 4′-0″. A flight of stairs for a commercial stairway should have a total rise not greater than 12′-0″ between landings or floors.

Circular and Spiral Stairways. Circular stairways are required to have the same minimum tread depth and maximum riser heights as straight-run stairways. The minimum tread depth, measured 12″ from the narrower end of the tread, must not be less than 11″. The minimum tread depth at the narrower end should not be less than 10″.

Spiral stairways are permitted in commercial buildings only within residential units, or from spaces not greater than 250 sq ft and serving not more than five occupants, or from overhead galleries or catwalks. Spiral stairways must have a minimum tread depth of 7½″, measured 12″ from the narrower end. A minimum headroom of 7′-6″ must be provided in a spiral stairway, and the maximum riser height should not be more than 9½″. The minimum stairway width for spiral stairways is 2′-2″.

The curved riser is a prefabricated component that is constructed by sawing kerfs in the back and forming the riser around a template.

Handrails. Commercial and ADA-compliant stairways are required to have handrails on each side. Aisle stairways that are provided with a center handrail do not require handrails along the sides. In addition, stairways within dwelling units and spiral stairways are only required to have a handrail on one side. Changes in elevation from one room to another with only one riser do not require handrails. Handrail height must not be less than 34″ or more than 38″, measured from the tread nosings. Intermediate handrails are required on wide stairways so that no part of the stairway is more than 30″ from a handrail. For ADA-compliant stairways, handrails should be continuous on each side of the stairway and the handrail on U-shaped stairways must also be continuous. If handrails are noncontinuous, the handrails must extend at least 12″ beyond the top riser and at least 12″ plus the depth of one tread beyond the bottom riser.

For optimum graspability, cylindrical handrails must have a minimum outside diameter of 1¼″ and a maximum outside diameter of 2″. For noncylindrical handrails, the handrails must have a perimeter dimension greater than 4″ and not more than 6¼″ with a maximum cross-section dimension of 2¼″. Handrails must extend at least 12″ beyond the top riser and at least one tread depth beyond the bottom riser. A clearance of at least 1½″ must be provided between the handrail and wall or other surface.

Stairwell Opening

A stairwell opening must be framed in the floor at the top of the stairway. In new construction, the width and length of a stairwell opening is shown in the prints. In remodeling work, a stairwell opening may have to be cut into an existing floor.

The width of a stairwell opening should be the same as the rough width of the stairs. The stairway opening length must be calculated to ensure the proper amount of headroom between the lower steps of the stairway and the end of the opening above. **See Figure 64-11.**

Straight-Flight Stairways

A straight-flight stairway is the simplest type of stairway to build. **Figure 64-12** shows a procedure for constructing a straight-flight stairway from the first to second story of a building.

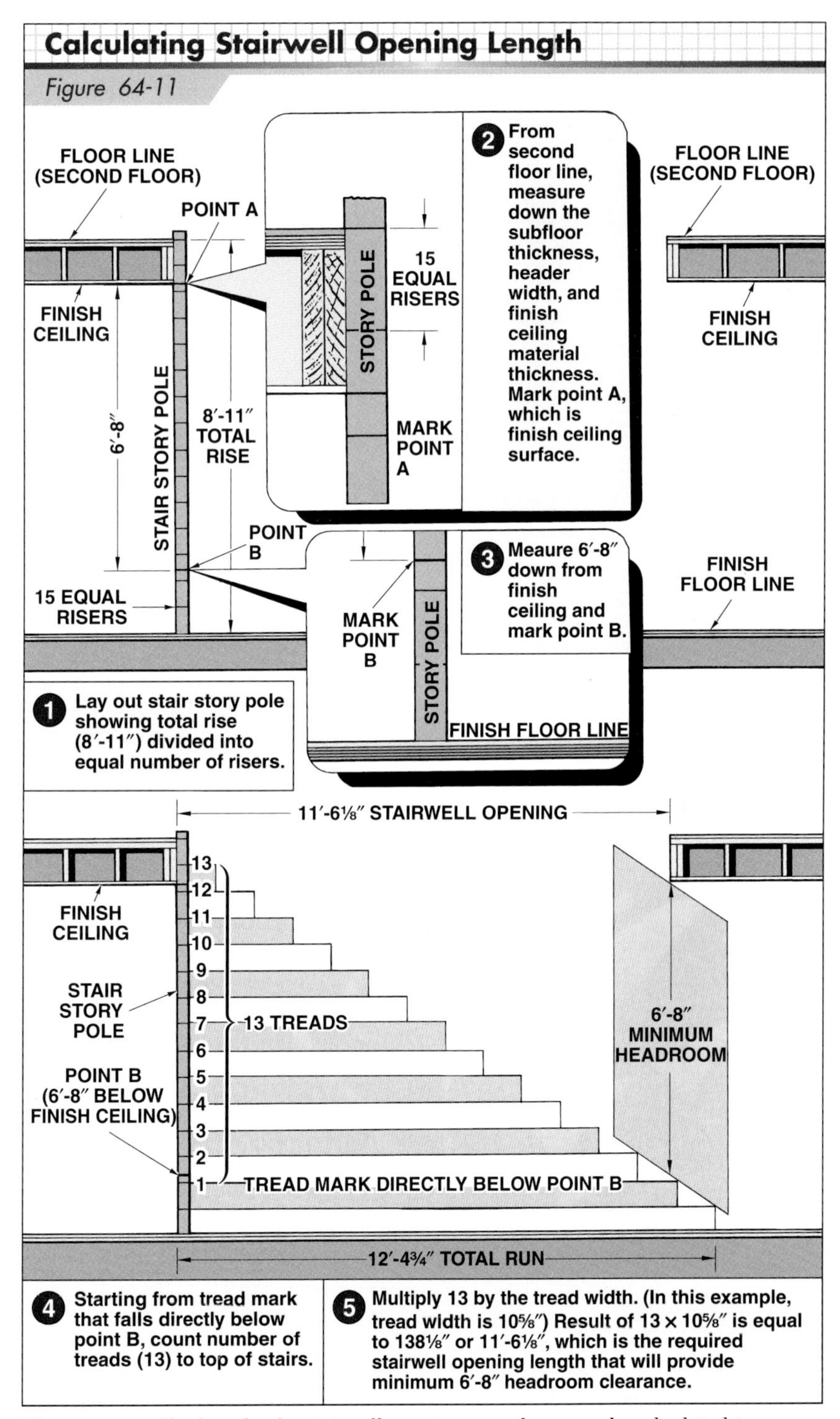

Figure 64-11. *The length of a stairwell opening must be properly calculated to ensure the proper amount of headroom.*

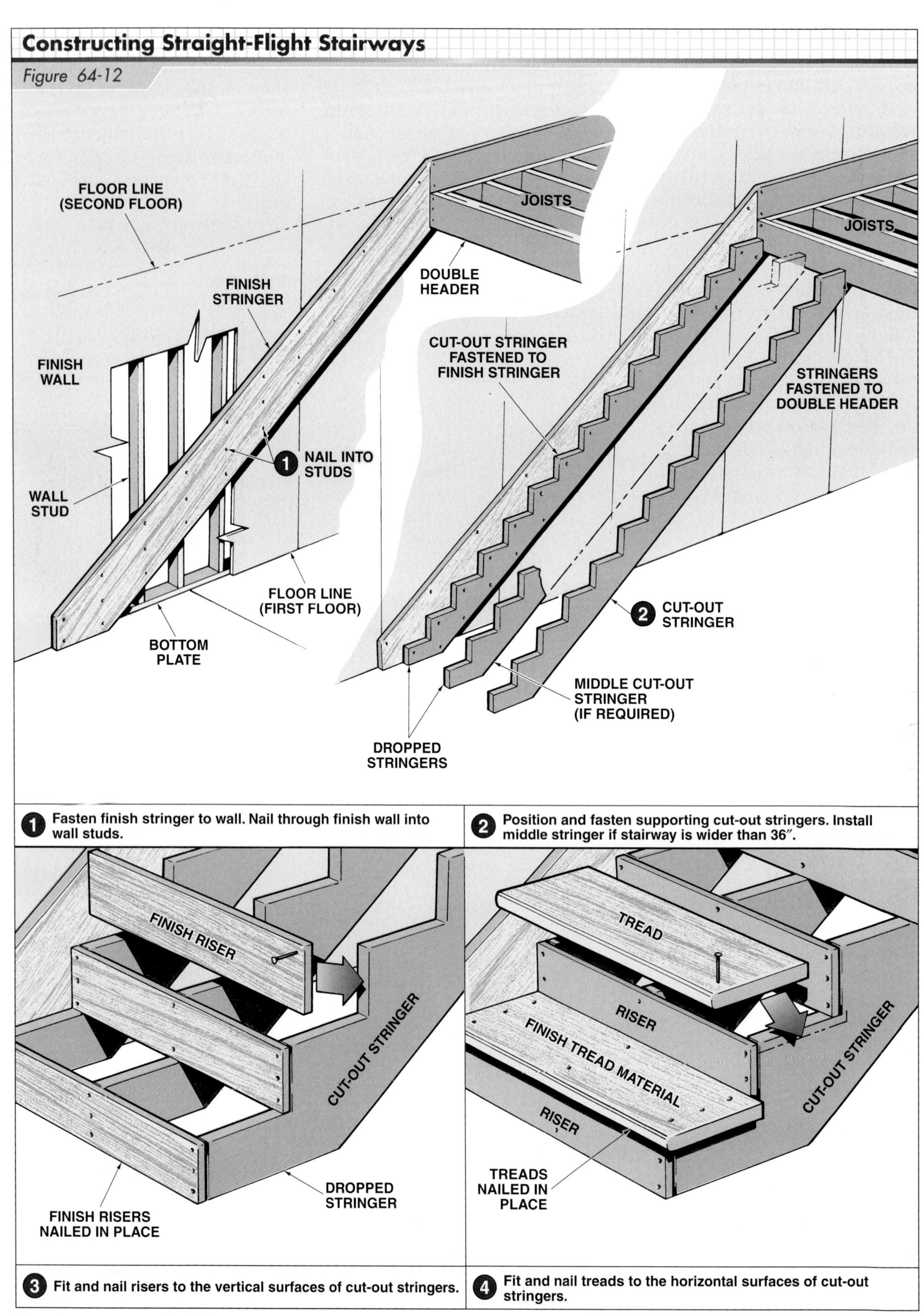

Figure 64-12. *A straight-flight stairway runs directly between different floor levels.*

For a straight-flight stairway with a landing, the first step is to frame the landing. The stairway is then constructed against the landing. **Figure 64-13** shows a procedure for constructing an L-shaped stairway with a landing. The stairway in this example is open on one side.

Stairways with Winders

Most stairways with winders are L-shaped. Instead of a landing separating the flights, a series of winder treads are used to make the turn. The winder section of most L-shaped stairways consists of three or four treads.

Stairways with winders are not as safe as straight-flight stairways. The depth of a winder tread varies from one end to the other. Stairways with winders are usually installed only where space does not permit a straight-flight stairway with a landing. **See Figure 64-14.**

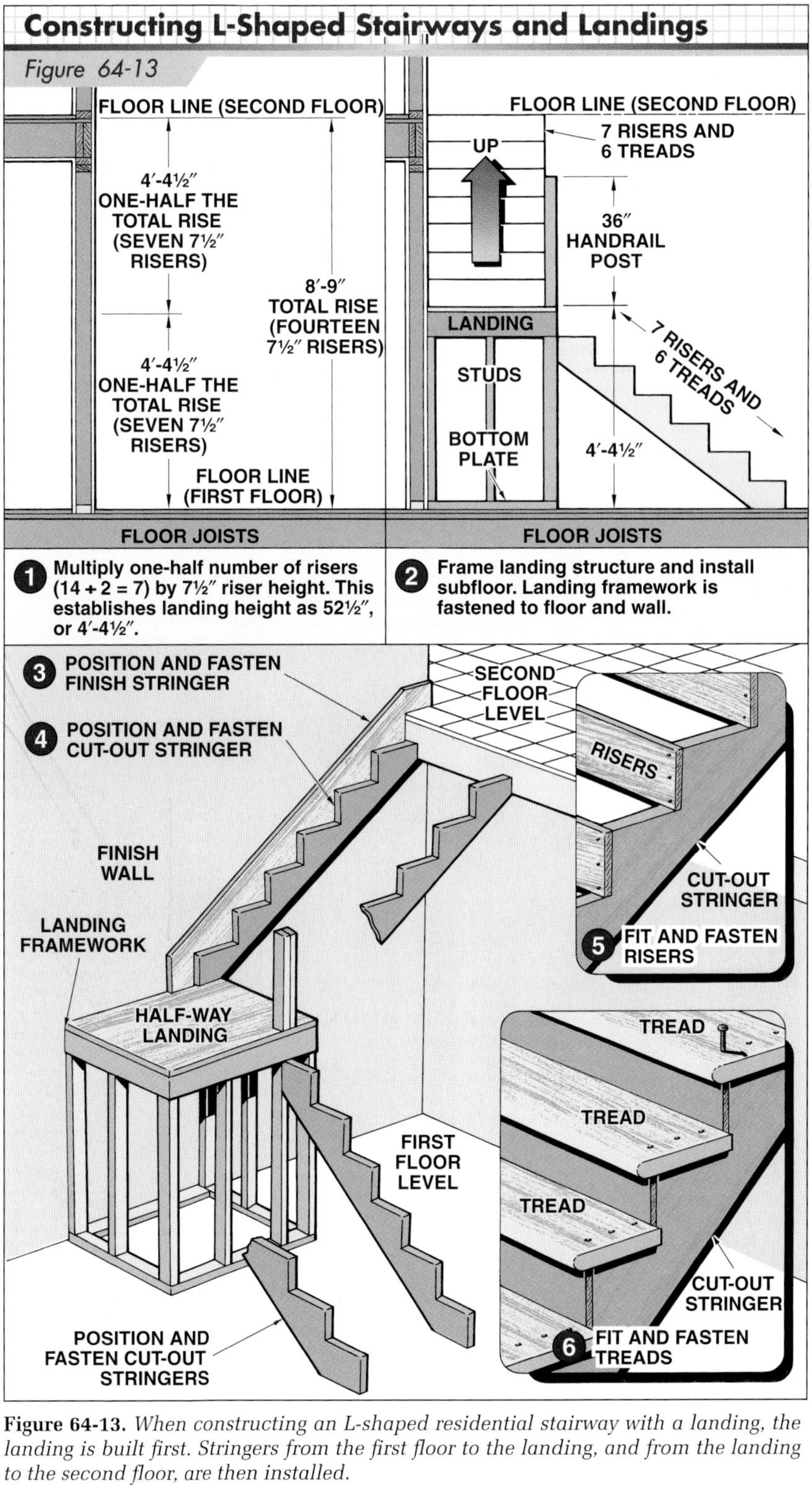

Circular and elliptical stairways are classified as either left- or right-handed stairways. Stairway direction is determined by the turn a stairway makes when facing it from the bottom.

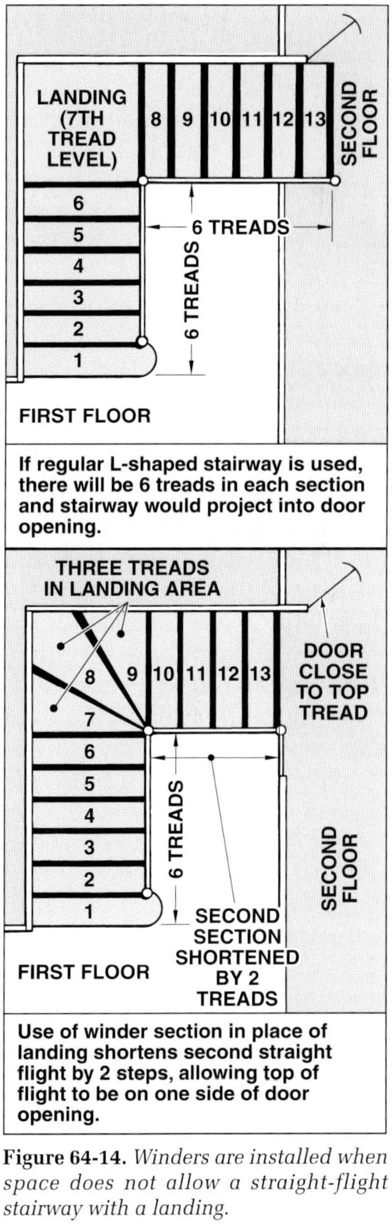

Constructing L-Shaped Stairways and Landings

Figure 64-13

1. Multiply one-half number of risers (14 ÷ 2 = 7) by 7½″ riser height. This establishes landing height as 52½″, or 4′-4½″.
2. Frame landing structure and install subfloor. Landing framework is fastened to floor and wall.
3. POSITION AND FASTEN FINISH STRINGER
4. POSITION AND FASTEN CUT-OUT STRINGER
5. FIT AND FASTEN RISERS
6. FIT AND FASTEN TREADS

Figure 64-13. *When constructing an L-shaped residential stairway with a landing, the landing is built first. Stringers from the first floor to the landing, and from the landing to the second floor, are then installed.*

If regular L-shaped stairway is used, there will be 6 treads in each section and stairway would project into door opening.

Use of winder section in place of landing shortens second straight flight by 2 steps, allowing top of flight to be on one side of door opening.

Figure 64-14. *Winders are installed when space does not allow a straight-flight stairway with a landing.*

An important dimension that must be established in designing a stairway with winders is the *line of travel.* **See Figure 64-15.** The line of travel is measured from the turn at the narrow ends of the winder steps, and is the place where a person is likely to walk when using the stairway. The line of travel is 12″ per the *International Residential Code* and *International Building Code.* The depth of a winder tread at the line of travel should be 10″ for residential construction and 11″ for commercial construction. The narrow ends of the winder treads should be at least 6″ deep.

The building codes specify a minimum width of 6″ at the narrow ends of winder treads. In this case, it is not possible to maintain a 10″ tread depth with a 12″ line of travel. Therefore, the winder treads at the line of travel will be deeper than the tread depth of the straight-flight section of the stairway.

If a stairway with winders is open on one side, the narrow ends must be mortised into a post. If the stairway is enclosed by walls on both sides, supporting stringers are required at the narrow and wide ends of the winder section. Cut-out or housed stringers can be used. Prefabricated winder stairways are usually built with housed stringers.

Before a stairway is installed, the winder section should be laid out to full scale in the floor area where the stairway is to be installed. Only from a full-scale layout can dimensions be obtained for the cuts of the stringer. General steps in the layout and construction of a stairway with a three-part winder section are shown in **Figure 64-16.**

The greatest winder tread depth at the line of travel must not exceed the smallest winder tread depth by more than ⅜″, measured at a right angle to the tread's leading edge.

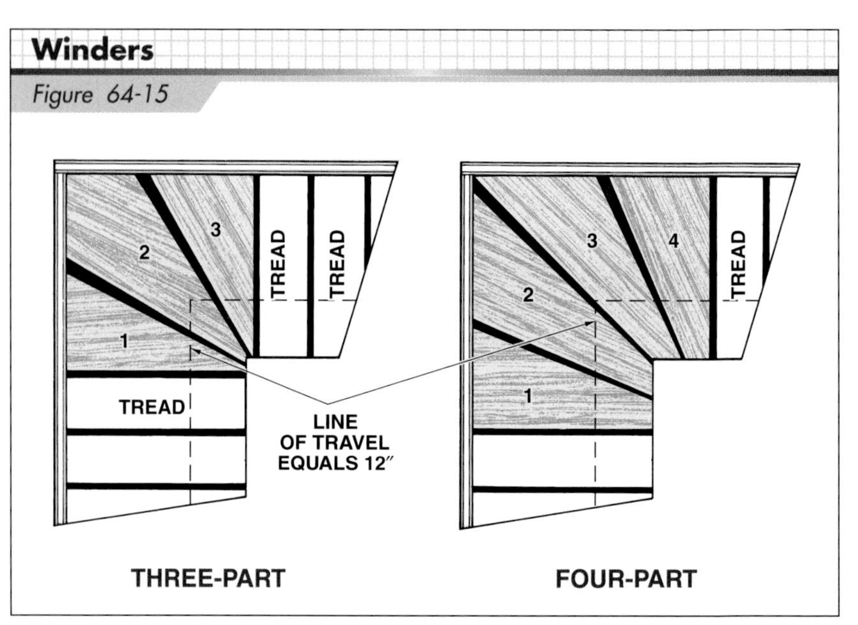

Figure 64-15. *The line of travel for a stairway with winders is 12″. Whenever possible, the depth of a winder tread at the line of travel should be the same as the tread depth along the straight-flight section.*

Curved stringers for custom circular stairways comprised of laminated layers of wood may require assembly by a specialty stair contractor.

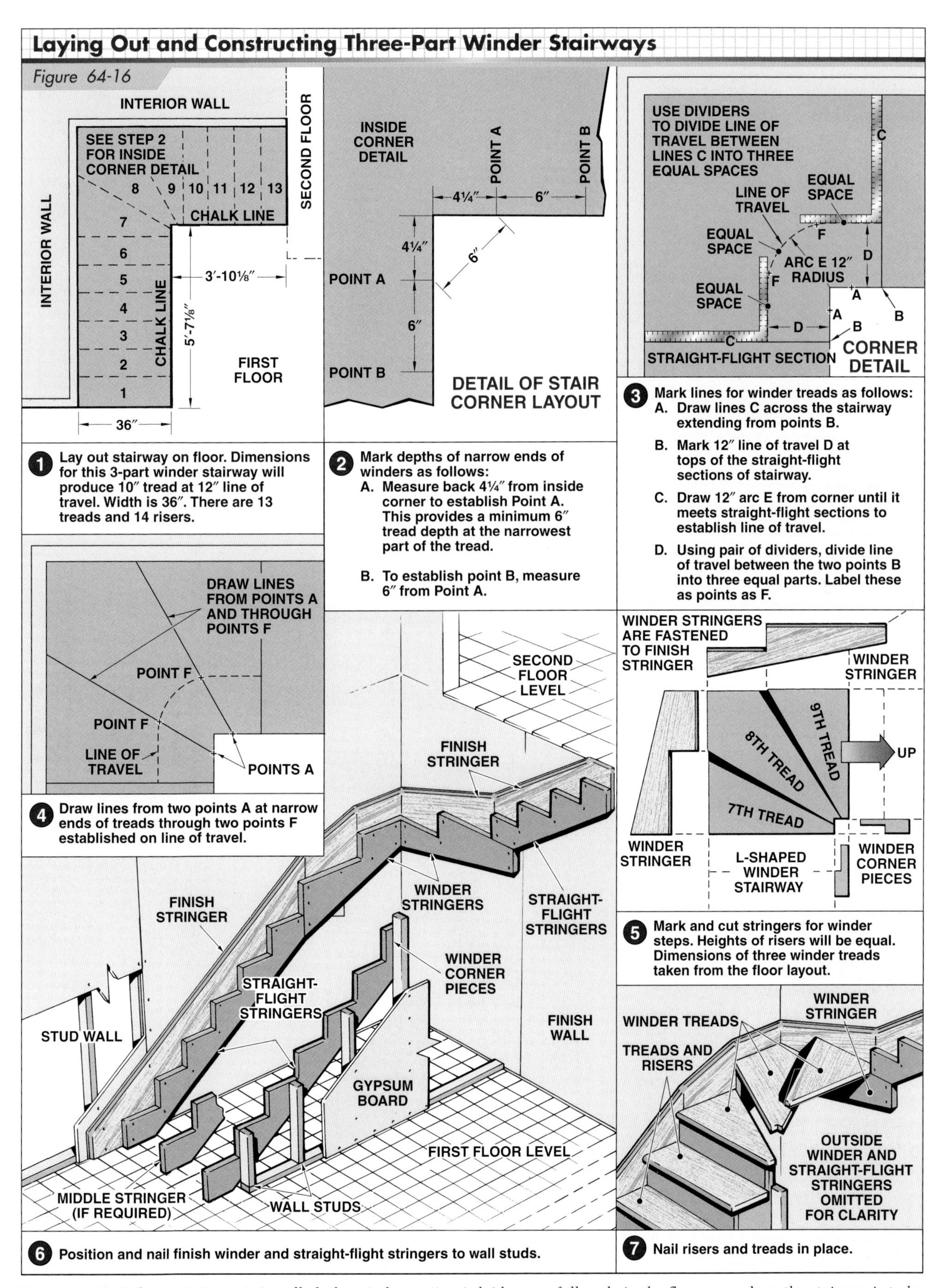

Figure 64-16. *Before a stairway is installed, the winder section is laid out to full scale in the floor area where the stairway is to be installed. In this example, the line of travel is 14″. The tread depth along the winder and straight-flight line of travel will be 10″. The total width of the stairway is 36″.*

PREFABRICATED AND CUSTOM STAIRWAYS

For prefabricated stairways, the individual parts of the stair system are cut and prebuilt in a stair contractor's shop that specializes in this kind of work. The parts are then delivered to the job site. **See Figure 64-17.** In certain cases, the stairway may be prefabricated in sections at the shop. The stair parts may be installed by carpenters already working on the job site, but they are more frequently assembled and installed by the carpenters working for the stair building contractor. Basic prefabricated stairways will include the finished treads and risers, railings, and housed stringers.

Stairway stringers are installed first. Next, the precut treads and risers are set into the grooves of the stringers and secured with glue and wedges. **See Figure 64-18.** Glue blocks fasten the bottoms of the risers to the edges of the treads. Glue blocks (instead of nails) are used with housed stringers to reduce the amount of squeaking. If one side of the stairway is open, a *mitered stringer* is used in which the corner joints of the risers and stringers are mitered. **See Figure 64-19.**

The term custom stairway describes a much higher quality of stairway compared to typical prefabricated stairways. They are usually made of high quality hardwood and designed for beauty and strength. Custom stairways often contain intricate details and custom handrails and balustrades. They are designed by specialty stair contractors and installed by highly skilled stair specialists. **See Figure 64-20.**

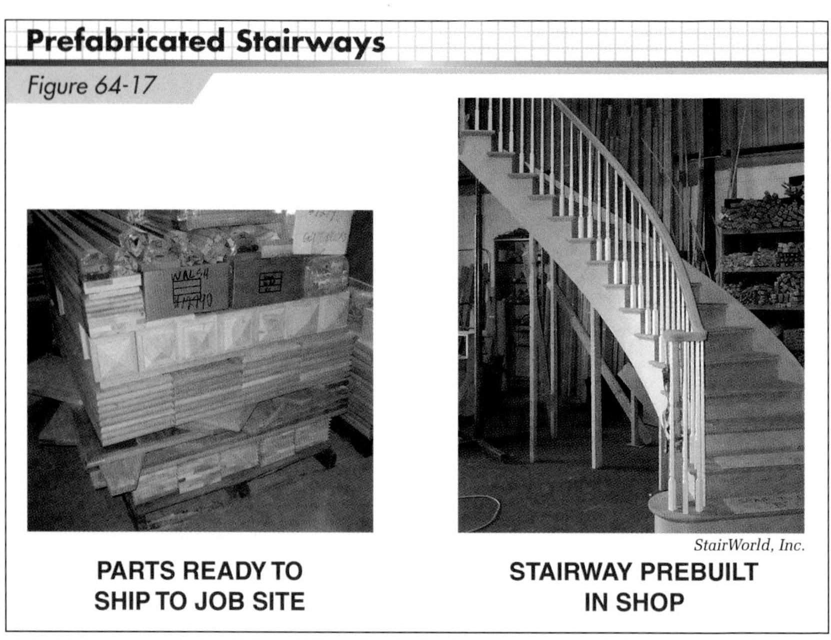

Figure 64-17. *Prefabricated stairways are delivered as a package and installed at the job site.*

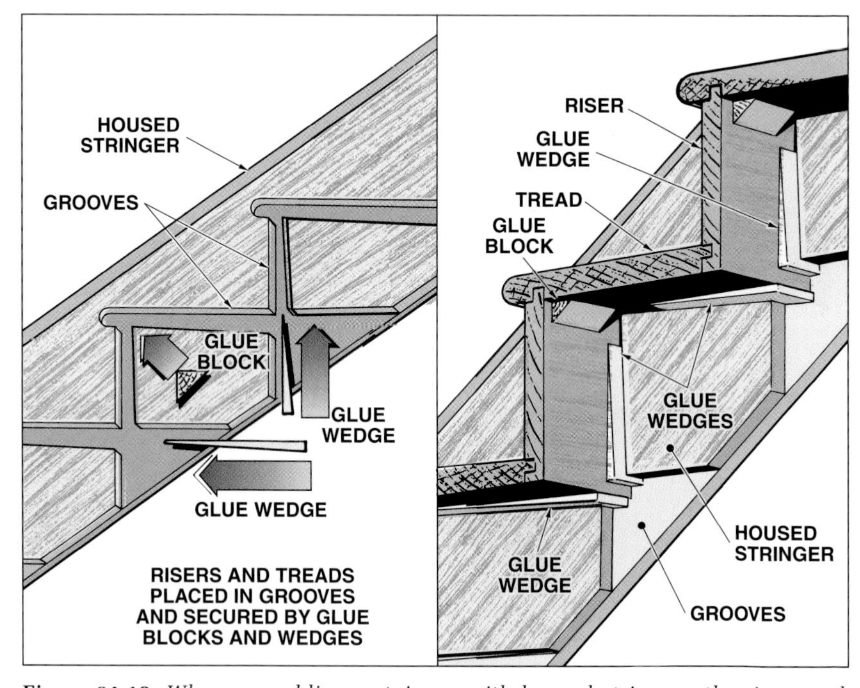

Figure 64-18. *When assembling a stairway with housed stringers, the risers and treads are placed in the grooves and secured with glue and wedges. Glue blocks fasten the bottoms of the risers to the edges of the treads.*

CONSTRUCTING EXTERIOR STAIRWAYS

A variety of exterior stairways are used for access to front and rear entrances, decks, and porches. An exterior stairway may be constructed entirely of wood, or its treads and risers may be finished with stone, tile, or other nonwood material. **See Figure 64-21.** Concrete is also a common material to use for exterior stairways due to its durability.

The basic layout methods for constructing wood exterior stairways are similar to those for interior stairways. One difference between exterior and interior stairways is that the riser height for exterior stairs is typically 6″ to 7″ rather than the 7″ to 7½″ common for interior stairs.

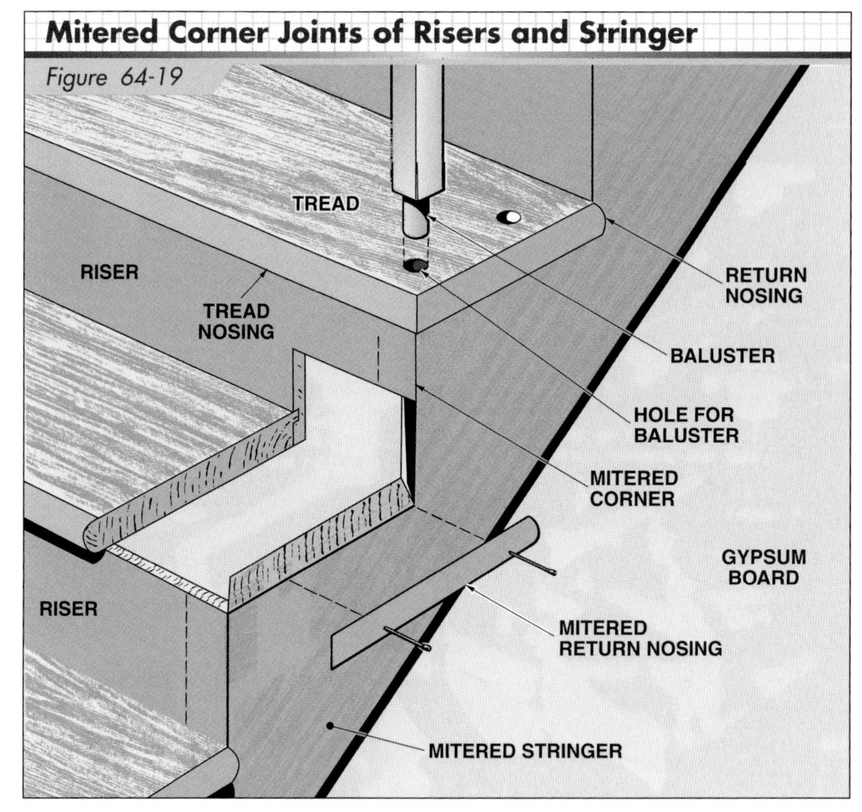

Figure 64-19. *In a mitered stringer, the corner joints of the risers and the stringer are mitered.*

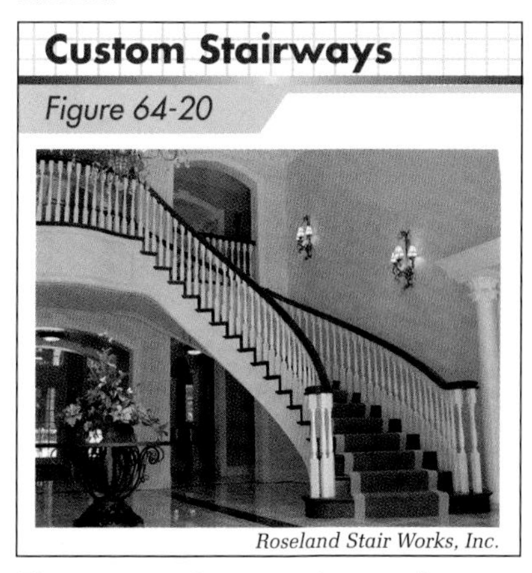

Figure 64-20. *Custom stairways often contain intricate details and custom handrails and balustrades.*

The steps of an exterior stairway may be finished with enclosed treads and risers, or the risers may remain open. **See Figure 64-22.** If an enclosed tread and riser design is used, the tread surfaces of the stringers are cut at a slight angle to provide a ⅛″ forward slope to provide proper water drainage.

A concrete bottom step is recommended for wood stairways if the bottom of the stringer does not rest on a concrete slab. If wood posts are used as part of the stairway structure, the posts should be supported by concrete piers that extend into firm soil.

One procedure for constructing an exterior stairway is shown in **Figure 64-23.** In this example, the stairway leads to a landing at the rear of the building.

Figure 64-21. *An exterior stairway may be constructed of wood or may be finished with nonwood materials.*

Ramps may be constructed adjacent to a building to provide access for individuals with disabilities. Section 1010 of the International Building Code outlines the requirements for ramps used as a means of egress. Items such as ramp slope, cross slope, vertical rise, minimum dimensions, landings, and changes in direction are addressed in this section.

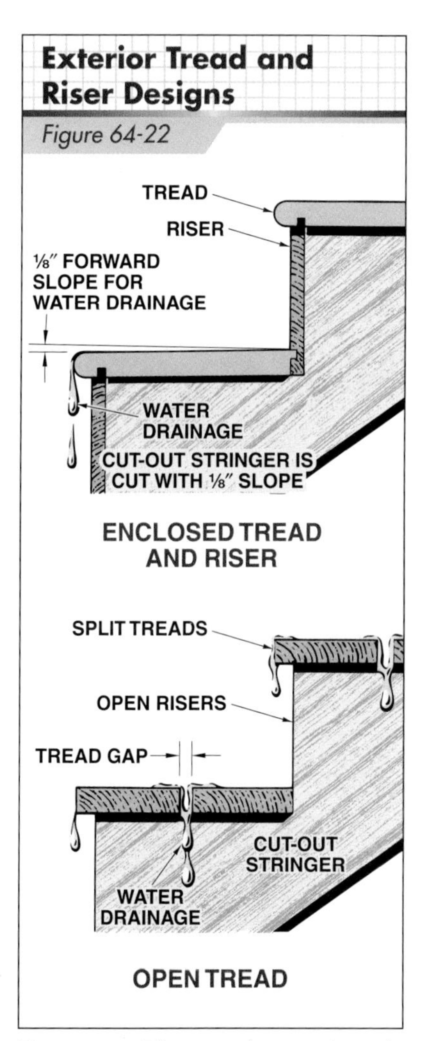

Figure 64-22. *The steps of an exterior stairway may be finished with enclosed treads and risers or the risers may remain open.*

Constructing Exterior Stairways with Landings

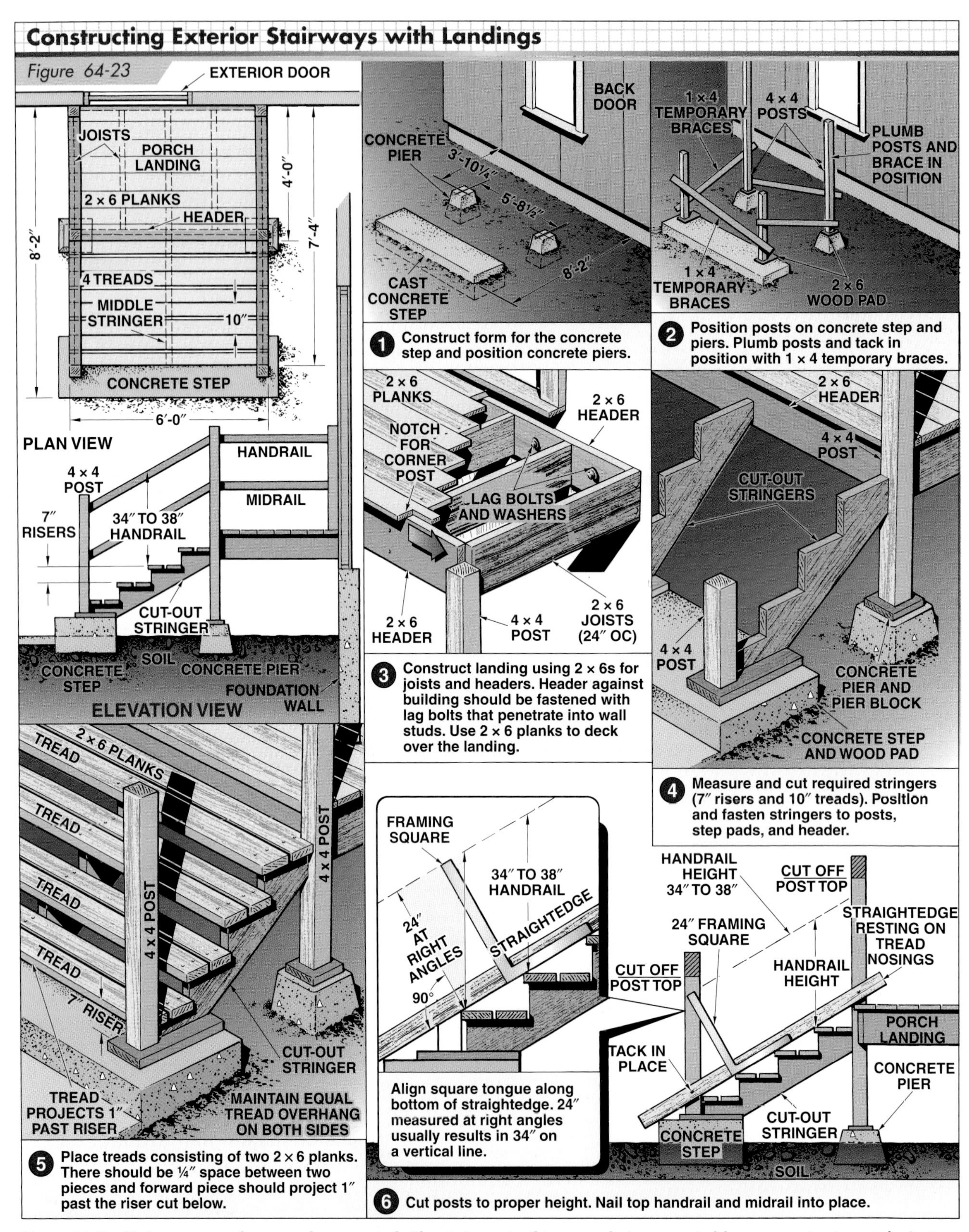

Figure 64-23. *Stairways must be properly supported. The stairway in this example is supported by a concrete step at the bottom, concrete piers, and a 2 × 6 header at the top.*

SECTION

15

Post-and-Beam Construction

CONTENTS

UNIT 65

Post-and-Beam Construction

OBJECTIVES

1. List common types of post-and-beam construction.
2. Describe the principles of post-and-beam construction.
3. Explain how post-and-beam walls are constructed.
4. Explain how post-and-beam roofs are constructed.
5. Describe how post-and-beam buildings are constructed.
6. Describe the differences between timber frame construction and post-and-beam construction.
7. Review post-frame construction.

Post-and-beam construction methods were made part of North American culture as early as 1620 when English settlers began arriving. Early post-and-beam construction consisted of a basic frame of posts and beams. Spaces between the posts and roof beams were filled with *wattle,* which is an intertwined mixture of twigs, reeds, and branches that was covered with mud. All materials for the building were found locally with the exception of square-cut nails, which were imported from England until the settlers were able to produce their own.

A revival of post-and-beam construction began in the 1940s and continues today. Three types of post-and-beam construction are *residential post-and-beam, timber frame,* and *post-frame.* While there are some similarities among the types, there are several important differences.

In the past, tongue-and-groove planks were frequently used for the floor and roof coverings of post-and-beam buildings. Thus, post-and-beam construction at one time was referred to as plank-and-beam construction. With the availability of new materials and methods today, planks are used less often.

Classic Wood & Beam

Roof and floor planks are kiln-dried to remove excess moisture from the lumber.

An informative source for post-and-beam construction is the Canadian Wood Council.

POST-AND-BEAM CONSTRUCTION

In *post-and-beam construction,* large wood members are used to frame the structure and create large, open areas within the building. The basic framework of a building consists of vertical posts and horizontal or sloping beams. The posts in the exterior walls of the house support the beams that are part of the roof and ceiling. Posts may also be used beneath beams that support the floor. **Figure 65-1** compares the framework of a post-and-beam house with a conventional, platform-framed house.

The exposed wood beams in post-and-beam construction provide an attractive, natural look to the structure interior. The beams may be solid wood, or engineered wood products such as built-up beams, glulam, or laminated veneer lumber beams. **See Figure 65-2.** Floor-to-ceiling windows can be set between the wall posts of exterior walls, providing a great amount of natural light to the interior.

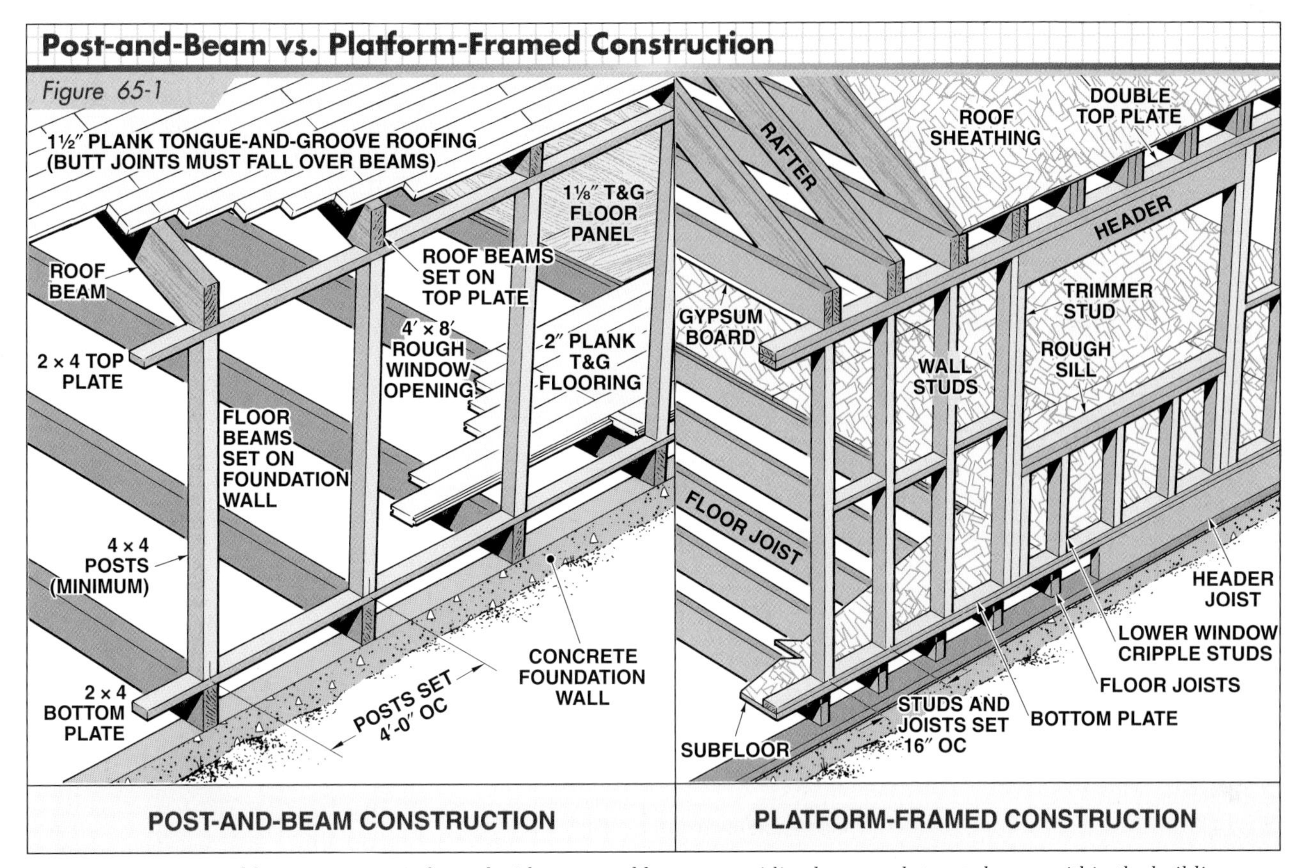

Figure 65-1. *A post-and-beam structure is framed with posts and beams, providing large unobstructed areas within the building.*

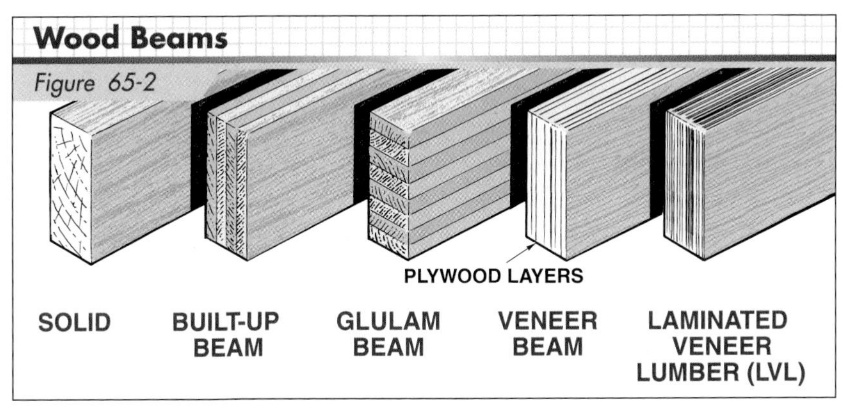

Figure 65-2. *Solid and engineered wood products are used as beams in post-and-beam construction.*

Metal connectors are used to fasten key structural members together. **See Figure 65-3.** Plain and ornamental connectors are available, and are attached with nails, lag bolts, or machine bolts, depending on the application. Holddowns and anchor tiedown systems are required in seismic and hurricane areas.

Post-and-beam houses are constructed over any conventional foundation system including full-basement, crawl-space, and slab-at-grade.

Post-and-Beam Floors

Although conventional floor joist systems can be used, solid or engineered wood beams can be placed over posts that are supported by concrete piers. The prints indicate the specified spacing between beams. Beams are tied to posts with metal connectors. The outside ends of the beams rest on top of the foundation walls. Wood planks or a panel subfloor is nailed to the tops of the beams. **See Figure 65-4.**

Post-and-beam systems are often used beneath the first floor of a platform-framed building. In some sections of the United States, posts and beams over piers may serve as the foundation support for the building.

The size of beams used for a floor depends on the live and dead loads and on the span between supports. Beams with 4″ × 8″ nominal dimensions are often used when the beams are spaced 4′ OC.

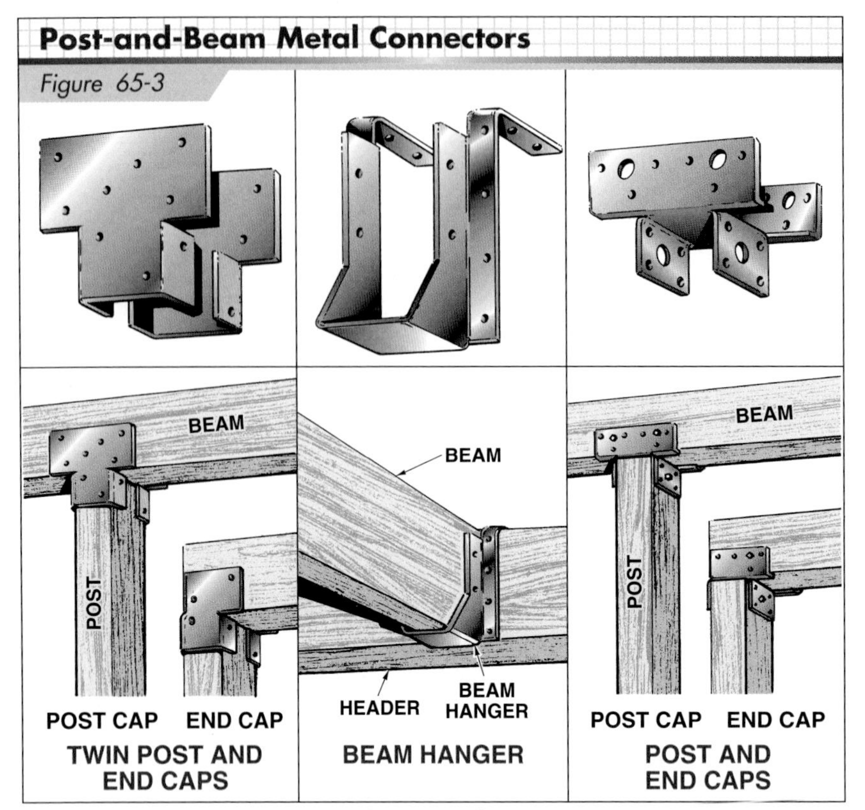

Figure 65-3. *Metal connectors provide additional strength to post-and-beam structures.*

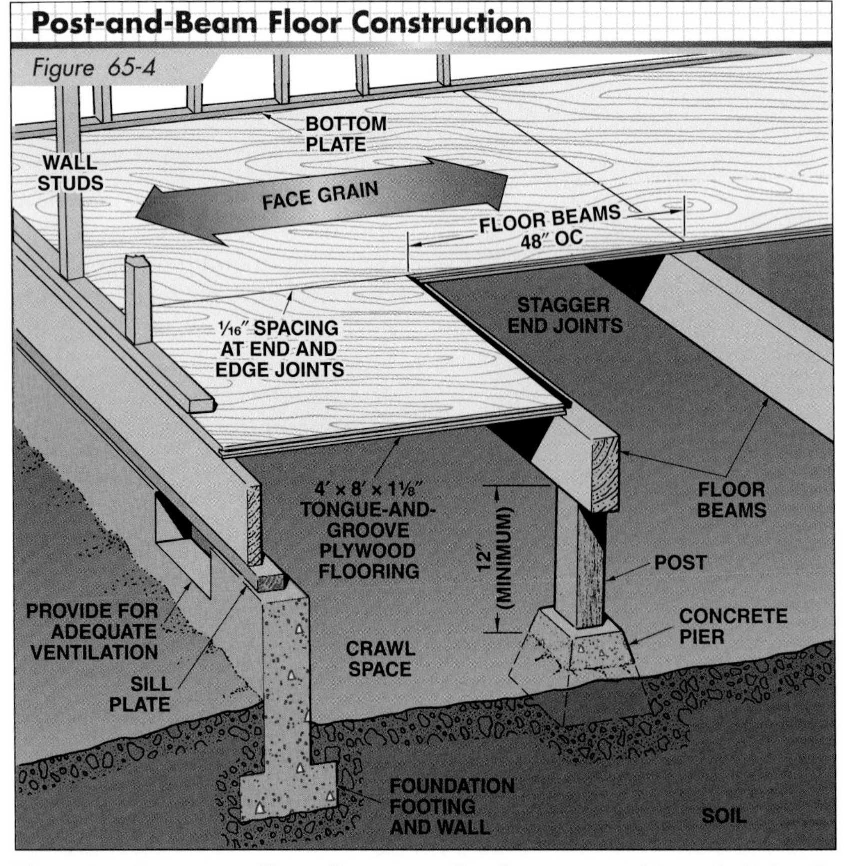

Figure 65-4. *In a post-and-beam floor system, floor beams are used instead of joists. In this example, the floor beams are spaced 4′ OC and are supported by posts resting on concrete piers. The deck material is 1⅛″ thick plywood.*

Covering Post-and-Beam Floors. Two-inch-thick tongue-and-groove planks have traditionally been used for post-and-beam floors. Plank flooring is installed so joints between butt ends are staggered. Each plank should span at least two openings between floor beams. **See Figure 65-5.** Post-and-beam floors may also be constructed using 1⅛″ thick tongue-and-groove OSB or plywood panels.

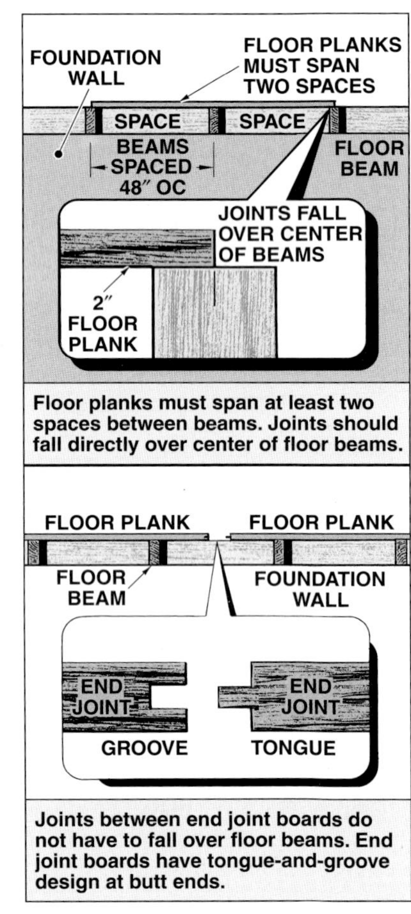

Figure 65-5. *Floor planks must span at least two openings between beams. Ends of end joint boards do not need to fall over beam centers. When installing plank flooring, joints between butt ends are staggered.*

Where heavier loads are anticipated, such as a bearing partition, bathtub, or refrigerator, additional framing should be installed beneath the floor to transmit the load to the beams.

Post-and-Beam Walls

The walls of a post-and-beam building are constructed of posts spaced 4′, 6′, or 8′ OC. Posts may be solid or engineered products, and should be no less than 4″ × 4″ nominal size. Post size is determined by the load it must support and by wall height. A long post without lateral bracing may buckle. As the post buckles, the top plate sags. Buckling can be reduced by using a wider post. A properly sized post supports the weight and keeps the top plate straight. **See Figure 65-6.**

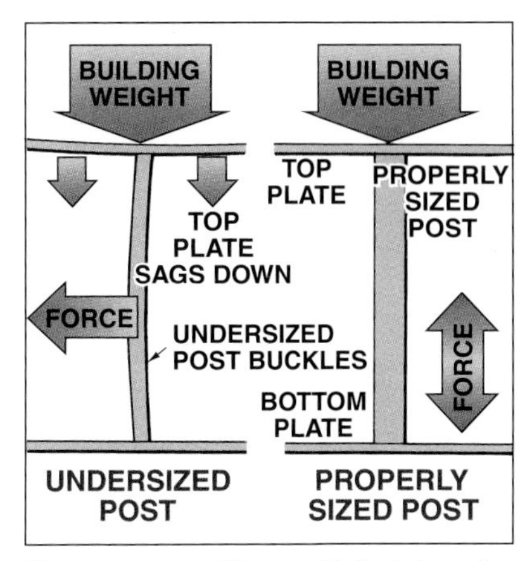

Figure 65-6. *The wall height of a post-and-beam building and the load it must support determine the size of the post that should be used. An undersized post may buckle.*

One method of constructing post-and-beam walls is to nail the posts to a bottom and top plate. **See Figure 65-7.** Another method is to anchor the post bottom directly to the floor with a metal bracket. Posts that butt directly to beams should be reinforced with a strap or bracket. **See Figure 65-8.**

Spaces between perimeter wall posts can be filled in using conventional framing and sheathing methods, or *structural insulated panels (SIPs)* can be used. The exterior surfaces are finished with the same materials as used for wood-framed construction.

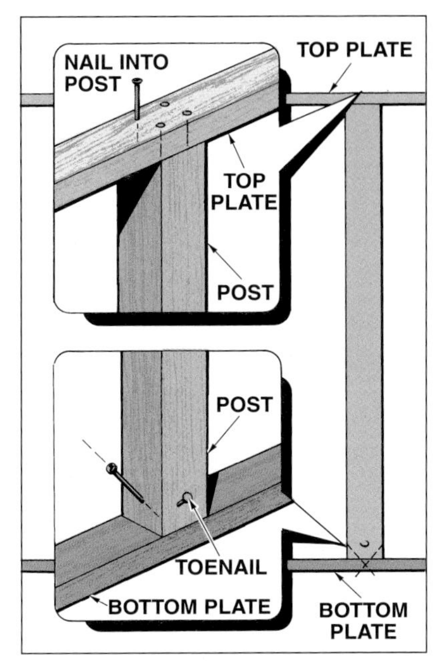

Figure 65-7. *Wall posts may be nailed to the top and bottom plates.*

Framing between Posts. When framing between posts, the studs may be held back the thickness of the exterior sheathing. When the sheathing is applied, the outside face of the post remains visible. The studs may be aligned flush with the post, in which case the post will be covered when sheathing is applied. **See Figure 65-9.**

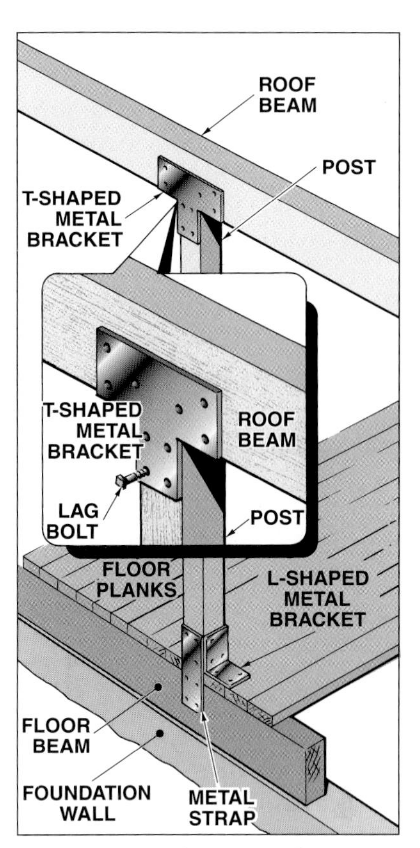

Figure 65-8. *Brackets or metal straps are used to fasten the posts directly to the floor and to the beam at the top of the posts.*

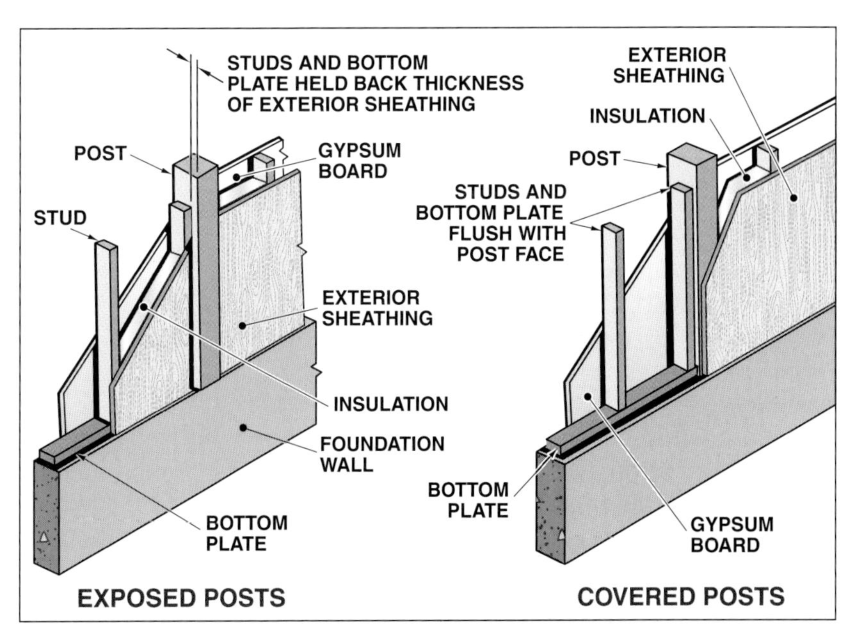

Figure 65-9. *Studs placed between posts are either held back the thickness of the exterior sheathing or are flush with the face of the posts.*

Gypsum board or other interior wall finish material is applied to the interior side of the wall. The space between the exterior sheathing and interior wall finish material is filled with insulation.

Kolbe & Kolbe Millwork Co., Inc.

Figure 65-10. *Exposed roof beams and plank decking materials are characteristic of post-and-beam construction.*

Post-and-Beam Roofs

The main structural members of a post-and-beam roof are the exposed ridge beam, roof beams, and planks or panels used for the roof covering. **See Figure 65-10.** This type of combined roof-ceiling design, called a *cathedral roof,* is also used over buildings that have conventional wood-framed walls.

Post-and-Beam Roof Designs. The basic post-and-beam roof designs are *longitudinal* and *transverse.* **See Figure 65-11.** A good grade of appearance lumber should be installed for roof beams. Solid lumber and engineered wood products, such as glulam or laminated veneer lumber (LVL), are commonly used. Roof trusses may also be used for transverse roofs.

Longitudinal roof beams run the full length of the building and are supported by posts at each end. In larger buildings, it may be necessary to provide intermediate supports between the ends.

Transverse roof beams extend from the exterior walls to the ridge beam. One end of each roof beam rests over a post in the exterior wall. The other end butts against or rests on top of the ridge beam as shown in **Figure 65-12.**

The procedure for calculating the length of a transverse beam and marking the angle cuts is basically the same as for common rafters. Reviews of these layout methods as they apply to transverse beams are shown in **Figures 65-13** and **65-14.**

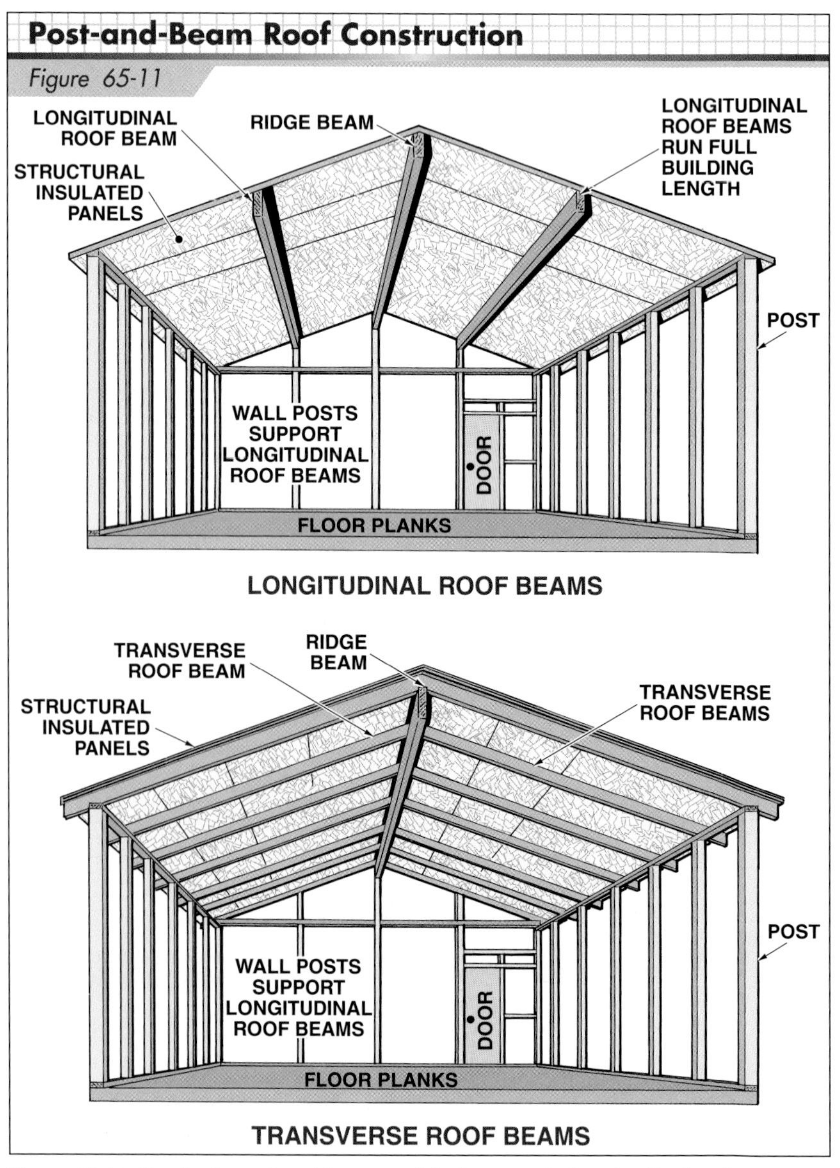

Figure 65-11. *Two basic post-and-beam roof designs are the longitudinal and transverse designs.*

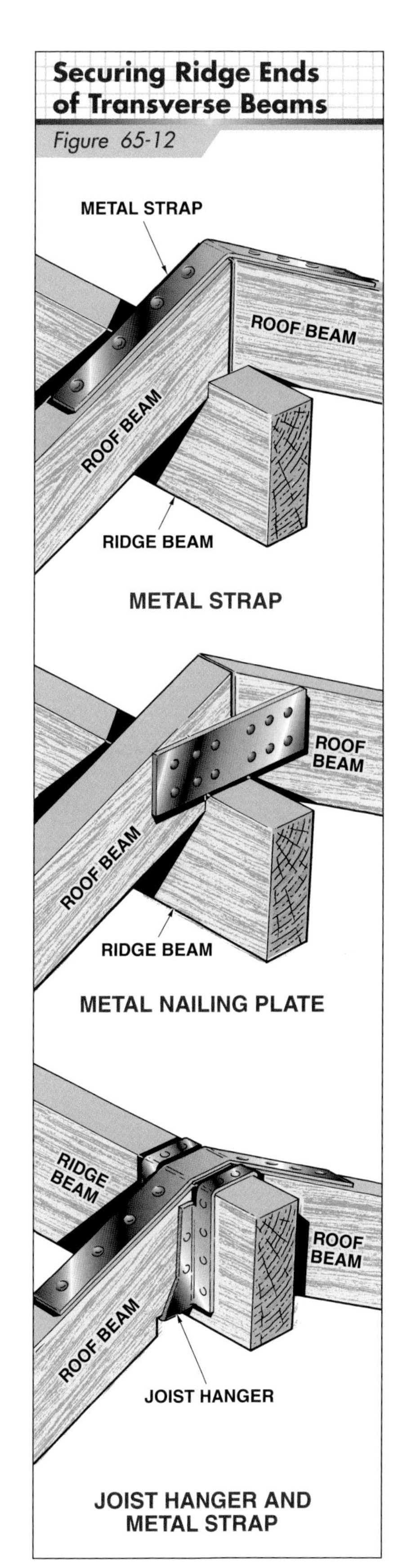

Figure 65-12. *The ridge ends of transverse beams are secured to the ridge beam with metal straps, metal nailing plates, or joist hangers.*

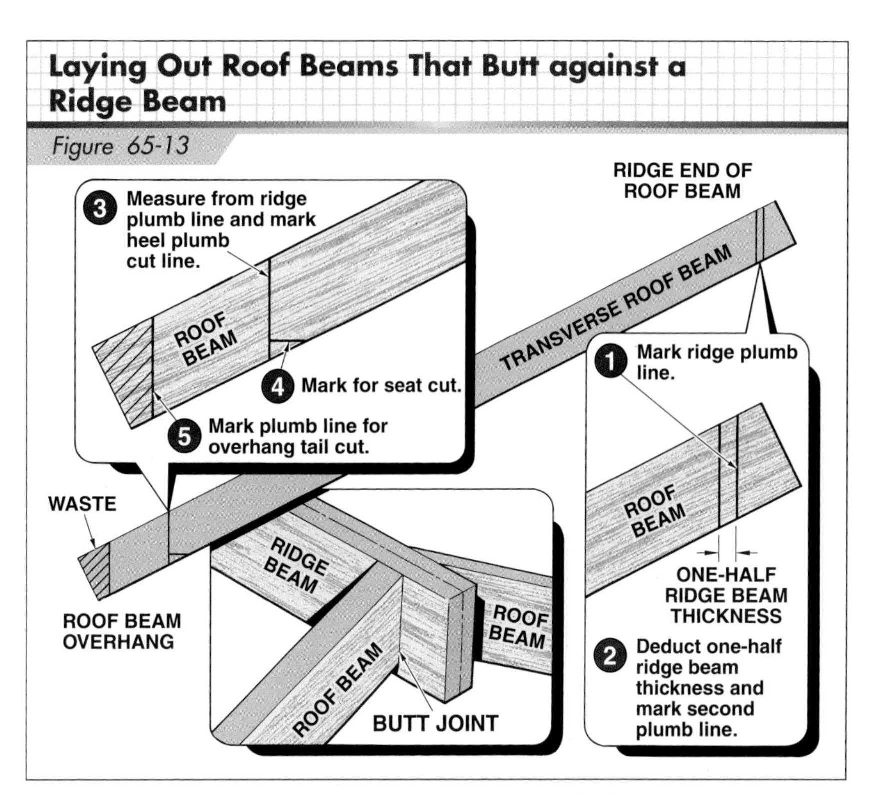

Figure 65-13. *Transverse roof beams may butt against a ridge beam.*

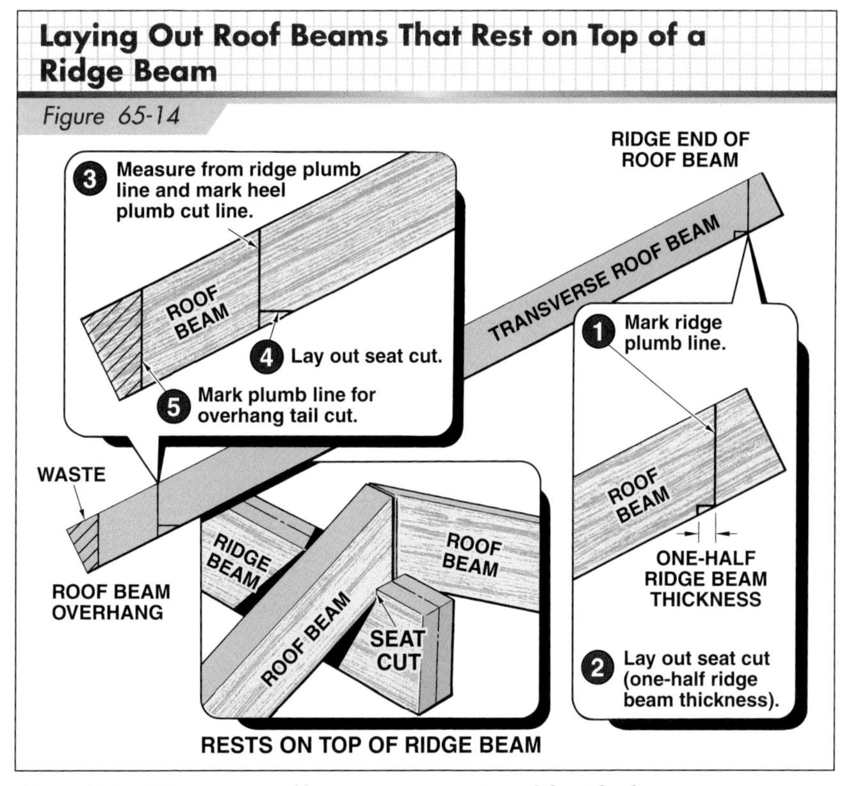

Figure 65-14. *Transverse roof beams may rest on top of the ridge beam.*

Roof Covering. Tongue-and-groove planks have been traditionally used for roof decking. The planks are available in 1½″, 2½″, and 3½″ thicknesses. **See Figure 65-15.** Another traditional material for roof covering has been 1⅛″ thick tongue-and-groove plywood panels. If the underside of roof decking is to be exposed, insulation must be placed on top of the deck. A standard procedure is to first place a vapor barrier and then rigid insulation. **See Figure 65-16.** After the insulation is installed, a variety of standard roof finishes can be applied.

Structural insulated panels, similar to those used for walls, have gained wide acceptance as a post-and-beam roof covering. SIPs offer the advantage of sheathing and insulation.

Constructing Post-and-Beam Buildings

Figure 65-17 shows a procedure for constructing a post-and-beam framework. However, the procedure may vary depending on the design of a particular structure.

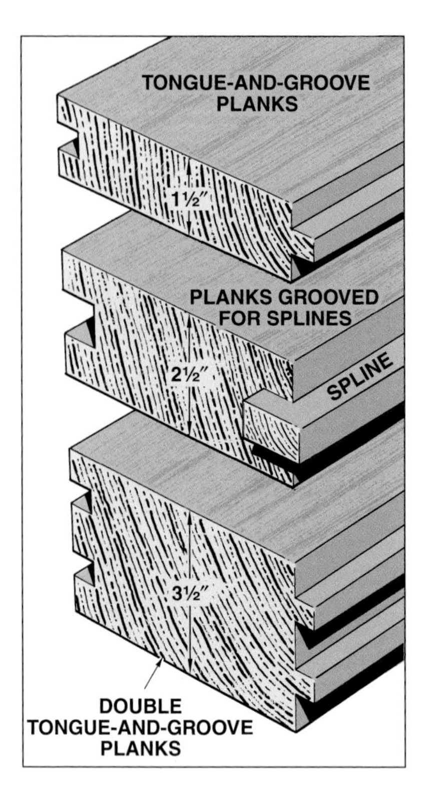

Figure 65-15. *Roof planks are 1½", 2½", or 3½" thick.*

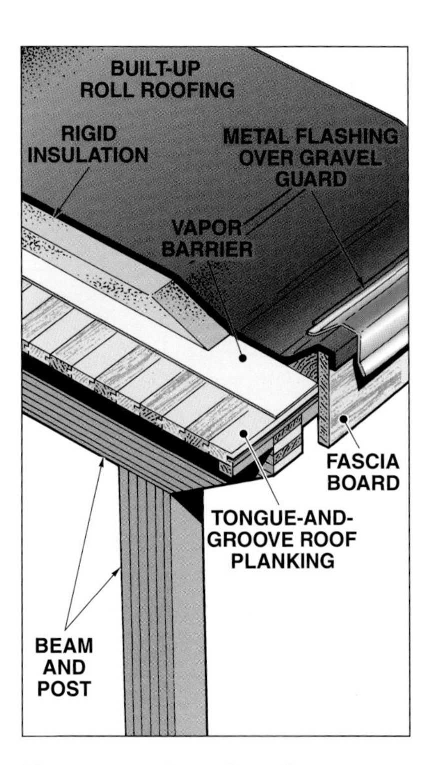

Figure 65-16. *Typical insulation over a post-and-beam roof consists of a vapor barrier and rigid insulation placed under the roofing material.*

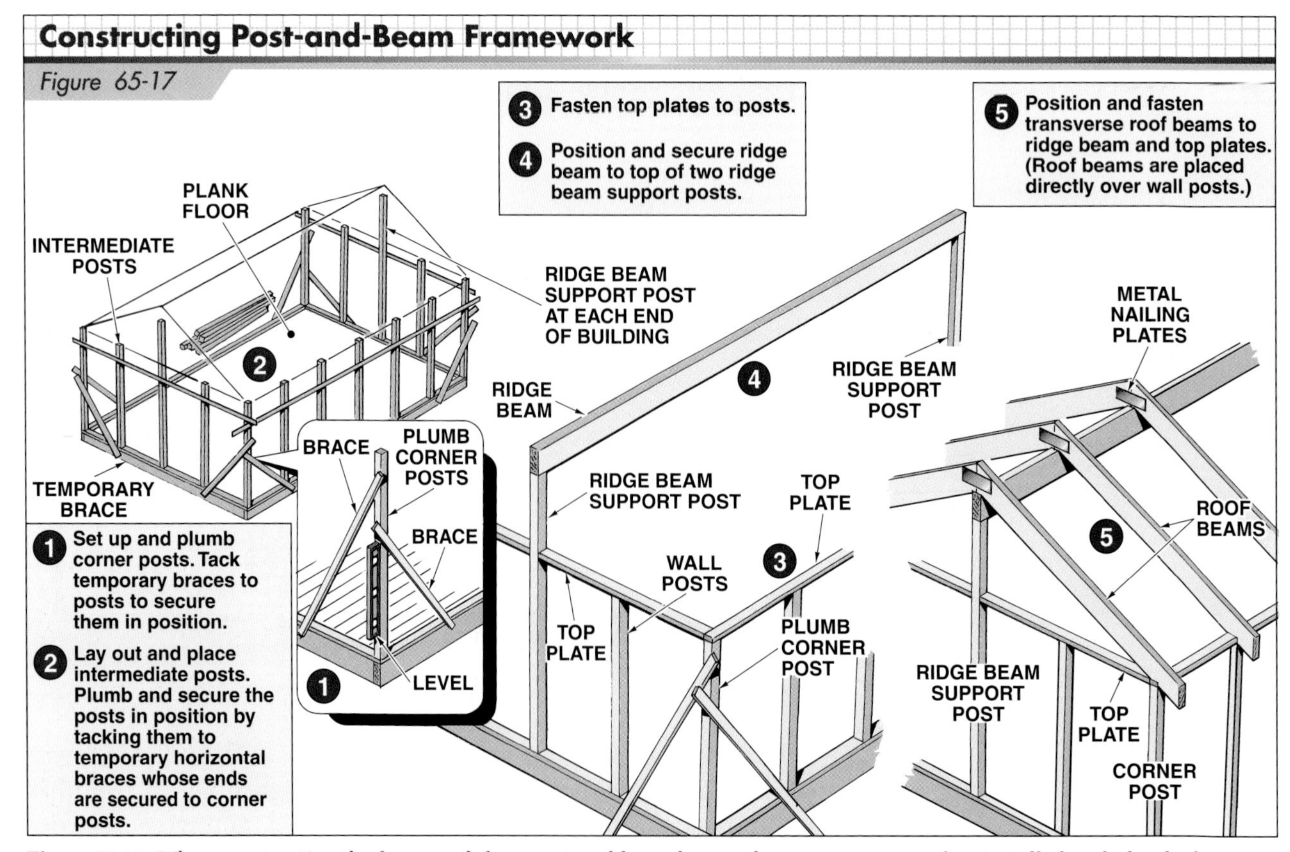

Figure 65-17. *When constructing the framework for a post-and-beam house, the corner posts are first installed and plumbed.*

Interior partitions are erected after the outside wall and roof construction has been completed. Standard framing procedures are used to erect interior partitions. The tops of partitions placed across the width of the house must be sloped to the shape of the roof. **Figure 65-18** shows two methods for constructing interior partitions. In one method, the lower wall section is framed on the floor and raised into place. Studs for the top section are cut to length and placed in position over studs in the lower section. In another method, the interior partition is framed with full-length studs running from the bottom plate to a top plate nailed below the roof. One or two rows of blocking may be required with this method.

Post-and-beam buildings are usually framed with tongue-and-groove decking and insulated with rigid foam insulation. Since the interior face of the roof deck is exposed, a vapor barrier is placed on top of the deck and at least two-thirds of the total insulation value of the roof must be above the vapor barrier.

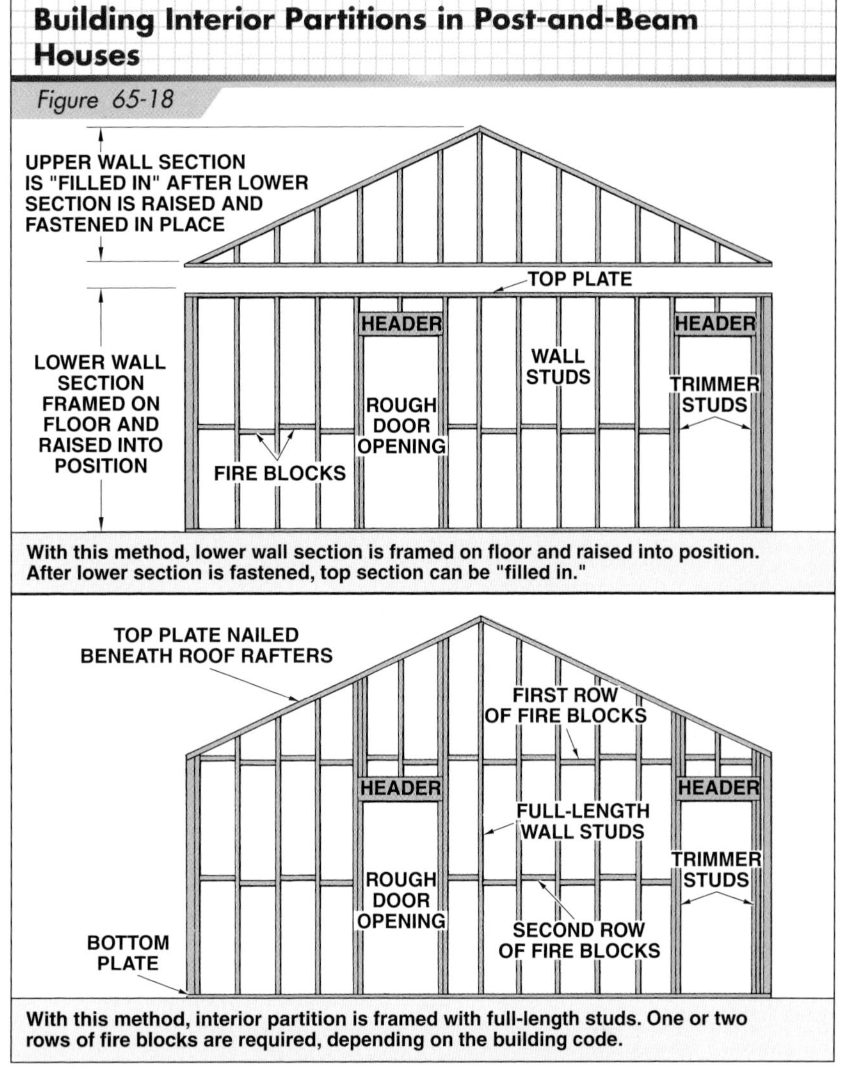

Figure 65-18. *Interior partitions in an exposed-roof post-and-beam house may be built as one or two sections.*

Classic Wood & Beam

Timber frame walls and other components are transported to the job site by truck.

Glulam timbers must be treated with care during storage at a job site. Place the glulam timbers on a firm, level surface to avoid warpage. Support glulam timbers with blocks to provide adequate and uniform support. If a covered storage area is not available, the timbers should be blocked well off the ground at a well-draining location. Glulam timbers should be separated with stickers arranged vertically over the block supports so air can circulate around all sides of each member. Long members should be stored on edge.

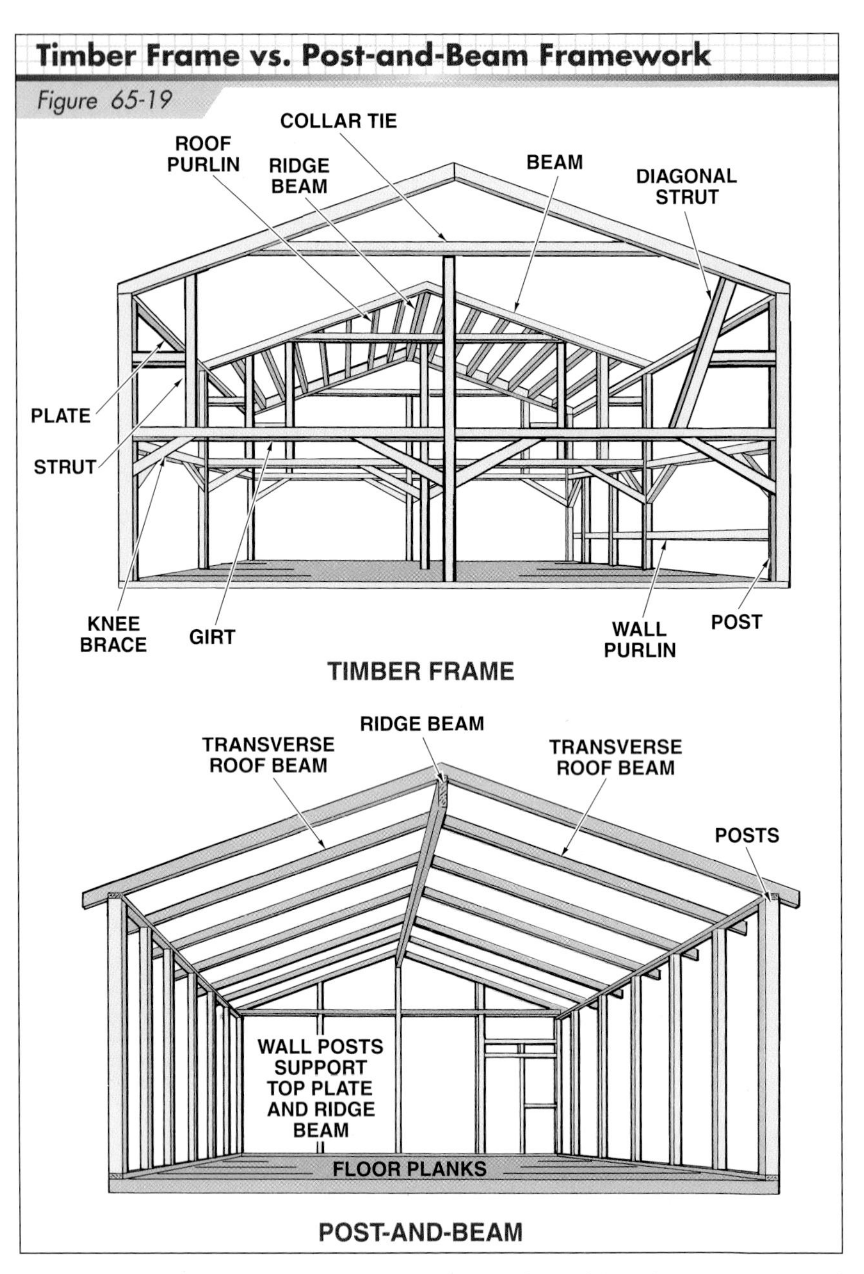

Figure 65-19. *The framework of a timber frame building differs from a conventional post-and-beam building.*

TIMBER FRAME CONSTRUCTION

Post-and-beam and timber frame houses appear very similar in that they both provide viewing of exposed beams and permit large open spaces within the building. In post-and-beam construction, vertical loads are carried by beams that rest on posts. In timber framing, vertical loads are conveyed to the posts through *bents.* A bent is a structural, interconnected system of timbers contained in a wall. The framework for a small timber frame building is shown in **Figure 65-19.** Several bent designs are shown in **Figure 65-20.**

Classic Wood & Beam

The top plate and posts must be leveled or plumbed, respectively, to ensure a properly constructed building.

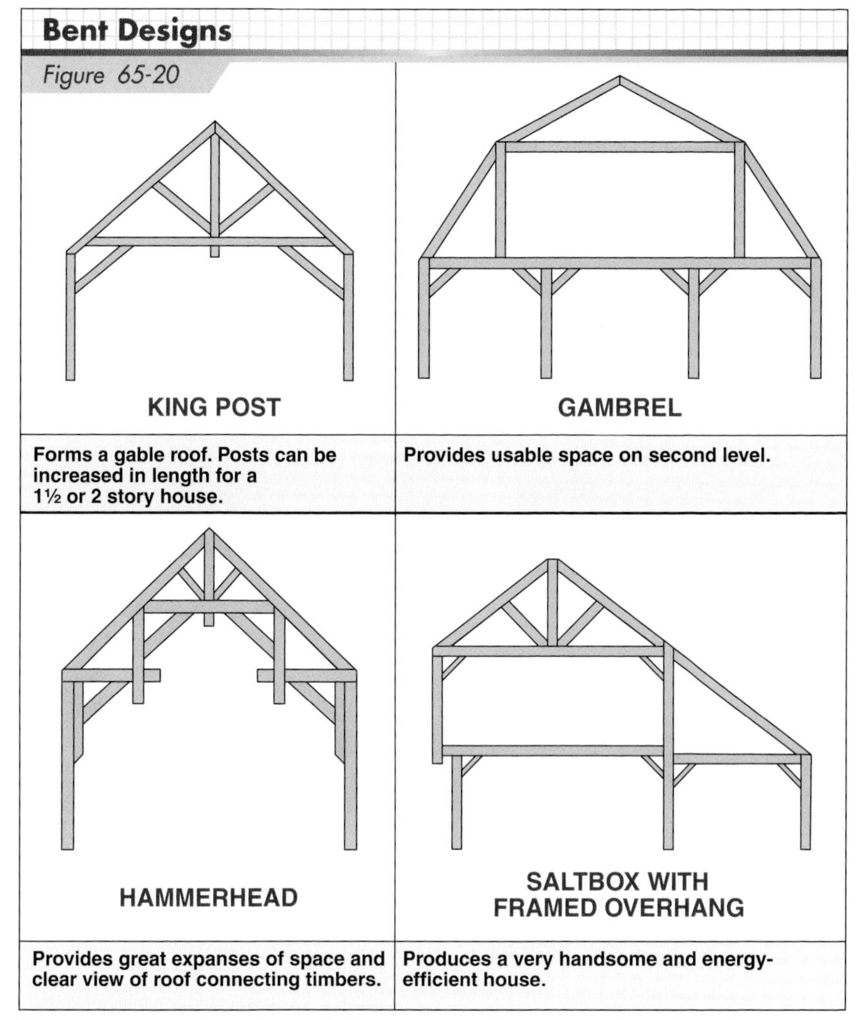

Figure 65-20. *A variety of bent designs are available for timber frame buildings.*

Another difference between post-and-beam and timber frame construction is the means by which the timbers are connected to one another. Joints in timber frame construction are connected using a variety of wood joints such as mortise and tenon, wood peg, and wedge. **See Figure 65-21.**

Due to the complex joinery involved, timber frame buildings are typically erected by highly skilled carpenters who are specially trained in timber frame construction. Timber frame walls are constructed in a horizontal position at the job site. If a builder has the proper facilities, the walls can be fabricated in a shop and transported by truck to the job site. **See Figure 65-22.** Larger walls are raised and set in place by crane. Wall and roof construction for timber frame houses is similar to post-and-beam houses.

The American Institute of Timber Construction (AITC) is the national technical trade association of the structural glued laminated (glulam) timber industry. AITC represents glued laminated timber manufacturers in the United States, in addition to a number of installers, suppliers, sales representatives, engineers, architects, designers, and researchers.

POST-FRAME CONSTRUCTION

Post-frame construction consists of wood roof trusses or rafters connected to vertical posts or timbers. Post-frame construction is used for small, medium, and large commercial and agricultural structures. In the past, post-frame construction was referred to as *pole construction* since round poles were used as posts.

Classic Wood & Beam

Concealed connectors in post-and-beam structures may be tightened with a ratchet and socket.

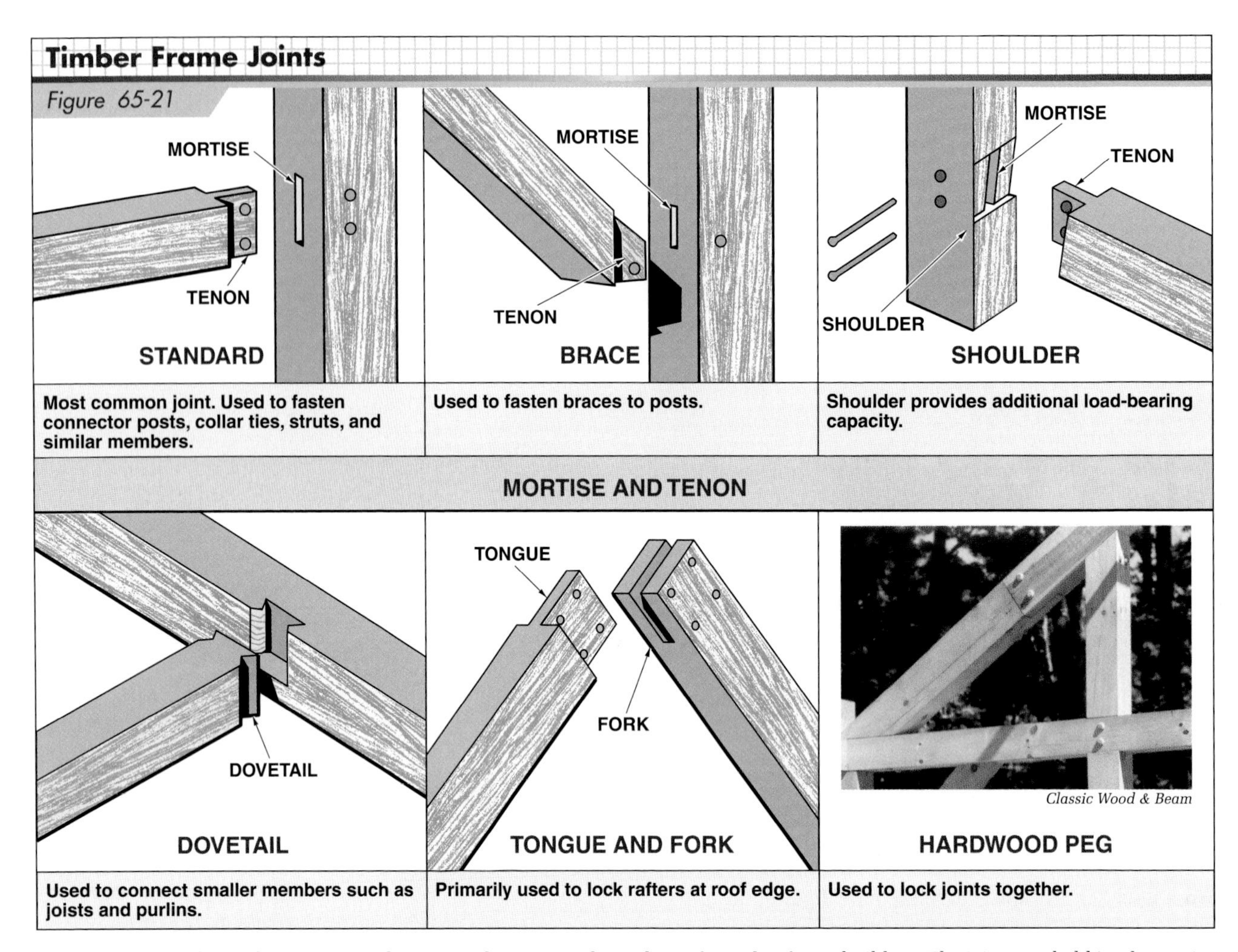

Figure 65-21. *Traditional joinery is used to secure the structural members of a timber frame building. The joints are held in place using hardwood pegs that are glued and driven into holes in the members.*

Classic Wood & Beam

Figure 65-22. *Timber frame walls can be constructed in a shop and transported to the job site by truck.*

Large-dimension timbers have an inherent fire resistance. Wood burns slowly — approximately 0.024″/min. The char on the wood surface as it burns protects and insulates the unburnt wood.

Modern post-frame structures use square pressure-treated posts as their primary support. The square post faces provide the bearing surface for other support members. The pressure-treated posts are typically embedded in concrete, although they may be buried in the ground or anchored to a concrete foundation. **See Figure 65-23.** Other framing members include roof purlins and wall *girts,* which support roof and wall coverings. The most common roof and wall covering for post-frame buildings is corrugated steel, which is commonly prepainted. Other wall coverings include exterior plywood and wood siding. Shingles and wood sheathing are also common on post-frame buildings.

Constructing Post-Frame Buildings

Post-frame buildings are commonly erected by specialized field crews employed by the post-frame building manufacturer. A common post-frame construction method is as follows:

1. Using an earth auger, bore holes for the posts. Holes should be a minimum of 48″ deep, and must extend below the frost line.
2. Deposit concrete in the holes to a depth of approximately 6″ and allow it to set sufficiently.
3. Place the pressure-treated poles in the holes, resting on top of the 6″ concrete pad. Plumb and brace the post and deposit concrete around the post until the hole is filled to the top.
4. If required, erect slab forms so that concrete can be placed for a floor slab. (For some projects, a floor slab is not required.) The concrete slab may also be placed at the completion of the building.
5. Construct the basic framework for the structure and install roof and wall coverings.

Post-Frame Framework. Horizontal girts, spaced 48″ OC, are attached to the posts using nails or lag screws. Door and window openings are framed where required. Roof trusses are set in place using a crane.

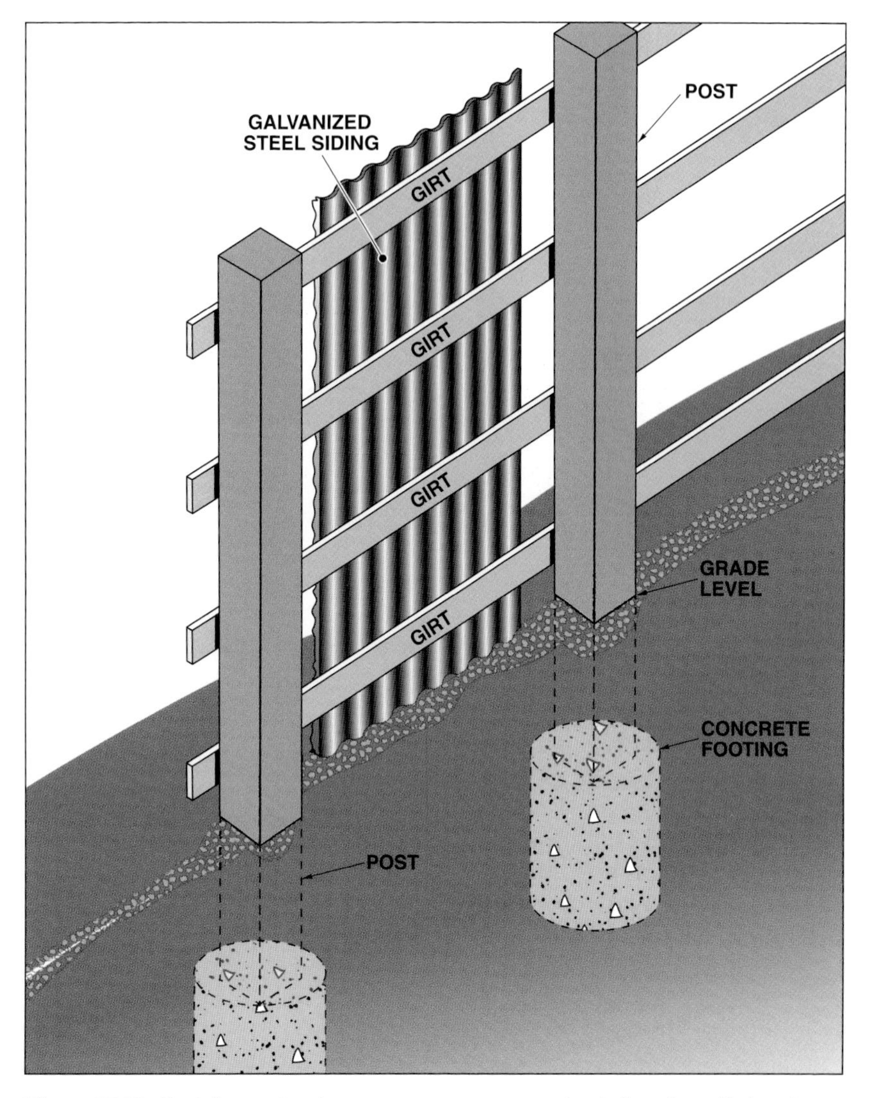

Figure 65-23. *Post-frame structures use square pressure-treated posts as their primary support. Girts support the wall coverings and tie the posts together.*

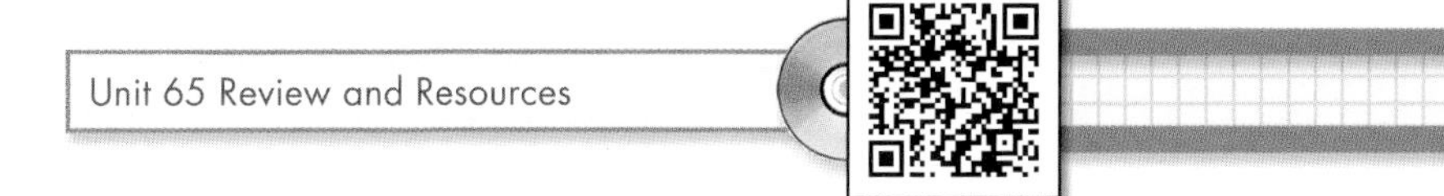

UNIT 66 Heavy Timber Construction

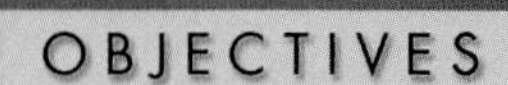

OBJECTIVES

1. List advantages of heavy timber construction.
2. List advantages of glulam timber design.
3. Describe the design of glulam timbers.
4. Define deflection and camber in a glulam beam.
5. List and describe the information given by the APA trademark on a glulam timber.
6. Discuss common types of connectors used with structural components.
7. Describe methods of heavy timber construction using glulams.
8. Discuss methods for fabricating timber domes.

Heavy timber construction is similar to post-and-beam construction, but is applied to larger structures. Commercial buildings, factories, bridges, warehouses, and sports facilities are erected using heavy timber construction methods. **See Figure 66-1.**

Timber construction was common in North America from the colonial period until the early 1900s, when there was a sharp decline in its use due to more productive balloon and platform framing methods. Heavy timber construction re-emerged in the 1980s but was faced with the diminishing availability of old-growth trees with the necessary cross sections and density for producing large-size timbers. Heavy timber construction now employs the use of engineered wood products such as glued laminated (glulam) timbers.

Timber roof trusses can be used on heavy timber structures provided the truss members meet minimum size requirements. Select a fastener and fastener arrangement that provides a pleasing appearance for trusses exposed to view.

FIRE AND SEISMIC RESISTANCE

Two important advantages of heavy timber construction are the fire resistance of the timbers and the ability of a heavy timber structure to resist seismic forces.

Heavy timbers maintain their structural integrity when subjected to the intense heat of fires in temperatures ranging from 1290°F to 1650°F. Unprotected steel beams exposed to a similar amount of heat will buckle and twist, causing the collapse of supporting walls, roof, and floors of a building. Large timbers burn slowly, at approximately 0.024″ per minute. The *char* formed on the wood surface as the timber burns helps to protect the wood below the charred surface. The unburned portion of a thick timber will retain 85% to 90% of its strength and continue to support its assigned loads.

Heavy timber structures have more seismic and wind resistance than most other wood-frame structures. Due to its

APA—The Engineered Wood Association

Glulam timbers are stronger than comparable sizes of solid dimension lumber.

flexible nature, a timber framework can absorb some movement without any breakage and distribute the stress over an entire structure.

APA—The Engineered Wood Association

Western Wood Structures, Inc.

APA—The Engineered Wood Association

Figure 66-1. *Large structures are frequently erected with heavy timber frames.*

Engineering design codes in North America provide design criteria for metallic connections used in heavy timber construction but do not provide specifications for joint connections. Joint connection performance is directly affected by workmanship.

BUILDING WITH GLULAM TIMBER

One of the most important factors in revitalizing heavy timber construction has been the development of glulam timbers. Glulam timbers are stronger than solid dimension lumber of equal size and can span greater distances without intermediate supports. Glulam offers the advantage and flexibility of creating a variety of shapes that retain their structural values.

Glulam Timber Design

Glulam is manufactured from wood laminations (lams) that are bonded together with adhesive. Smaller trees harvested from second- and third-growth forests are commonly used for the lams. Douglas fir, larch, and southern yellow pine are most commonly used to manufacture glulam timbers, although other lumber species are also acceptable.

Unlike plywood panels, in which each veneer is placed at a 90° angle to the adjoining layers, the grain of each lam runs parallel and in the same direction as the length of the glulam timber. Common thicknesses of individual lams are 1⅜″ and 1½″, with widths ranging from 2½″ to 10″. Wider glulam timbers can be special ordered. End and edge jointing makes it possible to produce longer and wider timbers than sizes normally available in solid lumber. **See Figure 66-2.**

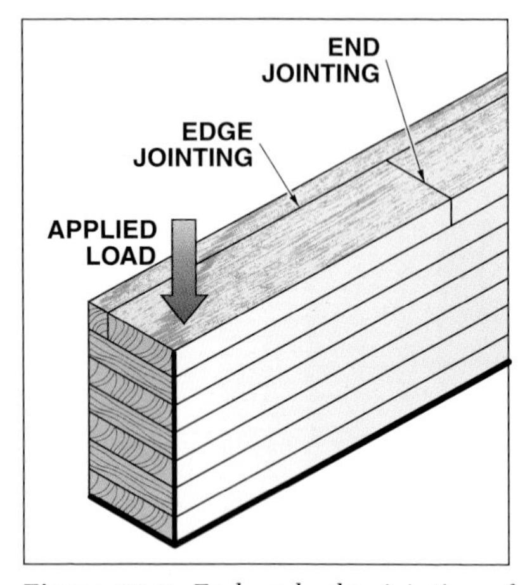

Figure 66-2. *End and edge jointing of lams makes it possible to produce longer and wider timbers than sizes available in solid lumber. Note the grain orientation of the lams.*

Glulam timbers are designed for different load stresses. Stresses placed upon a glulam beam vary depending on the clear span and the load the beam must support. The maximum load stresses occur at the top and bottom of a beam; therefore, the strongest lams are placed at the top and bottom of a beam.

The species and grade of an individual lam determine its strength. To make efficient use of the lumber, high-grade lams are placed in areas of the beam subjected to greater stresses, and lower grade lams are placed in lower stress zones. For maximum strength, glulam beams are usually placed with the width of the lams facing the applied load *(horizontally laminated members)*. A table comparing solid wood and glulam dimensions for heavy timber components is shown in **Figure 66-3.**

MINIMUM DIMENSIONS OF STRUCTURAL COMPONENTS IN HEAVY TIMBER CONSTRUCTION

Supported Assembly	Structural Component	Solid Wood*	Glulam*
Roofs only	Columns	5½ x 7½	5⅛ x 7½
	Arches supported on top of walls or abutments	3½ x 5½	3⅛ x 6
	Beams, girders, and trusses	3½ x 5½	3⅛ x 6
	Arches supported at or near floor line	5½ x 5½	5 x 6
Floors or floors and roofs	Columns	7½ x 7½	7 x 7½
	Beams, girders, trusses, and arches	5½ x 9½ or 7½ x 7½	5⅛ x 9 or 7 x 7½

* in in.

Figure 66-3. *A smaller glulam timber is equal in strength to a larger solid wood timber.*

Fiber-Reinforced Glulam Timbers

Fiber-reinforced glulam timbers are a recent development in glulam technology that provides a significant increase in strength of glulam members. Thin sheets of high-strength, resin-encased fibers, less than 1⁄10″ thick, are placed between the lams. Although fiber-reinforced glulam timbers may be used for all heavy timber operations, the timbers are primarily used where the weight of the timber (dead load) is a concern, such as in bridge and dome construction.

Deflection and Camber

Deflection is the movement of a structural component resulting from stress produced by a heavy applied load. Deflection is measured at a midpoint between the supports at the ends of a beam or at the end of a cantilevered beam. **See Figure 66-4.** Deflection is an important consideration in beam design as too much deflection will cause a beam to sag and weaken the beam, as well as affect its appearance.

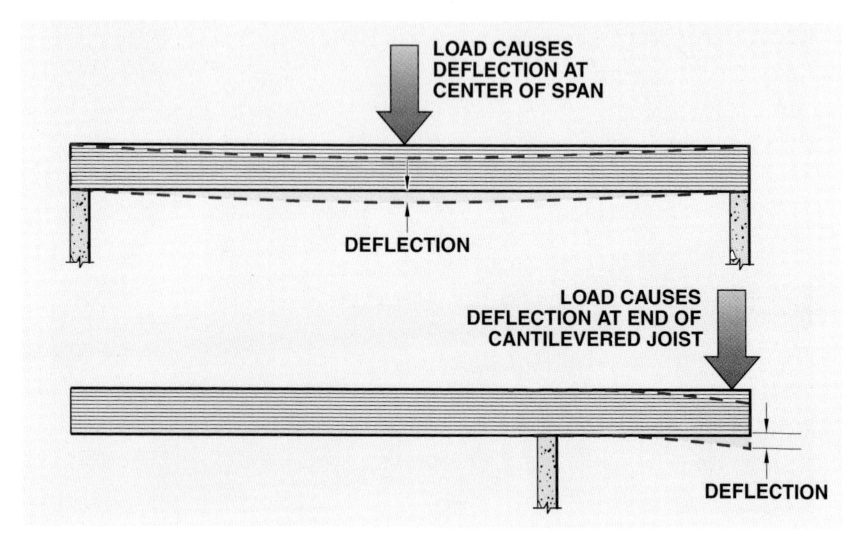

Figure 66-4. *Deflection is the movement of a structural component resulting from stress produced by a heavy applied load.*

APA—The Engineered Wood Association

Heavy timber trusses allow great distances to be spanned without intermediate supports.

To prevent beam deflection resulting from an applied load, an engineer must calculate the proper amount of *camber* in a beam design. Camber is the slight upward curve in a structural member. **See Figure 66-5.** Solid wood structural members, such as joists, have a natural crown that must be placed upward when installed. Glulam timbers can be produced with the exact required camber to compensate for deflection of the member under a load. Glulam manufacturers generally recommend that glulam roof beams be cambered for 1½ times the deflection caused by the dead load to ensure the beam will not sag after many years of supporting loads. Some conditions may require more or less camber.

The total load-carrying capacity of a solid or glulam member is based on the weight of the beam, plus the live and dead loads in pounds per square foot (PSF). **Figure 66-6** provides a table for determining glulam floor, roof beam, and purlin sizes.

Balanced and Unbalanced Beams. The most critical area of a glulam beam in controlling strength of the beam is the outermost tension zone. Glulam timbers are manufactured as *unbalanced* or *balanced* members. In unbalanced beams, the lams used on the tension side of the beam are of a higher grade than the lams used on the compression side. **See Figure 66-7.** This lam arrangement allows a more efficient use of lumber resources while maintaining the strength of the beam. Since unbalanced beams have different bending stresses assigned to the tension and compression zones, the beams must be installed accordingly. The top of an unbalanced beam is clearly stamped with the word "TOP." Unbalanced beams are usually used for simple span applications.

Balanced beams are symmetrical in lumber quality. Balanced beams are primarily used for cantilevered or continuous span applications where either the top or bottom of the member may be stressed in tension. Balanced beams may also be used for simple span applications, although unbalanced beams are best suited for these applications.

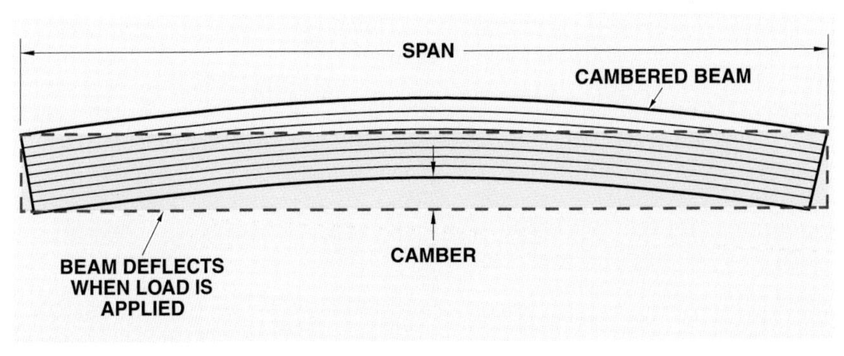

Figure 66-5. *Camber is the slight upward curve in a structural member and is designed to compensate for deflection of the member under a load.*

Glulam Trademark

A trademark is applied to one or more surfaces of a glulam timber. An example of an APA trademark is shown in **Figure 66-8.** Descriptions of the trademark elements are as follows:

- *Structural use.* Glulam timbers are used for a variety of structural applications, which are indicated with a letter. "B" indicates a simple span bending member, "C" indicates a compression member, "T" indicates a tension member, and "CB" indicates a continuous or cantilevered span bending member.
- *Glulam appearance classification.* Glulam appearance classifications (grades) are Framing, Industrial, Architectural, and Premium.
- *Mill number.* Identifies the mill where the glulam member was manufactured. Each mill where glulam members are manufactured is assigned its own mill number.
- *Applicable laminating specification.* Refers to the American Plywood Association

Engineered Wood System (APA EWS) specification EWS Y117, *Glulam Design Properties and Layup Combinations.*

- *Wood species.* Western wood (WW) species are indicated.
- *Applicable ANSI standard.* The American National Standards Institute (ANSI) publication ANSI A190.1-1992, *Structural Glued Laminated Timber,* is the ANSI standard for glulam beams.
- *Structural grade designation.* The APA EWS 24F-1.8E designation is a glulam grade commonly used in residential applications.

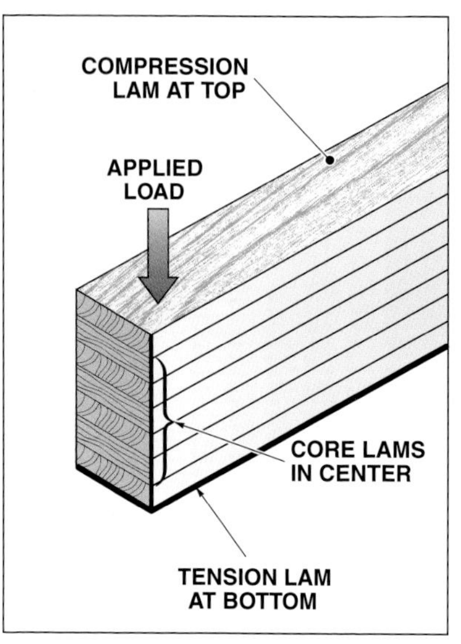

Figure 66-7. *In unbalanced beams, tension lams are of a higher grade than compression lams.*

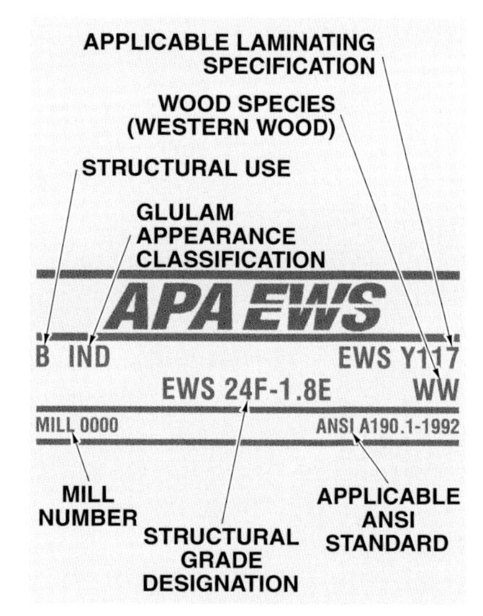

APA—The Engineered Wood Association

Figure 66-8. *Trademarks identifying glulam timbers are stamped on the timbers.*

GLULAM FLOOR, ROOF BEAM, AND PURLIN SIZES*

Span†	Spacing†	Total Load-Carrying Capacity— Roof Beams and Purlins						Floor Beams 50 PSF
		30 PSF	35 PSF	40 PSF	45 PSF	50 PSF	55 PSF	
12	6	3⅛ × 7½	3⅛ × 7½	3⅛ × 7½	3⅛ × 7½	3⅛ × 7½	3⅛ × 7½	3⅛ × 9
	8	3⅛ × 7½	3⅛ × 7½	3⅛ × 7½	3⅛ × 9	3⅛ × 9	3⅛ × 9	3⅛ × 10½
	10	3⅛ × 7½	3⅛ × 9	3⅛ × 9	3⅛ × 9	3⅛ × 9	3⅛ × 10½	3⅛ × 10½
	12	3⅛ × 9	3⅛ × 9	3⅛ × 9	3⅛ × 9	3⅛ × 10½	3⅛ × 10½	3⅛ × 12
16	8	3⅛ × 9	3⅛ × 10½	3⅛ × 10½	3⅛ × 10½	3⅛ × 12	3⅛ × 12	3⅛ × 13½
	12	3⅛ × 10½	3⅛ × 12	3⅛ × 12	3⅛ × 12	3⅛ × 13½	3⅛ × 13½	3⅛ × 15
	14	3⅛ × 12	3⅛ × 12	3⅛ × 13½	3⅛ × 13½	3⅛ × 15	3⅛ × 15	3⅛ × 15
	16	3⅛ × 12	3⅛ × 13½	3⅛ × 13½	3⅛ × 15	3⅛ × 15	3⅛ × 16½	3⅛ × 16½
20	8	3⅛ × 12	3⅛ × 12	3⅛ × 13½	3⅛ × 13½	3⅛ × 13½	3⅛ × 15	3⅛ × 16½
	12	3⅛ × 13½	3⅛ × 13½	3⅛ × 15	3⅛ × 15	3⅛ × 16½	5⅛ × 13½	5⅛ × 15
	16	3⅛ × 15	3⅛ × 16½	3⅛ × 16½	3⅛ × 18	5⅛ × 15	5⅛ × 16½	5⅛ × 16½
	20	3⅛ × 16½	3⅛ × 18	5⅛ × 15	5⅛ × 16½	5⅛ × 16½	5⅛ × 18	5⅛ × 18
24	8	3⅛ × 13½	3⅛ × 15	3⅛ × 15	3⅛ × 16½	3⅛ × 16½	3⅛ × 16½	5⅛ × 16½
	12	3⅛ × 16½	3⅛ × 16½	3⅛ × 18	5⅛ × 15	5⅛ × 16½	5⅛ × 16½	5⅛ × 18
	16	3⅛ × 18	5⅛ × 16½	5⅛ × 16½	5⅛ × 18	5⅛ × 18	5⅛ × 19½	5⅛ × 21
	20	5⅛ × 16½	5⅛ × 16½	5⅛ × 18	5⅛ × 19½	5⅛ × 21	5⅛ × 21	5⅛ × 22½
28	6	3⅛ × 16½	3⅛ × 16½	3⅛ × 18	3⅛ × 18	5⅛ × 16½	5⅛ × 16½	5⅛ × 19½
	8	3⅛ × 18	3⅛ × 16½	5⅛ × 18	5⅛ × 18	5⅛ × 18	5⅛ × 19½	5⅛ × 21
	10	5⅛ × 18	5⅛ × 18	5⅛ × 19½	5⅛ × 21	5⅛ × 19½	5⅛ × 22½	5⅛ × 24
	12	5⅛ × 18	5⅛ × 19½	5⅛ × 21	5⅛ × 22½	5⅛ × 21	5⅛ × 25½	5⅛ × 25½
32	8	3⅛ × 18	5⅛ × 16½	5⅛ × 18	5⅛ × 18	5⅛ × 18	5⅛ × 19½	5⅛ × 21
	12	5⅛ × 18	5⅛ × 19½	5⅛ × 19½	5⅛ × 21	5⅛ × 21	5⅛ × 22½	5⅛ × 24
	16	5⅛ × 19½	5⅛ × 21	5⅛ × 22½	5⅛ × 24	5⅛ × 25½	5⅛ × 25½	5⅛ × 27
	20	5⅛ × 21	5⅛ × 22½	5⅛ × 25½	5⅛ × 27	5⅛ × 28½	5⅛ × 30	6¾ × 27
40	12	5⅛ × 22½	5⅛ × 24	5⅛ × 24	5⅛ × 25½	5⅛ × 27	6¾ × 25½	6¾ × 28½
	16	5⅛ × 24	5⅛ × 27	5⅛ × 28½	5⅛ × 30	6¾ × 28½	6¾ × 28½	6¾ × 31½
	20	5⅛ × 27	5⅛ × 30	6¾ × 28½	6¾ × 30	6¾ × 31½	6¾ × 33	6¾ × 33
	24	5⅛ × 30	6¾ × 28½	6¾ × 30	6¾ × 33	6¾ × 34½	6¾ × 36	6¾ × 37½
48	12	5⅛ × 27	5⅛ × 28½	5⅛ × 30	6¾ × 28½	6¾ × 28½	6¾ × 30	6¾ × 33
	16	5⅛ × 30	6¾ × 28½	6¾ × 30	6¾ × 31½	6¾ × 34½	6¾ × 36	6¾ × 37½
	20	6¾ × 28½	6¾ × 31½	6¾ × 34½	6¾ × 36	8¾ × 33	8¾ × 36	8¾ × 36
	24	6¾ × 31½	6¾ × 34½	8¾ × 33	8¾ × 34½	8¾ × 37½	8¾ × 39	8¾ × 40½

* in in.
† in ft

Figure 66-6. *The total load-carrying capacity of a glulam member is based on the weight of the beam and the live and dead loads. For example, a 3⅛″× 16½″ beam is required to support a 50 PSF rated roof with a 20′ span and with the beams spaced 12′ OC.*

STEEL CONNECTORS

Concealed connectors are available for tying together structural components. Concealed connectors are corrosion-resistant steel plates that fit into machined slots in the timbers and are secured with small-diameter pins. **See Figure 66-9.** *Steel connectors* are the primary means for tying together the structural components of heavy timber construction. A variety of steel connectors are available and each type is designed for a specific use. Most steel connectors are bolted to the surface of the timbers. **See Figure 66-10.**

Pocopson Industries, Inc.

Figure 66-9. *Corrosion-resistant steel plates fit into machined slots in the timbers. Small-diameter steel pins are inserted into predrilled holes and through the steel plates.*

Expansion and Contraction

Wood expands and contracts due to changes in its moisture content. The expansion and contraction primarily occur across the wood grain. The movement caused by expansion and contraction not only weakens a connection, but may also cause splits running along the grain where connector hardware is installed in the wood members. **See Figure 66-11.**

Moisture penetration is also a concern at timber connections. Moisture accumulation can result in wood decay, especially in untreated wood. Steel connectors subjected to water penetration are designed with *weep holes* (drain holes) that allow moisture to drain from the connectors.

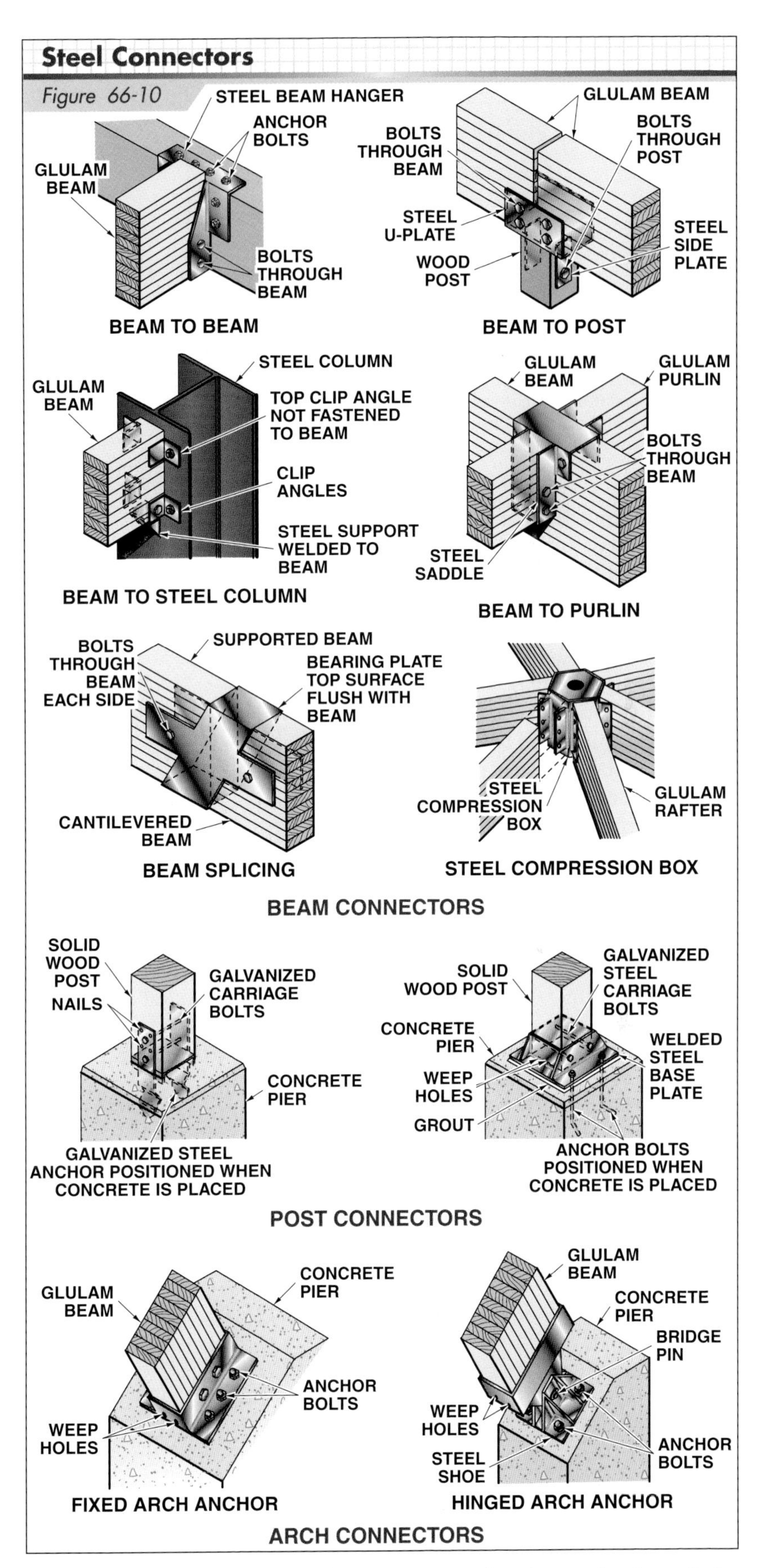

Figure 66-10. *Steel connectors are the primary means for tying together heavy timber components.*

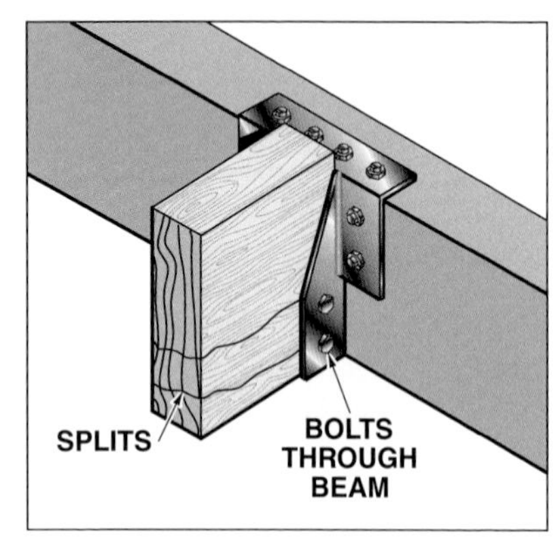

Figure 66-11. *Movement caused by expansion and contraction can cause splits along the grain at the fastening points of wood members.*

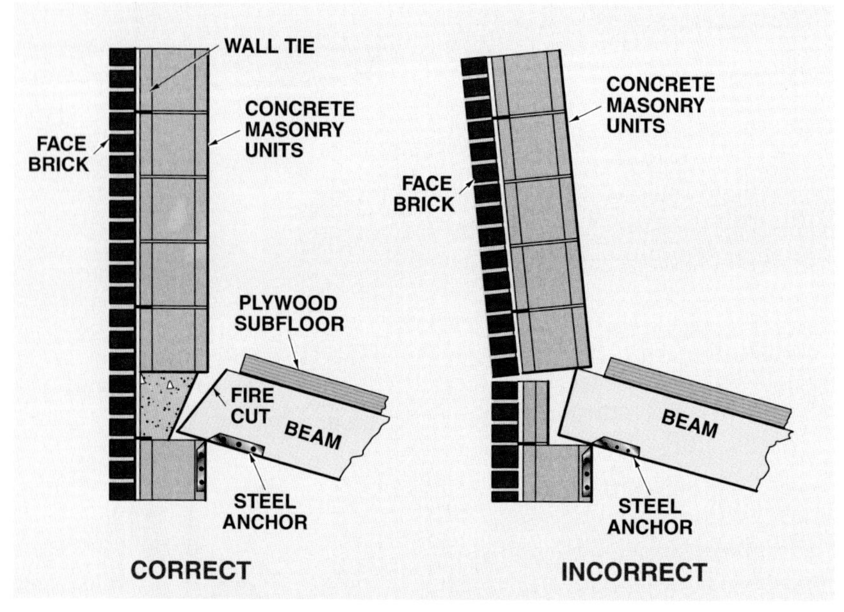

Figure 66-12. *The ends of timbers placed in masonry wall pockets should be cut at an angle (fire cut) to prevent damage to the wall in case of timber collapse from a prolonged fire.*

Timber Connections to Concrete and Masonry. Buildings may be designed with an internal timber framework and exterior concrete or masonry walls. Pockets are provided in the walls to support the ends of the timbers. A ½″ gap is allowed between the bearing ends of the timber and concrete or masonry wall because of possible water accumulation in the pocket. In addition, steel connectors should have drain holes. Ends of timbers placed in concrete or masonry walls must be cut at an angle *(fire cut)* to allow the beam to collapse without damaging the wall if the timbers burn through due to a prolonged fire. **See Figure 66-12.**

HEAVY TIMBER CONSTRUCTION

In glulam timber construction, beams and *purlins* are used for the main structural support. Beams support the ends of the purlins.

The sequence of glulam or solid timber construction is similar to post-and-beam construction described in Unit 65. Posts are plumbed and braced as they are being set in place. The beams are then positioned and steel connectors are attached. On residential and light construction projects, the timbers are raised into place with a forklift or material lift. Large construction projects may require a crane to raise the timbers. Walls are then enclosed with panel sheathing or structural insulated panels (SIPs).

Glulam beams support a plank roof for this building. Note that the bottoms of the posts are elevated from the concrete using post anchors.

Roof Construction

Various roofing materials and methods are used for heavy timber buildings. Roof frames may be covered with tongue-and-groove (T&G) planks or plywood. Roof finish is commonly built-up roofing except for truss roofs, which are often covered with metal or one of the various materials used on pitched roofs.

Tongue-and-Groove Planks. T&G planks are a traditional method of covering roof frames and are still popular today. Planks have an attractive appearance when viewed from the interior of a building.

Tongue-and-groove planks are available in solid wood and glulam designs in a variety of thicknesses and widths. **See Figure 66-13.**

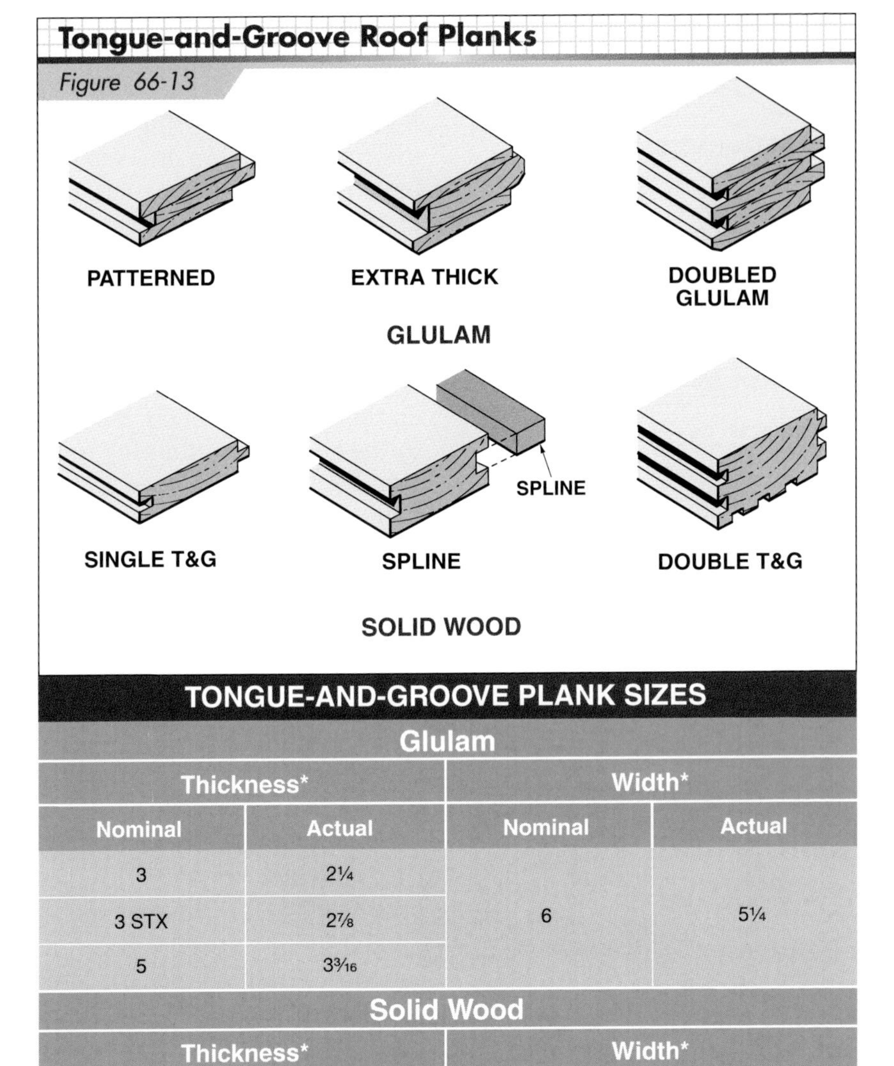

TONGUE-AND-GROOVE PLANK SIZES

Glulam			
Thickness*		Width*	
Nominal	Actual	Nominal	Actual
3	2¼	6	5¼
3 STX	2⅞		
5	3³⁄₁₆		

Solid Wood			
Thickness*		Width*	
2	1½	5, 6, 8, 10, 12	4, 5, 6¾, 8¾, 10¾
3	2½	6	5¼
4	3½	6	5¼

* in in.

Figure 66-13. *Glulam and solid wood roof planks are available in different thicknesses and widths.*

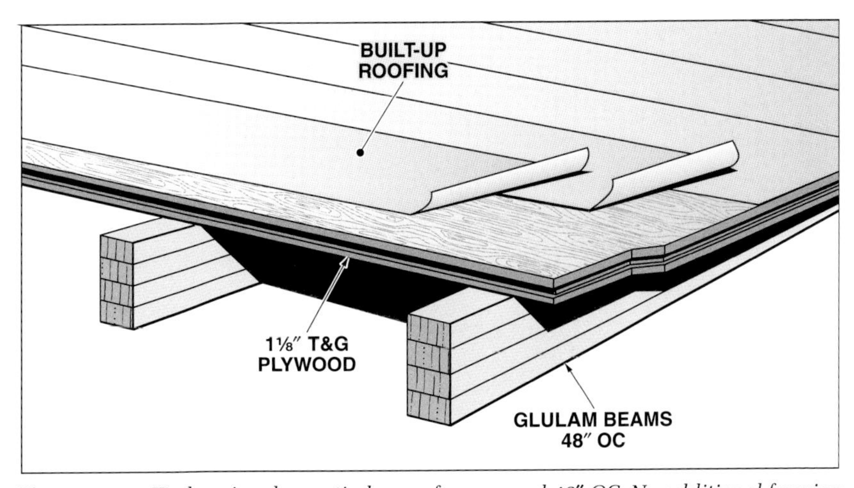

Figure 66-14. *Timbers in a heavy timber roof are spaced 48″ OC. No additional framing is required between the timbers.*

Heavy Timber Roofs. Heavy solid or glulam timbers, spaced 48″ OC, are used without additional framing and are covered with plywood or planks. **See Figure 66-14.**

Preframed Panelized Roofs. With preframed panelized roofs, 2 × 4 or 2 × 6 stiffeners (subpurlins) are nailed to the plywood panels before they are installed over the purlins. **See Figure 66-15.** The stiffeners are spaced 16″ or 24″ OC and are fastened to the purlins with metal joist hangers. Preframed panelized roof systems provide additional support for the panels and make it possible to space purlins as far apart as 8′ OC. The most common glulam purlin width used with preframed panelized systems is 2½″, although other widths are available.

Timber Roof Trusses

Timber roof trusses may be fabricated from solid timbers, parallel strand lumber (PSL), or glulam. Glulam members provide superior strength and appearance. Timber roof trusses are similar to lighter trusses used in residential construction.

Heavy timber roof trusses allow great distances to be spanned without intermediate support and are available in a variety of designs. Timber roof trusses can be used with timber frame structures, wood-frame buildings, and buildings with concrete or masonry exterior walls.

Truss Fabrication. The main components of timber trusses are the top and bottom chords and the webs. The components are fastened together with mechanical devices designed to overcome the stresses and weight exerted on and by the heavy timbers. When placed over the supporting walls, individual timber trusses are tied together with horizontal purlins. **See Figure 66-16.**

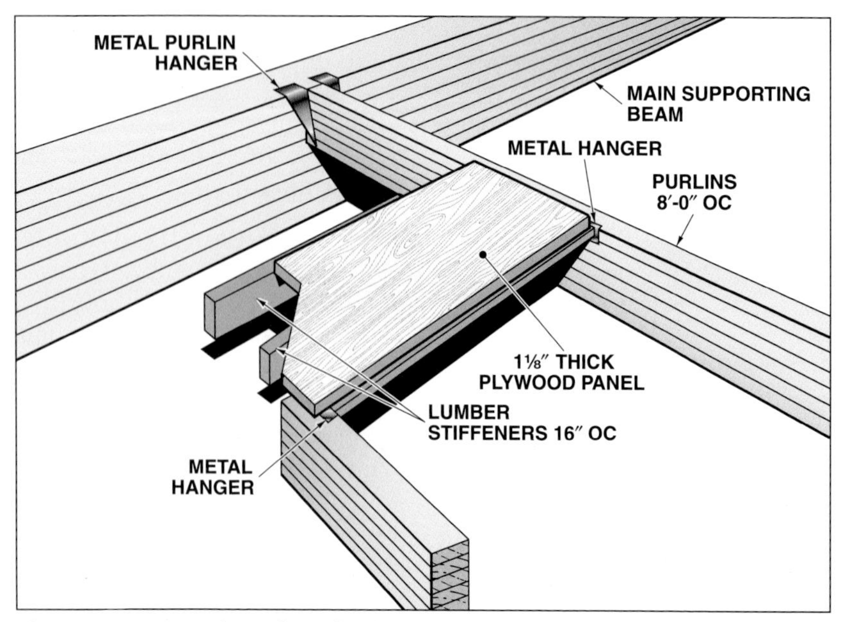

Figure 66-15. *In a preframed panelized roof system, stiffeners are attached to the panels before being placed over the purlins.*

Pocopson Industries, Inc.

Figure 66-16. *Heavy timber trusses are frequently combined with a heavy timber framework.*

In addition to metal plate connectors, *split-ring* and *shear plate connectors* are also used. **See Figure 66-17.** Split-ring connectors are installed in grooves that have been cut into the surfaces of truss members being joined using a special grooving tool. The grooves must be the same size as the split-ring connectors. A bolt passes through the timbers and split-ring connectors. As the nut on the bolt is tightened, the split in the ring allows the ring to be simultaneously tightened against both sides of the groove.

Shear plates are primarily used to fasten wood and steel members together. The shear plate is placed into a cutout in the wood member so it is flush with the surface of the truss member. A bolt passes through the wood and steel members and the shear plate. When tightened, the shear plate prevents any movement at the connection.

Large timbers are impractical for kiln drying since stresses would result from differences in moisture content between the interior and exterior of the timber. Timbers are usually dressed green with a moisture content above 19%.

Timber begins to shrink when its moisture content falls below about 28%. The amount of shrinkage depends on the climatic conditions, especially humidity. Timbers exposed to the outdoors shrink from 1.8% to 2.6% in width and thickness. Timbers used indoors shrink from 2.4% to 3.0% in width and thickness. Timber length is changed insignificantly in either case.

Timber Domes

A *dome* is a curved roof structure which is similar in appearance to one-half of a sphere.

Domes are designed to cover and span medium to large areas without intermediate supports. Historically, domes have served as major roof structures over churches, temples, and mosques. National and many state capitols have domes constructed over high vertical walls. Modern dome applications range from residential structures to commercial and public structures such as gymnasiums, stadiums, pools, ice rinks, and assembly halls. In many cases, domes are not just a roof, but are the entire structure. **See Figure 66-18.** The base of the dome may rest on low walls or footings at ground level.

Western Wood Structures, Inc.

Figure 66-18. *Domes are designed to cover wide span areas.*

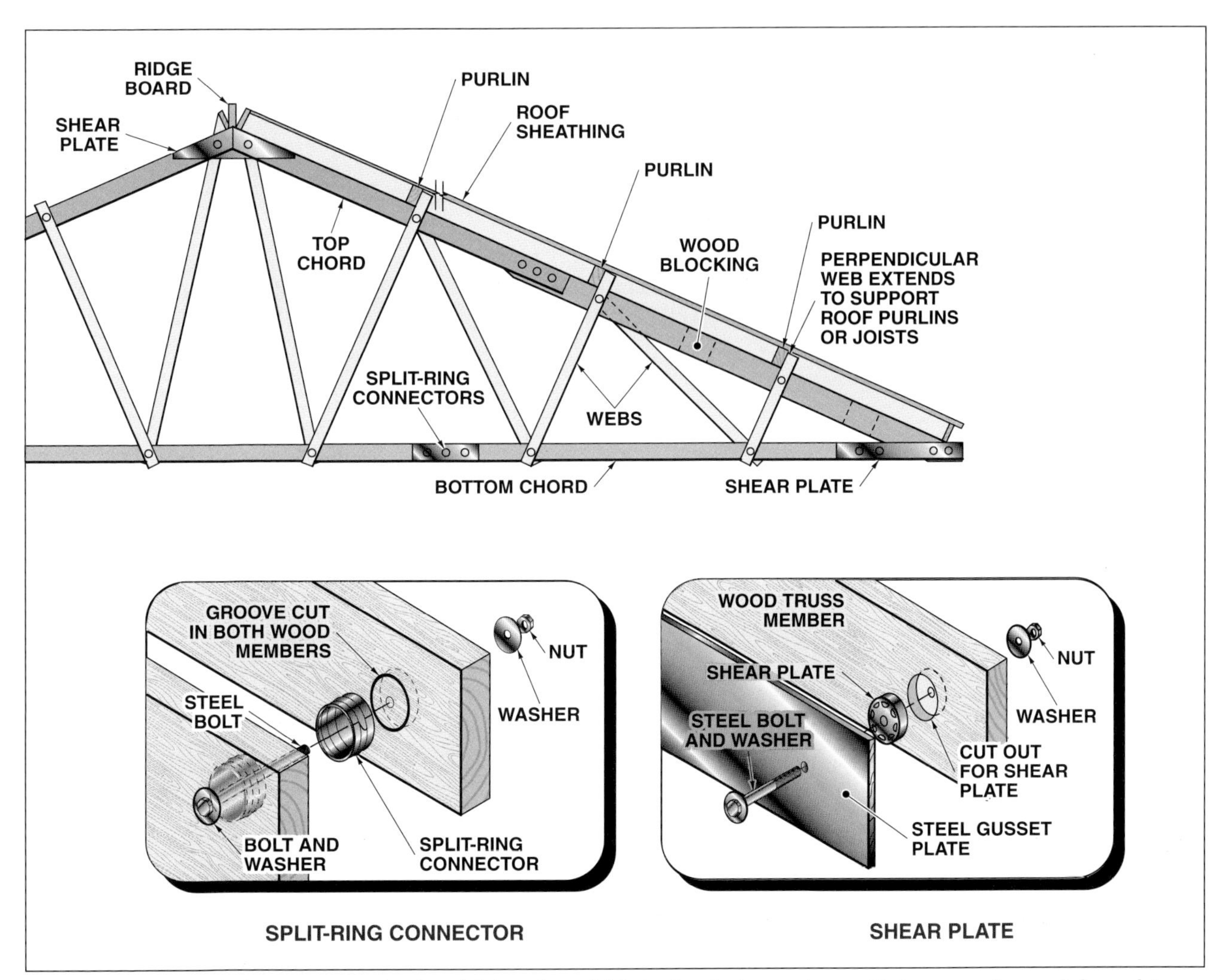

Figure 66-17. *Heavy timber trusses require the addition of special fastening devices such as split-ring connectors and shear plates.*

Glulam timbers are commonly used for dome construction, hence the term *timber domes.* Timber domes are constructed using a grid system of beams and purlins tied together with steel connecting devices and hubs. **See Figure 66-19.** One construction method is to fabricate triangular glulam units on the ground. The units are then lifted into place by crane and connected to steel hubs.

The domes are frequently sheathed with 2″ thick tongue-and-groove planks, although panel systems may also be used. **See Figure 66-20.** Climate control is achieved by installing vinyl-faced fiberglass insulation blankets to the interior surface of the decking. The fiberglass also enhances the acoustics and light reflection.

Western Wood Structures, Inc.

Figure 66-19. *The framework for a dome creates a grid system of beams and purlins tied together with steel connectors and hubs.*

Western Wood Structures, Inc.

Figure 66-20. *Dome roofs are frequently covered with 2″ plank sheathing.*

Unit 66 Review and Resources

SECTION

16

Heavy Concrete Construction

CONTENTS

UNIT 67

Foundation Design for Heavy Construction

OBJECTIVES

1. Explain the principle of piles and how they are placed.
2. Identify common types of piles.
3. Explain the advantages of using micropiles.
4. List and describe common types of caissons and their uses.
5. List and describe common types of spread foundations
6. List and describe the components of wind turbine foundations.
7. Explain the process of installing wind turbine foundations.

The major considerations in designing a foundation for a large concrete building are the size and weight of the building and the soil conditions on the job site. Some soil types can support smaller and lighter concrete buildings with little excavation. Larger and heavier buildings on the same soil type may require deep excavations before forms for foundation walls and footings can be constructed. **See Figure 67-1.**

PILES

Many tall concrete structures are designed with deep basements and heavy column loads. However, because of general soil conditions and the close proximity of existing buildings surrounding the new building, it may not be possible to excavate the entire site area to the depth required to reach load-bearing soil. Under these conditions, *piles* are often used to provide support for the foundation.

Piles are long structural members that penetrate deep into the soil. Piles are placed beneath grade beams that support load-bearing walls. **See Figure 67-2.** Piles may also be grouped and joined together with a concrete pile cap. **See Figure 67-3.** Grouped piles with a concrete cap are necessary when columns provide the main structural support for the building above and when the column load exceeds the capacity of an individual pile.

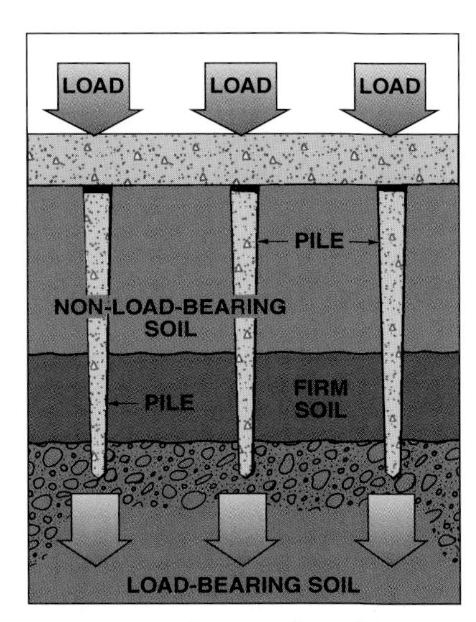

Figure 67-2. *Piles are column-like structural members that carry building loads through non-load-bearing soil to lower levels of bearing soil. Piles receive their loads from columns or grade beams.*

© Case Foundation

Figure 67-1. *Excavations for heavy construction foundations may require tiebacks and bracing to provide lateral support for exposed walls of adjacent buildings.*

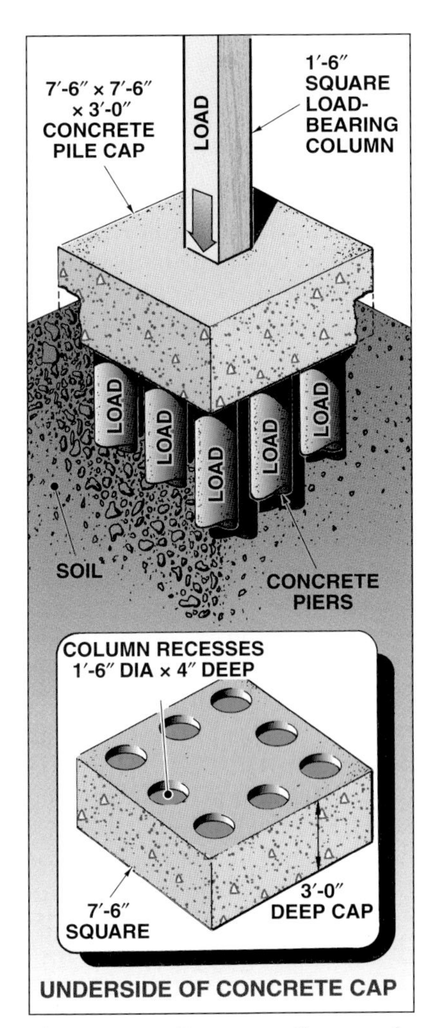

Figure 67-3. *Concrete piles may be grouped under a concrete cap when loads exceed the capacity of an individual pile. A load-bearing column rests on top of the cap.*

Piles are placed into the ground with a pile-driving rig. The rig has a drop, mechanical, or vibratory hammer that directly drives a complete pile or pile casing into the ground. Pile-driving rigs are powered by compressed air, steam, diesel pistons, or hydraulic fluid under pressure. Mobile cranes with special attachments can also be used to drive piles.

Pile Types

The three main types of piles are *bearing, friction,* and *sheet piles.* Piles are made of wood, steel, or concrete. Composite piles have a wood lower section and a concrete upper section.

An "extractor" is used to remove driven piles from the ground.

Bearing Piles. Bearing piles are the most common type of pile used for heavy construction. Bearing piles must penetrate completely through unstable soil layers to the firm, load-bearing soil below.

Friction Piles. Friction piles do not need to penetrate to load-bearing soil. Rather, friction piles only need to reach a point where soil resistance is sufficient to support a portion of the load. A major share of the load is supported by the surrounding soil pressing against the pile. A comparison between bearing and friction piles is shown in **Figure 67-4.**

Sheet Piles. Sheet piles are not intended to carry vertical loads; their primary purpose is to resist horizontal pressure. They are often used to restrain earth around the perimeters of deep excavations.

Wood Piles. Wood piles were used by the Romans over 2000 years ago and are still used today. Douglas fir, southern pine, red pine, western hemlock, and larch are commonly used for wood piles. Wood piles are pressure-treated for protection against decay and deterioration.

Wood piles are light, relatively inexpensive, and available in many areas. Wood piles have an indefinite life expectancy under water, so they are particularly suitable for wharves, docks, and other structures built over water, but are not often used for buildings constructed on dry land.

Concrete Piles. Concrete piles may be *precast* or *cast-in-place.* Precast piles are usually fabricated in a factory and are delivered to a job site, where they are driven into the ground with a pile-driving rig. Precast concrete piles must be properly protected while being driven to prevent damage to the pile. **See Figure 67-5.** A *driving head* is a steel device placed on the head of a pile to receive the blows from the pile-driving rig and protect the pile while it is being driven. A *pile shoe* is attached to a pile tip to prevent the pile from breaking while it is being driven. Some precast concrete pile shapes are shown in **Figure 67-6.**

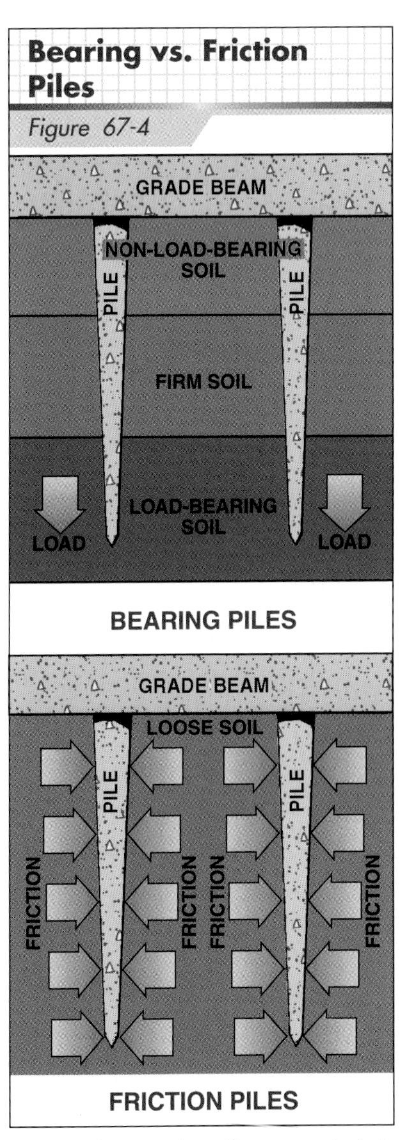

Figure 67-4. *Bearing piles transmit their loads directly to firm, load-bearing soil. Friction piles rely largely on pressure and friction from surrounding soil to help support the load.*

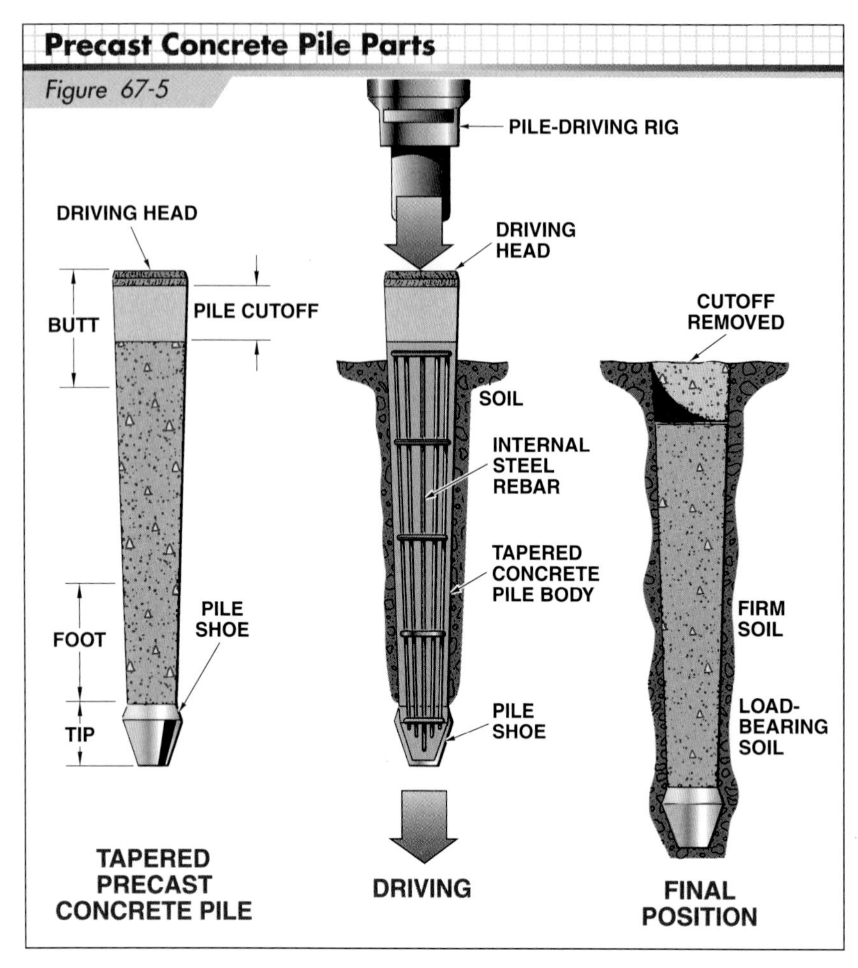

Figure 67-5. *Precast concrete piles are usually prefabricated in a factory. The driving head and pile shoe protect the pile against damage when driving the pile.*

A high-slump concrete mixture is used for cast-in-place piles.

Cast-in-place concrete piles are *shell* or *shell-less* piles. The shell type of pile is made by first driving a steel shell called a *casing* into the ground. Casings range from 20″ to 5′ diameter or larger. As the casing is driven into the ground, soil is removed within the casing with an auger or other special mechanism. Concrete is then placed in the casing, which acts as a form and prevents mud and water from mixing with the concrete. After the concrete hardens, the casing is left in place. Examples of shell concrete piles are shown in **Figure 67-7.**

For shell-less piles, the casing is withdrawn and the hole is filled with concrete. The casing can be completely withdrawn before the concrete is placed or the concrete can be placed as the casing is gradually being withdrawn. In either case, shell-less piles can be used only with firm, cohesive soil.

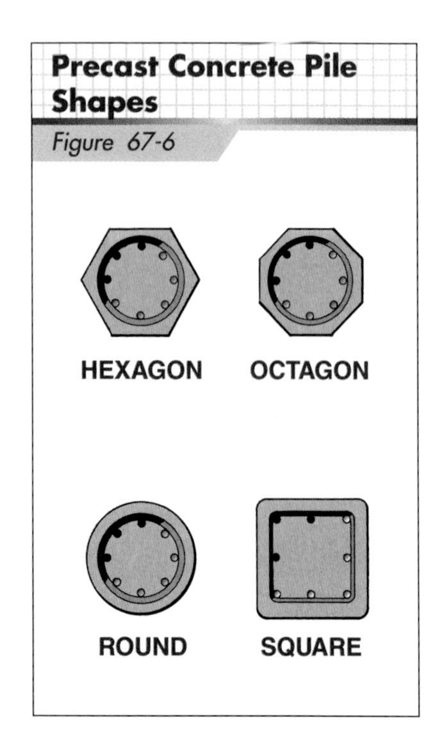

Figure 67-6. *Precast concrete piles are available in several cross-sectional shapes.*

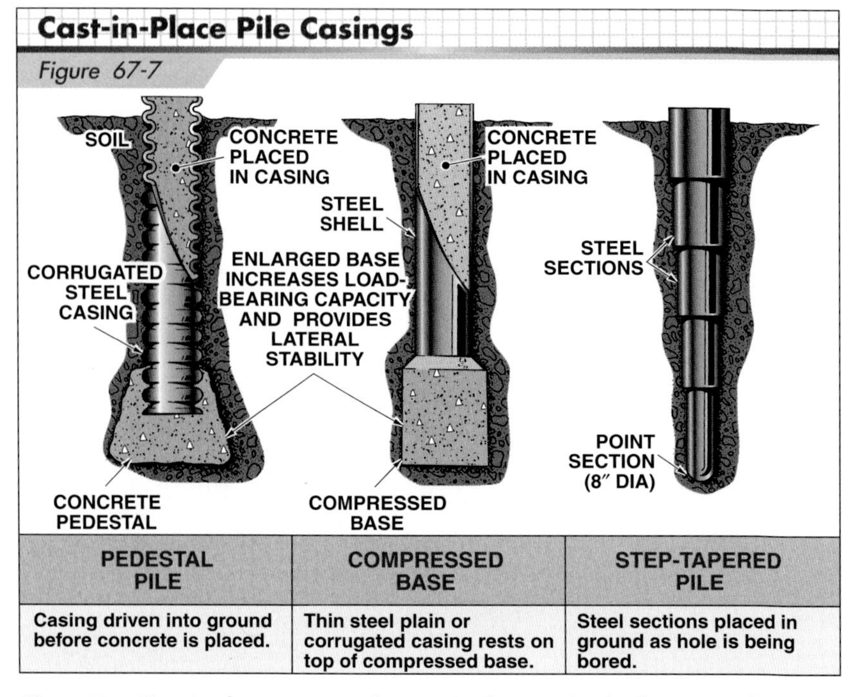

PEDESTAL PILE	COMPRESSED BASE	STEP-TAPERED PILE
Casing driven into ground before concrete is placed.	Thin steel plain or corrugated casing rests on top of compressed base.	Steel sections placed in ground as hole is being bored.

Figure 67-7. *Cast-in-place concrete piles require that a casing be driven into the ground before concrete is placed.*

Steel Piles. Steel piles may be *H-shaped* or *tubular*. H-shaped piles are used with wood lagging to restrain the earth around deep excavations. Tubular piles, also referred to as pipe piles, are normally filled with concrete after they have been driven. **See Figure 67-8.** When driven into the ground, the lower ends of tubular piles are closed off with a steel *boot* to prevent earth from filling the pile. Tubular piles may also be placed in prepared holes drilled by an earth auger.

MICROPILES

Micropiles are a recent development of conventional piles. Originally used in Europe, micropile technology spread to North America in the 1970s. The use of micropiles has grown significantly since the mid-1980s. Micropiles serve many purposes, but their most important functions are foundation support and soil retention and stabilization.

Micropiles, also called *minipiles,* are small diameter (3″ to 12″) drilled and grouted piles. **See Figure 67-9.** They consist of high-strength cement grout and steel reinforcement and have design loads ranging from 3 tons to over 500 tons. Micropiles can extend to depths of 200′. They may be used individually or grouped together. Single or multiple rebar, threaded rod, or post-tensioning strands are used as reinforcement. In some situations the reinforcement may be post-tensioned to increase the strength of the micropiles.

Micropile Installation

Micropiles require drilling machines that are smaller than conventional pile-driving rigs. The installation procedure causes minimal vibration and noise, which produces fewer disturbances to adjacent structures, soil, and the environment. **See Figure 67-10.** A high-pressure grouting machine is used to supply and pump the grout into the micropile hole.

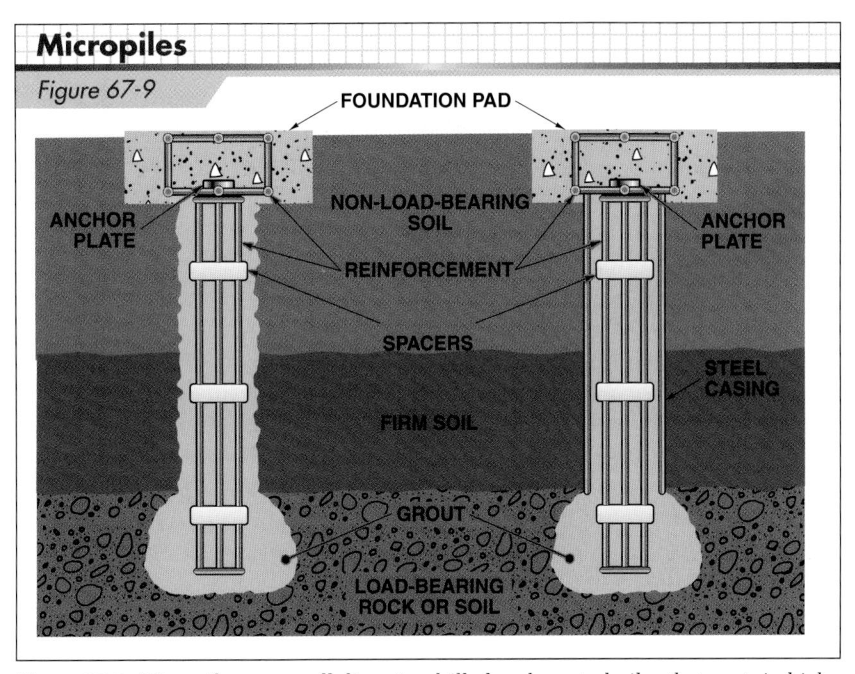

Figure 67-9. *Micropiles are small diameter drilled and grouted piles that contain high-strength cement grout and steel reinforcement.*

© Case Foundation

Figure 67-8. *Tubular piles are driven into the ground and the soil contents are removed with a long auger. The piles are then filled with concrete.*

Con-Tech Systems Ltd.

Figure 67-10. *Micropiles require pile-driving machines that are smaller than conventional pile-driving rigs.*

A typical micropile is installed similar to cast-in-place concrete piles. **See Figure 67-11.** A hole is drilled at the required angle (usually 90° to the ground) over a predetermined point while a steel casing is inserted as needed. Once the target depth is reached, the drill is removed and reinforcement with a tremie attached is placed into the hole.

Then, high-strength grout is pumped into the hole and the tremie is pulled up as grout fills the casing. The casing may be partially withdrawn and more grout introduced under pressure. The casing is then completely removed or cut off at the proper location. An anchor nut or plate is attached to the top of the reinforcement to provide connection between the micropile and the structure above. In some procedures, the reinforcement is introduced after the grout has been pumped.

Micropile Applications

Micropiles are preferred for difficult geological conditions that include the presence of boulders, underground caves, ground shift, slope stabilization, and earth retention. Micropiles can be used for conditions of low headroom and where there is limited access to the job site. They can be installed at any angle below horizontal that requires the same type of equipment used for ground anchor and grouting projects.

Foundation Support. In new construction, micropiles are a good solution for providing additional foundation support over difficult subsurface conditions. The micropiles are driven into the ground. A footing or grade beam can be attached and placed over the micropiles.

Micropiles can also be installed to provide additional support for already existing foundation systems. Holes are drilled through the existing footing to allow the pile holes to be bored below. Additional concrete and reinforcement is placed around the footing. **See Figure 67-12.**

In addition to buildings, micropiles are also used in many different areas of heavy construction. Micropiles are used to support highways, bridges, wharves, wind turbines, and transmission towers. **See Figure 67-13.** Micropiles can be used for new foundations or to underpin already existing structures.

Earth Retention and Slope Stabilization. Micropiles are an effective means to control earth retention where access and right-of-way is limited. One common procedure is to drive a combination of vertical and angled micropiles. This makes it possible to resist both tension and compression loads. The procedure for installing micropiles for slope stabilization is similar to the procedure for earth retention.

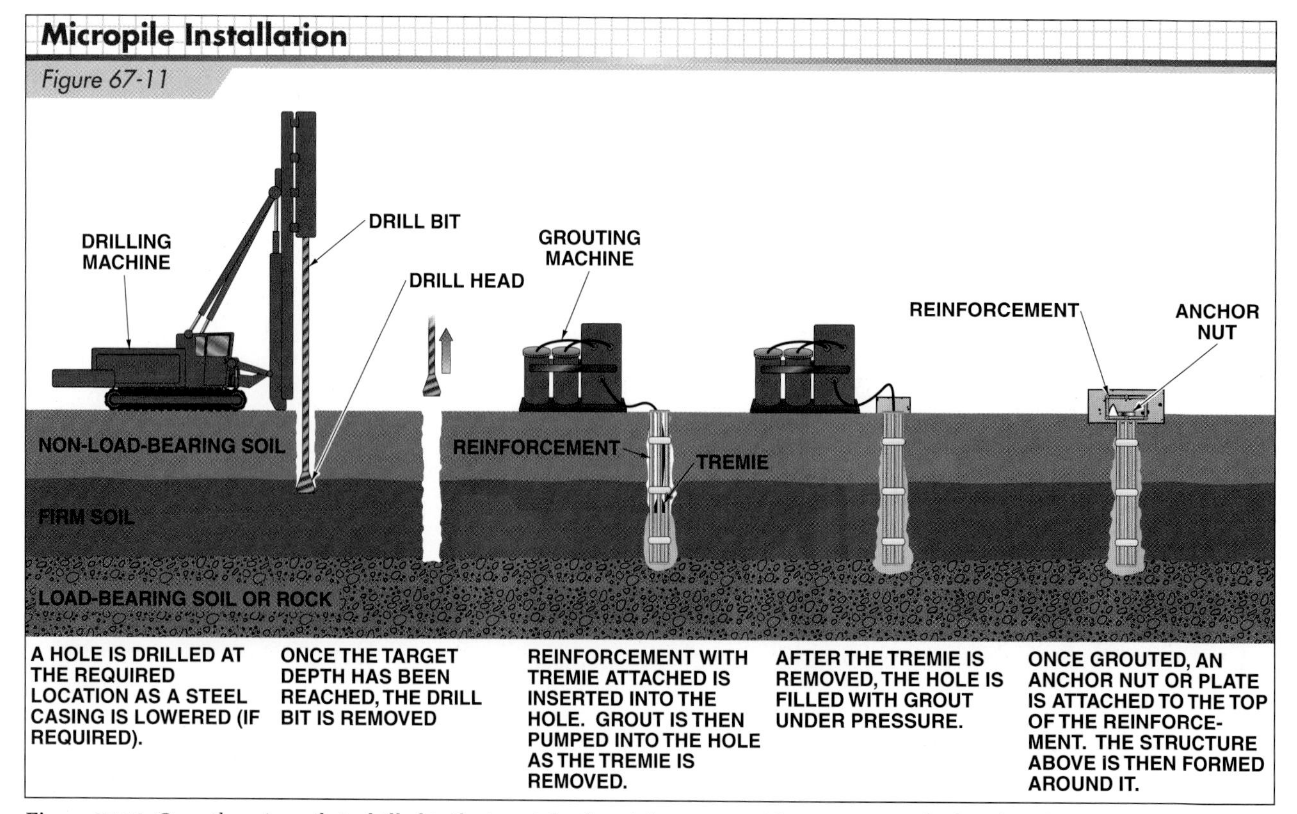

Figure 67-11. *Once the micropile is drilled to the target depth, reinforcement with a tremie attached is placed and high-strength grout is pumped into the hole.*

Micropile Foundation Supports

Figure 67-12

EXISTING COLUMN
EXISTING FOOTING
NEW FOOTING
MICROPILES
REINFORCEMENT

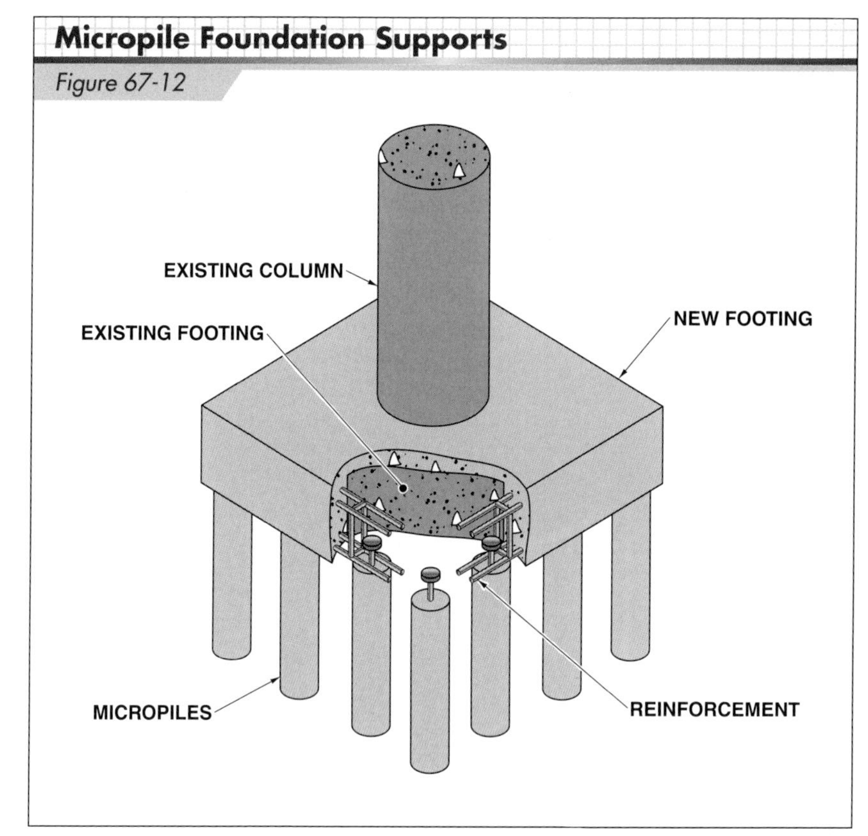

Figure 67-12. *Micropiles can be installed to provide additional support for already existing foundation systems.*

Micropile Applications

Figure 67-13

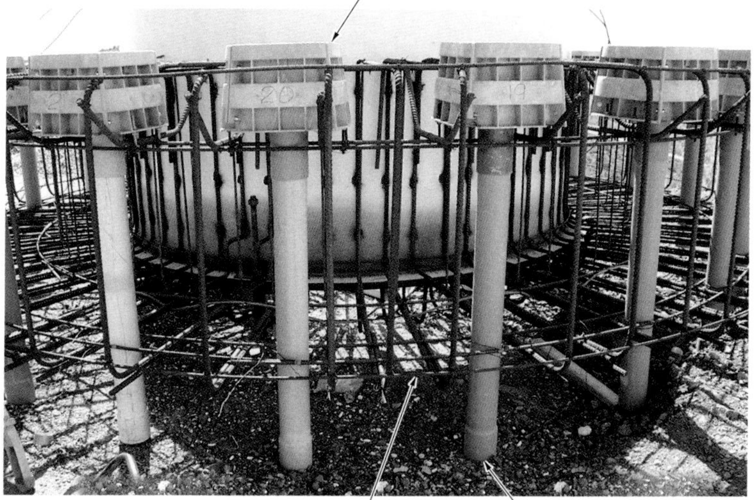

ANCHOR BOX EMBEDDED IN TOWER FOUNDATION ALLOWS ACCESS FOR POST-TENSIONING OF MICROPILE REINFORCEMENT
WIND TURBINE FOUNDATION REINFORCEMENT
MICROPILE CASINGS

Con-Tech Systems Ltd.

Figure 67-13. *Micropiles are used to support wind turbines, highways, bridges, wharves, and transmission towers.*

CAISSONS

In certain areas of the country, good load-bearing soil may not be reached with piles. Under these conditions, *caissons* are used. A caisson is a cylindrical or rectangular casing that is placed in the ground and filled with concrete. Caissons can extend deeper than piles and be constructed to much wider diameters than piles.

Bored Caissons

Bored caissons are commonly made by cranes fitted with earth auger attachments. The casing, which is usually a steel cylinder, is then placed in the hole. Modern drilling equipment and methods make it possible to bore holes 10′ in diameter to a depth of over 150′. The casing for caissons extending to greater depths is made up of sections that are added as the drilling proceeds. **See Figure 67-14.**

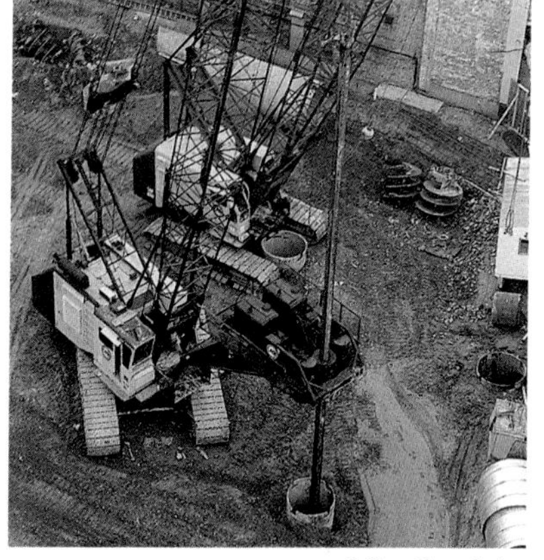

© Case Foundation

Figure 67-14. *Bored caissons are formed by an earth auger mounted on a crane drilling into the ground. Sections of casing are added as the drilling proceeds.*

Belled Caissons

Belled caissons provide greater load-bearing area at the bottom of the caisson. **See Figure 67-15.** After the caisson hole has been drilled to the desired depth, a belling tool is attached to the drilling head. The belling tool is lowered into the caisson in the closed position. At the proper depth, the belling tool is expanded to remove the bell-shaped portion of the caisson. **See Figure 67-16.**

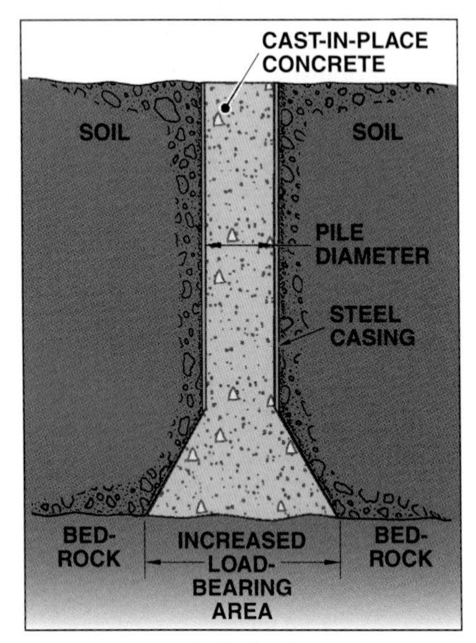

Figure 67-15. *A belled caisson has greater load-bearing area at its bottom.*

Figure 67-16. *A belling tool is expanded to create the bell-shaped portion of a belled caisson.*

SPREAD FOUNDATIONS

If building loads and soil conditions do not require a system of piles or caissons, a *spread foundation* is usually adequate to support concrete buildings. A spread foundation generally consists of a foundation wall resting on a wider base (footing). A spread foundation for heavy construction is similar to an inverted T-shaped foundation used for residential and other light construction foundations. The main difference between a spread foundation and inverted T-shaped foundation is the footing size. Footings are wider and thicker for heavy construction than for light construction. **See Figure 67-17.**

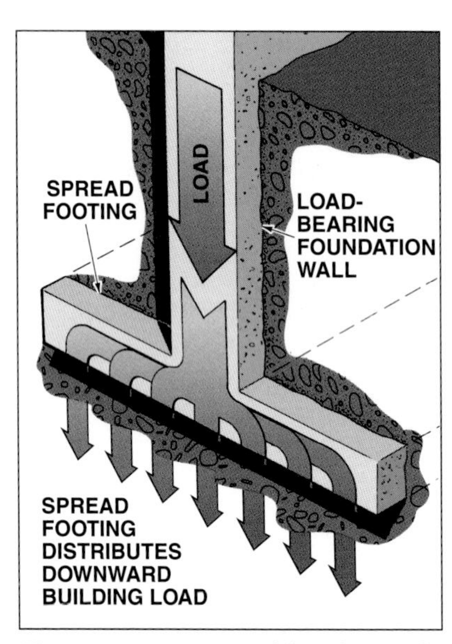

Figure 67-17. *In a spread foundation, the footing below the foundation wall distributes the building load over a wider area.*

Matt Foundations

Matt foundations are used in soils of low load-bearing strength. A matt foundation is a type of spread foundation that consists of a solid slab of heavily reinforced concrete placed beneath the entire building area. In some cases, the slab may be 3′ to 8′ thick.

WIND TURBINE FOUNDATIONS

A *wind turbine* is a machine that converts the kinetic energy of wind into rotating mechanical energy. Wind rotates rotor blades that capture the kinetic energy of the wind and convert it to rotating mechanical force. This mechanical force drives an electric generator.

A wind turbine system includes a reinforced concrete foundation, a tower, a nacelle, and a hub and blade assembly. **See Figure 67-18.** The total height of a wind turbine system can reach over 400′ from the ground to the tip of a blade. The tower alone can reach up to 344′. Together, the tower, nacelle, and hub and blade assembly can weigh over 300 tons.

Large wind turbines require concrete foundations that can support the weight of the turbine components and resist the forces of the spinning blades.

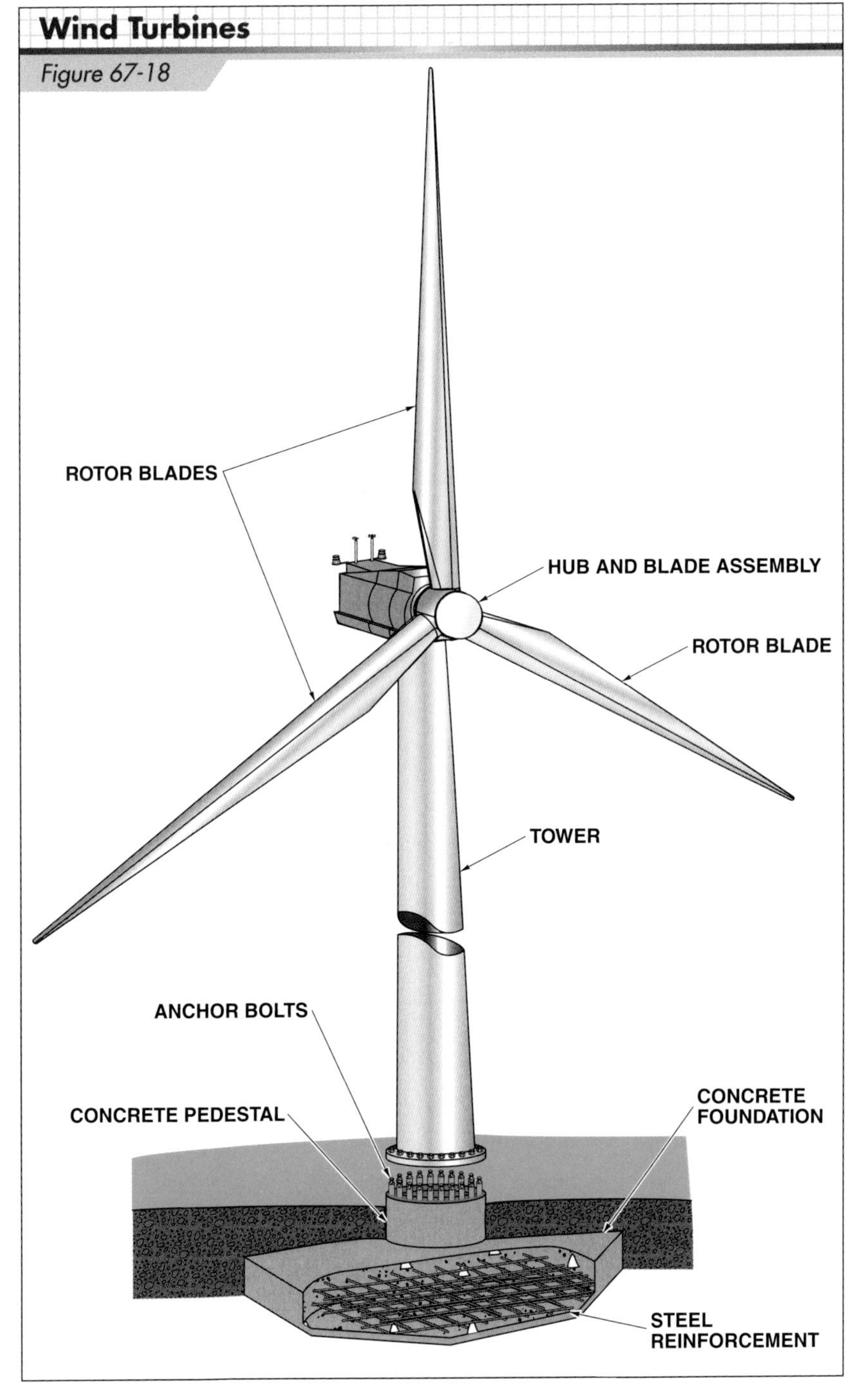

Figure 67-18. *A wind turbine system includes a reinforced concrete foundation, a tower, a nacelle, and a hub and blade assembly.*

A reinforced concrete foundation is designed to properly support the weight of the tower, nacelle, and hub and blade assembly. The foundation consists of a spread footing and pedestal. The reinforced concrete foundation also resists the forces produced by the spinning blades and the environment. The weight of the concrete foundation can be over 750 tons.

Foundation Reinforcement

Most land-based wind turbine foundations are octagonal in shape and can be up to 65′ wide and 10′ deep at the pedestal. After excavation, a mud slab is poured to support the foundation reinforcement. On this slab, a complex network of steel reinforcement, anchor bolts, and embed rings are secured in the required position. The steel reinforcement consists of a complex arrangement of various sizes of rebar. Wind turbine foundations can have over 20 tons of rebar. **See Figure 67-19.**

Attached to the network of rebar are at least two foundation rings that properly align the anchor bolts. **See Figure 67-20.** The two types of foundation rings are embed and template. Embed rings are made from thick steel (up to 2″) and may be over 20′ in diameter. An embed ring is permanently encased at the bottom of the foundation.

Template rings are similar to embed rings. Template rings have the same diameter as embed rings, but their thickness may be the same or slightly thinner. Depending on the design of the foundation, there may be more than one template ring. Any template rings other than the top ring are permanent parts of the foundation. The top template ring may be a permanent part of the foundation or it may be temporarily used to position the anchor bolts until the concrete hardens.

Anchor bolts are composed of high-strength hardened steel and can be over 12′ long. They range in diameter from 7⁄8″ to 2½″. Fully threaded anchor bolts can be plain steel, epoxy coated, or galvanized. In most situations, the anchor bolts are encased in a PVC sleeve to protect them from exposure to the concrete. There may be over 150 anchor bolts for a single wind turbine.

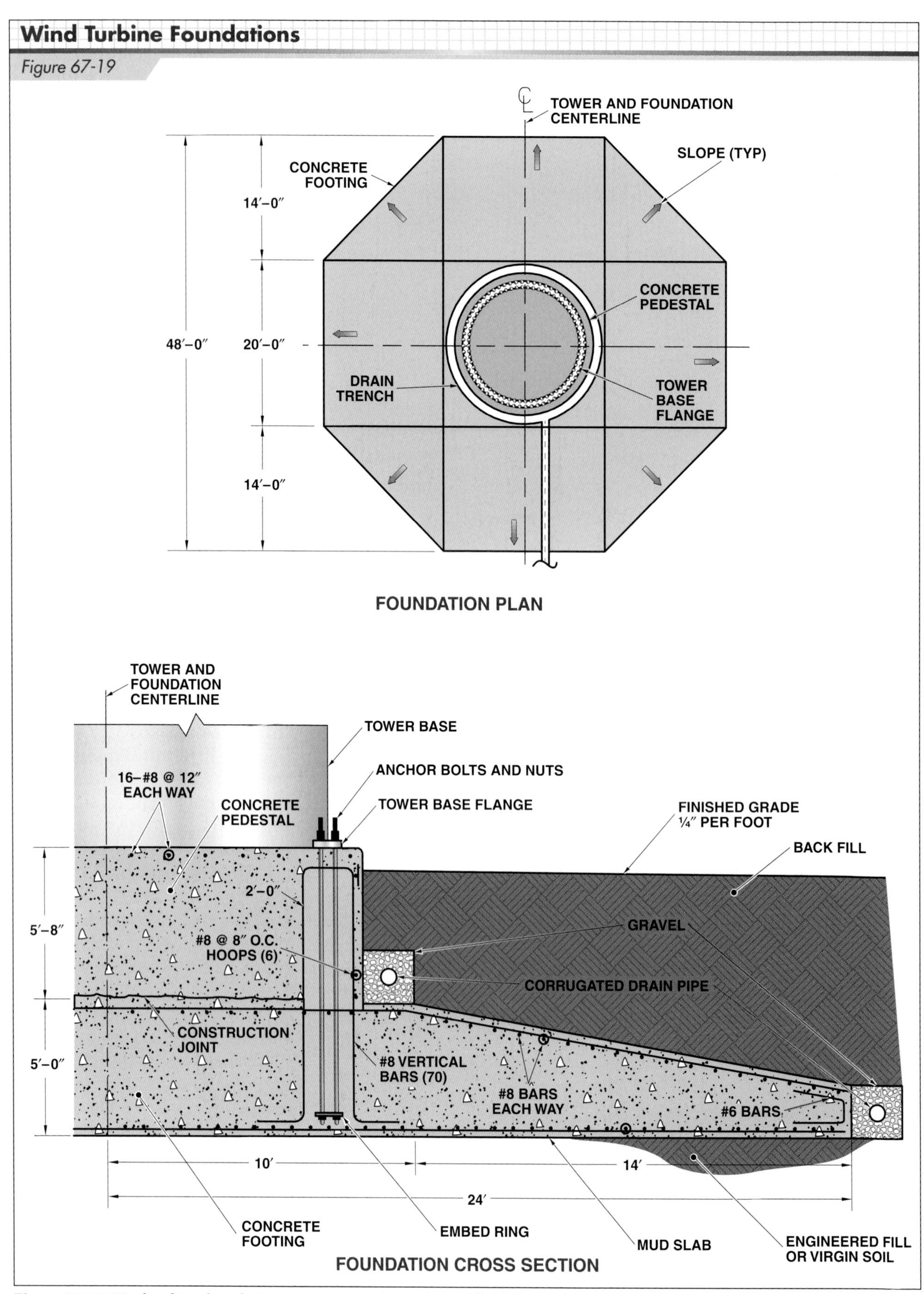

Figure 67-19. *Wind turbine foundations are composed many tons of high-strength concrete and a complex arrangement of steel reinforcement.*

Foundation Rings and Anchor Bolts

Figure 67-20

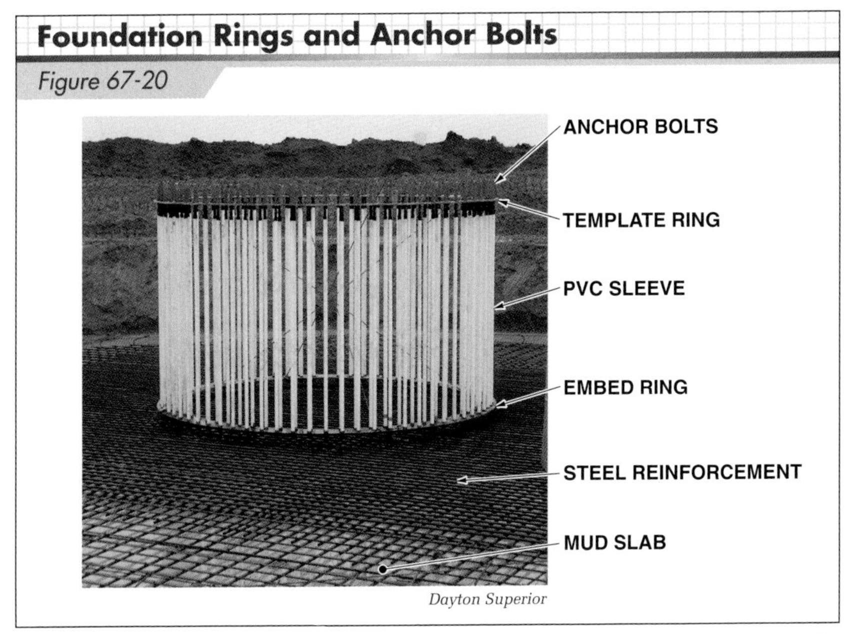

Figure 67-20. *Foundation rings and anchor bolts secure a wind turbine assembly to the foundation.*

The Deep Foundations Institute is a technical association of companies and professionals involved in the deep foundations sector of the construction industry. The institute serves as a primary means through which members may participate in improving the planning, design, and construction aspects of deep foundations and deep excavations. The institute develops and publishes various publications related to deep foundations and deep excavations, including inspector's manuals, technical guides, and references.

Foundation Formation

Once the reinforcement, anchor bolts, and rings are set, the sides of the footing are formed using prefabricated form panels. **See Figure 67-21.** When the forms are in place, high-strength concrete is placed into forms and allowed to harden. Then the pedestal is formed by using prefabricated form panels and placing more high-strength concrete.

After the concrete has cured to the required strength, the forms are removed and the foundation is backfilled, leaving only the pedestal exposed. When the concrete has reached its targeted design strength, a large crane is then used to lower the tower base section onto the pedestal. The tower base section is then secured and grouted in place as needed.

Wind Turbine Foundation Forms

Figure 67-21

Dayton Superior

Figure 67-21. *The sides of the footing and pedestal are formed with prefabricated form panels.*

Unit 67 Review and Resources

UNIT 68 Heavy Concrete Construction Formwork

OBJECTIVES

1. List materials used for plywood forms.
2. Describe procedures for laying out and constructing foundation and basement forms for large concrete buildings.
3. Identify common types of patented ties and waler systems.
4. Describe forms for pilasters, columns, and beams.
5. Describe the construction of floor and roof forms.
6. Explain how hand-set forms are placed.
7. Describe the procedure for laying out stairway forms.
8. Review prefabricated panel forming systems.

The basic structural parts of a concrete building are the walls, floors, beams, and columns. Forms are constructed to the shapes of the structural parts and concrete is then placed in the forms. The forms are removed after the concrete has properly set and hardened.

Forms must be rigid enough to resist bulging, twisting, or moving during concrete placement. In addition, forms must be tightly constructed to prevent concrete from leaking through the joints. Leaks will cause ridges on the surface of the hardened concrete. All debris such as wood scraps, sawdust, and nails should be removed from the inside of forms before concrete is placed. Cleanout holes are cut at the bottoms of forms during formwork construction. Compressed air is used to blow out debris from the forms. The clean-out holes are then blocked off.

Forms must also be designed for easy removal without damaging the concrete after the concrete has hardened. Wood is commonly used for building forms. Patented forming systems, consisting of metal frames combined with wood panels, are also frequently used. Steel, aluminum, and plastic panels are also available.

PLYWOOD FORMS

Most wood forms today are constructed of plywood backed by various combinations of studs, walers, strongbacks, and some type of bracing. Plywood and composite panels provide a large, smooth surface that can withstand rough treatment without splitting. Another advantage of plywood is its ability to bend when curved surfaces are required in the form design. **See Figure 68-1.** Textured plywood can also be used for a textured finish on the concrete.

Most exterior grades of plywood can be used for form construction. However, Plyform®, medium-density overlay (MDO) plywood, and high-density overlay (HDO) plywood are manufactured specifically for concrete forms. HDO plywood is used where the smoothest possible finish is required. Plyform, MDO, and HDO plywood panels can be reused many times if properly cared for and maintained.

Gates & Sons, Inc.

Figure 68-1. *An advantage of plywood in form construction is that it can be bent when a curved surface is required.*

Formwork Panel Maintenance

Formwork panels must be properly stored and maintained to ensure long life in the field. When stored for an extended period of time, the panels should be placed flat and level. Panel faces should not be exposed to direct sunlight for prolonged periods of time. The panels should be covered so that moisture cannot accumulate on the panels.

A *form-release agent* should be applied to formwork panels to make it easier to strip the panels from the hardened concrete. Form-release agents are available in a variety of forms, including pastes and spray-on agents. Water-soluble emulsions and environmentally friendly, biodegradable solutions are recommended. Environmentally friendly release agents are water-soluble and do not contain petroleum or other volatile organic compounds (VOCs). Form-release agents can be applied to a variety of formwork materials including wood, fiberglass, steel, and aluminum.

After the formwork panels have been stripped and cleaned of any sediment that has adhered to the panels, a form-release agent should be applied. Spraying is the most common means of applying form-release agents. **See Figure 68-2.** However, manufacturers may recommend that their form-release agents be applied using a mop or a cloth.

Form-release agents must not come in contact with rebar.

Symons Corporation

Figure 68-2. *A release agent is sprayed on panels to be used for concrete forms to facilitate form removal. Follow manufacturer instructions for the safe use and application of form-release agents.*

FOUNDATION AND BASEMENT FORMS

The procedure for laying out and constructing footing forms for large concrete buildings is similar to residential and other light construction work. The major difference is the greater width and depth required for footings for large concrete buildings.

In constructing basement wall forms, the first step is to place all the outside panel sections. **See Figure 68-3.** If possible, holes for snap ties should be drilled in the panels before they are erected.

Figure 68-3. *Outside form walls are erected and rebar is placed prior to constructing inside form walls.*

After all wall panels have been positioned and the walers and strongbacks (if required) have been attached, the wall is aligned and braced. Braces are usually placed 6′ to 8′ apart. Braces may be nailed to the wall and to wood stakes driven into the ground, or patented braces may be used. One type of patented wood brace features a turnbuckle that is welded to an anchor bracket, which is fastened to the ground with steel stakes. **See Figure 68-4.** A brace plate is used to fasten the brace to a waler or metal strongback on the wall. Another type of brace is made completely of steel. **See Figure 68-5.**

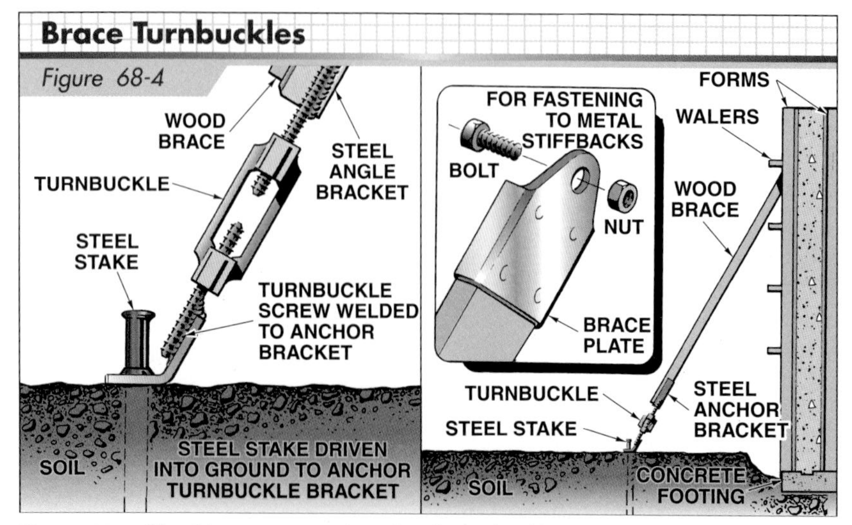

Figure 68-4. *Wood braces are equipped with turnbuckles at their lower ends to allow wall panels to be aligned. Turnbuckles are fastened to the ground with steel stakes. A brace plate is used to attach a brace to a waler or stiffback.*

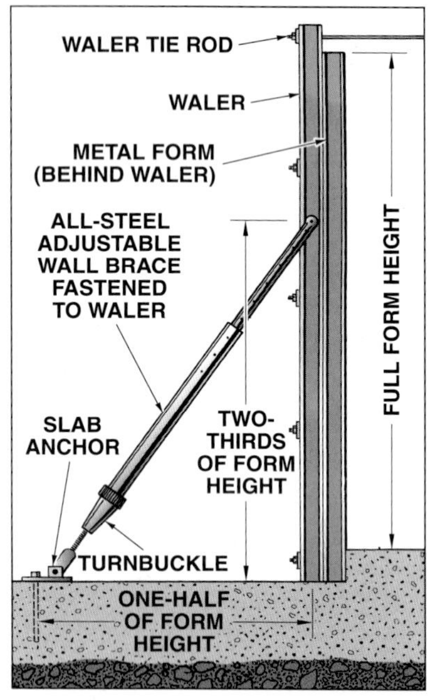

Figure 68-5. *All-steel wall braces may be used to support wall forms. Braces should be properly positioned to provide adequate support.*

Figure 68-6. *Metal window frames are secured to outside form walls as the walls are erected. The frames remain in place after the form walls are removed.*

Steel rebar assemblies may be formed on the ground by tying the individual rebar together with wire ties. The assemblies are raised into position with a crane and positioned in the wall forms.

Doors and Windows

After the outside wall form panels for the basement walls are set, provisions are made for door and window openings. Steel or aluminum window and door frames are commonly attached to the wall form panels. Brackets extending from the frames will be embedded in the concrete to secure the frame in position when the forms are stripped. **See Figure 68-6.**

A traditional method for providing for door and window openings is to set wood window and door bucks as shown in **Figure 68-7.** A rectangular opening at the bottom allows observation of the flow and consolidation of concrete beneath the buck. When concrete reaches the bottom of the buck, the piece that has been cut out is replaced and cleated down. The entire buck is removed when the forms are stripped from the hardened concrete walls. The wedge-shaped key strip remains in the concrete and serves as a nailing strip for the finish wood window frame.

Reinforcing Steel

Reinforcing steel bars (rebar) are placed by reinforcing-steel workers after the outside wall forms are set but before the inside wall forms are erected. **See Figure 68-8.** With some forming systems, rebar is placed first and the forms are built around the rebar.

Construction Joints

Concrete for tall buildings is placed one story at a time; therefore, a horizontal construction joint occurs at each floor level. The wall sections are structurally tied together by continuous rebar.

It may be necessary to place concrete for long walls in sections, creating vertical construction joints. When constructing long walls, carpenters must place a *bulkhead* inside or at the end of the form. **See Figure 68-9.** Depending on the wall thickness, bulkheads are made of 1 × 8s or 2 × 4s, which are notched around the rebar. In some situations, the bulkhead members are placed along the sides of the rebar and the spaces between the members are filled.

Bulkheads must be able to withstand a large amount of pressure as concrete is being placed.

Constructing Rough Window and Door Bucks

Figure 68-7

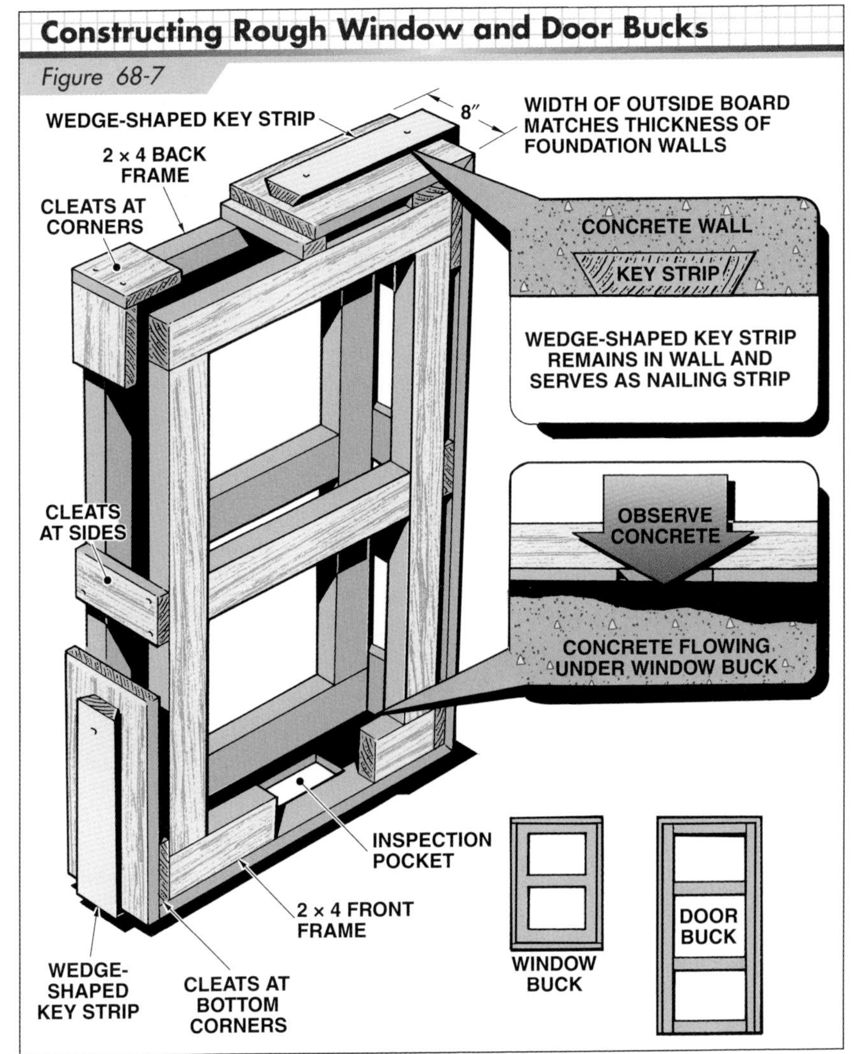

Figure 68-7. *Door and window bucks use similar construction techniques. Door bucks do not require a bottom piece with an inspection pocket.*

Utility Installation

Provision for utilities, such as water pipes, electrical conduit, outlet boxes, and sleeves for holes in the wall, must be placed after the outside form wall panels are set. Carpenters coordinate with other tradesworkers to ensure the utilities are placed in the proper position.

Doubling Up Walls

After the outside form walls are set in position, rebar is placed, and required door and window bucks and utilities are installed, carpenters *double up* the walls by setting the inside wall forms in position. The bottoms of the inside wall panels are placed in the proper position and the panel is slowly tilted into the vertical position. Form ties extending from the outside form wall are inserted through predrilled holes in the inside form wall panels. Inside form wall walers are then attached. Clamps, edges, or other devices are placed at both ends of the ties, spacing and fastening together the opposite form walls. **See Figure 68-10.** Some types of ties are pushed through the panel holes after the walls have been doubled up.

Figure 68-8. *Rebar is placed after the outside form walls are set and before the inside form walls are erected.*

Figure 68-9. *Bulkheads may be required in long walls. Bulkheads are made of 1 × 8s or 2 × 4s which are notched around the rebar.*

Figure 68-10. *Form walls are doubled up after the rebar, door and window frames, and utilities have been properly located.*

PATENTED TIES AND WALER SYSTEMS

Form ties combine with walers to secure and space the opposing form walls. A variety of patented form ties are available for different forming systems. Form ties are generally classified by their load-carrying capacity and method of use. The load-carrying capacity is *working load value,* which is the amount of pressure against the individual tie when placing concrete in the forms.

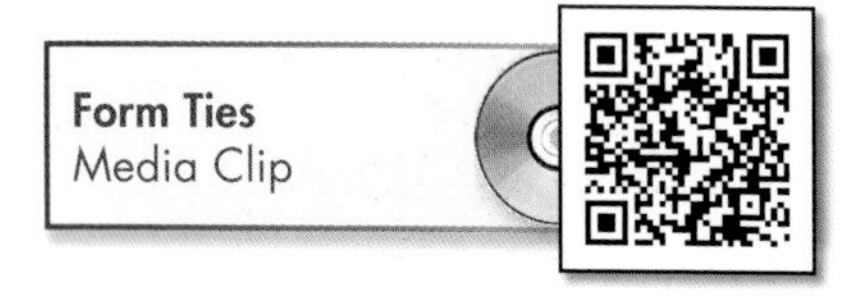

Light Form Ties

Light form ties have safe working load values of 3750 lb or less. Light form ties are also referred to as *through ties* since they are one continuous piece which extends through the wall cavity and both sides of the formwork. Light form ties are manufactured from heavy-gauge wire. Examples of light form ties are *snap, loop,* and *pencil rod ties.*

Snap Ties. Snap ties are commonly used with double- or single-waler systems and are available for wall thicknesses ranging from 6″ to 26″. The basic design of most snap ties allows a section of the tie to extend from the outside surfaces of opposing walers. Snap ties are tightened by driving slotted metal wedges behind the buttons at the ends of the ties. A hole is typically provided in the wedge so a duplex nail can be driven through the hole to prevent the wedge from loosening during concrete placement and vibration.

Small plastic cones or metal washers on snap ties act as spacers between the walls. A *breakback* consisting of a grooved section next to the tapered edge of the cone allows the tie to be broken off (snapped) after the concrete has hardened and the forms have been removed. A variation of a snap tie design is a system that features a nut button that enables the tie to be snapped by turning the button with a socket wrench while the forms and wedges are still in place. **See Figure 68-11.**

Loop and Pencil Rod Ties. Loop and pencil rod ties are two other types of light form ties. Loop ties have a loop at each end. Tapered wedges are driven through the loops to secure the ties in position. Pencil rod ties are designed for use on battered walls and on wider walls where snap ties are not practical.

High tensile strength fiberglass ties are also used. Tie rods are available in long lengths and are cut to length with an abrasive blade in a circular saw. Fiberglass ties are placed in predrilled holes after the walls have been doubled up. Self-gripping wedges placed against the form walers tighten the ties.

Medium to Heavy Ties

Medium to heavy ties are *internal disconnecting ties* and have safe working load values exceeding 3750 lb. Internal disconnecting ties, such as waler rods, coil ties, and taper ties are used for heavier and thicker walls where greater pressure is exerted during concrete placement. **See Figure 68-12.**

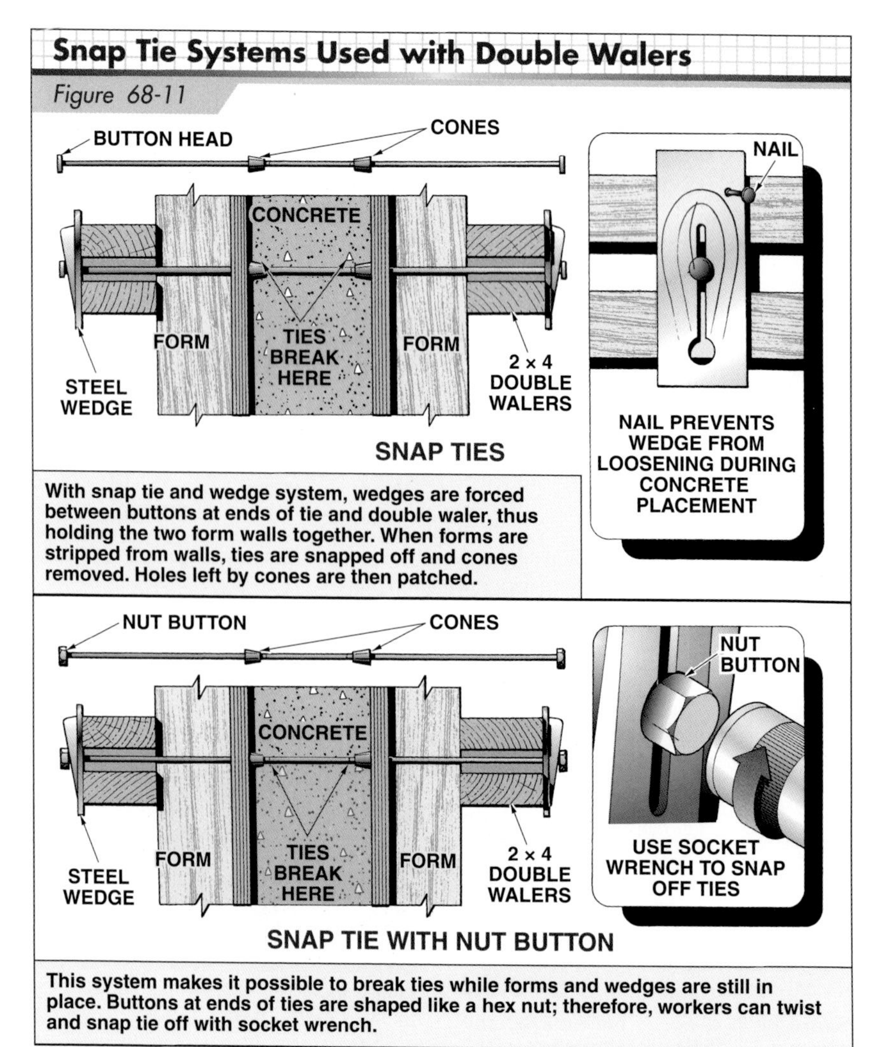

Figure 68-11. *Snap ties are used to secure and properly space opposing form walls.*

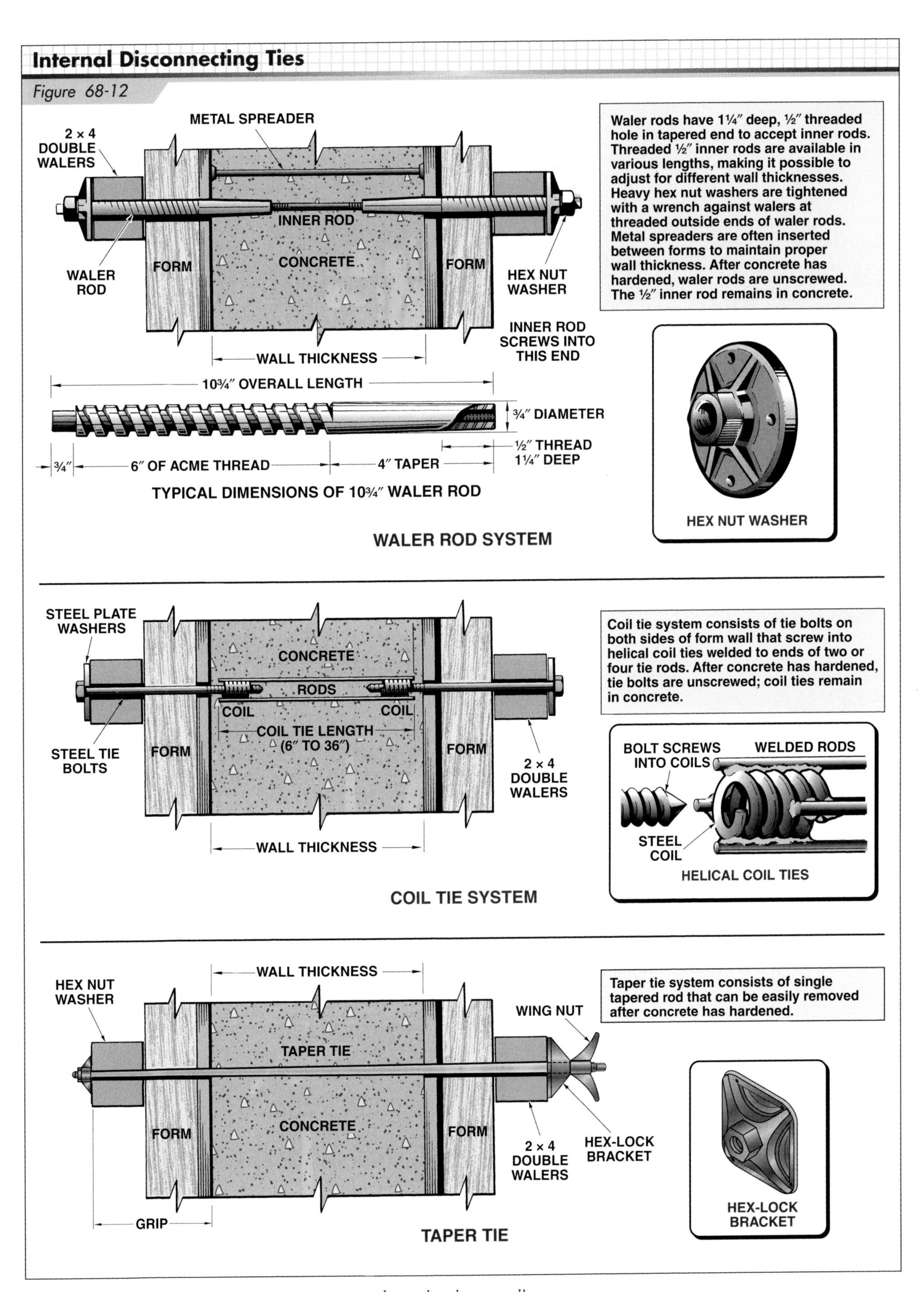

Figure 68-12. *Internal disconnecting ties are commonly used on heavy walls.*

Waler Rods. Waler rods consist of two external rods or bolts that screw into an internally threaded device. Waler rods are available in graduated lengths to accommodate wall thicknesses ranging from 8″ to 36″. The internally threaded device is inserted through holes in the wall form after the walls have been doubled up. Large hex nut washers are tightened against the walers to secure the walls. Metal spreaders are placed between the walls to maintain the proper width. The outer sections of the tie can be screwed out after the concrete has hardened, and the internal device remains in the concrete wall.

Coil Ties. When using coil ties, bolts pass through the double walers and screw into a helical coil welded to two or four rods. After the concrete has hardened, the bolts are withdrawn and the coils remain inside the wall.

Taper Ties. Taper ties consist of a single tapered rod terminated with a hex nut washer on one end and a bracket and wing nut on the other end. Taper ties should be lubricated prior to use. Taper ties are inserted through predrilled holes in the formwork panels, and the hex nut washer and wing nut are tightened to achieve the desired wall thickness. After the concrete has been placed and properly hardened, the formwork is stripped, leaving just the taper ties in place. A hammer is used to loosen and remove the ties for future use.

Waler Systems

Traditional heavy construction panel forming methods use vertical 2 × 4 studs spaced 12″ to 16″ OC. **See Figure 68-13.** The studs are reinforced by double or single walers and/or strongbacks, depending on design considerations. Snap ties, waler rods, and coil ties are commonly used with this system.

Figure 68-13. *Heavy construction panel forming methods commonly use 2 × 4 studs, which are reinforced by walers. Note the snap tie wedges along the walers.*

One forming method featuring a single waler system eliminates the need for studs beneath the walers. After the wall form panels are set in place, snap brackets and walers are attached. Wedges secure the walers in place. Strongbacks and braces are then placed at intervals. **See Figure 68-14.**

Form liners may be installed on form walls to provide the concrete with a wood grain, stone, brick, or other texture.

PILASTER, COLUMN, AND BEAM FORMS

Columns and beams are a major part of the basic structure of most concrete buildings. The structural design of many multistory buildings is a continuous system of columns supporting floor slabs, with the same floor plan at each level. Columns and

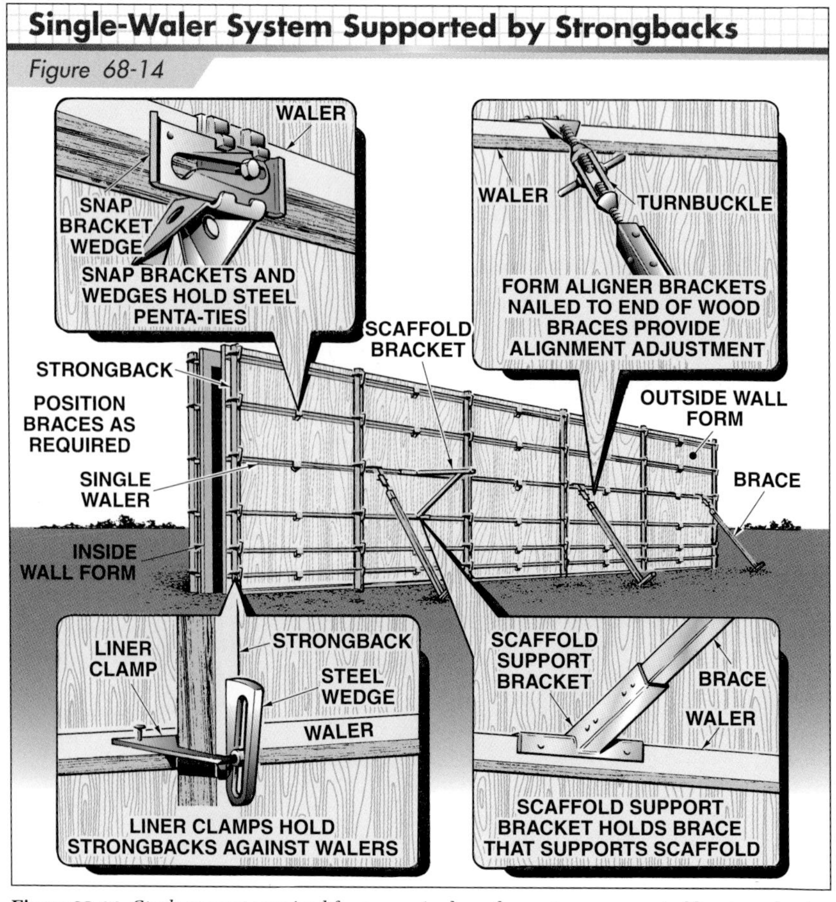

Figure 68-14. *Studs are not required for some single-waler systems supported by strongbacks.*

pilasters support the beams, which in turn tie into and support the floor and roof systems. Beams also tie together the walls of the building.

Pilasters

If a wall design includes *pilasters,* pilaster forms are erected as the wall forms are constructed. **See Figure 68-15.** Pilasters are projections from the face of a wall which add strength to the wall and may also support the ends of beams.

Columns

Column forms are constructed of plywood, prefabricated steel-framed plywood, tubular fiber, fiberglass, and metal. Column forms are subject to much greater pressure than wall forms when concrete is being placed. Tight joints and strong tie supports around the form are necessary.

Figure 68-15. *Pilasters provide additional strength to a wall and support the ends of beams. After the wall forms are positioned, walers and studs will be placed.*

When wood forms are used for rectangular columns, the sides are constructed of plywood and backed with 2 × 4 stiffeners. Stiffeners may be omitted for smaller columns that require a lighter form assembly. A cleanout hole is provided at the bottom of one of the sides. After the sides are assembled, they are placed in a template fastened to the column footing. **See Figure 68-16.**

The height of a column is typically three times its largest horizontal dimension.

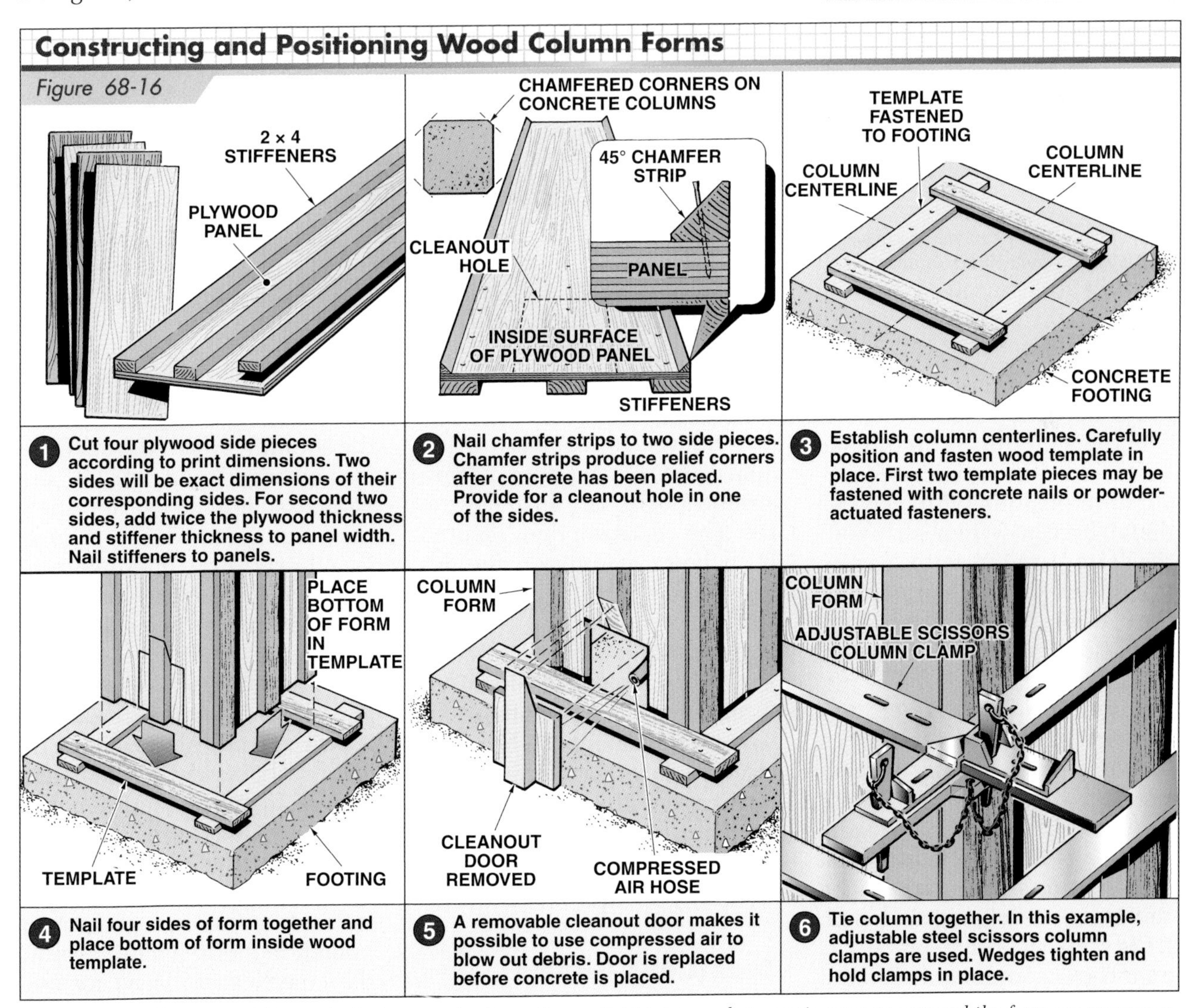

Figure 68-16. *When assembling and placing wood column forms, tight joints and strong tie supports around the form are necessary. A cleanout hole is provided at the bottom of one of the sides.*

Hinged metal *scissor clamps* are often used to tie together the column form sides. The scissor clamps are tightened and held in place by cam devices or wedges driven into slots located in the clamp. **See Figure 68-17.** After the concrete has hardened and the forms have been removed, the corners may be temporarily protected with wood strips.

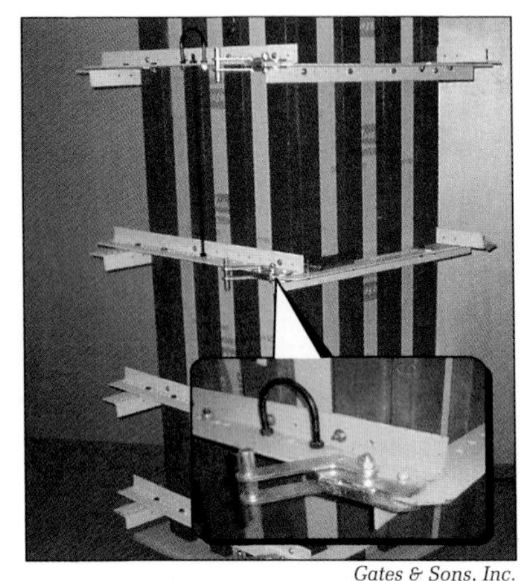

Gates & Sons, Inc.

Figure 68-17. *Scissor clamps are tightened and held in place by cam devices or wedges driven into slots in the clamp.*

Steel Column Forms

Figure 68-18

PERI Formwork Systems, Inc.

Figure 68-18. *Prefabricated round steel forms are used to form large columns for heavy construction projects.*

Tubular fiber forms are sometimes used to construct round concrete columns. The tubular forms are positioned and secured in position with wood or metal braces. When stripped, the forms are cut with a knife or saw and carefully pried from the concrete. Tubular fiber forms can be easily cut to length on a job site using a handsaw.

Round steel and fiberglass forms are used to form large columns for heavy construction projects. Round steel forms are available in diameters ranging from 14″ to 10′. **See Figure 68-18.** The sections that make up a round steel form range in length from 1′ to 10′. Since bracing is an integral component of round steel forms, additional bracing is not required.

A fiberglass form is pulled apart and placed in position around previously installed rebar. The edges are then secured with bolts at closure flanges reinforced with a predrilled steel bar. The form is plumbed and secured with braces tied to a steel bracing collar. Fiberglass forms provide a smooth architectural finish and can be easily stripped. Fiberglass forms can be combined with a two-piece capital form to construct a column with a capital. **See Figure 68-19.**

Beams and Girders

Forms for *beams* and *girders* rest on top of and are tied to the column forms. Although the terms "beam" and "girder" are often used interchangeably, they have distinct meanings. A beam is a horizontal member that supports a bending load over an opening, as from column to column. A girder is a heavy beam that supports other beams and girders. **See Figure 68-20.**

Two concrete placement methods can be used for beams and girders. In one method, concrete for the beams and girders is placed separately after concrete for the columns has hardened. In the other method, concrete is poured *monolithically*, meaning the concrete is placed for columns, beams, and girders at the same time.

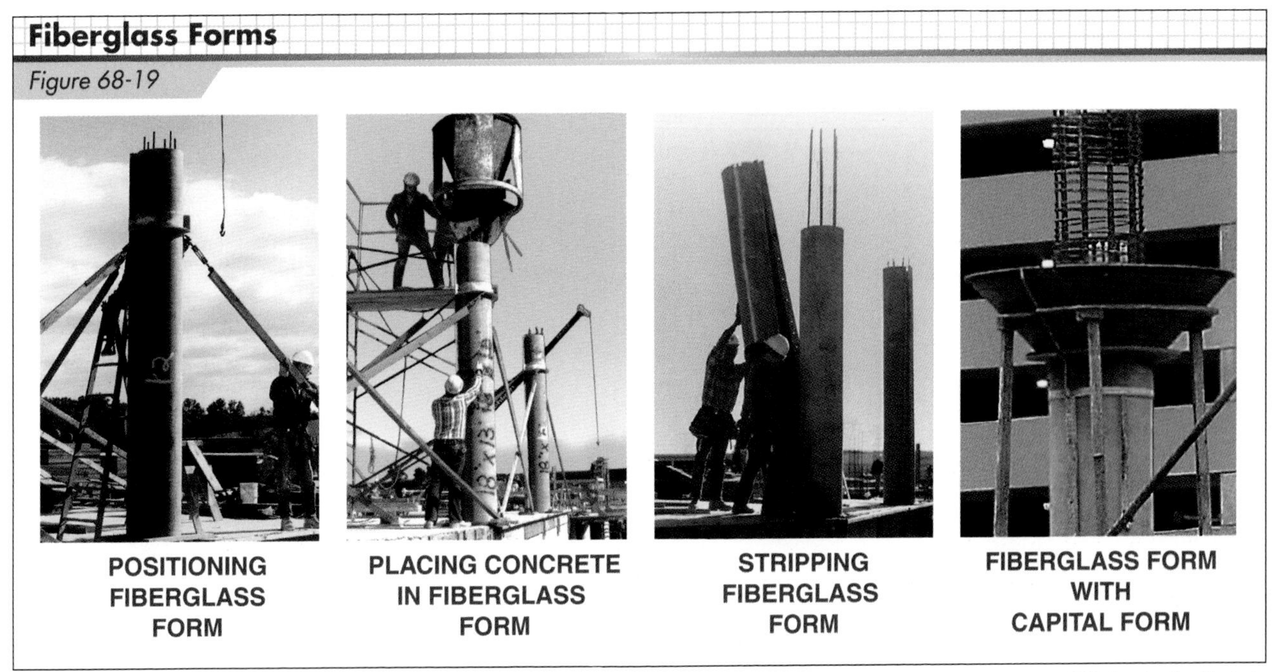

Figure 68-19. *A fiberglass form produces a smooth architectural finish with one vertical seam.*

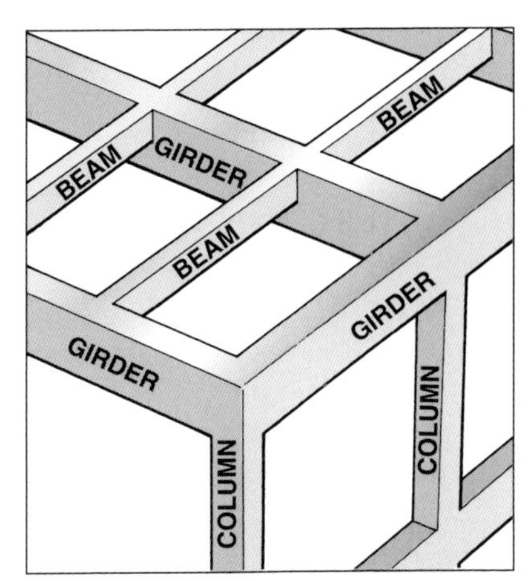

Figure 68-20. *A structural frame for a concrete building may include columns, beams, and girders.*

Beam and girder forms consist of a bottom piece (soffit) and sides. **See Figure 68-21.** The entire unit is supported by shores placed at intervals. Patented metal shores are commonly used to support the forms, although wood T-shores can also be used. Beams and girders are heavily reinforced with rebar that tie into the rebar of the column below and the floor above. An example of beam-and-column concrete construction is shown in **Figure 68-22.**

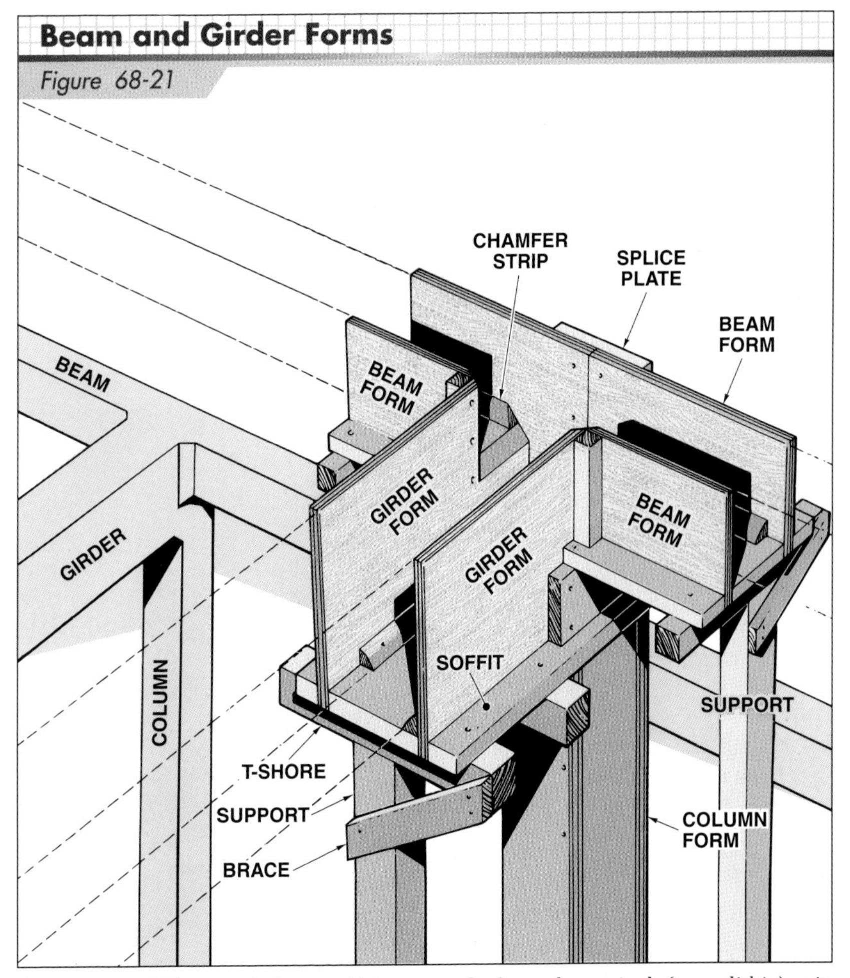

Figure 68-21. *Columns, girders, and beams may be formed as a single (monolithic) unit. Note that the beam and girder join over the column.*

Figure 68-22. *Concrete is placed at the same time for the walls, columns, beams, and girders of a monolithic unit.*

Portland Cement Association

Figure 68-24. *A flat-slab floor may be supported by drop panels over the columns.*

Portland Cement Association

Figure 68-25. *A flat-slab floor may be supported by drop panels and capitals over the columns.*

FLOOR AND ROOF FORMS

Several basic designs are used for concrete floors and roofs. All designs require formwork consisting of a deck supported by joists and shores. The formwork must be strong enough to support the weight of the form material, concrete, and load imposed by workers and equipment.

Beam-and-Slab Floors

Beam-and-slab systems are suitable for floors that will bear heavy loads. The floor slab rests on top of closely spaced beams that tie into girders supported by columns. A lighter design features concrete joists tied into girders. **See Figure 68-23.**

Flat-Slab Floors

Beams and girders are not used with a flat-slab system. The slab receives its main support from the columns and the thickened sections over the columns called *drop panels*. **See Figure 68-24.** A variation of the drop panel system uses column *capitals*, sometimes called *drop heads*, over the column. **See Figure 68-25.**

Portland Cement Association

Figure 68-23. *This concrete floor system features a slab resting on concrete joists that tie into girders.*

A stepped wall may be required on some construction projects. Note the guardrail along the top of the stepped wall.

Concrete Joist Systems

Concrete joist systems consist of concrete joists placed monolithically with a floor slab, beams, girders, and columns. Concrete joist systems allow for thinner and lighter floor slabs with high bearing capacities. Two basic concrete joist system designs are the *one-way* and *two-way* (waffle) joist systems. Both systems are formed with reusable metal or fiberglass pans.

Standard types and sizes of removable forms for one- and two-way joist systems are included in ANSI/CRSI A48.1, Forms for One-Way Joist Construction, and ANSI/CRSI A48.2, Forms for Two-Way Joist Construction, respectively.

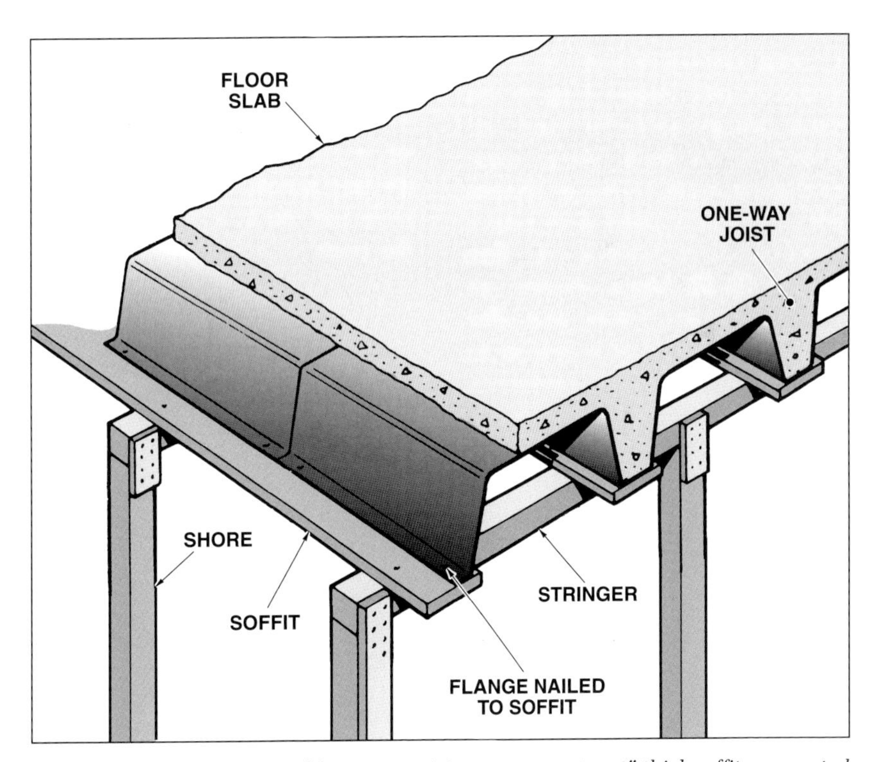

Figure 68-26. *Long pans used for one-way joist systems rest on 2″ thick soffits supported by shores and stringers.*

Long pans are used to form one-way joist systems. The pans are nailed to the tops or sides of 2″ thick soffits, which are supported by shores and stringers. **See Figure 68-26.** Long pans commonly frame into girder forms.

Dome pans are used to form two-way joist systems and are supported in the same manner as long pans. Since two-way joists do not include beams or girders, a solid area equal to the slab thickness and joist is formed around the supporting columns. **See Figure 68-27.**

Waffle Forms

Figure 68-27

Western Forms, Inc.

Figure 68-27. *Aluminum dome forms are used for a two-way joist system.*

Symons Corporation

Figure 68-29. *Metal scaffold shoring is used to support high form soffits.*

Slab Decks and Shoring

Slab decks are formed over metal, engineered wood, or solid wood shores, depending on the deck design and the load imposed on the deck. Metal (steel and aluminum) tubular shores are commonly used with heavy construction formwork. **See Figure 68-28.** Metal shores can be reused many times, and offer the convenience of screw jacks for making height adjustments. To support higher soffits, metal scaffold shores may be employed for greater stability. **See Figure 68-29.** Engineered wood products, such as veneered beams and I-joists, are also commonly used to support form soffits. **See Figure 68-30.** Aluminum beams are also frequently used.

A traditional method of shoring slab deck soffits involves the use of wood shoring. The shores are cut short to allow for wedges over a wood sill. **See Figure 68-31.** The wedges are used to drive the shores up tightly and to line up the floor above. Metal shore jacks can also be used for this purpose. Horizontal braces are nailed or screwed to the posts at mid-height to tie the posts together. Plywood cleats secure the stringers (beams) to the shores. Stringers are then set on top of the shores and joists are laid across the stringers. The plywood deck is then nailed to the joists.

When using engineered wood or metal shoring to support wood panel deck forms, the stringers are supported by metal shores that are adjusted with an adjusting screw at the top. Engineered wood and aluminum beams are available in 8′ to 30′ lengths, in 2′ increments.

Patent Construction Systems

Figure 68-28. *Tubular shores and metal beams are commonly used with heavy construction formwork.*

Safway Steel Products, Inc.

Figure 68-30. *Engineered wood products, such as laminated veneer lumber (LVL) beams and wood I-joists, provide proper support for form soffits.*

When the deck formwork has been completed, rebar is placed over the deck and properly positioned. The floor rebar are tied to rebar in the walls, beams, and columns. **See Figure 68-32.**

FORMWORK CONSTRUCTION

The general sequence for constructing hand-set forms for the first floor of a building is shown in **Figure 68-33.** The procedure for setting forms for subsequent floors is similar.

APA—The Engineered Wood Association

Figure 68-32. *Rebar is placed over the deck of the floor form. Note the opening for the beam extending the length of the deck.*

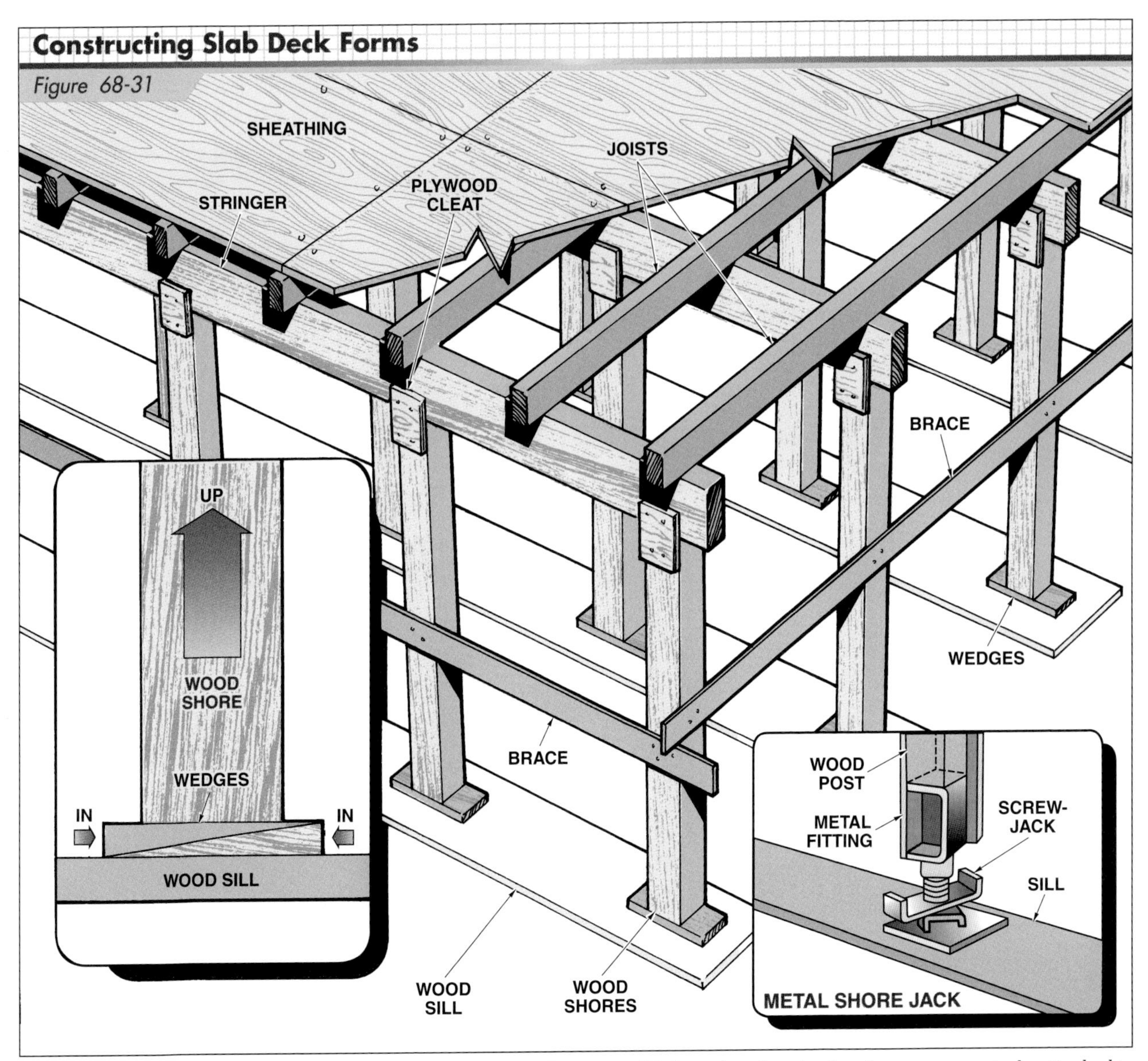

Figure 68-31. *Slab decks may be supported by wood shores and stringers. Wedges placed under the shores or screw jacks attached to the shores provide for vertical adjustment.*

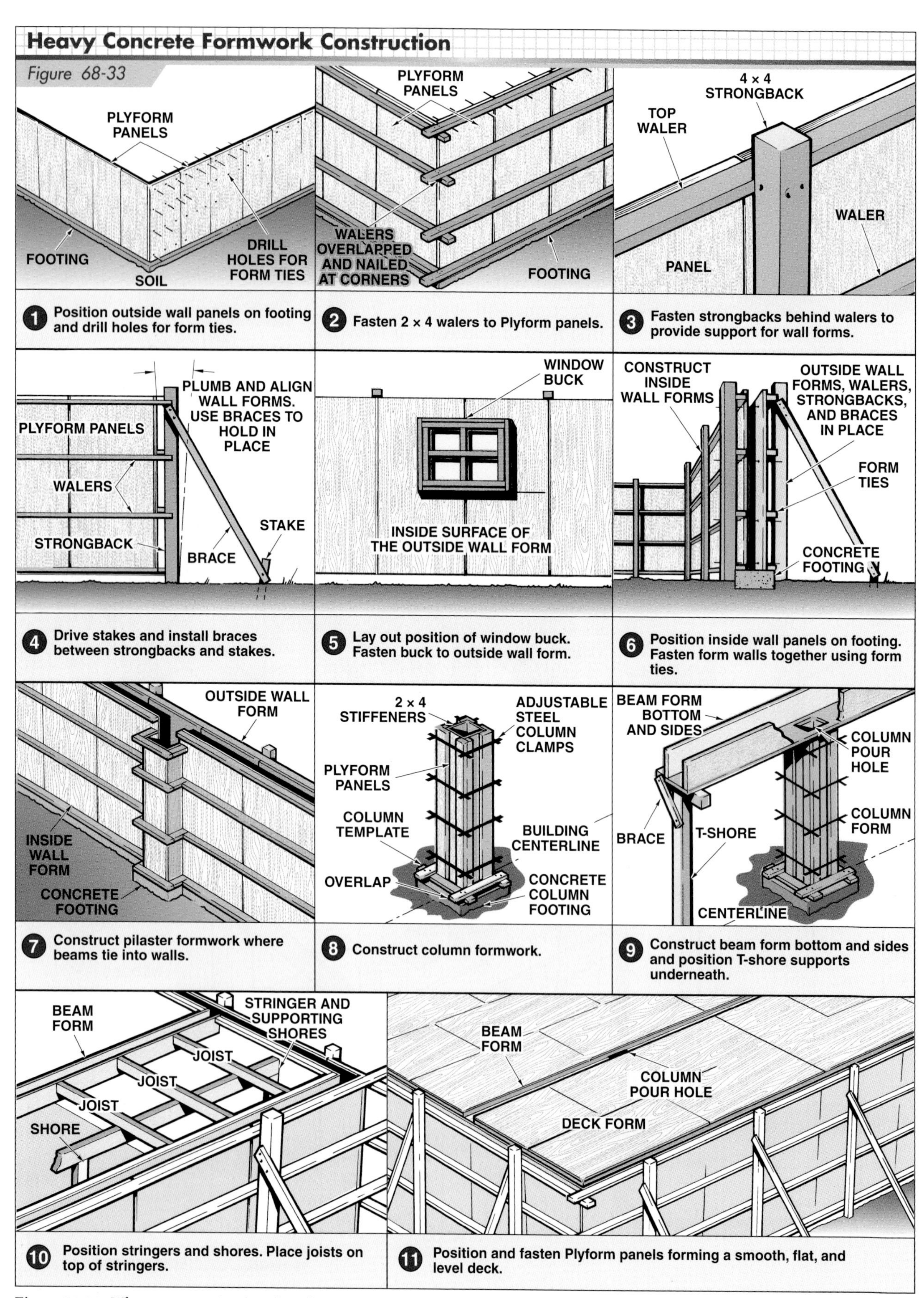

Figure 68-33. *When constructing first-floor heavy construction forms over foundation footings, wall forms are first placed, followed by column, beam, and deck formwork. Note reinforcing steel is not shown.*

When rebar is placed in walls and columns, it will extend above the next floor level. The rebar may project several feet above floor level or possibly to the top of the second-floor walls and columns. **See Figure 68-34.**

APA—The Engineered Wood Association

Figure 68-34. *Rebar for columns usually extends past the floor and beams that are supported by the columns. Rebar extending from this column will be tied to additional steel placed inside the form for the column of the floor above.*

Symons Corporation

Prefabricated reinforced floor panels are positioned with a forklift. Note the columns with the rebar extending from the top along the right side.

When forms are constructed for the outside walls of the second story, the bottoms of the outside form wall panels must be secured to the top of the first-story wall. Anchor bolts are embedded in the concrete of the previously placed wall. **See Figure 68-35.** The anchor bolts extend through walers at the bottom of the panels. Large washers and nuts are tightened over the walers.

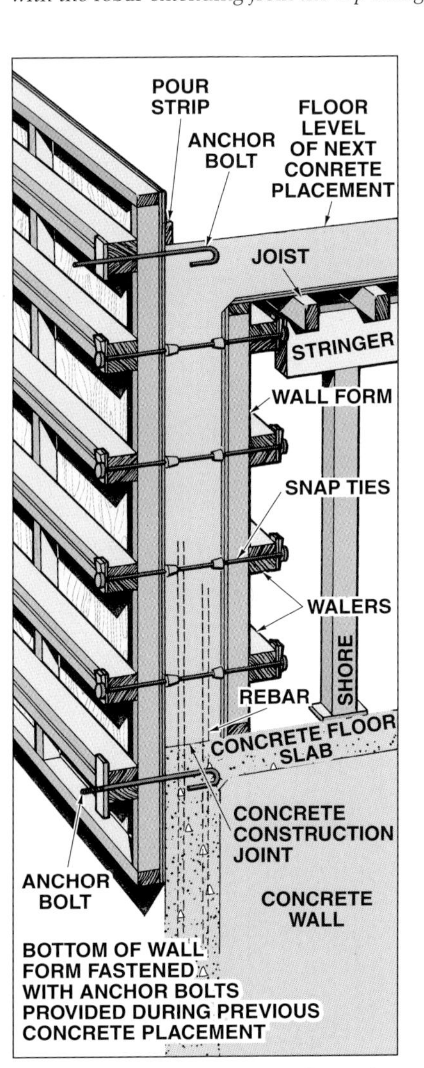

Figure 68-35. *An outside wall panel is fastened to a previously placed floor using anchor bolts which are embedded in the floor slab.*

STAIRWAY FORMS

Concrete stairway form layout is similar to that for a conventional wood stairway. For both concrete and wood stairways, the same procedure is used to calculate the tread and riser dimensions based on the total rise and run of the stairway.

Construction of a concrete stairway form, however, can be more complicated than construction of a wood stairway. On heavy construction projects, the architect may indicate the unit run and unit rise, as well as a permissible tolerance. The unit run and unit rise are then field verified on the job and can be adjusted as necessary. **Figure 68-36** shows a procedure for laying out and constructing an open stairway form. **Figure 68-37** shows a procedure for laying out closed concrete stairway forms.

PREFABRICATED PANEL FORMING SYSTEMS

Most large concrete construction projects use some type of prefabricated panel forming system. Prefabricated panel forming systems range from individual panels that can be placed by hand to large sections of panels that must be raised and placed by crane. Prefabricated panel forming systems are used when numerous reuses of the panels are anticipated. The use of prefabricated panel forms lowers labor and material costs.

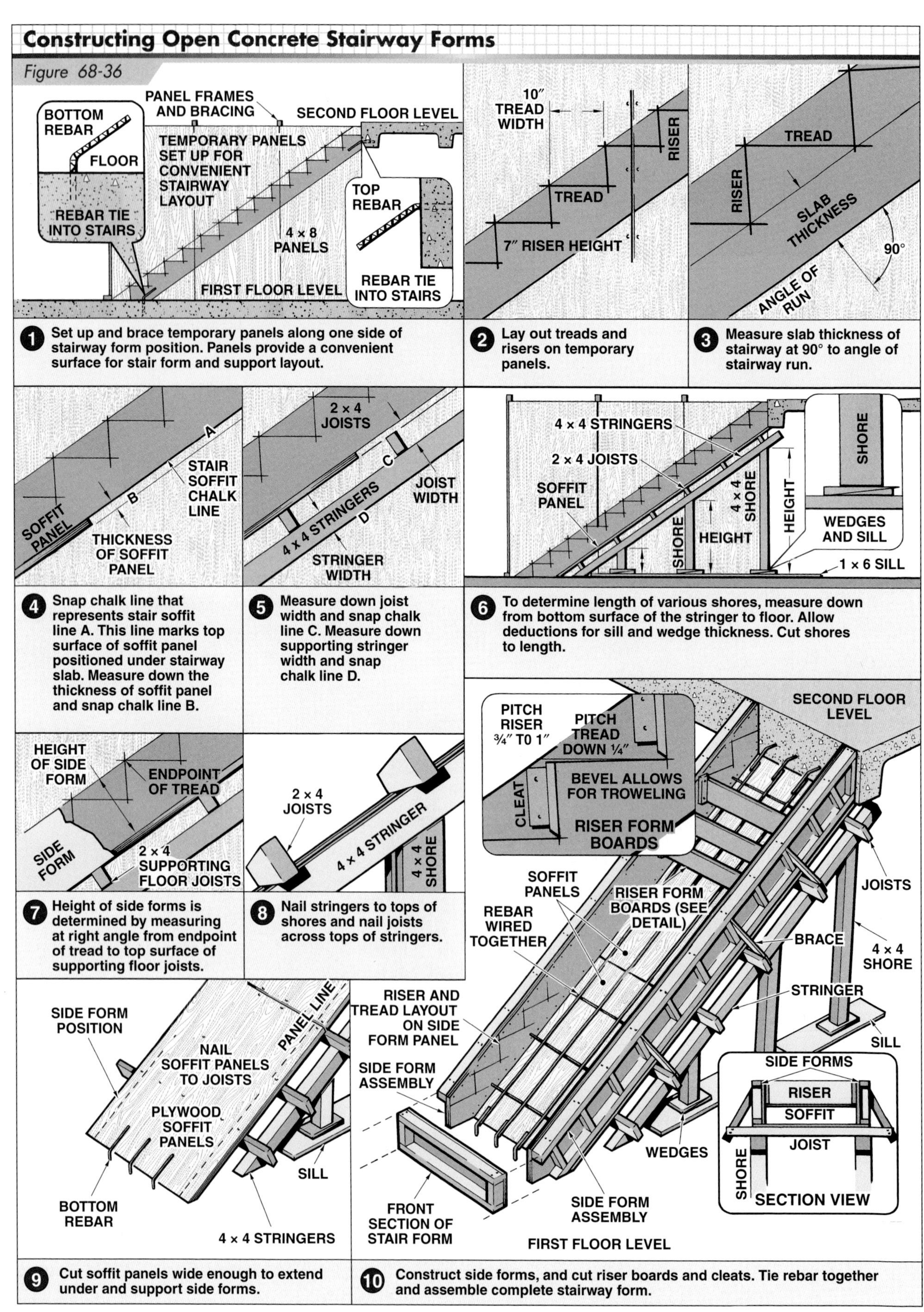

Figure 68-36. *Rebar extending from the floor level above and below is tied to rebar in the stairway.*

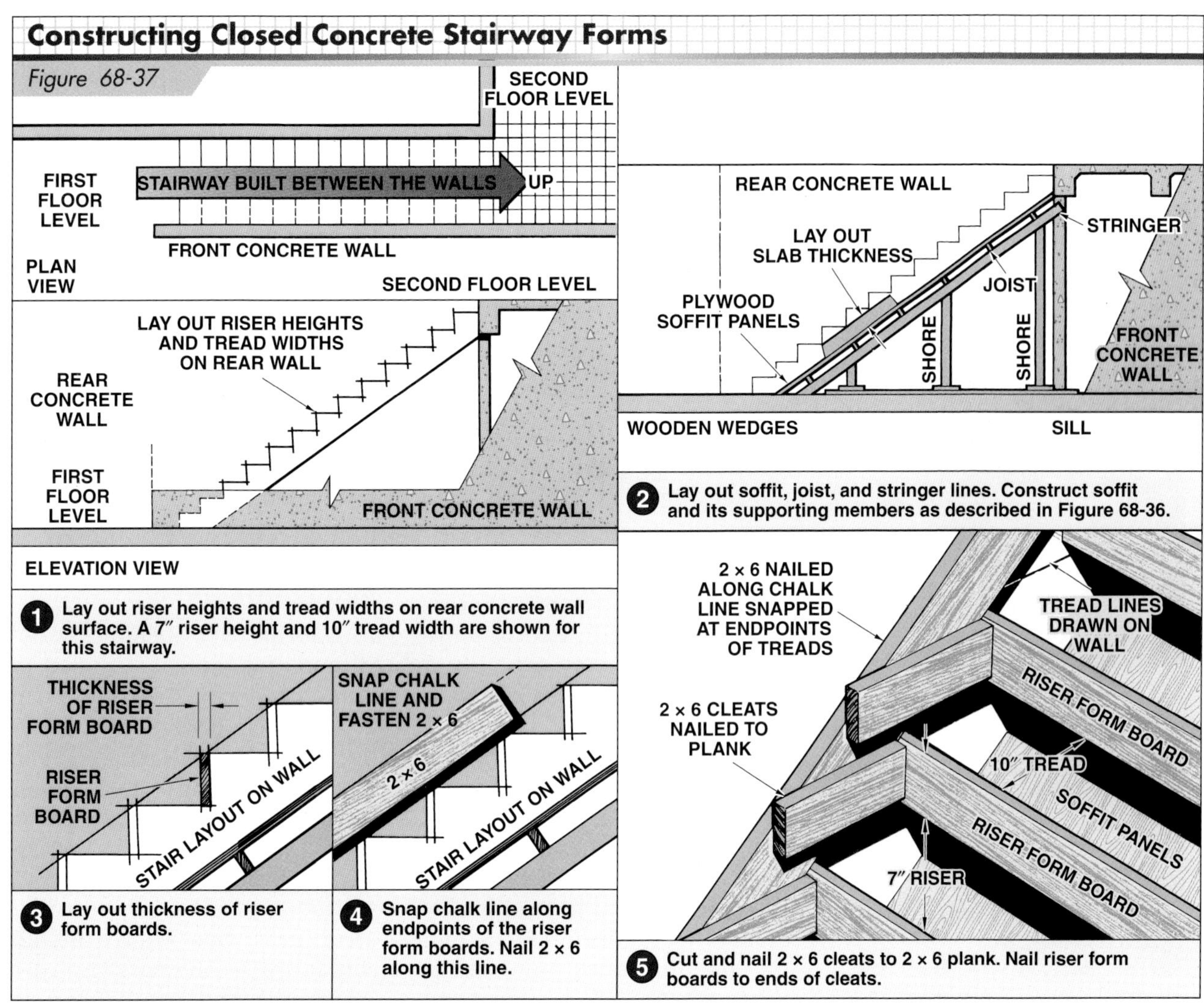

Figure 68-37. *A closed concrete stairway is built between two walls.*

Prefabricated panel forming systems can be rented or purchased from various manufacturers. Special-purpose custom-made panel forms are available for special forming operations. While many prefabricated panels forming systems are similar, each company has specific hardware, installation procedures, and maintenance guidelines. When installing prefabricated panel forms, manufacturer instructions must be followed closely to prevent form damage or worker injury.

Prefabricated Panel Forms

The most commonly used prefabricated panel forms consist of modular panel sections usually 2′ to 4′ in width and 2′ to 10′ in height. Smaller filler pieces of various sizes are also available. Manufacturers also provide the accessories (ties, walers, wedges, braces, etc.) for their particular prefabricated panel forms. Although metal-framed plywood panels are the most common type of prefabricated panel forms used today, the use of all-metal panel forms and plastic panel forms is growing. **See Figure 68-38.**

Metal-Framed Plywood Panel Forms. A metal-framed plywood panel form consists of ½″ or ⅝″ Plyform® set in an aluminum or steel frame. Horizontal metal stiffeners spaced approximately 1′ apart provide additional support. The frames are designed so the panels can be easily replaced when worn or damaged. The panel sections, often called hand-set forms, are light and easy to handle. Slots in the metal-side rails are used to join the panels together with wedge bolts. Walers are secured with metal waler ties. Wire, flat, or round ties space and hold the panels together. Braces are secured to the panels with wedge-shaped metal plates.

All-Metal Panel Forms. Prefabricated all-metal panel forms are made of aluminum or steel. Aluminum hand-set forms consisting of an aluminum face stiffened with an aluminum frame are frequently used in the construction of residential foundations. Steel forms are used in heavy construction. Steel forms can be combined into ganged panel forms, which are widely used in the precast industry. Accessories to assemble, align, and brace steel and aluminum forms are provided by the manufacturer. With proper care and maintenance, steel and aluminum forms can be used indefinitely.

Figure 68-38. *Metal-framed plywood panels are the most common type of prefabricated panel forms used today, but the use of all-metal panel forms and plastic panel forms is growing.*

Plastic Panel Forms. Panel forms made with a strong plywood core and plastic facing have been in use for many years. More recently, there has been a growing use of panel forms made primarily of plastic. Plastic panel forms can be nailed, screwed, and cut as easily as wood. Plastic panel forms also provide the following advantages compared to plywood panel forms:

- perfectly smooth concrete finish
- no swelling or shrinkage
- no loss due to rotting
- cheaper to produce
- longer lifespan
- easier to clean, repair, and recycle

Gang Forms

On structures where concrete must be placed in two or more stages, gang forms are often used. Gang forms are large panels constructed by fastening together a series of smaller panel forms. Walers, strongbacks, lifting brackets, and scaffold brackets are then fastened to the panels. The entire unit is lifted and placed in position by crane. **See Figure 68-39.** When the concrete has set sufficiently, the forms are then removed and moved to another location.

Flying Forms

Flying forms are prefabricated forming systems that usually consist of a wood deck and a metal support system. **See Figure 68-40.** Flying forms may be used to form floor slabs and other structural members such as beams. The forms are placed in position with a crane. **See Figure 68-41.** After placing concrete and allowing it to harden sufficiently, the entire flying form is removed and lifted to the next floor level. The use of flying forms greatly increases production on multistory buildings.

Slip Forms

Slip forms were originally developed for the construction of curved concrete structures such as silos and towers. Slip forming methods have expanded to the construction of rectangular buildings, caissons, building cores, underground shafts, shearwall buildings, communication towers, and a variety of other structures. **See Figure 68-42.** Slip forming can save significant labor, time, and material cost on construction projects.

MEVA Formwork Systems, Inc.

Figure 68-39. *Prefabricated gang forms commonly include walers, strongbacks, lifting brackets, and scaffold brackets. Note the toprails, midrails, and toeboards in place to protect workers against falls and prevent materials from falling from the platforms.*

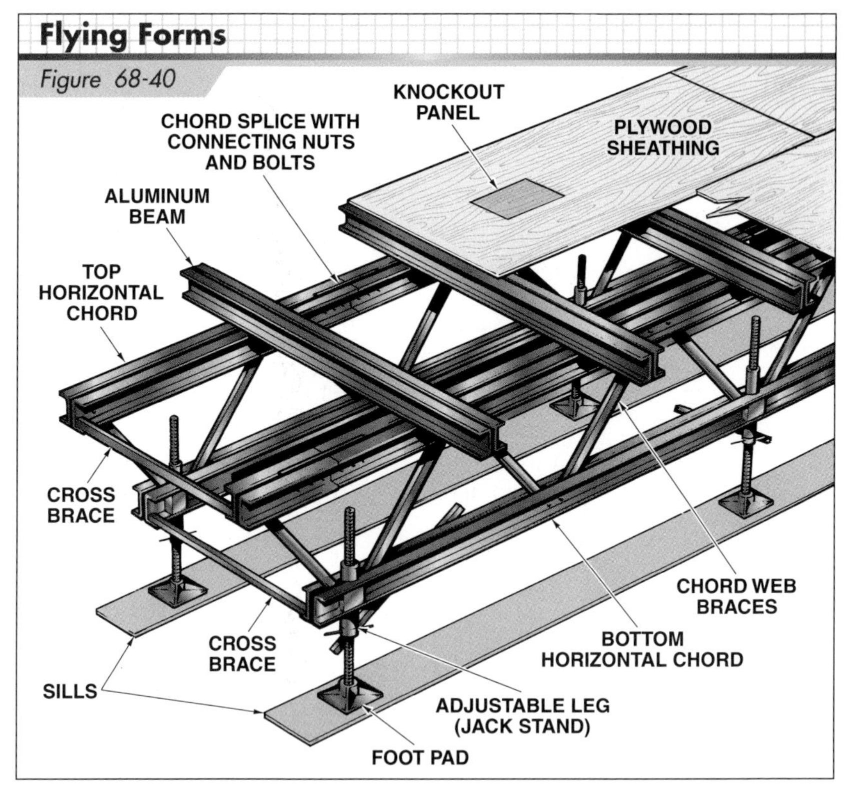

Figure 68-40. *The support for a flying floor form unit is provided by aluminum trusses placed on either side and aluminum beams placed across the trusses. A plywood deck is fastened on top of the beams. Adjustable jacks are used to raise the unit into position.*

Portland Cement Association

Figure 68-42. *Slip forms are commonly used in the construction of tall buildings. Concrete is transported to the upper level of the building and is placed in forms using a pumping apparatus.*

Patent Construction Systems Safway Steel Products, Inc.

Figure 68-41. *Flying forms used for floors or other structural members are set into place by crane.*

Most slip forms consist of 4′ high inner and outer walls of ⅜″ to ¾″ plywood panels supported by a 2 × 4 studs and 2 × 6 walers. Inner and outer form walls are slightly tapered outward at the bottom (⅛″ per foot) to reduce the amount of drag on the concrete as the form is raised. The walls are secured together with steel *cross beams* and *yoke legs.* Cross beams tie together the tops of opposing yokes and provide a mounting surface for the hydraulic jacks. The yoke legs are made of steel and are spaced approximately 6′ apart along the length of the slip form. The yoke legs are adjusted to the wall width and are fastened to the cross beams at the top end and to the walers along the bottom end. Hydraulic jacks are mounted on the cross beams. Slip forms are raised by electrically or pneumatically powered hydraulic *jacks* that climb *jackrods* extending into the form. Perfect coordination of the hydraulic jacks is essential for accurate alignment of the forms; all hydraulic jacks must be lifted at the same time and at the same rate. Jackrods are threaded at each end. Additional lengths of jackrod are fastened to the upper threaded end when required. **See Figure 68-43.**

Slip forms move up the jackrods as concrete is being placed, at a rate ranging from 2″ to 70″ per hour. The climbing speed depends on the type of structure, rate of concrete placement, and how quickly rebar and built-ins can be placed. *Built-ins* consist primarily of door and window bucks and beam pockets. Provision must also be made for the placement of brackets, anchors, utility (plumbing and electrical) recesses, and similar items. Slip forms do not operate 24 hours a day on most types of slip form construction. Therefore, construction joints occur on the surface of the structure.

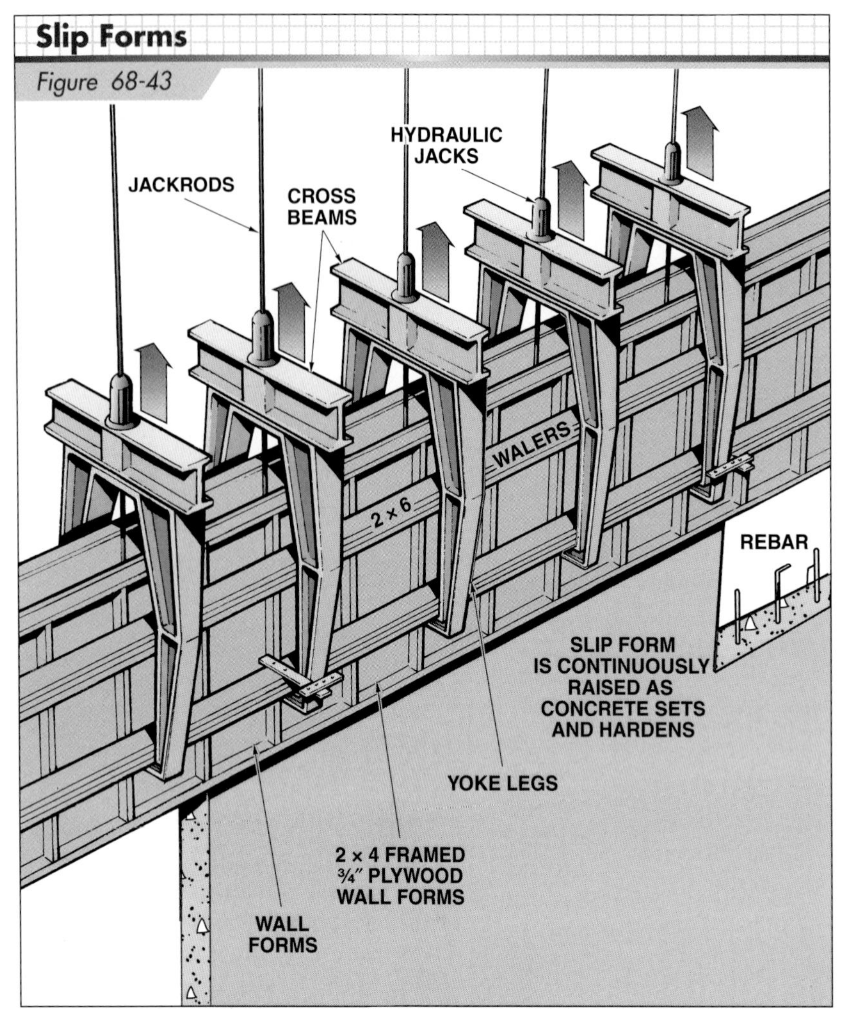

Figure 68-43. *The basic design of a standard slip form includes jackrods, hydraulic jacks, cross beams, and yoke legs. Additional features, such as scaffold, are custom made for the structure being erected.*

CONCRETE SLABS ON METAL DECKS

For some commercial buildings, corrugated metal decking is commonly topped with several inches of concrete to form floor slabs. The corrugated metal decking acts as the bottom form for the concrete and is left in place after the concrete has hardened, becoming an integral part of the floor system. Steel angles that act as pour stops make up the outer edge of the slab and are placed around any openings in the slab. Steel angles are also left in place after the concrete has been placed. The concrete thickness varies depending on the loads to be supported. A common concrete thickness is 6″.

Composite and Noncomposite Concrete Slabs

Concrete slabs on corrugated metal decking can be noncomposite or composite. A *noncomposite slab* is an elevated concrete slab independent of the structural steel in supporting loads after the concrete has hardened. Because the concrete is not connected physically to the steel structure, a noncomposite slab cannot support any tensile loads. For this reason, noncomposite slabs are not used as often as composite slabs.

A *composite slab* is an elevated concrete slab that is mechanically attached to the structural steel framing of a building. The term composite describes a method of combining different construction elements to support a load and resist outside forces. One important feature of a composite slab system is that it does not usually require any of the temporary formwork that is required by typical elevated slabs. Composite slabs are lower in weight than traditional elevated concrete slabs because they contain lower quantities of conventional reinforcement. The main components of a composite slab system are structural steel members, corrugated metal decking, shear studs, and reinforcement. **See Figure 68-44.**

Structural Steel Members. Structural steel members make up the framework for a composite slab. The most common members used to support the corrugated metal decking are steel beams. If shoring is not used, the steel beams may be slightly cambered during fabrication to offset the anticipated deflection of the beam under the weight of the concrete. *Camber* is the slight upward curve in a structural member designed to compensate for deflection of the member under load. After the concrete has been placed, the beams deflect to their intended position, which results in a flat and level floor surface.

Corrugated Metal Decking. Corrugated metal decking provides a work platform during construction and a base for the concrete slab. The metal decking is usually galvanized steel, stainless steel, or aluminum that has been corrugated to provide increased rigidity and strength. Corrugated metal decking is available in lengths from 6′ to 50′ and widths of 24″, 30″, and 36″. Corrugated metal decking may be manufactured with ribs, tabs, or louvers to increase its bond with the concrete. Corrugated metal decking is secured to the top of the structural steel members using self-tapping screws or by welding them together.

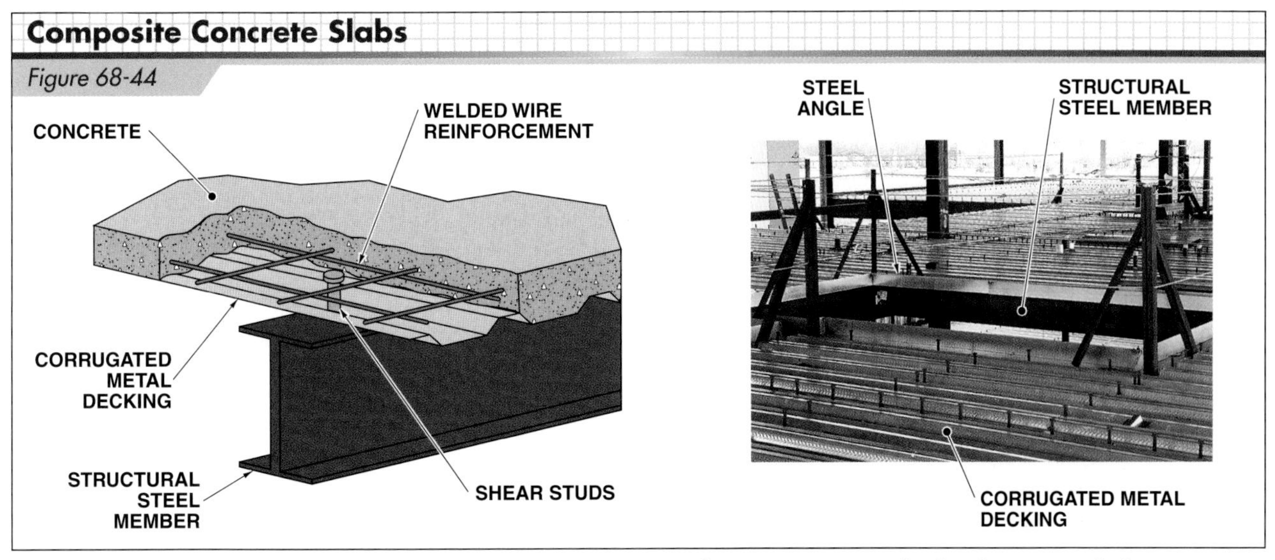

Figure 68-44. *Composite concrete slabs combine different construction elements to support a load and resist outside forces.*

Shear Studs. Shear studs play a critical role in composite concrete slabs. A *shear stud* is a metal post that directly connects the hardened concrete to the framework of structural steel members below. Most often, shear studs are welded through the corrugated metal decking to the top of the steel beams. However, some systems bolt or fasten the studs to the steel beams with powder-actuated drive pins. Common shear stud diameters are ⅜″, ½″, ⅝″, ¾″, and 1″. The minimum length of a shear stud should be at least 1½″ longer than the depth of the deck rib with at least ½″ of concrete over the top of the stud. **See Figure 68-45.**

Slab Reinforcement. Composite concrete slabs are reinforced with welded wire reinforcement and rebar. Welded wire reinforcement is used throughout the slab. Rebar is generally used to reinforce particular areas of the slab, such as around openings. Fiber reinforcement may be added to the concrete mix as well to increase the tensile strength and to control cracking.

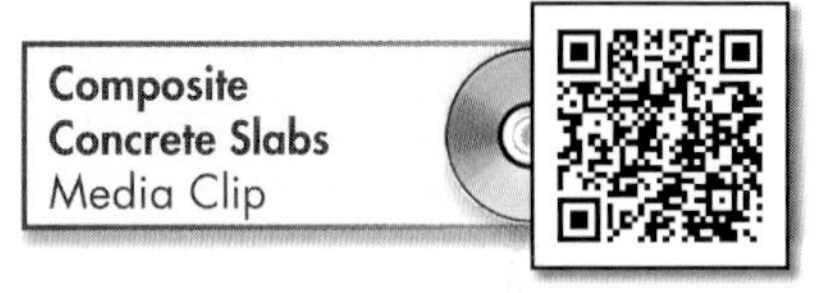

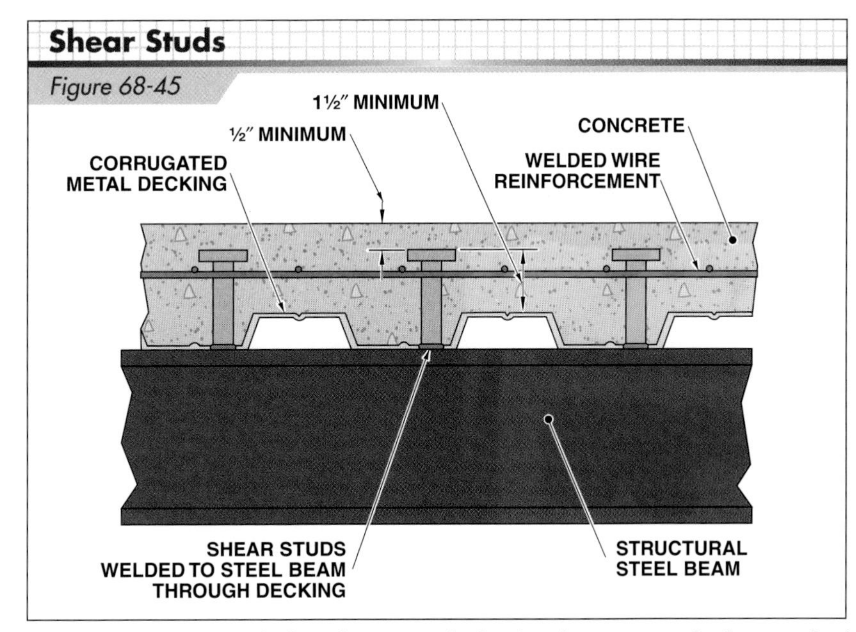

Figure 68-45. *Shear studs directly connect the hardened concrete to the framework of structural steel members below.*

UNIT 69 Concrete Placement for Heavy Construction

OBJECTIVES

1. List and describe methods of placing concrete.
2. Describe procedures for preventing segregation of concrete.
3. Describe how cold joints in concrete can be avoided.
4. List common concrete admixtures.
5. Describe common concrete tests.
6. Describe the effects of weather on concrete.
7. List and describe common types of concrete floor slabs.
8. Describe how concrete slabs are created.
9. Describe the function and formation of control joints and expansion joints.
10. Explain the function of wall expansion joints.
11. Describe the concrete curing process.

Concrete for heavy construction is prepared to specification at a batch plant and delivered to a job site with a transit-mix truck. **See Figure 69-1.** For projects in isolated areas, stationary mixers may be used to mix concrete on the job site. The composition of concrete and various formulas for concrete mixtures are discussed in Unit 37.

Figure 69-1. *Ready-mixed concrete is prepared in a batch plant and deposited in transit-mix trucks for delivery to a job site.*

CONCRETE PLACEMENT OPERATIONS

For the basement and lower floors of a tall concrete building, concrete is usually placed (poured) using a *concrete pump.* **See Figures 69-2** and **69-3.** A concrete pump is a truck-mounted piece of equipment that moves concrete, through a pipeline attached to a boom, to a distant location. Concrete is deposited into the pump hopper from a transit-mix truck and transported through the pipeline using a reciprocating or screw pump.

The ancient Romans were the first to use a form of concrete, in 27 BC. A volcanic ash known as pozzolana was mixed with slaked (crumbled) lime and sand. The mixture set hard and also set underwater.

As the building gains in height, either freestanding or climbing tower cranes are used to transport concrete to upper levels in *concrete buckets.* Concrete is deposited in the concrete buckets from a transit-mix truck.

ECO-Block, LLC

Figure 69-2. *A common method of placing concrete is pumping. Stabilizers on the concrete pump truck are positioned before the boom is extended to pump the concrete. Here, concrete is pumped into insulating concrete forms (ICFs).*

Concrete buckets are available in various shapes and designs and range from ⅓ cubic yard (cu yd) to 12 cu yd in capacity. Typical concrete buckets have a ¾ cu yd to 1 cu yd capacity.

Portland Cement Association

Figure 69-3. *Concrete is pumped through a pipeline, which is supported by a boom and cables. A flexible hose at the end of the pipeline allows the concrete to be placed where desired.*

A concrete bucket is raised by cable and positioned, if possible, directly over the wall form or floor slab where the concrete is to be placed. The concrete is then released by opening the gate at the bottom of the bucket. **See Figure 69-4.**

Power buggies may be used to transport concrete or other materials on a job site. Concrete is deposited in a power buggy and is then moved to its final location. The front-end bucket on the buggy is then tilted to discharge the concrete into the forms. **See Figure 69-5.**

Power buggies are available in various sizes with bucket payload capacities up to 12 cu ft. Power buggies are used to transport materials such as sand and gravel as well as fresh concrete.

Figure 69-4. *A concrete bucket is commonly used to place concrete in wall forms. The bucket being used on this job is equipped with a bottom dump chute that makes concrete placement convenient. The worker at the right is pulling down on the handle that opens the gate at the bottom of the bucket to release the concrete.*

Figure 69-5. *Power buggies are used to transfer concrete over short distances and deposit it in place.*

Concrete is a mixture of a small amount of cement added to a much larger quantity of fine and coarse aggregate (sand and gravel). When water is added to the mixture, the cement acts as the agent to bond together the aggregate. The cement and aggregate ingredients must be thoroughly mixed together for the concrete to properly set and harden. Incomplete mixing of the mixture can cause *segregation,* a condition in which the fine aggregate-cement ingredients of the concrete separate from the coarse aggregate. Improper concrete placement procedures can also cause segregation. If concrete has been properly mixed before being discharged, segregation can be avoided using the following procedures:

- The drop of concrete should always be vertical (straight down), and not angled.
- An *elephant trunk* should be attached to the lower end of a drop chute to deposit concrete into deep walls or tall columns. **See Figure 69-6.**
- *Drop chutes* should be used to prevent concrete from striking rebar or the side of the form above the level of placement. **See Figure 69-7.**
- The free-fall distance of concrete should be from 4′ to 6′ maximum.

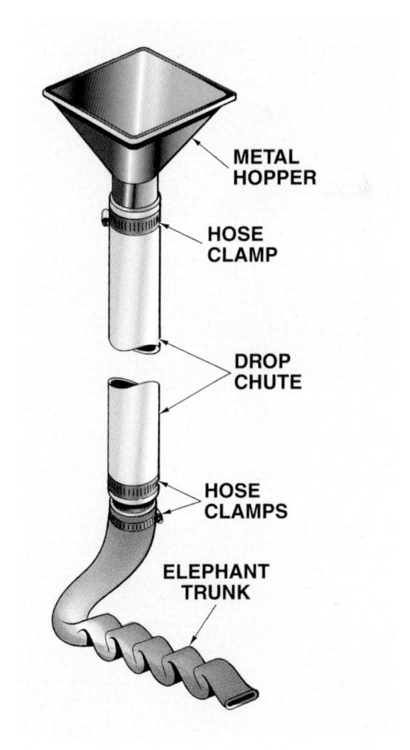

Figure 69-6. *An elephant trunk, when attached to a hopper and drop chute, allows concrete to be properly placed at the bottom of a deep form such as a tall column.*

HOPPER AND DROP CHUTE PREVENT CONCRETE FROM STRIKING FORM WALLS
CONCRETE HITS FORM WALLS
NARROW CONCRETE FORM
CORRECT
INCORRECT

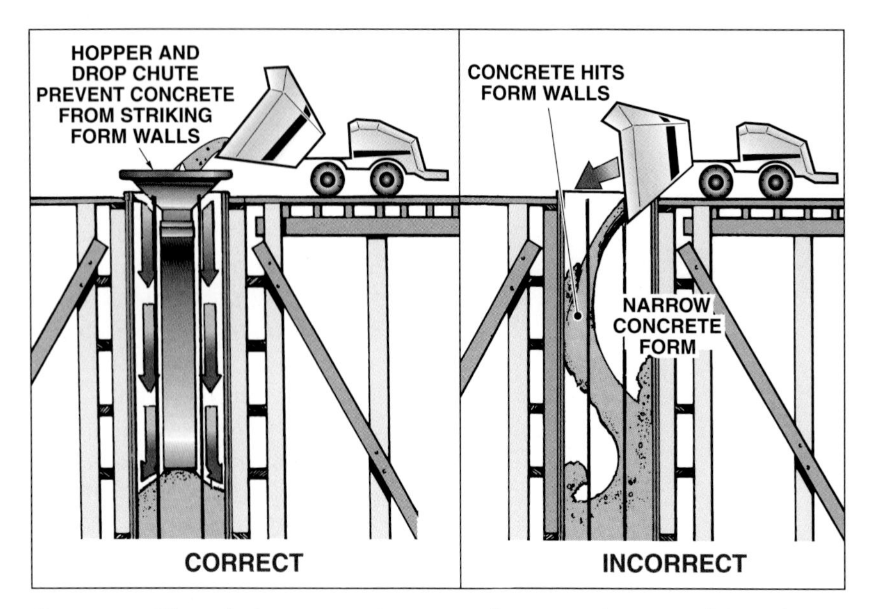

Figure 69-7. *When placing concrete in a narrow form, use a hopper and drop chute to prevent concrete from striking against the rebar and the form walls.*

Concrete for a form should not be deposited in one corner; rather, it should be discharged from different positions until an even *lift* (layer) has been placed in the form. **See Figure 69-8.** The procedure should then be repeated until enough lifts have been placed to fill the form. Thinner lifts are recommended in forms containing heavier or more closely spaced rebar.

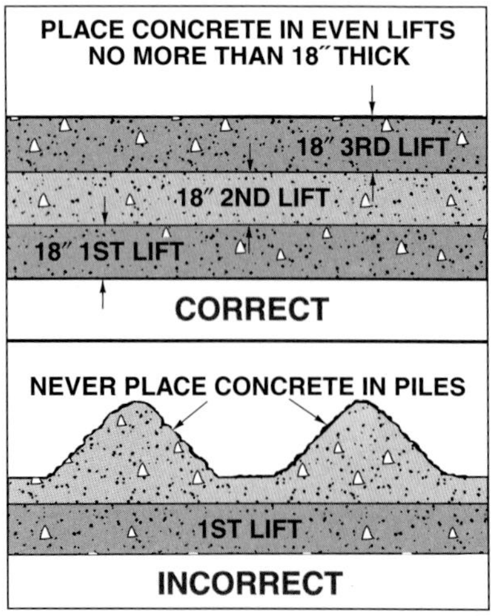

PLACE CONCRETE IN EVEN LIFTS NO MORE THAN 18″ THICK
18″ 3RD LIFT
18″ 2ND LIFT
18″ 1ST LIFT
CORRECT
NEVER PLACE CONCRETE IN PILES
1ST LIFT
INCORRECT

Figure 69-8. *Concrete should be placed in forms in even lifts.*

While concrete is being deposited, it must be *consolidated* to create a close arrangement of solid particles by reducing excess space between the particles. When multiple lifts of concrete are placed, each new lift is consolidated with the lift below it to prevent cold joints in the concrete. *Vibration* is an effective method for consolidating concrete. *Internal* (immersion) *vibrators* are used most often for heavy concrete construction. Internal vibrators are powered by electricity, gas, or compressed air. **See Figure 69-9.**

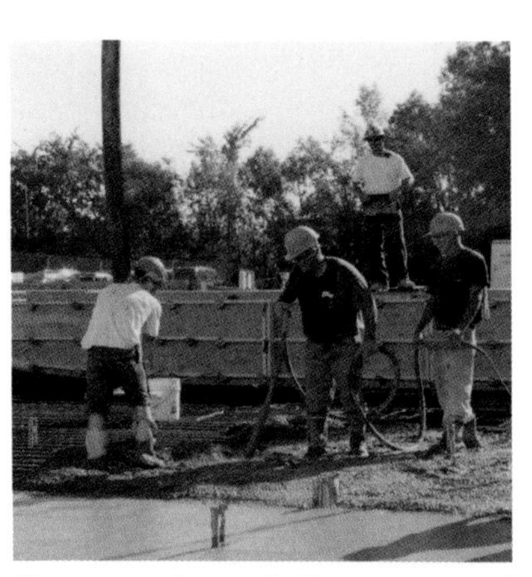

Figure 69-9. *Internal vibrators are used to consolidate concrete while it is being placed.*

Systematic vibration of each new lift of concrete is necessary to provide proper consolidation. The vibrator should penetrate a few inches into the previous lift at regular intervals for a period of 5 seconds to 15 seconds. If the concrete is not penetrated with the vibrator at various points and with sufficient depth, improper consolidation between the two lifts results. To aid consolidation in concrete placed in a sloping layer, start vibration at the bottom of the slope so compaction is increased by the weight of the newly added concrete. **See Figure 69-10.**

VIBRATOR HEAD
2ND LIFT
1ST LIFT OF CONCRETE
CORRECT
INCORRECT
VIBRATOR HEAD
CORRECT
INCORRECT

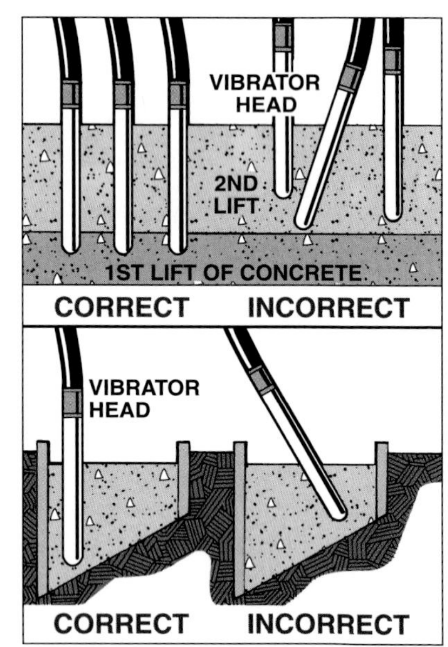

Figure 69-10. *The metal head of an internal vibrator is placed into the concrete for a period of 5 seconds to 15 seconds and should penetrate a few inches into the previous lift.*

External vibrators may be used to consolidate concrete in column or high wall forms. An external vibrator is attached to the form wall at predetermined locations. External vibrators generate and transmit vibration waves from the exterior to the interior of the concrete at a frequency of 6000 to 12,000 vibrations per minute. External vibrators are electrically, pneumatically, or hydraulically powered.

Compression test cylinders are filled at the job site. Note the cylinders in the background that were previously filled.

Typical surface defects resulting from improper consolidation are honeycombs, air pockets, and sand or gravel streaks. These defects can also be caused by faulty form construction or an improper concrete mixture.

Concrete Admixtures

Admixtures are ingredients that are added to the basic concrete mixture to modify its properties. Admixtures are added to the mixture prior to or during mixing to achieve the desired results.

Air-Entraining Admixtures. *Air-entraining admixtures* produce microscopic air bubbles in a concrete mixture. The air bubbles improve the workability of concrete as well as its resistance to freezing and thawing. Air-entraining admixtures are of particular benefit in concrete placed during cold weather.

Water-Reducing Admixtures. *Water-reducing admixtures* are added to a concrete mixture to reduce the amount of water required to produce a desired mixture. Concrete strength is increased when using water-reducing admixtures, as long as the water content in the mixture is reduced and the cement content and slump remain constant. Many water-reducing admixtures also retard the setting time of the concrete, which allows a slower curing process.

Accelerating Admixtures. Concrete gains strength very slowly at low temperatures. *Accelerating admixtures* reduce the setting time of concrete and improve the strength of concrete at an early stage. Accelerating admixtures are especially useful in cold weather conditions to allow concrete to set before it freezes.

Set-Retarding Admixtures. *Set-retarding admixtures* are added to concrete to extend its setting time. Set-retarding admixtures are useful in hot weather conditions, when concrete sets so rapidly that it cannot be finished properly. In addition, set-retarding admixtures may be added to delay the set in some applications such as when placing concrete in a large foundation or when more time is needed to complete finishing.

Pozzolans. *Pozzolans* used as concrete admixtures are permicite, volcanic ash, tuffs, diatomaceous earth, fly ash, calcined clays, and shale. By themselves, pozzolans would add little value to concrete. However, in the presence of moisture and under ordinary temperature conditions, pozzolans react with other cement materials to form compounds that have cementing properties. As a result, pozzolans may be used as a partial replacement for cement. In addition, pozzolans may be used to help reduce internal temperatures when concrete is being placed in massive structures such as dams.

Some pozzolans are used to reduce concrete expansion caused by alkali-reactive aggregate. Other pozzolans improve the sulfate resistance of concrete. However, pozzolans also may create a disadvantage. As a rule, concrete with pozzolans requires more mixing water to achieve the same slump (consistency) as concrete without pozzolans. In addition, when hardened, concrete with pozzolans has a tendency to crack because of greater contraction while drying.

Other Admixtures. Other admixtures are available to improve the workability of the concrete, to help waterproof it, and to increase bonding strength when new concrete is placed against existing concrete. Grouting agents are available to improve the quality of grout used for a variety of filling, patching, and repair purposes. Gas-forming agents are sometimes added to concrete and grout to cause a slight expansion before hardening. Coloring admixtures impart a desired color to concrete.

CONCRETE TESTS

Various test procedures are involved in the manufacture, mixing, and placement of concrete. Many test procedures are conducted in a laboratory. Two tests that are performed or prepared at a job site are the *slump test* and *compression test*. Carpenters are not usually involved in these tests, but they should understand why the tests are conducted.

Slump Tests

A slump test measures the *slump,* or consistency, of fresh concrete. Slump directly affects flowability of concrete during placement. ASTM C143, *Standard Test Method for Slump of Hydraulic Cement Concrete,* details the slump test procedure. Samples are taken at intervals during discharge of the concrete into the forms. A slump cone made of sheet metal, 4″ in diameter at the top, 8″ diameter at the bottom, and 12″ high, is used. The inside of the slump cone is dampened before use and placed on a smooth, nonabsorbent surface for the test. Concrete is placed in the cone in three equal layers and is rodded with a ⅝″ diameter by 24″ long rod by moving the rod up and down in the concrete. Excess concrete at the top of the slump cone is removed. The cone is then removed with a slow, even motion. The cone is then inverted and placed next to the concrete sample. A ruler is then used to measure the concrete slump. **See Figure 69-11.** Some allowable slump ranges for various types of construction are shown in **Figure 69-12.**

Portland Cement Assocation

A slump test is a field test that measures concrete consistency.

Compression Tests

A concrete mixture varies in strength according to the proportions of its ingredients. Ingredient proportions are determined for each job according to certain factors such as the shape and size of structural members to be built (walls, slabs, beams, columns), their required strength, and the environmental conditions they will face.

Engineers involved in designing the structure or concrete field specialists calculate the *compressive strength* required for its members. Compressive strength is the pounds per square inch (psi) of force the concrete can withstand 28 days after it has been placed. Twenty-eight days is the average time required for concrete to gain its full strength.

The *water-cement ratio* largely determines compressive strength. The water-cement ratio is calculated by dividing the weight of the water used in 1 cu yd of concrete by the weight of the cement used in 1 cu yd. Thus, if the weight of the water used in 1 cu yd is 10 lb, and the weight of the cement is 20 lb, the water-cement ratio is 0.50 (10 ÷ 20 = 0.50). **Figure 69-13** shows a table of several water-cement ratios and their relation to compressive strength.

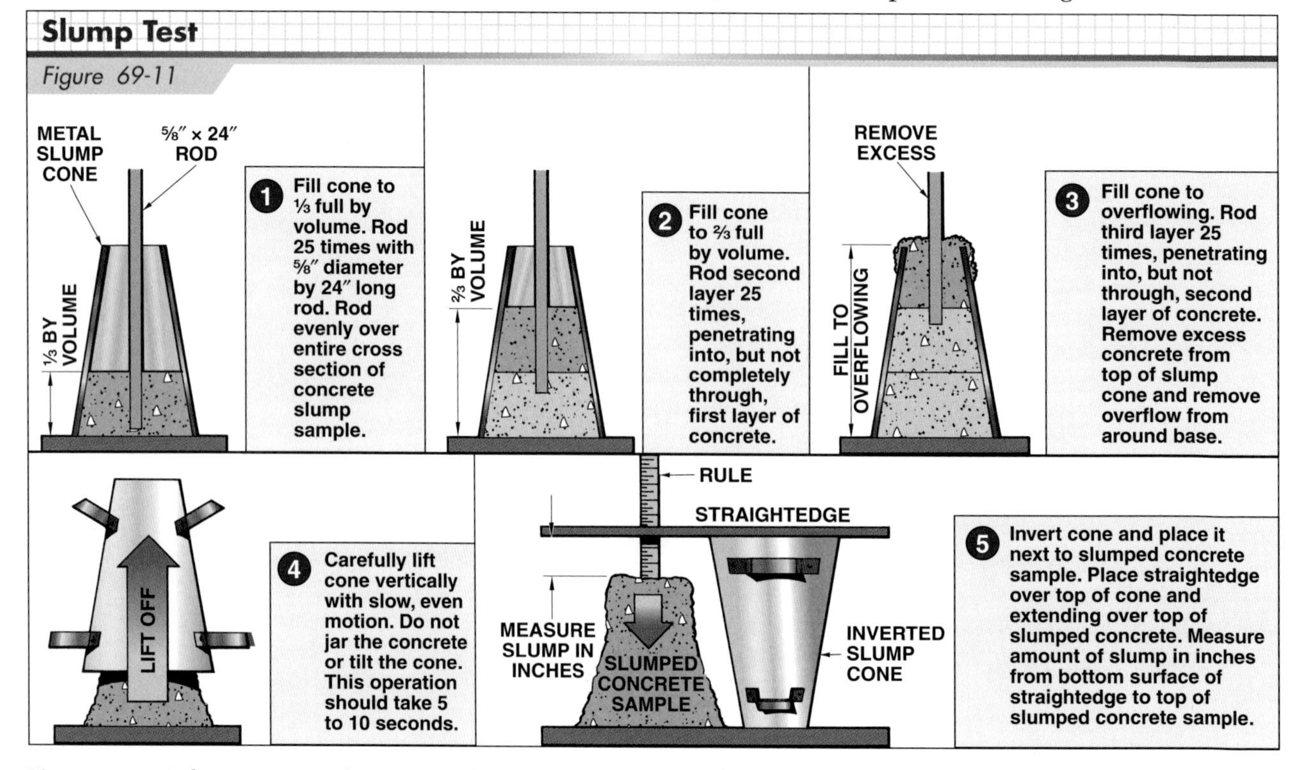

Figure 69-11. *A slump test is used to measure the consistency of concrete.*

ALLOWABLE SLUMP RANGES

Concrete Construction	Slump*	
	Maximum†	Minimum
Reinforced foundation walls and footings	3	1
Plain footings, caissons, and substructure walls	3	1
Beams and reinforced walls	4	1
Building columns	4	1
Pavements and slabs	3	1
Mass concrete	2	1

* in in.
† May be increased 1″ for consolidation by hand methods such as rodding and spacing

Figure 69-12. *Concrete slump varies for different types of construction.*

WATER-CEMENT RATIOS

Compressive Strength at 28 Days*	Water-Cement Ratio†	
	Non-Air-Entrained Concrete	Air-Entrained Concrete
6000	0.41	—
5000	0.48	0.40
4000	0.57	0.48
3000	0.68	0.59
2000	0.82	0.74

* by weight in lb
† in psi

Figure 69-13. *The compressive strength increases as the water-cement ratio decreases. Note that air-entrained concrete requires a lower water-cement ratio than non-air-entrained concrete.*

ASTM C172, *Standard Practice for Sampling Freshly-Mixed Concrete,* and ASTM C31, *Standard Practice for Making and Curing Concrete Test Specimens in the Field,* detail the sampling process, and handling, storing, and curing concrete. For the compression test, at least three samples are taken and placed into cylinders while the concrete is being discharged. The cylinders are covered with a lid and carefully stored on the job site in a place where they are protected from jarring. After 24 hours the cylinders are taken to a laboratory, where the concrete sample is removed from the cylinder and cured for 28 days.

After the 28-day curing period, the sample is capped with a thin layer of capping compound. After the cap has hardened, the sample is placed in a hydraulic press that is equipped with a pressure gauge. Pressure is exerted until the sample is broken. The pressure gauge registers the load required to break the sample. **See Figure 69-14.** ASTM C39, *Standard Test Method for Compressive Strength of Cylindrical Concrete Specimens,* describes the procedure for testing concrete samples.

EFFECTS OF WEATHER ON CONCRETE

Extreme hot or cold weather conditions pose special problems for placing concrete if proper precautions are not taken.

Effects of Hot Weather

Some challenges caused by hot weather are increased water demand, early slump loss, and increased rate of setting, which result in strength loss and the possibility of cracking. The ideal ambient (surrounding air) temperature during concrete placement is 50°F to 70°F. However, many specifications require that the ambient temperature be less than 90°F.

One way to maintain low concrete temperature is to keep the concrete ingredients as cool as possible before mixing. Water, before being placed in the mixture, can be cooled by refrigeration, liquid nitrogen, or crushed ice. Care must be taken to ensure that the ice is completely melted before placement to prevent weak spots in the concrete. Aggregate and cement can be stored in a shady place before use and kept moist by sprinkling with water.

Equipment such as mixers, chutes, hoppers, and pump lines should be shaded or covered with wet burlap prior to placing the concrete. The forms, rebar, and the subgrade should also be fogged with cool water before the concrete placement begins. Water-reducing admixtures may also be incorporated into the mixture.

ELE International, Inc.

Figure 69-14. *A concrete sample is placed in a hydraulic press equipped with a pressure gauge to determine the compressive strength. Pressure is placed on the sample until it breaks.*

During cold weather placement, the concrete and forms are covered with tarpaulins to retain the heat. Gravel is placed over the tarpaulin to hold it in position.

Effects of Cold Weather

Hydration takes place at a slower rate when temperatures are low. Hydration is the chemical reaction between cement and water that bonds molecules, resulting in the hardening of the concrete mixture.

When the temperature drops to 0°F or below, concrete ingredients should be heated. Raising the temperature of the mix water is the most efficient and effective method for increasing concrete temperature. Care should be taken to ensure the mix water does not exceed 140°F since contact with hot mix water can cause the cement to flash set, forming cement balls. Air-entraining or accelerating admixtures may also be used when placing concrete in cold weather conditions. After concrete placement, the forms and concrete are covered with tarpaulins or plastic film to retain the heat. If necessary, enclosed areas can be further warmed with heaters. In some cases, straw or hay may be spread over the tarps to retain concrete heat.

Concrete should not be placed on frozen soil; therefore, precautions must be taken to prevent frost from setting into the ground where concrete is to be placed. After the forms have been constructed, the insides of the forms and the rebar must be kept free of ice. Deicing chemicals are often used to prevent ice from forming.

CONCRETE FLOOR SLABS

A concrete floor slab may be laid directly on the ground (slab-at-grade) or it may be supported by walls, beams, and columns.

Superflat floors may be required for certain construction projects such as warehouse floors. Forklifts at these facilities may require near-perfect floor levelness to avoid spilling loads.

Slab-at-Grade Floor Slabs

Slab-at-grade concrete floor slabs are placed at the basement or first floor level and receive direct support from the ground. Concrete for slab-at-grade floor slabs may be placed at the initial stages of foundation work, in which case the foundation wall also acts as the perimeter form for the floor slab.

Ground preparation is important for a slab-at-grade floor slab, as the *subgrade* (ground below the slab) must be able to uniformly support the slab and any other loads carried by the slab. The subgrade must be properly graded, compacted, and moistened. Vegetation and other foreign matter must be removed. Soft or muddy spots must be dug out and replaced with well-compacted fill.

Before a concrete for a slab-at-grade floor slab is placed, a layer of granular rock (gravel) is laid down. A vapor barrier in the form of a waterproof membrane (polyethylene, butyl rubber, or asphalt-impregnated sheets) is placed on top of the gravel.

The perimeter form of a floor slab establishes the outside finish levels of the slab. To maintain the proper floor levels throughout, a system of *screed rails* and *screed supports* is set up. The screed system consists of a straightedge that moves along screed rails held by some type of support. **See Figure 69-15.** Screed rails are placed at intervals so the straightedge bottom is set to the desired level of the concrete slab. The straightedge resting on the screed rails *strikes off* (levels off) the concrete. Metal screed supports are commonly used. **See Figure 69-16.** The screed rails and supports are removed from the concrete as soon as the area has been struck off. Mechanical vibratory screeds that consolidate as well as strike off the concrete are also available.

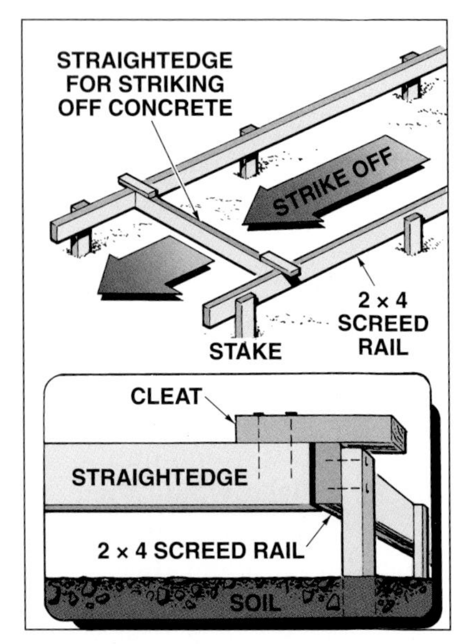

Figure 69-15. *Screed rails are placed at intervals so the bottom of the straightedge is set to the desired level of the concrete slab. The straightedge is moved back and forth to strike off the concrete. The screed rails are removed from the concrete as soon as the area has been struck off.*

Metal Screed Supports

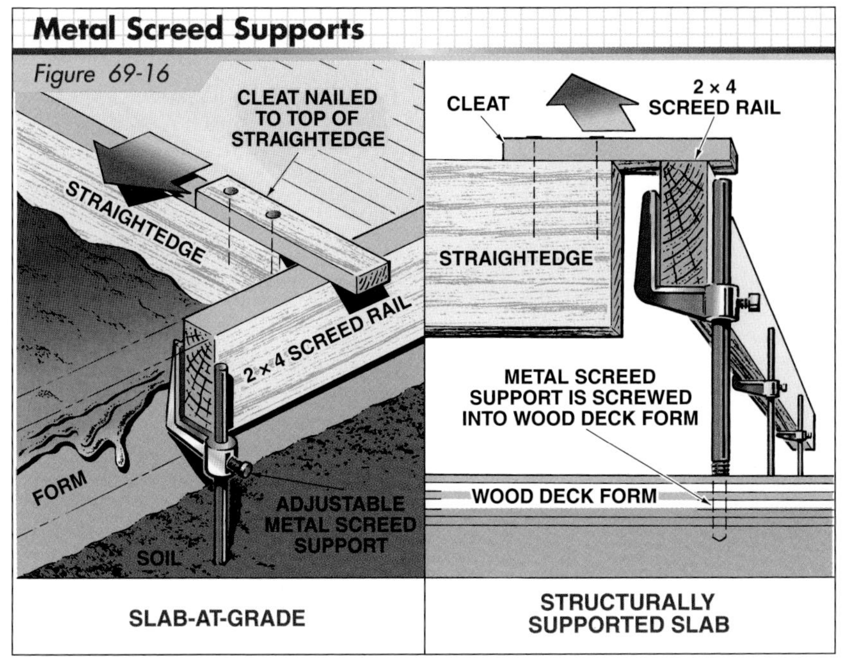

Figure 69-16. *Adjustable metal screed supports are driven into the ground for a slab-at-grade, or screwed into the wood deck form for structurally supported slabs.*

Fiber-reinforced concrete uses glass, metal, or plastic fibers to provide extra strength. Fibers in concrete reduce cracking caused by drying shrinkage and thermal expansion. The fibers also increase impact capacity, add abrasion resistance, provide tensile strength, reinforce the concrete against shattering, and provide an alternative to welded wire reinforcement.

As in all concrete construction, concrete for a slab-at-grade floor slab should be placed as close as possible to its final location to reduce unnecessary labor and to prevent segregation. Concrete deposited for floor slabs must be properly consolidated using a vibrator or manual means.

As the concrete is being placed, *cement finishers* strike off the concrete. When the screeds are removed, the concrete is finished. A *bullfloat* is used to eliminate high and low areas in the concrete. Power trowels and hand trowels produce the final finished surface of the concrete. **See Figure 69-17.**

Waterstops

A *waterstop* is a PVC, rubber, or stainless steel barrier used to prevent the passage of liquid or gas under pressure through a joint in a concrete slab or wall. Structures such as water-treatment facilities and water reservoirs require waterstops to prevent passage of liquids. As shown in **Figure 69-18,** split formwork must be constructed at waterstop joints. Split forms allow half of the waterstop to be embedded in the first section of concrete. The centerline of the waterstop should be aligned with the center of the slab joint. Properly secure the exposed flange of the waterstop to prevent displacement and folding over the waterstop during concrete placement.

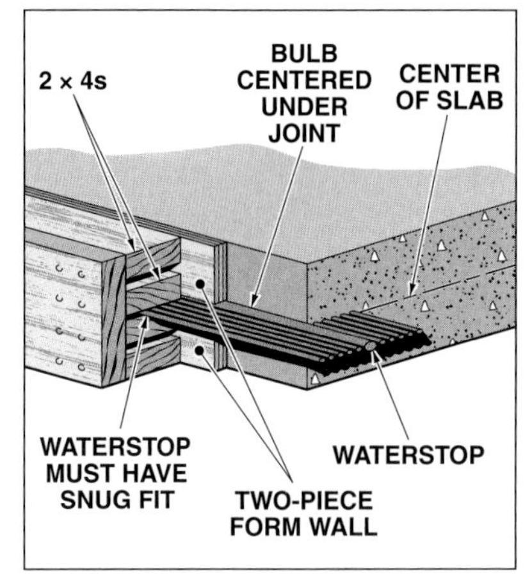

Figure 69-18. *Waterstops require special formwork to remain in position during concrete placement.*

Control Joints

Like other construction materials, concrete expands and contracts with moisture and temperature changes. To minimize cracking from expansion and contraction, *control joints* are provided in a concrete slab. A control joint is a groove in a concrete slab that creates a weakened plane and controls the location of cracking in a slab. **See Figure 69-19.**

Figure 69-17. *Concrete is finished by power troweling or hand troweling.*

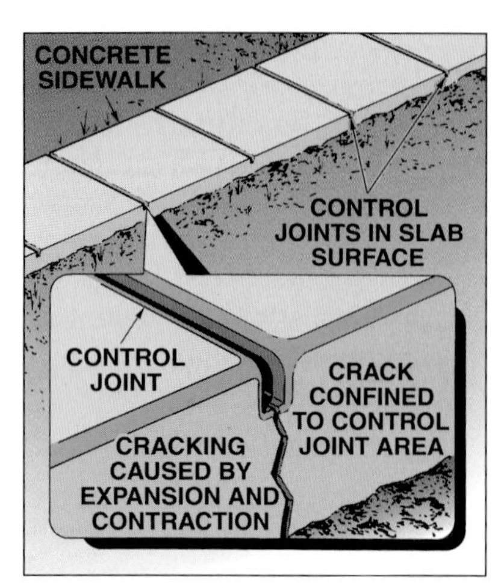

Figure 69-19. *If concrete cracks as a result of expansion and contraction, the crack will be confined to the area of the control joint.*

Figure 69-20. *Control joints are tooled into a concrete slab using a groover. The straightedge ensures a straight control joint.*

Figure 69-22. *An asphalt-impregnated strip is placed between adjoining sections of concrete to allow for movement caused by expansion and contraction of the slab.*

Control joints are formed immediately after finishing a concrete slab or the day after concrete is placed. When control joints are formed in newly finished concrete, the joints may be *tooled* or formed using metal or plastic T-strips. When tooling a control joint, a chalk line is snapped on the concrete surface and a straightedge is placed along the line as a guide for a *groover.* **See Figure 69-20.**

When forming a control joint with metal or plastic strips, a chalk line is snapped on the concrete surface and a cut is made in the surface with the edge of a trowel. The strips are inserted into the cut and the top edge of the T-strip is removed prior to troweling the surface.

Another method of creating a control joint is to cut into the slab with a concrete saw fitted with a masonry blade after the concrete has sufficiently hardened. **See Figure 69-21.** The cut is made to approximately one-fourth the slab thickness. If cracking later occurs, it will be confined to the area of the control joint.

Slab Expansion Joints

Expansion (isolation) *joints* are joints that separate adjoining sections of concrete to allow for movement caused by expansion and contraction of the slab. Expansion joints are required when a concrete slab is laid for a basement floor after the foundation walls, piers, and columns have been erected. Expansion joints are also placed where a driveway slab butts against a garage slab. **See Figure 69-22.**

Figure 69-21. *A control joint is cut into a concrete slab using a concrete saw.*

An expansion joint extends through the entire slab thickness. An expansion joint is commonly formed using an asphalt-impregnated strip around the slab perimeters before the concrete is placed. The strip remains in place after the concrete has set. **See Figure 69-23.**

Structurally Supported Slab Floors

The main difference between a slab-at-grade floor slab and a structurally supported slab floor is that concrete for a structurally supported slab is placed on top of the deck forms and is monolithically joined with the walls, columns, and beams. Concrete placement, screeding, and finishing operations are the same as described for slab-at-grade floor slabs.

Wall Expansion Joints

Expansion joints in walls serve a similar purpose to expansion joints in slabs—they separate adjoining concrete wall sections to allow for movement caused by expansion and contraction. Concrete for longer walls is often placed in sections, or concrete tilt-up panels are placed individually to create walls. The wall sections are subject to various stresses that cause movement, including temperature changes, settlement, seismic action, and wind load. Expansion joints help to account for this movement.

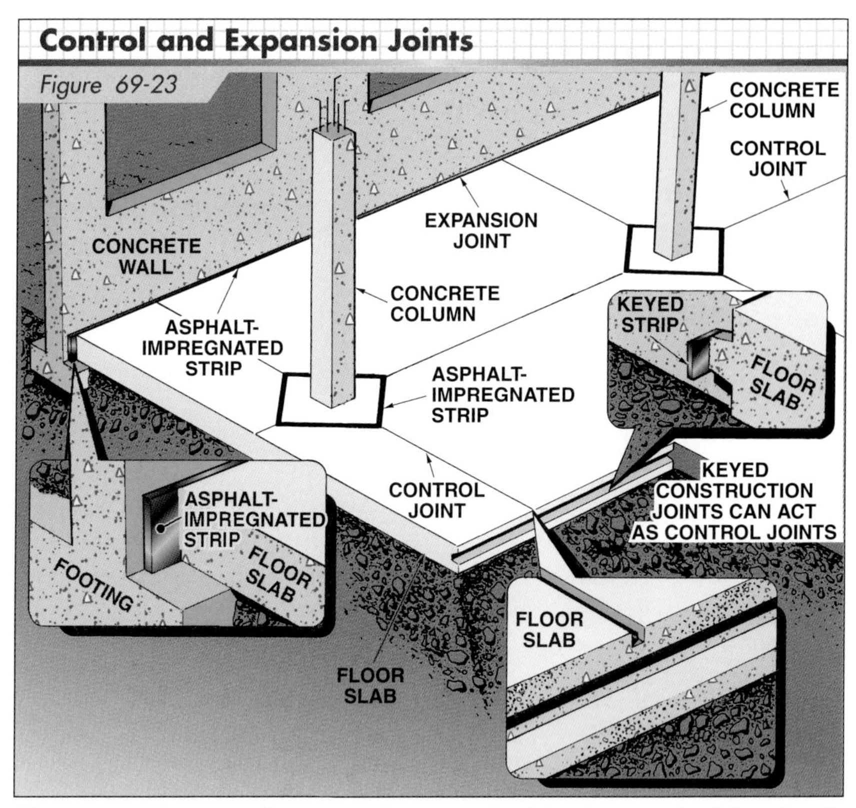

Figure 69-23. *A concrete floor system has expansion joints between the slab and walls and around columns. Control joints are placed at intervals in the slab to control cracking.*

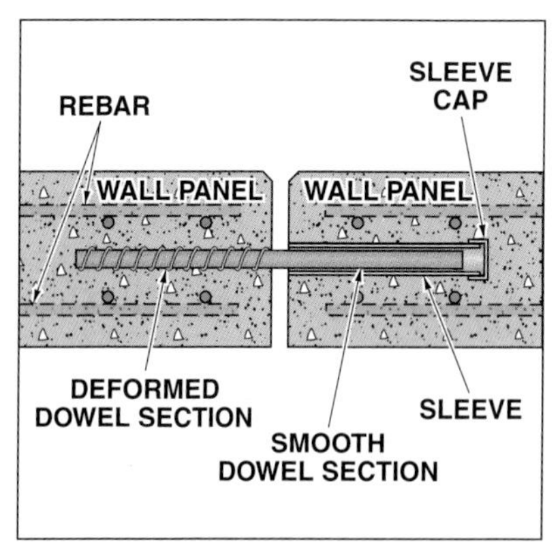

Figure 69-24. *A sleeved dowel system may be used to maintain wall alignment while allowing for movement of adjacent wall sections.*

Expansion joints extend the full height and width of a wall and may also include the foundation footing. The typical gap between walls is ½″. The construction prints typically include information about the location of expansion joints and the distance between joints. Expansion joints are commonly located at the ends of wall sections and where a wall butts against the face of another wall.

Preparation. The wall sections on each side of an expansion joint cannot be connected to one another, including horizontal rebar spanning the expansion joint gap. Rebar should terminate 2″ to 3″ from the expansion joint. When creating an expansion joint between adjacent wall sections, a ½″ thick piece of preformed bituminous fiber material can be temporarily tacked along the edge of the first wall using construction adhesive or nails. After concrete for the adjacent wall section has been placed and hardened, the bituminous material is removed, leaving a ½″ gap.

Sleeved Dowel System. In order to maintain wall alignment while allowing for the movement of adjacent wall sections, a *sleeved dowel system* may be used. The deformed end of a dowel is embedded in one of the wall sections with the smooth end extending from the edge of the wall section. A sleeve, which is capped on one end, is aligned with the smooth end and is embedded in the adjacent wall section. **See Figure 69-24.** The dowel spacing is indicated on the construction prints.

Joint Sealants. A *sealant* is used to prevent air and water from passing through an expansion joint and into a building. The sealant must be flexible enough to absorb movement without breaking its bond to the walls. Common sealants include silicones, acrylics, and ethylene vinyl acetate (EVA) sealants.

A *backer rod* is first forced into the expansion joint to limit the depth of the sealant. Backer rods are flexible, closed-cell polyethylene foam rods that are available in a variety of lengths. An exterior skin on the backer rod prevents the sealant from adhering to the rod. **See Figure 69-25.**

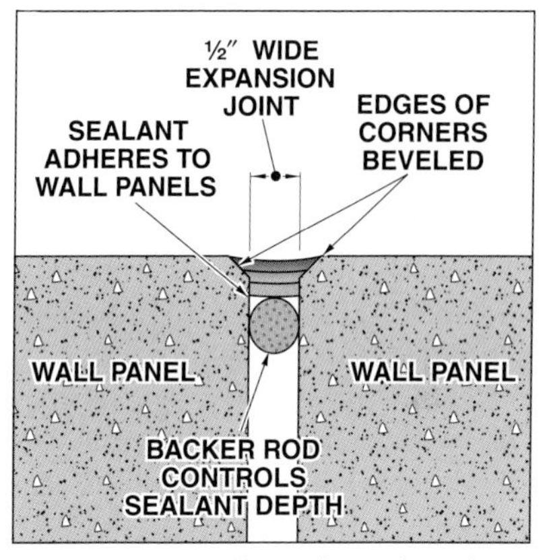

Figure 69-25. *Backer rods are forced into expansion joints to provide backing for the sealant.*

CURING CONCRETE AND REMOVING FORMS

Curing is the process of maintaining proper concrete moisture content and concrete temperature long enough to allow hydration to occur. Hydration begins as soon as water is added to the concrete mixture and continues as long as water is present in the concrete and the temperature conditions are favorable to maintain moisture content.

The first three days after concrete is placed are critical to the quality of concrete. During this period, concrete is most vulnerable to irreparable damage. Seven days after concrete placement the concrete attains approximately 70% of its strength, and at 14 days concrete attains about 85% of its strength. Under normal curing conditions, maximum concrete strength is reached at 28 days.

Curing floor slabs is more difficult than curing walls since more surface area is exposed for slabs. Greater surface exposure allows moisture in the concrete to evaporate more quickly if proper precautions are not taken.

A variety of methods are used to ensure that a concrete slab retains moisture for a minimum of three days. The slab may be misted or flooded with water. **See Figure 69-26.** The slab may also be covered with burlap or polyethylene film to help the concrete retain moisture and also to protect it from damage from frost, direct sun, traffic, and debris. **See Figure 69-27.** Chemical sealing compounds for curing can be applied by hand sprayers or by automatic self-powered sprayers. **See Figure 69-28.**

Forms must be *stripped* (removed) carefully to avoid injuring workers or damaging the concrete. In general, stripping forms from walls, columns, and beam sides is recommended no sooner than three days after the concrete has been placed. Forms for floor slabs should be stripped no sooner than seven days after concrete placement, and forms for arch centers and beam bottoms should be stripped no sooner than 14 days after concrete placement. On large construction projects, recommended times for stripping forms may be included in the print specifications.

Portland Cement Association

Figure 69-26. *A concrete slab may be misted with water to ensure the concrete has adequate moisture during curing.*

Portland Cement Association

Figure 69-27. *Burlap sheets are wetted and placed over concrete slabs to ensure the slabs have adequate moisture during hydration.*

Figure 69-28. *A liquid-membrane compound is sprayed onto fresh concrete to form a chemical barrier to prevent loss of moisture from the concrete.*

Portland Cement Association

During hot weather, concrete may need to be chilled to prevent premature setting of the concrete.

Unit 69 Review and Resources

UNIT 70

Precast Concrete Systems

OBJECTIVES

1. List and describe types of cast concrete structural members and their uses.
2. Explain the procedures for raising concrete members into position and fastening them together.
3. Identify common types of prestressed concrete and their manufacture.
4. Explain the manufacture and use of tilt-up members.
5. Describe lift-slab construction.

Many concrete structures are partially or completely constructed with *precast* (prefabricated) concrete systems instead of cast-in-place systems.

Some precast concrete members, such as wall and floor sections and beams and columns, are made in *casting beds (forms)* at precast plants. **See Figure 70-1.** The precast members are then transported by truck to a job site where they are placed in position using a crane. Precast members may also be *site cast* in forms that are constructed on the job. Site cast precast members are usually limited to wall and floor panels. **Figure 70-2** shows several types of precast structural members.

Hamilton Form Company, Ltd.

Figure 70-1. *Precast members may be made in casting beds at precast plants.*

Portland Cement Association

In the pretensioning process, rebar and high-tensile steel tendons encased in plastic sheathing are placed and secured in forms prior to concrete being placed. The steel tendons are stretched and concrete is placed into the form.

PRECAST CONCRETE

Some buildings are constructed with precast beam-and-column framework. One common type of construction is a steel-framed building with precast concrete wall panels, or *curtain walls*. Precast components are also used for sections of buildings that have complex architectural designs. **See Figure 70-3.**

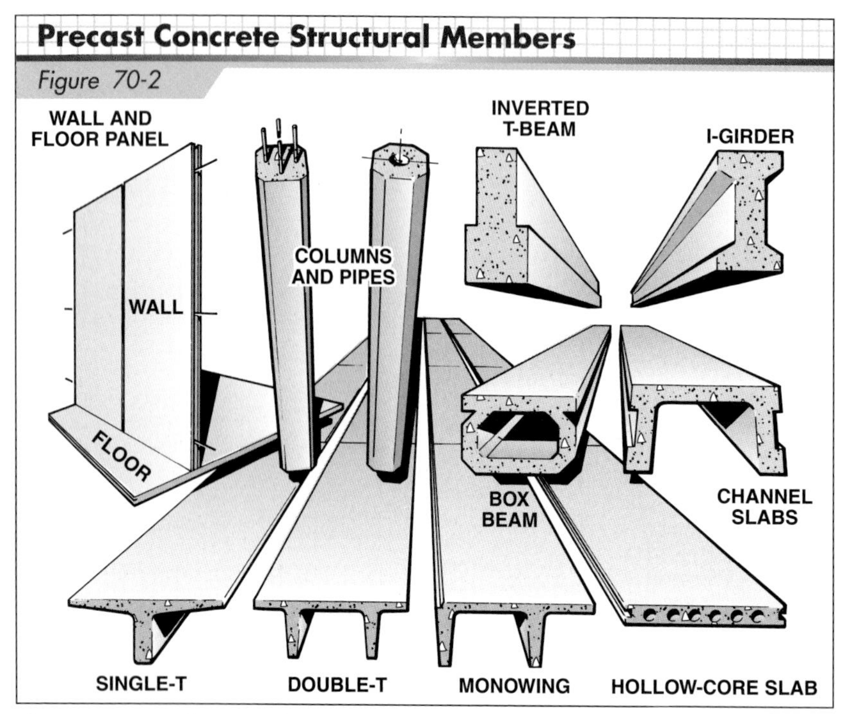

Figure 70-2. *A wide range of precast structural members is available.*

Figure 70-3. *Precast components are used for sections of a building with a complex architectural design.*

Precast members are raised and placed in position by crane. Bolts, clips, or other inserts are provided for the crane attachment. Hollow-core precast units, which combine strength with light weight, are quicker and easier to install. **See Figure 70-4.**

After the precast members are set in position and properly braced, various methods are used to connect the members. Most methods involve bolting or welding the members together. For example, the bottom of a column may have a metal anchor that fastens to bolts which were set in the foundation footing when the concrete was placed. Other examples are shown in **Figures 70-5** and **70-6.**

The National Precast Concrete Association (NPCA) is a trade association representing manufacturers of plant-produced concrete products and suppliers of the raw materials. The NPCA provides technical and production information about precast concrete.

PRESTRESSED CONCRETE

Prestressed concrete is precast concrete in which internal stress is introduced into the concrete to place it in a high state of compression. The internal stresses are placed on the concrete with high-tensile steel tendons (cables). Additional reinforcement with rebar may not be required. Prestressed concrete members have a greater resistance to lateral pressures than conventional reinforced members. As a result, lighter structural members can be used.

Two methods used to prestress concrete are *pretensioning* and *post-tensioning.* Pretensioning is commonly done in precast plants. Post-tensioning is more common with site cast precast members.

In the pretensioning process, the high-tensile steel tendons are placed in casting beds. The tendons are then stretched to the prescribed tension using powerful pretensioning jacks. **See Figure 70-7.** Concrete is then placed in the casting beds. The concrete bonds to the tendons as it hardens. When the concrete has set to its specified strength, tension on the tendons is released. The backward pull of the tendons places the concrete under compression. **See Figure 70-8.** The compressive force adds considerable strength to the concrete member. As a result, prestressed concrete beams and floor slabs can support heavier loads and have greater resistance to lateral pressure.

Figure 70-4. *Clips are attached to precast members to raise them into position.*

In the post-tensioning process, flexible metal or plastic ducts are placed in casting beds and concrete is placed around them. The ducts create hollow openings along the length of the concrete member. Tendons are then fed through the ducts. When a precast member is placed in position at the job site, the tendons are tensioned to the specified level and anchored. In some cases, the tendons are placed through ducts in adjoining structural members to tie the precast members together.

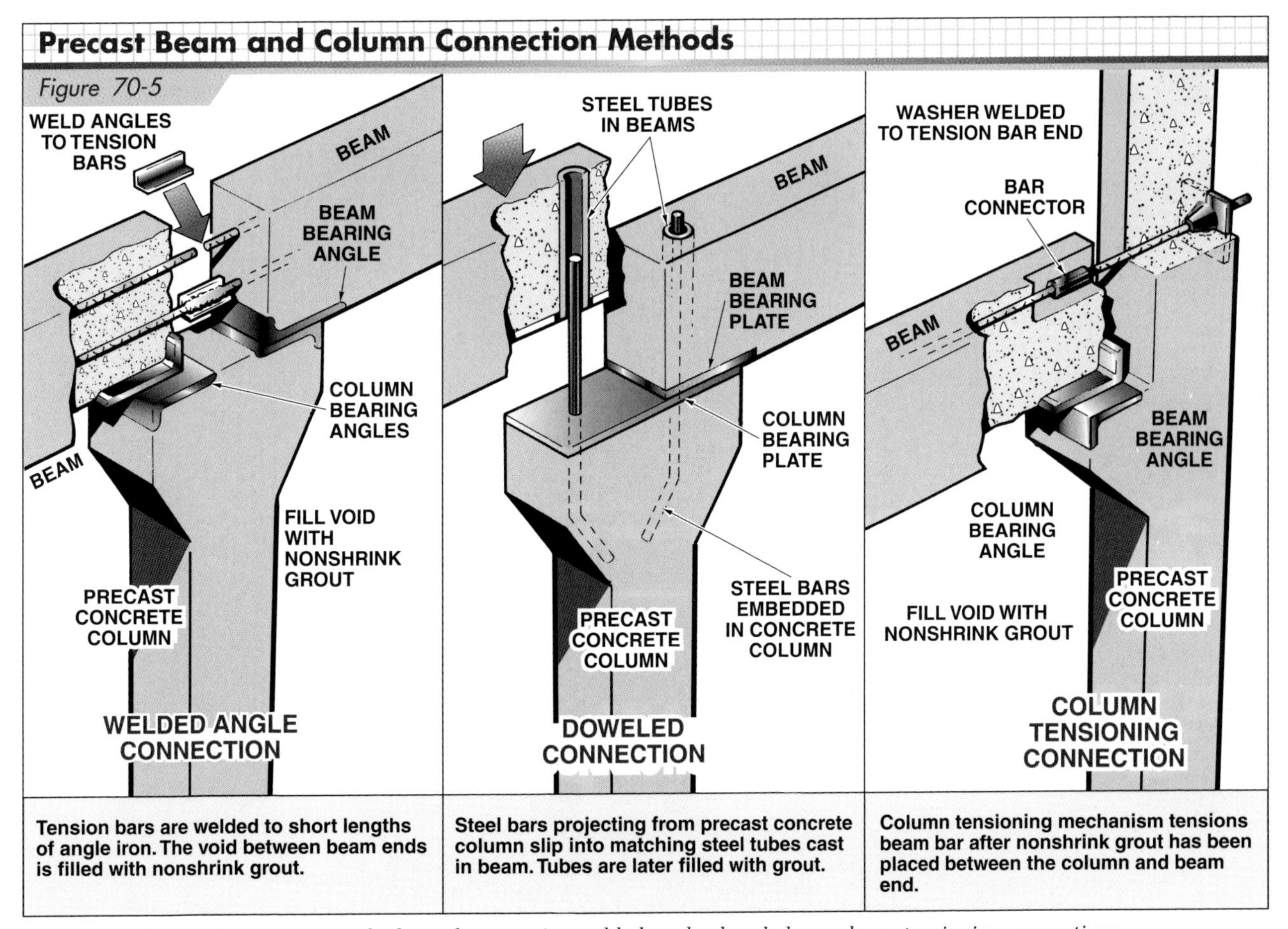

Figure 70-5. *Precast beams are attached to columns using welded angle, doweled, or column tensioning connections.*

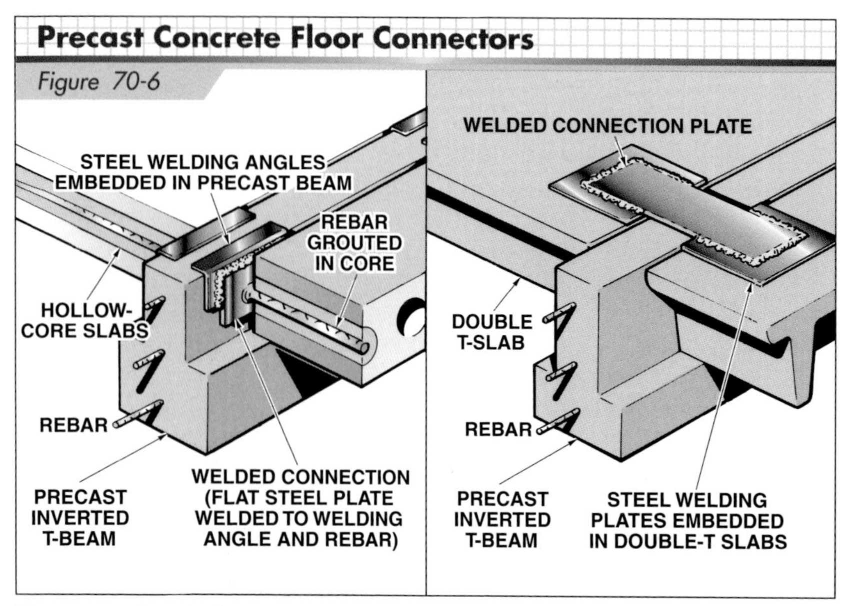

Figure 70-6. *Precast beams are commonly welded to precast floor slabs using welding angles or connection plates.*

Portland Cement Association

Figure 70-7. *Pretensioning jacks are used to stretch the cables extending from a precast structural member.*

At least 10,000 buildings enclosing more than 650 million sq ft are constructed annually. Tilt-up construction offers reasonable cost, low maintenance, durability, and speed of construction, especially in buildings of more than 10,000 sq ft, with 20′ or higher side walls, and using common panel sizes and appearance.

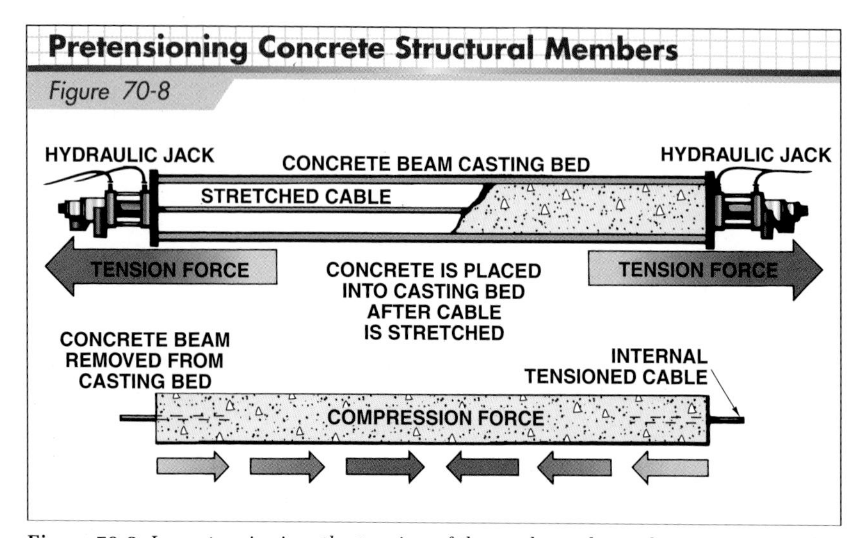

Figure 70-8. *In pretensioning, the tension of the tendons places the concrete member under compression.*

TILT-UP CONSTRUCTION

Tilt-up construction is a method of precast construction in which structural members (usually walls) are cast horizontally at a location adjacent to their eventual final position. Tilt-up members are usually constructed in casting panels on a floor slab and then raised in place by crane. Walls may be cast outside the building on a temporary concrete slab, wood platform, or well-compacted fill. Tilt-up construction is commonly used in one- and two-story buildings, although higher structures can also be erected using the tilt-up method.

Tilt-up formwork consists mainly of 2″ thick edge forms fastened to the slab. A traditional method of securing the edge forms consists of flat planks bolted or pinned down behind the edge form. **See Figure 70-9.** Short 1 × 4 or 2 × 4 braces secure the top of the edge forms to the flat planks. Another method of securing edge forms involves the use of triangular plywood brackets nailed to short 2 × 4 pads secured to the slab or triangular metal brackets fastened directly to the slab. The basic procedure for forming and raising tilt-up members is as follows:

1. Casting beds for wall panels are formed directly on the floor slab or adjacent to their final position. When forming tilt-up members on a concrete slab, two coats of *bond-breaker liquid* are applied to the slab to prevent precast wall panels from sticking to it.
2. Vertical and horizontal rebar are set in place. Horizontal rebar should extend at least 12″ beyond the edges of the panels if columns will be cast in place between the wall sections after they have been raised.
3. Steel-framed or wood bucks are placed for door and window openings.

4. Electrical conduit and outlet boxes are set in place. Inserts for the crane attachment are also provided.
5. Concrete is placed in the casting beds, vibrated for proper consolidation, and worked to the desired finish.
6. After the concrete has hardened and reached the proper strength, the precast wall panels are raised by crane and maneuvered over the wall footings. **See Figure 70-10.** Once the wall panels have been set in position on the footings, the panels are secured in position with temporary braces. **See Figure 70-11.**

Structural Connections

Tilt-up wall panels are secured to the floor slab and foundation after they have been raised. One method to secure tilt-up wall panels is to rest the ends of the panels on grout pads placed at intervals on the foundation footings. The grout pads are placed to level the wall. The space between the bottom of the wall and the footing is then filled with grout. Rebar protruding from the wall are tied to rebar extending from the concrete floor slab that has been placed to within a few feet of the foundation wall. The area between the foundation wall and floor slab is then filled with concrete.

ECO-Block, LLC

Figure 70-10. *A tilt-up wall section is carefully lifted into position. Note the braces attached to the face of the wall. The braces will be secured to the floor slab when the wall section is in position.*

Another method to secure a tilt-up wall panel to the foundation is to set the bottom of the wall in a keyway provided at the top of the foundation footing. A foam *backer rod* is placed in the keyway before raising the tilt-up wall panel to prevent moisture from seeping between the wall panel and footing. **See Figure 70-12.** The space between the edge of the keyway and wall panel is then filled with grout.

Figure 70-11. *As wall panels are raised, temporary braces are secured at the lower ends. Braces remain in position until all panels are in place and roof girders and beams are set in position.*

Figure 70-12. *The bottom of a tilt-up wall panel may be placed in a keyway in the footing. A backer rod positioned under the wall panel prevents moisture from seeping between the wall panel and footing.*

ECO-Block, LLC

Figure 70-9. *The form for this tilt-up wall section consists of 2 × 6s set on edge and resting against the concrete slab. Other 2 × 6s are fastened to the slab to secure the bottom of the form planks in position. Short pieces nailed to the top of the form and to the flat 2 × 6s will be installed to brace the top of the form. Rebar and lifting inserts are in position. The inserts will later receive bolts that secure the lift plates to the wall.*

Columns or *chord bars* may be used to tie together the vertical edges of wall panels. Cast-in-place columns that required a form to be built between the wall panels were common in early tilt-up construction. Contemporary designs use independent precast columns or columns cast monolithically with the walls. **See Figure 70-13.** Independent precast columns are formed with oversized recesses that accommodate the wall panel edges. Welding plates are used to secure the walls to the columns. When columns are cast as part of the wall panel, one half of a column is formed at both ends of each panel. The half columns at the panel ends are then joined with welding plates after the panels have been raised.

Columns are not required when using chord bars. The chord bars extend horizontally through the wall panels, and small pockets around the ends of the bars are blocked out when forming the panel sections. The bars are spliced and welded together after the walls have been raised, and the pockets are filled with concrete.

The second floor of a two-story tilt-up building is usually cast-in-place, requiring the construction of a suspended floor slab form. Other methods include wood or metal floor joists, or trusses supported by steel brackets and angles secured to the walls. Roofs of tilt-up structures are framed with open-web steel joists, glulam timbers, or trusses and sheathed with plywood, OSB, or other approved material.

Steel columns and open-web steel roof trusses are commonly combined with precast exterior walls.

LIFT-SLAB CONSTRUCTION

Lift-slab construction combines precast concrete or steel columns with floor slabs cast on the job site. After the foundation work has been completed and the ground floor slab has been placed, columns extending the entire height of the building are set up. On very tall structures, the columns cannot be set up for the entire height of the building at one time; therefore, columns are erected in sections as the floors are being raised.

Floor slabs for a lift-slab building are *stack-cast* (cast directly on top of each other). Resin-type bond-breaking compounds are applied between the slabs. Slab thicknesses vary from 7″ to 10″, depending upon span and load conditions. Maximum slab strength can be attained by post-tensioning the slabs after they have hardened.

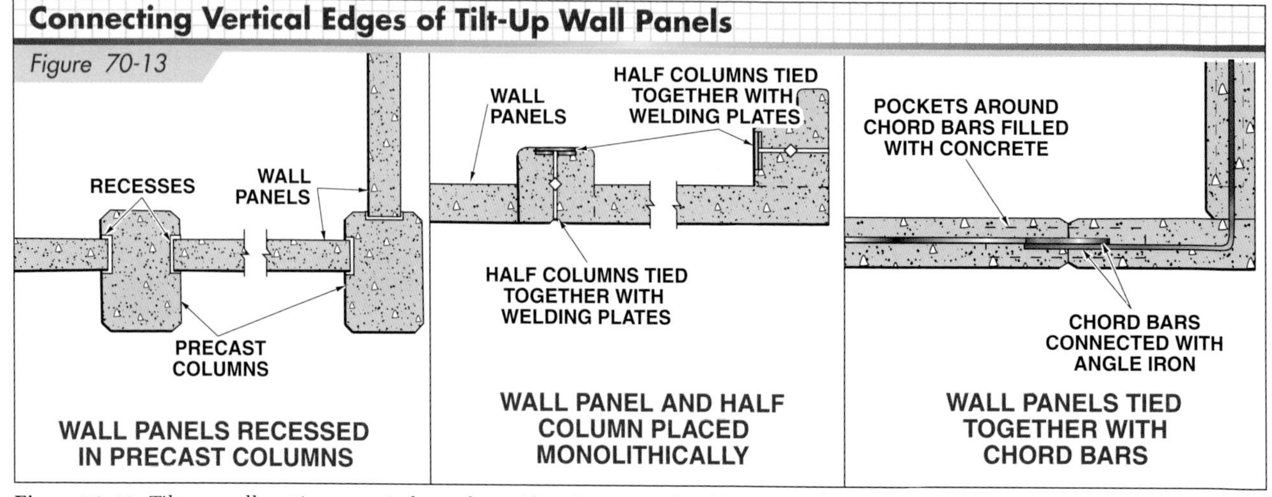

Figure 70-13. *Tilt-up wall sections are tied together with columns or chord bars.*

OSHA 29 CFR 1926.705, Requirements for Lift-Slab Operations, details the safety requirements for lifting floor slabs of a lift-slab building.

Hydraulic jacks are placed at the top of each column. Lifting rods extending from the jacks are connected to lifting collars that have been placed in the slabs around the columns. The lifting jacks are connected to a central, electrically controlled console that simultaneously operates all the jacks within a ¼″ tolerance. Normal lifting rates are 7′ to 10′ per hour. **Figure 70-14** reviews the general procedure for constructing a lift-slab building.

On smaller buildings, the roof slab is usually lifted to its final position first. The lifting sequence that follows varies according to the building height. For column stability, several stacked layers may be raised to a higher floor level.

To secure the floor slab in the proper position, steel *shear bars* may be inserted through the columns and under the slab after the slab has been elevated to its final position. Shear bars are designed to provide a permanent attachment between the slab and column. When steel columns are used, the floor slab may be secured in position by welding a steel collar under the slab to the steel column. After the slabs have been permanently attached to the column, gaps between the slab and the column are filled with grout. Exterior walls in lift-slab construction are curtain walls of various materials, usually metal.

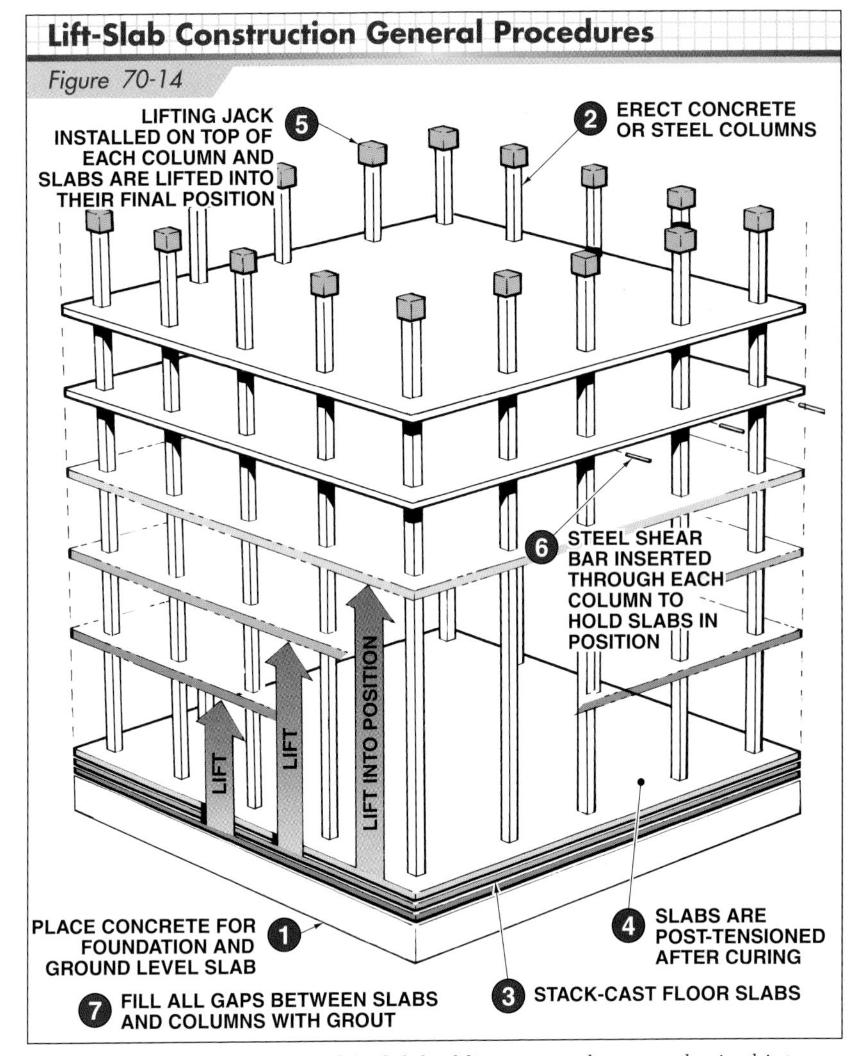

Figure 70-14. *Floor slabs for a lift-slab building are stack-cast and raised into position with lifting jacks installed at the tops of the columns.*

Unit 70 Review and Resources

APPENDIX

ABBREVIATIONS

Term	Abbreviation	Term	Abbreviation	Term	Abbreviation
Aluminum	AL	Flashing	FL	Roof Drain	RD
Anchor Bolt	AB	Floor	FL	Roofing	RFG
Asphalt Tile	AT.	Footing	FTG	Room	RM
At Finished Face	AFF	Foundation	FDN	Rough Opening	RO
Basement	BSMT	Furnace	FURN	Screen	SCR
Bathroom	B	Gauge	GA	Sewer	SEW.
Bathtub	BT	Galvanized Iron	GI	Shake	SHK
Beam	BM	Girder	GDR	Sheathing	SHTHG
Bedroom	BR	Glass	GL or GLS	Sheet Metal	SM
Benchmark	BM	Grade	GR	Shingle	SHGL
Block	BLK	Ground	GND or GRD	Shower	SH
Board	BD	Gypsum Board	GYP BD	Siding	SDG
Brick	BRK	Hardboard	HBD	Sill	SL
Building	BL or BLDG	Hardwood	HDWD	Sink	S or SK
Building Line	BL	Head	HD	Skylight	SLT
Cabinet	CAB.	Heat	H or HT	Sliding Door	SL DR
Casement	CSMT	Hose Bibb	HB	Soffit	SF or SOF
Cedar	CDR	Insulation	INS or INSUL	Soil Pipe	SP
Ceiling	CLG	Interior	INT	Solar Panel	SLR PAN.
Cement	CEM	Jamb	JB or JMB	South	S
Center	CTR	Joist	J or JST	Stack Vent	SV
Centerline	CL	Kitchen	K or KIT.	Stairs	ST
Chimney	CHM	Laundry	LAU	Stairway	STWY
Closet	CLO	Lavatory	LAV	Steel	STL
Column	COL	Light	LT	Stone	STN
Concrete	CONC	Linen Closet	LC or LCL	Street	ST
Concrete Block	CONC BLK	Linoleum	LINO	Tongue-and-Groove	T&G
Cornice	COR	Living Room	LR	Top of Concrete	TOC
Corrugated	CORR	Louver	LV or LVR	Top Hinged	TH
Detail	DET or DTL	Medicine Cabinet	MC	Top of Slab	TOS
Diameter	D or DIA	Metal	MET. or MTL	Top of Steel	TOS
Dining Room	DR	Noncombustible	NCOMBL	Tread	TR
Dishwasher	DW	North	N	Typical	TYP
Door	DR	On Center	OC	Unexcavated	UNEXC
Dormer	DRM	Opening	OPNG	Utility Room	URM
Double-Hung Window	DHW	Overhang	OH.	Vent	V
Douglas Fir	DF	Overhead Door	OH. DR	Ventilation	VENT.
Downspout	DS	Panel	PNL	Vent Stack	VS
Drain	DR	Partition	PTN	Vinyl Tile	VA TILE
Drywall	DW	Plate	PL	Washing Machine	WM
East	E	Plywood	PLYWD	Water	W
Electric	ELEC	Porch	P	Waterproof	WP
Elevation	EL	Pressure-Treated	PT	Water Closet	WC
Excavate	EXC	Rafter	RFTR	Water Heater	WH
Exterior	EXT	Redwood	RWD	Welded Wire Reinforcement	WWR
Face Brick	FB	Refrigerator	REF	West	W
Fill	F	Reinforced	REINF	White Pine	WP
Finish	FNSH	Reinforcement Bar	REBAR	Wide Flange	WF
Finish Floor	FNSH FL	Retaining Wall	RW	Window	WDW
Fireplace	FP	Ridge	RDG	Wood	WD
Fireproof	FPRF	Riser	R	Wood Blocking	WBL
Fixture	FXTR	Roof	RF	Yellow Pine	YP

ARCHITECTURAL SYMBOLS . . .

MATERIAL	ELEVATION	PLAN VIEW	SECTION
EARTH			
BRICK	WITH NOTE INDICATING TYPE OF BRICK (COMMON, FACE, ETC.)	COMMON OR FACE FIREBRICK	SAME AS PLAN VIEWS
CONCRETE		LIGHTWEIGHT STRUCTURAL	SAME AS PLAN VIEWS
CONCRETE MASONRY UNITS		OR	OR
STONE	CUT STONE RUBBLE	CUT STONE RUBBLE CAST STONE (CONCRETE)	CUT STONE CAST STONE CONCRETE RUBBLE OR CUT STONE
WOOD	SIDING PANEL	WOOD STUD REMODELING DISPLAY	ROUGH MEMBERS FINISHED MEMBERS PLYWOOD
PLASTER		WOOD STUD, LATH, AND PLASTER METAL LATH AND PLASTER SOLID PLASTER	LATH AND PLASTER
ROOFING	SHINGLES	SAME AS ELEVATION	
GLASS	OR GLASS BLOCK	GLASS GLASS BLOCK	SMALL SCALE LARGE SCALE

. . . ARCHITECTURAL SYMBOLS

MATERIAL	ELEVATION	PLAN VIEW	SECTION
FACING TILE	CERAMIC TILE	FLOOR TILE	CERAMIC TILE LARGE SCALE; CERAMIC TILE SMALL SCALE
STRUCTURAL CLAY TILE			SAME AS PLAN VIEW
INSULATION		LOOSE FILL OR BATTS; RIGID; SPRAY FOAM	SAME AS PLAN VIEWS
SHEET METAL FLASHING		OCCASIONALLY INDICATED BY NOTE	
METALS OTHER THAN FLASHING	INDICATED BY NOTE OR DRAWN TO SCALE	SAME AS ELEVATION	SMALL SCALE; STEEL; CAST IRON; ALUMINUM; BRONZE OR BRASS
STRUCTURAL STEEL	INDICATED BY NOTE OR DRAWN TO SCALE	OR	REBARS; SMALL SCALE; LARGE SCALE; **L-ANGLES, S-BEAMS, ETC.**

PLOT PLAN SYMBOLS

NORTH (N)	FIRE HYDRANT	WALK	ELECTRIC SERVICE (E OR)
POINT OF BEGINNING (POB)	MAILBOX	IMPROVED ROAD	NATURAL GAS LINE (G OR)
UTILITY METER OR VALVE	MANHOLE	UNIMPROVED ROAD	WATER LINE (W OR)
POWER POLE AND GUY	TREE	BUILDING LINE	TELEPHONE LINE (T OR)
LIGHT STANDARD	BUSH	PROPERTY LINE	NATURAL GRADE
TRAFFIC SIGNAL	HEDGE ROW	PROPERTY LINE	FINISH GRADE
STREET SIGN	FENCE	TOWNSHIP LINE	EXISTING ELEVATION (+ XX.00′)

DECIMAL EQUIVALENTS OF AN INCH

Fraction	Decimal	Fraction	Decimal	Fraction	Decimal	Fraction	Decimal
1/64	0.015625	17/64	0.265625	33/64	0.515625	49/64	0.765625
1/32	0.03125	9/32	0.28125	17/32	0.53125	25/32	0.78125
3/64	0.046875	19/64	0.296875	35/64	0.546875	51/64	0.796875
1/16	0.0625	5/16	0.3125	9/16	0.5625	13/16	0.8125
5/64	0.078125	21/64	0.328125	37/64	0.578125	53/64	0.828125
3/32	0.09375	11/32	0.34375	19/32	0.59375	27/32	0.84375
7/64	0.109375	23/64	0.359375	39/64	0.609375	55/64	0.859375
1/8	0.125	3/8	0.375	5/8	0.625	7/8	0.875
9/64	0.140625	25/64	0.390625	41/64	0.640625	57/64	0.890625
5/32	0.15625	13/32	0.40625	21/32	0.65625	29/32	0.90625
11/64	0.171875	27/64	0.421875	43/64	0.671875	59/64	0.921875
3/16	0.1875	7/16	0.4375	11/16	0.6875	15/16	0.9375
13/64	0.203125	29/64	0.453125	45/64	0.703125	61/64	0.953125
7/32	0.21875	15/32	0.46875	23/32	0.71875	31/32	0.96875
15/64	0.234375	31/64	0.484375	47/64	0.734375	63/64	0.984375
1/4	0.250	1/2	0.500	3/4	0.750	1	1.000

DECIMAL AND METRIC EQUIVALENTS

Fractions	Decimal Inches	Millimeters
1/16	.0625	1.58
1/8	.125	3.18
3/16	.1875	4.76
1/4	.250	6.35
5/16	.3125	7.97
3/8	.375	9.52
7/16	.4375	11.11
1/2	.500	12.70
9/16	.5625	14.29
5/8	.625	15.88
11/16	.6875	17.46
3/4	.750	19.05
13/16	.8125	20.64
7/8	.875	22.22
1	1.00	25.40

DECIMAL EQUIVALENTS OF A FOOT

Inches	Decimal Foot Equivalent	Inches	Decimal Foot Equivalent	Inches	Decimal Foot Equivalent
1/16	0.0052	4 1/16	0.3385	8 1/16	0.6719
1/8	0.0104	4 1/8	0.3438	8 1/8	0.6771
3/16	0.0156	4 3/16	0.3490	8 3/16	0.6823
1/4	0.0208	4 1/4	0.3542	8 1/4	0.6875
5/16	0.0260	4 5/16	0.3594	8 5/16	0.6927
3/8	0.0313	4 3/8	0.3646	8 3/8	0.6979
7/16	0.0365	4 7/16	0.3698	8 7/16	0.7031
1/2	0.0417	4 1/2	0.3750	8 1/2	0.7083
9/16	0.0469	4 9/16	0.3802	8 9/16	0.7135
5/8	0.0521	4 5/8	0.3854	8 5/8	0.7188
11/16	0.0573	4 11/16	0.3906	8 11/16	0.7240
3/4	0.0625	4 3/4	0.3958	8 3/4	0.7292
13/16	0.0677	4 13/16	0.4010	8 13/16	0.7344
7/8	0.0729	4 7/8	0.4063	8 7/8	0.7396
15/16	0.0781	4 15/16	0.4115	8 15/16	0.7448
1	0.0833	5	0.4167	9	0.7500
1 1/16	0.0885	5 1/16	0.4219	9 1/16	0.7552
1 1/8	0.0938	5 1/8	0.4271	9 1/8	0.7604
1 3/16	0.0990	5 3/16	0.4323	9 3/16	0.7656
1 1/4	0.1042	5 1/4	0.4375	9 1/4	0.7708
1 5/16	0.1094	5 5/16	0.4427	9 5/16	0.7760
1 3/8	0.1146	5 3/8	0.4479	9 3/8	0.7813
1 7/16	0.1198	5 7/16	0.4531	9 7/16	0.7865
1 1/2	0.1250	5 1/2	0.4583	9 1/2	0.7917
1 9/16	0.1302	5 9/16	0.4635	9 9/16	0.7969
1 5/8	0.1354	5 5/8	0.4688	9 5/8	0.8021
1 11/16	0.1406	5 11/16	0.4740	9 11/16	0.8073
1 3/4	0.1458	5 3/4	0.4792	9 3/4	0.8125
1 13/16	0.1510	5 13/16	0.4844	9 13/16	0.8177
1 7/8	0.1563	5 7/8	0.4896	9 7/8	0.8229
1 15/16	0.1615	5 15/16	0.4948	9 15/16	0.8281
2	0.1667	6	0.5000	10	0.8333
2 1/16	0.1719	6 1/16	0.5052	10 1/16	0.8385
2 1/8	0.1771	6 1/8	0.5104	10 1/8	0.8438
2 3/16	0.1823	6 3/16	0.5156	10 3/16	0.8490
2 1/4	0.1875	6 1/4	0.5208	10 1/4	0.8542
2 5/16	0.1927	6 5/16	0.5260	10 5/16	0.8594
2 3/8	0.1979	6 3/8	0.5313	10 3/8	0.8646
2 7/16	0.2031	6 7/16	0.5365	10 7/16	0.8698
2 1/2	0.2083	6 1/2	0.5417	10 1/2	0.8750
2 9/16	0.2135	6 9/16	0.5469	10 9/16	0.8802
2 5/8	0.2188	6 5/8	0.5521	10 5/8	0.8854
2 11/16	0.2240	6 11/16	0.5573	10 11/16	0.8906
2 3/4	0.2292	6 3/4	0.5625	10 3/4	0.8958
2 13/16	0.2344	6 13/16	0.5677	10 13/16	0.9010
2 7/8	0.2396	6 7/8	0.5729	10 7/8	0.9063
2 15/16	0.2448	6 15/16	0.5781	10 15/16	0.9115
3	0.2500	7	0.5833	11	0.9167
3 1/16	0.2552	7 1/16	0.5885	11 1/16	0.9219
3 1/8	0.2604	7 1/8	0.5938	11 1/8	0.9271
3 3/16	0.2656	7 3/16	0.5990	11 3/16	0.9323
3 1/4	0.2708	7 1/4	0.6042	11 1/4	0.9375
3 5/16	0.2760	7 5/16	0.6094	11 5/16	0.9427
3 3/8	0.2813	7 3/8	0.6146	11 3/8	0.9479
3 7/16	0.2865	7 7/16	0.6198	11 7/16	0.9531
3 1/2	0.2917	7 1/2	0.6250	11 1/2	0.9583
3 9/16	0.2969	7 9/16	0.6302	11 9/16	0.9635
3 5/8	0.3021	7 5/8	0.6354	11 5/8	0.9688
3 11/16	0.3073	7 11/16	0.6406	11 11/16	0.9740
3 3/4	0.3125	7 3/4	0.6458	11 3/4	0.9792
3 13/16	0.3177	7 13/16	0.6510	11 13/16	0.9844
3 7/8	0.3229	7 7/8	0.6563	11 7/8	0.9896
3 15/16	0.3281	7 15/16	0.6615	11 15/16	0.9948
4	0.3333	8	0.6667	12	1.0000

ENGLISH-TO-METRIC CONVERSION

Quantity	To Convert	To	Multiply by
Length	inches	millimeters	25.4
	inches	centimeters	2.54
	feet	centimeters	30.48
	feet	meters	.3048
	yards	centimeters	91.44
	yards	meters	.9144
Area	square inches	square millimeters	645.2
	square inches	square centimeters	6.452
	square feet	square centimeters	929.0
	square feet	square meters	.0929
	square yards	square meters	.8361
Volume	cubic inches	cubic millimeters	1639
	cubic inches	cubic centimeters	16.39
	cubic feet	cubic centimeters	2.832
	cubic feet	cubic meters	.02832
	cubic yards	cubic meters	.7646
Liquid Measure	pints	cubic centimeters	473.2
	pints	liters	.4732
	quarts	cubic centimeters	946.3
	quarts	liters	.9463
	gallons	cubic centimeters	3785
	gallons	liters	3.785
Weight	ounces	grams	28.35
	ounces	kilograms	.02835
	pounds	grams	453.6
	pounds	kilograms	.4536
	short tons (2000 lb)	kilograms	907.2
	short tons (2000 lb)	metric ton (1000 kg)	.9072
Pressure	inches of water column	kilopascals	.2491
	feet of water column	kilopascals	2.989
	pounds per square inch	kilopascals	6.895
Temperature	degrees Fahrenheit (°F)	degrees Celsius (°C)	$\frac{5}{9}$(°F − 32)

METRIC-TO-ENGLISH CONVERSION			
Quantity	**To Convert**	**To**	**Multiply By**
Length	millimeters	inches	.03937
	centimeters	inches	.3937
	meters	feet	3.281
	meters	yards	1.0937
Area	square millimeters	square inches	.00155
	square centimeters	square inches	.1550
	square centimeters	square feet	.0010
	square meters	square feet	10.76
	square meters	square yards	1.196
Volume	cubic centimeters	cubic inches	.06102
	cubic meters	cubic feet	35.31
	cubic meters	cubic yards	1.308
Liquid Measure	liters	pints	2.113
	liters	quarts	1.057
	liters	gallons	.2642
Weight	grams	ounces	.03527
	kilograms	pounds	2.205
	metric ton (1000 kg)	pounds	2205
Pressure	kilopascals	inches of water column	4.014
	kilopascals	feet of water column	.3346
	kilopascals	pounds per square inch	.1450
Temperature	degrees Celsius (°C)	degrees Fahrenheit (°F)	($\frac{9}{5}$ °F) + 32

NOMINAL AND MINIMUM DRESSED SIZES OF BOARDS, DIMENSION LUMBER, AND TIMBERS

	THICKNESS					WIDTH				
	Nominal Inch	Minimum Dressed				Nominal Inch	Minimum Dressed			
		Dry		Green			Dry		Green	
		inch	mm	inch	mm		inch	mm	inch	mm
Boards	¾	⅝	16	11/16	17	2	1½	38	1 9/16	40
	1	¾	19	25/32	20	3	2½	64	2 9/16	65
	1¼	1	25	1 1/32	26	4	3½	89	3 9/16	90
	1½	1¼	32	1 9/32	33	5	4½	114	4⅝	117
						6	5½	140	5⅝	143
						7	6½	165	6⅝	168
						8	7¼	184	7½	190
						9	8¼	210	8½	216
						10	9¼	235	9½	241
						11	10¼	260	10½	267
						12	11¼	286	11½	292
						14	13¼	337	13½	343
						16	15¼	387	15½	394
Dimension	2	1½	38	1 9/16	40	2	1½	38	1 9/16	40
	2½	2	51	2 1/16	52	2½	2	51	2 1/16	52
	3	2½	64	2 9/16	65	3	2½	64	2 9/16	65
	3½	3	76	3 1/16	78	3½	3	76	3 1/16	78
	4	3½	89	3 9/16	90	4	3½	89	3 9/16	90
	4½	4	102	4 1/16	103	4½	4	102	4 1/16	103
						5	4½	114	4⅝	117
						6	5½	140	5⅝	143
						8	7¼	184	7½	190
						10	9¼	235	9½	241
						12	11¼	286	11½	292
						14	13¼	337	13½	343
						16	15¼	387	15½	394
Timbers	5 & thicker			½ off	13 off	5 & wider			½ off	13 off

AREA—PLANE FIGURES

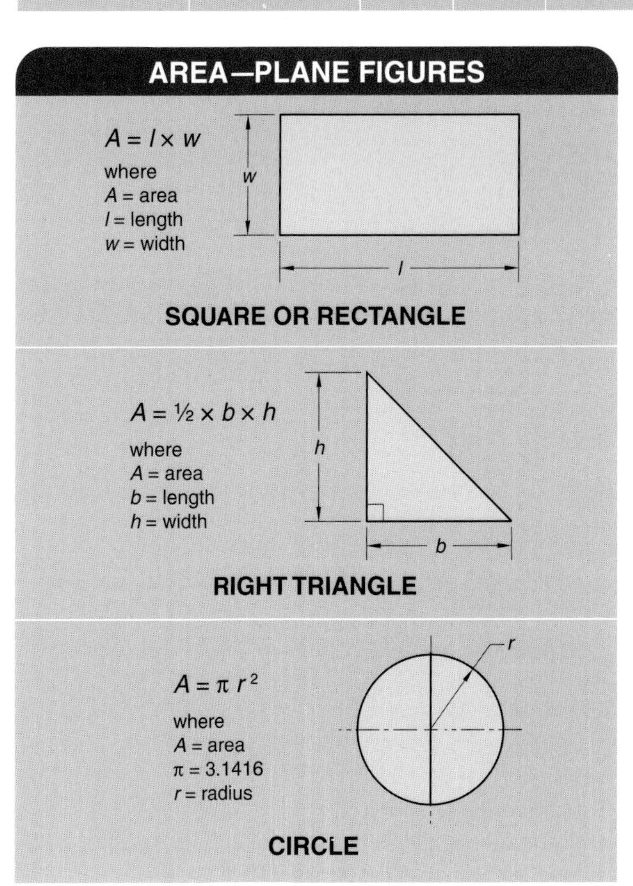

VOLUME—SOLID FIGURES

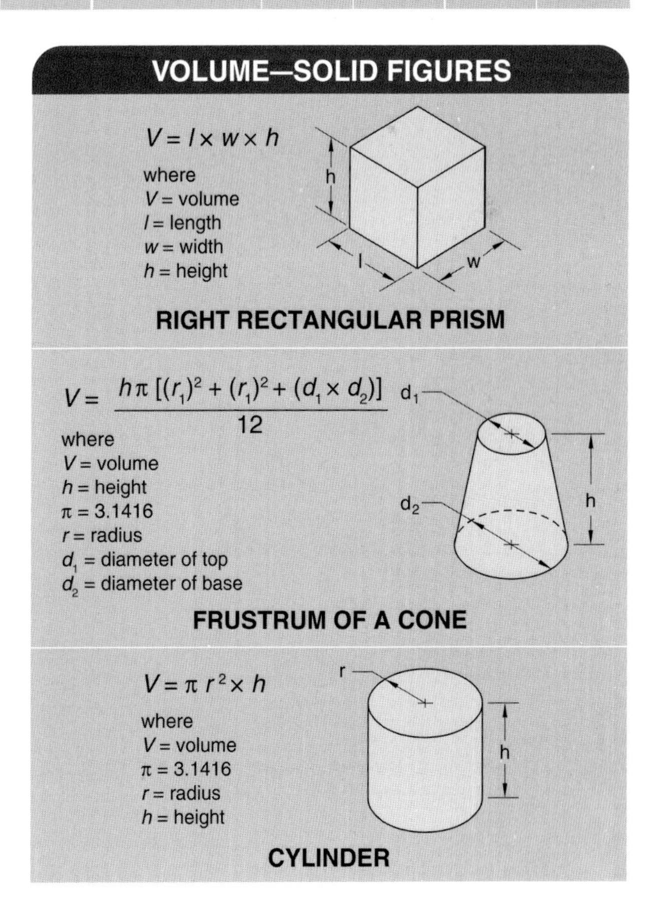

WESTERN WOOD PRODUCTS ASSOCIATION FRAMING GRADE LUMBER			
PRODUCT	**GRADES**	**WWPA GRADING RULES SECTION REFERENCE**	**APPLICATIONS**
Structural Light Framing (SLF) (2 × 2 through 4 × 4)	Select Structural No. 1 No. 2 No. 3	42.10 42.11 42.12 42.13	Structural applications where high-strength design values are required in light framing sizes.
Light Framing (LF) (2 × 2 through 4 × 4)	Construction Standard Utility	40.11 40.12 40.13	Basic framing lumber for wall framing, plates, sills, cripple studs, blocking, and similar applications.
Stud (2 × 2 through 4 × 18)	Stud	41.13	All-purpose grade designed primarily for stud uses, including load-bearing walls.
Structural Joists and Planks (SJ&P) (2 × 5 through 5 × 18)	Select Structural No. 1 No. 2 No. 3	62.10 62.11 62.12 62.13	Intended for engineering applications for lumber 5″ and wider, such as joists, rafters, headers, beams, trusses, and general framing.
SPECIAL DIMENSION LUMBER GRADES			
PRODUCT	**GRADES**	**WWPA GRADING RULES SECTION REFERENCE**	**APPLICATIONS**
Structural Decking (2 × 4 through 4 × 12)	Selected Decking	55.11	Used where the appearance of the best face is of primary importance.
	Commercial Decking	55.12	Customarily used when appearance is not of primary importance.
Machine Stress-Rated (MSR)	2850 Fb-2.3E 2700 Fb-2.2E 2550 Fb-2.1E 2400 Fb-2.0E 2250 Fb-1.9E 2100 Fb-1.8E 1950 Fb-1.7E 1800 Fb-1.6E 1650 Fb-1.5E 1500 Fb-1.4E 1450 Fb-1.3E 1350 Fb-1.3E 1200 Fb-1.2E 900 Fb-1.0E		Primarily used for trusses, but is also used for structural purposes where assured strength capabilities are required, such as rafters and floor and ceiling joists.
TIMBER GRADES			
PRODUCT	**GRADES**	**WWPA GRADING RULES SECTION REFERENCE**	**APPLICATIONS**
Beams and Stringers (5″ and thicker, width more than 2″ greater than thickness)	Dense Select Structural* Dense No. 1* Dense No. 2* Select Structural No. 1 No. 2	53.00 & 170.00 53.00 & 170.00 53.00 & 170.00 70.10 70.11 70.12	Designed for beam and stringer uses when sizes larger than 4″ nominal thickness are required.
Post and Timbers (5″ × 5″ and larger, width not more than 2″ greater than thickness)	Dense Select Structural* Dense No. 1* Dense No. 2* Select Structural No. 1 No. 2	53.00 & 170.00 53.00 & 170.00 53.00 & 170.00 80.10 80.11 80.12	Designed for vertically loaded applications where sizes larger than 4″ nominal thickness are required.

* Douglas fir or Douglas fir-larch only

WESTERN WOOD PRODUCTS ASSOCIATION APPEARANCE LUMBER GRADES

Product	Grades	Equivalent Grades in Idaho White Pine	WWPA Grading Rules Section Reference
HIGH-QUALITY APPEARANCE GRADES			
Select (all species)	B & BTR Select C Select D Select	Supreme Choice Quality	10.11 10.12 10.13
Finish (usually available only in Douglas fir and Hemlock-fir	Superior Prime E		10.51 10.52 10.53
Special Western Red Cedar Pattern Grades	Clear Heart A Grade B Grade		20.11 20.12 20.13
GENERAL-PURPOSE APPEARANCE GRADES			
Common Boards (WWPA Rules) (primarily in pines, spruces, and cedars)	1 Common 2 Common 3 Common 4 Common 5 Common	Colonial Sterling Standard Utility Industrial	30.11 30.12 30.13 31.14 30.15
Alternate Boards (WCLIB Rules) (primarily in Douglas fir and Hemlock-fir	Select Merchantable Construction Standard Utility Economy		WCLIB† 118-a 118-b 118-c 118-d 118-e
Special Western Red Cedar Pattern* Grades	Select Knotty Quality Knotty		WCLIB† 111-e 111-f

* Includes Finish, Paneling, Ceiling, and Siding grades.
† West Coast Lumber Inspection Bureau's *West Coast Lumber Standard Grading Rules*

SOUTHERN FOREST PRODUCTS ASSOCIATION FRAMING LUMBER GRADES . . .

DIMENSION LUMBER	
Grade	**Grade Characteristics and Applications**
Dense Select Structural* Select Structural Nondense Select Structural*	High quality, relatively free of characteristics which impair strength or stiffness. Recommended for applications where high strength, stiffness, and good appearance are desired.
No. 1 Dense* No. 1 No. 1 Nondense*	Recommended for general utility and construction where high strength, stiffness, and good appearance are desired.
No. 2 Dense* No. 2 No. 2 Nondense*	Recommended for most general construction applications where moderately high design values are required. Allows well-spaced knots of any quality.
No. 3	Assigned design values meet a wide range of design requirements. Recommended for general construction applications where appearance is not a controlling factor. Many pieces included qualify as No. 2, except for a single limiting characteristic.
Stud	Suitable for stud applications including use in load-bearing walls. Composite of No. 3 strength and No. 1 nailing edge characteristics.
Construction (2″ to 4″ wide only)*	Recommended for general framing applications. Good appearance, but graded primarily for strength and serviceability.
Standard (2″ to 4″ wide only)*	Recommended for general framing applications. Characteristics are limited to provide good strength and serviceability.
Utility (2″ to 4″ wide only)*	Recommended where a combination of economical construction and good strength is desired. Used for applications such as studs, blocking, plates, bracing, and rafters.
Economy	Usable lengths suitable for bracing, blocking, bulkheads, and other utility purposes where strength and appearance are not controlling factors.

TIMBERS	
Grade	**Grade Characteristics and Applications**
Dense Select Structural Select Structural	Recommended where high strength, stiffness, and good appearance are desired.
No. 1 Dense No. 1 No. 2 Dense No. 2	No. 1 and No. 2 are similar in appearance to corresponding grades of 2″ thick Dimension Lumber. Recommended for general construction applications.
No. 3	Nonstress rated, but economical for general construction applications such as blocking and fillers.

MACHINE STRESS-RATED (MSR) LUMBER	
Grade	**Grade Characteristics and Applications**
2700f-2.2E 2550f-2.1E 2400f-2.0E 2250f-1.9E 2100f-1.8E 1950f-1.7E 1800f-1.6E 1650f-1.5E	Lumber that has been evaluated by mechanical stress-rating equipment. MSR lumber is distinguished from visually stress-graded lumber in that each piece is nondestructively tested. MSR lumber is also required to meet certain visual grading requirements.

* Most mills do not manufacture all products and make all grade separations.

. . . SOUTHERN FOREST PRODUCTS ASSOCIATION FRAMING LUMBER GRADES

MACHINE-EVALUATED LUMBER (MEL)

Grade	Grade Characteristcs and Applications
M-23 **M-19** **M-10** **M-6**	Well-manufactured material evaluated by calibrated mechanical grading equipment which measures certain properties and sorts lumber into various strength classifications. MEL is also required to meet certain visual requirements.

FINISH

Grade	Grade Characteristics and Applications
B & B	Highest recognized grade of finish. Generally clear, although a limited number of pin knots are permitted. Finest quality for natural stain finish.
C	Excellent for painting or natural finish where requirements are less exacting. Reasonably clear, but permits limited number of surface checks and small, tight knots.
C & Btr	Combination of B & B and C grades; satisfies requirements for high-quality finish.
D	Economical, serviceable grade for natural or painted finish.

FLOORING, DROP SIDING, PANELING, CEILING AND PARTITION, OG BATTS, BEVEL SIDING, MISCELLANEOUS MILLWORK

Grade	Grade Characteristics and Applications
B & B **C** **C & Btr** **D**	See finish grades for face side; reverse side wane limitations are lower.
No. 1	No. 1 Flooring and Paneling not provided under Southern Pine Inspection Bureau Grading Rules as a separate guide, but if specified, will be designated and graded as D; No. 1 Drop siding is graded as No. 1 Boards.
No. 2	Graded as No. 2 Boards. High utility value where appearance is not a factor.
No. 3	More manufacturing imperfections allowed than in No. 2, but suitable for economical use.

SHOP AND MOLDING*

Grade	Grade Characteristic and Applications
No. 1 **No. 2** **No. 3**	Recommended for molding and millwork applications. Currently graded according to rules developed by the Western Wood Products Association.

* Most mills do not manufacture all products and make all grade separations.

APA TRADEMARKS . . .

PERFORMANCE RATED PANELS

Panel Type and Trademark			Description, Bond Classification, and Thickness
APA Rated Sheathing	APA THE ENGINEERED WOOD ASSOCIATION RATED SHEATHING 40/20 19/32 INCH SIZED FOR SPACING EXPOSURE 1 000 PS 2-92 SHEATHING PRP-108 HUD-UM-40	APA THE ENGINEERED WOOD ASSOCIATION RATED SHEATHING 24/16 7/16 INCH SIZED FOR SPACING EXPOSURE 1 000 PRP-108 HUD-UM-40	Specially designed for subfloor and wall and roof sheathing. Also good for broad range of other construction and industrial applications. Can be manufactured as OSB, plywood, or a composite panel. Bond Classifications: Exterior, Exposure 1, Exposure 2. Common Thicknesses: 5/16, 3/8, 7/16, 15/32, 1/2, 19/32, 5/8, 23/32, 3/4.
APA Structural 1 Rated Sheathing	APA THE ENGINEERED WOOD ASSOCIATION RATED SHEATHING STRUCTURAL I 32/16 15/32 INCH SIZED FOR SPACING EXPOSURE 1 000 PS 1-95 C-D PRP-108	APA THE ENGINEERED WOOD ASSOCIATION RATED SHEATHING 32/16 15/32 INCH SIZED FOR SPACING EXPOSURE 1 000 STRUCTURAL I RATED DIAPHRAGMS-SHEAR WALLS PANELIZED ROOFS PRP-108 HUD-UM-40	Unsanded grade for use where shear and cross-panel strength properties are of maximum importance, such as panelized roofs and diaphragms. Can be manufactured as OSB, plywood, or a composite panel. Bond Classifications: Exterior, Exposure 1. Common Thicknesses: 5/16, 3/8, 7/16, 15/32, 1/2, 19/32, 5/8, 23/32, 3/4.
APA Rated Sturd-I-Floor	APA THE ENGINEERED WOOD ASSOCIATION RATED STURD-I-FLOOR 24 oc 23/32 INCH SIZED FOR SPACING TAG NET WIDTH 47-1/2 EXPOSURE 1 000 PS 2-92 SINGLE FLOOR PRP-108 HUD-UM-40	APA THE ENGINEERED WOOD ASSOCIATION RATED STURD-I-FLOOR 20 oc 19/32 INCH SIZED FOR SPACING TAG NET WIDTH 47-1/2 EXPOSURE 1 000 PRP-108 HUD-UM-40	Specially designed as combination subfloor-underlayment. Provides smooth surface for application of carpet and pad and possesses high concentrated and impact load resistance. Can be manufactured as OSB, plywood, or a composite panel. Available square edge or tongue-and-groove. Bond Classifications: Exterior, Exposure 1, Exposure 2. Common Thicknesses: 19/32, 5/8, 23/32, 3/4, 1, 1 1/8.
APA Rated Siding	APA THE ENGINEERED WOOD ASSOCIATION RATED SIDING 24 oc 19/32 INCH SIZED FOR SPACING EXTERIOR 000 PRP-108 HUD-UM-40	APA THE ENGINEERED WOOD ASSOCIATION RATED SIDING 303-18-S/W 11/32 INCH 16 oc GROUP 1 SIZED FOR SPACING EXTERIOR 000 PS 1-95 PRP-108 FHA-UM-40	For exterior siding, fencing, etc. Can be manufactured as plywood, as a composite panel, or as an overlaid OSB. Both panel and lap siding available. Special surface treatment such as V-groove, channel groove, deep groove (such as APA Texture 1-11), brushed, rough sawn, and overlaid (MDO) with smooth- or texture-embossed face. Span rating (stud spacing for siding qualified for APA Sturd-I-Wall Applications) and face grade classification (for veneer-faced siding) indicated in trademark. Bond Classification: Exterior. Common Thicknesses: 11/32, 3/8, 7/16, 15/32, 1/2, 19/32, 5/8.

SANDED AND TOUCH-SANDED PANELS . . .

Panel Type and Trademark	Description, Bond Classification, and Thickness
APA A-A A-A • G-1 • EXPOSURE 1-APA • 000 • PS1-95	Use where appearance of both sides is important for interior applications such as built-ins, cabinets, furniture, partitions; and exterior applications such as fences, signs, boats, shipping containers, tanks, ducts, etc. Smooth surfaces suitable for painting. Bond Classifications: Interior, Exposure 1, Exterior. Common Thicknesses: 1/4, 11/32, 3/8, 15/32, 1/2, 19/32, 5/8, 23/32, 3/4.
APA A-B A-B • G-1 • EXPOSURE 1-APA • 000 • PS1-95	For use where appearance of one side is less important, but where two solid surfaces are necessary. Bond Classifications: Interior, Exposure 1, Exterior. Common Thicknesses: 1/4, 11/32, 3/8, 15/32, 1/2, 19/32, 5/8, 23/32, 3/4.

. . . APA TRADEMARKS . . .

. . . SANDED AND TOUCH-SANDED PANELS

Panel Type and Trademark	Description, Bond Classification, and Thickness
APA A-C APA THE ENGINEERED WOOD ASSOCIATION A-C GROUP 1 EXTERIOR 000 PS 1-95	For use where appearance of only one side is important in exterior or interior applications, such as soffits, fences, farm buildings, etc. Bond Classification: Exterior. Common Thicknesses: 1/4, 11/32, 3/8, 15/32, 1/2, 19/32, 5/8, 23/32, 3/4.
APA A-D APA THE ENGINEERED WOOD ASSOCIATION A-D GROUP 1 EXPOSURE 1 000 PS 1-95	For use where appearance of only one side is important in interior applications, such as paneling, built-ins, shelving, partitions, flow racks, etc. Bond Classifications: Interior, Exposure 1. Common Thicknesses: 1/4, 11/32, 3/8, 15/32, 1/2, 19/32, 5/8, 23/32, 3/4.
APA B-B B-B • G-2 • EXT-APA • 000 • PS1-95	Utility panels with two solid sides. Bond Classifications: Interior, Exposure 1, Exterior. Common Thicknesses: 1/4, 11/32, 3/8, 15/32, 1/2, 19/32, 5/8, 23/32, 3/4.
APA B-C APA THE ENGINEERED WOOD ASSOCIATION B-C GROUP 1 EXPOSURE 1 000 PS 1-95	Utility panel for farm service and work buildings, boxcar and truck linings, containers, tanks, agricultural equipment; as a base for exterior coatings; and other exterior uses or applications subject to high or continuous moisture. Bond Classification: Exterior. Common Thicknesses: 1/4, 11/32, 3/8, 15/32, 1/2, 19/32, 5/8, 23/32, 3/4.
APA B-D APA THE ENGINEERED WOOD ASSOCIATION B-D GROUP 2 EXPOSURE 1 000 PS 1-95	Utility panel for backing, sides of built-ins, industry shelving, slip sheets, separator boards, bins, and other interior or protected applications. Bond Classifications: Interior, Exposure 1. Common Thicknesses: 1/4, 11/32, 3/8, 15/32, 1/2, 19/32, 5/8, 23/32, 3/4.
APA Underlayment APA THE ENGINEERED WOOD ASSOCIATION UNDERLAYMENT GROUP 1 EXPOSURE 1 000 PS 1-95	For application over structural subfloor. Provides smooth surface for application of carpet and pad and possesses high concentrated and impact load resistance. For areas to be covered with resilient flooring, specify panels with "sanded face." Bond Classifications: Interior, Exposure 1. Common Thicknesses: 1/4, 11/32, 3/8, 15/32, 1/2, 19/32, 5/8, 23/32, 3/4.
APA C-C Plugged APA THE ENGINEERED WOOD ASSOCIATION C-C PLUGGED GROUP 2 EXTERIOR 000 PS 1-95	For use as an underlayment over structural subfloor, refrigerated or controlled atmosphere storage rooms, pallet fruit bins, tanks, boxcar and truck floors and linings, open soffits, and other similar applications where continuous or severe moisture may be present. Provides smooth surface for application of carpet and pad and possesses high concentrated and impact load resistance. For areas to be covered with resilient flooring, specify panels with "sanded face." Bond Classification: Exterior. Common Thicknesses: 11/32, 3/8, 15/32, 1/2, 19/32, 5/8, 23/32, 3/4.
APA C-D Plugged APA THE ENGINEERED WOOD ASSOCIATION C-D PLUGGED GROUP 2 EXPOSURE 1 000 PS 1-95	For open soffits, built-ins, cable reels, separator boards, and other interior or protected applications. Not a substitute for Underlayment or APA Rated Sturd-I-Floor as it lacks their puncture resistance. Bond Classifications: Interior, Exposure 1. Common Thicknesses: 3/8, 15/32, 1/2, 19/32, 5/8, 23/32, 3/4.

. . . APA TRADEMARKS

SPECIALTY PLYWOOD PANELS

Panel Type and Trademark	Description, Bond Classification, and Thickness
APA Decorative 	Rough sawn, brushed, grooved, or striated faces. For paneling, interior accent walls, built-ins, counter facing, exhibit displays. Can also be made by some manufacturers in Exterior for exterior siding, gable ends, fences, and other exterior applications. Use recommendations for Exterior panels vary with the particular product. Check with the manufacturer. Bond Classifications: Interior, Exposure 1. Common Thicknesses: 5/16, 3/8, 1/2, 5/8.
APA High Density Overlay (HDO) HDO • A-A • EXT-APA • 000 • PS1-95	Has a hard, semi-opaque resin-fiber overlay on both faces. Abrasion-resistant. For concrete forms, cabinets, countertops, signs, and tanks. Also available with skid-resistant screen-grid surface. Bond Classification: Exterior. Common Thicknesses: 3/8, 1/2, 5/8, 3/4.
APA Medium Density Overlay (MDO) 	Smooth, opaque, resin-fiber overlay on one or both faces. Ideal base for paint, both indoors and outdoors. For exterior siding, paneling, shelving, exhibit displays, cabinets, signs. Bond Classification: Exterior. Common Thicknesses: 11/32, 3/8, 15/32, 1/2, 19/32, 5/8, 23/32, 3/4.
APA Marine MARINE • A-A • EXT-APA • 000 • PS1-95	Ideal for boat hulls. Made only with Douglas fir or western larch. Subject to special limitations on core gaps and face repairs. Also available with HDO or MDO faces. Bond Classification: Exterior. Common Thicknesses: 1/4, 3/8, 1/2, 5/8, 3/4.
APA Plyform Class 	Concrete form grades with high reuse factor. Sanded both faces and mill-oiled unless otherwise specified. Special restrictions on species. Also available in HDO for very smooth concrete finish, and with special overlays. Bond Classification: Exterior. Common Thicknesses: 19/32, 5/8, 23/32, 3/4.
APA Plyron PLYRON • EXPOSURE 1 • APA • 000	Hardboard face on both sides. Faces tempered, untempered, smooth, or screened. For countertops, shelving, cabinet doors, and flooring. Bond Classifications: Interior, Exposure 1. Common Thicknesses: 1/2, 5/8, 3/4.

A

abbreviation: Letter or group of letters representing a term or phrase.

abrasive cutoff saw: Power saw equipped with an abrasive blade and used to cut metal framing members. The fence is adjusted from side to side for angled cuts.

abrasive paper: Paper or cloth with an abrasive material glued to one side. Used for smoothing wood or metal surfaces.

abut: To join members along an edge or end but not overlap.

accelerating admixture: Concrete admixture that reduces concrete setting time and improves strength of concrete at an early stage.

access flooring: Raised floor system used to create a chase for routing electrical conductors, computer wiring, ventilation ducts, and other utilities.

acclimate: To become used to new climates or conditions.

acoustical tile: Ceiling tile with small holes and fissures that act as sound traps to reduce the reflection of sound.

acoustic window: Triple-pane windows with the inner layer made of laminated glass; effective in minimizing noise infiltration into a building.

active door: Door of a double door unit used for normal traffic flow.

actual length of rafter: Length of a main rafter or valley jack rafter after it has been shortened because of ridge board and/or rafter thickness.

actual lumber size: Thickness and width of a piece of lumber after shrinkage resulting from drying and after surfacing.

adhesive: Substance used to bond two surfaces together. Available in solid, liquid, and semiliquid forms.

admixture: Substance other than water, portland cement, or aggregate that is added to concrete to modify its properties.

advanced framing: Engineered framing system that reduces material costs and construction waste and allows for increased insulation use.

aerial lift: Extendable or articulated equipment designed to position personnel and materials in elevated positions.

A-frame: Structure with a gable roof extending to the foundation to resemble the letter "A."

aggregate: Hard granular material, such as sand and gravel, that is mixed with cement to provide structure and strength in concrete.

air compressor: Equipment used to compress air for pneumatic tools and equipment.

air-dried lumber: Lumber that has been dried to an acceptable moisture content by being stacked for a period of time in the open air.

air duct: Pipe, usually rectangular and made of sheet metal, used to conduct hot or cold air in heating or cooling systems.

air-entraining admixture: Concrete admixture that produces microscopic air bubbles in concrete to improve its workability and its resistance to freezing and thawing.

air pocket: Void in hardened concrete caused by poor consolidation when the concrete was placed.

allowable span: Distance allowed between supporting points for various sizes of beams, joists, and roof rafters.

alteration work: Change or addition to an existing building. Also called *remodeling*.

anchorage point: Device that provides a secure point of attachment for lifelines, lanyards, or other deceleration devices.

anchor bolt: Bolt used to secure sill plates, columns, and beams to concrete or other masonry.

anchor clip: Strap-like device embedded into the top of a foundation wall and used to fasten sill plates.

Laying Out 45° Angles

A method for using a framing square to mark 45° angles on a wider piece of material.

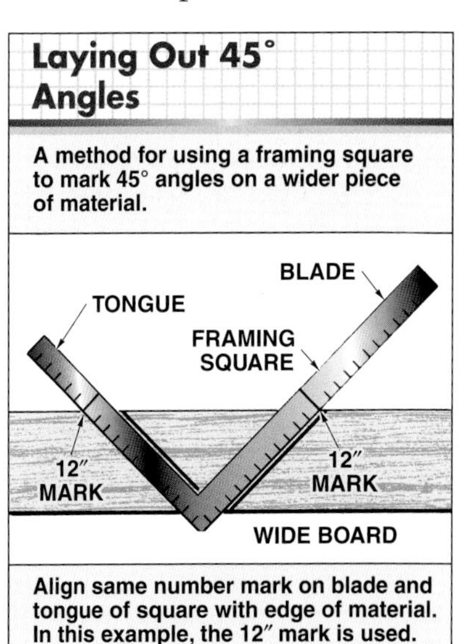

Align same number mark on blade and tongue of square with edge of material. In this example, the 12″ mark is used.

anchor tiedown system: Connection device used in areas where the most extreme uplift due to seismic conditions or high winds is anticipated.

angle grinder: A handheld tool that removes material by abrasive action.

angle iron: Structural steel bent to form a 90° angle and used for support or fastening purposes.

annual ring: Layer of wood seen in cross section of a tree trunk indicating a year's growth.

annular nail: A nail with circular ridges on the shank to provide greater holding power.

apprenticeship: Training for a trade or skilled craft that requires a wide variety of skills and knowledge. Provided with instruction and on-the-job experience in practical and theoretical aspects of the trade.

apron: Trim piece placed below the stool of a finished window frame.

arbor: Shaft on which a cutting tool is mounted, such as the blade of a circular saw.

arc: Part of the circumference of a circle.

arch: Curved structure designed to support the weight above an opening.

architect: Person qualified and licensed to design and oversee construction of a building.

archway: Passage area under an arch.

areaway: Small sunken area allowing light or air into a basement window that is partially below the grade level around the building. Also called *window well.*

argon gas: Odorless, colorless gas inserted between panes of glass to reduce heat transfer.

asphalt: Petroleum product obtained from crude oil. It is waterproof and is the basis for many products used for roof, wall, and floor covering.

asphalt shingle: Shingle made of asphalt-saturated felt and surfaced with mineral granules. Usually used as a finish roofing material.

astragal: Piece of molding attached to the edge of the inactive door of a pair of double doors. Serves as a stop for the active door.

attic: Space between a roof and a ceiling.

attic rafter vent: Foam plastic insert that is fit between the rafters or trusses in an attic to prevent insulation from blocking airflow from soffit vents.

awning window: Window that is hinged at the top and swings out at the bottom.

axis: Straight line around which a body or member rotates.

B

backed out: Refers to the back of casing material being hollowed out on the back side to give the molding flexibility, producing a tight fit even if there is unevenness between the jamb and the wall.

backer block: Wood blocks used with wood I-joists to provide a flat, flush surface for attachment of top- or face-mounted joist hangers or other structural elements by filling the space between the outside edge of the I-joist flange and the web of the adjoining I-joist.

backer rod: **1.** Cylindrical, flexible, closed cell polyethylene foam used as backing for expansion joint sealant or installed between a tilt-up wall panel and foundation. Used to prevent moisture from seeping between the adjacent members. **2.** A flexible polyethylene foam material that is used to control caulk joint depth and prevent three-sided adhesion.

backfill: Soil or gravel used to fill the space between a completed foundation wall and the excavated areas on one or both sides of the wall.

back framing: Secondary and nonstructural framing that is performed after structural framing is complete. Includes framing attic scuttles, drop ceilings, chases for utilities, etc.

backing: **1.** Pieces nailed over the top wall plates to provide a nailing surface for the edges of ceiling finish materials. **2.** Process of beveling each side of the top edge of a hip rafter so the ends of the roof sheathing will not be pushed up and out of line with the rest of the roof. **3.** Pieces nailed between studs to provide a surface for fastening plumbing fixtures. Also, pieces nailed behind studs to provide a surface for corner nailing.

back priming: Painting or treating with a sealer the backs of wood siding materials before they are nailed to walls.

backsplash: Pieces that extend up from a countertop and are fastened to the wall.

balanced beam: Glulam beams that are symmetrical in lumber quality and are primarily used for cantilevered or continuous span applications where either the top or bottom of the member may be stressed in tension.

ball-bearing butt hinge: Butt hinge with ball bearings at the knuckle joints to prevent wear. Used most often on heavy doors in public buildings.

balloon framing: Framing method in which studs extend from the sill plate to the roof. Second floor joists are fastened to the studs, but they receive their main support from a ribbon notched into and fastened to the studs.

ball-peen hammer: A hammer with a round, slightly curved face and round head.

baluster: Upright piece that extends between the handrails and the treads. Also called *banister.*

balustrade railing: Railing with newel posts at both ends and balusters in between.

band saw: A stationary power saw that has a flexible saw blade forming a continuous loop around two parallel wheels.

bargeboard: Finish board covering the projecting and sloping portion of a gable roof.

bark: Surface covering of a tree trunk. Consists of an outer layer of dead, dry tissue and a thin inner layer of living tissue.

bark pocket: Patch of bark nearly or wholly enclosed in wood.

barrel: Rounded hollow section located between the leaves of a hinge that holds the pin. Also called the *knuckle.*

barricade: Structure set up around a construction job to prevent unauthorized persons from entering working areas. Covered barricades also protect the public from falling objects.

baseboard: Molding placed at the base of a wall and fitted to the floor. Also called *base molding.*

base cabinet: Cabinet placed against a wall and resting on the floor.

base cap: Small decorative molding sometimes nailed at the top of a baseboard.

basement: Lowest story of a building, which is partly or completely below the ground.

base shoe: Strip of molding nailed at the joint between a baseboard and the floor.

base track: Bottom track of a metal-framed wall.

basket hitch: Sling hitch in which the sling is passed under a load and both ends of the sling are connected to a hoist hook.

batch plant: Facility where ready-mixed concrete is mixed to specification, then discharged into transit-mix trucks for delivery to a job site.

batten: Thin, narrow strip of lumber usually used to cover the joint between wider boards.

batterboard: Level board nailed to stakes driven into the ground. String is attached to batterboards to identify property lines, building lines, and pier locations.

Fastening Lines to Batterboards

A reliable method for preventing lines from moving out of their exact positions on batterboards.

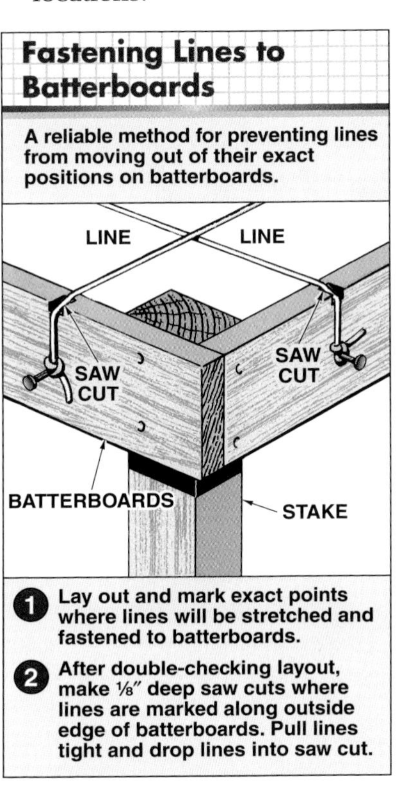

1. **Lay out and mark exact points where lines will be stretched and fastened to batterboards.**
2. **After double-checking layout, make 1/8″ deep saw cuts where lines are marked along outside edge of batterboards. Pull lines tight and drop lines into saw cut.**

battered foundation: Foundation with inside sloped walls to provide a wide base.

batt insulation: Sections of fiberglass material, usually 48″ long, placed between studs.

bay window: Window projecting from a wall. Creates a recessed area that is square, rectangular, or polygonal in shape.

beam hanger: Metal connector used to support a beam.

beam pocket: Opening in a concrete wall to receive the end of a wood or steel beam so the top of the beam is at the same level as the top of the wall.

bearing pile: Pile that penetrates through layers of unstable soil until it reaches firm, bearing soil.

bearing plates: Steel plates (3″ square) with an oblong hole in the middle to allow the plates to be adjustable in case an anchor bolt is not centered on the sill plate. Bearing plates significantly increase the strength of anchor bolt connections.

bearing wall: Wall that supports loads from above as well as its own weight.

bed molding: Molding placed over a frieze board and against the cornice soffit. May be placed between the rafters when the frieze board is notched into the rafters of an open cornice.

bedrock: Solid layer of rock beneath the earthen materials.

belled caisson: Concrete caisson flared at its bottom to provide a greater bearing area.

bents: Structural, interconnected system of timbers contained in a timber frame wall.

berm: Raised earth embankment.

bevel: Angled surface across the edge of a piece of material.

bevel siding: Tapered siding board that laps over the board below.

bibb: Threaded faucet for a hose attachment.

bid: Contractor's offer to a customer to perform construction work for a specified price.

bi-fold doors: Pair of folding doors in which only one door is hung from the jamb. The second door is hinged to the first door and is further supported by a roller guide that slides along a track mounted to the head jamb.

bind: To interfere with the cutting action of a saw as a result of wood fibers pressing against the blade.

bird's mouth: Triangular notch formed by the seat and heel plumb cut lines of a rafter where the lower end of the roof rafter fits over the top plates.

bit: Drilling device with a screw point that is held in the jaws of a drill.

bituminous: Containing tars, pitches, or asphalts. Commonly used for built-up roofing or waterproofing.

blank: Piece of wood that will produce the desired pattern of molding with the least amount of waste.

blanket insulation: Fiberglass or rock wool insulation usually backed by treated paper that serves as a vapor barrier.

blind nailing: Driving a nail so its head is concealed.

blind stop: Rectangular piece of molding placed toward the exterior of the head and side pieces of a window frame to serve as a stop for screens or storm windows.

blind valley construction: Method of building an intersecting roof. The main roof is sheathed and the intersecting roof section is constructed on top of the sheathing. Valley rafters are not required.

block flooring: Small squares made of narrow wood strips held together by splines or backing material. Usually installed with a mastic adhesive.

blocking: Wood piece fastened between structural members to tie together and strengthen the members.

blockout: Piece placed inside a form to provide a small recess or ledge at the top of the concrete wall to later support the end of a beam or other structural member.

bloodborne pathogen: Virus or bacteria of a disease in the blood that may be transmitted by another worker coming into contact with the infected worker's blood.

blown-in blanket insulation: Type of loose fill insulation in which a retention fabric made of spun polypropylene is fastened to wall studs using adhesive or staples before gypsum board or other interior wall covering is applied. A small opening is made in the fabric and loose fill fiberglass insulation is blown into the wall cavity.

blue stain: Nonharmful blue-gray discoloration of lumber caused by a fungus growth on unseasoned wood.

board: Lumber measuring no more than 1″ thick and 4″ to 12″ wide.

board-and-batten siding: Siding design that features wide vertical boards. The joints between the boards are covered with a narrow piece called the batten.

board foot: Volume of a piece of lumber measuring 1″ × 12″ × 12″ or its equivalent.

body harness: Personal fall-arrest device that protects internal body organs, the spine, and other bones in a fall.

bolt: 1. Cylindrical fastener, usually consisting of a piece of metal with a head or hooked end and a fully or partially threaded body. **2.** Portion of a lockset projecting from a door into a jamb to hold the door closed.

bond beam: Continuous reinforced course of concrete masonry units around the top of masonry walls. Typically contains steel reinforcement and is filled with grout.

bow: Distortion of a piece of lumber resulting in a curve from end to end across the flat plane of the wide side.

bow window: Series of windows set in a curved bow projecting from the building.

box beam: Hollow beam made of plywood sides nailed to a framework of 2 × 4 chords and stiffeners.

box nail: Flat-head nailed with a thinner shank and head than a common nail.

brace: Structural piece, either temporary or permanent, to hold members or assemblies in a plumb and/or fixed position.

brad: Small finish nail ranging from ½″ to 2″ in length.

breakback: Grooved section of a snap tie next to the tapered edge of the cone that allows the tie to be broken off after the concrete has hardened and the forms have been removed.

breezeway: Roofed passageway, open at ends or sides, extending from a house to a garage.

bridging: Solid wood blocking or other type of wood or metal pieces nailed between joists to stiffen and hold them in position.

bridle hitch: Sling hitch in which two or more slings share a common fitting as a means of attachment to the hoist hook.

British thermal unit (Btu): Measure of heat transfer based on the amount of heat required to raise the temperature of 1 lb of water 1°F.

buck: Frame placed inside a concrete form to provide an opening for a door or window.

bucket: Large container into which concrete is discharged from a mixer or transit-mix truck. The bucket is then raised by crane and moved to a location where the concrete is to be placed.

builder's level: Survey instrument used to establish and verify grades and elevations and to set up level points over long distances.

building code: Set of regulations that establish the required standards for the materials and methods of construction in a city, county, or state. Building codes are enforceable by law.

building envelope: **1.** Consists of the exterior wall finish and weather barrier applied over the wall sheathing. **2.** A continuous thermal and air boundary separating the conditioned space from any unconditioned space or from the outside.

Building information modeling (BIM): Integrated, electronically managed system that aligns all working drawings, structural drawings, and shop drawings into a consistent system.

building line: Line set up on batterboards to represent the outside face of the exterior wall of a building.

building paper: Asphalt-saturated felt material used as a moisture barrier and installed under siding and other exterior finish material.

building permit: Permit legally required by local or state building authorities for new construction and major renovations.

building science: Study of interaction between a building (including its HVAC, mechanical, and electrical systems), its inhabitants, and the surrounding environment. Also called *building dynamics* or *building physics.*

built-in: Refers to cabinets and other furniture that is built into the wall structure.

built-up beam: Beam made of two or more members nailed or bolted together.

built-up header: Header over a rough opening made up of two or more members fastened together.

built-up roofing: Type of roofing consisting of layers of roofing felt bonded together on the job site with hot bitumen. The final layer is covered with a layer of light-colored pebbles to reflect heat from the sun.

bulkhead: Vertical piece set inside a form to stop concrete at a certain place, thus providing a vertical construction joint.

butterfly roof: Roof with two sides sloping toward the central part of the building.

butt joint: Joint in which one piece butts squarely against another.

buttress: Projecting masonry structure built against a wall to give additional lateral strength and support.

bypass doors: Doors suspended from tracks with roller hangers, often used for closets. Two or three bypass sliding doors may be used to cover an opening, with the tracks designed to let the doors slide past one another.

C

cabinetmaker: Person who works in a cabinet shop and is skilled in the layout and construction of cabinetry.

caisson: Large cylindrical or box-like casing placed in the ground and filled with concrete. Caissons have wide diameters and usually extend deeper than piles.

callout: Note on drawing with a leader line pointing to the related feature.

camber: Slight upward curve in a structural member designed to compensate for deflection of the member under load.

cambium: Layer of tissue equal to a one-cell thickness and located between the bark and sapwood of a tree.

cam-out: Condition that occurs when the screwdriver bit slips out of the screw head when exerting pressure to turn the screw.

cantilevered joist: Joist projecting from the wall below to support a floor or upper level which extends past the wall below.

cant strip: Triangular piece placed where a flat roof deck intersects with a wall. Used to prevent a sharp angle for the roofing material.

capillary action: Natural process in which water rises from the water table to the surface of the ground.

capital: Flared section at the top of a concrete column that helps support the floor slab above. Also called *drop head.*

capture window: Part of a laser level detector that receives the laser beam from a laser level.

carbide tip: Tungsten carbide metal tip that is braised to the end of each tooth of a circular saw blade and to the cutting edges of masonry bits to prolong sharpness.

carcinogen: Cancer causing agent or material.

carpenter: Professional tradesworker engaged in the construction of wood and light-gauge steel building framework, concrete formwork, and interior and exterior finish.

carpenter's level: Hand tool used to establish level and plumb lines. Also called *hand level.*

carport: Roofed car shelter not enclosed at the sides.

casement window: Window hinged on one side and having the same swing action as a door.

casing: **1.** Molding that covers the space between finished door and window frames and the wall. **2.** Steel shell driven into the ground or placed in a prebored hole to act as a form to receive the concrete for a cast-in-place concrete pile.

casing nail: Nail with a flared head. Used for outside finish work and finish flooring.

casting bed: System of forms and supports for producing concrete members.

Marking Door Side Casings

A fast and accurate method for cutting pieces of side casing when several door openings are same height.

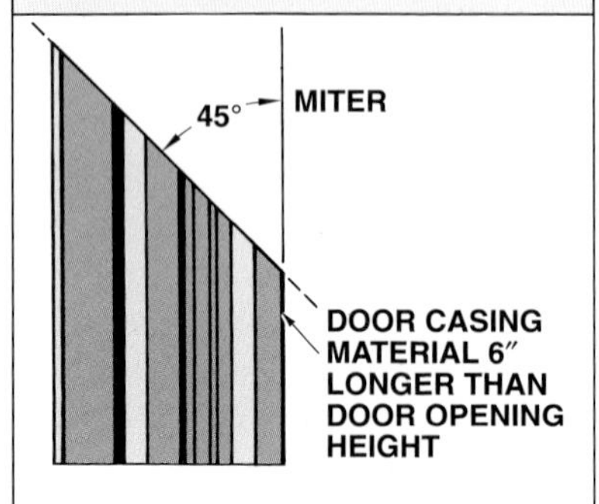

1. Cut a 45° miter at one end of casing material. Length from long point of miter to other end of casing material should be about 6″ longer than door opening height.

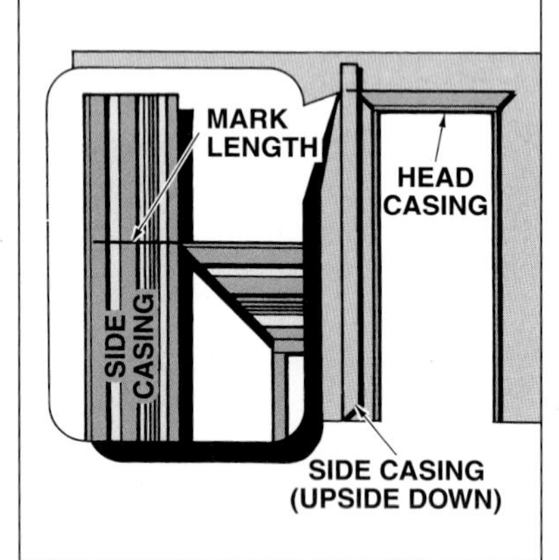

2. Place the side piece of casing in upside-down position next to door frame. Mark side casing even with the upper edge of head casing. Square line across side casing and cut it to length.

cast-in-place pile: A pile formed by placing concrete into a steel casing that has been driven into the ground.

caulk: A resilient construction material that incorporates synthetic polymers and is used to seal joints and cracks to prevent moisture and air leakage. Also called *joint sealant*.

caulking compound: Filler material applied at joints and cracks to prevent air and water infiltration.

ceiling grid: Light metal framework used to support the tiles of a suspended ceiling.

ceiling molding: Molding placed at the top of a wall close to or against the ceiling.

ceiling tile: Rectangular or square fibrous tile used to finish ceilings.

cellulose: Principal substance in the walls of wood cells.

cement: Ingredient that binds together sand and gravel in a concrete mixture after water is added.

center of gravity (CG): Point at which an object's total mass can be considered to be concentrated. Also know as *center of mass*.

center point test: Test used to determine the suitability of concrete for powder-actuated fasteners.

central heating: System in which heat is generated from a single source and is distributed by ducts to all parts of a building.

ceramic tile: Thin tile made of fired clay used as a finish material on floors, walls, and countertops.

chain: Series of metal links connected to one another to form a continuous line.

chain-of-custody: Path that raw materials harvested from Forest Stewardship Council-certified forests take through processing, manufacturing, and distribution until they are a final product and ready for sale to the consumer.

chair rail: Horizontal trim molding attached to a wall 32″ above the floor as ornamentation and protection against damage from chairs being pushed against the wall.

chamfer: Cut made from the face to the side of a board.

chamfer strip: Piece placed in the corners of beam and column forms to produce a beveled edge on the concrete member.

channel rustic siding: Rabbeted type of board siding that creates a grooved channel effect between the boards.

chase: Vertical enclosure within a structure that allows for placement of ductwork, pipes, and wiring.

check: Lumber defect caused by a lengthwise separation of the wood across the annual growth rings during seasoning.

check rail: Bottom rail of the top sash of a window and top rail of the bottom sash. Check rails are adjacent to each other when both window sections are in a closed position.

chimney: Noncombustible vertical structure surrounding a flue to convey smoke, flue gases, and fumes away from the combustion source.

choker hitch: Sling hitch in which one end of the sling is wrapped under or around the load, passed through the eye at the other end of the sling, and then connected to the hoist hook.

chord: A top or bottom member of a roof or floor truss.

circuit: Conductor or series of conductors through which electrical current flows.

circular saw: Portable power tool used to cut lumber to length or panels to length and width.

circular stairway: Type of winding stairway in which all stairs radiate from a common center.

circumference: Perimeter of a circle.

cladding: Exterior covering of a window.

clamp: Device used to hold or secure pieces together to prevent movement or separation.

clay: Fine-grained natural earth material that is plastic when wet and compact and brittle when dry.

cleanout hole: Small opening at the base of a wall or column form to allow debris to be removed. The opening is closed shut before concrete is placed.

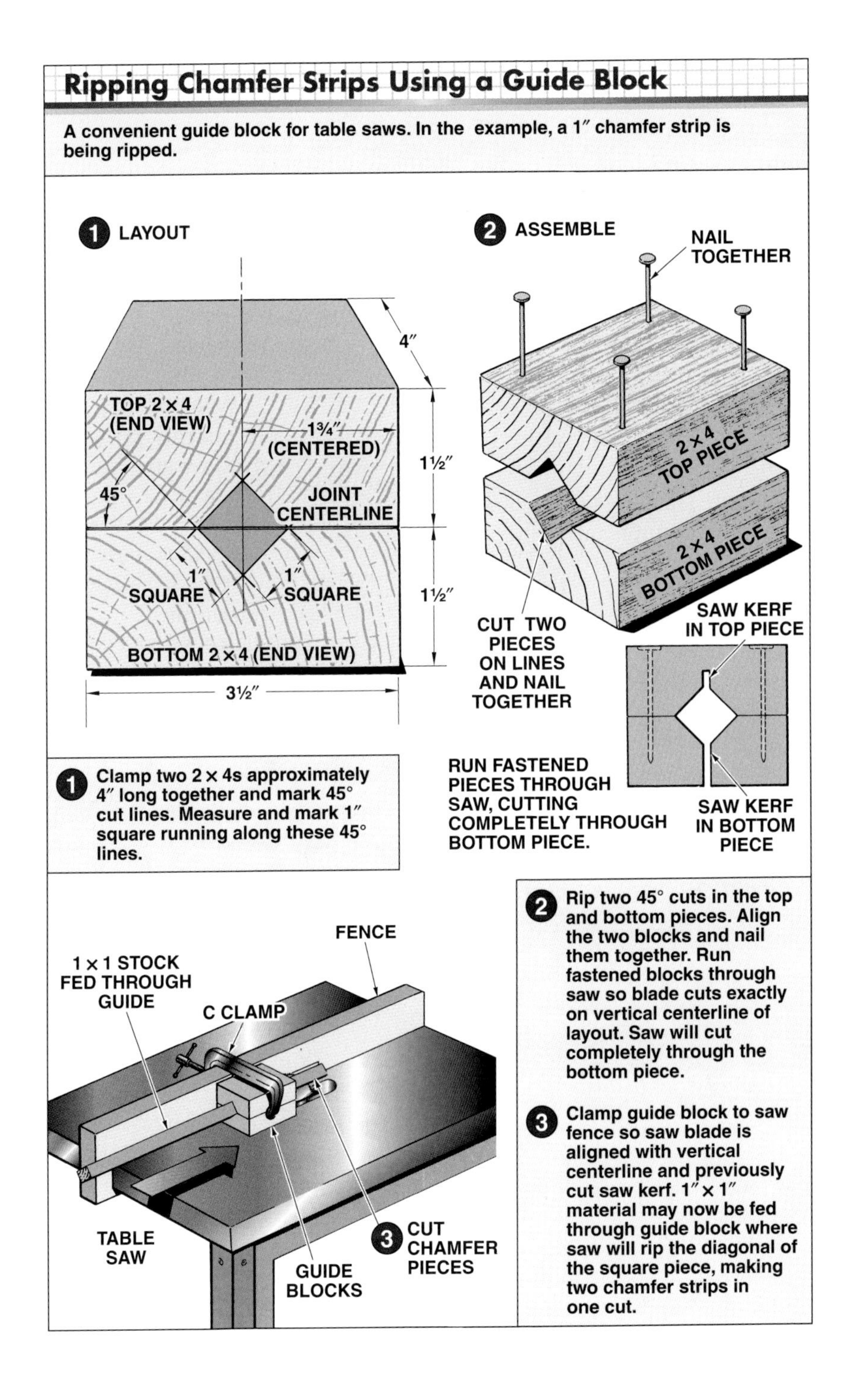

cleat: Narrow wood piece nailed across another board (or several boards) to strengthen it, or to provide support for a shelf or some other object.

cleated stringer: Stringer made of cleats nailed to a plank. The cleats support the ends of the treads.

climbing tower crane: Tower crane that is secured to the floor at the high-rise structure being erected and can be periodically raised as new floor levels are added to the structure.

clinching: Process of bending over the points of nails that have been driven through boards.

clip angle: Short L-shaped piece of metal (typically with 90° bend) commonly used as connector in metal-framed walls.

closed finish stringer: Finish board nailed against the wall side of a stairway with its top edge parallel to the slope of the stairs.

closed panel construction: Prefabricated panelized construction method in which interior and exterior wall sections are finished off and plumbing and wiring are installed at the manufacturing facility.

closed railing: Stairway railing in which the section below the rail is framed in and covered with gypsum board, plaster, or paneling.

coil tie: Device used to hold opposite form walls in position, which consists of bolts that extend through the walers on both sides and screw into a coil inside the wall.

cold bridging: Occurs when construction components (usually structural members or door or window frames), which have greater thermal conductivity than the rest of the building, extend from the interior face to the exterior face of the building and conduct thermal energy from one side of the wall to the other.

cold forming: Manufacturing process in which light-gauge steel members are shaped by rolling or shaping the steel without the addition of heat.

cold joint: Joint formed when concrete is placed on top of or next to a previous placement of concrete that has already hardened.

collar tie: Horizontal member connecting two opposite rafters. It is usually placed at every second or third pair of rafters.

column: Vertical structural member that supports the weight of other members.

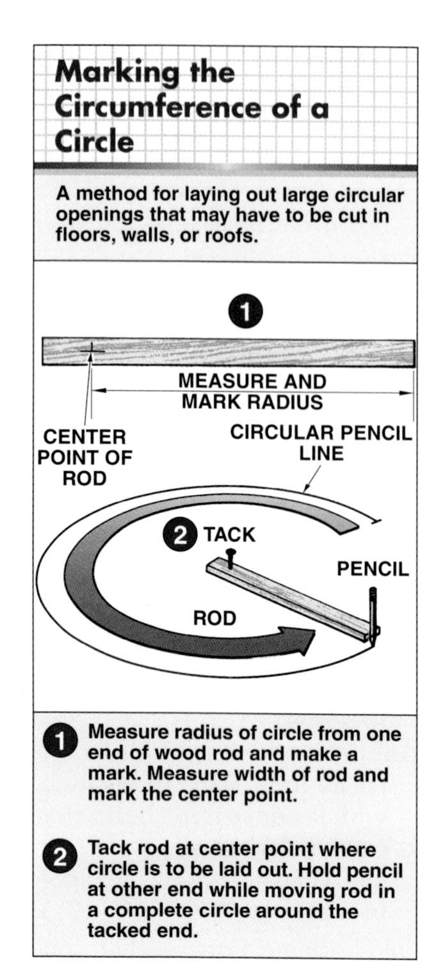

column clamp: Adjustable metal device used to tie together column forms.

combination blade: Multipurpose blade used with power saws. Suitable for crosscutting and ripping operations.

combustion-and-ventilation effect: Uncontrolled airflow created by wood-, oil-, or natural gas-burning appliances that require makeup air to support combustion. A draft is created by air being drawn into the building.

common length difference: Amount that jack rafters should increase or decrease in length when the same center-to-center measurements are maintained.

common nail: Flat-headed nail used most often in rough work.

common rafter: Rafter that extends from the top wall plates to the ridge of a gable roof.

compass direction: Direction based on the compass points north, south, east, and west.

compensator: Part of a laser level that automatically maintains the levelness of the laser level while it is being used.

component: Construction unit usually preassembled before being set in place.

composite panel: Engineered panel consisting of a reconstituted wood core with a face and back of softwood veneer.

composite slab: An elevated concrete slab that is mechanically attached to the structural steel framing of a building.

compound miter saw: A stationary power tool primarily used to cut molding at angles for finish work.

compression: Stress caused by a squeezing together or crushing force.

compression test: Quality control test that is used to determine the compressive strength of concrete.

compressive strength: Greatest amount of compression that a material can withstand before it fractures.

concrete: Construction material made of cement, sand, and gravel. Used for foundations, entire buildings, flatwork, and many other types of structures.

concrete masonry unit (CMU): A precast block, solid or hollow, used in the construction of walls. Also referred to as *concrete block.*

concrete mixture: Proportions of cement, sand, and gravel in a mixture of concrete.

concrete pour: Trade term for placing concrete.

concrete pump: Pumping apparatus used to place concrete at a remote location.

concrete saw: A power saw with an abrasive or diamond rotating blade that is used to score and cut concrete.

condensation: Formation of drops of water on the insides of exterior walls and windows. Occurs when the warm moisture-laden air within the building comes in contact with the cooler surfaces of the exterior walls.

condominium: Apartment building in which each apartment is owned individually.

conductance (C): Degree to which a nonhomogeneous material transfers heat.

conduction: Movement of heat through a solid or liquid substance.

conductivity (k): Degree to which a material transfers heat.

conduit: Tuber or pipe used to support and protect electrical conductors.

conifer: Species of softwood tree that bears cones and has thin, needle-shaped leaves.

consistency: Wetness and flowability of concrete at the time it is placed.

consolidating concrete: Working freshly placed concrete so that each layer is compacted with the layer below and voids caused by water or air pockets are eliminated.

construct: To build or put together.

construction forklift: Piece of equipment used for transporting building materials and equipment around a job site, usually relatively close to the ground.

construction-grade screw: Self-tapping screw with coarser and more steeply pitched threads, commonly used to fasten subfloor, roof sheathing, and wall sheathing to framing members.

construction joint: Joint that occurs between separate placements of reinforced concrete.

contact cement: Rubber- or butane-based liquid adhesive that adheres instantly on contact.

contour line: One of the lines drawn on a survey plan and some plot plans that pass through points having the same elevation on a lot.

control joint: Groove cut into the surface of concrete flatwork. It helps to control cracking due to shrinkage of the concrete as it hardens.

convection: Movement of heat from one area to another by air or water.

coped joint: Joint made by cutting the end of a piece of molding to the shape of the piece it will fit against. A coping saw is commonly used for this purpose.

core: Innermost layer of plywood. It may consist of hardwood or softwood sawed lumber, veneer, or reconstituted wood.

corner bead: Metal or plastic strip placed to reinforce the outside corner of a wall with a drywall or plastered finish.

corner post: Combination of studs or studs and blocks placed at the outside and inside corners of framed buildings. Also called *corner assembly.*

corner return: Trim member that covers the exposed grain on the end of the stool and apron.

cornice: Finish applied to the area under the eaves where the roof and sidewalls meet.

counterbore: To increase the diameter of a hole through part of its length to receive the head of a bolt or nut.

counterflashing: Flashing usually used where chimneys meet the roofline. It covers the shingle flashing and prevents infiltration of moisture.

countersink: To make a depression in wood or metal where a screw or bolt is to be placed so that the head of the screw or bolt will be flush with or below the surface of the material.

course: Layer or row of exterior finish material such as shingles, board siding, or brick.

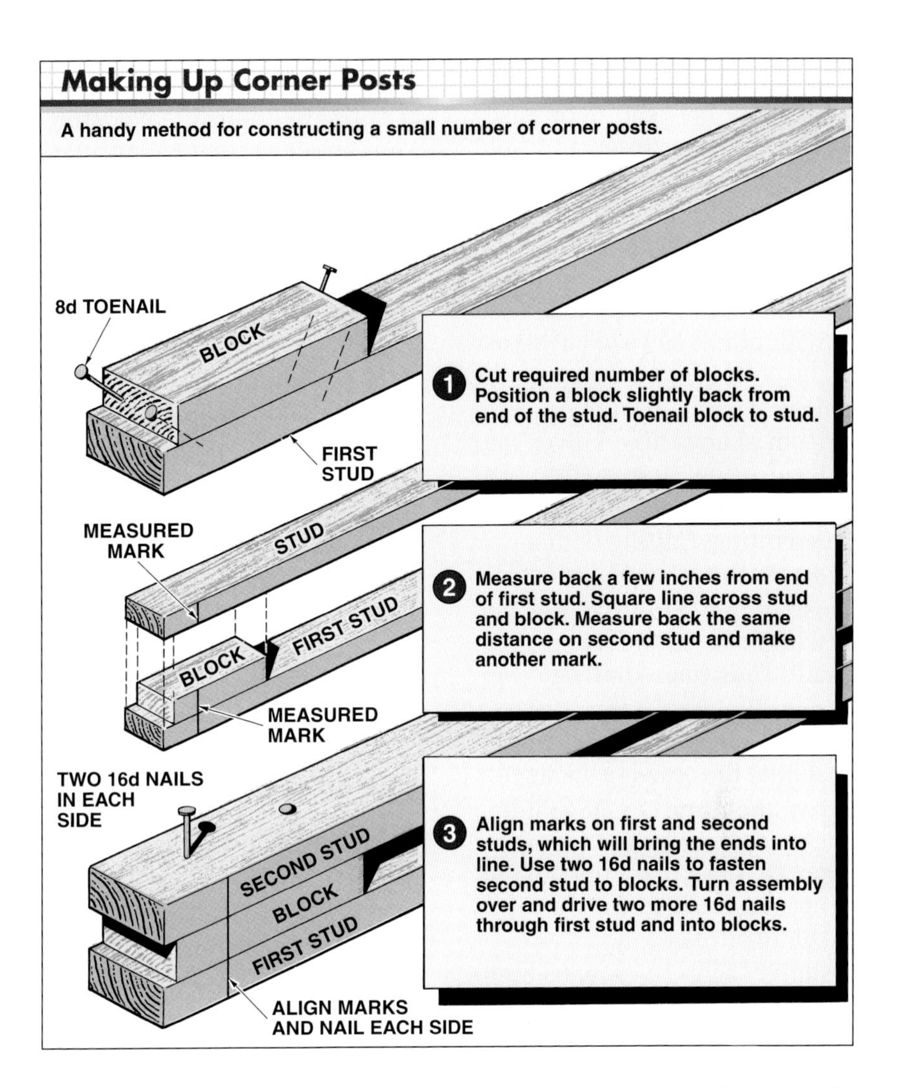

court: Open space partly or entirely surrounded by buildings.

cove base: Resilient material placed at the bottom of a wall next to a floor finished off with resilient tiles.

cove molding: Molding with a concave curve on its exposed side.

crane: A combination of a hoist and a supporting structure designed to move loads aerially.

crank operator: Device often used to open casement windows and hold them open.

crawl space: Narrow space between a floor unit and the ground.

crawl-space foundation: Type of foundation that features a narrow space between the bottom of a floor unit and ground. Also called *basementless foundation.*

cripple stud: One of the studs placed between headers and top plates or between rough sills and bottom plates.

cripple wall Used to extend the height of low foundation walls, usually for a full-basement house.

critical angle: Minimum and maximum stairway angles for walking safety.

crook: Distortion (warpage) resulting in a curve from end to end along the narrow edge of a piece of lumber.

crossband: Veneer layers placed between the core and the face plies of plywood.

cross beam: Slip form wall connector that ties together the tops of opposing yokes and provides a mounting surface for the hydraulic jacks.

cross bridging: Pair of narrow wood or metal pieces set diagonally between floor joists. They help to stiffen and hold the joists in position.

crosscut blade: Circular saw blade with teeth that form a knife-like edge with a face bevel of 10° to 15°. It is used with power saws to cut across the grain of materials.

crosscut saw: Handsaw with teeth shaped like knives and used to cut across the wood fibers and grain of the wood.

crosscutting: Cutting with a saw across the grain of lumber.

cross slope: Slope across the width of a sidewalk or patio to allow for water drainage.

crown: Bow along the edge of a joist. Joists should be placed so that the crown is facing up.

crown molding: Ornate molding characterized by a convex curve. Used in traditional roof cornice finish and at the top of interior walls.

C-shape: Light-gauge metal framing member with a C-shaped cross section and consisting of a web, flange, and lip.

cup: Distortion (warpage) resulting in a curve across the grain from edge to edge of a piece of lumber.

curing concrete: Process of retaining the moisture of freshly placed concrete to ensure proper hydration.

curtain wall: Light, non-bearing exterior wall set into and attached to the steel or concrete structural members of a building.

custom-built house: House designed by an architect for a specific customer.

cut-in brace: Older type of diagonal brace with pieces cut in between the studs of a wood-framed wall.

cut-out stringer: Stringer that has been cut to receive all the treads and risers of a stairway.

cutting plane line: Line (identified by letters) that cuts through a part of a structure on an elevation or plan view drawing. It refers to a separate section view or detail drawing given for that area.

cylinder: Part of a lockset that contains the keyhole and tumbler mechanism.

cylinder knob: Doorknob that contains the lock cylinder.

cylindrical lock: Bored lock with a cylinder that contains the tumbler mechanism into which the key fits. The cylinder is located in the knob section of the lock.

D

dado: Groove, rectangular in shape, cut across the grain of a board.

dadoed stringer: Stringer with dadoes cut to receive and support the ends of the treads of a stairway.

dampproofing: Material used to prevent the passage of water or water vapor.

datum point: Point of elevation reference, established by local authorities, from which other elevations in the area are measured. Often referred to as the *point of beginning (POB).*

daylighting: Method of capturing and redirecting natural light for use in the interior of a building using special design or equipment.

dead blow hammer: A striking tool with a head made of steel pellets encased in a plastc coating.

dead bolt: Locking device consisting of a solid metal bar that can be thrown or retracted with a knob or key. Considered a good security provision for any door.

dead load: Weight of the permanent structure of a building, including all materials that make up the units for walls, floors, ceilings, roofs, etc.

debarking: Process of stripping bark from a log.

decay: Disintegration and breakdown of wood caused by a wood-destroying fungus. Also called *dry rot.*

deceleration device: Personal fall-arrest device that dissipates a substantial amount of energy during a fall arrest or limits the energy imposed on a worker during a fall arrest.

deceleration distance: Additional vertical distance a falling worker travels, excluding lifeline elongation and free-fall distance, before stopping, from the point at which the deceleration device begins to operate.

decibel (dB): Unit used to measure sound intensity.

deciduous: Tree species that shed their leaves during the fall and remain leafless until the following spring.

deflection: Amount of bending that occurs in a wood member subjected to a load or stress.

degree: Unit by which angles are measured.

derrick: Hoisting equipment that consists of a vertical tower with a swinging boom hinged to its base.

detail: In prints, an enlarged picture of a part of the structure that cannot be adequately explained in more general plan or elevation view drawings.

detector: Device that serves as a target for laser beams from laser levels. Also called *sensor* or *receiver.*

dew point: Temperature at which condensation occurs in a given area.

diagonal brace: Brace placed in a framed wall to increase lateral strength.

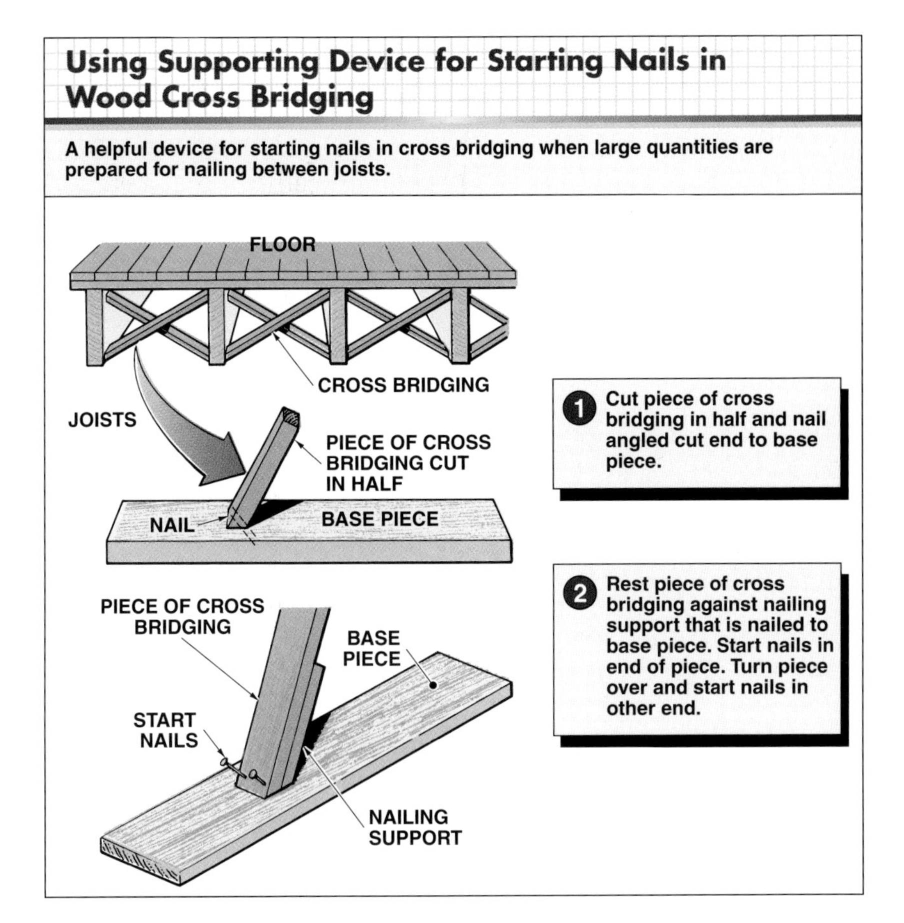

Using Supporting Device for Starting Nails in Wood Cross Bridging

A helpful device for starting nails in cross bridging when large quantities are prepared for nailing between joists.

dimension: Measurement; the distance between two points.

dimension lumber: Includes products that are 2″ to 4″ thick by 2″ and wider and includes studs, joists, planks, roof rafters, trusses, and other components that form the framework of a building.

dimple: Slight depression made by a hammer head when it is used to drive nails into gypsum board.

dome: Curved roof structure similar in appearance to one-half of a sphere.

dome form: Dome-shaped sections used for constructing waffle slab forms. Also called *pans.*

door closer: Device mounted on a door and its jamb to return the door to a closed position. It controls the speed and also prevents the door from slamming. Also called *door check.*

door hand: Direction in which a door swings.

door header: Wood member placed across the top of a rough door or window opening in a framed wall. It supports the weight from structures above the opening. Also called *lintel.*

door holder: Device fastened near the bottom of a door to hold the door in an open position.

door pull: Metal handle, not part of a lock, used to pull open a door.

door stop: Narrow piece nailed to a nonrabbeted door jamb to stop and hold the door in position when the door is closed.

dormer: Shed or gable framework projecting from the side of a roof to add light, ventilation, and space to an attic area.

double-acting floor hinge: Self-closing hinge which also allows a door to swing in both directions. The hinge mechanism fits into a notch at the bottom of the door.

double coursing: Shingles applied in two layers to a sidewall.

double-hung window: A window unit with upper and lower sash sections that move up and down.

double-ply application: Double layer of gypsum board applied to a wall.

doubler plate: Horizontal plate nailed to the upper surface of the top plate of a wood-framed wall. It strengthens the load-bearing capacity of the upper section of the wall. Also called *cap plate.* Together, the doubler plate plus the top plate are called a *double top plate.*

double-shear joist hanger: Metal joist hanger that provides greater strength than standard joist hangers. Includes dimples or domes to guide nails through joists and supporting member at a 45° angle, resulting in a stronger connection.

doubling up walls: Trade term for placing the second (opposite) wall of a concrete form.

dovetail joint: Wood joint similar to a mortise-and-tenon joint except that its interlocking parts are narrower at the heel.

dowel: Wood or metal pin used to strengthen a joint between two pieces.

doweled joint: Joint in which glued dowels extend into both pieces.

downspout: Pipe carrying water from a roof gutter to the ground.

drain tile: Pipe made of clay, plastic, or concrete sections. Placed alongside the foundation footing for the purpose of moving away water collecting around the foundation.

drill: Device with cutting edges to make holes in wood, metal, or plastic materials. The hand or power-operated tool for holding the device is also called a drill.

drill press: A stationary power tool for drilling or modifying holes of various depths.

Dividing Material into Two Equal Parts

A method for dividing a piece of material into two equal pieces. In this example, a 3½″ wide board is used.

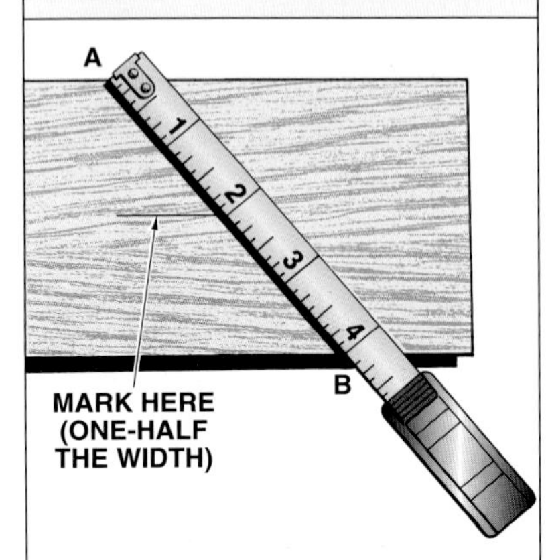

1. Hold rule to one edge of material at point A. Slant rule so a number that can be divided by two (in this example, 4) is even with the opposite edge at point B.
2. Make a pencil mark on the board at 2″ mark on rule. This is one-half the distance between points A and B and, therefore, one-half the width of the board.

Dividing Material into Three or Four Equal Parts

A method for dividing a piece of material into three or four equal pieces. In this example, a 5½″ wide board is used.

A To divide board into three equal parts, slant rule so a number that can be divided by 3 is even with two edges of board. In this example, 6 is used. Place marks at 2″ and 4″ numbers on rules.

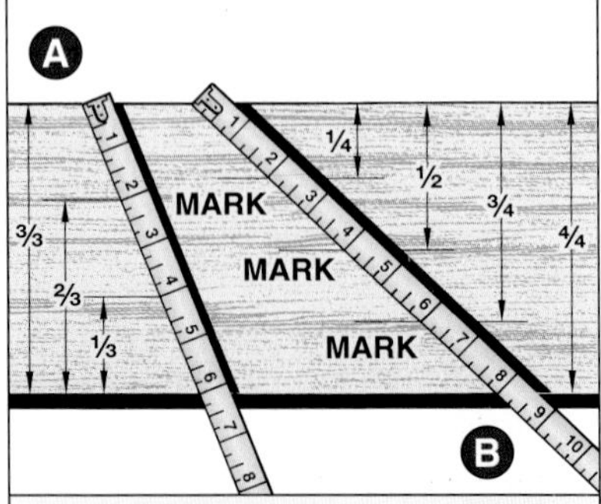

B To divide board into four equal parts, slant rule so a number that can be divided by 4 is even with two edges of board. In this example, 8 is used.

D-ring: Metal ring attached to a safety harness to which a lifeline is attached.

drip cap: Piece of molding placed at the top of an exterior wood frame to direct water away from the door and window openings.

drip edge: Metal piece bent over the exposed edges of roof sheathing. Placed after felt underlayment has been put down.

drive pin: Nail made of hardened steel that can be driven into concrete. It is usually shot into the concrete with a powder-actuated fastening tool.

driving head: Steel device placed on the head of a pile to receive the blows from the pile-driving rig and protect the pile while it is being driven.

drop chute: Long chute usually attached to the bottom of a hopper. Used to prevent concrete from striking against the reinforcing steel and form walls when it is being placed in higher walls.

drop panel: Thickened section over concrete columns that helps support the floor slab above.

dropping hip rafters: Process of deducting an amount from the seat cut so that the rafter drops. As a result, the ends of the roof sheathing can rest on the corners of the dropped rafter and be in line with the rest of the roof.

dropping stringers: Process of deducting an amount from the height of the first riser so that the first and last finished risers will equal the riser heights of the rest of the stairway.

drop siding: Board siding design that comes in tongue-and-groove or shiplap patterns.

drywall frame: Metal door jamb designed for walls finished off with gypsum board.

duplex nail: Double-headed nail designed to be pulled out easily. Used in temporary construction. Also called *double-headed nail.*

duplex receptacle: Electrical outlet with two plug-in receptacles.

E

earmuffs: Hearing protection device worn over the ears.

earplugs: Hearing protection device made of moldable rubber, foam, or plastic, and inserted into the ear canal.

easement: Legal right-of-way provision on another person's property.

eaves: Part of a roof that projects past the side walls of a building.

edge: The narrowest dimension of a piece of lumber.

edge-grained lumber: Lumber produced by quartering a softwood log lengthwise, then cutting boards out of each section.

electrical shock: Condition that results when a body becomes part of an electrical circuit.

electrode: Thin metal rod used in arc welding to provide filler at the welded joint between metals.

electronic distance measurement: Feature of a total station instrument that allows distances to be measured between points with a high degree of accuracy (within 0.001′) without the use of a measuring tape.

elephant trunk: Flexible hose-like device sometimes attached to the bottom of a drop chute. Used when concrete is being dropped from great heights.

elevation: One of the heights established for different levels of a building.

elevation view: Print drawing giving a view from the side of a structure.

embedded steel: Steel components, such as rebar and anchor bolts, that are placed within the outer faces of concrete surfaces.

end-matched lumber: Boards or panels with tongue-and-groove edges at their end sides.

energy audit: A comprehensive review of the energy use of a building.

engineered lumber: Building product manufactured using solid wood members or veneers, wood strands and fibers, or a combination of solid wood members and wood strand members.

engineered wood products: Class of building materials manufactured from solid and reconstituted wood products which are combined with waterproof adhesives, resins, and other binders and subjected to extreme heat and pressure to form panels, timbers, and other structural or nonstructural products.

English measurement: Measurement based on yards, feet, inches, and inch-fractions. Also called *customary measurement.*

epoxy: A two-part adhesive that consists of a resin and a hardner.

equilibrium moisture content: Point at which the moisture content in wood is at the same level as the moisture in the surrounding air.

escutcheon plate: Decorative metal plate placed against the door face behind the knob and lock.

evergreen: A softwood species of trees that retains its leaves during all seasons of the year.

excavation: Cavity dug in the ground.

excavator: Heavy construction equipment used to remove soil and deposit it in trucks for removal or deposit it elsewhere on the job site.

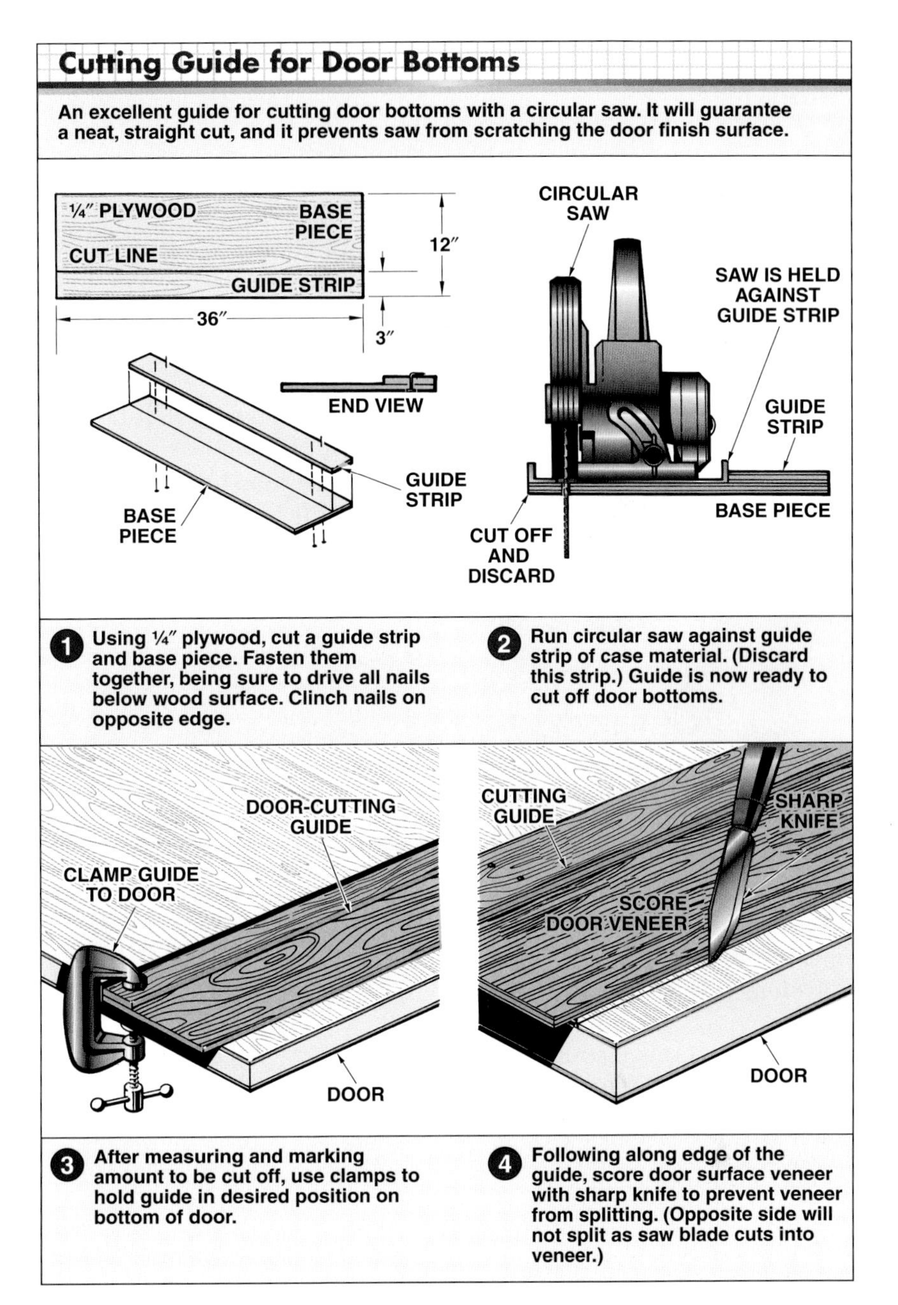

exit device: Mechanism with a horizontal bar that operates one or more door latches. When pushed in, the bar retracts the latches, enabling the door to be opened. Mainly used in theaters, schools, and public buildings. Sometimes called *panic-exit door bolt.*

expansion anchor: Device placed in concrete and other types of solid masonry walls for fastening purposes. As a screw or bolt is driven into the anchor, it expands and presses against the concrete.

expansion joint: Joint extending through the entire thickness of a concrete slab wherever the slab butts against walls, piers, or columns. Expansion joints allow for expansion of a slab due to temperature changes.

exposure: Distance between exposed edges of overlapping shingles.

exterior elevation: View from the side of an object used to clarify and provide additional details regarding information shown on a floor plan.

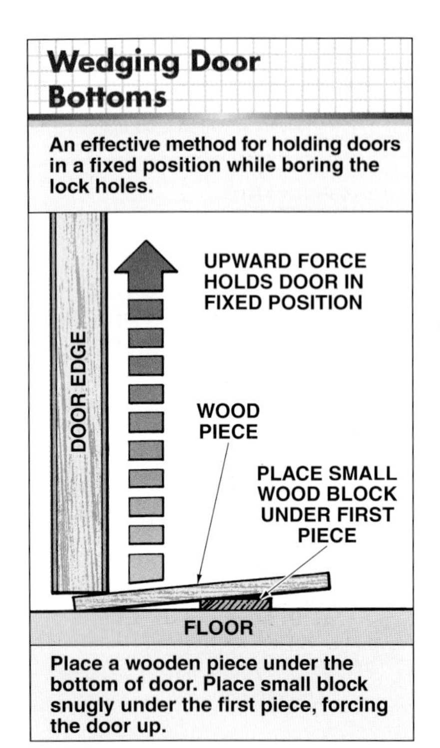

exterior finish: Materials such as roof shingles, wall siding, and window and exterior door frames used to finish the exterior face of a building.

exterior insulation and finish system (EIFS): Method used to provide exterior building protection through application of exterior sheathing, insulation board, reinforcing mesh, a base coat of acrylic copolymers and portland cement, and a finish of acrylic resins.

eyebolt: Bolt with a looped head that is fastened to a load to provide a lift point.

F

facade: Exterior of the front of a building.

face brick: High-quality brick used as exterior wall finish.

face frame: Frame attached to the front of a cabinet. It holds the doors and contains the drawer openings. Also called *front frame.*

face-mount joist hanger: Metal connector nailed to the face of a beam or ledger and used to support a joist.

face-nailing: Nailing on the surface of the lumber.

face shield: Transparent plastic device that covers and protects the entire face from the hazard of flying particles.

face side: Side of a board or other building material that has the best appearance.

factory and shop lumber: Lumber that will be further manufactured into millwork such as door jambs, window sashes and frames, and molding.

factory edge: Straight edge of a panel after it has been manufactured.

falsework: Structural supports and bracing required for temporary construction loads.

fascia board: Horizontal finish piece nailed to the tail end of the roof rafters.

fascia rafter: Rafter placed toward the outside face of the framework for a gable end overhang. Also called *barge rafter* or *bargeboard.*

fastener: General term for a metal device, such as a nail, bolt, or screw, to secure adjacent members together.

fenestration: An intentional opening in a building envelope.

fiber: Narrow tapered wood cell closed at both ends. Also called *tracheid.*

fiberboard: Engineered wood product made out of wood fibers. Includes particleboard and medium- and high-density fiberboard and hardboard. Commonly used in the furniture industry.

fiber-cement siding: Exterior finish material composed of cement, ground sand, and cellulose fiber that is autoclaved (cured with pressurized steam) to increase its strength and dimensional stability.

fiberglass: Insulating material made of spun glass fibers.

fiberglass-reinforced-plastic (FRP) plywood: Engineered panel product that consists of a tough glass fiber-reinforced overlay bonded to plywood.

fiber-reinforced glulam timber: Type of glulam beam in which thin sheets of high-strength resin-encased fibers, less than 1⁄10″ thick, are placed between the lams.

fiber saturation point: Point at which water has evaporated from the cell cavities of wood but the cell walls still retain water.

filler block: Wood piece used to fill the rectangular space between a pair of I-joists that are used as a single member. Filler blocks permit vertical load to be shared between the two I-joists and force each joist to absorb equal amounts of the load and bend the same amount under the applied load.

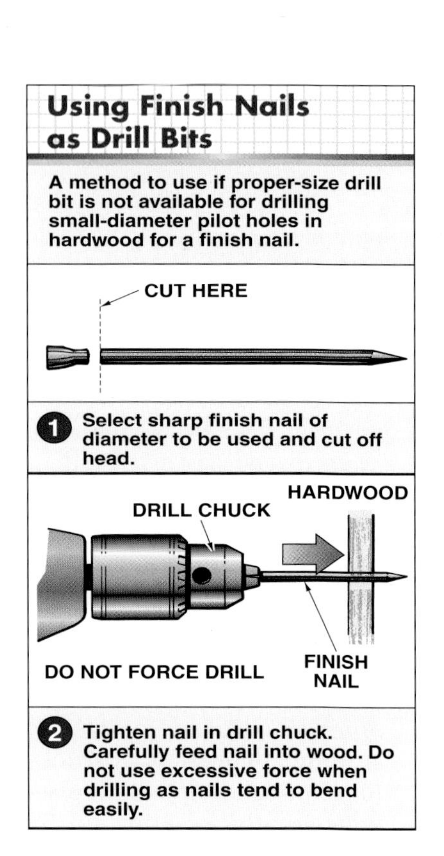

filler rod: In concrete formwork, a threaded rod made in various lengths that joins together the two outer sections of a waler rod assembly.

finger joint: Interlocking joint used to splice the ends of lumber pieces together.

finish floor elevation: Height of the first floor after finish materials have been applied, in relation to the benchmark established on the construction site.

finish flooring: Material used for the exposed, finished surface of a floor. Some examples are hardwood flooring, tile, and rugs.

finish grade: One of the various levels of the lot surface after grading work has been completed.

finish nail: Nail with a small barrel-shaped head that allows it to be set below the surface. Used for finish work.

finish stringer: Finish board placed against a wall and behind a cut-out stringer.

fink truss: Roof truss sloping in two directions with web members that form a W shape.

fireblocking: Installing wood members in concealed spaces to cut off vertical and horizontal draft openings and to form a barrier between stories, and between the top story and attic or roof space.

fireblocks: Horizontal pieces placed between the studs to slow down the passage of flames in case the structure catches on fire. Also called *fire stops.*

fire cut: Angled cut at the end of a joist or timber where it is set into a brick wall. Prevents the joist or timber from damaging the wall in the event it collapses due to fire.

fire-rated door: Fire-resistant door assembly, including the frame and hardware, commonly equipped with an automatic door closer.

fire-resistance rating: Measurement of the resistance of a material or component to failure when exposed to fire. Expressed as the number of hours or minutes that a material or component will maintain its integrity.

fire-retardant lumber: Lumber treated with a fire-retardant chemical.

firestopping caulk: Material commonly used to seal holes drilled through walls or structural members. Available as intumescent and endothermic caulk.

fire wall: Wall constructed of fire-resistant materials and subdividing a building to help retard the spread of fire.

fixed-sash window: Window that cannot be opened.

fixture: Electrical or plumbing device attached to a wall, floor, or ceiling. Examples of fixtures are lights, water closets, and sink basins.

flagstone: Flat, fine-grained, evenly split rocks. When set in mortar, flagstone can be used for the finished surface of outside walks and terraces and for sections of an interior floor.

flange: Projecting rim at the top and bottom webs of a steel beam or wood I-joist.

flashing: Strips of metal, plastic, or asphalt-saturated felt placed at roof areas vulnerable to water leakage and around window and door openings.

flat-grained lumber: Lumber produced by cutting a softwood log so that the annual growth rings are at an angle of 45° or less to the wide surface of the boards.

flat slab floor: Concrete slab supported mainly by drop panels over columns.

flat strapping: Sheet steel cut to a specified width without any bends; typically used for bracing.

Removing Finish Nails

A method for removing finish nails from materials without splitting or damaging surface finish.

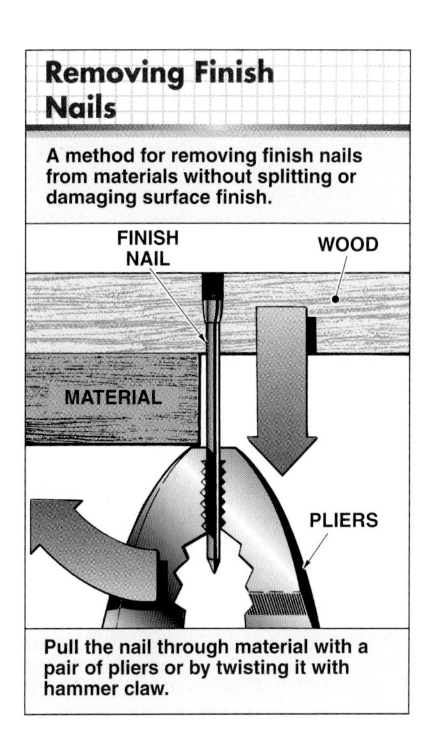

Pull the nail through material with a pair of pliers or by twisting it with hammer claw.

flatwork: Work connected with concrete slabs used for walks, driveways, patios, and floors.

flexible insulation: Insulating materials that come in blanket or batt form and are placed between studs and joists.

flexible membrane flashing: Door and window flashing consisting of a bituminous or butyl rubber adhesive core with release paper on one side and a metal foil or flexible plastic backing on the other side.

flight: Unbroken and continuous series of steps from one floor to another or from a floor to a landing.

floor beams: Used in a post-and-beam floor system. Beams are spaced apart 4′ OC or more. Flooring usually consists of 1½″ tongue-and-groove planks or 1⅛″ panels.

floor guide: Metal or plastic device that holds the bottom of a sliding door in correct position as it is moved.

flooring: Material used in the construction of floors.

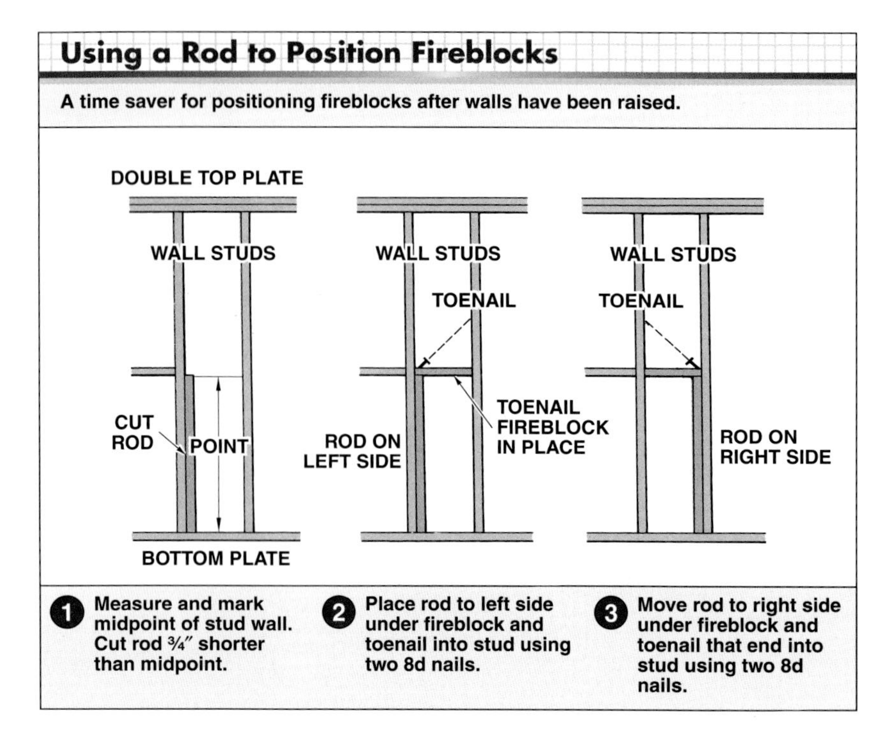

floor plans: Drawings in a set of prints that give a plan view of each floor of the building.

floor truss: Prefabricated structural unit made of top and bottom chords tied together with web members. Used for the same purpose as floor joists or beams.

flue: Passage through a chimney for gas, fumes, or smoke to rise.

flush bolt: Sliding bolt mechanism mortised into a door at its top and bottom edge. Usually used to hold in a fixed position the inactive door of a pair of double doors.

flush door: Door with a flat surface made of a frame covered with plywood or hardboard face panels.

flush pull: Recessed device mortised into sliding doors to provide a finger hold for moving the doors.

flying form: Complete form unit consisting of a wood deck and metal supports. The entire assembly is raised and set into place by crane.

foamed-in-place insulation: Chemical foam that is poured or sprayed into wall cavities. The foam then expands and completely fills the cavity.

footing: Base of a foundation system. It bears directly on the soil.

footing drain: Perforated pipes laid along the outside of foundation footings to collect rainwater and water from melting snow percolating down through the backfill and move it away from the foundation. Also called *drain tile.*

Forest Stewardship Council (FSC): Trade association that has established global standards for responsible forestry, including a chain-of-custody for certified forest products.

form: Braced structure built to the shape of the structural member and into which concrete is placed.

formaldehyde: Chemical used to produce glues used in the manufacture of particleboard, plywood, veneers, and other wood products.

form-release agent: Chemical agent that prevents the adhesion of fresh concrete to the forming surface.

formwork: Construction of concrete forms.

forstner bit: Wood-drilling bit used to bore flat-bottom holes.

foundation: The part of a building that rests on and extends into the ground. It provides support for the structural loads above.

foundation anchors: Metal connectors that provide a solid connection between the foundation and structural members to resist uplift and lateral forces.

foundation plan: Drawing in a set of prints that gives a plan view as well as section views of the foundation of a building.

foundation plates: Metal connector used to fasten the foundation to the side of the sill plate.

frame construction: Framed wall, floor, ceiling, and roof units.

framing anchor: A metal device used to strengthen the ties between structural members of a wood-framed building.

framing square: Valuable measuring and layout tool consisting of a 2″ wide by 24″ long blade and a 1½″ wide by 16″ long tongue. Also called *steel square* or *rafter square.*

free-fall distance: Vertical distance between the fall-arrest attachment point on the body harness before the fall and the attachment point when the personal fall-arrest system applies force to arrest the fall.

freestanding tower crane: Tower crane that is secured to a concrete foundation next to the structure being erected.

friction catch: Catch used to hold cabinet doors or other types of light doors in position. A metal tongue mounted inside the door engages a jaw device mounted inside the cabinet.

friction pile: Pile that relies on the pressure of the surrounding soil for a major part of its bearing capacity.

frieze board: Finish piece nailed below or notched between the rafters of an open cornice.

front setback: Distance from the property line to the front of a building.

front walk: Walkway that extends from the sidewalk or driveway to the front entrance of a building.

frost line: Depth to which soil freezes in a particular area.

furring strips: Narrow wood strips nailed to a wall or ceiling surface as a nailing base for finish materials.

fused: Melted together by the heat of an electric arc.

Cutting Quantities of Framing Members

A fast and accurate method of cutting large quantities of studs, trimmer studs, or joists with a circular saw.

2 × 4 END PIECE

2 × 4 FENCE PIECE

90°

¾″ PLYWOOD BOTTOM PIECE

1. Set up bench consisting of plywood panels supported by two sawhorses. Nail 2 × 4 strip to act as fence along long edge of bench. Nail second strip at one end of bench. This end piece must be nailed at an exact right (90°) angle to fence piece.

CUTTING BENCH

SNAP CHALK LINE

PUSH

2 × 4 STUDS

MEASURE AND MARK LENGTH

2. After squaring one end of all pieces to be cut, place them on top of bench. Push all pieces firmly against end piece. Measure desired length of pieces at opposite sides and make two marks. Snap chalk line across top of pieces and cut with circular saw.

G

gable end overhang: Triangular upper section of the end wall of a building with a gable roof.

gable roof: Roof that has a ridge at the center and slopes in two directions.

gable stud: Stud that extends from the top plate of an end wall to the bottom of a common rafter at the gable end of a gable roof.

gain: Recess cut into a piece of wood for door hinges and other hardware. Also a notch cut into lumber, such as a notch cut into a stud for a ribbon board.

galvanizing: Coating iron or steel with zinc to prevent rusting.

gambrel roof: Roof similar to a gable roof but with a break in each slope at an intermediate point between the ridge and the two exterior side walls of the building.

gang form: Large form panel made of smaller panel sections.

gas-filled window: Window unit, consisting of two or more panes filled with argon, krypton, or xenon gas, that reduces the overall heat transfer between the interior and exterior of a building.

gauge: Standard of measure for metal thickness and the diameter of wire. Also spelled *gage.*

general contractor: Licensed individual or firm that can enter into legal contracts to do construction work, and is in charge of the overall organization and supervision of a construction project. Also called *building contractor.*

geometric stairway: Winding stairway of circular or elliptical design.

girder: Large horizontal wood, steel, or concrete member used to help support a floor, ceiling, roof, or other load.

girt: Wood horizontal stiffener placed around the perimeter of a post-frame building to support the wall covering.

glazing: Installing glass in a window sash or door and applying putty to hold the panes in position.

glue block: Small piece of wood glued at the joint of two pieces of wood that meet at an angle.

glue bond: Bond between pieces of wood in products such as veneered panels and glued, laminated (glulam) timbers.

glued laminated timber: Heavy timber made up of planks joined together by a very strong bonding adhesive. Often referred to in the abbreviated form *glulam.*

glue-nailing: Fastening system that combines the use of an adhesive with nailing.

gooseneck: Curved or bent section on a handrail.

Dividing Improper Fractions

A shortcut for dividing improper fractions into two equal parts. In this example, 5⅝″ is used.

1. Determine largest whole number that can be divided into the whole number of the improper fraction (in this case, 5 ÷ 2 = 2). The answer is 2 (ignore the remainder).

$$\begin{array}{r} 2 \\ 2\overline{)5} \\ \underline{4} \\ 1 \text{ (ignore this remainder)} \end{array}$$

2. Add the numerator (5) and the denominator (8), which equals 13.

   5 numerator
   \+ 8 denominator
   13

3. Double denominator (8), which equals 16.

   8 denominator
   \+ 8 denominator
   16

4. The final answer is 2¹³⁄₁₆″.

   5⅝″ ÷ 2 = 2¹³⁄₁₆″

grade: The identification of the quality of a piece of lumber. The grade determines the lumber's price and what it can be used for.

grade beam: Reinforced concrete foundation wall placed at ground level. It rests on and is tied to deeply penetrating piers or piles that provide the main structural support for the beam and building loads above.

grader: Heavy construction equipment used in final grading operations on large construction sites to level the earth surface.

grading: Removing or adding soil to the surface of the lot so that there is enough slope for surface water to flow away from the building.

grain: Direction, size, and arrangement of the wood fibers in a piece of lumber.

gravel: Crushed rock. Particles range in size from ¼″ to 1½″ in diameter.

gravel streak: Surface defect in hardened concrete caused by poor consolidation when the concrete was placed.

gray water: Wastewater generated from domestic processes such as bathing, cleaning, and washing laundry.

green building: Refers to building design and construction methods that efficiently use materials, energy, water, and other natural resources. Also called *sustainable building.*

green lumber: Lumber with high moisture content; typically refers to freshly sawn unseasoned lumber.

grip-handle lock: Lockset with a grip handle on one or both sides. The latch is retracted with a thumbpiece.

grip span: Distance between the thumb and fingers when the tool jaws are open or closed.

ground: **1.** Safety feature to prevent shock due to a fault in an electrical system. It consists of an added ground wire running from a plug or equipment to the ground. **2.** Narrow piece of wood nailed to a framed wall to serve as a thickness gauge for plaster. Also provides backing and a nailing surface for wall molding.

ground clamp: In arc welding, a device for securing to the workpiece the end of the ground cable running from the welding machine.

ground fault circuit interrupter: Device used to protect against electrical shock by detecting an imbalance of current in the normal conductor pathway and then quickly opening the circuit (in as little as ¹⁄₄₀ of a second).

grout: Thin mixture of cement, sand, and water used for patching and leveling.

guardrail: Temporary railing placed around floor openings, across exterior door and window openings, and on scaffolds during construction.

guide strip: Narrow strip used to hold a cabinet drawer in position as it slides in and out. If the strip is nailed to the drawer, a dado is cut in the side of the cabinet. If the strip is nailed to the cabinet side, a dado is cut in the drawer.

gutter: Wood or metal trough attached to the eaves to receive water runoff from the roof. Also called *eaves trough.*

gypsum board: Panels made of a gypsum rock base sandwiched between specially treated paper. They have largely replaced plaster as a finish interior wall covering. Also called *wallboard* and *drywall.*

H

hammer: Striking tool used to drive and remove nails.

hammer-drill: Electric or pneumatic tool used to bore holes in hard surfaces such as concrete and masonry. The hardened bit is rotated and moved in a reciprocating motion.

handrail: Rail that is grasped by the hand to provide support when a person uses a stairway.

hanger wire: Wire used to hang grids or other objects from a ceiling.

hardboard: Nonstructural reconstituted wood panel product used for interior paneling, cabinets, underlayment, and exterior siding.

hardwood lumber: Lumber that comes from deciduous broad-leaved tree species.

harness: Fall protection equipment used to protect internal body organs, the spine, and other bones in a fall.

hazardous material: A substance capable of posing a risk to health, safety, or property.

header: **1.** Horizontal structural member placed over the top of a wall opening that distributes the load to either side of the opening. **2.** Joist or rafter perpendicular to trimmer joists or rafters and used to support and frame openings (Canadian usage).

header joists: **1.** Continuous pieces of lumber that are the same size as the floor joists and are nailed into the ends of the floor joists to prevent them from rolling or tipping. Also called *rim joists* and *band joists.* **2.** Doubled joists nailed between the trimmer joists of floor or ceiling openings.

headlap: Distance (in inches) from the lower edge of an overlapping shingle to the upper edge of the shingle in the second course below.

headroom: Minimum vertical clearance required from any tread on a stairway to any part of the ceiling structure above the stairway.

heartwood: Lumber from the core of a tree, which is darker in color and has higher decay resistance.

heat flow: Movement of heat toward colder air.

heavy construction: Concrete and heavy timber construction methods used to erect structures such as office and apartment buildings, factories, bridges, freeways, and dams.

heavy timber construction: System using heavy posts and girders for the basic structural members of buildings, bridges, trestles, waterfront docks, and piers.

heel plumb cut line: Mark indicating the rafter plumb cut at the building line.

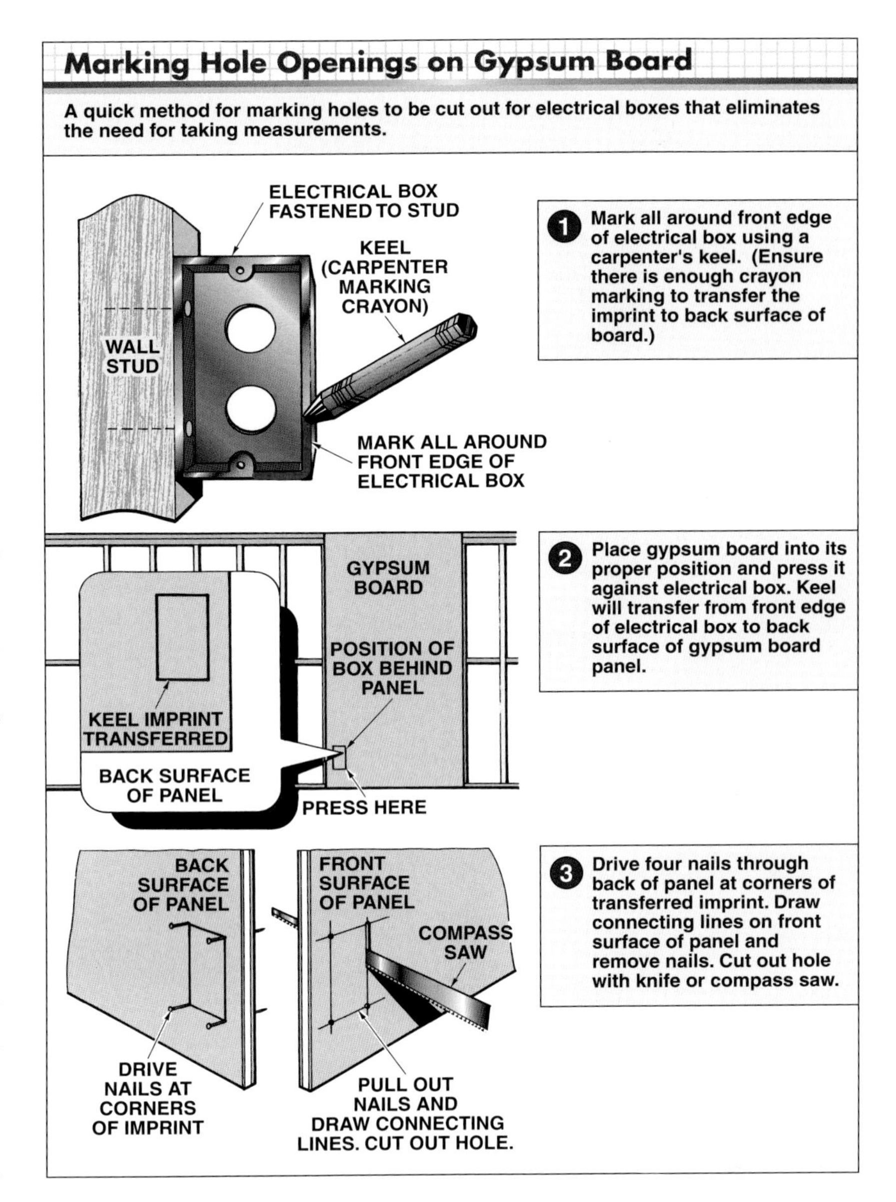

high-definition surveying (HDS): Surveying process that utilizes a 3D laser scanner to obtain three-dimensional data for a variety of civil, industrial, and construction applications.

high-density overlay (HDO) plywood: Exterior grade plywood with a resin-impregnated fiber veneer. Used for concrete forming and countertops.

high-pressure laminate (HPL): Laminate floor manufacturing process in which the components in the top layer are pressed together under intense heat and pressure to form a laminate. The laminate and bottom layer are glued to the carrier to form the finished laminate flooring.

hinge plates: Hinged, two-piece metal connectors used as an alternative to piggyback trusses.

hinges: Metal devices with movable joints attached to the door and jamb. They secure the door to the jamb and allow the door to swing back and forth. Also called *butt hinges.*

hip jack rafter: Jack rafter that extends from the top wall plate to a hip rafter.

hip rafter: Rafter that runs at a 45° angle from the corner of a building to the ridge of a hip roof.

hip roof: Roof that slopes in four directions from a central ridge.

hip-valley jack rafter: Short rafter that extends from a hip rafter to a valley rafter.

hitch: Binding of rope to another object, usually temporary.

holddown: Metal connector used to tie the foundation securely to the walls.

hollow-backed: A piece of wood molding or flooring material having a depression between the two edges of the back. The depression allows for a flatter, tighter fit against an irregular surface.

hollow-core door: Lightweight, less expensive type of flush door. With the exception of the outside frame, the space between the face panels is filled with a mesh or cellular material.

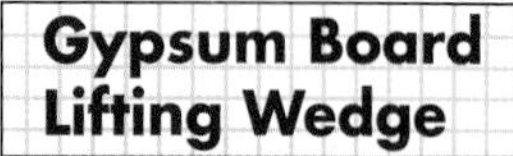

A simple aid for lifting gypsum board panels and holding them firmly against the ceiling while nailing them in place.

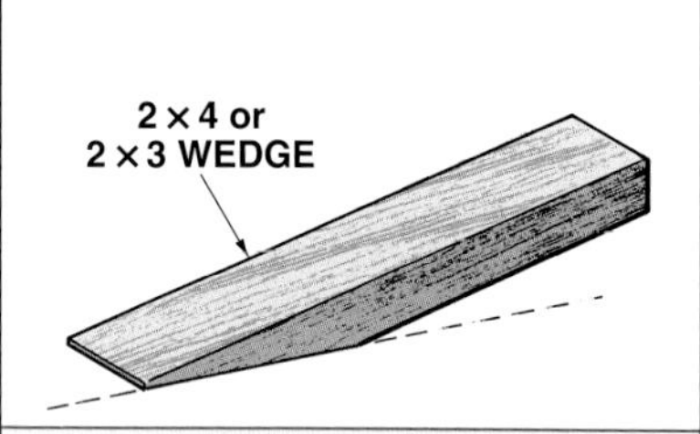

1. Cut a sharply pointed bevel at one end of short piece of 2 × 4 or 2 × 3 material.

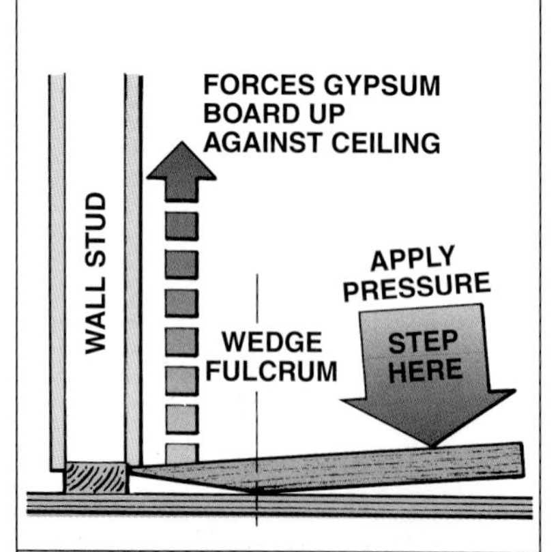

2. Place pointed end of lifting wedge under gypsum board panel. Raise panel tightly against ceiling by pressing down on other end of wedge with your foot. Drive three or four nails or screws into panel before releasing pressure on wedge.

hollow-core slab: Lightweight precast, prestressed concrete slab section used for floors, ceilings, and walls. Cores in the slab provide for air distribution and raceways for power and communication wiring.

honeycomb: 1. Surface defect in hardened concrete caused by poor consolidation when the concrete was placed. **2.** White pits and specks on the surface of lumber. Similar to defect called white speck, but pits are deeper or larger. It is caused by a tree fungus.

hook: Curved implement used for quickly and temporarily connecting rigging to loads or lifting equipment.

hopper: Funnel-shaped box used when placing concrete into a form.

hopper window: Window that hinges at the bottom and swings out at the top.

horizontal lifeline: Lifeline connected to fixed anchors at both ends.

horizontal sliding window: A window that opens and shuts by sliding horizontally in upper and lower tracks.

housed stringer: Stringer that has been routed out to receive treads and risers that are glued and wedged into place.

housewrap: Translucent spun-plastic sheet material that is tightly wrapped around a building to prevent water and air penetration into the structure.

housing tract: Residential development where many houses are constructed for sale on a common tract of land.

hub: Stake placed in the corner of a lot when the lot is being surveyed and its exact boundaries established.

humidity: Amount of dampness in the air.

hydration: Chemical reaction that takes place when water is combined with cement, sand, and gravel in a concrete mix. Hydration causes the concrete to harden.

I

ice and water guard: Flexible adhesive membrane used to prevent moisture penetration and interior damage from water backup due to an ice dam or wind-driven rain.

ice dams: Roof obstruction created when warm air rising through the attic is combined with melting snow, water, and cold outside temperature.

immersion vibrator: Tool used to consolidate freshly poured concrete. It consists of an electrically or pneumatically activated metal vibrating head that is dipped into the concrete.

impact driver: Cordless power screwdriver that uses both rotational force and a hammer mechanism to drive fasteners.

impact insulation class (IIC) rating: Rating system used to indicate the resistance of any part of a building structure to structure-borne sound transmission. A high number indicates more resistance to sound transmission than a low number.

impact wrench: Corded or cordless tool that applies a large amount of force to a small area, such as bolt heads, with little effort from the worker.

impermeable: Pertains to materials that are completely vapor-resistant, allowing no water to pass through.

inactive door: The door of a double door unit that is held in place by flush bolts and will not be used for normal traffic.

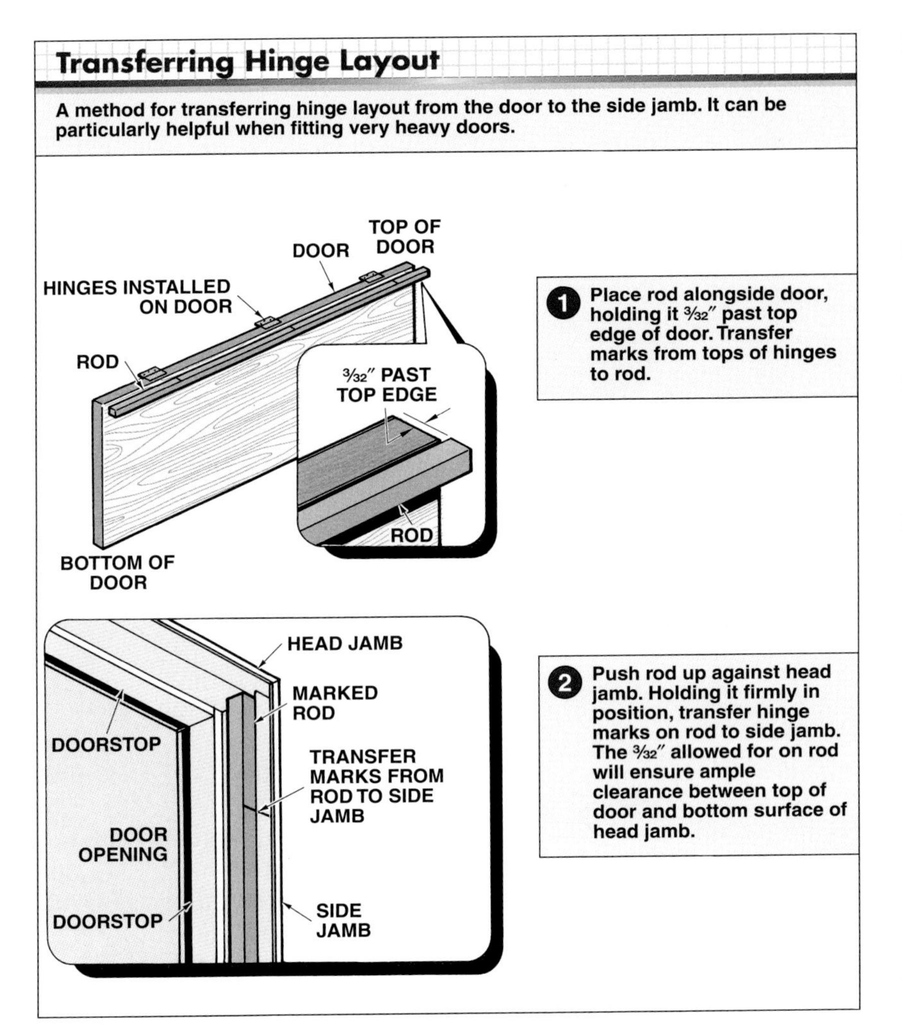

infiltration: Heat loss as a result of leakage through cracks around door and window frames.

in-line framing: Framing system in which all joists, studs, and roof rafters are in a direct line with one another. Also referred to as *stacking.*

insert: One of the bolt-like devices set into precast concrete wall sections to secure the lifting plates needed for crane hookups.

inspect: To check for deficiencies and to ensure structure is constructed per the building code.

inspector: Local official authorized to inspect and approve construction work.

install: To position a component or item so that it is ready for use.

insulating concrete forms (ICFs): Specialized forming system which consists of a layer of concrete sandwiched between layers of insulating foam material on each side.

insulating glass: Window unit comprised of two or more pieces of glazing separated by hermetically sealed airspace for improved thermal efficiency.

interior finish: Application of finish wallcovering, molding, cabinets, and interior door jambs. Also included are the hanging of doors and installation of finish hardware.

interior trim: Moldings and other decorative members installed inside a structure.

internal vibrator: Tool that consists of a motor, a flexible shaft, and an electrically or pneumatically powered metal vibrating head that is dipped into and pulled through concrete.

intersecting roof: Roof consisting of two or more roof sections sloping in different directions. Also called *combination roof.*

inverted T-shaped foundation: Type of foundation constructed with a stem wall supported by a spread footing.

invisible hinge: Hinge mortised into the door and jamb with no parts visible when the door is in a closed position.

isolation joint: Joint placed around the perimeters of concrete work where it butts up to pre-existing hardened concrete. The joint commonly is filled with an asphalt-impregnated strip.

J

jack rafter: One of the short rafters that extend from a main rafter to the ridge or top plate, or from one main rafter to another.

jalousie window: Window with a series of small glass slats that are opened and closed by a crank operator.

jamb: Finish frame of a door opening.

joint: The place where two pieces of material meet or are joined together.

Joint Apprenticeship and Training Committee (JATC): Labor-management organization responsible for the supervision and organization of the apprentice programs in a trade.

jointer plane: Hand tool approximately 22″ to 24″ long used to true edges of lumber and smooth the face of long boards.

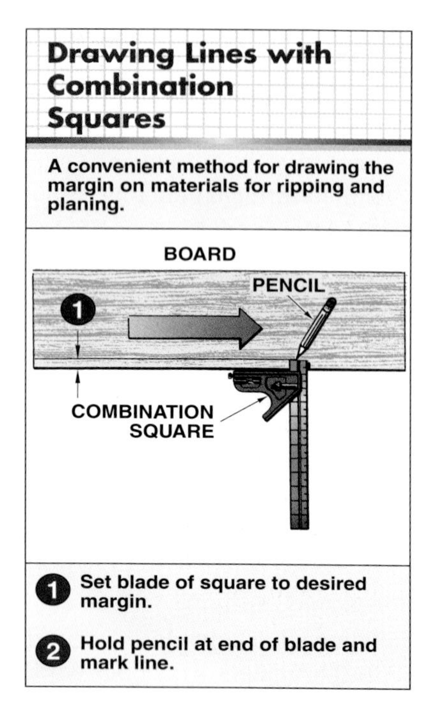

jointing lumber: Process of making the edges of boards straight and true before they are fitted together. Also, process of planing door edges when fitting and hanging doors.

joist: Horizontal plank placed on edge to support a floor or ceiling.

joist hanger: Metal strip used to support the end of a joist that is to be flush with and nailed against another joist or girder.

joist tie: Wood or metal piece notched into the tops of joists where they butt together over a wall or girder.

journeyman: Worker who has completed an apprenticeship training course and/or passed certification requirements for working in the trade.

K

kerf: Groove or notch made by a saw.

key strip: Chamfered pieces of 2 × 4s pressed into the concrete toward the center of the footing to create a keyway.

keyway: Groove formed in concrete at the top surface of a spread footing. It helps to secure the bottom of the foundation wall to be placed on top of the footing.

kicker: A 45° stake used to support concrete form boards.

kick plate: Metal or plastic plate mounted at the bottom of a door face to prevent damage from foot pressure against the door. Usually found on doors in public buildings.

kiln-dried lumber: Lumber dried in a temperature-controlled building called a kiln.

kingpost truss: Simple roof truss sloping in two directions with a vertical post extending from the ridge to the bottom chord.

knob: Projecting metal, plastic, or wood handle fastened to the face of a cabinet door or drawer for opening and closing purposes.

knocked-down unit: Unassembled structural unit. Components are prefabricated in a shop but are assembled on the job site.

knockout: Prefabricated opening provided in the webs of steel studs for utilities such as electrical wiring and plumbing pipes. Also referred to as *punchouts.*

knot: Lumber defect caused by a section of broken limb remaining embedded in the tree trunk.

krypton gas: Colorless, odorless, tasteless gas occurring in trace amounts in the atmosphere. Injected in the space between window panes to reduce the overall heat transfer between the interior and exterior of a building.

L

ladder: Structure for climbing consisting of two siderails joined at intervals by rungs or steps.

ladder jack scaffold: Metal device that hooks to the ladder rungs and provides support for planks that may support a single worker doing light work.

lag bolt: Threaded fastener with hexagonal head and a thread design for use in wood.

lagging: Heavy wood planks placed between soldier piles to form a shoring system around excavations.

Lally column: Steel pipe column used to support wood or steel girders.

laminated glass: Type of safety glass produced by placing a sheet of polyvinyl butral (PVB) between two sheets of glass and subjecting the assembly to intense heat and pressure. Lamination prevents glass from shattering and also provides a much higher sound insulation rating than standard glass.

laminate flooring: Composite flooring material which consists of a transparent top layer, decorative layer, carrier layer, and bottom layer.

landing: Horizontal platform in a stairway used to break a run of stairs.

landing newel post: Vertical member that supports a handrail at a landing or change of direction in a stairway.

lanyard: Flexible line of rope, wire rope, or strap that generally has a connector at each end for connecting a body harness to a deceleration device, lifeline, or anchorage point.

lap: Amount that materials extend over or past each other.

lap joint: Joint made by overlapping two pieces of material.

laser: Acronym formed from the initial letters of the words light amplification by stimulated emission of radiation.

laser diode: Solid-state device that produces a highly concentrated laser beam.

laser distance meter: A handheld measuring tool that uses a laser to measure the distance from the meter to an object or surface.

laser hand level: Hand level that can be used as standard carpenter's level or be switched on to emit a laser beam.

laser level: Survey instrument in which a concentrated beam of light is projected horizontally or vertically from a source and used as a reference for leveling or verifying horizontal or vertical alignment. Also called *construction laser.*

latch bolt: Retractable device that is part of a lockset. It engages the strike plate on the door jamb and holds the door in a closed position.

latch retractor: The part of a lock mechanism that engages the latch unit and retracts the bolt.

latch unit: The part of a lockset that contains a retractable latch bolt.

lateral force: Sideways pressure against a wall or other structural member.

lathe: A stationary power tool that rotates a piece of wood or metal to allow shaping and finishing of the material with specialized tools.

lattice: Grillework of narrow wood or metal strips fastened at 90° to one another to create an interwoven effect.

lattice-boom crane: Mobile crane with a boom constructed from one or more gridworks of thin steel members.

leaf: The flat pieces of a door hinge that screw into the jamb and door.

ledger strip: Strip of lumber nailed along the bottom edge of a girder to give support to joists butting against the girder.

LEED®: Leadership in Energy and Environmental Design. Organization that provides standards and guidance for green building.

let-in brace: Diagonal brace notched into the studs of a wood-framed wall.

level: 1. Line or plane that would be parallel to still water. **2.** Tool used for leveling and plumbing purposes.

lifeline: Fall protection device anchored above the work area, offering a free-fall path in case of a fall.

lift: Layer of fresh concrete or loose soil to be consolidated or compacted, respectively.

lifting bracket: Metal device attached to a gang form to provide for a crane hookup.

lift plate: One of the devices fastened to inserts embedded in precast concrete wall sections and used for a crane hookup.

lift point: Any point on a load where rigging can be attached.

lift-slab construction: Construction method in which floor slabs are stack-cast, then lifted into place by hydraulic jacks and anchored to preset columns.

light construction: Residential buildings and small to medium-sized commercial buildings.

light-gauge steel: Cold-formed, light-gauge metal framing members available in many shapes including C-shapes, tracks, U-channels, and furring channels.

lignin: Second most abundant substance found in wood. It covers the cell walls and cements them together.

lineal foot: Refers to a line one foot in length as differentiated from a square foot or cubic foot.

line of sight: An imaginary straight line extending from the telescope of a builder's level or transit-level to the object being sighted.

line of travel: Area that a person is likely to walk on when using a winding stairway.

link: Plain, rigid, closed loop used to connect multiple rigging components.

linoleum: Composite material used as a finish floor covering. It comes in tile or sheet form.

lintel: Wood, stone, or steel member placed across the top of a rough door or window opening. It supports the weight from above.

lip: Part of a C-shape that extends from the flange along the open edge. The lip increases the strength of the member and acts as a stiffener to the flange.

lipped door: Cabinet door with a rabbeted edge.

lite: Section of window glass. Also called *pane.*

live load: All moving and changing loads that may be placed on different sections of a building. Live load factors may be people, furnishings, snow, and wind.

load-bearing wall: Wall that supports loads from above as well as its own weight.

loader: Large wheeled tractor with hydraulic arms that control a bucket mounted in front.

lock block: Solid piece of wood placed between the face panels of a hollow-core door where the holes will be drilled for the door lock.

locking C-clamp: Tool used to hold metal framing members tightly together while they are being fastened.

locking latch: Device that locks horizontal sliding windows.

lockset: Entire lock unit, including the locks, strike plate, and all the trim pieces.

longitudinal roof beam: Roof beam that runs the length of a post-and-beam roof structure.

longitudinal section: Section view in a print drawing representing a cut along the length of the building.

lookout: Level pieces nailed between the wall and the ends of the rafters to provide a nailing base for the soffit material of a closed cornice.

loose fill insulation: Insulation material that is poured directly from a bag or is blown into place with a pressurized hose.

loose-pin hinge: Hinge used most often for hanging doors. It consists of a pair of leaves, a barrel or knuckle, and a removable pin.

lot: Piece of land or property having established boundaries.

lot survey: Survey of a piece of property, usually carried out by a qualified surveyor or engineer. Its main purpose is to stake out the corners of the property. Also, grades may be checked and recorded at the corners and at various points within the property.

louver: Slotted opening for ventilation.

low-emittance (low-E) coating: Surface coating applied to glass to reduce the passage of heat and ultraviolet rays through the windows.

L-shaped foundation: Type of foundation with the footing extending from only one side of the foundation wall.

L-shaped stairway: Stairway that runs along two walls and includes a landing where the stairway changes direction. Used when the space is not sufficient for a straight-flight stairway.

lumber: Any materials cut from a log (boards, planks, timbers) and used for construction purposes.

Measuring from 1′ Mark

Commonly referred to as "cutting a foot," this is a method for ensuring a more precise reading when taking longer layout measurements.

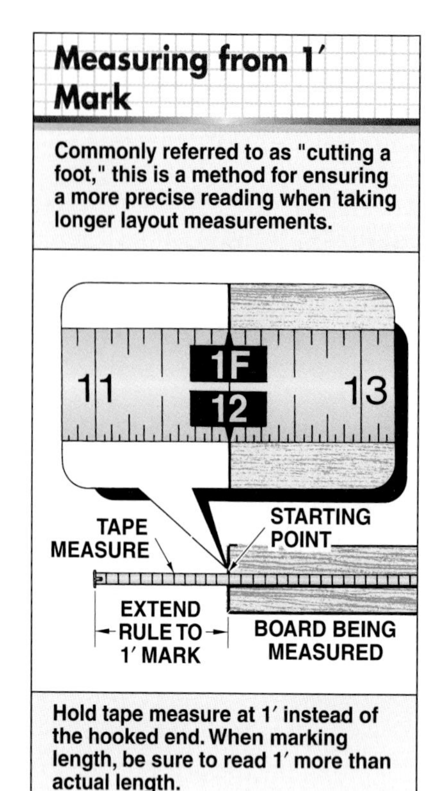

Hold tape measure at 1′ instead of the hooked end. When marking length, be sure to read 1′ more than actual length.

M

machine bolt: Bolt with a square or hexagonal head that is used to fasten together wood or metal pieces.

magnetic catch: Catch used to hold cabinet or other types of lightweight doors in position. It usually consists of a magnet fastened inside the cabinet and a metal plate fastened inside the door.

main entrance door: Exterior door on the front or street side of a building.

main stairway: Stairs that serve the inhabited (living) areas of a house. Also called *primary stairway.*

mallet: A soft, double-faced striking tool.

mansard roof: Roof similar to a hip roof but with a break in the slopes at an intermediate point between the ridge and the four outside walls of a building.

manufactured housing: General term for the prefabricated housing industry.

masonry: Molded or shaped construction materials such as concrete blocks, bricks, stones, and tiles.

masonry nail: Hardened nail with a fluted shank that can be driven into a concrete or masonry surface without deforming.

masonry veneer: Exterior finish cover of brick or stone, usually applied to a wood stud wall.

mastic: Thick, pasty adhesive used for fastening wallboard, paneling, ceiling and floor tile, and other finish materials.

material safety data sheet (MSDS): Right-to-know information used to relay hazardous material information from the manufacturer, importer, or distributor to the worker.

matt foundation: Foundation system consisting of a thick, heavily reinforced concrete slab placed beneath the entire building area.

mechanical core: Prefabricated modular unit that contains the hooked-up kitchen and bath fixtures as well as other mechanical equipment required for the house.

medium-density fiberboard (MDF): Nonstructural reconstituted panel product used for a variety of exterior and interior products including siding, molding, furniture, shelving, and cabinets.

medium-density overlay (MDO) plywood: Exterior grade plywood finish with a fiber resin overlay. Commonly used for furniture, signs, and displays.

medullary ray: One of the wood rays in a tree that extend radially from the pith toward the outside of the trunk. Their purpose is to store and transport food.

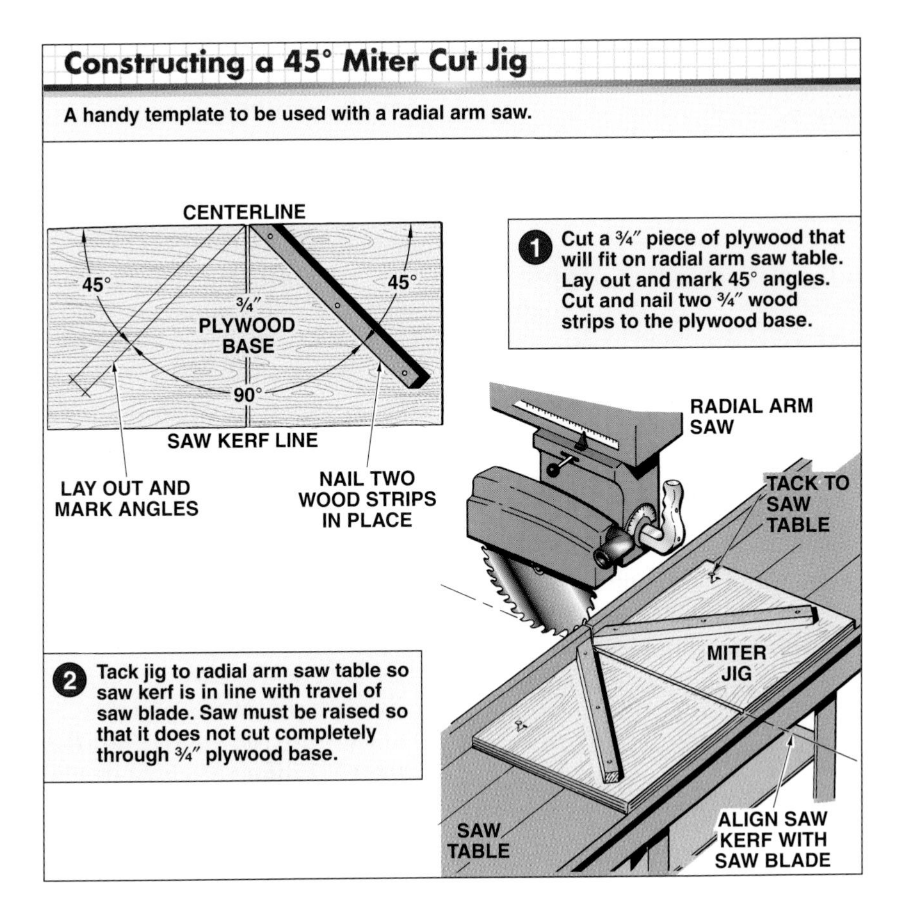

mesh core (door): Crisscross backing for the face panels used in hollow-core doors.

metal connector: Fastening device used in wood- and metal-framed construction to connect adjacent members together. Extremely important in areas subject to earthquakes, tornados, and hurricanes.

metal connector plate: Flat metal truss connection device with teeth punched from the base metal.

micropiles: Small diameter (3″ to 12″) drilled and grouted piles.

mil: Unit of thickness that is equal to 1⁄1000 of an inch (0.001″).

millwork: Interior trim stock that is produced in a mill and delivered to the job site.

millwright: Person who installs machinery and other mechanical equipment in mills and factories.

minute: Unit of measurement of angles equal to 1⁄60 of a degree.

mitered stringer: Stringer used on open sides of a stairway where the joints between the finish riser material and the stringer are mitered.

miter joint: Wood joint formed with two pieces cut at a 45° angle and fastened together to form a 90° angle.

miter protractor: A layout tool that is used to measure and transfer angles for mitered trim work.

mixture segregation: Condition that occurs when aggregate settles because the mixture is too thin to support the aggregate.

mobile crane Crane that can be moved within and between job sites.

mobile scaffold: Scaffold equipped with casters to allow the scaffold to be easily moved over a level paved or concrete surface.

model code: Building code developed through conferences between building officials and industry representatives from all parts of the country.

modular home: In manufactured housing, a building entirely constructed of prefabricated modular units. Also called *sectional home.*

modular unit: In manufactured housing, a completed three-dimensional section of a house.

moisture content: Amount of moisture present in wood at a given time. It is usually expressed as a percentage of the dry weight of the wood.

moisture meter: Instrument used in the field to give an instant reading of moisture content in wood.

molder: Machine that produces different patterns of molding from softwood and hardwood lumber. Also called *sticker.*

monolithic concrete: Concrete placed in forms in a continuous pour without construction joints.

mortise: Cavity cut into lumber to receive a similarly shaped tenon projecting from another piece of wood. Also, cavity for receiving hardware, such as a mortised lock.

mortise-and-tenon joint: Joint between two members where a tenon cut into the end of one piece fits into a cavity mortised in the other piece.

mortised lock: Older type of lock that fits into a mortised cavity in a door.

mounting plate: Plate placed on one side of a door. Holes are provided for two machine screws that engage and tighten into threaded holes in the cylindrical housing, thus securing the lock in place.

mounting rail: Wood strip provided at the top and bottom of the back interior of a wall cabinet. Wood screws that fasten the cabinet to the wall penetrate the mounting rails and are driven into the wall studs. Also called *hanging rail.*

mudshore: Timber used to brace and straighten a concrete form. One end is nailed to the form and the other end butts against a wood pad placed on the ground.

mudsill anchor: Metal connector used to fasten a sill plate to the foundation.

mullion: Vertical bar in a window frame that separates two windows.

multifolding door: Series of doors hinged to each other and supported by wheeled hangers that run across an overhead track.

multiple glazing: Two or three layers of glass with spaces in between set into a window sash to improve insulation.

muntin: Small pieces that separate window lights enclosed by a window sash.

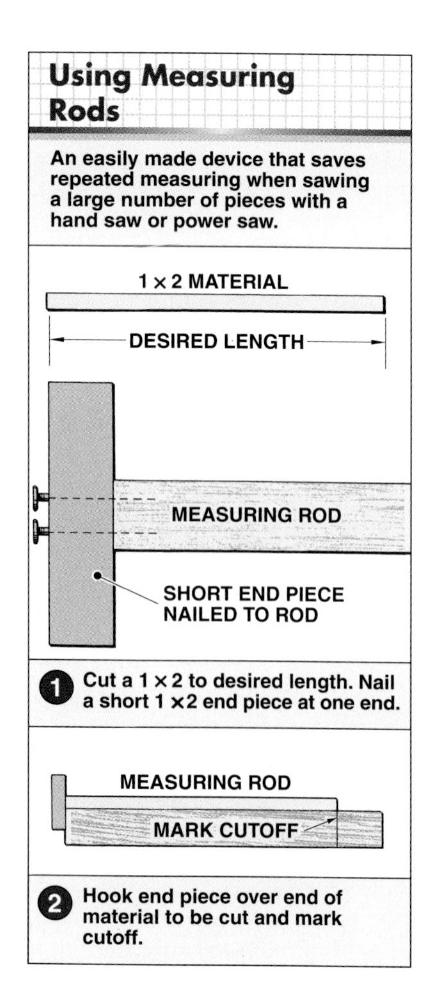

N

nailing plate: A metal plate nailed at the joint between truss members.

nail set: Hand tool, approximately 4″ long, with a slightly cupped tip. Used to drive nail heads below the surface of material.

natural grade: The various levels of the lot surface before any finish grading takes place. Also called *existing grade.*

newel post: One of the main upright members that support the handrails of a stairway.

nominal lumber size: Commercial size given to a piece of lumber based on its thickness and width at the time it was manufactured in the sawmill, before it was surfaced and seasoned.

noncomposite slab: Elevated concrete slab independent of the structural steel in supporting loads after the concrete has hardened.

non-load-bearing wall: Wall that supports no load other than its own weight.

nonveneered panels: Panel products manufactured from reconstituted wood.

nosing: Projection of the tread beyond the face of the riser.

notch: Rabbet-like cut made across the edge of a piece of lumber. An example is the notch made in a joist or stud to accommodate a pipe or conduit.

O

oil-borne preservative: Wood preservative best known in the form of pentachlorophenol. Highly toxic to fungi and insects.

on-center: Distance between center points of framing members or other building components.

one-way joist system: Floor slab system that has cast-in-place joists running in one direction.

open panel construction: Prefabricated panelized construction method. The outsides of the wall sections are finished off, but the insides are left open.

open stringer: Stringer that has been cut out to support the treads on the open side of a stairway.

orbital motion: Circular motion common to some types of electric sanders and jig saws.

orientation: Position of a building on a lot and the direction that the different walls will face.

oriented strand board (OSB): Panel product made of layers of wood strands bonded together with a phenolic resin.

oriented strand lumber (OSL): Engineered wood product made from flaked wood strands that have a high length-to-thickness ratio and are combined with waterproof adhesive.

orthographic drawing: A two-dimensional drawing used for making up prints.

oscillating motion: Back-and-forth movement commonly used in some electric finish sanders.

overhang: Part of the rafter that extends past the building line.

overhead garage door: A garage door that swings or rolls up when opened.

overlaid plywood: Plywood panel with factory-applied, resin-treated fibers on one or both sides of the panel.

overlay door: Cabinet door designed to cover the edges of the face frame.

oxyacetylene welding: Fusion process brought about by a welding flame applied to the edges and surfaces of metal pieces. Filler metal may also be added.

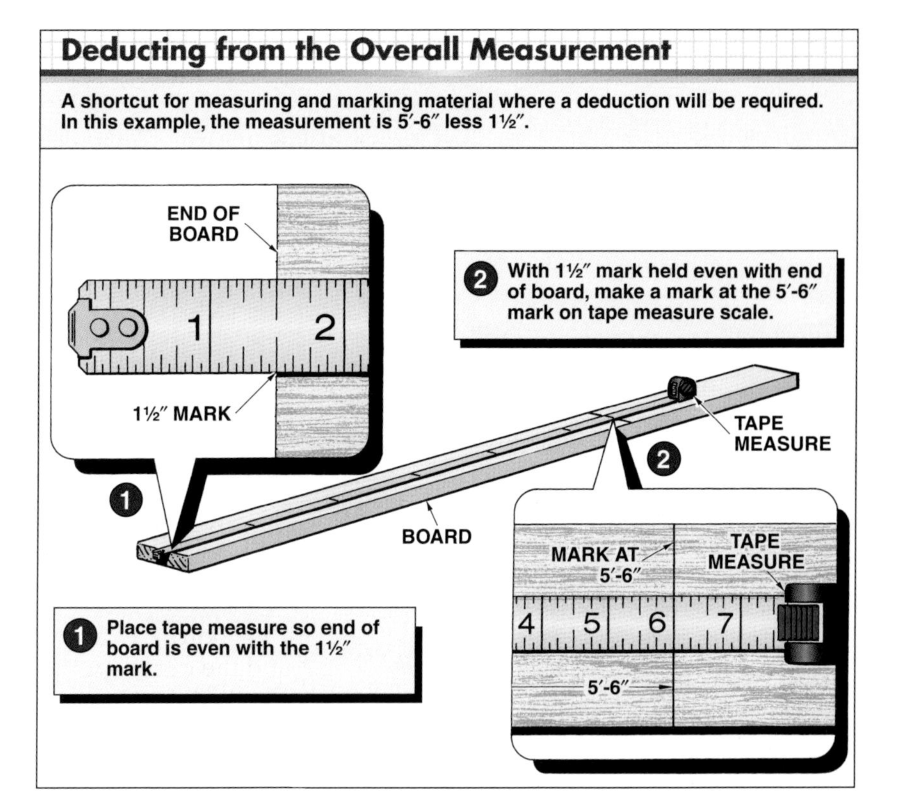

P

palm nailer: Compact pneumatic tool used to drive fasteners for metal connectors.

pane: Flat piece of glass cut to size for glazing a window.

panel clip: Device used at the unsupported edges of panel roof sheathing.

panel door: Door made of sunken panels fitted between vertical stiles and horizontal rails. Also called *stile-and-rail door.*

paneling: Panels applied as a finish for interior walls.

panelization: System for prefabricating walls, floors, and roofs in separate sections.

panelized construction: Method widely used in manufactured housing. The basic units of this system are framed wall sections built in a factory. They are delivered to the job site where they are to be installed.

paperless drywall: Panel product having a gypsum board core, and a front and back covered with fiberglass facing material.

parallel-strand lumber: Engineered lumber product manufactured from strands or elongated flakes of wood that are blended with a waterproof adhesive and cured under pressure.

parapet: Low wall at the edge of a roof.

parquet: Wood flooring design in which strips of wood are arranged in a square pattern with the grain running in the same direction. Grain patterns in adjacent squares run perpendicular to one another.

particleboard: Panel product produced by combining wood particles such as chips or flakes with a resin binder and hot-pressing them into panels. Also called *flakeboard* or *chipboard.*

parting strip: Thin vertical piece of wood that separates the upper and lower sash of a double-hung window. Also called *parting bead.*

partition: Interior wall.

patented ties: Devices used to hold together opposite wall forms during the placement of concrete.

patio: Outdoor paved area adjacent to a house, used for dining and recreation.

pattern: Anything that is used to mark the dimensions of an item on a material in order to produce duplicates. For example, a pattern is used to lay out common rafters.

payback period: The time elapsed until a project's resulting savings equal its cost.

peeler block: Section of a log from which strips of veneer are peeled off with a special lathe knife.

penny: Measure of nail length. Abbreviated by the letter *d.*

performance rated panel: Structural wood panel that conforms to performance-based standards such as dimensional stability, bond durability, and structural integrity.

perimeter: Outside boundary of an area.

perlite: Natural volcanic glass material used as a lightweight aggregate in concrete and as loose fill insulation.

perm rating: Measurement of water vapor flow through a material or a combination of materials.

personal fall-arrest system: Safety system used to arrest (stop) a worker in a fall from a working level.

personal protective equipment (PPE): Equipment used to protect against safety hazards on a job site. Includes protective clothing, head protection, eye and face protection, hearing protection, hand and foot protection, back protection, knee protection, and respiratory protection.

phenol formaldehyde: Synthetic resin commonly used as a binder in reconstituted wood panels.

Phillips-head screw: Wood screw with a cross slot in the head.

pictorial drawing: Three-dimensional view that shows three sides of an object or structure.

pier: Square, round, or battered concrete base set in the soil to directly support posts or columns. Also used to directly support grade beams.

pier block: Wood piece anchored to the top of a pier to provide a nailing surface for the bottom of a post.

pilaster: Column-like projection from a wall. It helps strengthen the wall and may also provide added support for a beam.

pile: Long, slender concrete, steel, or wood structural member that penetrates through unstable soil layers until it rests on firm soil. Piles provide support for grade beams or columns that carry the structural load of a building.

pile-driving rig: Machine used for driving piles.

pile shoe: Steel device attached to a pile tip to prevent the pile from breaking while it is being driven.

pilot hole: Hole drilled to facilitate driving a wood screw into wood. The diameter of the hole should be slightly smaller than the screw size.

pin: Bolt-shaped device placed into the barrel of a hinge. It holds together the two leaves of the hinge.

pitch: 1. Angle that a roof slopes from the outside walls of the building to the ridge of the roof. Expressed as a fraction. **2.** Resinous substance derived from the sap of various coniferous trees.

pitch pocket: Opening that extends parallel to the annual growth rings of a piece of lumber. It contains solid or liquid pitch.

pitch streak: Visible accumulation of pitch in a piece of lumber. Differs from pitch pocket in that the pitch saturates the section of wood fiber.

pith: Small, soft central core of a tree trunk, branch, or twig.

plainsawn lumber: Hardwood lumber cut out of a log so that the annual growth rings are at an angle of 45° or less to the wide surface of the boards.

plank: Lumber over 1″ thick and 6″ or more in width.

plank flooring: Wood finish floor pieces ranging in width from 3″ to 8″ and in thickness from 5/16″ to 25/32″. They give a more traditional effect than strip flooring materials.

planter strip: Area between the street curb and sidewalk where a lawn or other vegetation may be planted.

plan view: Drawing in a set of prints that presents a view looking down on an object.

plasma arc cutting (PAC): Cutting process that uses a constricted arc and high-pressure gas to form a high-velocity jet of plasma (ionized gas) for cutting metal.

plastic laminate: Product made of three or four layers of plastic material bonded together under high heat and pressure. Used to surface countertops, wall surfaces, shelving, cabinets, etc.

plate: One of the horizontal pieces at the top and bottom of framed walls to which the upright wall members are fastened.

plate level: Carpenter's level with an extendable arm to provide a longer reach for the level.

platform framing: Framing method in which each story of the building is framed as a unit consisting of walls, joists, and a subfloor. The subfloor acts as a "platform" for the construction of the story above. Also called *western framing.*

plot plan: Plan included in a set of prints showing the size of the lot, location of the building on the lot, grades, and all other information needed to perform work required before construction of the foundation begins.

plumb: Exact vertical and perpendicular line. It would be at a 90° angle to a level plane.

plumb bob: Cone-shaped metal weight fastened to a string to plumb an item. Force of gravity on the weight causes string to hang in true vertical plane (plumb).

ply: Single veneer sheet used to manufacture plywood.

plywood: Product made of wood layers (veneers) glued and pressed together under high heat and pressure.

pneumatic tool: A tool powered by compressed air.

pocket sliding (recessed) door: Sliding door suspended from a track by roller-type hangers. When opened, it slides into a cavity prepared in the wall.

pole construction: Method in which pressure-treated poles are sunk into the ground to provide the foundation and basic anchor for the framework of a building.

polyurethane: Chemical material used in construction adhesives and foamed-in-place insulation.

ponding: Curing a concrete slab by flooding its surface with water.

portable band saw: A handheld power saw that has a flexible metal saw blade forming a continous loop around two parallel pulleys.

portland cement: Ground and calcined (heated) mixture of limestone, shells, cement rock, silica sand, clay, shale, iron ore, gypsum, and clinker.

positive-placement nailer: Specialized pneumatic nailer designed specifically for driving nails for metal connectors.

post: Upright member used to support beams or girders.

post-and-beam construction: Type of construction in which the basic framework of the building consists of vertical posts and horizontal or sloping beams.

post base: Metal strap or other type of device set into the tops of concrete piers or walls. Its purpose is to provide an anchor for the bottoms of wood posts.

post cap: Metal device used to strengthen the tie between the top of a wood post and the girder above.

post-tensioning concrete members: Procedure of stretching and anchoring the cables used for prestressed concrete members after the concrete has hardened.

pour strip: Strip of wood tacked inside one of the form walls to indicate the height to which the concrete should be poured.

powder-actuated tool: Tool with a powder-filled round (shell) that shoots a drive pin or stud into hard surfaces such as concrete, masonry, and steel without predrilling holes for anchors.

powder load: Metal casing that contains a powder propellant and is crimped closed on the end and sealed, and used to drive powder-actuated fasteners into a surface.

power buggy: Power-driven vehicle driven by an operator to carry fresh concrete to the place on the job site where it is to be poured.

pozzolans: Admixtures that under certain conditions have cementing properties and can be used as a partial replacement for cement in the concrete mix.

precast concrete: Concrete structural members that have been cast and cured in a casting yard or factory and are then delivered to the job site.

Placing Subfloor Panels

A method for ensuring the first row of floor panels is set in a straight line even if exterior wall is slightly irregular.

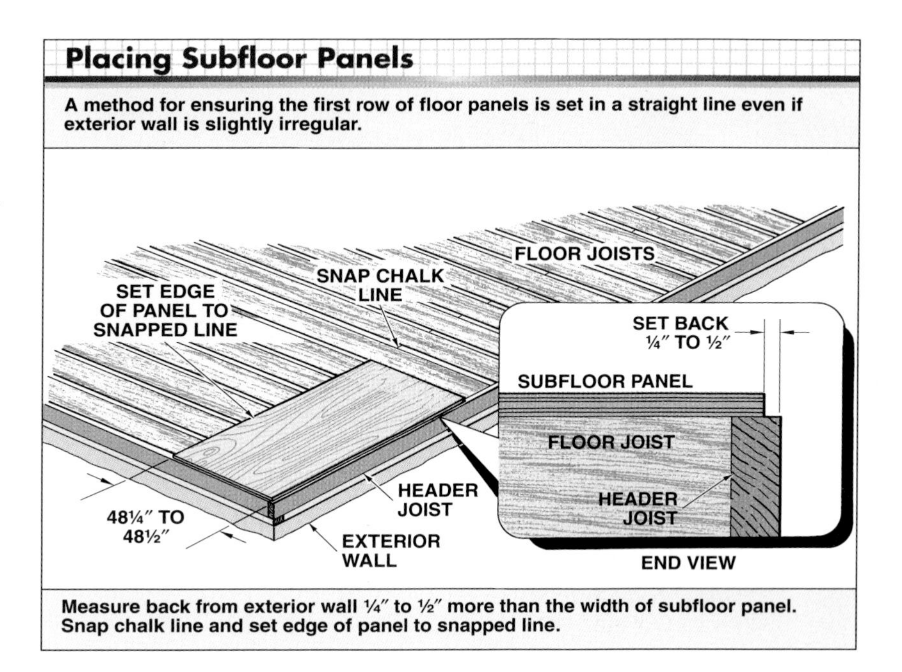

Measure back from exterior wall ¼″ to ½″ more than the width of subfloor panel. Snap chalk line and set edge of panel to snapped line.

prefabricated building unit: Building unit constructed or built in a manufacturing facility and delivered to a job site.

prefabrication: Construction of building members or units in a yard or factory.

preferred angle: Most comfortable walking angles of a stairway.

prefinished materials: Panels, boards, and other types of building materials that have received finish coats of paint, stain, or sealer on their appearance sides at the factory before delivery to the job site.

prehung door: Prefabricated door unit. It usually consists of a door already hung in the jamb, with bored lock holes and pre-fitted stops and casing.

preservative: Substance applied to or injected into wood to protect it from fungi and insects.

pressure differential: The difference between the pressures on either side of a barrier.

pressure treatment: Process of treating lumber with chemical preservatives by applying strong pressure to the pieces placed inside a tank.

prestressed concrete: Precast concrete that is reinforced by high-tensile steel cables placed under great tension.

pretensioning: Procedure of stretching the cables used for prestressed concrete members prior to placing the concrete in a casting bed.

prism: Reflective target for a total station instrument that receives the infrared beam from the instrument.

property lines: Recorded, legal boundaries of a piece of property.

protective helmet: Hat made of plastic or metal that protects a worker from injury caused by falling objects. Also called *hard hat.*

pueblo revival architecture: Architectural style which is a blend of Native American, Hispanic, and Anglo building traditions.

pump jack scaffold: Type of scaffolding consisting of 4 × 4 uprights and platforms supported by bracket devices that can be raised by a pumping action.

purlin: Horizontal timber held in place by braces. It is placed beneath the roof rafters at an intermediate point between the ridge and outside wall.

push bar: Device used to open a hopper window and hold it in a fixed position.

pushbutton: Device located inside the knob of a cylindrical lock. When pushed in, it locks the door.

push plate: Metal or plastic plate fastened to the lock side face of a door. Mounted at arm level, it protects the door from wear. Used most often on doors in public buildings.

putlog: Horizontal truss that extends between two separate scaffolds. Also called *bridge* or *trestle*.

Q

quarter round: Trim molding with a quarter-circle profile.

quartersawn lumber: Lumber produced by quartering a hardwood log lengthwise, then cutting boards out of each section.

R

rabbet: Groove cut along the edge of a piece of lumber.

rabbet joint: Joint formed at the ends of two pieces that have been rabbeted.

radial arm saw: Stationary power saw typically used to crosscut lumber to length, but can also be set up for ripping and angled cuts.

radiation: Transfer of heat through space.

rafter: Sloped structural roof member that supports roof sheathing and roof loads.

rafter tables: Tables found on the blade of a framing square. Used to compute the lengths of roof rafters.

rail: Horizontal piece of a panel door frame.

rainscreen wall: Moisture-management system that incorporates a vented or porous exterior finish, an air cavity, a drainage plane, and an airtight interior support wall.

rainwater harvesting system: On-site water collection and holding system for rainwater collected from roofs, impervious surfaces, and parking lots.

rake end: Overhang construction at the end of a gable roof.

ramp: Sloping runway from a lower to a higher level for passage of workers and materials on a construction job.

ready-mixed concrete: Concrete mixed at a batch plant and delivered by truck to the job site.

reamer and bucket: Device used for drilling large-diameter holes for caissons. The reamer is situated within the drilling bucket and digs into the ground with a revolving movement.

reciprocal motion: Up-and-down motion such as the action of the projecting blade of a reciprocating saw.

reciprocating saw: Multipurpose cutting tool in which the blade reciprocates (quickly moves back and forth) to create the cutting action.

reconstituted wood: Wood products made up of particles, flakes, or strands hot-pressed and bonded together into panel-sized sheets.

reflective insulation: Insulation materials that reflect heat rather than absorb it.

register: Grilled frame through which heated or cooled air is released into a room.

reinforced concrete: Concrete that contains steel reinforcement (rebar) or fiberglass reinforcing rod to strengthen it.

reinforcing bars: Deformed steel bars placed in concrete to increase its ability to withstand weight and pressure. They also help tie together structural concrete members. Also called *rebar.*

resawn lumber: Lumber pieces run through a special bevel saw to produce a coarse, textured pattern.

residential construction: Structures in which people live, such as houses, condominiums, and apartment buildings.

residue: Waste portion of a log (chips, bark, trimmings, shavings, sawdust) after the log has been cut into lumber at a sawmill.

resilient channel: One of the metal pieces nailed across wood studs or joists to which drywall material is fastened. They help break the sound vibration path through a wall or floor.

resilient tile: Type of floor tile that yields when pressure is applied, then returns to its original position.

resin: Sticky material obtained from the sap of certain trees, especially pine and fir species. Resins are often used in making varnishes, paints, plastics, and adhesives.

respiratory protection: Device worn over the mouth and nose to protect workers from inhaling dangerous dusts or fumes.

retaining wall: Masonry or wood wall constructed to hold back a bank of earth.

retarder: Admixture added to concrete to delay the stiffening of the concrete.

return on investment (ROI): The ratio of a project's financial gain to its cost after a certain length of time.

reveal: Space (1/8″ to 3/16″ typical) along the edge of a door casing. Creates a better appearance and puts the casing out of the way of the door hinges.

ribband: Piece of lumber, 1″ × 4″, laid flat and nailed to the tops of ceiling joists. It is placed at the center of the spans and helps prevent twisting and bowing of the joists.

ribbon: In balloon framing, narrow board notched into studs to support joists.

ridge: Highest point at the top of a roof where the roof slopes meet.

ridge beam: Beam that is placed between or below the roof beams in post-and-beam construction.

ridge board: Horizontal board or plank placed at the ridge of a roof, to which the upper ends of the rafters are nailed.

ridge plumb line: Marking for the plumb cut made at the ridge end of a roof rafter.

rift sawn: Sawing method in which a log is first quartered lengthwise. Lumber is cut from each quartered section, with the cuts made at a 30° or greater angle to the annual rings.

rigging: Equipment required to connect the hoist hook of a crane to attachment points on a load.

rigid foam insulation: Foam insulation consisting of polyurethane and polystyrene foam or fiberglass strands pressed into panels. Rigid foam insulation has a higher R-value per inch of thickness than any other type of insulation.

rim board: Plywood, oriented strand board, glulam timber, or laminated veneer lumber used to tie together wood I-joists.

rim track: Metal framing member used to tie together the outer ends of the metal joists. Serves the same purpose as a header joist in wood-frame construction.

rip: Cut made lengthwise in a wood member, parallel to the grain.

rip blade: Blade with flat-faced teeth shaped like a chisel. It is used with power saws to rip in the direction of the grain.

ripping: Cutting with a saw in the same direction as the wood grain.

ripsaw: Handsaw designed to cut with the wood grain. Ripsaw teeth are shaped like chisels, to be effective for cutting with the wood fibers.

rise: Vertical measurement.

riser: Piece forming the vertical face of a step.

roller hanger: Wheel-like device used to suspend and move sliding doors along tracks attached to the head jamb.

roof beam: Beam that is part of the roof structure of a post-and-beam building and to which the roof decking is nailed.

roof truss: Prefabricated structural roof unit made of top and bottom chords tied together with web members. The top chord acts as a roof rafter and the bottom chord serves as a ceiling joist.

rope grab: Deceleration device that travels on a lifeline.

rotary hammer: Electric tool with cutting bit that is rotated and moved with percussive action to bore into hard material.

rough lumber: Lumber hat has not been dressed (surfaced) but has been sawed, edged, and trimmed.

rough opening: Framed opening in a wall into which a finished door or window unit will be placed.

rough sill: Horizontal piece nailed across the bottom of a rough window opening.

rough-terrain forklift: Heavy-duty lift truck designed to traverse the rough terrain of construction sites and transport and place materials where they are required.

round sling: Continous loop sling consisting of unwoven synthetic fiber yarns enclosed in a protective cover.

rubber tile: Resilient floor tile of synthetic materials or natural rubber combined with mineral fillers.

run: Horizontal measurement.

runway: Temporary platform-like structure that provides passage for workers and materials on a construction job.

R-value: Measure of the effectiveness of a material to provide thermal insulation. Higher R-values indicate greater insulating capabilities.

S

saddle: Structure with a ridge sloping in two directions that is placed between the back side of a chimney and the roof sloping toward it. Its purpose is to divert water away from the chimney. Also called *cricket.*

safety goggles: Device made of specially treated glass or plastic worn over the eyes to protect the worker from eye hazards caused by flying particles.

safety net: A meshed net sometimes used when safety belts and lifelines are impractical, and placed below a work area that is 25′ or more above the ground, water surface, or a floor below.

sandpaper: Heavy paper with abrasive materials (garnet, emery, flint, etc.) glued to one side. Used with a hand sanding block or electric sander to smooth wood surfaces.

sap: Watery fluid that circulates through a wood plant carrying food and other substances to the tissues.

sapwood: Pale-colored living wood near the outside of a tree trunk.

Altering Dovetail Saw to Cut Holes in Thin Materials

A dovetail saw can be altered by cutting an angle at the end of the blade with a hacksaw. This makes it possible to start and finish neatly cut holes in thinner material such as 1/4″ paneling.

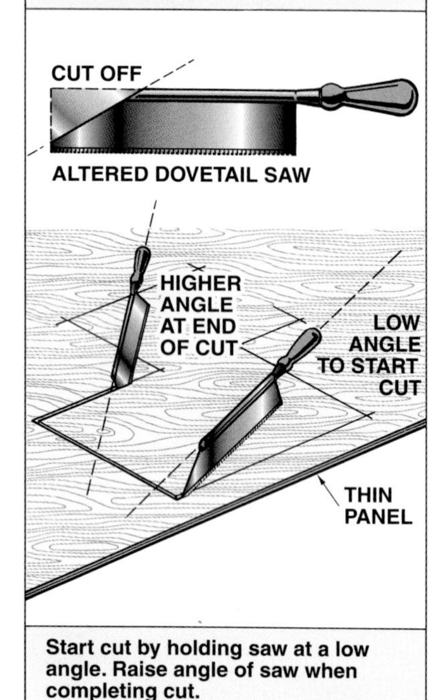

Start cut by holding saw at a low angle. Raise angle of saw when completing cut.

sash: Wood or metal frame into which glass panes are set.

sash lift: Hooked or bar device fastened to the lower window of a double-hung window unit. It acts as a handhold for lifting and lowering the window.

sash lock: Device attached to a window sash for locking purposes.

sash weight: Weight used to balance the sash so that it will remain in any vertical position when opened. Not used in new construction.

sawhorse: Portable work bench used by carpenters. It consists of a top piece supported by legs and tied together with end pieces.

sawmill: Plant where lumber is manufactured from wood logs.

scaffold: Temporary, braced platforms set up around buildings to enable carpenters to complete work that is out of reach from the floor or ground level.

scale drawing: Drawing in which inches or inch-fractions represent one foot of the actual measurement of a building.

scarf joint: Splice made by notching and lapping the ends of two pieces of timber. Also, end joint made by overlapping two pieces of molding with 45° angle cuts.

schedule: Detailed list on a print that provides information about building components such as doors, windows, and mechanical equipment. Numbers and/or letters on the print refer to the schedule.

scissors truss: Roof truss design featuring ceilings that slope up from the walls to the center of the span between the walls.

screeding concrete: Leveling newly placed concrete. Also called *striking off.*

screed: A piece placed across the tops of sidewall forms and used to strike off the concrete.

screed rail: Temporary wood or metal pieces positioned in an area where a concrete slab is to be placed. The tops of these pieces are set to the finish surface of the concrete.

screed support: Adjustable metal device that supports a screed rail.

screw anchor: Light-duty anchor for fastening materials to hollow walls.

scribe strip: Thin piece fitted against a wall to cover the space between an end cabinet and the wall.

scribing: Process of fitting the edge of a trim piece against an irregular surface.

scuttle: Small opening with a removable lid located in a ceiling below an attic to allow access into the attic.

sealant: Flexible material used to prevent air and water infiltration into a building. Common sealants include silicone, acrylic, and urethane.

seasoning lumber: Process of removing moisture from unseasoned lumber prior to marketing.

seat cut: Horizontal cut at the lower end of a rafter where the rafter rests on top of the wall plate.

second: Unit of measurement of an angle equal to 1/60 of a minute. (A minute is 1/60 of a degree.)

section view: Drawing of a vertical or horizontal cut made through a part of a building. It gives information about both the interior and exterior of the structure or object.

segregation: Separation of sand-cement ingredients from gravel due to improper placement of the concrete.

seismic risk zone: Area of the Earth's surface where the conditions for earthquakes exist.

self-drilling anchor: Heavy-duty anchoring device used to fasten material to concrete. It is driven by a rotary hammer. Sharp teeth at the end of the device drill the hole. The anchor is then secured with an expander plug.

self-drilling screw: Screw that, when driven into thinner metals, will drill a hole, cut threads, and fasten in one operation.

self-piercing screw: Screw with a sharp point capable of penetrating and tapping thin metal.

self-retracting lifeline: Type of vertical lifeline that contains a line that can be slowly extracted from or retracted to the drum under slight tension. When a worker falls, the drum automatically locks and arrests the fall.

self-tapping screw: Screw that, when driven into a prepared hole in thinner metals, will cut threads and fasten in one operation.

semiconcealed hinge: Small hinge similar in appearance to a loose-pin butt hinge. An offset on one leaf makes it possible to screw the leaf into the back of a door.

serrated-blade forming tool: A tool with a cutting action carried out with a steel blade containing hundreds of preset razor-sharp teeth that cut like chisels. Holes between the teeth permit the passage of shavings.

service door: Usually a rear-entry exterior door.

service stairway: Stairs that serve noninhabited areas of a house such as a basement or attic.

service walk: Walkway that extends from a driveway or sidewalk to the rear entrance of a building.

set: Offsetting of the teeth of a handsaw to one side or another to provide a kerf that is wider than the blade thickness.

set-retarding admixture: Substance added to concrete to extend its setting time. Also called *retarder.*

set of saw: Angle at which saw teeth are alternately bent from side to side. Helps prevent binding while cutting wood.

shackle: U-shaped metal connector with holes drilled into the ends for receiving a removable pin or bolt.

shake: **1.** Type of shingle split from red cedar logs. It gives a more rustic appearance than regular shingles. **2.** Lumber defect caused by a lengthwise separation of the wood, usually between or through the annual growth rings.

shank: Body of a nail or screw. The shank is the part driven into wood.

shear stud: Metal post that directly connects the hardened concrete to the framework of structural steel members below.

shear wall: Wall designed to resist lateral forces due to earthquake, wind, or other causes. A shear wall is often created by placing performance-rated panels on one or both sides.

sheathing: Panels or boards placed on the outside of an exterior framed wall or roof to provide greater insulation, strength, and a nailing base for finish materials.

sheathing paper: Layer of water-resistant paper applied over board sheathing or, where no sheathing is used, directly on stud walls. It is placed before the siding is nailed in place. Also called *building paper.*

shed roof: Roof that slopes in only one direction. Also called *lean-to roof.*

sheet piles: Piles that interlock with each other and are not intended to carry vertical loads. Their primary purpose is to resist horizontal pressure, and they are frequently used around excavations.

shielding: Use of a portable protective device capable of withstanding cave-in forces in trenches.

shim shingle: A narrow strip, usually cut from cedar shingles, used to plumb and straighten a door jambs or furring strips.

shingle panel: Panel to which one or two courses of shingles or shakes are bonded.

shiplap siding: Boards rabbeted on both edges so that when laid edge to edge they make a half-lap joint.

shock-absorbing lanyard: Personal fall-arrest lanyard with a specially woven, shock-absorbing inner core to reduce fall arrest forces.

shore: Wood timber or metal device placed in a vertical, horizontal, or angled position to provide temporary support.

shoring: System used to prevent the sliding or collapse of the earth banks around an excavation. Also, temporary bracing against a wall or beneath any type of structure that exerts weight from above.

shortened valley rafter: In an intersecting roof with unequal spans, the shorter valley rafter extending from an intersecting corner of the building. The top end butts against the supporting valley rafter.

shortening: Process of deducting an amount from the ridge plumb line of a common rafter equal to one-half the thickness of the ridge. In the case of hip and valley rafters, or rafter jacks butting against a rafter, the deduction is one-half the 45° thickness.

shutter: Louvered or flush rectangular panels located at each side of a window. They may be hinged in order to close over the windows as added protection against weather. Often they only serve a decorative purpose.

side: Widest dimension of a piece of lumber.

side cut: Angle cut where a hip or valley rafter comes up against the ridge of a roof. It is also formed where jack rafters are nailed against hip or valley rafters.

sidelap: Area where materials (shingles, underlayment felt, etc.) lap over each other at the side edges.

sidewalk: Walkway that extends along a street and borders the building lot.

sidewall: Outside wall of a building.

side yard: Distance from the property line to the side of a building.

siding: All types of exterior board, panel, and shingle wall covering applied by carpenters.

siding story pole: Rod upon which the layout of the siding boards can be determined in relation to the finish window openings.

sill: Slanted piece at the bottom of the finished window frame.

sill anchor: Anchor bolt positioned in a concrete form at the time the concrete is being placed. Its purpose is to fasten the sill plate to the top of the wall.

sill pan: Preformed polyvinyl chloride (PVC) plastic flashing used at the base of window and door openings to direct water that may have penetrated in or around the frames to the outside.

sill plate: Wood plate fastened to the top of a foundation wall. It provides a nailing base for floor joists or studs. Also called *mudsill*.

silt: Earth material consisting of fine mineral particles that are midway in size between sand and clay.

SI metric system: Decimal measurement system based on the kilogram and kilometer.

single coursing: Applying wood shingles to a sidewall in a single layer.

single-ply membrane: Roofing material made of flexible sheets of combined synthetic plastic materials. Attached to wood and steel decks with fasteners or specified adhesives.

site: Location of a construction project.

skylight: Opening in a roof or ceiling covered with glass or transparent plastic to provide light in the space below. Sometimes also used for ventilation. Some types are called *roof windows*.

slab-at-grade foundation: Foundation system that combines concrete foundation walls with a concrete floor slab that rests directly on a bed of gravel that has been placed over the ground. Also called *slab-on-grade*.

slat: Narrow strip of wood or metal.

slate: Gray-blue colored rock used as a finish roof covering.

sledgehammer: A heavy, double-faced striking tool used for heavy-duty striking work.

sleeper: Wood strips fastened to a concrete slab to provide a nailing base for wood finish flooring.

sleeve: Metal or fiber cylinder set inside a form before the concrete is placed. It provides a hole for the passage of pipes or other objects through the finished concrete wall.

sleeved dowel system: Self-adjusting ties between two concrete walls, allowing movement caused by expansion and contraction.

sling: Flexible length of load-bearing material that is used to rig a load.

sling angle: Acute angle between the horizontal plane and the sling leg.

sling hitch: Arrangement of one or more slings used for connecting a load to a hoist hook.

slip form: Forming system that moves continuously upward while the concrete is being placed.

slope diagram: Small triangle on a set of prints indicating the unit rise and unit run of a roof.

slump: Measure of the consistency of freshly mixed concrete.

slump test: Test taken at the time concrete is placed to measure the consistency (slump) of the concrete.

snap bracket: Wedge used with snap ties in a single-waler form system.

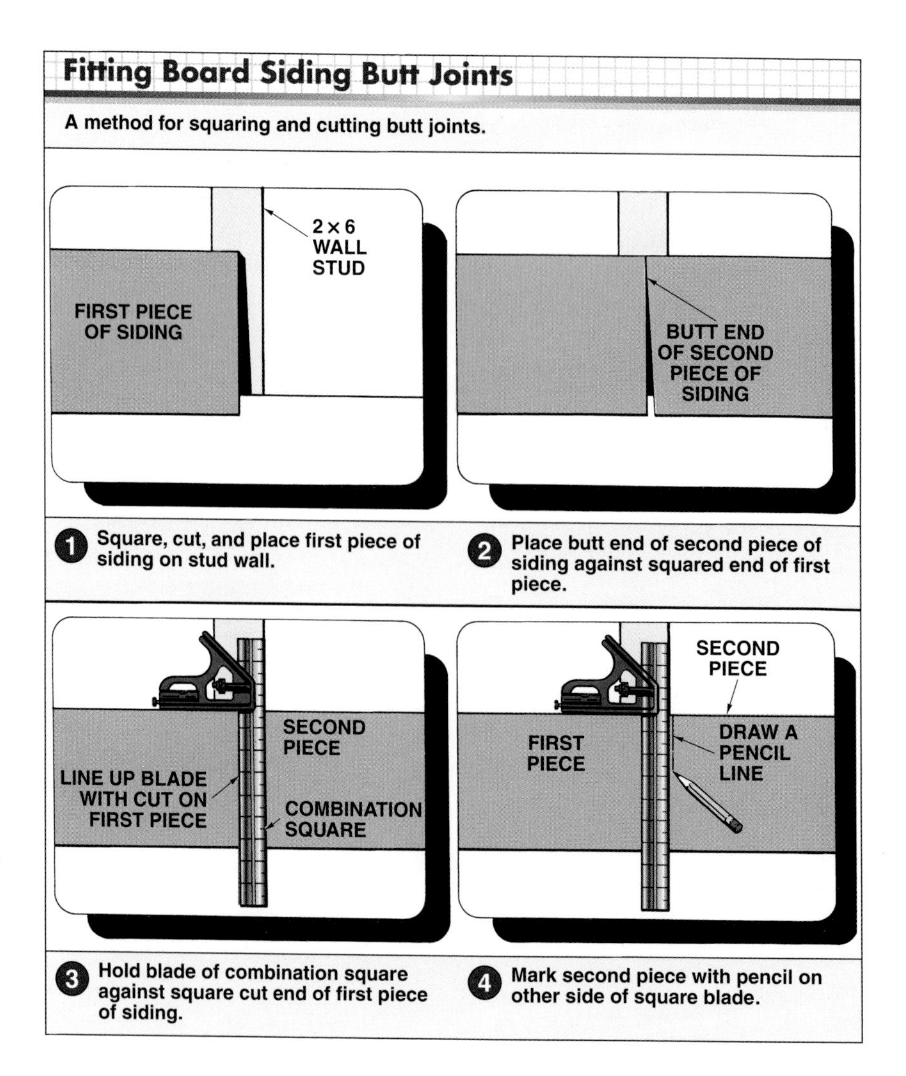

snap lock: Surface-mounted locking device sometimes used with hopper windows or bottom-hinged cabinet doors. It snaps into a locking position when the door or window is shut.

snap tie: Patented tie system with cones acting as spreaders. Grooved breakbacks are provided for breaking off the ends of the ties extending from the hardened concrete wall.

socket: Wire rope end attachment that allows connections to other hardware.

soffit: Underside of a building member such as a cornice, arch, beam, staircase, etc.

softwood lumber: Lumber that comes from coniferous (evergreen) tree species. Most of these trees have needle-shaped leaves.

soil auger: Large device powered by a drilling rig for the purpose of boring pier holes into the ground.

solar heat gain coefficient: Measure used to indicate how well a window unit blocks heat caused by sunlight.

solar heating system: Heating system that uses a collector to absorb solar energy. Distribution and storage components are also required.

solar house: Building designed with large south-facing glazed areas to collect and store solar energy.

soldier piles: Steel members driven into the ground for shoring around excavations. They are driven at intervals and wood planks called lagging are placed between the piles. Also called *soldier beams.*

sole plate: Horizontal piece at the base of a framed wall to which the bottoms of the upright members are nailed. Also called *bottom plate.*

solid beam: Beam made of one piece of timber.

solid blocking: Wood pieces, the same thickness and width as the joists, placed between the joists to stiffen and hold the joists in position.

solid board paneling: Solid wood boards, usually 3/4″ thick and 8″ to 12″ wide, applied as a finish for interior walls.

solid-core door: Flush doors with a solid core, usually consisting of staggered wood blocks.

solid-surface countertop: Countertop manufactured from composite resin material. Very durable and stain-resistant.

sound transmission coefficient (STC): Rating for the sound transmission performance of a wall or ceiling system.

spacer strip: Strip placed underneath the bottom edge of the first row of bevel siding. It should be the same thickness as the thin edge of the siding.

span: Horizontal distance between supports for joists, beams, and trusses.

specifications: Written document included with a set of prints clarifying the working drawing and supplying additional data.

Speed® Square: Triangular shaped layout tool used for roof rafter layout. Also referred to as a *pocket square.*

splash board: Board placed at the top of a form on the side opposite to where the concrete is being placed. Its purpose is to prevent the concrete from spilling over the opposite side.

spline: Thin strip of wood placed into mortises or grooves that have been cut in boards where they are being joined together.

split: Separation of wood fibers across the annual rings and extending entirely through a piece of lumber.

split jamb: Type of jamb used with adjustable pre-hung door units. The jamb comes in two halves and can be adjusted for the width of the wall.

spot clinching: Method for joining two metal framing members. Provides a strong connection without the use of mechanical fasteners.

spreader: Wood or metal piece placed between form walls to assure that the finished concrete wall will be the correct thickness.

spreader bar: Rigging device used to distribute weight while lifting wide frames, such as trusses or building walls, by crane.

spread footing: Wide base placed beneath a foundation wall. It spreads the load over a greater ground area.

spread foundation: Foundation consisting of a spread footing with a keyway and an independently placed foundation wall erected on it.

spring balance: Coiled spring device used to control the up-and-down movement of double-hung window sashes.

Making Square Cuts

A method for squaring and cutting posts that have been set in place.

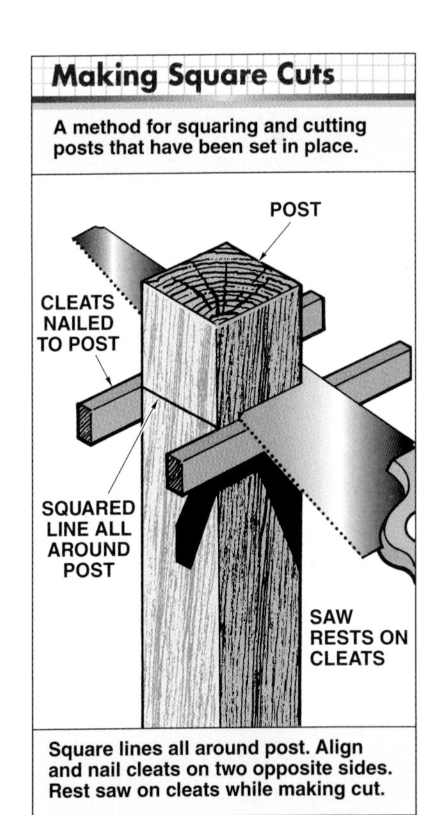

Square lines all around post. Align and nail cleats on two opposite sides. Rest saw on cleats while making cut.

spring hinge: Self-closing door hinge operated by a coil mechanism in the barrel. Used most often with doors installed in public buildings.

springwood: Portion of a tree trunk's annual growth ring that is formed during the early part of the season's growth. It is usually weaker and less dense than summerwood.

squash block: Dimension lumber of rim board used to carry a point load that would otherwise bear directly on a wood I-joist, such as when a post is installed directly over another post on a lower level. Typically installed in pairs to allow even distribution of a load.

stack-cast slabs: Series of concrete slabs cast directly on top of each other. A resin-type compound is used as a bond-breaker to prevent the slabs from adhering to each other.

stack effect: Uncontrolled airflow based on the principle that warm air rises.

stack framing: Method of construction where roof trusses, wall studs, and floor joists are constructed in a vertical line so that the load transfers directly to the structural members below it.

stair flight: Section of stairs going from one floor or landing to another.

stair landing: Platform between one flight of stairs and another.

stair story pole: Rod upon which the total number of risers for a stairway is marked off. It is checked to see if adjustments are necessary for the exact riser height.

stairway: Complete set of steps or series of flights extending from one story of a structure to another. Includes treads, risers, landings, handrail components, and structural supports.

stairwell: Space in which the flights of a stairway are placed.

standing part: Portion of a rope that is unaltered or not involved in making a knot or hitch.

staple: U-shaped metal fastener driven by manually operated or pneumatically powered stapling tools. Used for many fastening operations in construction work.

starter course: First course of sidewall wood shingles. It is usually doubled even if single coursing is used for the rest of the wall.

starting newel post: Post that supports a handrail at the bottom of a stairway.

stepped foundation: A foundation system used on sloped and hillside lots. The walls and footings are shaped like steps.

stick-built: Refers to a building that is constructed piece by piece on a job site rather than using modular or panelized construction.

stickers: Wood boards placed between layers of lumber in a lumber pile to allow air to circulate.

stiff leg: Piece used as a temporary shore to help hold an upper wall cabinet in position while it is being fastened to the wall.

stile: Vertical pieces of a panel door frame.

stock plans: Existing plans that can be purchased from concerns that produce a variety of working drawings for home construction.

stool: Trim piece usually notched over the interior sill edge of a traditionally finished window opening.

stop: Device with a soft tip that is fastened near the bottom of a door or into the wall base of the floor. It prevents the door from hitting against the wall when fully opened.

storm door: Additional, exterior door installed during the winter for added protection against cold and wet weather.

storm window: Additional, removable window installed during the winter for added protection against cold and wet weather.

story: Section of a building that runs from one floor to another floor or a roof.

stove bolt: Bolt with a slotted flat or round head. Usually used for lighter work.

straightedge: Long wood or metal piece with straight and parallel edges used to check and set surfaces to a straight line. Also used with a hand level for plumbing.

straight-flight stairway: Stairway that runs in a direct line from one floor to another. Also called *straight-run stairway.*

strap hinge: Hinge with long plates that is surface-mounted. Used with heavy doors or a gate.

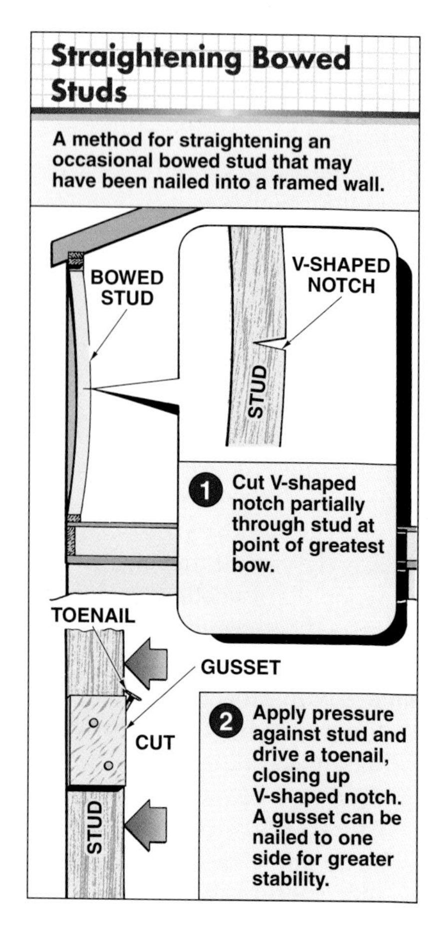

stressed-skin panels: Structural units used in floor, wall, and roof systems. They are made of plywood panels nailed and glued to 2 × 4s.

strike board: Straightedge used for screeding concrete.

strike plate: Metal piece mortised into a door jamb. It receives the latch bolt when the door is in a closed position.

stringer: **1.** Timber placed on top of shores to support the joists and panels of a deck form. **2.** Sloping members that provide the main support for the treads, risers, and other parts of a staircase. Also called *carriage* or *strings.*

strip flooring: Wood finish floor pieces ranging in width from 1½″ to 3½″ and in thickness from 5/16″ to 25/32″.

stripping forms: Removing forms from a structure after the concrete has adequately cured.

strongback: **1.** Piece of 2 × 4 nailed to a wider plank and also fastened down to the tops of ceiling joists at the center of their spans. The strongback keeps the joists in alignment and provides central support. **2.** Piece placed in back of and across walers to reinforce and stiffen a form. Also called *stiffback.*

structural insulation panel (SIP): A structural member consisting of a thick layer of rigid foam insulation pressed between two oriented strand board or plywood panels.

structural insulated sheathing (SIS): Wall panel sheathing that adds structural strength and insulation value to an exterior wall.

structural steel framing: Steel framing method that involves the use of bolted or welded connections to form the framework of a structure. Also referred to as *red iron steel framing.*

stub joist: One of the short joists running at right angles from a regular joist to an outside wall. This may occur under some roof sections when the rafters and ceiling joists do not run in the same direction.

stucco: Exterior finish material composed of cement, sand, lime, and water.

stud: **1.** Upright wood or steel member that extends from the bottom to the top plates of a framed wall. **2.** Device usually driven into concrete with a powder-actuated tool. One end is a nail that is embedded in the concrete. The other end is a threaded bolt that receives the nut used to secure the object being fastened.

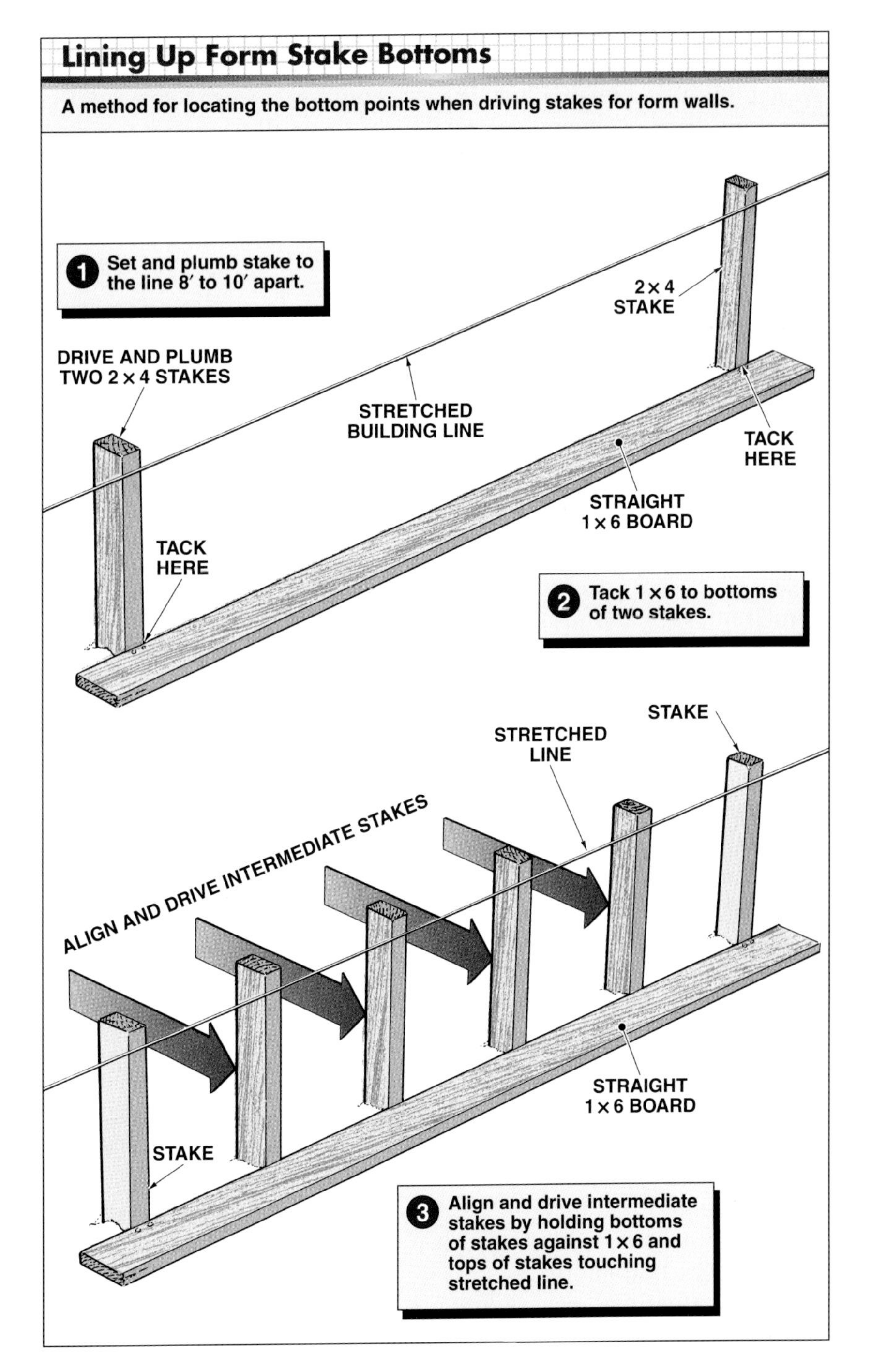

subcontractor: Licensed individual or firm that can enter into legal contracts to do work such as painting, electrical work, plumbing work, plastering, sheet-metal work, etc.

subdivision map: Map showing the established lots of a section of a city or township. The dimensions of each lot are given and the necessary property setbacks.

subfascia: Horizontal pieces the same thickness and width as the roof rafters. They are nailed to the tail ends of the rafter overhangs and covered with finish fascia material.

subfloor: Consists of structurally rated panels or boards fastened to the tops of the floor joists. It provides a base for the finish floor materials.

summerwood: Portion of a tree trunk's annual growth ring that is formed after the springwood is formed. It is usually stronger and denser than springwood.

sump: Pit located in the basement floor to collect and drain off water. Also called *sump well.*

sump pump: Water pump placed in a sump to pump out water collecting in the well.

supporting valley rafter: In an intersecting roof with unequal spans, the longer valley rafter extending from one intersecting corner of the building to the main ridge.

surfaced lumber: Lumber that has been smoothed in a planing mill. Also called *dressed lumber.*

surface hinge: Hinge with leaves that are mounted to the face frame and the face of the door.

surface water: Water from rain and melting snow that stays on the ground surface.

survey point: Small nail driven into the top of a corner stake (hub) to identify the exact corner of the property.

suspended ceiling: Ceiling made of a light metal grid hung by wire from the original ceiling. Ceiling tiles are dropped into the grid frame.

sustainable site: Building site that has reduced disruption of local plant and animal life, conserves existing natural areas, and restores damaged areas.

swale: One of the slopes required on a lot to ensure water drainage away from the building.

symbol: Pictorial representation of a structural or material component used on prints; commonly standardized.

systems scaffold: Type of scaffold with nodes welded to the posts and bearers for attachment of braces and railings.

T

table saw: Stationary power saw used for straight sawing. It is of great value in interior finish work such as cutting sheet goods and trim materials.

tack hammer: A lightweight hammer used to drive small nails into finish work.

tag line: Rope attached to a large member being placed by crane to help guide the member into position.

tail cut: Cut at the tail end of a rafter overhang.

tail joist: One of the members that run from the header to a supporting wall or girder in a floor or ceiling opening.

tangential cut: Type of cut produced in plainsawn or flat-grained lumber.

taper-ground saw blade: Saw blade that is thinner at the top edge than at the cutting edge. The taper helps prevent binding while sawing wood.

T-beam: Precast, prestressed concrete beam shaped like a T and suitable for spanning longer distances than conventionally shaped precast beams.

telehandler: Construction forklift designed to place materials up to a few stories in elevation and from a short distance.

telescopic-boom crane: Mobile crane with an extendable boom composed of nested sections.

tenon: Projecting part of a piece of lumber cut and shaped to fit into a mortise.

tensile strength: Resistance of a material to forces attempting to tear it apart.

tension: Pulling or stretching force.

tension strap: Flat metal piece used to brace studs against lateral movement.

tensioning jack: Hydraulic jack used for tensioning cables in prestressed concrete members.

termite shield: Metal shield placed under and extending out from the sill plate to prevent the passage of termites.

T-guide: Device used to hold the bottoms of sliding cabinet doors in position. It fastens to the bottom of the cabinet and fits into a groove at the bottom of the door.

theoretical length: The length of a rafter before it is shortened because of ridge or rafter thicknesses.

thermal bridging: Interruption of insulation by a framing member that has higher thermal conductivity than the insulation.

thermal conductor: Material capable of transmitting heat.

thermal imager: A device that detects heat patterns in the infrared wavelength spectrum without making direct contact with the targeted area.

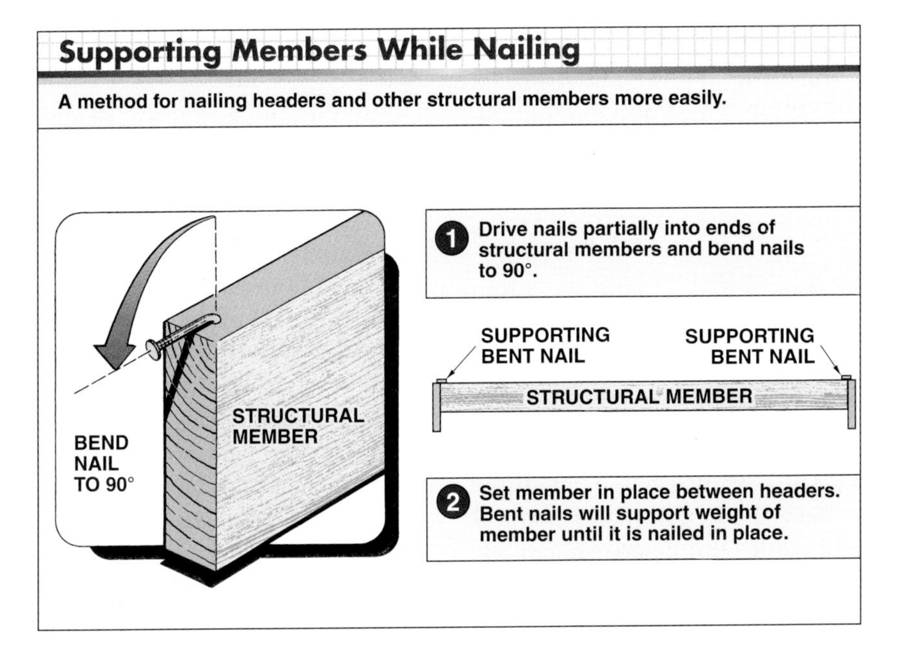

thermal insulation: Insulation materials placed in the walls, floors, ceilings, and roof to control heat flow and help maintain a comfortable temperature in the building.

thermal performance: Ability of a window to act as a barrier to heat transfer.

thermal storage unit: Heat storage unit required in an active solar heating system.

thermostat: Electrically operated instrument that can be set to automatically control the room temperatures produced by heating or cooling equipment.

threshold: Wood, metal, or stone piece set between the jamb and the bottom of a door opening. Also called *door sill.*

thimble: Curved piece of metal that supports a loop of rope and protects it from sharp bends and abrasion.

Thrombe wall: Masonry wall used to store heat in an indirect-gain passive solar heating system.

tiedown: A metal device bolted to a wood post in a wall and fastened to a bolt anchored in the concrete foundation.

tile: Building material made of cement, plastic, or other resilient material and used as a wall, floor, ceiling, and roof finish material.

tilt-up construction: Construction method in which the wall sections of a building are cast in place on the job site and lifted into position by crane.

timber: Lumber pieces that are no less than 5″ in their least dimension.

timber connector: Metal ring, plate, or grid embedded in adjacent timber members to increase the strength of the joint.

toeboard: Narrow board located at the back of the recessed area under a base cabinet to provide foot space when a person stands close to the cabinet. Also called *toe strip.*

toe holds: Boards temporarily nailed on top of panels as a safety measure when carpenters are sheathing steeper roofs.

toenail: To nail diagonally into two pieces.

toggle bolt: Anchoring device used to fasten light materials to hollow walls.

tongue-and-groove (T&G) lumber: Boards or planks with a groove in one edge and a tongue on the other edge. The tongue fits into the groove of a matching piece of lumber.

top-flange joist hanger: Metal joist hanger that is fastened to the tops of beams or ledgers.

toplap: Area where one shingle overlaps another shingle in the course below the first shingle.

top plate: One of the horizontal pieces to which the tops of the vertical members of a framed wall are nailed.

top track: Horizontal metal framing member that ties studs together at the top of a wall.

total rise: 1. Ridge height of a roof, measured from the top of the walls. **2.** Vertical distance of a stairway from a bottom floor to the floor above.

total run: 1. Distance equal to one-half the total roof span. **2.** Horizontal length of a stairway measured from the foot of the stairway to a point plumbed down from where the stairway ends at the deck or landing above.

total span: Measurement equal to width of the building.

total station instrument: Survey instrument that combines digital data processing and survey technology to perform leveling, plumbing, and horizontal and vertical measurement operations.

tower crane: Crane consisting of a high tower and jib. Sections can be added to the tower to achieve greater heights.

track: Metal channel attached to the head jamb. Roller hangers attached to the doors move along the track.

trademark: Information stamped on a lumber or panel product at the time of its manufacture to identify its grade, rating, mill number, and other pertinent information.

transit-mix truck: Truck equipped with a large drum concrete mixer for delivery of ready-mixed concrete to the job site.

transmittance (U): A factor that expresses the amount of heat in Btu transferred in 1 hour through 1 sq ft of the floor, ceiling, wall, or roof area of a building.

transom: Small opening over a door or window. It usually contains a movable sash or stationary louver.

transverse beam: Roof beam that extends from an outside wall to a ridge beam of a post-and-beam roof structure.

transverse section: Section view in a print representing a cut across the width of a building.

tread: Horizontal walking surface of the step of a stairway.

tremie: Device consisting of a funnel with a tube-like chute attached at the bottom. It is used to place concrete in forms under water.

trench: Ditch dug in the ground down to bearing soil for foundation footings.

trenching: Digging trenches for foundation footings.

trench shield: Movable shoring system consisting of steel plates and braces that are bolted or welded together and used to support trench walls while work is in progress.

trestle jack: Adjustable metal horse used to support low working platforms.

trim: Finish materials such as molding placed around doors and windows and at the top and bottom of walls.

trimmer joists: Doubled joists placed at floor or ceiling openings to which the header joists are fastened.

trimmer stud: One of the vertical members of a framed wall nailed to studs and supporting the ends of the headers placed at the top of rough openings.

trim ring: Finish piece placed against a door face or a plate before a cylinder unit is installed.

tripod: Three-legged adjustable support for survey instruments.

trolley: Roller device for a bi-fold door that fits into an overhead track. It is attached to the top of the second door for support.

truss spacer: Metal device that provides a fast and accurate method for spacing trusses and eliminates the need to mark the top plate before placing trusses.

T-shore: Shore with a braced head piece used to support forms for concrete beams and girders.

tube-and-clamp scaffold: Type of scaffold that consists of heavy-gauge galvanized steel tubes that are joined using coupling pins or interlocking ends. Also called *tube-and-coupler scaffolds.*

tubular lock: Less expensive type of cylindrical lock.

turnbuckle: **1.** Device often used with wood 2 × 4 or 2 × 6 braces. It consists of two rods held together with a coupling that can be turned to adjustable the tension of the rods. **2.** Rigid rigging attachment with an adjustable length.

turnbutton: Device located in the inside knob of a cylindrical lock. When turned in one direction it locks the door.

twist: Type of warping in which there is deviation from the flat planes of all four faces by a spiraling or torsional action. Usually the result of improper seasoning.

two-way joist system: Concrete floor slab system that has joists running perpendicular to each other.

U

U-factor: Measure of how well a window unit prevents heat from escaping from a building.

unbalanced beam: Glulam beam in which the lams used on the tension side of the beam are of a higher grade than the lams used on the compression side.

undercutting: Tilting a saw so that the back edge of the cut is slightly in from the face edge.

underlayment: **1.** Thin plywood or nonveneered panels fastened over the subfloor to provide a smooth surface for the placement of the finish floor material. **2.** Usually a layer of asphalt-saturated felt paper laid down over the roof sheathing before most types of roof shingles are placed.

United Brotherhood of Carpenters and Joiners of America: The carpenters' union, which was founded in 1881. The union negotiates agreements with contractors' associations concerning wages, fringe benefits, working conditions, provisions for apprenticeship training, the hourly wage scale, and the number of hours in the work week.

unit rise: **1.** Number of inches a common rafter will rise vertically for each foot of run. **2.** Riser height calculated by dividing the total rise by the number of risers in a stairway.

unit run: **1.** Unit of the total run, based on 12". **2.** Width of the tread calculated by dividing the total run by the number of treads in the stairway.

urea formaldehyde: Synthetic resin made by combining urea and formaldehyde. It is widely used as a binder in the manufacture of reconstituted wood panels. Also a basic ingredient in a type of foamed-in-place insulation.

Undercutting When Fitting Panels

A method for fitting panels when back of panel will not be visible.

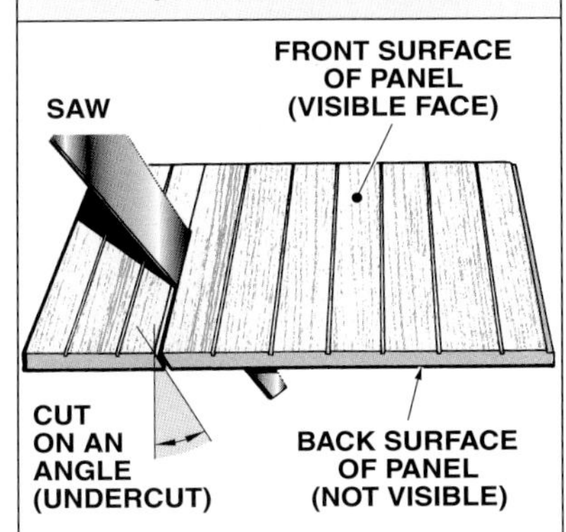

Angle saw slightly when making cut to prevent back surface of panel from touching and holding joint open. It will also reduce amount of planing that may be required for a better fit.

U-shaped stairway: Stairway consisting of two sets of stairways which run parallel to one another.

utilities: Services provided to the public, requiring electrical and plumbing hookups in a building.

V

valley: Sloped intersection of two roof surfaces.

valley cripple jack rafter: Jack rafter sometimes placed between the end sections past the point of intersection of a shortened and supporting valley rafter.

valley jack rafter: Jack rafter extending from a valley rafter to a ridge.

valley rafter: Rafter found on intersecting roofs and that runs at an angle from the intersecting corners of a building to the ridge. It is located at the bottom of the valley formed where different sloping roof sections come together.

vapor barrier: Thin, moisture-resistant material placed over the ground or in walls to retard the passage of moisture.

veneer: Thin layer of wood.

veneered beam: Beam made of glued veneers.

vent: Opening that provides air ventilation in any space of a building.

ventilation: Provisions made in a building to permit warm air to escape and to allow the circulation of air in enclosed areas.

vernier: Graduated scale that gives fractions of a degree on leveling instruments.

vertical hitch: Sling hitch in which one end of the sling connects to the hoist hook and the other end connects to the load.

vertical lifeline: Lifeline that is connected to a fixed anchor at an upper end that is independent of a work platform.

vibration: Consolidation of concrete through the use of mechanical vibration equipment.

vinyl-clad window units: Window units with a frame and sash consisting of a wood core covered with a thin layer of prefinished rigid vinyl.

vinyl siding: Prefinished plastic product that gives a similar effect to wood board siding.

vinyl tile: Resilient floor tile with basic ingredients of polyvinyl chloride resin binders and mineral fillers.

vise: Clamping device typically anchored to a solid base such as workbench.

visible transmittance value: Measure of the amount of light transmitted through a window.

V-joint: Slightly beveled joint between boards or panels that creates a shadow line.

volatile organic compound (VOC): Organic chemical compound with a high vapor pressure to significantly vaporize and enter the atmosphere. Considered a pollutant and can be dangerous to health.

W

waffle slab: Floor slab characterized by a waffle-like appearance at its underside. Its structural design makes possible the use of lighter concrete.

wainscot: Traditionally, wood panel work covering the lower section of a plastered or gypsum board wall. Wainscoting material may also be plastic laminate, tile, glass, etc.

wainscot cap: Piece of finish molding placed at the top of a wainscot.

waler rod: Device used to secure and hold together opposite form walls. It consists of waler rods (also called *she bolts)* extending past the walers on each side. They are joined inside the wall by a threaded filler rod. The whole assembly is tightened by nut washers on opposite sides.

walers: Horizontal pieces placed on the outsides of the form walls to strengthen and stiffen the walls. The form ties are also fastened to the walers. Also called *wales.*

wallboard anchor: Metal or plastic anchor designed for use in gypsum board. Does not require predrilling prior to installation.

wall cabinet: Cabinet not resting on the floor and fastened to the upper part of a wall.

wall framing story pole: Rod laid out according to information provided by the working drawings for a building. It gives lengths of the studs, trimmers, and top and bottom cripples for the walls.

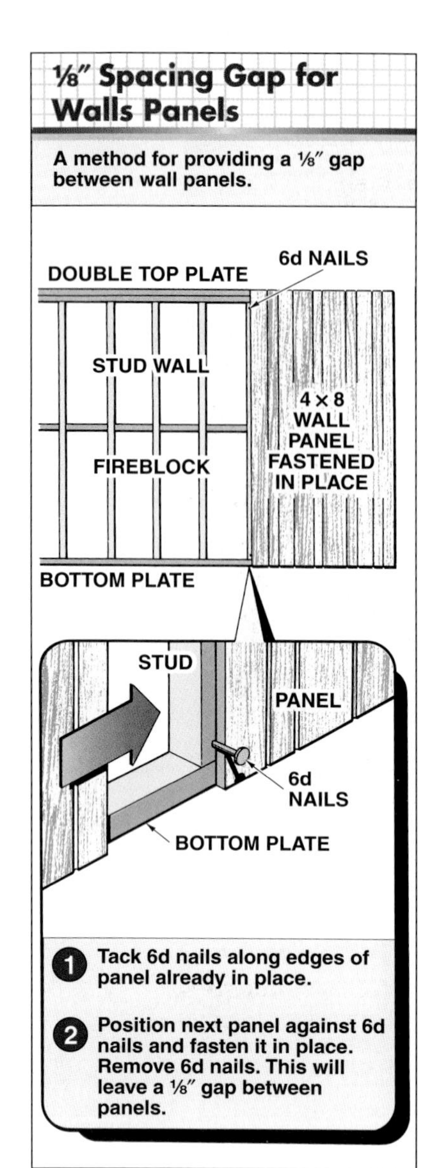

wall jack: Device used to raise walls when small carpentry crews are working on a job site. Typically used in pairs.

wane: Lumber defect caused by the presence of bark or a lack of wood on the edge or corner of a piece of lumber.

warm side: Inner surface of the outside shell (walls, roof, etc.) of a building.

warpage: Condition in which lumber has bent or twisted out of its original shape. Often caused by uneven shrinkage during seasoning.

waterborne preservative: Wood preservative used to treat lumber and plywood for residential construction.

water-cement (W/C) ratio: Amount of water used in a concrete mix in relation to the amount of cement. Major factor in the compressive strength of concrete.

water level: Accurate leveling tool that is based on the principle that water will find its own level in a system open to atmospheric pressure.

water-reducing admixture: Substance added to a concrete mixture to reduce the amount of water needed to produce a desired mix.

water table: **1.** Highest point below the surface of the ground that is normally saturated with water in a given area. **2.** Piece placed below the first course of siding to direct water drainage away from the foundation wall.

water wall: Water containers used to store heat as part of an indirect-gain passive solar system.

wind effect: Uncontrolled airflow created when wind blows against a building. A high-pressure area is created on the windward side and air is forced into the building.

weather barrier: Second line of protection for a building to prevent air and water infiltration into the building.

weatherstripping: Metal, felt, or plastic strips used to prevent air or moisture infiltration through the spaces between the outer edges of doors and windows and their finish frames.

web: Truss member that runs between and ties together the top and bottom chords.

web sling: Sling made from flat narrow strapping that is woven from yarns of strong synthetic fibers.

web stiffener: Dimension lumber, oriented strand board, or structural panel materials used to reinforce a wood I-joist web.

wedges: Devices driven against the walers and at the ends of the form ties. They hold together opposite form walls.

wedge ties: Patented ties consisting of a strap and wedges. Used with plank forming systems.

weep hole: Small hole extending through a concrete retaining wall to allow water to drain from behind the wall.

welded-frame scaffold: Type of scaffold that consists of heavy-gauge tubular steel bearers and posts that are welded together to form the frames.

welded wire reinforcement: Heavy steel wire welded together in a grid pattern and used to reinforce concrete slabs resting directly on the ground. Also called *wire mesh.*

white speck: Small white pits or spots on the surface of wood caused by a fungus in the living tree.

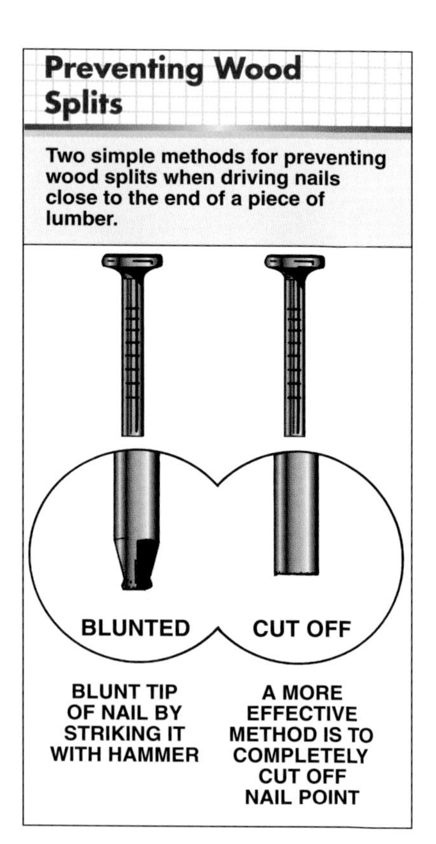

wide-flange beam: Steel beam with flanges at the top and bottom of the web. Also called *W-beam*.

windowsill: Finish horizontal member installed directly below a window.

wind turbine: Machine that converts the kinetic energy of wind into rotating mechanical energy.

wind uplift: Wind flowing over a pitched roof surface that creates an upward suction, pulling on the roof surface.

winder: Wedge-shaped step in a stairway.

winder stairway: Stairway that has a series of winding steps.

wood I-joist: Engineered wood product widely used in the construction of wood-framed floor units and flat roofs, and that is a high-performance alternative to solid dimension lumber.

wood preservative: Chemical in liquid form applied to wood for protection against fungi, decay, and insect attack.

working drawings: Set of plans drawn up by an architect. Contains all the dimensions and structural information needed to complete a construction project.

working load limit (WLL): Maximum linear force that a rigging component may be safely subjected to.

working load value: Amount of pressure against an individual form tie when placing concrete in forms.

working part: Portion of the rope involved in making the knot or hitch.

woven corner: Outside corner finish for sidewalls where alternate corner edges of the shingles are exposed.

Placing and Cutting Wood Wedges under Windowsills

A method for cutting and placing wood wedges under windowsills.

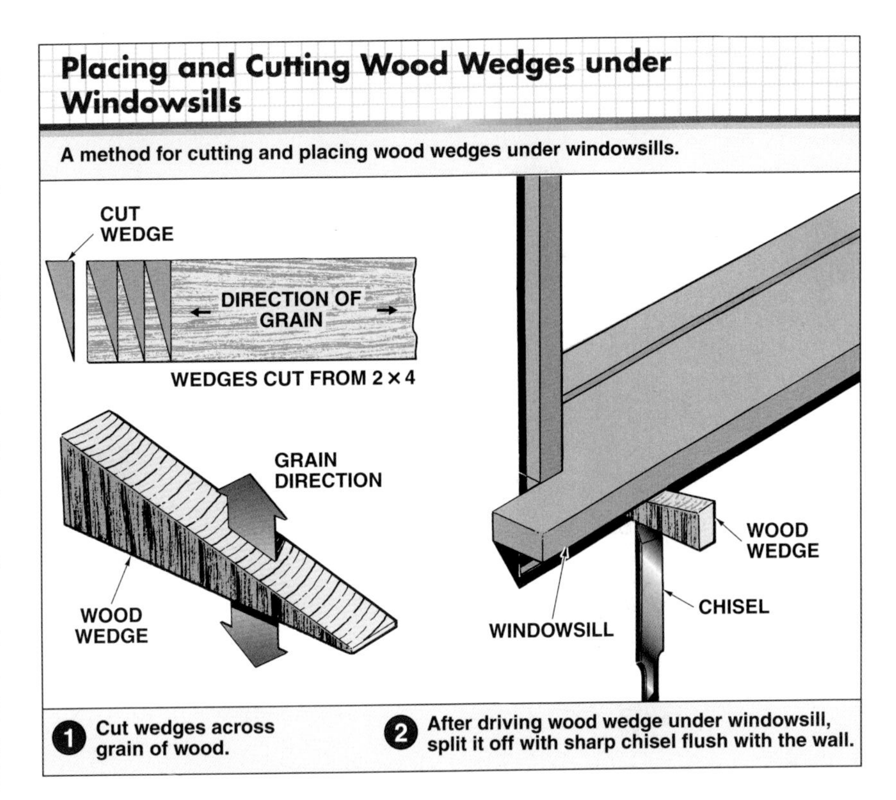

1. Cut wedges across grain of wood.
2. After driving wood wedge under windowsill, split it off with sharp chisel flush with the wall.

X

xerography process: Print reproduction process similar to the photocopying process. Provides high-quality reproduction with little loss of clarity of details.

Y

yard lumber: All lumber sold for structural building purposes.

yoke: Tie or clamping device used around column forms to prevent them from spreading due to pressure imposed by concrete.

Z

zoning regulations: Local regulations that govern the type of buildings and structures that may be erected in different areas of a community. Most zones come under the general categories of residential, commercial, and manufacturing.

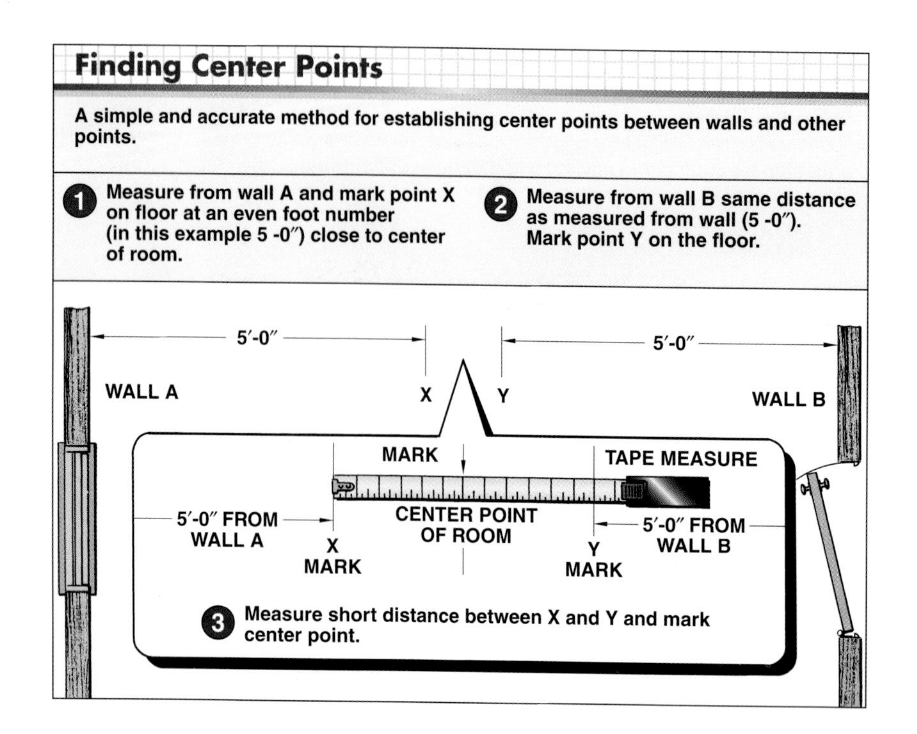
Finding Center Points
A simple and accurate method for establishing center points between walls and other points.
1 Measure from wall A and mark point X on floor at an even foot number (in this example 5 -0″) close to center of room.
2 Measure from wall B same distance as measured from wall (5 -0″). Mark point Y on the floor.
5′-0″
5′-0″
WALL A
X
Y
WALL B
MARK
TAPE MEASURE
5′-0″ FROM WALL A
CENTER POINT OF ROOM
5′-0″ FROM WALL B
X MARK
Y MARK
3 Measure short distance between X and Y and mark center point.

A

B

C

D

E

F

G

H

I

J

K

L

M

N

O

P

Q

R

S

T

U

V

W

Y

Z

USING THE *CARPENTRY* INTERACTIVE DVD

Before removing the Interactive DVD from the protective sleeve, please note that the book cannot be returned for refund or credit if the DVD sleeve seal is broken.

Windows System Requirements

To use this DVD on a Windows® system, your computer must meet the following minimum system requirements:

- Microsoft® Windows® 7, Windows Vista®, or Windows® XP operating system
- Intel® 1.3 GHz processor (or equivalent)
- 128 MB of available RAM (256 MB recommended)
- 335 MB of available hard disk space
- 1024 × 768 monitor resolution
- DVD drive (or equivalent optical drive)
- Sound output capability and speakers
- Microsoft® Internet Explorer® 6.0 or Firefox® 2.0 web browser
- Active Internet connection required for Internet links

Macintosh System Requirements

To use this DVD on a Macintosh® system, your computer must meet the following minimum system requirements:

- Mac OS® X 10.5 (Leopard) or 10.6 (Snow Leopard)
- PowerPC® G4, G5, or Intel® processor
- 128 MB of available RAM (256 MB recommended)
- 335 MB of available hard disk space
- 1024 × 768 monitor resolution
- DVD drive (or equivalent optical drive)
- Sound output capability and speakers
- Apple® Safari® 2.0 web browser or later
- Active Internet connection required for Internet links

Opening Files

Insert the Interactive DVD into the computer DVD drive. Within a few seconds, the home screen will be displayed allowing access to all features of the DVD. Information about the usage of the DVD can be accessed by clicking on Using This Interactive DVD. The Quick Quizzes®, Illustrated Glossary, Flash Cards, Master Math® Problems, Printreading Tests, Measurement Activity, Media Library, and ATPeResources.com can be accessed by clicking on the appropriate button on the home screen. Clicking on the ATP logo (www.atplearning.com) accesses information on related educational products. Unauthorized reproduction of the material on this DVD is strictly prohibited.